D0164495

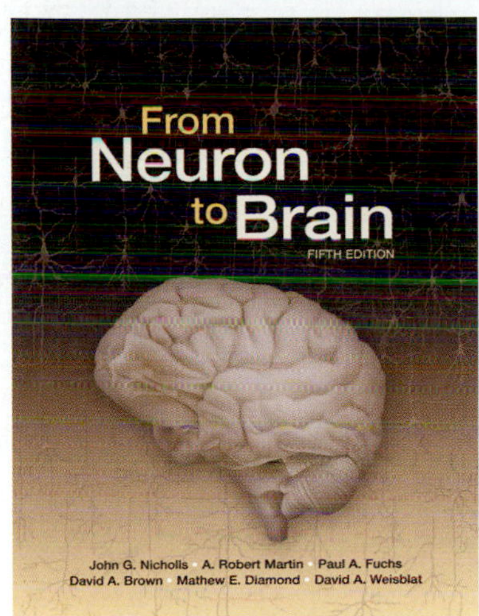

ABOUT THE COVER

Computer artwork of a human brain © Pasieka/Science Photo Library/Fotosearch. Neurons (background) are from the Third Edition cover, designed by Laszlo Meszoly.

From Neuron to Brain, Fifth Edition

Address inquiries and orders to
Sinauer Associates, Inc.
23 Plumtree Road
Sunderland, MA 01375 U.S.A.
FAX: 413-549-1118
E-mail: publish@sinauer.com
Internet: www.sinauer.com

Library of Congress Cataloging-in-Publication Data
From neuron to brain / John G. Nicholls ... [et al.]. -- 5th ed.
p. cm.
Includes bibliographical references and index.
ISBN 978-0-87893-609-0
1. Neurophysiology. 2. Brain. 3. Neurons. I. Nicholls, John G.
QP355.2.K83 2012
612.8--dc23
2011037528

8 7 6 5 4 3 2 Printed in U.S.A.

From Neuron to Brain

FIFTH EDITION

John G. Nicholls
International School for Advanced Studies, Trieste, Italy

A. Robert Martin
Emeritus, University of Colorado School of Medicine

Paul A. Fuchs
The Johns Hopkins University School of Medicine

David A. Brown
University College London

Mathew E. Diamond
International School for Advanced Studies, Trieste, Italy

David A. Weisblat
University of California, Berkeley

Sinauer Associates, Inc. • Publishers
Sunderland, Massachusetts • USA

This book is dedicated to the memory of our friend and colleague, Steve Kuffler.

In a career that spanned 40 years, Stephen Kuffler made experiments on fundamental problems and laid paths for future research to follow. A feature of his work is the way in which the right problem was tackled using the right preparation. Examples are his studies on denervation, stretch receptors, efferent control, inhibition, GABA and peptides as transmitters, integration in the retina, glial cells, and the analysis of synaptic transmission. What gave papers by Stephen Kuffler a special quality were the clarity, the beautiful figures, and the underlying excitement. Moreover, he himself had done *every* experiment he described. Stephen Kuffler's work exemplified and introduced a multidisciplinary approach to the study of the nervous system. At Harvard he created the first department of neurobiology, in which he brought together people from different disciplines who developed new ways of thinking. Those who knew him remember a unique combination of tolerance, firmness, kindness, and good sense with enduring humor. He was the J. F. Enders University Professor at Harvard, and was associated with the Marine Biological Laboratory at Woods Hole. Among his many honors was his election as a foreign member of the Royal Society.

Preface to the Fifth Edition

When the First Edition of our book appeared in 1976, its preface stated that our aim was "...to describe how nerve cells go about their business of transmitting signals, how the signals are put together, and how out of this integration higher functions emerge. This book is directed to the reader who is curious about the workings of the nervous system but does not necessarily have a specialized background in biological sciences. We illustrate the main points by selected examples...".

This new, Fifth Edition has been written with the same aim in mind but in a very different context. When the First Edition appeared there were hardly any books, and only a few journals devoted to the nervous system. The extraordinary advances in molecular biology, genetics, and immunology had not yet been applied to the study of nerve cells or the brain, and the internet was not available for searching the literature. The explosion in knowledge since 1976 means that even though we still want to produce a readable narrative, the topics that have to be addressed and the number of pages have increased. Inevitably, descriptions of certain older experiments have had to be jettisoned, even though they still seem beautiful. Nevertheless, our approach continues to be to follow ideas from their conception to the latest developments. To this end, in this edition we have retained descriptions of classical experiments as well as the newest findings. In this way we hope to present key lines of research of interest for practicing research workers and teachers of neurobiology, as well as for readers who are not familiar with the field.

For the Fifth Edition, new chapters have been added, others have been completely rewritten and all have been brought up to date. In addition there are two obvious, major changes from the last edition. First, as in the original text by Kuffler and Nicholls, a discussion of the visual system now comes in the first part of the book, with the same intent. Placing this information at the beginning of the book highlights its importance and immediately gives the reader a glimpse of the whole functioning "brain" that comprises part of our title. Experience shows that a beginning student who wants to know how the brain works is often drawn to the subject (as were the authors of this book) by fascination with higher functions. Thanks to the elegance of work done on the retina and visual cortex, it is possible to show the clear link between the properties of nerve cells and our visual perception. Moreover, this information can evoke curiosity about succeeding chapters, which deal with the cellular, biophysical, and molecular properties of nerve cells.

A second major change is in the number of authors. From two, then three and four, we have now grown to six authors. In the last years the amount of new information that is produced and the number of new techniques that must be described have grown faster than ever. As a result, the choices that have to be made about what to include or leave out are so sophisticated that it would be a full-time task beyond the capabilities of two or three neuroscientists (especially if they still wished to continue their research and get grants). Nevertheless, as in the past, we have aimed to maintain a consistent, continuous, narrative style. As in the past, we hope to make the whole book appear to the reader as if it were the work of a single author.

John G. Nicholls   A. Robert Martin   Paul A. Fuchs
David A. Brown   Mathew E. Diamond   David A. Weisblat
October, 2011

Acknowledgments

We owe special thanks to our friend and colleague, Bruce Wallace, who was an author of the Third and Fourth Editions of this book. Although it was not possible for him to collaborate for this Fifth Edition, material from his previous chapters has been used extensively in new chapters that have been rewritten and brought up to date. His work represents an essential contribution and we are extremely grateful that we are able to retain in the text much of his scholarly, clear, and authoritative expositions of key topics.

We are grateful to the numerous colleagues who have encouraged us and influenced our thinking. We particularly thank Dr. Denis Baylor, who read all of the chapters. For reading individual chapters we thank Drs. O. W. Hill, W. Kolbinger, G. Legname, F. F. de Miguel, K. J. Muller, U. J. McMahan, L. Szczupak, and C. Trueta. Helpful discussions and advice were provided by E. Arabzadeh, S.Bensmaïa, D. H. Jenkinson, W. B. Kristan, Jr., D.-H. Kuo, S. R. Levinson, A. Mallamaci, L. Mei, M. Ruegg, W. J. Thompson, and D. Zoccolan. We are most grateful to colleagues who provided original plates and figures, including: K. Briggman, A. Brown, R. Constanzo, J. Fernandez, D. Furness, M. S. Graziano, L. L. Iversen, A. Kaneko, J. Kinnamon, E. Knudsen, S. R. Levinson, A. Mallamaci, U. J. McMahan, R. Michaels, K. J. Muller, E. Newman, E. Puzuolo, R. Reed, S. Roper, M. Ruegg, M. M. Salpeter, P. Shrager, P. Sterling, M. Stryker, J. Taranda, R. Vassar, D. A. Wagenaar, and D. Zoccolan.

It is a pleasure to thank Andy Sinauer, our publisher and friend. His remarkable sense of style, keen interest, and sound judgment about the contents, illustrations, and appearance of the book have motivated and encouraged us at every stage. We were fortunate to have the guidance of Sydney Carroll who was with us from the inception of the book and provided valuable input to its design and layout. We are also grateful to Kathaleen Emerson, Chelsea Holabird, and Azelie Aquadro who edited our text with great care and courtesy; Chris Small, Joan Gemme, and Marie Scavotto, who played major roles in producing this new edition, and to artist Elizabeth Morales, who transformed our drawings and graphs into elegant and attractive figures.

From the Preface to the First Edition

Our aim is to describe how nerve cells go about their business of transmitting signals, how these signals are put together, and how out of this integration higher functions emerge. This book is directed to the reader who is curious about the workings of the nervous system but does not necessarily have a specialized background in biological sciences. We illustrate the main points by selected examples, preferably from work in which we have first-hand experience. This approach introduces an obvious personal bias and certain omissions.

We do not attempt a comprehensive treatment of the nervous system, complete with references and background material. Rather, we prefer a personal and therefore restricted point of view, presenting some of the advances of the past few decades by following the thread of development as it has unraveled in the hands of a relatively small number of workers. A survey of the table of contents reveals that many essential and fascinating fields have been left out: subjects like the cerebellum, the auditory system, eye movements, motor systems, and the corpus callosum, to name a few. Our only excuse is that it seems preferable to provide a coherent picture by selecting a few related topics to illustrate the usefulness of a cellular approach.

Throughout, we describe experiments on single cells or analyses of simple assemblies of neurons in a wide range of species. In several instances the analysis has now reached the molecular level, an advance that enables one to discuss some of the functional properties of nerve and muscle membranes in terms of specific molecules. Fortunately, in the brains of all animals that have been studied there is apparent a uniformity of principles for neurological signaling. Therefore, with luck, examples from a lobster or a leech will have relevance for our own nervous systems. As physiologists we must pursue that luck, because we are convinced that behind each problem that appears extraordinarily complex and insoluble there lies a simplifying principle that will lead to an unraveling of the events. For example, the human brain consists of over 10,000 million cells and many more connections that in their detail appear to defy comprehension. Such complexity is at times mistaken for randomness; yet this is not so, and we can show that the brain is constructed according to a highly ordered design, made up of relatively simple components. To perform all its functions it uses only a few signals and a stereotyped repeating pattern of activity. Therefore, a relatively small sampling of nerve cells can sometimes reveal much of the plan of the organization of connections, as in the visual system.

We also discuss "open-ended business," areas that are developing and whose direction is therefore uncertain. As one might expect, the topics cannot at present be fitted into a neat scheme. We hope, however, that they convey some of the flavor that makes research a series of adventures.

From Neuron to Brain expresses our approach as well as our aims. We work mostly on the machinery that enables neurons to function. Students who become interested in the nervous system almost always tell us that their curiosity stems from a desire to understand perception, consciousness, behavior, or other higher functions of the brain. Knowing of our preoccupation with the workings of isolated nerve cells or simple cell systems, they are frequently surprised that we ourselves started with similar motivations, and they are even more surprised that we have retained those interests. In fact, we believe we are working toward that goal (and in that respect probably do not differ from most of our colleagues and predecessors). Our book aims to substantiate this claim and, we hope, to show that we are pointed in the right direction.

Stephen W. Kuffler   John G. Nicholls

August, 1975

Media and Supplements

eBook

(www.coursesmart.com; ISBN 978-0-87893-632-8)

The *From Neuron to Brain* eBook is an economical option that includes convenient tools for searching the text, highlighting, and note-taking. Available via standard Web browsers as well as iPhone/iPad and Android devices.

Sylvius: An Interactive Atlas and Visual Glossary of Human Neuroanatomy

S. Mark Williams, Leonard E. White, and Andrew C. Mace

Sylvius is a unique digital learning environment for exploring and understanding the structure of the human central nervous system. Includes surface anatomy (photographic, MR, and brainstem model), sectional anatomy (photographic, MR, and brainstem cross-sections), pathways, and a visual glossary.

Instructor's Resource Library

(ISBN 978-0-87893-656-4)

Available to qualified adopters of *From Neuron to Brain*, Fifth Edition, the Instructor's Resource Library includes electronic versions of all of the textbook's figures and tables, in both JPEG and PowerPoint formats.

About the Authors

Mathew E. Diamond, David A. Weisblat, John G. Nicholls, Paul A. Fuchs, A. Robert Martin, David A. Brown

John G. Nicholls is Professor of Neurobiology and Cognitive Neuroscience at the International School for Advanced Studies in Trieste, Italy. He was born in London in 1929 and received a medical degree from Charing Cross Hospital and a Ph.D in physiology from the Department of Biophysics at University College London, where he did research under the direction of Sir Bernard Katz. He has worked at University College London, at Oxford, Harvard, Yale, and Stanford Universities and at the Biocenter in Basel. With Stephen Kuffler, he made experiments on neuroglial cells and wrote the First Edition of this book. He is a Fellow of the Royal Society, a member of the Mexican Academy of Medicine, and the recipient of the Venezuelan Order of Andres Bello. He has given laboratory and lecture courses in neurobiology at Woods Hole and Cold Spring Harbor, and in universities in Asia, Africa, and Latin America. His work concerns regeneration of the nervous system after injury and mechanisms that give rise to the respiratory rhythm.

A. Robert Martin is Professor Emeritus in the Department of Physiology at the University of Colorado School of Medicine. He was born in Saskatchewan in 1928 and majored in mathematics and physics at the University of Manitoba. He received a Ph.D. in Biophysics in 1955 from University College, London, where he worked on synaptic transmission in mammalian muscle under the direction of Sir Bernard Katz. From 1955 to 1957 he did postdoctoral research in the laboratory of Herbert Jasper at the Montreal Neurological Institute, studying the behavior of single cells in the motor cortex. He has taught at McGill University, the University of Utah, Yale University, and the University of Colorado Medical School, and has been a visiting professor at Monash University, Edinburgh University, and the Australian National University. His research has contributed to the understanding of synaptic transmission, including the mechanisms of transmitter release, electrical coupling at synapses, and properties of postsynaptic ion channels.

Paul A. Fuchs is Director of Research and the John E. Bordley Professor of Otolaryngology-Head and Neck Surgery, Professor of Biomedical Engineering, Professor of Neuroscience, and co-Director of the Center for Sensory Biology at the Johns Hopkins University School of Medicine. Born in St. Louis, Missouri in 1951, Fuchs graduated in biology from Reed College in 1974. He received a Ph.D. in Neuro- and Biobehavioral Sciences in 1979 from Stanford University where he investigated presynaptic inhibition at the crayfish neuromuscular junction under the direction of Donald Kennedy and Peter Getting. From 1979 to 1981 he did postdoctoral research with John Nicholls at Stanford University, examining synapse formation by leech neurons. From 1981 to 1983 he studied the efferent inhibition of auditory hair cells with Robert Fettiplace at Cambridge University. He has taught at the University of Colorado and the Johns Hopkins University medical schools. His research examines excitability and synaptic signaling of sensory hair cells and neurons in the vertebrate inner ear.

David A. Brown is Professor of Pharmacology in the Department of Neuroscience, Physiology and Pharmacology at University College London. He was born in London in 1936 and gained a B.Sc. in Physiology from University College London and a Ph.D. from St. Bartholomew's Hospital Medical College ("Barts") in London studying transmission in sympathetic ganglia in the laboratory of Peter Quilliam (where he first met John Nicholls). He then did a post-doc at the University of Chicago where he helped design an integrated neurobiology course for graduate medical students. After returning to a junior faculty position at Barts, he subsequently chaired departments of Pharmacology at the School of Pharmacy in London and finally at University College. In between times he has worked in several labs in the United States and elsewhere, including three semesters in A. M. Brown's Department of Physiology and Biophysics at the University of Texas in Galveston, and a split year as Fogarty Scholar-in-Residence at NIH in the labs of Michael Brownstein, Julie Axelrod, and Marshall Nirenberg. While at Galveston, he and Paul Adams (another ex-Barts colleague) discovered the M-type potassium channel, which provided new insight into the way neurotransmitters could alter nerve cell excitability by indirectly regulating a voltage-gated ion channel. He has since worked on the regulation of other ion channels by G protein–coupled receptors and previously on the actions and transport of GABA. He is a Fellow of the Royal Society, a recipient of the Feldberg Prize, and has an Honorary Doctorate from the University of Kanazawa in Japan.

Mathew E. Diamond is Professor of Cognitive Neuroscience at the International School for Advanced Studies in Trieste Italy (known by its Italian acronym, SISSA). He earned a Bachelor of Science degree in Engineering from the University of Virginia in 1984 and a Ph.D. in Neurobiology from the University of North Carolina in 1989. Diamond was a postdoctoral fellow with Ford Ebner at Brown University and then an assistant professor at Vanderbilt University before moving to SISSA to create the Tactile Perception and Learning Laboratory in 1996. His main interest is to specify the relationship between neuronal activity and perception. The research is carried out mostly in the tactile whisker system in rodents, but some experiments attempt to generalize the principles found in the whisker system to the processing of information in the human tactile sensory system.

David A. Weisblat is Professor of Cell and Developmental Biology in the Department of Molecular and Cell Biology at the University of California, Berkeley. He was born in Kalamazoo, Michigan in 1949, studied biochemistry as an undergraduate with Bernard Babior at Harvard College, where he was introduced to neurobiology in a course led by John Nicholls. He received his Ph.D. from Caltech for studies on the electrophysiology of *Ascaris* in 1976 with Richard Russell. He began studying leech development with Gunther Stent in the Department of Molecular Biology at Berkeley and was appointed to the Zoology Department there in 1983. As a postdoc, he developed techniques for cell lineage tracing by intracellular microinjection of tracer molecules. Current research interests include the evolution of segmentation mechanisms, D quadrant specification, and axial patterning. Work from his laboratory has helped to establish the leech *Helobdella* as a tractable representative of the super-phylum Lophotrochozoa, for the study of evolutionary developmental biology. He has assisted or organized courses in Africa, India, Latin America, and at Woods Hole, Massachusetts.

Brief Table of Contents

Contents

PART V SENSATION AND MOVEMENT 383

PART VI DEVELOPMENT AND REGENERATION OF THE NERVOUS SYSTEM 529

CHAPTER 25 Development of the Nervous System 531

PART VII CONCLUSION 613

PART I

Introduction to the Nervous System

This introduction provides a framework for approaching chapters that deal with signaling, development, and functions of the nervous system. Readers who are curious about the brain but unfamiliar with neurobiology often face difficulties in coming to grips with the subject. For example, the terminology of neurobiology, derived from anatomy, physics, biochemistry, and molecular biology, is disconcerting. But because of the elaborate structure of the nervous system and the specialized features of neural signaling, it is unavoidable.

Accordingly, the first three chapters in this book provide an introduction and overview of key concepts and definitions for readers who are approaching neurobiology for the first time. In Chapter 1 the principal morphological, physiological, and molecular properties of nerve cells and their connections are described. As an example of a well-defined structure in which the processing of signals is now largely understood, we use the retina. A major advantage is that from the outset the electrical signals generated by retinal cells can be directly correlated to perception; this enables features of the way in which we see the world to be understood at the cellular level. In Chapters 2 and 3 the signals are further followed from the eye to the cerebral cortex. There, it is shown how the meaning of the messages becomes transformed in a remarkable manner through precise interconnections along the pathway. The beauty and clarity of the experiments make it possible to understand the material with little background knowledge, to see where research on the brain is heading and to appreciate why the detailed study of cellular and molecular mechanisms described in the later chapters is so interesting and important.

Our principal objective at this stage is to enable the naive reader to think about higher functions from the outset and to see how they are dependent on and correlated to cellular mechanisms employed by nerve cells. To achieve this aim, the material is presented with only essential concepts and facts, the rest being dealt with in later chapters.

CHAPTER 1
Principles of Signaling and Organization

In this chapter we have chosen the retina (see also Chapter 18) as an example to describe signals in nerve cells (neurons) and to correlate such signals with our perception of the outside world. The orderly structure of the retina, where the initial steps that lead to vision occur, allows us to follow, in detail, the signals from neuron to brain and to our perception of the outside world. This chapter presents basic information that prepares the reader to deal with detailed descriptions of signaling and perception that follow.

Information is transmitted in nerve cells by electrical signals. A major goal of neurobiological research is to decode the content of the information that these signals transmit. The meaning of the signals is determined by where the nerve fibers arise and where they go, as well as on the frequency and regularity of the signals themselves. Signals in the optic nerve carry visual information from the retina. Similar impulses in a sensory nerve in the fingertip convey information about touch, while impulses in a motor nerve give rise to a muscular movement. Within the brain, each individual nerve cell receives inputs from thousands of others. By integrating this information, the cell creates a new message that can convey a complex meaning, such as the presence of a vertical bar of light in one's field of vision or the rough texture of sandpaper touched by one's finger.

One simplifying feature is that in many areas of the nervous system (including the retina), nerve cells with similar properties are grouped together, often in layers or clusters. Another is that the brain uses stereotyped electrical signals to transmit information. These consist of changes in voltage produced by electrical currents flowing across cell membranes. Neurons use only two types of electrical signals: local graded potentials, which spread over short distances, and action potentials, which are conducted rapidly over long distances.

Information is transmitted between one neuron and the next at specialized junctions, known as synapses. Here the incoming signals release chemical transmitter molecules that bind to specific chemoreceptor molecules in the membrane of the target cell. This interaction gives rise to a local graded potential that excites or inhibits the cell depending on the transmitter and the corresponding receptors. The efficacy of synaptic transmission is modified by activity, hormones, and drugs.

During development, neurons depend on molecular signals derived from other cells. These signals determine the shape and position of each neuron, its survival, its transmitter, and the targets to which it connects. Once mature, most nerve cells do not divide. Molecules in the environment of a neuron influence its capacity for repair after injury.

The central nervous system (CNS) is an unresting assembly of cells that continually receives information, analyzes it, perceives it, and makes decisions. The brain can also take the initiative and produce coordinated, effective muscle contractions for swimming, swallowing or singing. Throughout the following chapters, we attempt to explain these complex functions of the brain in terms of the underlying activity of nerve cells. A second aim is to understand the cellular and molecular mechanisms by which those signals arise. A third aim is to review the way in which structures and connections that subserve functions are established during development, become modified by experience, and repair themselves after injury. In this chapter, we summarize key concepts and essential background material.

Signaling in Simple Neuronal Circuits

Events that occur during the performance of simple reflexes can be followed and analyzed in detail. For example, when the tendon below the knee is tapped by a small hammer, thigh muscles are stretched and electrical impulses travel along sensory nerve fibers to the spinal cord, where motor neurons are excited, producing impulses and contractions in muscles. The end result is that the leg extends and the knee is straightened. Such simplified circuits are essential for regulating muscular contractions that control movements of the body. In simple reflex behavior, in which a stimulus leads to a defined output, the roles played by the signals in just two types of cells—sensory and motor—can be readily understood.

Complex Neuronal Circuitry in Relation to Higher Functions

Analyzing signaling in complex pathways that involve many different types of neurons is far more difficult than analyzing simple reflexes. The transmission of information to the brain for perception of a sound, a touch, an odor, a visual image, or the execution of a simple voluntary movement requires the sequential activation of a multitude of cells, neuron after neuron. Serious problems in the analysis of signaling and circuitry arise from the dense packing of nerve cells, the intricacy of their connections, and the profusion of cell types. The brain, unlike an organ such as the liver, consists of heterogeneous populations of cells. By contrast, if you have discovered how one region of the liver functions, you know a great deal about the liver as a whole. Knowledge of the cerebellum, however, gives you little idea about the workings of the retina or any other part of the central nervous system. Nevertheless, the retina can provide a well-defined structure for illustrating and explaining fundamental mechanisms that operate throughout the nervous system.

Despite the immense complexity of the nervous system, it is now possible to understand many aspects of the way in which neurons act as the building blocks for perception. The first three chapters of this book show how one can record the activity of neurons in the pathway from the eye to the brain, follow signals first in cells that specifically respond to light, and then step by step through successive relays. As you read this page, signaling within the eye itself ensures that the black letters stand out from the white page in either the dim light in a room or the brilliant sunshine on a beach (not perhaps the ideal place to concentrate on this book). Specific connections in the brain create a single image in the mind's eye, even though the two eyes are situated apart on the head and have somewhat different views of the outside world. Moreover, mechanisms exist to ensure that this image stays still (even though our eyes are continually making small movements) and to provide accurate information about the distance to the page so that you can turn it with your fingers.

How do the connections of nerve cells enable such phenomena to occur? Much is now known about how these attributes of vision stem from well-defined neuronal circuits in the eye and in the initial relays within the brain, but numerous questions remain about the relation between neuronal properties and behavior. New methods, such as **functional magnetic resonance imaging** of the human brain (abbreviated as fMRI), can reveal the locations of structures that become active as you read. But it is still not known how the neuronal signals in those regions cause you to perceive and understand the words in front of you. Moreover, while you are reading, you must maintain the posture of your body, head,

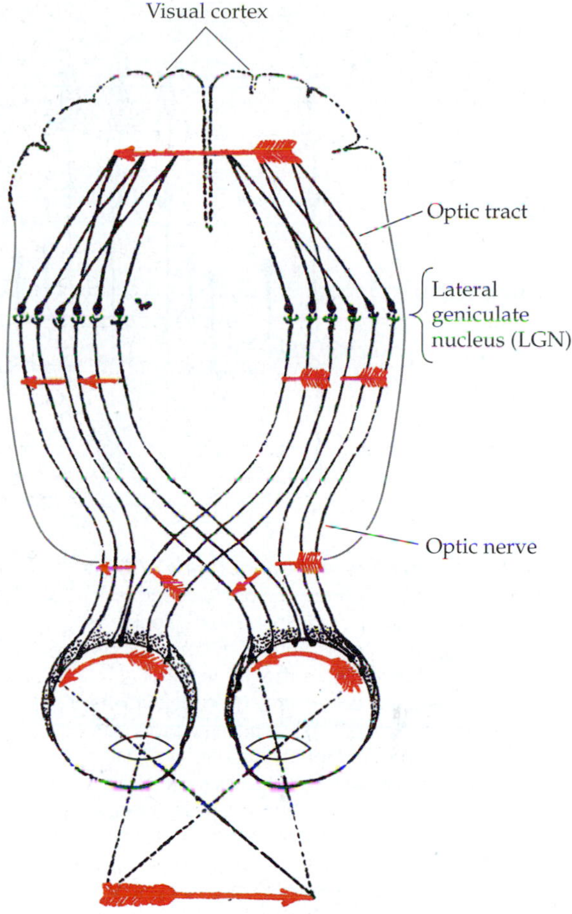

FIGURE 1.1 Pathways from the Eyes to the Brain. Pathways travel through the optic nerve and the optic tract. The interposed relay is the lateral geniculate nucleus. Arrows indicate how the lens reverses images and how the specific crossing of axons causes the right visual field to be represented in the left brain, and vice versa. The figure has been modified from an original drawing by Ramón y Cajal, which dates from 1892 ([1909–1911], Eng. trans., 1995).

and arms. The brain must further ensure that the eyeball remains moist through continuous secretion of tears, that breathing continues, and that untold other involuntary and unconscious bodily tasks are carried out. A full description of the mechanisms underlying such integrated behavior is well beyond the scope of this book or, indeed, of current knowledge.

The following discussion deals with questions of how neurons are organized and how electrical signals arise, propagate, and spread from nerve cell to nerve cell. The retina serves to introduce and illustrate general principles that operate throughout the nervous system.

Organization of the Retina

Analysis of the visual world depends on the information coming from the retina—the initial stage of processing. Figure 1.1 shows pathways from the eye to higher centers in the brain. The optical image that falls on the retina is reversed by the lens but is otherwise a faithful representation of the external world. How can this picture be translated into our visual experience by way of electrical signals that are initiated in the retina and then travel through the optic nerves? An essential concept, applicable at every level of the nervous system, is that for function to be understood one must know the anatomy of the structure involved.

Shapes and Connections of Neurons

Figure 1.2 shows the various types of cells and their arrangement in the vertebrate retina. Light entering the eye passes through the layers of transparent cells to reach the **photoreceptors**. Then, signals leaving the eye through the optic nerve fibers of **ganglion cells** provide the entire input for all of our vision.

The drawing of Figure 1.2 was made by Santiago Ramón y Cajal[1] before the turn of the twentieth century. Ramón y Cajal was one of the greatest students of the nervous system, selecting examples from a wide range of animals. He had an unfailing instinct for the essential. An important lesson from his work is that the shape and position of a neuron, as well as the origin and destination of its processes in the neural network, supply valuable clues to its function.

Ramón y Cajal, about 1914

In Figure 1.2, it is apparent that the cells in the retina, like those elsewhere in the CNS, are densely packed. Early anatomists had to tease nervous tissue apart to see individual cells. Staining methods that impregnate every neuron are virtually useless for investigating cell shapes and connections because a structure like the retina appears as a dark blur of intertwined cells and processes. Most of Ramón y Cajal's pictures were made with the Golgi staining method, which by a still unknown mechanism, stains just a few neurons at random out of the whole population, yet stains those few cells in their entirety. The electron micrograph of Figure 1.3 shows that the extracellular space surrounding neurons and their supporting cells is restricted to clefts only about 25 nanometers (25×10^{-9} meters) wide.

The schematic presentation in Figure 1.2 gives an idea of the orderly arrangement of the retinal neurons. It is possible to distinguish the photoreceptors, **bipolar cells**, and ganglion cells. The lines of transmission are from input to output, from photoreceptors through to ganglion cells. In addition, two other types of cells, horizontal and amacrine, make transverse connections linking the pathways. How do the cells in Ramón y Cajal's drawings contribute to the picture of the world we see?

[1] Ramón y Cajal, S. [1909–1911] 1995. *Histology of the Nervous System*, 2 vols. Translated by Neely Swanson and Larry Swanson. Oxford University Press, New York.

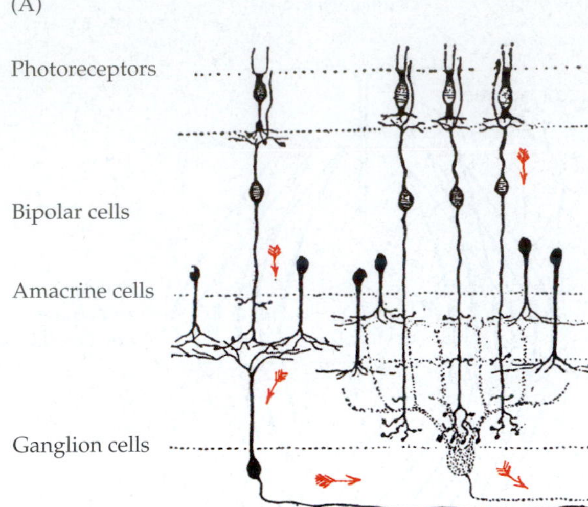

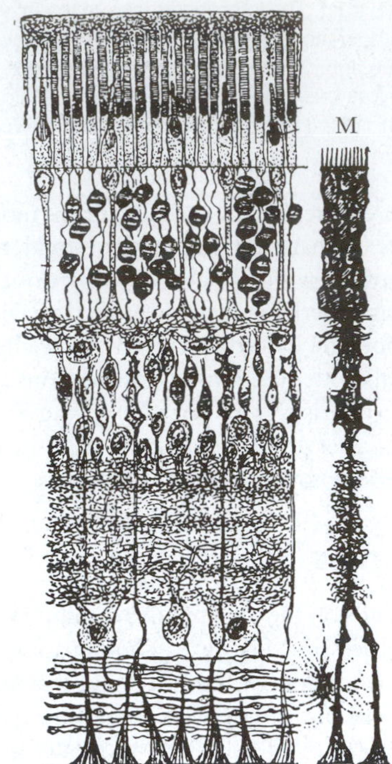

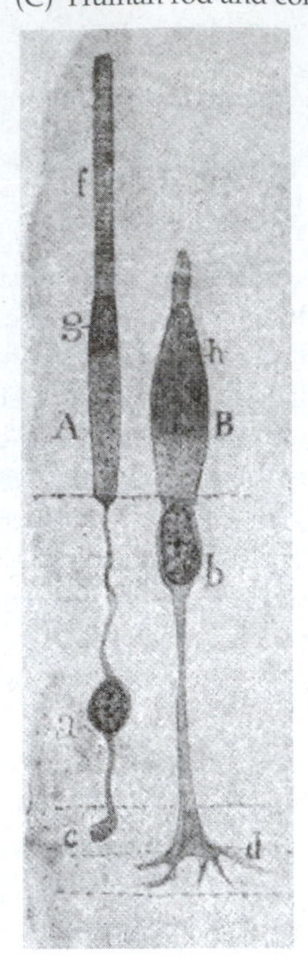

FIGURE 1.2 Structure and Connections of Cells in the Mammalian Retina. The photoreceptors (rods and cones) connect to bipolar cells. Bipolar cells in turn connect to ganglion cells, whose axons constitute the optic nerve. Horizontal cells (not shown) and amacrine cells make connections that are predominantly horizontal. (A) The scheme proposed by Ramón y Cajal for the direction taken by signals as they pass from receptors to the optic nerve fibers. This scheme still holds in general, but essential new pathways and feedback groups have been discovered since Ramón y Cajal's time. (B) Ramón y Cajal's depiction of the cellular elements of the retina and their orderly arrangement. The Müller (M) cell, shown on the right, is a satellite glial cell. (C) Drawings of a human rod (left) and cone (right) isolated from the retina. Light passes through the retina (in these drawings from bottom to top) to be absorbed by the outer segment (top) of the photoreceptor. There, it produces a signal that spreads to the terminal to influence the next cell in line. By recording electrically from each cell in the retinal circuit, we can follow signals step by step and understand how the meaning of the signals changes. (After Ramón y Cajal, 1995.)

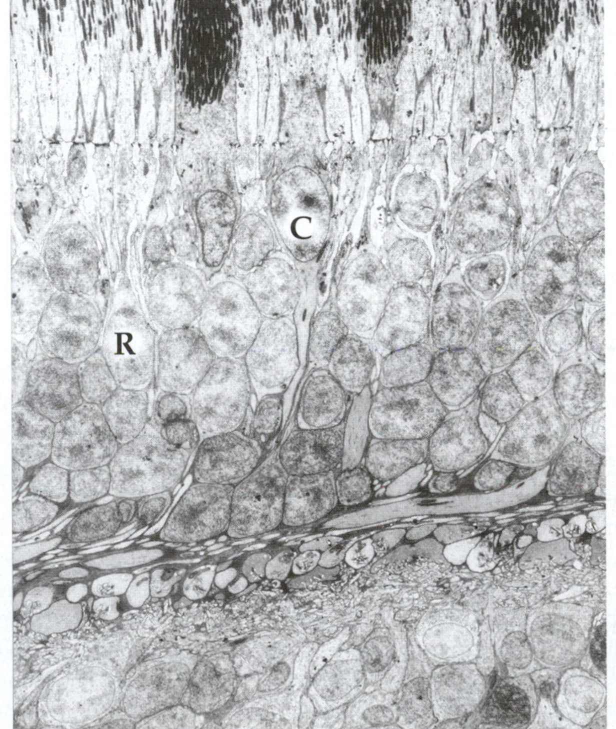

FIGURE 1.3 Dense Packing of Neurons in Monkey (Macaque) Retina. This electron micrograph shows a characteristic feature of the central nervous system in which the cell membranes are separated by narrow, fluid-filled clefts. The photoreceptors and their processes can be followed to the outer plexiform layer where their terminals contact bipolar and horizontal cells. One cone (C) and one rod (R) are labeled. (Micrograph kindly provided by P. Sterling and Y. Tsukamoto.)

Cell Body, Axons, and Dendrites

The ganglion cell shown in Figure 1.4 illustrates features shared by neurons throughout the nervous system. The **cell body** contains the nucleus and other intracellular organelles common to non-neuronal as well as neuronal cells. The long process that leaves the cell body to form connections with target cells is known as the **axon**. The term **dendrite** applies to branches upon which incoming fibers make connections and that act as receiving stations for excitation or inhibition. It will become apparent in later chapters that the terms for describing neuronal structures, particularly dendrites, are somewhat ambiguous but these terms are still convenient and widely used. In addition to the ganglion cell, Figure 1.4 shows other representative neurons.

Not all neurons conform to the simple plan of the cells shown in Figure 1.4. Certain nerve cells do not have an axon, while others have an axon onto which incoming connections are made. Still others have dendrites that can conduct impulses and transmit to target cells. While ganglion cells conform to the caricature of a stereotyped neuron with dendrites, a cell body, and an axon, others in the retina do not. For example, photoreceptors do not have an axon or obvious dendrites (see Figure 1.2C). Activity in photoreceptors does not arise through input from another neuron but from an external stimulus—light.

Techniques for Identifying Neurons and Tracing Their Connections

Although the staining technique devised by Golgi in 1873 is still widely used, many newer techniques have facilitated the functional identification of neurons and synaptic connections. Molecules that label a neuron in its entirety can be injected through a fine pipette. Fluores-

FIGURE 1.4 Shapes and Sizes of Neurons. Neurons have branches (called dendrites), on which other neurons form synapses, and axons that, in turn, make connections with other neurons. The motor neuron, drawn by the German neuroanatomist Otto Friedrich Karl Deiters in 1869, was dissected from a mammalian spinal cord. The other cells, stained by the Golgi method, were drawn by Ramón y Cajal. The pyramidal cell is from the cortex of a mouse, the mitral cell from the olfactory bulb (a relay station in the pathway concerned with smell) of a rat, the Purkinje cell from human cerebellum, and the ganglion cell from mammalian retina (animal not specified). (After Ramón y Cajal, 1995.)

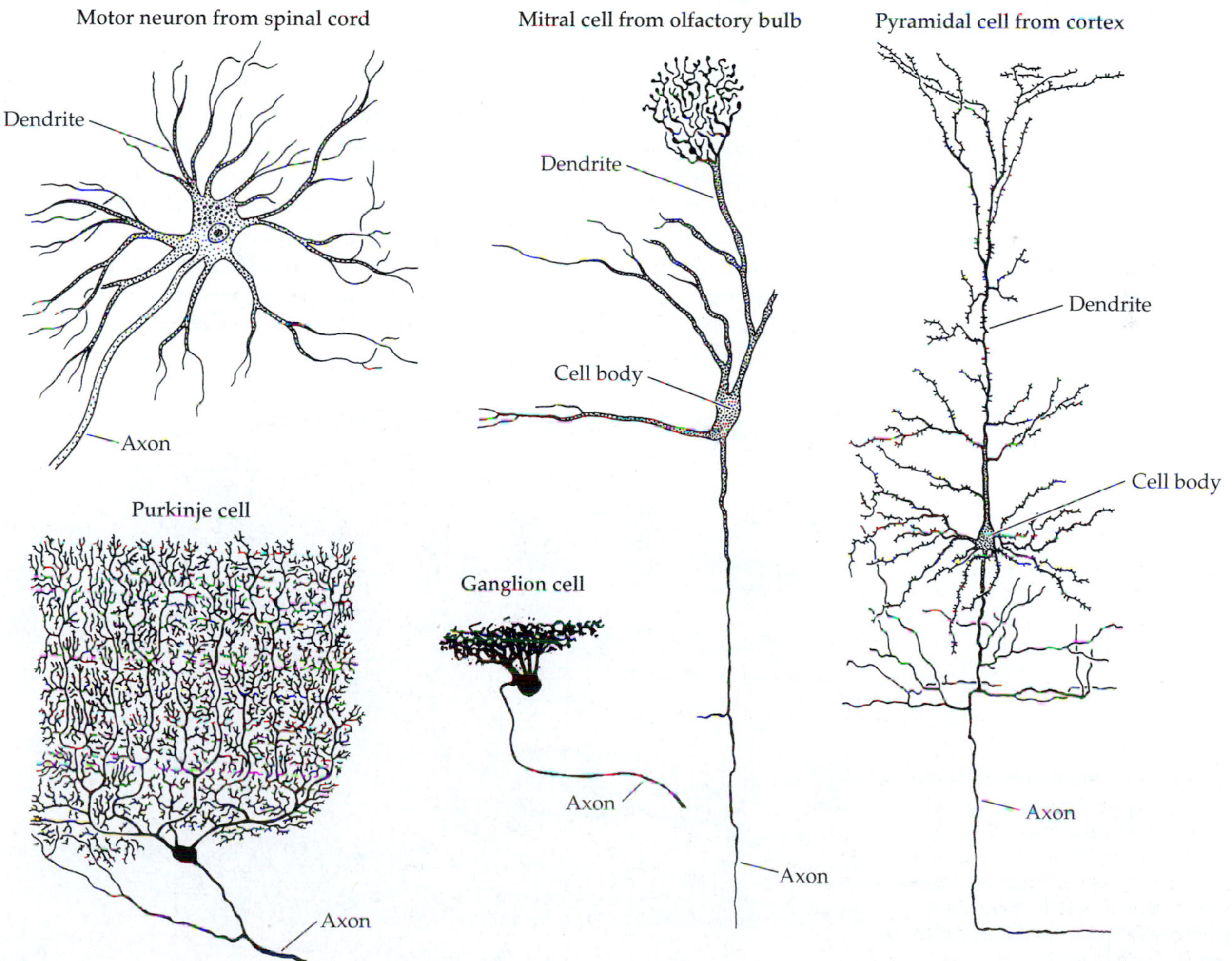

cent markers, such as Lucifer yellow, are visible as they spread through the fine processes of a living cell. Alternatively, other types of markers, such as the enzyme horseradish peroxidase (HRP) or biocytin, can be injected; after the tissue has been processed for histology, the marker in the cell appears as a dense product or bright fluorescence. Neurons can also be stained by extracellular application of horseradish peroxidase; the enzyme is taken up by axon terminals and transported to the cell body. Fluorescent carbocyanine dyes placed close to neurons dissolve in cell membranes and diffuse over the entire surface of the cell. These procedures are valuable for tracing the origins and destinations of axons from one region of the nervous system to another.

Antibodies have been made to characterize specific neurons, dendrites, axons, and synapses by selectively labeling intracellular or membrane components. Figure 1.5 shows a specific group of retinal cells (called bipolar cells because of their shape) that have been labeled by an antibody to the enzyme phosphokinase C. Antibody techniques also provide valuable tools for following the migration and differentiation of nerve cells during development. A complementary approach for characterizing neurons is by *in situ* hybridization. Specific tagged probes are used to label messenger RNA that codes for a channel, receptor, transmitter, or structural element of a neuron. In animals such as fruit flies, nematode worms, zebrafish, opossums and mice, specific cells, groups of cells, or the entire nervous system can be labeled by the introduction of marker genes, driven by appropriate control elements genes into the genome. Figure 1.6 shows living satellite glial cells (see below) in the cerebral cortex of an immature opossum. The cells were selectively labeled transgenically by introduction of a specific plasmid through the use of a technique known as electroporation: a weak current causes the gene to become incorporated into one particular type of dividing cell.[2] Alternatively, new strains of animals with stable transgenesis can be produced by introduction of genes into an oocyte, referred to as knock-out or knock-in animals.

Non-Neuronal Cells

The distinctive cell, labeled M in Figure 1.2B, represents a class of non-neuronal cells present in the retina. Such cells, known as glial cells (see Chapter 10), are abundant throughout the nervous system, and greatly outnumber neurons. Unlike neurons, they do not have axons or dendrites, and they have no specialized anatomical junctions with nerve cells. They play a number of roles in relation to neuronal signaling. For example, the axons of ganglion cells that run in the optic nerve conduct impulses rapidly because, once they leave the eye, the axons are surrounded by an insulating lipid sheath called **myelin.** Myelin is formed by glial cells that wrap themselves around axons during late stages of development. Retinal glial cells are known as Müller cells.

[2]Puzzolo, E., and Mallamaci, A. 2010. *Neural Development* 5:8.

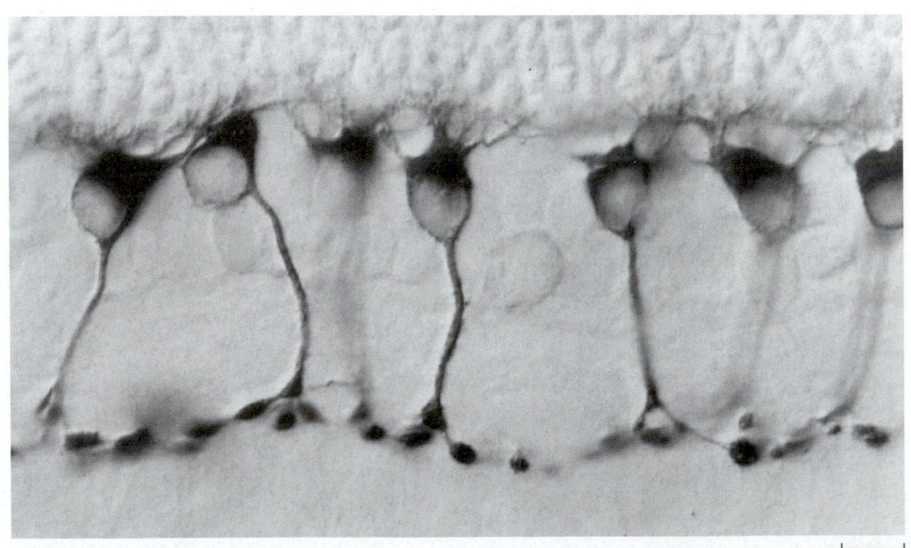

FIGURE 1.5 Population of Rod Bipolar Cells Stained by an Antibody against the Enzyme Phosphokinase C. Only bipolar cells that contain the enzyme are stained. Above are photoreceptors; below are ganglion cells. (Photograph kindly provided by H. M. Young and D. I. Vaney, University of Queensland.)

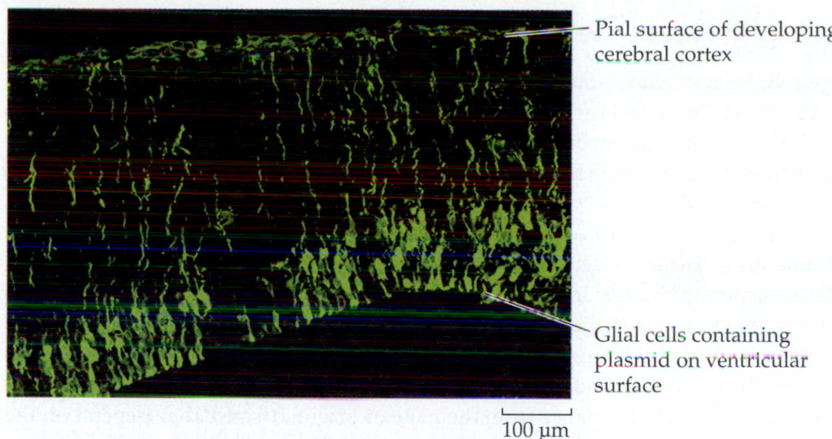

FIGURE 1.6 Glial Cells Selectively Labeled with Green Fluorescent Protein. The gene for the green fluorescent protein was incorporated through electroporation in the developing cortex of a neonatal opossum. To incorporate the gene, brief electrical currents are passed across the brain in the presence of plasmid DNA that has been injected into the ventricle. The enhanced green fluorescent protein gene (EGFP) was under the control of the tubulin a-1 (Ta1) promoter, which lights up glial cells, known as radial glia. They are analogous to the Müller glial cells in retina (see Figure 1.2). (Adapted from Puzzolo and Mallamaci, 2010, who kindly provided the photo).

Grouping of Cells According to Function

A remarkable feature of the retina is the clustering of cells according to function (see Figure 1.2). Photoreceptors and the horizontal, bipolar, amacrine, and ganglion cells all have their cell bodies and synapses situated in well-defined layers. Such layering is found in many regions of the brain. For example, the structure in which optic nerve fibers end (the lateral geniculate nucleus), consists of six readily distinguishable layers of cells. Throughout the nervous system, cells with similar functions are clustered in obvious circumscribed groups, known as nuclei (not to be confused with the cell nucleus) or ganglia (not to be confused with ganglion cells of the retina).

The simplified description of Figure 1.2 omits certain features of the organization of retinal cells. There are many distinctive types of ganglion cells, horizontal cells, bipolar cells, and **amacrine cells**, each with characteristic morphology, transmitters, and physiological properties. For example, the photoreceptors fall into two easily recognizable classes—the **rods** and the **cones**—which perform different functions. The elongated rods are extremely sensitive to small changes in illumination. As you read this page, the ambient light is too bright for the rods, which function only in dim light after a prolonged period in darkness. The cones respond to visual stimuli in bright ambient light. Moreover, the cones are further subdivided into red-, green-, or blue-sensitive photoreceptors. The amacrine cells provide an extreme example of cell type diversity: more than 20 types can be recognized by structural and physiological criteria. With few exceptions, the functional significance of this variety in cellular subtypes is not known. The properties and functions of these cells are described in Chapter 20.

Complexity of Connections

The arrows in Figure 1.2A indicate a through line of transmission from receptors to ganglion cells. Light falls on photoreceptors and generates electrical signals, which then influence bipolar cells. From bipolar cells, signals are conveyed to ganglion cells and then to higher centers in the brain that give rise to our perception of the outside world.

In reality, the picture is again far more complex. For example, there is a dramatic reduction in numbers from receptors to ganglion cells. More than 100 million photoreceptors provide input to just 1 million ganglion cells, by way of interposed cells. Each individual ganglion cell therefore receives inputs from many photoreceptors (convergence). Similarly, when the axon of a single ganglion cell reaches the next relay station in the lateral geniculate nucleus, it branches extensively to supply many target cells (divergence, see Figure 1.14).

In addition, other arrows could be inserted into Figure 1.2A that point sideways to indicate interactions among cells in the same layer (lateral connections) and even in the reverse direction—for example, from horizontal cells back toward photoreceptors (recurrent connections). Such convergent, divergent, lateral, and recurrent connections are consistent features of pathways elsewhere in the nervous system. Thus, the simple step-by-step processing of signals is dramatically influenced by parallel and feedback interactions.

Signaling in Nerve Cells

All nerve cells have a **resting potential**: the inside of the cell is negative with respect to the outside (extracellular) solution by a little less than a tenth of a volt. All electrical signals generated in nerve cells are superimposed on the resting potential. Some signals depolarize the cell membrane, making the inside less negative; others hyperpolarize it, making it more negative.

Electrical signals of nerve cells fall into two main classes. The first consists of **local graded potentials**. These are generated by extrinsic physical stimuli, such as light falling on a photoreceptor in the eye, sound waves deforming a hair cell in the ear, or a pressure of an object against a sensory nerve ending in the skin, or by activity at synapses (i.e. the junctions between nerve cells and their targets). Synapses are discussed later in the chapter. Local potentials vary in amplitude, depending on the strength of the activating signal. They usually spread only a short distance from their site of origin, because they depend on the passive electrical properties of the nerve cells.

Action potentials (also referred to as nerve impulses), constitute the second major category of electrical signals. Action potentials are initiated when localized graded potentials are sufficiently large to depolarize the cell membrane beyond a critical level (called the threshold). Once initiated, action potentials can propagate rapidly over long distances—for example, along the axons of ganglion cells in the optic nerve from the eye to the lateral geniculate nucleus on the way to the cortex, or from a motor cell in the spinal cord to a muscle in the leg. Unlike local graded potentials, action potentials are fixed in amplitude and duration, like the dots in a code. Signal transmission through the retina can be summarized by the following simplified scheme:

> Light → local graded signal in photoreceptor → local graded signaling bipolar cell → local graded potential in ganglion cell → action potential in ganglion cell → conduction to higher centers

Universality of Electrical Signals

An important feature of electrical signals is that they are virtually identical in all nerve cells of the body, whether they: (1) carry commands for movement, (2) transmit messages about colors, shapes, or painful stimuli, or (3) interconnect various portions of the brain. A second important feature of signals is that they are so similar in different animals that even a sophisticated investigator is unable to tell with certainty whether a record of an action potential is derived from the nerve fiber of whale, mouse, monkey, worm, tarantula, or professor. In this sense, action potentials can be considered to be stereotyped units. They are the universal coins for the exchange of information in all nervous systems that have been investigated. In the human brain, it is the great number of cells (probably 10^{10} to 10^{12} neurons) and the diversity of connections, rather than the types of signals, that accounts for the complexity of the tasks that can be carried out.

This idea was expressed in 1868 by the German physicist–biologist Hermann von Helmholtz. Starting from first principles, long before the facts as we know them were available, Helmholtz reasoned that:

> The nerve fibers have often been compared with telegraphic wires traversing a country, and the comparison is well fitted to illustrate the striking and important peculiarity of their mode of action. In the network of telegraphs we find everywhere the same copper or iron wires carrying the same kind of movement, a stream of electricity, but producing the most different results in the various stations according to the auxiliary apparatus with which they are connected. At one station the effect is the ringing of a bell, at another a signal is moved, at a third a recording instrument is set to work...In short, every one of the...different actions which electricity is capable of producing may be called forth by a telegraphic wire laid to whatever spot we please, and it is always the same process in the wire itself which leads to these diverse consequences...All the difference which is seen in the excitation of different nerves depends only upon the difference of the organs to which the nerve is united and to which it transmits the state of excitation.[3]

[3] Helmholtz, H. von. 1889. *Popular Scientific Lectures.* Longmans, London.

Techniques for Recording Signals from Neurons with Electrodes

For certain problems, it is useful to record the electrical activity from a single neuron or even a single channel in the cell membrane of a neuron; for other problems, one needs to survey the activity in many neurons, all at the same time. The following description summarizes briefly key techniques for recording neuronal activity discussed throughout this book.

Recordings of action potentials were first made in the early twentieth century with fine silver wires placed on the trunk of a peripheral nerve. Currents flowing between one pair of electrodes were used to stimulate nerve fibers, while a second pair of electrodes placed some distance away recorded an extracellular voltage signal produced by the resulting nerve cell activity. The voltage signal reflected the summed action potentials of many fibers (a compound action potential). Within the central nervous system, recordings from a neuron can be made with an **extracellular electrode**, which consists of a single wire insulated down to an exposed conducting tip or of a glass capillary filled with salt solution (Figure 1.7A). One can also record from many individual neurons simultaneously (on the order of 100) by positioning an array of electrodes in a particular area of the central nervous system.[4]

Intracellular microelectrodes can be used to measure the resting membrane potential of neurons and to record local potentials and action potentials generated in the cell membrane. Such microelectrodes are usually made of glass drawn to a fine-diameter tip (0.1 μm or smaller) and filled with a salt solution. The electrode is inserted through the cell membrane with the aid of a micromanipulator (Figure 1.7B). The cell membrane forms a tight seal around the glass so that the integrity of the cell is not disturbed. Microelectrodes are also used for passing electrical currents across the cell membrane or for injecting molecules into the cytoplasm.

An alternative is to measure the membrane potential by a procedure known as **whole-cell patch recording** (Figure 1.7C). A larger pipette with a polished tip of approximately 1 μm is applied to the surface of the cell, where it fuses with the membrane to form a tight seal. After the membrane within the pipette tip has been ruptured, the fluid in the pipette and the intracellular fluid become a single electrical compartment.

[4]Harris, J. A., Petersen, R. S., and Diamond, M. E. 1999. *Proc. Natl. Acad. Sci. USA* 96: 7587–7591.

Noninvasive Techniques for Recording and Stimulating Neuronal Activity

By the use of optical recording techniques, we can follow signaling in suitable preparations. Specially fabricated dyes that bind to the cell membrane change their light absorbance during changes in membrane potential. Other dyes measure changes in the level of

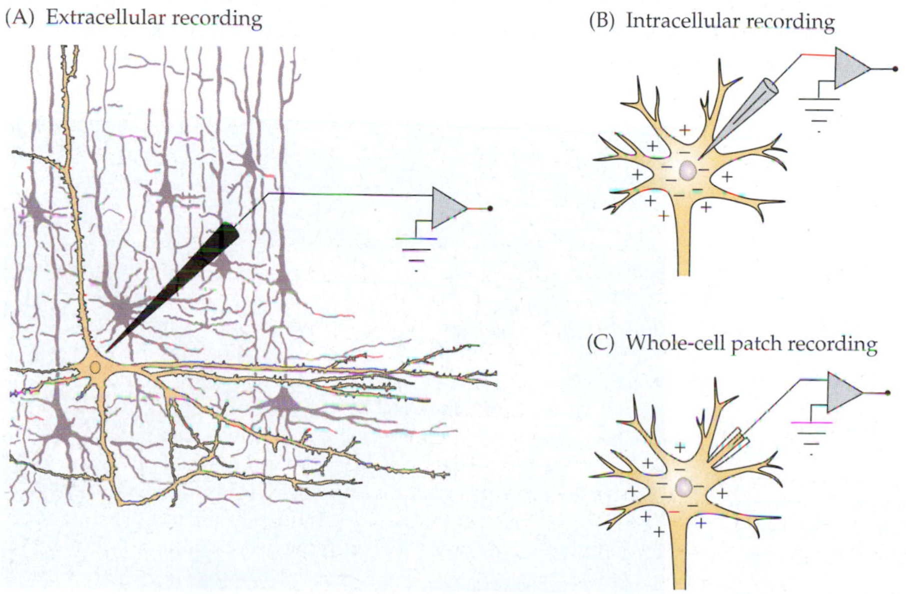

(A) Extracellular recording

(B) Intracellular recording

(C) Whole-cell patch recording

FIGURE 1.7 Electrical Recording Techniques. (A) The tip of a fine wire electrode is located close to a nerve cell in the cortex. (The wire above the tip is insulated.) Extracellular recording allows one to record from a single cell or from a group of cells. (B) Intracellular recordings are made with a fluid-filled glass capillary that has a tip of less than 1 μm in diameter, which is inserted into a neuron across the cell membrane. At rest, there is a potential difference of about 70 mV between the inside and outside of a cell, the inside being negative with respect to the outside. This difference is known as the resting potential. (C) Intracellular recordings are also made with patch electrodes. A patch electrode has a larger tip than that of an intracellular microelectrode; the tip makes an extremely tight seal with the cell membrane. If the seal is intact, the currents that flow as a single ion channel in the membrane opens or closes can be recorded. Alternatively, as shown here, the cell membrane can be ruptured to allow the diffusion of molecules between the pipette and the intracellular fluid of the cell (called whole-cell patch clamp).

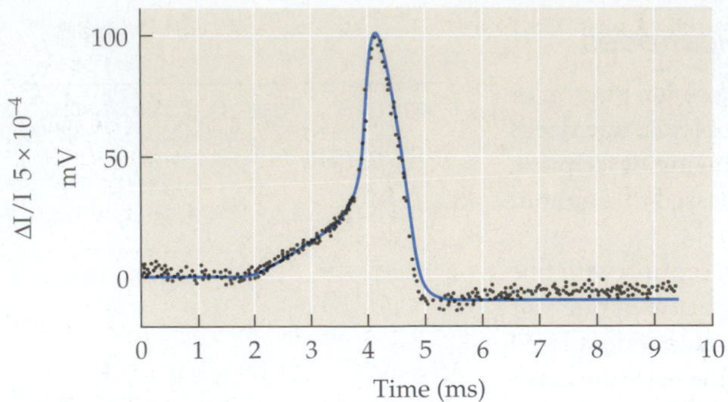

FIGURE 1.8 Comparison of Optical and Electrical Recordings. Faithful optical recording of an action potential (dots) from a squid giant axon compared to an electrical recording (solid line) with electrode. $\Delta I/I$ is a measure of the change in absorption by the voltage sensitive merocyanine dye. (After Homma et al., 2009.)

calcium inside nerve cells.[5] Changes in light emission by dyes that are fluorescent provide an index of activity occurring in a single neuron or in populations of neurons. A recording made from a large squid axon by a voltage-sensitive dye is shown in Figure 1.8. The optically recorded signal faithfully reproduces the voltage changes recorded by an electrode. With suitable dyes, microscopes and cameras, one can record the ongoing activity of hundreds of nerve cells simultaneously within a discrete region of the brain. By genetic engineering, it is now possible to introduce genes that will give rise to optical signals in active nerve cells without using dyes. Recordings that depend on changes in emission of light by active cells are most simply made in isolated tissues maintained alive in a culture dish or in intact CNS near the surface. Optical signals generated by deeply embedded cells are difficult to observe through overlying layers of tissue, unless one uses a laser technique, known as 2-photon microscopy.

Other noninvasive techniques are able to provide an indirect measure of activity in groups of neurons deep in the brain as well as in superficial neuronal layers. These techniques, known as positron emission tomography (PET) and functional magnetic resonance imaging (fMRI),[6] allow us to determine which regions of the awake, human brain are active when stimuli are presented, thoughts are pursued, or movements are initiated. They are particularly important in mapping brain regions involved in complex cognitive functions, such as reading, remembering, or imagining. The methods now available have a spatial resolution limited to hundreds of microns and seconds. Hence, present day imaging methods of the human brain cannot provide information about the exact cellular location or timing of messages that are being carried by active neurons. It is realistic to hope that this will be possible in the future since, year-by-year, the temporal and spatial resolution of imaging improve. The fMRI image in Figure 1.9 shows the location of activity moving from the eye to the cortex in response to a visual stimulus.

The overall activity of the eye and brain can also be observed in the electroretinogram and electroencephalogram. These techniques are whole-organ electrical signals detected with electrodes placed on the surface of the body. They have poor spatial resolution and are used mainly to diagnose disorders of function, such as epilepsy.[7]

Stimulation of an area in the brain or of a peripheral nerve is now practicable by a noninvasive technique, called **transcranial magnetic stimulation (TMS)**, which can be used in conscious patients or in animals. Stimulation of appropriate regions can cause the subject to see lights or induce movements. Strong magnetic fields are applied to the surface of the skull by means of a coil.[8] One can, for example, use fMRI to locate a discrete cortical region that becomes activated when the person perceives the direction of movement of dots on a computer monitor. When the same region is excited by transcranial magnetic stimula-

[5]Homma, Y. et al. 2009. *Philos. Trans. R. Soc. Lond. B Biol. Sci.* 364: 2453–2467.

[6]Ward, N. S., and Frackowiak, R. S. 2004. *Cerebrovasc. Dis.* 17(Suppl. 3): 35–38.

[7]Bandettini, P.A. 2009. *Ann. NY Acad. Sci.* 1156: 260–293.

[8]Bestmann, S. et al. 2008. *Exp. Brain. Res.* 191: 383–402.

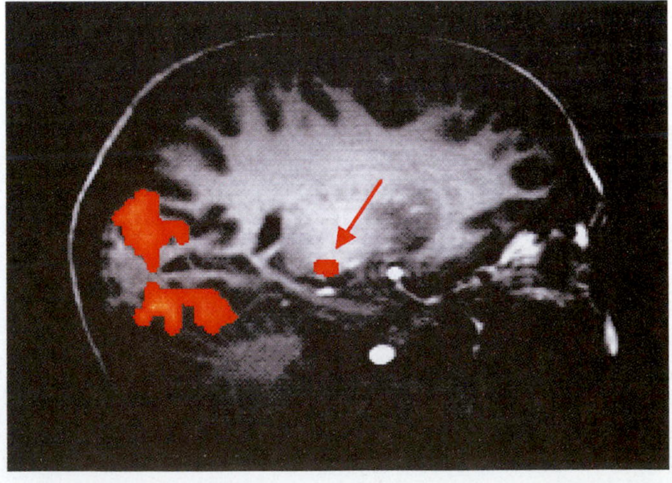

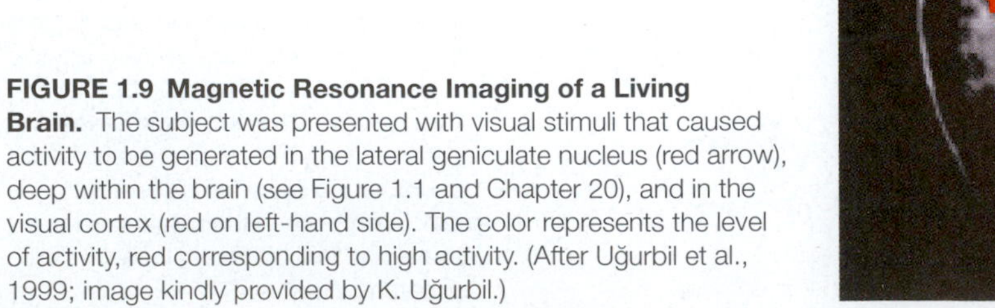

FIGURE 1.9 Magnetic Resonance Imaging of a Living Brain. The subject was presented with visual stimuli that caused activity to be generated in the lateral geniculate nucleus (red arrow), deep within the brain (see Figure 1.1 and Chapter 20), and in the visual cortex (red on left-hand side). The color represents the level of activity, red corresponding to high activity. (After Uğurbil et al., 1999; image kindly provided by K. Uğurbil.)

tion, the subject has difficulty in perceiving motion. In this way, the combination of imaging and TMS can be used to identify a brain area that is potentially involved in a perceptual task—and then to prove the point by silencing that specific brain area and simultaneously disrupting perception.

Spread of Local Graded Potentials and Passive Electrical Properties of Neurons

Implicit in the wiring diagram of Ramón y Cajal (see Figure 1.2A) is the idea that changes in illumination of the retina influence the activity of the photoreceptors and eventually the fibers leaving the eye. For this influence to take effect, signals must spread not only from cell to cell, but along a cell, from one end to the other. How, for example, does the electrical signal generated at one end of the bipolar cell (in contact with the photoreceptor), spread along its length to reach the terminal that is close to the ganglion cell?

To answer this question, it is useful to consider the relevant structural components that carry the signals. A neuron, such as a bipolar cell, can be considered to be a long tube filled with a watery solution of salts (dissociated into positively and negatively charged ions) and proteins, separated from the extracellular solution by an insulating membrane. The intracellular and extracellular solutions have the same osmolarity but different ionic compositions. The cell membrane, a lipid, is relatively impermeable to the ions on either side, but ions can move through specific ion channels formed by proteins that span the membrane. Electrical and chemical signals cause the various ion channels for sodium, potassium, calcium, and chloride to open or close. Detailed information about the molecular structure of ion channels, and the way in which they allow the flow of ions, is presented in Chapters 4 and 5.

The structure of a neuron in the retina or elsewhere limits its ability to conduct electrical signals. First, the intracellular fluid, or axoplasm, is about 10^7 times worse than a metal wire as a conductor of electricity because both the density of charge carriers (ions) and their mobility in the intracellular fluid are much lower than those of free electrons in a wire. Second, the movement of currents along the axon for any great distance is hampered by the fact that the membrane is not a perfect insulator. Consequently, any current flowing along the fiber is gradually lost to the outside by leakage through ion channels in the membrane.

Passively conducted electrical signals, then, are severely attenuated and limited to a short length of nerve fiber, 1 to 2 mm at most. For example, a local potential generated in a small sensory fiber in the big toe will spread only about one thousandth the distance that must be traversed for the signal to reach the spinal cord. In addition, when such a signal is brief, its time course may be severely distorted and its amplitude further attenuated by the electrical capacitance of the cell membrane. Nevertheless, graded localized potentials (Figure 1.10) provide the essential mechanisms for

FIGURE 1.10 Localized Graded Potentials. Intracellular recordings are made from (A) a bipolar cell and (B) a ganglion cell with microelectrodes. (A) When light is absorbed by the photoreceptors, it gives rise to a signal that in turn produces a localized, graded response in the bipolar cell. The resting potential across the membrane is reduced (the trace moves in an upward direction)—an effect known as a depolarization. The size of the signal in the bipolar cell depends on the intensity of illumination, hence the term graded. The depolarization spreads to the far end of the bipolar cell passively. As it spreads, it becomes smaller in amplitude owing to the poor conducting properties of neurons. At the terminal of the bipolar cell, depolarization causes the release of chemical transmitter. (B) The transmitter produces a local graded potential in the ganglion cell. Because it is localized, the potential cannot spread for more than 1 mm (at most) along the axon. Whereas the bipolar cell is short enough for a local potential to spread to its endings, the ganglion cell has an axon several centimeters long. In these illustrations, the local potentials were recorded from the cell bodies and were produced by transmitters acting on the dendrites. (A after Kaneko and Hashimoto, 1969; B after Baylor and Fettiplace, 1977.)

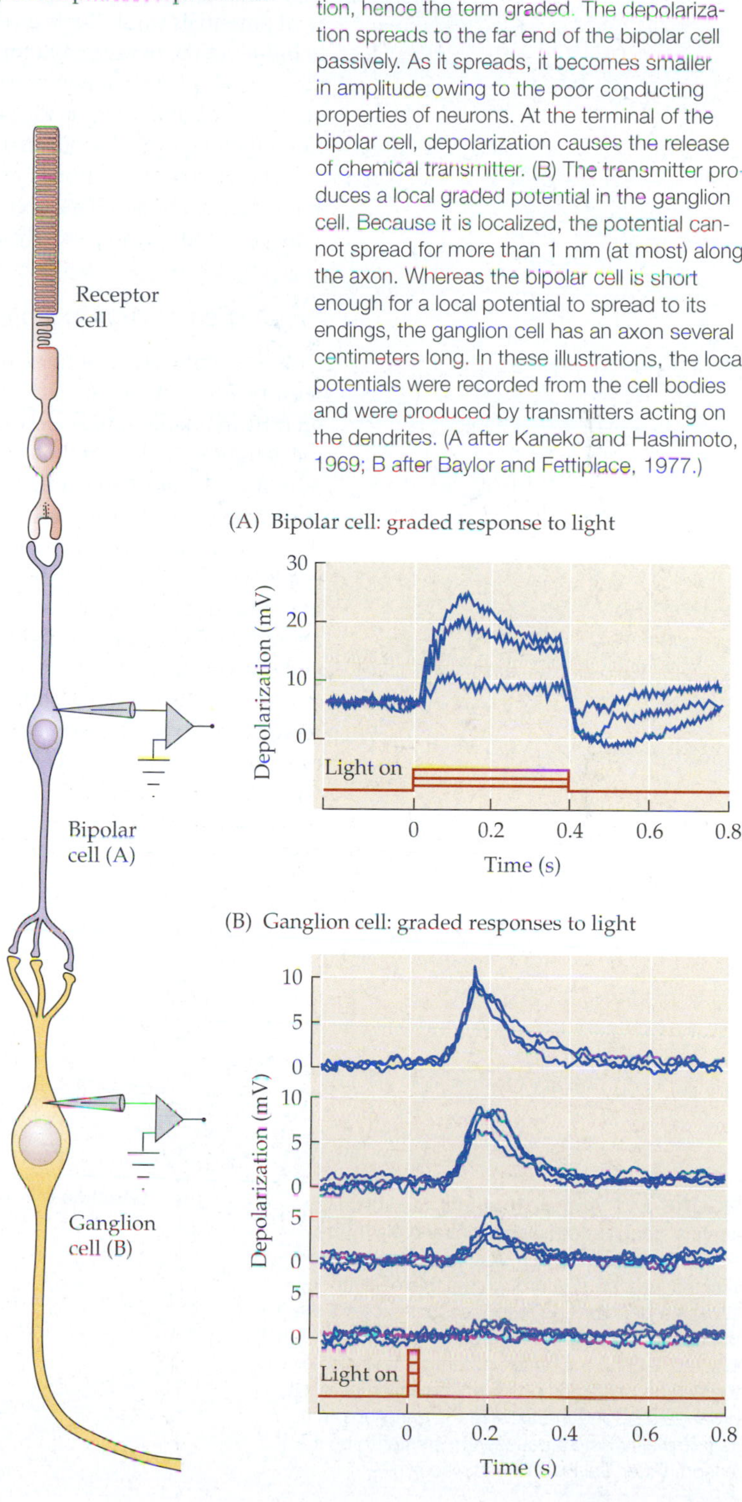

initiating propagated signals. The fact that nerve fibers are extremely small (between 1 and 20 μm in diameter in vertebrates) further reduces the amount of current they can carry.[9]

[9]Hodgkin, A. L. 1964. *The Conduction of the Nervous Impulse*. Liverpool University Press, Liverpool, England.

Spread of Potential Changes in Photoreceptors and Bipolar Cells

It is only because photoreceptor and bipolar cells are so short that local, graded signals can spread effectively from one end of the cell to the other. The electrical signal that results from illumination of the photoreceptor is generated in the outer segment of the rod or the cone. From there, it spreads passively along the cell to its terminal on the bipolar cell. If photoreceptor and bipolar cells were longer, centimeters or even millimeters in length, local potentials would fizzle out long before reaching the terminal and would not be able to influence the next cell. Photoreceptor and bipolar cells thereby constitute exceptions to the general rule that action potentials are necessary to carry information along the length of a cell. Ganglion cells, on the other hand, must generate action potentials to send signals along their elongated axons in the optic nerve.

The electrical recordings made from the neurons shown in Figure 1.10 were made from their cell bodies. The local potentials originate from synaptic actions on dendrites at a distance and spread passively to the recording site.

Properties of Action Potentials

One essential feature of the action potential is that it is a triggered, regenerative, all-or-nothing event. An action potential is initiated in a ganglion cell by the signals impinging on it from bipolar and amacrine cells, provided these signals are sufficient to depolarize the ganglion cell to threshold. Once initiated, the amplitude and duration of the action potential are not determined by the amplitude and duration of the stimulus. Larger stimulating currents do not give rise to larger action potentials, and stimuli of longer duration do not prolong the action potential. Figure 1.11shows that the action potential is a brief electrical pulse about 0.1 V in amplitude. At its peak, the voltage across the membrane reverses sign (i.e., the inside becomes positive). The action potential lasts for about 0.001 seconds (1 millisecond) and moves rapidly along the nerve fiber from one end to the other.

The entire action potential sequence must be completed before another action potential can be initiated at the same site. Thus, after each action potential, there is a period of enforced silence, usually lasting for a few milliseconds (the refractory period), during which a second impulse cannot be initiated. The maximal possible frequency of repeated action potentials is therefore limited by the refractory period.

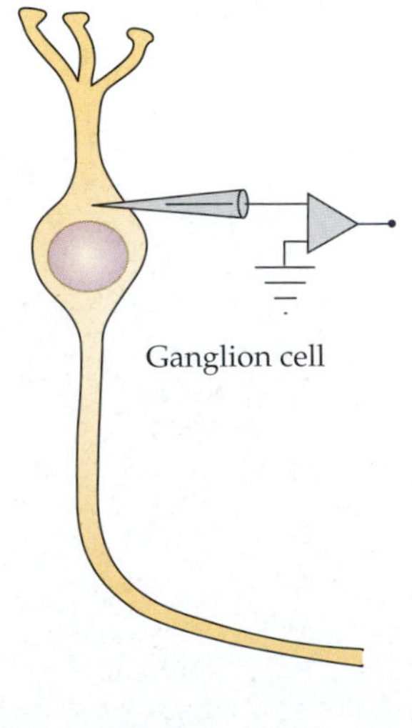

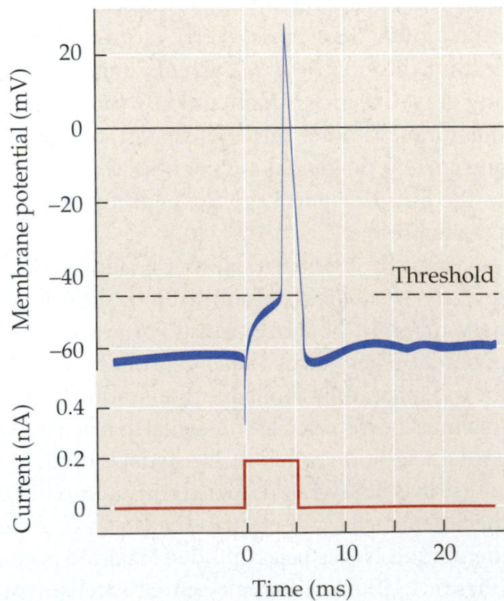

FIGURE 1.11 Action Potential recorded from a retinal ganglion cell with an intracellular microelectrode. When the stimulus, in this case current injected into the cell through the microelectrode, causes a depolarizing response that exceeds the threshold, the all-or-nothing action potential is initiated. During the action potential, the inside of the neuron becomes positive. The action potential propagates along the axon of the ganglion cell to its terminal, where it causes transmitter to be released. (After Baylor and Fettiplace, 1977.)

Propagation of Action Potentials along Nerve Fibers

The action potential itself causes electrical currents to spread passively ahead of itself along the axon. Although the resulting depolarization falls off steeply with distance, it nevertheless exceeds threshold. Hence, the action potential provides an electrical stimulus to the adjacent region of the axon. In this way, the impulse is reborn unchanged or "regenerated" as it propagates along the axon. The fastest action potentials in the human body travel in the largest fibers at a speed of about 120 meters/second (430 kilometers/hour or 270 miles/hour). They are therefore capable of conveying information rapidly over a long distance—for example it takes about 10 milliseconds for a motor fiber to conduct its action potentials over one meter from the spinal cord to a toe.

Action Potentials as the Neural Code

Given that the action potential is fixed in amplitude, how is information about the intensity of a stimulus conveyed? Intensity is coded by frequency of firing. A more effective visual stimulus produces a greater local potential and, as a consequence, a higher frequency of firing in ganglion cells (Figure 1.12). The phenomenon was first described by Adrian,[10] who showed that the frequency of action potential firing in a sensory nerve in the skin is a measure of the intensity of the stimulus. In addition, Adrian observed that stronger stimuli applied to the skin give rise to activity in a larger number of sensory fibers.

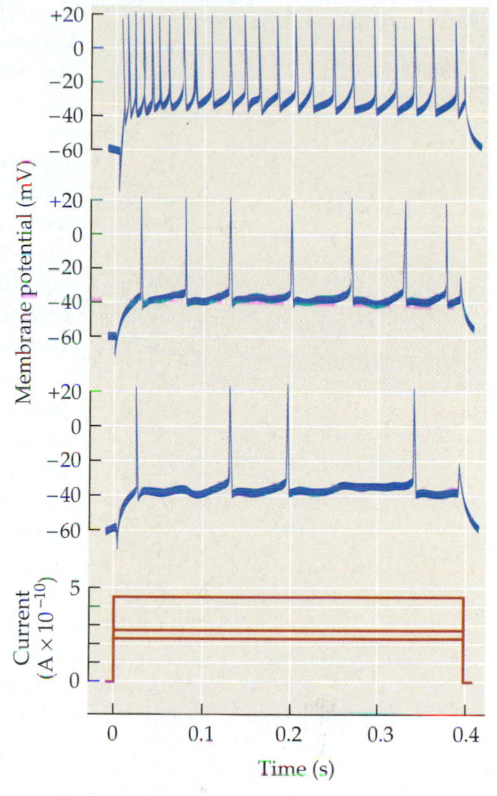

FIGURE 1.12 Frequency as a Signal of Intensity in a Retinal Ganglion Cell. Depolarizing current passed through the microelectrode produces local potentials. Larger currents produce larger local potentials and higher frequencies of firing. (After Baylor and Fettiplace, 1979.)

Synapses: The Sites for Cell-to-Cell Communication

The structure at which one cell hands its information to the next is known as a **synapse**. Through synaptic interactions, neurons such as ganglion cells take account of signals that arise from many photoreceptors and have fed onto horizontal, bipolar, and amacrine cells, thereby creating new messages. Study of the processes underlying synaptic transmission constitutes a major theme in modern neurobiology because these mechanisms are responsible for integration and plasticity and are the targets of many therapeutic drugs.

Chemically Mediated Synaptic Transmission

Figure 1.13 shows the highly organized structure at which a photoreceptor makes synaptic connections onto a bipolar cell. The **presynaptic terminal** of the photoreceptor is separated

[10] Adrian, E. D. 1946. *The Physical Background of Perception*. Clarendon, Oxford, England.

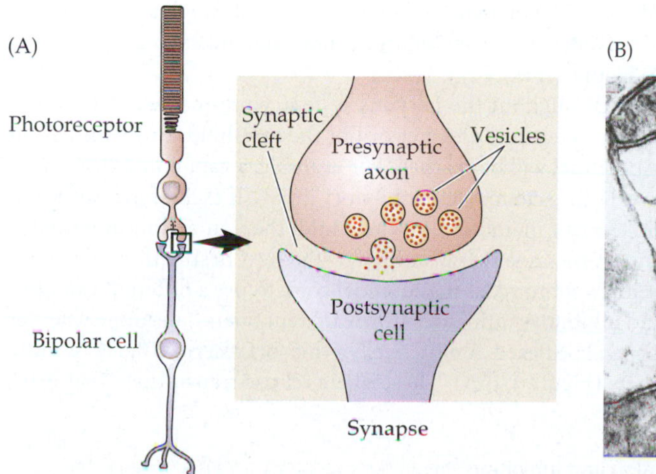

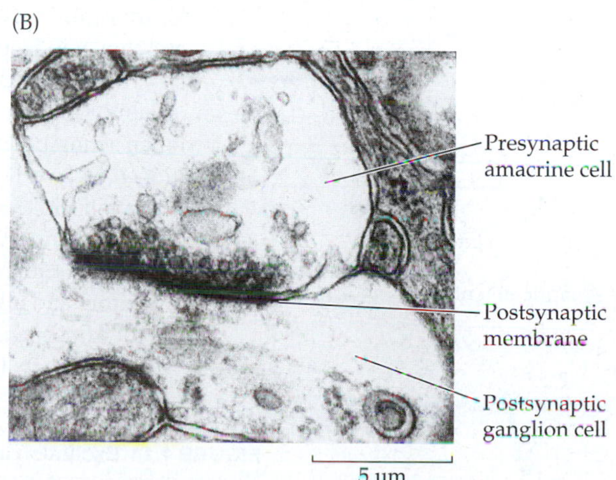

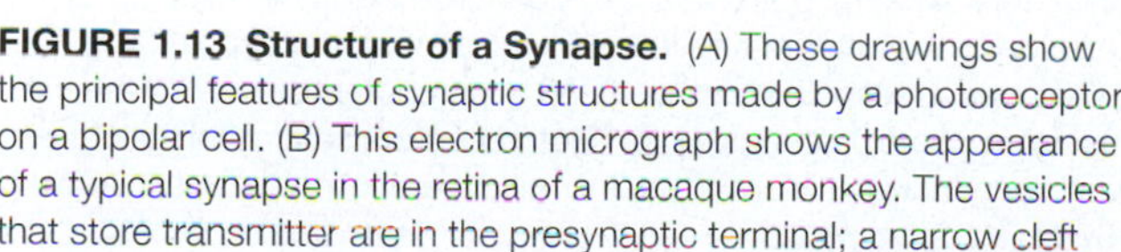

FIGURE 1.13 Structure of a Synapse. (A) These drawings show the principal features of synaptic structures made by a photoreceptor on a bipolar cell. (B) This electron micrograph shows the appearance of a typical synapse in the retina of a macaque monkey. The vesicles that store transmitter are in the presynaptic terminal; a narrow cleft separates the pre- and postsynaptic membranes. The postsynaptic membrane is densely stained. Transmitter released from the presynaptic amacrine cell diffuses across the cleft to interact with receptors on the postsynaptic ganglion cell. (Micrograph kindly provided by Y. Tsukamoto and P. Sterling.)

[11]Katz, B. 1971. *Science* 173: 123–126.

[12]Fillenz, M. 2005. *Neurosci. Biobehav. Rev.* 29: 949–962.

from the bipolar cells by a cleft that contains extracellular fluid. This space is cannot be traversed by currents generated in the photoreceptor. Instead, the photoreceptor terminal releases a neurotransmitter that is stored in presynaptic vesicles. The transmitter, the amino acid glutamate in this case, diffuses across the cleft to interact with specific protein molecules, known as receptors, which are embedded in the membranes of the postsynaptic bipolar cells. (This terminology can—but should not—lead to confusion: The term **receptor**, as used here, means a **chemoreceptor molecule** and is not the same concept as that of a **sensory receptor** cell that responds to external physical stimuli, such as a photoreceptor.) The transmitters in a neuron and the receptors on its surface can be identified and visualized by a variety of techniques, including antibody labeling and genetic transfection of so-called reporter constructs, which lead to the production of fluorescent proteins in cells expressing certain transcription factors.

Activation of the receptor molecules in a bipolar cell by glutamate sets up local graded potentials that spread to its terminals. The more transmitter released, the higher the concentration in the cleft, the larger the number of activated receptors, and the larger the local potential. Such events occur rapidly, with a delay of only about 1 millisecond between the arrival of the depolarizing signal in the presynaptic terminal and the appearance of the synaptic potential in the target. Essential features of synaptic transmission were first revealed by Katz, Kuffler, and their colleagues.[11] These investigators used the responses of receptors in muscle as a bioassay with extremely high sensitivity and time resolution for measuring transmitter release. Modern methods for measuring the transmitter directly include techniques such as voltammetry in which special carbon-coated probes produce graded electrical signals in response to a particular transmitter (such as serotonin), as it is released from a presynaptic terminal.[12]

Excitation and Inhibition

A feature of synaptic transmission, as exemplified by the interactions between photoreceptors and bipolar cells in the retina, is that the transmitter released by a presynaptic terminal can excite or inhibit the next cell, depending on the receptors that cell possesses. For example, one class of glutamate receptors localized to certain bipolar cells reacts to glutamate to cause an excitatory signal (i.e., depolarization of the cell membrane). This signal spreads passively to the bipolar cell terminals at the other end of the cell, where it causes liberation of transmitter. Other classes of bipolar cells contain different glutamate receptors that produce a signal of the opposite sign when activated by glutamate (i.e., hyperpolarization of the membrane). Again, the electrical signal spreads along the bipolar cell, but in this case, it suppresses the release of transmitter. **Excitatory and inhibitory synaptic potentials** in ganglion cells are shown in Figure 1.14.

In neurons throughout the nervous system, combined excitatory and inhibitory inputs determine whether or not the threshold for the initiation of an action potential will be reached. For example, a ganglion cell, as mentioned, receives both excitatory and inhibitory inputs. If excitation is sufficient to depolarize the cell membrane to threshold, then an action potential is generated and its message is transmitted to the next destination; if not, no message is sent. In motor cells in the spinal cord, to use a different example, excitatory and inhibitory influences from different fibers determine whether or not a finger will be flexed. A motor cell of this sort receives 10,000 or more incoming fibers (Figure 1.15A). These fibers release transmitters that drive

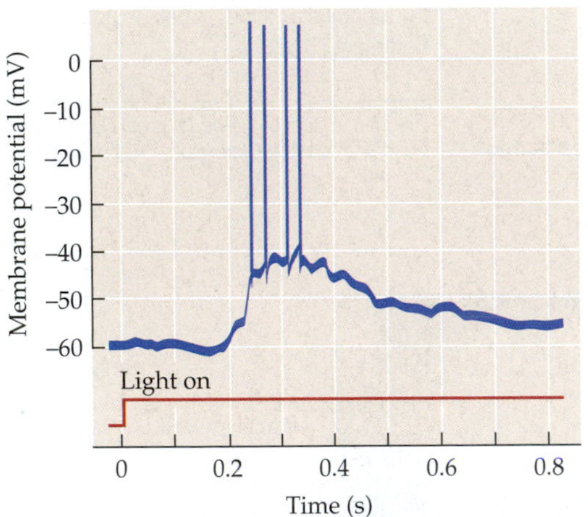

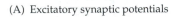

(B) Inhibitory synaptic potential

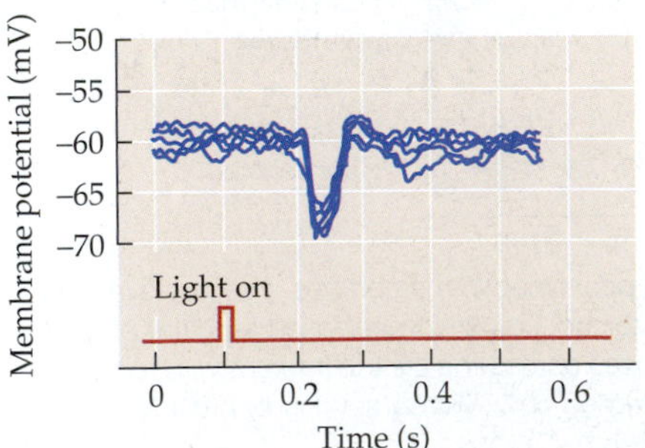

FIGURE 1.14 Excitation and Inhibition. Intracellular recordings from a ganglion cell showing excitatory and inhibitory synaptic potentials. (A) A ganglion cell is depolarized by continuous release of excitatory transmitter during retinal illumination. If the depolarizing synaptic potential is large enough, threshold is crossed and action potentials are initiated in the ganglion cell. (B) Illumination of a different group of photoreceptors causes inhibition. Hyperpolarization of the membrane makes it more difficult to initiate an action potential. (After Baylor and Fettiplace, 1979.)

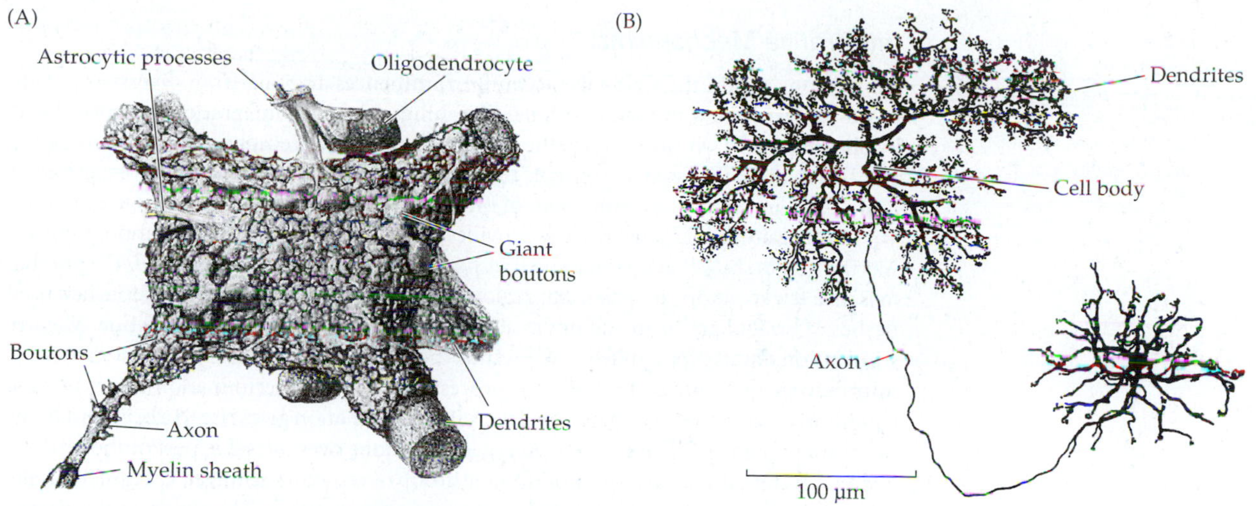

FIGURE 1.15 Multiple Connections of Individual Neurons. (A) Approximately 10,000 presynaptic axons converge to form endings that are distributed over the surface of a motor neuron in the spinal cord. The drawing is based on a reconstruction made from electron micrographs. (B) This drawing shows the divergence of the axon of a single horizontal cell that branches extensively to supply many postsynaptic target cells. (A from Poritsky, 1969; B after Fisher and Boycott, 1974.)

the membrane potential toward or away from the threshold for impulse initiation. An individual Purkinje cell in the cerebellum receives more than 100,000 inputs.

Electrical Transmission

Although synaptic transmission between most neurons involves the release of transmitter molecules, the membranes of many cells in the retina and in the rest of the nervous system are instead linked by specialized junctions. At such synapses electrical transmission occurs. The pre- and postsynaptic membranes are closely apposed and linked by channels that connect the intracellular fluids of the two cells. This close connection allows local electrical potentials and even action potentials to spread directly from cell to cell without a chemical transmitter and without delay. Metabolites and dyes can also spread from cell to cell. One important example in the retina is provided by the horizontal cells, which are electrically coupled in this way. By virtue of this property, graded depolarizing or hyperpolarizing potentials can spread from one horizontal cell to the next, with marked effect on the processing of visual information in the retina. Electrical synapses are found throughout the CNS of vertebrates and invertebrates. They also connect non-neuronal cells of other tissues in the body.

Modulation of Synaptic Efficacy

Chemically mediated synaptic transmission shows great plasticity. Dramatic changes occur in the amount of transmitter that is released by a signal—such as an action potential or a local potential—that invades a presynaptic terminal. The photoreceptors in the retina provide an example: The amount of the transmitter glutamate that is released by a rod or a cone in response to a standard light stimulus can be increased or decreased by feedback to the terminal from horizontal cells. The horizontal cells themselves are influenced by other photoreceptors. This feedback loop plays a critical role in the way the eye adapts to different levels of illumination.

Other mechanisms that influence transmitter release depend on the history of impulse activity. During and after a train of impulses in a neuron, the amount of transmitter it releases can increase or decrease dramatically, depending on the frequency and duration of the activity. Modulation of efficacy can also be postsynaptic in origin. Long- and short-term plasticity are the focus of intense contemporary research.

Integrative Mechanisms

All neurons within the CNS take account of influences arriving from diverse inputs to create their own new messages with new meanings. The term integration was introduced by Sherrington[13] (who also coined the words "synapse" and "receptive field"). Sherrington revealed many of the essential concepts that permeate modern neurobiology by experiments in which he measured the contractions of muscles, before electrical recordings were possible.

[13] Sherrington, C. S. 1906. *The Integrative Action of the Nervous System*. Reprint, Yale University Press, New Haven, CT, 1961.

[14] Kuffler, S. W. 1953. *J. Neurophysiol.* 16: 37–68.

Once again, retinal ganglion cells provide an excellent example of integration. Kuffler[14] was the first to show that a ganglion cell responds best to a small light spot or dark spot that falls on a few receptors in a particular region of the retina. Previously investigators had used flashes of bright light in an attempt to achieve maximal stimulation of the retina. Massive stimulation of this type applied to the retina, the ear, and other sensory systems gives little information about how information is processed with fine discrimination under normal conditions. Kuffler showed that a small region of illumination gives rise to a brisk discharge of action potentials (Figure 1.16A). A larger spot shone over the same part of the retina is far less effective: this is because an additional group of receptors arranged circumferentially around the first set also responds to the change in illumination. The action of these photoreceptors on bipolar cells gives rise to inhibition of ganglion cell firing (Figure 1.16B). Summation of the excitatory effect of a small central spot and the inhibitory effect from the surrounding region causes the ganglion cell to be relatively insensitive to diffuse light (Figure 1.16C). In a second major category of ganglion cells, the optimal visual stimulus consists of a small dark spot surrounded by light.

The meaning of the signal in a ganglion cell has thereby become more complex than information simply about light or dark. Instead, the action potentials report the presence of a contrasting pattern of light in a particular region of the visual field. This occurs because each ganglion cell is influenced, albeit indirectly, not by one photoreceptor but by many. For any given ganglion cell, the specific connections through bipolar, horizontal,

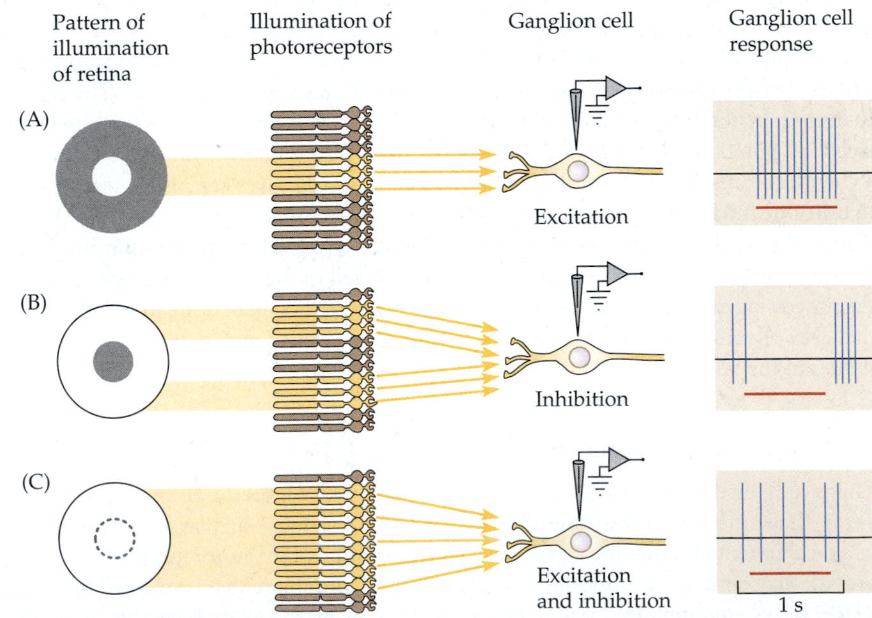

FIGURE 1.16 Integration by Ganglion Cells. Extracellular recordings made from a single ganglion cell in the retina of a lightly anesthetized cat, while patterns of light were presented to the eye (see Figure 1.17). (A) A small spot of light presented to a centrally located group of photoreceptors gives rise to excitation and a brisk discharge of action potentials. (B) Light presented as a ring, or annulus, to illuminate a circumferential group of photoreceptors gives rise to inhibition of the ganglion cell, which prevents the cell from firing. Removal of the inhibition at the end of illumination is equivalent to excitation, which gives rise to a burst of action potentials. (C) Illumination of both groups of receptors causes integration of excitation and inhibition and a weak discharge of action potentials. (After Kuffler, 1953.)

and amacrine cells determine the pattern of light that is required for it to discharge action potentials (see Chapter 20).

Complexity of the Information Conveyed by Action Potentials

At a distance of only three synaptic relays beyond the retina, even more sophisticated information about the visual world is provided by the action potentials in cortical nerve cells. Hubel and Wiesel[15] showed that cortical neurons do not respond simply to light or dark on the retina. Instead, their activation depends on the pattern of retinal illumination. Specific, distinctive patterns are the required and most effective stimuli for different types of cortical cells. For example, one type of cortical cell in the visual pathway responds selectively to a bar of light with a specific orientation (vertical, oblique, or horizontal), moving in a particular direction in a particular part of the visual field (vertical in the case of the cell illustrated in Figure 1.17). The firing of these cells is not influenced by diffuse light or by a bar of an inappropriate orientation or one moving in the wrong direction. Hence, its action potentials provide precise information about the visual stimulus to higher centers in the brain. This increase in the meaning attributed to a stereotyped action potential is explained by the precise connections of lower-order cells to the cortical cell and the way in which the cortical cell integrates incoming signals by summation of localized graded potentials.

The transformation of information can be simply summarized as follows:

- A signal in a photoreceptor indicates a change in light intensity in that area of the field of vision.
- A signal in a ganglion cell indicates the presence of contrast.
- A signal in a cortical neuron indicates the presence of an oriented bar or edge of light.

Complex integration occurs in other sensory systems. Thus, the position and direction of a mechanical stimulus, moving along a fingertip, act as selective stimuli for particular cells in the region of the cerebral cortex that are concerned with tactile stimuli. Mice, which use their whiskers much as we use our fingertips, have cells in the cortex that not only can distinguish the direction of motion but the roughness or smoothness of a surface.

Two important conclusions about signaling in the nervous system are: (1) nerve cells act as the building blocks for perception and (2) the abstract significance of the message can be extremely complex, depending on the number of inputs a neuron receives. It turns out that the progressive integration of information derived from lower order units can lead to the generation of highly complex and specific stimulus requirements for higher order central neurons. For example, later it will be shown (see Chapter 23) that specific cells exist in visual association areas that respond selectively to a face. In addition, the temporal grouping and patterning of impulses can provide information about the quality of the stimulus.

Reverse Traffic of Signals from Higher to Lower Centers

Implicit in the discussion of information transfer in this chapter is the concept of linear progression from receptors to perception or from motor commands to movement. It will be shown however that extensive signaling also occurs in the opposite direction: from the brain towards sensory receptors as well as from each higher center back towards the lower centers from which it received its inputs. In a few cases, the significance of descending information is understood, but in general, the functional role remains to be discovered.[16]

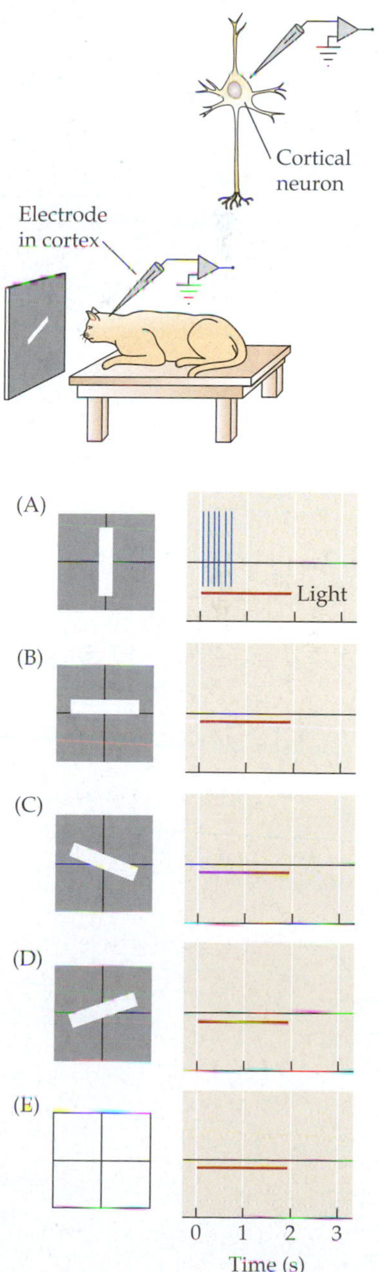

FIGURE 1.17 Information Conveyed by Action Potentials. Extracellular recordings from a neuron in the cerebral cortex of a lightly anesthetized cat. Action potentials in this cell indicate that a bar of light that is almost vertical shines on one particular part of the visual field. The small drawings to the left of the graphs show how visual stimuli such as bars or edges with different orientations and positions are presented to the eye. (A) The cortical cell fires a burst of action potentials when the light stimulus consists of a vertical bar of light in one particular part of the visual field. (B–E) Bars with different orientations or diffuse light fail to evoke action potentials. The cortical cell integrates information arriving by way of relays from a large number of photoreceptors, some of which (corresponding to those illuminated by the vertical bar) give rise to excitation on the cortical cell, the others giving rise to inhibition. (After Hubel and Wiesel, 1959.)

[15] Hubel, D. H., and Wiesel, T. N. 1977. *Proc. R. Soc. Lond. B, Biol. Sci.* 198: 1–59.

[16] Hawkins, J., and Blakeslee, S. 2004. *On Intelligence*. Times Books, New York.

Higher Functions of the Brain

In spite of the alarming flow of new treatises that appear day by day on consciousness, learning, and memory, only rudimentary information is available at present about the way in which the brain creates a complete image of the outside world, with its forms, colors, depth, and motion, or about the way in which it composes and executes complex, integrated movements of the body. Indeed, this open frontier is one of the most appealing aspects of research on the nervous system. We do not know how the tennis player runs to hit the ball in exactly the right place on the racket, so as to drive it to the far corner of the court, or how the coordinated finger and arm movements required for playing the violin are initiated or executed—let alone how we think and feel.

New approaches that shed light on mechanisms for the individual steps involved in higher functions often have their origin in psychophysical experiments. For example, tests on normal human subjects made with precise quantitative stimuli have shown that under suitable conditions, a person can detect the arrival of single quanta of light on photoreceptors of the eye. By carefully designed behavioral experiments made on rats and mice, one can produce symptoms of stress and anxiety resembling those seen in patients. As a further step, it then becomes possible to assess which brain structures and mechanisms play a part in such disorders of higher functions. In addition to the insights that these experiments provide about our emotions and our minds, they are essential for the development of new drugs that can mitigate the suffering of patients. Moreover, information flows back from the clinic toward basic research. Clinical observations, particularly on patients with discrete circumscribed lesions, provide unparalleled insights into mechanisms of perception, movement, and speech.

Cellular and Molecular Biology of Neurons

Like other types of cells, neurons possess the cellular machinery for metabolic activity for synthesizing intracellular and membrane proteins, and for distributing them to precise locations in the cell. Each type of neuron synthesizes, stores, and releases its characteristic transmitter(s). The receptors for specific transmitters are located at well-defined sites on the postsynaptic cell under the presynaptic terminals. In addition, other membrane proteins, known as pumps and transporters, maintain the constancy of the internal and external milieu of the cell. The presynaptic terminals of optic nerve fibers of ganglion cells (like those of photoreceptors and bipolar cells, and indeed like all presynaptic nerve terminals) contain in their membranes specific channels through which calcium ions can flow. Calcium entry triggers the release of transmitters and can activate intracellular cascades of enzymes and regulate numerous other cellular processes.

A major specialization in the cell biology of neurons, compared to other types of cells, arises from the presence of the axon. Axons do not have adequate machinery for synthesizing all the proteins they need. Hence, essential molecules are carried to the nerve terminals by a process known as axonal transport, often over long distances. Molecules required for maintenance of structure and function, as well as for the appropriate membrane channels, travel from the cell body in this way; similarly, molecules taken up at the ending are carried back to the cell body.

Neurons are different from most other cells in that, with few exceptions, they cannot divide after differentiation. As a result, in an adult human being, neurons in the central nervous system that have been destroyed usually cannot be replaced.

Signals for Development of the Nervous System

The high degree of organization in a structure such as the retina poses a fascinating problem. Whereas a computer requires a brain to wire it, the brain must establish and tune its own connections. What seems so puzzling is how the proper assembly of the parts endows the brain with its extraordinary properties.

In the mature retina, each cell type is situated in the correct layer—or even sublayer—and makes the correct connections with the appropriate targets. This arrangement is a prerequisite for function. For ganglion cells to develop, for example, precursor cells must

(A)

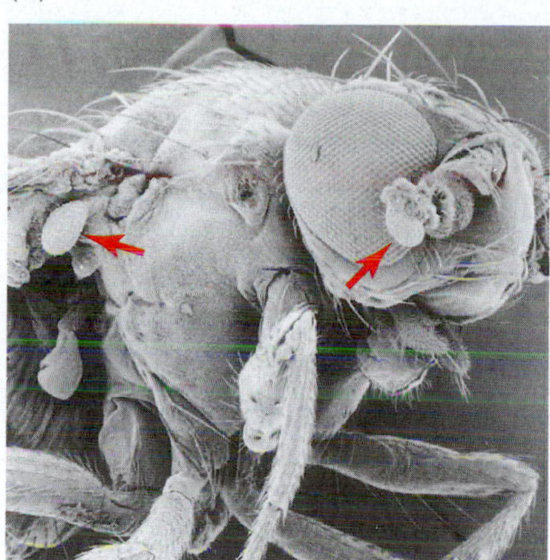

(B)

FIGURE 1.18 Genetic Influences on Development of the Eye in the Fruit Fly, *Drosophila*. A gene known as *eyeless* controls development of the eye in the fruit fly. After deletion of this gene, eyes fail to appear. Overexpression leads to the development of ectopic eyes that are morphologically normal. (A) This scanning electron micrograph shows ectopic eyes on the antenna (right arrow) and on the wing (left arrow). (B) Here, the wing eye is shown at higher magnification. A gene with strikingly similar sequence homology in the mouse can be inserted into the fly genome, and it also leads to the formation of ectopic eyes. (After Halder, Callaerts, and Gehring, 1995; micrographs kindly provided by W. Gehring.)

divide, migrate, differentiate into the appropriate shapes with the appropriate properties, and receive specific synapses. The axons must find their way over long distances through the optic nerve to end in the appropriate layer of the next relay station. Similar processes must occur for the various divisions of the nervous system so that complex structures required for function are formed.

Study of the mechanisms by which highly complex structures, such as the retina, are formed presents a key problem in modern neurobiology. An understanding of how intricate wiring diagrams are established in development often provides clues about function and about the genesis of functional disorders. In other words, if you know how an electrical circuit has been wired, you may be able to understand what the components are doing and, consequently, you may be able to repair it. Certain specific molecules are essential for differentiation, outgrowth of axons, pathfinding, synapse formation, and survival of neurons. Such molecules are now being identified at an ever-increasing rate, and their mechanisms of action are being studied. Interestingly, molecular signals that give rise to the outgrowth of axons and formation of connections can be regulated by electrical signals. Activity plays a role in determining the pattern of connections.

Genetic approaches have made it possible to identify genes that control the differentiation of entire organs, such as the eye as a whole. Gehring[17] and his colleagues have studied the expression of a gene in the fruit fly (*Drosophila*), known as *eyeless*, that controls the development of the eyes. After deletion of this gene in the germline, eyes fail to develop in the progeny for generation after generation. Homologous genes in mice and humans (known as *small eye* and *aniridia*, respectively) share extensive sequence identity and have similar developmental functions. If the fly *eyeless* gene or the mammalian homologue of the gene is introduced and overexpressed in the fly, it develops multiple ectopic eyes over its antennae, wings, and legs (Figure 1.18). The gene can therefore orchestrate the formation of an entire eye, in a mouse or a fly, even though the eyes themselves have completely different structures and properties.

Regeneration of the Nervous System after Injury

Not only does the nervous system wire itself when it is developing, but it can also restore certain connections after injury (again something your computer cannot do!). For example, axons in an arm can grow back after the nerve has been injured so that function can be restored; the hand can once again be moved, and sensation returns. Similarly, in a frog, fish, or an invertebrate like the leech, lesions in the central nervous system are followed by axon regeneration and functional recovery. After the optic nerve of a frog or a fish has been cut, fibers grow back to the brain and the animal can see again. However, in the adult mammalian CNS regeneration does not occur. The molecular signals that cause this failure are not yet known.

[17] Halder, G., Callaerts, P., and Gehring, W. J. 1995. *Science* 267: 1788–1792.

SUMMARY

- Neurons are connected to each other in a highly specific manner.
- At synapses, information is transmitted from cell to cell.
- In relatively simple circuits, such as those in the retina, it is possible to trace connections and understand the meaning of signals.
- Neurons within the eye and the brain act as building blocks for perception.
- Signals in neurons are highly stereotyped and similar in all animals.
- Action potentials conduct unfailingly over long distances.
- Local graded potentials depend on passive electrical properties of nerve cells and spread only over short distances.
- Owing to the peculiar structure of neurons, specialized cellular mechanisms are required for axonal transport of proteins and organelles to and from the cell body.
- During development, neurons migrate to their final destinations and become connected to their targets.
- Molecular cues provide guidance for growing axons.

Suggested Reading

All the experiments and concepts described in this introductory chapter are treated in more detail and fully referenced in later chapters. The following sources represent key reviews that show how essential concepts of neurobiology have developed over the years.

Currently in print and of great interest:

Hawkins, J., and Blakeslee, S. 2004. *On Intelligence*. Times Books, New York.

Hubel, D. H., and Wiesel, T. N. 2005. *Visual Perception*. Oxford University Press, Oxford.

Harder to buy new but still fascinating:

Adrian, E. D. 1946. *The Physical Background of Perception*. Clarendon, Oxford, England.

Helmholtz, H. 1962/1927. *Helmholtz's Treatise on Physiological Optics*. J. P. C. Southhall (ed.). Dover, New York.

Hodgkin, A. L. 1964. *The Conduction of the Nervous Impulse*. Liverpool University Press, Liverpool, England.

Katz, B. 1966. *Nerve, Muscle, and Synapse*. McGraw-Hill, New York.

Ramón y Cajal, S. [1909–1911] 1995. *Histology of the Nervous System*, 2 vols. Translated by Neely Swanson and Larry Swanson. Oxford University Press, New York.

Sherrington, C. S. 1906. *The Integrative Action of the Nervous System*. Reprint, Yale University Press, New Haven, CT, 1961.

CHAPTER 2
Signaling in the Visual System

Neuronal signals that are evoked by light begin in the retina. They are sent by ganglion cell axons to a relay, the lateral geniculate nucleus (LGN), and then to higher centers that produce our perception of scenes with objects and background, movement, shade, and color. Signaling at each level is best analyzed in terms of the receptive fields of neurons. A receptive field in the visual system is defined as the area of the retinal surface (or corresponding region of the visual field) that, upon illumination, enhances or inhibits the activity of a neuron. A useful strategy for analyzing the visual system is to define the optimal pattern of illumination and the receptive field for each neuron.

The receptive fields of most retinal ganglion cells and neurons in the lateral geniculate nucleus consist of small circular areas on the retina. The cells respond to contrast rather than diffuse illumination. Geniculate axons project to form a new map of the visual field in primary visual cortex. The receptive fields of neurons in the primary visual cortex for the most part consist of lines, bars, or edges with a particular orientation. Cortical neurons give no response to diffuse illumination. The optimal stimulus for a simple cell is an oriented edge or bar, which may be light or dark, with a defined width, shining on a precise place in the retina. Complex cells also respond to oriented bars but their discharges are evoked over a wider area than the simple cells. End inhibition, which is a decrease in the response of a neuron as the length of an image increases, gives rise to more elaborate stimulus requirements, such as a corner or a line that stops. Most cortical cells respond to appropriate illumination of both eyes. Receptive fields of simple cells result from convergence of a number of geniculate afferents with adjoining field centers. The response properties of complex cells depend on inputs from simple and other cortical cells. Cortical neurons detect only the edges of white or black patterns on a background with inverse contrast. The overall levels of illumination are measured by specialized retinal ganglion cells that project to areas other than the visual cortex.

The progression of receptive field properties from retina to complex cells suggests that inputs from one level are combined to produce more abstract requirements at the next. Information also flows in the opposite direction, for example from cortex to LGN, from one cortical layer to another and back. Throughout the visual pathways, the emphasis is on contrast, color, movement, depth, and boundaries, rather than on light detection. This distinction enables the nervous system to focus on what is important to the animal and to jettison irrelevant information in the visual fields.

This chapter describes the functional properties of neurons at successive stages in the visual pathways. We deal first with the output of the eye; second, with the next relay station, the lateral geniculate nucleus; and then with the primary visual cortex, the initial receiving center for visual information. Our aim is to show how neuronal activity is related to higher functions, such as visual perception, using as background knowledge only the basic information provided in Chapter 1. Chapter 3 shows in greater detail how structure and function are intimately related at every level (see also Chapter 22).

Experiments performed in recent years have produced an overwhelming body of work on psychophysics, color vision, dark adaptation, retinal pigments, transduction, transmitters, and the organization of the retina (see Chapter 20). Each of these topics can form the basis of a self-contained monograph (see the Suggested Reading section at the end of the chapter). The same applies to comparative aspects of the visual system in invertebrates, lower vertebrates, and mammals. Since a comprehensive account is not possible within the scope of this book, we have selected experiments that provide a continuous thread, extending from the properties of cells in the retina to mechanisms that underlie perception.

Pathways in the Visual System

The initial step in visual processing is the formation on each retina of a sharp image of the outside world. Essential for clear vision are: (1) correct focus of the image by adjustment of the curvature of the lens (accommodation), (2) regulation of light entering the eye by the diameter of the pupil, and (3) convergence of the two eyes to ensure that matching images fall on corresponding points of both retinas. Our vision depends critically on the region of the visual field that is being analyzed (Figure 2.1). We can read

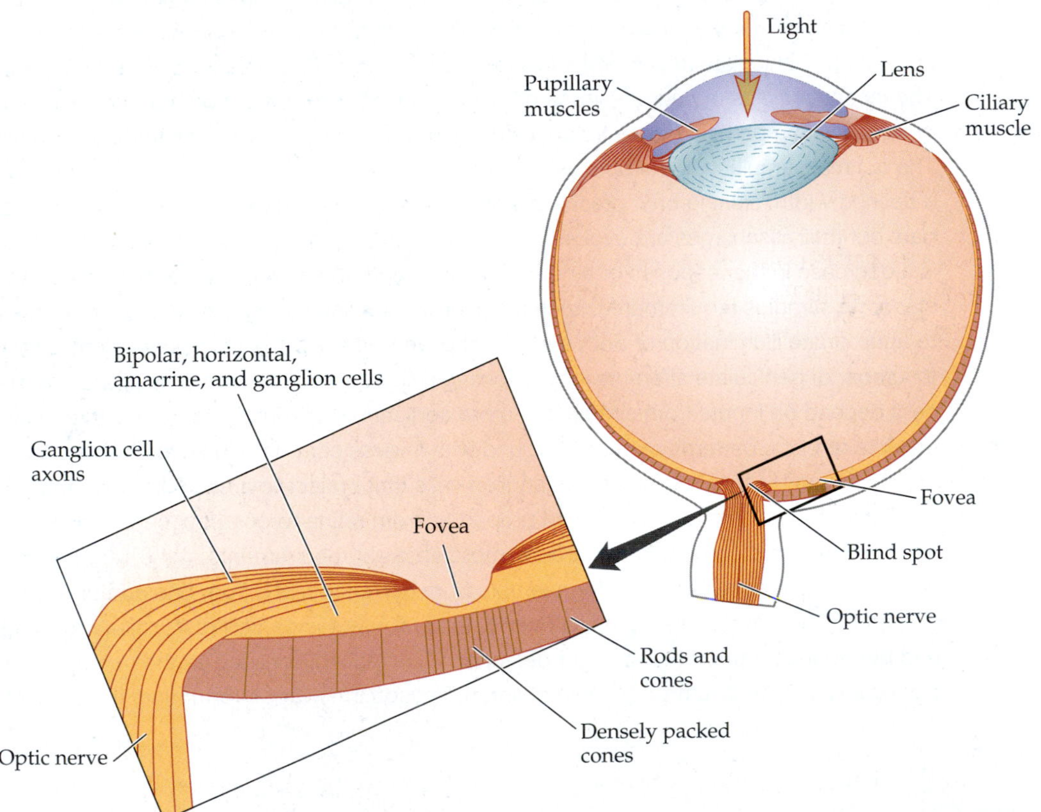

FIGURE 2.1 Pathways for Light and Arrangement of Cells in the Retina. Cross section through the eye. Light must pass through the lens and layers of cells in order to reach the rod and cone photoreceptors. The fovea is a specialized area, containing only densely packed, slender cones. It is used for fine discrimination. In the fovea, the superficial layers of cells are spread apart and this feature permits light to have more direct access to the photoreceptors than elsewhere in the retina. The point at which the optic nerve exits the eye has no photoreceptors and constitutes a blind spot.

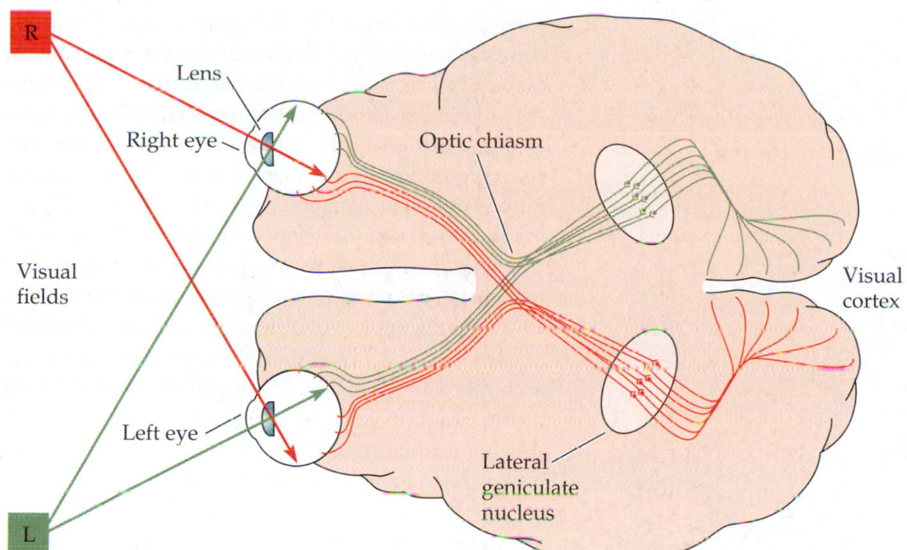

FIGURE 2.2 Visual Pathways. The right side of each retina, shown in green, projects to the right lateral geniculate nucleus. Thus, the right visual cortex receives information exclusively from the left half of the visual field.

small print at the center of gaze, where light falls on the fovea, but not in the peripheral field of vision. This loss of acuity arises from the way in which visual information is processed; it is not the result of blurred images or optical distortion outside the central region.

The pathways from the eye to the cerebral cortex are illustrated in Figure 2.2, which depicts some of the major landmarks of the visual system; this anatomical information constitutes the bare minimum for following the electrical signals as they pass from relay to relay.

The optic nerve fibers that arise from ganglion cells in the retina end on layers of cells in a relay station of the thalamus, which as mentioned is called the lateral geniculate nucleus (geniculate means *bent like a knee*). In each of the six principal layers of this structure (Figure 2.3), the outside world is represented as a coherent map of the field seen by one eye, either on the same or the opposite side. Geniculate axons in turn project through the optic radiation to the cerebral cortex. The six layers of the visual cortex and the arrangements of maps are dealt with in Chapter 3. For present purposes, it is sufficient to state that in the monkey, the optic radiation ends on a folded plate of cells about 2 mm thick (see Figure 2.7). This region of the brain is known as the primary visual cortex, or visual area 1, (also called V1), which lies posteriorly in the occipital lobe. Adjacent regions of cortex are also concerned with vision. From the primary visual cortex, the progression through the brain becomes ever more complex, with no end point in sight.

Figure 2.2 shows how the output from each retina divides in two at the optic chiasm. The right side of each retina projects to the right cerebral hemisphere. Because of optical reversal by the lens, the right side of each retina receives the image of the visual world on the left side of the head. Each cerebral hemisphere, therefore, sees the opposite side of the outside world. Accordingly, people with damage to the right cerebral hemisphere caused by trauma or disease become blind in the left visual field, and vice versa. Other pathways that branch off to the midbrain are not described here. They are concerned primarily with regulating eye movements, pupillary responses, and circadian rhythms (see Chapter 17).

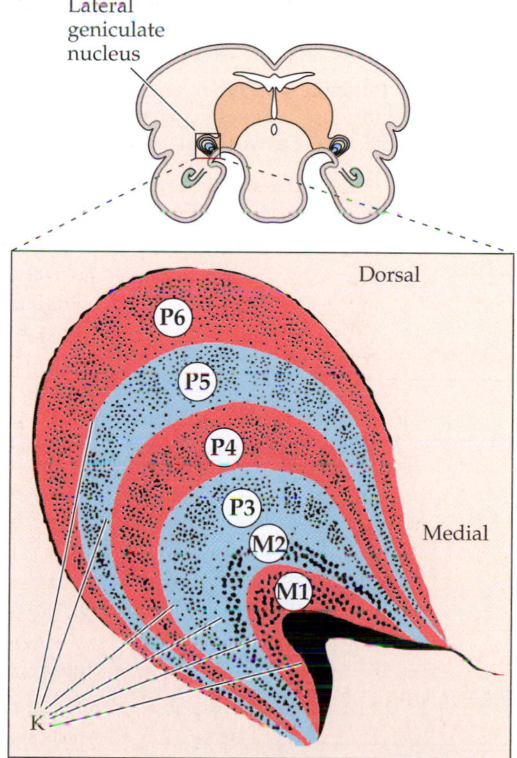

FIGURE 2.3 The Monkey Lateral Geniculate Nucleus (LGN) has six major layers designated parvocellular, or P (3, 4, 5, 6) and magnocellular, or M (1, 2), separated by the koniocellular (K) layers. In the monkey, each layer is supplied by only one eye and contains cells with specialized response properties. Red signifies input from the contralateral eye and blue from the ipsilateral eye. (After Hendry and Calkins, 1998.)

Convergence and Divergence of Connections

By examining the cellular anatomy of the various structures in the visual pathway, one can exclude the possibility that information is handed on un-

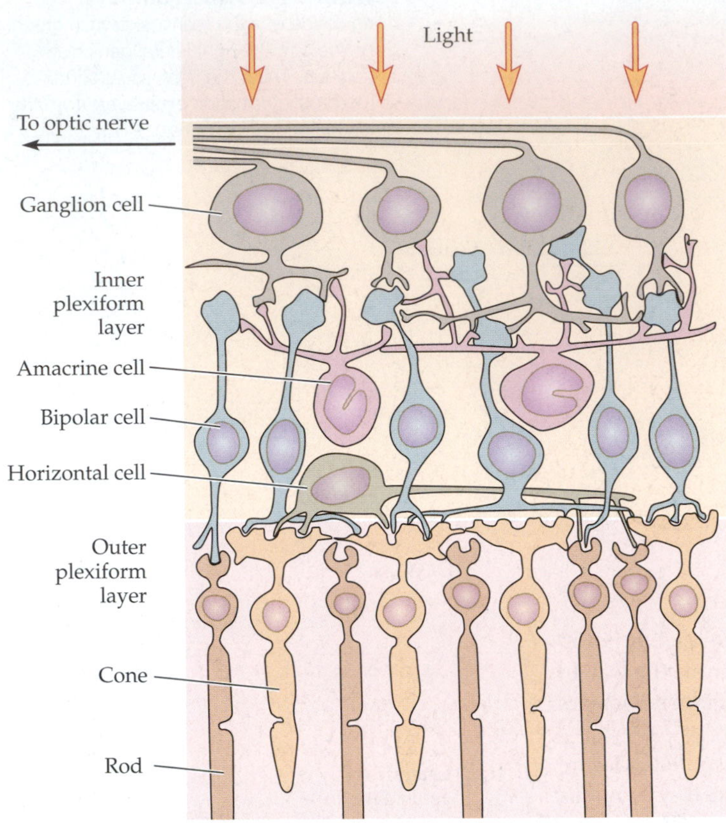

FIGURE 2.4 Principal Cell Types and Connections of Primate Retina to illustrate rod and cone pathways to ganglion cells (see Chapter 20). (After Dowling and Boycott, 1966; Daw, Jensen, and Brunken, 1990.)

changed from level to level (Figure 2.4). The neurons converge and diverge extensively at every stage; that is, each cell receives many inputs and makes connections with a number of other cells (see Chapter 1). Just as a ganglion cell is supplied (indirectly) from numerous rods and cones, so a neuron in the LGN receives its input from many ganglion cells and it in turn supplies many cortical neurons. Hence, as impulses travel to the cortex and within the cortex itself there occurs a funneling and, simultaneously, a dispersal of information. Converging impulses of different origin are integrated at each stage into an entirely new message that takes account of all the inputs. Moreover, except at the level of the ganglion cells, information is simultaneously flowing in the opposite direction, for example from cortex down to lateral geniculate nucleus.[1,2]

Receptive Fields of Ganglion and Geniculate Cells

Concept of Receptive Fields

Diffuse flashes of light are of little or no use for assessing function in the visual system. Instead, the technique of illuminating selected areas of the retina led to the concept of the **receptive field.** The concept has provided a key for understanding the significance of the signals, not only in the retina, but at successive stages in the cortex. As previously mentioned in Chapter 1, the term *receptive field* was coined originally by Sherrington in relation to reflex actions (see also Chapter 21); it was later introduced to the visual system by Hartline.[3] The receptive field of a neuron in the visual system can be defined as *the area of the retina from which the activity of a neuron can be influenced by light* (see also Chapter 20). Alternatively, one can define the receptive field as the region in the *visual field,* illumination of which influences the cell's activity. By definition, illumination outside a receptive field produces no effect on firing. The area itself can be subdivided into distinct regions, some of which increase activity and others of which suppress it.

The Output of the Retina

Many years before the electrical responses of photoreceptors or bipolar cells in the retina could be measured, important information was obtained by recording from ganglion cells. Thus, the first analysis of signaling in the retina was made at the output stage, the end result of synaptic interactions. It was a simplification and shortcut to go straight to the output.

As discussed in Chapter 1 (see Figure 1.16), retinal ganglion cells have circular receptive fields with concentric on and off regions. Stephen Kuffler first defined the organization of the receptive field in the cat visual system.[4] Synaptic inputs to ganglion cells that are responsible for receptive field organization are described in Chapter 20. Hubel has succinctly put Kuffler's achievement in perspective:

> What is especially interesting to me is the unexpectedness of the results, as reflected in the failure of anyone before Kuffler to guess that something like center–surround receptive fields could exist or that the optic nerve would virtually ignore anything so boring as diffuse light levels.[5]

The principal novelty in the study of the visual system was the use of discrete, circumscribed spots for stimulation of selected areas of the retina, instead of diffuse uniform

[1] Conley, M., Penny, G. R., and Diamond, I. T. 1987. *J. Comp. Neurol.* 256: 71–87.

[2] Ichida, J. M., and Casagrande, V. A. 2002. *J. Comp. Neurol.* 454: 272–283.

[3] Hartline, H. K. 1940. *Am. J. Physiol.* 130: 690–699.

[4] Kuffler, S. W. 1953. *J. Neurophysiol.* 16: 37–68.

[5] Hubel, D. H. 1988. *Eye, Brain and Vision.* Scientific American Library, New York.

illumination. A convenient way of illuminating particular portions of the retina is to anesthetize the animal and place it facing a screen or a computer, at a distance for which its eyes are properly refracted. When one then shines patterns of light onto the screen or displays computer-generated images, these will be well focused on the retinal surface (see Figure 1.16).

Such procedures had been foreshadowed by pioneering work on the eye of a simple invertebrate, the horseshoe crab *Limulus*,[3] and on the retina of the frog.[6,7] Kuffler's initial choice of the cat was a lucky one; in the rabbit, for example, the situation would have been more complicated. Rabbit ganglion cells have elaborate receptive fields that respond to such complex features as edges or to movement in a particular direction.[8,9] Equally complex are lower vertebrates, such as frogs and salamanders.[10] A general law seems to emerge: the dumber the animal, the smarter its retina (D. A. Baylor, personal communication).

Ganglion and Geniculate Cell Receptive Field Organization

When one records from a particular cell in the visual system, the first task is to find the location of its receptive field. Characteristically, most neurons throughout the visual system show discharges at rest even in the absence of illumination. Appropriate stimuli do not necessarily initiate activity but may modulate the resting discharge, causing either an increase or a decrease of frequency.

Figure 2.5 shows characteristic responses of two types of retinal ganglion cells to illumination in a lightly anesthetized cat. Ganglion cells with an "on" center, such as the one shown on the left, respond best to a small spot of light surrounded by darkness in a manner similar to the ganglion cells responses described in Chapter 1 (see Figure 1.16). The cell whose responses are shown on the right-hand side is an "off"-center cell, which responds best to a small dark spot surrounded by light. For the on-center receptive field in Figure 2.5, light produces the most vigorous response if it completely fills the center, whereas for most effective inhibition of firing, the light must cover the entire ring-shaped area. When the inhibitory annular light is turned off, the ganglion cell gives an exuberant off discharge. An off-center field has a converse organization, with inhibition arising in the circular center. For either cell, the spot-like center and its surround are antagonistic; therefore, if both center and surround are illuminated simultaneously, they tend to cancel each other's contribution.

The responses of cells in the LGN are similar to those of retinal ganglion cells, as shown in Figure 2.6.[11] As in the retina, a small spot of light, 2× or about 0.5 mm in diameter, shone onto a part of the receptive field is far more effective than diffuse illumination in producing

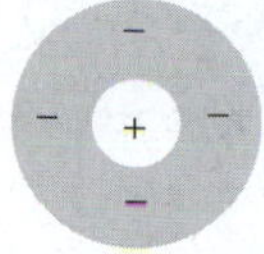

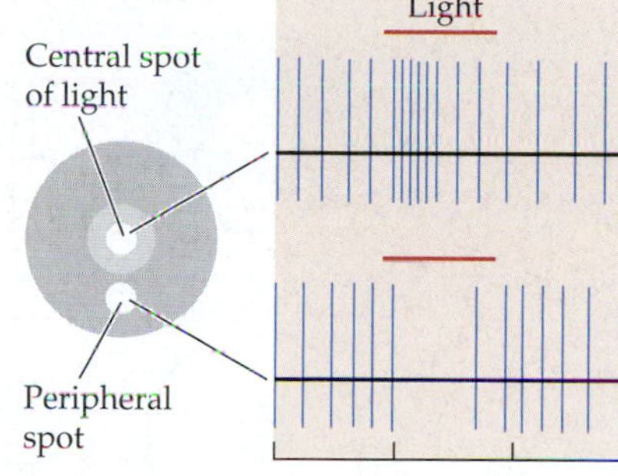

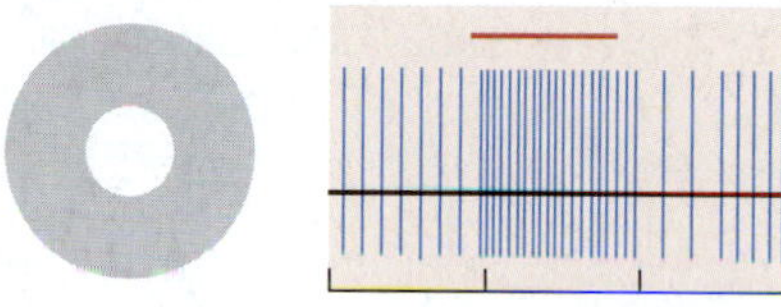

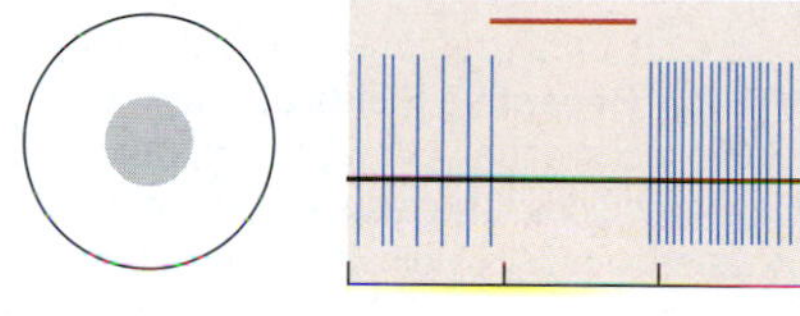

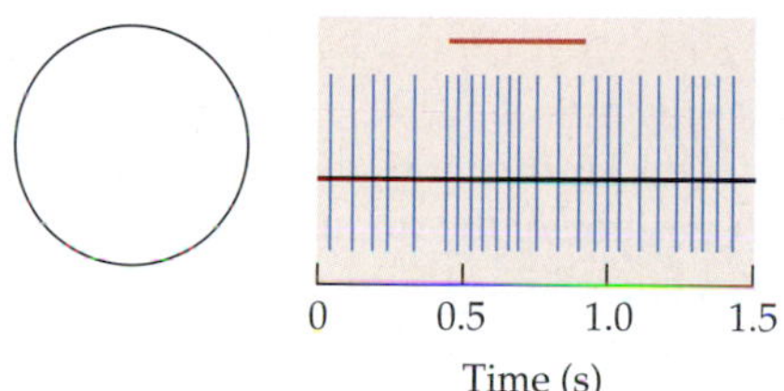

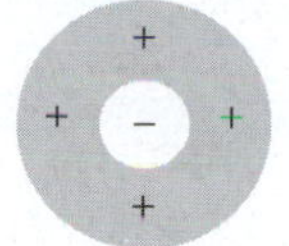

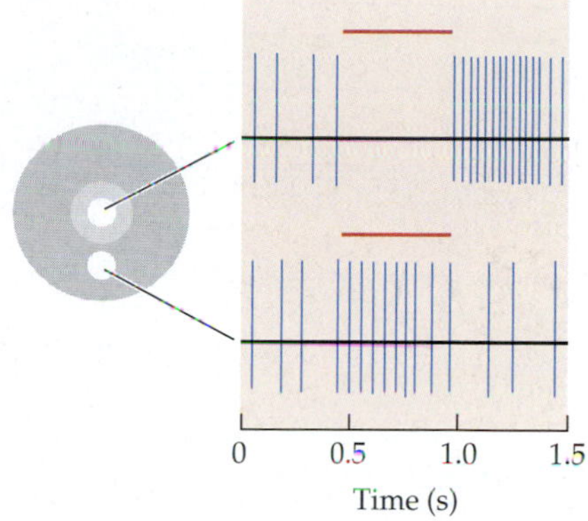

FIGURE 2.5 Receptive Fields of Ganglion Cells in the retinas of cats and monkeys are grouped into two main classes: on-center and off-center fields. On-center cells respond best to a spot of light shone onto the central part of the receptive field. Illumination (indicated by the red bar above records) of the surrounding area with a spot or a ring of light reduces or suppresses the discharges and causes responses when the light is turned off. Illumination of the entire receptive field elicits weak discharges because center and surround antagonize each other's effects, as with bipolar cells. Off-center cells slow down or stop signaling when the central area of their field is illuminated and accelerate when the light is turned off. Light shone onto the surround of an off-center receptive field causes excitation of the neuron. (After Kuffler, 1953.)

[6] Maturana, H. R. et al. 1960. *J. Gen. Physiol.* 43: 129–175.

[7] Barlow, H. B. 1953. *J. Physiol.* 119: 69–88.

[8] Barlow, H. B., Hill, R. M., and Levick, W. R. 1964. *J. Physiol.* 173: 377–407.

[9] Oyster, C. W., and Barlow, H. B. 1967. *Science* 155: 841–842.

[10] Baccus, S. A. et al. 2008. *J. Neurosci.* 28: 6807–6817.

[11] Hubel, D. H., and Wiesel, T. N. 1961. *J. Physiol.* 155: 385–398.

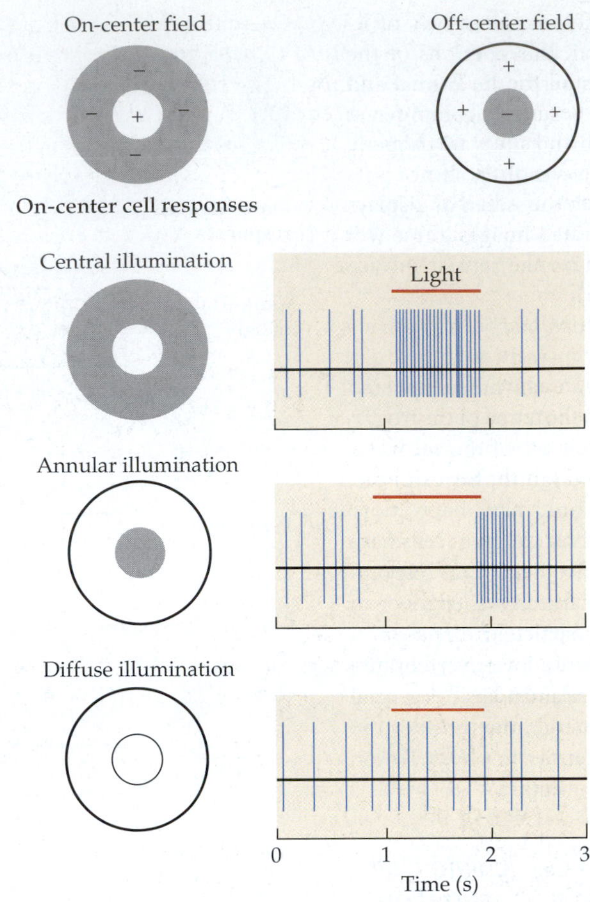

FIGURE 2.6 Receptive Fields of Lateral Geniculate Nucleus Cells. The concentric receptive fields of cells in the LGN resemble those of ganglion cells in the retina, consisting of on-center and off-center types. The responses illustrated are from an on-center cell in the cat LGN. The red bar above each record indicates illumination. The central and surround areas antagonize each other's effects, so diffuse illumination of the entire receptive field gives only weak responses (see bottom record) but is less pronounced than in retinal ganglion cells (compare to Figure 2.5). (After Hubel and Wiesel, 1961.)

excitation. Furthermore, the same spot of light can have opposite effects, depending on the exact position of the stimulus within the receptive field. In one area, a small spot of light excites the cell for the duration of illumination, while simply shifting the spot by 1 mm or less across the retinal surface gives rise to inhibition. Again, as in the retina, two basic receptive field types predominate, on-center and off-center geniculate cells. The receptive fields of both types are roughly concentric.

While ganglion and geniculate cells have very similar receptive field organization, they are not identical. For example, descending connections from layer 6 of the visual cortex project to geniculate neurons to modulate their firing; there is, however, no comparable descending input to ganglion cells. In addition there are subtle differences in receptive field properties, such as even greater failure of geniculate cells to respond to diffuse illumination. It is a general problem that the precise part played by thalamic structures (including the LGN) in transferring information to the cortex is still not fully understood[12,13] (see Chapter 21).

Sizes of Receptive Fields

Neighboring cells in the visual system collect information from very similar, but not quite identical, areas of the retina.[14] Even a small (0.1 mm) spot of light on the retina covers the receptive fields of many ganglion and geniculate cells. Some are inhibited,

[12] Sherman, S. M. 2007. *Curr. Opin. Neurobiol.* 17: 417–422.

[13] Guillery, R. W. 2005. *Prog. Brain Res.* 149: 235–256.

[14] Borghuis, B. G. et al. 2008. *J. Neurosci.* 28: 3178–3189.

others excited. Throughout the visual system, *neurons processing related information are clustered together*. In sensory systems this means that the central neurons dealing with a particular area of the surface can communicate with each other over short distances. This appears to be an economical arrangement, as it minimizes the need for long lines of communication and simplifies formation of connections (see Chapters 3 and 20). Since neighboring regions of the retina make connections with neighboring geniculate cells, the receptive fields of adjacent neurons overlap over most of their area.[15] Both the *area centralis* in the cat (the region of the cat retina with small receptive field centers) and the fovea in the monkey project onto the greater portion of each geniculate layer, and a similar distribution has been found in humans by the use of functional magnetic resonance imaging (fMRI).[16] There are relatively few cells devoted to the peripheral retina. This extensive representation of the fovea reflects the high density of foveal receptors necessary for high-acuity vision.

Moreover, the size of the receptive field of a ganglion, geniculate, or cortical cell depends on its location in the retina (or visual field). The receptive fields of cells situated in the central areas of the retina have much smaller centers than those at the periphery; receptive fields are smallest in the fovea, where the acuity of vision is highest.[17] The central on or off region of such a midget ganglion cell's receptive field can be supplied by a single cone and is accordingly only about 2.5 μm in diameter, subtending 0.5 minutes of arc—smaller than the period at the end of this sentence. Note that receptive fields can be described either as dimensions on the retina or as degrees of arc subtended by the stimulus. In human eyes, 1 mm on the retina corresponds to about 4°. For reference, the image of the moon has a diameter of 1/8 mm on the human retina, corresponding to 0.5° or 30 minutes of arc.

There are similar gradations of receptive field size and spatial dimension in the somatosensory system. A higher-order sensory neuron in the brain, responding to a fine touch applied to the skin of the fingertip, has a much smaller receptive field than that of a neuron whose field is on the skin of the upper arm (see Chapter 21). To discern the form of an object, we use our fingertips and fovea, not the less discriminating regions with poorer resolution.

Classification of Ganglion and Geniculate Cells

Superimposed on the general scheme of on- or off-center receptive fields, ganglion cells in the monkey retina can be grouped into two main categories denoted as M and P. The criteria are both anatomical and physiological. The M and P terminology is based on the anatomical projections of these neurons to the LGN and then to the cortex[18] (see Chapter 3). P ganglion cells project to the four dorsal layers of smaller cells in the LGN (the **parvocellular division**), whereas M ganglion cells project to the larger cells in two ventral layers (the **magnocellular division**). Chapter 3 describes how the characteristics of neurons in the M and P pathways are maintained at successive levels in the visual system. In brief, P ganglion cells have small receptive field centers, high spatial resolution, and are sensitive to color and *P*osition. P cells provide information about fine detail at high contrast. M cells have larger receptive fields than P cells and are more sensitive to small differences in contrast and to *M*ovement; they fire at higher frequencies and conduct impulses more rapidly along their larger-diameter axons. In the cat, which has no color vision, the classification of ganglion cells is different, with X, Y, and W groups.[15] Groups X and Y are in some respects parallel to P and M in their properties, but there are major differences and the two classifications are not interchangeable.

What Information Do Ganglion and Geniculate Cells Convey?

The most striking feature of ganglion and geniculate cells with their concentric fields is that they tell a different story from that provided by primary sensory receptors. They do not convey information about absolute levels of illumination, because they behave in a similar fashion at different background levels of light. They ignore much of the information of the photoreceptors, which work more like a photographic plate or a light meter. Rather, they measure differences within their receptive fields by comparing the degree

[15] Yeh, C. I. et al. 2009. *J. Neurophysiol.* 101: 2166–2185.

[16] Kastner, S., Schneider, K. A., and Wunderlich, K. 2006. *Prog. Brain Res.* 155: 125–143.

[17] Balasubramanian, V., and Sterling, P. 2009. *J. Physiol.* 587: 2753–2767.

[18] Malpeli, J. G., Lee, D., and Baker, F. H. 1996. *J. Comp. Neurol.* 375: 363–377.

of illumination between the center and the surround. They appear to be designed to encode simultaneous contrast and ignore gradual changes in overall illumination. They are exquisitely tuned to detect such contrast as the edge of an image crossing the opposing regions of a receptive field. Chapters 17 and 20 describe a class of retinal ganglion cells that do respond to diffuse illumination but do not project to the visual cortex or play a part in form perception.[19]

Experiments made in the salamander retina by Baylor, Meister, and their colleagues suggest that temporal aspects of firing by ganglion cells can also contribute to spatial resolution.[20,21] Throughout the previous discussion, the trains of impulses recorded from individual neurons have been treated as separate lines from which the analysis of visual input is made by the brain. Synchrony of firing by two cells, however, may be an additional variable. Analysis of the degree of synchrony can be used by higher centers to obtain information about light falling on the retina that cannot be deduced from looking at the firing of the two ganglion cells separately.

[19]Fu, Y. et al. 2005. *Curr. Opin. Neurobiol.* 15: 415–422.

[20]Meister, M., Lagnado, L., and Baylor, D. A. 1995. *Science* 270: 1207–1210.

[21]Gollisch, T., and Meister, M. 2008. *Science* 319: 1108–1111.

BOX 2.1
Strategies for Exploring the Cortex

In 1953, Stephen Kuffler pioneered the experimental analysis of the mammalian visual system by concentrating on receptive field organization and the meaning of signals in the cat optic nerve.[4] A clear, continuous thread from signaling to perception was subsequently provided through the beautiful experiments of Hubel and Wiesel and the large body of work they inspired.

The procedure used by Hubel and Wiesel, the monitoring of activity in single neurons, might seem an unprofitable way to study higher functions in which large numbers of cells take part. What chance do physiologists have of gaining insight into complex actions within the brain when they sample only one or a few of the billions of neurons in the brain, a hopelessly small fraction of the total number? A feature that simplifies the situation in the visual cortex is that the major cell types are laid out in an apparently well-ordered manner as repeating units: adjacent points in the retina project to adjacent points on the cortical surface. Thus, the visual cortex is designed to bring an identical set of neural analyzers to bear on each tiny segment of the visual field.

The problem faced by Hubel and Wiesel in 1958 was to find out how signals denoting small, bright, dark, or colored spots in the retina could be transmuted into signals that conveyed information about the shape, size, color, movement, and depth of objects. Techniques that are routinely used now—such as optical recording, horseradish peroxidase injection, or brain scanning—had not yet been thought of. At the outset, Hubel and Wiesel faced completely unanswered questions, which they tackled by assuming that visual centers in the cortex would perform their processing according to principles similar to those in the retina, but at a more advanced level. It is worth pointing out that they started their work at a time when not only was nothing known about how neurons functioned in the visual cortex, but far worse, the field abounded with misleading or frankly wrong hypotheses derived from experiments made by shining bright flashes of

David H. Hubel (left) and Torsten N. Wiesel during an experiment, about 1969. The cat, not shown, also faces the screen.

light into the eye (see Chapter 1) or by cortical lesions. For example, neurons in the visual cortex were described in 1955 as being on, off, on–off, or Type A neurons (which did not respond to anything at all). Pioneering work that reveals brand new concepts that stand the test of time often starts not from nothing but from a wealth of confused data.

One crucial strategy in Hubel and Wiesel's analysis was the use of stimuli that mimic those occurring under natural conditions. For example, edges, contours, and simple patterns presented to the eye revealed features of its organization that could never have been detected by using bright flashes without form. Another key to the success of Hubel and Wiesel's approach lay in asking not simply what stimulus evokes a response in a particular neuron, but rather what is the *most effective* stimulus. Pursuit of this question through the various stages of the visual system has elicited many surprising and remarkable results. Their early papers demonstrated that the receptive fields of simple and complex cells in the primary visual cortex constitute initial stages of pattern recognition.

Cortical Receptive Fields

Responses of cortical neurons, like those of the retinal ganglion and geniculate cells, tend to occur on a background of maintained activity. A consistent observation is that discharges of cortical neurons are not significantly influenced by diffuse illumination of the retina. Almost complete insensitivity to diffuse light is a more pronounced feature of the process already noted in the retina and the lateral geniculate nucleus; it results from equally matched antagonistic actions between the inhibitory and excitatory regions in the receptive fields of cortical cells. Thus, the neuronal firing rate is altered only when certain demands about the position and form of the stimulus on the retina are met. The receptive fields of most cortical neurons have configurations that differ from those of retinal or geniculate cells, so spots of light often have little or no effect. In his Nobel address, Hubel described the experiment in which Wiesel and he first recognized this essential property:

> Our first real discovery came about as a surprise. For three or four hours, we got absolutely nowhere. Then gradually we began to elicit some vague and inconsistent responses by stimulating somewhere in the midperiphery of the retina. We were inserting the glass slide with its black spot into the slot of the ophthalmoscope when suddenly, over the audio monitor, the cell went off like a machine gun. After some fussing and fiddling, we found out what was happening. The response had nothing to do with the black dot. As the glass slide was inserted, its edge was casting onto the retina a faint but sharp shadow, a straight dark line on a light background. That was what the cell wanted, and it wanted it, moreover, in just one narrow range of orientations. This was unheard of. It is hard now to think back and realize just how free we were from any idea of what cortical cells might be doing in an animal's daily life.[22]

By following a progression of clues, Hubel and Wiesel worked out the appropriate light stimuli for various cortical cells; initially they classified the receptive fields as **simple** or **complex**. Each of these categories includes a number of subgroups and important variables that bear on perceptual mechanisms. A major difference observed from ganglion and geniculate cells is that individual simple and complex cells are, for the most part, driven by both eyes.

Responses of Simple Cells

Most simple cells are found in layers 4 and 6 and deep in layer 3 (Figure 2.7). All these layers receive direct input from the LGN (although layer 4C is the most favored destination, as described in Chapter 3). Receptive fields of simple cells can be mapped with stationary spots of light, and they exhibit several variations.[23–25] One type of simple cell has a recep-

[22]Hubel, D. H. 1982. *Nature* 299: 515–524.

[23]Hubel, D. H., and Wiesel, T. N. 1959. *J. Physiol.* 148: 574–591.

[24]Hubel, D. H., and Wiesel, T. N. 1962. *J. Physiol.* 160: 106–154.

[25]Hubel, D. H., and Wiesel, T. N. 1968. *J. Physiol.* 195: 215–243.

(A)

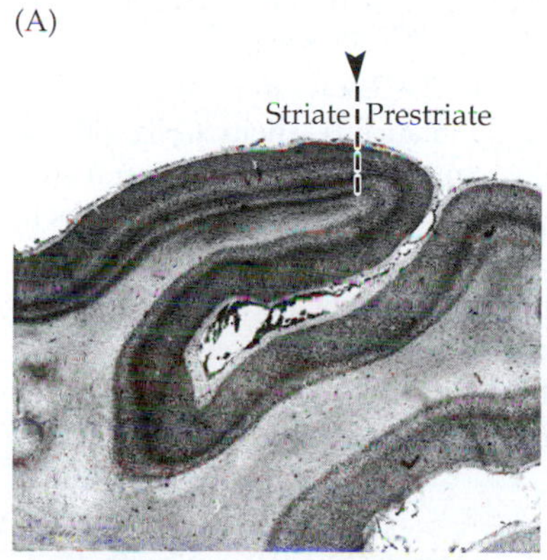

(B)

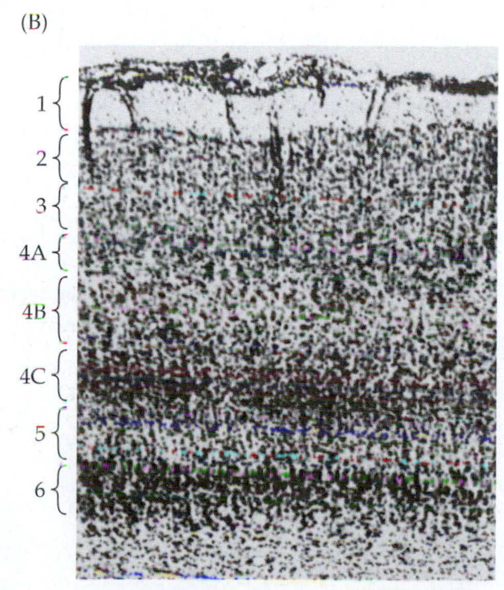

FIGURE 2.7 Architecture of Visual Cortex. (A) Section showing clear striation in area 17. (B) Distinct layering of cells in a section of striate cortex of the macaque monkey, stained to show cell bodies (Nissl stain). Fibers arriving from the LGN end in layers 4A, 4B, and 4C. (A after Hubel and Wiesel, 1972; B after Gilbert and Wiesel, 1979.)

FIGURE 2.8 Responses of a Simple Cell in Cat Striate Cortex to spots of light (A) and bars (C). The receptive field (B) has a narrow central on area (+) flanked by symmetrical antagonistic off areas (–). The best stimulus for this cell is a vertically oriented light bar in the center of its receptive field (see fifth record from the top in C). Other orientations are less effective or ineffective. Diffuse light does not stimulate. The bar above each record in A and C indicates the duration of stimulation. (After Hubel and Wiesel, 1959.)

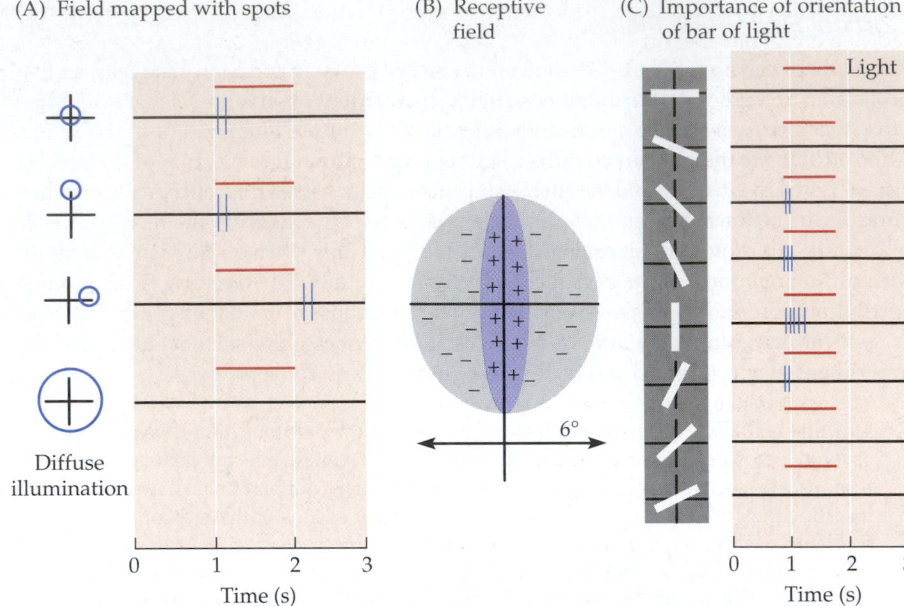

tive field that consists of an extended narrow central portion, flanked by two antagonistic areas. The center may be either excitatory or inhibitory. Figure 2.8 shows the receptive field of a simple cell in the striate cortex mapped out with spots of light that excited only weakly in the center (because the spots covered only a small fraction of the central on area).

The requirements of such simple cell are exacting, as illustrated in Figure 2.8C. For optimal activation, these cells need a bar of light that is not more than a certain width, that entirely fills the central area, and that is oriented at a certain angle. Illumination of the surrounding areas suppresses any ongoing activity or reduces the efficacy of a simultaneous center excitation. As predicted by mapping with spots of light, a vertically oriented bar is the most effective stimulus. Even small deviations from that pattern result in a diminished response. Various cells have receptive fields requiring a wide range of different orientations and positions. A new population of simple cells is therefore activated by rotating the stimulus or by shifting its position in the visual field. The distribution of inhibitory–excitatory flanks in various simple cell receptive fields may not be symmetrical or the field may consist of two longitudinal regions facing each other—one excitatory, the other inhibitory.

Figure 2.9 shows examples of four such receptive fields, all with a common axis of orientation but with differences in the distribution of areas within the field. For the receptive field in Figure 2.9A, a narrow slit of light oriented from 1 o'clock to 7 o'clock (assuming the visual field corresponds to a clock face with 12 o'clock high) elicits the best response. A dark bar in the same place but with light flanks suppresses ongoing spontaneous activity. Cells with the field shapes shown in Figure 2.9B and C fire optimally with a dark bar in the central area. For the field shown in Figure 2.9D, an edge with light on the left and darkness on the right produces the most effective "on" response, whereas reversing the dark and light areas is best for eliciting "off" discharges. In simple cells, the optimal width of the narrow light or dark bar is comparable to the diameters of the on- or off-center regions in the doughnut-shaped receptive fields of ganglion or lateral geniculate cells. Thus, cortical cells that have fields derived from the fovea are most excited by bars narrower than those that excite cells with fields in

FIGURE 2.9 Receptive Fields of Simple Cells in Cat Striate Cortex. In practice, all possible orientations are observed for each type of field. The optimal stimuli are a narrow slit or bar of light in the center for A; a dark bar for B and C; and an edge with dark on the right for D. Considerable asymmetry can be present, as in C. (After Hubel and Wiesel, 1962.)

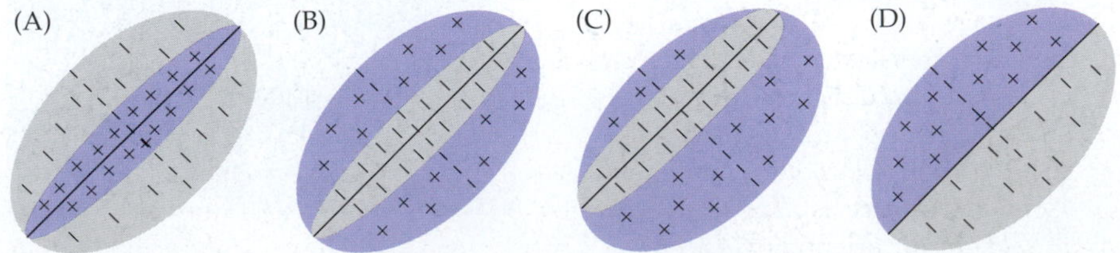

retinal periphery, corresponding to the smaller receptive fields of foveal ganglion cells. Thus, the common properties of all simple cells are (1) that they respond best to a properly oriented stimulus positioned so as not to encroach on antagonistic zones, and (2) that stationary slits or spots can be used to define on and off areas.

In addition to these features most simple cells respond to similar visual stimuli in either eye. **Binocular fusion**, which is not possible in the retina or the LGN, first appears in the visual cortex. Figure 2.10 shows the responses evoked in a simple cell in the cat visual cortex by a horizontally oriented bar, shone onto the same region of either the left or the right retina.

Another constant and remarkable feature is that in spite of all the different proportions of inhibitory and excitatory areas, the two contributions match exactly and cancel each other's effectiveness, so diffuse illumination of the entire receptive field produces a feeble response at best (see Figure 2.8). The off areas in cortical fields are not always able to initiate impulses in response to dark bars. Frequently (particularly in end inhibition and in the more elaborate fields to be described shortly), illumination of the off area can be detected only as a reduction in the discharge evoked from the on area. The movement of edges or bars of the appropriate orientation is a highly effective technique for initiating impulses. Once again, there is a specialization for detecting differences, but the spot-like contrast representation of ganglion cells has been transformed and extended into a line or an edge. Resolution has not been lost, but instead it has been incorporated into a more complex pattern.

For depth perception, which is a magnocellular pathway function, there exists another binocular specialization of receptive fields in which an object out of the plane of focus casts images on disparate parts of the two retinas.[23,26] Neurons with properties that fit the binocular cell's ability for depth perception have been found in primary and association visual cortex. For such cells, the best stimulus is an appropriately oriented bar in front of the plane of focus (for certain cells) or beyond it (for others). Impulses fail to be evoked by presenting the bar only to one eye or the other as well as to both eyes in the plane of focus. It is the disparity of the position on the two retinas that these cells require. Depth perception as such probably arises in higher cortical areas. For example, clusters of neurons with similar binocular disparity preferences are found in an area of visual association cortex known as V5 or middle temporal (MT; see Chapters 3 and 23).[27] The depth perception of a trained monkey was predictably altered when such a cluster was electrically stimulated.

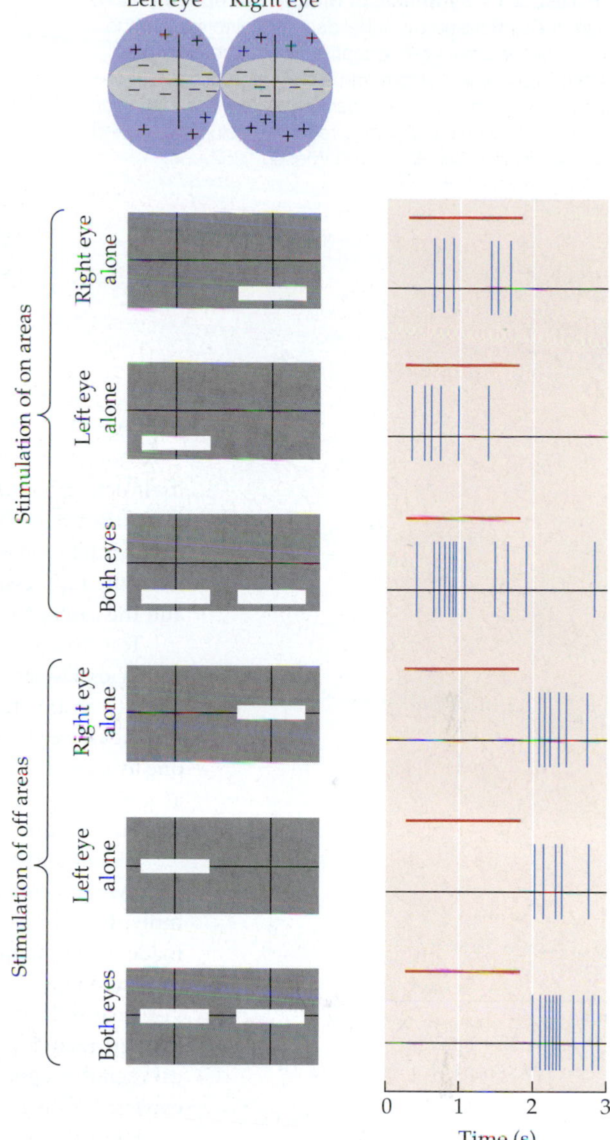

FIGURE 2.10 Binocular Activation of a Simple Cortical Neuron that has identical receptive fields in both eyes. Simultaneous illumination of corresponding on areas (+) of right and left receptive fields is more effective than stimulation of one alone (upper three records). In the same way, stimulation of off areas (–) in the two eyes reinforces off discharges (lower three records). In contrast, cells used for depth perception have receptive fields in both eyes but in disparate regions of the visual field. Such cells require that the bar be placed farther from or closer to the eye than the plane of focus. (After Hubel and Wiesel, 1959.)

Synthesis of the Simple Receptive Field

In 1962, Hubel and Wiesel provided a tentative hypothesis to explain the origin of cortical receptive fields.[24] Their scheme had the advantage of using known mechanisms to explain how a nerve cell can respond so selectively to a visual pattern—such as the oriented lines that excite simple cells. They suggested that in the cortex, simple cell receptive fields behave as if they are built from large numbers of geniculate fields.[24] This is illustrated in Figure 2.11, in which the fields of geniculate neurons connected to a cortical cell are lined up in such a way that a properly oriented bar of light, traversing their centers, would excite them all strongly. If the bar were widened or displaced slightly to either side, it would fall on the inhibitory surround of each cell and reduce or stop the excitatory output. Convergence of these geniculate neurons could produce a cortical cell whose optimal stimulus would be just such an oriented bar of light.

Connections such as these were postulated by Hubel and Wiesel as the simplest that could account for orientation selectivity. That is, the pattern of geniculate innervation

[26]Barlow, H. B., Blakemore, C., and Pettigrew, J. D. 1967. *J. Physiol.* 193: 327–342.

[27]DeAngelis, G. C., Cumming, B. G., and Newsome, W. T. 1998. *Nature* 394: 677–680.

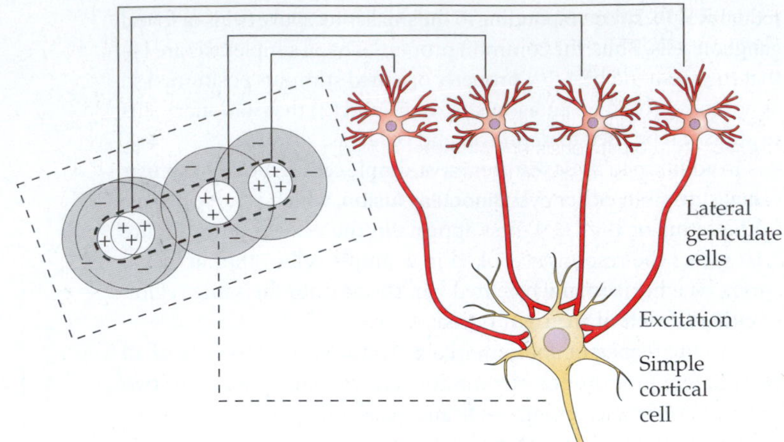

FIGURE 2.11 Synthesis of Simple Receptive Fields. Hypothesis devised by Hubel and Wiesel to explain the synthesis of simple-cell receptive fields. The elongated receptive field of a simple cell is produced by the convergence of inputs from many geniculate neurons (only four are shown) whose concentric receptive fields are aligned on the retina. (After Hubel and Wiesel, 1962.)

itself determines the response characteristics of cortical neurons. This would constitute a feed forward mechanism. An alternative hypothesis was that intracortical connections and lateral inhibition were responsible for the sharpness of receptive field orientation, stemming from the suppression of excitability by laterally placed or inappropriately oriented stimuli and the insensitivity to contrast.

Tests to distinguish between the various schemes could not be made directly by Hubel and Wiesel, since they made all their recordings with extracellular electrodes. Intracellular recordings are much more difficult to make and to maintain without damaging the cell. But they have an advantage: unlike extracellular recordings, an intracellular electrode allows one to assess the potency as well as the source of excitatory and inhibitory synapses arriving at a cell from various inputs. Thus, fibers coming to a neuron from the lateral geniculate nucleus or from the cortex can be stimulated selectively and their effects observed directly.

Ferster and his colleagues achieved a major advance toward a more detailed analysis of how a receptive field is built up from its inputs. They succeeded in recording from individual simple and complex cells with intracellular microelectrodes.[28–30] Recordings made from a simple cell that responds to a vertically oriented bar with high specificity are shown in Figure 2.12. In such records, one can observe the synaptic potentials activated by visual stimuli. They are produced by transmitter release from geniculate axons as well as from intracortical connections. These experiments have shown that direct geniculocortical excitatory potentials sum to greater amplitudes at preferred orientations, as would be expected if the arrangement illustrated in Figure 2.11 existed.

Figure 2.13 shows the effect of local cooling of the cortex.[31] The procedure abolishes all polysynaptic activity arising from intracortical connections, while leaving the geniculate

[28] Ferster, D., Chung, S., and Wheat, H. 1996. *Nature* 380: 249–252.

[29] Finn, I. M., Priebe, N. J., and Ferster, D. 2007. *Neuron* 54: 137–152.

[30] Priebe, N. J., and Ferster, D. 2008. *Neuron* 57: 482–497.

[31] Lampl, I. et al. 2001. *Neuron* 30: 263–274.

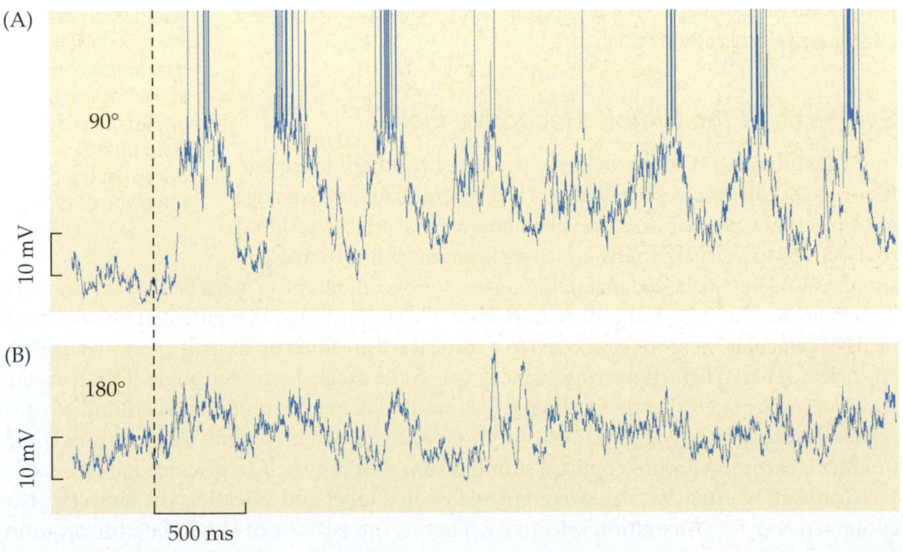

FIGURE 2.12 Intracellular Recording from a Simple Cell in Primary Visual Cortex of an Anesthetized Cat. With repeated presentations of a visual stimulus oriented vertically (90° in A), the cell shows strong depolarization, clear-cut synaptic excitatory potentials, and bursts of action potentials. In (B), similar stimuli applied at right angles (i.e., horizontally) fail to depolarize or stimulate the cell. Stimuli, which consisted of moving gratings, were applied at the time of the dotted line at the beginning of the traces. (After Lampl et al., 2001.)

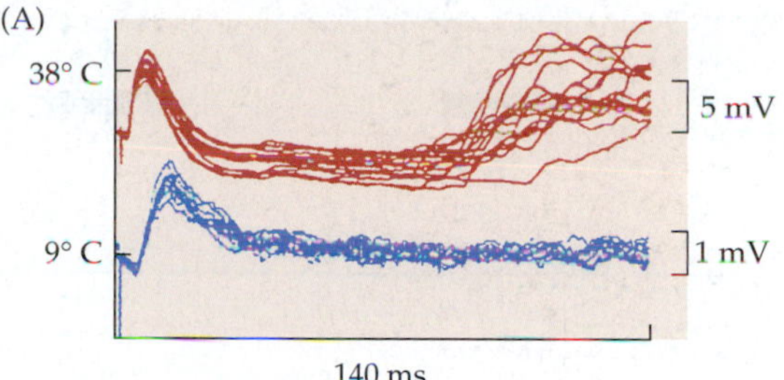

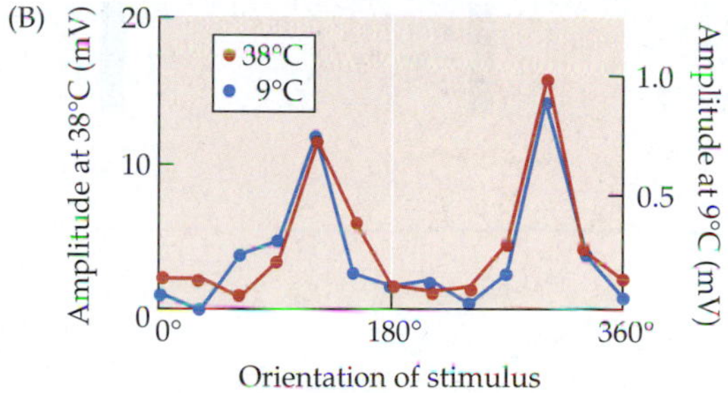

FIGURE 2.13 Effect of Cortical Cooling on the Response of Simple Cells (A) Electrical stimulation of the lateral geniculate nucleus causes short latency (monosynaptic) synaptic potentials at 38°C in a simple cell. These become smaller and slower at 9°C. Note that cooling eliminates long latency (polysynaptic) signals. (B) Orientation tuning to visual stimuli is comparable at both temperatures, consistent with the hypothesis that geniculate input specifies the simple cell receptive field and does not require intracortical feedback. (After Ferster, Chung, and Wheat, 1996.)

input intact so that only direct monosynaptic geniculate input persists (albeit slowed and reduced in amplitude). As shown in Figure 2.13B the orientation tuning remains even though the late (intracortical) response peak is absent. Although the pattern of geniculate inputs is sufficient for orientation tuning of simple cells in visual cortex, additional refinement is provided by both inhibitory and excitatory intracortical connections. Intracellular recordings from simple cells show that illumination of surrounding off areas produces inhibitory synaptic potentials. These can serve to sharpen orientation selectivity and to maintain tuning as visual contrast varies but are not, on their own, responsible for the formation of on and off areas.

Intracellular recordings[30] and cross correlations[32–34] have been made while the receptive field of a simple cell is mapped by small spots. These analyses make it possible to predict how the receptive field organization of a simple cell enables it to respond to a particular orientation and direction of movement (see below).

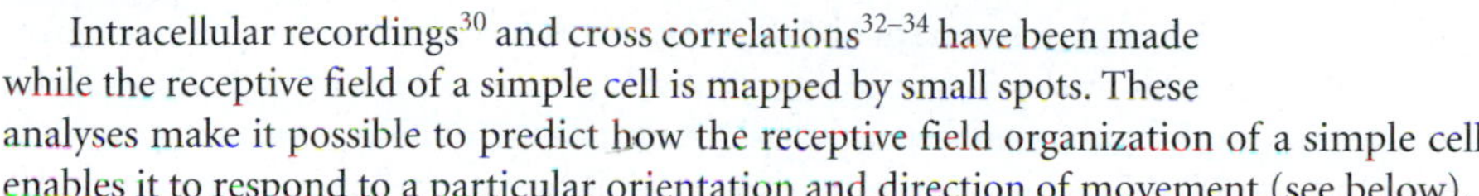

Responses of Complex Cells

In recordings made from individual neurons in the visual cortex, one finds, in addition to simple cells, other neurons, called complex cells, which behave quite differently. These complex cells, which are abundant in layers 2, 3, and 5, have two important properties in common with simple cells: (1) illumination of the entire field is ineffective, and (2) they require specific field axis orientation of a dark–light boundary. The responses of a complex cell are shown in Figure 2.14. Complex cells, like simple cells, have comparable receptive fields in corresponding regions of both retinas (see Figure 2.10). The demand, however, for precise positioning of the stimulus, observed in simple cells, is relaxed in complex cells. In addition, there are no longer distinct on and off areas that can be mapped with

[32] Alonso, J. M., et al. 2006. *Prog. Brain Res.* 154: 3–13.

[33] Alonso, J. M. 2009. *J. Physiol.* 587: 2783–2790.

[34] Alonso, J. M. 2002. *Neuroscientist* 8: 443–456.

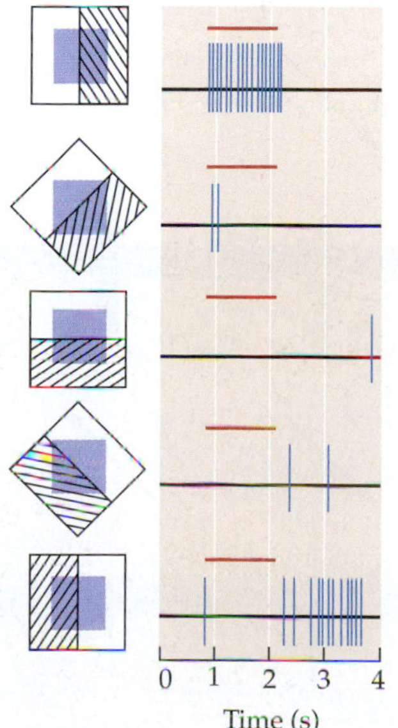

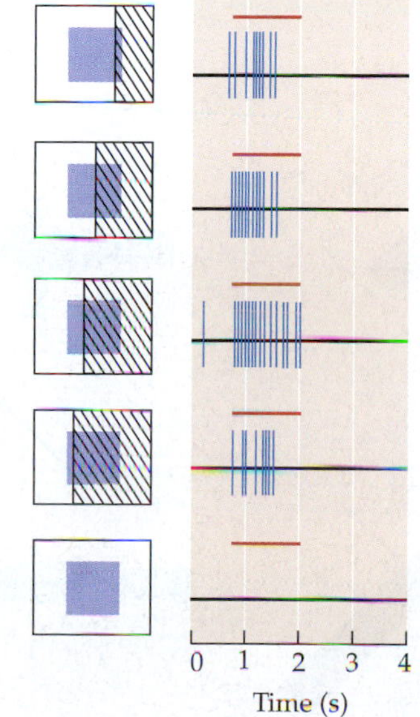

FIGURE 2.14 Responses of a Complex Cell in Cat Striate Cortex. The cell responds best to a vertical edge located within its receptive field (the blue square). (A) With light on the left and dark (hatching) on the right (first record), there is an on response. With light on the right (bottom record), there is an off response. Orientation other than vertical is less effective. (B) The position of the border within the field is not important. Illumination of the entire receptive field (bottom record) gives no response. (After Hubel and Wiesel, 1962.)

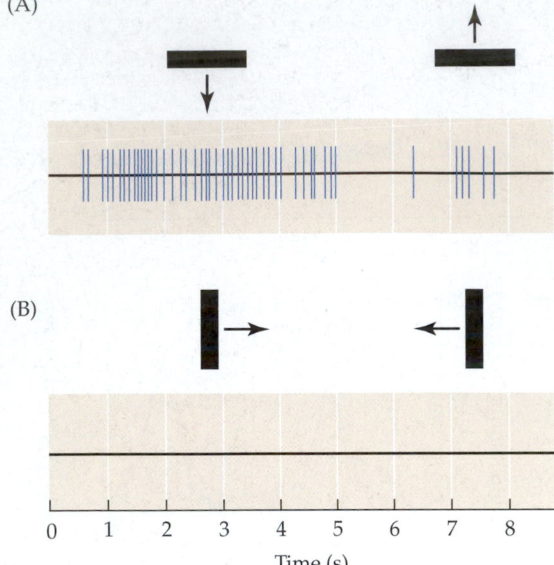

FIGURE 2.15 Preferred Direction of Movement for a Complex Cell in Cat Visual Cortex. In (A) the cell gives brisk responses to the downward movement of a horizontal bar but only a feeble response to upward movement. (B) A vertically oriented bar evokes no firing of action potentials in the same cell. (After Hubel and Wiesel, 1962.)

small spots of light. As long as a properly oriented stimulus falls within the boundary of the receptive field, most complex cells will respond, as in the examples illustrated in Figure 2.14. In this cell, the vertical edge causes nearly equivalent responses at any of four locations. Other orientations are ineffective. Extending the visual stimulus beyond the boundary of the field has no effect. The meaning of the signals arising from complex cells, therefore, differs significantly from that of simple cells. The simple cell localizes an oriented bar of light to a particular position within the receptive field, whereas the signals of a complex cell provide information about *orientation without strict reference to position.*

RESPONSES TO MOVING STIMULI Many simple and complex cells respond best to moving slits or bars of fixed width and precise orientation. Some cells respond only when the movement is in one direction. An example is shown in Figure 2.15. Downward movement of the bar evokes a far brisker response than upward movement; as the lower traces show, the orientation of the bar remains critical. Directional sensitivity is a feature commonly found in complex cells. Still more demanding complex cells can be found than the ones previously described, particularly in other cortical areas (see Chapters 3 and 23). Once again, intracellular recordings enable one to explain how such movement sensitivity is achieved in cortical neurons as the result of synaptic potentials evoked by visual stimuli.[30]

CORTICAL NEURONS THAT RESPOND TO LINES THAT STOP Another interesting feature of certain simple and complex cells is that they require bars or edges that do not extend beyond a fixed length.[25,35] Once again, the orientation of the stimulus and, often, the direction of movement are critical. Figure 2.16 shows recordings from a cell that responded best to an obliquely oriented bar of light. Stretching the bar beyond an optimal length reduced its effectiveness as a stimulus. It is as though there are additional off areas that exist on either side of the fields shown in dashed lines in Figure 2.16. Light with the appropriate pattern falling in these areas tends to suppress firing. However, *diffuse* illumination outside the field does not diminish the response. It is therefore not a conventional *off* area, as such. **End inhibition** or **end stopping** are the names given to this property.

[35] Pack, C. C. et al. 2003. *Neuron* 39: 671–680.

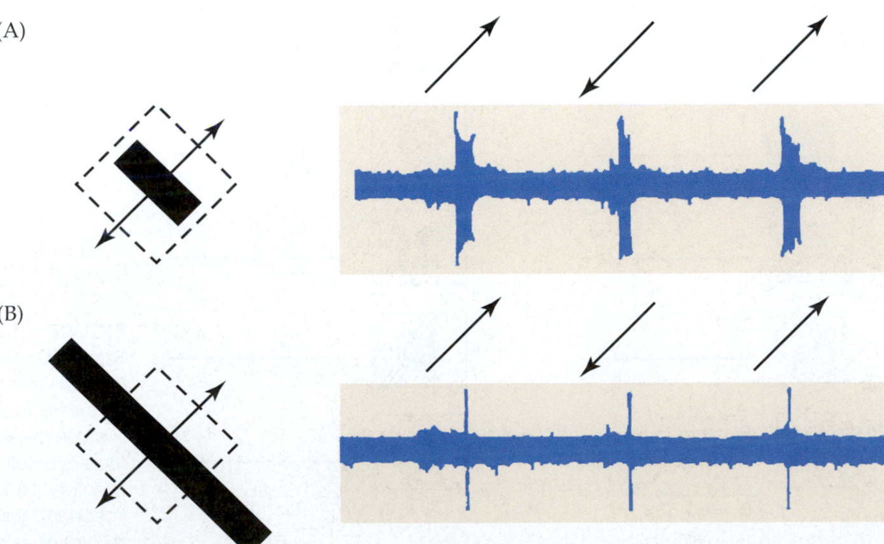

FIGURE 2.16 Action Potentials in a Complex Cell in layer 2 of cat visual cortex in response to a short bar. In (A) the neuron fires when the obliquely oriented bar is moved down or up. In this trial, the bar was short. In (B), however, with a longer bar of the same orientation and the same direction of movement, there was almost no response. Hence, this cell recognizes an edge that stops, a characteristic known as end stopping. (After Hubel and Wiesel, 1968.)

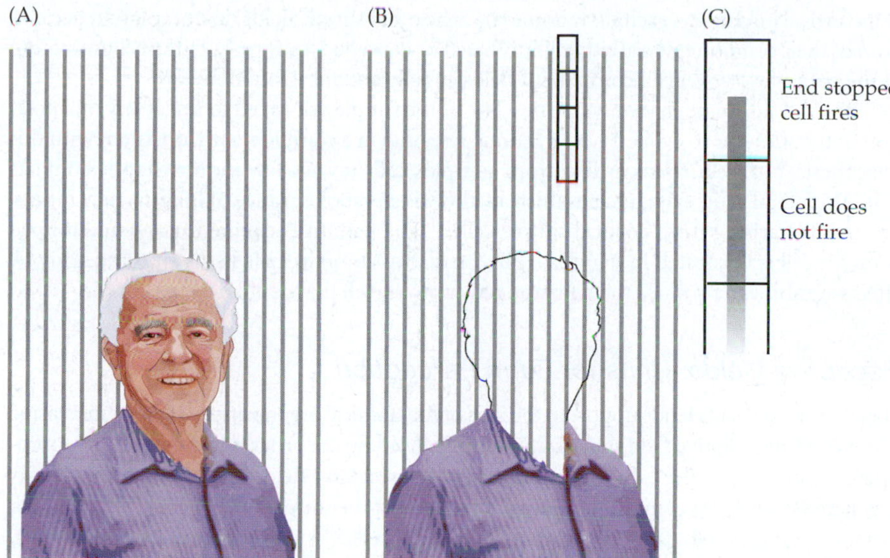

FIGURE 2.17 The Completion Phenomenon. (A) The field of a person viewing the head of his friend, who is sitting against striped wallpaper. (B) During a migraine attack, a small area of complete blindness occurs in the field, yet the stripes continue through the blind area. A possible explanation, at the cellular level, is that under normal conditions end-stopped neurons would be silent, except at the top and the bottom of the field where the line stops (C).

Hence, for a complex cell of this sort the best stimulus is an appropriately oriented bar or edge, of a defined length, that stops at a particular place. The cell requires a discontinuity in the visual pattern if it is to respond. It is worth emphasizing what a major achievement it was for Hubel and Wiesel to show unequivocally that a particular neuron requires *this* particular stimulus for it to fire.

The properties of the neurons that require a line to *stop* can provide a cue to a clinical observation in humans, known as the completion phenomenon, which occurs when small retinal or cortical lesions cause blind areas or scotomas. In this condition, forms or shapes projected onto the retina appear to contain an empty region in the visual field corresponding to the site of the lesion. Yet, if subjects look at striped wallpaper (Figure 2.17), a straight line pattern, or a zebra, they see the pattern continue through the blind area.

Synthesis of the Complex Receptive Field

In the same way that the receptive field of a simple cortical cell can be built up by the convergence of geniculate afferents, so too can the receptive field of a complex cortical cell be synthesized by combining those of simple cells.[24,36] Figure 2.18 presents a hypothetical complex cell that is excited by a vertical edge stimulus that falls anywhere within the area of the receptive field. This is so because wherever the edge falls, one of the simple fields is traversed

[36]Hirsch, J. A., and Martinez, L. M. 2006. *Trends Neurosci.* 29: 30–39.

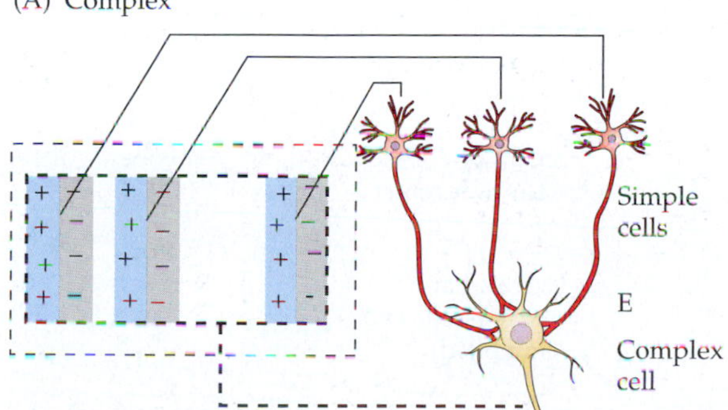

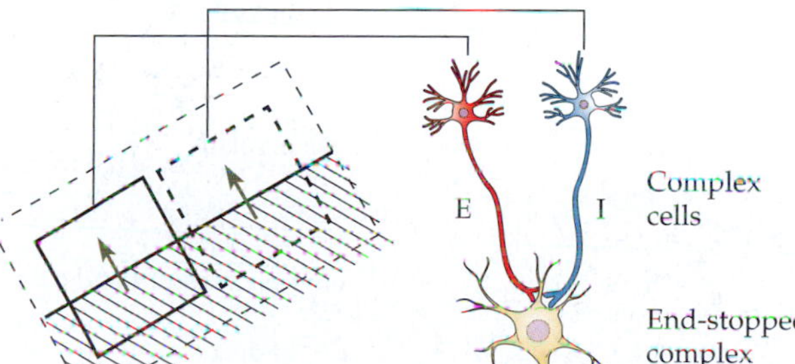

FIGURE 2.18 Synthesis of Complex Receptive Fields. (A) Convergent input from simple cells responding best to a vertically oriented edge at slightly different positions could bring about the behavior of a complex cell that responds well to a vertically oriented edge situated anywhere within its field. (B) Each of two complex cells responds best to an obliquely oriented edge. But, one cell is excitatory and the other is inhibitory to the end-stopped complex cell. Hence, an edge that covers both fields (as in the sketch) is ineffective, whereas a corner restricted to the left field would excite. E = excitation; I = inhibition. (After Hubel and Wiesel, 1962, 1965b.)

at its vertical inhibitory–excitatory boundary. The other simple fields do not respond because both of their components are either illuminated or darkened uniformly. Diffuse illumination of the entire field covers all component fields equally and therefore none fires.

One can postulate that only one or a few of the simple cells need to fire at any one position of the stimulus to evoke a near maximal response in a complex cell. Consistent with this hypothesis, intracellular recording from complex cells reveals few monosynaptic contacts from the lateral geniculate nucleus. Instead, there is a preponderance of long-latency inputs, presumably arising from cortical simple cells.[37] The tentative scheme for an end-stopped complex cell is illustrated in Figure 2.18B. There, two complex cells with opposing synaptic effects combine to produce the end-stopped complex cell that could detect a corner.

Receptive Fields: Units for Form Perception

Together these results lend support to the idea of hierarchical organization whereby increasing complexity in receptive field organization is produced by the orderly convergence of appropriate inputs. This feature does not mean that each receptive field of succeeding complexity is generated solely by combining inputs derived from the immediately preceding level. For instance, complex cells can receive inputs from LGN cells.[38] Moreover, numerous horizontal connections between visual neurons are prevalent throughout the cortex.[39] As mentioned previously, cortical input serves to sharpen the orientation tuning of simple cells. Nonetheless, the original working hypothesis proposed by Hubel and Wiesel in 1962 continues to provide a clear, elegant, and reasonable conceptual framework upon which to design new experimental tests.

Table 2.1 summarizes some of the key characteristics of receptive fields at successive levels of the visual system. Each eye conveys to the brain information collected from regions of various sizes on the retinal surface. The emphasis is not on diffuse illumination or the absolute amount of energy absorbed by photoreceptors. Rather the visual system extracts information about contrast by comparing the level of activity in cells with adjoining receptive fields. At each higher level, such neural computations define ever more elaborate spatial features.

This process can be appreciated by considering the types of signals generated by a square patch of light as shown in Figure 2.19. On-center retinal ganglion cells within the square will increase their discharge (at least initially), while off-center cells are suppressed. The

TABLE 2.1
Characteristics of receptive fields at successive levels of the visual system

Type of cell	Shape of field	What is best stimulus?	How good is diffuse light as a stimulus?
Photoreceptor	⊕	Light	Good
Ganglion		Small spot or narrow bar over center	Moderate
Geniculate		Small spot or narrow bar over center	Poor
Simple (layers 4 and 6)		Narrow bar or edge (some end-inhibited)	Ineffective
Complex (outside layer 4)		Bar or edge	Ineffective
End-inhibited complex (outside layer 4)		Line or edge that stops; corner or angle	Ineffective

[37] Finn, I. M., and Ferster, D. 2007. *J. Neurosci.* 27: 9638–9648.

[38] Martinez, L. M., and Alonso, J. M. 2001. *Neuron* 32: 515–525.

[39] Stettler, D. D. et al. 2002. *Neuron* 36: 739–750.

best-stimulated ganglion cells, however, are those subjected to the maximum contrast—that is, those having centers lying immediately adjacent to the boundary between the light and dark areas and consequently, having less activation of their inhibitory regions. Neurons in the LGN behave similarly. Cortical cells having receptive fields lying either completely within the square or outside it send no signals because diffuse illumination is not an effective stimulus. Only those simple cells with receptive fields oriented to coincide with the horizontal or vertical boundaries of the square will be stimulated.

Similar considerations apply to the stimulation of complex cells, which also require properly oriented bars or edges. End-inhibited simple or complex cells detect the corner of the square or a line that stops. There is an important difference between the two, however, which is related to the fact that the eyes continually make small saccadic eye movements (see Chapter 24). These are not perceived as motion but are essential for preventing photoreceptor adaptation as the eyes fixate. Each microsaccade causes a new population of simple cells with exactly the same orientation but with slightly different receptive field location to be thrown into action. For those complex cells that see the square, however, a boundary of appropriate orientation can be anywhere within the field. Thus, many of the same complex cells will continue to fire during eye movements, as long as the displacement is small and the pattern does not pass outside the receptive field of the cell.

If the preceding considerations are valid, the surprising conclusion is that the primary visual cortex receives little information about the absolute level of uniform illumination within the square. Signals arrive only from the cells with receptive fields situated close to the border. This hypothesis is supported by an easily replicated psychophysical experiment. A square that appears light when surrounded by a black border can be made to appear dark merely by increasing the brightness of the surround. In other words, we perceive the difference or contrast at the boundary, and it is by that standard that the brightness in the uniformly illuminated central area is judged.

This is not to say that general luminance is entirely ignored by the nervous system. For example, we do know if a room is dark or light and that the diameter of the eye's pupil varies with ambient light intensity. Such responses depend on a particular group of ganglion cells that send their output to a region of the brain called the suprachiasmatic nucleus, which regulates the day–night circadian rhythm (see Chapters 17 and 20).

Is orientation of stimulus important?	Are there distinct on and off areas within receptor fields?	Are cells driven by both eyes?	Can cells respond selectively to movement in one direction?
No	No	No	No
No	Yes	No	No
No	Yes	No	No
Yes	Yes	Yes (except in layer 4)	Some can
Yes	No	Yes	Some can
Yes	No	Yes	Some can

FIGURE 2.19 Responses of Neurons to a Pattern. When a square patch of light is presented to the retina, signals arise predominately from ganglion cells and lateral geniculate cells whose receptive fields lie close to the border of the square—and not from those subjected to uniform light or darkness. The cell whose fields are situated exactly at the four corners of the square will fire best of all. Simple and complex cells having receptive fields with the correct position (i.e., situated along the border or at a corner) and the correct orientation preference will also fire, while those not on the border or with an inappropriate orientation will remain silent. Activity level is indicated by the number of radiating lines around each field.

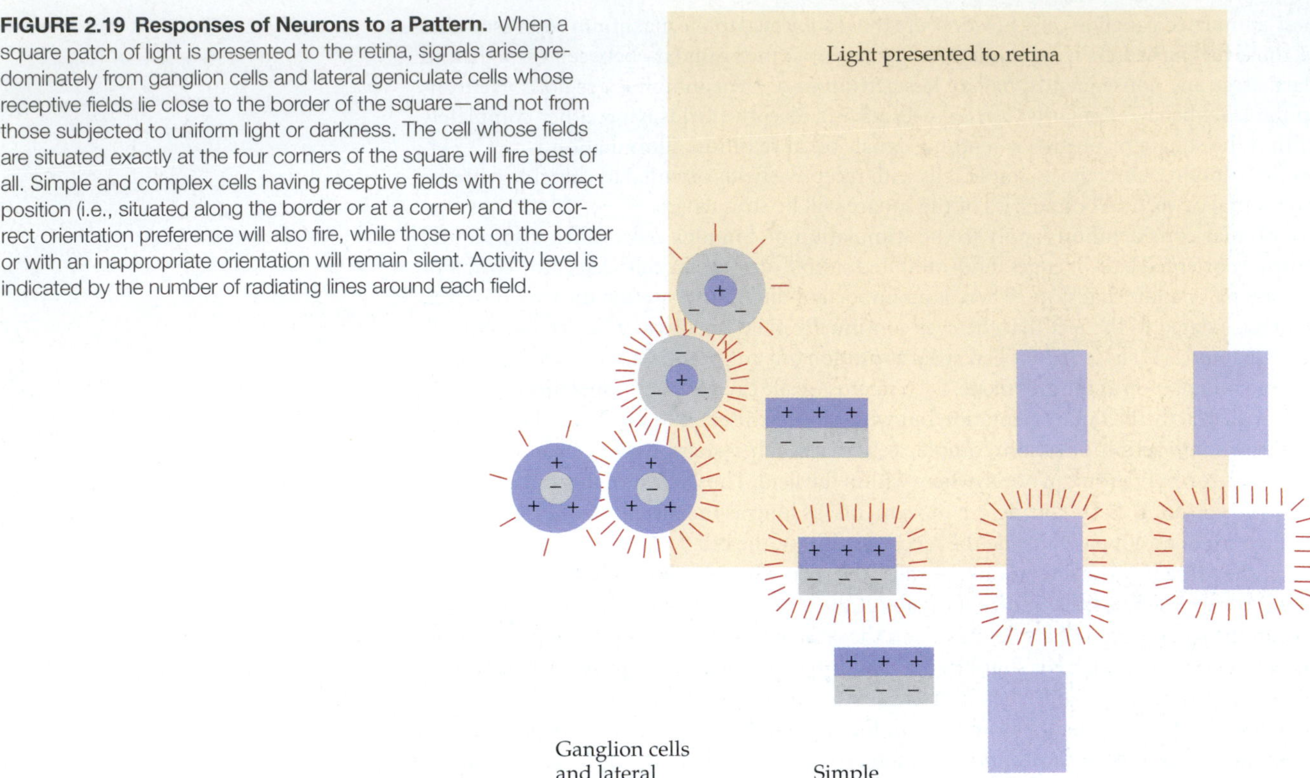

The work of Hubel and Wiesel and others has made clear that the first general step in visual analysis is to construct representations of lines or edges from the center–surround, spot-like receptive fields of the retina. In V1, the visual system begins to derive form from the retinal map. Finding these connections has given us our first glimpse into how the brain computes. But these first steps in the detection of a line or even a corner remain a long way from complete visual recognition during which shape, color, size, and motion are all combined, so that we can recognize a car, a cow, or the face of a friend. In Chapter 3, we describe how geniculate and cortical neurons are arranged spatially, in a highly orderly manner.

It is appropriate to close this chapter with a quotation from Sherrington, written long before receptive fields were mapped for single cells. Sherrington's somewhat opaque style (unlike that of Helmholtz) often makes it difficult to read his profoundly original papers and books. Yet the following paragraph reveals his poetic insight into the physiology of vision:

> The chief wonder of all we have not touched on yet. Wonder of wonders, though familiar even to boredom. So much with us that we forget it all the time. The eye sends, as we saw, into the cell-and-fibre forest of the brain throughout the waking day continual rhythmic streams of tiny, individual evanescent, electrical potentials. This throbbing streaming crowd of electrified shifting points in the spongework of the brain bears no obvious semblance in space-pattern, and even in temporal relation resembles but a little remotely the tiny two-dimensional upside-down picture of the outside world which the eye-ball paints on the beginnings of its nerve-fibres to electrical storm. And the electrical storm so set up is one which affects a whole population of brain-cells. Electrical charges having in themselves not the faintest elements of the visual—having, for instance, nothing of "distance," "right-side-upness," no "vertical," nor "horizontal," nor "color," nor "brightness," nor "shadow," nor "roundness," nor "squareness," nor "contour," nor "transparency," nor "opacity," nor "near," nor "far," nor visual anything—yet conjure up all these. A shower of little electrical leaks conjures up for me, when I look, the landscape; the castle on the height, or when I look at him, my friend's face and how distant he is from me they tell me. Taking their word for it, I go forward and my other senses confirm that he is there.[40]

[40] Sherrington, C. S. 1951. *Man on His Nature.* Cambridge University Press, Cambridge.

SUMMARY

- The receptive field of a neuron in the visual system is the area of the retina or visual field that, upon illumination, enhances or inhibits signaling.
- An important strategy for analyzing the visual system is to define the optimal light stimulus for each neuron.
- Ganglion cells in the retina and cells in the lateral geniculate nucleus respond best to contrast in the form of small spots of light surrounded by darkness or dark spots surrounded by light.
- Ganglion cells and lateral geniculate nucleus cells respond poorly to diffuse light.
- Simple cells in striate cortex, V1, respond to oriented light or dark bars. Their receptive fields can be mapped with spots of light, as though composed of adjoining lateral geniculate center–surround receptive fields.
- Complex cells in striate cortex also respond to oriented bars or edges. However, their receptive fields cannot be mapped with spots of light. They result from the convergence of multiple simple cells with adjoining receptive areas.
- Simple and complex cells respond to movement of bars or edges.
- End inhibition results when an additional suppressive zone specifies the optimal length for a simple or complex cell. These cells provide information about where a line ends or about the position of a corner.

Suggested Reading

General Reviews

Alonso, J. M. 2009. My recollections of Hubel and Wiesel and a brief review of functional circuitry in the visual pathway. *J. Physiol.* 587: 2783–2790.

Guillery, R. W. 2005. Anatomical pathways that link perception and action. *Prog. Brain Res.* 149: 235–256.

Hubel, D. H., and Wiesel, T. N. 2005. *Brain and Visual Perception.* Oxford University Press, NY, USA.

Priebe, N. J., and Ferster, D. 2008. Inhibition, spike threshold, and stimulus selectivity in primary visual cortex. *Neuron* 57: 482–497.

Original Papers

Finn, I. M., and Ferster, D. 2007. Computational diversity in complex cells of cat primary visual cortex. *J. Neurosci.* 27: 9638–9648.

Hubel, D. H., and Wiesel, T. N. 1959. Receptive fields of single neurones in the cat's striate cortex. *J. Physiol.* 148: 574–591.

Hubel, D. H., and Wiesel, T. N. 1968. Receptive fields and functional architecture of monkey striate cortex. *J. Physiol.* 195: 215–243.

Kuffler, S. W. 1953. Discharge patterns and functional organization of the mammalian retina. *J. Neurophysiol.* 16: 37–68.

Lampl, I., Anderson, J. S., Gillespie, D. C., and Ferster, D. 2001. Prediction of orientation selectivity from receptive field architecture in simple cells of cat visual cortex. *Neuron* 30: 263–274.

Stettler, D. D., Das, A., Bennett, J., and Gilbert, C. D. 2002. Lateral connectivity and contextual interactions in macaque primary visual cortex. *Neuron* 36: 739–750.

CHAPTER 3
Functional Architecture of the Visual Cortex

Vertically arranged clusters of neurons with similar properties are organized systematically through the thickness of the visual cortex. Each cluster receives its input from a small region of the visual field that is mapped onto the cortex. Maps on their own, however, do not provide insights into the functional architecture of the visual cortex; for that, one needs to know how the neurons that respond to specific visual stimuli are arranged in three dimensions. Detailed information about structure turns out to be essential for an understanding of how visual information is processed in the brain. The six layers of the primary visual cortex (V1) have specific organizational properties: incoming fibers from the lateral geniculate nucleus for the most part end in layer 4. Neurons in upper (2, 3) and deeper layers (5, 6) receive their inputs from the cortex and project to other layers or areas. Neurons that are preferentially driven by the right or the left eye are grouped in **ocular dominance columns**. Neurons whose line or edge preferences are at similar angles constitute **orientation columns**. The ocular dominance and orientation columns were first discovered by recording electrical activity, cell after cell, as electrodes traversed the cortical thickness. Ocular dominance and orientation columns are revealed directly in the living animal by optical techniques that display activated regions in the cortex. The word "column" is still used even though the arrangement consists of cortical slabs (for ocular dominance) or pinwheels (for orientation), rather than narrow columns. Each column of cortical cells functions as a module, operating on input from one location in visual space and forwarding the processed information to other areas. The location of the receptive fields of neurons in a column is the same throughout the depth of the cortex.

Neurons from the lateral geniculate nucleus with different properties supply distinct regions of the visual cortex. Thus, magnocellular axons (M, movement, contrast, depth) end more superficially in layer 4 than do parvocellular axons (P, position, color). On the other hand, koniocellular geniculate axons (K, short wavelength) project differentially to clusters outside layer 4, called blobs. These structures were first revealed by stains for the enzyme cytochrome oxidase. Although mixing of M, P, and K streams occurs throughout the visual pathways, features such as color and motion are analyzed separately. This phenomenon is illustrated not only by physiological recordings but also by the fact that a lesion in a discrete region of the brain can result in selective loss of a feature, such as color, rather than in an overall reduction in quality of visual images.

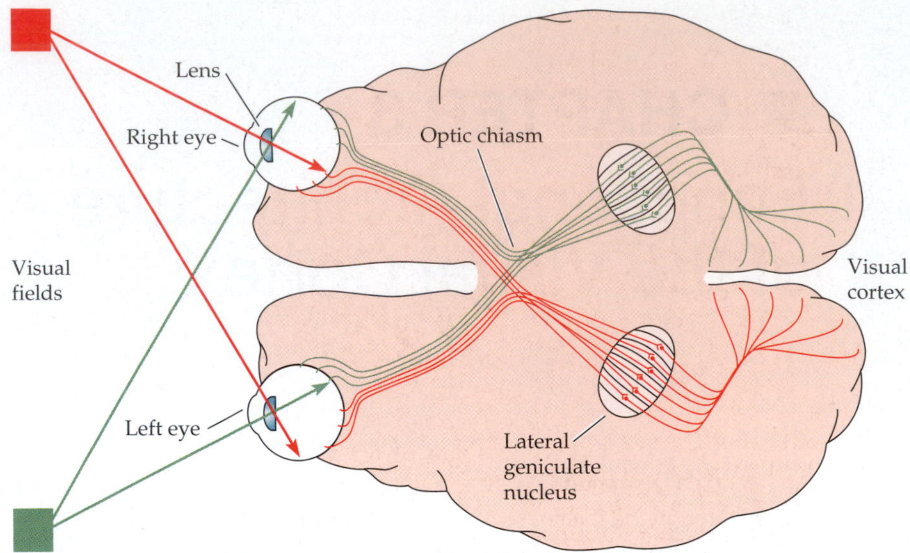

FIGURE 3.1 Visual Pathways. The left side of each retina, shown in red, projects to the left lateral geniculate nucleus. Thus, the left visual cortex receives information exclusively from the right half of the visual field.

In Chapter 2, we followed the flow of information from the retina to the primary visual cortex and described the effects of visual stimuli on successively higher-order cortical cells. This approach provided an understanding of the cellular mechanisms for the initial analysis of form and motion at each point in the visual field. Our task now is to describe how neurons are organized spatially to subserve their functions in the visual pathways (Figure 3.1). In this chapter, we show how inputs from the two eyes arrive at the visual cortex and how cells are grouped according to their physiological functions. We next examine the evidence that motion and color are analyzed in parallel channels through the cortex. Finally, we introduce examples of higher levels of processing in visual areas beyond the primary visual cortex, a topic that is dealt with in detail in Chapter 23.

From Chapters 1 and 2, it is apparent that knowledge of structure is not just a dry, didactic exercise that can be skipped over. Such knowledge turns out to be essential if one is to understand how different aspects of a scene (for example, *form, color, depth* and *motion*) are encoded in the cortex. A remarkable feature of the visual structure to be described in this chapter is that completely new and unexpected findings about how the brain is structured anatomically—came first from a series of *physiological* recordings made from single neurons, one at a time. Hubel and Wiesel have recalled that: "...to attack such a three-dimensional problem with a one-dimensional weapon is a dismaying exercise in tedium, like trying to cut the back lawn with a pair of nail scissors."[1]

It will be shown that in visual area 1 (V1) there is point-to-point representation of the retina: within this retinotopic map, inputs from the two eyes come together and are sorted into ocular dominance columns. These columns consist of vertically stacked groups of neurons that respond better either to the left eye or right eye. In addition, there are other major functional groupings, consisting of orientation columns of cells (all of which respond to a specific line orientation, such as vertical, oblique, or horizontal) and still other clusters of cells devoted to colors. Unusual in this chapter, as in Chapter 2, is the continuous reference to older experiments. The reason is that the work by Hubel and Wiesel represents a solid foundation, which has been confirmed, modified, and extended by others but not superseded.

Retinotopic Maps

Each area of cortex contains its own representation of the visual field, projected in an orderly, retinotopic manner.[2] Maps of the visual fields in adjacent areas are orientated in mirror fashion across cortical gyri (folds), a feature that facilitates communication.[3] Before the era of single-cell analysis, projection maps were made by shining light onto parts of the retina and recording with gross electrodes.[4] These maps, and those made by brain-imaging techniques, such as positron emission tomography (PET) and functional magnetic reso-

[1] Hubel, D. H., and Wiesel, T. N. 1977. *Proc. R. Soc. Lond. B, Biol. Sci.* 198: 1–59.

[2] Talbot, S. A., and Marshall, W. H. 1941. *Am. J. Ophthalmol.* 24: 1255–1264.

[3] Van Essen, D. C. 1997. *Nature* 385: 313–318.

[4] Daniel, P. M., and Whitteridge, D. 1961. *J. Physiol.* 159: 203–221.

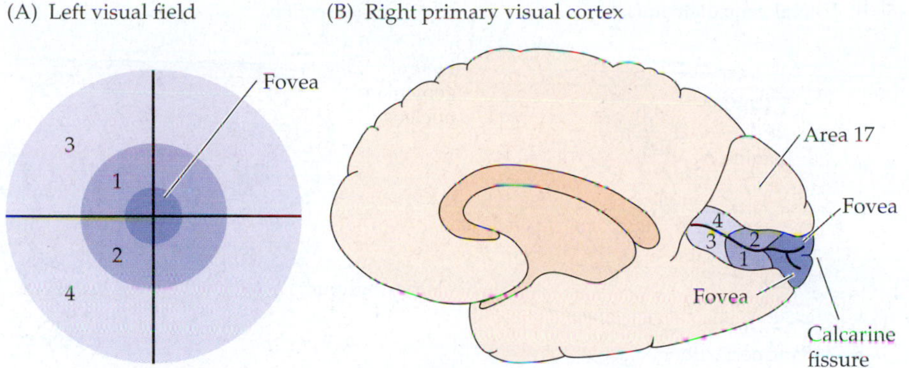

FIGURE 3.2 Visual Field Map of the Cortex. (A) The visual field is divided into a central zone (the fovea) and eight hemiquadrants. The left visual field maps onto the right primary visual cortex. (B) In humans, the primary visual cortex (area 17) lies almost entirely on the medial surface of the occipital lobe. The foveal region is situated in the farthest posterior portion, with the peripheral zones mapping anteriorly. The fovea claims a disproportionate share of the primary visual cortex.

nance imaging (fMRI), show that much more cortical area is devoted to representation of the fovea than to representation of the rest of the retina. This comes as no surprise, since form vision is served principally by the higher densities of photoreceptors in foveal and parafoveal areas and is analogous to the enlarged areas devoted to the hand and face in the primary somatosensory cortex (see Chapter 21). The retinal fovea is mapped onto the occipital pole of the cerebral cortex, while the map of the retinal periphery extends anteriorly, along the medial surface of the occipital lobe (Figure 3.2).[5] Because of image reversal by the eye, the upper visual field appears on the lower retina and is projected to V1 below the calcarine fissure; accordingly, lower visual fields map above the calcarine fissure.

From Lateral Geniculate Nucleus to Visual Cortex

Segregation of Retinal Inputs to the Lateral Geniculate Nucleus

Neurons in the six principal layers of the lateral geniculate nucleus (LGN) are grouped according to structure and function. In primates, including humans (as mentioned in Chapter 2), geniculate layers 6, 4, and 1 are supplied by the contralateral eye, while layers 5, 3, and 2 are supplied by the ipsilateral eye.[6] The complete segregation of endings from each eye into separate cell layers has been shown by electrical recording and a variety of anatomical techniques.[7,8] Particularly striking is the arborization of a single optic nerve fiber, shown in Figure 3.3, that had been injected with the enzyme horseradish peroxidase. The terminals are all confined to the layers supplied by that eye, with no spillover across the border.

Another major feature of the LGN is segregation according to the size of neurons, which in turn bears on their function. As described in Chapter 2, cells in the deep layers 1 and 2 of the lateral geniculate nucleus are larger than those in layers 3, 4, 5, and 6, and this gave rise to the terms magnocellular (**M**, large cells with large receptive fields) and parvocellular (**P**, small cells with small receptive fields; see Figure 3.3). In Chapter 2, it was shown that the same classification holds for ganglion cells in the retina. M neurons respond preferentially to movement, depth, and contrast, while P neurons respond to position and color. Between each of the M and P layers lies a zone of small cells: the interlaminar, or koniocellular (K), layers. K cells are distinct from M and P cells and provide the visual cortex with a third channel, which is concerned primarily with information about short-wavelength color (blue light).[9,10]

Cytoarchitecture of the Visual Cortex

Visual information passes to the cortex from the LGN through the optic radiation. In the monkey, the optic radiation ends mainly (but not exclusively[11]) in V1 (also called striate cortex or area 17, terms based on anatomical criteria developed at the beginning of the twentieth century). V1 lies posteriorly in the occipital lobe (see Figure 3.2) and can be recognized in cross section by its characteristic appearance. In sections of V1, incoming bundles of fibers form a clear stripe that can be seen by the naked eye (hence the name "striate"). The organization of geniculate projections to the striate cortex is illustrated in Figure 3.5.

In histological sections of the cortex, neurons can be classified according to their shapes. The two principal groups of neurons are known as stellate cells and pyramidal cells.

[5] Van Essen, D. C., and Drury, H. A. 1997. *J. Neurosci.* 17: 7079–7102.
[6] Guillery, R. W. 1970. *J. Comp. Neurol.* 138: 339–368.
[7] Hubel, D. H., and Wiesel, T. N. 1972. *J. Comp. Neurol.* 146: 421–450.
[8] Callaway, E. M. 2005. *J. Physiol.* 566: 13–19.
[9] Roy, S. et al. 2009. *Eur. J. Neurosci.* 30: 1517–1526.
[10] Casagrande, V. A. et al. 2007. *Cereb. Cortex* 17: 2334–2345.
[11] Sincich, L. C. et al. 2004. *Nat. Neurosci.* 7: 1123–1128.

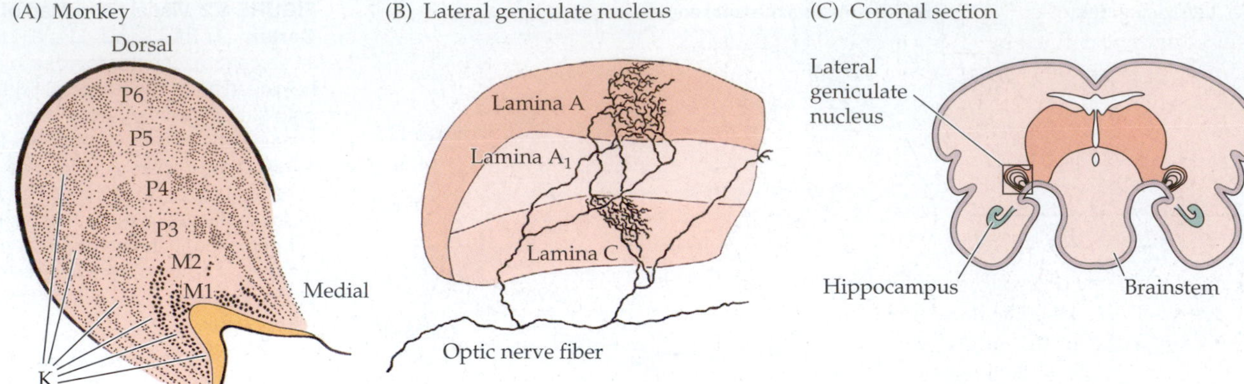

FIGURE 3.3 Organization of Lateral Geniculate Nucleus (LGN). (A) The monkey LGN has six major layers, designated parvocellular (P) 3, 4, 5, 6 and magnocellular (M) 1, 2, which are separated by the koniocellular (K) layers. Each layer is supplied by only one eye and contains cells with specialized response properties. (B) Termination of an optic nerve fiber in cat LGN, which has three principal layers. A single, on-center axon from the contralateral eye was injected with horseradish peroxidase. Branches end in layers A and C, but not in A_1. (C) Coronal section of the brain shows the position and orientation of the LGN in primate brain. (A after Hendry and Calkins, 1998; B after Bowling and Michael, 1980.)

Examples of these cells are shown in Figure 3.4B.[12] The axon of the pyramidal cell is long, descends into white matter, and leaves the cortex; the stellate cell axon tends to terminate locally. The two groups of cells exhibit other variations that bear on their functional properties, such as the presence or absence of spines on the dendrites. There are other fancifully named cortical neurons (double bouquet cells, chandelier cells, basket cells, and crescent cells) as well as neuroglial cells. Characteristically, processes of cells for the most part run vertically through the thickness of the cortex (at right angles to the surface, also referred to as "radially"). Connections between primary and higher-order visual cortices are made by axons that run in bundles through the white matter underlying the cellular layers.[13]

[12] Gilbert, C. D., and Wiesel, T. N. 1979. *Nature* 280: 120–125.

[13] Zeki, S. 1990. *Disc. Neurosci.* 6: 1–64.

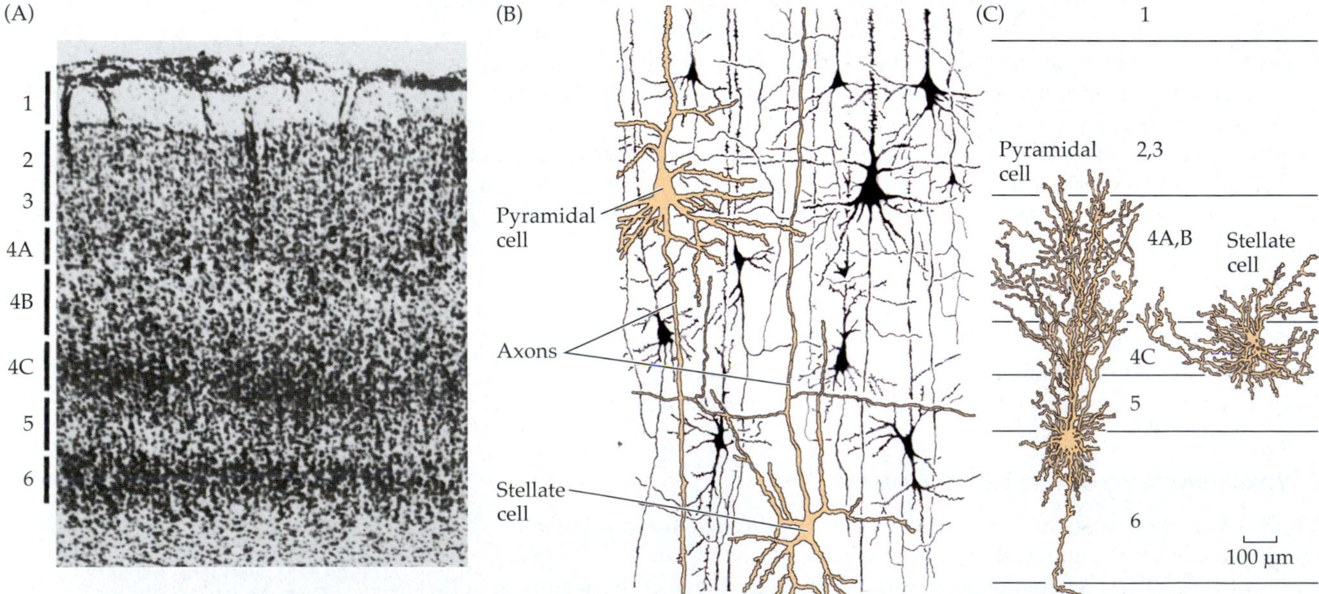

FIGURE 3.4 Architecture of Visual Cortex. (A) Distinct layering of cells in a section of striate cortex of the macaque monkey, stained to show cell bodies (Nissl stain). Fibers, arriving from the lateral geniculate nucleus, end in layers 4A, 4B, and 4C. (B) Drawing of pyramidal and stellate cells in cat visual cortex. The processes (stained with Golgi technique) for the most part run radially through the thickness of the cortex and extend for relatively short distances laterally. (C) Drawing from photographs of a pyramidal cell and a spiny stellate cell in cat cortex that had been injected with horseradish peroxidase after their activity had been recorded. Both were simple cells. (A from Hubel and Wiesel, 1972; B after Ramón y Cajal 1955; C from Gilbert and Wiesel, 1979.)

Other regions of cortex (V2–V5) are also concerned with vision. Their exact boundaries cannot be defined by simple inspection of the brain, but have been delineated by a number of techniques. These include imaging, experimental lesions, psychophysics, and physiology.[14–18] It will be shown that different visual areas communicate with one another in both directions and perform different types of analysis relating to form, color, depth, and movement (see Chapter 23).

Inputs, Outputs, and Layering of Cortex

Six layers of nerve cells are evident within the gray matter of V1, as in other regions of the brain (see Figure 3.4). The inputs are shown in the left-hand side of Figure 3.5. Incoming geniculate fibers for the most part end in layer 4, with some contacts also made in layer 6 (see below).[19] Superficial layers of the cortex receive inputs from the pulvinar a region of the thalamus. Numerous cortical cells, especially those in layer 2 and the upper portions of layers 3 and 5, receive inputs from neurons within the cortex.

The outputs from cortical layers 6, 5, 4, 3, and 2 are shown in the right-hand side of Figure 3.5. A single cell that sends efferent signals out of the cortex can also mediate

[14]Hubel, D. H., and Wiesel, T. N. 1965. *J. Neurophysiol.* 28: 229–289.

[15]Maunsell, J. H., and Newsome, W. T. 1987. *Annu. Rev. Neurosci.* 10: 363–401.

[16]Adams, D. L., and Zeki, S. 2001. *J. Neurophysiol.* 86: 2195–2203.

[17]Orban, G. A., Van Essen, D., and Vanduffel, W. 2004. *Trends Cogn. Sci.* 8: 315–324.

[18]Tsao, D. Y. et al. 2006. *Science* 311: 670–674.

[19]Hubel, D. H. 1988. *Eye, Brain and Vision*. Scientific American Library, New York.

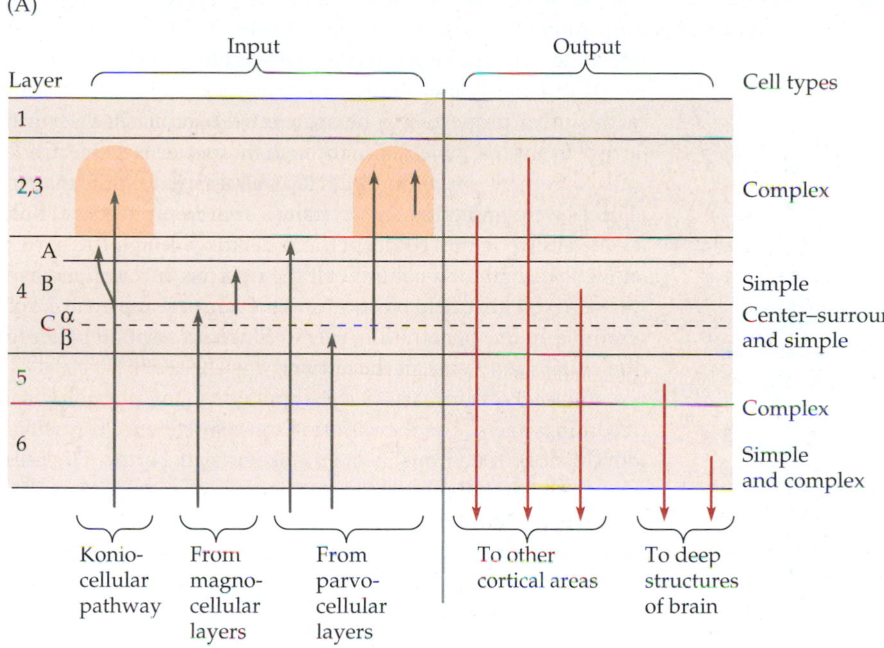

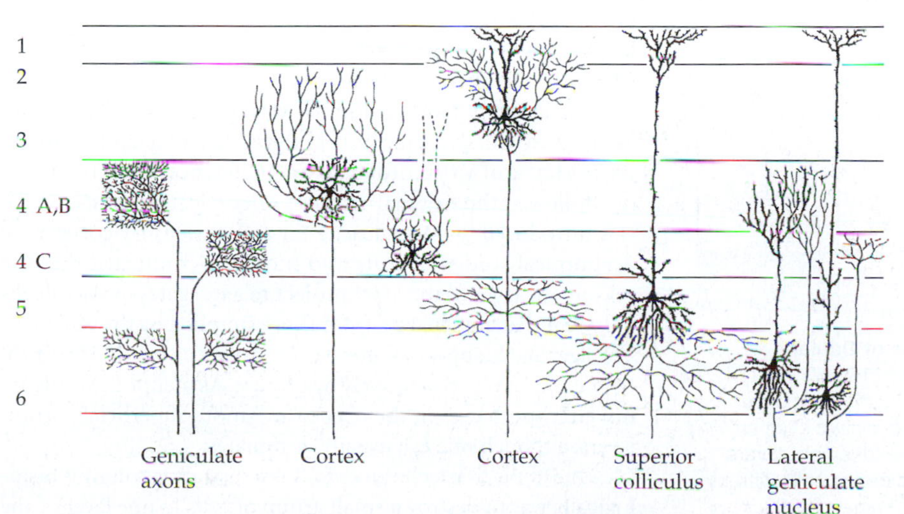

FIGURE 3.5 Connections of Visual Cortex. (A) The layers are shown with their various inputs, outputs, and types of cells. Note that inputs from the lateral geniculate nucleus end mainly in layer 4. Those arising from magnocellular layers end principally in 4Cα and 4B, whereas those from parvocellular layers end in 4A and 4Cβ. Simple cells are found mainly in layers 4 and 6, complex cells in layers 2, 3, 5, and 6. Cells in layers 2, 3, and 4B send axons to other cortical areas; cells in layers 5 and 6 send axons to the superior colliculus and the lateral geniculate nucleus. Koniocellular pathways end in superficial layers and in structures known as "blobs" (see Figure 3.11A). (B) Principal arborizations of geniculate axons and cortical neurons in the cat. In addition to these vertical connections, many cells have long horizontal connections running within a layer to distant regions of the cortex. (A after Hubel, 1988 and Sinich and Horton, 2005; B after Gilbert and Wiesel, 1981.)

[20] Hirsch, J. A. et al. 1998. *J. Neurosci.* 18: 8086–8094.

[21] Hubel, D. H., and Wiesel, T. N. 1959. *J. Physiol.* 148: 574–591.

[22] Mountcastle, V. B. 1957. *J. Neurophysiol.* 20: 408–434.

[23] Hubel, D. H., and Wiesel, T. N. 1962. *J. Physiol.* 160: 106–154.

[24] Hubel, D. H., and Wiesel, T. N. 1968. *J. Physiol.* 195: 215–243.

intracortical connections from one layer to another. For example, the axons of cells in layer 6, in addition to supplying the geniculate, can end in one of several other cortical layers, depending on the cell type.[20]

From this anatomy a general pattern emerges: information from the retina is transmitted to cortical cells (mainly in layer 4) by geniculate axons, handed on from neuron to neuron through the thickness of the cortex, and then sent out to other regions of the brain by fibers looping out through white matter. The radial, or vertical, organization of cortex suggests that *columns* of neurons serve as computational units, processing some feature of the visual scene and passing it on to other cortical regions. Lateral connections also exist, (i.e., parallel to the surface of the cortex); they enable neurons in neighboring areas to communicate with one another (see below). In addition, geniculate inputs project directly to areas of cortex beyond V1, for example to visual area 5 (see below), which is concerned with movement and depth perception.[11]

Ocular Dominance Columns

Whether we look at an object with one or both eyes, we see only one image, even though the positions of the object's projection are slightly different on the two retinas. Interestingly, well over 100 years ago Johannes Müller suggested that individual nerve fibers from the two eyes might fuse or become connected to the same cells in the brain. Thereby, he almost exactly anticipated Hubel and Wiesel's results (see Figure 2.9).[1,21]

Hubel and Wiesel's early experiments showed cortical cells with similar properties to be aggregated together in a vertical array.[1] In any one penetration through the cortex, as the electrode moved from the surface through layer after layer to white matter, all cells were found to share certain common properties. Thus, as the electrode moved deeper and deeper through the cortex along the length of a column, all the neurons encountered had the same receptive field position and the same eye preference. For example, in one penetration, every cell would respond better to the visual stimuli, say, in the left eye.

Figure 3.6 summarizes the results of many physiological recordings and shows the variation in eye preference (or ocular dominance) of neurons in the monkey striate cortex. The cells (total of 1116) are subdivided into seven groups. In groups 2, 3, 5, and 6, the effect of one eye is stronger than that of the other, while the cells in the middle group 4 are equally influenced. It is clear from the histogram that the majority of cells respond preferentially to one of the two eyes. Cells in groups 1 and 7, however, are driven exclusively by either the left or the right eye. These monocularly driven cells occur mainly in layer 4, the principal receiving station for geniculate inputs. Above and below layer 4, simple, complex, and end-stopped cortical neurons (see Chapter 2) are driven by both eyes.

"Ocular dominance columns" was the term that Hubel and Wiesel used to describe the vertical segregation of neurons through the thickness of V1 according to eye preference. Their terminology followed the original concept of cortical columns, which was introduced by Mountcastle for somatosensory cortex.[22–24] Anatomical studies on cats and monkeys confirmed that the cells in one layer of the LGN project to aggregates of target cells in layer 4 that are separate from those supplied by the other eye. The aggregates appear as alternating stripes or bands of cortical cells, each supplied exclusively by one eye. As mentioned, cells are driven by both eyes in the deeper and more superficial cortical layers, although one eye usually dominates.

One of the first techniques for demonstrating columns histologically was to destroy a small group of cells in one layer of the

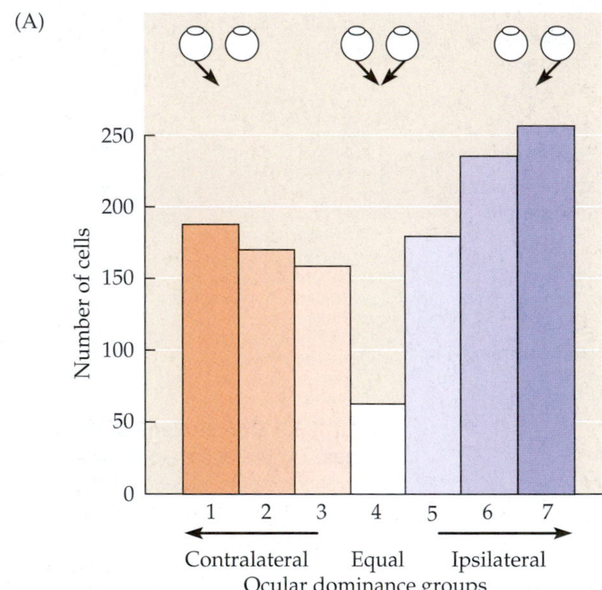

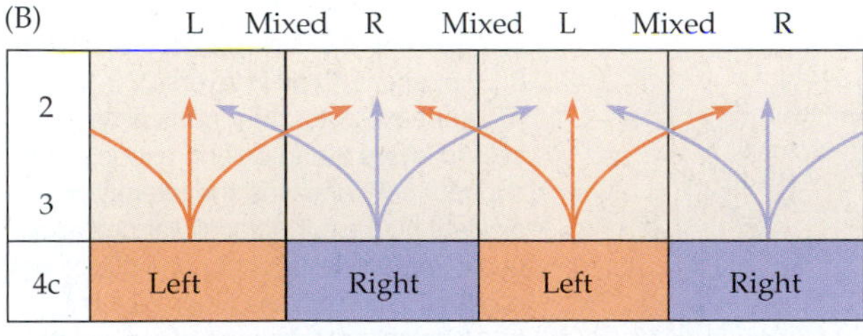

FIGURE 3.6 Physiological Demonstrations of Ocular Dominance Columns. (A) Eye preference of 1116 cells in V1 of 28 rhesus monkeys. Most cells (groups 2 through 6) are driven by both eyes. (B) Diagram to show how inputs from two eyes arriving in layer 4 of the cortex are combined in more superficial layers through horizontal or oblique connections to create cells with binocular fields. (After Hubel and Wiesel, 1968; Hubel, 1988.)

lateral geniculate nucleus and examine the cortex later for degenerating terminals. These terminals appeared in layer 4 in a characteristic pattern of alternating, well-demarcated regions[7] that corresponded to areas driven by the eye supplying the lesioned layer of the LGN.

Later, the ocular dominance pattern over the entire visual cortex was clearly demonstrated by injecting radioactive amino acids, such as leucine or proline, into one eye. The amino acid was taken up by ganglion cells in the retina and incorporated into protein. In time, the labeled protein was transported from ganglion cells along the optic nerve fibers to their terminals within the lateral geniculate nucleus. An extraordinary feature was that the label was then transferred to the geniculate cells and on to the endings of their axons in the cortex, revealing the striking ocular dominance pattern shown in Figure 3.7.[25–27] It is not yet known whether or how transsynaptic transport of protein is used physiologically.

[25] Specht, S., and Grafstein, B. 1973. *Exp. Neurol.* 41: 705–722.

[26] LeVay, S., Hubel, D. H., and Wiesel, T. N. 1975. *J. Comp. Neurol.* 159: 559–576.

[27] LeVay, S. et al. 1985. *J. Neurosci.* 5: 486–501.

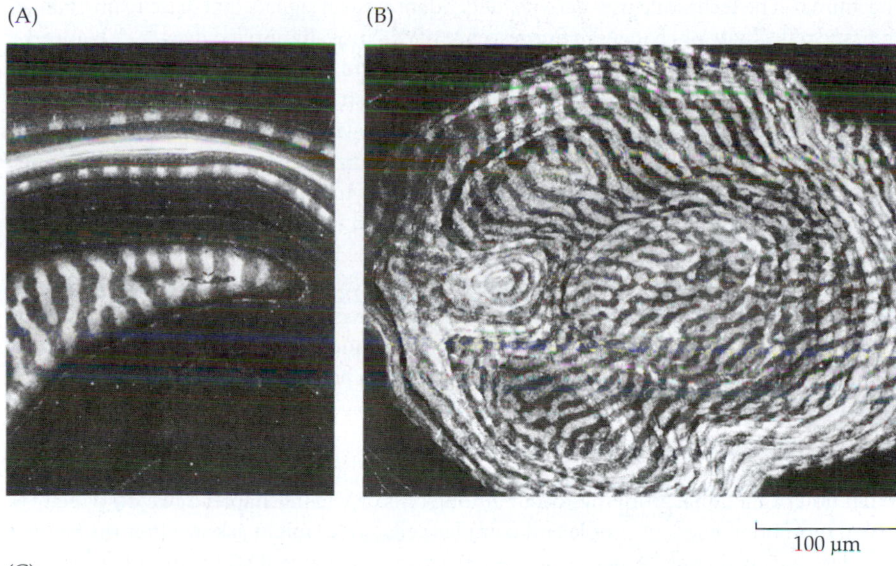

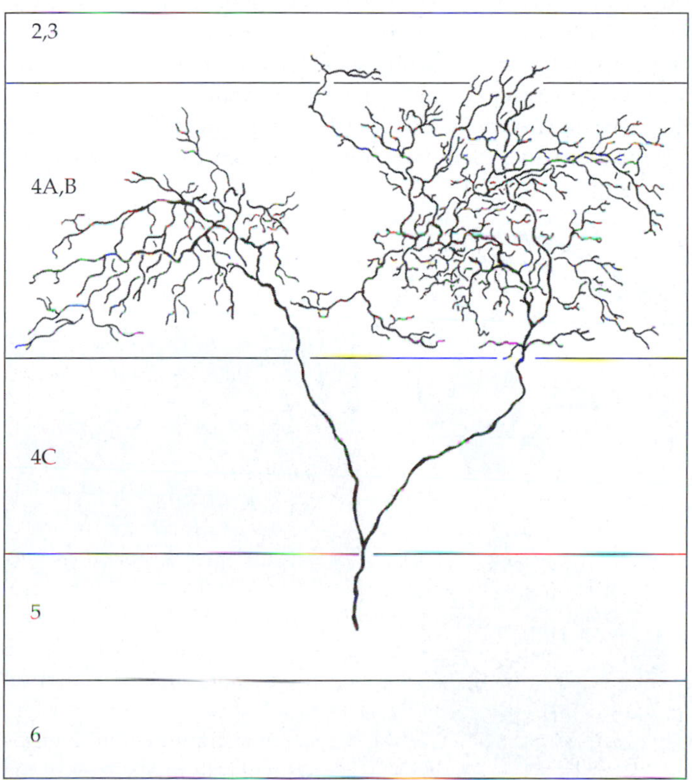

FIGURE 3.7 Ocular Dominance Columns (A,B) show the disposition in layer 4 of radioactive terminals of geniculate axons supplied by the injected eye (radioactivity causes deposition of silver grains in overlaid photographic emulsion, which then appear as white areas in dark-field photographs). The labeled stripes interdigitate with unlabeled areas supplied by the non-injected eye. The center-to-center distance between ocular dominance patches for one eye is approximately 1.0 mm. (A convenient way to visualize the pattern is to imagine that layer 4 corresponds to the layer of cream in a chocolate cake, seen from above after the upper layer of cake has been removed). (C) An off-center cell in the lateral geniculate nucleus that responded best to a small dark spot. It was injected with horse-radish peroxidase to show its terminals ending in layer 4 of the cat visual cortex. The terminals are grouped in two separate clusters, corresponding to columns supplied by the injected eye, separated by a vacant zone supplied by the other eye. (Autoradiographs kindly provided by S. LeVay; C after Gilbert and Wiesel, 1979.)

At the cellular level, a similar pattern has been revealed in layer 4 by injection of horseradish peroxidase into individual axons of lateral geniculate neurons as they approach the cortex.[8] The axon, shown in Figure 3.7C, is that of an off-center geniculate fiber (see Chapter 2) that gave transient responses to dark, moving spots. It ends in two distinct clusters of processes in layer 4. The processes are separated by a blank area corresponding in size to the territory supplied by the other eye. Such morphological studies have borne out and added depth to the original description of ocular dominance columns presented by Hubel and Wiesel in 1962.

Demonstration of Ocular Dominance Columns by Imaging

Ocular dominance columns in the visual cortex can be visualized directly in intact animals. Activity in a large area of cortex is monitored while visual stimuli are presented to the animal. The technique uses activity-dependent optical signals that arise from changes in tissue reflectivity or changes in fluorescence of extrinsically applied dyes.[28–30] Figure 3.8 shows ocular dominance columns detected by this kind of experiment, in which a visual stimulus was applied repeatedly to only one eye. The striped pattern is like that obtained after injection of radioactive amino acids into one eye (see Figure 3.7). The projection of these surface stripes through the depth of cortex confirms the concept of ocular dominance "slabs" (rather than columns) subdividing the retinotopic map. The representation of information from both eyes, regarding a stimulus at one position in the visual field, is integrated by neighboring cells in the visual cortex (see Figure 3.7B). The maps shown in Figures 3.8 and 3.10 were observed without the use of dyes; they are a reflection of altered blood flow and oxygenation in the active area. By use of a staining technique for the enzyme cytochrome oxidase (to be described below), Horton and his colleagues have mapped the complete arrangement of ocular dominance columns in *post mortem* human visual cortex.[31]

Orientation Columns

What other functional groupings occur among cells of V1? In Chapter 2, we described the orientation preferences of simple and complex cells, so we might ask whether this feature is systematically mapped across the cortex. A sample experiment is shown in Figure 3.9 in which a microelectrode is inserted perpendicular to the surface of the cortex in area V1 of the cat.[22] Each bar indicates the location of one cell and its preferred receptive field orientation in the progression through the cortical depth. Small lesions are made at significant points along the electrode track by passing current through the electrode. From these lesions and an end point (indicated by the circle at the end of each electrode track), the position of each recorded cell is reconstructed. In the left-hand track, the first 38 cells were optimally

[28] Grinvald, A. et al. 1986. *Nature.* 324: 361–364.

[29] Ts'o, D. Y. et al. 1990. *Science.* 249: 417–420.

[30] Ts'o, D. Y, Zarella, M., and Burkitt, G. 2009. *J. Physiol.* 587: 2791–2805.

[31] Adams, D. L., Sincich, L. C., and Horton, J. C. 2007. *J. Neurosci.* 27: 10391–10403.

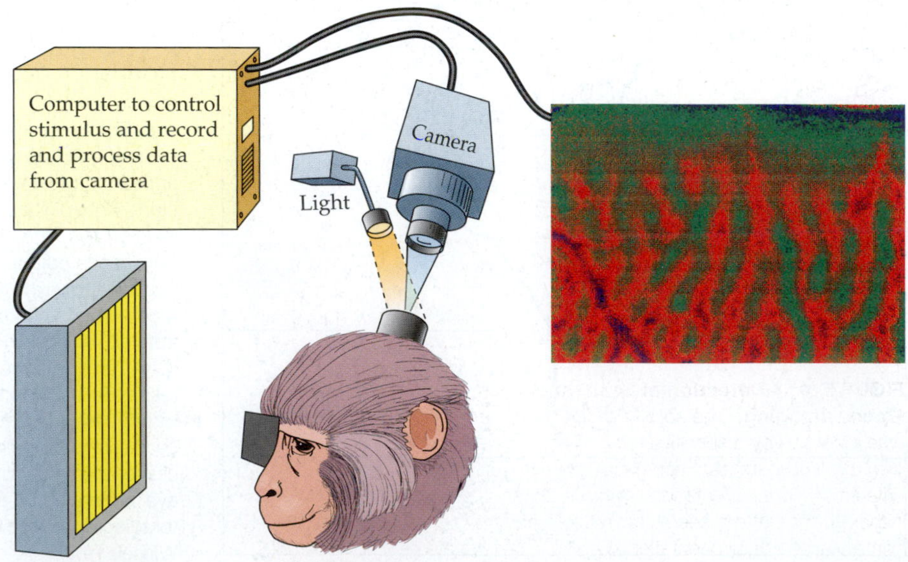

FIGURE 3.8 Display of Ocular Dominance Columns by Optical Imaging. A sensitive camera detects changes in light reflected from the monkey cortex following activity induced in just one eye. The intensity changes are color-coded so that active areas are red. The pattern of red stripes corresponds to ocular dominance columns revealed by anatomical labeling methods (see Figure 3.7). (After Ts'o et al., 1990.)

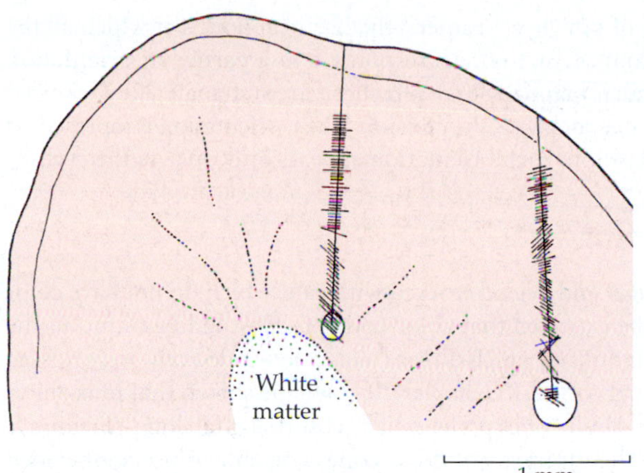

FIGURE 3.9 Axis Orientation of Receptive Fields of Neurons encountered as an electrode traverses the cortex of a cat. Cell after cell tends to have the same axis orientation, indicated by the angle of the bar to the electrode track. The penetration to the right is more oblique; consequently, the track crosses several columns and the axis orientations change frequently. The position of each cell is determined by making lesions repeatedly and at the end of the penetration (circle), and reconstructing the electrode track in serial sections of the brain. Such experiments have established that cat and monkey cells with similar axis orientation are stacked in columns running at right angles to the cortical surface. (After Hubel and Wiesel, 1962.)

driven by bars or edges at about 90° to the vertical, at one position in the visual field. After penetrating about 0.6 mm, the axis of the receptive field orientation changed to about 45°. In a second track (to the right) more tangential to the cortical surface than the first track, successive cells have slightly different receptive field positions and field axis orientations. With this oblique penetration, the field axes change in a regular manner, as though moving through a series of columns with different axis orientations. The orientation columns receive their input from cells with largely overlapping receptive fields on the retinal surface.

Information about how the orientation columns are spatially arranged in the visual cortex of monkeys and cats was first obtained by making tangential electrode penetrations through the cortex (see Figure 3.9).[1,23,32] Each 50 μm advance of the electrode horizontally along the cortex was accompanied by a change in field axis orientation of about 10°, sometimes in a regular sequence through 180°. The field axis orientation columns were narrower than those for ocular dominance—20 to 50 μm wide, compared with 250 to 500 μm.

After the initial anatomical demonstration of orientation columns based on the uptake of 2-deoxyglucose by active cells,[33] the organization was studied in detail by optical imaging techniques in living animals.[28,34–36] An example is the experiment shown in Figure 3.10. The presentation of visual stimuli in a variety of orientations produced activity in different cortical areas. The response to each orientation is represented by a different color (thus, yellow represents cells that prefer a horizontally oriented visual stimulus, while blue represents vertical). What is striking is the arrangement of the orientation columns with respect to one another. At first, the organization appears disorderly. However, careful

[32]Hubel, D. H., and Wiesel, T. N. 1974. *J. Comp. Neurol.* 158: 267–294.

[33]Sokoloff, L. 1977. *J. Neurochem.* 29: 13–26.

[34]Bonhoeffer, T., and Grinvald, A. 1991. *Nature* 353: 429–431.

[35]Hubener, M. et al. 1997. *J. Neurosci.* 17: 9270–9284.

[36]Ohki, K. et al. 2006. *Nature*. 442: 925–928.

(A)

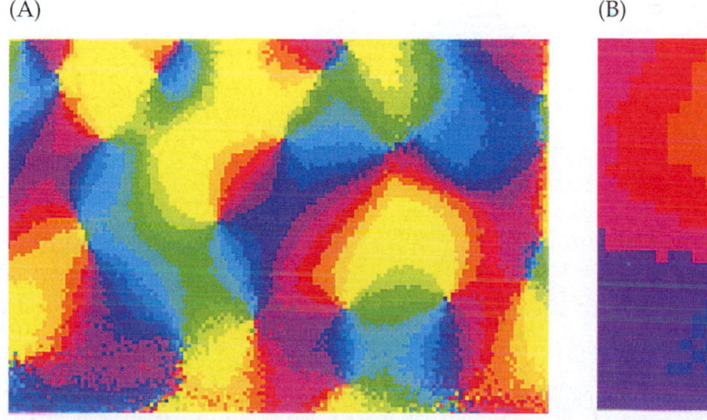

(B)

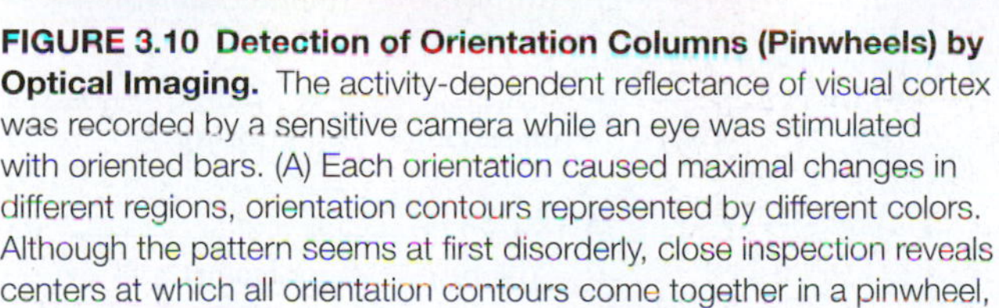

FIGURE 3.10 Detection of Orientation Columns (Pinwheels) by Optical Imaging. The activity-dependent reflectance of visual cortex was recorded by a sensitive camera while an eye was stimulated with oriented bars. (A) Each orientation caused maximal changes in different regions, orientation contours represented by different colors. Although the pattern seems at first disorderly, close inspection reveals centers at which all orientation contours come together in a pinwheel, as shown in (B) Note that each orientation is represented only once and that the sequence is beautifully precise. Such pinwheel centers occur at regular distances from each other. The bars on the right show eight different orientations for the visual stimuli, represented by colors (e.g., yellow represents horizontal, blue represents vertical). (After Bonhoeffer and Grinvald, 1991.)

inspection reveals the presence of pinwheel* centers; that is, focal points at which all the orientations come together. From there, the cells responsive to a particular orientation radiate in an extraordinarily regular manner. Some pinwheels are systematically organized with clockwise, others with counterclockwise, progression. Thus, orientation is represented in a radial rather than a linear fashion. Each orientation appears only once in the cycle.[37]

Cell Groupings for Color

In their initial experiments, Hubel and Wiesel worked with cats, which do not have color vision. It was therefore not to be expected that color-coded cells would be found in the visual cortex. However, even in monkeys, which do see color, color-coded cells were missed in early experiments on the visual cortex. In Chapter 20, it will be shown that in monkey and human retinas red, green, and blue cones preferentially absorb light in long-, medium-, and short-wavelength ranges. The inputs from those cones converge onto color-coded horizontal cells, ganglion cells, and lateral geniculate cells of the parvocellular division. The color-sensitive parvo cells were at first missed in monkey cortex because they are separately aggregated in small clusters, known as patches or "blobs," a name that has persisted in the literature. Blobs are roughly circular patches of cells (Figure 3.11) mainly in layers 2 and 3 but also in 5 and 6. They were first detected by Wong–Riley, who used stains to visualize cytochrome oxidase, an enzyme associated with areas of high metabolic activity.[38] Her experiments, devoted to a completely different problem, ended up revealing fundamental aspects of cortical structure. In primates, the patches of cytochrome oxidase stain are precisely arranged in parallel rows about 0.5 mm apart. Their positions correspond to the centers of ocular dominance columns.[39,40]

Electrical recordings show that many cells within the blobs are color-sensitive and have concentric fields with on and off regions.[41] These observations led to the suggestion that the cytochrome oxidase blobs represent a separate pathway for color, intermingled with the orientation and ocular dominance columns.[8,10,42,43] The blobs also receive inputs from M sublayers of layer 4C and contain neurons with M-like response properties.[44]

*Note: A pinwheel is a toy consisting of small brightly colored vanes pinned to a stick (hence the name) through their center point. The pinwheel rotates when moved by wind.

[37]Swindale, N. V., Matsubara, J. A., and Cynader, M. S. 1987. *J. Neurosci.* 7: 1414–1427.

[38]Wong-Riley, M. 1989. *Trends Neurosci.* 12: 94–101.

[39]Livingstone, M. S., and Hubel, D. H. 1984. *J. Neurosci.* 4: 309–356.

[40]Sincich, L. C., and Horton, J. C. 2005. *Ann. Rev. Neurosci.* 28: 303–326.

[41]Ts'o, D. Y., and Gilbert, C. D. 1988. *J. Neurosci.* 8: 1712–1727.

[42]Livingstone, M. S., and Hubel, D. 1988. *Science* 240: 740–749.

[43]Lu, H. D., and Roe, A. W. 2008. *Cereb. Cortex* 18: 516–533.

[44]Merigan, W. H., and Maunsell, J. H. R. 1993. *Annu. Rev. Neurosci.* 16: 369–402.

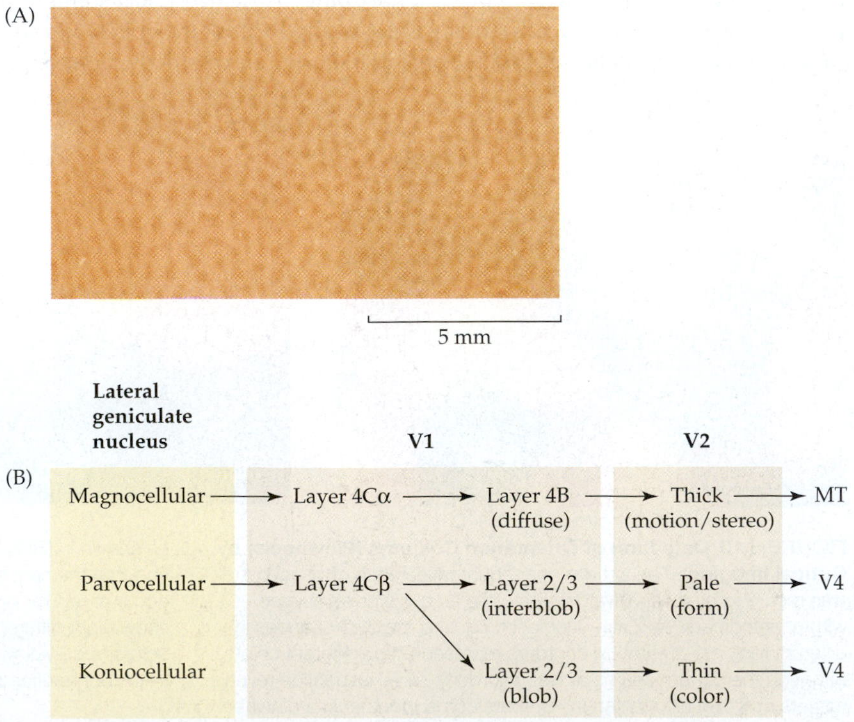

(A)

5 mm

(B)

Lateral geniculate nucleus	V1		V2	
Magnocellular →	Layer 4Cα →	Layer 4B (diffuse) →	Thick (motion/stereo) →	MT
Parvocellular →	Layer 4Cβ →	Layer 2/3 (interblob) →	Pale (form) →	V4
Koniocellular	(Layer 4Cβ ↘)	Layer 2/3 (blob) →	Thin (color) →	V4

FIGURE 3.11 Array of Small Blobs in Macaque Primary Visual Cortex (A) viewed in a horizontal section through layers 2–3. These small spots (which, according to Hubel, make the brain look as if it had measles) become apparent after the distribution of cytochrome oxidase has been revealed by histochemistry. (B) Diagram showing projections of M, P, and K geniculate fibers to V1 and the subsequent projection to V2. (After Sincich and Horton, 2005.)

The blob–interblob regions provide an as yet incompletely understood sorting and recombination of the M, P, and K pathways. Important aspects of color vision, including photoreceptor pigments, retinal connections, and color constancy, are dealt with in Chapter 20.

Connections of Magnocellular and Parvocellular Pathways between V1 and Visual Area 2 (V2)

Visual area 2 (V2, also known as area 18) immediately surrounds V1, receives inputs from it and feeds back onto it, as shown in Figure 3.12. In V2, the striate appearance is lost, large cells are found superficially, and coarse, obliquely running myelinated fibers are seen in the deeper layers. The technique of staining for cytochrome oxidase has made it possible to study columnar organization and to trace connections of magnocellular and parvocellular systems from one visual area to another. It has also been important for displaying features of cortical architecture in post mortem human brains. The pattern of connections from V1 to V2 is shown in Figure 3.12 in which it is clear that staining V2 with cytochrome oxidase produces a pattern different from that seen in V1.[40] The stain in V2 appears as a series of thick and thin stripes, alternating with paler areas having less enzyme activity. These parallel stripes run at right angles to the border between V1 and V2. After horseradish peroxidase is injected into blobs (parvo) of V1, it is taken up by axon terminals and transported retrogradely, revealing that nerve cells providing input to the blobs are located in the thin stripes in V2. The connections are reciprocal: injections of horseradish peroxidase into thin stripes label cells in V1 blobs.[38,45–48] In contrast, the interblob regions project to the pale stripes. The thick stripes receive primarily magnocellular information from layers 4B and 4Cα. Remarkably, this functional subdivision can even be distinguished at a molecular level; for instance, the monoclonal antibody Cat-301 preferentially labels magnocellular pathways throughout the monkey visual cortex.[49,50]

Magnocellular neurons that respond to depth or to movement are abundant in V2.[14,51] An example is shown in Figure 3.13. In this experiment, a dye that changes emission in response to changes in intracellular calcium concentration was injected into area V2 of a cat. Individual cells responded preferentially to movement of visual stimuli in one direction. This figure shows that resolution at the level of single cells can be achieved in living cortex.

[45] Livingstone, M. S., and Hubel, D. H. 1987. *J. Neurosci.* 7: 3371–3377.

[46] Ts'o, D. Y., Roe, A. W., and Gilbert, C. D. 2001. *Vision Res.* 41: 1333–1349.

[47] Sincich, L. C., Jocson, C. M., and Horton, J. C. 2007. *Cereb. Cortex* 17: 935–941.

[48] Shmuel, A. et al. 2005. *J. Neurosci.* 25: 2117–2131.

[49] Preuss, T. M., and Coleman, G. Q. 2002. *Cereb. Cortex* 12: 671–691.

[50] Deyoe, E. A. et al. 1990. *Vis. Neurosci.* 5: 67–81.

[51] Ohki, K. et al. 2005. *Nature* 433: 597–603.

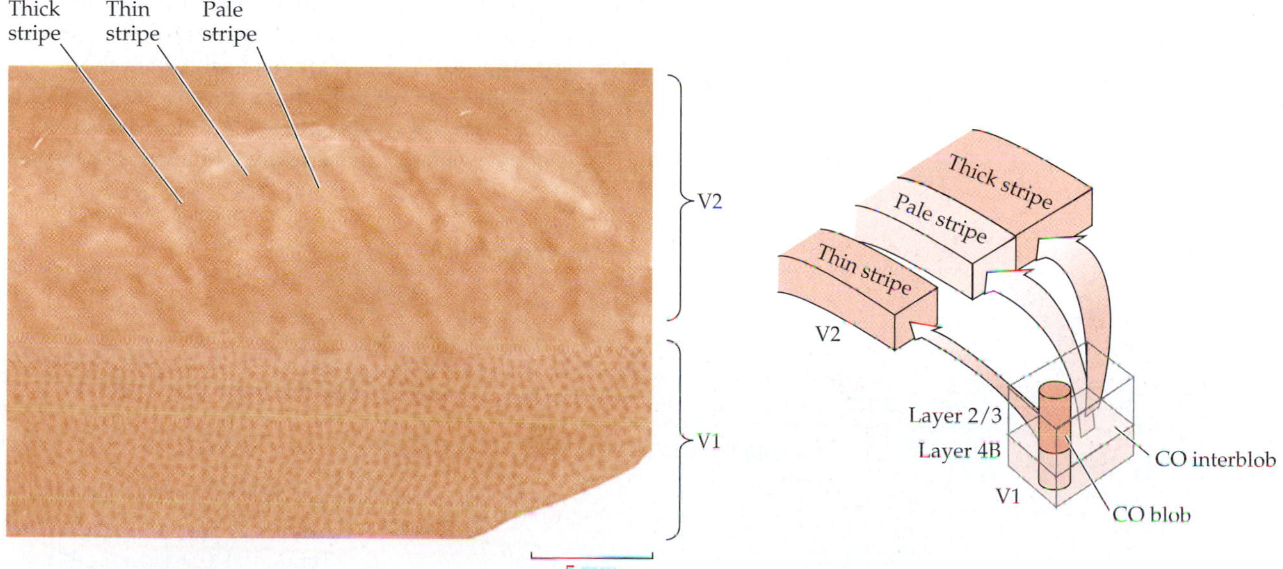

FIGURE 3.12 A Horizontal Section Through Cortex of Macaque MonkeyV1 stained to reveal the distribution of cytochrome oxidase blobs (also known as "patches") in V1 and stripes in V2. The border between the two areas is well defined and is in accord with characteristics of receptive field organization in V1 and V2. At the transition, the fine array of patches gives way to thick and thin stripes in repeating cycles. (After Sincich and Horton, 2005.)

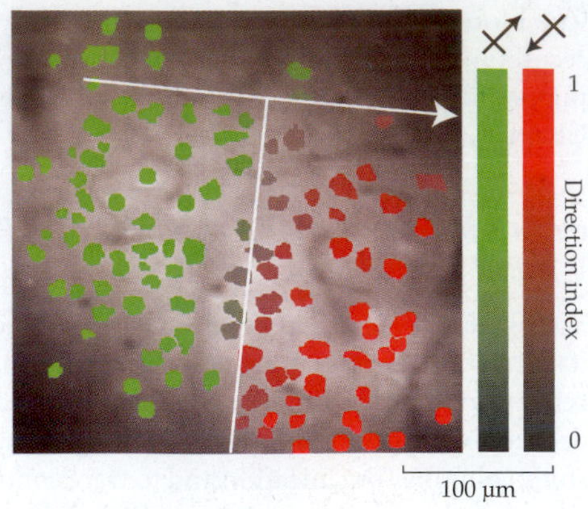

FIGURE 3.13 Neurons Responding to Moving Stimuli imaged by calcium dye in cat visual cortex (V2). Cells colored green responded to stimuli moving obliquely upwards, those colored red to the movement in the opposite direction. Gray cells responded to both movements. Cells are plainly arranged in columns with high resolution. (After Ohki et al., 2005.)

Relations between Ocular Dominance and Orientation Columns

Ocular dominance and orientation columns represent just two functional arrangements that apply to neurons in visual cortex. Direction of movement, spatial frequency (essentially a reflection of receptive field size), and image disparity (an important determinant in depth perception)[24,51–53] also appear in columnar arrangements in the cortex. The question then arises as to how all necessary aspects of image analysis can be carried out for each point on the retinotopically mapped cortex. The intermingling of functional columns shown in Figure 3.14 provides part of the answer. Indeed, long before optical imaging was used to observe such relationships, Hubel and Wiesel proposed a conceptual scheme they termed a hypercolumn (also known as "ice cube"). In the hypercolumn, all possible orientations for corresponding regions of the visual field for both eyes could be represented. An adjacent hypercolumn would analyze information in the same way for an adjacent but overlapping part of the visual field and so on, until the entire retina was mapped across the cortex.

Once the pinwheel arrangements for cells with particular orientation preferences were found, it was clear that the ice-cube hypercolumn was an oversimplification.[30] When optical imaging methods are used to reveal the relationship between orientation and ocular dominance columns, a complex picture is seen, such as that shown in Figure 3.15. In this experiment, cortical activity was imaged for eye-specific stimulation and for stimulation with a series of oriented bars. In Figure 3.15, each orientation contour is indicated by a separate color (an iso-orientation contour). Ocular dominance zones (or columns) are shown by dark lines and in B as clear or shaded. The orientation pinwheels are clearly seen as the convergence of iso-orientation contours, and a set of contour lines between pinwheels typically crosses an ocular dominance boundary at right angles (as in the original Hubel and Wiesel hypercolumn). That is, most orientation domains are split into ipsilateral and

[52] Barlow, H. B., Blakemore, C., and Pettigrew, J. D. 1967. *J. Physiol.* 193: 327–342.

[53] Westheimer, G. 2009. *J. Physiol.* 587: 2807–2816.

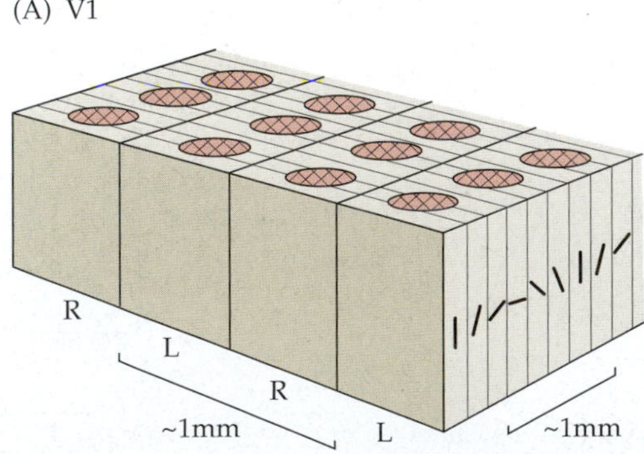

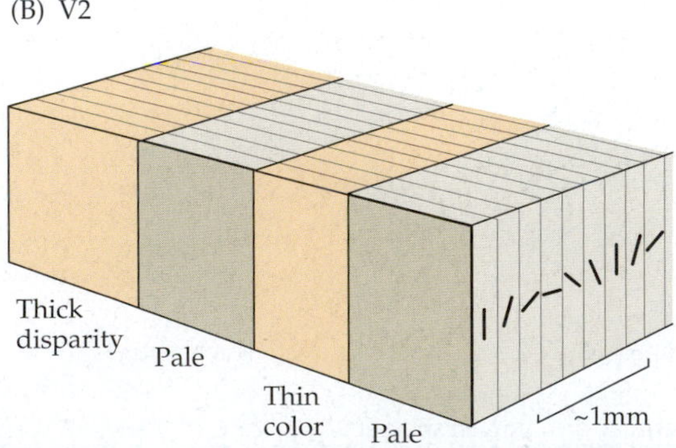

FIGURE 3.14 Tentative Scheme Proposed by Hubel and Wiesel for the arrangement of ocular dominance and orientation columns in V1. (A) Within a cube of cortical tissue (resembling an ice cube), the sets of columns for eye preference and orientation run at right angles to one another. Blobs are represented as circles. Similarly, in (B) (right-side drawing), the thick stripes and thin stripes revealed by stains for cytochrome oxidase are orthogonal to the orientation columns. (After Ts'o et al., 2009.)

(A)

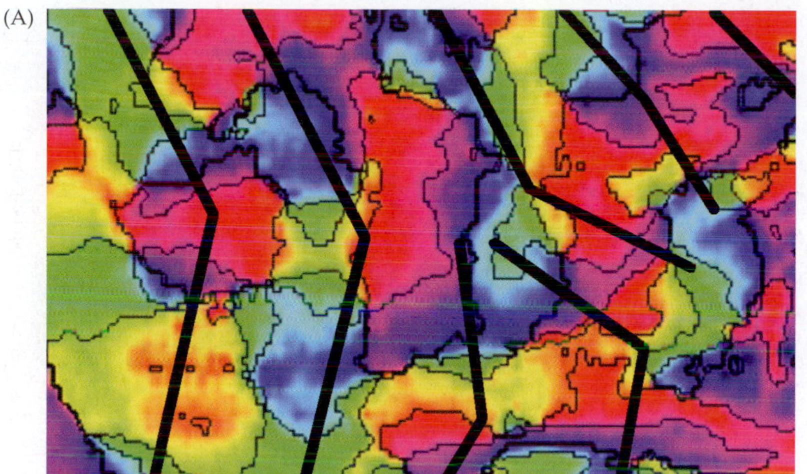

(B)

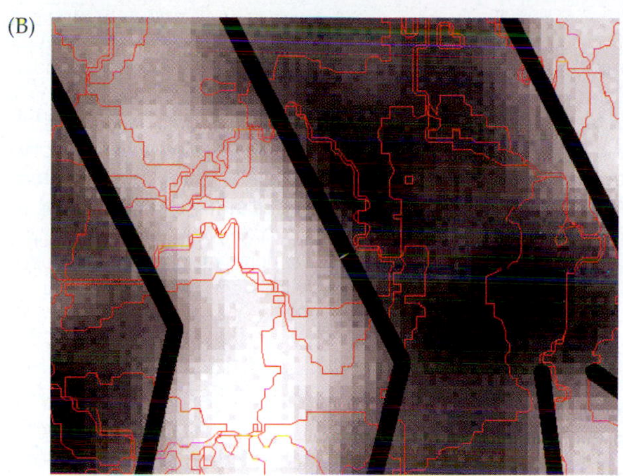

FIGURE 3.15 Spatial Relation of Orientation and Ocular Dominance Columns in primate cortex revealed by imaging. (A) Shows the broad, linear ocular dominance columns separated by thick, dark lines. The orientation columns (i.e., pinwheels) are color coded, as in Figure 3.10, with yellow representing neuron populations responding to horizontally oriented visual stimuli and blue representing vertical preference. (B) The same area of cortex with the ocular dominance columns tinted as dark (left eye) or pale (right eye). The centers of the pinwheels traced from (A) lie within the ocular dominance columns. Their contours, radiating around the clock, tend to cross the border of the ocular dominance column at right angles. While the picture is more complex than the original ice cube model, the principles of organization remain similar. (After Ts'o et al., 2009.)

contralateral halves, and thus serve the two eyes for that region of visual space. Each pinwheel center tends to occur near the center of an ocular dominance patch.[30,35]

Horizontal Intracortical Connections

In V1 additional connections have been described that suggest principles of organization in addition to those we have considered thus far. Whereas classical staining techniques (as described in Figure 3.4), such as Golgi impregnation, revealed a preponderance of neuronal processes that run perpendicularly from layer to layer, intracellular injections of single cells have demonstrated that cortical neurons also have long horizontal processes that extend laterally from column to column (Figure 3.16).[54–56] Connections such as these contribute to the synthesis of elongated receptive fields of simple cells in layer 6 of V1: The receptive fields of layer 5 cells are combined and added end-to-end on the layer 6 simple cell by way of long horizontal axons. Many simple and complex cells are found with long horizontal projections that extend more than 8 mm, corresponding to several hypercolumns. An individual neuron can therefore integrate information over an area of retina several times larger than the receptive field measured by conventional electrophysiological techniques.

Of particular interest is that connections are made only between columns that have similar orientation specificity. Evidence for such specific interconnections has been obtained by two additional methods. First, when labeled microbeads are injected into one column, they are transported to a distant hypercolumn with the same orientation preference (see Figure 3.16B). Second, by cross-correlation of the firing patterns of neurons with the same orientation preference in two widely separated columns, one can show that the neurons

[54] Gilbert, C. D., and Wiesel, T. N. 1989. *J. Neurosci.* 9: 2432–2442.

[55] Ts'o, D. Y., Gilbert, C. D., and Wiesel, T. N. 1986. *J. Neurosci.* 6: 1160–1170.

[56] Stettler, D. D. et al. 2002. *Neuron* 36: 739–750.

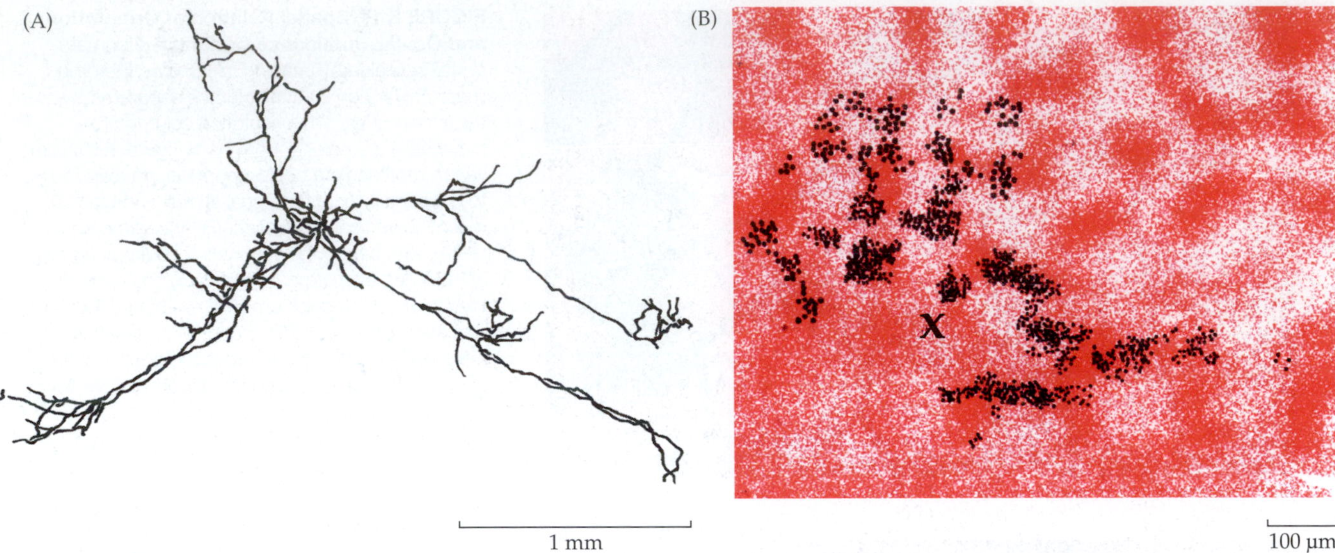

FIGURE 3.16 Horizontal Connections in Visual Cortex. (A) Surface view of pyramidal cell in cat V1, after labeling with horseradish peroxidase. The processes extend for nearly 3 mm across the cortical surface. Fine branches and synaptic boutons of this neuron occurred in several discrete clusters separated by 800 μm or more. (B) Labeled microspheres were injected into a region where cells had vertical orientation preference (large black "X"). The microspheres are taken up by axon terminals and transported back to the somata of cells projecting to the injection site. Vertical orientation columns were also labeled using 2-deoxyglucose during stimulation with vertically oriented bars of light. The labeled microspheres are found in 2-deoxyglucose-labeled areas, showing that horizontal connections occur between cells with the same orientation specificity. (A from Gilbert and Wiesel, 1983; B from Gilbert and Wiesel, 1989; kindly provided by C. Gilbert.)

are interconnected. Moreover, after a lesion has been made in the retina, cortical cells that have been deprived of input can show responses to distant stimuli that would be outside their normal receptive fields.[41]

In spite of its incompleteness, the original ice cube scheme shown in Figure 3.14 still represents the simplest way to imagine how orientation, color, and eye preference may be combined. There remains the challenge of incorporating additional features of visual analysis, such as the perception of depth and movement, into a comprehensive scheme.

Construction of a Single, Unified Visual Field from Inputs Arising in Two Eyes

A problem arises from the fact that each half of the brain deals only with the visual field on the opposite side of the world. How are the representations in the two cortices knitted together to produce the single image of the world that we perceive? That each hemicortex is wired to perceive one-half of the external world but not the other is a general property of the two hemispheres in relation to perception, not just for vision, but also for sensations of touch and position. What happens at the midline? How do the two sides of our brain knit the right world and the left world together with no hint of a seam or discontinuity?

The obvious way to preserve continuity is to join the left and the right visual fields together at the midline, in register. Such interactions would allow a complete picture to be formed with specific connections between the two hemispheres. There would, however, be little purpose in linking fields seen out of the corners of the two eyes that look on quite different parts of the world. Highly specific connections between neurons with receptive fields exactly at the midline have been found experimentally to run from cortex to cortex through the **corpus callosum**.[57] They are described in Box 3.1.

[57] Hubel, D. H., and Wiesel, T. N. 1967. *J. Neurophysiol.* 30: 1561–1573.

BOX 3.1
Corpus Callosum

The general question of transfer of information between the hemispheres has been studied in humans and in monkeys by Sperry, Gazzaniga, Berlucchi, and their colleagues.[58–62] Concentrating on the coordinating role of the corpus callosum, a bundle of fibers that runs between the two hemispheres, they have shown that the fibers are involved in the transfer of information and learning from one hemisphere to the other. Thus, the fusing or knitting together of the two fields of vision is mediated by fibers in the corpus callosum. Certain cells have receptive fields that straddle the midline and receive information about both sides of the visual world. These cells lie at the boundary of V1 and V2, and they combine inputs from both hemispheres by way of the corpus callosum. Interestingly, these cells lie in layer 3, which contains neurons known to send their axons to other regions of cortex. The organization and orientation preference of two receptive fields that are brought together in this way are similar. Projections from nerve cells in one hemisphere to the other have been shown anatomically by injecting horseradish peroxidase into the cortex at the boundary of V1 and V2.[63] Enzyme taken up by terminals is transported to neuronal cell bodies situated at exactly corresponding sites in opposite hemispheres. Cutting or cooling (which blocks conduction) the corpus callosum causes the receptive field to shrink and become confined to just one side of the midline (the usual arrangement for cortical cells). Furthermore, recordings from single fibers in the callosum show that they have receptive fields close to the midline, not in the periphery.

The role of callosal fibers is clearly demonstrated in an experiment of Berlucchi and Rizzolatti.[60] They made a longitudinal cut through the optic chiasm, thereby severing all direct connections to the cortex from the contralateral eye. Yet, provided the corpus callosum was intact, some cells in the cortex with fields close to the midline could still respond to appropriate visual stimulation of the contralateral eye.

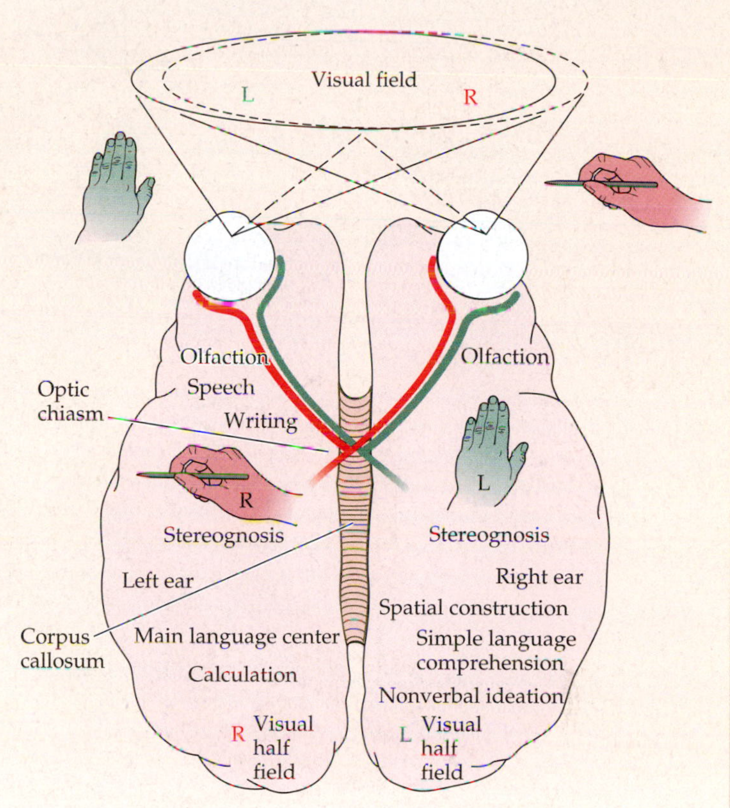

The lateralization of cortical function is shown schematically in the figure. The left hand and left visual field project to the right hemisphere. The right hand and right visual field project to the left hemisphere, in which specialized language function resides.

[58] Sperry, R. W. 1970. *Proc. Res. Assoc. Nerv. Ment. Dis.* 48: 123–138.
[59] Gazzaniga, M. S. 2005. *Nat. Rev. Neurosci.* 6: 653–659.
[60] Berlucchi, G., and Rizzolatti, G. 1968. *Science* 159: 308–310.
[61] Glickstein, M., and Berlucchi, G. 2008. *Cortex* 44: 914–927.
[62] Doron, K. W., and Gazzaniga, M. S. 2008. *Cortex* 44: 1023–1029.
[63] Shatz, C. J. 1977. *J. Comp. Neurol.* 173: 497–518.

Association Areas of Visual Cortex

In Chapter 23, we describe how psychophysical measurements, electrical recordings, brain imaging, and studies of patients with specific lesions have demonstrated the presence of physically separate brain regions beyond V1 (Figure 3.17), the visual association areas. These regions are concerned with different aspects of visual analysis, such as perception of depth, color,[64] movement, and faces.[18] It will be shown in Chapter 23, for example, that localized lesions of a small area (V4) lead to loss of all color vision in patients: they see everything in black and white, but have no defects in their memory—consciousness or intellect.[13] Similarly, it will be shown that in another area, the middle temporal cortex (MT; also known as V5), neurons are clustered that respond specifically to movements in specific directions and to depth perception.

MT is of particular interest since experiments made in this area bear on a general question regarding all the experiments described so far: does a neuron that responds to a particular

[64] Zeki, S. et al. 1991. *J. Neurosci.* 11: 641–649.

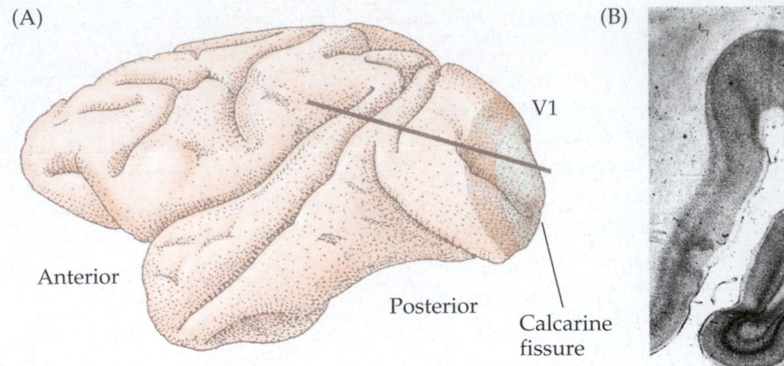

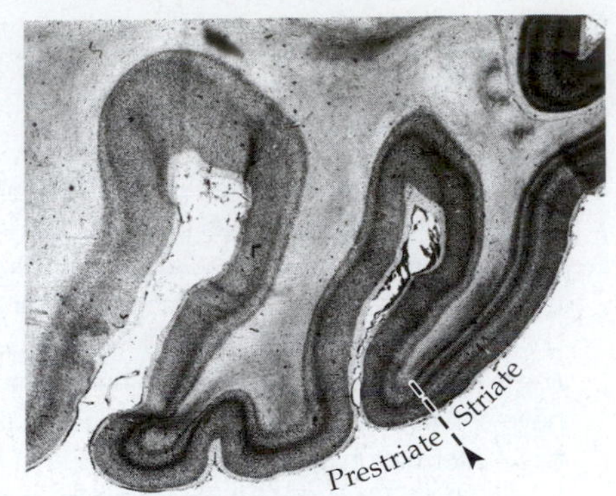

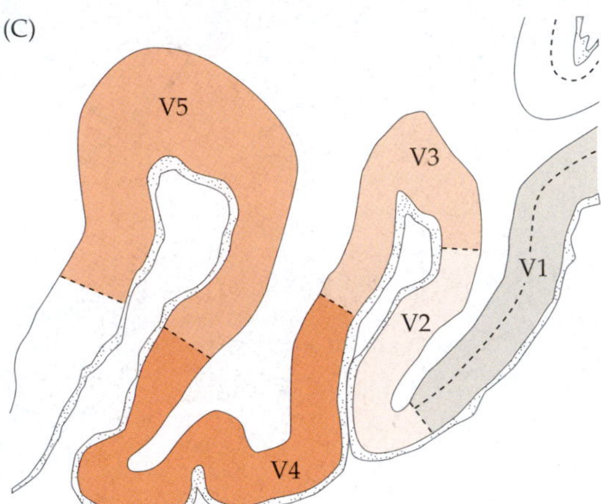

FIGURE 3.17 Relation of Primary Visual Cortex (V1) to V2, V3, V4, and V5 in the monkey. (A) The cortex and plane of section passing through V1 and V2. The boundary between V1 and V2 is unambiguous. (B,C) A section through the occipital cortex. In B, the boundary between V1 (striate) and V2 (prestriate) occurs at the dotted line, where the striped appearance is lost. The boundaries between V2, V3, V4, and V5 are revealed by a combination of physiological and anatomical studies. (After Zeki, 1990; micrograph kindly provided by S. Zeki and M. Rayan.)

stimulus, say a vertical bar moving towards the left, actually take part in the perception of the event by the animal? Or is it simply a concomitant along the sensory pathway? By stimulating and recording from single columns, Newsome and his colleagues have shown that individual columns and individual cells are indeed directly involved in the perception of movement in a specific direction.[65,66] Discussion of these elegant experiments requires more background knowledge and will be presented later in Chapter 23.

[65] Cohen, M. R., and Newsome, W.T. 2004. *Curr. Opin. Neurobiol.* 14: 169–177.

[66] Cohen, M. R., and Newsome, W. T. 2009. *J Neurosci.* 29: 6635–6648.

Where Do We Go from Here?

It has become possible to approach experimentally many longstanding questions about how the brain analyzes visual scenes falling on the retina. Results from anatomical, physiological, and psychophysical experiments have converged to reveal cortical circuits for perception. One important principle derived from such studies is that separate pathways beginning at the retina do extend through to consciousness. You can fool the systems used for assessing depth (Figure 3.18) and for detecting movement by using light and perspective conditions that bring in only the parvocellular or magnocellular channels; moreover, patients with lesions to specific areas of cortex can lose color vision with little impairment of their ability to recognize patterns.

An impressive aspect of the work on vision is the way in which redundant features of the image on the retina become jettisoned. *Discontinuities* in lines, angles, shading, and color are what the cortex deals with.

Still lacking is a comprehensive explanation of how the brain puts together the complete image that we see. Within V1 there is good evidence

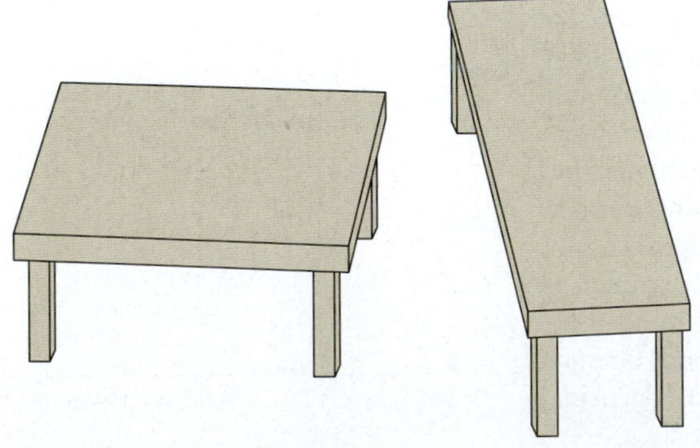

FIGURE 3.18 The Dimensions of Two Table Tops look very different but are identical. The artist forces the viewer to use the magnocellular M system for depth perception; this results in neglect of form recognition. It is pleasing that a highly complex perceptual event such as this can be described in terms of cellular mechanisms. (After Shepard, 1990.)

for a hierarchical principle, according to which inputs from lower order cells combine to create new and more complex messages in target cells. But the primary visual area is merely the first stage for cortical processing, sending visual input onward for further analysis and perception. You do not *see* with V1; it is only a way station on the way to *seeing*. Furthermore, the visual association areas are not merely successive, receiving stations. They feed back into V1 and interact among themselves. It is a sobering thought that we do not even know why so many fibers descend from V1 to the lateral geniculate nucleus. What seems to be a necessary next step is the development of novel imaging or recording techniques that do more than simply light up an area that is active at a particular instant. Rather, information needs to be obtained about the activity of large numbers of individual neurons at the cellular level, simultaneously millisecond by millisecond over prolonged periods and over large areas of the brain in conscious animals and human beings. Although intriguing speculations and books appear regularly about how the big picture is made, it must be said that we still do not have hard evidence or a comprehensive theory (but see Hawkins and Blakeslee, 2004 and Chapter 23).

SUMMARY

- Inputs from the two eyes segregate to different layers of the lateral geniculate nucleus.
- The layers are in retinotopic register. Neurons in each layer are functionally distinct, comprising magnocellular-, parvocellular-, or koniocellular-response types.
- The six layers of primary visual cortex serve as input and output stages of cortical processing.
- Geniculate afferents from the two eyes are segregated in layer 4C of striate cortex, establishing ocular dominance columns that can be detected physiologically and anatomically.
- Magnocellular and parvocellular layers of the LGN project specifically to different sublayers of cortical layer 4C.
- Neurons in primary visual cortex are organized according to eye preference and orientation selectivity.
- Most cortical neurons receive input from corresponding points in the visual field of the two eyes.
- The layout of ocular dominance slabs and orientation pinwheels can be visualized by the imaging of activity-dependent optical signals from the brain surface.
- Iso-orientation contours tend to intersect ocular dominance domains at right angles, and each orientation domain is shared between two ocular dominance columns.
- Magnocellular, parvocellular, and koniocellular pathways form parallel channels from the retina to V1 and beyond.
- Small areas, called blobs, in the center of each ocular dominance column are revealed by cytochrome oxidase staining and code for color.
- Imaging techniques allow one to map projections to primary and secondary visual areas as well as still more highly specialized regions concerned with the analysis of form, movement, color, and depth.

Suggested Reading

General Reviews

Callaway, E. M. 2005. Structure and function of parallel pathways in the primate early visual system. *J. Physiol.* 566: 13–19.

Cohen, M. R., and Newsome, W. T. 2004. What electrical microstimulation has revealed about the neural basis of cognition. *Curr. Opin. Neurobiol.* 14: 169–177.

Doron, K. W., and Gazzaniga, M. S. 2008. Neuroimaging techniques offer new perspectives on callosal transfer and interhemispheric communication. *Cortex* 44: 1023–1029.

Gilbert, C. D., and Wiesel, T. N. 1979. Morphology and intracortical projections of functionally characterised neurones in the cat visual cortex. *Nature* 280: 120–125.

Glickstein, M., and Berlucchi, G. 2008. Classical disconnection studies of the corpus callosum. *Cortex* 44: 914–927.

Hawkins, J., and Blakeslee, S. 2004. *On Intelligence*. Times Books, New York.

Hubel, D. H., and Wiesel, T. N. 1977. Functional architecture of macaque monkey visual cortex (Ferrier Lecture). *Proc. R. Soc. Lond. B, Biol. Sci.* 198: 1–59.

Hubel, D. H., and Wiesel, T. N. 2005. *Visual Perception*. Oxford University Press, Oxford. (This remarkable book contains all the papers by Hubel and Wiesel, as well as a stylish, informative, and witty commentary on how the work was done and what was found in later years by others.)

Mountcastle, V. B. 1997. The columnar organization of the neocortex. *Brain* 120: 701–722.

Sincich, L. C., and Horton, J. C. 2005. The circuitry of V1 AND V2: Integration of color, form, and motion. *Ann. Rev. Neurosci.* 28: 303–326.

Ts'o, D. Y, Zarella, M., and Burkitt, G. 2009. Whither the hypercolumn? *J. Physiol.* 587: 2791–2805.

Van Essen, D. C. 2005. Corticocortical and thalamocortical information flow in the primate visual system. *Prog. Brain Res.*149: 173–185.

Zeki, S. (ed.) 2005. Cerebral cartography 1905–2005. *Philos. Trans. R. Soc. Lond. B, Biol. Sci.* 360: 649–862.

Original Papers

Adams, D. L., Sincich, L. C., and Horton, J. C. 2007. Complete pattern of ocular dominance columns in human primary visual cortex. *J. Neurosci.* 27: 10391–10403.

Cohen, M. R., and Newsome, W. T. 2009. Estimates of the contribution of single neurons to perception depend on timescale and noise correlation. *J. Neurosci.* 29: 6635–6648.

Doron, K. W., and Gazzaniga, M. S. 2008. Neuroimaging techniques offer new perspectives on callosal transfer and interhemispheric communication. *Cortex* 44: 1023–1029.

Gilbert, C. D., and Wiesel, T. N. 1989. Columnar specificity of intrinsic horizontal and cortico-cortical connections in cat visual cortex. *J. Neurosci.* 9: 2432–2442.

Hubel, D. H., and Wiesel, T. N. 1959. Receptive fields of single neurones in the cat's striate cortex. *J. Physiol.* 148: 574–591.

Hubel, D. H., and Wiesel, T. N. 1968. Receptive fields and functional architecture of monkey striate cortex. *J. Physiol.* 195: 215–243.

LeVay, S., Hubel, D. H., and Wiesel, T. N. 1975. The pattern of ocular dominance columns in macaque visual cortex revealed by a reduced silver stain. *J. Comp. Neurol.* 159: 559–576.

Livingstone, M. S., and Hubel, D. 1988. Segregation of form, color, movement, and depth: Anatomy, physiology, and perception. *Science* 240: 740–749.

Ohki, K., Chung, S., Ch'ng, Y. H., Kara, P., and Reid, R. C. 2005. Functional imaging with cellular resolution reveals precise micro-architecture in visual cortex. *Nature* 433: 597–603.

Ohki, K., Chung, S., Kara, P., Hübener, M., Bonhoeffer, T., and Reid, R. C. 2006. Highly ordered arrangement of single neurons in orientation pinwheels. *Nature* 442: 925–928.

Tsao, D. Y., Freiwald, W. A., Tootell, R. B., and Livingstone, M. S. 2006. A cortical region consisting entirely of face-selective cells. *Science* 311: 670–674.

PART II

Electrical Properties of Neurons and Glia

This section deals with the functional and structural properties of the neuron that enable it to generate and transmit electrical signals, and sets the stage for following sections that deal with how neurons, once activated, transmit their signals from one to another in order produce integrated behavior.

We begin in Chapter 4 with the functional properties of ion channels in the nerve cell membrane. These channels determine the electrical characteristics of the neuron by allowing selective movement of ions into and out of the cell at rest and during signal generation. In Chapter 5 we discuss how the molecular structure of ion channels accounts for their ion selectivity and enables them to regulate ion movements in response to a variety of stimuli.

In Chapter 6 we move on to the behavior of the membrane as a whole and describe how its overall ion selectivity determines the resting membrane potential of the cell. Chapter 7 discusses in detail how selective activation of voltage-activated cation channels, principally sodium and potassium channels, leads to the generation of the action potential, which is the basis of all long-distance signaling in the nervous system. Chapter 8 describes how subthreshold "passive" signals spread in nerve fibers, and how the action potential, once initiated, is transmitted from one region of the neuron to the next.

In Chapter 9 we consider mechanisms for active transport of ions across the cell membrane, transport necessary to maintain the ionic composition of the cytoplasm in the face of constant leaks of ions into and out of the cell at rest and during electrical activity. Finally, in Chapter 10 we discuss the functional properties of glial cells and how glia and neurons interact during signal transmission.

CHAPTER 4
Ion Channels and Signaling

Electrical signals in nerve cells are all mediated by the flow of ions though aqueous pores in the nerve cell membrane. These pores are formed by transmembrane proteins called ion channels. It is possible to record and measure ionic currents through single channels. In this chapter, we discuss the functional properties of ion channels, such as their specificity for one ion species or another and how their activity is regulated. Most individual channels are selective for either cations or anions, and some cationic channels are highly selective for a single ion species, such as sodium.

Channels fluctuate between open and closed states, often with a characteristic open time. Their contribution to the flow of ionic current across the cell membrane is determined by the amount of time that they spend in the open state. Channel opening is regulated by a variety of mechanisms. Some of these are physical, such as changes in membrane tension or membrane potential; others are chemical, involving the binding of activating molecules (ligands) to sites on the extracellular mouth of the channel or of particular ions or molecules to the inner mouth.

In addition to kinetics of opening and closing, an important property of channels is the ability of an open channel to pass ionic current. One way in which ions can pass through a channel is by simple diffusion; another is by interacting with internal binding sites, hopping from one to the next as they pass through the pore. In either case, movement through the channel is passive, driven by concentration gradients and by the electrical potential gradient across the membrane. The ability of ions to move through the open channel down their electrochemical gradients depends on the permeability of the channel to the ion species involved and the concentration of the ions at either mouth of the channel. These two factors, permeability and concentration, determine the channel conductance.

In Chapter 1, we discussed how the transfer of information in the nervous system is mediated by two types of electrical signals in nerve cells: graded potentials, which are localized to specific regions of the nerve cell membrane, and action potentials, which are propagated along the entire length of a neuronal process. These signals are superimposed on a steady electrical potential across the cell membrane, called the **resting membrane potential.** Depending upon cell type, nerve cells at rest have steady membrane potentials ranging from about –30 mV to almost –100 mV. The negative sign indicates that the inside of the membrane is negative with respect to the outside. Signaling in the nervous system is mediated by changes in the membrane potential. For example, in sensory receptors, an appropriate stimulus, such as touch, sound, or light, causes local **depolarization** (making the membrane potential less negative) or **hyperpolarization** (membrane potential more negative). Similarly, neurotransmitters at synapses act by depolarizing or hyperpolarizing the postsynaptic cell. **Action potentials**, which are large, brief pulses of depolarization, propagate along axons to carry information from one place to the next in the nervous system.

All such changes in membrane potential are produced by the movement of ions across the nerve cell membrane. For example, inward movement of positively charged sodium ions reduces the net negative charge on the inner surface of the membrane or, in other words, causes depolarization. Conversely, outward movement of potassium ions results in an increase in net negative charge, causing hyperpolarization, as does inward movement of negatively charged chloride ions.

How do ions move across the cell membrane, and how is their movement regulated? The pathway for rapid movement of ions into and out of the cell is through **ion channels**, which are protein molecules that span the membrane and form pores through which ions can pass. Ion currents are regulated by controlling the rate at which these channels open and close. Thus, a variety of channel types in the membrane enables neurons to receive signals from the external world, or from other neurons, to carry signals over long distances, to modulate the activity of other neurons and effector organs, and to alter their own signaling properties in response to internal metabolic changes. All of the complexities of perception and analysis of neural signals depend ultimately on ion channel activity, as does the generation of complex motor outputs from the nervous system.

Properties of Ion Channels

The Nerve Cell Membrane

Cell membranes consist of a fluid mosaic of lipid and protein molecules. As shown in Figure 4.1A, the lipid molecules are arranged in a bilayer about 6 nm in thickness, with their polar, hydrophilic heads facing outward and their hydrophobic tails extending to the middle of the layer. The lipid is sparingly permeable to water and virtually impermeable to ions. Embedded in the lipid bilayer are protein molecules, some on the extracellular side, some facing the cytoplasm, and some spanning the membrane. Many of the membrane-spanning proteins form ion channels. Ions such as potassium, sodium, calcium, or chloride, move through such channels passively, driven by concentration gradients and by the electrical potential across the membrane.

Another set of membrane-spanning proteins function as **transport molecules** (**pumps** and **transporters**) that move substances across the membrane against their electrochemical gradients. Transport molecules maintain the ionic composition of the cytoplasm by pumping back across the cell membrane ion species that have leaked through channels into or out of the cell. They also perform the important function of carrying metabolic substances, such as glucose and amino acids, across cell membranes. Properties of transport molecules are discussed in Chapter 9.

What Does an Ion Channel Look Like?

The molecular composition of ion channels, and their configuration in the cell membrane are discussed in detail in later chapters, but it is useful at this point to have some idea of the general physical features of an ion channel protein. These are illustrated in Figure 4.1B. The protein spans the membrane, with a central water-filled pore open to both the intracellular and extracellular spaces. On each side of the membrane, the pore widens to

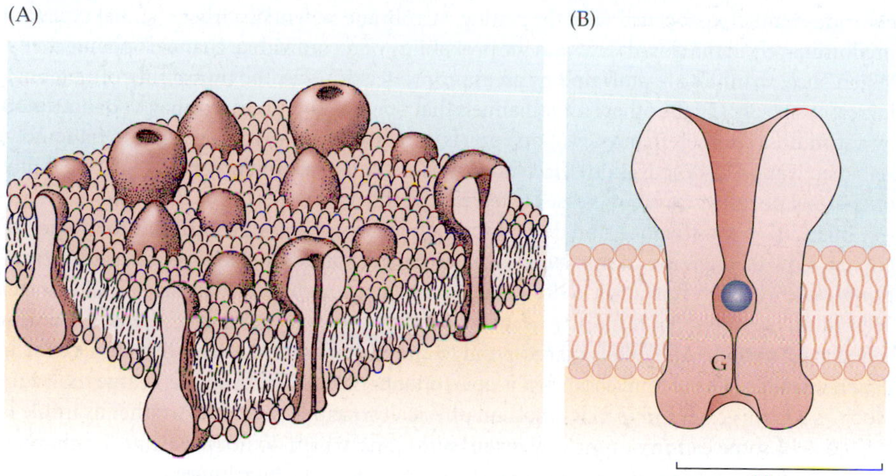

FIGURE 4.1 Cell Membrane and Ion Channel. (A) The cell membrane is composed of a lipid bilayer embedded with proteins. Some of the proteins traverse the lipid layer and some of these, in turn, form membrane channels. (B) Schematic representation of a membrane channel in cross section, with a central water-filled pore. The channel "gate" (G) opens and closes irregularly. Opening of the gate may be regulated by the membrane potential, the binding of a ligand to the channel, or other biophysical or biochemical conditions. A sodium ion, surrounded by a single shell of water molecules, is shown to scale in the pore for size comparison.

form a vestibule. Within the plane of the membrane, a segment of the pore is constricted and lined with a ring of negative charges to form a **selectivity filter** for cations. Conversely, a ring of positive charges would promote anion selectivity. Finally, the channel contains a **gate** that opens and closes to control ion movement through the channel. The size of the protein varies considerably from one channel type to the next, and some have additional structural features. Figure 4.1B represents a channel of medium dimensions.

Channel Selectivity

Membrane channels vary considerably in their selectivity: some are permeable to cations, some to anions. Some cation channels are selective for a single ion species, and allow permeation of sodium, potassium or calcium almost exclusively. Others are relatively nonspecific, allowing the passage of even small organic cations. Anion channels involved in signaling tend to have low specificity but are referred to as chloride channels, because chloride is the major permeant anion in biological solutions. There are also channels that connect adjacent cells (connexons) and allow the passage of most inorganic ions and many small organic molecules. These will be discussed in Chapter 8.

Open and Closed States

It is convenient to represent protein molecules as static structures; however, they are never still. Because of their thermal energies, all large molecules are inherently dynamic. At room temperature, chemical bonds stretch and relax, and twist and wave around their equilibrium positions. Although individual movements are only of the order of 10^{-12} meters in magnitude, with frequencies approaching 10^{13} Hz, such atomic trembling can underlie much larger and slower changes in conformation of the molecule. This is because numerous rapid motions of the atoms occasionally allow groups to slide by one another in spite of mutual repulsive interactions that would otherwise keep them in place. Such a transition, once achieved, can last for many milliseconds or even seconds. An example is hemoglobin, in which the binding sites for oxygen to the heme groups are buried inside the molecule and not immediately accessible. Oxygen binding, and its subsequent escape, can be accomplished only when a transient access pathway to the heme pocket is formed.[1]

In ion channel proteins, molecular transitions occur between open and closed states, with transitions between states being virtually instantaneous. If we examine the behavior of any given channel, we find that open times vary randomly—sometimes the channel opens for only a millisecond or less, sometimes for very much longer (see Figure 4.5). However, each channel has its own characteristic **mean open time** (τ) around which individual open times fluctuate.

Some channels in the resting cell membrane open frequently; thus, the probability of finding such channels in the open state is relatively high. Most of these are potassium and

[1] Karplus, M., and Petsko, G. A. 1990. *Nature* 347: 631–639.

chloride channels associated with the resting membrane potential. Other channel types are predominantly in the closed state, and the probability of an individual channel opening is low. When such channels are **activated** by an appropriate stimulus, the probability of openings increases sharply. On the other hand, channels that open frequently at rest may be **deactivated** by a stimulus (i.e., their frequency of opening is decreased). An important point to remember is that activation or deactivation of a channel means an increase or decrease in the probability of channel opening, not an increase or decrease in the mean channel open time.

In addition to activation and deactivation, there are two other conditions that regulate current flow through ion channels. The first is that certain channels can enter a conformational state in which activation no longer occurs, even though the activating stimulus is still present. In channels that respond to depolarization of the cell membrane, this condition is called **inactivation**; in channels that respond to chemical stimuli, the condition is known as **desensitization**. The second condition is **open channel block**. For example, a large molecule (such as a toxin) can bind to a channel and physically occlude the pore. Another example is the block of some cation channels by magnesium ions, which do not permeate the channel but bind in its inner mouth and prevent the permeation of other cations.

Modes of Activation

Figure 4.2 summarizes the modes of channel activation. Some channels respond specifically to physical changes in the nerve cell membrane. Most important in this group are the voltage-activated channels. An example is the **voltage-sensitive sodium channel**, which is responsible for the regenerative depolarization that underlies the rising phase of the action potential (see Chapter 7). Also in this group are stretch-activated channels, responding to mechanical distortion of the cell membrane. These are found, for example, in mechanoreceptors in the skin (see Chapter 18).

Other channels are activated when chemical agonists attach to binding sites on the channel protein. These **ligand-activated channels** are further divided into two subgroups, depending on whether the binding sites are extracellular or intracellular. Those responding to extracellular activation include, for example, cation channels in the postsynaptic membranes of skeletal muscle that are activated by the neurotransmitter acetylcholine (ACh), released from presynaptic nerve terminals (see Chapter 11). This activation allows sodium to enter the cell, causing muscle depolarization. Channels activated intracellularly

(A) Channels activated by physical changes in the cell membrane

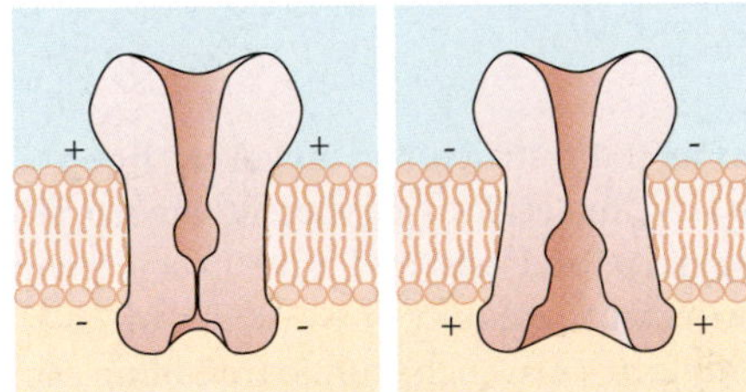

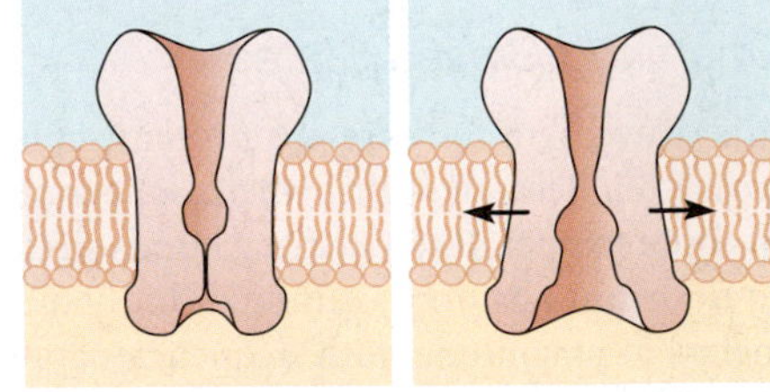

Voltage-activated

Stretch-activated

(B) Channels activated by ligands

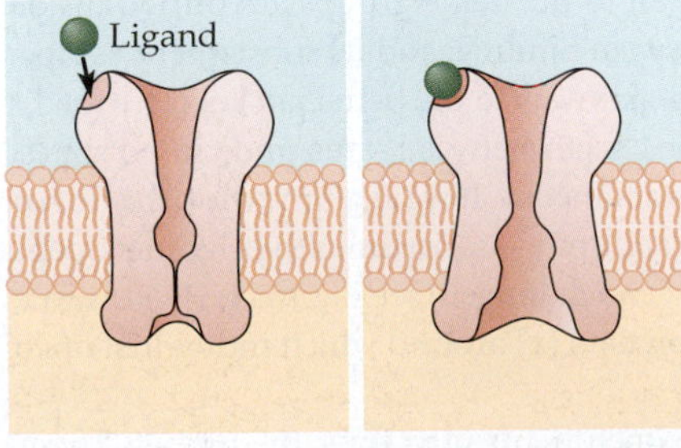

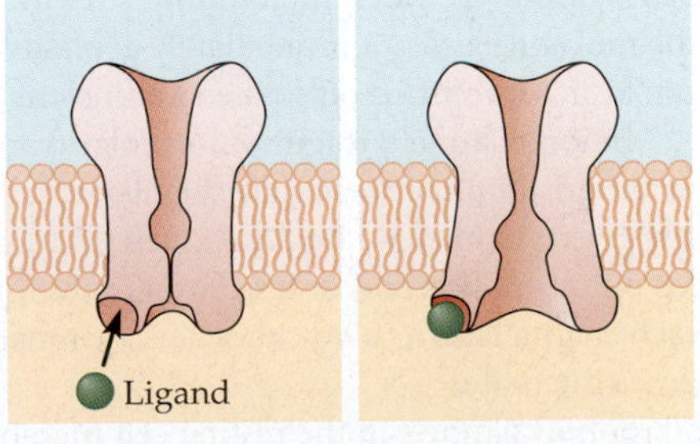

Extracellular activation

Intracellular activation

FIGURE 4.2 Modes of Channel Activation. The probability of channel opening is influenced by a variety of stimuli. (A) Some channels respond to changes in the physical state of the membrane, specifically: changes in membrane potential (voltage-activated), and mechanical distortion (stretch-activated). (B) Ligand-activated channels respond to chemical agonists, which attach to binding sites on the channel protein. Neurotransmitters, such as glycine and acetylcholine, act on extracellular binding sites. Included among a wide variety of intracellular ligands are calcium ions, subunits of G proteins, and cyclic nucleotides.

may be sensitive to local changes in concentration of a specific ion. For example, one type of potassium channel is activated by an increase in intracellular calcium concentration in the adjacent cytoplasm. In many neurons, these calcium-activated potassium channels play a role in repolarizing the membrane during termination of the action potential. Other intracellular ligands that modulate channel opening include the cyclic nucleotides: cyclic GMP is responsible for the activation of cation channels in retinal rods, thereby playing an important role in visual transduction (see Chapter 20).

It should be noted that these classifications are not rigid: some calcium-activated potassium channels are also voltage-sensitive, and some voltage-activated channels are sensitive to intracellular ligands.

Measurement of Single-Channel Currents

Intracellular Recording with Microelectrodes

The first experiments designed to examine the properties of membrane channels were done using glass microelectrodes to record membrane potentials or membrane currents from whole cells. In 1949, Ling and Gerard[2] adapted the glass microelectrode for intracellular recording from single cells. This achievement was a milestone as important as the introduction of patch clamp recording 3 decades later. The technique provided a method for accurate measurements of resting membrane potentials, action potentials, and responses to synaptic activation in muscle fibers and neurons.

The intracellular recording technique is illustrated in Figure 4.3A. A sharp glass micropipette, with a tip diameter of less than 0.5 μm and filled with a concentrated salt solution (e.g., 3M KCl), serves as an electrode and is connected to an amplifier to record the potential at the tip of the pipette. When the pipette is pushed against the cell membrane, penetration into the cytoplasm is signaled by the sudden appearance of the resting potential. If the penetration is successful, the membrane seals around the outer surface of the pipette, so that the cytoplasm remains isolated from the extracellular fluid.

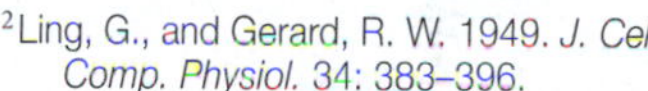

[2]Ling, G., and Gerard, R. W. 1949. *J. Cell Comp. Physiol.* 34: 383–396.

[3]Katz, B., and Miledi, R. 1972. *J. Physiol.* 224: 665–699.

Channel Noise

In the early 1970s, Katz and Miledi[3] did pioneering experiments on frog muscle fibers, in which they used intracellular recording techniques to examine the characteristics of the noise produced by acetylcholine (ACh) at the neuromuscular junction. At this synapse, ACh released from the presynaptic nerve terminal opens

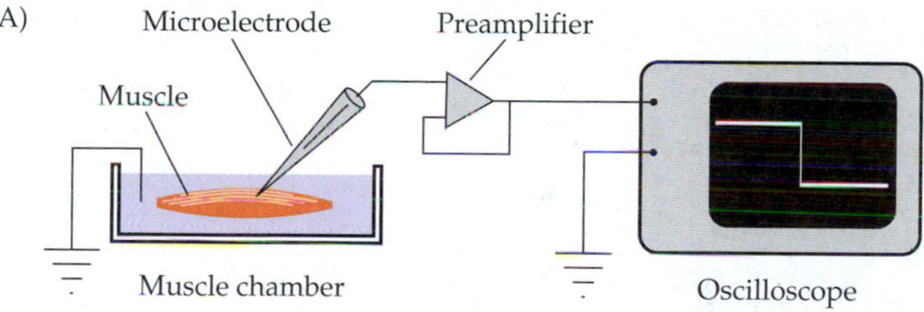

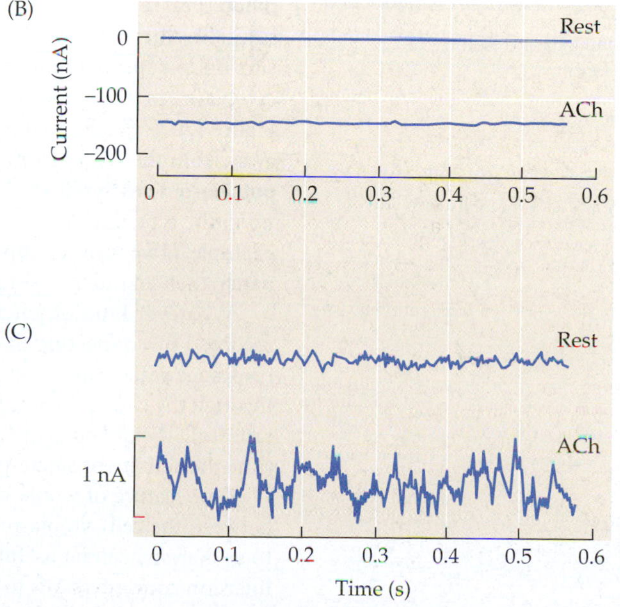

FIGURE 4.3 Intracellular Recording of Channel Noise. (A) Arrangement for recording membrane potentials of muscle fibers with a microelectrode. Electrode is connected to a preamplifier, and the signals are displayed on an oscilloscope or computer screen. Penetration of the electrode into a fiber is marked by sudden appearance of the resting potential (downward deflection on the screen). After penetration, changes in potential, resulting from channel activation, can be measured. (B) Intracellular records of the effect of acetylcholine (Ach). In this experiment, additional circuitry was used to record membrane current (rather than membrane potential). At rest (upper trace), there is no current across the membrane; application of acetylcholine (ACh) produces about 130 nanoamperes (nA) of inward current (lower trace). (C) Traces in (B) are shown at greater amplification. There is little fluctuation in the baseline at rest; the inward current produced by ACh shows relatively large fluctuations ("noise"), which is due to random opening and closing of ACh-activated channels. Analysis of the increased noise yields values for the single-channel current and mean open time of the channels. (Records modified from Anderson and Stevens, 1973.)

ligand-gated channels in the postsynaptic membrane. These channels allow cations to enter, thus depolarizing the muscle (see Chapter 11). Katz and Miledi applied ACh directly to the synaptic region and observed that the resulting depolarization was "noisy"; that is, during the depolarization, fluctuations in the electrical recording were larger than the normal baseline fluctuations at rest. This increase in noise was due to the random opening and closing of the ACh-activated channels. In other words, application of ACh resulted in the opening of a large number of channels, and this number fluctuated in a random way as ACh molecules bombarded the membrane.

By applying **noise analysis** techniques, Katz and Miledi were able to obtain information about the behavior of the individual channels activated by ACh. Subsequently, in similar experiments on the same preparation, Anderson and Stevens[4] used fluctuations in membrane current to deduce the size and duration of ionic currents through single channels (see Figure 4.3).

Although noise analysis involves moderately complex algebra, the principles underlying it are straightforward. First, if the single channel currents are relatively large, then the noise will be large as well. Second, channels that open for a relatively long time will produce only low-frequency noise; channels that open only briefly will produce higher frequency noise. Examination of the amplitude and the frequency composition of the noise produced by ACh-activated channels at the neuromuscular junction showed that about 10 million ions per second flowed through an open channel and that the mean open time of the channels (τ) was 1–2 milliseconds.

[4] Anderson, C. R., and Stevens, C. F. 1973. *J. Physiol.* 235: 665–691.

[5] Neher, E., Sakmann, B., and Steinbach, J. H. 1978. *Pflügers Arch.* 375: 219–228.

[6] Hamill, O. P., et al. 1981. *Pflügers Arch.* 391: 85–100.

Patch Clamp Recording

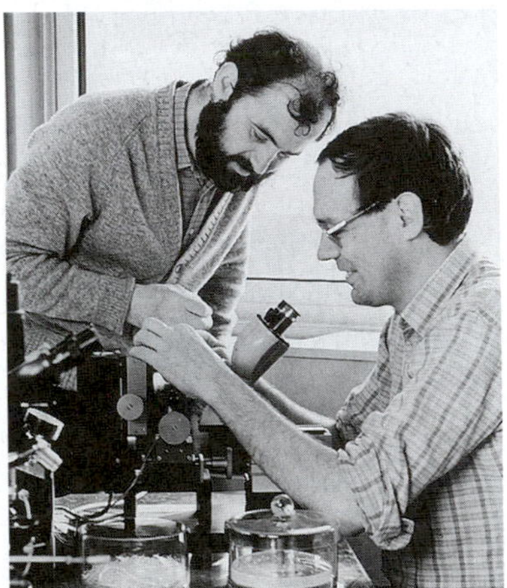

Erwin Neher (left) and Bert Sakmann (right)

Noise analysis has been largely supplanted by direct observation of channel currents, using **patch clamp recording** methods. These methods provide direct answers to questions of obvious physiological interest about channels; for example, how much current does a single channel carry? How long does a channel stay open? How do its open and closed times depend on voltage or on the activating molecule?

The development of the patch clamp by Erwin Neher, Bert Sakmann, and their colleagues,[5,6] has made an enormous contribution to our knowledge of the functional behavior of membrane channels. Patch clamp recording methods involve sealing the tip of a small (ca. 1 μm internal diameter) glass pipette to the membrane of a cell (Figure 4.4A). Under ideal conditions, with slight suction on the pipette, a seal resistance of greater than 10^9 ohms (hence a gigaohm seal) is formed around the rim of the pipette tip between the cell membrane and the glass (Figure 4.4B).

Patch clamp methods permit other recording configurations. Having made a seal to form a **cell-attached patch**, we can then pull the patch from the cell to form an **inside-out patch** (Figure 4.4C), with the cytoplasmic face of the patch membrane facing the bathing solution. Alternatively, after forming a cell-attached path, we can apply slight additional suction to rupture the membrane inside the patch and thereby provide access to the cell cytoplasm. In this condition, currents are recorded from the entire cell (**whole-cell recording**; Figure 4.4D). Finally, we may first obtain a whole-cell recording and then pull the electrode away from the cell to form a thin neck of membrane that separates and seals to form an **outside-out patch** (Figure 4.4E). Each of these configurations has an advantage, depending upon the type of channels studied and the kind of information one wishes to obtain. For example, if we want to apply a variety of chemical ligands to the outer membrane of the patch, then an outside-out patch is most convenient.

Whatever kind of patch is used, small currents across the patch membrane can be recorded by connecting the pipette to an appropriate amplifier (Figure 4.4F). The high-resistance seal ensures that such currents flow through the amplifier rather than escaping through the rim of the patch. The recorded events consist of rectangular pulses of current, reflecting the opening and closing of single channels. In other words, one observes in real time the activity of single-protein molecules in the membrane.

One feature of whole-cell recording with a patch pipette is that substances can move between the cell cytoplasm and the pipette solution. This exchange (sometimes referred to as dialysis) can be useful in providing a method for changing the preexisting intracellular ion concentrations to those in the pipette. On the other hand, particularly if the cell

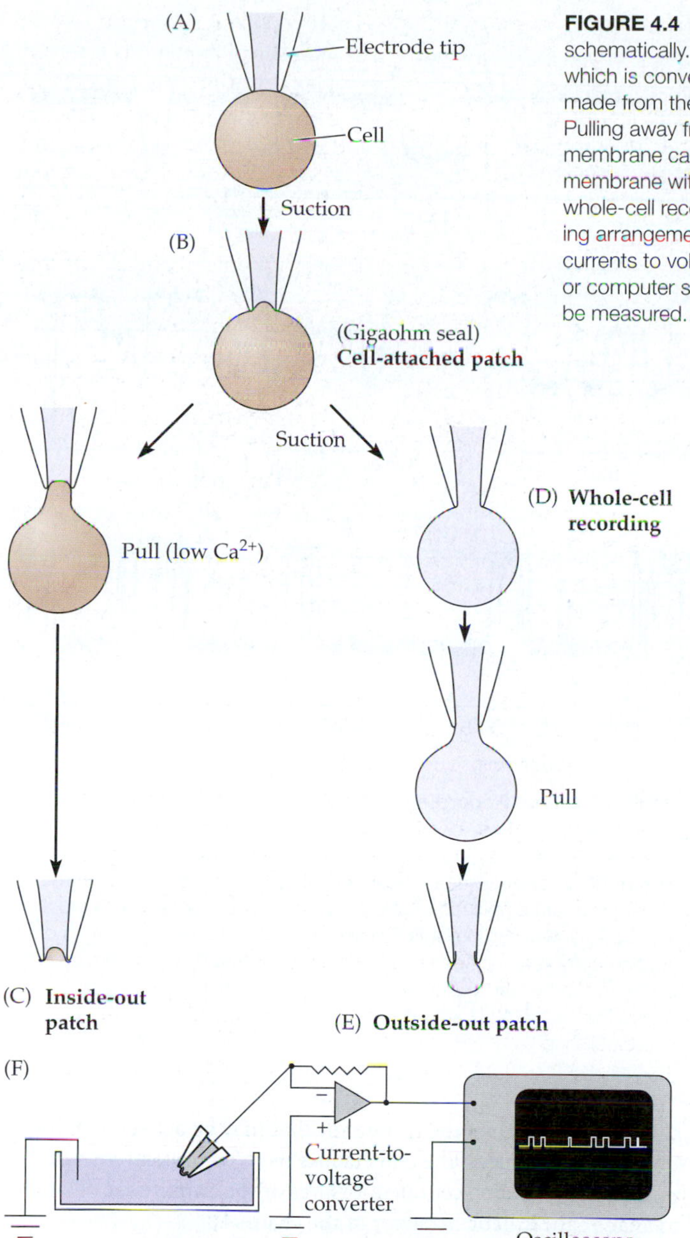

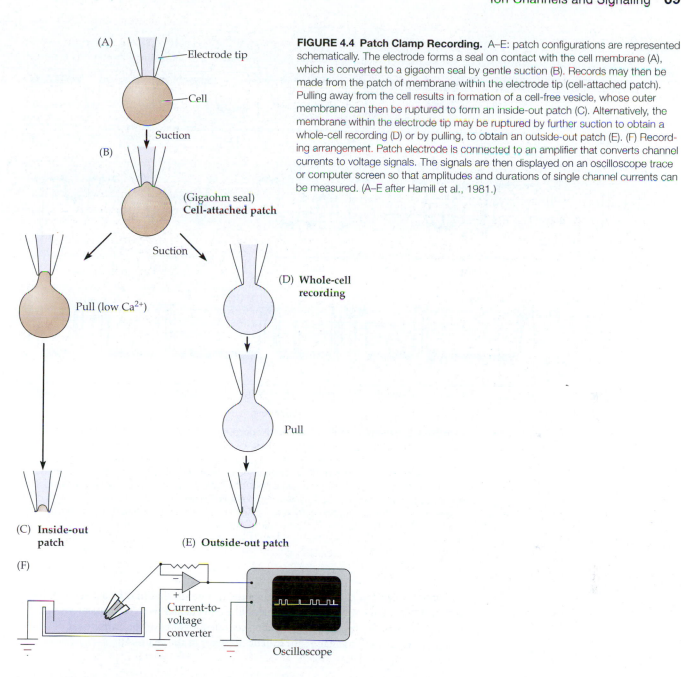

FIGURE 4.4 Patch Clamp Recording. A–E: patch configurations are represented schematically. The electrode forms a seal on contact with the cell membrane (A), which is converted to a gigaohm seal by gentle suction (B). Records may then be made from the patch of membrane within the electrode tip (cell-attached patch). Pulling away from the cell results in formation of a cell-free vesicle, whose outer membrane can then be ruptured to form an inside-out patch (C). Alternatively, the membrane within the electrode tip may be ruptured by further suction to obtain a whole-cell recording (D) or by pulling, to obtain an outside-out patch (E). (F) Recording arrangement. Patch electrode is connected to an amplifier that converts channel currents to voltage signals. The signals are then displayed on an oscilloscope trace or computer screen so that amplitudes and durations of single channel currents can be measured. (A–E after Hamill et al., 1981.)

is small, important cytoplasmic components can be lost rapidly into the pipette solution. Such loss can be avoided by use of a **perforated patch**.[7] The patch pipette is loaded with a pore-forming substance (e.g., the antibiotic nystatin) and a seal is formed to the cell. After a delay, pores are formed in the patch that allow whole-cell currents to be recorded without loss of intracellular macromolecules.

Single-Channel Currents

In their simplest form, single-channel current pulses appear irregularly, with nearly fixed amplitudes and variable durations (Figure 4.5A). However, in some cases, current records are more complex. For instance, channels exhibit open states with more than one current level, as in Figure 4.5B, where the open channels often close to smaller "substate" levels. In addition, channels may display complicated kinetics, such as channel openings that occur in **bursts** (Figure 4.5C).

In summary, patch clamp techniques offer two advantages for studying the behavior of channels. First, the isolation of a small patch of membrane allows allows us to observe the

[7] Horn, R., and Marty, A. 1988. *J. Gen. Physiol.* 92: 145–149.

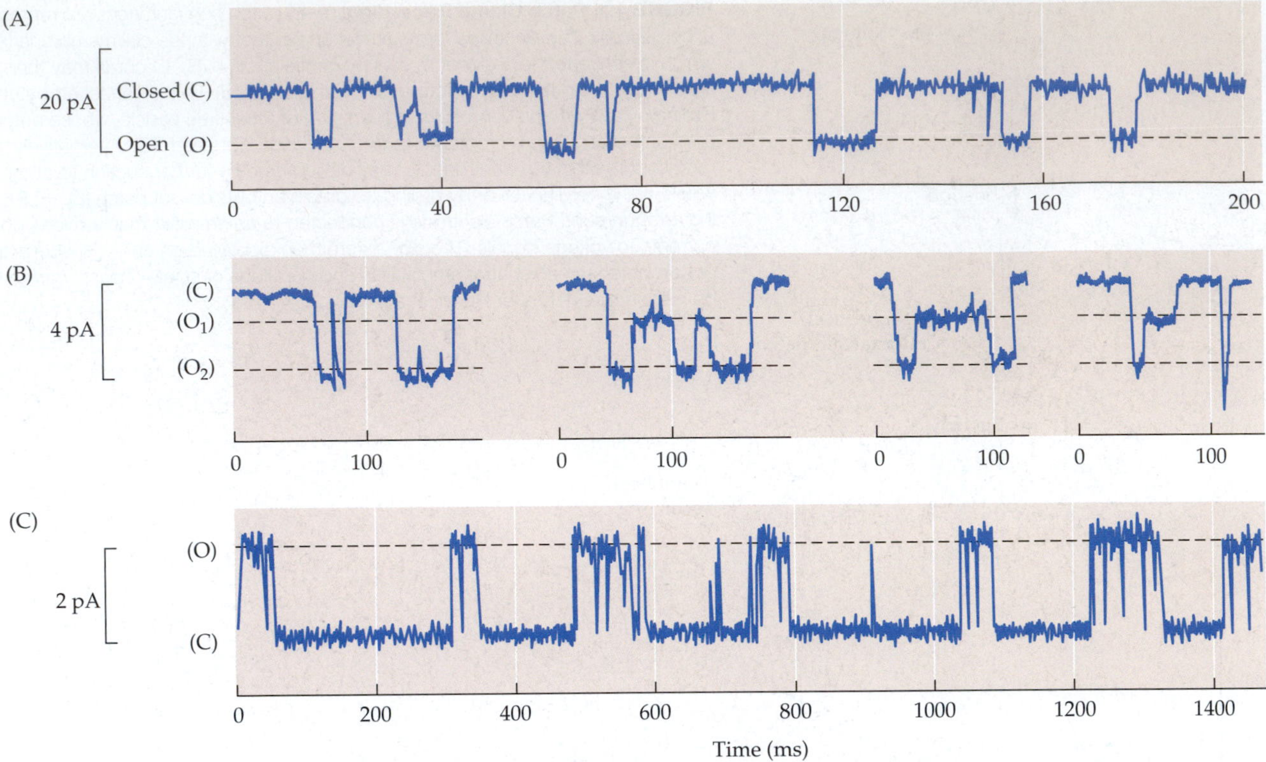

FIGURE 4.5 Examples of Patch Clamp Recordings. (A) Glutamate-activated channel currents recorded in a cell-attached patch from locust muscle occur irregularly, with a single amplitude and varied open times. Downward deflections indicate current flowing into the cell. (B) Acetylcholine-activated currents from single channels in an outside-out patch from cultured embryonic rat muscle reach a maximum amplitude of about 3 picoamperes (pA) and relax to a substrate current of about 1.5 pA. Downward deflections indicate inward current. (C) Pulses of outward current through glycine-activated chloride channels in an outside-out patch from cultured chick spinal cord cells are interrupted by fast closing and reopening transitions to produce bursts. (A after Cull-Candy, Miledi, and Parker, 1980; B after Hamill and Sakmann, 1981; C after A. I. McNiven and A. R. Martin, unpublished.)

activity of only a few channels, rather than the thousands that may be active in an intact cell; and second, the very high resistance of the seal enables us to record extremely small currents. As a result, we are able to obtain accurate measures of the amplitudes of single-channel currents and to analyze the kinetic behavior of the channels.

Channel Conductance

The kinetic behavior of a channel—that is, the durations of its closed and open states—can provide information about the steps involved in channel activation and the rate constants associated with these steps. The channel current, on the other hand, is a direct measure of how rapidly ions move through a channel. The current depends not only on channel properties but also on the transmembrane potential. Consider, for example, the outside-out membrane patch in Figure 4.6, which contains a single, spontaneously active channel that is permeable to potassium. The solutions in both the patch pipette and the bath contain 150 mM potassium. Potassium ions move in both directions through the open channel but, because the concentrations are equal, there is no net movement in either direction. Thus, no current is seen (Figure 4.6B). Fortunately, the patch clamp recording system allows us to apply a voltage to the pipette solution and thereby produce a voltage difference across the membrane patch. When a voltage of +20 mV is applied (Figure 4.6C), each channel opening results in a pulse of outward current, because positively charged potassium ions are driven outward through the channel by the electrical gradient between the pipette solution and the bath. On the other hand, when the inside

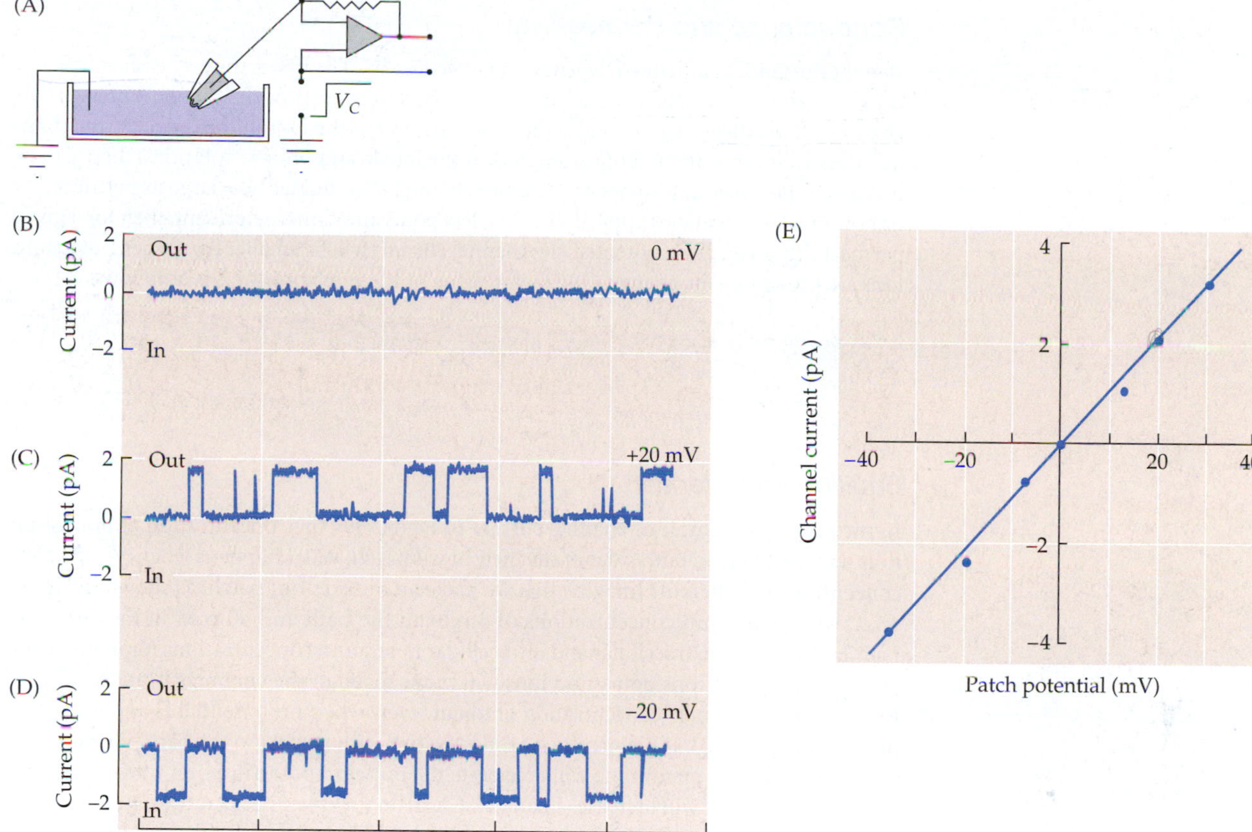

FIGURE 4.6 Effect of Potential on Currents through a single, spontaneously active potassium channel in an outside-out patch, with 150 mM potassium in both the electrode and the bathing solution. (A) The recording system. The output from the patch clamp amplifier is proportional to the current across the patch. A potential is applied to the electrode (and hence to the patch) by applying a command potential (V_C) to the amplifier as shown. Current flowing into the electrode is shown as negative. (B) When no potential is applied to the patch, no channel currents are seen, because there is no net flux of potassium through the channels. (C) Application of +20 mV to the electrode results in an outward current of about 2 pA through the channels. (D) A −20 mV potential results in inward channel currents of the same amplitude. (E) Channel currents as a function of applied voltage The slope of the line indicates the channel conductance (γ). In this case γ = 110 picosiemens (pS).

is made negative by 20 mV (Figure 4.6D), current flows in the other direction through the open channel into the pipette. The effect of voltage on the size of the current is plotted in Figure 4.6E. The relationship is linear; the current (I) through the channel is proportional to the voltage applied to it:

$$I = \gamma(V - V_0)$$

Many readers will recognize this as an expression of Ohm's law. V_0 is the voltage across the membrane at which the current is zero, and ($V - V_0$) is the **driving force** for current through the channel. In this example $V_0 = 0$, but that is not always the case (see Figure 4.7). The constant of proportionality, γ, is called the **channel conductance.** It is a measure of the ability of the channel to pass current. For a particular applied voltage, a high conductance channel will carry a lot of current, while a low conductance channel will carry only a little.

The unit of conductance is the siemens (S); 1S = 1 ampere/volt. In nerve cells, the potential across the membrane is usually expressed in units of millivolts (1 mV = 10^{-3}V), currents through single channels in picoamperes (1 pA = 10^{-12}A), and conductances of single channels in picosiemens (1 pS = 10^{-12}S). In Figure 4.6E, a potential of +20 mV produces a current of about 2.2 pA, so the channel conductance (γ = I/V) is 2.2 pA/20 mV = 110 pS.

Conductance and Permeability

The conductance of a channel depends upon two factors. The first is the ease with which ions can pass through the open channel; this is an intrinsic property of the channel known as the **channel permeability**. The second is the concentration of the ions in the region of the channel. Clearly, if there are no potassium ions in the inside and outside solutions, there can be no current flow through an open potassium channel—no matter how large its permeability or how great a potential is applied. If only a few potassium ions are present, then for a given permeability and a given potential, the channel current will be smaller than when potassium ions are present in abundance. One way to think of these relationships is as follows:

open channel → permeability

permeability + ions → conductance

Equilibrium Potential

In the previous example of channel current in Figure 4.6, the concentration of potassium ions was the same on both sides of the membrane patch. What happens when we make the concentrations different? Imagine that we make an outside-out patch, as shown in Figure 4.7A, with potassium concentrations of 3 mM in the bath and 90 mM in the electrode (similar to normal extracellular and intracellular potassium concentrations for many cells). Under those conditions potassium ions will move through the channels from the pipette to the bath down their concentration gradient, even when no potential is applied to the pipette (Figure 4.7B). If the pipette is made positive with respect to the bath, the potential gradient across the membrane will accelerate the outward potassium ion movement, and the channel current will increase (Figure 4.7C). If, on the other hand, if the pipette is made negative, outward movement of potassium will be retarded and the channel current will decrease (Figure 4.7D). With sufficiently large negativity, potassium ions will flow inward across the membrane against their concentration gradient (Figure 4.7E). If we make a number of such observations and plot channel current against applied voltage, the result is similar to that shown in Figure 4.7F.

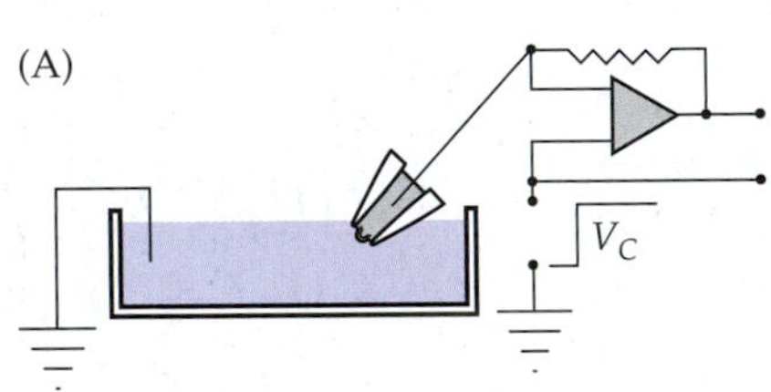

FIGURE 4.7 Reversal Potential for Potassium Currents in a hypothetical experiment using an outside-out patch with potasssium concentrations of 90 mM in the recording pipette ("intracellular") and 3 mM in the bathing solution ("extracellular"). (A) Recording arrangement. (B) With no potential applied to the pipette, flux of potassium from the electrode to the bath along its concentration gradient produces outward channel currents. (C) When a potential of +20 mV is applied to the pipette, outward currents increase in amplitude. (D) Application of –50 mV to the pipette reduces outward currents. (E) At –100 mV, currents are reversed. (F) The current–voltage relation indicates zero current at –85 mV, which is the potassium equilibrium potential (EK).

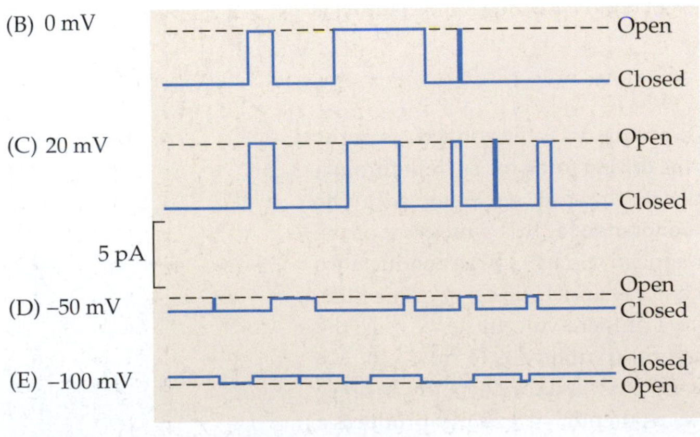

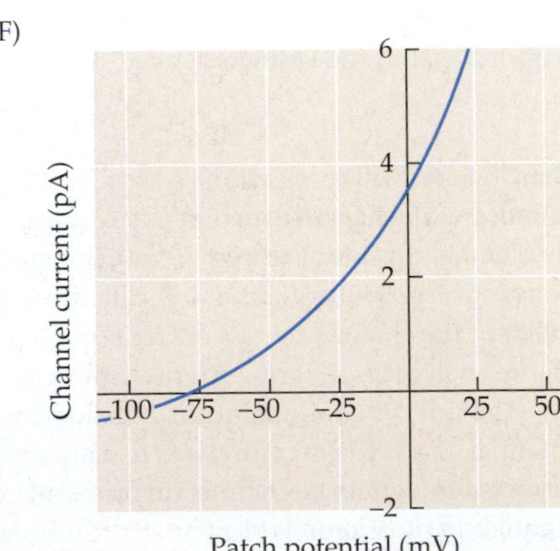

Figure 4.7F illustrates that the potassium current through the channel depends on both the electrical potential across the membrane and on the potassium concentration gradient—that is, on the **electrochemical gradient** for potassium. Unlike the result when the potassium concentration was the same on both sides of the membrane (Figure 4.6), the channel current is zero when the potential applied to the pipette is about –85 mV. In this condition, the concentration gradient, which would otherwise produce an outward flux of potassium through the channel, is balanced exactly by the electrical potential gradient that tends to move potassium inward. The potential that just balances the potassium concentration gradient is called the **potassium equilibrium potential**, E_K. When the membrane potential is at E_K, the driving force for potassium current is zero; at any other potential (V) the driving force is $(V–E_K)$. The equilibrium potential depends only on the ion concentrations on either side of the membrane and not on the properties of the channel or the mechanism of ion permeation through the channel.

The Nernst Equation

Exactly how large a potential is required to balance a given potassium concentration difference across the membrane? One guess might be that E_K would be proportional simply to the difference between the inside concentration $[K]_i$ and the outside concentration $[K]_o$; however, this assumption is not quite right. It turns out instead that the required potential depends on the difference between the *logarithms* of the concentrations:

$$E_K = k(\ln[K]_o - \ln[K]_i)$$

The constant k is given by RT/zF, where R is the thermodynamic gas constant, T the absolute temperature, z the valence of the ion (in this case +1), and F is the Faraday (the number of coulombs of charge in one mole of monovalent ion). The answer, then, is

$$E_K = \frac{RT}{zF}(\ln[K]_o - \ln[K]_i)$$

which is the same as

$$E_K = \frac{RT}{zF}\ln\frac{[K]_o}{[K]_i}$$

This is the **Nernst equation** for potassium. RT/zF has the dimensions of volts and is equal to about 25 mV at room temperature (20°C). It is sometimes more convenient to use the logarithm to the base 10 (log) of the concentration ratio, rather than the natural logarithm (ln). Then RT/zF must be multiplied by ln (10), or 2.31, which gives a value of 58 mV.

In summary,

$$E_K = 25\ln\frac{[K]_o}{[K]_i} = 58\log\frac{[K]_o}{[K]_i}$$

At mammalian body temperature (37°C), 58 mV increases to about 61 mV. For the cell shown in Figure 4.7, the value of E_K (–85 mV) agrees with the given concentration ratio (1/30).

It should be noted that the rate of diffusion of an ion down a concentration gradient is not strictly related to its concentration. In all but the very weakest solutions, ions are subject to interactions with one another, for example electrostatic attraction or repulsion. The result of such interactions is that the effective concentration of the ion is reduced. The effective concentration of an ion in solution is called its **activity**. In theory, the Nernst equation should be written with an activity ratio rather than a concentration ratio. However, because the total concentrations of ions inside and outside the cell are similar (see Chapter 6), the activity ratio for any particular ion is not significantly different from its concentration ratio.

Nonlinear Current–Voltage Relations

A second feature of the current–voltage relation in Figure 4.7F is that, unlike the one in Figure 4.6E, it is not linear. As we move away from the equilibrium potential in the depolarizing direction, the outward current increases more and more rapidly as the potential approaches zero. Conversely, when the membrane is made more negative (hyperpolarized), the inward current increases more slowly with hyperpolarization, This is because of the dependence of conductance on concentration. There is a much higher potassium concentration inside the pipette than in the external solution. Consequently there are more ions available to carry outward current than to carry inward current. The farther we move away from the equilibrium potential, the more prominent the effect becomes, so that the current–voltage relation has a marked upward curvature, even though, in this example, the channel permeability is quite independent of voltage.

Nonlinear current–voltage relations also occur in some channels because the channels themselves rectify (i.e., their permeability is voltage-dependent), allowing ions to move through the pores in one direction much more readily than in the other. One example is a voltage-sensitive potassium channel, called the **inward rectifier** channel, which allows potassium to move into the cell when the membrane potential is negative to the potassium equilibrium potential, but it permits little or no outward movement when the potential difference is reversed. Its voltage–current relation is similar to that shown in Box 4.1.

Ion Permeation through Channels

How do ions actually pass through channels? One way could be by diffusion through a water-filled pore. Diffusion formed the basis of early ideas about ion permeation, but for most channels diffusion does not provide an adequate description of the permeation process. This is because the channels themselves interact with the ions. For example, because they are charged, ions in solution are always accompanied by closely apposed water molecules. In the case of cations, the water molecules are oriented so that the oxygen molecules, which

BOX 4.1
Measuring Channel Conductance

Investigators often state that some channel or other has a particular conductance, say 100 pS. Because conductance depends upon concentration, such a statement tells us little unless we know the ionic conditions under which the measurement was made. For example, if the stated conductance is that of a potassium channel with a potassium concentration of 150 mM on either side of the membrane, then in a more physiological environment, with a potassium concentration of only 5mM on the extracellular side, the channel conductance can be expected to be five to ten times smaller.

A second problem arises when the current–voltage relation is not linear, either because the ion concentrations on either side of the channel are not symmetrical (see Figure 4.7F), or because the channel itself is heavily rectifying. Under these circumstances, the conductance is not constant, and it is necessary to specify the potential at which it is measured. One way to define the conductance is to use the relation $\gamma = I/(V-V_0)$. In the accompanying illustration, the equilibrium potential for the current (V_0) is –75 mV. When the patch is held at V = –25 mV, the driving force is 50 mV and the channel current is 0.6 pA. Thus the channel conductance is 0.6 pA/50 mV = 12 pS. This is the **chord conductance** of the channel at –25 mV, and is represented by the slope of the solid red line. The chord conductance at –25 mV is not the same as the chord conductance at –50 mV.

A second way of specifying conductance is to measure the slope of the current–voltage relation (dI/dV) at the point of interest. This is called the **slope conductance**. In this

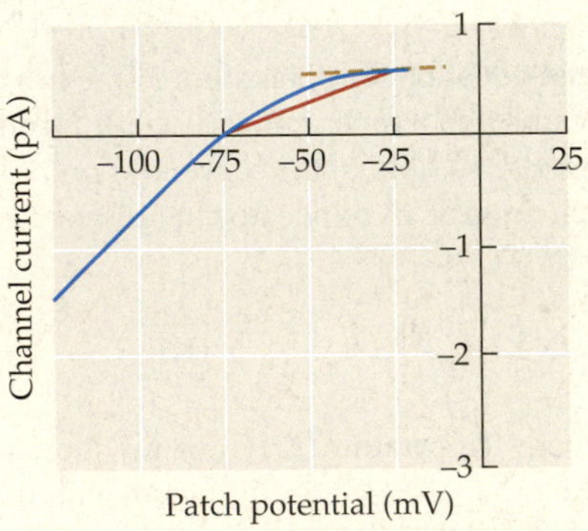

illustration, the slope of the current–voltage relation at –25 mV (dashed line) is about 3 pS. The measurement tells us that although the channel is passing a substantial current at –25 mV, it will not pass proportionately more as the driving force is increased.

In summary, the conductance characteristics of a channel can be specified precisely only by showing a complete current–voltage relation and by specifying the ionic conditions under which it was obtained. Single numbers given in the literature for channel conductance usually refer to chord conductance and, in the absence of additional information, provide only a rough indication of the channel characteristics.

carry net negative charges, lie closest to the ion. If the pore is relatively narrow, then an ion must acquire a certain amount of energy in order to escape from its associated waters of hydration and squeeze through the neck of the channel (see Chapter 5). Once in the channel, the ion may be attracted to or repelled by electrostatic charges lining the channel wall, or it may be bound to sites from which it must escape to continue its journey. Such interactions affect both ion selectivity and rate of ion flux through the channel. Channel models that deal with ion permeation in this way are called **Eyring rate theory** models.[8] In general, such models are more successful than simple diffusion models in describing channel selectivity and conductance.

An important point to remember is that all the ion fluxes that underlie signaling are due to ions moving passively through open channels along concentration and potential gradients. In other words, the neuron makes use of standing electrochemical gradients to generate ion movements, and hence to generate electrical signals. It is obvious that such fluxes would eventually dissipate the gradients. This does not happen, however, because cells use metabolic energy to maintain the ionic composition of the cytoplasm. Specialized mechanisms underlying the active transport of ions are discussed in Chapter 9.

[8] Johnson, F. H., Eyring, H., and Polissar, M. J. 1954. *The Kinetic Basis of Molecular Biology*. Wiley, New York.

SUMMARY

- Electrical signals in the nervous system are generated by the movement of ions across the nerve cell membrane. These ionic currents flow through the aqueous pores of membrane proteins known as ion channels.
- Channels vary in their selectivity: some cation channels allow only sodium, potassium, or calcium to pass, while others are less selective. Anion channels are relatively nonselective for smaller anions but pass mainly chloride ions because of the relative abundance of chloride in the extracellular and intracellular fluids.
- Channels fluctuate between open and closed states. Each channel has a characteristic mean open time. When channels are activated, their probability of opening increases. Deactivation reduces opening frequency. Channels may also be inactivated or blocked.
- Channels can be classified by their mode of activation: stretch-activated, voltage-activated, and ligand-activated.
- Ions move through channels passively in response to concentration and electrical gradients across the membrane.
- The net flux of ions through a channel down a concentration gradient can be reduced by an opposing electrical gradient. The electrical potential that reduces the net flux to exactly zero is called the equilibrium potential for that ion species. The relation between equilibrium potential and the concentration gradient is given by the Nernst equation.
- The driving force for movement of an ion across the membrane is the difference between its equilibrium potential and the actual membrane potential. The flow of ionic current through a channel depends on the driving force for the ion in question and on the conductance of the channel for that ion. The conductance depends, in turn, on the intrinsic ionic permeability of the channel and, in addition, on the inside and outside ion concentrations.

Suggested Reading

Hamill, O. P., Marty, A., Neher, E., Sakmann, B., and Sigworth, J. 1981. Improved patch-clamp techniques for high-resolution current recording from cells and cell-free membrane patches. *Pflügers Arch.* 391: 85–100.

Hille, B. 2001. *Ion Channels in Excitable Membranes,* 3rd ed. Sinauer Associates, Sunderland MA. pp. 347–375.

Pun, R. Y. K., and Lecar, H. 2001. Patch clamp techniques and analysis. In N. Sperelakis (ed.) *Cell Physiology Source Book,* 3rd ed. Academic Press, San Diego. pp. 441–453.

CHAPTER 5
Structure of Ion Channels

The molecular structure of ion channels has been resolved and related to their functional properties by a variety of experimental methods. These include biochemical isolation of channel proteins, molecular cloning to determine amino acid sequences of the proteins, site-directed mutagenesis to alter the sequences in selected locations, and expression of channel proteins in host cells to examine channel function. In addition, high-resolution electron microscopy and X-ray crystallography have revealed the three-dimensional structure of ion channels.

These combined experimental approaches have been applied most extensively to a ligand-activated channel, the nicotinic acetylcholine receptor (nAChR). The receptor is composed of five separate protein subunits, arranged around a central core. Two of these—the α-subunits—contain receptors for the ligand. Each subunit contains a large extracellular domain, a smaller intracellular domain, and four membrane-spanning regions (M1–M4), connected by intracellular and extracellular loops. The five M2 regions line the pore and form the channel gating structure. The ACh receptor is representative of a genetic superfamily known as Cys-loop receptors, which includes receptors for glycine, γ-aminobutyric acid (GABA), and 5-hydroxytryptamine (5-HT).

Voltage-activated channels form another superfamily. The voltage-activated sodium channel is a single large protein molecule with four repeating domains arranged around a central core, each with six transmembrane segments (S1–S6). In each domain, the loop of amino acids between S5 and S6 dips into the center of the structure to contribute to the pore lining. Voltage-activated calcium channels have a similar structure. Voltage-activated potassium channels are similar in molecular configuration, but with one important difference: instead of being single molecules, they are assembled from four separate subunits.

Detailed knowledge of the molecular structure of the Cys-loop receptors and voltage-activated channels has provided a firm basis for analyzing the structure and function of other channel types.

For the nervous system to function properly, neurons must perform a widely varied repertoire of electrical behavior. Thus, an impulse generated by one neuron may suppress the electrical activity of dozens of its neighbors, travel over long distances to produce excitation of other neuronal groups, and influence the responsiveness of still other target neurons in a variety of more subtle ways. All of these signals are mediated by the activation or deactivation of ion channels, thereby regulating the flow of ion currents across the nerve cell membranes. In this chapter we discuss experiments that have led to our current knowledge of the molecular structure of ion channels and how specific structural components are related to ion channel function.

Three experimental advances were instrumental in obtaining this information. The first was the development of techniques to isolate and sequence complementary DNA (cDNA) clones of channel proteins and thus obtain the corresponding amino acid sequences. These techniques also provide the opportunity to alter bases in the DNA (site-directed mutagenesis) and thereby substitute one amino acid for another at selected locations in the protein. The second advance was the development of techniques whereby messenger RNA (mRNA) derived from cDNA clones can be used to express channel proteins in host cells such as *Xenopus* oocytes. In this way, the functional properties of cloned channels can be measured. The combination of techniques makes it possible to tinker with specific regions of a channel molecule and then to determine how such tinkering affects its function. For example, changing even a single amino acid can have a marked effect on the ion selectivity of a channel. Finally, refined electron microscope imaging procedures and X-ray crystallography have provided a detailed physical view of channel structure at the molecular level. These approaches have yielded remarkably detailed information about two particular channel types: ligand-activated channels, represented by the nicotinic acetylcholine receptor, and voltage-activated cation channels.

Ligand-Activated Channels

The Nicotinic Acetylcholine Receptor

The first channel to be studied in detail was the **nicotinic acetylcholine receptor (nAChR)**. Note that for ligand-activated channels, the word *receptor* rather than *channel* is used routinely; this is because characterization of such molecules has relied primarily on the binding of activating molecules (agonists), antagonists, toxins, and antibodies, rather than on the specific channel properties of the protein. Nicotinic acetylcholine receptors are expressed in postsynaptic membranes of vertebrate skeletal muscle fibers, in neurons throughout the nervous systems of invertebrates and vertebrates, and in the neuroeffector junctions of electric organs of a number of electric fish. The receptors are activated by ACh released from presynaptic nerve terminals, and upon activation, open to form channels through which cations can enter or leave the postsynaptic cell. They are designated *nicotinic* because the actions of ACh are mimicked by nicotine, and also to distinguish them from the very different AChRs that can be activated by muscarine. Muscarinic AChRs (see Chapter 12) do not form ion channels; instead, their activation sets in motion intracellular messenger systems that, in turn, affect ion channel activity.

Biochemical isolation and characterization of the nAChR was facilitated by the availability of a concentrated source of receptors—the remarkably dense assembly of synapses on electrocyte membranes in the electric organ of the ray *Torpedo*. The electrocytes of this and other strongly electric fish are modified muscle cells. Large numbers of them are arrayed anatomically so that when depolarized simultaneously they can produce, as a group, voltages approaching 100 volts, with currents sufficient to stun nearby prey in the surrounding water.

After extraction from the electrocyte membranes, AChR molecules were separated from other membrane proteins by using their high affinity for α-bungarotoxin, a neurotoxin known to bind to channels with high specificity in intact electrocytes and in vertebrate muscle. Protein purification of the *Torpedo* AChR yielded a 250 kilodaltons (kD) complex, with two α-bungarotoxin binding sites. Separation on denaturing gels yielded four glycoprotein subunits (α, β, γ, and δ) of about 40, 50, 60, and 65 kD. Because toxin molecules bind to the α-subunit and to account for the molecular weight of the intact protein, a pentameric structure (α2βγδ) was proposed.[1] The isolated AChR was shown to retain the major functional properties of the native ionic channel when reincorporated into lipid vesicles.[2]

The size and orientation of the intact channel with respect to the lipid membrane has been determined by electron imaging, and by other physical techniques.[3–6] Its physical structure is shown in Figure 5.1. The five subunits form a circular array around a central pore. The extracellular domain extends about 8 nm above the plane of the membrane and is about 8 nm in diameter, with a central vestibule 2 nm in diameter leading into the transmembrane channel. The hatched area indicates the location of one of two binding sites. The second one is on the other α-subunit, adjacent to the δ-subunit. The intracellular domain extends about 4 nm below the membrane with five lateral openings between the subunits leading from the intracellular end of the channel to the cytoplasm. The pore of the open channel is about 0.7 nm in diameter, as predicted from earlier measurements of its selectivity to large cations.[7,8]

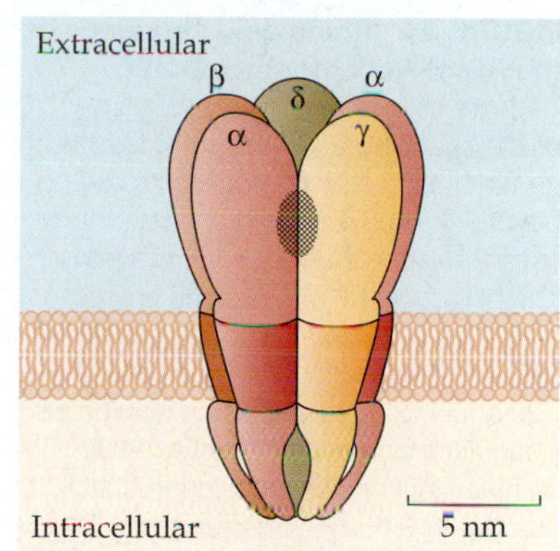

FIGURE 5.1 The ACh Receptor Consists of Five Subunits. There are two α, one β, one γ, and one δ-subunits spaced radially in increments of about 72 degrees around a central core. The extracellular domain extends about 8 nm above the plane of the membrane and contains a central vestibule of about 2 nm in diameter, which leads into the transmembrane channel. The hatched area indicates the location of one of two binding sites on the α-subunits. The intracellular domain extends about 4 nm below the membrane with five lateral openings between the subunits leading from the intracellular end of the channel to the cytoplasm. (After Unwin, 2005.)

Amino Acid Sequence of AChR Subunits

The cDNA for each subunit was cloned and sequenced by S. Numa and his colleagues in 1982.[9–11] Figure 5.2 shows the corresponding amino acid sequence for the α-subunit from *Torpedo*. The sequences of the other three subunits are highly similar (homologous), with various amino acid insertions and deletions, so that discussion of the structural configuration of any one subunit is generally applicable to the others. Sequences of subunits from human and bovine muscle are slightly different, and an additional subunit, similar to γ and designated ε, is found in fetal calf muscle.[12]

Higher Order Chemical Structure

Although the primary structure of the subunits does not provide unique information about how the polypeptide chains are arranged in the membrane, various models can be made, based on the characteristics of the amino acids in the sequence. As with any very large protein, segments of the molecule can be expected to fold into ordered secondary structures, such as α-helices or β-sheets. These secondary structures themselves fold to produce a tertiary structure in each subunit. Finally, five subunits join together to form the final quaternary structure (i.e., the complete channel). Models of the secondary and tertiary structures depend on a number of considerations, one being the identification of extended runs of nonpolar (and hence hydrophobic) amino acid residues in the primary sequence. Such sequences are capable of forming α-helices or other structures of sufficient length to span the membrane. In the original model proposed by Numa and his colleagues for the subunit structures[9] four such regions were identified (M1–M4; see Figure 5.2), and the model shown in Figure 5.3 was postulated. The reader may find it useful to use the hydropathy indices (Box 5.1) of the amino acid residues in these regions to verify the validity of these conclusions.

How do we know which parts of the molecule are extracellular and which are intracellular? To begin with, the NH_2 terminal is preceded by a relatively hydrophobic region of 24 amino acids (see Figure 5.2), which is taken to be the signal sequence necessary for insertion of the protein into the membrane. Consequently, the NH_2 terminal is placed extracellularly. Consistent with this orientation is the fact that the two adjacent cystein residues at positions 192 and 193 in the subunit are associated with the extracellular binding site for ACh. In addition, the initial extracellular segment constitutes about half of the entire molecule, which corresponds to the mass distribution of the intact receptor (see Figure 5.1). Given an even number of membrane crossings, the carboxylic acid (COOH) terminal is also extracellular. The general features of the model are in agreement with other observations. For example, when closely packed, AChRs aggregate in pairs (dimers) formed by disulfide cross-links extending between cysteine residues near the COOH terminals of the δ-subunits. These cross-links have been shown to be extracellular.[14,15]

Other Nicotinic ACh Receptors

After the nAChR was sequenced, similar isolation and sequencing was carried out for subunits of neuronal nAChR from autonomic ganglia and from vertebrate brain. Subunits

[1] Raftery, M. A. et al. 1980. *Science* 208: 1454–1457.

[2] Tank, D. W. et al. 1983. *Proc . Natl. Acad. Sci. USA* 80: 5129–5133.

[3] Wise, D. S., Schoenborn, B. P., and Karlin, A. 1981. *J. Biol. Chem.* 256: 4124–4126.

[4] Unwin, N., Toyoshima, C., and Kubalek, E. 1988. *J. Cell Biol.* 107: 1123–1138.

[5] Toyoshima, C. and Unwin, N. 1988. *Nature* 336: 247–250.

[6] Unwin, N. 2005. *J. Mol. Biol.* 346: 967–989.

[7] Maeno, T., Edwards, C., and Anraku, M. 1977. *J. Neurobiol.* 8: 173–184.

[8] Dwyer, T. M., Adams, D. J., and Hille, B. 1980. *J. Gen. Physiol.* 75: 469–492.

[9] Noda, M. et al. 1982. *Nature* 299: 793–797.

[10] Noda, M. et al. 1983b. *Nature* 301: 251–255.

[11] Noda, M. et al. 1983a *Nature* 302: 528–532.

[12] Takai, T. et al. 1985. *Nature* 315: 761–764.

[13] Kyte, J., and Doolittle, R. F. 1982. *J. Mol. Biol.* 157: 105–132.

[14] McCrea, P. D., Popot, J.-L., and Engleman, D. M. 1987. *EMBO J.* 6: 3619–3626.

[15] DiPaola, M., Czajkowski, C., and Karlin, A. 1989. *J. Biol.Chem.* 264: 15457–15463.

FIGURE 5.2 Amino acid sequence of the AChR Receptor α-Subunit. Sequences in orange indicate membrane-spanning α-helices (M1, M2, M3, and M4); blue-labeled segments extend beyond the membrane plane into the extracellular and intracellular spaces. The helix labeled (MA) resides entirely within the cytoplasm. The underlined cysteine (Cys) residues, at positions 128 and 142, are linked by a disulphide bond, so that the intervening thirteen residues form a Cys loop, characteristic of a large number of receptors. The initial underscored segment is the signal sequence. (Sequence after Numa et al., 1983; designation of helical segments from Unwin, 2005).

Met Ile Leu Cys

-20 −10
Ser Tyr Trp His Val Gly Leu Val Leu Leu Leu Phe Ser Cys Cys Gly Leu Val Leu Gly

10 20
Ser Glu His Glu Thr Arg Leu Val Ala Asn Leu Leu Glu Asn Tyr Asn Lys Val Ile Arg

30 40
Pro Val Glu His His Thr His Phe Val Asp Ile Thr Val Gly Leu Gln Leu Ile Gln Leu

50 60
Ile Ser Val Asp Glu Val Asn Gln Ile Val Glu Thr Asn Val Arg Leu Arg Gln Gln Trp

70 80
Ile Asp Val Arg Leu Arg Trp Asn Pro Ala Asp Tyr Gly Gly Ile Lys Lys Ile Arg Leu

90 100
Pro Ser Asp Asp Val Trp Leu Pro Asp Leu Val Leu Tyr Asn Asn Ala Asp Gly Asp Phe

110 120
Ala Ile Val His Met Thr Lys Leu Leu Leu Asp Tyr Thr Gly Lys Ile Met Trp Thr Pro

130 140
Pro Ala Ile Phe Lys Ser Tyr Cys Glu Ile Ile Val Thr His Phe Pro Phe Asp Gln Gln

150 160
Asn Cys Thr Met Lys Leu Gly Ile Trp Thr Tyr Asp Gly Thr Lys Val Ser Ile Ser Pro

170 180
Glu Ser Asp Arg Pro Asp Leu Ser Thr Phe Met Leu Ser Gly Glu Trp Val Met Lys Asp

190 200
Tyr Arg Gly Trp Lys His Trp Val Tyr Tyr Thr Cys Cys Pro Asp Thr Pro Tyr Leu Asp

210 220
Ile Thr Tyr His Phe Ile Met Gln Arg Ile Pro Leu Tyr Phe Val Val Asn Val Ile Ile

230 240
Pro Cys Leu Leu Phe Ser Phe Leu Thr Gly Leu Val Phe Tyr Leu Pro Thr Asp Ser Gly
M1

250 260
Glu Lys Met Thr Leu Ser Ile Ser Val Leu Leu Ser Leu Thr Val Phe Leu Leu Val Ile
M2

270 280
Val Glu Leu Ile Pro Ser Thr Ser Ser Ala Val Pro Leu Ile Gly Lys Tyr Met Leu Phe

290 300
Thr Met Ile Phe Val Ile Ser Ser Ile Ile Ile Thr Val Val Val Ile Asn Thr His His
M3

310 320
Arg Ser Pro Ser Thr His Thr Met Pro Gln Trp Val Arg Lys Ile Phe Ile Asp Thr Ile

330 340
Pro Asn Val Met Phe Phe Ser Thr Met Lys Arg Ala Ser Lys Glu Lys Gln Glu Asn Lys

350 360
Ile Phe Ala Asp Asp Ile Asp Ile Ser Asp Ile Ser Gly Lys Gln Val Thr Gly Glu Val

370 380
Ile Phe Gln Thr Pro Leu Ile Lys Asn Pro Asp Val Lys Ser Ala Ile Glu Gly Val Lys

390 400
Tyr Ile Ala Glu His Met Lys Ser Asp Glu Glu Ser Ser Asn Ala Ala Glu Glu Trp Lys
MA

410 420
Tyr Val Ala Met Val Ile Asp His Ile Leu Leu Cys Val Phe Met Leu Ile Cys Ile Ile
M4

430
Gly Thr Val Ser Val Phe Ala Gly Arg Leu Ile Glu Leu Ser Gln Glu Gly

[16] Millar, N. S., and Gotti, C. 2009. *Neuropharmacology* 56: 237–246.

analogous to the muscle receptor α-subunit are identified by the presence of adjacent cysteine residues near the proximal end of the amino terminal (positions 192 and 193 in Figure 5.2). The remaining subunits are designated as β. Together with the α, β, γ, δ, and ε-subunits from electroplax and muscle, they form a family of common genetic origin.[16] To date, 12 subunits have been isolated from chicken and rat nervous system: α2–α10 and β2,3,4. Injection into oocytes of mRNA for any one of α2–α6, or α10 with β2, β3, or β4 results in the formation of heteromultimeric channels. Expression of α7, α8, or α9 alone is sufficient for formation of homomultimeric channels. The existence of a large family of subunits available for channel formation enables them to combine selectively to form a diverse variety of channel isotypes with different

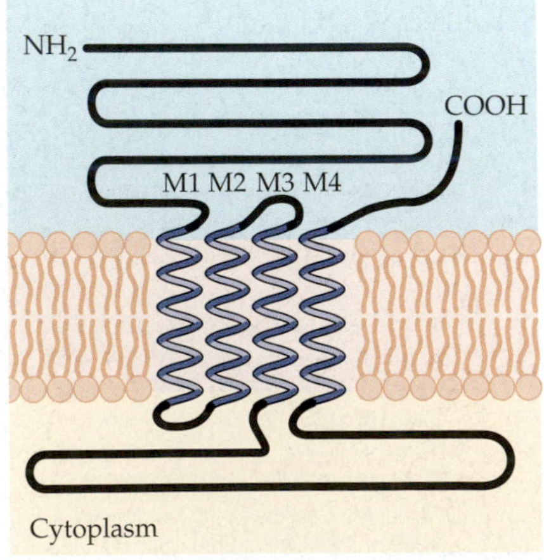

FIGURE 5.3 Model of the Tertiary Structure of an AChR Subunit, as proposed originally from amino-acid sequence analysis. Regions M1 through M4 each form membrane-spanning helices, and both the carboxy terminus and the amino terminus of the peptide lie in the extracellular space.

BOX 5.1
Classification of Amino Acids

Channel subunits, like all other peptides, are composed of amino acids, and it is the amino acid side chains that determine many of the local chemical and physical properties of the subunits. The twenty amino acids fall into three groups: neutral, acidic, and basic as shown (brackets contain three-letter and one-letter abbreviations). Acidic and basic amino acids are hydrophilic. The Neutral amino acids are arranged according to hydrophobicity indices (numbers below each amino acid) proposed by Kyte and Doolittle,[13] starting with the most hydrophobic (positive numbers) and progressing to the most hydrophilic (negative numbers). Sections of a peptide are candidates for membrane-spanning regions if they contain sequences of hydrophobic amino acids capable of forming an α-helix long enough to traverse the lipid bilayer (about 20 amino acid residues; see Figure 5.2).

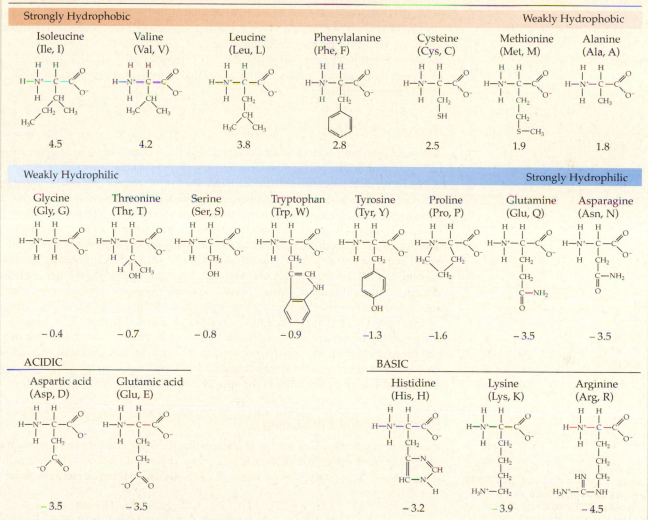

NEUTRAL

Strongly Hydrophobic → Weakly Hydrophobic

Isoleucine (Ile, I)	Valine (Val, V)	Leucine (Leu, L)	Phenylalanine (Phe, F)	Cysteine (Cys, C)	Methionine (Met, M)	Alanine (Ala, A)
4.5	4.2	3.8	2.8	2.5	1.9	1.8

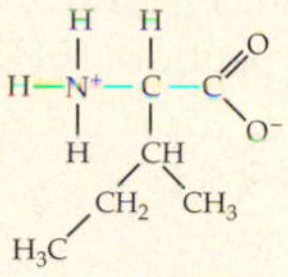

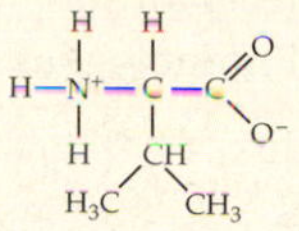

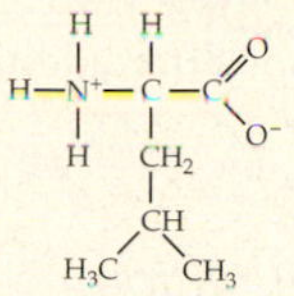

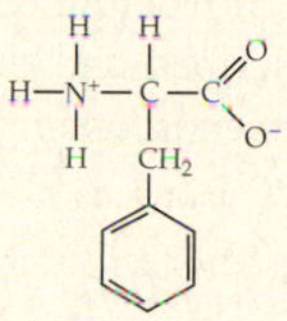

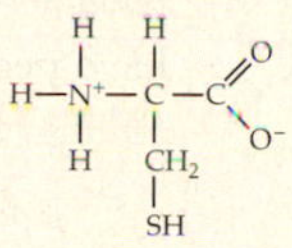

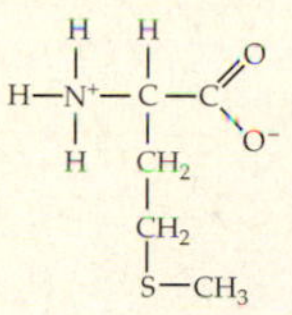

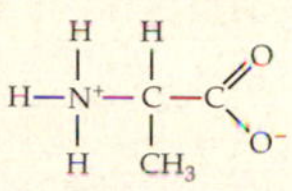

Weakly Hydrophilic → Strongly Hydrophilic

Glycine (Gly, G)	Threonine (Thr, T)	Serine (Ser, S)	Tryptophan (Trp, W)	Tyrosine (Tyr, Y)	Proline (Pro, P)	Glutamine (Glu, Q)	Asparagine (Asn, N)
−0.4	−0.7	−0.8	−0.9	−1.3	−1.6	−3.5	−3.5

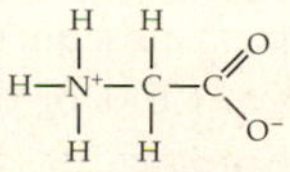

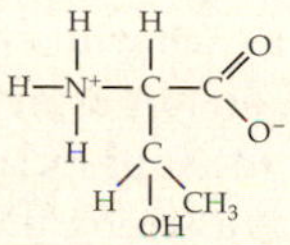

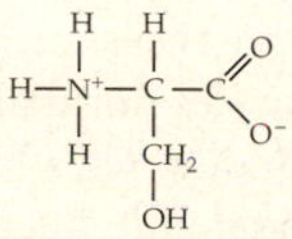

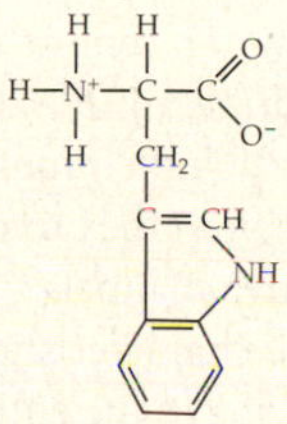

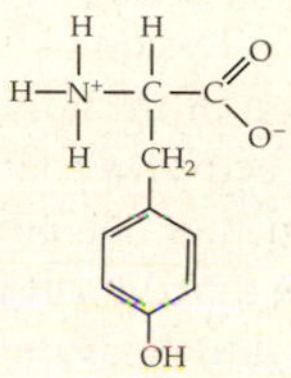

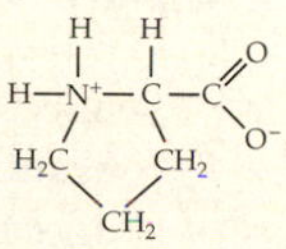

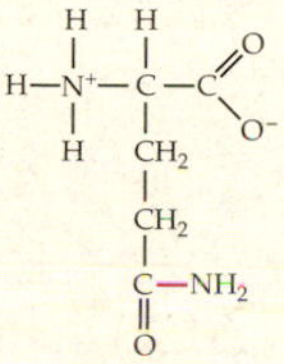

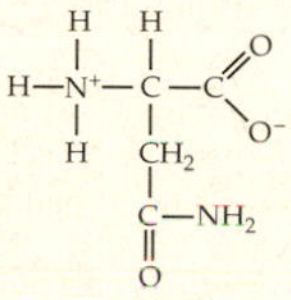

ACIDIC

Aspartic acid (Asp, D)	Glutamic acid (Glu, E)
−3.5	−3.5

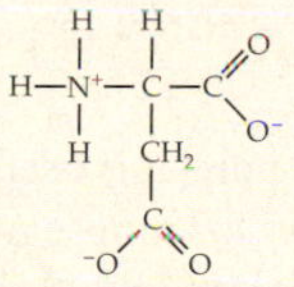

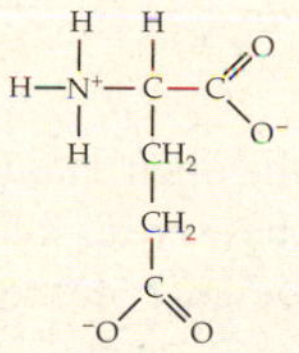

BASIC

Histidine (His, H)	Lysine (Lys, K)	Arginine (Arg, R)
−3.2	−3.9	−4.5

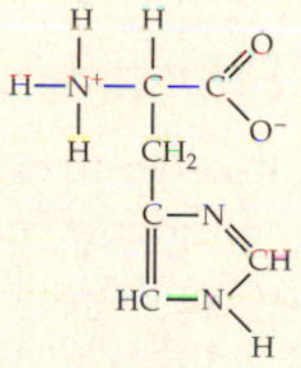

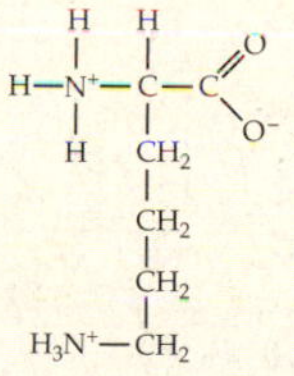

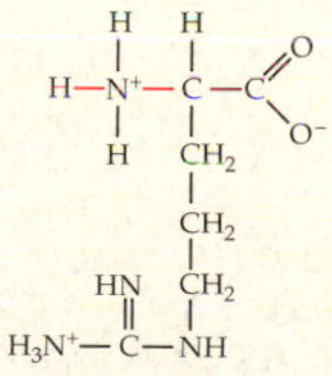

functional properties, such as ion selectivity, conductance, and kinetics. Specific examples related to functional properties of the nervous system will be discussed in later chapters.

A Receptor Superfamily

While the amino-acid sequence and chemical structure of the nAChR were being determined, it became apparent that other channel-forming receptors were similar in molecular composition. We now know that the nAChR is a member of a large **superfamily** of channel-forming receptors, known collectively as **Cys-loop receptors**, so-called because all subunits in the family have a pair of disulphide-bonded cysteines separated by 13 amino acid residues (Cys 128 and Cys 141 in Figure 5.2). Other members of the superfamily include receptors for serotonin (**5-HT3** and **MOD-1** receptors), the glycine receptor (**GlyR**), receptors for γ-aminobutyric acid (**$GABA_A$** and the invertebrate **EXP-1** receptors), an invertebrate glutamate-gated chloride receptor (**GluCl**), and a receptor activated by ionic zinc (**ZAC**).[17]

The 5-HT_3 receptors form cation channels with functional properties similar to those of nAChR.[18] Similarly, the GABA-activated EXP-1 channel[19] and the ZAC channel[20] are both cation selective. EXP-1 is found in invertebrates, and ZAC sequences have been identified in human and rat genomes. The remaining members of the family form channels that are anion selective. $GABA_A$ receptors,[21] together with GlyR,[22] subserve inhibitory synaptic transmission in vertebrate and invertebrate nervous systems (see Chapter 11). MOD-1 channels,[23] and GluCl channels[24] are found only in invertebrates.

Like the nAChR, each receptor type has a number of subunit isotypes. For example, 16 different $GABA_A$ subunit polypeptides have been identified in vertebrates by recombinant DNA techniques: α1–α6, β1–β3, γ1–γ3, δ, ε, π, θ. Three additional ρ = subunits contribute to $GABA_A$ variants, formerly specified separately as $GABA_C$ receptors.[21] Five GlyR subunits have been identified, four α and one β.[22] Two $5HT_3$ receptor subunits, $5HT_{3A}$, and $5HT_{3B}$, have been described and genes for three additional homologous polypeptides ($5HT_{3C}$–$5HT_{3E}$) have been isolated.[25]

Receptor Structure and Function

Two techniques have been essential in determining the relations between receptor structure and function. The first, site-directed mutagenesis, involves the construction of mutant cDNAs, with mutations directed at a particular site in the receptor protein, such that selected amino acids with particular properties (e.g., positively or negatively charged, highly polar or nonpolar) are replaced by others with different properties. The second technique is the expression of mutant receptors in host cells, such as *Xenopus* oocytes, subsequent to injection of the appropriate mRNA. Oocytes normally do not express nAChRs or other ligand-activated channels in their membranes. Yet, after the appropriate message has been injected, they not only express the protein subunits but also assemble them in the membrane to form functionally active channels.[26] In such experiments, electrical recording techniques are used to measure the characteristics of single-channel currents or whole-cell currents (representing the behavior of the entire population of inserted channels). Various mutations were found that affected ligand binding and thereby channel activation, while other mutations affected the ion selectivity of the channels and channel conductance. Some of the mutations affecting selectivity and conductance were located on the M2 helices, suggesting that these form the lining of the open channel (see Figure 5.5).

Structure of the Pore Lining

In order to examine the idea that the M2 helices line the open channel pore, it is useful to consider in more detail their amino acid sequences. These are as follows for mouse AChR α– and δ–subunits, going from cytoplasm (E 241, see Figure 5.2) to extracellular fluid:

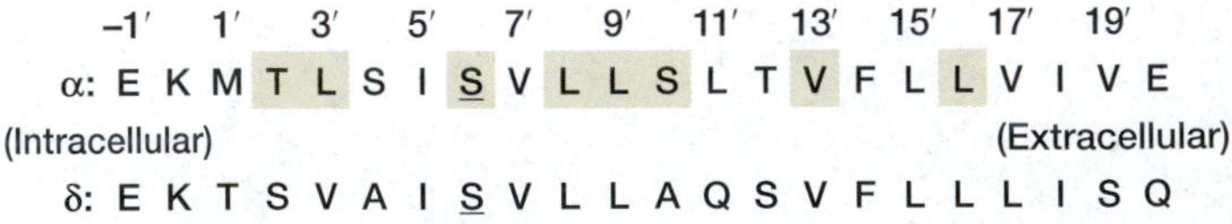

The primed numbering system facilitates comparison between M2 regions of different receptor subunits.[27] It starts at zero at the predicted cytoplasmic end of the helix and proceeds toward the extracellular region.

We might expect the relatively hydrophilic amino acids (see Box 5.1), such as the serines (S) and threonines (T), to be exposed to the aqueous pore, whereas the more hydrophobic isoleucines (I), for example, would be nestled against the membrane lipid or other parts of the protein. In accordance with this idea, replacing the serines in the underlined positions (6′) with alanines (which are weakly hydrophobic) produced a marked reduction in

[17] Lester, H. A. et al. 2004. *Trends Neurosci.* 27: 329–336.

[18] Peters, J. A., Hales, T. G., and Lambert, J. J. 2006. *Trends Pharmacol. Sci.* 26: 587–594.

[19] Beg, A. A., and Jorgensen, E. M. 2003. *Nat. Neurosci.* 6: 1145–1152.

[20] Davies, P. A. et al. 2003. *J. Biol. Chem.* 278: 712–717

[21] Olsen, R. W., and Sieghart, W. 2009. *Neuropharmacology* 56: 141–148

[22] Lynch, J. W. 2009. *Neuropharmacology* 56: 303–309.

[23] Ranganathan, R., Cannon, S. C., and Horvitz, H. R. 2000. *Nature* 408: 470–475.

[24] Cully, D. S., et al. 1994. *Nature* 371: 707–711.

[25] Niesler, B. et al. 2003. *Gene* 310: 101–111.

[26] Miledi, R., Parker, I. and Sumikawa, K. 1983. *Proc. R. Soc. Lond. B, Biol. Sci.* 218: 481–484.

[27] Charnet, P. et al. 1990. *Neuron* 4: 87–85.

channel conductance.[28] In addition, the binding affinity for the molecule QX222, which binds readily to the open channel in the native receptor, was greatly reduced. These effects are consistent with the idea that the serine residues are exposed to the aqueous channel.

Not all of the exposed amino acids are hydrophilic, however. The residues highlighted in color in the α-subunit sequence above, which include leucines and valine at positions 9′, 13′ and 16′, all appear to contribute to the pore lining. These were identified by Karlin and his colleagues with a technique known as the substituted-cysteine accessibility method (SCAM), in which residues are mutated, one at a time, to cysteine.[29] Mutant α-subunits were expressed in oocytes together with wild-type β, γ, and δ counterparts. Membrane currents produced by ACh were measured before and after exposure of the oocyte to the hydrophilic reagent methanethiosulfonate ethylammonium (MTSEA). The reagent reacts selectively with the cysteine sulfhydryl group, but it can only do so if the substituted cysteine is at a water-accessible position on the subunit (i.e., exposed to the aqueous pore). The reagent attenuated the responses to ACh only in channels with α-subunit mutations in the positions indicated. The pattern of exposure is consistent with the idea that the identified residues reside along one side of a helical sequence.

Analogous experiments were done by Changeux and his colleagues on native AChR by locating binding sites for chlorpromazine, a molecule that blocks ion flux through the channel.[30,31] Tritiated chlorpromazine was made to react with amino acid side chains within the open channel in response to an intense flash of ultraviolet light (photolabeling). The channel subunits were then isolated and scanned for radioactivity. Consistent with the substituted cysteine experiments, the radioactive label was present in all four subunits only on the M2 segments, and was located at positions 2′, 6′, and 9′.

High-Resolution Imaging of the AChR

A powerful approach to the question of channel topology is the use of high-resolution electron microscopy. Unwin and his colleagues have applied this technique to AChRs from *Torpedo* electric organs.[6] Isolated membranes from the electrocytes assemble readily into tubular vesicles, with the receptors themselves arrayed in an orderly lattice, as shown in the low-power micrographs of Figure 5.4A. Higher magnification images reveal the general shape and orientation of the receptor in the membrane (Figure 5.4 B,C).

By using very large numbers of images and combining digital averaging techniques with crystallographic analysis, Unwin and his colleagues were able to describe in detail the structure of the receptor at the molecular level, with a resolution of 0.4 nm. The results are summarized in Figure 5.5, which shows a view of two of the five subunits, on either side of the central core.

The extracellular domain of each subunit is built around a sequence of β-strands and their associated connecting loops, arranged into outer (red) and inner (blue) sheets. This arrangement is closely analogous to the crystal structure of another molecule, acetylcholine-binding protein (AChBP), described in detail by Brejc and his colleagues.[32] AChBP is a soluble protein that is secreted by snail glial cells at cholinergic synapses. Like nAChR, it is composed of five subunits, and it modulates synaptic transmission by binding ACh in the synaptic cleft.

[28] Leonard, R. J. et al. 1988. *Science* 242: 1578–1581.

[29] Karlin, A. 2002. *Nat. Rev. Neurosci.* 3: 102–114

[30] Giraudat, J. et al. 1986. *Proc. Natl. Acad. Sci. USA* 83: 2719–2723.

[31] Giraudat, J. et al. 1987. *Biochemistry* 26: 2410–2418.

[32] Brejc, K. et al. 2001. *Nature* 411: 269–276.

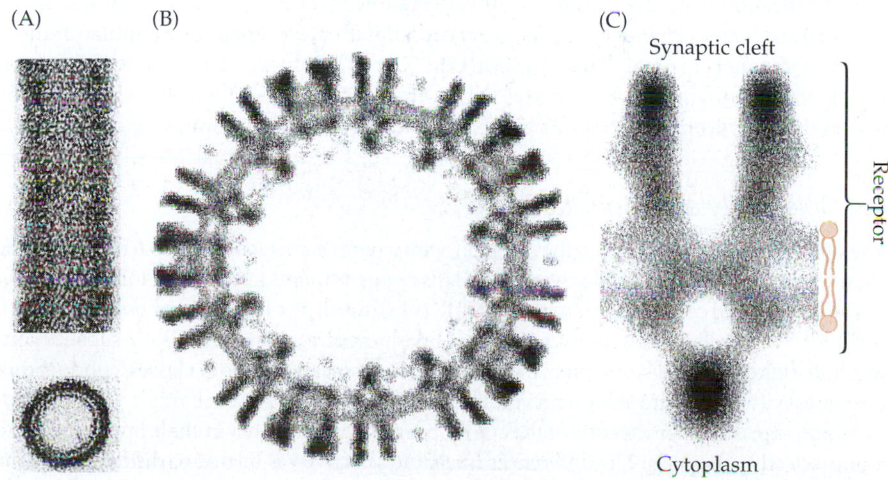

FIGURE 5.4 Electron Microscope Images of Acetylcholine Receptors. (A) Longitudinal and transverse images of cylindrical vesicles from postsynaptic membranes of *Torpedo*, showing closely packed ACh receptors. (B) Transverse section of the tube at higher magnification. (C) Further enlarged image of a single receptor, showing its position and size relative to the membrane bilayer. Dense blob under the receptor is an intracellular receptor-associated protein. (Images kindly provided by N. Unwin.)

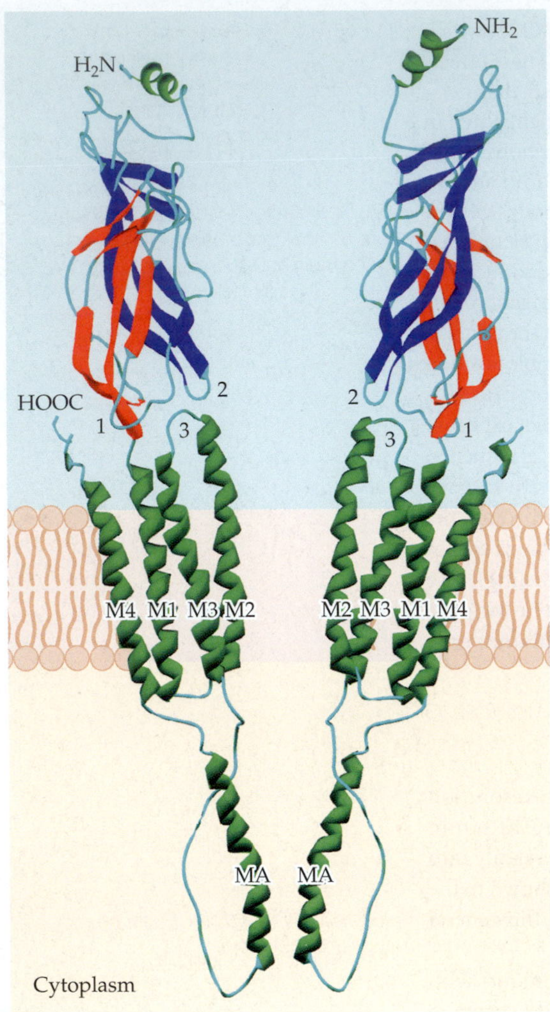

FIGURE 5.5 High-Resolution Structure of nAChR Subunits. Schematic view of two subunits, flanking the central cavity. The extracellular domain of each subunit is built around a sequence of β-strands and their associated connecting loops, arranged into outer (red) and inner (blue) sheets. It is connected to the first membrane-spanning α-helix, M1. The second α-helix, M2, lies central to the other three, so that the complete set of five M2 helices form the pore lining. The intracellular domain is formed by the sequence between M3 and M4, including the α-helix MA. Ligand binding to the extracellular domain causes reorientation of the inner and outer β-strands; this movement is transmitted to the pore region by interaction between one or more extracellular loops (1, 2) and the M2–M3 loop (3). (After Unwin, 2005.)

The transmembrane regions of the subunits are composed of the α-helices M1–M4 and are joined to the extracellular domains at the beginning of M1. The M2 helix lies central to the other three so that the five M2 segments together form the pore lining. The intracellular portions are formed by the sequences between M3 and M4, including the α-helix MA.

Receptor Activation

In an early series of lower-resolution experiments, Unwin examined the structure of the AChR in the presence and absence of ACh.[33] The results indicated that binding of ACh to the receptor resulted in a rotational displacement in the extracellular domain of the α-subunits. The subsequent high-resolution images showed that, in the absence of ACh, α and non-α-subunits have different conformations. The inner β-sheet in the α-subunit is rotated by about 10° with respect to its orientation in the non-α-subunits. In the presence of ACh, this difference is absent.[34] Unwin concluded that ACh binding results in rotation of the inner β-sheet, thereby bringing the α-subunit structure into conformation with that of the other subunits. The rotational movements of the β-sheets in the two α-subunits are transmitted to the M2 helices through interactions between one or more extracellular loops and the M2–M3 loops (see Figure 5.5), thereby causing the helices to move outward. These movements, together with coordinate outward movement of the other three M2 helices, then open the pore.[35]

This scheme is supported by observations of the X-ray structure (at 2.9 Å resolution) of a family of ligand-gated membrane channels in bacteria,[36–38] which are analogous in structure to mammalian cys-loop receptors and are activated by protons. In the closed state the channels are constricted by a ring of hydrophobic side chains in M2, equivalent to those at L9′ and V13′ in the mammalian channel. Upon activation, the constriction is removed by outward displacement of the M2 helices (Figure 5.6).

The M2 peptides, when expressed in lipid bilayers, form channels with selectivity and conductances similar to those of native nAChR channels from *Torpedo*.[39] In addition, spontaneously occurring channel currents are recorded with mean open times similar to those of native channel currents. Thus, it appears that current flow through the open channel is regulated by spontaneous fluctuations within the channel structure itself, perhaps in the region of the hydrophobic rings or at the more constricted cytoplasmic end of the pore.

Ion Selectivity and Conductance

The open channel is narrowest in the 2′ region, with a pore diameter of about 0.6 nm in anionic channels and 0.8 nm in cationic channels.[40] This region is therefore assumed to be critical for ion selectivity and conductance. In the nAChR, substituting the threonine at position 2′ with different residues illustrates the importance of both pore size and polarity. Polar substitutions result in higher channel conductance than nonpolar ones, and for both classes, conductance decreases with increasing side chain volume of the substituted residue.[41,42]

One surprising characteristic of the cys-loop receptors is that they include both cation and anion-selective channels. The differences in charge selectivity is related to differences in the

[33] Unwin, N. 1995. *Nature* 373: 37–43.
[34] Unwin, N. et al. 2002. *J. Mol. Biol.* 319: 1165–1176.
[35] Miyazawa, A., Fujiyoshi, Y., and Unwin, N. 2003. *Nature* 423: 949–955.
[36] Bocquet, N. et al. 2009. *Nature* 457: 111–114.
[37] Hilf, R. J., and Dutzler, R. 2008. *Nature* 452: 375–379.
[38] Hilf, R. J., and Dutzler, R. 2009. *Nature* 457:115–118.
[39] Montal, M. O. et al. 1993. *FEBS Lett.* 320: 261–266.
[40] Keramidas, A. et al. 2004. *Prog. Biophys. Mol. Biol.* 86: 161–204.
[41] Imoto, K. et al. 1991. *FEBS Lett.* 289: 193–200.
[42] Villarroel, A. et al. 1991. *Proc. R. Soc. Lond. B, Biol. Sci.* 243: 69–74.

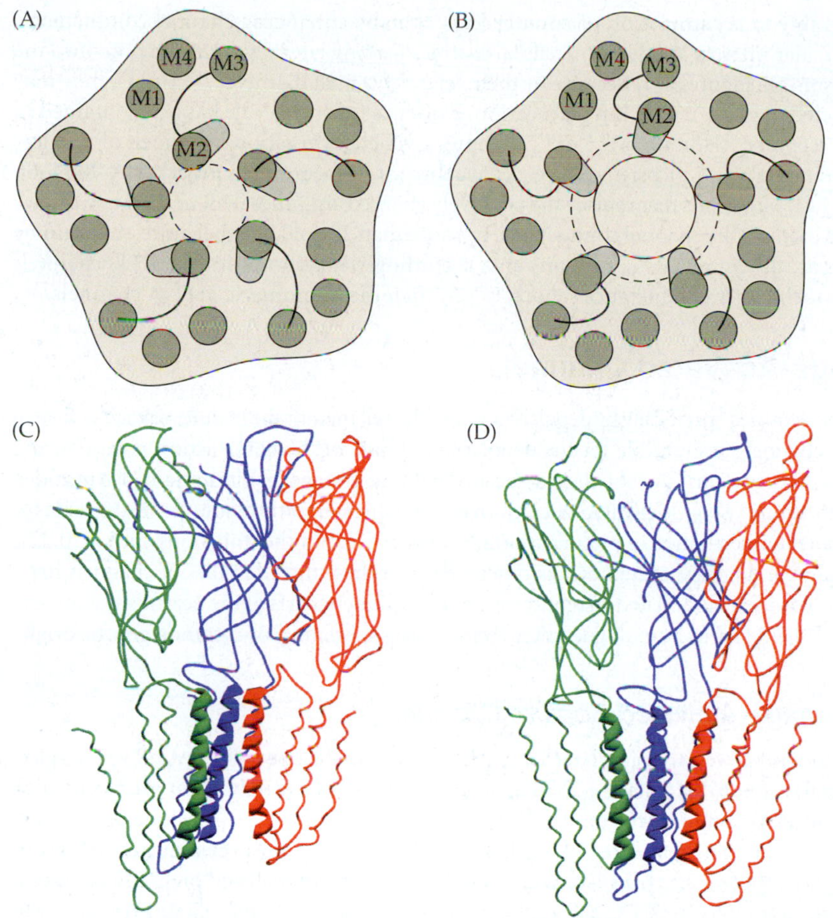

FIGURE 5.6 Receptor Channel Gating. Proposed mechanism for ACh receptor channel gating, looking into the pore from the extracellular side. (A) In the closed configuration, the pore is occluded by the ring of M2 helices. (B) Upon activation, the M2 helices swing outward to open the pore. (C,D) X-ray structure of an analogous bacterial ligand-gated channels at 3.1 Å, showing three of the M2 helices in the closed (C) and open (D) states. The transition from closed to open is the same as that postulated for the ACh receptor. (C and D after Isetlin and Hucho, 2009.)

sign of the charged residues along the ion pathway.[40] If we examine the amino acid sequence in the M2 subunits, we find that nAChR (and $5HT_3$ receptors) have a negatively charged residue (E^-) at position –1′, just at the cytoplasmic surface and another extracellularly at position 20′. Thus, the five M2 helices surround the pore with two distinct negatively charged rings that can be expected to promote cation selectivity. In glycine (and GABA) receptors, the corresponding residues at positions –1′ and 20′ are either positive or neutral (A^0):

		–1′	1′		19′	20′
AChR	α7	E^-	K^+	ISLGITVLLSLTVFMLLV	A^0	E^-
GlyR	α1	A^0	R^+	VGLGITTVLTMTTQSSGS	R^+	A^0

When the M2 residues in neuronal AChR homomultimeric α7 channels are altered to mimic those in GlyR channels the channel selectivity is changed from cationic to anionic.[43] Conversely, the reverse mutations in the GlyR produce cation-selective channels.[44] Charge selectivity of $GABA_C$ and $5HT_{3A}$ channels is also reversed by similar mutations.[45,46] Thus, anionic channels positively charged residues at position –1′, and cationic channels require negative residues in the same position.

Apart from determining charge selectivity of the channel, the charged rings have a marked effect on channel conductance.[47] In the nAChR, reducing the charge on the intracellular ring has the greatest effect, resulting in a reduction in both inward and outward current. Charge reduction in the outer ring also reduced conductance with a greater effect on inward current. Similar observations have been made on GlyR and GABA channels.[45]

Other segments of the ion pathway also contribute to ion selectivity and conductance. In the cation-selective channels, the walls of the channel vestibules have a net excess of negative charges, while in anion-selective channels the excess is positive. The outer vestibule is about 2 nm in diameter, and the effective radius for electrostatic interaction in physiological solutions is about 1 nm, so that excess charges in the vestibules might contribute

[43] Galzi, J.-L. et al. 1992. *Nature* 359: 500–505.

[44] Keramidas, A. et al. 2000. *Biophys. J.* 78: 247–259.

[45] Wotring, V. E., Miller, T. S., and Weiss, D. S. 2003. *J. Physiol.* 548: 527–540.

[46] Gunthorpe, M. J., and Lummis, S. C. R. 2001. *J. Biol. Chem.* 276: 10977–10983.

[47] Imoto, K. et al. 1998. *Nature* 335: 645–648.

considerably to accumulation of counter ions, thereby enhancing channel conductance. Of particular interest are the intracellular funnels leading laterally from the pore into the cytoplasm (see Figure 5.1). Because of their relatively small diameter (< 1 nm), they may be expected to play a role in ion permeation. Evidence for such a role has been obtained in 5-HT_3 receptors.[48] Homomeric 5-HT_{3A} receptors, which form cationic channels of very low conductance (about 1 picosiemens [pS]), have three arginine residues in each MA–M4 loop close to M4. Mutations that replace the positively charged arginines with neutral or negative residues increase the channel conductance by more than 20-fold. Furthermore, the arginine residues are absent in $5HT_{3B}$ subunits and, accordingly, heteromultimeric $5HT_3$ channels have a much larger conductance (about 16 pS) than the homomeric $5HT_3$A channels.

Voltage-Activated Channels

Channels activated specifically by depolarization of the cell membrane include voltage-activated sodium channels, responsible for the depolarizing phase of the nerve action potential, and voltage-activated potassium channels, associated with membrane repolarization. Also included in this group are voltage-activated calcium channels, which in some tissues are responsible for action potential generation or prolongation and subserve many other functions, such as muscle contraction and release of neurotransmitters. Each of these three families of channels has a number of isotypes found in different species and in different parts of the nervous system, and like the nAChR and its homologues, they constitute a superfamily of common genetic origin.

The Voltage-Activated Sodium Channel

The methods that were used to characterizing the molecular structure of the AChR were applied with equal success to the voltage-activated sodium channel. The essential steps were biochemical extraction and isolation of the protein,[49–51] followed by isolation of cDNA clones and deduction of the amino acid sequence.[52] As with the nAChR experiments, an electric fish—this time the eel, *Electrophorus electricus*—provided a rich source of material. and high-affinity toxins, principally tetrodotoxin (TTX) and saxitoxin (STX), were available to facilitate isolation of the protein. Both of these molecules block ion conduction in the native channels by occluding the pore of the open channel. Subsequently, sodium channels were isolated from brain and skeletal muscle. The sodium channel purified from eel consists of a single, large (260 kD) protein and is representative of a diverse family of structurally similar proteins.

In mammals, the functional sodium channel consists of the primary 260 kD protein (the α-subunit) acting in combination with one or more secondary structures (β-subunits). Genes have been identified that encode a family of nine α-subunits (Table 5.1), designated Na_v1.1–Na_v1.9.[53] One of the isoforms, Na_v1.4, has been found only in skeletal muscle and a second, Na_v1.5, in denervated skeletal muscle and heart muscle. Of the remainder, Na_v1.1–1.3 are found in the central nervous system; Na_v1.6–1.9 are found in the peripheral nervous system. Four β-subunits, β_1–β_4, are expressed in mammalian brain.[54,55] These range in size from 33–36 kD and have been shown to affect the sodium-channel kinetics and voltage dependence.[56]

[48]Kelley, S. P. et al. 2003. *Nature* 424: 321–324.

[49]Miller, J., Agnew, W. S., and Levinson, S. R. 1985. *Biochemistry* 22: 462–470.

[50]Hartshorn, R. P., and Catterall, W. A. 1984. *J. Biol. Chem.* 259: 1667–1675.

[51]Barchi, R. L. 1983. *J. Neurochem.* 40: 1377–1385.

[52]Noda, M. et al. 1984. *Nature* 312: 121–127.

[53]Goldin, A. L. et al. 2000. *Neuron* 28: 365–368.

[54]Yu, F. H., and Catterall, W. A, 2003. *Genome Biol.* 4: 207.1–207.7.

[55]Diss, J. K. J., Fraser, S. P., and Djamgoz, M. B. A. 2004. *Eur. Biophys. J.* 33: 180–195.

[56]Yu, F. H. et al. 2003 *J. Neurosci.* 23: 7577–7585.

TABLE 5.1
Voltage-activated sodium channels

Designation	Primary localization	Human gene
Na_v1.1–1.3	Central nervous system	*SNC1A–3A*
Na_v1.4	Skeletal muscle	*SNC4A*
Na_v1.5	Heart muscle	*SNC5A*
Na_v1.6	Central and peripheral nervous systems	*SNC8A*
Na_v1.7–1.9	Peripheral nervous system	*SNC9A–11A*

Amino Acid Sequence and Tertiary Structure of the Sodium Channel

The eel sodium channel is composed of a sequence of 1832 amino acids, containing four successive domains (I–IV) of 300 to 400 residues, with about a 50% sequence homology from one to the next. Each domain is architecturally equivalent to one subunit of the nAChR family of channel proteins. However, unlike nAChR subunits, the sodium channel domains are expressed together as a single protein. Within each domain are multiple hydrophobic or mixed hydrophobic and hydrophilic (amphipathic) sequences capable of forming transmembrane helices. As shown in Figure 5.7A, each domain has six such membrane-spanning segments,

(A) Sodium channel

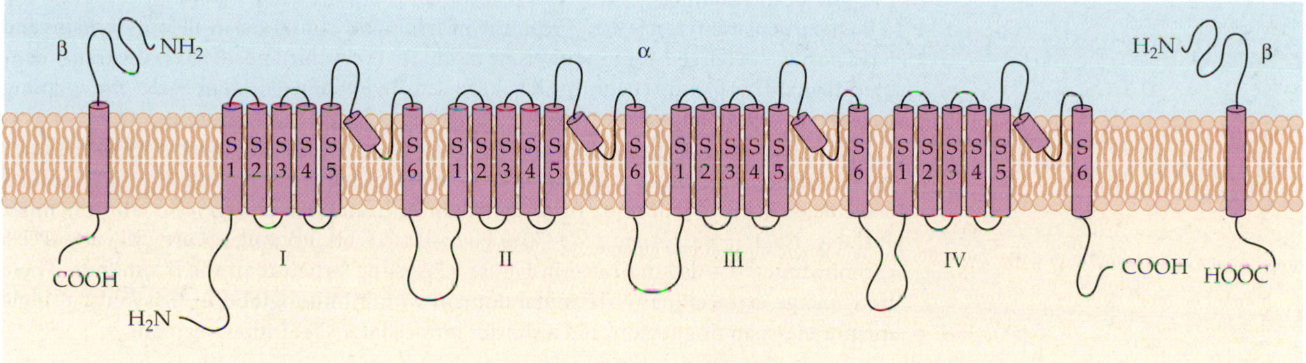

(B) Calcium channel

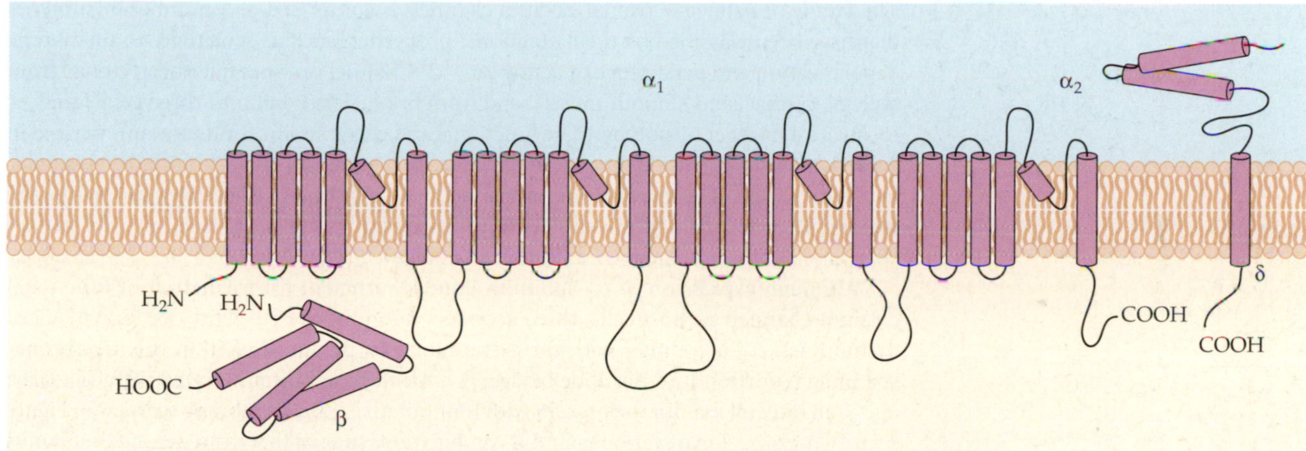

(C) Potassium channel

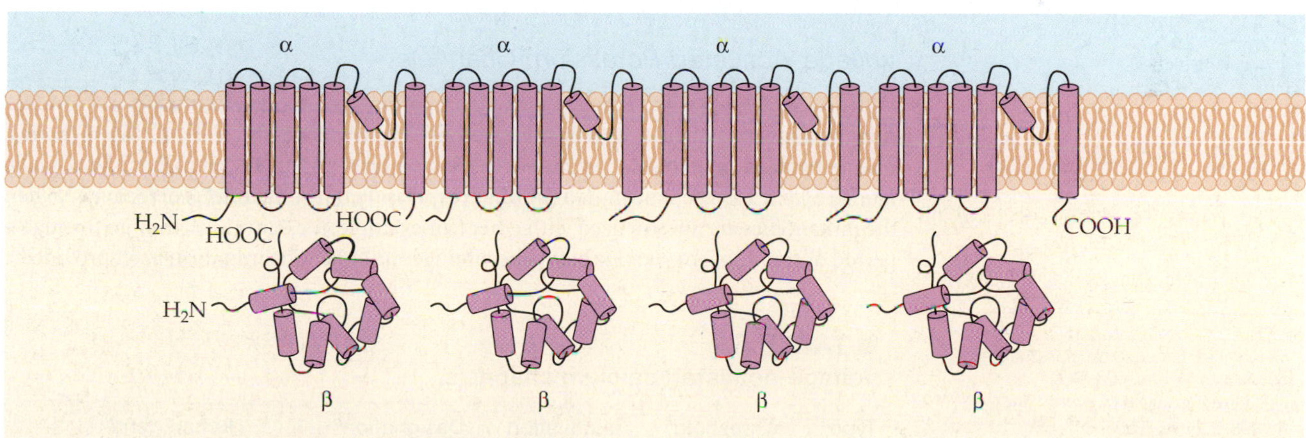

FIGURE 5.7 Voltage-Activated Channel Structure. (A) Primary (α) subunit of the voltage-activated sodium channel is a single protein with four domains (I–IV) connected by intracellular and extracellular loops. Each domain has six transmembrane segments (S1–S6), with a pore-forming structure (called a P-loop) between the fifth and sixth. The primary channel is accompanied by auxiliary (β) subunits. (B) The voltage-activated, calcium channel α_1-subunit is similar in structure and is accompanied by three auxiliary subunits: α_2, β, and δ. (C) The potassium channel is composed of four separate α-subunits, corresponding to one domain of the sodium channel α-subunit. Each is associated with a cytoplasmic β-subunit. (A and B after French and Zamponi, 2005; C after Long et al., 2005a.)

designated S1–S6. As with the AChR subunits, the domains are arranged radially around the pore of the channel.

Of particular interest is the S4 region, which is highly conserved in all four domains and has a positively charged arginine or lysine residue at every third position on the transmembrane helix. This feature occurs in all voltage-sensitive channels and provides the coupling between voltage changes across the membrane and channel activation.[57]

After translation, the channel protein is heavily glycosylated. About 30% of the mass of the mature eel channel consists of carbohydrate chains containing large amounts of sialic acid. In the α-subunits of mammalian sodium channels, glycosylation is more variable: $Na_v1.1$–$Na_v1.4$ are 15%–30% glycosylated, $Na_v1.5$ and 1.9 are only about 5% carbohydrate.[58,59] Also indicated in Figure 5.7A is the structure of the β-subunits. These have a large extracellular N-terminal domain with immunoglobulin-like folds, a single membrane-spanning region, and a shorter intracellular C-terminal segment.

Voltage-Activated Calcium Channels

The family of voltage-activated calcium channels is comprised of a number of subtypes that have been classified by their functional properties, such as sensitivity to membrane depolarization and persistence of activation.[60,61] Channel isotypes have been cloned from skeletal, cardiac, and smooth muscle, and from brain. These fall into three gene families, each with a number of isotypes. The functional and gene classifications are summarized in Table 5.2. The amino acid sequence of the primary channel-forming subunit (α_1) is similar to that of the voltage-activated sodium channel.[62] In particular, the putative transmembrane regions, S1 to S6, are highly homologous to those of the sodium channel, and the tertiary structure is entirely analogous, as indicated in Figure 5.7B.

Although expression of α_1-subunits alone is sufficient for formation of functional calcium channels in host cells, three accessory subunits are co-expressed in native cell membranes: α_2–δ, a dimer with the extracellular α_2 portion linked to the membrane-spanning δ portion by a disulfide bridge; β, a membrane protein located intracellularly; and γ, an integral membrane protein with four putative transmembrane segments (Figure 5.7B). Co-expression of various subunit combinations suggest that the α_2–δ and β-subunits influence both channel conductance and kinetics and that the γ-subunit plays a role in the voltage sensitivity of the channel.[63,64]

Voltage-Activated Potassium Channels

Voltage-sensitive potassium channels play an important role in nerve excitation and conduction. A number of distinct genetic messages give rise to a diverse family of these proteins. The first potassium to be sequenced was the potassium A-channel protein from *Drosophila*, named *Shaker* for a genetic mutant in which expression of the channel is defective.[65] When the mutant flies are anesthetized with ether (for example, for counting), they go through a period of trembling or *shaking* before becoming immobile. The mutation itself provided a

TABLE 5.2
Voltage-activated calcium channels

Type[a]	Threshold[b]	Inactivation	Designation[c]	Human gene
L	HV	Very slow	Ca_v 1.1, 1.2, 1.3, 1.4	*CACNA1S, 1C, 1D, 1F*
N	HV	Slow	Ca_v 2.2	*CACNA1B*
P/Q	HV	Slow	Ca_v 2.1	*CACNA1A*
R	HV	Very slow	Ca_v 2.3	*CACNA1E*
T	LV	Fast	Ca_v 3.1, 3.2, 3.3	*CACNA1G, 1H, 1I*

[a]Designation T, L, and N originally meant **T**ransient, **L**ong-lasting, and **N**either T nor L; P refers to **P**urkinje cells.
[b]HV and LV indicate high-voltage and low-voltage thresholds for activation.
[c]Classification according to Ertel et al. 2000.

[57] Hille, B. 2001. *Ion Channels in Excitable Membranes*, 3rd ed. Sinauer Associates, Sunderland MA, pp. 603–617.
[58] Marban, E., Yamagishi, T., and Tomaselli, G. F. 1998. *J. Physiol.* 508: 647–657.
[59] Tyrrell, L. et al. 2001. *J. Neurosci.* 21: 9629–9637.
[60] Hofmann, F., Biel, M., and Flockerzi, V. 1994. *Ann. Rev. Neurosci.* 17: 399–418.
[61] Randall, A., and Tsien, R. W. 1995. *J. Neurosci.* 15: 2995–3012.
[62] Tanabe, T. et al. 1987. *Nature* 328: 313–318.
[63] Walker, D., and De Waard, M. 1998. *Trends Neurosci.* 21: 148–154.
[64] French, R. J., and Zamponi, G. W. 2005. *IEEE Trans. Nanobiosci.* 4: 58–69.
[65] Papazian, D. M. et al. 1987. *Science* 237: 749–753.

different approach to cloning the channel—one that did not rely on prior identification of the protein. Genetic analysis indicated the approximate location of the *Shaker* gene on the *Drosophila* genome. Overlapping genomic clones from normal and mutant flies were isolated from that region, and comparison of normal and mutant sequences led to identification of the *Shaker* gene.

An unexpected finding was that the amino acid sequence of the protein was very much shorter than that of voltage-sensitive sodium or calcium channels. It contained a single domain, similar to domain IV of the eel sodium channel. Experimental evidence indicates that four of the single protein subunits assemble to form multimeric ion channels in the membrane (Figure 5.7C), thus mimicking the structure of the voltage-dependent sodium and calcium channels.[66]

Twelve distinct subfamilies of potassium channel proteins have been cloned. The relevant groups are listed in Table 5.3. Each subfamily is comprised of a number of isotypes (e.g., Kv11.1, 1.2, 1.3, etc.) Isotypes from the same subfamily, when expressed in host cells, combine to form heteromultimeric channels, but those from different subfamilies do not.[67] Like sodium and calcium channels, voltage-activated potassium channels are expressed with accessory (β) subunits.[68] Three subfamilies have been identified: Kvβ1, Kvβ2, and Kvβ3. When expressed with the primary subunits, the β-subunits affect the voltage-sensitivity and inactivation properties of the channels.

TABLE 5.3
Voltage-activated potassium channels

Subunit designation	Name	Human gene
Kv1.1–1.8	*Shaker*	*KCNA1–7, 10*
Kv2.1, 2.2	*Shab*	*KCNB1–2*
Kv3.1–3.4	*Shaw*	*KCNC1–4*
Kv4.1–4.3	*Shal*	*KCND1–3*
Kv7.1–7.5	*dKCNQ*	*KCNQ1–5*
Kv10.1, 10.2	*eag (ether-a-go-go)*	*KCNH1, 5*
Kv11.1–11.3	*erg*	*KCNH2, 6, 7*
Kv12.1–12.3	*elk*	*KCNH8, 3, 4*

Pore Formation in Voltage-Activated Channels

A consistent feature of all of the voltage-activated channel sequences is a moderately hydrophobic region in the extracellular S5–S6 loop. A segment of the loop contains an amino acid sequence that is highly conserved across all members of the potassium channel family (called the K^+ channel signature sequence).[69,70] As in the experiments described previously for the M2 region of the nAChR, mutations to this region of the *Shaker* potassium channel reduced the binding affinity for the blocking molecule tetraethylammonium (TEA) and altered the conductance properties of the channel.[71,72] It was concluded that this segment dips into the channel mouth to form the upper part of the pore region—a conclusion that is confirmed later by X-ray diffraction (Figure 5.8). S5–S6 is now known as the P-loop (*P* for "pore").

When the substituted cysteine accessibility method (SCAM) was used to identify residues exposed to the pore lining, the residues were found not only on the P-loop

[66] Timpe, L. C. et al. 1988. *Nature* 331: 143–145.

[67] Salkoff, L. et al. 1992. Trends Neurosci. 15: 161–166.

[68] Hanlon, M. R., and Wallace, B. A. 2002. *Biochemistry* 41: 2886–2894.

[69] Miller, C. 1992. *Current Biol.* 2: 573–575.

[70] MacKinnon, R. et al. 1998. *Science* 280: 106–109.

[71] Yool, A. J., and Schwarz, T. L. 1991. *Nature* 349: 700–704.

[72] Yellen, G. et al. 1991. *Science* 251: 939–942.

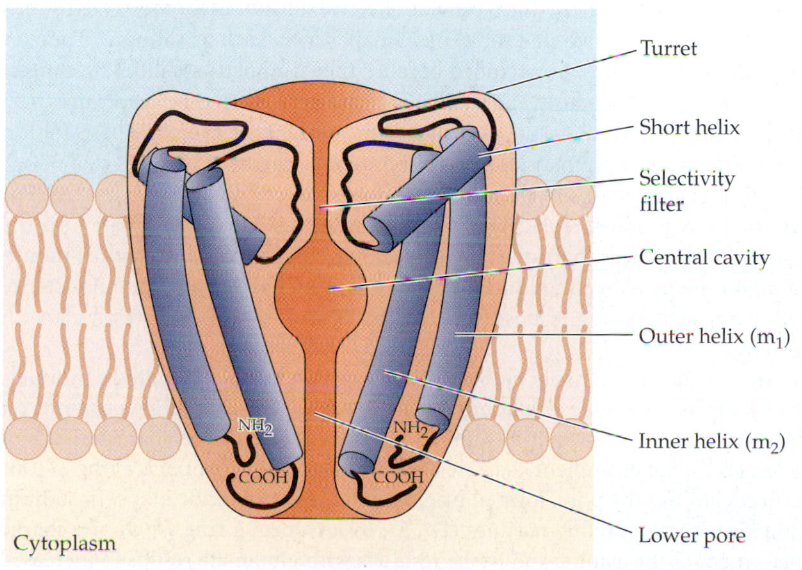

FIGURE 5.8. Structure of Potassium $K_{CS}A$ Channel. Sectional view of a $K_{CS}A$ potassium channel showing two of four subunits, one on either side of the central pore. Each subunit has two membrane-spanning helices and a short helix pointing into the pore. The connections between the outer helices and the short helices form four turrets that surround the pore entrance and contain binding sites for blocking molecules. The four connections between the short helices and the inner helices combine to form the selectivity filter, which allows the permeation of potassium, cesium, and rubidium but excludes smaller cations such as sodium and lithium (see text). (After Doyle et al., 1998.)

but also on the S6 transmembrane helix, indicating that S6 helices line the pore between the P-loop and the cytoplasm.[73,74]

High-Resolution Imaging of the Potassium Channel

The structure of potassium channels from *Streptomyces lividans* ($K_{CS}A$ channels) has been examined with X-ray crystallography at a resolution at 5.2 Å.[75] These bacterial channels belong to a class of potassium channels that are only weakly voltage sensitive and whose subunits have only two transmembrane segments, rather than six. The two segments, M1 and M2, are structurally equivalent to S5 and S6 in the voltage-activated channels, and the M1–M2 link represents the P-loop. The $K_{CS}A$ channel is a tetramer. Figure 5.8 is a sectional view of the channel, showing its major structural features. Near the amino terminus of each subunit is an outer helix (m_1) that spans the membrane from the cytoplasmic side of the membrane to the outer surface. The outer helix is followed by a short helix that points into the pore and then by an inner helix (m_2) that returns to the cytoplasmic side. The links between the outer and short helices form four turrets that surround the external opening of the pore and contain the binding sites for TEA and channel-blocking toxins. The link between the central end of the short helix and the inner helix contributes to the pore structure. These four inner links combine to form a restricted passage responsible for the ion selectivity of the channel—the selectivity filter. A relatively large central cavity and a smaller internal pore connect the selectivity filter with the cytoplasm.

The crystal structure of the complete *Shaker* voltage-dependent potassium channel is known as well.[76] The arrangement of the α-subunit is shown schematically in Figure 5.9A. The molecule has two distinct regions—the voltage-sensing region, consisting of S1–S4, and the pore-forming region formed by S5–S6.

Figure 5.9B shows a schematic view of the complete channel. The four S5–S6 helices of the α-subunits interlace to form a box-like structure around the pore. As in the $K_{CS}A$ channel, the S6 helices line the pore, and the S5–S6 linker dips into the extracellular opening to form the selectivity filter. The voltage-sensing regions sit somewhat apart at the four corners of the central structure.

Selectivity and Conductance

Selectivity for potassium is achieved by both the size and molecular composition of the selectivity filter,[77] which, in the $K_{CS}A$ channel, has a diameter of about 0.3 nm. The amino acids in its wall are oriented so that successive rings of four carbonyl oxygen atoms, one from each subunit, are exposed to the pore. The pore diameter is adequate to accommodate a dehydrated potassium ion (about 0.27 nm in diameter) but stripping waters of hydration from the penetrating ion requires considerable energy (see Chapter 4). This requirement is minimized by the exposed oxygens, which provide effective substitutes for the oxygen atoms that normally surround the hydrated molecule. Smaller ions, such as sodium (diameter 0.19 nm) or lithium (0.12 nm) are excluded because they cannot make intimate contact with all four oxygens simultaneously, and so remain hydrated. Ions larger than cesium (0.33 nm diameter) cannot penetrate the pore because of their size. This structural basis for ion selectivity is in accordance with traditional ideas about ion permeation through channels.[78]

Other regions of the potassium channel contribute to its conductance properties. For example, in the $K_{CS}A$ channel, replacing a neutral alanine residue by glutamate near the cytoplasmic end of M2, so as to form a ring of negative charges near the channel mouth results in a significant increase in conductance.[74] Similarly, in Kv2.1 channels, conductance is increased by replacing a ring of positively charged lysines in the outer vestibule with neutral residues.[79]

In sodium and calcium channels, both selectivity and conductance are altered by mutations in the P-loop. For example, mutations in the P-loops at corresponding positions on each of the four domains can result in sodium channels acquiring the characteristics of calcium channels.[80] One prominent characteristic of the calcium channel is a ring of four negative charges around the pore, formed by glutamate residues in the loops. In sodium channels, the four corresponding residues (D^-, E^-, K^+, A) form a ring with a net charge of –1. Replacement of the alanine and lycine residues with glutamate results in increased

[73] Liu, Y. et al. 1997. *Neuron* 19: 175–184.

[74] Lu, T. et al. 1999. *Neuron* 22: 571–580.

[75] Doyle et al. 1998. *Science* 280: 69–77.

[76] Long, S. B., Campbell, E. B., and MacKinnon, R. 2005a. *Science* 309: 897–903.

[77] Nimigean, C. M., Chappie, J. S., and Miller, C. 2005. *Biochemistry* 42: 9263–9268.

[78] Mullins, L. J. 1975. *Biophys. J.* 15: 921–931.

[79] Consiglio, J. F., Andalib, P., and Korn, S. J. 2003. *J. Gen. Physiol.* 121: 111–124.

[80] Heinemann, S. H. et al. 1992. *Nature* 356: 441–443.

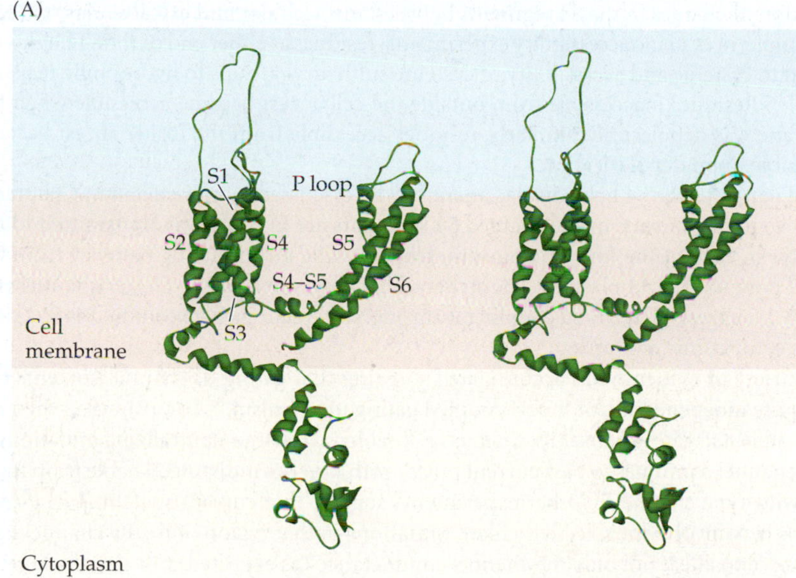

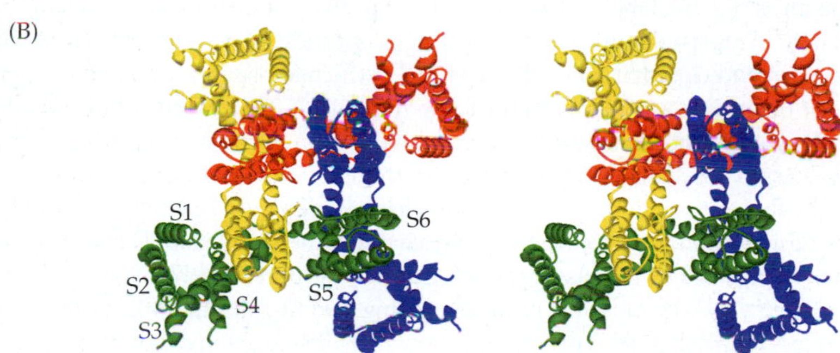

FIGURE 5.9 Stereo Images of Mammalian Kv1.2 Channels.* (A) Single subunit viewed parallel to the plane of the membrane, showing arrangement of helices S1 to S6, P-loop, S4–S5 linker, and cytoplasmic N-terminal region. (B) Complete structure, comprised of four subunits, looking into the pore from the extracellular side. S5 and S6 helices interlace to form the channel with P-loops dipping into the pore. Voltage-sensing regions S1 to S4 lie at the corners and are connected to the channel region by the S4–S5 linker. (After Long et al., 2005a.)

calcium permeability and block by calcium of permeation by sodium ions. Conversely, in calcium channels, the reverse mutations at the same positions reduced calcium permeability and allowed permeation of monovalent cations.[81]

Gating of Voltage-Activated Channels

Structural studies suggest that closed channels are occluded by the S6 segments coming together near the cytoplasmic end of the pore. This idea is supported by SCAM experiments in *Shaker* potassium channels. Probes presented to the cytoplasmic end of closed channels were able to penetrate only a very short distance.[71]

For voltage-activated gating to occur, there must be charged elements within the channel protein that are displaced by membrane depolarization. The S4 helix is an obvious candidate. As previously noted, it contains a string of positively charged lysine or arginine residues, located at every third position, and this feature is highly conserved within the superfamily of voltage-activated channels. These features suggest that the S4 helices comprise the voltage-sensing elements that link changes in membrane potential to the gating mechanism.[82] Application of a positive potential to the inside of the cell membrane (depolarization) would displace the positive charges so as to cause movement of the helix. Several techniques have provided support for the idea that activation is accompanied by

*VIEWING STEREOGRAMS. To see a three-dimensional representation from a pair of stereo images, hold the page at a normal viewing distance (about 12") and look at the space between the images, staring *through* the page as if you were trying to focus on a distant object beyond the book. Do *not* attempt to bring the images themselves into focus. "Ghosts" of the two images will drift together to form a third, three-dimensional image, lying between the other two. After the central image is stable you can then focus on its details.

[81] Yang, N. et al. 1993. *Nature* 366: 158–161.

[82] Sigworth, F. J. 1994. *Quart. Rev. Biophys.* 27: 1–40.

translation of charges in the S4 segments between intracellular and extracellular spaces.[83] For example, in cysteine accessibility experiments, residues at either end of the S4 helix were mutated to cysteine and accessibility of cysteine sulfhydryl groups to hydrophilic reagents tested.[84,85] Residues inaccessible from outside the cell at rest became accessible when the membrane was depolarized. Similarly, residues accessible from the inside at rest became inaccessible upon depolarization.

The nature of the S4 helix displacement will be discussed in more detail in Chapter 7. In any event, movements of the charged S4 segments are in some way transmitted to the S6 helices in each of the four domains, thereby opening a conducting pathway from the channel pore to the cytoplasm.[86] The structural arrangement of the *Shaker* channel (see Figure 5.3) suggests that S4–S6 coupling is through interactions between the S4–S5 linker and the cytoplasmic end of S6.

A variety of evidence has accumulated showing that gating of channel currents can occur quite independently of the S4 coupled gating mechanism.[87] For example, when the voltage-activated gate in *Shaker* Kv channels is disabled by charge-neutralizing mutations in S4, the channel continues to pass current pulses with kinetics indistinguishable from those in the wild-type channel.[88] Other experiments suggest that current switching in P-loop channels may involve the selectivity filter. Mutations in that region of the Kv channel have been shown to affect not only the channel conductance (as expected) but also the kinetics of the rapid gating transitions.[89] Similarly, mutations within the selectivity filter of inward rectifier channels (K_{ir}; see Table 5.3), which are constitutively active unless blocked by magnesium at their cytoplasmic end, alter their gating properties.[90,91] In addition, accessibility studies on open and closed SK and CNG channels, which are ligand-activated (see Table 5.3), suggest the presence of a gate in the same region.[92,93] These kinds of observations suggest the possibility of a dual gating system, with the voltage-coupled gating mechanism providing access to the channel and ionic currents though the open channel being switched on and off by fluctuations in the region of the selectivity filter.

Following activation, many voltage-sensitive channels then **inactivate**. That is, they enter a state that no longer allows the passage of ions through the pore. The principle mechanism of inactivation involves cytoplasmic residues moving into the mouth of the pore, thereby blocking access to the channel. This mechanism will be discussed in detail in Chapter 7.

Other Channels

There is a large number of other channels important for neuronal function, with a wide variety of structural configurations. Some are assembled from subunits that have as few as two transmembrane helices each. Others are formed by a single, large molecule. Some of these are catalogued in Table 5.4. An extensive list of channel subunits, their official names, and human and animal gene designations, can be found at the web site of *The Nomenclature Committee of the International Union of Basic and Clinical Pharmacology*: http://www.iuphar-db.org.

Glutamate Receptors

Glutamate is the most important and prevalent excitatory neurotransmitter in the central nervous system, activating three cation channel types (see Chapter 14). The three types have distinct functional properties and are distinguished experimentally by their different sensitivities to glutamate analogues.[94] One responds selectively to *N*-methyl-D-aspartate (NMDA). The other two are activated selectively by α-amino-3-hydroxy-5-methyl-4-isoxazole propionic acid (AMPA) and kainate. Because of this selectivity, the three chemical analogues are important experimental tools. Remember, however, that the native neurotransmitter for all three receptor types is glutamate, not one of the analogues.

So far, 16 cDNAs for glutamate receptor subunits have been identified by molecular cloning.[95] Seven of these—GluN1, GluN2A through 2D, and GluN3A and B—are involved in the formation of NMDA receptors. AMPA receptors are formed from another set, designated as GluA1–GluA4, and kainate receptors are assembled from subunits GluK1–3 combined with GluK4 or 5. Two homologous subunits, δ_1 and δ_2, remain unassigned to any particular receptor. The subunit proteins have four putative transmembrane seg-

[83] Bezanilla, F. 2008. *Nature Rev. Mol. Cell Biol.* 9: 323–332.

[84] Yang, N., George, A. L. and Horn, R. 1996. *Neuron* 16: 113–122.

[85] Larsson, H. P. et al. 1996. *Neuron* 16: 387–397.

[86] Long, S. B., Campbell, E. B., and MacKinnon, R. 2005b. *Science* 309: 903–908.

[87] Korn, S. J., and Trapani, J. G. 2005. *IEEE Trans Nanobiosci.* 4: 21–33.

[88] Bao, H. et al. 1999. *J. Gen. Physiol.* 113: 139–151.

[89] Liu, Y., and Joho, R. H. 1998. *Pflügers Arch.* 435: 654–661.

[90] Lu, T. et al. 2001. *Nature Neurosci.* 4: 239–246.

[91] So, I. et al. 2001. *J. Physiol.* 531.1: 37–50.

[92] Bruening-Wright, A. et al. 2002. *J. Neurosci.* 22: 6499–6506.

[93] Flynn, G. E., Johnson, J. P Jr., and Zagotta, W. N. 2001. *Nature Rev. Neurosci.* 2: 643–652.

[94] Lodge, D. 2009. *Neuropharmacol.* 56: 6–21.

[95] Dingledine, R. et al. 1999. *Pharmacol. Rev.* 51: 7–61.

TABLE 5.4
Other membrane channel types

Channel name	Ligand	Permeability	Subunits	Human genes
Ligand activated				
NMDA[a]	Glutamate	Cations	GluN1, 2A–D, 3A,B	*GRIN1, 2A–D, 3A,B*
AMPA[a]	Glutamate	Cations	GluA1–4	*GRIA1–4*
Kainate[a]	Glutamate	Cations	GluK1–5	*GRIK1–5*
P2X[b]	ATP	Cations	P2X1–P2X7	*P2RX1–7*
Intracellular activation				
CNG[c,d]	cAMP, cGMP	Cations	CNGA1–4, *CNGB1, B3*	*CNGA1–4, CNGB1, B3*
BK[e]	Calcium	Potassium	K_{Ca} 1.1	*KCNMA1*
SK[e]	Calcium	Potassium	K_{Ca} 2.1–2.3	*KCNN1–3*
IK[e]	Calcium	Potassium	K_{Ca} 3.1	*KCNN4*
Voltage-sensitive				
CLC[f]	—	Chloride	CLC0, CLC1,2	*CLCN0,1,2*
Kir[g]	—	Inward potassium	Kir1.1, Kir2.1–4, Kir3.1–4, Kir4.1–2, Kir5.1, Kir6.1–2, Kir7.1	*KCNJ1, J2, 12, 4, 14, J3, 6, 9, 5, J10, 15, J16, J8, 11, J13*
(2P)[h]	—	Potassium	K_{2P} 1.1–10.1, K_{2P}12.1, 13.1, K_{2P} 15.1–18.1	*KCNK1–10, 12, 13, 15–18*
TRP channels[i]				
TRPC	Various	Cations	TRPC1–7	
TRPV	Various	Cations	TRPV1–6	
TRPM	Various	Cations	TRPM1–8	
TRPML	Various	Cations	TRPML 1–3	
TRPP	Various	Cations	TRPP1–3	
TRPA1	Various	Cations	TRPA1	

[a]Collingridge et al. 2009; [b]Jarvis and Khakh, 2009; [c]Matulef and Zagotta, 2003; [d]Bradley et al. 2005; [e]Wei et al. 2005; [f]Pusch and Jentsch, 2005; [g]Kubo et al. 2005; [h]Goldstein et al. 2001; [i]Wu et al. 2010.

ments. However, the second segment, rather than crossing the membrane, enters from the cytoplasmic face to form a hairpin loop contributing to the pore lining.[96] The subunits are believed to be phylogenetically related to the potassium channel family, and their structure resembles that of an upside-down $K_{CS}A$ potassium channel subunit, with an additional transmembrane segment (Figure 5.10B).[97] Unlike nAChR subunits, their C-terminus

[96]Hollman, M., Maron, C., and Heinemann, S. 1994. *Neuron* 13: 1331–1343.

[97]Wollmuth, L. P., and Sobolevsky, A. I. 2004. *Trends in Neurosci.* 27: 321–328.

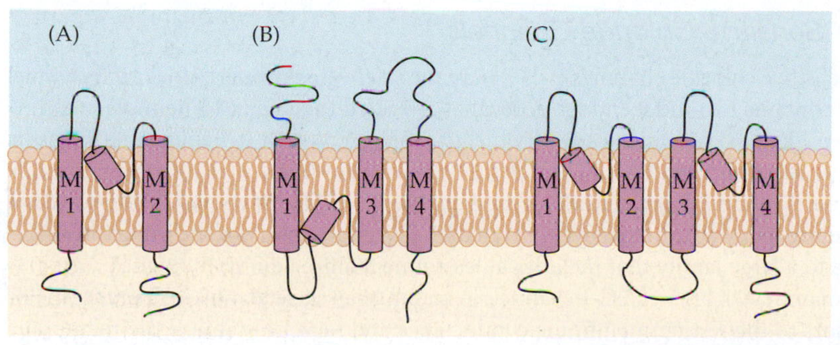

FIGURE 5.10 Other Channel Subunit Configurations. (A) Inward rectifier potassium channel subunit is similar to $K_{CS}A$ subunit, with two transmembrane segments and an intervening P-loop. (B) Glutamate receptor subunit has three transmembrane helices with an intramembranous loop entering from the cytoplasm. (C) The 2P potassium channel subunit has four transmembrane segments and two P-loops. Complete channel is a dimer, rather than a tetramer.

is intracellular. Both the N-terminus and the M3–M4 loop contribute to the ligand binding sites. The complete receptor is a tetramer.

ATP-Activated Channels

Adenosine-5′-triphosphate (ATP) acts as a neurotransmitter in smooth muscle cells, in autonomic ganglion cells, and in neurons of the central nervous system (see Chapter 14). Because ATP is a purine, its receptor molecules are known as purinergic (P) receptors. P2X receptors form ligand-gated cation channels that subserve a wide variety of functions.[98] P2Y receptors activate intracellular messenger systems. Seven P2X subunits (P2X1–P2X7) have been cloned.[99] Their proposed tertiary structure, with two membrane-spanning segments, is similar to that of the $K_{CS}A$ subunits, and subunits of inward-rectifier channels (Figure 5.10A).

Channels Activated by Cyclic Nucleotides

Receptors in the retina and olfactory epithelium are activated by intracellular cyclic AMP or cyclic GMP. The receptors form cation channels, with varying selectivities for potassium, sodium, or calcium. They are similar in structure to voltage-sensitive cation channels, being composed of four subunits, each with six membrane-spanning regions and a P-loop between S5 and S6.[100,101] The S4 helix contains a sequence of charged residues, but these are fewer in number than found in the voltage-sensitive family (usually four rather than six or seven). Consistent with its activation by intracellular ligands, most of the mass of the channel is on the cytoplasmic side of the membrane. In vertebrates, six members of the cyclic nucleotide-gated (CNG) gene family have been identified. CNGA1 and CNGB1 (later designated CNGB1a) were first found in bovine rod photoreceptors. Three CNGA1 subunits combine with one CNGB1 to form the native rod channel. In cones, CNG channels composed of two other subunits, designated CNGA3 and CNGB3, in a stoichiometric ratio of 2:2. Yet another pair of subunits, CNGA2 and CNGA4, is found in olfactory neurons. These combine with an alternatively spliced variant of CNGB1 (CNGB1b) to form the olfactory receptor channel, with stoichiometry $(A2)_2(A4)(B1b)$.

Calcium-Activated Potassium Channels

Calcium-activated potassium channels are activated by local changes in cytoplasmic calcium and are divided into three categories according to their potassium conductance: *big* (BK), *small* (SK), and *intermediate* (IK).[102,103] BK channels are also voltage sensitive and are activated by the concerted action of membrane depolarization and increased intracellular calcium. Their conductance is over 100 pS. SK channels are voltage-insensitive and have conductances of 10 pS or less. The IK channel conductance is in the order of 50 pS.

Structurally, SK and IK channel subunits share their overall transmembrane topology with voltage-activated potassium channels, but the only notable homology in amino acid sequence is in the P-loop. BK channel subunits are similar in topology but have seven, rather than six, membrane crossings, with the additional transmembrane segment (S0) carrying the amino terminus into the extracellular domain.

Voltage-Sensitive Chloride Channels

Voltage-sensitive chloride channels (CLC) have the interesting property that each channel molecule contains two independent pores that are gated together.[104] The molecule consists of two identical subunits, each of which is composed of 18 α-helices and forms its own independent channel.[105] CLC-0 was first cloned from the electric organ of *Torpedo*. The channel is found in high density on the noninnervated face of the cells and provides a low-resistance pathway for currents generated by electrical activity of the innervated face. It belongs to a large family that includes at least nine mammalian homologues. CLC-0 is found in mammalian brain. CLC-1 channels in mammalian skeletal muscle fibers are major contributors to the resting membrane conductance and serve to stabilize the membrane

[98] Abbracchino, M. P. et al. 2009. *Trends in Neurosci.* 32: 19–29.

[99] Jarvis, M. F., and Khakh, B. S. 2009. *Neuropharmacol.* 56: 208–215.

[100] Matulef, K., and Zagotta, W. N. 2003. *Annu. Rev. Cell. Dev. Biol.* 19: 23–44.

[101] Bradley, J., Reisert, J., and Frings, S. 2005. *Current Opin. Neurobiol.* 15: 343–349.

[102] Vergara, C. et al. 1998. *Curr. Opin. Neurobiol.* 8: 321–329.

[103] Falker, B., and Adelman, J. P. 2008. *Neuron* 59: 873–881.

[104] Miller, C., and White, M. M. 1984. *Proc. Nat. Aca. Sci. USA* 81: 2772–2775.

[105] Pusch, M., and Jentsch, T. J. 2005. *IEEE Trans. Nanobiosci.* 4: 49–57.

potential at its resting level. CLC-2 appears to be associated with cell volume regulation and therefore may be stretch-sensitive. Two other isotypes, CLC-K1 and CLC-K2, are associated with chloride reabsorption in the kidney. Other branches of the family, CLC-3 to CLC-7, are predominantly found in membranes of intracellular vesicles.

Inward-Rectifying Potassium Channels

Inward-rectifying potassium channels (K_{ir} channels) allow the movement of potassium ions into cells when the membrane potential is negative with respect to the potassium equilibrium potential, but allow little outward potassium movement when the driving force is in the opposite direction. The absence of outward potassium flux is associated with blockade of the channels by intracellular magnesium and/or by intracellular polyamines.[106] The channel is a tetramer with each subunit having only two membrane-spanning helices, similar to the $K_{CS}A$ channel (see Figure 5.10A). Seven subfamilies of the channel ($K_{ir}1$–$K_{ir}7$) have been cloned from brain, heart, and kidney.[107] The channels display a variety of functional properties. One family, $K_{ir}3$, forms channels that are activated by intracellular G proteins (see Chapter 12).

2P Channels

The family of 2P potassium channels is so called because each subunit contains two pore-forming loops.[108] The subunit has four transmembrane regions, M1–M4, with pore-forming loops M1–M2 and M3–M4 (Figure 5.10C). Two subunits assemble in the membrane to form the channel. 2P channels are open at normal cell resting potentials and are a major contributor to the resting potassium conductance, or "leak" pathways across cell membranes. The first 2P channel was cloned from *Drosophila,* and since then 16 additional mammalian subunits have been described.

Transient Receptor Potential (TRP) Channels

Transient receptor potential (TRP) channels constitute a very large family of cation-selective channels that are formed in the membrane by the assembly of four subunits. The first to be identified and cloned is associated with visual transduction in *Drosophila*. Photoreceptors in mutants lacking the *trp* gene are unable to maintain a sustained response to light. Instead, the receptor potential adapts rapidly (i.e., becomes transient). Since then, more than 50 TRP subunit genes have been identified, at least 28 of them in mammals.[109,110] The subunits are characterized by six transmembrane segments with a pore loop between the fifth and sixth segment, with extended cytoplasmic N- and C-terminal regions. Although the subunits lack charged residues in S4, some of the channels are voltage sensitive. The channels are generally non-selective, but their calcium permeability varies over a wide range. The mammalian TRP superfamily is divided into six groups (see Table 5.4), each with its own set of functional characteristics, which vary widely from one group to the next. Some channels respond to light, others to odorants, changes in osmolarity, or changes in temperature. For example, the TRPV family is involved in heat detection and TRPM8 in the detection of cold (see Chapter 19).

Diversity of Subunits

A feature of channel structure is the remarkably wide diversity of subunit isotypes. Thus, there are more than a dozen nAChR subunits and an even greater number of potassium channel and glutamate receptor subunits. How does this diversity arise? Generally, each channel or channel subunit is encoded by a separate gene, but two additional mechanisms have been identified. The first is called alternative splicing. Most proteins are encoded in several different segments of DNA known as exons. In some cases, instead of combining uniquely to form mRNA for a specific subunit, transcripts of the exons instead enter into various alternative combinations to generate mRNA for a variety of subunit isotypes. During transcription, an unknown regulatory mechanism determines which of the alternative RNAs are to be used. The remaining RNAs are excised from the transcript and the desired

[106] Lu, Z. 2004. *Annu. Rev. Physiol.* 66: 103–129.

[107] Nicholls, C. G., and Lopatin, A. N. 1997. *Annu. Rev. Physiol.* 59: 171–191.

[108] Goldstein, S. A. N. et al. 2001. *Nat. Rev. Neurosci.* 2: 175–184.

[109] Nilius, B., and Voets, T. 2005. *Pflügers Arch.* 451: 1–10.

[110] Wu, L.-J. et al. 2010. *Pharmacol. Rev.* 62: 381–404.

RNA segments spliced together to form the final mRNA. *Shaker* potassium channel isotypes are generated in this way.

The third method of obtaining subunit diversity is by **RNA editing**. Examples include the glutamate receptor subunits GluA2, GluK3, and GluK4 (formerly known as GluR2, GluR5, and GluR6). These carry either a glutamine or arginine residue within the pore. The presence of the arginine residue severely depresses the calcium permeability of the channel and alters its ionic conductance properties. It turns out that the genomic DNA for all three subunits harbors a glutamine codon (CAG) for that position, even though an arginine codon (CGG) can be found in the mRNAs.[111] The change in base sequence is accomplished by RNA editing in the cell nucleus. Virtually the entire message for GluA2 is edited in this way as well as some of the message for GluK3 and GluK4. In GluK4, additional A-G editing is found in a different region of the pore.[112] RNA editing at a single site in the glycine receptor subunit GlyRα_3 has been shown to underlie a marked increase in glycine sensitivity.[113]

[111] Sommer, B. et al. 1991. *Cell* 67: 11–19.

[112] Köhler, M. et al. 1993. *Neuron* 10: 491–500.

[113] Meier, J. C. et al. 2005. *Nat. Neurosci.* 8: 736–74.

Conclusion

The techniques used to examine the functional characteristics and structural details of the Cys-loop receptors on the one hand, and the voltage-activated cation channels on the other, have led to a remarkable increase in our detailed knowledge of channel structure and function. The two channel types, although appearing to be quite different at first glance, have a similar functional organization. Both are composed of three essential parts: a signal-sensing domain, a transmembrane pore-forming domain, and a cytoplasmic domain.

In Cys-loop receptors, the sensing domain is the ligand-binding region of the molecule, located appropriately in the extracellular space. The ligand-binding region is connected to M1 and, when activated, interacts with the connecting loop between M1 and M2 to shift M2 outward, thereby opening the pore. By necessity, voltage-activated channels have their sensing region in the membrane itself, with the S4 helix being the essential sensing element. Their activation is entirely equivalent to that of the Cys-loop receptors: the voltage sensor is connected to S5 and, upon depolarization, interacts with the S5–S6 loop to move S6 out of the pore.

The cytoplasmic region of the channels has a variety of functions. In the Cys-loop receptors and voltage-activated channels, the region plays a role in ion selectivity, and in voltage-activated channels, it plays a major role in inactivation. In addition, channels activated by intracellular ligands have cytoplasmic ligand-binding sites that are coupled with the transmembrane region so as to open the channel pore during activation.

As a general rule, channels that are relatively selective, such as the voltage-activated channels, are usually tetramers (one exception is the dimeric 2P channel); larger, less selective ligand-activated channels are pentamers. As an extension of this principle, the very largest and least selective channels—the gap junctions—have a hexameric structure (see Chapter 8).

The techniques now at hand will enable us to build on these general principles and provide even more intimate details of how the molecular organization of a given channel affects its functional properties, and hence, its role in nervous system function.

SUMMARY

- Nicotinic acetylcholine receptors (nAChR) from the electric organ of *Torpedo* consist of five subunits (two α and three others designated β, γ, and δ) arranged around a central pore.
- In each subunit of the AChR the string of amino acids folds to form four membrane-spanning helices (M1–M4) joined by intracellular and extracellular loops. The extracellular loops in the α-subunits contain the binding sites for ACh. The M2 helices form the central pore of the channel.
- Binding of ACh to the two α-subunits results in a change in configuration that is transmitted to the M2 helices, which move radially outward to widen the pore and allow ion flux through the channel.
- AChRs belong to a superfamily of Cys-loop receptors that includes receptors for γ-amino butyric acid (GABA), glycine, and 5-hydroxytryptamine (5-HT).
- The voltage-activated sodium channel from eel electric organs is a single molecule of about 1800 amino acids within which are four repeating domains (I–IV). The domains are arranged around the central pore and are architecturally equivalent to the subunits of other channels. Within each domain are six membrane-spanning helices (S1–S6) connected by intracellular and extracellular loops. The primary (α) subunits of mammalian sodium channels are homologous with the eel channel but are expressed in the membrane in concert with auxiliary (β) subunits.
- The family of voltage-activated calcium channel proteins is analogous in structure to the voltage-activated sodium channel. Voltage-activated potassium channels are structurally similar but with an important genetic difference—the four repeating units are expressed as individual subunits, not as repeating domains of a single molecule. Together, the three voltage-activated channels constitute a genetic superfamily.
- The voltage-activated sodium, calcium, and potassium channels are known as P-loop channels because of a loop of amino acids that enters the membrane from the extracellular side between S5 and S6 in each of the four domains. The P-loops form the outer part of the pore lining and play a major role in the ion selectivity of the channel. The remainder of the pore is lined by cytoplasmic ends of the S6 helices.
- Depolarization of the membrane displaces the S4 helices (that have a number of positively charged residues), which in turn, produces an outward radial movement of the S6 helices to open the pore.
- Inactivation of voltage-activated channels occurs when one of the intracellular loops of the amino acid swings into the cytoplasmic mouth of the open channel, thereby preventing ion flux through the pore.
- The techniques used to determine the detailed structure of these two ion channel families have led to a widespread understanding of the molecular organization of a number of other channel types.

Suggested Reading

General Reviews

Diss, J. K. J., Fraser, S. P., and Djamgoz, M. B. A. 2004. Voltage-gated Na^+ channels: multiplicity of expression, plasticity, functional implications and pathophysiological aspects. *Eur. Biophys. J.* 33: 180–195.

French, R. J., and Zamponi, G. W. 2005. Voltage-gated sodium and calcium channels in nerve, muscle and heart. *IEEE Trans. Nanobiosci.* 4: 58–69.

Karlin, A. 2002. Emerging structure of nicotinic acetylcholine receptors. *Nat. Rev. Neurosci.* 3: 102–114.

Keramidas, A., Moorhouse, A. J., Schofield, P. R., and Barry, P. H. 2004. Ligand-gated ion channels: mechanisms underlying ion selectivity. *Prog. Biophys. Mol. Biol.* 86: 161–204.

Korn, S. J., and Trapani, J. G. 2005. Potassium channels. *IEEE Trans. Nanobiosci.* 4: 21–35.

Lester, H. A., Dibas, M. I., Dahan, D. S., Leite, J. F., and Dougherty, D. A. 2004. Cys-loop receptors: new twists and turns. *Trends in Neurosci.* 27: 329–336.

Yu, F. H., and Catterall, W. A. 2005. Overview of the voltage-gated sodium channel family. *Genome Biol.* 4: 207.1–207.7.

Original Papers

Long, S. B., Campbell, E. B., and MacKinnon, R. 2005a. Crystal structure of a mammalian voltage-dependent *Shaker* family K^+ channel. *Science* 309: 897–903.

Long, S. B, Campbell, E. B., and MacKinnon, R. 2005b. Voltage sensor of Kv1.2: Structural basis of electromechanical coupling. *Science* 309: 903–908.

Miyazawa, A., Fujiyoshi, Y., and Unwin, N. 2003. Structure and gating mechanism of the acetylcholine receptor pore. *Nature* 423: 949–955.

Unwin, N. 2005. Refined structure of the nicotinic acetylcholine receptor at 4 Å resolution. *J. Mol. Biol.* 346: 967–989.

CHAPTER 6

Ionic Basis of the Resting Potential

At rest, a neuron has a steady electrical potential across its plasma membrane, the inside being negative with respect to the outside. In the neuron, the intracellular potassium concentration is high compared to that in the extracellular fluid, whereas the intracellular concentrations of sodium and chloride are low. As a result, potassium tends to diffuse out of the cell and sodium and chloride tend to diffuse in. The tendency for potassium ions to move out of the cell and chloride ions to move in is opposed by the membrane potential.

In this chapter, we discuss the relations between concentration and potential, first for a model cell whose membrane is permeable only to potassium and chloride. In this cell, the concentration gradients and the membrane potential can be balanced exactly, so that there is no net flux of either ion across the membrane. The membrane potential is then equal to the equilibrium potential for both potassium and chloride. In the model cell, changing the extracellular potassium concentration changes the potassium equilibrium potential, and hence the membrane potential. In contrast, changing extracellular chloride concentration eventually leads to an equivalent change in intracellular chloride so that the chloride equilibrium potential and the membrane potential are unchanged.

Real cells are also permeable to sodium. At rest, sodium ions constantly move into the cell, reducing the internal negativity of the membrane. As a result, potassium, being no longer in equilibrium, leaks out. If there were no compensation, these fluxes would lead to changes in the internal concentrations of sodium and potassium. However, the concentrations are maintained by the sodium–potassium exchange pump, which transports sodium out and potassium in across the cell membrane in a ratio of 3 parts sodium to 2 parts potassium. The resting membrane potential depends on the potassium equilibrium potential, the sodium equilibrium potential, the relative permeability of the cell membrane to the two ions, and the pump ratio. At the resting potential, the passive fluxes of sodium and potassium are exactly matched by the rates at which they are transported in the opposite direction. Because the sodium–potassium exchange pump transports more positive ions outward than inward across the membrane, it makes a direct contribution of several millivolts to the membrane potential.

The chloride equilibrium potential may be positive or negative relative to the resting membrane potential, depending on the chloride transport processes. Although the chloride distribution plays little role in determining the resting membrane potential, substantial chloride permeability is important in some cells for electrical stability.

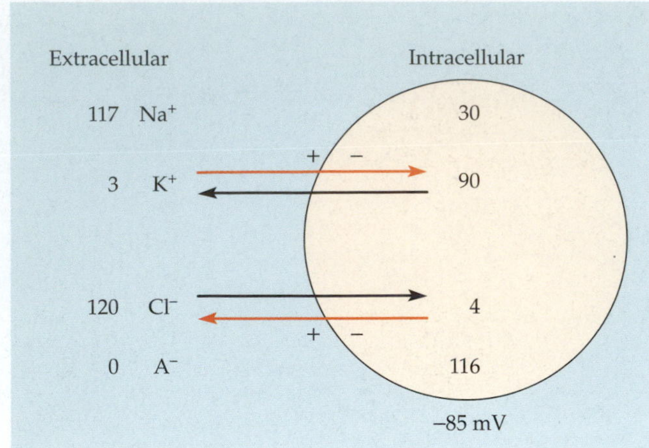

FIGURE 6.1 Ion Distributions in a Model Cell. The cell membrane is impermeable to Na^+ and the internal anion (A^-) and permeable to K^+ and Cl^-. The concentration gradient for K^+ tends to drive it out of the cell (black arrow), while the potential gradient tends to attract K^+ into the cell (red arrow). In a cell at rest, the two forces are exactly in balance. Concentration and electrical gradients for Cl^- are in the reverse directions. Ion concentrations are expressed in millimolar (m*M*).

Electrical signals are generated in nerve cells and muscle fibers primarily by changes in permeability of the cell membrane to ions such as sodium and potassium. Increases in permeability allow ions to move inward or outward across the cell membrane down their electrochemical gradients. As we discuss in Chapter 4, permeability increases are due to activation of ion channels. Ions moving through the open channels change the charge on the cell membrane, and hence, change the membrane potential. In order to understand how signals are generated, it is necessary first to understand how the standing ionic gradients across the cell membrane determine the resting membrane potential.

A Model Cell

It is useful to begin with the idealized model cell shown in Figure 6.1. This cell contains potassium, sodium, chloride, and a large anion species, and it is bathed in a solution of sodium and potassium chloride. Other ions present in real cells (e.g., calcium or magnesium) are ignored for the moment, as their direct contributions to the resting membrane potential are negligible. The extracellular and intracellular ion concentrations in the model cell are similar to those found in frogs. In birds and mammals, ion concentrations are somewhat higher, while in marine invertebrates (e.g., squid) they are very much higher (Table 6.1). The model cell membrane is permeable to potassium and chloride, but not to sodium or to the internal anion. There are three major requirements for a cell to remain in a stable condition:

1. The intracellular and extracellular solutions must each be electrically neutral. For example, a solution of chloride ions alone cannot exist; their negative charges must be balanced by an equal number of positive charges on cations such as sodium or potassium (otherwise electrical repulsion would literally blow the solution apart).
2. The cell must be in osmotic balance. Otherwise, water will enter or leave the cell, causing it to swell or shrink. Osmotic balance is achieved when the total concentration of solute particles inside the cell is equal to that on the outside.
3. There must be no net movement of any particular ion into or out of the cell.

Ionic Equilibrium

How are the concentrations of the permeant ions maintained in the model cell, and what electrical potential is developed across the cell membrane? Figure 6.1 shows that the two ions are distributed in reverse ratio—potassium is more concentrated on the inside of the cell, chloride on the outside. Imagine first that the membrane is permeable only to potassium. The question that arises immediately is why potassium ions do not diffuse out of the cell until the concentrations on either side of the cell membrane are equal. The answer is that the process is self-limiting. As the potassium ions diffuse outward, positive charges accumulate on the outer surface of the membrane and excess negative charges are left on the inner surface. As a result, an electrical potential develops across the membrane, with the inside being negative with respect to the outside. The electrical gradient slows the efflux of positively charged potassium ions, and when the potential becomes sufficiently large, further net efflux of potassium is stopped. The potential at which this occurs is called the **potassium equilibrium potential** (**E_K**). At E_K, the effects of the concentration gradient and the potential gradient on ion flux through the membrane balance one another exactly. Individual potassium ions still enter and leave the cell, but no *net* movement occurs. The potassium ion is in equilibrium.

TABLE 6.1
Concentrations of ions inside and outside freshly isolated axons of squid

	Concentration (m*M*)		
Ion	Axoplasm	Blood	Seawater
Potassium	400	20	10
Sodium	50	440	460
Chloride	60	560	540
Calcium	0.1 μ*M*[a]	10	10

Source: After Hodgkin, 1964.

[a] Ionized intracellular calcium from Baker, Hodgkin, and Ridgeway, 1971.

The conditions for potassium to be in equilibrium across the cell membrane are the same as those described in Chapter 4 for maintaining zero net flux through an individual channel in a membrane patch. There, a concentration gradient was balanced by a potential applied to the patch pipette. The important difference here is that the ion flux itself produces the required transmembrane potential. In other words, equilibrium in the model cell is automatic and inevitable. Recall from Chapter 4 that the potassium equilibrium potential is given by the Nernst equation:

$$E_K = \frac{RT}{zF} \ln \frac{[K]_o}{[K]_i} = 58 \log \frac{[K]_o}{[K]_i}$$

where $[K]_o$ and $[K]_i$ are the internal and external potassium ion concentrations. For the cell shown in Figure 6.1, E_K is 58 log (1/30) = –85 mV. Suppose now that, in addition to potassium channels, the membrane has chloride channels. Because for an anion $z = -1$, the equilibrium potential for chloride is:

$$E_{Cl} = -58 \log \frac{[Cl]_o}{[Cl]_i}$$

or from the properties of logarithmic ratios:

$$E_{Cl} = 58 \log \frac{[Cl]_i}{[Cl]_o}$$

In our model cell, the chloride concentration ratio is again 1/30 and E_{Cl} is also –85 mV. As with potassium, the membrane potential of –85 mV balances exactly the tendency for chloride to move down its concentration gradient, in this case *into* the cell.

In summary, both the tendency for potassium ions to leave the cell and for chloride ions to diffuse inward are opposed by the membrane potential. Because the concentration ratios for the two ions are of exactly the same magnitude (1:30), their equilibrium potentials are exactly the same. As potassium and chloride are the only two ions that can move across the membrane and both are in equilibrium at –85 mV, the model cell can exist indefinitely without any net gain or loss of ions.

Electrical Neutrality

The charge separation across the membrane of our model cell means that there is an excess of anions inside the cell and of cations outside. This appears to violate the principle of electrical neutrality but, in fact, does not. Potassium ions diffusing outward collect as excess cations against the outer membrane surface, leaving excess anions closely attracted to the inner surface. Both the potassium ions and the counter ions they leave behind are, in effect, removed from the intracellular bulk solution, leaving it neutral. Similarly, chloride ions diffusing inward add to the collection of excess anions on the inner surface of the membrane and leave counterions in the outer charged layer, so that the extracellular solution remains neutral as well. The outer layer of cations and inner layer of anions, of equal and opposite charges, are not in free solution but are held to the membrane surface by mutual attraction. Thus, the membrane acts as a capacitor, separating and storing charge. This does not mean that any given anion or cation is locked in position against the membrane. Ions in the charged layer interchange freely with those in the bulk solution. The point is that although the identities of the ions in the layer are constantly changing, their total number remains constant, and the bulk solution stays neutral.

Another question we might ask about charge separation is whether or not the number of ions accumulated in the charged layer represents a significant fraction of the total number of ions in the cell. The answer is that it does not. If we consider our model cell to have a radius of 25 μm, then at a concentration of 120 millimolar (m*M*), there are roughly 4×10^{12} cations and an equal number of anions in the cytoplasm. At a membrane potential of –85 mV the amount of charge separated by the membrane is about 5×10^{11} univalent ions/cm^2 (see Chapter 8). Our cell has a surface area of about 8×10^{-5} cm^2, so there are approximately 4×10^7 negative ions collected at the inner surface of the membrane, which is 1/100,000 the number in free solution. Thus, the movements of potassium and chloride ions into the charged layer as they establish the membrane potential have no significant effect on intracellular free ion concentrations.

The Effect of Extracellular Potassium and Chloride on Membrane Potential

In neurons and in many other cells, the steady-state resting membrane potential is sensitive to changes in extracellular potassium concentration but is relatively unaffected by changes in extracellular chloride. To understand how this comes about, it is useful to consider the consequences of such changes in the model cell. We will assume throughout this discussion that the volume of the extracellular fluid is infinitely large relative to the volume of the cell. Thus, movements of ions and water into or out of the cell have no significant effect on extracellular concentrations. Figure 6.2A shows the changes in intracellular composition and membrane potential that result from increasing extracellular potassium from 3 m*M* to 6 m*M*. The extracellular sodium concentration is reduced by 3 m*M* to keep the osmolarity unchanged. The increase in extracellular potassium reduces the concentration gradient for outward potassium movement, while initially leaving the electrical gradient unchanged. As a result, there will be a net inward movement of potassium ions. As positive charges accumulate on its inner surface, the membrane is depolarized. This in turn means that chloride ions are no longer in equilibrium and they move into the cell as well. Potassium and chloride entry continues until a new equilibrium is established, and both ions are at a new concentration ratio consistent with the new membrane potential—in this example –68 mV.

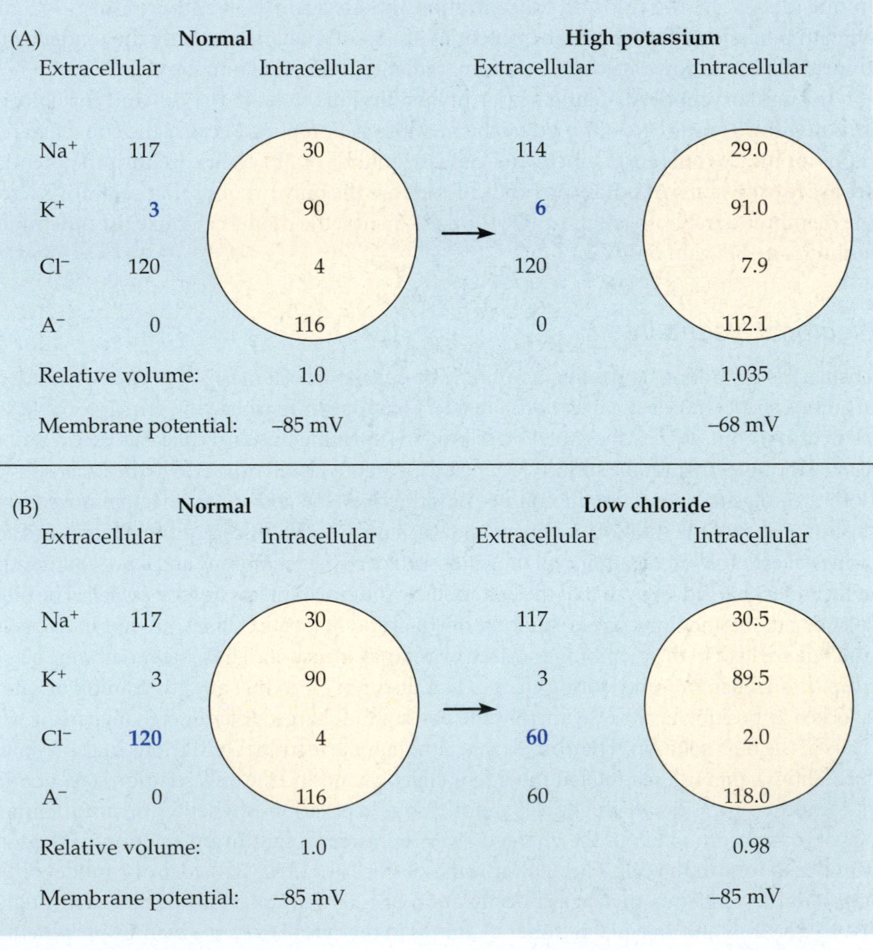

FIGURE 6.2 Effects of Changing Extracellular Ion Composition on intracellular ion concentrations and membrane potential. In (A) extracellular K^+ is doubled, with a corresponding reduction in extracellular Na^+ to keep osmolarity constant. In (B) half the extracellular Cl^- is replaced by an impermeant anion, A^-. Ion concentrations are m*M* and extracellular volumes are assumed to be very large with respect to cell volumes so that fluxes into and out of the cell do not change extracellular concentrations.

Potassium and chloride entry is accompanied by entry of water to maintain osmotic balance, resulting in a slight increase in cell volume. When the new equilibrium is reached, intracellular potassium has increased in concentration from 90 m*M* to 91 m*M*, intracellular chloride from 4 m*M* to 7.9 m*M*, and the cell volume has increased by 3.5%.

At first glance it seems that more chloride than potassium has entered the cell, but think what the concentrations would be if the cell did *not* increase in volume: the concentrations of both ions would be greater than the indicated values by 3.5%. Thus, the intracellular chloride concentration would be about 8.2 m*M* (instead of 7.9 m*M* after the entry of water), and intracellular potassium would be about 94.2 m*M*, both being 4.2 m*M* higher than in the original solution. In other words, we can think first of potassium and chloride entering in equal quantities (except for the trivial difference required to change the charge on the membrane) and then of water following to achieve the final concentrations shown in the figure.

Similar considerations apply to changes in extracellular chloride concentration, but with a marked difference: when the new steady state is finally reached the membrane potential is essentially unchanged. The consequences of a 50% reduction in extracellular chloride concentration are shown in Figure 6.2B, in which 60 m*M* of chloride in the solution bathing the cell is replaced by an impermeant anion. Chloride leaves the cell, depolarizing the membrane toward the new chloride equilibrium potential (–68 mV). Potassium, no longer being in equilibrium, leaves as well. As in the previous example, potassium and chloride leave the cell in equal quantities (accompanied by water). Because the intracellular concentration of potassium is high, the fractional change in concentration produced by the efflux is relatively small. However, the efflux of chloride causes a sizable fractional change in the intracellular chloride concentration, and hence in the chloride equilibrium potential. As chloride continues to leave the cell the equilibrium potential returns toward its original value. The process continues until the chloride and potassium equilibrium potentials are again equal and the membrane potential is restored.

Membrane Potentials in Squid Axons

The idea that the resting membrane potential is the result of an unequal distribution of potassium ions between the extracellular and intracellular fluids was first proposed by Julius Bernstein[1] in 1902. He could not test this hypothesis directly, however, because there was no satisfactory way of measuring membrane potential. It is now possible to measure membrane potential accurately and to see whether changes in external and internal potassium concentrations produce the potential changes predicted by the Nernst relation. The first such experiments were done on giant axons that innervate the mantle of the squid. The axons are up to 1 mm in diameter,[2] and their large size permits the insertion of recording electrodes into their cytoplasm to measure transmembrane potential directly (Figure 6.3A). Further, they are remarkably resilient and continue to function even when their axoplasm has been squeezed out with a rubber roller and replaced with an internal perfusate (Figure 6.3B)! Thus their internal, as well as external ion composition can be controlled. A. L. Hodgkin, who together with A. F. Huxley initiated many experiments on squid axon (for which they later received the Nobel prize), has said:

> It is arguable that the introduction of the squid giant nerve fiber by J. Z. Young in 1936 did more for axonology than any other single advance during the last forty years. Indeed a distinguished neurophysiologist remarked recently at a congress dinner (not, I thought, with the utmost tact), "It's the squid that really ought to be given the Nobel Prize."[3]

The concentrations of some of the major ions in squid blood and in the axoplasm of the squid nerves are given in Table 6.1 (several ions, including magnesium and internal anions, are omitted). Experiments on isolated axons are usually done in seawater, with the ratio of intracellular to extracellular potassium concentrations being 40:1. In these conditions, the membrane potential is –65 to –70 mV, considerably less negative than the potassium equilibrium potential of –93 mV, but more negative than the chloride equilibrium potential, which is about – 55 mV.

Bernstein's hypothesis was tested by measuring the resting membrane potential and comparing it with the potassium equilibrium potential at various extracellular potassium

[1] Bernstein, J. 1902. *Pflügers Arch.* 92: 521–562.

[2] Young, J. Z. 1936. *Q. J. Microsc. Sci.* 78: 367–386.

[3] Hodgkin, A. L. 1973. *Proc. R. Soc. Lond., B, Biol. Sci.* 183: 1–19.

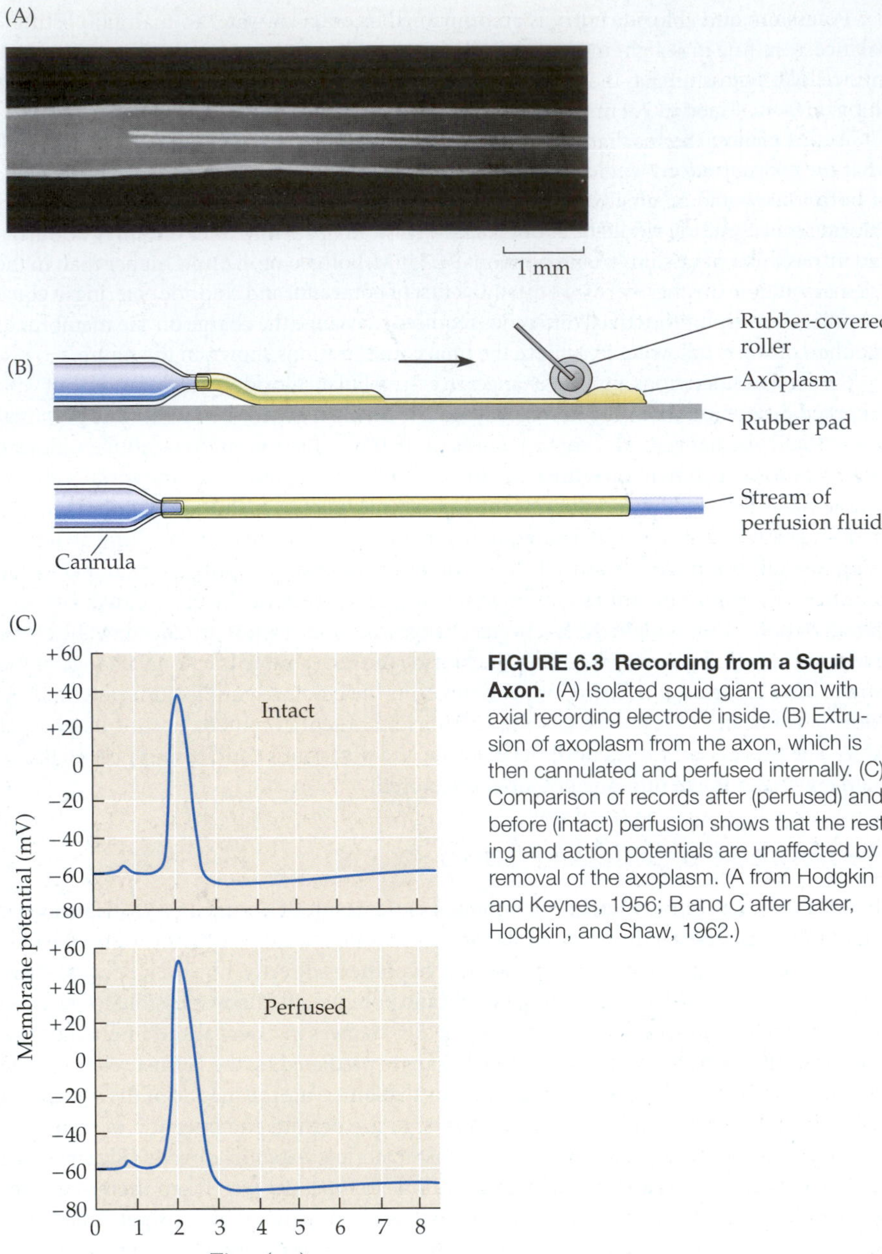

FIGURE 6.3 Recording from a Squid Axon. (A) Isolated squid giant axon with axial recording electrode inside. (B) Extrusion of axoplasm from the axon, which is then cannulated and perfused internally. (C) Comparison of records after (perfused) and before (intact) perfusion shows that the resting and action potentials are unaffected by removal of the axoplasm. (A from Hodgkin and Keynes, 1956; B and C after Baker, Hodgkin, and Shaw, 1962.)

concentrations. As with our model cell, these changes would not be expected to produce a significant change in internal potassium concentration. From the Nernst equation, changing the concentration ratio by a factor of 10 should change the membrane potential by 58 mV at room temperature. The results of an experiment on squid axon, in which the external potassium concentration was changed are shown in Figure 6.4. The external concentration is plotted on a logarithmic scale on the abscissa and the membrane potential on the ordinate. The expected slope of 58 mV per tenfold change in extracellular potassium concentration is realized only at relatively high concentrations (solid straight line), with the slope becoming less and less as external potassium is reduced. This result indicates that the potassium ion distribution is not the only factor contributing to the membrane potential.

The Effect of Sodium Permeability

From the experiments on squid axon we can conclude that the hypothesis made by Bernstein in 1902 is almost correct; the membrane potential is strongly but not exclusively dependent on the potassium concentration ratio. We can explain the deviation from the Nernst relation shown in Figure 6.4 simply by abandoning the notion that the membrane is impermeable to sodium. Real nerve cell membranes, in fact, have a permeability to sodium that ranges between 1% and 10% of their permeability to potassium.

To consider the effect of sodium permeability, we begin with our model cell (see Figure 6.1) and, for the moment, ignore any movement of chloride ions. The Nernst equation tells us that sodium would be in equilibrium at a membrane potential +34 mV (E_{Na}), far from the actual membrane potential of –85 mV. So, if we make the membrane permeable to sodium, both the concentration gradient and the membrane potential tend to drive sodium into the cell. As sodium ions enter the cell, they accumulate on the inner surface of the membrane, causing depolarization. As a result, potassium is no longer in equilibrium and potassium ions leave the cell. As the depolarization progresses, the membrane potential moves closer to the sodium equilibrium potential and farther from the potassium equilibrium potential. As it does so, the sodium influx decreases and the potassium efflux increases. The process continues until the influx of sodium is exactly balanced by the efflux of potassium. At that point there is no further charge accumulation, and the membrane potential remains constant. In summary, the membrane potential lies between the potassium and sodium equilibrium potentials and is the potential at which the sodium and potassium currents are exactly equal and opposite.

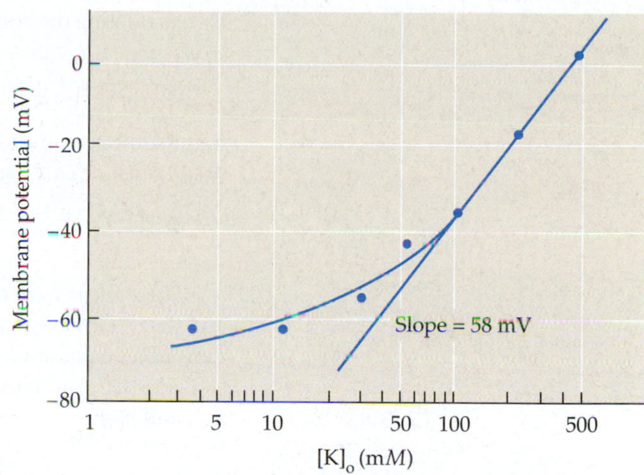

FIGURE 6.4 Membrane Potential versus External Potassium Concentration in squid axon, plotted on a semilogarithmic scale. The straight line is drawn with a slope of 58 mV per tenfold change in extracellular potassium concentration, according to the Nernst equation. Because the membrane is also permeable to sodium, the points deviate from the straight line, especially at low potassium concentrations. (After Hodgkin and Keynes, 1956.)

Chloride ions participate in the process as well but, as we have already seen, there is ultimately an adjustment in intracellular chloride concentration in the model cell, so that the chloride equilibrium potential matches the new membrane potential. As the cation fluxes gradually reach a balance, the intracellular chloride concentration increases until there is no net chloride flux across the membrane.

The Constant Field Equation

To determine the exact membrane potential in our model cell, we have to consider the individual ion currents across the membrane. The inward sodium current (i_{Na}) depends on the driving force for sodium, which is the difference between the membrane potential and the sodium equilibrium potential, $V_m - E_{Na}$, (see Chapter 2). The sodium current also depends on the sodium conductance of the membrane, or g_{Na}. The sodium conductance is a measure of the ease with which sodium can pass through the membrane and depends on the number of open sodium channels—the more open channels the greater the conductance. So, the sodium current is:

$$i_{Na} = g_{Na}(V_m - E_{Na})$$

The same relationship holds for potassium and chloride:

$$i_K = g_K(V_m - E_K)$$

$$i_{Cl} = g_{Cl}(V_m - E_{Cl})$$

If we assume that chloride is in equilibrium, so that $i_{Cl} = 0$, then for the membrane potential to remain constant, the potassium and sodium currents must be equal and opposite:

$$g_K(V_m - E_K) = -g_{Na}(V_m - E_{Na})$$

It is useful to examine this relationship in more detail. Suppose g_K is much larger than g_{Na}. Then, if the currents are to be equal, the driving force for potassium efflux must be much smaller than that for sodium entry. In other words, the membrane potential must be much closer to E_K than to E_{Na}. Conversely, if g_{Na} is relatively large, the membrane potential will be closer to E_{Na}.

By rearranging the equation we arrive at an expression for the membrane potential:

$$V_m = \frac{g_K E_K + g_{Na} E_{Na}}{g_K + g_{Na}}$$

If, for some reason, chloride is not at equilibrium, then chloride currents across the membrane must be considered as well and the equation becomes slightly more complicated:

$$V_m = \frac{g_K E_K + g_{Na} E_{Na} + g_{Cl} E_{Cl}}{g_K + g_{Na} + g_{Cl}}$$

These ideas were developed originally by Goldman,[4] and independently by Hodgkin and Katz.[5] However, instead of considering equilibrium potentials and conductances, they derived an equation for membrane potential in terms of ion concentrations outside ($[K]_o$, $[Na]_o$, $[Cl]_o$) and inside ($[K]_i$, etc.) the cell, and membrane *permeability* to each ion (p_K, p_{Na}, and p_{Cl}):

$$V_m = 58 \log \frac{p_K[K]_o + p_{Na}[Na]_o + p_{Cl}[Cl]_i}{p_K[K]_i + p_{Na}[Na]_i + p_{Cl}[Cl]_o}$$

Note that the chloride ratios are reversed, as occurred previously in the Nernst equation, because the chloride valence is –1.

As before, if chloride is in equilibrium, the chloride terms disappear. This equation is sometimes called the GHK equation for its originators, and is known also as the **constant field equation**, because one of the assumptions made in arriving at the expression was that the voltage gradient or "field" across the membrane is uniform. The constant field equation is entirely analogous to the previous equation and makes the same predictions: when the permeability to potassium is very high relative to the sodium and chloride permeabilities, the sodium and chloride terms become negligible and the membrane potential approaches the equilibrium potential for potassium: $V_m = 58 \log ([K]_o/([K]_i)$. Increasing sodium permeability causes the membrane potential to move toward the sodium equilibrium potential.

The constant field equation provides us with a useful general principle to remember: the membrane potential depends on the relative conductances (or permeabilities) of the membrane to the major ions, and on the equilibrium potentials for those ions. In real cells the resting permeabilities to potassium and chloride are relatively high. Hence, the resting membrane potential is close to the potassium and chloride equilibrium potentials. When sodium permeability is increased, as occurs during an action potential (see Chapter 7) or an excitatory postsynaptic potential (see Chapter 11), the membrane potential moves toward the sodium equilibrium potential.

[4]Goldman, D. E. 1943. *J. Gen. Physiol.* 27: 37–60.

[5]Hodgkin, A. L., and Katz, B. 1949. *J. Physiol.* 108: 37–77.

The Resting Membrane Potential

As useful as the constant field equation is, it does not provide us with an accurate description of the resting membrane potential because the requirement for zero net current across the membrane is not, in itself, adequate. Instead, for the cell to remain in a stable condition *each* individual ionic current must be zero. As a result, under the conditions of the constant field equation, the cell would gradually fill up with sodium and chloride and lose potassium. In real cells, intracellular sodium and potassium concentrations are kept constant by a sodium–potassium exchange pump (sodium–potassium ATPase; see Chapter 9). As sodium and potassium leak into and out of the cell, the pump transports a matching amount of each ion in the opposite direction (Figure 6.5). Thus, metabolic energy is used to maintain the cell in a **steady state**.

In order to have a more complete and accurate description of the resting membrane potential, we must consider both the passive ion fluxes and the activity of the pump. Again, we first consider the currents carried by passive fluxes of sodium and potassium across the membrane:

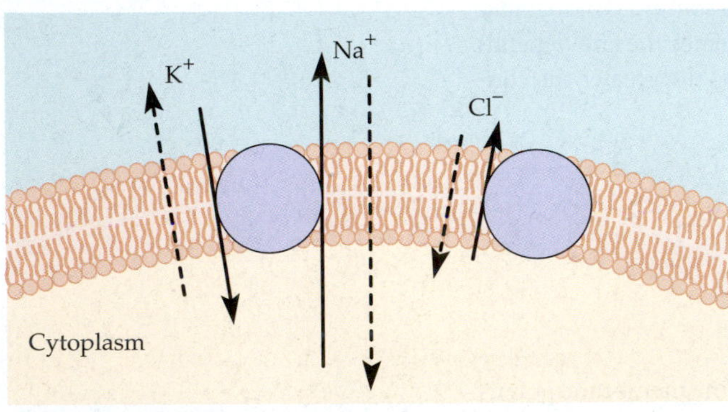

FIGURE 6.5 Passive Ion Fluxes and Pumps in a Steady State. (A) Net passive ion movements across the membrane are indicated by dashed arrows and transport systems by solid arrows and circles. Lengths of arrows indicate the magnitudes of net ion movements. Total flux is zero for each ion. For example, net inward leak of Na^+ is equal to rate of outward transport. Na:K transport is coupled with a ratio of 3:2. In any particular cell, Cl^- transport may be outward (as shown) or inward.

$$i_{Na} = g_{Na}(V_m - E_{Na})$$
$$i_K = g_K(V_m - E_K)$$

We no longer assume that the sodium and potassium currents are equal and opposite, but if we know how they are related we can, as before, obtain an equation for the membrane potential in terms of the sodium and potassium equilibrium potentials and their relative conductances. The relationship between the sodium and potassium currents is given by the characteristics of the pump. Because it keeps intracellular sodium and potassium concentrations constant by transporting the ions in the ratio of 3Na to 2K (see Chapter 9), it follows that the passive ion fluxes must be in the same ratio: i_{Na}:i_K = 3:2. So we can write:

$$\frac{i_{Na}}{i_K} = \frac{g_{Na}\left(V_m - E_{Na}\right)}{g_K\left(V_m - E_K\right)} = -1.5$$

The ratio is negative because the sodium and potassium currents are flowing in opposite directions. By rearranging the equation we get an expression for the membrane potential:

$$V_m = \frac{1.5g_K E_K + g_{Na} E_{Na}}{1.5g_K + g_{Na}}$$

The equation is similar to the expression derived previously for the model cell and makes the same kinds of predictions. As before, the membrane potential depends on the relative magnitudes of g_K and g_{Na}. The difference is that the potassium term is multiplied by a factor of 1.6. Because of this factor, the membrane potential is closer to E_K than would otherwise be the case. Thus, the driving force for sodium entry is increased and that for potassium influx reduced. As a result, the passive fluxes are in a ratio of 3Na to 2K rather than 1:1.

In summary, the real cell differs from the model cell in that the resting membrane potential is the potential at which the passive influx of sodium is 1.5 times the passive efflux of potassium, rather than the potential at which the two fluxes are equal and opposite. The passive inward and outward currents are determined by the equilibrium potentials and conductances for the two ions; the required ratio of 3:2 is determined by the transport characteristics of the pump.

The problem of finding an expression for the resting membrane potential of real cells, taking into account the transport activity, was first considered by Mullins and Noda,[6] who used intracellular microelectrodes to study the effects of ionic changes on membrane potential in muscle. Like Goldman, Hodgkin, and Katz, they derived an expression for membrane potential in terms of permeabilities and concentrations. The result is equivalent to the equation we have just derived using conductances and equilibrium potentials:

$$V_m = 58 \log \frac{rp_K[\mathrm{K}]_o + p_{Na}[\mathrm{Na}]_o}{rp_K[\mathrm{K}]_i + p_{Na}[\mathrm{Na}]_i}$$

where r is the absolute value of the transport ratio (3:2). The equation provides an accurate description of the resting membrane potential, provided all the other permeant ions (e.g., chloride) are in a steady state.

Chloride Distribution

How do these considerations apply to chloride? As for all other ions, there must be no net chloride current across the resting membrane. As already discussed (see Figure 6.2B), chloride is able to reach equilibrium simply by an appropriate adjustment in internal concentration, without affecting the steady-state membrane potential. In many cells, however, there are transport systems for chloride as well (see Chapter 9). In squid axon and in muscle, chloride is transported actively into the cells; in many nerve cells, active transport is outward (see Figure 6.5). The effect of inward transport is to add an increment to the equilibrium concentration such that there is an outward leak of chloride equal to the rate of transport in the opposite direction.[7] Outward transport has the reverse effect.

An Electrical Model of the Membrane

The characteristics of the nerve cell membrane endow it with the electrical properties illustrated in Figure 6.6. First, because the membrane is an insulating layer separating electrical

[6] Mullins, L. J., and Noda, K. 1963. *J. Gen. Physiol.* 47: 117–132.

[7] Martin, A. R. 1979. Appendix to Matthews, G. and Wickelgren, W. O. *J. Physiol.* 293: 393–414.

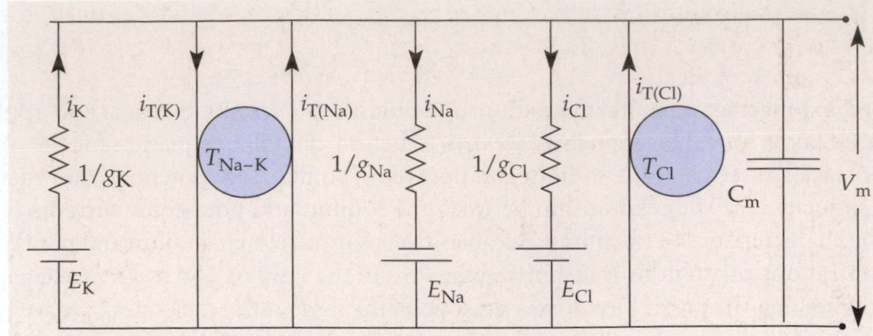

FIGURE 6.6 Electrical Model of the Steady-State Cell shown in Figure 6.2. (A) E_K, E_{Na} and E_{Cl} are the Nernst potentials for the individual ions. The individual ion conductances are represented by resistors, with a resistance 1/g for each ion. The individual ion currents, i_K, i_{Na}, and i_{Cl} are equal and opposite to the currents $i_{T(K)}$, $i_{T(Na)}$, and $i_{T(Cl)}$ supplied by the sodium–potassium exchange pump (T_{Na-K}) and the chloride pump (T_{Cl}), so that the net flux of each ion across the membrane is zero. The resulting membrane potential (V_m) determines the amount of charge stored on the membrane capacitor (C_m).

charges on its inner and outer surfaces, it has the properties of a capacitor. In parallel with the capacitor are conductance pathways, represented by resistors, that allow ion fluxes into and out of the cell. The electrical resistance in each pathway is inversely related to the conductance for the ion in question: the greater the ionic conductance the lower the resistance to current flow. Passive ion currents through the resistors are driven by batteries that represent the equilibrium potentials for each of the ions. The passive currents are equal and opposite to the corresponding currents generated by the pumps, so that the net current across the membrane for each ion is zero.

Predicted Values of Membrane Potential

How do these considerations explain the relation between potassium concentration and membrane potential shown in Figure 6.4? This can be seen by using real numbers in the equations. In squid axon, the permeability constants for potassium and sodium are roughly in the ratio 1.0:0.04.[5] We can use these relative values, together with the ion concentrations given in Table 6.1, to calculate the resting membrane potential in seawater:

$$V_m = 58 \log \frac{(1.5)(10)+(0.04)(460)}{(1.5)(400)+(0.04)(50)} = -73\,\text{mV}$$

Now we can see quantitatively why, when extracellular potassium is altered, the membrane potential fails to follow the Nernst potential for potassium. If, in the numerator of the equation, we look at the magnitude of the term involving extracellular potassium concentration ($1.5 \times 10 = 15$ m*M*) and the term that involves extracellular sodium concentration ($0.04 \times 460 = 18.4$ m*M*), we see that potassium contributes only about 45% of the total. Because of this, doubling the external potassium concentration does not double the numerator (as would happen in the Nernst equation), and as a consequence, the effect on the potential of changing extracellular potassium concentration is less than would be expected if potassium were the only permeant ion. When the external potassium concentration is raised to a high enough level (100 m*M* in Figure 6.4), the potassium term becomes sufficiently dominant for the voltage-current relation to approach the theoretical limit of 58 mV per tenfold change in concentration. This effect is enhanced by the fact that many potassium channels are voltage-activated (see Chapter 7). When the membrane is depolarized by increasing the extracellular potassium concentration, the voltage-activated channels open, thereby increasing the potassium permeability. As a result, the relative contribution of potassium to the membrane potential is still further increased.

In general, nerve cells have resting potentials of the order of –70 mV. In some cells, such as vertebrate skeletal muscle,[8] the resting potential can be –90 mV or larger, reflecting

[8] Fatt, P., and Katz, B. 1951. *J. Physiol.* 115: 320–370.

a low ratio of sodium-to-potassium permeability. Glial cells in particular have a very low resting permeability to sodium so that their membrane potential is nearly identical to the potassium equilibrium potential (see Chapter 10). Other cells, such as leech ganglion cells[9] and photoreceptors in the retina,[10] have relatively high membrane permeability to sodium and resting membrane potentials as small as –40 mV.

Contribution of the Sodium–Potassium Pump to the Membrane Potential

The sodium–potassium transport system is **electrogenic** because each cycle of the pump results in the net outward transfer of one positive ion, thereby contributing to the excess negative charge on the inner face of the membrane. How large is this contribution? An easy way to find out is to calculate what the membrane potential would be if the pump were *not* electrogenic or, in other words, if $r = 1$. Repeating the previous calculation with this condition gives the following:

$$V_m = 58 \log \frac{(1.0)(10) + (0.04)(460)}{(1.0)(400) + (0.04)(50)} = -67\,\text{mV}$$

The result is 6 mV less than the previous value, so the pump contributes –6 mV to the resting potential. In general, the size of the pump contribution depends on a number of factors, particularly the relative ion permeabilities. For a transport ratio of 3:2, the steady state contribution to the resting membrane potential is limited to a maximum of about –11 mV.[11] If the transport process is stopped, the electrogenic contribution disappears immediately and the membrane potential then declines gradually as the cell gains sodium and loses potassium.

It is interesting that the *rate* of transport does not appear in the equation, apart from the implicit requirement that it must match the passive ionic fluxes. Theoretical calculations indicate that once that requirement is met any further increase in sodium–potassium pump activity can be expected to have very little effect on the steady-state resting membrane potential.[12] This is largely because the transport system is dependent on intracellular sodium concentration (see Chapter 9). Any increase in pump rate results in hyperpolarization and depletion of intracellular sodium. As the sodium concentration falls, the pump rate and the membrane potential return toward their previous value.

What Ion Channels Are Associated with the Resting Potential?

The resting permeabilities of membranes to sodium, potassium, and chloride have been determined in many nerve cells. Underlying these is a wide variety of membrane channels that permit the passage of anions and cations into and out of the cell. However, the precise identification of the channels underlying these ionic **leak currents** in any specific cells is difficult. A significant fraction of the resting potassium current is likely to be through 2P potassium channels (see Chapter 5), which tend to be open at the resting membrane potential.[13,14] In addition, many nerve cells have M-type potassium channels that are open at rest and closed by intracellular messengers (see Chapter 17). M-currents, together with accompanying h-currents, are responsible for most of the potassium leak in sympathetic ganglion cells.[15] H-currents are carried by HCN (hyperpolarization-activated cyclic nucleotide-gated) channels that are activated by hyperpolarization—some of which are open at normal resting potentials.[16] HCN channels are cation channels with a sodium-to-potassium permeability ratio (p_{Na}/p_K) of about 0.25 and thus are responsible for a fraction of the sodium leak current as well. Other contributors to resting potassium permeability include channels activated by intracellular cations, namely sodium-activated and calcium-activated potassium channels. Finally, a few voltage-sensitive potassium channels associated with the action potential may be open at rest.

Apart from HCN channels, the major source of resting sodium permeability is the so-called NALCN (sodium leak channel, non-selective) channel that is open at rest.[17] NALCN channels are virtually non-selective for monovalent cations ($p_{Na}/p_K \approx 1.1$), so that at normal resting potentials the main ion movement through the channels is inward sodium flux. An additional sodium influx occurs not through channels, but rather through

[9]Nicholls, J. G., and Baylor, D. A. 1968. *J. Neurophysiol.* 31: 740–756.

[10]Baylor, D. A., and Fuortes, M. G. F. 1970. *J. Physiol.* 207: 77–92.

[11]Martin, A. R., and Levinson, S. R. 1985. *Muscle Nerve* 8: 354–362.

[12]Fraser, J. A., and Huang, C. L.-H. 2004. *J. Physiol.* 559: 459–478.

[13]Brown, D. A. 2000. *Curr. Biol.* 10: R456–459.

[14]Goldstein, S. A. N. et al. 2001. *Nat. Rev. Neurosci.* 2: 175–184.

[15]Lamas, J. A. et al. 2002. *Neuroreport* 13: 585–591.

[16]Biel, M. et al. 2009. *Physiol. Rev.* 89: 847–885.

[17]Lu, B. et al. 2007. *Cell* 129: 371–383.

[18]Sokolov, S., Scheuer, T., and Catterall, W. A. 2007. *Nature* 446: 76–78.

[19]Staley, K. et al. 1996. *Neuron* 17: 543–551.

[20]Meladinić, M. et al. 1999. *Proc. R. Soc. Lond., B, Biol. Sci.* 266: 1207–1213.

[21]Gold, M. R., and Martin, A. R. 1983. *J. Physiol.* 342: 99–117.

[22]Barchi, R. L. 1997. *Neurobiol. Dis.* 4: 254–264.

[23]Cannon, S. C. 1996. *Trends. Neurosci.* 19: 3–10.

sodium-dependent secondary active transport systems (see Chapter 9). Finally, tetrodotoxin has been shown to block a small fraction of the resting sodium conductance,[9] indicating a contribution by voltage-activated sodium channels.

An unusual pathological leak current in skeletal muscle cells is found in a neuromuscular disease known as hypokalaemic periodic paralysis. The current is unusual in that it does not involve ion movements through open channels. Instead, it is associated with cation fluxes through a normally occluded protein pore around the sodium channel S4 helix.[18] This so-called omega current (see Chapter 7) is associated with a mutation in which charge-carrying arginine residues on the helix are replaced by smaller neutral amino acids, thereby allowing cations to permeate the pore. The result is a constant inward leak of sodium into the muscle cell.

Chloride channels of the CLC family (see Chapter 5) are widely distributed in nerve and muscle. The presence of chloride channels is important in that they serve to stabilize the membrane potential (see below). The channels also interact with chloride transport systems to determine intracellular chloride concentrations.[19,20] For example, in embryonic hippocampal neurons CLC channel expression is low, and E_{Cl} is positive to the resting membrane potential because of inward transport and accumulation of chloride in the cytoplasm. In adult neurons, expression of CLC channels increases the chloride conductance of the membrane so that excess accumulation cannot occur, and E_{Cl} becomes equal to the membrane potential. In central nervous system neurons, chloride channels can account for as much as 10% of the resting membrane conductance.[21]

Changes in Membrane Potential

It is important to keep in mind that this discussion of resting membrane potential is always in reference to *steady state* conditions. For example, we have said that changing extracellular chloride concentration has little effect on membrane potential because the intracellular chloride concentration accommodates to the change. That is true in the long run, but the intracellular adjustment takes time, and while it is occurring, there will indeed be a transient effect.

The steady-state potential is the baseline upon which all changes in membrane potential are superimposed. How are such changes in potential produced? In general, transient changes, such as those that mediate signaling between cells in the nervous system, are the result of transient changes in membrane permeability. As we already know from the constant field equation, an increase in sodium permeability (or a decrease in potassium permeability) will move the membrane potential toward the sodium equilibrium potential, producing depolarization. Conversely, an increase in potassium permeability will produce hyperpolarization. Another ion of importance in signaling is calcium. Intracellular calcium concentration is very low, and in most cells, E_{Ca} is greater than +150 mV. Thus, an increase in calcium permeability results in calcium influx and depolarization.

The role of chloride permeability in the control of membrane potential is of particular interest. As we have noted, chloride makes little contribution to the resting membrane potential. Instead, intracellular chloride concentration adjusts to the potential and is modified by whatever chloride transport mechanisms are operating in the cell membrane. The effect of a transient increase in chloride permeability can be either hyperpolarizing or depolarizing, depending on whether the chloride equilibrium potential is negative or positive to the resting potential. The equilibrium potential, in turn, depends on whether intracellular chloride is depleted or concentrated by the transport system. In either case, the change in potential is usually relatively small. Even so, an increase in chloride permeability can be important for the regulation of signaling because it tends to hold the membrane potential near the chloride equilibrium potential and thus attenuates changes in potential produced by other influences.

Stabilization of the membrane potential in this way is important for controlling the excitability of many cells, such as skeletal muscle fibers, that have a relatively high chloride permeability at rest. In such cells, a transient influx of positive ions causes less depolarization than would otherwise be the case, because it is countered by an influx of chloride through already open channels. This mechanism is of some significance, as illustrated by the fact that chloride channel mutations that reduce chloride conductance are responsible for several muscle diseases. The diseased muscles are hyperexcitable (myotonic) due to loss of the normal stabilizing influence of a high-chloride conductance.[22,23]

SUMMARY

- Nerve cells have high intracellular concentrations of potassium and low intracellular concentrations of sodium and chloride so that potassium tends to diffuse out of the cell and sodium and chloride tend to diffuse in. The tendency for potassium and chloride to diffuse down their concentration gradients is opposed by the electrical potential across the cell membrane.
- In a model cell that is permeable only to potassium and chloride, the concentration gradients can be balanced exactly by the membrane potential, so that there is no net flux of either ion across the membrane. The membrane potential is then equal to the equilibrium potential for both potassium and chloride.
- Changing the extracellular potassium concentration changes the potassium equilibrium potential and, hence, the membrane potential. Changing extracellular chloride concentration, on the other hand, leads ultimately to a change in intracellular chloride, so that the chloride equilibrium potential and the membrane potential differ from their original values only transiently.
- In addition to being permeable to potassium and chloride, the plasma membrane of real cells is permeable to sodium. As a result, there is a constant influx of sodium into the cell and an efflux of potassium. These fluxes are balanced exactly by active transport of the ions in the opposite directions, in the ratio of 3 sodium to 2 potassium. Under these circumstances, the membrane potential depends on the sodium equilibrium potential, the potassium equilibrium potential, the relative conductance of the membrane to the two ions, and the sodium–potassium exchange pump ratio.
- Because the sodium–potassium exchange pump transports more positive ions outward than inward across the membrane, it can make a direct contribution of several mV to the membrane potential.
- The chloride equilibrium potential may be positive or negative to the resting membrane potential, depending on chloride transport processes. Although the chloride distribution plays little role in determining the resting membrane potential, high chloride permeability is important for electrical stability.

Suggested Reading

Hodgkin, A. L., and Katz, B. 1949. The effect of sodium ions on the electrical activity of the giant axon of the squid. *J. Physiol.* 108: 37–77. (The constant field equation is derived in Appendix A of this paper.)

Mullins, L. J., and Noda, K. 1963. The influence of sodium-free solutions on membrane potential of frog muscle fibers. *J. Gen. Physiol.* 47: 117–132.

CHAPTER 7
Ionic Basis of the Action Potential

The ionic mechanisms responsible for generating action potentials have been described quantitatively, largely through the use of the voltage clamp method to measure membrane currents. From such measurements, it is possible to determine which components of the currents are carried by each ion species, and to deduce the magnitude and time course of the underlying changes in ionic conductances. The experiments have shown that depolarization rapidly increases sodium conductance and, more slowly, potassium conductance. The activation of sodium conductance is transient, being followed by inactivation. The increase in potassium conductance persists for as long as the depolarizing pulse is maintained. The dependence of sodium and potassium conductances on membrane potential and their sequential timing account quantitatively for the amplitude and time course of the action potential as well as for other phenomena such as threshold and refractory period.

Other channels, in addition to the voltage-dependent sodium and potassium channels just described, can be involved in action potential generation. In some cells, voltage-activated calcium channels are responsible for the rising phase of the action potential, and repolarization can involve currents associated with activation of a number of additional potassium channel types. The action potential itself can be followed by afterpotentials, which are periods of hyperpolarization or depolarization mediated by prolonged changes in membrane conductance.

A. L. Hodgkin, 1949

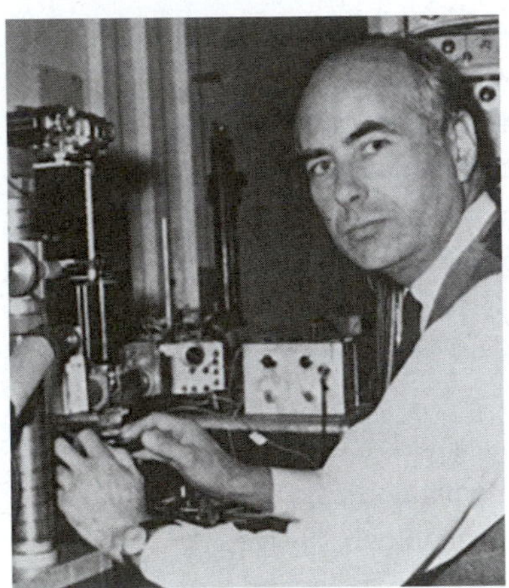

A. F. Huxley, 1974

In Chapter 6, we showed how the membrane potential of a nerve cell depends on its relative permeability to the major ions present in the extracellular and intracellular fluids, particularly sodium and potassium. The membrane potential of a cell at rest is near the **potassium equilibrium potential** because of the relatively high permeability of its membrane for potassium. It was discovered more than 100 years ago that sodium ions were necessary for nerve and muscle cells to produce an action potential, and evidence gradually accumulated to support the idea that the mechanism underlying action potential generation was an increase in sodium permeability that drove the membrane potential toward the sodium equilibrium potential.

Critical experiments were done in the early 1950s by Hodgkin, Huxley and Katz,[1] and Hodgkin and Huxley.[2–5] Two main features contributed to the success of their experiments: One was the availability of the giant axon of the squid, and the other a newly developed experimental tool called the **voltage clamp**. They found that the rising and falling phases of the action potential were accompanied by a large, transient influx of sodium ions, followed by an efflux of potassium. They then were able to deduce the underlying changes in membrane conductance and to show that these were of the correct magnitude and time course to account exactly for the magnitude and time course of the action potential. Since that time a great number of similar experiments have been carried out on other nerve cells and on muscle cells, and many new experimental techniques have been applied to the problem. Although some of the details of their theoretical treatment are still being tested and refined, after more than 50 years their basic conclusions still stand.

Voltage Clamp Experiments

The voltage clamp technique was devised by Cole and his colleagues[6] and developed further by Hodgkin, Huxley and Katz.[1] The aim of the experiments was to identify the nature of the ionic currents underlying the action potential and to determine their magnitude and time course. The experimental arrangement is described in Box 7.1. All we need to know to understand the experiments themselves is that the method permits us to set the membrane potential of the cell almost instantaneously to any level and hold it there (i.e., "clamp" it) while at the same time recording the current flowing across the membrane. Figure 7.1A shows an example of the currents that occur when the membrane potential is stepped suddenly from its resting value (in this example –65 mV) to a depolarized level (–9 mV). The current produced by the voltage step consists of three phases: (1) a brief outward surge lasting only a few microseconds (μs), (2) an early inward current, and (3) a late outward current.

Capacitative and Leak Currents

The initial brief surge of current is the **capacitative current**, which occurs because the step from one potential to another alters the charge on the membrane capacitance. If the clamp amplifier is capable of delivering a large amount of current, then the membrane can be charged rapidly and this current will last only a very short time. In practice, the surge of capacitative current lasts only about 20 μs and can be ignored when analyzing the subsequent ionic currents.

The capacitative current is followed by a small, steady outward current, known as **leak current**, which flows through the resting membrane conductances. Leak current is carried largely by potassium and chloride ions, varies linearly with voltage displacement from rest in either direction, and lasts throughout the duration of the voltage step. However, during most of the response, it is obscured by the much larger ionic currents associated with the action potential.

Ionic Currents Carried by Sodium and Potassium

Turning now to the second and third phases of membrane current, Hodgkin and Huxley showed that these currents were due to the entry of sodium across the cell membrane, followed by the exit of potassium. In addition, they were able to deduce the relative size and time course of the separate currents. One convenient procedure was to abolish the sodium current by replacing most of the extracellular sodium by choline (an impermeant cation). With an appropriate reduction in extracellular sodium concentration, the sodium equilibrium potential could be made equal to the potential during the depolarizing step. As a result, there was no driving force for sodium entry during the step and thus no net current

[1]Hodgkin, A. L., Huxley, A. F., and Katz, B. 1952. *J. Physiol.* 116: 424–448.

[2]Hodgkin, A. L., and Huxley, A. F. 1952. *J. Physiol.* 116: 449–472.

[3]Hodgkin, A. L., and Huxley, A. F. 1952. *J. Physiol.* 116: 473–496.

[4]Hodgkin, A. L., and Huxley, A. F. 1952. *J. Physiol.* 116: 497–506.

[5]Hodgkin, A. L., and Huxley, A. F. 1952. *J. Physiol.* 117: 500–544.

[6]Marmont, G. 1949. *J. Cell. Physiol.* 34: 351–382.

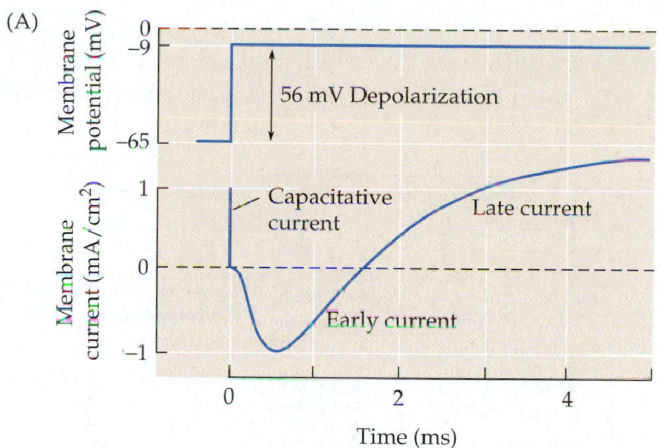

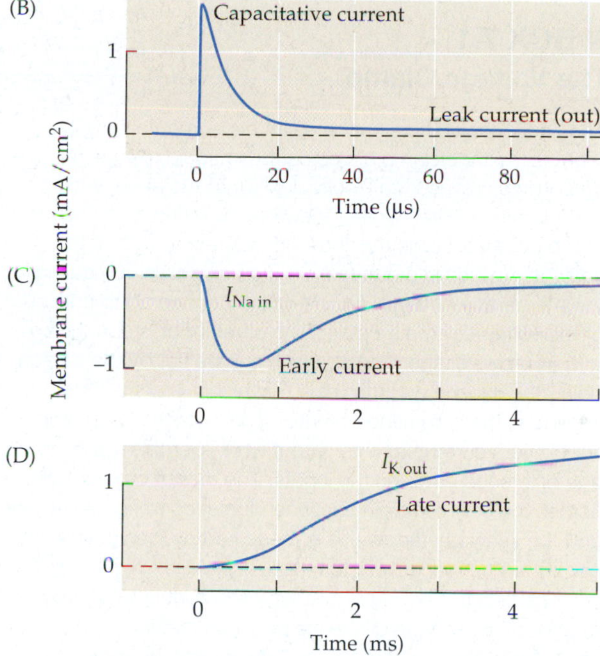

FIGURE 7.1 Membrane Currents Produced by Depolarization. (A) Currents measured by voltage clamp during a 56 mV depolarization of a squid axon membrane. The currents (lower trace) consist of a brief, positive capacitative current, an early transient phase of inward current, and a late, maintained outward current. These three currents are shown separately in B, C, and D, respectively. The capacitative current (B) lasts for only a few microseconds (note the change in time scale). The small outward leakage current is due to movement of potassium and chloride. The early inward current (C) is due to sodium entry and the late outward current (D) to potassium movement out of the fiber.

through the activated sodium channels. This left only the potassium current, as shown in Figure 7.1D. Subtraction of the potassium current from the total ion current (Figure 7.1A) then revealed the magnitude and time course of the sodium current (Figure 7.1C).

[7]Llewellyn, L. E. 2009. *Prog. Mol. Subcell. Biol.* 46: 67–87.

[8]Watters, M. R. 2005. *Semin. Neurol.* 25: 278–289.

[9]Terlau, H., and Olivera, B. M. 2004. *Physiol. Rev.* 84: 41–68.

Selective Poisons for Sodium and Potassium Channels

Pharmacological methods now exist for blocking sodium and potassium currents selectively. Two toxins that have been particularly useful for inactivating sodium channels are tetrodotoxin (TTX) and its pharmacological equivalent saxitoxin (STX).[7,8] TTX is a virulent poison, concentrated in the ovaries and other organs of puffer fish. STX is synthesized by marine dinoflagellates and concentrated by filter-feeding shellfish, such as the butter clam *Saxidomus*. Its virulence competes with that of TTX—ingestion of a single clam (cooked or not) can be fatal. Sodium channels are blocked by several snail toxins (conotoxins) as well.[9]

The great advantage of TTX for neurophysiological studies is its high specificity. When a TTX-poisoned axon is subjected to a depolarizing voltage step, no inward sodium current is seen, only the delayed outward potassium current (Figure 7.2B). The potassium current is unchanged in amplitude and time course by the poison. Application of TTX to the inside of the membrane by adding it to an internal perfusing solution has no effect. The actions of STX are indistinguishable from those of TTX. Both toxins

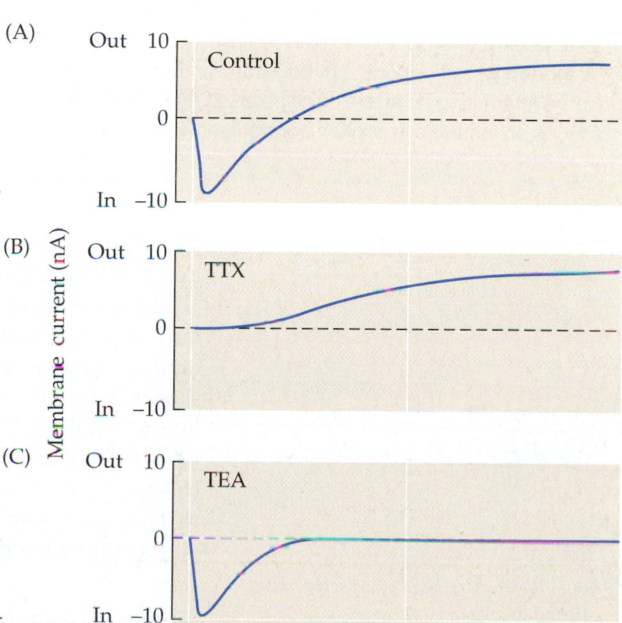

FIGURE 7.2 Pharmacological Separation of Membrane Currents into Sodium and Potassium Components. Membrane currents produced by clamping the membrane potential to 0 mV in a frog myelinated nerve. (A) is a control record in normal bathing solution. In (B), the addition of 300 nanomolar (nM) tetrodotoxin (TTX) causes the sodium current to disappear while the potassium current remains. In (C), addition of tetraethylammonium (TEA) blocks the potassium current, leaving the sodium current intact. (Modified from Hille, 1970.)

BOX 7.1
The Voltage Clamp

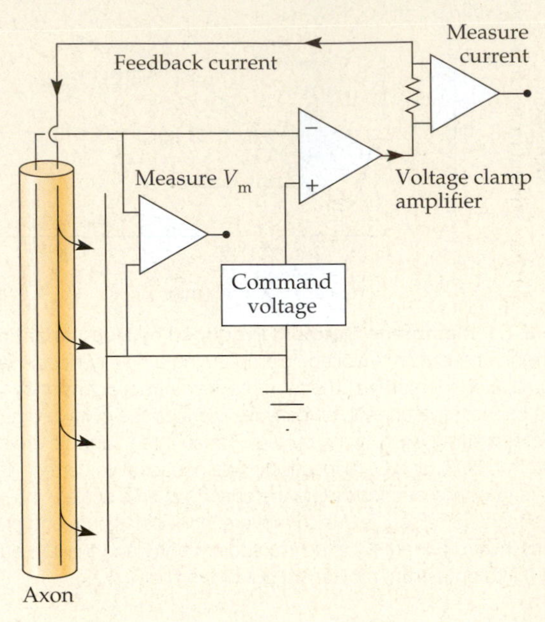

The figure shown here illustrates an experimental arrangement for voltage clamp experiments on squid axons. Two fine silver wires are inserted longitudinally into the axon, which is bathed in sea water. One of the wires provides a measure of the potential inside the fiber with respect to that of the seawater (which is grounded) or, in other words, a measure of the membrane potential (V_m). It is also connected to one input of the voltage clamp amplifier. The other input is connected to a variable voltage source, which can be set by the person doing the experiment; the value to which it is set is thus known as the command potential. The voltage clamp amplifier delivers current from its output whenever there is a voltage difference between the inputs. The output current flows across the cell membrane between the second fine silver wire and the seawater (arrows); it is measured by the voltage drop across a small series resistor.

The circuit is arranged so that the output current tends to cancel any voltage difference between the two inputs (negative feedback). It works as follows: first, suppose that the resting potential of the fiber is –70 mV and the command potential is set to –70 mV as well. Because the voltages at the two inputs of the amplifier are equal, there will be no output current. If the command potential is stepped to, say, –65 mV, then because of the 5 mV difference between the inputs, the amplifier delivers positive current into the axon and across the cell membrane. The current produces a voltage drop across the membrane, driving V_m to –65 mV and removing the voltage difference between the two inputs. In this way, the membrane potential is kept equal to the command potential. If the circuitry is properly designed, the change in V_m is achieved within a few microseconds.

Now suppose that the command potential is stepped from –70 mV to –15 mV. We would expect the amplifier to deliver positive current to the axon to drive V_m to –15 mV, which is indeed what happens, but only transiently. Then something more interesting occurs. The depolarization to –15 mV produces an increase in sodium conductance and there is a consequent flow of sodium ions *inward*, across the membrane. In the absence of the clamp, this influx would tend to depolarize the membrane still further (i.e., toward the sodium equilibrium potential); with the clamp in place, however, the amplifier provides just the correct amount of negative current to hold the membrane potential constant. In other words, the current provided by the amplifier is exactly equal to the current flowing across the membrane. Here, then, is the great power of the voltage clamp: in addition to holding the membrane potential constant, it provides an exact measure of the membrane current required to do so. Voltage clamp measurements can now be made in small nerve cells by using the whole-cell method of patch clamp recording (see Chapter 4).

appear to bind to the same site in the outer mouth of the channel through which sodium ions move, thereby physically blocking ion current through the channel.[10]

Isolation of the currents underlying the action potential has also been aided by the discovery of a number of substances that block voltage-activated potassium channels. For example, in squid axons and in frog myelinated axons, Armstrong, Hille, and others have shown that voltage-activated potassium currents are blocked by tetraethylammonium (TEA), in concentrations greater than 10 m*M*, as shown in Figure 7.2C.[11] In squid axons, TEA must be added to the internal solution and exerts its action at the inner mouth of the potassium channel; in other preparations, such as the frog node of Ranvier, TEA is effective when applied to the outside as well. Other compounds, such as 4-aminopyridine (4-AP), block potassium currents as well, as do the peptide toxins apamin, dendrotoxin, and charybdotoxin.[12]

Dependence of Ion Currents on Membrane Potential

Having established that the early and late currents were due to sodium influx followed by potassium efflux, Hodgkin and Huxley then determined how the magnitude and time

[10] Hille, B. 1970. *Prog. Biophys. Mol. Biol.* 21: 1–32.

[11] Armstrong, C. M., and Hille, B. 1972. *J. Gen. Physiol.* 59: 388–400.

[12] Jenkinson, D. H. 2006. *Brit. J. Pharmacol.* 147: S63–S71.

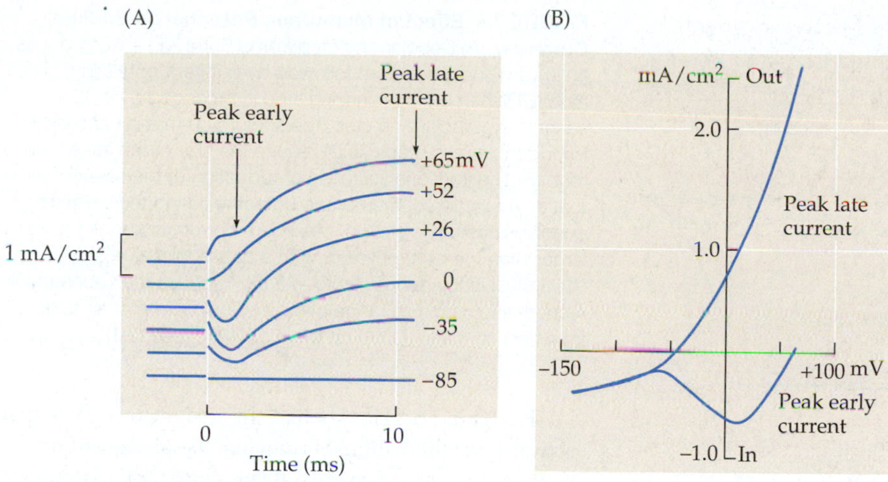

FIGURE 7.3 Dependence of Early and Late Currents on Potential. (A) Currents produced by voltage steps from a holding potential of –65 mV to a hyperpolarized level of –85 mV and then to successively increasing depolarized levels as indicated. The late potassium current increases as the depolarizing steps increase. The early sodium current first increases, then decreases with increasing depolarization, is absent at +52 mV and reversed in sign at +65 mV. (B) Peak currents plotted against the potential to which the membrane is stepped. Late outward current increases rapidly with depolarization. Early inward current first increases in magnitude, then decreases, reversing to outward current at about +52 mV, which is the sodium equilibrium potential. (After Hodgkin, Huxley, and Katz, 1952).

course of the currents depended on membrane potential. Currents produced by various levels of depolarization from a holding potential of –65 mV are shown in Figure 7.3A. First of all, a step hyperpolarization to –85 mV (lower record) produces only a small inward leak current, as would be expected from the resting properties of the membrane. As already shown in Figure 7.1, moderately depolarizing steps each produce an early inward current followed by a sustained outward current. With greater depolarizations, the early current becomes smaller, is absent at about +52 mV, and then reverses to become outward as the depolarizing step is increased still further to +65 mV.

The dependence of the two currents on membrane potential is shown in Figure 7.3B, in which the peak amplitude of the early current and the steady-state amplitude of the late current are plotted against the potential to which the membrane is stepped. With hyperpolarizing steps, there are no time-dependent (early and late) currents; the membrane simply responds as a passive resistor, with the expected inward current. The late current also behaves as one would expect of a resistor in the sense that depolarization produces outward current. However, as the depolarization is increased, the magnitude of the current becomes much greater than expected from the resting membrane properties. This is due to the activation of voltage-gated potassium channels. As the membrane potential becomes more positive, more and more potassium channels open, allowing additional current to flow through the membrane. The behavior of the early inward current is much more complex than that of the outward current. The early current first increases with increasing depolarization and then decreases, becoming zero at about +52 mV and then reversing in sign. The reversal potential is very near the equilibrium potential for sodium, as expected for a current carried by sodium ions.

The magnitude of the sodium current depends on the sodium conductance (g_{Na}) and on the driving force for sodium entry ($V_m - E_{Na}$). One might expect, therefore, that the peak inward current would *decrease* as the membrane potential step gets progressively closer to the sodium equilibrium potential and the driving force is reduced. However, between about –50 mV and +10 mV there is a marked increase in sodium conductance due to the activation of voltage-gated sodium channels, which greatly outweighs the decrease in driving force. Thus, the sodium current, $i_{Na} = g_{Na}(V_m - E_{Na})$, increases. In this voltage range, the current–voltage relation is said to have a region of negative slope conductance. Beyond about +10 mV there is no further increase in conductance and the current decreases toward zero as the step potential approaches the sodium equilibrium potential.

Inactivation of the Sodium Current

The time courses of the sodium and potassium currents are markedly different. The potassium current is much delayed relative to the onset of the sodium current, but once developed, it remains high throughout the duration of the step. The sodium current, on the other hand, rises much more rapidly but then decreases to zero, even though the membrane depolarization is maintained. This decline of the sodium current is called **inactivation**.

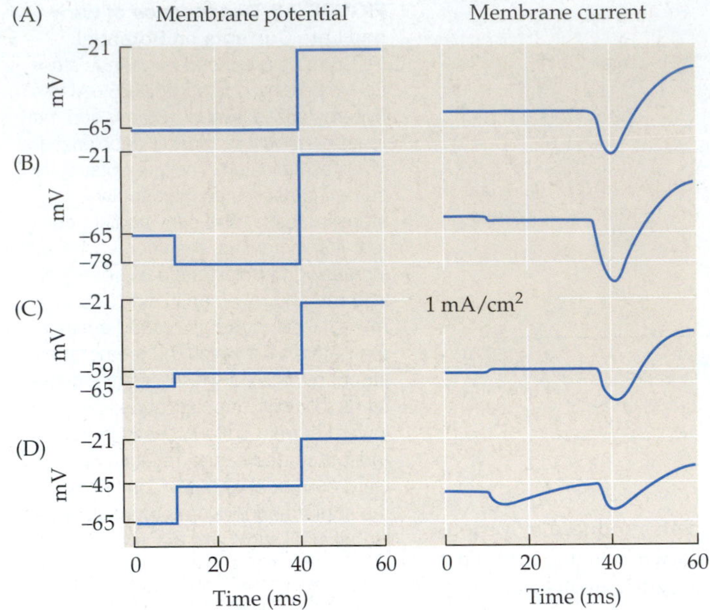

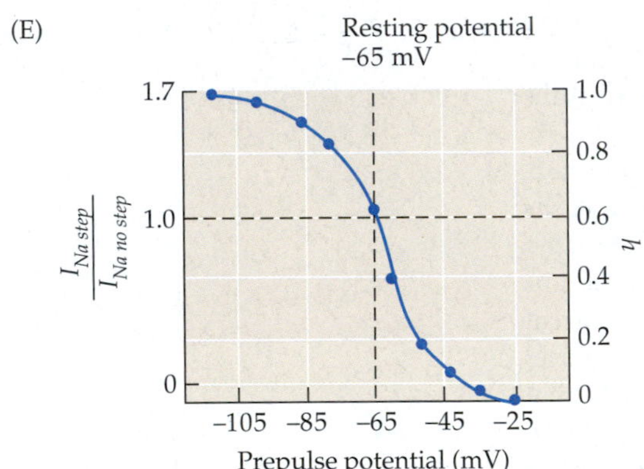

FIGURE 7.4 Effect of Membrane Potential on Sodium Currents. (A) Depolarizing step from –65 to –21 mV produces an inward sodium current followed by outward potassium current. (B) When the depolarizing step is preceded by a 30 millisecond hyperpolarizing step, the sodium current is increased. Prior depolarizing steps (C,D) reduce the size of the inward current. (E) The fractional increase or reduction of the sodium current as a function of membrane potential during the preceding conditioning step is shown. Maximum current is about 1.7 times larger than the control value, with a hyperpolarizing step to –105 mV. Depolarizing step to –25 mV reduces the subsequent response to zero. Full range of the sodium current is scaled from zero to unity (100%) on the *h* ordinate (see text).

Hodgkin and Huxley had the insight to realize that activation of the sodium current and its subsequent inactivation represented two separate processes and so designed experiments to study the nature of inactivation in detail. In particular, they investigated the effect of hyperpolarizing and depolarizing prepulses on the peak amplitude of the sodium current produced by a subsequent depolarizing step.

Records from such an experiment are shown in Figure 7.4. In Figure 7.4A, the membrane is stepped from a holding potential of –65 mV to –21 mV, producing a peak sodium current of about 1 milliampere/centimeter^2 (mA/cm^2). When the step is preceded by a hyperpolarizing prepulse to –78 mV, the peak sodium current is increased (Figure 7.4B). Depolarizing prepulses, on the other hand, cause a decrease in the sodium current (Figure 7.4C,D). The effects of hyperpolarizing and depolarizing prepulses are time-dependent; thus, brief pulses of only a few milliseconds (ms) duration have little effect. In the experiment shown here, the prepulses are of sufficient duration (30 ms) for the effects to reach their maximum. The results are shown quantitatively in Figure 7.4E, in which the peak sodium current is plotted against the potential during the prepulse. The peak current after a prepulse is expressed as a fraction of the control current. With a depolarizing prepulse to about –30 mV the subsequent sodium current was reduced to zero (i.e., inactivation was complete). Hyperpolarizing prepulses to –100 mV or beyond increased the sodium current by a maximum of about 70%. Hodgkin and Huxley represented this range of sodium currents from zero to their maximum value with a single parameter (h), varying from zero to 1, as indicated on the right-hand ordinate of Figure 7.4E. Note that zero means no activation (complete inactivation), and 1 means full activation (no inactivation). In these experiments, the peak sodium current reached only about 60% of its maximum value upon depolarization from the resting potential. Subsequent experiments have shown that all neurons show some degree of sodium channel inactivation at rest.

Sodium and Potassium Conductances as Functions of Potential

Having measured the magnitude and time course of sodium and potassium currents as a function of the membrane potential (V_m) and knowing the equilibrium potentials (E_{Na} and E_K), Hodgkin and Huxley were then able to deduce the magnitude and time courses of the sodium and potassium conductance changes, using the relations:

$$g_{Na} = \frac{I_{Na}}{(V_m - E_{Na})}$$

$$g_K = \frac{I_K}{(V_m - E_K)}$$

The results for five different voltage steps are shown in Figure 7.5A. Both g_{Na} and g_K increase progressively with increasing membrane depolarization. The relations between peak

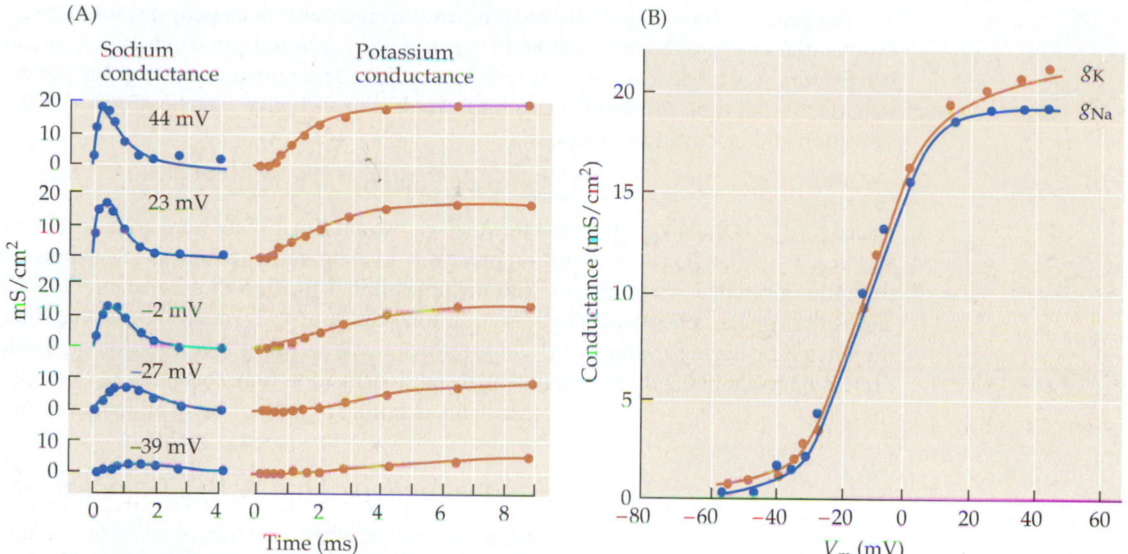

FIGURE 7.5 Sodium and Potassium Conductances. (A) Conductance changes produced by voltage steps from –65 mV to the indicated potentials. Peak sodium conductance and steady-state potassium conductance both increase with increasing depolarization. (B) Peak sodium conductance and steady-state potassium conductance plotted against the potential to which the membrane is stepped. Both increase steeply with depolarization between –20 and +10 mV. mS = millisiemens (After Hodgkin and Huxley, 1952a.)

conductance and membrane potential are shown for sodium and potassium in Figure 7.5B. The curves are remarkably similar, indicating that the two gating mechanisms have similar voltage sensitivities. In summary, the results obtained by Hodgkin and Huxley indicate that depolarization of the nerve membrane leads to three distinct processes: (1) activation of a sodium conductance mechanism; (2) subsequent inactivation of that mechanism; and (3) delayed activation of a potassium conductance mechanism.

Quantitative Description of Sodium and Potassium Conductances

After obtaining the experimental results, Hodgkin and Huxley proceeded to develop a mathematical description of the precise time courses of the sodium and potassium conductance changes produced by the depolarizing voltage steps. To deal first with the potassium conductance, one might imagine that the effect of a sudden change in membrane potential would be to cause the movement of one or more charges in a voltage-sensitive potassium channel that would then lead to channel opening. If a single process were involved, then the change in the overall potassium conductance might be expected to be governed by ordinary, first-order kinetics—that is, its rise after the onset of the voltage step would be exponential. Instead the onset of the conductance change was found to be S-shaped, with a marked delay. Because of this delay and because the potassium conductance increase occurred during depolarization but not hyperpolarization (i.e., it rectified), it was called the **delayed rectifier.** Hodgkin and Huxley were able to account for the S-shaped onset of the conductance by assuming that the opening of each potassium channel required the activation of four first-order processes, for example the movement of four charged particles in the membrane. In other words, the S-shaped time course of activation could be fitted by the product of four exponential distribution. The increase in potassium conductance for a given voltage step, then, was described by the relation:

$$g_K = g_{K(max)} n^4$$

where $g_{K(max)}$ is the maximum conductance reached for the particular voltage step and n is a rising exponential function varying between zero and unity, given by $n = 1 - e{-}(t/\tau_n)$.

The exponential time constant (τ_n) is also voltage dependent, with the increase in conductance becoming more rapid with larger depolarizing steps. At 10°C, τ_n ranges from about 4 ms for small depolarizations to 1 ms for depolarization to zero mV.

The time course of the rise in sodium conductance, also S-shaped, was fitted by an exponential raised to the third power. In contrast, the fall in sodium conductance due to inactivation was consistent with a simple exponential decay process. For a given voltage step, the overall time course of the sodium conductance change was the product of the activation and inactivation processes:

$$g_{Na} = g_{Na(max)}m^3h$$

where $g_{Na(max)}$ is the maximum level to which g_{Na} would rise if there were no inactivation and $m = 1 - e{-}(t/\tau_m)$. The inactivation process is a falling, rather than a rising, exponential and is given by $h = e{-}(t/\tau_h)$. As with the potassium activation time constant, both τ_m and τ_h are voltage dependent. The activation time constant τ_m is much shorter than that for potassium, having a value at 10°C of the order of 0.6 ms near the resting potential and decreasing to about 0.2 ms at zero potential. The inactivation time constant for τ_h, on the other hand, is similar to that for potassium activation.

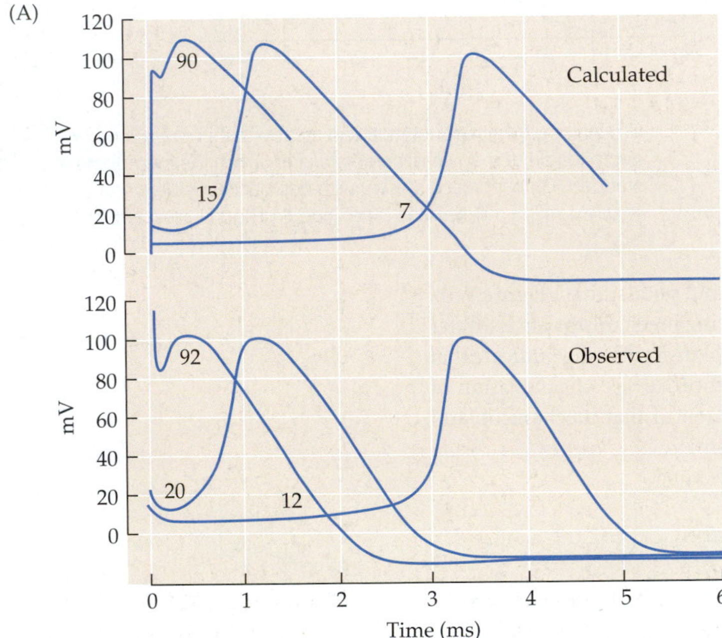

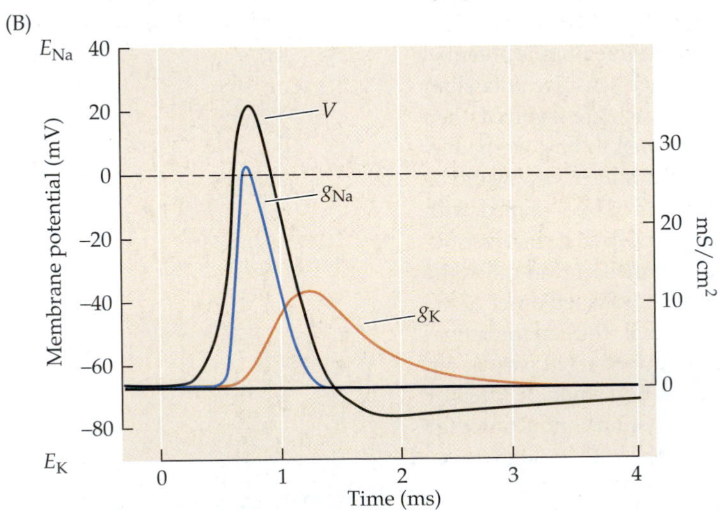

FIGURE 7.6 Reconstruction of the Action Potential. (A) Calculated action potentials produced by brief depolarizations of three different amplitudes (upper panel) are compared with those recorded under the same conditions (lower panel). (B) The relation between conductance changes (g_{Na} and g_K) and the action potential (V) are calculated for a propagated action potential in a squid axon. (After Hodgkin and Huxley, 1952d.)

Reconstruction of the Action Potential

Once the theoretical expressions were obtained for sodium and potassium conductances as a function of voltage and time, Hodgkin and Huxley were able to predict both the entire time course of the action potential and of the underlying conductance changes. Starting with a depolarizing step to just above threshold, they used their equations to calculate what the subsequent potential changes would be at successive intervals of 0.01 ms. Thus, during the first 0.01 ms after the membrane had been depolarized to say, –45 mV, they calculated how g_{Na} and g_K would change, what increments of I_{Na} and I_K would result, and then the effect of the net current on V_m. Knowing the change in V_m at the end of the first 0.01 ms, they then repeated the calculations for the next time increment, and so on, all through the rising and falling phases of the action potential. This was a laborious exercise in the days before computers or even electronic calculators; Huxley had to make do with a slide rule.

The calculations duplicated with remarkable accuracy the naturally occurring action potential in the squid axon. Calculated and observed action potentials produced by brief depolarizing pulses at three different stimulus strengths are compared in Figure 7.6A. In order to appreciate fully the magnitude of this accomplishment, it is necessary to keep in mind that the calculations used to duplicate the action potential were based on current measurements made under completely artificial conditions, with the membrane potential clamped first at one value and then at another. The mechanisms of action potential generation are summarized in Figure 7.6B, which shows the calculated magnitude and time course of a propagated action potential in a squid axon, together with the calculated changes in sodium and potassium conductance.

Threshold and Refractory Period

In addition to describing the action potential, Hodgkin and Huxley were able to use their results to explain, in terms of ionic conductance changes, many other properties of excitable axons, including the refractory period and the threshold for excitation (see Chapter 1). Further, their findings are applicable to a wide variety of other excitable tissues.

How do the findings predict the threshold membrane potential at which the impulse takes off? It would seem that a discontinuity, such as that which occurs at the **threshold for excitation**, would require a discontinuity in g_{Na} or g_K, yet both vary smoothly with membrane potential. The phenomenon can be understood if we imagine passing current through the membrane to depolarize it just to threshold and then turning the current off. Because the membrane is depolarized, there will be an increase in outward current over that at rest (through potassium and leak channels). We will also have activated some sodium channels, increasing the inward sodium current. At threshold, the inward and outward currents are exactly equal and opposite, just as they are at rest. However, there is an important difference: the balance of currents is now unstable. If a few extra sodium ions enter the cell the depolarization is increased, g_{Na} increases, and more sodium enters. The outward current can no longer keep up with the sodium influx, and the regenerative process explodes. If, on the other hand, a few extra potassium ions leave the cell, the depolarization is decreased, g_{Na} decreases rapidly, sodium current decreases, and the excess outward current causes repolarization. As the membrane potential approaches its resting level, the potassium current decreases until it again equals the resting inward sodium current. Depolarization above threshold results in an increase in g_{Na} sufficient for inward sodium movement to swamp outward potassium movement immediately. Subthreshold depolarization fails to increase g_{Na} sufficiently to override the resting potassium conductance.

And how is the refractory period explained? Two changes develop during an action potential that make it impossible for the nerve fiber to produce a second action potential immediately. First, inactivation of sodium channels is maximal during the falling phase of the action potential and requires several more milliseconds to be removed. During this time few, if any, channels are available to contribute to a new increase in g_{Na}. This results in an **absolute refractory period** that lasts throughout the falling phase of the action potential during which no amount of externally applied depolarization can initiate a second regenerative response. Second, because of activation of potassium channels, g_K is very large during the falling phase of the action potential and decreases slowly back to its resting level. Thus, a very large increase in g_{Na} is required to override the residual g_K and thereby initiate a regenerative depolarization. These mechanisms combine to produce a **relative refractory period**, during which the threshold gradually returns to normal as the potassium channels close and the sodium channels recover from inactivation.

It was an extraordinary achievement for Hodgkin and Huxley to have provided rigorous quantitative explanations for such complex biophysical properties of membranes. Their findings have since been shown to be generally applicable to action potential generation in both invertebrate and vertebrate neurons. However, the characteristics of action potentials vary greatly from one cell to another. For example, their durations can range from as little as 200 microseconds up to many milliseconds. Associated with these differences in duration is the rate at which neurons can fire repetitively. Some manage to generate only a few action potentials per second, while others can fire at frequencies approaching 1000/second. These differences are related to differences in the channel types underlying depolarization and repolarization.[13] For example, in central nervous system (CNS) neurons with brief spikes and high-frequency capability, the voltage-gated potassium channels associated with repolarization usually belong to the Kv3 family (see Chapter 5 for potassium channel classification). These channels have a very steep voltage dependence and activate and deactivate rapidly. They thereby promote rapid repolarization and, consequently, rapid removal of sodium channel inactivation.

Further differences from the squid axon relate to the fact that in most neurons, currents through voltage-activated sodium and potassium channels are not the only contributors to the action potential. For example, inward currents through voltage-activated calcium channels can enhance and prolong depolarization. At the same time, calcium entry into the cell can generate outward potassium currents through calcium-activated potassium channels and thereby shorten the action potential. Thus, the characteristics of any given action potential depend on which types of ion channels are present in the membrane and on how currents through these channels interact.

Although our knowledge of single-channel behavior has provided a new breadth to our understanding of the molecular mechanisms underlying the action potential, by no

[13]Bean, B. P. 2007. *Nat. Rev. Neurosci.* 8: 451–465.

stretch of the imagination would single-channel studies on their own—without the previous voltage clamp experiments and insights—have been able to account for how a nerve cell generates and conducts impulses. The older work has become enriched by, rather than supplanted by, the new.

Gating Currents

Hodgkin and Huxley suggested that sodium channel activation was associated with the translocation of charged structures or particles within the membrane. Such charge movements would be expected to add to the capacitative current produced by a depolarizing voltage step. After a number of technical difficulties were resolved, currents of the expected magnitude and time course were finally seen.[14,15] They are known as **gating currents.**

How is the gating current separated from the usual capacitative current expected with a step change in membrane potential (e.g., see Figure 7.1)? Briefly, currents associated merely with charging and discharging the membrane capacitance should be symmetrical. That is, they should be of the same magnitude for depolarizing steps as for hyperpolarizing steps. On the other hand, currents associated with sodium channel activation should appear upon a depolarization of 50 mV from a holding potential of –70 mV, but not upon 50 mV hyperpolarization. In other words, if the channels are already closed, there should be no gating current upon further hyperpolarization. Similarly, gating currents associated with channel closing might be expected at the termination of a brief depolarizing pulse but not after a hyperpolarizing pulse. One experimental way of recording gating currents is to sum the currents produced by two identical voltage steps of opposite polarity, as shown in Figure 7.7A. The asymmetry due to gating currents is shown in parts (a) and (b) of the figure. The current at the beginning of the depolarizing pulse is larger than that produced by the hyperpolarizing pulse because of the additional charge movement associated with gating of the sodium channel. When the two currents are added (c), the net result is the gating current. Gating currents are also known as **asymmetry currents.** An example of gating current in squid axon, obtained by cancellation of the capacitative current, is shown in Figure 7.7B. Voltage-sensitive potassium currents were blocked with TEA. A step depolarization of an internally perfused squid axon produced an outward gating current, followed by inward sodium current. The sodium current was much smaller than usual because extracellular sodium concentration was reduced to 20% of normal. In Figure 7.7C, the gating current is shown alone after tetrodotoxin was added to the solution to prevent sodium from entering

[14] Armstrong, C. M., and Bezanilla, F. 1974. *J. Gen. Physiol.* 63: 533–552.

[15] Keynes, R. D., and Rojas, E. 1974. *J. Physiol.* 239: 393–434.

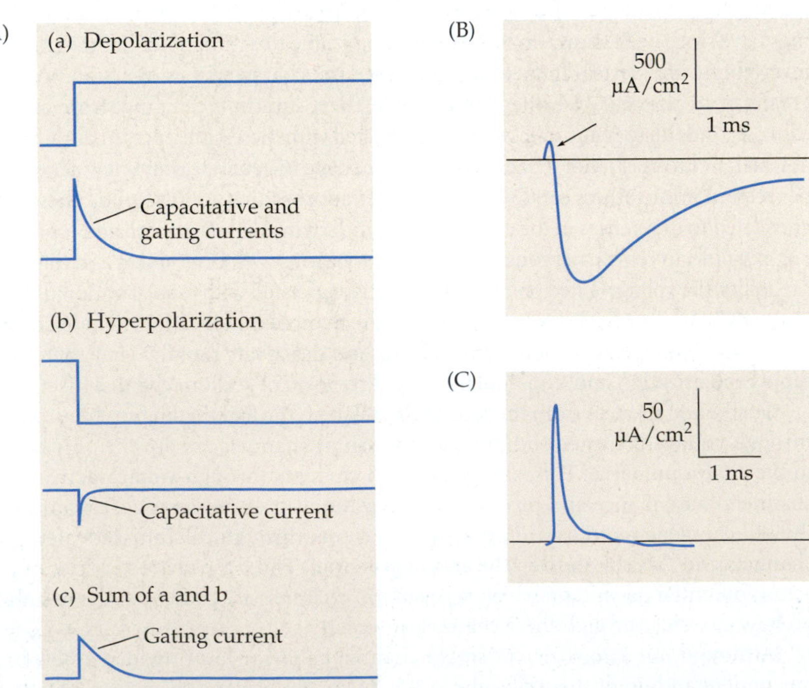

FIGURE 7.7 Sodium-Gating Current. (A) Method of separating gating current from capacitative current. Depolarizing pulse (a) produces a capacitative current in the membrane, plus a gating current. Hyperpolarizing pulse of the same amplitude (b) produces capacitative current only. When the responses to a hyperpolarizing and a depolarizing pulse are summed (c), capacitative currents cancel out and only the gating current remains. (B) Current record from a squid axon in response to depolarizing pulse, after cancellation of capacitative current. Inward sodium current is decreased by reducing extracellular sodium to 20% of normal. A small outward current (arrow) preceding the inward current is the sodium channel gating current. (C) Response to depolarization from same preparation after adding TTX to the bathing solution and recorded at higher amplification. Only gating current remains. (B and C after Armstrong and Bezanilla, 1977.)

the cell (note the change in scale). The evidence that asymmetry currents are, in fact, associated with sodium channel activation has been summarized by Armstrong.[16]

Mechanisms of Activation and Inactivation

It is well established that the gating currents are associated with translocation of positively charged residues in the S4 segments of the channel protein (see Chapter 5), and the total charge transfer has been estimated in *Shaker* potassium channels to be about 13 unit charges per channel or, if all four S4 segments participate, about $3e_0$ per segment.[17] Various schemes have been proposed in which the S4 helix responds to depolarization by translating and/or rotating within the membrane, thereby effecting a transfer of charge between the inner and outer aqueous compartments.[18] Examples include a paddle model, in which the helix moves within the membrane to provide an outward charge translocation of about 20 Å,[19] and a rotational model with charge translation over a relatively short distance.[20] An example of the rotational model is shown in Figure 7.8. When the membrane is at the resting potential, positive charges on the S4 helix are exposed to the cytoplasm by way of an aqueous crevice extending into the protein structure, and the cytoplasmic end of the channel is blocked by close apposition of the S6 helices (Figure 7.8A). Upon depolarization, the change in potential drives the charges on S4 into the extracellular region of the crevice, causing the helix to rotate (Figure 7.8B). The rotary movement is coupled to the S6 helix so as to open the pore.

The idea of a crevice or **gating pore** around S4 is supported by experiments on sodium and potassium channels in which the charged argenine residues on the outer segment of the S4 helix are replaced with much shorter residues, such as alanine or serine.[21–23] These substitutions give rise to cationic currents (known as **omega currents**) through the closed channels, presumably by way of the gating pore.

The identification of a channel structure associated with inactivation was made first on potassium A-channels from *Drosophila*. Experiments on this channel provided evidence that a particular intracellular string of amino acids is associated with inactivation, and revived the **ball-and-chain model** of sodium channel inactivation that had been proposed more than a decade earlier by Armstrong and Bezanilla.[24] In this model, a clump of amino acids (the ball) is tethered by a string of residues (the chain) to the main channel structure. Upon depolarization, the ball binds to a site in the inner vestibule of the channel, thereby blocking the pore (Figure 7.9). This model of inactivation was tested in potassium A-channels by examining the behavior of channels formed in oocytes from mutant subunits (recall that the A-channel is a tetramer rather than a single polypeptide). Mutations and deletions were made in the 80 or so amino acids between the amino terminus and the first (S1) membrane helix.[25] Channels formed by mutant subunits with deletion of residues 6–46 showed virtually no inactivation, suggesting that some or all of these residues were involved in the normal inactivation process. When a synthetic peptide matching the first 20 amino acids in the N-terminal chain was simply added to the solution bathing the cytoplasmic face of the membrane, inactivation was restored with a linear dose-dependence over the concentration range of 0 to 100 μ*M*.[26] This amazing observation provides unusually strong support for the idea that in potassium A-channel subunits, the first 20 or so amino acid residues constitute a blocking particle responsible for inactivation of the channel. Because it involves the N-terminal structure, this type of inactivation in potassium channels is often referred to as N-type inactivation.

(A) Closed channel

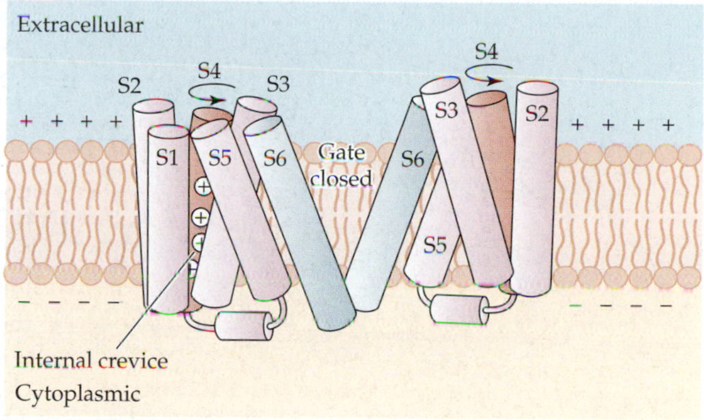

(B) Open channel

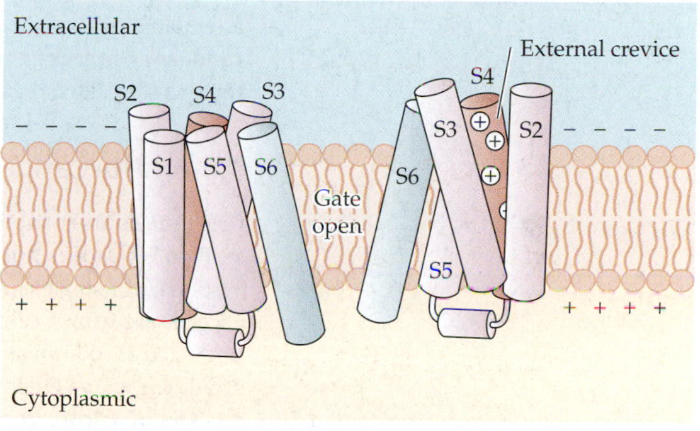

FIGURE 7.8 Proposed Scheme for Voltage Gating. Figure shows two apposed potassium channel subunits. Aqueous crevices connect S4 segments to the inner and outer solutions. (A) In the closed channel, positively charged residues on S4 are exposed to the intracellular solution via the internal crevice. S6 segments occlude the intracellular mouth of the channel. (B) Depolarization of the membrane causes rotation of S4, exposing the charges to the external crevice, and thus to the extracellular surface. S4 rotation interacts with the S6 segments to open the gate. (Based on Cha et al., 1999.)

[16] Armstrong, C. M. 1981. *Physiol. Rev.* 61: 644–683.

[17] Bezanilla, F. 2005. *IEEE Trans. Nanobiosci.* 4: 34–48.

[18] Bezanilla, F. 2008. *Nature Rev. Mol. Cell Biol.* 9: 323–332.

[19] Jiang, Y. et al. 2003. *Nature* 423: 42–84.

[20] Chanda, B. et al. 2005. *Nature* 436: 852–856.

[21] Tombola, F., Pathak, M. M., and Isacoff, E. E. 2005. *Neuron* 45: 379–388.

[22] Sokolov, S., Scheuer, T., and Catterall, W. A. 2005. *Neuron* 47: 183–189.

[23] Gamal El-Din, T. M. et al. 2010. *Channels* 4: 1–8.

[24] Armstrong, C. M., and Bezanilla, F. 1977. *J. Gen. Physiol.* 70: 567–590.

[25] Hoshi, T., Zagotta, W. N., and Aldrich, R. W. 1990. *Science* 250: 533–550.

[26] Zagotta, W. N., Hoshi, T. and Aldrich, R. W. 1990. *Science* 250: 568–571.

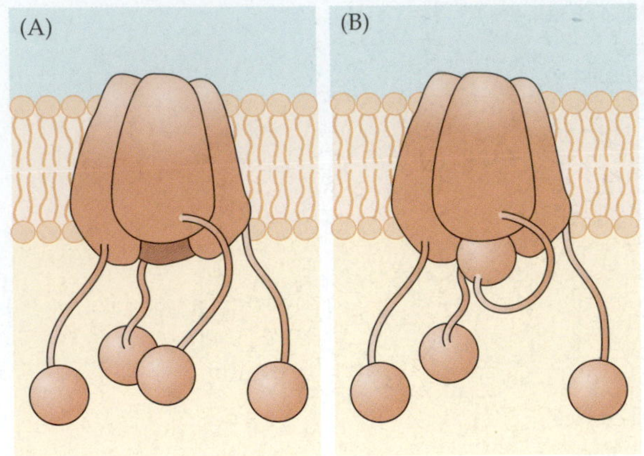

FIGURE 7.9 Ball-and-Chain Model of Inactivation of a Voltage-Activated Potassium Channel. One ball and chain element is chained to each of the four channel subunits. (A) Gating elements at the cytoplasmic end of the channel are open. (B) One of the four inactivation balls enters the open vestibule to block the open channel.

Some potassium channels also display a much slower C-type inactivation, originally suspected to involve the C-terminal but later found to be related to structures near the outer mouth of the pore.[27,28]

Experiments with sodium channels have identified the intracellular loop between domains III and IV as being closely involved in the inactivation process. The loop is about 45 residues in length, and is envisioned as a hairpin that swings into the inner vestibule of the channel to block the pore. In experiments with rat brain channels expressed in oocytes, three adjacent amino acid residues in the middle of the loop have been identified as essential for inactivation to occur.[29] When they were removed or replaced by site-directed mutagenesis, inactivation was severely attenuated or abolished. Similar experiments also identified groups of glycine and proline residues at either end of the loop involved in inactivation. These are assumed to be **hinge regions** that allow the hairpin to flip into the vestibule.[30]

Activation and Inactivation of Single Channels

The time course of activation and inactivation of the macroscopic sodium current shown in Figure 7.1 does not mirror the time course of current through a single channel. The responses of single channels to a depolarizing voltage step are shown in Figure 7.10. The records are from a cell-attached patch on a cultured rat muscle fiber that contained only a few active sodium channels.[31] To remove any resting inactivation of sodium channels, a steady command potential was applied to the electrode, hyperpolarizing the patch membrane to –100 mV or so. In a series of successive trials, a 40 mV depolarizing pulse was applied to the electrode for about 20 ms, as shown in Figure 7.10B trace (a). In about one-third of the trial traces in (b), no sodium channels were activated. In the remainder, one or more single channel currents appeared during the pulse, occurring most frequently near the onset of depolarization. The mean channel current was 1.6 picoamperes (pA). Assuming

[27]Hoshi, T., Zagotta, W. N., and Aldrich, R. W. 1991. *Neuron* 7: 547–556.

[28]Choi, K. L., Aldrich, R. W., and Yellen, G. 1991. *Proc. Natl. Acad. Sci. USA* 88: 5092–5095.

[29]Kellenberger, S. et al. 1997. *J. Gen. Physiol.* 109: 589–605.

[30]Kellenberger, S. et al. 1997. *J. Gen. Physiol.* 109: 607–617.

[31]Sigworth, F. J., and Neher, E. 1980. *Nature* 287: 447–449.

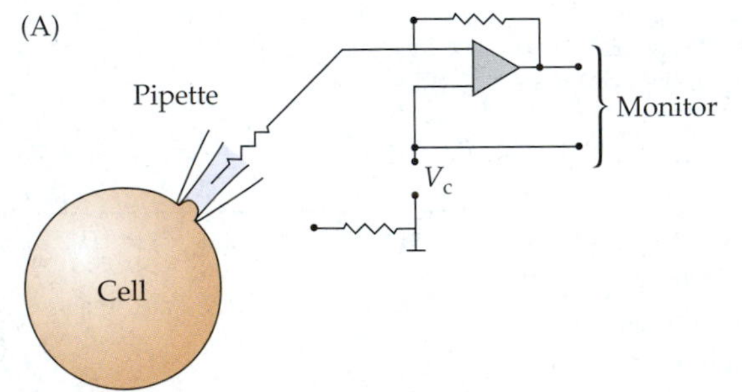

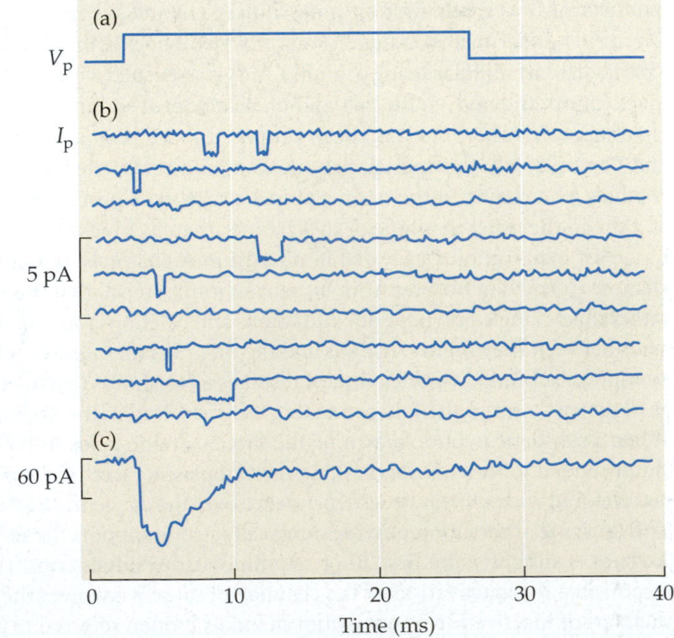

FIGURE 7.10 Sodium Channel Currents. Currents were recorded from cell-attached patch on cultured rat muscle cell. (A) Recording arrangement. (B) Repeated depolarizing voltage pulses applied to the patch, with waveform shown in (a), produce single-channel currents (downward deflections) in the nine successive records, as shown in (b). The sum of 300 such records (c) shows that channels open most often in the initial 1–2 millisecond after the onset of the pulse, after which the probability of channel opening declines with the time constant of inactivation. (After Sigworth and Neher, 1980.)

the sodium equilibrium potential to be +30 mV, the driving potential for sodium entry was about –90 mV; thus, the single-channel conductance was about 18 picosiemens (pS). This is comparable to sodium channel conductances measured in a variety of other cells. When 300 of the individual traces were added together (c), the summed current followed the same time course as that expected from the whole-cell sodium current.

A major point of interest in Figure 7.10 is that the mean channel open time (0.7 ms) is short relative to the overall time course of the summed current. Specifically, the time constant of decay of the summed current (about 4 ms) does not reflect the length of time that individual channels remain open. Instead, it indicates a slow decay in the probability of channel opening. The processes of activation (m^3) and inactivation (h) represent first an increase, and then a decrease in the *probability* that a channel will open for a brief period. Their product (m^3h) describes the time course of the overall probability change. The probability increases rapidly near the beginning of the pulse, reaches a peak, and then decreases with time. In any given trial an individual channel may open immediately after the onset of the pulse, at any subsequent time during the pulse, or not at all.

Afterpotentials

One of the characteristics of the action potential in Figure 7.6B is that it is followed by a transient **afterhyperpolarizing potential** (**AHP**) that persists for several milliseconds before the membrane potential returns to its resting level. The AHP occurs because delayed rectifier channels continue to open for a period that outlasts the action potential, and the resulting increase in potassium conductance drives the membrane toward the potassium equilibrium potential. As the channel openings gradually cease, the membrane potential returns to its resting level.

In addition to potassium channels associated with the delayed rectifier current, neurons have a number of potassium channel types whose activation contribute to the falling phase of the action potential and subsequent afterhyperolarization.[32] The most prominent of these are calcium-activated potassium channels (see Chapter 5). During the action potential, calcium ions enter the cell through voltage-activated calcium channels (see next section) and produce increases in potassium conductance that contribute to both the repolarization and the subsequent hyperpolarizing potentials. This effect is illustrated in Figure 7.11. In Figure 7.11A, an action potential produced in a frog spinal motoneuron by a brief depolarizing current pulse is followed by two separate phases of hyperpolarization. The first is due to the persistent activation of delayed rectifier channels, as seen in Figure 7.6. The second, slower phase is due to potassium efflux through calcium-activated potassium channels and disappears after the preparation has been soaked in Ca^{2+}-free solution (Figure 7.11B). The prolonged time course reflects the time taken for the cytoplasmic calcium concentration to return to its resting value. Short trains of action potentials have a cumulative effect on the size of the slow AHP, as successive impulses add additional increments to the intracellular calcium accumulation (Figure 7.11C).

The slow AHP is mediated primarily by potassium entry through so-called small K (SK) channels (see Chapter 5). Other potassium channels, known as big K (BK) are present in many neurons as well. These open very quickly and contribute to the rapid termination of the action potential in concert with delayed rectifier channels.

Potassium ion fluxes associated with repolarization and hyperpolarization play an important role in regulating the frequency of ongoing action potential activity. An example is shown in Figure 7.11D,E. A steady current pulse applied to a vagal motoneuron depolarizes the membrane and produces a train of action potentials that gradually decreases in frequency and finally stops (D). Although not apparent in the record, the frequency adaptation and eventual cessation of activity occur because the depolarization produced by the applied current declines from its initial value, finally falling below threshold for excitation. This decrease in magnitude of the depolarization is because of a steady increase in potassium conductance produced by the ongoing increase in intracellular calcium concentration during the train. The increase in potassium conductance is apparent when the events are observed on a slower time scale—a large, slow AHP is seen to persist for several seconds after the end of the depolarizing pulse.

When the same procedure is repeated with calcium entry blocked, the frequency adaptation and the prolonged AHP both disappear (Figure 7.11E). The frequency of the

[32] Hille, B. 2001. *Ion Channels of Excitable Membranes*, 3rd ed. Sinauer Associates, Sunderland, MA. pp. 136–147.

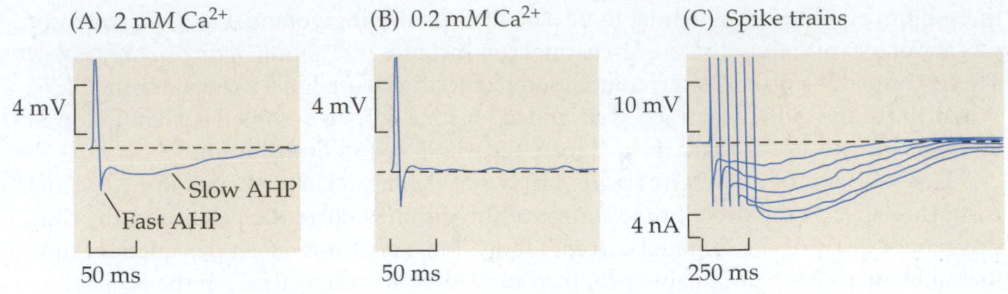

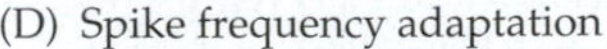

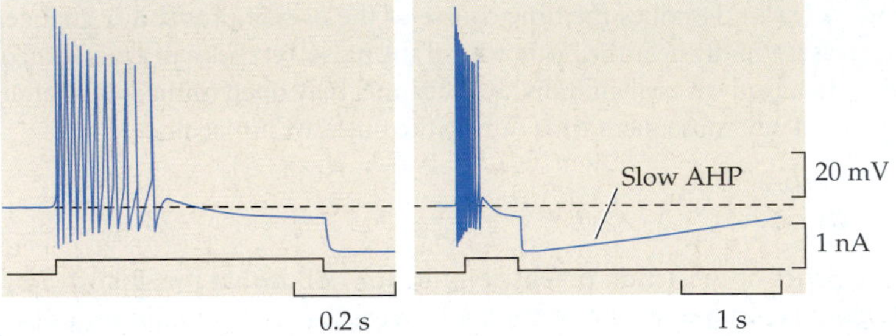

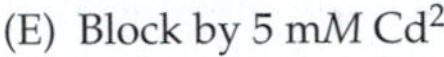

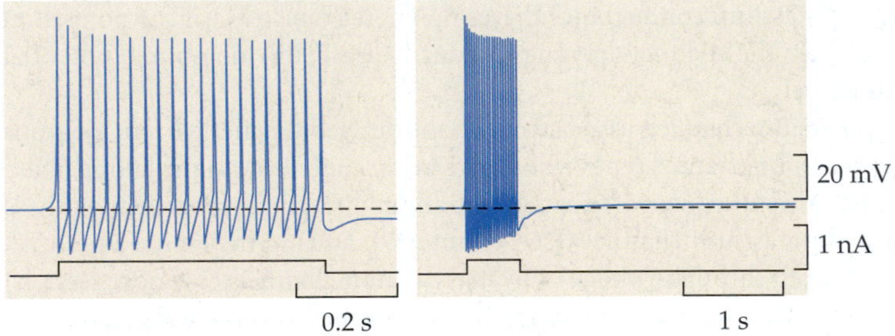

FIGURE 7.11 Hyperpolarizing Afterpotentials. (A) Action potential in a frog spinal motoneuron is followed by an afterhyperpolarization (AHP) with two phases: slow and fast. (B) After the preparation is soaked in low calcium bathing solution, the slow AHP disappears, suggesting that it depends on calcium influx during the action potential. (C) Superimposed records of trains of action potentials in a guinea pig vagal motoneuron. Size of the slow AHP increases as the number of successive action potentials in a train is increased from one to six. (D) A steady depolarizing current produces a train of action potentials that decreases in frequency (adapts) before dying out. Longer duration record reveals a large, slow AHP after the depolarizing current pulse is removed. (E) After block of voltage-sensitive calcium channels by the addition of cadmium to the bathing solution, the adaptation and slow AHP disappear. (A and B after Barrett and Barrett, 1976; B–E after Yarom, Sugimori, and Llinas, 1985.)

ongoing train of action potentials is now determined by the interaction between the early AHP and the depolarizing current pulse. After each action potential, the increased potassium conductance repolarizes the membrane below threshold for action potential initiation; as the potassium conductance returns toward its resting level, depolarization by the current pulse is restored and another action potential is initiated.

One additional factor can contribute to hyperpolarization and regulation of repetitive activity, namely the sodium–potassium exchange pump. The pump extrudes three sodium ions in exchange for two potassium ions and therefore contributes to the membrane potential (see Chapter 6). The rate of transport by the pump, and hence its contribution to the membrane potential, increases with increasing intracellular sodium concentration. Consequently, sodium accumulation during repetitive activity, particularly in small cells, can result in transient hyperpolarization that lasts until the excess sodium is extruded.[33,34]

Some neurons, for example cerebellar Purkinje cells, exhibit prolonged depolarization following the action potential.[35,36] This **afterdepolarizing potential** (**ADP**) is associated with activation of a TTX-sensitive sodium current (I_{NaP}), which has a low threshold of activation and very slow inactivation.[37]

Just as calcium-activated potassium currents determine the magnitude and time course of the AHP, they also participate in regulating the ADP. After a single action potential or a short train of action potentials, the slow sodium current competes with potassium currents for control of the membrane potential. This interaction is illustrated in Figure 7.12. With normal extracellular calcium concentration (2 m*M*) a brief depolarization applied to a rat hippocampal pyramidal cell produces a single action potential, followed by a prolonged ADP. When calcium influx during the action potential is decreased by reducing the external calcium concentration to 1.2 m*M* and then to zero, calcium influx and hence the calcium-activated potassium current is also reduced and the ADP becomes

[33] Jansen, J. K. S., and Nicholls, J. G. 1973. *J. Physiol.* 229: 635–655.

[34] Darbon, P. et al. 2003. *J. Neurophysiol.* 90: 3119–3129.

[35] Azouz, R., Jensen, M. S., and Yaari, Y. 1996. *J. Physiol.* 492: 211–223.

[36] Golomb, D., Yue, C, and Yaari, Y. 2006. *J. Neurophysiol.* 96: 1912–1926.

[37] French, C. R. et al. 1990. *J. Gen. Physiol.* 95: 1139–1157.

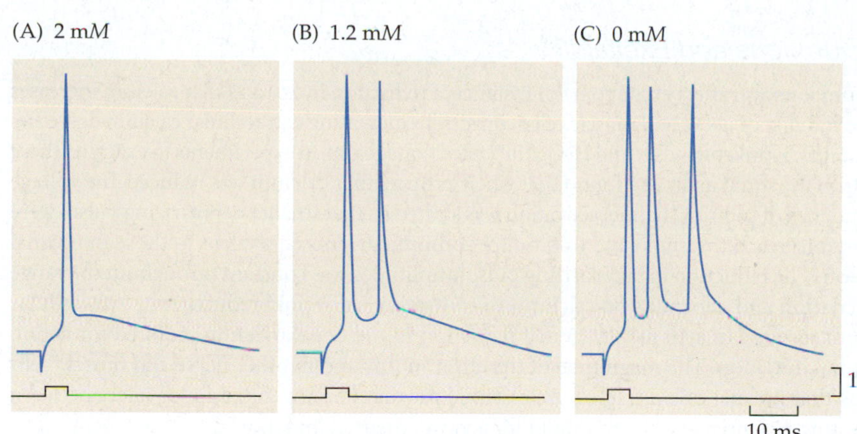

FIGURE 7.12 Depolarizing Afterpotentials. A single action potential in a rat hippocampal pyramidal cell, produced by a brief depolarizing pulse, is followed by a sustained afterdepolarization (ADP). When the external Ca^{2+} concentration is reduced from 2 m*M* to 1.2 m*M* the ADP becomes larger and initiates a second action potential. Reduction of external Ca^{2+} concentration to zero results in a further increase in the ADP. (After Golomb, Yue, and Yaari, 2006.)

larger. The increase in amplitude leads to a brief burst of action potentials in response to the single stimulus. Such bursting behavior, which is typical of a variety of cells in the central nervous system, is regulated by the interplay of sodium and potassium currents following the initial excitation.

There remains the question of how the afterdepolarizing potential is terminated, given the very slow inactivation of the sodium channels. Termination is accomplished by potassium current through voltage-activated M-channels, which activate and inactivate slowly. M-channels are open at the resting potential and further activated by depolarization (see Chapter 6). Depolarization by the prolonged sodium current increases M-channel activation, and the resulting increase in outward potassium current repolarizes the membrane until activation of the sodium channels is removed.[35]

The channel isotype responsible for the prolonged sodium current in CA3 cells has not been identified, but it is known that they contain an abundance of SCNA8 (Na_V6) channels.[38] Cerebellar Purkinje cells also exhibit both prolonged sodium currents that produce ADPs and burst responses to single stimuli. These characteristics are disrupted in Purkinje cells from mutant mice in which SCNA8 channels are not expressed.[39]

The Role of Calcium in Excitation

Calcium Action Potentials

The membranes of nerve and muscle fibers contain a variety of voltage-activated calcium channels (see Chapter 5 for calcium channel classifications and properties). Calcium ions enter the cell through these channels during the action potential, and this entry plays a key role in a variety of important processes. For example, a transient increase in intracellular calcium during the action potential is responsible for secretion of chemical transmitters by neurons and for contraction of muscle fibers.

In some muscle fibers and some neurons, calcium currents become sufficiently large to contribute significantly to, or even be solely responsible for, the rising phase of the action potential. Because g_{Ca} increases with depolarization, the process is a regenerative one, entirely analogous to that discussed for sodium. Calcium action potentials were first studied in invertebrate muscle fibers by Fatt and Ginsborg[40] and subsequently by Hagiwara.[41] Calcium action potentials occur in cardiac muscle in a wide variety of invertebrate neurons and neurons in the vertebrate autonomic and central nervous systems.[42] They also occur in immature neurons during development as well as in nonneural cells, such as endocrine cells and some invertebrate egg cells. The voltage-dependent calcium currents can be blocked by adding millimolar concentrations of cobalt, manganese, or cadmium to the extracellular bathing solution. Barium can substitute for calcium as the permeant ion, while magnesium cannot. A particularly striking example of the coexistence of sodium and calcium action potentials in the same cell is found in the mammalian cerebellar Purkinje cell, which generates sodium action potentials in its soma and calcium action potentials in the branches of its dendritic tree.[43,44]

[38] Schaller, K. L. et al. 1995. *J. Neurosci.* 15: 3231–3242.

[39] Raman, I. M. et al. 1997. *Neuron* 19: 881–891.

[40] Fatt, P., and Ginsborg, B. L. 1958. *J. Physiol.* 142: 516–543.

[41] Hagiwara, S., and Byerly, L. 1981. *Annu. Rev. Neurosci.* 4: 69–125.

[42] Hille, B. 2001. *Ion Channels of Excitable Membranes*, 3rd ed. Sinauer Associates, Sunderland, MA. pp. 95–98.

[43] Llinas, R., and Sugimori, M. 1980. *J. Physiol.* 305: 197–213.

[44] Ross, W. N., Lasser-Ross, N., and Werman, R. 1990. *Proc. R. Soc. Lond., B, Biol. Sci.* 240: 173–185.

Calcium Ions and Excitability

Calcium ions also affect excitation. For instance, a reduction in extracellular calcium increases the excitability of nerve and muscle cells; conversely, increasing extracellular calcium decreases excitability. Frankenhaeuser and Hodgkin[45] used voltage clamp experiments to examine these effects in the squid axon and found that when extracellular calcium was reduced, the voltage dependence of sodium channel activation was shifted so that smaller depolarizing pulses were required to reach threshold and to produce sodium currents equivalent to those in normal solution. The reduction in depolarizing pulse amplitudes was constant throughout the range of excitation and depended on calcium concentration. A fivefold reduction in extracellular calcium resulted in a 10 mV to 15 mV reduction in the depolarization required for action potential initiation. The magnitude of the effect in this and in other nerve and muscle cells means that normal calcium levels are essential for maintaining a margin of safety between the resting potential and the threshold for action potential initiation.

[45] Frankenhaeuser, B., and Hodgkin, A. L. 1957. *J. Physiol.* 137: 218–244.

SUMMARY

- The action potential in most nerve cell membranes is produced by a transient increase in sodium conductance that drives the membrane potential toward the sodium equilibrium potential, which is then followed by an increase in potassium conductance that returns the membrane potential to its resting level.
- The increases in conductance occur because sodium and potassium channels in the membrane are voltage dependent; that is, their opening probability increases with depolarization.
- Voltage clamp experiments on squid axons have provided detailed information about the voltage dependence and time course of the conductance changes. When the cell membrane is depolarized, the sodium conductance is activated rapidly and then inactivated. Potassium conductance is activated with a delay and remains high as long as the depolarization is maintained.
- The time and voltage dependence of the sodium and potassium conductance changes account precisely for the amplitude and time course of the action potential as well as for other phenomena, such as activation threshold and refractory period.
- Activation of sodium and potassium conductances by depolarization requires, in theory, charge movements within the membrane. Appropriate charge movements, called gating currents, have been measured.
- Calcium plays an important role in excitation. In some cells calcium influx, rather than sodium influx, is responsible for the rising phase of the action potential. Extracellular calcium also controls membrane excitability, which increases with decreasing calcium concentration.

Suggested Reading

General Reviews

Armstrong, C. M., and Hille, B. 1998. Voltage-gated ion channels and electrical excitability. *Neuron* 20: 371–380.

Bean, B. P. 2007. The action potential in mammalian central neurons. *Nat. Rev. Neurosci.* 8: 451–465.

Hille, B. 2001. *Ion Channels of Excitable Membranes,* 3rd ed. Sinauer Associates, Sunderland, MA. Chapters 2–5.

Original Papers

Frankenhaeuser, B., and Hodgkin, A. L. 1957. The action of calcium on the electrical properties of squid axons. *J. Physiol.* 137: 218–244.

Hodgkin, A. L., and Huxley, A. F. 1952. Currents carried by sodium and potassium ion through the membrane of the giant axon of *Loligo*. *J. Physiol.* 116: 449–472.

Hodgkin, A. L., and Huxley, A. F. 1952. The components of the membrane conductance in the giant axon of *Loligo*. *J. Physiol.* 116: 473–496.

Hodgkin, A. L., and Huxley, A. F. 1952. The dual effect of membrane potential on sodium conductance in the giant axon of *Loligo*. *J. Physiol.* 116: 497–506.

Hodgkin, A. L., and Huxley, A. F. 1952. A quantitative description of membrane current and its application to conduction and excitation in nerve. *J. Physiol.* 117: 500–544.

Hodgkin, A. L., Huxley, A. F., and Katz, B. 1952. Measurement of current-voltage relations in the membrane of the giant axon of *Loligo*. *J. Physiol.* 116: 424–448.

CHAPTER 8
Electrical Signaling in Neurons

In order to understand how signals are transmitted from one place to another in the nervous system it is necessary to know how they are generated in individual neurons and how they spread from one region of the cell to the next. Small depolarizing or hyperpolarizing signals spread passively along a nerve axon or dendrite, decreasing in amplitude over a short distance. The amplitude attenuation depends on a number of factors, principally the diameter and membrane properties of the fiber. Signals spread farther along a fiber with a large diameter and high membrane resistance. The electrical capacitance of the membrane influences both the time course and spatial spread of the electrical signal. Larger depolarizations give rise to action potentials. Once initiated, action potentials are regenerative and self-propagating, so they travel from their point of origin over the full length of the nerve process with no loss of amplitude.

The axons of many vertebrate nerve cells are covered by a high-resistance, low-capacitance myelin sheath. This sheath acts as an effective insulator and forces currents associated with the nerve impulse to flow through the membrane where the sheath is interrupted (nodes of Ranvier). The impulse jumps from one such node to the next and thereby its conduction velocity is increased. In addition, electrical activity can pass between neurons through specialized regions of close membrane apposition called gap junctions. Intercellular channels, called connexons, provide pathways for current flow through gap junctions.

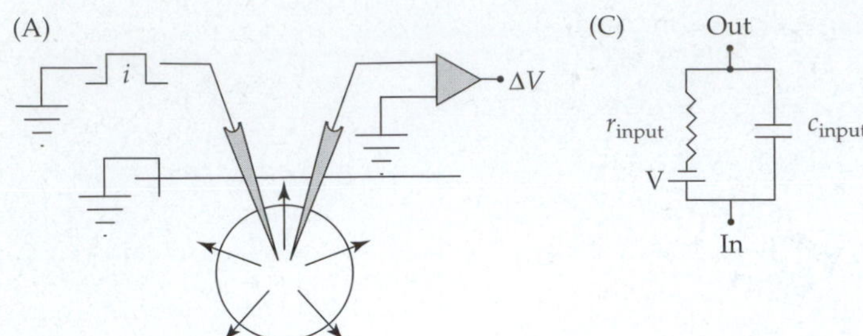

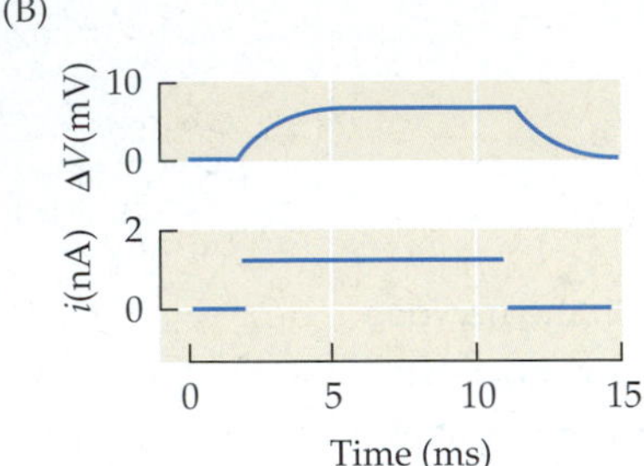

FIGURE 8.1 Response to Current Injection recorded from a spherical cell immersed in bathing solution. (A) Two electrodes are inserted into the cell, one to pass current and the other to record the resulting voltage change. Positive current injected into the cytoplasm flows outward across the cell membrane. (B) A pulse of positive current (i) produces a depolarization (ΔV) that rises gradually to its final value and then decays after the current is terminated. (C) An electrical model of the cell. r_{input} = the resistance of the cell membrane; c_{input} = membrane capacitance; V = resting membrane potential.

In this chapter we consider how electrical signals are transmitted along the processes of neurons. Small depolarizing or hyperpolarizing signals, generated locally in an axon or dendrite, die out over a relatively short distance. These signals occur in sensory receptors and in postsynaptic regions of neurons. If local depolarizations are sufficiently large, they produce action potentials that rise transiently to many times the amplitude of the initial signal. Action potentials are self-propagating and spread from their point of origin throughout the entire neuronal process without loss of amplitude. The spread of both local potentials and action potentials depend on the passive electrical properties of the cell.

We begin by considering current flow across the membrane of a simple spherical cell. In the experiment illustrated in Figure 8.1A, two microelectrodes are inserted into the cell, one to pass current across the membrane and the other to record the resulting change in membrane potential. A positive current pulse produces a depolarization that rises gradually to its final value and then dissipates after termination of the pulse (Figure 8.1B). Two questions arise from this kind of experiment: (1) What determines the size of the potential change? (2) What determines the rate of rise and fall of the potentials at the beginning and end of the current?

To answer these questions, it is useful to refer to the electrical model of the cell shown in Figure 8.1C. The **input resistance** of the cell, r_{input}, represents the pathway for current flow across the cell membrane. Input resistance is the inverse of the input conductance of the cell ($r_{input} = 1/g_{input}$), which in turn is made up of the sum of the membrane conductances to sodium, potassium, chloride, and other ions, lumped into a single pathway. The capacitor (c_{input}) represents the **input capacitance** of the cell, and is the electrical equivalent of the charged layers on the extracellular and intracellular surfaces of the lipid membrane. Its capacitance is determined by the electrical insulating properties and thickness of the membrane lipid. The battery, V, represents the resting membrane potential.

In considering the size of the potential change, it is useful to recall Ohm's law: a given amount of current, i, passed through a resistor, r, produces a voltage $v = ir$ (see Appendix A). In the case of our spherical cell, the input resistance of the cell determines the maximum amplitude finally reached during the voltage change (ΔV_{max}). So, we can write:

$$\Delta V_{max} = ir_{input}$$

The rise of the change in potential is slowed because the cell membrane has an input capacitance. To understand its effect, it is necessary to recall that the charge on the capacitor, q, varies with potential, V, according to the relation $q = cV$ (see Appendix A). This means that a membrane potential change must be accompanied by a corresponding change in the charge stored in the membrane capacitance. At the beginning of the current pulse, only a fraction of the current flows across the input resistance to produce a change in potential. The remainder is diverted to the capacitor in order to alter its charge. Initially, the rate of change of potential is relatively rapid and a capacitative current is relatively large. At the end of the pulse the potential change has reached its final value, the capacitor is fully charged to the new level, and all of the current flows through the resistor. After the current pulse is terminated the potential returns slowly to its original level as current drains off the capacitor through the resistor.

The rise of the membrane potential change follows an exponential time course, its time dependence being described by the relation $\Delta V_m = \Delta V_{max}(1 - e^{-t/\tau})$, where τ is the **input time constant** of the membrane, and the time, t, is measured from the onset of the pulse.

Similarly, when the current pulse is terminated, the potential change decays exponentially to zero: $\Delta V_m = \Delta V_{max} e^{-t/\tau}$. The input time constant is given by the product of the input resistance and the input capacitance: $\tau = r_{input} c_{input}$.

Specific Electrical Properties of Cell Membranes

The input resistance and capacitance of the spherical cell depend not only on the resistive and capacitive properties of the cell membrane but also on the size of the cell. The **specific capacitance** of the membrane (C_m) is the capacitance of one square centimeter of membrane. The input capacitance of a spherical cell is given simply by the capacitance per unit area times the surface area of the cell: $c_{input} = C_m (4\pi r^2)$. The larger the cell, the greater the input capacitance. Typically, cell membranes have capacitances of the order of 1 $\mu F/cm^2$ ($1 \mu F = 10^{-6}$ farads ; see Appendix A for definitions of electrical constants). This means that at a resting potential of –80 mV the amount of charge stored on the inner surface of the membrane ($q = C_m V$) is $(1 \times 10^{-6} F) \times (80 \times 10^{-3V}) = 8 \times 10^{-8}$ coulombs/cm^2, which is 5×10^{11} univalent ions (0.8 picomole [pmol]) for each square centimeter.

Similarly, the **specific resistance** of the membrane (R_m) is the resistance of one square centimeter of membrane. However, its relation to input resistance is reversed: In general, a large cell has more ionic channels in its membrane than a small cell, and therefore will have a lower resistance to the flow of ionic current. In other words, the input resistance of a cell is inversely related to membrane area. For a spherical cell: $r_{input} = R_m/(4\pi r^2)$. For this reason, R_m has the units Ωcm^2. Measurements of the specific resistance of a variety of neuronal membranes range from less than 1,000 Ωcm^2 for membranes with a large number of active ion channels to more than 50,000 Ωcm^2 for membranes with relatively few such channels. Suppose we select a value of 2000 Ωcm^2, which is equivalent to 5×10^8 pS/cm^2; then, if the average open membrane channel has a conductance of 50 picosiemens (pS), the number of open channels in the resting membrane at any given time would be 10^7/cm^2. The input time constant, being the product of the input resistance and the input capacitance, is independent of cell size and is equal to the time constant of the membrane: $\tau_m = R_m C_m$.

Flow of Current in a Nerve Fiber

The effects of injecting a pulse of current into a nerve fiber, such as an axon or a dendrite, are more complex than those seen in a simple spherical cell because the membrane is extended to enclose a cylinder and can no longer be represented by a single resistor and capacitor. In addition, because current injected into a fiber not only flows outward through the membrane but also spreads longitudinally along the core of the fiber, the response is affected by the electrical properties of the cytoplasm.

Figure 8.2A shows an experimental arrangement for measuring the response of a nerve fiber to current injection through a microelectrode. As with the spherical cell, a second electrode is used to measure the response, but this time it is withdrawn repeatedly and reinserted to measure the voltage change at increasing distances from the current-passing electrode. The results of such a set of measurements are shown in Figure 8.2B. The potential change near the point of current injection (0 mm from the current electrode) is relatively large and rises rapidly to its final value. As we move farther and farther from the current electrode, the potentials become smaller and slower. Potentials produced in this way are known as **electrotonic potentials.**

The final amplitudes of the electrotonic potentials are plotted against distance from the current-passing electrode in Figure 8.2C. The amplitudes fall exponentially with distance according to the relation:

$$\Delta V = \Delta V_0 e^{-x/\lambda}$$

where x is the distance from the current source and λ is the **length constant** of the fiber. The voltage change at zero separation is proportional to the size of the injected current:

$$\Delta V_0 - i r_{input}$$

where r_{input} is the input resistance of the fiber.

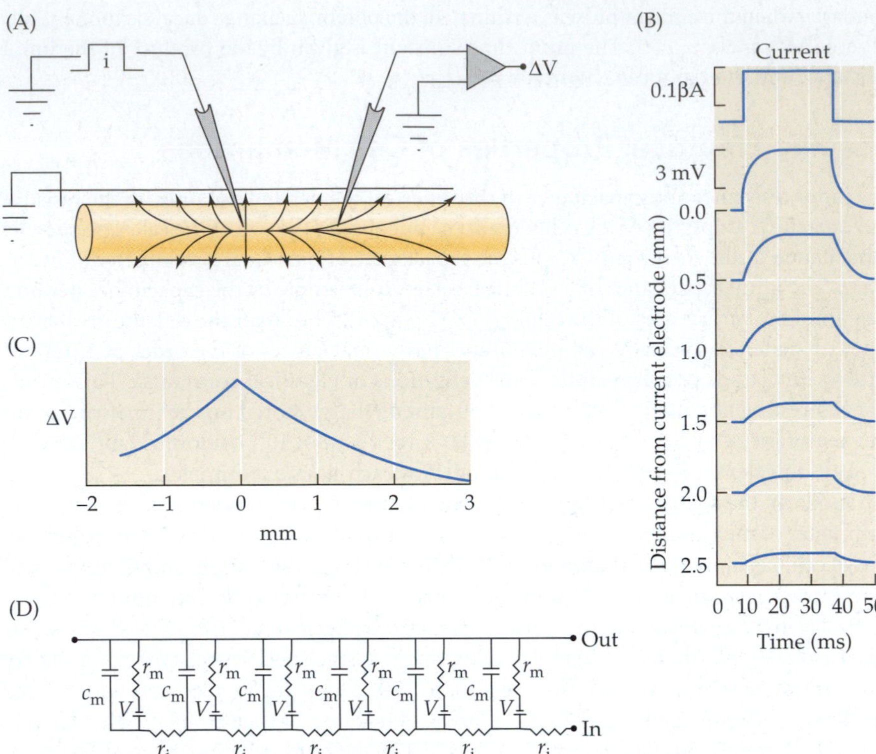

FIGURE 8.2 Response to Current Injection recorded from a nerve fiber immersed in bathing solution. (A) Two electrodes are inserted into the fiber—one to pass current and the other to record the resulting voltage change. Positive current injected into the cytoplasm flows longitudinally along the fiber and outward across the cell membrane. (B) Responses to a rectangular current pulse recorded at various distances from the current electrode. At zero electrode separation, the electrotonic potential rises rapidly to its maximum amplitude. As the recording electrode is moved to progressively increasing distances from the current source, the potentials become smaller and slower in time course. (C) Plot of response amplitude against distance from the current electrode. Decay of the response (Δ*V*) with distance is exponential. (D) An electrical model of the axon. V = membrane potential; r_m and c_m = resistance and capacitance per unit length of axon; r_i = internal longitudinal resistance per unit length of the cytoplasm. (Records in B based on Hodgkin and Rushton, 1946.)

In summary, two factors, r_{input} and λ, determine both the size of the response to current injection and how far the signal will spread along the fiber. To illustrate how these two factors depend on the properties of the fiber, it is again convenient to refer to an electrical model, shown in Figure 8.2D. The model is obtained by imagining that the fiber is cut along its length into a series of short cylinders. The membrane resistance of each cylindrical segment is represented by the resistor r_m, its capacitance by c_m, and the membrane potential by the battery V_m. The internal pathway for ion flow along the axoplasm from the midpoint of one cylinder to the next is represented by the resistor r_i. Because nerves in a recording chamber are normally bathed in a large volume of fluid, resistance along the outside of the cylinders is represented as being zero. This approximation is not always adequate in the central nervous system (CNS), where nerve axons, dendrites, and glial cells are closely packed, thereby causing pathways for extracellular current flow to be restricted. Any length can be selected for the cylinders themselves; however, by convention, the resistances r_m and r_i and the capacitance c_m are specified for a 1-cm length of axon. Their relations to the specific properties of the membrane and cytoplasm are given in Box 8.1. This type of analysis was first used to examine the electrical behavior of undersea cables, and the parameters r_m, c_m, and r_i are sometimes referred to as "cable properties."

BOX 8.1
Relation between Cable Constants and Specific Membrane Properties

The cable parameters r_m, c_m, and r_i refer to a 1-cm length of fiber. The first two parameters, r_m and c_m, depend on the specific resistance and capacitance of the fiber membrane and, in addition, on fiber diameter. The capacitance c_m has the units μF/cm. Its relation to the specific membrane capacitance is given by $c_m = C_m\,(2\pi a)$, with $2\pi a$ being the surface area of a unit length of fiber of radius a. The resistance r_m, which has the units Ωcm, is related to the specific membrane resistance by $r_m = R_m/(2\pi a)$.

The internal longitudinal resistance, r_i, is expressed as Ω/cm, and is related to the specific resistance of the cytoplasm, ρ, by $r_i = \rho/(\pi a^2)$. The specific resistance depends, in turn, on the intracellular ion concentrations and mobilities. The specific resistance of squid axoplasm is about 30 Ωcm at 20°C, or about 10^7 times that of copper. In mammals, in which the cytoplasmic ion concentration is lower, the specific resistance is about 125 Ω cm at 37°C; in frogs, with still lower ion concentration, the specific resistance is about 250 Ω cm at 20°C.

The relations given above can be used to deduce the dependence on fiber diameter of the input resistance and length constant. Thus, the input resistance is:

$$r_{input} = 0.5\left(r_m r_i\right)^{1/2} = \left(\frac{\rho R_m}{2\pi^2 a^3}\right)^{1/2}$$

and the length constant is:

$$\lambda = \left(\frac{r_m}{r_i}\right)^{1/2} = \left(\frac{aR_m}{2\rho}\right)^{1/2}$$

So, the input resistance decreases as fiber size increases, varying with the 3/2 power of the radius, and the length constant increases with the square root of the radius.

Current injected into the fiber flows spreads longitudinally along the fiber away from the tip of the electrode; as it does so, some of it is lost by ions moving laterally through the membrane. The distance the potential spreads depends on the relative ease with which ions carrying the current escape through the cell membrane, as compared with the ease of ion movement through the cytoplasm. If the membrane resistance is low relative to that of the cytoplasm, then current will leak outward through the membrane before it can spread very far. A higher resistance membrane, on the other hand, will allow a greater portion of the current to spread laterally before escaping to the external solution. So the length constant increases as membrane resistance increases, and decreases when the internal longitudinal resistance increases. The exact relation is given by:

$$\lambda = \left(\frac{r_m}{r_i}\right)^{1/2}$$

The input resistance depends on the resistance encountered by current flowing from the point of injection back into the extracellular fluid. Because the pathways for current flow include both r_i and r_m, both contribute to the input resistance:

$$r_{input} = 0.5\,(r_m r_i)^{1/2}$$

The factor 0.5 appears because the fiber extends in both directions from the point of current injection; each half has a resistance $(r_m r_i)^{1/2}$.

In general, large processes have greater length constants than small ones. This is because the membrane resistance depends on the surface area of the fiber and therefore varies inversely with fiber diameter, whereas the internal resistance depends on cross-sectional area and varies inversely with the square of the diameter. So as fiber diameter increases, r_i decreases more rapidly than r_m and the ratio r_m/r_i increases. Conversely, large processes have a lower input resistance than small processes because both the input resistance and longitudinal resistances are smaller (see Box 8.1). Other properties being equal, an excitatory synaptic potential (see Chapter 11) will be smaller in a large dendritic process (smaller r_{input}) than in a small one. On the other hand, in a larger dendrite the synaptic potential will spread farther toward the cell body (larger λ).

In axons and dendrites, the rates of rise and fall of electrotonic potentials depend on both the membrane time constant ($\tau = r_m c_m$) and the internal longitudinal resistance, but do not follow an exponential time course. At the onset of the pulse current flows rapidly into nearby membrane capacitors, but then slows as an increasingly larger fraction of the current is diverted to longitudinal flow through the cytoplasm. The potentials rise much more slowly at points distant from the current electrode than at points nearby because the internal longitudinal resistance reduces the current flowing into the more distant capacitors.

Action Potential Propagation

Once generated, action potentials typically propagate along the entire length of a nerve fiber without attenuation. This propagation depends on the passive spread of current ahead of the active region to depolarize the next segment of membrane to threshold. To illustrate the nature of the current flow involved in impulse generation and propagation, we can imagine the action potential frozen at an instant in time and plot its spatial distribution along a fiber, as shown in Figure 8.3. The distance occupied depends on the duration of the action potential and the rate of conduction along the fiber. For example, if the action potential duration is 2 milliseconds (ms) and if it is conducted at 10 meters per second (m/s), or 10 mm/ms, then the potential will be spread over a 20 mm length of axon (almost an inch). Near the leading edge of the action potential, the cell membrane is depolarized by a rapid influx of sodium ions along their concentration gradient. Just as when current is injected through a microelectrode, the inward current spreads longitudinally through the axoplasm. Current spread ahead of the active region causes a new segment of membrane to be polarized toward threshold. Behind the peak of the action potential the potassium conductance is high and current flows out through potassium channels, returning the membrane potential toward its resting level.

Normally, impulses arise at one end of an axon and travel to the other. However, there is no inherent directionality to propagation. Impulses produced in the middle of a muscle fiber at a neuromuscular junction travel away from the junction in both directions toward the tendons. However, except in unusual circumstances, an action potential cannot double back on itself, reversing its direction of propagation. This is because the peak depolarization is followed by a **refractory period** during which re-excitation cannot occur (see Chapter 7). During this period, sodium channels remain inactivated, and the potassium conductance is still high, so that even if sodium channels were active depolarization would be difficult. Because of the refractory period the active region of the membrane, as it moves along fiber, is followed by a refractory region that cannot be re-excited. As the membrane potential returns to its resting value, sodium channel inactivation is removed, potassium conductance returns to normal, and excitability recovers.

The rate of propagation of the action potential is influenced by both the space constant and time constant of the fiber. If the time constant is small, the membrane will depolarize to threshold quickly and the conduction velocity will be relatively high. If the space constant is large, the depolarization will spread a correspondingly large distance ahead of the active region, again speeding propagation of the signal. The result is that large fibers conduct more rapidly than small fibers because of their larger space constants. As already noted, the membrane time constant is independent of fiber size. Numerically, conduction velocity increases with the square root of fiber diameter.

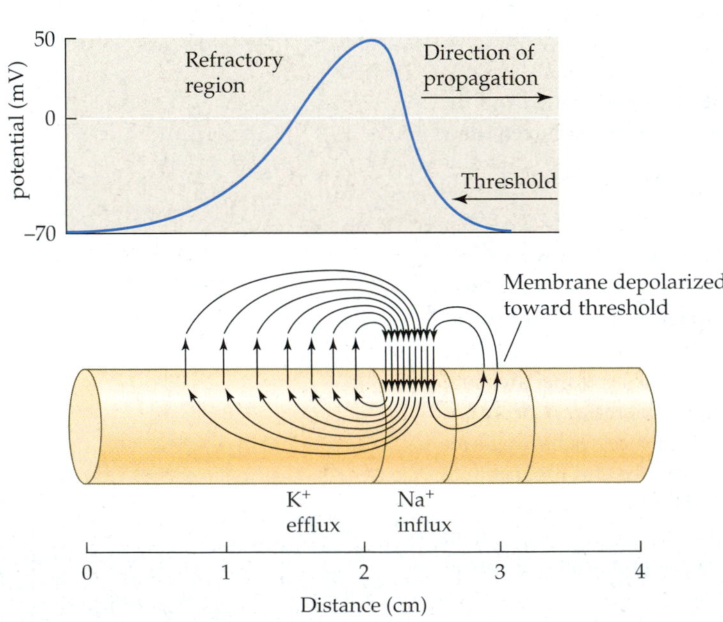

FIGURE 8.3 Current Flow during an Action Potential at an instant in time. Rapid depolarization on the rising phase of the action potential is due to influx of positively charged sodium ions. The positive current spreads ahead of the impulse to depolarize the adjacent segment of membrane toward threshold. Efflux of potassium ions behind the peak leads to repolarization.

Myelinated Nerves and Saltatory Conduction

In the vertebrate nervous system, the larger nerve fibers are myelinated. Myelin is formed in the periphery by Schwann cells and in the CNS by oligodendrocytes (see Chapter 10). The cells wrap themselves tightly around axons, and with each wrap the cytoplasm in between the membrane pair is squeezed out so that the result is a spiral of tightly packed membranes. The number of wrappings (lamellae) ranges from a low of between 10 and 20 to a maximum of about 160. In this way the effective membrane resistance is greatly increased and the membrane capacitance reduced. The myelin sheath usually occupies 20% to 40% of the overall diameter of the fiber and is interrupted periodically by nodes of Ranvier, exposing narrow patches of axonal membrane.[1] The internodal distance is usually about 100 times the external diameter of the fiber and ranges from 200 μm to 2 mm.

The effect of the myelin sheath is to restrict membrane current flow largely to the nodes, because ions cannot flow easily into or out of the high-resistance internodal region and the internodal capacitative currents are very small as well. As a result excitation jumps from node to node, thereby greatly increasing the conduction velocity. Such impulse propagation is called saltatory conduction (from Latin *saltare*, to jump, leap or dance). Saltatory conduction does not mean that the action potential occurs in only one node at time. While excitation is jumping from one node to the next on the leading edge of the action potential, many nodes behind are still active. Myelinated axons not only conduct more rapidly than unmyelinated ones but also are capable of firing at higher frequencies for more prolonged periods of time.

Conduction velocities of myelinated fibers vary from a few meters per second up to more than 100 m/s. The world speed record is held by myelinated axons of the shrimp, which conduct at speeds in excess of 200 m/s (447 miles/hour).[2] In the vertebrate nervous system, peripheral nerves have been classified into groups according to conduction, velocity, and function (Box 8.2). Theoretical calculations suggest that in myelinated fibers conduction velocity should be proportional to the diameter of the fiber. In mammals large myelinated fibers (>11 μm in diameter) have a conduction velocity in meters per second equal to

[1] Susuki, K., and Rasband, M. N. 2008. *Curr. Opin. Cell Biol.* 20: 616–623.

[2] Xu, K., and Terakawa, S. 1999. *J. Exp. Biol.* 202: 1979–1989.

BOX 8.2
Classification of Vertebrate Nerve Fibers

If we use external electrodes to stimulate all the fibers in a peripheral nerve at one end and then record its electrical response some distance away the record will have a series of peaks. The peaks occur because of dispersion of nerve impulses that travel at different velocities in different fibers, and therefore arrive at the recording electrode at different times after the stimulus. For example, a record taken from a rat sciatic nerve with 50 mm between the stimulating and recording electrode might look like the following (the rapid deflection at the beginning is an artifact due to current spread from the stimulating electrode):

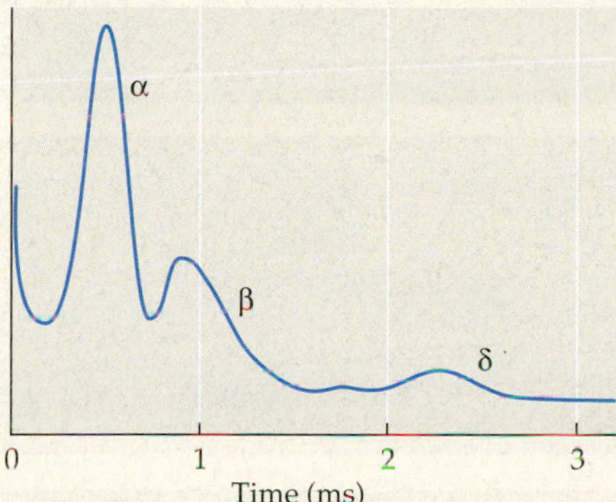

Vertebrate nerve fibers were classified into groups on the basis of differences in conduction velocity, combined with differences in function. Unfortunately, two such classifications arose independently. In the first system, group A refers to myelinated fibers in peripheral nerve; in mammals, these conduct at velocities ranging from 5 to 120 m/s. Group A fibers were further subdivided according to conduction velocity into α (80–120 m/s), β (30–80 m/s), and δ (5–30 m/s). These conduction velocity peaks are indicated in the record. The term γ fibers is reserved for motor nerves supplying muscle spindles (see Chapter 24), which have conduction velocities that span the β and lower part of the α range. Group B consists of myelinated fibers in the autonomic nervous system that have conduction velocities in the lower part of the A-fiber range. Group C refers to unmyelinated fibers, which conduct very slowly (less than 2 m/s).

The second nomenclature is used to classify to sensory fibers arising in muscle: Group I, corresponding to Aα, Group II to Aβ, and Group III to Aδ. Group I afferent fibers were further classified into two subgroups, depending on whether they conveyed information from muscle spindles (Ia) or from sensory receptors in tendons (Ib) (see Chapters 19 and 24).

Fiber size affects other electrical properties of nerves in addition to conduction velocity. When a nerve trunk is stimulated with external electrodes, large fibers are excited more readily than small ones (i.e., their excitation threshold is lower). This is fortunate for clinical purposes, as it allows testing of thresholds and conduction velocity in motor nerves, for example, without exciting much smaller pain fibers. Just as larger fibers are easier to excite, they are also harder to block, for example by cooling or by local anesthetic, which means that pain fibers can be blocked without interfering with conduction in larger motor and sensory fibers. One exception to this general rule is block by localized pressure for which large axons are affected first, then smaller ones, as the pressure is increased.

approximately six times their outside diameter in μm; for smaller fibers the constant of proportionality is about 4.5.[3]

Distribution of Channels in Myelinated Fibers

In myelinated fibers, voltage-sensitive sodium channels are highly concentrated in the nodes of Ranvier, with potassium channels more concentrated under the paranodal sheath.[4] The properties of the axon membrane in the paranodal regions normally covered by myelin were examined by Ritchie and his colleagues by loosening the myelin with enzyme treatment or osmotic shock.[5] Voltage clamp studies were then made of currents in the region of the node and compared with those obtained before the treatment. Previous experiments had suggested that, in rabbit nerve, nodes of Ranvier normally display only inward sodium current upon excitation. Repolarization after sodium channel inactivation appeared not to involve a transient increase in potassium conductance (as in other cells considered so far) but instead only passive current flow though a relatively large resting conductance. When the axon membrane adjacent to the nodes (the paranodal region) was exposed, excitation then produced a delayed outward potassium current, with no increase in inward current, indicating that the newly exposed membrane contained delayed rectifier channels but not sodium channels. Later immunocytochemical studies on rat myelinated nerve confirmed that delayed rectifier potassium channels (Kv1.1 and Kv1.2) were confined to the paranodal region (Figure 8.4A).[6,7]

In the meantime, further voltage clamp experiments revealed that the potassium current underlying repolarization at intact nodes had more complex features than expected of passive ion flux through resting channels—most notably a slowly inactivating component, labeled I_{Ks}.[8,9] A clue to the nature of the slowly inactivating current was provided by the observation that mutations that impair the function of the potassium channel subunit KCNQ2 cause neuronal hyperexcitability, manifested by epileptic seizures and by involuntary muscle contractions (myokymia caused by spontaneous discharges in peripheral motor axons).[10] KCNQ2 (Kv7.2) combines with other members of the KCNQ family to form the voltage-sensitive channels underlying the M-current (see Chapter 17). M-channels are activated by depolarization and inactivate very slowly. Their activation threshold is such

[3]Arbuthnott, E. R., Boyd, I. A., and Kalu, K. U. 1980. *J. Physiol.* 308: 125–157.

[4]Vabnick, I., and Shrager, P. 1998. *J. Neurobiol.* 37: 80–96.

[5]Chiu, S. Y., and Ritchie, J. M. 1981. *J. Physiol.* 313: 415–437.

[6]Wang, H. et al. 1993. *Nature* 365: 75–79.

[7]Rasband, M. N. et al. 1998. *J. Neurosci.* 18: 36–47.

[8]Dubois, J. M. 1983. *Prog. Biophys. Mol. Biol.* 42: 1–20.

[9]Roper, J., and Schwarz, J. R. 1989. *J. Physiol.* 416: 93–110.

[10]Dedek, K. et al. 2001. *Proc. Natl. Acad. Sci. USA* 98: 12272–12277.

(A)

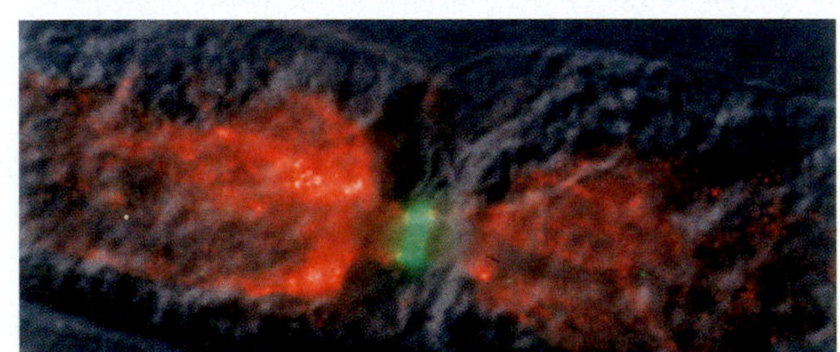

(B)

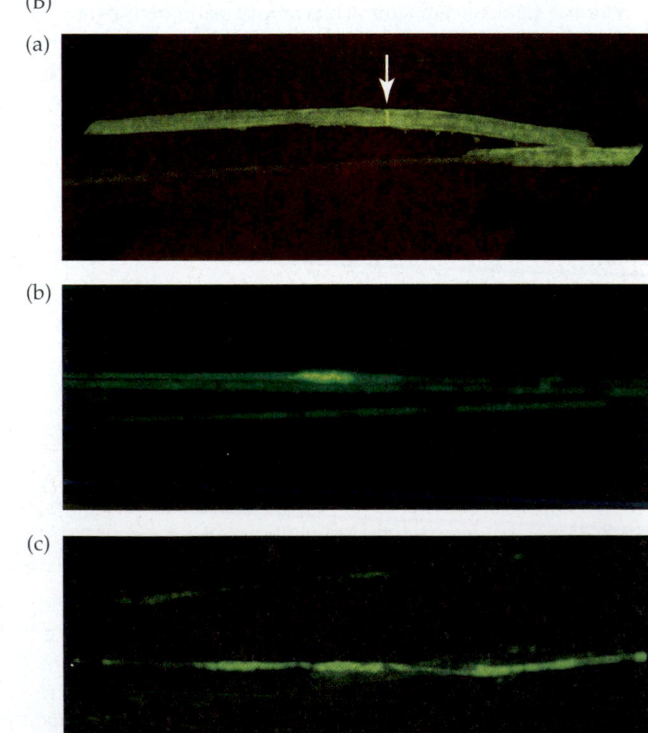

FIGURE 8.4 Distribution of Sodium and Potassium Channels in myelinated axons. (A) In rat sciatic nerve, sodium channels (green) are tightly clustered in the node of Ranvier and potassium channels (red) are sequestered in the paranodal region. Note the sharp decrease in axon diameter within the node. (B) Disruption of sodium channel distribution after demyelination of goldfish lateral line nerve. (a) In the myelinated axon, sodium channel staining (yellow–green) is restricted to the nodal region (arrow). (b) 14 days after the beginning of demyelination, sodium channels appear in irregular patches. (c) At 21 days, more patches have appeared and are distributed along the length of the nerve (A from Rasband et al., 1998, courtesy of Dr. P. Shrager; B from England, Levinson, and Shrager, 1996, courtesy of Dr. S. R. Levinson.)

that a fraction of the population is open at rest. Subsequent immunohistochemical studies confirmed that KCNQ2 subunits are expressed at nodes of Ranvier in rat sciatic nerve as well as at nodes and initial segments of myelinated CNS fibers.[11] In addition, voltage clamp studies on rat peripheral nerve fibers have shown that the I_{Ks} currents have all the biophysical and pharmacological properties of M-currents.[12] It is now apparent the M-channels are an important regulator of neuronal excitability throughout the nervous system. At nodes of Ranvier they serve to prevent spontaneous (ectopic) action potential discharges or repetitive discharges following invasion by a single action potential,[12] while at axon initial segments they regulate action potential threshold to suppress spontaneous firing.[13]

Mammalian axons that have been demyelinated chronically by exposure to diphtheria toxin can develop continuous conduction through a demyelinated region, suggesting that after demyelination, voltage-activated sodium channels appear in the exposed axon membrane.[14] Labeling of demyelinated nerves with antibodies to sodium channels shows that channels disappear from clusters in the former nodal regions and that new channels are distributed along previously myelinated regions (Figure 8.4B). Voltage-activated potassium channels are redistributed as well.[15] Upon remyelination there is restoration of normal clustering of sodium and potassium channels in newly formed nodes and paranodal regions.

Geometry and Conduction Block

A simple unmyelinated axon with uniform membrane properties is not representative of an entire neuron, with its cell body, elaborate dendritic arborization, and numerous axonal branches. The complex geometry of the neuron provides many possibilities for blocking action potential propagation. Specifically, propagation may fail wherever there is an abrupt expansion of membrane area. In such a situation the active membrane may not be able to provide enough current to depolarize the larger membrane area to threshold. For example, where an axon divides into two branches the safety factor for propagation is reduced because current from the active segment of the single axon must contribute sufficient current to depolarize both branches to threshold. Under normal circumstances an impulse will usually propagate into both branches, but after repeated firing propagation may become blocked at the branch point. Other factors contribute to such block. For instance, in leech sensory cells, block can occur because of persistent hyperpolarization induced by increased electrogenic activity of the sodium pump (see Chapter 9), and long lasting increases in potassium permeability, both of which increase the amount of current required for depolarization to threshold.[16,17]

In myelinated peripheral nerve the safety factor for conduction is about 5; that is, the depolarization produced at a node by excitation of a preceding node is approximately 5 times larger than necessary to reach threshold. Again, this safety factor is reduced where branching occurs. Similarly, when the myelin sheath terminates, for example near the end of a motor nerve, the current from the last node is then distributed over a large area of unmyelinated nerve terminal membrane and, as a consequence, provides less overall depolarization than would occur at a node. It is advantageous that the last few internodes before an unmyelinated terminal are shorter than normal, so that more nodes contribute to depolarization of the terminal.

Conduction in Dendrites

Apart from considerations of geometry, some regions of the neuron have a lower threshold for action potential initiation than others. This was first observed in spinal motoneurons by J. C. Eccles and his colleagues.[18] They found that upon depolarization action potentials were initiated first in the axon hillock, where the initial segment of the axon joins the cell body, and then propagated both outward along the axon and back into the soma and dendrites of the cell. At about the same time, Kuffler and Eyzaguirre found that depolarization of the dendrites in the crayfish stretch receptor initiated action potentials in or near the cell body, rather than in the dendrites themselves.[19] Observations of this kind led to the idea that dendrites were generally inexcitable and served only to transmit signals passively from dendritic synapses to the initial segment of the axon. This idea arose in spite of numerous observations to the contrary. For example, earlier extracellular recordings of electrical ac-

[11] Devaux, J. J. et al. 2004. *J. Neurosci.* 24: 1236–1244.

[12] Schwarz, J. R. et al. 2006. *J. Physiol.* 573: 17–34.

[13] Shah, M. M. et al. 2008. *Proc. Natl. Acad. Sci. USA* 22: 7869–7874.

[14] England, J. D. Levinson, S. R., and Shrager, P. 1996. *Microsc. Res. Tech.* 34: 445–451.

[15] Bostock, H., Sears, T. A., and Sherratt, R. M. 1981. *J. Physiol.* 313: 301–315.

[16] Gu, X. N., Macagno, E. R., and Muller, K. J. 1989. *J. Neurobiol.* 20: 422–434.

[17] Baccus, S. A. et al. 2000. *J. Neurophysiol.* 83: 1693–1700.

[18] Coombs, J. S., Eccles, J. C., and Fatt, P. 1955. *J. Physiol.* 130: 291–325.

[19] Kuffler, S. W., and Eyzaguirre, C. 1955. *J. Gen. Physiol.* 39: 87–119.

FIGURE 8.5 Spread of Action Potentials in Dendrites. (A) Records from a cerebellar Purkinje cell, obtained by impaling the cell at the indicated locations and passing a depolarizing current through the electrode. Near the end of the dendritic tree (a) depolarization produces long duration calcium action potentials. In the cell soma (d) a steady depolarizing current produced high-frequency sodium action potentials, interrupted periodically by calcium action potentials. At intermediate locations (b,c) depolarization produces calcium action potentials in the dendrite. Accompanying sodium action potentials generated in the soma spread passively into the dendritic tree and die out after a short distance. (B) Conduction in a cortical pyramidal cell. A cortical cell dendrite is depolarized by activating distal excitatory synapses. (a) Moderate depolarization of the dendrite spreads to the soma, where it initiates an action potential (blue record). In the dendrite, the initial depolarization is larger, and is followed by an action potential that spreads back from the soma (red record). (b) Larger depolarization produces a calcium action potential in the dendrite that precedes action potential initiation in the soma. (A after Llinás and Sugimori, 1980; B after Stuart, Schiller, and Sakmann, 1997.)

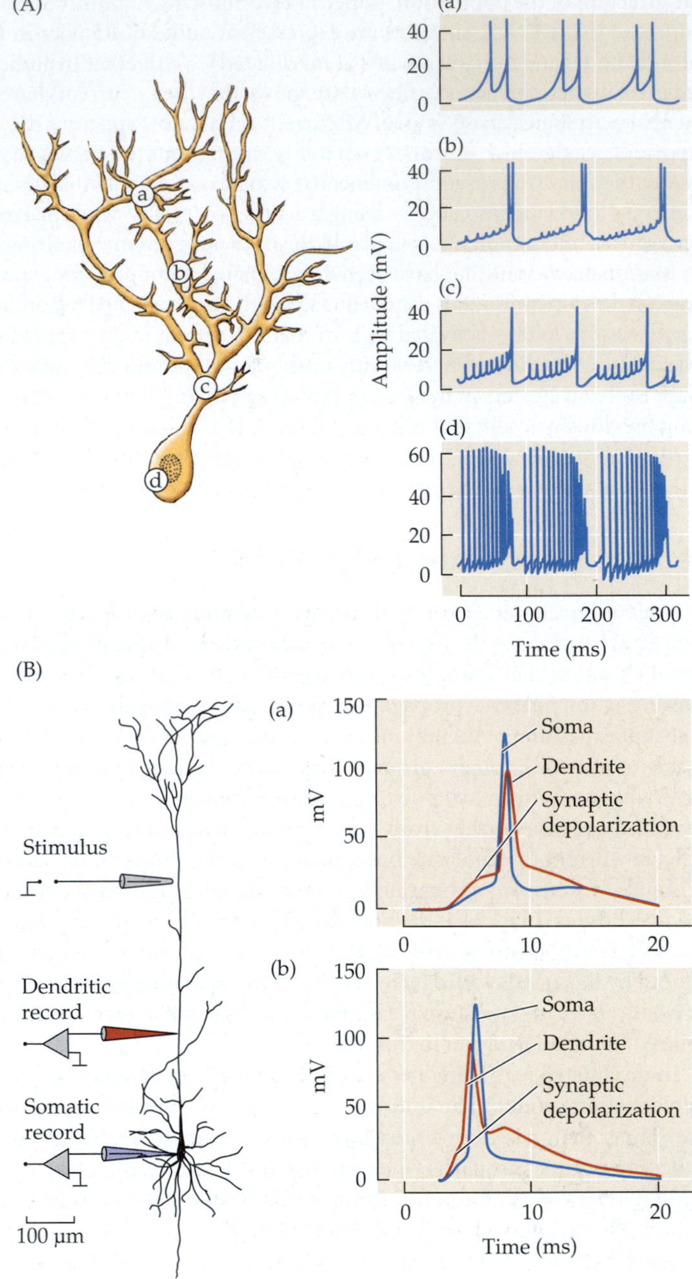

tivity within the mammalian motor cortex by Li and Jasper gave clear indication of action potentials traveling upward along pyramidal cell dendrites from their cell bodies to the cortical surface, with a conduction velocity of about 3 m/s.[20]

Dendritic action potentials are now known to occur in a variety of neurons, and are mediated by regenerative sodium and calcium currents. Cerebellar Purkinje cells, in addition to producing sodium action potentials in their somatic region, generate calcium action potentials in their dendrites.[21] As shown in Figure 8.5A, calcium action potentials generated in a dendrite spread effectively into the soma. Somatic action potentials, on the other hand, are not propagated into the dendrites but spread passively a short distance into the dendritic tree.

Like Purkinje cells, cortical pyramidal cells exhibit sodium action potentials in their somatic regions, usually arising in the initial segment of the axon. In addition, regenerative calcium potentials are observed in the distal dendrite.[22,23] Responses of a pyramidal cell to depolarization of the distal dendrite by activation of excitatory synapses (see Chapter 11) are shown in Figure 8.5B. Modest synaptic activation (a) produces dendritic depolarization

[20] Li, C.-L., and Jasper, H. H. 1953. *J. Physiol.* 121: 117–140.

[21] Llinás, R., and Sugimori, M. 1980. *J. Physiol.* 305: 197–213.

[22] Stuart, G., Schiller, J., and Sakmann, B. 1997. *J. Physiol.* 505: 617–632.

[23] Svoboda, K. et al. 1999. *Nat. Neurosci.* 2: 65–73.

that is attenuated as it spreads passively toward the soma. Upon arrival, the depolarization produces a somatic action potential that then spreads back into the dendrite. Stronger synaptic depolarization (b) results in direct activation of a dendritic calcium action potential that precedes the action potential generated in the soma.

Although there is now ample evidence of regenerative activity in dendrites, the general principle that the axon hillock is the most excitable region of the cell still holds for most types of neurons that have been examined.[24] Propagation of electrical signals in a dendrite is clearly more complex than in an axon.[25] Dendrites are likely to contain a variety of voltage-dependent channels in addition to those associated with the action potential, and the coexistence of action potentials and synaptic potentials in the dendritic tree add still more complexity. For example, the safety factor for backpropagation of action potentials will depend on the input resistance of the various branches; the input resistances, in turn, will depend on the extent of activity at excitatory and inhibitory synapses. Thus, whether or not backpropagation occurs depends on synaptic activity.[26–28] At the same time, synaptic channels that are voltage-dependent will behave differently from one moment to the next, depending on whether backpropagation has occurred.[29] These factors add new dimensions to signal processing that are only beginning be understood.

Pathways for Current Flow between Cells

In most circumstances an action potential arriving at a nerve terminal has little or no direct electrical influence on the next cell. Currents generated by the action potential flow preferentially through the low-resistance extracellular space. Contact between cells is usually limited to a small area of apposition, and the resistances of the small areas of membrane are far too high to allow any significant current flow between the cells. Certain cells, however, are **electrically coupled.** These include cardiac and smooth muscle cells, epithelial cells, gland cells, and a variety of neurons. Here we describe special intercellular structures that allow the processes of one neuron to be in electrical continuity with the next, thereby allowing current flow between them. The specific properties and functional role of electrical synapses are discussed in Chapter 11.

At sites of electrical coupling the intercellular current flows through gap junctions. The gap junction is a region of close apposition of two cells characterized by aggregates of particles distributed in corresponding arrays in each of the adjoining membranes (Figure 8.6A,B). Each particle, called a connexon, is comprised of six protein subunits arranged in a circle, about 10 nm in diameter, around a central core. Identical particles in the apposing cells are exactly paired to span the 3.5-nm gap in the region of contact.

About 20 connexon subunits (connexins) have been isolated, ranging in weight from 26 to 57 kilodaltons (kD).[30,31] Each was named according to its deduced weight; for example, Cx32 (32 kD), found in rat liver, Cx40, in heart muscle, and so on. Connexons can be assembled from one type of connexin (homomeric) or more than one (heteromeric), and an intercellular channel can be formed from two identical connexons (homotypic) or two different connexons (heterotypic). The primary amino acid sequence of the connexins indicates that the molecule is composed of four transmembrane helices, M1–M4, connected by one intracellular and two extracellular loops (Figure 8.6C). This structure has been confirmed by electron crystallographic studies on Cx43.[32] A study with high-resolution (3.5Å) crystallography has provided a detailed image of the molecular structure of the entire Cx26 intercellular channel.[33] The molecule has an outer diameter of 9 nm at the cytoplasmic ends, tapering to a minimum of 5 nm in the middle of the extracellular bridge. The open pore has a diameter of 3.5 nm at each entrance, narrowing to 1.4 nm at the outer faces of the membranes and widening to 2.5 nm in the intercellular region (Figure 8.6D). M1 is the major pore-lining helix. In the intercellular region, the E1 and E2 loops of each connexin form a tight, double-layered ring around the channel interior with six copies of the n-terminal half of E1 lining the pore. The two rings interdigitate in the center of the cleft with an overlap of about 6 nm.

Conductance and permeability of the connexons have been studied extensively, both as hemichannels expressed in individual host cells, such as *Xenopus* oocytes, and as complete gap junction channels expressed in paired cells.[23,24] Complete channels allow passage of molecules up to about 1 nm in diameter and 1,000 Daltons in molecular mass. However, molecular selectivity and conductance vary widely depending on subunit composition.[34]

[24] Stuart, G. et al. 1997. *Trends Neurosci.* 20: 125–131.

[25] Waters, J., Schaefer, A., and Sakmann, B. 2005. *Prog. Biophys. Mol. Biol.* 87: 145–170.

[26] Tsubokawa, H., and Ross, W. N. 1996. *J. Neurophysiol.* 76: 2896–2906.

[27] Sandler, V. M., and Ross, W. N. 1999. *J. Neurophysiol.* 81: 216–224.

[28] Larkum, M. E., Zhu, J. J., and Sakmann, B. 1999. *Nature* 398: 338–341.

[29] Markram, H. et al. 1997. *Science* 275: 213–215.

[30] Sosinsky, G. E., and Nicholson, B. J. 2005. *Biochim. Biophys. Acta* 1711: 99–125.

[31] Evans, W. H., and Martin, P. E. M. 2002. *Mol. Membr. Biol.* 19: 121–136.

[32] Fleishman, S. J. et al. 2004. *Mol. Cell* 15: 879–888.

[33] Maeda, S. et al. 2009. *Nature* 458: 597–604.

[34] Ma, M., and Dahl, G. 2006. *Biophys. J.* 90: 151–163.

(A)

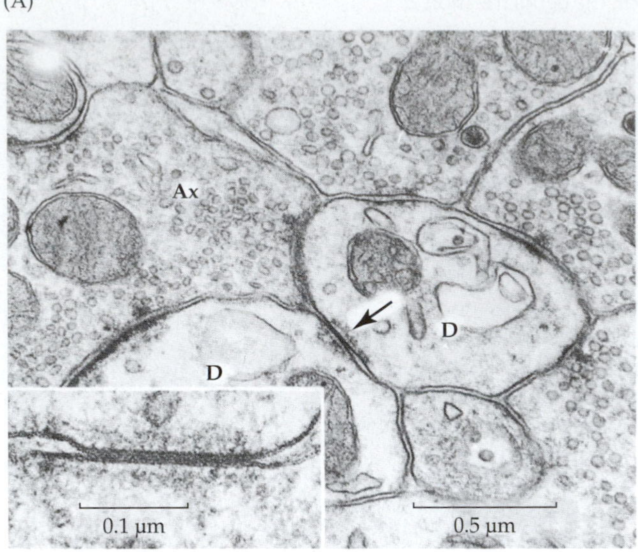

FIGURE 8.6 Gap Junctions between Neurons. (A) Two dendrites (labeled D) in the inferior olivary nucleus of the cat are joined by a gap junction (arrow), shown at higher magnification in the inset. The usual space between the cells is almost obliterated in the contact area, which is traversed by cross bridges. (B) Schematic view of a gap junction, showing close membrane apposition bridged by a group of paired connexons—each connexon is composed of six connexin subunits. (C) Connexin subunit consists of four membrane-spanning helices M1–M4. Amino acid loops E1 and E2 project into the extracellular space. (D) Cross section through a connexon pair, showing relative dimensions and based on high-resolution images of Cx26. Open channel is 3.5 nm in diameter at each entrance, narrowing to 1.4 nm at the extracellular membrane faces. Purple band represents the region of overlap of the extracellular loops. (A from Sotelo, Llinàs, and Baker, 1974; B after Purves et al., 2008; D based on Maeda et al., 2009.)

(B)

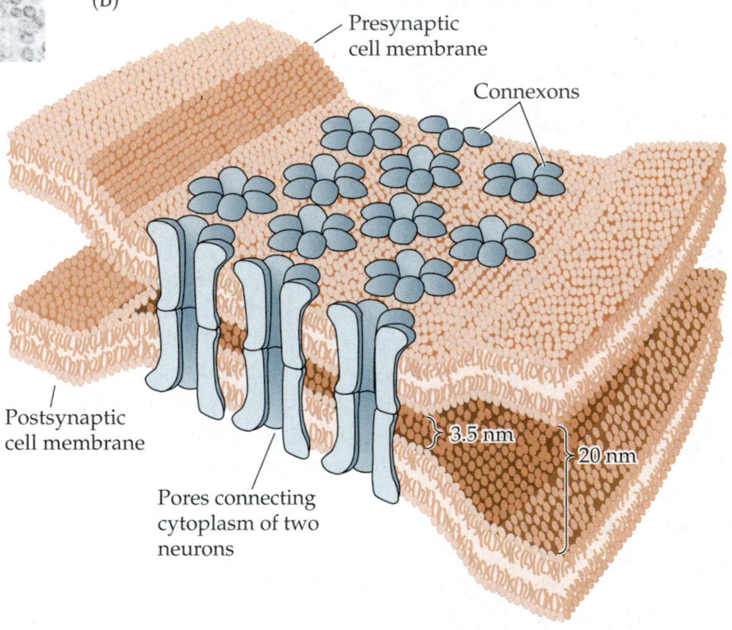

(C)

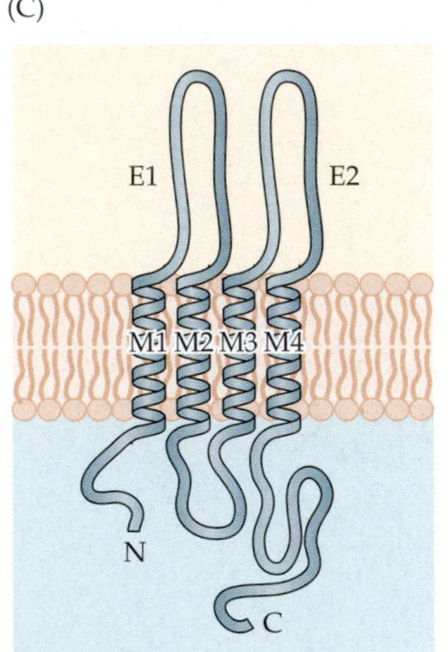

(D)

0 2 4 6 8 10
Diameter (nm)

For example, Cx43 channels are more than 100 times more permeable to adenosine triphosphate (ATP) and 40 times more permeable to glutamate than are Cx32 channels. Channel conductance is also highly variable, ranging from 15 pS for Cx36 channels to 300 pS for Cx43 in normal physiological solutions.

Even with the lowest single-channel conductance, individual gap junctions provide a significant pathway for current flow between cells because of the high packing density of the channels. For example, a tight junction 0.2 µm in diameter may contain more than 300 connexon pairs.[35] At 15 pS each, the total conductance through the junction would be 4.5 nanosiemens (nS). This is several thousand times greater than the conductance through two tightly apposed cell membranes with the same area and no connexons.

[35]Cantino, D., and Mugnani, E. 1975. *J. Neurocytol.* 4: 505–536.

SUMMARY

- The spread of local graded potentials in a neuron and the propagation of action potentials along a nerve fiber depend on the electrical properties of the cytoplasm and the cell membrane.
- When a steady current is injected into a cylindrical fiber, the size of a local graded potential is determined by the input resistance of the fiber (r_{input}) and the distance over which it spreads is determined by the length constant of the fiber (λ); r_{input} and λ depend, in turn, on the resistance of the fiber membrane (r_m) and the longitudinal resistance of the axoplasm (r_i).
- In addition to having a resistance, a nerve fiber membrane has a capacitance. The effect of the membrane capacitance (c_m) is to slow the rise and decay of signals. The magnitude of this effect is determined by the membrane time constant, $\tau = r_m c_m$.
- Propagation of an action potential along a fiber depends on the passive spread of current from the active region into the next segment of membrane. The conduction velocity depends on the time constant and length constant of the membrane and is proportional to fiber diameter.
- Large nerve fibers in vertebrates are wrapped in myelin sheaths, with regularly spaced gaps or nodes between ensheathing glia or Schwann cells. During action potential propagation, excitation jumps from one node to the next (saltatory conduction).
- Action potential propagation is influenced by geometrical factors that produce changes in membrane area. For example, block of propagation may occur at branch points in nerve terminal arborizations, and conduction may have a preferred direction in tapered dendrites.
- Transfer of electrical signals from one cell to the next requires special low-resistance connections called gap junctions. A gap junction is formed by a collection of connexons, which are proteins that form aqueous channels between the cytoplasms of adjacent cells.

Suggested Reading

General Reviews

Evans, W. H., and Martin, P. E. M. 2002. Gap junctions: structure and function. *Mol. Membr. Biol.* 19: 121–136.

Sosinsky, G. E., and Nicholson, B. J. 2005. Structural organization of gap junction channels. *Biochim. Biophys. Acta.* 1711: 99–125.

Susuki, K., and Rasband, M. N. 2008. Molecular mechanisms of node of Ranvier formation. *Curr. Opin. Cell Biol.* 20: 616–623.

Original Papers

Maeda, S., Nakagawa, S., Suga, M., Yamashita, E., Oshima, A., Fujiyoshi, Y., and Tsukihara, T. 2009. Structure of the connexin 26 gap junction channel at 3.5× resolution. *Nature* 458: 597–604.

Schwarz, J. R., Glassmeier, G., Cooper, E. C., Kao, T. C., Nodera, H., Tabuena, D., Kaji, R., and Bostock, H. 2006. KCNQ channels mediate I_{Ks}, a slow K^+ current regulating excitability in the rat node of Ranvier. *J. Physiol.* 573: 17–34.

Vabnick, I., and Shrager, P. 1998. Ion channel redistribution and function during development of the myelinated axon. *J. Neurobiol.* 37: 80–96.

Waters, J., Schaefer, A., and Sakmann, B. 2005. Backpropagating action potentials in neurones: measurement, mechanisms and potential functions. *Prog. Biophys. Mol. Biol.* 87: 145–170.

CHAPTER 9

Ion Transport across Cell Membranes

There is a constant flux of ions across the plasma membrane of neurons, both at rest and during electrical activity. The flux is driven by electrical and chemical concentration gradients. In the face of such movements, concentrations in the cytoplasm are kept constant by transport mechanisms that use energy to move ions back across the membrane against their electrochemical gradients. Maintenance of intracellular ion concentrations is essential to maintain both the resting membrane potential and the ability to generate electrical signals.

Primary active transport uses energy provided by the hydrolysis of adenosine triphosphate (ATP). The most prevalent transport mechanism of this kind is the sodium–potassium exchange pump. The molecule responsible for transport is an enzyme, sodium–potassium ATPase, which carries three sodium ions out of the cell and two potassium ions in, for every molecule of ATP hydrolyzed. Each cycle of the pump results in the net movement of one ionic charge across the membrane, so the pump is said to be electrogenic. Two different forms of calcium ATPase are responsible for calcium transport out of the cytoplasm: plasma membrane calcium ATPase pumps calcium out of the cell, and endoplasmic and sarcoplasmic reticulum ATPases pump calcium from the cytoplasm into intracellular compartments.

Secondary active transport uses the movement of sodium down its electrochemical gradient to transport other ions across the membrane either in the same direction (cotransport) or in the opposite direction (ion exchange). An example is sodium–calcium exchange, in which the influx of three sodium ions is used to transport a single calcium ion out of the cell. Like all other transport systems, this exchange system is reversible, and depending on the chemical and electrical gradients for the two ions, it can be made to run in a forward or backward direction. A second sodium–calcium exchange system, found in retinal cells, transports a single calcium ion plus one potassium ion out of the cell in exchange for four sodium ions. Sodium influx is also utilized to transport chloride and bicarbonate across the cell membrane. All such mechanisms provide pathways for sodium to enter the cell down its electrochemical gradient, and therefore depend on sodium–potassium ATPase to maintain that gradient.

Transport of neurotransmitters plays an important role in nervous system function. In presynaptic terminals transmitters are packaged in vesicles, ready for release. After release, they are removed from extracellular spaces by uptake mechanisms in the plasma membranes of neurons and glial cells. Transport into vesicles is by a hydrogen-transmitter exchange system. The hydrogen gradient is maintained, in turn, by hydrogen ATPase in the vesicle membrane. Uptake of the transmitters from extracellular spaces is by sodium cotransport systems.

A number of transport ATPases and ion exchangers have been isolated and cloned, and their configurations in the membrane have been determined. All appear to have 10 to 12 transmembrane segments, and are assumed to form channel-like structures through which substances are moved by alternate exposure of binding sites to the extracellular and intracellular spaces.

At rest and during electrical activity, ions move into and out of cells along their electrochemical gradients. Clearly, if such ion movements continued without compensation the system eventually would run down and both the concentration gradients and the potential across the cell membrane would disappear. Recovery of ions that leak into or out of the cell at rest or during electrical activity is accomplished by a variety of transport mechanisms that move ions back across the membrane against their electrochemical gradients. **Primary active transport** is driven directly by metabolic energy, specifically by the hydrolysis of ATP. **Secondary active transport** uses energy provided by the flux of an ion (usually sodium) down its established electrochemical gradient to transport other ions across the cell membrane, either in the same direction (cotransport) or in the opposite direction (ion exchange).

The Sodium–Potassium Exchange Pump

Most excitable cells have resting membrane potentials in the range of –60 to –90 mV, while the **equilibrium potential for sodium (E_{Na})** is usually of the order of +50 mV (see Chapter 6). Thus, there is a large electrochemical gradient tending to drive sodium into the cell, and it enters continuously through various pathways in the cell membrane. In addition, the **equilibrium potential for potassium (E_K)** is more negative than the resting potential, so that potassium ions continually move out of the cell. In order to maintain the viability of the cell, it is necessary to transport sodium back out and potassium back in, against their electrochemical gradients. This perpetual task is carried out by the **sodium–potassium exchange pump**, which in its usual mode of operation transports three sodium ions out across the cell membrane for every two potassium ions carried inward.

Early studies of sodium–potassium exchange were done by Hodgkin and Keynes and their colleagues on the giant axon of the squid,[1,2] where it was shown clearly that the source of energy for the transport process was hydrolysis of ATP. During the same period, Skou[3] demonstrated that an ATPase isolated from crab nerve had many of the biochemical properties that would be expected of a sodium–potassium exchange pump. Specifically, just as both sodium and potassium are required for the pump to work, the enzyme was stimulated by the simultaneous presence of sodium and potassium. Moreover, both sodium–potassium exchange and activity of the enzyme were inhibited by the poisonous glycoside ouabain. These observations led to the conclusion that the enzyme, **sodium–potassium ATPase**, is itself the transport molecule. It is one of a family of P-type ATPases (so-called because they form a phosphorylated intermediate) that derive energy for ion translocation from hydrolysis of ATP. Other members of the same family include calcium ATPases, which transport calcium out of the cytoplasm of cells, and vacuolar hydrogen ATPase, which transports protons into intracellular acidic vacuoles, including synaptic vesicles.

Biochemical Properties of Sodium–Potassium ATPase

The biochemical properties of sodium–potassium ATPase have been known for a number of years.[4,5] The stoichiometry of cation binding to the enzyme is as expected from the transport characteristics: Three sodium and two potassium ions are bound for each molecule of ATP hydrolyzed. The requirement for sodium is remarkably specific. It is the only substrate accepted for net outward transport; conversely, it is the only monovalent cation *not* accepted for inward transport. Thus, lithium, ammonium, rubidium, cesium, and thallium are all able to substitute for potassium in the external solution but not for sodium in the internal solution. The requirement for external potassium is not absolute. In its absence the pump will extrude sodium at about 10 percent of capacity in an uncoupled mode. The transport system is blocked specifically by digitalis glycosides (drugs used for treating congestive heart failure), particularly ouabain and strophanthidin. Although they block sodium and potassium transport by ATPase, these drugs have no effect on the passive movements of sodium and potassium through membrane channels.

Experimental Evidence that the Pump Is Electrogenic

Because sodium–potassium ATPase transports unequal numbers of sodium and potassium ions, each cycle of the pump results in the outward movement of one positive charge

[1] Hodgkin, A. L., and Keynes, R. D. 1955. *J. Physiol.* 128: 28–60.

[2] Caldwell, P. C. et al. 1960. *J. Physiol.* 152: 561–590.

[3] Skou, J. C. 1957. *Biochim. Biophys. Acta* 23: 394–401.

[4] Skou, J. C. 1988. *Methods Enzymol.* 156: 1–25.

[5] Jorgensen, P. L., Hakansson, K. O., and Karlish, S. J. D. 2003. *Annu. Rev. Physiol.* 65: 817–849.

across the membrane. For this reason the pump is said to be **electrogenic.** The electrogenic nature of the pump was tested experimentally in squid axons,[1,2,6] and in a remarkable set of experiments by Thomas, using snail neurons.[7,8] Snail neurons are sufficiently large to permit the insertion of several micropipettes through the cell membrane into the cytoplasm without damaging the cell. To examine how internal sodium concentration affected pump current and membrane potential, Thomas used two intracellular pipettes to deposit ions in the cell, one filled with sodium acetate and the other with lithium acetate (Figure 9.1A). A third intracellular pipette was used as an electrode to record membrane potential. A fourth pipette was used as a current electrode for voltage clamp experiments (see Chapter 7), and yet a fifth, made of sodium sensitive glass, monitored the intracellular sodium concentra-

[6]Baker, P. F. et al. 1969. *J. Physiol.* 200: 459–496.

[7]Thomas, R. C. 1969. *J. Physiol.* 201: 495–514.

[8]Thomas, R. C. 1972. *J. Physiol.* 220: 55–71.

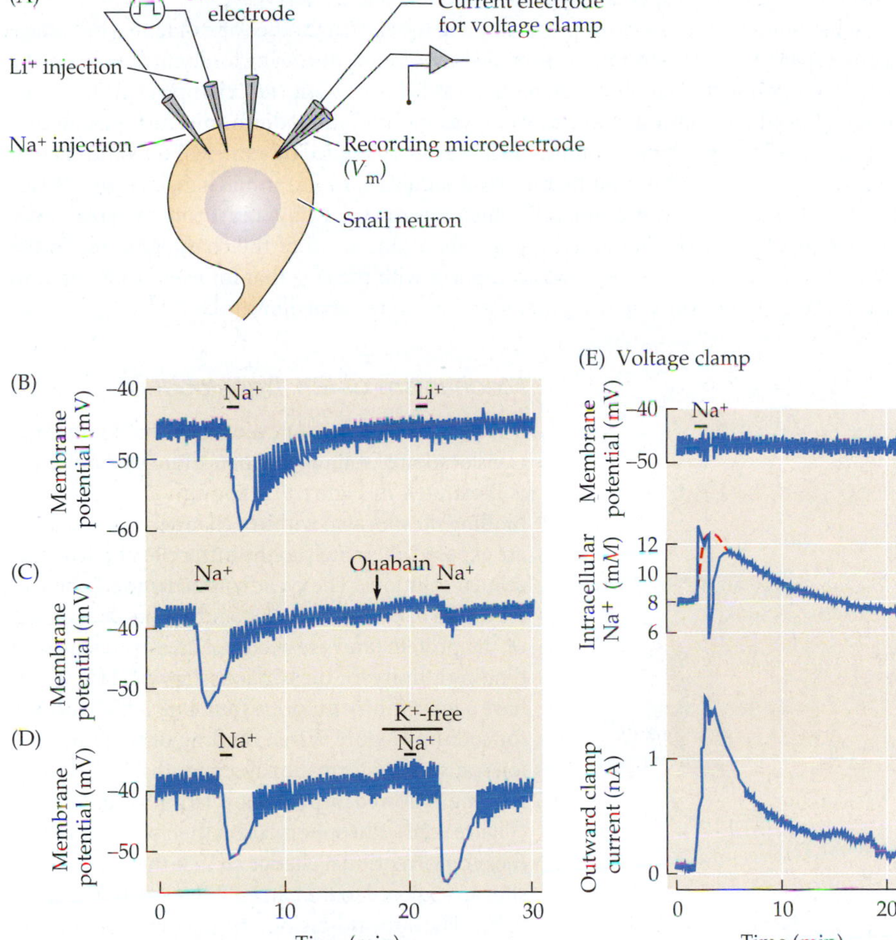

FIGURE 9.1 Effects of Sodium Injection into a snail neuron on intracellular sodium concentration, membrane potential, and membrane current. (A) Two micropipettes are used to inject either sodium or lithium; a sodium-sensitive electrode measures $[Na]_i$; another electrode measures membrane potential and is also used in combination with a current-passing electrode to hold the membrane potential steady while measuring membrane current (shown in E). (B) Hyperpolarization of the membrane following intracellular injection of sodium. Note that the small rapid deflections are spontaneously occurring action potentials, reduced in size because of the poor frequency response of the pen recorder. Injection of lithium does not produce hyperpolarization. (C) After application of ouabain (20 μg/ml), which blocks the sodium pump, hyperpolarization by sodium injection is greatly reduced. (D) Removal of potassium from the extracellular solution blocks the pump, so that sodium injection produces no hyperpolarization until potassium is restored. (E) Measurements of membrane current, with the membrane potential clamped at –47 mV. Sodium injection results in an increase in intracellular sodium concentration, and an outward current across the cell membrane. Sharp deflections on the sodium concentration record are artifacts from the injection system; the dashed line indicates the time course of the change in concentration. (After Thomas, 1969.)

tion! To inject sodium, the sodium-filled pipette was made positive with respect to the lithium pipette. Thus, current flow in the injection system was between the two pipettes, with none of the injected current flowing through the cell membrane.

The result of sodium injection is shown in Figure 9.1B. After a brief injection, the cell became hyperpolarized by about 15 mV, presumably because of increased pump activity. The potential recovered gradually over several minutes, as the excess sodium was extruded. Injection of lithium (by making the lithium pipette positive) produced no hyperpolarization.

Several lines of evidence showed that the potential change after sodium injection was due to the action of a sodium pump. For example, the hyperpolarization could be greatly reduced or abolished by addition of the transport inhibitor, ouabain, to the bathing solution (Figure 9.1C). Similarly, sodium injection had little effect on potential when potassium was absent from the external solution; reintroduction of potassium after sodium injection, however, resulted in immediate hyperpolarization (Figure 9.1D).

Quantitative estimates of the pump rate and the exchange ratio were obtained by voltage clamp experiments. This technique provided a means of measuring ionic current across the membrane, while the membrane potential was held constant (i.e., clamped). At the same time, intracellular sodium concentration was monitored. Sodium injection produced a transient rise in intracellular sodium, which was accompanied by a surge of outward current whose amplitude and duration followed the sodium concentration change (Figure 9.1E). The net charge carried out of the cell, which was calculated by measuring the area under the membrane current, amounted to only about one-third of the charge injected in the form of sodium ions. This result was consistent with the idea that for every three sodium ions pumped out of the cell, two potassium ions were carried inward.

Mechanism of Ion Translocation

The general sequence of events believed to underlie translocation of sodium and potassium by the enzyme is illustrated in Figure 9.2. Sodium and potassium binding sites, located within a channel-like structure, are exposed alternately to the intracellular and extracellular solutions. The cyclic conformational changes are driven by phosphorylation and dephosphorylation of the protein, and are accompanied by changes in binding affinity for the two ions. Inward-facing sites have a low affinity for potassium and a high affinity for sodium (Figure 9.2A). Binding of three sodium ions causes a conformational change that leads to ATP binding, followed by phosphorylation of the enzyme (Figure 9.2B). Phosphorylation then produces a further conformational change so that the ion-binding sites are exposed to the extracellular solution (Figure 9.2C). The outward-facing sites have low sodium and high potassium affinities, so that the sodium ions are replaced by two potassium ions (Figure 9.2D). Potassium binding leads to dephosphorylation of the

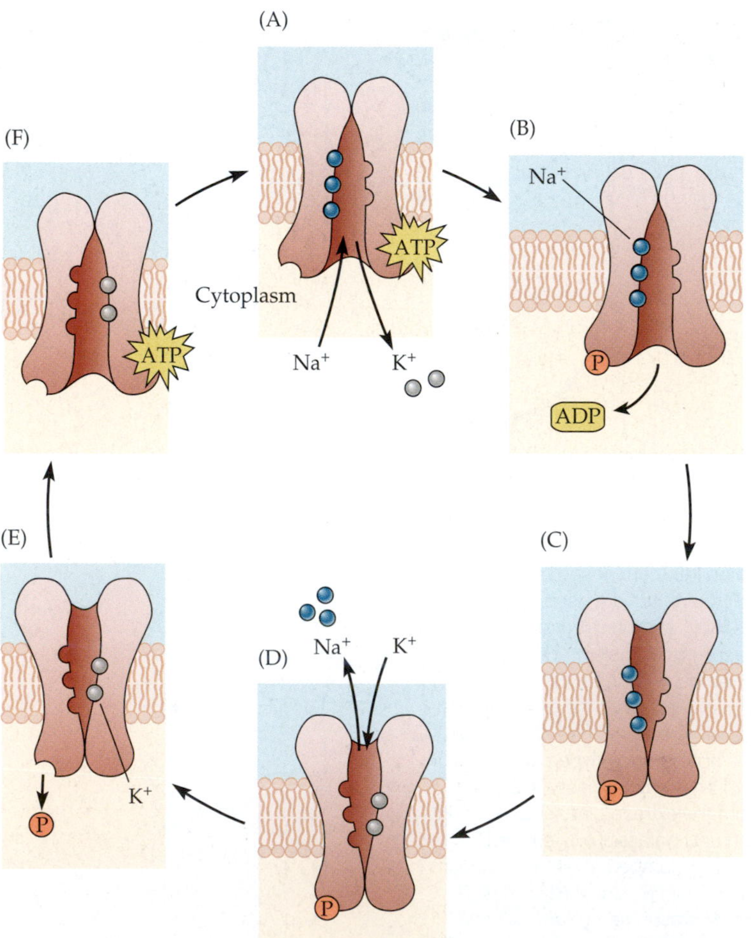

FIGURE 9.2 Ion Translocation by Na–K ATPase. (A) Inward-facing binding sites have a high affinity for sodium, a low affinity for potassium, and bind three sodium ions. (B) Sodium binding is followed by phosphorylation of the enzyme. (C) Phosphorylated enzyme undergoes a conformational change so that binding sites face the extracellular solution. (D) Outward-facing sites have a low affinity for sodium, a high affinity for potassium, and bind two potassium ions. (E) Potassium binding leads to dephosphorylation. (F) Dephosphorylation is followed by a return to the starting conformation. (After Horisberger, 2004.)

enzyme and a return to the starting conformation (Figure 9.2E,F). The potassium ions are then released into the cytoplasm.

Calcium Pumps

Changes in intracellular calcium are important for many neuronal functions. For example, increases in cytoplasmic calcium concentration mediate the release of neurotransmitters at synapses, activation of ion channels in the cell membrane, and regulation of a number of cytoplasmic enzymes. In muscle, increased intracellular calcium initiates contraction. Because these functional roles are associated with transient increases in cytoplasmic calcium concentration, it is important that the resting concentration be kept low. Otherwise, the various mechanisms would be activated continuously rather than in response to specific stimuli.

Transient increases in cytoplasmic calcium concentration are mediated either by entry of calcium through channels in the plasma membrane or by release from intracellular stores; return to resting concentration is achieved by activation of a number of transport proteins in the plasma membrane and in membranes of the intracellular storage compartments, principally the endoplasmic reticulum (sarcoplasmic reticulum in muscle) and mitochondria.[9,10] In most neurons cytoplasmic concentrations of free calcium range from 10 to 100 n*M*. The extracellular calcium concentration in vertebrate interstitial fluid is in the range of 2 to 5 m*M*, so that outward transport across the membrane is against a substantial electrochemical gradient. Transport systems within the cell maintain high concentrations of calcium within intracellular compartments. Calcium concentrations in the endoplasmic reticulum can reach 400 μ*M*, and in the sarcoplasmic reticulum of muscle can be as high as 10 m*M*. Transport of calcium out of the cytoplasm, across the plasma membrane, and into membrane-bound intracellular compartments is accomplished by **calcium ATPase**.[11] An additional mechanism for transport of calcium across the plasma membrane that does not involve enzyme activity is discussed later in this chapter.

Endoplasmic and Sarcoplasmic Reticulum Calcium ATPase

One family of calcium ATPases is concentrated in membranes of the endoplasmic reticulum of neurons and in the sarcoplasmic reticulum of muscle. These ATPases transport calcium from the cytoplasm into the membrane-bound intracellular compartments. In muscle, contraction is triggered by the release of calcium from the sarcoplasmic reticulum into the myoplasm. Rapid return of cytoplasmic calcium to its resting level, and hence prompt muscle relaxation, is accomplished by a high density of calcium ATPase in the sarcoplasmic reticulum membrane. The calcium transport cycle is analogous to that described for sodium–potassium ATPase and begins with the attachment of two calcium ions to high-affinity ($K_{m(Ca)} \approx 100$ n*M*) sites facing the cytoplasm. The enzyme is then phosphorylated and undergoes a conformational change, resulting in the release of the calcium ions to the reticular compartment. Release of the bound calcium is followed by dephosphorylation and return to the starting molecular configuration.

Plasma Membrane Calcium ATPase

Calcium ATPase is also found in the plasma membrane of all cells. The plasma membrane enzyme is similar in structure and function to its endoplasmic and sarcoplasmic reticular counterpart but differs in some details. The intracellular binding site has a high affinity for calcium, but only one calcium ion is bound during the transport cycle. In nerve and muscle the enzyme is only sparsely distributed in the plasma membrane, so its transport capacity is relatively low. Nonetheless, it is adequate to compensate for calcium influx into resting cells.

Sodium–Calcium Exchange

Many ion transport mechanisms make use of an entirely different principle for the uphill transfer of ions across the cell membrane: instead of relying on hydrolysis of ATP, ion movement is coupled to the inward flux of sodium down its electrochemical gradient. Sodium, entering the cell down its electrochemical gradient through transport molecules provides

[9] MacLennan, D. H., Abu-Abed, M., and Kang, C. H. 2002. *J. Mol. Cell. Cardiol.* 34: 897–918.

[10] Rizzuto, R., and Pozzan, T. 2006. *Physiol. Rev.* 86: 369–408.

[11] Carafoli, E., and Brini, M. 2000. *Curr. Opin. Chem. Biol.* 4: 152–161.

the energy required to carry other ions uphill against their electrochemical gradients, either into or out of the cell. A simple example is the 1:1 sodium–hydrogen exchange mechanism that contributes to the maintenance of intracellular pH. Protons are carried out of the cell against their electrochemical gradient in exchange for inward movement of sodium. Hydrogen, calcium, potassium, chloride, and bicarbonate are all transported in this way. These secondary transport systems account for a measurable fraction of the sodium entry into the resting cell. Ultimately, of course, the mechanisms depend on sodium–potassium ATPase to maintain the sodium gradient by pumping sodium ions back out of the cell. In some instances, potassium ions moving out of the cell down their electrochemical gradient also contribute energy to ion transport processes.

The NCX Transport System

At least two sodium–calcium exchange systems are found in plasma membranes. The most widely distributed of these is the sodium-calcium exchange (NCX) system.[12,13] The transport molecule carries one calcium ion outward for each three sodium ions entering the cell. The NCX exchanger has a lower affinity for calcium ($K_{1/2Ca} \approx 1.0\ \mu M$) than does calcium ATPase, but because the exchange molecules occur at a much higher density their transport capacity is about 50 times greater. The exchange system is called into play in excitable cells when calcium influx during electrical activity overwhelms the transport ability of the ATPase.

The experiment shown in Figure 9.3 illustrates the operation of the sodium–calcium exchange mechanism in a squid axon. The intracellular concentration of ionized calcium was measured by luminescence of the calcium indicator, aequorin. In the steady state, the influx of calcium along its electrochemical gradient is balanced by outward transport through the ion exchanger (Figure 9.3A). At the beginning of the experiment (Figure 9.3B), the intracellular calcium concentration is relatively high, as the axon is bathed in high (112 m*M*) levels of calcium. When the extracellular calcium concentration is reduced, the passive influx decreases. As a result, the intracellular calcium concentration falls, restoring the driving force for calcium entry until the passive influx again equals the rate of extrusion. Reducing the extracellular sodium concentration, on the other hand, increases intracellular calcium concentration. This is because the reduced driving force for sodium entry reduces the rate at which the exchanger extrudes calcium. Intracellular calcium concentration then rises until calcium influx is reduced by the same amount. Replacement of sodium by lithium (which does not enter the exchanger) results in a further increase in intracellular calcium.

[12]Blaustein, M. P., and Lederer, W. J. 1999. *Physiol. Rev.* 79: 763–854.

[13]Annunziato, L., Pignatoro, G., and DiRenzo, G. F. 2004. *Pharmacol. Rev.* 56: 633–654.

Reversal of Sodium–Calcium Exchange

Ion exchange mechanisms can be made to run backwards by altering one or more of the ionic gradients involved in the exchange. An interesting feature of the NCX family of exchangers is that such reversal can occur readily under physiological conditions,

(A)

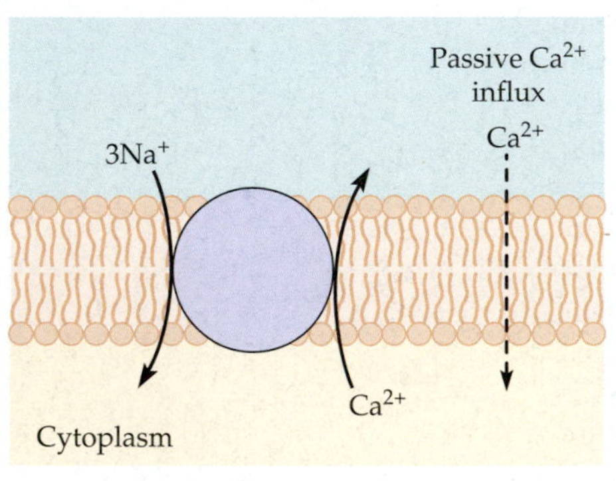

(B)

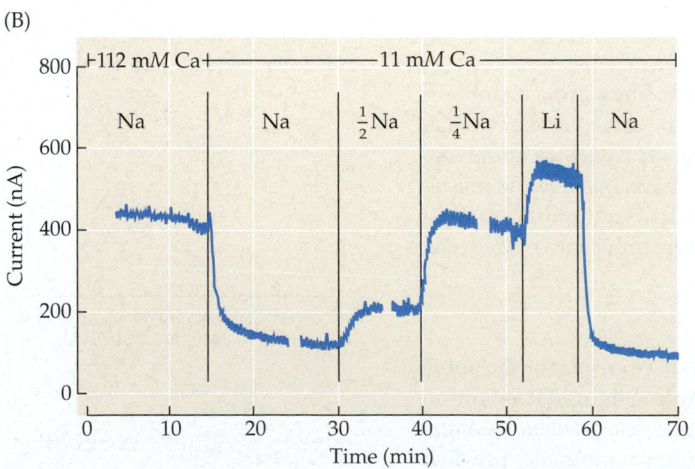

FIGURE 9.3 Transport of Calcium Ions out of a squid axon. (A) Scheme for sodium–calcium exchange. Influx of three sodium ions down their electrochemical gradient is coupled to the extrusion of one calcium ion. Calcium concentration reaches a steady state when outward transport through the exchanger is equal to the inward calcium leak. (B) Effect of changes in extracellular calcium and sodium on intracellular calcium concentration. Changes in intracellular calcium concentration are measured by changes in luminescence of injected aequorin, indicated by current from the photodetector. Increased readings mean increased intracellular free-calcium concentration. Reducing the extracellular calcium concentration from 112 m*M* to 11 m*M* reduces the intracellular concentration. Reducing extracellular sodium reduces outward calcium transport, and hence increases intracellular concentration. Lithium ions do not substitute for sodium in the transport system. (After Baker, Hodgkin, and Ridgeway, 1971.)

in which case calcium *enters* through the system and sodium is extruded. The direction of transport is determined simply by whether the energy provided by the entry of three sodium ions is greater than or less than the energy required to extrude one calcium ion. One factor determining this energy balance is the membrane potential of the cell. Such dependence on potential arises because the exchange is not electrically neutral; rather each forward cycle of the transport molecule results in a net transfer of one positive charge inward across the membrane. As a result, forward transport is facilitated by membrane hyperpolarization and impeded, or even reversed, by depolarization.

The energy dissipated by sodium entry (or required for extrusion) is simply the product of charge moved across the membrane and the driving force for this movement; in other words, it is the charge multiplied by the difference between the sodium equilibrium potential (E_{Na}) and the membrane potential (V_m). For three sodium ions, this is $3(E_{Na} - V_m)$. Similarly, for a single (divalent) calcium ion, the energy is $2(E_{Ca} - V_m)$. At some value of membrane potential the energies will be exactly equal and no exchange will occur. If we call this point the reversal potential, V_r, then:

$$3(E_{Na} - V_r) = 2(E_{Ca} - V_r)$$

and (by rearrangement)

$$V_r = 3E_{Na} - 2E_{Ca}$$

At potentials more negative than V_r, sodium moves into the cell and calcium is transported out. At more positive potentials, the sodium and calcium fluxes are reversed.

Now suppose a nerve cell has internal sodium and calcium concentrations of 15 m*M* and 100 n*M*, respectively, and is bathed in a solution containing 150 m*M* sodium and 2 m*M* calcium. These concentrations are reasonable physiological values for mammalian cells. Using the Nernst equation (see Chapter 4), the equilibrium potential for sodium is +58 mV and for calcium is +124 mV. Ion movement through the exchanger is zero at a membrane potential (V_r) of –74 mV. This is in the range of resting membrane potentials for many cells so that, in any given cell, ion movements through the exchanger may be in one direction or the other, depending on the exact membrane potential, and on whether or not there has been previous sodium or calcium accumulation.

Sodium–Calcium Exchange in Retinal Rods

It is evident that the NCX type of exchanger would be a poor system for calcium extrusion from cells with low resting potentials; rather than being extruded, calcium would be accumulated until a relatively high steady-state concentration was reached in the cytoplasm. An example of such a cell is the vertebrate retinal rod, which has a resting membrane potential of the order of –40 mV (see Chapter 20). These cells contain a second kind of sodium–calcium exchange molecule in the plasma membranes of their outer segments.[14] In addition to transporting sodium and calcium, the molecule transports potassium, and therefore is given the designation NCKX. The NCKX system is given an additional boost over the NCX exchanger by two differences in stoichiometry: (1) four (rather than three) sodium ions enter for each calcium ion extruded and (2) additional work is available from the extrusion of one potassium ion that, like sodium, moves down its electrochemical gradient during the exchange. The reversal potential for sodium–potassium–calcium exchange is:

$$V_r = 4E_{Na} - E_K - 2E_{Ca}$$

Using the previous assumptions about E_{Na} (58 mV) and E_{Ca} (124 mV) and assuming that $E_K = -90$ mV, then $V_r = +74$ mV. Clearly, transport through the RetX exchanger is unlikely to reverse.

Another way of viewing the exchange system is to ask what the intracellular concentration of calcium would be if V_r were equal to the resting potential of the cell. Using the same assumptions about sodium and potassium, we find that when $V_r = -40$ mV, $E_{Ca} = 181$ mV. With 2 m*M* extracellular calcium this is equivalent to an intracellular concentration of 1 n*M*. In other words, at a membrane potential of –40 mV it is energetically possible for the exchanger to reduce the cytoplasmic calcium concentration to 1 n*M*. However, this low concentration could not be achieved in practice because it is about two orders of magnitude smaller than the affinity of the exchange molecule for calcium.

[14]Schnetkamp, P. P. M. 2004. *Pflügers Arch.* 447: 683–688.

Chloride Transport

Intracellular chloride concentration is closely regulated in all cells. Such regulation is particularly important in neurons because direct synaptic inhibition (see Chapter 11) depends on the maintenance of low intracellular chloride. Although chloride-sensitive ATPases have been demonstrated in a number of tissues, including the brain,[15] the bulk of chloride transport across the plasma membrane of nerve cells is by secondary transport mechanisms. The most important are two cation-coupled transport systems: (1) a sodium–potassium–chloride transport mechanism that moves chloride into the cell, and (2) an outward potassium–chloride cotransport. In addition, all cells have one or more chloride–bicarbonate exchange systems that are concerned primarily with intracellular pH regulation.

Inward Chloride Transport

In many cells, such as skeletal muscle fibers, tubular cells of the kidney, and squid axons, chloride is actively accumulated. Chloride accumulation is dependent on the extracellular concentrations of both sodium and potassium, and the system transports all three ions inward across the cell membrane.[16] The movement of sodium down its concentration gradient supplies the required energy. In squid axons the Na–K–Cl stoichiometry has been shown to be 2:1:3. In cells of the kidney the Na–K–Cl stoichiometry is 1:1:2, and the renal transport system operates in parallel with a second, potassium-independent mechanism that transports sodium and chloride in a 1:1 ratio (Figure 9.4A).

Outward Potassium–Chloride Cotransport

A second major transport mechanism for chloride involves cotransport of chloride and potassium outward, across the cell membrane (Figure 9.4B). Potassium–chloride cotransport is best known for its role in cell volume regulation in a wide variety of tissues, but there is increasing evidence that it is important in regulation of intracellular chloride concentration in neurons.[17] As ion transport by the system is insensitive to extracellular sodium concentration, the energy required for outward chloride transport is supplied solely by the outward movement of potassium down its electrochemical gradient.

Both the inward and outward transport systems are blocked by the chloride transport inhibitors furosemide and bumetanide.

Chloride–Bicarbonate Exchange

Chloride–bicarbonate exchange systems, operating in parallel with sodium–hydrogen exchange, serve primarily to regulate intracellular pH.[18] Simple chloride–bicarbonate exchange is driven

[15] Gerencser, G. A., and Zhang, J. 2003. *Biochim. Biophys. Acta* 1618: 133–139.

[16] Russell, J. M. 2000. *Physiol. Rev.* 80: 211–276.

[17] Mercado, A., Mount, D. B. and Gamba, G. 2004. *Neurochem. Res.* 29: 17–25.

[18] Romero, M. F., Fulton, C. M., and Boron, W. F. 2004. *Pflügers Arch.* 447: 495–509.

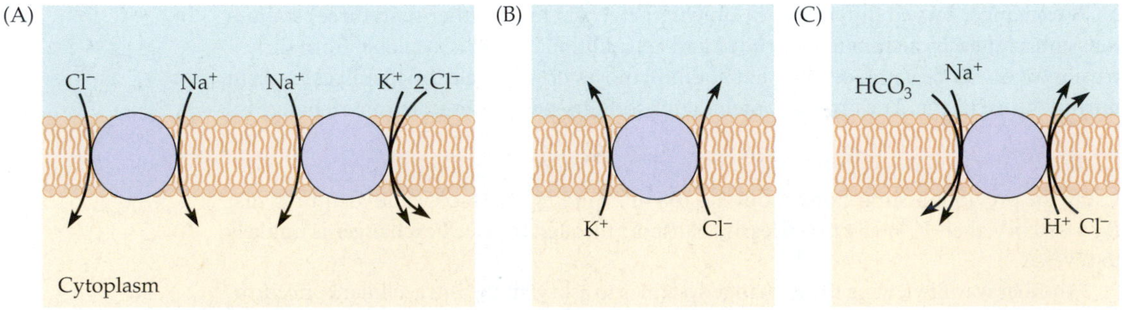

(A) Cl⁻ Na⁺ Na⁺ K⁺ 2 Cl⁻ Cytoplasm (B) K⁺ Cl⁻ (C) HCO_3^- Na⁺ H⁺ Cl⁻

FIGURE 9.4 Mechanisms of Chloride Transport. (A) In many cells, inward chloride transport is mediated by two independent mechanisms, both using the electrochemical gradient for sodium. One is sodium–chloride cotransport. The other is sodium–potassium–chloride cotransport, with a stoichiometry of 1:1:2. (B) Potassium–chloride cotransport uses the outward electrochemical gradient for potassium to transport chloride out of the cell. (C) Sodium dependent chloride–bicarbonate exchange uses the inward sodium gradient to move chloride out of the cell, at the same time exchanging bicarbonate with hydrogen ions. Note that all the chloride transport systems are electrically neutral.

by the outward gradient for bicarbonate produced by alkalinization of the cytoplasm and carries bicarbonate out of the cell and chloride inward on a 1:1 basis. Quantitatively, the role of this system is minimal in regulating intracellular chloride concentration in excitable cells.

Sodium-dependent chloride–bicarbonate exchange was first studied in squid axon and in snail neurons.[19,20] Recovery from acidification of the cytoplasm, by injection of hydrochloric acid (HCl), or exposure to CO_2, was prolonged when extracellular bicarbonate concentration was reduced or when intracellular chloride was depleted. In addition, recovery was virtually abolished when sodium was removed from the extracellular bathing solution. Thus, recovery involved inward movement of sodium down its electrochemical gradient, accompanied by bicarbonate, in exchange for chloride. Because the system is electrically neutral, it is assumed to include the outward transport of protons as well, as indicated in Figure 9.4C. The exchange mechanism is inhibited by 4-acetamido-4′-isothiocyanostilbene-2,2′-disulfonic acid (SITS) and a related compound, 4,4′-diisothiocyanostilbene-2,2′-disulfonate (DIDS).

Transport of Neurotransmitters

In addition to transporting inorganic ions, nerve cells have mechanisms for accumulating a variety of other substances, including those involved in synaptic transmission (see Chapter 11). Neurotransmitters are transported into organelles within the cytoplasm of presynaptic nerve terminals (synaptic vesicles), where they are stored ready for release. After release, transmitters, such as norepinephrine, serotonin, γ-aminobutyric acid (GABA), glycine, and glutamate, are recovered from the synaptic cleft by transporters in the plasma membranes of either the terminals themselves or of adjacent glial cells. All of these recovery processes involve secondary transport mechanisms that use energy from the movement of sodium, potassium, or hydrogen ions down their electrochemical gradients to power transmitter accumulation.

Transport into Presynaptic Vesicles

Neurotransmitters are synthesized within the cytoplasm of nerve terminals and then concentrated into presynaptic vesicles by secondary transport mechanisms coupled to proton efflux. This transport mechanism is analogous to sodium-driven secondary transport across the plasma membrane, but instead of a sodium gradient the system uses a proton gradient established by the transport of hydrogen ions from the cytoplasm into the vesicle by hydrogen ATPase.[21]

Three genetic families of proton-coupled transporters are expressed in the lipid bilayers of secretory vesicles (Table 9.1) designated as *SLC17* and *SLC18* and *SLC32* (SLC

[19] Russell, J. M., and Boron, W. F. 1976. *Nature* 264: 73–74.

[20] Thomas, R. C. 1977. *J. Physiol.* 273: 317–338.

[21] Edwards, R. H. 2007. *Neuron* 55: 835–858.

TABLE 9.1
Classification of neurotransmitter transporters

Gene family	Transporter	Substrate
Vesicle loading		
SLC18	VMAT1,2	Norepinephrine, dopamine, histamine, serotonin
SLC18	VAChT	Acetylcholine
SLC32	VIAAT, VGAT	Glycine, GABA
SLC17	VGLUT1,2,3	Glutamate
Transmitter uptake		
SLC1	EAAT1 (GLAST), EAAT2–5	Glutamate
SLC6	GAT1,2	GABA
	NET	Norepinephrine
	DAT	Dopamine
	SERT	Serotonin
	GLYT1,2	Glycine

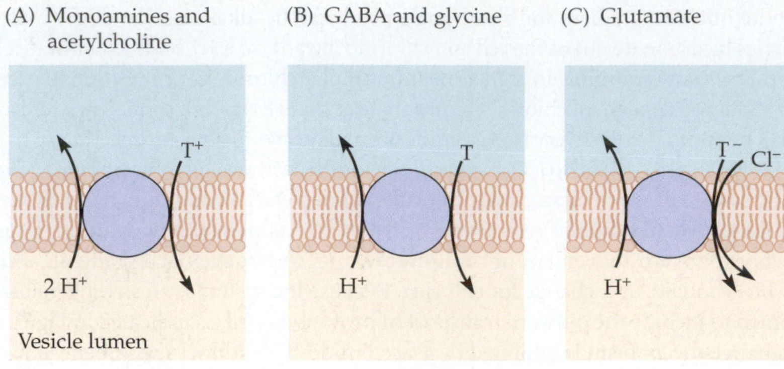

FIGURE 9.5 Transport of Neurotransmitters into Synaptic Vesicles. Hydrogen ATPase transports protons into the vesicle from the cytoplasm, creating an electrochemical gradient for proton efflux. Proton efflux through the secondary transport molecule provides the energy for accumulation of the neurotransmitter (T) in the vesicle. Proposed stoichiometry is 2:1 for monoamine and acetylcholine transport (A), 1:1 for glycine and GABA (B) and for glutamate (C). Glutamate transport is accompanied by chloride entry into the vesicle.

means solute carrier). *SLC18* has three isotypes.[22] Two of these are the monoamine transporters VMAT1 (*SLC18A1*) and VMAT2 (*SLC18A2*), which are responsible for vesicular accumulation of norepinephrine, dopamine, histamine, and serotonin (5-HT). The third, VAChT (*SLC18A3*), is responsible for accumulation of acetylcholine (ACh). GABA and glycine are concentrated by the product of a single gene (*SLC32*), referred to as the vesicular inhibitory amino acid transporter (VIAAT) or vesicular GABA transporter (VGAT).[23] Glutamate has its own transport family, *SLC17*, for which three isoforms have been identified: VGLUT1, 2, and 3 (*SLC17A7*, *A6*, and *A8*).[24] Glutamate transport is unusual in that it includes cotransport of chloride into the vesicle.[25] VGLUT1 and VGLUT2 are distributed separately to synapses in the brain.[26] VGLUT3 is found in various neuronal and non-neuronal structures, including non-glutamatergic synapses. It is responsible for vesicular glutamate uptake in cochlear hair cells. VGLUT3 knockout mice are profoundly deaf because their hair cells no longer release glutamate.[27]

There is a substantial electrochemical gradient for proton efflux across the vesicle membrane. For example, in filled monoaminergic and cholinergic vesicles, the pH gradient across the vesicle membrane is about –1.4 units and the electrical gradient is about +40 mV. The stoichiometry for the uptake systems is shown in Figure 9.5. Because the exchanges are not electrically neutral, all are dependent on the potential across the vesicle membrane as well as on the pH gradient.

Transmitter Uptake

After they are released from presynaptic nerve terminals into the synaptic cleft, most neurotransmitters are recovered either by the nerve terminals themselves or by adjacent glial cells (see Chapter 10). In general, such recovery serves two purposes: (1) The transmitter is removed from the extracellular space in the region of the synapse, which helps to terminate its synaptic action and prevents diffusion to other synaptic regions; (2) transmitter molecules recovered by the nerve terminal can be packaged again for rerelease. All uptake mechanisms use the electrochemical gradient for sodium to carry transmitter substances across the plasma membrane into the cytoplasm.

Two major transmitter uptake families are expressed in the cell membranes of neurons and glial cells.[28,29] Members of the *SLC1* family mediate the uptake of glutamate and neutral amino acids, while *SLC6* family members are responsible for uptake of dopamine, 5-HT, norepinephrine, glycine, and GABA. Each family has a number of isotypes (see Table 9.1). In the SLC1 system, inward transport of one glutamate ion is coupled to the influx of two sodium ions and the efflux of one potassium ion, coupled with either the extrusion of an hydroxyl ion or influx of a proton (Figure 9.6A).[30] In the SLC6 system, uptake of each transmitter molecule is accompanied by the entry of two sodium and one chloride ion (Figure 9.6B).[31]

One interesting aspect of the indirectly coupled transport mechanisms for transmitter uptake is that they are not, as a rule, electrically neutral. Thus, the direction of transport can be reversed by membrane depolarization, sometimes within the physiological range of membrane potentials.[32] Outward transport of GABA by this mechanism has been demon-

[22] Elden, L. E. et al. 2004. *Pflügers Arch.* 447: 636–640.

[23] Gasnier, B. 2004. *Pflügers Arch.* 447: 756–759.

[24] Takamnori, S. 2006. *Neurosci. Res.* 55: 343–351.

[25] Bellochio, E. E. et al. 2000. *Science* 289: 957–960.

[26] Fremeau, R. T. et al. 2004. *Science* 304: 1815–1819.

[27] Seal, R. P. et al. 2008. *Neuron* 24: 173–174.

[28] Gether, U. et al. 2006. *Trends Pharmacol. Sci.* 27: 375–383.

[29] Torres, G. E., and Amara, S. G. 2007. *Curr. Opin. Neurobiol.* 17: 304–312.

[30] Kanai, Y., and Hediger, M. A. 2004. *Pflügers Arch.* 447: 469–479.

[31] Chen, N-H, Reith, M. E. A., and Quick, M. W. 2004. *Pflügers Arch.* 447: 519–531.

[32] Attwell, D., Barbour, B., and Szatkowski, M. 1993. *Neuron* 11: 401–407.

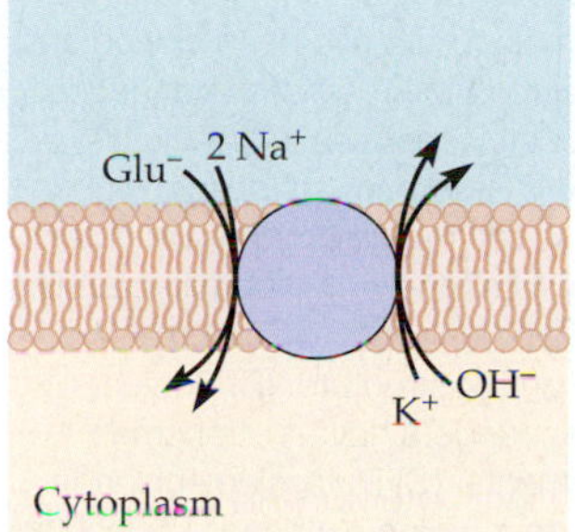

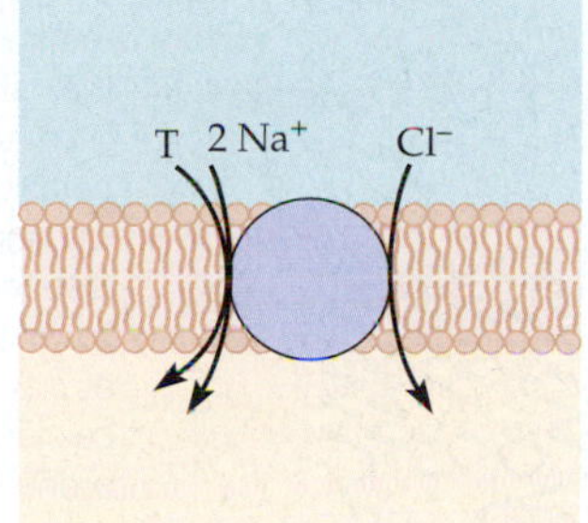

FIGURE 9.6 Uptake of Neurotransmitters. (A) Glutamate uptake is coupled to the influx of two sodium ions and the efflux of one potassium ion, and is accompanied by the extrusion of one hydroxyl (or one bicarbonate) ion. (B) In the GABA, glycine, and monoamine (norepinephrine, dopamine, and serotonin) uptake systems, recovery of the transmitter (T) is coupled to influx of two sodium ions and accompanied by the uptake of a single chloride ion. Choline from the hydrolysis of acetylcholine is recovered into the nerve terminal with the same stoichiometry as the GABA/glycine/monoamine systems.

strated in catfish retinal cells.[33] The transport system, then, not only mediates reuptake of the transmitter, but might also function as a mechanism for transmitter release. Reversal of transmitter uptake can have deleterious effects. After brain damage by stroke or trauma, outward transport of glutamate can lead to the accumulation of cytotoxic amounts of glutamate around neurons in the damaged area and thereby to further cell death.[34]

The *SLC1* family includes five high-affinity glutamate transporters: excitatory amino acid transporter or EAAT1, 2, 3, 4, and 5 (see Table 9.1). EAAT1, known as glial high affinity glutamate transporter (GLAST) in humans, is found in glia and is particularly abundant in the cerebellum. It also occurs in supporting cells in the retina, and it is responsible for uptake of glutamate released by hair cells in the cochlea.[35] EAAT2 is found in astrocytes, particularly in the cerebral cortex and hippocampus. EAAT3 is expressed in neurons throughout the brain, including those in cerebral cortex, hippocampus, superior colliculus, and thalamus. EAAT4 is found primarily in cerebellar Purkinje cells on postsynaptic dendritic spines and has a substantial chloride conductance associated with substrate transport. EAAT5 occurs mainly in rod photoreceptors and bipolar cells in the retina and exhibits a prominent chloride conductance.

The *SLC6* family includes transporters for GABA (GAT), norepinephrine (NET), dopamine (DAT), serotonin (SERT), and glycine (GLYT). The GABA transporter has three isoforms, GAT1, 2, and 3. GAT1 is the predominant neuronal transporter in the brain. It is found primarily along axons and around presynaptic nerve terminals. In the brain, GAT2 is found principally in meninges, ependyma, and choroid plexus and does not appear to be involved in neural signaling. GAT3 is expressed in glial cells. NET is found in neuroadrenergic neurons throughout the peripheral and central nervous systems and in adrenal chromaffin cells. DAT is found in dopaminergic neurons in the brain, localized around synaptic junctions. SERT is widely distributed in brain and is expressed in extrasynaptic axonal membranes. GLYT has two isoforms. GLYT2 is the neuronal transporter, found in association with glycinergic nerve terminals. It differs from other members of the *SLC6* family in that the transport stoichiometry is 3 sodium to 1 chloride to 1 glycine, suggesting that it is able to maintain extracellular glycine concentration at extremely low levels. GLYT1 is the predominant glial glycine transporter and has five splice variants: GLYT1a–e. GLYT1b and c are nervous system specific, while GLYT1e and f are found only in the retina.

ACh is recycled in a different way. It is synthesized in the cell cytoplasm from acetyl coenzyme A (acetyl-CoA) and choline (see Chapter 15). After release from the presynaptic terminal, its postsynaptic action is terminated by an enzyme (acetylcholinesterase) that hydrolyzes it to acetate and choline. Approximately half of the choline is recovered by a high-affinity (K_m 2μM) uptake mechanism and reused for ACh synthesis.[36] Like the monoamine and GABA/glycine transport systems, choline uptake is dependent on extracellular sodium and chloride concentrations.

Molecular Structure of Transporters

So far, we have dealt with functional properties of transporters with no reference to their molecular structure. As with membrane channels (see Chapter 5), each functional group is represented by a specific transport protein or, more commonly, a family of proteins.

[33] Cammack, J. N., and Schwartz, E. A. 1993. *J. Physiol.* 472: 81–102.

[34] Rossi, D. J., Oshima, T., and Attwell, D. 2000. *Nature* 403: 316–321.

[35] Glowatzki, E. et al. 2006. *J. Neurosci.* 26: 7659–7664.

[36] Iversen, L. L. et al. 2009. *Introduction to Neuropsychopharmacology*. Oxford University Press, New York, pp. 130–131.

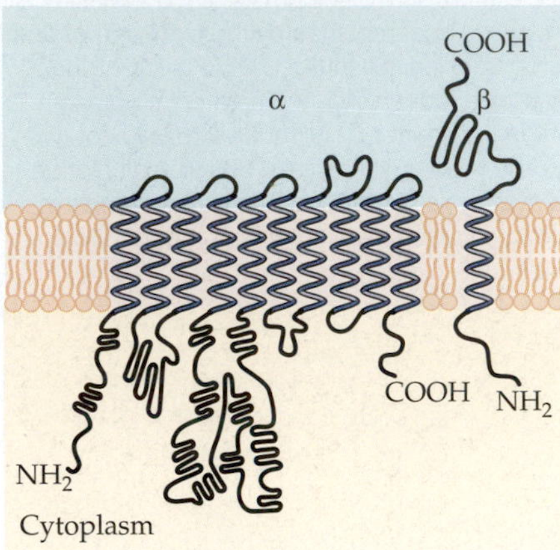

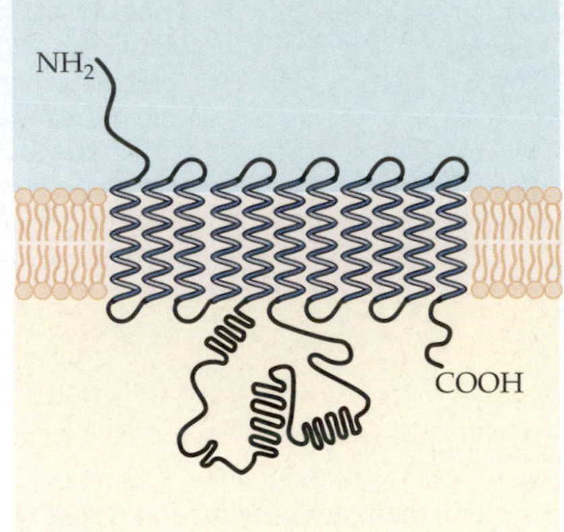

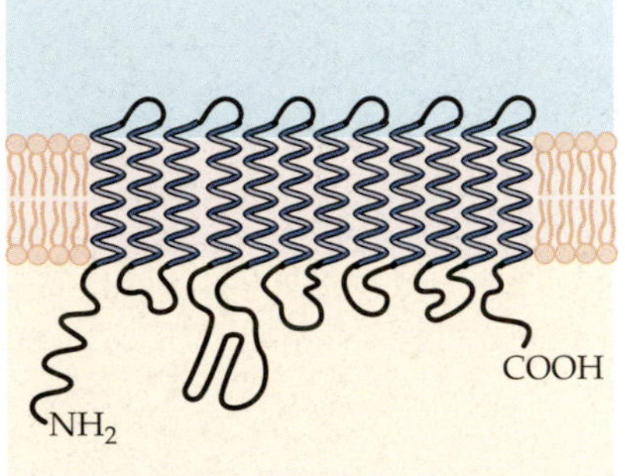

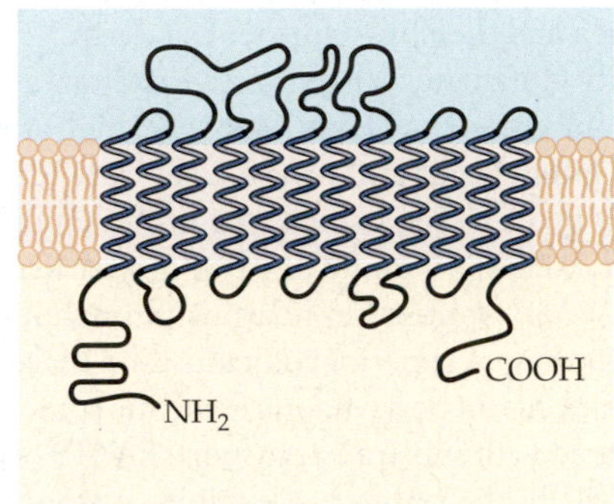

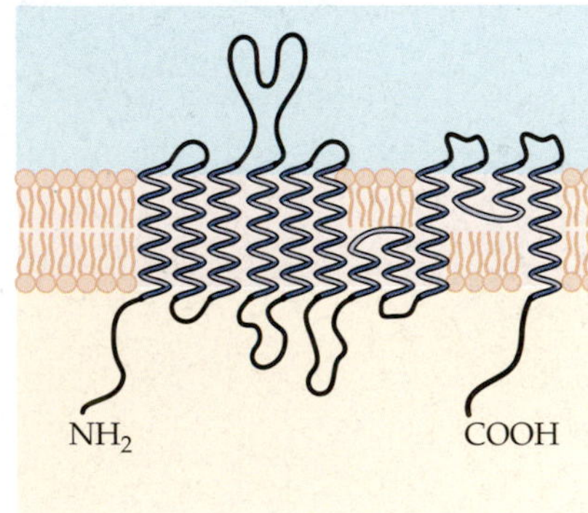

FIGURE 9.7 Molecular Configurations of Transport Molecules. A number of transport molecules have been cloned and their structure in the membrane deduced from hydropathy analyses. (A) Sodium–potassium ATPase consists of an α–subunit, with eight to ten transmembrane segments (see text) and a smaller β–subunit spanning the membrane only once. Calcium ATPases have a similar structure. (B) Sodium–calcium exchangers have 11 transmembrane segments. (C) Potassium–chloride cotransporters, anion exchange molecules, and sodium–potassium–chloride transporters all share the same membrane configuration, characterized by 12 transmembrane segments. (D) Monoamine uptake transporter subunit has 12 transmembrane segments. (E) Glutamate transport subunit has eight transmembrane segments and extracellular and intracellular membrane loops.

Many of these proteins have been isolated and cloned, and deductions have been made about their configurations in the membrane. Their structures are summarized below, and examples are shown in Figure 9.7. The proteins have nine to twelve transmembrane segments and are assumed to form channel-like structures through which substances are moved by alternate exposure of binding sites to the extracellular and intracellular spaces.

ATPases

The molecular structure of sodium–potassium ATPase is known in some detail.[37,38] It is assembled from two subunits, α and β. The α-subunit, with an apparent molecular weight of about 100 kD, is responsible for the enzymatic activity of the pump and contains all the substrate binding sites. The smaller (35 kD) β-subunit has a number of extracellular glycosylation sites and is necessary for pump function, but its precise role is not clear. Their proposed subunit structures are shown in Figure 9.7A. The α-subunit has ten transmembrane segments, four of which (M4, M5, M6 and M8) are believed to form the transmembrane pore containing the cation-binding sites. The nucleotide binding and phosphorylation sites have been localized to the large cytoplasmic region between M4 and M5. The β-subunit contains only one putative membrane-spanning region, with the bulk of the peptide in the extracellular space. There are four α-subunit isoforms (α_1–α_4), all expressed in the nervous system. Two (β_1 and β_2) of three known β-subunit isoforms are found in nervous tissue.

Calcium ATPases consist of a single polypeptide chain of about 100 kD, analogous in structure to the α-subunit of sodium–potassium ATPase but with an extended cytoplasmic segment at the carboxyl end.[11] Unlike sodium–potassium ATPase, it does not require a β-subunit for enzyme activity. The sarcoplasmic and endoplasmic reticulum calcium ATPase (SERCA) family consists of three basic gene products—SERCA1, SERCA2, and SERCA3—each containing two alternatively spliced transcripts. The molecular structure

[37] Horisberger, J. D. 2004. *Physiology* 19: 377–387.

[38] Martin, D. W. 2005. *Semin. Nephrol.* 25: 282–281.

of SERCA1a has been determined by X-ray crystallography at a resolution of 0.26 nm.[39] The family of plasma membrane calcium pumps is similar in structure to the SERCA enzymes, and is comprised of four basic gene products (PMCA1–4). Approximately 30 additional isoforms are generated by alternative splicing.

Sodium–Calcium Exchangers

Sodium–calcium exchange molecules are widely spread throughout the animal kingdom. The NCX family of exchangers includes three mammalian homologues, NCX1–3. NCX1 is found in heart muscle, while NCX2 and 3 are found in brain.[40] The cardiac exchanger contains 938 amino acids with nine transmembrane segments (Figure 9.7B). Homologous regions known as α-repeats are found within the molecule in segments 2, 3, and 7. The molecule has an extracellular reentry loop between segments 2 and 3, and a cytoplasmic reentry loop between segments 7 and 8. A large intracellular loop between segments 5 and 6 is the site of extensive alternative splicing. The loop contains important regulatory binding sites, but is not essential for the transport process.

The NCKX1 exchanger is somewhat larger, with 1199 amino acids. Other members of the family, NCKX2–4, are smaller in size, containing 600–660 residues. All have 11 transmembrane segments and, like the NCX exchangers, have a large intracellular loop in the middle of the molecule.[14]

Chloride Transporters

The K–Cl cotransporters, KCC1–KCC4, and the Na–K–2Cl transporters, NKCC1 and NKCC2, are both members of the same genetic family of electroneutral cation-coupled transport molecules and are similar in structure.[41] The KCC proteins consist of about 1100 amino acids with intracellular amino and carboxyl terminals separated by 12 transmembrane segments (Figure 9.7C). The NKCC molecule is slightly larger (about 1200 amino acids) with a similar topological arrangement. NKCC2 is absent from nervous tissue, being specific to kidney. KCC2 is unique in being specific to brain.

A sodium-dependent, chloride–bicarbonate exchanger (NDCBE) has been cloned from human brain and also from squid axon.[42,43] The squid axon protein consists of about 1200 amino acids, with intracellular amino and carboxyl terminals separated by 14 transmembrane segments. The human exchanger is slightly smaller.

Transport Molecules for Neurotransmitters

The monoamine and ACh transporters (VMAT 1 and 2, and VAChT) were the first vesicular transporters to be cloned.[44] The molecules are 520 to 530 amino acids in length and hydropathy analysis suggests that they contain 12 transmembrane segments (Figure 9.7D).[22] VAChT is about 40% identical to VMAT1 and VMAT2, which are about 65% identical to one another. The vesicular transport molecule for GABA and glycine has also been cloned.[45,46] Its structure is distinct from that of the monoamine and ACh transporters, with only ten putative transmembrane segments. The three members of the vesicular glutamate transport family are roughly 65 kD in mass and contain 560 to 590 amino acids, with ten potential transmembrane helices.[47]

The family of proteins associated with uptake of norepinephrine (NET), serotonin (SERT), dopamine (DAT), GABA (GAT), and glycine (GLYT) into axon terminals proteins have an apparent mass in the range of 80 to 100 kD.[48] The primary sequence of the choline transporter suggests that it may belong to the same superfamily.[49] The crystal structure of a bacterial homologue ($LeuT_{Aa}$) of the monoamine transporters has been analyzed at a resolution of 1.65 Å.[50] The molecule is a dimer with each subunit containing twelve transmembrane segments.

Five members of the family of proteins responsible for the uptake of glutamate have been isolated.[51] They are relatively small, containing 500 to 600 amino acids and having apparent masses of about 65 kD. The structure of a homologous bacterial transporter (Glt_{Ph}) has been resolved at a 3.5 Å resolution.[52] Each of its three subunits has eight transmembrane segments with two hairpin loops imbedded in opposite faces of the plasma membrane (Figure 9.7E).

[39] Toyoshima, C. et al. 2000. *Nature* 405: 647–655.

[40] Phillipson, K. D., and Nicoll, D. A. 2000. *Annu. Rev. Physiol.* 62: 111–133.

[41] Herbert, S. C., Mount, D. B., and Gamba, G. 2004. *Pflügers Arch.* 447: 580–593.

[42] Grichtchenko, I. I. et al. 2001. *J. Biol. Chem.* 276: 8358–8363.

[43] Virkki, L. V. et al. 2003. *Am. J. Physiol.* 285: C771–C780.

[44] Schuldiner, S., Schirvan, A., and Linial, M. 1995. *Physiol. Rev.* 75: 369–392.

[45] McIntire, S. L. et al. 1997. *Nature* 389: 870–876.

[46] Sagne, C. et al. 1997. *FEBS Lett.* 417: 177–183.

[47] Fremeau, R. T., Jr. et al. 2002. *Proc. Natl. Acad. Sci. USA* 99: 14488–14493.

[48] Nelson, N., and Lill, H. 1994. *J. Exp. Biol.* 196: 213–228.

[49] Mayser, W., Schloss, P., and Betz, H. 1992. *FEBS Lett.* 305: 31–36.

[50] Yamashita, A. et al. 2005. *Nature* 437: 215–223.

[51] Palacin, M. et al. 1998. *Physiol. Rev.* 78: 969–1054.

[52] Yernool, D. et al. 2004. *Nature* 431: 811–818.

Significance of Transport Mechanisms

Primary and secondary active ion transport molecules provide essential background mechanisms for maintaining cell homeostasis. However, their roles in nervous system function often extend well beyond such relatively mundane housekeeping duties, leading to an active role of transport in cell signaling. For example, activation of sodium–potassium ATPase by sodium accumulation during action potential activity in small nerve branches can produce a transient hyperpolarization and block conduction.

Another example is related to the effect of the neurotransmitter GABA on $GABA_A$ receptors, which when activated form chloride channels (see Chapter 11). In mature brain cells GABA causes inward chloride flux and hence hyperpolarization. However, in embryonic rat brain cells GABA produces depolarization. During postnatal development, the depolarizing response gradually disappears and eventually is replaced by hyperpolarization. This change in synaptic behavior occurs because intracellular chloride concentration is relatively high in embryonic neurons, and it decreases to the mature level during the postnatal period. The change in intracellular chloride concentration is the direct result of altered chloride transport across the cell membrane. In embryonic cells, chloride transport is dominated by the inwardly directed NKCC1 transporter, so that intracellular chloride concentration is relatively high. During postnatal development, expression of NKCC1 virtually disappears.[53] At the same time, expression of the outward chloride transporter KCC2, which is absent in embryonic cells, shows a marked increase.[54,55]

Finally, it is worth noting that monoamine transporters such as SERT are prime targets for drugs (e.g., fluoxetine, Prozac) used in the treatment of psychiatric disorders such as depression and anxiety.[56]

In summary, it is useful to think of ion channels as mediating electrical signaling, and transport molecules as maintaining the background conditions under which such signaling occurs. We should remember, however, that the two types of molecules often interact in ways more complicated than this to regulate nervous system function.

[53] Plotkin, M. D. et al. 1997. *Am. J. Physiol. Cell Physiol.* 272: C173–C183.

[54] Rivera, C. et al. 1999. *Nature* 397: 251–255.

[55] Lu, J., Karadsheh, M., and Delpire, E. 1999. *J. Neurobiol.* 39: 558–568.

[56] Iverson, L. L. et al. 2009. *Introduction to Neuropsychopharmacology*. Oxford University Press, New York, pp. 306–316.

SUMMARY

- A number of membrane proteins transport substances into and out of cells. One example is sodium–potassium ATPase, which transports three sodium ions outward across the cell membrane, and two potassium ions inward, per molecule of ATP hydrolyzed. The transport system maintains intracellular sodium and potassium concentrations at constant levels in spite of steady leakage into and out of the cells.
- Calcium concentrations in the cell cytoplasm are kept low by two classes of calcium ATPases. One, plasma membrane calcium ATPase, transports calcium out of the cell. The other, endoplasmic and sarcoplasmic reticulum ATPase, concentrates calcium into intracellular compartments.
- Another mechanism for calcium transport is sodium–calcium exchange: Sodium, entering the cell down its electrochemical gradient, provides energy for outward transport of calcium. This mechanism is an example of indirect transport, which relies on maintenance of the sodium gradient by sodium–potassium ATPase. In most neurons the transport molecule exchanges three sodium ions for one calcium ion. Under some physiological conditions, the exchanger can run in reverse. In retinal rods the transport molecule carries one calcium ion and one potassium ion out of the cell in exchange for the entry of four sodium ions.
- There are two main mechanisms for extrusion of chloride from cells. One is chloride–bicarbonate exchange, which is important for intracellular pH regulation and depends on the sodium electrochemical gradient for its operation. The second is potassium–chloride cotransport, which relies on the electrochemical gradient for outward potassium movement across the cell membrane. In some cells chloride is accumulated, rather than extruded. Chloride accumulation relies on the sodium electrochemical gradient and is accompanied by inward movement of potassium.
- In presynaptic vesicles a hydrogen-transmitter exchange system transports neurotransmitters from the cytoplasm into the vesicle lumen. Neurotransmitters released from presynaptic terminals are removed from extracellular spaces by uptake systems in neuron and glial cell membranes utilizing sodium-transmitter cotransport.
- The amino acid sequences of most transport molecules are known, and predictions have been made about their configuration in the membrane. Most have nine to twelve transmembrane segments and are assumed to form pore-like structures through which ions move by alternate exposure of binding sites to the extracellular and intracellular spaces.

Suggested Reading

Carafoli, E., and Brini, M. 2000. Calcium pumps: Structural basis for and mechanisms of calcium transmembrane transport. *Curr. Opin. Chem. Biol.* 4: 152–161.

Horisberger, J. D. 2004. Recent insights into the structure and mechanism of the sodium pump. *Physiology* 19: 377–387.

Martin, D. W. 2005. Structure–function relationships in the Na^+, K^+-pump. *Semin. Nephrol.* 25: 282–281.

Mercado, A., Mount, D. B., and Gamba, G. 2004. Electroneutral cation–chloride cotransporters in the central nervous system. *Neurochem. Res.* 29: 17–25.

Philipson, K. D., and Nicoll, D. A. 2000. Sodium–calcium exchange: A molecular perspective. *Annu. Rev. Physiol.* 62: 111–133.

Russell, J. M. 2000. Sodium–potassium–chloride cotransport. *Physiol. Rev.* 80: 211–276.

Schnetkamp, P. P. 2004. The *SLC24* Na^+/Ca^{2+}- K^+ exchanger family: Vision and beyond. *Pflügers Arch.* 447: 683–688.

Torres, G. E., and Amara, S. G. 2007. Glutamate and monoamine transporters: New visions of form and function. *Curr. Opin. Neurobiol.* 17: 304–312.

CHAPTER 10
Properties and Functions of Neuroglial Cells

Nerve cells in the central and peripheral nervous system are surrounded by satellite cells. These consist of Schwann cells in the periphery and neuroglial cells in the central nervous system. In this chapter, we discuss the structure and properties of the satellite cells, their interactions with neurons, and open questions regarding their functions.

Neuroglial cells make up about one-half of the volume of the brain and greatly outnumber neurons. The main classes of neuroglial cells are oligodendrocytes, astrocytes, and radial glial cells. Microglial cells constitute a separate population of wandering phagocytotic cells in the CNS. Neurons and glial cells are densely packed. Their membranes are separated from each other by narrow fluid-filled extracellular spaces that are about 20 nanometers (nm) wide. Glial cell membranes, like those of neurons, contain channels for ions, receptors for transmitters, ion transport pumps, and amino acid transporters. In addition, glial cells are linked to each other by low-resistance gap junctions that permit direct passage of ions and small molecules. As in kidney and epithelial cells, waves of increased intracellular calcium concentration spread rhythmically through networks of glial cells connected by electrical synapses. Glial cells have more negative resting potentials than neurons. They do not generate action potentials.

Among the essential functions of satellite cells are the following: Oligodendrocytes and Schwann cells form myelin around axons, which speeds up conduction of the nerve impulse. Glial cells and Schwann cells secrete neurotrophic molecules and guide growing axons to their targets during development. Astrocytes within the central nervous system cause capillaries to become impermeable to certain molecules, and thereby create the blood–brain barrier. Microglial cells invade regions of damage or inflammation and phagocytose debris.

By virtue of the close apposition of glial and neuronal membranes, dynamic interactions occur between the two types of cells. Neurons release K^+ and transmitters into narrow extracellular spaces and thereby depolarize glial membranes. This action in turn gives rise to waves of increased cytoplasmic calcium concentration that spread through electrically coupled glial networks. Activated glial cells liberate adenosine triphosphate (ATP), glutamate, and potassium into extracellular space. They also increase the circulation of blood locally in an area of brain with high neural activity. A well-established role of glia in the central nervous system is spatial buffering through uptake of transmitters and K^+. An area of current research concerns the way in which glial cells could influence signaling at neuron-to-neuron synapses.

Nerve cells in the brain are intimately surrounded by satellite cells called **neuroglial cells**, **glial cells**, or **glia**. It has been estimated that they outnumber neurons by at least 10 to 1 and make up about one-half of the bulk of the nervous system. From the time of their discovery, the problem of what the glial cells do has posed a challenging question for neurobiologists. In spite of the prevalence of glial cells, the physiological activities of the nervous system are often discussed in terms of neurons only, as if glia did not exist. This chapter emphasizes experiments dealing with the physiological properties of glial cells and their functional interactions with neurons. There exist now textbooks[1] and journals (*Glia, Neuron Glia Biology*) devoted to glial cells.

Historical Perspective

Glial cells were first described in 1846 by Rudolf Virchow, who thought of them as "nerve glue" and gave them their name. Excerpts from a paper by Virchow give the flavor of his thinking:

> Hitherto, considering the nervous system, I have only spoken of the really nervous part of it. But. . . it is important to have a knowledge of that substance...which lies between the proper nervous parts, holds them together and gives the whole its form...[this] has induced me to give it a new name, that of neuroglia...Experience shows us that this very interstitial tissue of the brain and spinal marrow is one of the most frequent sites of morbid change...Within the neuroglia run the vessels, which are therefore nearly everywhere separated from the nervous substance by a slender intervening layer, and are not in immediate contact with it.[2]

In subsequent years, neuroglial cells were studied intensively by neuroanatomists and pathologists, who knew them to be the most common source of tumors in the brain. This is not so surprising, because certain glial cells—unlike the vast majority of neurons—can still divide in the mature animal. Among early speculations about glial cell function in relation to neurons were structural support, the provision of nutrients to neurons, secretion of trophic molecules, and the prevention of "cross talk" by current spread during conduction of nerve impulses.[3–5]

Appearance and Classification of Glial Cells

A distinct structural feature of neuroglial cells compared with neurons is the absence of axons. A representative picture of mammalian neuroglial cells is shown in Figure 10.1. In the vertebrate central nervous system, glial cells are usually subdivided into several distinctive classes.[1,6]

Astrocytes make contact with capillaries and neurons. There are two principal subgroups: (1) fibrous astrocytes, which contain filaments and are prevalent among bundles of myelinated nerve fibers in the white matter of the brain; and (2) protoplasmic astrocytes, which contain less fibrous material and are abundant in the gray matter around nerve cell bodies, dendrites, and synapses.

Oligodendrocytes are predominant in the white matter, where they form myelin around larger axons (see Chapter 8).

Radial glial cells play an essential role in the developing central nervous system. They stretch through the thickness of the spinal cord, retina, cerebellum, or cerebral cortex to the surface, forming elongated filaments along which developing neurons migrate to their final destinations. In addition, cells resembling radial glia have been shown to generate neurons in the developing cerebral cortex. In adult CNS, radial glia are represented by Bergmann cells in the cerebellum and Müller cells in the retina.

Ependymal cells that line the inner surfaces of the brain, in the ventricles, are usually classified as glial cells.

Microglial cells are distinct from neuroglial cells in structure, properties, and lineage.[7,8] They resemble macrophages in the blood and probably arise from them.

In vertebrate peripheral nerves and ganglia, **Schwann cells** are analogous to glial cells. They form myelin around the fast-conducting axons. Schwann cells also enclose smaller axons (less than 1 micrometer [μm] in diameter) without forming a myelin sheath.

---

[1] Kettenmann, H., and Ransom, B. R. (eds.). 2005. *Neuroglia*, 2nd ed. Oxford University Press, New York.

[2] Virchow, R. 1959. *Cellularpathologie*. Hirschwald, Berlin. (Excerpts are from pp. 310, 315, and 317.)

[3] Golgi, C. 1903. *Opera Omnia*, Vols. 1 and 2. U. Hoepli, Milan, Italy.

[4] Ramón y Cajal, S. [1909–1911] 1995. *Histology of the Nervous System*, Vol. 1. Oxford University Press, New York.

[5] Webster, H., and Aström, K. E. 2009. *Adv. Anat. Embryol. Cell Biol.* 202: 1–109.

[6] Butt, A. M. et al. 2005. *J. Anat.* 207: 695–706.

[7] Ransohoff, R. M., and Perry, V. H. 2009. *Annu. Rev. Immunol.* 27: 119–145.

[8] Farber, K., and Kettenmann, H. 2005. *Brain Res. Brain Res. Rev.* 48: 133–143.

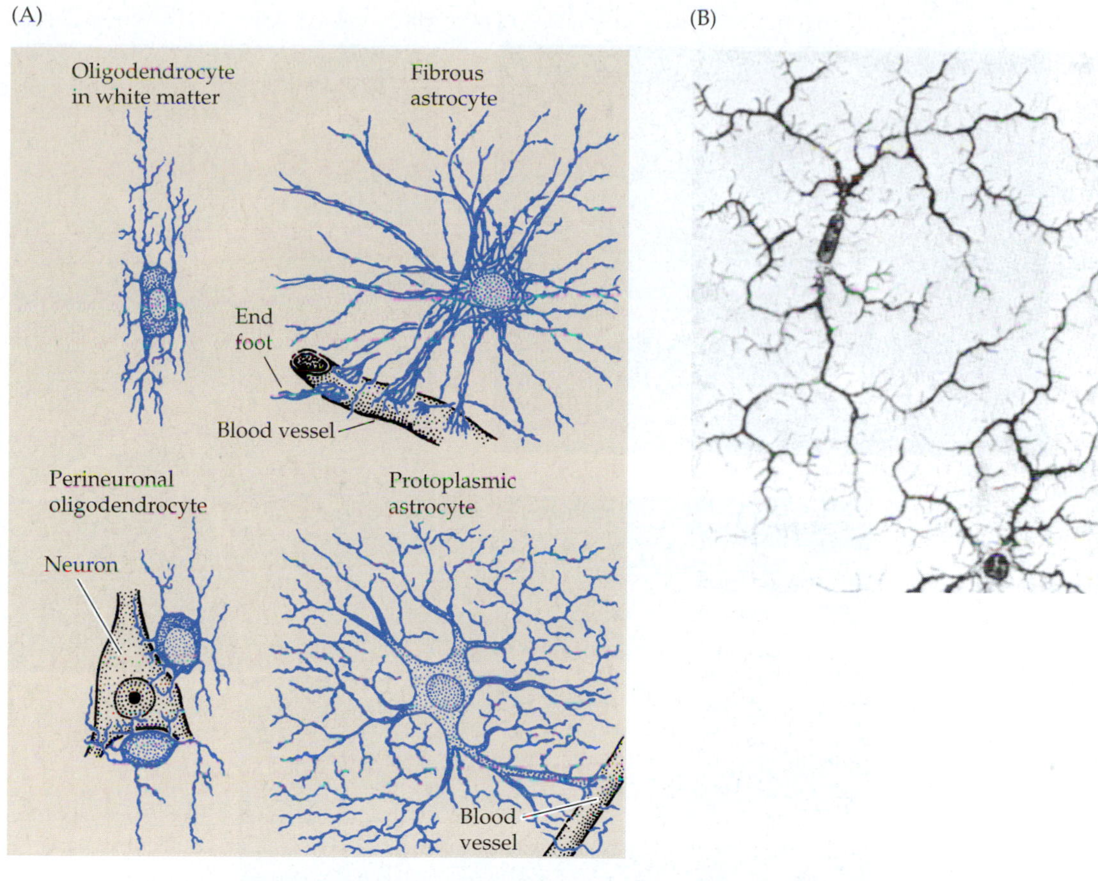

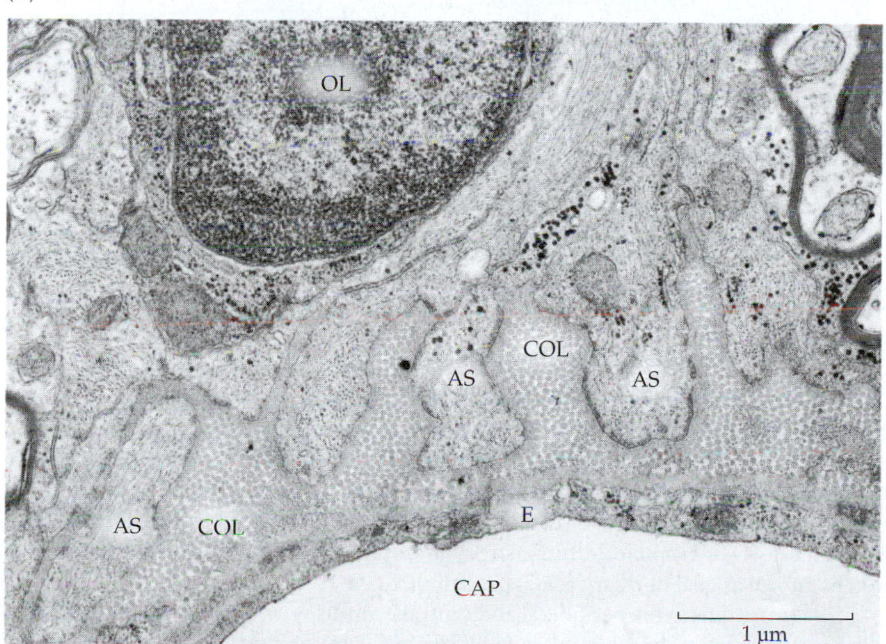

FIGURE 10.1 Neuroglial Cells in Mammalian Brain. (A) Oligodendrocytes and astrocytes, stained with silver impregnation, represent the principal types of neuroglial cells in vertebrate brain. They are closely associated with neurons. (B) Microglial cells are small, wandering, macrophage-like cells. (C) Electron micrograph of glial cells in rat optic nerve. In the lower portion is the lumen of a capillary (CAP) lined with endothelial cells (E). The capillary is surrounded by end feet formed by processes of fibrous astrocytes (AS). Between the end feet and the endothelial cells is a space filled with collagen fibers (COL). In the upper portion part of a nucleus of an oligodendrocyte (OL) is visible, and to the right axons are surrounded by myelin wrapping. (A and B after Penfield, 1932, and Del Rio-Hortega, 1920; C from Peters, Palay, and Webster, 1991.)

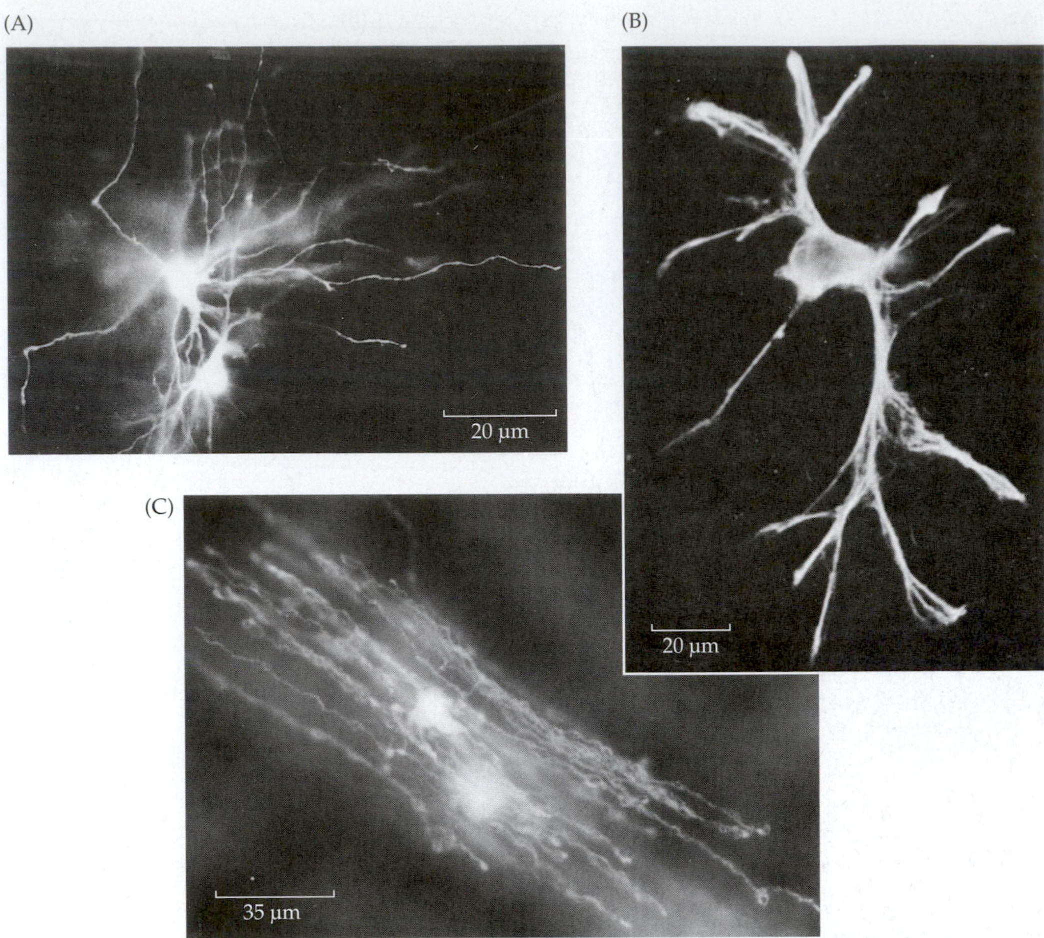

FIGURE 10.2 Glial Cells Labeled by Intracellular Injection and Antibodies. (A) Dye coupling of astrocytes in intact rat optic nerve. An astrocyte was filled with Lucifer yellow by intracellular injection. The dye spread to a neighboring astrocyte. (B) Fibrous astrocyte, freshly dissociated from salamander optic nerve, labeled by an antibody against glial fibrillary acidic protein (GFAP). The antibody staining and the shape unequivocally establish its identity. (C) Oligodendrocytes in rat optic nerve filled with Lucifer yellow by intracellular injection. The appearance of the longitudinal processes that run symmetrically in parallel is characteristic for oligodendrocytes. (A and C after Butt and Ransom, 1993, photos kindly provided by A. M. Butt; B from Newman, 1986.)

The term **satellite cell** is used in this chapter to refer collectively to non-neuronal cells in general; it includes glial cells in the CNS and Schwann cells in the periphery.

The various types of satellite cells can be distinguished by injecting them with labels, such as dyes in living preparations or by immunological techniques (Figure 10.2). Antibodies have been prepared that bind specifically to astrocytes, oligodendrocytes, microglia, or Schwann cells.[9] For example, fibrous astrocytes can be stained with an antibody against a protein known as glial fibrillary acidic protein or GFAP (see Figure 10.2B).[10]

Like the neurons in the peripheral and central nervous system, glial cells and Schwann cells have different embryological origins. Glial cells in the central nervous system are derived from precursor cells that line the neural tube, whereas Schwann cells arise from the neural crest (see Chapter 25). In animals such as the leech[11] and zebrafish[12] glial cell development can be observed directly in the living embryo. Precursor cells can be labeled by injection of a marker or by infection at an early stage with a virus encoding a labeled gene that is handed on to the descendants.[13] The labeled cells can then be identified as astrocytes or oligodendrocytes. In this way one can follow cell lineages and pinpoint the

[9] Zuo, Y. et al. 2004. *J. Neurosci.* 24: 10999–11009.

[10] Bignami, A., and Dahl, D. 1974. *J. Comp. Neurol.* 153: 27–38.

[11] Stent, G. S., and Weisblat, D. A. 1985. *Annu. Rev. Neurosci.* 8: 45–70.

[12] Lewis, K. E., and Eisen, J. S. 2003. *Prog. Neurobiol.* 69: 419–449.

[13] Luskin, M. B. 1998. *J. Neurobiol.* 36: 221–233.

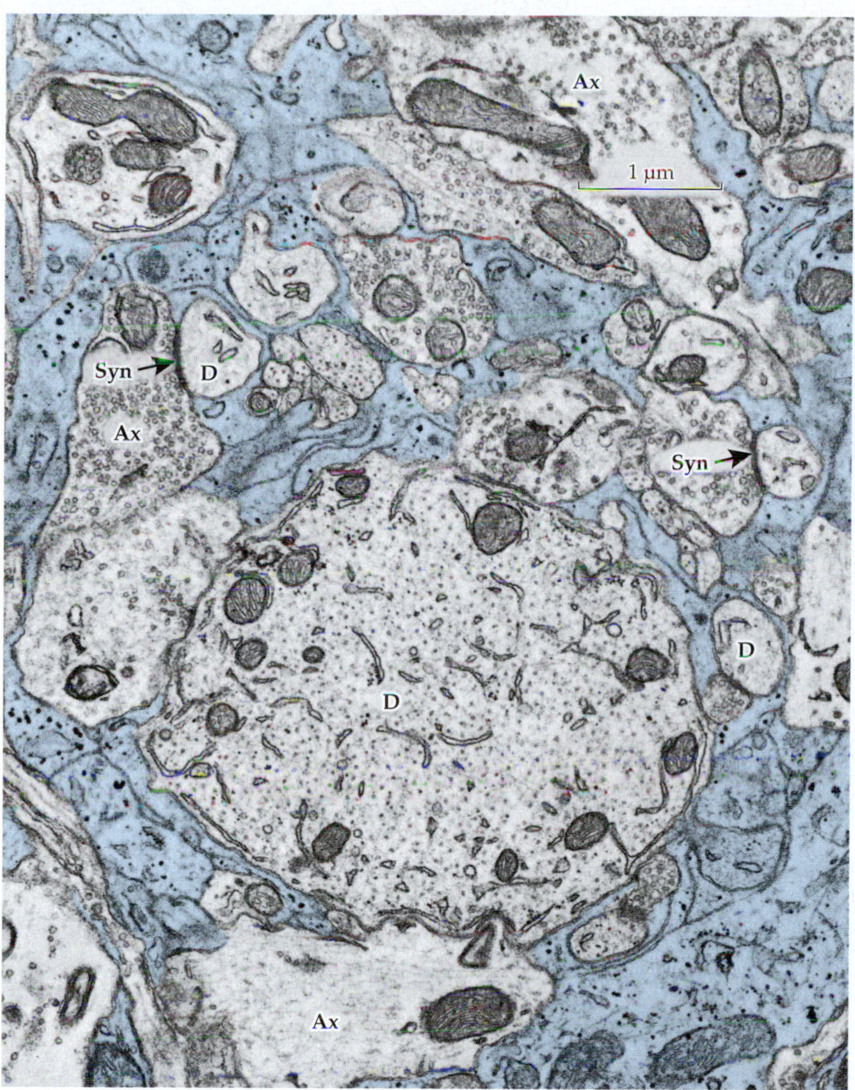

FIGURE 10.3 Neurons and Glial Processes in Rat Cerebellum. The glial contribution is lightly colored. The neurons and glial cells are always separated by clefts about 20 nm wide. The neural elements are dendrites (D) and axons (Ax). Two synapses (Syn) are marked by arrows. (From Peters, Palay, and Webster, 1991.)

stages at which glial cells diverge from neurons during development. Microglial cells arise from mesoderm and are able to move to sites of injury.

Structural Relations between Neurons, Glia, and Capillaries

A glance at electron micrographs of brain tissue brings home the close packing of neurons and glia. Figure 10.3 shows an example from the cerebellum of a rat. The section is filled with neurons and glial cells, which can be distinguished by a number of criteria. Glial processes tend to be thin, at times less than 1 μm thick. Larger volumes of glial cytoplasm appear only around the glial nuclei. The extracellular space is restricted to narrow clefts, about 20 nm wide that separate all cell boundaries. No special connections, such as synaptic structures or gap junctions, are seen between neurons and glia in adult CNS, and physiological tests fail to reveal direct low-resistance pathways between them.[14] During development, occasional structures resembling synapses have been observed between glial precursors and neurons.[15] Electrical synapses do, however, link glial cells to one another through gap junctions (see Chapter 7).[14,16] The relation between glial cells, neurons, capillaries, and extracellular space is diagrammed in Figure 10.4.

[14]Kuffler, S.W., and Potter, D. D.1964. *J. Neurophysiol.* 27: 290–320.

[15]De Biase, L. M. et al. 2010. *J. Neurosci.* 30: 3600–3611.

[16]Scemes, E. et al. 2007. *Neuron Glia Biol.* 3: 199–208.

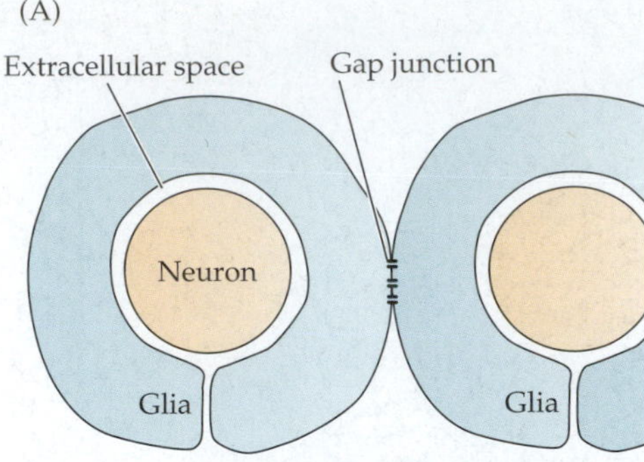

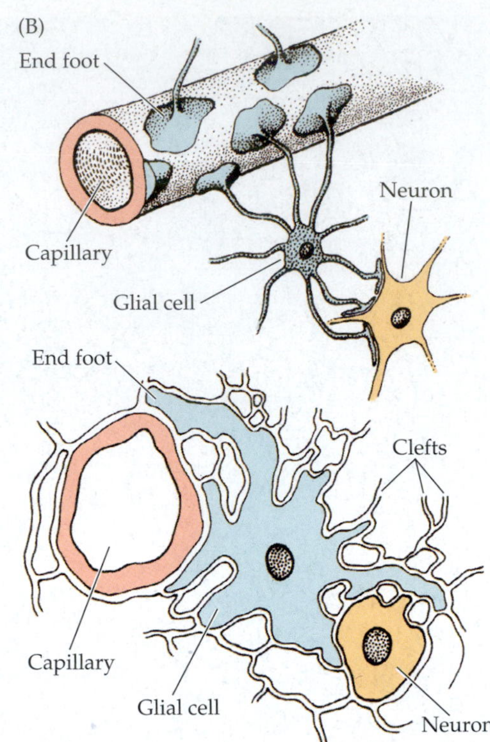

FIGURE 10.4 Neurons, Glia, Extracellular Space, and Blood. (A) Neuronal–glial and glial–glial relationships. While neurons are always separated from glia by continuous clefts, the interiors of glial cells are linked by gap junctions. (B) Relations of capillaries, glia, and neurons as seen by light and electron microscopy. The pathway for diffusion from the capillary to the neuron is through the aqueous intercellular clefts. Cell dimensions are not in proportion. (After Kuffler and Nicholls, 1966.)

Physiological Properties of Neuroglial Cell Membranes

For technical reasons, Kuffler and Potter[14] used the CNS of the leech to record from glial cells—at a time when nothing was known about their membrane properties. The glial cells in a leech ganglion are large and transparent (see Figure 18.12). They appear under the dissecting microscope as spaces between nerve cells and can be recorded from with sharp microelectrodes or by patch electrodes.[17] After one has established its physiological properties, the glial cell can be injected with a fluorescent marker, such as Lucifer yellow, and its form observed in the living preparation. Once leech glial cells had been described, it became practicable to record from and label amphibian and mammalian glial cells,[18] which were found to share many key properties with the leech glia. Far from being a roundabout approach, use of the leech turned out to be a shortcut to the study of glia in the vertebrate CNS.[19]

As in the leech, mammalian glial cells have resting potentials greater (more negative inside) than those of neurons. The largest membrane potentials recorded from neurons are about –75 mV, whereas the values for glial cells approach –90 mV. Another distinguishing feature of mature glial cells in situ is the absence of conducted action potentials.

The glial membrane behaves like a potassium electrode; that is, its behavior follows the Nernst equation in solutions containing different potassium concentrations (see Chapter 6). Though other channels are present, sodium and chloride make only a small contribution to the resting membrane potential.[20]

The distribution of potassium channels has been explored over the surface of Müller glial cells and astrocytes isolated from the retina and optic nerves of many species.[21] The potassium sensitivity is distributed in a characteristic pattern, being highest over the end feet and lower over the soma of the Müller cell. Figure 10.5 shows an isolated salamander Müller cell and its responses to a high concentration of potassium applied to different regions of the surface from a pipette. At early stages of development, potassium channels are distributed more uniformly over Müller cells. Thus, the high concentration at the end feet occurs as tadpoles become transformed into adult frogs.[22] The possible significance is discussed later in this chapter.

[17] Müller, M, and Schlue, W. R. 1998. *Brain Res.* 781: 307–319.

[18] Ransom, B. R., and Sontheimer, H. 1992. *J. Clin. Neurophysiol.* 9: 224–251.

[19] Kuffler, S. W., and Nicholls, J. G. 1966. *Ergeb. Physiol.* 57: 1–90.

[20] Kuffler, S. W., Nicholls, J. G., and Orkand, R. K. 1966. *J. Neurophysiol.* 29: 768–787.

[21] Brew, H. et al. 1986. *Nature* 324: 466–468.

[22] Rojas, L., and Orkand, R. K. 1999. *Glia* 25: 199–203.

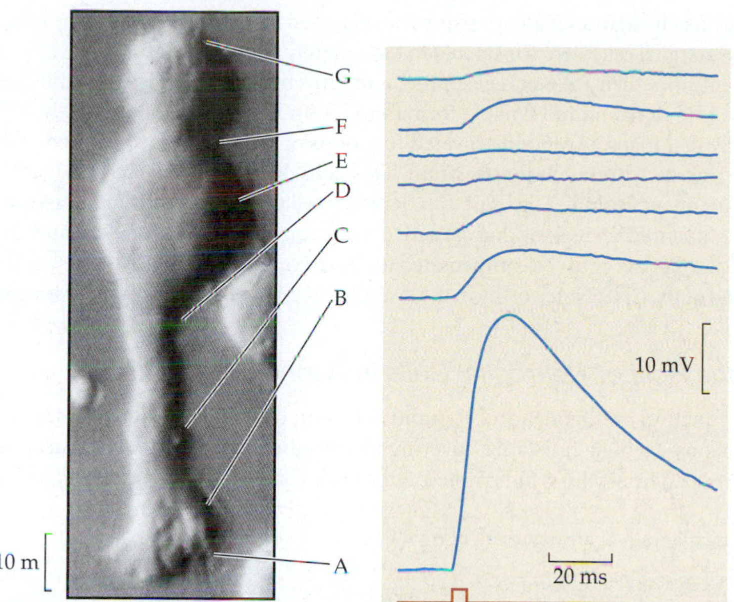

FIGURE 10.5 Responses to Potassium of a Müller Glial Cell isolated from salamander retina. Recordings were made with an intracellular electrode while potassium was applied to different sites. A is the end foot, and G is the distal part of the cell. The sensitivity to potassium is much greater at the end foot, suggesting a higher concentration of potassium channels in that region. (After Newman, 1987; micrograph kindly provided by E. A. Newman.)

Ion Channels, Pumps, and Receptors in Glial Cell Membranes

Glial cells and Schwann cells display a variety of ion channels and pumps in their membranes:

1. As already shown, potassium conductances predominate.[19]
2. Voltage-activated sodium and calcium channels are present in the membranes of Schwann cells and astrocytes.[23] In Müller cells, the overall ratio of potassium permeability to sodium permeability cells has been estimated to be approximately 100:1. As previously mentioned, in spite of occasional claims to the contrary,[24] the activation of sodium and calcium channels does not give rise to action potentials in mature glial cells. During development, action potentials have been seen in glial precursors. (see De Biase et al.[15])
3. Patch clamp recordings reveal the presence of chloride channels in Schwann cells and astrocytes.[25]
4. Ion pumps for transport of sodium and potassium as well as of bicarbonate and protons occur in glial cells.[26]
5. Transporters for glutamate, γ-aminobutyric acid (GABA), and glycine are abundant in glial membranes: They take up transmitter liberated by neurons.[27]
6. Oligodendrocytes, astrocytes, and Schwann cells display receptors for transmitters.[28,29]

Electrical Coupling between Glial Cells

Adjacent glial cells are linked to each other by gap junctions (see Chapter 11). In this respect they resemble epithelial and gland cells as well as smooth muscle and heart muscle fibers. Ions and small molecules are exchanged directly between cells without passing through the extracellular space, and such interconnections may be used for reducing concentration gradients.[16,19,30,31] Optical measurements of intracellular calcium show that waves and brief spikes of increased concentration occur in activated glial cells and spread from one coupled cell to the next (see below). As in the leech, no gap junctions between neurons and glial cells have been detected in vertebrate glial cells. Current flow across nerve membranes has no direct effect on the neighboring glial membrane.

Functions of Neuroglial Cells

Over the years, almost every nervous system task for which no other obvious explanation has been found has been attributed to glial cells. In the following sections, we first discuss well-established functions of glial cells and then intriguing questions about their functional role,

[23] Rose, C. R., Ransom, B. R., and Waxman, S. G. 1997. *J. Neurophysiol.* 78: 3249–3258.

[24] Barres, B. A., Chun, L. L., and Corey, D. P. 1988. *Glia* 1: 10–30.

[25] Ritchie, J. M. 1987. *J. Physiol. (Paris)* 82: 248–257.

[26] Blaustein, M. P. et al. 2002. *Ann. N.Y. Acad. Sci.* 976: 356–366.

[27] Marcaggi, P., and Attwell, D. 2004. *Glia* 47: 217–225.

[28] Karadottir, R. et al. 2005. *Nature* 438: 1162–1166.

[29] Verkhratsky, A., Krishtal, O. A., and Burnstock, G. 2009. *Mol. Neurobiol.* 39: 190–208.

[30] Loewenstein, W. R. 1999. *The Touchstone of Life*. Oxford University Press, New York.

[31] Iglesias, R. et al. 2009. *J. Neurosci.* 29: 7092–7097.

which require further elucidation. Modern papers on glia often contain sentences to the effect that "For many years, glial cells were considered to have merely a supporting function..."—a puzzling statement since their roles in development and myelin formation were established long ago. Their role in signaling and synapse formation is now being studied intensively.[32–35]

It seems likely that many as yet undiscovered functions of CNS glia are concerned with maintaining the microenvironment for the brain. This densely packed, delicate structure is situated within an inflexible skull. The brain can neither swell nor shrink without disastrous consequences for function. Yet regions that are active receive an increased blood flow, and any swelling that follows damage must be compensated for by decreased volume elsewhere within the bony surround of the CNS. Noteworthy in the brain is the absence of any lymphatic system.

[32] Perea, G., Navarrete, M., and Araque, A. 2009. *Trends Neurosci.* 32: 421–431.

[33] Halassa, M. M., and Haydon, P. G. 2010. *Annu. Rev. Physiol.* 72: 335–355.

[34] Allen, N. J., and Barres, B. A. 2005. *Curr. Opin. Neurobiol.* 15: 542–548.

[35] Barres, B. A. 2008. *Neuron* 60: 430–440.

Myelin and the Role of Neuroglial Cells in Axonal Conduction

One important function of oligodendrocytes and Schwann cells is to produce the myelin sheath around axons—a high-resistance covering akin to the insulating material around wires (see Chapter 8). The myelin is interrupted at the nodes of Ranvier (Figure 10.6), which

(A) Schematic diagram of arrangement of myelin

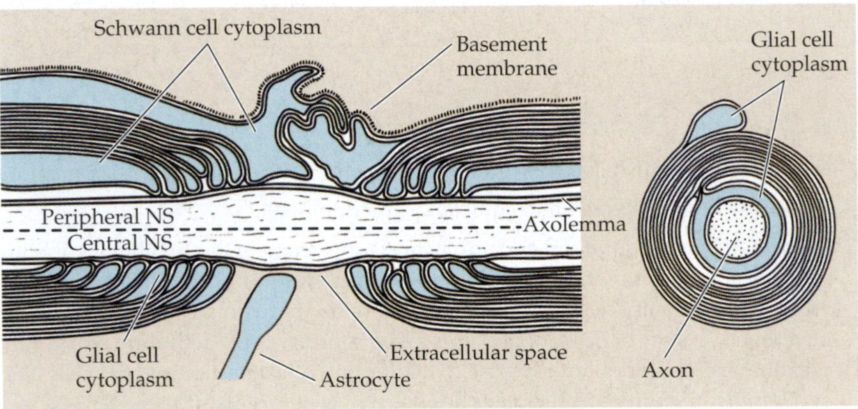

(B)

Ax

0.5 μm

(C)

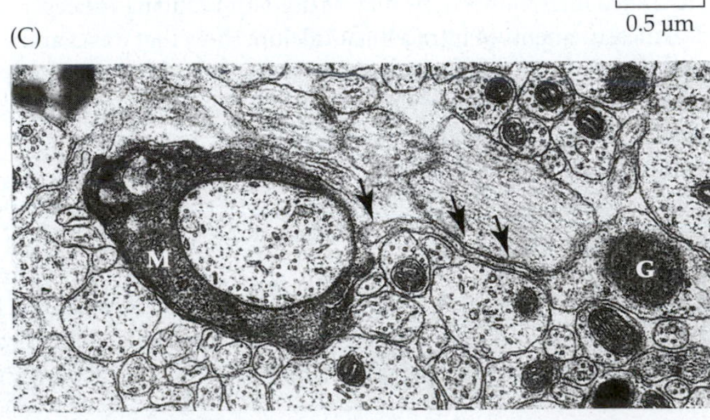

FIGURE 10.6 Myelin and Nodes of Ranvier. Oligodendrocytes and Schwann cells form the wrapping of myelin around axons. (A) At the nodes of Ranvier, like the one shown here on the left, the myelin is interrupted and the axon is exposed. The upper half of the nodal region, with a loose covering of processes, is typical of the arrangement in peripheral nerves. The lower part is representative of a node within the central nervous system. Here an astrocytic finger comes into close apposition with the nodal membrane. To the right is a transverse section through a myelin-covered axon. (B) Electron micrograph of a nodal region in a myelinated fiber in rat CNS. At the edge of the node is a specialized close-contact area between the membrane of the axon (Ax) and the membrane of the myelin wrapping (arrows). (C) Cross section of a myelinated axon at a node that is contacted by a process (marked with arrows) from a perinodal astrocytic glial cell (G). Myelin (M) is absent at the site of contact between the astrocyte and the node. (A after Bunge, 1968; B from Peters, Palay, and Webster, 1991; C from Sims et al., 1985; micrograph kindly provided by J. Black and S. Waxman.)

occur at regular intervals.[36] Since the ionic current associated with the conducted nerve impulse cannot flow across the myelin, ions move in and out at the nodes. As a result, the conduction velocity is increased. At nodes within the CNS, a characteristic feature is the presence of astrocytic processes that contact the axon.[37]

The association of Schwann cells or oligodendrocytes with axons to form myelin raises a number of interesting problems. For example, what are the genetic or environmental factors that enable glial cells to select the appropriate axons, to surround them at the right time, and to maintain myelin sheaths around them? What neurological disorders result from disease or genetic abnormalities of myelin? For the formation of the myelin sheath during development, complex and precise interactions occur between axons and satellite cells. The spacing of the nodes, the seals between the two types of cells at the paranodal areas, and the distribution of sodium and potassium channels must be matched and positioned for rapid conduction.

The dynamic interactions between neurons and Schwann cells have been analyzed by experiments made in living fish embryos and in tissue culture.[38–40] Schwann cell development, myelination and remyelination of axons, parallel the events that occur in vivo. At the molecular level, key proteins that play a part in Schwann cell–axon interactions have been identified. For example, Shooter and his colleagues have shown that when Schwann cells are cultured in a dish on their own, a peripheral myelin protein (known as PMP22) is synthesized.[41] Under these conditions, the turnover of PMP22 is rapid, and it is degraded in the endoplasmic reticulum. When neurons are introduced to the culture, as illustrated in Figure 10.7, the fate of the protein changes. After contact between Schwann cells and axons, PMP22 is translocated to the Schwann cell membrane. This is an essential step for myelin to be formed. The signals that pass between neurons and Schwann cells are not yet known, but it has been shown that a potent neurotrophin, a nerve growth factor (see Chapter 24), can regulate myelin formation.[42,43]

The exact amount of PMP22 that is produced is critical for proper myelination; with over- or underexpression of PMP22, disorders occur. Figure 10.8 shows that a change in a single amino acid of PMP22 (e.g., from leucine to proline) results in "trembler" mice, which exhibit deficient myelination and serious neurological defects. Hereditary human neuropathies arise in families with the same mutation.

[36]Bunge, R. P. 1968. *Physiol. Rev.* 48: 197–251.

[37]Black, J. A., and Waxman, S. G. 1988. *Glia* 1: 169–183.

[38]Buckley, C. E. et al. 2010. *Glia* 58: 802–812.

[39]Liu, N. et al. 2005. *J. Neurosci. Res.* 79: 310–317.

[40]Nave, K. A., and Trapp, B. D. 2008. *Annu. Rev. Neurosci.* 31: 535–561.

[41]Pareek, S. et al. 1997. *J. Neurosci.* 17: 7754–7762.

[42]Chan, J. R. et al. 2004. *Neuron* 43: 183–191.

[43]Xiao, J., Kilpatrick, T. J., and Murray, S. S. 2009. *Neurosignals* 17: 265–276.

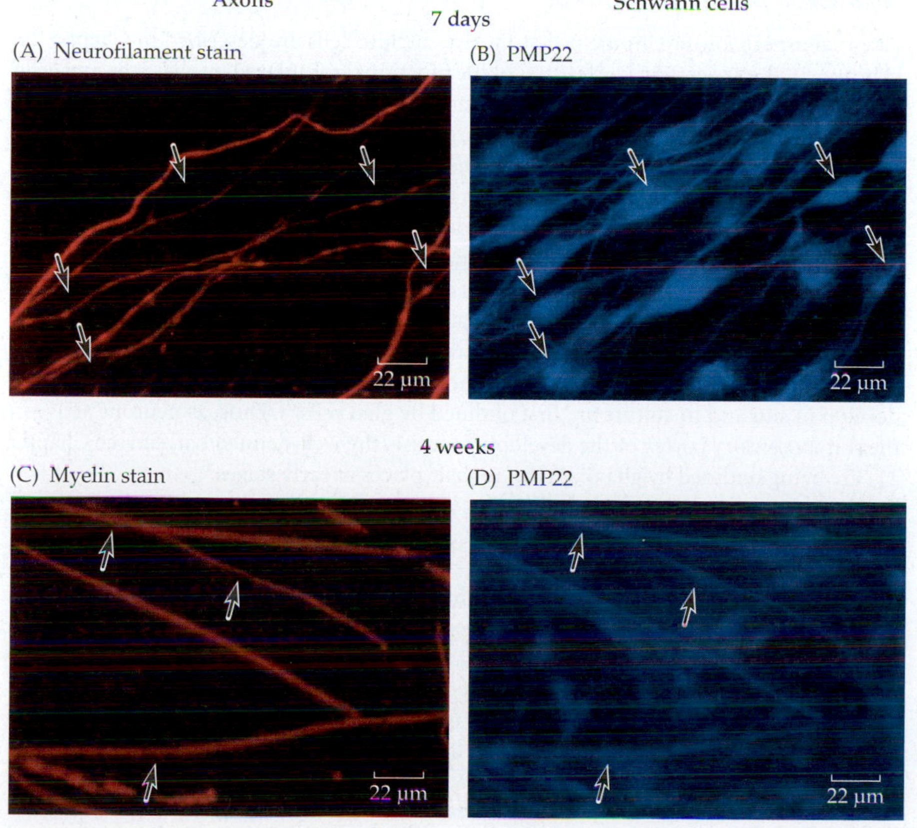

FIGURE 10.7 Localization of Myelin Protein (PMP22) in short- and long-term myelinating cultures of axons and Schwann cells. Changes in distribution of myelin protein PMP22 induced by co-culture of axons with Schwann cells. (A, B) A 1-week-old co-culture of neuronal axons (A) and Schwann cells (B) that are doubly stained with monoclonal antineurofilament and polyclonal PMP22 antiserum. Arrows point to Schwann cells that are in contact with neuronal processes. At this early stage the glial and neuronal proteins have different distributions, with PMP22 mainly in Schwann cell bodies. (C,D) After 4 weeks in medium that promotes myelination, PMP22 becomes co-localized with myelin segments (stained by antibody P0). Arrows point to axons (C) and to the cell bodies of elongated Schwann cells (D), with uniform PMP22 staining over the cell membrane. (From Pareek et al., 1997; photos kindly provided by E. Shooter.)

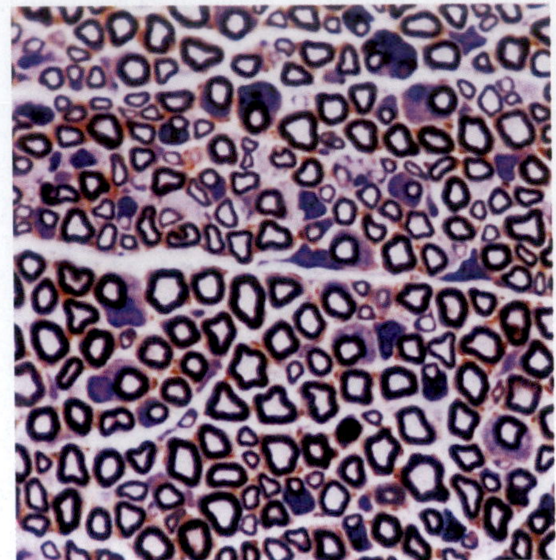

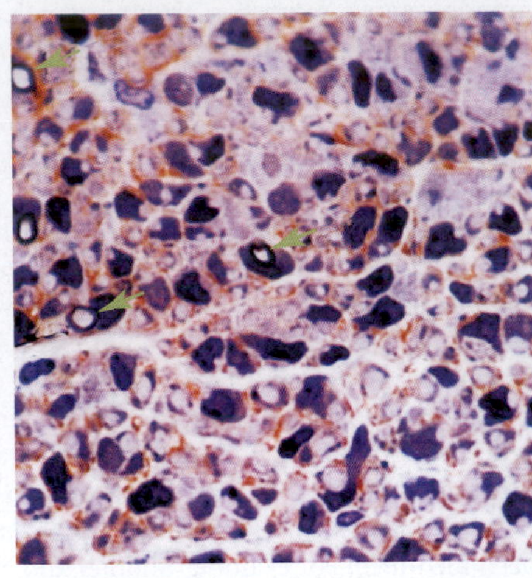

FIGURE 10.8 Deficient Myelination in "Trembler" Mutant Mice with a genetic defect in a myelin protein, PMP22. Morphological appearance of sciatic nerves in normal (A) and mutant trembler mice (B), aged 10 days. Note the marked differences in axon caliber and myelin thickness between normal and trembler mice (indicated by arrows in B) in microscopic sections at equivalent magnifications. Also note the severity of dysmyelination. A single amino acid mutation from leucine to proline produces trembler neuropathy in mice and in humans. (From Notterpek, Shooter, and Snipes, 1997.)

A number of experiments have provided evidence that glial cells can influence the clustering of sodium channels in myelinated nerve fibers. Changes occur in the distribution of ion channels in nodes, paranodal areas, and internodes as axons become myelinated, demyelinated, and remyelinated.[44,45] The process resembles the clustering of transmitter receptors at postsynaptic sites brought about by innervation. Astrocytic fingers in the nodal region themselves show intense labeling with saxitoxin (a toxin that binds to sodium channels; see Chapter 4), indicating a high density of sodium channels in the glial membrane.[46] It has been proposed that transfer of sodium channels might occur from astrocytes to nodes of Ranvier,[47] but there is no direct evidence for this interesting speculation.

Glial Cells and Development

Essential aspects of development that involve satellite cells are described in Chapter 25. Here we emphasize certain key features of the functional role played by glia, Schwann cells, and microglia. For example, glial and Schwann cells secrete molecules, such as laminin and nerve growth factor,[48,49] that promote the outgrowth of neurites, both in culture and in the animal. Glial cells can also act as repellents that inhibit neurite outgrowth.[50] An inhibitory protein, known as NOGO-A, is present in myelin and oligodendrocytes. This molecule stops neuronal growth cones from advancing and causes their collapse.[51] Growth-inhibiting proteins could help to delineate tracts and boundaries in the spinal cord: growing fibers would be stopped from inappropriately entering or wandering out of specified pathways. The possible effects of these proteins on regeneration after injury are discussed in Chapter 27.

During CNS development, glial cells have been shown to play a role during the aggregation of neurons into well-defined nuclei. Prospective nuclei and structures that develop in situ and in culture are first outlined by glial cells.[52] Thus, as neurons arrive in the somatosensory cortex of the developing mouse, the well-defined barrels (see Chapter 21) are being outlined by glia that take up their places at early stages.[53]

A mechanism by which radial glial cells guide neuronal migration during development has been demonstrated in experiments by Rakic,[54,55] Hatten,[56,57] and their colleagues (see also Chapter 25). During development of the cerebral cortex, hippocampus, and cerebellum in both monkeys and humans, nerve cells migrate to their destinations along radial glial processes. A time-lapse sequence of the movement of a hippocampal neuron along a radial glial cell is shown in Figure 10.9. Migrating neurons recognize surface molecules on the glial processes specific for their neuronal type. For example, hippocampal radial glial cells provide guidance cues for hippocampal but not for cortical neurons. The mature neurons send out axons and form connections with their targets.

An unexpected new finding is that during development neurons can be produced by division of cells that resemble radial glial cells. Thus, in addition to providing scaffolds for

[44] Susuki, K., and Rasband, M. N. 2008. *Curr. Opin. Cell Biol.* 20: 616–623.
[45] Feinberg, K. et al. 2010. *Neuron* 65: 490–502.
[46] Ritchie, J. M. et al. 1990. *Proc. Natl. Acad. Sci. USA.* 87: 9290–9294.
[47] Shrager, P., Chiu, S. Y., and Ritchie, J. M. 1985. *Proc. Natl. Acad. Sci. USA* 82: 948–952.
[48] Yu, W. M., Yu, H., and Chen, Z. L. 2007. *Mol. Neurobiol.* 35: 288–297.
[49] Bampton, E. T., and Taylor, J. S. 2005. *J. Neurobiol.* 63: 29–48.
[50] Caroni, P., and Schwab, M. E. 1988. *J. Cell Biol.* 106: 1281–1288.
[51] Schwab, M. E. 2004. *Curr. Opin. Neurobiol.* 14: 118–1124.
[52] Faissner, A., and Steindler, D. 1995. *Glia* 13: 233–254.
[53] Steindler, D. A. et al. 1989. *Dev. Biol.* 131: 243–260.
[54] Rakic, P. 1981. *Trends Neurosci.* 4: 184–187.
[55] Rakic, P. 2003. *Cereb Cortex.* 13: 541–549.
[56] Hatten, M. E. 1999. *Annu. Rev. Neurosci.* 22: 511–539.
[57] Solecki, D. J. et al. 2004. *Nat. Neurosci.* 7: 1169–1170.

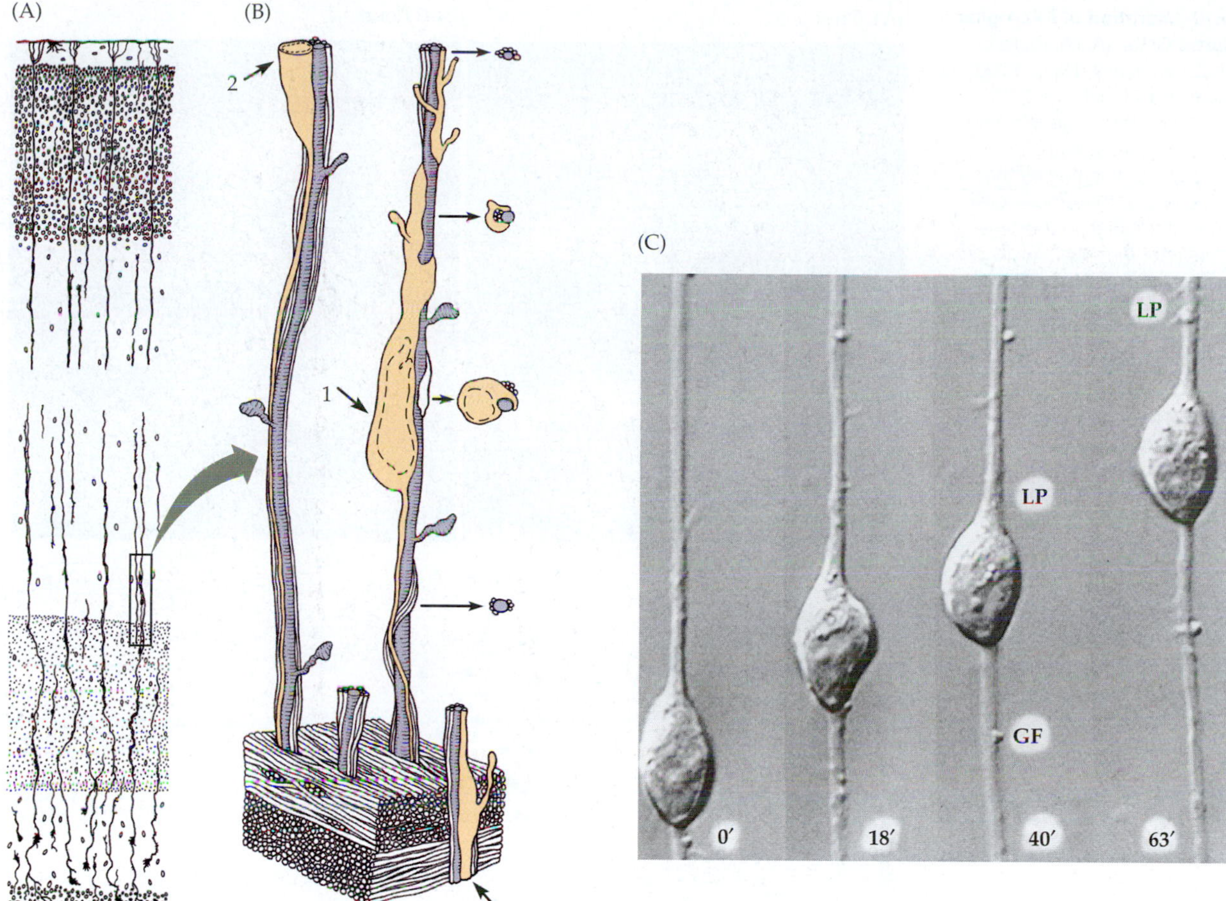

FIGURE 10.9 Neurons Migrating along Radial Glia during development. (A) Camera lucida drawing of the occipital lobe of developing cortex of a monkey fetus at mid-gestation. Radial glial fibers run from the ventricular zone below to the surface of the developing cortex above. (B) Three-dimensional reconstruction of migrating neurons. The migrating cell (1) has a voluminous leading process that follows the radial glia, using it as guideline. Cell 2, which has migrated farther, retains a process still connected to the radial glia. Cell 3 is beginning to send a process along the radial glia before migrating. (C) Migration of a hippocampal neuron along a radial glial fiber (GF) in vitro. As time progresses, the leading process (LP) moves farther up, with the neuronal cell body following. Times indicated at the bottom represent real time in minutes, taken from video photography. (A and B after Rakic, 1988; C from Hatten, 1990.)

neuronal migration, radial glia act as stem cells for neurogenesis.[58–60] In particular, radial glial cells in the developing cortex give rise to projection neurons, such as those that send axons in the pyramidal tracts to spinal cord motoneurons.[61]

Role of Microglial Cells in CNS Repair and Regeneration

Astrocytes, microglia, and Schwann cells react to neuronal injury by replication. They participate in the removal of debris and in scar formation (see also Chapter 27).[7,8,62] As a first step, resident microglial cells and macrophages, which invade damaged CNS from blood at the site of an injury divide, scavenge debris from dying cells.

In the leech, the role of microglial cells in regeneration has been studied by Muller and his colleagues.[63] (As an aside, it may be mentioned that it was in the CNS of the leech that such wandering cells first were given the name "microglia," by del Rio-Hortega.[64]) Normally microglial cells are evenly distributed in leech ganglia and in the bundles of axons that link them (Figure 10.10). Immediately after damage to the CNS, microglial cells migrate to the site of the lesion, at a rate of about 300 μm/hour. There they accumulate, phagocytose damaged tissue, and produce laminin—a molecule that promotes neurite outgrowth (see

[58] Hansen, D.V. et al. 2010. *Nature* 464: 554–561.

[59] Rakic, P. 2003. *Glia* 43: 19–32.

[60] Gotz, M., and Barde, Y. A. 2005. *Neuron* 46: 369–372.

[61] Hevner, R. F. 2006. *Mol. Neurobiol.* 33: 33–50.

[62] Slobodov, U. et al. 2009. *J. Mol. Neurosci.* 39: 99–103.

[63] Chen, A. et al. 2000. *J. Neurosci.* 20: 1036–1043.

[64] Del Rio-Hortega, P. 1920. *Trab. Lab. Invest. Biol. Madrid.* 18: 37–82.

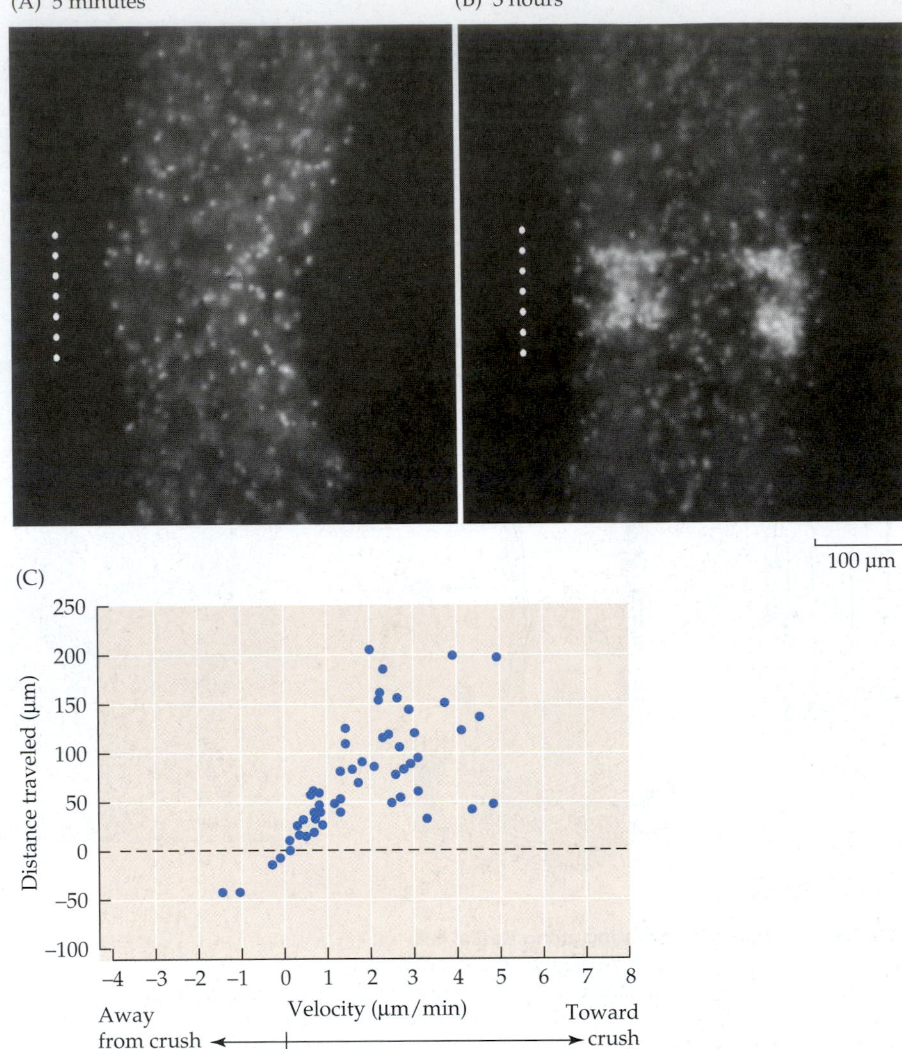

FIGURE 10.10 Migration of Microglial Cells in Injured CNS. (A) Microglia in the leech CNS were stained with a fluorescent nuclear dye (Hoechst 33342). The bundle of axons linking ganglia had been crushed 5 minutes earlier. The extent of the crush is indicated by the dotted line. The nuclei of microglial cells were still evenly distributed at this time. (B) Three hours after the injury, microglial cells had accumulated at the crush site. There they produced the growth-promoting molecule laminin. (C) Velocities and distances traveled by microglial cells as they moved toward a lesion in leech CNS. Microglial cells were tracked by video-microscopy at 10-minute intervals in injured leech preparations. In uninjured preparations, microglial cells make only short, random movements. (A and B from Chen et al., 2000; micrographs kindly provided by K. J. Muller; C after McGlade-McCulloh et al., 1989.)

Chapter 25). What causes the microglial cells to become activated and to move rapidly to the place where the nervous system was damaged? Several lines of evidence show that microglia become activated to move by extracellular accumulation of ATP, which is released after injury. At the same time attraction toward the lesion, as opposed to random movement, is mediated by the gas nitric oxide (NO). NO activates a soluble guanylate cyclase that acts as an attractant for directed microglial migration.[65,66] In mice, as in leeches, nitric oxide activates the migration of microglial cells towards spinal cord lesions.[67]

Schwann Cells as Pathways for Outgrowth in Peripheral Nerves

The way in which Schwann cells can guide axons and promote outgrowth has been demonstrated in a series of experiments by Thompson and his colleagues.[68–70] They took advantage of the visibility and accessibility of motor nerve terminals on skeletal muscle end plates (see Chapter 27). Figure 10.11A provides a diagrammatic summary of these experiments. In adult rats, the soleus muscle in the leg was partially denervated. In confirmation of earlier studies it was shown that uninjured motor axons sprouted to occupy denervated end plates. Under these circumstances, an axon can sprout to innervate up to five times the usual number of muscle fibers. Axon growth was visualized by an antibody against neurofilaments (red in Figure 10.11B). Schwann cells were stained with a different specific antibody (blue in Figure 10.11B). Direct observation showed that the first outgrowth consisted of Schwann cells that extended from denervated muscle fibers to reach

[65] Duan, Y., Sahley, C. L., and Muller, K. J. 2009. *Dev. Neurobiol.* 69: 60–72.

[66] Samuels, S. E. et al. 2010. *J. Gen. Physiol.* 36: 425–442.

[67] Dibaj, P. et al. 2010. *Glia* 58: 1133–1144.

[68] Son, Y. J., and Thompson, W. J. 1995. *Neuron* 14: 125–132.

[69] Son, Y. J., and Thompson, W. J. 1995. *Neuron* 14: 133–141.

[70] Son, Y. J., Trachtenberg, J. T., and Thompson, W. J. 1996. *Trends Neurosci.* 19: 280–285.

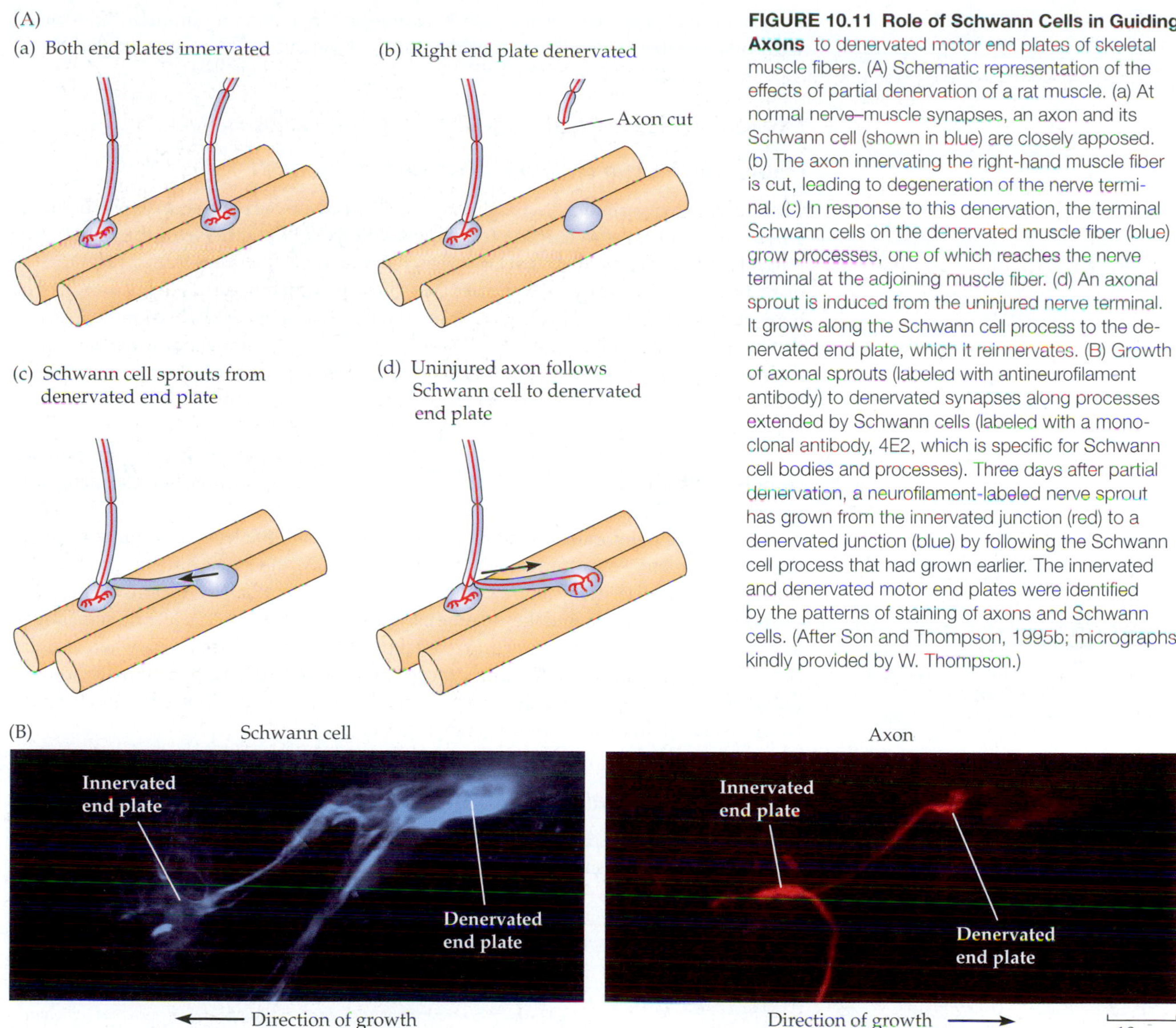

FIGURE 10.11 Role of Schwann Cells in Guiding Axons to denervated motor end plates of skeletal muscle fibers. (A) Schematic representation of the effects of partial denervation of a rat muscle. (a) At normal nerve–muscle synapses, an axon and its Schwann cell (shown in blue) are closely apposed. (b) The axon innervating the right-hand muscle fiber is cut, leading to degeneration of the nerve terminal. (c) In response to this denervation, the terminal Schwann cells on the denervated muscle fiber (blue) grow processes, one of which reaches the nerve terminal at the adjoining muscle fiber. (d) An axonal sprout is induced from the uninjured nerve terminal. It grows along the Schwann cell process to the denervated end plate, which it reinnervates. (B) Growth of axonal sprouts (labeled with antineurofilament antibody) to denervated synapses along processes extended by Schwann cells (labeled with a monoclonal antibody, 4E2, which is specific for Schwann cell bodies and processes). Three days after partial denervation, a neurofilament-labeled nerve sprout has grown from the innervated junction (red) to a denervated junction (blue) by following the Schwann cell process that had grown earlier. The innervated and denervated motor end plates were identified by the patterns of staining of axons and Schwann cells. (After Son and Thompson, 1995b; micrographs kindly provided by W. Thompson.)

the intact axons (part c of Figure 10.11A). Only later did those axons sprout and follow the track laid down by Schwann cells. Implantation of Schwann cells next to an uninjured axon also promoted sprouting without a denervated target.

In other experiments a peripheral nerve was cut; as expected, after a short delay axons grew out from the proximal stump. Once again, however, the first step was the outgrowth of Schwann cell processes that provided the substrate on which the axons could subsequently grow toward their targets. Direct electrical stimulation of the partially denervated muscle reduced the number of bridges formed by Schwann cells and the ability of nerve fibers to grow along them.[71]

A Cautionary Note

Although experiments made at neuromuscular synapses clearly implicate Schwann cells in directing neurite outgrowth during regeneration, it seems prudent to limit the scope of such generalizations. For example, in the CNS of the leech, lesioned axons grow back to re-form their original connections after all the large glial cells surrounding them have been killed. Abundant connections and synapses are formed in embryonic CNS in which glial cells are sparsely distributed. Moreover, in the complete absence of neuroglial cells, synapses with an

[71] Love, F. M., Son, Y. J., and Thompson, W. J. 2003. *J. Neurobiol.* 54: 566–576.

array of normal properties form rapidly in culture and in developing animals.[72] It therefore seems unlikely that "synapses might neither form nor function if there were no glia."[73]

Effects of Neuronal Activity on Glial Cells

Potassium Accumulation in Extracellular Space

That nerve activity can depolarize glial cells is illustrated by experiments shown in Figure 10.12. The recordings were made from a glial cell in the optic nerve of the mud puppy (*Necturus*). Action potentials that are initiated in the nerve fibers by electrical stimulation or by flashes of light travel past the impaled glial cell, which becomes depolarized.[74] The depolarization is graded. Similarly, in the mammalian cortex, glial cells become depolarized depending on the number of nerve fibers activated and on the frequency when neurons in their vicinity are activated by stimulation of neural tracts, peripheral nerves, the surface of the cortex, or sensory input.[75] Astrocytes within an orientation column of the visual cortex are depolarized by visual stimuli of the appropriate orientation.[76,77]

The cause of glial depolarization is potassium efflux from axons. When potassium accumulates in the intercellular clefts, it changes the $[K]_o/[K]_i$ ratio and alters the membrane potential of glial cells.

Changes in membrane potential in glial cells indicate the level of impulse traffic in their environment. Potassium signaling between neurons and glia is different from that in specific synaptic activity. Synaptic actions are confined to specialized regions on neuronal cell bodies and dendrites, and they may be excitatory or inhibitory. In contrast, signaling by potassium is not confined to structures containing receptors but occurs anywhere the glial cell is exposed to potassium. Neurons exposed to increased external potassium concentrations become less depolarized than glia because the neuronal membrane deviates from the Nernst equation in the physiological range (see Chapter 6).

72 Williams, P. R. et al. 2010. *J. Neurosci.* 30: 11951–11961.

73 Pfrieger, F. W., and Barres, B. A. 1996. *Curr. Opin. Neurobiol.* 6: 615–621.

74 Orkand, R. K., Nicholls, J. G., and Kuffler, S. W. 1966. *J. Neurophysiol.* 29: 788–806.

75 Ransom, B. R., and Goldring, S. 1973. *J. Neurophysiol.* 36: 869–878.

76 Van Essen, D, and Kelly, J.1973. *Nature* 241: 403–405.

77 Schummers, J., Yu, H., and Sur, M. 2008. *Science* 320:1638–1643.

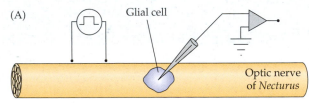

Potassium and Calcium Movement through Glial Cells

Currents flow between regions of a cell that are at different potentials. Nerve cells, of course, use this as the mechanism for conduction: current flows between inactive regions of an axon

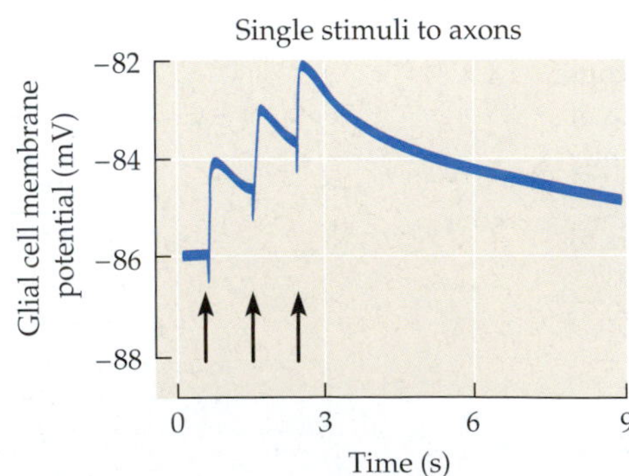

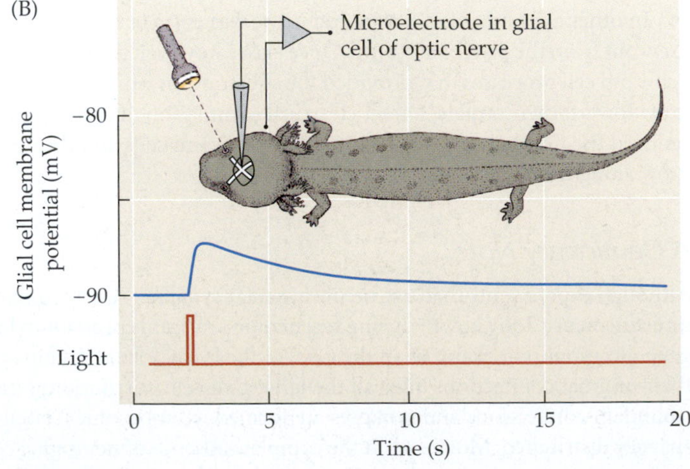

FIGURE 10.12 Effect of Action Potentials on Glial Cells in mud puppy optic nerve. (A) Synchronous impulses evoked by electrical stimulation of nerve fibers cause glial cells to become depolarized. The amplitude of the potentials depends on the number of axons activated and on the frequency of stimulation. (B) Illumination of the eye with a 0.1 second flash of light causes depolarization of a glial cell in the optic nerve of an anesthetized mud puppy with intact circulation. Lower trace monitors light stimulus. (After Orkand, Nicholls, and Kuffler, 1966.)

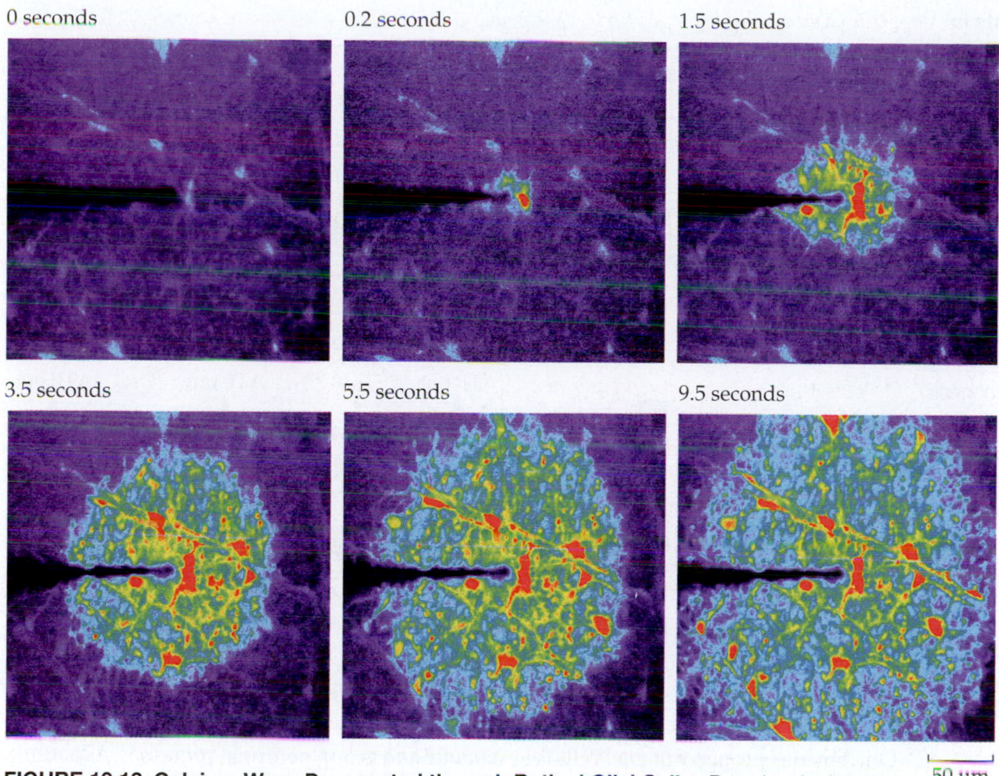

FIGURE 10.13 Calcium Wave Propagated through Retinal Glial Cells. Pseudocolor images of Ca^{2+} fluorescence within astrocytes (larger cells) and Müller cells (smaller spots) at the vitreal surface of the retina. Red represents the highest intensity and blue, the lowest. The Ca^{2+} wave is evoked by a mechanical stimulus to a single astrocyte. The wave is initiated at the stimulated cell (top panel, middle) and propagates outward through neighboring astrocytes and Müller cells. Elapsed times following stimulation are noted at the top of each panel. (Used with permission from E. A. Newman, unpublished.)

and the region that is occupied by a nerve impulse. Since glial cells are linked to each other by low-resistance connections,[14] their conducting properties are similar to those of a single, elongated cell. Consequently, if several glial cells become depolarized by increased potassium concentrations in their environment, they draw current from the unaffected cells. Similarly, an elongated Müller cell that extends through the thickness of the retina generates current when the potassium concentration increases locally over part of its surface (Figure 10.13; see also Figure 10.5). Inward current in the region of raised $[K]_o$, carried by potassium ions, spreads to other regions of the glial cell and through gap junctions to other glial cells. Currents generated by glial cells contribute to recordings made from the eye or the skull with extracellular electrodes. Such recordings, known as the electroretinogram (ERG) and the electroencephalogram (EEG) are valuable for the clinical diagnosis of pathological conditions.

Calcium Waves in Glial Cells

In networks of glial cells in culture or in situ, transient increases in cytoplasmic calcium concentration arise by release from intracellular stores (see Figure 10.13). Using fluorescent indicators, one can observe such oscillatory waves of increased calcium concentration as they propagate from glial cell to glial cell through intercellular junctions.[78] Pannexins, or hemi-junctions permeable to ATP, are present in extrajunctional glial cell membranes. As a result, ATP leaks out of activated glial cells into extracellular space.[31] Calcium waves occur spontaneously[79] or can be triggered by depolarization, by transmitters such as ATP,[29] or by mechanical stimulation. They resemble the calcium waves seen in neuronal networks and in epithelial cells.[80] Propagating intracellular calcium waves that trigger the release of ATP or glutamate can influence neuronal firing patterns (see below). There is evidence that calcium waves in cortical radial glia modulate the production of neurons during development.[81]

[78] Metea, M. R., and Newman, E. A. 2006. *Glia* 54: 650–655.

[79] Kurth-Nelson, Z. L., Mishra, A., and Newman, E. A. 2009. *J. Neurosci.* 29: 11339–11346.

[80] Oheim, M., Kirchhoff, F., and Stühmer, W. 2006. *Cell Calcium* 40: 423–439.

[81] Weissman, T. A. et al. 2004. *Neuron* 43: 647–661.

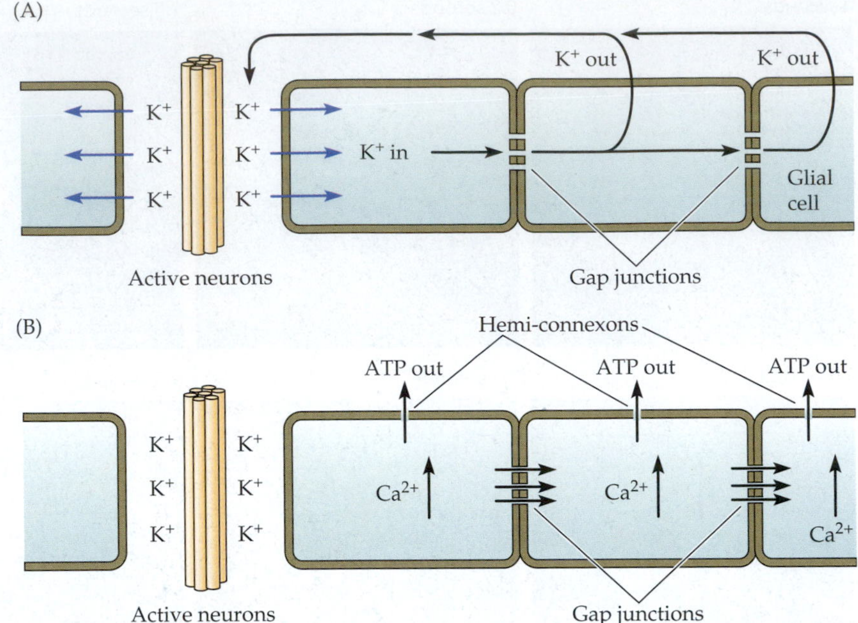

FIGURE 10.14 Potassium Currents in Glial Cells. (A) The glial cells in the diagram are linked by gap junctions. Potassium released by active axons in one region depolarizes the glial cell and enters it, causing current flow and outward movement of potassium through potassium channels elsewhere in the glial tissue. The concept of spatial buffering of potassium has been postulated as a mechanism for influencing neuronal function by glial cells. (B) Depolarization of the glial cell can cause calcium waves that spread through the network. The raised intracellular calcium concentration allows ATP to leak out from the glia through hemichannels (see Chapter 8).

Spatial Buffering of Extracellular Potassium Concentration by Glia

One obvious property of glial cells is to separate and group neuronal processes. As a result, the potassium concentration increases around some neurons while others in a separate compartment are protected. An attractive concept is that glial cells regulate the potassium concentration in intercellular clefts, a process known as **spatial buffering.**[19,82] According to this hypothesis, glial cells act as conduits for uptake of potassium from the clefts to preserve the constancy of the environment.[83] Since glial cells are coupled to each other, potassium enters in one region and leaves at another, as already described (Figure 10.14). That potassium will move through glial cells as a consequence of potassium buildup is inevitable. It is, however, not simple to estimate quantitatively how much potassium actually moves or how much these movements reduce the extracellular potassium concentration. For such calculations, numerous assumptions about geometry, conductance, diffusion, and active transport of potassium into neurons and glial cells must be made.[84]

Glial Cells and Neurotransmitters

Transmitters such as GABA, glutamate, glycine, purines, and acetylcholine act on glial membranes to produce depolarizing or hyperpolarizing responses.[1,28,29,85,86] Figure 10.15 shows activation of $GABA_A$ receptors by GABA in retinal Müller cells. These GABA receptors are similar to those of neurons in many respects. Similarly, glial cell membranes contain receptors for ATP and glutamate, which depolarize, allow calcium to enter, and initiate calcium waves.

Glial cells play a key role in transmitter uptake in the CNS, under normal and pathological conditions. The extracellular concentration of a transmitter such as glutamate, norepinephrine, or glycine that has been released at synapses is reduced in part by diffusion away from the site of release, but mainly by uptake into neurons and into glial cells.[87–89] As in neurons, glutamate transport in glial cells is coupled to inward movement of sodium along its electrochemical gradient (see Chapter 9). In the absence of a removal mechanism, excessively high levels of external glutamate can activate *N*-methyl-D-aspartate (NMDA) receptors in neurons, which in turn can lead to calcium entry and cell death. Quantitative estimates indicate that glial cell transport plays a key role in preventing such excessive rises in extracellular glutamate concentration.

Release of Transmitters by Glial Cells

If glial cells themselves become depolarized by raised extracellular potassium or by glutamate, or if intracellular sodium concentration is increased, their membranes transport

[82] Kofuji, P., and Newman, E. A. 2004. *Neuroscience* 129: 1045–1056.
[83] Kofuji, P. et al. 2000. *J. Neurosci*. 20: 5733–5740.
[84] Odette, L. L., and Newman, E. A. 1988. *Glia* 1: 198–210.
[85] D'Antoni, S. et al. 2008. *Neurochem. Res.* 33: 2436–2443.
[86] Qian, H. et al. 1996. *Proc. R. Soc. Lond. B, Biol. Sci.* 263: 791–796.
[87] Furness, D. N. et al. 2008. *Neuroscience* 157: 80–94.
[88] Takeda, H., Inazu, M., and Matsumiya, T. 2002. *Naunyn Schmiedebergs Arch Pharmacol.* 366: 620–623.
[89] Gomeza, J. et al. 2003. *Neuron* 40: 785–796.

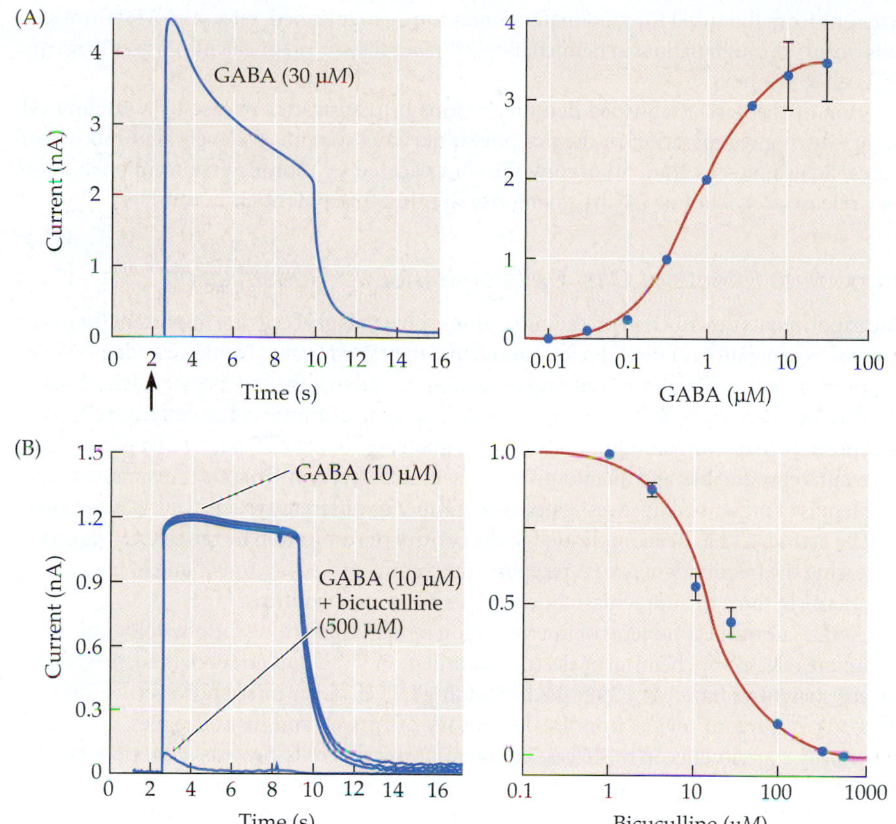

FIGURE 10.15 Responses of Glial Cells to GABA. Responses of Müller glial cells in skate retina to GABA. (A) Current induced by GABA (30 μ*M*) in a glial cell voltage clamped at 0 mV. The dose–response relation for the peak of the GABA current is shown on the right (error bars indicate standard error of the mean). (B) The effect of GABA was blocked by bicuculline, a $GABA_A$ antagonist. (After Qian et al., 1996.)

glutamate and serine out of the glial cell into the extracellular space.[90,91] This mechanism is similar to that for reversed transport described in Chapter 9. Figure 10.16 illustrates an experiment showing currents associated with glutamate release by glial cells. Such outward transport can exacerbate the deleterious effects of brain injury. Injured and dying nerve cells release glutamate and K^+ and depolarize glial cells, which in turn release more glutamate.

Another transmitter that is secreted by glial cells is ATP.[92] The mechanism for ATP however is

[90] Billups, B., and Attwell, D. 1996. *Nature* 379: 171–174.

[91] Henneberger, C. et al. 2010. *Nature* 463: 232–236.

[92] Pangrsic, T. et al. 2007. *J. Biol. Chem.* 282: 28749–28758.

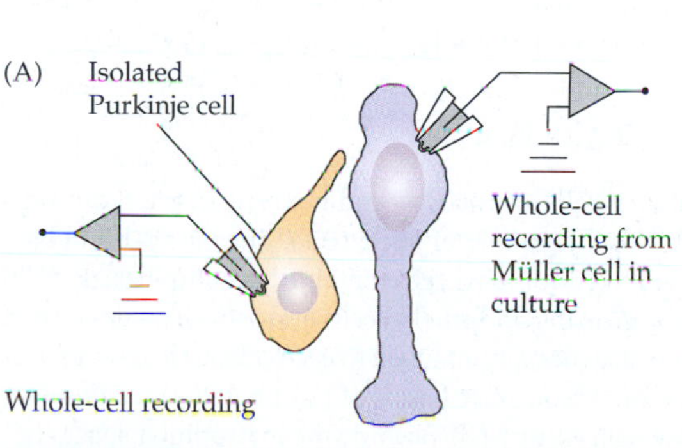

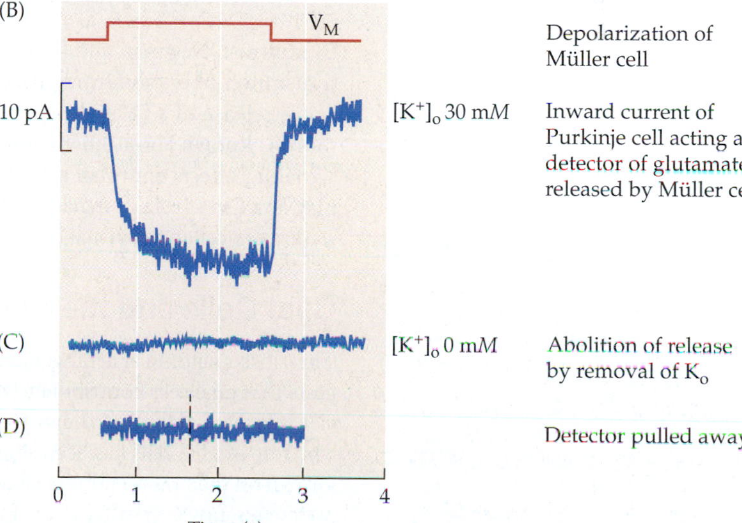

FIGURE 10.16 Release of Glutamate by Glial Cells. Release of glutamate generated by reversal of the glutamate uptake carrier in a Müller cell. (A) Depolarization-induced release of glutamate from a Müller cell (right) is monitored by recording glutamate-elicited currents from an adjacent Purkinje cell (left). The Purkinje cell acts as a detector with high sensitivity and time resolution. (B) Depolarization of the Müller cell from –60 to +20 mV (top trace) elicits an inward current in the nearby Purkinje cell. The Purkinje cell current is generated by activation of its glutamate receptors. The response to glutamate disappears when the Purkinje cell is moved away from the Müller cell (C) or when extracellular K^+ is omitted (D). In fluid containing 0 μ*M* K^+, reverse glutamate transport by the Müller cell is blocked. (After Billups and Attwell, 1996.)

different from that used for glutamate. As previously mentioned, release of ATP from glial cells occurs through pannexin hemichannels during the spread of calcium waves from one glial cell to the next.[31]

One of the best-established demonstrations of transmitter release by a satellite cell occurs during regeneration in the peripheral nervous system. At denervated motor end plates, Schwann cells come to occupy the sites vacated by motor nerve terminals. There they release acetylcholine (ACh), giving rise to miniature potentials in muscle.[93]

Immediate Effects of Glial Cells on Synaptic Transmission

The experiments described in preceding sections show that glial cells are important for development, regeneration, and myelin formation; that their membranes contain ion channels and receptors; that they liberate and take up transmitters; and that they propagate calcium waves.

A natural question that arises concerns the role of transmitters liberated from glial cells at neuronal synapses in the adult CNS. Do glutamate, serine, or ATP released by glia produce clear-cut, reproducible, and quantitatively measurable effects in synaptic transmission? This problem is at present being investigated in a number of different systems.[33,91,94] Secretion of ATP by astrocytes has been implicated in the control of respiration by raised CO_2[95]levels, in increasing the frequency of AMPA receptor currents in cortical neurons,[96]and in influencing circadian rhythms of suprachiasmatic nucleus neurons in culture.[97]

At the same time, problems of interpretation remain concerning the physiological role of glutamate released by chemical or electrical stimulation.[98,99] Thus, no specialized, presynaptic release structures have been identified in adult glial cells in situ (see however De Biase et. al. 2010).[15] A recent review from the laboratory of Attwell and his colleagues, who made pioneering experiments on the release of glutamate by glial cells, provides a cautionary note:

> In the past 20 years, an extra layer of information processing, in addition to that provided by neurons, has been proposed for the CNS. Neuronally evoked increases of the intracellular calcium concentration in astrocytes have been suggested to trigger exocytotic release of the 'gliotransmitters' glutamate, ATP and D-serine. These are proposed to modulate neuronal excitability and transmitter release, and to have a role in diseases as diverse as stroke, epilepsy, schizophrenia, Alzheimer's disease and HIV infection. However, there is intense controversy about whether astrocytes can exocytose transmitters *in vivo*. Resolving this issue would considerably advance our understanding of brain function. [100]

By contrast, Newman and his colleagues have been able to use natural stimuli to activate transmitter liberation from glial cells. In the retina, signals evoked by light in Müller cells cause release of ATP and thereby inhibit ganglion cells.[79,99,102] In summary, at present more is known about how neuronal activity affects glial signaling than about how glial signaling affects neuronal activity. There is as yet no compelling evidence that glial cells play any direct role in dynamic activities, such as the stretch reflex, or in the responses of a complex cell in the visual cortex under normal physiological conditions (see Chapter 2).

Glial Cells and the Blood–Brain Barrier

The close anatomical arrangement of glial cells, capillaries, and neurons in the brain suggests that glial cells contribute to the blood–brain barrier (Box 10.1). The blood–brain barrier is located at the junctions of specialized endothelial cells that line brain capillaries.[103,104] The role of glial cells has been shown by growing endothelial cells and astrocytes in culture. Grown on their own, endothelial cells are occasionally attached to each other. The presence of astrocytes, however, triggers the formation of complete bands of tight junctions resembling those seen in vivo.[105] These junctions, which completely occlude the intercellular spaces between endothelial cells, account for the impermeability of brain capillaries. Molecules must go through, rather than between, endothelial cells. Conversely, the presence of endothelial cells from brain capillaries in culture causes distinctive assemblies of membrane particles to appear in astrocytes. These interactions are specific for astrocytes and brain-capillary endothelial cells. No comparable results occur with fibroblasts or endothelial cells from peripheral vessels. Methods for promoting the uncoupling of endothelial cells to produce an increase in

93 Reiser, G., and Miledi, R. 1988. *Pflügers Arch.* 412: 22–28.

94 Perea, G., and Araque, A. 2010. *Brain Res. Rev.* 63: 93–102.

95 Gourine, A. V. et al. 2010. *Science* 329: 571–575.

96 Fiacco, A., and McCarthy, K. D. 2004. *J. Neurosci.* 24: 722–732.

97 Womac, A. D. et al. 2009. *Eur. J. Neurosci.* 30: 869–876.

98 Fiacco, T. A., Agulhon, C., and McCarthy, K. D. 2009. *Annu. Rev. Pharmacol. Toxicol.* 49: 151–174.

99 Agulhon, C., Fiacco, T. A., and McCarthy, K. D. 2010. *Science* 327: 1250–1254.

100 Hamilton, N. B., and Attwell, D. 2010. *Nat. Rev. Neurosci.* 11: 227–238.

101 Newman, E. A. 2004. *Neuron Glia Biol.* 1: 245–252.

102 Newman, E. A. 2003. *J. Neurosci.* 23: 1659–1666.

103 Brightman, M. W., and Reese, T. S. 1969. *J. Cell Biol.* 40: 668–677.

104 Wolburg, H. et al. 2009. *Cell Tissue Res.* 335: 75–96.

105 Tao-Cheng, J. H., Nagy, Z., and Brightman, M. W. 1987. *J. Neurosci.* 7: 3293–3299.

capillary permeability are now being investigated. These experiments could provide a method for circumventing the blood–brain barrier (see the next section) and allowing therapeutically useful substances to enter the brain that would otherwise not be able to do so.[106]

Astrocytes and Blood Flow through the Brain

We can now take account of three facts to suggest a possible role for astrocytes in the mammalian brain. First, they envelop the brain capillaries with their end feet. (Indeed, it was this feature that led Golgi and so many others to suggest that they provide materials to the neurons.) Second, as we show in Chapter 3, neuronal activity in a particular region of the brain causes a dramatic localized increase in the blood flow through the tissue, as

[106] Palmer, A. M. 2010. *Neurobiol. Dis.* 37: 3–12.

BOX 10.1
The Blood–Brain Barrier

A homeostatic system controls the fluid environment in the brain and prevents fluctuations in its composition. This constancy seems particularly important in a system in which the activity of so many cells is integrated and small variations may upset the balance of delicately poised excitatory and inhibitory influences.[107,108] Within the brain, there are three fluid compartments: (1) the blood supplied to the brain through a dense network of capillaries, (2) the cerebrospinal fluid (CSF) that surrounds the bulk of the nervous system and is contained in the internal cavities (ventricles), and (3) the fluid in the intercellular clefts (Figure I).

The blood–brain barrier depends on specialized properties of the endothelial cells of capillaries in the brain, which are far less permeable than those supplying organs in the periphery.[104,105] Proteins, ions, and hydrophilic molecules cannot pass through the blood–brain barrier, while lipophilic molecules (such as alcohol) and gases can. A second essential component of the blood–brain barrier is the choroid plexus: Specialized epithelial cells surround the choroid plexus capillaries and secrete CSF (Figure IIA).[109] CSF itself is almost devoid of protein, containing only about 1/200 of the amount present in blood plasma. Proteins, electrolytes, transmitters, and a variety of drugs, including penicillin, injected directly into the bloodstream act rapidly on peripheral tissues such as muscle, heart, or glands—but they have little or no effect on the CNS. When administered by way of the CSF, however, the same substances exert a prompt and strong action.

[107] Abbott, N. J. et al. 2010. *Neurobiol. Dis.* 37: 13–25.
[108] Saunders, N. R. et al. 2008. *Trends Neurosci.* 31: 279–286.
[109] Wolburg, H., and Paulus, W. 2010. *Acta. Neuropathol.* 119: 75–88.

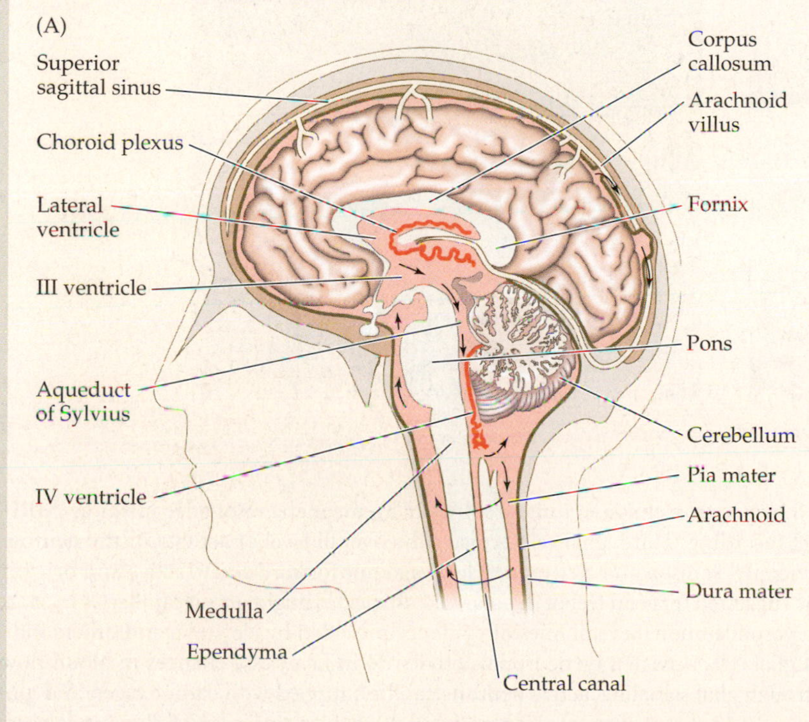

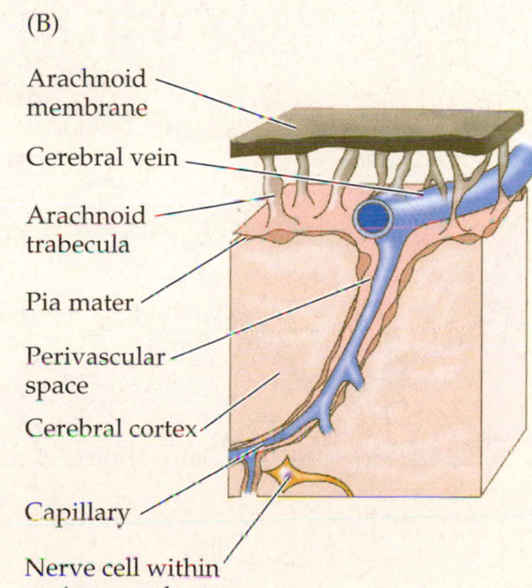

Figure I Distribution of Cerebrospinal Fluid and its relation to larger blood vessels and to structures surrounding the brain. (A) All spaces containing CSF communicate with each other. (B) CSF is drained into the venous system through the arachnoid villi.

(continued on next page)

BOX 10.1 *(continued)*

Within the brain, ions and small particles reach neurons by passing through the narrow 20 nm intercellular clefts and not through glia. In Figure IIB, after injection of microperoxidase into CSF, electron-dense molecules deposited by peroxidase reaction are lined up in clefts and fill extracellular spaces. This result shows that large molecules can pass between the ependymal cells that line the ventricles and through intercellular clefts. In contrast, the junctions between endothelial cells lining the blood capillaries in the brain provide a barrier. Tracers injected into CSF do not enter the capillaries. Figure IIC shows the opposite result. When enzyme was injected into the circulation, brain capillaries filled with enzyme but none entered the intercellular spaces. During development the barrier is already present, but the range of substances that can enter the CSF is qualitatively and quantitatively different.[110] Knowledge of blood–brain barrier properties is important for understanding pharmacological actions of drugs and their effects on the body.

[110] Johansson, P. A. et al. 2008. *Bioessays*. 30: 237–248.

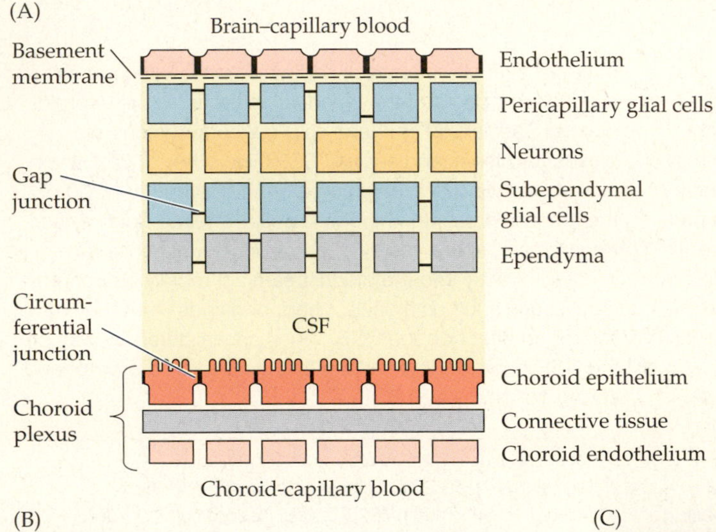

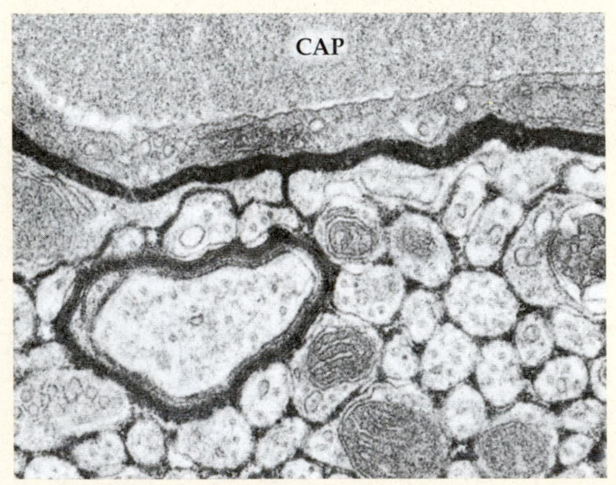

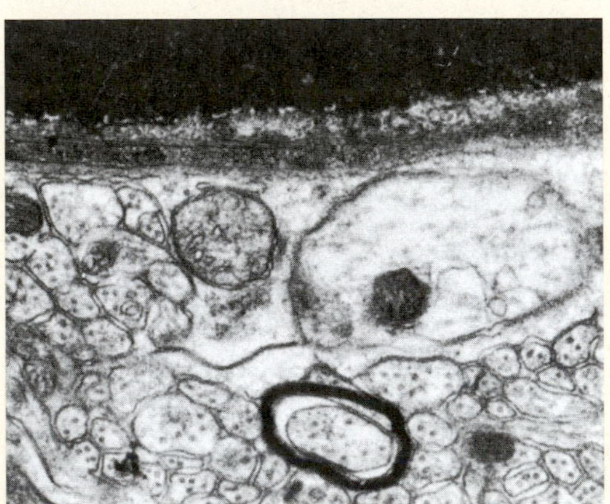

Figure II Pathways for Diffusion in the Brain. (A) Schematic presentation of cells involved in the exchange of materials between blood, CSF, and intercellular spaces. Molecules are free to diffuse through the endothelial cell layer that lines capillaries in the choroid plexus, which are not tightly linked. They are, however, restrained by circumferential tight junctions between the choroid epithelial cells, which secrete CSF. There are no barriers between the bulk fluid of CSF and the various cell layers, such as ependyma, glia, and neurons. The endothelial cells lining brain capillaries are joined by circumferential junctions. This prevents free diffusion of molecules out of the blood into the brain tissue or from the brain into the blood. (B) Demonstration in the mouse that the enzyme microperoxidase diffuses freely from cerebrospinal fluid into the intercellular spaces of the brain, which are filled with the dark reaction product. No enzyme is seen in the capillary (CAP). (C) When injected into the circulation, the enzyme fills the capillary but is prevented by the capillary endothelium from escaping into the intercellular spaces. (B and C from Brightman, Reese, and Feder, 1970.)

measured by positron emission tomography (PET), magnetic resonance imaging (MRI), or optical recording. Third, glial cells register the overall level of activity of the neurons in their vicinity. Paulson and Newman[111] long ago put forward an attractive and original idea: they suggested that end feet of depolarized astrocytes might act on capillaries to cause localized vasodilatation. Several lines of evidence provided by Newman and others indicate that glial cells activated by neuronal activity do in fact cause changes in blood flow. Thus, through glial signaling, active neurons can be supplied with extra oxygen and glucose. Figure 10.17 shows dilatation of blood vessels of the retina in situ following calcium waves in Müller radial glial cells.[112] As for mechanisms, recent experiments have provided

[111] Paulson, O. B., and Newman, E. A. 1987. *Science* 237: 896–898.

[112] Metea, M. R., and Newman, E. A. 2006. *J. Neurosci.* 26: 2862–2870.

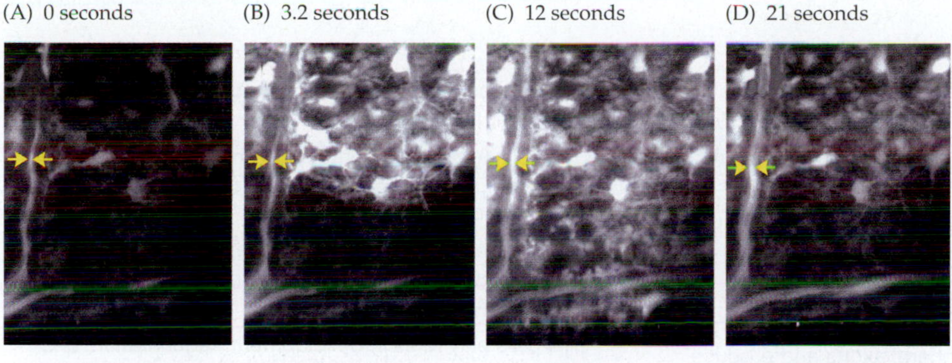

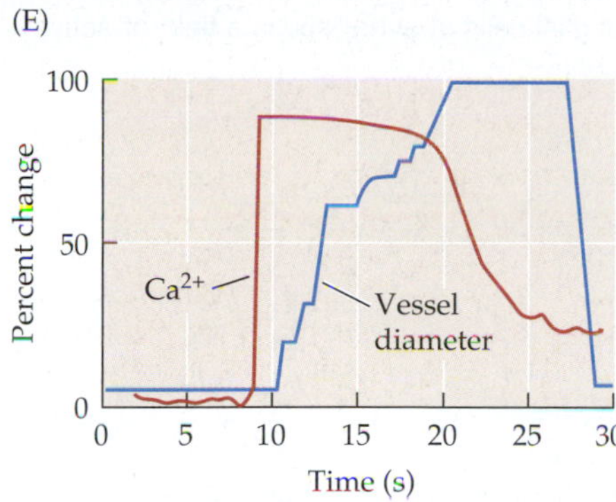

FIGURE 10.17 Propagated Glial Calcium Waves evoke changes of diameter in distant blood vessels in the retina. (A–D) Fluorescent waves representing an ATP-evoked calcium wave spreading through glial cells. The yellow arrows indicate the diameter of the vessel on the left of the photo, which dilates as the Ca^{2+} concentration rises in that area at about 12 seconds. The calcium wave was initiated by ATP micro-injection at a point just above the top right corner of the figures A–D. (E) Measurements showing increases of calcium and vessel diameter. (After Metea and Newman, 2006b; photo kindly supplied by E. A. Newman.)

conflicting results about whether it is ATP or potassium that acts on capillary endothelial cells.[113–115] Nevertheless, Paulson and Newman's suggestion has been confirmed; ironically it is reminiscent of Golgi's original idea, which it reproduces, but with signals flowing in the opposite direction. Instead of glial cells carrying nutrients through their cytoplasm from blood to neurons, neuronal activity leads to highly localized vasodilatation and increased blood flow just where it is needed.

Transfer of Metabolites from Glial Cells to Neurons

While the previous paragraph shows that glial cells influence cerebral blood flow, the hypothesis that nutrients are transferred to neurons through glial cells has been harder to prove by direct experiments. Several lines of evidence in invertebrates indicate that glial cells supply lactate to nerve cells. It has been proposed that lactate, rather than glucose, is taken up by active neurons to provide their prime source of energy.[116,117] The demonstration of such transfer in invertebrates and vertebrate neurons in culture has suggested a similar role for glial cells within the intact central nervous system under conditions of high neuronal activity or anoxia. However, in intact animals it is a major task to make direct measurements of the amount of lactate liberated, the increase in concentration, or the timing of release and uptake in relation to neuronal activity[118] (see review by Fillenz[119]).

Glial Cells and Immune Responses of the CNS

In the past it was generally accepted that the tissues of the central nervous system were not patrolled by the surveillance mechanisms of the immune system. The blood–brain barrier, the absence of a lymphatic system, and the comparative ease with which grafts can be accepted all suggest the absence of immune responses to foreign antigens. Thus, CNS functions are not disrupted by the massive allergic reactions to a bee sting or poison ivy. Astrocytes in culture and in situ, however, have been shown to react with T lymphocytes, whose activity they can either stimulate or suppress. Evidence has now accumulated to show that microglia and activated T lymphocytes do enter the brain and can mediate acute inflammation of brain tissue.[120–122] The role played by glial cells in interactions between the immune system and the nervous system still represents an interesting and challenging problem.

[113]Metea, M. R., Kofuji, P., and Newman, E. A. 2007. *J. Neurosci.* 27: 2468–2471.

[114]Girouard, H. et al. 2010. *Proc. Natl. Acad. Sci. USA* 107: 3811–3816.

[115]Koehler, R. C., Roman, R. J., Harder, D. R. 2009. *Trends Neurosci.* 32: 160–169.

[116]Brown, A. M., and Ransom, B. R. 2007. *Glia* 55: 1263–1271.

[117]Magistretti, P. J. 2009. *Am. J. Clin. Nutr.* 90: 875–880.

[118]Aubert, A. et al. 2005. *Proc. Natl. Acad. Sci USA* 102: 16448–16453.

[119]Fillenz, M. 2005. *Neurochem. Int.* 47: 413–417.

[120]Kaur, G. et al. 2010. *Neurosurg. Clin. N. Am.* 21: 43–51.

[121]Perry, V. H., Nicoll, J. A., and Holmes, C. 2010. *Nat. Rev. Neurol.* 6: 193–201.

[122]Rotshenker, S. 2009. *J. Mol. Neurosci.* 39: 99–103.

SUMMARY

- Glial cells in the brain and Schwann cells in the periphery envelop neurons.
- Oligodendrocytes have short processes and myelinate large axons.
- Astrocytes surround brain capillaries, determine their permeability, and regulate blood flow through them.
- The blood–brain barrier depends on interactions of astrocytes and capillary endothelial cells.
- Schwann cells myelinate peripheral axons and produce trophic molecules.
- Microglial cells remove debris after damage and are involved in inflammatory responses of the nervous system.
- Glial cells have more negative resting potentials than neurons and do not produce action potentials.
- Glial cells are electrically coupled to each other but not to neurons.
- Glial cell membranes contain ion channels for sodium, potassium, and calcium as well as receptors, pumps, and transporters.
- Waves of increased cytoplasmic calcium concentration, evoked in glial cells by depolarization or ATP, cause release of ATP.
- Glial cells play roles in development, regeneration, and homeostatic control of the fluid environment of neurons.
- The role of glial cells at synapses is a field of active investigation.

Suggested Reading

General Reviews

Barres, B. A. 2008. The mystery and magic of glia: a perspective on their roles in health and disease. *Neuron* 60: 430–440.

Brown, A. M., and Ransom, B. R. 2007. Astrocyte glycogen and brain energy metabolism. *Glia* 55: 1263–1271.

Kettenmann, H., and Ransom, B. R. (eds.) 2005. *Neuroglia.* 2nd Ed. Oxford University Press, New York.

Kuffler, S. W., and Nicholls, J. G. 1966. The physiology of neuroglial cells. *Ergeb. Physiol.* 57: 1–90.

Newman, E. A. 2004. A dialogue between glia and neurons in the retina: modulation of neuronal excitability. *Neuron Glia Biol.* 1: 245–252.

Rakic, P. 2003. Developmental and evolutionary adaptations of cortical radial glia. *Cereb. Cortex.* 13: 541–549.

Schwab, M. E. 2004. Nogo and axon regeneration. *Curr. Opin. Neurobiol.* 14: 118–124.

Webster, H., and Aström, K. E. 2009. Gliogenesis: historical perspectives, 1839–1985. *Adv. Anat. Embryol. Cell. Biol.* 202: 1–109.

Original Papers

Buckley, C. E., Marguerie, A., Alderton, W. K., and Franklin, R. J. 2010. Temporal dynamics of myelination in the zebrafish spinal cord. *Glia* 58: 802–812.

Duan, Y., Sahley, C. L., and Muller, K. J. 2009. ATP and NO dually control migration of microglia to nerve lesions. *Dev. Neurobiol.* 69: 60–72.

Fillenz, M. 2005. The role of lactate in brain metabolism. *Neurochem. Int.* 47: 413–417.

Girouard, H., Bonev, A. D., Hannah, R. M., Meredith, A., Aldrich, R. W., and Nelson, M. T. 2010. Astrocytic endfoot Ca^{2+} and BK channels determine both arteriolar dilation and constriction. *Proc. Natl. Acad. Sci. USA* 107: 3811–3816.

Hamilton, N. B., and Attwell, D. 2010. Do astrocytes really exocytose neurotransmitters? *Nat. Rev. Neurosci.* 11: 227–238.

Hansen, D. V., Lui, J. H., Parker, P. R., and Kriegstein, A. R. 2010. Neurogenic radial glia in the outer subventricular zone of human neocortex. *Nature* 464: 554–561.

Iglesias, R., Dahl, G., Qiu, F., Spray, D. C., and Scemes, E. 2009. Pannexin 1: the molecular substrate of astrocyte "hemichannels." *J. Neurosci.* 29: 7092–7097.

Johansson, P. A., Dziegielewska, K. M., Liddelow, S. A., and Saunders, N. R. 2008. The blood-CSF barrier explained: when development is not immaturity. *Bioessays* 30: 237–248.

Kuffler, S. W., and Potter, D. D. 1964. Glia in the leech central nervous system: Physiological properties and neuron-glia relationship. *J. Neurophysiol.* 27: 290–320.

Kurth-Nelson, Z. L., Mishra, A., Newman, E. A. 2009. Spontaneous glial calcium waves in the retina develop over early adulthood. *J. Neurosci.* 29: 11339–11146.

Love, F. M., Son, Y. J., and Thompson, W. J. 2003. Activity alters muscle reinnervation and terminal sprouting by reducing the number of Schwann cell pathways that grow to link synaptic sites. *J. Neurobiol.* 54: 566–576.

Marcaggi, P., and Attwell, D. 2004. Role of glial amino acid transporters in synaptic transmission and brain energetics. *Glia* 47: 217–225.

Metea, M. R., and Newman, E. A. 2006. Calcium signaling in specialized glial cells. *Glia* 54: 650–655.

Newman, E. A. 2003. Glial cell inhibition of neurons by release of ATP. *J. Neurosci.* 23: 1659–1666.

Rotshenker, S. 2009. The role of Galectin-3/MAC-2 in the activation of the innate-immune function of phagocytosis in microglia in injury and disease. *J. Mol. Neurosci.* 39: 99–103.

Verkhratsky, A., Krishtal, O. A., and Burnstock, G. 2009. Purinoceptors on neuroglia. *Mol. Neurobiol.* 39: 190–208.

Zuo, Y., Lubischer, J. L., Kang, H., Tian, L., Mikesh, M., Marks, A., Scofield, V. L., Maika, S., Newman, C., Krieg, P., and Thompson, W. J. 2004. Fluorescent proteins expressed in mouse transgenic lines mark subsets of glia, neurons, macrophages, and dendritic cells for vital examination. *J. Neurosci.* 24: 10999–11009.

PART III

Intercellular Communication

In Part II we described how ionic and electrical gradients are maintained across a nerve cell membrane, how these gradients are used to generate electrical impulses (action potentials), and how these impulses are conducted along the processes of the nerve cell (the axons). Now we have to consider what happens when the impulse gets to the end of the axon: how is this information transmitted to another neuron or to an effector cell such as a muscle cell?

In most cases the information is transmitted across specialized contact zones called synapses. The process is termed synaptic transmission. In some instances the presynaptic action potential is conducted electrotonically across the synapse to directly depolarize the postsynaptic neuron—electrical transmission. More commonly, however, the presynaptic action potential induces the release of a chemical transmitter substance that binds to specialized target proteins (receptors) on the postsynaptic membrane. This process is called chemical transmission.

In Chapters 11 and 12 we describe two forms of chemical transmission—one in which the transmitter substance actually binds to, and opens, an ion channel, to produce a very rapid, brief postsynaptic response (direct transmission), and one in which the transmitter binds to a different type of receptor to produce a slow, long-lasting postsynaptic effect by inducing a cascade of intracellular changes (indirect transmission). Then in Chapter 13 we go back to the presynaptic nerve endings and discuss how the transmitter is stored in them and how it is released by the presynaptic action potential.

In fact, there are a large number of different chemical transmitter substances in the nervous system, which makes it possible to selectively modify transmission at some synapses and not at others using specific chemicals and drugs. The pathways using the different transmitters and some of their actions on individual neurons and on overall brain function are summarized in Chapter 14. Chapter 15 gives more information about the synthesis, storage and inactivation of these transmitters.

Finally, in Chapter 16 we consider how the efficiency of transmission across synapses can show long-lasting increases or decreases depending on the amount of synaptic traffic (synaptic plasticity); this is important, for example, in the establishment of memory.

CHAPTER 11
Mechanisms of Direct Synaptic Transmission

Synapses are points of contact between nerve cells and their targets where signals are handed on from one cell to the next. This process of synaptic transmission may be mediated through the release of a chemical neurotransmitter substance from the nerve terminal by the incoming action potential (chemical transmission) or, at certain junctions, by the direct spread of electric current from the presynaptic neuron to the postsynaptic cell (electrical transmission).

At direct chemical synapses, the transmitter binds to receptors in the membrane of the postsynaptic cell that are themselves ion channels (ionotropic receptors). As a result, the conformation of the receptor changes, the channel opens, ions flow, and the membrane potential changes—all within a millisecond or so. A slower process, indirect chemical transmission, involving additional, intermediary steps, is described in Chapter 12.

The channels opened at excitatory synapses allow cations to enter, driving the membrane potential toward the action potential threshold. At inhibitory synapses, transmitters open channels that are permeable to anions, tending to keep the membrane potential negative to threshold. At both excitatory and inhibitory synapses, the direction of current flow is determined by the balance of concentration and electrical gradients acting on the permeant ions.

Synapses between motor nerves and skeletal muscle fibers provide important preparations for understanding the mechanisms of direct chemical synaptic transmission. In the mammalian central nervous system (CNS), directly mediated excitation occurs at synapses where the transmitter (usually glutamate) activates excitatory ionotropic receptors. Inhibitory synaptic transmission in the CNS is mediated by release of transmitters (usually γ-aminobutyric acid [GABA] or glycine) that activate inhibitory ionotropic receptors.

More than one type of transmitter may be released at a single chemical synapse, and many transmitters act both rapidly, by binding to and opening ion channels directly, and more slowly through indirect mechanisms. In another process known as presynaptic inhibition, a chemical transmitter acts on the presynaptic nerve ending to reduce the amount of neurotransmitter released. Electrical transmission occurs at synapses specialized for very fast reflex responses and also in the mammalian CNS, where it helps to coordinate nerve cell activity.

Synaptic Transmission

Action potentials travel down large motor axons or up large sensory axons, at speeds of up to 120 meters per second and at intervals down to 5 ms or so. It would seem rather pointless if there were then a long delay before the muscle fiber, or the next neuron, could be informed of the action potential's arrival. Accordingly, there are specialized junctions between one cell and another called **synapses**. These enable the recipient cell to respond within less than a millisecond of the action potential's arrival at the nerve terminal. For a long time it was unclear how such a fast transmission process could occur (Box 11.1). One possibility, widely held until the early 1950s, was that the electrical current in the presynaptic terminal spread passively to the postsynaptic cell (i.e., electrical transmission, Figure 11.1A). The alternative suggestion was that the information was transmitted by the local release and action of a chemical neurotransmitter (i.e., chemical transmission, Figure 11.1B). In fact, it is now clear that chemical transmission is the prevalent mode of synaptic communication in the vertebrate central and peripheral nervous system. However, electrical transmission is used at certain invertebrate and vertebrate synapses specialized for very fast responses, and it has become increasingly apparent that a form of supplementary electrical transmission is also quite widely used in the mammalian central nervous system as a mechanism for coordinating the electrical activity of groups of neurons. In this chapter we start by discussing chemical transmission, then consider examples of electrical transmission.

Chemical Synaptic Transmission

Certain obvious questions arise when one considers the elaborate scheme necessary for **chemical synaptic transmission**, which entails the secretion of a specific chemical by a nerve terminal and its interaction with specific postsynaptic receptors (see Figure 11.1B). How does the terminal liberate the chemical? Is there a special feature of the action potential mechanism that causes secretion? How is the interaction of a transmitter with its postsynaptic receptor rapidly converted into excitation or inhibition? The release process will be considered in detail in Chapter 13; the present discussion is concerned with the question of how transmitters act on the postsynaptic cell at direct chemical synapses.

Many of the pioneering studies of chemical synaptic transmission were done on relatively simple preparations, such as the skeletal neuromuscular junction of the frog. At the time, this particular preparation had the advantage that the neurotransmitter acetylcholine (ACh) had been definitively identified, while the transmitters at synapses in the CNS were completely unknown (see Chapter 14).

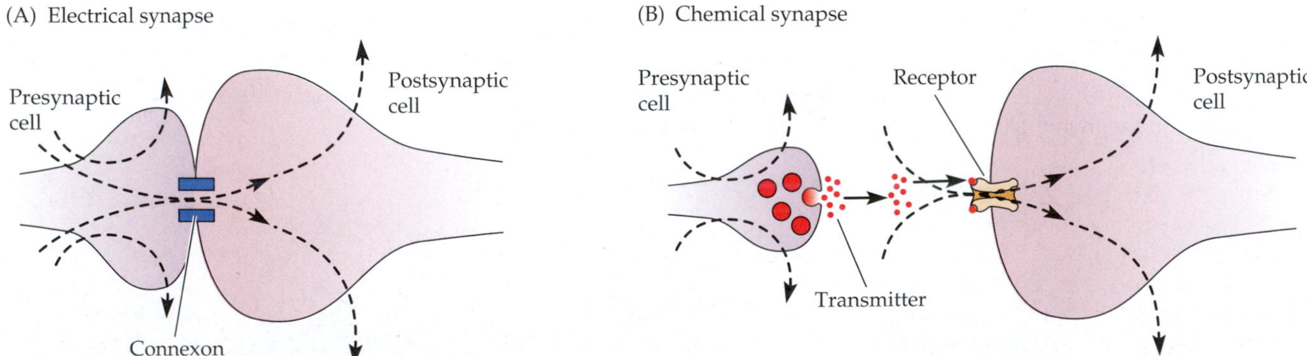

FIGURE 11.1 Electrical and Chemical Synaptic Transmission. (A) At electrical synapses, current flows directly from one cell to another through connexons (intercellular channels that cluster to form gap junctions). (B) At a chemical synapse, depolarization of the presynaptic nerve terminal triggers the release of neurotransmitter molecules, which open ion channel receptors on the postsynaptic membrane, causing excitation or inhibition.

BOX 11.1
Electrical or Chemical Transmission?

In the second half of the nineteenth century there was vigorous debate between proponents of the **cell theory**, who considered that neurons were independent units, and those who thought that nerve cells were a **syncytium** interconnected by protoplasmic bridges. Not until the late nineteenth century did it become generally accepted that nerve cells are independent units. This disagreement about synaptic structure was accompanied by a parallel disagreement about function. In 1843, Du Bois-Reymond showed that flow of electrical current was involved in both muscle contraction and nerve conduction, and it required only a small extension of this idea to conclude that transmission of excitation from nerve to muscle was also due to current flow (see Figure 11.1A).[1] Du Bois-Reymond himself favored an alternative explanation: the secretion by the nerve terminal of an excitatory substance that then caused muscle contraction (see Figure 11.1B). However, the idea of animal electricity had such a potent hold on people's thinking that it was more than 100 years before contrary evidence finally overcame the assumption of electrical transmission between nerve and muscle and, by extension, between nerve cells in general.

One reason why the idea of chemical synaptic transmission seemed unattractive is the speed of signaling between nerve cells or between nerve and muscle. The fraction of a second that intervenes between stimulation of a motor axon and contraction of the corresponding muscle did not appear to provide sufficient time for a chemical neurotransmitter to be released from the nerve terminal and interact with receptors on the postsynaptic target to cause excitation. This difficulty did not exist in the autonomic nervous system, which controls glands and blood vessels, where the effects of nerve stimulation are slow and prolonged (see Chapter 17). Thus, the first explicit suggestion of chemical transmission came from experiments on the sympathetic nervous system by T. R. Elliot in 1904.[2] He noted that an extract from the adrenal gland, adrenaline (epinephrine), mimicked the action of sympathetic nerve stimulation when applied directly to the target tissues and concluded with the words "Adrenalin might then be the chemical stimulant liberated on each occasion when the impulse arrives at the periphery." This was not pursued further at that time, but instead, in 1921 Otto Loewi did a direct and simple experiment that established the chemical nature of at autonomic synapses between the vagus nerve and the heart.[3] He perfused the heart of a frog and stimulated the vagus nerve, thereby slowing the heartbeat. When the fluid from the inhibited heart was transferred to a second unstimulated heart, it too began to beat more slowly. Apparently, stimulation of the vagus nerve had caused an inhibitory substance to be released into the perfusate. In subsequent experiments, Loewi and his colleagues demonstrated that the substance was mimicked in every way by acetylcholine (ACh).

Henry Dale (left) and Otto Loewi, mid-1930s. (Kindly provided by Lady Todd and W. Feldberg.)

It is an amusing sidelight that Loewi had the idea for his experiment in a dream, wrote it down in the middle of the night, but could not decipher his writing the next morning. Fortunately, the dream returned, and this time Loewi took no chances; he rushed to the laboratory and performed the experiment. Later he reflected:

> On mature consideration, in the cold light of morning, I would not have done it. After all, it was an unlikely enough assumption that the vagus should secrete an inhibitory substance; it was still more unlikely that a chemical substance that was supposed to be effective at very close range between nerve terminal and muscle be secreted in such large amounts that it would spill over and, after being diluted by the perfusion fluid, still be able to inhibit another heart.[3]

Subsequently, in the early 1930s, the role of ACh in synaptic transmission in **ganglia** in the autonomic nervous system was firmly established by Feldberg and his colleagues.[4] Highlights of such experiments and ideas from the beginning of the twentieth century are contained in the writings of Dale, who for several decades was one of the leading figures in British physiology and pharmacology.[5] Among his many contributions are the clarification of the action of acetylcholine at synapses in autonomic ganglia and the establishment of its role in **neuromuscular transmission**. Thus, in 1936, Dale and his colleagues demonstrated that acetylcholine was released by stimulating the motor nerves supplying skeletal muscle,[6] and then that application of ACh to the muscle caused contraction and that the effects of both ACh and motor nerve stimulation were blocked by curare (see Box 11.2).[5]

The idea of chemical transmission between one nerve cell and another, particularly in the mammalian central nervous system, took rather longer to become accepted. There were several reasons for this. First, there were no model synapses to study that were comparable to the frog neuromuscular junction (see below). Second, the chemical transmitters in the central nervous system were not known (acetylcholine is not the main transmitter), so there were no good pharmacological tools to use. Third, neuron-to-neuron transmission could only be studied using electrical recording techniques, which (one suspects) might have led the investigators to think in terms of electrical current flow. One of the most ardent propo-

(continued on next page)

BOX 11.1 *(continued)*

nents of the electrical transmission theory was J. C. Eccles. However, he became converted to the chemical hypothesis when he and his colleagues first used microelectrodes to record from inside a motor neuron in the spinal cord of an anesthetized cat. There, they found that stimulation of excitatory or inhibitory afferent nerves produced changes in the membrane potential of the motor neuron in *opposite* directions.[7] (The nature of these potentials is described in more detail below). They could not explain the opposite polarity of the inhibitory response on their previous electrical hypothesis, so they wrote:

> It may therefore be concluded that inhibitory synaptic action is mediated by a specific transmitter substance that is liberated from the inhibitory synaptic knobs and causes an increase in polarization of the subjacent membrane of the motoneurone.[7]

Interestingly, not long after chemical transmission was firmly established as the accepted form of synaptic activation, in 1959, Furshpan and Potter found electrical transmission of excitation between giant axons in the crayfish.[8] Then, in 1963, electrical synaptic transmission was reported in the avian ciliary ganglion.[9] Examples of electrical transmission have multiplied manyfold since then, and we now we know that both chemical and electrical synapses are abundant in both vertebrate and invertebrate nervous systems.

[1] Du Bois-Reymond, E. 1848. *Untersuchungen über thierische Electricität.* Reimer, Berlin.
[2] Elliot, T. R. 1904. *J. Physiol.* 31: (Proc.) xx–xxi.
[3] Loewi, O. 1921. *Pflügers Arch.* 189: 239–242.
[4] Feldberg, W. 1945. *Physiol. Rev.* 25: 596–642.
[5] Dale, H. H. 1953. *Adventures in Physiology*. Pergamon, London.
[6] Dale, H. H., Feldberg, W., and Vogt, M. 1936. *J. Physiol.* 86: 353–380.
[7] Brock, L. G., Coombs, J. S., and Eccles, J. C. 1952. *J. Physiol.* 117: 431–460.
[8] Furshpan, E. J., and Potter, D. D. 1959. *J. Physiol.* 145: 289–325.
[9] Martin, A. R., and Pilar, G. 1963. *J. Physiol.* 168: 443–463.

Synaptic Structure

As always, knowledge of structure is a prerequisite for understanding function. Figure 11.2 illustrates the principal morphological features of the neuromuscular junction of the frog. Individual axons branch from the incoming motor nerve, lose their myelin sheath, and give off terminal branches that run in shallow grooves on the surface of the muscle. The **synaptic cleft** between the terminal and the muscle membrane is about 30 nanometers (nm) wide. Within the cleft is the **basal lamina**, which follows the contours of the muscle fiber surface. On the muscle, **postjunctional folds** radiate into the muscle fiber from the cleft at regular intervals. The grooves and folds are peculiar to skeletal muscle and are not a general feature of chemical synapses. In skeletal muscle, the region of postsynaptic specialization is known as the **motor end plate.** Schwann cell lamellae cover the nerve terminal, sending fingerlike processes around it at regularly spaced intervals.

Within the cytoplasm of the terminal, clusters of **synaptic vesicles** are associated with electron-dense material attached to the presynaptic membrane, forming **active zones.** Synaptic vesicles are sites of ACh storage; upon excitation of the axon terminal, they fuse with the presynaptic membrane at the active zone to spill their contents into the synaptic cleft by **exocytosis** (see Chapter 13).

Synapses on nerve cells are usually made by nerve terminal swellings called **boutons,** which are separated from the postsynaptic membrane by the synaptic cleft. The presynaptic membrane of the bouton displays electron-dense regions with associated clusters of synaptic vesicles, forming active zones similar to, but smaller than, those seen in skeletal muscle (Figure 11.2C). At nerve–nerve synapses the postsynaptic membrane often appears thickened and has electron-dense material associated with it.

Synaptic Potentials at the Neuromuscular Junction

Early studies by Eccles, Katz, and Kuffler used extracellular recording techniques to study the **end-plate potential (EPP)** in muscle.[10–12] The EPP is the depolarization of the end-plate region of the muscle fiber following motor nerve excitation, produced by acetylcholine released from the presynaptic nerve terminals. Synaptic potentials similar to these are seen in nerve cells. A synaptic potential that excites a postsynaptic cell is usually referred to as an **excitatory postsynaptic potential (EPSP)**, and one that inhibits is called an **inhibitory postsynaptic potential (IPSP).**

Normally the amplitude of the end-plate potential in a skeletal muscle fiber is much greater than that needed to initiate an action potential. The amplitude can be reduced by

[10] Eccles, J. C., and O'Connor, W. J. 1939. *J. Physiol.* 97: 44–102.
[11] Eccles, J. C., Katz, B., and Kuffler, S. W. 1941. *J. Neurophysiol.* 4: 362–387.
[12] Eccles, J. C., Katz, B., and Kuffler, S. W. 1942. *J. Neurophysiol.* 5: 211–230.

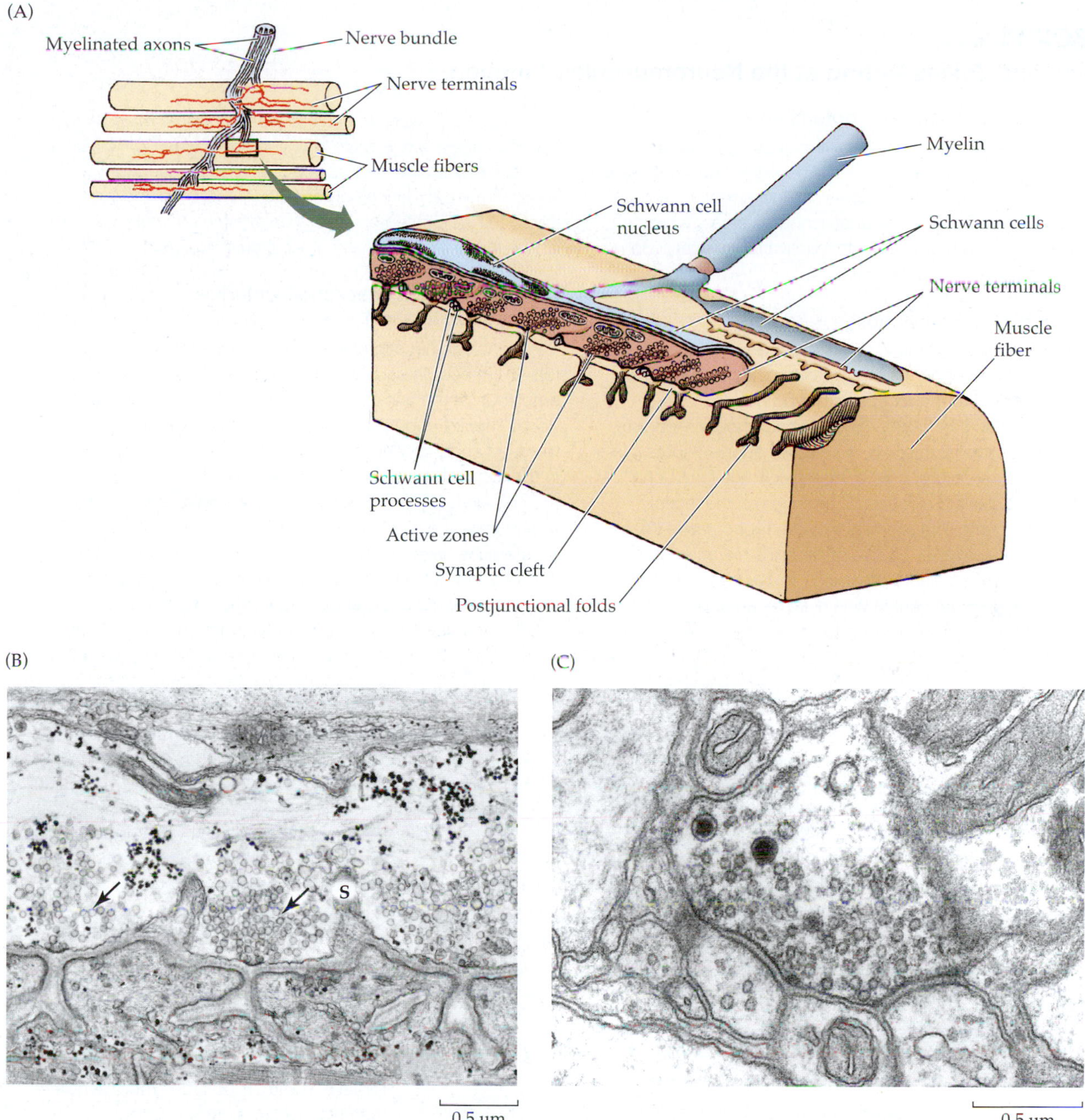

FIGURE 11.2 Structure of Chemical Synapses. (A) A three-dimensional sketch of part of the terminal arbor of a motor axon at the frog skeletal neuromuscular junction. Inset shows several skeletal muscle fibers and their innervation. Synaptic vesicles are clustered in the nerve terminal in special regions opposite the openings of the postjunctional folds. These regions, called active zones, are the sites of transmitter release into the synaptic cleft. Fingerlike processes of Schwann cells extend between the terminal and the postsynaptic membrane, separating active zones. (B) Electron micrograph of a longitudinal section through a portion of the neuromuscular junction. In the nerve terminal, clusters of vesicles lie over thickenings in the presynaptic membrane—the active zones (arrows). Schwann cell processes (S) separate the clusters. In the muscle, postjunctional folds open into the synaptic cleft directly under the active zone. The band of fuzzy material in the cleft, which follows the contours of the postjunctional folds, is the synaptic basal lamina. (C) Electron micrograph of synapses in the central nervous system of the leech. As at the frog neuromuscular junction, clusters of synaptic vesicles are focused on dense regions of the presynaptic membrane, forming active zones, and are juxtaposed to postsynaptic densities. (B kindly provided by U. J. McMahan; C kindly provided by K. J. Muller.)

adding curare, a blocker of the postsynaptic receptors, to the bathing solution (Boxes 11.2 and 11.3). With sufficient curare (about 1 μM), the amplitude of the end-plate potential is reduced to below threshold, so that it is no longer obscured by the action potential (Figure 11.3).

BOX 11.2
Drugs and Toxins Acting at the Neuromuscular Junction

Agonists, antagonists, and potentiators

Drugs have frequently been used to increase our understanding of neuromuscular transmission. These fall into two classes: drugs that act directly on the acetylcholine receptors and those that inhibit the enzyme acetylcholinesterase, which is responsible for the hydrolysis and inactivation of acetylcholine (see Appendix B).

Drugs acting on acetylcholine receptors

There are two types of these drugs: **agonists**, which stimulate the receptors and open the nicotinic channels; and **antagonists**, which bind to the receptors but do not open the ion channels, and so block the action of acetylcholine. Agonists include: acetylcholine itself (1), the natural agonist; some synthetic choline esters, such as carbamoylcholine (carbachol); and the plant alkaloid nicotine (2), from which the receptors get their name. Antagonists include: tubocurarine (3), a component of curare (a mixture of alkaloids extracted from the South American plants *Strychnos toxifera* and *Chondrodendron tomentosum*, which was used as an arrow poison to paralyze prey); and α-bungarotoxin, a component of the venom of the Taiwanese banded krait *Bungarus multicinctus*. Tubocurarine is a reversible blocking agent (see Box 11.3). Drugs with a similar mechanism of action are used to relax skeletal muscles during surgery. Bungarotoxin binds irreversibly to the ACh receptors and is used experimentally to count or see the receptors (as shown in Figure 11.7).

Drugs that inhibit acetylcholinesterase (anticholinesterases)

These include neostigmine (4) and physostigmine or eserine (5). At the neuromuscular junction, they do not alter the peak amplitude of the end-plate currents but rather slow their decay rate by about three times.[13,14] This is because the released acetylcholine is not hydrolyzed and so stays longer in the synaptic cleft, until it is cleared by diffusion; this allows the acetylcholine that is released by a single nerve impulse to stimulate the nicotinic receptors several times. Anticholinesterases have much more effect on the response of the end plate to ACh in the bathing solution, because a high proportion of the acetylcholine is normally hydrolyzed by the cholinesterase in the neuromuscular junction before it can access the receptors. The drugs can be used to reverse the blocking effect of tubocurarine or to improve the muscle response in diseases in which neuromuscular transmission is defective, such as myasthenia.

Anticholinesterase drugs have a more dramatic effect on both the amplitude and duration of the slow synaptic responses mediated by **muscarinic** acetylcholine receptors[15] (see Chapters 12 and 13). The reasons are that the muscarinic receptors are further away from the presynaptic nerve endings than the nicotinic receptors, so more of the released ACh is hydrolyzed before it reaches the receptors. In addition, the muscarinic receptors are 100 to 1000 times more sensitive than nicotinic receptors to acetylcholine so a low concentration of residual unhydrolyzed acetylcholine has a large effect. This underlies their limited use in treating Alzheimer's disease (see Chapter 14).

Agonists

(1) 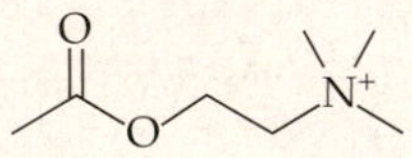 Acetylcholine

(2) 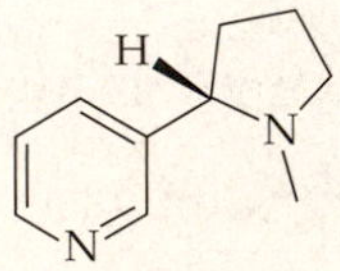 Nicotine

Antagonist

(3) 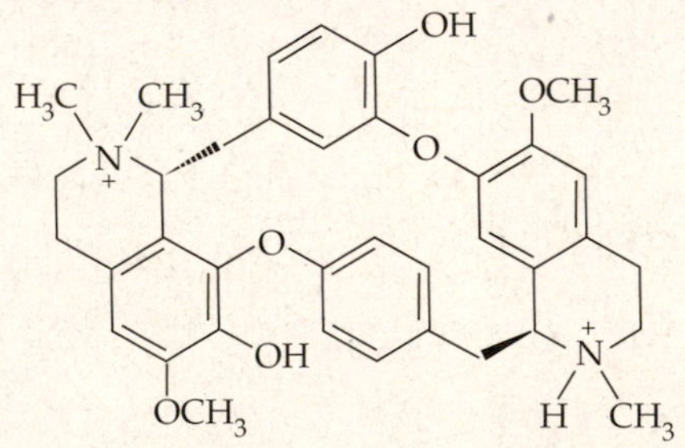 Tubocurarine

Potentiators – cholinesterase inhibitors

(4) 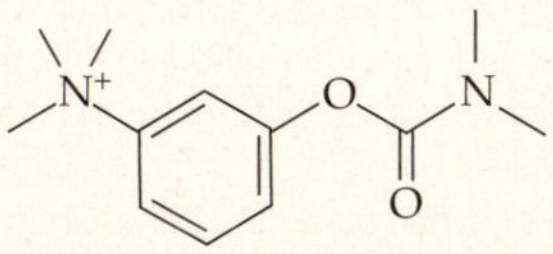 Neostigmine

(5) 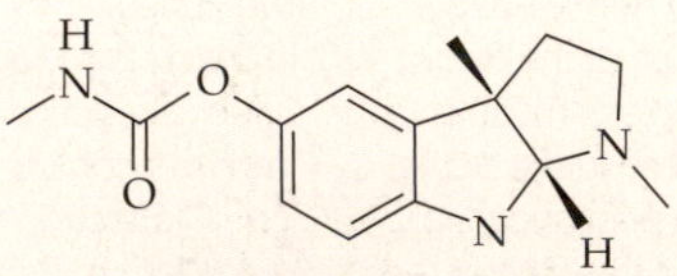 Physostigmine (eserine)

[13] Katz, B., and Miledi, R. 1973. *J. Physiol.* 231: 549–574.
[14] Magleby, K. L., and Terrar, D. A. 1975. *J. Physiol.* 244: 467–495.
[15] Brown, D. A., and Selyanko, A. A. 1985. *J. Physiol.* 365: 335–364.

BOX 11.3
Action of Tubocurarine at the Motor End Plate

Reversible competitive antagonism

Tubocurarine acts as a *reversible competitive antagonist* of acetylcholine at the motor end plate (see Box 11.2), which means that it readily dissociates from the acetylcholine receptors and that its blocking action can be overcome by increasing the concentration of acetylcholine. Competition occurs because both molecules only bind transiently to the receptors. So, when an ACh molecule detaches from the receptors, a tubocurarine molecule may take its place; and conversely, when the tubocurarine dissociates, ACh may take its place. The probability of one or the other substance occupying the receptor then depends on: (1) the relative number of molecules of each substance available (i.e., on their relative concentrations) and (2) on the relative times for which they occupy the receptor. Thus, tubocurarine binds to the receptor about 100 times longer than acetylcholine; so for equal concentrations, the competition is weighted in favor of tubocurarine, but this would be equalized if the concentration of ACh were 100 times greater. This competition can be quantified as shown below.

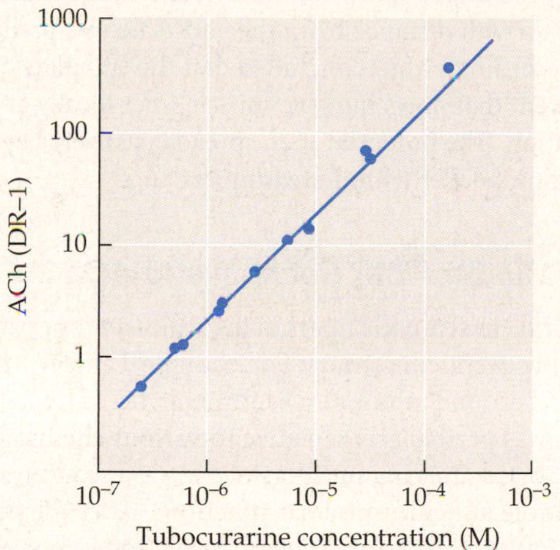

Take a simple reversible reaction between agonist (A) and receptor (R):

$$A + R \leftrightarrow AR \quad K_A$$

where K_A is the equilibrium dissociation constant. Then the proportion of receptors occupied by the agonist at equilibrium (P_{AR}) in the absence of antagonist may be given by:

$$P_{AR} = [A]/\{[A] + K_A\}$$

where [A] = concentration of A.

When the antagonist B is also present, the new receptor that is occupied by the agonist ($P_{AR(B)}$) is now given by:

$$P_{AR(B)} = [A]/\{[A] + K_A(1 + [B]/K_B)\}$$

where [B] is the concentration of B and K_B is the equilibrium dissociation constant for the reaction B + R ↔ BR. This is called the Gaddum equation[16] (see[17] for a full derivation). Thus, to get the same response to the agonist in the presence of the antagonist as that seen before adding the antagonist, the concentration of agonist must be increased from A to A_B such that $A_B / A = \{1 + [B] / K_B\}$. The term A_B / A is frequently called the **dose ratio**, or DR, and the relationship:

$$DR - 1 = [B]/K_B$$

is known as the Schild equation.[18]

In a classic piece of work, Donald Jenkinson[19] tested this out for tubocurarine, and acetylcholine antagonism in frog muscle. He measured the depolarization produced by ACh in the presence of increasing concentrations of tubocurarine and then plotted DR–1 against the concentration of tubocurarine ([B]) on a double-logarithmic scale. As shown in Figure 11.4, this followed a linear relation over a thousand-fold range of tubocurarine concentrations. This provides convincing evidence for true competitive inhibition. Subsequent work[20] revealed that tubocurarine could also block the nicotinic receptor ion channels. However, this only becomes significant at more hyperpolarized membrane potentials (–120 mV) than those in Jenkinson's experiments or those in muscle fibers in vivo.

[16] Gaddum, J. H. 1943. *Trans. Faraday Soc.* 39: 323–332.

[17] Jenkinson, D. H. 2011. In *Textbook of Receptor Pharmacology*, 3rd ed. CRC Press, London, U.K.

[18] Arunlakshana, O., and Schild, H. O. 1959. *Brit. J. Pharmacol. Chemother.* 14: 48–58.

[19] Jenkinson, D. H. 1960. *J. Physiol.* 152: 309–324.

[20] Colquhoun, D., Dreyer, F., and Sheridan, R. E. 1979. *J. Physiol.* 293: 247–284.

The intracellular microelectrode[21] was used by Fatt and Katz[22,23] to study in detail the time course and spatial distribution of the end-plate potential in muscle fibers treated with curare. They stimulated the motor nerve and recorded the end-plate potential intracellularly at various distances from the end plate (Figure 11.4). At the end plate the depolarization rose rapidly to a peak and then declined slowly over the next 10 to 20 ms. As they moved the recording microelectrode farther and farther away from the end plate, the end-plate potential amplitude became progressively smaller and its time to peak progressively longer. Fatt and Katz showed that after reaching its peak, the end-plate potential decayed at a rate that was

[21] Ling, G., and Gerard, R. W. 1949. *J. Cell Comp. Physiol.* 34: 383–396.

[22] Fatt, P., and Katz, B. 1951. *J. Physiol.* 115: 320–370.

[23] Nicholls, J. G. 2007. *J. Physiol.* 578: 621–622.

FIGURE 11.3 Synaptic Potentials Recorded with an Intracellular Microelectrode from a mammalian neuromuscular junction treated with curare. The curare concentration in the bathing solution was adjusted so that the amplitude of the synaptic potential was near threshold and hence, on occasion evoked an action potential in the muscle fiber. (From Boyd and Martin, 1956.)

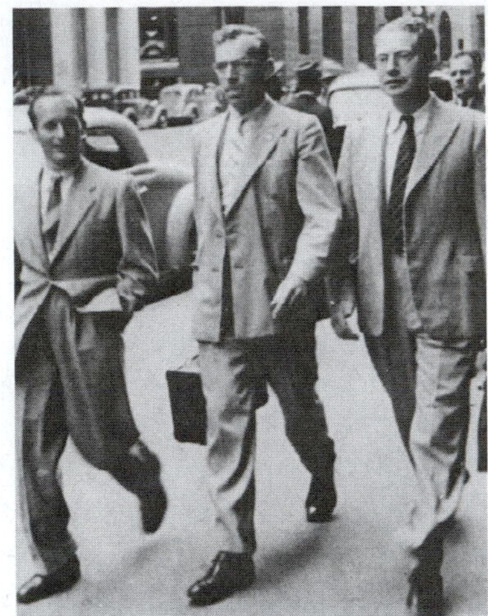

Stephen Kuffler, John Eccles, and Bernard Katz (left to right) in Australia, about 1941.

consistent with the time constant of the muscle fiber membrane, and that the decrement in end-plate potential peak amplitude with distance from the end plate was predicted by the muscle fiber cable properties. Accordingly, they concluded that the end plate potential is generated by a brief surge of current that flows into the muscle fiber locally at the end plate and causes a rapid depolarization. The potential then spreads passively beyond the end plate in both directions, becoming smaller with increasing distance.

Mapping the Region of the Muscle Fiber Receptive to ACh

The existence of special properties of skeletal muscle fibers in the region of innervation has been known since the beginning of the twentieth century. For example, Langley[24] assumed the presence of a "receptive substance" around motor nerve terminals, based on the finding that this region of the muscle fiber was particularly sensitive to various chemical agents, such as nicotine. This conclusion showed amazing insight since it was inconceivable that nicotine could play a physiological role at neuromuscular junctions. After the introduction of the glass microelectrode for intracellular recording, microelectrodes were also used for discrete application of ACh (and later other drugs as well) to the end plate region of muscle.[25] The technique is illustrated in Figure 11.5A.

A microelectrode is inserted into the end plate of a muscle fiber for recording membrane potentials, while an ACh-filled micropipette is held just outside the fiber. To apply ACh, a brief positive voltage pulse is applied to the top of the pipette, causing a spurt of positively charged ACh ions to leave the pipette tip. This method of ejecting charged molecules from pipettes is known as **ionophoresis.**[25] Using this method of application, del Castillo and Katz showed that ACh depolarized the muscle fiber only at the end-plate region and only when applied to the outside of the fiber.[26] When the ACh-filled pipette is placed in close apposition to the end-plate region, the response to ionophoresis is rapid (Figure 11.5B). Movement of the pipette by only a few micrometers results in a reduction in amplitude and slowing of the response.

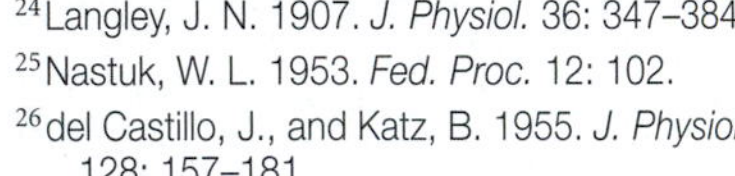

[24]Langley, J. N. 1907. *J. Physiol.* 36: 347–384.

[25]Nastuk, W. L. 1953. *Fed. Proc.* 12: 102.

[26]del Castillo, J., and Katz, B. 1955. *J. Physiol.* 128: 157–181.

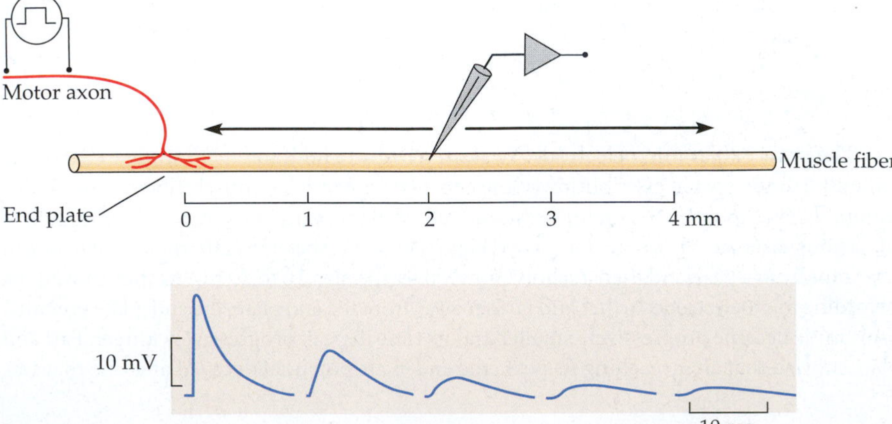

FIGURE 11.4 Decay of Synaptic Potentials with Distance from the End-Plate Region of a Muscle Fiber. As the distance from the end plate increases, synaptic potentials recorded by an intracellular electrode decrease in size and rise more slowly. (After Fatt and Katz, 1951.)

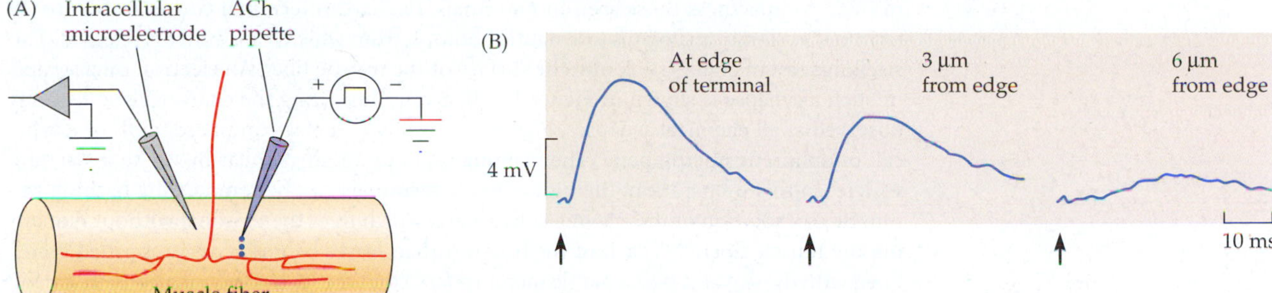

FIGURE 11.5 Mapping the Distribution of ACh Sensitivity by Ionophoresis at the Frog Neuromuscular Junction. (A) An ACh-filled pipette is placed close to the neuromuscular junction, and ACh is ejected from the tip by a brief, positive voltage pulse (ionophoresis). An intracellular microelectrode is used to record the response from the muscle fiber. (B) Responses to small ionophoretic pulses of ACh applied at different distances from the axon terminal (indicated by the blue dots in [A]). The amplitude and rate of rise of the response decrease rapidly as ACh is applied farther from the terminal. (After Peper and McMahan, 1972.)

The receptive substance postulated by Langley is now known to be the nicotinic ACh receptor. The technique of ionophoresis made it possible to map with high accuracy the distribution of postsynaptic ACh receptors in muscle fibers[27] and nerve cells.[28] This method is particularly useful with thin preparations in which the presynaptic and postsynaptic structures can be resolved with interference contrast optics,[29] and the position of the ionophoretic pipette in relation to the synapse can be determined with precision.

One such preparation is the neuromuscular junction of the snake, shown in Figure 11.6. The end plates in snake muscle are about 50 micrometers (μm) in diameter, resembling

[27] Miledi, R. 1960. *J. Physiol.* 151: 24–30.

[28] Dennis, M. J., Harris, A. J., and Kuffler, S. W. 1971. *Proc. R. Soc. Lond., B, Biol. Sci.* 177: 509–539.

[29] McMahan, U. J., Spitzer, N. C., and Peper, K. 1972. *Proc. R. Soc. Lond., B, Biol. Sci.* 181: 421–430.

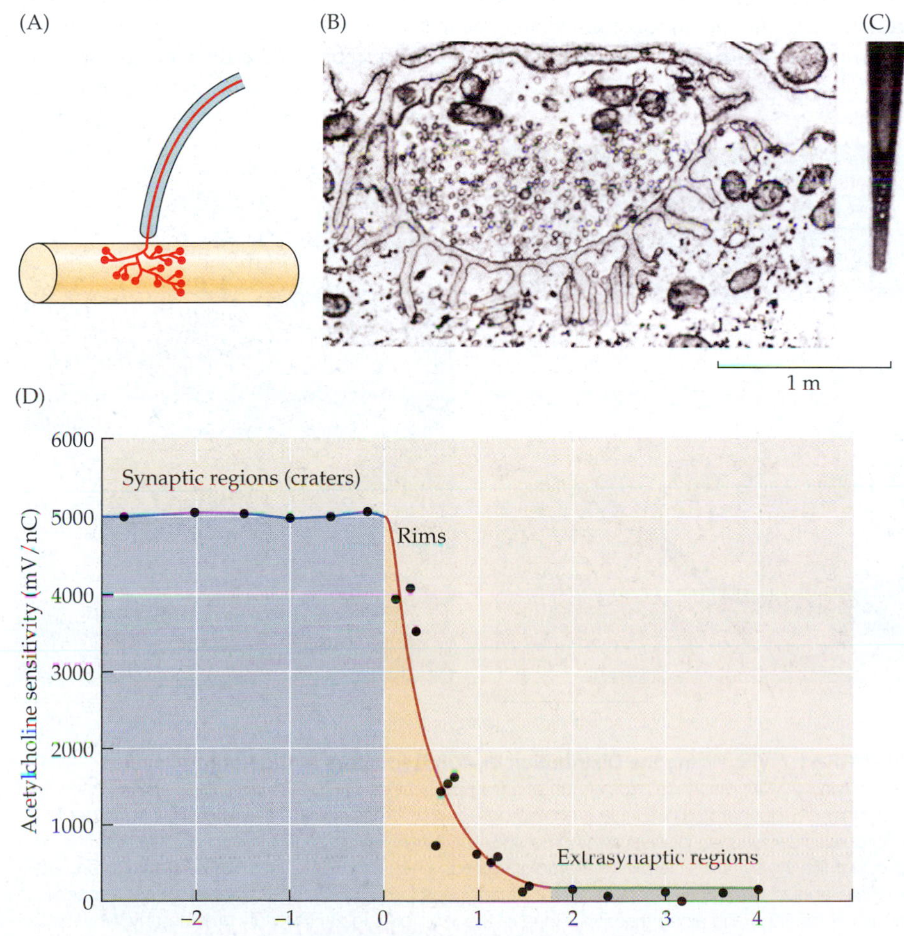

FIGURE 11.6 Acetylcholine Receptor Distribution at the Skeletal Neuromuscular Junction of the Snake. (A) An end plate on a skeletal muscle of a snake. The axon terminates in a cluster of boutons. (B) Electron micrograph of a cross section through a bouton. Synaptic vesicles, which mediate ACh release from the nerve terminal, are 50 nanometers (nm) in diameter. (C) Electron micrograph of the tip of a micropipette used for ionophoresis of ACh, shown at the same magnification as (B). The pipette has an outer diameter of 100 nm and an opening of about 50 nm. (D) ACh was applied by ionophoresis across the postsynaptic crater left after the nerve terminal was removed with collagenase. The craters have a uniformly high sensitivity to ACh (5000 mV/nC), with the sensitivity declining steeply at the rims of the craters. Extrasynaptic regions have a uniformly low ACh sensitivity (100 mV/nC). (After Kuffler and Yoshikami, 1975a.)

in their compactness those seen in mammals. Each axon terminal consists of 50 to 70 terminal swellings, analogous to synaptic boutons, from which transmitter is released. The swellings rest in craters sunk into the surface of the muscle fiber. An electron micrograph of such a synapse is shown in Figure 11.6B, again illustrating the characteristic features observed at all chemical synapses. Figure11.6C shows an electron micrograph of a typical ionophoretic micropipette. The opening is about 50 nm, similar in size to a synaptic vesicle. In this preparation, the motor nerve terminals can be removed by bathing the muscle in a solution of the enzyme collagenase, which frees the terminal without damaging the muscle fiber.[30,31] Each of the boutons then leaves behind a circumscribed crater lined with the exposed postsynaptic membrane, so that the ACh-filled micropipette can be placed directly on the postsynaptic membrane. Then 1 picocoulomb (pC) of charge passed through the pipette releases enough ACh to cause, on the average, a 5-mV depolarization. The sensitivity of the membrane is then said to be 5000 mV/nanocoulomb (nC) (Figure 11.6D). In contrast, at a distance of about 2 μm, just outside the crater, the same amount of ACh produces a response that is 50 to 100 times smaller. Along the rims of the craters, the sensitivity fluctuates over a wide range. Therefore, the conclusion from physiological mapping is that the ACh receptors are highly concentrated in the region of the synapse.

Morphological Demonstration of the Distribution of ACh Receptors

Another way to determine the distribution of ACh receptors is to use α-bungarotoxin, the snake toxin that binds highly selectively and irreversibly to nicotinic ACh receptors. The distribution of bound toxin can be visualized using histochemical techniques. For example, fluorescent markers can be attached to α-bungarotoxin and the distribution of receptors visualized by fluorescence microscopy (Figure 11.7A); or the enzyme horseradish peroxidase (HRP) can be linked to α-bungarotoxin and its dense reaction product visualized in the electron microscope (Figure 11.7B).[32] Such techniques confirm that receptors are highly restricted to the membrane immediately beneath the axon terminal. Even more precise quantitative estimates of the concentration of ACh receptors than these can be obtained

[30] Betz, W. J., and Sakmann, B. 1973. *J. Physiol.* 230: 673–688.

[31] Kuffler, S. W., and Yoshikami, D. 1975. *J. Physiol.* 244: 703–730.

[32] Burden, S. J., Sargent, P. B., and McMahan, U. J. 1979. *J. Cell Biol.* 82: 412–425.

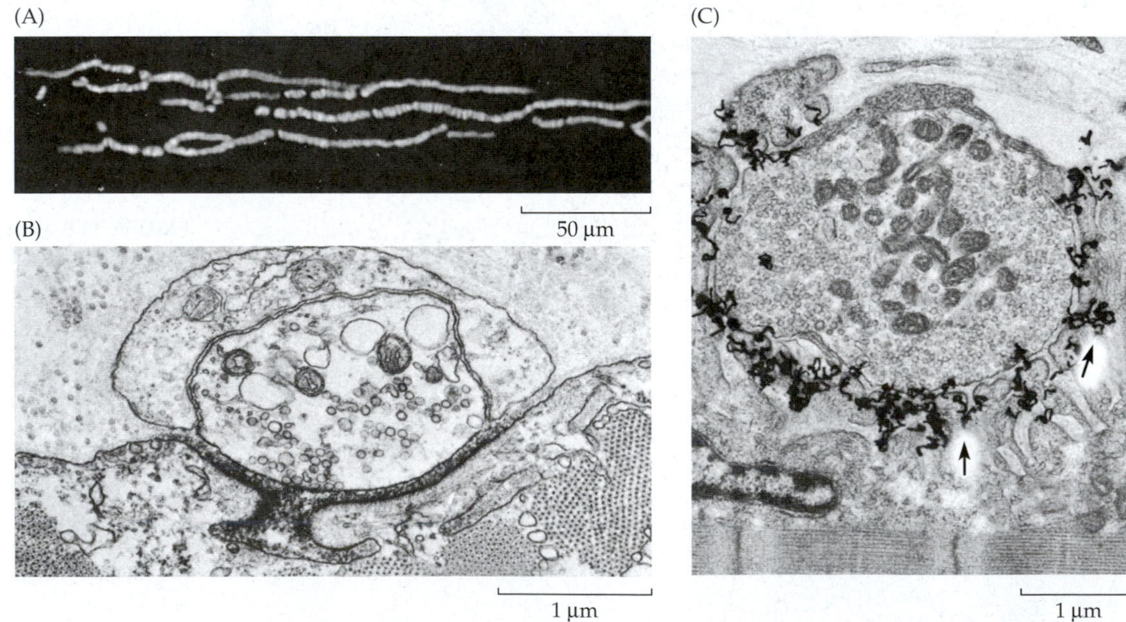

FIGURE 11.7 Visualizing the Distribution of ACh Receptors at the Neuromuscular Junction. (A) Fluorescence micrograph of a frog cutaneous pectoris muscle fiber stained with rhodamine α-bungarotoxin. (B) Electron micrograph of a cross section of a frog cutaneous pectoris neuromuscular junction labeled with horseradish peroxidase–α-bungarotoxin. A dense reaction product fills the synaptic cleft. (C) Autoradiograph of a neuromuscular junction in a lizard intercostal muscle labeled with [^{125}I]-α-bungarotoxin. Silver grains (arrows) show that receptors are concentrated at the tops and along the upper third of the junctional folds. (A kindly provided by W. J. Betz; B kindly provided by U. J. McMahan; C from Salpeter, 1987, kindly provided by M. M. Salpeter.)

using radioactive α-bungarotoxin and autoradiography (Figure 11.7C).[33] By counting the number of silver grains exposed in the emulsion, the density of receptors can be determined. In muscle the density is highest along the crests and upper third of the junctional folds (about $10^4/\mu m^2$); the density in extrasynaptic regions is much lower (about $5/\mu m^2$).[34] Transmitter receptors are highly concentrated in the postsynaptic membrane at synapses throughout the central and peripheral nervous systems.

Measurement of Ionic Currents Produced by ACh

How does ACh produce an inward current at the end plate? Experiments by Fatt and Katz led them to conclude that ACh produces a marked, nonspecific increase in permeability of the postsynaptic membrane to small ions.[22] Two techniques were subsequently used to assess the permeability changes produced by ACh. One involved the use of radioactive isotopes, which showed that the permeability of the postsynaptic membrane was increased to sodium, potassium, and calcium but not to chloride.[35] This experiment provided convincing evidence concerning the ion species involved but did not reveal the details of the conductance changes, their timing, or their voltage dependence. Precise information was provided by voltage clamp experiments, first performed by A. and N. Takeuchi, who used two microelectrodes to voltage clamp the end-plate region of muscle fibers.[36] The experimental arrangement is shown in Figure 11.8A. Two microelectrodes were inserted into the end-plate region of a frog muscle fiber—one for recording membrane potential (V_m), the other for injecting current to clamp the membrane potential at the desired level. The nerve was then stimulated to release ACh or, in later experiments, ACh was applied directly by ionophoresis. Subsequently, similar experiments were carried out by Magleby and Ste-

[33]Fertuck, H. C., and Salpeter, M. M. 1974. *Proc. Natl. Acad. Sci. USA* 71: 1376–1378.

[34]Salpeter, M. M. 1987. In *The Vertebrate Neuromuscular Junction*. Alan R. Liss, New York, pp. 1–54.

[35]Jenkinson, D. H., and Nicholls, J. G. 1961. *J. Physiol.* 159: 111–127.

[36]Takeuchi, A., and Takeuchi, N. 1959. *J. Neurophysiol.* 22: 395–411.

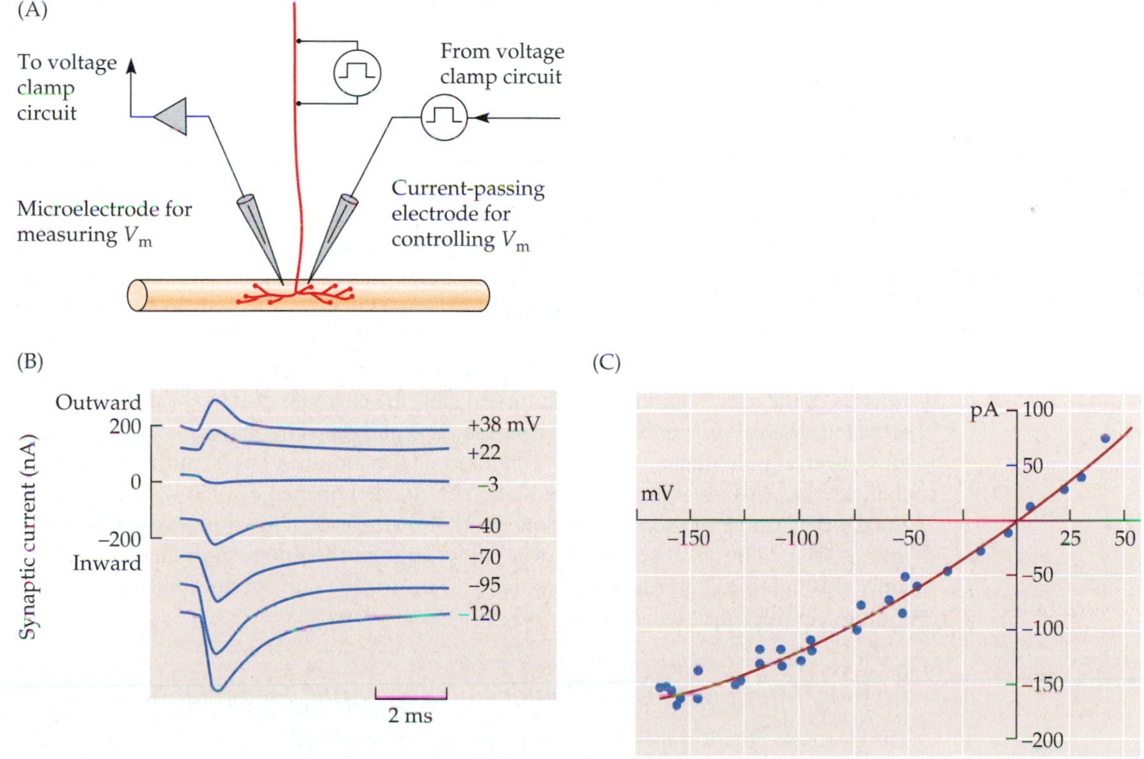

FIGURE 11.8 Reversal Potential for Synaptic Currents Measured by Voltage Clamp Recording. (A) Scheme for voltage clamp recording at the motor end plate. (B) Synaptic currents recorded at membrane potentials between –120 and +38 mV. When the muscle membrane potential is clamped below 0 mV, synaptic current flows into the muscle. Such inward current would depolarize the muscle if it were not voltage clamped. When the end-plate potential is clamped above 0 mV, synaptic current flows out of the cell. (C) Plot of peak end-plate current as a function of membrane potential. The relation is nearly linear, with the reversal potential close to 0 mV. (After Magleby and Stevens, 1972.)

vens[37] in muscle fibers treated with hypertonic glycerol. This stops the muscle fibers from contracting, though leaving the fibers in an artificially depolarized state.

Figure 11.8B illustrates results from such a glycerol-treated muscle fiber. With the muscle membrane potential clamped at –40 mV, nerve stimulation produced an inward current, which would have caused a depolarization if the fiber had not been voltage clamped. At more-negative holding potentials, the end-plate current increased in amplitude. When the membrane was depolarized, the end-plate current decreased in amplitude. With further depolarization, the current reversed direction and was outward.

Figure 11.8C shows a plot of the peak amplitude of the end-plate current as a function of holding potential. The current changed from inward to outward near zero membrane potential. This is called the **reversal potential, V_r**. In earlier experiments on intact muscle fibers, A. and N. Takeuchi estimated the reversal potential to be about –15 mV.

Significance of the Reversal Potential

The reversal potential for the end-plate current gives us information about the ionic currents flowing through the channels activated by ACh in the postsynaptic membrane. For example, if the channels were permeable exclusively to sodium, then current through the channels would be zero at the sodium equilibrium potential (about +50 mV). The other major ions, potassium and chloride, have equilibrium potentials near –90 mV, which is the normal resting membrane potential (see Chapter 6). The calcium equilibrium potential is approximately +120 mV. None of the ions has an equilibrium potential in the range of 0 to –15 mV. What ions, then, are involved in the response? Consistent with the results of radioactive tracer experiments,[35] A. and N. Takeuchi showed that changing the concentrations of sodium, potassium, or calcium in the bathing solution resulted in changes in the reversal potential, but changes in extracellular chloride did not.[38] They concluded that the effect of ACh was to produce a general increase in *cation* permeability.

Relative Contributions of Sodium, Potassium, and Calcium to the End-Plate Potential

ACh opens channels that, at the normal resting potential, allow sodium and calcium ions to leak inward and potassium ions to leak outward along their electrochemical gradients. Because the calcium conductance of the channels is small, the contribution of calcium to the overall synaptic current can be ignored, as can that of other cations such as magnesium. (It should be noted that the low calcium *conductance* is due to its low extracellular and intracellular concentrations; calcium *permeability* is about 20% of the sodium permeability.) The equivalent electrical circuit is shown in Figure 11.9A. The resting membrane consists of the usual sodium, potassium, and chloride channels. It is in parallel with ACh-activated channels for sodium and potassium, Δg_{Na} and Δg_K. The Takeuchis calculated that for a reversal potential V_r = –15 mV, the ratio of the sodium to potassium conductance changes, $\Delta g_{Na}/\Delta g_K$, is about 1.3 (Box 11.4). The channel opened by ACh is, in fact, nearly equally permeable to sodium and potassium.[39,40] However, taking the extracellular and intracellular solutions together, there are more sodium than potassium ions available to move through the channels (see Chapter 6). Thus, for the same permeability change, the sodium conductance change is slightly larger (see Chapter 4).

Resting Membrane Conductance and Synaptic Potential Amplitude

The electrical circuit shown in Figure 11.9A can be simplified by representing the resting membrane as a single conductance, g_{rest} (equal to the sum of all the ionic conductances), and a single battery, V_{rest} (equal to the resting membrane potential). Likewise, the synaptic membrane can be represented by a single conductance, Δg_s, and a battery whose voltage is equal to the reversal potential, V_r (Figure 11.9B). A feature of this electrical circuit is that the amplitude of a synaptic potential depends on both Δg_s and g_{rest}.

For simplicity, consider the steady-state membrane potential that would develop if the synaptic conductance were activated for a long period of time. If Δg_s were much larger than g_{rest}, then the membrane potential would approach V_r. However, if Δg_s were equal to g_{rest},

[37] Magleby, K. L., and Stevens, C. F. 1972. *J. Physiol.* 223: 151–171.

[38] Takeuchi, A., and Takeuchi, N. 1960. *J. Physiol.* 154: 52–67.

[39] Takeuchi, N. 1963. *J. Physiol.* 167: 128–140.

[40] Adams, D. J., Dwyer, T. M., and Hille, B. 1980. *J. Gen. Physiol.* 75: 493–510.

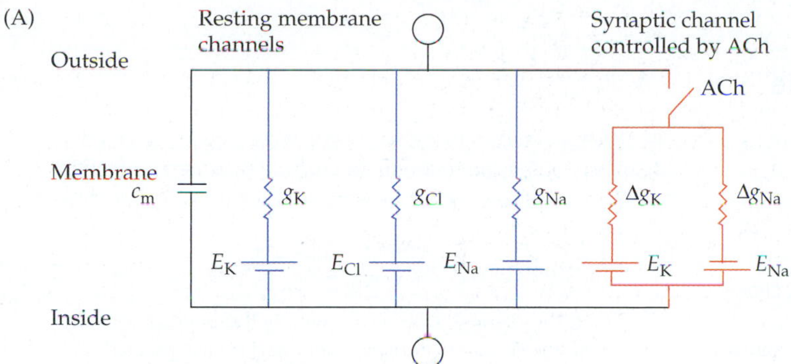

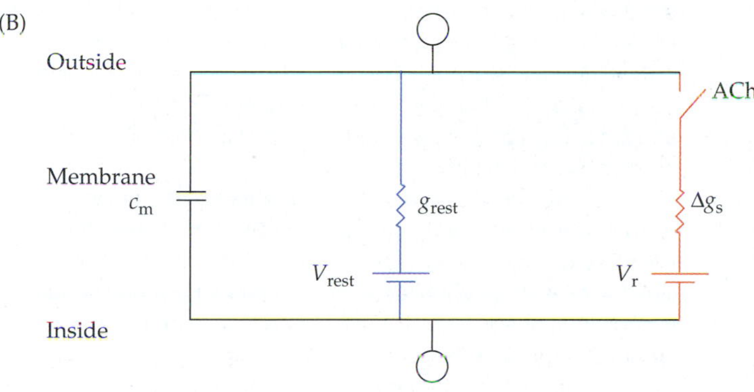

FIGURE 11.9 Electrical Model of the Postsynaptic Membrane. Channels activated by ACh are in parallel with the resting membrane channels and with the membrane capacitance, c_m. (A) The synaptic channel opened by ACh is electrically equivalent to two independent pathways for sodium and potassium. The resting membrane has channels for potassium, chloride, and sodium. (B) The synaptic channel can be represented as a single pathway with conductance Δg_s and a battery equal to the reversal potential V_r. The resting membrane can be represented as a single pathway with conductance g_{rest} and a battery equal to V_{rest}.

then the change in membrane potential produced by activating the synaptic conductance would be only one-half as great. Thus, the amplitude of a synaptic potential can be increased by either increasing the synaptic conductance (i.e., activating more synaptic channels) or by decreasing the resting conductance. Indeed, a reduction in membrane conductance is an important mechanism for modulating synaptic strength. For example, certain inputs to autonomic ganglion cells in the bullfrog can *close* potassium channels, thereby increasing the amplitude of excitatory synaptic potentials produced by other inputs to the cell (see Chapter 17). Similarly, an excitatory current of a given amplitude will give rise to a larger depolarization in a small neuron than in a large one that has a lower input resistance (see Chapter 7 and 24).

Kinetics of Currents through Single ACh Receptor Channels

To what extent does the time course of the end-plate current reflect the behavior of individual ACh channels? For example, do individual channels open and close repetitively during the end-plate current, with the probability of channel opening declining with time? Or do individual channels open only once, so that the time course of the current is determined by how long channels remain open?

Definitive answers to such questions came only with the advent of patch clamp techniques, by which the behavior of individual channels could be observed directly (see Chapter 4).[41] When ACh was applied continuously, ACh channels were shown to open instantaneously, in an all-or-nothing fashion, and then close at a rate that matched exactly the rate of decay of the end-plate current.[42,43] These observations can be interpreted according to the following scheme for the interaction between the transmitter molecule A (for agonist) and the postsynaptic receptor molecule R (for receptor):

$$\underset{shut}{A + R} \underset{k_{-1}}{\overset{k_1}{\longleftrightarrow}} \underset{shut}{AR + A} \underset{k_{-2}}{\overset{k_2}{\longleftrightarrow}} \underset{shut}{A_2R} \underset{\alpha}{\overset{\beta}{\longleftrightarrow}} \underset{open}{A_2R}$$

In this scheme, two ACh molecules sequentially combine with the channel (one on each α-subunit; see Chapter 5). With only one ACh molecule bound, the channel does not open (or opens very rarely). However, when the second ACh molecule binds, the channel

[41] Neher, E., and Sakmann, B. 1976. *Nature* 260: 799–802.

[42] Dionne, V. E., and Leibowitz, M. D. 1982. *Biophys. J.* 39: 253–261.

[43] Sakmann, B. 1992. *Neuron* 8: 613–629.

BOX 11.4
Electrical Model of the Motor End Plate

How did A. and N. Takeuchi calculate the ratio of sodium to potassium conductance for the channels opened by acetylcholine (ACh)? They proposed an electrical model of the muscle cell membrane similar to that shown in Figure 11.9A. Although ACh receptors do not form separate pathways for sodium and potassium, the two ions move through the channel independently. Therefore, the synaptic conductance and reversal potential can be represented by separate conductances (Δg_{Na} and Δg_K) and driving potentials (E_{Na} and E_K) for sodium and potassium. Accordingly, separate expressions can be written for the sodium and potassium currents (ΔI_{Na} and ΔI_K):

$$\Delta I_{Na} = \Delta g_{Na}(V_m - E_{Na})$$

$$\Delta I_K = \Delta g_K(V_m - E_K)$$

These equations provide a means of determining the relative conductance changes to sodium and potassium produced by ACh once the reversal potential (V_r) is determined. Since the Takeuchis considered only *changes* in current resulting from the action of ACh, they could ignore the resting membrane channels. The net synaptic current is zero at the reversal potential; therefore, at this potential the inward sodium current is exactly equal and opposite to the outward potassium current. So when $V_m = V_r$,

$$\Delta g_{Na}(V_r - E_{Na}) = -\Delta g_K(V_r - E_K)$$

It follows that

$$\frac{\Delta g_{Na}}{\Delta g_K} = \frac{-(V_r - E_K)}{-(V_r - E_{Na})}$$

We can rearrange the equations regarding synaptic sodium and potassium currents to predict the reversal potential when the relative conductances are known:

$$V_r = \frac{\Delta g_{Na}E_{Na} + \Delta g_K E_K}{(\Delta g_{Na} + \Delta g_K)}$$

Thus, the reversal potential is simply the average of the individual equilibrium potentials, weighted by the relative conductance changes. This relationship can be extended to include any number or variety of ions, so it is applicable at any synapse where transmitters produce a change in conductance of the postsynaptic membrane to one or more ions. This relationship was found to predict how changes in E_{Na} and E_K, produced by changes in extracellular concentrations of sodium and potassium, affected the reversal potential at the neuromuscular junction.[30]

Such predictions were accurate only for small changes in extracellular sodium and potassium, however, because channel conductance is determined in part by ion concentration (see Chapters 2 and 5). Therefore, the effect of a large change in sodium, potassium, or calcium concentration on reversal potential is predicted accurately only if the resulting change in conductance is taken into account. Alternatively, the analysis can be made in terms of permeabilities, using the constant field equation developed by Goldman, Hodgkin, and Katz (see Chapter 6).

undergoes a very fast (microseconds) change in conformation from the closed (A_2R) to the open (A_2R^*) state. The transitions between the open and closed states are characterized by the rate constants α and β, as indicated. Now consider the time course of the end-plate current, as illustrated in Figure 11.10. ACh arriving at the postsynaptic membrane opens a large number of channels almost simultaneously. Because ACh is lost rapidly from the synaptic cleft (through hydrolysis by the enzyme cholinesterase and by diffusion), each channel appears to open only once. As the channels close, the synaptic current declines. Thus, the time course of decay of the end-plate current reflects the rate at which individual ACh channels close. Channels close at the rate α $[A_2R^*]$; that is, many channels close very quickly, and fewer and fewer channels close at longer and longer times. As with all independent or random events, the open times are distributed exponentially, and the mean open time (τ) is equal to the time constant of the decay of the end-plate current, $1/\alpha$.

With improved time resolution it was discovered that many of these apparent single channel openings were interrupted by one or more brief closures.[44,45] In

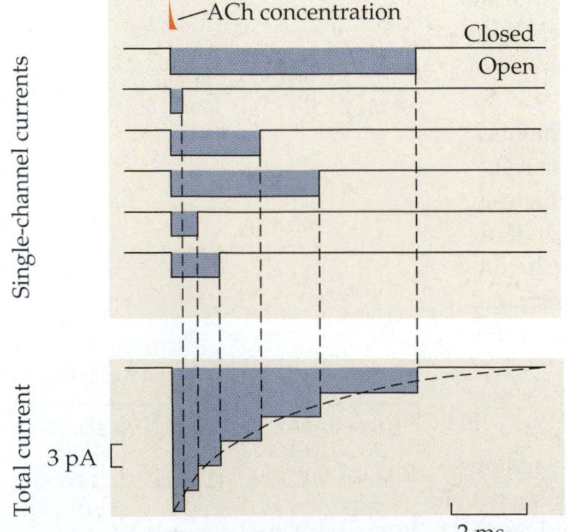

FIGURE 11.10 Total End-Plate Current Is the Sum of Individual Channel Currents. Current flow through six individual channels is depicted in the top panel. Channels open instantaneously in response to ACh (added at the red marker point). ACh is rapidly hydrolyzed, so its concentration falls quickly (red marker), preventing any further channel openings. Channel open times are distributed exponentially. The individual channel currents sum to give the total end plate current (lower panel). The time constant of the decay of the total current is equal to the mean open time of the individual channels.

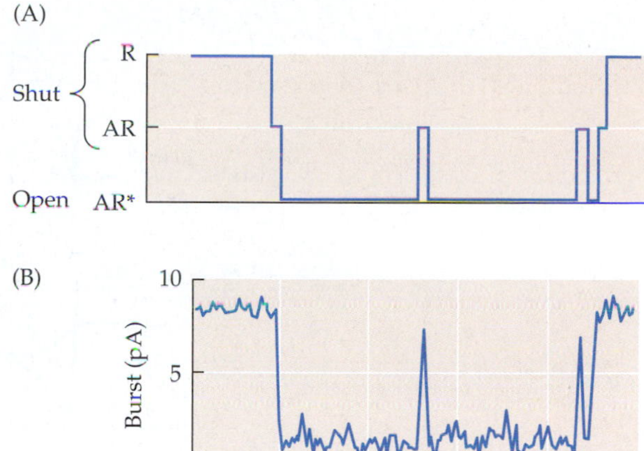

FIGURE 11.11 End-Plate Nicotinic Acetylcholine Receptors Open in Bursts. (A) Predicted burst of three openings (AR*) assuming the fastest possible binding rate (5 x 10^8 moles^{-1} s^{-1}). The channel flips briefly to the ACh-bound but returns to the shut (AR) state twice before the acetylcholine dissociates and the channel reverts to the R state. (B) High-resolution recording of an equivalent three-opening burst with two brief closures from a frog end-plate receptor activated by the nicotinic agonist suberyldicholine. (After Colquhoun, 2007.)

other words, instead of just a single opening, the ACh-bound channel can give a short burst of two or more openings (Figure 11.11). This may be explained by supposing that, when the channel has reverted to the shut (A_2R) state, instead of just going back to the mono-liganded shut state AR, it can flicker back and forth a few times to the open state A_2R^* before the second ACh molecule dissociates. The reason for this is that the conformational opening rate constant β is about the same as the dissociation rate constant k_{-2}. This means that the original open channel lifetime (as depicted in Figure 11.11) is actually the mean burst length. The mean time-constant for decay of this burst (and hence for the end plate current decay) is given by $\tau_{burst} = 1/(\alpha+k_{-2})$ – about double that for the closure of a single channel. It is worth noting that an ACh receptor channel open for a millisecond at –70 mV will conduct about 20,000 cations into the cell, i.e., 10,000 ions for each ACh molecule that binds to the receptor–an enormous signal amplification.

The properties of acetylcholine receptors change during development. There is a fetal form of the acetylcholine receptor, which has a low conductance and a long and variable open time, and an adult form, which has a higher conductance and shorter open time.[46] The switch from embryonic to adult receptors is caused by a change in subunit composition (see Chapters 5 and 25), and splice variants of one of the embryonic subunits may account for variation in mean channel open time early in development.[47] The change in properties is well adapted to the need for adequate current to stimulate the larger, fully developed muscle fibers.

Excitatory Synaptic Potentials in the CNS

The principles of synaptic excitation in the CNS follow those elucidated so beautifully at the neuromuscular junction: a pulse of transmitter released from the presynaptic bouton opens cation-selective channel receptors in the postsynaptic membrane. These receptors generate an inward **excitatory postsynaptic current** (**EPSC**) which, in turn depolarizes the postsynaptic membrane to produce an excitatory postsynaptic potential (EPSP).

However, there are important differences between neuromuscular and CNS synapses. The first essential difference is that the principal excitatory transmitter is L-glutamate, not acetylcholine. Second, glutamate activates two different types of ionotropic glutamate receptor, the α-amino-3-hydroxy-5-methyl-4-isoxazolepropionic acid (AMPA) receptor, so-named after its selective agonist, and the *N*-methyl-D-aspartate (NMDA) receptor[48,49] (Figure 11.12). Neither receptor is structurally homologous to the nicotinic receptor.[50] Nevertheless, the AMPA receptor has similar permeability characteristics and serves the same function as the nicotinic receptor, which is to mediate fast transient excitatory transmission throughout the CNS. Thus, fast synaptic potentials evoked by glutamate can be reconstructed from the kinetics of AMPA-receptor channels,[51] just as with end-plate nicotinic channels. The NMDA glutamate receptor has several unique properties, however. First, at the normal resting potential it is tonically blocked by magnesium ions in the extracellular fluid,[52,53] which means that, even when it is activated, no current flows until the cell is depolarized towards zero from the normal resting potential (see Figure 11.12). The reason is that the positive Mg^{2+} ions bind tightly in the channel when pulled into it by the normal transmembrane electrical gradient; however, this pull is reduced and the Mg^{2+} ions dissociate when the transmembrane gradient is reduced. The natural form of depolarization for unblocking the channel would be a train of preceding AMPA receptor-mediated synaptic potentials. This voltage-sensitivity also means that the NMDA receptor is not responsible for normal fast signaling between brain cells—only

44 Colquhoun, D., and Sakmann, B. 1981. *Nature* 294: 464–466.

45 Colquhoun, D., and Sakmann, B. 1985. *J. Physiol.* 369: 501–557.

46 Mishina, M. et al. 1986. *Nature* 321: 406–411.

47 Herlitze, S. et al. 1996. *J. Physiol.* 492: 775–787.

48 Watkins, J. C., and Evans, R. H. 1981. *Annu. Rev. Pharmacol. Toxicol.* 21: 165–204.

49 Sah, P., Hestrin, S., and Nicoll, R. A. 1990. *J. Physiol.* 430: 605–616.

50 Wollmuth, L. P., and Sobolevsky, A. I. 2004. *Trends Neurosci.* 27: 321–328.

51 Edmonds, B., Gibb, A. J., and Colquhoun, D. 1995. *Annu. Rev. Physiol.* 57: 495–519.

52 Nowak, L. et al. 1984. *Nature* 307: 462–465.

53 Mayer, M. L., Westbrook, G. L., and Guthrie, P. B. 1984. *Nature* 309: 261–263.

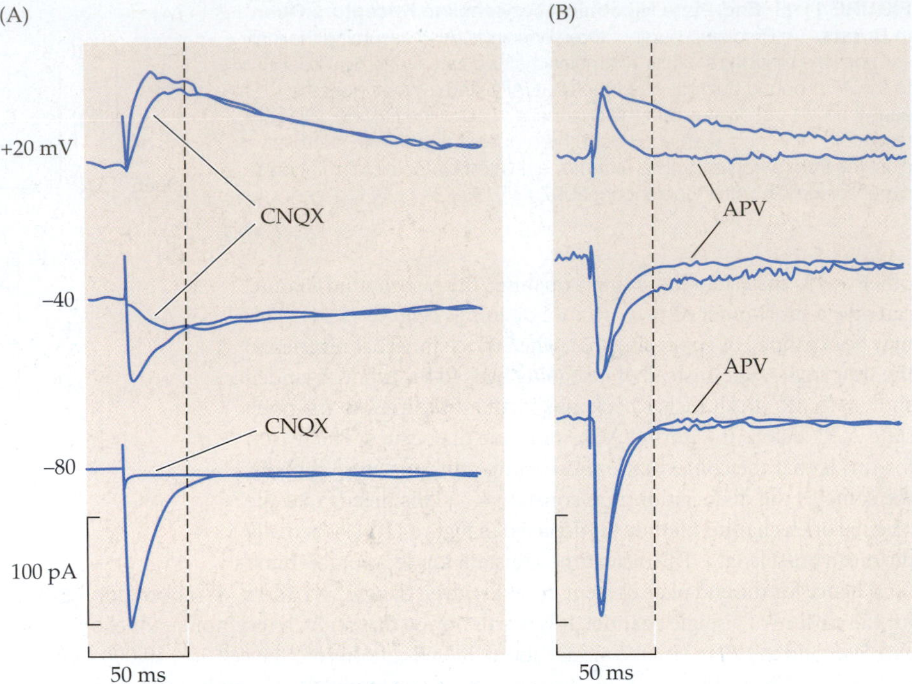

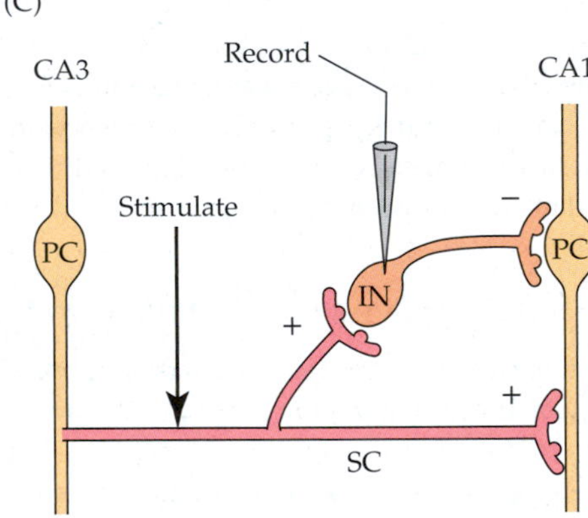

FIGURE 11.12 The Excitatory Neurotransmitter Glutamate Acts on Two Different Receptors to Produce Two Excitatory Postsynaptic Currents (EPSCs). EPSCs were recorded with a patch electrode from an interneuron in the CA1 region of a rat hippocampal slice preparation. The synaptic potentials were produced by stimulating the glutamate-releasing afferent fibers in the Schaeffer collaterals. The interneuron membrane potential was held at three different values, –80, –40, and +20 mV. (A) The traces show the effect of blocking the AMPA receptors with CNQX. This suppressed all of the EPSC at –80 mV but only the first part at +20 mV, leaving a large, slower component. The latter was due to the simultaneous activation of voltage-dependent NMDA receptors, since blocking the NMDA receptors with APV (B) left a pure, fast AMPA receptor-mediated EPSC. (C) Schematic of recording. PC = pyramidal cells; IN = interneuron; SC = Schaeffer collaterals; excitation (+); inhibition (–). (A and B after Sah et al., 1990).

the AMPA receptor does this. Second, the NMDA channels gate rather slowly,[51] which means that they open more during high-frequency synaptic activity; then, once opened, they stay open longer than the AMPA receptors. Third, the NMDA receptors have an unusually high permeability to calcium ions[54] and can thereby generate large increases in intracellular calcium concentration; this has a number of "second messenger" effects on neuronal function (see Chapter 12). All of these properties confer on the NMDA glutamate receptor a special role in synaptic development and plasticity (see Chapter 16).

In the CNS, the majority of excitatory synapses are located on the dendrites of the neurons. Synaptic potentials spread down the dendrites and through the soma to the normal site of action potential initiation at the axon hillock or axon initial segment (AIS).[55] In almost all neurons, a single synaptic EPSP is far too small to initiate an action potential; to reach threshold, the individual EPSPs have to be summated and integrated in the soma/dendritic region. In many neurons, such as the pyramidal cells of the hippocampus and cerebral cortex, and the Purkinje cells of the cerebellum, the excitatory synapses are located on small projections from the dendrites termed dendritic spines[56] (Figure 11.13). These spines play a special role in the long-term changes in brain function that accompany and follow synaptic activity (see Chapters 12 and 15).

[54]Burnashev, N. 1996. *Curr. Opin. Neurobiol.* 6: 311–317.

[55]Coombs, J. S., Curtis, D. R., and Eccles, J. C. 1957. *J. Physiol.* 139: 232–249.

[56]Spruston, N. 2008. *Nat. Rev. Neurosci.* 9: 206–221.

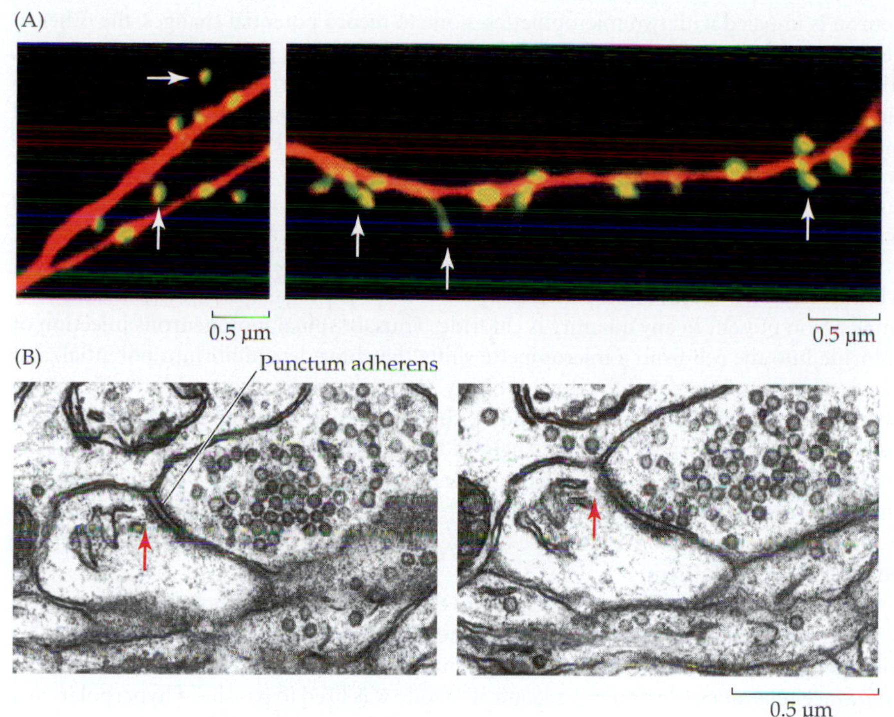

FIGURE 11.13 Dendritic Spines and Excitatory Spine Synapses in Hippocampal Pyramidal Neurons. (A) Dendrites labeled with red Alexa594 and counterstained for green fluorescent protein (GFP)-labeled, actin-binding protein. The latter is concentrated in the dendritic spines (arrows). (B) Electron micrographs of vesicle-filled synaptic boutons contacting dendritic spines. The arrows mark the postsynaptic spine apparatus projecting toward the punctum adherens (point of pre- and postsynaptic membrane adhesion). (A after Zito et al., 2004; B after Spacek and Harris, 1998.)

Direct Synaptic Inhibition

The principles that underlie direct chemical synaptic excitation at the neuromuscular junction also apply to **direct chemical inhibitory synapses.** Whereas excitation occurs by opening channels in the postsynaptic membrane whose reversal potential is *positive* to threshold, direct chemical synaptic inhibition is achieved by opening channels whose reversal potential is *negative* to threshold. Direct chemical synaptic inhibition occurs by activating channels permeable to chloride—an anion that typically has an equilibrium potential at or near the resting potential. Pioneering studies of direct chemical synaptic inhibition were made on the crustacean neuromuscular junction,[57,58] the crayfish stretch receptor,[59] and spinal motoneurons of the cat, at which gamma amino butyric acid (GABA) and glycine are the principal transmitters.[60]

Reversal of Inhibitory Potentials

Spinal motoneurons are inhibited by sensory inputs from antagonistic muscles, by way of inhibitory interneurons in the spinal cord. The effect of activation of inhibitory inputs can be studied by an experiment similar to that illustrated in Figure 11.14A. The moto-

[57] Dudel, J., and Kuffler, S. W. 1961. *J. Physiol.* 155: 543–562.

[58] Takeuchi, A., and Takeuchi, N. 1967. *J. Physiol.* 191: 575–590.

[59] Kuffler, S. W., and Eyzaguirre, C. 1955. *J. Gen. Physiol.* 39: 155–184.

[60] Coombs, J. S., Eccles, J. C., and Fatt, P. 1955. *J. Physiol.* 130: 326–373.

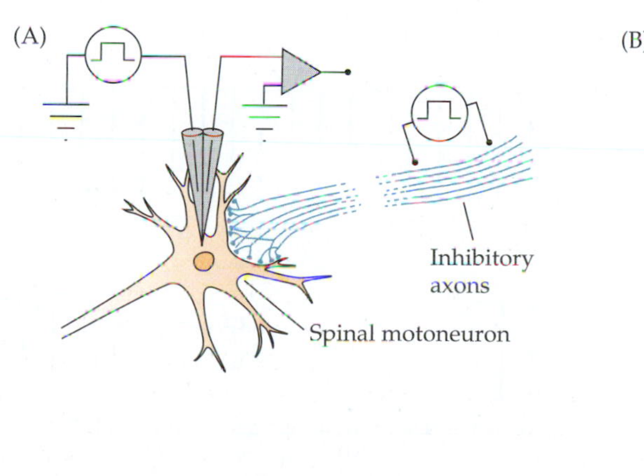

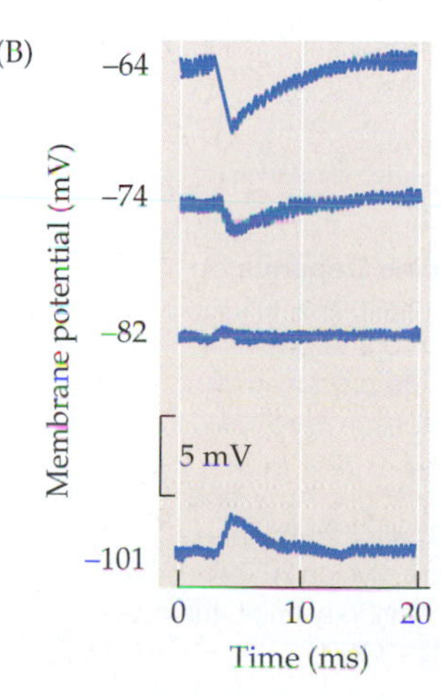

FIGURE 11.14 Direct Inhibitory Chemical Synaptic Transmission. (A) Scheme for intracellular recording from a cat spinal motoneuron and stimulation of inhibitory synaptic inputs. The membrane potential of the motoneuron is set to different levels by passing current through a second intracellular microelectrode. (B) Intracellular records of synaptic potentials evoked at membrane potentials between –64 and –101 mV. The reversal potential is between –74 and –82 mV. (After Coombs, Eccles, and Fatt, 1955b.)

neuron is impaled with two micropipettes—one to record potential changes, the other to pass current through the cell membrane. At the normal resting potential (about –75 mV), stimulation of the inhibitory inputs causes a slight hyperpolarization of the cell—the **inhibitory postsynaptic potential (IPSP)** (Figure 11.14B). When the membrane is depolarized by passing positive current into the cell, the amplitude of the IPSP is increased. When the cell is hyperpolarized to –82 mV, the inhibitory potential is very small and reversed in sign, and at –100 mV the reversed inhibitory potential is increased in amplitude. The reversal potential in this experiment is thus about –80 mV.

Inhibitory channels are permeable to anions, with permeabilities roughly correlated with the hydrated radius of the penetrating ion.[61] In physiological circumstances, the only small anion present in any quantity is chloride. Thus, in spinal motoneurons injection of chloride into the cell from a micropipette shifts the chloride equilibrium potential, and hence the reversal potential for the inhibitory synaptic potential toward zero (i.e., in the positive direction). In other preparations, changes in extracellular chloride have been shown to produce corresponding changes in the chloride equilibrium potential and the IPSP reversal potential, but such experiments often give ambiguous results. This is because changes in extracellular chloride concentration lead eventually to proportionate changes in intracellular concentration as well (see Chapter 5), so any change in chloride equilibrium potential is only transient.

One way around this difficulty is to remove chloride entirely, as shown in Figure 11.15. The records are from a reticulospinal cell in the brainstem of the lamprey, in which inhibitory synaptic transmission is mediated by glycine.[62] Membrane potential was recorded with an intracellular microelectrode. A second electrode was used to pass brief hyperpolarizing current pulses into the cell; the resulting changes in potential provided a measure of the cell's input resistance. Finally, a third micropipette was used to apply glycine to the cell close to an inhibitory synapse, using brief pressure pulses. Glycine application resulted in a slight hyperpolarization, with a marked reduction in input resistance (Figure 11.15A), as would be expected if glycine activated a large number of chloride channels. To test this idea, chloride was removed from the bathing solution and replaced by the impermeant ion isethionate. As a result, intracellular chloride was also removed by efflux, through chloride channels open at rest. After 20 minutes, glycine application produced no detectable change in membrane potential or input resistance (Figure 11.15B), indicating that no ions other than chloride pass through the inhibitory channels. The restoration of normal extracellular chloride concentration (Figure 11.15C) resulted in restoration of the response.

[61] Hille, B. 2001. *Ion Channels of Excitable Membranes*, 3rd ed. Sinauer, Sunderland, MA.

[62] Gold, M. R., and Martin, A. R. 1983. *J. Physiol.* 342: 99–117.

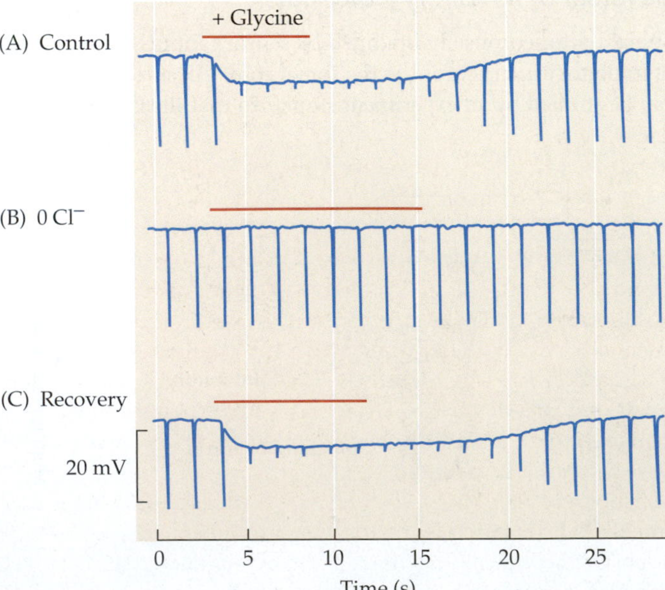

FIGURE 11.15 Inhibitory Response to Glycine Depends on Chloride. Intracellular microelectrode recordings from a neuron in the brainstem of the lamprey. (A) Resting membrane potential is –63 mV. Brief downward voltage deflections are produced by 10 nA current pulses from a second intracellular microelectrode; their amplitude indicates membrane resistance. On application of glycine (red bar), the cell is hyperpolarized by about 7 mV, and membrane resistance is reduced drastically. (B) After 20 minutes in chloride-free bathing solution, the response to glycine is abolished. (C) Five minutes after return to normal chloride solution the response has recovered. (From Gold and Martin, 1983.)

Given that the inhibitory response involves an increase in chloride permeability, the reversal potential for the inhibitory current will be equal to the chloride equilibrium potential (E_{Cl}) and the magnitude of the current will be given by:

$$\Delta i_{\text{inhibitory}} = \Delta i_{Cl} = \Delta g_{Cl}(V_m - E_{Cl})$$

At membrane potentials positive to E_{Cl}, the current is outward, resulting in membrane hyperpolarization. In this case outward current is carried by an influx of negatively charged chloride ions. At membrane potentials negative to E_{Cl}, inhibition causes an efflux of chloride ions, resulting in depolarization.

Early in postnatal development of the mammalian central nervous system, GABA and glycine paradoxically depolarize and thereby excite neurons in the central nervous system.[63] This effect is due not to differences in the properties of the channels opened by GABA and glycine, but to a difference in the nature of a transporter that regulates the intracellular chloride concentration[64] (see Chapter 9). In embryonic neurons, chloride is transported into the neuron by a sodium–potassium–2-chloride cotransporter (NKCC), thereby generating a high intracellular chloride concentration. Because E_{Cl} is positive to V_m, activation of chloride channels by GABA or glycine then results in outward movement of chloride and hence depolarization. At birth and during early postnatal development, another transporter, the potassium–chloride cotransporter (KCC2), becomes expressed. This extrudes chloride from the neuron. E_{Cl} then becomes negative to E_m, so GABA or glycine will now hyperpolarize the cell. This chloride switch is important in brain development because GABA-releasing synapses develop earlier than glutamate-releasing synapses in the brain, so the ability of GABA to depolarize and excite embryonic neurons (and thereby to increase intracellular calcium) is thought to be crucial for the proper embryonic development of synapses and circuits.[63] On the other hand, the postnatal switch from depolarization to hyperpolarization is essential for the normal function of inhibitory circuits after birth. Thus, mice in which both genes for the outward chloride transporter KCC2 have been deleted (and so do not express any KCC2 protein) die shortly after birth because their normal central respiratory circuits do not work (see Chapter 23) so they cannot breathe.[65] The chloride transporters are capable of maintaining different intracellular chloride concentrations in neuronal processes from those in the soma.[66–68] However, in mammalian peripheral neurons and nerve fibers (which have GABA receptors but no GABAergic synapses), E_{Cl} is always depolarized to E_m, even in adult animals.[69,70]

Stephen W. Kuffler in 1975

Presynaptic Inhibition

So far we have defined excitatory and inhibitory synapses on the basis of the effect of the transmitter on the postsynaptic membrane—that is, based on whether the postsynaptic permeability change is to cations or to anions. However, a number of early experiments indicated that in some instances it was difficult to account for inhibition in terms of postsynaptic permeability changes alone.[71,72] The paradox was resolved by the discovery of an additional inhibitory mechanism, **presynaptic inhibition**,[73] described in the mammalian spinal cord by Frank and Fuortes[72] and by Eccles and his colleagues[74] and at the crustacean neuromuscular junction by Dudel and Kuffler.[57] Presynaptic inhibition results in a reduction in the amount of transmitter released from excitatory nerve terminals.[76]

As shown in Figure 11.16, the action of the inhibitory nerve at the crustacean neuromuscular junction is exerted not only on the muscle fibers, but also on the excitatory terminals. The presynaptic effect is brief, reaching a peak in a few milliseconds and declining to zero after a total of 6 to 7 ms. For the maximum inhibitory effect to occur, the impulse must arrive in the inhibitory presynaptic terminal several milliseconds before the action potential arrives in the excitatory terminal. The importance of accurate timing is shown in Figure 11.16, in which parts A and B show the excitatory and inhibitory potentials following separate stimulation of the corresponding nerves. In Figure 11.16C, both nerves are stimulated, but the action potential in the inhibitory nerve follows that in the excitatory nerve by 1.5 ms, arriving too late to exert any effect. In Figure 11.16D, on the other hand, the action potential in the inhibitory nerve precedes that in the excitatory nerve and now strongly reduces the excitatory postsynaptic potential. The presynaptic effect, like that on the postsynaptic membrane, is mediated by γ-aminobutyric acid (GABA), and is associated with a marked increase in chloride permeability in the presynaptic terminals.[77,78]

[63] Ben-Ari, Y. et al. 2007. *Physiol. Rev*. 87: 1215–1284.

[64] Payne, J. A. et al. 2003. *Trends Neurosci*. 26: 199–206.

[65] Hübner, C.A. et al. 2001. *Neuron* 30: 515–524.

[66] Price, G. D., and Trussell, L. O. 2006. *J. Neurosci*. 26: 11432–11436.

[67] Khirug, S. et al. 2008. *J. Neurosci*. 28: 4635–4639.

[68] Trigo, F. F., Marty, A., and Stell, B. M. 2008. *Eur. J. Neurosci*. 28: 841–848.

[69] Adams, P. R., and Brown, D. A. 1975. *J. Physiol*. 250: 85–120.

[70] Gallagher, J. P., Higashi, H., and Nishi, S. 1978. *J. Physiol*. 275: 263–282.

[71] Fatt, P., and Katz, B. 1953. *J. Physiol*. 121: 374–389.

[72] Frank, K., and Fuortes, M. G. F. 1957. *Fed. Proc*. 16: 39–40.

[73] Rudomin, P. 2009. *Exp. Brain Res*. 196: 139–151.

[74] Eccles, J. C., Eccles, R. M., and Magni, F. 1961. *J. Physiol*. 159: 147–166.

[75] Spacek, J., and Harris, K. M. 1998. *J. Comp. Neurol*. 393: 58–68.

[76] Kuno, M. 1964. *J. Physiol*. 175: 100–112.

[77] Takeuchi, A., and Takeuchi, N. 1966. *J. Physiol*. 183: 433–449.

[78] Fuchs, P. A., and Getting, P. A. 1980. *J. Neurophysiol*. 43: 1547–1557.

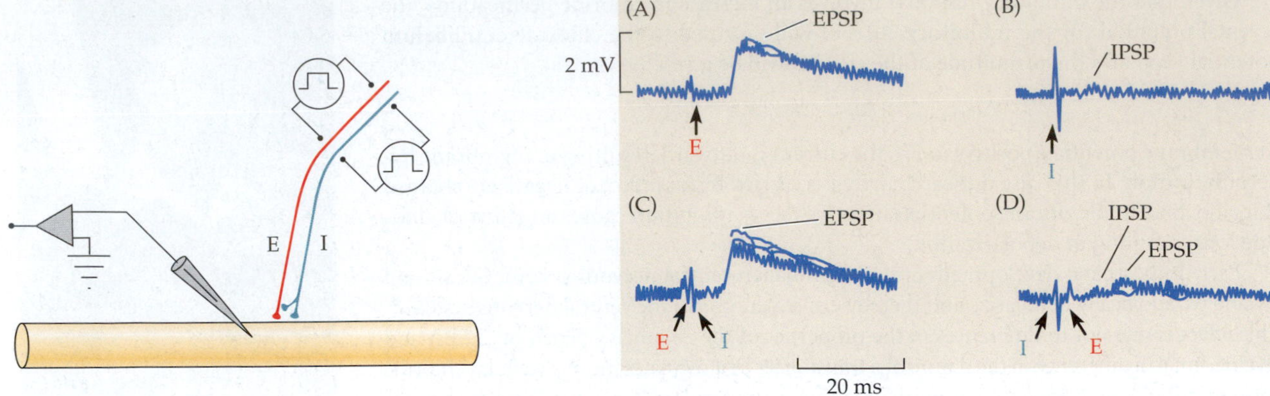

FIGURE 11.16 Presynaptic Inhibition in a Crustacean Muscle Fiber Innervated by One Excitatory and One Inhibitory Axon. (A) Stimulation of the excitatory axon (E) produces a 2 mV EPSP. (B) Stimulation of the inhibitory axon (I) produces a depolarizing IPSP of about 0.2 mV. (C) If the inhibitory stimulus follows the excitatory one by a short interval, there is no effect on the EPSP. (D) If the inhibitory stimulus precedes the excitatory one by a few milliseconds, the EPSP is almost abolished. The importance of precise timing indicates that the inhibitory nerve is having a presynaptic effect, reducing the amount of excitatory neurotransmitter that is released. (After Dudel and Kuffler, 1961.)

In the nervous system in general, presynaptic and postsynaptic inhibition serve quite different functions. Postsynaptic inhibition reduces the excitability of the entire cell, rendering it relatively less responsive to all excitatory inputs. Presynaptic inhibition is much more specific, aimed at a particular input and leaving the postsynaptic cell free to go about its business of integrating information from other sources.[79] Presynaptic inhibition implies that inhibitory axons make synaptic contact with axon terminals. Such axo–axonic synapses have been demonstrated directly by electron microscopy at the crustacean neuromuscular junction[80] and at numerous locations in the mammalian CNS.[81] Moreover, inhibitory nerve terminals themselves can be influenced presynaptically;[82] the requisite ultrastructural arrangement has been reported at inhibitory synapses on crayfish stretch receptors.[83]

Finally, there is an additional, quite different form of presynaptic inhibition in the mammalian nervous system that does not require the presence of specific presynaptic axo-axonal synapses, and can affect the release of transmitters from both excitatory and inhibitory terminals. This is due to the ability of transmitters (including glutamate[84] and GABA) to spill over from the synaptic cleft, back onto the presynaptic terminals where they activate metabotropic (G protein–coupled) receptors. These receptors, in turn suppress calcium entry into the terminals, and thereby reduce transmitter release (see Chapter 12). An example of this type of feedback inhibition (also called auto-inhibition) at a GABA-releasing synapse in the hippocampus is shown in Figure 11.17. Here, the authors[85] recorded the monosynaptic inhibitory postsynaptic currents from a hippocampal CA1 pyramidal neuron produced by GABA and released by stimulating fibers in the nearby stratum radiatum. A second stimulus, given 100 ms after the first, produced a much smaller current (a phenomenon called paired-pulse depression). (Figure 11.17A) This depression was due to the spread of the released GABA to the presynaptic terminals, where it activated the metabotropic $GABA_B$ receptor and so reduced the amount of transmitter released by the second stimulus, as shown schematically in Figure 11.17B. The paired-pulse depression was much reduced when the $GABA_B$ receptors were selectively blocked with the compound 2-hydroxy-saclofen or by CGP 35348. Presynaptic inhibition mediated by metabotropic receptors is slower in onset but lasts much longer than that mediated by ionotropic receptors (previously shown in Figure 11.16). It needs a minimal interval between the two stimuli of 20 to 30 ms to take effect, is maximal at around 100 ms, and lasts up to a second or more. The delayed onset reflects the time it takes for the receptor to activate a G protein and inhibit the calcium channels, while the slow offset is due to the slow kinetics of G protein recovery (see Chapter 12).

[79] Lomeli, J. et al. 1998. *Nature* 395: 600–604.

[80] Atwood, H. L., and Morin, W. A. 1970. *J. Ultrastruct. Res.* 32: 351–369.

[81] Schmidt, R. F. 1971. *Ergeb. Physiol.* 63: 20–101.

[82] Nicholls, J. G., and Wallace, B. G. 1978. *J. Physiol.* 281: 157–170.

[83] Nakajima, Y., Tisdale, A. D., and Henkart, M. P. 1973. *Proc. Natl. Acad. Sci. USA* 70: 2462–2466.

[84] Pinheiro, P. S., and Mulle, C. 2008. *Nat. Rev. Neurosci.* 9: 423–436.

[85] Davies, C. H., and Collingridge, G. L. 1993. *J. Physiol.* 472: 245–265.

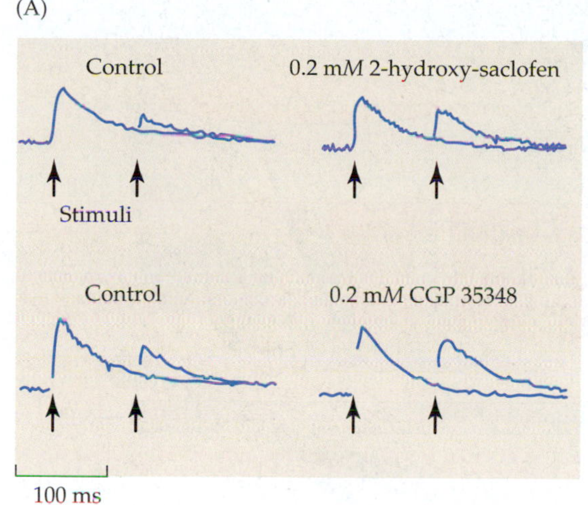

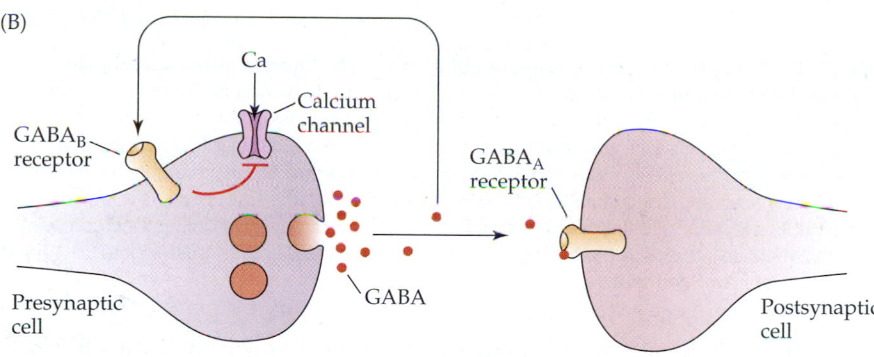

FIGURE 11.17 Presynaptic Auto-Inhibition at a GABA-Releasing Synapse in the Hippocampus. (A) Inhibitory synaptic currents (IPSCs) in a pyramidal neuron of a hippocampal slice preparation produced by two successive stimuli, delivered 100 ms apart, to inhibitory fibers in the adjacent stratum oriens. Excitatory currents were blocked with glutamate antagonists. The IPSC evoked by the second stimulus is much smaller than that produced by the first stimulus (paired pulse depression). This depression was largely due to activation of the presynaptic GABA$_B$ receptors, since it was reduced by blocking these receptors with GABA$_B$ antagonists (0.2 mM 2-hydroxy-saclofen or 0.2 mM CGP 35348; right-hand panels). (B) GABA released from the presynaptic terminal activates chloride-conducting GABA$_A$ receptors on the postsynaptic membrane to produce an inhibitory postsynaptic current (IPSC). It also activates metabotropic GABA$_B$ receptors on the presynaptic ending that inhibit calcium channels and reduce transmitter release. (A after Davies and Collingridge, 1993).

Transmitter Receptor Localization

Figures 11.6 and 11.7 show that the acetylcholine receptors (AChRs) are highly concentrated at the motor end plate. This is because they are aggregated into a specialized postsynaptic apparatus formed by cytoskeletal, membrane, and membrane-associated proteins (Figure 11.18).[86,87] At the mammalian neuromuscular junction the acetylcholine receptors are retained in the postsynaptic complex for several days (half-time 4.7 days);[88] by contrast, the residence half-time of the extrasynaptic receptors is about a day, which is similar to the turnover time of most membrane proteins. Within the postsynaptic apparatus, a 43 kilodaltons (kD) AChR-associated protein, called rapsyn, and components of the dystrophin complex play a key role in AChR localization. The dystrophin complex, which links together the myofiber cytoskeleton, membrane, and surrounding extracellular matrix, also provides structural support for the muscle cell.[89] Mutations in components of this complex give rise to Duchenne's muscular dystrophy, in which muscle fibers are damaged and degenerate.[90] The dystrophin complex is also involved in the maintenance of some other synapses in the central nervous system, and in the localization of aquaporin water channels, so its disruption can cause a variety of nervous system disorders.[91] When a motor nerve is cut and allowed to degenerate (see Chapter 26), the ACh receptor clustering mechanism at the end plate is disrupted such that the receptor lifetime is shortened.[92] At the same time, the number of extrajunctional receptors increases, causing denervation supersensitivity,[93] and their subunit composition reverts to the embryonic type as a result of a change in transcription.[94]

The postsynaptic apparatus at excitatory synapses in the central nervous system is also a very complex structure, containing more than 200 proteins as determined by mass spectrometry.[95,96] Three families of these proteins interact with glutamate receptors in the postsynaptic density (Figure 11.19).[97] Proteins in each of the families have one or more PDZ domains, which are conserved regions that mediate protein–protein interactions. PDZ is

[86]Sanes, J. R., and Lichtman, J. W. 2001. *Nat. Rev. Neurosci.* 2: 791–805.

[87]Banks, G. B. et al. 2003. *J. Neurocytol.* 32: 709–726.

[88]Akaaboune, M. et al. 2002. *Neuron* 34: 865–876.

[89]Blake, D. J. et al. 2003. *Physiol. Rev.* 82: 291–329.

[90]Davies, K. E., and Nowak, K. J. 2006. *Nat. Rev. Mol. Cell Biol.* 7: 762–773.

[91]Waite, A. et al. 2009. *Ann. Med.* 41: 344–359.

[92]Loring, R. H., and Salpeter, M. M. 1980. *Proc. Natl. Acad. Sci. USA* 77: 2293–2297.

[93]Axelsson, J., and Thesleff, S. 1959. *J. Physiol.* 147: 178–193.

[94]Witzemann, V., Brenner, H. R., and Sakmann, B. 1991. *J. Cell Biol.* 114: 125–141.

[95]Collins, M. O. et al. 2006. *J. Neurochem.* 97(Suppl. 1): 16–23.

[96]Cheng, D. et al. 2006. *Mol. Cell. Proteomics* 5: 1158–1170.

[97]Kim, E., and Sheng, M. 2004. *Nat. Rev. Neurosci.* 5: 771–781.

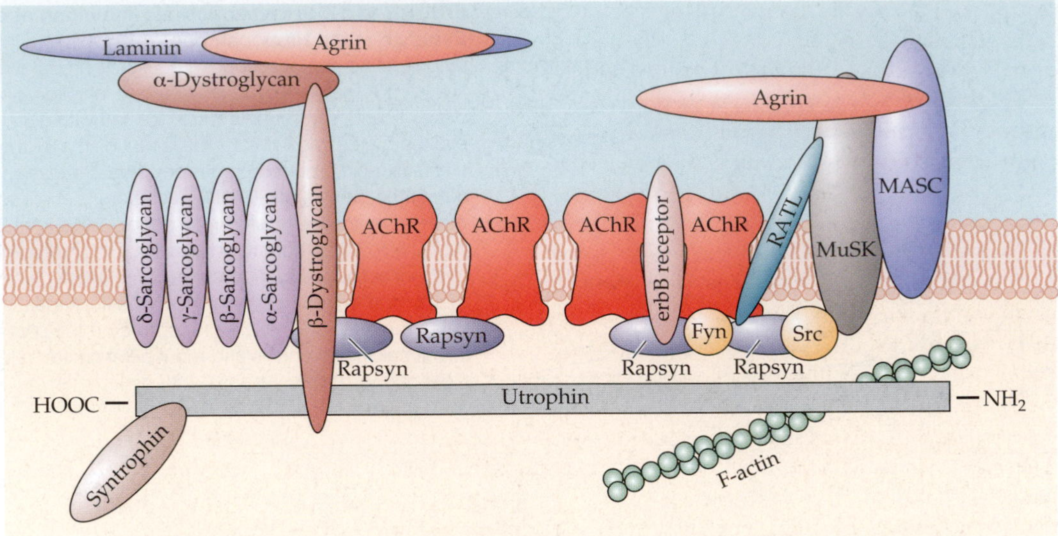

FIGURE 11.18 Postsynaptic Components of AChR-Rich Regions at the Vertebrate Skeletal Neuromuscular Junction. The dystrophin glycoprotein complex (utrophin, α- and β-dystroglycan, and the sarcoglycans) links together the actin cytoskeleton, the membrane, and the extracellular matrix. Agrin, secreted by the motor nerves (see McMahan[114]) binds to laminin and α-dystroglycan; it signals, through the receptor tyrosine kinase MuSK, to trigger formation of the postsynaptic apparatus during development (see Chapter 25). Rapsyn plays a key role in linking MuSK and AChRs to the cytoskeleton. RATL and MASC are as yet unidentified components that mediate interaction of MuSK with rapsyn and agrin, respectively. (See Figure 27.12 and Sanes and Lichtman[86] for further details.)

an acronym for three proteins with this common domain: **P**ostsynaptic density-95 (PDS-95); **D**rosophila disc protein DlgA; and **Z**onula occludens-1 protein (zo-1). NMDA-type glutamate receptors bind to proteins of the PSD-95 family, which are major components of the postsynaptic density. AMPA-type glutamate receptors bind to proteins of the GRIP and PICK families, and metabotropic glutamate receptors bind to members of the Homer protein family. Although they serve as scaffolding proteins to localize the receptors to subsynaptic sites, the AMPA-type receptors turnover quite rapidly and are free to move in

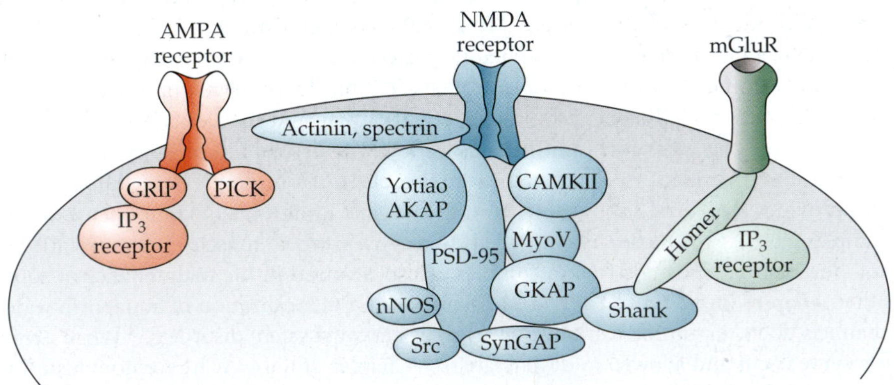

FIGURE 11.19 Glutamate Receptors Are Linked to a Postsynaptic Scaffold that includes proteins involved in intracellular signaling cascades. Metabotropic glutamate receptors, mGluR, are described in Chapter 12. GRIP, PICK, PSD-95 and Homer are PDZ-containing proteins that bind directly to the receptors. Actinin and Shank are proteins that bind to the cytoskeletal protein F-actin. Yotiao AKAP, GKAP, and synaptic Ras GTPases (SynGAP) are scaffold proteins that anchor and activate enzymes: AKAP binds protein kinases A and C and the protein phosphatase calcineurin; GKAP binds guanylate kinase; and SynGAP binds and activates the small GTPase Ras. nNOS = neuronal nitric oxide synthase; IP_3 = inositol-1,4,5-trisphosphate; CAMKII = calmodulin (CaM) kinase II; Src is a small GTPase protein. (After Sheng and Lee, 2000.)

and out of the PSD (as is necessary for the rapid changes required for synaptic plasticity, see Chapter 16). More importantly, they create an intracellular scaffold to recruit important signaling proteins that modify glutamate receptor function, dendritic architecture, and gene transcription downstream of receptor activation. These signaling proteins include nitric oxide synthase, calmodulin (CaM) kinase II, receptor tyrosine kinases, a guanosine triphosphate (GTP)ase-activating protein for small synaptic Ras GTPases (SynGAP), and inositol-1,4,5-trisphosphate (IP_3) receptors, and Ras-like small GTPases. Thus, these proteins determine not only receptor location, but also the consequences of receptor activation.

Localization of glycine and GABA receptors at inhibitory synapses in the central nervous system also requires ancillary subsynaptic proteins.[98,99] Postsynaptic clustering of glycine receptors requires the 93 kD subsynaptic protein gephyrin, which binds to the receptors and connects them to tubulin in the microtubules.[100,101] Thus, clustering is prevented when gephyrin synthesis is inhibited with an antisense nucleotide.[102] Gephyrin also is essential for localization of some $GABA_A$ receptors in postsynaptic membranes, although direct interactions between gephyrin and $GABA_A$ receptor subunits have not been demonstrated.[100] Gephyrin interacts with several intracellular components that mediate responses to activity and trophic factors.[98] Such interactions are thought to play a central role in the assembly and stabilization of postsynaptic specializations at inhibitory synapses.

Electrical Synaptic Transmission

Identification and Characterization of Electrical Synapses

In addition to the chemical transmission seen at most vertebrate synapses, there are places where true electrical transmission does occur. These places are either at synapses on reflex pathways where a particularly fast response is necessary or (in the retina and the CNS) where electrical transmission assists in coordinating or amplifying the activity of groups of neurons.

An example of a fast electrical synapse was provided in 1959 by Furshpan and Potter. Using intracellular microelectrodes to record from nerve fibers in the abdominal nerve cord of the crayfish, they discovered an **electrical synapse** between neurons that mediated the animal's escape reflex (Figure 11.20A).[8] They demonstrated that an action potential in a lateral giant fiber led (by direct intercellular current flow) to depolarization of a giant motor fiber leaving the cord (Figure 11.20B). The depolarization was sufficient to initiate an action potential in the postsynaptic fiber. The electrical coupling was in one direction only; depolarization of the postsynaptic fiber did not lead to presynaptic depolarization (Figure 11.20C). In other words, the synapse **rectified.**

Unlike the crayfish giant synapse, most electrical synapses do not exhibit rectification, but conduct equally well in both directions. The morphological specialization for electrical coupling at the crayfish giant synapse and other electrical synapses is the **gap junction.**[103,104] (see Chapter 8). Most gap junctions are formed by an assembly of **connexons,** each composed of six transmembrane connexin molecules[105,106] (or, in invertebrates, non-homologous but structurally similar innexon molecules[107]). The connexons (or innexons) in the membranes of the two connected neurons line up to form a conducting pathway that allow current to pass from one cell to the next. Rectification probably results from the contributions of different connexins or innexons to the pre- and postsynaptic connexons.[108] In the leech, electrical coupling between pairs of touch sensory neurons has the remarkable property that depolarization spreads readily from either cell to the other, but hyperpolarization spreads poorly; that is, the electrical connections are doubly rectifying.[109,110] This may result from the functional contribution of a rectifying innexon (*hm-inx2*) to a heteromeric assembly of innexin molecules in both pre- and postsynaptic membranes.[111]

Many connexins (especially connexin-36, Cx36) are strongly expressed in the vertebrate central nervous system,[104,112] and electrical transmission has now been demonstrated at a wide variety of central synapses.[113,114] Electrical connections are particularly prominent between inhibitory GABA-releasing interneurons in the hippocampus,[113] cerebral cortex,[115] and cerebellum.[116] In the rat somatosensory cortex, a single interneuron may be electrically coupled with up to 10 other interneurons at distances up to 200 μm away, forming mini-networks of 10 to 40 neurons.[117] Their effect is to synchronize interneuronal activity and contribute to some of the synchronized oscillations of electrical activity that can be

[98] Moss, S. J., and Smart, T. G. 2001. *Nat. Rev. Neurosci.* 2: 240–250.
[99] Kneussel, M., and Loebrich, S. 2007. *Biol. Cell* 99: 297–309.
[100] Sheng, M., and Lee, S. H. 2000. *Nat. Neurosci.* 3: 633–635.
[101] Fritschy, J. M., Harvey, R. J., and Schwarz, G. 2008. *Trends Neurosci.* 31: 257–264.
[102] Kirsch, J. et al. 1993. *Nature* 366: 745–748.
[103] Loewenstein, W. 1981. *Physiol. Rev.* 61: 829–913.
[104] Bennett, M. V. 1997. *J. Neurocytol.* 26: 349–366.
[105] Saez, J. C. et al. 2003. *Physiol. Rev.* 83: 1359–1400.
[106] Söhl, G., Maxeiner, S., and Willecke, K. 2005. *Nat. Rev. Neurosci.* 6: 191–200.
[107] Phelan, P. et al. 1998. *Trends Genet.* 14: 348–349.
[108] Phelan, P. et al. 2008. *Curr. Biol.* 18: 1955–1960.
[109] Baylor, D. A., and Nicholls, J. G. 1969. *J. Physiol.* 203: 591–609.
[110] Acklin, S. E. 1988. *J. Exp. Biol.* 137: 1–11.
[111] Dykes, I. M. et al. 2004. *J. Neurosci.* 24: 886–894.
[112] Bennett, M. V., and Zukin, R. S. 2004. *Neuron* 41: 495–511.
[113] Connors, B. W., and Long, M. A. 2004. *Annu. Rev. Neurosci.* 27: 393–418.
[114] McMahan, U. J. 1990. *Cold Spring Harb. Symp. Quant. Biol.* 55: 407–418.
[115] Hestrin S., and Galarreta, M. 2005. *Trends Neurosci.* 28: 304–309.
[116] Dugué, G. P. et al. 2009. *Neuron* 61: 126–139.
[117] Amitai, Y. et al. 2002. *J. Neurosci.* 22: 4142–4152.

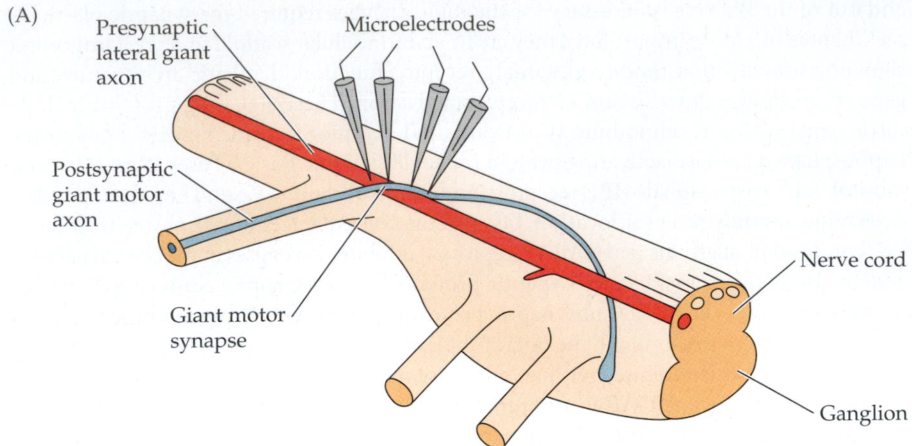

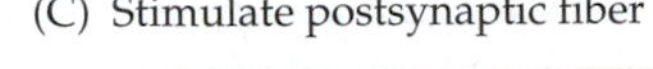

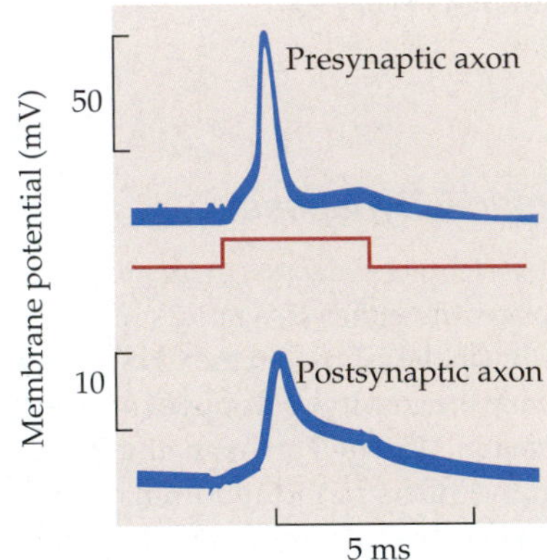

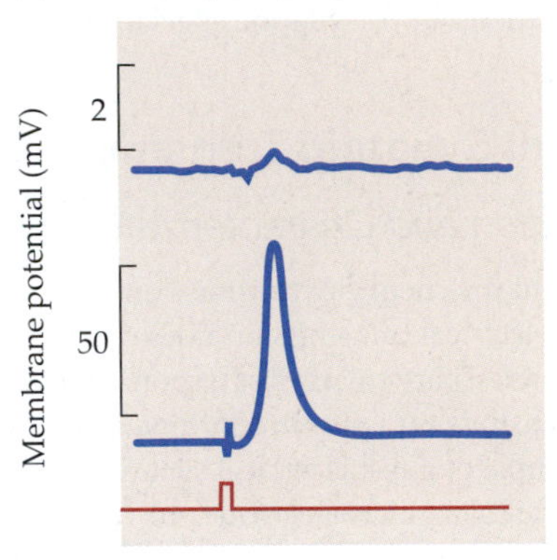

FIGURE 11.20 Electrical Synaptic Transmission at a Giant Synapse in the Crayfish Central Nervous System. (A) The experimental preparation. The presynaptic lateral giant axon makes an electrical synapse with the postsynaptic giant motor axon in the abdominal nerve cord. (B) Depolarization of the presynaptic axon spreads immediately to the postsynaptic fiber. In this case each cell reaches threshold and fires an action potential. (C) When the postsynaptic axon is stimulated directly to give an action potential, depolarization spreads poorly from the postsynaptic to the presynaptic axon. The synapse is said to rectify. (After Furshpan and Potter, 1959.)

recorded in these parts of the brain.[118] There is also a particularly rich network of electrical synapses in the retina.[119] The functions of these retinal electrical synapses are diverse. For example, those between horizontal cells greatly extend the cell's receptive field and enhance the signal/noise ratio for light detection under low-light conditions; while those between amacrine cells and rods are necessary for rod input to ON bipolar cells (see Chapter 20).

The degree of electrical coupling between cells is usually expressed as a **coupling ratio.** A ratio of 1:4 means that one-fourth of the presynaptic voltage change appears in the postsynaptic cell. For cells to be strongly coupled, the resistance of the junction between the cells must be very low and there must be a reasonable match between the sizes of the presynaptic and postsynaptic elements (see Chapter 7). Efficient coupling depends not only on the number of coupled connexons but also on their individual conductances. Thus, the channels can be closed by an increase in intracellular calcium, and their conductance is reduced when they are phosphorylated by cyclic adenosine monophosphate (cAMP)-dependent protein kinase. This means that electrical connectivity mediated by gap junctions can be altered by neurotransmitters that affect intracellular calcium or cAMP (see Chapter 12). This occurs in the retina, where light stimulates the release of dopamine from amacrine cells. This increases cAMP in horizontal cells, which in turn activates cAMP-dependent protein kinase. Through the action of the enzyme, the gap junction connexin becomes phosphorylated, resulting in a decrease in its conductance. As a result the receptive field of the horizontal cell shrinks. Conversely, reduced dopamine release at night has the opposite effect of increasing the horizontal cell's receptive field and enhancing the detection of dim objects.[119]

Comparison of Electrical and Chemical Transmission

Electrical and chemical transmission often coexist at a single synapse. Such combined electrical and chemical synapses were first found in cells of the avian ciliary ganglion, where a chemical

[118] Whittington, M. A., and Traub, R. D. 2003. *Trends Neurosci.* 26: 676–682.

[119] Bloomfield, S. A., and Völgyi, B. 2009. *Nat. Rev. Neurosci.* 10: 495–506.

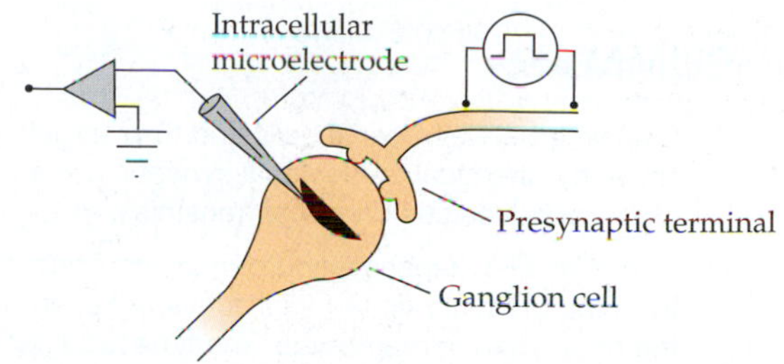

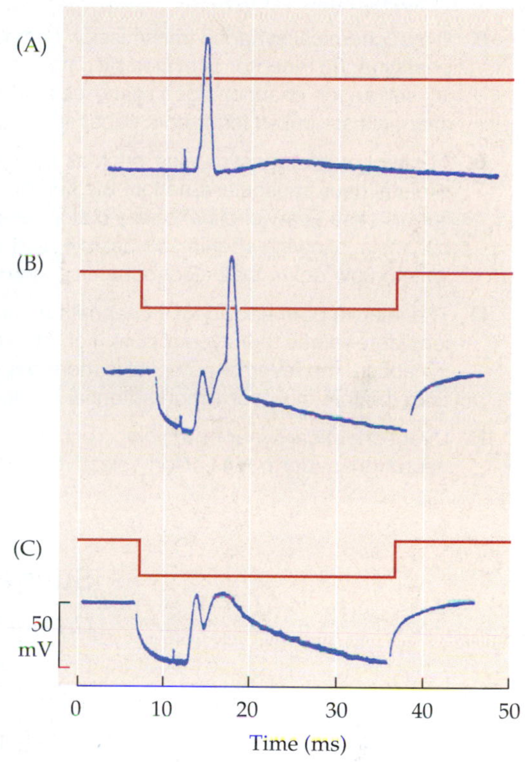

FIGURE 11.21 Electrical and Chemical Synaptic Transmission in a Chick Ciliary Ganglion Cell. (A) Stimulation of the preganglionic nerve produces an action potential in the ganglion cell (blue trace, recorded with an intracellular microelectrode). (B) When the ganglion cell is hyperpolarized by passing current through the recording electrode (red trace), the cell reaches threshold later, revealing an earlier, transient depolarization (blue trace). This depolarization is an electrical synaptic potential (coupling potential), caused by current flow into the ganglion cell from the presynaptic terminal. In A, the electrical synaptic potential depolarized the ganglion cell to threshold, initiating an action potential. (C) Slightly greater hyperpolarization prevents the ganglion cell from reaching threshold, exposing a slower chemical synaptic potential. The chemical synaptic potential follows the coupling potential with a synaptic delay of about 2 ms at room temperature. (After Martin and Pilar, 1963.)

synaptic potential (produced by ACh) is preceded by an electrical coupling potential (Figure 11.21).[9] Similar synapses occur widely in vertebrates—for example, onto spinal interneurons of the lamprey,[120] spinal motoneurons of the frog, and inhibitory cells in the cerebral cortex.[121] Postsynaptic cells may also receive separate chemical and electrical synaptic inputs from different sources. For example, in leech ganglia (see Chapter 18), motor neurons receive three distinct types of synaptic input from sensory neurons signaling three different modalities: one input is chemical, one electrical, and one combined electrical and chemical.[122]

One characteristic of electrically mediated synaptic transmission is the absence of a **synaptic delay.** At chemical synapses, there is a pause of approximately 1 ms between the arrival of an impulse in the presynaptic terminal and the appearance of an electrical potential in the postsynaptic cell. The delay is due to the time taken for the terminal to release transmitter (see Chapter 13). At electrical synapses, there is no such delay, and current spreads instantaneously from one cell to the next.

The presence of both electrical and chemical transmission at the same synapse provides a convenient means of comparing the two modes of transmission. This is illustrated in Figure 11.21, which shows intracellular records from a cell in the ciliary ganglion of the chick. Stimulation of the preganglionic nerve leads to an action potential in the postsynaptic cell, with very short latency (Figure 11.21A). When the cell is hyperpolarized slightly (Figure 11.21B), the action potential arises at a later time, revealing an early, brief depolarization that is now subthreshold because the cell has been hyperpolarized. This depolarization is an electrical coupling potential, produced by current flow from the presynaptic nerve terminal into the cell. Further hyperpolarization (Figure 11.21C) blocks the initiation of the action potential altogether, revealing the underlying chemical synaptic potential. These cells then have the property that, under normal conditions, initiation of a postsynaptic action potential by chemical transmission is preempted by electrical coupling. In this example the coupling potential precedes the chemical synaptic potential by about 2 ms, giving us a direct measure of the synaptic delay. Additional experiments on these cells have shown that the electrical coupling is bidirectional; that is, the synapses do not rectify.

There are several advantages to electrical transmission. One is that electrical synapses are more reliable than chemical synapses; transmission is less likely to fail because of synaptic depression or to be blocked by neurotoxins. A second advantage is the greater speed of electrical transmission. Speed is important in rapid reflexes involving escape reactions, in which the saving of a millisecond may be crucial for surviving an attack by a predator. Other functions include the synchronization of electrical activity of groups of cells[112,113] and intercellular transfer of key molecules, such as calcium, adenosine triphosphate (ATP), and cAMP.[103] Connexins have quite wide pores (12–14 Å),[105] which allow large molecules up to 1 kD to pass through. These include dyes such as Lucifer yellow and neurobiocytin, that can be used to visualize electrical coupling between neurons.[103,104,116] In the brain and retina, they are especially abundant during embryonic and early postnatal development and may play an important role in generating the rhythmic electrical activity necessary for neural development.[123] Finally, they are subject to biochemical and neurotransmitter regulation, in the retina (as previously noted) and elsewhere.[124,125] Thus, gap junctions do not function merely as passive connections but can be dynamic components of neuronal circuits.

[120] Rovainen, C. M. 1967. *J. Neurophysiol.* 30: 1024–1042.

[121] Shapovalov, A. I., and Shiriaev, B. I. 1980. *J. Physiol.* 306: 1–15.

[122] Nicholls, J. G., and Purves, D. 1972. *J. Physiol.* 225: 637–656.

[123] Peinado, A., Juste, R., and Kayz, L. C. 1993. *Neuron* 10: 103–114.

[124] O'Donnell, P., and Grace, A. A. 1993. *J. Neurosci.* 13: 3456–3471.

[125] Halliwell, J. V., and Horne, A. L. 1998. *J. Physiol.* 506: 175–194.

SUMMARY

- Signaling between nerve cells and their targets can occur by chemical or electrical synaptic transmission. Most synapses use chemical transmission.
- At chemical synapses a neurotransmitter released from the presynaptic terminal activates receptors in the postsynaptic membrane. The time required for transmitter release imposes a minimum synaptic delay of approximately 1 ms.
- Direct chemical synaptic transmission occurs when the postsynaptic receptor activated by a neurotransmitter is itself an ion channel. Such ligand-activated ion channels are called ionotropic transmitter receptors.
- At direct excitatory synapses, such as the vertebrate skeletal neuromuscular junction, the neurotransmitter (in this case acetylcholine) opens cation-selective channels, allowing sodium, potassium, and calcium ions to flow down their electrochemical gradients.
- The relative permeability of a channel for various ions determines the reversal potential. At excitatory synapses, the reversal potential is more depolarized than the threshold for action potential initiation.
- Direct chemical synaptic inhibition occurs when a neurotransmitter opens anion-selective channels, which allow chloride ions to flow down their electrochemical gradient. The reversal potential for such currents is the chloride equilibrium potential (E_{Cl}); inhibition occurs if E_{Cl} is hyperpolarized to threshold.
- The intracellular chloride concentration is set by inward or outward chloride pumps. During early development the pump is inward so the channels opened by the inhibitory transmitter allow an outward flow of chloride ions, producing a depolarization.
- Receptors for inhibitory transmitters may also be present on presynaptic terminals, where their activation reduces transmitter release.
- Receptors for excitatory and inhibitory transmitters are aggregated into specialized regions of the postsynaptic membrane by an array of scaffolding proteins.
- Electrical synaptic transmission is mediated by the direct flow of current from cell to cell, through ion channels called connexons that span the synaptic cleft. Electrical transmission is very rapid and used at specialized synapses mediating very fast reflexes. It is also used in the retina and brain to coordinate the activity of groups of neurons.

Suggested Reading

General Reviews

Bennett, M. V., and Zukin, R. S. 2004. Electrical coupling and neuronal synchronization in the mammalian brain. *Neuron* 41: 495–511.

Edmonds, B., Gibb, A. J., and Colquhoun, D. 1995. Mechanisms of activation of muscle nicotinic acetylcholine receptors and the time course of endplate currents. *Annu. Rev. Physiol.* 57: 469–493.

Engelman, H. S., and MacDermott, A. B. 2004. Presynaptic ionotropic receptors and control of transmitter release. *Nat. Rev. Neurosci.* 5: 135–145.

Hille, B. 2001. *Ion Channels of Excitable Membranes*, 3rd ed. Sinauer, Sunderland, MA. pp.169–199.

Hirsch, N. P. 2007. Neuromuscular junction in health and disease. *Brit. J. Anaesth.* 99: 132–138.

Katz, B. 1981. Electrical exploration of acetylcholine receptors. *Postgrad. Med. J.* 57(Suppl. 1): 84–88.

Kim, E., and Sheng, M. 2004. PDZ domain proteins of synapses. *Nat. Rev. Neurosci.* 5: 771–781.

Kneussel, M., and Loebrich, S. 2007. Trafficking and synaptic anchoring of ionotropic inhibitory neurotransmitter receptors. *Biol. Cell* 99: 297–309.

Moss, S. J., and Smart, T. G. 2001. Constructing inhibitory synapses. *Nat. Rev. Neurosci.* 2: 240–250.

Nicholls, J. G. 2007. How acetylcholine gives rise to current at the motor end-plate. *J. Physiol.* 578: 621–622.

Sakmann, B. 1992. Elementary steps in synaptic transmission revealed by currents through single ion channels. *Neuron* 8: 613–629.

Sanes, J. R., and Lichtman, J. W. 2001. Induction, assembly, maturation and maintenance of a postsynaptic apparatus. *Nat. Rev. Neurosci.* 2: 791–805.

Todman, D. 2008. John Eccles (1903–97) and the experiment that proved chemical synaptic transmission in the central nervous system. *J. Clin. Neurosci.* 15: 972–977.

Original Papers

Akaaboune, M., Grady, R. M., Turney, S., Sanes, J. R., and Lichtman, J. W. 2002. Neurotransmitter receptor dynamics studied in vivo by reversible photo-unbinding of fluorescent ligands. *Neuron* 34: 865–876.

Coombs, J. S., Eccles, J. C., and Fatt, P. 1955. The specific ionic conductances and the ionic movements across the motoneuronal membrane that produce the inhibitory post-synaptic potential. *J. Physiol.* 130: 326–373.

del Castillo, J., and Katz, B. 1955. On the localization of end-plate receptors. *J. Physiol.* 128: 157–181.

Dudel, J., and Kuffler, S. W. 1961. Presynaptic inhibition at the crayfish neuromuscular junction. *J. Physiol.* 155: 543–562.

Fatt, P., and Katz, B. 1951. An analysis of the end-plate potential recorded with an intra-cellular electrode. *J. Physiol.* 115: 320–370.

Furshpan, E. J., and Potter, D. D. 1959. Transmission at the giant motor synapses of the crayfish. *J. Physiol.* 145: 289–325.

Kirsch, J., Wolters, I., Triller, A., and Betz, H. 1993. Gephyrin antisense oligonucleotides prevent glycine receptor clustering in spinal neurons. *Nature* 366: 745–748.

Kuffler, S. W., and Yoshikami, D. 1975. The distribution of acetylcholine sensitivity at the post-synaptic membrane of vertebrate skeletal twitch muscles: Iontophoretic mapping in the micron range. *J. Physiol.* 244: 703–730.

Magleby, K. L., and Stevens, C. F. 1972. A quantitative description of end-plate currents. *J. Physiol.* 223: 171–197.

Martin, A. R., and Pilar, G. 1963. Dual mode of synaptic transmission in the avian ciliary ganglion. *J. Physiol.* 168: 443–463.

Neher, E., Sakmann, B., and Steinbach, J. H. 1978. The extracellular patch clamp: A method for resolving currents through individual open channels in biological membranes. *Pflügers Arch.* 375: 219–228.

Sah, P., Hestrin, S., and Nicoll, R. A. 1990. Properties of excitatory postsynaptic currents recorded in vitro from rat hippocampal interneurones. *J. Physiol.* 430: 605–616.

Takeuchi, A., and Takeuchi, N. 1960. On the permeability of the end-plate membrane during the action of transmitter. *J. Physiol.* 154: 52–67.

Takeuchi, A., and Takeuchi, N. 1966. On the permeability of the presynaptic terminal of the crayfish neuromuscular junction during synaptic inhibition and the action of γ-aminobutyric acid. *J. Physiol.* 183: 433–449.

Takeuchi, A., and Takeuchi, N. 1967. Anion permeability of the inhibitory post-synaptic membrane of the crayfish neuromuscular junction. *J. Physiol.* 191: 575–590.

CHAPTER 12
Indirect Mechanisms of Synaptic Transmission

In addition to opening ion channels, neurotransmitters bind to other membrane receptors, known as metabotropic receptors. Metabotropic receptors influence ion channels indirectly through membrane-associated or cytoplasmic second messengers. At many synapses in the central and autonomic nervous systems, excitatory and inhibitory transmission occurs solely by these indirect mechanisms. At other locations, indirect mechanisms serve to modulate direct transmission.

Most metabotropic receptors (G protein–coupled receptors) produce their effects by first interacting with G proteins in the cell membrane. G proteins, so called because they bind guanine nucleotides, are trimers of three subunits: α, β, and γ. When a G protein is activated by its receptor, the α- and $\beta\gamma$-subunits dissociate. The free subunits can diffuse, and then bind to and modulate the activity of intracellular targets. Some G protein subunits bind to ion channels, producing relatively brief effects. For example, when acetylcholine (ACh) binds to its muscarinic receptors in the heart atrium, a G protein is activated and the freed $\beta\gamma$-subunit then opens a potassium channel, thereby slowing the heart. A second mechanism of G protein action is through activation of enzymes that produce intracellular second messengers. An example is the activation of β-adrenergic receptors in the heart by norepinephrine. The α-subunit of the dissociated G protein stimulates the enzyme, adenylyl cyclase. The resulting increase in intracellular cyclic adenosine monophosphate (AMP), or cAMP, activates another enzyme, cAMP-dependent protein kinase, which modifies the activity of channels and enzymes through phosphorylation. Such responses may last for seconds, minutes, or hours—often persisting long after the transmitter interaction with the receptors has stopped. These mechanisms provide both amplification and radiation of signals.

Potassium and calcium channels are prime targets for such indirect transmitter action. Indirect action can cause channels to be opened, closed, or changed in their voltage sensitivity. Thus, indirectly acting transmitters open potassium channels in heart atrial cells; inhibit N-type calcium channels and M-type potassium channels in sympathetic neurons; and increase the probability that calcium channels will open in response to depolarization in cardiac muscle cells. Changes in channel activation in axon terminals modify transmitter release. In postsynaptic cells, such changes alter spontaneous activity and the responses to synaptic inputs.

In addition, there are tertiary messengers, which are generated by some forms of synaptic activation. These include endocannabinoids (lipid messengers) and a gas, nitric oxide (NO). These molecules diffuse freely and so have effects outside the synapse or cell in which they are made. Calcium ions constitute another tertiary messenger and produce both short-term and long-term changes in neuron excitability and synaptic function. The long-term changes include effects on gene transcription and synaptic wiring.

Direct Versus Indirect Transmission

In Chapter 11, we described the process of direct chemical transmission. In this process, a neurotransmitter released from a presynaptic ending binds to and within a millisecond, opens ion channels (ionotropic receptors) in the postsynaptic membrane (Figure 12.1A). Thus, direct transmission is extremely fast and is required for high-speed, integrated motor performance, such as playing a trill on the piano or for discriminating different notes at frequencies up to several kilohertz.

However, many of the essential human functions require a more gradual and longer-lasting form of communication. For example, when we are excited or frightened, the sympathetic nervous system is stimulated (see Chapter 17) and as a result, the heart rate gradually increases over many seconds. This action does not require ultra-fast transmission from the sympathetic nerves to the heart but instead requires a rather careful adjustment to the endogenous cardiac rhythm. This form of slow modulatory transmission is also an essential component of synaptic activity in the central nervous system (CNS), where it is superimposed on the faster type of transmission mediated by the ionotropic receptors. There it serves to adjust the efficiency of synaptic transmission, and the excitability of the neuron, over seconds or minutes (or even longer).

Slow modulatory transmission uses a completely different type of membrane receptor from the ionotropic receptor responsible for fast transmission. Slow transmission uses a so-called **metabotropic** receptor. These receptors are not ion channels and do not directly excite or inhibit a neuron. Instead, they interact with other membrane proteins to initiate a sequence of steps leading to a change in ion channel activity or other metabolic processes within the neuron (Figure 12.1B). For this reason, this form of transmission is referred to as **indirect transmission**. For the vast majority of metabotropic receptors, the first target

(A) Direct transmitter action

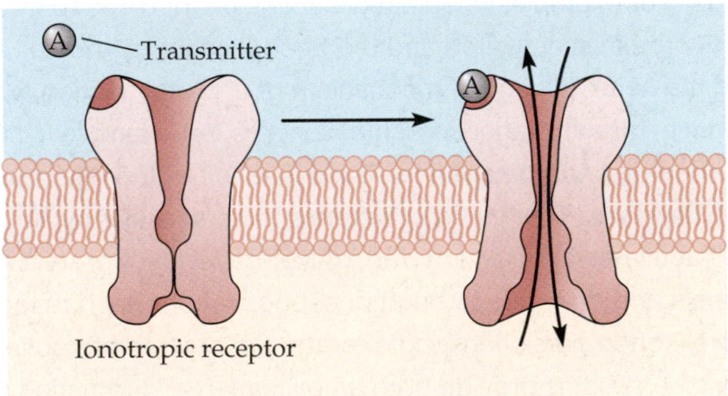

(B) Indirect transmitter action

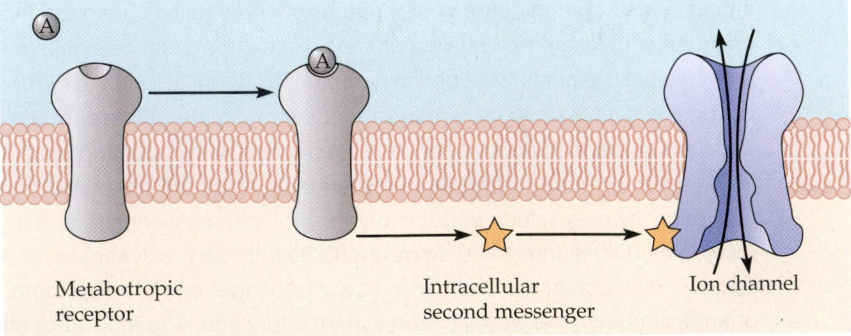

FIGURE 12.1 Direct and Indirect Transmitter Action. (A) At direct chemical synapses, the transmitter molecule A binds to an ionotropic receptor. Ionotropic receptors are ligand-activated ion channels. (B) Indirectly acting transmitters bind to metabotropic receptors. Metabotropic receptors are not themselves ion channels but rather activate intracellular second-messenger signaling pathways that influence the opening and closing of ion channels.

protein with which they interact is another membrane protein, called a **G protein**; hence, metabotropic receptors of this type are termed **G protein–coupled receptors** (or GPCRs for short).

In the rest of this chapter we describe G proteins and G protein–coupled receptors, give examples of how they modify ion channels, both directly and indirectly, through subsequent enzymatic cascades; and describe some effects of other downstream messengers of receptor activation, such as NO, endocannabinoids, and calcium ions.

G Protein–Coupled Metabotropic Receptors and G Proteins

Structure of G Protein–Coupled Receptors

GPCRs make up a superfamily of membrane proteins characterized by seven transmembrane domains, with an extracellular amino terminus and an intracellular carboxy terminus (Figure 12.2).[1–3] For this reason, they are sometimes called 7-transmembrane (7TM) or heptahelical receptors. More than a thousand different GPCRs have been identified (Box 12.1). Those activated by ACh are called muscarinic receptors; those that bind norepinephrine are known as adrenergic receptors. Others respond to γ-aminobutyric acid (GABA), serotonin (5-HT), dopamine, glutamate, purines, or neuropeptides; and some are activated by light (rhodopsin; see Chapter 20), by odorants (see Chapter 19), or by proteases.

Biochemical, structural, and molecular genetic experiments have indicated several distinct modes of ligand binding, each of which ultimately produces a similar rearrangement of the α-helical regions that form the transmembrane core of the receptor (see Figure 12.2). Portions of the second and third cytoplasmic loops, together with the membrane proximal region of the carboxy tail, mediate binding to and activation of the appropriate G protein.[1,3]

[1] Rosenbaum, D. M., Rasmussen, S. G. F., and Kobilka, B. 2009. *Nature* 459: 356–363.

[2] Ji, T. H., Grossmann, M., and Ji, I. 1998. *J. Biol. Chem.* 273: 17299–17302.

[3] Oldham, W. H., and Hamm, H. E. 2008. *Nat. Rev. Mol. Cell Biol.* 9: 60–71.

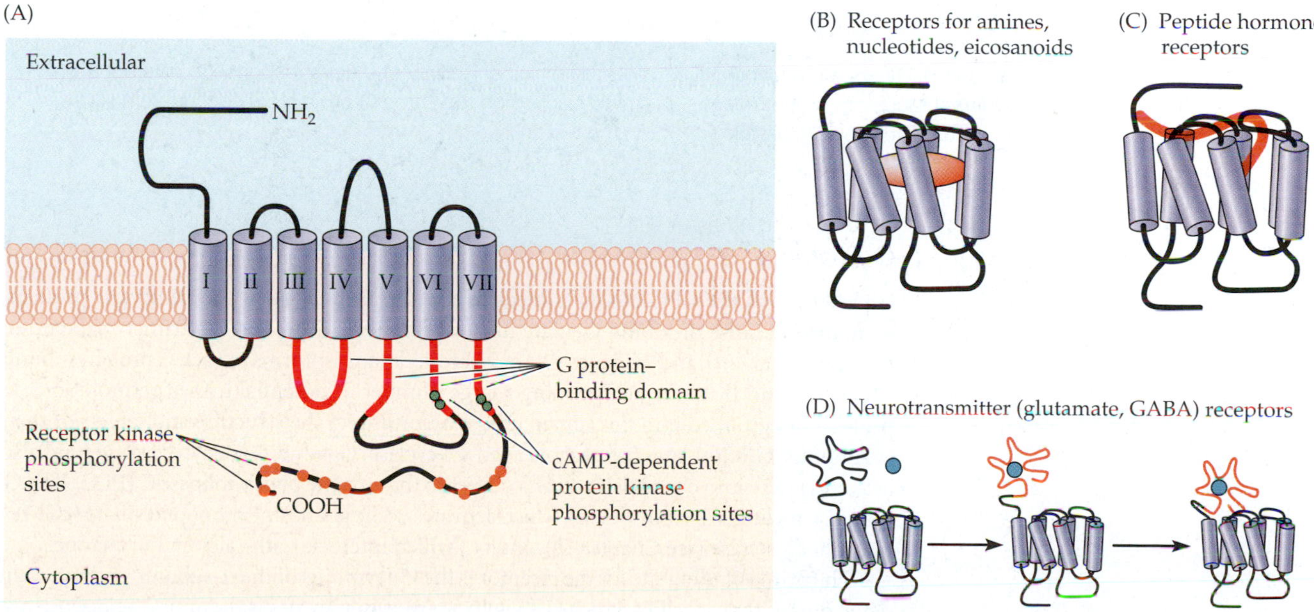

FIGURE 12.2 Metabotropic or G Protein–Coupled Transmitter Receptors. (A) Indirectly coupled transmitter receptors have seven transmembrane domains, an extracellular amino terminus, and an intracellular carboxy terminus. The second and third cytoplasmic loops, together with the amino terminal region of the intracellular tail, mediate binding to the appropriate G protein. Phosphorylation of sites on the third cytoplasm loop and the carboxy terminus by second-messenger–related kinases, such as cAMP-dependent protein kinase, causes receptor desensitization. Phosphorylation of sites on the carboxy terminus by G protein receptor kinases (GRKs), such as β-adrenergic receptor kinase (βARK), causes receptor desensitization, binding of the protein arrestin, and termination of the response. (B) Portions of the transmembrane domains form the ligand-binding sites of metabotropic receptors that bind amines, nucleotides, and eicosanoids. (C) Ligands bind to the outer portions of the transmembrane domains of peptide hormone receptors. (D) The amino terminal tail forms the ligand-binding domain of metabotropic receptors for glutamate and γ-aminobutyric acid (GABA)s. (After Ji, Grossmann, and Ji, 1998.)

BOX 12.1
Receptors, G Proteins, and Effectors: Convergence and Divergence in G Protein Signaling

There are more than 200 different metabotropic receptors that can couple to G proteins but a much smaller number of different G proteins. The primary receptor interaction site on the G protein is the C-terminus of the α-subunit. Hence, as viewed by the receptor, there are only three families of common G proteins: G_s, G_q and G_i/G_o. Although each family contains several members, their C-terminal sequences cannot be distinguished by the receptor. This means that there is substantial convergence of different transmitters or hormones, acting through different receptors, onto the same G protein. On the output side, while some G proteins have rather specific effects (e.g., G_o selectively interacts with certain types of neuronal calcium channels), others—particularly those that activate enzymes to produce second messengers, such as cyclic AMP—can produce multiple effects on any given cell, providing divergent signaling. G_s, G_q, and G_i are present in nearly all cells, and G_o is abundant in all nerve cells. In most cells, G protein molecules are about 10 times more plentiful than receptor molecules, so different receptors do not necessarily compete for the same G protein molecules. Possible ways in which selectivity of a cell's response to different transmitters acting on the same G protein may occur are discussed in the text.

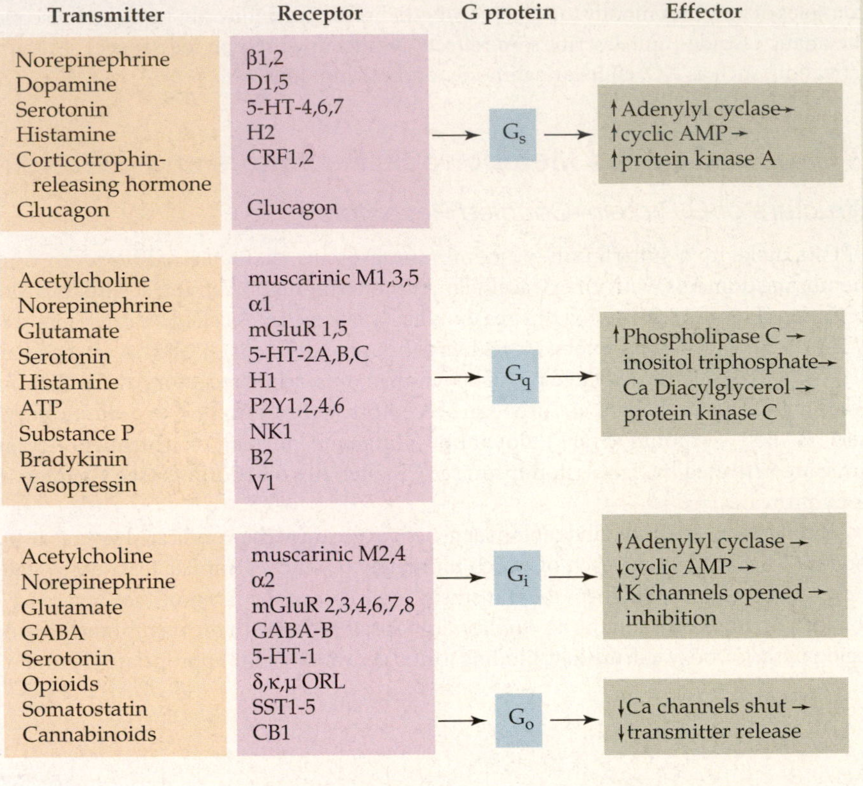

G Proteins

G proteins link the GPCR to its immediate effector protein. They are membrane proteins, so-named because they bind guanine nucleotides. Each G protein is a trimer made up of three subunits: α, β, and γ (Figure 12.3).[3] There are many isoforms of each G protein subunit (21 for α, 6 for β, 12 for γ), providing a large number of potential trimer permutations. G proteins are grouped into three main classes according to the structure and targets of their α-subunits: G_s stimulates the enzyme adenylyl cyclase; G_i inhibits adenylyl cyclase (and also activates potassium channels); and G_q couples to the enzyme phospholipase C (PLC). The G_i class also includes G_t (transducin), which activates cyclic guanosine monophosphate (cGMP) phosphodiesterase (see Chapter 19), and G_o, which interacts with calcium ion channels.

The prime binding site for the receptor is the C-terminus of the α-subunit, and the main determinant that specifies which G protein a receptor activates is the amino acid sequence of this C-terminus (although interactions with other subunits may facilitate binding). This sequence differs among the different classes of G protein (G_s, G_q and $G_{i/o}$) but is identical or very similar among different members of each class. Hence, while some receptors are promiscuous and can couple to more than one class of G proteins, most metabotropic neurotransmitter receptors show a preferential interaction with one or other of the three common classes of G protein, as shown in Box 12.1.

THE G PROTEIN CYCLE The way in which the G protein works is illustrated in Figure 12.3. In the resting state, guanosine diphosphate (GDP) is bound to the α-subunit, and

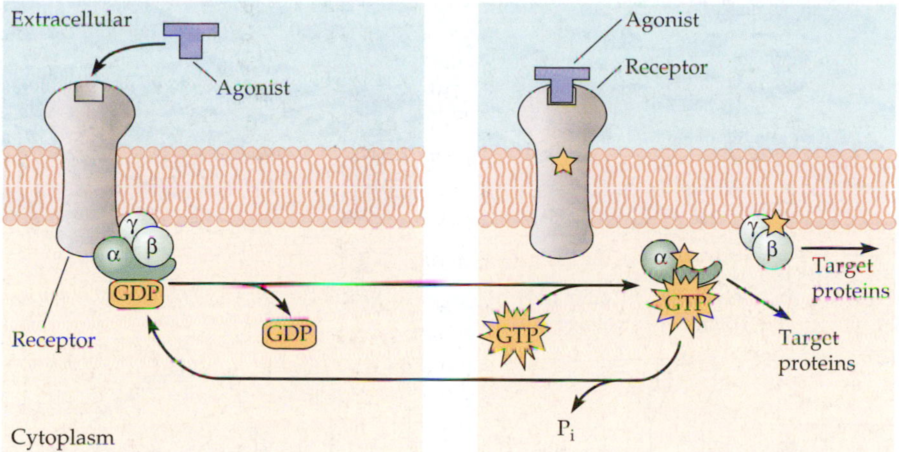

FIGURE 12.3 Indirectly Coupled Transmitter Receptors Act through G Proteins. G proteins are trimers of α-, β-, and γ-subunits. Activation of a metabotropic receptor by agonist binding promotes the exchange of guanosine triphosphate (GTP) for guanosine diphosphate (GDP) on the α-subunit of the G protein. This action activates the α-subunit and the βγ-complex, causing them to dissociate from the receptor and from one another. The free activated α-GTP subunit and βγ-complex each interacts with target proteins. Hydrolysis of GTP to GDP and inorganic phosphate (P_i) by the endogenous GTPase activity of the α-subunit leads to re-association of the αβγ-complex, terminating the response.

the three subunits are associated as a trimer. Interaction with an activated receptor allows guanosine triphosphate (GTP) to replace GDP on the α-subunit, resulting in dissociation of the α- and βγ-subunits. (The β- and γ-subunits remain together under physiological conditions.) Free α- and βγ-subunits then diffuse, bind to and modulate the activity of target proteins.[3–5] The free α-subunit has intrinsic GTPase activity that results in the hydrolysis of the bound GTP to GDP. This permits the re-association of the α- and βγ-subunits into the G protein complex and terminates their activity.

The lifetime of the activated G protein subunits is modulated by proteins called GTPase-activating proteins, or GAPs, which influence the rate at which GTP, bound to the α-subunit, is hydrolyzed.[6] One family of GAPs, the RGS proteins—Regulators of **G** protein **S**ignaling—play an important role in determining the time course of transmitter action.[7] The details of the interactions of the G protein subunits with each other and with their receptors and targets have been explored using biomolecular X-ray crystallography[1,3] and live-cell imaging[8,9] techniques. In addition, probes have been developed for identifying responses mediated by G proteins (Box 12.2).

Modulation of Ion Channel Function by Receptor-Activated G Proteins: Direct Actions

G proteins affect ion channels in two different ways. First, a subunit of the activated G protein (usually the βγ-complex) may interact **directly** with the ion channel; or, second, the G protein (usually the GTP-bound α-subunit) may activate one or more enzymes to alter ion channel function **indirectly** through one or more **second messengers**.

G Protein Activation of Potassium Channels

Much of our knowledge about the direct interaction of G proteins with ion channels comes from experiments concerning the way in which stimulation of the vagus nerve slows or stops the heart beat. As first shown by Loewi (see Box 11.1), this is due to the release of ACh from the vagus nerve endings. The ACh then binds to a G protein–coupled receptor—the M2 **muscarinic** receptor (see Box 12.1)—so-called because it is selectively activated by the drug muscarine,[10] an alkaloid in the fly agaric (*Amanita muscaria*) mushroom. Figure 12.4 illustrates what happens when ACh is applied to the sinoatrial node of the rabbit heart. In Figure 12.4A a brief application of ACh causes a temporary cessation of spontaneous action potentials and a hyperpolarization of the heart muscle cell (as first shown by Burgen and Terroux[11] and Hutter and Trautwein[12]). This hyperpolarization is due to the opening of potassium channels,[13] generating the outward potassium current seen in Figure 12.4B.

The role of G proteins in coupling the muscarinic receptors to these ion channels was established in a series of experiments by Breitwieser, Szabo, Pfaffinger, Trautwein, Hille, and their colleagues. They found that intracellular GTP is required;[14] that activation of potassium channels by muscarinic agonists is greatly prolonged by intracellular application

[4] Dascal, N. 2001. *Trends Endocrinol. Metab.* 12: 391–398.

[5] Clapham, D. E., and Neer, E. J. 1997. *Annu. Rev. Pharmacol. Toxicol.* 37: 167–203.

[6] Berman, D. M., and Gilman, A. G. 1998. *J. Biol. Chem.* 273: 1269–1272.

[7] Doupnik, C. et al. 1997. *Proc. Natl. Acad. Sci. USA* 94: 10461–10466.

[8] Hein, P. et al. 2005. *EMBO J.* 24: 4106–4114.

[9] Raveh, A, Riven, I. and Reuvenny, E. 2009. *J. Physiol.* 587: 5331–5335.

[10] Dale, H. H. 1914. *J. Pharmac. Exp. Ther.* 6: 147–190.

[11] Burgen, A. S. V., and Terroux, K. G. 1953. *J. Physiol.* 120: 449–464.

[12] Hutter, O. F., and Trautwein, W. 1956. *J. Gen. Physiol.* 39: 715–733.

[13] Sakmann, B., Noma, A., and Trautwein, W. 1983. *Nature* 303: 250–253.

[14] Pfaffinger, P. J. et al. 1985. *Nature* 317: 536–538.

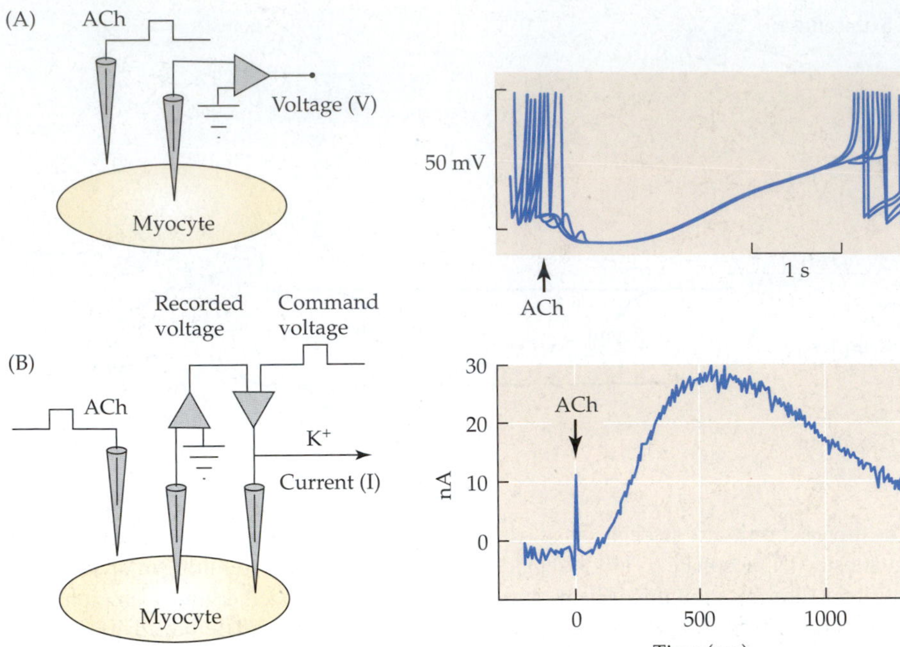

FIGURE 12.4 Acetylcholine (ACh) Opens Potassium Channels in Sinoatrial Cells of the Rabbit Heart. (A) A brief ionophoretic ejection of ACh from a micropipette transiently hyperpolarizes the cell membrane and inhibits spontaneous action potentials for ~3 seconds. (B) When membrane current is recorded using voltage clamp (see Box 7.1), a similar ACh application produces an outward K^+ current, I_K, which starts after about 50 milliseconds (ms), and lasts about 1.5 seconds. (A after Trautwein et al., 1982; B after Trautwein et al., 1981.)

BOX 12.2
Identifying Responses Mediated by G Proteins

Several tests can be used to identify responses mediated by G proteins. For example, activation of the α-subunit requires that bound GDP be replaced by GTP. Accordingly, G protein mediated events have an absolute requirement for cytoplasmic GTP and will be blocked by intracellular perfusion with solutions lacking GTP. Two analogues of GTP—GTPγS and Gpp(NH)p—are useful because they cannot be hydrolyzed by the endogenous GTPase activity of the α-subunit. Like GTP, they can replace GDP on the α-subunit and activate it. However, because they cannot be hydrolyzed, these analogues activate the α-subunit permanently. Thus, intracellular perfusion with either of these analogues enhances and greatly prolongs agonist-induced activation of G protein mediated responses and may even initiate responses in the absence of agonist. On the other hand, GDPβS, an analogue of GDP, binds strongly to the GDP site on the α-subunit and resists replacement by GTP. Thus, GDPβS inhibits G protein mediated responses by maintaining the αβγ-complex in the inactive state.

Two bacterial toxins are useful for characterizing G protein mediated processes. Each is an enzyme that catalyzes the covalent attachment of ADP-ribose to an arginine residue on the α-subunit. Cholera toxin acts on α_s, irreversibly activating it; pertussis toxin acts on members of the α_i family, irreversibly blocking their activation and so inhibiting responses mediated by the corresponding G proteins.

GTP
Guanosine 5′-triphosphate

GDP
Guanosine 5′-diphosphate

GTPγS
Guanosine 5′-*O*-[γ-thio] triphosphate

Gpp(NH)p
Guanosine 5′-[β,γ-imido] triphosphate

GDPβS
Guanosine 5′-*O*-[β-thio] diphosphate

(A)

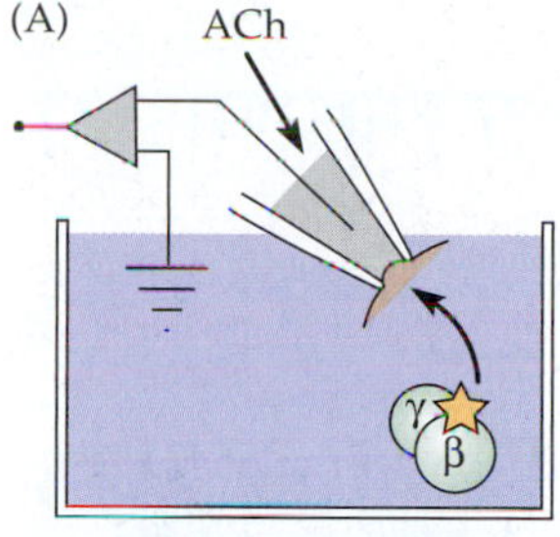

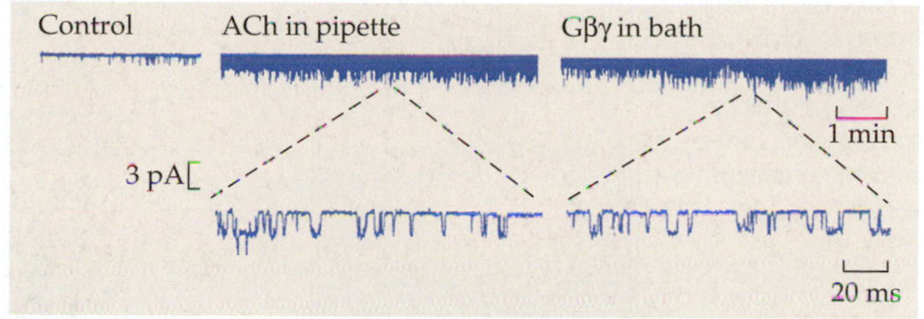

(B)

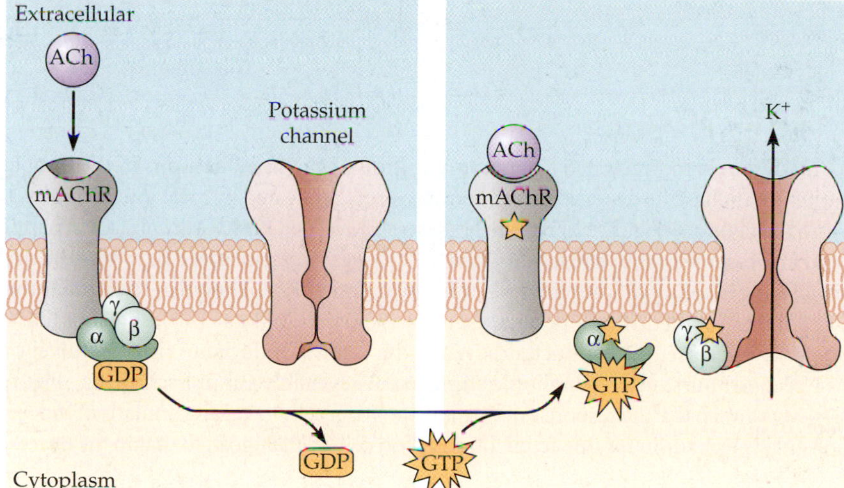

FIGURE 12.5 Direct Modulation of Channel Function by G Proteins. (A) Application of the Gβγ complex to the intracellular surface of an isolated patch of membrane from a rat atrial muscle cell (Gβγ in bath) results in an increase in potassium channel activity similar to that seen when acetylcholine (ACh) is added to the extracellular side of the patch (ACh in pipette). (B) Schematic representation of events in an intact cell. Binding of ACh to muscarinic receptors (mAChR) activates a G protein (indicated by a star); activated βγ-complex binds directly to and opens a potassium channel. (A after Wickman et al., 1994.)

of the non-hydrolyzable analogue of GTP, known as Gpp(NH)p;[15] and that muscarinic activation of potassium channels is blocked by pertussis toxin,[14] which inactivates G_i proteins (see Box 12.2). An important advance came from experiments by David Clapham and his colleagues using inside-out excised membrane patches (Figure 12.5). They showed that the cardiac potassium channels were opened when pure recombinant βγ-subunit was applied to the intracellular side of the patch.[16] This demonstrated that the βγ-subunit, rather than the α-subunit, is responsible for potassium channel activation. Subsequent experiments with a cloned muscarinic potassium channel (originally named GIRK1, for G protein activated inward rectifier K^+ channel, but later renamed Kir3.1) indicated that βγ-subunits interact directly with potassium channels (Figure 12.5B).[17,18]

Using muscle cells dissociated from the atrium of the heart, Soejima and Noma found that potassium channel activity in cell-attached patches was increased when muscarinic agonists were added to the patch pipette solution, but not when agonists were added to the bath (Figure 12.6).[19] Thus, activated βγ-subunits appear unable to traverse the region of the pipette–membrane seal to influence channels on the other side. This so-called **membrane-delimited** mechanism of G protein action reflects the limited range over which βγ-subunits can act. In fact, some evidence suggests that the G protein, GTPase-activating protein RGS4,[20] and GIRK channel are closely associated and may then form a receptor, the G protein–GIRK channel complex.[9,21]

Similar G protein activated inward rectifier potassium channels (GIRK channels)[22] are also present in many **nerve cells**. There, they can be activated by a number of different neurotransmitters through receptors coupled to G_i/G_o proteins—for example, ACh, norepinephrine, GABA, and dopamine.[23] Their activation induces a postsynaptic hyperpolarization (inhibitory synaptic potential) very similar to the cardiac response to vagal stimulation. Figure 12.7 shows an example from an experiment on bullfrog ganglion cells.[24] When the nicotinic receptors in the ganglion had been blocked with curare, a few electrical stimuli, applied to the preganglionic sympathetic nerves in the descending lum-

[15] Breitwieser, G. E., and Szabo, G. 1985. *Nature* 317: 538–540.

[16] Wickman, K. D. et al. 1994. *Nature* 368: 255–257.

[17] Reuveny, E. et al. 1994. *Nature* 370: 143–146.

[18] Huang, C-L. et al. 1995. *Neuron* 15: 1133–1143.

[19] Soejima, M., and Noma, A. 1984. *Pflügers Arch.* 400: 424–431.

[20] Fowler C. E. et al. 2007. *J. Physiol.* 580: 51–65.

[21] Benians A. et al. 2005. *J. Biol. Chem.* 280: 13383–13394.

[22] Bichet, D., Haase F. A., and Jan, L. Y. 2003. *Nat. Rev. Neurosci.* 4: 957–967.

[23] North, R. A. et al. 1987. *Proc. Natl. Acad. Sci. USA* 84: 5487–5491.

[24] Dodd, J., and Horn, J. P. 1983. *J. Physiol.* 334: 271–291.

FIGURE 12.6 Direct, or Membrane-Delimited, Effects of G Proteins Operate over Short Distances. (A) Effects of acetylcholine (ACh) were assayed by cell-attached, patchclamp recording. ACh could be perfused into either the patch pipette or the bath. (B) Recordings of single-channel currents before and during addition of ACh. Compared with the control, channel activity increased only when ACh was added to the patch pipette. (After Soejima and Noma, 1984.)

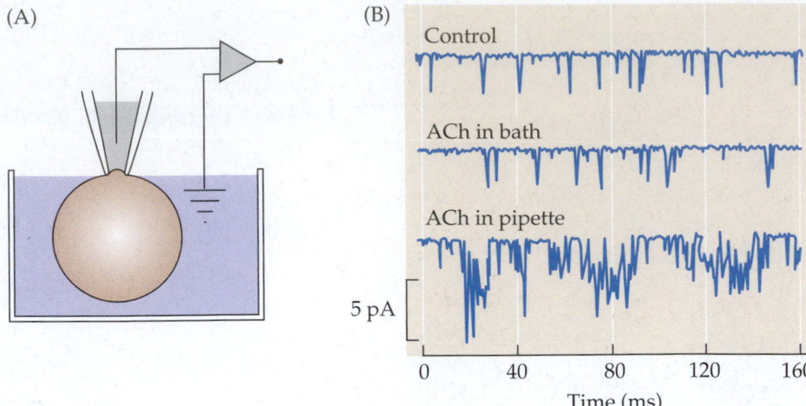

bar sympathetic chain, generated a hyperpolarization of the small neurons in the lumbar sympathetic ganglia. The hyperpolarization was closely replicated by local ionophoresis of ACh onto the neurons (Figure 12.7B). Further tests showed that both the inhibitory synaptic potential and response to ACh reversed when the cell was hyperpolarized beyond –102 mV in normal Ringer solution (containing 2 millimolar [m*M*] K^+). They also found that this reversal potential shifted 58 mV per tenfold increase in extracellular $[K^+]$, showing that it was due to an increased K^+ conductance. When the cell was induced to fire repetitively, a short (1 second) burst of preganglionic stimulation reversibly suppressed firing (Figure 12.7C)—very much like the response of cardiac sinoatrial cells to vagal stimulation. Indeed, the molecular mechanism for this action on ganglion cells is analogous to that in the heart.[25]

[25] Fernandez-Fernandez, J. M. et al. 2001. *Eur. J. Neurosci.* 14: 283–292.

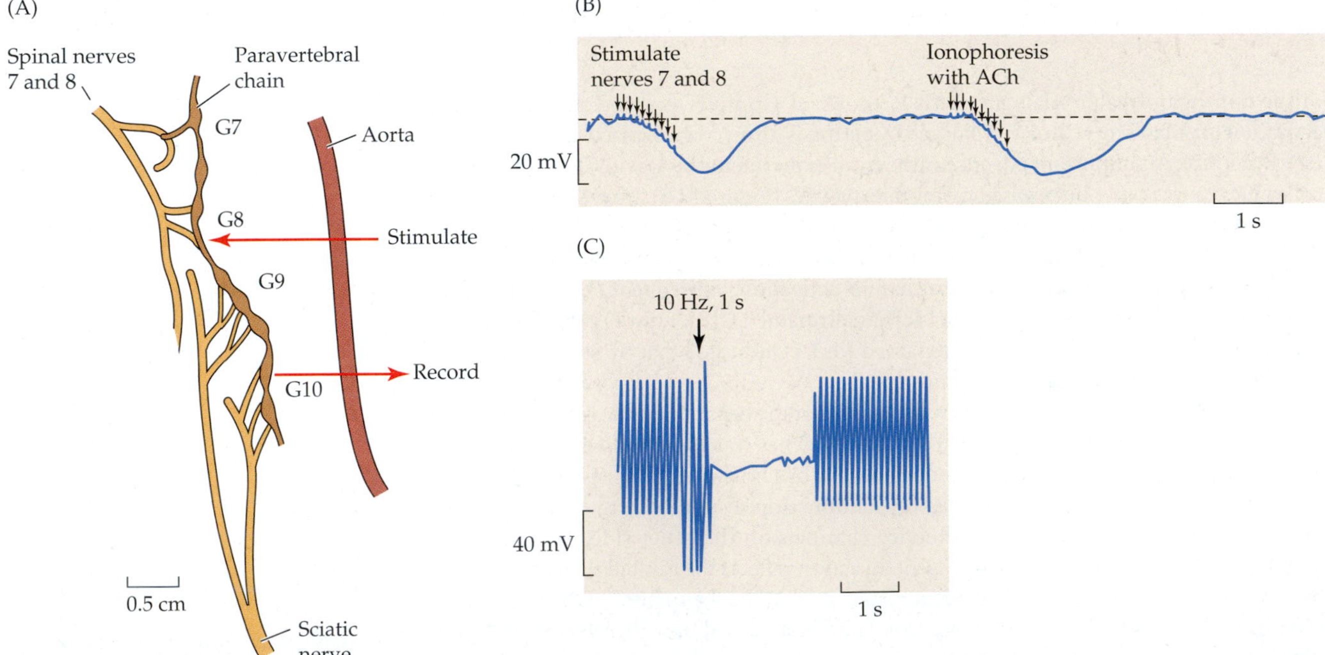

FIGURE 12.7 Cholinergic Synaptic Inhibition in the Nervous System Results from the Opening of Potassium Channels. (A) Drawing of the frog lumbar sympathetic nervous system. Neurons in the 9th and 10th sympathetic ganglia are innervated by descending cholinergic preganglionic fibers in spinal nerves 7 and 8 and send their postganglionic axons out in the sciatic nerve. (B) Intracellular recordings from a small neuron in the 9th sympathetic ganglion. A series of stimuli to the descending preganglionic fibers produces a slow hyperpolarization, which is imitated by electrophoretically applying acetylcholine to the neuron through a micropipette. This action results from stimulating muscarinic receptors because the nicotinic receptors were blocked using tubocurarine. (C) Preganglionic stimulation also suppresses action potential discharges of the postganglionic neuron. The neuron was made to fire repetitively by 4 minutes of preganglionic stimulation at 60 impulses per minute. This releases luteinizing hormone releasing hormone (LHRH; see Chapter 17), which depolarizes the cell because it produces a prolonged inhibition of the M-current. (A after Dodd and Horn, 1983a; B after Dodd and Horn, 1983b.)

FIGURE 12.8 Presynaptic Autoreceptors Reduce Transmitter Release. (A) Norepinephrine (NE) released from sympathetic neurons combines with α_2-adrenergic receptors (called autoreceptors) in the terminal membrane, activating a G protein. The activated $\beta\gamma$-complex binds to calcium channels, decreasing calcium influx and so limiting further transmitter release. (B) Norepinephrine reduces the release of transmitter from sympathetic ganglia. Ganglia were loaded with radioactive norepinephrine and then enclosed in a perfusion chamber. Transmitter release was evoked by depolarization with a solution containing 50 m*M* potassium (green bars). Addition of 30 μ*M* unlabeled norepinephrine to the perfusion solution (red bar) reduced the amount of radiolabeled transmitter released in response to potassium-induced depolarization. (B after Lipscombe, Kongsamut, and Tsien, 1989.)

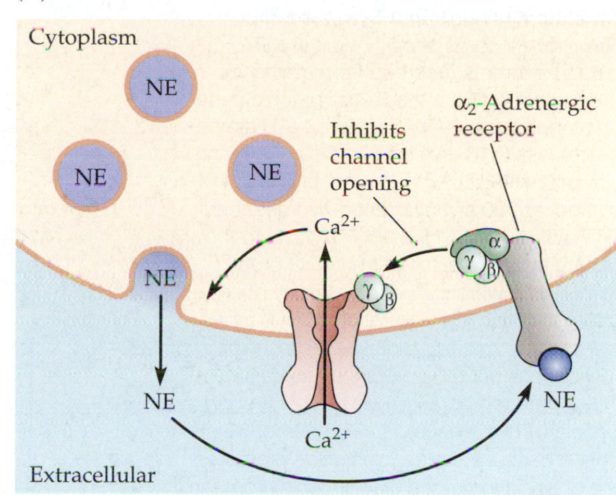

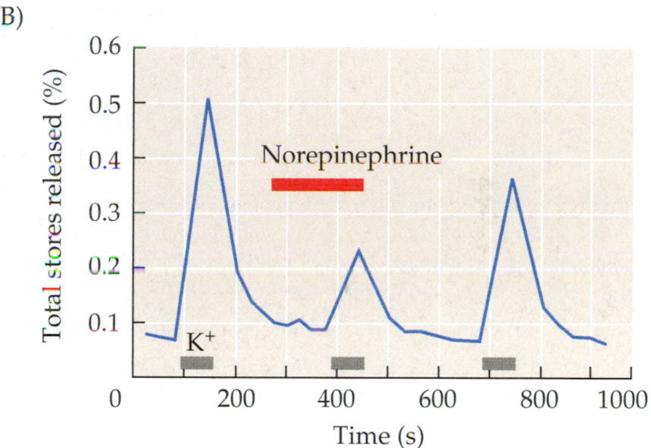

G Protein Inhibition of Calcium Channels Involved in Transmitter Release

When released, many neurotransmitters (including ACh, glutamate, GABA, mono-amines such as norepinephrine, and many peptides) not only stimulate postsynaptic receptors but also activate metabotropic receptors on the presynaptic terminals to reduce their own release (see Chapter 11). An example of this feedback inhibition (or autoinhibition) of the release of norepinephrine from nerve terminals in frog sympathetic ganglia[26] is illustrated in Figure 12.8. As shown schematically in Figure 12.8A, the transmitter acts not only on postsynaptic target cells, but also on the nerve terminals themselves. Figure 12.8B shows the release of radioactively tagged norepinephrine from the terminals in response to depolarization by elevated external potassium (green bar). Addition of norepinephrine to the bathing solution (red bar) produced a marked reduction of transmitter release from the terminals.

This decrease in release is due to activation of presynaptic metabotropic α2-adrenergic receptors (see Box 12.1). These activate G_o proteins, which in turn directly inhibit activation of N-type ($Ca_V2.2$) calcium channels (see Chapter 5). This effect is shown in the records of Figure 12.9. Single-channel calcium currents from a cell-attached patch, produced by membrane depolarization, are reduced in frequency when norepinephrine is added to the patch solution. As with the effect of ACh on potassium channels (see Figure 12.6), there was no response of the channels to the application of norepinephrine outside the patch (the response is membrane-delimited), suggesting again a close association between the G proteins and the channels.

Results of experiments in which α- or βγ-subunits or βγ-sequestering peptides were over-expressed or injected into cells indicated that it is the βγ-subunits that inhibit N-type calcium channels.[27–29] This is illustrated in Figure 12.10A. The inward calcium current produced by depolarizing a sympathetic ganglion cell (a), was reduced strongly by adding norepinephrine (NE) to the bathing solution. A similar effect was observed when a non-hydrolysable GTP analog was added to the bath (b) or when βγ-subunits were over-expressed in the cell by prior injection of their cDNAs (c). The βγ-subunits bind directly to the calcium channel α-subunit, primarily at the intracellular linker between domains I and II[30] (see Figure 5.7B).

[26] Lipscombe, D., Kongsamut, S., and Tsien, R. W. 1989. *Nature* 340: 639–642.
[27] Ikeda, S. R. 1996. *Nature* 380: 255–258.
[28] Herlitze, S. et al. 1996. *Nature* 380: 258–262.
[29] Delmas, P. et al. 1998. *J. Physiol.* 506: 319–329.
[30] De Waard, M. et al. 2005. *Trends Pharmacol. Sci.* 26: 427–436.

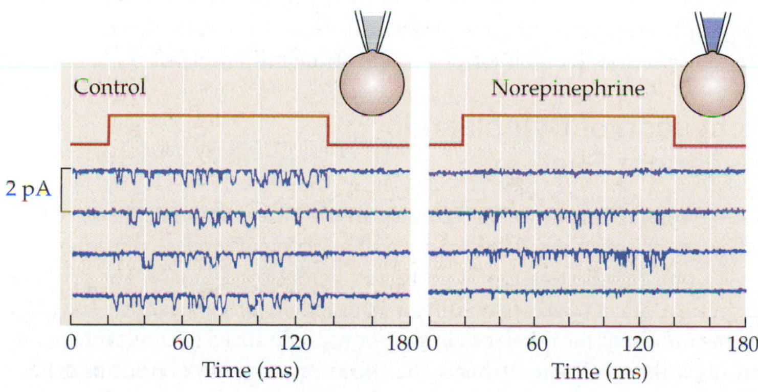

FIGURE 12.9 Norepinephrine Inhibits Calcium Channel Activity. Single-channel currents were recorded in cell-attached patches; channels were activated with a depolarizing pulse (top trace). When 30 μ*M* norepinephrine was included in the patch electrode, the unitary currents did not change in size, but channel openings were less frequent and of shorter duration. (After Lipscombe, Kongsamut, and Tsien, 1989.)

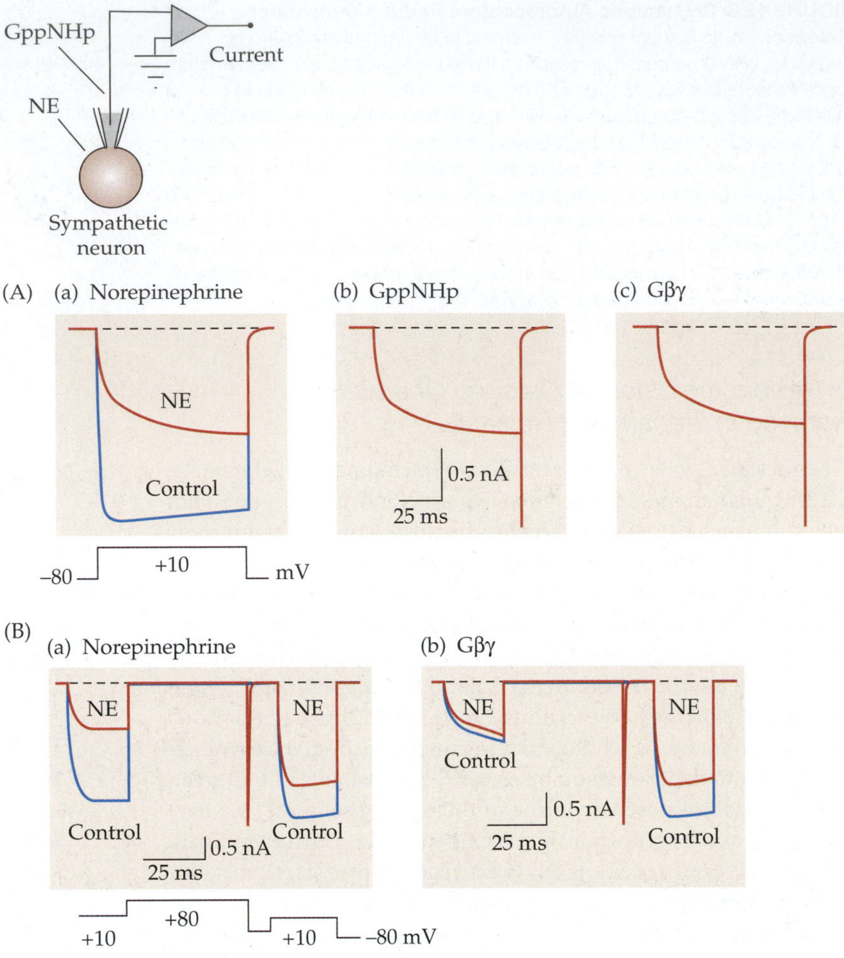

FIGURE 12.10 G Protein Regulation of Calcium Currents in a Sympathetic Neuron. Records show N-type ($Ca_V2.2$) calcium currents recorded from a rat sympathetic neuron with a whole-cell patch pipette on stepping from –80 to +10 mV. (A) Norepinephrine (NE, 10 μ*M*) inhibits the current and slows its activation (a). This effect is imitated by adding 500 μ*M* of the non-hydrolyzable GTP-analog GppNHp (see Box 12.2) to the pipette solution (b), or when free G protein βγ-subunits were over-expressed by prior cDNA injection (c). Calibration bars: 0.5 nA (vertical), 20 ms (horizontal). (B) The inhibitory actions of both norepinephrine (a) and Gβγ (b) are temporarily reversed by strongly depolarizing the neuron to +80 mV for 50 ms. This is because the depolarization promotes the dissociation of the Gβγ-subunits from the calcium channel. (After Ikeda, 1996.)

This interaction is voltage-dependent, with inhibition being reduced by depolarization; hence, inhibition diminishes as the depolarizing pulse is maintained.[31,32] In Figure 12.10A, this voltage-dependence appears as a slowing of the calcium current; this suggests that the channels have been converted from their normal "willing to open" state to a "reluctant to open" state.[33] The voltage-dependence is shown in another way in Figure 12.10B, by delivering a very large depolarization for 50 ms between two calcium current test pulses. This accelerates the dissociation of the G protein βγ-subunit so that the inhibition produced by norepinephrine (a) or by the expressed βγ-subunits (b), seen during the first test pulse, is completely reversed when the second test pulse is given. On repolarization, channel inhibition is restored quite quickly, over 100 ms or so (not shown), as the released Gβγ-subunits re-associate with the channel, suggesting that they have not moved far away. It has been calculated that the receptor, G protein, and channel must all be within less than 1 μm distance.[34] Recent experiments have shown that the primary response of the G protein–coupled receptor to a transmitter is also sensitive to membrane voltage[35,36] and that this property may also modify their effect on neurotransmitter release.[37]

G Protein Activation of Cytoplasmic Second Messenger Systems

Many G proteins do not bind directly to ion channels. Instead they modulate the activity of enzymes involved in cytoplasmic second messenger systems: adenylyl cyclase, phospholipase C, phospholipase A_2, phosphodiesterase, and phosphatidylinositol 3-kinase. The products of these enzymes, in turn, affect targets that influence the activity of ion channels and other cellular processes. In contrast to the rapid localized responses produced by direct interaction of G protein subunits with membrane channels, the effects produced by G proteins that ac-

[31] Tsunoo, A., Yoshii, M., and Narahashi, T. 1986. *Proc. Nat. Acad. Sci. USA* 83: 9832–9836.

[32] Grassi, F., and Lux, H. D. 1989. *Neurosci. Lett.* 105: 113–119.

[33] Bean, B. P. 1989. *Nature* 340: 153–157.

[34] Zhou, J., Shapiro, M. S., and Hille, B. 1997. *J. Neurophysiol.* 77: 2040–2048.

[35] Parnas, H. and Parnas, I. 2007. *Trends Neurosci.* 30: 54–61.

[36] Mahaut-Smith, M. P., Martinez-Pinna, J., and Gurung, I. S. 2008. *Trends Pharmacol. Sci.* 29: 421–429.

[37] Kupchik, Y. M. et al. 2008. *Proc. Natl. Acad. Sci. USA* 105: 4435–4440.

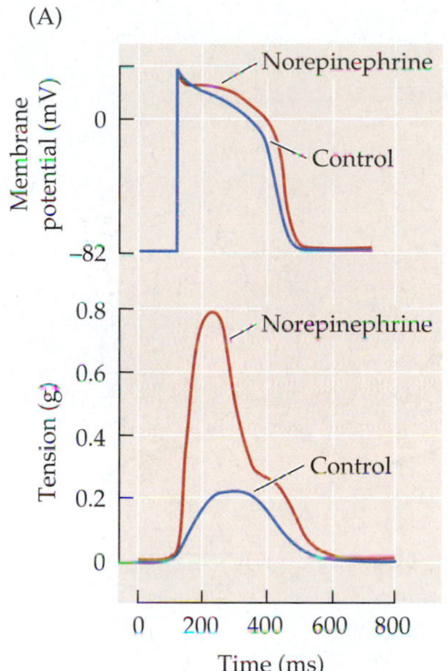

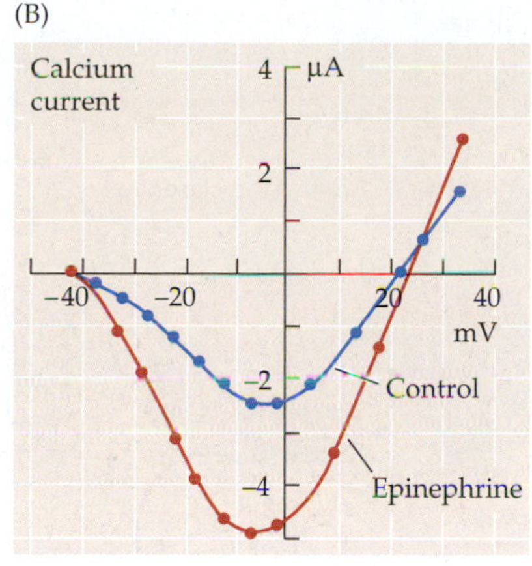

FIGURE 12.11 Activation of β-Adrenergic Receptors in Cardiac Muscle Increases Calcium Current. (A) The increase in calcium current produced by activation of β-adrenergic receptors, in this case by addition of 10^{-6} *M* norepinephrine, increases action potential amplitude and duration and the tension produced by cardiac muscle cells. (B) The current–voltage relationship of calcium current in a myocardial cell is measured under voltage clamp conditions in the absence and presence of 0.5 μ*M* epinephrine—a β-adrenergic receptor agonist. (A after Reuter et al., 1983; B after Reuter, 1974.)

tivate cytoplasmic second messenger systems are slower and more widespread. They include contractions of smooth and cardiac muscles, and glandular secretion, among other effects.

β-Adrenergic Receptors Activate Calcium Channels via a G Protein—the Adenylyl Cyclase Pathway

One of the most thoroughly studied examples of indirect synaptic transmission mediated by an intracellular second messenger is the activation of β-adrenergic receptors in cardiac muscle cells by norepinephrine.[38,39] This produces an increase in the rate and force of contraction of the heart, accompanied by an increase in the amplitude and duration of the cardiac action potential (Figure 12.11A). Voltage clamp studies by Reuter, Trautwein, Tsien, and others indicate that these effects are accompanied by a marked increase in the calcium current associated with the action potential (Figure 12.11B). This increased calcium current is responsible for the increased size of the action potential and the increased calcium influx is the cause of the marked increase in contraction.

Single-channel recording from cardiac muscle cells, using the cell-attached mode of the patch clamp technique, confirms that stimulation with a β-adrenergic receptor agonist, such as norepinephrine or isoproterenol, produces an increase in calcium channel activity (Figure 12.12). Moreover, it is not necessary that the agonist be added to the pipette solution to observe the response. Adding isoproterenol to the medium bathing the cell causes an increase in activity of calcium channels *within* the patch—a diagnostic test for responses mediated by diffusible cytoplasmic second messengers.[38]

[38] Tsien, R. W. 1987. In *Neuromodulation: The Biochemical Control of Neuronal Excitability*. Oxford University Press, New York, pp. 206–242.

[39] McDonald, T. F. et al. 1994. *Physiol. Rev.* 74: 365–507.

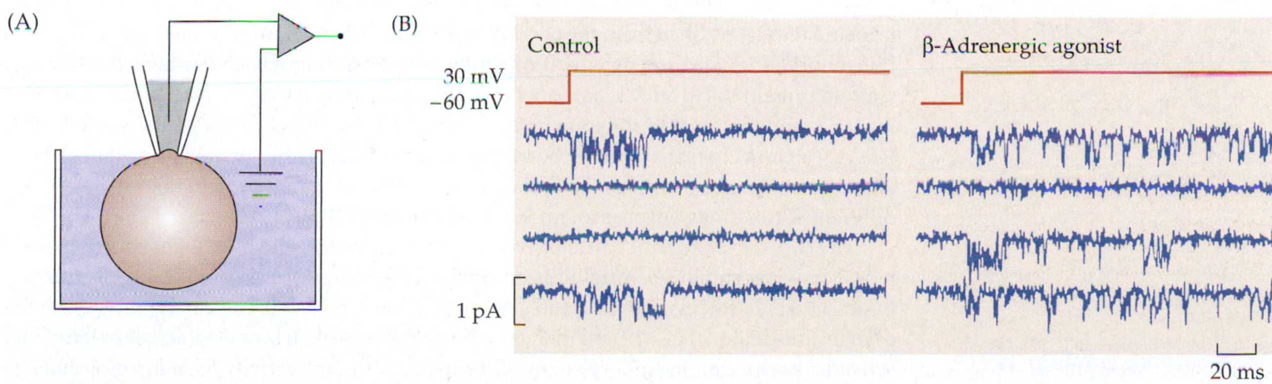

FIGURE 12.12 β-Adrenergic Agonists Cause an Increase in Calcium Channel Activity during a depolarizing pulse. (A) Recordings are from a voltage clamped cell-attached patch. (B) Consecutive records of the activity of a patch containing two calcium channels. Addition of 14 μ*M* isoproterenol, a β-adrenergic agonist, to the bath causes an increase in calcium channel activity during depolarization. (After Tsien, 1987.)

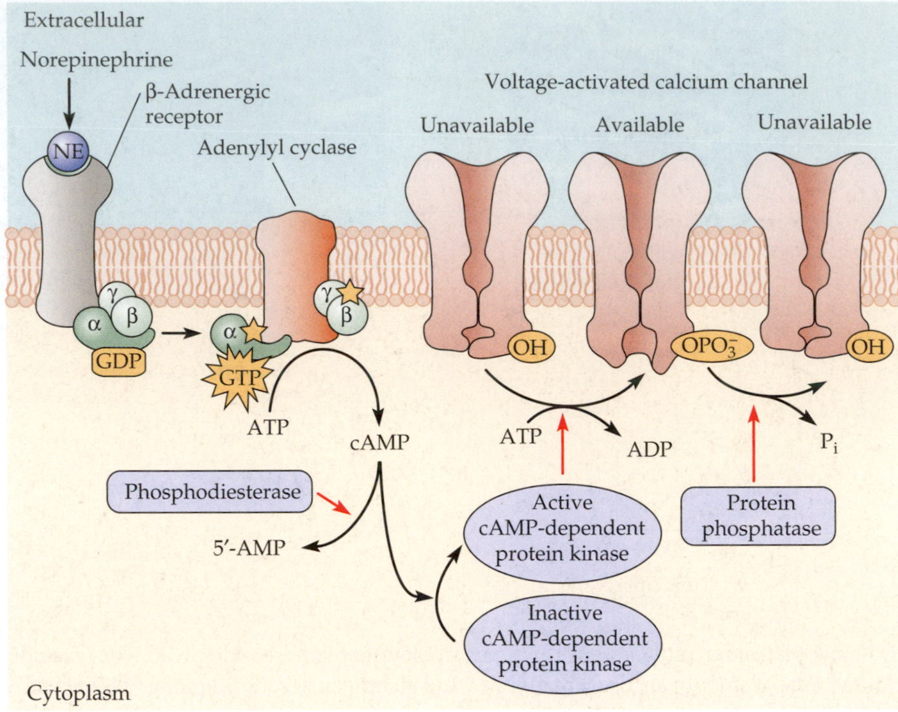

FIGURE 12.13 β-Adrenergic Receptors Act through the Intracellular Second Messenger Cyclic AMP to Increase Calcium Channel Activity. Binding of norepinephrine to β-adrenergic receptors activates, through a G protein, the enzyme adenylyl cyclase. Adenylyl cyclase catalyzes the conversion of ATP to cyclic AMP (cAMP). As the concentration of cAMP increases, it activates cAMP-dependent protein kinase, an enzyme that phosphorylates proteins on serine and threonine residues (—OH). The response to norepinephrine is terminated by the hydrolysis of cAMP to 5′-AMP and the removal of protein phosphate residues by protein phosphatases. In cardiac muscle cells, norepinephrine causes phosphorylation of voltage-activated calcium channels, converting them to a form that can be opened by depolarization (available).

Activation of β-adrenergic receptors is coupled to the increase in calcium conductance through the intracellular second messenger cyclic AMP (cAMP) (Box 12.3). As illustrated in Figure 12.13, binding of norepinephrine to β-adrenergic receptors on heart cells activates a G protein, G_s. The α-subunit of G_s then binds to and activates the enzyme adenylyl cyclase. Adenylyl cyclase converts ATP to cAMP, a readily diffusible intracellular second messenger that activates another enzyme, cAMP–dependent protein kinase (protein kinase A or PKA). The catalytic subunits of this protein kinase mediate the transfer of phosphate from ATP to the hydroxyl groups of serine and threonine residues in a variety of enzymes and channels, thereby modifying their activity. In this case, phosphorylation of cardiac calcium channels increases their probability of opening.

Several lines of evidence are consistent with this scheme as outlined in Box 12.3. For example, calcium channel activity is increased by forskolin, by membrane-permeable derivatives of cAMP, by inhibitors of phosphodiesterase, and by direct intracellular injection of cAMP itself. Similarly, intracellular injection of the catalytic subunit of cAMP–dependent protein kinase leads to an increase in calcium current, while injection of excess regulatory subunit or inhibitors of protein kinase block adrenergic stimulation of calcium currents. ATPγS, an analogue of ATP, augments adrenergic activation of calcium channels by forming stably phosphorylated proteins, while intracellular injection of protein phosphatases prevents or reverses adrenergic stimulation of calcium currents by rapidly removing protein phosphate residues.

Subsequent experiments established that the increased channel activity produced by protein kinase A resulted from the phosphorylation of the calcium channels themselves.[40,41] Phosphorylation and enhancement of cardiac calcium channel activity is facilitated by a scaffold protein called an A-kinase anchoring protein, or AKAP79,[42] which binds to protein kinase A and targets it to the ion channel. Modulation of the activity of the same L-type (Ca_V1) calcium channels by other hormones is also mediated by channel phosphorylation, either through effects on adenylyl cyclase and cAMP-dependent protein kinase or through different second messenger–protein kinase signaling pathways.[39]

The two-step enzymatic cascade involving adenylyl cyclase and cAMP-dependent protein kinase provides significant amplification compared to direct opening or closing of channels by activated G proteins. Each activated adenylyl cyclase molecule can catalyze the synthesis of many molecules of cAMP and thereby activate many protein kinase molecules—and each activated kinase can phosphorylate many proteins. Thus, the activity of many molecules of

---

[40] Curtis, B. M., and Catterall, W. A. 1986. *Biochemistry* 25: 3077–3083.

[41] Flockerzi, V. et al. 1986. *Nature* 323: 66–68.

[42] Gao, T. et al. 1997. *Neuron* 19: 185–196.

BOX 12.3
Cyclic AMP as a Second Messenger

Experiments by Sutherland, Krebs, Walsh, Rodbell, Gilman, and their colleagues, initially aimed at understanding how the hormones epinephrine and glucagon elicit breakdown of glycogen in the liver, led to the discovery of cyclic AMP (cAMP) and the concept of intracellular second messengers.[43–45] They showed that binding of the hormone to its receptor activates a G protein, which, in turn, stimulates the enzyme adenylyl cyclase. Adenylyl cyclase catalyzes the synthesis of cAMP from ATP. The increase in cAMP concentration activates cAMP-dependent protein kinase, an enzyme that phosphorylates its target proteins on serine and threonine residues. The cAMP is subsequently degraded by phosphodiesterase to AMP, and the phosphate residues on the target proteins are removed by protein phosphatases (see Figure 12.13).

Some tests used to determine if the response to a transmitter or hormone is mediated by cAMP depend on activating adenylyl cyclase or elevating cAMP directly. For example, intracellular injection of cAMP and addition of membrane-permeable derivatives of cAMP, such as 8-bromo-cAMP or dibutyryl-cAMP, mimic cAMP-mediated responses. Similarly, direct activation of adenylyl cyclase by forskolin mimics the response. Inhibitors of phosphodiesterase, such as the methylxanthines, theophylline, and caffeine, either mimic or enhance the response, depending on the endogenous level of cyclase activity. Other procedures test the involvement of cAMP-dependent protein kinase (also known as protein kinase A, or PKA). This enzyme is composed of two regulatory and two catalytic subunits. In the absence of cAMP, the four subunits exist as a complex, with the regulatory subunits blocking the activity of the catalytic subunits. When cAMP binds to the regulatory subunits, the complex dissociates, freeing active catalytic subunits. Thus, intracellular injection of purified catalytic subunit will mimic responses mediated by increased cAMP, while injection of excess regulatory subunits will be inhibitory. Additional inhibitors of this enzyme have been developed, including H-8 (which also inhibits several other protein serine kinases), specific peptide inhibitors, and derivatives of ATP that cannot be used by the kinase as a source for phosphate residues. These inhibitors block responses mediated by cAMP.

On the other hand, treatments that inhibit protein phosphatases augment and prolong responses mediated by cAMP. These include injection of specific phosphatase inhibitors and of ATPγS, an analogue of ATP that can be used as a cosubstrate by cAMP-dependent protein kinase, forming phosphoproteins with thiophosphate linkages, which are resistant to hydrolysis by protein phosphatases.

[43] Sutherland, E. W. 1972. *Science* 177: 401–408.
[44] Schramm, M., and Selinger, Z. 1984. *Science* 225: 1350–1356.
[45] Gilman, A. G. 1987. *Ann. Rev. Biochem.* 56: 615–649.

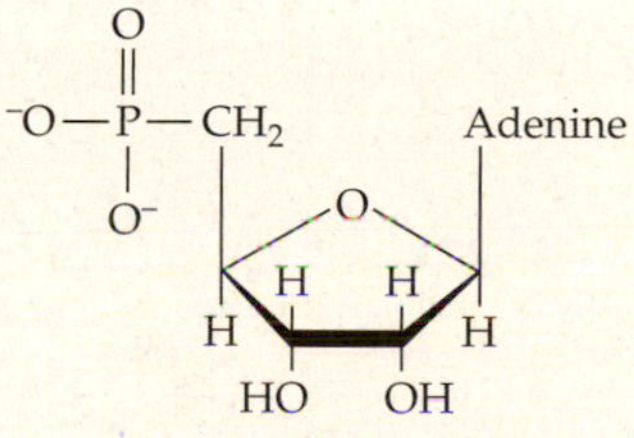

AMP
Adenosine 5′-monophosphate

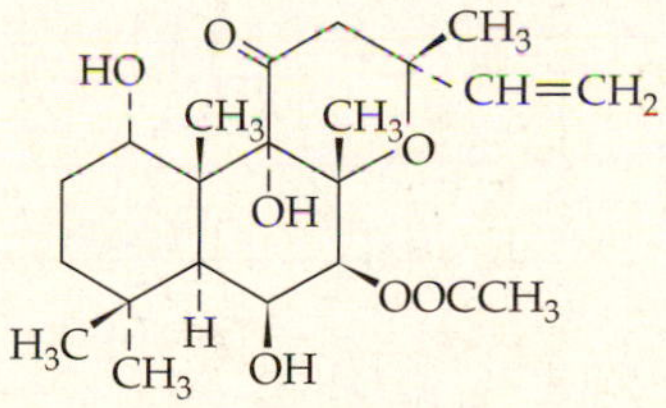

Forskolin

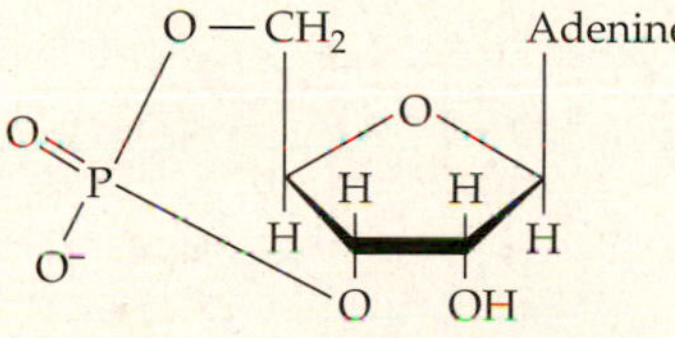

Cyclic AMP (cAMP)
Adenosine 3′, 5′-monophosphate

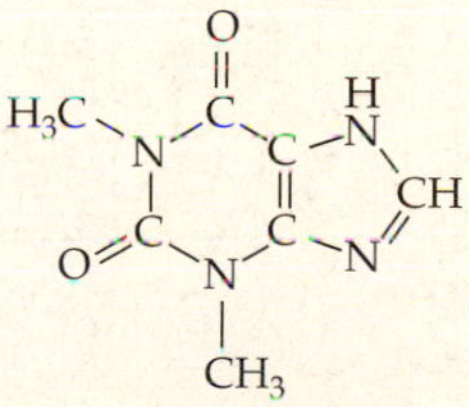

Theophylline
1, 3-Dimethylxanthine

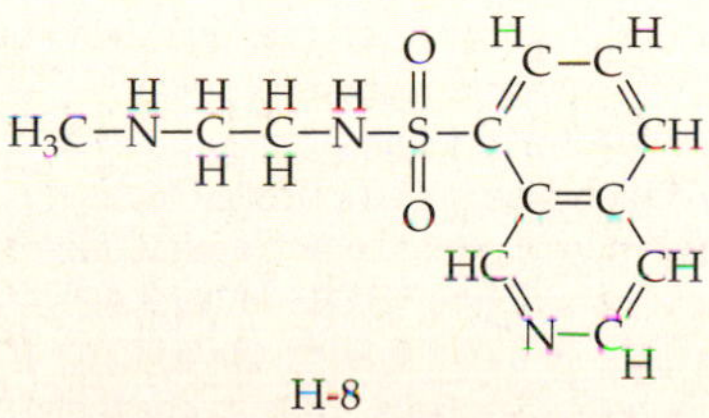

H-8
N-2-[(methylamino)ethyl]-5-isoquinolinesulfonamide

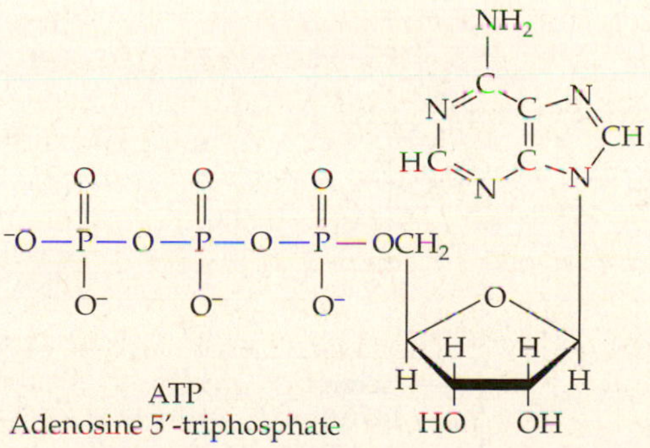

ATP
Adenosine 5′-triphosphate

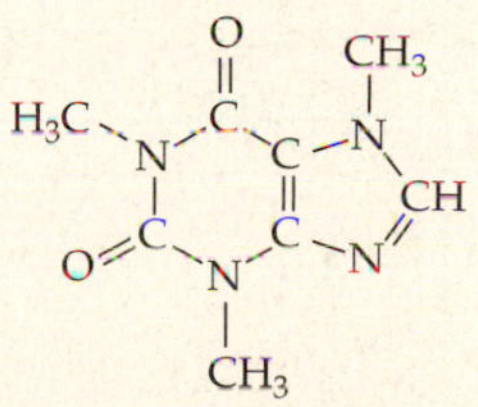

Caffeine
1, 3, 7-Trimethylxanthine

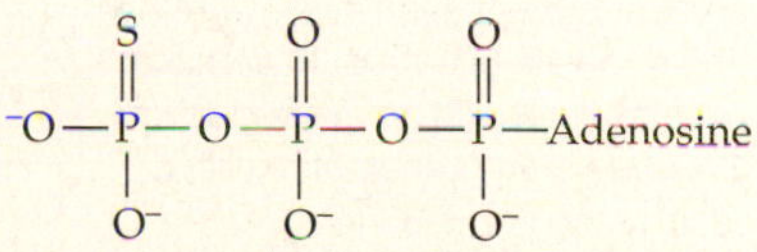

ATPγS
Adenosine 5′-*O*-[γ-thio] triphosphate

a target protein at widespread sites may be modulated by the occupation of a few receptors. Moreover, cAMP-dependent protein kinase can phosphorylate a variety of proteins and so modulate a broad spectrum of cellular processes. However, all of this takes time. Thus, in frog atrial muscle fibers, it takes about 5 seconds after activating the adrenergic β-receptors before the calcium current begins to increase. Most of this is taken up by the time needed to generate sufficient cAMP. In experiments in which a sudden increase in cAMP was generated by flash photolysis of the precursor o-nitrobenzyl cAMP, an increase in calcium current was seen within 150 ms.[46]

Once generated, cAMP is metabolized to AMP by cyclic nucleotide phosphodiesterase enzymes (PDE) (Box 12.4). Phosphodiesterase activity plays a crucial role in determining the duration of cAMP action and in limiting the spread of cAMP from its site of generation; this has the effect of compartmentalizing the response of cardiac cell calcium channels to nearby β-receptor stimulation.[47]

CYCLIC AMP DOES MORE THAN ACTIVATE ADENYLATE CYCLASE Not all of the effects of cAMP result from activation of adenylyl cyclase. Thus, the increase in **heart rate** produced by norepinephrine is due to a **direct** action of the cAMP on the channels responsible for the pacemaker current in the sinoatrial node (which is where the cardiac rhythm in the mammalian heart is generated).[48,49] This is an inward current through cation-permeant, hyperpolarization-activated cyclic nucleotide-gated (HCN) channels that are opened by membrane hyperpolarization.[50] Thus, as shown in Figure 12.14A, HCN

[46] Nargeot, J. et al. 1983. *Proc. Natl. Acad. Sci. USA* 80: 2385–2399.

[47] Fischmeister, R. et al. 2006. *Circ. Res.* 99: 816–828.

[48] Brown, H. F., DiFrancesco, D., and Noble, D. 1979. *Nature* 280: 235–236.

[49] DiFrancesco, D., and Tortura, D. P. 1991. *Nature* 351: 145–147.

[50] Accili, E. A. et al. 2002. *News Physiol. Sci.* 17: 32–37.

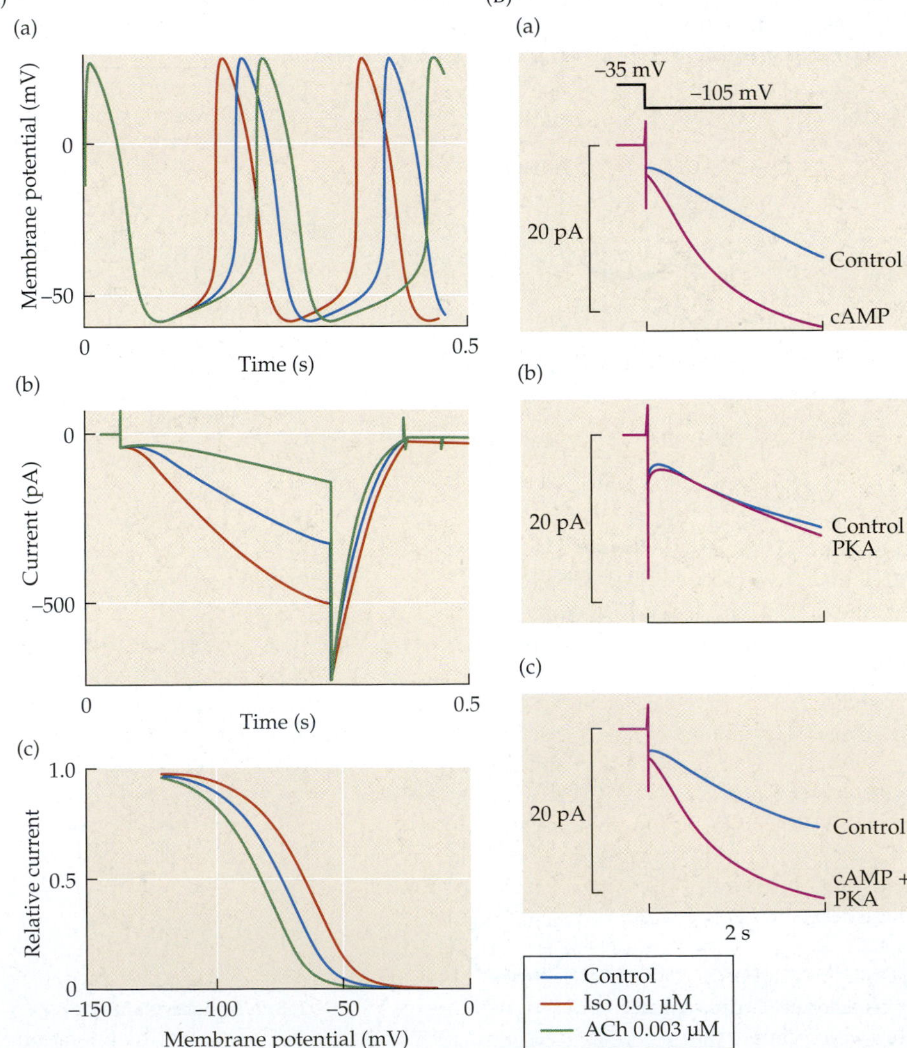

FIGURE 12.14 The Sinoatrial (SA) Node Heart Rate Is Regulated by β-Receptors through a Direct Effect of cAMP on the SA Pacemaker Current I_h. (A) Recordings from an intact rabbit SA node fiber. (a) The β-receptor agonist isoprenaline (Iso) increases the rate of spontaneous action potentials (red trace), whereas acetylcholine (ACh) reduces it (green trace). (b) ACh and Iso have opposite effects on the inward pacemaker current I_h (activated by stepping to –85 mV). (c) ACh and Iso shift the activation curves for I_h in opposite directions. Thus, h-channels open more readily with hyperpolarization in Iso solution, giving the faster-rising pacemaker current seen in (a), while ACh has the opposite effect. (B) Pacemaker h-current recorded in an excised inside-out SA node membrane patch with a 2-second step from –35 to –105 mV. Current is increased by cAMP (a) but not by the catalytic subunit of protein kinase A (PKA) (b), and PKA has no further effect after cAMP (c). (A after Accili et al., 2002; B after DiFrancesco and Tortora, 1991.)

BOX 12.4
Phosphatidylinositol-4,5-bisphosphate (PIP_2) and the phosphoinositide (PI) Cycle

Phosphatidylinositol-4,5-bisphosphate (PIP_2) only makes up about 1% of the phospholipids in neuronal cell membranes[51] but plays an important role in transmitter action. It is rapidly hydrolyzed through activation of the enzyme phospholipase C (PLC) when G_q coupled receptors are stimulated. PIP_2 itself, and the two products of its hydrolysis, inositol-1,4,5-trisphosphate (IP_3) and diacylglycerol (DAG) then act as second messengers that alter ion channel function and nerve cell activity (see Figure 12.15).

Most of the PIP_2 resides in the inner leaflet of the outer cell membrane.[51] It is composed of two fatty acyl chains (arachidonic acid and a fatty acid) which insert in the membrane, linked through the 1-phosphate to IP_3, which is negatively-charged and hydrophilic, so projects into the cytoplasm.

Synthesis of PIP_2 starts from inositol (in the cytosol) and phosphatidic acid (in the membrane), which are combined to form phosphatidylinositol (PI). This is then sequentially phosphorylated at the 4- and 5-positions on the inositol ring by PI4-kinase and PI5-kinase respectively to yield PIP_2. Phospholipase C (PLC) cleaves off the inositol ring with its three phosphates in the 1,4 and 5 positions to give IP_3, which goes into the cytoplasm, leaving diacylglycerol (DAG) in the membrane. PLC works very slowly at rest but is strongly activated by the GTP-bound α-subunit of G_q, and hence by metabotropic receptors that activate G_q. IP_3 is dephosphorylated by inositol phosphatase, eventually to inositol, while DAG is phosphorylated by DAG kinase to generate phosphatidic acid—thus, completing the PI cycle.

[51] Gamper, N. S., and Shapiro, M. S. 2007. *Nat. Rev. Neurosci.* 8: 1–14.

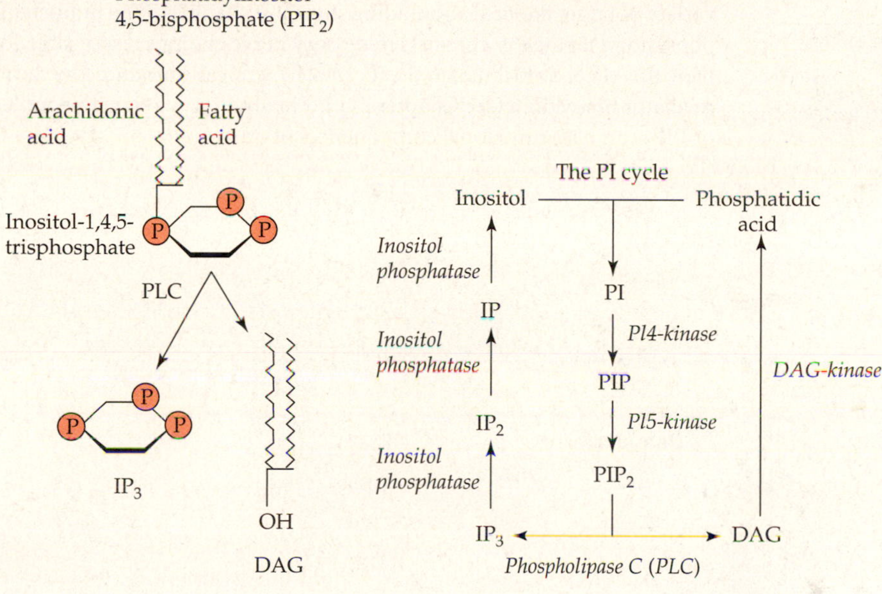

channels open during the hyperpolarization after an action potential in the cardiac sinoatrial node, and the resulting depolarization (the pacemaker potential) triggers the next action potential. An adrenergic agonist (in this case isoproterenol) causes an increase in the rate of pacemaker depolarization, and hence an increase in action potential frequency (a). This is because of an increase in the magnitude of the pacemaker current (b). Graph (c) shows that the increased current is due to a shift in voltage sensitivity: in isoproterenol, less hyperpolarization is required to activate the current. Records from a membrane patch, in Figure 12.14B, show that a similar increase in pacemaker current is produced by cAMP, but not by protein kinase A. Thus, cAMP increases the current by acting directly on the channels, rather than by phosphorylation. HCN channels are also present in many neurons. In some of these, they generate pacemaker currents like those in the heart, but they also have other functions.[52] Because of differences in channel subunit composition, not all of these neuronal HCN channels are affected by cAMP.[52]

SOME RECEPTORS INHIBIT ADENYLYL CYCLASE An important property of adenylyl cyclase is that it can be **inhibited** by transmitters that act on those metabotropic receptors which couple to the G protein G_i (i = inhibitory). Thus, in the heart, the activation

[52] Wahl-Schott, C., and Biel, M. 2009. *Cell. Mol. Life Sci.* 66: 470–494.

of G_i by ACh (via M2 muscarinic receptors) inhibits the activation of adenylyl cyclase by β-adrenoceptor stimulation. This action prevents the increase in cAMP and so inhibits the increase in calcium current amplitude produced by activating the β-receptors with isoprenaline.[53] By lowering cAMP, acetylcholine also has the opposite effect of β-adrenoceptor stimulation on the cardiac pacemaker current and action potential frequency, as shown in Figure 12.14A. This probably accounts for the opposing effects of sympathetic and vagal nerve activity on the heart rate at normal physiological rates of stimulation.[50]

G Protein Activation of Phospholipase C

Activation of phospholipase C (PLC) is the first step in another important G protein signaling pathway. This enzyme is preferentially activated by metabotropic receptors that couple to the G protein G_q (see Box 12.1). PLC catalyzes the hydrolysis of the membrane phospholipid phosphatidylinositol-4,5-bisphosphate (PIP_2). This yields two potential second messengers: inositol-1,4,5-trisphosphate (IP_3), which is water-soluble and enters the cytoplasm; and diacylglycerol (DAG), which stays in the membrane (Figure 12.15 and Box 12.4). IP_3 releases calcium ions from the endoplasmic reticulum, and hence contributes to various Ca^{2+}-dependent processes (see "Calcium as a Second Messenger," below). DAG remains in the membrane where it activates protein kinase C (PKC) to phosphorylate a variety of target molecules including some ion channels. One important role for PKC is to phosphorylate the ion channels in sensory nerve endings responsible for sensing burning pain; this effect contributes to the thermal hyperalgesia produced by the local inflammatory mediator bradykinin (see Chapter 21).[54] Finally, PIP_2 is a signaling molecule, so hydrolysis of PIP_2 can have functional consequences of its own.

[53] Fischmeister, R., and Hartzell, H. C. 1986. *J. Physiol.* 376: 183–202.

[54] Cesare, P., and McNaughton, P. A. 1996. *Proc. Natl. Acad. Sci. USA* 93: 15435–15439.

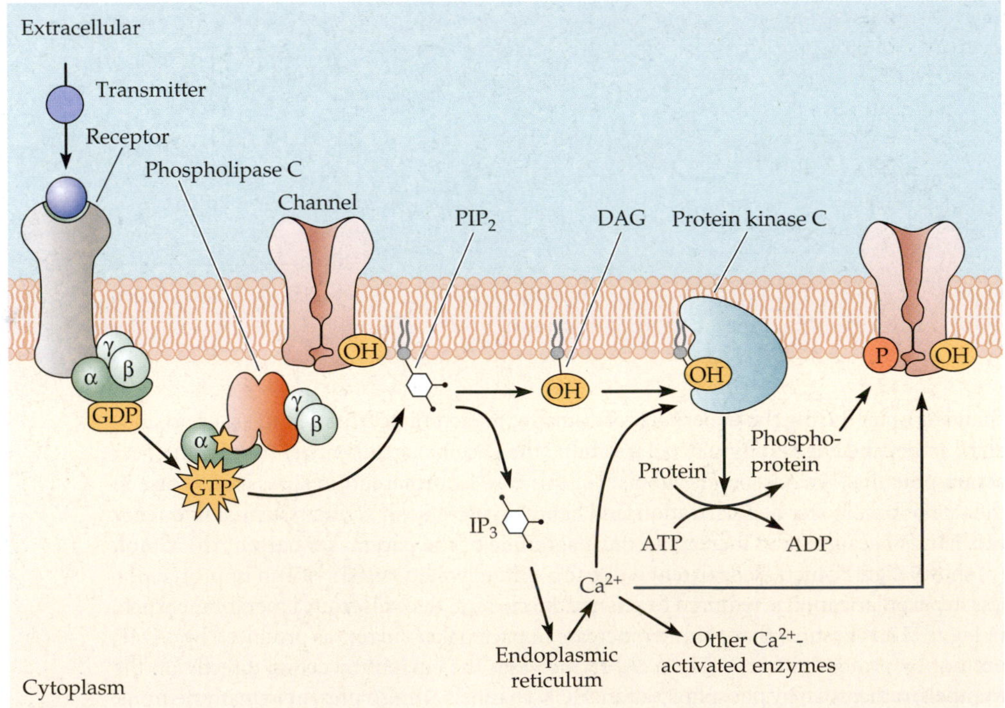

FIGURE 12.15 Signaling by PIP_2 and by Products of PIP_2 Hydrolysis. When a transmitter stimulates a receptor coupled to G_q, the GTP-bound G_q protein activates phospholipase C to induce the hydrolysis of phosphatidylinositol-4',5'-bisphosphate (PIP_2). This releases two intracellular second messengers: diacylglycerol (DAG) and inositol 1,4,5-trisphosphate (IP_3). IP_3 releases calcium from the endoplasmic reticulum into the cytoplasm. DAG and calcium together activate protein kinase C (PKC). PKC catalyzes increased protein phosphorylation. Ion channel function may be modified (1) by loss of PIP_2, (2) by PKC-induced phosphorylation (direct or indirect) or (3) by an effect of calcium (direct or indirect).

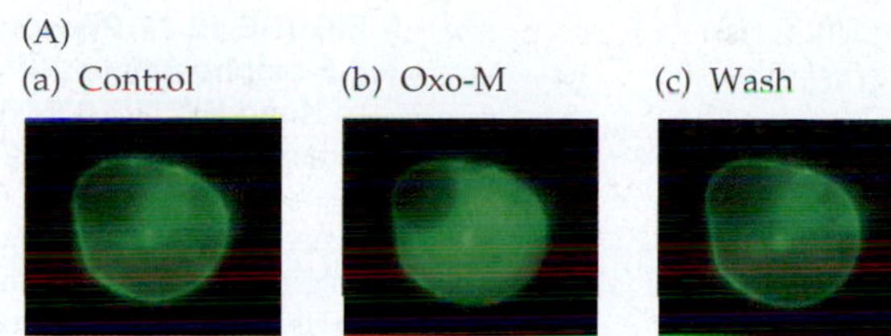

FIGURE 12.16 Hydrolysis of Phosphatidylinositol-4,5-bisphosphate (PIP_2). PIP_2 hydrolysis accompanies cholinergic inhibition of M-current in an isolated rat sympathetic neuron. (A) The GFP-tagged PH-domain of phospholipase Cδ (GFP-PLCδ-PH) is used as a fluorescent probe to observe PIP_2 hydrolysis. At rest (a), the probe binds to PIP_2 in the membrane. The muscarinic agonist oxotremorine-M (Oxo-M) (b) stimulates PIP_2 hydrolysis and membrane PIP_2 falls, so the probe leaves the membrane and goes into the cytoplasm where it binds to free inositol-1,4,5-trisphosphate (IP_3). On washing out the Oxo-M (c), PIP_2 is resynthesized so the probe returns to the membrane. (B) The time course of GFP-PLCδ-PH movement, registered as the change of fluorescence in a region of the cytoplasm. Fluorescence is expressed as fractional increase over original baseline, $\Delta F/F_O$. (C) The time-course of M-current inhibition and recovery, recorded as the loss and recovery of outward K^+ current at −20 mV. The membrane was hyperpolarized to −50 mV for 2 seconds every 15 seconds to check the change of conductance (given by downward current deflections). Note that the change of M-current closely follows the fluorescence change. (After Winks et al., 2005.)

Direct Actions of PIP_2

PIP_2 is required for the proper function of many membrane proteins, including a number of ion channels.[51] One such PIP_2-regulated ion channel is the **M-channel**, which is a voltage-gated K^+ channel that regulates the excitability of many central and peripheral neurons.[55] M-channel currents were first described in sympathetic neurons;[56] in these cells, activation of muscarinic receptors results in closure of M-channels and membrane depolarization (see Chapter 17). Channel closure following stimulation of muscarinic receptors is tightly coupled to the hydrolysis of PIP_2. This effect was shown using a fluorescent probe that binds to PIP_2 in the membrane (Figure 12.16). When a sympathetic neuron expressing the probe is challenged with a muscarinic agonist, the probe leaves the membrane and moves into the cytosol at the same time as the M-current is reduced. It then moves back to the membrane as the M-current recovers.[57] This particular probe—the pleckstrin-homology (PH) domain of phospholipase-Cδ, tagged with the jellyfish green fluorescent protein (GFP), so that it can be seen—binds to both PIP_2 and to its hydrolysis product IP_3. Hence, translocation might result from the loss of PIP_2 from the membrane or gain of IP_3 in the cytoplasm. However, other evidence shows that channel closure results from the loss of PIP_2, not the formation of IP_3. Hence, current inhibition is reduced when the amount of PIP_2 in the membrane is increased, so that it is more difficult to deplete as a consequence of overexpressing its synthetic enzyme phosphatidylinositol-4-phosphate-5-kinase.[57,58] Similarly, a probe that binds to PIP_2 but not to IP_3 also shows a dissociation from the membrane.[59]

The direct effect of PIP_2 on M-channel activity is illustrated in Figure 12.17. The constituent potassium channel subunits that make up M-channels (Kv7.2 and Kv7.3;[60] see Chapter 5) were expressed in Chinese Hamster Ovary (CHO) cells.[61] When the membrane potential in a cell-attached patch electrode was set to 0mV, there was modest channel activity. After pulling the patch from the cell, thereby exposing the inner membrane surface to the bathing solution, activity declined but was restored by adding a PIP_2 analogue to the bathing solution. Using this expression system, the kinetics of all of the steps linking the muscarinic receptor to the closure of M-channels have now been examined in detail by B. Hille and his colleagues.[62–64]

A similar signaling pathway appears to be responsible for the muscarinic cholinergic inhibition of M-current (and consequent increased excitability) in mammalian central neurons.[65] Indeed, loss of PIP_2 contributes to the inhibition of a wide variety of PIP_2-dependent

[55] Brown, D. A., and Passmore, G. M. 2009. *Brit. J. Pharmacol.* 156: 1185–1195.

[56] Brown, D. A., and Adams, P. R. 1980. *Nature* 283: 673–676.

[57] Winks, J. S. et al. 2005. *J. Neurosci.* 25: 3400–3413.

[58] Suh, B. C. et al. 2006. *Science* 314: 1454–1457.

[59] Hughes, S. et al. 2007. *Pflügers Arch.* 455: 115–124.

[60] Wang, H-S. et al. 1998. *Science* 282: 1890–1893.

[61] Li, Y. et al. 2005. *J. Neurosci.* 25: 9825–9835.

[62] Suh, B. C. et al. 2004. *J. Gen. Physiol.* 123: 663–683.

[63] Falkenburger, B. H., Jensen, J. B., and Hille, B. 2010. *J. Gen. Physiol.* 135: 81–97.

[64] Falkenburger, B. H., Jensen, J. B., and Hille, B. 2010. *J. Gen. Physiol.* 135: 99–114.

[65] Shen, W. et al. 2005. *J. Neurosci.* 25: 7449–7458.

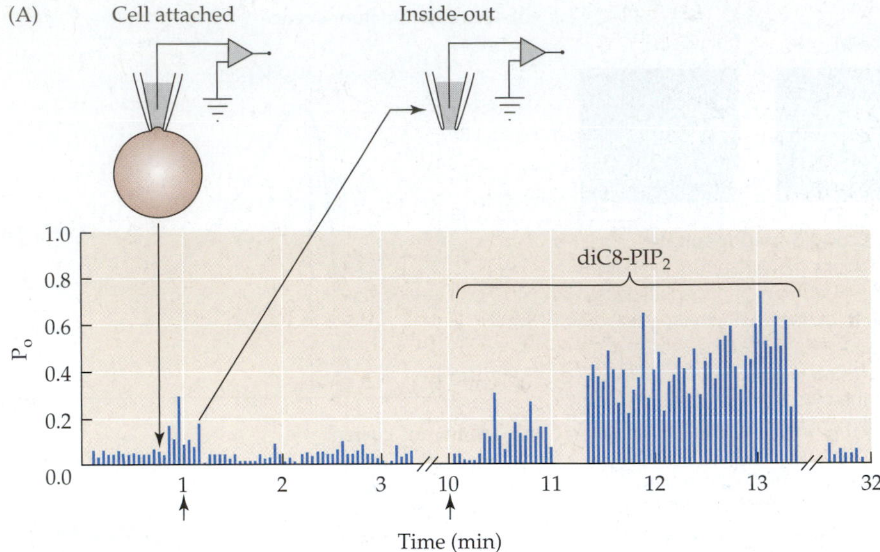

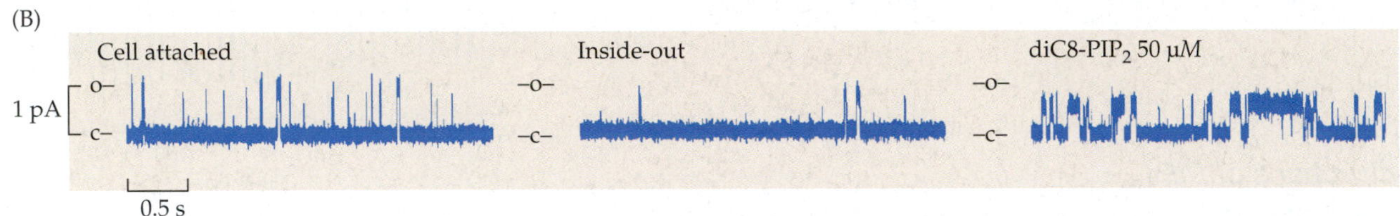

FIGURE 12.17 Phosphatidylinositol-4,5-bisphosphate (PIP_2) Is Necessary to Keep Kv7.2/7.3 (M-Type) Potassium Channels Open. Kv7.2 and Kv7.3 mRNAs were co-expressed in a Chinese hamster ovary (CHO) cell and single-channel activity recorded using a cell-attached pipette set to 0 mV membrane potential. With the pipette on-cell, the channel shows a modest open probability (P_o) of 0.1–0.2. When the membrane patch was excised into inside-out mode (inside surface facing the bath solution), channel activity was lost. Addition of 50 μ*M* dioctanoyl-phosphatidylinositol-4,5-bisphosphate (diC8-PIP_2) to the bath solution restored and increased activity. (A) A continuous time-plot of open probability recorded in 3-second runs is shown, while (B) shows sample currents at a faster speed. (After Li et al., 2005.)

ion channels by muscarinic (and other) receptors that stimulate its hydrolysis, including inward rectifier K^+ channels, twin-pore K^+ channels, Ca^{2+} channels, and Trp channels.[51]

G Protein Activation of Phospholipase A_2

Another target of G protein action is phospholipase A_2. This enzyme acts on DAG and on certain membrane phospholipids, such as PIP_2, to release the fatty acid arachidonic acid.[66] Arachidonic acid modulates neuronal signaling by direct effects on ion channels,[67] indirectly through activation of protein kinase C,[68] and through the actions of its metabolites.[69] In *Aplysia* sensory neurons, for example, arachidonic acid is produced in response to receptor activation by the peptide Phe-Met-Arg-Phe-NH2 (FMRFamide) and is metabolized to 12-hydroperoxy-5,8,10,14-eicosatetraenoic acid (12-HPETE). 12-HPETE, in turn, acts to open S-current potassium channels.[70,71]

Convergence and Divergence of Signals Generated by Indirectly Coupled Receptors

As noted earlier (see Box 12.1), a large number of different G protein–coupled receptors converge on a much smaller number of G proteins. Accordingly, they can all potentially produce the same effects on a neuron. Conversely, activation of one type of G protein can, through various pathways, potentially affect several different ion channels and other proteins in the cell. For example, the activity of a single rat superior cervical sympathetic neuron can be modulated by at least nine transmitters acting through five G protein–coupled pathways that influence two calcium channels and at least one potassium channel;[72] while in different guinea-pig sympathetic neurons, muscarinic receptors can affect five different potassium currents.[73]

Selectivity and variation in the cell's response depends on: (1) what receptors are present in that particular cell; (2) which transmitters or hormones the cell actually sees; (3) what

[66] Bazan, N. G. 2006. In *Basic Neurochemistry: Molecular, Cellular and Medical Aspects*, 7th ed. Lippincott-Raven, Philadelphia, pp. 731–741.

[67] Meves, H. 2008. *Brit. J. Pharmacol.* 155: 4–16.

[68] Majewski, H., and Iannazzo, L. 1998. *Prog. Neurobiol.* 55: 463–476.

[69] Piomelli, D. 2001. *Trends Neurosci.* 22: 17–19.

[70] Piomelli, D. et al. 1987. *Nature* 328: 38–43.

[71] Buttner, N., Siegelbaum, S. A., and Volterra, A. 1989. *Nature* 342: 553–555.

[72] Hille, B. 1994. *Trends Neurosci.* 17: 531–536.

[73] Cassell, J. F., and McLachlan, E. M. 1987. *Brit. J. Pharmacol.* 91: 259–261.

effectors are present; and (4) how the receptors, G proteins, and effectors are arranged in the cell membrane. As an example of selectivity in response, GABA (acting on $GABA_B$ receptors) can activate both G_i and G_o G proteins, so as to cause potassium channels to open and calcium channels to close. Both effects occur postsynaptically, but only calcium channel inhibition is seen presynaptically. The reason is that the presynaptic nerve endings do not possess the necessary G protein activated potassium channels.[74] An example of selectivity between similar receptors is provided again by the sympathetic ganglion. In mammals, each sympathetic neuron contains both M2 and M4 muscarinic acetylcholine receptors, each of which can potentially activate both G_i and G_o G proteins. However, in the rat, only the M4 receptor activates G_o (to inhibit calcium channels), while only the M2 receptor activates G_i (to open potassium channels).[25,75] These cells also show segregation between different receptors that couple to G_q and hydrolyze PIP_2. Thus, the hydrolysis resulting from activation of bradykinin receptors is accompanied by an increase in IP_3 and the release of calcium from the endoplasmic reticulum,[76] whereas no release of calcium occurs after stimulating the muscarinic ACh receptors even though they hydrolyze just as much PIP_2 and produce as much IP_3.[57,59,76] This difference is because the bradykinin receptor is held in very close association with the inositol trisphosphate receptor through the actin cytoskeleton, to form what may be termed a signaling microdomain.[77] The assembly of receptors, G proteins, and their immediate effectors into multi-protein complexes—aided by cytoskeletal or scaffolding proteins—seems to be quite a common feature of signaling through G protein–coupled receptors.[78] Likewise, ion channels are frequently assembled with signaling protein complexes.[79] Such complexes serve both to increase signaling efficiency and to segregate different receptor and ion channel signaling pathways.

Divergence of G protein signaling enables a transmitter to generate an integrated response of a neuron or an effector cell to stimulation of a metabotropic receptor. For example, sympathetic nerve stimulation releases norepinephrine, which acts on β-adrenergic receptors in the heart and activates adenylate cyclase (see above). This not only modifies the calcium channels to promote an increased entry of calcium during each heart beat; it also affects various processes involved in the intracellular storage and release of calcium, and in the contractile mechanism itself, to produce a fully coordinated response of the heart to fear or excitement.[39] In the long term, with persistent or repeated stimulation, responses to metabotropic receptor stimulation can extend to changes in gene transcription, to produce long-lasting structural and functional changes in the innervated neurons or effectors.

Retrograde Signaling via Endocannabinoids

Endocannabinoids (Box 12.5) are further products of phospholipid metabolism in nerve cell membranes. They are of obvious interest in view of widespread recreational use of cannabis but—more importantly for our purposes—are now known to mediate a widespread form of physiological inhibition in the CNS.[80] This contributes to several aspects of CNS function, including the sensitization to sensory stimulation that occurs in chronic or neuropathic pain,[81] and forms the basis for the therapeutic applications of cannabinoid drugs.[82]

Unlike diacylglycerol (DAG) or inositol trisphosphate, endocannabinoids are released from nerve cells into the extracellular space following stimuli that increase intracellular calcium (including action potentials) or activate phospholipase C (such as G_q-coupled metabotropic receptors). When released in the brain they induce retrograde inhibition of transmitter release by stimulating CB1 cannabinoid receptors (see Box 12.5) on presynaptic nerve terminals.[83,84] Thus, they introduce two new concepts: **retrograde synaptic signaling**, and the release of a transmitter-like messenger that is not stored in synaptic vesicles but is synthesized on demand.

Figure 12.18A illustrates this retrograde action in a hippocampal pyramidal cell.[85] The records show ongoing inhibitory synaptic currents (downward deflections) due to spontaneous release of GABA from the presynaptic terminals. Following a train of action potentials in the cell the spontaeous inhbitory activity is almost totally suppressed, and takes several seconds to recover. Two additional features characterize this phenomenon, known as **depolarization-induced suppression of inhibition (DSI)**. First, the postsynaptic response to extrinsically applied GABA was unaffected, so the inhibition of transmission

[74] Takahashi, T., Kajikawa, Y., and Tsujimoto, T. 1998. *J. Neurosci.* 18: 3138–3146.

[75] Fernandez-Fernandez, J. M. et al. 1999. *J. Physiol.* 515: 631–637.

[76] Delmas, P. et al. 2002. *Neuron* 34: 209–220.

[77] Delmas, P., Crest, M., and Brown, D. A. 2004. *Trends Neurosci.* 27: 41–47.

[78] Bockaert, J. et al. 2010. *Annu. Rev. Pharmacol. Toxicol.* 50: 89–109.

[79] Levitan, I. B. 2006. *Nat. Neurosci.* 9: 305–310.

[80] Hashimotodani, Y., Ohno-Shosaku, T., and Kano, M. 2007. *Neuroscientist* 13: 127–137.

[81] Pernia-Andrade, A. J. et al. 2009. *Science* 325: 760–764.

[82] Iversen, L. 2003. *Brain* 126: 1252–1270.

[83] Wilson, R. I., and Nicoll, R. A. 2002. *Science* 296: 678–682.

[84] Kano, M. et al. 2009. *Physiol. Rev.* 89: 309–380.

[85] Pitler, T. A., and Alger, B. E. 1992. *J. Neurosci.* 12: 4122–4132.

(A)

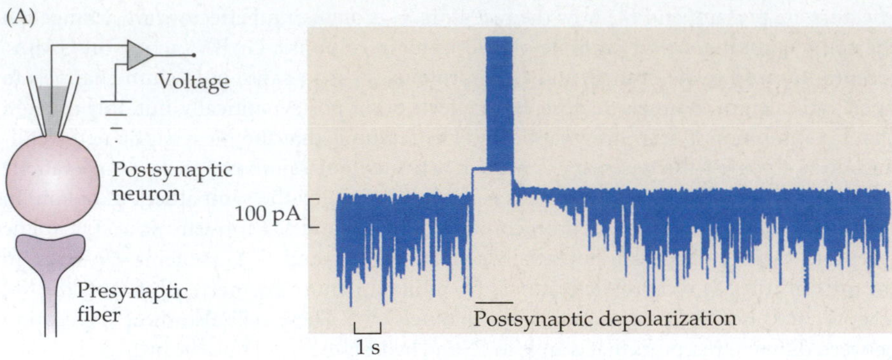

(B)

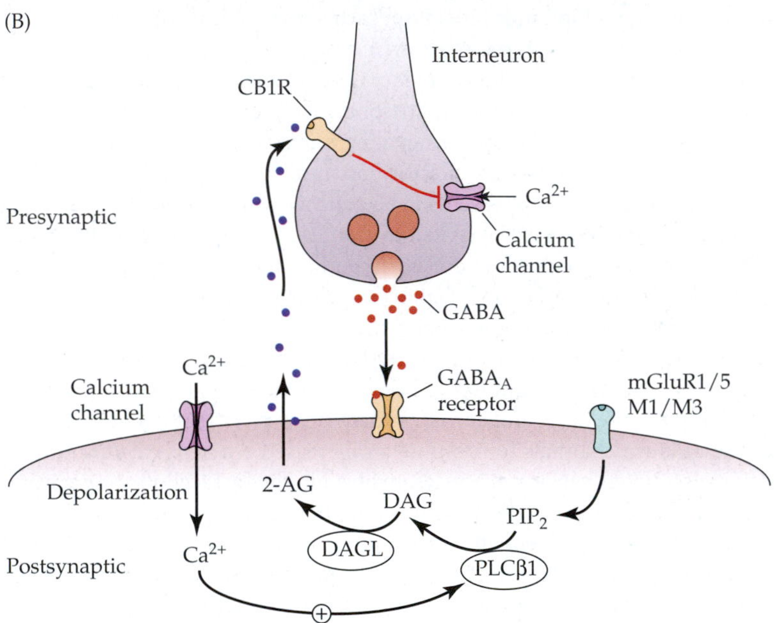

FIGURE 12.18 An Endocannabinoid Acts as a Retrograde Messenger for Depolarization-Induced Suppression of Inhibition (DSI). (A) Recording of spontaneous GABA-mediated inhibitory postsynaptic currents (IPSCs) in a hippocampal pyramidal neuron held at –80 mV. The cell was depolarized for 1 second to induce a train of action potentials, which transiently suppresses the IPSCs. This effect is called depolarization-induced suppression of inhibition, or DSI. (B) Suggested explanation for hippocampal DSI. Postsynaptic depolarization (or action potential activity) opens Ca^{2+} channels to produce an influx of Ca^{2+} ions. This activates phospholipase Cβ1 (PLCβ1) to hydrolyze phosphatidylinositol-4'5'-bisphosphate (PIP_2), yielding diacylglycerol (DAG). DAG is converted to 2-arachidonoyl glycerol (2-AG) by diacylglycerol lipase (DAGL). 2-AG is then released into the interstitial space and activates the CB1 cannabinoid receptor (CB1R) on the presynaptic terminals. The CB1 receptor activates G_i/G_o and their βγ-subunits then inhibit Ca^{2+} entry through presynaptic Ca^{2+} channels; it may also inhibit synaptic vesicle fusion. Metabotropic glutamate receptors (mGluR1 or mGluR5) or M1 or M3 muscarinic acetylcholine receptors can also independently activate phospholipase C via the G protein G_q, to generate 2-arachidonoyl glycerol through the same biochemical pathway. (A after Pitler and Alger, 1992; B after Hashimotodani et al., 2006.)

is presynaptic in origin. Second, the response was suppressed by injecting an intracellular calcium buffer, showing that the effect depended on elevation of calcium concentration in the cytosol. It has been suggested that the retrograde effect is initiated by calcium entry into the cell through voltage-sensitive calcium channels. A similar retrograde inhibition of both inhibitory and excitatory transmission, depolarization-induced suppression of excitation (DSE), has been described in cerebellar Purkinje cells.[86,87]

The identity of the retrograde messenger as an endocannabinoid was resolved some 10 years later by three independent groups working on different central synapses.[88–90] It was found that retrograde inhibition could be imitated by applying cannabinoids and blocked

[86] Llano, I., Leresche, N., and Marty, A. 1991. *Neuron* 6: 564–674.

[87] Kreitzer, A. C., and Regehr, W. G. 2001. *Neuron* 29: 717–727.

[88] Ohno-Shosaku, T., Maejima, T., and Kano, A. 2001. *Neuron* 29: 729–738.

[89] Wilson, R. I., and Nicoll, R. A. 2001. *Nature* 410: 588–592.

[90] Varma, N. et al. 2001. *J. Neurosci.* 21: RC188 (1–5).

BOX 12.5
Formation and Metabolism of Endocannabinoids

Two endocannabinoids are known—anandamide and 2-arachidonoyl glycerol (2-AG). They both contain arachidonic acid (see Box 12.4) linked to ethanolamine and glycerol, respectively, and yield arachidonic acid upon metabolism by fatty acid amide hydrolase or monoacylglycerol lipase (MGL). They are formed from Ca^{2+}-dependent hydrolysis of membrane phospholipids (N-arachidonoyl phosphatidylethanolamine or phosphatidylinositol) by phospholipase D or phospholipase C, respectively.

The principal endocannabinoid in the mammalian nervous system is 2-AG. It is released into the interstitial space where it can act on CB1 cannabinoid receptors. These are G protein–coupled receptors (GPCRs) that preferentially couple to the G proteins G_i and G_o. As a result, they can inhibit adenylyl cyclase, activate GIRK (Kv3) potassium channels, and inhibit Ca_V2 calcium channels. Effects of endocannabinoids are replicated by Δ^9-tetrahydrocannabinol (the active principle of cannabis or marijuana) and the synthetic water-soluble compound WIN-5122, and are antagonized by rimonabant.

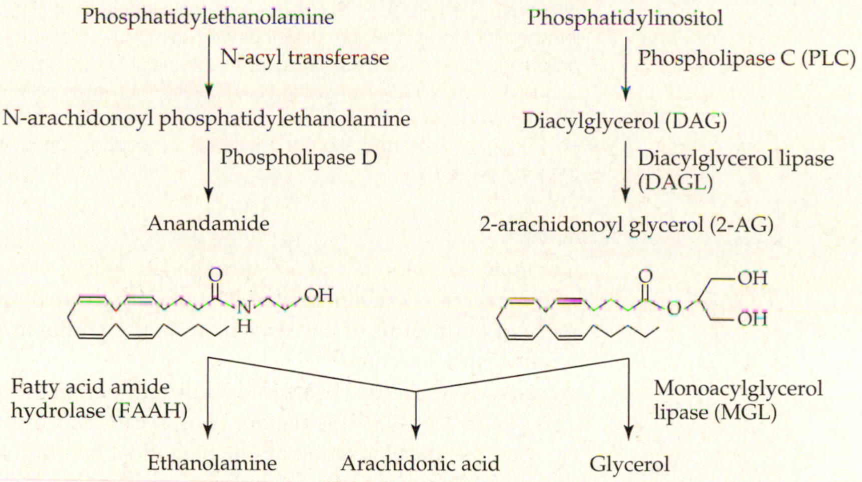

Phosphatidylethanolamine
N-acyl transferase
N-arachidonoyl phosphatidylethanolamine
Phospholipase D
Anandamide

Phosphatidylinositol
Phospholipase C (PLC)
Diacylglycerol (DAG)
Diacylglycerol lipase (DAGL)
2-arachidonoyl glycerol (2-AG)

Fatty acid amide hydrolase (FAAH)
Monoacylglycerol lipase (MGL)
Ethanolamine Arachidonic acid Glycerol

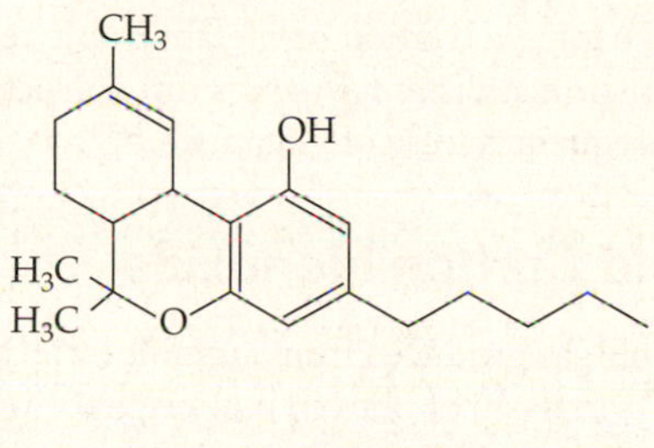

Δ9-tetrahydrocannabinol 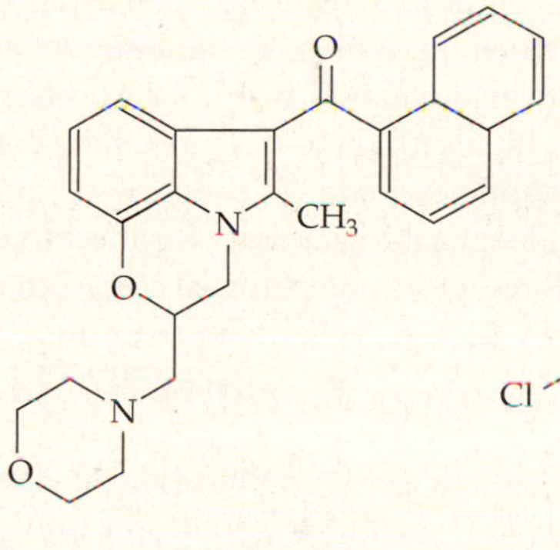 WIN 55212-2 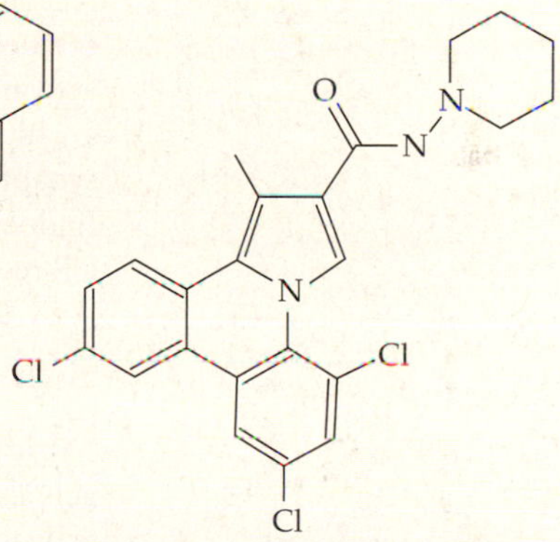 Sr141716A (rimonabant)

by CB1 receptor antagonists, and that it was also prevented in mice in which the CB1 receptor has been genetically deleted. The mechanism proposed for DSI in the hippocampus is summarized in Figure 12.18B. Calcium entry enhances the activity of PLCβ1, which, in turn, leads to hydrolysis of PIP_2 and sequential generation of DAG and 2-arachidonoyl glycerol (2-AG) (Box 12.5). 2-AG released from the cell then binds to CB1 receptors[91] on the presynaptic terminal to inhibit transmitter release. CB1 receptors interact with G_i and G_o proteins to cause inhibition of calcium channels.[92–94] Because calcium influx into the presynaptic terminals is reduced, transmitter release is attenuated.[94]

The release of the endocannabinoid 2-AG is also increased when PLCβ1 is activated by stimulating postsynaptic G_q-coupled metabotropic M1-muscarinic or metabotropic

[91] Matsuda, L. et al. 1990. *Nature* 346: 561–564.

[92] MacKie, K., and Hille, B. 1992. *Proc. Natl. Acad. Sci. USA* 89: 3825–3829.

[93] Caulfield, M. P., and Brown, D. A. 1992. *Brit. J. Pharmacol.* 106: 231–232.

[94] Kushmerick, C. et al. 2004. *J. Neurosci.* 24: 5955–5965.

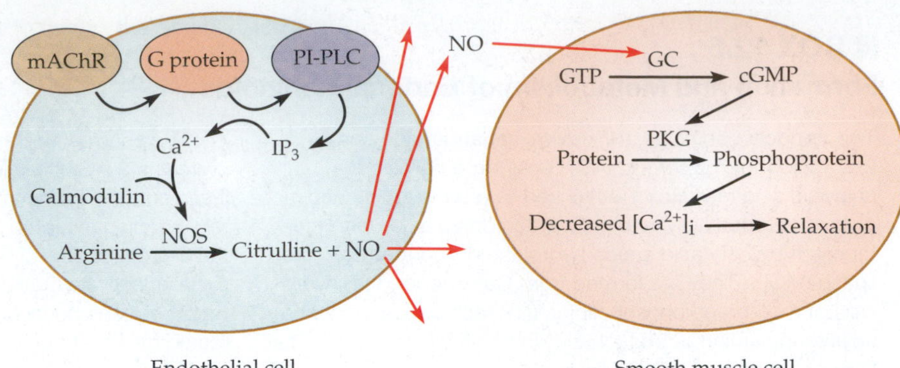

FIGURE 12.19 Paracrine Signaling by Release of Nitric Oxide. ACh binds to muscarinic receptors (mAChR) on vascular endothelial cells, activating phosphatidylinositide-specific phospholipase C (PI-PLC). PI-PLC forms inositol trisphosphate (IP_3), which releases calcium from intracellular stores. Calcium, together with calmodulin, activates nitric oxide synthase (NOS), producing nitric oxide (NO). NO diffuses into neighboring smooth muscle cells and stimulates guanylyl cyclase (GC), increasing cGMP. cGMP activates cGMP-dependent protein kinase (PKG). The resulting increases in protein phosphorylation leads to a decrease in intracellular calcium concentration, causing relaxation. NO is rapidly degraded, so that it affects only nearby cells—hence the term paracrine.

glutamate receptors 1 and 5,[84] through IP_3-induced calcium release. Since stimulation of these receptors reinforces the effect of Ca^{2+} channel opening, the simultaneous occurrence of both forms of postsynaptic response reinforce each other, providing a form of "coincidence detection."[95]

2-AG is synthesized and released from any part of the nerve cell membrane that detects a rise in Ca^{2+} or PLC stimulation. Thus, it can exert diffuse effects. In the hippocampus, for example, depolarization of one pyramidal neuron can suppress inhibitory synapses on a neighboring neuron.[89] Its effect is primarily on inhibitory terminals, where CB1 receptors are most concentrated, rather than on excitatory terminals.[96] In contrast, in the cerebellum, retrograde inhibition of excitatory transmission is normally confined to the activated synapses, because the postsynaptic calcium signals in the Purkinje cell dendrites are highly localized.[97] A more diffuse effect may occur during intense stimulation or when synaptic excitation is coupled with postsynaptic activation of metabotropic receptors.[84] Endocannabinoids may also affect transmission at distant synapses through activation of CB receptors on neuroglial cells, with consequent release of glutamate.[98]

Signaling via Nitric Oxide and Carbon Monoxide

Nitric oxide (NO), a water- and lipid-soluble gas produced from arginine by NO synthase, acts as a transmitter by diffusing from the cytoplasm of one cell into neighboring cells and activating guanylyl cyclase.[99] NO was first characterized as an important regulator of blood pressure, mediating the vasodilatation caused by acetylcholine.[100,101] Production of NO is initiated by the interaction of ACh with muscarinic receptors on vascular endothelial cells, which in turn leads to activation of phospholipase C, formation of IP_3, and release of calcium from intracellular stores (Figure 12.19). Calcium combines with calmodulin and activates NO synthase, producing NO. NO diffuses into neighboring smooth muscle cells and stimulates the soluble form of guanylyl cyclase, causing an increase in cGMP. The cGMP, in turn, activates a cGMP-dependent protein kinase. The resulting increases in protein phosphorylation modulate the activity of potassium and calcium channels and calcium pumps, leading to a decrease in intracellular calcium concentration and reduced Ca^{2+} sensitivity of the contractile proteins, which causes relaxation. NO is inactivated within seconds by reaction with superoxides and by formation of complexes with heme-containing proteins such as hemoglobin. The role of NO in relaxing blood vessel smooth muscles explains the vasodilator and hypotensive action of glyceryl trinitrate and other nitro-compounds, which release free NO and so act as NO donors. It also explains why

95 Hashimotodani, Y. et al. 2005. *Neuron* 45: 257–268.

96 Katona, I. et al. 2000. *Neuroscience* 100: 797–804.

97 Brown, S. P., Brenowitz, S. D., and Regehr, W. D. 2003. *Nat. Neurosci.* 10: 1047–1058.

98 Navarrete, M., and Araque, A. 2010. *Neuron* 68: 113–126.

99 Ignarro, J. 1990. *Ann. Rev. Physiol.* 30: 535–560.

100 Furchgott, R. F., and Zawadzki, J. V. 1980. *Nature* 288: 373–376.

101 Palmer, R. M. J., Ferrige, J., and Moncada, S. 1987. *Nature* 324: 524–526.

phosphodiesterase-5 (PDE-5) inhibitors such as sildenafil (trade name, Viagra) enhance and prolong penile erections: erection results from NO-induced cGMP-mediated penile vasodilation, and cGMP is selectively degraded to 5′-GMP by PDE-5, so PDE-5 inhibitors reduce the rate of cGMP inactivation.

The role of NO as a **neurotransmitter** was first established in the peripheral autonomic nervous system,[102,103] where it is released from **non-adrenergic, non-cholinergic (NANC)** fibers to relax smooth muscle in the same manner as NO released from endothelial cells. In this case, however, the stimulus for NO synthetase is Ca^{2+} entry into the nerve terminals during the presynaptic action potential. A role in **brain function** was first suggested from experiments on the cerebellum, in which NO was shown to be responsible for the increase in cGMP produced by stimulating the NMDA subtype of ionotropic glutamate receptors (see Chapter 5).[104,105] The NMDA receptor in cerebellar granule cells is closely associated with NO synthase, so that the enzyme is activated by Ca^{2+} ions entering through the NMDA channels. The released NO then stimulates guanylyl cyclase in the Purkinje cells or in adjacent astroglial cells.[106] In Purkinje cells, the cGMP thus formed activates the cGMP-dependent protein kinase, leading to phosphorylation and endocytosis of the postsynaptic AMPA receptor that subserves normal excitatory transmission from the parallel fibers, thereby complementing the effect of endocannabinoid in suppressing transmitter release from the presynaptic terminals. Conversely, in the hippocampus, postsynaptically generated NO may act as a retrograde messenger to contribute to long-term potentiation of excitatory transmission.[106] Because it is a freely-diffusible gas, NO formed in one neuron can affect the function of neighboring neurons and synapses, up to distances of 100 μm away.[107]

As with cAMP, not all of the effects of cGMP are due to activation of cGMP-dependent protein kinase. For example, the cGMP formed in the retina following the release of NO from the illuminated retina directly opens cyclic nucleotide-gated (CNG) cation channels in cone photoreceptors; the influx of calcium through these cation channels then increases glutamate release from the cone cells.[108] Also, NO itself may produce other effects than stimulation of guanylate cyclase, such as S-nitrosylation of the ion channel protein by free NO.[109]

Another endogenously produced gas with properties similar to those of NO is carbon monoxide (CO).[110] CO is formed from the degradation of heme by the enzyme heme oxygenase and, like NO, can activate guanylate cyclase to produce cGMP. In the intestinal nervous system, release of CO seems to co-operate with NO in generating the NANC-relaxation of intestinal smooth muscle since relaxation produced by stimulating the enteric neurons is equally reduced in mice lacking NO synthetase or heme oxygenase. CO may also have a role in signaling in the CNS, since inhibition of heme oxygenase (like inhibition of NO synthetase) can prevent or reverse the induction of long-term potentiation in the hippocampus.[111,112]

Like the endocannabinoids, NO and CO cannot be stored in synaptic vesicles, but are synthesized and released on demand. They then diffuse indiscriminately from the site at which they are produced into neighboring cells, their spread being limited only by their short lifespan. This type of signaling, which is an intermediate between neurotransmission and the release of hormones into the bloodstream by endocrine organs, has been termed **paracrine transmission** or, in the brain, **volume transmission.** Clearly, specificity in the effects of such signals depends on the distribution and properties of enzymes activated or inhibited by NO and CO.

Calcium as an Intracellular Second Messenger

Calcium is a ubiquitous second messenger.[113–115] The concentration of free Ca^{2+} ions in the cytoplasm at rest is around 100 nM (i.e., about 1/10,000th of that in the extracellular fluid). One reason for this low concentration of free calcium ions is that most of the cytoplasmic calcium is reversibly bound to (buffered by) calcium-binding proteins. Around 98% to 99.8% of the total calcium in the cytoplasm is buffered in this way.[116–118] Cytoplasmic calcium is in dynamic equilibrium with extracellular calcium, and with intracellular calcium stores, principally within the endoplasmic reticulum (ER) and mitochondria. The different mechanisms for regulating cytoplasmic calcium concentration are summarized in Figure 12.20.

Since the intracellular concentration of Ca^{2+} ions is so low, there is plenty of scope for increasing it. From the viewpoint of transmitter action, two mechanisms are predominant. First,

[102] Gillespie, J. S., Liu, X. R., and Martin, W. 1989. *Brit. J. Pharmacol.* 98: 1080–1082.

[103] Bult, H. et al. 1990. *Nature* 345: 346–347.

[104] Garthwaite, J., Charles, S. L., and Chess-Williams, R. 1988. *Nature* 336: 385–388.

[105] Bredt, D. S., and Snyder, S. H. 1989. *Proc. Natl. Acad. Sci. USA* 86: 9030–9033.

[106] Garthwaite, J. 2008. *Eur. J. Neurosci.* 27: 2783–2802.

[107] Steinert, J. R. et al. 2008. *Neuron* 60: 642–656.

[108] Savchenko, A., Barnes, S., and Kramer, R. H. 1997. *Nature* 390: 694–698.

[109] Ahern, P., Klyachko, V. A., and Jackson, M. B. 2002. *Trends Neurosci.* 25: 510–517.

[110] Snyder, S. H., Jaffrey, S. R., and Zakhary, R. 1998. *Brain Res. Brain Res. Rev.* 26: 167–175.

[111] Stevens, C. F., and Wang, Y. 1993. *Nature* 364: 147–149.

[112] Zhuo, M. et al. 1993. *Science* 260: 1946–1950.

[113] Ghosh, A., and Greenberg, M. E. 1995. *Science* 268: 239–247.

[114] Berridge M. J., Lipp, P., and Bootman, M. D. 2000. *Nat. Rev. Mol. Cell Biol.* 1: 11–21.

[115] Clapham, D. E. 2007. *Cell* 131: 1047–1058.

[116] Neher, E., and Augustine, G. J. 1992. *J. Physiol.* 450: 273–301.

[117] Trouslard, J., Marsh, S. J., and Brown, D. A. 1993. *J. Physiol.* 481: 251–271.

[118] Fierro, L., and Llano, I. 1996 *J. Physiol.* 496: 617–625.

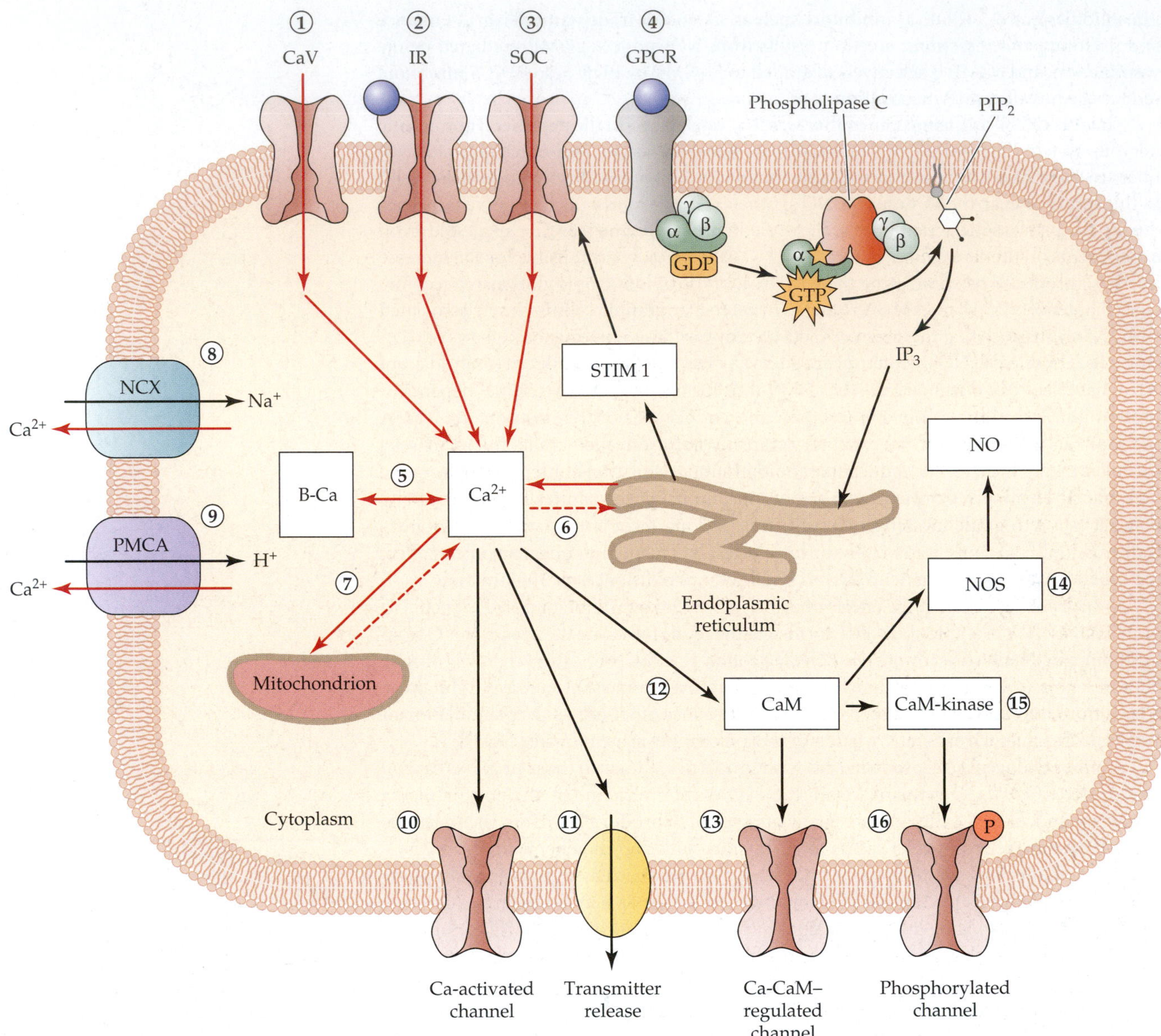

FIGURE 12.20 Calcium as an Intracellular Second Messenger. Intracellular Ca^{2+} ions in neurons are increased by (1) entry through voltage-gated calcium channels (CaV) during electrical activity; (2) activation of calcium-permeable ionotropic receptors (IR) by neurotransmitters such as glutamate and acetylcholine; (3) opening of store-operated calcium channels; and (4) activation of G protein–coupled receptors (GPCR) that stimulate phospholipase C (PLC) and generate inositol-1,4,5-trisphosphate (IP_3; see Figure 12.15). The IP_3 releases calcium from the endoplasmic reticulum (ER). Depletion of the ER calcium stores by IP_3 releases another signaling molecule STIM1 that couples to store-operated calcium channels (SOCs, principally Orai) in the outer membrane (4) to generate a second wave of calcium entry. (5) Most of the Ca^{2+} ions (98% to 99%) are rapidly but reversibly buffered by calcium buffers (B-Ca). Cytosolic $[Ca^{2+}]$ is subsequently restored by uptake into intracellular organelles—the ER (6) and, at high calcium loads, mitochondria (7)—and by extrusion via (8) the plasma membrane sodium–calcium exchange pump (NCX) or (9) the plasma membrane calcium ATPase (PMCA). Messenger functions include: (10) direct activation of calcium-dependent potassium, chloride, and cation channels; (11) stimulation of vesicular transmitter and hormone release; and (12) binding to and activation of, calmodulin (CaM). Ca-CaM may (13) modulate other ion channels (e.g., activation of small (SK) calcium-dependent potassium channels), and activate several enzymes. These activated enzymes include (14) nitric oxide synthase (NOS) to generate another messenger nitric oxide (NO; see Figure 12.19), and (15) calcium-calmodulin (CaM)–dependent protein kinases (CaM-kinase). This can phosphorylate ion channels (16) and many other molecules, to produce a variety of long-term changes in nerve cell function.

the high transmembrane Ca^{2+} gradient generates a large Ca^{2+} influx when calcium-permeable ionotropic receptors are stimulated. These include the NMDA class of ionotropic glutamate receptors[119] and nicotinic ACh receptors, especially those containing α7- or α9-subunits.[120] Second, metabotropic receptors that couple to G_q generate IP_3, which acts on ionotropic IP_3 receptors[121] to release Ca^{2+} from the endoplasmic reticulum (see Figure 12.10).[122]

[119] Burnashev, N. et al. 1995. *J. Physiol.* 485: 403–418.

[120] Fucile, S. 2004. *Cell Calcium* 35: 1–8.

[121] Mikoshiba, K. 2007. *J. Neurochem.* 102: 1426–1446.

[122] Streb, H. et al. 1983 *Nature* 306: 67–69.

Optical methods for imaging Ca^{2+} transients using fluorescent Ca^{2+}-binding compounds (Box 12.6) have revealed other properties of the Ca^{2+} transients produced by nerve activity and transmitter action. For example, increases in $[Ca^{2+}]$ are often confined to particular small regions of the neuron, creating calcium microdomains.[123–126] This results, in part, from the very slow diffusion of Ca^{2+} ions in the cytoplasm (about one-fiftieth of that in free solution[127]) because of binding to buffers and intracellular uptake. Thus, following synaptic activation of glutamate receptors, postsynaptic calcium transients may be confined for some time to individual dendrites[128] or even to a single dendritic spine.[129]

Calcium signals in these microdomains often appear as elementary events (variously called sparks, puffs, syntillas, etc.), reflecting the opening of single calcium-carrying channels or clusters of channels in the membrane or endoplasmic reticulum.[130] Though most frequently studied in non-neural cells, such events have also been seen in neurons.[131,132] Indeed, even before calcium imaging, they were detected in sympathetic neurons as spontaneous miniature outward currents (SMOCs), signifying the release of packets of calcium from the submembrane ER and consequent opening of calcium-activated potassium channels.[133] Rises in calcium concentration may also appear as oscillations or travelling waves.[114] The latter result from regenerative calcium-induced calcium release from intracellular stores. In neuronal dendrites, the calcium wave can appear as the discontinuous (saltatory) propagation of a calcium spike, rather like (but much slower than) the conduction of an action potential along a myelinated nerve fiber, as the calcium jumps from one cluster of IP_3 receptors to the next.[134]

Actions of Calcium

A rise in intracellular calcium has many effects on neuronal activity and function, as indicated in Figure 12.20. One direct effect is the activation of calcium-dependent potassium channels. An interesting example of this is the inhibition of cochlear hair cells by efferent cholinergic fibers in the auditory nerves.[135] Cholinergic activation would ordinarily be excitatory, as at the neuromuscular junction. However, calcium entering through the ACh-activated channels opens adjacent calcium-activated potassium channels, thereby producing inhibition (Figure 12.21A). As shown in Figure 12.21B, direct application of a

[123] Ross, W. N., Arechiga, H., and Nicholls, J. G. 1988. *Proc. Natl. Acad. Sci. USA* 85: 4075–4078.

[124] Llinás, R., Sugimori, M., and Silver, R. B. 1992. *Science* 256: 677–679.

[125] Oheim, M., Kirchhoff, F., and Stühmer, W. 2006. *Cell Calcium* 40: 423–439.

[126] Parekh, A. B. 2008. *J. Physiol.* 586: 3043–3054.

[127] Hodgkin, A. L., and Keynes, R. D. 1957. *J. Physiol.* 138: 253–281.

[128] Eilers, J., Plant, T., and Konnerth, A. 1996. *Cell Calcium* 20: 215–226.

[129] Denk, W., Sugimori, M., and Llinas, R. 1995. *Proc. Natl. Acad. Sci. USA* 92: 8279–8282.

[130] Cheng, H., and Lederer, W. J. 2008. *Physiol. Revs.* 88: 1491–1545.

[131] Ouyang, K. et al. 2005. *Proc. Natl. Acad. Sci. USA* 102: 12259–12264.

[132] Manita, S., and Ross, W. N. 2009. *J. Neurosci.* 29: 7833–7845.

[133] Brown, D. A., Constanti, A., and Adams, P. R. 1983. *Cell Calcium* 4: 407–420.

[134] Fitzpatrick, J. S. et al. 2009. *J. Physiol.* 587: 1439–1459.

[135] Fuchs, P. A., and Murrow, B. W. 1992. *J. Neurosci.* 12: 800–809.

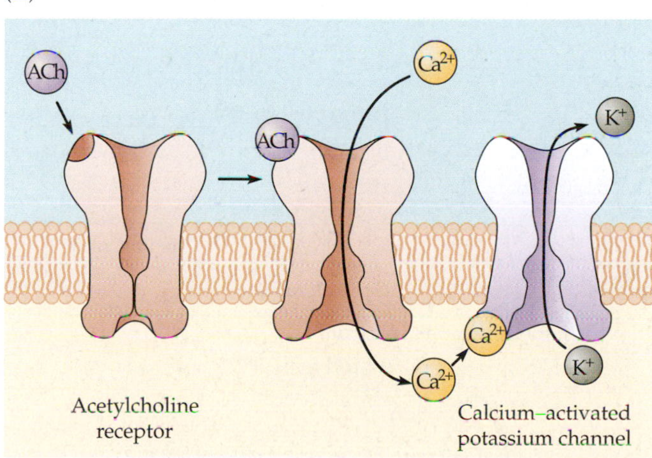

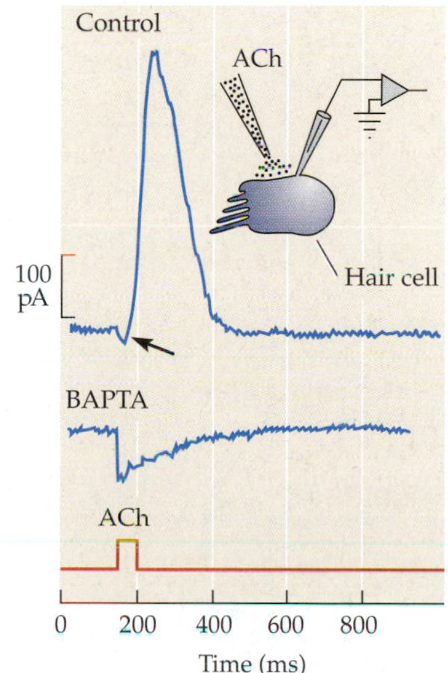

FIGURE 12.21 Inhibition by Ach-Activated Cation Channels. (A) In chick cochlea hair cells, ACh binds to nicotinic ionotropic receptors that allow cations, including calcium, to flow into the cell. Intracellular calcium causes calcium-activated potassium channels to open, leading to outward potassium current and hyperpolarization. (B) In a whole-cell recording (inset), application of ACh near the base of a hair cell produces a small, transient inward current (arrow) followed by a large outward current. In the intact cell, the outward current would be inhibitory. (C) If the calcium chelator BAPTA is added to the recording electrode, and hence to the cell cytoplasm, ACh application produces only inward current. No outward current is seen because incoming calcium ions are chelated and so prevented from activating potassium channels. (Records provided by P. A. Fuchs.)

BOX 12.6
Measuring Intracellular Calcium

Changes in the concentration of intracellular calcium ions are usually measured using luminescent or fluorescent calcium-binding compounds. One of the first such compounds was the luminescent jellyfish protein aequorin.[136] This comprises a 22kD apo-aequorin protein attached via oxygen to a fluorophore coelenterazine. On binding three Ca^{2+} ions, coelenterazine is converted to the amide and the complex splits, giving off a blue light (peak 460 nm).

Because it binds three Ca^{2+} ions, aequorin responds to a wide range of Ca^{2+} concentrations (from 0.1 to >100 μM) but suffers from the disadvantage that it is broken down after binding Ca^{2+}, so the concentration rapidly declines. However, because apo-aequorin is a protein, it can be expressed in cells from its complementary DNA (cDNA) by transfection. Cells readily take up coelenterazine, and can form aequorin from the expressed apo-aequorin. Then, by adding a targeting sequence to the cDNA, the aequorin can be expressed, and Ca^{2+} monitored in subcellular organelles such as the endoplasmic reticulum.[137]

Two very widely used calcium indicators are Fura-2 and Indo-1, derived from the calcium chelator 1,2-bis(o-aminophenoxy)ethane-N,N,N′,N′-tetraacetic acid (BAPTA) by Roger Tsien and his colleagues.[138] Fura-2 and Indo-1 are excited by ultraviolet light and show a shift in their excitation (Fura-2) or emission (Indo-1) spectrum when they bind Ca^{2+} ions. This is useful, since the concentration of Ca^{2+} can be calculated from the change in the ratio of the fluorescence at two excitation or emission wavelengths. This ratiometric method avoids artifacts arising from changes in concentration or fluorescence quenching.

Individually, these indicators cover a narrower range of calcium concentrations than aequorin, but many different fluorescent indicators are available to cover different ranges of [Ca^{2+}]. They are not proteins, so they cannot be expressed or directed to specific subcellular organelles. Instead, for this, Roger Tsien has devised another type of protein-based calcium reporter termed a cameleon.[139] These use calmodulin as the calcium-sensor, coupled to a calmodulin binding protein and two different fluorescent proteins from a jellyfish (based on green fluorescent protein, GFP) with different emission wavelengths. When the calmodulin binds calcium, it induces a conformational change in the calmodulin-binding protein, which brings the two GFP-derivatives closer together so that their fluorescences interact (Förster resonance energy transfer, or FRET). FRET can be recorded using two different emission wavelengths to calculate the concentration of Ca^{2+} ions.

[136] Shimomura, O., and Johnson, F. H. 1970. *Nature* 227: 1356–1357.
[137] Montero, M. et al. 1995. *EMBO J.* 14: 5467–5475.
[138] Grynkiewicz, G., Poenie, M., and Tsien, R. Y. 1985. *J. Biol Chem.* 260: 3440–3450.
[139] Miyawaki, A. et al. 1997. *Nature* 388: 882–887.

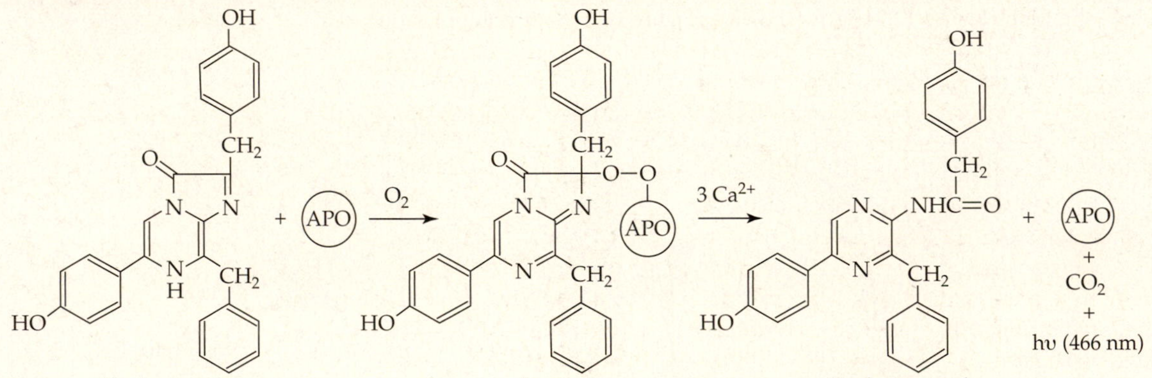

Coelenterazine + APO $\xrightarrow{O_2}$ Aequorin $\xrightarrow{3\,Ca^{2+}}$ Coelenteramide + APO + CO_2 + hυ (466 nm)

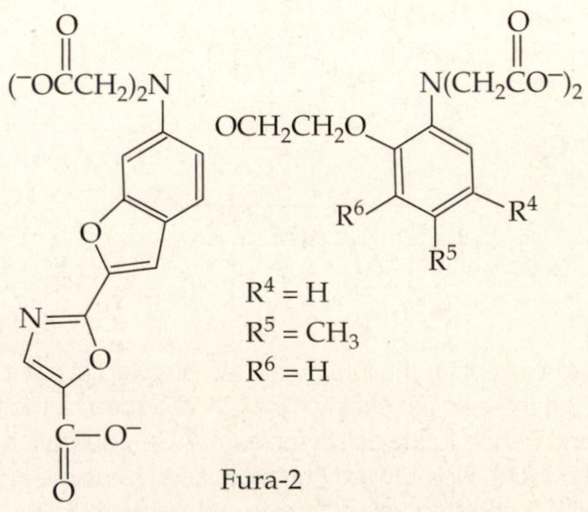

Fura-2

$R^4 = H$
$R^5 = CH_3$
$R^6 = H$

Fura-2 excitation spectrum

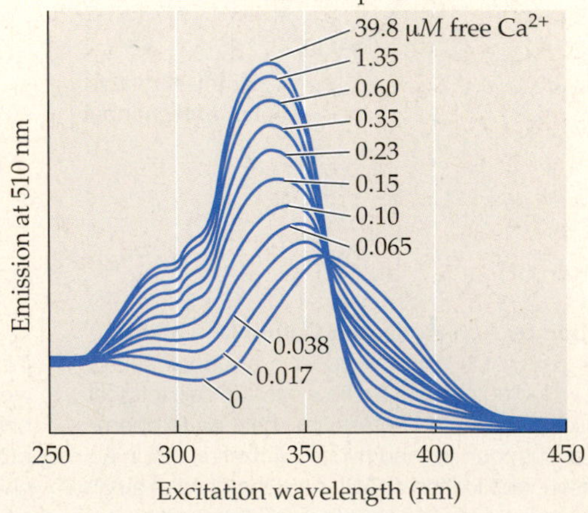

brief pulse of ACh produces a small transient inward current, which is due to activation of nicotinic receptors and is followed by a much larger outward current due to activation of potassium channels. After the calcium chelator 1,2-bis(o-aminophenoxy)ethane-N,N,N′,N′-tetraacetic acid (BAPTA) has been added to the bathing solution, activation of the potassium channels is blocked and only the inward synaptic current is seen. This mechanism is particularly effective because the cochlear nicotinic receptors are composed of α9/α10-subunits,[140] which have an unusually high calcium permeability.[120] A similar mechanism of inhibition may occur in some neurons in the brain.[141] In both brain and cochlea,[142,143] the responsive potassium channels belong to the small-conductance (SK, KCa2) family; for these channels, the calcium is not sensed by the channel protein itself but by closely associated calmodulin molecules.[144] Other calcium-activated channels that participate in second messenger-mediated responses include calcium-activated chloride channels, which contribute to olfactory sensing,[145] and cation channels that are activated by calcium-releasing metabotropic receptors,[146] which may be members of the Trp family of cation channels (see Chapter 5).

The principal molecular transducer of calcium actions is the calcium-binding protein calmodulin.[115] Calmodulin can bind four calcium ions. This induces a conformational change in the calmodulin molecule, which allows it to activate or inhibit other proteins. These include ion channels, such as the SK-type of calcium-dependent potassium channel mentioned above, and the IP_3-activated endoplasmic reticulum calcium channels, but the main targets of calcium–calmodulin (Ca-CaM) are enzymes, such as calcium-calmodulin–dependent protein kinases (CaM kinase), calcineurin (a protein phosphatase), and NO synthase (NOS). Activation of NOS produces the third messenger, NO (see Figure 12.19). CaM kinases can phosphorylate several ion channels[147] and play a major role in promoting long-term changes in nerve cell function such as long-term potentiation (see Chapter 15) and inducing changes in transcription.[113,148] Calcium also activates important calcium-binding enzymes independently of calmodulin, such as phospholipases, protein kinase C, and proteases such as calpain.

Prolonged Time Course of Indirect Transmitter Action

Synaptic interactions mediated by indirect mechanisms typically develop more slowly and last much longer than those mediated by direct mechanisms. At the skeletal neuromuscular junction, only 1 or 2 milliseconds are required for ACh to be released, diffuse across the synaptic cleft, and bind to and open ionotropic ACh receptors. These events are much too fast to be mediated by enzymes such as adenylyl cyclase or phospholipase C, which take many milliseconds to catalyze the synthesis of a single molecule of cAMP or the hydrolysis of a membrane lipid. Even activation of a membrane channel by binding of a G protein subunit to the channel itself tends to have a time course of hundreds of milliseconds, reflecting the lifetime of the activated α-subunit. Responses mediated by enzymatic production of diffusible cytoplasmic second messengers such as cAMP or IP_3 are slower still, lasting seconds to minutes and reflecting the slow time course of changes in second messenger concentration.

Yet experience tells us that changes in signaling in the nervous system can last a lifetime. How can such long-lasting changes in synaptic efficacy be produced? One answer comes from the properties of several of the protein kinases discussed in this chapter. These enzymes are themselves targets for phosphorylation. For example, when activated by calcium, CaM kinase II phosphorylates itself.[148] It can then become constitutively active and no longer requires the presence of the calcium–calmodulin complex for activity. This mechanism is one way by which a transient increase in calcium concentration can be translated into long-lasting activation of the kinase, which in turn can cause sustained changes in the activity of its other target proteins. Activated CaM kinase may be highly restricted in its location, to a single synapse or dendritic spine.[149]

For changes to persist for days or longer, protein synthesis is usually required. Many of the second messenger systems described in this chapter have been shown to produce changes in protein synthesis.[150] Such changes typically occur as a result of activation of one or more protein phosphorylation signaling cascades, which lead to phosphorylation of transcription factors and, consequently, altered gene expression. The most rapid effects that have been

[140] Lustig, L. R. 2006. *Anat. Rec.* 288A: 424–234.

[141] Gulledge, A. T., and Stuart, G. J. 2005. *J. Neurosci.* 28: 10305–10320.

[142] Oliver, D. et al. 2000. *Neuron* 26: 595–601.

[143] Kong, J-H., Adelman, J. P., and Fuchs, P. A. 2008. *J. Physiol.* 586: 5471–5485.

[144] Maylie, J. et al. 2004. *J. Physiol.* 554: 255–261.

[145] Stephan, A. B. et al. 2009. *Proc. Natl. Acad. Sci. USA* 106: 10776–10781.

[146] Congar, P. et al. 1997. *J. Neurosci.* 17: 5366–5379.

[147] Levitan, I. B. 1994. *Annu. Rev. Physiol* 56: 193–212.

[148] Soderling, T. 2000. *Curr. Opin. Neurobiol.* 10: 375–380.

[149] Lee, S. J. et al. 2009. *Nature* 458: 299–304.

[150] Flavell, S. W., and Greenberg, M. E. 2008. *Annu. Rev. Neurosci.* 31: 563–590.

[151] Wang, D. O. et al. 2009. *Science* 324: 1536–1540.

measured occur in expression of immediate early genes, such as c-*fos*, which encodes the inducible transcription factor Fos. Upon translation, this protein enters the nucleus, where it regulates further gene expression, ultimately producing metabolic or structural changes that permanently alter the response of the cell.[151] The final translational step from messenger RNA (mRNA) to protein may then be restricted to individual synapses,[151] allowing such changes to be confined to specific neural pathways or patterns of neural activity.

SUMMARY

- Neurotransmitters activate metabotropic receptors that are not themselves ion channels, but instead modify the activity of ion channels, ion pumps, or other receptor proteins by indirect mechanisms.
- Examples of metabotropic receptors include muscarinic ACh receptors; α- and β-adrenergic receptors; some of the receptors for GABA, 5-HT, dopamine, and glutamate; and receptors for neuropeptides, light, and odorants. They produce their effects through G proteins.
- G proteins are αβγ-heterotrimers. In the resting state, GDP is bound to the α-subunit, and the three subunits are associated as a trimer. When activated by a metabotropic receptor, GDP is replaced with GTP, the α- and βγ-subunits dissociate, and the free subunits activate one or more intracellular targets. The activity of G protein subunits is terminated by hydrolysis of GTP to GDP both by the endogenous GTPase activity of the α-subunit followed by the recombination of α- and βγ-subunits into a trimer.
- Some G protein βγ-subunits bind directly to ion channels and stimulate or inhibit their activity. Other G protein α- or βγ-subunits activate adenylyl cyclase, phospholipase C, or phospholipase A_2, generating intracellular second messengers that can have widespread effects. Changes in membrane phospholipids that result from phospholipase C activation also affect ion channel function.
- Indirectly acting transmitters influence the activity of potassium and calcium channels. The changes in potassium and calcium channel activity in turn influence the resting potential, spontaneous activity, the response to other inputs, or the amount of calcium entering during an action potential—and thereby, the amount of transmitter release.
- Endocannabinoids act as tertiary messengers. They are synthesized and released in response to a rise in intracellular calcium. They serve as retrograde messengers, inhibiting transmitter release from presynaptic endings.
- NO also acts as a tertiary messenger that is synthesized and released in response to a rise in calcium. It stimulates cGMP formation to affect ion channels in the same or neighboring neurons.
- Changes in intracellular calcium or calcium–calmodulin concentration regulate ion channels, phospholipases C and A_2, protein kinase C, calpain, adenylyl cyclase, cyclic nucleotide phosphodiesterase, and NO synthase.
- Both the distribution of changes in intracellular calcium, which can be highly localized, and their dynamics (calcium waves and oscillations) are important determinants of calcium action.
- Transmitter actions mediated by indirect mechanisms have time courses that vary from milliseconds to years. Rapid effects are produced by direct changes in ion channel activity, effects of intermediate duration by activation and phosphorylation of enzymes and other proteins, and very long-lasting effects by regulation of protein synthesis.

Suggested Reading

General Reviews

Clapham, D. E. 2007. Calcium signaling. *Cell* 131: 1047–1058.

Delmas, P., and Brown, D. A. 2005. Pathways modulating neural KCNQ/M (Kv7) potassium channels. *Nat. Rev. Neurosci.* 6: 850–862.

Evans, R. M., and Zamponi, G. W. 2006. Presynaptic Ca^{2+} channels—integration centers for neuronal signaling pathways. *Trends Neurosci.* 29: 617–624.

Gamper, N. S., and Shapiro, M. S. 2007. Regulation of ion transport proteins by membrane phosphoinositides. *Nat. Rev. Neurosci.* 8: 1–14.

Garthwaite, J. 2008. Concepts of neural nitric oxide-mediated transmission. *Eur. J. Neurosci.* 27: 2783–2802.

Kano, M., Ohno-Shosaku, T., Hashimotodani, Y., Uchigashima, M., and Watanabe, M. 2009. Endocannabinoid-mediated control of synaptic transmission. *Physiol. Rev.* 89: 309–380.

Meves, H. 2008. Arachidonic acid and ion channels: an update. *Brit. J. Pharmacol.* 155: 4–16.

Oldham, W. H., and Hamm, H. E. 2008. Heterotrimeric G protein activation by G-protein-coupled receptors. *Nat. Rev. Mol. Cell Biol.* 9: 60–71.

Parekh, A. B. 2008. Ca^{2+} microdomains near plasma membrane Ca^{2+} channels: impact on cell function. *J. Physiol.* 586: 3043–3054.

Wayman, G. A., Lee, Y. S., Tokumitsu, H., Silva, A. J., and Soderling, T. R. 2008. Calmodulin-kinases: modulators of neuronal development and plasticity. *Neuron* 59: 914–931.

Wettschureck, N., and Offermanns, S. 2005. Mammalian G proteins and their cell type specific functions. *Physiol. Rev.* 85:1 159–1204.

Original Papers

DiFrancesco, D., and Tortura, D. P. 1991. Direct activation of cardiac pacemaker channels by intracellular cyclic AMP. *Nature* 351: 145–147.

Doupnik, C. A., Davidson, N., Lester, H. A., and Kofuji, P. 1997. RGS proteins reconstitute the rapid gating kinetics of gbetagamma-activated inwardly rectifying K^+ channels. *Proc. Natl. Acad. Sci. USA* 94: 10461–10466.

Falkenburger, B. H., Jensen, J. B., and Hille, B. 2010. Kinetics of M1 muscarinic receptor and G protein signaling to phospholipase C in living cells. *J. Gen. Physiol.* 135: 81–97.

Fuchs, P. A., and Murrow, B. W. 1992. Cholinergic inhibition of short (outer) hair cells of the chick's cochlea. *J. Neurosci.* 12: 800–809.

Ikeda, S. R. 1996. Voltage-dependent modulation of N-type calcium channels by G-protein beta gamma subunits. *Nature* 380: 255–258.

Lipscombe, D., Kongsamut, S., and Tsien, R. W. 1989. β-Adrenergic inhibition of sympathetic neurotransmitter release mediated by modulation of N-type calcium-channel gating. *Nature* 340: 639–642.

Steinert, J. R., Kopp-Scheinpflug, C., Baker, C., Challiss, R. A., Mistry, R., Haustein, M. D., Griffin, S. J., Tong, H., Graham, B. P., and Forsythe, I. D. 2008. Nitric oxide is a volume transmitter regulating postsynaptic excitability at a glutamatergic synapse. *Neuron* 60: 642–656.

Suh, B. C., Horowitz, L. F., Hirdes, W., Mackie, K., and Hille, B. 2004. Regulation of KCNQ2/KCNQ3 current by G protein cycling: the kinetics of receptor-mediated signaling by G_q. *J. Gen. Physiol.* 123: 663–683.

Wickman, K. D., Iñiguez-Lluhi, J. A., Davenport, P. A., Taussig, R., Krapivinsky, G. B., Linder, M. E., Gilman, A. G., and Clapham, D. E. 1994. Recombinant G-protein βγ-subunits activate the muscarinic-gated atrial potassium channel. *Nature* 368: 255–257.

Wilson, R. I., and Nicoll, R. A. 2001. Endogenous cannabinoids mediate retrograde signalling at hippocampal synapses. *Nature* 410: 588–592.

Winks, J. S., Hughes, S., Filippov, A. K., Tatulian, L., Abogadie, F. C., Brown, D. A., and Marsh, S. J. 2005. Relationship between membrane phosphatidylinositol-4,5-bisphosphate and receptor-mediated inhibition of native neuronal M channels. *J. Neurosci.* 25: 3400–3413.

CHAPTER 13
Release of Neurotransmitters

The stimulus for neurotransmitter release is depolarization of the nerve terminal. Release occurs as a result of calcium entry into the terminal through voltage-activated calcium channels. Invariably a delay of about 0.5 milliseconds intervenes between presynaptic depolarization and transmitter release. Part of the delay is due to the time taken for calcium channels to open; the remainder is due to the time required for calcium to cause transmitter release.

Transmitter is secreted in multimolecular packets (quanta), each containing several thousand transmitter molecules. In response to an action potential, anywhere from 1 to as many as 300 quanta are released almost synchronously from the nerve terminal, depending on the type of synapse. At rest, nerve terminals release quanta spontaneously at a slow rate, giving rise to spontaneous miniature synaptic potentials. At rest, there is also a continuous, nonquantal leak of transmitter from nerve terminals.

One quantum of transmitter corresponds to the contents of one synaptic vesicle and comprises several thousand molecules of a low-molecular-weight transmitter. Release occurs by the process of exocytosis, during which the synaptic vesicle membrane fuses with the presynaptic membrane, and the contents of the vesicle are released into the synaptic cleft. The components of the vesicle membrane are then retrieved by endocytosis, sorted in endosomes, and recycled into new synaptic vesicles.

A number of questions arise concerning the way in which presynaptic neurons release transmitter. Experimental answers to such questions require a highly sensitive, quantitative, and reliable measurement of the amount of transmitter released, with a time resolution in the millisecond range. In many of the experiments described in this chapter, this measurement is obtained by recording changes in the membrane potential of the postsynaptic cell. As discussed in Chapter 11, the vertebrate neuromuscular junction, where the transmitter is known to be acetylcholine (ACh), offers many advantages. However, to obtain more complete information about the release process, it is useful to be able to record from the presynaptic endings as well. For example, such recordings are needed to establish how calcium and membrane potential affect transmitter release. The presynaptic terminals at vertebrate skeletal neuromuscular junctions are typically too small for electrophysiological recording (but see Morita and Barrett, 1990[1]); however, this can be done at a number of synapses, such as the giant fiber synapse in the stellate ganglion of the squid,[2] giant terminals of goldfish retinal bipolar cells,[3] and calyciform synapses in the avian ciliary ganglion[4] and the rodent brainstem.[5] Moreover, new techniques allow transmitter release to be monitored by means that do not require electrical recording from the postsynaptic cell. In this chapter we discuss electrophysiological and morphological experiments that characterize the release process.

Characteristics of Transmitter Release

Axon Terminal Depolarization and Release

The stellate ganglion of the squid was used by Katz and Miledi to determine the precise relation between presynaptic membrane depolarization and the amount of transmitter released.[6] Simultaneous records were made of the action potential in the presynaptic terminal and the response of the postsynaptic fiber, as shown in Figure 13.1A. When tetro-

[1] Morita, K., and Barrett, E. F. 1990. *J. Neurosci.* 10: 2614–2625.

[2] Bullock, T. H., and Hagiwara, S. 1957. *J. Gen. Physiol.* 40: 565–577.

[3] Heidelberger, R., and Matthews, G. 1992. *J. Physiol.* 447: 235–256.

[4] Martin, A. R., and Pilar, G. 1963. *J. Physiol.* 168: 443–463.

[5] Borst, J. G. G., and Sakmann, B. 1996. *Nature* 383: 431–434.

[6] Katz, B., and Miledi, R. 1967. *J. Physiol.* 192: 407–436.

(A) Stellate ganglion of squid

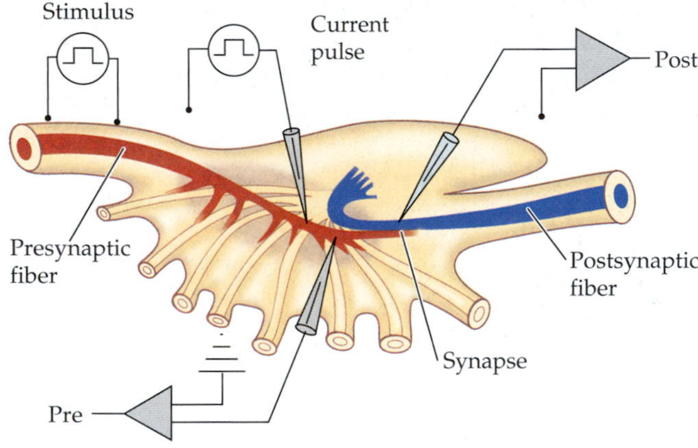

(B) Tetrodotoxin (TTX) paralysis

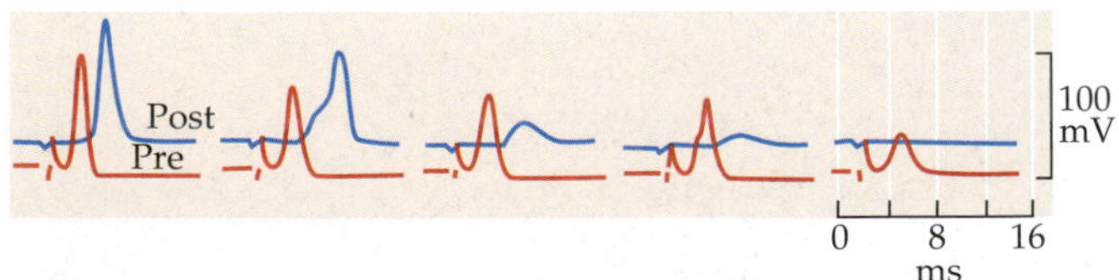

(C) Pre- and postsynaptic potential changes

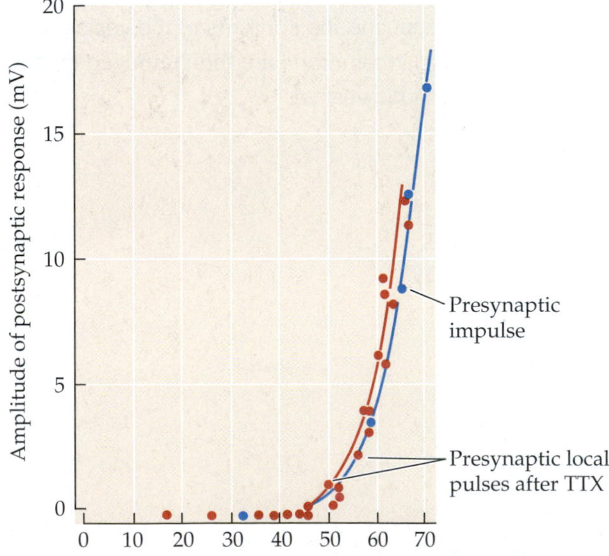

FIGURE 13.1 Presynaptic Impulse and Postsynaptic Response at a squid giant synapse. (A) Sketch of the stellate ganglion of the squid, illustrating the two large axons that form a chemical synapse. Pre- and postsynaptic axons are impaled with microelectrodes to record membrane potential, and an additional microelectrode is used to pass depolarizing current into the presynaptic terminal. (B) Simultaneous recordings from the presynaptic axons (red records) and postsynaptic axon (blue records) during the development of conduction block by tetrodotoxin (TTX). As the amplitude of the presynaptic action potential decreases, so does the size of the postsynaptic potential. Note that the two largest presynaptic action potentials evoke postsynaptic action potentials. (C) The relation between the amplitude of the presynaptic action potential and the postsynaptic potential. Blue circles represent results in B; red circles represent results obtained by applying depolarizing current pulses to the presynaptic terminals after complete TTX block. (A after Bullock and Hagiwara, 1957; B and C after Katz and Miledi, 1967b.)

dotoxin (TTX) was applied to the preparation, the presynaptic action potential gradually decreased in amplitude over the next 15 minutes (Figure 13.1B). The postsynaptic action potential also decreased in amplitude, but then abruptly disappeared because the excitatory postsynaptic potential (EPSP) failed to reach threshold. From this point on, the size of the synaptic potential could be used as a measure of the amount of transmitter released.

When the amplitude of the EPSP is plotted against the amplitude of the failing presynaptic impulse, as in Figure 13.1C (blue circles), the synaptic potential decreases rapidly as the presynaptic action potential amplitude falls below about 75 mV, and at amplitudes less than about 45 mV, there are no postsynaptic responses. TTX has no effect on the sensitivity of the postsynaptic membrane to transmitter, so the fall in synaptic potential amplitude indicates a reduction in the amount of transmitter released from the presynaptic terminal. Thus, there is a threshold for transmitter release at about 45 mV depolarization, after which the amount released, and hence the EPSP amplitude, increases rapidly with presynaptic action potential amplitude.

Katz and Miledi used an additional procedure to explore further the relation between the potential amplitude and transmitter release They placed a second electrode in the presynaptic terminal, through which they applied brief (1–2 ms) depolarizing current pulses, thereby mimicking a presynaptic action potential. The relationship between the amplitude of the artificial action potential and that of the synaptic potential was the same as the relation obtained with the failing action potential during TTX poisoning (Figure 13.1C, red circles). This result indicates that the normal fluxes of sodium and potassium ions responsible for the action potential are not necessary for transmitter release; only depolarization is required.

Synaptic Delay

One characteristic of the transmitter release process evident in Figure 13.1B is that there is a lag time between the onset of the presynaptic action potential and the beginning of the synaptic potential. This lag time is known as the **synaptic delay** (see Chapter 11). In these experiments on the squid giant synapse, which were done at about 10°C, the delay was 3 to 4 ms. Detailed measurements at the frog neuromuscular junction show a synaptic delay of 0.5 ms at room temperature (Figure 13.2).[7] The time is too long to be accounted for by diffusion of ACh across the synaptic cleft (a distance of 50 nanometers [nm]), which should take no longer than about 50 microseconds (μs). When ACh is applied to the junction ionophoretically from a micropipette, delays of as little as 150 μs can be achieved, even though the pipette is much farther from the postsynaptic receptors than are the nerve terminals. Furthermore, synaptic

[7]Katz, B., and Miledi, R. 1965. *J. Physiol.* 181: 656–670.

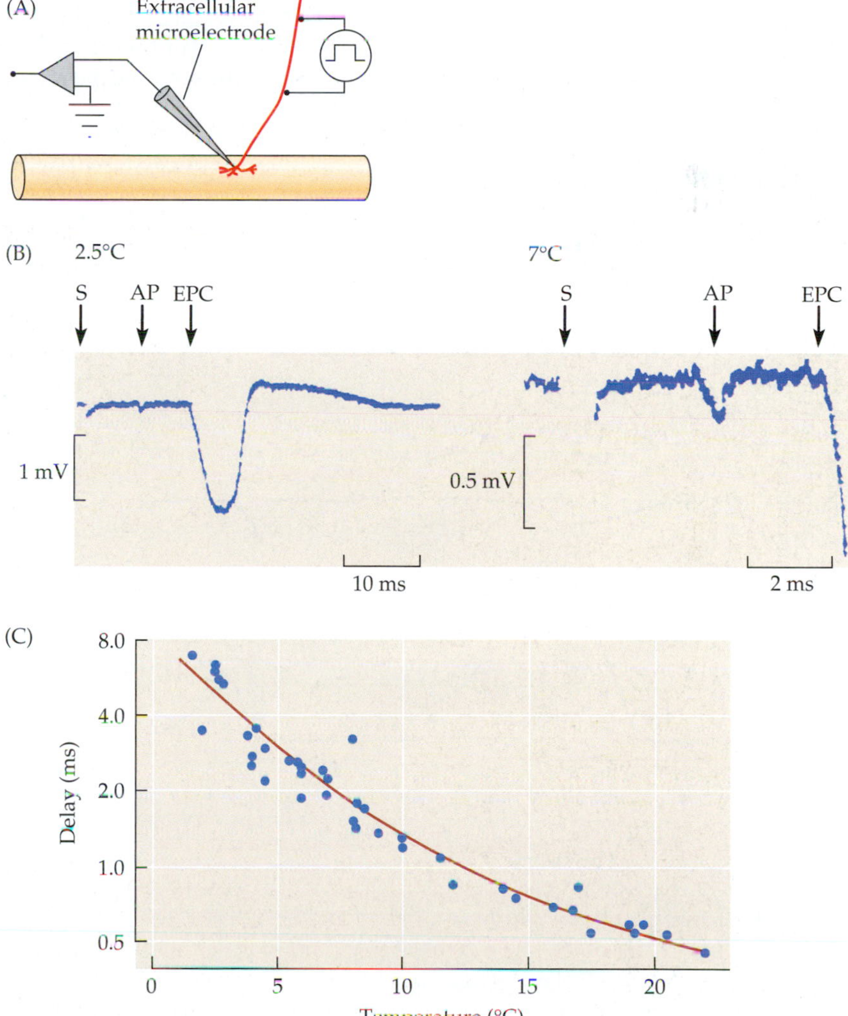

FIGURE 13.2 Synaptic Delay at a Chemical Synapse. (A) The motor nerve is stimulated while recording with an extracellular microelectrode at the frog neuromuscular junction. With this recording arrangement, current flowing into the nerve terminal or the muscle fiber is recorded as a negative potential. (B) Extracellular recordings of the stimulus artifact (S), the axon terminal action potential (AP), and the end-plate current (EPC) at 2.5°C and 7°C. The synaptic delay is the time between the action potential in the nerve terminal and the beginning of the end-plate current. (C) A plot of synaptic delay as a function of temperature , showing the decrease in synaptic delay with increasing temperature. (After Katz and Miledi, 1965.)

Ricardo Miledi

delay is much more sensitive to temperature than would be expected if it were due to diffusion. Cooling the frog nerve–muscle preparation to 2.5°C increases the delay to as long as 7 ms (Figure 13.2B), whereas the delay in the response to ionophoretically applied ACh is not perceptibly altered. Thus, the delay is largely in the transmitter release mechanism.

Evidence that Calcium Is Required for Release

Calcium has long been known as an essential link in the process of synaptic transmission. When its concentration in the extracellular fluid is decreased, release of ACh at the neuromuscular junction is reduced and eventually abolished.[8,9] The importance of calcium for release has been established at synapses, irrespective of the nature of the transmitter. (One exception is the release of GABA from horizontal cells in the fish retina.[10]) The role of calcium has been generalized further to other secretory processes, such as liberation of hormones by cells of the pituitary gland, release of epinephrine from the adrenal medulla, and secretion by salivary glands.[11,12] As discussed in the next section, evoked transmitter release is preceded by calcium entry into the terminal and is antagonized by ions that block calcium entry, such as magnesium, cadmium, nickel, manganese, and cobalt. Transmitter release can be reduced, then, either by removing calcium from the bathing solution or by adding a blocking ion. For transmitter release to occur, calcium must be present in the bathing solution at the time of depolarization of the presynaptic terminal.[13]

[8] del Castillo, J., and Stark, L. 1952. *J. Physiol.* 116: 507–515.

[9] Dodge, F. A., Jr., and Rahamimoff, R. 1967. *J. Physiol.* 193: 419–432.

[10] Schwartz, E. A. 1987. *Science* 238: 350–355.

[11] Penner, R., and Neher, E. 1988. *J. Exp. Biol.* 139: 329–345.

[12] Kasai, H. 1999. *Trends Neurosci.* 22: 88–93.

[13] Katz, B., and Miledi, R. 1967. *J. Physiol.* 189: 535–544.

Measurement of Calcium Entry into Presynaptic Nerve Terminals

Entry of calcium into the nerve terminal is through voltage-sensitive calcium channels of the Ca_V2 family (see Chapter 5) that are activated upon depolarization by the presynaptic action potential. Using voltage clamp techniques, Llinás and his colleagues measured the magnitude and time course of the calcium current produced by presynaptic depolarization at the squid giant synapse. An example is shown in Figure 13.3A. The sodium and potassium conductances associated with the action potential were blocked by TTX and tetraethylammonium (TEA) so that only the voltage-activated calcium channels remained. Depolarizing the presynaptic terminal to –18 mV (upper record in

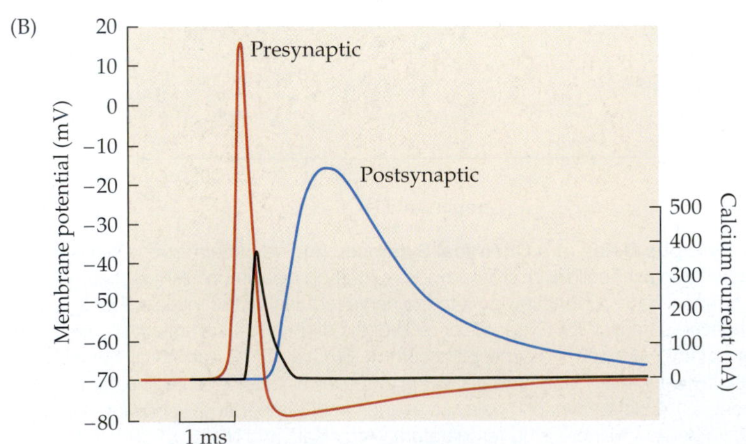

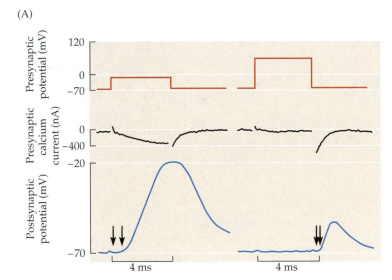

FIGURE 13.3 Presynaptic Calcium and Transmitter Release at the Squid Giant Synapse. The presynaptic terminal is voltage clamped and treated with TTX and TEA to abolish voltage-activated sodium and potassium currents. (A) Records show potentials applied to the presynaptic fiber (upper trace), presynaptic calcium current (middle trace), and EPSP in the postsynaptic fiber (lower trace). A voltage pulse from –70 to –18 mV (left panel) results in a slow inward calcium current and, after a delay of about 1 ms (arrows), an EPSP. A larger depolarization, to +60 mV (right panel), suppresses calcium entry. At the end of the pulse, a surge of calcium current is followed within about 0.2 ms (arrows) by an EPSP. (B) If a voltage change identical in shape to a normal action potential is produced by the voltage clamp (labeled Presynaptic), then the EPSP is indistinguishable from that seen normally (labeled Postsynaptic). The black curve gives the magnitude and time course of the calcium current. The synaptic delay between the beginning of the presynaptic depolarization and the beginning of the postsynaptic response is due in part to the time required to open calcium channels and in part to the time for calcium entry to trigger transmitter release. (After Llinás, 1982.)

left panel) produced an inward calcium current in the terminal that increased slowly in magnitude to about 400 nA (middle record), and a large synaptic potential in the postsynaptic cell (lower record). When the terminal was depolarized to +60 mV, approximating the calcium equilibrium potential, the calcium current was suppressed during the pulse (right panel) and no synaptic potential was seen. This demonstrates that depolarization of the terminal is not sufficient on its own to trigger release; calcium entry must also occur. On repolarization, there was a brief, inward calcium current through channels remaining open after the depolarization, accompanied by a small postsynaptic potential.

The effect of an artificial action potential is shown in Figure 13.3B. A presynaptic action potential, recorded before addition of TTX and TEA to the preparation, was played back through the voltage clamp circuit to produce exactly the same voltage change in the terminal. The postsynaptic potential is indistinguishable from that produced by a normal presynaptic action potential, confirming that the sodium and potassium currents that normally accompany the action potential are not necessary for transmitter release.

The experiment shown in Figure 13.3B also enabled Llinás and his colleagues to measure the magnitude and time course of the calcium current produced by the artificial action potential (black curve). The calcium current begins about 0.5 ms after the beginning of the presynaptic depolarization, and the postsynaptic potential begins about 0.5 ms later. Thus, the time required for the presynaptic terminal to depolarize and the calcium channels to open accounts for the first half of the synaptic delay; the time required for the calcium concentration to rise within the terminal and evoke transmitter release accounts for the remainder.

An experimental technique important for characterizing the role of calcium transmitter release is the use of calcium indicator dyes to estimate intracellular calcium concentrations (see Box 12.6).[14,15] The first dye to be used extensively was aequorin, a calcium-sensitive luminescent protein extracted from the jellyfish *Aequorea victoria* (see Chapter 12). It has now been largely replaced by synthetic compounds that change their fluorescence in the presence of calcium, based on calcium chelators such as ethylene glycol-bis[2-aminoethyl ether]-N,N,N′,N′-tetraacetic acid (EGTA).

An example of the use of luminescent dye to reveal changes in presynaptic calcium concentration is shown Figure 13.4. Aequorin injected into the resting presynaptic terminal of the squid giant synapse (A) revealed discrete microdomains of free calcium, some with relatively high concentrations (indicated by red and yellow peaks). After a brief train of presynaptic action potentials (B), the intracellular calcium concentration reached 100 to 200 μ*M*.

[14]Tsien, R. Y. 1989. *Annu. Rev Neurosci.* 12: 227–253.

[15]Rudolf, R. et al. 2003. *Nat. Rev. Mol. Cell Biol.* 4: 579–586.

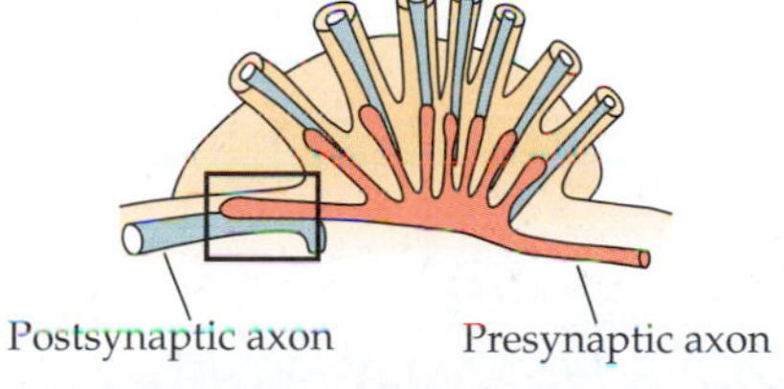

FIGURE 13.4 Microdomains of Calcium within the Presynaptic Terminal at the Squid Giant Synapse. (A) Distribution of calcium within the presynaptic axon terminal at rest, determined by intracellular injection of a calcium-sensitive dye (box in illustration above [A] shows the region imaged). (B) A brief train of presynaptic action potentials results in the appearance of microdomains of high calcium concentration within the axon terminal. (After Llinás, Sugimori, and Silver, 1992; micrographs kindly provided by R. Llinás.)

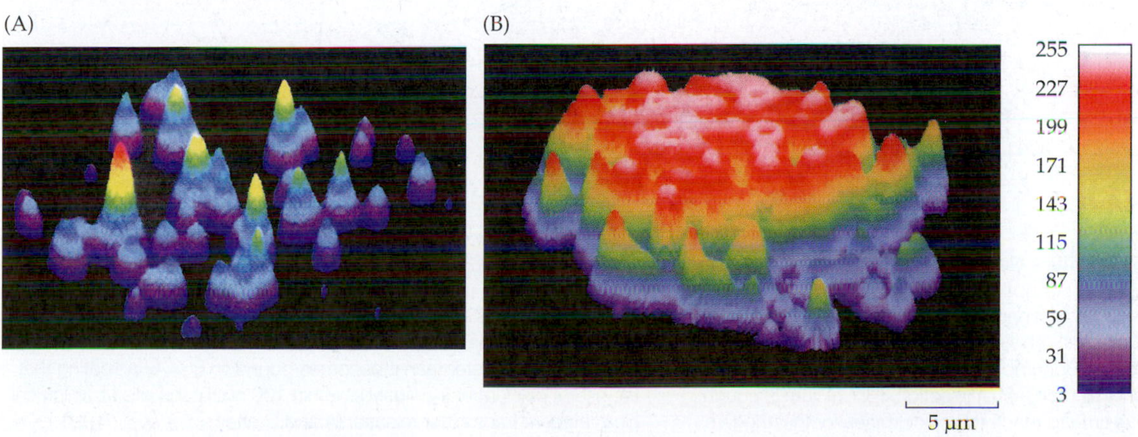

Localization of Calcium Entry Sites

Figure 13.4 illustrates an important point about the distribution of free calcium in the cytoplasm, namely that it is not at all uniform.[16] Calcium entering the terminal through a single channel collects briefly in a small **nanodomain** with the concentration falling rapidly over a radius of a few tens of nm from the channel as the ions diffuse into the bulk solution or are bound by intrinsic calcium chelators. Calcium entering through a group of closely apposed channels occupies a **microdomain** that can spread over a distance of a few hundred nm from the channel cluster. Because of the restricted spread of incoming ions, the spatial relation between calcium channels and their associated transmitter release sites is of critical importance.

Experiments on the squid giant synapse using calcium buffers have provided information about the proximity of calcium channels to the sites of transmitter secretion.[17] In these experiments, injection of 1,2-bis(o-aminophenoxy)ethane-N,N,N′,N′-tetraacetic acid (BAPTA), a potent calcium buffer, into the presynaptic terminal resulted in a severe attenuation of transmitter release, without affecting the presynaptic action potential (Figure 13.5A). On the other hand, EGTA, a calcium buffer of equal potency, had little effect on release (Figure 13.5C). This disparity is due to the fact that calcium is bound hundreds of times faster by BAPTA than by EGTA. Thus calcium ions have little opportunity to diffuse from their site of entry before being bound by BAPTA, but can traverse some distance before being captured by EGTA (Figure 13.5B and D). From the rates of calcium diffusion and binding to EGTA, it can be calculated that the calcium-binding site associated with the release process must lie within 100 nm or less of the site of calcium entry.

[16] Augustine, G. J., Santamaria, F., and Tanaka, F. 2003. *Neuron* 40: 331–346.

[17] Adler, E. M. et al. 1991. *J. Neurosci.* 11: 1496–1507.

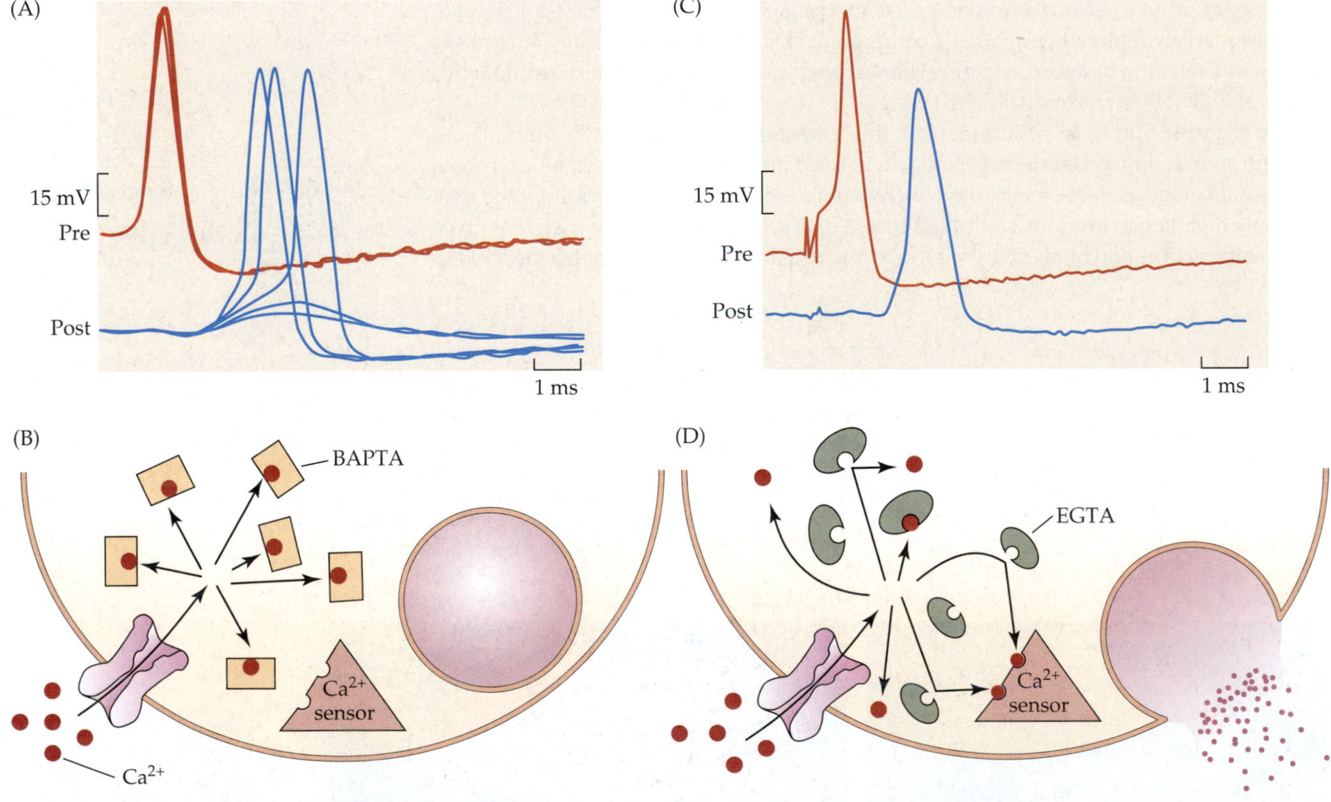

FIGURE 13.5 Calcium Enters Near the Site of Transmitter Release at the squid giant synapse. (A) Intracellular recordings from the presynaptic (Pre) and postsynaptic (Post) axons following injection of the fast calcium chelator 1,2-bis(o-aminophenoxy)ethane-N,N,N′,N′-tetraacetic acid (BAPTA). Superimposed traces show the reduction in the EPSP during a 4-minute BAPTA injection. (B) Calcium is bound to BAPTA before it has time to reach the calcium sensor that triggers transmitter release. (C) Superimposed intracellular recordings during a 4-minute injection of ethylene glycol-bis[2-aminoethyl ether]-N,N,N′,N′-tetraacetic acid (EGTA), a chelator that binds calcium more slowly. No change in EPSP amplitude is seen. (D) Calcium reaches the sensor that triggers release faster than it becomes bound to EGTA, indicating that the site of calcium entry must be within 100 nm of the site at which calcium triggers transmitter release. (A and C after Adler et al., 1991.)

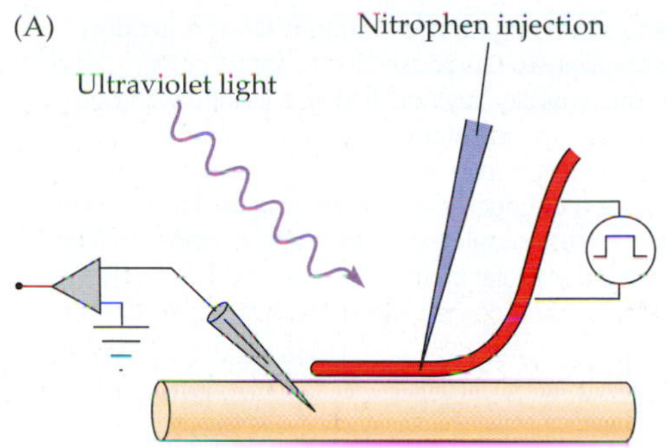

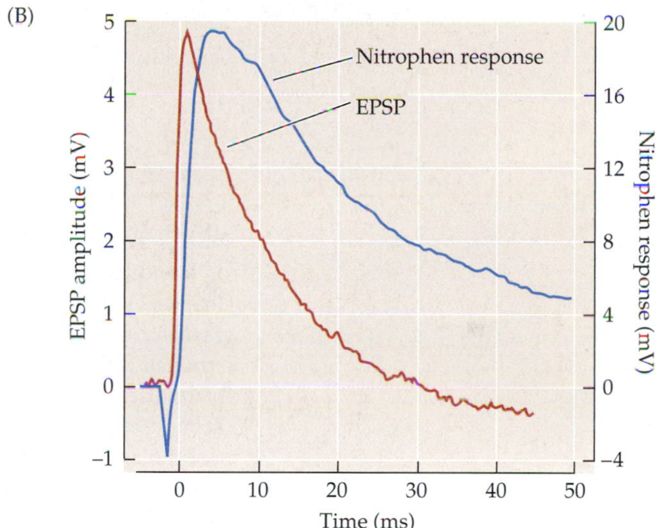

FIGURE 13.6 An Increase in Intracellular Calcium Is Sufficient to Trigger Rapid Transmitter Release at the squid giant synapse. (A) Nitrophen, a form of caged calcium, is injected into the presynaptic terminal. Transmitter release is monitored by recording intracellularly from the postsynaptic axon. (B) Intracellular records show the postsynaptic response to nerve stimulation (EPSP) and to release of calcium from nitrophen by a flash of ultraviolet light (nitrophen response). An abrupt increase in intracellular calcium causes an increase in transmitter release that is nearly as rapid as that produced by a presynaptic action potential. The decay of the nitrophen response is slower and incomplete because the photolyzed nitrophen buffers calcium to a concentration higher than the normal level at rest. (B after Zucker, 1993.)

On the other hand, similar experiments at some neuronal synapses have shown an effect of EGTA on release, suggesting that in these cells calcium may diffuse some distance from calcium channels to sites that trigger or modulate release.[5]

Transmitter Release by Intracellular Concentration Jumps

Another important technique for exploring the role of calcium in transmitter release is the use of photolabile calcium chelators that release "caged" calcium upon illumination.[18,19] This provides a means of producing a transient increase in intracellular calcium in the presynaptic terminal, divorced from any change in membrane potential. An example is shown in Figure 13.6. Nitrophen, an EDTA-based calcium cage, was injected into the presynaptic terminal of the squid giant synapse. Illumination of the terminal with a brief flash of ultraviolet light produced a transient increase in intracellular calcium concentration and release of transmitter. The jump in intracellular calcium concentration, which was estimated to be about 100 μ*M*, produced a postsynaptic depolarization that closely resembled the excitatory potential produced by nerve stimulation. Similar experiments on terminals of bipolar cells from the goldfish retina gave comparable results.[20]

More detailed examination of the role of intracellular calcium in the transmitter release process has been made possible by the utilization of a unique presynaptic structure in the vertebrate auditory pathway–the calyx of Held, which forms a glutamatergic synapse with neurons in the medial nucleus of the trapezoid body. The calyx is sufficiently large to allow the attachment of a patch clamp electrode for electrical recording,[21] and simultaneous patch clamp records can be obtained from the calyx and the postsynaptic cell.[22]

Using rat brain slices, Bollmann and Sakmann were able to combine the unique properties of the calyx of Held synapse, the use of calcium indicator dyes, and laser photolysis of caged calcium to obtain precise measures of the relation between intracellular calcium concentration transients and transmitter release.[23] A whole-cell patch pipette was used to load the calyx with caged calcium and a low-affinity calcium indicator dye. A second patch pipette recorded postsynaptic currents, and electrodes were placed on the slice for presynaptic nerve stimulation. As shown in Figure 13.7, a brief calcium transient of appropriate amplitude and time course could produce an excitatory postsynaptic current that was indistinguishable from that produced by presynaptic nerve stimulation. Thus, a transient increase in cytoplasmic calcium concentration can account completely for the magnitude and time course of transmitter release.

Other Factors Regulating Transmitter Release

From the evidence presented so far, we can conclude that transmitter release is triggered by an increase in intracellular calcium concentration, brought about by depolarization of the presynaptic terminal and opening of voltage-activated calcium channels. However, at

[18] Adam, S. R. et al. 1988. *J. Am. Chem. Soc.* 110: 3212–3220.

[19] Ellis-Davies, G. C. R. 2008. *Chem. Rev.* 108: 1603–1613.

[20] Heidelberger, R. et al. 1994. *Nature* 371: 513–515.

[21] Forsythe, I. D. 1994. *J. Physiol.* 479: 381–387.

[22] Borst, J. G. G., Helmchen, F., and Sakmann, B. 1995. *J. Physiol.* 489: 825–840.

[23] Bollman, J. H., and Sakmann, B. 2005. *Nat. Neurosci.* 8: 426–434.

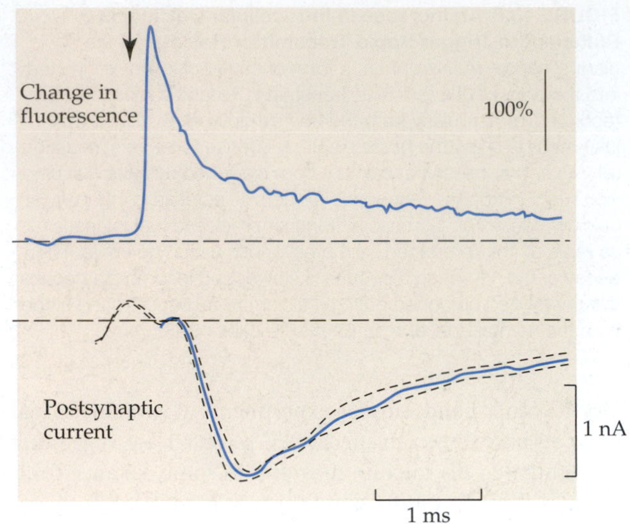

FIGURE 13.7 Change in Presynaptic Calcium Concentration and Excitatory Postsynaptic Currents at a synapse in the trapezoid body of the rat. The presynaptic calyx of Held was loaded with both caged calcium and a calcium indicator dye. Upper record: fluorescent signal indicating a transient increase in presynaptic calcium concentration, evoked by photolysis of caged calcium with a laser flash. Lower traces: postsynaptic currents produced by the calcium transient (solid curve) and by presynaptic stimulation (dashed curves). The postsynaptic currents are identical in time course. (Modified from Bollmann and Sakmann, 2005.)

some synapses additional regulatory factors have been shown to be present. In particular, I. and H. Parnas and their colleagues have shown that presynaptic autoreceptors (see Chapter 11) participate in the release process at neuromuscular junctions of the mouse and crayfish.[24,25] These are G protein–coupled receptors (see Chapter 12) on the presynaptic terminals which, when activated by released transmitter, act in turn to inhibit further release. As we discuss later in this chapter, presynaptic muscarinic cholinergic (M2) receptors in the mouse are continually exposed to a resting concentration of ACh in the range of 10–20 nanomolar (n*M*). This concentration is sufficient to block the receptors and thereby prolong the decaying phase of voltage-activated transmitter release. An additional feature of the M2 receptors is that their acetylcholine binding affinity is voltage-dependent.[26] Given these two observations, the authors postulated that depolarization of the nerve terminal, in addition to opening calcium channels, reduces the binding affinity of the M2 receptors, thereby relieving tonic inhibition of the release mechanism and facilitating its activation by the incoming calcium. Upon repolarization, the inhibitory action of the M2 receptors is restored, so that the release is terminated even though the calcium concentration in the region may still be elevated. The exact nature of the coupling between the voltage-sensitive autoreceptor and the release machinery is not known. Similar observations on the crayfish neuromuscular junction, which is glutamatergic, suggest that the proposed mechanism may have more general applicability. However, it does not contribute to the release process at the calyx of Held synapse, where depolarization has no effect on transmitter release triggered by uncaged calcium.[27] As shown in Figure 13.7, the time course of evoked transmitter release can be mimicked by a transient increase in cytoplasmic calcium with no accompanying depolarization.

Quantal Release

So far, the general scheme for transmitter release can be summarized as follows:

presynaptic depolarization → calcium entry → transmitter release

Now that this general framework has been established, it remains to be shown how transmitter is secreted from the terminals. In experiments on the frog neuromuscular junction, Fatt and Katz showed that ACh can be released from terminals in multimolecular packets, which they called **quanta.**[28] Later experiments by Kuffler and Yoshikami showed that each quantum corresponds to approximately 7000 molecules of ACh.[29] Quantal release then means that any response to stimulation will consist of roughly 7000 molecules, or 14,000 and so on, but not 4250 or 10,776. At any given synapse, the number of quanta released from the nerve terminal in response to an action potential (the **quantum content** of the synaptic potential) may vary considerably from trial to trial, but the mean number of molecules in each quantum (**quantal size**) is fixed (with a variance of about 10%).

[24] Kupchik, Y. M. et al. 2008. *Proc. Natl. Acad. Sci. USA* 105: 4435–4440.

[25] Parnas, I., and Parnas, H. 2010. *Pflügers Arch.* 460: 975–990.

[26] Ben-Chaim, Y. et al. 2006. *Nature* 444: 106–109.

[27] Felmy, F., Neher, E., and Schneggenberger, R. 2003. *Proc. Natl. Acad. Sci. USA* 100: 15200–15205.

[28] Fatt, P., and Katz, B. 1952. *J. Physiol.* 117: 109–128.

[29] Kuffler, S. W., and Yoshikami, D. 1975. *J. Physiol.* 251: 465–482.

Spontaneous Release of Multimolecular Quanta

The first evidence for packaging of ACh in multimolecular quanta was the observation by Fatt and Katz[28] that at the motor end plate, but not elsewhere in the muscle fiber, spontaneous depolarizations of about 1 mV occurred irregularly (Figure 13.8). They had the same time course as the potentials evoked by nerve stimulation. The spontaneous miniature end-plate potentials (MEPPs) were decreased in amplitude and eventually abolished by increasing concentrations of the ACh receptor antagonist curare, and were increased in amplitude and time course by acetylcholinesterase inhibitors, such as prostigmine (Figure 13.8C). These two pharmacological tests indicated that the potentials were produced by the spontaneous release of discrete amounts of ACh from the nerve terminal and ruled out the possibility that they might be due to single ACh molecules. Subsequently, patch electrode recordings demonstrated directly that the amount of current that flows through an individual ACh receptor will produce a potential change in the muscle fiber of approximately 1 μV. Thus, a spontaneous miniature potential is produced by the opening of about a thousand ACh receptors.Additional evidence confirmed, in a variety of different ways, that the spontaneous miniature potentials are indeed due to multimolecular packets of ACh liberated by the nerve terminal. For example, depolarization of the nerve terminal by passing a steady current through it causes an increase in frequency of the spontaneous activity, whereas muscle depolarization has no effect on frequency.[30] Botulinum toxin, which blocks release of ACh in response to nerve stimuli, also abolishes the spontaneous activity.[31] Shortly after denervation of a muscle, as the motor nerve terminal degenerates, the miniature potentials disappear.[32] Surprisingly, after an interim period, spontaneous potentials reappear in denervated frog muscle; these arise because of ACh released from Schwann cells that have engulfed segments of the degenerating nerve terminals by phagocytosis.[33]

Bernard Katz, 1950

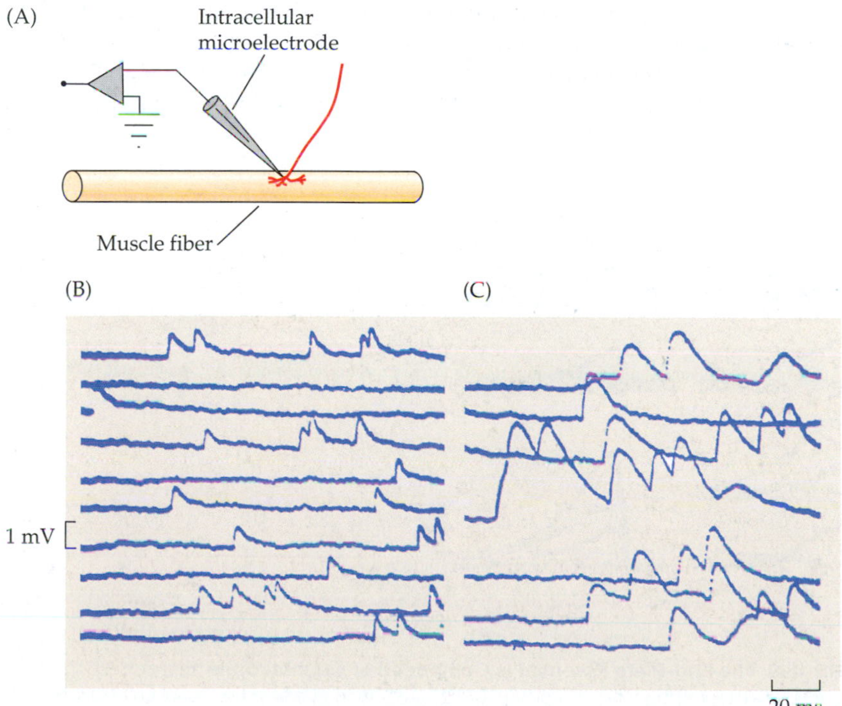

FIGURE 13.8 Miniature End-Plate Potentials at the Frog Neuromuscular Junction. (A) Intracellular recording from a muscle fiber in the region of the motor end plate. (B) Miniature end-pate potentials (MEPPs, about 1 mV in amplitude, occur spontaneously and are confined to the end-plate region of the muscle fiber. (C) After addition of prostigmine, which prevents acetylcholinesterase from hydrolyzing acetylcholine (ACh), miniature synaptic potentials are increased in amplitude and duration, but the frequency at which they occur is unchanged. This observation indicates that each miniature potential is due to a quantal packet of ACh, rather than to a single ACh molecule. (After Fatt and Katz, 1952.)

[30] del Castillo, J., and Katz, B. 1954. *J. Physiol.* 124: 586–604.

[31] Brooks, V. B. 1956. *J. Physiol.* 134: 264–277.

[32] Birks, R., Katz, B., and Miledi, R. 1960. *J. Physiol.* 150: 145–168.

[33] Reiser, G., and Miledi, R. 1989. *Brain Res.* 479: 83–97.

Fluctuations in the End-Plate Potential

A typical synaptic potential at the skeletal neuromuscular junction depolarizes the postsynaptic membrane by 50 mV to 70 mV, many times greater than the depolarization produced by a single quantum. In order to find out how this response to stimulation was related to the spontaneously released quanta, Fatt and Katz reduced the amplitude of the evoked synaptic potential by lowering the extracellular calcium and adding extracellular magnesium. Under these conditions the responses fluctuated in a stepwise manner, as shown in Figure 13.9A. Some stimuli produced no response at all—a failure of transmission. Some stimuli produced a response of about 1 mV in amplitude, similar in size and shape to an MEPP; others evoked responses that appeared to be two, three, or four times larger.

This remarkable observation led Fatt and Katz to propose the **quantum hypothesis:** that the single quantal events observed to occur spontaneously also represented the building blocks for the synaptic potentials evoked by stimulation. Normally the end-plate potential is made up of about 200 quantal units, and variations in its size are not obvious. In low calcium concentrations, the quantal *size* remains the same, but the quantum *content* is small—perhaps 1 to 3 quanta—and fluctuates randomly from trial to trial, resulting in stepwise fluctuations in the amplitude of the end-plate potential.

Statistical Analysis of the End-Plate Potential

Del Castillo and Katz realized that to test the quantum hypothesis adequately would require a statistical analysis.[34] Accordingly, they proposed that the motor nerve terminal contains a very large number of quantal packets of ACh (n), each of which has a probability (p) of being released in response to a nerve impulse, and that quanta are released independently—that is, the release of one has no influence on the probability of release of the next. Then, in a large number of trials, the mean number of quanta released per trial (m) would be given by np, and the number of times the response consisted of 0, 1, 2, 3, 4…, or x quanta would be given by the **binomial distribution** (Box 13.1). However, del Castillo and Katz could not test their expectation of a binomial distribution experimentally because they had had no

[34] del Castillo, J., and Katz, B. 1954. *J. Physiol.* 124: 560–573.

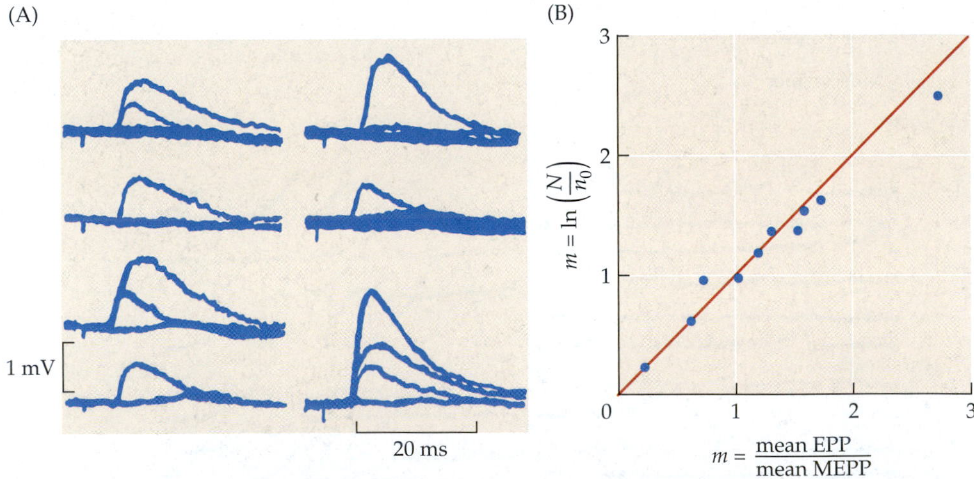

FIGURE 13.9 The End-Plate Potential Is Composed of Quantal Units that correspond to spontaneous miniature potentials. Presynaptic release of ACh at a frog neuromuscular junction was reduced by lowering the calcium concentration in the bathing solution. (A) Sets of intracellular records, each showing two to four superimposed responses to nerve stimulation. The amplitude of the end-plate potential (EPP) varies in a stepwise fashion; the smallest response corresponds in amplitude to a spontaneous miniature end-plate potential (MEPP). (B) Comparison of the mean quantal content (m) of the EPP, determined in two ways: by applying the Poisson distribution, $m = \ln(N/n_0)$ (ordinate), and by dividing the mean EPP amplitude by the mean MEPP amplitude (abscissa). Agreement of the two estimates supports the hypothesis that the EPP is composed of quantal units that correspond to spontaneous MEPPs. (A after Fatt and Katz, 1952; B after del Castillo and Katz, 1954a.)

BOX 13.1
Statistical Fluctuation in Quantal Release

When del Castillo and Katz saw fluctuations in the quantum content of the end-plate potential, they proposed that the release was a statistical process and that, consequently, it would be possible to predict the variations from one trial to the next by the binomial distribution. How does a statistical process lead to fluctuations, and how is it that these are described by the binomial equation? It is useful to look at a simple numerical example.

Suppose that a nerve terminal contains 3 quanta (a, b, and c), each with a 10% chance of being released upon arrival of an action potential, and suppose further that each time any one is released it is immediately replaced. If we call the number of available quanta n, and the release probability p, then in our example $n = 3$, $p = 0.1$. We will also define q as the probability that a quantum will *not* be released. So,

$$q = 1 - p = 0.9$$

In any trial, what is the likelihood that no quantum is released? We will call this p_0. The probability of one not being released is $q = 0.9$; and, for all three to not be released:

$$p_0 = q^3 = 0.729$$

To see a single response requires that one quantum be released and the other two not. The chance of this is pq^2, and there are three ways for it to happen: either a, b, or c is released, and the remaining two not.

So,

$$p_1 = 3pq^2 = 0.243$$

By similar reasoning:

$$p_2 = 3p^2q = 0.027$$

and

$$p_3 = p^3 = 0.001$$

The sum of all the probabilities ($q^3 + 3pq^2 + 3p^2q + p^3$) is 1.0, which means that we have accounted correctly for all possible release combinations.

In 1000 trials, then, we would expect to see 729 failures, 243 single releases, 27 doubles (yielding 54 quanta) and one response with 3 quanta, for a total of 300 quanta.

The average number of quanta released per trial, which we will call m, is 300/1000 = 0.3.

So,

$$m = np$$

The distribution is called a binomial distribution because $q^3 + 3q^2p + 3qp^2 + p^3$ are the terms we get when we multiply out the binomial (two variables) expression $(q + p)^3$.

If there are n quanta in the terminal, instead of 3, then the probabilities of seeing 0, 1, 2, etc. releases are given by the successive terms of the expansion of $(q + p)^n$.

The probability that x quanta will be released (p_x) is given by the relation:

$$p_x = \frac{p^x q^{n-x}(n!)}{(n-x)!\,x!}$$

The Poisson distribution is based on completely different reasoning. It simply describes how the occurrences of random (independent) events in time depend on their average over time, (i.e., on m). When the quantal release probability, p, is small (for practical purposes < 0.1) binomial predictions are not significantly different from those of the Poisson distribution, in which:

$$p_x = \frac{e^{-m}(m^x)}{x!}$$

It should be noted that while results that conform to the Poisson distribution support the hypothesis of a binomial distribution, they do not confirm it. However, experiments have shown that when the release probability is relatively high the fluctuations in quantum content are, indeed, described by the binomial equation.

way of measuring n or p. The only available measure was m. In order to deal with this difficulty, they reasoned as follows:

> Under normal conditions, p may be assumed to be relatively large, that is a fairly large part of the synaptic population responds to an impulse. However, as we reduce the Ca and increase the Mg concentration, the chances of responding are diminished and we observe mostly complete failures with an occasional response of one or two units. Under these conditions, when p is very small, the number of units x which make up the e.p.p. in a large series of observations should be distributed in the characteristic manner described by Poisson's law.[34]

The **Poisson distribution** approximates the binomial distribution when p is very small. The crucial difference is that to predict a Poisson distribution it is necessary to know only m, the mean number of quanta released per trial. In practice, this means measuring the average response amplitude and the average miniature potential amplitude. Then:

$$m = \frac{\text{mean amplitude of evoked potentials}}{\text{mean amplitude of miniature potentials}}$$

For a Poisson distribution, in N trials the expected number of responses containing x quanta is given by

$$n_x = (N)\frac{e^{-m}(m^x)}{x!}$$

If the end-plate potential amplitudes are distributed according to the Poisson equation, then m can also be determined from the number of failures, n_0. When $x = 0$ in the Poisson equation, $n_0 = Ne^{-m}$ (since both m^0 and $0! = 1$). Rearranging this result gives

$$m = \ln\left(\frac{N}{n_0}\right)$$

Del Castillo and Katz bathed a neuromuscular junction in a solution containing low calcium and high magnesium concentrations and recorded a large number of end-plate potentials evoked by nerve stimulation, as well as a large number of MEPPs. When they calculated m in these two entirely different ways, they found the estimates in excellent agreement, providing strong support for the idea of a Poisson distribution (Figure 13.9B).

A more stringent test of the applicability of the Poisson equation is to predict the entire distribution of response amplitudes, using only m and the mean amplitude of the unit potential (Figure 13.10). To do this, m is calculated from the ratio of the mean evoked potential amplitude to that of the mean MEPPs, as before. Then the number of expected responses containing 0, 1, 2, 3, ...units is calculated. To account for the slight variation in size of the unit, the expected number of responses containing one unit is distributed about the mean unit size, with the same variance as the spontaneous events (Figure 13.10, inset). Similarly, the predicted number of responses, containing 2, 3, or more units, are distributed about their means with proportionately increasing variances. The individual distributions are then summed to give the theoretical distribution shown by the continuous curve. The agreement with the experimentally observed distribution (bars) provides additional support for the hypothesis.

At many synapses, the probability of transmitter release is sufficiently high that it is not necessary to rely on the Poisson distribution to test the quantum hypothesis. Under such conditions, the binomial distribution can be tested directly. As before, if we assume that the terminal contains n units, each with an average probability (p) of being released by a nerve stimulus, then the relative occurrence of multiple events predicted by the binomial distribution is:

$$n_x = (N)\frac{p^x q^{n-x}(n!)}{(n-x)!\,x!}$$

where n_x is the number of responses containing x quanta, N is the number of trials, and $q = 1 - p$. Adherence of the release process to binomial statistics was first demonstrated at the crayfish neuromuscular junction.[35]

[35] Johnson, E. W., and Wernig, A. 1971. *J. Physiol.* 218: 757–767.

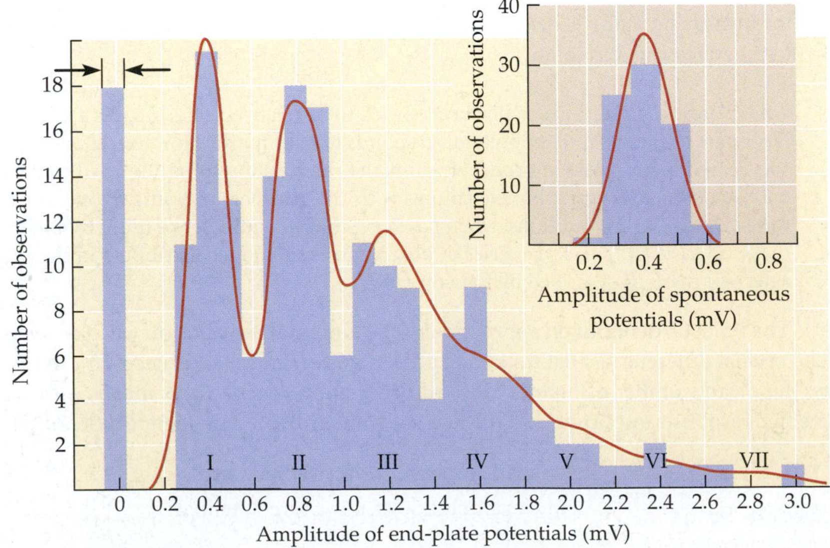

FIGURE 13.10 Amplitude Distribution of End-Plate Potentials at a mammalian neuromuscular junction in high (12.5 m*M*) magnesium solution. The histogram shows the number of end-plate potentials observed at each amplitude. The peaks of the histogram occur at 0 mV (failures) and at one, two, three, and four times the mean amplitude of the spontaneous MEPPs (inset), indicating responses comprising 1, 2, 3, and 4 quanta. The solid line represents the theoretical distribution of end-plate potential amplitudes calculated according to the Poisson equation and allowing for the spread in amplitude of the quantal size. The arrows indicate the predicted number of failures. (From Boyd and Martin, 1956.)

In summary, there is now ample evidence that transmitter is released in packets, or quanta. When the release probability (p) is very low, as in a low-calcium medium, the Poisson distribution provides a useful means of analyzing fluctuations. The applicability of the binomial hypothesis has been confirmed when the probability of release is high. In addition, binomial statistics can provide information as to whether changes in the amount of transmitter released arise from changes in the number of available quanta or in the probability of their release.

Quantum Content at Neuronal Synapses

One striking feature of the vertebrate nervous system is the reduction in mean quantum content as one moves from the neuromuscular junction, where there is little integration (m = 200–300), to autonomic ganglia (m = 2–20)[36,37] to synapses in the central nervous system (CNS; at which m can be as low as 1),[38,39] where postsynaptic cells are concerned with integrating myriad incoming signals. At the synapse between a primary afferent fiber from a muscle spindle and a spinal motor neuron, for example, the mean quantum content is about 1.[40] This does not mean, however, that transmission fails most of the time, as would be expected for a Poisson distribution. Rather, release conforms to binomial statistics, with a high probability ($p \sim 0.9$) and a low number of available quanta ($n \sim 1$).

Number of Molecules in a Quantum

Although it was clear from the experiments of Katz, Fatt, and del Castillo that at the neuromuscular junction one quantum contained more than one ACh molecule, the question of how many molecules were in a quantum remained. The first accurate determination was made by Kuffler and Yoshikami, who used very fine pipettes for ionophoresis of ACh onto the postsynaptic membrane of snake muscle.[29] By careful placement of the pipette, they were able to produce a response to a brief pulse of ACh that mimicked almost exactly the MEPP (Figure 13.11). To measure the number of molecules released by the pipette, ACh was released by repetitive pulses into a small (about 0.5 μl) droplet of saline under oil (Figure 13.12). The droplet was then applied to the end plate of a snake muscle fiber and the resulting depolarization measured. The response was compared with responses to droplets of exactly the same size containing known concentrations of ACh. In this way, the concentration of ACh in the test droplet was determined and the number of ACh molecules released per pulse was calculated. The pulse of ACh required to mimic an MEPP contained approximately 7000 molecules.

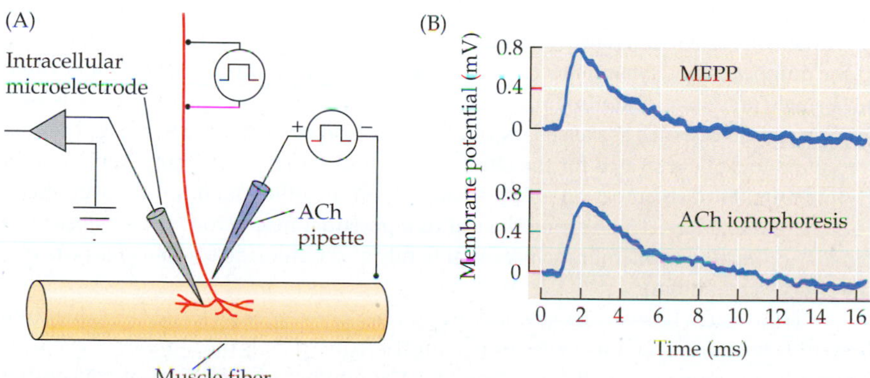

FIGURE 13.11 The Number of ACh Molecules in a Quantum, determined by mimicking an MEPP with an ionophoretic pulse of ACh. (A) An intracellular microelectrode records spontaneous MEPPs and the response to ionophoretic application of ACh. (B) An MEPP is mimicked almost exactly by an ionophoretic pulse of ACh. The rate of rise of the ionophoretic ACh pulse is slightly slower because the ACh pipette is further from the postsynaptic membrane than is the nerve terminal. (B after Kuffler and Yoshikami, 1975b.)

[36] Blackman, J. G., and Purves, R. D. 1969. *J. Physiol.* 203: 173–198.

[37] Martin, A. R., and Pilar, G. 1964. *J. Physiol.* 175: 1–16.

[38] Redman, S. 1990. *Physiol. Rev.* 70: 165–198.

[39] Edwards, F. A., Konnerth, A., and Sakmann, B. 1990. *J. Physiol.* 430: 213–249.

[40] Kuno, M. 1964. *J. Physiol.* 175: 81–99.

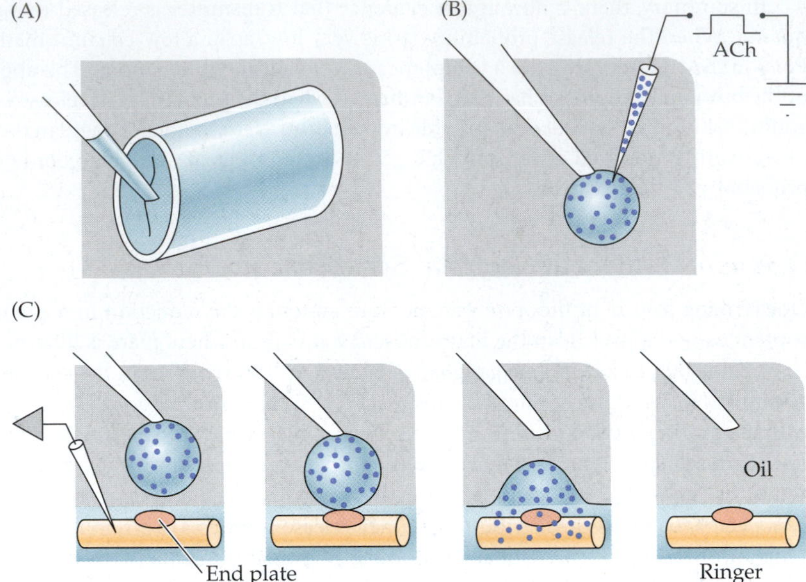

FIGURE 13.12 Assay of ACh Ejected from a Micropipette by Ionophoresis. (A) A droplet of fluid is removed from the dispensing capillary under oil. (B) ACh is injected into the droplet by a series of ionophoretic pulses, each identical to that used to mimic a spontaneous MEPP; (see Figure 13.10B). (C) After its volume is measured, the ACh-loaded droplet is touched against the oil–ringer interface at the end plate of a snake muscle, discharging its contents into the aqueous phase. The depolarization of the end plate is measured (not shown) and compared with that produced by droplets with known ACh concentration. Once the concentration in the test droplet is determined, the amount of ACh released per pulse from the electrode can be calculated. (After Kuffler and Yoshikami, 1975b.)

Number of Channels Activated by a Quantum

Given that a quantum of ACh consists of about 7000 molecules, one might expect that only a few thousand of these would actually combine with postsynaptic receptors at the neuromuscular junction—the remainder being lost to diffusion out of the cleft or hydrolysis by cholinesterases. This expectation is correct. The number of receptors activated by a quantum can be determined by comparing the conductance change that occurs during a miniature potential with that produced by a single ACh-activated channel.[41] Measurements of miniature end-plate currents by voltage clamp in frog muscle indicate a peak conductance change on the order of 40 nanosiemens (nS). A single frog ACh receptor has a conductance of about 30 picosiemens (pS) Thus, a miniature end-plate potential is produced by about 1300 open channels. (This corresponds to 2600 molecules of ACh, since it takes two molecules of ACh to open a channel; see Chapters 5 and 11.) This is similar to the number calculated by Katz and Miledi, who estimated the contribution of a single channel to the end-plate potential from noise measurements.[42] A similar value for the number of channels opened by a quantum of transmitter was obtained at glycine-mediated inhibitory synapses in lamprey brainstem cells.[43] Lower values are observed at other synapses. For example, at synapses on hippocampal cells, a quantal response corresponds to activation of 15 to 65 channels.[39,44]

Why are there such differences among synapses? A little thought leads to the conclusion that the number of postsynaptic receptors activated by a quantum of transmitter released from a single presynaptic bouton must be tailored to the size of the cell. In large cells with low input resistances, such as skeletal muscle fibers or lamprey Müller cells, a large number of receptors must be activated for the effect of a quantum to be significant. Activation of the same number of receptors on a very small cell, on the other hand, would overwhelm all other conductances, depolarizing the cell to a potential near zero if the synapse were excitatory or locking its membrane potential firmly at the chloride equilibrium potential if the effect were inhibitory.

How is the match between cell size and the number of receptors activated by a quantum achieved? Is the number of molecules in a quantum reduced, or is the number of available postsynaptic receptors lower? Precise values for the number of molecules of transmitter in a CNS synaptic vesicle are not available. However, the reported estimate for glutamate-containing vesicles is 4000,[45] which is the same order of magnitude as the number of ACh molecules in vesicles at the neuromuscular junction. On the other hand, analysis of quantal fluctuations at excitatory and inhibitory synapses on hippocampal cells suggests that the number of available postsynaptic receptors is much lower than at the neuromuscular junction. The amplitude of these quantal events at hippocampal synapses shows remarkably little

[41] Magleby, K. L., and Weinstock, M. M. 1980. *J. Physiol.* 299: 203–218.

[42] Katz, B., and Milei, R. 1972. *J. Physiol.* 244: 665–699.

[43] Gold, M. R., and Martin, A. R. 1983. *J. Physiol.* 342: 85–98.

[44] Jonas, P., Major, G., and Sakmann, B. 1993. *J. Physiol.* 472: 615–663.

[45] Villanueva, S., Fiedler, J., and Orrego, F. 1990. *Neuroscience* 37: 23–30.

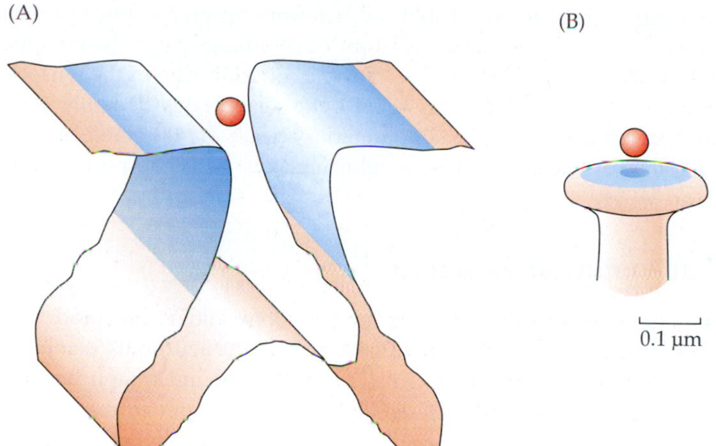

FIGURE 13.13 The Area of Postsynaptic Membrane Relative to the Size of a Synaptic Vesicle. (A) At the frog neuromuscular junction acetylcholine (Ach) receptors are packed at high density (~10,000/μm^2) over a large postsynaptic area (shaded blue). Accordingly, receptors outnumber ACh molecules, and the size of the quantal event varies with the variation in the number of molecules per quantum. (B) At a typical hippocampal synapse, postsynaptic receptors are packed less densely (~2800/μm^2) over a very small area (0.04 μm^2). As a result, the number of transmitter molecules in a quantum is sufficient to saturate the available receptors, and quantal events show very little fluctuation in amplitude.

variance, suggesting that the number of molecules released in a single quantum is always more than sufficient to activate all the available receptors. Conversely, at the neuromuscular junction, an increase in the number of transmitter molecules in a quantum can result in a larger quantal event.[46,47] The difference in available postsynaptic receptors deduced from quantal fluctuations is consistent with the difference in synaptic morphology (Figure 13.13). Thus, at the neuromuscular junction, receptors are packed at high density (~10,000/μm^2) throughout a large expanse of postsynaptic membrane, providing an essentially limitless sea of receptors for each quantum of transmitter. At a typical hippocampal synapse, the estimated postsynaptic receptor density is lower (~2800/μm^2),[48] and the area occupied by postsynaptic membrane is very small (0.04 μm^2);[49] thus, fewer than 100 postsynaptic receptors may be present.

Changes in Mean Quantal Size at the Neuromuscular Junction

Although the size of miniature synaptic potentials at any particular synapse tends to remain constant, exceptions occur under certain circumstances. For example, at the tadpole neuromuscular junction, various treatments that induce high rates of spontaneous transmitter release are followed by the appearance of small-mode MEPPs, which are a fraction of the normal quantal size.[50] At neuromuscular junctions in neonatal mice, a dominant fraction of the overall MEPP amplitude distribution consists of small-mode MEPPs, so that the overall distribution is heavily skewed toward the baseline, with no discernable quantal peaks.[51] Similar skewed amplitude distributions are seen during regeneration of nerve terminals following denervation.[52] The origin of these subminiature potentials is not clear. They may occur because of incomplete filling or emptying of synaptic vesicles (see the next section).

Conversely, spontaneous synaptic potentials, larger than the usual miniature potentials, are seen occasionally.[53] In some instances, these appear to be due to the spontaneous release of two or more quanta simultaneously; in other instances, their size shows no clear relation to normal quantal amplitude. Finally, in some myoneural diseases that afflict humans, such as myasthenia gravis, spontaneous miniature and evoked synaptic potentials are reduced in amplitude owing to a reduction in the number of receptors in the postsynaptic membrane.[54]

Nonquantal Release

In addition to being released by the motor nerve terminal in the form of individual quanta, ACh leaks continuously from the cytoplasm into the extracellular fluid. In other words, there is a steady nonquantal "ooze" of ACh from the presynaptic terminal.[55] Indeed, the amount of ACh that leaks from the nerve terminal in this way is about 100 times greater than that released in the form of spontaneous quanta. Its magnitude can be determined by comparing the total amount of ACh released from a muscle, measured biochemically, to the amount released as quanta, which is calculated from MEPP frequency and the total number of end plates in the muscle.

[46]Hartzell, H. C., Kuffler, S. W., and Yoshikami, D. 1975. *J. Physiol.* 251: 427–463.

[47]Salpeter, M. M. 1987. In *The Vertebrate Neuromuscular Junction*. Alan R. Liss, New York, pp. 1–54.

[48]Harris, K. M., and Landis, D. M. M. 1986. *Neuroscience* 19: 857–872.

[49]Schikorski, T., and Stevens, C. F. 1997. *J. Neurosci.* 17: 5858–5867.

[50]Kriebel, M. E., and Gross, C. E. 1974. *J. Gen. Physiol.* 64: 85–103.

[51]Erxleben, C., and Kriebel, M. E. 1988. *J. Physiol.* 400: 659–676.

[52]Denis, M. J., and Miledi, R. 1974. *J. Physiol.* 239: 571–594.

[53]Vautrin, J., and Kriebel, M. E. 1991. *Neuroscience* 41: 71–88.

[54]Drachman, D. B. 1994. *New England J. Med.* 330: 1797–1810.

[55]Vyskočil, F., Malomouzh, A. I., and Nikolsky, E. E. 2009. *Physiol. Res.* 58: 763–784.

Under normal circumstances, the slow dribble of ACh from the presynaptic terminal does not produce a postsynaptic response; the amount of cholinesterase in the synaptic cleft is sufficient to hydrolyze most of the ACh, so that the concentration in the synaptic cleft is not more than a few tens of n*M*. Its postsynaptic effect can be detected only when cholinesterase is inhibited. In contrast, the simultaneous release of 7000 molecules of ACh in a quantum locally overwhelms the enzyme, allowing ACh to reach its postsynaptic receptors and cause an MEPP.

Vesicles and Transmitter Release

Shortly after Katz and his colleagues demonstrated by electrophysiological methods that transmitter release was quantal, the first electron micrographs of the neuromuscular junction revealed that axon terminals contain many small membrane-bound synaptic vesicles (Figure 13.14).[56,57] Thus, it was suggested that a quantum of transmitter corresponds to the contents of one vesicle and that release occurs by a process of **exocytosis**, in which a vesicle fuses with the presynaptic plasma membrane and releases its contents into the synaptic cleft.[58]

Ultrastructure of Nerve Terminals

Ultrastructural studies provided support for the vesicle hypothesis of release. Many were first made at the neuromuscular junction. Subsequent experiments demonstrated that the principal morphological features of chemical synapses are similar throughout the nervous system, suggesting that at most chemical synapses, release occurs by exocytosis of transmitter-containing vesicles. A schematic view of a portion of the frog neuromuscular junction is shown in Figure 13.15, as it might appear if both the pre- and postsynaptic membranes were split open by the technique of freeze-fracturing. However, in practice a fracture would occur in one membrane or the other, not both at the same time.

In the upper portion of Figure 13.15, vesicles are lined up on the cytoplasmic face of the presynaptic membrane along regions of membrane thickening called **active zones**. Some are represented in the process of exocytosis. Along the active zones, intramembraneous particles protrude from the exposed fracture face of the cytoplasmic leaflet, and matching pits are seen on the fracture face of the outer leaflet. Sites of vesicle exocytosis are visualized as

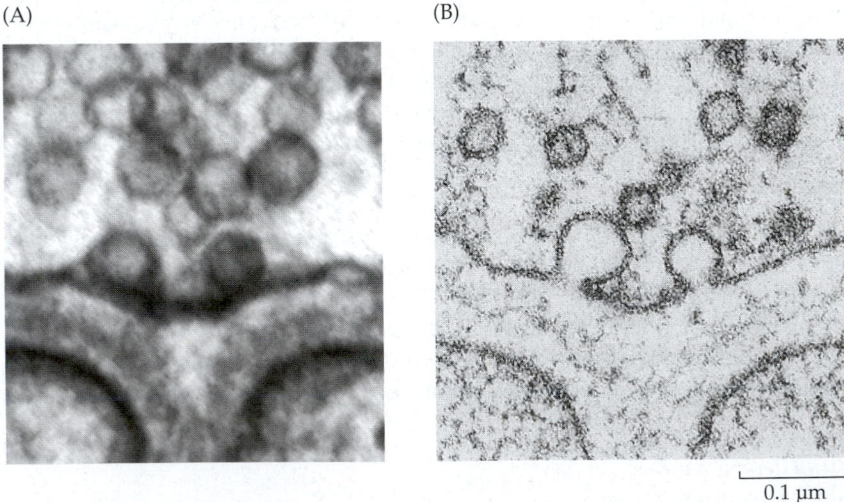

FIGURE 13.14 Release of Neurotransmitter by Synaptic Vesicle Exocytosis. High-power electron micrographs of frog neuromuscular junctions. (A) A cluster of synaptic vesicles within the presynaptic terminal contacts an electron-dense region of the presynaptic membrane, forming an active zone. (B) A single stimulus was applied to the motor nerve in the presence of 4-aminopyridine, a drug that greatly increases transmitter release by prolonging the action potential, and the tissue was frozen within milliseconds. Vesicles docked at the active zone have fused with the presynaptic membrane and released their contents into the synaptic cleft by exocytosis. (A micrograph kindly provided by U. J. McMahan; B from Heuser, 1989).

[56] Reger, J. F. 1958. *Anat. Rec.* 130: 7–23.

[57] Birks, R., Huxley, H. E., and Katz, B. 1960. *J. Physiol.* 150: 134–144.

[58] del Castillo, J., and Katz, B. 1956. *Prog. Biophys.* 6: 121–170.

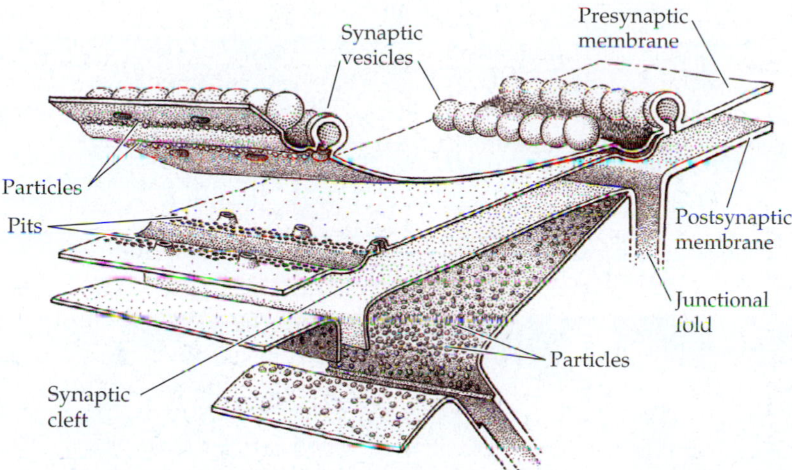

FIGURE 13.15 Synaptic Membrane Structure at the Frog Neuromuscular Junction. A Three-dimensional view of presynaptic and postsynaptic membranes, with each membrane split along its intramembranous plane as might occur in freeze-fracture. The cytoplasmic half of the presynaptic membrane at the active zone shows protruding particles on its fracture face whose counterparts are seen as pits on the fracture face of the outer membrane leaflet. Vesicles fusing with the presynaptic membrane give rise to pores and protrusions on the two fracture faces. The fractured postsynaptic membrane, in the region of the folds, shows a high concentration of particles on the fracture face of the cytoplasmic leaflet; these are ACh receptors. (Kindly provided by U. J. McMahan.)

large indentations in the cytoplasmic portion of the membrane and as fractured vesicle stalks in the outer portion. Particles on the inner fracture face of the postsynaptic membrane represent acetylcholine receptors.

Figure 13.16A shows a conventional transmission electron micrograph of a horizontal section through an active zone. It is equivalent to looking down onto an active zone from the nerve terminal cytoplasm. An orderly row of synaptic vesicles is lined up along either side of the band of dense material. Figure 13.16B shows a corresponding image of the fracture face of the cytoplasmic portion of the presynaptic membrane, which is equivalent to viewing the same region from the synaptic cleft. Rows of particles, each about 10 nm in diameter, flank the active zone on each side. Severed stalks, believed to indicate exocytotic openings, appear more laterally. As described earlier, electrophysiological experiments in which calcium buffers were injected into presynaptic terminals indicated a close association between calcium channels and release sites. Thus, at least some of the pits seen in Figure 13.16B might correspond to the voltage-activated calcium channels that trigger exocytosis. Results obtained with toxin-binding studies at the neuromuscular junction of the frog and the mouse are consistent with this idea.[59–61] Omega-conotoxin, which blocks neuromuscular transmission irreversibly by binding to presynaptic calcium channels,[62] was coupled to a fluorescent molecule. Upon microscopic examination, the fluorescence was found to be concentrated in narrow bands at 1-µm intervals, the same spacing as that of the active zones in the terminal.

A low-power freeze-fracture image of a frog neuromuscular junction is shown in Figure 13.16C. At the upper left, the first fracture face is that of the outer portion of the presynaptic membrane. The fracture then breaks across the synaptic cleft and exposes the face of the cytoplasmic portion of the postsynaptic membrane. Clusters of particles are seen along the sides of the postsynaptic folds. These are believed to correspond to the ACh receptors that are concentrated in this region of the end plate (see Chapter 11).[63–65]

Morphological Evidence for Exocytosis

An important experimental innovation developed by Heuser and Reese and their colleagues enabled frog muscle to be quick-frozen within milliseconds after a single shock to

[59] Robitaille, R., Adler, E. M., and Charlton, M. P. 1990. *Neuron* 5: 773–779.

[60] Cohen, M. W., Jones, O. T., and Angelides, K. J. 1991. *J. Neurosci.* 11: 1032–1039.

[61] Sugiura, Y. et al. 1995. *J. Neurocytol.* 24: 15–27.

[62] Olivera, B. M. et al. 1994. *Annu. Rev. Biochem.* 63: 823–867.

[63] Heuser, J. E., Reese, T. S., and Landis, D. M. D. 1974. *J. Neurocytol.* 3: 109–131.

[64] Peper, K. et al. 1974. *Cell Tissue Res.* 149: 437–455.

[65] Porter, C. W., and Barnard, E. A. 1975. *J. Membr. Biol.* 20: 31–49.

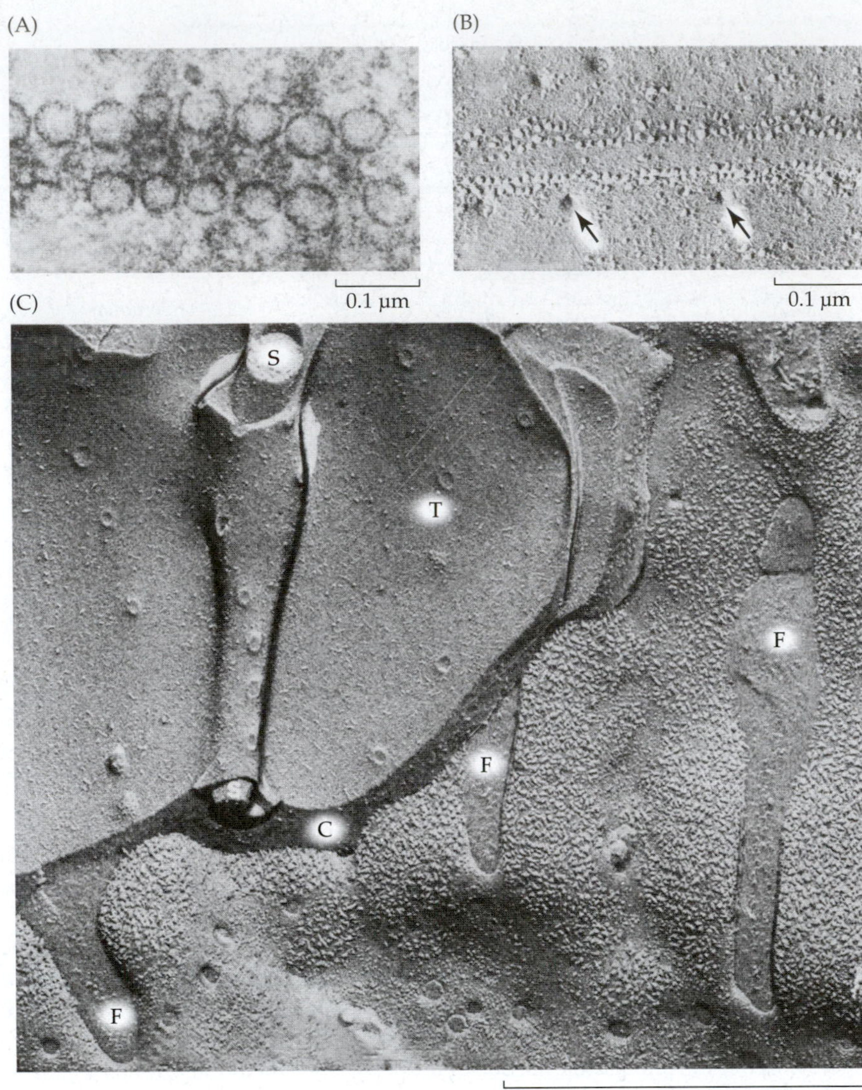

FIGURE 13.16 Structure of the Frog Neuromuscular Junction. (A) A transmission electron micrograph of a section through the nerve terminal, parallel to an active zone, shows two lines of vesicles. (B) Fracture face of the cytoplasmic half of the presynaptic membrane in an active zone The active zone is delineated by particles about 10 nm in diameter and flanked by pores (arrows) caused by fusion of synaptic vesicles with the membrane. (C) A low-power view of freeze-fractured synaptic region. The fracture passes through the presynaptic terminal (T), showing the fracture face of the outer leaflet, then crosses the synaptic cleft (C) to enter the postsynaptic membrane. The cytoplasmic leaflet on the fracture face between the postsynaptic folds (F) contains aggregates of ACh receptors. A Schwann cell process (S) passes between the nerve terminal and the muscle. (A from Couteaux and Pecot-Déchavassine, 1970; B and C from Heuser, Reese, and Landis, 1974.)

the motor nerve and then to be prepared for freeze-fracture.[66] With such an experiment it was possible to obtain scanning electron micrographs of vesicles caught in the act of fusing with the presynaptic membrane and to determine, with some accuracy, the time course of such fusion. To do this, the muscle is mounted on the undersurface of a falling plunger, with the motor nerve attached to stimulating electrodes. As the plunger falls, a stimulator is triggered, shocking the nerve at a selected interval before the muscle smashes into a copper block that is cooled to 4 K with liquid helium. An essential part of the experiment is that the duration of the presynaptic action potential is increased by addition of 4-aminopyridine (4-AP) to the bathing solution. This treatment greatly increases the magnitude and duration of quantal release evoked by a single shock and hence the number of vesicle openings seen in the electron micrographs (Figure 13.17A and B).

[66] Heuser, J. E. et al. 1979. *J. Cell Biol.* 81: 275–300.

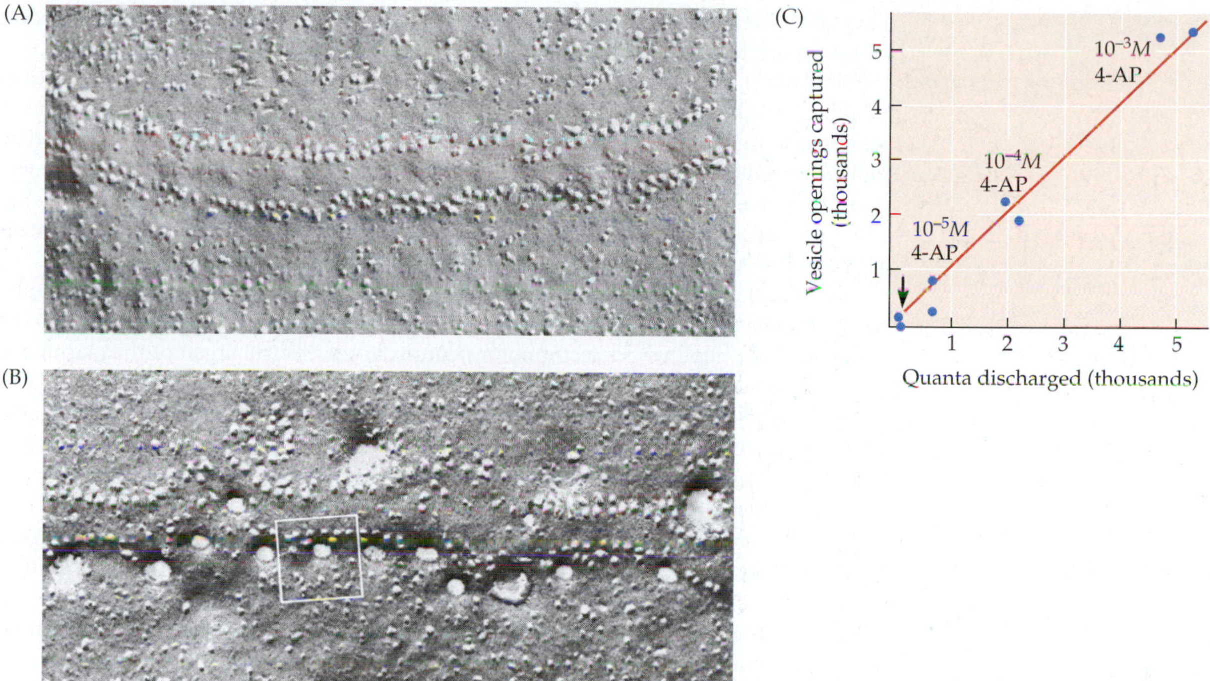

FIGURE 13.17 Vesicle Exocytosis Corresponds to Quantal Release. (A) Freeze-fracture electron micrograph of the cytoplasmic half of the presynaptic membrane in a frog nerve terminal (as if observed from the synaptic cleft). The region of the active zone appears as a slight ridge delineated by membrane particles (about 10 nm in diameter). (B) Similar view of a terminal that was frozen just at the time the nerve began to discharge large numbers of quanta (5 ms after stimulation). Holes (box) are sites of vesicle fusion. (C) Comparison of the number of vesicle openings (counted in freeze-fracture images) and the number of quanta released (determined from electrophysiological recordings). The diagonal line is the 1:1 relationship expected if each vesicle that opened released 1 quantum of transmitter. Transmitter release was varied by adding different concentrations of 4-AP (arrow indicates control, without 4-AP). (From Heuser et al., 1979; micrographs kindly provided by J. E. Heuser.)

Two important observations were made: First, the maximum number of vesicle openings occurred when stimulation preceded freezing by 3 to 5 ms, which corresponded to the peak of the postsynaptic current recorded from curarized, 4-AP–treated muscles in separate experiments. In other words, the maximum number of vesicle openings coincided in time with the peak postsynaptic conductance change determined physiologically. Second, the number of vesicle openings increased with 4-AP concentration, and the increase was related linearly to the estimated increase in quantum content of the end-plate potentials by 4-AP—again, obtained from separate physiological experiments (Figure 13.17C). Thus, vesicle openings were correlated both in number and in time course with quantal release. In later experiments, Heuser and Reese characterized the time course of vesicle openings in greater detail, showing that openings first increase during a 3 to 6 ms period after stimulation and then decrease over the next 40 ms.[67]

Release of Vesicle Contents by Exocytosis

A prediction of the hypothesis that neurotransmitter release occurs by vesicle exocytosis is that stimulation will release the total soluble contents of synaptic vesicles. This prediction was first tested, not in neurons, but in adrenal medullary cells, from which chromaffin granules could be purified and their contents analyzed.[68] Chromaffin granules are organelles analogous to but much larger than synaptic vesicles; they contain epinephrine, norepinephrine, ATP, the synthetic enzyme dopamine β-hydroxylase, and proteins called chromogranins. All of these components are released in response to stimulation of the

[67] Heuser, J. E., and Reese, T. S. 1981. *J. Cell Biol.* 88: 564–580.

[68] Kirshner, N. 1969. *Adv. Biochem. Psychopharmacol.* 1: 71–89.

adrenal medulla, and they appear in the perfusate in exactly the same proportions as are found within the purified granules.

Similarly, there is good correspondence in neurons between vesicle contents and release, although it is difficult to isolate pure populations of synaptic vesicles from nerve terminals in order to determine their contents. For example, small synaptic vesicles in sympathetic neurons contain norepinephrine and ATP; the larger dense-core vesicles contain, in addition, dopamine β-hydroxylase and chromogranin A. Stimulation of sympathetic axons results in the release of all of these vesicle constituents.[69] Similarly, vesicles isolated from cholinergic neurons contain ATP as well as ACh, and both are released by stimulation of cholinergic nerves.[70]

The idea that one quantum of transmitter corresponds to the contents of one synaptic vesicle has been examined quantitatively for cholinergic neurons. Vesicles purified from the terminals of the cholinergic electromotor neurons in the electric organ of the marine ray *Narcine brasiliensis* (a relative of *Torpedo californica*) were found to contain about 47,000 molecules of ACh.[71] If synaptic vesicles at the frog neuromuscular junction had the same intravesicular ACh concentration, then, making allowance for their smaller size, they would contain 7000 molecules of ACh. This is in excellent agreement with electrophysiological estimates of the number of ACh molecules in a quantum.[29]

In summary, there is now much evidence that synaptic vesicles are the morphological correlate of the quantum of transmitter, each vesicle containing a few thousand transmitter molecules. Vesicles can release their contents by exocytosis both spontaneously at a low rate (producing miniature synaptic potentials) and in response to presynaptic depolarization. As mentioned earlier, there is evidence that at some specialized synapses in the retina, depolarization can release transmitter through transport proteins in the presynaptic membrane, a mechanism that is nonquantal, not mediated by vesicle exocytosis, and not dependent on calcium influx.[72,73]

[69]Smith, A. D. et al. 1970. *Tissue Cell* 2: 547–568.

[70]Silinsky, E. M., and Redman, R. S. 1996. *J. Physiol.* 492: 815–822.

[71]Wagner, J. A., Carlson, S. S., and Kelly, R. B. 1978. *Biochemistry* 17: 1199–1206.

[72]Cammack, J. N., and Schwartz, E. A. 1993. *J. Physiol.* 472: 81–102.

[73]Cammack, J. N., Rakhilin, S. V., and Schwartz, E. A. 1994. *Neuron* 13: 949–960.

[74]Penner, R., and Neher, E. 1989. *Trends Neurosci.* 12: 159–163.

[75]Angleson, J. K., and Betz, W. J. 1997. *Trends Neurosci.* 20: 281–287.

Monitoring Exocytosis and Endocytosis in Living Cells

The first quantitative studies of exocytosis were made on dissociated non-neuronal secretory cells, such as mast cells and chromaffin cells, in which the discharge of large, dense-cored secretory granules could be followed with a variety of techniques. These techniques included light microscopy of granules labeled with fluorescent dye, measurement of the increase in membrane capacitance produced by incorporation of vesicle membrane into the plasma membrane of the cell, and amperometry, which detects the amines released in response to stimulation.[74,75] One such experiment is illustrated in Figure 13.18. Exocytosis was observed directly in cultured chromaffin cells by evanescent-wave microscopy, a fluorescence microscopy technique that greatly reduces background fluorescence by exciting only a 300-nm–thick layer of cytosol. Chromaffin granules within the cells were labeled

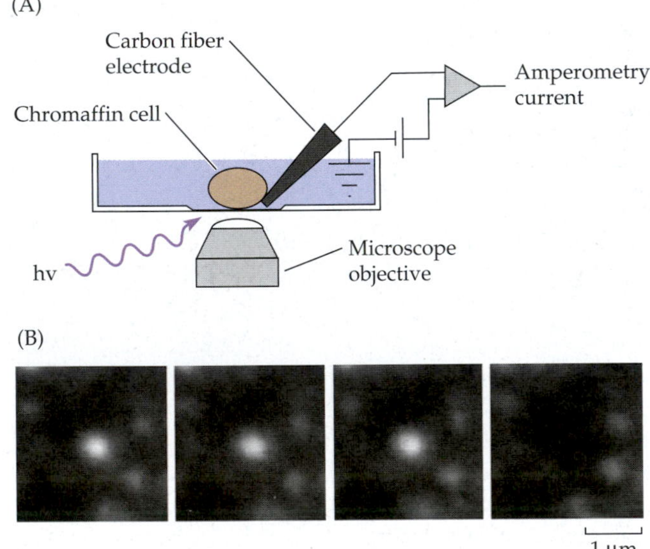

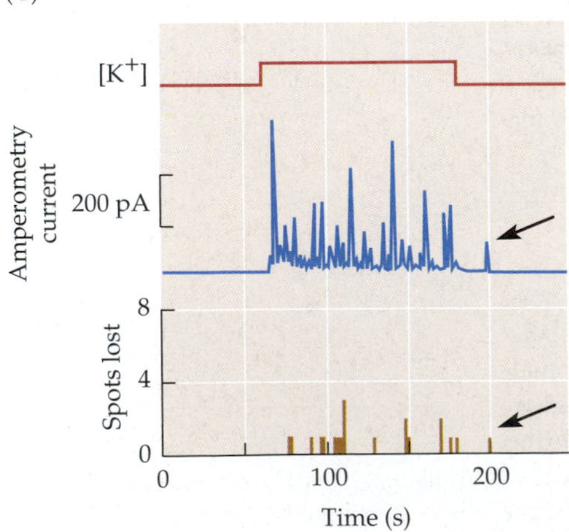

FIGURE 13.18 Exocytosis Observed in Living Cells. (A) Chromaffin cells growing on a glass cover slip in cell culture were labeled with a fluorescent dye, which becomes concentrated in chromaffin vesicles. Individual vesicles docked at the plasma membrane were visualized by evanescent-wave microscopy. At the same time, release of catecholamines was detected by amperometry. (B) High-power images of a single chromaffin vesicle at 2-second intervals after the cell was stimulated with high potassium. The spot disappears abruptly and permanently as the vesicle undergoes exocytosis and releases its fluorescent contents. (C) The time course of exocytosis in response to an increase in extracellular potassium concentration, as recorded by amperometric detection of catecholamine release and the disappearance of fluorescent spots. Note the coincidence of release and spot disappearance (the arrows mark one example). More events are recorded by amperometry than by fluorescence because the amperometric electrode detects exocytosis over a large part of the cell, while only a small portion of the cell surface is imaged by evanescent-wave microscopy. (After Steyer, Horstmann, and Almers, 1997; micrographs kindly provided by W. Almers.)

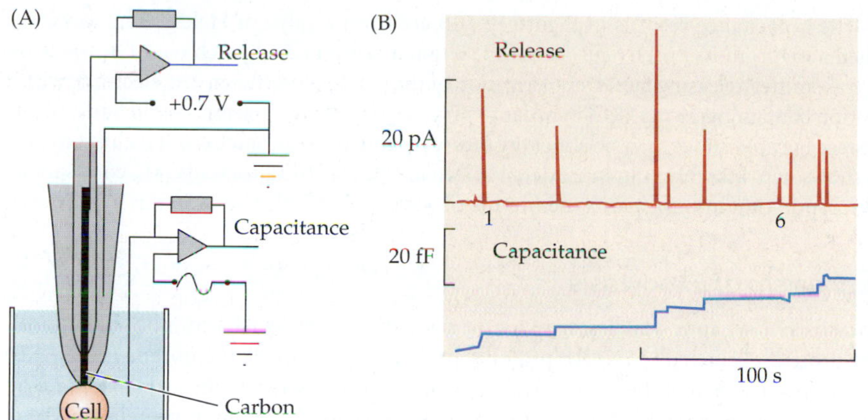

FIGURE 13.19 Coincident Increases in Membrane Capacitance and Release of Catecholamines from chromaffin cells. (A) A carbon fiber electrode inside the patch pipette measures catecholamine release by amperometry, while at the same time the electrode is used to measure capacitance within the patch. (B) Simultaneous recording of catecholamine release (top trace) and capacitance (bottom trace). All exocytotic events detected by catecholamine release coincide with increases in capacitance. Capacitance units are in femtofarads (1fF = 10^{-15} Farads). (After Albillos et al., 1997.)

with a fluorescent dye. The release of catecholamines was measured by amperometry—a very sensitive method in which a carbon fiber microelectrode is used to detect transmitters by the current they produce when they are oxidized. With evanescent-wave microscopy, individual fluorescent vesicles could be seen to dock at the plasma membrane and then disappear as they released their fluorescent contents by exocytosis.[76,77] Each time a fluorescent vesicle disappeared, the release of a quantum of transmitter was detected by amperometry.

The experiment shown in Figure 13.19 illustrates the use of capacitance measurements to monitor exocytosis. A cell-attached patch is made on a chromaffin cell, and inside the patch pipette a carbon fiber electrode detects catecholamine release. A sinusoidal signal applied to the bathing solution is used to measure changes in capacitance across the patch. Catecholamine release by exocytosis is accompanied by stepwise increases in capacitance arising from the addition of the granule membrane to the surface of the patch.[78]

Dye release and capacitance increases associated with exocytosis of vesicles have been made in CNS nerve terminals as well. Two examples are shown in Figure 13.20. Figure 13.20A shows records of dye release from individual boutons on cultured hippocampal neurons in response to presynaptic nerve stimulation.[79] Vesicles were first loaded with a fluorescent lipid marker, with roughly five vesicles/bouton being stained. Stimulation of the presynaptic nerve resulted in step-wise drops in bouton fluorescence, indicating the exocytosis of individual vesicles and consequent dispersal of the dye.

[76] Lang, T. et al. 1997. *Neuron* 18: 857–863.

[77] Steyer, J. A., Horstmann, H., and Almers, W. 1997. *Nature* 388: 474–478.

[78] Albillos, A. et al. 1997. *Nature* 389: 509–512.

[79] Richards, D. A. 2009. *J. Physiol.* 587: 5073–5080.

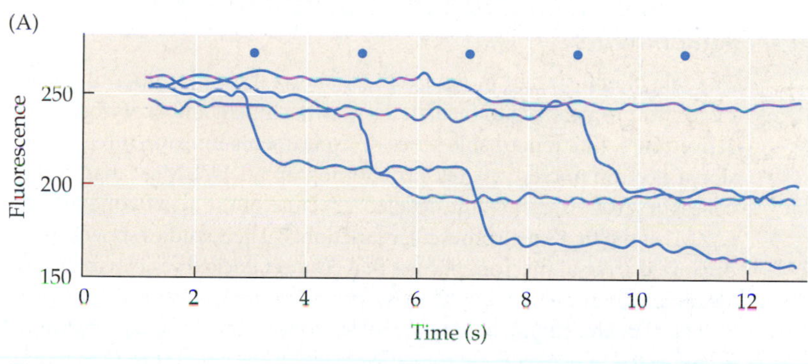

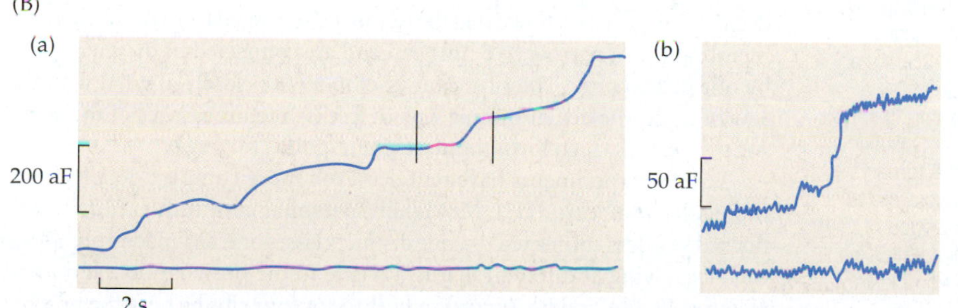

FIGURE 13.20 Release of Synaptic Vesicles from Presynaptic Nerve Terminals, monitored by loss of fluorescent dye (A) and increase in membrane capacitance (B). (A) Fluorescence records from four terminal boutons on a cultured hippocampal neuron. Vesicles in the bouton had been loaded previously with a fluorescent lipid marker. Stimulation of the presynaptic nerve (dots) caused step-wise drops in fluorescence, signaling single vesicle discharges. (B) Capacitance of a cell-attached membrane patch on the transmitter-releasing face of a calyx of Held. The patch was depolarized by perfusing the electrode with 25 m*M* KCl, starting at the beginning of the record. Membrane capacitance (upper trace) increased as vesicle membranes were incorporated into the patch (a). (b) is a magnified record of the segment between the two vertical lines in a, showing step-wise jumps. Capacitance calibration is in attofarads (1 aF = 10^{-18}F). (Records in A from Richards, 2009; in B from He et al., 2006.)

[80]He, L. et al. 2006. *Nature* 444: 102–105.

[81]Pang, Z. P., and Südhoff, T. C. 2010. *Curr. Opin. Cell Biol.* 22: 496–505.

[82]Pavlos, N. J. et al. 2010. *J. Neurosci.* 30: 13441–13453.

[83]Han, G. A. et al. 2010. *J. Neurochem.* 115: 1–10.

[84]Chen, X. et al. 2002. *Neuron* 33: 397–409.

[85]Harlow, M. L. et al. 2001. *Nature* 409: 479–484.

The capacitance records in Figure 13.20B are from a calyx of Held.[80] The calyx was pulled away from its postsynaptic neuron so that a cell-attached patch could be made on the transmitter-releasing face. Depolarization of the patch, by perfusing the electrode with a solution containing 25 m*M* KCl in order to produce transmitter release, resulted in a steady increase in capacitance as vesicle membranes were added to the patch (a). The inset in part (b) shows step-wise jumps in more detail. A step increase of 100 attofarads (aF) corresponds to incorporation into the patch membrane of a synaptic vesicle about 50 nm in diameter.

Mechanism of Exocytosis

Exocytosis of synaptic vesicles involves the cooperative action of a number of intracellular proteins, principally SNARE proteins attached to the vesicle membrane (v-SNARE) and the *target* region of the nerve terminal membrane (t-SNARE). The term SNARE is an abbreviation of **S**oluble ***N***-ethylmaleimide-sensitive factor **A**ttachment protein **RE**ceptor.

The interaction between the vesicle and membrane snares leading to exocytosis is illustrated in Figure 13.21.[81] Attached to the vesicle membrane is the SNARE protein synaptobrevin, together with a calcium-sensor, synaptotagmin. Two SNARE proteins are attached to the synaptic membrane, syntaxin, and SNAP-25. Figure 13.21A shows the vesicle in the *docked* position over the active zone, held in place by rab3 tethers (not shown). Rab proteins are a family of GTPases involved in a number of cytoplasmic functions, including targeting synaptic vesicles to appropriate membrane sites.[82]

The t-SNARE syntaxin is held in a closed state by the regulatory protein Munc18-1, and in that configuration it is unable to interact with other SNARE proteins.[83] Docking is followed by a priming stage (13.21B) in which the configuration of the Munc18-1/syntaxin complex is altered so that syntaxin is in its open state and assembles with synaptobrevin and SNAP 25 to form a four-helix bundle that binds the vesicle in close contact with the plasma membrane. Attached to the bundle is a cytoplasmic regulatory protein complexin, which serves to stabilize the interaction between the synaptobrevin and syntaxin helices.[84] Upon calcium influx to the nerve terminal, the incoming calcium binds to the synaptotagmin, which in turn, binds to the snare complex, displacing complexin and initiating pore formation (Figure 13.21C).

(A) Docked vesicle

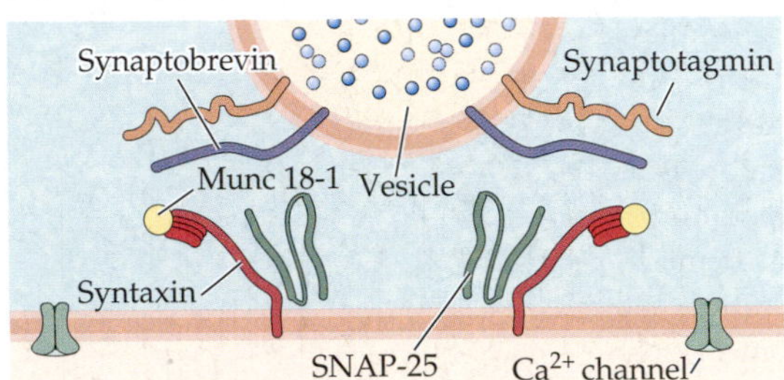

(B) Vesicle in primed position

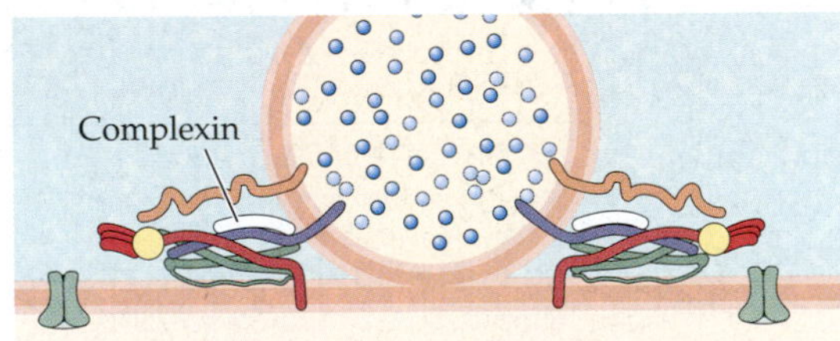

(C) Entering Ca^{2+} binds to synaptotagmin

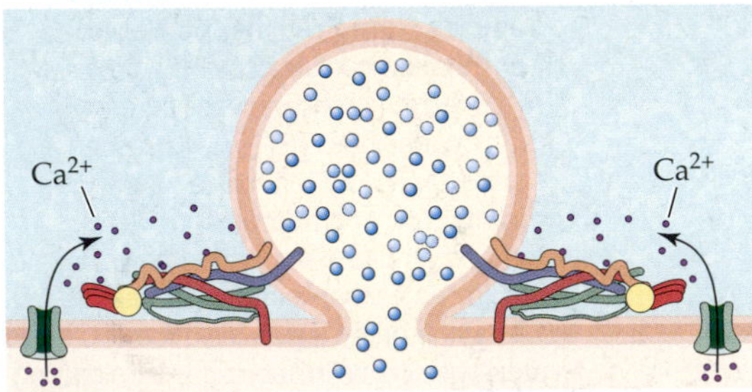

FIGURE 13.21 Mechanism of Exocytosis. (A) Docked vesicle. The vesicle membrane contains the SNARE protein synaptobrevin and the calcium sensor synaptotagmin. Two additional SNARE proteins, syntaxin and SNAP-25, are anchored to the nerve terminal plasma membrane at the active zone. Syntaxin is held in a folded, inactive configuration by Munc18-1. (B) Vesicle in the primed position. Syntaxin has entered its open configuration and formed a ternary SNARE complex with synaptobrevin and the two arms of SNAP-25. The complex is stabilized by the presence of complexin. (C) Calcium binds to synaptotagmin, which in turn binds to the SNARE complex, displacing complexin and inducing pore formation.

High-Resolution Structure of Synaptic Vesicle Attachments

The scheme illustrated in Figure 13.21 shows how a number of protein molecules are arranged to mediate vesicle attachment and exocytosis in the active zones. In a remarkable series of experiments utilizing three-dimensional electron microscopy (electron tomography), U. J. McMahan and his colleagues have examined the detailed structure of the active zone material (AZM) at the frog's neuromuscular junction.[85] Their studies show a highly organized array of macromolecules that connect docked vesicles to the presynaptic membrane. Images of these structures are shown in Figure 13.22. A central **beam**, running along the active zone in the presynaptic terminal is connected to two flanking rows of vesicles by a series of **ribs**. Other, more detailed, images have shown that the beams are separated from the presynaptic membrane by a narrow (≈ 5 nm) gap and are connected to the membrane by one or two pegs. The rib-beam assemblies are closely aligned with the row of macromolecular bumps seen in freeze-fracture images of the active zone (panel i), which form the attachment points for pegs.

Further experiments have revealed even more complex details of the molecular structure (U. J. McMahan, personal communication). Each docked vesicle is connected to three distinct classes of AZM macromolecules, each class lying at different depths vertical to the presynaptic membrane (Figures 13.23A and B). Accordingly, three or four ribs lie near the presyn-

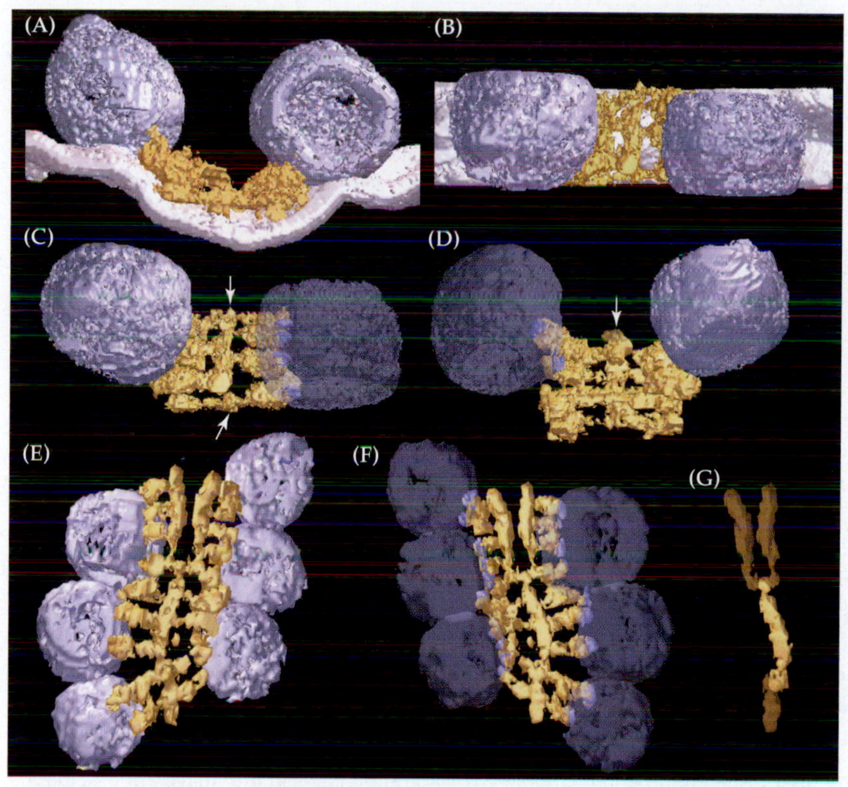

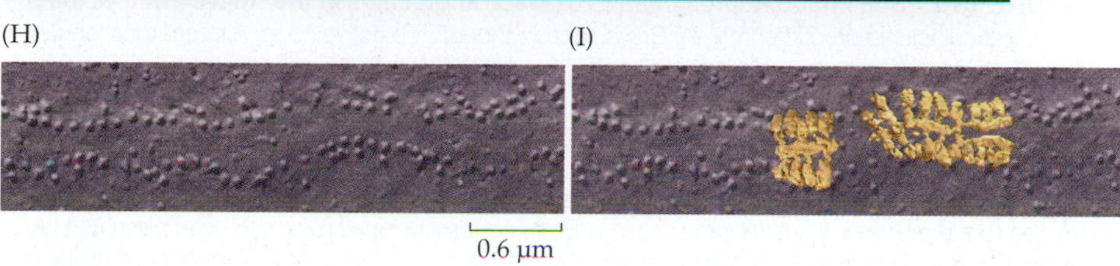

FIGURE 13.22 Organization of Active Zone Material at the frog's neuromuscular junction. Three-dimensional images obtained by electron microscope tomography. (A,B): Transverse and horizontal views of a short section of an active zone, showing two vesicles (blue) and active zone material (yellow). (C–F): Additional views with the presynaptic membrane removed. Active zone material is organized into central ribs, with a series of transverse beams connecting them to the vesicles. Vesicles in (C) and (D) are made transparent to reveal the points of beam attachment. Not visible are pegs that connect the beams to macromolecules in the presynaptic membrane. (G): Only the beams are shown. (H,I): Size and spacing of the rib-beam structure closely matches macromolecular bumps seen in a freeze-fracture replica of an active zone. (From Harlow, et al. 2001; image kindly provided by Dr. U. J. McMahan.)

aptic membrane, two **spars** are deep to the ribs, and five to seven **booms** are deep to the spars. Whereas the peripheral ends of the ribs, spars, and booms are connected to distinct domains on the membrane of docked vesicles, the central ends of each class are attached to distinct classes of macromolecules in a vertical assembly at the midline of the AZM, thus: ribs to beams, spars to **steps**, and booms to **masts**. Another class of core macromolecules, the **top masts**, link some of the nearby undocked vesicles to the masts; pegs link the ribs to the macromolecules in the presynaptic membrane thought to include calcium channels; and pins connect the vesicle membrane to the presynaptic membrane away from the main body of the AZM.

When the neuromuscular junctions are fixed while the axons are being electrically stimulated at high frequency (10 Hz) different stages of vesicle recycling can be captured. Thus is it possible to see former docked vesicles in various stages of merging with the presynaptic membrane, and undocked vesicles in stages of replacing previously docked vesicles. The varied images that are observed suggest a sequence of events like that shown in Figure 13.23C. Docked vesicles fuse with the presynaptic membrane to release transmitter while still associated with the AZM (a). As the fused vesicles flatten into the presynaptic membrane, they first dissociate from the booms, then the spars, and finally the ribs (b–e). Undocked vesicles that come to occupy the vacated docking sites on the presynaptic membrane initially form associations with the booms, then the spars, and finally the ribs and pins before contacting the presynaptic membrane.

Initial electron tomography studies on active zones of the mouse neuromuscular junction have shown that the superficial portion of the AZM is also made up of ribs, beams, and pegs linked to docked vesicles and presynaptic membrane macromolecules. The active zones in mouse are much smaller than those in the frog, and the gross positioning of the

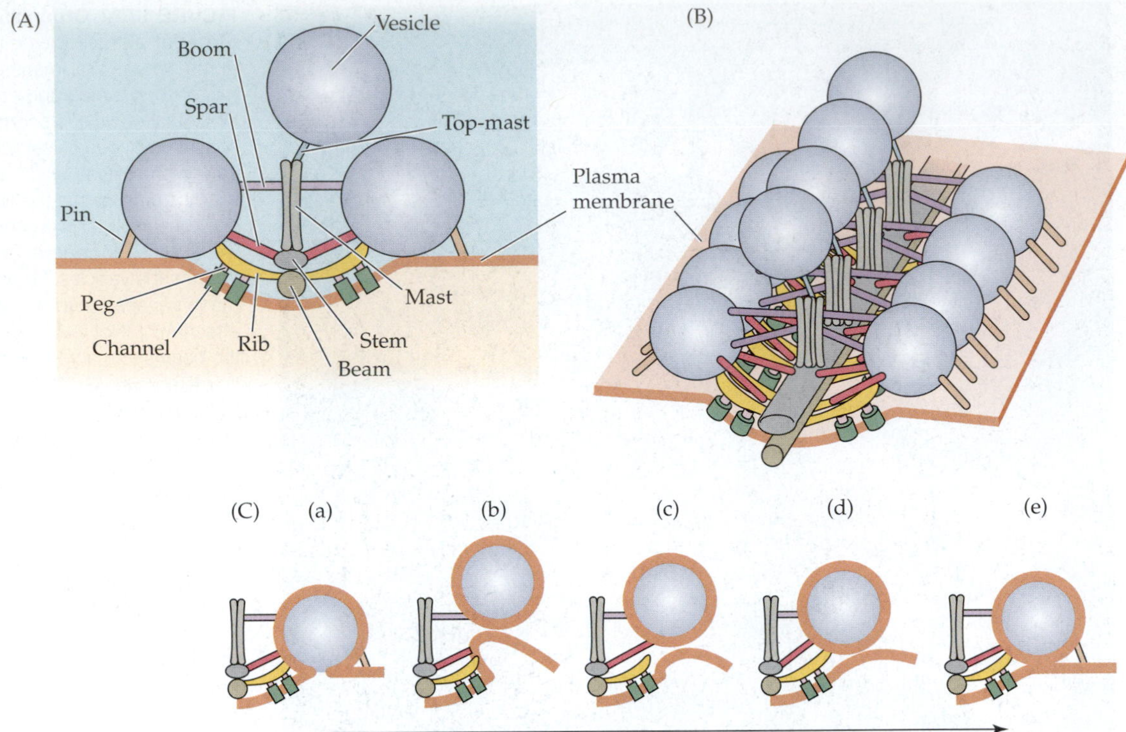

FIGURE 13.23 Schematic Representation of Macromolecules in the Active Zone at the neuromuscular junction of the frog. (A) Cross section though the active zone. A beam runs longitudinally along the active zone membrane and is connected laterally to adjacent docked synaptic vesicles (SV) by ribs. Ribs are connected to channels in the presynaptic membrane (PM) by short pegs. At the next level, a step gives rise to spars that extend out to the vesicles. Above the step, a multimolecular mast extends into the cytoplasm and is connected to the docked vesicles by booms. A topmast tethers an undocked vesicle to the mast. Lateral to the active zone the vesicles are connected to the presynaptic membrane by pins. (B) Three-dimensional view of the components in A, showing the multiplicity of molecular attachments. (C) Scheme for replacement of docked vesicles after exocytosis. (a): vesicle fuses with the membrane and releases transmitter while still associated with the active zone material. (b–e): As the fused vesicle collapses into the postsynaptic membrane, it dissociates first from the pins, then the booms, spars, and ribs, while the replacement vesicle associates first with booms, then spars, ribs, and pins. (Courtesy Dr. U. J. McMahan.)

AZM relative to the docked vesicles is somewhat different.[86] Altogether, the results indicate the AZM at synapses is a multifunctional organelle that directs all phases of exocytosis. It seems likely that the proteins revealed by biochemistry to be involved in vesicle docking and fusion with the presynaptic membrane (see Figure 13.21) will turn out to be components of AZM macromolecules.

Reuptake of Synaptic Vesicles

At neuromuscular, ganglionic, and CNS synapses, periods of intense stimulation have been shown to deplete synaptic vesicles and increase the surface area of the axon terminal, indicating that after releasing their contents, empty vesicles flatten out and become part of the terminal membrane.[87–89] An example is shown in the electron micrographs (Figure 13.24). At the resting synapse (A), synaptic vesicles are clustered over a presynaptic membrane thickening (release site). In a specimen fixed after prolonged stimulation (B), release sites are devoid of vesicles. When fixation is delayed for an hour after stimulation (C), synapses have recovered their initial morphological features.

How is the vesicle population restored? Heuser and Reese found that components of the vesicle membrane are retrieved and recycled into new synaptic vesicles.[90] They studied recycling of vesicles in frog motor nerve terminals by stimulating nerve–muscle preparations in the presence of horseradish peroxidase (HRP), an enzyme that catalyzes the formation

[86] Sharuna, N. et al. 2009. *J. Comp. Neurol.* 513: 457–468.

[87] Ceccarelli, B., and Hurlbut, W. P. 1980. *Physiol. Rev.* 60: 396–441.

[88] Dickinson-Nelson, A., and Reese, T. S. 1983. *J. Neurosci.* 3: 42–52.

[89] Wickelgren, W. O. et al. 1985. *J. Neurosci.* 5: 1188–1201.

[90] Heuser, J. E., and Reese, T. S. 1973. *J. Cell Biol.* 57: 315–344.

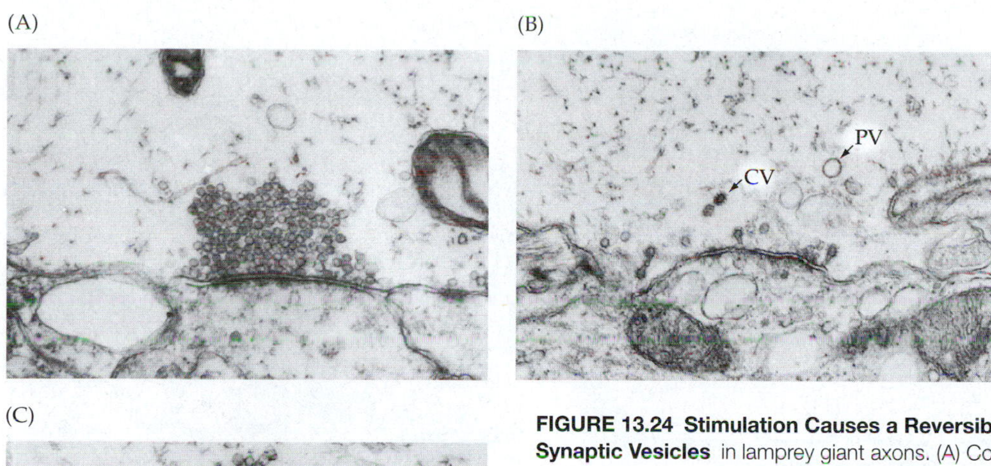

FIGURE 13.24 Stimulation Causes a Reversible Depletion of Synaptic Vesicles in lamprey giant axons. (A) Control synapse fixed after 15 minutes in saline. Synaptic vesicles are clustered at the presynaptic membrane. (B) Synapse fixed after stimulation of the spinal cord for 15 minutes at 20 Hz. Note the depletion of synaptic vesicles, the presence of coated vesicles (CV), pleomorphic vesicles (PV), and the expanded presynaptic membrane. (C) Synapse fixed 60 minutes after cessation of stimulation. Note similarities to the control synapse in A. (Micrographs kindly provided by W. O. Wickelgren.)

of an electron-dense reaction product. When electron micrographs of terminals fixed after short periods of electrical stimulation were examined, HRP was found primarily in coated vesicles around the outer margins of the synaptic region, suggesting that these vesicles had been formed from the terminal membrane by endocytosis and, in the process, had captured HRP from the extracellular space (Figure 13.25A). HRP also appeared, after a delay, in synaptic vesicles (Figure 13.25B). Synaptic vesicles loaded in this way with HRP could then be depleted of the enzyme by a second period of stimulation in HRP-free medium (Figure 13.25C), thus supporting the idea that the previously recaptured membrane and enclosed HRP had been recycled into the vesicle population from which release occurs.

After particularly intense stimulation in the presence of HRP, large, uncoated pits and cisternae containing HRP are seen (see Figure 13.25A). Such uncoated pits and cisternae appear to represent bulk, nonselective invaginations of excess presynaptic membrane.[88] Presumably, coated vesicles then remove specific components from such cisternae to be recycled, directly or through endosomes (Figure 13.26).

Vesicle Recycling Pathways

Proposed pathways for vesicle recycling are illustrated in Figure 13.26.[91] The best characterized of these is the classic endocytotic pathway, in which vesicles flatten completely into the presynaptic membrane following exocytosis, and components are retrieved by endocytosis via clathrin-coated pits.[92] Other proteins in addition to clathrin serve to identify the appropriate constituents for recycling. After retrieval, vesicles lose their coat to reform synaptic vesicles and refill with transmitter (orange cores). Following intense stimulation, a second pathway is activated in which vesicles pass through endosome-like structures that appear transiently in the cytoplasm. The endosomes are probably the result of direct retrieval of large pieces of membrane (bulk endocytosis).[93]

Finally, experiments, such as those shown in Figure 13.20, have suggested that during the release process, vesicles do not always collapse fully into the terminal membrane but instead form a transient fusion pore, after which they are recovered directly into the cytoplasm. The vesicle contents are discharged partially or completely, depending on the size

[91] Miller, T. M., and Heuser, J. E. 1984. *J. Cell Biol.* 98: 685–698.

[92] Doherty, G. J., and McMahon, H. T. 2009. *Annu. Rev. Biochem.* 78: 815–902.

[93] Wu, W., and Wu, L-G. 2007. *Proc. Natl. Acad. Sci. USA* 104: 10234–10239.

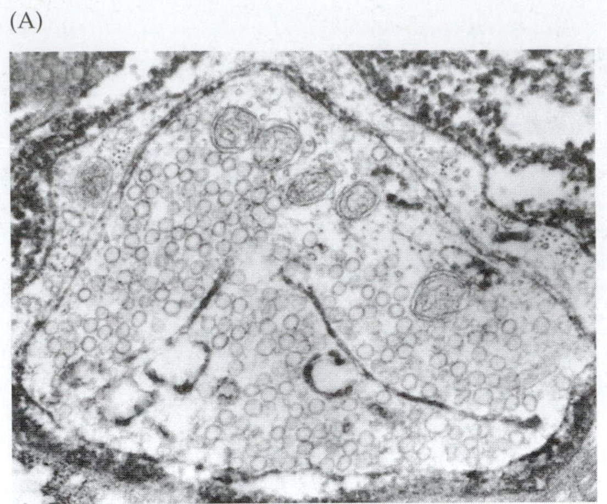

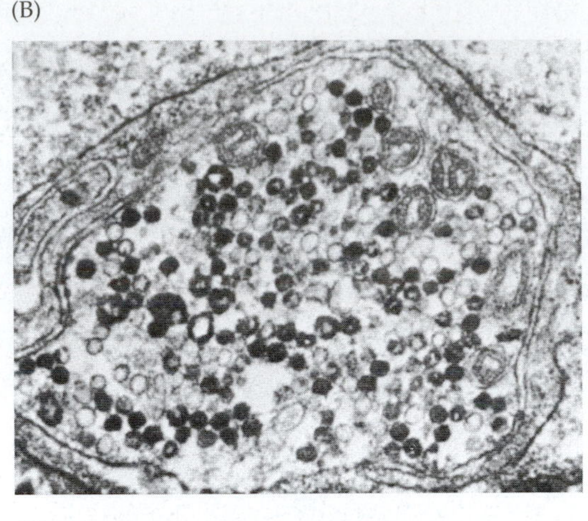

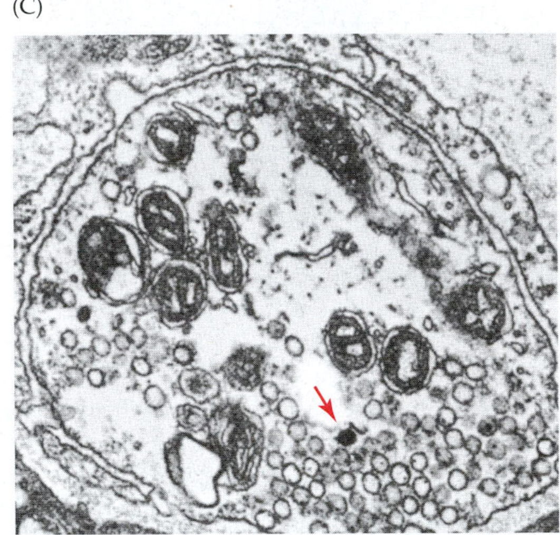

FIGURE 13.25 Recycling of Synaptic Vesicle Membrane. Electron micrographs of cross sections of frog neuromuscular junctions stained with horseradish peroxidase (HRP). (A) The nerve was stimulated for 1 minute in saline containing HRP; electron-dense reaction product can be seen in the extracellular space and in cisternae and coated vesicles. (B) The nerve was stimulated for 15 minutes in HRP, then allowed to recover for 1 hour while the HRP was washed out of the muscle. Many synaptic vesicles contain HRP reaction product, indicating that they have been formed from membrane retrieved by endocytosis. (C) The axon terminal was loaded with HRP and allowed to rest, as in B, then stimulated a second time and allowed to recover an additional hour. Few vesicles are labeled (arrow), indicating that the previously recaptured membrane and enclosed HRP had been recycled into the vesicle population from which release occurs. (From Heuser and Reese, 1973; micrographs kindly provided by J. E. Heuser.)

of the pore and duration of the opening. For example, dye releases often exhibit smaller than average jumps, indicating a restricted loss of dye into the extracellular space.[79] Perhaps more convincing is the observation that up steps in capacitance are often followed within a few hundred milliseconds by down steps of equal amplitude, suggesting that vesicle fusion is followed within a short period by retrieval.[80] The capacitance "flicker" is often accompanied by an increase in membrane conductance, which is indicative of the formation of

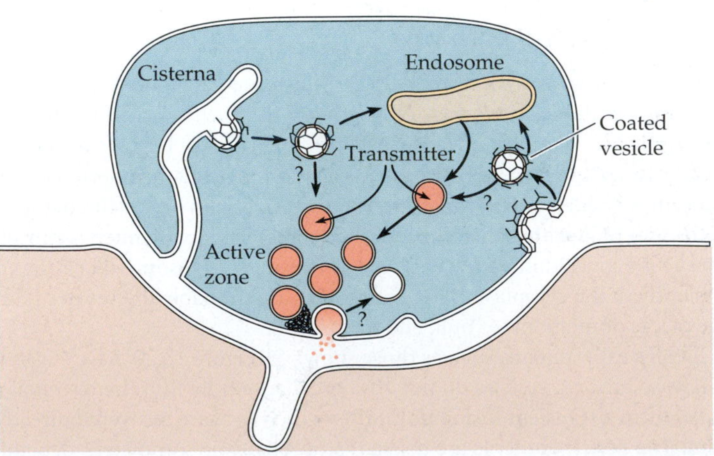

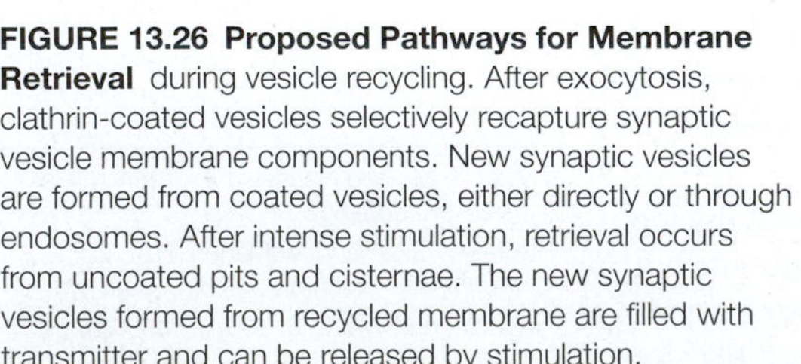

FIGURE 13.26 Proposed Pathways for Membrane Retrieval during vesicle recycling. After exocytosis, clathrin-coated vesicles selectively recapture synaptic vesicle membrane components. New synaptic vesicles are formed from coated vesicles, either directly or through endosomes. After intense stimulation, retrieval occurs from uncoated pits and cisternae. The new synaptic vesicles formed from recycled membrane are filled with transmitter and can be released by stimulation.

a constricted fusion pore between the vesicle lumen and the extracellular space. These, and similar experiments on endocrine cells, have been interpreted as evidence for "**kiss-and-run" exocytosis.** [94] Kiss-and-run is well established as a mode of exocytosis in endocrine cells and, in some circumstances, may underlie a substantial fraction of release.[95] Its role in release from nerve terminals is less clear. In the calyx of Held, about 20% of the capacitance up steps in the membrane patch were followed by down steps within 2 seconds.[80] In nerve terminals in the posterior pituitary gland, the proportion of fusion events represented by capacitance flickers is about 5%.[96] These nerve terminals end in free space, rather than forming synapses, and contain both large, dense-core vesicles and smaller synaptic vesicles. Accordingly, capacitance jumps of two distinct sizes were observed.

Ribbon Synapses

Short sensory receptors (see Chapter 19), and some second-order sensory cells, do not generate action potentials. So, transmitter release from these cells is ongoing, rather than occurring intermittently in response to action potential activity, and is modulated by graded changes in membrane potential. For example, photoreceptors and bipolar cells in the retina secrete glutamate continuously at rates that vary with changes in illumination. (see Chapter 20). Similarly, release of glutamate by hair cells in the auditory and vestibular systems is graded with polarity and intensity of the stimulus (see Chapter 22). Both systems are able to transmit graded information accurately and continuously over a wide range of stimulus intensities, often at very high rates of release.

This ongoing release of transmitter is associated with specialized machinery not seen at phasic synapses–the **synaptic ribbon.**[97] Synaptic ribbons are intracellular organelles that tether large numbers of vesicles near the presynaptic active zones (see Chapter 20). Their structure varies considerably from one type of synapse to another, but in general, they are anchored to the active zone and extend from the synaptic membrane into the cytoplasm (Figure 13.27A).

[94] He, L., and Wu, L-G. 2007. *Trends Neurosci.* 30: 447–455.

[95] Elhamdani, A., Azizi, F., and Artalejo, C. R. 2006. *J. Neurosci.* 26: 3030–3036.

[96] Klyachko, V. A., and Jackson, M. B. 2002. *Nature* 418: 89–92.

[97] Matthews, G., and Fuchs, P. 2010. *Nat. Rev. Neurosci.* 11: 812–822.

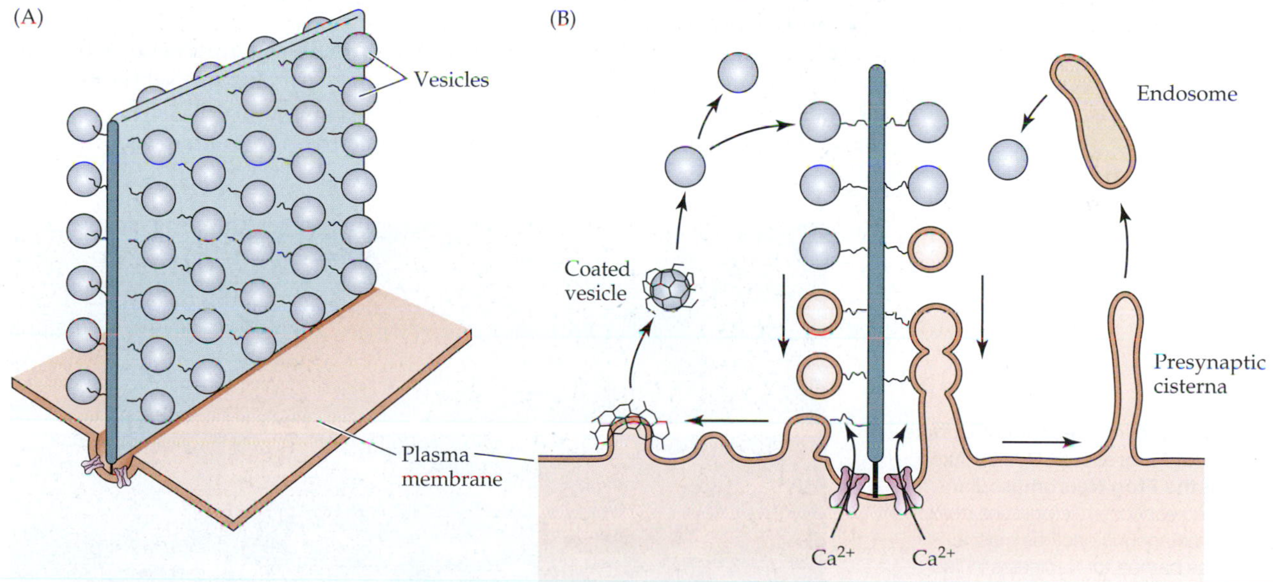

FIGURE 13.27 Schematic Representation of a Synaptic Ribbon. (A) Ribbon is anchored along the active zone of the presynaptic membrane and extends into the cytoplasm. Vesicles are tethered to the ribbon in vertical rows. (B) Possible modes of vesicle discharge and recovery. Calcium entering through channels at the base of the ribbon triggers release of single vesicles (left side of diagram), which are recovered by clathrin-mediated endocytosis. Vesicles move down the queue as release continues. Upon losing their coats, vesicles either remain in the cytoplasm or reattach to the ribbon. Alternatively (right side), several vesicles may discharge *en masse* by fusing serially with one another and with the membrane. Recovery may be through large endosomes and presynaptic cisterns.

The mechanisms of transmitter release at ribbon synapses are similar in principle to those at conventional synapses, with a number of differences in detail. Calcium entry is mediated by $Ca_V1.3$ and $Ca_V1.4$ channels in hair cells and retinal cells, respectively, rather than $Ca_V1.2$ in conventional synapses. The calcium channels are concentrated in the active zone near the bases of the ribbons. The intracellular machinery consists of many of the same components as in non-ribbon synapses, but utilizes different isoforms of some of the intracellular proteins associated with vesicle exocytosis and recovery.[98] The ribbon itself is composed largely of the protein RIBEYE.

The role of the ribbon in relation to the release process and vesicle recycling has not yet been defined precisely. A summary of proposed schemes for the sequence of events during vesicle fusion and recovery is shown in Figure 13.27B.

Vesicle Pools

One way to follow the sequence of events during recycling is by monitoring the uptake of highly fluorescent dyes to mark recycled vesicles. This technique, developed by W. J. Betz and his colleagues,[99] offers the advantage that vesicle recycling can be observed in living preparations by monitoring the stimulation-dependent accumulation and subsequent release of dye (Figure 13.28).

Experiments of this nature have revealed the presence of two distinct vesicle pools. One, the readily releasable pool (RRP), contains the first vesicles to be released upon stimulation, and is refilled rapidly by endocytosis. The second, the reserve pool, comes into play during depletion of the terminal and refills over a longer period.[100,101]

The existence of two such pools is supported by a variety of other evidence. For example, in the electric organ of the marine ray *Torpedo* it was found that during stimulation, newly synthesized ACh was not spread uniformly among synaptic vesicles, but was localized to those vesicles recently formed by recycling.[102,103] Studies on mammalian motor and sympathetic axon terminals have shown that newly synthesized transmitter molecules are preferentially released.[104,105] Such results suggest that a subpopulation of vesicles recycles rapidly, while most of the vesicles are held in reserve.

The time courses of vesicle recycling vary widely from one synapse to the next. At the neuromuscular junction, the RRP is able to refill in about 1 minute through the clathrin-mediated cycle, while the reserve pool takes about 15 minutes to recover from depletion and appears to refill through surface membrane infoldings and cysternae.[98] In contrast, endocytosis in

[98] Zanazzi, G., and Matthews, G. 2009. *Mol. Neurobiol.* 39: 130–148.

[99] Cochilla, A. J., Angleson, J. K., and Betz, W. 1999. *Annu. Rev. Neurosci.* 22: 1–10.

[100] Richards, D. A. et al. 2003. *Neuron* 39: 529–541.

[101] Wu, L-G., Ryan, T. A., and Lagnado, L. 2007. *J. Neurosci.* 27: 11793–11802.

[102] Zimmermann, H., and Denston, C. R. 1977. *Neuroscience* 2: 695–714.

[103] Zimmermann, H., and Denston, C. R. 1977. *Neuroscience* 2: 715–730.

[104] Potter, L. T. 1970. *J. Physiol.* 206: 145–166.

[105] Kopin, I. J. et al. 1968. *J. Pharmacol. Exp. Ther.* 161: 271–278.

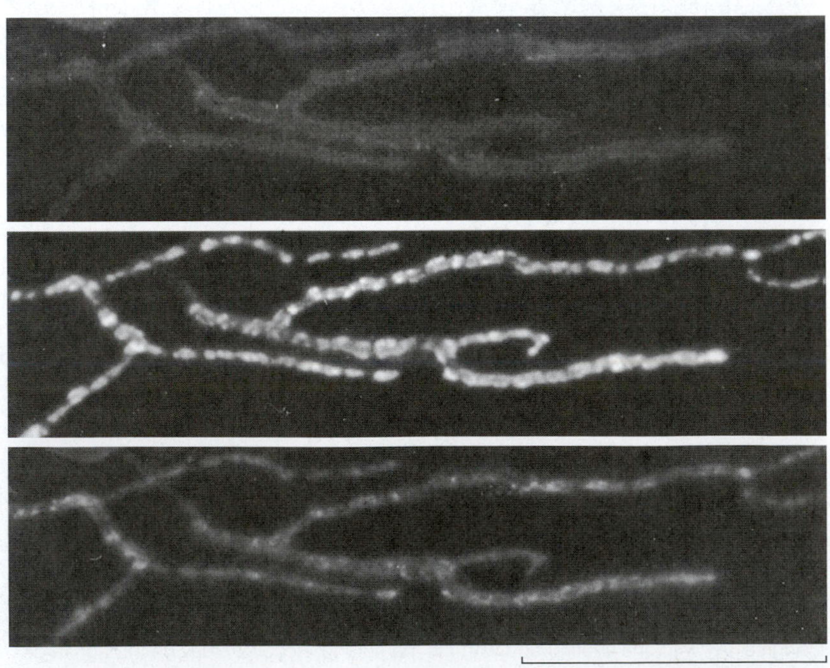

FIGURE 13.28 Activity-Dependent Uptake and Release of Fluorescent Dye by Axon Terminals at the Frog Neuromuscular Junction. Fluorescence micrographs of axon terminals in a cutaneous pectoris muscle. (A) Muscle was bathed for 5 minutes in fluorescent dye (2μ*M* FM1-43) and washed for 30 minutes. Only small amounts of dye remain associated with the terminal membrane. (B) The same muscle was then bathed in dye for 5 minutes while the nerve was stimulated (10 Hz) and washed for 30 minutes. The fluorescent patches are clusters of synaptic vesicles that were filled with dye during recycling. (C) The same muscle was then stimulated at 10 Hz for 5 minutes and washed for 30 minutes. Stimulation released most of the dye. (Micrographs kindly provided by W. J. Betz.)

retinal bipolar cells occurs in two separate phases with time constants of 1–2 seconds and 10–15seconds.[100] The slower component has been shown to be clathrin dependent. The fast component is consistent with kiss-and-run recycling, but direct observation of individual vesicles with high-resolution imaging techniques have indicated that in these cells vesicle fusion is always accompanied by complete loss of FM dye,[106] and the vesicle can merge completely into the surface membrane and retract with a time constant of 1 second.[107]

In synaptic terminals on cultured hippocampal neurons the major component of refilling has a time constant of the order of 15 seconds [108–110] A faster component, attributed to kiss-and-run, had a recovery time of about 500 milliseconds.[108] At these synapses, pH-sensitive fluorescent markers reveal brief exposures of the vesicle lumen to the extracellular space during exocytosis, suggestive of kiss-and-run exocytosis, as well as complete washout.[111]

At the calyx of Held, endocytosis has three components.[100] The first has a time constant of a few seconds and the second about 20 seconds. After intense stimulation, a third, rapid component is activated with a time constant of 1–2 seconds. The first component can be attributed to kiss-and-run, on the basis that transient capacitance flickers lasting a few hundred milliseconds are relatively common in the terminals.[80] The second component is consistent with clathrin-mediated endocytosis, and the third appears to involve bulk membrane recovery.

At ribbon synapses, dye-labeled vesicles have been shown to recycle selectively to the ribbons, suggesting that the attached vesicles represent the readily releasable pool. At synapses in the retina, good agreement has been found between the readily releasable pool size and the estimated total number of vesicles attached to the ribbons. Vesicles in the cytoplasm are highly mobile and presumably constitute the reserve pool.

[106] Zenisesek, D. et al. 2002. *Neuron* 1085–1097.

[107] Llobet, A., Beaumont, V., and Lagnado, L. 2003. *Neuron* 40: 1075–1086.

[108] Sankaranarayanan, S., and Ryan, T. A. 2000. *Nat. Cell Biol.* 2: 197–204.

[109] Gandhi, S. P., and Stevens, C. F. 2003. *Nature* 423: 607–613.

[110] Granseth, B. et al. 2006. *Neuron* 51: 773–786.

[111] Zhang, Q., Li, Y., and Tsien, R.W. 2009. *Science* 323: 1449–1453.

SUMMARY

- When an axon terminal is depolarized, voltage-activated calcium channels open, increase the intracellular calcium concentration, and cause transmitter release.
- Transmitter is released in multimolecular packets, or quanta, that arise when transmitter-containing synaptic vesicles fuse with the plasma membrane and release their contents by exocytosis. There is also a continuous, nonquantal leak of transmitter from axon terminals at rest.
- The synaptic delay between the beginning of the presynaptic depolarization and the beginning of the postsynaptic potential is due to the time required for the nerve terminal to depolarize, calcium channels to open, and increased intracellular calcium to cause exocytosis.
- In response to an action potential, anywhere from 1 to 300 quanta are released nearly simultaneously, depending on the synapse. When the nerve terminal is at rest, single quanta are released spontaneously at a low rate, producing miniature synaptic potentials.
- Synaptic vesicles contain several thousand molecules of transmitter. The number of postsynaptic receptors activated by a quantum of transmitter varies considerably, from about 15 to 1500, depending on the synapse.
- The distribution of amplitudes of spontaneous miniature and evoked postsynaptic potentials can be analyzed by statistical methods to determine the quantal size and quantum content of the response. Neuromodulatory influences that act presynaptically tend to influence quantum content; those that act postsynaptically tend to influence quantal size.
- After exocytosis, synaptic vesicles may flatten out into the plasma membrane. Components of the vesicle membrane are then specifically retrieved by endocytosis of coated vesicles and recycled into new synaptic vesicles. Under certain circumstances, vesicles may pinch back off without ever becoming incorporated into the surface membrane.

Suggested Reading

General Reviews

Cochilla, A. J., Angleson, J. K., and Betz, W. J. 1999. Monitoring secretory membrane with FM1-43 fluorescence. *Annu. Rev. Neurosci.* 22: 1–10.

Doherty, G. J., and McMahon, H. T. 2009. Mechanisms of endocytosis. *Annu. Rev. Biochem.* 78: 815 –902.

He, L., and Wu, L-G. 2007. The debate on the kiss-and-run fusion at synapses. *Trends Neurosci.* 30: 447–455.

Katz, B. 2003. Neural transmitter release: From quantal secretion to exocytosis and beyond. *J. Neurocytol.* 32: 437–446.

Parnas, H., Segel, L., Dudel, J., and Parnas, I. 2000. Autoreceptors, membrane potential and the regulation of transmitter release. *Trends Neurosci.* 23: 60–68.

Walmsley, B., Alvarez, F. J., and Fyffe, R. E. W. 1998. Diversity of structure and function at mammalian central synapses. *Trends Neurosci.* 21: 81–88.

Vyskoĉil, F., Malomouzh, A. I., and Nikolsky, E. E. 2009. Non-quantal acetylcholine release at the neuromuscular junction. *Physiol. Res.* 58: 763–784.

Zucker, R. S. 1993. Calcium and transmitter release. *J. Physiol.* (*Paris*) 87: 25–36.

Original Papers

Albillos, A., Dernick, G., Horstmann, H., Almers, W., Alvarez de Toledo, G., and Lindau, M. 1997. The exocytotic event in chromaffin cells revealed by patch amperometry. *Nature* 389: 509–512.

Betz, W. J., and Bewick, G. S. 1993. Optical monitoring of transmitter release and synaptic vesicle recycling at the frog neuromuscular junction. *J. Physiol.* 460: 287–309.

Bollman, J. H. and Sakmann, B. 2005. Control of synaptic strength and timing by the release-site Ca^{2+} signal. *Nat. Neurosci.* 8: 426–434.

Boyd, I. A., and Martin, A. R. 1956. The end-plate potential in mammalian muscle. *J. Physiol.* 132: 74–91.

del Castillo, J., and Katz, B. 1954. Quantal components of the end-plate potential. *J. Physiol.* 124: 560–573.

Edwards, F. A., Konnerth, A., and Sakmann, B. 1990. Quantal analysis of inhibitory synaptic transmission in the dentate gyrus of rat hippocampal slices: A patch-clamp study. *J. Physiol.* 430: 213–249.

Fatt, P., and Katz, B. 1952. Spontaneous subthreshold potentials at motor nerve endings. *J. Physiol.* 117: 109–128.

Fernandez, J. M., Neher, E., and Gomperts, B. D. 1984. Capacitance measurements reveal step-wise fusion events in degranulating mast cells. *Nature* 312: 453–455.

Harlow, M. L., Ress, D., Stoschek, A., Marshall, R. M., and McMahan, U. J. 2001. The architecture of active zone material at the frog's neuromuscular junction. *Nature* 409: 479–484.

Heuser, J. E., and Reese, T. S. 1973. Evidence for recycling of synaptic vesicle membrane during transmitter release at the frog neuromuscular junction. *J. Cell Biol.* 57: 315–344.

Heuser, J. E., Reese, T. S., Dennis, M. J., Jan, Y., Jan, L., and Evans, L. 1979. Synaptic vesicle exocytosis captured by quick freezing and correlated with quantal transmitter release. *J. Cell Biol.* 81: 275–300.

Katz, B., and Miledi, R. 1967. A study of synaptic transmission in the absence of nerve impulses. *J. Physiol.* 192: 407–436.

Katz, B., and Miledi, R. 1967. The timing of calcium action during neuromuscular transmission. *J. Physiol.* 189: 535–544.

Kuffler, S. W., and Yoshikami, D. 1975b. The number of transmitter molecules in a quantum: An estimate from ionophoretic application of acetylcholine at the neuromuscular synapse. *J. Physiol.* 251: 465–482.

Llinás, R., Sugimori, M., and Silver, R. B. 1992. Microdomains of high calcium concentration in a presynaptic terminal. *Science* 256: 677–679.

Matthews, G., and Fuchs, P. 2010. The diverse roles of ribbon synapses in sensory neurotransmission. *Nat. Rev. Neurosci.* 11: 812–822.

Miller, T. M., and Heuser, J. E. 1984. Endocytosis of synaptic vesicle membrane at the frog neuromuscular junction. *J. Cell Biol.* 98: 685–698.

Ryan, T. A., Reuter, H., and Smith, S. J. 1997. Optical detection of a quantal presynaptic membrane turnover. *Nature* 388: 478–482.

Steyer, J. A., Horstmann, H., and Almers, W. 1997. Transport, docking and exocytosis of single secretory granules in live chromaffin cells. *Nature* 388: 474–478.

Wu, L-G., Ryan, T. A., and Lagnado, L. 2007. Modes of synaptic vesicle retrieval at ribbon synapses, calyx-type synapses, and small central synapses. *J. Neurosci.* 27: 11793–11802.

Zenisesek, D., Steyer, J. A., Feldman, M. E., and Almers, W. 2002. A membrane marker leaves synaptic vesicles in milliseconds in retinal bipolar cells. *Neuron* 35: 1085–1097.

Zhang, Q., Li, Y., and Tsien, R.W. 2009. The dynamic control of kiss-and-run and vesicle reuse probed with single nanoparticles. *Science* 323: 1449–1453.

CHAPTER 14
Neurotransmitters in the Central Nervous System

A large number of different chemical substances act as neurotransmitters in the nervous system. Each transmitter subserves one or more particular functions in the brain; since we cannot discuss all of them in detail, we have selected certain of the major transmitters and describe their particular roles in relation to generating and modulating the activity of the central nervous system (CNS) and behavior.

In Chapter 11 we described the events underlying direct (or fast) synaptic transmission. In the vertebrate CNS, direct excitatory transmission is mediated by the amino acid glutamate, which acts on ionotropic glutamate receptors to open cation-permeable channels. Two other amino acids, γ-aminobutyric acid (GABA) and glycine, act as fast inhibitory neurotransmitters. They activate anion-permeable ionotropic $GABA_A$ and glycine receptors.

Superimposed on the fast action of these amino acids are the slower, modulatory actions of a wide range of transmitters and chemical messengers, mediated primarily through the activation of the metabotropic G protein–coupled receptors (see Chapter 12). The amino acids glutamate and GABA are included as modulatory transmitters, because, as well as activating ionotropic receptors, they also interact with metabotropic $GABA_B$ and glutamate receptors. Other important modulatory transmitters include: acetylcholine (ACh); the monoamines norepinephrine, dopamine, 5-hydroxytryptamine (5-HT or serotonin), and histamine; and adenosine triphosphate (ATP) and its dephosphorylated product, adenosine. ACh, 5-HT, and ATP also activate separate subsets of ionotropic receptors. However, these ionotropic receptors are mostly located on presynaptic nerve endings, not at postsynaptic sites, so (except for 5-HT) they rarely contribute to postsynaptic responses.

Neurons that secrete norepinephrine, dopamine, 5-HT, histamine, and ACh are for the most part confined to special nuclei in the midbrain and brainstem and are relatively few in number. Nevertheless, their axons innervate wide areas of the brain and spinal cord, so that these transmitters regulate important global functions of the brain, such as alertness, arousal, sleep–wake patterns, and memory recall; their loss causes neurodegenerative diseases such as Alzheimer's disease and Parkinson's disease.

Finally, the CNS contains a wide variety of neuropeptide transmitters and hormones, all of which seem to act exclusively on metabotropic receptors. These are generally confined to particular neural pathways. As examples we discuss substance P and the opioid peptides and their role in controlling pain; the orexin/hypocretin system and its function in sleep and feeding behavior; and the hypothalamic peptides vasopressin and oxytocin and their influence on social interaction.

Chemical Transmission in the CNS

The most fully investigated example of chemical transmission in the vertebrate nervous system is that between motor nerves and the skeletal muscle end-plate described in Chapter 11. At this synapse, ACh is the transmitter substance and the motor nerves that release it are termed **cholinergic.** ACh is also the excitatory transmitter at synapses between preganglionic and postganglionic neurons in the ganglia of the autonomic nervous system (see Chapter 17). Hence, it is not surprising that, in seeking excitatory transmitters in the CNS, investigators' first thoughts turned to ACh—especially since it was well known that ACh, choline acetyltransferase, and ACh esterase were all abundant in the brain.[1] Indeed, early work by Eccles and his colleagues did identify one such ganglion-like cholinergic synapse that lies between recurrent collateral branches of the cholinergic motor axons and inhibitory interneurons (or Renshaw cells) in the spinal cord.[2] However, this (plus a few other localized central synapses) proved the exception rather than the rule—indeed even at the Renshaw cell ACh has to share transmitter duties with another fast transmitter, glutamate.[3] (Acetylcholine *is* an important transmitter in the brain but has a different function from that at the neuromuscular junction, as described later in this chapter.)

Instead, the role of ACh as the direct excitatory transmitter in the vertebrate CNS is replaced by the amino acid glutamate; while two other amino acids, GABA and glycine act as direct inhibitory central transmitters (see Chapter 11). For these transmitters, the model synapses are not those in the vertebrate peripheral nervous system (PNS) but instead those in invertebrates like the crustacean neuromuscular junction and stretch receptor or the insect neuromuscular junction (see Box 14.1). It is as though, by some curious evolutionary change, the excitatory and inhibitory control mechanisms undertaken in the PNS of invertebrates has been brought inboard into the CNS in vertebrates.

The effects of direct excitatory and inhibitory transmission are modulated by a wide range of indirect transmitters that enhance or reduce transmission through the synaptic pathways in the CNS. These indirect transmitters include both small molecules, such as ACh, ATP, and the monoamines (i.e., norepinephrine, dopamine, 5-HT, and histamine), and also a variety of neuropeptides. All of these modulators act on G protein–coupled receptors as indirect transmitters (see Chapter 12), though some (ACh, ATP, and 5-HT) also activate ligand-gated ion channels at specific loci. Many of these modulators were first identified as transmitters or local messengers in the PNS. The enteric nervous system (in the gut wall) has been a particularly rich source of information, especially about peptide transmitters[4] (Box.14.2). Most of these modulatory transmitters can act on several different subtypes of G protein–coupled receptors (see Chapter 12), so their postsynaptic effects on different neurons can vary and can be quite subtle and complex. Hence, it is difficult to predict their overall contribution to brain function from actions at individual synapses—their functions are often first inferred from the overall behavioral effects of pharmacologically blocking their receptors or by deleting the genes required for the synthesis of the transmitter or its receptors.

[1] Feldberg, W. 1950. *Br. Med. Bull.* 6: 312–321.

[2] Eccles, J. C., Fatt, P., and Koketsu, K. 1954. *J. Physiol.* 126: 524–562.

[3] d'Incamps, B. L., and Ascher, P. 2008. *J. Neurosci.* 28:14121–14131.

[4] Furness, J. B. et al.1992. *Trends Neurosci.* 15: 66–71.

Mapping Neurotransmitter Pathways

An aid to understanding the physiological role of individual neurotransmitters in the CNS is to identify the particular neurons, nerve fibers, and synapses that contain them. Identification can be achieved using a variety of histochemical or immunohistochemical methods (Figure 14.1). For example, the distribution of neurons and fibers containing norepinephrine, dopa-

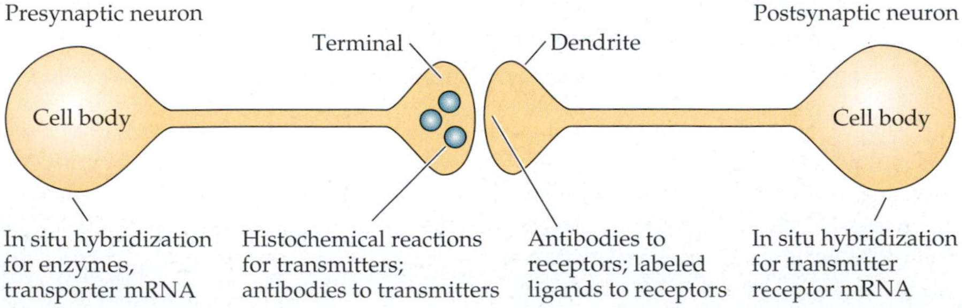

FIGURE 14.1 Methods for Identifying Neurotransmitters in the CNS. Labeled antibodies or nucleotide probes can be used to detect the expression of enzymes involved in synthetic and degradative pathways in the presynaptic neuron. The neurotransmitter itself can be detected by chemical reaction or by antibodies to a conjugated form of the transmitter. Specific uptake of radiolabeled transmitter can identify some neurons. Ligands or antibodies to postsynaptic receptors as well as nucleotide probes for receptor mRNA provide means to identify cells sensitive to a particular transmitter.

BOX 14.1
The Discovery of Central Transmitters: I. The Amino Acids

Gamma-aminobutyric acid (GABA)

Although GABA was originally discovered in the mammalian brain,[5–7] its inhibitory transmitter function was first established in the peripheral nervous system (PNS) of the crayfish. In this and other crustacea there are separate, identifiable inhibitory nerves that, when stimulated, inhibit the muscle fibers and the sensory stretch receptors by increasing their chloride conductance (see Chapter 11). In 1954, Ernst Florey[8] found that an inhibitory factor (Factor I) he extracted from brain could block the discharges of the stretch receptor, and he subsequently identified GABA as the main inhibitory ingredient.[9] Two sets of researchers, Kuffler and Edwards[10] and Boistel and Fatt,[11] then showed that GABA precisely imitated the action of inhibitory nerve stimulation, respectively, on the stretch receptor and the muscle. Through much painstaking work, Kravitz and his colleagues[12] showed that GABA was highly concentrated in crustacean inhibitory axons and (crucially) could be released from them when stimulated in sufficient amounts to be a transmitter.

It took longer to establish the role of GABA in the mammalian central nervous system (CNS). One reason is that there are no discrete inhibitory fiber bundles like those in the crayfish PNS. One approach used by Curtis and his colleagues[13] was to eject GABA ionophoretically around a neuron in the CNS while stimulating excitatory and inhibitory inputs. However, they concluded that GABA was not the inhibitory transmitter but instead had a general depressant action on the somatodendritic membrane of all central neurons, irrespective of their innervation. Using the same techniques, Krnjevic[14] and his colleagues took a more optimistic view. Using intracellular recordings, they subsequently showed that GABA hyperpolarized cortical neurons and increased their membrane conductance, just like stimulating inhibitory afferents,[15] and that both effects were due to an increase in chloride conductance.[15,16] One reason why Curtis and his colleagues rejected a role for GABA in spinal cord inhibition was that its action was not antagonized by strychnine,[17] which was known to block spinal cord inhibitory postsynaptic potentials (IPSPs). Spinal inhibition was not blocked by strychnine because the principal inhibitory transmitter there is glycine, as subsequently shown by Werman and his colleagues.[18]

Glutamate

It took even longer for glutamate to be recognized as the main excitatory transmitter in the CNS. There were several reasons why: although it is abundant in the brain, it is not unique to the brain (unlike GABA); it excited every central neuron tested; its action was replicated by other dicarboxylic amino acids;[19] and there were no antagonists of synaptic excitation to help identify the transmitter. Again, the strongest evidence came from experiments on invertebrates, particularly those of Takeuchi and Takeuchi[20] on lobster muscle, in which they found glutamate-sensitive hotspots that coincided with the excitatory postsynaptic potentials (EPSPs) initiated by motor nerve stimulation. As they stated, *"the receptors which respond to L-glutamate are identical with normal neuroreceptors."*[20] Glutamate eventually gained acceptance as the major excitatory transmitter in the mammalian CNS through demonstration of its calcium-dependent release from neurons, the development of specific blocking agents, the cloning of the receptors, and the identification of glutamate-containing synaptic vesicles.[21]

Inactivation of amino acid transmitters

One initial obstacle to the acceptance of amino acids as transmitters was that there seemed to be no enzyme like acetylcholinesterase that was capable of rapidly inactivating the transmitters. This issue was resolved with the discovery by Iversen[22] and others that there are special uptake systems by which the released amino acids (GABA, glutamate, and glycine) are rapidly re-accumulated into the nerves that release them as well as into neighboring neuroglial cells (see Chapters 9 and 15).

[5] Awapara, J. et al. 1950. *J. Biol. Chem.* 187: 35–39.
[6] Roberts, E., and Frankel, S. 1950. *J. Biol. Chem.* 187: 55–63.
[7] Udenfriend, S. 1950. *J. Biol. Chem.* 187: 65–69.
[8] Florey, E. 1954. *Arch. Int. Physiol.* 62: 33–53.
[9] Bazemore, A., Elliott, K. A., and Florey, E. 1956. *Nature* 178: 1052–1053.
[10] Kuffler, S. W., and Edwards, C. 1958. *J. Neurophysiol.* 21: 589–610.
[11] Boistel, J., and Fatt, P. 1958. *J. Physiol.* 144: 176–191.
[12] Otsuka, M. et al. 1966. *Proc. Natl. Acad. Sci. USA* 56: 1110–1115.
[13] Curtis, D. R., and Phillis, J. W. 1958. *Nature* 182: 323.
[14] Krnjevic, K., and Phillis, J. W. 1963. *J. Physiol.* 165: 274–304.
[15] Krnjevic, K., and Schwartz, S. 1967. *Exp. Brain Res.* 3: 320–336.
[16] Obata, K. et al. 1967. *Exp. Brain Res.* 4: 43–57.
[17] Curtis, D. R., Phillis, J. W., and Watkins, J. C. 1959. *J. Physiol.* 146: 185–203.
[18] Werman, R., Davidoff, R. A., and Aprison, M. H. 1968. *J. Neurophysiol.* 31: 81–95.
[19] Curtis, D. R., Phillis, J. W., and Watkins, J. C. 1960. *J. Physiol.* 150: 656–682.
[20] Takeuchi, A., and Takeuchi, N. 1964. *J. Physiol.* 170: 296–317.
[21] Lodge, D. 2009. *Neuropharmacology* 56: 6–21.
[22] Iversen, L. L. 1971. *Br. J. Pharmacol.* 41: 571–591.

mine, and 5-HT can be individually mapped using a fluorescent method devised by Hillarp and Falck,[23] in which each of these monoamines emits light of a characteristic wavelength under ultraviolet illumination (Figure 14.2). For cholinergic neurons, George Koelle and his colleagues[24] introduced a histochemical method for staining cholinesterase, the enzyme that hydrolyzes ACh. This method was used to map the distribution of cholinergic fibers in the brain,[25] but was subsequently replaced by a more specific immunohistochemical method for localizing the synthetic enzyme, choline acetyltransferase (ChAT).[26]

[23] Falck, B. et al. 1962. *J. Histochem. Cytochem.* 10: 348–354.
[24] Koelle, G. B., and Friedenwald, J. S. 1949. *Proc. Soc. Exp. Biol. Med.* 70: 617–622.
[25] Lewis, P. R., and Shute, C. C. 1967. *Brain* 90: 521–540.
[26] Wainer, B. H. et al. 1984. *Neurochem. Int.* 6: 163–182.

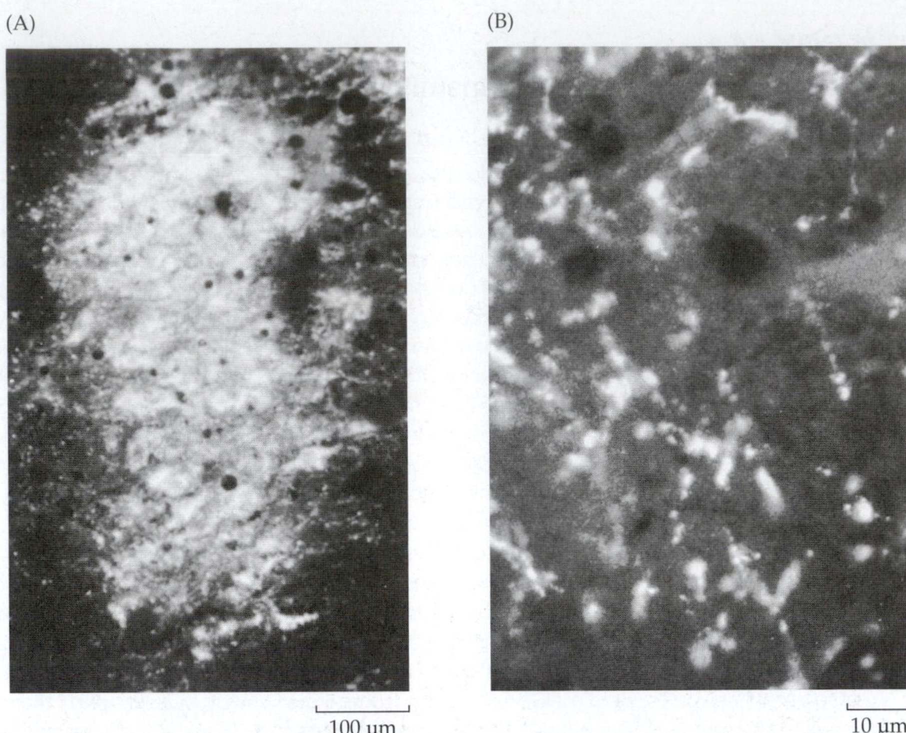

FIGURE 14.2 Visualization of Biogenic Amine-Containing Cells and their Terminal Arborizations by Formaldehyde-Induced Fluorescence. (A) Norepinephrine-containing cells in the locus coeruleus. (B) The terminal arborizations of locus coeruleus cells in the hippocampus. (From Harik, 1984.)

Immunohistochemical methods have now become the method of choice for localizing most transmitters and their receptors. By using fluorescently tagged secondary antibodies, neurons and fibers containing transmitter-related proteins (such as synthetic enzymes or membrane transporters, or, in the case of peptides, the transmitter itself or its precursor) can be readily seen and co-localization with other transmitters detected using different wavelength fluorophores on the secondary antibodies. Figure 14.3A shows an example in which nerve terminals in the dorsal raphe nucleus are tagged with an antibody against the

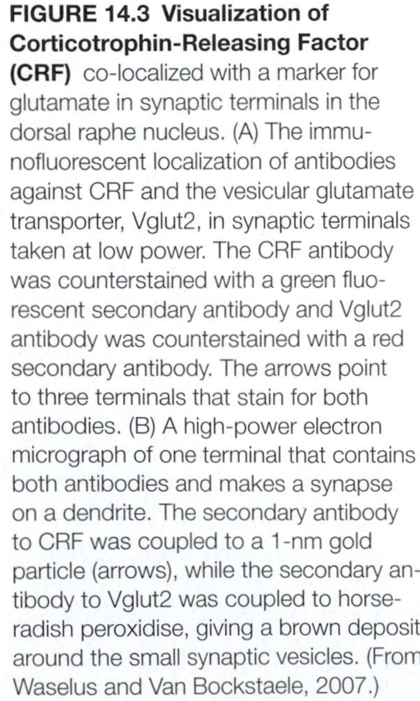

FIGURE 14.3 Visualization of Corticotrophin-Releasing Factor (CRF) co-localized with a marker for glutamate in synaptic terminals in the dorsal raphe nucleus. (A) The immunofluorescent localization of antibodies against CRF and the vesicular glutamate transporter, Vglut2, in synaptic terminals taken at low power. The CRF antibody was counterstained with a green fluorescent secondary antibody and Vglut2 antibody was counterstained with a red secondary antibody. The arrows point to three terminals that stain for both antibodies. (B) A high-power electron micrograph of one terminal that contains both antibodies and makes a synapse on a dendrite. The secondary antibody to CRF was coupled to a 1-nm gold particle (arrows), while the secondary antibody to Vglut2 was coupled to horseradish peroxidise, giving a brown deposit around the small synaptic vesicles. (From Waselus and Van Bockstaele, 2007.)

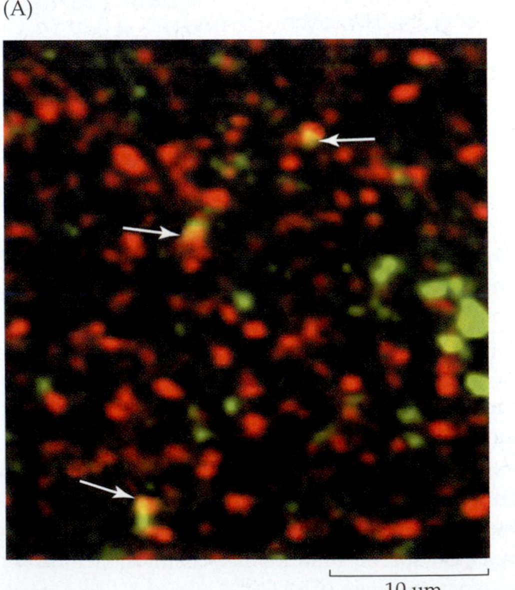

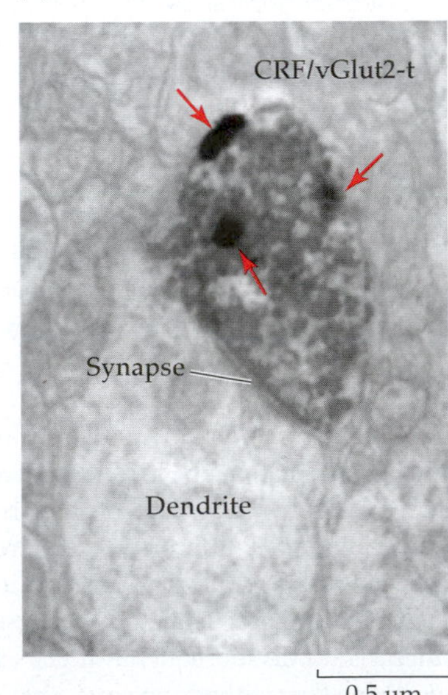

BOX 14.2
The Discovery of Central Transmitters: II. Neuropeptides

The first "neuropeptides" in the brain identified and sequenced were the hypothalamic hormones oxytocin and vasopressin, in 1955.[27] Since then, more than 60 other central neuropeptides have been identified.[28] How were they all discovered? The classical approach is to find an appropriate assay, test extracts from brain, and then purify and sequence the active peptides. The modern approach is to go for the gene or its messenger RNA (mRNA) product. These two approaches are well illustrated by the discovery of the **enkephalins** and the **orexins**, respectively.

The enkephalins

Enkephalins are two pentapeptides (5 amino acids) that are naturally released stimulants of opiate (morphine) receptors. They were discovered by John Hughes, Hans Kosterlitz, and their colleagues in 1975.[29] Morphine itself is a plant alkaloid that comes from the opium poppy *Papaver somniferum*; the idea that there might be a morphine-like substance in the mammalian brain stemmed from the fact that morphine seemed to act on specific receptors. So, there must be an endogenous ligand for them. In order to find this ligand, Hughes used the mouse vas deferens as a bioassay tissue. This tissue contracts when its afferent sympathetic nerves are stimulated; morphine reduces this contraction by inhibiting the release of norepinephrine. Using this tissue Hughes[30] detected morphine-like activity in mammalian brain extracts, which seemed to be due to a peptide because it was abrogated by peptidases. The amino-acid composition of the peptide was determined by chemical methods.[29] This showed that there were actually two peptides: one with the sequence Tyr-Gly-Gly-Phe-**Met**, which they termed met-enkephalin, the other with the sequence Tyr-Gly-Gly-Phe-**Leu**, which was called leu-enkephalin. Both were shown to replicate the effect of morphine on the vas deferens and to be antagonized by naloxone, a known morphine antagonist.

The orexins

Orexins are peptides that are released from nerves of hypothalamic origin, stimulate wakefulness, and enhance appetite; their absence causes the human disease narcolepsy. They were discovered independently by two groups of scientists starting from what are termed orphan genes—that is, genes or gene products that have no known function. One group began with the gene for the transmitter itself, the other from the gene for its receptor.

The first group (de Lecea and colleagues)[31] constructed a complementary DNA (cDNA) library from the rat hypothalamus and then made a systematic survey of which messenger RNA (mRNA) products of these cDNAs were selectively concentrated in the hypothalamus. The cDNA derived from one of these mRNA products (clone 35) encoded two peptides that were concentrated in neurons and synapses in the dorsolateral hypothalamus. These peptides excited hypothalamic neurons when applied to them.[32] They called the peptides hypocretins because they were similar to the gut hormone secretin (*hypothalamic secretin*).

The second group (Sakurai and colleagues)[33] started from the genes for orphan G protein–coupled receptors. They expressed the receptors in cell lines and then looked for substances in brain extracts that stimulated them and increased the cell's intracellular calcium. Again, they purified and sequenced two active proteins, and termed them orexin-A and orexin-B (because they had an appetite-stimulating or *orexigenic* action in rats). The researchers also determined the amino acid sequences of the responsive receptors, which they called OX1 and OX2. This work shows the power of studying orphan receptors. There are many such receptors, and they have led to the identification of several other neurotransmitters.[34,35]

[27] Du Vigneaud, V. 1955. *Harvey Lect.* 50: 1–26.
[28] Salio, C. et al. 2006. *Cell Tissue Res.* 326: 583–598.
[29] Hughes, J. et al. 1975. *Nature* 258: 577–580.
[30] Hughes, J. 1975. *Brain Res.* 88: 295–308.
[31] Gautvik, K. M. et al. 1998. *Proc. Natl. Acad. Sci. USA* 93: 8733–8738.
[32] De Lecea, L. et al. 1998. *Proc. Natl. Acad. Sci. USA* 95: 322–327.
[33] Sakurai, T. et al. 1998. *Cell* 92: 573–585.
[34] Wise, A. Jupe, S. C., and Rees, S. 2004. *Annu. Rev. Pharmacol. Toxicol.* 44: 43–66.
[35] Civelli, O. et al. 2006. *Pharmacol. Ther.* 110: 525–532.

vesicular glutamate transporter Vglut2 (indicating that the vesicles contain—and presumably release—glutamate) and counterstained with a red fluorescent secondary antibody, while an antibody tagging the peptide neurotransmitter corticotrophin-releasing factor (CRF) is counterstained with a green fluorescent secondary antibody. Terminals where both are present then show up as yellow.

Immunohistochemical techniques can also be applied at the level of the electron microscope, to see which transmitter or transmitter-specific marker protein is present in individual synapses or synaptic vesicles. Thus, the vesicular glutamate transporter Vglut2 in the glutamate-containing synaptic terminals shown in Figure 14.3A can be seen under the electron microscope in Figure 14.3B by using a secondary antibody coupled to the enzyme immunoperoxidase, which gives a brown deposit. The presence of CRF immunoreactivity can then be marked separately using another secondary antibody coupled to electron-dense gold particles.

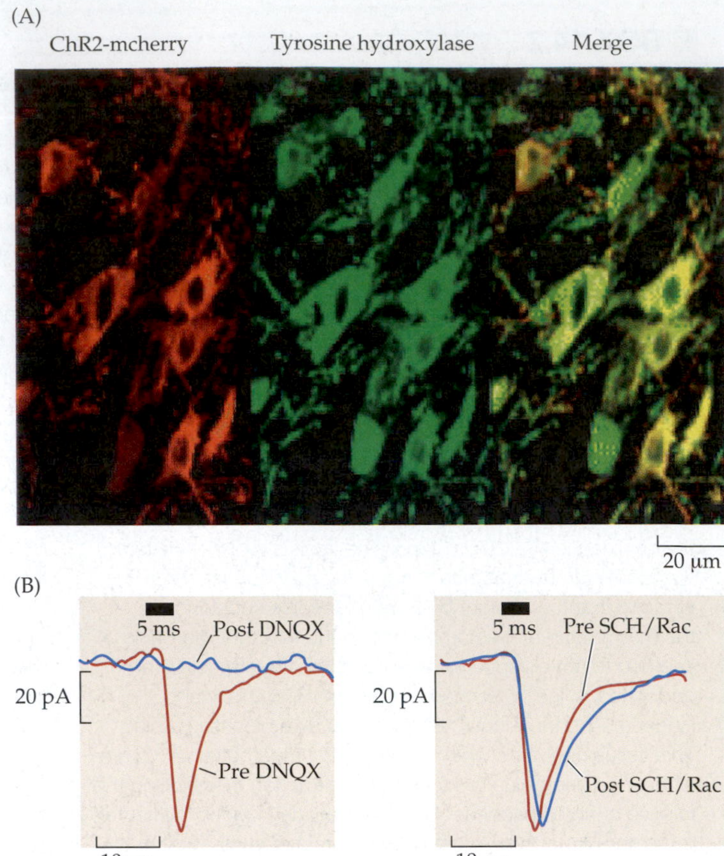

FIGURE 14.4 Transgenic Insertion of a Photo-Activatable Ion Channel that allows the selective stimulation of dopaminergic neurons and their processes. The light-sensitive cation channel, channelrhodopsin-2, tagged with the red fluorescent protein mcherry (ChR2-mcherry) was transgenically expressed from its cDNA in dopaminergic neurons in the mouse brain under the control of the promoter for the dopamine transporter. (A) Dopaminergic neurons in the ventral tegmental area (VTA) of the midbrain expressing the red ChR2-mcherry (left) and an antibody to the dopamine synthetic enzyme tyrosine hydroxylase (TH), counterstained with a green fluorescent secondary antibody (middle). The image on the right shows the left and center pictures merged, and it indicates that ChR2 is restricted to TH-positive neurons. (B) These records were made from a neuron in the nucleus accumbens (which is innervated by dopaminergic fibers from the VTA) in a slice of the midbrain taken from a ChR2-expressing mouse brain. They show short-latency excitatory postsynaptic currents generated when the presynaptic fibers are stimulated by activating the ChR2 with a 5-msec pulse of blue light. The current was inhibited by the glutamate antagonist (DNQX, left side, blue line) but not by the dopamine antagonists SCH23390 + raclopride (SCH/rac, right side). This means that the postsynaptic current did not actually result from the release of dopamine itself, but from the release of glutamate that is stored in the same nerve endings as the dopamine and that activates ionotropic glutamate receptors. (After Stuber et al., 2010).

Visualizing Transmitter-Specific Neurons in Living Brain Tissue

Certain neurons that contain a transmitter of interest can be color-coded in vivo by using transgenic methods. This can be very helpful for subsequent electrophysiological experiments, especially for neurons that are not very numerous. For example, dopamine-releasing (dopaminergic) neurons in the mesencephalon can be color-coded with the jellyfish green fluorescent protein (GFP, see Box 12.6) by tagging the GFP cDNA onto the cDNA that codes for the promoter region of the synthesizing enzyme tyrosine hydroxylase, then expressing it in mice using transgenic techniques.[36] Subsequently, when the mesencephalon is dissected and the cells are isolated and cultured, the dopamine-secreting cells are easily identifiable by their green fluorescence, even though they contribute only a minority of cells in the mixed-cell culture.

A similar approach has been used to express a fluorescently tagged, light-activated, cation-conducting protein, channelrhodopsin-2, in the dopaminergic neurons[37] (Figure 14.4A). Then, dopaminergic neurons in an isolated midbrain slice could be stimulated selectively by a brief flash of blue light, which opened the channelrhodopsin channels to depolarize the neuron (Figure 14.4B). Selectively expressed channelrhodopsins and related proteins are now widely used as a means of stimulating (or inhibiting) neurons,[38] particularly for analyzing neural circuits in vivo,[39] since light stimulation is easier and more versatile than electrical stimulation of individual neurons through microelectrodes. This field of research has been dubbed "optogenetics."[40,41]

Key Transmitters

In this section, we highlight some properties and functions of features for a selection of the transmitters previously referred to in Boxes 12.1 and 12.2. Details of chemical structures and metabolic pathways of the small-molecule transmitters are summarized in Appendix B. More detailed information can be obtained in the book by Cooper et al.[42]

[36] Jomphe, C. et al. 2005. *J. Neurosci. Methods* 146: 1–12.

[37] Stuber, G. D. et al. 2010. *J. Neurosci.* 30: 8229–8233.

[38] Gradinaru, V. et al. 2010. *Cell* 141: 154–165.

[39] Kravitz, A. V. et al. 2010. *Nature* 466: 622–626.

[40] Pastrana, E. 2011. *Nat. Methods* 8: 24–25.

[41] Diesseroth, K. 2011. *Nat. Methods* 8: 26–29.

[42] Cooper, J. R., Bloom, F. E., and Roth, R. H. 2003. *The Biochemical Basis of Neuropharmacology,* 8th ed. Oxford University Press.

Glutamate

This is the fast, excitatory transmitter that is released from neurons throughout the CNS. It activates two types of postsynaptic ionotropic receptors: fast-opening α-amino-3-hydroxy-5-methyl-4-isoxazolepropionic acid (AMPA) receptors and slower-opening *N*-methyl-D-aspartate (NMDA) receptors (see Figure 11.12) The AMPA receptors are responsible for normal fast transmission. They are made up of GluA1–4 subunits (see Chapter 5). The speed and duration of the evoked synaptic currents varies substantially between different synapses, depending on the subunit composition of the channel and on the splice-variant (*flip* or *flop*[43]) of the subunit.[44] The presence or absence of the GluA2 subunit, and the extent of its mRNA editing[45] also affect the calcium permeability of the AMPA receptors.[46]

One important feature of AMPA receptors is that they are not held in the postsynaptic membrane so tightly as are the nicotinic receptors at the neuromuscular junction, but are highly mobile.[47] The lifetime of individual AMPA receptors has been estimated at 10–30 minutes as the receptors recycle between the cell surface and the subsynaptic apparatus.[48] Several auxilliary synaptic proteins regulate this trafficking.[49] It allows rapid changes in the numbers and subunit composition of synaptic receptors during neural activity, providing the basis for many forms of synaptic plasticity (see Chapter 16).

The NMDA receptors are made up of GluN1–3 subunits (see Chapter 5). They differ from AMPA receptors in three main respects: the channels open more slowly; they are blocked by Mg^{2+} ions and only allow ions to pass when the membrane is depolarized; and they have a much higher permeability to Ca^{+} ions than most AMPA channels (see Chapter 11). Their voltage-dependence means that they act as coincidence detectors, only allowing current to pass when the neuron is simultaneously depolarized by, for example, high-frequency activation of AMPA channels, ongoing action potential activity, or co-activation of metabotropic glutamate receptors (see below). The consequent entry of Ca^{2+} ions then induces a variety of downstream effects (see Figure 12.20)—most notably, the induction of long-term potentiation (see Chapter 16). However, there is also a downside, which is that excessive Ca^{2+} entry through NMDA channels is also neurotoxic (excitotoxic, to give it its original name) and can cause neurodegeneration and ischemic neuron death (stroke).[50]

Glutamate also activates a distinctive class of G protein–coupled receptors called **metabotropic glutamate receptors (mGluRs)**.[51] There are eight of these receptors, mGluR1–8, which are divided into three groups: group I (mGluR1,5), group II (mGluR2,3), and group III (mGluR4,6,7,8) The group I receptors are the main postsynaptic receptors. They couple to the G protein G_q and hence to the phospholipase C–phosphoinositide pathway (see Figure 12.15). Their activation causes a membrane depolarization and increased excitability, mainly by inhibiting Ca^{2+}-activated and M-type, voltage-gated potassium currents. They also release calcium ions from intracellular stores by producing IP_3. IP_3 synergizes with NMDA receptor activation in inducing long-term potentiation. Thus, mice in which the gene for mGluR1 has been deleted[52] show defects in long-term potentiation and learning behavior, while mGluR5 gene deleted mice show deficits in various forms of learning and habituation.[53] In contrast, mGluR2, 3, and 4 receptors are more strongly (though not completely) represented in presynaptic glutamatergic fiber terminals. There, they couple to the G protein G_o, inhibit the Ca_V2 Ca^{2+} channels and so reduce transmitter release. In this way, they mediate a negative feedback control of glutamate release, much like the feedback inhibition of GABA release by presynaptic $GABA_B$ receptors (see Figure 11.17). When present postsynaptically, they activate inward rectifier potassium channels, hyperpolarizing the neuron—the opposite effect to that of stimulating mGluR1 or 5 receptors. The mGluR6 receptor hyperpolarizes retinal bipolar cells in a different way: it stimulates cGMP phosphodiesterase, producing a fall in cGMP and closure of cyclic nucleotide-gated cation channels (see Chapter 20).

GABA (γ-Aminobutyric acid) and glycine

These are the two main inhibitory neurotransmitters in the CNS. They activate chloride-conducting ionotropic GABA or glycine receptors (see Chapter 5) to produce an inhibitory postsynaptic potential (IPSP) in the postsynaptic neuron (see Chapter 11). The increase in chloride conductance tends to inhibit the generation of action potentials by the excitatory postsynaptic potentials (EPSPs). GABA is the predominant inhibitory transmitter

[43] Hollman, M., and Heinemann, S. 1994. *Annu. Rev. Neurosci.* 17: 31–108.

[44] Geiger, J. R. et al. 1997. *Neuron* 18: 1009–1023.

[45] Seeburg, P. H., and Hartner, J. 2003. *Curr. Opin. Neurobiol.* 13: 279–283.

[46] Cull-Candy, S., Kelly, L. and Farrant, M. 2007. *Curr. Opin. Neurobiol.* 17: 277–280.

[47] Tardin, C. et al. 2003. *EMBO J.* 22: 4656–4665.

[48] Choquet, D., and Trille, A. 2003. *Nat. Rev. Neurosci.* 4: 251–265.

[49] Nicoll, R. A., Tomita, S., and Bredt, D. S. 2006. *Science* 311: 1253–1256.

[50] Papardia, S., and Hardingham, G. E. 2007. *Neuroscientist* 13: 572–579.

[51] Niswender, C., and Conn, P. J. 2010. *Ann. Rev. Pharmacol. Toxicol.* 50: 295–322.

[52] Aiba, A. et al. 1994. *Cell* 79: 365–375.

[53] Bird, M. K., and Lawrence, A. J. 2009. *Trends Pharmacol. Sci.* 30: 617–623.

released by neurons in the cortex and midbrain, whereas glycine plays a prominent role in the brainstem and spinal cord. Some neurons have receptors for both transmitters, which act on separate and distinguishable chloride channels,[54] and both transmitters can sometimes be released from the same neuron.[55]

GABA and glycine are concentrated in interneurons. Their primary function is to control the output of the principal excitatory neurons by negative feedback through recurrent inhibitory collaterals from the axons of these neurons. This prevents excessive excitatory discharges from the principal neurons. One example (already mentioned) is the negative feedback inhibition of motoneuron discharges by recurrent axons that innervate glycinergic interneurons (Renshaw cells) in the spinal cord.[3] The importance of inhibitory interneurons in the brain is illustrated by the fact that blocking the inhibitory GABA or glycine receptors (with bicuculline or strychnine, respectively) produces convulsions. However, inhibitory interneurons in the cortex also have more subtle functions. Because each interneuron innervates several principal neurons, they serve to coordinate the output of the principal neurons and to synchronize activity within networks of neurons.[56–58] In addition to receiving recurrent collaterals from principal neuron axons, these GABAergic interneurons also receive a direct input from collaterals of the afferent fibers to principal neurons, producing what is termed feed-forward inhibition.[59] This helps to set the timing of principal neuron's response to afferent stimuli.[60]

RECEPTORS FOR GABA AND GLYCINE Ionotropic $GABA_A$ and glycine receptors are pentameric (five subunit) receptors homologous with nicotinic ACh receptors but with the difference that their structure favors permeation of the anion chloride instead of cations (see Chapter 5). The glycine receptor is made up of two α-subunits and three β-subunits, arranged as a ring in the order α-β-β-α-β, with four binding sites for glycine at the α-β interfaces.[61] Three genetic variants of the α-subunit are known (α_1 to α_3), but only one form of the β-subunits (β_1). Sivilotti eloquently discusses the mechanism whereby glycine opens these channels, as deduced from single channel analysis.[62]

The GABA receptor has a more complex structure and repertoire of subunits. There are 19 different subunits: 6 α-subunits, 3 β, and 3 γ, one each of δ, ε, π, θ, and three variants of a special ρ-subunit.[63] The most common combinations at inhibitory synapses are $\alpha_1\beta_2\gamma_2$ (60%), $\alpha_2\beta_3\gamma_2$ (15%–20%) and $\alpha_3\beta_n\gamma_2$ (10%–15%).[64] Because different combinations differ in rates of activation, deactivation and desensitization, their expression is tuned to the optimum requirements for inhibition at different synapses.[64] GABA receptors containing the δ-subunit are particularly interesting because these do not desensitize and are activated by unusually low (sub-micromolar) concentrations of GABA.[65] Delta subunits, along with α6-subunits, are present in receptors on cerebellar and hippocampal dentate gyrus granule cells. They are tonically activated by resting interstitial GABA concentrations (estimated at 300–600 nM) and so produce a steady component of resting membrane current.[66] The ρ-subunits, assembled as a homomeric pentamer, make up another unique type of $GABA_A$ receptor present in the retina.

The $GABA_A$ receptor is also the prime target for a number of important drugs that act allosterically to enhance GABA-mediated currents and IPSPs. These include benzodiazepine compounds, barbiturates, and certain anesthetics such as propafol, etomidate and the steroid anesthetic alphaxolone.[64,67] These three groups of drugs bind at different sites on the receptor and have different actions. For example, barbiturates[68,69] and alphaxolone[70] prolong the postsynaptic current, by slowing the closure of the channels,[71] whereas benzodiazepines increase the sensitivity of the channels to GABA (by slowing dissociation) without changing open time.[72] Enhancement of $GABA_A$ receptor currents is also responsible for some of the effects of ethanol,[73,74] which is why one should not mix alcohol with benzodiazepines. Naturally occurring regulators of $GABA_A$ receptor activity in the brain include endogenous neurosteroids[75,76] and zinc ions.[77]

$GABA_B$ RECEPTORS Like glutamate, GABA can also activate a G protein–coupled receptor. This is termed the $GABA_B$ receptor. It was discovered by Norman Bowery and his colleagues in 1980.[78] They found that GABA reduced the release of transmitters from peripheral and central neurons by activating a receptor that was not antagonized by $GABA_A$ receptor antagonists but could be selectively activated by the GABA analog, baclofen (β-chlorophenyl-GABA). They subsequently identified it as a G protein–coupled

[54] Gold, M. R., and Martin, A. R. 1984. *Nature* 308: 639–641.

[55] Jonas, P., Bischofberger, J., and Sandkühler, J. 1998. *Science* 281: 419–424.

[56] Cobb, S. R. et al. 1995. *Nature* 378: 75–78.

[57] Whittington, M. A., and Traub, R. D. 2003. *Trends Neurosci.* 26: 676–682.

[58] Klausberger, T., and Somogyi, P. 2008. *Science* 321: 53–57.

[59] Buszaki, G. 1984. *Prog. Neurobiol.* 22: 131–153.

[60] Pouille, F., and Scanziani, M. 2001. *Science* 293: 1159–1163.

[61] Betz, H., and Laube, B. 2006. *J. Neurochem.* 97: 1600–1610.

[62] Sivilotti, L. 2010. *J. Physiol.* 588: 45–58.

[63] Olsen, R. W., and Sieghart, W. 2008. *Pharmacol. Rev.* 60: 243–260.

[64] Möhler, H. 2006. *Cell Tissue Res.* 326: 505–516.

[65] Brown, N. et al. 2002. *Brit. J. Pharmacol.* 136: 965–974.

[66] Farrant, M., and Nusser, Z. 2005. *Nat. Rev. Neurosci.* 6: 215–229.

[67] Franks, N. P. 2008. *Nat. Rev. Neurosci.* 9: 370–386.

[68] Nicoll, R. A. et al. 1975. *Nature* 258: 625–627.

[69] Scholfield, C. N. 1978. *J. Physiol.* 275: 559–566.

[70] Scholfield, C. N. 1980. *Pflügers Arch.* 383: 249–255.

[71] Steinbach, J. H., and Akk, G. 2001. *J. Physiol.* 537: 715–733.

[72] Bianchi, M. T. et al. 2009. *Epilepsy Res.* 85: 212–220.

[73] Wallner, M., Hanchar, H. J., and Olsen, R. W. 2003. *Proc. Natl. Acad. Sci. USA* 100: 15218–15223.

[74] Kumar, S. et al. 2009. *Psychopharmacology (Berl.)* 205: 529–564.

[75] Hosie, A. M. et al. 2006. *Nature* 444: 486–489.

[76] Mitchell, E. A. et al. 2008. *Neurochem Int.* 52: 588–595.

[77] Smart, T. G., Hosie, A. M., and Miller, P. 2004. *Neuroscientist* 10: 432–442.

[78] Bowery, N. G. et al. 1980. *Nature* 283: 92–94.

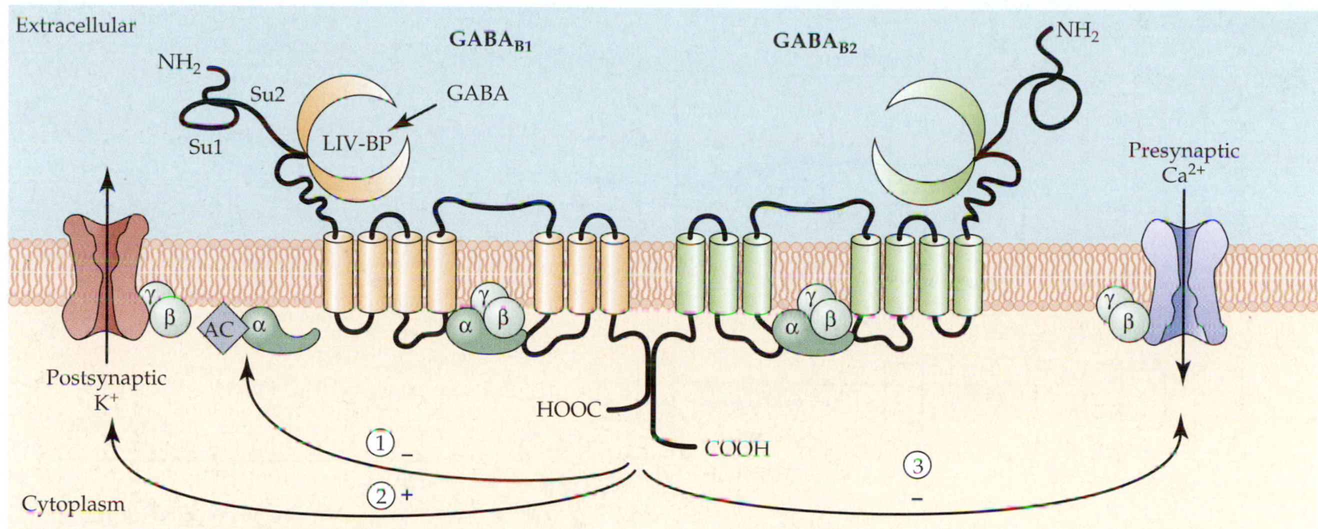

FIGURE 14.5 Schematic Diagram of the GABA$_B$ Receptor Dimer. The functional receptor is a dimer composed of GABA$_{B(1)}$ and GABA$_{B(2)}$ monomers joined through a coiled–coil domain at their C-termini. GABA binds to a part of the extracellular domain of the B(1) monomer at a region homologous to the bacterial periplasmic leucine–isoleucine–valine binding protein, or LIV-BP. The B(2) monomer responds by activating the G proteins G_i and G_o. The α-subunit of G_i inhibits adenylate cyclase (pathway 1), while the βγ-subunit activates the postsynaptic G protein gated inward rectifier channel (pathway 2). The βγ-subunit of G_o also inhibits the Ca_V2 class of calcium channels in the presynaptic terminals (pathway 3). (After Couve et al., 2000.)

receptor because (among other properties) its action was suppressed by Pertussis toxin. In fact, it turned out to be a very unusual receptor because the active receptor is a dimer of two closely similar receptors termed GABA$_{B(1)}$ and GABA$_{B(2)}$, conjoined through their intracellular C-termini[79,80] (Figure 14.5). The B1-subunit has the binding site for GABA whereas the B2-subunit is necessary to activate the G protein. Hence, both subunits have to be expressed and assembled to make a functional receptor.[81]

GABA$_B$ receptors are located both postsynaptically and presynaptically, and have different effects at the two sites.[82] Activation of *postsynaptic* receptors opens inwardly rectifying G protein–gated potassium channels[83] (Figure 14.6), which produces a membrane hyperpolarization and delayed postsynaptic inhibition.[82] These postsynaptic GABA$_B$ receptors are located around the periphery of the synapse (perisynaptically) and do not readily respond to GABA released by a single afferent stimulus; instead, they are activated by spillover GABA escaping from the synapse during repetitive stimulation or following simultaneous activation of several GABAergic inputs.[84]

Presynaptic GABA$_B$ receptors inhibit voltage-gated calcium channels in the nerve terminal and thereby reduce transmitter release. When GABA is released, it may act on receptors on its own terminal (autoinhibition; see Figure 11.17) or on adjacent terminals that release GABA or another transmitter (i.e., heterosynaptic inhibition). Inhibition of the presynaptic calcium current and its consequent effect on transmitter release is illustrated in the experiment on calyx of Held terminals shown in Figure 14.7A. These terminals, located in the medial nucleus of the trapezoid body, are large enough to be patch clamped.[85] When the presynaptic terminal was depolarized by injecting current through the patch electrode, application of baclofen reduced both the resulting inward calcium current and the EPSP (Figure 14.7B). The reduced EPSP could be explained solely by the reduced calcium current since the relationship between inward current and EPSP amplitude was unaltered (Figure 14.7C), However, at some other synapses there may be an additional effect of the GABA$_B$-released G protein βγ-subunits on the transmitter release mechanism itself.[86,87]

Acetylcholine

Cholinergic neurons are present throughout the brainstem, midbrain, and telencephalon and provide a widespread innervation to diverse areas of the brain.[88] An important source of cholinergic axons to the cerebral cortex and hippocampus is provided by groups of neu-

[79] Couve, A., Moss, S. J., and Pangalos, M. N. 2000. *Mol. Cell. Neurosci.* 16: 296–312.

[80] Bettler, B. et al. 2004. *Physiol. Rev.* 84: 835–867.

[81] Filippov, A. K. et al. 2000. *J. Neurosci.* 20: 2867–2874.

[82] Nicoll, R. A. 2004. *Biochem. Pharmacol.* 68: 1667–1674.

[83] Gahwiler, B. H., and Brown, D. A. 1985. *Proc. Natl. Acad. Sci. USA* 82: 1558–1562.

[84] Scanziani, M. 2000. *Neuron* 25: 673–681.

[85] Takahashi, T., Kajikawa, Y., and Tsujimoto, T. 1998. *J. Neurosci.* 18: 3138–3146.

[86] Wu, L. G., and Saggau, P. 1997. *Trends Neurosci.* 20: 204–212.

[87] Blackmer, T. et al. 2005. *Nat. Neurosci.* 8: 421–425.

[88] Woolf, N. J. 1991. *Prog. Neurobiol.* 37: 475–524.

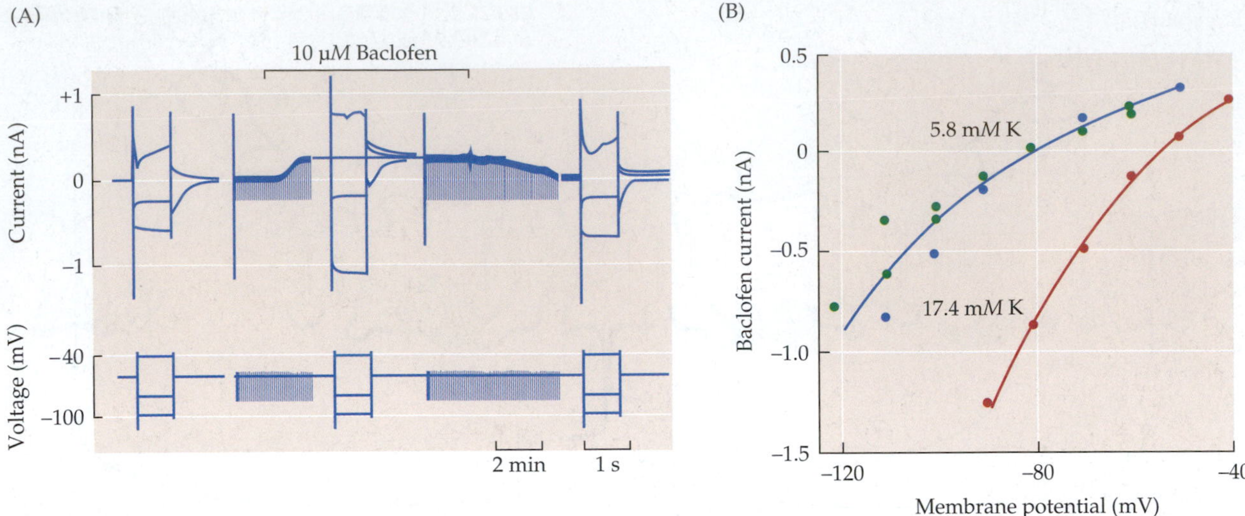

FIGURE 14.6 The $GABA_B$-Selective GABA Analog Baclofen Activates a Potassium Current in a Hippocampal Pyramidal Neuron. Records from an experiment on a slice of rat hippocampus cultured in vitro for 3 weeks. (A) A continuous record of membrane current recorded at –61 mV, with one-second steps to –81 mV applied every 5 seconds (downward deflections). Just before, during, and after application of baclofen, three voltage steps, to –41, –81 and –101 mV, were applied and the currents recorded at a faster speed. Baclofen produced an outward current at –61 mV and increased the membrane conductance as shown by the increased response to the voltage steps. (B) The extra current produced by baclofen (after subtracting the resting current) recorded from this cell in 5.8- and 17.4-m*M* external [K^+]. The currents were measured at the end of one-second steps from –61 mV to the voltages shown. The current reversal potential (that is, the potential at which the current–voltage curve crossed the zero-current line) shifted +26 mV on increasing the K^+ concentration threefold, which is near that (+29 mV) expected for a K^+ current (see Chapters 6 and 11). Note that the curve shows inward rectification—that is, the current is larger at potentials negative to the reversal potential (when the direction of net K^+ flow is into the cell) than at potentials more positive to the reversal potential, when it is outward. (After Gahwiler and Brown, 1985b.)

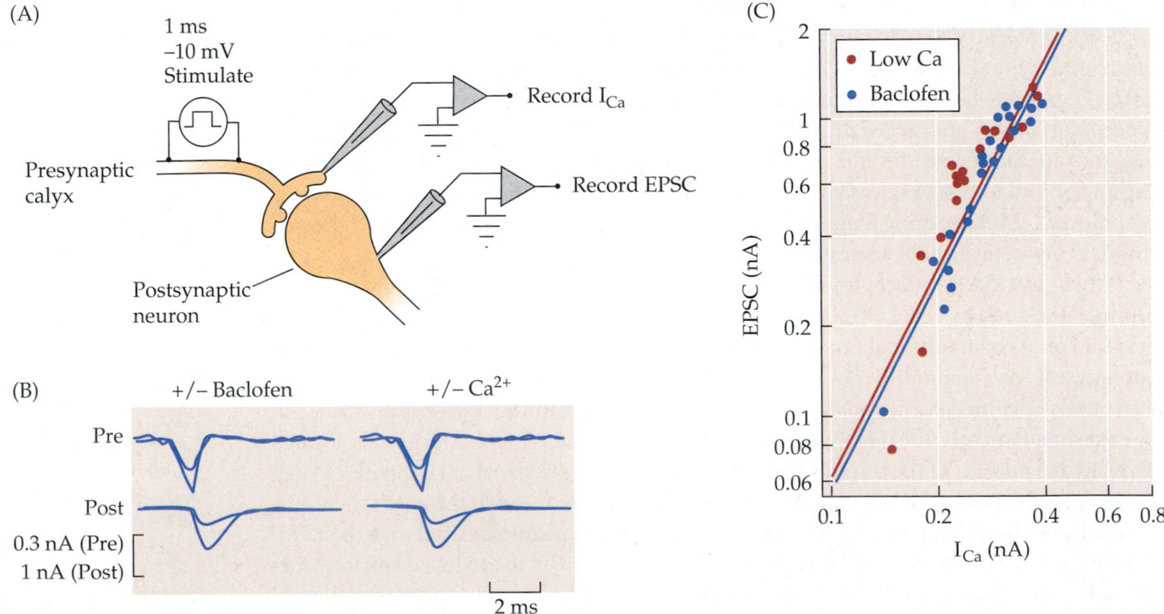

FIGURE 14.7 Presynaptic $GABA_B$ Receptors Reduce Transmitter Release by reducing terminal Ca^{2+} current. (A) The experiments were performed by simultaneously recording from the presynaptic calyx of Held in the medial nucleus of the trapezoid body and the innervated postsynaptic neuron. A 1-ms step from –70 mV to –10 mV was delivered to the calyx, and the resultant presynaptic Ca^{2+} current (I_{Ca}) and glutamatergic excitatory postsynaptic current (EPSC) were recorded. In (B), both presynaptic I_{Ca} (Pre) and postsynaptic EPSC (Post) were reduced by the $GABA_B$ receptor stimulant, baclofen (20 μ*M*), or by reducing external Ca^{2+} (replaced with Mg^{2+}). (C) is a plot of EPSC amplitude against presynaptic I_{Ca} in the presence of different concentrations of baclofen or Ca^{2+}. Note that baclofen did not alter the relation between Ca^{2+} current and transmitter release as measured by the EPSC, implying that its effect on transmitter release at this terminal can be explained wholly by the inhibition of the Ca^{2+} current. (After Takahashi et al., 1998.)

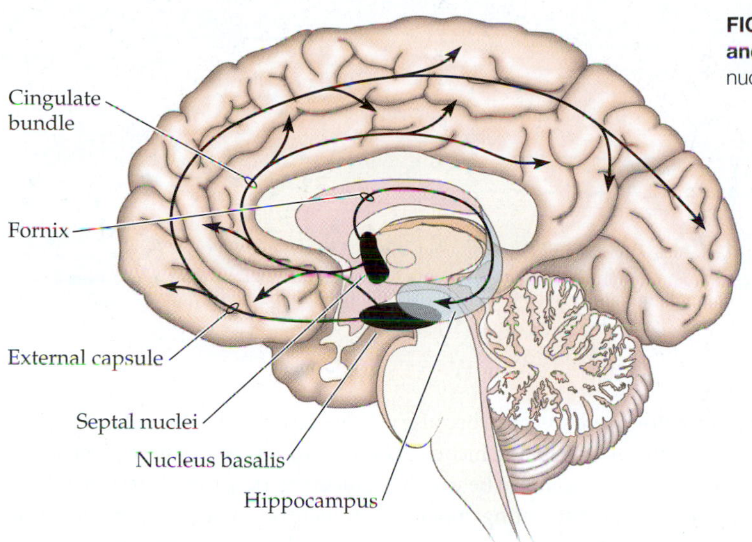

FIGURE 14.8 Cholinergic Innervation of the Cortex and Hippocampus by neurons in the septal nuclei and nucleus basalis.

rons in the nucleus basalis and medial septum of the basal forebrain (Figure 14.8). Neurons in the tegmental area provide a second source of cholinergic axons to the thalamus and midbrain dopaminergic areas. These projections are broad and diffuse, and are composed of unmyelinated axons. A third, more clearly defined tract, composed of myelinated fibers, is the habenulo–interpeduncular tract from the medial habenular nucleus to the interpeduncular nucleus. A fourth important group is cholinergic interneurons in the striatum, which provide a substantial innervation within the striatum and in the olfactory tubercle.[89]

One important difference between the central and peripheral cholinergic systems concerns the sites at which ACh is released from the cholinergic fibers. Thus, while some of the cholinergic axons to the cortex end at morphologically defined synapses onto pyramidal cell dendrites,[90] akin to those described in Chapter 11, most release sites in the cortex consist of swellings (or varicosities) along the axons that do not make direct contact with the postsynaptic neuron.[91] Localized release of ACh from these varicosities has been recorded in tissue culture.[92] The release process is as rapid as at true synapses but may occur some distance from its target receptors, so ACh acts diffusely, rather like a local hormone.[93] This action is also true for most of the other transmitters described below and is sometimes referred to as volume transmission[94] or (less flatteringly) the soup theory of transmission.[95] Once released, the ACh can then act on both ionotropic nicotinic and G protein–coupled muscarinic receptors.

NICOTINIC RECEPTORS AND THEIR FUNCTIONS Neuronal nicotinic receptors[96,97] are structurally homologous to those at the neuromuscular synapse, but the pentamer is made up of only two types of subunit, α and β. There are nine neuronal α-subunits (α2 to α10) and three β-subunits (β2 to β4). The different combinations vary in sensitivity to ACh and in their rates of desensitization.[98] The main types in the brain are α4β2-heteromers and homomers of α7, but there are many others, with varying α-β stoichiometries.[99] The α7-channels are important because they have a high permeability to Ca^{2+} ions; they are also the only neuronal channels blocked by α-bungarotoxin (see Chapter 11).

Although there are a few synapses where the receptors are in the postsynaptic membrane and generate a true EPSP (including the spinal Renshaw cell discussed earlier), these are quite exceptional: the vast majority of the receptors are on presynaptic axons and terminals.[97,98] Here, they serve to increase the release of other transmitters such as glutamate. This effect has been clearly shown at synapses between fibers in the interpeduncular tract and neurons in the interpeduncular nucleus (Figure 14.9).[100] At this synapse, synaptic transmission is mediated by glutamate,[101] even though the fibers themselves contain ACh. Nicotine increased both the glutamatergic EPSC following afferent stimulation and the frequency of spontaneous EPSCs. These effects were due to an increase in Ca^{2+} entry into the interpeduncular fibers. They were blocked by α-bungarotoxin, indicating that the Ca^{2+} entry was mediated by α7-nicotinic receptors.

[89] Zhou, F. M., Wilson, C. J., and Dani, J. A. 2002. *J. Neurobiol.* 53: 590–605.

[90] Turrini, P. et al. 2001. *Neuroscience* 105: 277–285.

[91] Umbriaco, D. et al. 1994. *J. Comp. Neurol.* 348: 351–373.

[92] Allen, T. G., and Brown, D. A. 1996. *J. Physiol.* 492: 453–466.

[93] Descarries, L., Gisiger, V., and Steriade, M. 1997. *Prog. Neurobiol.* 53: 603–625.

[94] Lendvai, B., and Vizi, E. S. 2008. *Physiol. Rev.* 88: 333–349.

[95] Sivilotti, L., and Colquhoun, D. 1995. *Science* 269: 1681–1682.

[96] Sargent, P. B. 1993. *Annu. Rev. Neurosci.* 16: 403–443.

[97] Dani, J. A., and Bertrand, D. 2007. *Annu. Rev. Pharmacol. Toxicol.* 47: 699–729.

[98] Role, L. W., and Berg, D. K. 1996. *Neuron* 16: 1077–1085.

[99] Millar, N. S., and Gotti, C. 2009. *Neuropharmacology* 56: 237–246.

[100] McGehee, D. S. et al. 1995. *Science* 269: 1692–1696.

[101] Brown, D. A., Docherty, R. J., and Halliwell, J. V. 1983. *J. Physiol.* 341: 655–670.

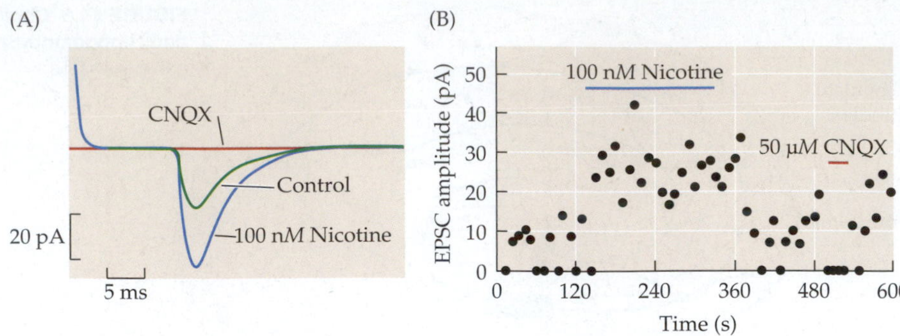

FIGURE 14.9 Presynaptic Nicotinic Receptors Enhance Excitatory Transmission. Stimulating the habenulo-interpeduncular tract in an in vitro brain slice produces an excitatory postsynaptic current (EPSC) in a neuron in the interpeduncular nucleus. (A) The EPSC results from the release of glutamate, since it is suppressed by a glutamate antagonist (CNQX). Nicotine (100 n*M*) increases the amplitude of the EPSC. (B) Other tests showed that this effect is due to enhanced release of glutamate, resulting from the entry of Ca^{2+} ions through activated presynaptic α7-containing nicotinic receptors and consequent rise in the intracellular Ca^{2+} concentration. (After McGehee et al., 1995.)

Nicotinic ACh receptors were so-called because they are activated by nicotine. The question then arises: do the effects of inhaling nicotine result from activating these presynaptic nicotinic receptors? The answer seems to be yes. Thus, nicotine increases glutamate release from excitatory afferent fibers innervating dopaminergic neurons in the ventral tegmental area.[102] This causes a persistent increase in the activity of these neurons and increased release of dopamine from their terminals in the nucleus accumbens and prefrontal cortex, which is thought to be responsible for the pleasurable (rewarding) aspect of nicotine's action[103] (see the section on dopamine below). Alpha4/beta2-containing nicotinic receptors are particularly involved in some of the effects of nicotine,[104] and a partial agonist of these receptors, varenicline,[105] is used to treat nicotine addiction.

MUSCARINIC RECEPTORS AND THEIR EFFECTS Most of the effects of ACh released from the cholinergic nerves to the cerebral cortex, hippocampus, and basal ganglia, as depicted in Figure 14.8, are mediated by muscarinic receptors.[106] There are five subtypes of this receptor, M1 through M5. The odd-numbered ones (mainly M1 and M3) link to the G protein G_q; their immediate effect is to stimulate hydrolysis of the membrane phospholipid phosphatidylinositol-4,5-bisphosphate. They are predominantly postsynaptic and their principal effect is to produce a form of postsynaptic excitation. The receptors are not actually located within the subsynaptic membrane itself but just outside the synapse,[107] and respond to diffusively released ACh, as discussed above. The even-numbered receptors (M2 and M4) link to the G proteins G_o and G_i. They are primarily presynaptic where they inhibit transmitter release, but some are postsynaptic and cause a form of postsynaptic inhibition.

- ***Postsynaptic excitation*** A characteristic form of muscarinic excitation is that seen in hippocampal pyramidal neurons after stimulating cholinergic afferents from the medial septum (Figure 14.10). This consists of a slow depolarization and enhanced action potential discharges.[108] It results from the inhibition of two types of K^+ channel: (1) a calcium-activated K^+ channel that normally generates a long-lasting ("slow") afterhyperpolarization (sAHP) following an action potential[109] and (2) the M-channel.[110,111] Inhibition of the slow afterhyperpolarization increases the frequency and duration of action potential discharges; inhibition of M-channel causes depolarization and enhances excitability by lowering the threshold for action potential generation in the axon initial segment.[112] Similar effects are seen in cerebral cortical pyramidal neurons[113,114] (innervated from the basal forebrain); inhibition of the M-current also accounts for the cholinergic excitation of medium spiny neurons in the striatum.[115] Cholinergic afferents also excite inhibitory interneurons in the hippocampus and cortex that indirectly inhibit the principal neurons. This has a strong influence on cortical network behavior.[116]
- ***Postsynaptic inhibition*** Activation of muscarinic receptors directly inhibits some cortical neurons. There are two ways in which this can happen. First, in some cortical pyramidal cells, activation of postsynaptic M1 receptors can produce inhibition, probably by releasing Ca^{2+} from intracellular stores, which in turn activates calcium-dependent K^+ channels.[117] Second, a minority of neurons have postsynaptic G_i-coupled M2 or M4 receptors that hyperpolarize neurons by activating inward rectifier (Kir) K^+ channels[118] (see Figure 12.7).

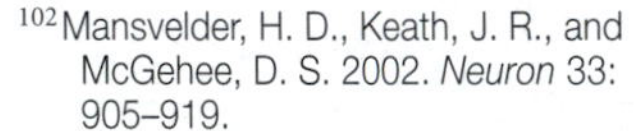

[102] Mansvelder, H. D., Keath, J. R., and McGehee, D. S. 2002. *Neuron* 33: 905–919.

[103] Livingstone, P. D., and Wonnacott, S. 2009. *Biochem. Pharmacol.* 78: 744–755.

[104] Xiao, C. et al. 2009. *J. Neurosci.* 29: 12428–12439.

[105] Rollema, H. et al. 2007. *Trends Pharmacol. Sci.* 28: 316–325.

[106] Brown, D. A. 2010. *J. Mol. Neurosci.* 41: 340–346.

[107] Yamasaki, M., Matsui, M., and Watanabe, M. 2010. *J. Neurosci.* 30: 4408–4418.

[108] Cole, A. E., and Nicoll, R. A. 1984. *J. Physiol.* 352: 173–188.

[109] Madison, D. V., Lancaster, B., and Nicoll, R. A. 1987. *J. Neurosci.* 7: 733–741.

[110] Halliwell, J. V., and Adams, P. R. 1982. *Brain Res.* 250: 71–92.

[111] Gahwiler, B. H., and Brown, D. A. 1985. *Nature* 313: 577–579.

[112] Shah, M. M. et al. 2008. *Proc. Natl. Acad. Sci. USA* 105: 7869–7874.

[113] McCormick, D. A., and Williamson, A. 1989. *Proc. Natl. Acad. Sci. USA* 86: 8098–8102.

[114] Benardo, L. S. 1993. *Neuroscience* 53: 11–22.

[115] Shen, W. et al. 2005. *J. Neurosci.* 25: 7449–7458.

[116] Lawrence, J. J. 2008. *Trends Neurosci.* 31: 317–327.

[117] Gulledge, A. T. et al. 2009. *J. Neurosci.* 29: 9888–9902.

[118] Eggermann, E., and Feldmeyer, D. 2009. *Proc. Natl. Acad. Sci. USA* 106: 11753–11758.

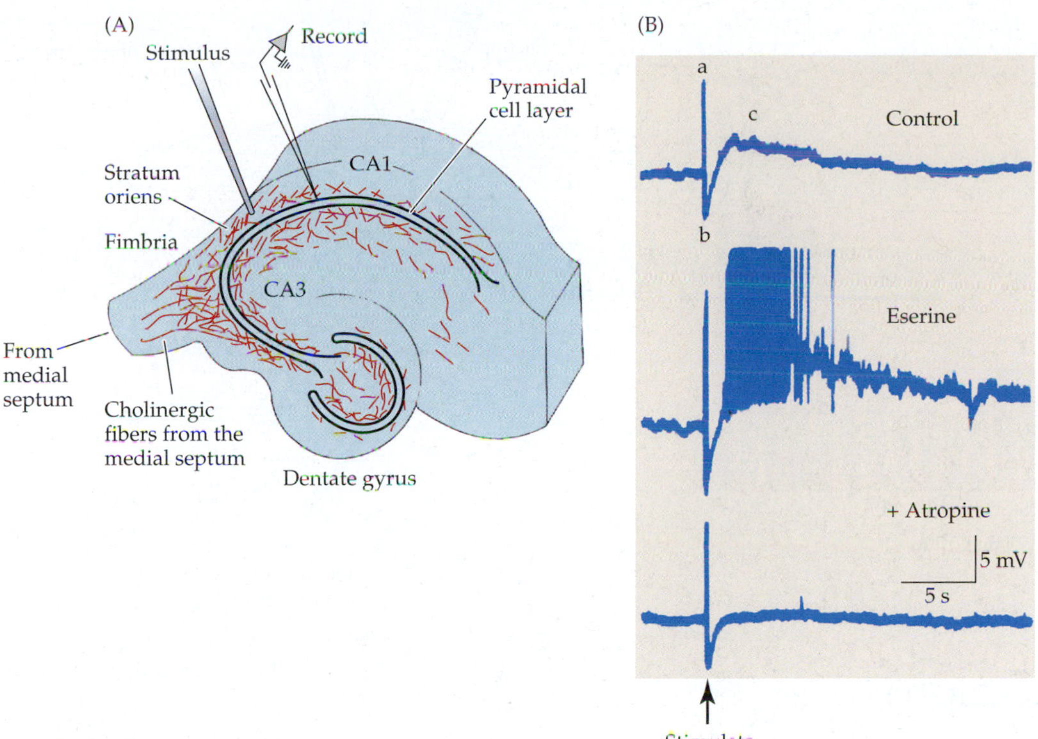

FIGURE 14.10 Stimulation of Cholinergic Fibers from the Medial Septum Excites Hippocampal Pyramidal Cells. (A) A drawing of a transverse hippocampal slice. Cholinergic fibers from the medial septum enter the slice through the fimbria and course through the stratum oriens. A stimulating electrode is placed in the stratum oriens and an intracellular pipette is used to record from a neuron in the pyramidal cell layer. (B) Stratum oriens stimulation for 5 seconds at 20 Hz produces: (a) summed fast excitatory postsynaptic potentials due to glutamate release (probably from the cholinergic fibers; see Chapter 15); (b) summed inhibitory synaptic potentials due to GABA release from interneurons; and (c) a slow excitatory postsynaptic potential. The latter is enhanced by the anticholinesterase eserine (physostigmine) and suppressed by the muscarinic receptor antagonist atropine. This effect of atropine shows that the slow excitatory postsynaptic potential results from activation of muscarinic acetylcholine receptors. (A after Nicoll, 1985; B after Cole and Nicoll, 1984.)

- ***Presynaptic inhibition*** The release of ACh from cholinergic fibers is subject to profound feedback inhibition by presynaptically located M2 and M4 receptors. Dudar and Szerb[119] first noted this phenomenon in 1969 when they were measuring the release of ACh from the surface of the cat cerebral cortex following activation of the ascending reticular formation. They found that the amount of ACh collected was increased threefold when the muscarinic antagonist atropine was added to the collecting cup (Figure 14.11A). The mechanism responsible for this effect cannot be easily studied in the brain itself but has been explored in tissue-cultured cholinergic basal forebrain neurons where the release of ACh from individual varicosities along their axons can be recorded with a nicotinic receptor detector patch (see Figure 15.9). The muscarinic agonist muscarine dramatically reduced this release[92] (Figure 14.11B). This effect is due to the activation of M4 receptors and probably results from the inhibition of the axonal calcium current by the activated G protein.[120] These presynaptic M2 and M4 receptors are very widely distributed in the nervous system. For example, they inhibit ACh release from cholinergic nerve endings in the hippocampus and striatum.[121]

OVERALL CONTRIBUTIONS TO BRAIN FUNCTION The ascending cholinergic system to the cortex and hippocampus plays a crucial role in attention[122] and memory.[123] In the cortex, ACh focuses attention to sensory information by selectively augmenting the responses of cortical neurons to a specific stimulus without increasing their background activity.[124–126]

[119] Dudar, J. D., and Szerb, J. C. 1969. *J. Physiol.* 203: 741–762.
[120] Allen, T. G., and Brown, D. A. 1993. *J. Physiol.* 466: 173–189.
[121] Zhang, W. et al. 2002. *J. Neurosci.* 22: 1709–1717.
[122] Sarter, M. et al. 2005. *Brain Res. Brain Res. Rev.* 48: 98–111.
[123] Hasselmo, M. E. 2006. *Curr. Opin. Neurobiol.* 16: 710–715.
[124] Murphy, P. C., and Sillitoe, A. M. 1991. *Neuroscience* 40: 13–20.
[125] Herrero, J. L. et al. 2008. *Nature* 454: 1110–1114.
[126] Goard, M., and Dan, Y. 2009. *Nat. Neurosci.* 12: 1444–1449.

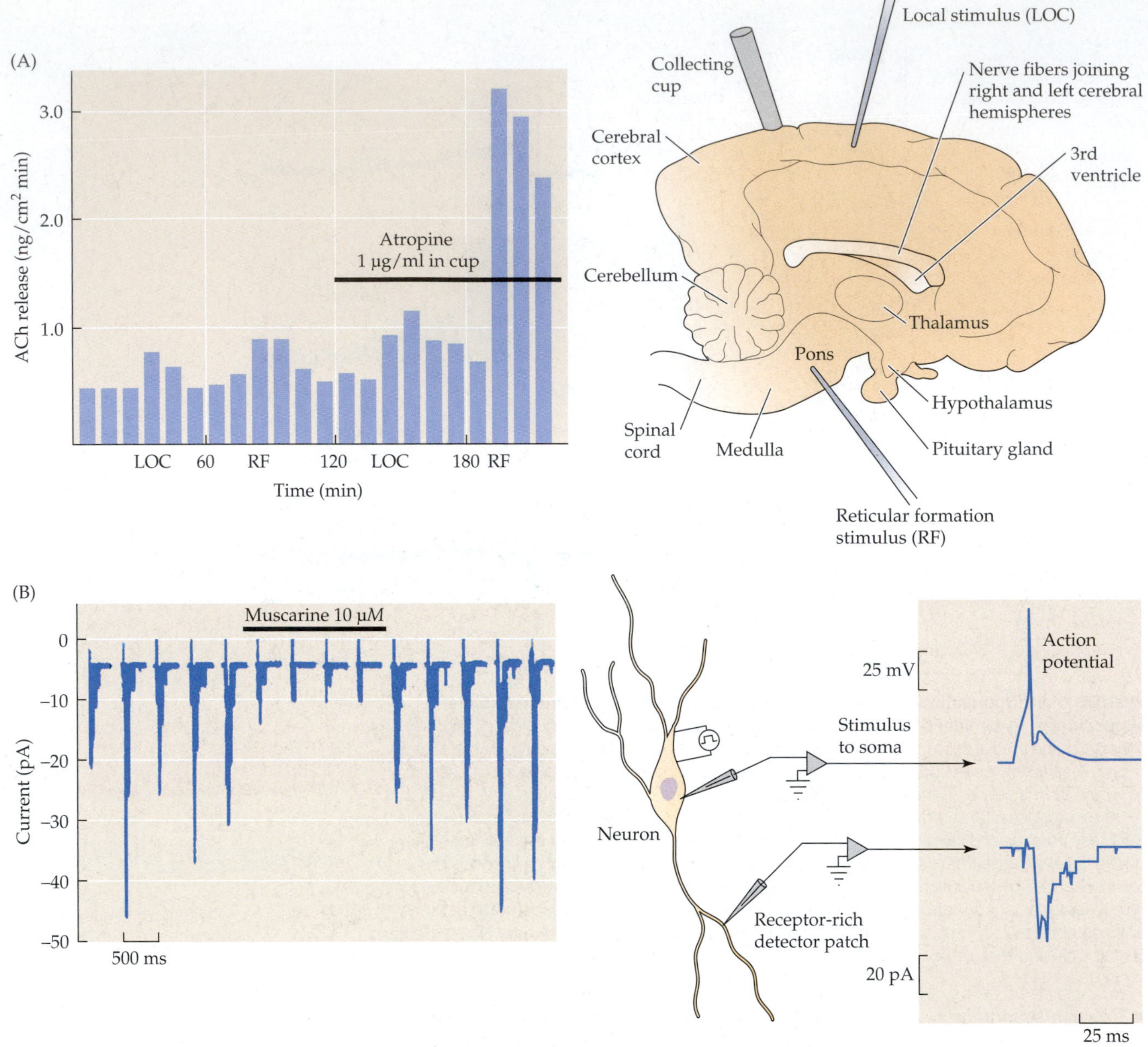

FIGURE 14.11 Muscarinic Auto-Receptors Inhibit Acetylcholine Release from Cholinergic Forebrain Afferents. (A) Acetylcholine (ACh) released from the surface of the parietal cortex of an anesthetized cat was collected into a cortical cup every 10 minutes and subsequently assayed. Local stimulation (LOC) of cholinergic afferent fibers and stimulation of midbrain reticular formation (RF) are shown. The muscarinic receptor antagonist, atropine (1 µg/ml), added to the cup enhanced evoked release. (B) Currents recorded from a nicotinic receptor detector patch placed adjacent to an ACh release site on a neurite from a tissue-cultured basal forebrain neuron. The neuron was stimulated to give an action potential once per minute, which travelled down the neurite. The ACh released following each action potential triggered a burst of nicotinic channel openings in the detector patch. Muscarine (10 µ*M*) strongly reduced the amount of ACh released. (A after Dudar and Szerb, 1965; B after Allen and Brown, 1996.)

Importantly, degeneration of the basal forebrain cholinergic neurons and of their cortical and hippocampal terminals is the earliest and most profound neurodegenerative change seen in **Alzheimer's disease**, and it makes a major contribution to the cognitive deficits that occur in this condition.[127,128] The cholinergic interneurons in the basal ganglia have a different role in regulating motor output. Thus, muscarinic receptor antagonists were once used to reduce the tremor in Parkinson's disease before being replaced by more effective dopaminergic agonists.

[127] Mesulam, M. 2004. *Learn. Mem.* 11: 43–49.

[128] Schliebs, R., and Arendt, T. 2006. *J. Neural Transm.* 113: 1625–1644.

Biogenic Amines

Biogenic amines include norepinephrine, 5-HT, dopamine, and histamine. They are notable for the fact that they are restricted to a very few neurons contained in discrete loci. However, from these nuclei, ramifying unmyelinated axons project to innervate large areas of the brain, through which they influence many states of brain function such as attention, arousal, sleep, and mood. In most cases, these axons do not terminate at discrete synapses; instead, like ACh, the transmitters are released from varicosities along the course of the axons and act in a diffuse manner called volume transmission.[129] These actions correlate with the fact that, with the one exception of 5-HT, their target receptors are all indirect G protein–coupled receptors that are adapted to respond to relatively low concentrations of transmitter on a slow time scale (see Chapter 12). Further, there are often many subtypes of receptor for each amine capable of affecting several different ion channels, so their effects on neuronal activity can be complex and varied.

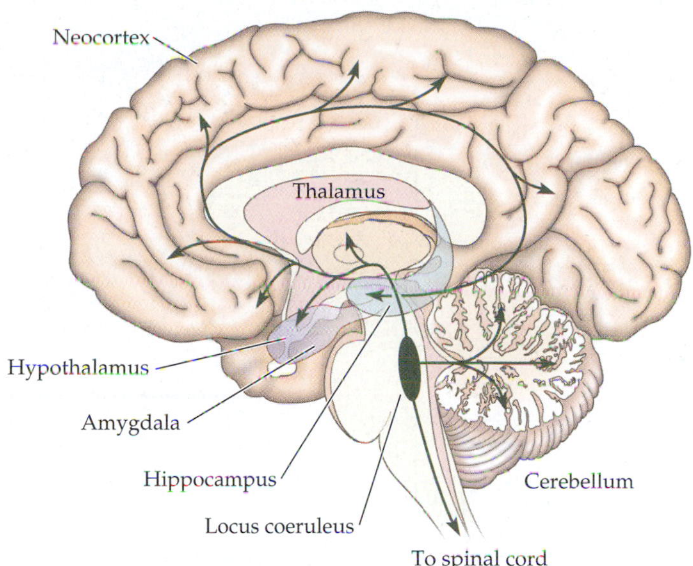

FIGURE 14.12 Projections of Norepinephrine-Containing Neurons in the Locus Coeruleus. The locus coeruleus lies in the pons just beneath the floor of the fourth ventricle. Neurons projecting from it innervate widespread regions of the brain and spinal cord.

NOREPINEPHRINE The noradrenergic neurons are concentrated in a nucleus in the dorsal part of the pons called the **locus coeruleus** (or blue spot, the blue being from melanin).[130–132] In the rat, each locus (either side of the midline) contains only about 1500 neurons (in humans, about 12,000); however, the axons of these neurons innervate wide areas of the CNS, including large parts of the cerebral cortex (via the medial forebrain bundle), hippocampus, hypothalamus, and amygdala, and they even run down the spinal cord[133] (Figure 14.12).

The locus coeruleus neurons fire spontaneously both **in vivo** and **in vitro** at a steady low rate, which can then be modified by the release of glutamate, GABA, enkephalins, and other transmitters from their afferent inputs.[131,134] Their firing rate can also be inhibited by norepinephrine, through α2-adrenoceptors; these hyperpolarize the neuron by opening potassium channels.[135,136] By this means, release of norepinephrine within the locus coeruleus exerts a negative-feedback control of the neurons' intrinsic firing rate.[137]

The primary effect of norepinephrine on its target neurons in the hippocampus and cerebral cortex is to inhibit the slow calcium-activated potassium current.[131,138] This action is mediated by β2-adrenoceptors and involves activation of adenylate cyclase.[139] Its effect is to increase the firing rate of the neuron. Norepinephrine also hyperpolarizes the target neurons by activating α2-receptors, as it does to the locus coeruleus neurons. The hyperpolarization reduces background activity while the potassium current inhibition facilitates responses to strong afferent stimulation. These actions enable afferents from the locus coeruleus to augment and sharpen the response of cortical pyramidal cells to the sensory stimulus of interest without increasing noise, as illustrated in Figure 14.13. By such means, the locus coeruleus can increase attention and facilitate learning and memory.[140]

5-HYDROXYTRYPTAMINE (5-HT OR SEROTONIN) Like the noradrenergic neurons, serotonergic neurons are localized to a few nuclei in the brainstem. These are the **raphe** nuclei that lie directly along the midline from the midbrain to the medulla (Figure 14.14). (The term *raphe* comes from the French for "seam"). In the rat, these nuclei contain only about 20,500 neurons, but their axons innervate wide areas of the brain and extend down the spinal cord.[141] The raphe neurons fire spontaneously at a slow, steady rate of around 1–3 Hz, but this rate can be modified by a variety of afferent inputs and is accelerated when an animal is awake or aroused and active.[142] The raphe neurons have 5-HT receptors and are subject to autoinhibition[143] by locally released 5-HT through activation of inwardly rectifying potassium channels.[144,145]

There is a remarkable number and variety of 5-HT receptors.[146,147] One of these, the 5-HT_3 receptor, is a ligand-gated cation channel, highly homologous to the nicotinic

[129] Fuxe, K. et al. 2010. *Prog. Neurobiol.* 90: 82–100.
[130] Dahlstom, A., and Fuxe, K. 1964. *Acta Physiol. Scand. Suppl.* 232: 1–55.
[131] Foote, S. L, Bloom, F. E., and Aston-Jones, G. 1983. *Physiol. Rev.* 63: 844–914.
[132] Berridge, C. W., and Waterhouse, B. D. 2003. *Brain Res. Brain Res. Rev*. 42: 33–84.
[133] Ungerstedt, U. 1971. *Acta Physiol. Scand. Suppl.* 367: 1–49.
[134] Williams, J. T. et al. 1984. *Neuroscience* 13: 137–156.
[135] Aghajanian, G. K., and VanderMaelen, C. P. 1982. *Science* 215: 1394–1396.
[136] Egan, T. M. et al. 1983. *J Physiol.* 345: 477–488.
[137] Aghajanian, G. K., Cedarbaum, J. M., and Wang, R. Y. 1977. *Brain Res*. 136: 570–577.
[138] Madison, D. V., and Nicoll, R. A. 1986. *J. Physiol*. 372: 221–244.
[139] Madison, D.V., and Nicoll, R. A. 1986. *J. Physiol.* 372: 245–259.
[140] Sara, S. J. 2008. *Nat. Rev. Neurosci.* 10: 211–223.
[141] Jacobs, B. L., and Azmitia, E. C. 1992. *Physiol. Rev.* 72: 165–229.
[142] Vandermaelen, C. P., and Aghajanian, G. K. 1983. *Brain Res.* 289: 109–119.
[143] Piñeyro, G., and Blier, P. 1999. *Pharmacol. Rev.* 51: 533–591.
[144] Wang, R. Y., and Aghajanian, G. K. 1977. *Brain Res.* 132: 186–193.
[145] Penington, N. J., Kelly, J. S., and Fox, A. P. 1993. *J. Physiol.* 469: 387–405.
[146] Barnes, N. M., and Sharp, T. 1999. *Neuropharmacology* 38: 1083–1152.
[147] Filip, M., and Bader, M. 2009. *Pharmacol. Rev*. 61: 761–777.

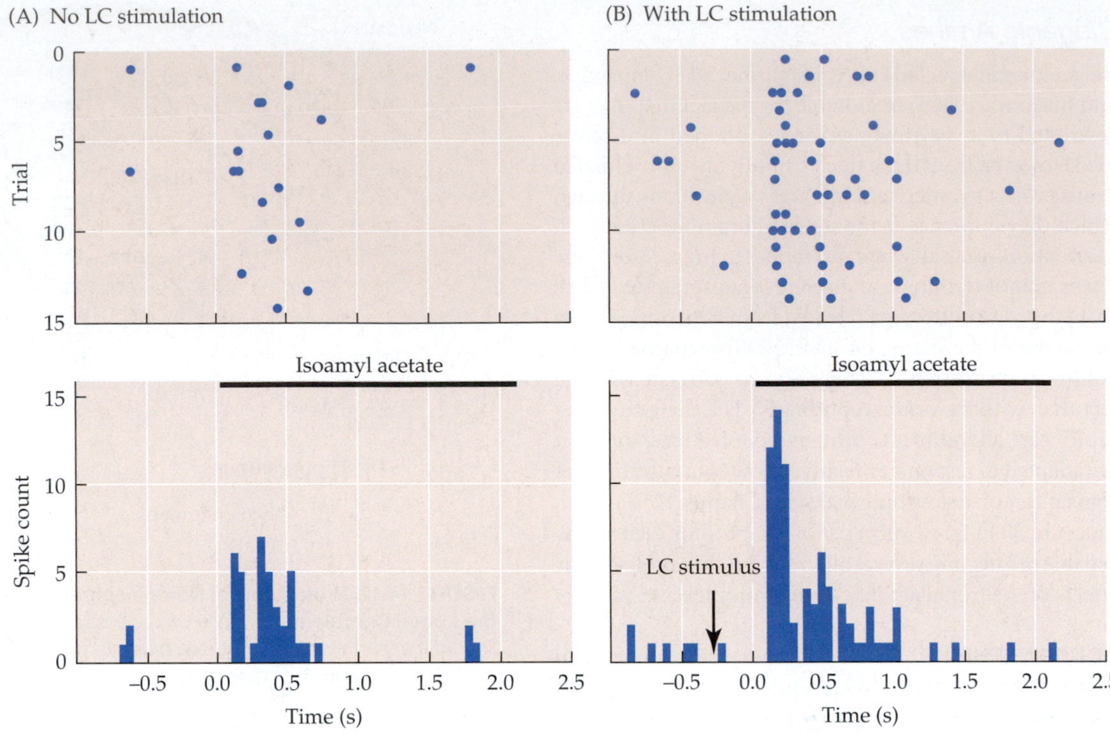

FIGURE 14.13 Locus Coeruleus Stimulation Sharpens and Enhances Sensory Processing in the Cerebral Cortex. Action potentials were recorded from a single neuron in the piriform cortex of an anesthetized rat for one second before and then during a 2-second olfactory stimulus with isoamyl acetate (bars at bottom) applied 14 times (i.e., 14 trials). Each dot in the upper plot represents one action potential. The bottom histogram shows the numbers of action potentials recorded, grouped into 50 ms bins. (A) Without locus coeruleus (LC) stimulation. (B) Stimulation of locus coeruleus for 250 ms before the olfactory stimulus. Such stimulation increased the reliability of the response (there are fewer blank horizontal lines in the upper plots) and sharpened it, by increasing the clustering of the action potentials within the first few hundred ms and reducing the jitter in the delay to the first action potential. (After Sara, 2008.)

[148]Barnes, N. M. et al. 2009. *Neuropharmacology* 56: 273–284.

[149]van Hooft, J. A. et al. 1998. *Proc. Natl. Acad. Sci. USA* 95: 11456–11461.

[150]Chameau, P., and van Hooft, J. A. 2006. *Cell Tissue Res.* 326: 573–581.

receptor[148] (so homologous, in fact, that it can actually form a functional chimeric channel with the nicotinic α4-subunit[149]). These ionotropic 5-HT receptors are widely distributed in the brain, both presynaptically and postsynaptically.[150] However, only a couple of true synaptic currents produced by serotonergic fiber stimulation have been reported, so presumably the receptors respond to diffusively released 5-HT.

All of the other 5-HT receptors are G protein–coupled, differentially activating G_o/G_i, G_q, or G_s (see Chapter 12). These receptors are found both postsynaptically and presynaptically, and some neurons express more than one type of receptor. Thus, the overall actions of 5-HT on individual neurons and synaptic pathways are complex and variable, and they cannot readily be extrapolated to

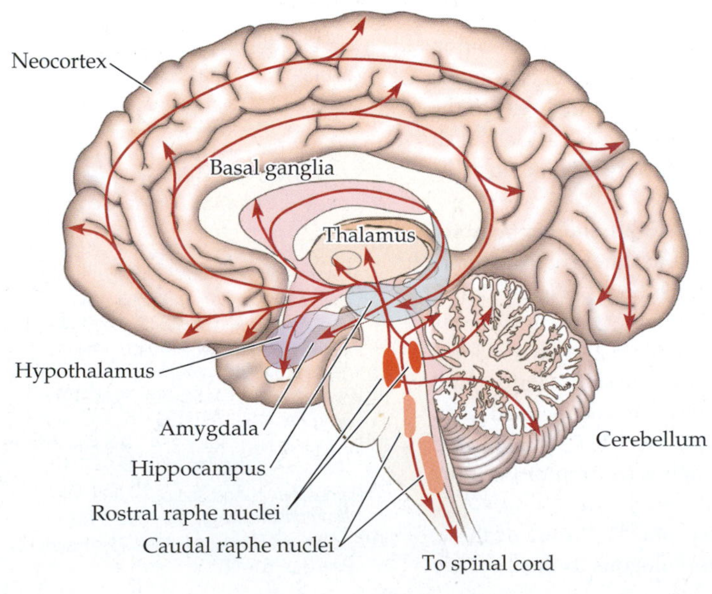

FIGURE 14.14 Neurons Containing 5-Hydroxytryptamine (5-HT) Form a Chain of Raphe Nuclei Lying along the Midline of the Brainstem. More caudal nuclei innervate the spinal cord, while more rostral nuclei innervate nearly all regions of the brain.

effects on brain function.[147,148] The overall actions can be more readily deduced from the actions of drugs that mimic 5-HT, block its action on specific receptors, or block its reuptake, or from the effects of individual 5-HT receptor knock-outs.[147] For example, projections to the hypothalamus help to regulate the sleep–wake cycle[142] and food intake,[151] so that drugs that inhibit 5-HT uptake reduce food intake. These same drugs, which include fluoxetine (or Prozac), are used as antidepressants, probably reflecting a general role for 5-HT in controlling affective states.[152] A more focused effect of descending serotonergic fibers is their role in reducing the release of transmitters from nociceptive spinal and trigeminal sensory afferents in the substantia gelatinosa.[153,154] Thus, the drug sumatriptan, which selectively activates these (5-HT_{1D}) receptors, can suppress the symptoms of migraine and trigeminal neuralgia. 5-HT also contributes to the expression of aggression in a range of animal species, including crustacea.[155]

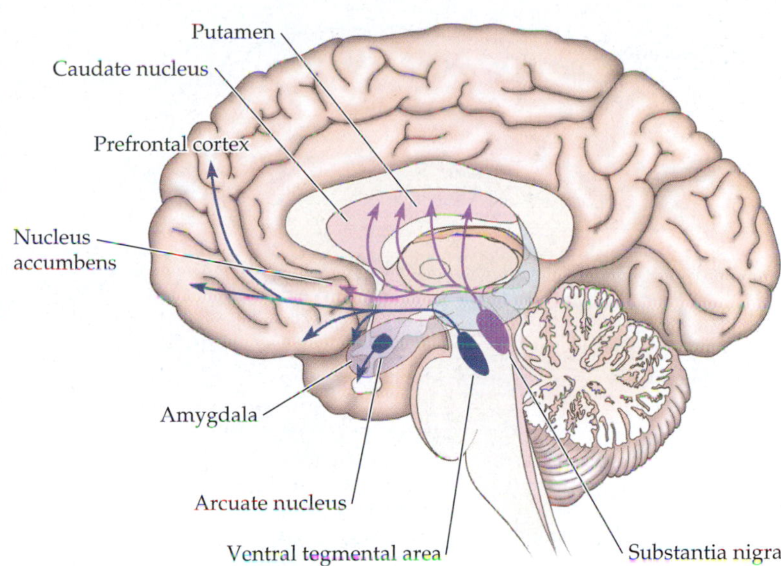

FIGURE 14.15 Neurons Containing Dopamine Are Found in Nuclei in the Hypothalamus and Midbrain. Those in the arcuate nucleus project to the median eminence of the hypothalamus, forming the tuberoinfundibular system. Dopamine neurons in the substantia nigra project to the caudate nucleus and putamen (collectively called the striatum) of the basal ganglia, forming the nigrostriatal pathway. Dopamine neurons in the ventral tegmental area project to the nucleus accumbens, amygdala, and prefrontal cortex, forming the mesolimbic and mesocortical systems.

DOPAMINE Dopamine (3-hydroxytyramine) is an intermediary in the synthetic pathway to norepinephrine (see Appendix B). However, some important groups of neurons do not have the enzyme dopamine β-hydroxylase that converts dopamine to norepinephrine, so dopamine is the end product.[156] These **dopaminergic** neurons are concentrated in four nuclei in the brainstem (Figure 14.15). One group in the arcuate nucleus sends axons to the hypothalamus, where they control prolactin secretion from pituitary lactotrophs.[157] The other three groups are in the ventral tegmental area and substantia nigra (so-called because the neurons contain neuromelanin and stain black). Axons from these neurons project to the neostriatum in the basal ganglia, limbic structures such as the nucleus accumbens, and the prefrontal cortex. As with other catecholaminergic systems, a very few dopaminergic neurons reach out to many target cells—there are about 7000 neurons in the substantia nigra but each neuron gives rise to an estimated 250,000 varicosities in the basal ganglia.[158]

Also, like other catecholamine-releasing neurons, dopaminergic neurons show spontaneous firing, the rate and rhythm of which are controlled by afferent inputs.[159] The released dopamine may act on one or more of five receptors (D_1 through D_5). Activating D_1 or D_5 receptors stimulates adenylate cyclase and phospholipase C, while $D_{2,3,4}$ receptors inhibit adenylate cyclase, open potassium channels, and inhibit calcium currents (see Chapter 12). Thus, the postsynaptic action of released dopamine varies with different target neurons and can be complex.[160] Nearly all dopaminergic neurons have autoreceptors, on both their terminals and soma–dendrite regions. These are mostly D_2 receptors whose activation inhibits neuron firing (if on the soma or dendrites) or inhibits dopamine release (if at the terminals).

The physiological importance of the dopaminergic projection to the basal ganglia became apparent when it was discovered that it was much reduced in patients with **Parkinson's disease** (a disease affecting motor control, see Chapter 24).[161] In Parkinson's disease, there is a selective degeneration of the dopaminergic neurons in the substantia nigra, resulting in the loss of dopaminergic terminals in the striatum. This finding led to one of the most astonishing advances in medical treatment, namely, that giving the precursor of dopamine, L-dihydroxyphenylalanine (L-DOPA), increased the amount of dopamine in the residual terminals and dramatically reversed the motor deficits.[162,163]

Other dopaminergic projections serve different functions. The projection from the ventral tegmental area to the nucleus accumbens is a reward signaling system, with the neurons increasing their activity in anticipation of or in response to pleasurable events.[103,164] This area is the principal site of action of the addictive drug cocaine, which inhibits the dopamine

[151] Garfield, A. S., and Heisler, L. K. 2009. *J. Physiol*. 587: 49–60.

[152] Cowen, P. J. 2008. *Trends Pharmacol. Sci.* 29: 433–436.

[153] Jennings, E. A., Ryan, R. M., and Christie, M. J. 2004. *Pain* 111: 30–37.

[154] Yoshimura, M., and Furue, H. 2006. *J. Pharmacol. Sci.* 101: 107–117.

[155] Kravitz, E. A. 2000. *J. Comp. Physiol. A* 186: 221–238.

[156] Carlsson, A. et al. 1958. *Science* 127: 471.

[157] van den Pol, A. N. 2010. *Neuron* 65: 147–149.

[158] Yurek, D. M., and Sladek, J. R., Jr. 1990. *Annu. Rev. Neurosci.* 13: 415–440.

[159] Diana, M., and Tepper, J. M. 2002. In *Handbook of Experimental Pharmacology*, volume 154, part 1. Springer-Verlag, Berlin. pp. 1–62.

[160] Nicola, S. M., Surmeier, D. J., and Malenka, R. C. 2000. *Annu. Rev. Neurosci*. 23: 185–215.

[161] Ehringer, H., and Hornykiewicz, O. 1960. *Klin. Wochenschr.* 38: 1236–1239.

[162] Birkmayer, W., and Hornykiewicz, O. 1962. *Arch. Psychiatr. Nervenkr*. 203: 560–574.

[163] LeWitt, P. A. 2008. *New England J. Med.* 359: 2468–2476.

[164] Schultz, W., Dayan, P., and Montague, R. R. 1997. *Science* 275: 1593–1599.

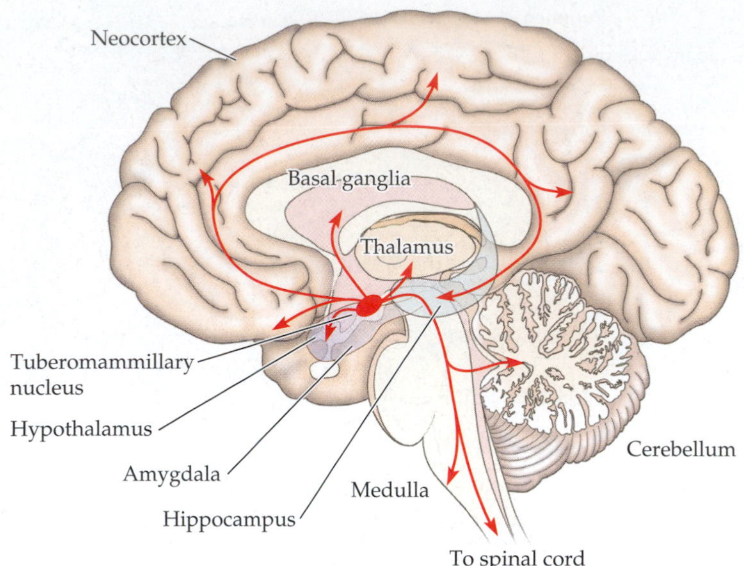

FIGURE 14.16 Histamine-Containing Neurons Are Localized to the Tuberomammillary Nucleus in the Hypothalamus. These neurons have diffuse projections throughout the brain and spinal cord.

transporter and increases the amount of dopamine released in the nucleus accumbens.[165] This effect, and the accompanying behavioral response, is abolished in transgenic mice in which the normal dopamine transporter is replaced with one that is insensitive to cocaine.[166] As noted previously, another addictive drug, nicotine, also increases dopamine release in the nucleus accumbens, though by a different mechanism. The projection to the prefrontal cortex is thought to participate in cognitive functions[167] and to control affective states such as schizophrenia;[168] nearly all drugs currently used to treat the latter condition work by blocking dopamine receptors.

HISTAMINE Histamine was first identified as a mediator of inflammation. However, when the first antihistamines were introduced (in the 1930s, to treat hay fever, rashes, and insect bites), their main side effect was to make the recipient sleepy. This provided the first clue that histamine might also be important in the brain—a role that has been amply confirmed.[169] There is a dense aggregation of histamine-containing neurons in the tuberomammillary nucleus of the dorsal hypothalamus (Figure 14.16). From there, histaminergic fibers project to wide areas of the brain, including other parts of the hypothalamus, the cerebral cortex, hippocampus, amygdala, and basal ganglia as well as down the spinal cord. Histamine also acts as a transmitter in some invertebrates, for example, at arthropod photoreceptor synapses.[170]

There are three subtypes of histamine receptor in the brain, all G protein-coupled: H_1, H_2, and H_3. H_1 receptors (the original antihistamine receptors) are the principal postsynaptic receptors. They activate the G protein G_q to stimulate phospholipase C (see Chapter 12) and depolarize neurons by inhibiting potassium currents (including M-currents[113]) (Figure 14.17A). H_2 receptors are also postsynaptic—their main effect is to inhibit the slow calcium-activated potassium current and increase action potential discharges[171] (Figure 14.17B), which they do by activating adenylate cyclase and increasing cyclic AMP. In contrast, H_3 receptors act as autoinhibitory receptors.[172] They are located on the somata and dendrites of the histaminergic neurons, where they reduce their natural firing rate (Figure 14.17C), and on their terminals, where they reduce transmitter release.

As might be anticipated from the wide distribution of their efferent fibers, tuberomammillary histaminergic neurons affect many brain functions. One notable role is in regulating sleep. Their activity is closely coupled to the sleep–wake cycle (Figure 14.18). They are more active during the awake state and induce cortical arousal through their direct projections to the thalamus and cortex and by activating other ascending arousal systems.[173] Blockade of the histamine H_1 receptors on their target neurons explains the somnolent effect of the original antihistamines. A projection to the suprachiasmatic nucleus and to other regions also affects the sleep–wake cycle through an alteration in circadian rhythm.[174]

Histaminergic neurons also control motor behavior. Mice in which histamine levels have been reduced by deleting the synthetic enzyme histidine decarboxylase show exaggerated responses to locomotor stimulants;[175] and a mutation in the gene encoding this enzyme has recently been found in a human family exhibiting Tourette's syndrome,[176] a neurological disorder in which the affected person shows uncontrolled jerky spontaneous movements and verbal expostulations.

Adenosine Triphosphate (ATP)

ATP is an essential transmitter and chemical messenger in the PNS. This discovery stems from the pioneering work of Pamela Holton, who first showed that ATP was released from peripheral nerve endings when their axons were stimulated,[177] and by Geoffrey Burnstock,

[165] Kalivas, P. W., and Duffy, P. 1990. *Synapse* 5: 48–58.
[166] Chen, R. et al. 2006. *Proc. Natl. Acad. Sci. USA* 103: 9333–9338.
[167] Seamans, J. K., and Yang, C.R. 2004. *Prog. Neurobiol.* 74: 1–58.
[168] Howes, O. D., and Kapur, S. 2009. *Schizophr. Bull.* 35: 549–562.
[169] Haas, H. L., Sergeeva, O. A., and Selbach, O. 2008. *Physiol. Rev.* 88: 1183–1241.
[170] Stuart, A. E, Borycz, J., and Meinertzhagen, I. A. 2007. *Prog. Neurobiol.* 82: 202–227.
[171] Haas, H. L., and Konnerth, A. 1983. *Nature* 302: 432–434.
[172] Arrang, J. M., Garbarg, M., and Schwartz, J. C. 1983. *Nature* 302: 832–837.
[173] Lin, J. S. 2000. *Sleep Med. Rev.* 4: 471–503.
[174] Abe, H. et al. 2004. *Brain Res. Mol. Brain Res.* 124: 178–187.
[175] Kubota, Y. et al. 2002. *J. Neurochem.* 83: 837–845.
[176] Ercan-Sencicek, A. G. et al. 2010. *New England J. Med.* 362: 1901–1908.
[177] Holton, P. 1959. *J. Physiol.* 145: 494–504.

(A) H_1 action

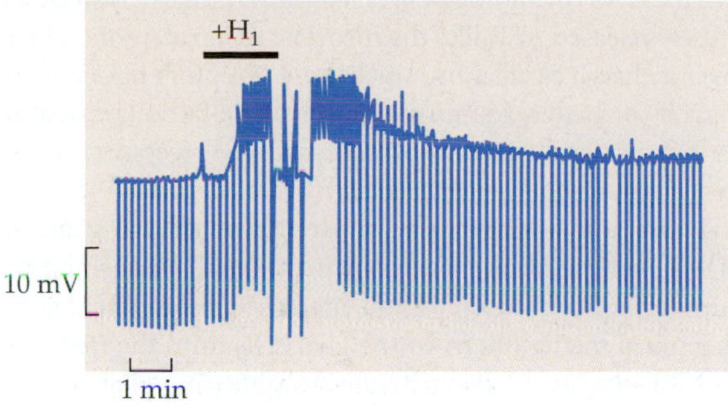

(C) H_3 action

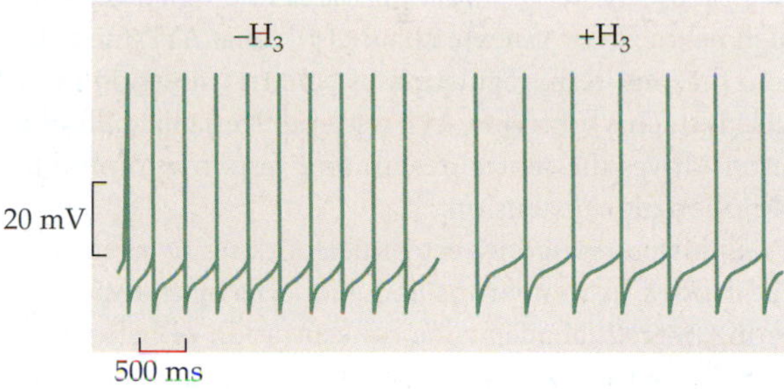

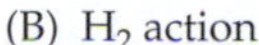

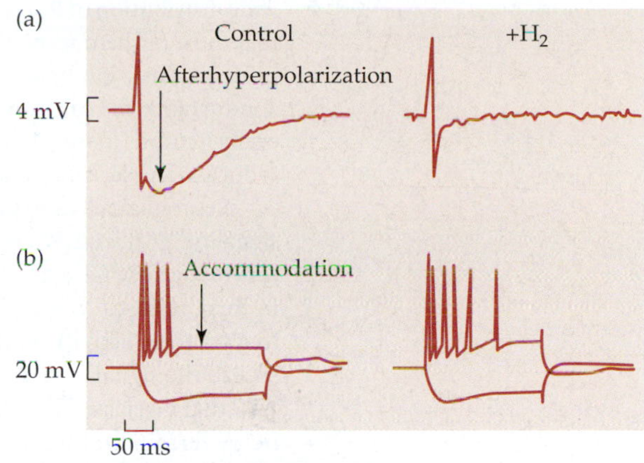

FIGURE 14.17 Effects of Stimulating Three Different Histamine Receptors. (A) Record from a cell in the pontine reticular formation. Downward deflections are transient voltage responses to brief hyperpolarizing constant current injections. Application of histamine (bar) caused depolarization and induced action potential firing (truncated by the recorder). When the cell was repolarized during the response to histamine, the brief current injections produced a larger voltage deflection, showing that the membrane conductance was reduced. This finding was attributed to the reduction of a resting potassium conductance. (B) Stimulating H_2 receptors in a human hippocampal pyramidal cell (a) eliminates the long-lasting calcium-dependent afterhyperpolarization following an action potential, and (b) reduces accommodation during action potential discharges produced by a 250-millisecond depolarizing current. (C) Stimulating H_3 autoreceptors slows action potential discharges in a tuberomammillary histaminergic neuron. These records show the spontaneous activity of a neuron recorded when H_3 receptors were blocked with the selective H_3 antagonist thioperamide ($-H_3$) and then when the receptors were unblocked by omitting the thioperamide ($+H_3$). (After Haas and Panula, 2003.)

who established its transmitter function in the autonomic nervous system (see Chapter 17) and dubbed ATP-releasing nerves "purinergic."[178] It is now recognized that ATP also plays a significant role in intercellular communication within the CNS.[179] There are two types of receptors for ATP: P_{2X} receptors, which are ligand-gated cation channels,[180] and G protein–coupled receptors, called P_{2Y} receptors.[181] Seven types of P_{2X} receptor subunits and eight different P_{2Y} receptors are known. Many of the $P2_{2X}$ receptors and some of the P_{2Y} receptors (most widely, the $P2Y_1$ receptor) are expressed on neurons and glial cells in the brain.

ATP acts as a fast transmitter on P_{2X} receptors at a few synapses in the brain.[182,183] At some synapses, it acts as the

[178] Burnstock, G. 1972. *Pharmacol. Rev.* 24: 509–581.

[179] Abbracchio, M. P. et al. 2009. *Trends Neurosci.* 32: 19–29.

[180] Surprenant, A., and North, R. A. 2009. *Annu. Rev. Physiol.* 71: 333–359.

[181] Abbracchio, M. P. et al. 2006. *Pharmacol. Rev.* 58: 281–341.

[182] Edwards, F. A., Gibb, A. J., and Colquhoun, D. 1992. *Nature* 359: 144–147.

[183] Pankratov, Y. et al. 2006. *Pflügers Arch.* 452: 589–597.

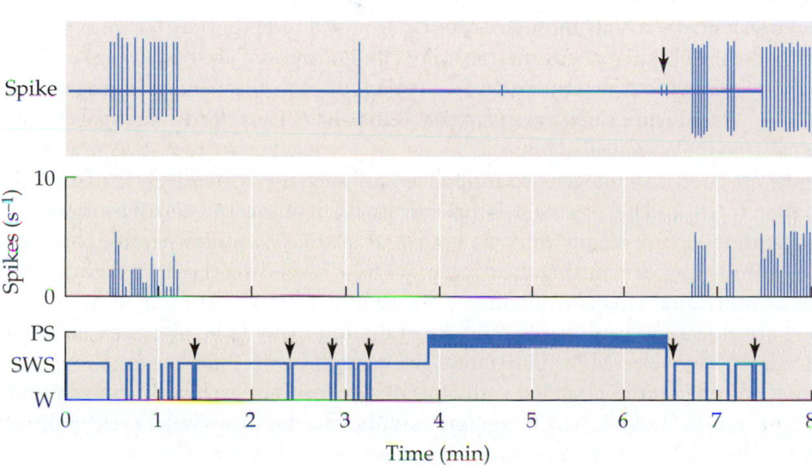

FIGURE 14.18 Histaminergic Neuron Activity Is Related to the Awake State. The top record shows action potential discharges (labeled "spike") from a single histaminergic neuron in the tuberomammillary nucleus of an unanaesthetized mouse with an extracellular recording electrode. The trace below is a ratemeter output of action potential frequency (spikes s^{-1}). The bottom diagram shows the awake-sleep state of the mouse, as judged from the electroencephalogram (EEG). W = awake; SWS = slow-wave sleep; PS = paradoxical sleep. The neuron only fired when the mouse was awake, and was completely silent when the mouse was asleep. It also stayed silent when the mouse woke up for just very short periods (arrowheads). (Adapted from Takahashi et al., 2006.)

sole transmitter[184] but more often it is co-released with another transmitter.[185] However, the wide distribution of P_{2X} receptors in the CNS (including in presynaptic terminals[181]) suggests that most of them respond to diffusely released ATP, like the nicotinic ACh receptors. Their principal function may be to deliver a charge of calcium, since these receptors have a high conductance for calcium ions.[179] ATP (or better, adenosine diphosphate [ADP]) can also excite neurons in the hippocampus by stimulating $P2Y_1$ G protein–coupled receptors, which induce a depolarizing cation current[186] or inhibit the M-current.[187]

Neuroglial cells also possess ATP receptors, which may be involved in neuron-to-glial cell signaling.[188] It is envisaged that: (1) ATP released from neurons activates glial cell purinoceptors to increase intracellular calcium ion concentration; (2) the increased intraglial calcium in turn, releases ATP and other chemical mediators from the glial cells onto the neurons to alter their activity; and (3) the ATP so released also activates neighboring glial cells to spread the signal.[188] Thus, recent research has shown that glial cells can be depolarized by ATP that is released from specialized release sites along axons,[189] as well as from purinergic nerve terminals; that ATP released following nerve injury activates microglial cells and causes them to release ATP through pannexin or innexin channels;[190] that ATP-mediated microglial activation contributes to the long-term regulation of pain transmission in the spinal cord;[191] and that, in the nucleus tractus solitarius, ATP released from glial cells onto neighboring chemosensitive neurons drives the central respiratory response to produce changes in arterial blood pH and carbon dioxide tension.[192]

Extracellular ATP is rapidly dephosphorylated by ecto-nucleotidases to adenosine, which joins a pool of adenosine extruded from neurons and glial cells by transporters, producing a constant but fluctuating interstitial adenosine concentration of between 25 and 250 nanomolar (nm).[193] Adenosine can activate another family of G protein–coupled receptors on neurons called adenosine receptors. Some of these (A_1 and A_{2A} receptors) are sufficiently sensitive to adenosine to be activated at the normal levels of interstitial adenosine, so producing a tonic reduction of neuron excitability and transmitter release.[194] This is important in the regulation of sleep.[195] For example, accumulation of adenosine during periods of wakefulness depresses the activity of neurons in the tuberomammillary nucleus, thus precipitating sleepiness. This explains the stimulant action of caffeine, since caffeine blocks adenosine receptors. Thus, mice in which the A_{2A} gene is deleted are unusually aggressive and no longer respond to the stimulant effect of caffeine.[196]

Peptides

There are differences between peptide transmitters (or **neuropeptides**) and small molecule transmitters like amino acids and monoamines. First, peptides are not synthesized de novo in nerve endings but instead are split off from larger precursor molecules by peptidase enzymes. The precursors are made in the cell bodies, transported down the axons to the terminals, and then packaged (with their peptidase enzymes) into large secretory (i.e., dense-core) synaptic vesicles (see Figure 15.8). Second, they are often present in the same nerve terminals (though in different vesicles) that release small-molecule transmitters like the amino acids and so may be co-released.[24] Third, in the mammalian nervous system at least, the peptides act exclusively as **indirect transmitters**, mainly through G protein–coupled receptors (see Chapter 12), and tend to straddle the boundary between transmitter and hormone. The best example of true peptidergic transmission is that mediated by a gonadotrophin-like peptide in frog sympathetic ganglia[197] (see Figure 17.2); most peptides seem to have a less clearly identifiable, more complex effect in the mammalian brain.

Peptides are good immunogens, so antibodies have been used extensively to identify the neurons, fiber tracts, and nerve terminals that contain them or their precursor molecules.[198] However, peptides do not readily enter the brain and relatively few non-peptide chemicals capable of stimulating or blocking their receptors have been discovered, so their specific functions at individual synapses are often difficult to determine. As pointed out in Box 14.2, there are at least 60 neuropeptides in the brain, too many to be discussed individually. Instead, we have selected for illustration five peptides with different functions: two, substance P and the opioid peptides, with a particular (but not exclusive) connection to pain perception; the orexins, which regulate feeding and the sleep–wake cycle; and two

[184] Robertson, S. J., and Edwards, F. A. 1998. *J. Physiol.* 508: 691–701.

[185] Jo, Y. H., and Role, L. W. 2002. *J. Neurosci.* 22: 4794–4804.

[186] Bowser, D. N., and Khakh, B. S. 2004. *J. Neurosci.* 24: 8606–8620.

[187] Filippov, A. K. et al. 2006. *J. Neurosci.* 26: 9340–9348.

[188] Fields, R. D., and Burnstock, G. 2006. *Nat. Rev. Neurosci.* 7: 423–436.

[189] Thyssen, A. et al. 2010. *Proc. Natl. Acad. Sci. USA* 107: 15258–15263.

[190] Samuels, S. E. et al. 2010. *J. Gen. Physiol.* 136: 425–442.

[191] Tsuda, M., Tozaki-Saitoh, H., and Inoue, K. 2010. *Brain Res. Rev.* 63: 222–232.

[192] Gourine, A. V. et al. 2010. *Science* 329: 571–575.

[193] Dunwiddie, T. V., and Masino, S. A. 2001. *Annu. Rev. Neurosci.* 24: 31–55.

[194] Haas, H. L., and Selbach, O. 2000. *Naunyn Schmiedebergs Arch. Pharmacol.* 362: 375–381.

[195] Basheer, R. et al. 2004. *Prog. Neurobiol.* 73: 379–396.

[196] Ledent, C. et al. 1997. *Nature* 388: 674–678.

[197] Kuffler, S. W. 1980. *J. Exp. Biol.* 89: 257–286.

[198] Hökfelt, T. et al. 2000. *Neuropharmacology* 39: 1337–1356.

closely related peptides, vasopressin and oxytocin, which appear to play an interesting role in the way in which animals (and humans) interact with each other.

Substance P

Substance P is a peptide with 11 amino acids, having the composition Arg-Pro-Lys-Pro-Gln-Gln-Phe-Phe-Gly-Leu-Met. It is a member of the family of tachykinins. There are two other tachykinins in the mammalian CNS, the neurokinins (NKs) A and B, but many more in non-mammalian vertebrates and invertebrates.[199] Substance P is so called because in 1931, von Euler and Gaddum first detected its biological activity in a *powder* made from extracts of horse intestine and brain.[200] Substance P and other neurokinins act on a group of G_q-G protein–coupled receptors called NK receptors. Substance P preferentially targets NK_1 receptors but can also activate NK_2 and NK_3 receptors at high concentrations.

Substance P is highly concentrated in primary afferent pain-sensing neurons and their unmyelinated axons, and in the central endings of these neurons in the substantia gelatinosa in the dorsal horn of the spinal cord.[201] When primary afferent fibers are stimulated, substance P (along with glutamate) is co-released as a neurotransmitter and then depolarizes the target dorsal horn neurons.[202] Its role in nociceptive (pain) transmission has been examined using NK_1 receptor antagonists and mice in which the NK_1 receptor has been genetically deleted (NK_1 knock-outs). It is not involved in the transmission of acutely painful stimuli, such as a tail-pinch or burn, but NK_1 knock-out mice showed a reduced response to the prolonged pain produced by injecting a painful inflammatory substance.[203,204] The gradual increase in the firing of dorsal horn neurons during repeated C-fiber stimulation (termed "wind-up") that is associated with sensitization to painful stimuli was also strongly reduced in these animals.[205] Rather disappointingly, however, NK_1 antagonist drugs seem not to have any significant effects on inflammatory pain in humans.[206] Substance P is also strongly expressed in some supraspinal regions, such as the substantia nigra.[203] It may be involved in the control of affective states, since mice in which the NK_1 receptor gene has been deleted show behavioral changes similar to mice treated with antidepressant drugs.[207]

Opioid Peptides

Opioid peptides are a family of endogenous neuropeptides that interact with opioid receptors (as do exogenous substances such as morphine). They include the pentapeptides leu-enkephalin and met-enkephalin (Tyr-Gly-Gly-Phe-Leu and Tyr-Gly-Gly-Phe-Met), which were originally discovered by Hughes, Kosterlitz, and their colleagues (see Box 14.2), plus a number of larger peptides containing the enkephalin sequences, such as the α-endorphins, β-endorphins, and the dynorphins (dynorphin A and B, neo-dynorphin).[208] They are formed by peptidase cleavage from three large precursor proteins (Figure 14.19) and stimulate one or more of three opioid receptors labeled μ, δ, and κ (sometimes called MOR, DOR, and KOR for mu-, delta- and kappa-opioid receptor). These are G protein–coupled receptors that activate the pertussis toxin-sensitive G proteins, G_o and G_i. Like others of this type of receptor, they inhibit calcium channels to reduce transmitter release, activate inwardly rectifying potassium channels to hyperpolarize neuron, and inhibit adenylate cyclase (see Chapter 12).

[199] Severini, C. et al. 2002. *Pharmacol. Rev.* 54: 285–322.

[200] von Euler, U. S., and Gaddum, J. H. 1931. *J. Physiol.* 72: 74–87.

[201] Hökfelt, T. et al. 1975. *Science* 190: 889–890.

[202] Otsuka, M., and Yoshioka, K. 1993. *Physiol. Rev.* 73: 229–308.

[203] Cao, Y. Q. et al. 1998. *Nature* 392: 390–394.

[204] De Felipe, C. et al. 1998. *Nature* 392: 394–397.

[205] Suzuki, R., Hunt, S. P., and Dickenson, A. H. 2003. *Neuropharmacology.* 45: 1093–1100.

[206] Hill, R. 2000. *Trends Pharmacol. Sci.* 21: 244–246.

[207] Yan, T. C., Hunt, S. P., and Stanford, S. C. 2009. *Neuropharmacology* 57: 627–635.

[208] Weber, E., Evans, C. J., and Barchas, J. D. 1983. *Trends Neurosci.* 6: 333–336.

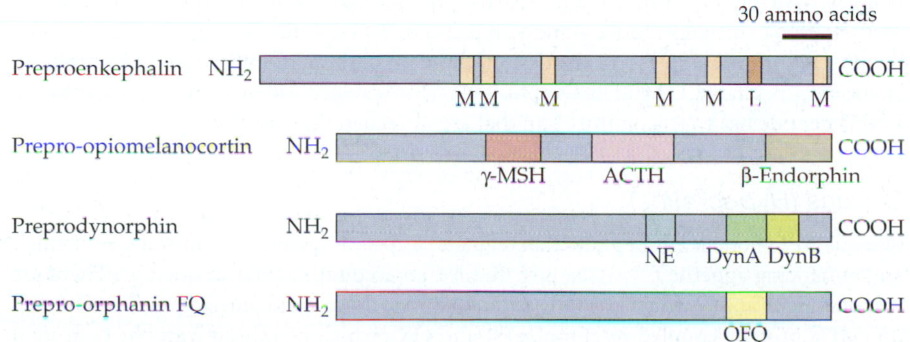

FIGURE 14.19 Four Precursor Proteins that Generate the Opioid Peptides and other peptide hormones by peptidase cleavage. M = met-enkephalin; L = leu-enkephalin; γ-MSH = γ-melanocyte stimulating hormone; ACTH = adrenocorticotrophic hormone; NE = α-neoendorphin; DynA and DynB = dynorphins A and B; OFQ = orphanin FQ/nociceptin. (After Darland et al., 1998.)

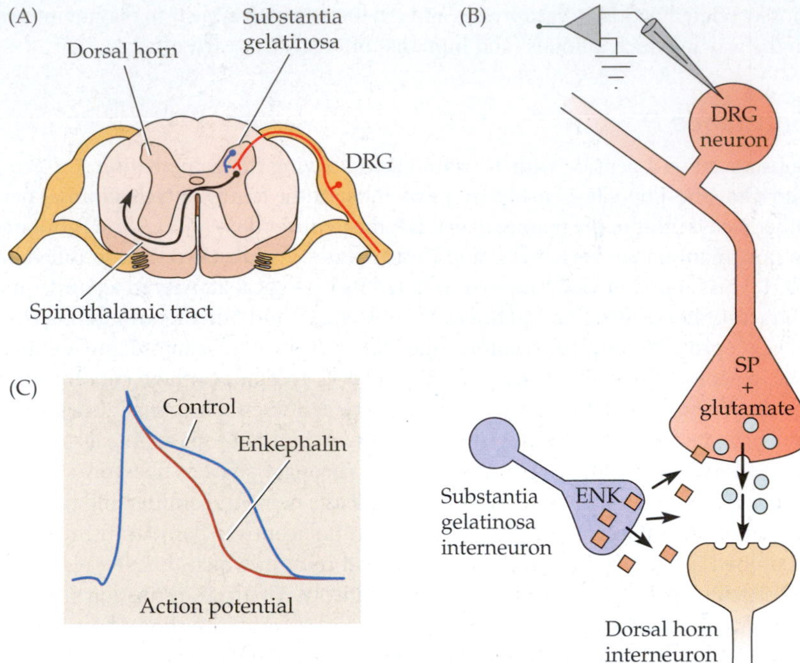

FIGURE 14.20 Pathway for Transmission of Pain Sensation in the Spinal Cord. (A,B) Dorsal root ganglion (DRG) cells responding to noxious stimuli release substance P (SP) and glutamate at their synapses with interneurons in the dorsal horn of the spinal cord. Interneurons containing enkephalin (ENK) in the substantia gelatinosa of the dorsal horn block transmission partly by inhibiting transmitter release from terminals of the DRG cells. (C) Intracellular recordings from the dorsal root ganglion cell demonstrate that enkephalin acts by causing a decrease in the duration of the action potential, reflecting reduced calcium current, which would reduce the amount of transmitter released from sensory nerve terminals. (C after Mudge, Leeman, and Fischbach, 1979.)

Opioid peptides are closely involved in the control of pain. Thus, they are strongly expressed in the periaqueductal gray matter of the midbrain and in the dorsal horn of the spinal cord.[209,210] At the latter site, enkephalin-containing fibers innervate nociceptive dorsal horn neurons[211] and inhibit them by activating an inwardly rectifying potassium current.[212] Enkephalin released from descending fibers in the brainstem or from spinal interneurons also inhibits the release of substance P from nociceptive afferent fibers.[213] This probably results from the inhibition of the calcium current required for transmitter release from afferent nerve endings, as is indicated by the experiments on sensory neurons depicted in Figure 14.20.[214] Opioid receptor antagonists and genetic deletion of the opioid receptors[215] both induce hyperalgesia (increased sensitivity to pain), demonstrating that the opioid peptides play a physiological role in controlling pain. However, these peptides and their receptors are also widely expressed in other parts of the brain not directly concerned with pain, such as the basal ganglia and subcortical limbic areas,[216] suggesting their involvement in other functions.

Another interesting opioid peptide has the dual name orphanin FQ and nociceptin.[217,218] This is because orphanin was identified as the previously unknown ligand for an *orphan* receptor called NOP (formerly called ORL1), hence orphanin,[219] and nociceptin because it *increased* the response to painful stimuli in animals.[220] The "FQ" is added to orphanin because it is identical to dynorphin A (see Figure 14.9) except for two amino acid substitutions: phenylalanine (F) and glutamine (Q). The NOP receptor is homologous to the opioid receptors, couples to their same G proteins, and has the same effects on neurons when stimulated as do opioid receptors.[219] The reason it *increases* pain, rather than reducing it like other opioids, is probably because the NOP receptors are located on different neurons than those with enkephalin or dynorphin receptors. It has been suggested that the NOP neurons normally reduce activity in pain pathways, so that when they are inhibited by nociceptin/orphanin FQ transmission, the pain pathway is enhanced (disinhibited). Nociceptin/orphanin FQ and its receptors are also expressed widely throughout the brain, so this peptide has effects on the brain that are unrelated to pain.[216,217]

Orexins (Hypocretins)

Orexins are hypothalamic peptides that regulate sleep and feeding by promoting awakening and enhancing appetite[221,222] (see Box 14.2 for an account of their discovery). There are two orexins, A and B, which are both generated from the same precursor protein. They act on two G protein–coupled receptors, OX_1 and OX_2, which mainly activate the G protein

[209] Hökfelt, T. et al. 1977. *Proc. Natl. Acad. Sci. USA* 74: 3081–3085.
[210] Marvizón, J. C., Chen, W., and Murphy, N. 2009. *J. Comp. Neurol.* 517: 51–68.
[211] Ma, W. et al. 1997. *Neuroscience* 77: 793–811.
[212] Marker, C. L. et al. 2006. *J Neurosci.* 26: 12251–12259.
[213] Collin, E. et al. 1991. *Neuroscience* 44: 725–731.
[214] Mudge, A. W., Leeman, S. E., and Fischbach, G. D. 1979. *Proc. Natl. Acad. Sci. USA* 76: 526–530.
[215] Kieffer, B. L., and Gavériaux-Ruff, C. 2002. *Progr. Neurobiol.* 66: 285–306.
[216] Mansour, A. et al. 1988. *Trends Neurosci.* 11: 308–314.
[217] Darland, T., Heinricher, M. M., and Grandy, D. K. 1998. *Trends Neurosci.* 21: 215–221.
[218] Meis, S. 2003. *Neuroscientist* 9: 158–168.
[219] Reinscheid, R. K. et al. 1995. *Science* 270: 792–794.
[220] Meunier, J. C. et al. 1995. *Nature* 377: 532–535.
[221] Tsujino, N., and Sakurai, T. 2009. *Pharmacol. Rev.* 61: 162–176.
[222] Bonnavion, P., and de Lecea, L. 2010. *Curr. Neurol. Neurosci. Rep.* 10: 174–179.

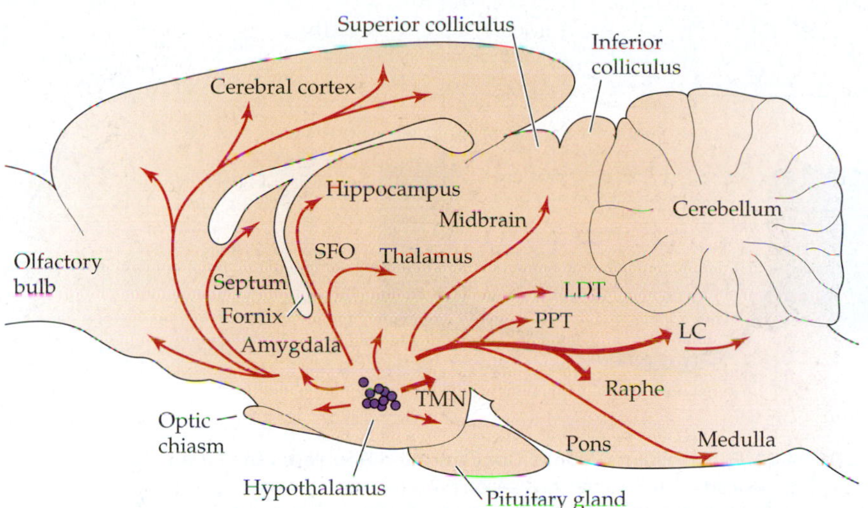

FIGURE 14.21 Orexin-Containing Neurons in the Hypothalamus and Their Axonal Projections. LC = locus coeruleus; LDT = laterodorsal thalamus; PPT = pedunculopontine tegmental area; TMN = tuberomammillary nucleus; SFO = subfornical organ. (After Tsujino and Sakurai, 2009.)

G_q and, when stimulated, increase the activity of the target neuron. Orexin-containing neurons are localized to the posterior lateral hypothalamus, but their axons innervate large areas of the brain (Figure 14.21).

The role of orexins in regulating sleep was established by two genetic observations. First, mice in which the DNA sequences encoding the orexins had been deleted showed excessive nighttime sleeping.[223] (Mice are nocturnal so this is equivalent to daytime sleeping in humans.) The second clue came from work on dogs with the sleep disorder narcolepsy—a disorder that also quite commonly affects people (about 1 in 2000 in the United States). The victim suffers from repeated periods of extreme daytime sleepiness, suddenly falling asleep at the most inconvenient times (sleep attacks); in extreme cases, these episodes are coupled with loss of muscle tone, so that the person collapses (cataplexy). It was found that dogs that had this disease (Doberman pinschers and Labradors) had specific mutations in the gene encoding the OX_2 orexin receptor.[224] Human narcolepsy patients do not necessarily have this mutation but instead lose orexin-containing neurons so they do not secrete enough orexin.[222] Orexin neurons induce awakening via pathways to the locus coeruleus, dorsal raphe, and tuberomammillary nuclei, activating noradrenergic, serotonergic, and histaminergic neurons, respectively (Figure 14.22). Stimulation of histaminergic neurons could play a role because the awakening effect of orexin injected into the lateral ventricles of mice is abolished when the

[223] Chemelli, R. M. et al. 1999. *Cell* 98: 437–451.

[224] Lin, L. et al. 1999. *Cell* 98: 365–376.

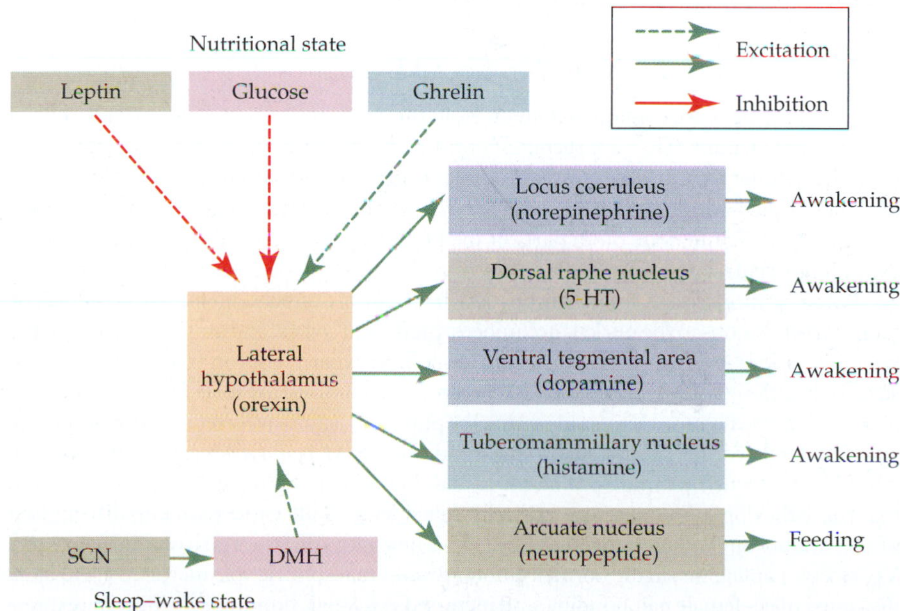

FIGURE 14.22 Integration of Awakening and Feeding Signals by the Orexinergic Neurons in the Lateral Hypothalamus. Orexinergic neurons receive information about the sleep–wake state from the suprachiasmatic nucleus and about the nutritional state from extracellular glucose levels and from the levels of the hormones leptin and ghrelin. They signal awakening and feeding through the various pathways indicated on the right. SCN = suprachiasmatic nucleus; DMH = dorsal medial hypothalamus; dashed lines = afferent signals; full lines = efferent signals. (After Tsujino and Sakurai, 2009.)

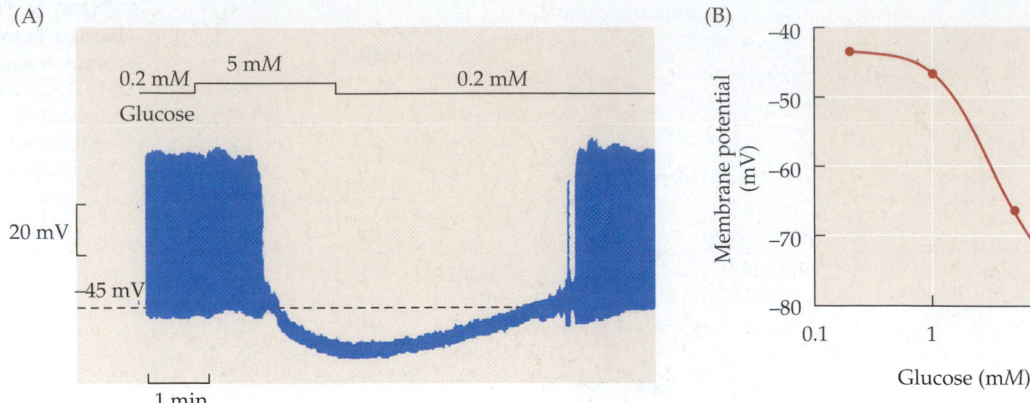

FIGURE 14.23 Extracellular Glucose Concentration Regulates the Firing of Orexin-Containing Neurons. (A) Spontaneous action potentials recorded from an orexin-expressing neuron in a slice of rat hypothalamus in vitro, recorded first in a solution containing 0.2 m*M* glucose and then on raising the glucose to 5 m*M*. Glucose hyperpolarized the neuron and inhibited action potentials. (B) Here, it is shown that the membrane potential is progressively increased as the glucose concentration is raised from 0.2 m*M* to 10 m*M*. (After Burdakov et al., 2005.)

histamine H_1-receptor gene is deleted.[225] In turn, the orexinergic neurons receive information about the state of the sleep–wake cycle from the suprachiasmatic nucleus.

Orexinergic neurons also act as sensors of nutritional state and respond to changes in extracellular glucose levels. As shown in Figure.14.23, orexinergic neurons are spontaneously active when the concentration of glucose in the extracellular medium is low (0.2 m*M* in this experiment). However, increasing the glucose concentration progressively silences the neurons, by activating twin-pore potassium channels and hyperpolarizing them.[226] Orexinergic neurons are also inhibited by the hormone leptin (secreted by adipocytes, or fat cells, in proportion to the amount of fat they contain; see also Chapter 17) but are activated by another hormone, ghrelin (secreted from the stomach and pancreas as a hunger signal) (see Figure 14.22). Hence, the orexinergic neurons signal low glucose and hunger, which leads to increased food intake, partly through activation of hypothalamic neuropeptide Y (NPY)-secreting neurons but mainly by triggering orexin's awakening action. In other words, you have to be awake to eat, and hunger wakes you up. Hence, orexin-containing neurons help to coordinate the brain's response to information about the body's nutritional and sleep–wake balance.

Vasopressin and Oxytocin: The Social Brain

These two peptides are contained in neurons in the supraoptic and paraventricular nuclei of the hypothalamus. They are secreted into the bloodstream from nerve endings in the posterior pituitary gland. Vasopressin promotes water reabsorption from the kidney (it was originally called antidiuretic hormone); oxytocin induces lactation. However, processes of these neurons innervate other parts of the brain (Figure 14.24); thus, the peptides can excite many central neurons.[227]

Both peptides strongly influence the social behavior of animals and humans via direct or indirect connections to the nucleus accumbens, prefrontal cortex, and amygdala. Some of the key observations leading to this conclusion came from experiments on voles.[228] Some voles, such as the prairie vole (*Microtus ochrogaster*) are monogamous: the male forms a strong pair bond with the female, cooperates in rearing the pups, and aggressively repels a strange prairie vole. Other related vole species, such as the meadow vole (*M. pennsylvanicus*) or the montane vole (*M. montanus*), are more promiscuous and do not form tight pair bonds. It was found that this behavioral difference was critically dependent on the expression and distribution of the vasopressin V_{1A} receptor in the male.[229] Thus, intracerebroventricular injection of a V_{1A} receptor antagonist in the normally monogamous male prairie vole increased aggression, disrupted male–female pair bonding, and increased extra-pair bonding,[230] while overexpres-

[225] Huang, Z. L. et al. 2001. *Proc. Natl. Acad. Sci. USA* 98: 9965–9970.

[226] Burdakov, D. et al. 2006. *Neuron* 50: 711–722.

[227] Raggenbass, M. 2001. *Prog. Neurobiol.* 64: 307–326.

[228] McGraw, L. A., and Young, L. J. 2010. *Trends Neurosci.* 33: 103–109.

[229] Lim, M. M. et al. 2004. *Nature* 429: 754–757.

[230] Winslow, J. T. et al. 1993. *Nature* 365: 545–548.

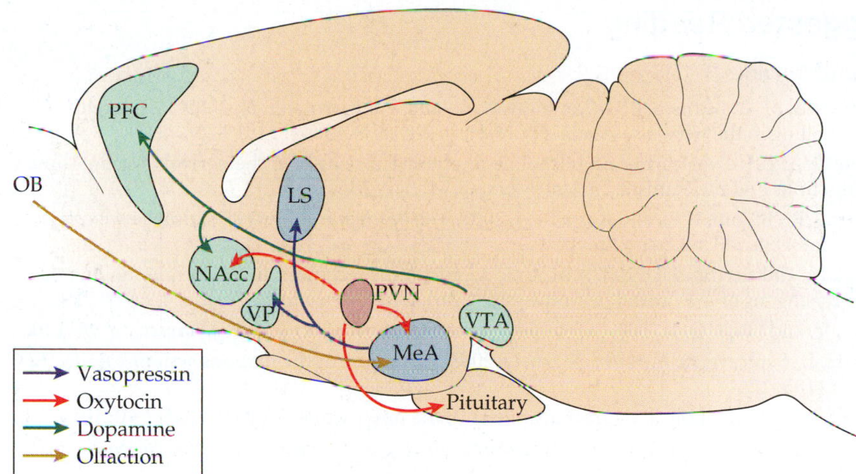

FIGURE 14.24 Proposed Neural Circuits Underlying the Effects of Vasopressin and Oxytocin on Social Interaction. LS = lateral septum; MeA = medial nucleus of the amygdala; NAcc = nucleus accumbens; OB = olfactory bulb; PFC = prefrontal cortex; PVN = paraventricular nucleus; VP = ventral pallidum; VTA = ventral tegmental area. (After McGraw and Young, 2010.)

sion of the V_{1A} receptor gene *AVPR1A* dramatically increased pair bonding in the more promiscuous meadow vole.[228] Differences in V_{1A} receptor expression in the two species are related to the presence of a long insert in the noncoding region of the *AVPR1A* gene in the meadow vole.[231] Interestingly, the human gene shows polymorphisms in this region, which have been correlated with the strength of pair bonding in human males.[232]

Oxytocin has the complementary effect of strengthening pair bonding in female voles. Thus, oxytocin release in the nucleus accumbens is greatly increased after mating, while infusion of oxytocin facilitates pair bonding and blocking oxytocin receptors hinders it.[229] Pair bonding and maternal–pup bonding is also reduced in mice in which the oxytocin receptor or the enzyme CD38 (which facilitates oxytocin release) are disrupted.[233] Facilitation of pair bonding involves a suppression of the normal fear response mediated by the amygdala, which is where the olfactory recognition signal required for social interaction is processed (see Figure 14.24). In humans, this effect translates to an increase in trust and empathy,[234,235] essential prerequisites for interpersonal bonding. Indeed, oxytocin and vasopressin may provide the long-sought chemical basis for at least part of that human emotion termed love.[236,237]

[231]Young, L. J. et al. 1999. *Nature* 400: 766–768.

[232]Walum, H. et al. 2008. *Proc. Natl. Acad. Sci. USA* 105: 14153–14156.

[233]Higashida, H. et al. 2010. *J. Neuroendocrinol.* 22: 373–379.

[234]Baumgartner, T. et al. 2008. *Neuron* 58: 639–650.

[235]Hurlemann, R. et al. 2010. *J. Neurosci.* 30: 4999–5007.

[236]Zeki, S. 2007. *FEBS Lett.* 581: 2575–2579.

[237]Young, L. J. 2009. *Nature* 457:148.

SUMMARY

- There is a wide variety of chemical transmitters in the CNS. Neurons that use a particular transmitter can be identified by immunohistochemical and other visual methods.
- Many neurons release more than one transmitter, and each transmitter can act on a variety of target receptors. Hence, their postsynaptic effects on individual neurons are often complex.
- The main excitatory transmitter is glutamate. This acts on two types of ionotropic receptor (AMPA and NMDA receptors) to generate excitatory postsynaptic potentials. It also acts on a metabotropic receptor, mGluR, to produce pre-and postsynaptic modulatory effects.
- GABA and glycine are inhibitory transmitters in the brain and spinal cord. Glycine acts on ionotropic receptors. GABA acts on both ionotropic ($GABA_A$) and metabotropic ($GABA_B$) receptors.
- Cholinergic neurons are concentrated in the basal forebrain. Their axons innervate wide areas of the cerebral cortex and many subcortical regions. They play a role in cortical arousal, attention, and memory. ACh released from these nerves acts on ionotropic nicotinic receptors, which are mainly located on presynaptic nerve endings, and on metabotropic muscarinic receptors.
- The monoamines norepinephrine, 5-HT, dopamine, and histamine are contained in a relatively few neurons clustered in discrete nuclei in the brainstem and midbrain. Axons of noradrenergic, serotonergic, and histaminergic neurons project to wide areas of the forebrain and affect numerous brain functions. Dopaminergic axons project to the basal ganglia and limbic system, and play a specific role in the control of motor function
- ATP is released as a cotransmitter from many neurons. It acts on ionotropic P_{2X} and metabotropic P_{2Y} receptors on both neurons and neuroglial cells.
- A number of different neuropeptides are released from central neurons. Examples described are: substance P and the enkephalins, which control pain; orexin, which affects sleep and feeding; and oxytocin and vasopressin, which affect social behavior.

Suggested Reading

General Reviews

Abbracchio, M. P., Burnstock, G., Verkhratsky, A., and Zimmermann, H. 2009. Purinergic signalling in the nervous system: an overview. *Trends Neurosci.* 32: 19–29.

Brown, D. A. 2010. Muscarinic acetylcholine receptors (mAChRs) in the nervous system: some functions and mechanisms. *J. Mol. Neurosci.* 41: 340–346.

Cooper, J. R., Bloom, F. E., and Roth, R. H. 2003. *The Biochemical Basis of Neuropharmacology*, 8th ed. Oxford University Press.

Dani, J. A., and Bertrand, D. 2007. Nicotinic acetylcholine receptors and nicotinic cholinergic mechanisms of the central nervous system. *Annu. Rev. Pharmacol. Toxicol.* 47: 699–729.

Dayan, P., and Huys, Q. J. 2009. Serotonin in affective control. *Annu. Rev. Neurosci.* 32: 95–126.

Haas, H. L., Sergeeva, O. A., and Selbach, O. 2008. Histamine in the nervous system. *Physiol. Rev.* 88: 1183–1241.

Hokfelt, T. 2010. Looking at neurotransmitters in the microscope. *Prog. Neurobiol.* 90: 101–118.

Iversen,S. D., and Iversen, L. L. 2007. Dopamine: 50 years in perspective. *Trends Neurosci.* 30: 188–193.

Krnjevic, K. 2010. When and why amino acids? *J. Physiol.* 588: 33–44.

Livingstone, P. D., and Wonnacott, S. 2009. Nicotinic acetylcholine receptors and the ascending dopamine pathways. *Biochem. Pharmacol.* 78: 744–755.

Lodge, D. 2009. The history of the pharmacology and cloning of ionotropic glutamate receptors and the development of idiosyncratic nomenclature. *Neuropharmacology* 56: 6–21.

McGraw, L. A., and Young, L. J. 2010. The prairie vole: an emerging model organism for understanding the social brain. *Trends Neurosci.* 33: 103–109.

Mesulam, M. 2004. The cholinergic lesion of Alzheimer's disease: pivotal factor or side show? *Learn. Mem.* 11: 43–49.

Sakurai, T. 2007. The neural circuit of orexin (hypocretin): maintaining sleep and wakefulness. *Nat. Rev. Neurosci.* 8: 171–181.

Salio, C., Lossi, L., Ferrini, F., and Merighi, A. 2006. Neuropeptides as synaptic transmitters. *Cell Tissue Res.* 326: 583–598.

Sara, S. J. 2008. The locus coeruleus and noradrenergic modulation of cognition. *Nat. Rev. Neurosci.* 10: 211–223.

Surmeier, D. J. 2009. A lethal convergence of dopamine and calcium. *Neuron* 62: 163–164.

Original Papers

Edwards, F. A., Gibb, A. J., and Colquhoun, D. 1992. ATP receptor-mediated synaptic currents in the central nervous system. *Nature* 359: 144–147.

Gourine, A.V., Kasymov, V., Marina, N., Tang, F., Figueiredo, M. F., Lane, S., Teschemacher, A. G., Spyer, K. M., Deisseroth, K., and Kasparov, S. 2010. Astrocytes control breathing through pH-dependent release of ATP. *Science* 329: 571–575.

Kravitz, A. V., Freeze, B. S., Parker, P.R.L., Kay, K., Thwin, M. T., Deisseroth, K., and Kreitzer, A. C. 2010. Regulation of parkinsonian motor behaviours by optogenetic control of basal ganglia circuitry. *Nature* 466: 622–626.

Lin, L., Faraco, J., Li, R., Kadotani, H., Rogers, W., Lin, X., Qiu, X., de Jong, P. J., Nishino, S., and Mignot E. 1999. The sleep disorder canine narcolepsy is caused by a mutation in the hypocretin (orexin) receptor 2 gene. *Cell* 98: 365–376.

Shen, W., Hamilton, S. E., Nathanson, N. M., and, Surmeier, D. J. 2005. Cholinergic suppression of KCNQ channel currents enhances excitability of striatal medium spiny neurons. *J. Neurosci.* 25: 7449–7458.

Stell, B. M., Brickley, S. G., Tang, C. Y., Farrant, M., and Mody, I. 2003. Neuroactive steroids reduce neuronal excitability by selectively enhancing tonic inhibition mediated by δ subunit-containing GABAA receptors. *Proc. Natl. Acad. Sci. USA.* 100: 14439–14444.

Whim, M. D., and Moss, G. W. 2001. A novel technique that measures peptide secretion on a millisecond timescale reveals rapid changes in release. *Neuron* 30: 37–50.

Winslow, J. T., Hastings, N., Carter, C. S., Harbaugh, C. R., and Insel, T. R. 1993. A role for central vasopressin in pair bonding in monogamous prairie voles. *Nature* 365: 545–548.

CHAPTER 15

Transmitter Synthesis, Transport, Storage, and Inactivation

At chemical synapses, neurons release neuropeptides as well as low-molecular-weight transmitters. Low-molecular-weight transmitters, such as acetylcholine (ACh), γ-aminobutyric acid (GABA), and glutamate, are synthesized in the axon terminal. A number of mechanisms ensure that their supply is adequate to meet the demands of release; these include storage of transmitters in synaptic vesicles, rapid changes in the activity of enzymes mediating transmitter synthesis, and long-term changes in the number of enzyme molecules in the terminal. Neuropeptides are synthesized and incorporated into vesicles in the cell body, then shipped down the axon for storage and release. More than one neurotransmitter may be stored in and released from the same nerve terminal.

Synaptic vesicles and other organelles move by fast axonal transport toward the terminal (anterograde transport) and back to the cell body (retrograde transport). Slow axonal transport moves cytoplasmic proteins and components of the axonal cytoskeleton from the cell body toward the terminal.

The final step in chemical synaptic transmission is removal of the transmitter from the synaptic cleft. Low-molecular-weight transmitters are either degraded after release or taken up into glial cells or axon terminals where they are repackaged into vesicles and released again. Neuropeptides are removed by diffusion. Drugs that interfere with transmitter degradation or uptake can have profound effects on signaling, indicating that such removal processes play an important role in synaptic function.

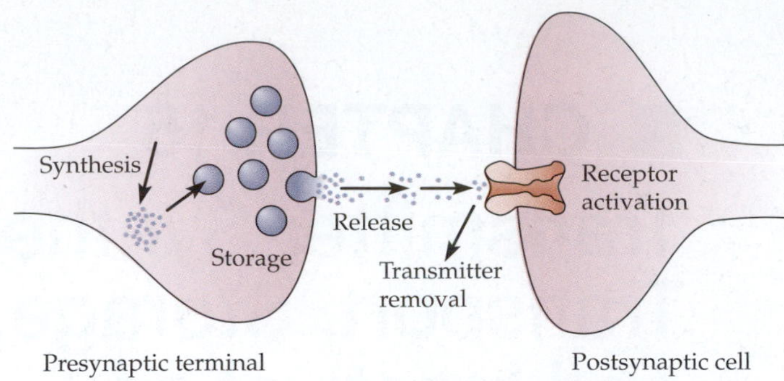

FIGURE 15.1 Chemical Synaptic Transmission. At chemical synapses, neurotransmitters are synthesized, stored in synaptic vesicles, and released by exocytosis. Transmitters diffuse across the synaptic cleft, activate receptors on the postsynaptic cell, and then are removed by diffusion, uptake, or degradation.

Chapters 11, 12, and 14 have shown that there are many different chemical neurotransmitters. Some are small molecules (less than 1 kilodalton in mass); these include acetylcholine, the amino acids glutamate and GABA, and several monoamines (norepinephrine [NE], dopamine, 5-hydroxytryptamine [5-HT] and histamine). Others (peptides) can be several kilodaltons in size. In this chapter, we consider how some of the key transmitters are made (or synthesized); how they are stored in the nerve endings; how they (or their synthetic enzymes) get to the nerve ending (by axoplasmic flow); and how, after their release, the transmitters are degraded or their postsynaptic action is otherwise terminated (Figure 15.1). Mechanisms of release are described in Chapter 13.

Neurotransmitter Synthesis

Where are transmitter molecules synthesized, and how are transmitter stores maintained and replenished? Are transmitters shipped ready-made to the nerve terminals, or are they assembled there from precursors provided by the cell body? The answers to such questions are different for different transmitters. Classical low-molecular-weight transmitters are produced within the axon terminal from common cellular metabolites and are incorporated into small synaptic vesicles (50 nanometers [nm] in diameter) for storage and release. Nitric oxide (NO), carbon monoxide (CO), and the endocannabinoids are also synthesized within terminals. However, since they cannot be packaged in vesicles, they immediately diffuse out of nerve terminals to act on their targets (see Chapter 12). Neuropeptide transmitters, on the other hand, are synthesized in the cell body, packaged in large dense-core vesicles (100–200 nm in diameter), and shipped down the axon.

Principal biochemical pathways for the synthesis and degradation of acetylcholine, GABA, glutamate, catecholamines (dopamine, norepinephrine, and epinephrine), and histamine are shown in Appendix B.

Synthesis of ACh

Birks and MacIntosh performed one of the first thorough investigations into how transmitters are accumulated in nerve terminals and how transmitter stores are maintained during periods of activity in their studies on ACh in the terminals of preganglionic axons in the superior cervical ganglion of the cat (Figure 15.2A,B; see also Chapter 17).[1] They cannulated the carotid artery and the jugular vein, perfused the ganglion with solutions containing anticholinesterase, and analyzed the perfusate for ACh. A small amount of ACh was continually released from the ganglion at rest, amounting to 0.1% of the total contents each minute (Figure 15.2C). The fact that the level of ACh in the ganglion remained constant meant that ACh was synthesized continually at rest. Subsequently it was shown that the ongoing rate of synthesis of ACh at rest, determined by monitoring the incorporation of radioactively labeled choline into ACh, is so high that an amount equal to the entire store of ACh is degraded and resynthesized within the axon terminals every 20 minutes.[2]

Birks and MacIntosh then stimulated the preganglionic nerve with long trains of impulses and found that the quantity of ACh released from the ganglion increased 100-fold, so that an amount corresponding to 10% of the original content was released each minute (see Figure 15.2C). Remarkably, this rate of release was maintained for over an hour with no

[1] Birks, R. I., and MacIntosh, F. C. 1961. *Can. J. Biochem. Physiol.* 39: 787–827.
[2] Potter, L. T. 1970. *J. Physiol.* 206: 145–166.

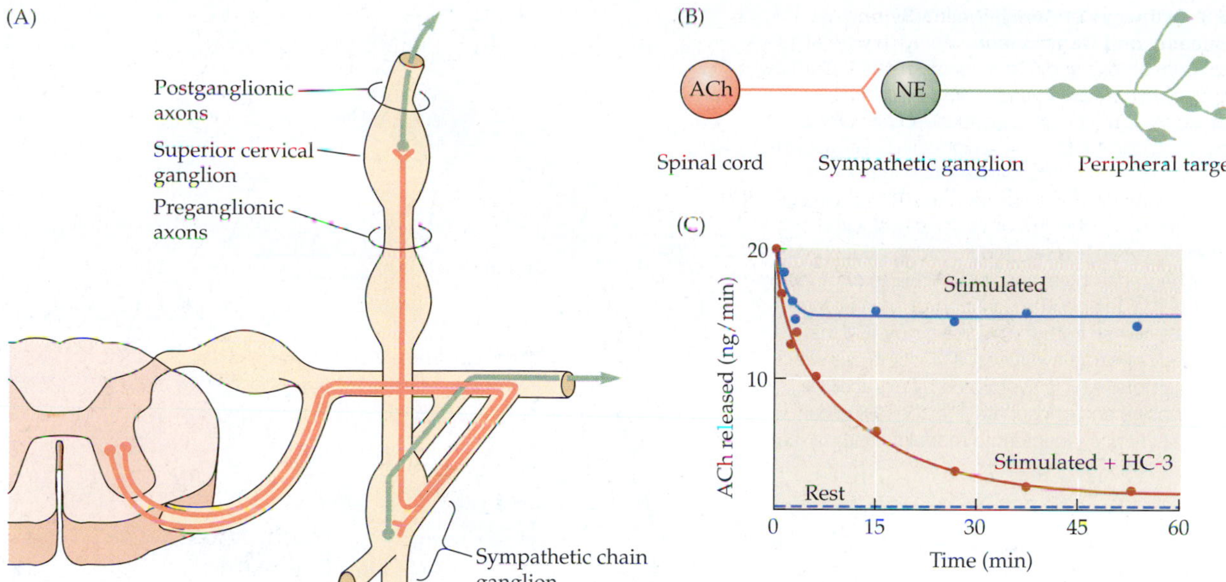

FIGURE 15.2 Measuring the Release of ACh from the terminals of preganglionic axons in the cat superior cervical ganglion. (A) Preganglionic axons reach the superior cervical ganglion from more posterior ganglia in the sympathetic chain. (B) Preganglionic neurons, whose cell bodies lie in the spinal cord, release acetylcholine (ACh) as a transmitter at synapses in sympathetic ganglia. Ganglion cells release norepinephrine (NE) from varicosities along their processes in the periphery. (C) Release of ACh from a cat sympathetic ganglion perfused with oxygenated plasma containing 3×10^{-5} *M* eserine (physostigmine) to inhibit acetylcholinesterase. In control medium, preganglionic stimulation at 20/s causes a sustained, 100-fold increase in the rate of ACh release compared to release at rest. Release decreases rapidly during stimulation in the presence of 2×10^{-5} *M* hemicholinium (HC-3), which blocks choline uptake. (After Birks and MacIntosh, 1961.)

change in the level of ACh in the ganglion. Thus, during an hour of stimulation an axon terminal can release an amount of ACh equal to many times its original content without having its stores depleted.

The only exogenous ingredient the nerve terminals need to maintain their stores of ACh under such conditions is choline, which is taken up from the surrounding fluid via an active transport process (Figure 15.3). The requirement for extracellular choline was demonstrated both by perfusing the preparation with solutions lacking choline and by blocking choline uptake into the axon terminals with hemicholinium (HC-3). In both cases, the level of ACh in the ganglion and the amount released by stimulation fell rapidly (see Figure 15.2C).

How is ACh synthesis controlled to meet the demands of release? Our understanding of the mechanisms regulating ACh synthesis and storage in cholinergic nerve terminals is surprisingly limited. The enzymatic reactions are summarized in Figure 15.3 and shown in detail in Appendix B. Acetylcholine is synthesized from choline and acetyl coenzyme A (acetyl-CoA) by the enzyme choline acetyltransferase (ChAT) and is hydrolyzed to choline and acetate by acetylcholinesterase (AChE). Both enzymes are found in the cytosol. Because the reaction catalyzed by ChAT is reversible, one factor controlling the level of ACh is the **law of mass action**. For example, a fall in ACh concentration caused by release would favor net synthesis until equilibrium was reestablished. However, the regulatory mechanisms at work within cholinergic axon terminals are more complex than this. For example, under resting conditions the accumulation of ACh is limited by ongoing hydrolysis by intracellular acetylcholinesterase; inhibition of AChE within nerve terminals causes the ACh content to increase.[1,2] Thus, the level to which ACh accumulates represents a steady state between ongoing synthesis and degradation. This is a common feature of the metabolism of low-molecular-weight transmitters. Although it seems wasteful, such constant turnover may be an unavoidable consequence of the mechanisms that ensure an adequate supply of transmitter is always available.

Much of the ACh in nerve terminals is sequestered in synaptic vesicles, whereas ACh synthesis and degradation occur in the cytosol. Thus, to have an effect on the rate of synthesis, the release of ACh must reduce the cytoplasmic concentration of ACh, which presumably

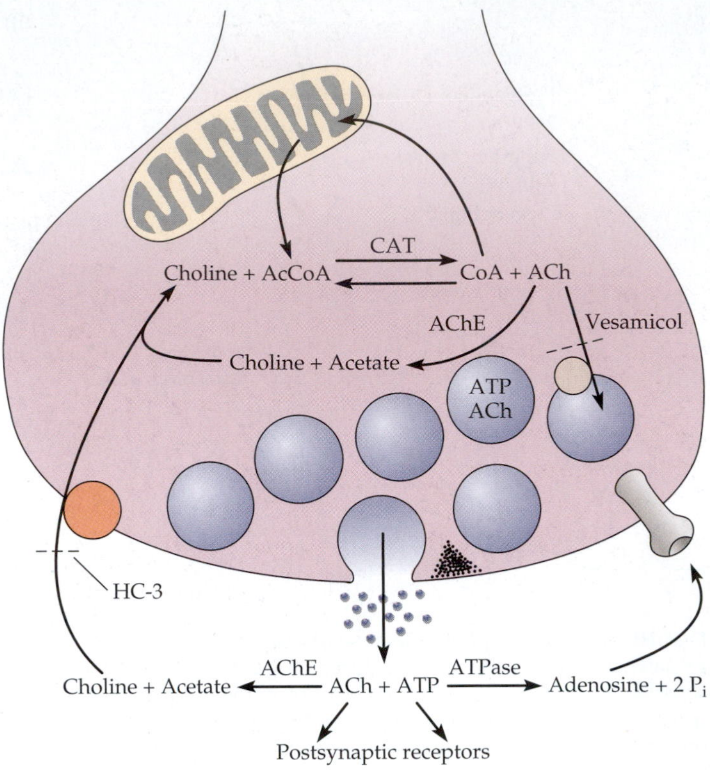

FIGURE 15.3 Pathways of Acetylcholine Synthesis, Storage, Release, and Degradation. Acetylcholine (ACh) is synthesized from choline and acetyl coenzyme A (AcCoA) by choline acetyltransferase (ChAT) and is degraded by acetylcholinesterase (AChE). AcCoA is synthesized primarily in mitochondria; choline is supplied by a high-affinity active transport system that can be inhibited by hemicholinium (HC-3). ACh is packaged into vesicles together with ATP for release by exocytosis. Transport of ACh into vesicles is blocked by vesamicol. Vesicular ACh is protected from degradation. After release, ACh is degraded by extracellular AChE to choline and acetate. About half of the choline transported into cholinergic axon terminals comes from the hydrolysis of ACh that has been released. At some synapses, ATP combines with postsynaptic receptors. ATP is hydrolyzed by extracellular ATPases to adenosine and phosphate (P_i); adenosine can combine with presynaptic receptors to modulate release (see Chapter 14).

occurs from the movement of cytoplasmic ACh into newly formed vesicles. Similar interplay between cytoplasmic synthesis and vesicular storage and release is a common feature of the metabolism of low-molecular-weight transmitters. Maintaining the correct balance between ACh synthesis and vesicular uptake is facilitated by the fact that the gene for the vesicular transporter VAChT is contained in the first intron of the gene for the synthetic enzyme ChAT[3] so that the entire gene forms a cholinergic neuron gene.[4] This process ensures a coordinated expression of the two proteins for synthesis and vesicular uptake through a common gene promoter.[5] In cholinergic nerve terminals in the central nervous system, the supply of choline and co-substrate acetyl-CoA (made in mitochondria) and the activity of ChAT have also been shown to regulate the rate of ACh synthesis.[6,7]

Synthesis of Dopamine and Norepinephrine

Another mechanism by which the rate of synthesis of substances in cells is controlled is **feedback inhibition**, in which the rate-limiting step in a biosynthetic pathway is inhibited by the final product. A good example comes from studies by von Euler, Axelrod, Udenfriend, and their colleagues on the synthesis, storage, and release of norepinephrine in sympathetic neurons and in secretory cells of the adrenal medulla.[8] Adrenal medullary cells resemble sympathetic neurons in many ways. They share the same embryonic origin, are innervated by cholinergic axons that originate in the central nervous system, and release a catecholamine in response to stimulation. The term **catecholamine** is used to designate collectively the substances 3,4-dihydroxyphenylalanine (DOPA), dopamine, norepinephrine, and epinephrine—all of which contain a catechol nucleus (i.e., a benzene ring with two adjacent hydroxyl groups) and an amino group (see Appendix B.) Mammalian sympathetic neurons release norepinephrine (those in frog release epinephrine); adrenal medullary cells release epinephrine as well as norepinephrine.

Norepinephrine is synthesized from the common cellular metabolite tyrosine in a series of three steps. First, tyrosine is converted to DOPA by the enzyme tyrosine hydroxylase, then DOPA to dopamine by aromatic L-amino acid decarboxylase (AAAD), and finally dopamine to norepinephrine by dopamine β-hydroxylase (Figure 15.4; see also Appendix B). The conversions of tyrosine to DOPA and DOPA to dopamine occur in the cytoplasm.

[3] Erickson, J. D. et al. 1994. *J. Biol. Chem.* 269: 21929–21932.
[4] Eiden, L. E. 1998. *J. Neurochem.* 70: 2227–2240.
[5] Prado, M. A. et al. 2002. *Neurochem. Int.* 41: 291–299.
[6] Jope, R. 1979. *Brain Res. Rev.* 1: 313–344.
[7] Parsons, S. M., Prior, C., and Marshall, I. G. 1993. *Int. Rev. Neurobiol.* 35: 279–390.
[8] Axelrod, J. 1971. *Science* 173: 598–606.

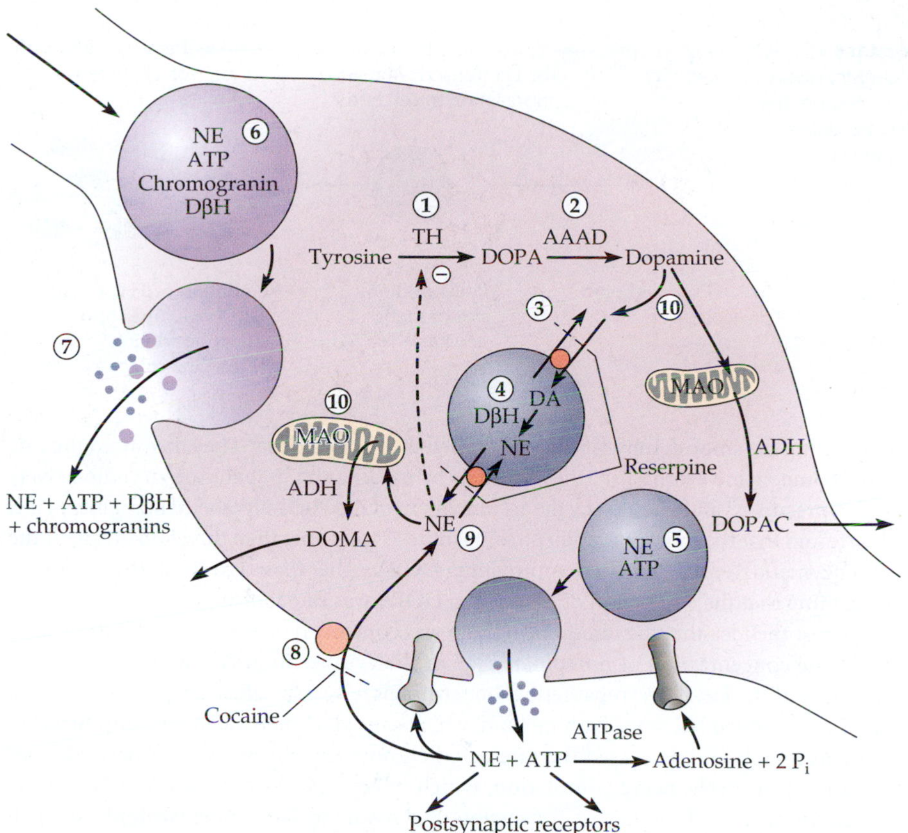

FIGURE 15.4 Pathways of Norepinephrine Synthesis, Storage, Release, and Uptake. (1) Tyrosine is converted to DOPA by tyrosine hydroxylase (TH). (2) DOPA is converted to dopamine (DA) by aromatic L-amino acid decarboxylase (AAAD). (3) Dopamine is transported into vesicles, where it is converted to norepinephrine (NE) by dopamine β-hydroxylase (DβH) (4). Norepinephrine inhibits TH, thus regulating synthesis by feedback inhibition. Transport of dopamine and norepinephrine into vesicles is blocked by reserpine. (5) Vesicles also contain adenosine triphosphate (ATP); (6) large dense-core vesicles contain soluble DβH and chromogranins as well. (7) All soluble components of vesicles are released together. NE, ATP, adenosine, and peptides derived from chromogranins can bind to pre- or postsynaptic receptors. (8) After release, norepinephrine is transported back into the varicosity by an uptake mechanism that is blocked by cocaine. (9) Norepinephrine in the cytoplasm can be repackaged into vesicles for release. (10) Within the varicosity, monoamine oxidase (MAO) and aldehyde dehydrogenase (ADH) degrade norepinephrine to 3,4-dihydroxymandelic acid (DOMA) and dopamine to 3,4-dihydroxyphenyl-acetic acid (DOPAC).

Dopamine is then transported into synaptic vesicles, where it is converted to norepinephrine by dopamine β-hydroxylase, which is associated with the vesicle membrane. Much of the norepinephrine is stored within vesicles; some escapes into the cytoplasm, where it is susceptible to degradation by monoamine oxidase.

Neurons that release dopamine as a transmitter contain tyrosine hydroxylase and AAAD but they lack dopamine β-hydroxylase. Other neurons, as well as adrenal medullary cells, release epinephrine, which is derived from norepinephrine by the action of phenylethanolamine *N*-methyltransferase.

Typically the first enzyme in a multiple-step pathway is rate limiting and is inhibited by the final product. In extracts of the adrenal medulla, the activity of tyrosine hydroxylase was found to be two orders of magnitude lower than that of an AAAD and dopamine β-hydroxylase, suggesting that tyrosine hydroxylation was the rate-limiting step. Moreover, tyrosine hydroxylase was shown to be inhibited by norepinephrine (and by dopamine and epinephrine as well). Thus, as dopamine, norepinephrine, or epinephrine accumulates, further synthesis will be inhibited until a steady state is reached, at which point the rate of synthesis is equal to the rate of degradation and release (see Figure 15.4).

Evidence that feedback inhibition regulates the synthesis of norepinephrine in neurons came from experiments by Weiner and his colleagues on terminals of sympathetic axons

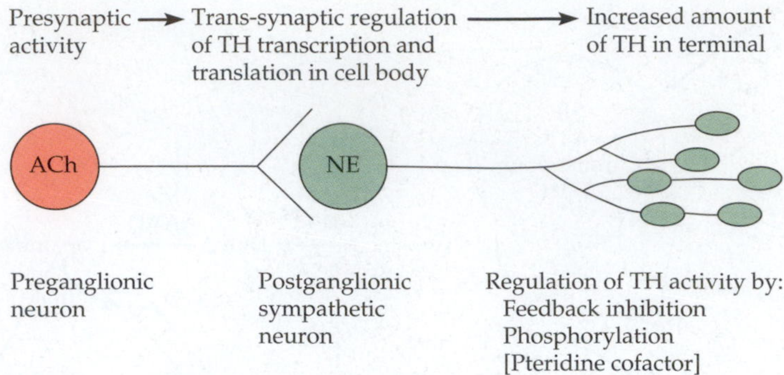

FIGURE 15.5 Regulation of Tyrosine Hydroxylase in sympathetic neurons. The expression of tyrosine hydroxylase (TH) is influenced by the activity of the presynaptic neuron, a process referred to as trans-synaptic regulation. This determines the amount of TH present in the cell and nerve terminal. Within nerve terminals, there is local control of tyrosine hydroxylase activity.

innervating the smooth muscles of a duct called the vas deferens.[9] They measured the rate of norepinephrine synthesis in the terminals by bathing the preparation in radioactively labeled precursors and monitoring the accumulation of radioactively labeled norepinephrine. They found that the rate of norepinephrine synthesis was more than threefold greater if the first enzymatic step was bypassed by providing DOPA rather than tyrosine as the precursor, confirming that the conversion of tyrosine to DOPA was rate limiting.

To test the idea that the rate-limiting step was controlled by feedback inhibition, they varied the concentration of norepinephrine in the cytoplasm in two ways. First, taking advantage of the fact that sympathetic axon terminals have a specific transport mechanism for NE, they added NE to the bathing fluid, which caused an increase in NE concentration in the terminals and decreased the rate at which norepinephrine was synthesized from tyrosine. Conversely, nerve stimulation, which lowers the concentration of NE in the cytoplasm, increased the rate of conversion of tyrosine to NE almost twofold. No such increase was seen, however, if NE was added to the bath during nerve stimulation. Apparently, uptake from the medium was sufficient to maintain the level of norepinephrine in the axon terminals and so limit its biosynthesis.

Additional factors affect catecholamine synthesis (Figure 15.5). When axon terminals are stimulated to release norepinephrine, tyrosine hydroxylase acquires a higher affinity for its cofactor tetrahydrobiopterin (see Appendix B) and becomes less sensitive to inhibition by NE.[10] These changes are associated with a reversible phosphorylation of the enzyme by kinases activated by the influx of calcium ions.[11,12] An additional factor that regulates tyrosine hydroxylase activity is the concentration of tetrahydrobiopterin, which is synthesized from guanosine triphosphate.[13] Thus, a variety of mechanisms act to ensure that the rate of synthesis of norepinephrine meets the demands of release.

Synthesis of 5-Hydroxytryptamine (5-HT)

Serotonin is synthesized from tryptophan. The first step, conversion of tryptophan to 5-hydroxytryptophan (5-HTP) by the enzyme tryptophan hydroxylase, is rate limiting (see Appendix B).[14] 5-HTP is decarboxylated to serotonin (also called 5-HT) by AAAD, the same enzyme that converts DOPA to dopamine. Stimulation of neurons releasing 5-HT causes an increase in the rate of conversion of tryptophan to 5-HT. It has been suggested that this is due to changes in the properties of tryptophan hydroxylase caused by calcium-dependent phosphorylation,[15] similar to the effects of stimulation on tyrosine hydroxylase. Like tyrosine hydroxylase, tryptophan hydroxylase requires the cofactor tetrahydrobiopterin, and serotonin synthesis is thought to be regulated by the availability of this cofactor.

Neurons cannot synthesize tryptophan. Thus, the initial event leading to 5-HT synthesis is the facilitated transport of tryptophan from blood into cerebrospinal fluid (see Chapter 8). Other neutral amino acids (phenylalanine, leucine, and methionine) are transported from the blood into the brain by the same carrier. Thus, an important determinant of the level of 5-HT in serotonergic neurons is the relative amount of tryptophan compared to other neutral amino acids in the diet. As a result, behaviors associated with 5-HT function (see Chapter 14) are particularly sensitive to dietary influences.[16] For example, volunteers fed a low-protein diet for a day and then given a tryptophan-free amino acid mixture showed an increase in aggressive behavior[17] and changes in sleep cycle.[18]

[9]Weiner, N., and Rabadjija, M. 1968. *J. Pharmacol. Exp. Ther.* 160: 61–71.

[10]Joh, T. H., Park, D. H., and Reis, D. J. 1978. *Proc. Natl. Acad. Sci. USA* 75: 4744–4748.

[11]Zigmond, R. E., Schwarzchild, M. A., and Rittenhouse, A. R. 1989. *Annu. Rev. Neurosci.* 12: 415–461.

[12]Nagatsu, T. 1995. *Essays Biochem.* 30: 15–35.

[13]Nagatsu, T., and Ichinose, H. 1999. *Mol. Neurobiol.* 19: 79–96.

[14]Boadle-Biber, M. C. 1993. *Prog. Biophys. Mol. Biol.* 60: 1–15.

[15]Hamon, M. et al. 1981. *J. Physiol. (Paris)* 77: 269–279.

[16]Sandyk, R. 1992. *Int. J. Neurosci.* 67: 127–144.

[17]Moeller, F. G. et al. 1996. *Psychopharmacology (Berl)* 126: 96–103.

[18]Voderholzer, U. et al. 1998. *Neuropsychopharmacology* 18: 112–124.

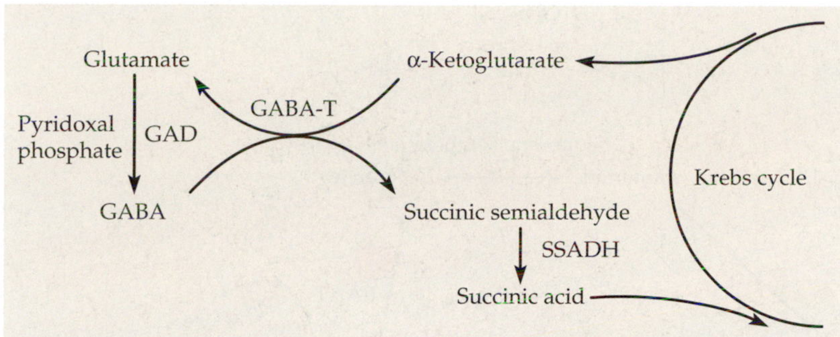

FIGURE 15.6 GABA Synthesis and Metabolism. GABA is synthesized from glutamate by the enzyme glutamic acid decarboxylase (GAD), which requires pyridoxal phosphate as a cofactor. Glutamate is synthesized from α-ketoglutarate by the enzyme GABA α-oxoglutarate transaminase (GABA-T) or from glutamine (see Figure 15.7). GABA is metabolized to succinic acid by GABA-T and succinic semialdehyde dehydrogenase (SSADH).

Synthesis of GABA

GABA is synthesized from glutamate by the enzyme glutamic acid decarboxylase (GAD). This reaction was first characterized as part of the so-called GABA shunt—a series of reactions by which α-ketoglutarate could be converted to succinate (Figure 15.6). The GABA shunt was originally considered to be a brain-specific pathway for glucose metabolism that bypassed part of the Krebs cycle (hence the term "shunt"). The discovery that GABA is the major inhibitory transmitter in the brain, together with the finding that glutamic acid decarboxylase is found only in neurons releasing GABA (see Chapter 14), suggests that the GABA shunt is not of general importance in glucose metabolism.

Kravitz and his colleagues showed that in crustacean inhibitory neurons, physiological levels of GABA inhibit glutamic acid decarboxylase, indicating that feedback inhibition regulates the accumulation of GABA.[19] Several additional regulators of GABA synthesis have been identified in the mammalian brain, including adenosine triphosphate (ATP), inorganic phosphate, and the cofactor pyridoxal phosphate.[20] Two forms of glutamic acid decarboxylase (GAD_{67} and GAD_{65}) are present in brain.[21,22] GAD_{67} has a high affinity for pyridoxal phosphate and so may be constitutively active. GAD_{65} has a lower affinity, and its activity may be rapidly regulated by cofactor availability. Mutant mice lacking GAD_{65} have normal behavior and levels of GABA but are slightly more susceptible to seizures. GAD_{67} knockout mice show a substantial reduction in brain GABA and die shortly after birth from severe cleft palate.[23]

Synthesis of Glutamate

Glutamate is the major excitatory transmitter in the brain. There is more than one pathway for glutamate synthesis in cells. In neurons, glutamate destined for release as a transmitter is derived primarily from glutamine by a phosphate-activated form of the enzyme glutaminase[24] (Figure 15.7). Much of the glutamate released by neurons is taken up by glial cells and converted to glutamine. Glutamine, in turn, is released from the glial cells, taken up by neurons, and converted back to glutamate.[25–27]

Short- and Long-Term Regulation of Transmitter Synthesis

The regulatory mechanisms described so far operate rapidly to change the rate of synthesis within nerve terminals. In addition to such short-term effects, there are long-term regulatory mechanisms. A good example comes from the response of the sympathetic nervous system to prolonged exposure of an animal to stress. When the body is stressed, sympathetic neurons are activated. With prolonged activation, the levels of tyrosine hydroxylase and dopamine β-hydroxylase in the cell bodies and terminals of sympathetic neurons increase as much as three- to fourfold.[28,29] The increase is due to the synthesis of new enzyme molecules. Other enzymes of norepinephrine synthesis and degradation, such as AAAD and monoamine oxidase, are not affected.

The increase is triggered by synaptic activation of sympathetic neurons (see Figure 15.5). Such **trans-synaptic regulation** provides a mechanism whereby the synthetic capability of the neurons can be matched to the rate of transmitter release.[30] Experiments on human sympathetic ganglia have demonstrated that electrical stimulation of preganglionic fibers induces a marked increase in the level of the mRNAs for tyrosine hydroxylase and dopa-

[19]Hall, Z. W., Bownds, M. D., and Kravitz, E. A. 1970. *J. Cell Biol.* 46: 290–299.

[20]Martin, D. L. 1987. *Cell. Mol. Neurobiol.* 7: 237–253.

[21]Erlander, M. G. et al. 1991. *Neuron* 7: 91–100.

[22]Soghomonian, J. J., and Martin, D. L. 1998. *Trends Pharmacol. Sci.* 19: 500–505.

[23]Asada, H. et al. 1997. *Proc. Natl. Acad. Sci. USA* 94: 6496–6499.

[24]Albrecht, J. et al. 2007. *Front Biosci.* 12:332-343.

[25]Palmada, M. and Centelles, J. J. 1998. *Front. Biosci.* 3: 701–718.

[26]Hertz L. 2004. *Neurochem. Int.* 45: 285–296.

[27]Torres, G. E., and Amara, S. G. 2007. *Curr. Opin. Neurobiol.* 17: 304–312.

[28]Thoenen, H., Mueller, R. A., and Axelrod, J. 1969. *Nature* 221: 1264.

[29]Thoenen, H., Otten, U., and Schwab, M. 1979. In *The Neurosciences: Fourth Study Program*. MIT Press, Cambridge, MA. pp. 911–928.

[30]Comb, M., Hyman, S. E., and Goodman, H. M. 1987. *Trends Neurosci.* 10: 473–478.

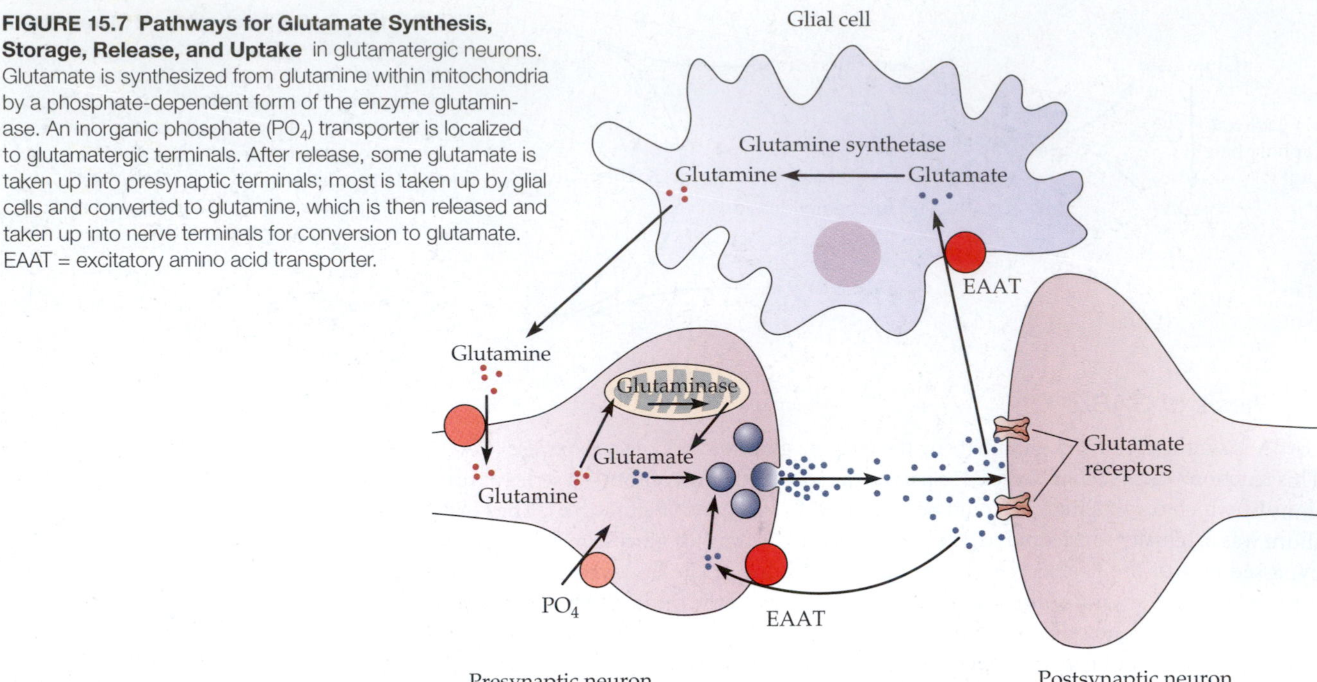

FIGURE 15.7 Pathways for Glutamate Synthesis, Storage, Release, and Uptake in glutamatergic neurons. Glutamate is synthesized from glutamine within mitochondria by a phosphate-dependent form of the enzyme glutaminase. An inorganic phosphate (PO_4) transporter is localized to glutamatergic terminals. After release, some glutamate is taken up into presynaptic terminals; most is taken up by glial cells and converted to glutamine, which is then released and taken up into nerve terminals for conversion to glutamate. EAAT = excitatory amino acid transporter.

mine β-hydroxylase within 20 minutes, suggesting that the regulation of genes involved in norepinephrine synthesis is very rapid and sensitive.[31]

Synthesis of Neuropeptides

Regulation of the stores of peptide transmitters is complicated by the separation between the sites of synthesis and release. Peptides are synthesized on ribosomes, which are found predominantly (but not exclusively[32]) in neuronal cell bodies. This arrangement has two consequences. First, the rate of synthesis of peptides is regulated in the cell body, and the peptides must then be moved to the terminal by axonal transport (which will be discussed in the next section). This is a slow process compared to the rapid local control of the synthesis and storage of low-molecular-weight transmitters within the axon terminal. Second, the amount of a peptide available for release is limited to the amount on hand in the terminal. However, the binding of peptides to their receptors occurs at a much lower concentration (in the range of 10^{-10} to 10^{-8} *M*) than the binding of low-molecular-weight transmitters, such as ACh, to their receptors (10^{-7} to 10^{-4} *M*). In addition, the mechanisms by which they are removed from the synaptic cleft are generally slower. Moreover, neuropeptide receptors, like other metabotropic receptors, act indirectly through intracellular pathways that can provide great amplification (see Chapter 12). As a consequence, only a few molecules of a peptide are needed to influence a postsynaptic target, so the demands of release can be met by the supply of molecules transported from the cell body.

Peptides are synthesized as part of larger precursor proteins, which often contain the sequences for more than one biologically active peptide[33] (Figure 15.8). The initial steps in neuropeptide precursor synthesis are those typical of secreted proteins: synthesis in the endoplasmic reticulum, signal peptide cleavage, processing in the Golgi apparatus, and incorporation into large (100–200 nm) dense-core vesicles. Later steps are unique to neurons and endocrine cells. They are catalyzed sequentially by: (1) specific endoproteases that cleave the precursor protein into the appropriate peptide molecules; (2) exopeptidases, which remove C-terminal basic residues; and (3) a bifunctional, amidating enzyme that converts the C-terminal peptidyl glycine to the corresponding peptide amide[34,35] (see Figure 15.8B). Proteolytic processing begins in the *trans*-Golgi network and continues within large dense-core vesicles as they are transported down the axon and stored in the terminal. Some cells synthesize more than one transmitter peptide; these can be differentially sorted into vesicles and targeted to different terminals.[36]

[31]Schalling, M. et al. 1989. *Proc. Natl. Acad. Sci. USA* 86: 4302–4305.

[32]Giuditta, A. et al. 2008. *Physiol. Rev.* 88: 515–555.

[33]Mains, R. E., and Eipper, B. A. 1999. In *Basic Neurochemistry: Molecular, Cellular, and Medical Aspects*, 6th ed. Lippincott-Raven, Philadelphia. pp. 363–382.

[34]Seidah, N. G., and Chretien, M. 1997. *Curr. Opin. Biotechnol.* 8: 602–607.

[35]Steiner, D. F. 1998. *Curr. Opin. Chem. Biol.* 2: 31–39.

[36]Salio C. et al. 2006. *Cell Tissue Res.* 326: 583–598.

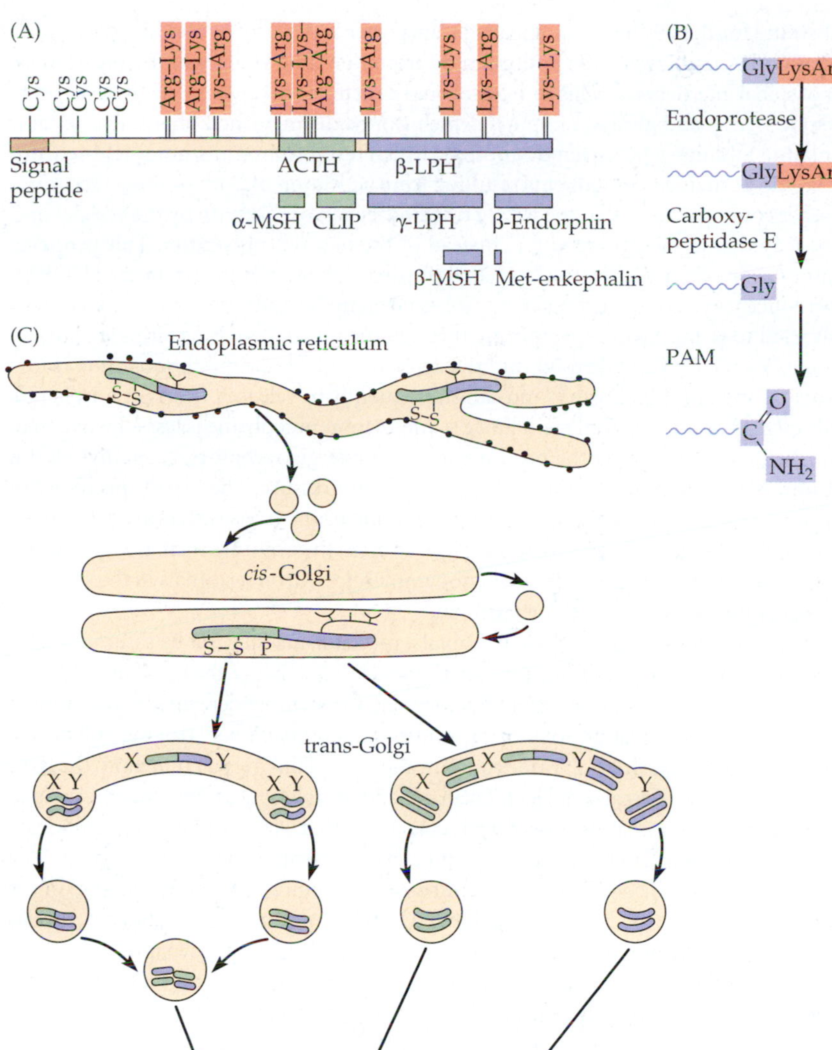

FIGURE 15.8 Synthesis of Neuropeptides. (A) Structure of bovine pro-opiomelanocortin. The locations of known peptide components are shown by colored boxes. ACTH = adrenocorticotropic hormone; LPH = lipotropic hormone; MSH = melanocyte stimulating hormone; CLIP = corticotropin-like intermediate peptide. Paired basic amino acid residues—common targets for processing enzymes—are indicated. (B) Processing of neuropeptide precursors usually begins with cleavage on the carboxy-terminal side of the recognition site by an endoprotease. The basic residues are trimmed by carboxypeptidase E. If the peptide ends in glycine (Gly), the enzyme peptidylglycine α-amidating monooxygenase (PAM) converts the carboxy terminus to an amide. (C) Neuropeptide precursors are directed into the lumen of the endoplasmic reticulum by a signal sequence. In the endoplasmic reticulum, disulfide bonds are formed and N-linked glycosylation occurs. The propeptide is then transported through the Golgi apparatus, where further modifications, such as sulfation and phosphorylation, take place. Two packaging schemes are illustrated. On the left, a propeptide is packaged into vesicles budding from the Golgi; as the vesicle matures, the propeptide is cleaved, resulting in two peptides (X and Y) packaged in the same vesicle. On the right, a propeptide is cleaved within the Golgi, followed by sorting of peptides into separate vesicles. (After Sossin, Fisher, and Scheller, 1989.)

Storage of Transmitters in Synaptic Vesicles

Low-molecular-weight transmitters, such as ACh and NE, are synthesized and packaged into vesicles in the axon terminal. In electron micrographs, such synaptic vesicles tend to be small (50 nm in diameter) and can appear clear (e.g., ACh, amino acid transmitters) or have dense cores (e.g., biogenic amines). The concentration of low-molecular-weight transmitters in vesicles is approximately 0.5 *M*—much greater than that in the surrounding cytoplasm.

The accumulation of transmitters in synaptic vesicles is mediated by specific transport proteins. They are members of a large superfamily of solute carrier (SLC) transporters[37] and are described in Chapter 9 (see Figure 9.5 and Table 9.1). An important point to note is that there are fewer types of vesicular transporters than there are neurotransmitters, so some SLCs transport more than one transmitter. Thus, while there are individual vesicular ACh transporters (VAChTs) and three vesicular glutamate transporters (VGLUT1, VGLUT2, and VGLUT3), there is only one vesicular inhibitory amino acid transporter (VIAAT) for both of the two inhibitory transmitters GABA and glycine and only one vesicular monoamine transporter (VMAT, with two variants VMAT1 and 2) for all of the monoamines (i.e., norepinephrine, dopamine, serotonin, and histamine). A vesicular transporter for ATP has recently been identified, the vesicular nucleotide transporter (VNUT, from the *SLC17A* gene family), which also transports adenosine diphosphate (ADP) and guanosine triphosphate (GTP).[38]

This poor substrate specificity has some interesting consequences. Thus, it means that some vesicles can store and release more than one transmitter. For example, GABA and glycine are

[37] Hediger, M. A. et al. 2004. *Pflügers Arch.* 447: 465–468.

[38] Sawada, K. et al. 2008. *Proc. Natl. Acad. Sci. USA* 105: 5683–5686.

both substrates for the inhibitory amino acid transporter VIAAT. Jonas and his colleagues[39] found that GABA and glycine are both released from the same vesicles in the presynaptic boutons of spinal interneurons. This co-release could be detected because they activate separate postsynaptic GABA and glycine receptors, which can be distinguished pharmacologically. The miniature inhibitory postsynaptic currents, which result from transmitter release from a single vesicle, contained components resulting from activation of both of these receptors.

Another consequence is that the wrong transmitter can be taken up by the vesicles and then be released as a "false transmitter" instead of the natural transmitter. This property can be useful. For example, the drug α-methyldihydroxyphenylalanine (α-methyl DOPA), which was once used to reduce blood pressure, is taken up into adrenergic nerve terminals and converted to α-methylnorepinephrine through the normal NE biosynthetic pathway (see Figure 15.4); it is then taken up into the synaptic vesicles on the monoamine transporter (displacing norepinephrine) and can subsequently be released from the adrenergic terminals instead of norepinephrine.[40] Since α-methylnorepinephrine is less effective than norepinephrine in stimulating the postjunctional adrenergic receptors, α-methyl DOPA inhibits noradrenergic transmission and reduces blood pressure. The broad specificity of the monoamine transporters also means that the monoamine transmitters are potentially capable of replacing each other. Thus, when 5-HT levels in the brain are increased, serotonin can be taken up into the synaptic vesicles of dopaminergic neuron terminals in the striatum and subsequently released from these terminals.[41]

A further point to note is that vesicular uptake mechanisms may not be saturated. This means that the concentration of transmitter in the vesicles (and hence the quantum of transmitter released) is not necessarily constant but may vary in a manner dependent on a number of factors including the availability of the transmitter, the proton gradient driving uptake (see Chapter 9), and the intracellular chloride concentration.[42,43] Thus, by patching onto the large presynaptic terminals of the calyx of Held, Takahashi and his colleagues found that the quantal size measured from the miniature postsynaptic currents varied with the amount of glutamate (between 1 and 100 m*M*) in the patching pipette.[44] The activity of VGLUT is regulated by the concentration of chloride ions in the terminals, which binds to an allosteric site on the transporter protein; this chloride effect is inhibited by some products of metabolism, such as ketone bodies, with a consequent reduction in hippocampal miniature synaptic currents.[45] The reduction in glutamatergic transmission was suggested to explain the antiepileptic effect of a ketogenic diet, which is high in fat and low in protein and carbohydrate.

Co-Storage and Co-Release

Following a suggestion by Sir Henry Dale,[46] neurons are customarily characterized by the transmitter they release—for example, cholinergic neurons (releasing ACh), adrenergic neurons (releasing norepinephrine or epinephrine), glutamatergic (releasing glutamate), and so on. However, it is now clear that many neurons (perhaps the majority) can release more than one transmitter: this effect has been dubbed "cotransmission."[47] In some cases cotransmission occurs because two transmitters are stored in the same vesicle—for example, the co-release of co-stored glycine and GABA[39] described above. Another example is ATP, which is frequently co-stored with catecholamines or ACh in synaptic vesicles and released as a cotransmitter.[48,49] Alternatively, co-released transmitters are stored in different vesicles, but in the same nerve terminal; they may then be differentially released, depending on the frequency and pattern of action potential activity.[50,51] Thus, effective release of peptides usually requires repetitive action potentials or bursts of action potentials, unlike co-released amino acids or amines (see Chapter 13). A similar distinction has been observed between the pattern of activity required for the release of the same transmitter (5-HT) from clear vesicles at the nerve terminals and dense-core vesicles in the soma of the same neuron.[52]

In the mammalian central nervous system, the fast transmitters (glutamate or GABA) are often contained in and released from the same nerve terminals as those that store ACh, a monoamine, or a peptide.[53] Thus, the recurrent collaterals of motor axons that excite Renshaw interneurons in the spinal cord, which were originally deduced to be cholinergic,[54] have recently been shown to also release glutamate.[55] Co-release of glutamate and ACh has also been recorded from single axons of cholinergic neurons from the basal forebrain[56] (Figure 15.9). Previous immunocytochemical work had indicated that VGLUT and the ACh

[39] Jonas, P., Bischofberger, J., and Sandkuhler, J. 1998. *Science* 281: 419–424.

[40] Kopin, I. J. 1968. *Annu. Rev. Pharmacol.* 8: 377–394.

[41] Zhou, F. M. et al. 2005. *Neuron* 46: 65–74.

[42] Erickson, J. D. et al. 2006. *Neurochem. Int.* 48: 643–649.

[43] Edwards, R. H. 2007. *Neuron* 55: 835–858.

[44] Ishikawa, T., Sahara, Y., and Takahashi, T. 2002. *Neuron* 34: 613–621.

[45] Juge, N. et al. 2010. *Neuron* 68: 99–112.

[46] Dale, H. H. 1933. *J. Physiol.* 80: 10–11.

[47] Burnstock, G. 1976. *Neuroscience* 1: 239–248.

[48] Dowdall, M. J., Boyne, A. F., and Whittaker, V. P. 1974. *Biochem. J.* 140: 1–12.

[49] De Potter, W. P., Smith, A. D., and De Schaepdryver, A. F. 1970. *Tissue Cell* 2: 529–546.

[50] Kupfermann, I. 1991. *Physiol. Rev.* 71: 683–732.

[51] Whim, M. D., Church, P. J., and Lloyd, P. E. 1993. *Mol. Neurobiol.* 7: 335–347.

[52] De-Miguel, F. F., and Trueta, C. 2005. *Cell. Mol. Neurobiol.* 25: 297–312.

[53] Seal, R. P., and Edwards, R. H. 2006. *Curr. Opin. Pharmacol.* 6: 114–119.

[54] Eccles, J. C., Fatt, P., and Koketsu, K. 1954. *J. Physiol.* 126: 524–562.

[55] Lamotte d'Incamps, B., and Ascher, P. 2008. *J. Neurosci.* 28: 14121–14131.

[56] Allen, T. G., Abogadie, F. C., and Brown, D. A. 2006. *J. Neurosci.* 26: 1588–1595.

synthesizing enzyme ChAT were sometimes both present in basal forebrain neurons, but their co-release at cortical synapses could not be tested directly using electrophysiological methods because the postsynaptic (muscarinic) action of ACh is slow and temporally dissociated from the afferent nerve impulses (see Chapters 12 and 14). The solution adopted was to generate an autaptic glutamate-transmitting neuronal synapse in single cell culture (Figure 15.9A,B) and then to use a skeletal muscle myoball containing nicotinic receptors as a fast-responding ACh detector. In this way, it could be shown that both transmitters were released simultaneously from the same axon (though not necessarily from the same bouton) (Figure 15.9C). A similar autapse approach has been used to show co-release of glutamate with dopamine from single dopaminergic neurons of the ventral midbrain.[57] In such cases, glutamate subserves the basic function of synaptic excitation while the co-released amine (or peptide) provides a more long-lasting modulatory function. Since the latter effect relates

[57] Sulzer, D. et al. 1998. *J. Neurosci.* 18: 4588–4602.

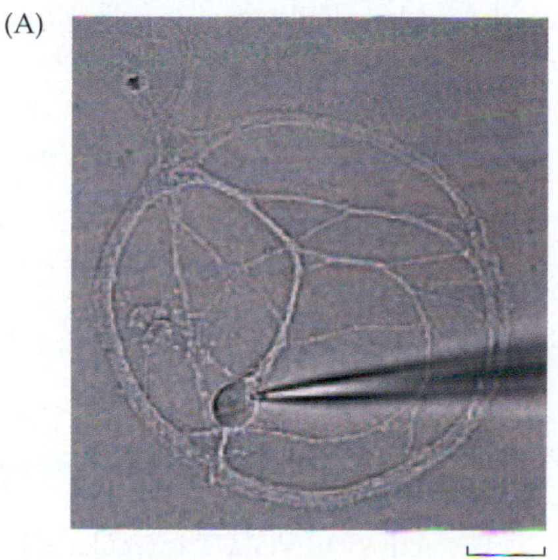

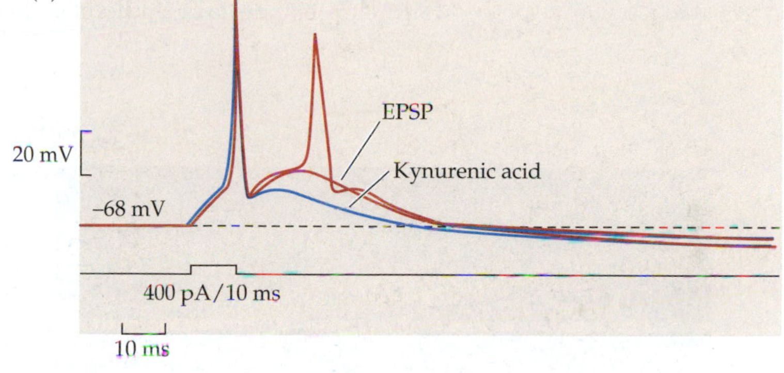

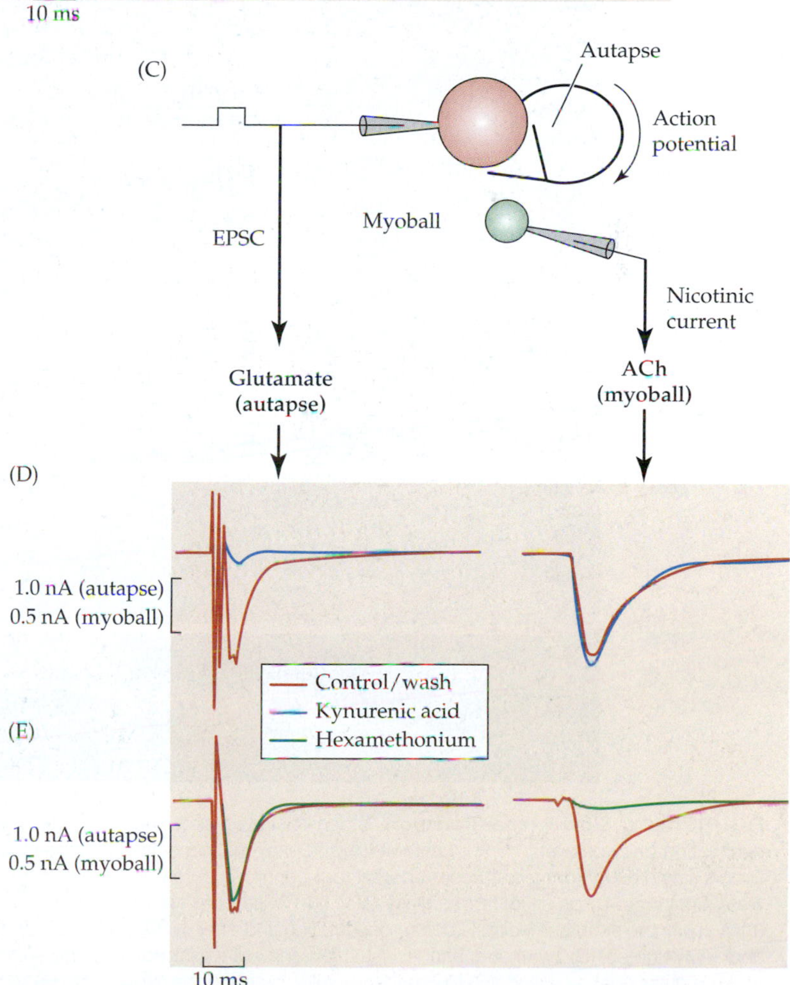

FIGURE 15.9 Simultaneous Release of Two Neurotransmitters from a Single Neuron. (A) A single cholinergic basal forebrain neuron dissociated from a rat brain was cultured in isolation for 3 weeks. During that time it sent out an axon which branched and went back to innervate the same neuron forming an autapse. (B) An action potential was initiated by passing current into the neuron soma through a patch electrode. The action potential was conducted along the axon to the autapse, resulting in an excitatory postsynaptic potential (EPSP) and a second action potential in the soma, which is due to the release of glutamate, since it is reduced by the glutamate receptor antagonist kynurenic acid (1 m*M*). (C) Experimental arrangement. Stimulus was applied through the patch clamp electrode, which also recorded the excitatory EPSC produced by release of glutamate from the autapse (D and E, left columns). A second patch electrode bearing a myoball (i.e., a small sphere of muscle membrane containing nicotinic receptors that can detect acetylcholine) was placed close to the axon forming the autapse. Responses of the myoball to ACh released from the autapse are shown in the right hand columns of (D) and (E).(D,E) Simultaneous recordings of autaptic and myoball currents generated by a single action potential. The autaptic current is blocked by 1 m*M* kynurenic acid (D), showing that it is due to the release of glutamate, whereas the myoball current is blocked by 100 μM of the nicotinic blocker hexamethonium (E), showing that it is due to the release of acetylcholine. (After Allen et al., 2006).

most closely to the overall function of the neurons (see Chapter 14), it is useful to preserve their amine or peptide-based labels, such as cholinergic, etc., as originally suggested by Dale.[46]

Axonal Transport

Proteins found in axon terminals must have been shipped there from the cell body, where they are synthesized. The first evidence for movement of material along axons came from experiments by Weiss and his colleagues, who ligated peripheral nerves and described the ballooning out of axons just proximal to the site of constriction, and the subsequent movement of the accumulated material along the axons after the constriction was removed.[58] These effects suggested that normally there is a continuous bulk movement of axoplasm along the axon at the rate of 1–2 mm/day, which was given the name **axoplasmic flow**. This idea was buttressed by later experiments using radioactively labeled amino acids[59] and fluorescent proteins[60] to follow the movement of proteins from neuronal cell bodies along peripheral and central axons. Such movement has even been observed in single axons in cell culture (Figure 15.10)[61] and in organelles along living nerve fibers, followed by cinematography.[62]

[58]Weiss, P., and Hiscoe, H. B. 1948. *J. Exp. Zool.* 107: 315–395.

[59]Droz, B., and Leblond, C. P. 1963. *J. Comp. Neurol.* 121: 325–346.

[60]Dahlstrom, A. B. 2010. *Prog. Neurobiol.* 90: 119–145.

[61]Koehnle, T. J., and Brown, A. 1999. *J. Cell Biol.* 144: 447–458.

[62]Forman, D. S., Padjen, A. L., and Siggins, G. R. 1977. *Brain Res.* 136: 197–213.

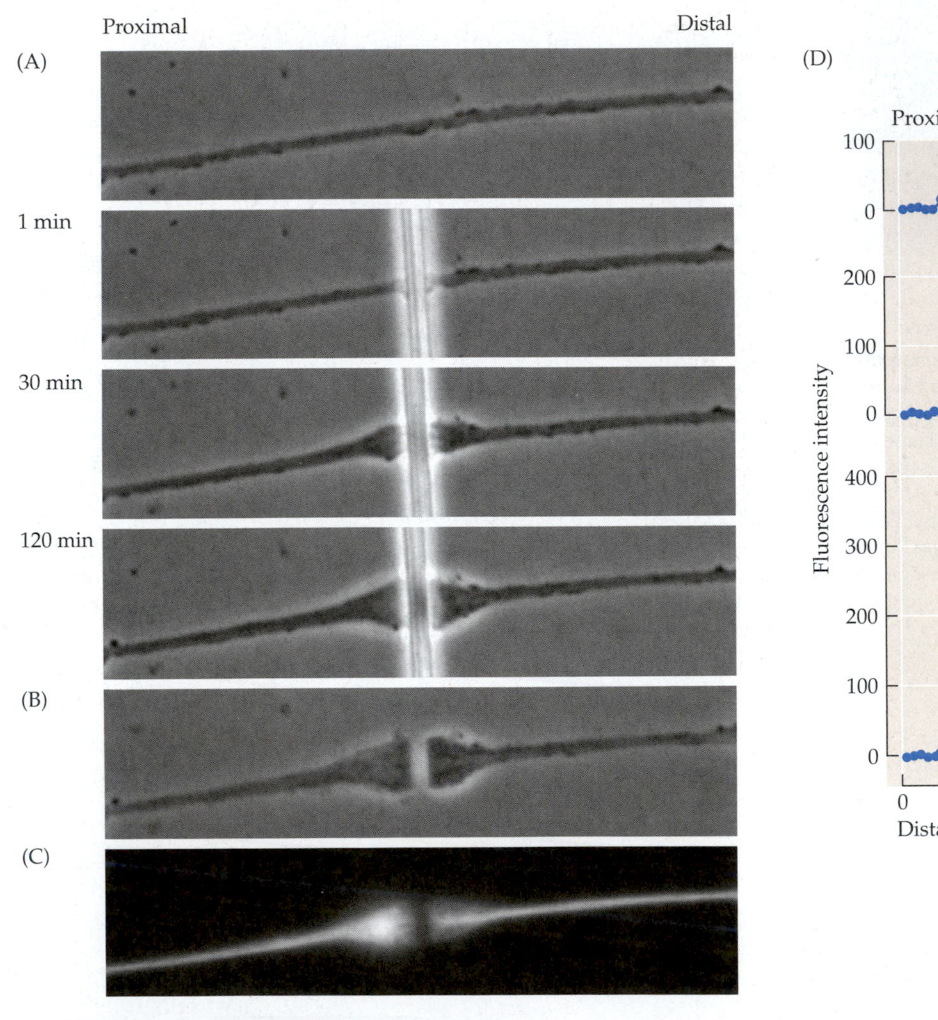

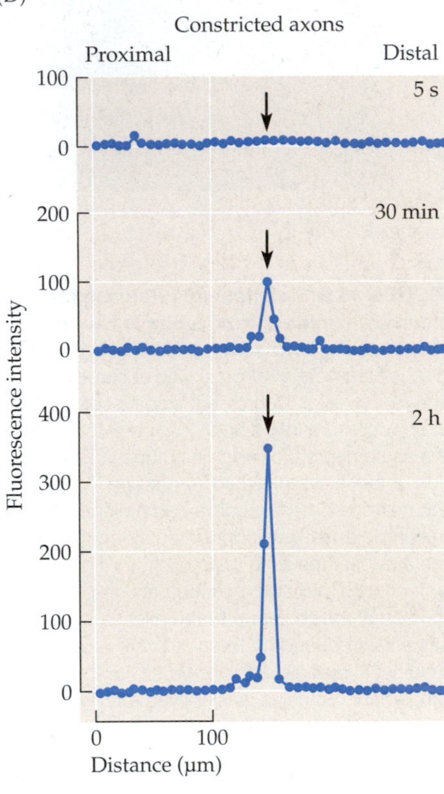

FIGURE 15.10 Slow Axonal Transport is demonstrated by the accumulation of cytoskeletal components at the site of axonal constriction. A single axon from a cultured rat dorsal root ganglion neuron was constricted by pressure from a glass fiber. (A) Phase-contrast images show the axon immediately before constriction and after 1, 30, and 120 minutes. (B) Two hours after constriction, the cell was fixed and the glass fiber removed. (C) Fluorescence micrograph of the axon labeled with anti-neurofilament protein antibodies. (D) Graphs of fluorescence intensity as a function of distance along axons constricted for 5 seconds, 30 minutes, or 2 hours. The time course of accumulation of neurofilament protein at the site of the constriction (arrow) indicated that the average transport rate was approximately 3 mm/day. (After Koehnle and Brown, 1999; micrographs kindly provided by A. Brown.)

Rate and Direction of Axonal Transport

Measurement of the time course of accumulation of material proximal to a constriction or in axon terminals demonstrated characteristic differences in the rates of movement within the broad spectrum of components being transported. Structural proteins, such as tubulin and neurofilament proteins, move at the slowest rates (1–2 mm/day), while membrane-enclosed organelles, such as mitochondria and vesicles (including synaptic vesicles packed with transmitter), move much faster (up to 400 mm/day).[63] Such rapid movement could not be accounted for by the bulk flow of cytoplasm, and thus the general term **axonal transport** was adopted.

Some proteins and organelles move toward the axon terminal (**anterograde transport**) and others from the terminal to the cell body (**retrograde transport**).[62,64] Retrograde transport of membrane-enclosed organelles returns material to the cell body for recycling or degradation, and it has been shown to be crucial for the movement of trophic molecules, such as nerve growth factor, from axon terminals back to cell bodies (see Chapter 25).[65]

Neuroanatomists have developed tracers, such as horseradish peroxidase, fluorescently labeled beads, and even viruses that are carried in anterograde and retrograde directions by axonal transport. Using tracers such as these, it is possible to map synaptic connections, even over long distances, by visualizing individual axons, their terminal arborizations, and their cell bodies.[66,67]

Microtubules and Fast Transport

Although early experiments demonstrated that axonal transport required metabolic energy and relied on intact microtubules, for 30 years little progress was made in understanding its mechanism. Then two technological advances triggered very rapid progress: (1) the development of microscopic techniques that allowed direct visualization of single vesicles within cells,[68,69] and (2) the finding that vesicle movements persisted in cell-free systems, such as extruded squid axoplasm.[70] Studies by Reese, Sheetz, Schnapp, Vale, Block, and their colleagues have demonstrated that transport occurs by the attachment of organelles, such as mitochondria and vesicles, to microtubules. Mechanochemical enzymes, or motors, hydrolyze ATP and use the energy to carry organelles along the microtubule track (Figure 15.11).[71,72]

Microtubules have an inherent polarity; in axons, the plus (i.e., positive) end points toward the distal axon terminal. Anterograde transport is powered by kinesin, which moves organelles toward the plus end, while retrograde transport is powered by cytoplasmic dynein, which moves organelles toward the minus (i.e., negative) end (Figure 15.12).[73] Specific receptors on the surface of organelles mediate the attachment of either kinesin or cytoplasmic dynein and thus regulate the direction of organelle movement (Figure 15.13).[74] Remarkably, a single kinesin motor has been shown to pull an organelle along at speeds equivalent to fast axonal transport;[75] each molecule of ATP hydrolyzed produces a step of approximately 8 nm, corresponding to the distance from one $\alpha\beta$ tubulin dimer to the next along the microtubule protofilament.[76,77] Differences in the rate of transport of different components arise from differences in the proportion of time they remain on track and in the resistance they encounter trying to penetrate the dense network of cytoskeletal and cross-bridging elements within the axon. In synaptic regions, actin forms the major cytoskeletal protein and myosins are the main molecular motors.[78] Analogous stepping transport by myosin has recently been examined in detail using high-speed atomic force microscopy.[79]

Mechanism of Slow Axonal Transport

Soluble proteins of intermediary metabolism and cytoskeletal elements, such as microtubules and neurofilaments, move from the cell body toward the axon terminal by slow transport.[62]

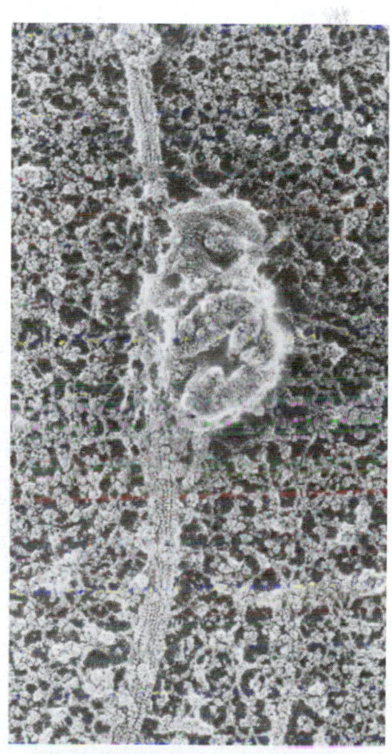

FIGURE 15.11 Identifying the Organelles and Tracks Mediating Fast Axonal Transport. Electron micrograph of a vesicle attached to a microtubule in extruded squid axoplasm. Before fixation this organelle was observed by light microscopy moving along a filamentous track at a rate corresponding to fast axonal transport. The electron micrograph shows that the organelle is a synaptic vesicle, and the track is a microtubule. A layer of granular and finely filamentous material coats the glass substrate. (From Schnapp et al., 1985.)

[63]Grafstein, B., and Forman, D. S. 1980. *Physiol. Rev.* 60: 1167–1283.

[64]Vallee, R. B., and Bloom, G. S. 1991. *Annu. Rev. Neurosci.* 14: 59–92.

[65]Bartlett, S. E., Reynolds, A. J., and Hendry, I. A. 1998. *Immunol. Cell Biol.* 76: 419–423.

[66]Kuypers, H. G. J. M., and Ugolini, G. 1990. *Trends Neurosci.* 13: 71–75.

[67]Teune, T. M. et al. 1998. *J. Comp. Neurol.* 392: 164–178.

[68]Inoue, S. 1981. *J. Cell Biol.* 89: 346–356.

[69]Allen, R. D., Allen, N. S., and Travis, J. L. 1981. *Cell Motil.* 1: 291–302.

[70]Brady, S. T., Lasek, R. J., and Allen, R. D. 1982. *Science* 218: 1129–1131.

[71]Vale, R. D., and Fletterick, R. J. 1997. *Annu. Rev. Cell Dev. Biol.* 13: 745–777.

[72]Vallee, R. B., and Gee, M. A. 1998. *Trends Cell Biol.* 8: 490–494.

[73]Hirokawa, N. 1998. *Science* 279: 519–552.

[74]Sheetz, M. P. 1999. *Eur. J. Biochem.* 262: 19–25.

[75]Howard, J., Hudspeth, A. J., and Vale, R. D. 1989. *Nature* 342: 154–158.

[76]Svoboda, K. et al. 1993. *Nature* 365: 721–727.

[77]Mandelkow, E., and Hoenger, A. 1999. *Curr. Opin. Cell Biol.* 11: 34–44.

[78]Hirokawa, N., Niwa, S. and Tanaka, Y. 2010. *Neuron* 68: 610–638.

[79]Kodera, N. et al. 2010. *Nature* 468: 72–76.

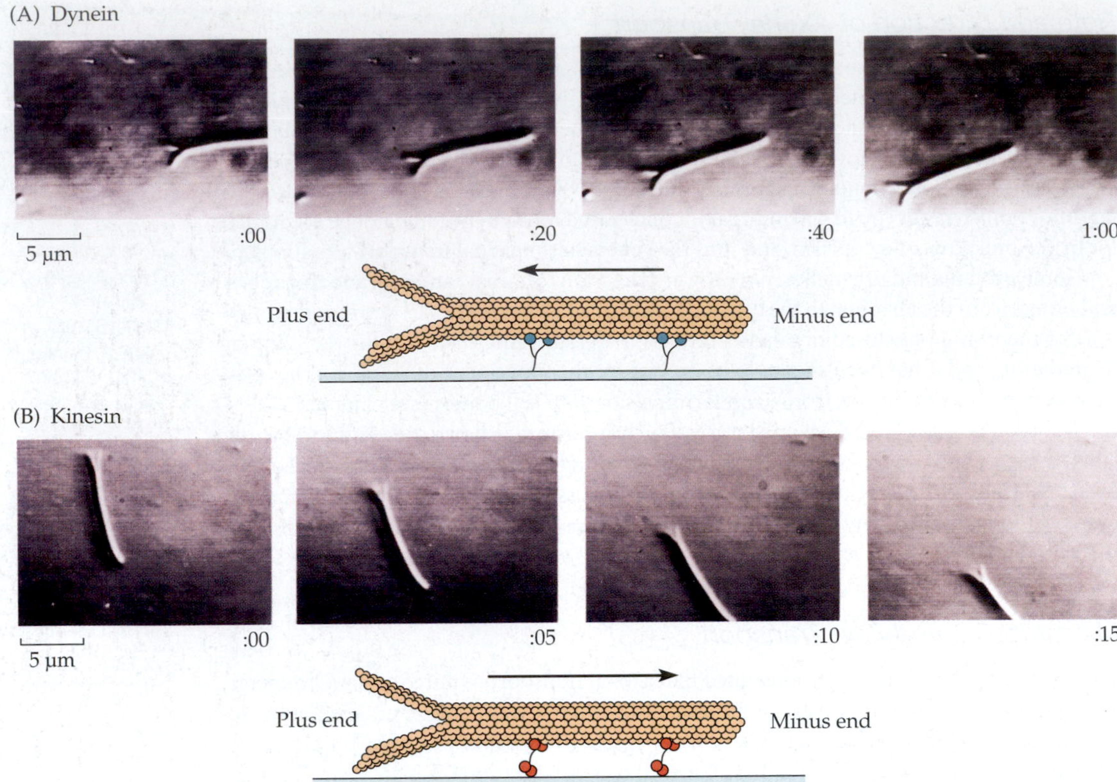

FIGURE 15.12 The Molecular Motors Dynein and Kinesin Propel Microtubules in Opposite Directions. (A,B) Sequential images of the movement of microtubule fragments on purified fast-transport motors. Time is indicated in minutes. Purified cytoplasmic dynein (A) or kinesin (B) was adsorbed to a cover slip, and fragments of microtubules were added. When the fragments contacted the surface, they were propelled toward their frayed (distal or +) end on dynein and toward their compact (proximal or –) end on kinesin, as illustrated. (After Paschal and Vallee, 1987; micrographs kindly provided by R. Vallee.)

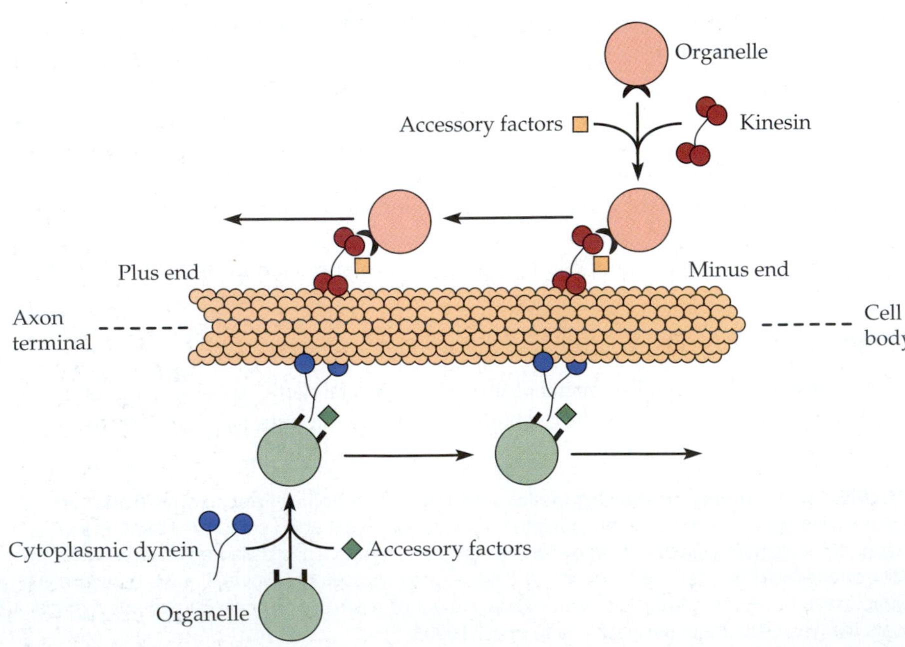

FIGURE 15.13 Fast Axonal Transport Powered by Kinesin and Dynein. In the axon, microtubules are stationary and have a polarity: The plus end points toward the axon terminal, the minus end toward the cell body. Kinesin and cytoplasmic dynein, together with accessory factors, attach to organelles and propel them toward the axon terminal and cell body, respectively. (After Vallee, Shpetner, and Paschal, 1989.)

There is considerable debate about whether microtubules and neurofilaments move as intact polymers[80] or if polymerized filaments are stationary and tubulin and neurofilament monomers or oligomers are transported.[81] What is clear is that diffusion cannot account for the axonal transport of cytoskeletal proteins and that an active process is involved.

Removal of Transmitters from the Synaptic Cleft

The final step in chemical synaptic transmission is the removal of transmitter from the synaptic cleft. The mechanisms for transmitter removal include diffusion, degradation, and uptake into glial cells or nerve terminals.

Removal of ACh by Acetylcholinesterase

As described in Chapter 11, the action of ACh is terminated by the enzyme acetylcholinesterase (AChE), which hydrolyzes ACh to choline and acetate. Much of the choline is transported back into the nerve terminal and reused for ACh synthesis. At the vertebrate skeletal neuromuscular junction, AChE is bound to the synaptic basal lamina—that portion of the muscle fiber's sheath of extracellular matrix material that occupies the synaptic cleft and junctional folds (Figure 15.14).[82] There are 2600 catalytic subunits of AChE per square micrometer of synaptic basal lamina[83] (compared to the 10^4 AChRs per square micrometer in the postsynaptic membrane).

It might seem inefficient to have acetylcholinesterase situated between the axon terminal and the postsynaptic membrane, forcing molecules of ACh to traverse a minefield of degradative enzymes before having an opportunity to interact with their postsynaptic receptors. However, if the dimensions of the cleft and the rates of ACh diffusion, binding, and hydrolysis are taken into consideration, a simple scheme emerges, which is called

[80] Baas, P. W., and Brown, A. 1997. *Trends Cell Biol.* 7: 380–384.

[81] Hirokawa, N. et al. 1997. *Trends Cell Biol.* 7: 382–388.

[82] McMahan, U. J., Sanes, J. R., and Marshall, L. M. 1978. *Nature* 271: 172–174.

[83] Salpeter, M. M. 1987. In *The Vertebrate Neuromuscular Junction*. Alan R. Liss, New York. pp. 1–54.

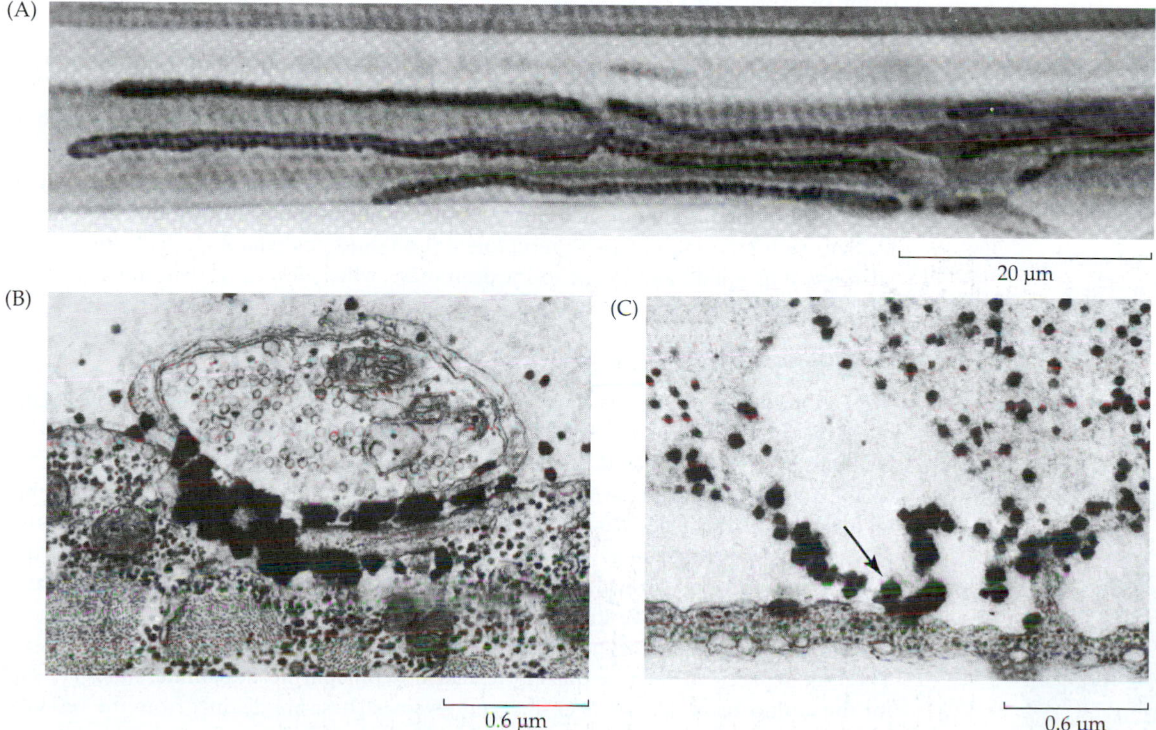

FIGURE 15.14 Acetylcholinesterase Is Concentrated in the synaptic basal lamina at the skeletal neuromuscular junction. (A) Light micrograph of a neuromuscular junction in a frog cutaneous pectoris muscle stained by a histochemical procedure for acetylcholinesterase. The dark reaction product lines the synaptic gutters and junctional folds. (B) Electron micrograph of a cross section of an axon terminal from a muscle stained for acetylcholinesterase as in part A. The electron-dense reaction product fills the synaptic cleft and the junctional folds. (C) Electron micrograph of a damaged muscle in which the nerve terminal, Schwann cell, and muscle fiber have degenerated and been phagocytized, leaving only empty basal lamina sheaths (see Chapter 27). The damaged muscle was stained for acetylcholinesterase; reaction product is associated with the synaptic basal lamina (arrow). (Micrographs kindly provided by U. J. McMahan.)

the **saturated disk.**[84] Following the release of one quantum, the concentration of ACh increases almost instantaneously (within microseconds) across the width of the cleft to a level high enough (0.5 mM) to saturate both ACh receptors and esterase within a disk approximately 0.5 μm in diameter centered on the release site. Binding of ACh to its receptors and to AChE is rapid compared to the rate at which AChE can hydrolyze ACh (it takes AChE 0.1 milliseconds to hydrolyze one molecule of ACh). Therefore, the fraction of ACh molecules released that bind initially to postsynaptic receptors is determined by the ratio of receptors to esterase; thus, approximately 20% of the ACh molecules bind to AChE and 80% to ACh receptors.

Binding causes a precipitous fall in ACh concentration. The concentration then remains low because AChE can hydrolyze ACh molecules much faster (10 molecules per ms) than they are released from receptors as the channels close ($\tau = 1$ ms; see Chapter 11). Thus, by 0.1 ms or so after release, the concentration of ACh in the cleft has fallen to levels that make the probability of two ACh molecules being available to bind and open another receptor negligible.

Such an analysis predicts that inhibition of acetylcholinesterase should have a more pronounced effect on the duration of the synaptic potential than on its amplitude, which is the case. Amplitude is increased 1.5-fold to 2-fold, and the duration is increased 3-fold to 5-fold.[85,86] Thus, the organization of the neuromuscular junction and the density and kinetic properties of ACh receptors and AChE combine to produce a synapse that is capable of very rapid responses and efficient use of ACh.

Acetylcholinesterase seems to have a similar function in cholinergic nerve–nerve synapses, such as those in autonomic ganglia, in that its inhibition prolongs the synaptic current.[87,88] However, much of the AChE at these sites is located on the presynaptic fibers,[89,90] where it probably has the additional function of reducing the ACh overspill during repetitive nerve activity. Cholinesterase inhibitors produce a much more dramatic enhancement and prolongation of the slow synaptic current produced by repetitive afferent stimulation, which results from stimulation of muscarinic ACh receptors outside the immediate area of the synapse[91] (see Chapters 14 and 17).

Removal of ATP by Hydrolysis

The action of ATP like that of ACh is rapidly terminated by hydrolysis.[92] Ecto-ATP diphosphohydrolase (ecto-ADPase or ecto-apyrase) hydrolyzes ATP to ADP and ADP to AMP. AMP is converted to adenosine by ecto-5′-nucleotidase. Both enzymes are found on glial cells, and at synaptic sites on neurons. At many synapses, adenosine modulates synaptic transmission by combining with receptors on pre- and postsynaptic cells.[93] Its action is terminated by uptake and by adenosine deaminase, which degrades adenosine to inosine.

Removal of Transmitters by Uptake

The actions of dopamine, norepinephrine, glutamate, 5-HT, glycine, and GABA are terminated by uptake of the transmitter into presynaptic nerve terminals, postsynaptic cells, or glial cells by specific transport proteins.[94,95] Transmitters transported into nerve terminals can be repackaged and released again. This process was first shown for norepinephrine uptake into sympathetic nerve endings by Julius Axelrod and his colleagues.[96,97] They found that radioactively labeled NE was accumulated in sympathetically innervated tissues, that this was prevented by sympathetic denervation, and that the labeled NE could be released again by sympathetic nerve stimulation.

The properties of these membrane transporters are described in detail in Chapter 9. There are separate membrane transporters for norepinephrine, dopamine, and 5-HT, and for the amino acids glutamate, GABA, and glycine. These are distinct from the vesicular transporters and show less substrate overlap (though there is some overlap between the norepinephrine and dopamine transporters). The transporters for the monoamines are important targets for a number of drugs used to treat psychiatric disorders, such as depression, and also for some recreational drugs, such as cocaine and the amphetamines[94] (see Chapter 14).

There appear to be no specific uptake mechanisms for peptide transmitters. Their transmitter action is terminated by diffusion, and some by peptidase hydrolysis.[98]

[84] Bartol, T. M. et al. 1991. *Biophys. J.* 59: 1290–1307.
[85] Fatt, P., and Katz, B. 1951. *J. Physiol.* 115: 320–370.
[86] Katz, B., and Miledi, R. 1973. *J. Physiol.* 231: 549–574.
[87] MacDermott, A. B. et al. 1980. *J. Gen. Physiol.* 75: 39–60.
[88] Rang, H. P. 1981. *J. Physiol.* 311: 23–55.
[89] Weitsen, H. A., and Weight, F. F. 1977. *Brain Res.* 128: 197–211.
[90] Davis, R., and Koelle, G. B. 1978. *J. Cell Biol.* 78: 785–809.
[91] Brown, D. A., and Selyanko, A. A. 1985. *J. Physiol.* 365: 335–364.
[92] Zimmermann, H. 2006. In *Novartis Foundation Symposium 276: Purinergic Signalling in Neuron-Glial Interactions.* Wiley, Chichester, UK. pp. 113–128.
[93] Dunwiddie, T. V., and Masino, S. A. 2001. *Annu. Rev. Neurosci.* 24: 31–55.
[94] Gether, U. et al. 2006. *Trends Pharmacol. Sci.* 27: 375–383.
[95] Iversen, L. 2006. *Brit. J. Pharmacol.* 147(Suppl. 1): S82–S88.
[96] Hertting, G. et al. 1961. *Nature* 189: 66.
[97] Hertting, G., and Axelrod, J. 1961. *Nature* 192: 172–173.
[98] Isaac, R. E., Bland, N. D., and Shirras, A. D. 2009. *Gen. Comp. Endocrinol.* 162: 8–17.

Julius Axelrod, 1970
(Kindly provided by L. L. Iversen.)

SUMMARY

- ACh, norepinephrine, epinephrine, dopamine, 5-HT, histamine, ATP, GABA, glycine, and glutamate are low-molecular-weight transmitters. Neuropeptides form a second group of transmitters.
- Many neurons release more than one transmitter, typically a low-molecular-weight transmitter and one or more neuropeptides.
- Low-molecular-weight transmitters are synthesized in axon terminals, packaged in small synaptic vesicles, and stored for release. Feedback mechanisms control the number and activity of enzymes that catalyze transmitter synthesis so as to maintain an adequate supply of transmitter.
- Neuropeptides are synthesized in the cell body, processed and packaged in large dense-core vesicles in the Golgi apparatus, and transported to the axon terminal.
- Slow axonal transport moves soluble proteins and components of the cytoskeleton from the cell body to the axon terminal at a rate of 1 to 2 mm/day.
- Fast axonal transport moves vesicles and other organelles at speeds of up to 400 mm/day either toward the terminal (anterograde transport) or toward the cell soma (retrograde transport). Fast transport is mediated by molecular motors that move organelles along microtubules.
- The final step in chemical synaptic transmission is the removal of transmitter from the synaptic cleft by diffusion, degradation, or uptake. Prompt transmitter removal is important for normal synaptic function.

Suggested Reading

General Reviews

Axelrod, J. 1971. Noradrenaline: fate and control of its biosynthesis. *Science* 173: 598–606.

Chen, N. H., Reith, M. E., and Quick, M. W. 2004. Synaptic uptake and beyond: the sodium- and chloride-dependent neurotransmitter transporter family SLC6. *Pflügers Arch.* 447: 519–531.

Cooper, J. R., Bloom, F. E., and Roth, R. H. 2003. *The Biochemical Basis of Pharmacology*, 8th ed. Oxford University Press, New York.

Edwards, R. H. 2007. The neurotransmitter cycle and quantal size. *Neuron* 55: 835–858.

Hediger, M. A., Romero, M. F., Peng, J. B., Rolfs, A., Takanaga, H., and Bruford, E. A. 2004. The ABCs of solute carriers: physiological, pathological and therapeutic implications of human membrane transport proteins. Introduction. *Pflügers Arch.* 447: 465–468.

Hirokawa, N., Niwa, S., and Tanaka, Y. 2010. Molecular motors in neurons: transport mechanisms and roles in brain function, development, and disease. *Neuron* 68: 610–638.

Hökfelt, T., Broberger, C., Xu, Z. Q., Sergeyev, V., Ubink, R., and Diez, M. 2000. Neuropeptides—an overview. *Neuropharmacology* 39: 1337–1356.

Iversen, L. 2006. Neurotransmitter transporters and their impact on the development of psychopharmacology. *Brit. J. Pharmacol.* 147, Suppl. 1: S82–88.

Prado, M. A., Reis, R. A., Prado, V. F., de Mello, M. C., Gomez, M. V., and de Mello, F. G. 2002. Regulation of acetylcholine synthesis and storage. *Neurochem. Int.* 41: 291–299.

Seal, R. P., and Edwards, R. H. 2006. Functional implications of neurotransmitter co-release: glutamate and GABA share the load. *Curr. Opin. Pharmacol.* 6: 114–119.

Torres, G. E., and Amara, S. G. 2007. Glutamate and monoamine transporters: new visions of form and function. *Curr. Opin. Neurobiol.* 17: 304–312.

Vallee, R. B., and Bloom, G. S. 1991. Mechanisms of fast and slow axonal transport. *Annu. Rev. Neurosci.* 14: 59–92.

Original Papers

Allen, T. G., Abogadie, F. C., and Brown, D. A. 2006. Simultaneous release of glutamate and acetylcholine from single magnocellular "cholinergic" basal forebrain neurons. *J. Neurosci.* 26: 1588–1595.

Birks, R. I., and MacIntosh, F. C. 1961. Acetylcholine metabolism of a sympathetic ganglion. *Can. J. Biochem. Physiol.* 39: 787–827.

Brady, S. T., Lasek, R. J., and Allen, R. D. 1982. Fast axonal transport in extruded axoplasm from squid giant axon. *Science* 218: 1129–1131.

Forman, D. S., Padjen, A. L., and Siggins, G. R. 1977. Axonal transport of organelles visualized by light microscopy: cinemicrographic and computer analysis. *Brain Res.* 136: 197–213. [An accompanying movie, *Movement of Organelles in Living Nerve Fibers*, is available at the following website: http://www.alp.mcgill.ca/Pub/Pub_Main_Display.asp?LC_Docs_ID=4911]

Howard, J., Hudspeth, A. J., and Vale, R. D. 1989. Movement of microtubules by single kinesin molecules. *Nature* 342: 154–158.

Jonas, P., Bischofberger, J., and Sandkühler, J. 1998. Corelease of two fast neurotransmitters at a central synapse. *Science* 281: 419–424.

Kodera, N., Yamamoto, D., Ishikawa, R., and Ando, T. 2010. Video imaging of walking myosin V by high-speed atomic force microscopy. *Nature* 468: 72–76.

Kuromi, H., and Kidokoro, Y. 1998. Two distinct pools of synaptic vesicles in single presynaptic boutons in a temperature-sensitive *Drosophila* mutant, *shibire*. *Neuron* 20: 917–925.

McMahan, U. J., Sanes, J. R., and Marshall, L. M. 1978. Cholinesterase is associated with the basal lamina at the neuromuscular junction. *Nature* 271: 172–174.

Schnapp, B. J., Vale, R. D., Sheetz, M. P., and Reese, T. S. 1985. Single microtubules from squid axoplasm support bi-directional movement of organelles. *Cell* 40: 455–462.

CHAPTER 16
Synaptic Plasticity

The efficacy of transmission at a synapse is not fixed, but can vary as a consequence of patterns of ongoing activity. Short trains of presynaptic action potentials can produce either facilitation of transmitter release from the presynaptic terminal that persists for several hundred milliseconds, or depression of release lasting for seconds, or a combination of both. More prolonged trains of presynaptic action potentials produce post-tetanic potentiation (PTP), an increase in transmitter release that can last for several minutes. An intermediate phase of enhancement, classified as augmentation, decays with a time course similar to that of synaptic depression. These changes in synaptic efficacy are closely linked to accumulation of calcium in the presynaptic cytoplasm during activity, and its subsequent extrusion.

At many synapses repetitive activity can produce not only short-term changes, but also alterations in synaptic efficacy that last for hours or even days. The two phenomena of this type are known as long-term potentiation (LTP) and long-term depression (LTD). LTP is mediated by an increase in calcium concentration in the postsynaptic cell that sets in motion a series of second-messenger systems that recruit additional receptors into the postsynaptic membrane and, in addition, increase receptor sensitivity. LTD appears to be associated with smaller increases in postsynaptic calcium concentration and is accompanied by a reduction in the number and sensitivity of postsynaptic receptors. LTP and LTD appear to involve presynaptic mechanisms as well. Both LTP and LTD have been postulated to be substrates for various forms of learning and memory formation.

So far, we have discussed excitatory and inhibitory synaptic transmission in terms of a single action potential arriving at the presynaptic nerve terminal, causing depolarization, calcium entry, and transmitter release, followed by a postsynaptic potential change. Under such circumstances, except for statistical variations, the postsynaptic responses at any given synapse are relatively regular and stereotyped. During everyday activity, however, synapses in the nervous system are not usually activated by the arrival of an occasional presynaptic action potential. Instead, constant streams of action potentials invade the terminals—sometimes regularly with clock-like intervals, other times in bursts of varying frequency and duration. Such ongoing activity can have marked effects on the efficacy of synaptic transmission.

Short-Term Changes in Signaling

Facilitation and Depression of Transmitter Release

When a brief train of stimuli is applied to a presynaptic nerve, the amplitude of the resulting postsynaptic potentials may progressively increase (synaptic **facilitation**), decrease (synaptic **depression**), or undergo a combination of both. This process is illustrated in Figure 16.1A, which shows end-plate potentials recorded from a frog neuromuscular junction, produced by a short train of impulses to the motor nerve. The end-plate potentials were reduced in amplitude by lowering calcium in the bathing solution so that the initial quantum content of the potentials was low (less than 10). The amplitudes of the potentials (measured from the starting point of each rising phase) increase progressively during the train. Furthermore, the effect outlasts the stimulus train, so that the response to a test stimulus occurring 230 milliseconds (ms) later is still larger than the first response in the sequence.

Transmitter release can also be subject to synaptic depression if the number of quanta released by a train of stimuli is large. A similar experiment on a muscle in higher calcium concentration is shown in Figure 16.1B. Here, the quantal release is very large, but the responses have been reduced in amplitude by blocking the postjunctional acetylcholine (ACh) receptors with curare. During repetitive stimulation, the responses become progressively smaller in amplitude. As with facilitation, depression outlasts the stimulus train (not shown) and can persist for several seconds. After a long train of repetitive stimulation, depression can be severe, reducing the amplitude of the synaptic potential to less than 20% of its previous value.

Figure 16.1C illustrates how facilitation and depression can interact at intermediate levels of release. During the train of impulses, the initial facilitation is overridden by depres-

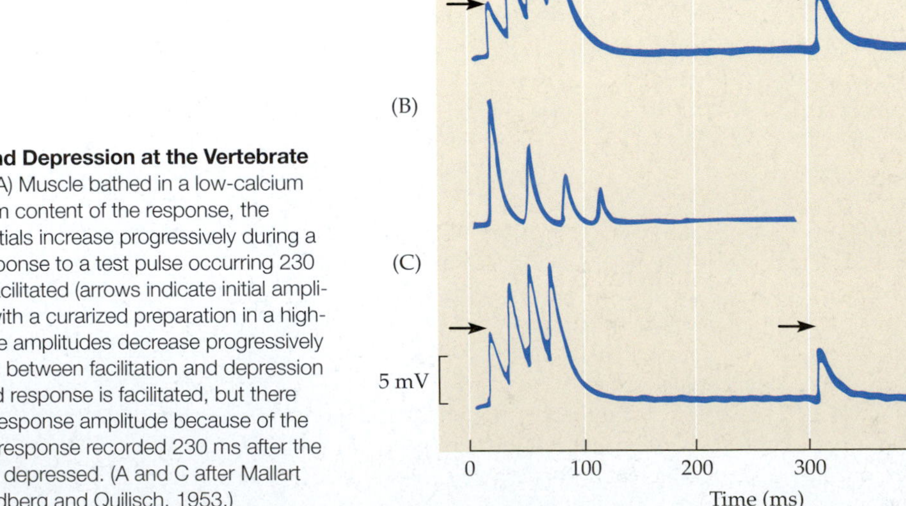

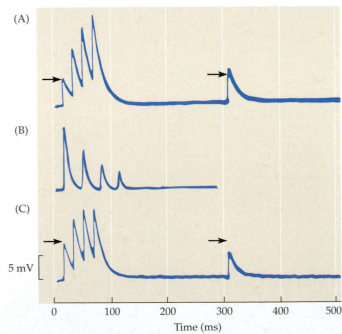

FIGURE 16.1 Facilitation and Depression at the Vertebrate Neuromuscular Junction. (A) Muscle bathed in a low-calcium solution to reduce the quantum content of the response, the amplitudes of end-plate potentials increase progressively during a train of four impulses. The response to a test pulse occurring 230 milliseconds (ms) later is still facilitated (arrows indicate initial amplitude). (B) Similar experiment with a curarized preparation in a high-calcium solution. The response amplitudes decrease progressively during the train. (C) Interaction between facilitation and depression in normal calcium. The second response is facilitated, but there is no subsequent increase in response amplitude because of the onset of depression. The test response recorded 230 ms after the end of the stimulus train is still depressed. (A and C after Mallart and Martin, 1968; B after Lundberg and Quilisch, 1953.)

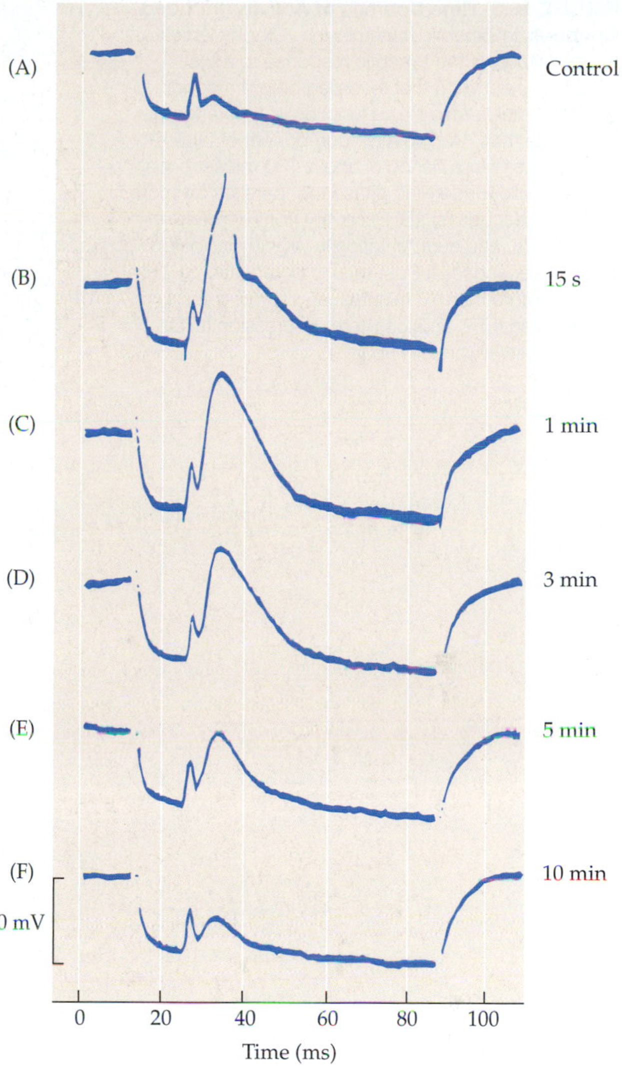

FIGURE 16.2 Post-Tetanic Potentiation (PTP) of the excitatory postsynaptic potential (EPSP) in a chick ciliary ganglion cell, produced by preganglionic nerve stimulation. Potentials were recorded with an intracellular microelectrode. To prevent action potential initiation, the EPSP amplitude was reduced with curare and a hyperpolarizing pulse applied though the recording electrode before each stimulus. (A) The control record shows electrical coupling potential (brief depolarization) followed by a small EPSP. (B) Response recorded 15 seconds after the end of a train of 1500 stimuli applied to the preganglionic nerve. The EPSP amplitude is more than six times greater than control, giving rise to an action potential. Amplitude of coupling potential is unchanged. (C–F) Test stimuli at 1, 3, 5, and 10 minutes after the tetanus show slow decline of potentiation, with the EPSP in the last record still more than twice the control value. (From Martin and Pilar, 1964b.)

sion, so that there is no increase in response amplitude after the second response. Later, when the test pulse is given, facilitation has largely worn off and only depression remains.

Post-Tetanic Potentiation and Augmentation

A relatively long, high-frequency train of stimuli (commonly called a "tetanus" because such a train of stimuli applied to a muscle or to its motor nerve produces a tetanic muscle contraction) usually results in synaptic depression but is followed a few seconds later by an increase in synaptic potential amplitude that can persist for tens of minutes. This is called **post-tetanic potentiation (PTP)**. An example is shown in Figure 16.2 from an experiment on a cell in a ciliary ganglion from a chicken, treated with curare to reduce the amplitude of the excitatory postsynaptic potential (EPSP). In addition, in order to prevent the synaptic potential from triggering an action potential, the cell was hyperpolarized (long downward deflection) before stimulating the presynaptic nerve. The first upward deflection is an electrical coupling potential (see Chapter 11), the second, slower depolarization is the EPSP, produced by release of ACh from the presynaptic terminal. It is the EPSP that is of interest. Initially the EPSP was only about 4 mV in amplitude (because of curarization). The presynaptic nerve was then stimulated at 100 pulses/second for 15 seconds (1500 stimuli), which caused a transient depression of the EPSP (not shown). Fifteen seconds later, however, a single test stimulus produced an EPSP well over 20 mV in amplitude (Figure 16.2B)—so large, in fact, that it exceeded threshold and produced an action potential! The EPSP amplitudes produced by subsequent test shocks then declined (Figure 16.2C–E), but the response was still twice the pretetanic amplitude 10 minutes after the end of the tetanic stimulation (Figure 16.2F).

PTP is also associated with an increase in the rate of spontaneous, or tonic, transmitter release from the nerve terminal. The ongoing appearance of individual miniature synaptic potentials is accelerated after the tetanus and then declines to resting level over the same time course as the potentiation itself.[1,2]

An intermediate phase of enhancement of transmitter release by repetitive stimulation has been designated **augmentation.**[3] It is produced by stimulus trains of moderate duration, comes on more slowly than facilitation, and decays over a period of several seconds. At the frog neuromuscular junction, augmentation and facilitation together can increase synaptic potential amplitude by a factor of more than five.

Facilitation, depression, augmentation, and PTP, although first observed at the vertebrate neuromuscular junction, occur throughout the vertebrate and invertebrate nervous systems. Their characteristic time courses are as summarized in Figure 16.3A–C. The decay of facilitation is not a simple exponential process, but instead has two components. A small component decays over a relatively long period with a time constant of about 250 ms. Superimposed on that is

[1] Liley, A. W. 1956. *J. Physiol.* 133: 571–587.
[2] Martin, A. R., and Pilar, G. 1964. *J. Physiol.* 175: 16–30.
[3] Magleby, K. L., and Zengel, J. E. 1976. *J. Physiol.* 257: 449–470.

FIGURE 16.3 Time Courses of Activity-Induced Changes in synaptic transmission. Graphs indicate the amplitude of the synaptic response to a test stimulus—relative to that recorded before a conditioning stimulus train—as a function of time after the end of the train. (A) The main component of facilitation decays over a period of about 100 milliseconds, with a smaller underlying phase that persists for more than 0.5 seconds (s). (B) Recovery from depression is complete, and augmentation is largely dissipated after 10 seconds. (C) Post-tetanic potentiation (PTP) lasts for more than 10 minutes, and (D) long-term potentiation (LTP) and long-term depression (LTD) can last well beyond 10 hours.

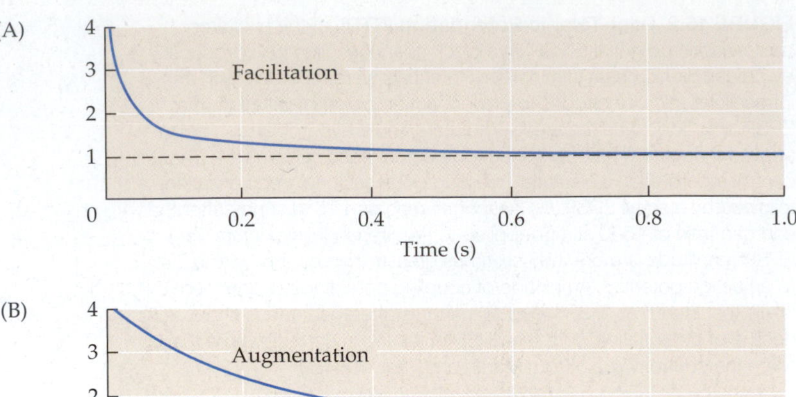

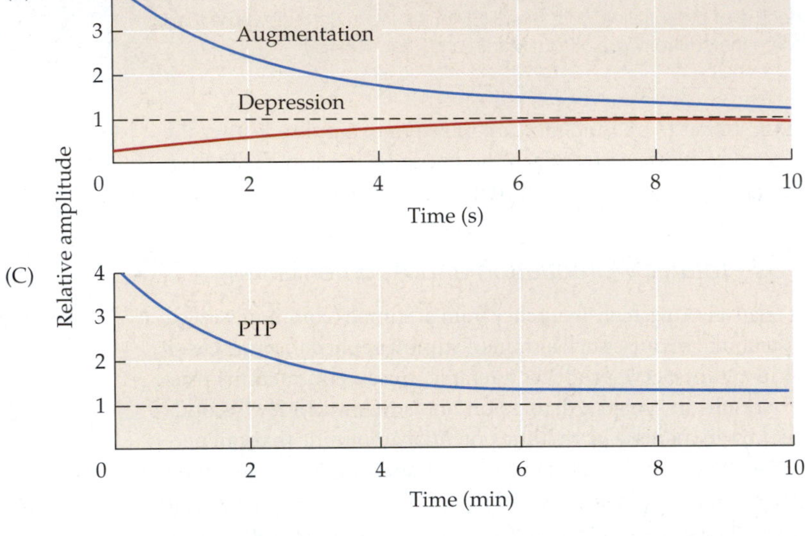

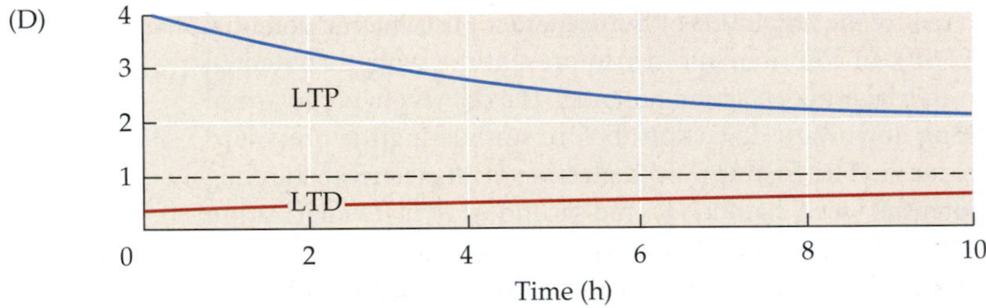

a much larger and briefer component with a decay time constant of about 50 ms.[4] Decay of augmentation and recovery from depression both occur over periods of several seconds. PTP usually decays with a time constant of several minutes. These four phenomena are referred to as short-term changes in synaptic efficacy to distinguish them from two changes that last much longer, found at synapses in the central nervous system (CNS)—they are **long-term potentiation (LTP)** and **long-term depression (LTD)**. As indicated in Figure 16.3D, LTP and LTD persist for hours. They will be discussed in detail later in this chapter.

Mechanisms Underlying Short-Term Synaptic Changes

Short-term changes in synaptic efficacy during and following repetitive activity are related to increases or decreases in the number of quanta of transmitter released from the presynaptic terminals and not, for example, to a change in the size of an individual quantum or to a change in the sensitivity of the postsynaptic membrane.[1,5,6] As we discuss in Chapter 13, the amount of transmitter released by any one stimulus depends on the size of the readily releasable pool of synaptic vesicles within the terminal and on the fractional release from the pool (i.e., the average probability that any vesicle will be released). The fractional release, in turn, is dependent on the amount of calcium that enters the terminal through voltage-activated calcium channels. The release probability varies with the third or fourth power of the calcium influx.[7–9] Traditionally, facilitation and potentiation have been ascribed to increases in release

[4] Mallart, A., and Martin, A. R. 1967. *J. Physiol.* 193: 679–694.

[5] del Castillo, J., and Katz, B. 1954. *J. Physiol.* 124: 574–585.

[6] Kuno, M. 1964. *J. Physiol.* 175: 100–112.

probability mediated by increases in available calcium, and depression to depletion of the releasable pool of vesicles. Recent experiments have revealed a rather more complicated picture.

FACILITATION Experimental evidence obtained by Katz and Miledi in 1968 suggested that facilitation of transmitter release during a train of action potentials in the presynaptic terminal might be related to a progressively increasing residue of calcium left over from the previous stimuli.[10] The residual amounts, themselves insufficient to trigger release, would add to the next bolus of incoming calcium, thereby increasing the release probability. The decay of facilitation over time would then represent the return of intracellular calcium to its resting level. This idea that calcium accumulation is involved in some way is supported by observations at a variety of synapses that buffering intracellular calcium concentration with calcium chelators attenuates facilitation.[11]

Subsequent studies indicate that increased cytoplasmic calcium has additional effects that can contribute to facilitation. For example, experiments on glutamatergic synapses in the auditory brainstem of the rat have shown that facilitation is also accompanied by a progressive increase in calcium influx with each impulse during a conditioning train.[12,13] These synapses have calyciform presynaptic terminals (calyces of Held), and whole-cell patch clamp recordings can be made simultaneously from the terminals and the postsynaptic cells.[14] Transmitter release is mediated by P-type ($Ca_V2.1$) calcium channels.[15] The calcium currents facilitate, and subsequently decay, with a time course similar to that of facilitation of transmitter release. Like facilitation, the increase in calcium current is dependent on calcium accumulation and is due to a shift in the voltage sensitivity of calcium channel activation. Although increased calcium influx is clearly a major contributor to facilitation, other factors, such as a direct effect of residual calcium on release, still play a role.[9,16]

DEPRESSION Depression of synaptic potentials is seen only after a relatively large quantal release, which suggests that one underlying factor is depletion of vesicles from the nerve terminal during the conditioning train.[1,17] However, large quantal release is also accompanied by large calcium entry, so calcium accumulation may contribute in some way as well. Indeed, early experiments on the frog neuromuscular junction indicated that depression was accompanied by a reduction in release probability.[18] Similar experiments on cultured rat hippocampal neurons have led to the same conclusion.[19]

Experiments on calyx of Held synapses have shown depression of presynaptic calcium currents by prior conditioning pulses.[15,20] As with facilitation, the amount of depression depends on the amount of calcium entering the cytoplasm during the conditioning period. During repetitive activation at low frequencies, this reduction in calcium current and the consequent reduction in release probability are the major factors underlying depression. However, at frequencies of 100/s or more and with normal levels of transmitter release, depletion of the available pool contributes as well, accounting for as much as 50% of the reduction in release. At these frequencies, calcium currents are first facilitated and then depressed, and recovery from depression occurs over a period of several tens of seconds. In spite of the early facilitation of calcium current during the train, the corresponding excitatory postsynaptic currents show immediate depression due to depletion. Subsequent recovery occurs in two phases, the first over a period of seconds, reflecting recovery from depletion, and the second with a time course parallel to that of recovery from depression of the presynaptic calcium current.

AUGMENTATION AND PTP Like facilitation and depression, augmentation and PTP are associated with increased intracellular calcium concentration. Experiments on the neuromuscular junction of the frog showed that if calcium is removed from the bathing solution during application of the conditioning train, then no potentiation occurs.[21] At the crayfish neuromuscular junction, PTP is reduced in magnitude and duration by treatments that interfere with calcium uptake and release by mitochondria, suggesting that calcium influx during the tetanus may be accompanied by rapid mitochondrial calcium loading.[22] In that event, excess mitochondrial calcium might be released slowly, thereby prolonging the elevation of cytoplasmic calcium concentration and maintaining the potentiation.

On the other hand, PTP is not dependent on sodium entry, as it can be produced by trains of artificial depolarizing pulses applied to the nerve terminal in the presence of

[7] Dodge, F. A., Jr., and Rahamimoff, R. 1967. *J. Physiol.* 193: 419–432.

[8] Catterall, W. A., and Few, A. P. 2008. *Neuron* 59: 882–901.

[9] Neher, E., and Sakaba, T. 2008. *Neuron* 59: 861–872.

[10] Katz, B., and Miledi, R. 1968. *J. Physiol.* 195: 481–492.

[11] Zucker, R. S., and Regehr, W. G. 2002. *Annu. Rev. Physiol.* 64: 355–405.

[12] Cuttle, M. F. et al. 1998. *J. Physiol.* 512: 723–729.

[13] Borst, J. G. G., and Sakmann, B. 1998. *J. Physiol.* 513: 149–155.

[14] Borst, J. G. G., Helmchen, F., and Sakmann, B. 1995. *J. Physiol.* 489: 825–840.

[15] Forsythe, I. D. et al. 1998. *Neuron* 20: 797–807.

[16] Xu, J., He, L., and Wu, L-G. 2007. *Curr. Opin. Neurobiol.* 17: 352–359.

[17] Mallart, A., and Martin, A. R. 1968. *J. Physiol.* 196: 593–604.

[18] Betz, W. J. 1970. *J. Physiol.* 206: 629–644.

[19] Sullivan, J. M. 2007. *J. Neurophysiol.* 97: 948–950.

[20] Xu, J., and Wu, L-G. 2005. *Neuron* 46: 633–645.

[21] Rosenthal, J. L. 1969. *J. Physiol.* 203: 121–133.

[22] Tang, Y-G., and Zucker, R. F. 1997. *Neuron* 18: 483–491.

tetrodotoxin (TTX).[23] In that circumstance, the magnitude of the potentiation is increased with increasing extracellular calcium concentration, and, in very high calcium (83 m*M*), PTP can last for more than two hours. While sodium is not necessary for potentiation, sodium entry nevertheless contributes to its duration. At the rat neuromuscular junction, potentiation is prolonged by treatments that block extrusion of sodium by Na-K ATPase, such as adding ouabain or removing potassium from the bathing solution.[24] Prolongation may occur because increased intracellular sodium reduces the rate at which accumulated calcium is extruded by sodium–calcium exchange.

In the calyx of Held, high-resolution fluorescent dye techniques have been used to measure calcium transients produced by single presynaptic stimuli. After induction of PTP, presynaptic calcium influx increased by about by about 15%.[25] At the peak of PTP, average cytoplasmic calcium concentrations increased from a resting value of about 50 n*M* to over 200 n*M*. The calcium concentration then declined to its resting level with a time course that paralleled that of the decay of PTP.[26,27]

The increase in presynaptic calcium influx suggests that PTP and augmentation are due, at least in part, to an increase in transmitter release probability. The sustained increased in intracellular calcium concentration may also result in an increase in the size of the releasable pool. One technique for estimating the releasable pool size is to induce rapid, ongoing transmitter release until the pool is exhausted and then estimate how much has been released. For example when a train of stimuli, say at 100 Hz, is applied to the presynaptic nerve the resulting synaptic potentials (or currents) decrease steadily in amplitude toward zero. The summed amplitudes of all the synaptic responses then provide an estimate the total amount of transmitter that was in the pool. A minor complication is that the responses never go to zero because of ongoing replenishment of the pool from back-up stores, but appropriate allowance can be made to accommodate this factor.

Depletion experiments on synapses between cultured hippocampal neurons indicate that augmentation is due to an increase in release probability, with no apparent increase in the size of the releasable pool of transmitter.[28] On the other hand, depletion experiments on the calyx of Held synapses have shown that during PTP the releasable pool does increase in size.[29] The magnitude of the PTP depended on the frequency and duration of the conditioning tetanus, with synaptic currents reaching a maximum of about four times the control amplitude following a 6000-shock tetanus. The potentiation was accompanied by a maximum increase of about 70% in the size of the releasable pool, and an increase in release probability by a factor of about 2.5. PTP decayed in two phases. The first phase, with a time-constant of about one minute, was due to a return of the release probability toward its resting value and probably represented an underlying contribution of augmentation to the overall process. The second phase extended over a period of 10 to 20 minutes and represented the return of the releasable pool to its resting size.

The various factors underlying short-term synaptic plasticity seem at first to be somewhat bewildering, so it is useful at this point to summarize what we have learned. First, all forms of short-term plasticity are related to presynaptic calcium accumulation. This accumulation results in calcium-activated alterations in influx though presynaptic calcium channels. At modest levels of activation a progressive increase in intracellular calcium concentration first facilitates channel currents; then, as the calcium concentration continues to build up, facilitation gives way to depression. The changes in channel currents follow the same time course as facilitation and depression of transmitter release. At higher levels of release, depression is also mediated in part by depletion of the releasable pool. During prolonged stimulation, calcium accumulation eventually leads to augmentation which, like facilitation, is associated with increased calcium flux through the voltage-activated channels. The early stages of PTP are attributable, at least in part, to augmentation, and the more prolonged component reflects an increase in size of the releasable pool of vesicles.

The next question to be answered is how increased intracellular calcium acts to modulate calcium channel function. Catterall and his colleagues have approached this question by looking at the effects on facilitation and depression of mutations in $Ca_V2.1$ channels.[30] The channels were expressed in cultured superior cervical ganglion (SCG) neurons, and the transfected cells formed cholinergic synaptic connections with adjacent SCG cells in the culture. Mutations were made on the channel α_1-subunits at two nearby locations on the C terminus. At one

[23]Weinreich, D. 1970. *J. Physiol.* 212: 431–446.

[24]Nussinovitch, I., and Rahamimoff, R. 1988. *J. Physiol.* 396: 435–455.

[25]Habets, R. L., and Borst, J. G. 2006. *J. Neurophysiol.* 96: 2868–2876.

[26]Korogood, N., Lou, X., and Schneggerburger, R. 2005. *J. Neurosci.* 25: 5127–5137.

[27]Habets, R. L., and Borst, J. G. 2005. *J. Physiol.* 564: 173–187.

[28]Stevens, C. F., and Wesseling, J. F. 1999. *Neuron* 22: 139–146.

[29]Habets, R. L., and Borst, J. G. 2007. *J. Physiol.* 581: 467–478.

[30]Mochida, S. et al. 2008. *Neuron* 57: 210–216.

location, an isoleucine and a methionine residue at adjacent positions (IM) were replaced by two alanines (AA). At the other, a calmodulin-binding domain (CBD) was deleted.

When two stimuli were applied to a presynaptic cell containing native channels, the second EPSP in the postsynaptic cell showed depression at short, inter-shock intervals (< 50 ms) and then showed facilitation as the interval between the stimuli was increased to 100 ms. Facilitation then declined toward zero as the inter-shock intervals approached 200 ms. At synapses with IM–AA mutations, facilitation was virtually absent. Deletion of the CBD segment, on the other hand, removed almost all of the depression. The authors conclude that changes in calcium channel conductance observed during facilitation and depression of transmitter release are mediated by conformational changes in the channel produced by selective binding of calcium–calmodulin to the C-terminal sites.

Although the experiments that we have discussed here provide a general picture of factors underlying short-term plasticity, the actual details may vary considerably from one type of synapse to the next. For one thing, most of the experiments in which changes in calcium channel currents were examined were done on calyx of Held synapses, which contain $Ca_V2.1$ channels with P/Q-type currents. Release of transmitter at synapses in the peripheral nervous system and in some parts of CNS is subserved predominantly by calcium influx through $Ca_V2.2$ channels with N-type currents, which may have quite different properties.[8] Another factor to consider is the progressive effects of increased calcium concentration inside the presynaptic terminal membrane. In the calyx of Held synapse, the increase in intracellular calcium first mediates facilitation of channel currents; then, as the concentration builds up further, facilitation switches to inhibition. With this scenario, the overall effect of calcium influx can be expected to depend markedly on physical factors, one example being the size of the terminal. In very small terminals with a large surface-to-volume ratio, calcium concentration could build up rapidly so that the overriding effect is inhibitory, whereas in very large terminals the build up could be much smaller, yielding only facilitation. Similarly, differences between presynaptic terminals in the cytoplasmic concentration and distribution of calcium chelators could produce marked differences in behavior.

Long-Term Changes in Signaling

In the CNS, repetitive activity can produce changes in synaptic efficacy that last much longer than those seen at peripheral synapses. These longer lasting changes have been found in a variety of brain locations and are particularly intriguing because their long duration suggests that they may be associated in some way with memory. Two basic changes can be induced: long-term potentiation (LTP) and long-term depression (LTD).

Long-Term Potentiation

Long-term potentiation (LTP) was first described by Bliss and Lømo in 1973 at glutamatergic synapses in the hippocampal formation.[31,32] This structure, which lies within the temporal lobe of the brain, consists of two regions known as the hippocampus and the dentate gyrus, which in cross section appear as interlocking C-shaped strips of cortex, plus the neighboring subiculum (Figure 16.4). Its orderly arrangement of cells and input pathways enables recording electrodes to be inserted into the brain of the intact animal and placed in close proximity to known cell types, or even intracellularly, to record synaptic potentials. Similarly, stimulating electrodes can be located in specific input pathways. Bliss and Lømo demonstrated that high frequency stimulation of inputs to cells in the dentate gyrus produced a subsequent increase in the amplitude of excitatory synaptic potentials that lasted for hours or even for days (Figure 16.5). This is now known as **homosynaptic LTP.** Although LTP has been shown to occur in other regions of the brain, including several neocortical areas, and even in autonomic ganglia,[33] it has been studied most extensively in CA1 pyramidal cells in hippocampal slices in vitro.[34]

Associative LTP in Hippocampal Pyramidal Cells

Experiments by T. H. Brown and his colleagues revealed that repetitive activity at one synaptic input to a cell could potentiate synaptic potentials generated by stimulation of

[31] Bliss, T. V. P., and Lømo, T. 1973. *J. Physiol.* 232: 331–356.

[32] Lømo T. 2003. *Philos. Trans. R. Soc. Lond., B, Biol. Sci.* 358: 617–620.

[33] Alkadhi, K. A., Alzoubi, K. H., and Aleisa, A. M. 2005. *Prog. Neurobiol.* 75: 83–108.

[34] Malenka, R. C., and Nicoll, R. A. 1999. *Science* 258: 1870–1874.

FIGURE 16.4 The Hippocampal Formation lies buried in the temporal lobe and consists of two interlocking C-shaped strips of cortex, the dentate gyrus and hippocampus, together with the neighboring subiculum. Granule cells in the dentate gyrus (black) are innervated by the perforant fiber pathway (red) from the subiculum and, in turn, send axons (mossy fibers) to make synapses on CA3 pyramidal cells (green). CA3 pyramidal cells project axons (Schaffer collaterals) to pyramidal cells in CA1 (black).

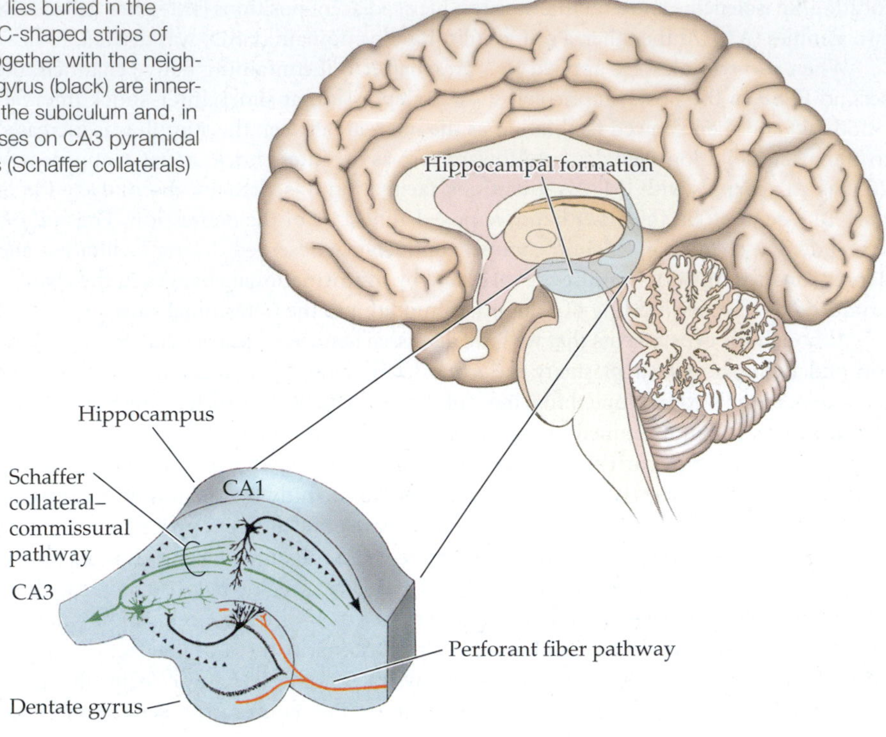

another input to the same cell.[35] This is called **associative LTP**. An example is shown in Figure 16.6. Intracellular recordings were made from pyramidal cells in area CA1 of the hippocampus, and two extracellular stimulating electrodes were located in the input pathway (the Schaffer collateral/commissural tract) in such a way as to stimulate subpopulations of axons innervating two different regions on the dendritic arbors of the pyramidal cells (Figure 16.6A). The stimulus intensities were adjusted so that electrode I evoked a large synaptic potential in a pyramidal cell, while electrode II evoked a much smaller one. Records of EPSPs produced by electrode II are shown in Figure 16.6B. Brief trains of stimuli (100 Hz for 1 second repeated once again after 5 seconds) applied to electrode I resulted in LTP of the synaptic potentials at that input (not shown), as in the experiments of Bliss and Lømo. The stimuli applied to electrode I had no effect on the

[35]Barionuevo, G., and Brown, T. H. 1983. *Proc. Natl. Acad. Sci. USA* 80: 7347–7351.

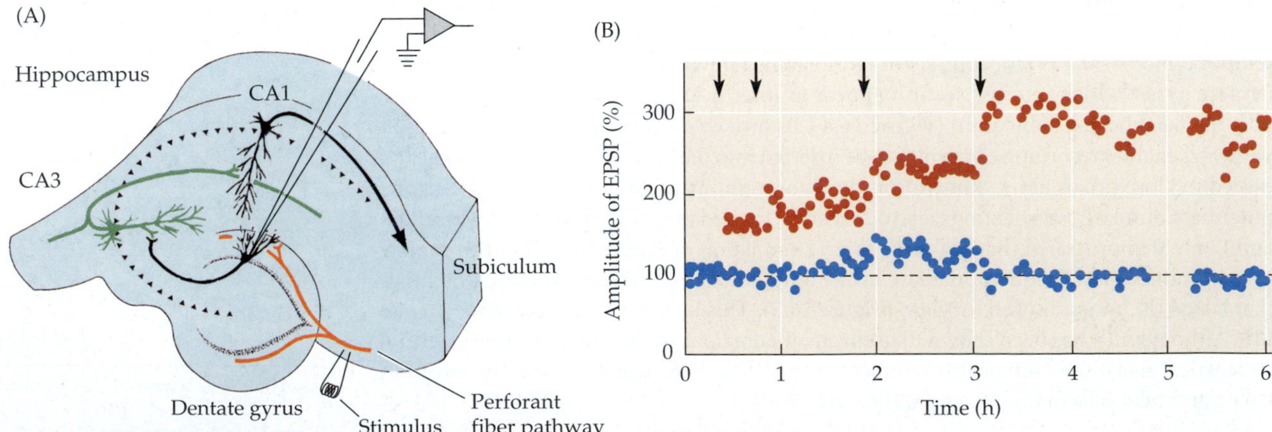

FIGURE 16.5 Long-Term Potentiation (LTP) in the hippocampus of an anesthetized rabbit. (A) Synaptic responses to perforant pathway stimulation were recorded from granule cells in the dentate gyrus. (B) Brief tetanic stimuli (15/s for 10 s) were given at times marked by the arrows. Each tetanus caused an increase in the amplitude of the synaptic response (red circles), eventually lasting for hours. Responses in a control pathway not receiving tetanic stimulation (blue circles) were unchanged. (After Bliss and Lømo, 1973.)

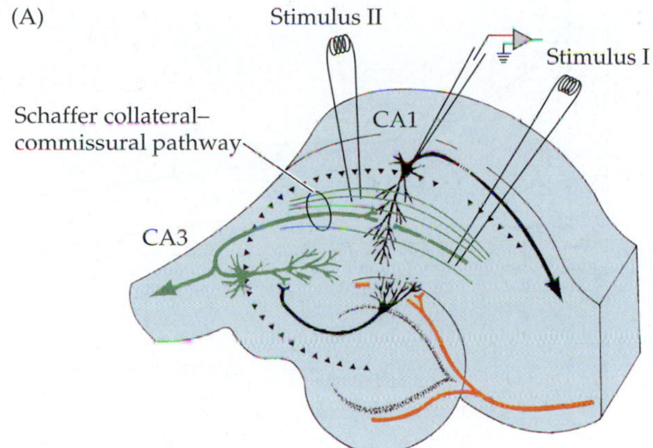

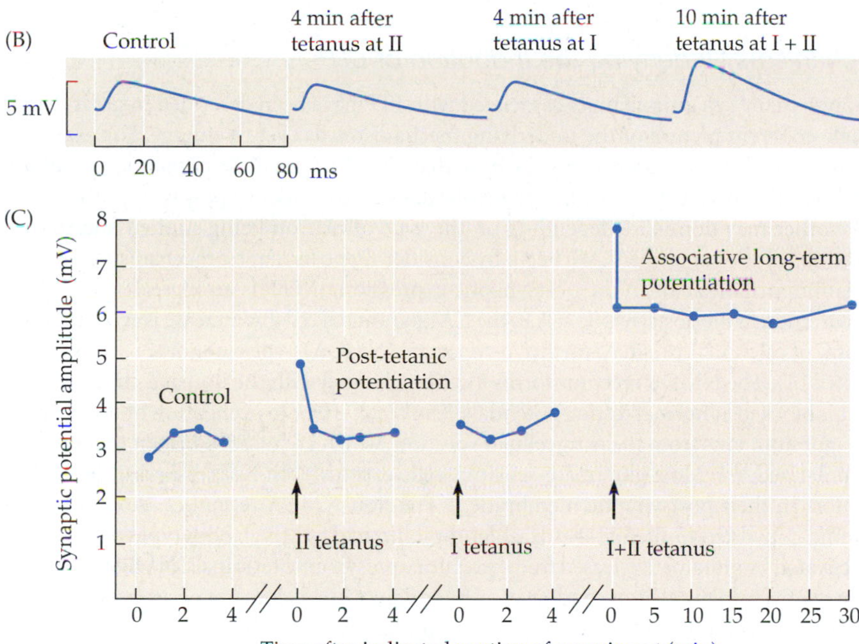

FIGURE 16.6 Associative LTP in a Rat Hippocampal Slice. (A) Intracellular records were made from CA1 pyramidal cells while stimulating two distinct groups of presynaptic fibers in the Schaffer collateral/commissural pathway (Stimulus I, and Stimulus II). The stimuli were adjusted so that responses to stimulation at site I were five times greater than those to stimulation at site II. (B) Averaged responses to stimulation at site II in control condition, after tetanic stimulation at site I (100/s for 1 s), after a similar tetanus at site II, and after a combined tetanus at I and II. Only the combined tetanus produced potentiation; test shock 10 minutes later indicated a two-fold increase in response amplitude. (C) Summary of the results in part B showing the time course of the changes in response amplitude. Stimulation at site I had no effect, stimulation at II produced a brief potentiation of the response, and combined stimulation produced LTP. (After Barrionuevo and Brown, 1983.)

smaller EPSP produced by electrode II (Figure 16.6B, second record). Also, high-frequency stimulation with electrode II did not produce LTP of the smaller response (third record). However, after high-frequency stimulation by I and II together, there was an increase in the size of the EPSPs produced by electrode II (fourth record), lasting for tens of minutes (Figure 16.6C). This was called associative LTP because the prolonged increased response to input II was produced only when repetitive stimulation of that input was associated with simultaneous stimulation of input I.

Associative LTP is particularly interesting because it illustrates a possible mechanism for conditioned reflexes, the classic example being conditioning a dog to salivate in response to the sound of a bell.

Like short term changes in signaling, LTP can be separated into three components, based on their time course of decay.[36] Their induction depends in part on the duration and intensity of the conditioning stimulus. The response of CA1 pyramidal neurons to three different conditioning protocols is summarized in Figure 16.7. A single 100 Hz conditioning train increased the size of the EPSP by about 130%, and the increase decayed back to zero over the next 2 hours. Four conditioning trains produced about the same increase, but the potentiation decayed less rapidly—after 2 hours the EPSPs were still potentiated by 30%. With eight conditioning trains, the decay was still slower, with 80% of the potentiation persisting after 2 hours.

[36]Raymond, C. R. 2007. *Trends Neurosci.* 30: 167–175.

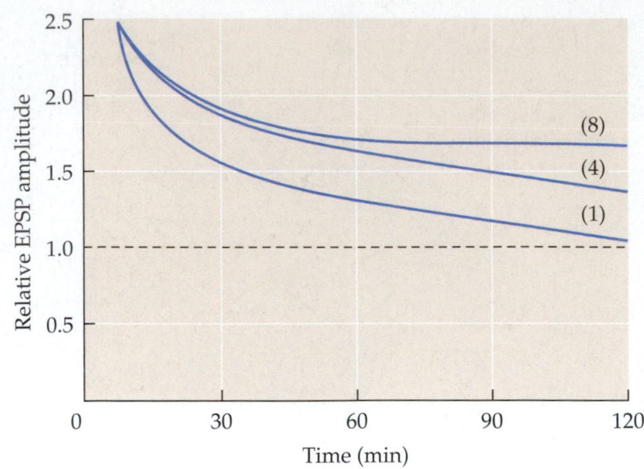

FIGURE 16.7 Three Phases of Decay of LTP in CA1 neuron. A single brief 100-Hz conditioning train applied to the Schaeffer collateral pathway (1) increases EPSP amplitude to more than 2 times normal. The potentiation decays to half its initial value in about 30 minutes. Four conditioning trains (4) produce about the same potentiation, but the decay time is much prolonged. When eight conditioning trains are applied (8), potentiation shows an early decay and then remains at about 1.7 times normal throughout the rest of the recording period. (After Raymond, 2007.)

Mechanisms Underlying the Induction of LTP

Although many synaptic changes associated with LTP have been described in great detail, a single coherent picture of the underlying mechanisms has yet to emerge. For example, some experiments have shown conclusively that the phenomenon is presynaptic, others that that only postsynaptic factors are involved. Current evidence points to both.[37] One or the other may dominate depending on the type of neuron being studied and on the conditions of the experiment. We begin by considering postsynaptic mechanisms for LTP.

An important factor in the postsynaptic expression of LTP is an increase in calcium concentration in the postsynaptic cell. In the CA1 pyramidal cell this increase is accomplished by entry of calcium through *N*-methyl-D-aspartate (NMDA)-type glutamate receptors (see Chapter 14). The NMDA receptor forms a cation channel with the unusual characteristic that it is blocked at normal resting potentials. The block is due to occupation of the channel by magnesium ions from the extracellular solution, which are removed when the receptor is depolarized.[38,39] Most glutamate-sensitive cells express both NMDA and non-NMDA receptors in their postsynaptic membrane.[40] The non-NMDA receptors are sensitive to α-amino-3-hydroxy-5-methyl-4-isoxazolepropionic acid (AMPA). Both types of receptors are activated by glutamate released from excitatory presynaptic terminals. NMDA receptors have a relatively high calcium conductance, but calcium entry does not occur until synaptic depolarization is sufficient to remove magnesium block of the channels.

Involvement of NMDA receptors in LTP is indicated by the fact that NMDA antagonists block the induction of LTP, but do not prevent LTP if they are applied after it has been induced.[41,42] However, recent experiments indicate that entry of calcium through the synaptic channels is not in itself adequate for the induction of LTP, and that the three different phases of LTP (see Figure 16.7) are mediated by different calcium pathways.[43] In dendritic spines and distal dendrites, calcium entry into the cytoplasm through NMDA receptors serves to trigger additional calcium release from the endoplasmic reticulum (ER). Calcium-induced calcium release from the ER depends on activation of ryanodine (Ry) receptors and/or inositol triphosphate (IP_3) receptors. Block of Ry receptors selectively inhibits calcium release in synaptic spines and abolishes short-lasting LTP. Block of IP_3 receptors inhibits calcium release along dendrites and abolishes LTP of intermediate duration. Long-lasting LTP that persists after the first two phases does not depend on NMDA receptors at all. Instead, it is mediated by calcium entry through L-type voltage-sensitive calcium channels in the cell soma.

Silent Synapses

One of the major factors in the postsynaptic expression of LTP is the activation of **silent synapses.**[44] In the first few postnatal days, CA1 synapses in rats and mice are virtually devoid of AMPA receptors, so that most do not respond to presynaptic stimulation. As dendritic spines develop, NMDA receptors are expressed first, followed by AMPA receptors, which over the next few weeks appear at roughly half the synapses. Suppose that in a mature synapse some presynaptic excitatory boutons overlie postsynaptic regions on dendritic spines

[37] Lisman, J. E. 2009. *Neuron* 63: 261–264.

[38] Nowak, L. et al. 1984. *Nature* 307: 462–465.

[39] Mayer, M. L., Westbrook, G. L., and Guthrie, P. B. 1984. *Nature* 309: 261–263.

[40] Takumi, Y. et al. 1999. *Ann. NY Acad. Sci.* 868: 474–481.

[41] Collingridge, G. L., Kehl, S. J., and McClennan, H. 1983. *J. Physiol.* 334: 33–46.

[42] Muller, D., Joly, M., and Lynch, G. 1988. *Science* 242: 1694–1697.

[43] Raymond, C. R., and Redman, S. J. 2006. *J. Physiol.* 570: 97–111.

[44] Kerchner, G. A., and Nicoll, R. A. 2008. *Nat. Neurosci.* 9: 813–825.

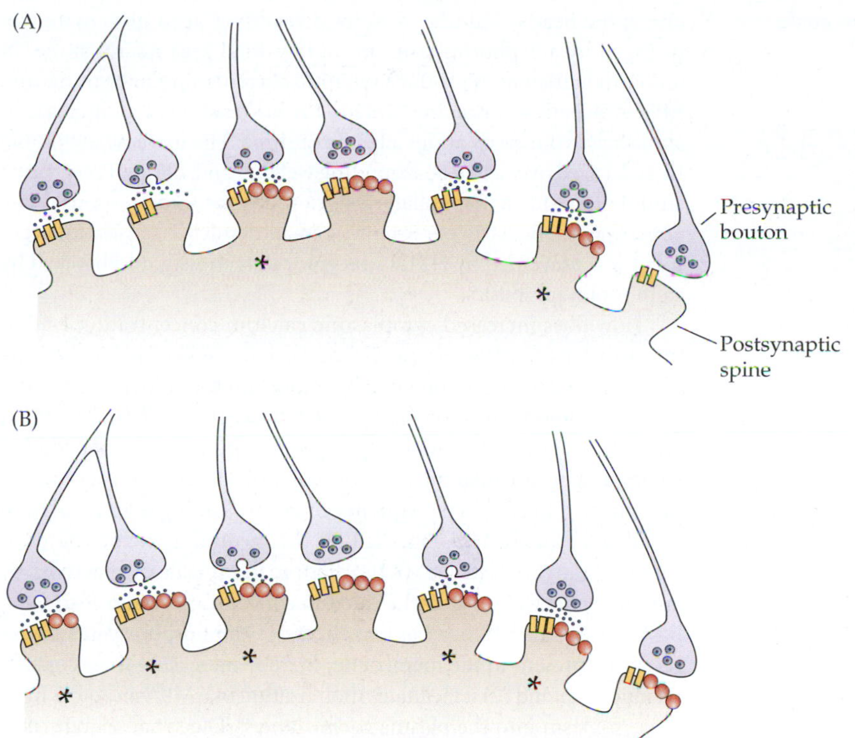

FIGURE 16.8 Proposed Mechanism for Increased Postsynaptic Response during LTP. (A) Release of five quanta of glutamate from presynaptic boutons (indicated by omega figures in the bouton membranes) activates only two postsynaptic spines (asterisks) because many spines contain no AMPA receptors (red circles) and are "silent." Thus the quantum content of the response is two, even though five quanta are released. NMDA receptors (yellow rectangles) do not respond because depolarization is insufficient to remove the magnesium block. (B) During potentiation, AMPA receptors are inserted into the postsynaptic membranes of the spines and the quantum content of the response is increased to five.

that contain only a few AMPA receptors, or perhaps none at all. Under resting conditions, release of a quantum of glutamate from these boutons will produce little or no response; thus, these synaptic contacts will be silent and only a fraction of the synaptic apparatus will respond to presynaptic excitation (Figure 16.8A). Now suppose that the induction of LTP leads to insertion of AMPA receptors into the postsynaptic membranes of the silent synapses. These will now respond to quanta released from the presynaptic terminal, and the quantum content of the response will increase (Figure 16.8B).

There is now substantial evidence that AMPA receptors in the postsynaptic membrane, which turn over quite rapidly at rest (see Chapter 14), are upregulated during the expression of LTP. The most direct evidence is the demonstration that AMPA receptor subunits are delivered to dendritic spines after repetitive stimulation accompanied by NMDA receptor activation.[45] The AMPA receptor subunit GluR1 (now called GluA1; see Chapter 5) was tagged with green fluorescent protein (GFP) and expressed transiently in hippocampal CA1 pyramidal cells. When cell dendrites were examined with laser-scanning and electron microscopy, most of the protein (GluA1–GFP) in the dendrites was found in intracellular compartments and only about half of the dendritic spines showed fluorescence. After stimulation, tagged receptors were delivered rapidly to dendritic spines and, in addition, to clusters on the dendrites. Almost all of the spines were fluorescent, even those nearly devoid of label before stimulation. The results indicate that many of the excitatory dendritic spines are silent and receive a full complement of AMPA receptors after repetitive stimulation.

Additional evidence for upregulation of AMPA receptors has been summarized by Nicoll and his colleagues.[32,44] For example, it has been shown by immunohistochemistry that all Schaffer collateral–commissural synapses contain NMDA receptors but that only a fraction of these contain co-localized AMPA receptors.[46] Correspondingly, electrophysiological experiments reveal many synapses in CA1 pyramidal cells that are activated only by NMDA; these synapses acquire AMPA responses during LTP.[47,48] Still other experiments have demonstrated that after induction of LTP new spines appear on the CA1 pyramidal cell dendrites.[49] Potentiation of the synaptic response by about 80% was accompanied by roughly a 13% increase in measured spine density. A particularly interesting experiment involved the induction of LTP by artificial application of glutamate to dendritic spines of CA1 cells.[50] The cells were transfected with green fluorescent protein so that the spines

[45] Shi, S. H. et al. 1999. *Science* 284: 1811–1816.

[46] Takumi, Y. et al. 1999 *Nat. Neurosci.* 2: 618–624.

[47] Liao, D., Hessler, N. A., and Malinow, R. 1995. *Nature* 375: 400–404.

[48] Isaac, J. T. R., Nicoll, R. A., and Malenka, R. C. 1995. *Neuron* 15: 427–434.

[49] Engert, F., and Bonhoeffer, T. 1999. *Nature* 399: 66–70.

[50] Matsuzaki, M. et al. 2004. *Nature* 429: 761–766.

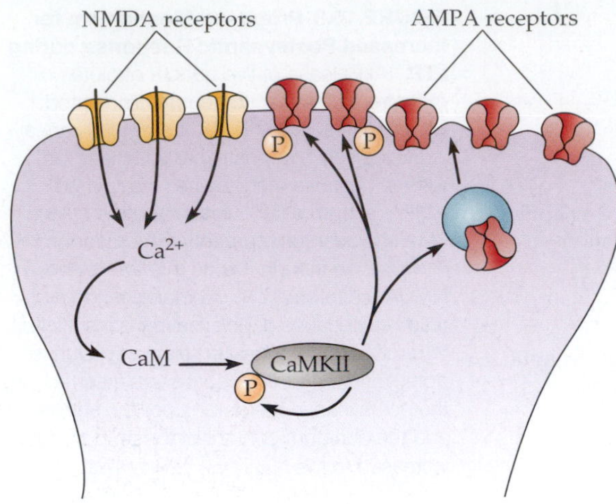

FIGURE 16.9 Proposed Mechanism for LTP. Activation of NMDA receptors allows calcium entry into the spine, activating calcium–calmodulin-dependent protein kinase II (CaMKII), which undergoes autophosphorylation, thereby maintaining its own activity after the calcium concentration has returned to normal. CaMKII phosphorylates AMPA receptors already present in the postsynaptic membrane and promotes the insertion of new receptors from a reserve pool. (After Malenka and Nicoll, 1999.)

and spine heads could be visualized readily. Caged glutamate was uncaged by a 2-photon laser, in a very small area near a spine, in amounts sufficient to produce synaptic currents similar to miniature EPSCs. Repetitive uncaging (1/s for 1 min) resulted in an increase in spine head volume by about a factor of three. The increase in volume then decayed over a period extending well beyond 1 hour. Presynaptic stimulation of Schaffer collaterals produced the same response. The effect was particularly persistent in small dendritic spines. Enlargement was prevented by NMDA receptor antagonists, and blocked by calmodulin inhibitors.

How does increased cytoplasmic calcium concentration lead to upregulation of AMPA receptors? One calcium-dependent biochemical pathways that has been shown to be important for induction of LTP is calcium–calmodulin-dependent protein kinase II (CaMKII).[51] CaMKII is found in high concentrations in the postsynaptic densities of dendritic spines, and intracellular injection of inhibitors of CaMKII prevents the induction of LTP.[52,53] The sequence of events leading to upregulation is shown in Figure 16.9. During LTP, the incoming calcium binds to calmodulin to activate CaMKII, which maintains its own activity by autophosphorylation after the calcium concentration has returned to basal levels. CaMKII then has two effects: (1) it phosphorylates AMPA receptors present in the membrane, thereby increasing their channel conductance, and (2) it facilitates mobilization of AMPA receptors from the cytoplasm into the plasma membrane.[54] The synapses return to their previous resting state over time because of constitutive recycling of AMPA receptors into and out of the membrane.[55]

The third phase of LTP, which can last for very long times, is dependent on gene transcription and protein synthesis.[56,57] As we have already noted, its induction is dependent on calcium accumulation in the cell soma by influx through voltage-sensitive calcium channels. Several biochemical pathways are involved in linking calcium accumulation to transcription. For example, activation of modulatory receptors may be linked to adenyl cyclase, leading to phosphorylation of the cAMP response element-binding protein (CREB). The resulting gene expression involves not only receptor proteins, but also other structural and functional proteins, suggesting that an essential long-term consequence of LTP is growth and remodeling of the synapse. The mechanisms whereby transcripts and proteins are transported from the soma to specific synaptic regions are not known.

Presynaptic LTP

LTP is not only expressed postsynaptically. For example, at mossy fiber synapses on CA3 pyramidal cells, LTP occurs even when a postsynaptic increase in calcium concentration is completely prevented by blocking both NMDA receptors and voltage-sensitive calcium channels, and loading the postsynaptic cell with a high concentrations of a calcium chelator. This observation led to the conclusion that LTP at that synapse was entirely presynaptic.[58] The underlying mechanisms are not at all clear, but one step may involve the recurrent action of glutamate on presynaptic metabolic glutamate receptors (mGluR; see Chapter 12) or on presynaptic kainate receptors (see Chapter 5).

A number of experiments have indicated that presynaptic LTP occurs at CA1 pyramidal cell synapses as well.[59] In these experiments, LTP was associated not with an increase in the amplitude of miniature EPSPs (quantal size), as would be expected from an increase in postsynaptic sensitivity, but rather with an increase in the number of quanta released from the terminal (quantum content). An example is shown in Figure 16.10, where the amplitude distribution of the synaptic potentials is plotted before and after potentiation. In this and other experiments, statistical analysis of the amplitude distributions indicated that the mean quantum content of the potentiated synaptic potential was increased.[60,61]

The results shown in Figure 16.10 could be accounted for in theory by the scheme shown in Figure 16.8, where quantal release is the same before and after conditioning, but the observed quantum content is increased because additional postjunctional spines are

[51] Schulman, H. 1995. *Curr. Opin. Neurobiol.* 5: 375–381.

[52] Malenka, R. C. et al. 1989. *Nature* 340: 554–557.

[53] Malinow, R., Schulman, H., and Tsien, R. W. 1989. *Science* 245: 862–866.

[54] Malinow, R., and Malenka, R. C. 2002. *Annu. Rev. Neurosci.* 25: 103–126.

[55] Shi, S. et al. 2001. *Cell* 105: 331–343.

[56] Abraham, W. C., and Williams, J. M. 2003. *Neuroscientist* 9: 463–474.

[57] Lynch, M. A. 2004. *Physiol. Rev.* 84: 87–136.

[58] Nicoll, R. A., and Schmitz, D. 2005. *Nat. Rev. Neurosci.* 6: 863–876.

[59] Veronin, L. L., and Cherubini, E. 2004. *J. Physiol.* 557: 3–12.

[60] Malinow, R., and Tsien, R. W. 1990. *Nature* 346: 177–180.

[61] Beckers, J. M., and Stevens, C. F. 1990. *Nature* 346: 724–729.

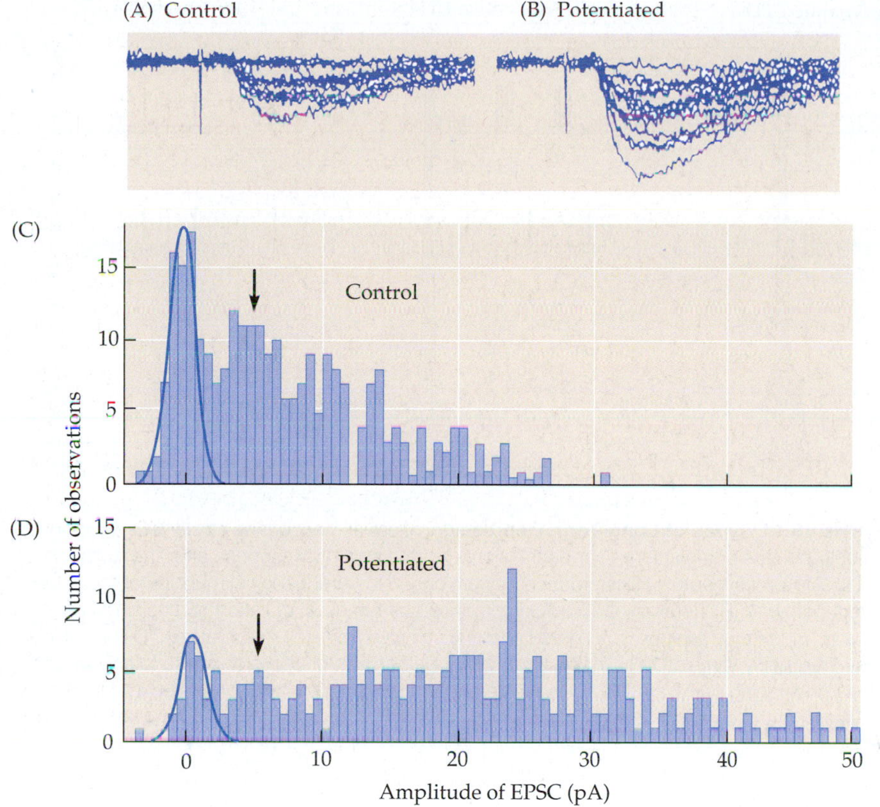

FIGURE 16.10 Change in Quantum Content of synaptic response during LTP. Records from rat hippocampal slice. (A,B) Sixteen superimposed, whole-cell records of synaptic currents in CA1 pyramidal cell before (A) and after (B) conditioning train of stimuli. Note quantal steps in the current amplitudes. After conditioning, the fraction of failures is decreased and there are many more multi-quantal responses. (C,D) Distribution of current amplitudes before (C) and after (D) conditioning. Normal curve is fitted to baseline noise (i.e., failures); arrows indicate mean current produced by a single quantum. After potentiation, the number of failures is reduced and the mean current is increased in amplitude by a factor of almost three, while the single quantum current is unchanged. (After Malinow and Tsien, 1990.)

made responsive by insertion of AMPA receptors. Subsequent experiments indicate that this is not entirely the case. Transmitter release after induction of LTP was examined at single synaptic contacts on CA1 pyramidal cell dendrites.[62] The cells were injected with a calcium-sensitive fluorescent dye, and simultaneous recordings were made of EPSP amplitudes in the cell soma and of changes in fluorescence in single dendritic spines. Single quanta of transmitter released from the presynaptic terminal by presynaptic (Schaeffer collateral) stimulation produced brief calcium transients. The release occurred irregularly, with numerous failures interspersed between single- or multiquantal release. Typically, the probability of release on any one trial was about 20%. After repetitive stimulation, the probability roughly tripled and then declined over a period of more than an hour. During the period of potentiation, the amplitude of the fluorescent signal remained constant, indicating no change in postsynaptic receptor sensitivity at individual release sites.

In summary, LTP in CA1 pyramidal cells has three temporal components. During the conditioning stimulus, calcium influx across the synaptic membrane increases calcium concentration in the cytoplasm and releases additional calcium from the endoplasmic reticulum (ER). Two pathways appear to be involved in release from the ER: one mediated by Ry receptors in dendritic spines, the second by IP_3 receptors along the dendrites. These pathways are associated, respectively, with the first and second components of LTP. The increased intracellular calcium then activates CaMKII, as shown in Figure 16.9, leading to increased conductance of existing receptors and insertion of AMPA receptors in the membrane. The third, long-lasting component of LTP appears to depend primarily on calcium entry into the cell soma through voltage-dependent calcium channels and is associated with gene transcription and expression. Finally, LTP at CA1 synapses also appears to involve an increase in transmitter release from the presynaptic nerve terminal, possibly lasting throughout the first two phases. Details of the mechanisms associated with the presynaptic changes have yet to be revealed.

Long-Term Depression

One of the earliest demonstrations of long-term synaptic depression (LTD), the inverse of LTP, was in relation to the gill-withdrawal reflex in the sea slug *Aplysia*. Experiments by

[62] Enoki, R., Hu, Y., Hamilton, D., and Fine, A. 2009. *Neuron* 62: 242–253.

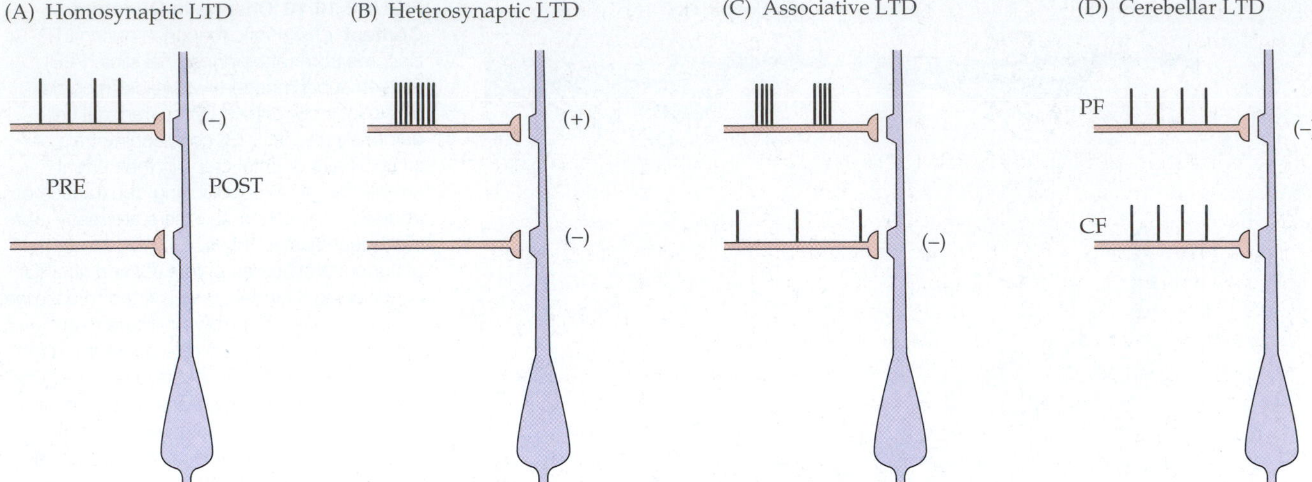

FIGURE 16.11 Types of Long-Term Depression, classified according to stimulus conditions. Symbols indicate potentiation (+) or depression (–) of the synaptic response after the conditioning stimuli. (A) Homosynaptic LTD is produced by prolonged low-frequency stimulation of the same afferent pathway. (B) Heterosynaptic LTD is produced by tetanic stimulation of a neighboring pathway, which may itself be potentiated after the stimulus train. (C) Associative LTD is produced by low frequency stimulation of the test pathway together with brief out-of-phase tetani applied to the conditioning pathway. (D) LTD in the cerebellum is produced by coordinated low-frequency stimulation of the climbing-fiber (CF) and parallel-fiber (PF) inputs to Purkinje cells. (After Linden and Connor, 1995.)

E. R. Kandel and his colleagues indicated that habituation of the reflex by repeated tactile stimulation of the siphon or mantle was with depression of synaptic transmission in an identified gill motor neuron in the abdominal ganglion.[63]

In the vertebrate central nervous system, LTP was first reported at the Schaffer collateral input to CA1 pyramidal cells[64] and has since been studied in a number of other regions of the brain.[65] **Homosynaptic LTD** is a prolonged depression of synaptic transmission produced by previous repetitive activity in the same pathway (Figure 16.11A). It can be induced by a variety of stimulus protocols, such as prolonged low-frequency stimulation (1–5 stimuli/s for 5–15 min), low frequency stimulation with paired pulses, or brief high frequency stimulation (50–100 stimuli/s for 1–5 s). In Schaffer collaterals homosynaptic LTD is blocked by NMDA receptor antagonists[66,67] as well as by pyramidal cell hyperpolarization and postsynaptic injection of calcium chelators. However, in other brain locations NMDA antagonists have no effect. Instead, induction of LTD appears to involve metabotropic glutamate receptors (mGluR).

Heterosynaptic LTD is a prolonged depression of synaptic transmission produced by previous activity in a different afferent pathway to the same cell (Figure 16.11B). This form of LTD was first reported as a correlate of homosynaptic LTP induced in Schaffer collateral input to CA1 pyramidal cells (i.e., induction of LTP in one pathway resulted in depression of transmission at nearby synapses). Later experiments, using chronic recordings from perforant pathway dentate gyrus synapses, showed that the depression could persist for days.[68] The phenomenon is accompanied by a rise in postsynaptic calcium concentration, which in the hippocampus is mediated by NMDA receptor activation.[69] However, it can also be induced by direct postsynaptic depolarization without activation of NMDA receptors. In the dentate gyrus, it is blocked by L-type calcium channel blockers.[70]

Associative LTD has been reported to occur with stimulus protocols similar to those used for associative LTP. Combined weak and strong stimulation of two inputs results in depression of the weakly stimulated input (Figure 16.11C). One difference between the protocols for associative LTP and associative LTD is that during induction of LTD the two stimuli are delivered out of phase. As in associative LTP, postsynaptic depolarizing pulses can substitute for the strong synaptic stimulation.[71] Overall, however, experiments made to demonstrate associative LTD in the hippocampus have yielded inconsistent results, and in

[63] Castellucci, V. et al. 1970. *Science* 167: 1745–1748.

[64] Lynch, G. S., Dunwiddie, T., and Gribkoff, V. 1977. *Nature* 266: 737–239.

[65] Massey, P. V., and Bashir, Z. I. 2007. *Trends Neurosci.* 30: 176–184.

[66] Dudek, S. M., and Bear, M. F. 1992. *Proc. Natl. Acad. Sci. USA* 89: 4363–4367.

[67] Mulkey, R. M., and Malenka, R. C. 1992. *Neuron* 9: 967–975.

[68] Krug, M. et al. 1985. *Brain Res.* 360: 264–272.

[69] Barry, M. F. et al. 1996. *Hippocampus* 6: 3–8.

[70] Christie, B. R., and Abraham, W. C. 1994. *Neurosci. Lett.* 167: 41–45.

[71] Debanne, D., and Thompson, S. M. 1996. *Hippocampus* 6: 9–16.

some cases have required special stimulating protocols, such as prior priming by so-called theta (5 stimuli/s) stimulation.

LTD in the Cerebellum

One important site of LTD is the cerebellar cortex. There, Purkinje cells receive excitatory input from two sources: parallel fibers (PF) that arise from granule cells and form synapses on secondary and tertiary dendrites; and climbing fibers (CF) from the inferior olive nucleus, which make strong synaptic connections on the soma and proximal dendritic tree. Parallel fibers use glutamate as a transmitter and their synapses contain both mGluR and AMPA receptors. The transmitter used by the climbing fibers has not been definitely identified. No NMDA receptors are found in the adult cerebellum.[72]

Cerebellar LTD was first studied by Ito and his colleagues.[73] They applied a series of low-frequency (1–4/s) paired stimuli to parallel fiber and climbing fiber pathways (Figure 16.11D) for about 5 minutes. Subsequent responses to parallel fiber stimulation were depressed for several hours. In addition, when application of glutamate to the dendritic field was paired with climbing fiber stimulation, subsequent responses to glutamate were depressed, suggesting that the phenomenon was mediated by postsynaptic changes. Reliable demonstration of cerebellar LTD proved somewhat difficult in vivo, but was established as an unambiguous phenomenon in cerebellar slice preparations[74] and in culture.[75] It was also shown, in both slice preparations and cerebellar cultures, that Purkinje cell depolarization, producing calcium action potentials in the dendrites (see Chapter 7), could substitute for climbing fiber stimulation in inducing LTD.[76,77] However, neither climbing fiber stimulation nor depolarization alone is effective in inducing LTD; co-activation of glutamate receptors, either by parallel fiber stimulation or by direct application of glutamate, is required. In the case of parallel fiber stimulation, LTD is input specific, that is, only the stimulated inputs are depressed. LTD induction is blocked by postsynaptic calcium chelators,[78] and there is a large calcium accumulation following climbing fiber stimulation.[79]

Mechanisms Underlying LTD

Conditions under which LTD can be produced vary considerably, depending on which type is being studied and in what location. One consistent feature is that LTD, like LTP, depends on postsynaptic accumulation of calcium.[63] In the hippocampus, calcium entry appears to be primarily through NMDA receptors, although heterosynaptic LTD can be induced by depolarization alone without receptor activation and is attenuated by L-type calcium channel blockers. This suggests that while local depolarization and calcium accumulation through NMDA receptors produces LTP at the activated input, spread of depolarization to adjacent synaptic regions can produce LTD by calcium entry through voltage-gated calcium channels. In other brain regions free calcium concentration may be elevated by release from intracellular stores by IP_3 after activation of metabotropic glutamate receptors (see Chapter 12). In cerebellar Purkinje cells, where NMDA receptors are absent, calcium entry is through voltage-sensitive calcium channels that generate dendritic action potentials.

Why does calcium accumulation produce LTP in some circumstances and LTD in others? At the moment, there is no clear answer to that question except that the difference appears to be related to the increase in concentration. A relatively large increase results in LTP, a smaller increase in LTD. In accordance with this idea is the observation that a stimulus barely able to induce homosynaptic LTP in CA1 cells with normal extracellular calcium concentration produces LTD when extracellular calcium is reduced.[65]

The mechanisms whereby increased cytoplasmic calcium concentration lead to LTD expression appear to be the exact reverse of those responsible for LTP. LTD expression involves a reduction in postsynaptic sensitivity to applied glutamate and a reduction in amplitude of miniature excitatory synaptic potentials.[80,81] Furthermore, during LTD in cultured hippocampal cells, there is a decrease in the number of AMPA receptors clustered in the postsynaptic membrane.[82] These changes are accompanied by dephosphorylation of the GluA1 subunit of the AMPA receptors.[83] As a general rule, during LTP trafficking of AMPA receptors into the membrane, and increased receptor sensitivity is associated with

[72] Levenes, C., Daniel, H., and Crepel, F. 1998. *Prog. Neurobiol.* 55: 79–91.

[73] Ito, M., Sakurai, M., and Tongroach, P. 1982. *J. Physiol.* 324: 113–134.

[74] Sakurai, M. 1987. *J. Physiol.* 394: 463–480.

[75] Hirano, T. 1990. *Neurosci. Lett.* 119: 141–144.

[76] Crepel, F., and Jaillard, D. 1991. *J. Physiol.* 432: 123–141.

[77] Hirano, T. 1990. *Neurosci. Lett.* 119: 145–147.

[78] Sakurai M. 1990. *Proc. Natl. Acad. Sci. USA* 87: 3383–3385.

[79] Ross, W. N., and Werman, R. 1987. *J. Physiol.* 389: 319–336.

[80] Oliet, S., Malenka, R. C., and Nicoll, R. A. 1996. *Science* 271: 1294–1297.

[81] Murashima, M., and Hirano, T. 1999. *J. Neurosci.* 19: 7326–7333.

[82] Carrol, R. C. et al. 1999. *Nat. Neurosci.* 2: 454–460.

[83] Lee, H-K. et al. 1998. *Neuron* 21: 1151–1162.

protein kinase activity, whereas during LTD trafficking of receptors out of the membrane and reduced sensitivity involves activation of protein phosphatase.

Presynaptic LTD

Habituation of the gill withdrawal reflex in *Aplysia* was initially shown to be accompanied by a reduction in the quantum content of the EPSP in the gill motor neuron.[84] This finding bolstered the idea, widely held at the time, that long-term plasticity, like its short-term counterpart, was essentially a presynaptic phenomenon. However, later experiments on LTD in *Aplysia* revealed that postsynaptic mechanisms were involved as well, mimicking those seen in the vertebrate CNS.[85]

In the hippocampus, there is evidence that supports presynaptic involvement in LTD in both CA1 and CA3 neurons.[86,87] One indication of such involvement is provided by the experiments described previously in relation to LTP, in which transmitter release was examined at single synaptic contacts on CA1 pyramidal cell dendrites.[61] As we have already discussed, during LTP the probability of release at any given contact increased markedly during LTP. Similarly, after induction of LTD, the probability of release at the dendritic synaptic contacts was reduced by an amount commensurate with the reduction in amplitude of EPSPs recorded from the cell soma.

Significance of Changes in Synaptic Efficacy

It is a matter of basic belief amongst those who study nervous system function that learning and memory involve long-term changes in synaptic efficacy and, for this reason, the mechanisms underlying LTP and LTD are of particular interest. This interest is strengthened further because both phenomena exhibit a characteristic postulated by Donald Hebb to be required for associative learning,[88] namely that increases in synaptic strength should occur when the presynaptic and postsynaptic elements are coactive. Synapses that exhibit this property are known as "Hebbian," and it is sometimes assumed that if the requirement has been satisfied, then learning has occurred. At any rate, a number of correlations have been established between changes in behavior and long-term changes in synaptic efficacy in several regions of the brain.[56,89]

For example, spatial learning in intact animals and LTP in hippocampal slices display a number of similarities.[90–92] Both can be blocked by antagonists of NMDA receptors or metabotropic glutamate receptors and also by inhibitors of calcium–calmodulin protein kinase. However, the nature of the behavioral deficit associated with the block is not always clear. For example, rats under NMDA antagonists have general sensorimotor disturbances that interfere with negotiating a water maze (suggesting that learning ability has been compromised), but they can learn it readily if they first become familiar with the general requirements of the task.[93] Thus, NMDA receptor-mediated LTP does not seem to be an essential requirement. Similar ambiguities have been encountered with gene deletions—some that eliminate LTP produce a deficit in the spatial learning ability, others do not.[94]

There is growing evidence for the idea that LTP in the amygdala might be a substrate for aversive (or fear) conditioning. Rats trained to associate foot shock with an auditory tone exhibit an exaggerated auditory startle reflex, and cells in the amygdala show an LTP-like increase in their synaptic response to electrical stimulation of the auditory pathway from the medial geniculate nucleus.[95,96] Conversely, induction of LTP at the same synapses by electrical stimulation results in an increase in the response to auditory stimuli.[69] Both effects are blocked by NMDA receptor antagonists.[97,98] In conclusion, while an unequivocal relation between LTP and spatial learning tasks has not been established, LTP may play a role in more discrete learning paradigms such as classical conditioning.

---

[84] Castellucci, V., and Kandel, E. R. 1974. *Proc. Natl. Acad. Sci. USA* 71: 5004–5008.

[85] Glanzman, D. L. 2009. *Neurobiol. Learn. Mem.* 92: 147–154.

[86] Berretta, N., and Cherubini, E. 1998. *Eur. J. Neurosci.* 10: 2957–2963.

[87] Domenici, M. R., Berretta, N., and Cherubini, E. 1998. *Proc. Natl. Acad. Sci. USA* 95: 8310–8315.

[88] Hebb, D. O. 1949. *The Organization of Behavior.* Wiley, New York.

[89] Kessels, H. W., and Malinow, R. 2009. *Neuron* 61: 340–350.

[90] Izquierdo, I., and Medina, J. H. 1995. *Neurobiol. Learn. Mem.* 63: 19–32.

[91] Martinez, J. L., Jr., and Derrick, B. E. 1996. *Annu. Rev. Psychol.* 47: 173–203.

[92] Elgersma, Y., and Silva, A. J. 1999. *Curr. Opin. Neurobiol.* 9: 209–213.

[93] Cain, D. P. 1998. *Neurosci. Biobehav. Rev.* 22: 181–193.

[94] Holscher, C. 1999. *J. Neurosci. Res.* 58: 62–75.

[95] Rogan, M. T., Staubil, U. V., and LeDoux, J. E. 1997. *Nature* 390: 604–607.

[96] McKernan, M. G., and Shinnick-Gallagher, P. 1997. *Nature* 390: 607–611.

[97] Miserendino, M. J. et al. 1990. *Nature* 345: 716–718.

[98] Fanselow, M. S., and Kim, J. J. 1994. *Behav. Neurosci.* 108: 210–212.

SUMMARY

- Short periods of synaptic activation can result in facilitation, depression, or augmentation of transmitter release, or a combination of these effects.
- Facilitation decays gradually over a few hundred milliseconds, while synaptic depression and augmentation persist for several seconds.
- Facilitation is related to a persistent increase in cytoplasmic calcium concentration in the presynaptic terminal.
- Longer periods of repetitive stimulation result in post-tetanic potentiation (PTP) of transmitter release, which can last for tens of minutes and, like facilitation, is mediated by an increase in presynaptic terminal calcium concentration.
- In various parts of the central nervous system repetitive stimulation can result in long-term potentiation (LTP) or long-term depression (LTD) of synaptic strength.
- The change in synaptic efficacy during LTP or LTD may be homosynaptic, involving only the stimulated input, or heterosynaptic, affecting adjacent synapses on the same dendrite. In addition, heterosynaptic effects may be associative, requiring the coordinate activation of both synapses.
- LTP is produced by an increase in calcium concentration in the postsynaptic cell and involves both the insertion of new receptors into the postsynaptic membrane and an increase in receptor sensitivity.
- LTD also requires an increase in postsynaptic calcium concentration and is mediated by a decrease in receptor number and sensitivity.
- Both LTP and LTD can also involve changes in transmitter release from the presynaptic terminal.
- Although there are some correlations between LTP and LTD and behavioral tasks involved in learning, no unequivocal relation between these long-term synaptic changes and memory formation have been established.

Suggested Reading

General Reviews

Lynch, M. A. 2004. Long-term potentiation and memory. *Physiol. Rev.* 84: 87–136.

Malenka, R. C., and Nicoll, R. A. 1999. Long-term potentiation—A decade of progress? *Science* 285: 1870–1874.

Massey, P. V., and Bashir, Z. I. 2007. Long-term depression: multiple forms and implications for brain functions. *Trends Neurosci.* 30: 176–184.

Raymond, C. R. 2007. LTP forms 1, 2, and 3: Different mechanisms for the "long" in long-term potentiation. *Trends Neurosci.* 30: 167–175.

Südhof, T., and Malenka, R. C. 2008. Understanding synapses: Past, present, and future. *Neuron* 60: 468–476.

Original Papers

Abraham, W. C., and Williams, J. M. 2003. Properties and mechanisms of LTP maintenance. *Neuroscientist* 9: 463–474.

Barrionuevo, G., and Brown, T. H. 1983. Associative long-term potentiation in hippocampal slices. *Proc. Natl. Acad. Sci. USA* 80: 7347–7351.

Bliss, T. V. P., and Lømo, T. 1973. Long-lasting potentiation of synaptic transmission in the dentate of the anesthetized rabbit following stimulation of the perforant path. *J. Physiol.* 232: 331–356.

del Castillo, J., and Katz, B. 1954. Statistical factors involved in neuromuscular facilitation and depression. *J. Physiol.* 124: 574–585.

Enoki, R., Hu, Y., Hamilton, D., and Fine, A. 2009. Expression of long-term plasticity at individual synapses in hippocampus is graded, bidirectional, and mainly presynaptic: Optical quantal analysis. *Neuron* 62: 242–253.

Ito, M., Sakurai, M., and Tongroach, P. 1982. Climbing fibre induced depression of both mossy fibre responsiveness and glutamate sensitivity of cerebellar Purkinje cells. *J. Physiol.* 324: 113–134.

Kessels, H. W. and Malinow, R. 2009. Synaptic AMPA receptor plasticity and behavior. *Neuron* 61: 340–350.

Mallart, A., and Martin, A. R. 1967. Analysis of facilitation of transmitter release at the neuromuscular junction of the frog. *J. Physiol.* 193: 679–697.

Shi, S. H., Hayashi, Y., Petralia, R. S., Zaman, S. H., Wenthold, R. J., Svoboda, K., and Malinow, R. 1999. Rapid spine delivery and redistribution of AMPA receptors after synaptic NMDA receptor activation. *Science* 284: 1811–1816.

Weinrich, D. 1970. Ionic mechanisms of post-tetanic potentiation at the neuromuscular junction of the frog. *J. Physiol.* 212: 431–446.

Zengel, J. E., and Magleby, K. L. 1982. Augmentation and facilitation of transmitter release. A quantitative description at the frog neuromuscular junction. *J. Gen. Physiol.* 80: 582–611.

PART IV

Integrative Mechanisms

Beyond the mechanisms that individual neurons, glial cells, and synapses use to communicate, of prime interest for neurobiology is the behavior of an animal as a whole. The first of the two chapters in this section, Chapter 17, deals with an essential function of the nervous system: the constant performance of "housekeeping tasks." The cardiovascular, respiratory, and intestinal needs of an animal must be regulated by the autonomic nervous system as the animal flies, swims, runs, feeds, or walks. In this chapter, the biophysical, molecular, and chemical mechanisms described previously for neurons come together. Indeed, many key discoveries, such as the identification of transmitters, were first made at sympathetic and parasympathetic synapses. In addition, it will be shown that the autonomic nervous system is not truly autonomous, and interacts with emotions and hormonal aspects of the functions of the brain.

Similarly, Chapter 18 describes how the complex behavior of relatively simple animals is brought about by the integrated actions of nerve cells. Examples include the way in which ants and bees forage for food and then, in an extraordinary manner, find their way home by using receptors that respond to ultraviolet light or to magnetic fields. In the leech, the circuits between identified individual sensory neurons, interneurons, and motor neurons have been traced and their biophysical properties analyzed. Such information provides a basis for explaining how complex movements are initiated and carried out by the animal.

CHAPTER 17
Autonomic Nervous System

The autonomic nervous system controls essential functions of the body. Thus, neurons of the autonomic nervous system supply smooth muscles in the eye, lung, gut, blood vessels, bladder, genitalia, and uterus. They regulate glandular secretion, blood pressure, heart rate, cardiac output, and body temperature, as well as food and water intake. In contrast to speedy conduction and muscle contractions required for limb movements, these housekeeping or vegetative functions are slower, last longer and are not under the direct control of the will.

Four distinct groupings of neurons make up the autonomic nervous system. The **sympathetic division** consists of neurons with myelinated axons that leave the spinal cord through ventral roots from thoracic and lumbar segments. They form synapses on nerve cells in sympathetic ganglia situated alongside and at a distance from the spinal cord, and on chromaffin cells in the adrenal medulla. Sympathetic postganglionic axons are unmyelinated and extend over long distances to target areas. The **parasympathetic division** consists of axons leaving through certain cranial and sacral nerves. They form synapses in ganglia situated within or close to the target organs. Parasympathetic postganglionic axons are, in general, shorter than those of the sympathetic nervous system. A third, highly complex division consists of millions of nerve cells in the intestinal wall, the **enteric nervous system**. The fourth division comprises neurons in the spinal cord, hypothalamus, and brainstem. Within the central nervous system (CNS), boundaries between the autonomic and somatic nervous systems are not sharply defined.

Synaptic transmission in the autonomic nervous system is extraordinary in its diversity, and makes use of all the known transmitters. Principles of transmission and integration that were first revealed at autonomic synapses include the chemical nature of synaptic transmission, reuptake of transmitter, autoreceptors on presynaptic terminals, co-release of more than one transmitter at a single terminal, and the role of second messengers. In autonomic ganglia, the transmitters include acetylcholine (ACh), peptides, and dopamine. Parasympathetic postganglionic nerve terminals release acetylcholine as the primary transmitter, which acts on muscarinic receptors in the target organs, and, in addition, they release nitric oxide (NO) and peptides. Postganglionic sympathetic neurons release norepinephrine, epinephrine, acetylcholine, purines, or peptides as primary transmitters. Autonomic neurons co-release peptides together with adenosine triphosphate (ATP).

Whereas much is known about the regulation of activity in smooth muscle and gland cells, less information is available about integrative mechanisms within the CNS that regulate autonomic functions. The periodic 24-hour cycle of activity, known as circadian rhythm, influences many autonomic functions. Experiments in which recordings were made from specific neurons in the retina and in the hypothalamus have revealed cellular mechanisms that generate the rhythm.

The name autonomic implies an independent system that runs on its own. In part, this is true. The autonomic nervous system controls blood vessels, the heart, glands, and smooth muscle throughout the gut, bronchi, bladder, and spleen, without our having to make conscious decisions. By a simple act of will one cannot increase the diameter of the pupil or the blood flow through one's little finger. It is possible of course to cheat the system, to some extent, by the use of tricks; thus, the generation of emotion by deliberately thinking of an exam, a dental appointment, or a film starlet can stimulate the sympathetic nervous system to increase the heart rate.

In practice, the performance of the autonomic nervous system is closely linked to voluntary movements. Exercise results in appropriate diversion of blood to the muscles and in stimulation of sweat glands; the action of standing up from a recumbent position requires circulatory adjustments so as to maintain blood flow to the brain. Ingestion of a meal reroutes blood to the stomach and intestines. By turning activity on or off in a widespread group of target cells, the autonomic nervous system deals with the housekeeping and maintenance work of the body. The brain establishes the priorities, setting in motion digestion, reproduction, micturition, defecation, or focusing in dim light, through mechanisms that are not decided by our conscious will. Of key concern for human beings are disorders of the autonomic nervous system that lead to conditions such asthma, constipation, diarrhea, ulcers, hypertension, heart disease, stroke, and retention of urine (or lack thereof). We simply take for granted the regulation of essential bodily functions. For instance, it is remarkable that all the readers of this book have body temperatures of about 37°C and blood pressure values of about 120/80 mm Hg, in spite of their very different metabolic rates (and states of mind while reading).

Recent experiments and classical work on the autonomic nervous system represent such an extensive and varied field that a comprehensive review is impossible in this chapter. Indeed, entire textbooks[1,2] and specialized journals[3–5] are devoted to important functions of the autonomic nervous system. A large amount of information is available about mechanisms that control the enteric nervous system and bladder, the diameter of the pupil, secretion by glands, and that regulate respiration, temperature, body weight, appetite, and reproduction.[6–8]

In this chapter, as in others, the main emphasis is on a few selected examples that illustrate cellular, molecular, and integrative mechanisms. It will be shown that although much is now known about the autonomic nervous system, many open questions remain, particularly about integrative mechanisms within the CNS. There the distinction between autonomic and somatic systems has no hard-and-fast boundaries. It is convenient to begin with a brief description of the principal features of the peripheral autonomic nervous system.

Functions under Involuntary Control

Sympathetic and Parasympathetic Nervous Systems

The principal anatomical features are shown in Figure 17.1. Virtually all the organs of the body are supplied by autonomic neurons. Even skeletal muscle fibers, which receive no direct innervation, are dependent on the autonomic nervous system—their blood supply is regulated according to need. Sympathetic preganglionic neurons are situated in the intermediolateral horn of the spinal cord of segments T1 to L3. Their myelinated axons pass through ventral roots to form synapses in ganglia situated alongside the vertebral column and peripheral to it (Figure 17.1A). From these ganglia, unmyelinated axons run to the tissues. By contrast, the parasympathetic outflow is restricted to cranial nerves III, VII, IX, and X and sacral roots S2, S3, and S4 (Figure 17.1B). The parasympathetic ganglia are located close to or in the tissues themselves. Hence, the parasympathetic myelinated preganglionic axon is long, whereas the unmyelinated postganglionic axon is short.

The actions of the two systems are often, but not always, antagonistic (Table 17.1). For example, excitation of sympathetic neurons leads to dilatation of the pupil, increased heart rate, and decreased gut motility. Parasympathetic excitation produces opposite effects, such as pupillary constriction, slowed heart rate, and increased gut motility. On the other hand, glandular secretion can be increased by activation of either system. Both systems can cause smooth muscles to contract or to relax, depending on the transmitter that is released and the types of receptors that are present on the muscle.

[1] Burnstock, G., ed. 1990–19. *The Autonomic Nervous System* 8 vols. Harwood Academic, New Jersey.

[2] Robertson, D. ed. 2004. *Primer on the Autonomic Nervous System*, Academic Press, London.

[3] *J. Autonomic Nervous System*

[4] *Autonomic Neuroscience*

[5] *J. Autonomic Pharmacology*

[6] Cooper, J. R., Bloom, F. E., and Roth, R. H. 2002. *The Biochemical Basis of Pharmacology*. Oxford University Press, New York.

[7] Fowler, C. J., Griffiths, D., and de Groat, W. C. 2008. *Nat. Rev. Neurosci.* 25: 7324–7332.

[8] Spyer, K. M., and Gourine, A. V. 2009. *Philos. Trans. R. Soc. Lond., B, Biol. Sci.* 364: 2603–2610.

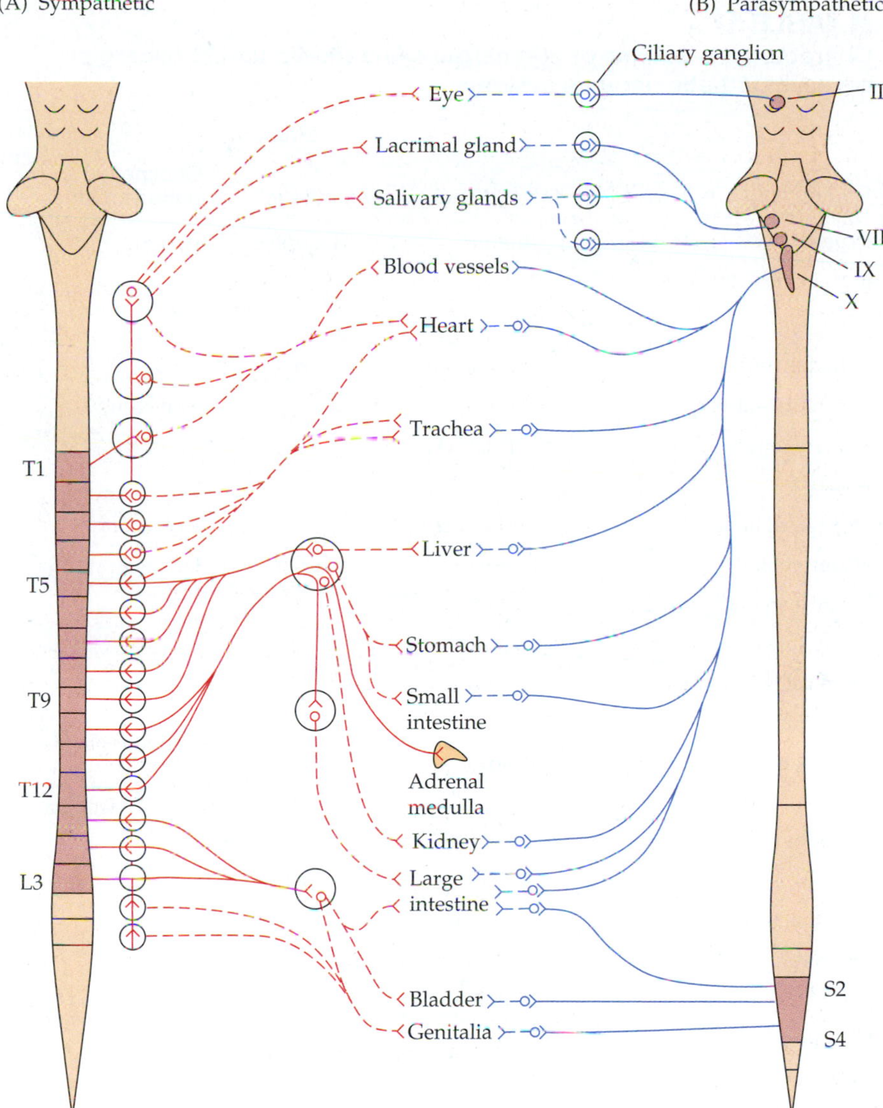

FIGURE 17.1 The Autonomic Nervous System and the target organs that it innervates. (A) Principal features of the sympathetic nervous system, including the paravertebral ganglia, peripheral ganglia, and adrenal medulla. (B) The parasympathetic nervous system has more restricted output and targets that it innervates in comparison to the sympathetic nervous system. Solid lines are myelinated preganglionic axons; dashed lines are unmyelinated postganglionic axons.

A major difference between the two autonomic systems is that, while the parasympathetic nervous system functions in a focused manner, the sympathetic nervous system tends to be thrown into action as a whole, with widespread generalized consequences for the body. The sympathetic system is typically activated under conditions of *fright, fight,* and *flight* as well as during intense exercise. The symptoms are familiar; they include dilated pupils, dry mouth, pounding heart, sweating, and enhanced emotions. The systemic effects of sympathetic neuronal activity are enhanced further by chromaffin cells in the adrenal medulla. These cells are modified ganglionic neurons. They receive cholinergic input from preganglionic axons and secrete epinephrine, norepinephrine, peptides, and ATP as hormones into the bloodstream.[9,10] Epinephrine in the blood reinforces and extends sympathetic activity. It can reach and bind to receptors in smooth muscle of bronchi, far from sympathetic nerve endings; epinephrine also binds to receptors in blood vessels that are insensitive to norepinephrine (see Box 17.1). Unlike norepinephrine, epinephrine produces vasodilatation as well as contraction of blood vessels.

By contrast, the parasympathetic nervous system is more focused in its activity. It is surely a considerable advantage that the pupil can constrict in bright light and that the lens of the eye accommodate for viewing nearby objects selectively, without concomitant and ill-timed arousal of bladder contractions or even more embarrassing parasympathetic effects.

[9]Crivellato, E., Nico, B., and Ribatti, D. 2008. *Anat. Rec. (Hoboken)* 291: 1587–1602.

[10]Fulop, T., Radabaugh, S., and Smith, C. 2005. *J. Neurosci.* 25: 7324–7332.

TABLE 17.1
Characteristic actions of adrenergic sympathetic and cholinergic parasympathetic nervous systems

	Effect of		
	Adrenergic sympathetic		Cholinergic parasympathetic
Organ	Action[a]	Receptor[b]	Action
Eye			
Iris			
Radial muscle	Contracts	α_1	—
Circular muscle	—	—	Contracts
Ciliary muscle	(Relaxes)	β	Contracts
Heart			
Sinoatrial node	Accelerates	β_1	Decelerates
Contractility	Increases	β_1	Decreases (atria)
Vascular smooth muscle			
Skin, splanchnic vessels	Contracts	α	—
Skeletal muscle vessels	Relaxes	β_2	—
Nerve endings	Inhibits release	α_2	—
Bronchiolar smooth muscle	Relaxes	β_2	Contracts
Gastrointestinal tract			
Smooth muscle			
Walls	Relaxes	α_1, β_2	Contracts
Sphincters	Contracts	α_1	Relaxes
Secretion	—	—	Increases
Myenteric plexus	Inhibits	α	Activates
Genitourinary smooth muscle			
Bladder wall	Relaxes	β_2	Contracts
Sphincter	Contracts	α_1	Relaxes
Metabolic functions			
Liver	Gluconeogenesis	α/β_2	—
	Glycogenolysis	α/β_2	—

[a]Accounts of the actions of the autonomic nervous system on target organs listed in this table that are not dealt with in this chapter are given in reviews and textbooks of physiology and pharmacology (see references in text).

[b]Not all the adrenergic receptors or effector cells are included; purinergic, peptidergic, and cholinergic mechanisms are dealt with in the text. Whereas epinephrine acts on all the adrenergic receptors, norepinephrine is effective on α_1, α_2, and β_1-receptors but only weakly on β_2. The various types of adrenergic and muscarinic receptors are characterized by the specific agonists and antagonists that bind to them and by their molecular structures.

Synaptic Transmission in Autonomic Ganglia

Certain mechanisms of transmission in the autonomic nervous system have already been described in earlier chapters (see Chapters 11, 12, and 14). These include the co-release of multiple transmitters from nerve endings, modulatory actions of autonomic transmitters, the properties of receptors that use second messengers, and the effects of acetylcholine and epinephrine on cardiac muscle. The way in which such mechanisms interact to influence signaling is well illustrated by experiments made on synaptic transmission in autonomic ganglia. These synapses also serve to demonstrate integrative processes occurring within the central nervous system that are even more complex.

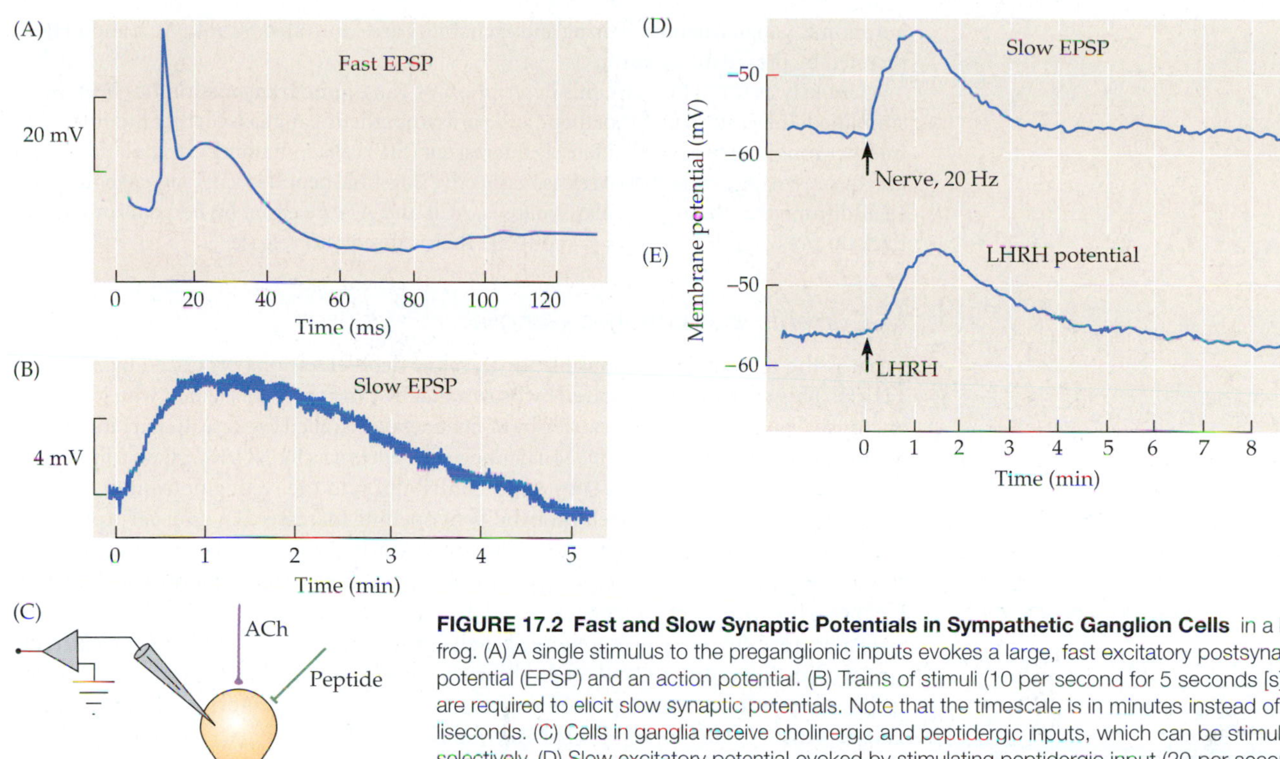

FIGURE 17.2 Fast and Slow Synaptic Potentials in Sympathetic Ganglion Cells in a bullfrog. (A) A single stimulus to the preganglionic inputs evokes a large, fast excitatory postsynaptic potential (EPSP) and an action potential. (B) Trains of stimuli (10 per second for 5 seconds [s]) are required to elicit slow synaptic potentials. Note that the timescale is in minutes instead of milliseconds. (C) Cells in ganglia receive cholinergic and peptidergic inputs, which can be stimulated selectively. (D) Slow excitatory potential evoked by stimulating peptidergic input (20 per second for 5 s). As in (B), the depolarization lasts several minutes. (E) Application of luteinizing hormone-releasing hormone (LHRH) to the same neuron by pressure from a micropipette. The peptide mimics the action of the naturally released transmitter. (After Kuffler, 1980.)

Autonomic ganglia constitute relay stations, the functional significance of which is not immediately obvious. At first glance the mechanism of direct, rapid transmission is similar to that at the skeletal neuromuscular junction.[11] Each presynaptic impulse releases acetylcholine, which acts on nicotinic receptors in the postsynaptic cell to open channels and produce a fast depolarization (see Chapter 11).[12] As at the nerve–muscle junction, a single presynaptic action potential is followed by one in the postsynaptic cell (Figure 17.2). The nicotinic receptors, however, are different in their subunit composition from those at the motor end plate, being made up of alpha and beta subunits but without gamma, delta, or epsilon subunits.[13]

Transmission at the synapse between the preganglionic axon and the ganglion cell is far more elaborate than would at first appear. Thus, a very different picture emerges with repetitive stimulation of the presynaptic axon at frequencies comparable to those occurring normally in the animal. Under these conditions, the ganglion is not simply a throughway, but is a site of complex interactions. With trains of impulses, prolonged depolarizing and hyperpolarizing, long-latency synaptic potentials arise in the ganglion cell.[14,15] They summate to produce a steady, subthreshold depolarization maintained for seconds, minutes, or even hours. During the depolarization a single presynaptic action potential can now give rise to multiple postsynaptic impulses. Both the fast and the slow synaptic potentials are evoked by the release of acetylcholine from the presynaptic nerve terminals. As before, the fast, direct synaptic potential results from activation of nicotinic ACh receptors. The slow potential is due to activation of muscarinic ACh receptors that are coupled to G proteins[16,17] (see Chapter 12). The low level of cholinesterase activity that is present in the ganglion does not greatly enhance nicotinic transmission, unlike its effect at neuromuscular synapses. It does however facilitate the activation of muscarinic receptors.

Kuffler and his colleagues found that a second transmitter also contributes to slow depolarizing and hyperpolarizing synaptic potentials. Certain presynaptic axons release a decapeptide that resembles luteinizing hormone-releasing hormone (LHRH). (LHRH is also known as gonadotropin-releasing hormone [GnRH]; see Figure 17.8.) Hence, in

[11] Gibbins, I. L., and Morris, J. L. 2006. *Cell Tissue Res.* 326: 205–220.

[12] McLachlan, E. M., ed. 1995. *Autonomic Ganglia.* Gordon and Breach, London.

[13] Ullian, E. M., McIntosh, J. M., and Sargent, P. B. 1997. *J. Neurosci.* 17: 7210–7219.

[14] Kuffler, S. W. 1980. *J. Exp. Biol.* 89: 257–286.

[15] Jan, Y. N., Jan, L. Y., and Kuffler, S. W. 1980. *Proc. Natl. Acad. Sci. USA* 77: 5008–5012.

[16] Brown, D. A. et al. 2007. *J. Physiol.* 582: 917–925.

[17] Suh, B. C., and Hille, B. 2008. *Annu. Rev. Biophys.* 37: 175–195.

autonomic ganglia, neuronal firing and excitability are controlled by both ACh and LHRH, secreted by presynaptic neurons.

Unlikely as it might seem, this description of ganglionic transmission has been oversimplified. Integration in sympathetic and parasympathetic ganglia is further modulated by interneurons known as small intensely fluorescent (SIF) cells containing catecholamines and also by presynaptic endings that release vasoactive intestinal peptide (VIP) and enkephalins.[18] In addition, as at neuromuscular synapses, ACh and ATP released by nerve terminals act back on presynaptic receptors to modulate their subsequent release.[19]

M-Currents in Autonomic Ganglia

What is the mechanism responsible for the slow depolarizations produced by ACh and LHRH? This question was resolved by Brown, Adams, and their colleagues, who identified an unusual potassium current carried by **M-channels**—so called because they are influenced by muscarinic ACh receptors.[20,21] M-channels, also known as KCNQ/K_V7, have a high open probability at rest and make a substantial contribution to the resting potassium conductance. With depolarization, their probability of opening increases. An unusual property of the M potassium channels is that activation of muscarinic receptors causes these channels to *close* (Figure 17.3). As a consequence, the resting influx of sodium ions is no longer in balance with potassium efflux and the cell depolarizes.

After their discovery in autonomic ganglia, M-channels were found in neurons in the spinal cord, hippocampus, and cerebral cortex.[22,23] M-channel closure produced by transmitters results from activation of phospholipase C.[17,23,24] This in turn leads to hydrolysis and depletion of membrane phosphatidylinositol-4,5-bisphosphate—a molecule that is required for channel opening (see Chapter 12).

What is the physiological importance of M-currents in autonomic ganglia? The principal effect of the M-current is to raise the threshold for firing an action potential. Thus, when

[18]Prud'homme, M. J. et al. 1999. *Brain Res.* 821: 141–149.

[19]Rogers, M., and Sargent, P. B. 2003. *Eur. J. Neurosci.* 18: 2946–2956.

[20]Adams, P. R., and Brown, D. A. 1980. *Brit. J. Pharmacol.* 68: 353–355.

[21]Delmas, P., and Brown, D. A. 2005. *Nat. Rev. Neurosci.* 6: 850–862.

[22]Hansen, H. H. et al. 2008. *J. Physiol.* 586: 1823–1832.

[23]Brown, D. A., and Passmore, G. M. 2009. *Brit. J. Pharmacol.* 156: 1185–1195.

[24]Hernandez, C. C. et al. 2008. *J. Physiol.* 586: 1811–1121.

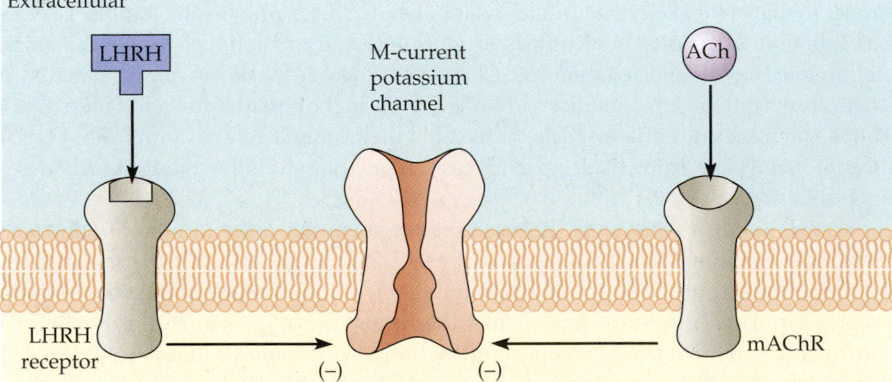

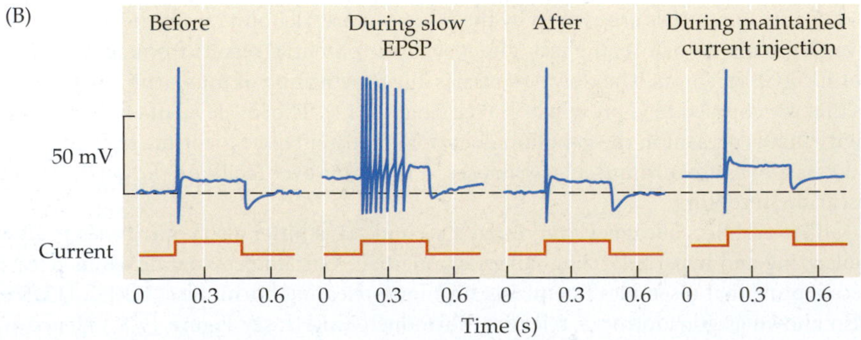

FIGURE 17.3 Inhibition of Potassium Currents in Sympathetic Ganglion Cells modulates responses to presynaptic stimulation. (A) Binding of ACh to muscarinic receptors (mAChR) and binding of LHRH to its receptor both inhibit M-current potassium channels. (B) The effect of the decrease in the M-current during the slow synaptic potential is to increase the excitability of the ganglion cell. Depolarizing current pulses applied through the microelectrode (lower traces) before and after a slow synaptic potential produce a single action potential. During the slow potential, the same current pulse elicits a burst of action potentials. Depolarizing the ganglion cell (to the same extent as occurs during closure of M-channels) by injecting a maintained current has no such effect on the responsiveness of the cell. (After Jones and Adams, 1987.)

a neuron with an M-current is depolarized, it only fires one or two action potentials then becomes silent. The steps that cause this to happen are:

> initial depolarization opens M-channels→M-channels generate an outward potassium current→potassium current counteracts inward sodium current in initial phase of action potential→prevents full action potential from developing

When the M-current is suppressed through closure of the K channels by the transmitter, this series of steps does not happen and the neuron fires repetitively (see Figure 17.3). In other words, the neuron switches from **phasic-firing** to **tonic-firing** mode when the M-current is suppressed. Some sympathetic neurons (e.g., in prevertebral ganglia) do not have M-currents and they do fire tonically rather than phasically.[25]

M-channels have a major effect on firing patterns in the autonomic nervous system. In cells with large M-currents, such as those that cause dilatation of the pupil, presynaptic inputs do not fire tonically and the output is roughly one to one.[26] By contrast, cells in lumbar ganglia that cause vasoconstriction receive a continuous bombardment from presynaptic inputs. This inhibits their M-currents through the muscarinic effect of acetylcholine. Accordingly, they fire tonically at varying frequencies, depending on the input, and produce greater or reduced tonic vasoconstriction. These results fit with the special requirement of discontinuous, episodic dilatation of the pupil on demand and maintained control of blood vessel diameter. Tonic and phasic discharges have additional effects; they can determine which types of transmitter are to be released by the terminals of a ganglion cell onto its targets.

Transmitter Release by Postganglionic Axons

Although acetylcholine is the principal transmitter used by postganglionic parasympathetic axons, they can co-release nitric oxide (NO)[27] and peptides (Figures 17.4 and 17.5) For example, ACh released by parasympathetic axons causes salivary glands to secrete by acting on muscarinic receptors. With high-frequency stimulation, the same axons also liberate a peptide called vasoactive intestinal peptide (VIP). VIP, originally found in gut and brain, causes vasodilatation, increased intracellular calcium concentration, and increased secretion of saliva that is not blocked by atropine, which is an antagonist for muscarinic receptors.[28]

For sympathetic postganglionic neurons, norepinephrine is the principal transmitter. Sympathetic axons innervating sweat glands and blood vessels in skeletal muscle, however, secrete acetylcholine instead of norepinephrine.[29] Sympathetic nerve fibers also secrete ATP and peptides, which are co-released with the conventional transmitters. The locations of some of the transmitters used in intestinal reflexes are shown in Figure 17.4 and 17.5.

Table 17.1 shows the principal locations of adrenergic receptors in the body and their mechanisms of action. With the advent of molecular biology, it became clear that the

[25] Wang, H-S., and McKinnon, D. 1995. *J. Physiol.* 485: 319–325.

[26] Janig, W., and McLachlan, E. M. 1992. *Trends Neurosci.* 15: 475–481.

[27] Toda, M., and Okamura, T. 2003. *Pharmacol. Rev.* 55: 271–324.

[28] Burnstock, G. 2006. *Trends Pharmacol. Sci.* 27: 166–176.

[29] Guidry, G. et al. 2005. *Auton. Neurosci.* 123: 54–61.

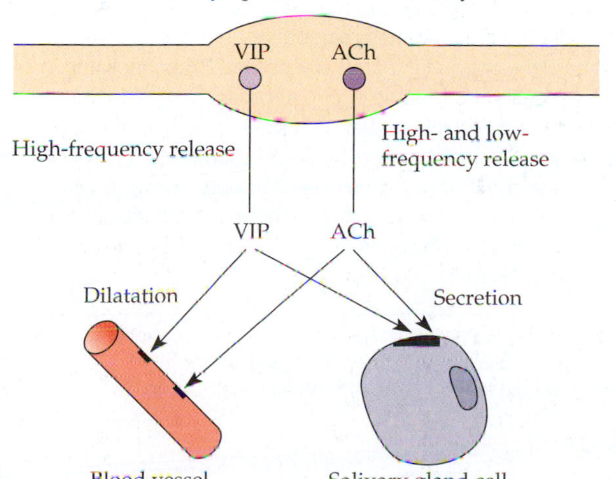

FIGURE 17.4 Cotransmission by Parasympathetic Postganglionic Neurons that release both acetylcholine and vasoactive intestinal peptide (VIP). Parasympathetic nerve fibers supplying the salivary gland secrete both transmitters, which are stored in separate vesicles. Stimulation at low frequencies releases acetylcholine but not VIP, while at higher frequencies both transmitters are released, causing vasodilatation and secretion of saliva. (After Burnstock, 1995.)

FIGURE 17.5 Localization of Transmitters and neuropeptides in neurons of the sympathetic nervous system. All known transmitters are found in the intestine. LHRH = luteinizing hormone-releasing hormone; NE = norepinephrine; VIP = vasoactive intestinal peptide. (After Hökfelt et al., 1980.)

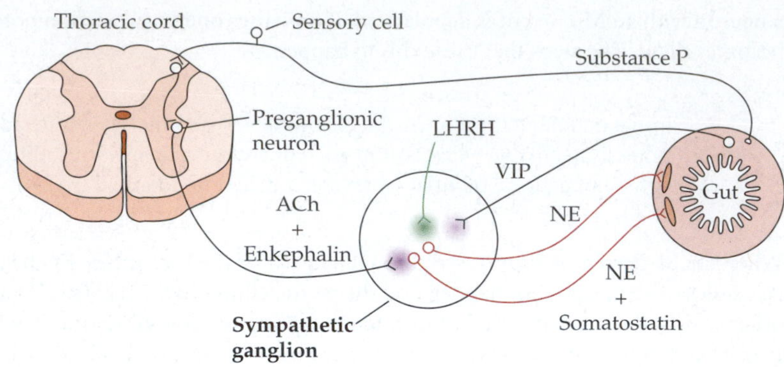

Geoff Burnstock

amino acid sequences are similar in the α_1 and α_2, as well as the β_1, β_2 and β_3 adrenergic receptors, and also in the muscarinic receptors activated by ACh. All these receptors contain seven transmembrane segments, are coupled to G proteins, and use second messengers (see Chapters 12 and 14; Box 17.1).

Purinergic Transmission

In a remarkable series of experiments, Burnstock and his colleagues demonstrated the existence of a major class of sympathetic transmitters, the purines: ATP and adenosine.[30] Certain sympathetic nerve fibers secrete ATP from their terminals, either as the principal transmitter or together with norepinephrine or acetylcholine. The experiments showing that ATP is a sympathetic transmitter were originally designed for a quite different purpose. Burnstock and Holman[31] recorded intracellularly from smooth muscle fibers of the reproductive system to investigate sympathetic synaptic transmission, which at the time was supposedly mediated by norepinephrine. Their recordings of spontaneous miniature potentials constituted an important finding, since, before this time, quantal release had

[30] Burnstock, G. 1995. *J. Physiol. Pharmacol.* 46: 365–384.

[31] Burnstock, G., and Holman, M. E. 1961. *J. Physiol.* 155: 115–133.

BOX 17.1
The Path to Understanding Sympathetic Mechanisms

The development of concepts about mechanisms of synaptic transmission in the autonomic nervous system spanned many years. The original work of Henry Dale in the 1930s and 1940s showed that excitatory and inhibitory transmission in target organs is extremely complicated. One source of confusion arose from the idea that epinephrine might be the transmitter molecule liberated by sympathetic nerves (as originally suggested by Elliott in 1904). It was not until von Euler[32] discovered that norepinephrine, not epinephrine, is the principal transmitter released by sympathetic nerves that the distinction between hormonal and transmitter actions could be accounted for.

Epinephrine released from the adrenal medulla reaches receptors in cells that are not innervated by sympathetic axons—for example, on smooth muscle fibers of bronchioles in the lung. It can also act on receptors that are insensitive to norepinephrine released from sympathetic nerve fibers. A major advance was made by Ahlquist,[33] who devised the scheme for classifying α and β-adrenergic receptors by comparing specific agonists and antagonists. This classification was essential for explaining the varied excitatory and inhibitory sympathetic actions that occur on blood pressure, smooth muscle, the gut, bronchi, and glands. The discovery that some sympathetic axons release acetylcholine (ACh) or purines was also essential.

An understanding of the transmitters and receptors used by the autonomic nervous system has allowed new drugs to be developed for treatment of diseases. One example is provided by the control of bronchioles in the lung. Bronchoconstriction occurs in asthma or severe anaphylactic reactions of the immune system. To initiate relaxation of smooth muscle in the bronchi, β_2-receptors are activated by giving epinephrine or a more specific agonist (such as salbutamol, a β_2-agonist used for the treatment of asthma), but not norepinephrine, which has little or no effect on β_2-receptors. Bronchodilatation allows the patient to breathe again. Another example is provided by the beta-blockers that are widely used and highly effective in the treatment of high blood pressure, coronary artery disease, and glaucoma. As their name implies, these drugs act by blocking the actions of norepinephrine and epinephrine on β-receptors in the heart, in the smooth muscle of blood vessels, in the kidney, and in the eye.[34]

[32] von Euler, U. S. 1956. *Noradrenaline*. Charles Thomas, Springfield, IL.

[33] Ahlquist, R. P. 1948. *Am. J. Physiol.* 153: 586–600.

[34] Black, J. W., and Prichard, B. N. 1973. *Br. Med. Bull.* 29: 163–167.

been observed only at the skeletal neuromuscular junction. Later, however, Burnstock and his colleagues[35] showed that norepinephrine is not the sole transmitter and that the sympathetic neurons co-release a purine (i.e., ATP) and a peptide (neuropeptide Y). The miniature potentials in the smooth muscle are, in fact, due to ATP rather than norepinephrine acting directly on ion channel receptors.

Two main families of receptors for purines have been identified, sequenced, and cloned (see Chapters 4 and 12). A family of eight or more P1 receptors is situated in pre- and postsynaptic structures in the periphery and in brain. They are activated preferentially by adenosine and are G protein coupled. Although ATP is the transmitter liberated from presynaptic endings, enzymes rapidly break it down to adenosine, which is the natural agonist for the P1 receptors. ATP does act directly on a family of P2 receptors[36] in the central and the peripheral nervous systems, where it regulates endocrine secretion, smooth muscle contractility, and activates nociceptive C-fibers.[37]

Sensory Inputs to the Autonomic Nervous System

This description of the autonomic nervous system has failed to mention essential components: sensory inputs and reflex regulation. Indeed, in textbooks of physiology and pharmacology, the autonomic nervous system is often treated as though it functioned as a purely motor system for smooth muscle, cardiac muscle, and glands.

One reason for neglect of autonomic regulation is the paucity of our knowledge. Mechanisms of smooth muscle contraction, such as blood vessel constriction and dilatation, the forward propulsion of material through the gut, and bladder emptying seem relatively straightforward compared to, say, playing tennis. And yet the integration that is required is far from simple. On the afferent side, sensory receptors in the eye, the lungs, the blood vessels, the viscera, and other target tissues provide information about the organs concerned[8,38] (see Figure 17.5). These include nociceptive receptors that signal information about painful stimuli.

A deceptively simple and well-studied reflex is the response of the circulation to changes in body position. With the human body lying flat, the brain is supplied by blood without differences in pressure between the legs and the head. The assumption of a vertical stance causes a drop in blood pressure above the level of the heart, as blood accumulates in the gut and the legs. In the absence of autonomic regulation, loss of consciousness results from standing up, owing to diminished blood flow through the brain.[2] The receptors that signal the need for a change in the pattern of circulation are situated in a large artery in the neck, the carotid artery. The endings are stretch receptors embedded in a swelling of the arterial wall, known as the carotid sinus. Distension of the wall causes increased firing, as shown in Figure 17.6.

The sensory axons run to the brainstem and terminate in a well-defined nucleus (the nucleus of the solitary tract). These neurons project to neurons in the brainstem reticular formation, which in turn project to the autonomic preganglionic neurons. In the horizontal position, a high rate of sensory firing gives rise to inhibition of cardiovascular sympathetic outputs (see Figure 17.6). Blood pressure, heart rate, and cardiac output are depressed, and blood vessels in the skin and gut are dilated, when one is lying down. With assumption of vertical posture, the pressure in the artery falls and the rate of firing of the carotid sinus axons is reduced dramatically, removing the central inhibition. The resulting release of sympathetic activity causes blood vessels in the skin and gut to constrict, and cardiac output and heart rate to rise. The increase in pressure maintains blood flow through the brain.

The ancient recordings in Figure 17.6C, made by Anrep and Starling using a pointer that scratches the surface of a rotating smoked drum, still provide good illustrations of these effects.[39] It was the same E. Starling who, with W. M. Bayliss in 1902, first coined the word "hormone" and proposed the concept of hormonal action while they worked on the control of secretion in the gut.

The description of the reflex presented in Figure 17.6 appears simple. But the central sympathetic and parasympathetic integrative mechanisms for rerouting blood where it is urgently needed remain a black box.[8] This is also the case for other autonomic reflexes for which the sensory and motor limbs are known—for example, enteric, excretory, and respiratory reflexes.[2,40]

[35] Kasakov, L. et al. 1988. *J. Auton. Nerv. Syst.* 22: 75–82.

[36] Soto, F., Garcia-Guzman, M., and Stühmer, W. 1997. *J. Membr. Biol.* 160: 91–100.

[37] Giniatullin, R., Nistri, A., and Fabbretti, E. 2008. *Mol. Neurobiol.* 37: 83–90.

[38] Saper, C. B. 2002. *Annu. Rev. Neurosci.* 25: 433–469.

[39] Starling, E. H. 1941. *Starling's Principles of Human Physiology*. Churchill, London.

[40] Cameron, O. G. 2009. *Neuroimage* 47: 787–794.

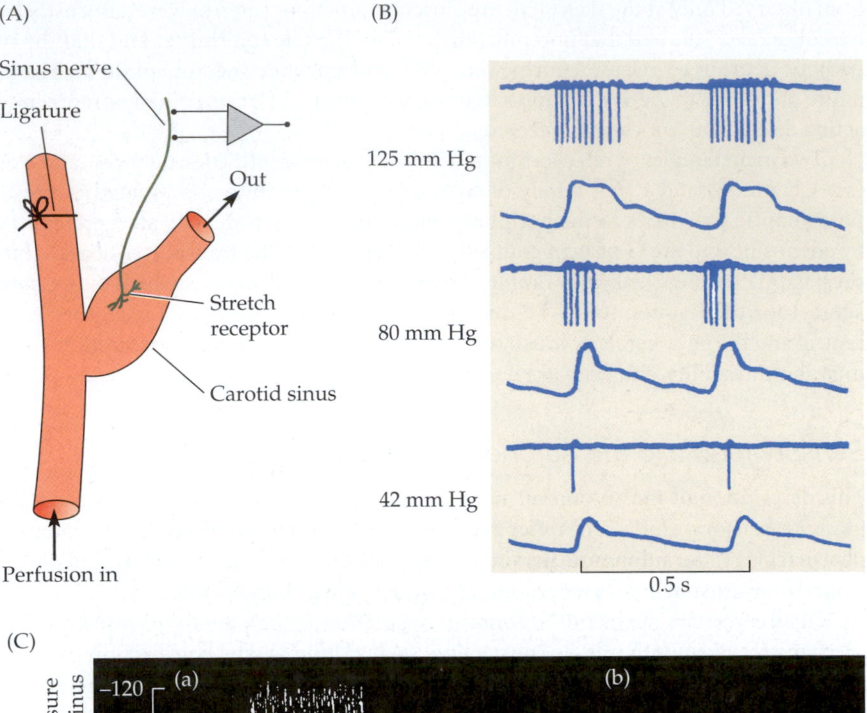

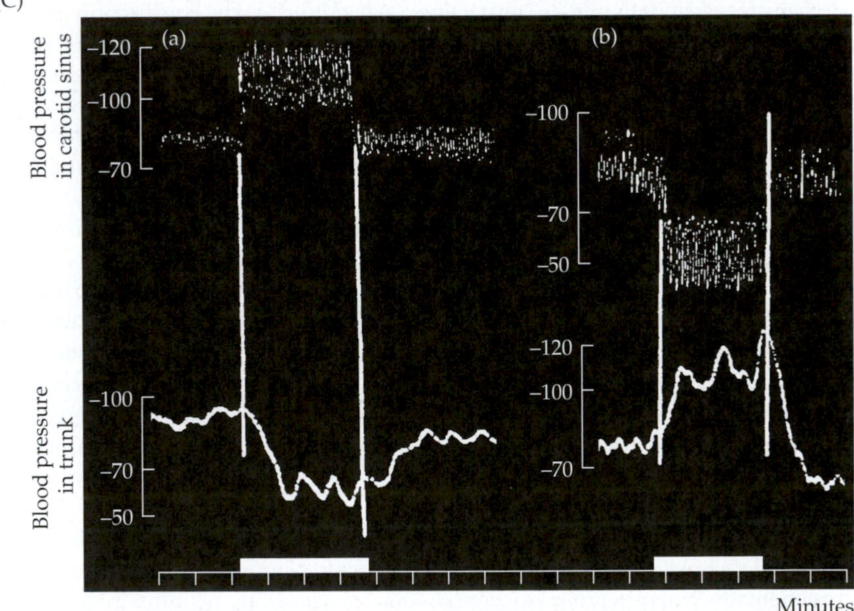

FIGURE 17.6 Firing of Carotid Sinus Stretch Receptors in response to raised blood pressure. (A) Experimental arrangement for recording from sensory nerve fibers in the carotid sinus while it is distended by the circulation or, as in the diagram, artificially perfused. (B) Relationship between blood pressure (lower trace) and the firing of a single afferent fiber from the carotid sinus at different levels of mean arterial pressure (from top down: 125, 80, and 42 mm of mercury, measured with a manometer). (C) A classic record made in 1924. The head of this animal was supplied with blood from a different animal so that blood pressure in the head arteries could be controlled separately by the experimenters. (a) Increased pressure in the head caused a fall in systemic blood pressure in the trunk of the animal. (b) Decreased pressure in the head caused an increase in systemic pressure. Such records were made before electrical recordings were possible; experimenters determined blood pressure using a mercury manometer and registered the movements with a fine pointer on a smoked drum. (B redrawn from Neil, 1954; C after Starling, 1941.)

The Enteric Nervous System

Local regulatory reflexes in the gut (see Figure 17.5) are extremely complex and are brought about by vast numbers of neurons. The enteric nervous system contains more than 10 million nerve cells arranged in the wall of the intestine as sensory neurons, interneurons, and motor neurons. Every known transmitter is represented there (and many of them were first discovered in the gut). To analyze the intrinsic circuits is difficult because of the profuse local reflexes and numbers of connections.[41,42] Functional analysis has been a major challenge even in simpler systems, such as the viscera of the lobster. When Selverston and his colleagues[43] began to study the stomatogastric ganglion, with its complement of only 30 neurons, it seemed that it could perhaps be worked out completely. Yet, although great progress has been made by electrical recordings from identified neurons and although principles of general significance for neurobiology have been discovered, a complete understanding is still not at hand. What appeared at first to be a simple circuit for regulating gut functions turned out to be plastic and modifiable rather than static and hardwired.

[41] Obaid, A. L. et al. 2005. *J. Exp. Biol.* 208: 2891–3001.

[42] Altaf, M. A., and Sood, M. R. 2008. *Dev. Disabil. Res. Rev.* 14: 87–95.

Regulation of Autonomic Functions by the Hypothalamus

Hormones provide an essential aspect of control of the autonomic nervous system. The secretion of hormones by glands (such as the thyroid, the ovary, the adrenal cortex) is regulated by releasing factors secreted in the CNS (discussed in the sections that follow). The hormones in turn act back on the CNS to regulate the secretion of releasing factors, creating a feedback loop.

The hypothalamus (Figures 17.7 and 17.8) is a brain area that controls integrative autonomic functions, including body temperature, appetite, water intake, defecation, micturition, heart rate, arterial pressure, sexual activity, lactation, and, on a slower timescale, growth.[44] The precision of these homeostatic mechanisms enables us to keep our body temperature at about 37° C, our blood pressure at about 120/80 mm Hg, our heart rate at 70 beats/minute, and our intake and output of water at 1.5 liters/day. In addition, the hypothalamus makes it possible for food to be propelled inexorably along the alimentary tract with appropriate secretions for digestion and absorption at every level. The hypothalamus is also a brain area in which appetite, emotions, and responses to bodily exercise are regulated. Emotions are coupled to autonomic responses. Even the thought of food leads to secretion of saliva, and the anticipation of exercise gives rise to increased sympathetic activity.

Appetite is regulated by a remarkable series of steps involving cells that store fat.[45,46] When adipocytes are activated by β_3-receptors they secrete a 167-amino acid protein known as leptin, which acts on cytokine receptors situated in the membranes of neurons in a variety of hypothalamic nuclei. Leptin gene expression is increased by over feeding and reduced by starvation (see Figure 17.7). Other delicately controlled functions are the extraordinarily precise and regular rhythms generated by the hypothalamus. Slow rhythms include those that control endocrine secretion.

For example, sexual and reproductive functions oscillate with periods of weeks that depend on secretion of peptide hormones by hypothalamic cells. These act on the anterior pituitary gland to stimulate secretion of other hormones into the bloodstream.

Hypothalamic Neurons That Release Hormones

A well-studied example of hormonal release is provided by neurons within the hypothalamus that secrete gonadotropin-releasing hormone (GnRH, which is the same as LHRH).[47] A prime action of these neurons is to secrete GnRH into a portal system of blood vessels that flow directly from the hypothalamus to the anterior pituitary gland (see Figure 17.8). Neurally released GnRH thereby acts selectively on a gland that is not directly innervated, enabling the central nervous system to control hormonal secretion.

[43] Selverston, A. I., and Ayers, J. 2006. *Biol. Cybern.* 95: 537–554.

[44] Eikeles, N., and Esler, M. 2005. *Exp. Physiol.* 90: 673–682.

[45] Rahmouni, K., Haynes, W. G., and Mark, A. L. 2004. In *Primer on the Autonomic Nervous System*. Academic Press, London. pp. 86–89.

[46] Williams, K. W., Scott, M. M., and Elmquist, J. K. 2009. *Am. J. Clin. Nutr.* 89: 985S–990S.

[47] Lee, V. H., Lee, L. T., and Chow, B. K. 2008. *FEBS J.* 275: 5458–5478.

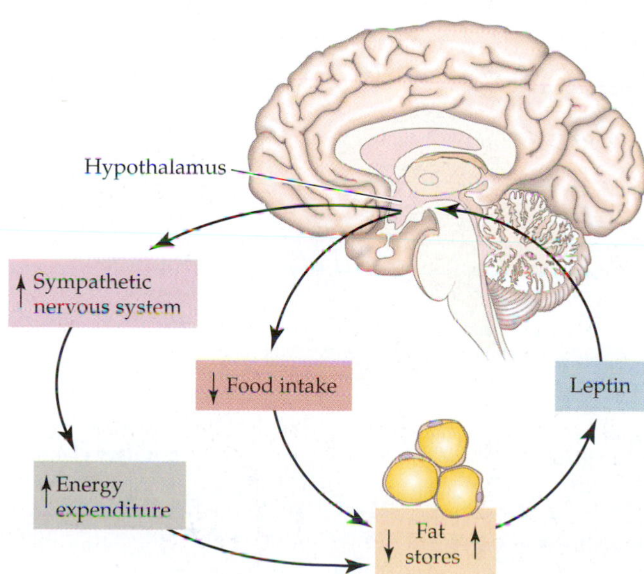

FIGURE 17.7 Mechanism by Which Leptin Regulates Body Weight. Leptin peptide is secreted into the circulation by adipose cells in fat depots. It acts on neurons in the hypothalamus that inhibit the intake of food and increase the expenditure of energy by activating the sympathetic nervous system. As a result, the fat content of the body decreases. (After Rahmouni, Haynes, and Mark, 2004.)

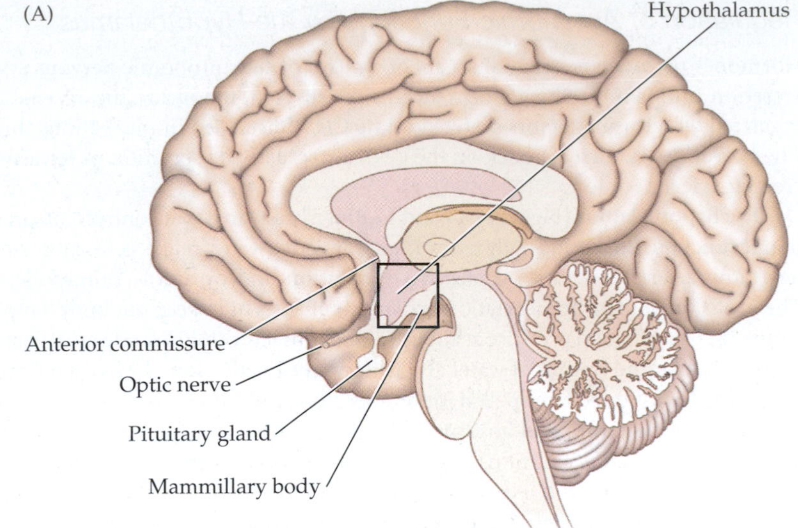

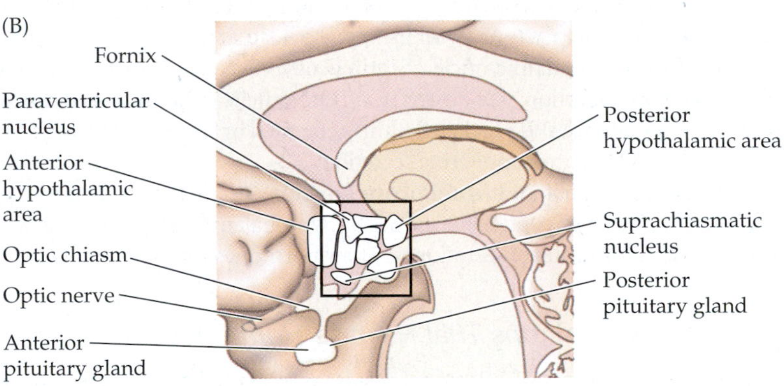

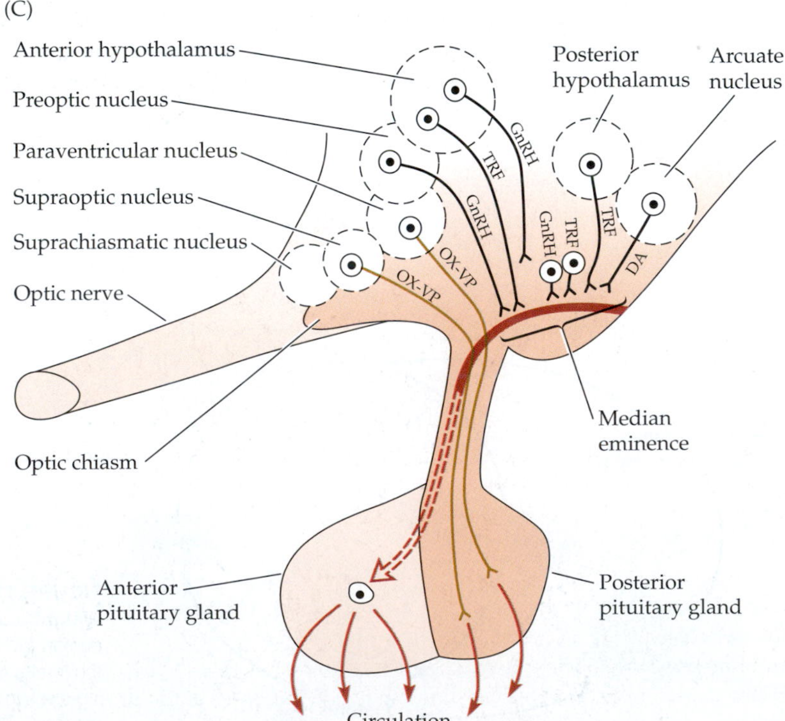

FIGURE 17.8 Hypothalamus and Pituitary Gland in the human brain. (A) Sagittal section of brain, with the area shown in B outlined. (B) Nuclei of the hypothalamus and adjacent structures. (C) Connections of hypothalamic neurons with the neurohypophysis (posterior pituitary gland) and adenohypophysis (anterior pituitary gland). Axons run directly to the neurohypophysis. There the terminals secrete hormones into the circulation. By contrast, releasing hormones released by neurons in the hypothalamus reach the adenohypophysis in high concentration through a dedicated group of portal vessels (red dashed lines). There they activate secretory cells, which liberate hormones into the circulation. DA = dopamine, GnRH = gonadotropin-releasing hormone, TRF = thyroid hormone-releasing factor, OX–VP = oxytocin–vasopressin.

Thereafter, the releasing hormone is diluted in the major vessels of the circulation and cannot, for example, influence synaptic transmission in autonomic ganglia. In the anterior pituitary gland (adenohypophysis), GnRH stimulates specific cells to secrete gonadotropin—a hormone that is essential for sexual and reproductive rhythms and functions.

This brief, oversimplified account cannot do justice to the beautiful experiments of G. W. Harris, who first demonstrated that the local release of a releasing hormone from the hypothalamus could provide an essential control mechanism.[48] His description of delivery of a chemical message through a system of blood vessels by highly localized transport was a revolutionary concept.

Distribution and Numbers of GnRH Cells

GnRH cells are dispersed throughout the hypothalamus, with no clearly defined nucleus or aggregate. The previous section dealt only with those GnRH cells close to the anterior pituitary gland (in the median eminence) that promote its gonadotropin secretion (see Figure 17.8). The release of the releasing hormone itself is also influenced by hormones, such as those secreted by the ovary that feed back into the brain and by synaptic inputs mediated by a variety of transmitters, including norepinephrine, dopamine, histamine, glutamate, and γ-aminobutyric acid (GABA).[49]

One extraordinary feature of the GnRH cells is their small number: 1300 in rats and 800 in mice.[50] Rats and mice (and human beings) would become extinct without these few, scattered cells in the brain. A second remarkable feature is their development (see Chapter 23). During embryonic days 10 to 15 in rats, the precursor cells first appear in a region known as the olfactory placode. This is the region destined to be the future olfactory mucosa. After dividing, the cells migrate along axons of the olfactory nerve and end up in the hypothalamus.[51] The pathways and molecular mechanisms of GnRH cell migration have been studied in embryos, in newborn opossums, and in culture systems.[52] Since all the cells can be reliably marked by specific antibodies to GnRH, they can be counted quantitatively at the site of origin and as they migrate. Other types of neurons migrate along the same axonal pathway as the GnRH cells. Before reaching the hypothalamus, however, they branch off along other axons to reach distinctively different destinations.[53]

Figure 17.8 shows that in addition to the GnRH cells in the hypothalamus, there exist specific populations of neurons that secrete other hormones required for autonomic functions. Metabolism, thyroid function, absorption of salts by the kidney, and growth all depend on releasing hormones that are secreted into the portal system and that act on the anterior pituitary gland.

Specific hypothalamic neurons in the supraoptic and paraventricular nuclei (see Figure 17.8) innervate the posterior pituitary gland directly. Their endings release antidiuretic hormone (ADH, also known as vasopressin) and oxytocin into the blood.[54–56] Hence, the control of water absorption by the kidney and the contractions of the uterus depend directly on the firing of hypothalamic neurons.

Circadian Rhythms

Of particular importance in the life of an animal are the circadian rhythms that control the day–night or wakefulness–sleep cycle. In the absence of all external synchronizing cues, 24-hour rhythmical cycles are maintained by an internal clock for prolonged periods (weeks or months) in invertebrates as well as vertebrates,[57–59] and even in explants or isolated neurons in culture.[60] The internal timing mechanism can be altered (or entrained) by providing regularly spaced light and dark stimuli. Autonomic functions are strongly influenced by biological clocks that act on the pineal gland and the secretion of melatonin.[61,62]

In mammals, a key structure in the hypothalamus for generating the rhythm of the internal clock is the suprachiasmatic nucleus (SCN). An important input to this nucleus is from the eye.[63,64] After destruction of the suprachiasmatic nucleus in rats, light and dark entrainment of endogenous rhythms becomes lost. Locomotor activity, drinking, sleep–wake cycles, as well as rhythms of hormone secretion become disrupted. If fetal hypothalamic tissue containing the SCN is transplanted to a host previously rendered arrhythmic by a

[48] Harris, G. W., and Ruf, K. B. 1970. *J. Physiol.* 208: 243–250.

[49] Bhattarai, J. P. et al. 2011. *Endocrinology* 152: 1551–1561.

[50] Wray, S., Grant, P., and Gainer, H. 1989. *Proc. Natl. Acad. Sci. USA* 86: 8132–8136.

[51] Cariboni, A., Maggi, R., and Parnevalas, J. G. 2007. *Trends Neurosci.* 30: 638–644.

[52] Tarozzo, G. et al. 1998. *Ann. N Y Acad. Sci.* 839: 196–200.

[53] Tarozzo, G. et al. 1995. *Proc. R. Soc. Lond., B, Biol. Sci.* 262: 95–101.

[54] Burbach, J. P. et al. 2001. *Physiol. Rev.* 81: 1197–1267.

[55] Amar, A. P., and Weiss, M. H. 2003. *Neurosurg. Clin. N. Am.* 14: 11–23.

[56] Kosterin, P. et al. 2005. *J. Membr. Biol.* 208: 113–124.

[57] Blau, J. et al. 2007. *Cold Spring Harb. Symp. Quant. Biol.* 72: 243–250.

[58] Saper, C. B., and Fuller, P. M. 2007. *Cold Spring Harb. Symp. Quant. Biol.* 72: 543–550.

[59] Colwell, C. S. 2011. *Nat. Rev. Neurosci.* 12: 553–569.

[60] Kononenko, N. I. et al. 2008. *Neurosci. Lett.* 436: 314–316.

[61] Piggins, H. D., and London, A. 2005. *Curr. Biol.* 15: 455–457.

[62] Pandi-Perumal, S. R. et al. 2006. *FEBS J.* 273: 2813–2838.

[63] Ralph, M. R. et al. 1990. *Science* 247: 975–978.

[64] Davidson, A. J., Yamazaki, S., and Menaker, M. 2003. *Novartis Found. Symp.* 253: 110–121.

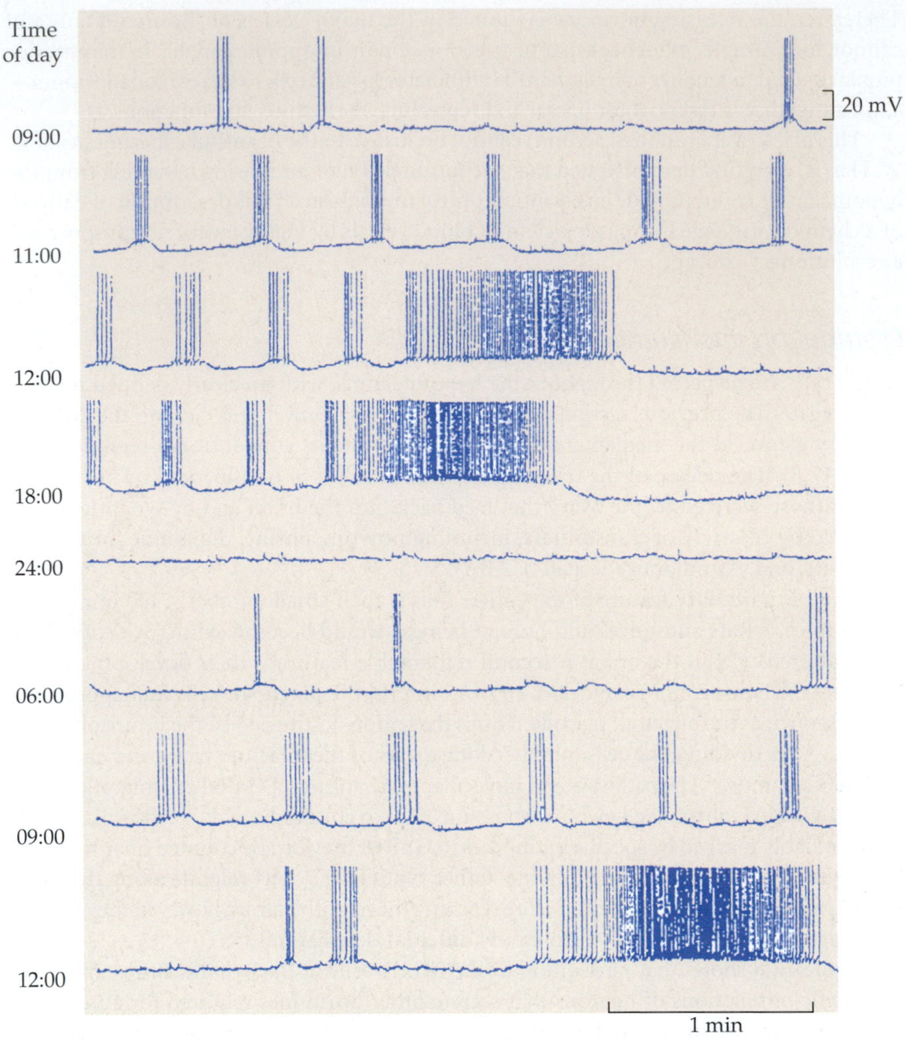

FIGURE 17.9 Circadian Firing of Neurons of crayfish eyestalk maintained in culture. During the 24-hour cycle, the action potential activity recorded intracellularly from a single neuron undergoes cyclical changes. Regular bursts occur at 10.00 A.M., followed by irregular firing between 12:00 noon and 6:00 P.M., and then silence at midnight; the next morning the regular bursts begin again. (From records kindly provided by H. Arechiga and U. Garcia.)

complete lesion of the SCN, then rhythmicity is restored with a free running period corresponding to the donor genotype.[65] Information about cellular and molecular mechanisms that produce regular night and day cycles has been obtained in both invertebrates and vertebrates.[66–68] Examples of circadian rhythms produced by cells in culture are shown in Figures 17.9 and 17.11.

Until recently, there was an apparent paradox concerning the way in which light and dark could entrain the day–night cycle. This paradox arose from the properties of neurons in the visual cortex, which do not respond at all to changes in the level of illumination, but rather only to bars, edges, and moving patterns. Even in the retina, recordings from ganglion cells had shown that the best stimulus for them to fire is not the level of illumination but contrast (see Chapters 1, 2, 3, and 20). This problem has been resolved by the discovery of a small number of specialized ganglion cells that are themselves photoreceptors. These photoreceptive ganglion cells contain the pigment melanopsin and respond to diffuse illumination (Figure 17.10). Their axons project to the suprachiasmatic nucleus rather than to the lateral geniculate nucleus. If all the rod

[65] Kaufman, C. M., and Menaker, M. J. 1993. *J. Neural Transplant. Plast.* 4: 257–265.

[66] Numano, R. et al. 2006. *Proc. Natl. Acad. Sci. USA* 103: 3716–3721.

[67] Siepka, S. M. et al. 2007. *Cold Spring Harb. Symp. Quant. Biol.* 72: 251–259.

[68] Lowrey, P. L., and Takahashi, J. S. 2004. *Annu. Rev. Genomics Hum. Genet.* 5: 407–441.

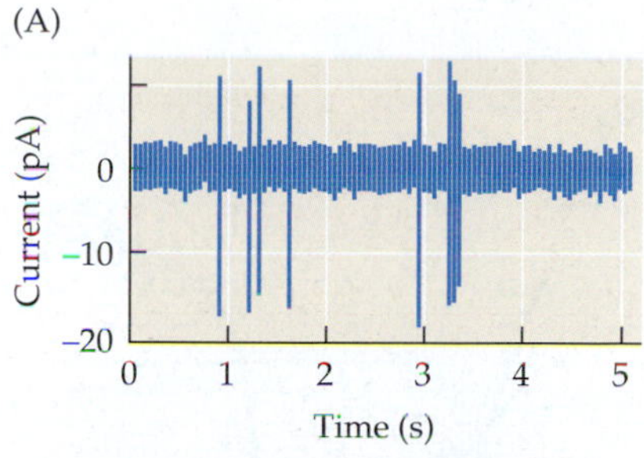

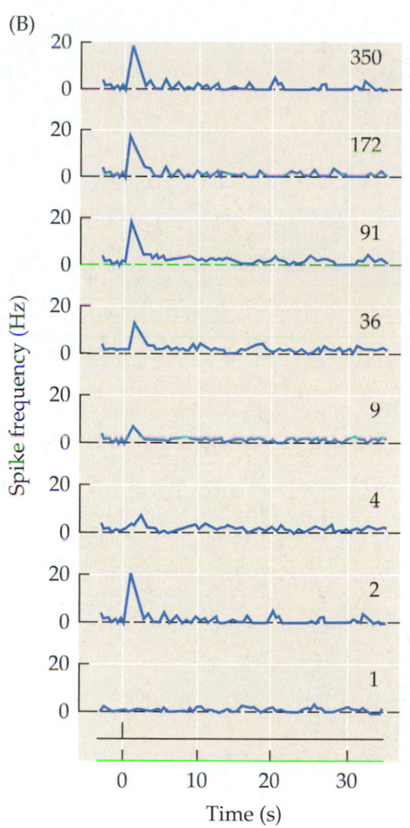

FIGURE 17.10 Recordings from an In Situ Retinal Ganglion Cell in response to illumination by a small spot shone onto the cell soma. (A) Spontaneous firing in darkness. (B) Effect of different-intensity flashes (50 ms duration, 480-nm wavelength, 40-μm diameter). The flash monitor is shown in the bottom trace and the relative intensity is indicated by numbers on the right of each trace. Responses were averaged over several trials. (After Do et al., 2009.)

and cone receptor responses are abolished, the pupil still constricts in bright light and the day–night rhythm can still be entrained.[69–72]

In neurons of the suprachiasmatic nucleus, the frequency of spontaneous action potentials increases during the day and decreases at night, as shown in Figure 17.11. By what mechanism is the rhythm produced? This problem has been investigated in slices of rat SCN in culture where GABA has been shown to be a major transmitter.[73] Yarom and his colleagues[74] have shown that certain groups of suprachiasmatic neurons in a slice respond to GABA with depolarization and increased rates of firing during the day (Figure 17.11A). The same concentration of GABA applied at night causes hyperpolarization and a decrease in rate (Figure 17.11B). Hence, as in developing central nervous system,[75] GABA can be an excitatory or an inhibitory transmitter[76] (see Chapters 11 and 25). The type of response depends on the level of internal chloride. As described in Chapter 11, when the internal chloride concentration is low, the chloride equilibrium potential (E_{Cl}) is more negative than the resting potential. Opening of channels by GABA allows chloride ions to enter and the membrane to hyperpolarize. With raised internal chloride concentrations, E_{Cl} shifts to a value positive with respect to the resting membrane potential. As a result, GABA causes chloride ions to move out of the cell and gives rise to a depolarization. The change from inhibition to excitation is actively regulated by the activity of NKCC1and KCC2 chloride transporters[77,78] (see Chapter 9). The mechanisms by which those transporters are differentially regulated during the day–night cycle are not yet known.

Common proteins that are associated with periodicity throughout the animal kingdom have been revealed by genetic techniques. Genes and proteins that control circadian rhythms have been identified and cloned in *Drosophila*.[79] In many species, these proteins, one of which is known as per (period), have been observed in pacemaker regions such as the suprachiasmatic nucleus.[66] In flies, deletion of the *per* gene abolishes circadian rhythm. Reintroduction of the *per* gene reestablishes the rhythm.[80] Although no link has been found between regulatory proteins and intracellular chloride concentration, it is gratifying that one can now begin to explain circadian rhythms in terms of genes and ion concentrations in well-defined groups of neurons.

[69] Berson, D. M., Dunn, F. A., and Takao, M. 2002. *Science*. 295: 1070–1073.

[70] Hattar, S. et al. 2002. *Science* 295: 1065–1070.

[71] Güler, A. D. et al. 2008. *Nature* 453: 102–105.

[72] Do, M. T. et al. 2009. *Nature* 457: 281–287.

[73] Choi, H. J. et al. 2008. *J. Neurosci.* 28: 5450–5459.

[74] Wagner, S. et al. 1997. *Nature* 387: 598–603.

[75] Cherubini, E., Gaiarsa, J. L., and Ben-Ari, Y. 1991. *Trends Neurosci.* 14: 515–519.

[76] Albus, H. et al. 2005. *Curr. Biol.* 15: 886–893.

[77] Wagner, S., Sagiv, N., and Yarom, Y. 2001. *J. Physiol.* 537: 853–869.

[78] Belenky, M. A. et al. 2010. *Neuroscience* 165: 1519–1537.

[79] Saez, L., Meyer, P., and Young, M. W. 2007. *Cold Spring Harb. Symp. Quant. Biol.* 72: 69–74.

[80] Zehring, W. A. et al. 1984. *Cell* 39: 369–376.

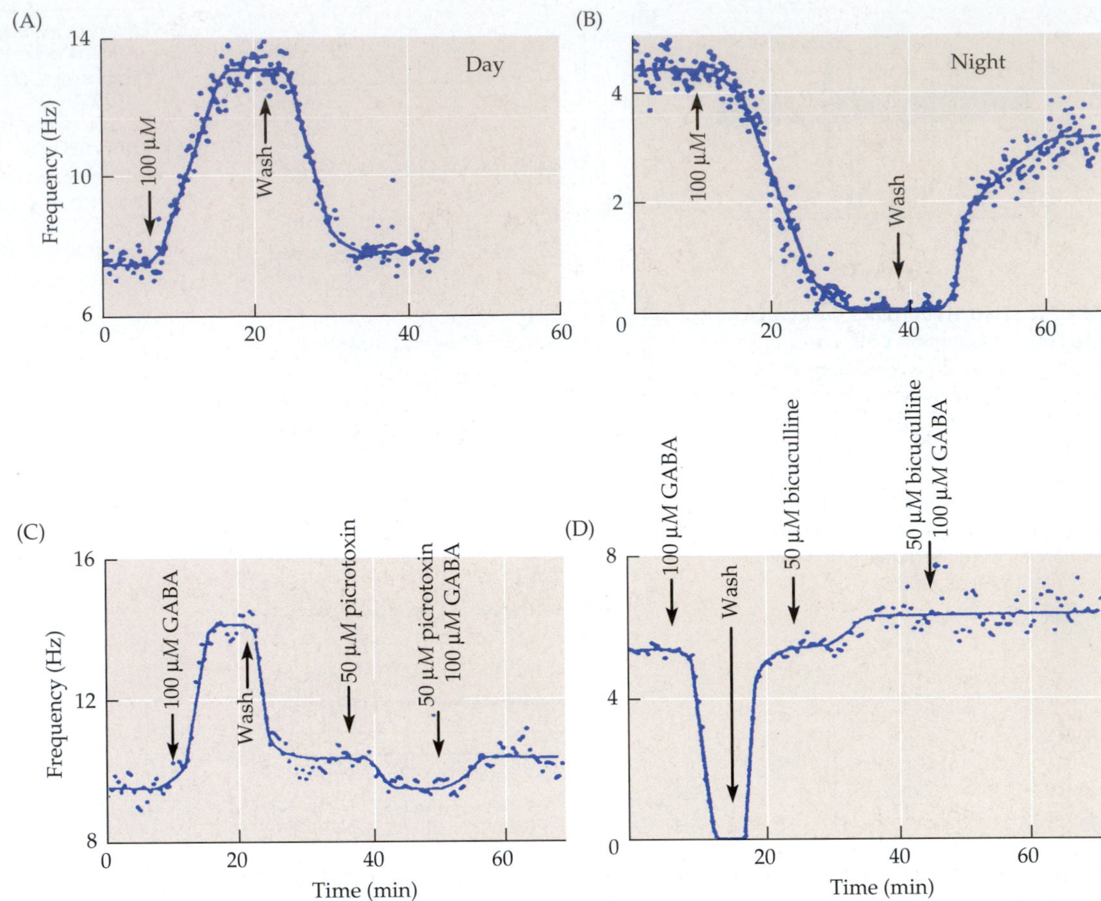

FIGURE 17.11 Circadian Rhythm of Slice of Rat Suprachiasmatic Nucleus maintained in culture. GABA was applied at different times while extracellular recordings were made from neurons. GABA gave rise to increases of action potential frequency in the daytime (A) and decreases at night (B). The recordings in C and D show that the effects of γ-aminobutyric acid (GABA) were blocked by GABA antagonists (bicuculline and picrotoxin). The change from excitation to inhibition can be accounted for in terms of changed intracellular chloride concentrations, which were assessed by whole-cell patch recordings (not shown). The conductance change produced by GABA remains unchanged during the day–night cycle. (After Wagner et al., 1997.)

SUMMARY

- The autonomic nervous system regulates essential functions of all internal organs and is itself regulated by hormonal and sensory feedback.
- Parasympathetic effects are focused, compared to the widespread effects of sympathetic activation.
- ACh is the principal transmitter used for transmission in autonomic ganglia, at parasympathetic nerve endings, and at certain sympathetic nerve endings.
- Norepinephrine is the principal transmitter for most sympathetic endings. Other transmitters include acetylcholine, peptides, and ATP.
- A single molecule—for example, LHRH, which is also known as GnRH—can act as a transmitter at synapses and a hormone within the brain.
- Analysis of effects mediated by the autonomic nervous system is complex, owing to the variety of receptors and the large numbers of peptide and nonpeptide transmitters.
- Epinephrine released as a hormone into the circulation from the adrenal medulla reaches receptors in target cells that are not affected by transmitter released from nerve endings.
- The hypothalamus is the region of the brain that controls the overall activities of the autonomic nervous system and also regulates the secretion of hormones.
- The hypothalamus, in turn, is influenced by higher centers of the central nervous system and by hormones.

Suggested Reading

General Reviews

Delmas, P., and Brown, D. A. 2005. Pathways modulating neural KCNQ/M (K_v7) potassium channels. *Nat. Rev. Neurosci.* 6: 850–862.

Burnstock, G. 2008. The journey to establish purinergic signalling in the gut. *Neurogastroenterol. Motil.* 20 (Suppl 1): 8–19.

Cooper, J. R., Bloom, F. E., and Roth, R. H. 2002. *The Biochemical Basis of Pharmacology*. Oxford University Press, New York.

Fu, Y., Liao, H. W., Do, M. T., and Yau, K. W. 2005. Non-image forming ocular photoreception in vertebrates. *Curr. Opin. Neurobiol.* 15: 415–422.

Robertson, D. 2004. *Primer on the Autonomic Nervous System*. Academic Press, London.

Siepka, S. M., Yoo, S. H., Park, J., Lee, C., and Takahashi, J. S. 2007. Genetics and neurobiology of circadian clocks in mammals. *Cold Spring Harb. Symp. Quant. Biol.* 72: 251–259.

Suh, B. C., and Hille, B. 2008. PIP_2 is a necessary cofactor for ion channel function: how and why? *Annu. Rev. Biophys.* 37: 175–195.

Williams, K. W., Scott, M. M., and Elmquist, J. K. 2009. From observation to experimentation: leptin action in the mediobasal hypothalamus. *Am. J. Clin. Nutr.* 89: 9855–9905.

Original Papers

Banks, F. C., Knight, G. E., Calvert, R. C., Thompson, C. S., Morgan, R. J., and Burnstock, G. 2006. The purinergic component of human vas deferens contraction. *Fertil. Steril.* 85: 932–939.

Brown, D. A., and Passmore, G. M. 2009. Neural KCNQ (Kv7) channels. *Brit. J. Pharmacol.* 156: 1185–1195.

Güler, A. D., Ecker, J. L., Lall, G. S., Haq, S., Altimus, C. M., Liao, H. W., Barnard, A. R., Cahill, H., Badea, T. C., Zhao, H., Hankins, M. W., Berson, D. M., Lucas, R. J., Yau, K. W., and Hattar, S. 2008. Melanopsin cells are the principal conduits for rod-cone input to non-image forming vision. *Nature* 453: 102–105.

Kuffler, S. W. 1980. Slow synaptic responses in autonomic ganglia and the pursuit of a peptidergic transmitter. *J. Exp. Biol.* 89: 257–286.

Merlin, C. Gegear, R. J., and Reppert, S. M. 2009. Antennal circadian clocks coordinate sun compass orientation in migratory monarch butterflies. *Science* 325: 1700–1704.

Numano, R., Yamazaki, S., Umeda, N., Samura, T., Supino, M., Takahashi, R., Ueda, M., Mori, A., Yamada, K., Sakaki, Y., Inouye, S. T., Menaker, M., and Tei, H. 2006. Constitutive expression of the Period1 gene impairs behavioral and molecular circadian rhythms. *Proc. Natl. Acad. Sci. USA.* 103: 3716–3721.

Spyer, K. M., and Gourine, A. V. 2009. Chemosensory Pathways in the Brainstem Controlling Cardio-Respiratory Activity. *Philos. Trans. R. Soc. Lond.* 364: 2603–2610.

Wagner S., Sagiv, N., Yarom, Y. 2001. GABA-induced current and circadian regulation of chloride in neurones of the rat suprachiasmatic nucleus. *J. Physiol.* 537: 853–869.

CHAPTER 18

Cellular Mechanisms of Behavior in Ants, Bees, and Leeches

Experiments on invertebrates have provided crucial insights into cellular and molecular mechanisms of signaling. Because of their simplified nervous systems and wide diversity, animals such as flies, bees, ants, worms, snails, lobsters, and crayfish offer advantages for studying how nerve cells integrate information to produce coordinated behavior.

Ants and bees are social insects whose complex behaviors exemplify an important principle: measurements of behavior provide insights into integrative mechanisms. Desert ants walk and bees fly over long distances, taking meandering paths while foraging for food. Once it has found a food source, however, the ant or bee orients toward its nest and returns there directly, in a straight line. Somehow, the individual ant or bee calculates the position from which it started, and heads straight for home. This navigation requires the integration of information provided by polarized light from the sun. The desert ant's eyes contain specific groups of photoreceptors for polarized light. These photoreceptors supply information to the CNS, which calculates and keeps track of movements in space to create a new vector. By contrast, bees, though closely related to ants, rely more on other navigational cues, including magnetic fields.

If one's goal is to work downward from behavior via the brain to the properties and connections of individual neurons, neither the ant nor the bee provides an experimentally favorable preparation, owing to the relatively small size and large number of cells in the insect CNS. The leech, by contrast, provides a convenient preparation for the analysis of how behavior is generated by networks of neurons. This simple animal swims, crawls, selects suitable victims to feed on and makes love to other leeches of the same species, all directed by a CNS composed of 32 experimentally accessible ganglia. Each ganglion contains only about 400 neurons; individual sensory cells, motoneurons, and interneurons can be recognized by visual inspection and by direct electrical recording. Thus, in the leech, it is possible to: identify individual neurons from ganglion to ganglion and from one specimen to another; measure their biophysical properties; trace their connections; and define their roles in integrative actions, such as reflexes, swimming, and avoidance behavior. Each mechanosensory cell in the ganglion responds selectively to touch, pressure, or noxious stimulation and innervates a well-defined area of skin. Sensory cells transmit information to interneurons and motoneurons by electrical and chemical synapses. After repetitive stimulation at natural frequencies, transmission in sensory neurons becomes blocked at branch points where small axons feed into larger processes. This action temporarily alters both the receptive field of the sensory neuron and its interactions with synaptic partners. With circuits consisting of identified neurons, it is also possible to study how individual neurons form connections during development and during regeneration after injury. Ants, bees, and leeches exemplify the ability of invertebrate nervous systems containing relatively few neurons to make complex computations. Thus, these invertebrates provide appealing preparations for studying cellular mechanisms that underlie behavior.

Throughout this book, invertebrate neurons are used to illustrate fundamental mechanisms that are significant for understanding the workings of the nervous system. For example, principles of general validity for conduction of action potentials and for synaptic transmission were derived from the squid giant axon and giant synapse. Cells of other invertebrates were used to illustrate passive electrical properties and the mechanisms responsible for synaptic inhibition.

The reasons for choosing a particular animal to work on are often technical. Certain problems can be solved more easily in invertebrate nervous systems. First, although invertebrate behavior may be elaborate, it is often highly stereotyped, and thus more readily analyzed. Second, given the relative simplicity and accessibility of many invertebrate nervous systems compared to vertebrates, it is often possible to recognize individual nerve cells and to study them by electrophysiological and molecular techniques. Third, the opportunities for genetic manipulations in the fruit fly (*Drosophila melanogaster*) and the nematode worm *Caenorhabditis elegans*[1,2] have been highly advantageous for studies on the development and function of the nervous system (see Chapter 25). Despite the divergent morphologies of the various kinds of animals, most of their genes, including those encoding developmental regulators, receptors, and channels are strikingly similar (see Chapter 5 and 25).[3]

From Behavior to Neurons and Vice Versa

Invertebrates provide opportunities to follow the thread of physiological mechanisms from the level of single cells, with their particular branching patterns and intercellular connections, to coordinated animal behaviors. In suitable species with large neurons, one can analyze the biophysical properties of single cells and observe how they give rise to the properties of networks of cells and to the behavior of the whole animal. Similarly, one can follow molecular events occurring in that cell as the animal modifies its behavior in response to outside influences and internal programs.[4,5] Such analyses are possible because stereotyped behavioral responses in invertebrates are performed by relatively few neurons, whereas analogous responses in mammals require many thousands of neurons.

In ants and bees, quantitative observation of behavior permits one to infer and then define underlying cellular and integrative mechanisms. The central nervous systems of insects have been used to study a variety of problems, including development and plasticity, flight, walking, navigation, and communication by sound.[6–12] The animals themselves and the scope of the problems are so varied that a comprehensive review is impossible. In subsequent chapters that deal with sensory mechanisms, perception, and motor coordination in the mammalian CNS, we shall see again that it is behavior (e.g., the detection of where a sound has arisen in space) that provides the starting point for revealing cellular mechanisms.

In addition to its advantages, each system also has its own set of disadvantages for analyzing the neural basis of behavior—indeed, beware the scientist who claims to have the ideal system! In ants and bees, individual neurons are small and hard to study physiologically. Happily, cellular neurobiology of other preparations provides a complementary approach to the analysis that goes from behavior to brain. Here too, examples abound; the neural circuits for coordinated elementary units of behavior (e.g., postural reflexes, feeding, heartbeat, circadian rhythms, escape reactions, and swimming) have been traced in crustaceans, annelids, and mollusks.

The literature on crustaceans, insects, and particularly the sea slug *Aplysia* (which has been extensively studied by Kandel and his colleagues), is far too extensive to be dealt with in a book of this size. Indeed, entire monographs are available on these systems.[13–18] Accordingly, as in other chapters, we use selected examples for detailed discussion of circuitry and behavior. We have singled out for discussion the nervous systems of ants, bees, and leeches. In ants and bees, behavioral analysis is the starting point. This has revealed neuronal mechanisms that provide the animals with information about polarized light and magnetic fields in the outside world. In contrast, and of particular relevance for this text, the leech features more limited behaviors than do the ant or bee, and have a highly stereotyped CNS with fewer neurons. Thus, the leech provides preparations in which one can study in great detail the properties, connections, and functions of individual identified nerve cells.

---

[1] Dittman, J. 2009. *Adv. Genet.* 65: 39–78.

[2] Leyssen, M., and Hassan, B. A. 2007. *EMBO Rep.* 8: 46–50.

[3] Copley, R. R. 2008. *Philos. Trans. R. Soc. Lond., B, Biol. Sci.* 363: 1453–1461.

[4] Bailey, C. H., and Kandel, E. R. 2008. *Prog. Brain Res.* 169: 179–198.

[5] Leonard, J. L., and Edstrom, J. P. 2004. *Biol. Rev. Camb. Philos. Soc.* 79: 1–59.

[6] Lichtneckert, R., and Reichert, H. 2008. *Adv. Exp. Med. Biol.* 628: 32–41.

[7] Jacobs, G. A., Miller, J. P., and Aldworth, Z. 2008. *J. Exp. Biol.* 211: 1819–1828.

[8] Frye, M. A., and Dickinson, M. H. 2001. *Neuron* 32: 385–388.

[9] Edwards, J. S. 1997. *Brain Behav. Evol.* 50: 8–12.

[10] Arthur, B. J. et al. 2010. *J. Exp. Biol.* 213: 1376–1385.

[11] Cator, L. J. et al. 2009. *Science* 323: 1077–1079.

[12] Engel, J. E., and Wu, C. F. 2009. *Neurobiol. Learn. Mem.* 92: 166–175.

[13] Kandel, E. R. 1979. *Behavioral Biology of Aplysia*. W. H. Freeman, San Francisco.

[14] Wiese, K., ed. 2002. *The Crustacean Nervous System*. Berlin, Germany.

[15] Muller, K. J., Nicholls, J. G., and Stent, G. S., eds. 1981. *Neurobiology of the Leech*. Cold Spring Harbor Laboratory, Cold Spring Harbor, NY.

[16] Shain, D. H., ed. 2009. *Annelids in Modern Biology*. Wiley-Blackwell, New Jersey.

[17] Atwood, H. L., ed. 1982. *Biology of Crustacea*. Academic Press, New York.

[18] Beadle, D. J., Lees, G., and Kater, S. B. 1988. *Cell Culture Approaches to Invertebrate Neuroscience*. Academic Press, London.

Navigation by Ants and Bees

An essential technique for understanding the workings of the nervous system is the quantitative analysis of behavior. In humans, key concepts that explain color vision and dark adaptation were first established through psychophysical experiments, in which mechanisms of perception were deduced from the responses of subjects exposed to different wavelengths or intensities of lights (see Chapters 1, 2, and 20). Similarly, in invertebrates, observations of animals under natural conditions have led to valuable insights into the role of receptors and of integrative mechanisms of the CNS.

The extraordinary performance of invertebrate nervous systems can be appreciated by considering complex navigation by ants,[19–21] which meander over long distances searching for food and then unerringly find their way home. A host of sensory cues are needed for such navigation to be successful. Because of the small sizes of the nerve cells in these insects, it is often hard to analyze directly membrane properties and synaptic transmission. And yet, as the following sections show, sensory mechanisms in these animals can be inferred and then analyzed at the cellular level through insights derived from behavioral experiments.

FIGURE 18.1 The Desert Ant, *Catalglyphis bicolor*, which navigates successfully by means of detection of ultraviolet light. It is able to return home in a direct path after having searched for food. (After Wehner, 1994b.)

The Desert Ant's Pathway Home

Wehner and his colleagues have conducted experiments to analyze how desert ants, *Cataglyphis bicolor* and related species (Figure 18.1), are able to wander for long distances in search of food and then return toward the nest in a straight line. The principle of their experiments is illustrated in Figures 18.2 and 18.3.[22,23] An area of desert in Tunisia around the nest and

[19] Wehner, R., and Muller, M. 2006. *Proc. Natl. Acad. Sci. USA* 103: 12575–12579.

[20] Merkle, T., and Wehner, R. 2008. *J. Exp. Biol.* 211: 3370–3377.

[21] Muller, M., and Wehner, R. 2010. *Curr. Biol.* 20: 1368–1371.

[22] Wehner, R. 1997. In *Orientation and Communication in Arthropods*. Birkhauser, Basel, Switzerland. pp. 145–185.

[23] Collett, M., Collett, T. S., and Wehner, R. 1999. *Curr. Biol.* 9: 1031–1034.

(A)

(B)

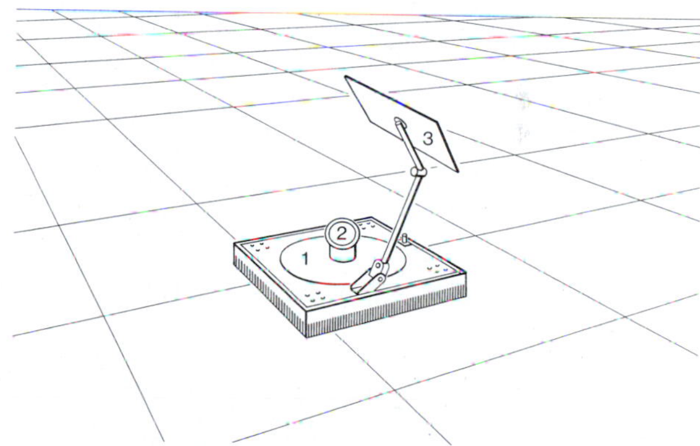

FIGURE 18.2 To Measure the Movements of an Ant it is placed on a grid marked out on the desert floor. (A) A desert ant is tracked as it migrates along the desert floor using a rolling optical laboratory cart to follow the ant and control the portion of sky it can observe. (B) The experimenter moves the cart, keeping the ant centered within the optical setup. The horizontal aperture (1) can be fitted with filters that cut out all light except ultraviolet light or light polarized in just one direction. A small aperture (2) sits atop a circular tube, and the screen (3) is used to prevent the ant from seeing sunlight directly. Since the cart has a frame, the little ant cannot see the skyline or markers on the ground and is also shielded from the wind. The white lines are 1 meter apart and are painted on the desert floor to enable the observers to track the ant's progress accurately. (After Wehner, 1994b; photograph kindly provided by R. Wehner.)

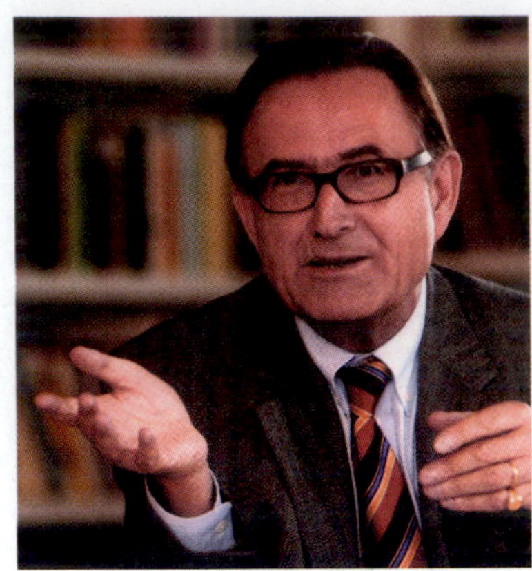

Rüdiger Wehner

the food supply of the ants is marked out in squares as shown in Figure 18.2. A single ant is then followed as it walks to the food source and back. Figure 18.3 illustrates the tortuous, 19-minute outward path from A (nest) to B (food; each dot marks 1 minute). The journey home by contrast is direct, unhesitating, and accurate, and takes only about 6 minutes.

Experiments show that the ant has integrated all the information about its movements, keeping track of both the angles turned (as with a compass) and the distances traveled (as with an odometer) as it wanders out over an area of more than 19,000 square meters (m^2). An ant can wander for hundreds of meters in search of food, then return to within 1 meter (m) of its nest—an error of less than 1% (in this respect the ant is doing far better than at least one author of this book, who does not have the mathematical skills for making computations of this sort).

How does the ant do this? A first guess might be that it uses pheromones or other chemical signals to leave a trail that would require neither compass nor odometer to follow. Chemical cues are not used, however, and would in any case be inoperative at the extremely high (up to 45° C) desert temperatures. Spatial cues and useful landmarks do not abound in the desert. Although objects near the nest do indeed provide information for finding the nest hole at short ranges,[24] objects along the pathway are not the principal cues that the ant uses for long-range navigation. This finding was shown by experiments in which the ant made the round trip while able to view only a portion of the sky.[25] To eliminate the sun and all other cues, the experimenter walked along with the ant, pushing a cart that kept the ant centered under an aperture to the sky (see Figure 18.2B). Inserted into the aperture were filters that determined the direction, wavelength, or angle of polarization of the light seen by the ant through the aperture to the sky. In the absence of the sun, landmarks, and odors, and with only the polarization of light from the sky to guide it, the ant headed straight for home, as the experimenter trundled along, screening it with the little cart.

[24] Bregy, P., Sommer, S., and Wehner, R. 2008. *J. Exp. Biol.* 211: 1868–1873.

[25] Muller, M., and Wehner, R. 1994. *J. Comp. Physiol. A* 175: 525–530.

[26] Muller, K. J. 1973. *J. Physiol.* 232: 573–595.

[27] Wehner, R. 1989. *Trends Neurosci* 12: 353-359.

Polarization of electromagnetic radiation refers to the situation in which the electrical vector of an electromagnetic wave is restricted to a single plane (and the magnetic vector is accordingly restricted to the orthogonal plane). Light from the sun becomes polarized as it traverses the atmosphere, as we can appreciate when a properly oriented polarizing filter darkens the sky by blocking light polarized in a different direction. Light reflected from the uneven surface of clouds, however, is no longer polarized. Hence, the brightness of the clouds is much less affected by such a filter. When the sun is directly overhead, the pattern of polarized light is the same in all directions, so the pattern could not be used as a compass. If, however, the sun shines at any angle other than vertical, an asymmetric pattern of polarization orientations occurs, as shown in Figure 18.4. This pattern could in principle be used as a map for navigation. Although human eyes cannot detect polarized light, those of the desert ant can (as can other arthropods, such as bees, wasps, and crustaceans).[26,27] The pattern of polarized light defines the position of the sun, whether or not it can be seen directly, and thus provides the required compass.

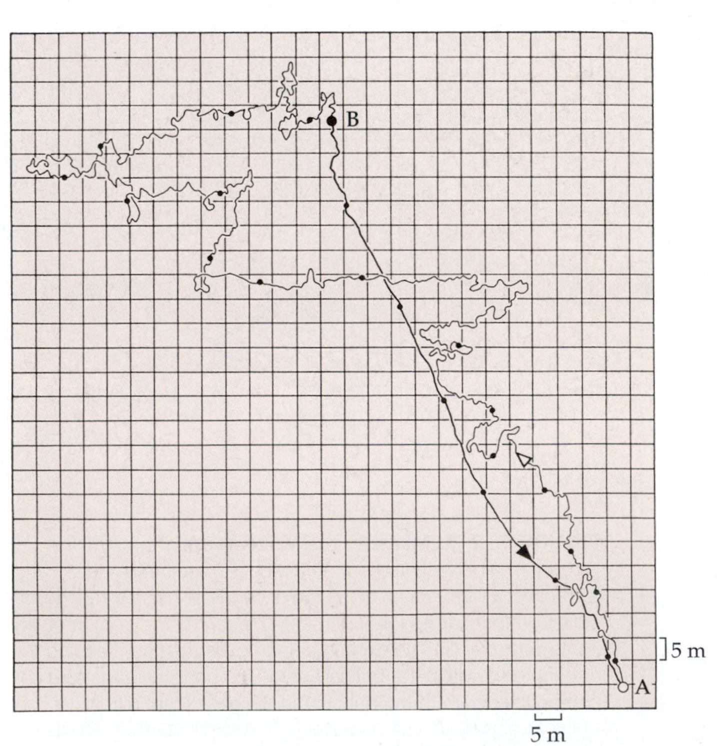

FIGURE 18.3 Pathways Taken by an Ant from its nest (A) to a source of food (B) and back. The distance between sequential dots on the trace represents the distance traveled in 1 minute. The ant follows a tortuous path (over 592 meters [m]) until it happens upon the food. Then, it heads straight for home (140 m) with no deviation and amazing accuracy. (After Wehner, 1994a.)

The odometer function for the pathfinding process appears to be generated by summing proprioceptive information associated with the locomotory movements of the ant's legs (so it is a pedometer rather than an odometer!), in which each step is associated with covering a certain distance. This integration of locomotory information was shown by ingenious experiments in which the stride length of individual ants was altered, just as they were about to start their return journey to the nest. Some ants had their legs (and thus their strides) lengthened by gluing on stilts made of boar's hairs; these ants routinely overshot the nest on the homeward journey before initiating the local searching movements that told the experimenters that the ants *thought* they were near home.

(A)

(B)

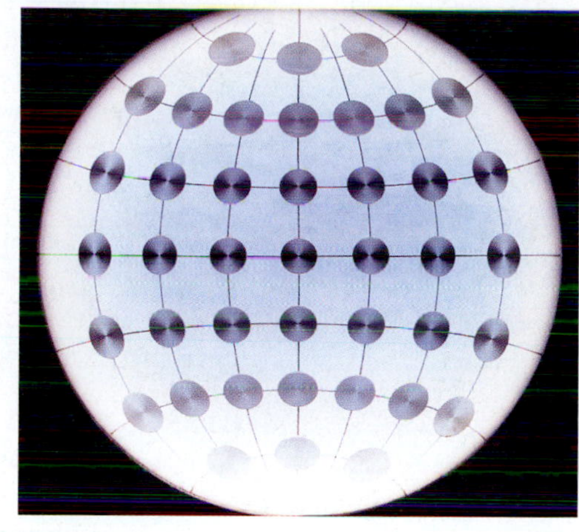

(C) (D)

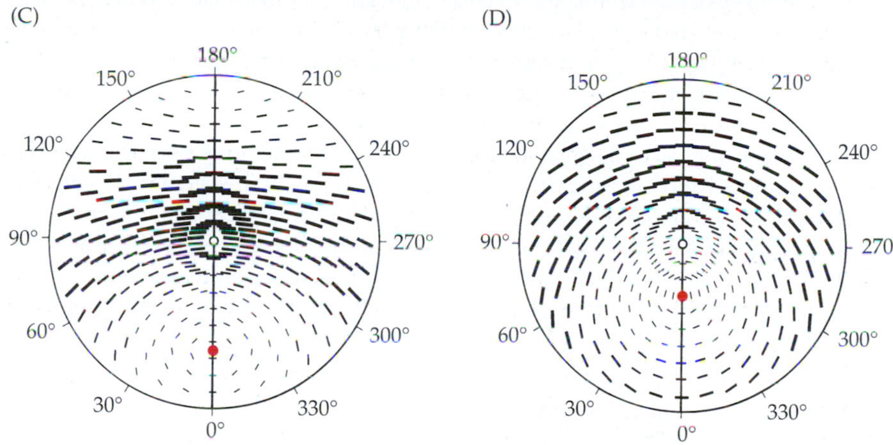

FIGURE 18.4 Patterns of Polarized Light. (A) An array of polarizing filters is mounted within a transparent dome that is oriented parallel to the horizon. (B) Thus, each filter is oriented toward, and reveals the polarization of light from, a different part of the sky. For a person standing at the equator at midday, with the sun shining directly overhead, the patterns would be completely symmetrical. (C,D) Different patterns of polarization are produced with the sun at two different positions (indicated by the red dot in each panel); thus, by observing the pattern of polarization, an ant need observe only a small part of the heavens to compute the position of the sun and thereby navigate successfully. (A and B after Wehner, 1994a; C and D after Wehner, 1997.)

Conversely, ants whose legs and strides were shortened by amputation of distal leg segments initiated the local searching movements before they reached the vicinity of the nest.[28,29]

Polarized Light Detection by the Ant's Eye

The dome-shaped compound eye of an insect consists of a multifaceted array of radially oriented, photo-detecting units called ommatidia—each ommatidium has its own nerve tract and sees the world from its own perspective.[30] In ants, there are nine distinct, elongated photoreceptor cells (rhabdomeres) per ommatidium, some whose peak sensitivity is for green light and others whose peak sensitivity is for UV light. As with vertebrate photo-detection (see Chapter 20), the visual response in ants and other insects starts when light is absorbed by membrane-bound pigment molecules (rhodopsin) within closely packed arrays of subcellular organelles in each rhabdomere. But in insect rhabdomeres, these organelles take the form of closely packed sausages (i.e., microvilli), oriented perpendicular to the long axis of the cell, instead of the stacks of pancakes (i.e., discs) seen in vertebrate rods and cones. This difference enables certain specialized ommatidia to respond preferentially to light that is polarized in a particular orientation. In the desert ant, these specialized ommatidia lie within the dorsal rim area of the eye, so they are normally oriented skyward (Figure 18.5).

[28]Wittlinger, M., Wehner, R., and Wolf, H. 2006. *Science* 312: 1965–1967.

[29]Wittlinger, M., Wehner, R., and Wolf, H. 2007. *J. Exp. Biol.* 210: 198–207.

[30]Zollikofer, C., Wehner, R., and Fukushi, T. 1995. *J. Exp. Biol.* 198: 1637–1646.

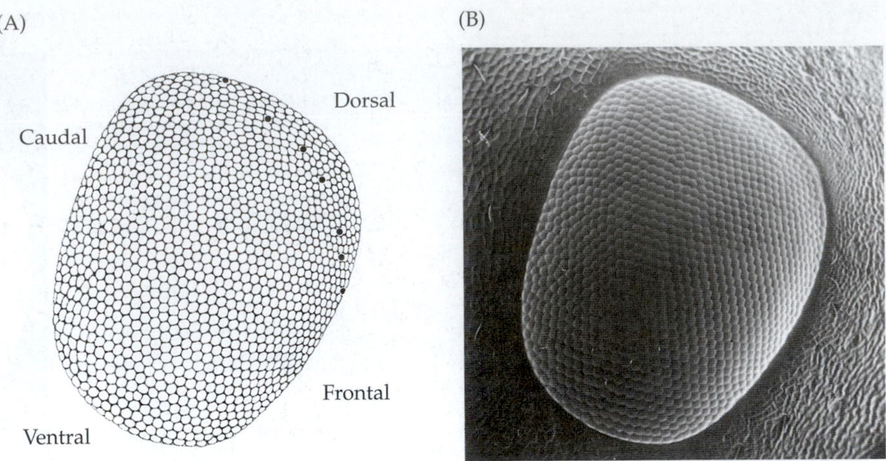

FIGURE 18.5 Polarized Light Detectors in the Eye of the Desert Ant. (A) The arrangement of photoreceptors (ommatidia) within a single, compound eye. The receptors for polarized light, used by the ant to navigate, lie in the dorsal rim of the eye, within the area bounded by black dots. (B) Scanning electron micrograph of the eye of the ant. (After Wehner, 1994a.)

The sensitivity of these ommatidia to polarization arises from the fact that the microvilli are stacked precisely parallel to one another along the entire length of their UV-sensitive rhabdomeres (Figure 18.6). As a result, the rhodopsin molecules in the microvilli are arranged in uniform register with respect to the electrical vector of the incident light.[31]

Since rhodopsin absorbs light optimally along the long axis of the molecule, one particular plane of polarized light will be most effective in generating electrical signals in that rhabdomere. Moreover, the orientations of the microvilli in different UV-sensitive rhabdomeres of the dorsal rim ommatidia are aligned precisely at 90° to one another as shown in Figure 18.6, an arrangement seen only in those photoreceptors concerned with polarized light. This orthogonal arrangement of receptors in one ommatidium is optimal for sensing the angle of polarization, by comparing the outputs of the orthogonally oriented

[31] Goldsmith, T. H., and Wehner, R. 1977. *J. Gen. Physiol.* 70: 453–490.

(A)

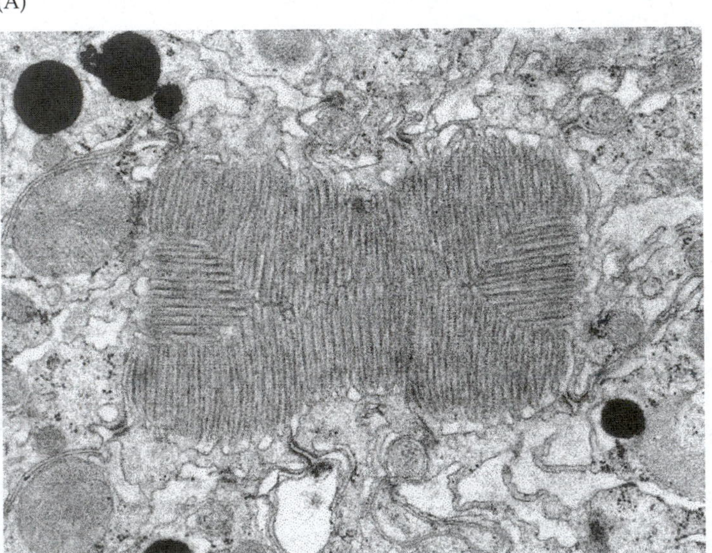

(B)

FIGURE 18.6 Arrangement of Photoreceptors Responsive to Polarized Light. (A) Electron micrograph of an ommatidium in the dorsal region of the eye of an ant. (B) The microvilli are arranged in a precisely orthogonal manner, summarized schematically. Heavy lines separate the eight numbered photoreceptors. Photoreceptors 1 and 5 respond preferentially to light polarized at right angles to that preferred by the other photoreceptors. Elsewhere in the eye, the microvilli are not at right angles like this. (Electron micrograph kindly provided by R. Wehner; B after Wehner, 1996.)

receptors—a single photoreceptor on its own could not differentiate between differences of intensity, wavelength, and polarization. Similar arrangements of polarized light receptors are found in the dorsal rim of bee eyes, and in crustacean eyes.

Evidence that polarized light is essential for the ant's navigation has been provided by the following experiments. First, if the eye is covered by a contact lens leaving only the dorsal rim exposed, the ant can still take a direct way home. Second, if the dorsal rim of the eye is blocked, pathfinding becomes disturbed. Third, if the pattern of polarization reaching the ant's eye is shifted by appropriate filters (placed on the cart), the return course becomes deviated by a precise and calculable extent.

For navigation to be successful, information arriving at the eye must be correlated with a celestial map in which the position of the sun determines the orientations of polarized light. As the ant walks away from the nest, constant reference is made to the complex but regular pattern in the sky by the ommatidia. This provides the nervous system with information about the direction of travel. The compound eye is hemispherical, and this allows for a faithful spatial representation of orientations like those shown in Figure 18.4. Moreover, when the orientation preferences of the ommatidia are analyzed, one finds a coherent arrangement that is suited to a representation in neural terms of the polarized light patterns of the sky. As a result, the amount of match or mismatch of the pattern of polarized light can be used to determine compass direction in a predictable manner (predictable for ants and for ant researchers).

Strategies for Finding the Nest

A further complication arises because the sun is not stationary. Hence, the ant has to compensate for the shifting pattern of polarization during the day. That it can do so has been shown in the experiments in which an ant is placed at a spot removed from the nest at one hour, kept there, and then released at a later time. An ant that has become familiar with the sun's rate of movement for at least one day is able to correct its trajectory in an appropriate manner, as though it had learned the patterns of polarized light at different times of day. In addition to the polarized light compass, the sun itself and external objects situated along the path can be used to aid navigation.[32] Distinctive features of the terrain and objects are of principal importance in the last part of the return to the nest, which constitutes a tiny hole in the desert. If the homing vector has led to an error, such that the ant has not arrived precisely at its nest, a new strategy is introduced.[25] A series of exploratory loops is made, increasingly larger but always returning to the starting point. This represents an optimal strategy for reconnoitering without getting lost.

Polarized Light and Twisted Photoreceptors

Ants and bees are evolutionarily related—together with wasps, they comprise the order of insects called Hymenoptera. Thus, it is not surprising that both ants and bees navigate using receptors for polarized light to provide a compass. The presence of such receptors could, in principle, be a mixed blessing, however. On the one hand, the precise arrays of microvilli allow the heavens to be scanned for the orientation of polarized light, as we have seen. But for discriminating shapes and colors, polarization can cause difficulties. For example, as bees fly, they need to identify flowers by their colors. Leaves and petals, however, vary in the way they reflect light in a manner that depends on how waxy the surfaces are. Leaves that are shiny reflect polarized light more than those that are matte. Consequently, the angle at which a leaf or a petal is illuminated and viewed will affect the amount and the direction of polarized light that is reflected.

The photopigments necessary for color vision in the bee are contained in microvilli (like those of the ant) of specific receptors sensitive to green, blue, or ultraviolet light. In the bee eye, as in the ant, the photopigments are arranged in precise, parallel series of rhabdomeres. With variable, uncalibrated contributions by polarized light, the signals regarding color would be ambiguous because the appreciation of color depends not only on wavelength but also on the relative absorption by different classes of color receptors. As Wehner and Bernard[33] put it, "This means that for the bee the hue of a given part of a plant would change, whenever an approaching bee changed its direction of flight and

[32] Collett, M. et al. 1998. *Nature* 394: 269–272.

[33] Wehner, R., and Bernard, G. D. 1993. *Proc. Natl. Acad. Sci. USA* 90: 4132–4135.

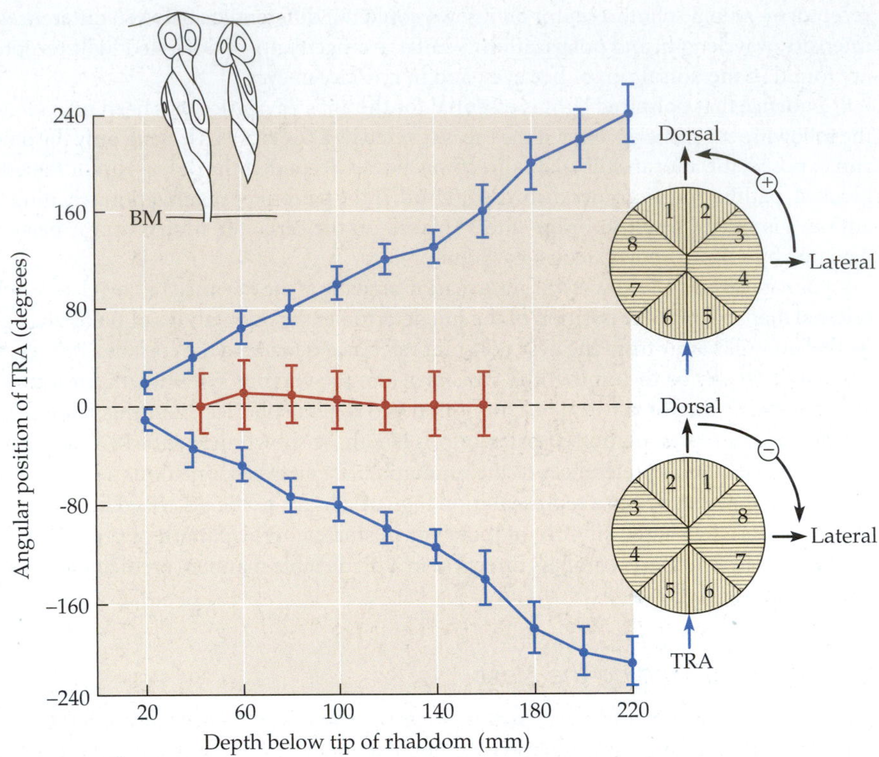

FIGURE 18.7 Twists in Most Bee Photoreceptors Minimize the Influence of Polarized Light. The inset illustrates the helical twisting of the rhabdomeres in two developing ommatidia (viewed from the side) as one progresses from the tips of the rhabdomeres near the surface of the eye (top) toward the basement membrane (BM). The graph plots the angular position of the transverse axis (TRA; defined in the schematics at right) of various photoreceptors against the depth beneath their tips. Receptors in the dorsal rim of the eye, which are sensitive to polarized light, show no twist (red circles). Elsewhere in the eye, photoreceptors twist either clockwise or counterclockwise (blue circles), as a result of which, they do not respond selectively to any particular orientation of polarized light. (After Wehner and Bernard, 1993.)

thus, its direction of view—a completely unwanted phenomenon. For example, when zigzagging over a meadow with all its differently inclined surfaces of leaves, the bee would experience pointillistic fireworks of false colors that would make it difficult to impossible to detect the real colors of the flowers." To avoid this problem, the ommatidia of the bee outside of the dorsal rim area contain so-called twisted receptors. By light and electron microscopy, it was found that the rhabdomeres were twisted along their long axes. This twist produces a progressive change in the orientation of their microvilli (Figure 18.7), so that they are not in a parallel array throughout the depth of the rhabdomere. Thus, the receptor no longer responds selectively to polarized light. Similar irregularities of the microvilli arrangement are seen in ommatidia outside the dorsal rim in the ant's eye. As in the bee, these ommatidia do not sense the plane of polarized light and can therefore be used for landmark detection.

Additional Mechanisms for Navigation by Ants and Bees

No seasoned explorer would be comfortable venturing out with just one navigation system—Steve Wozniak, a cofounder of the Apple computer company, is reported to drive about with four or more GPS devices—which, in turn, requires procedures for weighing conflicting information. Ants are no exception to this rule. In addition to the coupled polarization compass and pedometer system, the desert ant *Cataglyphis* can also use direct observations of the sun and wind in finding its path through a visually sparse environment, and it weighs

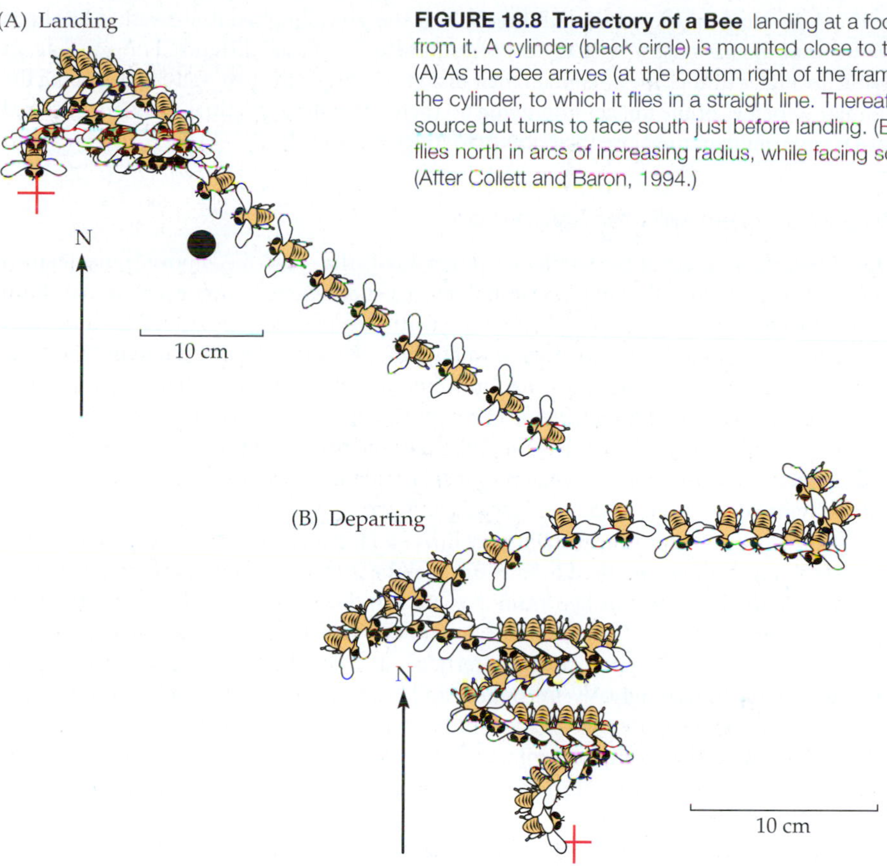

FIGURE 18.8 Trajectory of a Bee landing at a food source and departing from it. A cylinder (black circle) is mounted close to the food source (cross). (A) As the bee arrives (at the bottom right of the frame), it is oriented toward the cylinder, to which it flies in a straight line. Thereafter, it flies to the food source but turns to face south just before landing. (B) As the bee departs, it flies north in arcs of increasing radius, while facing south (flying backward). (After Collett and Baron, 1994.)

these two inputs differently depending on how high the sun is in the sky.[34] Similarly, ants inhabiting different visual environments use different sets of strategies for pathfinding.[35,36]

In addition to visual cues and polarized light,[37,38] bees can use a magnetic compass to orient themselves while searching for a target. Collett and his colleagues demonstrated this phenomenon in experiments in which they trained bees to collect sugar from a small bottle cap on a board.[39] For orientation, a black cylinder was placed in a constant compass direction at a fixed distance from the bottle cap. The cylinder and the sucrose were moved to different places on the board between trials. Periodically, the bottle cap was removed, leaving only the cylinder. The exploration of the board by a trained bee was monitored by video recording as it hunted for the sucrose and then returned home. Figure 18.8 shows the trajectory of a bee as it approached and later left the cylinder (black circle) and the food source (cross). What is clear is that the bee turned so as to face south before landing and again faced south shortly after taking off. In this way, it viewed the visual cue and the attractant (sugar) each time from a constant direction. Observation of the sky alone was not a sufficient explanation for this behavior. Bees faced south in the rain, under a completely overcast sky, or when the sky compass was eliminated. From such observations, one can conclude that somehow the animal can distinguish south from north, east, or west.

The fact that bees are sensitive to magnetic fields was shown by training bees under a tarpaulin, using imposed magnetic fields that shifted the magnetic north. The bees oriented themselves again to the *south*, but this direction was the south of the imposed magnetic field. How changes in imposed magnetic fields produce changes in behavior and pattern recognition is not known. These are sensory mechanisms not represented in our nervous system but present in animals such as birds, turtles, and certain invertebrates.

Without access to proprioceptive information of walking, how do bees judge the distance they have traveled on each *leg* of their flights? By analogy with the proprioceptive memory of the ants, one might imagine that bees count wing beats, but this would lead to

[34] Muller, M., and Wehner, R. 2007. *Naturwissenschaften* 94: 589–594.

[35] Collett, T. S., and Graham, P. 2004. *Curr. Biol.* 14: R475–477.

[36] Cheng, K. et al. 2009. *Behav. Processes* 80: 261–268.

[37] Lehrer, M., and Collett, T. S. 1994. *J. Comp. Physiol. A* 175: 171–177.

[38] Dacke, M., and Srinivasan, M. V. 2007. *J. Exp. Biol.* 210: 845–853.

[39] Collett, T. S., and Baron, J. 1994. *Nature* 368: 137–140.

errors depending on the speed and direction of the prevailing wind. Instead, it has been shown that the bees use a "visually driven odometer" to judge distance. For example, in flying through a tunnel, they maintain their distance between the two walls by balancing the angular velocities of the images impinging on their left and right retinas. In an uncluttered visual environment, they lose their ability to judge distance.[40]

Neural Mechanisms for Navigation

One satisfactory aspect of the studies on insect navigation described above is the detailed information now available about the initial sensory mechanisms. Since each ommatidium of the dorsal rim contains photoreceptors sensitive to planes of polarized light at right angles to one another, and since each ommatidium views the sky at a different angle, the array of ommatidia provides information to the brain about the spatial distribution of polarized light vectors. Moreover, one can compute how the system behaves when challenged. In behavioral studies, objects are placed in the path so that detours must be made and corrected for. Another challenge is to displace the insects at different times of day or with different distortions of the polarized light input.

At the same time, technical challenges have so far prevented scientists from making the detailed recordings required to fully unravel the integrative steps performed by neurons within the insect ant or bee brain. Neurons in the ant's optic lobe are particularly hard to record because of their small size and the surrounding sheath is very tough. Thus, Labhart and colleagues studied interneurons in crickets that receive inputs from receptors for polarized light.[41,42] As in ants and crustaceans, cricket ommatidia that are sensitive to the plane of polarized light contain rhabdomeres with orthogonally arranged microvilli—these rhabdomeres project to the interneuron, which computes information about the vector of polarization. Electrical recordings from such cells are shown in Figure 18.9. Their responses are just what would be predicted from behavioral studies with polarized light. More recently, similar neurons have been recorded from the more difficult ant preparation—only six polarization-sensitive neurons could be recorded successfully from 40 preparations.[43] Intriguingly, by a computational approach in which

[40] Srinivasan, M. et al. 1996. *J. Exp. Biol.* 199: 237–244.

[41] Labhart, T. 1988. *Nature* 331: 435–437.

[42] Labhart, T., Petzold, J., and Helbling, H. 2001. *J. Exp. Biol.* 204: 2423–2430.

[43] Labhart, T. 2000. *Naturwissenschaften* 87: 133–136.

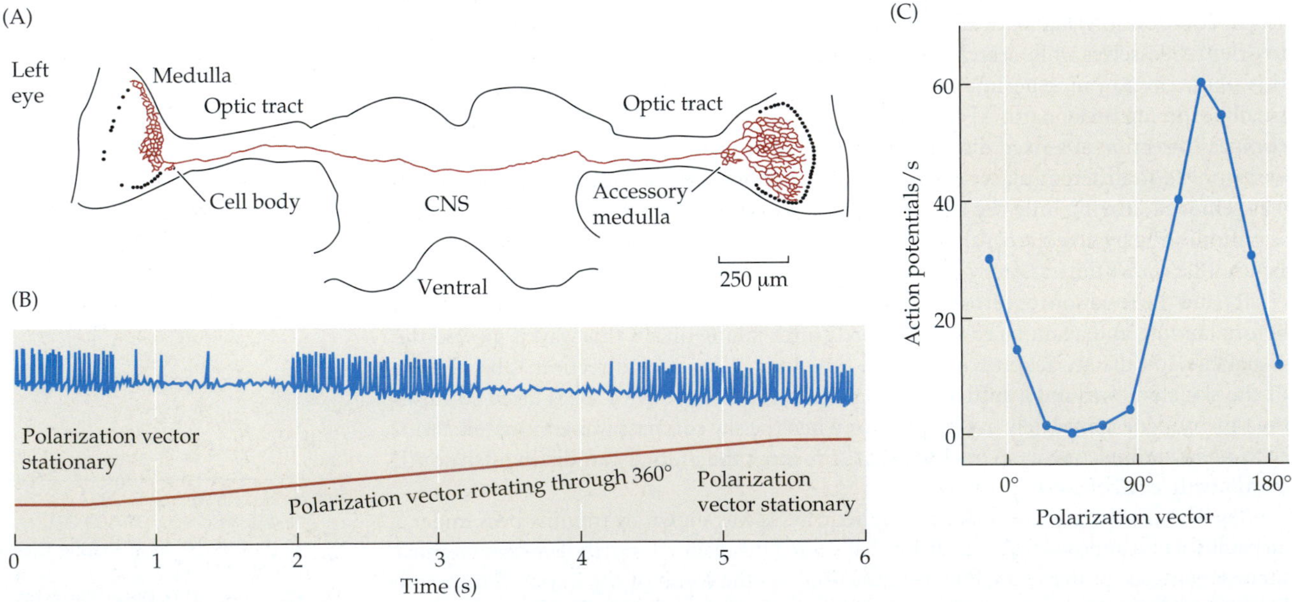

FIGURE 18.9 Electrical Responses of Polarization-Sensitive Interneurons in cricket CNS. The cricket interneurons are large enough to be impaled by microelectrodes. (A) Reconstruction of the interneuron stained by intracellular injection of a histological marker, neurobiotin. This neuron receives its input from the left eye. (B) Responses of a neuron of this type to polarized light as the polarization vector is rotated through 360°. (C) Graphical representation of response intensity plotted against the angle of the polarization vector. (After Labhart, 1988.)

FIGURE 18.10 Mobile Robot Known as Sahabot devised by Wehner and his associates. This robot is equipped with six polarized light sensors arranged in pairs. Each pair forms a polarization-detecting unit, with properties resembling those of the neuron shown in Figure 18.9. Each unit can be tuned to a specific direction of polarization. The robot is able to navigate with this polarized light compass, successfully recreating the behavior of the ant. For example, it can be driven along a tortuous path and then compute the shortest way back to the starting point. Such models illustrate the possibility of computing trajectories in this way but do not, of course, provide evidence that this is the system the ant uses. (After Wehner, 1997; photograph kindly provided by R. Wehner.)

the known properties of neurons are used, one can produce models or even robots that mimic accurately the navigational behavior of the desert ant using cues provided by polarized light (Figure 18.10). Still, to investigate the detailed links between sensory input and motor performance, we must turn to another system.

Behavioral Analysis at the Level of Individual Neurons in the CNS of the Leech

Since the days of ancient Greece and Rome, physicians have applied leeches to patients suffering from anything from epilepsy, angina, tuberculosis, and meningitis to black eyes and hemorrhoids—an unpleasant treatment that almost certainly did more harm than good to the unfortunate victims. Nor was it all fun and games for the leeches—by the nineteenth century, use of the medicinal leech *Hirudo* was so prevalent that the animal became almost extinct in Western Europe, forcing Napoleon to import about 6 million leeches from Hungary in one year to treat his soldiers. This mania for leeching had one benefit for contemporary biology—the ready availability of medicinal leeches facilitated basic research on their reproduction, development, and anatomy. In the late nineteenth century, the leech nervous system was extensively studied by a roster of distinguished anatomists, including Ramón y Cajal, Sanchez, Gaskell, Del Rio Hortega, Odurih, and Retzius.[44] At about the same time, C. O. Whitman, one of the founders of experimental embryology, used another leech species to follow the fates of early embryonic cells[45] (see Chapter 25). Interest in the leech thereafter declined, to be rekindled in 1960 when Stephen Kuffler and David Potter first applied modern neurophysiological techniques to its nervous system.[46] This research set the stage for extensive studies of its circuitry and behavior,[47–49] which in turn led to subsequent studies of leech development[50,51] and regeneration,[52] and of the cell biology or biophysics of its isolated neurons.[53] Strange though it seems, in recent decades, leeches have reentered the realm of medicine, both as a source of novel anticoagulants and, reprising their traditional role in blood letting, for the minimally invasive postsurgical treatment of venous congestion.[54,55]

With a nervous system containing only about one tenth the number of neurons of the insects discussed in the preceding sections, leeches still exhibit reflexes such as shortening and local bending in response to mechanical stimuli, along with behaviors involving coordination of the whole body, such as swimming and crawling. Leeches orient by means of sensory cues that allow them to access potential food sources with an efficiency that is the bane of those who swim or trek in leech habitat. Like other animals, leeches also modulate their behaviors in response to changes in physiological state and must also choose among mutually incompatible behaviors, such as feeding, mating, swimming, or crawling. Remarkable progress has been made in understanding how these processes are carried out, starting with the properties and connections of individual nerve cells.

Leech Ganglia: Semiautonomous Units

The body and the nervous system of the leech are segmented. That is, various tissues and organ systems are organized along the anterior–posterior axis into repeating units (segments) that are similar, though not identical, throughout the animal. The leech body contains 32

[44] Payton, W. B. 1981. In *Neurobiology of the Leech.* Cold Spring Harbor Laboratory, Cold Spring Harbor, NY. pp. 27–34.

[45] Maienschein, J. 1978. *J. Hist. Biol.* 11: 129–158.

[46] Kuffler, S. W., and Potter, D. D. 1964. *J. Neurophysiol.* 27: 290–320.

[47] Friesen, W. O., and Kristan, W. B. 2007. *Curr. Opin. Neurobiol.* 17: 704–711.

[48] Kristan, W. B., Jr., Calabrese, R. L., and Friesen, W. O. 2005. *Prog Neurobiol.* 76: 279–327.

[49] Li, Q., and Burrell, B. D. 2009. *J. Comp. Physiol. A* 195: 831–841.

[50] Weisblat, D. A., and Kuo, D.-H. 2009. In *Emerging Model Organisms, a Laboratory Manual.* Cold Spring Harbor Laboratory, Cold Spring Harbor, NY. pp. 245–274.

[51] Marin-Burgin, A., Kristan, W. B., Jr., and French, K. A. 2008. *Dev. Neurobiol.* 68: 779–787.

[52] Duan, Y. et al. 2005. *Cell Mol. Neurobiol.* 25: 441–450.

[53] Trueta, C., Mendez, B., and De-Miguel, F. F. 2003. *J. Physiol.* 547: 405–416.

[54] Weinfeld, A. B. et al. 2000. *Ann. Plast. Surg.* 45: 207–212.

[55] Mineo, M., Jolley, T., and Rodriguez, G. 2004. *Urology* 63: 981–983.

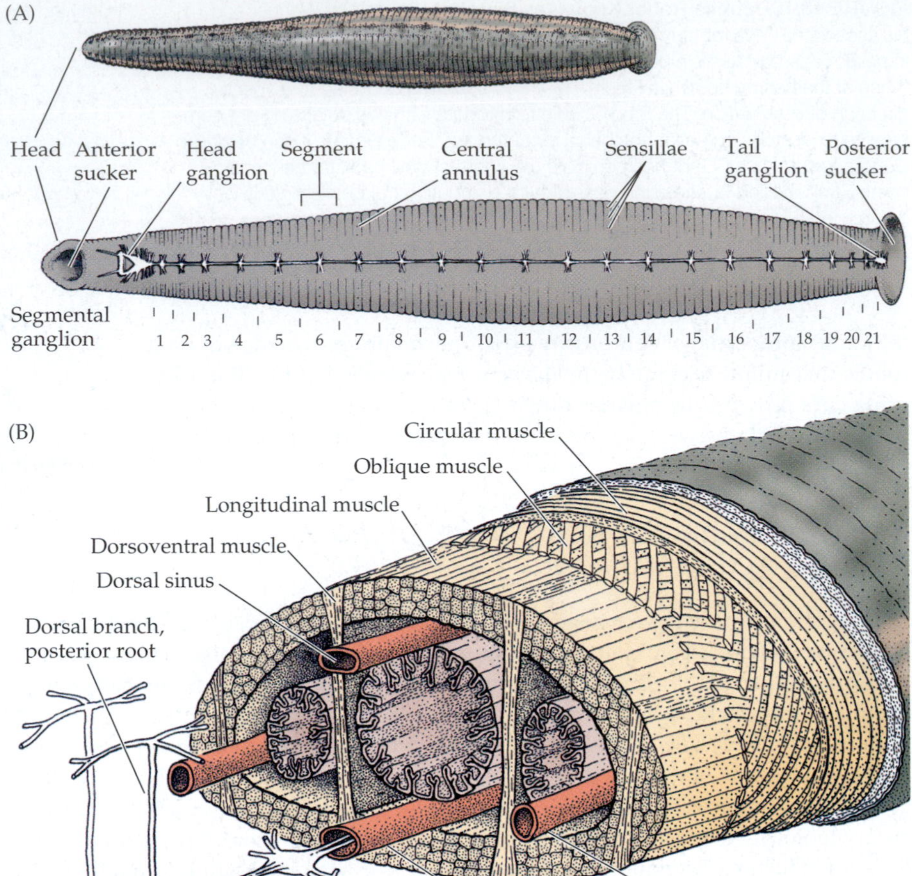

FIGURE 18.11 Central Nervous System of the Leech. (A) The segmented CNS of the leech includes a chain of 21 discrete midbody ganglia plus head and tail ganglia made up of fused segmental units. Over most of the body, five circumferential annuli make up each segment; the central annulus is marked by sensory organs (sensillae) responding to light and touch. (B) The nerve cord lies in the ventral part of the body within a blood sinus. Ganglia, which are linked to each other by bundles of axons (connectives), innervate the body wall by paired roots. The muscles are arranged in three principal layers: circular, oblique, and longitudinal. In addition, there are dorsoventral muscles that flatten the animal and fibers immediately under the skin that raise it into ridges.

segments; 21 standard segments make up the midbody of the animal, 4 fused segments make up most of the head, and 7 fused segments make up the tail. Each midbody segment is innervated by a morphologically stereotyped ganglion. Even the brain-like head and tail ganglia shown in Figure 18.11 consist primarily of fused segmental ganglia, in which many characteristic features of midbody ganglia are still recognizable.[56]

Each ganglion contains only about 400 nerve cells,[57] which have distinctive shapes, sizes, positions, and branching patterns, as do neurons in the periphery.[58] A ganglion receives sensory information from, and innervates muscles and other targets in, a well-defined territory of the body by way of paired axon bundles (**roots**), and it communicates with neighboring and distant parts of the nervous system through another set of bundles (**interganglionic connectives**). The coordinated operation of the whole nerve cord is also influenced by the brains at each end of the animal.

Perhaps the main appeal of the leech as a neurobiological preparation is the beauty of the ganglion as it appears under the microscope, with its neurons so recognizable and so familiar from segment to segment, specimen to specimen, and even species to species (Figure 18.12). As one looks at these limited aggregates of cells laid out in an orderly pattern, one cannot but marvel at how they, on their own, being the brain of the creature, are responsible for all its movements, hesitations, avoidance, mating, feeding, and sensations. In addition to the esthetic pleasure provided by the preparation, there is the intellectual excitement of trying to solve the circuitry and logic of a finite, well-organized nervous system. But, before one can work out how the animal performs its movements, it is necessary to know about the individual cells: their properties, connections, and functions.

[56] Coggeshall, R. E., and Fawcett, D. W. 1964. *J. Neurophysiol.* 27: 229–289.

[57] Macagno, E. R. 1980. *J. Comp. Neurol.* 190: 283–302.

[58] Huang, Y. et al. 1998. *J. Comp. Neurol.* 397: 394–402.

(A)

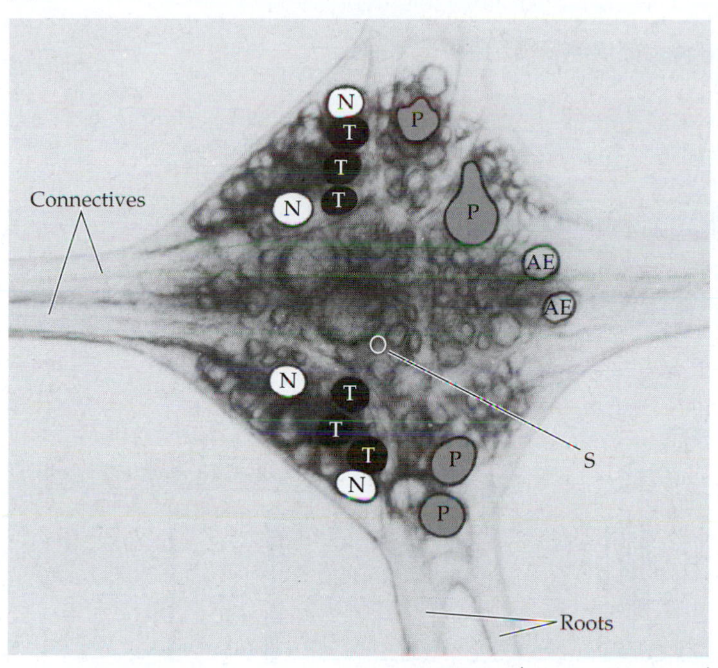

(B)

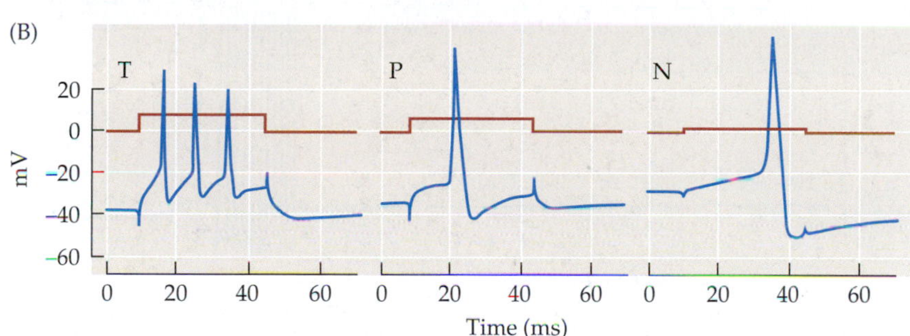

FIGURE 18.12 Leech Segmental Ganglia Contain Individually Identified Neurons. (A) Ventral view of a midbody ganglion. Some individual cells are clearly recognized by their size and location within the ganglion. For example, the pairs of sensory cells responding to touch (T), pressure (P) and noxious (N) mechanical stimulation of the skin are labeled, as are the annulus erector (AE) motoneurons outlined in the posterior part of the ganglion. Identification of other cells in the ganglion is based on more subtle but equally clear-cut physiological or morphological criteria. For example, the S cell is a small, unpaired neuron connected to its homologs in neighboring ganglia; S cells play a role in dishabituation and sensitization. (B) Each of the sensory cell types give distinctive action potentials. Impulses in T cells are briefer and smaller than those in P or N cells. N cell impulses have a larger afterhyperpolarization than do T or P cells. Current injected into cells through the microelectrode is monitored on the upper traces. (After Nicholls and Baylor, 1968; Stuart, 1970.)

Sensory Cells in Leech Ganglia

When one strokes, presses, or pinches the skin of a leech, a sequence of movements follows. One segment or more shortens abruptly, and the skin becomes raised into a series of distinct ridges. Subsequently, the animal bends, writhes, or swims away. One can reliably identify the individual sensory and motor cells that mediate these reflexes according to their shapes, sizes, positions, and electrical characteristics.[59–62] Figure 18.12 shows the distribution of identified sensory cells, motoneurons, and interneurons in a leech ganglion. The 14 neurons labeled T, P, and N in Figure 18.12 are sensory cells and represent three sensory modalities. Each cell responds selectively to touch (T), pressure (P), or noxious (N) mechanical stimulation of the skin. Figure 18.13 illustrates the responses of sensory cells to various forms of cutaneous stimuli.

T cells give transient responses to light touch of the skin surface (Figure 18.13A). Their sensory endings consist of small dilatations situated between epithelial cells on the surface of the skin. T cells adapt rapidly to a small, maintained indentation (i.e., they cease firing within a fraction of a second). The P cells respond only to a marked pressure or deformation of the skin and show a slowly adapting discharge (Figure 18.13B). The N cells require still stronger mechanical stimuli, such as pinching the skin with blunt forceps (Figure 18.13C,D). Similar to polymodal nociceptors in mammals, one or both N cells also respond selectively to acid, heat, and capsaicin.[63] The modalities and responses of these neurons in leech resemble those of mechanoreceptors in the human skin (see Chapter 21), which distinguish among touch, pressure, and noxious or painful stimuli. In the leech, however, individual neurons do the job of many equivalent neurons in a densely innervated region of our own skin, such as the fingertip.

[59] Nicholls, J. G., and Baylor, D. A. 1968. *J. Neurophysiol.* 31: 740–756.

[60] Stuart, A. E. 1970. *J. Physiol.* 209: 627–646.

[61] Lockery, S. R., and Kristan, W. B., Jr. 1990. *J. Neurosci.* 10: 1816–1829.

[62] Rodriguez, M. J., Perez-Etchegoyen, C. B., and Szczupak, L. 2009. *J. Comp. Physiol. A* 195: 491–500.

[63] Pastor, J., Soria, B., and Belmonte, C. 1996. *J. Neurophysiol.* 75: 2268–2279.

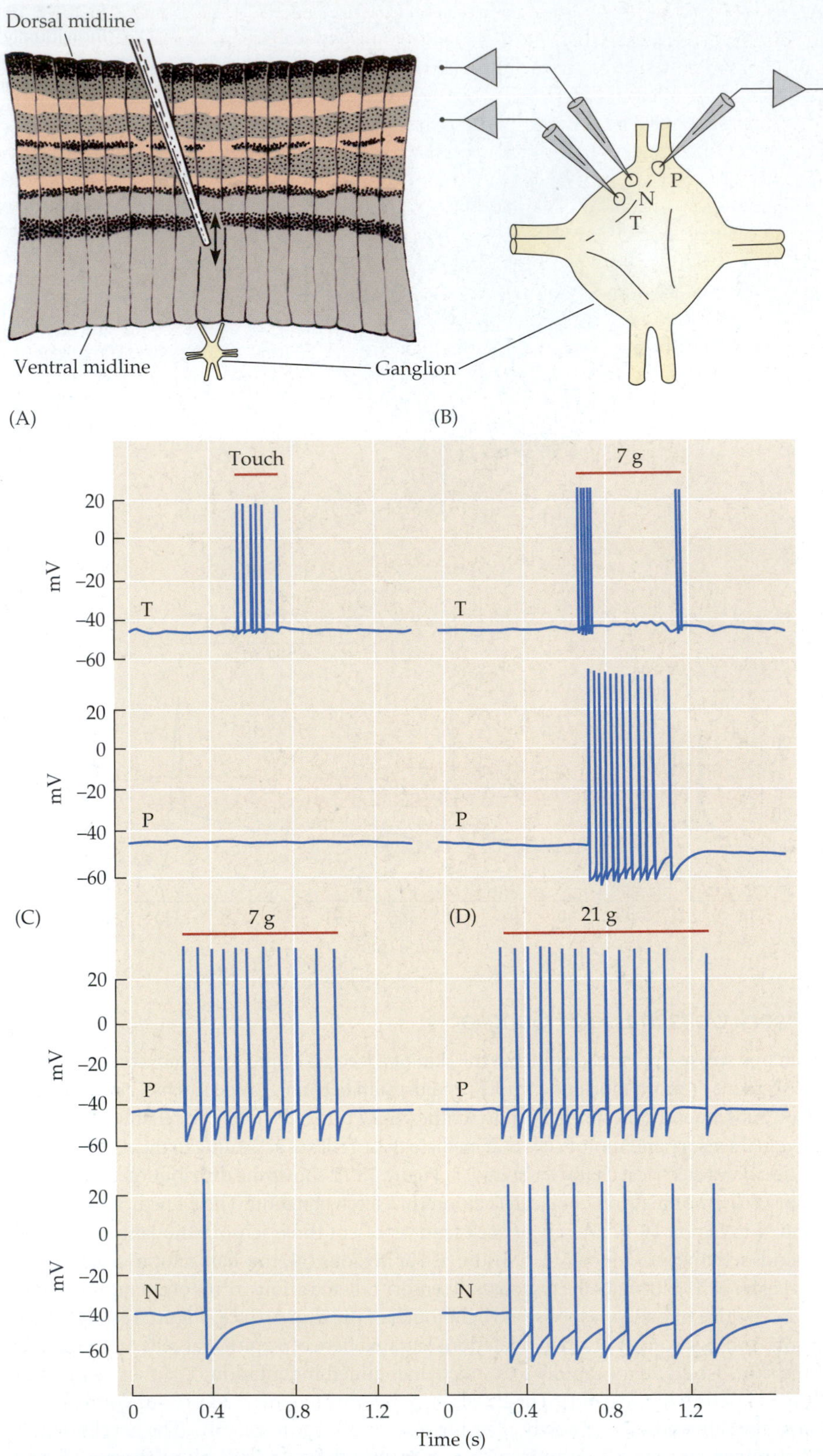

FIGURE 18.13 Sensory Neurons Respond to Skin Stimulation. Intracellular record of T, P, and N sensory cells (see Figure 18.12). The preparation consists of a piece of skin and the ganglion that innervates it. Cells are activated by touching or pressing their receptive fields in the skin. (A) A T cell responds to light touch that is not strong enough to stimulate the P cell. (B) Stronger, maintained pressure evokes a prolonged discharge from the P cell and a rapidly adapting on and off response from the T cell. (C,D) Still stronger pressure is needed to activate the N cell mechanically. (After Nicholls and Baylor, 1968.)

Figure 18.14 shows that each sensory cell innervates a defined territory, which is mapped by recording from a cell while applying mechanical stimuli to the skin or by labeling the cell and its axons with a marker, such as horseradish peroxidase.[64] The boundaries of receptive fields can be conveniently identified in relation to landmarks such as segment borders or the coloring of skin, so that one can predict reliably which cells will fire when a particular area is touched, pressed, or pinched. One of the touch-sensitive T cells innervates dorsal skin, another ventral skin, and a third laterally situated skin. Similarly, the two P sensory cells divide the skin into roughly equal dorsal and ventral areas. The elaborate and stereotyped branching pattern of a P sensory cell injected with horseradish peroxidase is shown in Figure 18.14; like P, the T and N cells also send axons along the connectives to neighboring ganglia, and then out to minor receptive fields on either side of the major region of innervation.[65] Curiously, while the terminal branches of an individual cell each supply a circumscribed area of skin with no overlap between them, endings from other sensory neurons of the same modality may encroach on the same territory to a limited extent within each segment (e.g., the dorsal and ventral P cells on one side of the ganglion overlap

[64]Blackshaw, S. E. 1981. *J. Physiol.* 320: 219–228.

[65]Yau, K. W. 1976. *J. Physiol.* 263: 489–512.

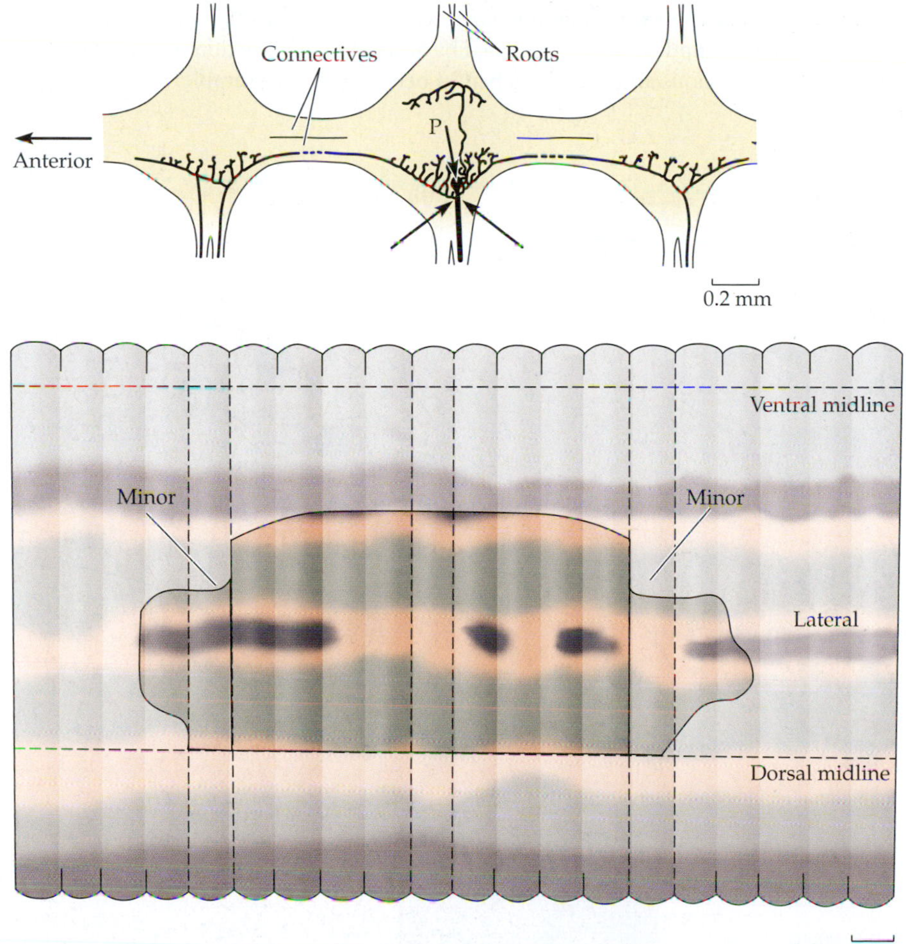

FIGURE 18.14 Receptive Field of a Pressure Sensory (P) Cell. This P cell has an axon that runs out through the root of its own ganglion to supply the skin of the segment in which it is situated. Other axons of smaller diameter pass along the connectives to neighboring ganglia. These axons pass through the appropriate roots to innervate additional (minor) territories in adjacent segments. The second P cell (not shown) innervates more medial territory (i.e., closer to the ventral midline) but with similar longitudinal extent. There is considerable overlap between the receptive fields of the two cells. Hence, pressure applied to dorsal skin will activate the P cell shown in this figure, pressure applied to ventral skin will activate the other P cell, and pressure applied on lateral skin will activate both P cells. The fact that axons with a small diameter supply the minor fields in adjacent segments has implications for conduction. Although pressure applied anywhere within the receptive field will activate action potentials, conduction can become blocked where an axon of small diameter feeds into a larger-diameter axon, as at the points marked by arrows. (After Gu, 1991.)

in lateral territory) and extensively between segments (e.g., the ipsilateral ventral P cells from adjacent segments overlap across segment borders). We will see in the following discussion that this seemingly sloppy pattern of innervation is important for processing sensory information.

In this system, which has such clear-cut boundaries in the periphery, it has also been possible to determine how the receptive fields become established during development and regeneration.[58,66,67] Additional sensory cells that respond specifically to light, chemical stimuli, vibration, and stretch of the body wall have been found in the head and in the periphery of the leech.[68,69]

Motor Cells

The criterion for showing that a cell is indeed a motor cell is that each impulse in the cell gives rise to a conducted action potential in axons leading to the muscle and then to a synaptic potential in the muscle fiber. More than 20 pairs of motor cells (motoneurons) supplying the various muscles that flatten, lengthen, shorten, and bend the body as well as others that control the heart have been identified in the segmental ganglia (Figure 18.15A,B).[60] Muscles receive inhibitory GABAergic and modulatory peptidergic inputs as well as excitatory inputs that are mediated by acetylcholine. Deletion of a single cell can give rise to an obvious and specific deficit in behavior.[70] For example, in each ganglion

[66]Wang, H., and Macagno, E. R. 1997. *J. Neurosci.* 17: 2408–2419.

[67]Wang, H., and Macagno, E. R. 1998. *J. Neurobiol.* 35: 53–64.

[68]Blackshaw, S. E., and Nicholls, J. G. 1995. *J. Neurobiol.* 27: 267–276.

[69]Blackshaw, S. E., and Thompson, S. W. 1988. *J. Physiol.* 396: 121–137.

[70]Bowling, D., Nicholls, J., and Parnas, I. 1978. *J. Physiol.* 282: 169–180.

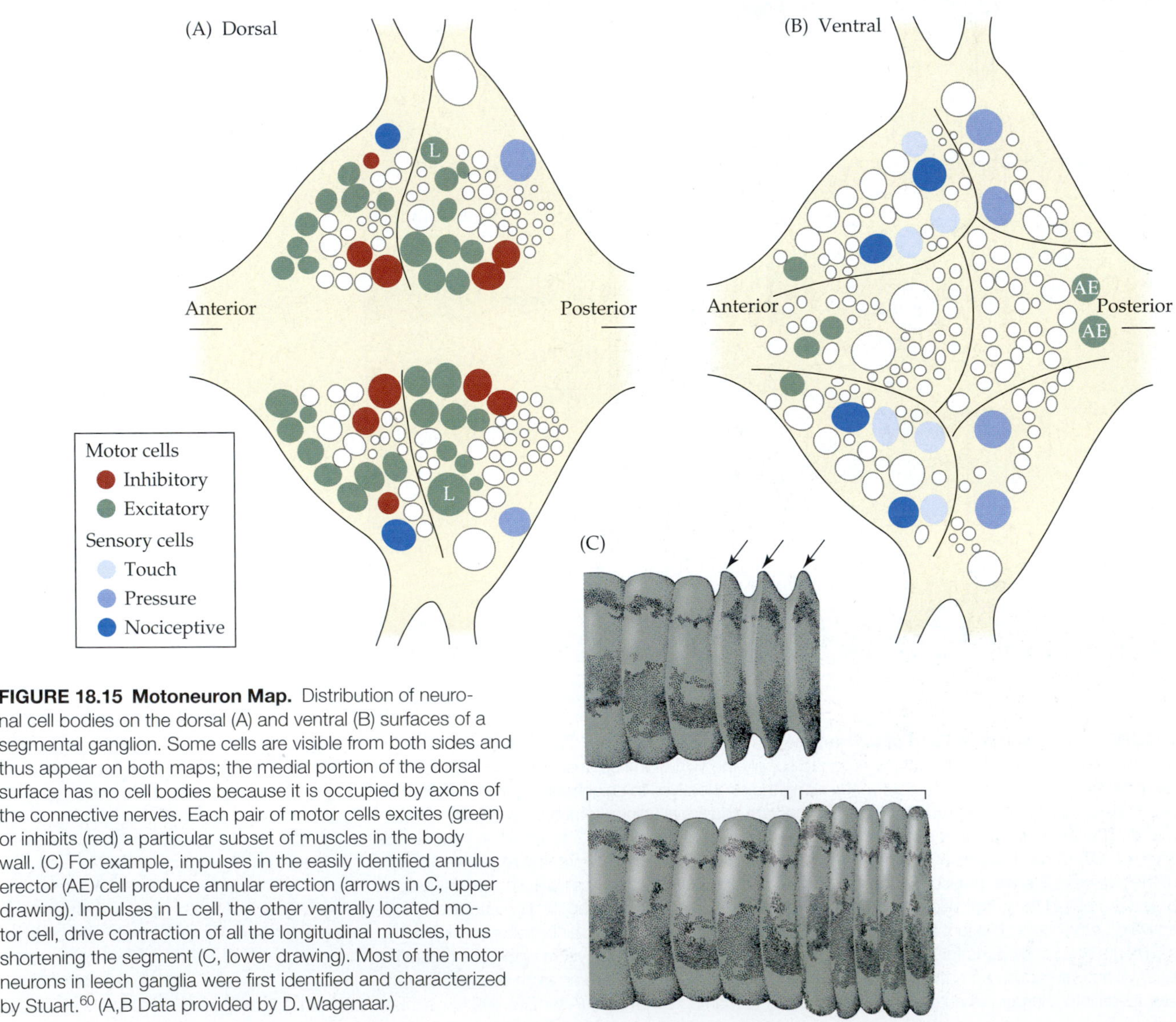

FIGURE 18.15 Motoneuron Map. Distribution of neuronal cell bodies on the dorsal (A) and ventral (B) surfaces of a segmental ganglion. Some cells are visible from both sides and thus appear on both maps; the medial portion of the dorsal surface has no cell bodies because it is occupied by axons of the connective nerves. Each pair of motor cells excites (green) or inhibits (red) a particular subset of muscles in the body wall. (C) For example, impulses in the easily identified annulus erector (AE) cell produce annular erection (arrows in C, upper drawing). Impulses in L cell, the other ventrally located motor cell, drive contraction of all the longitudinal muscles, thus shortening the segment (C, lower drawing). Most of the motor neurons in leech ganglia were first identified and characterized by Stuart.[60] (A,B Data provided by D. Wagenaar.)

there is only one annulus erector motor cell on each side (labeled AE in Figures 18.12 and 18.15B). Impulses in this cell cause the skin of the leech to be raised into ridges like an accordion (Figure 18.15C). When one AE motor cell is killed by injecting it with a mixture of proteolytic enzymes (pronase) in an otherwise intact animal, the region of skin that was innervated exclusively by the killed cell fails to become erect in response to appropriate sensory stimuli. This deficit is not permanent, however. Eventually branches of other AE cells come to supply the denervated territory.[71]

Connections of Sensory and Motor Cells

In the nervous systems of invertebrates, synapses between neurons are usually situated not on the cell bodies but on fine processes within a central region of the ganglion (the neuropil).[56,72–75] Synaptic potentials originating in the neuropil spread into the cell body, where they are recorded as excitatory and inhibitory potentials. Currents injected into the cell body can influence synaptic potentials and the release of transmitter. Despite its complexity, the neuropil is organized in an orderly manner. This fact was revealed by the experiments of Muller and McMahan, who devised the technique of intracellular injection of horseradish peroxidase, using identified nerve cells in the leech.

The branching patterns of sensory and motor cells within the neuropil are characteristic, each cell displaying its own configuration. Examples of the typical ramifications of identified neurons are shown in Figure 18.16 (see also Figure 18.14). A single sensory cell supplies many postsynaptic targets, and its presynaptic endings are themselves supplied by

[71] Blackshaw, S. E., Nicholls, J. G., and Parnas, I. 1982. *J. Physiol.* 326: 261–268.

[72] Muller, K. J., and McMahan, U. J. 1976. *Proc. R. Soc. Lond., B, Biol. Sci.* 194: 481–499.

[73] French, K. A., and Muller, K. J. 1986. *J. Neurosci.* 6: 318–324.

[74] Macagno, E. R., Muller, K. J., and Pitman, R. M. 1987. *J. Physiol.* 387: 649–664.

[75] Baker, M. W. et al. 2003. *J. Neurobiol.* 56: 41–53.

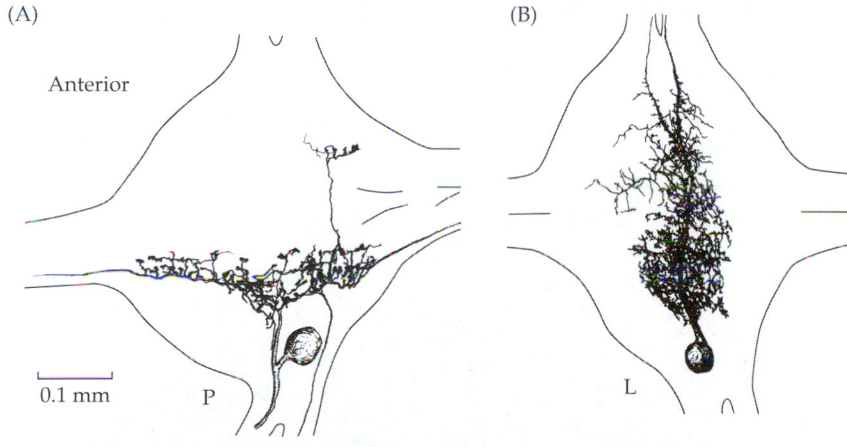

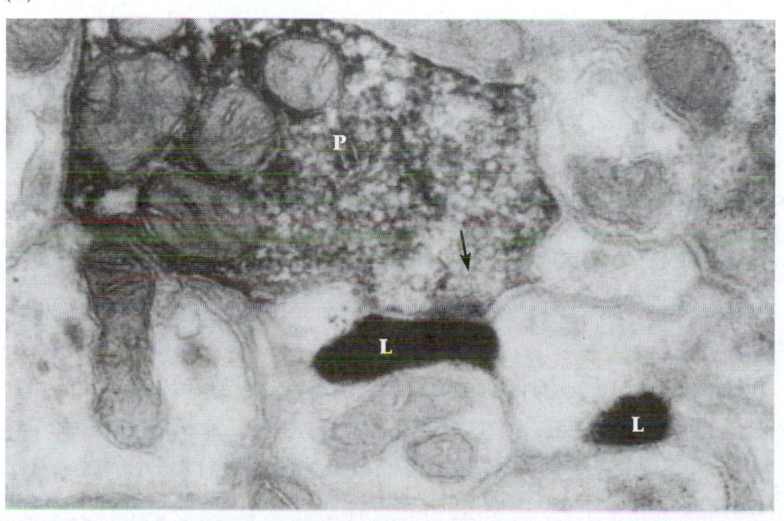

FIGURE 18.16 Structures of Pre- and Postsynaptic Cells labeled by intracellular injection of horseradish peroxidase. (A) The arborization of a pressure (P) cell (shown in Figure 18.14) is profuse, with numerous varicosities. The varicosities represent sites of presynaptic endings that release transmitter. (B) The L motoneuron sends its axons out through contralateral roots. Its processes within the ganglion are smooth and represent postsynaptic sites upon which synapses are made. (C) A synapse (arrow) made by a P cell onto an L motoneuron in the neuropil. Both cells were injected with horseradish peroxidase. (After Muller, Nicholls, and Stent, 1981.)

numerous terminals arising from other neurons that modulate its release of transmitters. Electrical synapses are observed when gap junctions form a narrow, direct cytoplasmic connection between neurons (see Chapter 11). Fluorescent dyes or other markers injected into one cell will usually cross into those cells with which it is electrically coupled (if, as with Lucifer yellow, the injected markers are small enough to pass through the gap junctions).

The T, P, and N sensory cells responding to mechanical stimuli make excitatory connections on the L motoneurons (called "L" because they innervate longitudinal muscles) used for shortening the leech. Several lines of evidence, including electron microscopy, have shown that the connections are direct (i.e., that there are no known intermediary cells; see Figure 18.16).[76] This fact is important because only if each constituent of a circuit and its properties are known can one pinpoint the sites at which any interesting modifications in signaling take place.

For example, the mechanism of transmission onto the L motor cell is different for each type of sensory cell. The N cells act through chemical synapses (with only a hint of electrical coupling), the T cells through rectifying electrical synapses, and the P cells by combination of both mechanisms.[76] The same P and N cells also make direct chemical synapses on another cell, the AE motoneuron. As illustrated in Figure 18.17, the differing properties of synapses linking specific pairwise combinations of neurons mean that the same cell may be affected differently by otherwise similar presynaptic inputs and that a given output from one cell can affect two postsynaptic targets differently. Transmitters used by leech neurons include acetylcholine, γ-aminobutyric acid (GABA), glutamate, dopamine, serotonin, and peptides.[77–81]

In *Hirudo* and in *Aplysia*,[82,83] as in the more complex nervous systems of vertebrates, there are delayed lines through interneurons that parallel the direct route and serve to coordinate more complex directional movements evoked by mechanical stimuli. For example, Kristan and his colleagues have shown that pressure applied to one side of the leech causes the animal to bend. The direction of the bend depends on the position of the mechanical stimulus on the body wall.[48,84] The reflex causes the bend to be directed away from the stimulus. Due to the partial overlap of their receptive fields, a single stimulus will activate multiple P neurons

[76]Nicholls, J. G., and Purves, D. 1972. *J. Physiol.* 225: 637–656.

[77]Cline, H. T. 1986. *J. Neurosci.* 6: 2848–2856.

[78]Thorogood, M. S., Almeida, V. W., and Brodfuehrer, P. D. 1999. *J. Comp. Neurol.* 405: 334–344.

[79]De-Miguel, F. F., and Trueta, C. 2005. *Cell Mol. Neurobiol.* 25: 297–312.

[80]Calvino, M. A., and Szczupak, L. 2008. *J. Comp. Physiol. A* 194: 523–531.

[81]Glover, J. C. et al. 1987. *J. Neurosci.* 7: 581–594.

[82]Hickie, C., Cohen, L. B., and Balaban, P. M. 1997. *Eur. J. Neurosci.* 9: 627–636.

[83]Walters, E. T., and Cohen, L. B. 1997. *Invert. Neurosci.* 3: 15–25.

[84]Lewis, J. E., and Kristan, W. B., Jr. 1998. *Nature* 391: 76–79.

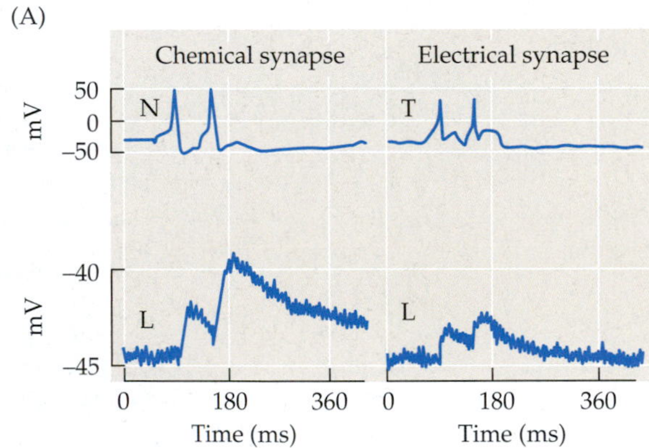

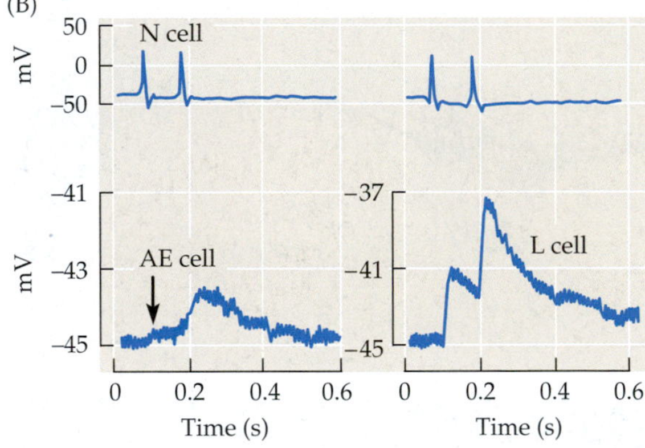

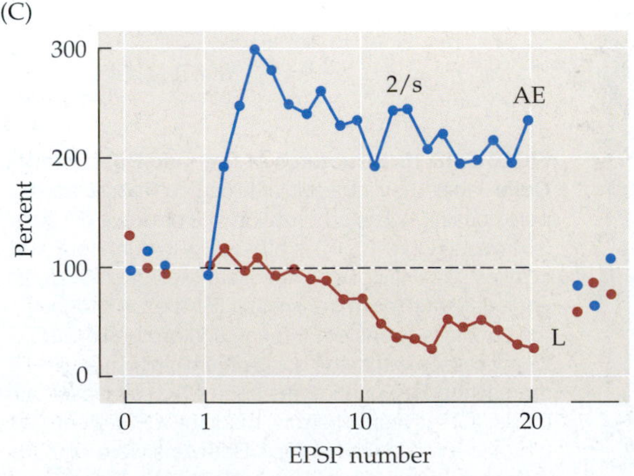

FIGURE 18.17 Short-Term Changes in Synaptic Plasticity between Sensory and Motor Neurons. (A) Chemical and electrical transmission. A nociceptive (N) or T cell is stimulated twice in succession, and its impulses are recorded (upper trace). Facilitation occurs at the chemical synapse between N and L cells, so the second impulse leads to a larger synaptic potential (facilitation; bottom left). In contrast, at the T–L synapse, no facilitation is seen, which is typical of electrical synapses. (B) Characteristics of transmitter release at different synapses made by a single presynaptic neuron. An N cell is stimulated and responses are recorded in L and annulus erector (AE) cells. Facilitation is greater at the N–AE synapse. (The small, first synaptic potential in the AE cell is marked by an arrow.) (C) When a train of impulses is evoked by stimulating the N cell at a rate of 2 per second (in elevated Ca^{+2}), the synaptic potentials in the AE cell are facilitated to more than double their original size, whereas those in the L cell decrease in amplitude (depression). The *x*-axis indicates the position number of the synaptic potential in the train. The *y*-axis gives the height of the synaptic potentials at each position relative to the average size of potentials recorded before the train (set at 100 %). (A after Nicholls and Purves, 1972; B and C after Muller and Nicholls, 1974.)

and information about the stimulus position can be obtained by comparing their relative frequency of firing. This process provides much greater spatial resolution than could be obtained by an equivalent number of sensory neurons with strictly non-overlapping fields, and it provides a compelling rationale for the sloppiness of the overlapping fields alluded to above. The information coming from the various P cells in each segment is processed by a network of two dozen or so interneurons and feeds to about that number of excitatory and inhibitory motoneurons, so that bending in the appropriate direction by contracting certain muscles and relaxing others can occur (Figure 18.18). This motion is achieved by a distributed processing mechanism: each sensory cell makes excitatory inputs to many of the interneurons; each of those makes excitatory inputs to many of the motoneurons. Connections among the interneurons themselves are minimal and a generalized inhibition appears critical to the proper functioning of the reflex.

Higher Order Behavior in the Leech

Analyzing how complex behavioral acts are built up from simple, elementary reflexes is a major goal of neurobiology and is one scientific aim for which tractable invertebrates are particularly well suited. Another major aim is to understand, at the cellular and molecular levels, how experience leads to behavioral changes of the sort we commonly refer to as learning and memory. For example, in most animals, repetitive stimuli lead to changes in the behavioral responses they evoke. Thus, the reflexive withdrawal from a mild tactile stimulus becomes weaker if it is repeated enough times (habituation), while a strong stimulus can produce an overall increase in sensitivity (sensitization). The sea slug *Aplysia* has been useful for elucidating

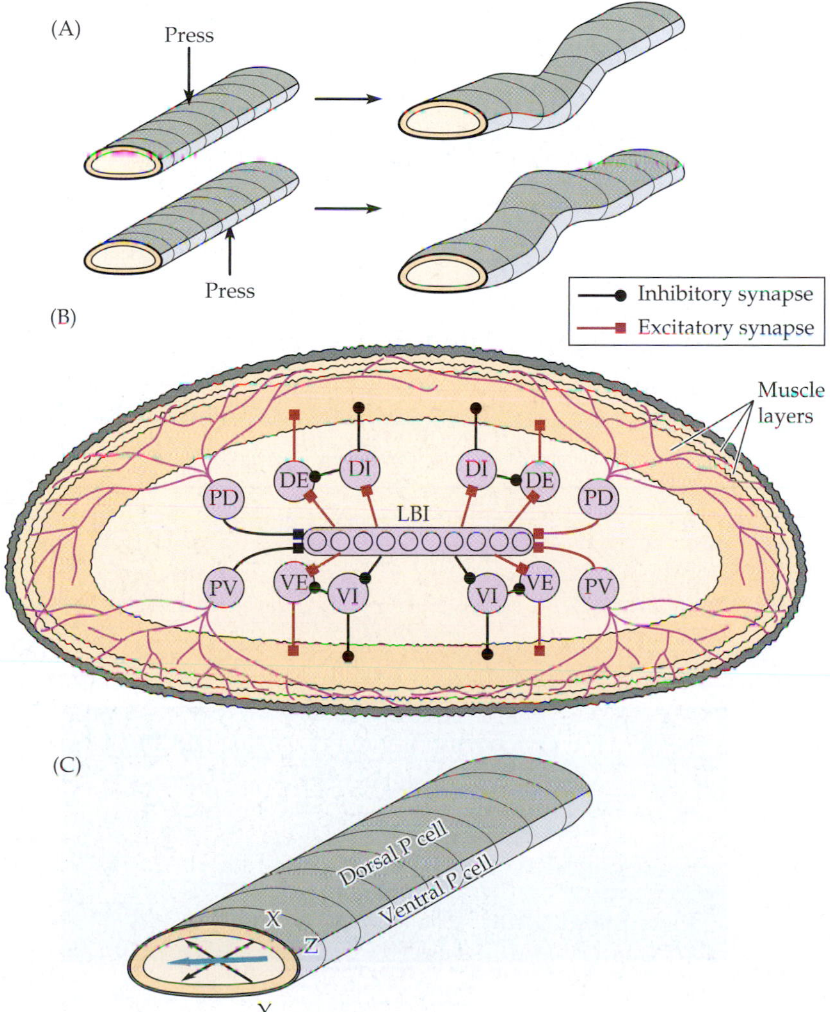

FIGURE 18.18 Integration by Interneurons in the leech bending reflex. When the skin of a leech is pressed at one point, a reflex bend occurs in the body so as to withdraw that region from the stimulus. Touch and pressure (P) cells convey signals to an array of interneurons. These in turn make excitatory and inhibitory connection on motoneurons. (A) When the P cell innervating dorsal skin is activated by pressure (arrows), longitudinal muscles contract near the point of contact (from excitatory motoneuron inputs), while those on the opposite, ventral side of the body relax (due to inhibitory inputs). Conversely, when ventral skin is pressed, the ventral muscles contract while those on the dorsal side relax. (B) Scheme for connections of the four P cells that connect to the interneurons that, in turn, excite motoneurons used for bending (PD and PV designate the P neurons innervating dorsal and ventral fields, respectively; LBI designates the group of local bend interneurons; DE, DI, VE, and VI designate excitatory and inhibitory motoneurons innervating dorsal and ventral muscles, respectively). (C) When both P cells are activated, by pressing simultaneously at X and Y, or by pressing in the zone where the dorsal and ventral sensory fields overlap (Z), the interneurons integrate the information provided by the relative firing frequencies of the dorsal and ventral P cells (black vector arrows), and drive the motoneurons so as to produce an appropriate bend (blue vector arrows). Thus, the direction of the bend can by predicted from the firing frequencies of the two P cells. (A and C, after Lewis and Kristan, 1998; B after Baca et al., 2008.)

molecular mechanisms by which transmission can be modified at individual synapses.[4,85–87] The leech *Hirudo* has been valuable for tracing the circuits by which individually identified cells act in concert to produce and maintain the rhythms of its heartbeat and swimming.

Habituation, Sensitization, and Conduction Block

In a simple nervous system, it is possible to identify a single neuron that plays a coordinating role in the behavior of an animal as a **command neuron**, which can initiate or orchestrate a particular behavioral response of the animal. This concept has been studied extensively in insects and crustaceans. In the leech, a complex function of this sort is well illustrated by a single, unpaired interneuron in each ganglion called the S cell (Figure 18.19; see also Figure 18.12).[88] The S neuron receives excitatory inputs from touch and pressure sensory cells and, in turn, excites the L motoneuron, which causes shortening (as already discussed). Each S cell is electrically connected by a large, rapidly conducting axon to the axon of its homologue in the neighboring ganglion. These S-to-S junctions occur in the midregion of the connective, as is revealed by injecting one S cell with horseradish peroxidase or another high molecular marker, which cannot cross gap junctions (see Figure 18.19).

The chain of S cells is essential for adaptive types of behavior. If repeated touch stimuli are applied to a segment of the leech's body wall, a reflex shortening is produced. The response becomes progressively weaker with each repeated touch—a process known as **habituation** (Figure 18.20).[89] After a stronger stimulus that activates P (pressure) and N (nociceptive) cells, as well as T (touch) cells, has been applied to a different region of the body wall, the shortening response to touch once again becomes evident. This recovery

[85] Si, K., Lindquist, S., and Kandel, E. 2004. *Cold Spring Harb. Symp. Quant. Biol.* 69: 497–498.

[86] Kandel, E. R. 2001. *Science* 294: 1030–1038.

[87] Glanzman, D. L. 2009. *Neurobiol. Learn. Mem.* 92: 147–154.

[88] Muller, K. J., and Carbonetto, S. 1979. *J. Comp. Neurol.* 185: 485–516.

[89] Sahley, C. L. et al. 1994. *J. Neurosci.* 14: 6715–6721.

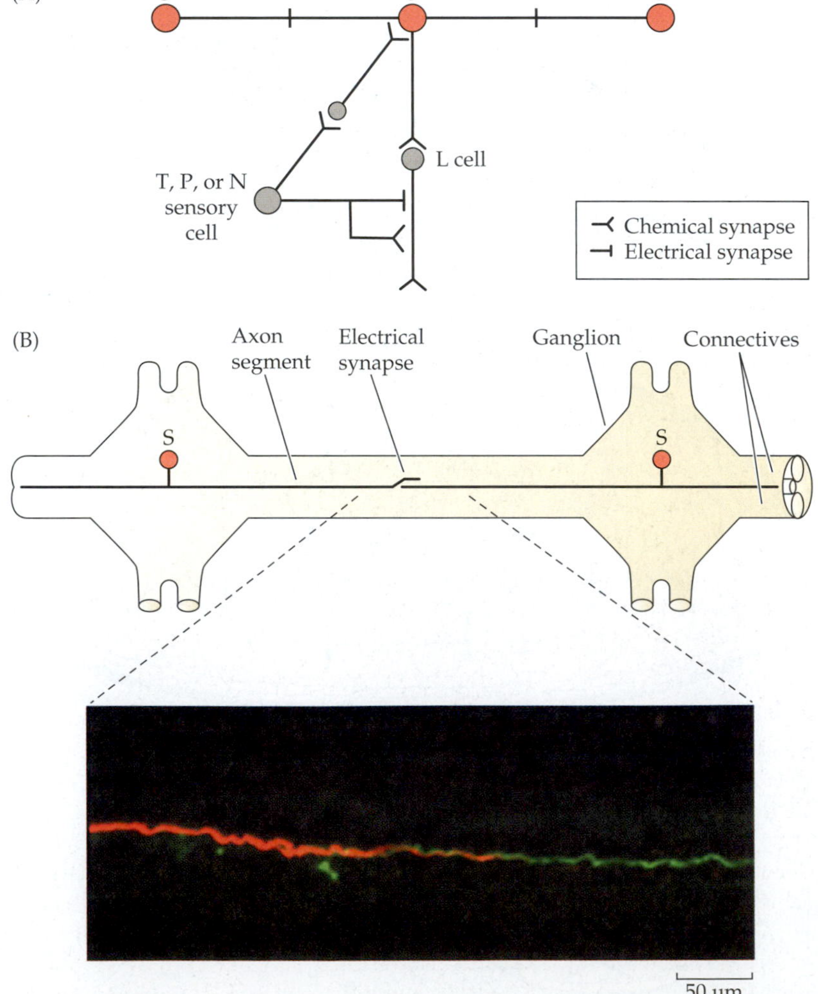

FIGURE 18.19 Known Connections of the Rapidly Conducting S Cells. (A) Electrical and chemical connections of sensory touch (T), pressure (P), and nociceptive (N) cells as well as S cells onto the L motoneuron. Each ganglion contains only one S cell, the axons of which make electrical synapses with its homologs. The S cell chain modulates (but is not required for) rapid shortening in response to sensory stimuli. (B) S cell processes connect to processes emanating from their homologs in neighboring ganglia midway along the interganglionic connective—by electrical synapses. The photomicrograph shows an interganglionic connective from a preparation in which the S cell whose process enters from the left (red) was injected with a mixture of Lucifer yellow (LY; yellow-green) and rhodamine-dextran (RD; red). RD (MW greater than 10 kDA) cannot cross from cell to cell through electrical junctions, whereas LY (MW ~0.5 kDa) can. Accordingly, the axon of the right-hand S cell became labeled with LY but not with RD. S cell axons regenerate and re-form their connections with very high specificity after injury. The micrograph, which resembles that of a normal animal, was in fact taken from a preparation in which the axon of one S cell had been severed and had re-formed its connections after regeneration. (After Mason and Muller, 1996, and Burrell et al., 2003; micrograph kindly provided by K. Muller.)

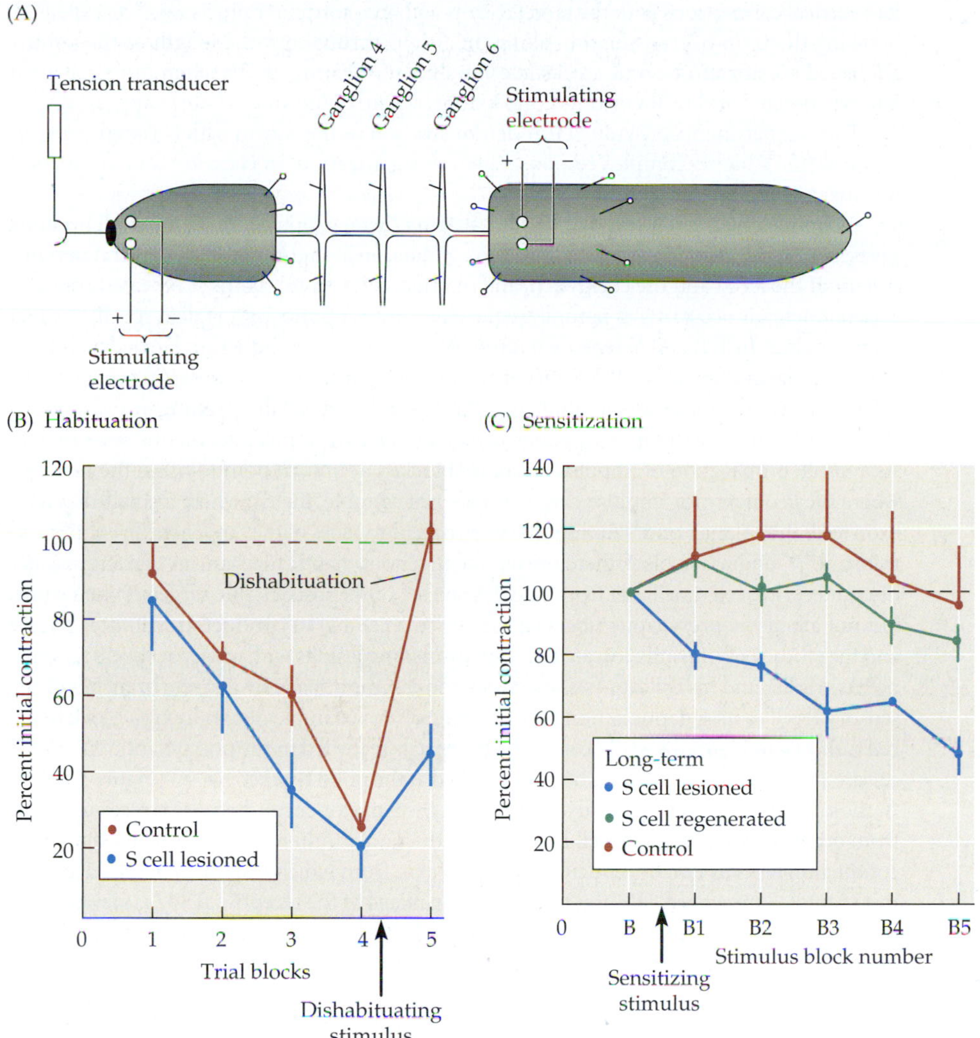

FIGURE 18.20 Habituation, Dishabituation and Sensitization of leech reflexes, and the role of the S cell. (A) Exposed ganglia connected to the anterior and posterior parts of the body. Stimuli are applied to anterior or posterior skin by electrodes or by mechanical stimuli that activate touch (T), pressure (P), and/or nociceptive (N) cells. Intracellular recordings are made from S cells, sensory cells and motoneurons, and muscle contractions are measured with a tension transducer. Weak, repetitive stimuli applied to the posterior portion of the animal give rise to habituation. A stronger stimulus gives rise to sensitization and to dishabituation. (B) Responses to weak electrical shocks to posterior skin. In repeated trials, the responses became smaller (habituation, red circles). After the fourth trial, a dishabituating strong stimulus was given that produced a bigger response. After killing an S cell in one segment, only habituation occurred (blue circles). (C) When a strong sensitizing stimulus was given (sensitization, red circles), the responses became stronger than normal instead of habituating. After elimination of the S cell, by killing it in one segment or by cutting its axon, sensitization could no longer be elicited (blue circles). Repeated stimuli gave only habituation. After a severed S cell axon had regenerated as shown in Figure 18.19, a strong stimulus once again gave rise to sensitization (green circles) and dishabituation (not shown). (B,C) The control animals were subjected to sham operations, i.e., the body wall was opened but the S cell was left intact. This experiment demonstrates that a single cell is essential for these complex responses. (After Sahley et al., 1994, and Modney, Sahley, and Muller, 1997.)

process is known as **dishabituation**. Similarly, if a strong stimulus is applied without previous habituation, **sensitization** occurs; that is, the strength of the response to touch becomes greater than normal. The firing patterns of S cells do not change during habituation. But S cell connectivity is required during both sensitization and dishabituation (see Figure 18.20). In technically difficult experiments, an S cell axon was cut or an S cell was killed in its entirety by injection of pronase. Killing the S cell did not interfere with shortening or habituation as such, but it did abolish dishabituation and sensitization.

A related series of experiments confirmed the importance of the S cell in these processes; in this work, the axon of a single S cell was lesioned and then allowed to regenerate. A remarkable property of the S cell is that after its axon is severed, it grows back to re-form its electrical connections with the target S axon with extraordinary precision.[90] As expected, breaking the train of transmission along the S cells throughout the length of the animal abolished sensitization. Some weeks later, as shown in Figure 18.20, when the S cell axon had regenerated and re-formed its connections, sensitization once again reappeared.[91]

These experiments provide a clear demonstration of the way in which a single cell can play a part in a highly complex response, such as sensitization. In *Hirudo*[92] and in *Aplysia*,[4] sensitization involves serotonin.

A completely different mechanism for altering the synaptic action of one cell upon its postsynaptic targets is the failure of impulse conduction along axons. In the central nervous system of the leech and the cockroach, and in crustacean motor axons,[93] repeated trains of action potentials occurring at natural frequencies cause conduction to fail at specific axonal branch points. In T, P, and N sensory neurons of the leech, the conduction block depends on hyperpolarization caused by the electrogenic sodium pump and by long-lasting changes in a calcium-activated potassium conductance. For example, repeatedly stroking or pressing the skin of the leech causes trains of impulses and a prolonged hyperpolarization of P sensory cells. As a result, propagation of impulses becomes blocked at branch points within the neuropil where the geometry for impulse conduction is unfavorable, that is, where a small-diameter axon feeds into a larger one.[94] At the same time, other branches of the same neuron continue to transmit.[95] Conduction block therefore represents a nonsynaptic mechanism that temporarily disconnects a P cell or a T cell from one defined set of its postsynaptic targets. When some, but not all, of the presynaptic fibers connected to a cell fail to conduct, transmitter release and the efficacy of transmission are reduced. By making lesions with a laser at specific branch points, Muller and his colleagues assessed the contribution made by discrete branches of the sensory P cell to its synaptic action on motor cells.[96] An example of conduction block in a P cell is shown in Figure 18.21. The synaptic potentials recorded in the postsynaptic AE motor cell are abolished when impulses become blocked at the point marked by "X" (Figure 18.21B). At the same time, synaptic potentials still arise in the L motoneuron, because it receives inputs from other branches of the P cell in which conduction continues. Another consequence of conduction block is that by reducing the area of skin from which responses in the AE cell (in this example) are evoked, it causes temporary shrinkage of the receptive field (as viewed from the cell body) and thereby has the effect of improving the spatial resolution of the sensory neurons under conditions of prolonged stimulation.

[90] Elliott, E. J., and Muller, K. J. 1983. *J. Neurosci.* 3: 1994–2006.

[91] Modney, B. K., Sahley, C. L., and Muller, K. J. 1997. *J. Neurosci.* 17: 6478–6482.

[92] Burrell, B. D., and Crisp, K. M. 2008. *J. Neurophysiol.* 99: 605–616.

[93] Grossman, Y., Parnas, I., and Spira, M. E. 1979. *J. Physiol.* 295: 283–305.

[94] Yau, K. W. 1976. *J. Physiol.* 263: 513–538.

[95] Gu, X. N. 1991. *J. Physiol.* 441: 755–778.

[96] Gu, X. N., Muller, K. J., and Young, S. R. 1991. *J. Physiol.* 441: 733–754.

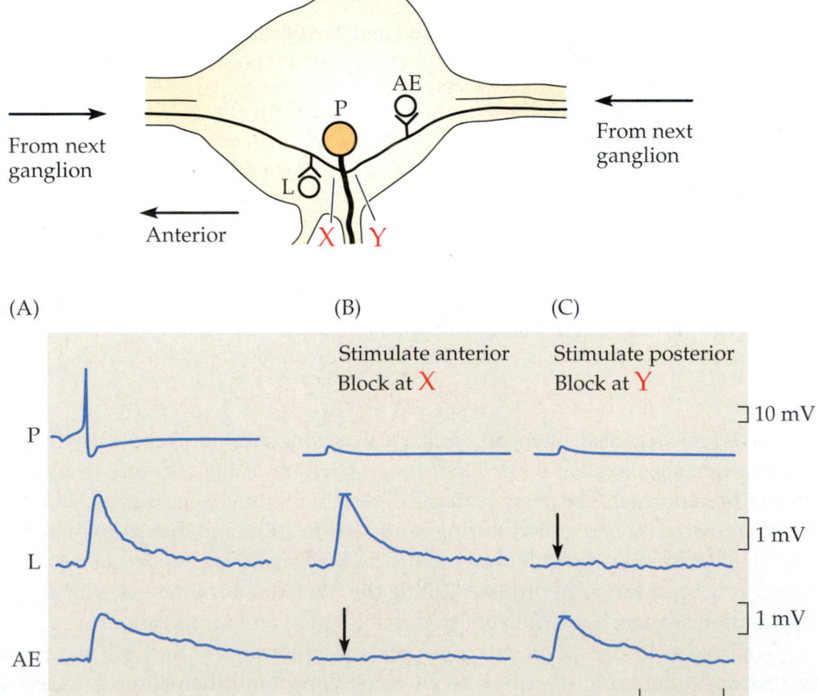

FIGURE 18.21 Effects of Conduction Block on synaptic transmission and integration. The medially situated pressure (P) cell (innervating dorsal skin and similar to that shown in Figure 18.14) sends presynaptic axons to motor cells that cause the longitudinal muscles to contract (L motoneuron cell) and the annuli to erect (annulus erector [AE] cell). (A) The synaptic potentials in the L cell and the AE cell were evoked by stimulating the P cell intracellularly. Similarly, when impulses originate in the neighboring segments at very low frequency, they travel along the small axons in the connective, invade all the branches of the P cell, and produce synaptic potentials in the AE and L cells. (B) When the skin in the next most anterior segment is pressed so that P-cell impulses arise there at higher frequency, the P cell becomes hyperpolarized and impulse conduction fails at the point labeled X. When this occurs, the L cell still receives its input and produces a synaptic potential, but the AE cell is temporarily disconnected (arrow). (C) Conversely, when the posterior field is stimulated so as to generate higher frequency impulses, they become blocked at Y, where the small axons encounter the larger axon. Once, again cells become disconnected (arrow). This time, however, it is the L cell that fails to show a synaptic potential, while the AE cell continues to respond. (After Gu, 1991.)

Conduction block represents an example of how work on a mammalian nervous system can be applied to understand an invertebrate—the opposite of what is so often claimed for research on simpler animals. The first description of conduction block was made by Barron and Matthews in 1935,[97] in their experiments on dorsal root axons of the mammalian spinal cord. Their results were largely forgotten, and only much later was conduction block rediscovered in invertebrates. It is now apparent that conduction block is a key feature of integrative mechanisms in dendrites in mammalian CNS.[98]

Circuits Responsible for the Production of Rhythmical Swimming

Leeches swim forward in a serpentine manner, with their dorsal surface uppermost (Figure 18.22A). As in vertebrates (see Chapter 14), biogenic amines play a role in modulating motor activity.[80,99–101] Quiescent (sluggish!), non-swimming leeches have lower blood levels

[97] Barron, D. H., and Matthews, B. H. 1935. *J. Physiol.* 85: 73–103.

[98] Tsubokawa, H., and Ross, W. N. 1997. *J. Neurosci.* 17: 5782–5791.

[99] Glover, J. C., and Kramer, A. P. 1982. *Science* 216: 317–319.

[100] Willard, A. L. 1981. *J. Neurosci.* 1: 936–944.

[101] Brodfuehrer, P. D. et al. 1995. *J. Neurobiol.* 27: 403–418.

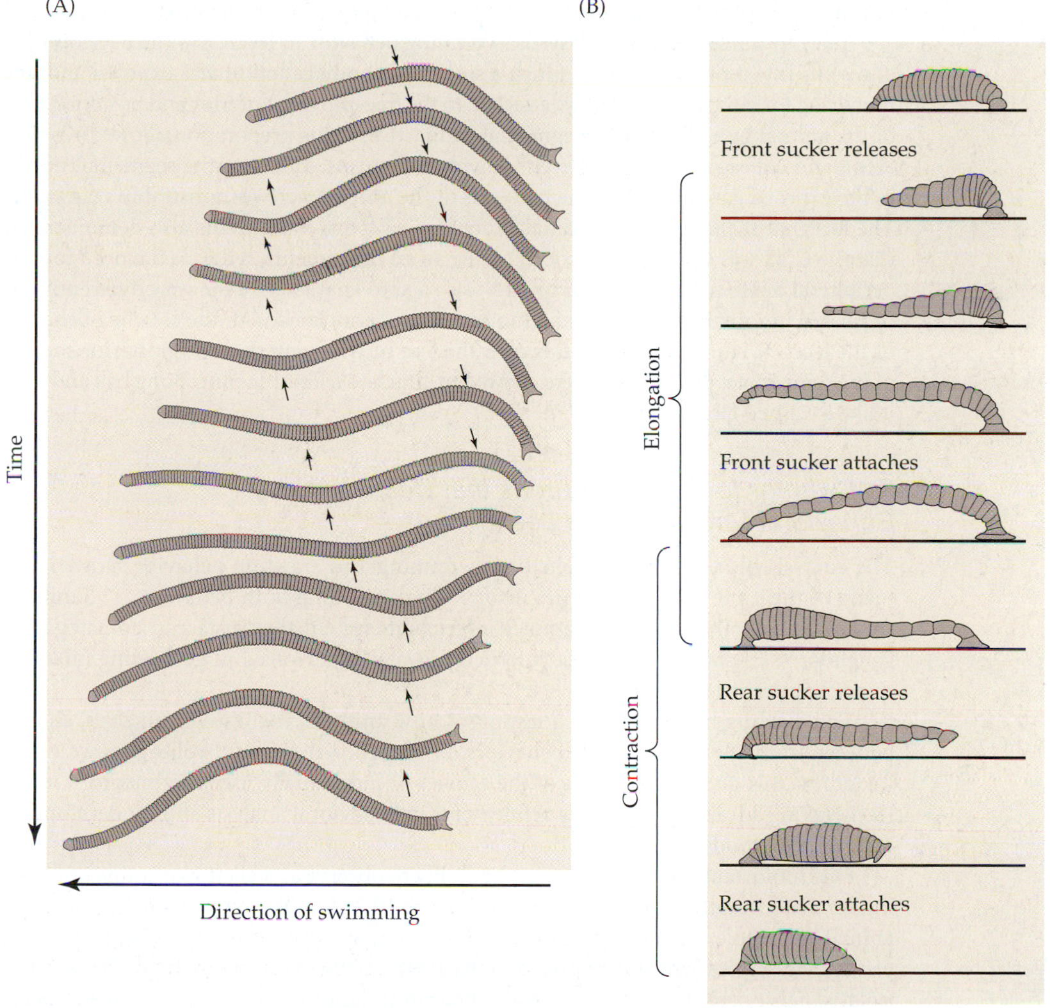

FIGURE 18.22 Locomotory Behaviors of the Leech. (A) In swimming, the leech releases its anterior sucker from the substrate, extends the body forward, and flattens dorsoventrally to generate a blade-shaped profile; then, it releases the posterior sucker and propels itself through the water by passing a wave of rapidly alternating dorsal and ventral flexures from front to back. (B) In vermiform (as opposed to inchworm style) crawling, the animal releases its front sucker and then reaches forward by simultaneously relaxing its longitudinal muscles and contracting its circular muscles to extend the hydrostatic skeleton. After extending, the front sucker attaches and the rear sucker detaches. The animal then shortens by contracting its longitudinal muscles and relaxing its circular muscles, and finally reattaches the rear sucker to complete the cycle. The alternating cycles of extension and contraction also progress from front to back. (A after Stent et al., 1978; B after Kristan, Calabrese, and Friesen, 2005.)

of 5-hydroxytryptamine (5-HT or serotonin) than do active leeches. Moreover, stimulation of identified cells (known as Retzius cells) that secrete 5-HT promotes an increase in its concentration in the blood and promotes swimming in the animal as well. The level of 5-HT can be lowered in embryos by means of a specific chemical (5,7-dihydroxytryptamine or 5,7-DHT) that selectively destroys the 5-HT neurons in the developing ganglia. After development, adult leeches treated in this way, and thereby depleted of serotonin, do not swim spontaneously, but they will do so if immersed in a weak solution of 5-HT.

Which neurons initiate, coordinate, and maintain the swim rhythm? This problem was first studied by Stent, Kristan, Friesen, and their colleagues.[102] The head and tail ganglia of the leech are not essential for generating the swim rhythm, which can occur in a few isolated segments or even a single segment of the animal. As in other invertebrates (e.g., the cockroach, locust, and cricket), in which central motor programs involving a small number of individual cells have been shown to control complex patterns of movement, the swim rhythm in the leech is established by inhibitory and excitatory synaptic interactions within the CNS. Peripheral receptors serve to trigger, enhance, depress, or halt leech swimming. Similarly, mammalian CNS preparations isolated from the body generate the respiratory rhythm in vitro (see Chapter 24).

A key to the successful analysis of swimming behavior in leech was the development of a semi-intact preparation, in which a single, midbody ganglion was exposed and immobilized for electrical recordings, attached to the free-moving anterior and posterior ends of the animal by only the intersegmental connectives. This preparation allowed Kristan, Stent, and colleagues to record from individual neurons and from the segmental nerves in the exposed ganglion while the two ends of the animal were demonstrably swimming. The fact that the ends could maintain coordinated swim movements also demonstrated that the CNS was sufficient for coordinating swim movements, without the need for any peripheral nerve net. A similar approach was used to characterize the very differently coordinated interneuron circuitry and motor outputs associated with the crawling behavior of the leech (Figure 18.22B). In this case, the free moving anterior and posterior suckers of the semi-intact preparation were allowed to attach to a floating ping-pong ball and turn it like a trained circus elephant.[103]

To Swim or to Crawl? Neurons that Determine Behavioral Choices in the Leech

The analysis of neural circuits eliciting swimming and crawling behavior showed that many of the same interneurons are involved in controlling both behaviors.[104] Similarly, studies from another carefully examined invertebrate system, the crustacean stomatogastric ganglion,[105–108] show that specific interneurons may be involved in generating multiple, distinct patterned movements.

This finding highlights the question of how animals, including ourselves, choose between mutually incompatible behaviors. Briggman, Kristan, and colleagues were able to address this question in terms of the activity of individually identified neurons in the leech, taking advantage of further refinements in behavioral analysis and modern optical recording methods.[109]

One important breakthrough was the ability to observe and elicit swimming or crawling *behavior* in fully isolated nerve cords (Figure 18.23A). Of course, no behavior of the animal as a whole occurs in these experiments, since most of the body has been removed. Nevertheless, from semi-intact preparations, it is possible to categorize fictive behavior of isolated nerve cords by recording the interneuron and motor patterns in various segments and how they are coordinated between segments. Moreover, it is possible to elicit these behaviors by administering mild shocks to a single segmental nerve, equivalent to stimulating the axons of a defined set of T and P neurons. A particular regimen was established in which an identical stimulus would elicit either fictive swimming or fictive crawling with roughly equal probabilities across multiple trials. How did the leech, in essence, *decide* which behavior to undertake?

This question was addressed by the combined use of optical methods and a systems approach to analyze the electrical activity patterns of many individual neurons in parallel. In multiple trials, shocks were applied to a segmental nerve extending from one ganglion in

[102] Stent, G. S. et al. 1978. *Science* 200: 1348–1357.

[103] Baader, A. P., and Kristan, W. B., Jr. 1995. *J. Comp. Physiol. A* 176: 715–726.

[104] Briggman, K. L., Abarbanel, H. D., and Kristan, W. B., Jr. 2006. *Curr. Opin. Neurobiol.* 16: 135–144.

[105] Bucher, D., Taylor, A. L., and Marder, E. 2006. *J. Neurophysiol.* 95: 3617–3632.

[106] Marder, E., and Bucher, D. 2007. *Annu. Rev. Physiol.* 69: 291–316.

[107] Hooper, S. L., and DiCaprio, R. A. 2004. *Neurosignals* 13: 50–69.

[108] Prinz, A. A., Bucher, D., and Marder, E. 2004. *Nat. Neurosci.* 7: 1345–1352.

[109] Briggman, K. L., Abarbanel, H. D., and Kristan, W. B., Jr. 2005. *Science* 307: 896–901.

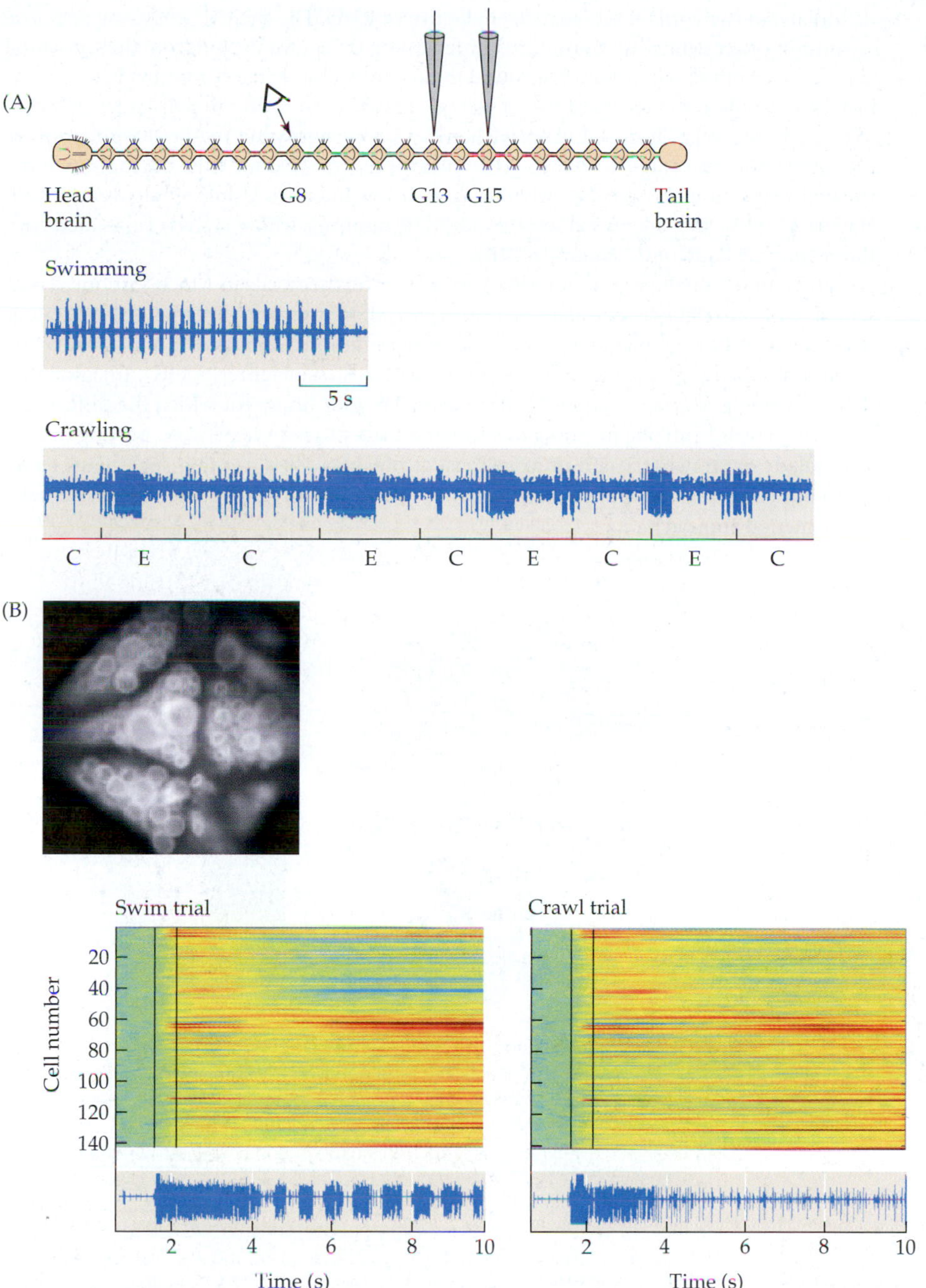

FIGURE 18.23 To Swim or to Crawl? Choosing between mutually exclusive behaviors in leech. (A) Schematic shows an isolated nerve cord. The interconnected, segmental organization of the leech nerve cord makes it possible to stimulate electrically in one segment (G15) to elicit behavior, monitor activity of one or more neurons by recording in another segment (G13), and then use optical methods to monitor the activity of large populations of neurons in a third segment (G8). Fictive behavior in isolated nerve cord preparations (validated by intermediate studies of semi-intact preparations) allowed Kristan and colleagues to study how the CNS generates appropriate patterns of motoneuron activity for different behaviors. In the examples shown here, the activity of the dorsal excitor (DE) motoneuron in G13 is detected as the largest spikes in extracellular recordings from the segmental nerve innervating the dorsal body wall. In the top trace, bursts of DE activity correspond to what would be a dorsal flexure of segment 13 during an episode of swimming. In the bottom trace, longer lasting, less frequent bursts correspond to what would be seen as the phases of longitudinal contraction (C) alternate with phases of extension (E) during an episode of crawling. Stimulation elicited either fictive swimming or fictive crawling with equal probability over multiple trials. (B) In the same preparation, the simultaneous activity of large numbers of neurons in G8 was monitored optically, using a charge-coupled device (CCD) camera and voltage-sensitive fluorescent dyes. The changes in membrane potential for each neuron were color-coded (red = depolarized; blue = hyperpolarized), and the data for elicited swim and crawl behavior were pooled separately for further analysis. Black vertical lines indicate the duration of the stimulus. (A after Mesce, Esch, and Kristan, 2008; B after Briggman, Abarbanel, and Kristan, 2005.)

the isolated nerve cord (as schematized in Figure 18.23A). The elicited behavioral response (swimming or crawling) was monitored by recording the motor output from the segmental nerve of a second ganglion. Simultaneously, the activity of more than 100 individual neurons in a third ganglion was analyzed (using voltage-sensitive dyes and a digital camera), focusing on the interval of jumbled neuronal activity after stimulation but before initiation of the motor program (Figure 18.23B). This interval represented the time during which the isolated nerve cord was *deciding* whether to swim or to crawl. Using a systems approach (Figure 18.24A), they identified several candidate neurons, whose activity correlated with the swim/crawl outcome (Figure 18.24B).

Then, in the same preparation they carried out further trials in which just one representative of one class of candidate neuron (cell 208, in just a single ganglion) was either hyperpolarized or depolarized at the time of stimulation. Remarkably, this minor perturbation biased the outcome of the swim/crawl trials (see Figure 18.24C). Imagine that there is a single neuron or ensemble of neurons in your brain, for which the differences in activity levels from one morning to another affects your *choice* between putting orange marmalade or strawberry jam on your toast at breakfast. It is not that such neurons do or do not exist, but one cannot imagine being able to identify them through a systematic experimental approach.

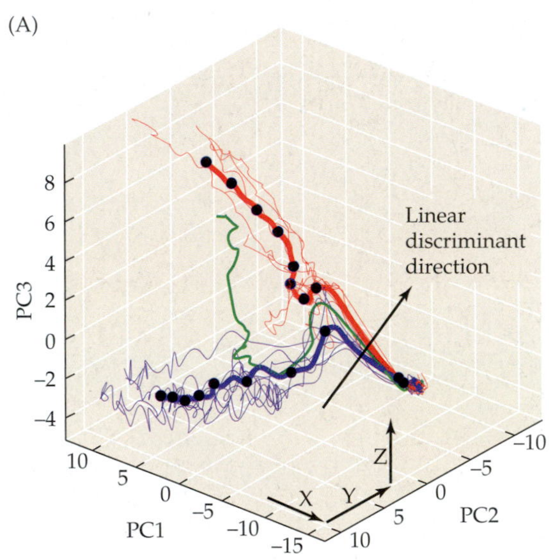

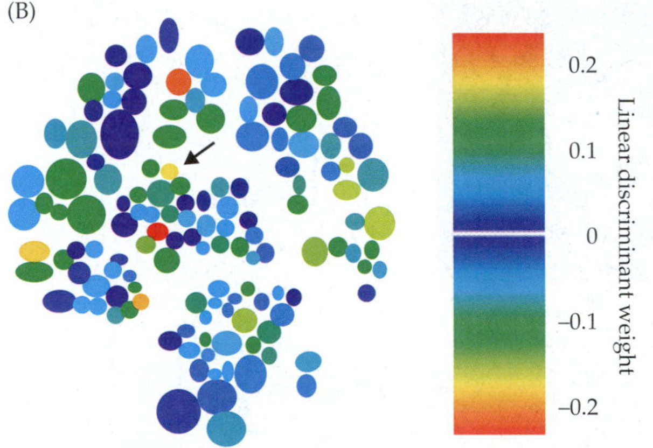

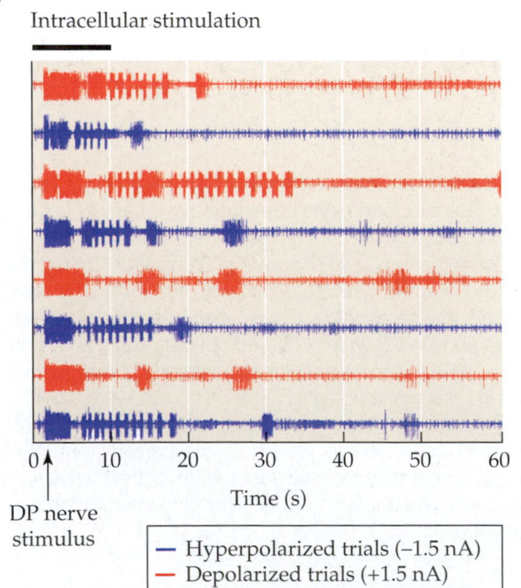

FIGURE 18.24 Biasing the Swim-Crawl Decision (A) Viewed in abstraction, the overall state of activity of the ganglion at each instant during a swim/crawl trial can be represented by a point along a trajectory in an N-dimensional space, with each dimension corresponding to the activity of one of the optically recorded neurons (143 in this experiment). Using a mathematical process known as principal component analysis (PCA), a new space is formulated in which as much of the trajectory as possible is restricted to just three of the 143 possible dimensions (principal components 1–3 = PC1, PC2, PC3; left panel), by taking different linear combinations of the activities of different neurons. Thin lines plot the individual trajectories of the swim (blue) and crawl (red) trials, and thick lines indicate the average of the swim and crawl trajectories, respectively. The green line highlights an exceptional case in which the preparation started out along a typical swim trajectory, but then deviated to exhibit fictive crawling. Further analysis permitted the calculation of a linear discriminant (long black arrow), corresponding to the point at which the average swim and crawl average trajectories diverged. (B) Mapping the weighting factors for each neuron in the linear discriminant back onto the ganglia data set revealed cell 208 (arrow) as a candidate for biasing the choice between swimming and crawling in response to stimulation. (C) Consistent with the mapping, hyperpolarizing or depolarizing a single cell 208 in the chain of ganglia sufficed to bias the response to stimulation to a statistically significant degree. (After Briggman, Abarbanel, and Kristan, 2005.)

Why Should One Work on Invertebrate Nervous Systems?

Throughout this book, and from the examples described in this chapter, it is evident that invertebrate nervous systems have been essential for approaching problem after problem relating to the biophysics, cell biology, and development of nerve cells. Particularly striking is the conservation of fundamental mechanisms in species after species through evolution. Often, experiments on an invertebrate have produced the very insight needed for starting the investigation of similar problems in a mammal. For example, a great impetus for the development of the slice technique for mammalian brain came from the work on invertebrate ganglia in which identified neurons could be seen directly in the microscope as they were impaled. Hartline's work on the horseshoe crab eye[110] provided the key stimulus for Kuffler's experiments on the cat retina.[111] At the same time it would be futile to hope to understand how the visual cortex of a monkey carries out its functions from studies made on an invertebrate.

What then is the point of studying, say, navigation by ants or bees? First, one can guess that even though we do not see polarized light or sense magnetic fields, the underlying principles for analyzing sensory information and translating it to an effective output will be used again in higher nervous systems, in one way or another. Second, the work on invertebrates illustrates a fundamental attitude toward biology—neurobiology is not restricted to the study of *the* brain (i.e., ours). Rather, there is an inherent fascination in trying to understand how the tiny, finite brain of an invertebrate can do wonderful and sophisticated computations that are essential for its survival. The way in which a leech swims, an ant navigates, a bee dances, a sea slug learns, a cricket sings, or a fly flies are all problems of interest in their own right.[112]

[110] Hartline, H. K. 1940. *Am. J. Physiol.* 130: 690–699.

[111] Kuffler, S. W. 1953. *J. Neurophysiol.* 16: 37–68.

[112] Baca, S. M. et al. 2008. *Neuron* 57: 276–289.

SUMMARY

- The properties of individual neurons and glia are similar between vertebrates and invertebrates. However, invertebrate nervous systems generally have far fewer cells and correspondingly simpler organizations than those of vertebrates.
- Correlated with their comparative simplicity, invertebrate behaviors, though tremendously sophisticated, are often more stereotyped than those of vertebrates and thus more amenable to study. Quantitative measurements of these behaviors shed light on fundamental principles in neurobiology.
- Ants and bees routinely carry out remarkable navigational feats using multiple sensory systems in parallel, including visual cues, the direction of polarization of light, and the ability to orient with respect to an external magnetic field. For measuring distance, however, ants rely on proprioceptive processing to measure the steps traveled, while bees do so by integrating the flow of visual information across their retinas.
- In the leech *Hirudo* and some other invertebrate preparations, much or all of the nervous system consists of relatively large, identified, experimentally accessible cells. In such preparations, it is possible to analyze behaviors—ranging from simple reflexes to higher order behaviors, such as learning and decision making—at the level of individual cells and synapses.
- The study of the invertebrate central nervous systems is fascinating in its own right and need not be directed toward understanding mechanisms in human brain.

Suggested Reading

General Reviews

Bailey, C. H., and Kandel, E. R. 2008. Synaptic remodeling, synaptic growth and the storage of long-term memory in Aplysia. *Prog. Brain Res.* 169: 179–198.

Cheng, K., Narendra, A., Sommer, S., and Wehner, R. 2009. Traveling in clutter: navigation in the Central Australian desert ant *Melophorus bagoti*. *Behav. Processes* 80: 261–268.

Collett, T. S., and Waxman, D. 2005. Ant navigation: reading geometrical signposts. *Curr Biol.* 15: R171–173.

Kristan, W. B., Jr., Calabrese, R. L., and Friesen, W. O. 2005. Neuronal control of leech behavior. *Prog. Neurobiol.* 76: 279–327.

Leonard, J. L., and Edstrom, J. P. 2004. Parallel processing in an identified neural circuit: the Aplysia californica gill-withdrawal response model system. *Biol. Rev. Camb. Philos. Soc.* 79: 1–59.

Marder, E., and Bucher, D. 2007. Understanding circuit dynamics using the stomatogastric nervous system of lobsters and crabs. *Annu. Rev. Physiol.* 69: 291–316.

Srinivasan, M. V. 2010. Honey bees as a model for vision, perception, and cognition. *Annu. Rev. Entomol.* 55: 267–284.

Walters, E. T., and Cohen, L. B. 1997. Function of the LE sensory neurons in *Aplysia. Invert. Neurosci.* 3: 15–25.

Original Papers

Bao, L., Samuels, S., Locovei, S., Macagno, E. R., Muller, K. J., and Dahl, G. 2007. Innexins form two types of channels. *FEBS Lett.* 581: 5703–5708.

Briggman, K. L., Abarbanel, H. D. and Kristan, W. B., Jr. 2005. Optical imaging of neuronal populations during decision-making. *Science* 307: 896–901.

Fuchs, P. A., Henderson, L. P. and Nicholls, J. G. 1982. Chemical transmission between individual Retzius and sensory neurones of the leech in culture. *J. Physiol.* 323: 195–210.

Gu, X. 1991. Effect of conduction block at axon bifurcations on synaptic transmission to different postsynaptic neurones in the leech. *J. Physiol.* 441: 755–778.

Lewis, J. E., and Kristan, W. B., Jr. 1998. Quantitative analysis of a directed behavior in the medicinal leech: Implications for organizing motor output. *J. Neurosci.* 18: 1571–1582.

Modney, B. K., Sahley, C. L., and Muller, K. J. 1997. Regeneration of a central synapse restores nonassociative learning. *J. Neurosci.* 17: 6478–6482.

Muller, M., and Wehner, R. 2007. Wind and sky as compass cues in desert ant navigation. *Naturwissenschaften* 94: 589–594.

Nicholls, J. G., and Wallace, B.G. 1978. Quantal analysis of transmitter release at an inhibitory synapse in the CNS of the leech. *J. Physiol.* 218: 171–185.

Srinivasan, M., Zhang, S., Lehrer, M., and Collett, T. 1996. Honeybee navigation en route to the goal: visual flight control and odometry. *J. Exp. Biol.* 199: 237–244.

Wang, H., and Macagno, E. R. 1997. The establishment of peripheral sensory arbors in the leech: In vivo time-lapse studies reveal a highly dynamic process. *J. Neurosci.* 17: 2408–2419.

Wittlinger, M., Wehner, R., and Wolf, H. 2006. The ant odometer: stepping on stilts and stumps. *Science* 312: 1965–1967.

PART V

Sensation and Movement

In this section we examine the mechanisms by which the brain acquires, analyzes, and acts upon information from the environment. Different forms of stimulus energy are converted (transduced) into electrical signals by specific sensory organs, receptor cells, and molecular mechanisms. The general principles of sensory transduction are described in Chapter 19. Mechanotransduction by sensory hair cells provides an example of relatively direct gating of ion channels by a stimulus to generate receptor potentials. In contrast, olfactory neurons are excited when chemicals from the environment bind to a G protein–coupled receptor, opening cyclic-nucleotide-gated ion channels by way of the enzyme adenylyl cyclase to synthesize cyclic AMP. Enzymatic transduction of stimulus energy also takes place in retinal photoreceptors, as described in Chapter 20, with the striking difference that photoreceptors are hyperpolarized when light causes a drop in the level of cyclic nucleotides. This chapter also explains the functional organization of the retina that leads from the relatively straightforward response of the photoreceptor to the requirements of retinal ganglion cells for contrasting patterns of light to be transmitted centrally.

Touch, pain, and texture sensation result from the activity of specific classes of receptor neurons that innervate skin, as described in Chapter 21. The tactile whiskers on the muzzle of rodents provide a prominent example of somatotopic mapping of the cortex, as well as the segregated analysis of stimulus features. The same organizational principles apply to sensory signaling from the inner ear (Chapter 22), although here the specialized construction of the end-organ confers unique stimulus qualities, such that the primary auditory cortex is mapped as an acoustic frequency gradient.

A central challenge in neurobiology is to understand the connection between transduction in the sensory periphery and the resulting behavior. Chapter 23 moves from primary sensory cortex to higher-order areas where *meaning* might emerge from the appropriate combination of stimulus features. Finally, in Chapter 24 the mechanisms of movement are described, ranging from the organization of stretch reflexes, to the genesis of anticipatory motor programs that establish sensory set points appropriate for the upcoming movement. While much between sensation and movement remains a black box, these chapters show that the box is getting smaller.

CHAPTER 19
Sensory Transduction

Stimulus intensity and timing are encoded in receptor potentials arising in the receptive endings of sensory cells. Receptor potentials can be depolarizing or hyperpolarizing; they increase in amplitude with increased stimulus intensity and saturate at higher stimulus levels. During sustained stimulation, the receptor potential adapts to a lower level. Adaptation can occur quickly or slowly. It arises from mechanical, electrical, or biochemical processes in different cell types. Receptors that adapt slowly can better encode stimulus duration. Rapidly adapting receptors are specialized to detect changes in the stimulus.

Transduction can be very direct, as in the conversion of sound or head movement by mechanosensory hair cells. Here, the stimulus energy directly alters the gating probability of mechanosensitive ion channels in the sensory hair bundle. Such direct transduction is fast and makes possible hearing at high frequencies. In comparison, indirect mechanisms of transduction rely on biochemical cascades. These are inherently slower than channel gating, but can greatly increase sensitivity through amplification at each step. For example, odor molecules (olfactants) bind to a large family of G protein–binding receptors in ciliated olfactory sensory neurons in the nose. A subsequent rise in cyclic adenosine monophosphate (cAMP) causes many cation channels to open, and the resulting depolarization generates action potentials.

Stimulus modality depends on a combination of direct and indirect transduction mechanisms. For example, certain taste stimuli (amino acids, sugar, and bitter compounds) are transduced by G protein–coupled membrane receptors in taste buds. Again, a rise in cAMP causes cation channels to open, and action potentials are initiated. But also, salts and acids (sour taste) can act directly on ion channels in some taste receptor cells. The transduction of painful stimuli in tissues results from both direct mechanical and indirect biochemical mechanisms of transduction as well as from sensitization by substances released from damaged cells, such as adenosine triphosphate (ATP).

We know the physical world through our senses. We reach out to touch nearby objects, and we receive other signals transmitted from afar. Neuronal sensory receptors are the gateways through which these signals pass. Right at the outset, receptors set the stage for all the analyses of sensory events that are subsequently made by the central nervous system (CNS). They define the limits of sensitivity and determine the range of stimuli that can be detected and acted upon. With rare exceptions, each type of neuronal receptor is specialized to respond preferentially to only one type of stimulus energy, called the **adequate stimulus**. For instance, rods and cones in the eye respond to light (see Chapter 20); nerve endings in the skin to touch, pressure, or vibration; and receptors in the tongue to chemical tastants. The stimulus, whatever its modality, is always converted (or transduced) to an electrical signal—the **receptor potential**. In general, the strength and duration of a stimulus are coded in electrical signals; recognition by the CNS of the modality of the stimulus and its position depend on the nature of the sensory ending and its anatomical location. Thus, a temperature receptor in the foot has its own pathway into the nervous system that is quite distinct from that of a vibration receptor in the hand, but in both axons, the communication is by action potential trains of variable frequency and duration.

For sensory stimuli, there is a great deal of amplification at the receptor level, so that very small external changes provide a trigger to release stored charges that appear as electrical potentials. For example, only a few molecules of specific odorant substances (pheromones) are sufficient to act as sex attractants for moths. Similarly, a few quanta of light trapped by receptors in the retina are sufficient to produce a visual sensation. This extreme sensitivity extends to the inner ear, where mechanical displacements of 10^{-10} meters (m) are detectable.[1] Equally remarkable are electroreceptors in some fish that can detect electrical fields of a few nanovolts per centimeter.[2,3] This signal is similar in magnitude to the field that would be produced if two wires connected to either pole of an ordinary flashlight battery could be dipped into the Atlantic Ocean—one at Bordeaux, the other at New York!

Sensory receptors have a well-defined range of stimuli to which they respond. For example, our auditory hair cells respond to sounds only within a bandwidth of about 20 to 20,000 Hz. The response by receptors in our retina to electromagnetic radiation is restricted to wavelengths between about 400 and 750 nanometers (nm). Shorter (near-ultraviolet) and longer (near-infrared) wavelengths go undetected. Restrictions of this kind are not usually because of unavoidable physical limitations. Instead, each system is tuned to the particular needs of the organism—whales and bats can hear much higher frequencies; snakes can detect infrared, and bees ultraviolet, radiation. For dogs and pigs, the sense of smell is more refined than in humans.

What mechanisms provide such great sensitivity and selectivity to sensory receptor cells? These vary according to the stimulus and receptor type. We will describe **mechanotransduction** by stretch receptors of muscle, touch receptors in skin, and hair cells of the inner ear in which mechanosensitive ion channels directly transduce the stimulus. In distinction to the direct impact of the physical stimulus on ion channels in hair cells, **chemotransduction** in olfactory neurons operates through G protein–coupling to an olfactory receptor protein, while **gustation** employs multiple mechanisms, some G protein–coupled and others that are more direct. We conclude with a discussion of **nociception** that underlies pain sensation and combines the transduction of chemical and mechanical stimuli. The detailed mechanisms underlying phototransduction are treated separately in Chapter 20.

Stimulus Coding by Mechanoreceptors

Short and Long Receptors

The receptor potential generated by the transduction process reflects the intensity and duration of the original stimulus. In some receptors, such as retinal rods and inner ear hair cells that do not have long axons, receptor potentials spread passively from the sensory region to the synaptic pole of the cell (Figure 19.1A). Such receptors are known as **short receptors**. The passage of information from the receptor end to the synaptic end of the cell does not require the intervention of action potentials. In some cells, passive spread of the receptor potential can reach a surprisingly distant point. For example, in some crustacean[4] and

[1] Bialek, W. 1987. *Annu. Rev. Biophys. Biophys. Chem.* 16: 455–478.

[2] Kalmijn, A. J. 1982. *Science* 218: 916–918.

[3] Heiligenberg, W. 1989. *J. Exp. Biol.* 146: 255–275.

[4] Roberts, A., and Bush, B. M. 1971. *J. Exp. Biol.* 54: 515–524.

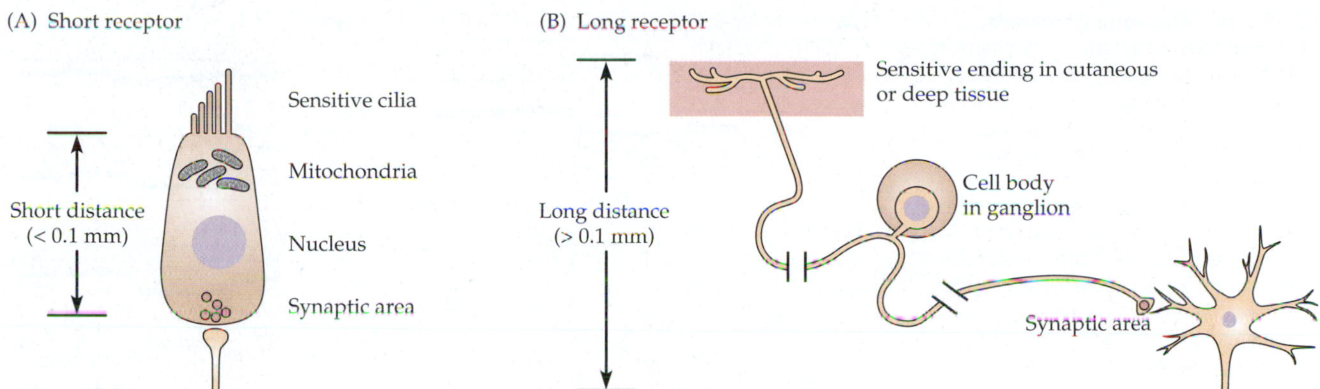

FIGURE 19.1 Short and Long Sensory Receptor Cells. (A) Short receptors—such as retinal rods and cones, and mechanosensory hair cells of the inner ear—are less than 100 μm in length. Thus, receptor potentials generated within the sensitive ending spread effectively throughout the cell, altering transmitter release at synaptic areas. (B) Long receptors, such as muscle spindle afferents and cutaneous mechanoreceptors, employ action potentials to conduct their signals to a distant second-order neuron. The amplitude of the receptor potential is encoded in the frequency of action potentials it generates.

leech[5] mechanoreceptors, and in photoreceptors in the barnacle eye,[6] the receptor potential spreads passively over a distance of several millimeters. In such cells the membrane resistance, and hence the length constant for spread of passive depolarization, is unusually high.

Whereas receptor potentials are usually depolarizing, certain short receptors respond to their adequate stimulus with a hyperpolarizing potential change. This action occurs, for example, in photoreceptors of the vertebrate retina and in cochlear hair cells that have both hyperpolarizing and depolarizing responses. Whatever the polarity of the receptor potential, short receptors release neurotransmitter tonically from their synaptic regions, with depolarization increasing and hyperpolarization decreasing the rate of release.

In **long receptors** (Figure 19.1B), such as those in skin or muscle, information from single receptors must be sent over a much greater distance to reach the second-order sensory cell (e.g., from the big toe to the spinal cord). In order to accomplish this relay, the receptor performs a second transformation process in which receptor potentials give rise to trains of action potentials whose duration and frequency code information about the duration and intensity of the original stimulus. These impulses then carry the information to the synaptic terminals of the cell.

The **frequency code** for stimulus intensity is established through the interaction of the maintained receptor current from sensory terminals and the conductance changes associated with the action potential. At the end of each action potential, the increased potassium conductance that occurs on the recovery phase drives the membrane in the hyperpolarizing direction, toward E_K (the potassium equilibrium potential). This increase in potassium conductance is transient, while the sustained transduction current depolarizes the membrane once more to the firing level. The stronger the receptor current, the sooner the firing level again is reached, and the higher the impulse frequency. Similar considerations apply to all neurons in which synaptic input, rather than a receptor potential, sums to alter the frequency of action potentials.

E. D. Adrian

Encoding Stimulus Parameters by Stretch Receptors

The way in which sensory receptors generate electrical signals was studied early on by Adrian and Zotterman[7] using extracellular recording from sensory nerve fibers arising in vertebrate muscle stretch receptors. Katz was the first to demonstrate the link between sensory stimuli and electrical signals in a mechanoreceptor,[8] when he recorded receptor potentials and showed that stretch caused a depolarization of the **sensory ending**. When the receptor potential was observed in isolation by blocking the nerve discharge with procaine (a local anesthetic), its amplitude could be seen to increase in a graded fashion with muscle stretch.

[5] Blackshaw, S. E., and Thompson, S. W. 1988. *J. Physiol.* 396: 121–137.

[6] Hudspeth, A. J., Poo, M. M., and Stuart, A. E. 1977. *J. Physiol.* 272: 25–43.

[7] Adrian, E. D., and Zotterman, Y. 1926. *J. Physiol.* 61: 151–171.

[8] Katz, B. 1950. *J. Physiol.* 111: 261–282.

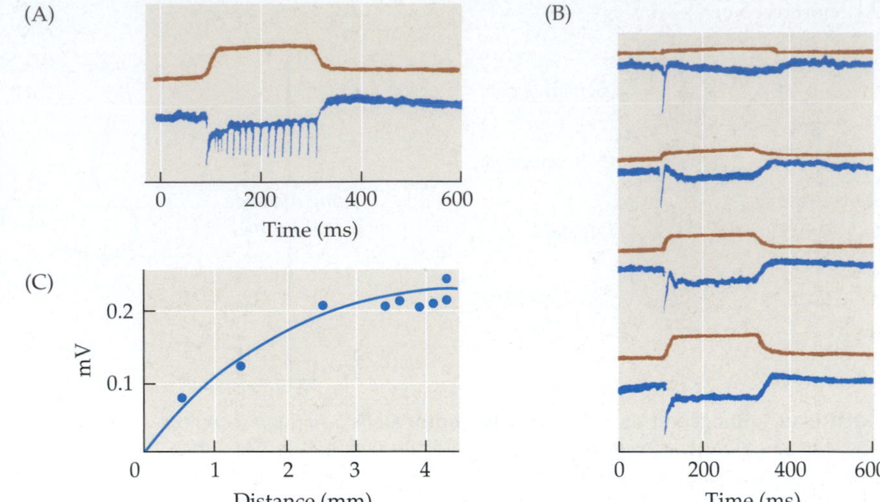

FIGURE 19.2 Receptor Potentials Recorded Extracellularly from a sensory nerve fiber supplying a muscle spindle. The recording electrode is placed as close as possible to the receptor. Downward deflection of the voltage record (lower traces) indicates receptor depolarization. (A) Stretching the muscle (upper trace) produces a receptor potential, upon which is superimposed a series of action potentials (lower trace). (B) Four stretches of increasing magnitude applied to the muscle after procaine has been added to the bathing solution. Action potentials (except for the first) are abolished by procaine, but the receptor potentials remain. (C) Plot of receptor potential amplitude against increase in muscle length. (After Katz, 1950.)

Figure 19.2 shows the relationship between receptor potential amplitude and stretch. The function begins with a slope of roughly 0.1 mV (extracellular recording) per millimeter of stretch, but flattens out at higher levels. Thus, the *sensitivity* of the sensory ending, in millivolts per millimeter of stretch, decreases as the stimulus grows. Many sensory receptors take advantage of this nonlinear relationship to provide amplitude coding over a wider range of stimulus intensities. In these receptors the response amplitude continues to increase, but in proportion to the logarithm of stimulus intensity. This effect is of great utility in receptors, such as hair cells and photoreceptors, that respond to stimuli whose amplitudes vary by several orders of magnitude.

These relationships between stimulus intensity and sensitivity correspond to those first described in 1846 by Weber. He measured the ability of subjects to discriminate weights held in their two hands and showed that it varied in proportion to the size of the weights. That is, the subjects were just able to detect a 3 gram (g) difference between weights of about 100 g each, whereas kilogram weights had to differ by 30 g (in each case, the detectable difference was about 3% of each object's weight). Fechner formalized this relationship by pointing out that this result implied a logarithmic relationship between stimulus and response. The Weber–Fechner law[9] is one formulation of the nonlinear relationship between stimulus strength and sensation. Although the exact form of the relationship depends on stimulus modality, this process applies to many aspects of perception and behavior. (For example, we might pinch pennies when choosing a pencil, but happily spend hundreds of dollars more for a classier computer.)

The Crayfish Stretch Receptor

Stimulus coding was analyzed in detail in crayfish stretch receptors by Eyzaguirre and Kuffler.[10] This preparation is particularly useful because the cell body of the stretch receptor lies in isolation—not in a ganglion, but on its own in the periphery, where it can be seen in live preparations (Figure 19.3A). It is large enough for penetration by intracellular microelectrodes. The cell inserts its dendrites into a nearby muscle strand and sends an axon centrally to a segmental ganglion (Figure 19.3B). In addition, the receptor receives inhibitory innervation from the ganglion; the muscle fibers into which it inserts receive excitatory and inhibitory innervation. Thus, receptor sensitivity is regulated by the central nervous system.

There are two types of crustacean stretch receptors with distinct structural and physiological characteristics, and their dendrites are embedded in different types of muscle. One responds well at the beginning of a stretch, but its response quickly wanes. This decrease in response to a steady stimulus is called **adaptation.** In contrast to the **rapidly adapting** receptor, the second type is **slowly adapting**; that is, its response is well maintained during prolonged stretch. The responses of a rapidly adapting and a slowly adapting stretch receptor are shown in Figure 19.4. In the slowly adapting receptor (Figure 19.4A), mild

[9]Boring, E. G. 1942. *Sensation and Perception in the History of Experimental Psychology*. Appleton-Century, New York.

[10]Eyzaguirre, C., and Kuffler, S. W. 1955. *J. Gen. Physiol.* 39: 87–119.

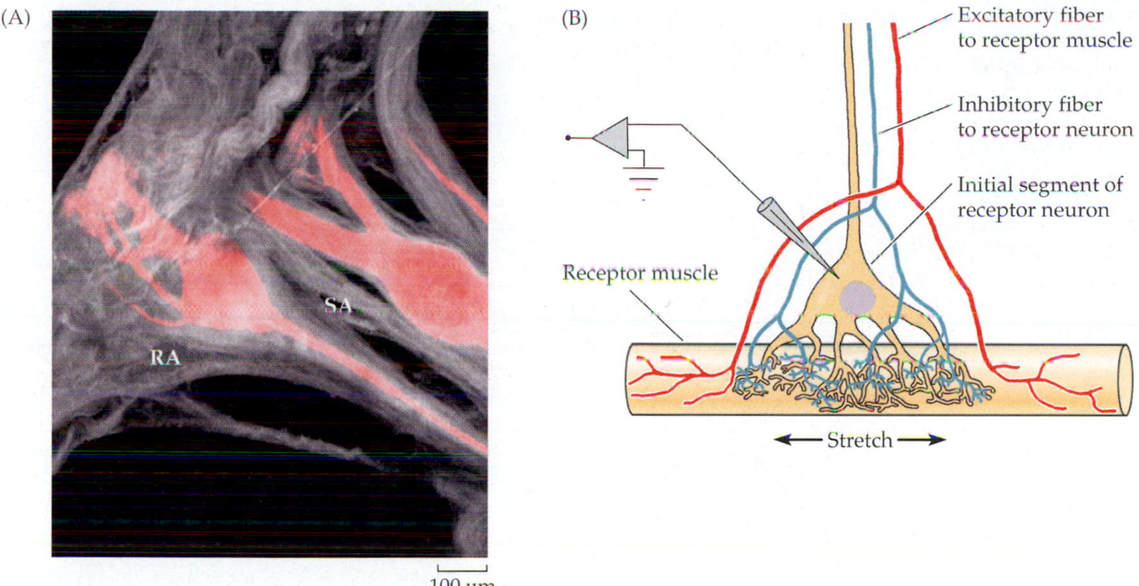

FIGURE 19.3 Crustacean Stretch Receptor. (A) Superimposed picture of tubulin filaments in muscle (gray) and the receptor neurons (red). (B) Relation between stretch receptor neuron and muscle, indicating the method of intracellular recording. The excitatory fiber to the muscle produces contraction; the inhibitory fiber innervates the neuron. Two additional inhibitory fibers are not shown. (A after Purali, 2005; B after Eyzaguirre and Kuffler, 1955.)

stretch of the muscle produces a depolarizing receptor potential of about 5 mV, lasting for the duration of the stretch. A larger stretch produces a larger potential that depolarizes the cell to above threshold and produces a train of action potentials that propagate centrally along the axon. A similar stretch of the muscle produces only transient responses in the rapidly adapting receptor (Figure 19.4B).

Muscle Spindles

Stretch receptors in mammalian skeletal muscles show mechanisms of action similar to those in crustaceans. Such stretch receptors were called **muscle spindles** by early anatomists because of their resemblance to the spindles used by weavers. (Muscle fibers within

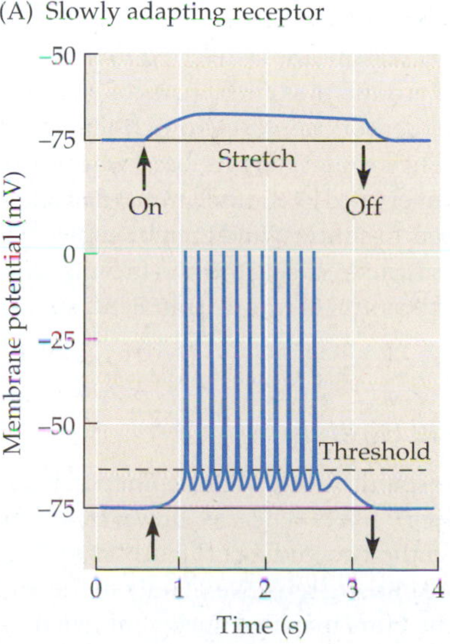

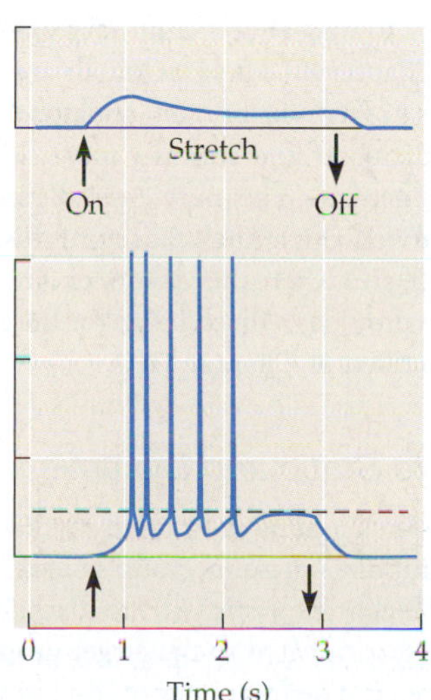

FIGURE 19.4 Responses of Stretch Receptor Neurons to increases in muscle length, recorded intracellularly as indicated in Figure 19.3B. (A) In a slowly adapting receptor, a weak stretch for about 2 seconds produces a subthreshold receptor potential that persists throughout the stretch (upper record). With a stronger stretch, a larger receptor potential sets up a series of action potentials (lower record). (B) In a rapidly adapting receptor, the receptor potential is not maintained (upper record), and during the large stretch, the action potential frequency declines (lower record). (After Eyzaguirre and Kuffler, 1955.)

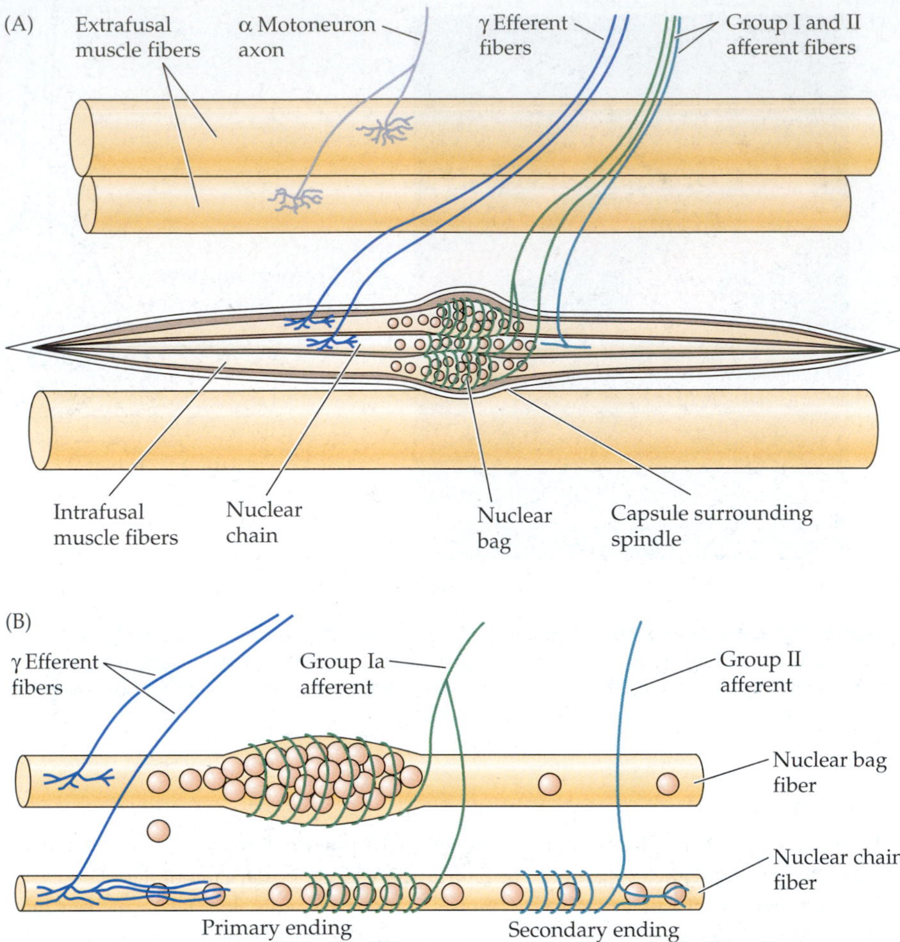

FIGURE 19.5 Mammalian Muscle Spindle. (A) Scheme of mammalian muscle spindle innervation. The spindle, composed of small intrafusal fibers, is embedded in the bulk of the muscle, which is made up of large muscle fibers supplied by α-motoneurons. γ-motor (fusimotor) fibers supply the intrafusal muscle fibers, and group I and group II afferent fibers carry sensory signals from the muscle spindle to the spinal cord. (B) Simplified diagram of intrafusal muscle types and their innervation. (B after Matthews, 1964.)

the spindle are called intrafusal fibers after the Latin word for "spindle," *fusus*.) Figure 19.5 illustrates schematically the sensory apparatus of spindles in leg muscles of the cat. The spindle consists of a capsule containing 8 to 10 intrafusal fibers. In the central, or equatorial, region there is in each fiber a large aggregation of nuclei. Their arrangement provides the basis for the classification of intrafusal fibers as bag or chain fibers, depending on whether the nuclei are grouped together centrally or are arranged linearly.

Two types of sensory neurons innervate each muscle spindle. The larger nerve fibers, group Ia afferents, have diameters of 12 to 20 micrometers (μm) and conduct impulses at velocities up to 120 m/s. (For a summary of the fiber classifications referred to here and elsewhere, see Chapter 8.) Their terminals are coiled around the central parts of both bag and chain fibers to form the **primary endings**. Smaller sensory nerves (Group II fibers) are 4 to 12 μm in diameter and conduct more slowly. They contact chain fibers, where they form **secondary endings**. The muscle spindle also is innervated by motoneurons (**fusimotor fibers**, or γ-motoneurons). They cause intrafusal fibers to contract and thereby stretch the central nuclear region where the sensory endings are situated, causing them to fire impulses. This interaction provides a mechanism for the efferent control of muscle spindle sensitivity that will be described in Chapter 24.

Responses to Static and Dynamic Muscle Stretch

When a rapid stretch is applied to a muscle and to the spindles within it, receptor potentials and bursts of impulses arise in group Ia and II sensory fibers. There is, however, a clear difference in the characteristics of the discharges in the two endings (Figure 19.6). The primary endings, connected to the larger group Ia axons, are sensitive mainly to the rate of change of stretch. The frequency of discharge is therefore maximal during the dynamic

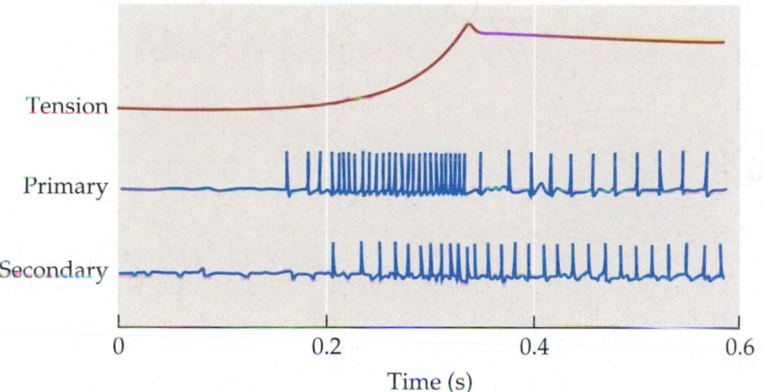

FIGURE 19.6 Specific Muscle Spindle Responses. Recordings of action potentials from single primary (group Ia) and secondary (group II) sensory afferent fibers originating in a cat muscle spindle. The primary fiber greatly increases its discharge rate as tension develops during the stretch; during the maintained phase of the stretch, it quickly adapts to a lower rate. The secondary fiber increases its firing rate more slowly as tension develops and maintains its discharge during the steady stretch. (After Jansen and Matthews, 1962.)

phase, while stretch is increasing, and it subsides to a lower steady level while the stretch is maintained. The secondary endings, connected to the smaller group II fibers, are relatively unaffected by the rate of stretch but are sensitive to the level of static tension.[11] The group Ia (dynamic) and group II (static) afferents are analogous to the rapidly adapting and slowly adapting receptors in the crayfish muscle and in other sensory systems.

Mechanisms of Adaptation in Mechanoreceptors

In muscle spindles, the viscoelastic properties of intrafusal fibers allow a gradual decrease in deformation of the sensory terminals.[12] A variety of processes have been shown to contribute to adaptation of crustacean stretch receptors.[11,13–15] In the slowly adapting stretch receptor, trains of impulses lead to an increase in internal sodium concentration and activation of the sodium pump. The net outward transport of positive charges by the pump reduces the amplitude of the receptor potential and hence the discharge frequency. Yet another factor contributing to adaptation is an increase in potassium conductance. For example, during a train of impulses in crayfish stretch receptors, calcium entry through voltage-activated channels causes the opening of calcium-activated potassium channels. The effect of this increase in potassium conductance is to "short out" the receptor potential, reducing its amplitude and the frequency of the sensory impulses. The rapidly adapting crayfish stretch receptor shows prompt adaptation of its firing rate even when a steady depolarizing current is applied experimentally. During imposed stretch, calcium influx through the transduction channels activates nearby calcium-dependent potassium channels to hyperpolarize the cell.[16]

Adaptation in the Pacinian Corpuscle

The Pacinian corpuscle is a rapidly adapting cutaneous mechanoreceptor[17] whose sensory terminal is enclosed in an onion-like capsule. Pressure applied slowly to the capsule produces no response at all; more rapidly applied pressure produces only one or two action potentials. However, the receptors are exquisitely sensitive to vibration up to frequencies of 1000/s. Although they are found generally in subcutaneous tissue, they are particularly common around footpads and claws of mammals, and in the interosseus membranes bridging the bones of the leg and forearm, where they act as sensitive detectors of ground vibration.[18] A similar structure, the Herbst corpuscle, is found in the legs, bills, and cutaneous tissue of birds (and in the tongues of woodpeckers!). Speculation about their physiological function includes detection by the duck's bill of aquatic vibrations due to small prey and, in soaring birds, detection of vibration of flight feathers due to improper aerodynamic trim.[19] In Chapter 21, we will describe the role of Pacinian corpuscles in the fingertip in sensing texture.

The mechanism of adaptation in the Pacinian corpuscle was studied in detail by Werner Loewenstein and his associates, who showed that it was due, in part, to the dynamic mechanics of the capsule.[20] When a mechanical pulse was applied to an isolated, intact corpuscle, a brief receptor potential appeared at the onset and withdrawal of the pulse (Figure 19.7A). The responses to sustained pressure are transient because compression of the sensitive ending is relieved by redistribution of fluid within the capsule. After the

[11] Matthews, P. B. 1981. *J. Physiol.* 320: 1–30.

[12] Fukami, Y., and Hunt, C. C. 1977. *J. Neurophysiol.* 40: 1121–1131.

[13] Nakajima, T., and Takahashi, K. 1966. *J. Physiol.* 187: 105–127.

[14] Nakajima, S., and Onodera, K. 1969. *J. Physiol.* 200: 187–204.

[15] Sokolove, P. G., and Cooke, I. M. 1971. *J. Gen. Physiol.* 57: 125–163.

[16] Erxleben, C. F. 1993. *Neuroreport* 4: 616–618.

[17] Bell, J., Bolanowski, S., and Holmes, M. H. 1994. *Prog. Neurobiol.* 42: 79–128.

[18] Quilliam, T. A., and Armstrong, J. 1963. *Endeavour* 22: 55–60.

[19] McIntyre, A. 1980. *Trends Neurosci.* 3: 202–205.

[20] Loewenstein, W. R., and Mendelson, M. 1965. *J. Physiol.* 177: 377–397.

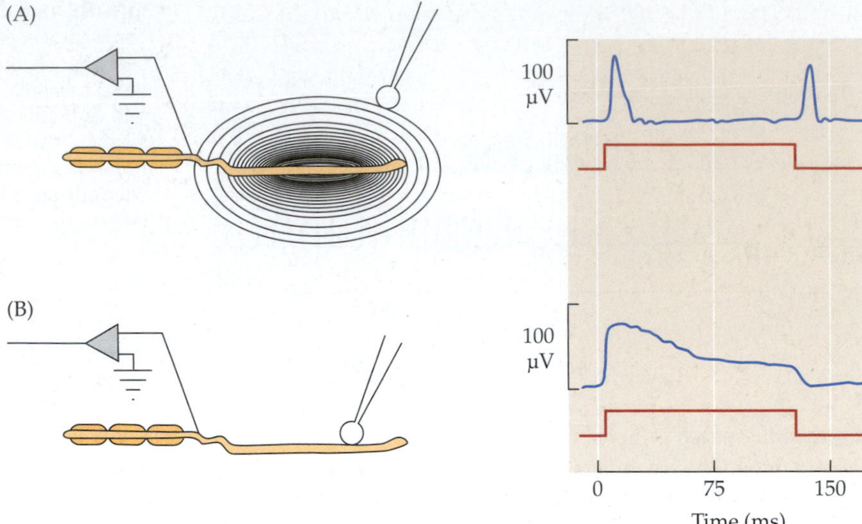

FIGURE 19.7 Adaptation in a Pacinian Corpuscle. (A) A pressure step applied to the body of the corpuscle (lower trace) produces a rapidly adapting receptor potential (upper trace), as a result of a transient wave of deformation that travels through the capsule to the nerve terminal. A similar response occurs on removal of the pulse. (B) After removal of the capsule layers, pressure applied to the nerve terminal produces a receptor potential that lasts for the duration of the pulse. (After Loewenstein and Mendelson, 1965.)

capsule was stripped carefully from the nerve ending, the receptor potential decayed only slowly during the pulse (Figure 19.7B). Nonetheless, even when the receptor potential was prolonged, there was still only a brief burst of action potentials in the afferent axon (not shown); that is, the properties of the axon itself are matched to those of the intact receptor. Such intrinsic adaptation is likely to involve ionic mechanisms like those described in the crayfish stretch receptor.

Direct Transduction by Mechanosensory Hair Cells

It is apparent that there must be stretch-sensitive ion channels localized to mechanosensory nerve terminals—the sites of receptor potential generation. Furthermore, those transduction channels must differ from the channels that support the action potential, since stretch-evoked receptor potentials continue in the presence of a local anesthetic that blocks conduction. Mechanosensitive ion channels are found in a wide variety of cell types and organs, including endothelial cells of blood vessels, baroreceptors in the carotid sinus, touch and pressure receptors in the skin, muscle stretch receptors, and mechanosensitive hair cells of the inner ear.[21–23]

By voltage clamp it has been shown that the current underlying the receptor potential in crayfish stretch receptors is associated with an increase in permeability to sodium and potassium,[24,25] as well as to divalent cations[26] and to larger organic cations such as tris (tris [hydroxymethyl] amino methane) and arginine. The increase in conductance produced by stretch is unaffected by tetrodotoxin[27] but can be altered by some local anesthetics.[28] Receptor potentials in vertebrate muscle spindles also are due to increased cation permeability.[29]

Single channels activated by membrane distortion were first observed in membrane patches from embryonic chick muscle cells[30] and other cell membranes having nothing to do with sensory transduction.[31] Patch recordings made from similar channels in the primary dendrites of the crayfish stretch receptor[32] show that their relative permeabilities to sodium, potassium, and calcium are consistent with previous observations of the whole cell. How membrane deformation causes these channels to open is not known.

Mechanosensory Hair Cells of the Vertebrate Ear

Our understanding of mechanotransduction has advanced furthest in studies of vertebrate hair cells. Mechanosensitive **hair cells** of the inner ear respond to acoustic vibration or head motion that causes fluid movements within the chambers of the inner ear. The exact form of the fluid movement depends on the shape and composition of the particular end organ involved. We discuss in Chapter 22 the structural and functional specializations that support auditory and vestibular signals. For now it will suffice to point out that hair cells in

[21] Ingber, D. E. 2006. *FASEB J.* 20: 811–827.
[22] Garcia-Anoveros, J., and Corey, D. P. 1997. *Annu. Rev. Neurosci.* 20: 567–594.
[23] Chalfie, M. 2009. *Nat. Rev. Mol. Cell. Biol.* 10: 44–52.
[24] Brown, H. M., Ottoson, D., and Rydqvist, B. 1978. *J. Physiol.* 284: 155–179.
[25] Rydqvist, B., and Purali, N. 1993. *J. Physiol.* 469: 193–211.
[26] Edwards, C. et al. 1981. *Neuroscience* 6: 1455–1460.
[27] Nakajima, S., and Onodera, K. 1969. *J. Physiol.* 200: 161–185.
[28] Lin, J. H., and Rydqvist, B. 1999. *Acta Physiol. Scand.* 166: 65–74.
[29] Hunt, C. C., Wilkinson, R. S., and Fukami, Y. 1978. *J. Gen. Physiol.* 71: 683–698.
[30] Guharay, F., and Sachs, F. 1984. *J. Physiol.* 352: 685–701.
[31] Sachs, F. 1988. *Crit. Rev. Biomed. Eng.* 16: 141–169.
[32] Erxleben, C. 1989. *J. Gen. Physiol.* 94: 1071–1083.

the **cochlea** are stimulated by fluid movements within the acoustic frequency range—in humans, from 20 to 20,000 Hz. The vestibular end organs of the inner ear are constructed quite differently and respond to the much lower frequencies generated during head movement. Mass loading of the saccule and utricle by an overlying **otolithic membrane** makes these epithelial sheets sensitive to linear acceleration. The hair cells in the semicircular canals are stimulated by angular acceleration during head rotation. Whatever the form of the fluid movement, in each case it causes deflection of a bundle of modified microvilli, or stereocilia, that project from the apical surface of the hair cell. Bundle deflection results directly in the opening of mechanosensitive ion channels.

Structure of Hair Cell Receptors

Hair cells and the surrounding supporting cells form epithelial sheets that separate dissimilar fluid spaces of the inner ear. The basolateral membranes of hair cells are bathed by perilymph, similar in composition to ordinary extracellular fluid containing high sodium and low potassium (Figure 19.8). The apical, hair-bearing surface of the hair cell faces endolymph, a solution similar in some ways to cytoplasm, having high potassium and low sodium and calcium concentrations. Hair cells make synaptic contact with afferent fibers on their basolateral surfaces. Many hair cells also receive synaptic input from efferent neurons from the brainstem.

(A)

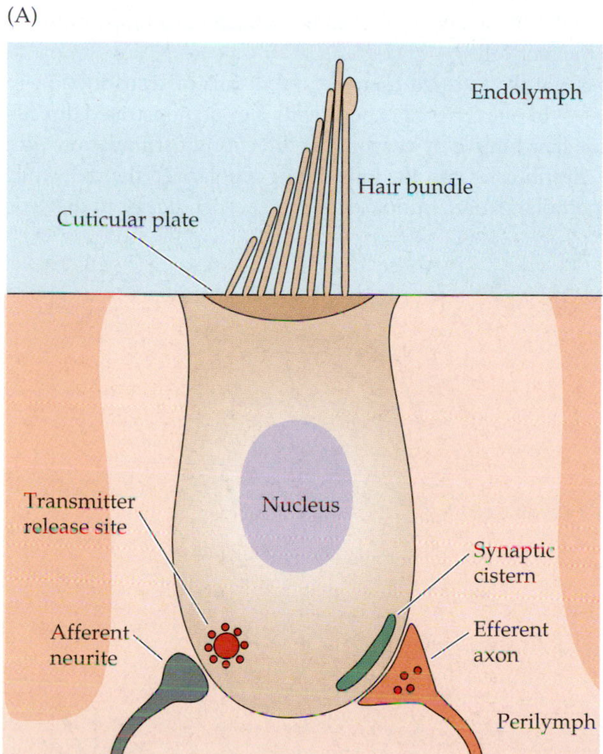

(B)

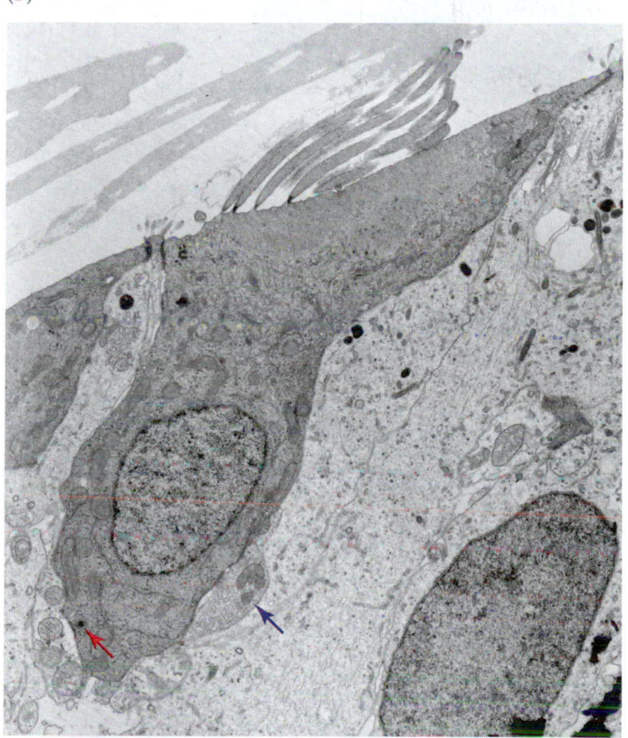

FIGURE 19.8 The Mechanosensory Hair Cell. (A) Schematic drawing highlighting the functional specializations of the hair cell. A bundle of specialized microvilli called stereocilia projects from the cuticular plate into the endolymphatic space. In some hair cells, a true cilium, the kinocilium, is found at one side of the hair bundle. Below the nucleus, hair cells form synapses with afferent and efferent neurons. Synaptic vesicles surround a dense body opposite the ending of an afferent neuron. Efferent neurons projecting from the brainstem form cholinergic synapses. Inside the hair cell, a synaptic cistern lies in close apposition to the plasma membrane underlying the efferent contact. (B) Transmission electron micrograph of a hair cell from chick inner ear. The hair bundle was bent over during fixation. This type of hair cell has an expanded cuticular surface. Afferent contact, on left, is associated with a ribbon (red arrow). A synaptic cistern associated with the efferent contact, (blue arrow) is not visible at this magnification. (Micrograph kindly provided by R. Michaels.)

[33]Flock, A., Flock, B., and Murray, E. 1977. *Acta Otolaryngol.* 83: 85–91.

[34]Flock, A. 1965. *Cold Spring Harb. Symp. Quant. Biol.* 30: 133–145.

[35]Lowenstein, O., and Wersall, J. 1959. *Nature* 184: 1807–1808.

[36]Crawford, A. C., and Fettiplace, R. 1985. *J. Physiol.* 364: 359–379.

[37]Hudspeth, A. J., and Corey, D. P. 1977. *Proc. Natl. Acad. Sci. USA* 74: 2407–2411.

[38]Hudspeth, A. J., and Jacobs, R. 1979. *Proc. Natl. Acad. Sci. USA* 76: 1506–1509.

[39]Corey, D. P., and Hudspeth, A. J. 1979. *Nature* 281: 675–677.

There are anywhere from a few dozen to hundreds of stereocilia (i.e., modified microvilli containing polymerized actin filaments), of graduated length, on different types of hair cells. The longest stereocilia are found on hair cells of the semicircular canals, the shortest in the high-frequency region of the cochlea. Within any one bundle, the stereocilia occur in an organ pipe or staircase array of ascending height. In many hair cells, a single true cilium, the kinocilium (containing a 9 + 2 microtubule array), is found near the middle of the tallest row of stereocilia. Cochlear hair cells have kinocilia early in development, but these disappear later. The stereocilia narrow to insert into a cuticular plate. During bundle deflection the stereocilia behave like rigid rods and bend at this insertion point.[33] A variety of lateral linkages cause the assembly of stereocilia to move as a unified hair bundle.

Transduction by Hair Bundle Deflection

It has been known for a number of years that electrical responses in hair cells are produced by deformation of the hair bundle;[34,35] however, direct experimental confirmation required the development of sensitive techniques for producing and measuring very small movements while recording from hair cells. Auditory and vestibular epithelia from cold-blooded vertebrates, such as turtles and frogs, have proved particularly advantageous for these experiments. Procedures used for nanostimulation of hair cells in the turtle inner ear are shown in Figure 19.9A (see also Figure 19.12). A microelectrode records the hair cell's membrane potential while a glass fiber attached to a piezoelectric manipulator pushes the hair bundle. Movements as small as 1 nm can be detected by projecting the image of the glass fiber, or the hair bundle itself, onto a pair of photodiodes. Such a stimulus produces a voltage change of 0.2 mV in the hair cell.[36]

A. J. Hudspeth (left), demonstrating force-feedback by the sensory hair bundle.

Hudspeth and colleagues elegantly revealed many of the details of transduction in vestibular hair cells of the frog.[37–39] In one series of experiments they demonstrated directly the functional orientation of the hair bundle by varying the direction of stimulation with a piezoelectric manipulator. Deflections toward the kinocilium depolarized the cell, while movement away resulted in hyperpolarization. Bundle deflection perpendicular to that axis

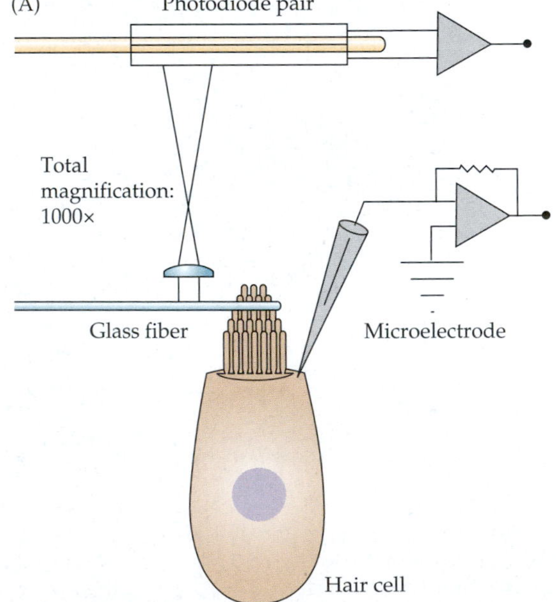

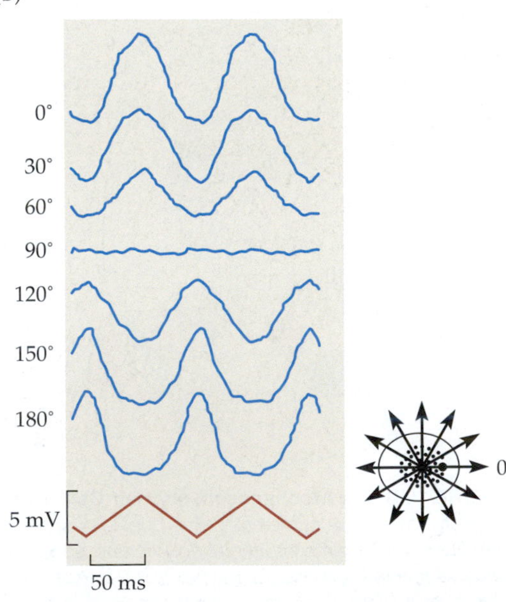

FIGURE 19.9 Recording Mechanotransduction in Hair Cells. (A) Microelectrodes are inserted into hair cells in an excised epithelium mounted on the stage of a compound microscope. The hair bundle is displaced by a glass fiber attached to a piezoelectric manipulator. The image of the glass fiber is enlarged and projected onto a photodiode pair so that motion causes a differential signal between them. Movements as small as 1 nm can be detected with this method. (B) Receptor potentials recorded from a hair cell in the excised saccule of a frog during bundle deflection at various angles. The kinocilium at the center of the tallest row of hairs lies at 0°. Maximal responses occur for motion toward and away from the kinocilium; no response is seen during motion at right angles to that line. (A after Crawford and Fettiplace, 1985; B after Shotwell, Jacobs, and Hudspeth, 1981.)

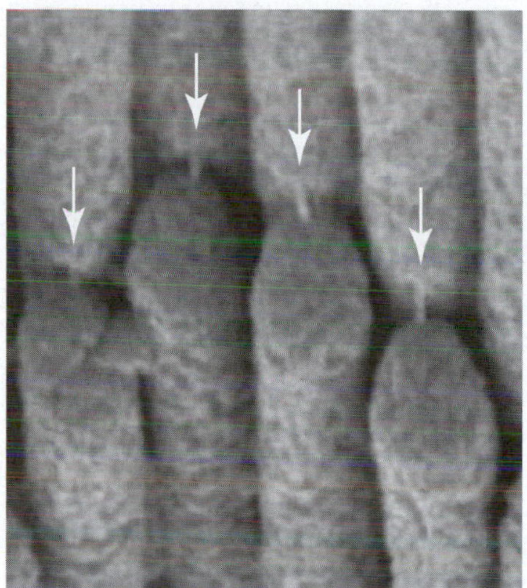

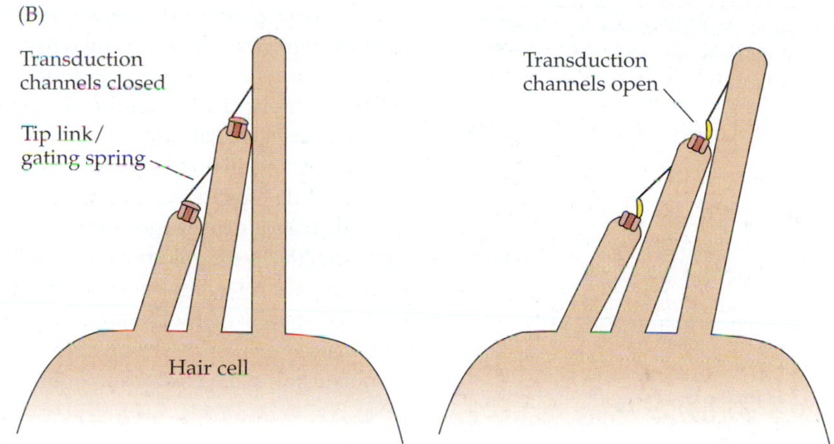

FIGURE 19.10 Tip Links on Hair Cell Stereocilia. (A) Scanning electron micrograph of a rat cochlear hair cell showing extracellular fibers that run from the tips of shorter stereocilia to the sides of adjacent taller stereocilia. (B) Tip links are positioned so that deflection of the hair bundle in the excitatory direction extends the tip link and pulls open the transduction channel (right), while the opposite motion shortens the tip link (left), allowing the channel to close. Current models include a spring-like element within the transduction apparatus (but not the tip link itself), with mechanosensitive channels located at the tops of stereocila. (Micrograph kindly provided by D. Furness.)

caused no change in membrane current.[40] The results of such an experiment are seen in Figure 19.9B, where the magnitude of the voltage change generated in the hair cell varies with the angle of bundle deflection.

Tip Links and Gating Springs

What structural feature of the hair bundle might underlie the directional selectivity of transduction? Pickles and colleagues used the scanning electron microscope to describe a unique class of extracellular linkages connecting the top of one hair with the side of the adjacent taller hair.[41] These **tip links** (Figure 19.10A) were observed only along the axis of mechanical stimulation (i.e., oriented up and down the staircase). The position of the tip links suggested that they might be involved in mechanotransduction, and treatments that break the tip link abolish transduction.[42,43] Indeed, extracellular recordings[44,45] as well as calcium imaging[46,47] indicated that the channels activated by mechanical stimuli are located at the very top of the stereocilia.

Quantitative measures of transduction and the identification of tip links are combined in the **gating spring** hypothesis of mechanotransduction in hair cells. Deflection of the hair bundle in the positive direction (i.e., toward the taller hairs) separates the tips to stretch a gating spring, thus pulling open the transduction channel (Figure 19.10B). When the bundle is pushed away from the tallest hairs, the spring is compressed and the channels close. Although such a scheme might seem somewhat fanciful, a direct physical connection between bundle mechanics and channel gating is required by the great speed at which transduction occurs in hair cells, with time constants of opening of about 40 μs.[48,49] Further, the energetics and mechanics of transduction are consistent with this model. For example, it is possible to measure a decrease in bundle stiffness as the transduction channels open, as though this molecular motion relieves some tension on the gating spring.[50] While the mechanism requires a spring-like element, the tip-link itself is composed of cadherin-23 and protocadherin-15, making it too stiff to serve this function;[51] thus, the molecular identity of the gating spring remains under investigation.

Transduction Channels in Hair Cells

What types of channels are opened in the tips of hair cells? These appear to be nonselective cation channels that have considerable calcium permeability. The single-channel conduc-

[40] Shotwell, S. L., Jacobs, R., and Hudspeth, A. J. 1981. *Ann. N Y Acad. Sci.* 374: 1–10.

[41] Pickles, J. O., Comis, S. D., and Osborne, M. P. 1984. *Hear. Res.* 15: 103–112.

[42] Crawford, A. C., Evans, M. G., and Fettiplace, R. 1991. *J. Physiol.* 434: 369–398.

[43] Assad, J. A., Shepherd, G. M., and Corey, D. P. 1991. *Neuron* 7: 985–994.

[44] Hudspeth, A. J. 1982. *J. Neurosci.* 2: 1–10.

[45] Jaramillo, F., and Hudspeth, A. J. 1991. *Neuron* 7: 409–420.

[46] Beurg, M. et al. 2009. *Nat. Neurosci.* 12: 553–558.

[47] Lumpkin, E. A., and Hudspeth, A. J. 1995. *Proc. Natl. Acad. Sci. USA* 92: 10297–10301.

[48] Corey, D. P., and Hudspeth, A. J. 1983. *J. Neurosci.* 3: 962–976.

[49] Crawford, A. C., Evans, M. G., and Fettiplace, R. 1989. *J. Physiol.* 419: 405–434.

[50] Howard, J., and Hudspeth, A. J. 1988. *Neuron* 1: 189–199.

[51] Kazmierczak, P. et al. 2007. *Nature* 449: 87–91.

[52]Ohmori, H. 1985. *J. Physiol.* 359: 189–217.
[53]Denk, W. et al. 1995. *Neuron* 15: 1311–1321.
[54]Vollrath, M. A., Kwan, K. Y., and Corey, D. P. 2007. *Annu. Rev. Neurosci.* 30: 339–365.
[55]Kwan, K. Y. et al. 2006. *Neuron* 50: 277–289.
[56]Gillespie, P. G., and Muller, U. 2009. *Cell* 139: 33–44.
[57]Hudspeth, A. J., and Gillespie, P. G. 1994. *Neuron* 12: 1–9.
[58]Libby, R. T., and Steel, K. P. 2000. *Essays Biochem.* 35: 159–174.
[59]Gillespie, P. G., Wagner, M. C., and Hudspeth, A. J. 1993. *Neuron* 11: 581–594.
[60]Ricci, A. J., and Fettiplace, R. 1997. *J. Physiol.* 501: 111–124.
[61]Martin, P. Mehta, A. D., and Hudspeth, A. J. 2000. *Proc. Natl. Acad. Sci. USA* 97: 12026–12031.

tance is large, greater than 100 pS.[42,52,53] Strikingly, single-channel conductance increases as one progresses from low to high frequency hair cells. From conductance measurements and measurements of total transducer current, it is possible to calculate that each hair cell has only about 100 transduction channels. This corresponds to perhaps as few as two channels per stereocilium!

The very small number of channels in each hair cell makes biochemical and molecular biological investigation especially challenging. Hair cell transduction channels have no intrinsic voltage dependence, nor are they ligand gated in any traditional sense. It is unlikely, therefore, that strong homologies exist between these and most voltage- or ligand-gated ion channels. A large number of mechanosensitive ion channels have been cloned from bacteria, yeast, nematodes, and flies,[21] providing candidate genes for future study.[54] As yet, however, even one very strong candidate was not elected, since genetic knockout failed to eliminate hair cell transduction in mice.[55]

Adaptation of Hair Cells

Hair cells are extremely sensitive, with threshold responses to bundle motion of less than 10^{-9} m.[56] It seems likely, then, that some type of adaptive process exists to maintain sensitivity in the face of a so-called background stimulus. For example, vestibular hair cells in the saccule and utricle must remain sensitive to subtle head movements while continuously being subject to gravitational force acting on the overlying otolithic membrane (see Chapter 22). Indeed, it has been shown by direct measurement that hair cells adapt to a prolonged displacement by establishing a new set point for their operating range, with no loss of sensitivity. This form of set point adaptation is thought to arise from the action of a non-muscle myosin that exerts tension on the transduction channels by pulling against the actin core of the stereocilium.[57] Myosins have been cloned from the inner ear,[58] and specific antibodies have been used to show that myosin 1C is located near the tips of stereocilia in frog hair cells.[59] This myosin-based adaptation depends on calcium influx and is relatively slow. It may be most prominent in vestibular hair cells. The more rapid adaptation seen in auditory hair cells may arise from calcium ions acting directly on the transduction channel, causing it to close.[60]

R. Fettiplace

The tight coupling between mechanical input and transduction channel gating implies that feedback will occur during transduction.[61] Thus, when the hair bundle is deflected, the calcium that enters through the open transduction channels can produce a further change in hair bundle stiffness or position. Similarly, alteration of hair cell membrane potential can move the hair bundle by changing the driving force for calcium entry. These feedback processes can give rise to mechanical resonance (or ringing) in bundle motion. Fettiplace and Crawford explored this phenomenon in auditory hair cells of the turtle using flexible glass fibers to deflect the hair bundle directly.[36] A 75-nm step to the butt end of the glass fiber caused a smaller deflection of the tip attached to the bundle (Figure 19.11). The tip

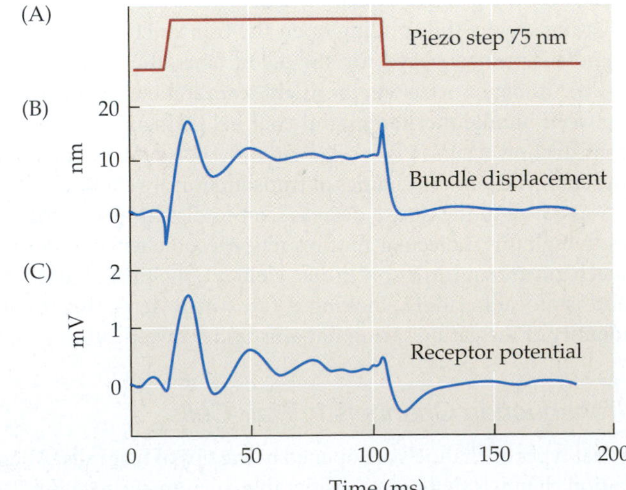

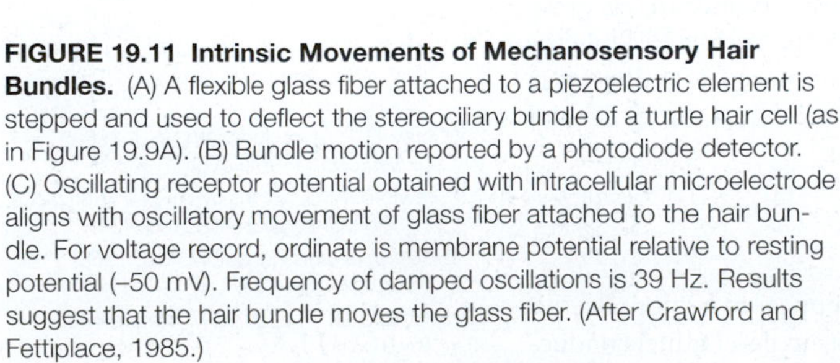

FIGURE 19.11 Intrinsic Movements of Mechanosensory Hair Bundles. (A) A flexible glass fiber attached to a piezoelectric element is stepped and used to deflect the stereociliary bundle of a turtle hair cell (as in Figure 19.9A). (B) Bundle motion reported by a photodiode detector. (C) Oscillating receptor potential obtained with intracellular microelectrode aligns with oscillatory movement of glass fiber attached to the hair bundle. For voltage record, ordinate is membrane potential relative to resting potential (–50 mV). Frequency of damped oscillations is 39 Hz. Results suggest that the hair bundle moves the glass fiber. (After Crawford and Fettiplace, 1985.)

motion was smaller because the bundle was relatively stiff compared to the wispy glass fiber. Further, although the imposed movement was square, the fiber tip showed a small oscillation that was coincident with an oscillatory receptor potential recorded from the hair cell. In other words, the glass fiber pushed the bundle, and the bundle *pushed* back! This oscillatory motion of the hair bundle results from transducer channel gating. The reciprocal interaction between bundle deflection and membrane potential contributes to the production of what are called ear sounds (i.e., otoacoustic emissions) that can be generated in most species, including humans.[62] The ability to elicit otoacoustic emissions from the ear has provided a way for audiologists to assess hair cell function directly, even in infants or comatose patients.[63]

Olfaction

Mechanotransduction in the ear attains high sensitivity by tightly coupling the stimulus energy to the hair cell's membrane potential. In contrast, great sensitivity is obtained in olfaction (smell) and vision, and in some forms of taste by chemical amplification—that is, second messenger pathways in which enzymatic cascades produce large numbers of intermediate products, thereby increasing by a thousand-fold the effect of one activated receptor molecule.

Olfaction is poorly developed in humans compared to dogs or pigs or butterflies. But at the same time, considerable effort (and advertising dollars) goes into human olfactory behavior (consider the numbers of soaps, deodorants, and perfumes that are aimed at securing a socially acceptable personal bouquet). In fact, olfactory signals are essential to human survival, stimulating feeding, reproduction, and mother–infant bonding.[64,65] Detection and discrimination of the unique blend of odors linked to those behaviors begins with a large family of molecular receptors in the olfactory receptor neurons.

Olfactory Receptors

Mammals detect odors with a patch of about 100,000 olfactory receptor neurons whose axons project through a thin portion of the frontal skull (the cribriform plate) to the olfactory bulb (Figure 19.12). The long cilia of the olfactory receptors extend into the nasal cavity, where they lie in a layer of mucus, approximately 50 μm thick in humans, that is entirely replaced every 10 min. The mucous layer protects the sensory epithelium, wash-

[62]Kemp, D. T. 1978. *J. Acoust. Soc. Am.* 64: 1386–1391.

[63]Lonsbury-Martin, B. L., and Martin, G. K. 2003. *Curr. Opin. Otolaryngol. Head Neck Surg.* 11: 361–366.

[64]Stern, K., and McClintock, M. K. 1998. *Nature* 392: 177–179.

[65]Brennan, P. A., and Kendrick, K. M. 2006. *Philos. Trans. R. Soc. Lond., B, Biol. Sci.* 361: 2061–2078.

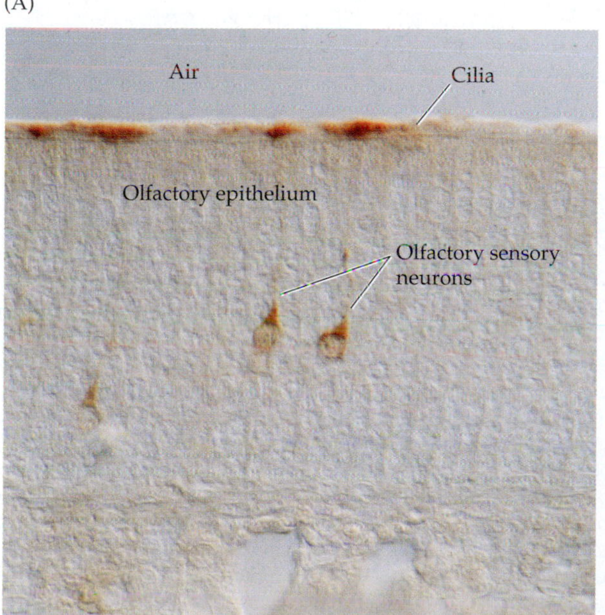

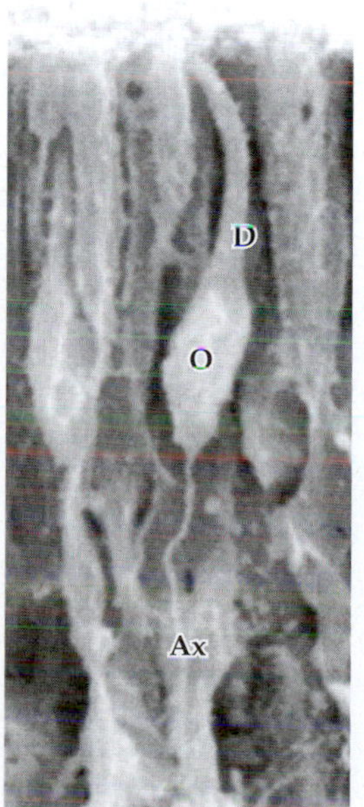

FIGURE 19.12 Olfactory Epithelium at Light Level and Scanning EM. (A) Section through the olfactory epithelium of a mouse. Individual olfactory neurons labeled with antibodies to a single molecular receptor. (B) Scanning electron micrograph (EM) of the olfactory epithelium of a hamster. The receptor neurons (O) have a long dendrite (D) that extends to the surface, and an axon (Ax) that projects from the epithelium to the olfactory bulb. The long sensory cilia at the tip of the dendrite form a dense mat in this preparation and are not individually resolved. (A courtesy of R. Reed; B micrograph courtesy of R. Costanzo.)

ing out potentially toxic airborne compounds, and all odorants must dissolve through this to reach the sensory cilia. An odorant-binding protein helps to concentrate hydrophobic odorants in this aqueous layer.[66] Olfactory receptors are continuously replaced throughout the animal's lifetime. Each receptor lives for a month or two, and new receptors arise from a layer of basal cells in the olfactory epithelium.[67]

The Olfactory Response

Early measurements of olfactory responses were made by Adrian[68] and Ottoson.[69] Since then, evidence has accumulated that odorant molecules interact with receptors in the ciliary membrane to produce an increase in conductance, resulting in depolarization. Action potentials then travel along the olfactory receptor axon into the CNS. Patch clamp techniques have been used to record both odorant-induced currents from isolated olfactory cells[70] and to record the precise time course and localization of the odor response.

An example of such an experiment on a cell isolated from the olfactory mucosa of the salamander[71] is shown in Figure 19.13. The membrane potential of the cell is held at –65 mV, and a solution containing a mixture of odorant molecules (approximately 0.1 m*M*) in 100 m*M* KCl is applied from a second pipette by a brief (35 ms) pressure pulse—first to the soma, and then to the distal portion of the dendrite and the cilia. Pipette solution applied to the soma produces a rapid inward current, which is due to the local increase in potassium concentration. The time course of the potassium response provides a measure of the speed of application and subsequent dissipation of the solution by diffusion into the surrounding bath. A second, smaller and slower inward current appears when the odorants reach the apical dendrite.

Solution applied to the apical dendrite and cilia produces only a small potassium response; presumably, there are fewer potassium channels in that portion of the cell. However, the odorant itself produces a large inward current that outlasts the time of application by several seconds. The experiment clearly indicates that the region of sensitivity to the odorants is the distal dendrite and cilia, and the prolonged time course of the dendritic response is consistent with the idea that the conductance change is produced by a second messenger system whose activity outlasts the initial odorant binding reaction.

[66] Pevsner, J. et al. 1988. *Science* 241: 336–339.

[67] Farbman, A. I. 1994. *Semin. Cell Biol.* 5: 3–10.

[68] Adrian, E. D. 1953. *Acta Physiol. Scand.* 29: 5–14.

[69] Ottoson, D. 1956. *Acta Physiol. Scand.* 35: 1–83.

[70] Maue, R. A., and Dionne, V. E. 1987. *J. Gen. Physiol.* 90: 95–125.

[71] Firestein, S., Shepherd, G. M., and Werblin, F. S. 1990. *J. Physiol.* 430: 135–158.

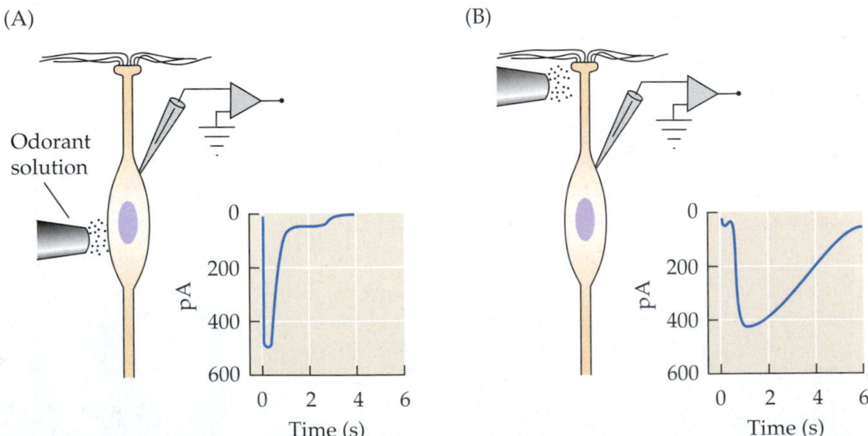

FIGURE 19.13 Responses of Isolated Olfactory Cells from the salamander. A patch clamp electrode is used to record whole-cell current. A solution containing 0.1 m*M* odorant mixture in 100 m*M* KCl is applied to the cell by a brief (35 ms) pressure pulse. (A) When solution is applied to the cell body, there is a rapid, transient inward current due to the increased KCl concentration, followed by a smaller, slower current as the odorant reaches the apical end of the dendrite. The time course of the fast inward current is indicative of the time course of application and dissipation of the electrode solution. (B) When the solution is applied to the dendrite, there is only a small, rapid current due to the KCl, but a large current, due to the odorant, lasts for several seconds after the electrode solution has washed away. (After Firestein, Shepherd, and Werblin, 1990.)

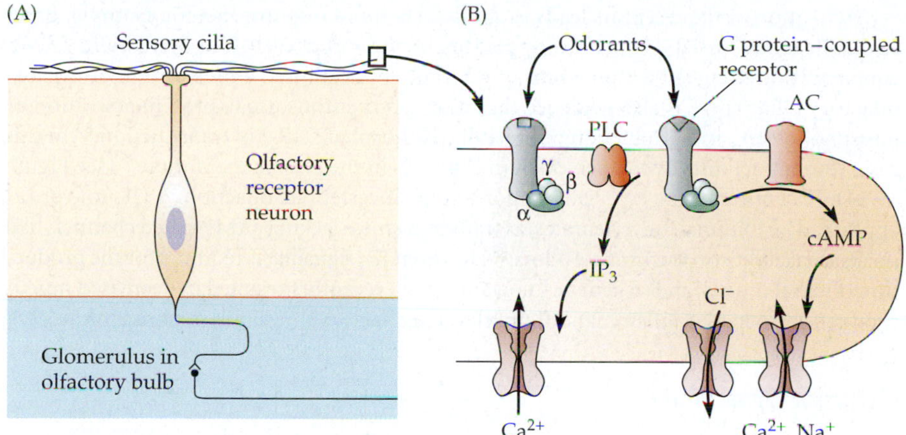

FIGURE 19.14 Transduction in Olfactory Cilia. (A) The molecular receptors for olfactants are found in sensory cilia that project into the mucous layer of the olfactory epithelium. Depolarizing receptor potentials in these long receptors give rise to action potentials that propagate along the olfactory receptor neuron's axon into the central nervous system. (B) Odorant molecules bind to specific G protein–coupled receptors in the plasma membrane of the olfactory cilia. This frees the α-subunit to activate adenylyl cyclase (AC) and raise the concentration of cyclic adenosine monophosphate (cAMP), which causes nonselective cation channels to open, depolarizing the membrane. Calcium-gated chloride current can enhance this effect. Other pathways may involve the activation of phospholipase C (PLC) and the consequent rise in IP_3 to act directly on plasma membrane calcium channels.

Cyclic Nucleotide-Gated Channels in Olfactory Receptors

The depolarization produced by odorants arises from the opening of nonselective cation channels, permeable to Na^+, K^+, and Ca^+.[72,73] The channels, which are activated by intracellular cAMP (Figure 19.14), are closely related to the cation channels opened by cyclic guanosine monophosphate (cGMP) in rod photoreceptors.[74] Indeed, several features of this transduction cascade are similar to that operating in photoreceptors, with the striking difference that an increase in light intensity leads to a *decrease* in cyclic nucleotide concentration and consequent hyperpolarization (see Chapter 20). Like rod photoreceptor channels, olfactory receptor channels are insensitive to changes in membrane potential. The influx of calcium, in turn, opens calcium-activated chloride channels, further contributing to the receptor potential.[75–77] Enhancement of the receptor potential by activation of chloride channels seems like an anomaly, as activation of chloride channels in most cells would be expected to produce hyperpolarization, or at least little change in membrane potential. However, the receptor regions of olfactory cells appear to have little resting chloride conductance and an intracellular chloride concentration similar to that of the mucus in which they are bathed. Thus, the chloride equilibrium potential, rather than being close to the resting membrane potential, is near 0 mV, and opening chloride channels results in outward chloride flux (i.e., inward, positive current).[78] Only a few open channels are required to initiate action potentials because of the very high input resistance in olfactory receptors, suggesting that even single odorant molecules can be detected.[79]

Coupling the Receptor to Ion Channels

How is odorant binding coupled to the gating of cAMP-dependent cation channels? The mechanism of activation is outlined in Figure 19.14. Receptors coupled to G proteins first bind the odorant. The activated G protein releases the α-subunit, which stimulates adenylyl cyclase to produce cAMP. An extensive family of candidate odorant receptor genes has been identified[80] that encode seven-transmembrane, G protein–coupled proteins related in structure to metabotropic neurotransmitter receptors (see Chapter 12). The greatest variability between encoded proteins occurs in the sequence of three of the transmembrane domains that form the ligand-binding pocket. A G protein specifically expressed in olfactory epithelia (G_{olf}) has also been identified,[81] as has an olfactory adenylyl cyclase.[82,83]

---

[72] Nakamura, T., and Gold, G. H. 1987. *Nature* 325: 442–444.

[73] Pifferi, S., Boccaccio, A., and Menini, A. 2006. *FEBS Lett*. 580: 2853–2859.

[74] Dhallan, R. S. et al. 1990. *Nature* 347: 184–187.

[75] Kleene, S. J., and Gesteland, R. C. 1991. *J. Neurosci.* 11: 3624–3629.

[76] Kurahashi, T., and Yau, K. W. 1993. *Nature* 363: 71–74.

[77] Pifferi, S. et al. 2009. *J. Physiol.* 587: 4265–4279.

[78] Kleene, S. J. 2008. *Chem. Senses*. 33: 839–859.

[79] Menini, A., Picco, C., and Firestein, S. 1995. *Nature* 373: 435–437.

[80] Buck, L., and Axel, R. 1991. *Cell* 65: 175–187.

[81] Jones, D. T., and Reed, R. R. 1989. *Science* 244: 790–795.

[82] Pace, U. et al. 1985. *Nature* 316: 255–258.

[83] Bakalyar, H. A., and Reed, R. R. 1990. *Science* 250: 1403–1406.

Activation of the receptors leads to a relatively rapid response. So, for example, Breer and colleagues[84] used a stop-flow apparatus to show that a 10-fold increase in cAMP concentration occurred within 50 ms of odorant application in a preparation of isolated olfactory cilia. There is also evidence that olfactory neurons use G protein activation of phospholipase C and production of inositol trisphosphate (IP_3) in transduction.[85] In this case, IP_3 may act directly to open calcium channels in the plasma membrane[86] (see Figure 19.14). IP_3 appears to be especially important for invertebrate olfaction.[87] The role of IP_3 appears to be minimal for vertebrates, as transgenic mice lacking cAMP-gated channels had no residual ability to discriminate odors.[88] However, IP_3 signaling is required for the production of nasal mucus; and so genetic knockout of IP_3 receptors resulted in decreased mucus production, increased inflammation, and elevated olfactory thresholds in transgenic mice.[89]

Odorant Specificity

Mammals can discriminate a very large number of odors, and the existence of hundreds, or possibly thousands, of olfactory receptor proteins are the substrate for this capability. A remaining difficulty is the lack of odorant specificity of individual olfactory receptor neurons (ORNs), each of which recognize a spectrum of odors rather than being highly selective.[90] One approach to further understanding this issue has been to determine the expression pattern of cloned receptor molecules among olfactory receptor neurons using in situ hybridization or expression of recombinant proteins. Each particular odorant receptor was found in a restricted area of the olfactory epithelium.[91,92] Different families of receptor genes appear to be expressed in zones extending along the length of the epithelium (Figure 19.15A,B), with any given gene expressed in only a small number of olfactory receptor neurons. Each ORN expresses only one of the olfactory receptor genes, a remarkable result of feedback regulation by the olfactory receptor protein.[93,94] Each molecular receptor has rather broad binding affinities, accounting for the limited specificity of individual ORNs. In fact, there is no 1:1 correspondence known between specific odorants and individual receptor proteins.

Remarkably, those ORNs expressing the same molecular receptor project to a single pair of medial and lateral glomeruli in the first central relay station, the olfactory bulb[95] (Figure

[84]Breer, H., Boekhoff, I., and Tareilus, E. 1990. *Nature* 345: 65–68.

[85]Boekhoff, I. et al. 1990. *EMBO J.* 9: 2453–2458.

[86]Restrepo, D. et al. 1990. *Science*. 249: 1166–1168.

[87]Ache, B. W., and Zhainazarov, A. 1995. *Curr. Opin. Neurobiol.* 5: 461–466.

[88]Brunet, L. J., Gold, G. H., and Ngai, J. 1996. *Neuron* 17: 681–693.

[89]Fukuda, N. et al. 2008. *Eur. J. Neurosci.* 27: 2665–2675.

[90]Reed, R. R. 2004. *Cell* 116: 329–336.

[91]Ressler, K. J., Sullivan, S. L., and Buck, L. B. 1993. *Cell* 73: 597–609.

[92]Vassar, R., Ngai, J., and Axel, R. 1993. *Cell* 74: 309–318.

[93]Serizawa, S. et al. 2003. *Science* 302: 2088–2094.

[94]Lewcock, J. W., and Reed, R. R. 2004. *Proc. Natl. Acad. Sci. USA* 101: 1069–1074.

[95]Mombaerts, P. et al. 1996. *Cell* 87: 675–686.

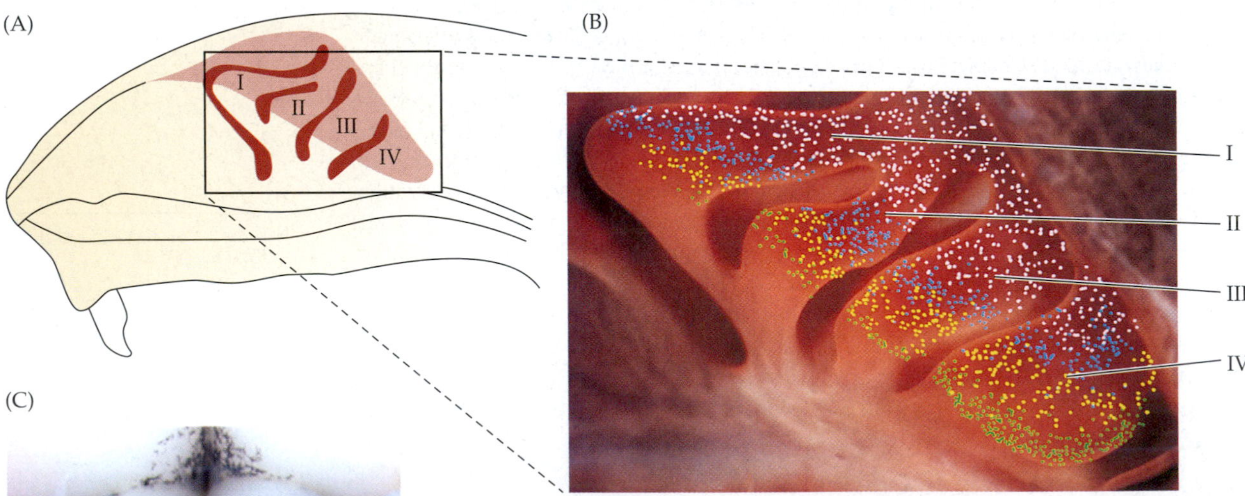

FIGURE 19.15 Expression of Specific Odorant Receptor Genes by Subsets of Olfactory Receptor Neurons. (A) The olfactory epithelium of a rat lies on a series of convolutions of the nasal cavity called turbinates—labeled I to IV. Bands of olfactory receptor neurons label positively for olfactory receptor mRNAs. (B) Probes for different mRNAs label non-overlapping populations of neurons (depicted as green, yellow, blue, and white dots). (C) Olfactory receptor neurons expressing one receptor gene project to unique glomeruli in each olfactory bulb. (A after Vassar, Ngai, and Axel, 1993; B kindly provided by R. Vassar; C from Tadenev et al., 2011.)

19.15C). This highly specific innervation pattern includes fasciculation of "like" axons prior to reaching the bulb, and also depends on expression of the molecular receptor protein.[96] These features of olfactory organization are still more intriguing when one recalls that olfactory receptor neurons turn over every few weeks throughout the lifetime of the organism![90] The mechanisms producing this selective innervation pattern remain largely unknown.

In addition to the main olfactory epithelium, mammals possess a vomeronasal organ used for the detection of pheromones that stimulate mating and other social behaviors. Vomeronasal receptor neurons (VRNs) project to an accessory olfactory bulb, which in turn, projects to the limbic system. VRNs express additional families of molecular receptors. G protein–coupled vomeronasal receptors 1 and 2 (VR1 and VR2)[97–99] each include over 100 genes in mice, some of which specifically bind known pheromones.[100] G protein–coupled formyl peptide receptors mediate immune cell responses to bacteria but also appear to be expressed specifically in VRNs,[101] as are "trace-amine associated receptors."[102] Formyl peptides and trace amines are present in urine and so could contribute to assessment of gender, social status, or the state of health of individuals.[103] Each vomeronasal neuron may express just one type of molecular receptor, and the pattern of expression differs between male and female rats.[104]

Mechanisms of Taste (Gustation)

Discussions of taste and smell are often combined because both senses are activated by chemical stimuli arriving from the outside world. Indeed, some tastants (i.e., taste stimuli) act on G protein–coupled receptors (GPCRs) in ways quite similar to those discussed for olfaction. However, other tastants, principally salts and acids, act directly on membrane conductances, and taste receptor cells differ anatomically from olfactory receptor neurons.

Taste Receptor Cells

Taste receptors are ciliated neuroepithelial cells and are found in taste buds on the tongue surface (Figure 19.16). Like olfactory receptors, taste cells are regenerated throughout life. Unlike olfactory receptors, taste cells do not have axons, but form chemical synapses with afferent neurites within the taste bud. Microvilli project from the apical pole of the taste cell into the open pore of the taste bud, where they come into contact with tastants dissolved in saliva on the tongue's surface. Strikingly, only a subset of cells within the taste bud actually respond to tastants[105] (Figure 19.16B). A second class of cells fails to respond to tastants. Instead, there is ultrastructural evidence that they receive chemical synaptic input from the taste cells. Communication between the two cell types may be mediated by ATP[106] and serotonin.[107]

[96] Bozza, T. et al. 2002. *J. Neurosci.* 22: 3033–3043.
[97] Dulac, C., and Axel, R. 1995. *Cell* 83: 195–206.
[98] Matsunami, H., and Buck, L. B. 1997. *Cell* 90: 775–784.
[99] Ryba, N. J., and Tirindelli, R. 1997. *Neuron* 19: 371–379.
[100] Boschat, C. et al. 2002. *Nat. Neurosci.* 5: 1261–1262.
[101] Riviere, S. et al. 2009. *Nature* 459: 574–577.
[102] Liberles, S. D., and Buck, L. B. 2006. *Nature* 442: 645–650.
[103] Tirindelli, R. et al. 2009. *Physiol. Rev.* 89: 921–956.
[104] Herrada, G., and Dulac, C. 1997. *Cell* 90: 763–773.
[105] DeFazio, R. A. et al. 2006. *J. Neurosci.* 26: 3971–3980.
[106] Finger, T. E. et al. 2005. *Science* 310: 1495–1499.
[107] Huang, Y. J. et al. 2005. *J. Neurosci.* 25: 843–847.

(A)

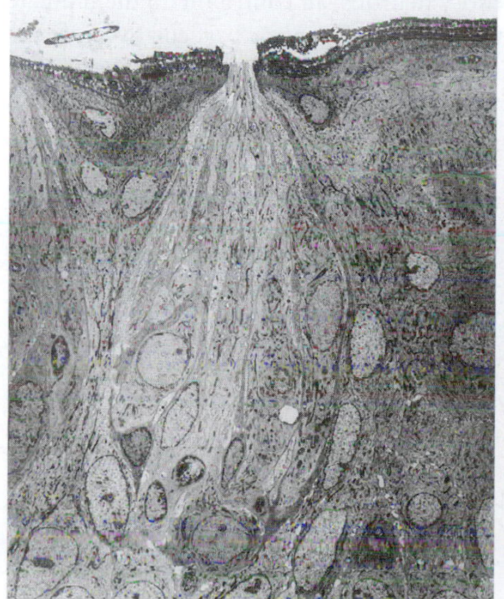

(B)

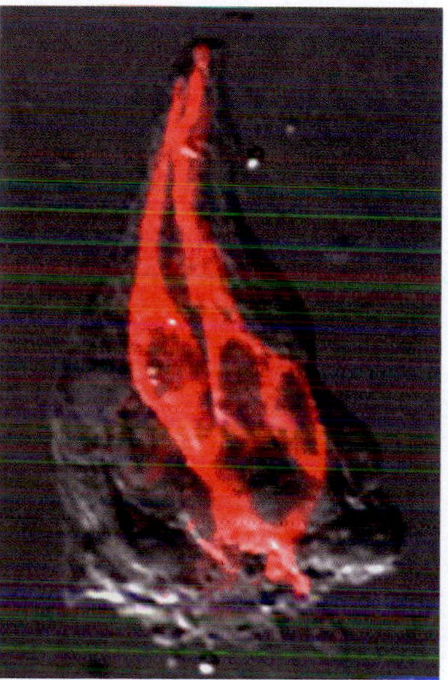

FIGURE 19.16 Taste Receptor Cells Are Found in Taste Buds within the lingual epithelium. (A) A transmission electron micrograph of a taste bud in the tongue of a rat. Individual taste receptor cells have microvilli that project into the taste pore to sample the saliva. (B) An individual taste bud dissected from the tongue of a rat. The taste receptor cells are labeled with an antibody to gustducin, a G protein involved in taste transduction. (A courtesy of R. Yang and J. Kinnamon; B courtesy of I. Wanner and S. D. Roper.)

Taste Modalities

Taste stimuli are usually subdivided into five categories: salt, sour, bitter, sweet, and umami, the last being a Japanese word for the taste of monosodium glutamate (MSG), or more generally, amino acid (i.e., meat) taste. Each category has its own transduction mechanism (Figure 19.17), which falls into one of two classes, both of which result in depolarization of the taste receptor membrane: (1) direct action of the tastant on ion channels, or (2) coupling of tastant receptors to ion channels through second messenger pathways involving G protein–coupled receptors (GPCRs).[108]

There is good agreement that the taste of salt is mediated by direct flux of sodium (or other monovalent cations) through channels in the apical membrane of the taste cell that are open at rest.[109] Sodium is present at higher concentration in salty foods (>100 m*M*) than in saliva, and it simply diffuses into the cells down its electrochemical gradient, thereby producing depolarization. The candidate channels are similar to epithelial sodium channels (ENaCs) found in frog skin and kidney. They are voltage insensitive and are blocked by the diuretic compound amiloride. The functional channels consist of three subunits.[110] The α-subunit has been detected in lingual epithelium,[111] and mice in which the ENaC α-subunit was genetically ablated lose their taste for sodium.[112]

Sour taste is produced by the high concentration of protons in acidic foods. Protons may act by entering taste cells through the same amiloride-blockable sodium channels.[113] Alternatively, protons may depolarize the cells by blocking normally open potassium channels.[114] A third mechanism is seen in frog taste cells, which have cation channels that are activated by protons.[115] In addition to acting at the taste cell's cilia, salts and protons may percolate through the taste pore (a paracellular pathway) to act on the same or other ion channels (including some that are amiloride-insensitive) in the basolateral membranes of the cell.[108] This illustrates what seems to be a general principle of gustation—that is, several parallel transduction pathways can exist for any one class of tastants.

Sweet, bitter, and umami tastes are mediated by GPCRs, probably acting through phospholipase C and IP_3 to release calcium from internal stores,[116] (see Chapter 12) and thereby activating calcium-dependent cation channels[117] In addition, the intracellular messenger cAMP is altered through the activity of gustducin, a G protein specific to taste cells.[118] Sensitivity to sweet, bitter, and umami is reduced in gustducin knockout mice.[108] A growing number of molecular receptors for sweet, bitter, and umami have been identified in recent years. The family of GPCR molecules, T1R1, T1R2, and T1R3 may be expressed in sweet receptor cells. A large family of GPCRs, the T2Rs[119,120] serve the taste of the widely varying class of bitter compounds. Finally, unusual forms of metabotropic glutamate receptors[121] have been implicated in mediating umami taste.

It is important to note that taste, as a percept, results from sensations in addition to those reported by the taste buds. The aromas detected by olfaction, as well as by texture and temperature, which are sensed by somatosensory neurons, all contribute to the ultimate identification and appreciation of food. The hot taste of chili peppers particularly reinforces this concept. Hot chilies are not sensed by taste cells per se, but rather by pain fibers in the

[108] Roper, S. D. 2007. *Pflügers Arch.* 454: 759–776.
[109] Avenet, P., and Lindemann, B. 1991. *J. Membr. Biol.* 124: 33–41.
[110] Canessa, C. M. et al. 1994. *Nature* 367: 463–467.
[111] Li, X. J., Blackshaw, S., and Snyder, S. H. 1994. *Proc. Natl. Acad. Sci. USA* 91: 1814–1818.
[112] Chandrashekar, J. et al. *Nature* 464: 297–301.
[113] Gilbertson, T. A., Roper, S. D., and Kinnamon, S. C. 1993. *Neuron* 10: 931–942.
[114] Kinnamon, S. C., Dionne, V. E., and Beam, K. G. 1988. *Proc. Natl. Acad. Sci. USA* 85: 7023–7027.
[115] Okada, Y., Mitamoto, T., and Sato, T. 1994. *J. Exp. Biol.* 187: 19–32.
[116] Zhang, Y. et al. 2003. *Cell* 112: 293–301.
[117] Zhang, Y. et al. 2007. *J. Neurosci.* 27: 5777–5786.
[118] McLaughlin, S. K., McKinnon, P. J., and Margolskee, R. F. 1992. *Nature* 357: 563–569.
[119] Adler, E. et al. 2000. *Cell* 100: 693–702.
[120] Matsunami, H., Montmayeur, J. P., and Buck, L. B. 2000. *Nature* 404: 601–604.
[121] Chaudhari, N., Landin, A. M., and Roper, S. D. 2000. *Nat. Neurosci.* 3: 113–119.

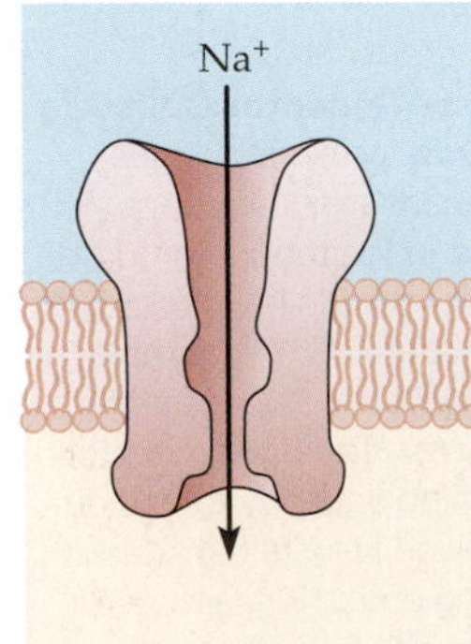

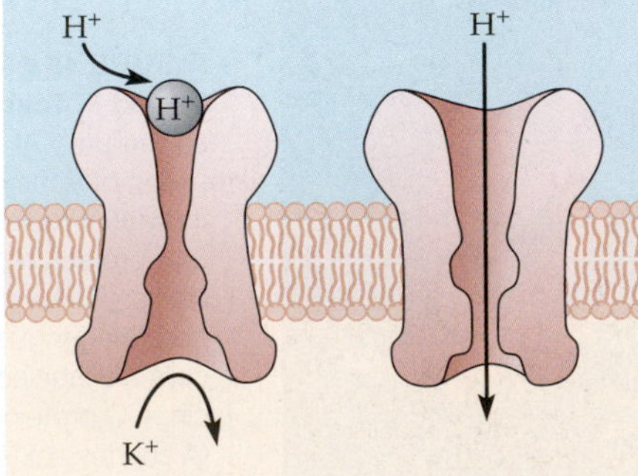

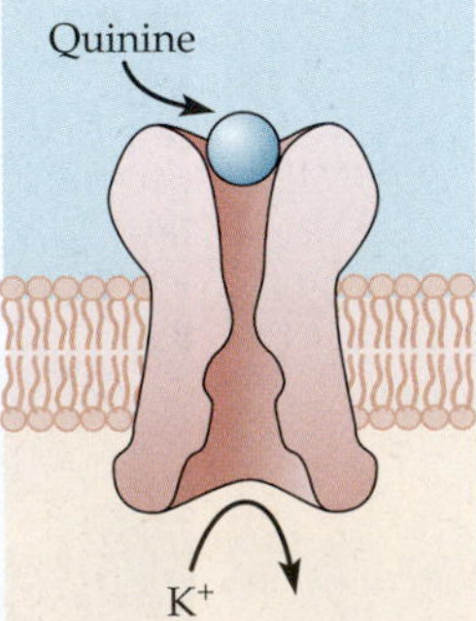

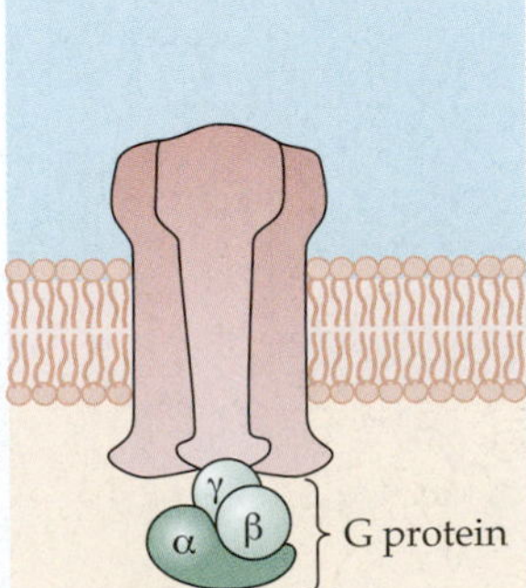

FIGURE 19.17 Mechanisms of Taste Transduction. Tastant molecules range from protons (acids) to simple salts to complex organic compounds. This wide range of chemical stimuli is transduced by a multiplicity of mechanisms. (A) Salts and (B) acids can permeate ion channels in the sensitive ending or block normally open potassium channels. (C) Some bitter compounds also block potassium channels to cause depolarization. (D) Sugars and amino acids (umami) interact with G protein–coupled receptors to initiate second-messenger cascades. All these mechanisms lead eventually to a depolarization, voltage-gated calcium influx, and increased release of transmitter onto associated afferent dendrites.

tongue that are activated by the compound capsaicin. A capsaicin receptor has been cloned and shown to be a calcium-selective cation channel.[122] The hot receptor, known as transient receptor potential (subfamily V, member 1), or TRP-V1, is also found in small diameter sensory fibers (C fibers) responding to noxious temperatures (see following section). Thus, nature has provided chili peppers with a chemical targeted to this receptor, possibly to discourage herbivores by activating pain fibers—a not entirely successful strategy in the case of humans with a preference for spicy foods.

Pain and Temperature Sensation in Skin

The somatosensory system includes a rich variety of encapsulated and free nerve endings that provide input from the body, the surface skin, as well as deeper tissues. Specialized receptors for fine touch and vibration are discussed in Chapter 21. Here we will examine the neural basis of pain and temperature sensation. These percepts arise largely from the activity of small caliber C fibers and Aδ fibers. The stimuli affect free nerve endings, without any accessory structures, and act largely through indirect mechanisms of transduction. One class of endings is activated selectively by noxious stimuli—mechanical injury, excessive heat or cold, or chemical damage. It is natural, then, to postulate a direct connection between activity in these fibers and the sensation of pain. The discovery of this separate population of nociceptors, together with the finding that low-threshold mechanoreceptors do not respond to painful stimuli, ruled out an earlier theory that pain results from the excessive mechanoreceptor stimulation. Indeed there is growing evidence for nociceptor submodalities; for example, a population of itch-specific afferent neurons.[123]

Temperature changes are transduced by free nerve endings through the activation of TRP ion channels. Channel permeability varies with changes in skin temperature. Four different TRP channels, TRPV1 to TRPV4, are activated over different warm temperature ranges[124] (Figure 19.18). As already noted, TRPV1 is activated by noxious heating and is sensitive to capsaicin. TRPM8 is a Ca^{2+}-permeable channel activated by lowering temperature.[125] Menthol and eucalyptol also activate this channel, which explains the cooling sensation evoked by these compounds. Painful (or noxious, <17°C) cold requires the additional participation of TRPA1 channels.[126]

[122] Caterina, M. J. et al. 1997. *Nature* 389: 816–824.

[123] Liu, Q. et al. 2009. *Cell* 139: 1353–1365.

[124] Lumpkin, E. A., and Caterina, M. J. 2007. *Nature* 445: 858–865.

[125] Reid, G. 2005. *Pflügers Arch.* 451: 250–263.

[126] Kwan, K. Y., and Corey, D. P. 2009. *J. Gen. Physiol.* 133: 251–256.

(A)

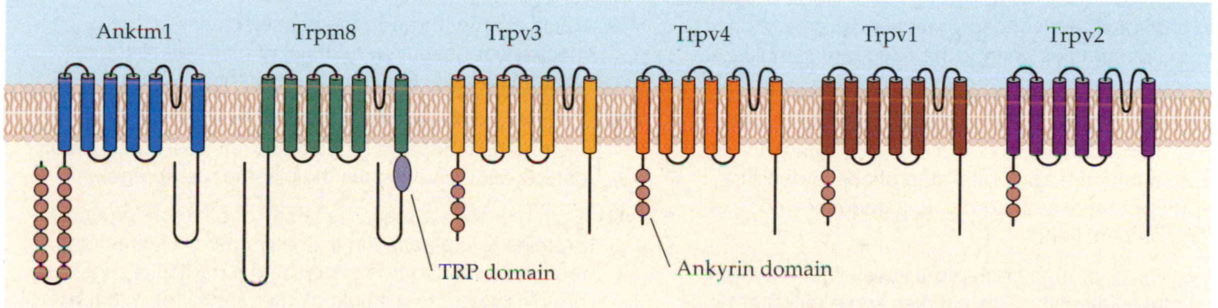

(B)

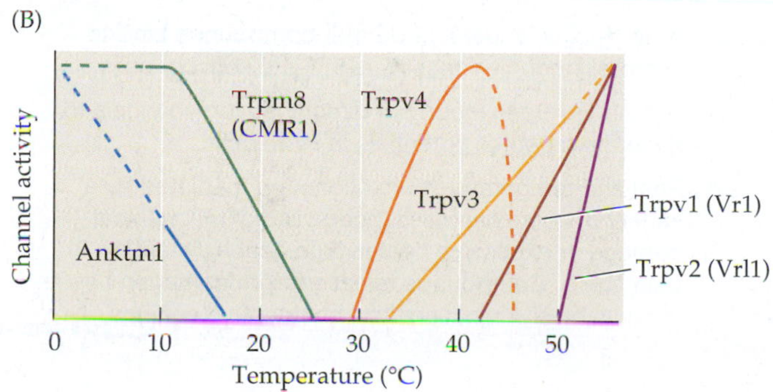

FIGURE 19.18 TRP Channels and Temperature Coding. (A) TRP channels are composed of six putative membrane-spanning units and cytoplasmic amino and carboxyl termini. (B) ThermoTRPs have different thermal activation ranges. In some cases chemical compounds also activate the receptor, producing a sensation of cooling (menthol on Trpm8) or heating (chili on Trpv1). Activation curves averaged across multiple studies; dashed portions extrapolated. (After Patapoutian et al., 2003.)

Activation and Sensitization of Nociceptors

Nociception (the perception of noxious or damaging stimuli) arises from a combination of direct and indirect actions on peripheral sensors. Painful heat (hotter than about 43° C) causes nonspecific cation channels (TRPV1) to open in C fiber endings.[127,128] Calcium and sodium ions enter and depolarize the cell, causing action potential generation. Prolonged exposure of these endings to capsaicin eventually causes calcium accumulation and cell death. For this reason capsaicin is used as a long-term analgesic, presumably relieving chronic pain by killing C fiber afferents.[129] Acids also may act to open cation channels directly, and an acid-sensitive ion channel (ASIC) has been cloned from nociceptive neurons.[130] Mechanical stimuli leading to skin damage can also produce direct activation of nociceptive endings.

In addition to being activated directly by painful stimuli, nociceptors respond to chemical activators, such as ATP, released from damaged cells. One ATP receptor subunit (P2X3) occurs specifically in C fiber somata in dorsal root ganglia and may contribute to the structure of nociceptive ATP receptors in the sensory terminals.[131–133] Cellular damage also leads to the release of cytoplasmic proteases, which then cleave serum proteins. In this manner the nine–amino acid peptide, bradykinin, is produced from kininogen, a ubiquitous inactive precursor. Bradykinin is a potent activator of C fiber endings. Unlike ATP, its effects are mediated by metabotropic receptors, rather than by direct action on membrane channels.[134] Bradykinin and other chemicals in damaged skin also act to increase the excitability of (i.e., sensitize) nociceptive endings activated by other stimuli. For example, responses to noxious heat stimuli are larger and occur at a lower temperature than normal in the presence of bradykinin.[128] Other inflammatory mediators include prostaglandins, serotonin, histamine, and substance P. Prostaglandin E_2 and serotonin increase sensitivity by lowering the threshold for activation of voltage-gated sodium currents.[135] Activated pain fibers release substance P not only from their synapses within the spinal cord (see Chapter 14), but also from their terminals in the skin. In the periphery, substance P may increase the excitability of C fibers by blocking K^+ channels.[136] The process of sensitization is accompanied by local vasodilatation and edema. The affected area becomes hyperalgesic, having a reduced threshold for pain.

[127] Bevan, S., and Yeats, J. 1991. *J. Physiol.* 433: 145–161.

[128] Cesare, P., and McNaughton, P. 1996. *Proc. Natl. Acad. Sci. USA* 93: 15435–15439.

[129] Szallasi, A., and Blumberg, P. M. 1996. *Pain* 68: 195–208.

[130] Waldmann, R. et al. 1997. *Nature* 386: 173–177.

[131] Chen, C. C. et al. 1995. *Nature* 377: 428–431.

[132] Lewis, C. et al. 1995. *Nature* 377: 432–435.

[133] Cook, S. P. et al. 1997. *Nature* 387: 505–508.

[134] Burgess, G. M. et al. 1989. *J. Neurosci.* 9: 3314–3325.

[135] Gold, M. S. et al. 1996 *Proc. Natl. Acad. Sci. USA* 93: 1108–1112.

[136] Adams, P. R., Brown, D. A., and Jones, S. W. 1983. *Brit. J. Pharmacol.* 79: 330–333.

SUMMARY

- Each type of sensory receptor responds preferentially to one type of stimulus energy, the adequate stimulus.
- Short and long receptors differ morphologically and functionally. Short receptors encode stimulus intensity directly in the amplitude of the receptor potential. Long receptors take the additional step of converting the receptor potential amplitude into a frequency code of action potential firing.
- The response of many receptors varies with the log of the stimulus intensity. This enables some receptor types to have a wide dynamic range.
- Most sensory receptors adapt during maintained stimuli. Adaptation arises from both mechanical and electrical factors. In some receptors, very rapid adaptation makes them tuned to rapidly varying stimuli, such as vibration.
- Mechanosensory hair cells of the inner ear couple movement directly to the gating of ion channels by physical connection. The tip link that connects adjacent stereocilia is stretched by deflection of the hair bundle and so pulls open an ion channel.
- Calcium entry through the nonselective mechanotransduction channel of hair cells leads to adaptation and closure of the channel.
- Olfactory neurons employ G protein–coupled membrane receptors that lead to the opening of cAMP-gated cation channels in the plasma membrane.
- Each member of the large family of olfactory receptor proteins is expressed in a small number of olfactory receptors. All neurons expressing a particular receptor protein project to a single glomerulus in the olfactory bulb.
- Amino acids, sugars, and bitter compounds bind to G protein–coupled receptors in taste sensory cells.
- Salt and protons (sour) act directly on ion channels to generate receptor potentials in taste cells.
- Pain and temperature sensations are mediated by a variety of chemical messengers. Direct mechanical damage or excessive heat initiate action potentials in pain fibers. Compounds released from damaged tissue, such as bradykinin, sensitize nociceptive endings.

Suggested Reading

General Reviews

Dussor, G., Koerber, H. R., Oaklander, A. L., Rice, F. L., and Molliver, D. C. 2009. Nucleotide signaling and cutaneous mechanisms of pain transduction. *Brain Res. Rev.* 60: 24–35.

Fettiplace, R. 2009. Defining features of the hair cell mechanoelectrical transducer channel. *Pflügers Arch.* 458: 1115–1123. Review.

Frings, S. 2009. Primary processes in sensory cells: current advances. *J. Comp. Physiol. A* 195: 1–19.

Gillespie, P. G., and Müller, U. 2009. Mechanotransduction by hair cells: models, molecules, and mechanisms. *Cell* 139: 33–44.

Hudspeth, A. J. 2008. Making an effort to listen: mechanical amplification in the ear. *Neuron* 59: 530–545. Review.

Kleene, S. J. 2008. The electrochemical basis of odor transduction in vertebrate olfactory cilia. *Chem. Senses* 33: 839–859.

Lumpkin, E. A., and Caterina, M. J. 2007. Mechanisms of sensory transduction in the skin. *Nature* 445: 858–865.

Roper, S. D. 2007. Signal transduction and information processing in mammalian taste buds. *Pflügers Arch.* 454: 759–776.

Roper, S. D., and Chaudhari, N. 2009. Processing umami and other tastes in mammalian taste buds. *Ann. N Y Acad. Sci.* 1170: 60–65.

Touhara, K., and Vosshall, L. B. 2009. Sensing odorants and pheromones with chemosensory receptors. *Annu. Rev. Physiol.* 71: 307–332.

Vega, J. A., García-Suárez, O., Montaño, J. A., Pardo, B., and Cobo, J. M. 2009. The Meissner and Pacinian sensory corpuscles: revisited new data from the last decade. *Microsc. Res. Tech.* 72: 299–309.

Vollrath, M. A., Kwan, K. Y., and Corey, D. P. 2007. The micromachinery of mechanotransduction in hair cells. *Annu. Rev. Neurosci.* 30: 339–365.

Original Papers

Beurg, M., Fettiplace, R., Nam, J. H., and Ricci, A. J. 2009. Localization of inner hair cell mechanotransducer channels using high-speed calcium imaging. *Nat. Neurosci.* 12: 553–558.

Buck, L., and Axel, R. 1991. A novel multigene family may encode odorant receptors: A molecular basis for odor recognition. *Cell* 65: 175–187.

Caterina, M. J., Schumacher, M. A., Tominaga M., Rosen T. A., Levine J. D., and Julius D. 1997. The capsaicin receptor: A heat-activated ion channel in the pain pathway. *Nature* 389: 816–824.

Chaudhari, N., Landin, A. M., and Roper, S. D. 2000. A metabotropic glutamate receptor variant functions as a taste receptor. *Nat. Neurosci.* 3: 113–119.

Crawford, A. C., and Fettiplace, R. 1985. The mechanical properties of ciliary bundles of turtle cochlear hair cells. *J. Physiol.* 364: 359–379.

DeFazio, R. A., Dvoryanchikov, G., Maruyama, Y., Kim, J. W., Pereira, E., Roper, S. D., and Chaudhari, N. 2006. Separate populations of receptor cells and presynaptic cells in mouse taste buds. *J. Neurosci.* 26: 3971–3980.

Dhallan, R. S., Yau, K. W., Schrader, K. A., and Reed, R. R. 1990. Primary structure and functional expression of a cyclic nucleotide-activated channel from olfactory neurons. *Nature* 347: 184–187.

Eyzaguirre, C., and Kuffler, S. W. 1955. Processes of excitation in the dendrites and soma of single isolated sensory nerve cells of the lobster and crayfish. *J. Gen. Physiol.* 39: 87–119.

Howard, J., and Hudspeth, A. J. 1988. Compliance of the hair bundle associated with gating of mechanoelectrical transduction channels in the bullfrog's saccular hair cell. *Neuron* 1: 189–199.

Hudspeth, A. J., and Corey, D. P. 1977. Sensitivity, polarity and conductance change in the response of vertebrate hair cells to controlled mechanical stimuli. *Proc. Natl. Acad. Sci. USA* 74: 2407–2411.

Kazmierczak, P., Sakaguchi, H., Tokita, J., Wilson-Kubalek, E. M., Milligan, R. A., Muller, U., and Kachar, B. 2007. Cadherin 23 and protocadherin 15 interact to form tip-link filaments in sensory hair cells. *Nature* 449: 87–91.

Loewenstein, W. R., and Mendelson, M. 1965. Components of adaptation in a Pacinian corpuscle. *J. Physiol.* 177: 377–397.

Nakamura, T., and Gold, G. H. 1987. A cyclic nucleotide-gated conductance in olfactory receptor cilia. *Nature* 532: 442–444.

Ricci, A. J., Crawford, A. C., and Fettiplace, R. 2003. Tonotopic variation in the conductance of the hair cell mechanotransducer channel. *Neuron* 40: 983–990.

Riviere, S., Challet, L., Fluegge, D., Spehr, M., and Rodriguez, I. 2009. Formyl peptide receptor-like proteins are a novel family of vomeronasal chemosensors. *Nature* 459: 574–577.

Vassar, R., Ngai, J., and Axel R. 1993. Spatial segregation of odorant receptor expression in the mammalian olfactory epithelium. *Cell* 74: 309–318.

Zhang, Y., Hoon, M. A., Chandrashekar, J., Mueller, K. L., Cook, B., Wu, D., Zuker, C. S., and Ryba, N. J. 2003. Coding of sweet, bitter, and umami tastes: different receptor cells sharing similar signaling pathways. *Cell* 112: 293–301.

CHAPTER 20
Transduction and Transmission in the Retina

The way in which neuronal signals are evoked by light to produce our perception of scenes with objects and background, movement, shade, and color begins in the retina. Responses to light start at receptors known as rods and cones that contain visual pigments. Rods are highly sensitive and can be activated by a single quantum of light. Color and daylight vision depend on cones. Absorption of light by the visual pigment of a photoreceptor activates a G protein, leading to a cascade of biochemical reactions. As a result, nucleotide-gated cation channels in the membrane close, causing the photoreceptor to become hyperpolarized. Light thereby reduces ongoing transmitter release onto postsynaptic bipolar and horizontal cells. Signals from photoreceptors finally reach ganglion cells, whose axons enter the optic nerve and constitute the sole output from the eye.

The connections between photoreceptors and ganglion cells involve bipolar, horizontal, and amacrine cells. Like rods and cones, bipolar and horizontal cells produce graded local potentials, not action potentials. Signaling by individual neurons in the retina and at successive levels of the visual system is best analyzed in terms of receptive fields, which are the building blocks for perception. Receptive field of a neuron in the visual system refers to the restricted area of the retinal surface that, upon illumination, enhances or inhibits the signaling of that cell. The receptive field of a retinal ganglion cell is a small circular area on the retina. Action potentials are evoked either in "on" ganglion cells by small spots of light shone onto the center of the receptive field, or in "off" cells by reducing the light (like a shadow) in the center of the receptive field. Two groups of ganglion cells are functionally important. Known as parvocellular (P) and magnocellular (M), they are distinguished by their sizes, positions, connections, and physiological responses. Smaller P ganglion cells exhibit fine spatial discrimination and some have color sensitivity. Larger M ganglion cells respond better to moving stimuli and to small changes in contrast. These distinctive properties of M and P divisions are maintained through successive stages in the brain. A small number of specialized ganglion cells respond directly to illumination and provide information about the level of background illumination.

The performance of nerve cells in the retina is described more fully here than in Chapters 1 and 2, which served to illustrate principles of signaling and organization. We begin with a brief review of the visual pathways to put in perspective the details of retinal organization that follow. Because photoreceptors produce unusual signals that give rise to functional transmission through the retina, their responses to light are dealt with in this chapter rather than in Chapter 19.

The Eye

The retina in the eye acts as a self-contained outpost of the brain. It collects information, analyzes it, and hands it on to higher centers through the well-defined optic nerve pathway for further processing. The initial step in visual processing is the formation on each retina of a sharp, inverted image of the outside world. Essential for clear vision are: (1) correct focus of the image by adjustment of the thickness of the lens (accommodation); (2) regulation of light entering the eye by the pupil diameter; (3) convergence of the two eyes to ensure that matching images fall on corresponding points of both retinas; and (4) eye movements that compensate for self-generated or forced movements of the head. Our vision is not uniformly detailed across the visual field but is most acute in the center. We can read small print at the center of gaze, but not in the peripheral field of vision. This loss of acuity arises from decreased receptor density and changes in connectivity in the peripheral retina more than from optical effects. We describe first the principal anatomical features of the visual pathway and then the stepwise transformation of signals through the retina as light is trapped by visual pigments to generate changes in electrical currents.

Anatomical Pathways in the Visual System

The pathways from the eye to the cerebral cortex are illustrated in Figure 20.1, which depicts some of the major landmarks of the visual system in a rat brain. The retinal ganglion cell axons enter the optic nerve, about half cross at the optic chiasm, and end in a part of the thalamus called the lateral geniculate nucleus. Because of optical reversal by the lens, which projects the left visual field onto the right side of each retina, each cerebral hemisphere *sees* the visual field on the opposite side.

Layering of Cells in the Retina

Among the reasons the retina is inviting for physiological research is the neat layering and stereotyped morphology of the relatively few nerve cell types—there are only five main

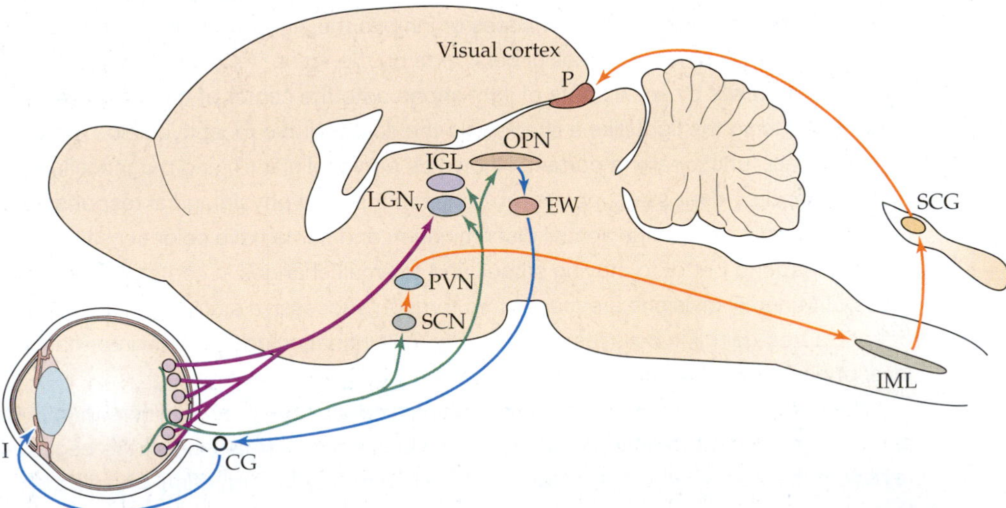

FIGURE 20.1 Central Targets of Retinal Ganglion Cells. The great majority of retinal ganglion cells respond to light transduced by photoreceptors and transmitted via bipolar cells, and send their axons to the lateral geniculate nucleus of the thalamus. From there information flows to visual cortex (see Chapter 2). These classic retinal ganglion cells (thick violet lines) carry information about color, shape, movement, all the details that give us form vision and the ability to analyze the visual world. A small minority of retinal ganglion cells are intrinsically photosensitive due to the expression of the photopigment melanopsin. These intrinsically photosensitive retinal ganglion cells (ipRGCs; thin green lines) project to the suprachiasmatic nucleus (SCN) (see Chapter 17) through the paraventricular nucleus (PVN) to form the retinohypothalamic tract (orange lines) that drives circadian rhythms. This circuit involves the intermediolateral nucleus (IML) of the spinal cord, and the superior cervical ganglion (SCG) that ultimately stimulates melatonin release from the pineal gland (P). Other ipRGCs project to the olivary pretectal nucleus (OPN) and from there to the Edinger-Westphal nucleus (EW) to activate motoneurons in the ciliary ganglion (CG) that control muscles of the iris (blue lines). (After Berson, 2003).

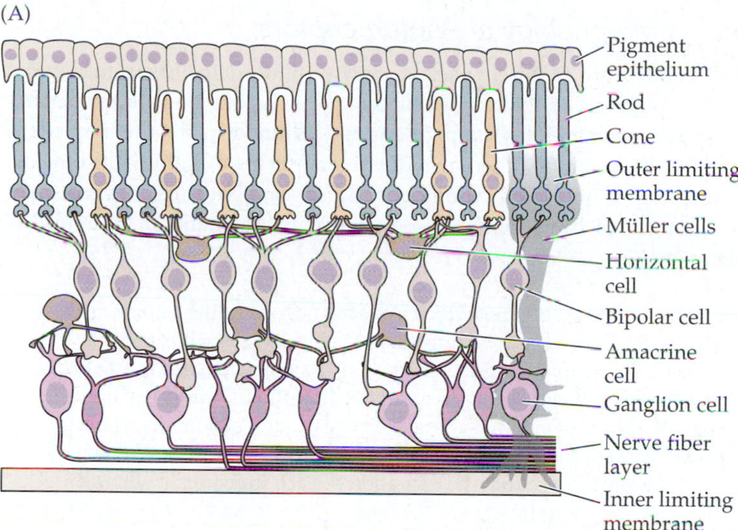

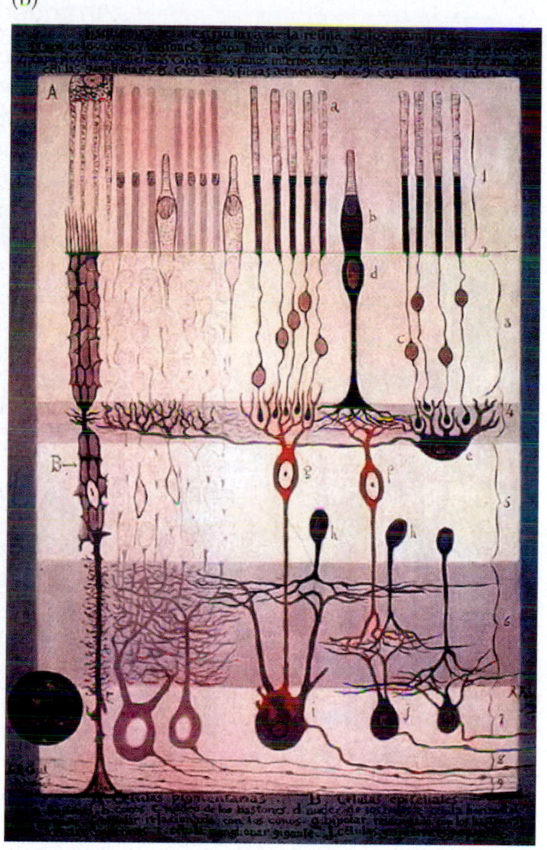

FIGURE 20.2 Schematized Views of the Retina showing the five principal cell types arranged in layers. (A) Light enters the retina, from below in this picture, and reaches the rods and cones, where it is absorbed, initiating signals in the outer segments. Photoreceptors make synaptic connections onto bipolar cells and horizontal cells in the outer plexiform layer. Bipolar cells make synapses with ganglion cells and amacrine cells in the inner plexiform layer. (B) Original drawing of retinal cell types from Ramón y Cajal. (B from Ramón y Cajal, c.1900.)

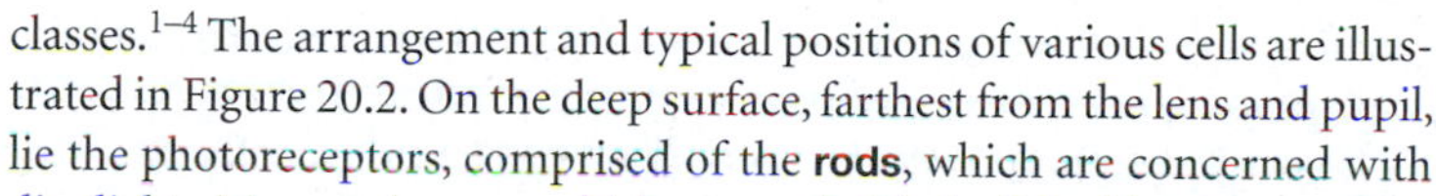

classes.[1–4] The arrangement and typical positions of various cells are illustrated in Figure 20.2. On the deep surface, farthest from the lens and pupil, lie the photoreceptors, comprised of the **rods**, which are concerned with dim light vision, and **cones**, which are used with daylight illumination when we can perceive color. Cones are connected to the **bipolar cells** that in turn connect to the **ganglion cells**, whose axons are the optic nerve fibers. The connectivity of rods is more complex, connecting to an intervening interneuron (amacrine cell) that piggybacks onto the cone circuitry.

Apart from this through-line, there are other cells that make predominantly lateral (i.e., side-to-side) connections. These are the **horizontal cells** and the **amacrine cells**. Only ganglion cells and some amacrine cells generate action potentials. Photoreceptors, horizontal cells, and bipolar cells give only locally graded signals. Within each of these major classes, there are subgroups exhibiting important differences in structure and function. Müller cells and astrocytes, the properties of which are described in Chapter 10, constitute the glial cells of the retina.

Phototransduction in Retinal Rods and Cones

Photoreceptors set the stage for vision and define how the outside world can be perceived; their spectral range varies between species. For example, many invertebrates can detect ultraviolet light, and some can use the polarized properties of skylight to navigate (see Chapter 18). Human photoreceptors normally detect neither of these signals, but they can out-perform some fellow mammals. Cats, lacking appropriate receptors, are color blind (though we are too at night, when all cats are gray). On the other hand, the sensitivity of mammalian rods in darkness is such that a single quantum of light can give rise to a measurable signal, and only seven or so rods need to be activated by single quanta for a conscious sensation.[5] Yet with less-sensitive cone photoreceptors, we can detect subtle tints and differences in contrast or color on a bright day, when the light intensity is 100 million times greater.

In terms of transduction mechanisms, it will be apparent that there are striking similarities between the generation of responses to light by photoreceptors compared to the responses to odors by olfactory receptors described in Chapter 19.

[1] Boycott, B. B., and Dowling, J. E. 1969. *Philos. Trans. R. Soc. Lond. B, Biol. Sci.* 255: 109–184.

[2] Sterling, P., and Demb, J. B. 2003. In *Synaptic Organization of the Brain*. Oxford University Press, New York.

[3] Balasubramanian, V., and Sterling, P. 2009. *J. Physiol.* 587: 2753–2767.

[4] Masland, R. H. 2001. *Nat. Neurosci.* 4: 877–886.

[5] Hecht, S., Shlaer, S., and Pirenne, M. H. 1942. *J. Gen. Physiol.* 25: 819–840.

Hermann von Helmholtz (1821–1894), together with one of his drawings and the frontispiece of his book on vision. Helmholtz made equally important and original contributions to the study of medicine, hearing, neurophysiology, and thermodynamics. It seems refreshing to read his prose today. (Photomontage courtesy of Dr. Rolf Boch.)

Arrangement and Morphology of Photoreceptors

The rods and cones constitute a densely packed array of photodetectors in the layer of retina adjacent to the pigment epithelium (Figure 20.3; see Figure 20.2) that is farthest from the cornea and incoming light. The pigment epithelium absorbs the light that is not absorbed by the photoreceptors, thereby reducing scattering of light within the eye. Thus, with the exception of one small retinal area, the fovea, light must traverse layers of cells and fibers before reaching the light absorbing outer segments of receptors. As Helmholtz wrote in 1867,

> There is in the retina a remarkable spot which is placed near its center...and which... is called the fovea or pit...[It] is of great importance for vision since it is the spot where the most exact discrimination is made. The cones are here packed most closely together and receive light which has not been impeded by other semi-transparent parts of the retina. We may assume that a single nervous...connection...runs from each of these cones through the trunk of the optic nerve to the brain...and there produces its special impression so that the excitation of each individual cone will produce a distinct and separate effect upon the sense.[6]

It is extraordinary that Helmholtz could write this paragraph before the word "synapse" or even the cell doctrine existed.

Counts made at the fovea reveal that cones are closely packed, with a density of 200,000/mm^2, and that rods are excluded. Moreover, the cones at the fovea are more slender than those in peripheral parts of the retina.[2] Since the fovea contains no rods, it constitutes a blind spot in very dim light, such as at night. A different blind spot corresponds to the region of the retina through which the optic nerve fibers leave the eye; at this spot, the optic disc, there are no photoreceptors at all.

Figure 20.4 shows three principal features of photoreceptor structure: (1) an outer segment within which light is absorbed by visual pigments; (2) an inner segment containing the nucleus, ion pumps, transporters, ribosomes, mitochondria, and endoplasmic reticulum; and (3) the synaptic terminal, which releases glutamate onto second-order cells and which also receives synaptic inputs. The release site of the synaptic ending is highly characteristic with one or more ribbon structures, along which are aligned the vesicles containing transmitter[7] (see Figure 20.17).

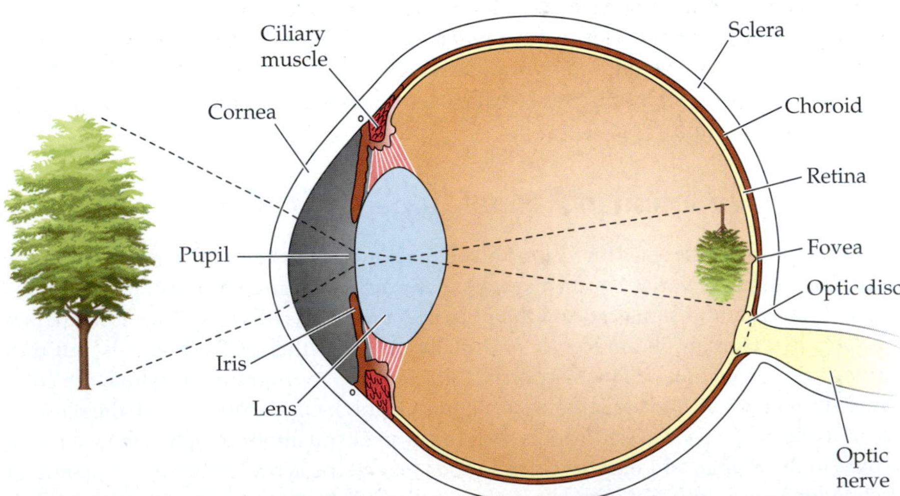

FIGURE 20.3 Structure of the Eye. Cross section through the eye showing the projection of a visual object onto the retina. Activation of ciliary muscle changes the thickness of the lens, and the opening of the pupil. The photosensitive retinal epithelium is interrupted by the exit of ganglion cell axons to form the optic nerve. Nearby lies the retinal fovea where visual acuity is best due to the high density of photoreceptors, and the thinning of the overlying cellular layers.

[6] Helmholtz, H. V. 1962/1924. *Helmholtz's Treatise on Physiological Optics*. Dover, New York.

[7] Zanazzi, G., and Matthews, G. 2009. *Mol. Neurobiol.* 39: 130–148.

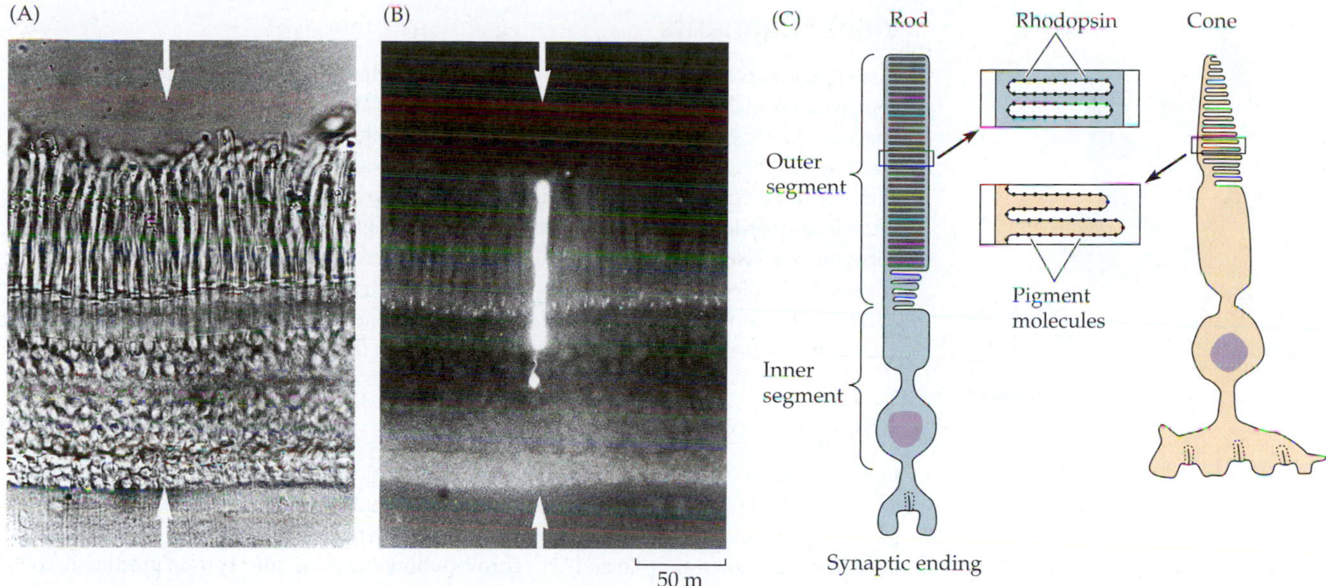

FIGURE 20.4 Photoreceptors in Retina. (A,B) Rod in toad retina injected with a fluorescent dye, Lucifer yellow, as seen in visible (A) and ultraviolet (B) light. Arrows mark identical points on the retina. (C) Diagram of a rod and a cone. In the rod, the pigment rhodopsin (black dots) is embedded in membranes arranged in the form of disks, which are not continuous with the outer membrane of the cell. In the cone, the pigment molecules are on infolded membranes that are continuous with the surface membrane. The outer segment is connected to the inner segment by a narrow stalk. The synaptic endings continually release transmitter in the dark. (A and B kindly provided by the late B. Nunn, unpublished; C after Baylor 1987.)

Electrical Responses of Vertebrate Photoreceptors to Light

As described earlier, sensory receptors typically respond to appropriate stimuli by graded local depolarization that initiates action potentials. Although the majority of invertebrate photoreceptors behave in this way (Figure 20.5A),[8] the responses of most vertebrate photoreceptors to light are very different. Figure 20.5B shows the responses of a turtle rod recorded with an intracellular microelectrode.[9] In the dark (at rest), the photoreceptor is depolarized by a continuous inward current flowing into the outer segment. Light turns off the ongoing inward current, leaving the always-present outward potassium current to hyperpolarize the cell. The following paragraphs deal with the way in which light is absorbed by photoreceptors and the mechanisms that underlie the production of electrical signals.

[8]Fuortes, M. G., and Poggio, G. F. 1963. *J. Gen. Physiol.* 46: 435–452.

[9]Baylor, D. A., Fuortes, M. G., and O'Bryan, P. M. 1971. *J. Physiol.* 214: 265–294.

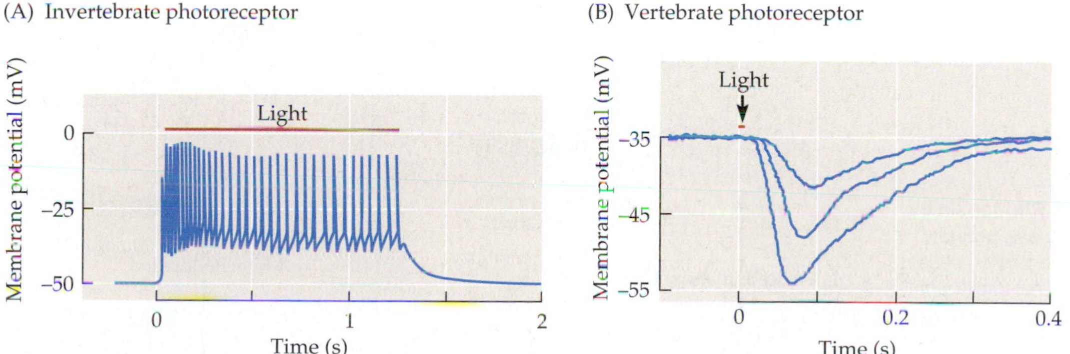

FIGURE 20.5 Responses of Photoreceptors. (A) Photoreceptors of an invertebrate (a horseshoe crab) respond to light with a depolarization that gives rise to impulses. This is the usual type of response elicited from sensory receptors activated by various stimuli, such as touch, pressure, or stretch (see Chapter 19). (B) Photoreceptors of a vertebrate (a turtle) respond with a hyperpolarization that is graded according to the intensity of the flash. (A after Fuortes and Poggio, 1963; B after Baylor, Fuortes, and O'Bryan, 1971.)

Wilhelm Kühne (1837–1900). Under the portrait, on the left, is shown the view from his room that was presented to the retina of a rabbit, causing bleaching and the clearly discernible image of the window arrangement (shown on the right). Kühne isolated visual purple for the first time. (Photomontage courtesy of Dr. Rolf Boch.)

Visual Pigments

Visual pigments are concentrated in membranes of the outer segments. Each rod contains approximately 10^8 pigment molecules. They are aggregated on several hundred discrete disks (approximately 750 in a monkey rod) that do not make contact with the outer membrane (see Figure 20.4C). By contrast, cones have pigment-laden infoldings that are continuous with the cell membrane. Pigment molecules make up about 80% of the total disk protein. The visual pigment is so closely packed on outer segment membranes that the distance between two visual pigment molecules in a rod is less than 10 nm.[10] This dense packing of sensitive molecules in serial layers of membranes traversed by light enhances the probability that a photon will be trapped on its way through the outer segment. The following question then arises: How are signals generated when light is absorbed by visual pigments?

Absorption of Light by Visual Pigments

Light is absorbed by the visual pigment **rhodopsin** in rods and a related pigment in cones. The events that occur when this happens have been studied by psychophysical, biochemical, physiological, and molecular techniques. Visual pigment molecules consist of two moieties: (1) a protein, known as opsin and (2) a chromophore (i.e., a chemical group producing color in a compound), 11-*cis* vitamin A aldehyde, known as **retinal**. The absorption characteristics of the visual pigments have been measured quantitatively by spectrophotometry.[11,12] When different wavelengths of light are shone through a solution of the rod visual pigment rhodopsin, blue–green light, at a wavelength of about 500 nanometers (nm), is absorbed most effectively. The absorption spectrum is similar when small spots of light of different wavelengths are shone onto single rods under the microscope. An elegant correspondence has been demonstrated between the absorption characteristics of rhodopsin and our perception in dim light. Quantitative psychophysical measurements made in human subjects show that blue-greenish lights at about 500 nm are optimal for perception of a dim light in the dark. In daylight, when rods are inactive and cones are used, we are more sensitive to red light, corresponding to the absorption spectra of cones (which will be discussed shortly).[10]

Once a photon has been absorbed by rhodopsin, retinal undergoes photoisomerization and changes from the 11-*cis* to an all-*trans* configuration. This transition is extremely rapid; it takes place in approximately 10^{-12} seconds. The protein then undergoes a series of transformational changes through various intermediates.[13] One conformation of the protein, metarhodopsin II, is of crucial importance for transduction (discussed later in the chapter). Figure 20.6 shows the sequence of changes occurring in bleaching and in regeneration of active rhodopsin. Metarhodopsin II appears after about 1 millisecond (ms). Regeneration of pigments after bleaching is slow, taking many minutes; it entails transfer of retinal between the photoreceptors and the pigment epithelium.[14]

[10] Dowling, J. E. 1987. *The Retina: An Approachable Part of the Brain.* Harvard University Press, Cambridge, MA.

[11] Brown, P. K., and Wald, G. 1963. *Nature* 200: 37–43.

[12] Marks, W. B., Dobelle, W. H., and Macnichol, E. F., Jr. 1964. *Science* 143: 1181–1183.

[13] Matthews, R. G. et al. 1963. *J. Gen. Physiol.* 47: 215–240.

[14] Pepperberg, D. R. et al. 1993. *Mol. Neurobiol.* 7: 61–85.

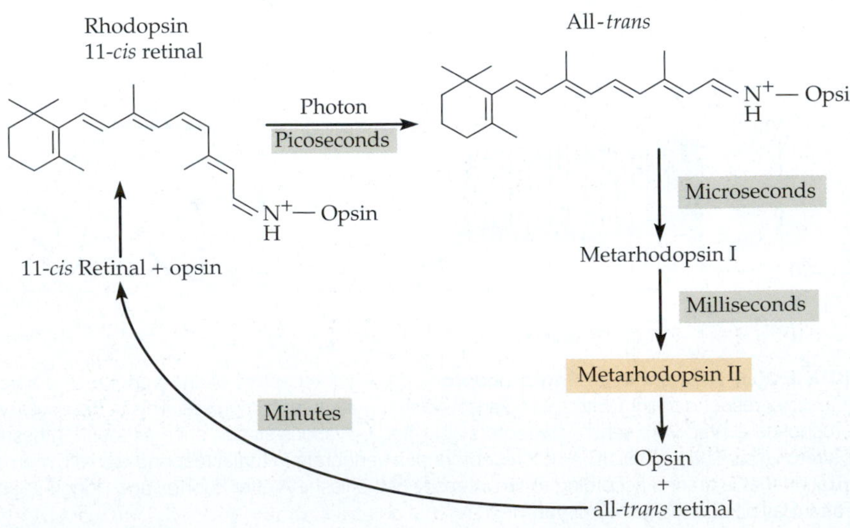

FIGURE 20.6 Bleaching of Rhodopsin by Light. In the dark, 11-*cis* retinal is bound to the protein opsin. Capture of a photon causes isomerization of the 11-*cis* retinal to all-*trans* retinal. The opsin–all-*trans* molecule, in turn, is rapidly converted to metarhodopsin II, which is dissociated to opsin and all-*trans* retinal. The regeneration of rhodopsin depends on interactions between photoreceptors and cells of the retinal pigment epithelium. Metarhodopsin II is the trigger that sets in motion activation of the second messenger system. (After Dowling, 1987.)

Structure of Rhodopsin

The protein opsin consists of 348 amino acid residues with seven hydrophobic regions of 20 to 25 amino acids, comprising seven transmembrane helices.[15] The amino terminus is located in the extracellular space (i.e., within the disk in rods) and the carboxy terminus in the cytoplasm (Figure 20.7). Retinal is attached to opsin through a lysine residue in the seventh membrane-spanning segment. Opsin is a member of the family of seven-transmembrane domain proteins that includes metabotropic neurotransmitter receptors, such as adrenergic and muscarinic receptors. Like rhodopsin, those receptors exert their effects through second messengers by activating G proteins (see Chapter 12). The stability of rhodopsin in the dark is extraordinarily high. Baylor calculated that spontaneous thermal isomerization of a single rhodopsin molecule should occur once every 3000 years or 10^{23} times more slowly than photoisomerization.[16]

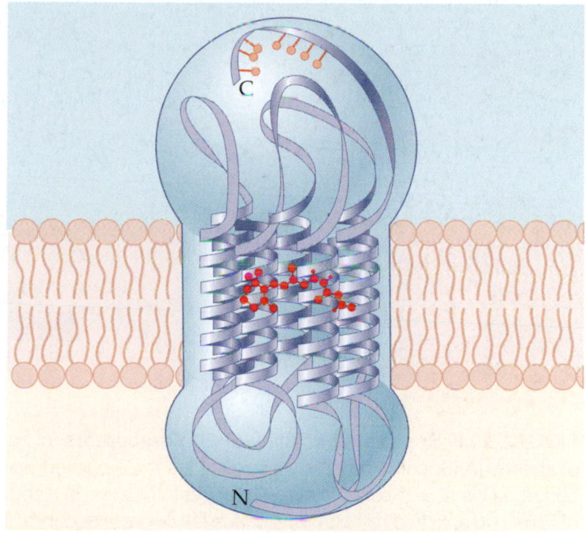

FIGURE 20.7 Structure of Vertebrate Rhodopsin in the Membrane. The helix is partly opened to show the position of retinal (red). C = carboxy terminus; N = amino terminus. (After Stryer and Bourne, 1986.)

Cones and Color Vision

Extraordinary insights and experiments by Young and Helmholtz in the nineteenth century defined crucial questions for color vision and at the same time provided clear, unequivocal explanations. Their conclusion that there must be three types of sensory photoreceptors for color has stood the test of time and has been confirmed at the molecular level. To set the stage, we quote again from Helmholtz, who compares the perception of light and sound as well as color and tone. One envies the clarity, force, and timeless beauty of his thinking, especially in view of the confusing, vitalistic concepts that were current at the time:

> All differences of hue depend upon combinations in different proportions of the three primary colors...red, green and violet...Just as the difference of sensation of light and warmth depends...upon whether the rays of the sun fall upon nerves of sight or nerves of feeling, so it is supposed in Young's hypothesis that the difference of sensation of colors depends simply upon whether one or the other kind of nervous fibers are more strongly affected. When all three kinds are equally excited, the result is the sensation of white light...If we allow two different colored lights to fall at the same time upon a white screen...we see only a single compound, more or less different from the two original ones. We shall better understand the remarkable fact that we are able to refer all the varieties in the composition of external light to mixtures of three colors if we compare the eye with the ear...In the case of sound...we recognize the long waves as low notes, the short as high-pitched, and the ear may receive at once many waves of sound, that is to say many notes. But here these do not melt into compound notes in the same way that colors...melt into compound colors. The eye cannot tell the difference if we substitute orange for red and yellow; but if we hear the notes C and E sounded at the same time, we cannot put D instead of them...if the ear perceived musical tones as the eye colors, every accord might be completely represented by combining only three constant notes, one very low, one very high and one intermediate, simply changing the relative strength of these three primary notes to produce all possible musical effects...But we find a continuous transition of colors into one another through numberless intermediate gradations...the way in which (colors) appear...depends chiefly upon the constitution of our nervous system...It must be confessed that both in man and in quadrupeds we have at present no anatomical basis for this theory of colors.[6]

Denis Baylor, 1991

These farsighted, accurate predictions were validated by quite different sets of observations. Using spectrophotometry, Wald, Brown, MacNichol, Dartnall, and their colleagues[11,12,17] showed the existence of three types of cones with different pigments in human retina (Figure 20.8A). Second, Baylor and his colleagues recorded currents from monkey and human cones. The results are shown in Figure 20.8B.[18] Three populations of cones were found with distinct but overlapping sensitivities in the blue, green, or red part of the spectrum. The wavelengths of light optimal for initiating electrical signals coincided precisely with the

[15] Nathans, J., and Hogness, D. S. 1984. *Proc. Natl. Acad. Sci. USA* 81: 4851–4855.

[16] Baylor, D. A. 1987. *Invest. Ophthalmol. Vis. Sci.* 28: 34–49.

[17] Dartnall, H. J., Bowmaker, J. K., and Mollon, J. D. 1983. *Proc. R. Soc. Lond. B, Biol. Sci.* 220: 115–130.

[18] Schnapf, J. L. et al. 1988. *Vis. Neurosci.* 1: 255–261.

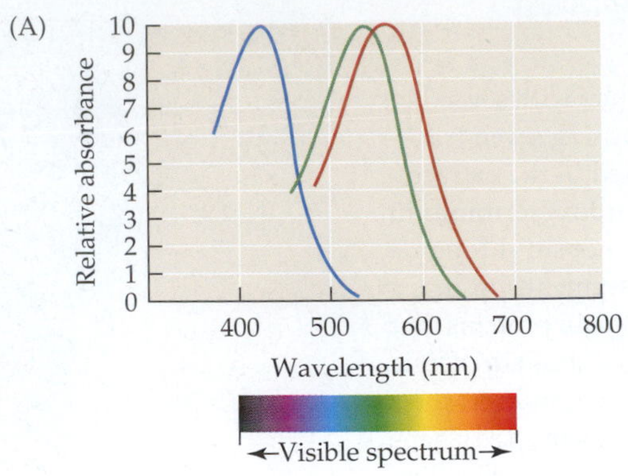

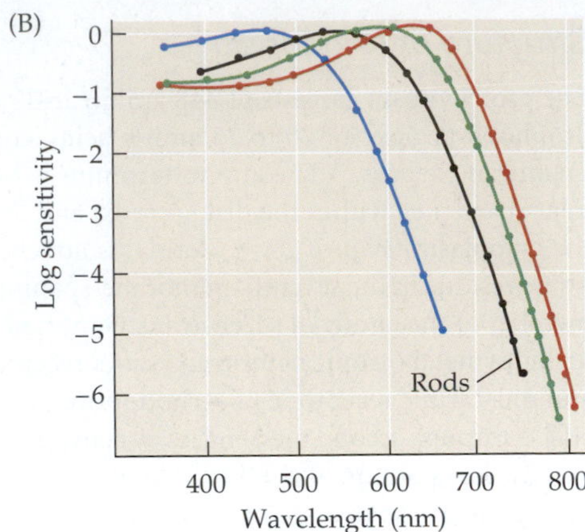

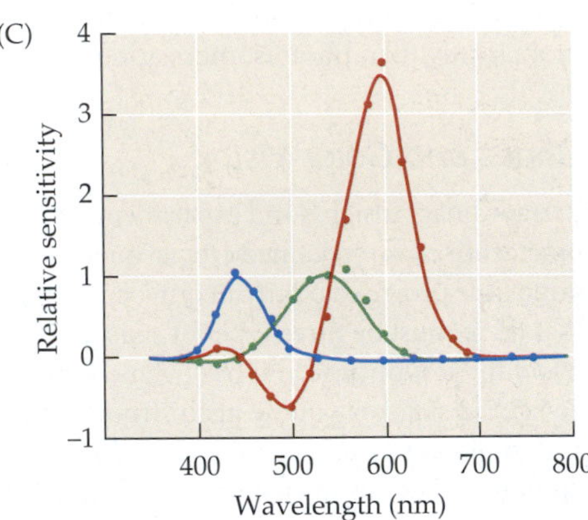

FIGURE 20.8 Spectral Sensitivity of Photoreceptors of human subjects and of visual pigments. (A) Spectral sensitivity curves of the three colored visual pigments showing absorbance peaks at wavelengths corresponding to blue, green, and red. (B) Spectral sensitivities of blue-, green-, and red-sensitive cones (as colored) and rods (black) from macaque monkeys. The responses were recorded by suction electrodes, then averaged and normalized. The curve through the rod spectrum was obtained from visual pigments in human subjects. (C) Comparison of spectral sensitivity of monkey cones with those obtained by human color matching. The continuous curves represent color-matching experiments in which the sensitivity at various wavelengths was determined in human subjects. The dots show results predicted from electrical measurements made by recording currents from single cones, after correcting for absorption in the lens and by pigments on the path to the outer segment. The correspondence between results obtained on single cells and by color matching is extraordinarily good. (A after Schnapf and Baylor, 1987; B after Baylor, 1987; C after Dowling, 1987.)

absorbance peaks of the visual pigments, which were demonstrated by spectrophotometry as well as by psychophysical measurements of spectral sensitivity (Figure 20.8C). Moreover, the genes for the blue, green, or red cone opsin pigments as well as the gene for rhodopsin were cloned and sequenced by Nathans.[19–21]

What accounts for the ability of different visual pigment molecules to trap specific wavelengths of light preferentially? It turns out that rhodopsin, the rod visual pigment, and all three cone visual pigments contain the same chromophore: 11-*cis* retinal. However, the amino acid sequences of the various pigment proteins differ from one another (Figure 20.9). Differences

[19] Nathans, J. 1987. *Annu. Rev. Neurosci.* 10: 163–194.

[20] Nathans, J. 1989. *Sci. Am.* 260: 42–49.

[21] Nathans, J. 1999. *Neuron* 24: 299–312.

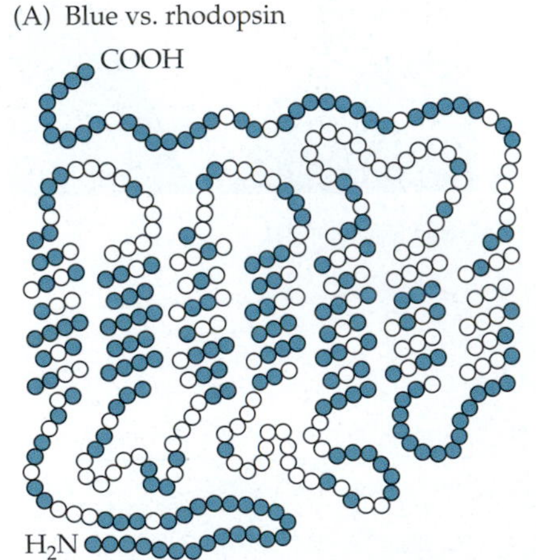

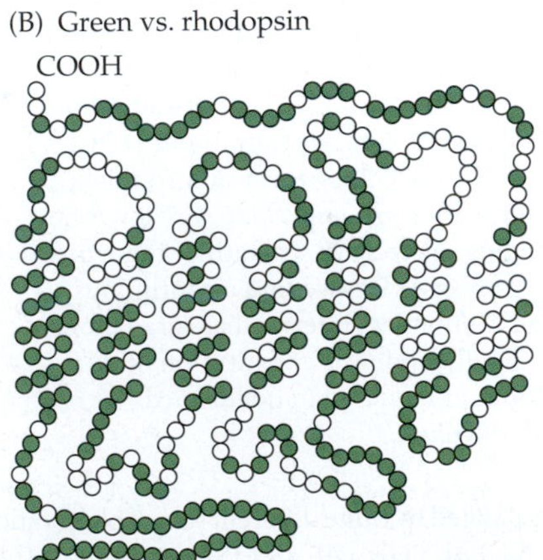

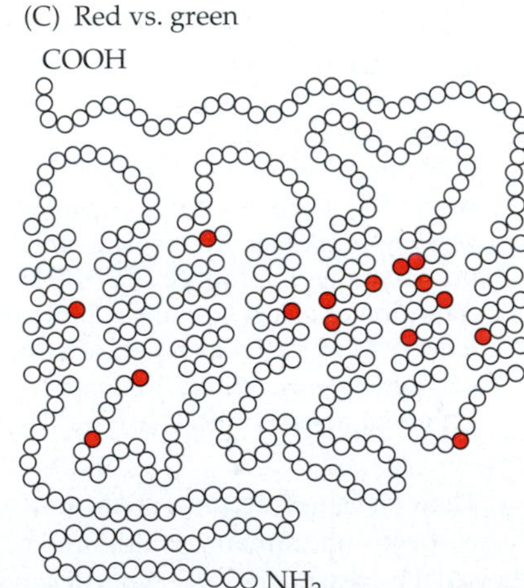

FIGURE 20.9 Comparisons of Amino Acid Sequences of red, green, and blue pigments with each other and with rhodopsin. Each colored dot represents an amino acid difference. (A, B) Blue and green pigments compared with rhodopsin. (C) Green and red pigments compared. The sequences of red and green pigments are highly similar. (From Nathans, 1989.)

in just a few amino acids account for the differences in spectral sensitivity. Insights into the evolution of color vision have been gained from comparisons of opsin genes in vertebrates, and among old and new world primates in particular,[22] while studies of inherited defects in humans[23,24] provide additional understanding of the molecular basis of color vision.

[22] Jacobs, G. H. 2008. *Vis. Neurosci.* 25: 619–633.

[23] Deeb, S. S. 2006. *Curr. Opin. Genet. Dev.* 16: 301–307.

[24] Deeb, S. S., and Kohl, S. 2003. *Dev. Ophthalmol.* 37: 170–187.

[25] Baylor, D. A., and Fuortes, M. G. 1970. *J. Physiol.* 207: 77–92.

[26] Luo, D. G., Xue, T., and Yau, K. W. 2008. *Proc. Natl. Acad. Sci. USA* 105: 9855–9862.

[27] Baylor, D. A., Lamb, T. D., and Yau, K. W. 1979. *J. Physiol.* 288: 589–611.

[28] Fesenko, E. E., Kolesnikov, S. S., and Lyubarsky, A. L. 1985. *Nature* 313: 310–313.

[29] Ruiz, M. L., and Karpen, J. W. 1997. *Nature* 389: 389–392.

[30] Taylor, W. R., and Baylor, D. A. 1995. *J. Physiol.* 483 (Pt 3): 567–582.

Color Blindness

Although a single type of photoreceptor cannot on its own provide information about color, three types of cones with properties shown in Figure 20.8 can. In principle, two types of cones with different pigments might be sufficient to distinguish colors of light, but many different mixtures of wavelengths would then appear identical. This is the situation in certain color-blind people, in whom Nathans has shown that a genetic defect results in an absence of one of the pigments. From our present perspective, one can only marvel that explanations at the molecular level so beautifully confirm the brilliant but rigorous speculations of Young and Helmholtz. Their idea that major attributes of color vision and color blindness are to be found within the receptors themselves has now been confirmed by direct physiological measurements and corresponding differences in genes and protein structure.[21]

Transduction

How does photoisomerization of rhodopsin give rise to a change in membrane potential? For many years, it was clear that some sort of internal transmitter was required for the generation of electrical signals in rods and cones. One reason is that information about the capture of photons in a rod outer segment must somehow be conveyed from rhodopsin, located in the disk, across the cytoplasm to the outer membrane. A second reason is the enormous amplification of the response. Baylor and his colleagues,[25] working on turtle photoreceptors, showed that decreases in membrane conductance and measurable electrical signals were produced when a single photon was absorbed and activated one pigment molecule out of about 10^8.

The sequence of events through which activated photopigment molecules change membrane potential has since been elucidated by patch clamp recordings from rod and cone outer segments and by molecular techniques.[26] The scheme for transduction from light to electrical signals is shown in Figure 20.10.

In darkness, a continuous *dark* current flows into the outer segment of rods and cones.[27] As a result, they have membrane potentials of approximately –40 mV, which is positive to the potassium equilibrium potential, E_K (–80 mV). The inward current in the dark is carried mainly by sodium, moving down its electrochemical gradient through cation channels in the outer segment. Hyperpolarization of the photoreceptor by light is brought about by closure of the channels, allowing the membrane potential to move toward E_K.

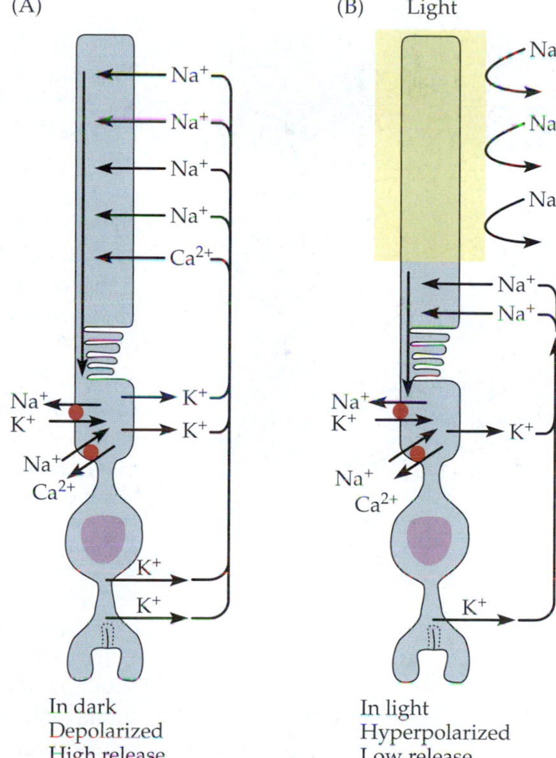

FIGURE 20.10 Dark Current in a Rod. (A) In darkness, sodium ions flow through cation channels of the rod outer segment, causing a depolarization; calcium ions also enter through the cation channels. The current loop is completed through the neck of the rod, with the outward movement of potassium through the inner segment membrane. (B) When the outer segment is illuminated, the channels close because of a decrease in intracellular cyclic guanosine monophosphate (cGMP), and the rod then becomes hyperpolarized. This hyperpolarization reduces transmitter release. Sodium, potassium, and calcium concentrations of the rod are maintained by pumps and exchangers in the inner segment (red circles); calcium exchangers are also present in the outer segment (see Box 20.1). (After Baylor, 1987.)

Properties of the Photoreceptor Channels

The cation channels in the outer segment, under normal physiological conditions, have calcium/sodium/potassium permeability ratios of 12.5:1.0:0.7, and a very low single-channel conductance of around 25 femtosiemens (fS).[28] However, that value depends in part on the activation level[29] and in part on the blocking effect of divalent cations.[30] Because the sodium concentration is much higher than that of calcium, about 85% of the inward current is carried by sodium. The driving force for potassium movement is, of course, outward. As calcium ions move through the channels, they are tightly bound to sites within the pore and thus interfere with the passage of other cations. Because of

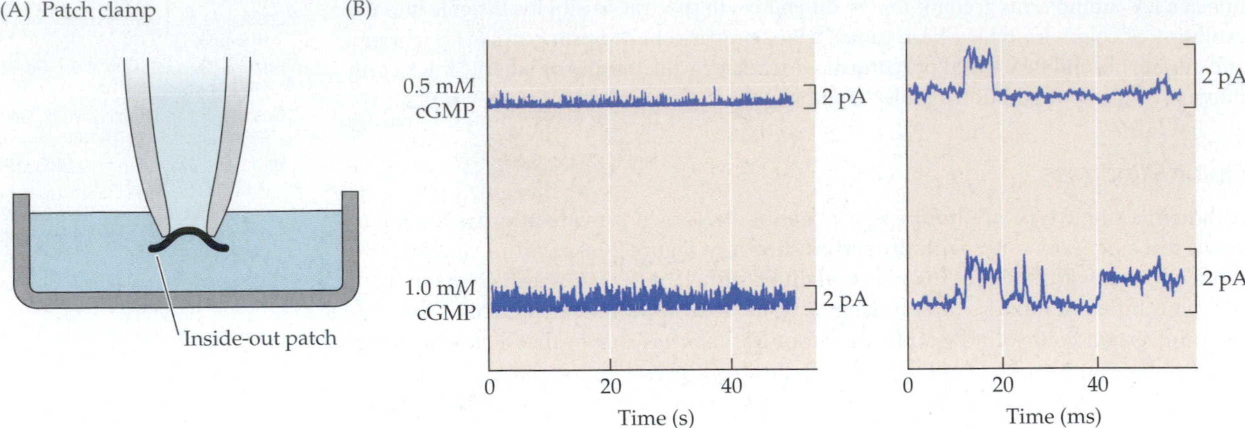

FIGURE 20.11 Role of Cyclic GMP in Opening Sodium Channels in rod outer segment membranes. Single-channel recordings made from inside-out patches bathed in various concentrations of cGMP. Channel opening causes deflections in the upward direction. The frequency of channel opening is extremely low in control recordings. Addition of cGMP causes single-channel openings, the frequency of which increases with increased concentration. (After Baylor, 1987.)

King Wai Yau

this property, removal of calcium from the solution around the cell allows sodium and potassium to move much more freely through the channel, increasing its conductance to about 25 picosiemens (pS).

Fesenko, Yau, Baylor, Stryer, and their colleagues[28,31,32] have shown that cyclic guanosine monophosphate (cGMP) acts as the internal transmitter from disk to surface membrane and fulfills the requirements of appropriate kinetics and great amplification. As shown in Figure 20.11, a high cytoplasmic concentration of cGMP keeps the cation channels in an open state. When the cGMP concentration of the fluid facing the inside of the membrane is reduced, channel openings become rare events. Thus, the membrane potential of the photoreceptors is a reflection of the cytoplasmic cGMP concentration—the higher the concentration, the more the cell is depolarized. The cGMP concentration in turn is inversely related to the intensity of ambient light. Increasing light intensity reduces cGMP concentration and reduces the fraction of open channels. In the absence of cGMP, almost all the channels are closed and the resistance of the outer segment membrane approaches that of a channel-free lipid bilayer.

Molecular Structure of Cyclic GMP–Gated Channels

Complementary DNAs for rod outer segment channels have been isolated and the amino acid sequences determined for channel subunits from human, bovine, mouse, and chicken retinas. There is a pronounced sequence similarity between cDNAs of outer segment channel subunits and the subunits of other cyclic-nucleotide–gated channels—for example, those found in the olfactory system.[33] Their membrane regions share structural similarities with other cation-selective channels, particularly in the S4 region and the region of the pore (see Chapter 5). The photoreceptor channels are tetramers made up of at least two different subunit proteins, α and β, with apparent molecular sizes of 63 and 240 kilodaltons (kD), respectively. The intracellular nucleotide-binding site is near the carboxy terminus of the α and β-subunits.[34]

The cGMP Cascade

The sequence of events leading to reduction in cGMP concentration and consequent closing of the cation channels is shown in Figure 20.12. The decrease in internal cGMP concentration triggered by light is brought about by metarhodopsin II, an intermediary in the bleaching process in which the chromophore separates from the protein (see Figure 20.6). Metarhodopsin II acts on the G protein, **transducin**, which consists of α, β, and γ polypeptide chains.[35,36]

[31] Stryer, L., and Bourne, H. R. 1986. *Annu. Rev. Cell Biol.* 2: 391–419.

[32] Yau, K. W., and Nakatani, K. 1985. *Nature* 317: 252–255.

[33] Torre, V. et al. 1995. *J. Neurosci.* 15: 7757–7768.

[34] Kaupp, U. B., and Seifert, R. 2002. *Physiol. Rev.* 82: 769–824.

[35] Stryer, L. 1987. *Sci. Am.* 257: 42–50.

[36] Chen, C. K. 2005. *Rev. Physiol. Biochem. Pharmacol.* 154: 101–121.

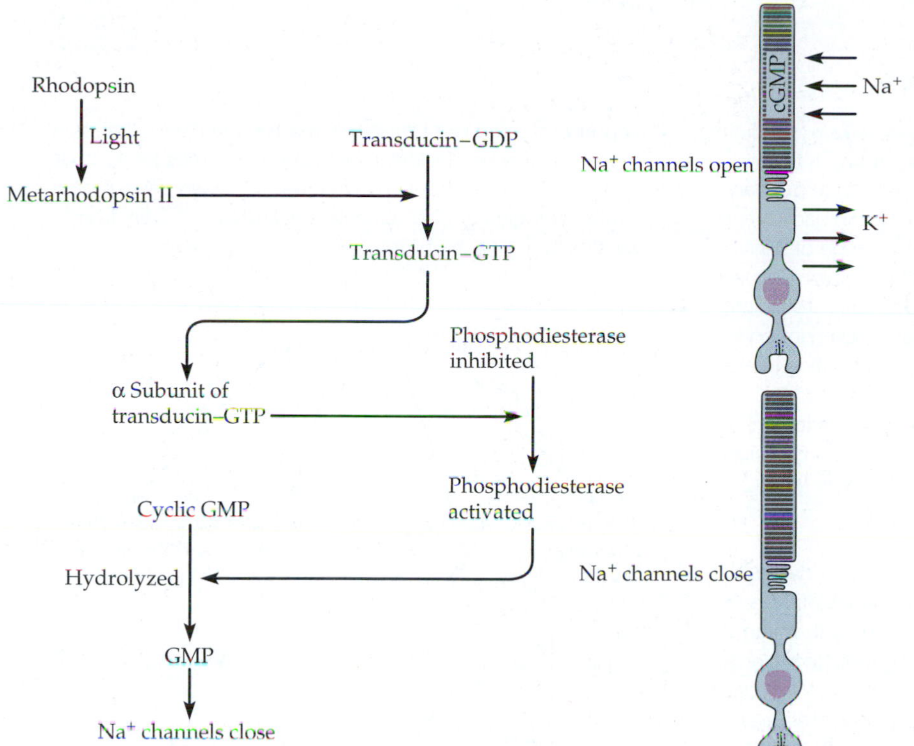

FIGURE 20.12 Coupling of Photopigment Activation to G Protein Activation. The G protein transducin binds guanosine triphosphate (GTP) in the presence of metarhodopsin II, leading to activation of phosphodiesterase, which in turn hydrolyzes cyclic GMP. With the reduced concentration of cGMP, sodium channels close. GDP = guanosine diphosphate (After Baylor, 1987.)

The transient interaction of metarhodopsin II and transducin causes guanosine diphosphate (GDP) bound to the α-subunit to be exchanged for guanosine triphosphate (GTP). This activates the α-subunit, which separates from the β and γ-subunits, and in turn activates a membrane-bound phosphodiesterase, which is the enzyme that hydrolyzes cGMP. The cGMP concentration falls, fewer sodium channels open, and the rods hyperpolarize. The cascade is terminated by phosphorylation of the carboxy-terminal region of the active metarhodopsin II (Box 20.1). The key role of cGMP in controlling the channels during the response is supported by biochemical experiments. Illumination of photoreceptors can cause a 20% decrease in the internal cGMP concentration.[16]

Amplification through the cGMP Cascade

Two steps of the cGMP cascade provide the great amplification that accounts for the exquisite sensitivity of rods to light. First, a single molecule of active metarhodopsin II catalyzes the exchange of many molecules of transducin GDP for GTP and thereby liberates hundreds of G protein α-subunits.[36] Second, each α-subunit activates a molecule of phosphodiesterase in the disk that can hydrolyze a large number of cytoplasmic cGMP molecules and thereby allow many channels to close.

Responses to Single Quanta of Light

The finding that single quanta of light can give rise to a conscious sensation raises a number of tantalizing questions. How large is the unitary response; how is it distinguished from ongoing noise; and how is such minimal information transferred faithfully through the retina to higher centers? To measure unitary responses to single quanta, Baylor and his colleagues recorded photo-activated currents of individual rods in retinas from toad, monkey, and human.[37] These experiments provide a rare example of the way in which a process as complex as seeing the dimmest possible flashes of light can be correlated with the events that occur in single molecules.[38]

The procedure was to isolate a piece of retina from the animal or from a cadaver and maintain it in darkness. To measure currents, the outer segment of the rod was sucked

[37]Schnapf, J. L., and Baylor, D. A. 1987. *Sci. Am.* 256: 40–47.

[38]Rieke, F., and Baylor, D. A. 1998. *Biophys. J.* 75: 1836–1857.

BOX 20.1
Adaptation of Photoreceptors

The light intensities that humans use for vision range from the detection of single photons to bright sunlight on a tropical beach, a dynamic range of more than 10^{13}. Since pupillary constriction reduces light by only 10-fold, adaptation by photoreceptors accounts for most of the wide range of sensitivity[39] and is achieved largely by retinal cones, which are used in daylight and starlight. Although our rods saturate at light levels that we associate with daytime vision, even indoors, our cones can adjust to very bright lights, saturating just before light intensities become damaging.

Thus, it is essential to ensure that responses to light can occur at different background levels of illumination. If bright, ambient light were able to close all the nucleotide-gated channels, the receptor would be unable to register any further increase in intensity.

Calcium is one factor of key importance for adaptation of photoreceptors.[40,41] In the dark, the nucleotide-gated channels are open and calcium ions continually flow into the photoreceptor. Calcium is extruded by ion pumps and exchangers in the outer segment (see Figure 20.10).[42] Under conditions of steady illumination, the channels close and calcium entry is reduced. Because calcium is still actively extruded (see Figure 20.10), the intracellular calcium concentration falls. The reduced intracellular calcium concentration opposes closure of the nucleotide-gated channels by several mechanisms.

First, calcium reduces the affinity of channels to cGMP through the calcium-binding protein, calmodulin. Lowered calcium concentration therefore increases the affinity of the channels to cGMP and, thereby, potentiates channel opening and current flow during illumination. Second, lowered intracellular calcium favors intracellular cGMP accumulation by increasing the activity of guanylate cyclase (i.e., promotes cGMP synthesis) and inhibiting the activation of phosphodiesterase (i.e., slows cGMP hydrolysis). Third, lowered intracellular calcium causes phosphorylation of metarhodopsin II and thereby speeds up its inactivation. The rate of termination of the catalytic activity of metarhodopsin II is of importance for transduction, since, while it is active, the cascade continues to generate a signal. The phosphorylation of activated rhodopsin is mediated by recoverin,[43] a calcium-binding molecule that is also involved in the inhibition of phosphodiesterase. Dark adaptation, which is the reverse of light adaptation, is principally limited by the regeneration of photopigment, which for full adaptation can take up to 10 minutes for cones and 30 minutes for rods.

Molecular mechanisms for adaptation were analyzed by Baylor and his colleagues[41], who measured the rate of adaptation of photoreceptor responses in normal and transgenic mice in which 15 amino acids were deleted from the C terminus of rhodopsin, the presumptive site for phosphorylation of activated rhodopsin.[44,45] In a normal rod (part B of the figure), a flash of light produces an outward current that adapts (i.e., decreases) as expected. Part C of the figure shows the response of a rod in a transgenic mouse to a flash—the response, instead of declining, lasts about 20 times longer than normal. Thus, the 15 amino acids that had been deleted constitute the region of the molecule required for recovery from the effects of a flash of light.

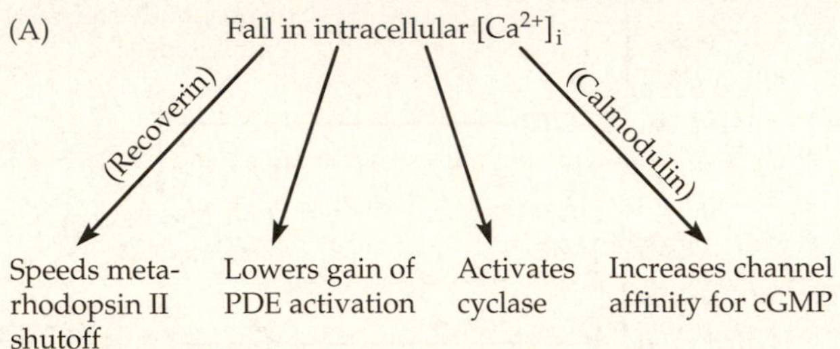

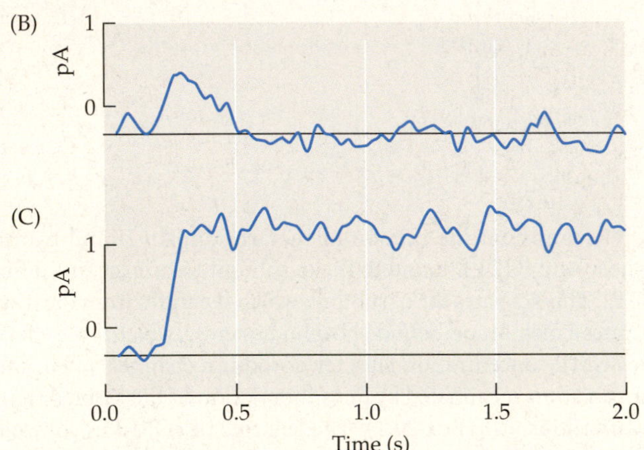

Mechanisms of Adaptation. (A) Diagram showing effects of a fall in the intracellular calcium concentration on mechanisms that influence adaptation of photoreceptors to steady light. (B) Responses to flashes delivered at time zero were recorded from a rod by suction electrodes. The response elicited by activation of a normal rhodopsin molecule declined as expected. (C) The trace was recorded from a rod in a transgenic mouse in which rhodopsin molecules had been truncated by the deletion of 15 amino acids from the C terminus. The duration of the response was prolonged through failure of the altered rhodopsin molecule to become shut off after the flash. PDE = phosphodiesterase (After Baylor, 1996).

[39] Fain, G. L., Matthews, H. R., and Cornwall, M. C. 1996. *Trends Neurosci.* 19: 502–507.

[40] Baylor, D. 1996. *Proc. Natl. Acad. Sci. USA* 93: 560–565.

[41] Koutalos, Y., and Yau, K. W. 1996. *Trends Neurosci*. 19: 73–81.

[42] Morgans, C. W. et al. 1998. *J. Neurosci.* 18: 2467–2474.

[43] Baylor, D. A., and Burns, M. E. 1998. *Eye (Lond).* 12 (Pt 3b): 521–525.

[44] Chen, J. et al. 1995. *Science* 267: 374–377.

[45] Fu, Y., and Yau, K. W. 2007. *Pflügers Arch.* 454: 805–819.

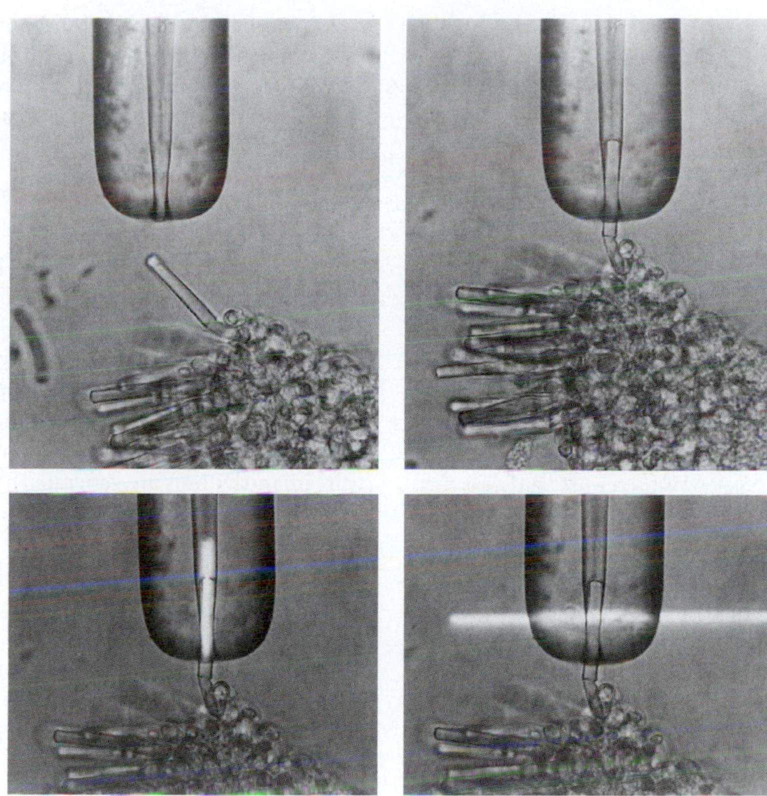

FIGURE 20.13 Method for Recording Membrane Currents of a Rod Outer Segment. A suction electrode with a fine tip is used to suck up the outer segment of a rod that protrudes from a piece of toad retina. Slits of light illuminate the receptor with precision. Since the electrode fits tightly around the photoreceptor, current flowing into it or out of it is recorded. (From Baylor, Lamb, and Yau, 1979.)

into a fine pipette (Figure 20.13). As expected, in darkness a current flowed continuously into the outer segment. Flashes of light closed channels in the outer segment, causing a decrease in the dark current. Figure 20.14A shows responses to very dim light flashes, corresponding to one or two quanta of light on the outer segment. The currents are small and quantal in nature. That is, sometimes a dim flash evokes a unitary response, sometimes a doublet, and sometimes nothing at all.

In monkey rods, the current reduction caused by transduction of a single photon is about 0.5 picoamperes (pA). This corresponds to the closure of approximately 300 channels, or of about 3% to 5% of the rod channels that are open in the dark, and is the result of the large amplification through the cGMP cascade. Moreover, because of the extreme stability of visual pigments mentioned earlier, random isomerizations and spurious channel closings are rare events. This allows the effects

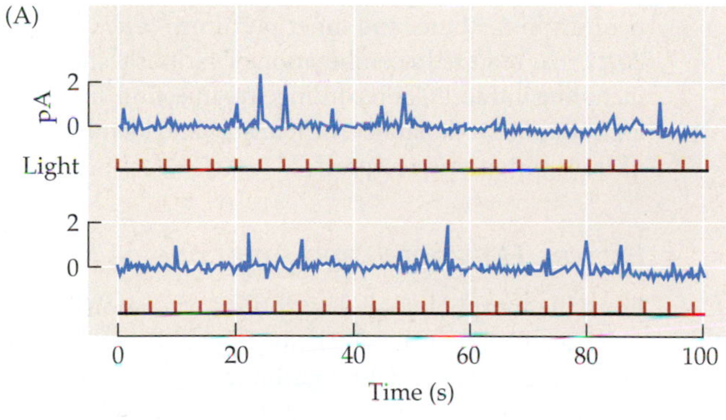

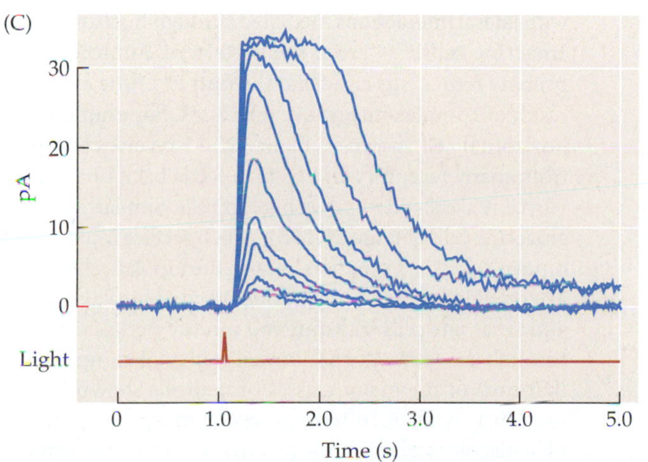

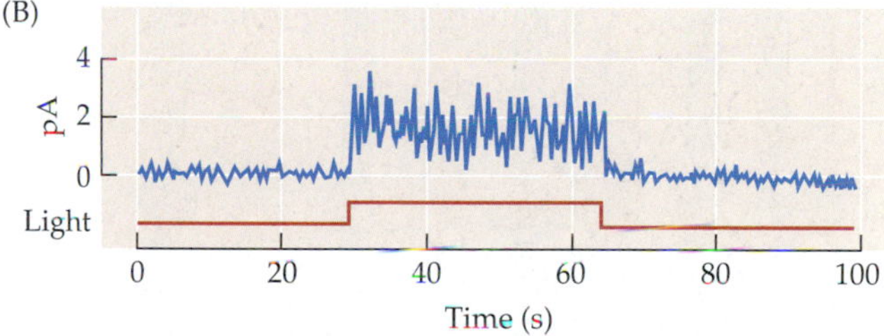

FIGURE 20.14 Recordings Made by Suction Electrode from Monkey Rod Outer Segment. (A) Responses to dim flashes, applied as indicated in the red traces (labeled "Light"), are shown in the two current traces. The currents fluctuate in a quantal manner. Smaller deflections are the currents generated by single photons interacting with visual pigments. Photoisomerizations often failed to occur. (B) Steady, more intense illumination (bottom trace) gives rise to a burst of signals. (C) Records from a rod in a monkey retina with flashes of increasing intensity. These currents are the counterpart of voltage traces shown in Figure 20.5B. (From Baylor, Nunn, and Schnapf, 1984.)

of single-light quanta to stand out against an extremely quiet background. It has been shown that electrical coupling through gap junctions between photoreceptors provides an additional smoothing effect that reduces the background noise and improves the signal-to-noise ratio for responses of rods to single quanta.[46]

Circadian Photoreceptors in the Mammalian Retina

Form and color vision in mammals are served by rods and cones, each with their specific photopigment. In contrast, other light-dependent behaviors, such as pupillary reflexes and entrainment of circadian rhythms depend on entirely different sensory receptors and visual pigments. Blind mice that fail to produce either rods or cones still exhibit circadian rhythms, and these can be shifted by light exposure during subjective night (entrained).[47,48] This behavior depends on a tiny population of intrinsically photosensitive retinal ganglion cells (ipRGCs)[49]—about 3% of the total ganglion cells—that express the photoprotein melanopsin (see also Chapter 17) and project axons to the suprachiasmatic nucleus (where circadian rhythms are generated)[50] (see Figure 20.1). Melanopsin was first identified in melanocytes of frog skin and is both structurally related to rhodopsin and has similar light absorption characteristics.[51] Melanopsin-containing retinal ganglion cells slowly depolarize in response to bright light. Accumulating evidence suggests that melanopsin is coupled through a non-transducin G protein to phospholipase C that generates IP_3 and diacyl-glycerol (DAG), finally to gate cation current through transient receptor potential (TRP)-like channels.[52] This process is similar to the depolarizing phototransduction cascade known for invertebrates[53,54] (see Figure 20.5A), but obviously differs from the process by which rhodopsin couples to the hyperpolarization of rods and cones.

[46] Bloomfield, S. A., and Volgyi, B. 2009. *Nat. Rev. Neurosci.* 10: 495–506.

[47] Foster, R. G. et al. 1991. *J. Comp. Physiol. A* 169: 39–50.

[48] Freedman, M. S. et al. 1999. *Science* 284: 502–504.

[49] Berson, D. M., Dunn, F. A., and Takao, M. 2002. *Science* 295: 1070–1073.

[50] Hattar, S. et al. 2002. *Science* 295: 1065–1070.

[51] Provencio, I. et al. 1998. *Proc. Natl. Acad. Sci. USA* 95: 340–345.

[52] Peirson, S. N., Halford, S., and Foster, R. G. 2009. *Philos. Trans. R. Soc. Lond., B, Biol. Sci.* 364: 2849–2865.

[53] Wang, T., and Montell, C. 2007. *Pflügers Arch.* 454: 821–847.

[54] Berson, D. M. 2007. *Pflügers Arch.* 454: 849–855.

[55] Boycott, B., and Wässle, H. 1999. *Invest. Ophthalmol. Vis. Sci.* 40: 1313–1327.

[56] Schiller, P. H. 2010. *Proc. Natl. Acad. Sci. USA* 107: 17087–17094.

Synaptic Organization of the Retina

Numerous questions arise about the transmission of signals in the retina. How do rods and cones influence bipolar cells? And, how are horizontal and amacrine cells involved in signaling? The analysis of signal processing by these neurons in the outer and inner plexiform layers (Figure 20.15) has required a combination of various techniques, including intracellular recording, dye injection, morphological studies, cellular neurochemistry, and identification of transmitters and receptors.

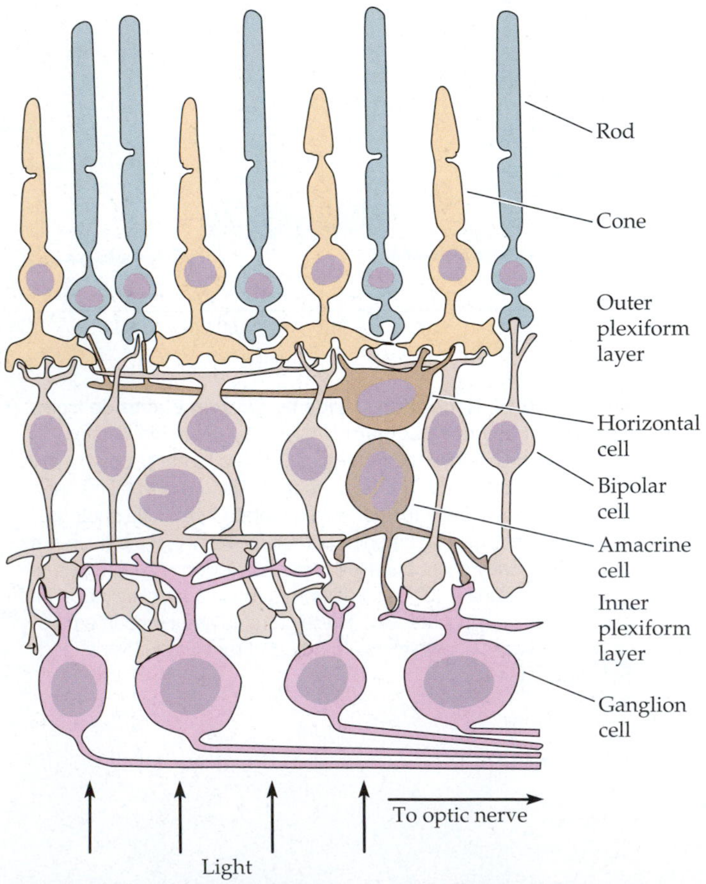

FIGURE 20.15 Principal Cell Types and Connections of Primate Retina to illustrate rod and cone pathways to ganglion cells. (After Dowling and Boycott, 1966 and Daw, Jensen, and Brunken, 1990.)

Bipolar, Horizontal, and Amacrine cells

From the morphological descriptions of Ramón y Cajal (see Figure 20.2) a general scheme for the wiring diagram of the retina emerges: a through-line of transmission from photoreceptors to ganglion cells by way of bipolar cells, with lateral interactions mediated through horizontal and amacrine cells.[55,56] From the pattern of connections in primate retina, it is clear that the output of the eye is the result of complex integrative processes. For example, the horizontal cell shown in Figure 20.15 receives synapses from many receptors and in turn feeds back onto them. Horizontal cells also end on bipolar cells. Similarly, certain amacrine cells (Figure 20.16B), which receive inputs from bipolar cells, send synapses back to the bipolar cells as well as to ganglion cells. One can conclude that horizontal cells and amacrine cells transmit and modify signals traveling through the retina. An additional source of complexity is that each of the major classes of neurons shown in Figures 20.15 and 20.16 has numerous morphological and pharmacological subtypes. By physiological, biochemical,

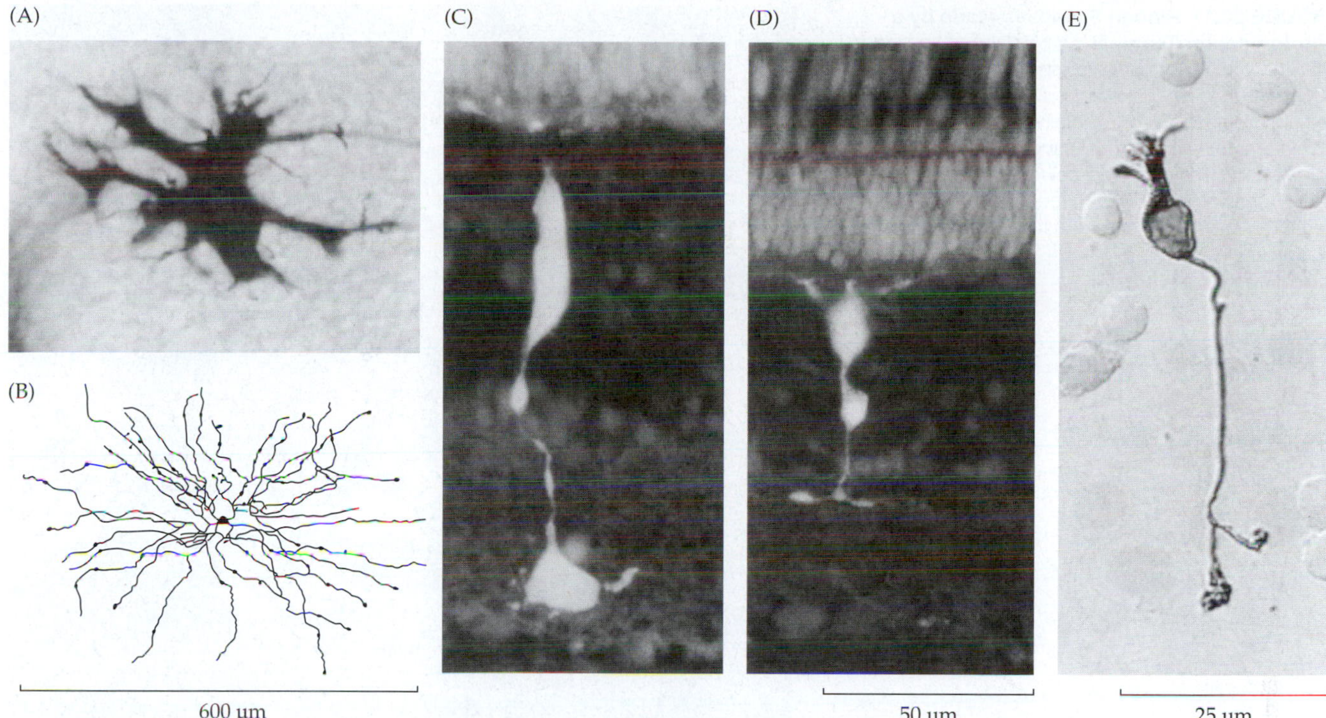

FIGURE 20.16 Horizontal, Amacrine, and Bipolar Cells. (A) A horizontal cell in dogfish retina injected with horseradish peroxidase. (B) Indolamine-accumulating amacrine cell from rabbit retina injected with Lucifer yellow. (C) A depolarizing on-center bipolar cell of goldfish injected with fluorescent dye. (D) A hyperpolarizing off-center goldfish bipolar cell. (E) A bipolar cell isolated from rat retina stained for protein kinase C. (A, B, and D kindly provided by A. Kaneko, unpublished; C from Yamashita and Wässle, 1991; E from Masland, 1988.)

and anatomical criteria, five major classes of bipolar cells, two or more types of horizontal cell, at least 20 types of amacrine cells, and more than 10 types of ganglion cells have been described.[57,58] It is thought that all morphologically defined retinal cell types have been identified, with distinct functional roles assigned to nearly half of these[59] making the retina arguably the most completely known component of the nervous system.

Molecular Mechanisms of Synaptic Transmission in the Retina

Virtually all the known neurotransmitters have been found in the retina.[60–63] Glutamate is the transmitter liberated by photoreceptors and bipolar cells. Horizontal cells secrete γ-aminobutyric acid (GABA). Some amacrine cells secrete dopamine, others indolamines, others acetylcholine. There are also less conventional transmitters, such as nitric oxide, that have been identified in retinal signaling. The distributions and functional significance of the transmitters, receptors, receptor subunits, and transporters have been explored in detail by immunohistochemistry, in situ hybridization, and pharmacology.

Peptide transmitters also are found in the retina,[64] where they are implicated in aspects of retinal development and disease. Vasoactive intestinal peptide (VIP) for example, is expressed in a subset of amacrine cells under the control of brain-derived neurotropic factor (BDNF)[65] and VIP expression has been related to the development of myopia (nearsightedness).[66] VIP, glucagon, and other neurotransmitters mediate communication between the neural retina and the retinal pigment epithelium that may participate in activity-dependent eyeball growth.[67]

The continuous, quantized release of glutamate from retinal photoreceptors and bipolar cells is achieved by specialized active zones known as ribbon synapses.[7] An example from a retinal cone is shown in Figure 20.17. The cone terminal contains vesicles that are tethered to a long, flat electron-dense organelle—the ribbon. The ribbon structure is thought to reflect its specialized function: the ability to sustain high rates of vesicular release modu-

[57]Kolb, H. 1997. *Eye (Lond).* 11 (Pt 6): 904–923.

[58]MacNeil, M. A. et al. 1999. *J. Comp. Neurol.* 413: 305–326.

[59]Masland, R. H. 2001. *Curr. Opin. Neurobiol.* 11: 431–436.

[60]Wässle, H. et al. 1998. *Vision Res.* 38: 1411–1430.

[61]Mora-Ferrer, C., and Neumeyer, C. 2009. *Vision Res.* 49: 960–969.

[62]Brandstatter, J. H. 2002. *Curr. Eye Res.* 25: 327–331.

[63]Shen, Y., Liu, X. L., and Yang, X. L. 2006. *Mol. Neurobiol.* 34: 163–179.

[64]Herbst, H., and Thier, P. 1996. *Exp. Brain Res.* 111: 345–355.

[65]Cellerino, A. et al. 2003. *J. Comp. Neurol.* 467: 97–104.

[66]Tkatchenko, A. V. et al. 2006. *Proc. Natl. Acad. Sci. USA* 103: 4681–4686.

[67]Rymer, J., and Wildsoet, C. F. 2005. *Vis. Neurosci.* 22: 251–261.

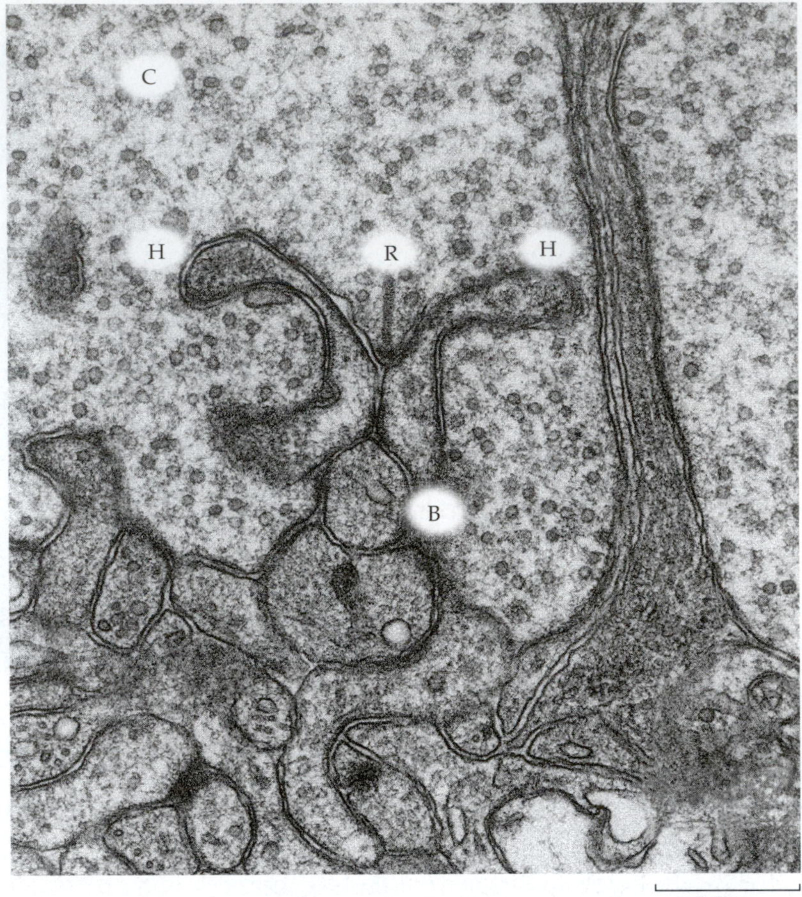

FIGURE 20.17 Ribbon Synapses Made by a Photoreceptor Terminal on bipolar and horizontal cell endings. Presynaptic vesicles in a cone terminal (C) are aligned along the ribbon (R). This type of synapse is adapted for maintained release of quanta of glutamate onto bipolar (B) and horizontal (H) cells in darkness. The horizontal cell, which releases γ-aminobutyric acid (GABA), feeds back onto the receptor terminal. (Micrograph kindly provided by P. Sterling.)

lated by changes in presynaptic voltage but without benefit of action potentials increases its bandwidth or frequency response. Analogous structures are found in sensory hair cells and may serve to concentrate vesicles near to membrane docking sites.[68] In keeping with their functional specialization, ribbons have their own variants of the molecular elements in the canonical soluble N-ethylmaleimide-sensitive factor attachment protein receptor, or SNARE, complex of neuronal synapses[69] (see Chapter 13). Unique to ribbon synapses is the protein RIBEYE[70] with unknown function but comprising perhaps two-thirds of the ribbon volume.[71]

A host of other proteins has been localized to the synaptic ribbon (Figure 20.18), also with limited certainty about their function. One whose identity and function is certain is the voltage-gated calcium channel. Ribbon synapses release continuously but also require rapid modulation with changes in stimulus intensity (especially for auditory hair cells, see Chapter 22). In contrast to standard chemical synapses where transmitter release is mediated by N,P/Q and R type calcium channels (see Chapter 13), ribbons are served by so-called L-type calcium channels with rapid gating kinetics, relatively little inactivation, and an activation range that spans the resting potential of the cell.[72] Stochastic gating of the L-type channels supports the release of hundreds of vesicles per second from each ribbon, although how that rapid release takes place and what mechanisms ensure a steady supply of vesicles remain unknown.

Receptive Fields of Retinal Neurons

The receptive field of a neuron in the visual system can be defined as the area of the retina for which light influences the activity of that neuron. As described in Chapter 2, our understanding of retinal receptive fields was obtained first by studying the output carried by the axons of retinal ganglion cells. Extracellular recording of action potentials was possible

[68] Matthews, G., and Fuchs, P. 2010. *Nat. Rev. Neurosci.* 11: 812–822.

[69] tom Dieck, S., and Brandstatter, J. H. 2006. *Cell Tissue Res.* 326: 339–346.

[70] Schmitz, F., Konigstorfer, A., and Sudhof, T. C. 2000. *Neuron* 28: 857–872.

[71] Zenisek, D. et al. 2004. *J. Neurosci.* 24: 9752–9759.

[72] Heidelberger, R., Thoreson, W. B., and Witkovsky, P. 2005. *Prog. Retin. Eye Res.* 24: 682–720.

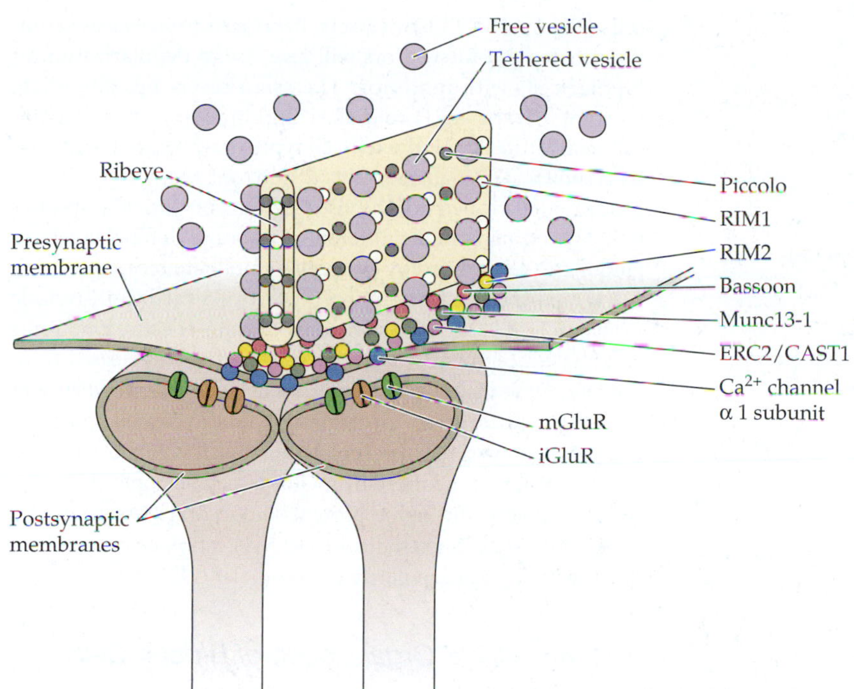

FIGURE 20.18 Molecular Components of Ribbon Synapses. The electron-dense body of the ribbon is made up largely of the Ribeye (CtBP2) protein, with contributions from CtBP1/BARS and KIF3A. Piccolo and Rab3-interacting molecule 1 (RIM1) are ribbon-associated proteins. L-type calcium channels cluster in the plasma membrane beneath the ribbon and associate with Bassoon, RIM2, Munc13-1, and ERC2/CAST1 proteins. Metabotropic (mGluR) and ionotropic (iGluR) glutamate receptors are located in the postsynaptic membranes of bipolar and horizontal cell. (After Dieck and Brandstatter, 2006.)

long before intracellular electrodes were inserted into individual retinal neurons. Although a product of necessity, beginning with the output signal was nonetheless an excellent starting point for analyzing retinal processing. Two important features were established by such recordings. The first was that the receptive field of a retinal ganglion cell is organized in a circular, **center-surround** manner. That is, light in the receptive field surround area produces an opposite effect to that in the center (see Figure 2.5).

The second essential feature obtained from ganglion cell recordings was that some retinal ganglion cells were excited by light in their receptive field center, the on retinal ganglion cells; in contrast, other ganglion cells were inhibited by light in their receptive field center, the off retinal ganglion cells. This dichotomy of on and off pathways is now known to be a fundamental property of all vertebrate retinas and establishes an organizing principle for higher visual centers.[56] From these recordings of retinal ganglion cells, it is clear that the eye tells the brain about *patterns* of light and dark. Since photoreceptors can only report light intensity, it must fall to retinal interneurons to provide the analytical steps leading to the more structured retinal output.

Responses of Bipolar Cells

Each bipolar cell receives its direct input either from rods or cones. Rod bipolar cells are typically supplied by 15 to 45 receptors. One type of cone bipolar, the midget bipolar, receives its input from a single cone. As one might expect, midget bipolar cells are found in the through-line from the fovea, where acuity is highest. They end on specialized ganglion cells. Other bipolar cells are supplied by a convergent input from 5 to 20 adjacent cones. H bipolar cells, like photoreceptors, are hyperpolarized by an increase in light. However, D bipolar cells are depolarized by light, laying the basis for the off and on output carried by retinal ganglion cells.

The responses and receptive fields of bipolar cells depend on two mechanisms. First, the continuous release of glutamate from photoreceptors in the dark keeps some bipolar cells depolarized and others hyperpolarized, depending on whether the cells have excitatory or inhibitory glutamate receptors. Second, light causes photoreceptors to be hyperpolarized, thereby reducing glutamate release. Accordingly, decreased tonic release from illuminated photoreceptors will reduce excitation of bipolar cells with excitatory (i.e., sign preserving) receptors, giving rise to hyperpolarization.[73] These are called H (hyperpolarizing) bipolar

[73]Kaneko, A., and Hashimoto, H. 1969. *Vision Res.* 9: 37–55.

(A) Central illumination

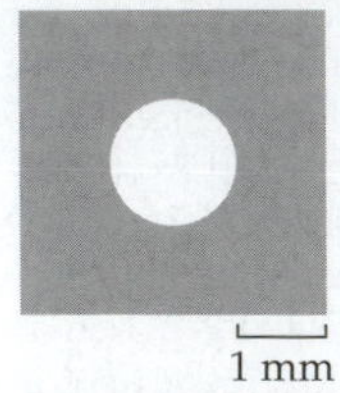

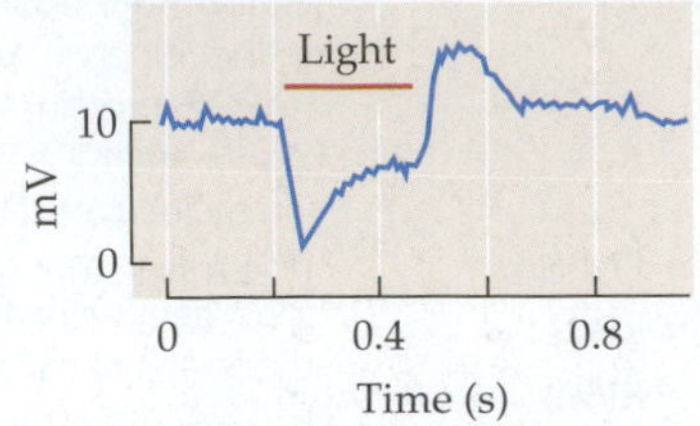

(B) Annular illumination

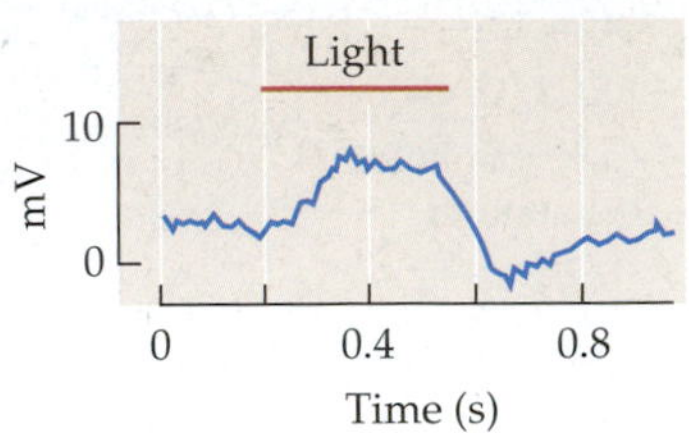

FIGURE 20.19 Receptive Field Organization of a Hyperpolarizing (H) Bipolar Cell. (A) Records made from the bipolar cell in the goldfish retina show a hyperpolarization in response to illumination of the center of the receptive field. Annular illumination causes the cell to respond with a depolarization (B). Diffuse light would have little effect on the cell. For a D bipolar cell, illumination of the center would produce depolarization, while the annulus would produce hyperpolarization. (After Kaneko, 1970.)

cells (Figure 20.19). Conversely, decreased tonic release from illuminated photoreceptors will give rise to depolarization of bipolar cells with inhibitory (i.e., sign reversing) glutamate receptors. These are D (depolarizing) bipolar cells. D bipolar cells constitute one of the few cell types in which glutamate has been shown to have an inhibitory action (see Chapter 14). Thus, a fundamental property of the visual system, on and off responses to light, originates in the differential response of bipolar cells to glutamate, determined by each cell's glutamate receptors.

H bipolar cells have α-amino-3-hydroxy-5-methyl-4-isoxazole propionic acid (AMPA) or kainate-type cation-selective ionotropic glutamate receptors.[74] Depolarized by glutamate released from photoreceptors in the dark, H bipolar cells hyperpolarize when light reduces glutamate release. In contrast, as Kaneko, his colleagues and others have shown, the hyperpolarizing transmission to D bipolar cells in the dark is mediated by mGluR6 metabotropic glutamate receptors that act through G proteins and second messengers.[75,76] These close a TRP melastatin 1 (TRPM1) cation channel, which opens with light to depolarize the D bipolar cell.[77,78]

Receptive Field Organization of Bipolar Cells

The receptive field of a hyperpolarizing bipolar cell is shown in Figure 20.19. A small spot of light shone onto the central part of the field causes a sustained hyperpolarization. Illumination by an annulus, leaving the center dark, causes depolarization. Thus, the central area, driven directly by photoreceptors, is enveloped by an antagonistic surround. The H bipolar cell of Figure 20.19 can be described as having an off-center receptive field, since it is depolarized when the spot of light goes off.

D bipolar cells have similarly shaped concentric fields, except that illumination of the center causes depolarization and illumination of the surround causes hyperpolarization. Because it is depolarized when the light goes on, the D bipolar cell has an on-center receptive field. The terminology of on and off responses is used extensively to describe receptive field properties at successive levels of the visual system. An important principle is that a single photoreceptor can contribute to the receptive field centers of both on and off bipolar cells and to the surrounds of others.

Rod Bipolar Cells

Rod bipolar cells depart from the straightforward pattern described for cone bipolar cells. All rod bipolar cells are on type, expressing sign-reversing inhibitory mGluR6 metabotropic glutamate receptors.[79] Rather than contacting retinal ganglion cells directly, rod bipolar cells synapse onto AII amacrine cells.[80] Each AII amacrine cell depolarizes due to the summed input of many rod bipolar cells, and in turn forms synapses onto the axon terminals of cone bipolar cells to influence the activity of retinal ganglion cells.[81] It has been suggested that this indirect arrangement results from the later evolutionary arrival of rod photoreceptors to the vertebrate retina, piggybacking onto the preexisting cone circuitry.[82] AII amacrine cells collect input from large numbers of rod photoreceptors, and they form electrical synapses onto the synaptic pedicle of on cone bipolars and make inhibitory glycinergic synapses with off cone bipolars.[80] This connectivity suggests that rods provide a wide-field measure of background light levels.[81] The retinal processing of rod photoreception is an emerging story, since there is now evidence for electrical synapses of rods with cones[83] and for glutamatergic synapses of rods and cones onto some of the same bipolar cells.[84,85]

Horizontal Cells and Surround Inhibition

Rods and cones make synapses with bipolar cells and horizontal cells. The responses of D and H bipolar cells to surround illumination are mediated by horizontal cells. Each horizontal cell

[74] DeVries, S. H. 2000. *Neuron* 28: 847–856.
[75] Kikkawa, S. et al. 1993. *Biochem. Biophys. Res. Commun*. 195: 374–379.
[76] Masu, M. et al. 1995. *Cell* 80: 757–765.
[77] Nakanishi, S. et al. 1998. *Brain Res. Brain Res. Rev.* 26: 230–235.
[78] Koike, C. et al. 2010. *Cell Calcium* 48: 95–101.
[79] Nomura, A. et al. 1994. *Cell* 77: 361–369.
[80] Famiglietti, E. V., Jr., and Kolb, H. 1975. *Brain Res*. 84: 293–300.
[81] Wässle, H. 2004. *Nat. Rev. Neurosci.* 5: 747–757.
[82] Lamb, T. D. 2009. *Philos. Trans. R. Soc. Lond. B, Biol. Sci.* 364: 2911–2924.
[83] Deans, M. R. et al. 2002. *Neuron* 36: 703–712.
[84] Pang, J. J. et al. 2010 *Proc. Natl. Acad. Sci. USA* 107: 395–400.
[85] Soucy, E. et al. 1998. *Neuron* 21: 481–493.

receives inputs from a large number of photoreceptors. Horizontal cells, like H bipolar cells, respond to illumination of photoreceptors by hyperpolarization (because glutamate release from photoreceptor terminals opens depolarizing ionotropic receptors and release decreases with illumination). Another feature of horizontal cells is that they are electrically coupled to each other.[86,87] Lucifer yellow dye injected into one horizontal cell spreads readily to others through gap junctions. Any one horizontal cell, therefore, is influenced by light shone on a large area of retina because of current flow from its neighbors. An interesting feature of the electrical synapses between horizontal cells is that they become uncoupled by dopamine.[88]

Horizontal cells make inhibitory synaptic connections; they release GABA back onto the photoreceptors and bipolar cells.[89–92] Thus, depolarization of photoreceptors in the dark is antagonized by inhibitory input from horizontal cells. Receptor illumination results in horizontal cell hyperpolarization and a reduction in GABA release. Hence, hyperpolarization of photoreceptors by diffuse light is countered by a reduction in the GABA inhibition coming from horizontal cells. It has been suggested that nitric oxide, which is synthesized by photoreceptors and horizontal cells, also contributes to the inhibition of glutamate release by photoreceptors.[93] In summary, there is negative feedback onto the photoreceptors through the horizontal cells:

illumination→ photoreceptor hyperpolarization→ horizontal cell hyperpolarization→ photoreceptor depolarization

The connections involved in the off-center/on-surround responses of an H bipolar cell are shown schematically in Figure 20.20. For simplicity, the center is represented by a single photoreceptor and the surround by a few neighboring receptors connected to a single horizontal cell. The response to illumination of the central photoreceptor is straightforward (Figure 20.20A). Photoreceptor activation results in hyperpolarization and therefore a

[86] Kaneko, A. 1971. *J. Physiol.* 213: 95–105.

[87] Liu, C. R. et al. 2009. *Neuroscience* 164: 1161–1169.

[88] Tornqvist, K., Yang, X. L., and Dowling, J. E. 1988. *J. Neurosci.* 8: 2279–2288.

[89] Kaneko, A., and Tachibana, M. 1986. *J. Physiol.* 373: 443–461.

[90] Schwartz, E. A. 1987. *Science* 238: 350–355.

[91] Yang, X. L., Gao, F., and Wu, S. M. 1999. *Vis. Neurosci.* 16: 967–979.

[92] Deniz, S. et al. 2011. *J. Neurochem.* 116: 350–362.

[93] Savchenko, A., Barnes, S., and Kramer, R. H. 1997. *Nature* 390: 694–698.

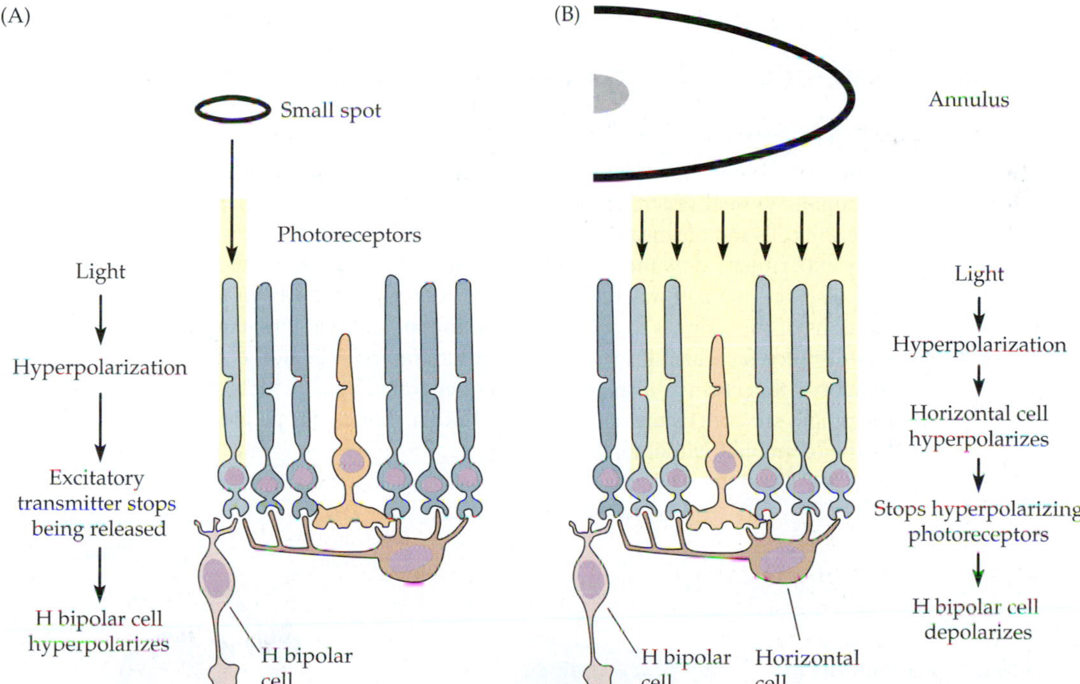

FIGURE 20.20 Connections of Photoreceptors, Bipolar Cells, and Horizontal Cells. The figure illustrates connections required to elicit responses in bipolar cells. (A) Light falling on a single photoreceptor causes it to become hyperpolarized. As a result, glutamate stops being released and the H bipolar cell, as in Figure 20.19, becomes hyperpolarized through loss of excitation. (B) Light falling on the surrounding area in the form of an annulus again prevents glutamate from being released by photoreceptors. As a result, the horizontal cell becomes hyperpolarized; this hyperpolarization prevents the horizontal cell from releasing its inhibitory transmitter, γ-aminobutyric acid (GABA), onto the photoreceptor. The photoreceptor that is connected to the H bipolar cell therefore becomes depolarized (through removal of inhibition). It once again releases glutamate and depolarizes the bipolar cell. With diffuse light, the depolarizing and hyperpolarizing effects cancel each other out. Thus, horizontal cells play an essential part in the construction of the receptive field properties of bipolar cells.

reduction in glutamate release. As a result, the bipolar cell is hyperpolarized. The horizontal cell receives a hyperpolarizing input as well, but it is from only one photoreceptor and the effect is small, as is the negative feedback onto the central photoreceptor.

The response to surround illumination involves an extra step (Figure 20.20B). The horizontal cell, which receives input from several photoreceptors in the surround, is hyperpolarized by illumination. The hyperpolarization reduces GABA release by the horizontal cell. Reduced inhibition of the photoreceptors tends to produce depolarization. The depolarizing feedback effect is minimal on the surround receptors, which are being strongly hyperpolarized by illumination. The central receptor, however, is receiving no illumination; its only input is removal of horizontal cell inhibition. Consequently, the central receptor is depolarized, release of glutamate is increased, and the H bipolar cell is depolarized. Comprehensive reviews and papers describe the morphology and the properties of photoreceptor terminals, bipolar cells, and the feedback synapses of horizontal cells onto bipolar cells.[2,10,94,95]

Significance of Receptive Field Organization of Bipolar Cells

What are the physiological implications of bipolar cell receptive fields? D and H bipolar cells do not simply respond to light. Rather, they begin to analyze information about patterns. Their signals convey information about small spots of light surrounded by darkness or about small dark spots surrounded by light. They respond to contrasting patterns of light and dark over a small area of retina. In addition to the broad categories of D and H bipolar cells mentioned in the list that follows, approximately 11 types of cone bipolar cells have been distinguished by morphological and immunohistochemical criteria, presumably underlying additional functional specificity.[4,96] Here we summarize three principal types.

1. D and H cone bipolar cells respond best to small light or dark spots.
2. The centers of D and H midget bipolar cells are supplied by single cones.
3. Rod bipolar cells are D, or on-center.

Receptive Fields of Ganglion Cells

The Output of the Retina

As described in Chapter 2, Stephen Kuffler pioneered the experimental analysis of the mammalian visual system by concentrating on the receptive field organization of retinal ganglion cells and the meaning of their signals in the cat. An important feature of Kuffler's early experiments was the use of the intact, undissected eye—the normal refraction of which served as a pathway for stimulation.[97] A convenient way of illuminating particular portions of the retina is to anesthetize the animal and place it facing a screen or monitor at a distance for which its eyes are properly refracted. When one then shines patterns of light onto the screen or displays computer-generated images, these will be well focused on the retinal surface (Figure 20.21). In fact, Kuffler shone light patterns into the eye directly by use of an ophthalmoscope.

[94] Thoreson, W. B. 2007. *Mol. Neurobiol.* 36: 205–223.

[95] Yang, X. F. et al. *Neuroscience* 173: 19–29.

[96] Lee, B. B., Martin, P. R., and Grunert, U. 2010. *Prog. Retin. Eye Res.* 29: 622–639.

[97] Kuffler, S. W. 1953. *J. Neurophysiol.* 16: 37–68.

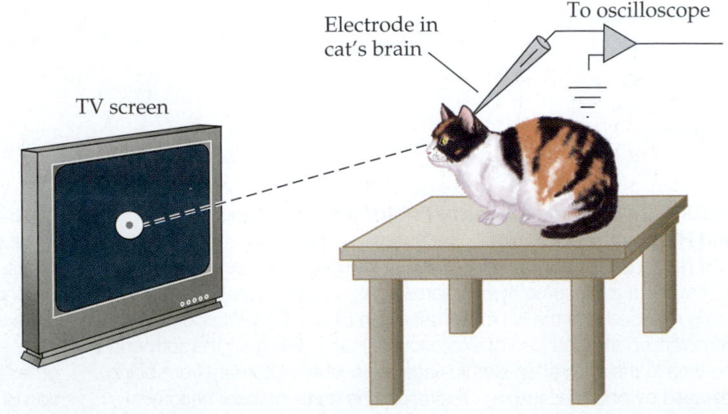

FIGURE 20.21 Stimulation of Retina with Patterns of Light. The eyes of an anesthetized, light-adapted cat or monkey focus on a movie or a television screen with various patterns of light generated by a computer or shone by a projector. An electrode records the responses from a single cell in the visual pathway. Light or shadow falling onto a restricted area of the screen may increase or decrease the frequency of signals given by the neuron. One can delineate the receptive field of the cell by determining the areas on the screen from which the neuron's firing is influenced.

Ganglion Cell Receptive Field Organization

When one records from a particular ganglion cell, the first task is to find the location of its receptive field. Characteristically, most ganglion cells and other neurons at higher levels in the visual system fire at rest even in the absence of patterned illumination. Appropriate stimuli do not necessarily initiate activity but may modulate the resting discharge; responses of ganglion cells can consist of either an increase or a decrease of action potential frequency.

As is the case for bipolar cells, which were first studied many years later, there are two basic receptive field types: on-center and off-center ganglion cells. The receptive fields of both types are roughly concentric, with the ganglion cell soma in the geometrical center of its field. Figure 20.22, adapted from a paper by Kuffler, shows the response of an on retinal ganglion cell to increased light in either the center or the surround of its receptive field. Light produces the most vigorous response if it completely fills the center, whereas for most effective inhibition of firing, the light must cover the entire ring-shaped surround. When the inhibitory annular light is turned off, the ganglion cell gives an exuberant off discharge (not shown). An off-center field has a converse organization, with inhibition arising in the circular center. For either cell, the spot-like center and its surround are antagonistic; therefore, if both center and surround are illuminated simultaneously, they tend to cancel each other's contribution, although the center response prevails.

(A) On-center field

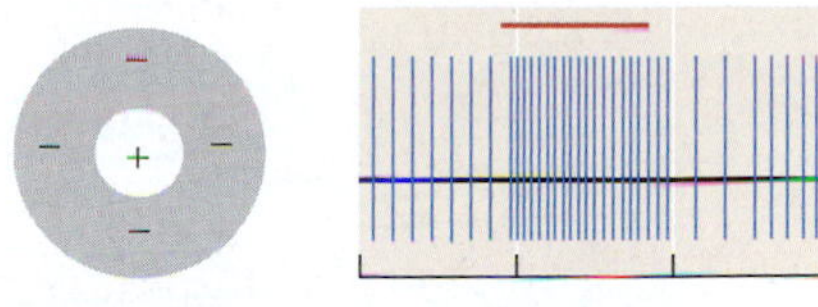

(B) Off-center field

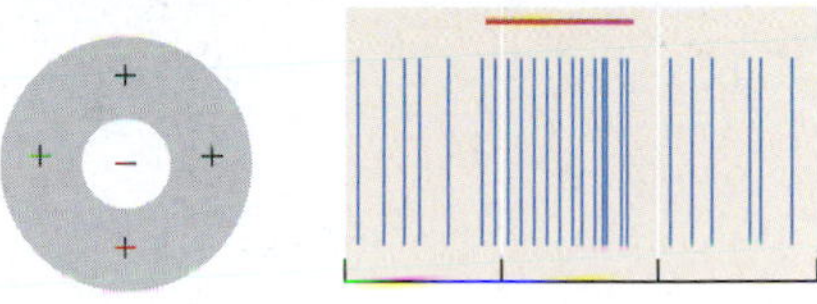

(C) Diffuse illumination

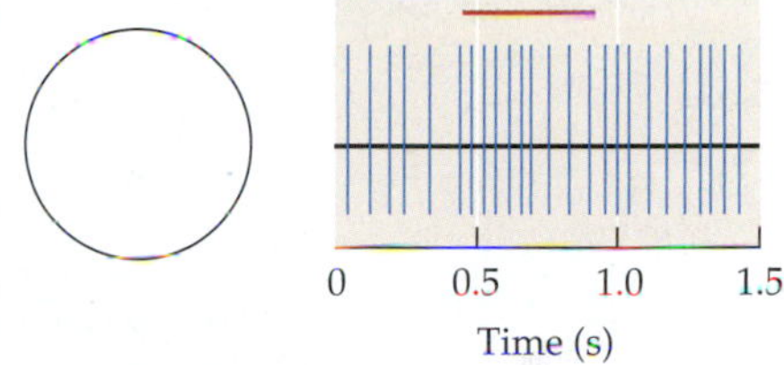

FIGURE 20.22 Receptive Fields of Retinal Ganglion Cells. Retinal ganglion cells have center-surround receptive fields in which light produces opposite effects depending on its location. (A) For on-center retinal ganglion cells a spot of light in the center (indicated by the plus sign) causes increased firing of action potentials (stylized recording on the right, timing of light indicated by the red bar). The same spot of light falling on the inhibitory surround would reduce activity in these cells. (B) A central spot of light in an off-center retinal ganglion cell slows the rate of action potential generation, and conversely, if directed to the surround would increase firing. (C) For both on-center and off-center retinal ganglion cells, diffuse light that covers both center and surround produces little or no change in firing. (After Kuffler, 1953.)

Sizes of Receptive Fields

Neighboring ganglion cells collect information from very similar, but not quite identical, areas of the retina. Even a small (0.1 mm) spot of light on the retina covers the receptive fields of many ganglion cells. Some ganglion cells are inhibited, others excited. This characteristic organization, with neighboring groups of receptors projecting onto neighboring ganglion cells in the retina, is retained at all levels in visual pathways. Throughout the visual system, the systematic analysis of the positions of cells and their receptive fields demonstrates the general principle that neurons processing related information are clustered together. In sensory systems, this organization means that the central neurons dealing with a particular area of the surface can communicate with each other over short distances. This avoids the necessity for long lines of communication and simplifies the making of connections.

The size of the receptive field of a ganglion cell depends on its location in the retina. The receptive fields of cells situated in the central areas of the retina have much smaller centers than those at the periphery; receptive fields are smallest in the fovea, where the acuity of vision is highest.[98,99] The central on or off region of such a midget ganglion cell's receptive field can be supplied by a single cone and is accordingly only about 2.5 micrometers (μm) in diameter, subtending 0.5 minutes of arc—smaller than the period at the end of this sentence. Note that receptive fields can be described either as dimensions on the retina or as degrees of arc subtended by the stimulus. In our eyes, 1 millimeter (mm) on the retina corresponds to about 4°. For reference, the image of the full moon has a diameter of 1/8 mm on our retina, corresponding to 0.5° or 30 minutes of arc.

Classification of Ganglion Cells

Superimposed on the general scheme of on- or off-center receptive fields, ganglion cells in the monkey retina can be grouped into two main categories denoted as M and P. The criteria are both anatomical and physiological. The M and P terminology is based on the anatomical projections of these neurons to the lateral geniculate nucleus and from there to the cortex (see Chapter 2). P ganglion cells project to the four dorsal layers of smaller cells in the lateral geniculate nucleus (the parvocellular division—*parvo* is Latin for small). M ganglion cells project to the larger cells in the two ventral layers (the magnocellular division—*magno* is Latin for large). The separate characteristics of neurons in the M and P pathways are maintained at successive levels in the visual system through and beyond the primary visual cortex. In brief, P ganglion cells (also known as midget ganglion cells)

[98] Kier, C. K., Buchsbaum, G., and Sterling, P. 1995. *J. Neurosci.* 15: 7673–7683.

[99] Croner, L. J., and Kaplan, E. 1995. *Vision Res.* 35: 7–24.

have small receptive field centers, high spatial resolution, and most are sensitive to color. P cells provide information about fine detail but require higher contrast stimuli.[100,96] M cells (also known as parasol ganglion cells because of the shapes of their more extensive dendritic arbors, which collect wider inputs) have larger receptive fields than P cells and are more sensitive to small differences in contrast and to movement. M cells fire phasically at higher frequencies and conduct impulses more rapidly along their larger-diameter axons than do P cells along their smaller-diameter axons. In the cat, which has no color vision and has a single type of cone, the classification of ganglion cells is different, with X, Y, and W groups.[101] X, Y, and W systems of the cat are also concerned with discrimination of patterns and movement but show marked differences from the primate M and P classification.[102] In rabbits, ganglion cells show directional sensitivity to movement.[103] The magnocellular/parvocellular classification in the primate visual system provides a convenient and useful framework for studying pathways and properties, and reflects functionally distinct processing.

Synaptic Inputs to Ganglion Cells Responsible for Receptive Field Organization

Inputs to ganglion cells from bipolar and amacrine cells occur in the inner plexiform layer (see Figure 20.15), which has been further subdivided into at least 10 layers according to the terminations of bipolar and amacrine cells on the ganglion cell dendrites. A fuller description of the intricate connections from photoreceptors to ganglion cells is given in lucid and complementary reviews by Sterling[2] and a more recent summary by Wu.[104] As might be expected, depolarizing on-center and hyperpolarizing off-center cone bipolar cells make excitatory glutamatergic synapses with corresponding on- and off-center ganglion cells. Consequently, the change in membrane potential of a bipolar cell causes the ganglion cell to which it is connected to change its membrane potential in the same direction. The axons of on-center bipolar cells terminate closer to the ganglion cell layer than do the off-center bipolar cells, and bipolar cells for the P pathway are separate from those for the M pathway. As previously described, transmission from rod bipolar cells to ganglion cells is indirect, by way of a special type of amacrine cell (known as the AII amacrine cell).[4,85] The connections are so precise that both rods and cones in the same part of the retina supply the same ganglion cell appropriately, but generally by way of different interposed cells. The precision of these connections also is reflected in the arborization pattern of each ganglion cell type (Figure 20.23). On and off ganglion cells each establish specific dendritic patterns so that each functional class optimally covers the retinal surface.[81]

Amacrine Cell Control of Ganglion Cell Responses

Because there are nearly 30 different types of amacrine cells in the mammalian retina, the rod amacrine cells represent only a subset. Although the relative proportions of the various types differ between species, amacrine cells mediate or contribute to a variety of

[100] Kaplan, E., and Shapley, R. M. 1986. *Proc. Natl. Acad. Sci. USA* 83: 2755–2757.

[101] Enroth-Cugell, C., and Robson, J. G. 1966. *J. Physiol.* 187: 517–552.

[102] Benardete, E. A., Kaplan, E., and Knight, B. W. 1992. *Vis. Neurosci.* 8: 483–486.

[103] Levick, W. R. 1967. *J. Physiol.* 188: 285–307.

[104] Wu, S. M. *Invest. Ophthalmol Vis. Sci.* 2010 51: 1263–1274.

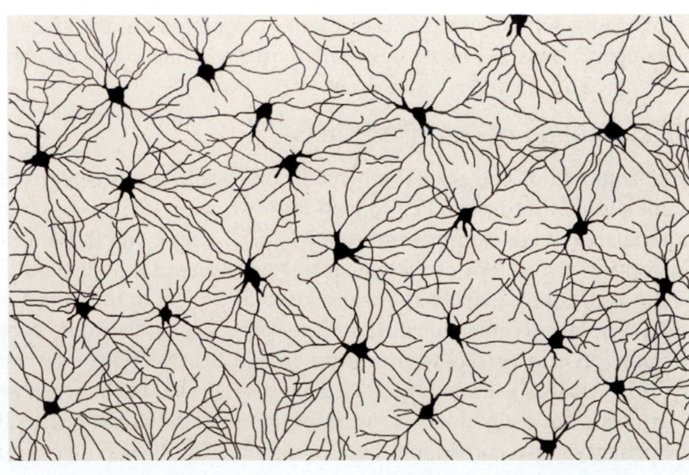

FIGURE 20.23 Dendritic Arborization Pattern of On Ganglion Cells in the cat retina labeled with antibody to neurofilament. The cell bodies form a regular mosaic, and their dendritic trees cover the retinal field without leaving gaps and without extensive overlap. Other anatomically and functionally defined classes of ganglion cells provide equivalent retinal coverage, so that multiple parallel channels process information about light at each location on the retina. (After Wässle, 2004.)

distinctive types of responses and receptive field properties of retinal ganglion cells.[105,106] Many of these effects are what might be called dynamic, or changing, and depend upon the variable range of stimulus intensity, stimulus movement relative to the background scene, or spatial asymmetries in that pattern. For example, together with bipolar cells, amacrine cells show an early stage of processing called **contrast adaptation** or **contrast gain control**, which rapidly affects the contrast sensitivity of the ganglion cell, so that the cell can give a full range of responses to the changes in illumination intensity it experiences within a timeframe of seconds.[107] This works a little like the contrast adjustments that can be made to digital photographs but on a second-by-second basis, over time.

A retinal ganglion cell's receptive field can also change dynamically in a way that maximizes the cell's sensitivity to spatial features that differ from a surrounding pattern.[108] An example is a cell that can improve its ability to detect an object in a patterned field, such as among tall tree trunks in a forest, as compared to a less regular background. Experiments and modeling have shown that amacrine cells, with their lateral projections, and bipolar cells contribute to the effect, which has been termed **predictive coding**.

Although in primates most sensitivity to the direction of movement of stimuli is thought to arise through cortical or tectal connections, there are retinal ganglion cells (as shown particularly in rabbits) that detect movement direction. In rabbits it is the starburst amacrine cells that account for sensitivity to direction of movement,[109,110] and starburst amacrine cells with similar properties are found in primates including humans.[111,112] Eye movements cause the entire visual field to move. As a compensatory mechanism, some retinal ganglion cells are wired to respond preferentially to the movement of a small object relative to overall movement of the visual field. Experiments and related modeling indicate that certain amacrine cells provide the processing and connections for sensing such relative motion of objects in a moving field of vision.[113,114] For the larger movements called **saccades**, which are associated with shifting gaze, visual perception is suppressed in part because of central control that is exerted by oculomotor centers, but in addition, some retinal ganglion cell responses are suppressed,[115] apparently by the circuitry that detects global eye movements. Finally, signaling to retinal ganglion cells by a moving stimulus reaches the edge of each ganglion cell's dendritic field before illuminating the cell body; a change in cell sensitivity as a result of the stimulus causes the population of firing ganglion cells to lead the stimulus in what is called motion anticipation.[116]

What Information Do Ganglion Cells Convey?

It should be clear that ganglion cell signals tell a different story from that of primary sensory receptors. The sizes, shapes, and dynamic properties of their receptive fields are tuned to detect spatial, temporal, or color contrast, and like bipolar cells from which they derive input, half of the ganglion cells depolarize in response to light. Their synaptic inputs tune them to detect stimuli like the edge of an image crossing the opposing regions of their receptive field. As detailed in Chapter 2, from this starting point, higher order neurons in the visual system construct lines, corners, and all the features that underlie visual object recognition. And a very few ganglion cells provide information about the level of ambient illumination.

Throughout the previous discussion, the spike trains from individual neurons have been treated as separate lines from which the brain analyses visual input. Since the firing of ensembles of retinal ganglion cells is crucial for the next levels of analysis into the cortex, some studies have compared activity within groups of ganglion cells and identified additional roles of amacrine cells.[117,118] Experiments made in the salamander retina by Baylor, Meister, and their colleagues suggest that the temporal aspects of firing by ganglion cells also can contribute to retinal analysis.[119,120] Thus, when recordings were made simultaneously from ganglion cells with closely adjacent fields, action potentials occurred synchronously for some visual stimuli. For example, a spot of light that straddled the border of two off-center retinal ganglion cells produced synchronous firing. Electrical synapses between amacrine and ganglion cells or between ganglion cells are thought to contribute to coordinated firing, referred to as concerted signaling of ganglion cells.[121,122]

It is appropriate to close this chapter with a quotation from Sherrington, written long before receptive fields were mapped for single cells. Unlike that of Helmholtz, Sherrington's somewhat opaque style often makes it difficult to read his profoundly

[105] Baccus, S. A. 2007. *Annu. Rev. Physiol.* 69: 271–290.

[106] Grimes, W. N. et al. *Neuron* 65: 873–885.

[107] Baccus, S. A., and Meister, M. 2002. *Neuron* 36: 909–919.

[108] Hosoya, T., Baccus, S. A., and Meister, M. 2005. *Nature* 436: 71–77.

[109] Fried, S. I., Munch, T. A., and Werblin, F. S. 2002. *Nature* 420: 411–414.

[110] Munch, T. A., and Werblin, F. S. 2006. *J. Neurophysiol.* 96: 471–477.

[111] Rodieck, R. W. 1989. *J. Comp. Neurol.* 285: 18–37.

[112] Rodieck, R. W., and Marshak, D. W. 1992. *J. Comp. Neurol.* 321: 46–64.

[113] Olveczky, B. P., Baccus, S. A., and Meister, M. 2007. *Neuron* 56: 689–700.

[114] Baccus, S. A. et al. 2008. *J. Neurosci.* 28: 6807–6817.

[115] Roska, B., and Werblin, F. 2003. *Nat. Neurosci.* 6: 600–608.

[116] Berry, M. J., 2nd, et al. 1999. *Nature* 398: 334–338.

[117] Shlens, J., Rieke, F., and Chichilnisky, E. 2008. *Curr. Opin. Neurobiol.* 18: 396–402.

[118] Maffei, L., and Galli-Resta, L. 1990. *Proc. Natl. Acad. Sci. USA* 87: 2861–2864.

[119] Meister, M. Lagnado, L., and Baylor, D. A. 1995. *Science* 270: 1207–1210.

[120] Meister, M., and Berry, M. J., 2nd. 1999. *Neuron* 22: 435–450.

[121] Brivanlou, I. H., Warland, D. K., and Meister, M. 1998. *Neuron* 20: 527–539.

[122] Schnitzer, M. J., and Meister, M. 2003. *Neuron* 37: 499–511.

original papers and books. Yet, the following paragraph reveals his poetic insight into the physiology of vision:

> The chief wonder of all we have not touched on yet. Wonder of wonders, though familiar even to boredom. So much with us that we forget it all the time. The eye sends, as we saw, into the cell-and-fibre forest of the brain throughout the waking day continual rhythmic streams of tiny, individual evanescent, electrical potentials. This throbbing streaming crowd of electrified shifting points in the sponge-work of the brain bears no obvious semblance in space-pattern, and even in temporal relation resembles but a little remotely the tiny two-dimensional upside-down picture of the outside world which the eye-ball paints on the beginnings of its nerve-fibres to electrical storm. And the electrical storm so set up is one which affects a whole population of brain-cells. Electrical charges having in themselves not the faintest elements of the visual—having, for instance, nothing of "distance," "right-side-upness," no "vertical," nor "horizontal," nor "color," nor "brightness," nor "shadow," nor "roundness," nor "squareness," nor "contour," nor "transparency," nor "opacity," nor "near," nor "far," nor visual anything—yet conjure up all these. A shower of little electrical leaks conjures up for me, when I look, the landscape; the castle on the height, or when I look at him, my friend's face and how distant he is from me they tell me. Taking their word for it, I go forward and my other senses confirm that he is there.[123]

[123] Sherrington, C. S. 1951. *Man on His Nature.* Cambridge University Press, Cambridge, UK.

SUMMARY

- Rod and cone photoreceptors respond to illumination in dim and bright light, respectively.
- The visual pigments are densely packed on rod and cone membranes.
- Transduction occurs in a series of steps involving a G protein and cyclic GMP.
- In darkness, photoreceptors are depolarized and continuously release glutamate.
- Light causes the closing of nucleotide-gated cation channels, hyperpolarization, and reduction of glutamate release.
- H bipolar cells are depolarized by glutamate released from photoreceptors in the dark and so, like the photoreceptor, hyperpolarized by light.
- D bipolar cells are hyperpolarized by glutamate released from photoreceptors in the dark and so are depolarized by light.
- Receptive field refers to the area of the visual field or retina, illumination of which influences the signals of a cell in the visual system.
- Photoreceptors, horizontal cells, and bipolar cells do not produce action potentials.
- Ganglion cells and amacrine cells generate action potentials.
- Bipolar cells and ganglion cells have concentric receptive fields with on or off centers and antagonistic surrounds.
- Large ganglion cells, known as magnocellular or M cells have large receptive fields and respond well to movement.
- Smaller ganglion cells, known as parvocellular or P cells have smaller receptive fields and respond to color and fine detail.
- A small subset of ganglion cells contain pigments and respond to illumination

Suggested Reading

General Reviews

Berson, D. M. 2007. Phototransduction in ganglion-cell photoreceptors. *Pflügers Arch.* 454: 849–855.

Chen, C. K. 2005. The vertebrate phototransduction cascade: amplification and termination mechanisms. *Rev. Physiol. Biochem. Pharmacol.* 154: 101–121.

Dowling, J. E. 1987. *The Retina: An Approachable Part of the Brain.* Harvard University Press, Cambridge, MA.

Lamb, T. D. 2009. Evolution of vertebrate retinal photoreception. *Philos. Trans. R. Soc. Lond. B, Biol. Sci.* 364: 2911–2924.

Lee, B. B., Martin, P. R., and Grunert, U. 2010. Retinal connectivity and primate vision. *Prog. Retin. Eye Res.* 29: 622–639.

Schiller, P. H. 2010. Parallel information processing channels created in the retina. *Proc. Natl. Acad. Sci. USA* 107: 17087–17094.

Sterling, P., and Demb, J. B. 2003. Retina. In G. M. Shepherd (ed.), *Synaptic Organization of the Brain*. Oxford University Press, New York.

Wässle, H. 2004. Parallel processing in the mammalian retina. *Nat. Rev. Neurosci.* 5: 1–11.

Yau, K. W., and Hardie, R. C. 2009. Phototransduction: motifs and variations. *Cell* 139: 246–264.

Original Papers

Baylor, D. A., Lamb, T. D., and Yau, K. W. 1979. The membrane current of single rod outer segments. *J. Physiol.* 288: 589–611.

Berson, D. M., Dunn, F. A., and Takao, M. 2002. Phototransduction by retinal ganglion cells that set the circadian clock. *Science* 295: 1070–1073.

Boycott, B. B., and Dowling, J. E. 1969. Organization of primate retina: Light microscopy. *Philos. Trans. R. Soc. Lond. B, Biol. Sci.* 255: 109–184.

Chen, J., Makino, C. L., Peachey, N. S., Baylor, D. A., and Simon, M. I. 1995. Mechanisms of rhodopsin inactivation in vivo as revealed by a COOH-terminal truncation mutant. *Science* 267: 374–377.

Croner, L. J., and Kaplan, E. 1995. Receptive fields of P and M ganglion cells across the primate retina. *Vision Res.* 35: 7–24.

Hattar, S., Liao, H. W., Takao, M., Berson, D. M., and Yau, K-W. 2002. Melanopsin-containing retinal ganglion cells: architecture, projections, and intrinsic photosensitivity. *Science* 295: 1065–1070.

Heidelberger, R., Thoreson, W. B., and Witkovsky, P. 2005. Synaptic transmission at retinal ribbon synapses. *Prog. Retin Eye Res.* 24: 682–720.

Kaneko, A. 1970. Physiological and morphological identification of horizontal, bipolar and amacrine cells in goldfish retina. *J. Physiol.* 207: 623–633.

Kaneko, A., Delavilla, P., Kurahashi, T., and Sasaki, T. 1994. Role of L-glutamate for formation of on-responses and off-responses in the retina. *Biomed. Res.* 15(Suppl. 1): 41–45.

Kuffler, S. W. 1953. Discharge patterns and functional organization of the mammalian retina. *J. Neurophysiol.* 16: 37–68.

Meister, M., Lagnado, L., and Baylor, D. A. 1995. Concerted signaling by retinal ganglion cells. *Science* 270: 1207–1210.

Schnapf, J. L., Kraft, T. W., Nunn, B. J., and Baylor, D. A. 1988. Spectral sensitivity of primate photoreceptors. *Vis. Neurosci.* 1: 255–261.

Trong, P. K., and Rieke, F. 2008. Origin of correlated activity between parasol retinal ganglion cells. *Nat. Neurosci.* 11: 1343–1351.

Zenisek, D., Horst, N. K., Merrifield, C., Sterling, P., and Matthews, G. 2004. Visualizing synaptic ribbons in the living cell. *J. Neurosci.* 24: 9752–9759.

CHAPTER 21
Touch, Pain, and Texture Sensation

The entire body surface is covered with tactile receptors. In this way, the somatosensory system differs from sensory systems whose receptors are clustered within organs, such as the eye, nose, or ear. The sensory system of the skin is essential for identifying shapes and textures; for recognizing, grasping, and manipulating objects; for registering temperature; for directing visual attention; and for avoiding dangerous objects signaled by the pain they evoke. In this chapter, we describe the processing that leads from contact of an object with the skin to recognition of the physical properties of that object. The functional organization of the somatosensory system is addressed both in rats and mice, where the whiskers are particularly important, and in primates, where the fingertips are particularly important.

Most somatosensory nerve fibers terminate in tiny accessory structures distributed in different positions within the skin—hair follicles, or Meissner, Merkel, Pacinian, or Ruffini's corpuscles. But, there is another class of fibers that ends with no accessory structure—the so-called free nerve endings. Each type of fiber is activated by a specific stimulus and leads to a characteristic sensation. The terminations in accessory structures are associated with touch, pressure, stretch, or vibration; free nerve endings are associated with hot or cold, painful or itchy sensations. The connection between receptor activity and sensation was first posited from studies in anesthetized animals, but it has been confirmed by recording from single nerve fibers in the arm of awake, human subjects, who describe the feeling evoked by electrical stimulation of the fiber.

Signals travel to the spinal cord, where they are sorted out into distinct ascending pathways according to the functional category of the receptor. All pathways reach the cerebral cortex after multiple relays. Though stimuli that cause damage lead to activation of a specific pathway, the subjective experience of pain is more complex than the mere perceptual consequence of such activity. For this reason, the psychological state of a person, including the expectation of reduced pain (placebo) or of intense pain (nocebo) truly alters the degree of suffering. The modulation occurs through endogenous opioid mechanisms.

At all levels of the somatosensory pathways, neurons are arranged into topographic maps where the spatial relationships of the body are conserved: adjacent stimuli activate adjacent neurons. These maps are grossly distorted because larger territories in the central nervous system (CNS) are dedicated to smaller, more densely innervated skin areas. The representation of the whiskers in mice and rats is one notable example. The whiskers are represented in layer IV of cortex by an oversized grid of columns, called barrels. All the neurons in one barrel respond to movement of one whisker on the opposite side of the face. In humans and other primates, the fingertips are densely innervated, providing the tactile input necessary for object manipulation. The corresponding magnification of the fingertips (and lips) in somatosensory cortex gives rise to distorted cortical representations. Cortical map topography can be altered by manipulations during development, but also throughout life by training in sensory discrimination tasks.

A detailed account of sensory processing, from the skin receptor to the somatosensory cortex, has been established to explain the perception of texture. To identify textures, rats palpate an object with brief touches of their whiskers. The way that the whiskers skip along the surface varies according to the surface's roughness. Whisker motion is converted to spike trains by receptor neurons—the coarser the surface, the greater the rate of neuronal firing. The signals ascend to the barrel cortex, where the neuronal firing rate determines the animal's judgment of texture. In primates, there are two channels for sensing texture. The perception of coarse textural features relies on a spatial mechanism—at any given instant, the degree of roughness can be decoded by a "snapshot" of the contrast in firing among slowly adapting neurons with nearby receptive fields. The perception of finely textured surfaces is determined by the efficacy with which finger motion along them excites the nerve terminations in the Pacinian corpuscles. The vibration mechanism in primates has much in common with the mechanism by which the whiskers extract surface properties.

From Receptors to Cortex

Receptors in the Skin

Touch receptors are mechanoreceptors: they transduce mechanical energy applied to the skin into the language of the nervous system; that is, sequences of action potentials. The key to understanding this first stage of the somatic sensation is to know that each type of touch receptor is sensitive to particular features of mechanical energy and insensitive to all other features. This is because morphological specializations—usually the structures encapsulating the nerve fiber termination—make that ending respond selectively to one specific type of stimulus. The capsular structures are deformed by mechanical stimuli, and thereby transmit skin deformation to the nerve terminal (Figure 21.1). A single sensory fiber arriving from the spinal cord can branch to innervate from a few to tens of separate capsules in the skin. All capsules innervated by one fiber are of the same type.

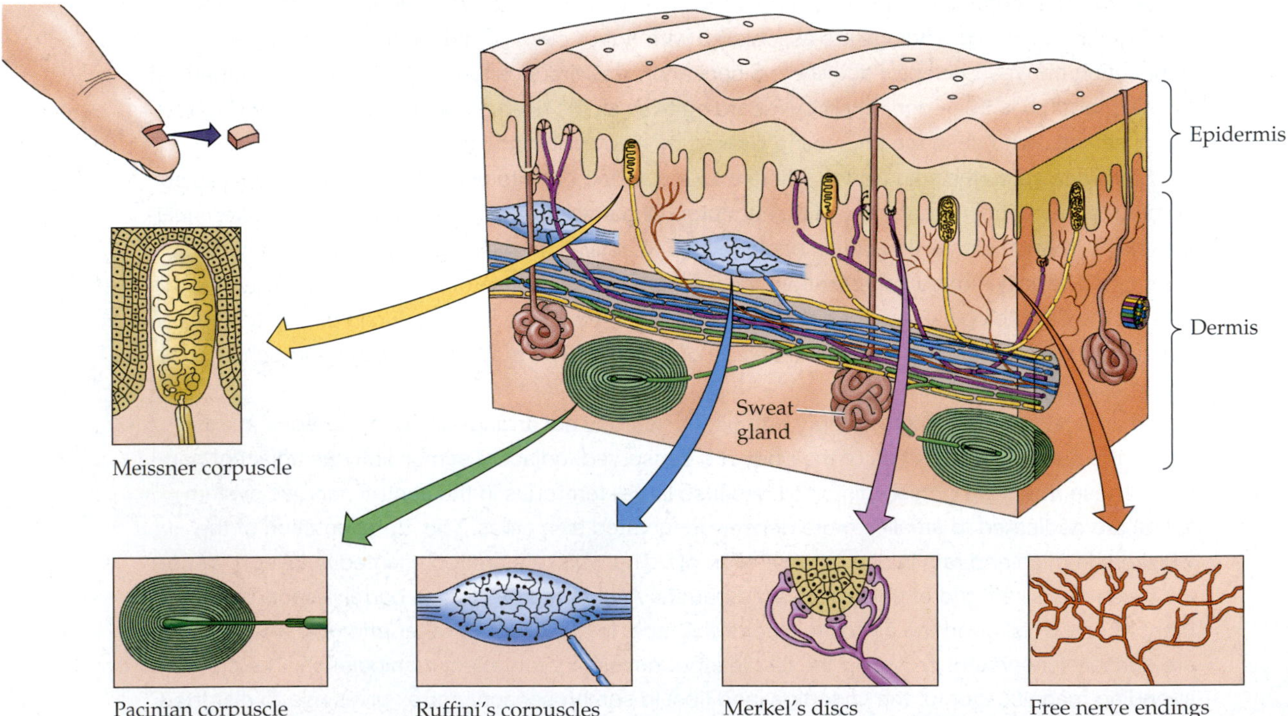

FIGURE 21.1 Section of Skin Showing the Morphology and Position of Tactile Receptors. This figure refers to glabrous skin (meaning hairless, such as the skin of the palms) in primates. The Meissner, Pacinian, and Ruffini's corpuscles as well as Merkel's disc are all specialized structures that filter the mechanical energy that excites the nerve termination. The free nerve endings are—as the name implies—uncovered, which exposes them to substances released by the skin tissues.

The **Pacinian corpuscle** is an intriguing example of how the capsular structure selects the features that reach the nerve termination through the surrounding tissue, and it is described in detail in Chapter 19. **Merkel's disc** is another structure that filters mechanical energy. It is a small epithelial cell found under the fingerprint ridges. It transmits compressive strain to the sensory nerve ending. The responses are sustained during the period of pressure, with firing frequency proportional to the pressure applied to the skin.[1] Another structure, **Ruffini's corpuscle**, is activated by pressure or compression.[2] These endings are located deeper in the skin and have the greatest density at the base of the fingernails, in the tissue overlying joints and ligaments, as well as in the palm.[3] Both the Ruffini complex (the corpuscle together with the nerve termination) and the Merkel complex (the disc together with the nerve termination) are slowly adapting structures and transmit information about how firmly your hand is grasping an object or how hard your foot is pressing on the floor.

Meissner's corpuscle in the skin of the lips, palm, fingers, and sole of the foot confers upon the nerve terminals extraordinary mechanical sensitivity to initial contact and to motion. The receptor complex is a rapidly adapting structure and is strongly activated by low-frequency vibrations, up to 50 Hz.[4] Application of local anesthesia to the superficial skin layers, which affects nerve terminations in Meissner's corpuscles but not in Pacinian corpuscles, diminishes the sense of low-frequency vibration but not high-frequency vibration.[5]

The receptive fields—the area of skin in which stimulation evokes a response—of Meissner's and Merkel's receptors are small, a few millimeters (mm) in diameter, while those of Pacinian and Ruffini receptors are several centimeters (cm) in diameter. Most stimuli activate many kinds of receptors at once, so that impulses from different classes signal different aspects of the stimulus.[6] For example, if you grasp a vibrating cell phone in your hand, pressure and vibration sensory channels will be activated in parallel.

Hair follicles are innervated by a network of nerve terminals with different morphologies and different positions along the hair shaft and the wall of the follicle. A small hair follicle, like that present in the mammalian skin, is illustrated in Figure 21.2A, while a large whisker follicle,[7] like that on the snout of a rat or mouse, is illustrated in Figure 21.2B. The whisker is a special hair that will be discussed later in the chapter. The nerve endings are excited when the hair is bent, vibrated, or pulled. It is not yet known how the selectivity of individual endings varies according to morphology and position within the follicle.

[1] Iggo, A., and Muir, A. R. 1969. *J. Physiol.* 200: 763–796.

[2] Bolanowski, S. J., Jr. et al. 1988. *J. Acoust. Soc. Am.* 84: 1680–1694.

[3] Pare, M., Smith, A. M., and Rice, F. L. 2002. *J. Comp. Neurol.* 445: 347–359.

[4] LaMotte, R. H., Mountcastle, V. B. 1975. *J. Neurophysiol.* 38: 539–559.

[5] Mahns, D. A. et al. 2006. *J. Neurophysiol.* 95: 1442–1450.

[6] Birder, L. A., and Perl, E. R. 1994. *J. Clin. Neurophysiol.* 11: 534–552.

[7] Rice, F. L., Mance, A., and Munger, B. L. 1986. *J. Comp. Neurol.* 252: 154–174.

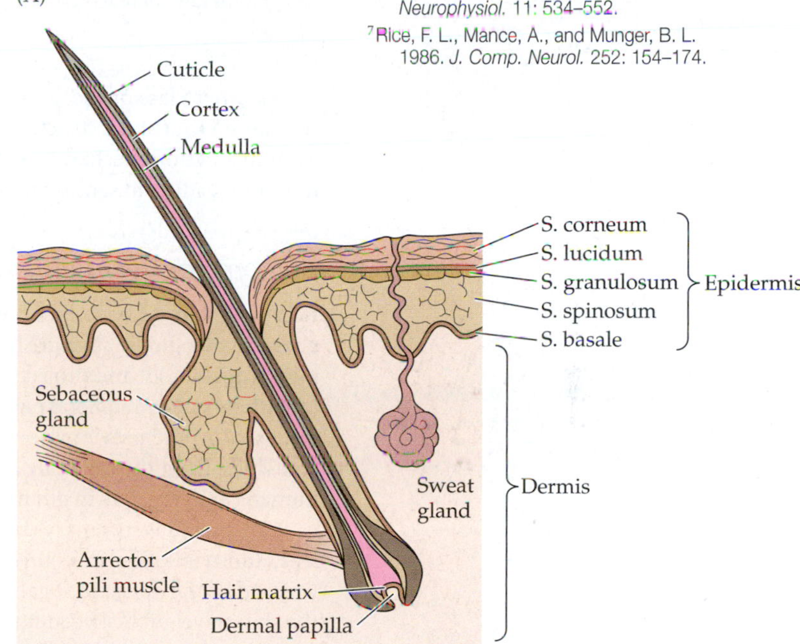

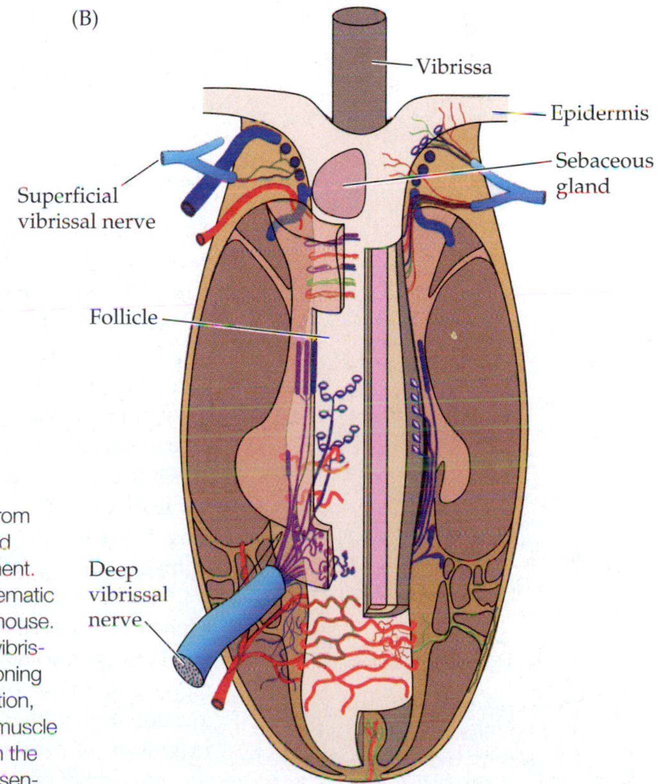

FIGURE 21.2 Hair Follicles. (A) Schematic view of a hair shaft and follicle from the skin of a primate. Nerve terminals are distributed throughout the follicle and wrapped around the base of the hair, generating impulses for any hair movement. The muscle contracts to erect the hair in response to cold or arousal. (B) Schematic view of a special hair follicle, that of the large whisker of the snout of a rat or mouse. Nerve terminations enter through the superficial vibrissal nerve and the deep vibrissal nerve to occupy many different locations within the follicle, and their positioning is likely to be closely related to type of hair movement that excites them (vibration, bending, pulling, etc.). As in the case of the unspecialized skin hair in (A), the muscle wrapped around the follicle (not shown) contracts to erect the hair; however in the case of the whisker, the action is a rhythmic sweeping motion used to collect sensory signals. (B courtesy of Frank Rice.)

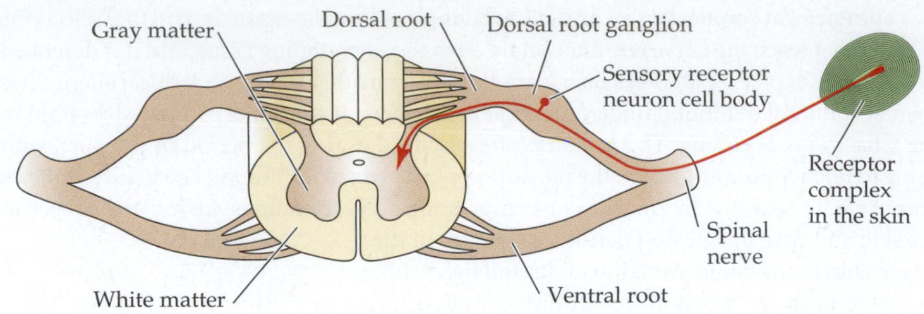

FIGURE 21.3 Morphology of Sensory Receptor Neurons. The cell body lies in a dorsal root ganglion situated in the aperture between the vertebrae of the spine. A central branch joins a dorsal root and projects into the spinal cord; a peripheral branch joins other fibers in a peripheral nerve and then terminates as a specialized cutaneous receptor complex; in this example, a Pacinian corpuscle.

Another class of receptor is a **free nerve ending**—one that is not connected to any specialized capsule or structure (see Figure 21.1). Through a variety of specialized chemical transduction mechanisms, described in Chapter 19, the free nerve endings generate impulses leading to sensation of pain, temperature, itch, or tickle.[6,8,9]

Anatomy of Receptor Neurons

Receptor neurons have three components: (1) a cell body that lies in a dorsal root ganglion situated in the aperture between the vertebrae of the spine, (2) a central branch that joins a dorsal root and projects into the spinal cord, and (3) a peripheral branch that joins other fibers in a peripheral nerve and then terminates as the specialized receptor complex discussed in the previous section (Figure 21.3). The longest of these peripheral axons, stretching from the cell body to receptor endings in the toe, exceeds 1 meter in length in humans and 3 meters in giraffes.

The previous summary applies to the innervation of the skin of the hands, feet, arms, legs, and trunk. For the skin of the face, the cell bodies are in the trigeminal ganglion (instead of the dorsal root ganglion) situated lateral to the brainstem; fibers projecting to the skin travel in the trigeminal nerve—the fifth cranial nerve—and the central processes from these ganglion cells terminate in the trigeminal nuclei of the brainstem.

Because sensory neurons carry signals over long distances, transmission of information to second-order neurons cannot be accomplished by a graded potential like that of the photoreceptor (see Chapter 19). Instead, action potentials are generated very close to the nerve terminal and they travel past the ganglion cell soma and into the spinal cord or brainstem.

Sensations Evoked by Afferent Signals

An increase in the understanding of touch sensation has come from the method of **microneurography** in human subjects.[10] A fine metal microelectrode is advanced through the skin of the arm and into a nerve, which contains fibers traveling from the hand to the spinal cord. This set-up allows the investigator to record the impulses in conscious subjects while applying a stimulus to the skin and, at the same time, to monitor the subject's sensations (Figure 21.4). Most fibers have little or no spontaneous activity and fire only when the skin is stimulated. Both the receptive field of a single human nerve fiber as well as the type of stimulus that is most effective in evoking an action potential conforms exactly to those previously identified in laboratory animals.[11] A single fiber is sensitive to one or a few distinct skin spots, and their size varies according to receptor submodality. A fiber is sensitive to just one submodality—for instance, light pressure, or low-frequency vibration, or high-frequency vibration.[12]

The advantage over animal neurophysiology lies in the fact that the subject can describe what he or she feels. From this came the discovery that electrical stimulation of a single fiber innervating a mechanoreceptor can evoke an elementary sensation that is experienced as tapping, vibration, or pressure. The sensation feels as if it is occurring at the receptive field location of that fiber. Varying the frequency of impulse generation causes a change in the magnitude of the sensation. For example, increasing the stimulation frequency increases the intensity of perceived pressure.

[8] Ikoma, A. et al. 2003. *Arch. Dermatol.* 139: 1475–1478.

[9] Schmelz, M. et al. 2003. *J. Neurophysiol.* 89: 2441–2448.

[10] Vallbo, A. B., and Hagbarth, K. E. 1968. *Electroencephalogr. Clin. Neurophysiol.* 25: 407.

[11] Jarvilehto, T., Hamalainen, H., and Laurinen, P. 1976. *Exp. Brain Res.* 25: 45–61.

[12] Mano, T., Iwase, S., and Toma, S. 2006. *Clin. Neurophysiol.* 117: 2357–2384.

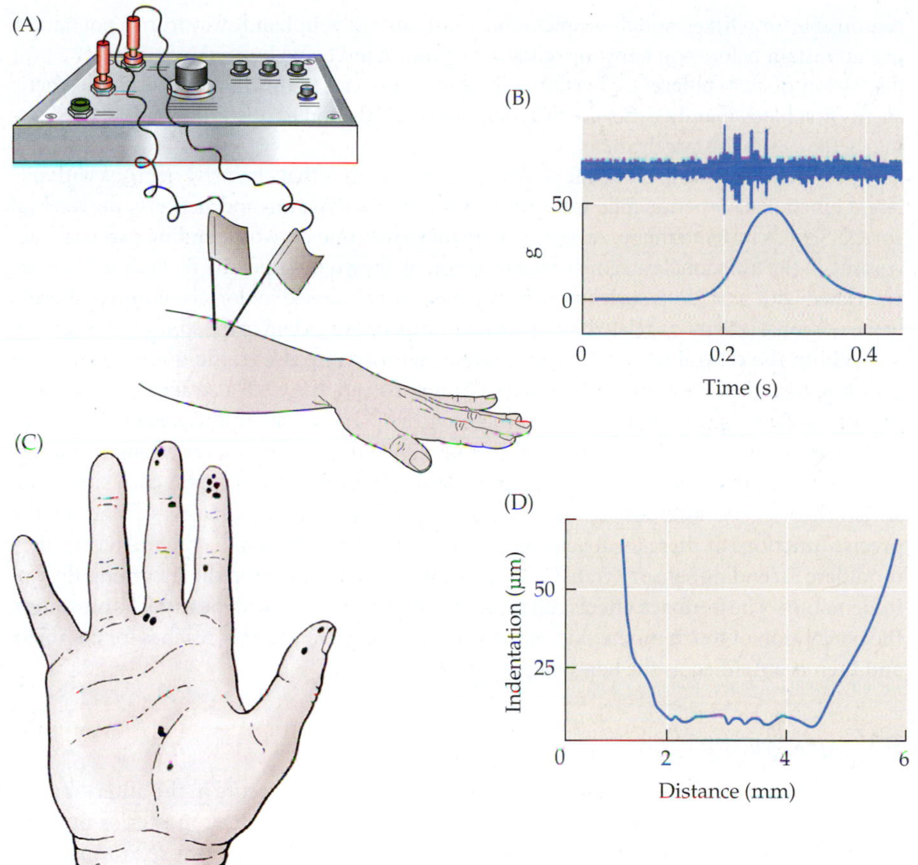

FIGURE 21.4 Microneurography of a Human Nerve. (A) Setup of the experiment. Two microelectrodes penetrate the radial nerve and the potentials are led by wires to electrical amplifiers. Use of two electrodes allows measurement of conduction velocity. (B) Top: responses to light pressure of multiple neurons recorded from a peripheral nerve through the same electrode. The units had overlapping receptive fields at the stimulation point. Bottom: the time course of the skin indentation force in grams (g). (C) Receptive fields of rapidly adapting touch receptors (presumably Meissner's corpuscles) in a human hand. Fields were mapped by recording discharges from single fibers in the median nerve. Each dot represents the receptive field of one fiber. (D) Indentation (in micrometers) required to produce a response in one unit is plotted against location within the receptive field. The region of maximum sensitivity is about 3 mm in diameter, within which indentations of only a few micrometers are sufficient to produce a response. Points of maximum sensitivity within the region are believed to correspond to the position of individual endings of branches of the same afferent. (A after McGlone et al., 2007; B after Wessberg et al., 2003; C and D after Johanson and Vallbo, 1983.)

In order for the subject to feel a sensation of pain, impulses are required from many fibers arising from nociceptive free nerve endings, an integrative process known as spatiotemporal summation. Above threshold, the intensity of pain is correlated with the number of action potentials.[13]

Microneurography studies confirm the stimulus-specificity of afferent fibers. In addition, they confirm that the activity of a single primary afferent fiber can produce a conscious sensation.[14,15] The threshold of single fibers (i.e., the strength of skin indentation required to excite impulses) matches the subjects' threshold for detecting the same stimulus.[15] We can deduce that transmission from skin to cerebral cortex is highly reliable, as in vision, and that impulses are delivered from the periphery into a low-noise central processing network: if the network were active in the absence of stimuli, the arrival of impulses from one fiber could not be readily detected. Of course, single-fiber activation is not a normal mode of operation—contact with objects produces activity patterns that are complex with respect to time and space. But the signals carried by single fibers constitute the elemental building blocks for meaningful somatic sensations.

Ascending Pathways

Receptor type, and therefore submodality, is correlated with the diameter of the afferent fiber and its myelination. As discussed in Chapter 8, diameter and myelination determine conduction velocity. Fibers that carry signals conveying information about light (i.e., non-painful) touch are myelinated and have conduction velocities up to about 60 meters/second (m/s); this class of fiber is called A-β. Fibers that carry signals about painful stimuli and temperature have conduction velocities from 2–10 m/s (A-δ fibers; myelinated) down to 0.2–2 m/s (C fibers; unmyelinated). Microneurography shows that activity in A-δ fibers is associated with a rapid stinging or pricking sensation called first pain, whereas activity in C fibers is associated with a burning or aching sensation called second pain. The faster pain fibers (A-δ) are also

[13] Torebjork, E. 1985. *Philos. Trans. R. Soc. Lond., B, Biol. Sci.* 308: 227–234.

[14] Johansson, R. S., and Vallbo, A. B. 1979. *J. Physiol.* 297: 405–422.

[15] Ochoa, J., and Torebjork, E. 1983. *J. Physiol.* 342: 633–654.

responsible for reflexes, and this explains how you can jerk your hand away from a pot handle just an instant before you have any sensation of your hand being burned (see Chapter 24 for discussion of such reflexes). Thermoreceptors are innervated by C fibers. Local anesthetic agents first block mainly C-fiber activity, suppressing pain and temperature sense while leaving tactile sensibility largely intact.

A fundamental principle of the somatosensory system is that chains of neurons with different functional properties follow distinct pathways to the thalamus and, hence, to the cerebral cortex. Signals travel to the spinal cord, where they enter one of two ascending pathways, according to the functional category of the receptor. As shown in Appendix B, neuronal activity correlated with sensations such as touch and pressure are carried in dorsal column pathways, while neuronal activity correlated with pain and temperature ascend in the spinothalamic tract.

Within the central nervous system, somatosensory impulse traffic does not move in only one direction, from nerve terminal to brain; instead, the CNS also sends messages in the reverse direction. Thus, ascending somatosensory pathways are influenced by *descending* pathways—projections from each level cascade down to preceding levels. Somatosensory cortical areas project densely to the same thalamic, brainstem, and spinal cord sites from which they receive afferent signals. As in the visual system (see Chapters 2 and 3), the precise functions of these feedback connections are not yet known. In general terms, they modulate ascending sensory transmission, facilitating or inhibiting the incoming flow of information. The feedback effects can be highly focused in time and location. For instance, the perception of touch on the skin of the hand is reduced as the arm reaches for an object and then is amplified as the hand palpates its target.[16]

Somatosensory Cortex

The first stage of cortical processing of somatosensory signals occurs in the anterior parietal region, that is, the area targeted by axons from the thalamic ventroposterior nucleus. In primates, input from low-threshold cutaneous receptors—light pressure and vibration—projects to a medial-lateral strip along the crown of the postcentral gyrus, area 3b in Brodmann's terminology (see Appendix C). A second target of cutaneous inputs is area 1, a strip immediately posterior to area 3b. Input from proprioceptors projects to a medial-lateral strip at the anterior border of the somatosensory field (Brodmann's area 3a, just behind motor cortex) and a strip at the posterior border (area 2). These areas together are known as **primary somatosensory cortex (SI)**, though Kaas[17] argues that only area 3b should be considered primary somatosensory cortex. Their arrangement is illustrated in the Figure 21.5. There is disagreement as to whether the four areas carry out hierarchical processing or parallel processing. In the first view, areas 1 and 2 integrate inputs from areas 3a and 3b to generate complex, higher-order response properties, while in the second view each of the four areas operates, independently, on its own thalamic input.

Lesions to SI in humans result in a loss of touch sensation on the contralateral side of the body. Although some sensations of touch are eventually restored, the ability to discriminate shape and texture is disrupted permanently.[18–20] Spatially restricted lesions can be made in animals to provide clearer insights into the sensory functions of specific cytoarchitectonic fields. In monkeys, lesions confined to area 3b produce the severest

[16] Chapman, C. E., and Beauchamp, E. 2006. *J. Neurophysiol.* 96: 1664–1675.

[17] Kaas, J. H. 1983. *Physiol. Rev.* 63: 206–231.

[18] Weinstein, S. et al. 1958. *J. Comp. Physiol. Psychol.* 51: 269–275.

[19] Schwartzman, R. J., and Semmes, J. 1971. *Exp. Neurol.* 33: 147–158.

[20] Bohlhalter, S., Fretz, C., and Weder, B. 2002. *Brain* 125: 2537–2548.

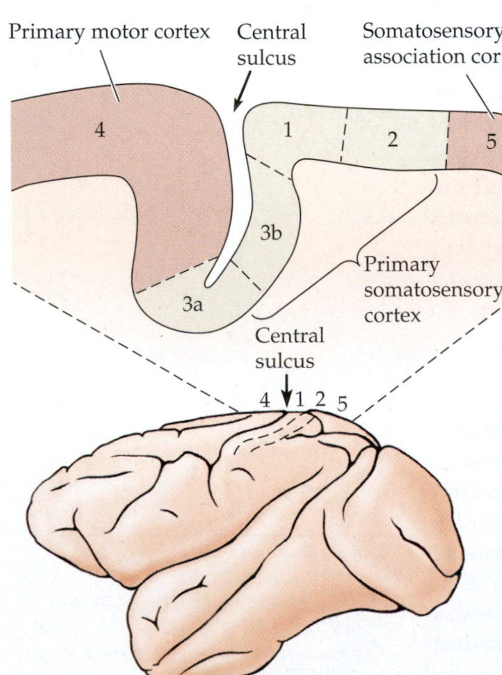

FIGURE 21.5 Organization of Primate Somatosensory Cortex. The primary somatosensory region is located in the anterior parietal cortex, just behind the central sulcus. The figure depicts the brain of a monkey. The cytoarchitectonic areas (seen in the sagittal section, upper plot) are 3a, 3b, 1, and 2. Area 3b receives dense input from the core of the thalamic ventral posterior nucleus, receiving signals related to light touch. The other somatosensory areas receive less dense thalamic input, arising from shell areas surrounding the ventral posterior nucleus core. Area 1, like area 3b, receives signals related to light touch, while areas 3a and 2 receive signals related to deeper tissues and joints. Areas 4 and 5 are regions that receive dense projections from the adjacent somatosensory cortex and use the information for different functions. Area 4 is part of motor cortex (see Chapter 24), while area 5 integrates somatosensory information with that of other sensory modalities to create a percept of the body in relation to surrounding space.

sensory deficits, spanning multiple tactile functions. Ablation of area 1 results in deficits in texture discrimination, whereas lesions in area 2 impair the ability to discriminate the size and shape of objects.[21,22]

Pain Perception and its Modulation

The various forebrain regions that play a part in processing nociceptive signals have been identified, but their specific roles have been debated since the early 1900s. Most investigators distinguish between lateral and medial networks. The **lateral system**, whose main components are the lateral regions of thalamus and primary (SI) and **secondary somatosensory cortices (SII)**, is believed to be involved in processing the sensory discriminative aspects of pain. This system specifies where on the body the sensation arises and, according to the nature of the sensation, what might be the cause. A nociceptive pathway to SI has been traced and some of its neurons do respond to noxious stimuli.[23] Also, SI activation increases in relation to the subjective strength of a painful stimulus, according to functional magnetic resonance imaging (fMRI) studies.[24–26] However, neurosurgical ablation of SI has effects on the perception of pain that vary widely across subjects.

The **medial system**, whose main components are the medial regions of the thalamus, and the insular, anterior cingulate, and prefrontal cortices, is believed to be involved in processing the affective, motivational aspects of pain.[27] Ablation of these regions of cortex profoundly reduces the subjective unpleasantness of noxious stimuli. Moreover, the strength of activation of these areas, observed by brain imaging, is correlated with the judgment of how distressing the pain sensation is.[26]

Although "pain pathways" from the skin have been identified,[28] it is incorrect to suppose that a specified quantity of nociceptor impulses travels in an uninterrupted stream to the cerebral cortex to produce a proportional quantity of pain. Rather, a given level of nociceptor activation can produce dramatically different subjective experiences under different conditions. A soldier charging into battle may not even notice injuries that would, under other circumstances, produce exceptional pain. At the opposite end of the spectrum, a chronic pain syndrome can exist even without abnormal activity in skin nociceptors.[29] It is the central processing of nociceptor signals, not their mere presence, that gives rise to the sensory and affective experience. Understanding pain thus requires understanding how the stream of impulses arriving from the nociceptors is processed by the rest of the nervous system. And, as pointed out by Melzack,[30] the subjective experience of pain—more than any other experience originating in a sensory organ—is modulated by social, cultural, and personal factors.

The discovery in the 1970s of endogenous opioid receptors in the brain[31–33] implied the existence of brain systems whose normal function would be to modulate the processing of nociceptive signals. The endogenous system acts by the release of peptides (known as **endorphins**) in the spinal cord and in other regions of the central nervous system, where they inhibit neuronal activity in the pain pathway. The name *endorphin* comes from the fact that they are **endogenous** and that their chemical structure resembles that of the opiate painkiller **morphine**. Endorphins act through three major receptors: μ, δ, and κ. The receptor types can be distinguished according to the chemical substance for which they have greatest affinity; for instance, the most effective agonist for the μ-opioid receptors is the opium alkaloid, morphine.

Can we control the operation of the endogenous pain modulation system and thereby suppress chronic pain that has outlived its usefulness as an alarm bell? Can we reduce dependence on addictive painkillers? Recent research shows that **placebo** treatment—the suggestion of pain reduction without medication or functional treatment—can effectively reduce even a postoperative pain experience.[34] Placebo functions by activating opioid receptors[35,36] and placebo-evoked pain suppression is eliminated by blocking μ-opioid receptors.[37] Functional magnetic resonance imaging experiments show that the effectiveness of suggestion in reducing subjective pain is positively correlated with the degree of suppression of activity in anterior cingulate and insular cortex.[38,39] Thus, placebo acts directly on the networks that normally generate the unpleasant experience of pain. A fascinating finding is that the expectation of a *negative* outcome can cause an intensification of pain sensations and can cause even a normally innocuous stimulus to be felt as painful.[40] This **nocebo** effect occurs by driving the endogenous opioidergic system in the opposite direction of the placebo effect.

[21] Randolph, M., and Semmes, J. 1974. *Brain Res*. 70: 55–70.

[22] Semmes, J., Porter, L., and Randolph, M. C. 1974. *Cortex* 10: 55–68.

[23] Kenshalo, D. R. et al. 2000. *J. Neurophysiol.* 84: 719–729.

[24] Davis, K. D. 2000. *Neurol. Res*. 22: 313–317.

[25] Apkarian, A. V. et al. 2005. *Eur. J. Pain* 9: 463–484.

[26] Christmann, C. et al. 2007. *Neuroimage* 34: 1428–1437.

[27] Wiech, K., Preissl, H., and Birbaumer, N. 2001. *Anaesthesist* 50: 2–12.

[28] Ochoa, J., and Torebjork, E. 1989. *J. Physiol.* 415: 583–599.

[29] Namer, B., and Handwerker, H. O. 2009. *Exp. Brain Res.* 196: 163–172.

[30] Melzack, R. 1973. *The Puzzle of Pain*. Harmondsworth: Penguin Books.

[31] Pert, C. B., and Snyder, S. H. 1973. *Science* 179: 1011–1014.

[32] Terenius, L. 1973. *Acta Pharmacol. Toxicol.* (*Copenh.*) 32: 317–320.

[33] Simon, E. J., Hiller, J. M., and Edelman, I. 1973. *Proc. Natl. Acad. Sci. USA* 70: 1947–1949.

[34] Pollo, A. et al. 2001. *Pain* 93: 77–84.

[35] Zubieta, J. K. et al. 2005. *J. Neurosci.* 25: 7754–7762.

[36] Zubieta, J. K., and Stohler, C. S. 2009. *Ann. N Y Acad. Sci.* 1156: 198–210.

[37] Amanzio, M., and Benedetti, F. 1999. *J. Neurosci.* 19: 484–494.

[38] Wager, T. D. et al. 2004. *Science* 303: 1162–1167.

[39] Petrovic, P. et al. 2005. *Neuron* 46: 957–969.

[40] Benedetti, F. et al. 2006. *J. Neurosci.* 26: 12014–12022.

Somatosensory System Organization and Texture Sensation in Rats and Mice

The Whiskers of Mice and Rats

How do impulses from skin receptors eventually lead to the sensation (i.e., the registration of the elemental properties) and perception (i.e., recognition of the identity and meaning of the sensation) of the things that we contact? As a first example, we use the sensory pathways originating in the whiskers of mice and rats to illustrate the main principles of the organization and function of touch. We then show how these principles apply to sensation and perception in primates.

Mice and rats are active in dark environments and have poor vision; their survival depends on the sense of touch. Through the whiskers, the brain builds up percepts of objects: their position, size, shape, and texture.[41] The heart of the tactile system is an array of some 35 long whiskers on each side of the face, thick hairs that the animal flicks forward and backward (there are also shorter, and less mobile hairs packed densely around the nose and lips that the animal can use to collect additional information).[42] Contact with an object activates mechanoreceptor terminations located in the whisker follicles, giving rise to neuronal signals (see Figure 21.2B). The first study to utilize the whiskers as an object of research was carried out by S. Vincent in 1912 for her Ph.D thesis at the University of Chicago (using rats captured in the city!).[43] She built an elevated labyrinth and found that the time required to run from the start to the end of the maze and the number of errors (for example, stepping into dead-end arms) increased dramatically when whiskers were clipped. The rats were slowed by the need to feel the edge of the platforms with their feet rather than with their whiskers.

Magnification Factor

The behavioral importance of the rodents' whiskers goes hand in hand with the richness of the skin innervation. The whisker pad is the most densely supplied skin area of the body studied so far—about 200 trigeminal ganglion neurons innervate each whisker follicle.[44] In Figure 21.6A, it is evident that the primary somatosensory cortex occupies a greater territory

[41] Diamond, M. E. et al. 2008. *Nat. Rev. Neurosci.* 9: 601–612.

[42] Brecht, M., Preilowski, B., and Merzenich, M. M. 1997. *Behav. Brain. Res.* 84: 81–97.

[43] Vincent, S. B. 1912. *Behav. Monogr.* 1-82.

[44] Welker, E., and van der Loos, H. 1986. *J. Neurosci.* 6: 3355–3373.

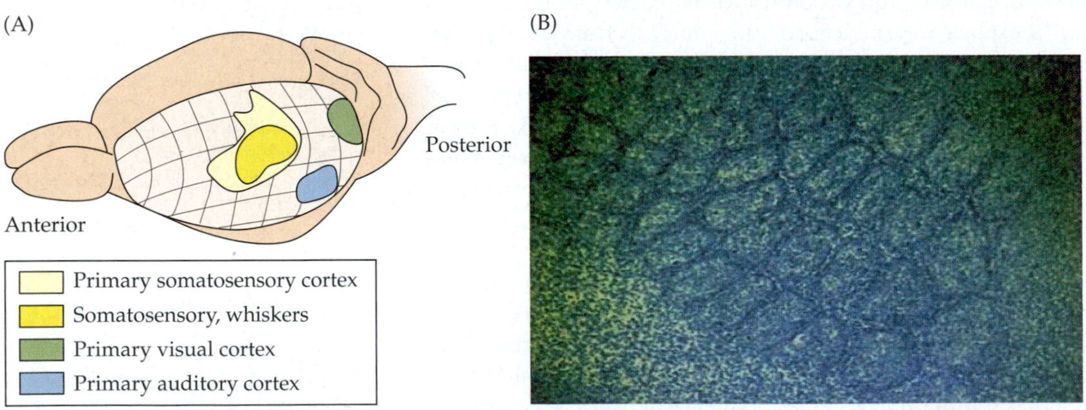

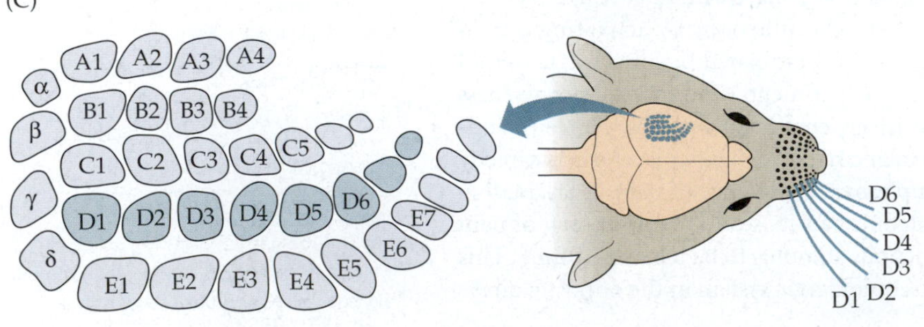

FIGURE 21.6 Cortical Representation of the Whiskers in Rats and Mice. (A) Relative positions and sizes of primary sensory cortical fields. (B) Barrels in the mouse somatosensory cortex visible in a tangential section through layer IV. The dark rings are the cell-dense walls of the barrels. (C) Arrangement of the barrels in the left hemisphere of a rat, with each barrel labeled by its corresponding whisker. In the sketch on the right, whiskers D1 to D6, which project to barrels D1 to D6 are shown full-length. The follicles of other whiskers are depicted as dots. (B after Blakemore, 1977.)

than do the primary sensory fields for vision and audition. And within primary somatosensory cortex, the whisker field is disproportionately large. Although this can be interpreted as a greater magnification of the whiskers' cortical representation with respect to that of other body parts (i.e., in terms of square millimeters of skin projecting to square millimeters of cortical surface), it can also be interpreted as conservation, across body areas, of the relationship between receptor density and cortical territory. In other words, the number of tactile receptors projecting to each square millimeter of cortical surface is held constant. As a result, the richly innervated whisker follicles gain a large territory.

Topographic Map of the Whiskers and Columnar Organization

To visualize the organization of the mouse somatosensory cortex, Woolsey and van der Loos[45] separated the cerebral cortex from the underlying structures. Then, they flattened the curved cortex between two slides and sectioned the slab in a tangential plane that yielded a large horizontal expanse of layer IV in a single section. Once the tissue was stained, distinct clusters of densely packed neurons were visible. The grid-like arrangement of the 35 clusters resembled wine casks, calling out for the name **barrels** (Figure 21.6B). The insight of Woolsey and van der Loos was that the spatial arrangement of the barrels replicates exactly the spatial arrangement of the whiskers on the snout. The authors concluded that the barrel field is a map of the whiskers on the contralateral snout. Every whisker on the snout was given a label (e.g., D3) and could be readily matched with one cortical barrel (again, D3). Within a few years, the prediction of a connection from each whisker to its own cortical barrel was extended from mice to rats and confirmed by many physiological studies.[46–49] Thus, the receptive fields of the neurons in barrel D3 are centered on the so-called *principal* whisker, D3. Whiskers surrounding the principal whisker may also excite cells to a lesser degree. Neurons in the layers above and below layer IV, which receive their main sensory input from layer IV, also receive a strong input from the principal whisker.[46] So, the neuronal population extending through all layers forms a cortical column associated with a single whisker, a functional unit anchored by the layer IV barrel.

Because two whiskers that are adjacent to each other on the animal's face are represented in adjacent cortical barrels, the barrel field constitutes a **topographic map** (Figure 21.6C). Beyond the special case of whiskers, it is common to refer to a brain representation as a "map" whenever the spatial relationship among sensory receptors is conserved in the central representation of the sense organ. A historical note on the discovery of cortical somatosensory maps is given in Box 21.1.

The map of cortical columns serves to organize the storage of sensory experiences. One demonstration is provided by an experiment in which, to begin, all whiskers except one were trimmed from the snout of the rats (Figure 21.7A). Then, the animals were trained, in the dark, to perch on one platform, extend the intact whisker across a gap to touch and, thus, localize a second platform, and jump across the gap to reach a reward. Next, the *trained* whisker was clipped off and a previously clipped whisker was fixed with glue to a different whisker stub (Figure 21.7B). When retested, the number of trials necessary to reacquire the task was found to increase as a function of the distance between the site of the trained whisker and the new (i.e., attached) whisker (Figure 21.7C). Moreover, the transfer of learning between any two whisker locations was perfectly explained by the spatial distribution of physiological activity within the barrel field map (Figure 21.7D). This experiment suggests the presence of a sensory memory trace governed by the precise topography of the sensory receptors.[50]

Map Development and Plasticity

When a strain of mice is bred to have different numbers of whiskers, the barrel field differs in precise correspondence with the face; thus, extra whiskers (or missing whiskers) are accompanied by extra (or missing) barrels in the matching part of the barrel field.[51,52] Destruction at birth of a whisker follicle will cancel that whisker's module from all maps in the ascending sensory pathway.[53] But once the anatomical connection between one whisker and its corresponding cortical barrel is established a few days after birth, that barrel remains fixed throughout life.

[45] Woolsey, T. A., and van der Loos, H. 1970. *Brain. Res.* 17: 205–242.

[46] Armstrong-James, M., Fox, K., and Das-Gupta, A. 1992. *J. Neurophysiol.* 68: 1345–1358.

[47] Simons, D. J. 1978. *J. Neurophysiol.* 41: 798–820.

[48] Welker, C. 1976. *J. Comp. Neurol.* 166: 173–189.

[49] Petersen, R. S., and Diamond, M. E. 2000. *J. Neurosci.* 20: 6135–6143.

[50] Harris, J. A., Petersen, R. S., and Diamond, M. E. 1999. *Proc. Natl. Acad. Sci. USA* 96: 7587–7591.

[51] van der Loos, H., and Dorfl, J. 1978. *Neurosci. Lett.* 7: 23–30.

[52] van der Loos, H., Dorfl, J., and Welker, E. 1984. *J. Hered.* 75: 326–336.

[53] Jeanmonod, D., Rice, F. L., and van der Loos, H. 1977. *Neurosci. Lett.* 6: 151–156.

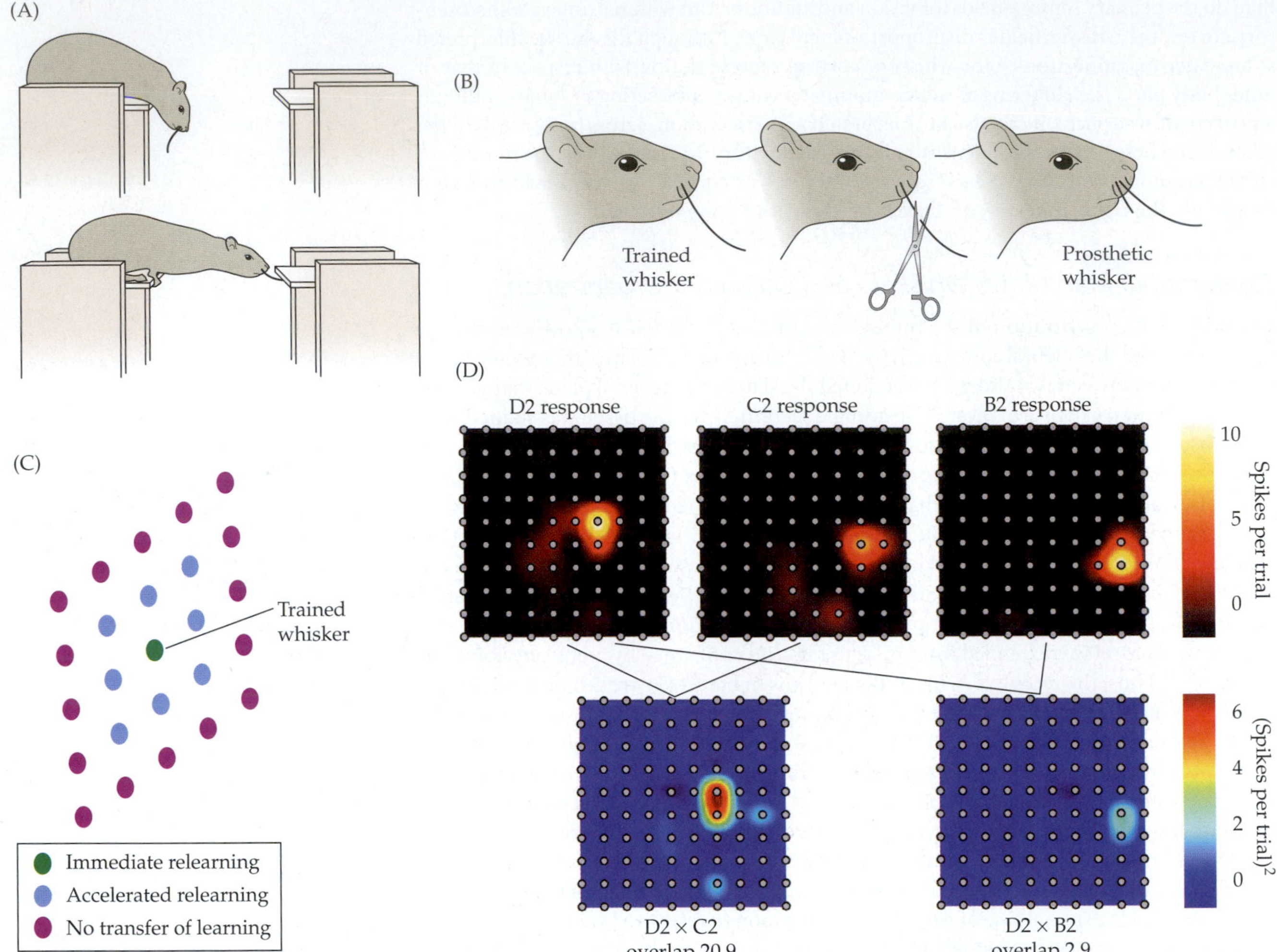

FIGURE 21.7 Role of the Cortical Whisker Map in Sensory Learning. (A) In the gap-crossing task, the rat learned to identify, in the dark, the location of a reward platform by making contact using a single whisker, referred to as the "trained whisker." In this drawing, first it leans into the gap (upper sketch) and then it extends forward to reach the target (lower sketch). (B) The trained whisker (left) was clipped off (middle) and a different whisker was fixed to a whisker stump some specific distance from the stump of the trained whisker (prosthetic whisker, right). (C) Schematic depiction of the whisker grid. The speed at which the rat relearned the gap-crossing task depended on the distance from the site of the trained whisker (T) to the site of attachment of the new whisker, indicating that learning was localized. (D) In the upper three plots, the number of action potentials (spikes per trial) evoked by separate, brief movements of three different whiskers (D2, C2, and B2) is illustrated. The responses were measured across a 10 × 10 grid of microelectrodes placed in the barrel cortex; electrode positions are designated by gray dots. The lower two plots indicate the degree of overlap in the cortical representation between whisker pairs (D2 × C2, and D2 × B2). This is calculated by multiplying the separate responses to the two whiskers at each electrode; thus the response scale is (spikes squared per trial) squared. Cortical overlap is higher for the closer whisker pair (D2 × C2) than for the more widely separated whisker pair (D2 × B2). The key result is that the exact quantity of cortical overlap between the representations of any two whiskers perfectly accounts for the degree of learning transfer between those two whisker sites. (After Harris et al.,1999.)

The topographic map at the level of cortex is not generated ad hoc from a disordered input—rather, all along the pathway, the neighboring relations are received in an orderly manner from the preceding level (the descending feedback projections mentioned earlier in the chapter respect the same topographic order). Each barrel receives input from a discrete **barreloid** in the ventroposterior medial nucleus of the thalamus; each thalamic barreloid receives input from a **barrelette** in the brainstem trigeminal nucleus; and each barrelette receives input from a whisker. During development the modular clustering appears in a temporal sequence from periphery to cortex.[54]

Although the principal input to a given barrel is its topographically matched whisker, neurons do respond to stimulation of neighboring whiskers, particularly in the column of neurons above and below the layer IV barrel. These inputs form the **surround receptive**

[54] Andres, F. L., and van der Loos, H. 1985. *Anat. Embryol. (Berl.)* 172: 11–20.

BOX 21.1
Variation across Species in Cortical Maps

The expanded representations of particularly sensitive body areas and their locations within distorted maps were described by Adrian beginning in the 1920s in work that opened the modern epoch in the study of somatosensory cortex. The experiments were accomplished by stimulating skin sites while picking up potentials, summated across millions of neurons, through electrodes (tiny spheres of copper) placed on the cortical surface. The line of research was continued by Clinton Woolsey through the 1950s, until microelectrode methods took over. Comparing the cortical skin representation across numerous species, these pioneers found fascinating species-specific maps. Recently, a particularly remarkable cortical map has been discovered: the representation of the pink fleshy appendages, called rays, ringing the snout of the star-nosed mole. These densely innervated tactile probes, which the animal uses to find worms, have an enormous topographically organized cortical representation.

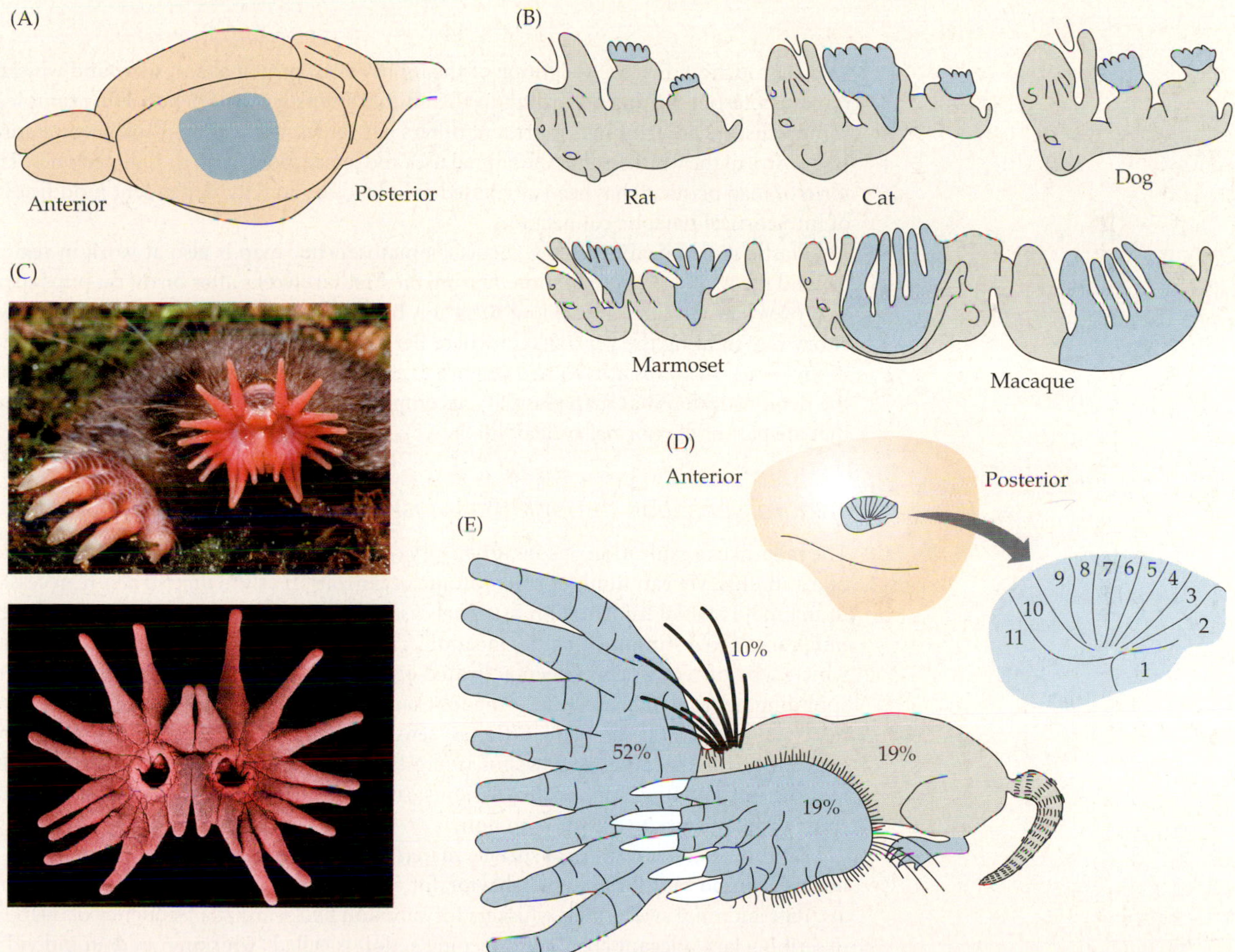

Cortical somatosensory maps. The skin areas that are most important to natural behavior, and are most richly innervated, are associated with expanded cortical representations. (A) View from above of the rodent brain. The location of somatosensory cortex on the cerebral hemisphere, shown here in blue shading, is typical of the general mammalian plan. (B) With the same orientation as in (A), the cortical maps for five species are illustrated. Forelimb and hindlimb are shaded, from which it can be noted that these regions are expanded in the primates, marmoset, and macaque. The maps are not shown in scale to each other. (C) A star-nosed mole. Top: Note the forelimb shaped for digging and the prominent finger-like "rays" around the snout. Bottom: Photomicrograph of the nose, showing the 11 rays per side (22 total) distributed about the central holes—the nostrils. (D) Again, a star-nosed mole. Left: view from the side of the left hemisphere with the nose representation shown in blue. Right: Closer view of the nose representation, in which each ray possesses a separate area. Adjacent rays have adjacent cortical territories. (E) Schematic depiction of the mole's body representation in somatosensory cortex. Percentages indicate the proportion of primary somatosensory cortex responding to the given body part. (B after Harlow and Woolsey, 1958; C photos courtesy of K. Catania; D,E after Catania, 1999, Catania and Kaas, 1997, and Sachdev and Catania, 2001.)

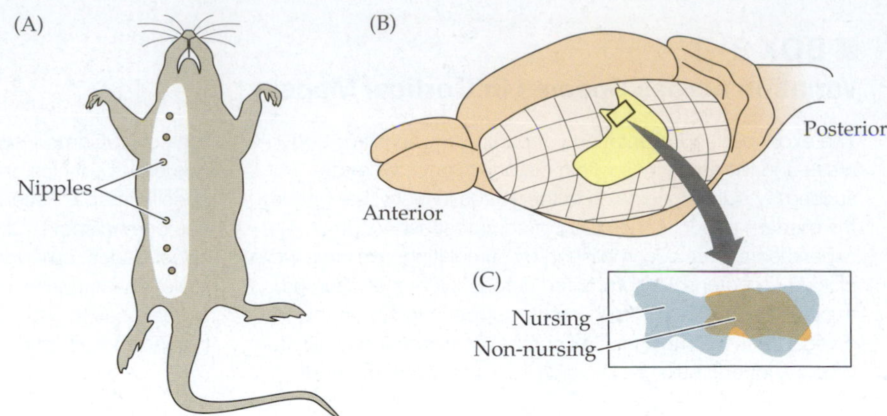

FIGURE 21.8 Somatosensory Cortical Plasticity Accompanying Natural Tactile Experience. (A) Mammary region of a female rat. (B) Rat brain, showing the somatosensory cortical map in yellow. Rectangle indicates the area in which responses to stimulation of the mammary region were sampled by microelectrodes. (C) Outline of responding areas within the sampled region in non-nursing (orange) and nursing (gray) rats. Cortical area devoted to input from the mammary region of the skin is greatly enlarged during nursing.

field. Throughout life, the weighting of the inputs from principal and surround whiskers remains plastic, shifting according to the animal's sensory experience. For example, if some whiskers are trimmed short and others left intact, cortical neurons become more responsive to the whiskers that the animal uses and less responsive to the clipped ones. This form of map plasticity has been attributed mainly to the activity-dependent modification of intracortical synaptic connections.[55–57]

Plastic reorganization of the cortical somatosensory map is also at work in regions beyond the whisker representation. During the first few weeks after birth, rat pups spend well over half the time in suckling behavior. In the nursing mother rat, the cortical territory representing the ventral skin where her pups suckle increases many fold, only to retract when the pups are weaned (Figure 21.8).[58] The significance of this experiment is the demonstration that map plasticity accompanies the variations in sensory experience that are part of the normal cycle of life.

Texture Sensation through the Whiskers: Peripheral Mechanisms

The map-like organization of somatosensory cortex is essential for detecting the location of a stimulus. We can think of somatotopic organization as the *infrastructure* necessary to maintain spatial information as signals ascend to cerebral cortex. However, there is much more to a stimulus than its location. The *quality* of touch sensation is related to which type of receptor has been activated and the details of the activity in the corresponding neurons. This section provides examples of how touch sensations arise from neuronal activity. Of the many different sensations that originate in the skin, we have selected texture because it allows for interesting comparisons between the rat whisker and the primate fingertip sensory systems; some mechanisms are unique to each, and some mechanisms are common to both.

Texture is defined by local surface material and microgeometry. The perception of texture is important for animal behavior, for example in the selection of nesting materials. Rats and mice sweep their whiskers forward and backward at a frequency of 5–15 Hz, describing large elegant arcs. The sweeping action is called "whisking" and an individual cycle is called a "whisk." Whisking is an example of active sensing—that is, generation of receptor motion to create sensory signals. When the animals set out to identify the texture of an object, they approach it and then palpate its surface with brief, light whisks (Figure 21.9A). To find out how well rats can identify textures, investigators have trained them in the dark to take one action upon contact with one texture and a different action upon contact with a second texture (e.g., turn to the left or right reward location according to the presented stimulus). The rats' performance is remarkably fast and accurate. On an easy task, for instance when one texture is pebbly and the other perfectly smooth, they can extract texture identity from just one to three touches per whisker;[59] this translates to a total time between initial contact and texture identification as short as 100–200 ms. If the task is made more difficult by using textures that are more alike, rats require more training and more contact time per trial. But after many practice sessions, they are able

[55] Diamond, M. E., Armstrong-James, M., and Ebner, F. F. 1993. *Proc. Natl. Acad. Sci. USA* 90: 2082–2086.

[56] Diamond, M. E., Huang, W., and Ebner, F. F. 1994. *Science* 265: 1885–1888.

[57] Celikel, T., Szostak, V. A., and Feldman, D. E. 2004. *Nat. Neurosci.* 7: 534–541.

[58] Xerri, C., Stern, J. M., and Merzenich, M. M. 1994. *J. Neurosci.* 14: 1710–1721.

[59] von Heimendahl, M. et al. 2007. *PLoS Biol.* 5: e305.

(A)

(B)

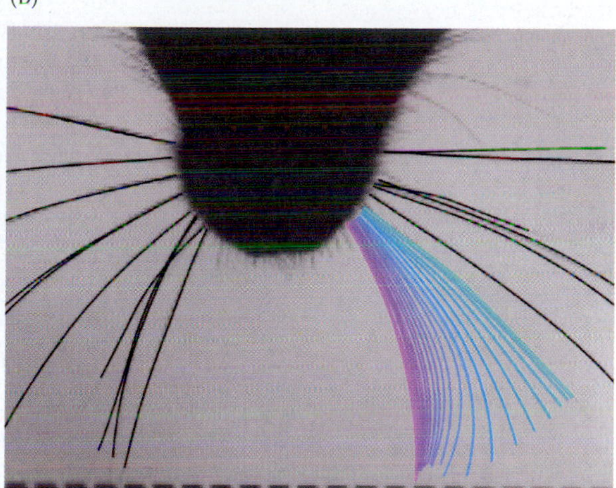

FIGURE 21.9 Texture Discrimination Behavior. (A) This sketch depicts the rat as it leans across a gap to touch a textured plate with its whiskers. The cable on the head of the animal carries neuronal signals to the computer. (B) One frame from a high-speed (1000 frames/second) video is shown. The orientation of the snout of the rat and the texture is similar to that in (A). This texture contains notch-like grooves. One whisker was tracked through a sequence of frames and the traces, from violet to light blue, shows the whisker position over 1 ms timesteps. The whisker tip was blocked in a groove and then sprung free as the rat whisked in the posterior direction. Irregular, fast events, like this one, provide information to the sensory receptors about texture. (A Drawn by Marco Gigante; B after Zuo, Perkon, and Diamond, 2011.)

to make discriminations that are difficult even for human subjects, for example selecting between a smooth surface and one with grooves that are 30 μm deep and spaced at 90 μm.[60]

The capacity to discriminate between textures begins with the signal transmitted by the whiskers. On any given whisk, many whiskers touch the surface, but they do so in a seemingly disorderly manner, passing in front of and behind their neighbors. This observation suggests that the sensory system does *not* utilize the specific spatial arrangement of the whiskers in the sensing of textures. Indeed, if only a few whiskers are left intact (the others clipped), rats can still discriminate textures. So, each whisker, by itself, appears to transmit a highly informative message. When moving in air, whisker motion is continuous and fluid. But along a surface, the whisker's trajectory is characterized by an irregular, skipping motion made up of intermixed high and low velocities.[61] The underlying process is known as stick and slip: the whisker tip tends to get fixed in place (i.e., sticks) and bends until the force behind it overcomes the resistance caused by friction; at this instant, the whisker springs loose (i.e., slips) only to get stuck again.[62] The movement of whiskers on a surface containing grooves resembles their movement along irregular, coarse textures. For instance, Figure 21.9B shows the trajectory of a single whisker as it is released from the groove of a plate. Discrimination can occur because each texture is associated with a distinct trajectory of sticks and slips, and the coarser the texture or the greater the spatial frequency of grooves, the greater is the probability of such slips and the greater their velocity when they occur.

Among sensory receptor neurons, spiking probability increases with progressively higher whisker speed and acceleration.[63] Since whisking along coarse textures leads to higher-velocity slips than does whisking along smooth textures,[61,62] rough texture is translated to a greater rate of neuronal firing.[59,64] As we will see next, the animals' judgment of roughness varies according to the firing rate in the sensory pathway.

Texture Sensation through the Whiskers: Cortical Mechanisms

The barrel cortex is critical for texture discrimination. After rats are trained to use their whiskers to discriminate between sandpapers with different grain sizes, the capacity is lost after a lesion of barrel cortex.[65] Instead, the rats begin to use their forepaw to touch and identify textures.

[60] Carvell, G. E., and Simons, D. J. 1990. *J. Neurosci.* 10: 2638–2648.

[61] Arabzadeh, E., Zorzin, E., and Diamond, M. E. 2005. *PLoS Biol.* 3: e17.

[62] Wolfe, J. et al. 2008. *PLoS Biol.* 6: e215.

[63] Shoykhet, M., Doherty, D., and Simons, D. J. 2000. *Somatosens. Mot. Res.* 17: 171–180.

[64] Lottem, E., and Azouz, R. 2009. *J. Neurosci.* 29: 11686–11697.

[65] Guic-Robles, E., Jenkins, W. M., and Bravo, H. 1992. *Behav. Brain. Res.* 48: 145–152.

In a direct test of the proposal that neuronal firing rate distinguishes between different texture sensations, activity was measured in cortical barrels while rats identified a contacted texture as rough or smooth, indicating their choice on each trial by moving to a water reward spout on either the left or right.[59] Success rate was over 80%. On the set of trials when a rat correctly identified the stimulus, the average firing rate of neurons in barrel cortex was higher for rough trials than for smooth trials. The texture-specific firing rate appeared immediately preceding the instant of choice, consistent with the notion that this activity is the signal that leads the rat to identify the stimulus. On error trials, the firing-rate code was reversed (i.e., lower for rough than for smooth), meaning that the rat makes its decision (right or wrong) based upon the magnitude of whisker-evoked activity in barrel cortex. The temporal pattern of spikes might also contribute to the discrimination of texture,[61,66] but this idea has proven hard to test in behaving animals.

Somatosensory System Organization and Texture Sensation in Primates

Magnification Factor

The whiskers are the critical touch organ in rats and mice, whereas in humans and other primates the fingertips are their equivalent. Each fingertip is innervated by axons from 250–300 sensory neurons (about the same number as the whisker). Because individual axons branch out and terminate in multiple receptor structures, the density of mechanoreceptors reaches the remarkably high value of over 1000 per square cm. The large number of receptors is responsible for the fine tactual acuity of the fingertips, which enables Braille experts to read at speeds in excess of 100 words a minute or Venetian glass blowers to assess the smoothness of their vases.

The rich fingertip innervation density is, like the whisker innervation of rodents, connected with an expanded cortical representation in which the representation of a 1 cm^2 area of skin on the fingertip occupies a somatosensory cortical area 100 times greater than the representation of a 1 cm^2 area of skin on the shoulder. In humans, innervation density and cortical magnification are high for the lips as well—crucial in the sensory aspects of eating, speaking, and kissing.

Topographic Map of the Skin and Columnar Organization

The magnification of certain areas of skin gives rise to the famous distorted cortical maps. In humans, it was discovered in the 1930s in patients undergoing brain operations. As the surgeon electrically stimulated a restricted cortical locus, the patient reported a sensation referred to a particular position on the body; touching that place on the skin led to an evoked potential at the same cortical site.[67] The original scheme, the **homunculus**, was correct concerning the relative locations of the major parts of the body in the coronal plane. Later, details such as the fine organization of the face[68] and the medial-to-lateral sequence of fingers[69,70] were corrected by using fMRI. Like the cortical representation of the whiskers in the mouse and rat, the homunculus is a topographic map because neighboring sites on the skin are represented at neighboring sites in the cortex. Processing of painful stimuli also exhibits a topographic organization within the ascending sensory pathways.[71]

In the 1950s, the use of microelectrodes to record the responses of neuronal clusters to well-controlled stimuli in anesthetized animals led to two crucial advances beyond the basic cortical body map. First, columnar organization[72] was discovered and described for the first time by Mountcastle and his colleagues in primate somatosensory cortex. When the electrode penetrated cortex at an angle normal to the surface, all neurons responded to the same receptor class and shared overlapping receptive fields on the skin (thus, barrels in mouse and rat somatosensory cortex are a special case of columnar organization). Second, map topography was worked out in much greater detail. Areas 3b and 1 (Figure 21.10) each contain a map arranged in an elongated medial-lateral strip along the postcentral gyrus (also see Appendix C); the two maps are arranged as mirror images reflected about the cytoarchitectonic border.[73] Because a three-dimensional surface (the body) is projected onto a two-dimensional surface (the cortex), map discontinuities are inevitable.

[66] Arabzadeh, E., Panzeri, S., and Diamond, M. E. 2006. *J. Neurosci.* 26: 9216–9226.

[67] Jasper, H., and Penfield, W. 1954. *Epilepsy and the Functional Anatomy of the Human Brain*, 2nd ed. Boston: Little, Brown and Co.

[68] Moulton, E. A. et al. 2009. *Hum. Brain Mapp.* 30: 757–765.

[69] Blankenburg, F. et al. 2003. *Cereb. Cortex* 13: 987–993.

[70] van Westen, D. et al. 2004. *BMC Neurosci.* 5: 28.

[71] DaSilva, A. F. et al. 2002. *J. Neurosci.* 22: 8183–8192.

[72] Powell, T. P., and Mountcastle, V. B. 1959. *Bull. Johns Hopkins Hosp.* 105: 133–162.

[73] Merzenich, M. M. et al. 1978. *J. Comp. Neurol.* 181: 41–73.

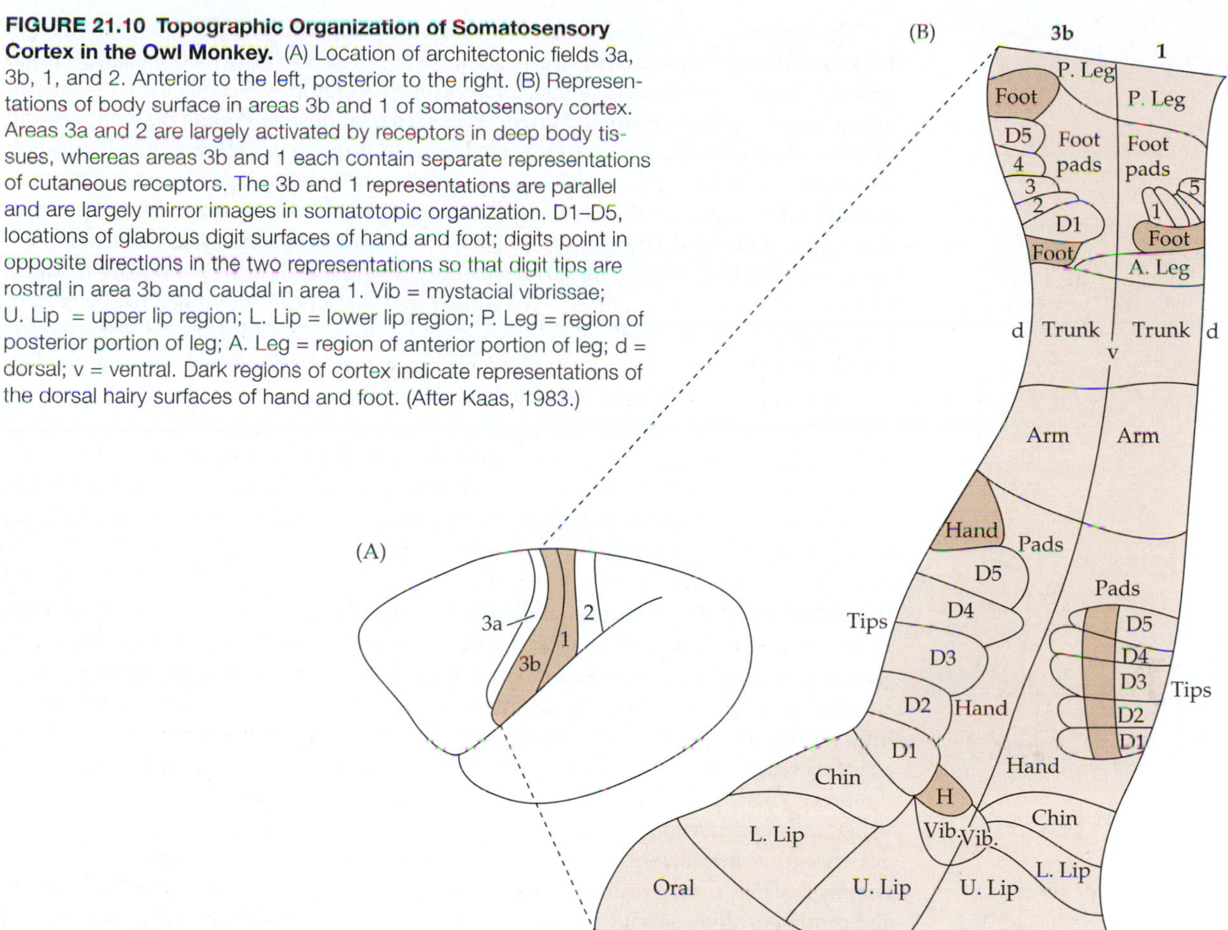

FIGURE 21.10 Topographic Organization of Somatosensory Cortex in the Owl Monkey. (A) Location of architectonic fields 3a, 3b, 1, and 2. Anterior to the left, posterior to the right. (B) Representations of body surface in areas 3b and 1 of somatosensory cortex. Areas 3a and 2 are largely activated by receptors in deep body tissues, whereas areas 3b and 1 each contain separate representations of cutaneous receptors. The 3b and 1 representations are parallel and are largely mirror images in somatotopic organization. D1–D5, locations of glabrous digit surfaces of hand and foot; digits point in opposite directions in the two representations so that digit tips are rostral in area 3b and caudal in area 1. Vib = mystacial vibrissae; U. Lip = upper lip region; L. Lip = lower lip region; P. Leg = region of posterior portion of leg; A. Leg = region of anterior portion of leg; d = dorsal; v = ventral. Dark regions of cortex indicate representations of the dorsal hairy surfaces of hand and foot. (After Kaas, 1983.)

Map Plasticity

Sensory discrimination capacities can improve with training. Is cortical map plasticity involved? In one study, monkeys were trained to discriminate the frequency of vibrations applied to one fingertip. Primary somatosensory cortex showed a clear change in the representation of stimuli applied to the trained skin. The spatial representation expanded by a factor of 1.5–3, and the temporal precision of neuronal responses to vibrations improved as well.[74,75] In humans, violin players have an enlarged representation of the fingertips of the left hand—the fingers that are used to manipulate the strings.[76] Blind people who are expert Braille readers show heightened tactile acuity with the fingertip used to read the raised dots, accompanied by an expanded cortical representation of the same finger.[77] To summarize, cortical topographic maps show a remarkable capacity throughout life to make fine adjustments in their representation of the external world. In general, selected parts of the sensory apparatus are allocated an expanded cortical representation when their behavioral significance is amplified by shifts in sensory demands or by training on specific tasks. The increased number of cortical cells processing the relevant stimuli may be associated with improved sensory capacities, though a causal relationship is hard to prove.

Texture Sensation through the Fingertip: Peripheral Mechanisms

One important way in which fingertip touch differs from whisker touch is that we (primates) manipulate objects with our hands, whereas rodents do *not* manipulate objects with their whiskers. The fingertips provide tactile feedback necessary for the motor control of the hand, such as peeling a banana, holding a pencil, or buttoning a button. While we handle objects, texture is sensed quickly and provides essential information both for identifying the object and for generating the appropriate motor plan. For instance, texture is a large part of the sensation that tells us how tightly we must grasp an object to keep it from slipping through our hands.

[74]Recanzone, G. H., Merzenich, M. M., and Schreiner, C. E. 1992. *J. Neurophysiol.* 67: 1071–1091.

[75]Recanzone, G. H. et al. 1992. *J. Neurophysiol.* 67: 1031–1056.

[76]Elbert, T. et al. 1995. *Science* 270: 305–307.

[77]Van Boven, R. W. et al. 2000. *Neurology* 54: 2230–2236.

The mechanisms for sensing texture through the primate fingertip differ in a number of ways from those of the whiskers. Receptors in the 35 whisker follicles form a discontinuous grid of 35 exquisitely sensitive points, whereas touch receptors in the fingertips are distributed in a continuous, overlapping sheet. The spatial relationships among the whiskers shift from one contact to the next, whereas the layout of fingertip touch receptors is constant over time. Several lines of evidence suggest that the perception of coarse textural features, through the fingertip, does rely on a spatial mechanism—at any given instant, the degree of roughness can be decoded by a "snapshot" (a spatial profile in a short time window) of the contrast in firing among slowly adapting neurons with nearby receptive fields. In contrast, the perceived roughness of finely textured surfaces is determined by the efficacy with which finger motion along the surface excites the rapidly adapting Pacinian system. We can view this latter process as a vibration mechanism, and it seems not to depend on comparing the distribution of excitation across separate receptors (a spatial mechanism). The reader will have noted that, of the two channels, the vibration mechanism has much in common with the prevailing view of how whiskers extract surface properties. This formulation, with its distinct morphological and functional channels for smooth and rough, has been named the **duplex theory** of texture perception. Evidence for the duplex theory is given in the following sections.

COARSE TEXTURES The spatial distribution of receptors allows the primate sensory system to gather signals which turn out to be particularly important for sensing coarse textures. Textures of this sort—for which the center-to-center distance between raised elements is larger than about 200 microns—can be discriminated simply by pressing the finger against the surface.[78] The fact that finger movement is not necessary points to slowly adapting receptors, presumably of the Merkel type, as the crucial peripheral termination; these receptors do not require motion across the receptive field but rather show an increased firing rate in proportion to the pressure applied to their tiny receptive field.[2] More direct evidence for slowly adapting coding of coarse textures comes from physiological and psychophysical experiments using stimuli formed by raised dots that vary in height, diameter, and center-to-center spacing.[79,80] Human subjects rated the roughness of the stimuli, and their estimates were compared to the responses of monkey slowly adapting neurons for the same set of stimuli. The stimuli judged as roughest were the ones that evoked the greatest contrast in firing rate of nearby slowly adapting receptors. Why might *contrast* in firing among adjacent receptors be correlated with roughness? As illustrated in Figure 21.11, if the surface is rough or grainy, adjacent receptors absorb different forces. In contrast, if the surface is smooth, or if there is little space between adjacent grains, force is distributed equally across adjacent receptors.

[78] Harris, J. A., Harris, I. M., and Diamond, M. E. 2001. *J. Neurosci.* 21: 1056–1061.

[79] Connor, C. E., and Johnson, K. O. 1992. *J. Neurosci.* 12: 3414–3426.

[80] Connor, C. E. et al. 1990. *J. Neurosci.* 10: 3823–3836.

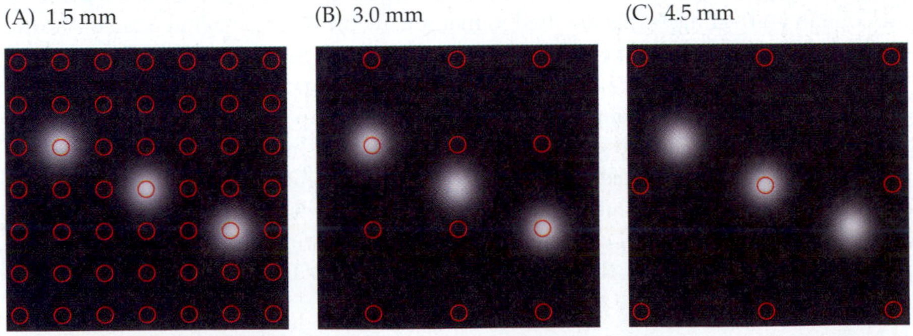

(A) 1.5 mm (B) 3.0 mm (C) 4.5 mm

FIGURE 21.11 Scheme for Detection of Coarse Features: Peripheral Mechanisms. Red circles represent regular, raised features on the surface of an object (e.g., grains, dots, pegs). Center-to-center spacing of features is in (A) 1.5 mm, in (B) 3.0 mm, and in (C) 4.5 mm. In each panel, the receptive fields of three slowly adapting primary afferent neurons are represented by the three light-colored spots. In (A), all three neurons are excited by surface features, so contact evokes little contrast in the firing rate among the three. In (B), two neurons are excited and one is not, causing a contrast in firing rate. In (C), there is also contrast in firing rate, but only one of three neurons is excited. Due to the contrast in firing rates, (B) would feel roughest. (After Connor and Johnson, 1992.)

The division of labor between rapidly and slowly adapting receptors in sensing rough texture is not absolute. There is evidence that rapidly adapting receptors of the Meissner type also contribute to the perception of coarse textures, but the mechanisms are not yet clear. In response to vibration, such receptors fire in phase with each displacement for frequencies up to about 50 Hz.[11,81–83] Some subjects may use vibrations of this kind to identify texture as they move the fingertip along a surface.[84,85]

FINE TEXTURES Slowly adapting receptors only contribute if the separation between pebbles, dots, or ridges is more than about 200 microns; if spacing is less than 200 microns, each slowly adapting receptor averages over the different bumps within its receptive field, and the mechanism of firing contrast no longer works. Therefore, to sense the quality of a fine-grained texture, a finger must move along its surface. The mechanism depends on the firing rate of Pacinian receptors. This conclusion was derived from many different types of experiment.[86]

The first kind of experiment studied the effect of high-frequency adaptation on fine texture sensation. Pacinian receptors (see Chapter 19) register vibrations in the frequency range of 50–400 Hz with peak sensitivity for vibrations in the range of about 250–300 Hz.[2] If sinusoidal vibrations centered on the Pacinian frequency band are continuously applied to a fingertip, high-frequency vibration sensitivity is lost, and such adaptation causes subjects to lose the ability to discriminate between fine textures but not between coarse textures. Adaptation of the Meissner rapidly adapting receptor with prolonged vibrations of 10 Hz does not affect the sensing of fine textures.

To collect more evidence, one experiment used a device to measure skin vibrations near the contact point between fingertip and the surface (Figure 21.12). On each trial, subjects were asked to compare the roughness of two surfaces. A higher degree of perceived roughness was associated with an increasing intensity of skin vibration *within the range of maximum Pacinian sensitivity*; vibrations at frequencies below or above the Pacinian peak sensitivity led to lower judgments of roughness.[87] Remarkably, when subjects were presented the same texture twice, as if they were two separate stimuli, they judged the texture as rougher on the presentation that evoked stronger vibrations in the Pacinian range.[86]

As a final test, artificial vibrations were added. When the plate surface was made to vibrate in the preferred Pacinian frequency range while the subject was palpating it, the texture felt rougher, even when subjects were not aware of the vibration itself.[88]

[81] Freeman, A. W., and Johnson, K. O. 1982. *J. Physiol.* 323: 21–41.

[82] Coleman, G. T. et al. 2001. *J. Neurophysiol.* 85: 1793–1804.

[83] Lundstrom, R. J. 1986. *Scand. J. Work Environ. Health.* 12: 413–416.

[84] Gamzu, E., and Ahissar, E. 2001. *J. Neurosci.* 21: 7416–7427.

[85] Cascio, C. J., and Sathian, K. 2001. *J. Neurosci.* 21: 5289–5296.

[86] Hollins, M., and Bensmaia, S. J. 2007. *Can. J. Exp. Psychol.* 61: 184–195.

[87] Bensmaia, S. J., and Hollins, M. 2003. *Somatosens. Mot. Res.* 20: 33–43.

[88] Hollins, M., Fox, A., and Bishop, C. 2000. *Perception* 29: 1455–1465.

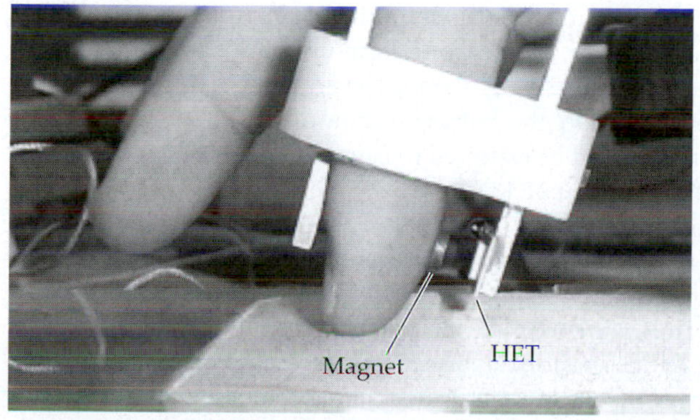

(B)

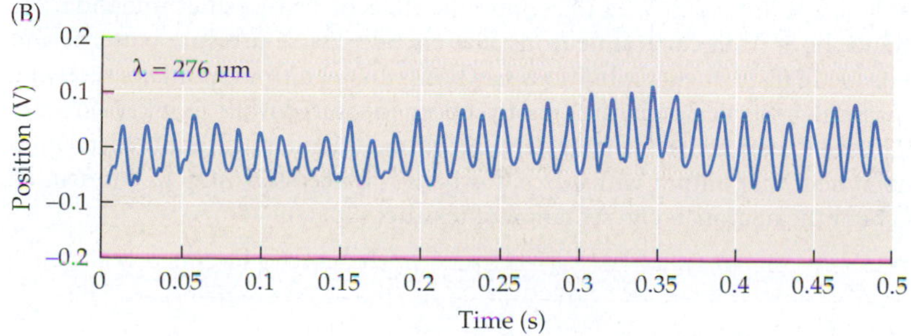

FIGURE 21.12 Skin Vibrations Evoked by Motion across a Fine Texture. (A) Experimental set-up. As the fingertip moves across the surface, small vibrations are produced in the skin. These vibrations travel a short distance up the finger and cause the magnet to vibrate, which, in turn, produces a current in the Hall Effect Transducer (HET); then, the HET current can be amplified and stored in the computer. (B) Transducer reading associated with movement along a surface with a spatial period of 276 micrometers (μm). Both the frequency and amplitude of the vibrations varied with spatial period of the probed surface. Surfaces that produced vibrations with a large amount of power in the frequency range of Pacinian receptors were perceived as roughest. (After Bensmaia and Hollins, 2003; photo courtesy of S. Bensmaia.)

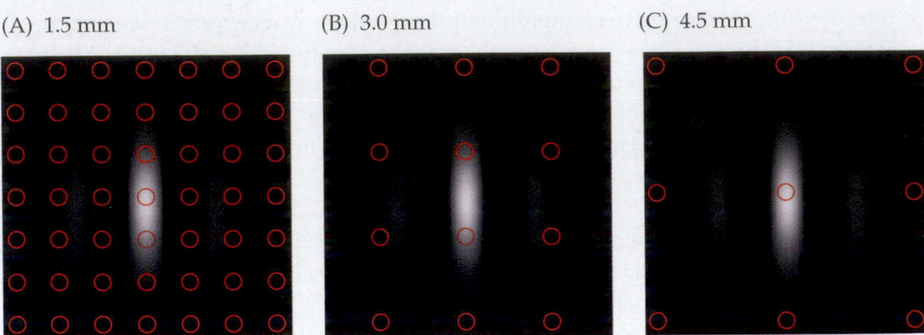

FIGURE 21.13 Scheme for Detection of Coarse Features: Cortical Mechanisms. As in Figure 21.11, red circles represent regular, raised features on the surface of an object (e.g., grains, dots, pegs). Center-to-center spacing of features is in (A) 1.5 mm, in (B) 3.0 mm, and in (C) 4.5 mm. The object surfaces are depicted in relation to the receptive field of a single, slowly adapting cortical neuron. Cortical neurons receive inputs from many peripheral receptors and have complex receptive fields, in this case an elongated excitatory central region (light area) flanked by alternating inhibitory (dark) and excitatory (light) subfields. The texture in (A) evokes a weak response in this neuron because the dots fall on both excitatory and inhibitory regions of the receptive field. The pattern in (B) produces strong excitation of the neuron because two of the dots fall on the central excitatory region and none on the flanking inhibitory subfield. The pattern in (C) excites the neuron to a lesser extent because only one dot falls on the excitatory portion. Thus, this cortical neuron would respond most strongly to a texture with a 3-mm spacing, which would be perceived as rough. (After Connor and Johnson, 1992.)

Texture Sensation through the Fingertip: Cortical Mechanisms

The evidence described above concerns the coding of texture by the peripheral nervous system. In the somatosensory cortex, the distinction between rapidly adapting and slowly adapting neurons is less sharp. In area 3b, the somatosensory cortical area with the strongest thalamic input, the responses of neurons are determined by three components within the receptive field: (1) a single, central region of a few millimeters diameter which, when touched, produces excitation at short latency, (2) surrounding regions which, when touched produce inhibition at nearly the same short latency as the central excitatory field, and (3) a region that overlaps both the short-latency excitatory and inhibitory regions and which, when touched, gives rise to inhibition but at a longer latency.[89] Clearly, whenever object contact activates all three components, for example as the fingertip moves along a surface, the neuronal responses are hard to predict by simply summing the three receptive field components. Nevertheless, neurons have been found whose properties could support the representation of texture. First, neurons, whose receptive field properties would be expected to cause them to fire at higher rates for contact with coarser textures (Figure 21.13), have been identified; their output could give rise to the percept of roughness.[89–91]

Signals from Pacinian neurons also reach the somatosensory cortex, and some neurons are essentially tuned to give the greatest response to skin vibrations aligned with the peak of Pacinian sensitivity.[4,82,92–99] Thus, vibrations like those illustrated in Figure 21.12B, which occur from movement across a fine surface, can be expected to excite cortical neurons.

It is certain that, in primates, the cerebral cortex does have a role in texture perception, and, as in rats, lesions in SI lead to severe impairments in roughness discrimination. In particular, ablation of Brodmann area 1 has a specific effect on texture discrimination.[21,22] Yet the evidence both from cortical neurons that encode coarse pressure patterns and those that encode high-frequency vibrations has been obtained from monkeys receiving the stimuli passively—cortical activity has not yet been measured while monkeys identify surface roughness. Experiments involving explicit texture judgment will allow correlation of trial-to-trial neuronal output with percept, which is a necessary step to confirm all hypotheses about the neuronal representation of texture.

[89] DiCarlo, J. J., and Johnson, K. O. 2000. *J. Neurosci.* 20: 495–510.

[90] DiCarlo, J. J., Johnson, K. O., and Hsiao, S. S. 1998. *J. Neurosci.* 18: 2626–2645.

[91] DiCarlo, J. J., and Johnson, K. O. 2002. *Behav. Brain. Res.* 135: 167–178.

[92] Hyvarinen, J. et al. 1968. *Science* 162: 1130–1132.

[93] Ferrington, D. G., and Rowe, M. 1980. *J. Neurophysiol.* 43: 310–331.

[94] Hyvarinen, J., Poranen, A., and Jokinen, Y. 1980. *J. Neurophysiol.* 43: 870–882.

[95] Burton, H., and Sinclair, R. J. 1991. *Brain Res.* 538: 127–135.

[96] Gardner, E. P. et al. 1992. *J. Neurophysiol.* 67: 37–63.

[97] Sinclair, R. J., and Burton, H. 1993. *J. Neurophysiol.* 70: 331–350.

[98] Lebedev, M. A., and Nelson, R. J. 1996. *Exp. Brain Res.* 111: 313–325.

[99] Zhang, H. Q. et al. 2001. *J. Neurophysiol.* 85: 1805–1822.

SUMMARY

- Low-threshold mechanoreceptors form one class of skin receptor. There are several types, each defined by the specialized structure surrounding the nerve termination. The morphology of the structure makes the membrane sensitive to particular properties of mechanical stimuli.
- Another class of skin receptor is the free nerve ending, a terminal that is activated by strong mechanical, thermal, or chemical (painful) stimuli.
- The method of microneurography—recording and stimulating single nerve fibers in the arm of human subjects—elucidates the connection between individual sensory receptor neurons and elemental sensations.
- Different submodalities of skin sensation (light touch, painful touch, temperature) are relayed by separate anatomical routes to the contralateral thalamic nuclei.
- Primary somatosensory cortex (SI) is the main target of the somatosensory thalamic nuclei. Cortical neurons are excited by skin stimulation, and lesions cause profound tactile deficits (touch, shape, texture, etc.).
- The sense of pain and its emotional effect are influenced by social, cultural, and personal factors. These factors interact with the processing of painful stimuli by modulating the release of endorphins.
- In rodents, the organization of the whisker sensory system introduces several fundamental principles shared with the primate touch system.
- Individual whisker follicles lead, through a chain of synaptic relays, to individual cortical columns, called barrels. Barrels are arranged as a topographic map of the corresponding whiskers; the whisker map occupies a large territory of cerebral cortex.
- Rats can distinguish between different textures through whisker contact. Contact with a surface causes an irregular whisker trajectory of stops and starts: the pattern of motion is distinctive for each texture.
- In primates, the fingertips are crucial to many behaviors. They are densely innervated and are associated with a greatly expanded cortical representation.
- The skin is represented by a mosaic of columns in somatosensory cortex. The columns are arranged to produce a distorted map of the skin (more columns for the fingertips and lips). The layout of cortical maps and the properties of single neurons within cortical maps are shaped by sensory experience.
- In primates, there are two channels for sensing texture. The perception of coarse textures is based on the difference in firing rate between adjacent slowly adapting receptors. The perception of fine surfaces is based on vibrations in the skin, transduced by rapidly adapting Pacinian receptors.

Suggested Reading

General Reviews

Apkarian, A. V., Bushnell, M. C., Treede, R. D., and Zubieta, J. K. 2005. Human brain mechanisms of pain perception and regulation in health and disease. *Eur. J. Pain* 9: 463–484.

Bolanowski, S. J., Jr., Gescheider, G. A., Verrillo, R. T., and Checkosky, C. M. 1988. Four channels mediate the mechanical aspects of touch. *J. Acoust. Soc. Am.* 84: 1680–1694.

Diamond, M. E., Petersen, R. S., and Harris, J. A. 1999. Learning through maps: functional significance of topographic organization in primary sensory cortex. *J. Neurobiol.* 41: 64–68.

Hollins M., and Bensmaia, S. J. 2007. The coding of roughness. *Can. J. Exp. Psychol.* 61: 184–195.

Mano T., Iwase S., and Toma, S. 2006. Microneurography as a tool in clinical neurophysiology to investigate peripheral neural traffic in humans. *Clin. Neurophysiol.* 117: 2357–2384.

Melzack, R. 1973. *The Puzzle of Pain*. Harmondsworth: Penguin Books.

Torebjork, E. 1985. Nociceptor activation and pain. *Philos. Trans. R. Soc. Lond., B, Biol. Sci.* 308: 227–234.

Original Papers

Arabzadeh E., Zorzin E., and Diamond, M. E. 2005. Neuronal encoding of texture in the whisker sensory pathway. *PLoS Biol.* 3:e17.

Albert T., Pantev C., Wienbruch C., Rockstroh B., and Taub, E. 1995. Increased cortical representation of the fingers of the left hand in string players. *Science* 270: 305–307.

Pollo, A., Amanzio M., Arslanian A., Casadio C., Maggi G., and Benedetti, F. 2001. Response expectancies in placebo analgesia and their clinical relevance. *Pain* 93: 77–84.

Van Boven, R. W., Hamilton, R. H., Kauffman T., Keenan, J. P., and Pascual-Leone, A. 2000. Tactile spatial resolution in blind Braille readers. *Neurology* 54: 2230–2236.

van der Loos H., and Dorfl, J. 1978. Does the skin tell the somatosensory cortex how to construct a map of the periphery? *Neurosci. Lett.* 7: 23–30.

von Heimendahl, M., Itskov, P. M., Arabzadeh E., and Diamond, M. E. 2007. Neuronal activity in rat barrel cortex underlying texture discrimination. *PLoS Biol.* 5: e305.

Wager, T. D., Rilling, J. K., Smith, E. E., Sokolik A., Casey, K. L., Davidson, R. J., Kosslyn, S. M., Rose, R. M., and Cohen, J. D. 2004. Placebo-induced changes in fMRI in the anticipation and experience of pain. *Science* 303: 1162–1167.

CHAPTER 22
Auditory and Vestibular Sensation

The meaning of a sound is derived from analysis of its frequency components and its temporal sequence. The frequency selectivity of auditory hair cells depends on their electrical characteristics and on their position along the mechanically tuned basilar membrane. Therefore, the sensory epithelium is tonotopically arranged. In the mammalian cochlea, each sensory fiber of the eighth cranial nerve innervates a single inner hair cell and so responds best to the acoustic frequency that best activates that cell. Efferent feedback from the central nervous system reduces hair cell sensitivity and frequency selectivity in the inner ears of reptiles, birds, and mammals. Hair cells of certain vertebrates are electrically tuned to provide acoustic frequency selectivity.

Afferent fibers from the cochlea form synapses within brainstem nuclei. Second-order neurons project to the superior olivary complex or into pathways that ascend through the inferior colliculus to the medial geniculate nucleus of the thalamus. Neurons in primary auditory cortex receive input from both ears and encode features of sound that are more complex than those detected in the periphery. The localization of a sound arises from neural computations involving comparisons between inputs to the two ears. Accordingly, the central auditory pathway includes a complex set of synaptic relays and feedback connections at which binaural comparisons are made or other aspects of timing and frequency composition are determined.

Position and motion of the head are sensed and transduced by the vestibular apparatus: three semicircular canals, the utricle, and saccule. Hair cells within each of these structures are activated by fluid flow induced by rotation or displacement of the head. Vestibular signals travel to the brain in sensory fibers of the eighth nerve but do not normally reach consciousness; they mediate automatic behaviors that maintain body posture and stability. An important mechanism for maintaining stability of gaze is the vestibulo-ocular reflex (VOR). This reflex is mediated by a 3-neuron arc that includes sensory neurons, interneurons, and motoneurons. The VOR is capable of impressive plasticity through modulation by the cerebellum.

Mechanosensory hair cells of the inner ear respond to a wide range of mechanical stimuli, in terms of both intensity and frequency. Hair cells in the ear encode stimulus intensities that vary over six orders of magnitude. In addition, hair cells cover an equally impressive kinetic range, from vestibular coding of head movements that occur at 0.05 Hz to acoustic coding of sounds at more than 100 kHz (in whales and bats). And yet, at the molecular level, all vertebrate hair cells operate in essentially the same way—a deflection of their stereocilia causes a receptor potential and subsequent activation of postsynaptic neurons by release of glutamate (see Chapter 19). The specific range of stimulus frequencies and intensities to which particular hair cells respond is determined in large part by the accessory structures in which they are imbedded. In this chapter we explore the structural adaptations that confer stimulus specificity onto hair cell signaling and follow that signaling through higher levels of analysis. Our focus will be largely on the inner ear of mammals, but important lessons also have been learned by the study of other vertebrates, particularly at the level of cellular physiology.

As seen in Figure 22.1, the labyrinth of the inner ear includes several elaborately shaped fluid-filled ducts. The tightly wound spiral of the cochlea (from the Greek *kochlias*, snail) lies ventral to 3 orthogonally positioned loops of the semicircular canals, part of the vestibular system. In addition, two relatively flat (macular) epithelia, the saccule and utricle, lie within the central vestibular chamber. A comparison of the labyrinth in reptiles, birds, and mammals provides an interesting correlation of structure and function. This point is made in drawings (Figure 22.1B,C) from von Bekesy,[1] whose studies of inner ear function in the early twentieth century established the basis of cochlear frequency selectivity. As seen in Figure 22.1, the disposition of semicircular canals is similar in reptiles, birds and mammals, implying that the function (detecting rotation of the head in space) is similar among vertebrates. In contrast, the spiraled cochlea of the mammal replaced the shorter, and still shorter auditory end organs of birds and reptiles. This evolutionary progression in auditory structure can be correlated with the frequency range of hearing,

[1] von Bekesy, G. 1960. *Experiments in Hearing*. Mcgraw-Hill, New York.

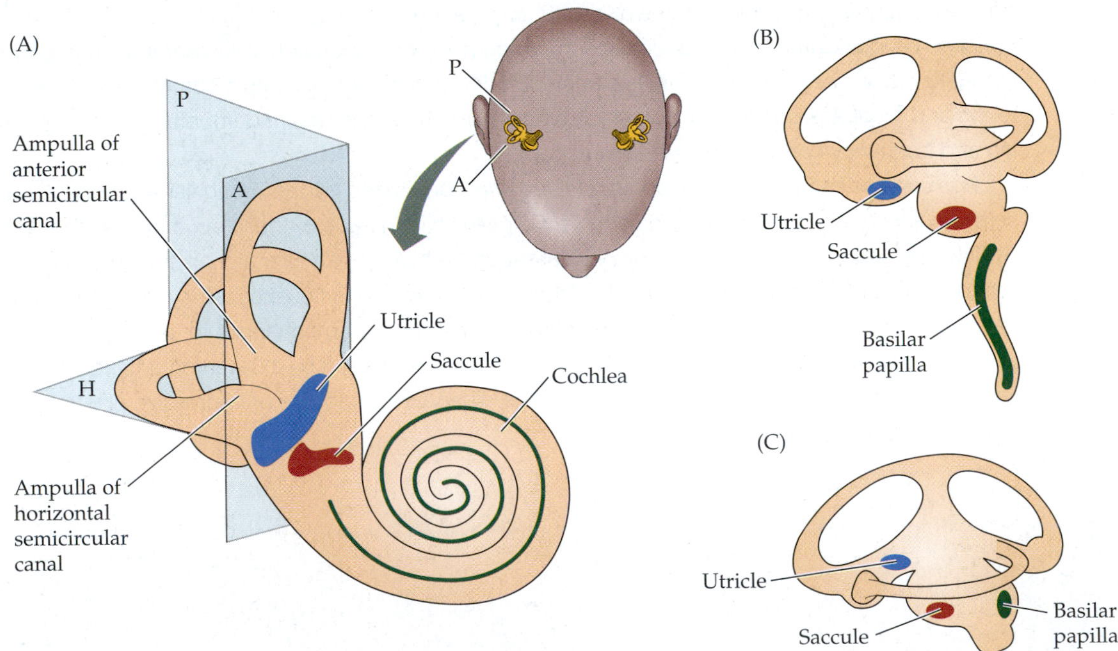

FIGURE 22.1 Sensory Structures of the Inner Ear. The inner ears of (A) mammals (human), (B) birds (chicken), and (C) reptiles (turtle) have in common the orthogonal arrangement of semicircular canals and two macular epithelia, the utricle and saccule. These mechanisms provide essentially identical information in all vertebrates. In contrast, the auditory end organ (the cochlea in mammals) increases progressively in length from the 1-mm long basilar papilla of turtles, to the 5-mm long basilar papilla of chicken, to the 30-mm long cochlea of mammals. The length increase serves as the substrate for expansion into higher frequency hearing, from 1 to 5 to 20 kHz. (B,C after von Bekesy, 1960.)

up to 1 kHz in turtles, 5 kHz in chickens, and 30 kHz in rats. The longer, coiled cochlea of mammals provides a mechanism to expand the high frequency range of hearing. We shall see that a number of cellular specializations also contribute to the ability to hear higher frequencies.

How is sound frequency encoded by cochlear hair cells, and how are these signals processed at higher levels of the auditory system? How is the spatial location of a sound determined? What are the neural computations that relate fluid movement of the inner ear to head position and movement? We begin by examining the structure and function of the sensory organs.

The Auditory System: Encoding the Frequency Composition of Sound

In terrestrial vertebrates, sound waves enter the outer ear, strike the **tympanic membrane** (eardrum), and via mechanical coupling through the **ossicles** of the middle ear, are converted to fluid waves in the mammalian cochlea (Figure 22.2A). The fluid waves, in turn,

(A)

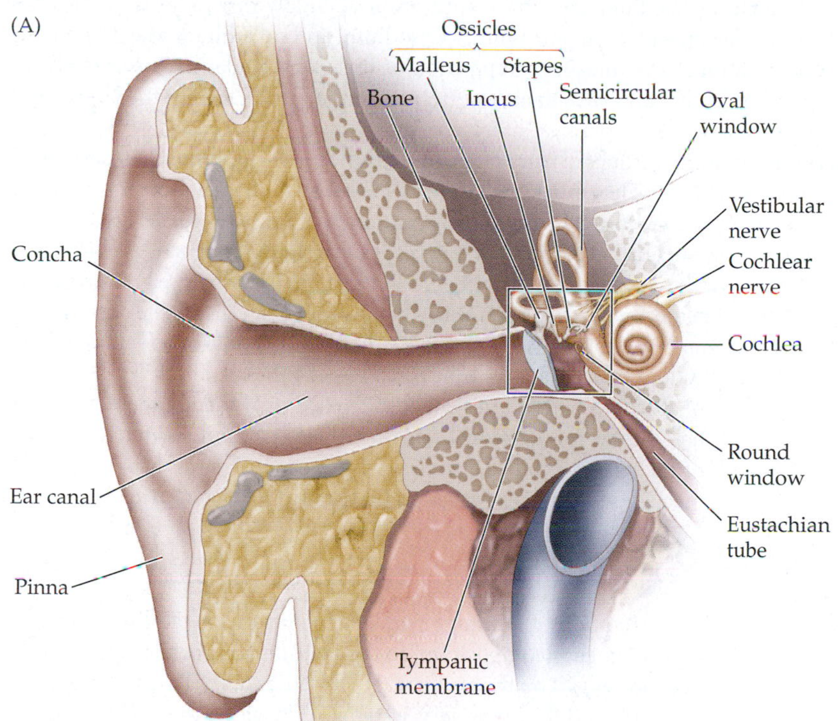

(B)

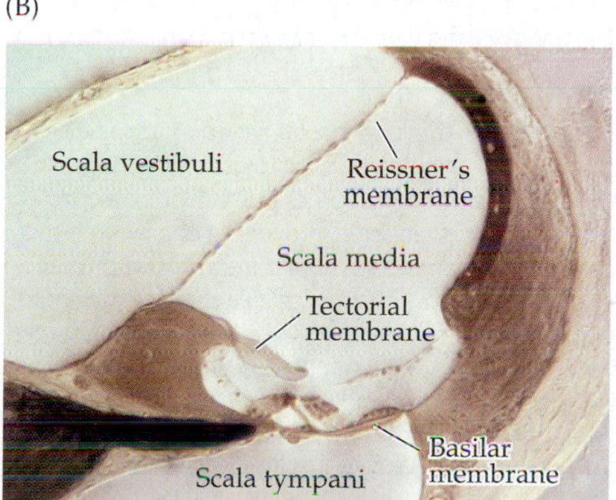

(C)

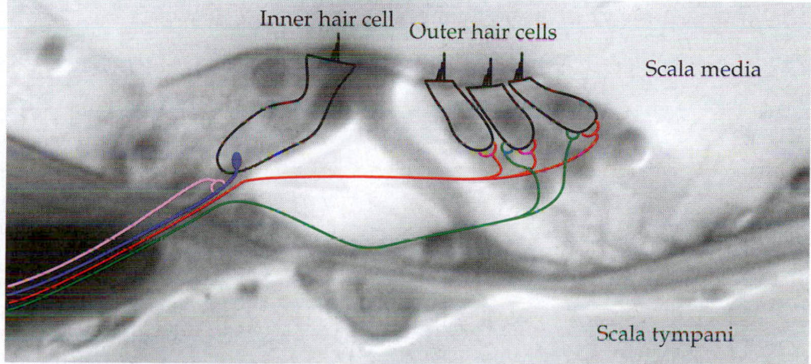

FIGURE 22.2 Structure of the Cochlea (A) The external, middle, and inner ear, showing the eardrum and its bony connections to the oval window. (B) The cochlea, showing the scala media bounded by Reissner's membrane (which contains endolymph, a high-potassium solution) and the structural relations among the basilar membrane, inner hair cells, outer hair cells, and tectorial membrane. (C) Hair cells form synapses with the terminals of auditory nerve fibers that have their cell bodies in the spiral ganglion. Approximately 95% of afferent fibers are postsynaptic to IHC. As many as 20 afferent fibers are in contact with a single IHC. OHCs have few afferent contacts but instead are the postsynaptic targets of cholinergic efferent neurons that project from the olivary complex in the brainstem.

cause vibration of the **basilar membrane**, on which sit sensory **hair cells** in the **organ of Corti**. This process has been summarized poetically by Aldous Huxley:

> Pongileoni's blowing and the scraping of the anonymous fiddlers had shaken the air in the great hall, had set the glass of the windows looking on to it vibrating; and this in turn had shaken the air in Lord Edward's apartment on the further side. The shaking air rattled Lord Edward's membrana tympani; the interlocked malleus, incus and stirrup bones were set in motion so as to agitate the membrane of the oval window and raise an infinitesimal storm of fluid in the labyrinth. The hairy endings of the auditory nerve shuddered like weeds in a rough sea; a vast number of obscure miracles were performed in the brain, and Lord Edward ecstatically whispered "Bach"![2]

The Cochlea

The cochlear duct is divided into three compartments: The **scala media** contains a high-potassium solution, the endolymph. It is separated from the overlying **scala vestibuli** by Reissner's membrane and from the fluid space of the **scala tympani** by intercellular tight junctions between the apical ends of the hair cells and their surrounding supporting cells (Figure 22.2B). The scala tympani and scala vestibuli contain perilymph, similar in composition to cerebrospinal fluid; the ionic composition of endolymph in the scala media is like that of cytoplasm, with high potassium, low sodium, and calcium ions held to micromolar concentration.[3] The unusual composition of this extracellular fluid is established by ion transport in the **stria vascularis**, a secretory epithelium that lines the lateral wall of the scala media.

There are two distinct groups of hair cells in the mammalian cochlea: **inner hair cells** (**IHC**) and **outer hair cells** (**OHC**). These reside in the organ of Corti on the basilar membrane (Figure 22.2B,C). Their bundles of sensory hairs project up into an overlying acellular gelatinous sheet, the **tectorial membrane.** Inner and outer hair cells differ by position (inner hair cells are closer to the central axis of the cochlear coil) and innervation pattern. Inner hair cells receive more than 90% of the afferent contacts to the cochlea;[4–6] while outer hair cells are the postsynaptic targets of the efferent nerve supply (see Figure 22.2C).[7] This differential innervation pattern raises interesting questions concerning the functional roles of these two cell types that will be discussed later in this chapter. The tectorial membrane overlies the hair bundles of both inner and outer hair cells, and differential motion of the tectorial and basilar membranes causes lateral shear to gate the mechanotransduction channels (as detailed in Chapter 19).

[2] Huxley, A. 1928. *Point Counter Point*. Harper Collins, New York.

[3] Wangemann, P. 2006. *J. Physiol.* 576: 11–21.

[4] Brown, M. C. 1987. *J. Comp. Neurol.* 260: 591–604.

[5] Kiang, N. Y. et al. 1982. *Science* 217: 175–177.

[6] Spoendlin, H. 1969. *Acta Otolaryngol.* 67: 239–254.

[7] Warr, W. B. 1975. *J. Comp. Neurol.* 161: 159–181.

Frequency Selectivity: Mechanical Tuning

Interpretation of auditory signals depends on the ability to analyze their frequency composition. In the mammalian cochlea, this analysis depends in part on the mechanical properties of the basilar membrane. Its width and thickness vary systematically along the cochlea (Figure 22.3). At the base of the cochlea, near the oval window, the membrane is narrow and rigid, while at the cochlear apex, it is wide and flexible. The consequences of this progressive variation in structure were exam-

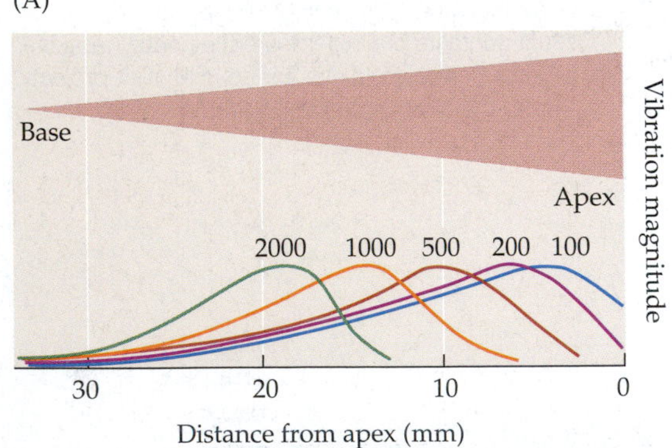

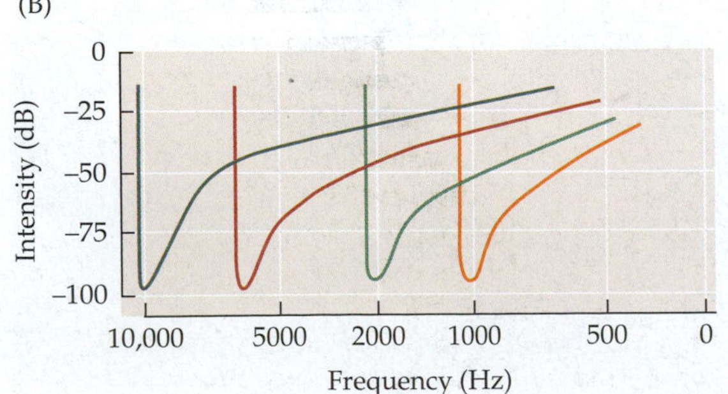

FIGURE 22.3 Cochlear Tuning (A) The location of maximum displacement of the basilar membrane in the cochlea by sound waves depends on frequency. The curves represent relative displacement at the indicated frequencies (100–2000 Hz). At the low frequencies, maximum displacement is near the wider (more flexible) apical membrane; higher frequencies produce maximal displacement near the narrower (stiffer) base. (B) Idealized tuning curves of four individual eighth-nerve fibers innervating different locations on the cochlea. Sound intensity, in decibels (dB), needed to produce discharges in a fiber is plotted against frequency of the auditory stimulus. The best frequencies for the fibers (i.e., the frequencies requiring the least stimulus intensity) are 1000, 2000, 5000, and 10,000 Hz. (A after von Bekesy, 1960; B after Katsuki, 1961.)

ined by von Bekesy,[1] who used stroboscopic illumination of reflective particles scattered on the basilar membrane to visualize the pattern of vibration. What he observed was that high-frequency sounds cause maximal vibration of the thicker, stiffer basal end of the membrane and that low-frequency sounds cause maximal vibration of the more flexible apical membrane.

As a consequence of mechanical tuning of the basilar membrane, hair cells at the cochlear base (nearest the stapes footplate) are stimulated preferentially by high-frequency sound, whereas hair cells progressively further along the cochlear coil (nearer to the apex) respond best to progressively lower-frequency sound (Figure 22.3A). The resulting **tonotopic map** is reflected in the stimulus requirements of cochlear afferent neurons that selectively innervate single cochlear hair cells. The frequency selectivity of cochlear afferents can be determined by recording action potentials during systematic presentation of pure tones of various frequencies and intensities. The resulting **tuning curve** is V-shaped, with a best or characteristic frequency defined as the pure tone to which the fiber is most sensitive (Figure 22.3B). The characteristic frequency of each sensory fiber is determined by the position along the cochlear duct of the hair cell it contacts.

Electromotility of Mammalian Cochlear Hair Cells

The frequency-dependent vibration of cochlear membranes described by von Bekesy provided an elegant basis for cochlear tuning. However, when examined closely it was found that the physical properties of those membranes were not adequate to explain the great sensitivity and sharp tuning of individual cochlear afferent neurons. Furthermore, it was demonstrated that sharp tuning was physiologically vulnerable,[8,9] suggesting that some energy-requiring biological process participated as well. The resolution lies in the fact that outer hair cells of the cochlea actively contribute to cochlear vibration through a process called **electromotility**. Outer hair cells shorten in response to depolarization and lengthen during hyperpolarization of their membranes, as though they were tiny muscles (Figure 22.4).[10,11] However, these movements are *not* generated by actin–myosin but result from direct effects of voltage on a charged motor protein, prestin, which is expressed at high levels in the basolateral membrane.[12] It is thought that the voltage-dependent conformational changes undergone by prestin are somehow coupled to the cytoskeleton. Electromotility of outer hair cells adds to the vibration amplitude of the cochlear partition during stimulation by sound, and thus increases the deflection of the stereocilia of the inner hair cell.[9] In this way, outer hair cells enhance cochlear tuning by adding mechanical energy to movement of the basilar membrane (Figure 22.5). Given this role as cochlear amplifiers, the puzzling efferent innervation pattern of the cochlea now makes more sense. The central nervous system can regulate cochlear sensitivity by adjusting the *gain* of the amplifier, that is, by altering outer hair cell motility.

[8] Ashmore, J. 2008. *Physiol. Rev.* 88: 173–210.

[9] Dallos, P. 2008. *Curr. Opin. Neurobiol.* 18: 370–376.

[10] Brownell, W. E. et al. 1985. *Science* 227: 194–196.

[11] Ashmore, J. F. 1987. *J. Physiol.* 388: 323–347.

[12] Zheng, J. et al. 2000. *Nature* 405: 149–155.

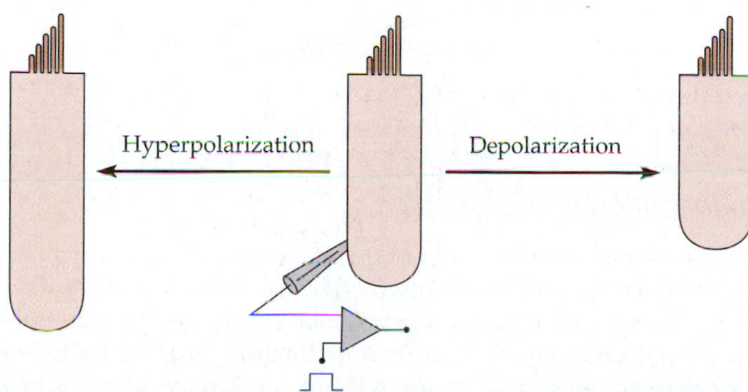

FIGURE 22.4 Electromotility of Outer Hair Cells Whole-cell patch pipettes are used to clamp the membrane potential of outer hair cells isolated from the mammalian cochlea. Depolarization causes the cells to shorten; hyperpolarization makes them longer. Such length changes can be as large as 30 nm/mV.

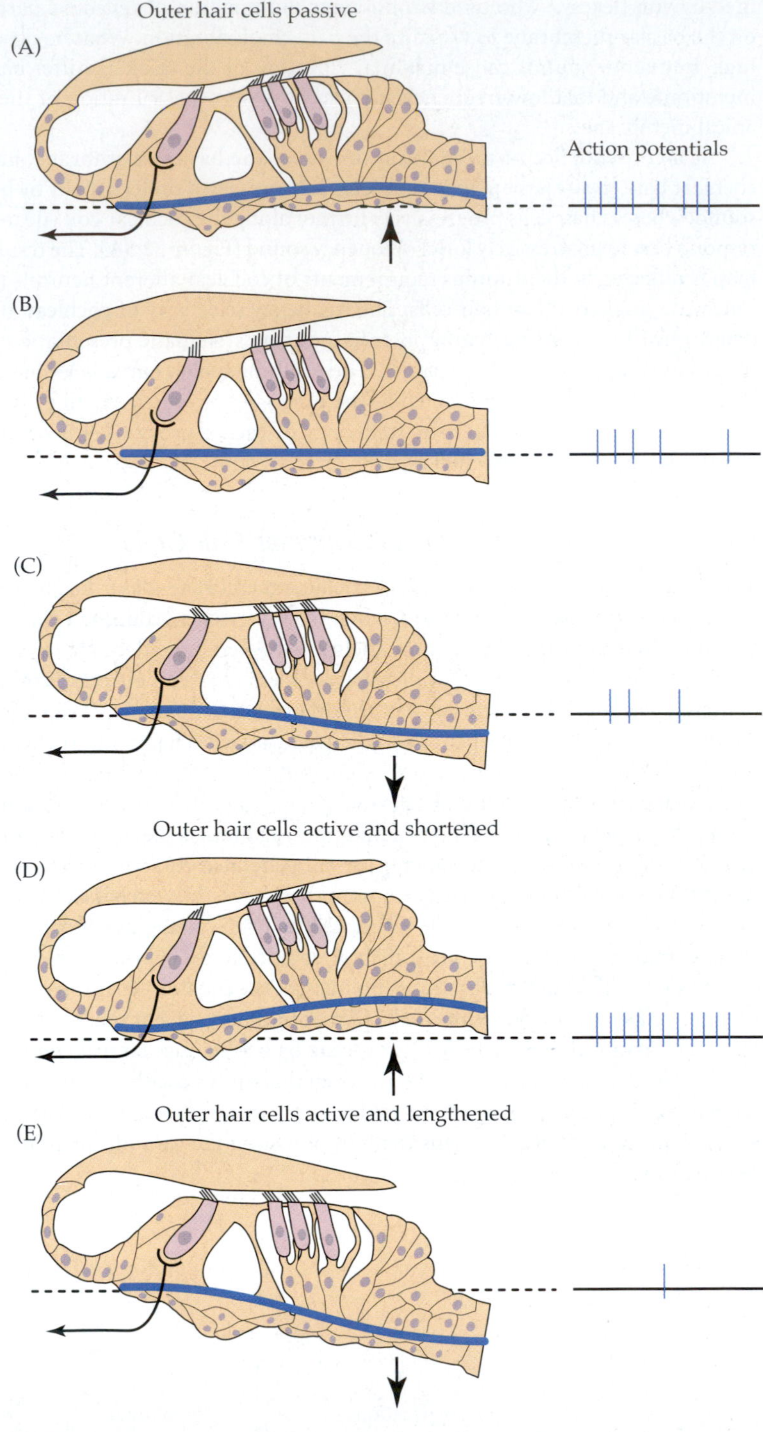

FIGURE 22.5 Motion of the Cochlear Partition During sound-evoked fluid waves, the cochlear partition (basilar membrane and organ of Corti) moves up and down. (A) During the rarefaction phase of a sound wave, the basilar membrane and tectorial membrane move upward, resulting in a shearing motion toward the longest stereocilia of the hair bundle—depolarizing the hair cells relative to the resting condition (B). (C) Downward motion during the compressive phase of a sound produces the opposite displacement of the hair bundles, and hair cells are hyperpolarized relative to rest. (D) Active contraction of outer hair cells during depolarization accentuates the upward motion of the basilar membrane. (E) Hyperpolarization of outer hair cells causes them to elongate. The net effect is that motion of the basilar membrane is greater because of the so-called electromotility of the outer hair cells, and this is reflected in the activity pattern of afferent fibers contacting the inner hair cells. The extent of motion is *greatly* exaggerated here for the purpose of illustration.

Efferent Inhibition of the Cochlea

In mammals, neurons of the superior olivary complex in the brainstem project to ipsilateral and contralateral cochleae (Figure 22.6A).[13] Activation of this pathway causes the release of acetylcholine at efferent synapses onto hair cells and suppresses the response of cochlear afferent fibers to sound.[14,15] An example of the reduction in sensitivity of cochlear afferents by efferent inhibition is shown in Figure 22.6B.[16] Of interest is that suppression is maximal near the best frequency for the afferent, reinforcing the role of outer hair cells in cochlear frequency selectivity.

Efferent fibers respond to sound and innervate restricted portions of the cochlea; as a result, feedback inhibition is frequency specific.[17,18] The effect is to reduce the

[13] Rasmussen, G. 1946. *J. Comp. Neurol.* 84: 141–219.

[14] Galambos, R. 1956. *J. Neurophysiol.* 19: 424–437.

[15] Jasser, A., and Guth, P. S. 1973. *J. Neurochem.* 20: 45–53.

[16] Winslow, R. L., and Sachs, M. B. 1987. *J. Neurophysiol.* 57: 1002–1021.

[17] Wiederhold, M. L., and Kiang, N. Y. 1970. *J. Acoust. Soc. Am.* 48: 950–965.

[18] Liberman, M. C. 1988. *J. Neurophysiol.* 60: 1779–1798.

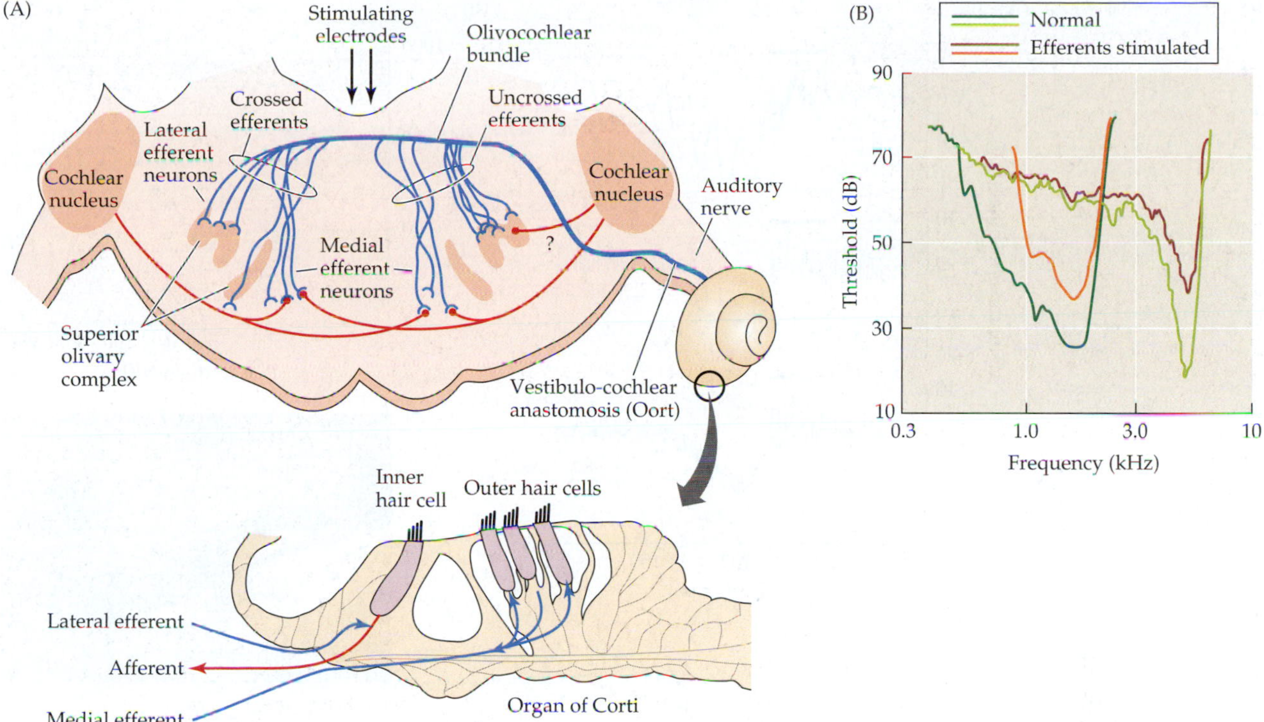

FIGURE 22.6 Efferent Inhibition of the Cochlea (A) Efferent neurons reside near the superior olivary complex. Medial efferents give rise to axons that synapse on outer hair cells. The axons of lateral efferents synapse onto Type I afferent dendrites beneath inner hair cells. (B) Tuning curves of single cochlear afferent fibers and the effect of inhibition. During inhibition (electrical shocks delivered to efferent axons in the floor of the IVth ventricle) the afferent fiber becomes less sensitive (requires louder sound to elicit a threshold response). This inhibitory effect is maximal at the center frequency in sharply tuned fibers, so that frequency selectivity is reduced. (A,B after Warr, Guinan, and White, 1986.)

cochlear response at the relevant frequency, which allows the cochlea to detect differences in intensity of loud sounds that would otherwise saturate the response. In other words, inhibition increases the dynamic range of perception at that frequency. Efferent feedback also plays a role in protecting the cochlea from loud sound damage.[19] Indeed, the strength of efferent feedback is inversely correlated with the degree of acoustic injury from loud sound.[20]

All vertebrate hair cell organs are subject to efferent regulation, and intracellular recordings were first achieved in experimentally tractable preparations, such as the fish lateral line.[21] Detailed studies in the turtle ear showed that activation of the efferent pathway caused large hyperpolarizing inhibitory postsynaptic potentials (IPSPs) in hair cells[22] that would be expected to suppress the hair cell's response to sound. In fact, the inhibition of acoustic sensitivity was even more interesting, as shown by the intracellular recordings in Figure 22.7A. The hair cell was stimulated with pure tones at three frequencies: one that corresponded to the best or characteristic frequency, one at a higher frequency, and one at a lower frequency. The intensities of the tones were adjusted so that each evoked an oscillating receptor potential of the same amplitude. A short train of shocks to the efferent axons hyperpolarized the cell and severely attenuated its response to a tone at 220 Hz (the characteristic frequency). At lower and higher frequencies of acoustic stimulation, activation of the efferents still resulted in hyperpolarization, but the voltage oscillations at high frequencies were unchanged, while those at low frequencies were actually enhanced in amplitude. This differential effect of inhibition results in broadening of the hair cell's frequency response (Figure 22.7B), qualitatively similar to the *detuning* of cochlear afferents in mammals (see Figure 22.6B). The basis for frequency tuning in the turtle ear is described in a later section.

[19] Rajan, R. 1995. *J. Neurophysiol.* 74: 598–615.

[20] Taranda, J. et al. 2009. *PLoS Biol.* 7: e18.

[21] Flock, A., and Russell, I. 1976. *J. Physiol.* 257: 45–62.

[22] Art, J. J., Fettiplace, R., and Fuchs, P. A. 1984. *J. Physiol.* 356: 525–550.

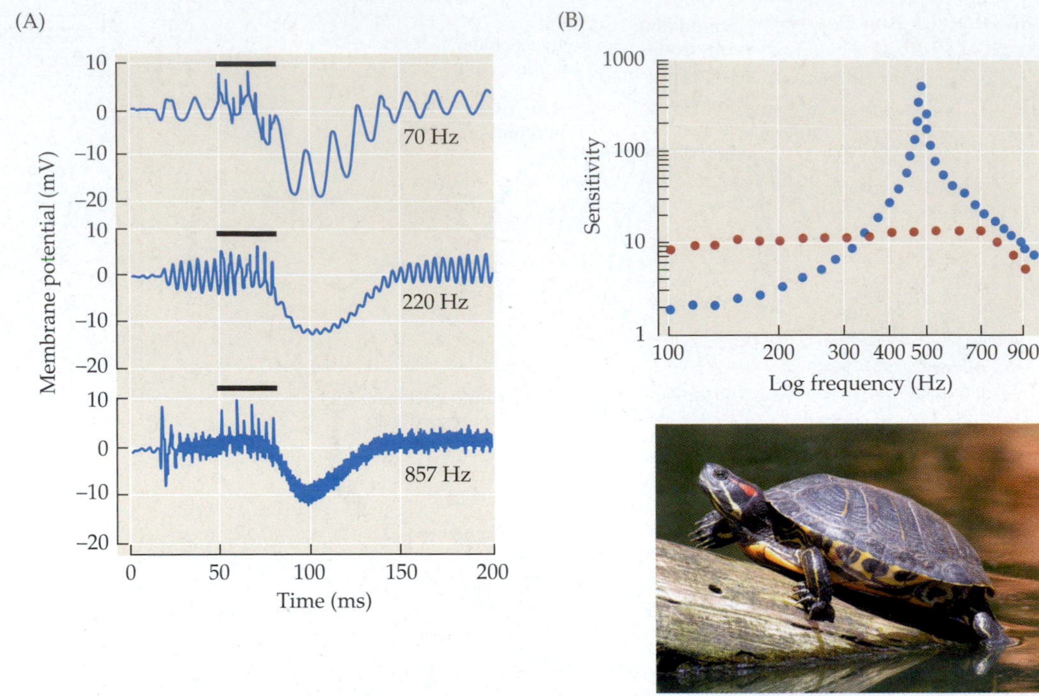

Trachemys scripta elegans

FIGURE 22.7 Effect of Efferent Stimulation on cochlear hair cell responses to acoustic stimuli. (A) Center record: The oscillatory response of a cell to an acoustic stimulus of 220 Hz (its resonant frequency) is inhibited by a brief train of efferent stimuli (indicated by the bar), and the cell is hyperpolarized. Upper record: The response to an acoustic stimulus at 70 Hz. The stimulus intensity was adjusted so that the oscillatory response was similar in magnitude to that at 220 Hz. Efferent stimulation produces a hyperpolarization, but an increase, rather than a decrease, in the oscillatory response. Lower record: The oscillatory response to an 857 Hz stimulus (again adjusted to produce a response similar to that at the resonant frequency) is unchanged by efferent stimulation. (B) Sensitivity of another cell (in millivolts per unit of sound pressure) as a function of frequency in the absence (blue) and presence (red) of efferent stimulation. Efferent inhibition reduces the response at the resonant frequency and increases the sensitivity at low frequencies, thereby degrading tuning. (A,B after Art et al., 1985; photograph by Neil Hardwick/Alamy.)

The ionic mechanism of cholinergic inhibition was established in chicken hair cells.[23,24] The acetylcholine receptors (AChRs) are ligand-gated cation channels through which sodium and calcium enter the cell, and this leads to the activation of calcium-dependent potassium channels. This two-channel mechanism produces a biphasic change in membrane potential, dominated by the much larger and longer-lasting hyperpolarization. A similar cholinergic inhibitory mechanism is found in mammalian hair cells.[25–27] The hair cell's response to ACh is antagonized by α-bungarotoxin, and there is good evidence that two unusual members of the nicotinic receptor family of genes, α9 and α10, form the hair cell AChR.[28,29]

This synaptic inhibition can explain how efferent synapses on outer hair cells suppress cochlear sensitivity measured as the response of inner hair cells and their associated afferent neurons. Since outer hair cell electromotility is driven by rapidly alternating acoustic stimuli, the prolonged inhibitory hyperpolarization interrupts that motility. As a result, inhibition reduces the active mechanical contribution of outer hair cells to vibration of the cochlear partition,[30] and so reduces excitation of inner hair cells. In this way the sensitivity and tuning of cochlear afferent neurons is reduced.[17]

It is interesting that electromotility is found only in outer hair cells of the mammalian cochlea. Both prestin (the hair cell motor protein) and α10 (a component of the hair cell AChR) have undergone purifying selection during evolution of the vertebrate ear, resulting in isoforms unique to the mammalian cochlea.[31] Nonetheless, efferent inhibition in the turtle produces a similar loss of tuning,[32] implying that some other voltage-dependent process enhances frequency selectivity in nonmammalian hair cells.

[23] Fuchs, P. A., and Murrow, B. W. 1992. *J. Neurosci.* 12: 800–809.

[24] Martin, A. R., and Fuchs, P. A. 1992. *Proc. R. Soc. Lond., B, Biol. Sci.* 250: 71–76.

[25] Blanchet, C. et al. 1996. *J. Neurosci.* 16: 2574–2584.

[26] Evans, M. G. 1996. *J. Physiol.* 491: 563–578.

[27] Housley, G. D., and Ashmore, J. F. 1991. *Proc. R. Soc. Lond., B, Biol. Sci.* 244: 161–167.

[28] Elgoyhen, A. B. et al. 1994. *Cell* 79: 705–715.

[29] Elgoyhen, A. B. et al. 2001. *Proc. Natl. Acad. Sci. USA* 98: 3501–3506.

[30] Russell, I. J., and Murugasu, E. 1997. *J. Acoust. Soc. Am.* 102: 1734–1738.

[31] Franchini, L. F., and Elgoyhen, A. B. 2006. *Mol. Phylogenet. Evol.* 41: 622–635.

[32] Art, J. J., and Fettiplace, R. 1984. *J. Physiol.* 356: 507–523.

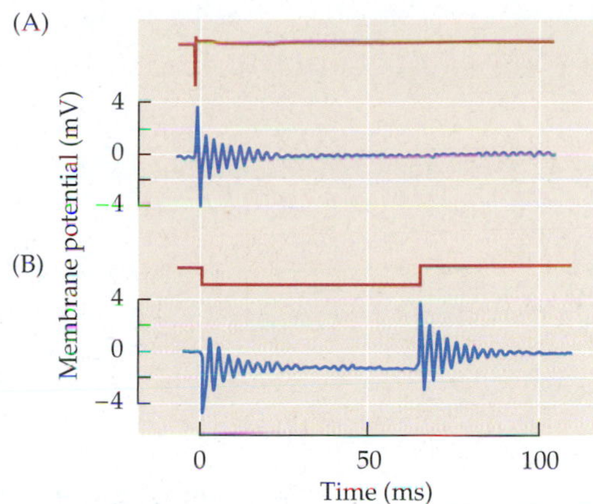

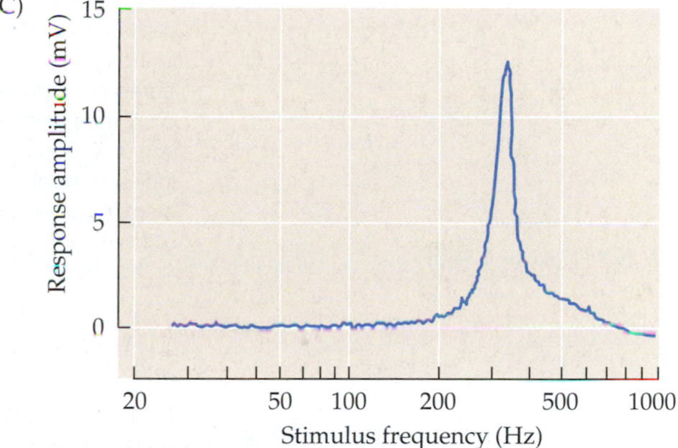

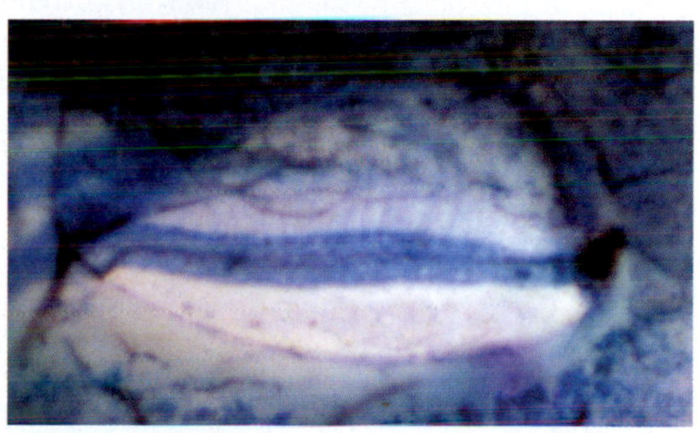

FIGURE 22.8 Hair Cell Tuning in the Turtle Cochlea. (A) The effect of an acoustic click (indicated by the upper trace) on the membrane potential of a hair cell (lower trace, relative to the resting membrane potential of −50 mV), recorded with an intracellular microelectrode. The click produces a damped oscillation in membrane potential at a frequency of about 350 Hz, with an initial peak-to-peak amplitude of about 8 mV. (B) A hyperpolarizing current pulse (upper trace) applied to the same cell produces similar oscillations at both the onset and termination of the pulse, indicating that the frequency of oscillation is an intrinsic electrical property of the hair cell. (C) When a hair cell with oscillatory responses, such as those seen in A and B, is stimulated with pure tones ranging from 25–1000 Hz, the peak-to-peak amplitude of the receptor potential has a sharp maximum at about 350 Hz. (D) Excised turtle inner ear stained with methlyene blue. The clear ellipse is the basilar membrane along which the sensory hair cells (horizontal blue band) are tonotopically arrayed from low to high frequencies, left to right. (C after Fettiplace, 1987; D courtesy of Paul Fuchs.)

Frequency Selectivity in Nonmammalian Vertebrates: Electrical Tuning of Hair Cells

We have seen that the expanded frequency range of mammalian hearing is based on cellular and structural specializations that confer sharp and sensitive mechanical tuning to an extended cochlear duct. And yet, within their range of hearing, nonmammalian vertebrates achieve good acoustic frequency selectivity despite having much shorter basilar membranes that are not mechanically tuned, as in mammals.[33] Crawford and Fettiplace undertook studies in the turtle that showed that the mechanosensory hair cells were frequency selective through a mechanism of *intrinsic* electrical tuning.[34] Intracellular recordings from a hair cell in the turtle's basilar papilla (the auditory sensory epithelium) are shown in Figure 22.8. When a brief acoustic stimulus (a click) was presented, the hair cell's membrane potential underwent a damped oscillation or ringing that occurred at a frequency of about 350 Hz (Figure 22.8A). This frequency is the same as that of pure tones to which the hair cell was most sensitive, as shown by the frequency sweep experiment in Figure 22.8C. Here, a constant intensity tone was presented to the external ear, and its frequency was gradually swept from 20–1000 Hz. The voltage response in the hair cell peaked near 350 Hz. The voltage ringing produced by an acoustic transient and the tuning curve produced by frequency sweeps are equivalent demonstrations of the hair cell's frequency selectivity.

Figure 22.8B shows the important result that occurs when a microelectrode was used to inject a rectangular current pulse (an electrical transient) into the cell. Voltage ringing was produced with frequency and rate of decay that were identical to those caused by an acoustic transient. The conclusion from these experiments is that the hair cell's frequency selectivity depends on the electrical properties of the plasma membrane. Electrical and acoustic tuning frequencies of hair cells are equivalent and vary systematically along the length of the turtle's basilar papilla, producing a tonotopic array of tuned detectors.

Hair Cell Potassium Channels and Electrical Tuning

What properties of the cell membrane provide this electrical tuning, and how do these properties vary to determine different tuning frequencies? Recordings from hair cells in the frog's saccule revealed that interactions between voltage-gated calcium channels and **calcium-sensitive** (or **Ca^{2+}-gated**), **voltage-gated potassium** (**BK**) channels can produce voltage ringing.[35,36] Studies of hair cells isolated from the turtle ear[37,38] demonstrated that the characteristic frequency of each cell is determined in a remarkably elegant and straightforward way—namely, by the density and kinetic properties of the BK potassium channels in each cell (Figure 22.9).

[33] Manley, G. A. 2000. *Proc. Natl. Acad. Sci. USA* 97: 11736–11743.

[34] Crawford, A. C., and Fettiplace, R. 1981. *J. Physiol.* 312: 377–412.

[35] Hudspeth, A. J., and Lewis, R. S. 1988. *J. Physiol.* 400: 237–274.

[36] Lewis, R. S., and Hudspeth, A. J. 1983. *Nature* 304: 538–541.

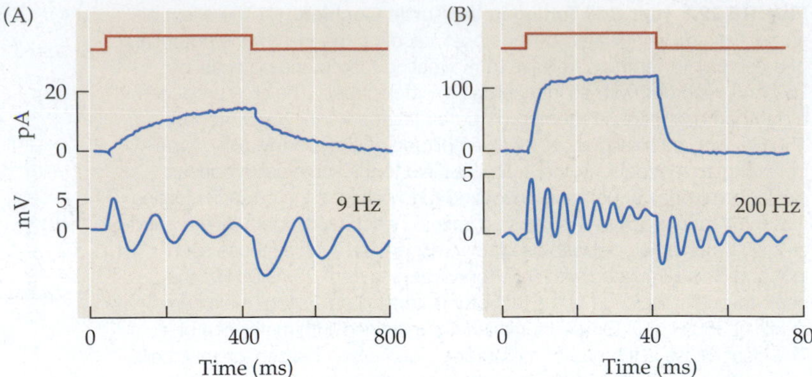

FIGURE 22.9 Tuning Frequency and Potassium Conductance in isolated turtle hair cells measured by whole-cell patch clamp recording. (A) The middle record shows outward current, carried mainly by potassium, produced by a depolarizing voltage command (duration indicated in the top record). Current rises slowly to a maximum of 15 pA. A small current step of the same duration produces oscillatory voltage responses at the beginning and end of the pulse (lower record) with a resonant frequency of 9 Hz. (B) In another cell, a depolarizing pulse produces a much larger, rapidly rising outward current (middle record; note the changes in current and timescales), indicating a greater density of potassium channels with faster kinetics. The oscillatory response to a small current pulse reveals a concomitant increase in tuning frequency to 200 Hz (lower record). (After Fettiplace, 1987.)

In hair cells tuned to lower frequencies, the total potassium conductance is smaller and slower to activate, and thus gives rise to relatively slow voltage oscillations. In higher-frequency cells, the potassium conductance is larger and more rapidly activating. These distinctions extend to the single channel level (Figure 22.10), where it can be shown that slow BK currents are produced by channels with longer mean open times. Channels with shorter mean open times carry faster currents. Single channel conductance is the same, so that high- and low-frequency cells also must express different numbers of channels.

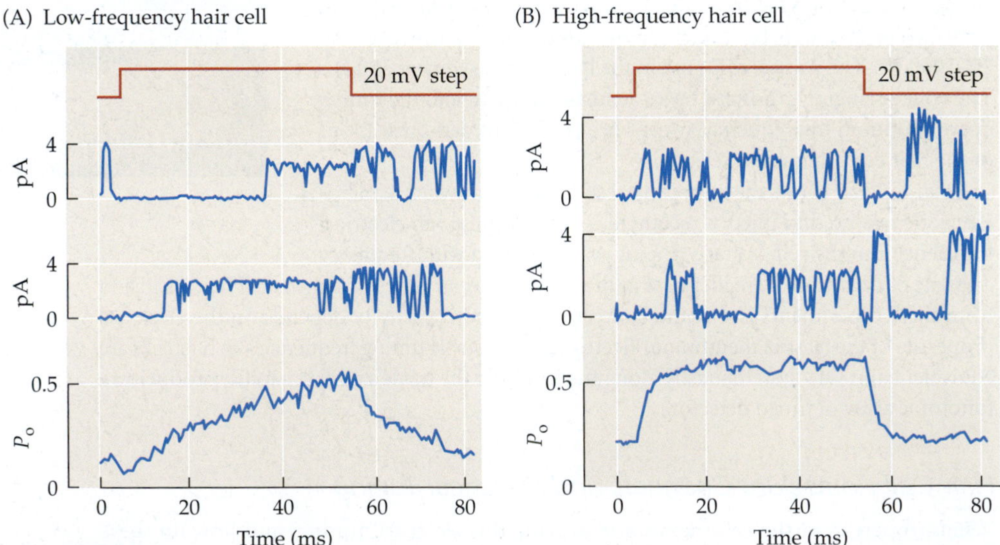

FIGURE 22.10 Single Ca^{2+}- and Voltage-Gated Potassium (BK) Channel Currents in turtle auditory hair cells. (A) Cell attached patch recording from a low-frequency hair cell. Channel opening upward. Lowest record is average compiled from large numbers of records as seen in middle two traces. Activation and deactivation occurs over the course of ~40 ms. (B) Cell attached patch recording from a high-frequency hair cell. Average record shows activation and deactivation rates that are more rapid for the single BK channel from this cell than the cell in (A). (A,B from Art and Fettiplace, 1987.)

The BK channels in hair cells of frogs,[39] turtles,[40] and chickens[41,42] are encoded by a gene whose mRNA is subject to alternative splicing of its composite exons. Some differentially spliced isoforms of the channel are kinetically distinct.[43] Additional variability may be provided by an accessory β-subunit that combines with the channel and slows its gating kinetics.[44] Although BK channels also are expressed in the mammalian cochlea, there is little alternative splicing,[45] and no evidence for electrical resonance in inner or outer hair cells. Nonetheless, as in birds,[46] BK channels appear in rodent cochlear hair cells near the onset of hearing[47] and are more numerous in higher-frequency hair cells.[48,49] The expression of mRNA significantly precedes that of functional channels, and developmental and tonotopic expression patterns appear to be regulated at the level of membrane localization of protein.[50,51]

The Auditory Pathway: Transmission between Hair Cells and Eighth Nerve Fibers

Depolarizing and hyperpolarizing receptor potentials alter the open probability of voltage-gated calcium channels in the hair cell's basolateral membrane. Calcium entry in turn alters the rate of release of neurotransmitter (glutamate) onto the terminal of a postsynaptic afferent neuron. Hair cells, like retinal photoreceptors and bipolar cells, employ so-called **ribbon synapses** for tonic transmitter release[52] (Figure 22.11; see also Chapters 13 and 20). Even in the absence of a stimulus, the membrane potential of the hair cell is positive to the threshold for gating of voltage-activated calcium channels; consequently, release of glutamate is ongoing and excites the afferent fibers, giving rise to spontaneous action potentials. At frequencies below ~5 kHz, the sinusoidal receptor potential in cochlear hair cells alternately increases and decreases the rate of transmitter release, producing phase-locking in the postsynaptic firing pattern. At frequencies greater than 5 kHz, the hair cell's membrane time constant prevents rapid changes in membrane potential and the resulting afferent activity simply increases above the spontaneous rate for the duration of the tone, without cycle-by-cycle phase locking.

A type I afferent neuron in the mammalian cochlea has a single dendrite that is postsynaptic to a single ribbon in a single inner hair cell.[53] Afferent action potential activity often exceeds 100 Hz, requiring that the ribbon synapse have an impressive capacity to marshal and release vesicles.[54–56] Given that each inner hair cell ribbon tethers only 100–200 vesicles,[57] it is still mysterious how this occurs. A further surprise is that spontaneous

[37] Art, J. J., and Fettiplace, R. 1987. *J. Physiol.* 385: 207–242.

[38] Art, J. J., Wu, Y. C., and Fettiplace, R. 1995. *J. Gen. Physiol.* 105: 49–72.

[39] Rosenblatt, K. P. et al. 1997. *Neuron* 19: 1061–1075.

[40] Jones, E. M., Laus, C., and Fettiplace, R. 1998. *Proc. R. Soc. Lond., B, Biol. Sci.* 265: 685–692.

[41] Jiang, G. J. et al. 1997. *Proc. R. Soc. Lond., B, Biol. Sci.* 264: 731–737.

[42] Navaratnam, D. S. et al. 1997. *Neuron* 19: 1077–1085.

[43] Jones, E. M., Gray-Keller, M., and Fettiplace, R. 1999. *J. Physiol.* 518: 653–665.

[44] Ramanathan, K. et al. 1999. *Science* 283: 215–217.

[45] Langer, P., Grunder, S., and Rusch, A. 2003. *J. Comp. Neurol.* 455: 198–209.

[46] Li, Y. et al. 2009. *BMC Dev. Biol.* 9: 67.

[47] Kros, C. J. 2007. *Hear. Res.* 227: 3–10.

[48] Engel, J. et al. 2006. *Neuroscience* 143: 837–849.

[49] Wersinger, E. et al. 2010. *PLoS One.*

[50] Kim, J. M. et al. 2010. *J. Comp. Neurol.* 518: 2554–2569.

[51] Sokolowski, B. et al. 2009. *Biochem. Biophys. Res. Commun.* 387: 671–675.

[52] Matthews G., and Fuchs, P. 2010. *Nat. Rev. Neurosci.* 11: 812–822.

[53] Liberman, M. C. 1982. *Science* 216: 1239–1241.

[54] Frank, T. et al. 2009. *Proc. Natl. Acad. Sci USA* 106: 4483–4488.

[55] Griesinger, C. B., Richards, C. D., and Ashmore, J. F. 2005. *Nature* 435: 212–215.

[56] Rutherford, M. A., and Roberts, W. M. 2006. *Proc. Natl. Acad. Sci USA* 103: 2898–2903.

[57] Sobkowicz, H. M. et al. 1982. *J. Neurosci.* 2: 942–957.

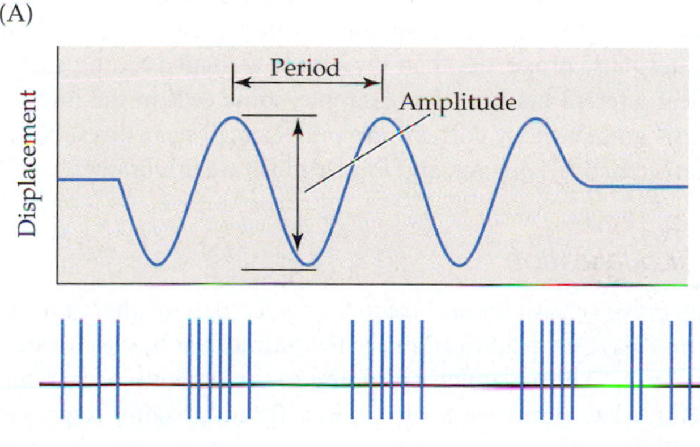

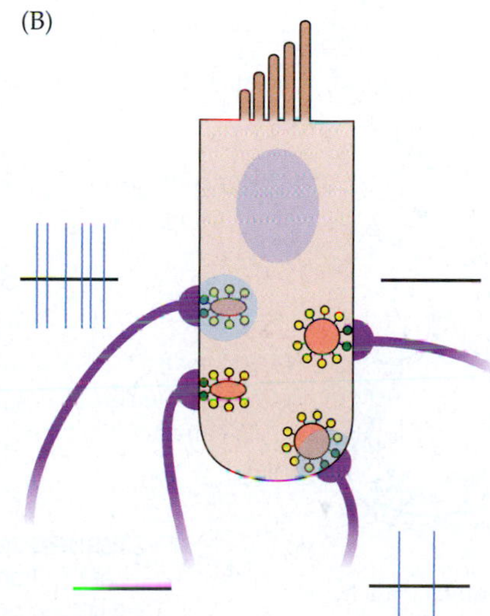

FIGURE 22.11 Hair Cell to Afferent Signaling (A) Type I spiral ganglion neurons are postsynaptic to single ribbon synapses of a single inner hair cell. The ribbon synapse has an electron-dense core to which are tethered ~100 synaptic vesicles. Sinusoidal stimulation of the hair cell gives rise to phase-locked activity in afferent neurons. (B) Individual afferent fibers contacting a single inner hair cell can have different firing rates, both spontaneous and evoked.

transmitter release from ribbon synapses appears to be multivesicular; that is, it can be composed of several vesicles released simultaneously.[58] It is likely that ribbon function may involve unique proteins differing from those serving release at other chemical synapses.[59,60]

Stimulus Coding by Primary Afferent Neurons

Neuronal signaling in the auditory pathway begins with the spiral ganglion neurons that receive transmitter released from hair cells and send their central axons to the cochlear nucleus of the brainstem. Many decades of single fiber recordings have catalogued the acoustic responses of these primary afferents.[61] Each spiral ganglion neuron responds selectively to the frequency of sound that is optimal for the inner hair cell to which it is attached. Each inner hair cell is the sole presynaptic partner of a group of type I afferent neurons, numbering from 10 to 30, depending on location in the cochlea (see Figure 22.11). Both acoustic threshold and spontaneous firing rate vary among this pool of afferents[53] helping to extend the dynamic range of the cochlea. Presumably, individual ribbon synapses of an inner hair cell can have different release properties.[62,63] The selective innervation of inner hair cells on the mechanically tuned basilar membrane produces an array of 10,000 or so afferent neurons that serve as **frequency-labeled lines**—the first stage of the tonotopically organized auditory pathway. In addition to this *pitch-is-place* mechanism, phase-locked firing of afferent action potentials to acoustic sinusoids can be used to encode frequency up to about 3 kHz in the guinea pig[64] and up to 10 kHz in the barn owl.[65]

Brainstem and Thalamus

The main auditory pathways are illustrated schematically in Figure 22.12. Auditory fibers of the eighth nerve travel centrally and send branches to both the dorsal and ventral cochlear nuclei.[66] Second-order axons ascend in the contralateral lateral lemniscus to innervate cells in the inferior colliculus (the nucleus of the lateral lemniscus is a synaptic way station for some of these fibers). Neurons in the ventral cochlear nucleus also provide collateral branches to both the ipsilateral and contralateral superior olivary nuclei. Third-order cells in the olivary nuclei, in turn, send ascending fibers to the inferior colliculus. The ascending pathway continues through the medial geniculate nucleus of the thalamus to the auditory region on the transverse surface of the temporal lobe of the cerebral cortex.

Each level in the auditory pathway is tonotopically mapped. However, as one ascends higher in the auditory system, individual cells have more complex response properties than the simple V-shaped tuning curves of cochlear afferent neurons. For example, some cells in the inferior colliculus[67] and auditory cortex[68] are only excited near threshold at their characteristic frequency, and louder tones are inhibitory.

Sound Localization

The impressive sensitivity and frequency selectivity of the auditory system may have evolved to improve the animal's ability to locate a sound in space. The advantages of doing so are obvious; long-range signals emitted as sound waves can reveal a distant predator or prey in

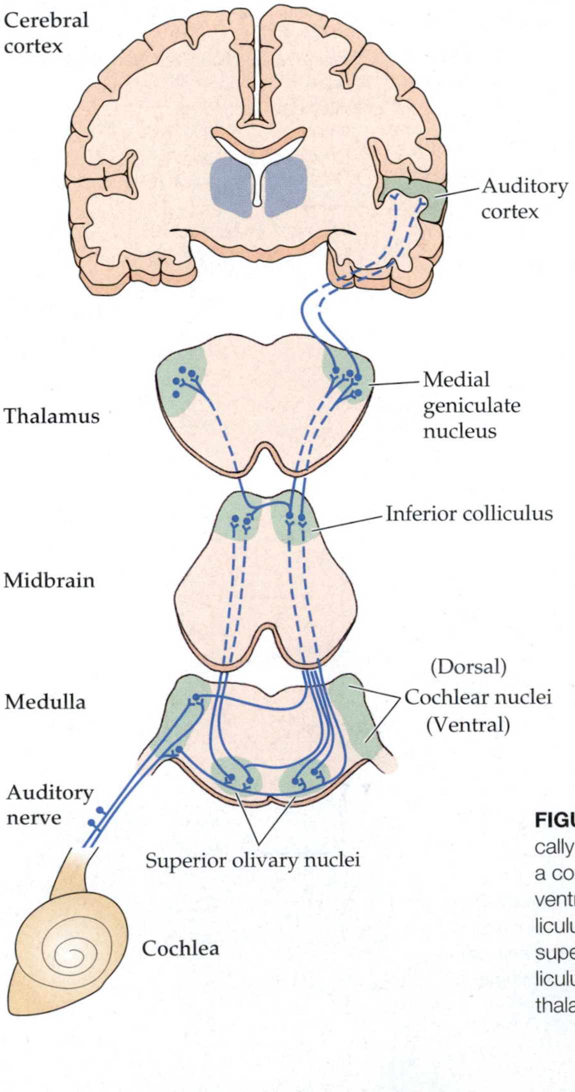

FIGURE 22.12 Auditory Pathways Central auditory pathways are shown schematically on transverse sections of the medulla, midbrain, and thalamus, as well as on a coronal section of the cerebral cortex. Auditory nerve fibers end in the dorsal and ventral cochlear nuclei. Second-order fibers ascend to the contralateral inferior colliculus; those from the ventral cochlear nucleus also supply collaterals bilaterally to the superior olivary nuclei. Further bilateral interaction occurs at the level of the inferior colliculus. Neurons of the inferior colliculus project to the medial geniculate nucleus of the thalamus, which in turn projects to the auditory cortex. (After Berne and Levy, 1988.)

[58] Glowatzki, E., and Fuchs, P. A. 2002. *Nat. Neurosci.* 5: 147–154.

[59] Safieddine, S., and Wenthold, R. J. 1999. *Eur. J. Neurosci.* 11: 803–812.

[60] Roux, I. et al. 2006. *Cell* 127: 277–289.

[61] Young, E. D., and Sachs, M. B. 1979. *J. Acoust. Soc. Am.* 66: 1381–1403.

[62] Meyer, A. C. et al. 2009. *Nat. Neurosci.* 12: 444–453.

[63] Grant, L., Yi, E., and Glowatzki, E. 2010. *J. Neurosci.* 30: 4210–4220.

[64] Palmer, A. R., and Russell, I. J. 1986. *Hear. Res.* 24: 1–15.

[65] Koppl, C. 1997. *J. Neurosci.* 17: 3312–3321.

[66] Fekete, D. M. et al. 1984. *J. Comp. Neurol.* 229: 432–450.

[67] Ramachandran, R., Davis, K. A., and May, B. J. 1999. *J. Neurophysiol.* 82: 152–163.

[68] Sadagopan, S., and Wang, X. 2008. *J. Neurosci.* 28: 3415–3426.

the absence of visual or other cues. However, in contrast to the visual or somatosensory systems, the auditory neuroepithelium cannot code location directly. Instead, sound location is computed from binaural comparisons of timing and intensity that occur within the central auditory system. The auditory pathway, therefore, is correspondingly complex and involves numerous subcortical synaptic relays and extensive crossing over at nearly every level.

The **dorsal cochlear nucleus** is used largely for monaural frequency analysis[69] and provides a relatively direct, tonotopically organized projection onto the contralateral primary auditory cortex, or A1, via the inferior colliculus and medial geniculate nucleus of the thalamus. In contrast, second-order neurons in the **ventral cochlear nucleus** project both ipsilaterally and contralaterally to the superior olivary complex in the brainstem. Most neurons in the **medial superior olive** (**MSO**) are excited by stimulation of either ear[70] (and so are designated EE neurons) but respond best when a tone is presented to the two ears with a characteristic delay, corresponding to arrival first at one ear, then the other. The velocity of sound in air is 340 m/s, so the maximal time difference imparted by the human head (about 18 cm in diameter) is 0.5 ms for a sound arriving along the axis of the two ears, and successively more frontal locations impart smaller time differences. In addition to differences in arrival time, continuous sound sources give rise to phase differences at the two ears.[71,72]

Cells in the **lateral superior olive** receive excitation from the ipsilateral ventral cochlear nucleus (Figure 22.13). Cells in the contralateral ventral cochlear nucleus project across the midline to form synapses in the **medial nucleus of the trapezoid body** (**MNTB**). The cells of the MNTB *inhibit* neurons in the lateral superior olive.[73] Thus, neurons in the lateral superior olive are excited by ipsilateral but inhibited by contralateral sound and so are designated EI neurons. Such an interaction would be useful for detecting differences in the intensity of sound at the two ears. As much as a 10-fold difference in intensity is found at high frequencies, for which the head serves as an effective sound shadow.[74]

Psychophysical studies have shown that localization is accomplished by combining differences between the two ears in time of arrival and/or intensity of the incoming sound.[75,76] Thus, if clicks are presented through earphones with different delays, the sound is localized toward the ear in which the click arrives first. Humans can detect interaural time differences of as little as 5 μs—remarkable resolution considering that action potentials are approximately 1 ms in duration, which once again emphasizes the importance of precise timing for auditory function. If the clicks are made simultaneously but are of different intensities, localization occurs to the side with the loudest click. Both phase and intensity differences vary as a function of frequency. For the human head, phase differences between the two ears are more significant below 2 kHz, whereas intensity differences become more prominent at higher frequencies.

[69] Malmierca, M. S. 2003. *Int. Rev. Neurobiol.* 56: 147–211.

[70] Cant, N. B., and Casseday, J. H. 1986. *J. Comp. Neurol.* 247: 457–476.

[71] Buell, T. N., Trahiotis, C., and Bernstein, L. R. 1991. *J. Acoust. Soc. Am.* 90: 3077–3085.

[72] Kuwada, S. et al. 2006. *J. Neurophysiol.* 95: 1309–1322.

[73] Moore, M. J., and Caspary, D. M. 1983. *J. Neurosci.* 3: 237–242.

[74] Brungart, D. S., Durlach, N. I., and Rabinowitz, W. M. 1999. *J. Acoust. Soc. Am.* 106: 1956–1968.

[75] Blauert, J. 1982. *Scand. Audiol. Suppl.* 15: 7–26.

[76] Buell, T. N., and Hafter, E. R. 1988. *J. Acoust. Soc. Am.* 84: 2063–2066.

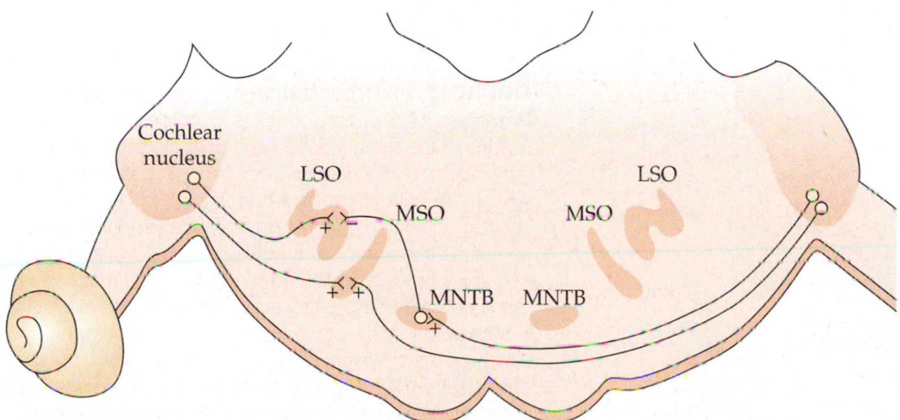

FIGURE 22.13 Binaural Connections in the Olivary Complex. Neurons in the ventral cochlear nucleus project to the ipsilateral LSO (lateral superior olive) as well as to the contralateral (MNTB) medial nucleus of the trapezoid body and the contralateral MSO (medial superior olive). Thus, MSO neurons are excited by both ears, and LSO neurons are excited ipsilaterally but are inhibited contralaterally by way of the intervening inhibitory interneuron in the MNTB.

The ability to localize the vertical position of a sound depends heavily on its frequency composition.[77] The external ear and head are not in mirror symmetry above and below. As a consequence, frequency components are differentially reflected depending on whether a sound rises or falls toward the listener.[78]

Auditory Cortex

Auditory input processed through both dorsal and ventral cochlear nuclei ascends to the auditory cortex. The **primary auditory cortex (A1)** is located on the superior bank of the temporal lobe, corresponding to Brodmann's areas 41 and 42 (see Appendix C) (Figure 22.14). In cats, A1 is exposed on the lateral surface of the brain; many combined anatomical–physiological studies have been performed in this species.[79] Microelectrode recordings in anesthetized animals have shown that A1 has a columnar organization, with cells along a vertical track all having the same best frequency.[80,81] The auditory periphery (organ of Corti) is essentially a one-dimensional map of frequency, so auditory cortex is laid out in isofrequency slab-like formation, along which other acoustic dimensions can be mapped.[82]

The auditory cortex in monkeys contains multiple cochleotopic maps, with parallel projections from the medial geniculate nucleus to all these areas.[83] The most posterior of the three areas corresponds to A1.[84] Surrounding this central core are secondary auditory areas that interconnect with primary cortex but also with subdivisions of the medial geniculate nucleus. Thus, pathways exist for both serial and parallel processing in the auditory cortex.[85]

By analogy to other sensory cortices, one would expect the tonotopic map of A1 to be subdivided into different functional zones. What acoustic features will be mapped onto cortex? Neurons within an isofrequency slab are clustered according to whether inputs from the two ears summate or suppress.[86] This clustering is thought to reflect binaural computations of sound location carried out in olivary nuclei, although there is no contiguous cortical map of auditory space.[82] And, while evidence also exists for systematic variations in intensity and bandwidth coding in A[1,87–89] their functional significance remains obscure.

Environmental sounds, including vocalizations, consist of constant frequency elements, frequency-modulated elements, and noise bursts.[90] That is, recognition of acoustic objects must depend on analysis not just of frequency composition but also of how that composition changes in time and the relative timing between different acoustic elements. There is growing appreciation that many cortical neurons are particularly sensitive to the time-varying components of species-specific vocalizations or calls.[91] Over 20 years ago, Suga and colleagues established this principle in studies of echo-locating bats,[92] and more recently, it has been examined in other species.[93,94] Human vocalizations (i.e., speech sounds) are determined largely by the resonant frequencies of the vocal tract, called **formants**. Vowels are identified by the relative frequencies of the first two formants. Consonants depend more heavily on rapid temporal transitions.[91]

[77] May, B. J., and Huang, A. Y. 1996. *J. Acoust. Soc. Am.* 100: 1059–1069.
[78] Rice, J. J. et al. 1992. *Hear. Res.* 58: 132–152.
[79] Winer, J. A., and Lee, C. C. 2007. *Hear. Res.* 229: 3–13.
[80] Merzenich, M. M., Knight, P. L., and Roth, G. L. 1975. *J. Neurophysiol.* 38: 231–249.
[81] Reale, R. A., and Imig, T. J. 1980. *J. Comp. Neurol.* 192: 265–291.
[82] Schreiner, C. E., and Winer, J. A. 2007. *Neuron* 56: 356–365.
[83] Hackett, T. A., Preuss, T. M., and Kaas, J. H. 2001. *J. Comp. Neurol.* 441: 197–222.
[84] Romanski, L. M., and Averbeck, B. B. 2009. *Annu. Rev. Neurosci.* 32: 315–346.
[85] Lee, C. C., and Winer, J. A. 2005. *Cereb. Cortex* 15: 1804–1814.
[86] Middlebrooks, J. C., Dykes, R. W., and Merzenich, M. M. 1980. *Brain Res.* 181: 31–48.
[87] Clarey, J. C., Barone, P., and Imig, T. J. 1994. *J. Neurophysiol.* 72: 2383–2405.
[88] Heil, P., Rajan, R., and Irvine, D. R. 1994. *Hear. Res.* 76: 188–202.
[89] Sutter, M. L., and Schreiner, C. E. 1995. *J. Neurophysiol.* 73: 190–204.
[90] Kanwal, J. S., and Rauschecker, J. P. 2007. *Front. Biosci.* 12: 4621–4640.
[91] Young, E. D. 2008. *Philos. Trans. R. Soc. Lond., B, Biol. Sci.* 363: 923–945.
[92] Tsuzuki, K., and Suga, N. 1988. *J. Neurophysiol.* 60: 1908–1923.
[93] Wang, X., and Kadia, S. C. 2001. *J. Neurophysiol.* 86: 2616–2620.
[94] Schnupp, J. W. et al. 2006. *J. Neurosci.* 26: 4785–4795.

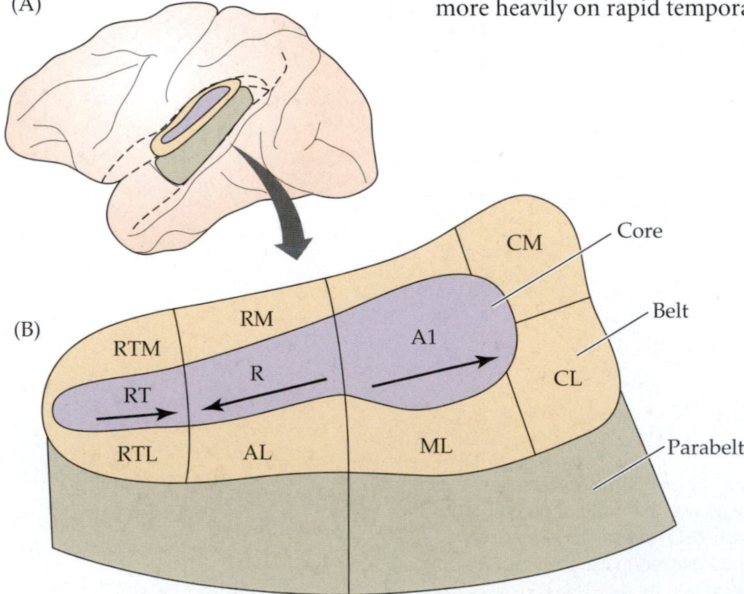

FIGURE 22.14 Organization of Auditory Cortex in Primates. (A) Auditory cortex is found on the superior surface of the temporal lobe and consists of a central core region, surrounded by "belt" and "parabelt" regions that are thought to correspond to higher level association areas. (B) The core region of primate auditory cortex contains three complete tonotopic maps (low- to high-frequency gradient indicated by arrows). The most caudal map is considered primary auditory cortex (A1). Surrounding the core is the belt region divided into geographic zones, some of which are tonotopically mapped, but less distinctly than A1. Parabelt is still higher order auditory association cortex. A1 = primary or core auditory area; R = rostral core auditory area; RM = rostromedial; RT = rostral temporal core auditory area; RTM = RT medial; RTL = RT lateral; AL = anterolateral belt auditory cortex; ML = middle-lateral belt auditory cortex; CM = caudal-medial belt auditory cortex; CL = caudal-lateral belt auditory cortex. (After Romanski and Averbeck, 2009.)

Understanding cortical representation of these complex acoustic objects requires more comprehensive analytical methods, such as reverse correlation. In this approach the response of a single cortical neuron is used to examine the features of a broadband, randomly modulated stimulus leading up to that response. A neuron's selectivity for some combination of spectrotemporal features (called the spectrotemporal response function, or STRF) is defined by its preferred rate (i.e., modulation in time), and scale (i.e., breadth of frequency selectivity). Using these parameters, a hypothetical neuron selective for a vowel sound would have a high scale (narrow spectral bandwidth) and low rate (little temporal modulation). A neuron encoding consonants would have low scale (broad bandwidth) and high rate (rapid transients).[95]

The processing of auditory signals is complex.[96,97] Behaviorally important sounds must be extracted from a rich and variable acoustic environment.[98] Not only the frequency content but also the sequence in time of incoming sounds must be analyzed in some way[99] (e.g., playing a tape recording of human speech backward produces gibberish). In humans, the basic elements of speech, called **phonemes**, are common to all languages and are the sounds first babbled by babies, before particular sounds are selected to be combined into words.[100] These basic sounds are composed of the formants and temporal transitions previously described. One evident challenge for auditory perception is to identify a particular temporal pattern of frequency changes as an acoustic object, even as the absolute frequencies vary with different speakers (male versus female voices, for example). By analogy with the visual system, which contains cells that respond to slits, corners, edges, and other geometrical forms (see Chapter 3), we might expect to find higher-order cells in the human auditory cortex that respond to particular formants or, perhaps, phonemes. However, in contrast to the other senses, auditory information is extensively processed at subcortical levels, as suggested by its more complex subcortical circuitry. For example, neurons of the inferior colliculus signal interaural time disparities equivalent to those of behavioral threshold.[101]

There is an emerging realization that primary auditory cortex may encode relatively complex auditory features, or even complete auditory percepts,[102] rather than performing the hierarchical assembly of features as in other primary sensory cortices. This organizational plan may reflect the greater challenge of encoding stimuli whose *sense* is not primarily spatial, as it is for vision and touch. It remains to be seen whether auditory cortex analyzes component features of acoustic stimuli, or more complex assemblies of features that might be regarded as auditory objects.[96]

The Vestibular System: Encoding Head Motion and Position

The inner ear contains the machinery for detecting movement of the head. Three semicircular canals, one horizontal and two vertical, are designed to detect angular motion of the head (rotation), while the macular epithelia, saccule, and utricle (surmounted by otolithic crystals) all signal linear acceleration (see Figure 22.1). The arrangement of three orthogonally disposed vestibular canals is conserved throughout all vertebrates (with the exception of jawless fishes, such as the lamprey with only two canals and a single otolithic macula).[103] Even invertebrate cephalopod molluscs (e.g., squid, octopi) have vestibular-like organs, called statocysts.[104]

Vestibular information is essential to maintenance of posture, efficient locomotion, and stability of gaze. Sometimes referred to as "the silent sense," vestibular function remains largely subconscious, at least, until something goes wrong. Motion sickness is one temporary but common experience of inappropriate vestibular input, while more serious, longer-lasting deficits can significantly degrade the quality of life. One easily appreciated example arises from consideration of the vestibulo-ocular reflex (VOR) that enables one's gaze to remain fixed on a target as the head moves. Without this reflex, activities such as driving a car become impossible, since gaze fixation is required to read signs, avoid obstacles, etc., while the car, and hence the head, is in motion.[105]

Vestibular Hair Cells and Neurons

As in the cochlea, the central chamber of the vestibular labyrinth is filled with potassium-rich endolymph. The stereociliary bundles of sensory hair cells project into this space.

[95] Mesgarani, N. et al. 2008. *J. Acoust. Soc. Am.* 123: 899–909.

[96] Nelken, I., and Bar-Yosef, O. 2008. *Front. Neurosci.* 2: 107–113.

[97] Sachs, M. B. 1984. *Annu. Rev. Physiol.* 46: 261–273.

[98] Nelken, I., Rotman, Y., and Bar Yosef, O. 1999. *Nature* 397: 154–157.

[99] Griffiths, T. D. et al. 1998. *Nat. Neurosci.* 1: 422–427.

[100] De Boysson-Bardies, B. et al. 1989. *J. Child Lang.* 16: 1–17.

[101] Shackleton, T. M. et al. 2003. *J. Neurosci.* 23: 716–724.

[102] Nelken, I. 2004. *Curr. Opin. Neurobiol.* 14: 474–480.

[103] Lowenstein, O., Osborne, M. P., and Thornhill, R. A. 1968. *Proc. R. Soc. Lond., B, Biol. Sci.* 170: 113–134.

[104] Williamson, R., and Chrachri, A. 2007. *Philos. Trans. R. Soc. Lond., B, Biol. Sci.* 362: 473–481.

[105] Land, M. F. 2009. *Vis. Neurosci.* 26: 51–62.

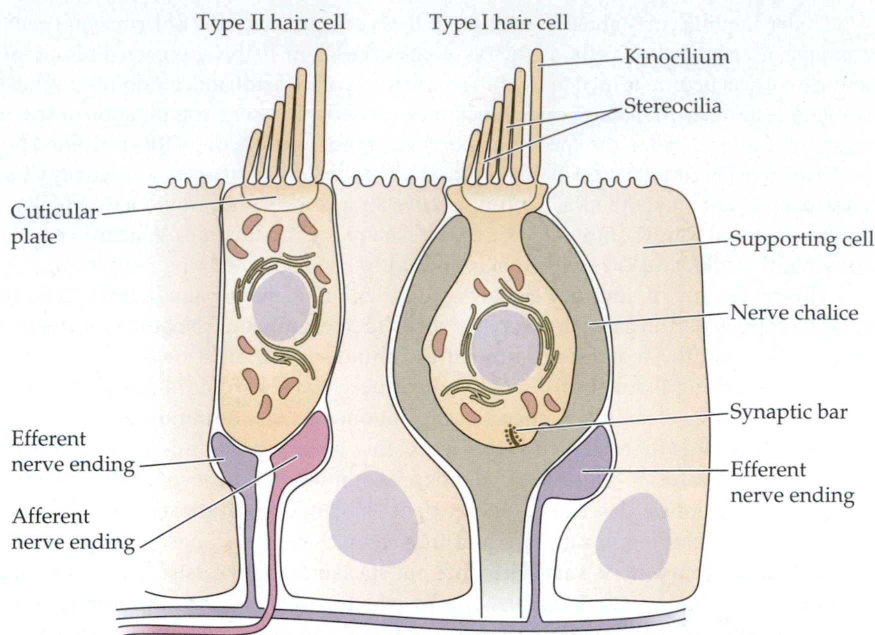

FIGURE 22.15 Vestibular Hair Cells and Neurons. Two types of hair cell are found in all vestibular epithelia. Type I is amphora-shaped and entirely engulfed by the highly specialized calyx afferent ending. Type II hair cells are more "typical," as they are columnar and contacted by bouton-type afferent endings. Vestibular afferent neurons also can be classified by morphology. Calyx-only afferents have the largest diameter axons, only make calyx endings, and tend to be found near the center of each sensory epithelium. Bouton-only afferents have smaller diameter axons and tend to be found in the periphery of the epithelium. Dimorphic afferents make both calyx and bouton endings and are scattered throughout the epithelium. Efferent neurons also innervate vestibular epithelia and make contacts directly onto type II hair cells, and onto the calyx ending surrounding type I hair cells. Efferent activation can result in either excitation or inhibition of vestibular afferent activity.

Consequently, potassium ions enter through open transducer channels to depolarize the hair cell. In contrast to the cochlea, the vestibular labyrinth has only a small negative endolymphatic potential.[106] Hair cell receptor potentials are additionally shaped by a variety of voltage-gated potassium channels whose distribution varies by cell type.[107] As in the cochlea, resting tension within the hair bundle keeps a small fraction of transducer channels open at rest—that is, in the absence of explicit stimulation. Thus, head movements that deflect vestibular hair bundles toward the tallest stereocilia depolarize by opening additional transducer channels, while oppositely oriented deflections hyperpolarize by closing those transducer channels that are open at rest.

Responses of vestibular hair cells to movements are determined largely by their location and orientation within the vestibular apparatus. Neither a high-resolution micrograph of a vestibular hair cell nor an intracellular electrical recording would identify a particular cell as being one from the cristae of the semicircular canals or from the maculae of the utricle and saccule.[108] Rather, hair cell structure and function vary topographically in each end organ.[109] The hair cells are of two types (Figure 22.15). Type I vestibular hair cells are amphora-shaped and enclosed by a specialized afferent nerve terminal, the calyx. The hair cells release glutamate onto the calyx from ribbon type synapses. Type II hair cells are columnar in shape, have ribbon synapses in apposition to bouton-like endings of afferent neurons, and once again, employ glutamate as a neurotransmitter.

Afferent fibers contact multiple hair cells and are distinguished by the types of synaptic endings they make: calyx-only, bouton-only, and dimorphic (i.e., calyx and bouton).[110] In addition, electrical recordings from vestibular afferents show that they can be divided into regular and irregular subtypes on the basis of their firing patterns.[111] Irregularly firing afferents tend to have larger diameter axons with calyx or dimorphic endings, and their

[106] Marcus, D. C. et al. 2002. *Am. J. Physiol. Cell Physiol.* 282: C403–407.

[107] Brichta, A. M. et al. 2002. *J. Neurophysiol.* 88: 3259–3278.

[108] Eatock, R. A. et al. 1998. *Otolaryngol. Head Neck Surg.* 119: 172–181.

[109] Holt, J. C. et al. 2007. *J. Neurophysiol.* 98: 1083–1101.

[110] Goldberg, J. M. 2000. *Exp. Brain Res.* 130: 277–297.

[111] Eatock, R. A., Xue, J., and Kalluri, R. 2008. *J. Exp. Biol.* 211: 1764–1774.

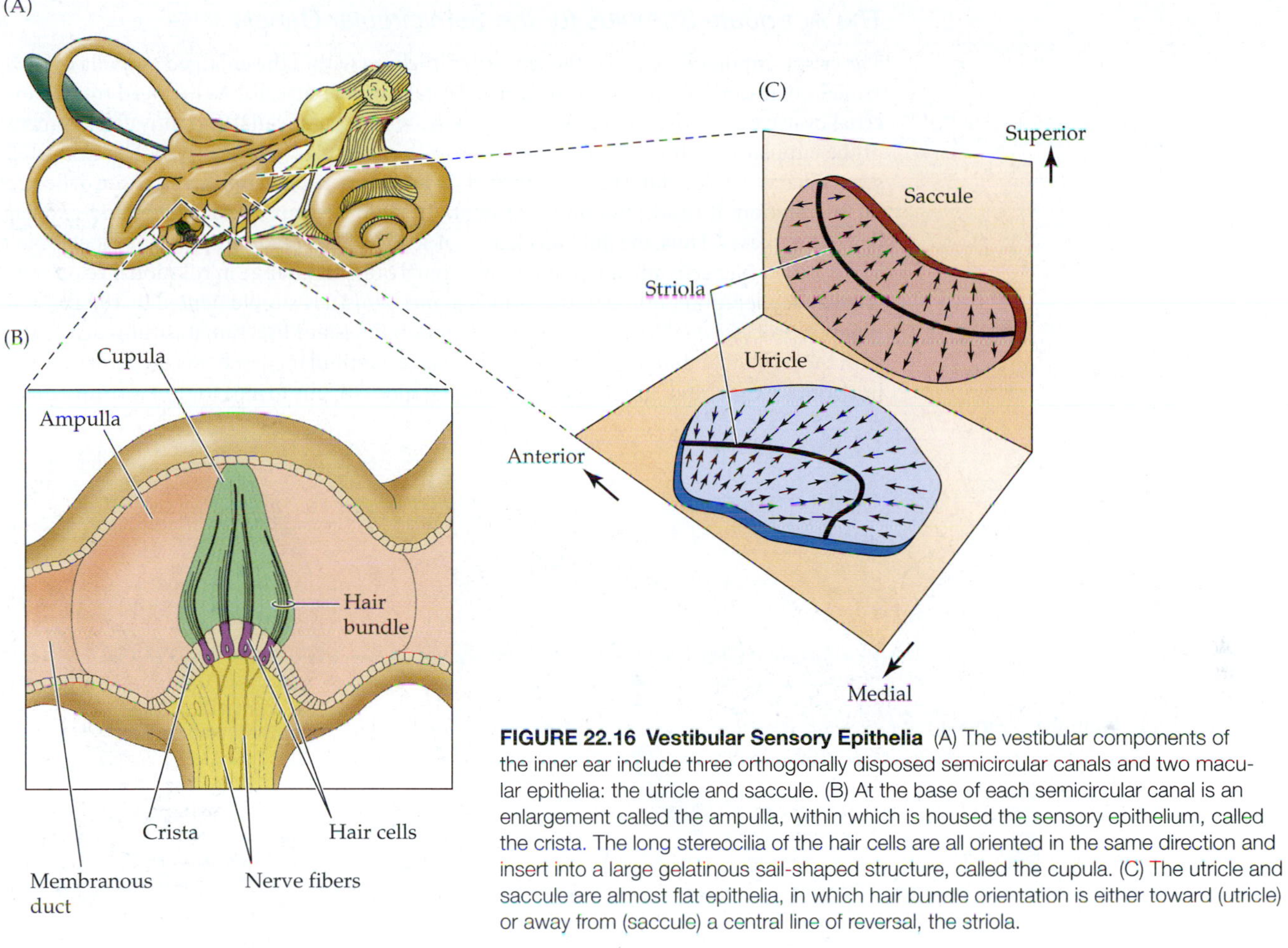

FIGURE 22.16 Vestibular Sensory Epithelia (A) The vestibular components of the inner ear include three orthogonally disposed semicircular canals and two macular epithelia: the utricle and saccule. (B) At the base of each semicircular canal is an enlargement called the ampulla, within which is housed the sensory epithelium, called the crista. The long stereocilia of the hair cells are all oriented in the same direction and insert into a large gelatinous sail-shaped structure, called the cupula. (C) The utricle and saccule are almost flat epithelia, in which hair bundle orientation is either toward (utricle) or away from (saccule) a central line of reversal, the striola.

responses to head movement are more transient, or phasic.[112] Vestibular afferent fibers can have high spontaneous firing rates, so that head movements are encoded by both increasing and decreasing firing rate.

Cholinergic efferent innervation is directed to the calyces on type I hair cells and directly onto type II hair cells. Depending upon the species and end organ under study, efferent feedback can be inhibitory, excitatory, or both.[113]

The Adequate Stimulus for the Saccule and Utricle

The macular epithelia, saccule and utricle, are commonly referred to as otolithic organs because each is overlaid by a mass of calcium carbonate crystals encased in a gelatinous matrix (Figure 22.16). These otoliths (or ear stones) reside in the endolymphatic space in contact with hair cell stereocilia. Head motion results in stereociliary deflection because the inertial load of the otoliths delays their motion relative to that of the head. Because the utricle lies approximately in the horizontal plane when the head is upright, it signals horizontal motion. The saccule hangs nearly vertically in the upright head, so it responds maximally to vertical motion. In each instance the effective stimulus is linear acceleration. Constant linear velocity is not a stimulus (as we experience during vehicular travel). Hair cell orientations vary systematically within each macular epithelium (see Figure 22.16). In the utricle, the tall edge of each hair bundle (and the position of the kinocilium) points toward a central line of reversal. In the saccule hair, bundles point away. Thus, analysis of linear acceleration must involve integration of both negative and positive changes in afferent activity from each end organ as well the combination of inputs from both saccule and utricle from both sides of the head.

[112] Baird, R. A. et al. 1988. *J. Neurophysiol.* 60: 182–203.

[113] Highstein, S. M. 1991. *Neurosci. Res.* 12: 13–30.

The Adequate Stimulus for the Semicircular Canals

The crista ampullaris, which is the sensory epithelium within the enlarged ampulla of each semicircular canal, is comprised of hair cells and supporting cells. As the head rotates, inertial drag causes movement of fluid within the semicircular canal and thus displacement of the cupular membrane that stretches across the ampulla (Figure 22.17). The elongated stereociliary bundles of the hair cells project up into the acellular cupula and so are deflected by this motion. If head rotation is constant, the fluid eventually catches up and cupular deflection ceases. Thus, the fluid mechanics of semicircular canals impart a particular sensibility; their hair cells and afferent neurons signal only the change in rotational velocity or angular acceleration (although this principal mechanism is supplemented by cellular and synaptic processes).[114] The orientation of hair bundles is unidirectional within each crista, with the result that fluid motion toward the central vestibular chamber is excitatory for the horizontal canals, and inhibitory (i.e., reducing spontaneous firing) for the vertical canals.

[114] Highstein, S. M. et al. 2005. *J. Neurophysiol.* 93: 2359–2370.

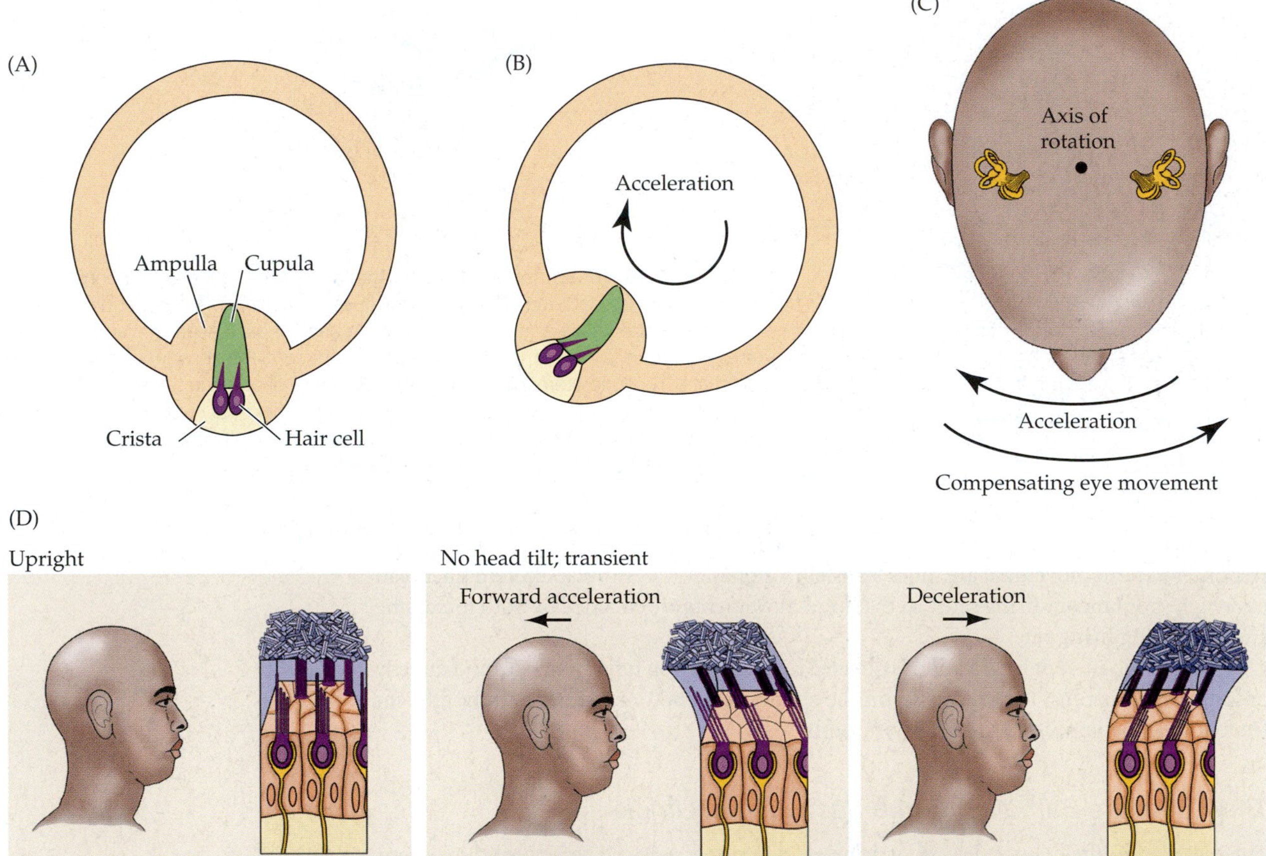

FIGURE 22.17 Stimulation of Vestibular Epithelia (A–C) Fluid motion within each semicircular canal deflects the cupula and the inserted stereocilia, causing receptor potentials in the hair cells. Horizontal canals in the two ears provide opposite signals for rotation on the horizontal plane. The right anterior and left posterior canals as well as the left anterior–right posterior canals provide opposing signals about those planes of rotation, approximately 45° from head vertical (see Figure 21.1). (D) The hair bundles of macular hair cells project into an overlying gelatinous mass containing calcium carbonate crystals (i.e., the otoliths or ear stones). The otolith provides an inertial load, so that linear acceleration of the head results in differential motion of the epithelium and the overlying otolith, and thus, causes receptor potentials in hair cells of each macula. The utricle lies approximately in the horizontal plane of the upright head and the saccule approximately vertically to it, making each relatively sensitive to motion in those planes. The arrows show the direction of utricular hair bundle deflection for horizontal acceleration and deceleration. However, all head motion produces components of motion that affect both maculae. Further, the oppositely oriented hair bundles in each macula give still greater differentiation to the peripheral signal.

The Vestibulo-Ocular Reflex

The vestibulo-ocular reflex (VOR) depends on a simple (3-neuron) circuit by which head motion elicits equal and opposite motion of the eyeballs so that the fovea can remain fixated on an object of interest in visual space.[115] We will describe one particular example, horizontal head rotation, but in fact, all directions of movement activating both cristae and macular epithelia participate in foveation by this mechanism.[116] A rotation of the upright head to the left is compensated by an equal rightward rotation of the eyes (Figure 22.18). Leftward head rotation is excitatory to the left semicircular canal and suppresses activity in the right canal. Left-side canal afferents excite second-order neurons in the vestibular nucleus of the brainstem, which excite motor neurons in the oculomotor nucleus that in turn activate the medial rectus muscle of the left eye, causing it to rotate rightward. Second-order vestibular neurons also excite motor neurons (in the abducens nucleus) to the lateral rectus muscle of the right eye, causing it to rotate rightward an equal amount. Thus leftward head rotation is accompanied by coordinate, conjugate, and opposing motions that keep the two eyes fixated on the original visual target. Reduced activity in the right horizontal semicircular produces complementary reduction of activity and relaxation of opposing muscles: left lateral rectus and the right medial rectus.

Under normal conditions the gain of this reflex is one, and it can operate near normally even with input from only one ear. Obviously, complete loss of the vestibulo-ocular reflex can severely limit normal life activities. Somewhat less devastating is a condition called superior semicircular canal dehiscence, a thinning or complete loss of bone in the floor of the cranial vault at the top of the arch of the posterior (superior) semicircular canal.[117] Such a bony defect allows for transmission of mechanical energy from the cranial fluids to those of the superior semicircular canal, thereby exciting reflex eye movements in the plane of the canal. One disturbing consequence is that loud sounds or impacts absorbed through the head can give rise to eye movements and vertigo. For example, the action and sound of water from a shower results in the shower stall appearing to move! Fortunately, the defect can be repaired surgically.

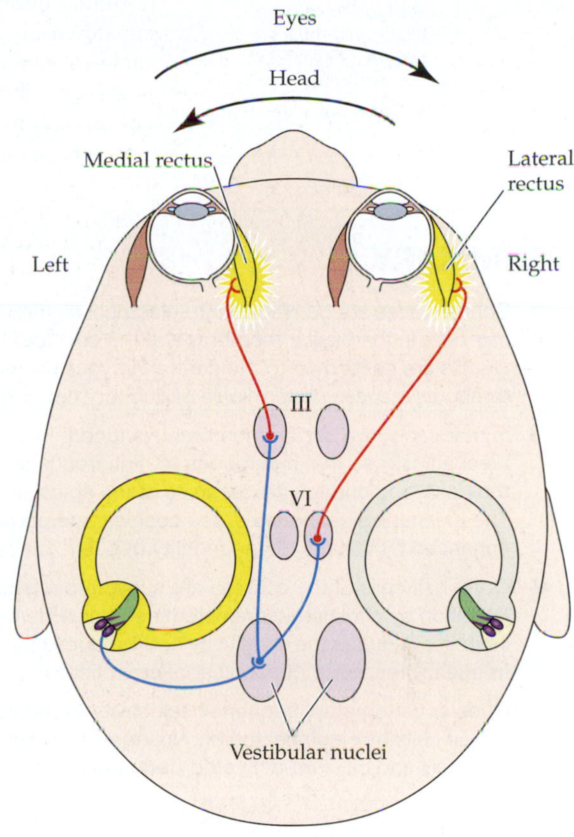

FIGURE 22.18 The Vestibulo-Ocular Reflex. Leftward rotation of the head causes excitation of left horizontal canal and suppression of its contralateral twin. Increased afferent activity causes excitation of principal cells of the vestibular nucleus in the brainstem, which in turn excites motorneurons of the ipsilateral medial rectus and contralateral lateral rectus, causing rightward rotation of the eyes. The net effect is to maintain gaze on a point in space, even as the head is deflected.

Higher Order Vestibular Function

As mentioned, vestibular function is largely subconscious, driving or modulating motor activity but not usually giving rise to conscious sensation. This state can change however, with vertigo and nausea arising from unusual vestibular input (motion sickness), or dysfunction. Happily, in most circumstances one can adapt to that altered condition, gaining sea legs, as it were.

Adaptation of the vestibulo-ocular reflex (VOR) has been studied extensively. The VOR is highly plastic, and adapts readily to altered sensory input, as wearers of bifocal glasses can attest. In fact, experimental manipulation of the VOR often involves magnifying or reducing lenses that alter the eye's focal length and thus the rate at which the visual scene shifts during eye movement. Subsequently, the VOR gain (i.e., the ratio of head to eye movement) changes or adapts with experience so that the appropriate compensation is made. This plasticity results from cerebellar modulation of the reflex.[118,119] Purkinje cells in the vestibular portion of the cerebellar cortex (flocculus and nodulus) project directly to inhibit principal neurons of the vestibular nuclei. Changes in synaptic efficacy may take place in the cerebellar Purkinje cells or in their target neurons in the vestibular nuclei.[120,121] As predicted, cerebellar lesions prevent VOR adaptation.[122]

How does vestibular sensation reach consciousness? The answer is confounded by the complication that vestibular sensation usually includes visual, proprioceptive, and tactile stimulation.[123] Further, there is little evidence for a specific vestibular neocortex. Rather,

[115] Lorente de No, R. 1933. *Arch. Neurol. Psychiatry* 30: 245–291.

[116] Cohen, B., Maruta, J., and Raphan, T. 2001. *Ann. N Y Acad. Sci.* 942: 241–258.

[117] Minor, L. B. 2005. *Laryngoscope* 115: 1717–1727.

[118] Ito, M. 1972. *Brain Res.* 40: 81–84.

[119] Lisberger, S. G. 2009. *Neuroscience* 162: 763–776.

[120] Cullen, K. E. et al. 2009. *J. Vestib. Res.* 19: 171–182.

[121] Blazquez, P. M., Hirata, Y., and Highstein, S. M. 2004. *Cerebellum* 3: 188–192.

[122] Luebke, A. E., and Robinson, D. A. 1994. *Exp. Brain Res.* 98: 379–390.

[123] Angelaki, D. E., and Cullen, K. E. 2008. *Annu. Rev. Neurosci.* 31: 125–150.

[124]Angelaki, D. E., Klier, E. M., and Snyder, L. H. 2009. *Neuron* 64: 448–461.

[125]Yates, B. J., and Miller, A. D. 1994. *J. Neurophysiol.* 71: 2087–2092.

[126]Balaban, C. D. 1999. *Curr. Opin. Neurol.* 12: 29–33.

vestibular information reaches many cortical association areas, combining with visual, somatosensory, and motor signals.[124] This fact by no means diminishes the importance of the so-called silent sense. In addition to being essential for effective locomotion and visual analysis, vestibular input plays a role in autonomic reflexes, helping maintain blood pressure and respiration with postural changes.[125] It has been proposed that the nausea arising from vertigo might serve to condition avoidance of potentially dangerous situations.[126]

SUMMARY

- Sound waves are converted into electrical signals by hair cells in the basilar membrane of the cochlea. These signals are passed on to the central nervous system through synapses on terminals of auditory nerve fibers.
- In mammals, the basilar membrane is tuned mechanically so that it responds to high frequencies at the basal end and low frequencies at the apical end. The mechanical response of the cochlear membrane is enhanced by voltage driven motility of outer hair cells.
- Outer hair cells of the cochlea are subject to efferent inhibition by cholinergic brainstem neurons. Efferent inhibition reduces the sensitivity and broadens the frequency response of cochlear afferent fibers.
- In lower vertebrates, frequency selectivity is imparted to hair cells by electrical tuning. Voltage-gated calcium channels and calcium-activated potassium channels interact to enhance the voltage response at a frequency that corresponds to the characteristic acoustic frequency for each hair cell.
- The central auditory pathway, including cortex, is tonotopically mapped. Response properties of cells in auditory cortex are complex, showing binaural interactions and dependence on temporal combinations of tones. Binaural comparisons of sound intensity and timing are used to compute the locations of sounds in space. These computations are made with synaptic connections in nuclei of the superior olive.
- Stimulus specificity for vestibular hair cells is conferred by the mechanics of their surrounding tissues. Semicircular canals detect angular acceleration, while the macular organs, saccule, and utricle detect linear acceleration.

Suggested Reading

General Reviews

Ashmore, J. 2008. Cochlear outer hair cell motility. *Physiol. Rev.* 88: 173–210. Review.

Dallos P. 2008. Cochlear amplification, outer hair cells and prestin. *Curr. Opin. Neurobiol.* 18: 370–376.

Eatock, R. A., Xue, J., and Kalluri, R. 2008. Ion channels in mammalian vestibular afferents may set regularity of firing. *J. Exp. Biol.* 211 (Pt 11): 1764–1774. Review.

Goldberg, J. M. 2000. Afferent diversity and the organization of central vestibular pathways. *Exp. Brain. Res.* 130: 277–297. Review.

Nelken, I. 2008. Processing of complex sounds in the auditory system. *Curr. Opin. Neurobiol.* 18: 413–417. Review.

Robles, L., and Ruggero, M. A. Mechanics of the mammalian cochlea. *Physiol. Rev.* 81: 1305–1352. Review.

Schreiner, C., and Winer, J. 2007. Auditory cortex mapmaking: principles, projections and plasticity. *Neuron* 56: 356–364.

Young, E. 2008. Neural representation of spectral and temporal information in speech. *Philos. Trans. R. Soc. Lond., B, Biol. Sci.* 363: 923–945.

Original Papers

Crawford, A. C., and Fettiplace, R. 1981. An electrical tuning mechanism in turtle cochlear hair cells. *J. Physiol.* 312: 377–412.

Grant L, Yi, E., and Glowatzki, E. 2010. Two modes of release shape the postsynaptic response at the inner hair cell ribbon synapse. *J. Neurosci.* 30: 4210–4220.

Highstein, S. M., Rabbitt, R. D., Holstein, G. R., and Boyle, R. D. 2005. Determinants of spatial and temporal coding by semicircular canal afferents. *J. Neurophysiol.* 93: 2359–2370.

Holt, J. C., Chatlani, S., Lysakowski, A., and Goldberg, J. M. 2007. Quantal and nonquantal transmission in calyx-bearing fibers of the turtle posterior crista. *J. Neurophysiol.* 98: 1083–1101.

Hudspeth, A. J., and Lewis, R. S. 1988. Kinetic analysis of voltage- and ion-dependent conductances in saccular hair cells of the bull-frog, *Rana catesbeiana*. *J. Physiol.* 400: 237–274.

Nelken, I., Rotman, Y., and Bar Yosef, O. 1999. Responses of auditory cortex neurons to structural features of natural sounds. *Nature* 397: 154–157.

Palmer, A. R., and Russell, I. J. 1986. Phase-locking in the cochlear nerve of the guinea-pig and its relation to the receptor potential of inner hair-cells. *Hear. Res.* 24: 1–15.

Ramanathan, K., Michael, T. H., Jiang, G. J., Hiel, H., and Fuchs, P. A. 1999. A molecular mechanism for electrical tuning of cochlear hair cells. *Science* 283: 215–217.

Winslow, R. L., and Sachs, M. B. 1987. Effect of electrical stimulation of the crossed olivocochlear bundle on auditory nerve response to tones in noise. *J. Neurophysiol.* 57: 1002–1021.

Zheng, J., Shen, W., He, D. Z., Long, K. B., Madison, L. D., and Dallos, P. 2000. Prestin is the motor protein of cochlear outer hair cells. *Nature* 405: 149–155.

CHAPTER 23
Constructing Perception

All the data that sensory receptors collect from the outside world are present in the peripheral sensory system, but neuronal activity does not lead to conscious experience until it can be further elaborated in the cerebral cortex. There, ongoing streams of sensory signals are transformed from representations of basic elements into more complex combinations of features; current experiences become meaningful once compared, within cortex, to recent and distant memories as well as to expectations. The knowledge gained from the external world is also used to prepare the appropriate motor output. How this occurs is the subject of this chapter. The cerebral cortex is one of the most studied parts of the nervous system and so our aim here is neither to survey every line of research nor to supply conclusions in an encyclopedic manner. Instead, we consider in detail two forms of processing: the first is perception of vibration applied to the skin, the second is recognition of an object in the visual field.

Monkeys can be trained to compare the frequencies, f_1 and f_2, of a base and a comparison tactile vibration, separated in time by a delay of one or more seconds. They must produce different motor responses according to whether the first or the second frequency is higher: $f_1 > f_2$ or $f_2 > f_1$. Their performance is similar to that of human subjects performing the same task. Neurons in the primary somatosensory cortex (SI) encode vibration frequency by their firing rate, with higher vibration frequency leading to higher firing rate. When a recording electrode is centered in a cortical column made up of rapidly adapting neurons, electrical microstimulation through the electrode can substitute for skin deflections in the behavioral task, and the monkeys' performance remains good. Thus, sensations can be inserted artificially into the cerebral cortex.

During the comparison task, the activity of neurons in secondary somatosensory cortex (SII) and in frontal cortex differs from that in SI in several important ways. Cortical neurons beyond SI can have either ascending or descending firing rates as vibration frequency increases. Whereas the activity of SI neurons does not persist after presentation of the base stimulus, SII neurons continue to fire for a brief period and stimulus-related firing of neurons in frontal cortex continues throughout the entire delay interval.

During the comparison stimulus, SI neurons show firing related only to that stimulus; in contrast, the neuronal activity in SII and frontal cortex depends on whether the frequency is greater than or less than that of the base stimulus. This comparison is transformed into the subsequent motor act in premotor cortex and motor cortex.

In primates, recognition of objects within a visual scene is mediated by a processing stream in the cerebral cortex running in an anterior direction from the occipital lobe to the inferior temporal lobe. Disruption of tissue along this pathway leads to a deficit in object recognition known as agnosia. If the lesion involves a specific region on the ventral surface of the temporal lobe, patients can selectively lose the ability to recognize faces. Measurements of human brain activity, using functional magnetic resonance imaging (fMRI), and of single-neuron activity in monkeys both indicate that neurons at progressively more anterior stages of processing are excited by progressively more complex images until neurons are excited explicitly by objects rather than elemental forms.

In both monkeys and humans, a region has been found that seems to be dedicated to the detection and recognition of faces.

We can recognize a familiar object under a variety of viewing angles and lighting conditions. Physiological correlates of such invariance have been found in the form of neurons that fire according to the identity of an object, independently of the details of how the object is viewed. Progressive increases—both in the complexity of the images that activate neurons as well as in the invariance of neuronal response to viewing conditions—suggest a hierarchical transformation along the posterior to anterior axis.

The posterior-to-anterior transformation is accompanied by a flow of signals in the opposite direction. Those inputs (known as top–down) serve to direct attention to an expected stimulus location or to salient object features, to activate the recall of images, and to learn associations between visual images that tend to occur in sequence. The processing streams of vision, touch, hearing, and other senses reach the hippocampus, where experiences are stored as episodes involving all the sensory modalities contributing to that event.

What Is the Function of Cortical Processing?

Sensory pathways lead from receptors across synapses to the thalamus and from there to the cerebral cortex. What operations are carried out on sensory signals once they reach the cortex? Clearly, the answer will depend on the sensory modality, and even on the type of stimulus within a modality. For example, speech and music are processed in cortex in very different ways.[1] But one general answer holds up for any sensory input: intracortical processing serves to integrate and distribute elemental sensory signals in such a way that those signals can gain meaning by linkage with stored knowledge and can be acted on through motor behavior.

This proposal for constructing perception was put forward in a convincing manner by Whitfield.[2] Through analysis of the behavioral effects of lesions in the auditory system, he noted that animals can carry out surprisingly fine sensory discrimination tasks even after ablation of sensory cortex and its connected regions, provided that the task does not require the transformation of "sensory data" into "objects."[2] He therefore postulated that the information present in subcortical centers related to the elemental physical characteristics of a stimulus (tone, wavelength, vibration frequency) can be accessed even by an animal with sensory cortex ablated. However, a deficit appears when the subject is required to endow simple sensations with the quality of belonging to objects. On page 146 of his seminal article, Whitfield concludes that it is the cortex that transforms physical characteristics into the percept of real things that are "out there" in the world. Once intracortical processing converts elemental sensory attributes into percepts of known objects, he continues, the animal appears to be able to use the result of one problem to generalize—that is, to solve a closely related problem. This capacity emerges because, in the late stages of cortical processing, information is organized as objects and concepts, rather than as a set of more or less elaborate features.

This chapter begins with a discussion of how tactile stimuli are encoded, stored in memory, and acted on in the course of a behavioral task. The sensory input employed in these experiments is comparatively simple, allowing us to gain a complete picture of its representation at multiple levels of cortex. We then proceed to the visual system and present evidence about how intracortical processing transforms sensory data about images into percepts of objects and movement in the real world. The visual experiments deal with more complex stimuli so it is difficult to gain a complete quantification of their neuronal representation. Yet the research is compelling because it offers the possibility to connect brain function to perceptual experiences—like recognizing a face—that are critical to our lives.

[1] Peretz, I. 2006. *Cognition* 100: 1–32.

[2] Whitfield, I. C. 1979. *Brain Behav. Evol.* 16: 129–154.

Ranulfo Romo

Tactile Working Memory Task and its Representation in Primary Somatosensory Cortex

Behavioral Task

This section on tactile processing focuses on experiments by Ranulfo Romo and colleagues, who have provided insights into how stimuli are encoded by neuronal activity and how

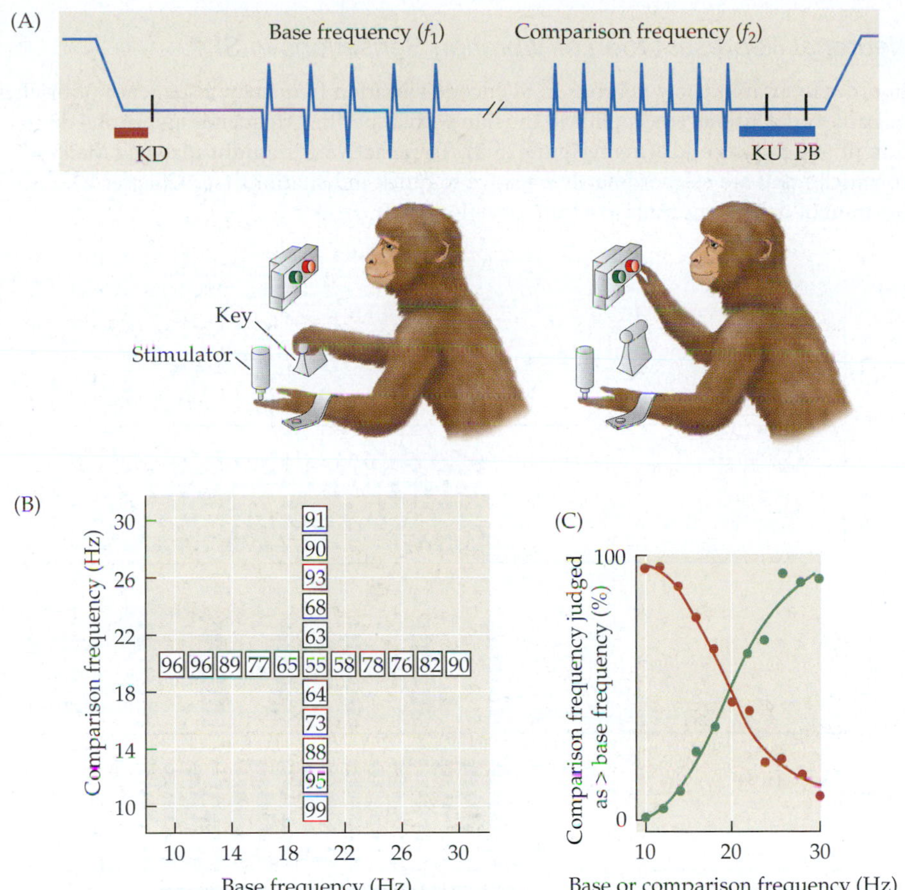

FIGURE 23.1 Vibration Discrimination Task and Performance. (A) The upper plot shows the probe position (*y*-axis) across time in one trial. The lower drawings show the monkey receiving the stimulus (left) and responding at the conclusion of the trial (right). Experimental sequence is as follows. First, the mechanical stimulator is lowered, indenting one fingertip of the restrained hand. The monkey uses its free hand to press a key down (KD; red bar). Then the monkey receives the first vibration from the stimulator, which oscillates vertically at the base stimulus frequency (f_1). After a delay, the monkey receives the second vibration at the comparison frequency (f_2). The monkey lets the key up (KU) and presses the medial or lateral (green and red, respectively) response button (PB) to indicate whether f_2 was lower or higher than f_1 (blue bar). (B) Results from a typical session. Horizontal row of boxes indicates percentage of correct responses at various base frequencies that precede a comparison frequency of 20 Hz. Vertical row indicates percentage of correct responses at varying comparison frequencies after a base frequency of 20 Hz. Note that the larger differences between f_1 and f_2 led to better performance. The complete experiment included all possible combinations of the base and comparison frequency (not shown). (C) Results of another typical session presented as curves that illustrate the percent of trials in which the monkey judged the comparison stimulus frequency, f_2, as higher than the base stimulus frequency, f_1. For the red points, f_2 was 20 Hz, and the *x*-axis gives varying values of f_1. The green points show the reverse: f_1 was 20 Hz and the *x*-axis gives varying values of f_2. Measurement of performance in this manner is known as the psychometric curve; steep curves like these indicate proficient performance of the task and provide a direct way of comparing the animal's performance to neuronal responses.

such activity leads to assessment of the stimulus. Monkeys were trained to discriminate the difference in frequency between two mechanical vibrations applied sequentially to one fingertip (Figure 23.1A).[3,4] The first vibration frequency, f_1, and the second vibration frequency, f_2, vary across trials. The first and second stimuli are referred to as the base and comparison, respectively. At the end of the comparison stimulus, the monkeys use their free hand to press a response button, expressing one of two decisions: $f_1 > f_2$ or $f_2 > f_1$. The delay between the base and comparison is 1 second (s) in most experiments, but monkeys can perform the task even with a 10 s delay. The greater the difference between f_1 and f_2, the better the performance (Figure 23.1B and C).

[3] Romo, R. et al. 1997. *Cereb. Cortex* 7: 317–326.

[4] Hernandez, A. et al. 1997. *J. Neurosci.* 17: 6391–6400.

[5]LaMotte, R. H., and Mountcastle, V. B. 1975. *J. Neurophysiol.* 38: 539–559.

[6]Powell, T. P., and Mountcastle, V. B.1959. *Bull. Johns Hopkins Hosp.* 105: 133–162.

Neuronal Representation of Vibration Sensations in SI

In order to examine how neurons in SI encode vibration frequency, a sequence of brief, pulsatile indentations was applied to the skin with a specified frequency and a total duration of 500 milliseconds (ms) (Figure 23.2). These activated rapidly adapting Meissner corpuscles that are extraordinarily sensitive to quick indentations (see Chapter 21),[5] so the train of deflections leads to a train of action potentials.[6]

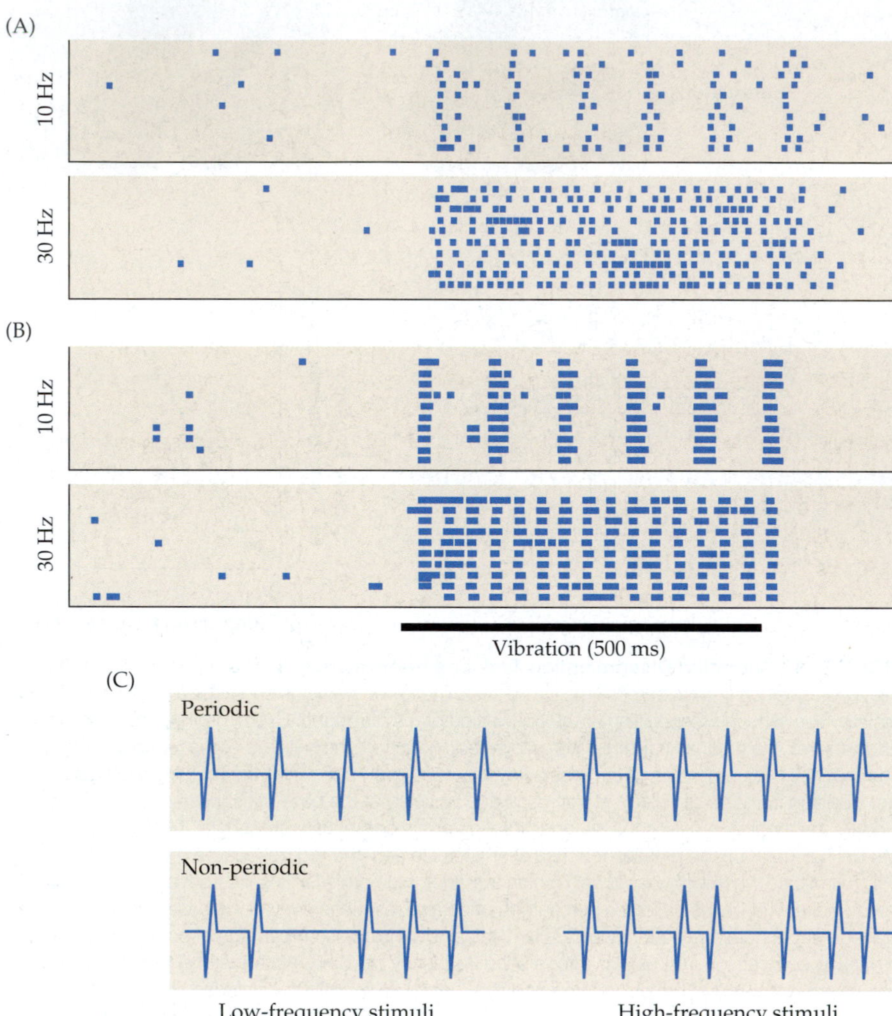

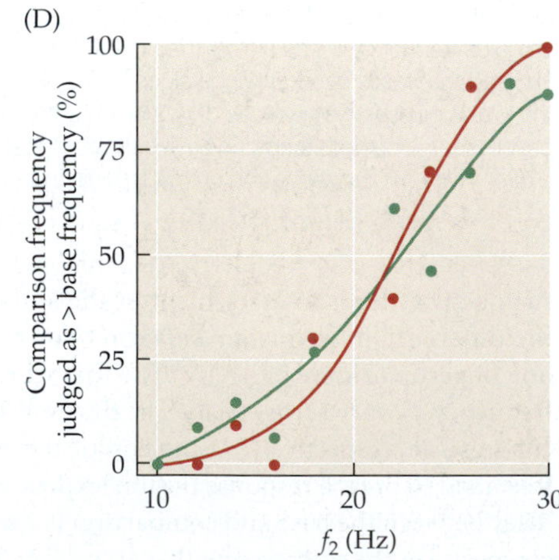

FIGURE 23.2 Neuronal Coding of Vibration Frequency in Primary Somatosensory Cortex (SI). (A) Responses to ten trials with stimulus at 10 Hz (upper plot) and 30 Hz (lower plot). Spike times are blue squares. Black bar indicates the 500-millisecond (ms) period in which the stimulus was presented. This neuron had a low degree of firing periodicity but encoded vibration frequency by firing rate. (B) Responses of a neuron that encoded vibration frequency by periodicity but not by firing rate (since at low vibration frequency each deflection evoked more spikes). (C) Low-frequency (left) and high-frequency (right) stimuli in a periodic (upper) and non-periodic (lower) stimulus train. (D) The red dots and line are the psychometric curve for one session derived from the trials in which the base stimulus was 20 Hz (see Figure 23.1C for derivation of such curves). The green dots and line are a so-called neurometric curve and represent the most accurate performance that an observer of neuronal activity could achieve. The neuron analyzed here encoded vibration frequency by firing rate. For non-periodic stimuli, performance of the monkey and the performance available in SI neuronal activity were similar, but only for neurons that encoded vibration frequency by firing rate. Neurons that were phase-locked could not encode non-periodic stimuli. (After Salinas et al., 2000.)

Figure 23.2 shows the response of individual SI neurons to a stimulus train. Action potentials were recorded with an array of microelectrodes inserted into the cortical area receiving input from the stimulated fingertip. Over 50% of neurons exhibited frequency-dependent modulation in firing rate: as vibration frequency increased, firing rate increased (Figure 23.2A). A large proportion of neurons fired in phase with each indentation; for instance, a vibration of 10 Hz led to a response with intervals of about 100 ms between spikes (Figure 23.2B). Both kinds of signal—the total number of spikes in the train and the periodicity within the train—carry information about vibration frequency. In theory, the periodicity of spikes within the train provides more information about the base frequency than does the firing rate. It may seem intuitive that cortical areas central to SI use the more informative, phase-locked signal, but two kinds of evidence lead to the surprising conclusion that vibration frequency is extracted from the firing rate instead.

The first kind of evidence comes from the comparison between SI neuronal activity on single trials and the choice made by the monkey on that same trial. The likelihood of the monkey making a correct discrimination on a given trial was unrelated to whether the spike trains were more periodic or less periodic than average on that trial. In contrast, in a small number of single neurons the firing rate fluctuated together with the monkeys' decisions.[7] Since firing rate seems to contribute to the monkeys' judgment, it is the better candidate for the critical code.

A second kind of evidence is convincing. Having practiced the discrimination task for months using periodic stimuli, the monkeys were then presented with non-periodic stimuli in which indentations were separated by random time intervals (Figure 23.2C). Phase-locked neurons now fired with irregular spike trains. If the brain required periodic spike trains to discriminate frequency, performance would degenerate dramatically. Yet, the performance was equally strong for periodic and non-periodic stimuli. The choices of the monkey were predictable from the overall firing rate, not the periodicity, of SI neurons (Figure 23.2D).[8]

Temporal jitter in a train of skin deflections certainly affects the central processing of the stimulus.[9–11] Under the conditions presented here, it is safe to conclude that the percept of vibration frequency is constructed from the firing rates of SI neurons.

Replacement of Vibrations by Artificial Stimuli

The investigators then tested the effect of stimulating the SI cells directly, thereby bypassing the pathway from skin to cortex.[12] An artificial stimulus consisted of current injections[13] delivered at the same frequency as the mechanical stimulus it substituted (Figure 23.3). Stimulation sites in the SI cortex were selected with receptive fields on the fingertip at the location of the mechanical stimulating probe. Remarkably, the monkeys were able to discriminate between the frequencies of the tactile base stimulus and the electrical comparison stimulus with the same accuracy as that obtained with two tactile stimuli (Figure 23.3A and B). The accuracy was also maintained when the base stimulus was electrical (Figure 23.3C), indicating that the subjects could form the working memory of the base stimulus from an artificial input. Even with artificial inputs used as *both* the base and comparison stimuli, monkeys could achieve discrimination levels close to those measured when mechanical stimuli were delivered to the fingertips (Figure 23.3D).

The use of direct cortical stimuli allowed a further insight into the role of firing periodicity. When the electrical stimulation was non-periodic, the performance remained good and confirmed that firing rate, not temporal pattern, was the feature used by the brain to compute vibration frequency. The monkey responded to electrical stimulation when the stimulus site was in cortical columns formed by rapidly adapting neurons (see Chapter 21); stimulation of columns formed by slowly adapting neurons produced no response.

To summarize, the full cognitive operation may be triggered by the signal emanating from a limited number of neurons in one cortical column. The demonstration of artificial sensation has been of importance for the new field of brain–machine interfaces,[14] where one long-term goal is to insert sensations directly inserted into the brain of patients with nonfunctional sensory pathways.

[7] Salinas, E. et al. 2000. *J. Neurosci.* 20: 5503–5515.

[8] Hernandez, A. et al. 2000. *Proc. Natl. Acad. Sci. USA* 97: 6191–6196.

[9] Lak, A. et al. 2010. *Proc. Natl. Acad. Sci. USA* 17: 7981–7986.

[10] Lak, A., Arabzadeh, E., and Diamond, M. E. 2008. *Cereb. Cortex* 18: 1085–1093.

[11] Godde, B., Diamond, M., and Braun, C. 2010. *Neurosci. Lett.* 480: 143–147.

[12] Romo, R. et al. 2000. *Neuron* 26: 273–278.

[13] Stoney, S. D., Jr., Thompson, W. D., and Asanuma, H 1968. *J. Neurophysiol.* 31:659–669.

[14] Fagg, A. H. et al. 2007. *J. Neurosci.* 27: 11842–11846.

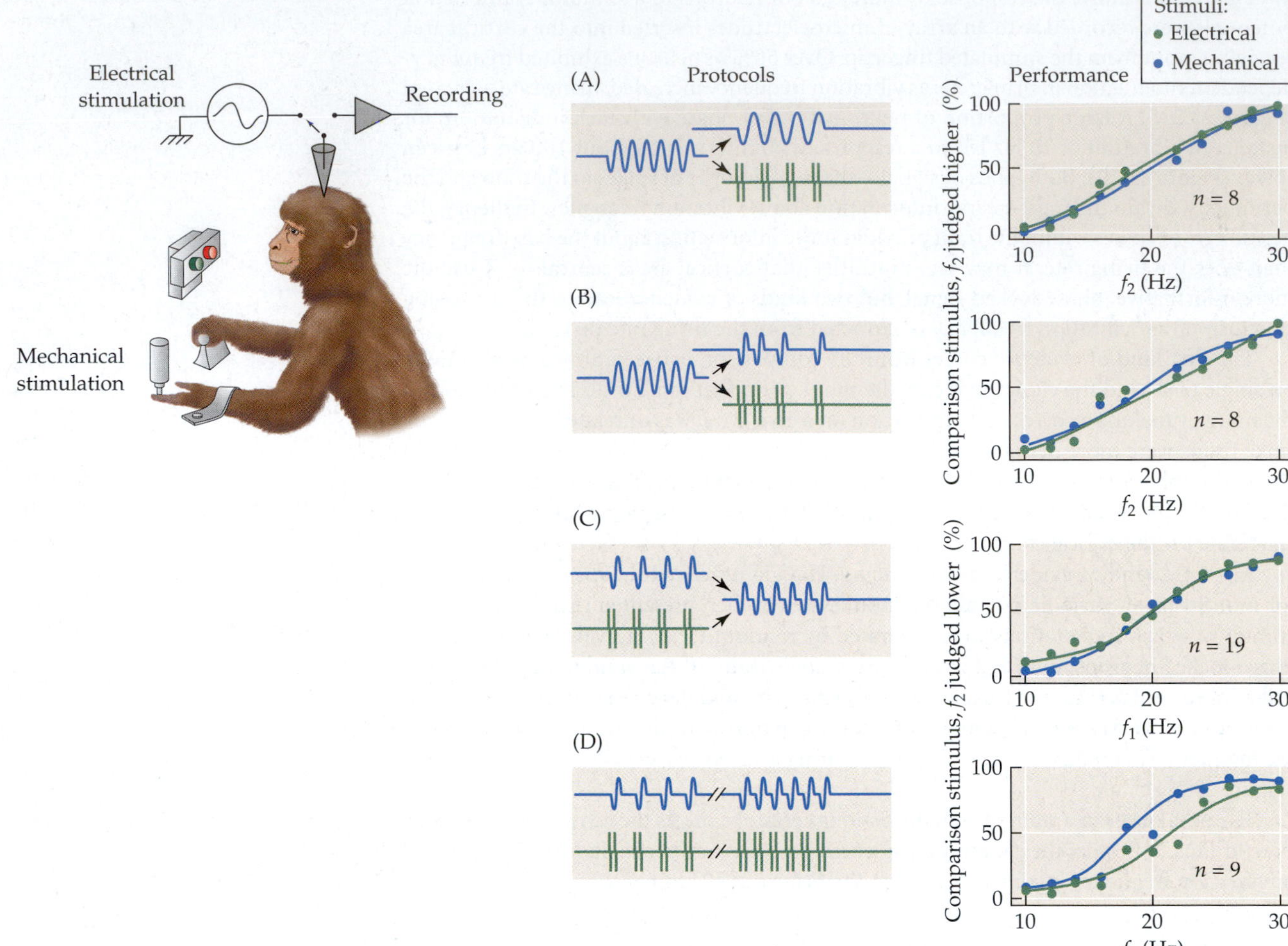

FIGURE 23.3 Replacement of Vibrations by Artificial Stimuli. After training with mechanical stimuli, skin stimuli were replaced by trains of electrical-current bursts (green) injected into primary somatosensory cortex (SI) at the same frequencies as natural stimuli. In half of the trials, the monkeys compared two mechanical vibrations delivered to the skin. In the other half, one or both of the stimuli were replaced by electrical stimulation in the cortex. The diagrams on the left show four of the protocols used. The curves on the right show the monkey's performance as psychometric functions (constructed in the manner explained in Figure 23.1C), in the corresponding session with only skin stimuli (blue dots and curve) or electrical stimuli (green dots and curve). (A) All stimuli were periodic. The comparison stimulus could be either mechanical or electrical. (B) The base stimulus was periodic and the comparison stimulus was non-periodic. The comparison could be either mechanical or electrical. (C) All stimuli were periodic. The base stimulus could be either mechanical or electrical. (D) All stimuli were periodic. In electrical stimulation trials, both base and comparison stimuli were artificial. Tactile stimuli were either sinusoids or trains of short mechanical pulses. (After Romo and Salinas, 2003.)

Transformation from Sensation to Action

Activity in SI across Successive Stages of the Task

In the same set of experiments described above, the investigators devised a way of characterizing the relationship between the stimuli and neuronal firing during presentation of the base stimulus, the comparison stimulus, and the delay between the two. Under the assumption that neuronal firing rate has a linear relationship to the stimuli, the firing rate (r) within any phase of the task can be described by:

$$r = b + a_1 f_1 + a_2 f_2$$

where b is the background firing rate (unrelated to any stimulus), f_1 is the base stimulus frequency, and f_2 is the comparison stimulus frequency. For a given neuron, a_1 and a_2 are coefficients that describe how the neuronal discharge rates throughout a trial depend upon stimulus frequencies in the base and comparison periods. The parameters a_1 and a_2 change

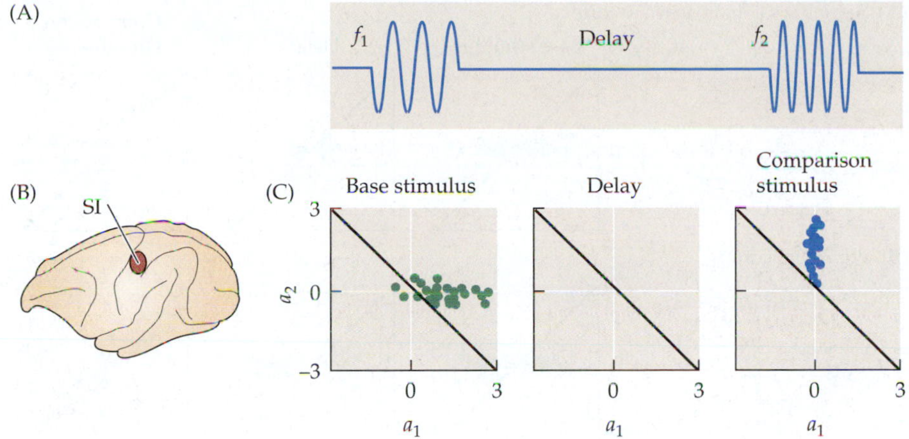

FIGURE 23.4 Neuronal Activity in SI Cortex Characterized by Firing-Rate Coefficients a_1 and a_2. (A) Time course of the behavior. (B) Location of recordings sites in SI cortex. (C) Within the f_1 interval, the delay interval, and the f_2 interval, firing rate can be expressed as $(a_1 \times f_1) + (a_2 \times f_2) + b$. Each point gives the values of a_1 or a_2 for one neuron. Green dots correspond to neurons with significant a_1 coefficients (regardless of a_2); blue dots indicate neurons with significant a_2 coefficients only. During f_1, most values of a_1 are positive and values of a_2 are zero. During the delay, both a_1 and a_2 are zero. During f_2, most values of a_2 are positive and values of a_1 are zero. Points are not plotted if both a_1 and a_2 are zero (as is the case in the middle graph).

over the course of the trial. To understand their meaning an example is useful. Consider first the firing of a neuron in the base period during presentation of the stimulus with frequency f_1. Suppose that across a set of trials with a variety of stimulus frequencies the neuron's firing rate increases by two spikes per second for each 1 hertz (Hz) increase in f_1. Then $a_1 = 2$ for that phase of the task. During this period, the comparison stimulus (with frequency f_2) can have no effect on the firing rate because the stimulus has not yet been presented, so $a_2 = 0$ (unless, of course, the monkey is reading the computer that runs the experiment and predicts the upcoming stimulus). Consider now the firing of the same neuron during the comparison period. If stimulation during the comparison period itself increases the firing rate by 3 Hz for every 1 Hz increase in f_2, then $a_2 = 3$. If prior stimulation during the base period has no residual effect on the firing rate, then $a1 = 0$. However, in a working memory task such as this, the brain must carry a trace of recent events. Therefore, we could predict that in some regions of the cortex, neuronal firing during the comparison stimulus is affected by the value of the past stimulus, f_1. In that case, a_1 will have a non-zero value.

Figure 23.4 illustrates how the neurons in one cortical region can be characterized by the values of a_1 and a_2. Figure 23.4A presents the events of one trial: the base stimulus defined by frequency f_1, the delay, and the comparison stimulus defined by frequency f_2. Figure 23.4B indicates the first cortical region examined, the fingertip representation of SI. Figure 23.4C shows the activity of one set of SI neurons during different stages of the task. In all plots, the x-axis gives the value of a_1, and the y-axis gives the value of a_2. During presentation of the base stimulus (left graph of Figure 23.4C), most neurons had a positive value of a_1 because they encoded f_1 by increasing their firing rate as frequency was increased. The values of a_2 were zero, since sensory neurons cannot vary their firing rate according to a future stimulus, f_2. During the delay (middle graph), a_1 values returned to zero, indicating that the base stimulus had no residual effect on neuronal discharge. During presentation of the comparison stimulus (right graph), the neurons had a positive value of a_2, while a_1 remained zero.

Activity in Regions beyond SI

The frequency comparison task requires more than the coding of stimulus features in SI cortex. In the following sections, we examine the behavior of cortical regions that operate on tactile signals successive to SI. Figure 23.5 shows the projection of SI to SII, which then projects to the inferior convexity of the prefrontal cortex (PFC), a region known to have a role in sensory working memory tasks.[15] Medial premotor cortex (MPC) receives input from SII and from PFC; it is believed to be involved with motor planning. M1 is involved in execution of motor commands (see Chapter 24).

During the base stimulus, neurons in SII showed more complex responses than those in SI (Figure 23.6A, left). Many of the responsive

[15] Constantinidis, C., Franowicz, M. N., and Goldman-Rakic, P. S. 2001. *Nat. Neurosci.* 4: 311–316.

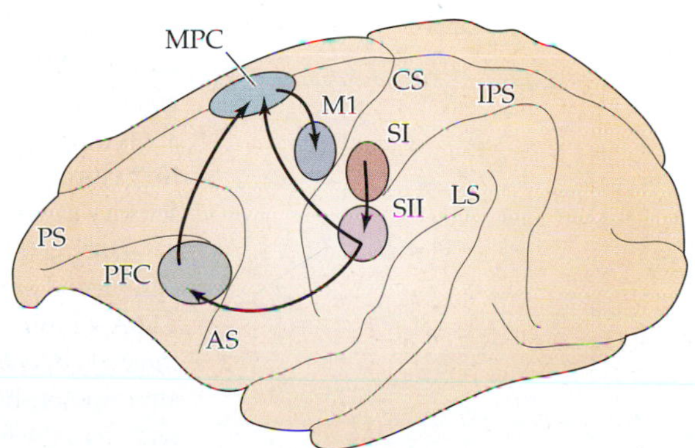

FIGURE 23.5 Cortical Regions Involved in the Vibration Comparison Task. The drawing shows the location of each cortical region. Arrows indicate the expected flow of tactile information. AS = arcuate sulcus; CS = central sulcus; IPS = intraparietal sulcus; LS = lateral sulcus; M1 = primary motor cortex; MPC = medial premotor cortex; PFC = prefrontal cortex; PS = principal sulcus; SI = primary somatosensory cortex; SII = secondary somatosensory cortex.

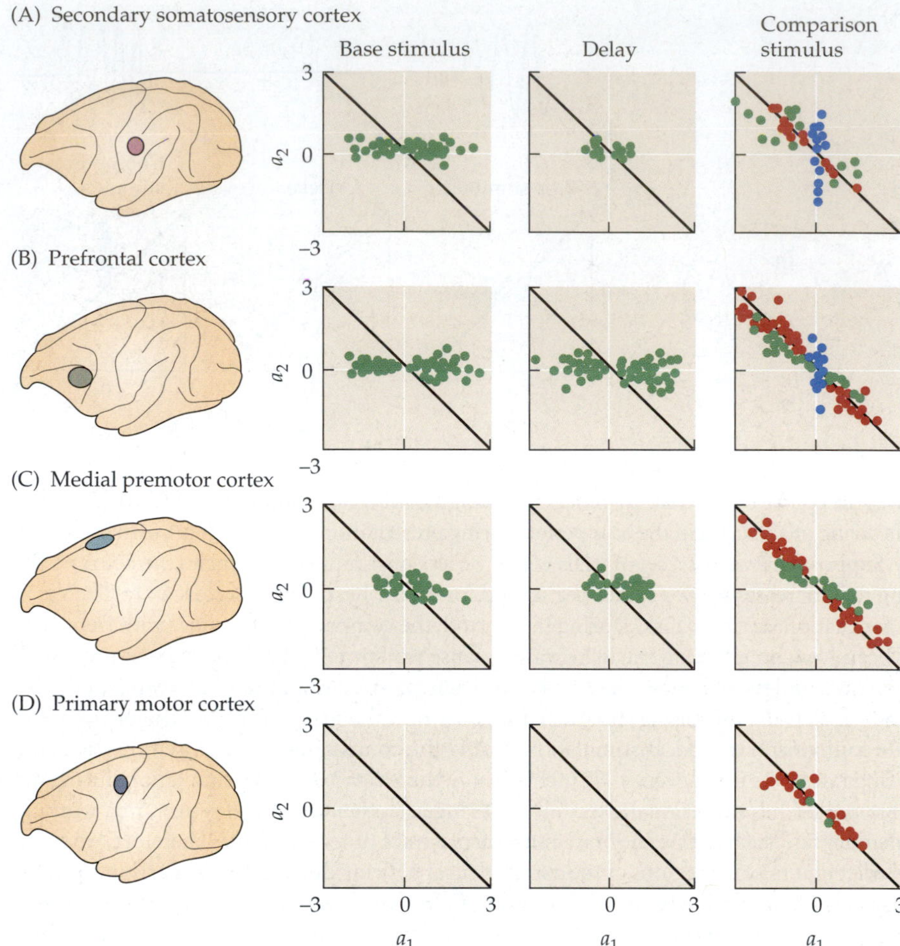

FIGURE 23.6 Neuronal Activity beyond SI Characterized by Firing Rate Coefficients a_1 and a_2. As in Figure 23.4, green dots indicate neurons with significant a_1 coefficients (regardless of a_2); blue dots indicate neurons with significant a_2 coefficients only. Left set of graphs refers to the base stimulus, middle set of graphs refers to the delay, and right set of graphs refers to the comparison stimulus. Red dots indicate coefficients computed during the last 300 milliseconds (ms) of the comparison stimulus. (A) Responses recorded from secondary somatosensory cortex SII. (B) Responses recorded from prefrontal cortex (PFC). (C) Responses recorded from medial premotor cortex (MPC). (D) Responses recorded from primary motor cortex (M1). All areas except M1 show stimulus coding during the base stimulus, but (unlike in SI) a_1 can have negative as well as positive values. Neurons continue to represent the base stimulus during the delay, also unlike in SI. During the comparison stimulus, many neurons have firing characterized by the property $a_2 = -a_1$. This is especially so during the final part of the second stimulus, as illustrated by the red dots. (After Romo and Salinas, 2003.)

neurons in SII actually *decreased* their firing rate as stimulus frequency increased. The opposing dependence of frequency—some firing rates rising and others falling as vibration frequency increased—reflects a fundamental transformation of the signal received from SI.

During the base stimulus, a mix of neurons with positive and negative values of $a1$ were found in all the frontal lobe areas except primary motor cortex (M1; Figure 23.6B–C, left column of plots). Latencies (i.e., time from stimulus onset until response onset) were longer in SII than in SI and progressively longer in PFC and MPC (not shown). Among these areas, no sharp anatomical boundary can be drawn to separate sensory from non-sensory areas during presentation of the base stimulus. In M1, very few neurons responded to the first vibration (Figure 23.6D, left), so this region can be considered not to participate in sensory processing during this behavior.

During the delay following the base stimulus, many neurons in SII, prefrontal cortex, and medial premotor cortex maintained activity related to the preceding vibration. (Figure 23.6A–C, middle column of plots). During presentation of the second stimulus, the monkey must (1) decide whether f_2 is greater or less than f_1, and (2) proceed to press the appropriate response button. Thus, changes in firing rate during the comparison period that are related to the frequency difference, $f_2 - f_1$, are of particular importance. No such responses were seen in SI (see Figure 23.4), but were clearly present in all subsequent cortical areas (Figure 23.6, third column of plots). The dependence on $f_2 - f_1$ was particularly evident during the last 300 msec of the comparison period (red dots). The parameters a_1 and a_2 fall closely along the line $a_2 = -a_1$. Referring back to the earlier equation and substituting $-a_1$ for a_2, this gives us the relation:

$$r = b + a_1 f_1 + a_2 f_2 = b + a_1 f_1 - a_1 f_2$$

$$\text{or } r = b + a_1 (f_1 - f_2)$$

In other words, when points lie close to the line $a_2 = -a_1$, it means that the neuron is responding as a function of $f_1 - f_2$; such neurons appear to *compare* f_2 and f_1.

Since the task is composed of distinct phases, the time course of signals is particularly interesting. Figure 23.7 plots the proportion of neurons in different cortical regions that had significant (positive or negative) values of a_1 (green line) and a_2 (red line) over time. The proportion of neurons whose firing was described by $a_2 = -a_1$ is illustrated by the blue lines. SI firing rates were characterized by a_1 during the base stimulus and by a_2 during the comparison stimulus. SII firing rates were characterized by a_1 during the base stimulus and showed a memory of f_1 during the first 500 ms of the delay period. During the comparison stimulus, many SII neurons were characterized by a_2 alone but others were described by $a_2 = -a_1$; such neurons encode the difference between the base and comparison stimulus frequencies.

In comparison to SII neurons, PFC neurons showed a more pronounced working memory for the base stimulus and a more prominent representation of the stimulus difference during the comparison period. MPC neurons were characterized by a less strong representation of the current stimulus but by a very strong representation of the stimulus difference. Neurons in motor cortex, M1, showed no stimulus representation or working memory. Their activity was related only to the outcome of the stimulus difference, f_2 versus f_1. As would be expected, the responses began later in the comparison period and extended into the interval when the monkey actually pressed the response button.

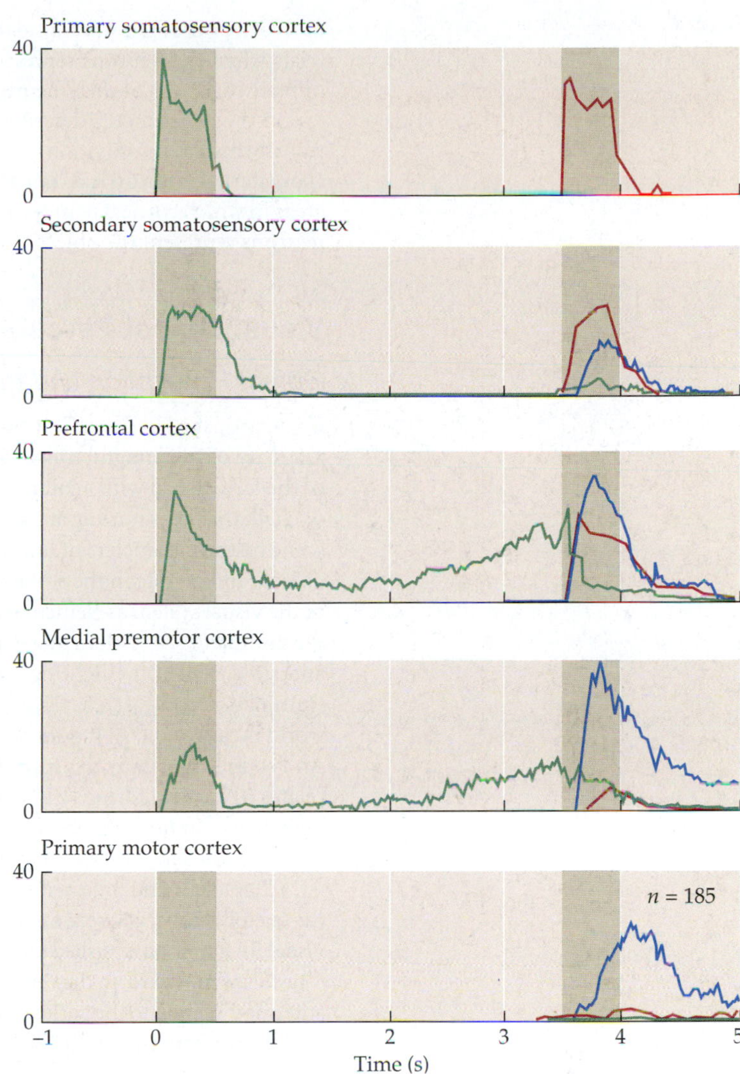

FIGURE 23.7 Time Course of Response Parameters throughout the Trial Period. Magnitudes of a_1 (green) and a_2 (red) vary from one cortical location to the next. Blue traces indicate neurons whose activity directly represents the comparison between f_1 and f_2 for all different values of the two stimuli (that is, either firing more for $f_2 > f_1$, or for $f_2 < f_1$). Responses are expressed as a percentage of total number of responding neurons. (After Hernandez et al., 2000, 2002; Romo et al., 2002, 2004.)

Neurons Associated with Decision Making

Monkeys make errors in detecting differences in the frequency of vibration, especially when the difference is small. What kinds of neuronal activity are associated with errors? Insights can be gained by comparing activity of an individual neuron on correct versus incorrect trials and, more specifically, by determining to what extent any differences in activity could predict an error. This is referred to as the **choice probability index** for the neuron. A high value means that on trials when the neuron's activity is different from its average activity in correct trials, there is a high probability of the monkey making an error.[16] An intuitive way of thinking of this is that if a neuron's choice probability index is very high, then when that neuron makes a mistake, so to speak, the whole brain is likely to make a mistake. Neurons in SI were found to have a low choice probability index. Typically, they encoded the stimuli accurately, whether the final action turned out to be right or wrong, although the trial-to-trial variability in firing rate of a small number of SI neurons did predict errors. Neurons in frontal cortex showed much higher choice probability index. In MPC, for instance, the choice probability index grew during the delay period and peaked during the comparison period. So, errors usually do not depend on a faulty representation of the stimulus in the ascending pathway through SI. Most errors seem to arise through a declining persistence of the representation of f_1 and, consequentially, a faulty comparison between the f_1 and f_2.

Earlier, it was noted that a set of neurons showed firing late in the comparison period, which can be represented by $a_1 = -a_2$. If the activity of these neurons is so closely related to the monkey's task (because their firing rate depends on the difference between f_2 and f_1), you might expect that they contribute disproportionately to the final decision. In support

[16] Parker, A. J., and Newsome, W. T. 1998. *Annu. Rev. Neurosci.* 21: 227–277.

of this hypothesis, it was found that the monkeys were most likely to make errors on those trials when the neurons representing "$a_2 = -a_1$" did not accurately encode the frequency difference. Errors of these neurons were predictive of errors by the monkey.

In the experiments described above, we have examined the neuronal processing that transforms tactile sensation into a decision. In the next section, our goal is once again to characterize intracortical transformations, but now in the visual system. We will examine areas in which cortical neurons represent the features of objects and areas in which cortical neurons represent the object's identity, such as a face.

Visual Object Perception in Primates

Object Perception and the Ventral Visual Pathway

If you are asked to describe what you see, the description will inevitably center on a world made up of meaningful things—for instance, "I see a badly dressed professor at the front of the lecture hall with an illustration of the brain projected behind him," or, "I see a piece of chalk that he is hurling at an inattentive student." It is unlikely that you will describe the scene by the parameters of luminance, contrast, spatial frequency, wavelength, etc. Yet, the objects and people in the scene could not be perceived unless encoded by the early stages of the visual system as elementary features. If pressed to do so, you could perhaps catalogue the complete set of elementary features making up the scene, but it would be laborious and slow. And if the lighting conditions changed, the catalog would have to be altered (luminance, contrast, etc.). Describing the same scene according to your knowledge of the world is easy and fast. It does not change if the lighting or the angle of viewing is changed. And so the visual world is formed of basic physical features but our subjective experience relates to a world formed of people and things that have significance due to accumulated knowledge. In this section of the chapter, we present evidence about the processing along the occipito-temporal axis that leads from perception of features to perception of objects.

Chapters 2 and 3 described visual processing from the retina to the posterior pole of the occipital lobe, where the primary visual cortex (V1) is situated. The brain's treatment of visual images is not finished at that point. In humans and other primates, information from V1 radiates outward, in the ventral-anterior direction to the temporal lobe and in the dorsal-anterior direction to the parietal lobe.[17] In this chapter, we focus on the **ventral pathway**—that is, the processing stream that underlies how we perceive, identify, and remember objects that we see—but also describe one key function of the **dorsal pathway**, motion perception.

V1 is the point of departure for a series of visual processing steps that take place in the secondary visual cortex (V2), V4, and from there to the ventrolateral surface of the temporal lobe where a posterior to anterior stream continues in a large area known as IT (inferotemporal cortex). Inferotemporal cortex consists of a number of subregions; commonly the posterior part is called TEO and the anterior part TE; TE can be further subdivided into posterior and anterior regions, TEp and TEa, respectively (Figure 23.8). As we will see, neurons in this intracortical stream, at locations progressively farther anterior from V1, are selectively activated by increasingly complex visual images. From the 1980s

[17] Fellman, D. J., and Van Essen, D. C. 1991. *Cereb. Cortex* 1: 1–47.

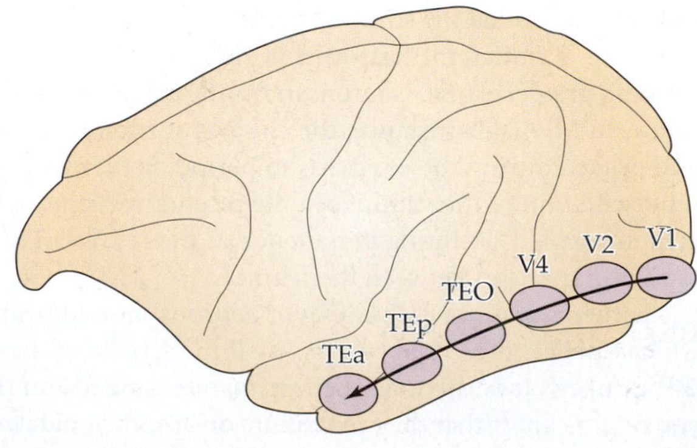

FIGURE 23.8 Ventral Intracortical Visual Pathways. Visual processing areas of the macaque brain are shown. The arrow indicates the posterior to anterior flow of information. V1 = primary visual cortex; V2 = secondary visual cortex; V4 = fourth visual cortical area; TEO = posterior part of inferotemporal cortex; TE = anterior part of inferotemporal cortex; TEa = anterior part of TE; TEp = posterior part of TE. Parts of the inferotemporal cortex extend to the ventral surface of the temporal lobe, not seen here. Ovals indicate the relative positions, but not the boundaries, of the cortical region. Analogous processing streams exist in the human brain.

onward, much effort has been dedicated to understanding neuronal processing in the ventral visual pathway, and we shall present some crucial findings.

Deficits in Object Perception

The difference between seeing elementary forms and perceiving objects was first proposed by Hermann Munk in 1881.[18] (Munk was the first scientist to perform precise ablations and careful behavioral observations in laboratory conditions, which led to his sharp insights.) His subjects were dogs that received either a lesion restricted to the posterior pole of the occipital lobe or else a lesion elsewhere, including regions farther anterior and lateral (angular gyrus). Those with bilateral occipital lobe ablation showed complete blindness, bumping into tables and walls. Those with more anterior lesions, sparing the occipital pole, showed what Munk called "psychic blindness"—they did not collide with furniture, yet they did not recognize by vision previously familiar objects.[18] Psychic blindness was interpreted as a loss of visual memory, or loss of visual images stored in the occipital lobe.

The psychic blindness described by Munk has much in common with the **visual agnosia** syndromes occurring in human patients (*agnosia* means loss of knowledge in Greek). A noted case is patient DF, a young woman who suffered brain damage in 1988 as a result of anoxia from carbon monoxide poisoning.[19] DF had no problem grasping objects placed in front of her or moving through the world without bumping into things because the dorsal visual pathway was intact. She was unable however to indicate the size, shape, and orientation of an object, either verbally or manually: she could perform actions related to objects, but she had no explicit knowledge about the object's identity. Brain scans reveal that her agnosia was the result of damage to the ventral processing stream.

The most remarkable form of human visual agnosia, **prosopagnosia**, is the selective loss of the ability to recognize people's faces. Although prosopagnosic patients are aware that faces are faces—that is, they know the category of the visual image—they fail to identify faces reliably or to achieve a sense of familiarity from seeing faces of family members, famous persons, and other individuals they previously knew well. Although they have trouble forming memories of new faces, other new objects can be learned. However, prosopagnosic patients may identify individuals by salient details such as clothing and hairstyle or by nonvisual features such as voice.

The question has been raised as to whether a deficit in face recognition is really specific to faces or else might be general to other sorts of images that are scanned by the fovea in the same way. The bulk of evidence indicates that a lesion restricted to a specific location in the temporal lobe can produce an agnosia that is genuinely selective to faces.[20] Later, we shall see that the area whose destruction causes prosopagnosia can be defined as a "face area" by criteria other than the lesion effects.

Images that Activate Neurons in the Ventral Stream

Discovery of Responses to Complex Stimuli in Monkeys

In landmark studies, neurons were found in the inferior temporal lobe of macaque monkeys that responded selectively to images of behaviorally significant objects, like hands,[21,22] and additional explorations of the inferior temporal cortex revealed a set of neurons that responded selectively to face images.[23,24] Thus, there exists a class of temporal lobe neurons that seem to *see* not bars or spots of light but rather faces and hands—they are excited by *objects* not elemental forms.

In humans and monkeys, face identity ("who is it?") and face expression ("what is the intention of this person?") are forms of perception essential to social interaction and survival. Even newborn infants preferentially look at face-like arrangements of features as compared to jumbles of face features or non-biological stimuli, suggesting that our interest in faces is at least partly built into brain circuitry.[25]

The Special Case of Faces

In the early studies in which face-responsive neurons were found, territories *dedicated* to this category of stimulus were not detected; only about 30% of neurons were face-responsive.[21,24,26,27]

[18] Munk, H. 1881. *Ueber die Functionen der Grosshirnrinde; gesammelte Mittheilungen aus den Jahren 1877-80.* Hirschwald, Berlin.

[19] Milner, A. D. et al. 1991. *Brain* 114: 405–428.

[20] Wada, Y., and Yamamoto, T. 2001. *J. Neurol. Neurosurg. Psychiatry* 71: 254–257.

[21] Desimone, R. et al. 1984. *J. Neurosci.* 4: 2051–2062.

[22] Gross, C. G., Rocha-Miranda, C. E., and Bender, D. B. 1972. *J. Neurophysiol.* 35: 96–111.

[23] Rolls, E. T. 1984. *Hum. Neurobiol.* 3: 209–222.

[24] Perrett, D. I. et al. 1984. *Hum. Neurobiol.* 3: 197–208.

[25] Johnson, M. H. 2005. *Nat. Rev. Neurosci.* 6: 766–774.

[26] Perrett, D. I., Rolls, E. T., and Caan, W. 1982. *Exp. Brain Res.* 47: 329–342.

[27] Baylis, G. C., Rolls, E. T., and Leonard, C. M. 1987. *J. Neurosci.* 7: 330–342.

(A)

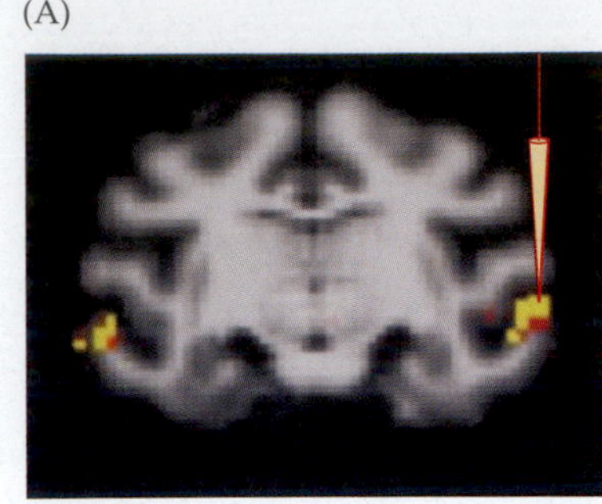

(B)

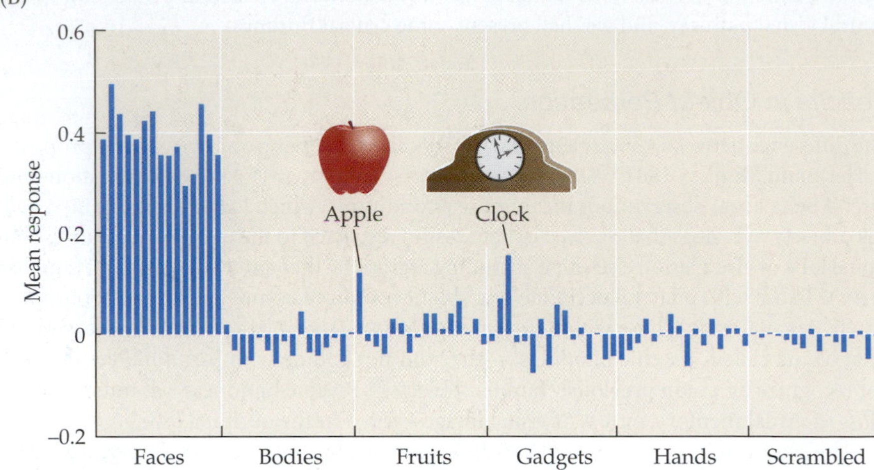

FIGURE 23.9 Face-Selective Neurons in Temporal Lobe. (A) Presentation of face images produced a large response in selected areas of the temporal lobe of macaque monkeys. High levels of activity detected in fMRI experiments are colored yellow–red. Electrodes were directed to such areas, as indicated schematically on the right side. (B) Once the electrodes reached the target area, neuronal responses were measured. Six categories of stimulus were presented (pictures of faces, bodies, fruits, gadgets, hands, and scrambled patterns), with 16 individual images per category. The bars give the mean strength of response of all 286 cells tested to all 96 stimuli, and it is clear that most neurons in the selected brain region produced a strong response to face images and much weaker responses to all other categories. The responses to some non-face stimuli, like the apple and the clock, might have occurred because they contained some face-like features. Mean response (*y*-axis) was calculated as a proportion of each neuron's strongest response to any stimulus. (After Freiwald, Tsao, and Livingstone, 2009.)

So, what can we conclude about the existence of an area whose principal function is to process such images? The issue was solved first by identifying with functional magnetic resonance imaging (fMRI) (Box 23.1) a region in the anterior temporal lobe in which signals were elevated selectively by faces. Once the coordinates were registered, recording electrodes were directed to the same region, where it was found that 97% of neurons respond preferentially to faces (both monkey and human faces) compared to other objects (Figure 23.9).[28] Thus, when correctly targeted, a truly face-selective region can be identified. An elegant experiment established a causal relationship between the activity of face-selective neurons and face perception. The investigators excited small clusters of inferotemporal neurons by means of electrical microstimulation while the monkeys performed the task of judging whether noisy visual images belonged to "face" or "non-face" categories. Microstimulation of face-selective sites (but not other sites) increased the likelihood of the monkey reporting the presence of a face. The magnitude of the effect depended on the degree of face selectivity of the stimulation site, the size of the stimulated cluster of face-selective neurons, and the exact timing of microstimulation.[29]

In humans, the physiological evidence for the involvement of the occipito-temporal pathway in object recognition depends largely on brain imaging rather than on electrical recordings. In fMRI experiments, a set of regions in the ventral pathway from occipital to temporal cortex was found to be activated more strongly when subjects viewed real objects as compared to scrambled objects, textures, stationary dot patterns, coherently moving dots, or gratings.[30] When subjects viewed images of objects (faces and cars) broken in a stepwise manner into an increasing number of blocks, regions that were progressively farther from the occipital pole in the posterior-to-anterior dimension showed less activation as the degree of scrambling increased. These results suggest the existence of a hierarchical axis along which the neuronal properties shift in sensitivity from local object features to a more global and holistic representation.[31]

Presentation of different categories of images activates neurons in different locations.[32] The most category-specific regions are found at the anterior extreme of the ventral pathway in an area known as the **parahippocampal place area** (PPA—activated by observation of places, buildings, landscapes) and in the fusiform face area (FFA—activated, as the name implies, by observation of faces).[33,34] This is presumably the homologous center in humans to the face-selective region found in monkeys.[28]

[28] Tsao, D. Y. et al. 2006. *Science* 311: 670–674.

[29] Afraz, S., Kiani, R., and Esteky, H. 2006. *Nature* 442: 692–695.

[30] Grill-Spector, K. et al. 1998. *Hum. Brain Mapp.* 6: 316–328.

[31] Lerner, Y. et al. 2001. *Cereb. Cortex* 11: 287–297.

[32] Grill-Spector, K., and Malach, R. 2004. *Annu. Rev. Neurosci.* 27: 649–677.

[33] Kanwisher, N., and Yovel, G. 2006. *Philos. Trans. R. Soc. Lond., B, Biol. Sci.* 361: 2109–2128.

[34] Downing, P. E. et al. 2006. *Cereb. Cortex* 16: 1453–1461.

BOX 23.1
Functional Magnetic Resonance Imaging

Functional magnetic resonance imaging (fMRI) has changed neuroscience by allowing investigators to visualize the activation of the brain in healthy human subjects during cognitive tasks. The signal on which the image is based is related to changes in blood volume, oxygen consumption, and blood flow during brain activity. The method relies primarily on measurement of the **blood oxygen level-dependent** (**BOLD**) signal. As neurons do not have internal reserves for glucose and oxygen, an increase in activity requires more glucose and oxygen to be delivered rapidly through the blood stream. Through a process called the **hemodynamic response**, blood flows to the active neurons and astrocytes at a greater rate than to inactive neurons (see Chapter 10). This extra supply of blood results in a surplus of oxygen-bound hemoglobin in the veins of the active area. The ratio between oxygen-bound and oxygen-free hemoglobin is the BOLD signal obtained by the fMRI machine.

The physical principle for detecting the BOLD signal depends on magnetism: oxygenated and deoxygenated hemoglobin react differently to the magnetic field that is applied inside the fMRI apparatus. The differential reaction to the magnetic field makes it possible to detect the heightened supply of oxygenated blood present in active brain tissue.

The change in blood flow induced by neuronal activity is not instantaneous; indeed, there is a lag of approximately 1–5 seconds. The delay between electrical activity and BOLD signal is known as the **hemodynamic lag**, and it limits the temporal precision with which fMRI can detect neuronal responses to stimuli. The spatial resolution is about 1 millimeter.

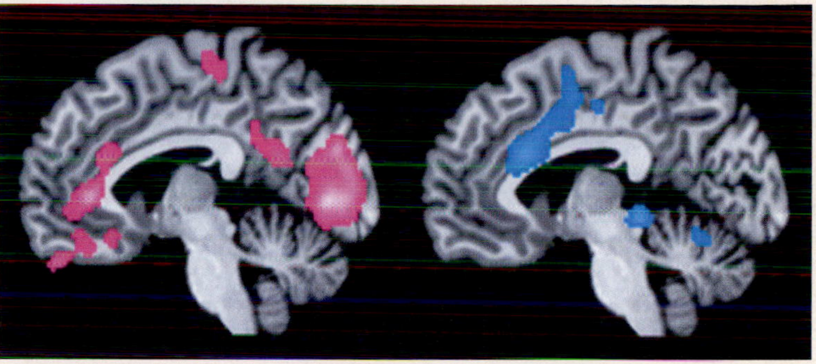

Brain Activity as Revealed by Averaging fMRI Scans across a Group of 20 Subjects. The scans are a view of the midline structures of the right side of the brain in a sagittal plane. To produce the activation shown on the left, subjects received pleasant visuo-tactile stimulation; they looked at pictures of agreeable things like silk and roses while touching the same objects. To produce the activation shown on the right, the subjects received unpleasant visuo-tactile stimulation; they looked at pictures of disagreeable things like worms and slugs while touching objects with tactile properties similar to the pictured objects. Areas in which the BOLD signal was significantly higher than that under a control condition (experience of neutral emotions) are denoted in color. Note that the two types of stimulation activated different brain regions, providing some insight into the areas involved with feeling pleasure and disgust. (Scans courtesy of G. Silani.)

A fascinating fMRI experiment suggested that activation in face-selective areas in humans is related to the act of perceiving a face, not just to the processing of images with a face-like form or contour. Subjects viewed the Rubin vase–face illusion for periods of 9 seconds and indicated whether they perceived the vase or the face. In those intervals when subjects perceived a face, there was a larger signal in the face-selective regions of the ventral stream as compared to intervals when they perceived a vase (Figure 23.10). All the while, the physical stimulus remained constant, so the variation in cortical activation could only be correlated with what the subjects perceived.[35]

Perceiving the presence of a face and recognizing the identity of the particular face are related but not equivalent operations. There is evidence that the fusiform face area contributes to both operations. In this region, the magnitude of the fMRI signal is correlated on a trial-by-trial basis with successful face identification.[36]

Perceptual Invariance and Neuronal Response Invariance

Although we see a familiar object many times and recognize it each time, its image as it falls on the retina is never exactly the same. If the activity of object-responsive neurons in the temporal lobe truly underlies our identification of objects, one would expect their responses to be to some degree tolerant to changes in viewing conditions that affect object appearance but not identity. Can we confirm such invariance in the human temporal lobe using brain-imaging methods? Each voxel (1 unit of brain volume scanned) in an fMRI experiment reports the strength of activation of many thousands of neurons, giving a spatial resolution far too poor to investigate individual inferotemporal neurons.

[35] Hasson, U. et al. 2001. *J. Cogn. Neurosci.* 13: 744–753.

[36] Grill-Spector, K., Knouf, N., and Kanwisher, N. 2004. *Nat. Neurosci.* 7: 555–562.

(A)

(B)

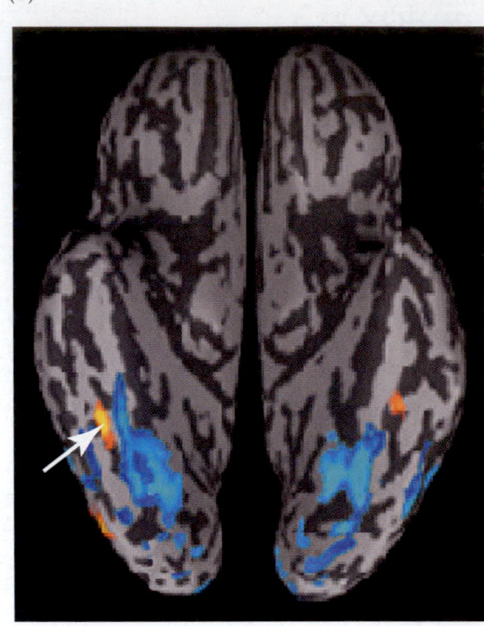

FIGURE 23.10 Brain Activation Directly Correlated with Face Perception. (A) Face–vase illusion. Subjects alternately perceive two faces gazing inward at each other, or else one ornate vase in the middle. It is impossible to perceive the faces and the vase simultaneously. (B) The brain viewed from below. Regions of the ventral surface of the temporal lobe where the signal was stronger when subjects perceived the vase are in blue; regions where the signal was stronger when subjects perceived the faces are in yellow–orange. The yellow–orange hot spot, indicated by the arrow, is the fusiform face area. (B from Hasson, et al., 2001.)

As a way around the spatial constraints, investigators have used an **adaptation procedure** to look for response invariance. The idea is that if the excitation of a neuronal population is related specifically to the viewing of a particular object, then repetition of that image at a high rate will excite the neurons progressively less until their response reaches some low, steady-state level; at that point, the neurons have *adapted*. Next, a new image is presented. If it portrays the same object but is made up of altered elemental features (e.g., size, contrast, or color), then that object will be processed by these same, adapted neurons and the overall response will be suppressed. If the new image portrays a different object, it will activate a different set of neurons; these, being in the non-adapted state, will give a large response.

As a test for invariance to viewing conditions, the image of one object was shown repetitively to human subjects. fMRI signals in a temporal lobe region known as **posterior fusiform** (homologous to inferotemporal in monkeys) continued to adapt even when size, visual field position, direction of illumination, or viewing angle varied from one single image to the next. If viewing conditions remained constant but object identity varied from one image to the next, responses of neurons in this region did not adapt. These results suggest that the neuronal populations in the posterior fusiform region fired according to their selectivity for the *identity* of the object, independently of the viewing conditions.[37]

Electrophysiological recordings in monkeys have the single-neuron resolution necessary to investigate response invariance more directly (Figure 23.11). Certain classes of inferotemporal neurons in monkeys respond with a similar strength when an image is placed before the eyes independently of size, contrast, angle of lighting, blurring, spatial frequency, and position in the visual field.[38] Some of these neurons are excited by a particular face, even at different viewing angles. Invariance is learned by exposure to the same object across changing views.[39,40] Theoretical accounts for the generation of invariance have been proposed,[41,42] but collecting adequate experimental data is difficult.

[37] Grill-Spector, K. et al. 1999. *Neuron* 24: 187–203.

[38] Ito, M. et al. 1995. *J. Neurophysiol.* 73: 218–226.

[39] Logothetis, N. K. et al. 1994. *Curr. Biol.* 4: 401–414.

[40] Li, N., and DiCarlo, J. J. 2008. *Science* 321: 1502–1507.

[41] DiCarlo, J. J., and Cox, D. D. 2007. *Trends Cogn. Sci.* 11: 333–341.

[42] Booth, M. C., and Rolls, E. T. 1998. *Cereb. Cortex* 8: 510–523.

Dorsal Intracortical Visual Pathways and Motion Detection

While visual perception has been discussed until now in relation to the ventral pathway processing involved in the recognition of objects, another essential function of vision is the analysis of motion. Motion is analyzed by the dorsal, magnocellular-parietal visual pathway (Figure 23.12A; see also Chapter 3). Magnocellular pathway neurons are sensitive to moving stimuli, and this trait is maintained through V1 and V2. From there, the dorsal stream is directed toward the parietal lobe. A key region in the dorsal pathway is the middle temporal

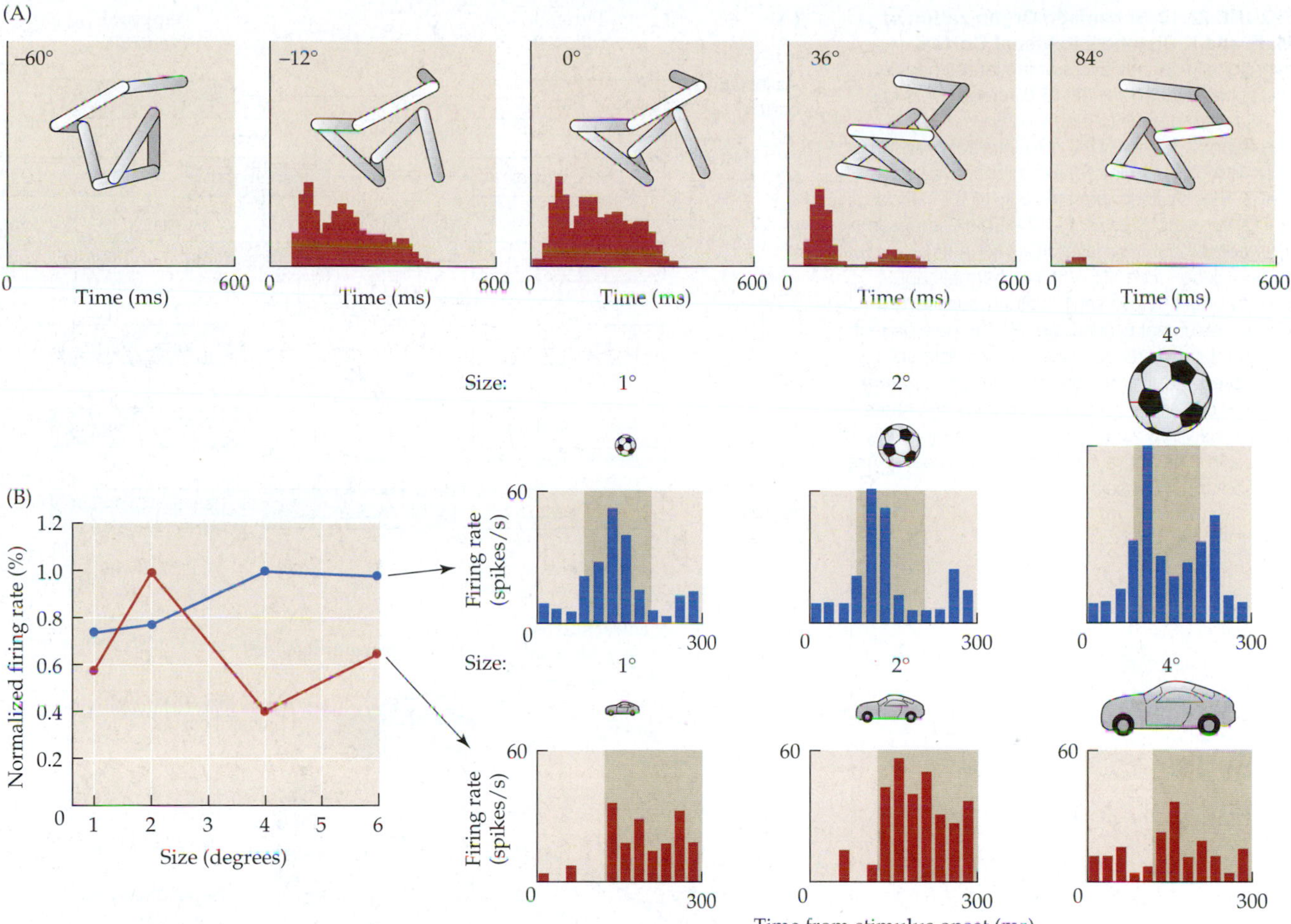

FIGURE 23.11 Responses of Macaque Monkey Inferotemporal Neurons to Changes in Viewing Angle and Stimulus Size. (A) Each form resembles a wire paper clip bent into an arbitrary three-dimensional shape. During training, the monkey learned to identify specific wire shapes, with each shape seen from one specific viewing angle. During testing, wire forms were projected onto the screen at the angle familiar to the monkey but also at angles never seen before. The histograms show the activity of one neuron, in firing rate across time, summated over several trials, when a familiar target was presented at different angles (angles of rotation in the horizontal or vertical plane are indicated on each panel). Stimulus onset is at 0 milliseconds (ms). The neuron's response was strong when the form was presented at the familiar viewing angle (central panel; 0°) and at angles ranging from –12° to +36°. For larger rotations, the neuron no longer fired, its response resembling that to unfamiliar objects. Thus, we could say the neuron *recognized* the form unless it was rotated so much as to be unfamiliar. (B) Reponses of two inferotemporal neurons, one designated in blue and one in red, to presentation of images of a soccer ball and a car. The sizes of the images were changed randomly across presentations, from 1° to 6° of visual field. The plot on the left shows that both neurons fired for the stimulus at levels of 40–100% of maximum firing rate even when image size differed. The plots on the right show the two objects at differing sizes. Below each image, the neuronal firing rate, averaged across trials, is plotted for the 300 ms interval immediately after stimulus presentation. (A after Logothetis, Pauls, and Poggio, 1995; B after Zoccolan, et al., 2007.)

(MT) cortex, also known as visual area 5 (V5). Area MT is located in the posterior bank of the dorsal part of the superior temporal sulcus (Figure 23.12B).

In a comprehensive and elegant series of experiments, Newsome and his colleagues have analyzed how activity of neurons in area MT allows a monkey to assess the direction of movement. In many respects, the results are consistent with those described above for perception of vibration frequency.

The procedure is to teach monkeys to respond to the direction in which a visual stimulus is moving, while recordings are made with a microelectrode from the active neurons. Area MT is retinotopically mapped.[43] Neurons selective for the speed and direction of a moving stimulus are clustered together in columns with a similar preferred direction.[44–47] Such neurons respond poorly or not at all to motion in the opposite, or null, direction. When small regions of MT are chemically lesioned with a neurotoxin, a monkey's ability to detect a moving pattern of dots in a corresponding region of the visual field is impaired.[48]

[43] Maunsell, J. H., and Newsome, W. T. 1987. *Annu. Rev. Neurosci.* 10: 363–401.

[44] Zeki, S. M. 1974. *J. Physiol.* 236: 549–573.

[45] Maunsell, J. H. R., and Van Essen, D. C. 1983. *J. Neurophysiol.* 49: 1127–1147.

[46] Albright, T. D. 1984. *J. Neurophysiol.* 52: 1106–1130.

[47] Malonek, D., Tootell, R. B. H., and Grinvald, A. 1994. *Proc. R. Soc. Lond. B, Biol. Sci.* 258: 109–119.

[48] Newsome, W. T., and Pare, E. B. 1988. *J. Neurosci.* 8: 2201–2211.

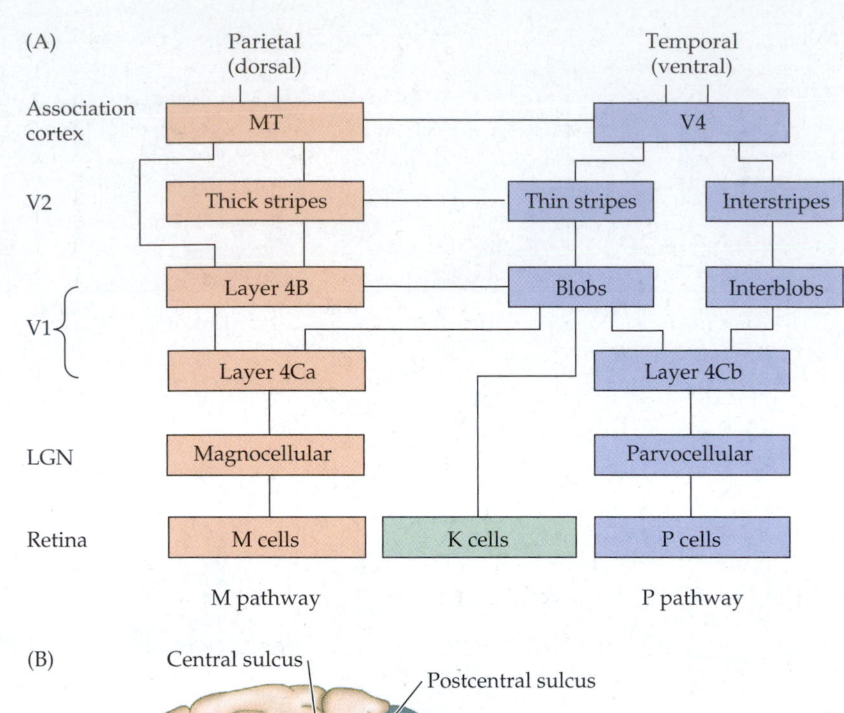

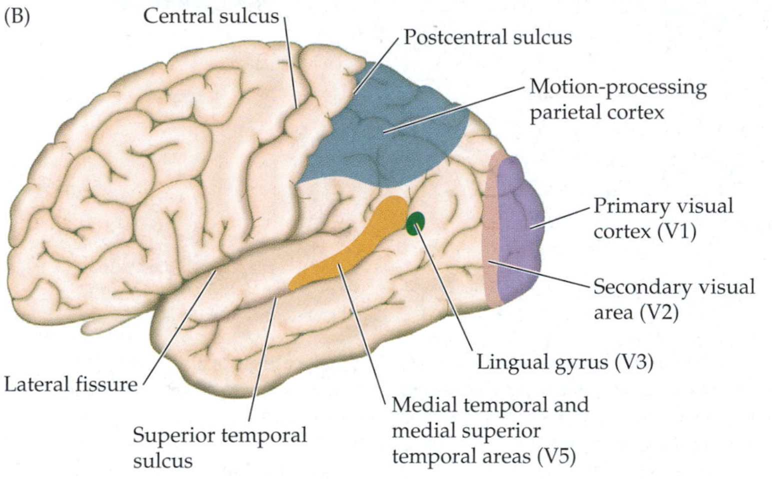

FIGURE 23.12 Schematic Organization of M, P, and K Channels to Visual Cortex. (A) Functionally distinct layers of the lateral geniculate nucleus (LGN) project to different layers in V1. K layers project to blobs in layers 2 and 3. The M and P layers of 4C may interact preferentially with blob and interblob regions in layers 2 and 3. Blobs project preferentially to thin stripes in V2. Thin stripes project to V4. Thick stripes in V2 receive input from layer 4B in V1 and project to association area MT (V5). M-channels appear to project to dorsal (parietal) visual cortex, where movement is analyzed. P-channels project preferentially to area V4, where color vision is processed. Additional details on the architecture of visual cortex are given in Chapter 3. (B) Areas of cortex involved in the dorsal stream of visual processing, as portrayed on a human brain. (After Merigan and Maunsell, 1993.)

Area MT is involved in visual tracking. This was shown by experiments in which a monkey was trained to track a moving target with its eyes (Figure 23.13).[49] The normal pattern of eye movements is seen in the upper record in Figure 23.13 in which the moving target (trajectory begins at time 0) was acquired by a rapid saccade (the downward deflection occurring 200 ms later) and then retained on the fovea by an accurate tracking or smooth pursuit process. After a small injection of neurotoxin (ibotenic acid) into the foveal region of MT cortex, the monkey's ability to track the moving target was markedly impaired. In particular, after the initial saccade, the subsequent tracking velocity was much slower than the target velocity. The deficit is visible in the traces in the lower record in Figure 23.13. That underestimation also applies to the initial saccade made by the lesioned animal. It positioned the eye as though the estimated velocity were lower

[49]Dursteler, R. M., Wurtz, R. H., and Newsome, W. T. 1987. *J. Neurophysiol.* 57: 1262–1287.

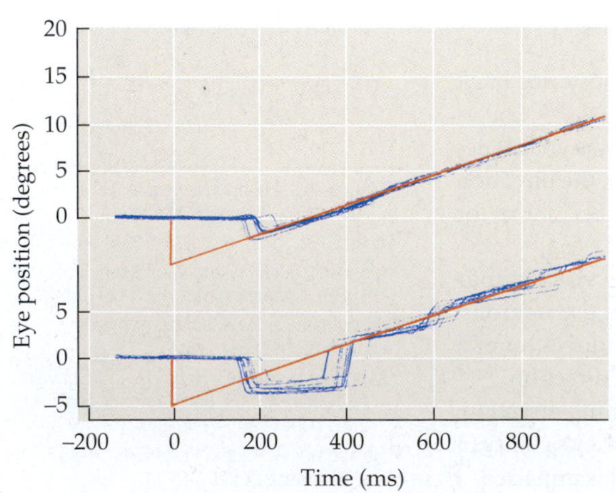

FIGURE 23.13 Involvement of MT in Tracking Visual Motion. A monkey was trained to track a moving target (stimulus path shown by red line), and its eye position relative to the target is shown in the upper record. After an initial saccade to center the target on the fovea (the rapid downward eye deflection), the eye closely followed the target path. After injection of a neurotoxin in area MT (lower record), the initial saccade was too large and overshot the target, and subsequent tracking was slower than required, as though the computation of target speed were faulty. (From Newsome and Wurtz, 1988.)

than the actual velocity. Somehow, the lesion perturbed the ability of area MT to compute an accurate estimate of target velocity.

How is motion computed in area MT? As stated earlier, cells are clustered into columns of similar preferred directions across the retinotopic map. Thus, the movement of a target across the retina ought to activate those columns aligned with the direction of movement. But moving visual targets will not activate only one such column, being more likely to exhibit a complex patterns of motion that activate many sets of directionally tuned neurons to varying degrees. Thus, it will require some form of neural computation to derive an average movement vector.

Newsome and his colleagues studied the neural arithmetic performed by multiple direction columns, using electrical microstimulation to alter eye movements in trained monkeys.[50] The columnar organization means that the cells affected by microstimulation had similar functional properties. The microelectrode that recorded the preferred direction of a column of cells was then used to pass amounts of current to activate that same column during a target-tracking eye movement (Figure 23.14). Then, the tracking eye movement was compared with and without electrical stimulation to ask how components of the visual motion map sum. Eye position tracked target position closely in the control condition (Figure 23.14A). When electrical stimulation activated an MT column whose directional preference was different from that of the moving target, the resulting eye movement lay somewhere between the actual direction of the target and the preferred direction of the stimulated cells. The stimulation "pulled" the monkey's eye movement towards the direction associated with the stimulated cells (Figure 23.14B).

The conclusion is that the vectorial average of the activated direction columns ultimately sets the eye movement direction. An appealing feature of these experiments is that the monkey's behavior (eye movements) is a direct measure of the analysis made by higher centers in the cortex. Similar vector averaging has long been known to occur during the generation of saccadic eye movements and will be seen in motor cortex in Chapter 24.[51]

Additional insights into the function of area MT came from experiments in which the monkey viewed sequential pairs of random dot stimuli moving in one of four possible directions.[52] This first stimulus was called the sample; it was analogous to what we referred to as the base stimulus in the earlier fingertip vibration experiments. It was followed after a brief delay by a second stimulus referred to as the test, analogous to the comparison stimulus in the fingertip vibration experiment (Figure 23.15A).

[50] Groh, J. M., Born, R. T., and Newsome, W. T. 1997. *J. Neurosci.* 17: 4312–4330.

[51] Robinson, D. A., and Fuchs, A. F. 1969. *J. Neurophysiol.* 32: 637–648.

[52] Cohen, R., and Newsome, W. T. 2004. *Curr. Opin. Neurobiol.* 14: 1–9.

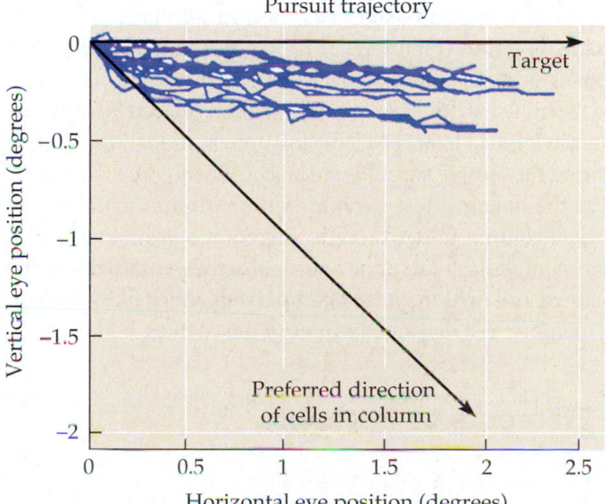

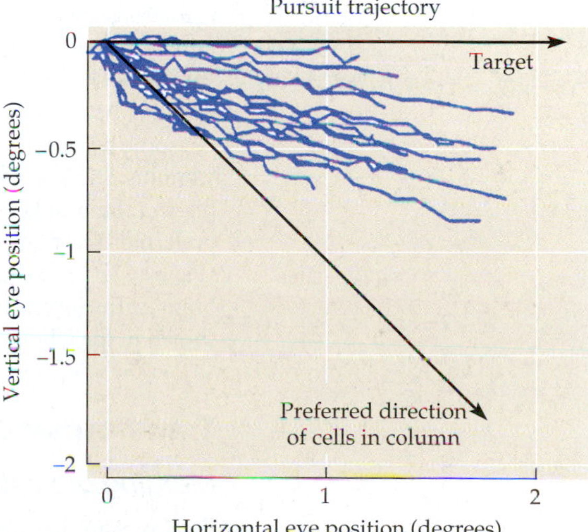

FIGURE 23.14 The Direction of Eye Movements Can Be Altered by Electrical Stimulation in Area MT. (A) Eye movements were recorded in response to a moving visual target. Earlier, an electrode had been inserted into area MT, and the preferred direction of cells in that location was noted. This preferred direction differed from that of the moving target. (B) When this location in area MT was stimulated electrically, the resulting eye movements were biased in the preferred direction of cells in the stimulated region. These results suggest that visual motion is computed as the vector sum of several preferred directions in area MT. (After Groh, Born, and Newsome, 1997.)

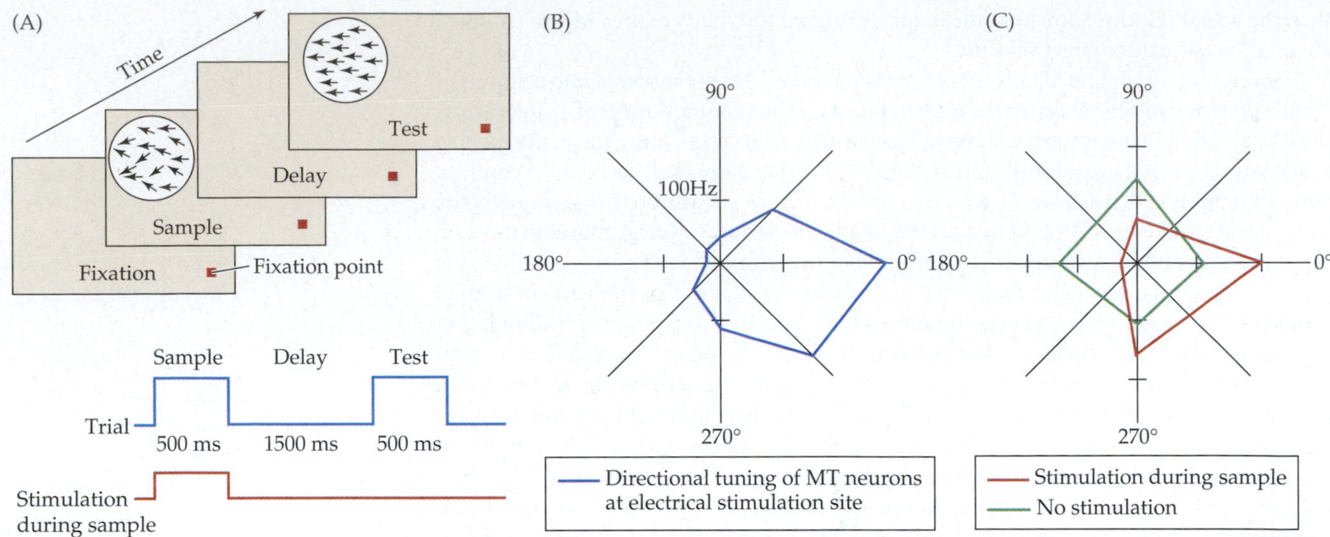

FIGURE 23.15 Area MT Microstimulation Affects a Working Memory Task. (A) Schematic illustration of the task. After monkeys fixated on the small red square, they were presented with a random dot stimulus for 500 milliseconds (ms; sample). The stimulus consisted of dots moving in a direction determined by a probability distribution whose mean was in one of the four cardinal directions (up, down, left, right). Following a 1500-ms delay period with no image, the test stimulus was presented for 500 ms. The test stimulus consisted of dots moving coherently in either the same direction as in the sample, or in the opposite direction. The monkeys indicated whether the test stimulus matched the direction of the sample by pressing a button. For each sample direction, half of the test stimuli were the same direction as the sample and half were in the opposite direction. On half of the trials, microstimulation was applied during the sample period. (B) Direction tuning of a sample MT multi-unit site at which electrial stimulation was applied. Firing rate (action potentials/second) is plotted in relation to moving dot direction. The preferred direction of the cells at this site was for motion to the right or down and to the right. (C) Behavioral data from one experiment. The plot shows the percentage of trials on which the monkey reported that the test matched the direction of the sample for each of the four possible sample directions. In the no stimulation condition (green line), the monkey reported a match roughly 50% of the time for all four directions, leading to an average of about 90% correct, as the test actually matched the sample exactly half the time. When microstimulation was applied (red line) to the same site whose tuning is depicted in (B), the monkey reported a match for nearly every trial in which the test stimulus was rightward and for most trials when the test stimulus was downward. By contrast, the monkey almost never reported a match when the test was leftward. This outcome indicates that microstimulation during presentation of the sample biased the monkey's perception of the motion toward the preferred direction of the stimulated cells. (After Cohen and Newsome, 2004.)

At the end of each trial, the monkey pressed one of two buttons to indicate whether the direction of motion of the test stimulus was the same as or different from that of the sample. A microelectrode for recording and stimulating was placed in a column in area MT containing neurons tuned to a specific direction (Figure 23.15B). On trials with no electrical stimulation of area MT, the monkey performed nearly perfectly because the difference in direction between the sample and the non-matching test stimuli was large (at least 90 degrees). On some trials, microstimulation was applied in area MT during presentation of the sample stimulus. Stimulation applied during the sample stimulus strongly influenced performance, biasing the monkey to choose, as the match, a test stimulus whose motion matched the preferred direction of the stimulated column rather than the direction of sample stimulus (Figure 23.15B and C). This experiment indicates that the neuronal activity inserted into the brain by the investigator can be stored and used for future comparison, much like what was shown in the vibration working memory task described earlier in the chapter.

Transformation from Elements to Percepts

Merging of Features

Having seen that the neuronal mechanisms underlying motion perception in the dorsal stream are to some degree akin to those in the somatosensory system, we now return to the discussion of the ventral stream. In a recent set of studies, investigators tried to understand how neurons in inferotemporal cortex build up responses to complex shapes.[53,54] They focused on the selectivity of many of these neurons for two-dimensional boundary shape. The experiments examined posterior processing stages in inferotemporal cortex (areas TEO

[53] Brincat, S. L., and Connor, C. E. 2004. *Nat. Neurosci.* 7: 880–886.

[54] Yamane, Y. et al. 2008. *Nat. Neurosci.* 11: 1352–1360.

and posterior TEp; see Figure 23.8). For each neuron studied, the monkey was presented in rapid succession with ~1,000 stimuli in which shape characteristics varied in small steps to provide a rich and quantifiable dataset. From the responses to the full stimulus set, neurons were fitted with tuning curves. The main result is that combinations of elementary forms, such as oriented curves and contour fragments, formed the tuning curve of these neurons.

In a further set of experiments, three-dimensional shapes were used. Since this experiment entailed an even larger set of potential stimuli, the investigators developed a clever technique whereby new generations of stimuli evolved continuously according to neuronal responsiveness to the stimuli in the previous generation. The new generation contained combinations of features that previously were found to excite the neuron. The method allowed the experimenters to quickly converge upon the optimal stimulus shape for every neuron. The main result is that the way the inferotemporal neurons encoded both two- and three-dimensional images was consistent with a buildup of complex properties based on the integration of simpler properties.

Speed of Processing

The operations necessary to extract meaning from an image—feature selection, recombination, comparison to memory—invoke the idea of an arduous and time-consuming process, yet visual analysis can be effortless and fast. When presented with images that are flashed for as little as 20–80 ms, human subjects can make complex judgments about the content of the scene before them. They can reliably determine whether or not the scene contains an animal (motor response: "go") or does not ("no-go"). In a different task, they can determine the presence or absence of food. How much time does the brain need to extract this information? Reaction measures, like "go" versus "no-go," are imprecise because they include the time required for response execution.

A better approach is to look for the first sign in the brain's electrical activity that distinguishes between go versus no-go trials. It has been found from such experiments that a scalp potential related to response inhibition on no-go trials becomes evident roughly 150 ms after stimulus onset.[55] Considering that responses begin in V1 at 30–100 ms after stimulus onset,[56] we can conclude that the intracortical processing that leads to extraction of high-level image content can be accomplished in some tens of milliseconds.

Fast neuronal processing has been seen in the monkey ventral stream, consistent with human reaction times. For instance, in the face-selective area of anterior temporal lobe, responses distinguish the presence of a face versus another category of image at 130 ms after stimulus onset. The same population of responding neurons identifies the current face from among all faces in the stimulus library 60 ms later.[28]

Normally, inspection of a visual scene is more extended in time, so the claim cannot be made that analysis always involves just one fast pass through the cortex. The timing measures reveal the capacity for incoming signals to rapidly access stored knowledge (for instance the go/no-go task requires comparison of the new image to the storehouse of what animals look like) and show that any given stage of processing can operate rapidly.

Forms of Coding

We have referred to neuronal coding in a simplified manner, implying that a neuron fires or does not fire for any given stimulus. Inspection of neuronal firing shows that this simplification is not accurate. What are the detailed firing patterns of individual neurons, and how do neurons work together to represent the visual world? Are objects represented by activity in a relatively small number of neurons that are each selective for the shape or identity of a specific object (i.e., a sparse code), or are they represented by a pattern of activity across a large number of less selective neurons (i.e., a population code)?

The problem is intriguing and challenging. Intuition would suggest that at posterior levels, close to V1, information is encoded by the activity of a very large set of neurons. Since, at anterior levels, neurons are feature selective, then the coding of an image would be expected to become sparser—a smaller proportion of neurons would be active, with each neuron's activity specifying a larger amount of object information. But what happens when the current stimulus matches the preferred stimulus of an entire population?

[55] Thorpe, S. J., Fize, D., and Marlot, C. 1996. *Nature* 381: 520–522.

[56] Schmolesky, M. T. et al. 1998. *J. Neurophysiol.* 79: 3272–3278.

The experiment described earlier in which the fusiform face area of monkeys was explored proves informative in this regard.[28] The region of interest in the anterior temporal lobe was first identified using fMRI responses to face images. Electrodes were advanced to the responding region. Individual neurons were found to respond with different magnitudes to a wide variety of faces so that the exact identity of the presented face was encoded by a population of neurons. By the relative magnitude of response in the population of 94 neurons, it was possible for the experimenter to "decode" which of 96 possible faces was presented on a given trial with an accuracy of 74%.[28]

The distribution of response magnitudes across the population is not the only coding mechanism. One study found that neurons in the monkey inferotemporal cortex carried information about the category of stimulus according to *when* they fire.[57] Many cells responded to human and non-primate animal faces with comparable magnitudes but responded significantly more quickly to human faces than to non-primate animal faces. Differences in onset latency may be used to increase the coding capacity as well as to enhance or suppress information about particular object groups by time-dependent modulation.

Top–Down Inputs

Our description of functional processing in the ventral pathway portrayed a cascade of signals from V1 towards the anterior temporal lobe, a so-called downstream flow. But, information travels in the anterior-posterior direction as well, that is, from the frontal cortex and the hippocampus to the temporal lobe and from the temporal lobe back to the occipital lobe. Thus, neurons at every processing center receive feedback from neurons in downstream centers. Understanding of the roles of these **top–down inputs** (to introduce yet more jargon) is still incomplete, but recent work has pointed to a number of functions.

Imagining or recalling a visual stimulus that is not currently being viewed can produce activation in V1, as revealed by fMRI.[58,59] Since disruption of V1 activity by transcranial magnetic stimulation disrupts visual recall, this activity constitutes one component of the recall of visual memories.[60] Given that the activation cannot originate in the retina, investigators argue that responsibility for re-evoking V1 activity must lie with top-down signals.

Modulation of cortical processing by attention also is believed to be a top-down function. Directing subjects' attention to different locations in the visual field enhances fMRI signals in cortical areas that are retinotopically aligned to the attended location and inhibits signals in cortical areas that are aligned to the unattended location.[61–63] Attention to objects and faces enhances activation, measured by fMRI, in object- and face-selective regions, respectively.[64–66]

People as well as monkeys can readily learn that the presence of one specific image predicts the appearance of a second specific image a short time later (e.g., a plate, then food). Top-down inputs seem to be critical to the formation of associations between image pairs. In one experiment performed in monkeys, recordings were made simultaneously from neurons in two inferotemporal cortex areas—(1) area TE and (2) a region in perirhinal cortex just anterior to TE—while the animals learned associations between pairs of shapes.[67–69]

Even in naive animals, neurons normally fired for specific stimuli, which allowed the experimenters to select two images, image A, "preferred" by the perirhinal neuron, and image B, "preferred" by the TE neuron. The non-preferred stimuli evoked little response before training. Then, the animal began to learn the association: image A was used as a cue for the appearance of image B. As the monkey learned that A predicted B, TE neurons began to respond to image A, the cue. The key finding is the relative timing of spikes. When image A appeared, perirhinal neurons responded *before* TE neurons. When image B appeared soon thereafter, perirhinal neurons responded *after* the TE neurons. Thus, TE neurons received signals about their preferred stimulus, B, through a posterior-to-anterior flow of visual sensory information, while they received signals that cued the future appearance of stimulus A by an anterior-to-posterior flow of information. Additional experiments identified the frontal cortex as the source of the top-down information flow during cued recall.

Further Processing

The hippocampal formation, a structure lying medial to the cerebral cortex, receives input from all sensory modalities through their cortical processing streams (see Chapter 16). It

---

57 Kiani, R., Esteky, H., and Tanaka, K. 2004. *J. Neurophysiol.* 94: 1587–1596.

58 Ress, D., Backus, B. T., and Heeger, D. J. 2000. *Nat. Neurosci.* 3: 940–945.

59 Kastner, S. et al. 1999. *Neuron* 22: 751–761.

60 Kosslyn, S. M. et al. 1999. *Science* 284: 167–170.

61 Brefczynski, J. A., and DeYoe, E. A. 1999. *Nat. Neurosci.* 2: 370–374.

62 Macaluso, E., Frith, C. D., and Driver, J. 2000. *Science* 289: 1206–1208.

63 Tootell, R. B., and Hadjikhani, N. 2000. *Nat. Neurosci.* 3: 206–208.

64 Avidan, G. et al. 2003. *Neuroimage* 19: 308–318.

65 O'Craven, K. M., Downing, P. E., and Kanwisher, N. 1999. *Nature* 401: 584–587.

66 Wojciulik, E., Kanwisher, N., and Driver, J. 1998. *J. Neurophysiol.* 79: 1574–1578.

67 Naya, Y., Sakai, K., and Miyashita, Y. 1996. *Proc. Natl. Acad. Sci. USA* 93: 2664–2669.

68 Tomita, H. et al. 1999. *Nature* 401: 699–703.

69 Naya, Y., Yoshida, M., and Miyashita, Y. 2001. *Science* 291: 661–664.

encodes and then stores events, experiences, and episodes, and is also essential to the process of recalling events. Information from the ventral visual pathway reaches this structure after further processing in the entorhinal cortex.

It is interesting to consider in what form a visual stimulus is represented at this level. Different pictures of Marilyn Monroe, even if greatly modified as in Andy Warhol's famous portraits, can evoke retrieval of stored knowledge about the subject. Interestingly, the same concept of Marilyn Monroe (or Pamela Anderson, in the experiment in Figure 23.16) can be evoked with other stimulus modalities, for instance by reading her name or hearing it spoken. Is there a neuronal substrate for the representation of an identity independent of the sensory modality by which it is evoked?

By using presentations of pictures and of spoken and written names, it was shown that (1) single neurons in the human medial temporal lobe responded selectively to representations of the same individual across different sensory modalities; (2) the degree of multimodal invariance increases along the hierarchical structure within the medial temporal lobe; and (3) such neuronal representations could be generated within less than a day or two.

These results demonstrate that single neurons can encode percepts in an explicit, selective, and invariant manner, even if evoked by different sensory modalities. In short, if we take the neuronal response as a report of the presence of a given thing in our environment, it is tempting to believe that such neurons are the basis of subjective experience.

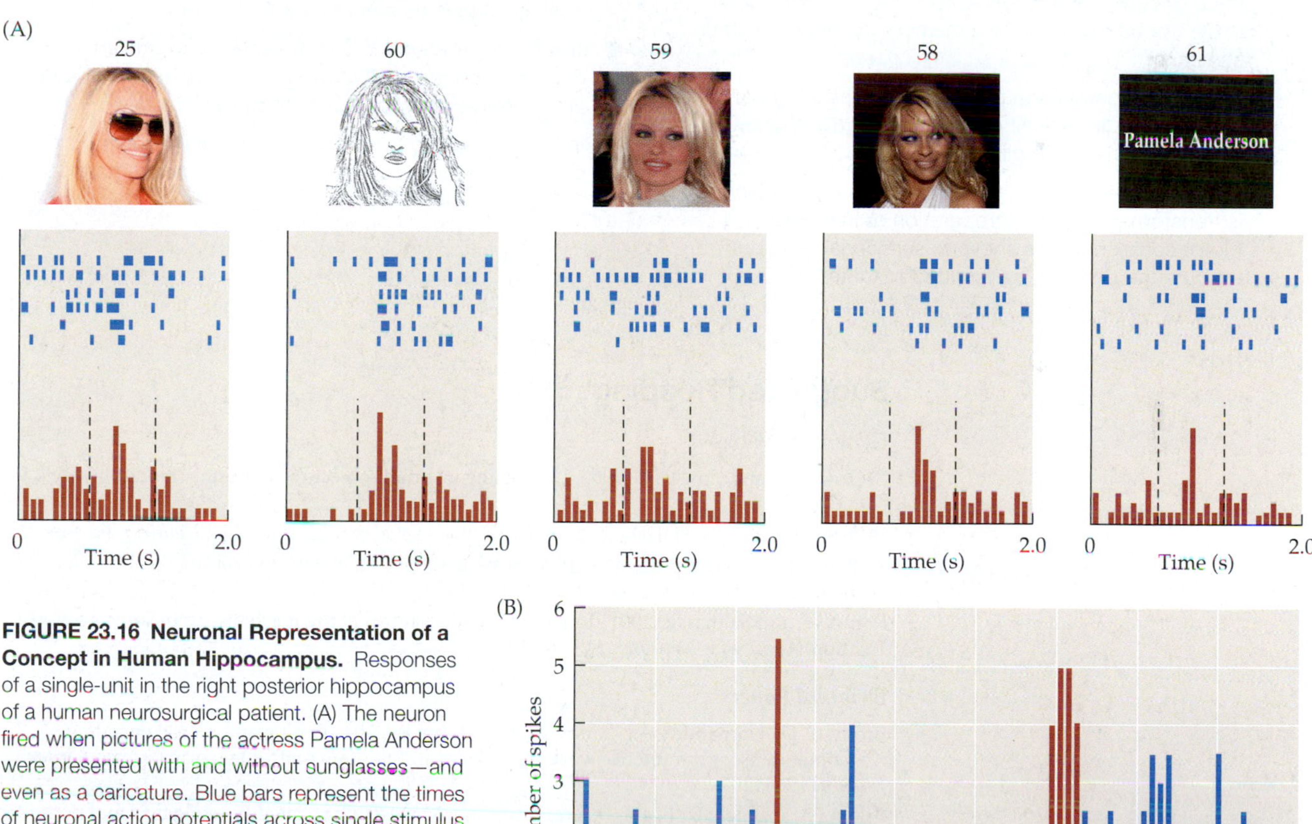

FIGURE 23.16 Neuronal Representation of a Concept in Human Hippocampus. Responses of a single-unit in the right posterior hippocampus of a human neurosurgical patient. (A) The neuron fired when pictures of the actress Pamela Anderson were presented with and without sunglasses—and even as a caricature. Blue bars represent the times of neuronal action potentials across single stimulus presentations and the histograms below are the action potentials summated in time bins. Vertical dashed lines are the start and end of stimulus presentation. The fact that the neuron fired when the written letter string was presented as a stimulus indicates that the neuron's activity is part of the representation of the concept of Pamela Anderson, not just the visual likeness. (B) The vertical bars give the number of spikes fired by the same neuron for 87 different stimuli. Those stimuli related to Pamela Anderson (red bars) evoked the largest response. The neuron fired for other stimuli, so its selectivity was not absolute. (After Quiroga et al., 2005.)

SUMMARY

- All information available about the external world is present in sensory receptors, but sensory signals are perceived as meaningful real-world objects only after elaboration of these signals within the cerebral cortex.
- The sensing, perception, and judgment of a vibration applied to the fingertip has proven to be a useful inquiry. Monkeys can be trained to compare the frequency of two sequential vibrations, the base and comparison stimulus.
- In SI cortex, vibration frequency is encoded by neuronal firing rate.
- Either the base or comparison stimulus, or both, can be substituted by a train of electrical pulses delivered to somatosensory cortex. The monkey senses the artificial stimuli as being natural.
- Neurons in SI encode the base stimulus and the comparison stimulus separately, showing no memory trace or comparison mechanism. Neurons in SII and in the frontal cortex carry a memory trace for the base stimulus.
- During the comparison stimulus, neurons in frontal cortex show an explicit computation of the relation between the two stimuli.
- In the premotor cortex, the result of the comparison is transformed into the preparation of the motor act. Motor cortex neurons show no sensory activity but execute the decision transmitted to them.
- Object recognition in the visual modality depends on processing in the ventral pathway, which courses along the inferior temporal lobe.
- Lesions in the ventral pathway can produce highly specific losses in the ability to identify objects.
- In monkeys and humans there are regions dedicated to the processing of faces. Activity in such areas is necessary and sufficient to produce the percept of a face.
- An object may activate a neuron in the inferotemporal cortex even when viewing conditions change, a property known as invariance.
- The dorsal intracortical pathway proceeds from the occipital lobe to the parietal lobe. Along this pathway, the region MT encodes the direction of motion of objects in the visual scene.
- Although computationally complex, processing through multiple stages occurs remarkably rapidly.
- Top-down inputs course in the anterior to posterior direction and are involved with attention, learning, and stimulus-free recall of earlier images.
- In the hippocampus, successive to the inferotemporal cortex, objects are encoded in a supramodal manner.

Suggested Reading

General Reviews

DiCarlo. J. J., and Cox, D. D. 2007. Untangling invariant object recognition. *Trends Cogn. Sci.* 11: 333–341.

Grill-Spector, K., and Malach, R. 2004. The human visual cortex. *Annu. Rev. Neurosci.* 27: 649–677.

Parker, A. J., and Newsome, W. T. 1998. Sense and the single neuron: probing the physiology of perception. *Annu. Rev. Neurosci.* 21: 227–277.

Romo, R., and Salinas, E. 2001. Touch and go: decision-making mechanisms in somatosensation. *Annu. Rev. Neurosci.* 24: 107–137.

Original Papers

Brody, C. D., Hernandez, A., Zainos, A., and Romo, R. 2003. Timing and neural encoding of somatosensory parametric working memory in macaque prefrontal cortex. *Cereb. Cortex* 13:1196–1207.

Lak, A., Arabzadeh, E., Harris, J., and Diamond, M. 2010. Correlated physiological and perceptual effects of noise in a tactile stimulus. *Proc. Natl. Acad. Sci. USA* 17: 7981–7986.

Quiroga, R. Q., Reddy, L., Kreiman, G., Koch, C., and Fried, I. 2005. Invariant visual representation by single neurons in the human brain. *Nature* 435: 1102–1107.

Romo, R., Hernandez, A., Zainos, A., Brody, C. D., and Lemus, L. 2000. Sensing without touching: psychophysical performance based on cortical microstimulation. *Neuron* 26: 273–278.

Tsao, D. Y., Freiwald, W. A., Tootell, R. B., and Livingstone, M. S. 2006. A cortical region consisting entirely of face-selective cells. *Science* 311: 670–674.

CHAPTER 24
Circuits Controlling Reflexes, Respiration, and Coordinated Movements

As in sensory systems, the neural organization of motor control is hierarchical. Smaller, simpler elements are integrated into more complex circuits at higher levels of the nervous system. Sensory input, feedback loops, and descending motor commands control spinal motoneurons, which innervate skeletal muscles and constitute the final common path of the motor system. Stretch reflexes that originate in muscle spindle afferents represent a basic automatic type of movement. Nevertheless, even at this level, multiple connections of spinal interneurons are necessary for coordinated reflex contractions of flexors and extensors on both sides of the body. Central pattern generators within the central nervous system can generate complex behaviors without feedback from the periphery. Thus, respiratory movements, which are unfailing, regular, and automatic, arise by use of a motor program that depends on rhythmical activity of command neurons within the brainstem. At the same time, the rate and depth of respiration are modulated by sensory inputs as well as by voluntary and involuntary commands from higher centers, including the cortex. Walking and running are mediated by programmed neuronal interactions in the spinal cord and in higher centers. These ensure that limbs move appropriately, with the correct phase relations.

Purely voluntary movements, such as bringing a cup of tea to one's mouth, involve the recruitment of collateral components of motor control, namely anticipation and planning. The destination must be defined, the trajectory designed, synergistic muscles contracted appropriately, and movements initiated in parts of the body, apart from the arm, to compensate for shifts in balance and the effect of gravity. Similarly, if the head and eyes are moved, say to scan the horizon, corollary commands have to be made to allow the image of the world in the mind's eye to remain stationary. For smooth and goal-directed actions to be accomplished, for the head to stay upright and the body to fight against gravity, complex feedback loops are required. They comprise interactions of structures such as the cerebellum, basal ganglia, vestibular apparatus, thalamus, and cortex, as well as stretch reflexes. Sensorimotor integration is carried out in the motor cortex, premotor cortex, and parietal association cortex. Individual cortical neurons code for movement of the arm in a particular direction and are arranged in columns.

Cerebellar deficits cause a loss of coordination and balance, but little change in strength or sensation. Diseases that affect the basal ganglia give rise to spontaneous, disruptive motor outputs or a reduction of voluntary movement. Many aspects of motor control still remain obscure, owing to the extraordinary complexity of the circuitry and the synaptic mechanisms.

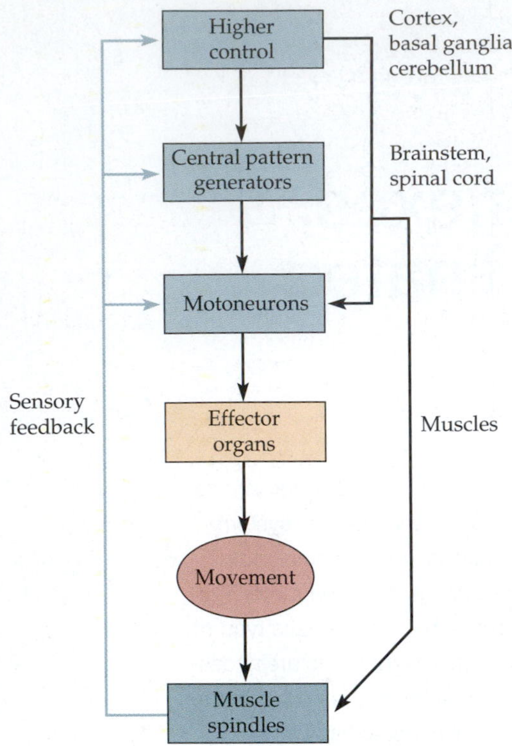

FIGURE 24.1 Scheme for Motor System Organization. Limb muscles are controlled by motoneurons and interneurons of the spinal motor apparatus. Interneurons within the spinal cord and brainstem make up central pattern generators that direct the motor apparatus. Motor output is planned and refined by the motor cortex, basal ganglia, and cerebellum. At every level of motor control, sensory input serves to initiate, inform, and modulate output and is itself influenced by commands from higher levels.

For a leech or a fish to swim, for an owl or a cat to catch a mouse, for a bear or a child to ride a bicycle, a blackbird to trill, or for Dietrich Fischer Dieskau to sing "Die Winterreise" muscles of the body must be brought into play in rapid succession, in a coordinated manner. The intricate mechanisms required for execution of even a simple movement, let alone riding a bicycle, pose a major challenge to neurobiologists. How you point your finger is far harder to analyze than how you perceive a moving object. The reason is that in a sensory system such as vision, one can follow signals from their inception as they spread from receptors to second, third, or fourth order synapses and through the cortex. There, neurons encode in their impulses abstract information about the stimulus, such as a face in the visual field (see Chapter 23). The progression is from simple to ever more complex. By contrast, in the motor system, one cannot simply follow signals downward from the top (the cortex) to find out how a voluntary movement of the body is carried out, let alone initiated. The ultimate events in the motor system can be readily studied in terms of muscles, motoneurons, and simple reflexes. But the descending influences that play upon them arise from a multitude of diverse centers above the level of the spinal cord that interact with one another through intricate servo loops. Central mechanisms that underlie our ability to appose our thumb to the digits of the same hand in rapid succession require motor programs and coordinated activity of many separate descending systems Our future ability to understand such events depends on knowledge of how the component parts of the motor system, spinal cord, brainstem, vestibular apparatus, cerebral cortex, cerebellum, and basal ganglia work together to plan and execute movements.

As in sensory systems, principles of organization have emerged to simplify the task. The first is that motor control is arranged hierarchically. Increasingly complex motor tasks are organized in successively higher centers (Figure 24.1). A second important principle is that motor output is continually updated and adjusted by feedback. Both negative and positive feedback loops through the basal ganglia, vestibular apparatus, and cerebellum are essential to the timing and coordination of cortical motor programs. An extra complication of motor systems is that every movement of limbs, trunk, or head occurs in the gravitational field and alters the perception of the body's configuration in space. Hence, even before the initiation of a voluntary movement (for example, the raising of one leg), contractions must be set in motion to compensate for the predicted effects of gravity.

In this chapter, we begin with the synaptic organization responsible for spinal reflexes and the properties of spinal motoneurons, and then go on to describe how groups of neurons within the spinal cord and brainstem produce rhythmic, coordinated movements of respiration and locomotion. Next, we discuss descending influences from motor cortex, vestibular apparatus, cerebellum, basal ganglia, and motor cortex. How, in the brain of a higher animal, the decision to move an arm is made, or how voluntary actions are initiated, remain as important questions for future work.

C. S. Sherrington with one of his pupils (J. C. Eccles) in the mid-1930s.

Adrian, whose name has cropped up again and again in earlier chapters, long ago posed problems regarding motor systems, in sentences that could not be bettered: (1) "The chief function of the central nervous system is to send messages to the muscles which will make the body move effectively as a whole and for this to take place the contraction of each muscle must be capable of delicate adjustment,"[1] and (2) "We may learn a skilled movement by employing certain muscles and therefore certain groups of nerve-cells in the motor area of the brain, but when we have learnt it we can carry out the movement with an entirely different set of muscles and nerve-cells—we can write our name with a pencil held between the toes when we have learnt to do it with the fingers. We can draw a triangle small or large when we have learned its shape…"[2]

The Motor Unit

Sherrington called the spinal motoneuron the **final common path** because all the neural influences that concern movement or posture converge upon it. The major motor nerve cells

[1] Adrian, E. D. 1959. *The Mechanism of Nervous Action*. University of Pennsylvania Press, Philadelphia.

[2] Adrian, E. D. 1946. *The Physical Background of Perception*. Clarendon, Oxford, England.

of the spinal cord are called α motoneurons. Smaller motoneurons, called γ, that regulate the sensitivity of muscle spindles, will be discussed later. (Note: the terms α and γ arose from an early classification of conduction velocity; see Chapter 8). A single α motoneuron innervates a group of muscle fibers, and together the motoneuron and its target fibers form a functional element known as the **motor unit**. The number of muscle fibers in a motor unit ranges from a few—for example, in muscles used to extend or flex the fingers—to several thousand in the large proximal muscles of the limbs.

When a motor neuron discharges, all the muscle fibers to which it is connected contract. The smoothness and precision of our movements are brought about by varying the number and timing of motor units brought into play.[1] Contractions of single motor units are not apparent as small twitches when the whole muscle contracts because the individual activations are asynchronous and are smoothed out by the elastic properties of the muscles. For example, the 25,000 muscle fibers in the cat soleus are supplied by 100 α motoneurons. A contraction of the whole muscle can therefore be graded in 100 unequal steps by **recruitment** of motor units.

Skeletal muscle fibers are not homogeneous: some are faster in their contractions than others and their contractile mechanisms fatigue more rapidly. The slowly conducting, fatigue-resistant fibers (also called red muscles) depend on oxidative metabolism for energy production, while the fast, rapidly contracting fibers (or white muscles) depend on glycolysis. The pattern of activation of any given motoneuron is matched to the properties of its muscle fibers.[3]

[3]Kanning, K. C., Kaplan, A, and Henderson, C. E. 2010. *Annu. Rev. Neurosci.* 33:409–440.

[4]Brannstrom, T. 1993. *J. Comp. Neurol.* 330: 439–454.

[5]Eccles, J. C. 1981. *Appl. Neurophysiol.* 44: 5–15.

[6]Hultborn, H. 2006. *Prog. Neurobiol.* 78: 215–242.

[7]Mendell, L. M., and Henneman, E. 1971. *J. Neurophysiol.* 34: 171–187.

[8]Lucas, S. M., and Binder, M. D. 1984. *J. Neurophysiol.* 51: 50–63.

[9]Brown, A. G., and Fyffe, R. E. W. 1981. *J. Physiol.* 313: 121–140.

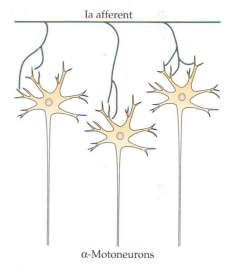

Synaptic Inputs to Motoneurons

The recruitment and fine control of motoneurons to produce coordinated movements requires that influences from a variety of sources should play upon them in the appropriate sequence and with appropriate balance. It is therefore not surprising that the average motoneuron receives many thousands of synaptic inputs[4] (see Figure 1.13A). These inputs relay instructions from higher centers and from sensory receptors in the periphery. Inputs to the cell produce excitatory and inhibitory postsynaptic potentials (EPSPs and IPSPs), and presynaptic inhibition selectively regulates the efficacy of the incoming signals. Each time the motoneuron is sufficiently depolarized, an impulse originates in a particular region of the cell, called the **axon hillock** (see Chapter 8).

Much is now known about the mechanisms of synaptic transmission onto this cell and the interaction of excitatory and inhibitory synapses.[5,6] One important excitatory input is from the muscle spindles (see Chapter 19): The group Ia afferent fibers make monosynaptic excitatory connections on motoneurons. By painstakingly recording from all the motor neurons supplying a particular muscle (its **motor pool**), Mendell and Henneman[7] have shown that each Ia afferent fiber from a muscle sends an input to as many as 300 motoneurons—virtually all those supplying that muscle.[8]

The arborization pattern of an individual Ia afferent can be seen directly after intracellular injection of the enzyme horseradish peroxidase (HRP). The labeled Ia afferent axon branches extensively along the rostro-caudal axis of the spinal cord to contact members of the motor pool (Figure 24.2A). Close examination of HRP-filled afferent fibers allows their contacts with individual motoneurons to be mapped (Figure 24.2B). The convergence of Ia afferents onto motoneurons is anatomically discrete, each afferent making two to five contacts on the dendritic tree.[9] All the Ia contacts onto a given motoneuron tend to occur within the same general region of the dendritic tree, although not necessarily at the same electrical

(A)

Ia afferent

α-Motoneurons

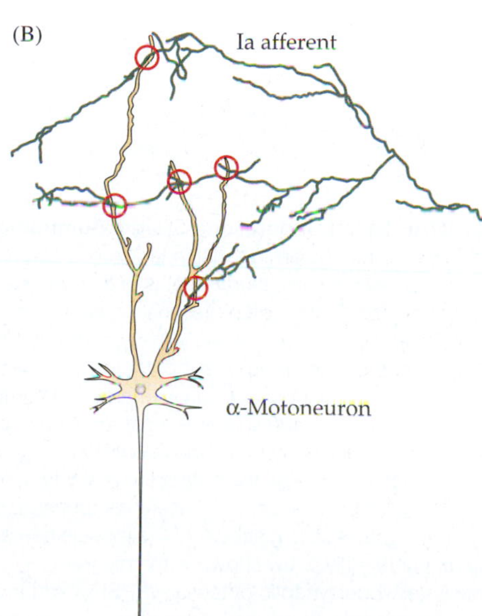

FIGURE 24.2 Contacts between Stretch Receptor Afferents and Spinal Motoneurons. (A) A single muscle spindle fiber (Ia) afferent sends branches to several motoneurons. (B) A more detailed view shows the afferent fiber passing over multiple dendritic branches, indicating possible points of synaptic contact (red circles). Innervation patterns of this kind can be seen experimentally by labeling afferent fibers and motoneurons with histological markers, such as horseradish peroxidase (HRP). (After Burke and Glenn, 1996.)

distance from the soma.[10] It is remarkable that a single axon collateral provides all these contacts, while other branches of the same axon pass by to supply other motoneurons.

Unitary Synaptic Potentials in Motoneurons

An impulse in a single Ia afferent fiber gives rise in a motoneuron to only a very small (about 200 microvolts [μV]) monosynaptic excitatory potential. This potential corresponds, on average, to the release of a total of one or two quanta of transmitter from the four to seven synaptic contacts (i.e., not every bouton releases a quantum with every impulse). Quantitative measurements of release were first made by Kuno,[11] who dissected small sensory nerve bundles in dorsal roots and recorded the potentials produced by stimulation of single Ia afferent fibers.[12] Single potentials on the order of 200 μV can be expected to have little influence on the firing pattern of a motoneuron, but they sum during brief bursts of activity to produce a buildup of depolarization in a process called **temporal summation** (Figure 24.3A). In addition, sufficient stretch of a muscle like the soleus can activate of all 50 of its Ia stretch receptors, resulting in **spatial summation** of all the inputs contacting different regions in the motoneuron's dendritic tree (Figure 24.3B). Integration of a multitude of excitatory and inhibitory synaptic inputs determines whether or not the motoneuron will reach threshold for the action potential.

The Size Principle and Graded Contractions

How are motor units recruited to produce smoothly graded movements? As already described, the force of contraction can be increased by bringing in additional motoneurons, and by increasing their rate of firing. However, further refinement is provided by recruiting the motoneurons according to their size. A motoneuron with a small cell body and a small axon innervates relatively few muscle fibers, so their activation causes only modest increases in muscle tension. Large motoneurons with large-diameter axons contact many muscle fibers, and an impulse from such a motoneuron gives rise to a large increase in muscle tension. When a contraction occurs, small motor units fire first, producing small increments in tension. As the strength of the contraction increases, larger units are recruited, each contributing progressively more tension.[13] Finely graded control is thereby achieved, enabling small or large movements to be produced efficiently. The orderly recruitment of motoneurons is referred to as the **size principle**.

For example, in the soleus muscle of the cat, the firing of a small motoneuron may give rise to an increase in tension of about 5 grams (g), whereas a larger unit may contribute more than 100 g, and the maximum contraction brought about by all the motor units firing may

[10] Burke, R. E., and Glenn, L. L. 1996. *J. Comp. Neurol.* 372: 465–485.

[11] Kuno, M. 1971. *Physiol. Rev.* 51: 647–678.

[12] Kirkwood, P. A., and Sears, T. A. 1982. *J. Physiol.* 322: 287–314.

[13] Henneman, E., Somjen, G., and Carpenter, D. O. 1965. *J. Neurophysiol.* 28: 560–580.

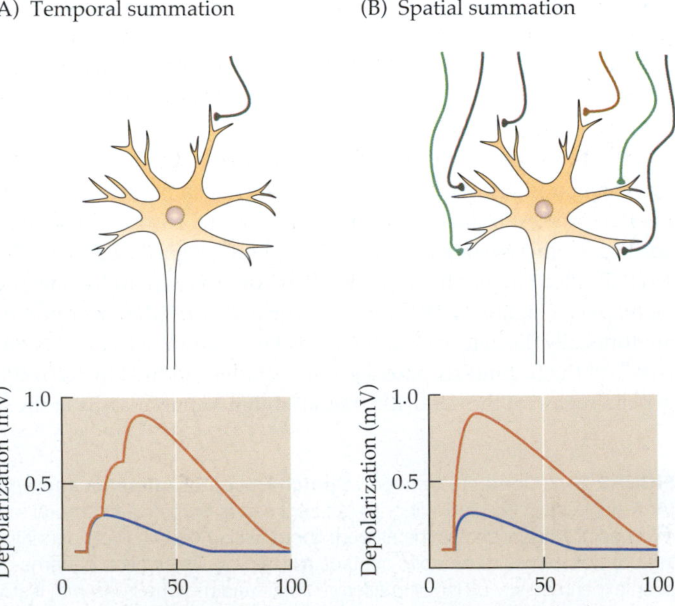

FIGURE 24.3 Temporal and Spatial Summation. (A) A single action potential in a single Ia afferent produces a synaptic potential in a motoneuron that is only a fraction of a millivolt (mV; blue trace). When the presynaptic fiber fires three action potentials rapidly in succession, the synaptic potentials (brown traces) ride on the falling phase of the previous one, so they build up to a larger depolarization—temporal summation. (B) A muscle, such as the soleus in a cat, may have as many as 50 muscle spindles, and an equivalent number of Ia afferent fibers. These all diverge to contact the majority of motoneurons in the motor pool. Thus, 50 Ia afferents converge onto each motoneuron. A strong stretch of the muscle can activate all the Ia afferents (only a few shown in the diagram): the individual excitatory postsynaptic potentials (EPSPs) add to depolarize the motoneuron by spatial summation.

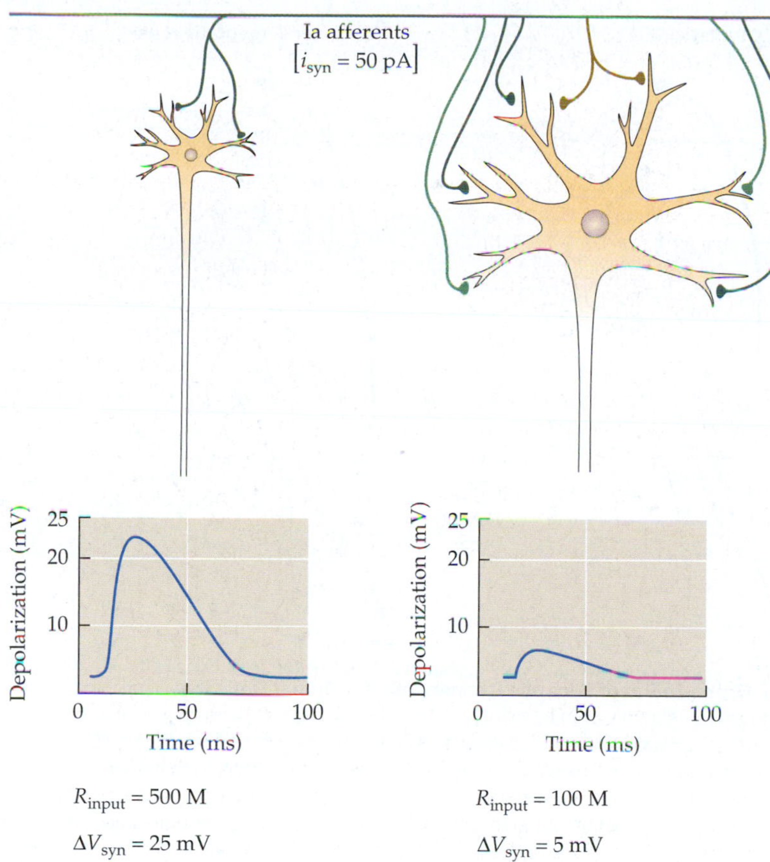

FIGURE 24.4 The Size Principle. Current flow into a motor nerve cell produces a change in membrane potential that is proportional to the input resistance (r_{input}). Input resistance is inversely proportional to the radius of the cell, so equivalent synaptic currents (i_{syn}) produce greater depolarization ($\Delta V_{syn} = i_{syn} r_{input}$) of smaller motoneurons (see Chapter 8). The small motor neuron on the left and the large one on the right, both receive the same input from Ia afferent fibers. The synaptic currents, which are the same in both cells (50 pA), produce a larger depolarization in the smaller motoneuron.

reach over 3.5 kilograms (kg). Plainly, a small motor unit would be relatively ineffective if brought in when the contraction is near its maximum, and a large unit firing in the lower range would perturb fine movements. The fact that the motor units are recruited in order of increasing size means that each additional unit adds a relatively fixed fraction (about 5%) to the existing tension.

The principle that motor unit recruitment adds a fixed fraction, rather than an absolute force increment, to existing muscle tension is also reflected in the sensory aspects of our motor activity. We judge weights by the muscular force needed to support them, and we can easily distinguish the difference between 2 and 3 g, but not between 2002 and 2003 g. Again, it is the *relative* change that is important, as enunciated in the Weber–Fechner relationship (see Chapter 19). Indeed much of our perception of the world is determined in the same way, and we act accordingly. You would not mind paying $2003 for an item normally costing $2002, but would be outraged at paying $3 for a $2 postage stamp.

How do the cellular properties of motoneurons help to explain the size principle? Suppose that all the motoneurons innervating a muscle receive the same synaptic input. The voltage change produced by the synaptic current in each motoneuron depends on its input resistance, which is a function of cell size (Figure 24.4). As shown in Chapter 8, the input resistance varies inversely with the cell radius. Thus, any given synaptic current will produce a larger voltage change in smaller motoneurons, making them more likely to reach threshold than the large motoneurons. As the strength of sensory input increases, larger and larger motoneurons will be brought to threshold.

Spinal Reflexes

Reciprocal Innervation

Limb movements are produced by the coordinated contraction of groups of muscles that work together, referred to as **agonists**. Opposing muscles are called **antagonists**. **Extensor**

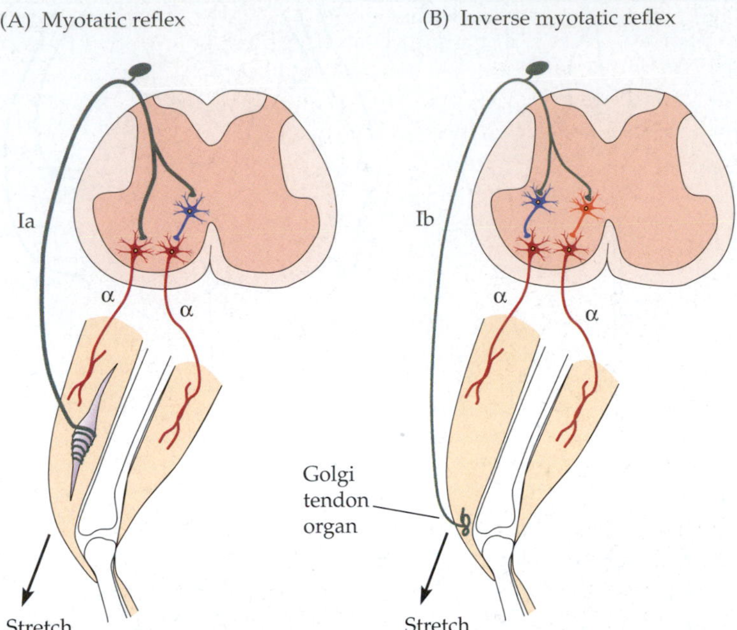

FIGURE 24.5 Organization of Synaptic Connections for Spinal Reflexes. The spinal cord is shown in transverse section, with inhibitory interneurons in blue. (A) In the myotatic reflex, stretch of the muscle spindle generates impulses that travel along group Ia afferent fibers to the spinal cord and produce monosynaptic excitation of α-motoneurons to that same muscle. Impulses also excite interneurons that, in turn, inhibit motoneurons supplying the antagonist muscles. (B) Stretch or contraction of the muscle pull on the tendon and generate impulses in the Golgi tendon organ's Ib afferent fiber. Ib fibers inhibit motoneurons that supply the same muscle.

muscles open or extend the joints and oppose the force of gravity; **flexor** muscles close or flex the joints and pull the limbs toward the body. When a myotatic reflex (see Chapter 1) is activated by muscle stretch (e.g., by tapping the patellar tendon to produce a knee jerk), the primary sensory endings in the muscle spindles of the extensors are deformed and initiate impulses in group Ia afferent fibers going to the spinal cord. These impulses produce monosynaptic excitation of the α-motoneurons projecting back to the muscle that has been stretched, resulting in a reflex contraction.

Extensor muscle contraction is accompanied by simultaneous inhibition of the α-motoneurons that innervate antagonistic flexor muscles. This occurs because Ia afferents activate spinal interneurons that inhibit the antagonist α-motoneurons (Figure 24.5A). The principle of one group of muscles being excited while its antagonists are inhibited was first described by Sherrington, who called it **reciprocal innervation.**[14] We have emphasized the knee jerk in this analysis because of its simplicity; but reflexes that play a part in bodily movements normally occur as the result of increases in tension that are graded, rather than the abrupt and synchronous input to all the sensory endings for stretch produced by a hammer.

For the sake of simplicity, a number of pathways are omitted from Figure 24.5. For example, discharges in the smaller group II afferents from muscle spindles reinforce the reflex largely by way of interneurons.[15] Such intraspinal connections have been worked out in detail by Lundberg, Jankowska, and their colleagues.[16] Inhibitory interneurons are also activated by **Golgi tendon organs** whose sensitive endings are encapsulated near the tendon–muscle junctions[17] (Figure 24.5B). Their afferent fibers are designated type Ib, to distinguish them from the primary spindle afferent fibers. These stretch receptors are in series with contracting skeletal muscles. They can be made to discharge impulses by passive stretch, but muscle contraction, to which they are more sensitive, is the principal stimulus that activates firing. A contraction of one or two muscle fibers, leading to a tension increase

[14]Sherrington, C. S. 1906. *The Integrative Action of the Nervous System*, 1961 ed. Yale University Press, New Haven, CT.

[15]Marchand-Pauvert, V. et al. 2005. *J. Physiol.* 566: 257–271.

[16]Bannatyne, B. A. et al. 2009. *J. Physiol.* 587: 379–399.

[17]Crago, P. E., Houk, J. C., and Rymer, W. Z. 1982. *J. Neurophysiol.* 47: 1069–1083.

of less than 100 milligrams (mg), can cause a brisk discharge. Axons from tendon organs activate interneurons that, in turn, inhibit α-motoneurons supplying their muscle of origin; thus, their action is the opposite to that of spindle afferents.[18–20]

In summary, the stretch reflex has several underlying mechanisms. First, impulses in Group Ia and II afferent fibers from the spindles activate motoneurons supplying their own muscles, causing contraction. At the same time, the afferent impulses are transmitted through interneurons to inhibit antagonist muscles. Finally, the muscle contraction itself activates Group Ib afferent fibers that inhibit the motoneurons to terminate ongoing activity.

Central Nervous System Control of Muscle Spindles

Sensory responses from muscles are complicated by the fact that the spindles themselves contain specialized contractile elements, called **intrafusal muscle fibers** (see Chapter 19). These fibers contract in response to excitation by a dedicated group of motor nerve fibers (Figure 24.6) that are 2–8 micrometers (μm) in diameter; they are also known as fusimotor or γ efferent fibers.[21,22] It is important to note that the contractile elements within the spindle do not run uninterruptedly from end to end (see Figure 24.6). As they contract, they stretch the central gelatinous region of the spindle in which the sensory nerve endings are embedded, thereby initiating sensory impulses in the group I and II afferent nerve fibers. Note also that the intrafusal muscle fibers, unlike the extrafusal muscle fibers, bear no load even if the muscle is lifting many kilograms. This enables them to contract reliably in a graded manner when they are stimulated by the γ-motoneurons. In principle, the spinal cord could command an intrafusal muscle fiber to shorten its length by 50% and be sure to obtain the desired result. By contrast, a simple motor command to extrafusal muscle fibers via α-motoneurons could not guarantee shortening to a precise length. The extent of contraction of extrafusal fibers making up the muscle mass will depend not only on the command but also on the load that has to be lifted.

The role of fusimotor fibers was established in a series of technically difficult and definitive experiments on anesthetized cats by Kuffler, Hunt, and Quilliam.[23] They recorded the electrical activity of an individual dorsal root sensory fiber coming from a muscle spindle while simultaneously stimulating a single fusimotor γ-fiber in the ventral root that supplied the same spindle. Fusimotor stimulation produced an increase in sensory activity, but no increase in tension of the muscle as a whole. Trains of impulses in γ-fusimotor neurons either accelerated the sensory discharge produced by stretching the muscle or initiated a sensory discharge in the relaxed muscle.

What is the function of the γ motor system and what is achieved by the efferent regulation of muscle spindle discharges? When the entire muscle is stretched, the intrafusal fibers

[18]Matthews, P. B. C. 1972. *Mammalian Muscle Receptors and Their Central Action.* Edward Arnold, London.

[19]Edin, B. B., and Vallbo, A. B. 1990. *J. Neurophysiol.* 63: 1307–1313.

[20]Windhorst, U. 2007. *Brain. Res. Bull.* 73: 155–202.

[21]Eccles, J. C., and Sherrington, C. S. 1930. *Proc. R. Soc. Lond. B* 106: 326–357.

[22]Leksell, L. 1945. *Acta Physiol. Scand.* 10(Suppl. 31): 1–84.

[23]Kuffler, S. W., Hunt, C. C., and Quilliam, J. P. 1951. *J. Neurophysiol.* 14: 29–51.

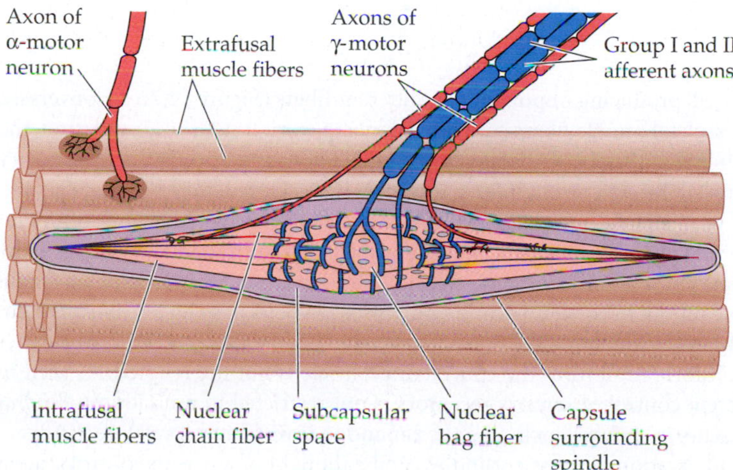

FIGURE 24.6 Mammalian Muscle Spindle. Scheme of mammalian muscle spindle innervation. The spindle, composed of small intrafusal fibers, is embedded in the bulk of the muscle. The main muscle is made up of large muscle fibers supplied by α-motoneurons. Gamma motor (or fusimotor) fibers supply the intrafusal muscle fibers, and group I and group II afferent fibers carry sensory signals from the muscle spindle to the spinal cord. Note that the central region of the intrafusal muscle fiber is made up of nuclei and that the contractile material does not run from end to end. As a result, when the intrafusal fiber at each end of the spindle contracts, the sensory endings are stretched (see also Figure 19.5).

FIGURE 24.7 Efferent Regulation of Muscle Spindles. Records of extrafusal muscle tension (red) and sensory discharge from muscle spindles (blue). (A) Stimulation of γ-efferent fibers causes contraction of muscle spindles, producing discharge in the afferent fibers. (B) Stimulation of α-fibers supplying the main muscle causes it to contract, reducing the stretch on the intrafusal fibers, and the sensory fibers stop firing. (C) When both α and γ-motor fibers are stimulated tension on the muscle spindles remains unchanged and the sensory discharge is undisturbed.

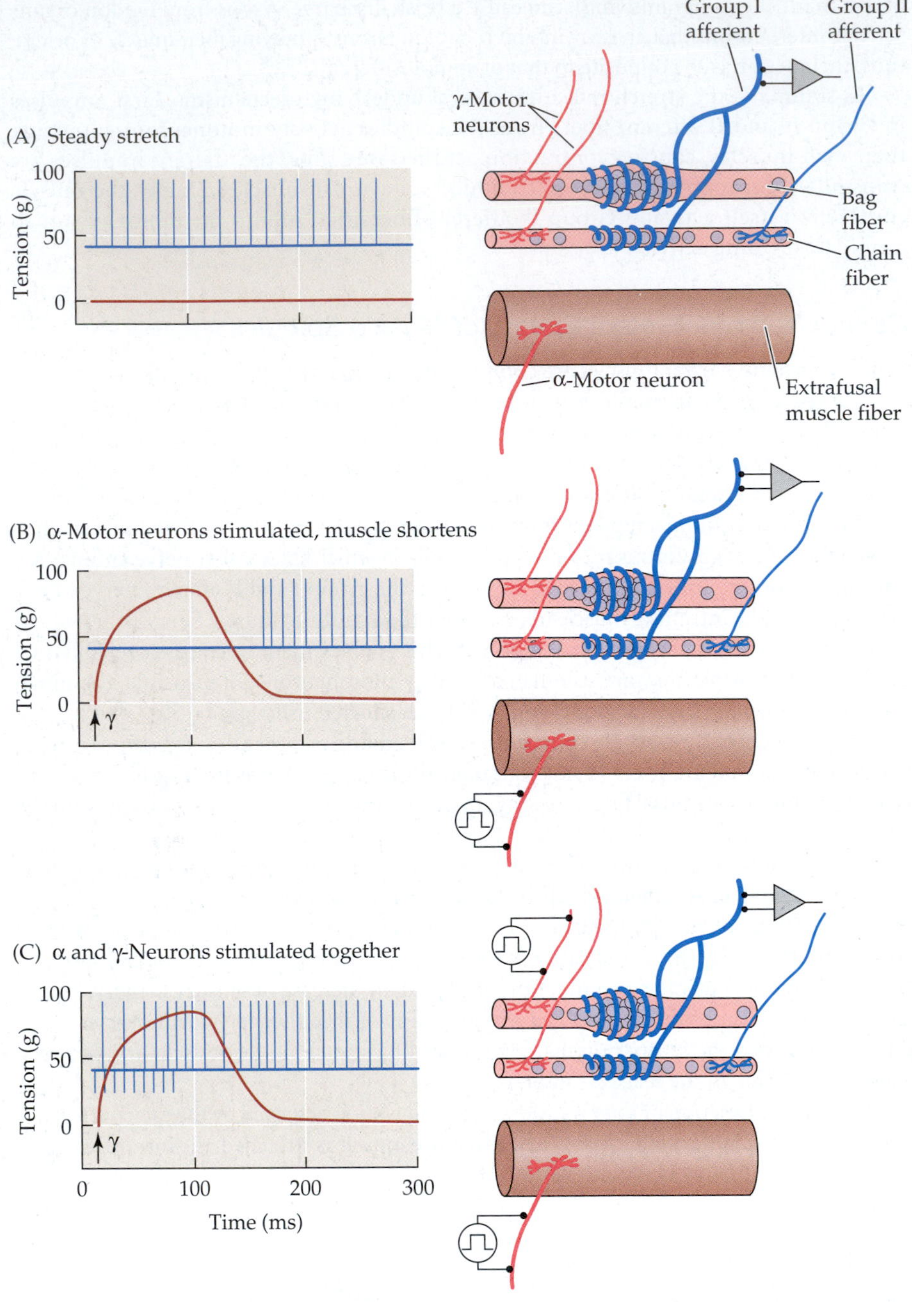

are stretched as well, producing impulses in the afferent fibers (Figure 24.7A). Conversely, when extrafusal skeletal muscle fibers contract following α-motoneuron stimulation, the intrafusal fibers go slack, thereby unloading the central regions and terminating the sensory discharge (Figure 24.7B). As a result, information about muscle length is no longer sent to the CNS. If this were all that occurred, the brain could not know whether the command to contract had been properly executed. Accordingly, when a contraction is produced, the spinal cord activates both γ motoneurons and α-motoneurons.[24,25] This causes the intrafusal muscle fibers to contract in concert with the muscle as a whole, thereby preventing them from going slack, and maintaining the sensory discharge (Figure 24.7C). The CNS ensures that the flow of information from the contracting muscle is not interrupted. As seen in Chapter 19, efferent control of sensory receptors is not restricted to muscle spindles but occurs in other sensory systems, including the ear and in invertebrate stretch receptors.

The motor innervation of muscle spindles can be thought of as a gain control system that continually adjusts the length of the spindle in order to maintain its sensitivity over

[24] Sears, T. A. 1964. *J. Physiol.* 174: 295–315.

[25] Allen, T. J., Ansems, G. E., and Proske, U. 2008. *Exp. Physiol.* 93: 391–398.

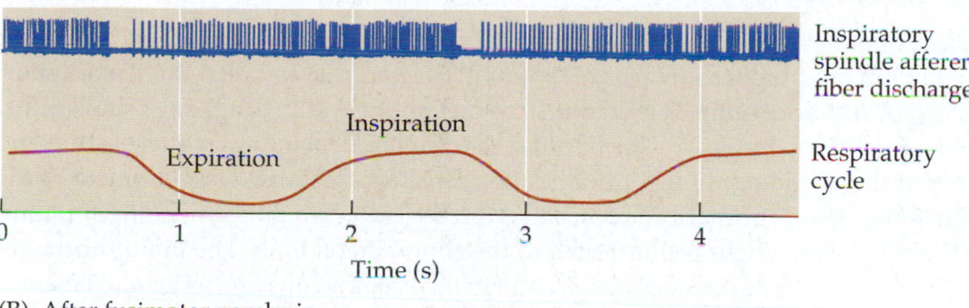

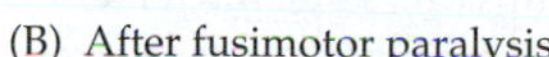

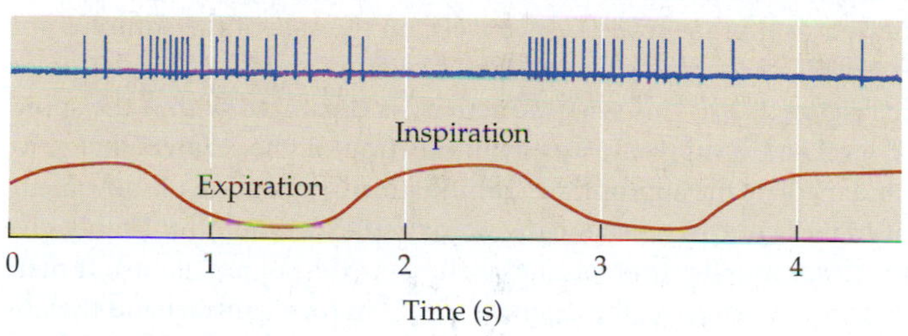

FIGURE 24.8 Coactivation of α and γ-Respiratory Motoneurons. (A) Extracellular recording of action potentials arising from a spindle in an inspiratory muscle during the respiratory cycle (lower trace, red). It is at first surprising that the sensory discharge from the inspiratory muscle spindle is highest during inspiration, while the muscles are shortening rather than being stretched. This phenomenon is due to simultaneous activation of γ fusimotor fibers to the spindle. (B) After the fusimotor fibers are blocked selectively by procaine, the spindles behave passively. The sensory discharge frequency increases discharges during expiration, when the main mass of the muscle is stretched; as expected, sensory firing stops during inspiratory movements when the main mass of the muscle is shortened. (After Critchlow and von Euler, 1963.)

all lengths of the muscle. At present, what is lacking is detailed information about how the firing of γ motoneurons is regulated by descending influences from higher centers.

An example of co-activation of α and γ-motoneurons is shown in Figure 24.8A[26] (see also Figure 24.10). The recordings show discharges in spindle afferents of inspiratory muscles during respiration. Figure 24.6B shows that sensory discharges in a spindle from an inspiratory muscle are, in fact, highest during *inspiration*, when the muscle is contracted and short, not during expiration when it is stretched. This apparent paradox is explained by the fact that α and γ-motoneurons are activated together, so that intrafusal contraction more than compensates for muscle shortening. When the γ-motoneurons are selectively blocked by a local anesthetic (Figure 24.8B), the sensory fibers fire only during expiration, while inspiratory muscles are being stretched.

Evidence for **co-activation** of large α and small γ-motoneurons has also been obtained from experiments on finger movements, where spindle afferents increase their firing, even during voluntary isometric contractions when the joint does not move.[19] An additional complication of spindle mechanisms is that more than one type of γ efferent system has been identified. Certain γ-motoneurons increase dynamic responses of spindle afferents (i.e., their responses to the *rate* of stretch), while other γ-motoneurons increase their static responses (i.e., to the degree of stretch).[27]

Clearly, the motor innervation of spindles prevents the afferent volley of action potentials from acting simply as indicators of muscle length (see Figure 24.8). A high rate of spindle discharge from an extended muscle with no fusimotor activity can be the same as that from a shortened muscle with active fusimotor fibers. That spindle discharges do, in fact, provide information about the position of limbs and fingers in space is possible only because the central nervous system continually monitors and modulates the outgoing fusimotor activity.[28] Imagine that your hand and your arm are commanded to pick up a beaker and bring it to your mouth without spilling a drop. If the beaker is heavier than expected, stronger contractions will be required. At every moment, the nervous system must calibrate how well the muscles have compensated for the load and must monitor speed and position in the trajectory as well as the remaining distance to be covered. The servo mechanism of the γ system, with its calibrated output to the intrafusal fibers, allows the muscle contractions to be regulated perfectly, even with the eyes closed. Note also that the spindle discharge throughout the contraction enhances the reflex drive onto the α-motoneurons and thereby aids the purposeful movement.

It is of interest that certain skeletal muscles contain no spindles and are controlled solely by α-motoneurons. An example is provided by the extraocular muscles that move the eye.[29] Feedback is visual, and the load on the muscles is constant and irrelevant (apart from minor effects of gravity[30]).

[26] Critchlow, V., and von Euler, C. 1963. *J. Physiol.* 168: 820–847.

[27] Durbaba, R. et al. 2003. *J. Physiol.* 550: 263–278.

[28] Smith, J. L. et al. 2009. *J. Appl. Physiol.* 106: 950–958.

[29] Daniel, P. 1946. *J. Anat.* 80: 189–193.

[30] Pierrot-Deseilligny, C. 2009. *Ann. N Y Acad. Sci.* 1164: 155–165.

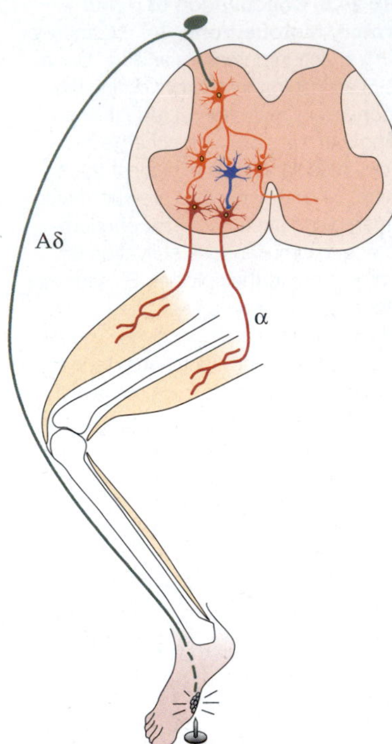

FIGURE 24.9 The Flexor Reflex is a limb-withdrawal reflex, produced in this example by stepping on a tack. Excitation of Aδ pain fibers results in elevation of the thigh (synaptic connections not shown) and flexing of the knee joint by polysynaptic excitation of flexor motoneurons and inhibition of extensors (blue interneurons are inhibitory). Also not shown are contralateral connections that subserve extension of the opposite leg for support.

Flexor Reflexes

Complex combinations of muscle activity involving multiple joints and sometimes more than one limb are produced by painful stimuli. The simplest is called the **flexor reflex**, which is activated, for example, when one steps on a sharp object, bangs one's shin against a bench, or touches a hot stove. The response is complex, depending on the location and intensity of the offending stimulus, but has two consistent features: (1) movement of the affected limb is always primarily flexion and is directed away from the offending stimulus; and (2) if necessary, weight is transferred to the contralateral limb. The input for the reflex arises from responses of nociceptive and tactile receptors in skin.[31] The movement is determined by the interplay of networks of excitatory and inhibitory spinal interneurons acting on flexor and extensor motoneurons, respectively, on the side of the stimulus and (if weight transfer is needed) by simultaneous extensor excitation and flexor inhibition on the contralateral side (Figure 24.9). This synaptic activity is organized within the spinal cord at the segmental level and is supplemented by inputs from higher centers that serve to maintain balance and mediate the appropriate continuation or cessation of movement.

In the absence of higher control mechanisms, spinal reflexes would inevitably give rise to oscillations: as a flexor muscle contracts it stretches its extensor antagonist. If there were no overriding control mechanisms, the extensor would in turn contract and thereby re-stretch the flexor, and so on. Indeed, a major function of higher centers, such as the basal ganglia (see below), is to guarantee smooth movements and to prevent spasticity, tremor, and clonus from occurring. Such symptoms appear in patients with lesions of descending tracts.

Generation of Coordinated Movements

There is an important difference between simple reflexes and coordinated, rhythmical movements. For several patterned motor acts, sensory feedback is not a prerequisite. For example, during birdsong, movements of the muscles follow each other in a rapid, orderly sequence without sufficient time for a feedback loop to be completed.[32] The next instructions are sent out from the central nervous system before it can analyze the preceding sound. After the CNS of a leech or a cockroach has been dissected out of the body and deprived of all sensory inputs, it continues to generate impulses in patterns that would normally result in swimming or walking. The presence of autonomous **central pattern generators** does not mean that feedback from the periphery is ignored entirely during behavior of the intact animal.[33] For example, if the dorsum of the foot of a walking cat touches a small twig during the swing phase (see Figure 24.14), the foot will be lifted elegantly over the twig. The role of sensory feedback is to modulate ongoing motor programs in accordance with the organism's needs and in response to unpredicted challenges imposed by the external world.

Two examples—respiration and walking—will be used to illustrate how central pattern generators within the mammalian CNS produce coordinated movements.

Neural Control of Respiration

Ceaseless rhythmical contractions of respiratory muscles ensure that oxygen and carbon dioxide can be exchanged between the blood and the lungs until the moment that one dies. To ensure that normal and premature babies can breathe as soon as they are born, programmed respiratory movements of the rib cage are already made in embryos (obviously without intake of oxygen). The respiratory rhythm is relentless and unfailing: you can commit suicide by not eating, but you cannot decide not to breathe any more. Nevertheless, this automatic behavior is modulated by sensory feedback and, within limits, can be controlled by the will.

The diaphragm and two antagonistic sets of muscles are responsible for drawing air into the lungs and expelling it. During inspiration, the diaphragm contracts and the rib cage is raised by the external intercostal muscles (Figure 24.10). As a result, the volume of the chest is increased, the lungs expand, and air enters. Expiration is achieved by relaxation of the diaphragm and contraction of the internal intercostal muscles. Other muscles of the thorax and abdomen also contribute to a variable extent, depending on the posture of the animal and the rate and depth of respiration.[34] An example of the respiratory rhythm in muscles of

[31] Sherrington, C. S. 1910. *J. Physiol.* 40: 28–121.

[32] Konishi, M. 2004. *Ann. N Y Acad. Sci.* 1016: 463–475.

[33] Briggman, K. L., and Kristan, W. B. 2008. *Annu. Rev. Neurosci.* 31: 271–294.

[34] Da Silva, K. M. C. et al. 1977. *J. Physiol.* 266: 499–521.

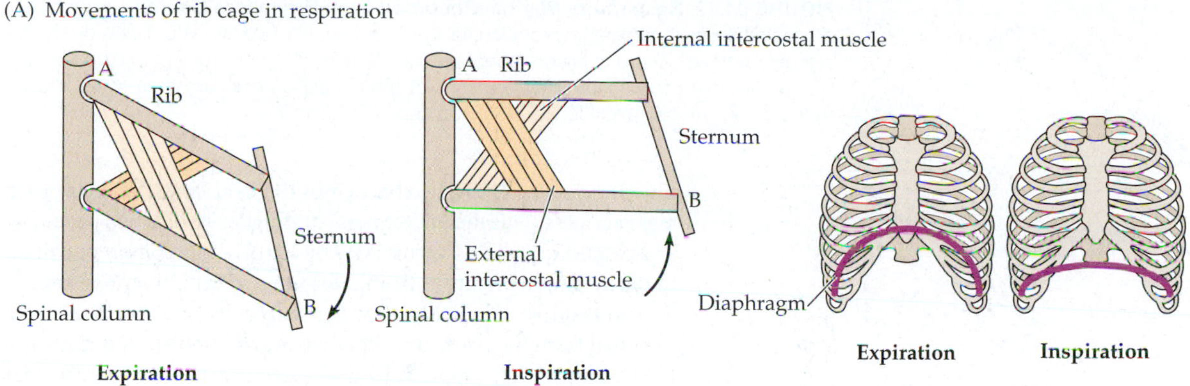

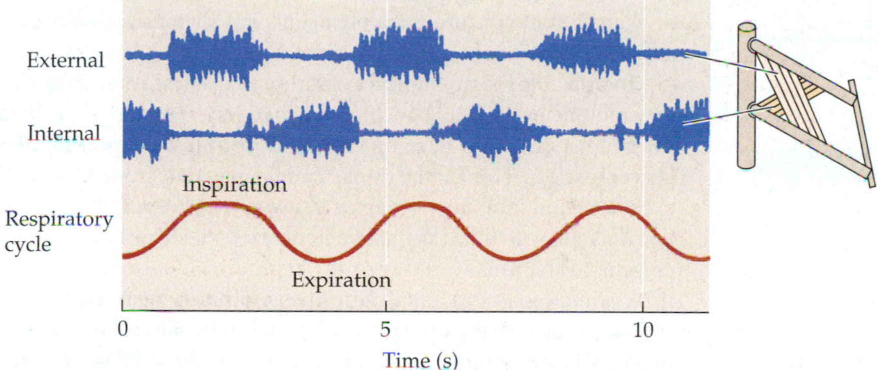

FIGURE 24.10 Movements of Rib Cage and Respiratory Muscles during expiration and inspiration. (A) Actions of the internal intercostal muscles (depressing the ribs during expiration) and external intercostal muscles (raising the ribs during inspiration). As the diaphragm contracts it expands the lungs. (B) Activity of respiratory muscles in the cat recorded with needle electrodes. Discharges of the external and internal intercostal muscles are out of phase.

an anesthetized cat is shown in Figure 24.10B. Activity of each muscle is registered by strain gauges and by recording its electrical activity with wire electrodes embedded in the body of the muscle; that is, by **electromyography** (**EMG**). Figure 24.10B shows that inspiratory and expiratory muscle contractions are accompanied by bursts of potentials, indicating motor unit discharges; it is apparent that the two sets of muscles contract out of phase.

As in limb muscles, the stretch reflex contributes to movement by maintaining the excitability of motoneurons. When the internal and external intercostal muscles are stretched alternately by commands from the central nervous system, their Ia afferent fibers fire at high frequencies. Those impulses contribute excitatory synaptic potentials to the homonymous motor neurons (i.e., motor neurons supplying the same muscle) and, through interneurons, contribute inhibition to the motor neurons of antagonist muscles. Figure 24.11 shows the effect on muscle fiber discharge activity, as an inspiratory muscle (the levator costae) is stretched by pulling on its tendon. With each inspiration of the animal, the electromyogram shows a burst of spikes. The activity is enhanced when the muscle is lengthened, but the basic rhythm is hardly altered.

As shown earlier in Figure 24.8, the afferent discharge from muscle spindles is maintained at a high frequency, even when the muscles are actively contracting, owing to activity of the fusimotor γ-efferent fibers.[26] Fusimotor activity presets muscle spindles for the length change expected during that half cycle. If an unexpected obstruction in the airway were to prevent the expected movement, the sensory endings in the muscle spindle would be stretched and would fire and increase excitation to the respiratory motoneurons.

Where is the pattern generator for respiration situated and how does it generate the rhythm?[35] Within the pons and medulla, there are pools of neurons that fire during inspiration or expiration and produce excitation and inhibition of respiratory motor neurons.[35] For example, during inspiration, the motoneurons supplying external intercostal (inspiratory)

[35]Feldman, J. L., Mitchell, G. S., and Nattie, E. E. 2003. *Annu. Rev. Neurosci.* 26: 249–266.

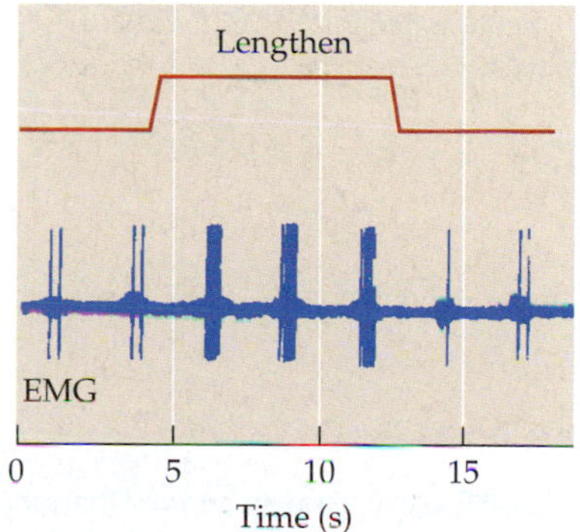

FIGURE 24.11 Stretch Reflex of an Inspiratory Muscle. During each inspiration the electromyogram (EMG) from a small muscle, the levator costae, shows bursts of action potentials, and the muscle contracts. Stretching the muscle by pulling on its tendon increases the number of impulses in each burst by increasing the reflex drive on the motor unit, without affecting the respiratory rate. (After Hilaire, Nicholls, and Sears, 1983.)

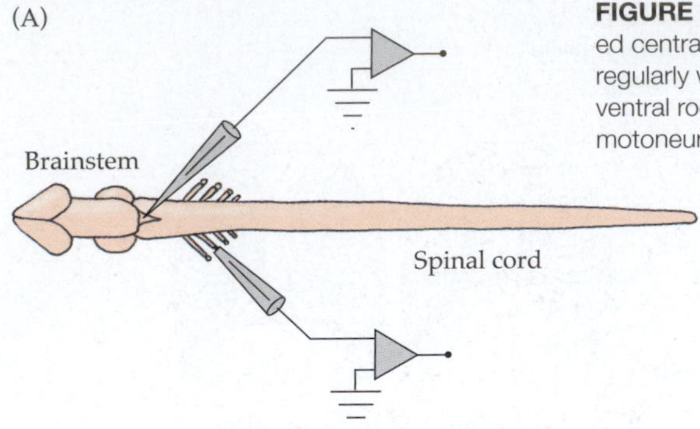

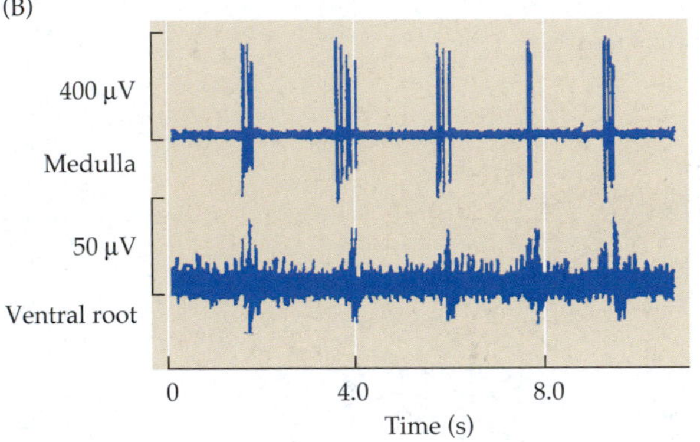

FIGURE 24.12 Respiratory Rhythm Recorded from Brainstem Neurons in an isolated central nervous system from a neonatal opossum (A). (B) Two brainstem cells discharge regularly with an interburst interval of about 2 seconds (s). Simultaneous recording from a ventral root that supplies the diaphragm shows the corresponding discharge of respiratory motoneurons. (Kindly provided by D. J. Zou and J. G. Nicholls.)

muscles are depolarized by a barrage of EPSPs arising from neurons in higher centers of the medulla or pons, causing bursts of action potentials. The inspiratory phase is terminated by a burst of inhibitory potentials to the inspiratory neurons from intermingled expiratory neurons.[36]

An example of individual rhythmically active brainstem neurons, recorded from the isolated CNS of a neonatal opossum, is shown in Figure 24.12.[37] The upper trace in Figure 24.12B is an extracellular record from a single neuron in the medulla, showing short bursts of impulses with a period of about 2 seconds (s). In the lower trace, recordings from a thoracic ventral root show corresponding rhythmic discharges of motoneurons supplying inspiratory muscles, occurring at the same frequency and with a slight delay.

In principle the rhythmical activation of inspiratory and expiratory motoneurons could be achieved in two ways: (1) neurons within the brainstem might have inherently rhythmical properties, like heart muscle cells, or (2) the rhythm might originate from activity of excitatory and inhibitory synapses between a network of brainstem neurons. In a rhythmically active network of neurons, inspiratory command neurons would turn off expiratory neurons, which in time would break free and inhibit the inspiratory neurons, and so on. Endogenously bursting pacemaker neurons have been found in neonatal rats within a region of the ventral medulla called the pre-Bötzinger complex.[35] Experiments made by St. John suggest that those pacemaker neurons may function in relation to gasping rather than to normal respiration.[38]

A major technical problem, as of now, is how to record from the entire population of neurons in an area of the brain. With electrodes, one can measure the activity of a small proportion of the population or the average activity of the whole population, which blurs fine details. An alternative is to survey simultaneously the activity of large populations of brainstem neurons, one by one, during inspiration and expiration by optical methods[39] (see also Chapters 1 and 3). An example of firing by respiratory neurons obtained by calcium imaging is shown in Figure 24.13. Results such as these have shown that inspiratory and expiratory neurons are intermingled rather than in completely separate pools. Several lines

[36]Davies, J. G., Kirkwood, P. A., and Sears, T. A. 1985. *J. Physiol.* 368: 63–87.

[37]Nicholls, J. G. et al.1990. *J. Exp. Biol.* 152: 1–15.

[38]St. John, W. M. 2009. *Philos. Trans. R. Soc. Lond. B, Biol. Sci.* 364: 2625–2633.

[39]Muller, K. J. et al. 2009. *Philos. Trans. R. Soc. Lond. B, Biol. Sci.* 364: 2485–2491.

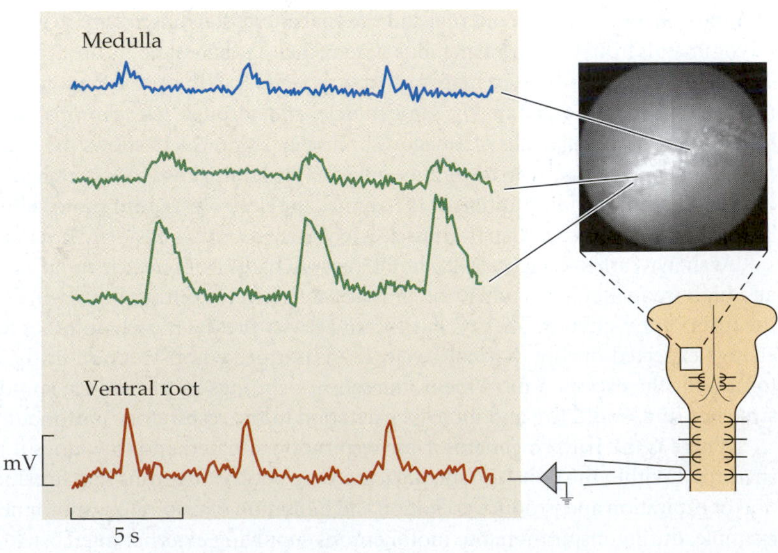

FIGURE 24.13 Fluorescence Changes in rostral medulla (top three traces) and integrated ventral root burst activity to the diaphragm (bottom trace) in an isolated brainstem–spinal cord preparation (stained with calcium-sensitive dye) of a 20-day-old embryonic mouse. On the right are a photograph and a drawing of such a preparation with an electrode attached to a cervical ventral root. The approximate region of view is indicated with a black square. Inspiratory- (top trace, blue) and expiratory-related (second and third traces, green) fluorescence changes are recorded optically from neurons in the rostral medulla. Colored lines join cells in the medulla, with their respective optical recordings. (From Eugenin et al., 2006.)

of evidence further suggest that rhythmicity might arise from properties of the network rather than being driven solely by pacemaker cells. As yet, the excitatory and inhibitory connections of the respiratory command neurons in the brainstem have not been traced.

Of key importance for the rhythmicity of breathing is the level of carbon dioxide in the arterial blood. Under conditions of reduced CO_2, the rate and depth of respiration are reduced; conversely, raised levels of CO_2 increase respiration. This effect depends on inputs from chemoreceptors in the carotid arteries and aorta and from neurons and glial cells in the medulla that are sensitive to the CO_2 levels.[40,41] The firing patterns of individual medullary neurons that control expiratory and inspiratory motoneurons have been shown to be influenced critically by CO_2. Changes in the steady level of CO_2 are translated into pronounced changes in the frequencies of firing of the interneurons and, therefore, of the respiratory motoneurons.[42] By genetic manipulations, Champagnat,[43] Brunet,[44] and their colleagues have shown that a particular gene, known as *Phox 2b*, is responsible during development for the generation of neurons that respond to increased levels of CO_2. Deletion of the gene in mice gives rise to animals that no longer breathe faster and deeper in a high CO_2 environment.

We know from everyday experience that modulation of breathing depends on voluntary commands as well as sensory input. You can inhale more deeply when you smell the odor of Kentucky Fried Chicken, maintain expiration for prolonged periods as you sing "Celeste Aida," or hold your breath in an offensive toilet. While the central pattern generator guarantees an unfailing rhythm, it is not fully autonomous. At present, although we know that glutamate and γ-aminobutyric acid (GABA) play important roles,[45] we do not yet know whether the respiratory rhythm arises from inherent rhythmicity of brainstem neurons, network properties, or combinations of both.

Locomotion

A striking feature of locomotion is that in vertebrates there is a consistent, highly stereotyped pattern of limb movements. In the walking cat shown in Figure 24.14, the left hind limb is lifted off the ground first, then the left forelimb, the right hind limb, and the right

[40] Spyer, K. M., and Gourine, A. V. 2009. *Philos. Trans. R. Soc. Lond. B, Biol. Sci.* 364: 2603–2610.

[41] Huckstepp, R. T. et al. 2010. *J. Physiol.* 588: 3901–3920.

[42] Eugenin, J., and Nicholls, J. G. 1997. *J. Physiol.* 501: 425–437.

[43] Champagnat, J. et al. 2009. *Philos. Trans. R. Soc. Lond. B, Biol. Sci.* 364: 2469–2476.

[44] Dubreuil, V. et al. 2009. *Philos. Trans. R. Soc. Lond. B, Biol. Sci.* 364: 2477–2483.

[45] Cifra, A. et al. 2009. *Philos. Trans. R. Soc. Lond. B, Biol. Sci.* 364: 2493–2500.

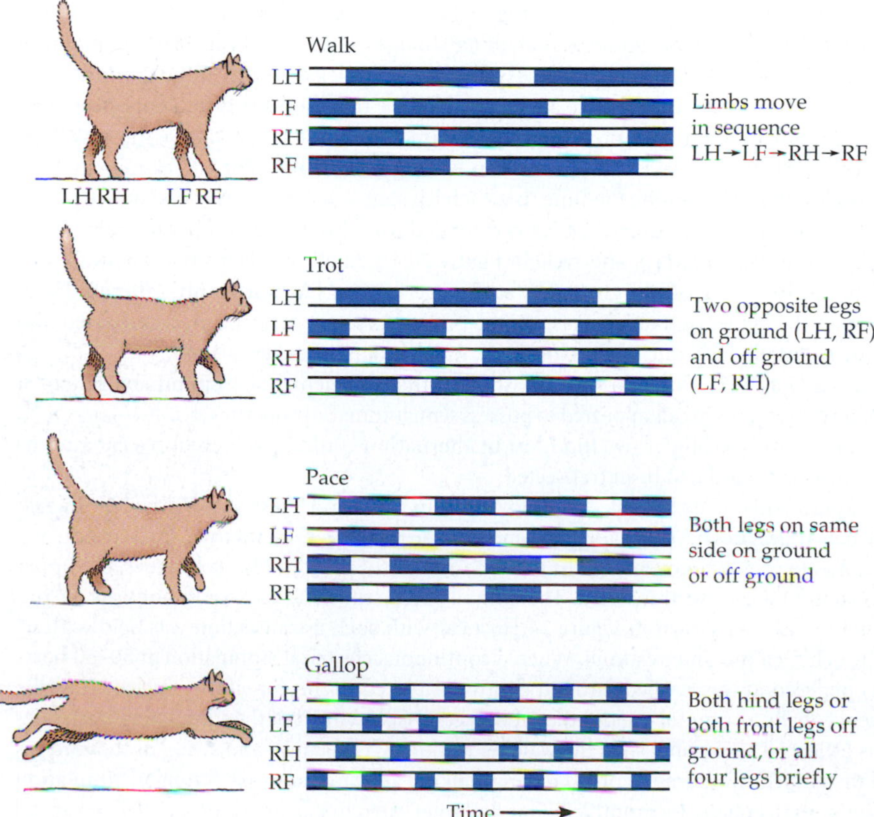

FIGURE 24.14 The Stepping Pattern of a Cat. Cat patterns seen during four different gaits of locomotion: walk, trot, pace, and gallop. The white bars show the time that a foot is off the ground (the swing phase, during which flexor motoneurons are active); the blue bars show the time a foot is on the ground (the stance phase, during which extensor motoneurons are active). During walking, the legs are moved in sequence, first on one side, then on the other. During a trot, a different pattern of interlimb coordination is used, in which diagonally opposite limbs are raised together. In a pace, the rhythm changes again, the limbs on the same side being raised together. Faster still is the gallop in which the hind limbs and then the front limbs leave the ground. LF = left foreleg, LH = left hind leg, RF = right foreleg, RH = right hind leg. (After Pearson, 1976.)

FIGURE 24.15 Constancy of Swing Phase during Locomotion. As the animal moves more and more rapidly (abscissa), the time spent by each foot on the ground (stance phase) becomes progressively shorter (ordinate). The time that each foot spends in the air (swing phase) is almost the same in a walk and a gallop. (After Pearson, 1976.)

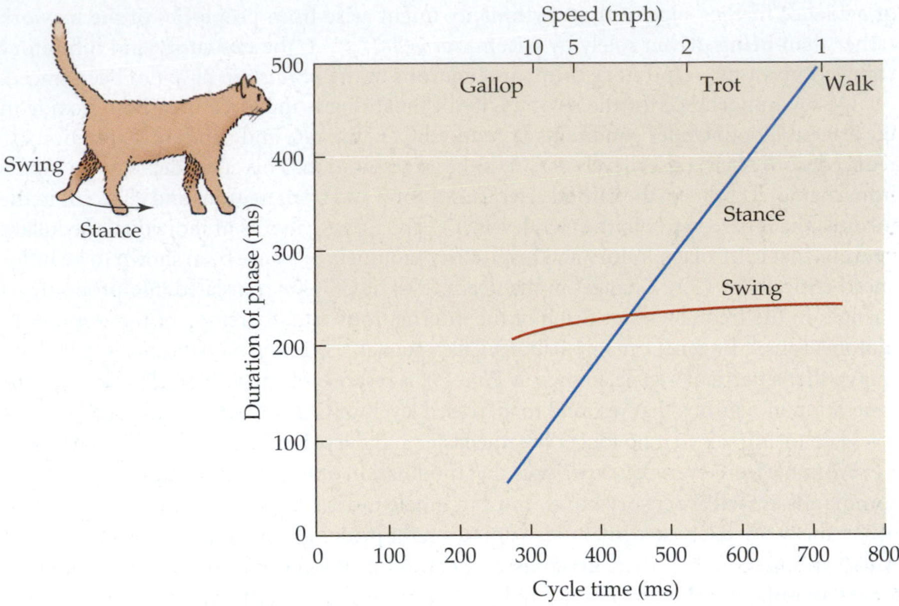

forelimb. This sequence provides stabilization by the forelimbs, while the hind limbs propel the animal. The tendency to turn produced by the hind limb is counteracted by the forelimbs, thereby preventing rotation and enabling movement to proceed straight forward. Such a sequence is common in vertebrates including crocodiles, rats, cats, and elephants (but not fish). Even in invertebrates with six legs, such as cockroaches, a similar sequential pattern is observed. During locomotion, each leg executes an elementary stepping movement that consists of two phases:[46] (1) a swing phase, during which the leg (having been extended to the rear) is flexed, raised off the ground, swung forward, and extended again to contact the ground; and (2) a stance phase, during which the leg is in contact with the ground, moving backward in relation to the direction taken by the body.

That the gait of a cat undergoes striking changes as its speed increases is shown in Figures 24.14 and 24.15. While the cat is walking, a single leg is raised off the ground at any one time. As the speed increases to a trot, two legs are raised off the ground at once—one front and one back on opposite sides of the animal. Still faster, at a gallop, the two front legs and then the two back legs alternate in leaving the ground. The increase in speed is accomplished by shortening the time that each leg stays on the ground—the stance phase. Thus, as the cat moves faster, each leg is extended for a briefer period before being bent, raised, and moved forward, one cycle in Figure 24.15. At all speeds from a slow walk to a gallop, the time spent off the ground by each leg as it swings forward is little altered. There is substantial evidence that intrinsic motor programs are genetically predetermined and appear spontaneously during development, independent of experience.[47]

As early as 1911, Graham Brown[48] showed that the elementary circuits required for walking movements in cats appeared to possess semiautomatic properties (see also Guertin[49]). The raising and placing of two hind feet in alternation could be achieved in a cat after its thoracic spinal cord had been transected.

Experiments made by Shik and Orlovsky[50] (in the then USSR) and Nistri,[51] Grillner,[52] and their colleagues have provided evidence for the role of central mechanism in producing coordinated walking movements. In the experiments of Shik and his colleagues, the upper brainstem of a cat was transected. The animal could still stand but could not walk or run spontaneously. As shown in Figure 24.16, a cat with such a transection was held with its feet touching a moving treadmill. When a continuous electrical stimulation at 30–60 hertz (Hz) per second was applied through electrodes placed in the cuneiform nucleus (i.e., the mesencephalic locomotor region) the cat walked. Displacement of the stimulating electrodes by as little as 0.3 mm abolished the walking response. The stance and swing of the forelegs and the electromyograms recorded during walking appeared normal. Stronger stimulation of the mesencephalic locomotor region by larger currents at the same frequency caused

[46] Pearson, K. 1976. *Sci. Am.* 245: 72–86.

[47] Dasen, J. S., and Jessell, T. M. 2009. *Curr. Top. Dev. Biol.* 88: 169–200.

[48] Brown, T. G. 1911. *Proc. R. Soc. Lond. B*, 84: 308–319.

[49] Guertin, P. A. 2009. *Brain Res. Rev.* 62: 45–56.

[50] Shik, M. L., and Orlovsky, G. N. 1976. *Physiol. Rev.* 56: 465–501.

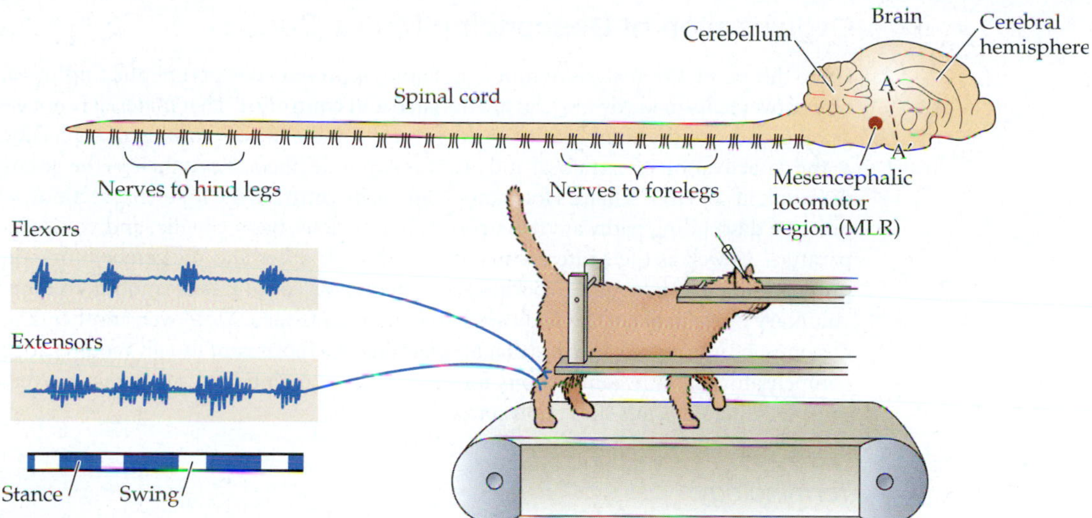

FIGURE 24.16 Locomotion by a Cat on a Treadmill after section of the brainstem (A–A′ in the diagram). Such an animal does not walk spontaneously. Electrical stimulation of the mesencephalic locomotor region (MLR in the diagram) causes the animal to walk on the treadmill. Electromyographs associated with locomotor activity are recorded by electrodes in limb muscles. The speed of walking or galloping depends on the rate of the treadmill. Increasing the strength or the rate of stimulation increases the strength of limb movements (as though the animal were walking uphill) but not their speed. (After Pearson, 1976.)

the propulsive forces of the leg muscles to be increased. But, while the strength of electrical stimulation could influence the force of the walking movements, the frequency of stepping was not altered provided that the speed of the treadmill remained constant. Rhythmic discharges of motoneurons that are appropriate for swimming or walking in the intact animal can also be elicited in isolated spinal cord segments of the turtle[53] and slices of rat spinal cord.[54] Although such spinal cord preparations in vitro have interneurons with properties necessary for producing locomotor patterns,[55] the switching from walk to trot to gallop is not understood.

Sensory Feedback and Central Pattern Generator Programs

Clear parallels emerge in the automaticity of breathing and of walking. For both types of movements, a central program orders the contractions of appropriate groups of muscles in a predestined sequence. For respiratory neurons in the brainstem, the drive depends on the CO_2 in their environment. The drive for the alternation of leg movements during walking or running depends on a descending stimulus supplied by glutamatergic, noradrenergic, dopaminergic and serotonergic neurons in the mesencephalic locomotor region.[57]

Although sensory input from sensory receptors in the periphery is not required for respiratory or walking movements to take place, it does play a major role in their regulation. Sensory feedback modulates the frequency and extent of the rhythms to be consistent with the moment-to-moment requirements of the animal. Thus, the rate and depth of respiration depend on the level of blood pH and carbon dioxide, which is sensed by receptors in the aortic and carotid bodies; stretch receptors in the lungs also modulate the rhythm.[56] Similarly, muscle and joint receptors play a key role in modulating the speed of locomotion. For example, the experiment in Figure 24.16 showed that when the treadmill was accelerated during constant electrical stimulation of the mesencephalic locomotor region, the locomotion of the cat changed from walking to trotting and then to galloping. Sensory feedback plainly controls the rate of stepping at greater treadmill speeds.[57] As the stance phase shortens, the leg moving to the rear on the treadmill becomes extended more rapidly. It therefore takes less time to reach the point at which afferent signals initiate the swing phase in which the leg is lifted and swung forward. As, expected, cutting dorsal roots abolishes the response to different treadmill speeds but not the walking evoked by electrical stimulation. Afferent feedback has been shown also to control human gait.[58]

[51] Taccola, G., and Nistri, A. 2006. *Crit. Rev. Neurobiol.* 18: 25–36.

[52] Kozlov, A. et al. 2009. *Proc. Natl. Acad. Sci. USA* 106: 20027–20032.

[53] Guertin, P. A., and Hounsgaard, J. 1998. *Neurosci. Lett.* 245: 5–8.

[54] Ballerini, L. et al. 1999. *J. Physiol.* 517: 459–475.

[55] Dougherty, K. J., and Kiehn, O. 2010. *Ann. N Y Acad. Sci.* 1198: 85–93.

[56] Mörschel, M., and Dutschmann, M. 2009. *Philos. Trans. R. Soc. Lond. B, Biol. Sci.* 364: 2517–2526.

[57] Pearson, K. G. 2008. *Brain Res. Rev.* 57: 222–227.

[58] Nielsen, J. B., and Sinkjaer, T. 2002. *J. Electromyogr. Kinesiol.* 12: 213–217.

Organization of Descending Motor Control

Up to this point, the analysis of motor systems has presented experimental findings about how movements made by muscles are initiated and controlled. That material is not easy to grasp because of the complexities of the muscle spindle, to say nothing of such concepts as the co-activation of extrafusal and intrafusal muscle fibers. Nevertheless, the discussion did result in coherent schemes for motor control. By contrast, the following sections, which describe descending pathways from cortex, red nucleus, basal ganglia, and vestibular apparatus—as well as their interconnections with each other and the cerebellum—do not provide a testable, adequate hypothesis to explain how coordinated, smooth, tremor-free, voluntary movements of the body are initiated or controlled. Moreover, another complex idea must be introduced, which is that the planning of a movement usually requires adequate compensatory muscle contractions for balance to be maintained, and those movements must be initiated *before* the action starts.

Terminology

It has been emphasized in this book that knowledge of anatomy is a prerequisite for understanding the physiology of the nervous system; anatomy, imaging, and lesions are not sufficient on their own, however, to explain complex functions, such as those involved in coordinated voluntary movements. Nevertheless, it is necessary to be able to recognize the names of structures and have some idea of their locations if one is to discuss function. For readers not familiar with the nervous system, terms like *medial vestibulospinal tract* may seem as bewildering as those used by high-energy physicists. Fortunately, anatomists have maintained some rigor in naming fiber tracts. A pathway is named according to its origin first (e.g., the vestibular nucleus in the brainstem) and then its termination (spinal cord). If two or more pathways from the same source run in the spinal cord, their individual locations are specified as medial or lateral, ventral or dorsal. Appendix C lists terms that indicate direction with respect to main axes of the CNS, such as *rostral*, *anterior*, and *ventral*.

Supraspinal Control of Motoneurons

The major descending pathways to motoneurons are shown in Figure 24.17 (see also Appendix C). These pathways can be classified into two groups, lateral and medial, according to their anatomy and physiological functions.[59]

Lateral Motor Pathways

Laterally situated pathways in the spinal cord are primarily concerned with phasic movements and fine manipulations, such as grasping or playing the piano. The **lateral corticospinal tract** originates in the motor and premotor areas of the cerebral cortex, in front of the central sulcus (Brodmann's areas 4 and 6; see Appendix C), as well as from a small strip of the postcentral somatosensory region (area 3) of the cerebral cortex (see Figure 24.18). Fibers originating from pyramidal nerve cells in those regions pass downward through the internal capsule and cerebral peduncles to the medullary pyramids, after which most cross the midline (decussate) and continue their descent laterally in the spinal cord. Axons of the lateral corticospinal tracts (also known as **pyramidal tracts**) terminate predominantly on motoneurons and interneurons in the lateral gray matter. An important feature is that many of the fibers descending from the cortex have terminal branches that end directly on the motoneurons controlling muscles of the digits.[60,61] In humans and other primates, interruption of the lateral corticospinal tract results primarily in a loss of the ability to move the fingers independently and a deficit in the ability to perform fine, precise tactile movements.[62]

The **rubrospinal tracts** originate in the red nuclei (see Figure 24.17) and cross the midline before descending in the spinal cord to end on interneurons and occasional motoneurons associated with the lateral motor system. Cells in the red nucleus are arranged somatotopically and receive excitatory inputs from the motor cortex and from the cerebellum. Although the precise functional role of the rubrospinal tract is unclear, it is thought to duplicate many

[59] Lemon, R. N. 2008. *Annu. Rev. Neurosci.* 31: 195–218.

[60] Cheney, D. P., and Fetz, E. E. 1980. *J. Neurophysiol.* 44: 773–791.

[61] Rouiller, E. M. et al. 1996. *Eur. J. Neurosci.* 8: 1055–1059.

[62] Lawrence, D. G., and Kuypers, H. G. J. M. 1968. *Brain* 91: 1–14.

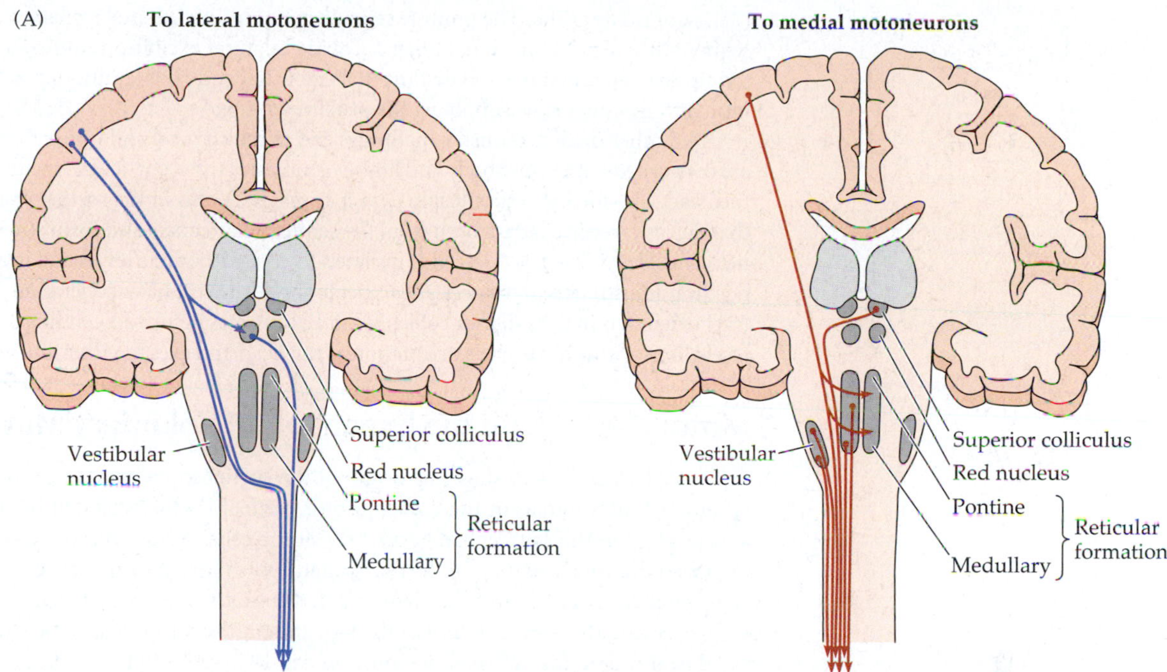

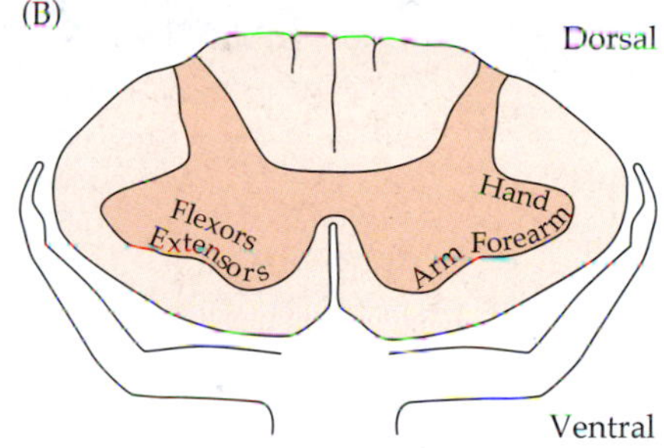

FIGURE 24.17 Major Motor Pathways in the vertebrate central nervous system supplying lateral motoneurons (blue) and medial motoneurons (orange) are shown schematically on coronal sections of the cerebral hemispheres, continuing to a longitudinal section of the brainstem and spinal cord. (A) Cells in the primary motor area of the cerebral cortex send axons to the contralateral spinal cord to form the lateral corticospinal tract, with collateral connections to the red nucleus. Axons from cells in the red nucleus cross the midline and descend in the rubrospinal tract. The lateral tracts supply monosynaptic and polysynaptic innervation largely to lateral motoneurons (which supply distal musculature; see [B]). Other cortical fibers descend without crossing to form the ventral corticospinal tract, supplying collaterals to brainstem nuclei. The postural muscles of the body are supplied predominantly by the motor regions of the brainstem through the reticulospinal tracts, originating in the pontine and medullary reticular formations. The vestibulospinal tracts originate in the vestibular nuclei, and the tectospinal tract originates in the superior colliculus. (B) Organization of motoneurons supplying the upper extremities, shown in a transverse section of the spinal cord in the cervical region. Muscles of the shoulder and arm are represented most medially, those of the hand most laterally. Extensor motoneurons are located near the margin of the gray matter; flexor motoneurons are more central.

of the functions of the corticospinal tract and to constitute a parallel pathway from the cortex.[63] In primates, lesions of the rubrospinal tract have little obvious effect, but after interruption of both the rubrospinal and corticospinal tracts, coordinated positioning of the hands and feet is severely impaired.[64]

Medial Motor Pathways

Medially situated descending tracts provide input to extensor motoneurons that are concerned mainly with sustained activities, such as stance and the adjustment of posture. Except for a small component from the uncrossed ventral corticospinal tract, the descending axons originate primarily in the brainstem (see Figure 24.17). The medial pathways include the **ventral corticospinal tract**, the lateral and medial **vestibulospinal tracts**, the **pontine** and medullary **reticulospinal tracts**, and the **tectospinal tract.** The cells of origin of the lateral vestibulospinal tract lie (as the name indicates) in the lateral vestibular nucleus. Each lateral vestibular nucleus receives input from the ipsilateral vestibular apparatus, in particular from the utricles of the labyrinth (see Chapter 19). The tract descends uncrossed in the spinal cord to provide input to the medial motoneurons supplying postural muscles, with monosynaptic excitatory inputs to extensor muscles and disynaptic inhibitory inputs to flexors. The tract is involved in the maintenance of posture and the regulation of extensor

[63] Zelenin, P. V. et al. 2010. *J. Neurosci.* 30: 14533–14542.

[64] Kennedy, P. R. 1990. *Trends Neurosci.* 13: 474–479.

(i.e., antigravity) tone. The pontine reticulospinal tract descends ipsilaterally and ends on segmental interneurons that, in turn, provide bilateral excitation to medial extensor motoneurons. The medullary reticulospinal tract descends bilaterally to provide inhibitory inputs to motoneurons supplying the proximal limbs.

Two other medial brainstem pathways end in the cervical and upper thoracic levels and are concerned with upper body and limb posture, and most particularly with the position of the head. The medial vestibulospinal tract arises from cells in the medial vestibular nucleus that, in turn, receive inputs both from the semicircular canals and from stretch receptors in neck muscles.[65] The tract descends ipsilaterally to midthoracic levels and is concerned with postural adjustments of the neck and upper limbs during angular acceleration. The tectospinal tract originates in the superior colliculus and decussates before descending to upper cervical levels. This pathway mediates orientation of the head and eyes to visual and auditory targets.

Motor Cortex and the Execution of Voluntary Movement

Figures 24.18 and 24.19 show the origin of the corticospinal tracts in primary (M1) and secondary motor areas of the cortex. Motor control is also provided by the somatosensory cortex on the postcentral gyrus.[66,67] Motor cells in the precentral gyrus motor are arranged in orderly manner to form a somatotopic pattern of muscle organization. As in the somatosensory cortex (see Chapter 21), the motor map in humans is distorted, with disproportionate representation of the face and hands compared to the trunk.

The motor maps were first demonstrated in 1870 by Fritsch and Hitzig, who produced movements of the body by stimulating the cerebral cortex of animals.[68] The somatotopic representation in humans was later mapped on the brains of patients during neurosurgery by Penfield and his colleagues.[69] Localized stimulation of the cortical surface with brief electrical shocks produced movements of a restricted region of the body, for example a finger. The position of the contracting muscle depended on the position of the stimulating electrode. Noninvasive recording techniques such as functional magnetic resonance imaging (fMRI) provide similar maps of the motor cortex.[70,71]

The secondary, or association, motor cortex consists of the premotor cortex (Brodmann's area 6), which lies anterior and somewhat lateral to M1, and the supplemental motor

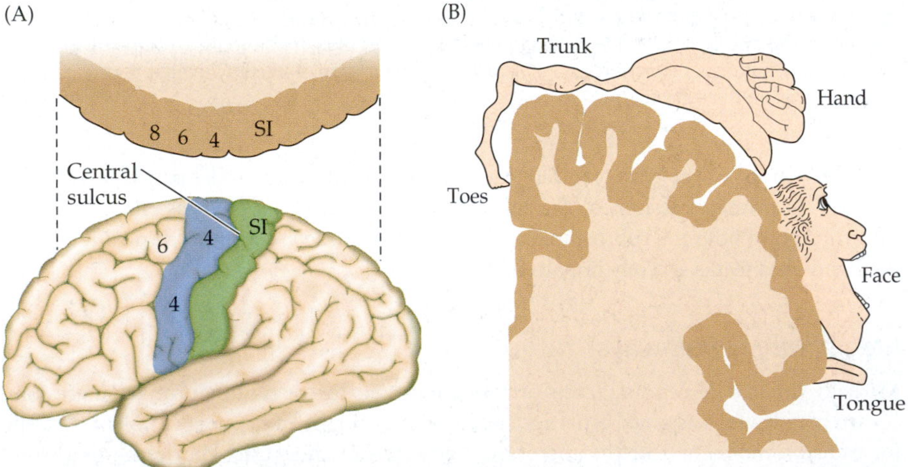

FIGURE 24.18 Motor Representation on the Cerebral Cortex. (A) Lateral view of the surface of the cerebral cortex. Motor movements result from activation of cells in area 4 of the cerebral cortex (including the cells of origin of the corticospinal tract), which is the primary motor area (M1). The motor system also includes area 6 (premotor area), extending onto the medial surface of the hemispheres. The green strip is the primary somatosensory cortex (SI; see Chapter 21). (B) Sketch of a coronal section through the cerebral hemisphere anterior to the central sulcus. The musculature of the human body is represented in an orderly but distorted fashion, with the leg and foot on the medial surface of the hemisphere and the head most lateral. The very large area devoted to the hand is indicative of the number of neurons involved in control of manipulations by the digits.

[65] Kaspar, J., Schor, R. H., and Wilson, V. J. 1988. *J. Neurophysiol.* 60: 1765–1768.

[66] Rizzolatti, G, and Wolpert, D. M. 2005. *Curr. Opin. Neurobiol.* 15: 624–625.

[67] Fogassi, L., and Luppino, G. 2005. *Curr. Opin. Neurobiol.* 15: 626–631.

[68] Fritsch, G., and Hitzig, E. 1870. *Arch. Anat. Physiol. Wiss. Med.* 37: 300–332.

[69] Penfield, W., and Rasmussen, T. 1950. *The Cerebral Cortex of Man. A Clinical Study of Localization of Function*. Macmillan, New York.

[70] Porro, C. A. et al. 1996. *J. Neurosci.* 16: 7688–7698.

[71] Rijntjes, M. et al. 1999. *J. Neurosci.* 19: 8043–8048.

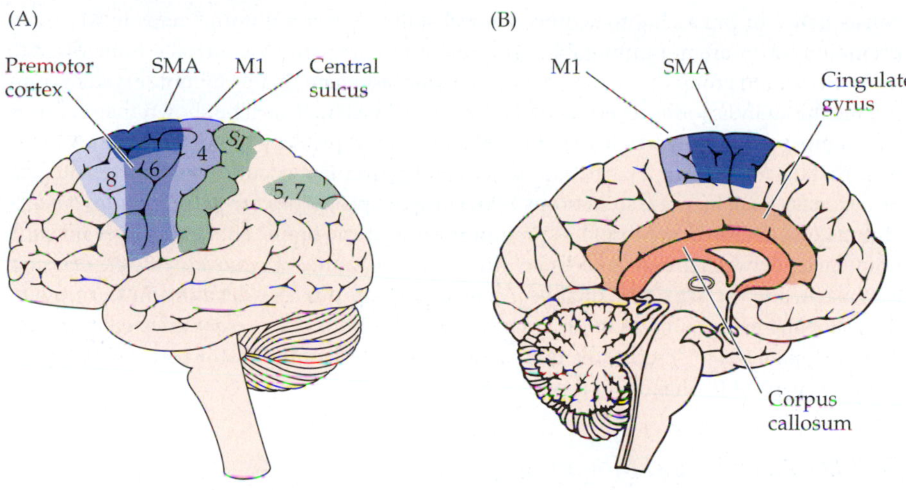

FIGURE 24.19 The Association Motor Cortices. (A) The primary and association motor cortices lie anterior to the central sulcus in Brodmann's areas 4 (primary motor cortex) and 6 (premotor cortex and supplemental motor area [SMA]). Frontal eye fields are found in area 8. Primary somatic cortex, SI (areas 3, 1, and 2), and especially association somatosensory cortex (areas 5 and 7), generate commands used in motor planning. (B) Medial surface of the cerebral hemisphere. The cingulate motor area lies between the cingulate gyrus proper and the medial extension of the primary motor and supplementary motor areas.

area, also anterior to M1 (see Figure 24.19). Both these areas, like M1, are somatotopically organized and receive input from sensory association cortex (posterior parietal areas 5 and 7).[72] Premotor cortex is strongly influenced by the cerebellum, and the supplementary motor area is connected with the basal ganglia (see below). Movements elicited by electrical stimulation in premotor and supplementary motor cortex are complex—for example, a reaching and grasp action—and are often bilateral (see Chapter 21). Motor-related activity is observed in both these areas of human brains using fMRI (Figure 24.20). Both areas project somatotopically to the primary motor cortex.

What Do Motor Maps Mean?

When motor maps were first discovered they revolutionized thinking about how movements are initiated and controlled. Movement at first seemed to be controlled by a distorted homunculus, who represented a fixed plan for the contractions of individual muscles. Yet, in time that idea turned out to be an oversimplification. The concept had arisen from the use of electrical currents to stimulate neurons artificially. There are parallels with the visual system. Thus, knowledge of the projection of the visual field onto area 17 was obtained by flashing bright lights. While this was a prerequisite for understanding the first steps in visual processing, it did not on its own reveal columnar architecture, functional groupings of neurons, or the importance of movement and orientated edges.

Unit recordings, fMRI, and transcranial magnetic electrical stimulation have shown that the map of motor cortex is plastic and changes following peripheral lesions or as a

[72] Preuss, T. M., Stepniewska, I., and Kaas, J. H. 1996. *J. Comp. Neurol.* 371: 649–676.

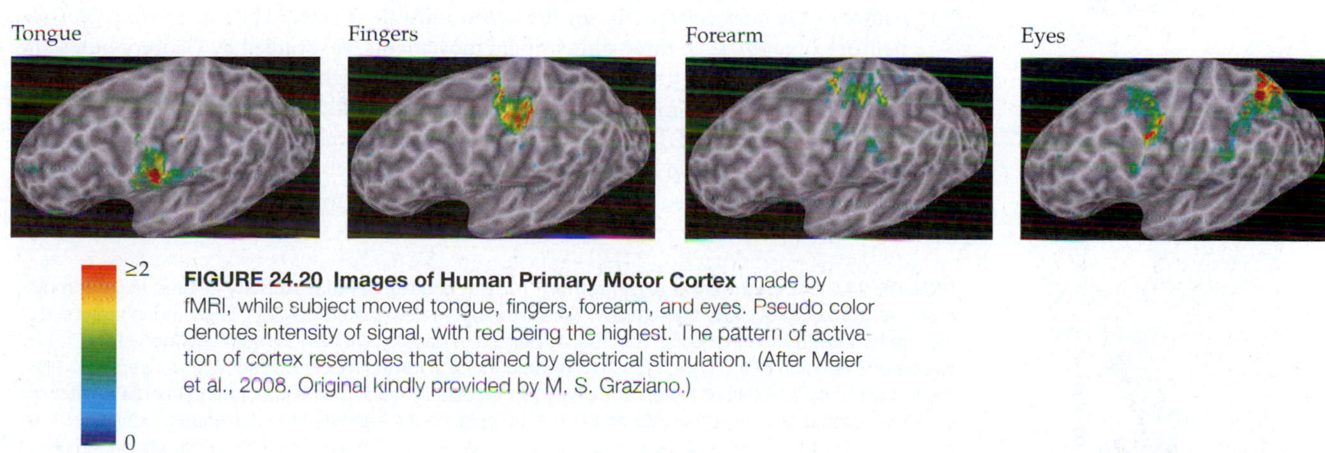

FIGURE 24.20 Images of Human Primary Motor Cortex made by fMRI, while subject moved tongue, fingers, forearm, and eyes. Pseudo color denotes intensity of signal, with red being the highest. The pattern of activation of cortex resembles that obtained by electrical stimulation. (After Meier et al., 2008. Original kindly provided by M. S. Graziano.)

[73]Ward, N. S., and Frackowiak, R. S. 2004. *Cerebrovasc. Dis.* 17(Suppl 3): 35–38.

[74]Blake, D. T., Byl, N. N., and Merzenich, M. M. 2002. *Behav. Brain Res.* 135: 179–184.

[75]Scott, S. H. 2008. *J. Physiol.* 586: 1217–1224.

[76]Evarts, E. V. 1965. *J. Neurophysiol.* 28: 216–228.

[77]Evarts, E. V. 1966. *J. Neurophysiol.* 29: 1011–1027.

[78]Krüger, J. et al. 2010. *Front. Neuroeng.* 3: 6.

[79]Georgopoulos, A. P., Schwartz, A. B., and Kettner, R. E. 1986. *Science* 243: 1416–1419.

[80]Georgopoulos, A. P. et al. 2007. *Proc. Natl. Acad. Sci. USA* 104: 11068–11072.

[81]Merchant, H., Naselaris, T., and Georgopoulos, A. P. 2008. *J. Neurosci.* 28: 9164–9172.

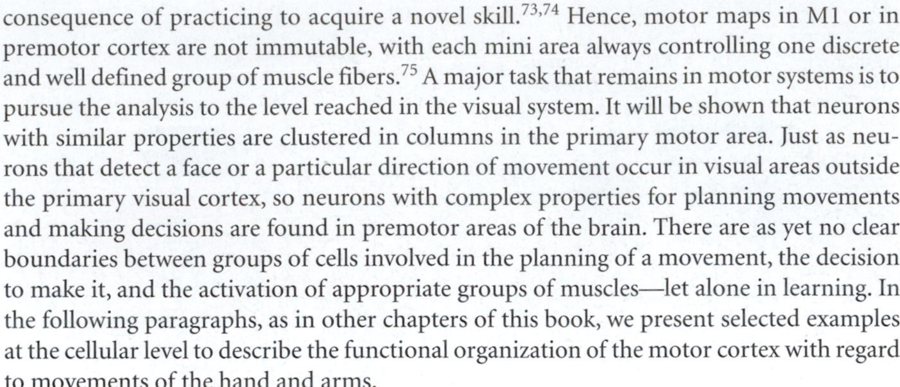

consequence of practicing to acquire a novel skill.[73,74] Hence, motor maps in M1 or in premotor cortex are not immutable, with each mini area always controlling one discrete and well defined group of muscle fibers.[75] A major task that remains in motor systems is to pursue the analysis to the level reached in the visual system. It will be shown that neurons with similar properties are clustered in columns in the primary motor area. Just as neurons that detect a face or a particular direction of movement occur in visual areas outside the primary visual cortex, so neurons with complex properties for planning movements and making decisions are found in premotor areas of the brain. There are as yet no clear boundaries between groups of cells involved in the planning of a movement, the decision to make it, and the activation of appropriate groups of muscles—let alone in learning. In the following paragraphs, as in other chapters of this book, we present selected examples at the cellular level to describe the functional organization of the motor cortex with regard to movements of the hand and arms.

Cellular Activity and Movement

How is the activity of neurons in motor cortex related to the initiation and performance of a movement? Do individual neurons in M1 cause contraction of a single muscle, direct the strength of contraction of specific muscle groups, control the magnitude of displacement around a joint, or move parts of the body in a particular direction? These questions were asked by Evarts,[76,77] who recorded the activity of pyramidal tract cells in the motor cortex during the performance of trained wrist movements by awake monkeys (Figure 24.21).

In this and other experiments described in the following sections (see also Chapters 21 and 23) electrodes are chronically implanted into the cortex, often for many weeks or even years,[78] and the animal is rewarded for making a correct choice. By loading the wrist to oppose either flexion or extension, Evarts could dissociate the force required for a movement from its direction. Early results were straightforward. Individual cortical cells fired impulses in association with either extension or flexion of the wrist. The discharge frequency, however, was related to the force required to execute the movement. This behavior of the cortical cells was not unlike the behavior of the spinal motoneurons to which they projected. Subsequent experiments showed that this particular kind of behavior is characteristic of corticospinal cells that end directly on spinal motoneurons.[60] Other classes of cells exhibit more complex behavior, depending on the imposed load or starting position of the limb.

Cortical Cell Activity Related to Direction of Arm Movements

Reaching out to grasp a desired object requires an elaborate series of neural computations. The object to be grasped and its spatial location are first identified visually. The position of the target must then be compared with that of the hand and a trajectory computed to join them. Finally, that spatial trajectory must be converted into coordinated contractions of muscles that will move the hand to that location in space. Note that there may be many trajectories that achieve the same goal—think of shoulder, elbow, wrist combinations. The choice of the most comfortable or the fastest must be selected. The way in which cortical neurons organize such three-dimensional movements was studied by Georgopoulos and his colleagues.[79–81] They recorded the activity of many single units with multiple electrodes in the motor cortex of monkeys that were making visually guided arm movements. Neurons in the arm area of the primary motor cortex discharged at maximum rates when the movement was made in a particular direction, toward left, right, up, or down. Preferred directions were not absolute; the discharges of such neurons fell off as the angle of reach

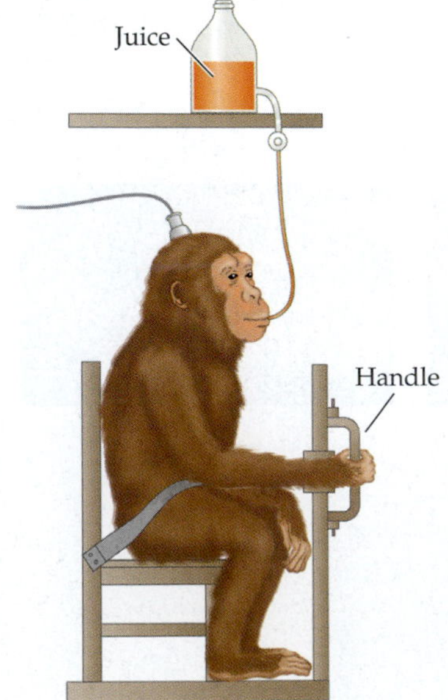

FIGURE 24.21 Experimental Arrangement for Recording Cellular Activity related to wrist movement. A monkey, previously trained to move a handle to a designated position, is seated in the chair with its forearm placed in a cuff. The monkey deflects a handle to the left or right between stops, by flexion or extension of the wrist. A system of weights, or a torque motor (not shown), is used to load the handle to oppose either flexion or extension. For visually guided movements, the handle position is indicated on a display screen. When the monkey places the handle in the designated position, it receives a reward of fruit juice. Single-unit activity is recorded with a microelectrode positioned in an appropriate area of the brain, by means of a microdrive fixed to the skull.

was altered. Further, the preferred direction varied with the initial position or posture of the limb. In one series of experiments, of 2385 recording sites that were analyzed, 985 (41.3%) were directionally tuned.

Figure 24.22 shows the intensity firing and preferred directions of individual units as the arm moved toward its target. It is the activity within an ensemble of neurons that determines the trajectory. The outcome, shown as arrows, encodes a direction that is equivalent to the vector sum of the preferred directions of all the active neurons. Particularly interesting is that neurons with similar preferred directions were clustered together in columns running through the depth of the cortex (Figure 24.23B). Cells situated at the center of the column fired according to the precise angle in which the movement was made (blue arrow in Figure 24.23B), whereas those nearer the periphery of the column showed greater variation in the preferred angular movement. Columns were approximately 50–100 μm wide and spaced regularly in a lattice 200 μm apart from each other. Georgopoulos and his colleagues have proposed that the preferred, bell-shaped directional tuning curve of a neuron results from excitatory thalamic input to a particular column, followed by a mixture of excitatory and inhibitory inputs mediated by local interneurons. These results obtained in motor cortex are reminiscent of the way in which moving visual targets are mapped in the middle temporal (MT) area of the visual cortex (see Chapters 3 and 22).

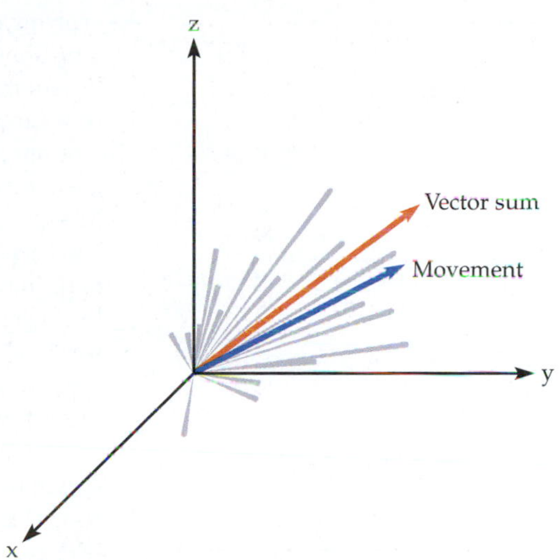

FIGURE 24.22 Encoding Movement in the Motor Cortex. The preferred direction of cortical neurons is shown in three dimensions. The activity of each neuron during a trained limb movement is plotted as a scalar whose length is proportional to its rate of firing. The vector sum for that population of neurons is an arrow whose direction is similar to that of the movement. (After Georgopoulos, Schwartz, and Kettner, 1986.)

Higher Control of Movement

Parallels with the visual system (see Chapters 2 and 3) are once again apparent in higher motor organization. Thus, while much is known about how columns of neurons in the visual cortex respond selectively to the orientation and movement of a stimulus, we do not know how the whole picture is put together. Similarly, while we have information about how columns of cells in the motor cortex move an arm in a certain direction, cellular mechanisms that enable an action to be planned, learned, or decided on are still not understood. The description below of feedback mechanisms, eye movements, anticipatory mechanisms, and mirror cells provides a short overview of

(A)

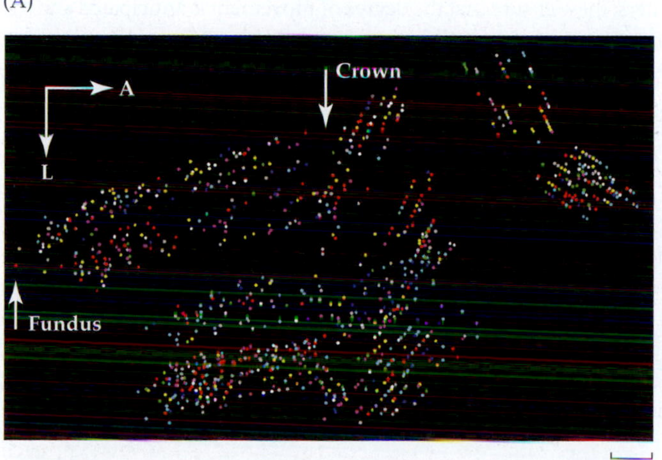

(B)

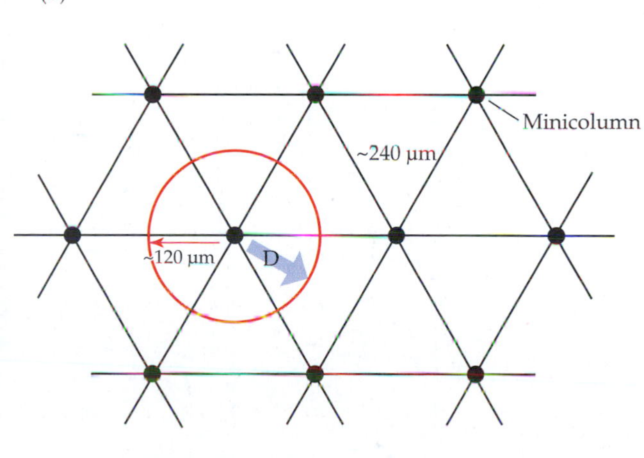

FIGURE 24.23 Mapping of Neurons in Motor Cortex Responding to Direction of Movement. (A) Each point represents a site at which activity is evoked by a particular direction along an *x*-, *y*-, or *z*-axis as the monkey makes a movement in 3D. The colors indicate the vector in a sphere as a cosine of the angle of movement. Dots of blue, magenta, green, gray, turquoise, red, white, and yellow, each corresponding to a specific direction of movement as in Figure 24.22. (B) The diagram shows the center of each column (i.e., the minicolumn) as a small point and the vector as a blue arrow. The center of each column is approximately 30 μm in diameter, while the columns are approximately 240 μm in diameter. Note that the columns for movement are clustered rather than random in their distribution. Minicolumns with similar preferred directions occur in doublets or triplets. (From Georgopoulos et al., 2007. Originals kindly provided by A. P. Georgopoulos.)

evolving concepts of higher control mechanisms employed by motor areas and their interactions with sensory systems.

An initial question concerns the role of feedback from the periphery in the execution of voluntary movements. Experiments made in conscious subjects show that it is important but not essential. Thus, a monkey can still move its arm after the dorsal roots have been sectioned, but the limb tends not to be used even though the motor control is intact. Marsden and his colleagues[82] have studied the manual motor performance of a man who had been deafferented by a severe peripheral neuropathy affecting dorsal roots. The power of the hand was almost unaffected and accurate voluntary movements could be made with his fingers, hand, and forearm. He could not, of course, judge weights of objects. Yet, he was unable to do simple tasks in daily life, such as holding a cup, writing with a pen, or buttoning up his shirt. Nor could he maintain muscle contractions for more than 1or 2 seconds without visual feedback.

A type of motor performance for which feedback is essential is the movement of the eyes in fixating and tracking objects. Eye movements represent a behavioral system of motor control relating to perception, with a controlled input and an output that can be measured with precision. Sensory information is derived from the vestibular apparatus and from the eyes themselves. To retain an image when the head rotates or to follow moving objects, the eyes move in a coordinated manner. Since the eyes can make only limited movements while the head moves or rotates, they turn slowly in the opposite direction and flick back rapidly. The abrupt deflections are known as **saccades.** (The role of the vestibular apparatus in saccadic movements, in particular the semicircular canals, has been discussed in Chapter 22.) An essential feature of saccades, and other movements, is that the motor system makes sure that the effects of an action on the body will be neutralized. Thus, we do not perceive saccadic eye movements as visual motion. The image on the retina is stable, and we are unaware of the visual stimulus generated by a saccade.[83–85] If you turn your head, the world does not seem to rotate. But, if you gently push the corner of one eye with your finger, the image moves (do this carefully, we shall not pay damages!). What is the difference between the two eye movements? When the nervous system sends commands to extraocular muscles to contract, they execute the commands faithfully. (Remember, as mentioned earlier, the loads that extraocular muscles need to overcome are virtually constant and relatively unaffected by gravity[29]). At the same time, corollary signals are sent from centers that plan the action to visual centers; these ensure that the degree of movement is anticipated and the perception of motion on the retina cancelled. When a finger is used to push the eye no such anticipatory information is provided to the visual system.

Anticipatory information also plays a major role in the control of body movements. If you raise one leg, your center of gravity shifts and without a correction you would fall. To apply the correction after the movement has been made would be inefficient and perhaps useless.

Figure 24.24 shows that what happens is that muscles on the opposite side of the body contract appropriately to maintain balance *before* the arm is raised by an act of will.[86] In the experiments shown in Figure 24.24, normal subjects were asked to raise one leg by 45° in response to a sound. Movements over the whole body were tracked, photographed, and analyzed and electromyograms were made of muscle contractions. The lower traces in Figure 24.24C recorded from the opposite side of

[82] Rothwell, J. C. et al. 1982. *Brain* 105: 515–542.

[83] Wurtz, R. H. 2008. *Vision Res.* 48: 2070–2089.

[84] Land, M. F. 2009. *Vis. Neurosci.* 26: 51–62.

[85] Watson, T. L., and Krekelberg, B. 2009. *Curr. Biol.* 19: 1040–1043.

[86] Lee, R. G. et al. 1995. *Can. J. Neurol. Sci.* 22: 126–135.

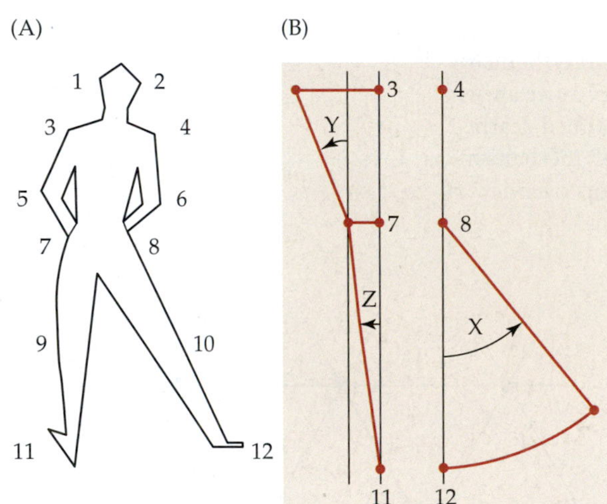

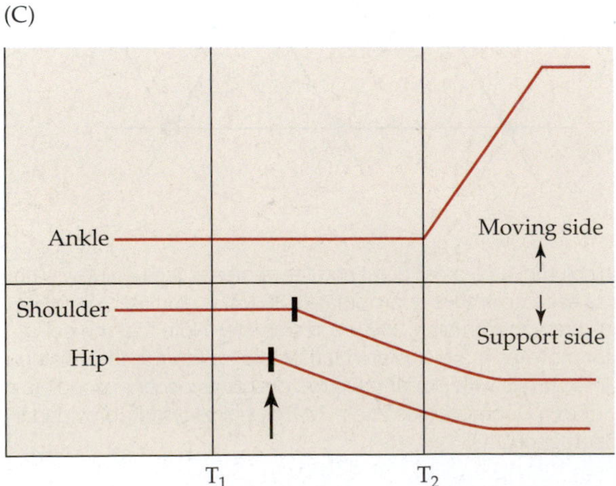

FIGURE 24.24 Compensatory Movements in Response to Intention of Raising the Leg on One Side of the Body. The drawing in (A) shows the raising of the leg, and the numbers indicate the points at which movements were analyzed by kinematic (pre-video!) analysis. B) Postural adjustments of the trunk in schematic form. (C) The first movements occur in anticipation on the opposite side of the body (see arrow), *before* the leg is raised. (After Lee et al., 1995.)

the body start to rise before the leg is raised. The importance of anticipatory compensations and corollary discharges has been described by Sommer and Wurtz:

> Each movement we make activates our own sensory receptors, thus causing a problem for the brain: the spurious, movement-related sensations must be discriminated from the sensory inputs that really matter, those representing our environment... Such circuits convey a copy of each motor command, known as a corollary discharge...to brain regions that use sensory input.[87]

As one considers higher aspects of motor control, the boundary between sensory and motor systems becomes lost. For instance, in a task in which an animal must execute the correct movement in response to a particular frequency of vibration (see Chapter 23), neurons in premotor cortex encode both the vibration itself as well as the upcoming motor response. As another example, the activity of certain cells in prefrontal cortex of monkeys and humans, known as **mirror neurons**, is related to both the observation and the planning of movements. Thus, Rizzolati and his colleagues have shown that the same neuron fires when a monkey performs an action, such as grasping a stick, and also when it observes another monkey performing the same function.[88] Such mirror neurons seem to play a part in learning how to move by copying the actions of others (see also Chapter 23). Another example of neurons whose activity depends on sensory discrimination, learning, and memory is provided by the work of Romo and his colleagues,[89] discussed in detail in Chapter 23.

Cerebellum and Basal Ganglia

This section deals with two structures (the cerebellum and the basal ganglia) that coordinate the movements of the body. They are essential for accuracy, prevent tremor and spasticity, and contribute to motor learning.

The Cerebellum

The principal anatomical features of the cerebellum are shown in Figure 24.25 and Appendix C. The cerebellum participates in motor control through extensive interconnections with the premotor cortex and the vestibular system.[90,91] It influences the spinal motor apparatus by way of brainstem motor nuclei. Although recent experiments have provided valuable information about molecular and cellular mechanisms of cerebellar function, the lucid summary given by Adrian (again!) more than 50 years ago[2] still seems elegant, clear, and accurate:

> The cerebellum has the...immediate and quite unconscious task of keeping the body balanced whatever the limbs are doing and of insuring that the limbs do whatever is required of them. Its actions show what complex things can be done by the mechanism of the nervous system in carrying out the decisions of the mind. If I decide to raise my arm, a message is dispatched from the motor area of one cerebral hemisphere to the spinal cord and a duplicate of that message goes to the cerebellum. There, as a result of interactions with other sensory impulses, supplementary orders are sent out to the spinal cord so that the right muscles come in at the exact moment when they are needed, both to raise my arm and keep my body from falling over. The cerebellum has access to all the information from the muscle spindles and pressure organs and so can put in the staff work needed to prevent traffic jams and bad coordination. If it is injured the timing breaks down, muscles come in too early or too late and with the wrong force. The staff work needs to be elaborate, particularly when the body has to be balanced on two legs and uses its arms for all manner of movement, but it is done by the machinery of the nervous system after the mind has given its orders. The cerebellum has nothing to do with formulating the general plan of the campaign. Its removal would not affect what we feel or think, apart from the fact that we should be aware that our limbs were not under full control and so should have to plan our activities accordingly.

Connections of the Cerebellum

The cerebellum receives proprioceptive, vestibular, and other sensory inputs from the entire body as well as a massive projection from motor and association cortex. These multiple

[87] Sommer, M. A., and Wurtz, R. H. 2008. *Annu. Rev. Neurosci.* 31: 317–338.

[88] Cattaneo, L., and Rizzolatti, G. 2009. *Arch. Neurol.* 66: 557–560.

[89] Romo, R., Hernández, A., and Zainos, A. 2004. *Neuron* 41: 165–173.

[90] Glickstein, M., Strata, P., and Voogd, J. 2009. *Neuroscience* 162: 549–559.

[91] Glickstein, M., Sultan, F., and Voogd, J. 2011. *Cortex* 47: 59–80.

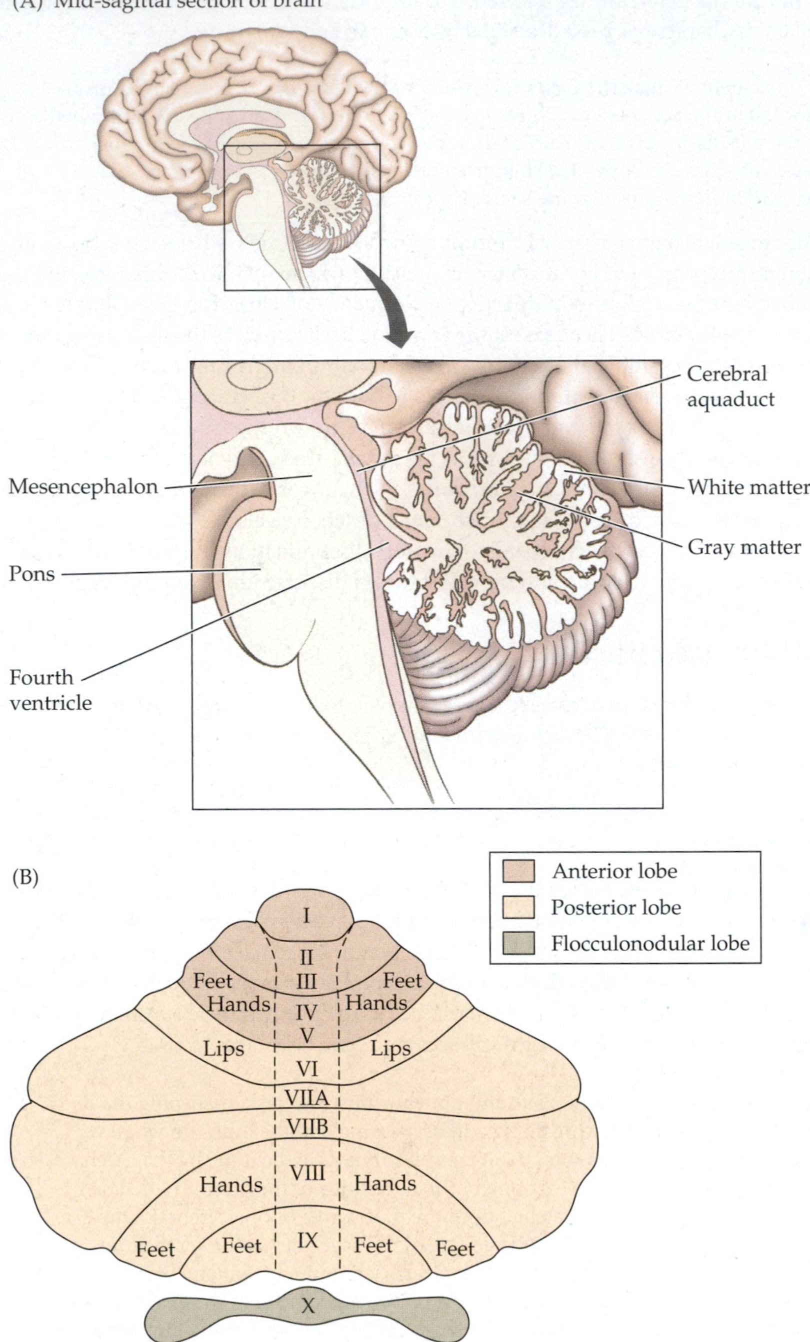

FIGURE 24.25 Position and Primary Architecture of the Cerebellum. (A) Shows the location of the cerebellum in relation to the midbrain and brainstem. (B) Shows the somatotopic representation of the body in the anterior and posterior lobes.

connections allow actual or intended movements to be compared during their execution with plans provided by the cortex.[92,93] Figure 24.26 shows the afferent and efferent pathways to and from the cerebellum.

The lateral hemispheres receive inputs from a wide area of cerebral cortex (via relay nuclei in the pons) and from the red nucleus (via the inferior olive). The flocculonodular lobe receives inputs from the vestibular nucleus.[94] The medial zone of the cerebellar cortex receives proprioceptive and cutaneous input from all levels of the spinal cord. For this reason, Sherrington referred to the cerebellum as "the head ganglion of the proprioceptive system."[95]

An essential feature of the cerebellum is that the projections are arranged in a highly orderly manner. Thus, the sensory inputs form multiple somatotopic representations on the cerebellar cortex, and these overlie motor representations in the same regions. It will

[92] Thach, W. T., Goodkin, H. G., and Keating, J. G. 1992. *Annu. Rev. Neurosci.* 15: 403–442.

[93] Llinás, R., Leznik, E., and Makarenko, V. I. 2002. *Ann. N Y Acad. Sci.* 978: 258–272.

[94] Tan, J., Epema, A. H., and Voogd, J. 1995. *J. Comp. Neurol.* 356: 51–71.

[95] Sherrington, C. S. 1933. *The Brain and Its Mechanism*. Cambridge University Press, London.

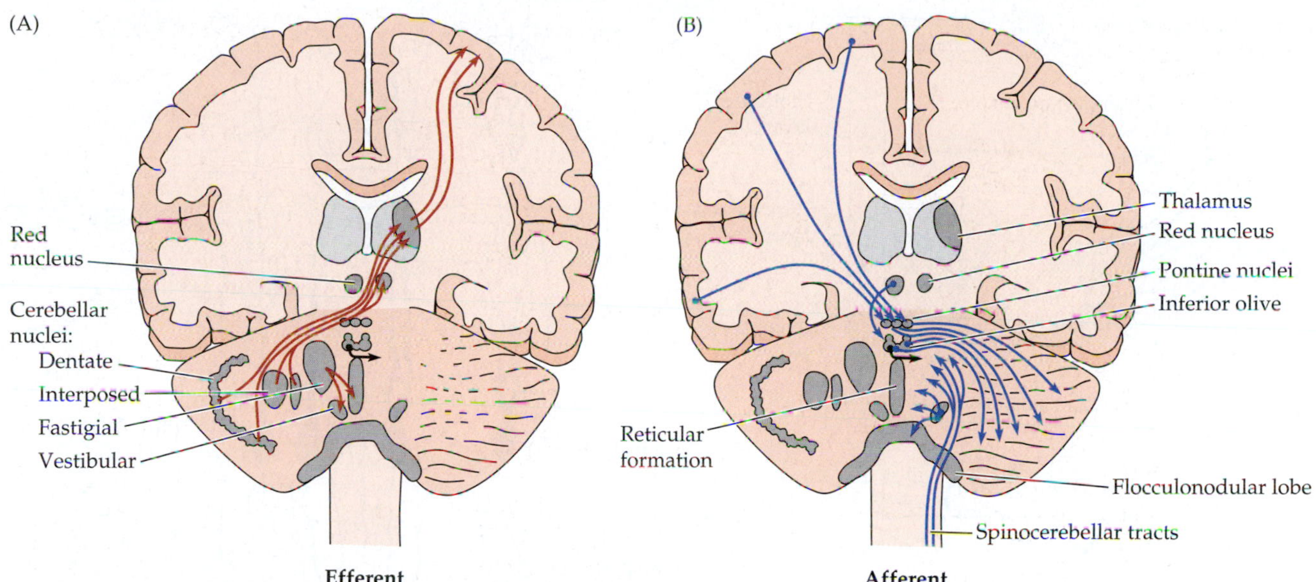

FIGURE 24.26 Efferent and Afferent Pathways of the Cerebellum shown in relation to the underlying nuclei, cerebral hemispheres, brainstem and spinal cord. (A) Outputs from the cerebellum (red) are through the dentate, interposed, and fastigial nuclei. Fibers from the dentate nucleus supply the contralateral motor cortex through the ventrolateral nuclei and parts of the ventroposterolateral nuclei of the thalamus. The interposed nuclei project to the contralateral red nucleus. The fastigial nucleus projects to the vestibular nucleus and the pontine and medullary reticular formations, contributing to the medial motor system. (B) Inputs (blue) to the lateral hemispheres of the cerebellum are from wide areas of the cerebral cortex, through the pontine nuclei. Afferent input from the red nucleus is relayed through the inferior olive. More medially, the cerebellum receives extensive input from the spinocerebellar tracts. The flocculonodular lobe is supplied by the vestibular nucleus.

become evident from the connections that posture and stance are influenced mainly by the output from cells situated in the midline of the cerebellum, whereas fine movements depend on those situated more laterally.

Output from the cerebellar cortex is solely through the axons of **Purkinje cells**, all of which are inhibitory. The Purkinje cell axons project onto cells of the deep cerebellar nuclei and onto vestibular nuclei, again in an orderly fashion (see Figure 24.26). Those from the flocculus and nodulus (vestibulocerebellum) project directly to the vestibular nuclei. The remainder project to the deep nuclei in a regular progression from medial to lateral across the cerebellum. Purkinje cells situated in the midline of the cerebellar cortex project to the fastigial nucleus, which projects to the vestibular nucleus and the reticular formation to influence the vestibulospinal and reticulospinal tracts—that is, the medial motor system concerned with stance and balance. Purkinje cells situated more laterally project to the interposed nucleus, and the most lateral project to the dentate nucleus.[96] Those nuclei send their outputs to the motor cortex via the ventrolateral nucleus of the thalamus. They exert their primary influence on the lateral motor system and thereby influence the musculature of the body concerned with fine movements. The interposed nucleus also projects to the red nucleus. Somatotopic order is maintained in each cerebellar cortical region and carried on through each nuclear projection.

Synaptic Organization of the Cerebellar Cortex

In few, if any, regions of the central nervous system have the functional connections of the incoming axons and the output been traced as fully as in the cerebellum.[97] In intact preparations and in slices, a variety of techniques, including anatomical and molecular studies as well as optical and electrical recordings, have been used to reveal the pattern of connections and their physiological mechanisms.

[96] Habas, C. 2010. *Cerebellum* 9: 22–28.

[97] Ito, M. 1984. *The Cerebellum and Neural Control*. Raven, New York.

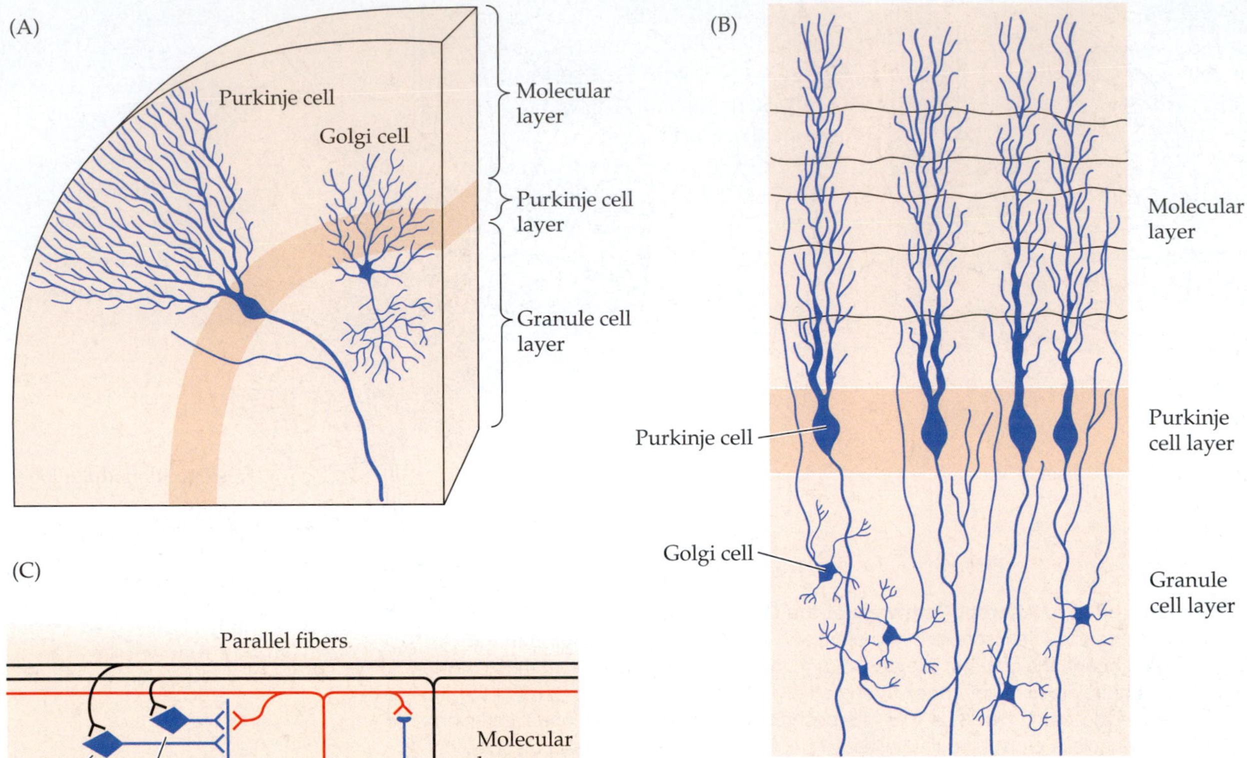

FIGURE 24.27 Synaptic Organization of the Cerebellum. (A) and (B) show the arborizations of Purkinje cells seen from two perspectives: along the parallel fibers and at right angles to them The cerebellar cortex is subdivided into the granule cell, Purkinje cell, and molecular layers. Red cells are excitatory, and blue cells are inhibitory. The sole output from the cortical layer to cells of the deep cerebellar nuclei is by inhibitory axons from Purkinje cells. (C) The axons of deep nuclear neurons form the output paths of the cerebellum. Mossy fiber inputs excite granule cells, whose axons ascend to the molecular layer to form a parallel fiber network. Parallel fibers form excitatory synapses on Purkinje cells, stellate cells, basket cells, and dendrites of Golgi cells. Climbing fibers form excitatory synapses on Purkinje cells. Both climbing fibers and mossy fibers make excitatory connections with cells in the deep cerebellar nuclei.

The cytoarchitecture of the cerebellum was revealed by Ramón y Cajal[98] and studied later in detail by light and electron microscopy. The cerebellar cortex is composed of three layers (Figure 24.27). The innermost layer is packed with 10^{10} to 10^{11} **granule cells**—approximating the sum of all other cells in the nervous system! They send axons to the outermost (molecular) layer to form a system of **parallel fibers**, each extending several millimeters along the folium. Also in the granule cell layer are **Golgi cells**, which make inhibitory synapses onto granule cells.

The second cortical layer is occupied by Purkinje cells, whose axons, as already mentioned, constitute the sole output from the cerebellar cortex. The Purkinje cell dendrites extend into the outer molecular layer of the cortex, with their planar arborizations oriented at right angles to the streams of parallel fibers. The parallel fibers make excitatory synaptic contacts onto spiny processes of the distal dendrites of the Purkinje cells. Figure 24.26 shows the way in which the Purkinje cells are stacked in a row along a folium, with the parallel fibers extending through them, rather like the wires laid on telephone poles. It is estimated that each Purkinje cell receives inputs from more than 200,000 parallel fibers.

Each parallel fiber engages a beam-like set of Purkinje cells, extending along the folium and projecting in an orderly manner to the underlying cerebellar nucleus. The significance

[98] Ramón y Cajal, S. 1995. *Histology of the Nervous System*. 2 vols. Oxford University Press, New York.

of this arrangement is that such a beam of Purkinje cells can span all the inputs from muscle spindles and joints in a region (for example, the shoulder, elbow, and wrist joints of the arm), thereby providing a possible representation of the body for coordinating complex movements. The length of the parallel fibers is sufficient to connect Purkinje cells that project to adjacent deep nuclei. Hence, different nuclei can function in a coordinated manner. The second cortical layer also contains **stellate** and **basket cells**, which provide inhibitory inputs to Purkinje cells from remote parallel fibers. The arrangement resembles the lateral inhibition seen in sensory systems.

Information flowing into the cerebellum from cortico-pontine relays and sensory systems is carried by **mossy fibers** that make excitatory synapses with granule cells, Golgi cells, and deep-nuclear neurons.[99] Mossy-fiber excitation of parallel fibers (the axons of granule cells) causes simple spike generation in Purkinje cells. These occur continuously at rates of 50 to 150/s and resemble conventional brief action potentials seen in other neurons. By contrast, a single **climbing fiber** arising in the inferior olive makes extensive connections onto the soma and proximal dendrite of 1 to 10 Purkinje cells. Climbing fibers cause powerful excitation of Purkinje cells,[100] and produce large plateau potentials that lead to complex spikes.[101] This activity involves calcium action potentials in the dendrites, leading to a calcium influx and long-term depression.[102,103] Several lines of evidence suggest that during classical conditioning the conditioned stimulus is transmitted by simple spikes (evoked by mossy fibers), while transmission of the unconditioned stimulus is mediated by complex spikes (evoked by climbing fibers).

What Does the Cerebellum Do and How Does It Do It?

Entire books and journals are devoted to the cerebellum, so it is a sobering thought that in 2009 a review of cerebellar structure and function should have to contain the following sentence:

> Although there is no unanimous agreement about what the cerebellum does or how it does it, some principles of its structure and function are well understood.[90]

That the cerebellum is required for the planning and performance of coordinated movements of the eyes and the body as a whole is clear from electrical and optical recordings made in conscious monkeys while they perform trained movements. Moreover, everyday actions that we take for granted, such as bringing a cup to one's mouth or sitting down, have to be learned in the first place. The idea that the cerebellum plays a part in motor learning is supported by the effects of lesions. Long-term depression and potentiation (see Chapter 16), which are prevalent at cerebellar synapses (such as those between parallel fibers and Purkinje cells or between mossy fibers and cerebellar nuclei), provide possible mechanisms for motor learning[104] (see also Welsh et al.[105]). Lesion experiments and clinical disorders support the idea that the cerebellum plays a key role in learning, in controlling movements, and even in higher cognitive functions. The literature is, however, too extensive to be described in this chapter (see Strick et al.[106] and Thach[107]).

Most lesions of the cerebellum in patients, whether due to disease or development, are diffuse rather than focal, with signs and symptoms that relate to balance and motor control.[108] As one might expect from their connections to the vestibular system, localized lesions of the nodulus and flocculus give rise to disturbances in equilibrium and eye movements as well as to disordered movements of the trunk. A common feature of cerebellar lesions is an intention tremor during a movement, say of the hand, in which small, rhythmical, and purposeless trembling begins and becomes stronger. This tremor is different from that occurring at rest with disorders of the basal ganglia (see below).

Earlier in this chapter, we have shown that a major function of descending control to spinal interneurons and motoneurons is to prevent oscillations. Without descending control, oscillation would, in principle, be inevitable since every muscle contraction activates stretch receptors in its antagonists. The cerebellum must make the distinction between true sensory input from outside the body and secondary inputs from movements generated by the CNS.[87] It predicts the sensory outcome of motor commands; then, through internal feedback, appropriate motor commands can be delivered to the muscles.

[99] Rokni, D., Llinas, R., and Yarom, Y. 2008. *Front. Syst. Neurosci.* 2: 192–198.

[100] Ito, M., and Simpson, J. I. 1971. *Brain Res.* 31: 215–219.

[101] Foust, A. et al. 2009. *Neuroscience* 162: 836–851.

[102] Miyakawa, H. et al. 1992. *J. Neurophysiol.* 68: 1178–1189.

[103] Zagha, E. et al. 2010. *J. Neurophysiol.* 103: 3516–3525.

[104] Coesmans, M. et al. 2004. *Neuron* 44: 691–700.

[105] Welsh, J. P. et al. 2005. *Proc. Natl. Acad. Sci. USA* 102: 17166–17171.

[106] Strick, P. L., Dum, R. P., and Fiez, J. A. 2009. *Annu. Rev. Neurosci.* 32: 413–434.

[107] Thach, W. T. 2007. *Cerebellum* 6: 163–167.

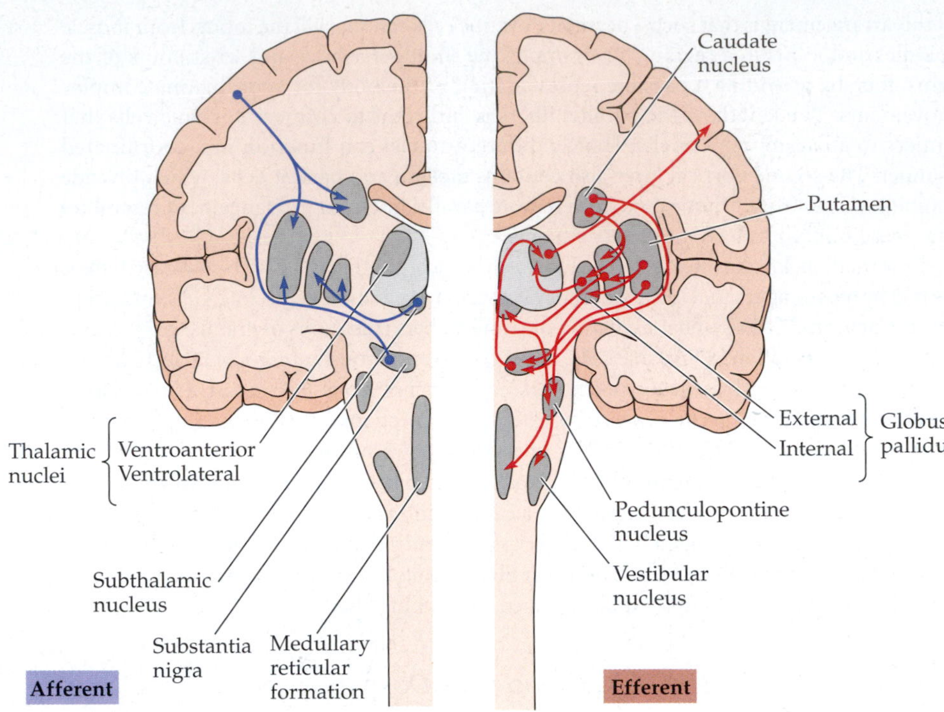

FIGURE 24.28 The Basal Ganglia. Coronal section through the cerebral hemispheres, continued as a longitudinal section of the brainstem and spinal cord. Basal ganglia include the caudate nucleus, putamen, and globus pallidus (external and internal divisions). Two additional nuclei, the substantia nigra and the subthalamic nucleus, have extensive interconnections with the basal ganglia and are sometimes included with them. The predominant input to the basal ganglia is from the cortex (left side, blue). Outputs from the basal ganglia go to the ventroanterior and ventrolateral nuclei of the thalamus, which in turn project to cortex (right side, red), completing a cortical feedback circuit. Additional output pathways project to the vestibular nucleus and medullary reticular formation through the pedunculopontine nucleus.

The Basal Ganglia

The nuclei that constitute the basal ganglia play an essential role in motor control. Whereas the cerebellum is primarily concerned with phasic movements of the body (for example, eye saccades and finger pointing), a principal function of the basal ganglia is to regulate posture, counteract tremor, and maintain steady muscular contractions. Neurons in basal ganglia can activate agonists and antagonists together—a mechanism appropriate for stabilizing a joint, such as the knee.[109] Basal ganglia also cooperate in terminating movements and in motor learning. It has even been suggested by Kandel and his colleagues[110] that the basal ganglia might play a role in the genesis of schizophrenia. Here, we provide only a brief summary and guide to the vast literature that deals with the structure and functions of the basal ganglia (see for example[111–113]).

The basal ganglia consist of nuclear masses situated beneath the outer cortical layers of the cerebral hemispheres. Key structures are the **caudate nucleus** and the **putamen** (known together as the **neostriatum**), and the external and internal divisions of the **globus pallidus** (Figure 24.28). Two midbrain structures, the **substantia nigra** and the **subthalamic nucleus,** have afferent and efferent connections with the basal ganglia and are part of the circuit. Dopaminergic neurons in the substantia nigra (see Chapter 14) project to the striatum (the nigrostriatal pathway). Nigral neurons release dopamine, which inhibits some neurons and excites others; the overall effect of dopamine on the striatum is excitatory.[114] The striatum receives widespread inputs from the cerebral cortex, particularly from the precentral gyrus. Major outputs of the basal ganglia from the globus pallidus are directed to the ventrolateral and ventroanterior nuclei of the thalamus (with overlapping with regions receiving input from the cerebellum) and back to the cortex. The basal ganglia modulate motor output through this complex feedback circuitry.[115]

Synaptic transmission and integration in the basal ganglia are highly plastic. Long term potentiation (LTP) and depression (LTD) occur at glutamatergic synapses on medium spiny projection neurons of the striatum.[116] Transmission in the basal ganglia is also modulated

[108] Dietrichs, E. 2008. *Acta Neurol. Scand. Suppl.* 188: 6–11.

[109] Mink, J. W., and Thach, W. T. 1991. *J. Neurophysiol.* 65: 330–351.

[110] Simpson, E. H., Kellendonk, C., and Kandel, E. 2010. *Neuron* 65: 585–596.

[111] Graybiel, A. M. 2008. *Annu. Rev. Neurosci.* 31: 359–387.

[112] Kreitzer, A. C., and Malenka, R. C. 2008. *Neuron* 60: 543–554.

[113] Kreitzer, A. C. 2009. *Annu. Rev. Neurosci.* 32: 127–147.

[114] Surmeier, D. J. et al. 2007. *Trends Neurosci.* 30: 228–245.

[115] Mink, J. W., and Thach, W. T. 1993. *Curr. Opin. Neurobiol.* 3: 950–957.

[116] Lovinger, D. M. 2010. *Neuropharmacology* 58: 951–961.

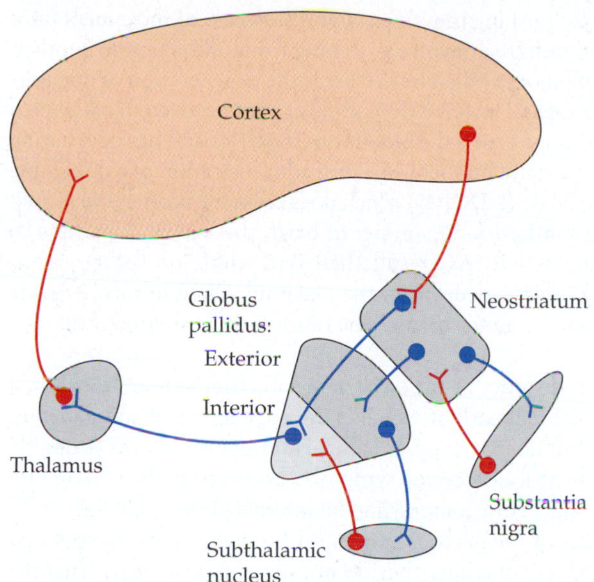

FIGURE 24.29 Functional Circuits of the Basal Ganglia. Glutamatergic neurons in cortex excite GABAergic cells of the neostriatum (caudate and putamen). Striatal neurons project to the external globus pallidus (the indirect pathway) and internal globus pallidus (the direct pathway) to inhibit GABAergic neurons in those nuclei. GABAergic neurons of the internal globus pallidus project to and inhibit the thalamus. Dopaminergic neurons of the substantia nigra produce net excitation of the striatum. Glutamatergic neurons of the subthalamic nucleus are inhibited by the projection from the external globus pallidus and excite GABAergic neurons of the internal globus pallidus. Excitatory neurons are shown in red and inhibitory neurons in blue. The gas, nitric oxide, is an important modulatory transmitter in the striatum, accumbens nuclei, subthalamic nuclei, and substantia nigra.

by the gas nitric oxide (NO), acting as a local transmitter released by neurons that diffuses locally.[117] Since it is a gas, release of NO depends on the rate of synthesis in neurons that make up approximately 2% of the total in the striatum.[118] In the basal ganglia, the principal action of NO is excitatory modulation of synaptic efficacy through second messenger systems (see Chapter 14).

Circuitry of the Basal Ganglia

The caudate and putamen function as the input stage of the basal ganglia, (Figure 24.29). The putamen receives its input from the sensorimotor strip surrounding the central sulcus, so its activity relates most directly to motor activity. The caudate is innervated by frontal cortex and is involved in higher-order cognitive processing. This parallel arrangement underlies the role of the basal ganglia in cognition and affect as well as in motor processing. GABAergic neurons of the caudate and putamen project to the globus pallidus and inhibit its activity. Neurons of the globus pallidus are also inhibitory;[119] they release GABA onto thalamic neurons in the ventroanterior and ventrolateral nuclei. Since neurons in the globus pallidus fire tonically, they continuously inhibit the flow of excitation from thalamus to cortex.[109] The firing rate of individual neurons in the globus pallidus concerned with wrist movements are relatively unaffected by its position, the velocity of movement, or the load on the wrist. The strongest stimulus for such neurons instead is a sudden movement, which causes a pronounced increase in frequency after a delay. Mink and Thach[109] have proposed that pallidal cell discharges are associated with the release of holding mechanisms responsible for joint fixation, thereby allowing the movement to occur. An analogy for the delayed firing is that of starting a car on a hill. You release the hand brake only after power has been applied to the wheels. The following is an oversimplified scheme of the functional connections of the basal ganglia:

> increased cortical activity → excitation of caudate nucleus and putamen → inhibition of globus pallidus → disinhibition of the thalamus → increased activity of the thalamocortical pathway → activation of cortex

Diseases of the Basal Ganglia

The importance of the basal ganglia in motor control is emphasized by the devastating consequences of neurodegenerative diseases that affect their function.[120] James Parkinson described the "shaking palsy" in 1817. **Parkinson's disease** is characterized by: a continuous tremor at rest ("pill rolling"), an increased tone due to simultaneous activation of antagonist

[117] Garthwaite, J. 2008. *Eur. J. Neurosci.* 27: 2783–2802.

[118] Del Bel, E. A. et al. *Cell Mol. Neurobiol.* 25:371-392.

[119] Rav-Acha, M. et al. 2005. *Neuroscience* 135: 791–802.

[120] DeLong, M., and Wichmann, T. 2009. *Parkinsonism Relat. Disord.* 15(Suppl. 3): S247–240.

[121] Gil, J. M., and Rego, A. C. 2008. *Eur. J. Neurosci.* 27: 2803–2820.

muscles, difficulty in initiating or finishing movements, and slowness of movement once begun. Deficits are prominent in tasks that involve switching from one action to another.

A key factor in the development of Parkinson's disease is the degeneration of dopaminergic neurons of the substantia nigra. However, dopamine cannot be given to patients to counteract the deficit because it cannot cross the blood–brain barrier and has devastating peripheral side effects. Instead, the standard therapy is oral administration of a dopamine precursor, L-amino acid decarboxylase (L-DOPA), which does cross the barrier and is used by substantia nigra neurons to synthesize dopamine. In brief, the loss of dopamine in Parkinson's disease reduces striatal activity. As a result, there is less inhibition of the globus pallidus, and its increased firing inhibits neurons in the thalamus. They, in turn, provide a lower level of excitatory input to the motor cortex. The result is the hypokinesis that is a dominant deficit of this disease.

Another clinically important disorder of the basal ganglia is **Huntington's disease**, a genetically determined disease, the hallmark of which is the appearance of spontaneous, disruptive movements that give this disease its other name, Huntington's chorea (from the Greek word for *dance*).[121] Patients typically develop symptoms and signs in their 20s or 30s, as a result of degeneration of striatal neurons supplying the external globus pallidus.

Although lesions and diseases of the nervous system lead to behavioral changes that provide essential clues about normal functions, they do not reveal mechanisms. That the basal ganglia and cerebellum organize, coordinate, and take part in the learning of movements is certain—but, how they do so is still not known.

SUMMARY

- A motor unit consists of a single α motor neuron and the skeletal muscle fibers it innervates.
- Muscle spindle afferents diverge to make synaptic contacts onto all the motoneurons innervating the muscle of origin.
- Muscular contraction begins with small motor units and progresses to large (the size principle of motor recruitment).
- The stretch reflex excites agonist muscles, and it inhibits antagonists through inhibitory interneurons.
- The muscle spindle's sensitivity to stretch is modulated by activation of γ-motoneurons (fusimotor) that cause the intrafusal muscle fibers to contract.
- α and γ-motor neuron co-activation continuously adjusts the spindle to maintain its sensitivity during programmed movements.
- Flexor reflexes and crossed extensor reflexes initiated by painful stimuli comprise elements of interlimb coordination essential for locomotion.
- Medial and lateral pools of spinal motoneurons innervate the muscles of the trunk and distal limbs, respectively.
- Respiration and locomotion provide examples of motor programs arising from pattern generators in the central nervous system.
- The primary motor cortex, or M1, lies anterior to the central sulcus and is somatotopically mapped.
- Many neurons in M1 are grouped in columns according to their direction selectivity.
- The planning of coordinated movements includes anticipatory adjustments to take account of the effects of muscular contractions on feedback mechanisms.
- The cerebellum plans and executes motor commands through feedback with the cortex and via descending commands through the red nucleus and brainstem nuclei. Lesions of the cerebellum disrupt coordination.
- The basal ganglia provide negative feedback to the cerebral cortex. The consequences of basal ganglia disease reflect the complex pattern of feedback loops that underlie their function.

Suggested Reading

General Reviews

Briggman, K. L., and Kristan, W. B. 2008. Multifunctional pattern-generating circuits. *Annu. Rev. Neurosci.* 31: 271–294.

Cattaneo, L., and Rizzolatti, G. 2009. The mirror neuron system. *Arch. Neurol.* 66: 557–560.

Georgopoulos, A. P., and Stefanis, C. N. 2007. Local shaping of function in the motor cortex: motor contrast, directional tuning. *Brain Res. Rev.* 55: 383–389.

Glickstein, M., Strata, P., and Voogd, J. 2009. Cerebellum: history. *Neuroscience* 162: 549–559.

Hultborn, H. 2006. Spinal reflexes, mechanisms and concepts: from Eccles to Lundberg and beyond. *Prog. Neurobiol.* 78: 215–242.

Konishi, M. 2004. The role of auditory feedback in birdsong. *Ann. N Y Acad. Sci.* 1016: 463–475.

Kreitzer, A. C. 2009. Physiology and pharmacology of striatal neurons. *Annu. Rev. Neurosci.* 32: 127–147.

Lemon, R. N. 2008. Descending pathways in motor control. *Annu. Rev. Neurosci.* 31: 195–218.

Pearson, K. G. 2008. Role of sensory feedback in the control of stance duration in walking cats. *Brain Res. Rev.* 57: 222–227.

Sherrington, C. S. 1906. *The Integrative Action of the Nervous System*, 1961 Ed. Yale University Press, New Haven, CT.

Sommer, M. A., and Wurtz, R. H. 2008. Brain circuits for the internal monitoring of movements. *Annu. Rev. Neurosci.* 31: 317–338.

St John, W. M. 2009. Noeud vital for breathing in the brainstem: gasping—yes, eupnoea—doubtful. *Philos. Trans. R. Soc. Lond. B, Biol. Sci.* 364: 2625–2633.

Taccola, G., and Nistri, A. 2006. Oscillatory circuits underlying locomotor networks in the rat spinal cord. *Crit. Rev. Neurobiol.* 18: 25–36.

Thach, W. T. 2007. On the mechanism of cerebellar contributions to cognition Cerebellum 6: 163–167.

Original Papers

Bannatyne, B. A., Liu, T. T., Hammar, I., Stecina, K., Jankowska, E., and Maxwell, D. J. 2009. Excitatory and inhibitory intermediate zone interneurons in pathways from feline group I and II afferents: differences in axonal projections and input. *J. Physiol.* 587: 379–399.

Coesmans, M., Weber, J. T., De Zeeuw, C. I., and Hansel, C. 2004. Bidirectional parallel fiber plasticity in the cerebellum under climbing fiber control. *Neuron* 44: 691–700.

Durbaba, R., Taylor, A., Ellaway, P. H., and Rawlinson, S. 2003. The influence of bag_2 and chain intrafusal muscle fibers on secondary spindle afferents in the cat. *J. Physiol.* 550: 263–278.

Eugenin, J., Nicholls, J. G., Cohen, L. B., and Muller, K. J. 2006. Optical recording from respiratory pattern generator of fetal mouse brainstem reveals a distributed network. *Neuroscience* 137: 1221–1227.

Georgopoulos, A. P., Merchant, H., Naselaris, T., and Amirikian, B. 2007. Mapping of the preferred direction in the motor cortex. *Proc. Natl. Acad. Sci. USA* 104: 11068–1072.

Henneman, E., Somjen, G., and Carpenter, D. O. 1965. Functional significance of cell size in spinal motoneurons. *J. Neurophysiol.* 28: 560–580.

Kuffler, S. W., Hunt, C. C., and Quilliam, J. P. 1951. Function of medullated small-nerve fibers in mammalian ventral roots: Efferent muscle spindle innervation. *J. Neurophysiol.* 14: 29–51.

Meier, J. D., Aflalo, T. N., Kastner, S., and Graziano, M. S. A. 2008. Complex organization of human primary motor cortex: a high-resolution fMRI study. *J. Neurophysiol.* 100: 1800–1812.

Rochat, M. J., Caruana, F., Jezzini, A., Escola, L., Intskirveli, I., Grammont, F., Gallese, V., Rizzolatti, G., and Umiltà, M. A. 2010. Responses of mirror neurons in area F5 to hand and tool grasping observation. *Exp. Brain Res.* 204: 605–616.

Romo, R., Hernández, A., and Zainos, A. 2004. Neuronal correlates of a perceptual decision in ventral premotor cortex. *Neuron* 41: 165–173.

Smith, J. L., Crawford, M., Proske, U., Taylor, J. L., and Gandevia, S. C. 2009. Signals of motor command bias joint position sense in the presence of feedback from proprioceptors. *J. Appl. Physiol.* 106: 950–958.

Welsh, J. P., Yamaguchi, H., Zeng, X. H., Kojo, M., Nakada, Y., Takagi, A., Sugimori, M., and Llinás, R. R. 2005. Normal motor learning during pharmacological prevention of Purkinje cell long-term depression. *Proc. Natl. Acad. Sci. USA* 102: 17166–17171.

Suggested Reading

PART VI

Development and Regeneration of the Nervous System

A question that naturally arises from reading about neuronal signaling, sensory perception, and motor functions concerns the way in which the nervous system is formed during development and how it is modified by experience and injury. Chapter 25 deals with the way in which the nervous system develops in an embryo, in terms of cellular and molecular mechanisms that give rise to highly complex structures. Selected examples are used to show how key molecules play crucial roles in the formation of the central and peripheral nervous systems.

A logical question concerns the role of experience in the formation of the brain after an animal is born. In Chapter 26 we describe the extraordinary progress that has been made in understanding how visual, auditory, and olfactory experiences in the first months of life influence not merely the performance of the nervous system but its very structure.

A related question of obvious importance concerns the ability of the nervous system to repair itself after injury. In Chapter 27 we show that nerve fibers in the periphery can regenerate to restore function after they have been damaged. By contrast, axons in the central nervous system are unable to do so, for reasons that are not yet known.

CHAPTER 25
Development of the Nervous System

During development, cells acquire specific neuronal identities, and establish orderly and precise synaptic connections, under the influence of factors intrinsic to the cells (their genotype, cytoplasmic inheritance and their changing patterns of gene expression), and factors in the embryonic environment (inductive and trophic interactions among cells, cues that guide cell migration and axon outgrowth, specific cell–cell recognition, and activity-dependent refinement of connections). Studies on invertebrates, such as the fruit fly, have been used for elucidating many widely shared developmental mechanisms. In this chapter we discuss how vertebrate structures (e.g., the spinal cord, autonomic nervous system, and cerebral cortex) form and describe selected experiments to illustrate underlying mechanisms of neurodevelopment at the cellular and molecular level. In addition to the inherent fascination of understanding how such complex structures assemble, knowledge of the development of the nervous system is useful for the understanding and treatment of many diseases.

The development of the vertebrate nervous system begins with the formation of the neural plate in the dorsal ectoderm of the early embryo. The neural plate then folds to produce the neural tube and the neural crest. Neurons and glial cells of the central nervous system (CNS) are produced by division of precursor cells, mainly in the ventricular zone of the neural tube. Postmitotic neurons migrate away from the ventricular surface to form the gray matter of the adult nervous system. Within each region of the developing nervous system, the fates of cells become restricted according to anteroposterior, dorsoventral, and local patterns. The identity of a segment is determined by expression of a series of patterning genes along the anteroposterior and the dorsoventral axes.

Neural crest cells form the peripheral nervous system (PNS). The phenotype adopted by a neural crest cell is determined by signals from its environment. Thus, a neural crest cell transplanted from one region to another early in development assumes the fate appropriate to its new location.

To reach their targets, neurons extend axons tipped with growth cones that explore the environment. Two classes of molecules have been identified as important substrates for growth cone movements: (1) cell adhesion molecules (CAMs), which allow one neuron to grow along another, and (2) extracellular matrix molecules (ECMs), which allow adhesion of axons to a substrate. Growth cone navigation is controlled by long- and short-range attractive and repulsive cues. Chemoattractants guide axons either to their ultimate synaptic partners or to an intermediate target, such as a guidepost cell. Chemorepellents prevent axons from entering inappropriate territories. Axonal projections made during development are often more extensive than those seen in the adult. Synapses form rapidly and are then trimmed to the adult pattern by trophic and activity-dependent mechanisms.

A common feature of central nervous system development is an initial overproduction of neurons followed by a period of cell death. Neuronal death is regulated by competition for trophic substances, which act alone or in combinations to sustain particular neuronal populations.

Development: General Considerations

The orderliness of the connections made by nerve cells with one another and with tissues in the periphery is a prerequisite for normal function. To create the precise neural architecture of the nervous system during development, the correct numbers and types of neurons must be generated, assume their appropriate positions, and then make synapses on the proper target cells. For example, for the stretch reflex to work, the Ia afferent sensory neuron in a dorsal root ganglion must send one axon to end on the appropriate region of a muscle spindle, and another axon centrally to make synapses on those motoneurons that innervate the muscle that contains the spindle. Other branches of the same central axon end on spinal interneurons or run in the dorsal columns to innervate cells in the dorsal column nuclei. In addition, the number of sensory and motor neurons must be matched to the size of the muscle and the number of spindles it contains.

A variety of questions arise when one considers the example just described. How do precursor cells acquire their identities as neurons or glial cells? What cues guide neurons to their correct positions? Which mechanisms enable a neuron to extend an axon to a particular target, among myriad possible choices, and form a synapse? First of all, the assemblage of 10^{10}–10^{12} cells and an almost uncountable number of synapses is established with a very much smaller number of genes, only 30,000 or so, which confirms that genes act in a combinatorial manner to specify cell identities. Moreover, the overall wiring plan requires that flexibility be maintained during critical periods in development, and even in the adult, so that synapses can be formed, modified, or removed—for example by altered patterns of activity (see Chapter 26).

The scope of all the problems relating to development, synapse formation, neural specificity, and changes in efficacy is too great to be described in a single chapter (for more comprehensive reviews see[1–3]). Our aim is to provide a concise, coherent account of neural development of the sort that might be delivered in two or three lectures, instead of in a complete course devoted to the subject. There is however an inherent difficulty: how is one to present the development of the most complex structure in the body for readers unfamiliar with basic principles of embryonic development? Instead of trying to discuss all the structures as they form and all the responsible molecules that have been identified, we have chosen to deal with a few issues in development at structural, cellular, and molecular levels. After two brief asides (dealing with the question of how cells become different from one another at the genetic level and the need to describe *what* happens during development before one attacks the problem of *how* it happens), selected topics are discussed in developmental sequence, beginning with the induction of neuroectoderm and early neural morphogenesis. Then the regional specification of neural tissue and the factors that determine the identity of individual neurons and glial cells are described. Finally, the mechanisms of axon outgrowth, target innervation, and synapse formation are considered, together with the role of growth factors and competitive interactions in shaping the final form of the nervous system.

Genomic Equivalence and Cell Type Diversity

The ability to clone whole animals from individual cells[4,5] provides dramatic evidence that the cells in a developing embryo all contain the same genes. It follows that the biochemical differences underlying the wide variety of neurons, glia, and other cells reflect qualitative and quantitative differences in the activity of specific genes among the various cell types. These differences are achieved at many levels, ranging from **transcriptional regulation** (the extent to which specific genes are transcribed into mRNAs) to **post-translation modifications** (e.g., phosphorylation, glycosylation, ubiquitination) of proteins that the various genes encode.[6–8] Which genes are transcribed in any given cell depends on what combinations of **transcription factors** are present and also on the state of the chromatin (DNA plus associated proteins). Transcription factors are proteins that bind to specific regulatory DNA sequences associated with each gene. They may either activate or repress transcription, depending in part on which other transcription factors are present.

Two major advances have led to a rapid increase in our ability to explain phenomena of development in terms of molecular mechanisms. First is the development of techniques to monitor and manipulate gene expression precisely and specifically. Second is the discovery that mechanisms and molecules that mediate neural development are remarkably similar

[1] Gilbert, S. F. 2010. *Developmental Biology*, 9th Ed. Sinauer, Sunderland, MA.

[2] Sanes, D. H., Reh, T. A., and Harris, W. A. 2011. *Development of the Nervous System*, 3rd Ed. Academic Press, Burlington, VT.

[3] Price, D. J. et al. 2011. *Building Brains: An Introduction to Neural Development*, Wiley Blackwell, Oxford, UK.

[4] Gurdon, J. B., and Melton, D. A. 2008. *Science* 322: 1811–1815.

[5] Takahashi, K., and Yamanaka, S. 2006. *Cell* 126: 663–676.

[6] Port, F., and Basler, K. 2010. *Traffic* 11: 1265–1271.

[7] Mabb, A. M., and Ehlers, M. D. 2010. *Annu. Rev. Cell. Dev. Biol.* 26: 179–210.

[8] Nguyen, M. D., Mushynski, W. E., and Julien, J. P. 2002. *Cell. Death Differ.* 9: 1294–1306.

throughout the animal kingdom. For example, as mentioned in Chapter 1, the gene that directs the formation of the eye in a developing chick, mouse, or human is similar to that controlling eye formation in the fruit fly *Drosophila melanogaster*,[9] and genes that function in patterning the developing embryo and nervous systems of flies or nematode worms often have homologs that function in chicks, zebrafish, and mice. This conservation of developmental regulatory genes across species unfortunately leads to confusion for students and researchers alike, because homologous genes discovered independently in different experimental systems receive a bewildering variety of names (Box 25.1).

Cell Fate Maps Provide a Description of Normal Development

A good starting point for many studies of development is the question of when and where the cells of interest arise during embryogenesis. The process of answering such questions is called **cell fate mapping**. In experimentally favorable species, one can follow patterns of proliferation, migration, and differentiation by cells that can be identified from one embryo to the next. For example, the nematode *Caenorhabditis elegans* contains only about 300 neurons. Its embryo is transparent and develops rapidly, so that each cell division can be visualized under the microscope.[10,11] For more complex embryos, including vertebrates, it is possible to mark individual cells early on and see what types of progeny they produce later in development. This approach for cell fate mapping was introduced using leech embryos; intracellular tracers, such as fluorescent dextran or the enzyme horseradish peroxidase (HRP), were micro-injected into individual cells whose lineages were then followed either in living embryos or after staining the embryos to visualize marked cells (Figure 25.1).[12] Now, comparable experiments can be done by injecting or electroporating DNA constructs encoding fluorescent protein reporter genes (such as **green fluorescent protein** [**GFP**]), or by creating transgenic animals that express such proteins.[13–16] Experiments of this kind allow one to identify the precursor cells in embryos that will give rise to sensory cells, interneurons, and motor cells in the CNS of the adult.

Other approaches that have proved useful for the study of cell lineage in vertebrate central nervous systems include mapping the fate of genetically marked cells in embryonic and adult chimeras produced by surgical and microsurgical techniques, including transplantation of

[9]Gehring, W. J., Kloter, U., and Suga, H. 2009. *Curr. Top. Dev. Biol.* 88: 35–61.

[10]Brenner, S. 1974. *Genetics* 77: 71–94.

[11]Hobert, O. 2010. *WormBook.* 4:1–24.

[12]Weisblat, D. A. et al. 1980. *Science* 209: 1538–1541.

[13]Chalfie, M. et al. 1994. *Science* 263: 802–805.

[14]Giepmans, B. N. et al. 2006. *Science* 312: 217–224.

[15]Sive, H. L., Grainger, R. M., and Harland, R. M. 2010. *Cold Spring Harb. Protoc.* Dec 1.

[16]Puzzolo, E., and Mallamaci, A. 2010. *Neural Dev.* 5: 8.

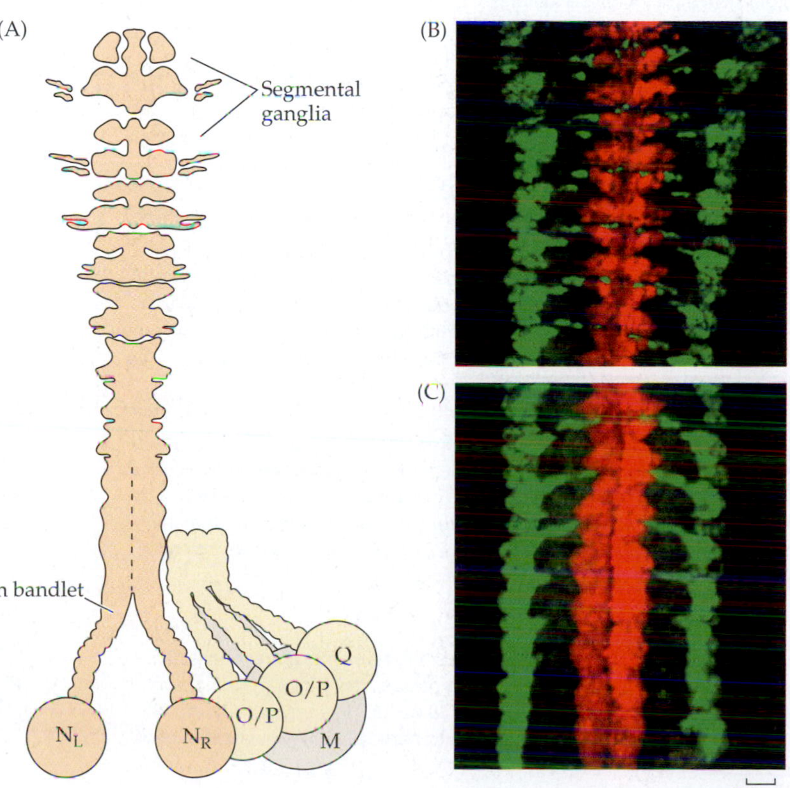

FIGURE 25.1 Lineage Tracing in Leech Development (A) Schematic representation of gangliogenesis in leech embryo. Stereotyped cell divisions in early development generate a posterior growth zone containing five bilateral pairs of segmentation stem cells called M, N, O/P, and Q teloblasts (only the N teloblast is shown on the left side). Through repeated unequal divisions, the teloblasts produce columns of cells (bandlets) that coalesce along the ventral midline and form segments arise in anteroposterior progression. To establish a fate map for segmental tissues, including the identified neurons of the segmental ganglia, teloblasts can be injected with lineage tracers such as fluorescent dyes conjugated to high molecular weight dextrans (to prevent the tracers from diffusing through gap junctions). (B,C) Views of an embryo in which both N teloblasts had been injected with rhodamine-conjugated dextran (red) and both Q teloblasts with fluorescein-conjugated dextran (green). The result shows that: (1) as in vertebrate embryos, leech segments differentiate in anteroposterior progression; (2) the cell fate distributions in each lineage are highly conserved from segment to segment; (3) most cells in the ventral ganglia arise from the ventral most N teloblast lineages, but a few neurons in each segment migrate ventrally from lateral and dorsal lineages. (After Shain et al., 2000.)

BOX 25.1
Conserved Signaling Pathways for Early Development and Neurogenesis

The signaling pathways listed here are only some of the best known of the many that function in neurogenesis and other aspects of development and physiology. The same pathways are used over and over again in development, often with different functional roles at different times and places in the embryos of diverse species. Often there are multiple versions of each pathway within a single species.

This list is far from complete—other molecules essential for various aspects of neurodevelopment are named as they appear in the text—moreover, in addition to their transcriptional effects, the Wnt and TGFβ pathways may also exert transcription-independent effects, for example, by affecting the state of protein phosphorylation in the target cell. Other signaling molecules not in this list may act entirely at the cytoplasmic level.

Terminology is a major problem for students interested in development at the molecular level. In recent years, scientists have identified many genes and gene products important for cell survival, growth, and differentiation, for axon extension and navigation, and for synapse formation and modification. These proteins and genes are named based on the history of their discovery and on the quirks of the investigators. The same molecules may receive multiple names from different investigators working on the same or different organisms. For example, *t*ransforming *g*rowth *f*actor-*beta* (TGFβ) and related molecules were first identified as proteins affecting cell proliferation and oncogenesis in vertebrates. Another family of molecules called bone morphogenetic proteins (BMPs), also discovered in vertebrates, turned out to be closely related to the TGFβs. Other vertebrate members of this TGFβ superfamily in vertebrates are called inhibin, activin and Vg-1, and in *Drosophila*, two genes are named *d*eca*p*entaplegic and *g*lass *b*ottom *b*oat, respectively!

For greater detail, the student should consult comprehensive books and reviews (see references[1,2]).

LIGAND(S)	RECEPTOR(S)
Retinoic Acid (RA)	*Retinoic Acid Receptors (RARs and RXRs)*
As a derivative of vitamin A (retinaldehyde), RA is a small, hydrophobic molecule that can readily diffuse across cell membranes. Thus, the receptors are nuclear rather than membrane proteins. In the absence of RA, RAR-RXR heterodimers bind to specific DNA elements and recruit other proteins that repress transcription of the target genes. When RA binds to the heterodimer, the repressor is replaced by a transcriptional activator.	
Hedgehog (Hh or Sonic)	*Patched-Smoothened Complex*
Hedgehog is a secreted, diffusing protein that binds to a transmembrane receptor, indirectly preventing the cleavage of a target cytoplasmic protein called Cubitus interruptus (Ci or Gli). In the absence of Hh signaling, Ci is cleaved and represses transcription of its target genes. In the presence of Hh, Ci is not cleaved and activates transcription of its targets.	
TGFß/BMP/Activin, etc.	*Receptor Serine/Threonine Kinases (RS/TKs)*
Binding of the secreted, diffusing protein ligands cause one transmembrane RS/TK to phosphorylate another, which in term phosphorylates one of a family of cytoplasmic proteins called SMADs. The phosphorylated SMAD binds a related protein, and the resultant complex is transported into the nucleus to regulate the transcription of target genes.	
Fibroblast growth factor (FGF), etc.	*Receptor Tyrosine Kinases (RTKs)*
A very large family of diffusible ligands and transmembrane receptors. Like the RS/TK receptors, ligand binding dimerizes the receptors. In this pathway, however, receptor activation triggers a kinase cascade, in which phosphorylation of one protein activates it to phosphorylate the next protein in the cascade and so on down the cascade. In addition to the final activation-by-phosphorylation of one or more transcription factors, the intermediate kinases in the cascade may also phosphorylate other cytoplasmic proteins and thus exert effects independent of any transcriptional regulation.	
Wingless/Int-1 (Wnt)	*Frizzled*
Wnts are also a family of secreted, diffusible proteins that act indirectly by binding to a transmembrane receptor, thereby regulating the cytoplasmic levels of a downstream target in the receiving cell. Here the key target is a protein called β-catenin, which also plays a structural role at the cell surface by interacting with a cell adhesion protein (cadherin). In the absence of Wnt signal, any free β-catenin in the cytoplasm is phosphorylated by a protein complex and thereby targeted for degradation. Wnt signaling indirectly inhibits the activity of the complex. Cytoplasmic β-catenin increases and is free to enter the nucleus to participate in the regulation of gene transcription.	
Delta/Serrate/Lag-2 (DSL)	*Notch*
In contrast to the other pathways, both the DSL ligands and the Notch receptor are transmembrane proteins. Thus, DSL–Notch signaling requires direct contact between the signaling and receiving cells. In this pathway, activation of the receptor by ligand enables the last of several proteolytic processing steps to occur. As a result, the *n*otch *in*tracellular *d*omain (NICD) is freed from the membrane and can enter the nucleus to participate in regulation of gene transcription.	

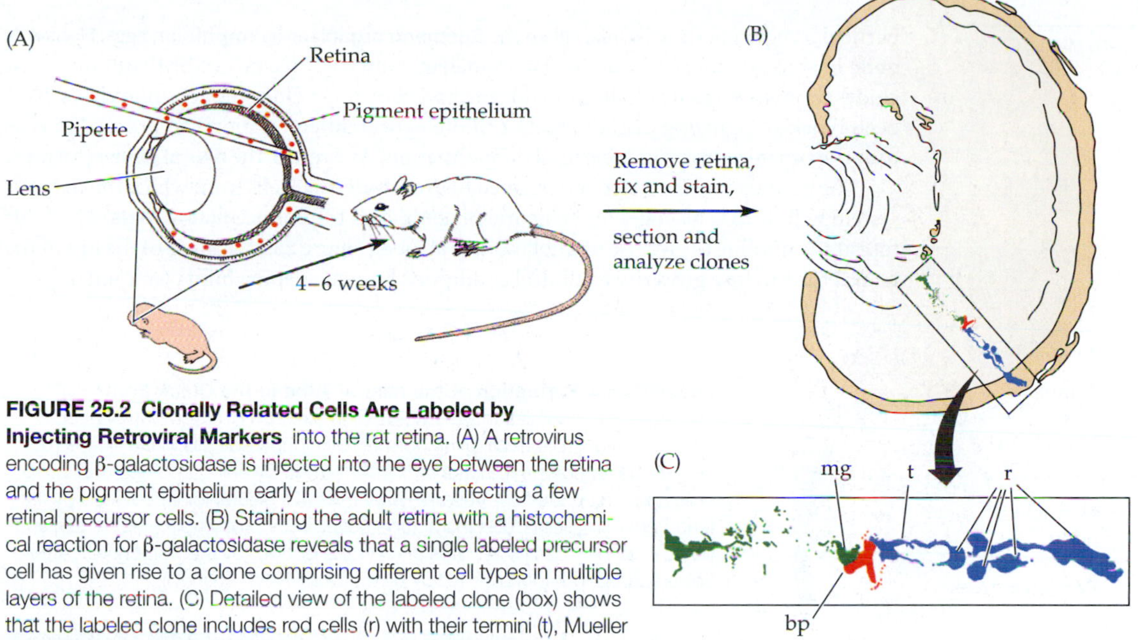

FIGURE 25.2 Clonally Related Cells Are Labeled by Injecting Retroviral Markers into the rat retina. (A) A retrovirus encoding β-galactosidase is injected into the eye between the retina and the pigment epithelium early in development, infecting a few retinal precursor cells. (B) Staining the adult retina with a histochemical reaction for β-galactosidase reveals that a single labeled precursor cell has given rise to a clone comprising different cell types in multiple layers of the retina. (C) Detailed view of the labeled clone (box) shows that the labeled clone includes rod cells (r) with their termini (t), Mueller glia (mg) and bipolar cells (bp). (After Turner and Cepko, 1987.)

individual cells,[17,18] and infecting cells in the CNS of developing animals with specially engineered viruses (Figure 25.2). For this latter purpose, viruses are constructed that will become permanently incorporated into the chromosomes of the host cell and replicate with them during mitosis.[19] Thus, the information encoded by the virus is not diluted during successive cell divisions and a suitably encoded marker, for example GFP, can be detected at any stage. Provided the number of cells infected is low, one can conclude that a cluster of stained cells found later in development is a clone, the progeny of a single parent cell. For example, Figure 25.2 shows the distribution of progeny cells in the adult retina after injection of a retroviral marker early in development.

Early Morphogenesis of the Nervous System

Early in vertebrate embryogenesis, the region of the gastrula that will give rise to the nervous system is a sheet of cells on the outer surface of the embryo called **ectoderm** (Figure 25.3). The ectodermal cells are under the influence of growth factors—including two proteins of the bone morphogenic protein (BMP) family, BMP-2 and BMP-4—that are secreted by the ectodermal cells themselves. Despite their name, in this context they promote the formation of epidermal tissue, while at the same time suppressing neural differentiation.[20,21] Next, a cocktail of diffusible signaling proteins that block the action of BMPs is released from a

[17] Rossant, J. 1985. *Philos. Trans. R. Soc. Lond., B, Biol. Sci.* 312: 91–100.

[18] Ho, R. K. 1992. *Dev. Suppl.* 65–73.

[19] Cepko, C., and Pear, W. 1996. *Curr. Protoc. Mol. Biol.* Sec. III, Unit 9.9. J. Wiley & Sons, Inc.

[20] Sasai, Y. 1998. *Neuron* 21: 455–458.

[21] Little, S. C., and Mullins, M. C. 2006. *Birth Defects Res. C Embryo Today* 78: 224–242.

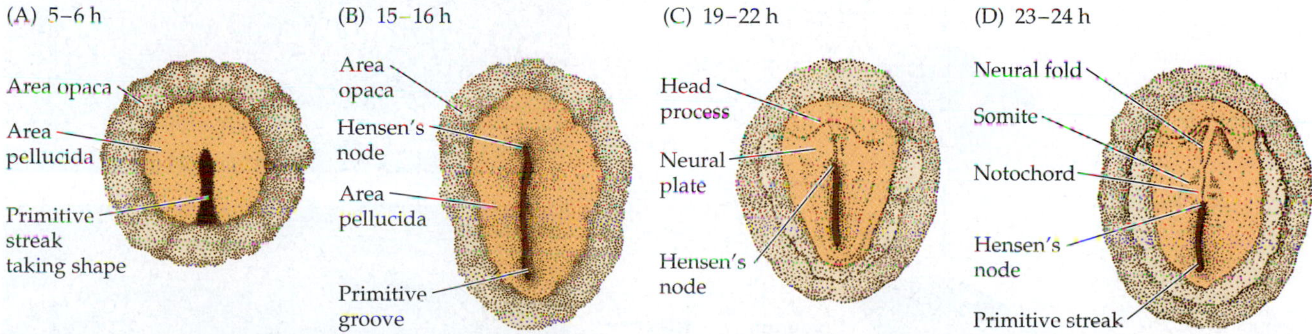

FIGURE 25.3 Early Morphogenesis in the Vertebrate Embryo. Dorsal views of the first day in the development of a chick embryo. (A) 5–6 h: formation and elongation of the primitive streak. (B) 15–16 h: formation of the primitive groove and Hensen's node. (C) 19–22 h: formation of the head process and neural plate. (D) 23–24 h: formation of the neural fold, notochord, and mesodermal somites. (After Gilbert, 2000.)

particular region of the gastrula, called the **Spemann organizer** in amphibian eggs, **Hensen's node** in chick, or simply the **node** in mammalian embryos. This mix of BMP inhibitors includes proteins known as follistatin, noggin, and chordin.[22] Once the actions of BMP have been blocked, signaling cascades that promote neural differentiation can proceed in cells near the organizer to promote neural differentiation. As a result, the **neural plate** is formed.

The neural plate is a sheet of elongated neuroectodermal cells from which the nervous system will form. The edges of the neural plate thicken to produce **neural folds.** The folds fuse at the midline to give rise to a hollow **neural tube** (Figure 25.4). Failure of the lips of the neural tube to fuse gives rise to clinical conditions known as spina bifida (in caudal spinal

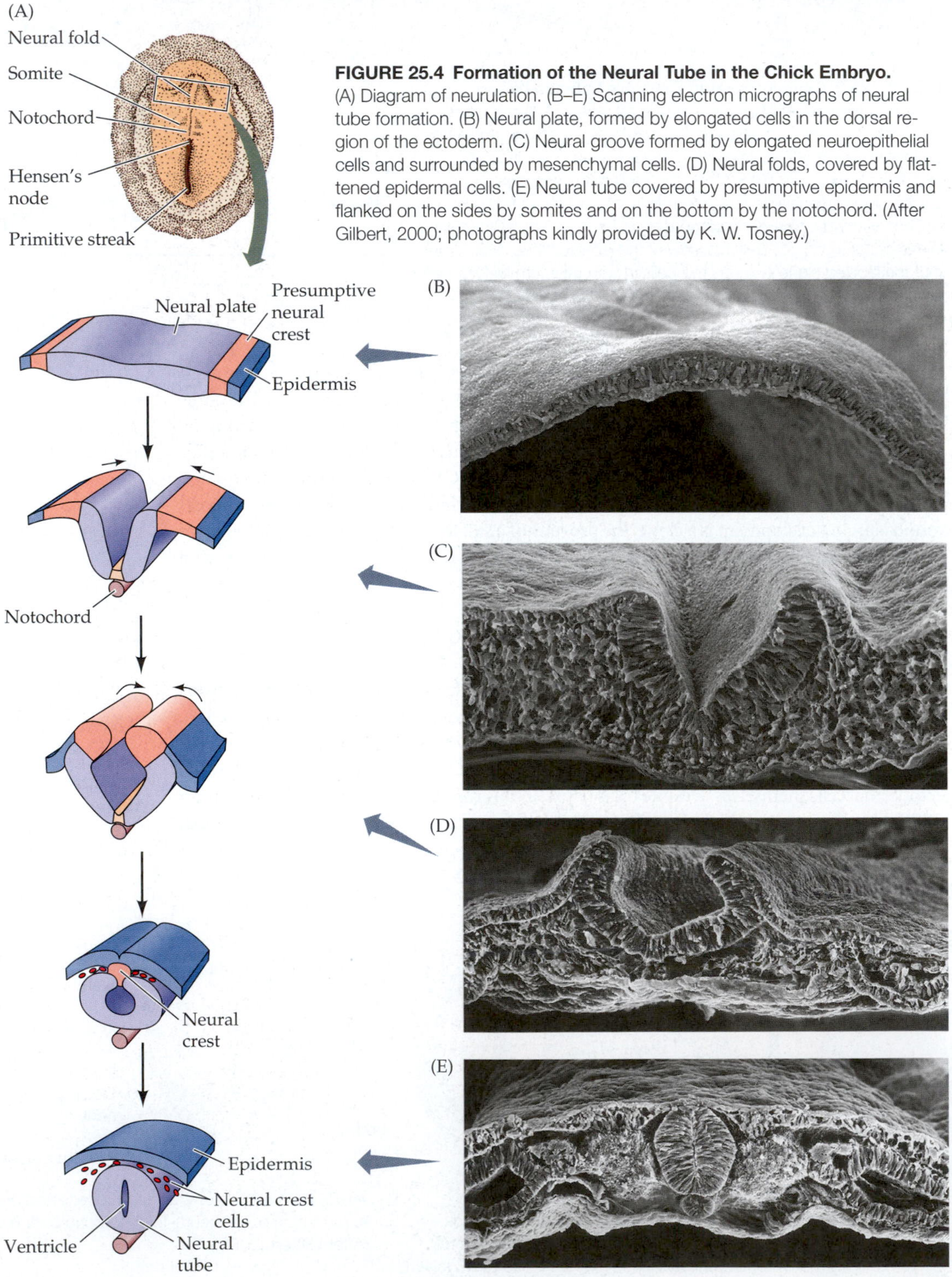

[22] Hemmati-Brivanlou, A., and Melton, D. 1997. *Annu. Rev. Neurosci.* 20: 43–60.

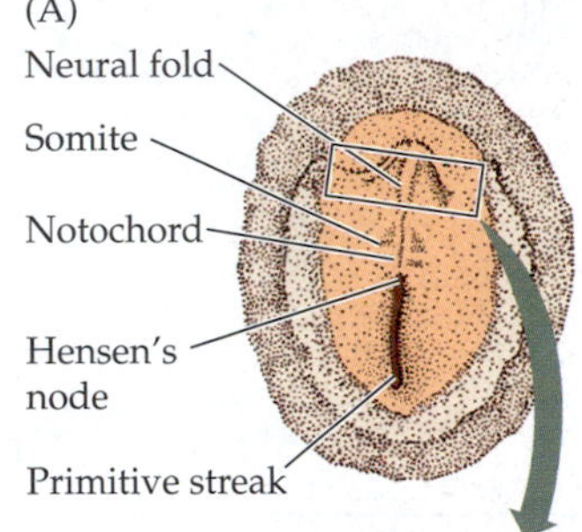

FIGURE 25.4 Formation of the Neural Tube in the Chick Embryo. (A) Diagram of neurulation. (B–E) Scanning electron micrographs of neural tube formation. (B) Neural plate, formed by elongated cells in the dorsal region of the ectoderm. (C) Neural groove formed by elongated neuroepithelial cells and surrounded by mesenchymal cells. (D) Neural folds, covered by flattened epidermal cells. (E) Neural tube covered by presumptive epidermis and flanked on the sides by somites and on the bottom by the notochord. (After Gilbert, 2000; photographs kindly provided by K. W. Tosney.)

cord) and anencephaly (failure of forebrain development). Some of the cells at the lips of the neural fold come to lie between the neural tube and the overlying ectoderm. These cells form the **neural crest**. Neural crest cells migrate away from the neural tube and give rise to several types of peripheral cells: neurons and satellite cells of the sensory, sympathetic, and parasympathetic nervous systems; cells of the adrenal medulla; pigmented cells of the epidermis; and bones and connective tissue in the head.

The observations described above illustrate two key principles of neural development. First, cells in the node secrete factors that dramatically affect the fates of neighboring cells and structures. In this way, one group of cells induces changes in an adjacent group by secretion of a signal. Such interactions are possible in an embryo because the distances for diffusion are short. Second, a single molecule such as BMP can function in the development of different systems: bone and nervous system. These features of development will be seen again and again in this chapter.

Patterning along Anteroposterior and Dorsoventral Axes

A key feature of neural development is the extraordinary variety of structures that are produced. Understanding how one region of the kidney or liver develops provides insights about how the whole organ arises. This is not so for the brain, which shows dramatic regional differences along both the anteroposterior and dorsoventral axes, and even less so for the nervous system as a whole. Although the basic principles are similar, different steps are needed to create an eye, a cerebellum, and a spinal cord. As development proceeds, the anterior (cephalic or rostral) portion of the neural tube undergoes a series of swellings, constrictions, and flexures that form anatomically defined regions of the brain (Figure 25.5). The caudal portion of the neural tube retains a relatively simple tubular structure and forms the spinal cord. Investigations of the regional specification of neural tissue in the vertebrate brain were aided immensely by the finding that most genes in the embryo of the fruit fly, *Drosophila melanogaster* (including many **developmental regulatory genes**), have vertebrate **homologs** (descended from the same gene in the ancestor of insects and vertebrates) that usually serve similar functions. One vertebrate preparation that offers many of the advantages of invertebrates such as *Drosophila*

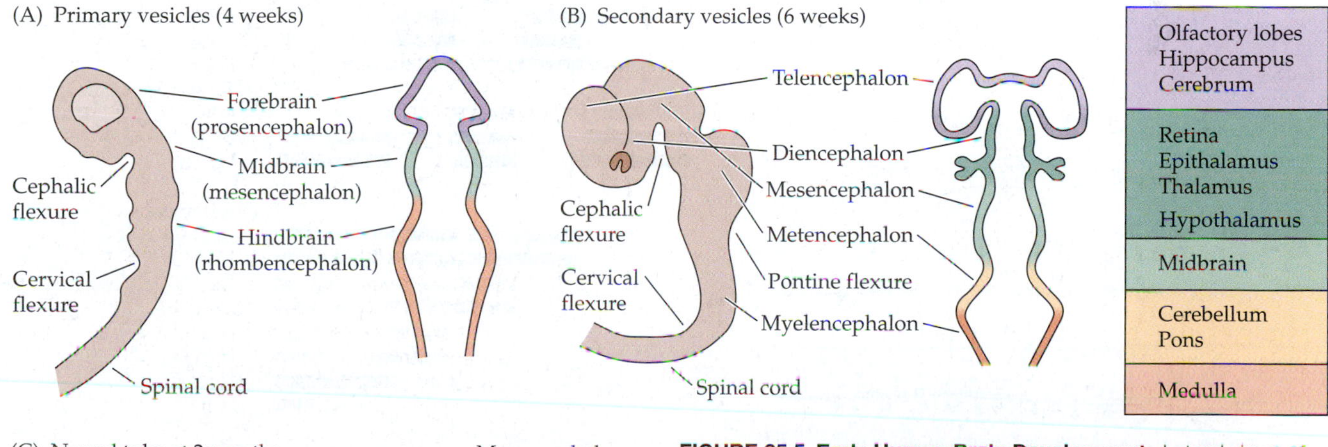

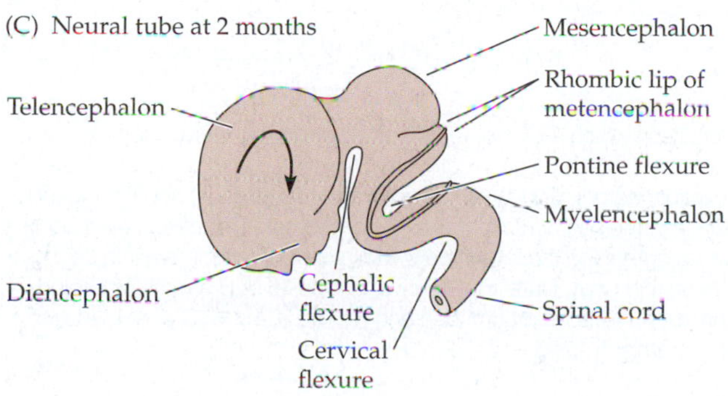

FIGURE 25.5 Early Human Brain Development. Lateral views of the developing brain and corresponding schematic horizontal sections through the vesicles. (A) At 4 weeks of development, the human CNS consists of three primary vesicles. (B) At 6 weeks of development, five secondary vesicles can be distinguished. Colored regions give rise to structures listed in the accompanying table. (C) By 2 months, a series of flexures, constrictions, and swellings form the various regions of the brain. Further development is dominated by rapid growth of the telencephalon in a "C" shape (arrow). (After Nolte, 1988.)

is the zebrafish *Danio rerio*, which was introduced by Streisinger.[23] The animal is transparent, and reproduces rapidly; throughout embryogenesis, individual cells can be observed under the microscope as they divide and grow. Mutations can be readily induced and maintained.[24]

Developmental regulatory genes often encode **transcription factors**, proteins that bind to DNA and thereby affect the level of expression of other genes, or components of **signaling pathways** by which cells influence one another. Development is characterized by the sequential and hierarchical expression of many different transcription factors at different times and in different parts of the embryo, often under the control of one or more signaling pathways. Each transcription factor influences the expression of (usually many) other genes, including other transcription factors. The main factor in specifying the properties of specific types of neurons, glia, and other cells is the particular combinations of transcription factors they express.

Anteroposterior Patterning and Segmentation in Hindbrain

The vertebrate hindbrain and forebrain provide examples of how developing nervous systems are patterned along the anteroposterior axis. Unlike the rest of the vertebrate brain, the embryonic hindbrain (**rhombencephalon**) has a conspicuously segmented structure. Each segment exhibits the same general pattern of neuronal differentiation, but from segment to segment the pattern is modified in specific ways (Figure 25.6). Several genes have

[23] Streisinger, G. et al. 1981. *Nature* 291: 293–296.

[24] Viktorin, G. et al. 2009. *Dev. Dyn.* 238: 1984–1998.

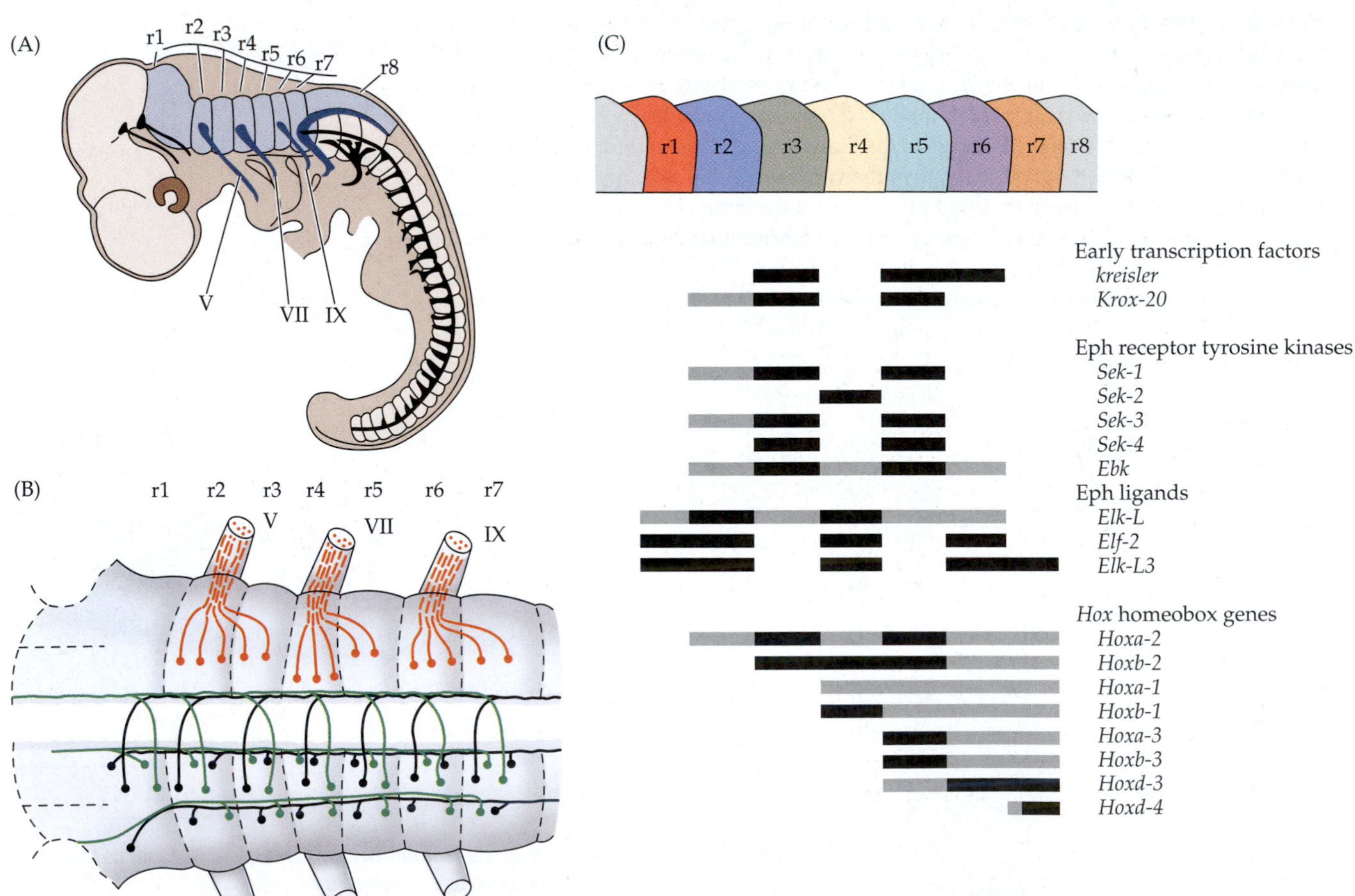

FIGURE 25.6 The Vertebrate Hindbrain Develops as a Conspicuously Segmented Structure. (A) Diagram of a 3-day chick embryo (lateral view), illustrating the segmental arrangement of rhombomeres (r1–r8) in the hindbrain. (B) Pattern of cell organization in rhombomeres r1 to r7 of the 3-day chick embryonic hindbrain (dorsal view). Reticular neurons (green and black) and branchiomotor neurons (orange) occur in a segmentally repeating pattern. Motor neurons send their axons into cranial nerves V, VII, and IX. (C) Segmental expression of genes in rhombomeres r1 to r8 of the vertebrate hindbrain. Gray bars indicate in which rhombomeres and how strongly individual genes are expressed; black bars indicate a high level of expression. Early transcription factors, Eph family receptor tyrosine kinases, and Eph ligands establish the segmental pattern of rhombomeres. The *Hox* homeobox genes determine the fate of cells within each rhombomere in a segmentally specific way. Data from chick and mouse. (A,B after Keynes and Lumsden, 1990, and Lumsden and Krumlauf, 1996; C after Lumsden and Krumlauf, 1996.)

been identified whose pattern of expression at early stages in development correlates with segmental boundaries of the hindbrain (see Figure 25.6).[25] These genes fall into two categories: (1) The first consists of genes that determine the fate of each segment. Many of these genes belong to the highly conserved *Hox* family. (2) Certain genes in the second category establish the overall architecture of repeating segmental units. Some genes in this group encode transcription factors (*kreisler*, *Krox-20*); others encode receptor tyrosine kinases or their ligands. A receptor tyrosine kinase is a transmembrane protein whose intracellular kinase domain is activated to phosphorylate tyrosine residues on intracellular protein targets when a ligand binds to its extracellular domain.

Hox genes were first characterized in *Drosophila* as examples of **homeotic genes**—genes which, when mutated, cause one body part to be changed so as to resemble another body part (**homeosis**[26]; for a more recent discussion on this see also[27]). For example, one famous homeotic mutation in *Drosophila* called *proboscipedia* causes an antenna to resemble a leg. Studies in *Drosophila* revealed that several homeotic genes occur as a tandem array along the chromosome, reflecting ancient gene duplications. Another indication that these genes arose from a single ancestral gene is that they all share a 60 amino acid DNA-binding motif called the **homeodomain**, encoded by a DNA sequence called a **homeobox**. It turns out that there are many *homeobox*-containing transcription factor genes in the genome, most of which are not part of the conserved cluster of homeotic genes and which have nothing to do with homeosis. The term *Hox* gene is used to specify the genes in the original *Drosophila* cluster and their close homologs in other organisms. An intriguing discovery is that these arrays of *Hox* genes, collectively called *Hox* clusters, are broadly conserved among animals and play a key role in assigning regional identities along the anteroposterior axis. Remarkably, there is a co-linear relationship between the anterior boundaries of the expression patterns of the various *Hox* genes in *Drosophila* and their arrangement within the cluster along the chromosome; this co-linearity is also conserved across species.[28,29]

The segmental pattern of *Hox* gene expression observed in the chick, zebrafish, and rodent hindbrain indicates that *Hox* genes act as patterning master genes in vertebrate development, and create structures appropriate to particular anteroposterior positions in the embryonic hindbrain. Evidence from cell transplantation, gene knockout, and gene misexpression studies are consistent with this idea.[30–32] Additional support is provided by the finding that mutations of *Hox* and other patterning genes in humans give rise to malformations of the corresponding CNS regions.[33,34] The next obvious question is, what determines the pattern of *Hox* gene expression? The answer is not fully known. One factor is a combination of signaling by **retinoic acid** from the paraxial mesoderm and by **fibroblast growth factor** (**FGF**) from a signaling center in rhombomere 4.[35,36] Equally interesting is the emerging evidence that changes in the state of the chromatin (such as reversible, covalent modifications of the DNA and its associated histone proteins), spread along the chromosome in the region of the *Hox* cluster to sequentially activate and repress transcription of the *Hox* genes.[37]

Apart from the Hox proteins, two other homeodomain transcription factors, encoded by genes called *ems* (empty spiracles) and *otd* (orthodenticle), also regulate head segment development in *Drosophila*. Their vertebrate homologues, *emx* and *otx*, play major roles in the development of the forebrain. Of particular interest are the functional consequences of mutations in *emx* and *otx*. Mutations in *emx* give rise to gross defects in the structure of the cortex, while defects in *otx* give rise to epilepsy.[38] The distribution of *emx* in developing mouse brain is shown in Figure 25.7. Note that the expression of the gene is restricted to the forebrain region and stops abruptly at its boundary with the midbrain.

Dorsoventral Patterning in the Spinal Cord

The vertebrate nervous system differentiates along the dorsoventral as well as the anteroposterior axis. Along the ventral midline of the developing neural tube lies a band of specialized glial cells, called the **floor plate** (see below). Adjacent, **basal** regions of the neural tube give rise to motoneurons, more dorsal **alar** regions give rise to interneurons, and the region that is the most dorsal forms the **roof plate** and the **neural crest**.

The specializations in ventral territory depend in large part on a signaling protein known as **Sonic hedgehog**.[39] Sonic hedgehog is initially produced and secreted by cells of the notochord, just ventral to the neural tube. Cells at the ventral portion of the neural tube

[25] Lumsden, A., and Krumlauf, R. 1996. *Science* 274: 1109–1115.

[26] Bateson, W. 1894. *Materials for the Study of Variation: Treated with Special Regard to Discontinuity in the Origin of Species*. Macmillan and Co., New York.

[27] Schmitt, S. 2003. *Hist. Philos. Life Sci.* 25: 193–210.

[28] Soshnikova, N., and Duboule, D. 2009. *Epigenetics* 4: 537–540.

[29] Duboule, D. 2007. *Development* 134: 2549–2560.

[30] Capecchi, M. R. 1997. *Cold Spring Harb. Symp. Quant. Biol.* 62: 273–281.

[31] Morrison, A. D. 1998. *BioEssays* 20: 794–797.

[32] Rhinn, M. et al. 2005. *Development* 132: 1261–1272.

[33] Boncinelli, E., Mallamaci, A., and Broccoli, V. 1998. *Adv. Genet.* 38: 1–29.

[34] Mallamaci, A. 2011. *Prog. Brain Res.* 189: 37–64.

[35] Hernandez, R. E. et al. 2004. *Development* 131: 4511–4520.

[36] Moens, C. B., and Prince, V. E. 2002. *Dev. Dyn.* 224: 1–17.

[37] Tschopp, P., and Duboule, D. 2011. *Dev. Biol.* 351: 288–296.

[38] Cipelletti, B. et al. 2002. *Neuroscience* 115: 657–667.

[39] Roelink, H. et al. 1995. *Cell* 81: 445–455.

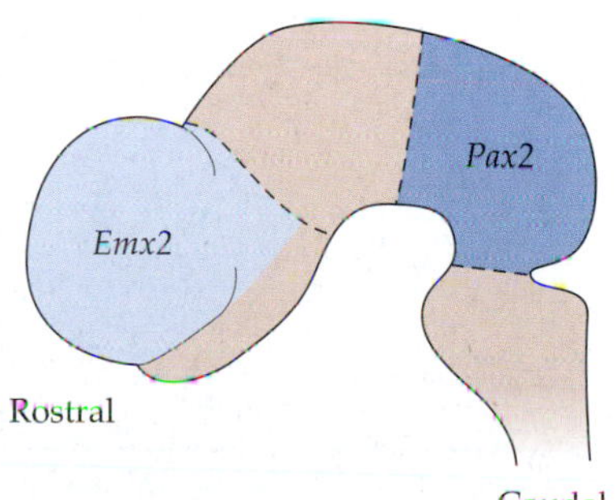

FIGURE 25.7 Expression of Patterning Genes *Emx2* and *Pax2* in Developing Mouse Brain. *Emx2* is present in the region that will give rise to forebrain, while *Pax2* is restricted to hindbrain. The expression of the genes is restricted and stops exactly at the boundary. Defects in the *Emx* gene give rise to gross deformations of the forebrain. (After a figure kindly provided by A. Mallamaci.)

(A)

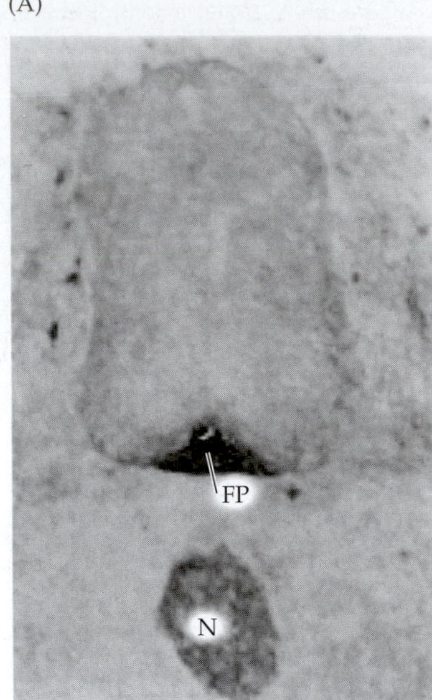

(B)

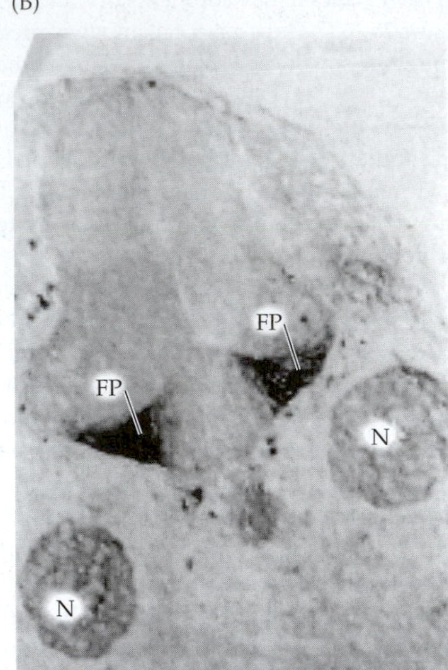

(C)

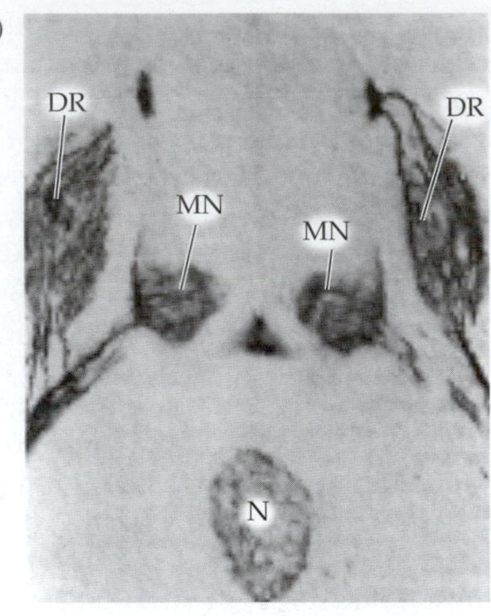

(D)

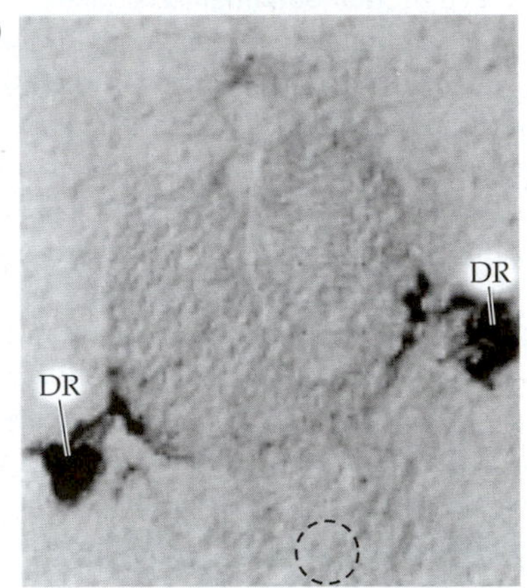

FIGURE 25.8 The Notochord (N) Induces Formation of the Floor Plate (FP) and Motor Neurons (MN) during development of the spinal cord. (A,B) Specific labeling with an antibody that recognizes floor plate cells (FP). (A) Normal chick embryo. (B) Addition of a second notochord (N) induces a second floor plate. (C,D) Specific labeling with an antibody that recognizes floor plate cells, motor neurons, and dorsal root ganglion cells (DR). In absence of notochord (dashed circle), the floor plate and motor neurons are absent, and dorsal roots are displaced ventrally. (After Placzek et al., 1991.)

experience higher levels of Sonic hedgehog protein than those elsewhere and are induced to become floor plate cells, which then make and secrete Sonic hedgehog on their own.[40] As a result, a ventrodorsal gradient of Sonic hedgehog is established in the neural tube with highest concentrations in the ventral regions. Within the neural tube, cells experiencing different levels of Sonic hedgehog protein are induced to express different homeodomain transcription factors, and these in turn specify the differentiation of different classes of cells along the ventrodorsal extent of the spinal cord.[41,42] Thus, an extra notochord produces an extra floor plate (Figure 25.8A and B), and in the absence of a notochord, the floor plate and the motoneurons, which require the highest levels of Sonic hedgehog, fail to develop (Figure 25.8C and D).[43]

In parallel to the establishment of the ventrodorsal gradient of Sonic hedgehog emanating from the floor plate cells, other processes in early development lead to the formation of roof plate cells at the dorsal side of the neural tube. The roof plate generates dorsoventral gradients of BMP and Wnt proteins.[44] The dorsoventral gradient determines the differentiation of dorsal, (i.e., sensory), regions of developing spinal cord. (For detailed information about the Sonic hedgehog, Wnt and BMP signaling pathways see refs.[37,42,45])

The examples of hindbrain segmentation and neural tube patterning illustrate a general principle: different combinations of cues (often secreted signaling molecules) along the anteroposterior and dorsoventral axes of the embryo specify cells to assume specific fates appropriate to their locations (Figure 25.9). Specialized differentiation is achieved through the expression of particular combinations of transcription factors, which determine regional identity throughout the CNS.[46] Hence, any one factor can induce quite different effects, depending on where in the embryo it is expressed. For example, Sonic hedgehog signaling induces the production of motor neurons in the spinal cord, serotonergic and dopaminergic neurons in the hindbrain, and oculomotor neurons in the anterior midbrain.[47,48]

[40] Yamada, T. et al. 1991. *Cell* 64: 635–647.

[41] Wilson, L., and Maden, M. 2005. *Dev. Biol.* 282: 1–13.

[42] Briscoe, J. 2009. *EMBO J.* 28: 457–465.

[43] Placzek, M. et al. 1991. *Development* 113(Suppl. 2): 105–122.

[44] Simon, H., Hornbruch, A., and Lumsden, A. 1995. *Curr. Biol.* 5: 205–214.

[45] Port, F., and Basler, K. 2010. *Traffic* 11: 1265–1271.

[46] Danjo, T. et al. 2011. *J. Neurosci.* 31: 1919–1933.

[47] Ericson, J. et al. 1995. *Cell* 81: 747–756.

[48] Ye, W. et al. 1998. *Cell* 93: 755–766.

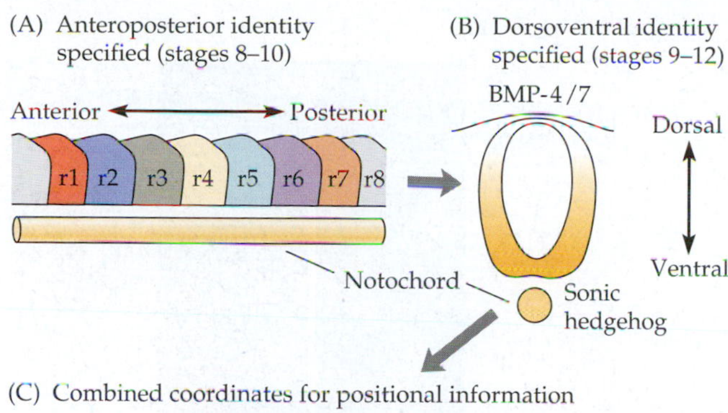

FIGURE 25.9 A Coordinate System of Positional Information in the Vertebrate Hindbrain is established in two steps. (A) First, anteroposterior position is encoded—for example, by *Hox* gene expression. (B) Subsequently, dorsoventral position is encoded by gradients of midline signals, such as Sonic hedgehog and bone morphogenetic protein (BMP)-4/7. (C) The resulting two-dimensional coordinate system of positional information restricts the repertoire of cell fates available to pluripotent precursor cells. (After Simon, Hornbruch, and Lumsden, 1995.)

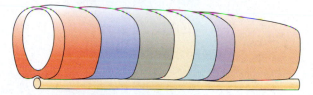

Cell Proliferation

The **cerebral cortex** contains billions of neurons and glial cells, organized into multiple layers as described in Chapter 3. In the following sections we describe important features for the development of the cortex, namely cell **proliferation** and **migration**.[49]

Cell Proliferation in the Ventricular Zone

One technique to determine where and when a cell was born is to introduce an analog of thymidine (such as tritiated thymidine or bromodeoxyuridine—BrdU) transiently into the live embryo.[50,51] The molecule is incorporated into the DNA (not RNA) only of cells that are in the DNA synthesis phase (S phase) of the cell cycle at the time of exposure. Thus, by examining tissues soon after the injection, it is possible to detect the label and to determine where and when cell divisions are taking place; and that information, combined with observations made at later times after the injection (see below), permits one to infer how the labeled cells have migrated.[52,53] Figure 25.10 illustrates the reconstructed sequence of events as neurons and glial cells are generated. The wall of the neural tube is initially composed of a single, rapidly dividing layer of cells; each cell extends a process from the luminal, or **ventricular**, edge to the external, or **pial**, surface (Figure 25.10A). As a cell progresses through the cell cycle, its nucleus moves back and forth through its cytoplasm between the ventricular and pial surfaces, a process known as *interkinetic nuclear migration*. Once the nucleus arrives back at the ventricular surface, the pial connection is lost and the cell divides. The two daughter cells may then go through further cycles of division. Alternatively, one (or both) of the daughter cells may migrate away from the ventricular zone to assume a neuronal or glial fate.

Most cells destined to become neurons are postmitotic, meaning they do not divide again (but see below). Glial cell precursors, on the other hand, can divide even after they have reached their final locations. As more and more postmitotic cells are produced, proliferation slows and the neural tube thickens, assuming a three-layered configuration: an innermost **ventricular zone** (where proliferation continues), an intermediate **mantle zone** containing the cell bodies of the migrating neurons, and a superficial **marginal zone** composed of the elongating axons of the underlying neurons (Figure 25.10B). This three-layered structure persists in the spinal cord and medulla. In other regions, such as the cerebrum, neurons migrate toward the pial surface into the marginal zone to form a **cortical plate**, which matures into the adult cortex (Figure 25.10C and D, and see below).

Cell Proliferation via Radial Glia

Another mechanism by which neurons are generated in the developing cortex involves the participation of glial cells. As the three-layered configuration of the neural tube takes shape,

[49] Clowry, G., Molnár, Z., and Rakic, P. 2010. *J. Anat.* 217: 276–288.

[50] Angevine, J. B., Jr., and Sidman, R. L. 1961. *Nature* 192: 766–768.

[51] Puzzolo, E., and Mallamaci, A. 2010. *Neural Dev.* 5: 5–8.

[52] Rakic, P. 1974. *Science* 183: 425–427.

[53] Luskin, M. B., and Shatz, C. J. 1985. *J. Comp. Neurol.* 242: 611–631.

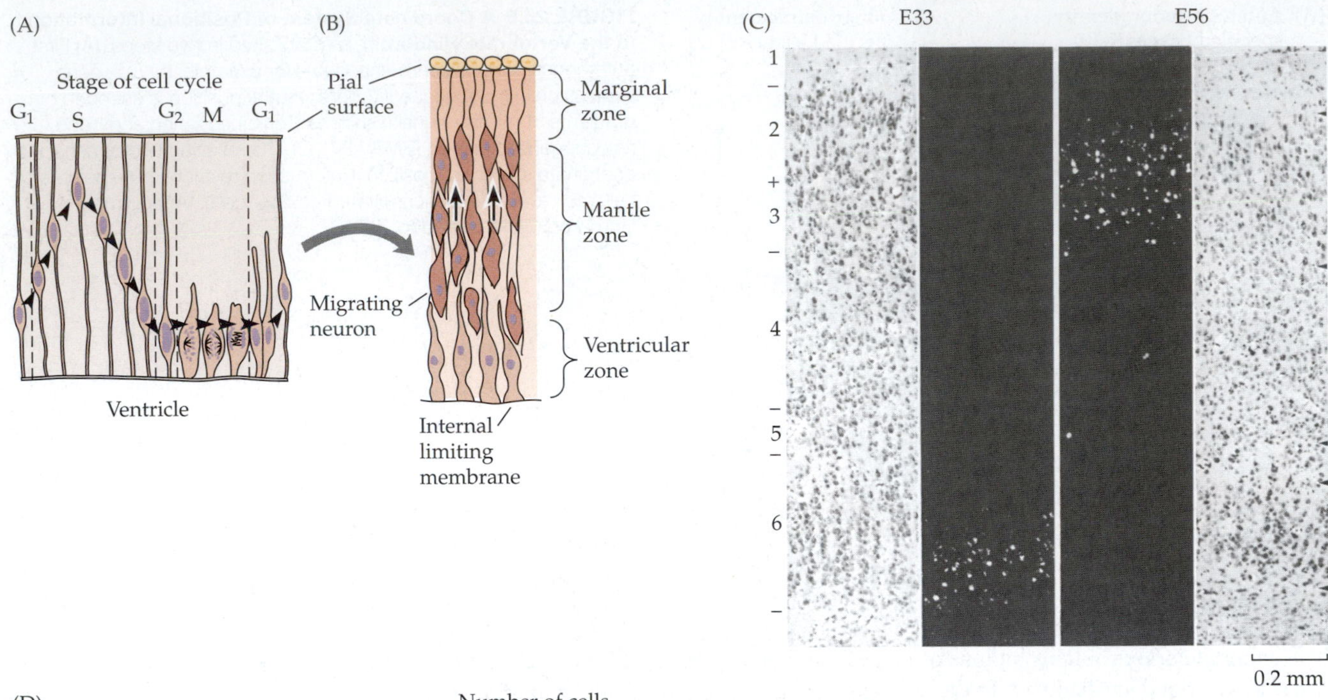

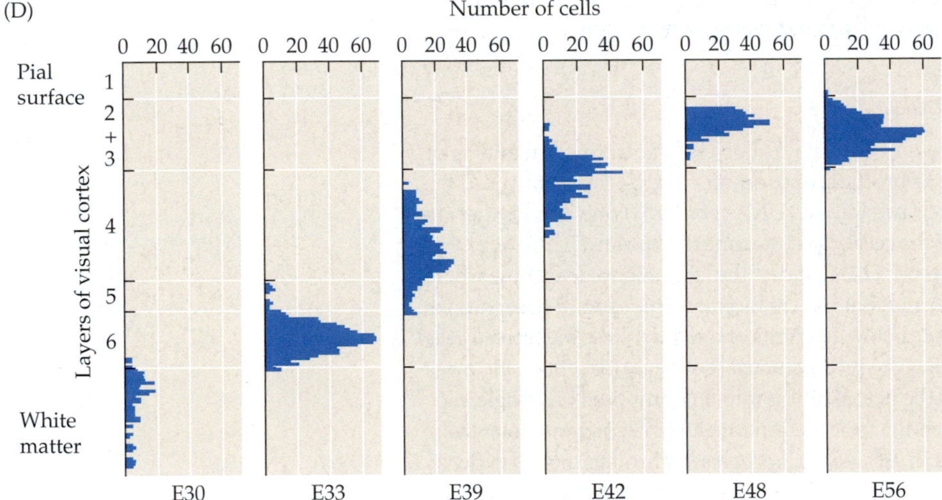

FIGURE 25.10 Differentiation of the Walls of the Neural Tube and Neurogenesis of Cat Cortex. (A) The position of the nuclei in cells in the primitive neural tube varies during the cell cycle (G_1 = first gap phase of cell cycle; S = DNA synthesis phase; G_2 = second gap phase; M = mitosis). (B) Cells become postmitotic, migrate away from the ventricular zone, and form the mantle zone. Their processes make up the marginal zone. (C) Autoradiographs of sections of the adult visual cortices of cats that had been injected with [^{3}H]-thymidine on embryonic day 33 (E33) or 56 (E56). Bright-field micrographs of the same sections stained with cresyl violet indicate that heavily labeled cells are located in layer 6 after the E33 injection and in layers 2 and 3 after injection on E56. (D) Histograms showing the distribution of cells labeled on various days between E30 and E56 illustrate the inside-out pattern of neurogenesis in the visual cortex. (After Luskin and Shatz, 1985a; micrograph kindly provided by M. B. Luskin.)

the majority of the ventricular zone cells lose the processes that originally reached to the pial surface. But a sub-class of cells known as radial glia (identified by their expression of specific molecular markers, see Chapter 10) maintain processes that extend through the thickness of developing cortex from ventricular to marginal zones. Kriegstein and his colleagues have shown that radial glial cells produce neuronal precursors.[54] Thus, cells destined to be neurons bud off from immature radial glia (Figure 25.11). (Radial glial cells should not be confused with the cytoplasmic processes of neuronal precursors shown in Figure

[54]Noctor, S. C. et al. 2001. *Nature* 409: 714–720.

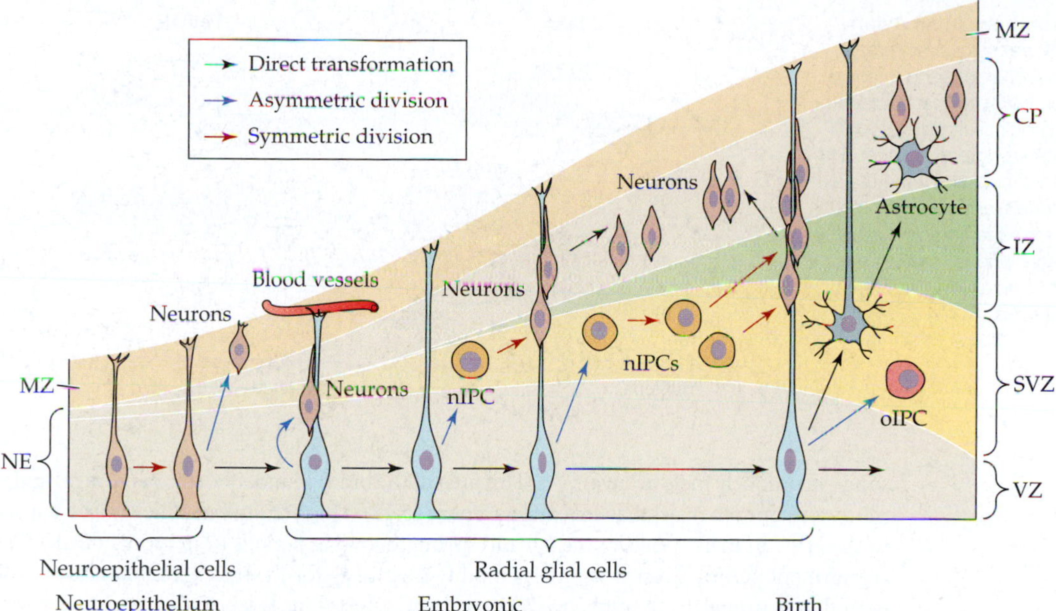

FIGURE 25.11 Production of Cortical Cells by Radial Glia. Following the early period of development (left) during which neurons arise primarily from neuroepithelial cells undergoing interkinetic nuclear migration (brown cells, left), additional cells are produced by radial glia (light blue cells, center). Clonal analysis reveals that some neurons arise as the direct descendants of radial glia. In other cases, the immediate progeny of the radial glia acts as a neuron intermediate progenitor cell (nIPC), which undergoes additional rounds of division. In mammals, most of the radial glia eventually differentiates into astrocytes (right). CP = cortical plate; IZ = intermediate zone; MZ = marginal zone; oIPC, oligodendrocyte intermediate progenitor cell; SVZ = subventricular zone; VZ = ventricular zone; NE = neuroepithelium. (After Kriegstein and Alvarez-Buylia, 2009.)

25.10). Cells that resemble radial glia also give rise to neurons in human cerebral cortices obtained from embryos postmortem.[55]

In mature mammalian CNS, radial glial cells (with certain exceptions, such those in retina and cerebellum) become transformed into astrocytes. In non-mammalian vertebrates, however, radial glial cells remain abundant into adulthood. This difference correlates with dramatic differences in the capacity for adult cortical neurogenesis (see below) and regeneration between mammalian and non-mammalian vertebrates (see Chapter 26).[56]

When Do Neurons Stop Dividing? Adult Neurogenesis

Unlike glial cells, the vast majority of neurons formed during development never divide again in the mature animal. Nevertheless, pluripotent stem cells can give rise to neurons,[57] and new neurons are continuously added to the hippocampus, olfactory bulb, and olfactory mucosa in some species of adult mammals.[58] It appears that these new cells arise from so-called self-renewing **neural stem cells** that have been isolated not only from the walls of the ventricles but also from the hippocampus of the adult brain.[59,60] Such cells can generate neurons, oligodendrocytes, and astrocytes (and even vascular endothelial cells[61]). Neurons that are continuously added to the olfactory bulb in vivo arise from slowly dividing stem cells in the innermost ependymal layer of the walls of the lateral ventricles (the vestige of the original ventricular zone).[62] After each division, one of the resulting cells enters the subependymal zone to become a **progenitor cell**. The progenitor cell divides, forming immature neurons that migrate rostrally into the bulb, where they differentiate as interneurons and integrate into the existing circuitry. Hopes have been expressed that stem cells could eventually be harvested to generate neurons or glia for treatment of nervous system diseases or spinal cord lesions.

In songbirds, Nottebohm and his colleagues have demonstrated that proliferation of important sets of neurons occurs throughout adult life. In canaries, a brain region known as the high vocal center (HVC) nucleus plays a crucial role in the acquisition and retention of

[55]Hansen, D. V. et al. 2010. *Nature* 464: 554–561.

[56]Campbell, K., and Gotz, M. 2002. *Trends Neurosci.* 25: 235–238.

[57]Okita, K., Ichisaka, T., and Yamanaka, S. 2007. *Nature* 448: 313–317.

[58]Kornack, D. R., and Rakic, P. 1999. *Proc. Natl. Acad. Sci. USA* 96: 5768–5773.

[59]Reynolds, B. A., and Weiss, S. 1996. *Dev. Biol.* 175: 1–13.

[60]Kelsch, W., Sim, S., and Lois, C. 2010. *Annu. Rev. Neurosci.* 33: 131–149.

[61]Wurmser, A. E. et al. 2004. *Nature* 430: 350–356.

[62]Luskin, M. B. 1993. *Neuron* 11: 173–189.

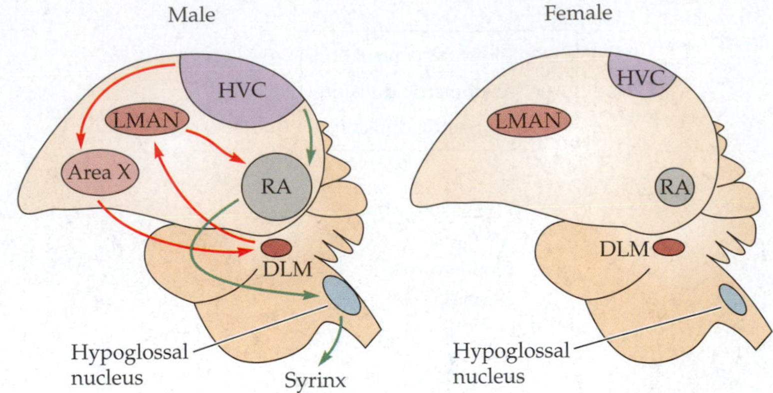

FIGURE 25.12 Sexual Dimorphism in an Avian Brain. Schematic diagram of the major brain areas and pathways involved in production of song in songbirds. The higher vocal center (HVC), robust nucleus of the archistriatum (RA), and hypoglossal nucleus form the posterior, vocal motor pathway (green arrows). The HVC, area X, medial dorsolateral nucleus of the thalamus (DLM), and lateral magnocellular nucleus of the anterior neostriatum (LMAN) form the anterior pathway (red arrows). HVC, hypoglossal nucleus, and RA are significantly larger in male birds; area X has not been observed in brains of female finches.

song—a uniquely male behavior.[63,64] This area of the brain is under hormonal control and is more developed in males than in females (Figure 25.12). The recruitment of newborn neurons to the HVC in males peaks in the fall and spring following periods of neuronal death. This recruitment period is just when males modify their song for the next breeding season. The period of neuronal death coincides with a drop in testosterone levels, and recruitment peaks when testosterone levels rise. Administration of testosterone to females causes an increase in the recruitment of new HVC neurons and induces the females to sing. The HVC nucleus and other structures associated with song production become enlarged in such androgenized females. New HVC neurons in males and females receive appropriate synaptic input and project axons to the proper targets. These remarkable observations indicate not only that new neurons arise in the adult songbird brain, but also that they can be assimilated into complex circuits so as to provide the substrate for the remodeling of a behavior as intricate as birdsong.

In light of the results showing that neurons are added to adult mammalian hippocampus,[65] a question that arises is whether new neurons are also added to the adult human cerebral cortex. Evidence that new cortical neurons *do* arise in rodents and even in monkeys gave rise to widely publicized speculation about the possible functions of newly born neurons in adult human cerebral cortex, but the basic observations were subsequently called into question.[66,67] An innovative and decisive approach to this problem was undertaken in two sets of experiments.[68,69] In the first set of experiments (Figure 25.13), Bhardwaj and colleagues made use of the fact that ^{14}C incorporated into the DNA of dividing cells during replication persists for the life of the cells. During the nuclear weapons tests of 1955–1963, ^{14}C in the environment increased sharply and was incorporated into the food supply. Consequently, one would expect to find elevated levels of ^{14}C in cortical neurons of people born during that time. Measurement of ^{14}C cortical labeling made by accelerator mass spectroscopy in human brains obtained postmortem showed that this was the case. But after the test ban treaty of 1963, levels of ^{14}C in the environment fell sharply. (Note that this rapid decline reflects dilution through the biogeochemical carbon cycle, not radioactive decay—the radioactive half life of ^{14}C is > 5000 years.) Thus, if new (unlabeled) cortical cells were to be produced in adulthood, one would expect the levels of cortical ^{14}C to decline year by year in people born during 1955–1963. Indeed, this was the case for glial cells and for the endothelial cells in blood vessels, as a result of the replacement of original ^{14}C-labeled cells by cells with progressively lower ^{14}C levels. No such decline was found in the neurons, however, as levels of ^{14}C in the cortical neurons of adults born before 1963 remained high throughout their lives. This showed that few if any new neurons had been produced in the adult cortex.

A second series of quite different experiments produced a similar result. Terminal patients afflicted with localized carcinomata, with no spread to the brain and no medication that could affect the cortex, volunteered to have bromodeoxyuridine (BrdU) injected into the ventricle. As mentioned above, BrdU is incorporated into the DNA of a cell while it is replicating its DNA in preparation for mitosis, but not in mature, postmitotic cells, and can be detected by antibody labeling. On postmortem examination, it was found that glial cells and endothelial cells of the volunteers were labeled, but neurons were not. Thus, rather few new neurons are produced in adult human cortex, unlike other regions such as hippocampus or in the cortex of a rodent.

[63] Nottebohm, F. 1989. *Sci. Am.* 260: 74–79.

[64] Adar, E., Nottebohm, F., and Barnea, A. 2008. *J. Neurosci.* 28: 5394–5400.

[65] Lledo, P. M., Alonso, M., and Grubb, M. S. 2006. *Nat. Rev. Neurosci.* 7: 179–193.

[66] Rakic, P. 2002. *J. Neurosci.* 22: 614–618.

[67] Vessal, M., and Darian-Smith, C. 2010. *J. Neurosci.* 30: 8613–8623.

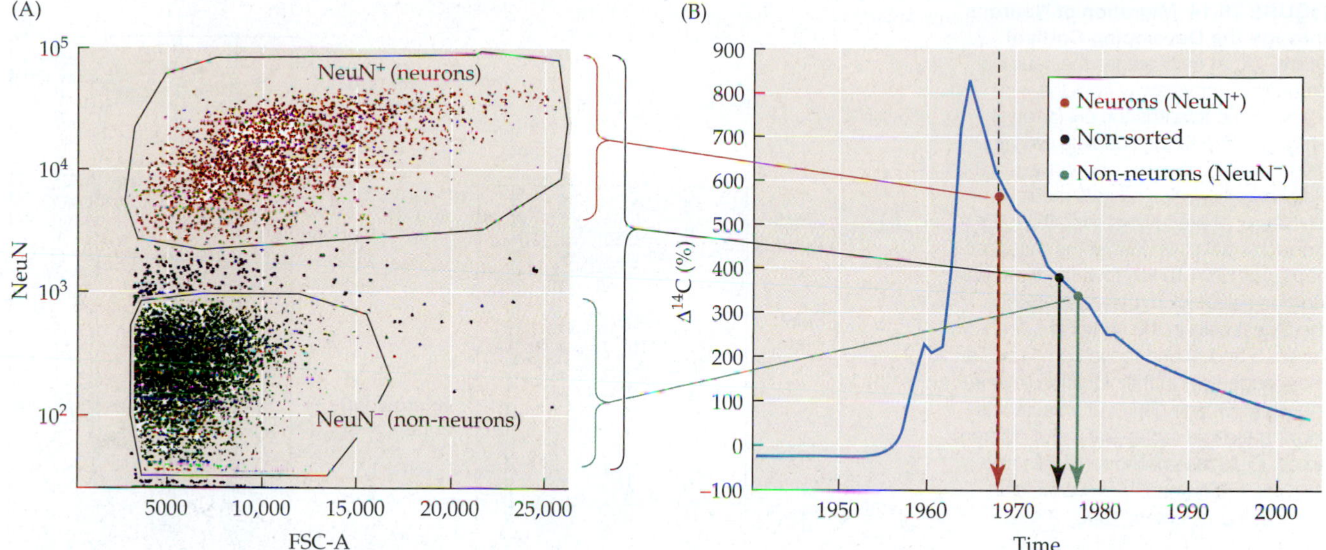

FIGURE 25.13 Failure of Neurons to Divide in Adult Human Cerebral Cortex. ^{14}C radioactivity was measured in cell nuclei of neurons and non-neuronal cells in human cortex in relation to the time of birth. (A) Neuronal (NeuN-positive, red) and nonneuronal (NeuN-negative, green) cell nuclei from the adult human cerebral neocortex were identified, separated, and isolated by flow cytometry. (B) In 1955–1963 the levels of ^{14}C in the atmosphere increased dramatically as a result of nuclear weapons tests and subsequently declined (blue line). One can infer the time of birth of cell populations by first relating the level of ^{14}C in their DNA to that in the atmosphere (guidelines from A to B), and then reading their age off the *x* axis (vertical arrows). The age of the individual is given by the dashed line. The average age of all cells in the prefrontal cortex is less than the individual (black circle and arrow), indicating cell turnover. Dating of nonneuronal cells (green circle and arrow) indicates that they are younger, whereas neurons (red circle and arrow) are approximately as old as the individual. These results and others (see text) indicate that glial and endothelial cells continue to divide while cortical neurons did not. FSC-A = forward scatter channel-A. (After Bhardwaj et al., 2006.)

Migration

Migration of Cortical Neurons

A cortical neuronal precursor that has undergone its final division migrates through the thickness of the developing CNS to reach its final destination.[70] As mentioned previously, migration through the cortex can be studied by applying a pulse of ^{3}H thymidine or BrdU on a particular day during development to mark cells undergoing DNA synthesis at that time. Postmitotic cells are not labeled, and in cells that continue to divide after the pulse, the label will be diluted during subsequent rounds of DNA synthesis and cell division. However, any cells that undergo their final round of DNA synthesis and then become born, so to speak, as postmitotic neurons during the pulse will remain heavily labeled. Thus, the fate of neurons born on a particular day can be determined by exposing an embryo to a pulse of label on that day, and then determining the locations of the labeled cells at a later date by autoradiography or immunohistochemistry.[50]

Such experiments reveal a systematic relationship between the time a neuron is born and its final position in the developing cerebral cortex; characteristically, development proceeds in an inside-out fashion.[52] Neurons of the deepest cortical layers are born first. Neurons that occupy layers that are more superficial are born later and migrate past the cells of the deeper layers to assume their final positions within the cortex. These neuronal precursors migrate through the cortex along **radial glial cells** (see Figure 25.10). As the walls of the neural tube thicken, through the continued division of cells in the ventricular layer and the accumulation of neurons in the mantle zone and cortical plate, the processes of the radial glial cells become elongated. From light- and electron-microscopic studies of the development of the cerebrum and cerebellum, Rakic and his colleagues[71] deduced that neurons move along this scaffolding of radial glial cells to reach their appropriate positions in the cortex (Figure 25.14). Observations in mutant mice and on cells maintained in culture[72] support this conclusion. The proteins mediating such neuronal migration include a neural glycoprotein, known as astrotactin,[73,74] and extracellular matrix adhesion molecules (discussed later). The cell biological mechanisms underlying cell migration

[68] Bhardwaj, R. D. et al. 2006. *Proc. Natl. Acad. Sci. USA* 103: 12564–12568.

[69] Rakic, P. 2006. *Science* 313: 928–929.

[70] Marín, O., and Rubenstein, J. L. 2003. *Annu. Rev. Neurosci*. 26: 441–483.

[71] Rakic, P. 1981. *Trends Neurosci*. 4: 184–187.

[72] Hatten, M. E., Liem, R. K. H., and Mason, C. A. 1986. *J. Neurosci.* 6: 2676–2683.

[73] Zheng, C., Heintz, N., and Hatten, M. E. 1996. *Science* 272: 417–419.

[74] Wilson, P. M. et al. 2010. *J. Neurosci.* 30: 8529–8540.

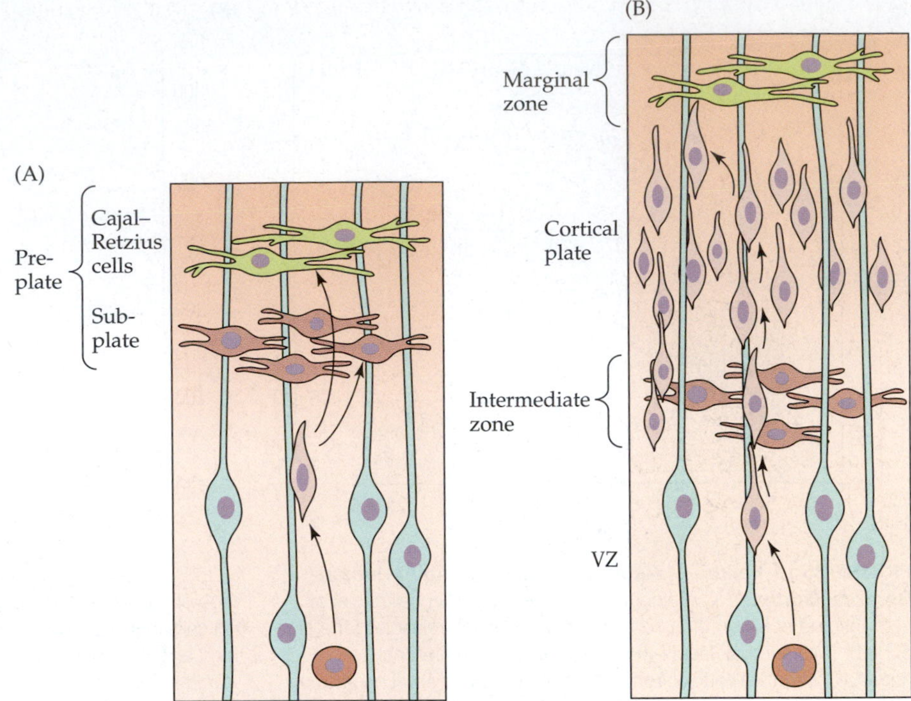

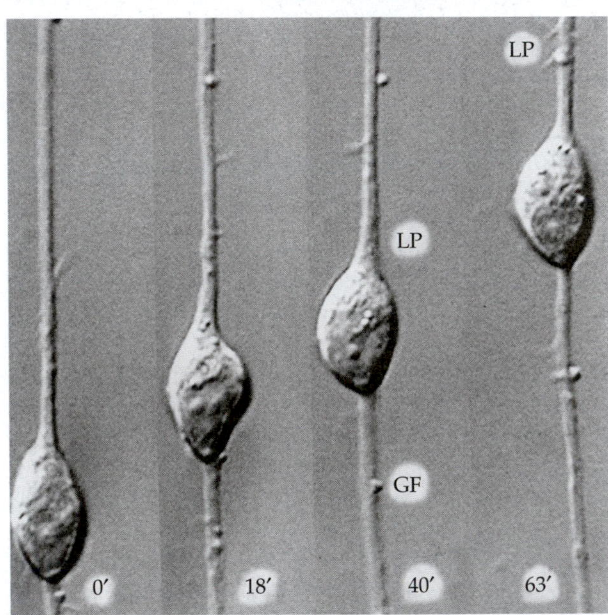

FIGURE 25.14 Migration of Neurons through the Developing Cortical Plate. (A) At early stages the neuronal precursors move away from the ventricular zone towards the pia (arrows). They give rise to the subplate cells (in the intermediate zone) and to the Cajal–Retzius cells, which secrete reelin. (B) Neuronal precursors migrate along the radial gial cells past all the cells that had been born earlier, towards the Cajal-Retzius cells, where they stop. Hence, the oldest cells in the cortex are deep (layers 5 and 6) and those born later are more superficial in their location (layers 2 and 3). (C) The neuronal precursor is moving along a radial glial cell at different times. Cells resembling radial glia also give rise to neurons. VZ = ventricular zone. (C from Hatten, 1990.)

and axon outgrowth (see below) are similar.[75] Both depend primarily on the actin-based microfilament cytoskeleton.

Neurons can also migrate without crawling along radial glial cells. For example, the formation of the cerebral cortex involves tangential migration of large numbers of neurons, (especially GABA-secreting stellate cells), laterally from one cortical region into neighboring areas during development.[76,77]

Another remarkable example of migration without radial glia is provided by a population of neurons expressing gonadotropin-releasing hormone (GnRH). These precursors migrate from the periphery *into* the CNS (see also Chapter 17).[78] The GnRH cells travel from the olfactory pit (an ectodermal derivative, or **placode**, that gives rise to the nasal epithelium), into the hypothalamus along a previously established axon tract. Patients suffering from

[75] Marin, O. et al. 2010. *Cold Spring Harb. Perspect. Biol.* 2: a001834.

[76] O'Rourke, N. A. et al. 1995. *Development* 121: 2165–2176.

[77] Govek, E. E., Hatten, M. E., and Van Aelst, L. 2010. *Dev. Neurobiol.* Dec 7.

[78] Wray, S., Grant, P., and Gainer, H. 1989. *Proc. Natl. Acad. Sci. USA* 86: 8132–8136.

a condition known as Kallmann's syndrome, (a genetic disorder of the olfactory placode) display a defective sense of smell and also sterility.[79,80] The cause is a failure of migration of olfactory precursors and GnRH cells to their destinations within the hypothalamus.

Genetic Abnormalities of Cortical Layers in Reeler *Mice*

The mutant mouse *reeler* (so called because of the uneven gait it displays) demonstrates how a neuron's final position can be determined by an extracellular signal. In the developing cortex of *reeler* mice, neurons fail to migrate past one another.[81] Therefore, their relative positions in the adult are inverted: neurons born at early times end up in the most superficial layers, whereas those born later end up in deeper layers. In spite of their aberrant positions, the misplaced neurons acquire the morphological appearance and make connections appropriate to their time of birth. Thus, the morphology of cortical neurons and the nature of their synaptic interactions are determined at the time the neuron is born and can be expressed independently of position. The *reeler* phenotype results from multiple consequences of the mutation in the cortex and elsewhere in the brain. The product of the *reeler* gene is a large extracellular matrix glycoprotein called reelin.[82,83] It is expressed in **Cajal–Retzius cells** that are situated superficially in the marginal zone of the cortex (see Figure 25.14).[84] There is evidence that reelin produced by these superficial cells provides a signal that stops migration ("All passengers, step down from the radial glia here, please"); at the same time, it is clear that reelin also has other functions.[85]

Determination of Cell Phenotype

The following paragraphs deal with the sequence of steps that establish the properties of a cell, that is, its **phenotype**, once it has reached its final destination. What determines whether a precursor cell will become a neuron or a glial cell? For a neuron, what factors determine its transmitter and the receptors on its surface? It will be shown that inductive interactions between cells (extrinsic factors), as well as the lineage from which the cells arise (intrinsic factors) are important for determining the fates of neurons.

Lineage of Neurons and Glial Cells

To determine which precursor cells give rise to glial cells or different types of neurons, experiments were done in which a virus was injected into the eye of a newborn rat. After the animal had reached adulthood, the retina contained clusters of stained cells that frequently included both glial cells and several types of neurons.[86] Thus, single cells in the retina at the time of birth had subsequently given rise to both neurons and glial cells (see Figure 25.2), and the rodent retina shows no evidence for specific lineages that give rise to particular types of neurons.

Extensive work shows that the various cell types in the highly regular insect eye also arise by indeterminate lineages (see Chapter 18).[87,88] Rather, progenitor cells respond to external cues. Both the cues present in the environment and the intrinsic responses of the cells to such cues change with time, generating a succession of different cell types.[89,90]

By contrast, when similar experiments were made in cerebral cortex, clones containing both neurons and glial cells were rare, suggesting that at the time of viral infection separate populations of progenitors for neurons and glial cells were already present in the cortical ventricular zone. Moreover, clones tended to contain exclusively pyramidal cells or non-pyramidal cells, suggesting that these lineages diverge early in neurogenesis.[91] In a general sense, this result also parallels studies of the ventral nerve cord (as opposed to the eye) in simpler systems, such as leeches and insects, in which individually identified neurons have been shown to arise by highly stereotyped lineages.[92–94]

Control of Transmitter Choice in the Peripheral Nervous System

In the peripheral nervous system, cell lineage, neuronal birthday, and local cues all affect the fate of sympathetic neurons, including the decision to secrete a particular transmitter, for example acetylcholine (ACh) or norepinephrine. Such questions have been studied in chick and quail by Le Douarin, Weston, and others.[95–97]

[79] Wray, S. 2010. *J. Neuroendocrinol.* 22: 743–753.

[80] Sarfati, J., Dodé, C., and Young, J. 2010. *Front. Horm. Res.* 39: 121–132.

[81] Caviness, V. S., Jr. 1982. *Dev. Brain Res.* 4: 293–302.

[82] D'Arcangelo, G. et al. 1997. *J. Neurosci.* 17: 23–31.

[83] Förster, E. et al. 2010. *Eur. J. Neurosci.* 31: 1511–1518.

[84] Meyer, G. 2010. *J. Anat.* 217: 334–343.

[85] Zhao, S., and Frotscher, M. 2010. *Neuroscientist* 16: 421–434.

[86] Turner, D. L., and Cepko, C. L. 1987. *Nature* 328: 131–136.

[87] Ready, D. F. et al. 1976. *Dev. Biol.* 53: 217–240.

[88] Sanes, J. R., and Zipursky, S. L. 2010. *Neuron* 66: 15–36.

[89] Cepko, C. L. et al. 1996. *Proc. Natl. Acad. Sci. USA* 93: 589–595.

[90] Randlett, O., Norden, C., and Harris, W. A. 2011. *Dev. Neurobiol.* 71: 567–583.

[91] McConnell, S. K. 1995. *Neuron* 15: 761–768.

[92] Weisblat, D. A., and Shankland, M. 1985. *Philos. Trans. R. Soc. Lond., B, Biol. Sci.* 312: 39–56.

[93] Kramer, A. P., and Weisblat, D. A. 1985. *J. Neurosci.* 5: 388–407.

[94] Doe, C. Q., Kuwada, J. Y., and Goodman, C. S. 1985. *Philos. Trans. R. Soc. Lond., B, Biol. Sci.* 312: 67–81.

[95] Weston, J. 1970. *Adv. Morphogenesis* 8: 41–114.

[96] Dupin, E., Ziller, C., and Le Douarin, N. M. 1998. *Curr. Top. Dev. Biol.* 36: 1–35.

[97] Teillet, M. A., Ziller, C., and Le Douarin, N. M. 2008. *Methods Mol. Biol.* 461: 337–350.

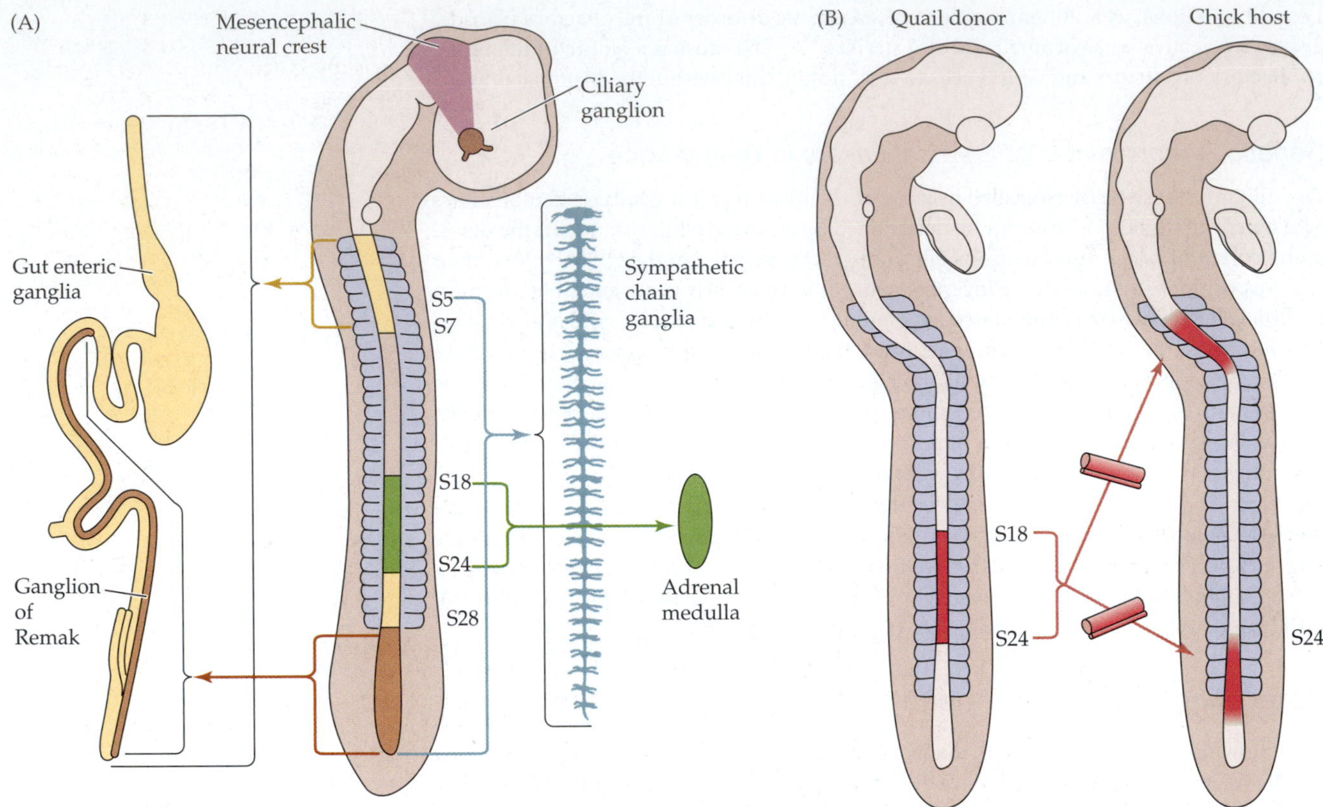

FIGURE 25.15 The Fate of a Neural Crest Cell is determined by environmental cues. (A) Neural crest cells give rise to a variety of peripheral ganglia. The ciliary ganglion is formed by migrating cells from the mesencephalic neural crest (beige triangle). The ganglion of Remak and the enteric ganglia of the gut are formed by cells from the vagal (somites 1–7) and lumbosacral (caudal to S28) regions of the neural crest. The ganglia of the sympathetic chain are derived from all regions of the neural crest caudal to S5. The adrenal medulla is populated by crest cells from S18–S24. (B) If crest cells from S18–S24, which are destined to form the adrenal medulla, are transplanted from a quail donor to the vagal or lumbosacral region of a host chick embryo, they will adopt the fate appropriate to their new location and populate the ganglion of Remak and the enteric ganglia of the gut. (After Le Douarin, 1986.)

Neural crest cells at different positions along the neuraxis give rise to different cell types of the peripheral nervous system (Figure 25.15A). To investigate whether the phenotype of cells derived from the neural crest was fixed early in development or could be altered at a later date, Le Douarin and her colleagues transplanted cells from quail embryos into host chick embryos (Figure 25.15B). The transplanted cells could then be recognized by cytological differences between quail and chick cells. When donor cells from one region of the neural crest were transplanted to a different region of a host embryo the donor cells assumed the fate appropriate to their new position in the host. For example, the normal destiny of a cell situated in segment 12 is to become a sympathetic ganglion cell and secrete norepinephrine as its transmitter. When transplanted to a region of segments 18–24, however, cells from segment 12 became chromaffin cells and secreted epinephrine.

Changes in the choice of neurotransmitter also occur during normal development. For example, sympathetic neurons that innervate sweat glands initially synthesize norepinephrine, but during the second and third weeks of life they are induced to synthesize acetylcholine by factors associated with their target.[98] Sympathetic ganglion cells in culture have been used to explore the mechanism by which neurotransmitters are switched. Neurons dissociated from the superior cervical ganglia of newborn rats and grown in culture in the absence of other cell types, contain tyrosine hydroxylase and synthesize catecholamines.[99] However, neurons that are grown together with heart muscle cells or sweat glands cease synthesizing catecholamines and begin synthesizing choline acetyltransferase and acetylcholine.

[98] Francis, N. J., and Landis, S. C. 1999. *Annu. Rev. Neurosci.* 22: 541–566.

[99] Mains, R. E., and Patterson, P. H. 1973. *J. Cell Biol.* 59: 329–345.

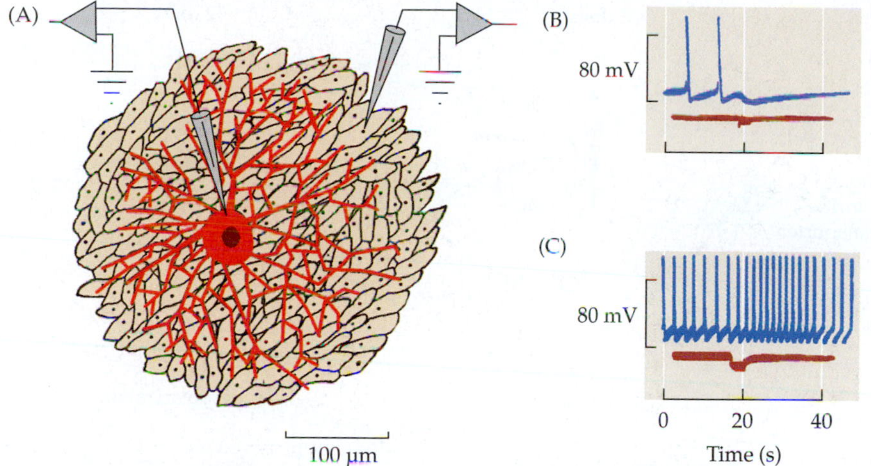

FIGURE 25.16 Single Neurons from Sympathetic Ganglia Can Release Both Acetylcholine (ACh) and Norepinephrine at synapses on heart cells in culture. (A) A microculture containing a single sympathetic neuron is grown on an island of cardiac muscle cells. (B) A brief train of impulses in the neuron (10 Hz, deflection of lower trace) produced inhibition of spontaneous myocyte activity due to release of ACh (upper trace). (C) Addition of atropine (10^{-7} M) blocked the inhibitory cholinergic response, revealing an excitatory effect, which is due to the release of norepinephrine. (After Furshpan et al., 1976.)

To establish unequivocally that this change occurs in individual cells, single neurons were cultured on micro-islands of heart cells (Figure 25.16).[100] The neuron rapidly extended neurites and established synaptic contact with the heart cells. Initially, these synapses were purely adrenergic, but over several days the cell began to release both ACh and norepinephrine. Finally, transmission became purely cholinergic. A factor that induces cholinergic differentiation of sympathetic neurons was identified from heart-conditioned medium.[101] The active molecule proved to be *l*eukemia *i*nhibitory *f*actor (LIF), a protein that had been characterized previously on the basis of its ability to induce differentiation of cells in the immune system. There is evidence that LIF, together with related molecules, mediates the switch in transmitter from norephinephrine to acetylcholine in the sympathetic neurons that innervate sweat glands in vivo.[102]

Changes in Receptors during Development

In addition to switching transmitters, cells may also change their expression of receptors during development. One example (already mentioned in Chapter 5) is provided by the nicotinic ACh receptors (made up of 5 subunits, ααβγδ) that are distributed over the surface of immature muscle fibers. Arrival of the motor nerve induces clustering of ACh receptors at the motor end plate and a change in their subunit composition: the γ-subunit is replaced by an ε. This switch to a new protein entails a change in transcription induced by arrival of the motor nerve (see also Chapter 27).

Another way of bringing about a developmental change in receptor function is a process called **RNA editing**. This phenomenon is exemplified by the class of excitatory glutamate receptors known as α-amino-3-hydroxy-5-methyl-4-isoxazole propionic acid (AMPA) receptors (see Chapter 11). There are four AMPA receptor genes (GluA1 through GluA4), each of which codes for an asparagine residue at a critical site in the loop that contributes to the ion channel. When AMPA receptors are assembled entirely from GluA subunits encoded by unediteds mRNAs, they show significant calcium permeability.[103] However, beginning during embryonic development and continuing into adulthood, an **RNA editing enzyme**, called adenosine deaminase (ADA) on RNA (ADAR2), modifies the GluA2 mRNA so that it is translated with an arginine residue instead of asparagine at the critical site.[104,105] Moreover, the presence of even one such asparagine-to-arginine-edited GluA subunit drastically reduces the calcium permeability of the receptor complex.[102] Curiously, it is not clear that the unedited version of the receptor is required during embryogenesis—mice that are genetically engineered to express a pre-edited version of GluR2 subunit developed with no obvious defects.[106] On the other hand, editing the GluA2 to encode a calcium-impermeant channel is essential, as deficient editing of GLuA2 is associated with postnatal lethality, early-onset epilepsy, and motoneuron death in amyotrophic lateral sclerosis (ALS).[107,108]

Changes in transmitter action also occur during development without changes in receptors or transmitters. Cherubini and his colleagues[109,110] have shown that at early postnatal stages GABA secreted by hippocampal interneurons acts as an excitatory transmitter. Only at about

[100] Furshpan, E. J. et al. 1976. *Proc. Natl. Acad. Sci. USA* 73: 4225–4229.

[101] Yamamori, T. et al. 1989. *Science* 246: 1412–1416.

[102] Stanke, M. et al. 2006. *Development* 133: 141–150.

[103] Hollmann, M., Hartley, M., and Heinemann, S. 1991. *Science* 252: 851–853.

[104] Seeburg, P. H. et al. 2001. *Brain Res.* 907: 233–243.

[105] Rueter, S. M. et al. 1995. *Science* 267: 1491–1494.

[106] Kask, K. et al. 1998. *Proc. Natl. Acad. Sci. USA* 95: 13777–13782.

[107] Brusa, R. et al. 1995. *Science* 270: 1677–1680.

[108] Kwak, S., and Kawahara, Y. 2005. *J. Mol. Med.* 83: 110–120.

[109] Cherubini, E., Gaiarsa, J. L., and Ben-Ari, Y. 1991. *Trends Neurosci.* 14: 515–519.

[110] Lagostena, L. et al. 2010. *J. Neurosci.* 30: 885–893.

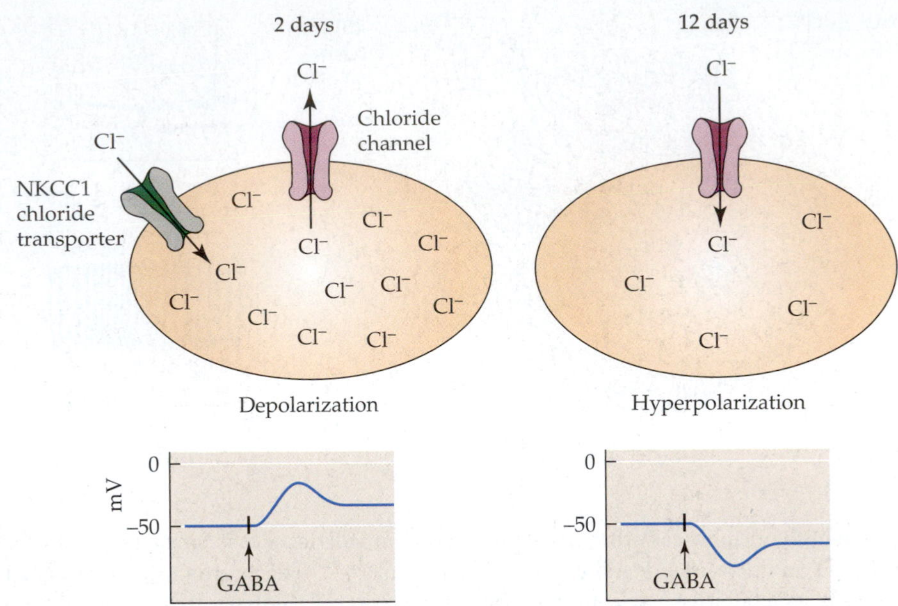

FIGURE 25.17 GABA as a Depolarizing Transmitter during Development. Two days after birth, CA3 neurons in the rat hippocampus are depolarized by GABA and by inhibitory synaptic inputs. At early stages, the opening of chloride channels by γ-aminobutyric acid (GABA) allows the negatively charged ion to exit. By 12 days, the effect of GABA and inhibitory inputs have reversed and, as in the adult, give rise to hyperpolarization (inward movement of Cl^-) at the normal resting potential. The change to hyperpolarization is caused by a decrease in intracellular chloride concentration. At 2 days, the concentration is high owing to the activity of an inward chloride transporter (NKCC1), which virtually disappears by day 12. (After Cherubini, Gaiarsa, and Ben-Ari, 1991.)

12 days in rats does GABA cause hyperpolarization and inhibition as in the adult (Figure 25.17). In this case, neither the receptors nor the transmitter change. Instead, the critical difference is that the intracellular chloride level is higher in immature postsynaptic neurons than in mature neurons, owing to the presence of an *inwardly* directed chloride transporter NKCC1 in the immature neurons (see Chapter 9).[111] With high intracellular Cl^-, opening GABA-gated channels leads to outward movement of Cl^- and depolarization of the cell. Later in development, NKCC1 in these neurons is replaced by an *outward* chloride transporter, KCC2. As a result, intracellular Cl^- become lower inside the cell and the hyperpolarizing response to GABA appears. There is evidence that excitation and firing produced by GABA plays a part in the formation of connections while circuits are being established.[112]

Axon Outgrowth and Growth Cone Navigation

Growth Cones, Axon Elongation, and the Role of Actin

Ramón y Cajal was the first to recognize the **growth cone** as the region of an axon responsible for navigation and elongation toward a target (Figure 25.18). Growth cones are situated at the tips of axons; they extend and retract broad membranous sheets, called lamellipodia, and slender, spike-like protrusions, termed filopodia, for distances of tens of micrometers, as if sampling the substrate in every direction.[113–115] Filopodia may adhere to the substrate, in which case the retraction serves to pull the growth cone in that direction. Dynamics of the actin cytoskeleton (i.e., polymerization, depolymerization, and interactions with non-muscle myosins) are responsible for producing protrusion and retraction of lamellipodia and filopodia as well as for the forward movement of the body of the growth cone (Figure 25.19). Both lamellipodia and filopodia are rich in filamentous actin; agents that inhibit actin polymerization, such as the fungal toxin cytochalasin B, immobilize growth cones. Much information is now available about mechanisms of neurite outgrowth. As we have seen for transmitter release (see Chapter 13) and for many other cellular processes, transient, localized increases in cytoplasmic calcium levels are important in regulating events in the growth cone.[116,117]

Cell and Extracellular Matrix Adhesion Molecules and Axon Outgrowth

Cell adhesion molecules (CAMs) are transmembrane or membrane-associated glycoproteins that mediate axon outgrowth by providing a favorable environment for growth cone extension. They are characterized by structural motifs in their extracellular portions that

[111] Tyzio, R. et al. 2011. *J. Neurosci.* 31: 34–45.
[112] Cherubini, E. et al. 2011. *Mol. Neurobiol.* 43: 97–106.
[113] Lowery, L. A., and Van Vactor, D. 2009. *Nat. Rev. Mol. Cell. Biol.* 10: 332–343.
[114] Grumbacher-Reinert, S., and Nicholls, J. 1992. *J. Exp. Biol.* 167: 1–14.
[115] Gallo, G., and Letourneau, P. C. 2004. *J. Neurobiol.* 58: 92–102.
[116] Roehm, P. C. et al. 2008. *Mol. Cell. Neurosci.* 37: 376–387.
[117] Hutchins, B. I., and Kalil, K. 2008. *J. Neurosci.* 28: 143–153.

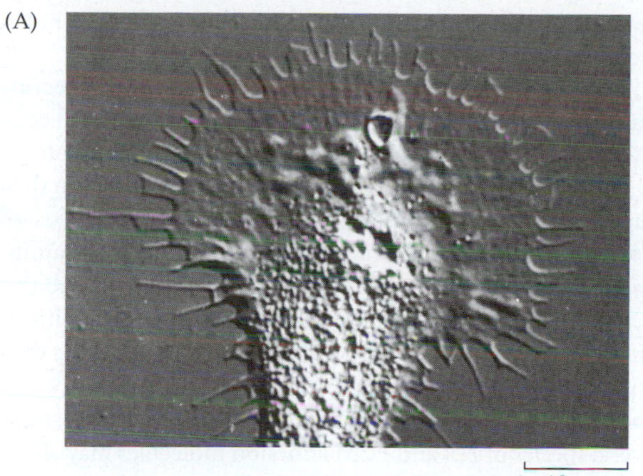

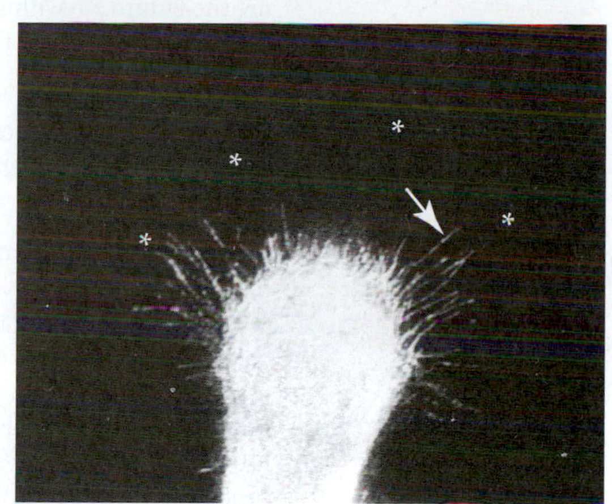

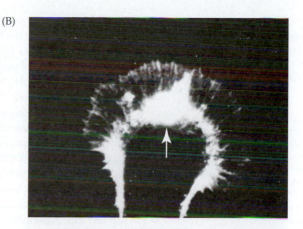

FIGURE 25.18 The Morphology of Growth Cones. (A) Growth cone observed by differential interference contrast microscopy. (B) Fluorescence micrograph showing the distribution of filamentous actin visualized with rhodamine-conjugated phalloidin. Actin filaments align with filopodia, or microspikes, in the periphery of the growth cone; randomly oriented filaments are often concentrated near the central domain (arrow). (C) Microtubule distribution visualized with antitubulin antibodies and fluorescein-conjugated secondary antibodies. Microtubules are concentrated in the axon. Most terminate in the central domain of the growth cone; some (arrow) extend toward the growth cone margin (asterisks). (After Forscher and Smith, 1988; micrographs kindly provided by S. J. Smith.)

resemble the constant-region domains of immunoglobulin (Ig) and the type III domains of fibronectin. Members of this Ig superfamily include neural CAM (N-CAM), neuroglial CAM (NgCAM), transient axonal glycoprotein-1 (TAG-1), myelin-associated glycoprotein (MAG), and Deleted in Colorectal Carcinoma (DCC).[118–120] These molecules mediate cell–cell adhesion, either through **homophilic binding** (e.g., N-CAM on one cell binding to N-CAM on another cell) or through **heterophilic binding** (e.g., neuronal cell adhesion molecule [NrCAM] on one cell binding to transient axonal glycoprotein-1 [TAG-1] on another cell). Another family of CAMs is exemplified by N-cadherin, which mediates homophilic, calcium-dependent cell adhesion.[120] In culture, expression of N-CAM and N-cadherin in

[118] Shapiro, L., Love, J., and Colman, D. R. 2007. *Annu. Rev. Neurosci.* 30: 451–474.

[119] Togashi, H., Sakisaka, T., and Takai, Y. 2009. *Cell Adh. Migr.* 3: 29–35.

[120] Hansen, S. M., Berezin, V., and Bock, E. 2008. *Cell. Mol. Life. Sci.* 65: 3809–3821.

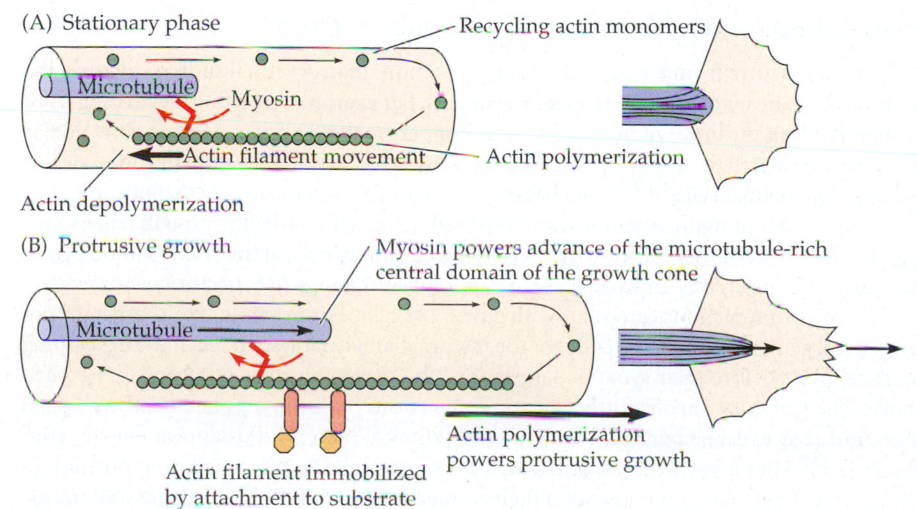

FIGURE 25.19 Model for Actin-Based Motility of Growth Cones. (A) Cross section (left) and top view (right) of a growth cone in stationary phase. Microtubule-attached myosin powers rearward movement of actin filaments, while filaments are undergoing continuous polymerization at the leading edge of the growth cone and depolymerization centrally. (B) Similar views of a growth cone during protrusive growth. Actin filaments are immobilized by attachment to the substrate. Actin polymerization now causes protrusion of the growth cone, while myosin cycling moves microtubules forward, advancing the central domain of the growth cone. (After Suter et al., 1998.)

cells promotes their aggregation, the extension of axons (on cellular but not extracellular matrix substrates), and the binding together of growing axons into fascicles.

Extracellular matrix (ECM) adhesion molecules, for example laminin, fibronectin, tenascin, cytotactin, and perlecan, are glycoproteins secreted by glial and mesodermal cells during development and by Schwann cells in adult peripheral nervous system. ECM proteins also provide favorable substrates for neurite extension.[121] For example, a leech neuron that grows only a few microns when plated on plastic can extend processes over hundreds of microns on a dish coated with laminin.[122] ECM molecules interact with cells via a family of receptors known as integrins, which span the membrane to link ECM proteins and the intracellular actin cytoskeleton, and thereby regulate cell shape and migration. In addition, the activation of integrin receptors by ECM activates intracellular signaling pathways that control cell growth, proliferation, and differentiation.[123]

Antibody perturbation studies indicate that growth cones rarely rely on a single substrate to support their movements; several species of cell and ECM adhesion molecules may affect neurite outgrowth within a particular molecular ecosystem of the embryo. Cues that support the growth of one type of migrating cell or growth cone may be ignored or avoided by others.[124]

Growth Cone Guidance: Target-Dependent and Target-Independent Navigation

The axon of a cell can extend up to a meter or more (far more in a giraffe or a whale) to form synapses on appropriate cells in a region that contains many potential targets. During development, cues in the environment of an axon guide it along specific pathways.

How do extracellular cues guide growth cones?[125] Ramón y Cajal originally proposed a chemoattractant model of axon guidance. According to this model, the growth cone navigates along a gradient of molecules released by its target. This can account for directed axon outgrowth when the distance from the nerve cell body to its target is short. For example, Lumsden and Davies[126] studied the growth of axons from the trigeminal ganglion in the head of the mouse into the adjacent epithelial tissue, a distance of less than 1 millimeter. (These axons ultimately give rise to the sensory innervation of the whiskers; see Chapter 21.) If the developing trigeminal ganglion is placed in culture near explants from several peripheral tissues, neurites grow from the ganglion toward their appropriate target, ignoring other tissues. Explants of target epithelium have this effect on axon outgrowth only if they are taken from embryos at the stage when innervation normally occurs.

In contrast, the ability of spinal motor axons to grow to the appropriate region in the limb does not depend on the presence of their target muscles. This was shown by removing the tissues that give rise to limb musculature, early in development.[124,127] Motor axons extended normally from the spinal cord, grew into the limb, and formed the appropriate pattern of muscle nerves—all in the absence of muscle. Thus, the factors that guide motor axons to their correct destinations in the limb are not supplied by the muscles that the axons ultimately innervate.

Target-Dependent Navigation via Guidepost Cells

When the distance from a neuron to its target is more than several hundred microns, the pathway may be marked with intermediate targets. For example, growth cones arising from sensory cells in the limbs of developing grasshoppers make abrupt changes in direction as they extend toward the CNS (Figure 25.20). The turns occur when the growth cones contact so-called **guidepost cells.**[128,129] Such behavior suggests that interactions with guidepost cells, which are often immature neurons, are responsible for redirecting the growth cones. This phenomenon can be demonstrated by removing the guidepost cells by laser ablation before the growth cone arrives, in which case the appropriate change in trajectory is not made.

Synaptic interactions can occur with guidepost cells. For example, axons arising from the lateral geniculate nucleus (LGN) of the mammalian visual system reach the developing cortical plate before their synaptic targets (i.e., the pyramidal cells of layer 4) have been born. The geniculate axons form synaptic connections with the subplate neurons, which are produced early, lie beneath the developing cortical plate, and disappear shortly after birth.[130,131] After a few weeks, when layer 4 pyramidal cells have reached their position in the cortex, geniculate axons abandon their connections with subplate neurons and invade

[121] Durbeej, M. 2010. *Cell Tissue Res.* 339: 259–268.

[122] Grumbacher-Reinert, S., and Nicholls, J. G. 1992. *J. Exp. Biol.* 167: 1–14.

[123] Denda, S., and Reichardt, L. F. 2007. *Methods Enzymol.* 426: 203–221.

[124] Bonanomi, D., and Pfaff, S. L. 2010. *Cold Spring Harb. Perspect. Biol.* 2: a001735.

[125] Kolodkin, A. L., and Tessier-Lavigne, M. 2010. *Cold Spring Harb. Perspect. Biol.* 3: 10.1101/cshperspect.a001727.

[126] Lumsden, A. G. S., and Davies, A. M. 1986. *Nature* 323: 538–539.

[127] Phelan, K. A., and Hollyday, M. 1990. *J. Neurosci.* 10: 2699–2716.

[128] Bentley, D., and Caudy, M. 1983. *Cold Spring Harb. Symp. Quant. Biol.* 48: 573–585.

[129] Shen, K., Fetter, R. D., and Bargmann, C. I. 2004. *Cell* 116: 869–881.

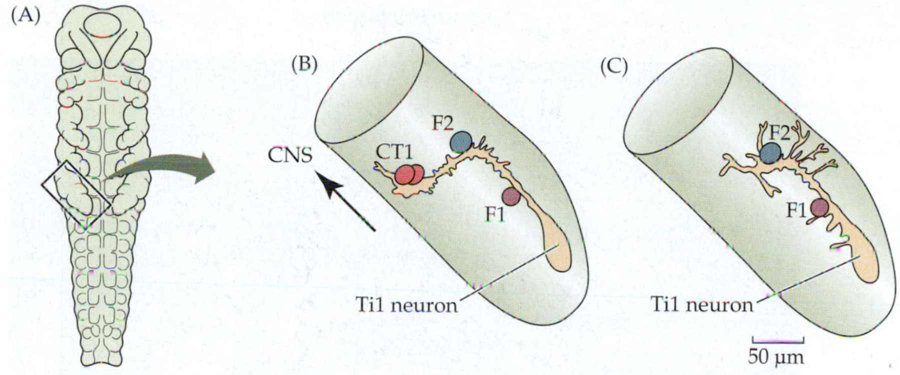

FIGURE 25.20 Growth Cones of Peripheral Neurons Rely on Guidepost Cells to navigate through the limb of the grasshopper (A). (B) In normal embryos, the axon of the Ti1 neuron encounters a series of guidepost cells on its route to the central nervous system: F1, F2, and two CT1 cells. (C) If the CT1 cells are killed early in development, the Ti1 neuron forms several axonal branches at the site of cell F2, with growth cones extending in abnormal directions. (After Bentley and Caudy, 1983.)

the cortex to establish the adult pattern of innervation. If the subplate neurons are eliminated early in development by local application of neurotoxins, LGN axons grow past the developing visual cortex, and ocular dominance columns fail to form.[132]

Growth Cone Navigation along Gradients

Studies of growth cone behavior in cell culture indicate that cell and extracellular matrix adhesion molecules do not provide long-range guidance cues. For example, growth cones neither extend up or down gradients of cell and matrix adhesion molecules, nor do they navigate using differences in the strength of adhesion of different surfaces. Adhesive molecules are either permissive for growth or not.[133,134] Instead, long-range guidance involves growth cone navigation along gradients of diffusible factors.

A good example of mechanisms and molecules that guide growth cones to their targets is provided by commissural interneurons in the vertebrate spinal cord. Early in development these sensory interneurons, which lie in the dorsal part of the spinal cord, extend axons that grow ventrally cross the midline, and then run longitudinally along the spinal cord toward their targets in the brainstem and thalamus (see Chapter 21).[135]

Axons of such commissural interneurons are initially attracted to the ventral midline by the protein **netrin-1**, a diffusible chemoattractant released by the floor plate cells that form the ventral midline of the nerve cord (Figure 25.21A).[136,137] Netrin-1 is a member of

[130] Shatz, C. J., and Luskin, M. B. 1986. *J. Neurosci.* 6: 3655–3658.

[131] Luskin, M. B., and Shatz, C. J. 1985. *J. Neurosci.* 5: 1062–1075.

[132] Kanold, P. O., and Shatz, C. J. 2006. *Neuron* 51: 627–638.

[133] McKenna, M. P., and Raper, J. A. 1988. *Dev. Biol.* 130: 232–236.

[134] Isbister, C. M., and O'Connor, T. P. 1999. *J. Neurosci.* 19: 2589–2600.

[135] Tessier-Lavigne, M. et al. 1988. *Nature* 336: 775–778.

[136] Kennedy, T. E. et al. 1994. *Cell* 78: 425–435.

[137] Dickson, B. J., and Gilestro, G. F. 2006. *Annu. Rev. Cell Dev. Biol.* 22: 651–675.

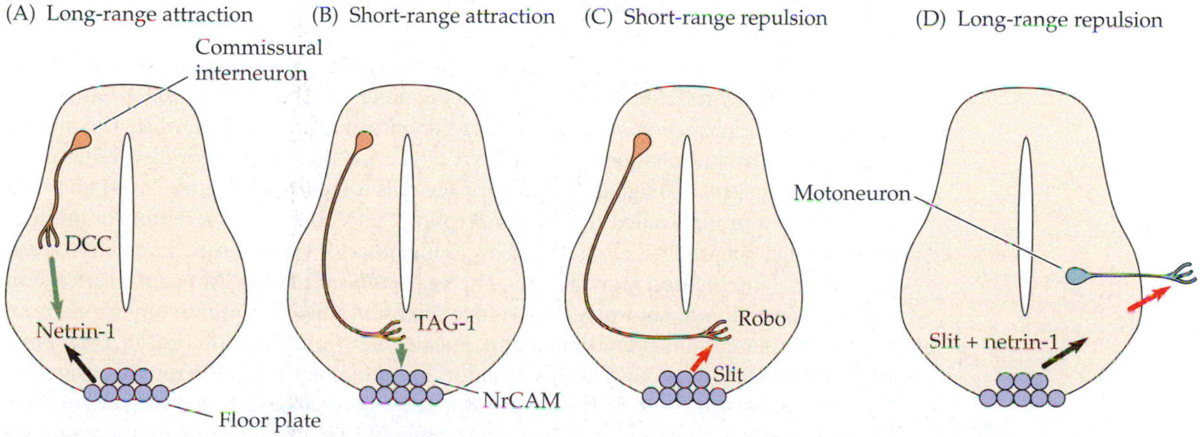

FIGURE 25.21 Long- and Short-Range Chemoattraction and Chemorepulsion guide developing axons in the vertebrate spinal cord. (A) Netrin-1, acting as a long-range chemoattractant, secreted by cells of the floor plate (blue), diffuses (black arrow), and binds to its receptor, Deleted in Colorectal Carcinoma (DCC) receptor on the growth cones of commissural neurons (orange), attracting them (green arrow). (B) Transient axonal glycoprotein-1 (TAG-1) on commissural axon growth cones binds to neuronal cell adhesion molecule (NrCAM) on floor plate cells. This short-range chemoattraction (green arrow) facilitates extension of commissural axon growth cones across the floor plate. (C) As they cross the midline, TAG-1 on commisural growth cones is replaced by Roundabout (robo), which binds to slit on floor plate cells. This short-range chemorepulsion (red arrow) prevents the axons from recrossing the floor plate. (D) Slit and netrin-1 diffuse from the floor plate (black arrow), interact with receptors on growth cones of motor neurons (blue), and repel them (red arrows). This long-range chemorepulsion helps direct the growth of motor axons away from the cord.

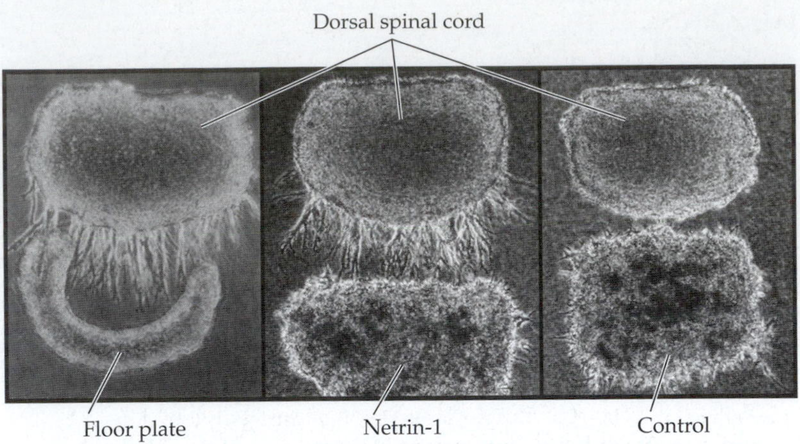

FIGURE 25.22 Netrins and Netrin Receptors function in long-range attraction and repulsion. (A) Micrographs of pieces of dorsal spinal cord from embryonic rats (top in each panel) cultured with a piece of floor plate tissue (left), an aggregate of COS cells secreting recombinant netrin-1 (middle), or control COS cells (right). The floor plate and netrin-1 both elicit the profuse and directed outgrowth of bundles of commissural axons from the dorsal spinal cord tissue. (After Tessier-Lavigne et al., 1988 and Kennedy et al., 1994; micrographs kindly provided by M. Tessier-Lavigne.)

a family of secreted proteins characterized by domains homologous to the amino-terminal domains of the γ chain of laminin-1.[136,138] Homologues of netrin, which can be repulsive as well as attractant under various circumstances, are also involved in axon guidance in *Drosophila* and in *C. elegans*, where it was first discovered and named unc-6.[139] The dorsoventral concentration gradient of netrin-1 attracts growing axons by interacting with a receptor expressed on them, which as mentioned above is called Deleted in Colon Cancer (DCC) and is a cell adhesion molecule.

The existence of a chemoattractant produced by the floor plate was demonstrated by culturing explants of dorsal spinal cord either alone or with pieces of floor plate. As shown in Figure 25.22, axons of commissural neurons specifically grew toward the floor plate, even when the tissues were separated by several hundred microns. This distance is too great to be spanned by a filopodium from a growth cone. Commissural neurons were similarly attracted to cell aggregates that secreted netrin-1 (see Figure 25.22, middle panel) but not to non-secreting control cells (right panel).

After being attracted in the direction of the floor plate, the commissural axon growth cones cross the ventral midline to the contralateral side (Figure 25.21B and C). Crossing is facilitated by the interaction of two cell surface-adhesive molecules: TAG-1, which is expressed on the surface of commissural axons, and NrCAM, expressed on the floor plate cells. As the axons cross, they receive signals from floor plate cells to stop synthesizing TAG-1 and start synthesizing a protein called roundabout (robo).[137,138] Robo is the receptor for another protein, called slit, which is also released by floor plate cells. The slit–robo interaction *repels* commissural interneuron growth cones. The loss of the TAG-1–NrCAM contact attraction and the acquisition of slit–robo short-range repulsion prevent commissural axons from re-crossing the midline. (The gene name *Robo* comes from the mutation in *Drosophila* that knocks out the function of this receptor. In mutant embryos, the repulsive signal is lost and the growth cones in the developing nerve cord cross and re-cross the midline. The thickened commissural nerves come to resemble a traffic circle or *ro*und-*abo*ut.[140]) Netrin-1, while attracting commissural neuron growth cones, acts in concert with slit to *repel* growth cones of motoneurons. These long-range repulsive interactions direct motor axons away from the cord toward the periphery (Figure 25.21D).[141]

Another family of proteins that act primarily as chemorepellents is the semaphorins.[142] Originally identified and implicated in axon guidance in grasshoppers, semaphorins were recognized as chemorepellents when a vertebrate homologue, collapsin-1, was purified and shown to cause retraction or collapse of growth cones in cell culture and to mediate long-range repulsion of axons in explant cultures. Neuropilins are highly conserved receptors for

[138] Chen, Z. et al. 2008. *Neuron* 58: 325–332.
[139] Hedgecock, E. M. et al. 1990. *Neuron* 4: 61–85.
[140] Seeger, M. et al. 1993. *Neuron* 10: 409–426.
[141] Brose, K. et al. 1999. *Cell* 96: 795–806.
[142] Derijck, A. A., Van Erp, S., and Pasterkamp, R. J. 2010. *Trends Cell Biol.* 20: 568–576.

the semaphorins.[143] Neurite outgrowth inhibitor-A (Nogo-A),[144] another growth inhibitory protein associated with oligodendrocytes and CNS myelin, is discussed in Chapter 27.

Rita Levi-Montalcini

Growth Factors and Survival of Neurons

Cell Death in the Developing Nervous System

A conspicuous feature of the nervous system development is that many cells are born to die. In invertebrates, for example, the sweeping changes that occur during metamorphosis include the programmed cell death (**apoptosis**) of many neurons.[145] However, in developing invertebrate and vertebrate nervous systems, cell death also occurs in the absence of such gross morphological changes.[146] Overproduction of neurons followed by a period of cell death is a common pattern throughout development. Some of the neurons that die may not have made any synapses or may have innervated an inappropriate target. Most, however, appear to have reached and innervated their correct targets. An important finding was that inhibitors of mRNA or protein synthesis prevented the normal death of neurons.[147] Thus, apoptosis is a process that activates intrinsic *suicide* machinery in a cell. DNA and proteins within the cell break down as a result of synthesis or activation of families of proteolytic enzymes called **caspases**.[148]

The extent of neuronal apoptosis in vertebrate embryos depends on the size of the target tissue.[149] Hamburger and his colleagues showed that when synaptic connections were first being formed on myofibers in developing limbs, 40% to 70% of the motoneurons that had sent axons into the limb died. Implantation of a supernumerary limb reduced the fraction of motoneurons that died, while removal of the limb bud exacerbated the death of motoneurons. These results suggested that motoneurons were competing for some **trophic substance** supplied by their target tissue.

In addition to cell death, another process, called synaptic reorganization, refines the pattern of innervation after axons reach their targets and make synaptic connections. Synaptic reorganization entails a reduction in the number of axons and synapses, accompanied by a reorganization of surviving connections, to achieve the adult pattern. As is true for cell death, synaptic reorganization also involves competition for limited supplies of trophic factors, such as those described in the following sections.

Nerve Growth Factor

The first trophic substance to be identified was nerve growth factor (NGF), acting on sensory and sympathetic neurons. In pioneering experiments, Levi-Montalcini, Cohen, and their colleagues first demonstrated that NGF stimulated the outgrowth of neurites neurons in culture.[150] They went on to show that NGF was also required for neuronal survival. When the action of NGF was blocked with antibodies in newborn mice, almost all of the sympathetic neurons died. The parasympathetic nervous system was not affected, and the dorsal root ganglia were only slightly smaller than normal. In subsequent experiments, it was shown that at an earlier, fetal stage in development, dorsal root ganglion sensory neurons likewise required NGF for survival. In adults, antibodies to NGF were much less effective on either cell population, even though NGF was still required for survival.[151–153] These observations set the stage for a molecular analysis of the mechanism of action of nerve growth factor by a number of research groups, including those of Levi-Montalcini, Shooter,[154] and Barde.[155] Their studies revealed the characteristics of NGF, the receptors to which it binds, and the subsequent metabolic events.

The fact that NGF is required for cell survival suggests that it acts either directly or indirectly on the cell body. When embryonic sympathetic neurons are grown in three-compartment culture chambers (Figure 25.23), the central compartment in which the neurons are placed must initially contain NGF for the cells to survive (Figure 25.23A).[156] However, after neurites have reached the side compartments, NGF can be removed from the central compartment and the cells will remain alive, so long as the side compartments contain NGF (Figure 25.23B). Removal of NGF from one side compartment causes neurites on that side to degenerate (Figure 25.23C). These results indicate that the trophic effects of NGF are mediated by signals evoked in growing terminals that influence the nucleus.

[143] Fujisawa, H. 2004. *J. Neurobiol.* 59: 24–33.

[144] Liu, B. P. et al. 2006. *Philos. Trans. R. Soc. Lond., B, Biol. Sci.* 361: 1593–1610.

[145] Truman, J. W., Thorn, R. S., and Robinow, S. 1992. *J. Neurobiol.* 23: 1295–1311.

[146] Buss, R. R., Sun, W., and Oppenheim, R. W. 2006. *Annu. Rev. Neurosci.* 29: 1–35.

[147] Martin, D. P. et al. 1988. *J. Cell Biol.* 106: 829–844.

[148] Ryan, C. A., and Salvesen, G. S. 2003. *Biol. Chem.* 384: 855–861.

[149] Hollyday, M., and Hamburger, V. 1976. *J. Comp. Neurol.* 170: 311–320.

[150] Levi-Montalcini, R. 1982. *Annu. Rev. Neurosci.* 5: 341–362.

[151] Gorin, P., and Johnson, E. M. 1979. *Proc. Natl. Acad. Sci. USA* 76: 5382–5386.

[152] Crowley, C. et al. 1994. *Cell* 76: 1001–1011.

[153] Smeyne, R. J. et al. 1994. *Nature* 368: 246–249.

[154] Shooter, E. M. 2001. *Annu. Rev. Neurosci.* 24: 601–629.

[155] Tucker, K. L., Meyer, M., and Barde, Y. A. 2001. *Nat. Neurosci.* 4: 29–37.

[156] Campenot, R. B. 1977. *Proc. Natl. Acad. Sci. USA* 74: 4516–4519.

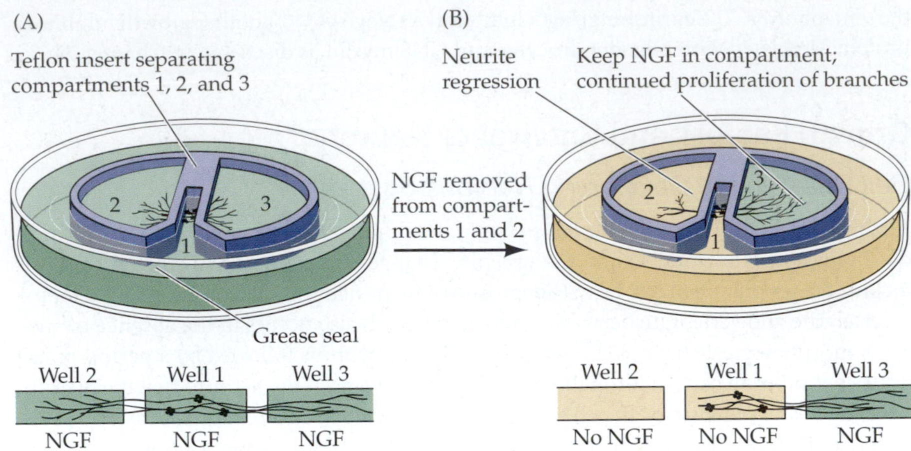

FIGURE 25.23 Nerve Growth Factor (NGF) and the Survival of Axon Branches from sympathetic ganglion cells grown in cell culture. (A) Neurons dissociated from neonatal sympathetic ganglia plated in the central compartment send neurites under a Teflon divider and into the adjacent compartments; all compartments contain NGF. (B) After initial outgrowth has occurred, removal of NGF from compartment 1 for 20 days has no effect, as neurons are maintained by NGF transported retrogradely from their terminals in the side compartments. Removal of NGF from compartment 2 causes the neurites entering it to degenerate, while those in the compartment containing NGF remain. (After Campenot, 1982.)

Retrograde transport of NGF occurs from nerve terminals to the cell body, but it is not clear whether this transport is required for survival (see Campenot[157]).

Like many secreted proteins, nerve growth factor is initially translated as a longer polypeptide (pro-NGF) that is subsequently cleaved to the mature form.[158] It had been generally assumed that only the mature forms of such signaling factors are active. However, it is now clear that not only are both forms of NGF active, but that they exhibit opposing biological activities, by acting on pairwise combinations of three different co-receptors (TrkA, p75NTR, and sortilin).[159–161] Mature NGF binds with high affinity to TrkA, either alone or in combination with p75NTR, to promote cell survival and differentiation. ProNGF (the NGF precursor), by contrast, binds in a complex with sortilin and p75NTR to promote apoptosis. Intracellular pathways used by two growth factor receptors (TrkA and p75NTR) to trigger survival, apoptosis, and growth are shown in Figure 25.24. Thus, in the words of Feng and colleagues,[159] "...the choice between life and death of neurons may be a delicate balance between cleavage or non-cleavage of proNGF, and expression of TrkA or p75NTR/sortilin receptor complexes on the cell surface."

NGF in the Central Nervous System

Nerve growth factor is not unique to the peripheral nervous system. A population of NGF-sensitive cells has also been found within the CNS.[162,163] These are cholinergic neurons located in the basal forebrain, particularly in the nucleus basalis. Their axons innervate the cerebral cortex and hippocampus (see Chapters 12 and 14). If the axons are cut in the adult rat, the cells die. If, however, NGF is infused into the CNS, then the cholinergic neurons survive axotomy. The number of cells that stain with markers for cholinergic function declines with age, as does the ability of rats to learn a maze or other spatial memory tasks. If NGF is infused into aged rats, the number of cells that can be stained increases and the rat's performance in spatial memory tasks improves.[164] These observations indicate that survival and growth of neurons within the CNS are likely to rely on factors that are similar to those in peripheral neurons. Much research is now in progress to assess, in molecular terms, defects that might give rise to mental deficits, such as Alzheimer's disease, in which one of the first signs is death of cholinergic neurons that supply the cortex.[165]

The Neurotrophins and other Families of Growth Factors

Other neurotrophic factors exist in the brain. One example, brain-derived neurotrophic factor (BDNF), has been shown to promote the survival of dorsal root ganglion neurons in culture and to rescue them in vivo from natural neuronal death (see below).[166–168] BDNF has a high degree of homology to NGF. Together NGF and BDNF defined a protein family called **neurotrophins.**[2] Other members of the family are NT-3, NT-4/5, NT-6, and glial derived neurotrophic factor (GDNF).[169,170] Early in development, before sensory neurons innervate their peripheral targets, neural crest cells and sensory neurons in dorsal root and trigeminal ganglia require BDNF or NT-3 for proliferation, differentiation, and survival.[171] At such early stages, neurotrophins

[157] Campenot, R. B. 2009. *Results Probl. Cell. Differ.* 48: 141–158.
[158] Greene, L. A., and Shooter, E. M. 1980. *Annu. Rev. Neurosci.* 3: 353–402.
[159] Feng, D. et al. 2010. *J. Mol. Biol.* 396: 967–984.
[160] Schecterson, L. C., and Bothwell, M. 2010. *Dev. Neurobiol.* 70: 332–338.
[161] Reichardt, L. F. 2006. *Philos. Trans. R. Soc. Lond., B, Biol. Sci.* 361: 1545–1564.
[162] Gage, F. H. et al. 1988. *J. Comp. Neurol.* 269: 147–155.
[163] Mufson, E. J. et al. 2003. *J. Chem. Neuroanat.* 26: 233–242.
[164] Fischer, W. et al. 1991. *J. Neurosci.* 11: 1889–1906.
[165] Capsoni, S., and Cattaneo, A. 2006. *Cell. Mol. Neurobiol.* 26: 619–633.
[166] Barde, Y-A. 1989. *Neuron* 2: 1525–1534.
[167] Ernsberger, U. 2009. *Cell Tissue Res.* 336: 349–384.
[168] Tettamanti, G. et al. 2010. *Gene* 450: 85–93.
[169] Kalb, R. 2005. *Trends Neurosci.* 28: 5–11.
[170] Airaksinen, M. S., and Saarma, M. 2002. *Nat. Rev. Neurosci.* 3: 383–394.
[171] Farinas, I. et al. 1996. *Neuron* 17: 1068–1078.

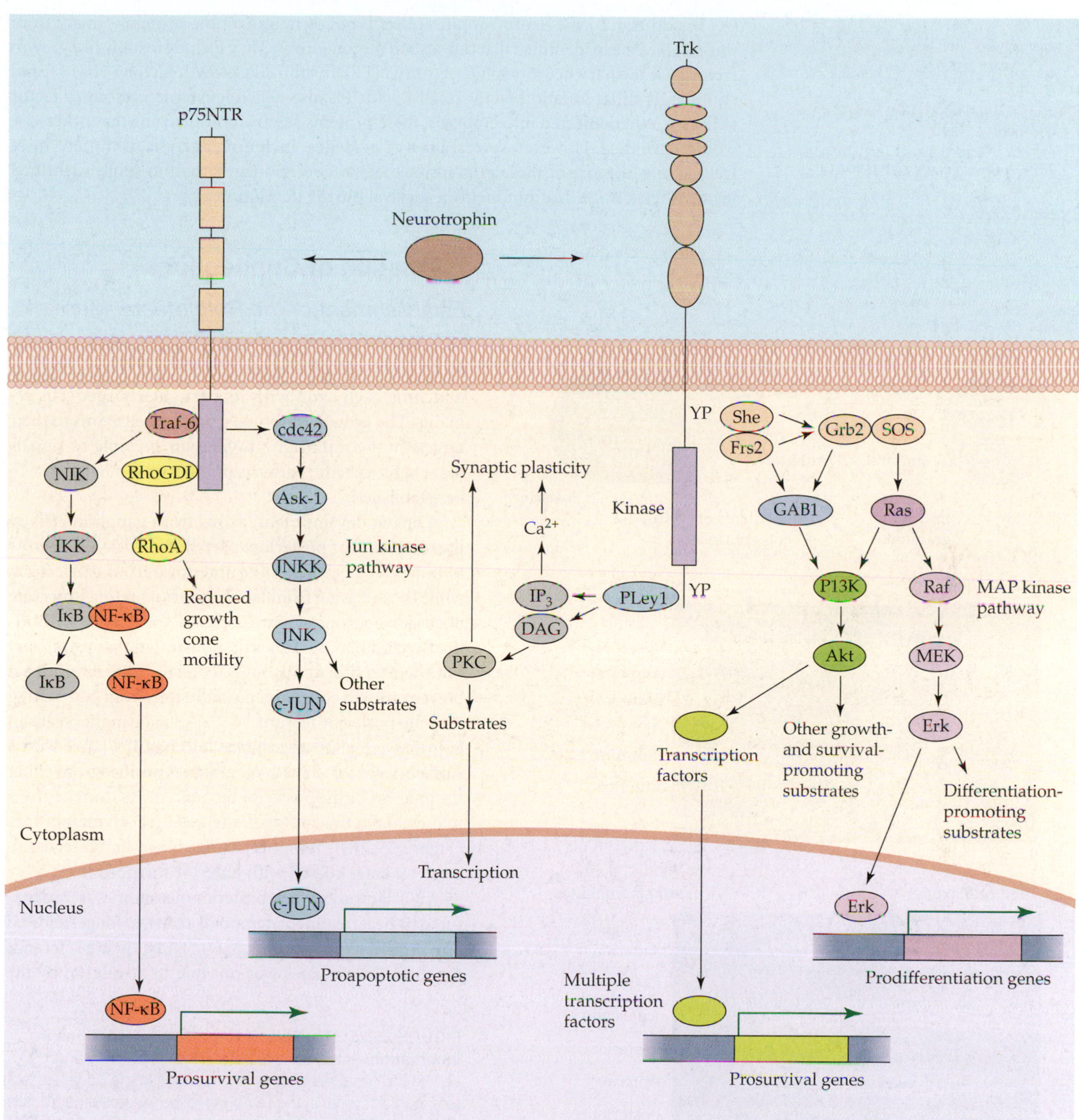

FIGURE 25.24 Interactions of Neurotrophins with Trk and p75NTR Receptors and complex intracellular signaling pathways, some of which are illustrated in this figure. The two receptors influence cell survival, apoptosis, and differentiation. The p75NTR pathway controls activation of certain genes that promote neuronal apoptosis and others that control growth cone mobility. Trk receptors controls major signaling pathways that promote neuronal differentiation, neurite outgrowth, and survival. (After Reichardt, 2006.)

appear to be provided by the neurons themselves and by the mesenchymal tissues through which the sensory axons grow. Later, after their axons reach their targets, sensory neurons begin to express NGF receptors and become dependent on target-derived NGF for survival;[172] neurotrophins, particularly BDNF, have also been shown to potentiate synaptic transmission and increase or decrease neuronal excitability by regulating ion channel expression.[173–175]

[172]Davies, A. M., and Lumsden, A. 1990. *Annu. Rev. Neurosci.* 13: 61–73.

[173]Lagostena, L. et al. 2010. *J. Neurosci.* 30: 885–893.

[174]Tongiorgi, E. 2008. *Neurosci. Res.* 61: 335–346.

[175]Kovalchuk, Y., Holthoff, K., and Konnerth, A. 2004. *Curr. Opin. Neurobiol.* 14: 558–563.

[176]Wu, W. et al. 2003. *J. Neurotrauma* 20: 603–612.

[177]Taylor, A. R. et al. 2007. *J. Neurosci.* 27: 634–644.

[178]Walter, J. et al. 1987. *Development* 101: 685–696.

[179]Walter, J., Henke-Fahle, S., and Bonhoeffer, F. 1987. *Development* 101: 909–913.

In addition to the neurotrophins, other types of neurotrophic proteins (many from muscle) have been identified that can sustain motoneurons. They include insulin-like growth factor (IGF), ciliary neurotrophic factor (CNTF), insulin-like growth factor 1 (IGF1), and cholinergic differentiation factor (CDF), which is also called leukemia inhibitory factor (LIF) .[176] When injected into embryos, these proteins rescue motoneurons that otherwise would have died. However, several lines of evidence, including analysis of mutant mice lacking one or more of these proteins or their receptors, suggest that no single identified factor is responsible for motoneuron survival during development.[177]

Formation of Connections

Establishment of the Retinotectal Map

Even after the growth cone guidance mechanisms described earlier guide axons to their destinations, the problem of matching each axon with its particular target cells remains. The growth of retinal ganglion cell axons to their targets in the midbrain provides an example of signals that enable specific patterns of innervation in a tissue to be established.

During development, axons from ganglion cells in the temporal (or posterior) part of the chick retina grow to innervate neurons in the anterior part of the tectum, while those arising from nasal (anterior) retina innervate posterior neurons (Figure 25.25). Experiments by Bonhoeffer and his colleagues demonstrated that axons sort out their territories through repulsive interactions that prevent growth cones of temporal axons from penetrating into the posterior tectum.[178,179] Retinal ganglion cells in culture were allowed to grow into parallel lanes whose surfaces were coated with membranes purified from either the anterior or the posterior tectum. Axons from neurons in the nasal retina grew equally well on both membranes (Figure 25.25B). Temporal axons, however, preferred to grow on lanes coated with anterior membranes (Figure 25.25C). However, when posterior membranes were denatured by heat treatment, temporal axons no longer showed any preference (Figure 25.25D). Thus, their preference for the anterior lanes was not due to attraction by the

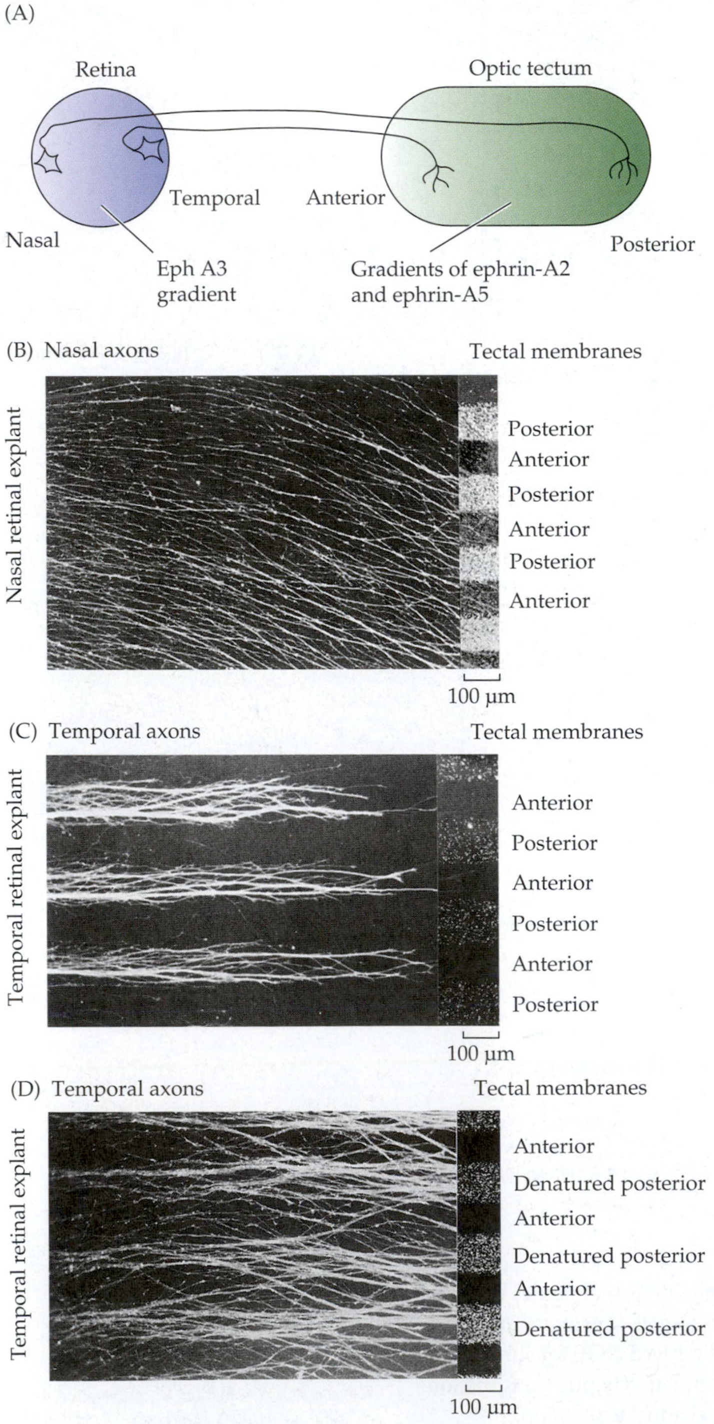

FIGURE 25.25 The Role of Repulsive Interactions in Innervation of the optic tectum in the chick. (A) Ganglion cells in the nasal retina innervate neurons in the posterior tectum; ganglion cells in the temporal retina innervate neurons in the anterior tectum. There is a nasotemporal gradient of the Eph A3 receptor tyrosine kinase in retinal ganglion cells and anteroposterior gradients of the Eph receptor ligands ephrin-A2 and ephrin-A5 in the tectum. Axons of temporal ganglion cells are prevented from entering the posterior tectum by the repulsive interaction of Eph receptors and ligands. (B) In cell culture, axons from neurons in the nasal retina grow equally well on lanes coated with membranes isolated from anterior or posterior tectum. (C) Axons from neurons in the temporal retina prefer to grow on anterior membranes. (D) Axons from temporal retina grow equally well on intact anterior membranes and denatured posterior membranes, indicating that normally they are repelled by heat-sensitive components of the posterior membranes. (B and D from Walter, Henke-Fahle, and Bonhoeffer, 1987; C from Walter et al., 1987; micrographs kindly provided by F. Bonhoeffer.)

anterior membranes, but rather to repulsion by the posterior membrane. Curiously, when not offered a choice of membranes, retinal axons elongated rapidly on either substrate.

The molecules responsible for this repulsive interaction in the optic tectum (or **superior colliculus** in mammals) are members of a family of receptor tyrosine kinases (known as Eph kinases) and their ligands (called ephrins).[180,181] Ephrin-A2 and ephrin-A5 are expressed in the tectum during the time retinotectal connections are being formed, and their concentration increases in a graded manner from anterior to posterior. The Eph A3 receptor is expressed on retinal axons in a corresponding nasotemporal gradient. When incorporated into lipid vesicles and added to the medium in which temporal retinal axons are growing, ephrins A2 and A5 cause the growth cones to detach from the substrate and retract.[182]

The other axis of the topographic projection maps the dorsoventral positions in the retina to mediolateral positions in the tectum. Here, patterning is achieved by a different family of ligands and receptors, the EphB/ephrin-B family. The cellular and molecular logic that underlies the dorsoventral/mediolateral mapping is more complex. Both attractive and repulsive interactions are involved, depending on the levels of ligands and receptors. Perturbation of these main mapping mechanisms reveals the importance of these and other ligand receptor signaling systems (e.g., other ephrins and Ephs; Wnt ligands and their Frizzled receptors) in fine-tuning the formation of patterns of innervation of a tissue. Later refinement of patterning is brought about by activity-dependent mechanisms[183] (see also Chapter 26).

Synapse Formation

The vertebrate skeletal neuromuscular junction provides a favorable preparation for studying the cellular and molecular mechanisms of synapse formation. The first steps in synapse formation occur rapidly. Mu-ming Poo and his colleagues have shown that as the growth cone of a motor axon approaches a myotube (i.e., an immature muscle fiber), depolarizing potentials arise due to the release of ACh from the growth cone (Figure 25.26).[184] Upon contact, the rate of spontaneous release of quanta of ACh rapidly increases, as does the size of the synaptic potential evoked by stimulating the axon. Thus, within minutes a functional synaptic connection is established. In Chapter 27, we give an account of the properties and functions of a protein known as **agrin**, which is released from the motor nerve axon as it approaches the muscle fiber. Agrin causes postsynaptic structures to be formed and ACh receptors to accumulate at the motor end plate.[185]

Factors involved in formation of synapses in the CNS are not well understood, but two more families of interacting transmembrane proteins, **neuroligins** and **neurexins**, may play key roles in the functional development of chemical synapses.[186,187] Neuroligins presented

[180] Drescher, U. et al. 1995. *Cell* 82: 359–370.

[181] Feldheim, D. A., and O'Leary, D. D. 2010. *Cold Spring Harb. Perspect. Biol.* 2: a001768.

[182] Cox, E. C., Muller, B., and Bonhoeffer, F. 1990. *Neuron* 4: 31–37.

[183] Baker, M. W., Peterson, S. M., and Macagno, E. R. 2008. *Dev. Biol.* 320: 215–225.

[184] Evers, J. et al. 1989. *J. Neurosci.* 9: 1523–1539.

[185] Reist, N. E., Werle, M. J., and McMahan, U. J. 1992. *Neuron* 8: 865–868.

[186] Craig, A. M., and Kang, Y. 2007. *Curr. Opin. Neurobiol.* 17: 43–52.

[187] Sudhof, T. 2008. *Nature* 455: 903–911.

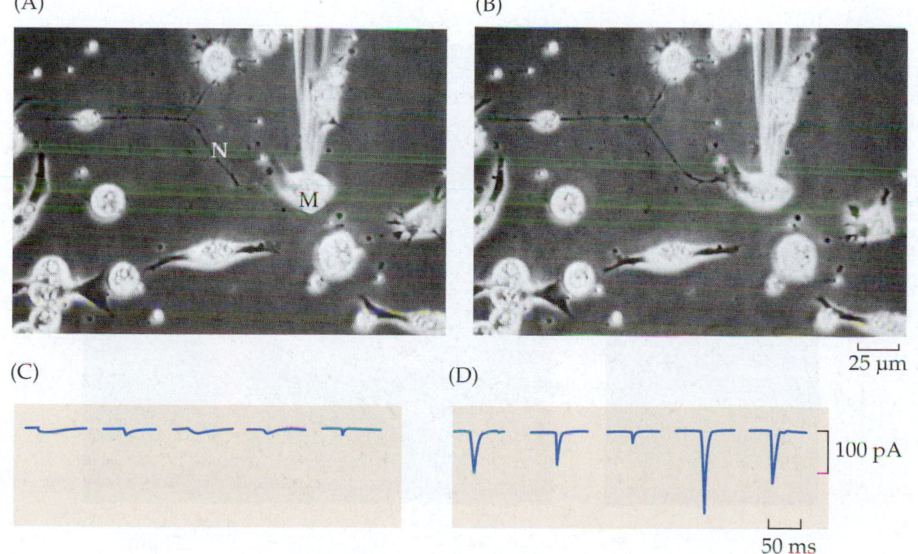

FIGURE 25.26 Rapid Formation of Functional Synaptic Connections between motor axons and muscle cells. (A,B) Phase-contrast photographs of a growing neurite (N) and a spindle-shaped myocyte (M) in a *Xenopus* neuron–muscle cell culture at the beginning (A) and end (B) of electrical recording. (C,D) Whole-cell patch clamp records from the myocyte. Spontaneous synaptic currents can be recorded within 1 minute of contact (C) and have increased in strength several-fold by 18 minutes (D). (From Evers et al., 1989.)

to neurons by expression in non-neuronal cells or linked to beads induce the formation of presynaptic specializations, and similar presentation of neurexins (which share extracellular domains similar to those of agrin) induce the formation of postsynaptic specializations.

Pruning and the Removal of Polyneuronal Innervation

Once synapses have formed and the population of neurons innervating a target has been restricted through cell death, axons of surviving neurons compete with one another for synaptic territory. This competition typically results in the loss of some of the terminal branches and synapses made initially—a process referred to as **pruning**.[188] Pruning ensures appropriate and complete innervation of a target by a particular population of neurons. In some cases, pruning also provides a mechanism for correcting mistakes; in other instances, it appears to reflect a strategy for establishing pathways.

A clear example of competitive pruning occurs in developing skeletal muscle. In the adult, each motor neuron innervates a group of up to 300 muscle fibers, forming a motor unit (see Chapter 24), but each muscle fiber is innervated by only one axon. In developing muscle, however, motor neurons branch extensively so that each muscle fiber comes to be innervated by axons from several motor neurons (Figure 25.27)—a phenomenon called **polyneuronal innervation**.[189,190] On each developing muscle fiber, the synaptic endings of multiple axons are interspersed at a single site and are juxtaposed to aggregates of ACh receptors and other components of the postsynaptic apparatus. As development progresses, axon branches are eliminated until the adult pattern is formed. This process does not involve

[188] Luo, L., and O'Leary, D. D. 2005. *Annu. Rev. Neurosci.* 28: 127–156.

[189] Redfern, P. A. 1970. *J. Physiol.* 209: 701–709.

[190] Brown, M. C., Jansen, J. K., and Van Essen, D. 1976. *J. Physiol.* 261: 387–422.

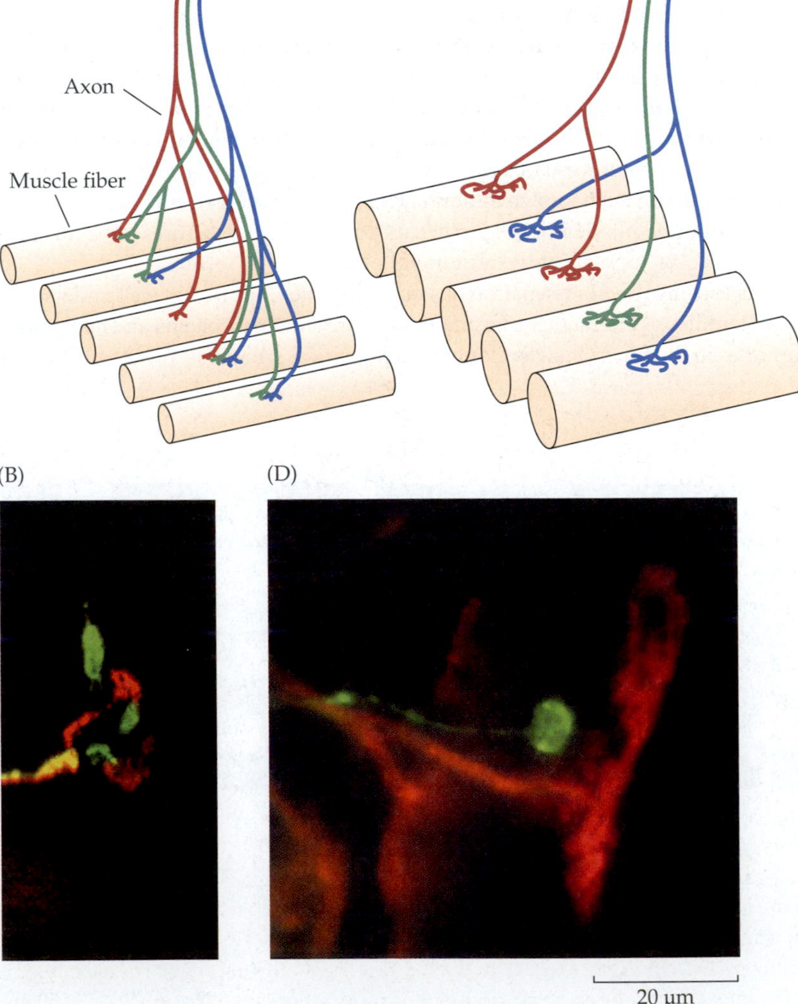

FIGURE 25.27 Polyneuronal Innervation and its Elimination at the vertebrate skeletal neuromuscular junction. (A) During embryonic development, motor axons branch to innervate many muscle fibers, and each muscle fiber is innervated by several motor axons (polyneuronal innervation). (B) Fluorescence micrograph of a neuromuscular junction of an E18 mouse showing the distribution of terminals of two axons, each labeled with a lipophilic probe (diI in red, diA in green). During the period of polyneuronal innervation, the terminal arbors of all motor axons innervating a particular muscle fiber interdigitate at a single synaptic site. (C) After birth, polyneuronal innervation is eliminated as axon branches retract, leaving each muscle fiber innervated by a single motor axon. (D) Fluorescence micrograph of a mouse neuromuscular junction during the period of removal of polyneuronal innervation. Two axons innervating the junction were labeled as in B. All terminals of one axon (green) have been eliminated, and the axon is being withdrawn. (Micrographs kindly provided by J. W. Lichtman.)

neuronal cell death (which occurs at an earlier developmental stage), only a reduction in the number of muscle fibers innervated by each motor neuron.

The removal of polyneuronal innervation is mediated by competition between axons of different motor neurons for synaptic space on the muscle cell. A clear example comes from studies of a small muscle in the toe of the rat.[191] When all but one of the motor axons innervating this muscle were cut early in development, the remaining axon spread to innervate many fibers within the muscle. During the period when elimination of polyneuronal innervation would normally occur, however, no synapses were lost. In the absence of competition, the surviving motor neuron maintained contacts with every myofiber it had innervated.

Neuronal Activity and Synapse Elimination

Neuronal activity plays a role in synapse elimination, influencing both the rate and outcome of the competition between axon terminals. Stimulation of the nerve to a muscle, via implanted electrodes, increases the rate of synapse elimination.[192] If activity is reduced, by applying tetrodotoxin in a cuff around the nerve to block action potentials or by inhibiting synaptic transmission, synapse elimination is slowed.[193] In muscles that receive input from axons that run in two different nerves, one can block impulses in one nerve and not the other.[194] In such cases, inactive axons are at a competitive disadvantage, as axons in the blocked nerve innervate smaller than normal motor units and those in the active nerve innervate more fibers than usual.

Activity-dependent competition extends to the level of branches of a single motor axon.[195] If a small region of an adult junction is inactivated by focal application of α-bungarotoxin, the inactive region of the junction is eliminated. If the entire junction is silenced, no elimination occurs.

Similar competition for synaptic targets occurs during the development of CNS pathways. One example is the formation of the ocular dominance columns in visual cortex (see Chapter 26), in which axons from the LGN conveying information from the two eyes initially overlap extensively in layer 4 of the cortex and then sort out into left-eye and right-eye columns.[196] Here the pattern of activity in the terminals from the two eyes plays a decisive role in determining the outcome of the competition. It is tempting to speculate that pruning may occur through competition for a limited amount of target-derived trophic factor, such as NGF or BDNF.

General Considerations of Neural Specificity and Development

Considerable progress has been made in understanding how nerve cells differentiate, find their targets, and establish synaptic contacts. However, when one contemplates the number and specificity of the fate decisions that must be made and connections that must be formed when wiring up the nervous system, the problem seems daunting. A commonplace analogy may be encouraging. Let us assume that we are ignorant about the workings and design of the postal system. A chapter from a book on the nervous system, without its illustrations, is posted in Trieste, Italy, and addressed to Sunderland, Massachusetts, USA, where it arrives a few days later. How does it get there? The writer knows only the closest mailbox and is unaware even of the post office in his district. The postal worker who empties the mailbox knows the post office; there the clerk who handles the mail may not know where Sunderland is but does know how to direct the package to the airport, and so on, to the right country, city, street, building, and eventually the correct person. If this were not enough, the illustrations that complement the chapter are posted separately from Berkeley and Baltimore to the same destination, where they arrive almost simultaneously with the chapter from Trieste. All the while, other mail is moving through the same mailboxes and post offices in different directions to different destinations.

The comforting feature of this analogy is that, although the problem seems altogether baffling at first sight, one can solve the postal puzzle by following the mail step by step to its destination. This would reveal some of the logic and design of postal organization (albeit without disclosing the identity of the designer). At any one step, only a limited number of instructions are followed and a limited number of mechanisms operate.

[191] Betz, W. J., Caldwell, J. H., and Ribchester, R. R. 1980. *J. Physiol.* 303: 265–279.

[192] Thompson, W. 1983. *Nature* 302: 614–616.

[193] Brown, M. C., Hopkins, W. G., and Keynes, R. J. 1982. *J. Physiol.* 329: 439–450.

[194] Ribchester, R. R., and Taxt, T. 1983. *J. Physiol.* 344: 89–111.

[195] Balice-Gordon, R. J., and Lichtman, J. W. 1994. *Nature* 372: 519–524.

[196] Wiesel, T. N. 1982. *Nature* 299: 583–591.

Some aspects of neural specificity may not be too different. A retinal ganglion cell sends its axon toward the back of the eye, where it makes a turn to enter the optic nerve together with fibers from other regions of the retina. The optic chiasm presents the next choice point, where the decision to enter the optic tract, leading either to the left of the right LGN may be made based on local chemical signals. Within the LGN, retinal axons may arrange themselves and innervate their targets according to gradients of repulsive molecules. Axons of geniculate neurons likewise follow a fairly simple path to their targets in the cortex, stopping along the way to form transient connections with subplate neurons. Thus, the seemingly complex task of forming specific connections between retinal ganglion cells and neurons in the visual cortex can be broken down into a series of relatively simple, independent events, each of which shapes the next. This general principle also applies to the even more complex question of how retinal ganglion cells, and all the many other types of neurons and non-neuronal cells, are generated in the proper numbers and places as the organism develops from the fertilized egg.

SUMMARY

- During vertebrate embryogenesis, proteins diffuse from the organizer, to induce formation of the neural plate, the edges of which fold upward to form the neural tube.
- In the vertebrate CNS, the fates of developing neurons are specified according to anteroposterior position and dorsoventral position.
- *Hox* genes determine anteroposterior identity in the hindbrain.
- Along the length of the neural tube, gradients of Sonic hedgehog protein (produced by the notochord and floor plate) and bone morphogenetic protein (emanating from the roof plate) regulate ventral and dorsal neural fates, respectively.
- Cells divide rapidly at the ventricular surface of the neural tube. Postmitotic neurons and glial progenitor cells migrate away from the ventricular surface of the neural tube to form the CNS.
- In mammalian cerebral cortex, development proceeds in an inside-out fashion. Neurons of the deepest cortical layers are born first.
- Division of neural stem cells in the central nervous system of adult birds and mammals continually produces new neurons. In the adult human, new cells are produced in hippocampus but not in cerebral cortex.
- Neuronal migration occurs along radial glial cells and pathways marked by cell surface and extracellular matrix components.
- The ultimate identity of a cell is determined by cell lineage and inductive interactions.
- Neural crest cells arise from the edge of the neural fold and migrate away from the neural tube to form the peripheral nervous system, pigment cells, and bones of the head.
- The transmitter synthesized by a sympathetic neuron is determined by its environment.
- The tip of a growing axon expands to form a growth cone.
- Cell surface and extracellular matrix adhesion molecules guide growth cones by short-range attractive and repulsive mechanisms.
- Netrins and semaphorins act as long-range chemoattractant or chemorepellents for different types of axons depending on what receptors are expressed on their growth cone surfaces.
- When a motoneuron growth cone contacts a muscle cell, functional synaptic transmission is established within minutes.
- The ephrins and the Eph family of receptor tyrosine kinases influence pathfinding, cell migration, and cell intermingling by a chemorepellent mechanism.
- Neurons rely on trophic factors for survival and differentiation.
- Programmed neuronal death (apoptosis) is a common feature of neural development.
- Synaptic connections, once established, are pruned to ensure appropriate and complete innervation of the target. Pruning occurs through activity-dependent competition among axon terminals for target-derived trophic substances.

Suggested Reading

General Reviews

Brenner, S. 1974. The genetics of *Caenorhabditis elegans*. *Genetics* 77: 71–94.

Briscoe, J. 2009. Making a grade: Sonic hedgehog signalling and the control of neural cell fate. *EMBO J.* 28: 457–465.

Buss, R. R., Sun, W., and Oppenheim, R. W. 2006. Adaptive roles of programmed cell death during nervous system development. *Annu. Rev. Neurosci.* 29: 1–35.

Clowry, G., Molnár, Z., and Rakic, P. 2010. Renewed focus on the developing human neocortex. *J. Anat.* 217: 276–288.

Ernsberger, U. 2009. Role of neurotrophin signalling in the differentiation of neurons from dorsal root ganglia and sympathetic ganglia. *Cell Tissue Res.* 336: 349–384.

Gehring, W. J., Kloter, U., and Suga, H. 2009. Evolution of the *Hox* gene complex from an evolutionary ground state. *Curr. Top. Dev. Biol.* 88: 35–61.

Gilbert, S. F. 2010. *Developmental Biology*, 6th ed. Sinauer, Sunderland, MA.

Kalb, R. 2005. The protean actions of neurotrophins and their receptors on the life and death of neurons. *Trends Neurosci.* 28: 5–11.

Kiecker, C., and Lumsden, A. 2005. Compartments and their boundaries in vertebrate brain development. *Nat. Rev. Neurosci.* 6: 553–564.

Le Douarin, N. M. 2008. Developmental patterning deciphered in avian chimeras. *Dev. Growth Differ.* 50(Suppl. 1): S11–28.

Levi-Montalcini, R. 1982. Developmental neurobiology and the natural history of nerve growth factor. *Annu. Rev. Neurosci.* 5: 341–362.

Lowery, L. A., and Van Vactor, D. 2009. The trip of the tip: understanding the growth cone machinery. *Nat. Rev. Mol. Cell Biol.* 10: 332–343.

Mallamaci, A. 2011. Molecular bases of cortico-cerebral regionalization. *Prog. Brain Res.* 189: 37–64.

Marín, O., and Rubenstein, J. L. 2003. Cell migration in the forebrain. *Annu. Rev. Neurosci.* 26: 441–483.

Mufson, E. J., Ginsberg, S. D., Ikonomovic, M. D., and DeKosky, S. T. 2003. Human cholinergic basal forebrain: chemoanatomy and neurologic dysfunction. *J. Chem. Neuroanat.* 26: 233–242.

Price, D. J., Jarman, A. P., Mason, J. O., and Kind, P. C. 2011. *Building Brains: An Introduction to Neural Development,* Wiley Blackwell, Oxford, UK.

Rakic, P. 2006. Neuroscience. No more cortical neurons for you. *Science* 313: 928–929.

Randlett, O., Norden, C. and Harris, W. A. 2010. The vertebrate retina: A model for neuronal polarization *in vivo*. *Dev. Neurobiol.* 6: 567–583.

Rubenstein, J. L. 2010. Three hypotheses for developmental defects that may underlie some forms of autism spectrum disorder. *Curr. Opin. Neurol.* 23: 118–123.

Sanes, D. H., Reh, T. A., and Harris, W. A. 2011. *Development of the Nervous System*, 3rd ed. Academic Press, Burlington, VT.

Schecterson, L. C., and Bothwell, M. 2010. Neurotrophin receptors: Old friends with new partners. *Dev. Neurobiol.* 70: 332–338.

Shooter, E. M. 2001. Early days of the nerve growth factor proteins. *Annu. Rev. Neurosci.* 24: 601–629.

Truman, J. W., Thorn, R. S., and Robinow, S. 1992. Programmed neuronal death in insect development. *J. Neurobiol.* 23: 1295–1311.

Original Papers

Adar, E., Nottebohm, F., and Barnea, A. 2008. The relationship between nature of social change, age, and position of new neurons and their survival in adult zebra finch brain. *J. Neurosci.* 28: 5394–5400.

Bhardwaj, R. D., Curtis, M. A., Spalding, K. L., Buchholz, B. A., Fink, D., Björk-Eriksson, T., Nordborg, C., Gage, F. H., Druid, H., Eriksson, P. S. et al. 2006. Neocortical neurogenesis in humans is restricted to development. *Proc. Natl. Acad. Sci. USA* 103: 12564–12568.

Campenot, R. B. 2009. NGF uptake and retrograde signaling mechanisms in sympathetic neurons in compartmented cultures. *Results Probl. Cell. Differ.* 48: 141–158.

Cherubini, E., Griguoli, M., Safiulina, V., and Lagostena, L. 2011. The depolarizing action of GABA controls early network activity in the developing hippocampus. *Mol. Neurobiol.* 43: 97–106.

Cox, E. C., Muller, B., and Bonhoeffer, F. 1990. Axonal guidance in the chick visual system: Posterior tectal membranes induce collapse of growth cones from the temporal retina. *Neuron* 4: 31–47.

Drescher, U., Kremoser, C., Handwerker, C., Loschinger, J., Noda, M., and Bonhoeffer, F. 1995. In vitro guidance of retinal ganglion cell axons by RAGS, a 25 kDa tectal protein related to ligands for Eph receptor tyrosine kinases. *Cell* 82: 359–370.

Ericson, J., Muhr, J., Placzek, M., Lints, T., Jessell, T. M., and Edlund, T. 1995. Sonic hedgehog induces the differentiation of ventral forebrain neurons: A common signal for ventral patterning within the neural tube. *Cell* 81: 747–756.

Evers, J., Laser, M., Sun, Y-A., Xie, Z-P., and Poo, M-M. 1989. Studies of nerve–muscle interactions in *Xenopus* cell culture: Analysis of early synaptic currents. *J. Neurosci.* 9: 1523–1539.

Gallo, G., and Letourneau, P. C. 2004. Regulation of growth cone actin filaments by guidance cues. *J. Neurobiol.* 58: 92–102.

Hansen, D. V., Lui, J. H., Parker, P. R., and Kriegstein, A. R. 2010. Neurogenic radial glia in the outer subventricular zone of human neocortex. *Nature* 464: 554–561.

Kennedy, T. E., Serafini, T., de la Torre, J. R., and Tessier-Lavigne, M. 1994. Netrins are diffusible chemotropic factors for commissural axons in the embryonic spinal cord. *Cell* 78: 425–435.

Lumsden, A. G. S., and Davies, A. M. 1986. Chemotropic effect of specific target epithelium in the developing mammalian nervous system. *Nature* 323: 538–539.

Puzzolo, E., and Mallamaci, A. 2010. Cortico-cerebral histogenesis in the opossum *Monodelphis domestica*: generation of a hexalaminar neocortex in the absence of a basal proliferative compartment. *Neural Development.* 5: 5–8.

Rakic, P. 1974. Neurons in rhesus monkey visual cortex: Systematic relationship between time of origin and eventual disposition. *Science* 183: 425–427.

Seeburg, P. H., Single, F., Kuner, T., Higuchi, M., and Sprengel, R. 2001. Genetic manipulation of key determinants of ion flow in glutamate receptor channels in the mouse. *Brain Res.* 907: 233–243.

Thompson, W. 1983. Synapse elimination in neonatal rat muscle is sensitive to pattern of muscle use. *Nature* 302: 614–616.

Wray, S., Grant, P., and Gainer, H. 1989. Evidence that cells expressing luteinizing hormone-releasing hormone mRNA in mouse are derived from progenitor cells in the olfactory placode. *Proc. Natl. Acad. Sci. USA* 86: 8132–8136.

Zheng, C., Heintz, N., and Hatten, M. E. 1996. CNS gene encoding astrotactin, which supports neuronal migration along glial fibers. *Science* 272: 417–419.

CHAPTER 26

Critical Periods in Sensory Systems

This chapter describes the effects of use and disuse on the visual systems of newborn kittens and monkeys. Experiments made by Wiesel and Hubel have shown that visual input in early life plays an essential role in developing structure and function of neurons in the visual cortex. At the time of birth, the receptive fields of neurons in the retina, lateral geniculate nucleus (LGN), and visual cortex resemble those of adults. There is, however, a marked difference in layer 4 of the primary visual cortex (V1) of the newborn animal, in which incoming geniculate axons from the two eyes overlap in their territory instead of being separated. As a result, neurons in layer 4 of neonatal cortex are driven by both eyes. During the first 6 weeks, the adult pattern is established. Lateral geniculate axons arising from the two eyes segregate so that cells in layer 4 respond to visual stimuli in only one eye. If one eye receives greater visual input, it takes over the territory of the other eye in the cortex. Closure of the lids of one eye during the first 3 months of life leads to blindness in that eye, and loss of its ability to drive cortical cells. Columns in the visual cortex and the geniculate axons supplied by the deprived eye shrink, while those supplied by the normal, undeprived eye expand. These results suggest that there is competition for territory in early life. Lid closure in adult animals has no effect on columnar architecture or responses to visual stimuli.

Evidence for competition between the two eyes is provided by experiments in which both eyes were deprived by lid closure shortly after birth or when an ocular muscle was cut in a newborn so as to produce a squint (strabismus). When neither eye has an advantage, normal columnar structure develops; however, each neuron in the cortex is driven by only one eye. The role of impulse activity in competition was shown by blocking impulse traffic in both optic nerves by application of tetrodotoxin (TTX). Under these conditions, ocular dominance columns failed to segregate. Appropriate sensory input is also essential in the development of the auditory system. Barn owls localize their prey by sight and sound. The visual and auditory maps of the outside world are normally aligned in the optic tectum at birth. When the visual input is displaced by applying prisms to the two eyes, young owls adapt to this mismatch in auditory and visual maps of the outside world. The receptive fields of auditory neurons in the tectum shift over days and weeks to fit the displaced visual map. Once the prisms are removed, the visual fields again shift abruptly, while the auditory fields move back slowly over time. After a critical period, such shifts in auditory receptive fields no longer occur. In owls brought up in an enriched environment with enhanced sensory experience, the critical period during which maps can be brought into register is prolonged. Sensory deprivation experiments are significant for considering the development of higher brain functions.

In previous chapters, we have emphasized that highly specific connections are necessary for the nervous system to function properly. This chapter shows that development and fine-tuning of connections between neurons are not complete at birth. For example, kittens—unlike monkeys that are able to see at birth—are born with their eyes closed. If the lids are opened and light is shone into an eye, the pupil constricts, but the animal appears to be completely blind.[1] By 10 days, the kitten shows evidence of vision and thereafter begins to recognize objects and patterns. When kittens are brought up in total darkness, the pupillary reflex continues to function, but they remain blind. On the other hand, adult cats that have been kept in darkness for prolonged periods can see immediately when exposed to light. Such results suggest that there is a critical period of vulnerability in early life. One might intuit that circuits, such as those used for spinal cord or pupillary reflexes, should be genetically determined, whereas the development of circuits controlling complex functions might require continuous interaction with the environment. What are the relative contributions of genetic factors and experience for cortical development? What mechanisms ensure that cortical circuits become stabilized when development is complete?

In this chapter, we show that the visual systems of the kitten and newborn monkey are first established in utero by intrinsic, genetically driven neuronal differentiation. Then, for a brief, well-defined period, visual experience refines neuronal circuits for optimal operation in the growing animal. We start by focusing on work that follows logically from the material presented in Chapters 2 and 3. Studies of the role of experience in forming the immature visual system set the stage for analyzing plasticity in other sensory circuits, notably the auditory system in birds and mammals.

The Visual System in Newborn Monkeys and Kittens

A good deal is known about the organization of the connections that underlie visual perception in the adult cat and monkey. A simple cell in the cortex selectively recognizes one well-defined type of visual stimulus, such as the movement of a narrow bar of light oriented vertically, in a particular region of the visual field of either eye. Such responses are possible because of the precise and orderly connections made to cortical cells from the retina by way of the lateral geniculate nucleus (see Chapters 2 and 3). It is natural to wonder whether cells and connections of this type are already present in the newborn animal or whether they develop as a result of visual experience, in which case, visual stimuli in early life would direct or refine a set of preexisting connections.

To study visually naive animals, monkeys were taken immediately after natural birth or after delivery by cesarean section, with care being taken to avoid exposure to light. To prevent form vision until animals were old enough to be studied, the lids were either sutured or the cornea was covered by a translucent occluder, which blurs images but allows light to pass. Other visually naive animals, such as kittens, rats, and ferrets, have also been examined during the first weeks after birth.[2–5]

Receptive Fields and Response Properties of Cortical Cells in Newborn Animals

A newborn monkey appears visually alert and is able to fixate. The responses of cortical neurons in many ways resemble those of the adult animal. For example, recordings made from individual cells in V1 show that the cells are not driven by diffuse illumination. As in a mature animal, they fire best when light or dark bars with a particular orientation are shone onto a particular region of the retina.[2] In recordings made in animals lacking prior visual experience, the range of orientations cannot be distinguished from that in adults. The receptive fields are also organized into antagonistic "on" and "off" areas that are driven by both eyes. Moreover, with oblique penetrations, the preferred orientation changes in a regular sequence as the electrode moves through the cortex (Figure 26.1). Figure 26.1 also shows that orientation preference maps with characteristic pinwheels (see Chapter 3) are already evident, with all orientations represented equally at the time of birth.[4]

[1]Riesen, A. H., and Aarons, L. 1959. *J. Comp. Physiol. Psychol.* 52: 142–149.

[2]Wiesel, T. N., and Hubel, D. H. 1974. *J. Comp. Neurol.* 158: 307–318.

[3]Hubel, D. H., and Wiesel, T. N. 1963. *J. Neurophysiol.* 26: 994–1002.

[4]Crair, M. C., Gillespie, D. C., and Stryker, M. P. 1998. *Science* 279: 566–570.

[5]Chapman, B., Stryker, M. P., and Bonhoeffer, T. 1996. *J. Neurosci.* 16: 6443–6453.

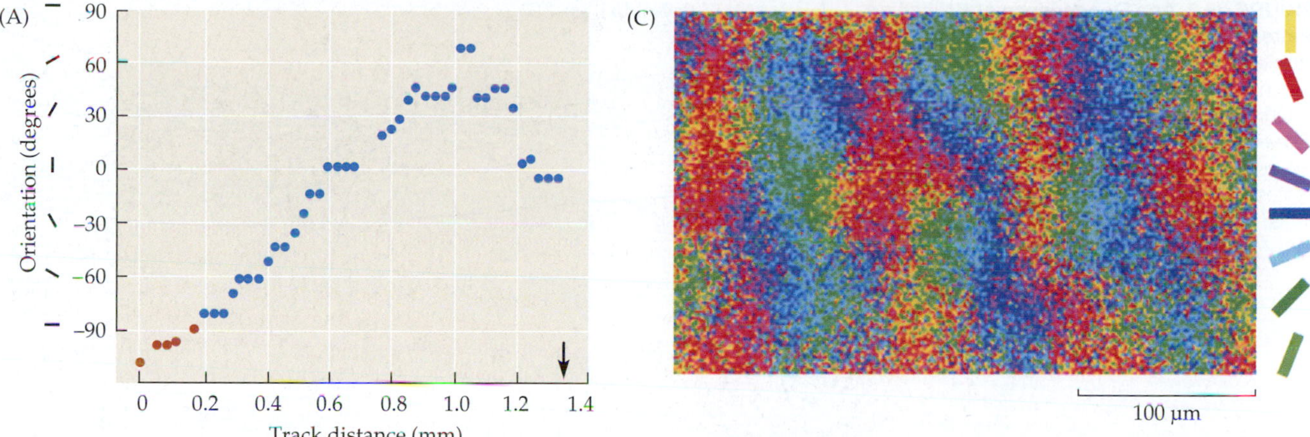

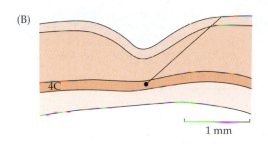

FIGURE 26.1 Orientation Columns in the absence of visual experience. (A) Axis orientation of receptive fields encountered by an electrode during an oblique penetration through the cortex of a 17-day-old baby monkey whose eyes had been sutured closed on the second day after birth. The receptive field orientation changes progressively as columns are traversed, indicating that normal orientation columns are present in the visually naive animal. Red dots are from the ipsilateral eye, blue dots from the contralateral eye. (B) The black dot marks the lesion made at the end of the electrode track in layer 4. (C) Orientation columns displayed by imaging in a 14-day-old kitten with lids sutured at birth. Colored bars (right) represent the orientation of the stimulus. Note that pinwheels are already present. (A,B after Wiesel and Hubel, 1974; C from Crair, Gillespie, and Stryker, 1998; micrograph kindly provided by M. C. Crair and M. P. Stryker.)

Ocular Dominance Columns in Newborn Monkeys and Kittens

At birth, most cells in all layers of V1 are already driven by both eyes—some better by one eye, some by the other, and some equally well by both. Figure 26.2 shows the distribution of responses of neurons distributed throughout all cortical layers according to eye preference in immature and adult monkeys.

The degree of ocular dominance is expressed in the histograms by grouping neurons into seven categories according to the discharge frequency with which they responded to stimulation of one or the other eye (see Chapter 3). The majority of cells respond to appropriate illumination of either eye. Cells in groups 1 and 7 in Figure 26.2 were driven only by visual stimuli applied to one eye, while those in groups 2 through 6 responded to both eyes.

The histograms of Figure 26.2A and B appear similar, with a range of eye preferences for cells throughout the cortical layers. What these data fail to show however is a striking and important difference in the properties of cells in layer 4. In that layer, cells in the newborn monkey are driven by both eyes; after 6 weeks of development, cells in layer 4 are driven by one eye only.[6] Outside of layer 4, cortical cells in newborn monkeys appear similar in their responses to those in adults, except that in some cells discharges are less vigorous or absent.

[6]Wiesel, T. N. 1982. *Nature* 299: 583–591.

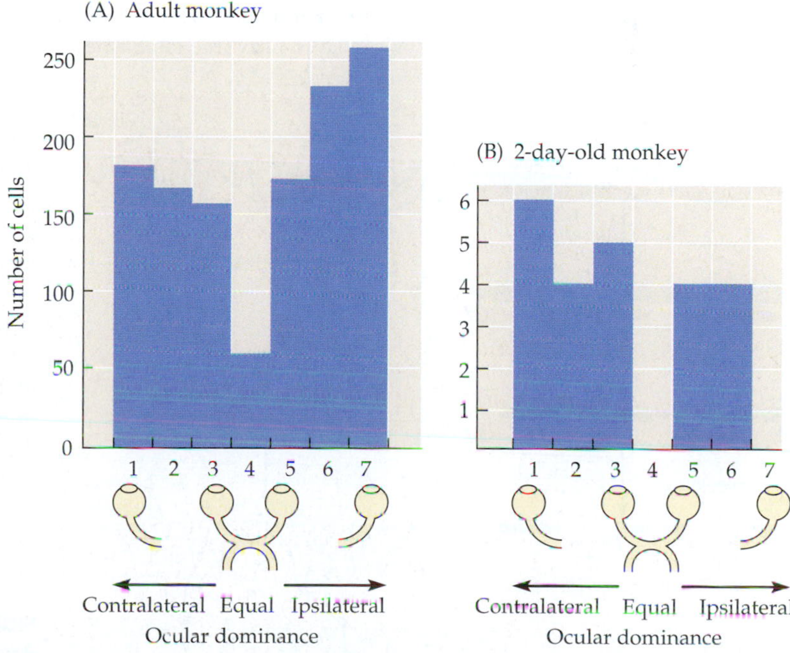

FIGURE 26.2 Ocular Dominance Distribution in the visual cortex of a newborn monkey. Cells in groups 1 and 7 of the histograms are driven by one eye only (ipsilateral or contralateral). All other cells have input from both eyes. In groups 2, 3, 5, and 6, one eye predominates. In group 4, both eyes have equal influence. (A) Normal adult monkey. (B) Normal 2-day-old monkey. (After Wiesel and Hubel, 1974.)

FIGURE 26.3 Age Dependence of Branching Patterns of Axons from Lateral Geniculate Nucleus. (A) Ending in layer 4, labeled by injection with horseradish peroxidase, an axon of a 17-day-old kitten. The axon spreads over a large uninterrupted territory in layer 4 of the visual cortex. (B) In the adult cat, the geniculate axon ends in two discrete tufts, interrupted by unlabeled fibers coming from the other eye. (After Wiesel, 1982.)

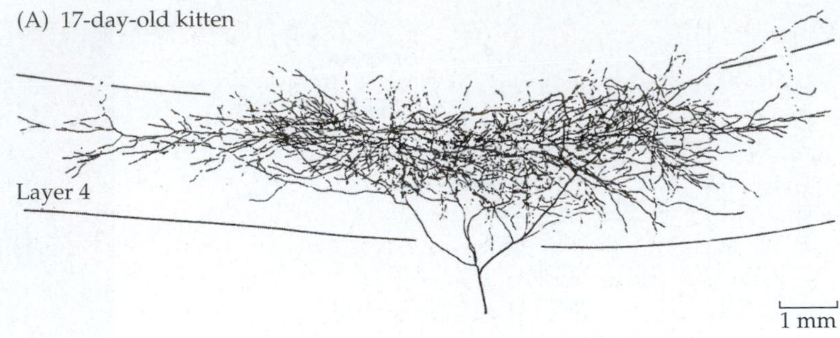

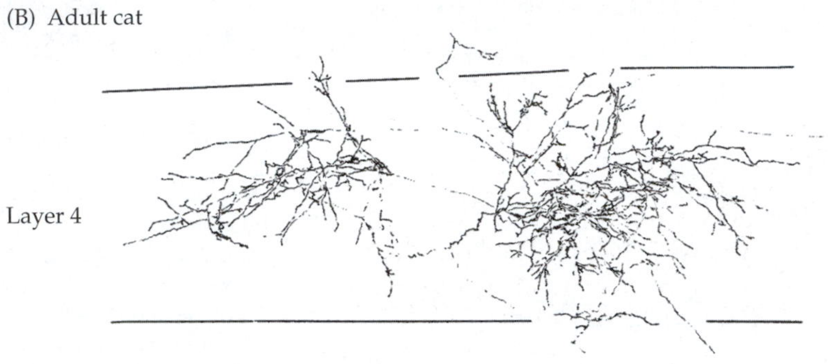

Postnatal Development of Ocular Dominance Columns

LeVay, Wiesel, and Hubel[7] (also Rakic;[8,9] see the next section) found a striking and important anatomical difference between adult and newborn in the geniculate innervation of layer 4 in V1. In contrast to the well-segregated geniculate terminals in the adult, arborizations of geniculate fibers ending in layer 4 overlap extensively in newborn kittens (Figure 26.3). As a result of the overlap in territories supplied by each eye, the immature neurons in layer 4 respond to stimulation of both eyes rather than just one.[7] Similarly, during the first 6 weeks of a monkey's life, the axons of geniculate cells in layer 4 retract to form smaller arborizations, as though pruned. In this way, separate domains of cortex in layer 4 become established, each supplied exclusively by one or the other eye (Figure 26.4). Comparable changes during development occur at the preceding stage in the lateral geniculate nucleus of the visual pathway.[10,11] As optic nerve fibers from the two eyes grow into the nucleus, their arborizations overlap extensively before they separate into distinct layers. The early postnatal development of ocular dominance columns and geniculate layers that starts before birth, proceeds in animals reared in total darkness. The retraction of geniculate axons in layer 4 is reminiscent of the events occurring during the development of nerve–muscle synapses in neonatal rats. At birth, each motor end plate is supplied by numerous motoneurons, but in a few weeks, most of the axons retract, leaving each muscle fiber supplied by just one motoneuron[12] (see Chapter 25).

[7]LeVay, S., Wiesel, T. N., and Hubel, D. H. 1980. *J. Comp. Neurol.* 191: 1–51.

[8]Kuljis, R. O., and Rakic, P. 1990. *Proc. Natl. Acad. Sci. USA* 87: 5303–5306.

[9]Rakic, P. 2006. *Cereb. Cortex.* 16(Suppl 1): i3–i17.

[10]Rakic, P. 1977. *Philos. Trans. R. Soc. Lond., B, Biol. Sci.* 278: 245–260.

[11]Shatz, C. J. 1996. *Proc. Natl. Acad. Sci. USA* 93: 602–608.

[12]Bernstein, M., and Lichtman, J. W. 1999. *Curr. Opin. Neurobiol.* 9: 364–370.

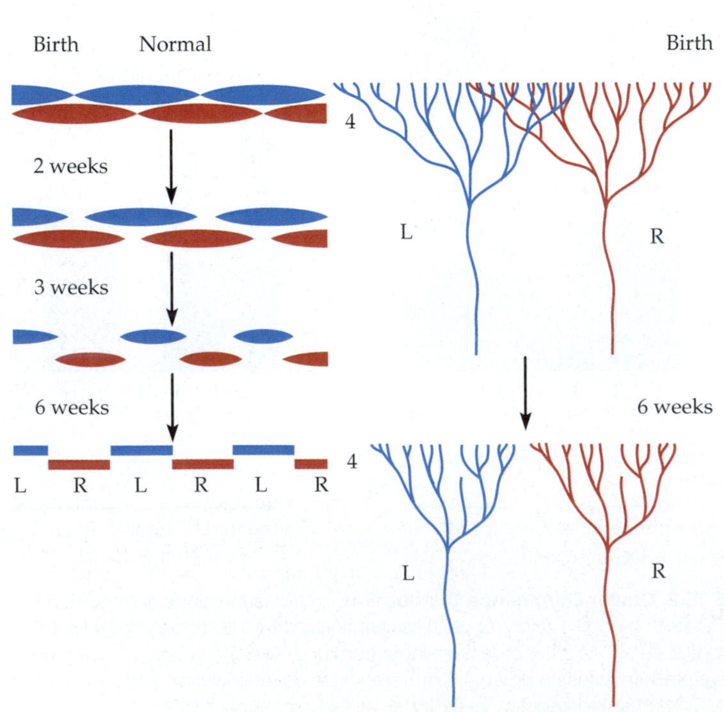

FIGURE 26.4 Retraction of Lateral Geniculate Nucleus Axons. Ending in layer 4 of the cortex during the first 6 weeks of life, the figure shows the overlap of inputs from the right (R) and left (L) eyes present at birth and the subsequent segregation into separate clusters corresponding to ocular dominance columns. The overlap at birth is greater in kittens than in monkeys. (After Hubel and Wiesel, 1977.)

Effects of Abnormal Visual Experience in Early Life

This section describes three types of experiments, mostly performed by Hubel and Wiesel in the first instance, in which animals were deprived of normal visual stimuli.[6,13] They studied the effects on the physiological responses of nerve cells and the structure of the visual system after (1) closing the lids of one or both eyes; (2) preventing form vision, but not access of light to the eye; and (3) leaving light and form vision intact, but producing an artificial strabismus (squint) in one eye. These procedures cause remarkable abnormalities in the function and anatomy of the cortex.

[13]Hubel, D. H. 1988. *Eye, Brain and Vision*. Scientific American Library, New York.

[14]Wiesel, T. N., and Hubel, D. H. 1963. *J. Neurophysiol.* 26: 1003–1017.

[15]Wiesel, T. N., and Hubel, D. H. 1963. *J. Neurophysiol.* 26: 978–993.

[16]LeVay, S., Stryker, M. P., and Shatz, C. J. 1978. *J. Comp. Neurol.* 179: 223–244.

[17]Wiesel, T. N., and Hubel, D. H. 1965. *J. Neurophysiol.* 28: 1029–1040.

Blindness after Lid Closure

When the lids of one eye were sutured during the first 2 weeks of life, monkeys and kittens still developed normally and used their unoperated eye. At the end of 1 to 3 months, however, when the operated eye was opened and the normal one closed, it was clear that the animals were practically blind in the previously deprived eyes. For example, kittens would bump into objects and fall off tables.[6,14] There was no gross evidence of a physiological defect within the eyes—pupillary reflexes appeared normal and so did the electroretinogram, which is an index of the average electrical activity of the eye. Records made from retinal ganglion cells in deprived animals showed no changes in their responses, and their receptive fields appeared normal.

Responses of Cortical Cells after Monocular Deprivation

Although responses of cells in the lateral geniculate nucleus appeared relatively unchanged after monocular deprivation,[15] there were major changes in the responses of cortical cells.[6,16,17] When electrical recordings were made in the visual cortex, very few cells could be driven by the eye that had been closed. The majority of those that did respond had abnormal receptive fields. Responses of cells driven by the undeprived eye were normal. Figure 26.5 shows ocular dominance histograms obtained from the cells examined in monkeys and kittens raised with closure of one eye during the first weeks of life.

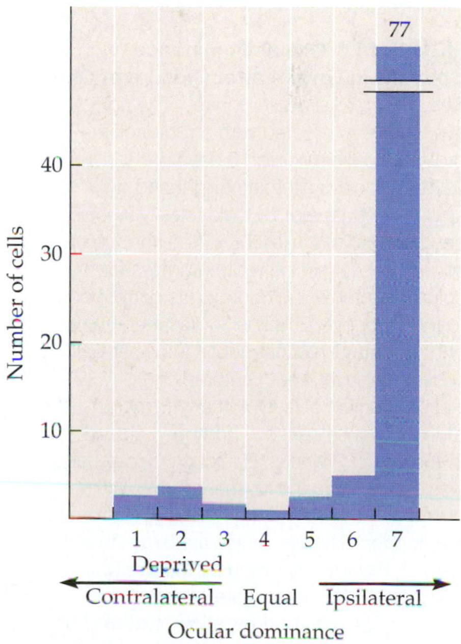

FIGURE 26.5 Damage Produced by Closure of One Eye Ocular dominance distribution in a monkey whose right eye was closed from 21 to 30 days of age. In spite of subsequent 4 years of binocular vision, most cortical neurons were unresponsive to stimulation of the deprived eye. (After LeVay, Wiesel, and Hubel, 1980.)

Relative Importance of Diffuse Light and Form for Maintaining Normal Responses

The results described so far indicate that if one eye is not used normally in the first weeks of life, its power wanes and it ceases to be effective in the visual cortex. These far-reaching changes are produced by the relatively minor procedure of closing the lids, without cutting any nerves. What is the important condition for maintaining and developing proper visual responses? Is diffuse light adequate?

Lid closure reduces the level of light that reaches the retina but does not exclude it. A series of experiments in newborn kittens demonstrated that form vision, rather than the mere presence of light, is required to prevent abnormal development of cortical connections. A plastic occluder (like a ping-pong ball) was placed over the cornea instead of closing the eyelids; the occluder prevented form vision but admitted light. All these cats were blind in the deprived eye.[14] Furthermore, cortical cells were no longer driven by the deprived eye. Neither retinal nor geniculate responses were noticeably changed under such conditions.

Morphological Changes in the Lateral Geniculate Nucleus after Visual Deprivation

Cells in the lateral geniculate nucleus of cat and of monkey are arranged in layers, each supplied predominantly by one or the other eye (see Chapter 2). In the same animals that showed marked abnormalities in the cortex after lid closure, the geniculate cells seemed at first to be normal. Nevertheless, it was shown that marked changes in morphology occurred after lid closure of one eye, resulting in cells that were noticeably smaller than in the layers supplied by the undeprived eye.[15] The reduction in size depended on the duration of lid closure. Surprisingly, the shrunken cells showed little physiological deficit. Several lines of evidence suggest that the size of the LGN may cells in reflect the extent of their arborization in the cortex.[18,19]

Morphological Changes in the Cortex after Visual Deprivation

The morphological consequences of eye closure are particularly conspicuous in layer 4 of the visual cortex V1, where geniculate fibers terminate in an orderly manner.[7,20] Changes in ocular dominance columns following lid closure in monkeys have been revealed by autoradiography of the cortex after injection of radioactive materials into one eye. After lid closure, there is a marked reduction in the width of ocular dominance columns receiving projections from the occluded eye. At the same time, the columns with inputs from the normal eye show a corresponding increase in width compared with that seen in a normal adult monkey. The shrinkage of ocular dominance columns is evident in Figure 26.6, in which the normal columns can be compared with columns in animals in which one eye had been closed at 2 weeks and left closed for 18 months. The changes indicate that geniculate axons activated by the normal eye retained or captured territory in the cortex lost by their weaker, visually deprived neighbors. These results are consistent with physiological observations made by recording from cells in layer 4. Almost all cells were driven only by the eye that had *not* been deprived. Certain features of the cortex, such as the alternating striped pattern of cytochrome oxidase staining in secondary visual cortex (V2),[21] are less vulnerable to deprivation than layer 4 in V1.[22]

[18] Guillery, R. W., and Stelzner, D. J. 1970. *J. Comp. Neurol.* 139: 413–421.

[19] Humphrey, A. L. et al. 1985. *J. Comp. Neurol.* 233: 159–189.

[20] Hubel, D. H., Wiesel, T. N., and LeVay, S. 1977. *Philos. Trans. R. Soc. Lond., B, Biol. Sci.* 278: 377–409.

[21] Horton, J. C., and Hocking, D. R. 1998. *Vis. Neurosci.* 15: 289–303.

[22] Hensch, T. K. 2004. *Annu. Rev. Neurosci.* 27: 549–579.

Critical Period for Susceptibility to Lid Closure

When the lids of one eye are closed in an adult cat or monkey, no abnormal consequences are seen.[6,7] Even if an eye is closed for over a year, the cells in the cortex continue to be driven

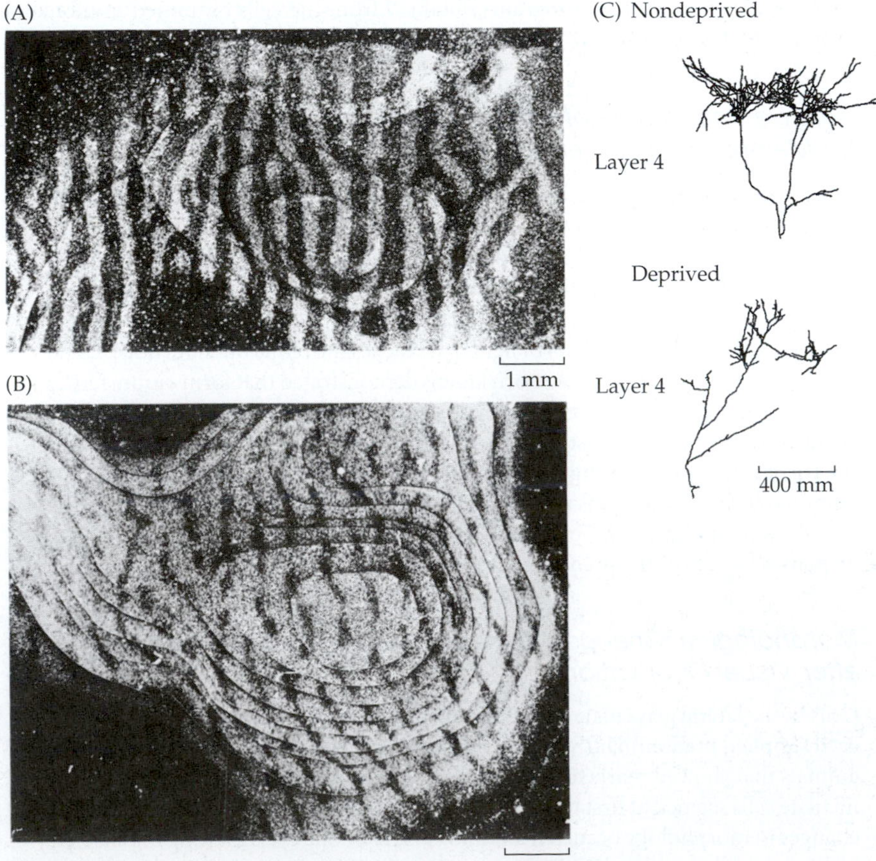

FIGURE 26.6 Ocular Dominance Columns in Layer 4 after Closure of One Eye. (A) Normal adult monkey. The right eye had been injected with a radioactive proline–fucose mixture 10 days previously. Layer 4 displays alternating light and dark stripes of equal width. Radioactively labeled geniculate axons in layer 4 of the right hemisphere appear as fine white granules forming columns. Intervening dark bands correspond to the other eye. This image was made as a photomontage reconstruction from parallel sections of layer 4 with autoradiography. (B) Reconstruction of layer 4 in an 18-month-old monkey whose right eye had been closed at the age of 2 weeks. Radioactive material was injected into the normal left eye. White grains demonstrate columns in layer 4 from the nondeprived eye, which are larger than normal. Columns supplied by the eye that had been closed (black) are narrower than normal. (C) Cortical arborization of labeled geniculate axons ending in layer 4 of a kitten in which one eye had been closed for 33 days. The terminal arborization of the geniculate axon from the deprived eye shows a dramatic reduction of branches compared to that from the nondeprived eye. (A,B from Hubel, Wiesel, and LeVay, 1977; C after Antonini and Stryker, 1993b.)

normally by both eyes and display the normal ocular dominance histogram. Moreover, after one eye has been completely removed in an adult monkey, the structure of layer 4 remains normal when observed with autoradiography or other staining methods, even though there is atrophy in the LGN. This finding indicates a remarkable resistance to change in layer 4 of the adult animal when compared with the changes seen in the immature animal.

In monkeys, the greatest sensitivity to lid closure is during the first 6 weeks of life.[6,7,20] At any time during that period, with a peak at 1 week of age,[23] substantial changes in eye preference and columnar architecture develop if one eye is closed for a few days. During the subsequent months (up to about 12 to 16 months), several weeks of closure are required to produce obvious changes in ocular dominance histograms or the width of columns in layer 4. At later times, changes do not develop even after surgical removal of one eye.

The period of greatest susceptibility to lid closure in kittens has been narrowed down to the fourth and fifth weeks after birth.[24,25] During the first 3 weeks or so of life, eye closure has little effect. This is not surprising, since the kittens' eyes are normally closed for the first 10 days. But abruptly, during weeks 4 and 5, sensitivity increases. Closure at that age for as little as 3 to 4 days leads to a sharp decline in the number of cells that can be driven by the deprived eye. An experiment in which littermates are compared is shown in Figure 26.7. In this example, 6- and 8-day closures starting at the age of 23 and 30 days (Figure 26.7A,B) caused about as great an effect as 3 months of monocular deprivation from birth. The susceptibility to lid closure declines after the critical period has passed and eventually disappears by about 3 months of age (Figure 26.7C). The critical period can, however, be prolonged by rearing kittens in the dark.[26,27] Then, susceptibility to monocular closure can still be demonstrated at 6 months of age, as though stimulus-driven activity is required to complete eye-specific cortical contacts. However, there is evidence that even a brief exposure of the kitten to light for a few hours may be sufficient to prevent such extension of the critical period.

Recovery during the Critical Period

To what extent is recovery possible after lid closure during the critical period? Even if the deprived eye in an adult cat or monkey is subsequently opened for months or years, the damage remains, with little or no recovery. Thus, the animal continues to be blind in

[23] Horton, J. C., and Hocking, D. R. 1997. *J. Neurosci.* 17: 3684–3709.

[24] Malach, R., Ebert, R., and Van Sluyters, R. C. 1984. *J. Neurophysiol.* 51: 538–551.

[25] Hubel, D. H., and Wiesel, T. N. 1970. *J. Physiol.* 206: 419–436.

[26] Cynader, M., and Mitchell, D. E. 1980. *J. Neurophysiol.* 43: 1026–1040.

[27] Daw, N. W. et al. 1995. *Ciba Found. Symp.* 193: 258–276; discussion 322–254.

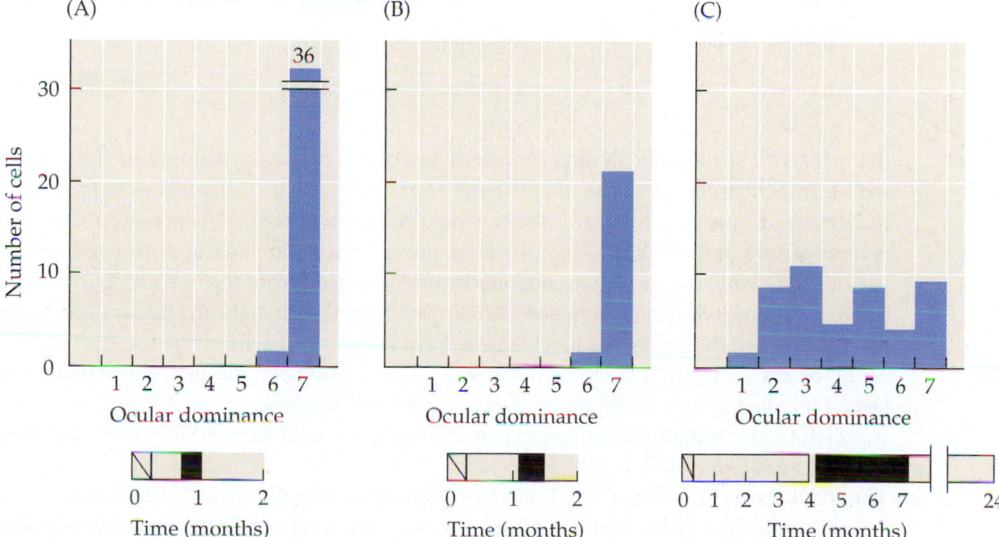

FIGURE 26.7 Critical Period in Kittens. Histograms showing eye preference in the visual cortex of kittens that were littermates, in which the right eye was closed at different ages. The period during which the eye was closed is indicated under the histograms. (A) Eyelids sutured for 6 days at 23 days of age. (B) Eyelids sutured for 9 days at 30 days of age. (C) The right eye was open the first 4 months, closed for 3 months, and then kept open until 2 years of age, when the recordings were made. (After Hubel and Wiesel, 1970.)

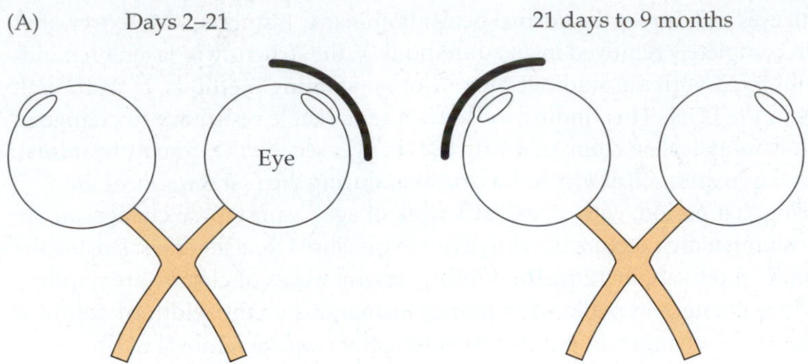

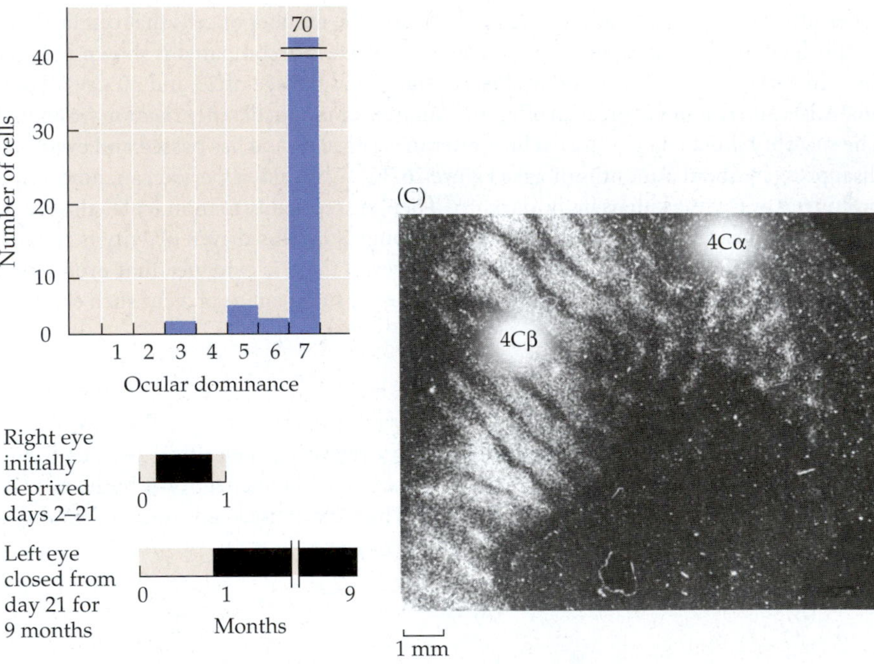

FIGURE 26.8 Effects of Reverse Suture on Ocular Dominance. (A) Histograms showing columnar organization in a monkey. The procedure was as follows. The right eye was closed from days 2 to 21 after birth, after which it was opened while the left eye was closed from day 21 for 9 months. (B) The ocular dominance histogram shows that almost all cells were driven exclusively by the right eye, which had been initially deprived. Virtually no cortical cells were driven by the left eye. Had both eyes been kept open at 21 days, the histogram would be reversed. Accordingly, fibers driven by the right eye recaptured cortical cells they had previously lost. (C) Tangential section of cortex passing through layer 4Cβ and 4Cα. The bands labeled by the right eye are expanded in layer 4Cβ even though it had been deprived of light for 19 days. During those first days, the columns supplied by the right eye had shrunk before expanding. Recovery did not occur equally well in other layers, such as 4Cα. (After LeVay, Wiesel, and Hubel, 1980.)

that eye, with shrunken columns and skewed ocular dominance histograms. In animals with monocular closure, experiments have been made in which the lids were opened in the deprived eye and closed over the normal eye (Figure 26.8). This procedure, termed **reverse suture**, leads to a recovery of vision, provided it is carried out during the critical period.[7,28,29] Monkeys and kittens not only begin to see again with the initially deprived eye, but they become blind in the other eye. Accompanying these changes, the ocular dominance histograms switch; that is, the newly opened eye drives most cells, while the eye that had been opened for the first weeks (now closed) cannot. Moreover, the anatomical pattern in layer 4 revealed by autoradiography shows a corresponding change: The shrunken regions supplied by the initially closed eye expand at the expense of the initially open (but then closed) eye (Figure 26.8C).

The conclusions from these experiments are that: (1) during the critical period in a normal animal, geniculate fibers supplying layer 4 of the cortex retract so that each eye supplies areas of comparable extent; (2) lid closure of one eye during the critical period leads to unequal retraction; and (3) reverse suture during the critical period produces *sprouting* of the geniculate axons so that an eye can recapture the cells it had lost (Figure 26.9).[30] If postponed until adulthood, reverse suture is without effect. For example, in a monkey in which the reverse suture was performed at 1 year of age, the labeled columns for the initially deprived eye remained shrunken.

[28] Blakemore, C., and Van Sluyters, R. C. 1974. *J. Physiol.* 237: 195–216.

[29] Kim, D. S., and Bonhoeffer, T. 1994. *Nature* 370: 370–372.

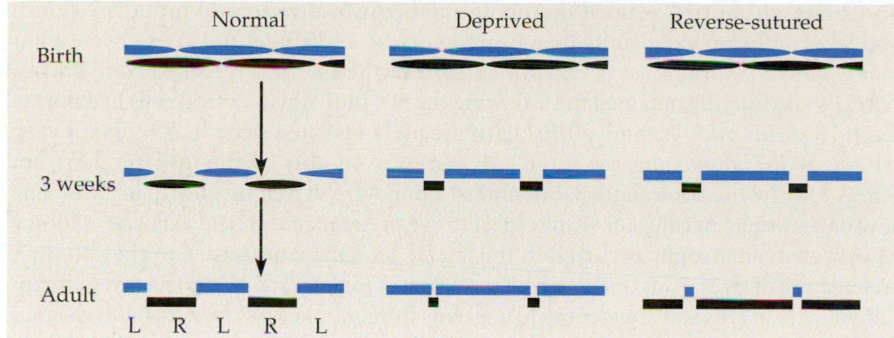

FIGURE 26.9 Summary of Effects of Eye Closure. In a normal monkey, ocular dominance columns have become well defined in layer 4 of the cortex by 6 weeks. Lid closure causes excessive retraction of geniculate fibers supplied by the deprived eye. Those supplied by the open eye retract less than usual, so their columns in layer 4 of the cortex are larger than normal in the adult. After reverse suture during the critical period, the initially deprived eye can recapture the territory it had lost in layer 4. (After Hubel and Wiesel, 1977.)

Requirements for Maintenance of Functioning Connections in the Visual System

At this stage one might be tempted to conclude that simple loss of activity in the visual pathways is the main factor that disrupts normal responses of cortical neurons. After all, cortical cells are driven not by diffuse illumination but by shapes and forms, so in the absence of form vision, cortical activity ceases. The following discussion shows that there must be additional causes that are more subtle than loss of activity. In particular, impulse activity from the two eyes interacts competitively to determine cortical connectivity.

Binocular Lid Closure and the Role of Competition

The first clue that loss of visually evoked activity cannot on its own account for the changed performance of neurons is shown by the following experiments. Both eyes were closed in monkeys that were either newborn at term or delivered prematurely by cesarean section.[2,6] From the preceding discussion, one might guess that cells in the cortex would subsequently be driven by neither eye. Surprisingly, however, after binocular closure for 17 days or longer, most cortical cells could still be driven by appropriate illumination, and the receptive fields of simple and complex cells appeared largely normal. The columnar organization for orientation was similar to that in controls (see Figure 26.1). The principal abnormality was the appearance of a substantial fraction of the cells that could not be driven binocularly. In addition, some spontaneously active cells could not be driven at all, and others did not require specifically oriented stimuli. However, the areas of cortex supplied by each eye were equal, and the pattern resembled that seen in normal adult monkeys—that is, in layer 4, cells were driven by one eye only, and columns were well defined when marked by autoradiography or by cytochrome oxidase. Binocular closure in kittens led to similar effects, except that more cortical cells continued to be binocularly driven.[17] The arborizations of lateral geniculate axons in layer 4 were not shrunken.[30] At the same time, cells in the lateral geniculate body showed atrophy (a decrease in size of approximately 40%) in all layers.

The conclusion from these experiments is that some, but not all, of the ill effects expected from closing one eye are reduced or averted by closing both eyes. It is as though inputs from the two eyes are in competition for representation in cortical cells, and with one eye closed, the contest becomes unequal.

Effects of Strabismus (Squint)

The abnormal effects described in the preceding discussion were produced by suturing eyelids or by using translucent diffusers, implicating loss of form vision. Following the clue that cross-eyed children (i.e., those with strabismus or squint) or wall-eyed children can become blind in one eye, Hubel and Wiesel produced artificial strabismus in kittens and newborn monkeys by cutting an eye muscle.[6,31] The optical axis of that eye was thereby deflected from normal. Under such conditions, illumination and pattern stimulation for each eye remained unchanged but the visual fields were not aligned.

[30] Antonini, A. et al. 1998. *J. Neurosci.* 18: 9896–9909.

[31] Hubel, D. H., and Wiesel, T. N. 1965. *J. Neurophysiol.* 28: 1041–1059.

The results at first seemed disappointing because after several months vision in both eyes of the operated animals appeared normal, and Hubel and Wiesel were about to abandon a laborious set of experiments. Nevertheless, they recorded from cortical cells and consistently obtained the following results. Individual cortical cells had normal receptive fields and responded briskly to precisely oriented stimuli. *But almost every cell responded only to one eye*; some cells were driven only by the ipsilateral eye and others only by the contralateral, but almost none were driven by both. The cells were, as usual, grouped in columns with respect to eye preference and field axis orientation.[32] As expected, no atrophy occurred in the lateral geniculate nucleus, and the columnar architecture of layer 4 was unchanged. The almost complete lack of binocular representation in cortical cells was apparent in histograms from monkeys with artificial strabismus. The critical period for displaced images to produce changes was comparable to that for monocular deprivation.

These experiments provide an example in which all the usual parameters of light are normal—the amount of illumination and form and pattern stimuli. The only change consists of a failure of the images to fall on corresponding regions of the two retinas. The factor that seems important for the loss of binocular convergence of geniculate projections to the cortex is lack of congruity of input from the two eyes. It is as though the homologous receptive fields in both eyes must be in register with one another, so that excitation will be simultaneous. The following experiments further support this idea.

During the first 3 months of its life, the eyes of a kitten were occluded with an opaque plastic cover that was switched on alternate days from one eye to the other, so that the two eyes received the same total experience, but at different times.[31] The result was the same as in the cross-eyed experiments, demonstrating that cells were driven predominantly by either one eye or the other, but not by both. The maintenance of normal binocularity, therefore, depends not only on the amount of impulse traffic but also on the appropriate spatial and temporal overlap of activity in the different incoming fibers.

Changes in Orientation Preference

A logical question to ask is whether raising animals in an environment in which they see only one orientation can change the orientation preference of cortical cells. An experimental approach that involved competition as well as deprivation was used by Carlson, Hubel, and Wiesel.[33] The lids of one eye were sutured in a newborn monkey. The animal was kept in darkness except when it placed its head in a holder (Figure 26.10). Then, with the head held vertically, it would see vertical stripes with the unsutured eye. Since the monkey received orange juice each time it placed its head in the holder correctly, it performed this maneuver frequently. Thus, during the critical period, one eye received no visual input, while the other saw only vertical stripes. After 57 hours of experience between 12 and 54 days after birth, normal levels of cortical activity were found, with cells of all orientations arranged as usual in columns. As expected, the open eye tended to dominate.

When tests were made for orientation preference, the results shown in Figure 26.10 were obtained. Both eyes could drive cells well when horizontal lines were the stimulus. However, the left eye (the open eye) was considerably more effective for vertical stripes. The probable explanation for this result is that neither eye saw horizontal bars or edges during the critical period. Hence, the stimulation for horizontality is analogous to binocular closure; that is, the competition is equal. For the vertical input, however, the open eye had an enriched experience and in a sense captured cells in vertical orientation columns that had previously been supplied by the deprived eye. Similar results have been obtained by Bonhoeffer and his colleagues in kittens reared in a striped environment.[34]

Segregation of Visual Inputs without Competition

In the experiments described so far, an underlying principle has been that the two eyes compete for connections and territory in the LGN and in layer 4 of V1, starting off with roughly equal opportunity. Rakic and his colleagues have used a different approach to study how neighboring groups of cells with defined properties sort out their terminals and their targets as they develop, *without* competition.[35]

[32] Lowel, S., and Singer, W. 1992. *Science* 255: 209–212.

[33] Carlson, M., Hubel, D. H., and Wiesel, T. N. 1986. *Brain Res.* 390: 71–81.

[34] Sengpiel, F., Stawinski, P., and Bonhoeffer, T. 1999. *Nat. Neurosci.* 2: 727–732.

[35] Meissirel, C. et al. 1997. *Proc. Natl. Acad. Sci. USA* 94: 5900–5905.

(A)

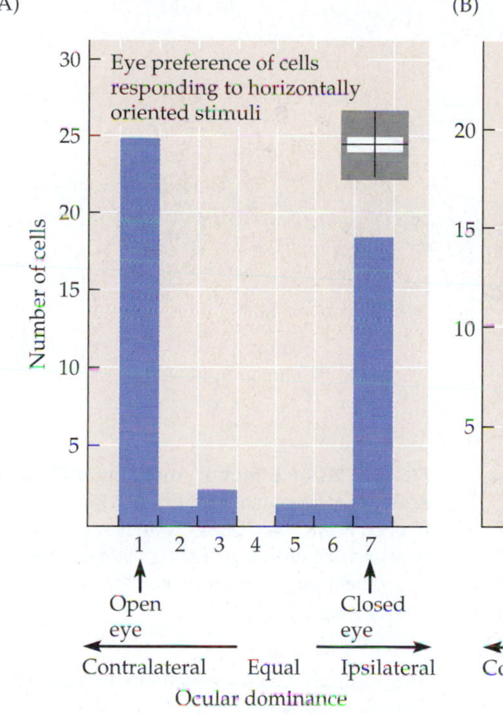

(B)

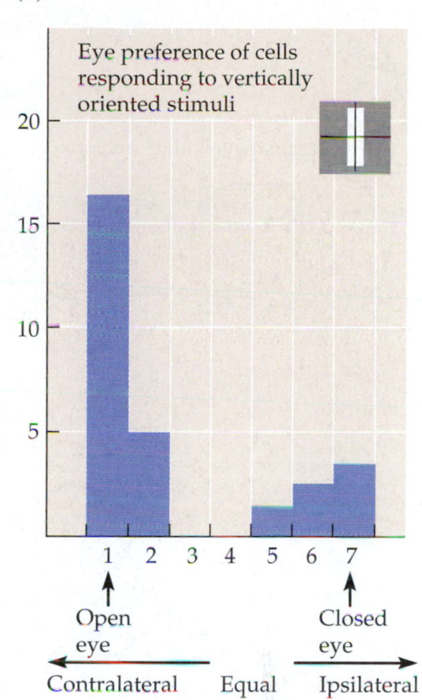

FIGURE 26.10 Orientation Preferences of Cortical Cells in a monkey with altered visual experience. The monkey was kept in a dark room. At 12 days, the right eye was closed. Whenever the monkey placed its head in a holder (to ensure that the head was not tilted), it received orange juice. At that time, it also saw vertical stripes with its left eye. For a total of 57 hours of exposure between 12 and 54 days, the only visual experience was an image of vertical lines seen by the left eye. The right eye saw nothing. (A) After 54 days the right eye was re-opened and horizontally oriented light stimuli shone onto the screen stimulated cortical cells driven by the left eye or the right eye equally well. In this histogram, no deprivation is apparent for horizontal orientation, except for a lack of binocular cells. (B) With vertically oriented stimuli, the left eye, which had been kept open to view vertical stripes, was much more effective in driving cortical cells. The histogram resembles that seen after monocular deprivation. The results suggest that competition was equal for horizontal stimuli that neither eye had ever seen, and unequal for vertical stimuli (favored by the left, open eye). (After Carlson, Hubel, and Wiesel, 1986.)

As described in Chapter 3, the magnocellular (M) and parvocellular (P) systems occupy distinct layers in the LGN and in visual cortex. By staining individual M and P axons during development, as they grow into the lateral geniculate nucleus, Rakic and his colleagues have shown that axons from the outset arrive in the correct M and P layers, where they form characteristic, non-overlapping arborizations. M fibers end only in geniculate layers 1 and 2, while P fibers end only in layers 3, 4, 5, and 6, with no spillover. Hence, when the two eyes develop their connections, one can suppose that competition plays a role for inputs that provide similar information about the visual world. By contrast, the M and P systems carry quite different types of information. Their connections (like those involved in the formation of blobs and or the stripes in visual area 2)[21] develop according to principles in which competition is not of the essence. A further example of connections that form without signs of competition is the alignment of orientation maps in the visual cortex of the kitten.[36]

Effects of Impulse Activity on the Developing Visual System

The role of impulse activity in shaping synaptic connections has been studied in experiments on kittens. When the lids are closed or the animal is brought up in complete darkness, impulse traffic in the visual pathways does not stop entirely. Neurons continue to fire spontaneously, and ocular dominance columns develop as segregated areas in layer 4.

Experiments by Stryker, Shatz, and their colleagues[37,38] have shown that this presumably equal spontaneous activity from the two eyes is important for normal development.

[36] Godecke, I., and Bonhoeffer, T. 1996. *Nature* 379: 251–254.

[37] Stryker, M. P., and Harris, W. A. 1986. *J. Neurosci.* 6: 2117–2133.

[38] Sretavan, D. W., Shatz, C. J., and Stryker, M. P. 1988. *Nature* 336: 468–471.

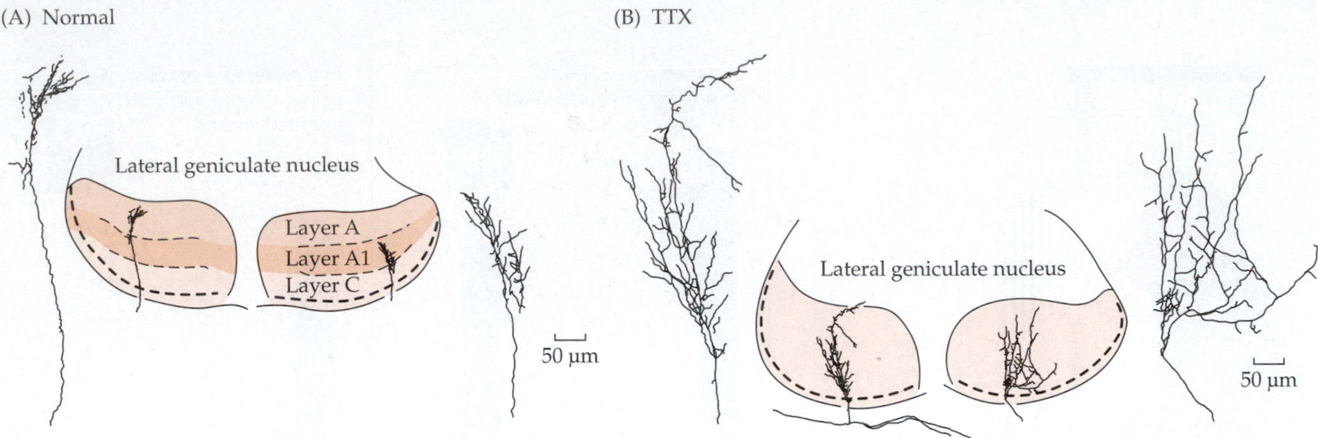

FIGURE 26.11 Effect of Abolition of Electrical Activity by Tetrodotoxin on arborization of optic nerve fibers terminating in the lateral geniculate nucleus. (A) In a normal kitten, the terminals of optic nerve fibers labeled with horseradish peroxidase are restricted to the single layer where they end. (B) After application of tetrodotoxin for 16 days during embryonic life, labeled axons show much larger arborizations that are not restricted to individual layers. (After Sretavan, Shatz, and Stryker, 1988.)

Tetrodotoxin (TTX), which blocks action potentials, was injected into both eyes of newborn kittens. Several days after removal of the toxin, impulse conduction in the visual pathways from retina through the geniculate to the cortex was restored. However, in the lateral geniculate nucleus, inputs from the two eyes had failed to segregate into separate layers (Figure 26.11). Moreover, cells in layer 4 of the visual cortex were still driven by both eyes as in the newborn animal, and the ocular dominance columns revealed by autoradiography resembled the neonatal pattern, with extensive overlap and no clear boundaries. Thus, in the absence of *all* firing, optic nerve fibers failed to segregate in the geniculate, and geniculate fibers failed to retract normally in layer 4 of the cortex.

The effects of lid closure on ocular dominance columns can also be modified by blocking impulse activity within the cortex itself (Figure 26.12). TTX was infused into the visual cortex of a kitten during the critical period while one eye was deprived of form and light.[39] After the TTX was removed, cortical cells were responsive to stimuli in both eyes, even though one had been deprived. Again, in the absence of activity, retraction failed to occur. Stryker and his colleagues performed similar experiments, using agents that inhibited firing of cortical neurons but left geniculate axons functional.[40] Their results indicated that postsynaptic activity also played a role in retraction. Thus, it is not simply the amount of incoming activity that is important but also whether that activity successfully drives cortical neurons.

[39]Reiter, H. O., Waitzman, D. M., and Stryker, M. P. 1986. *Exp. Brain Res*. 65: 182–188.

[40]Hata, Y., Tsumoto, T., and Stryker, M. P. 1999. *Neuron* 22: 375–381.

[41]Maffei, L., and Galli-Resta, L. 1990. *Proc. Natl. Acad. Sci. USA* 87: 2861–2864.

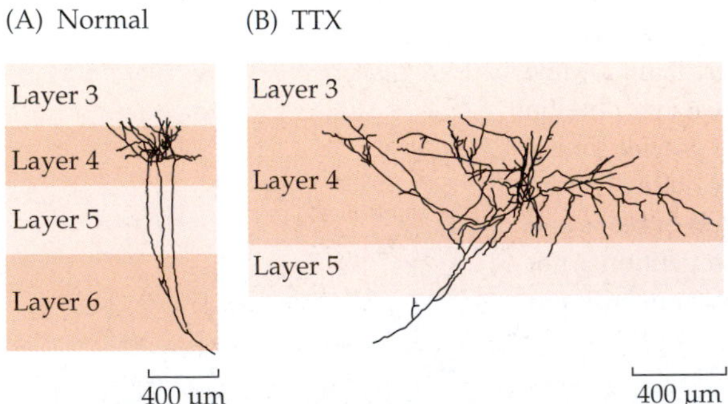

FIGURE 26.12 Increased Arborization of Lateral Geniculate Fibers. (A) Ending in layer 4 of visual cortex after application of tetrodotoxin to both eyes, normal arborization of a labeled geniculate axon in layer 4 (30-day-old animal). (B) Labeled geniculate axon in a kitten in which tetrodotoxin had been applied to the eyes for 12 days (29-day-old animal). The axons of this neuron cover a much larger area of cortex. (After Antonini and Stryker, 1993a.)

Synchronized Spontaneous Activity in the Absence of Inputs during Development

The experiments with TTX suggest that action potential activity in the visual pathway is necessary for the sorting of axons to their appropriate targets. Without ongoing activity, axons remain spread across layers in the LGN and across boundaries of ocular dominance columns in the cortex. Yet, as we have seen, much of the development has already proceeded by the time of birth. In the darkness of the womb, before a kitten or a monkey has seen anything and before photoreceptors have become functional, the layers of the LGN and the cortical columns are recognizable. Does this mean that early development proceeds without action potential activity, or is there intrinsic impulse activity in the system that guides development? Maffei and his colleagues demonstrated that synchronous bursts of action potential traffic do, in fact, propagate along the optic nerve in utero.[41]

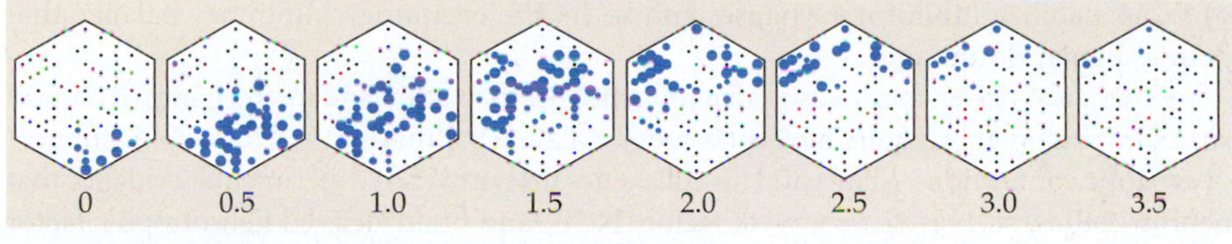

FIGURE 26.13 Wave of Impulse Activity Spreading across Isolated Retina of a neonatal ferret. The isolated retina was placed on recording electrodes, embedded in a regular array in the dish. The position of each of 82 retinal neurons is represented by a small black spot. Electrically active neurons are marked by larger blue spots, the sizes of which are proportional to the firing rates. Each frame represents the activity averaged over successive 0.5-second (s) intervals. During the time represented by the eight frames (3.5 s), action potentials begin with one small group of cells and spread slowly across the retina. A new wave begins shortly after, and then another, each spreading in a different direction. At this stage of development, photoreceptors in the ferret are not responsive to light. (After Meister et al., 1991.)

Meister, Baylor, and their colleagues showed that there are periodic, synchronized waves of firing of neighboring ganglion cells in retinas isolated from immature ferrets and fetal kittens.[42] In these experiments, the retina was placed in a chamber over an array of 61 electrodes; from each electrode, it was possible to identify the discharges of as many as four different ganglion cells. Recordings of this type revealed an ordered pattern of activity sweeping across the retina from ganglion cell to ganglion cell. An example is shown in Figure 26.13. Small black dots represent the positions of retinal ganglion cells, larger blue dots show the location of action potential discharges, and the size of the dot indicates discharge frequency. Successive frames were taken at 0.5-second intervals. The wave of activity spread across the retina over a period of about 3 seconds. Typically, such waves recurred repeatedly, separated by silent periods on the order of 2 seconds in duration. Wong[43] has reported a similar wave-like spread of transient changes in intracellular calcium concentration and suggested that these play a role in synchronization of the electrical activity. There is evidence that cholinergic neurons, starburst amacrine cells, and electrical coupling play a part in the generation of the coordinated firing of ganglion cells in the immature retina.[44–46] It will be shown below that waves of intrinsic activity also occur in the developing cochlea.

Role of γ-Aminobutyric Acid (GABA) and Trophic Molecules in Development of Columnar Architecture

The refinement of ocular dominance columns involves interactions between excitatory thalamocortical inputs and GABAergic inhibitory cortical interneurons. Chronic infusion of cat visual cortex with diazepam, a positive modulator of $GABA_A$ receptors, reduced the binocularity of cortical neurons and broadened anatomically defined ocular dominance columns.[47] Downregulation of $GABA_A$ receptors produced opposite effects. Similarly, knockout mice lacking the GABA synthesizing enzyme glutamic acid decarboxylase 65 kDa (GAD65) lacked ocular dominance plasticity, and plasticity was restored by cortical infusion of diazepam.[48] Cortical plasticity may involve a specific subtype of GABAergic interneuron. Synapse formation on pyramidal cells in the visual cortex by parvalbumin-positive GABAergic neurons is influenced by visual experience during early postnatal life.[49] Quite remarkably, transplantation of embryonic inhibitory neurons into mouse visual cortex (at postnatal day 10) extended the period of ocular dominance plasticity to postnatal day 43; this is 2 weeks beyond the normal close of the critical period on postnatal day 28.[50] These findings indicate that inhibitory interneurons are important for maintaining the period for development of ocular dominance. This hypothesis receives additional support from the role of extracellular matrix (ECM) in synapse stabilization and brain plasticity.[51] Large perineuronal nets of ECM develop around parvalbumin-positive cells (thought to be GABAergic inhibitory interneurons) at the end of the ocular dominance critical period. Enzymatic degradation of ECM can restore ocular dominance plasticity to visual cortex in adult rats.[52] It is postulated that growing perineuronal nets

[42] Meister, M. et al. 1991. *Science* 252: 939–943.

[43] Wong, R. O. 1999. *Annu. Rev. Neurosci.* 22: 29–47.

[44] Feller, M. B. 2009. *Neural Dev.* 4: 24.

[45] Zhou, Z. J. 1998. *J. Neurosci.* 18: 4155–4165.

[46] Brivanlou, I. H., Warland, D. K., and Meister, M. 1998. *Neuron* 20: 527–539.

[47] Hensch, T. K., and Stryker, M. P. 2004. *Science* 303: 1678–1681.

[48] Hensch, T. K. et al. 1998. *Science* 282: 1504–1508.

[49] Chattopadhyaya, B. et al. 2004. *J. Neurosci.* 24: 9598–9611.

[50] Southwell, D. G. et al. 2010. *Science* 327: 1145–1148.

[51] Fawcett, J. 2009. *Prog. Brain Res.* 175: 501–509.

[52] Pizzorusso, T. et al. 2002. *Science* 298: 1248–1251.

of ECM stabilize inhibitory synapses and so fix the excitatory/inhibitory balance that defines ocular dominance.

Experience-driven mechanisms that influence the formation and maintenance of cortical structure presumably require molecular cues for axonal pathfinding and for trophic support of synaptic connections. Maffei and his colleagues first provided experimental evidence that neurotrophins, such as nerve growth factor (NGF) and brain-derived neurotrophic factor (BDNF), can prevent the effects of monocular lid closure in the developing rat visual system.[53,54] The supposition is that column shrinkage during deprivation occurs when axons fail to receive sufficient trophic molecules from their target cell and withdraw their connections (see Chapter 25). Introduction of extrinsic NGF into the region allows synaptic connections to be maintained, thereby preventing column shrinkage. The introduction of anti-NGF antibody blocks this effect. Moreover, in the developing visual system, NGF antibodies cause shrinkage of cells and prolongation of the critical period, as though a normal trophic action of NGF were being antagonized.[55] BDNF also may play a role in cortical plasticity. Transgenic overexpression of cortical BDNF in mice hastened the formation of inhibitory synapses and accelerated the onset and closure of the ocular dominance critical period.[56]

Critical Periods in Somatosensory and Olfactory Systems

Whisker barrels constitute a prominent feature of the somatosensory cortex in rodents and correspond exactly in number and arrangement to the large whisker follicles on the rodent snout (see Chapters 21 and 23). Barrels fail to develop if a whisker is ablated in early life.[57] However, such peripheral lesions no longer alter cortical barrels after postnatal day 4,[58] revealing a sensitive period for structural plasticity in the trigeminal pathway.[59] One might expect that such fundamental processes as the formation of somatotopic maps would somehow be driven by intrinsic cues and would rely less on sensory experience for direction. Indeed, blockade of peripheral action potentials with TTX does not disrupt cortical barrel formation.[60] At the same time, although overall cortical structure appears normal, rats that have had their whiskers trimmed for the first 3 days of life (but not later) show changes in dendritic structure within the barrels and have both impaired sensory discrimination and reduced social interaction as adults.[61] It is not yet known how such early changes might relate to experience-dependent plasticity seen in the synaptic organization of the adult barrel cortex.[62]

Yet another system in which sensory input is required for the development of normal structure is olfaction. Closure of a nostril in the immature rat prevents the normal development of olfactory glomeruli.[63] In immature mice, each glomerulus of the olfactory bulb is supplied by the axons of olfactory receptor neurons expressing a variety of olfactory receptor proteins. In the adult, however, each glomerulus is supplied by axons that express the same unique odorant receptor protein (see Chapter 19). Blockage of one nostril in an immature mouse prevents such sorting out of the axonal supply to olfactory glomeruli. In contrast to the refinement of connections in other sensory systems that depends on electrical activity, olfactory bulb formation depends on other aspects of the G protein–coupled signaling cascade. Olfactory glomeruli develop normally in mice in which the odorant-activated, cyclic-nucleotide–gated channel is missing and whose olfactory receptor neurons are electrically inactive.[64] Instead, glomerular development depends on cyclic adenosine monophosphate (cAMP) production and was severely disturbed when the synthetic enzyme adenylyl cyclase 3 was disrupted.[65]

Sensory Deprivation and Critical Periods in the Auditory System

The auditory system provides familiar examples of the role of sensory experience in development. It is common knowledge that language is most easily acquired in the first years of life. In fact, there is an early loss of perceptual sensitivity to some phonetic distinctions that are not part of the native language; for example, Japanese-speaking infants lose their initial ability to distinguish between "R" and "L" sounds.[66] From clinical studies there is a clear correlation between speech or language delays in children and the severity of their hearing loss.[67] The profound importance of human language gives further impetus to studies of developmental plasticity in the auditory system.

[53] Berardi, N., and Maffei, L. 1999. *J. Neurobiol.* 41: 119–126.

[54] Pizzorusso, T. et al. 1994. *Proc. Natl. Acad. Sci. USA* 91: 2572–2576.

[55] Capsoni, S. et al. 1999. *Neuroscience* 88: 393–403.

[56] Huang, Z. J. et al. 1999. *Cell* 98: 739–755.

[57] Van der Loos, H., and Woolsey, T. A. 1973. *Science* 179: 395–398.

[58] Rebsam, A., Seif, I., and Gaspar, P. 2005. *J. Neurosci.* 25: 706–710.

[59] Erzurumlu, R. S. 2010. *Exp. Neurol.* 222: 10–12.

[60] Henderson, T. A., Woolsey, T. A., and Jacquin, M. F. 1992. *Brain Res. Dev. Brain Res.* 66: 146–152.

[61] Lee, L. J. et al. 2009. *Exp. Neurol.* 219: 524–532.

[62] Lebedev, M. A. et al. 2000. *Cereb. Cortex.* 10: 23–31.

[63] Zou, D. J. et al. 2004. *Science* 304: 1976–1979.

[64] Lin, D. M. et al. 2000. *Neuron* 26: 69–80.

[65] Chesler, A. T. et al. 2007. *Proc. Natl. Acad. Sci. USA* 104: 1039–1044.

[66] Werker, J. F., and Tees, R. C. 2005. *Dev. Psychobiol.* 46: 233–251.

[67] Schonweiler, R., Ptok, M., and Radu, H. J. 1998. *Int. J. Pediatr. Otorhinolaryngol.* 44: 251–258.

FIGURE 26.14 Plasticity of the Tonotopic Map in Auditory Cortex of Young Rats. Representative cortical tonotopic maps from a naive rat and a rat that had been exposed to 7.1-kHz tone pips from postnatal day 9 to day 30. Each polygon corresponds to a recording site and to the color codes for the characteristic frequency of the recorded neurons. The gray areas had characteristic frequencies in a range of 7.1 kHz ± 0.2 octaves. Note the enlarged representation of frequencies near 7.1 kHz in the tone-exposed animal. (Sanes and Bao, 2009, modified from Han et al., 2007.)

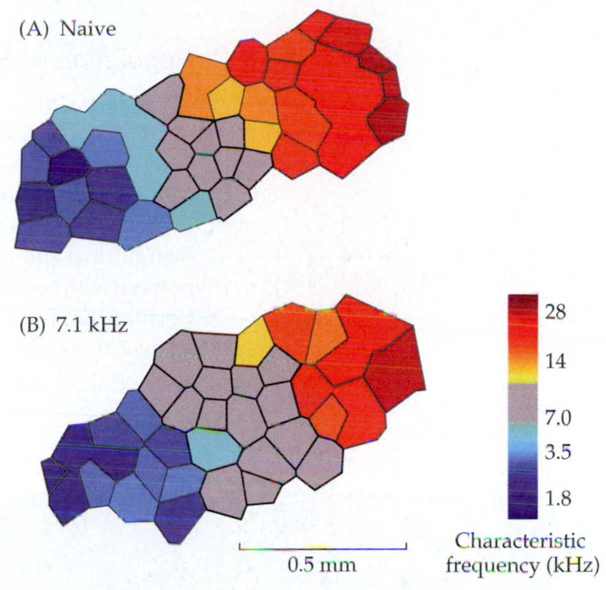

Studies in animals have focused on manipulation of the tonotopic map of cortex by over- or under exposure to selected sound frequencies, with expansion or contraction of related cortical representations.[68] For example, exposure of young rats to a continuous pure tone expanded cortical representation for that frequency[69] (Figure 26.14). Plasticity of the tonotopic cortical map (for near threshold tones) is restricted to a brief period from postnatal days 11–13,[97] but sensitive periods for more complex acoustic features, such as tuning bandwidth and temporal coding, occur progressively later.[70,71]

Studies on the development of lower levels of the auditory pathway have revealed similarities to events in the visual system. As in the developing retina, spontaneous bursting of primary afferent neurons occurs in the immature cochlea before the onset of hearing.[72–74] Bergles and colleagues have delineated the cellular mechanisms underlying this early spontaneous activity and shown that it is depends on transmitter release from sensory hair cells (Figure 26.15).[75] The sensory hair cells in turn are depolarized not by sound (the middle ear remains blocked until postnatal day 12), but rather by release of ATP from nearby supporting cells. As in the retina, waves of excitation spread along the cochlea, causing intracellular calcium to rise and trigger ATP release to excite neighboring supporting cells. Thus, ATP-mediated excitation causes neighboring hair cells and their postsynaptic afferents to be coordinately activated. The spontaneously active supporting cells make up a transient epithelium called Kolliker's organ that disappears by the onset of hearing, as does the spontaneous bursting of cochlear afferents.[76]

[68] Sanes, D. H., and Bao, S. 2009. *Curr. Opin. Neurobiol.* 19: 188–199.
[69] Han, Y. K. et al. 2007. *Nat. Neurosci.* 10: 1191–1197.
[70] Insanally, M. N. et al. 2009. *J. Neurosci.* 29: 5456–5462.
[71] Razak, K. A., Richardson, M. D., and Fuzessery, Z. M. 2008. *Proc. Natl. Acad. Sci. USA* 105: 4465–4470.
[72] Lippe, W. R. 1994. *J. Neurosci.* 14: 1486–1495.
[73] Jones, T. A. et al. 2007. *J. Neurophysiol.* 98: 1898–1908.
[74] Walsh, E. J., and McGee, J. 1987. *Hear Res.* 28: 97–116.
[75] Tritsch, N. X. et al. 2007. *Nature* 450: 50–55.
[76] Tritsch, N. X., and Bergles, D. E. 2010. *J. Neurosci.* 30: 1539–1550.

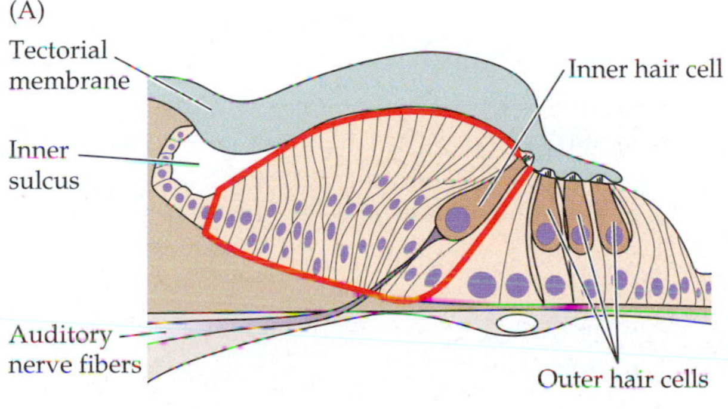

FIGURE 26.15 Spontaneous Purinergic Signaling in the Developing Cochlea. (A) Drawing of the organ of Corti in a young rat (postnatal day 7). Kölliker's organ (outlined in red) temporarily fills much of the space that will become the inner sulcus. Inner hair cells are activated by ATP released from surrounding cells. (B) Slow changes in membrane potential recorded from a supporting cell in the postnatal day 7 rat cochlea. (C) ATP-driven slow waves in supporting cells (upper trace, arbitrary units [AU] optical density) excite hair cells to release glutamate onto afferent nerve fibers. Lower trace is a voltage clamp recording from an afferent dendrite showing bursts of glutamatergic synaptic currents elicited from the inner hair cell by ATP released from supporting cells. Inset (*): Details of synaptic currents from region marked by asterisk. (After Tritsch et al., 2007.)

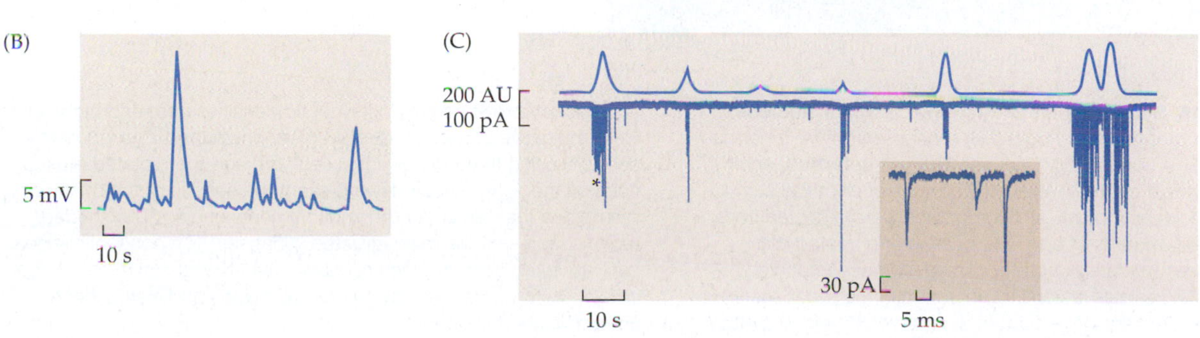

Regulation of Synapse Formation by Activity in the Cochlear Nucleus

The influence of sensory activity on synaptic structure in brainstem neurons has been studied in congenitally deaf cats by Ryugo, Niparko and colleagues using cochlear implants (Box 26.1) to restore hearing.[77] They focused their attention on the end bulb of Held—a large contact made by cochlear afferent fibers on cells in the cochlear nucleus in the brainstem. In young deaf cats (less than 6 months of age), the end bulbs have fewer synaptic vesicles than normal and exhibit expanded postsynaptic densities. When functional hearing was restored with cochlear implants, these morphological changes were reversed (Figure 26.16). Cochlear implants also have been shown to reverse the shrinkage of cells in the cochlear nucleus that occurs in deaf animals.[78]

[77] Ryugo, D. K., Kretzmer, E. A., and Niparko, J. K. 2005. *Science* 310: 1490–1492.

[78] Lustig, L. R. et al. 1994. *Hear Res.* 74: 29–37.

(A) Normal hearing cat

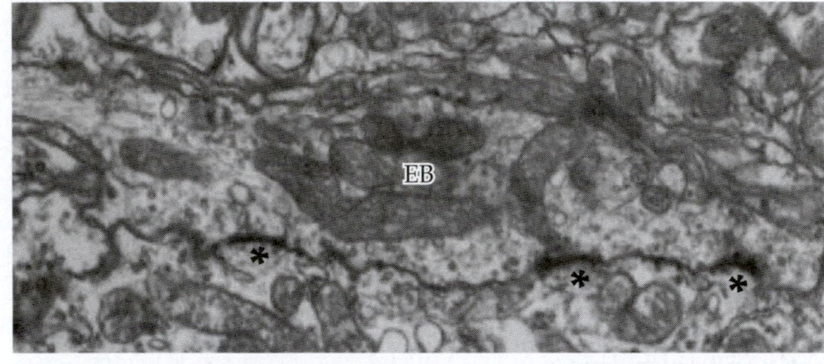

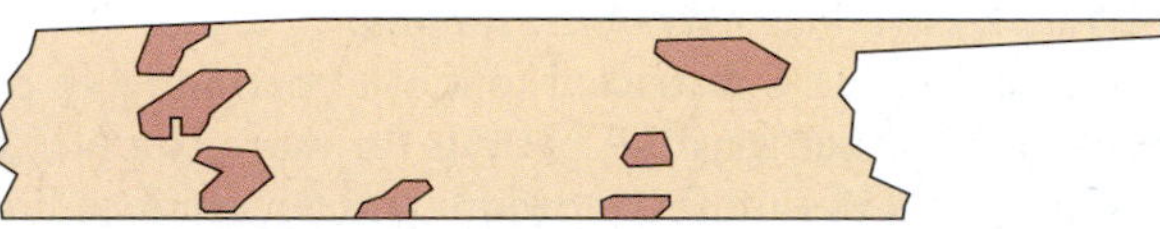

(B) Congenitally deaf cat, untreated

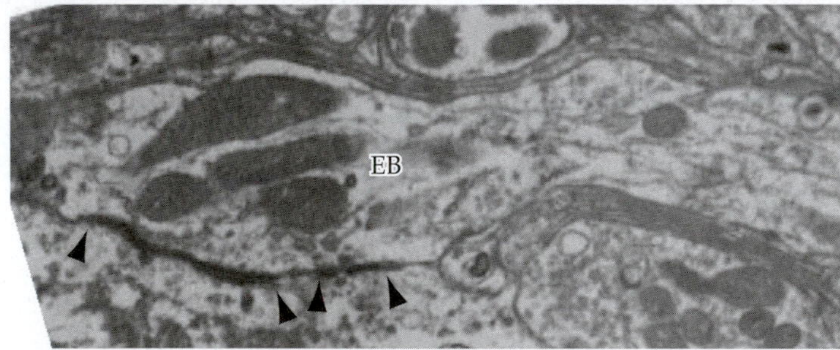

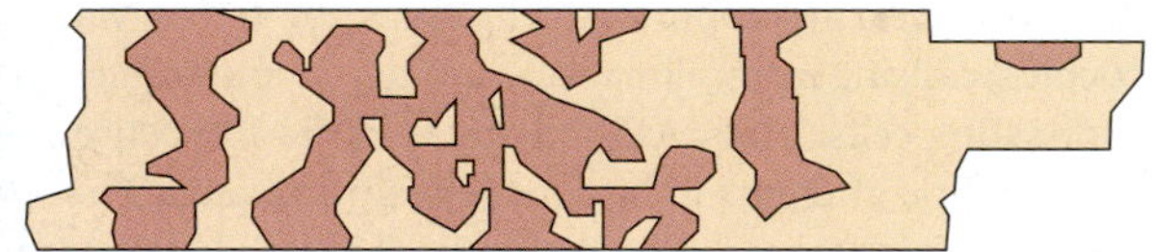

(C) Congenitally deaf cat, cochlear implant

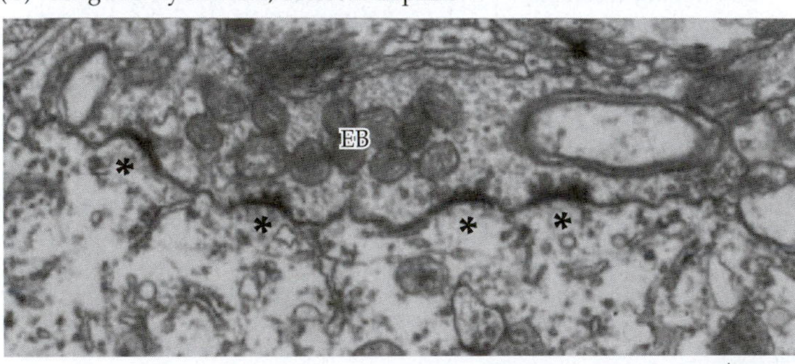

FIGURE 26.16 Synaptic plasticity in the cochlear nucleus. Electron micrographs of end bulb (EB) synapses from (A) a normal hearing cat, (B) a congenitally deaf cat that was untreated, and (C) a congenitally deaf cat that received 3 months of electrical stimulation from a cochlear implant. All micrographs were collected from cats that were 6 months of age. The hearing and treated cats exhibit synapses that are punctate, curved, and accompanied by nearby synaptic vesicles (asterisks). In contrast, the synapses from untreated deaf cats are large, flattened, and mostly void of synaptic vesicles (arrowheads). Postsynaptic densities from end bulbs of Held, indicative of release sites, were reconstructed through serial sections using three-dimensional software and then rotated (illustrations on the right). These views present the surface of the bushy cell membrane that lies under the auditory nerve ending. The shaded regions represent the reconstructed synapses. The congenitally deaf cats exhibited hypertrophied synapses, whereas the stimulated deaf cats exhibited synapses having normal shapes and distributions. (From Ryugo et al., 2005.)

BOX 26.1
The Cochlear Implant

Hearing loss is the most common sensory deficit experienced by humans.[79] Exposure to ototoxic compounds, damaging levels of sound (in the workplace or elsewhere), and age-related hearing loss (presbycusis) contribute to significantly elevated hearing thresholds in approximately 25% of people over 50 years of age. Also, one to three children in 1000 are born with significant hearing loss, 40% of which qualify as profoundly deaf,[80] with a genetic basis for half of these.[81] External hearing aids can relieve partial hearing loss, but until the late twentieth century, profound deafness was essentially untreatable. This picture changed dramatically with the advent of the cochlear implant.[82] By April 2009, approximately 200,000 people had received cochlear implants worldwide.[83] In the best cases, a cochlear implant can provide impressive functional recovery, such as telephone use for those who lost hearing as adults or age-appropriate schooling for deaf children implanted in the first year or two of life.

The cochlear implant consists of an array of electrodes inserted into the cochlear coil. An external microphone and analyzer decomposes sound into its frequency components and converts energy within each frequency band into a series of electrical pulses on one of the implant's electrodes. Each electrode activates nearby spiral ganglion neurons—mimicking the frequency-labeled lines provided in the hearing cochlea by synapses with sensory hair cells. The figure shows activation of a cochlear implant by two distinct syllables: "ah" and "choo." "Ah," a vowel sound, activates electrodes further along the implant, toward low-frequency locations in the cochlea. "Choo" begins with high-frequency frictives (i.e., consonant sounds) and so activates electrodes nearer to the insertion point at the high-frequency base of the cochlea.

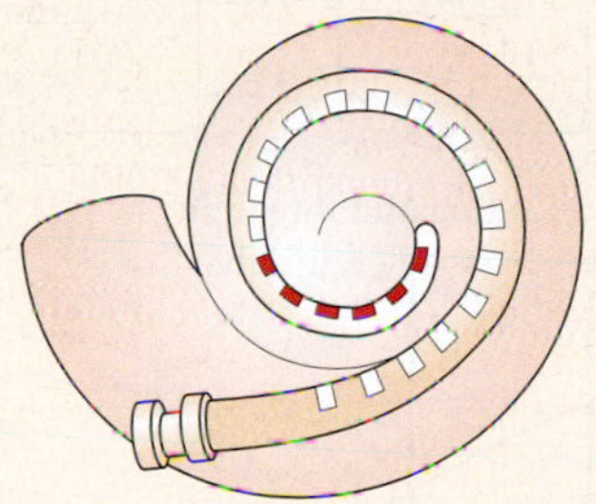

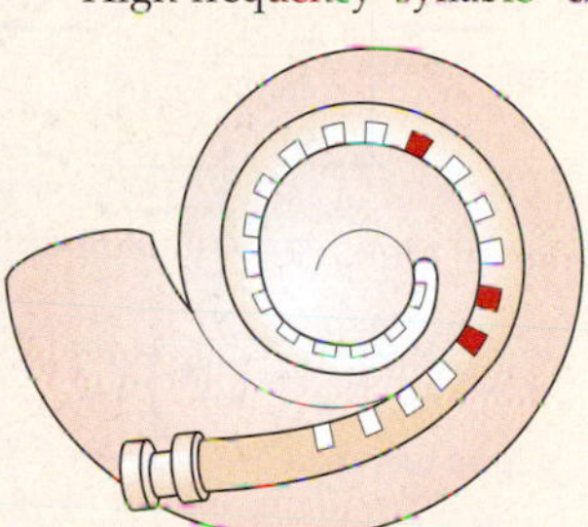

[79] Lustig, L. R. 2010. In *The Oxford Handbook of Auditory Science: The Ear*. Oxford University Press, New York. pp. 15–47.

[80] Smith, R. J. et al. 2005. *Lancet* 365: 879–890.

[81] Morton, N. E. 1991. *Ann. N Y Acad. Sci.* 630: 16–31.

[82] Moller, A. R. 2006. *Adv. Otorhinolaryngol.* 64: 1–10.

[83] NIDCD. 2011. *NIDCD Fact Sheet: Cochlear Implants*. National Institute on Deafness and Other Communication Disorders. Publication No. 11-4798. //www.nidcd.nih.gov/

Although information is becoming available about local differences in the molecular environment and about the inherent properties of neurons, we neither have a comprehensive picture of the molecular mechanisms that regulate cortical critical periods, nor can we explain why connections in the visual cortex should be so much more vulnerable than those in the retina or the spinal cord. Neurotrophins, a host of neuromodulators, inhibitory interneurons, and components of the extracellular matrix are all under consideration as factors that control cortical plasticity.[84] Still more candidates are emerging as DNA arrays are used to examine gene expression during development and after altered experience.[85,86] An additional motivation is to determine whether the adult cortex retains or can regain elements of plasticity for the purpose of repair, or to correct developmental disorders. Comparative studies of birds have provided considerable insight regarding brain plasticity, ranging from learning by songbirds[87] to determination of conditions for central plasticity in adult barn owls.

Critical Periods in the Auditory System of Barn Owls

During the critical period, neural organization is particularly sensitive to environmental experience, beyond that time equivalent sensory exposure does not produce an equivalent change in brain circuitry. The critical period serves to form an animal's nervous system in a way that is optimally adapted to the environment and, accordingly, one would expect the changes to be relatively permanent. However, emerging evidence indicates that even fundamental aspects of established circuitry can be modified if required by an altered environment.[88] This adaptability has been illustrated by the experiments of Knudsen and his colleagues on barn owls.[89] Early experience shapes the auditory spatial localization of neurons in the barn owl's optic tectum in a frequency-dependent manner.[90] The resulting auditory space map must be aligned with the visual world for the owl to locate and cap-

[84] Tropea, D., Van Wart, A., and Sur, M. 2009. *Philos. Trans. R. Soc. Lond., B, Biol. Sci.* 364: 341–355.

[85] Lyckman, A. W. et al. 2008. *Proc. Natl. Acad. Sci. USA* 105: 9409–9414.

[86] Majdan, M., and Shatz, C. J. 2006. *Nat. Neurosci.* 9: 650–659.

[87] Mooney, R. 2009. *Curr. Opin. Neurobiol.* 19: 654–660.

[88] Keuroghlian, A. S., and Knudsen, E. I. 2007. *Prog. Neurobiol.* 82: 109–121.

[89] Knudsen, E. I. 1999. *J. Comp. Physiol. A* 185: 305–321.

[90] Gold, J. I., and Knudsen, E. I. 1999. *J. Neurophysiol.* 82: 2197–2209.

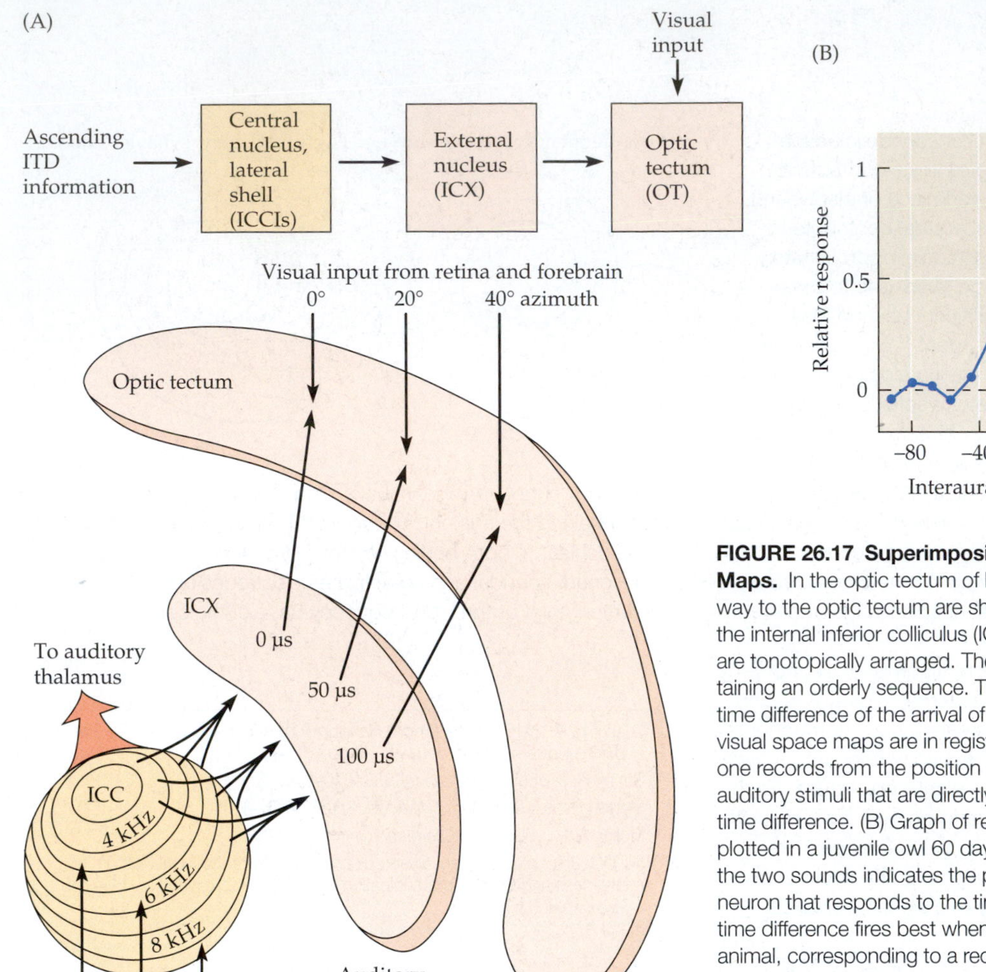

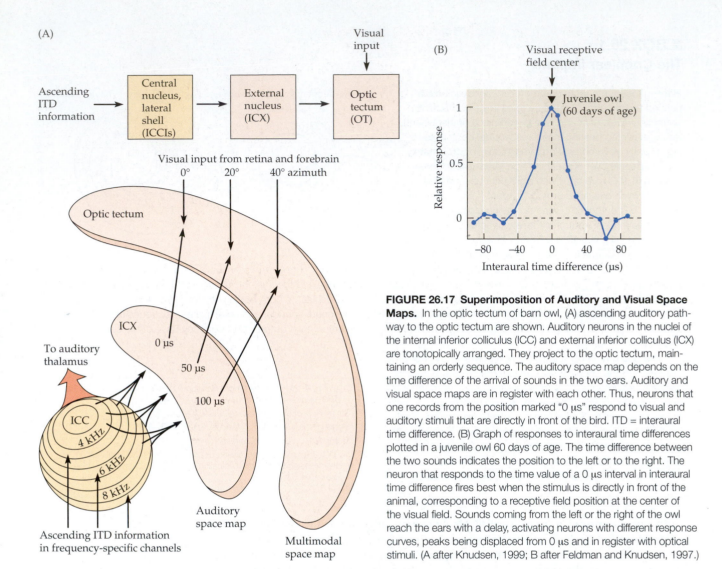

FIGURE 26.17 Superimposition of Auditory and Visual Space Maps. In the optic tectum of barn owl, (A) ascending auditory pathway to the optic tectum are shown. Auditory neurons in the nuclei of the internal inferior colliculus (ICC) and external inferior colliculus (ICX) are tonotopically arranged. They project to the optic tectum, maintaining an orderly sequence. The auditory space map depends on the time difference of the arrival of sounds in the two ears. Auditory and visual space maps are in register with each other. Thus, neurons that one records from the position marked "0 μs" respond to visual and auditory stimuli that are directly in front of the bird. ITD = interaural time difference. (B) Graph of responses to interaural time differences plotted in a juvenile owl 60 days of age. The time difference between the two sounds indicates the position to the left or to the right. The neuron that responds to the time value of a 0 μs interval in interaural time difference fires best when the stimulus is directly in front of the animal, corresponding to a receptive field position at the center of the visual field. Sounds coming from the left or the right of the owl reach the ears with a delay, activating neurons with different response curves, peaks being displaced from 0 μs and in register with optical stimuli. (A after Knudsen, 1999; B after Feldman and Knudsen, 1997.)

ture mice with optimum efficiency. The following account shows how altered visual input influences the representation of the auditory system in the owl's brain.

Barn owls are able to locate sound sources with exceptional accuracy.[91] Horizontal location is achieved by turning the head until the interaural time difference (see Chapter 22) is zero. Vertical localization is enabled by the fact that asymmetrical earflaps collect sound from above for one ear and from below for the other. So the head is tilted up or down until the sound intensity is equal in the two ears. The owl's eyes provide another index of the position and path taken by mice. Figure 26.17A shows that in normal adult owls the neural maps for visual and auditory space are precisely aligned in one layer of the optic tectum (which corresponds to the superior colliculus in mammals). Such maps were produced by measuring responses evoked in individual neurons in the tectum by sound stimuli from different locations and by light presented in different parts of the visual field.

Baby owls were raised with visual fields displaced by 23° to the left or the right by prisms placed over their eyes (Figure 26.18A).[92,93] As a result, the image of the visual field on the retina and hence on the tectum was shifted so that the visual and auditory maps were no longer in alignment (Figure 26.18B,C). Over the next 6 to 8 weeks, the auditory space map shifted until it once again came into exact register with the new, displaced, visual map. As a result, the owl became able to orient its eyes toward sounds in spite of the prisms. When the prisms were removed, the visual and auditory maps were again out of register. Provided that the owls were less than 200 days old, the auditory map shifted a second time to put its

[91] Moiseff, A. 1989. *J. Comp. Physiol. A* 164: 637–644.
[92] Knudsen, E. I., and Knudsen, P. F. 1990. *J. Neurosci.* 10: 222–232.
[93] Feldman, D. E., and Knudsen, E. I. 1997. *J. Neurosci.* 17: 6820–6837.

(A)

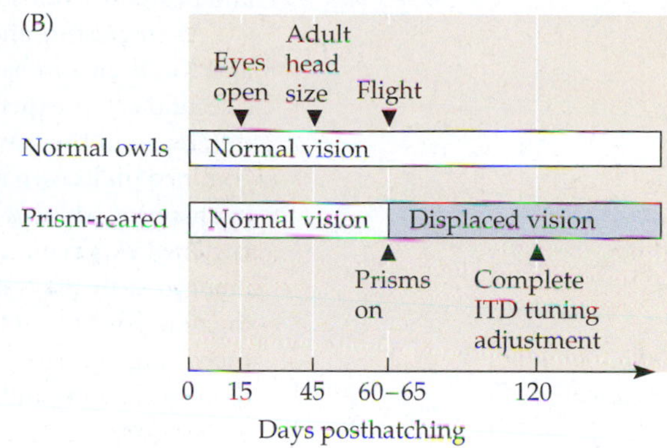

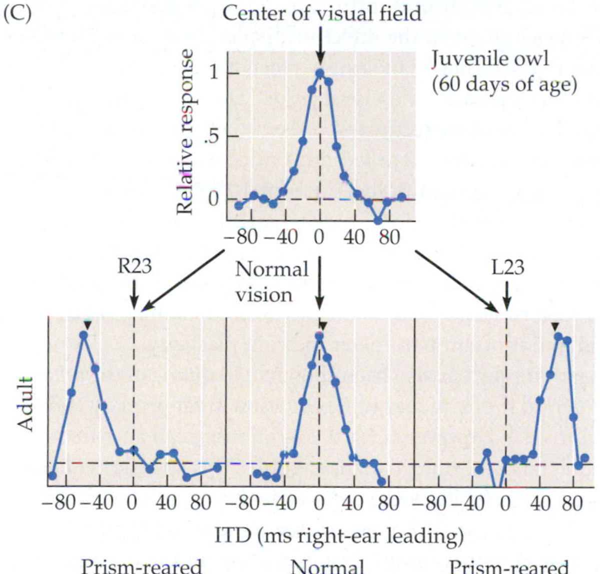

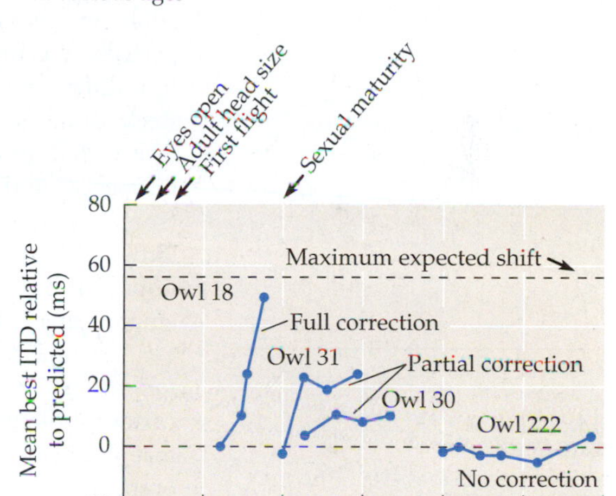

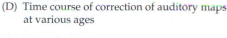

FIGURE 26.18 Shift of Auditory Receptive Fields. After application of prisms during the critical period, (A) a baby owl with glasses consisting of prisms that offset its visual field by 23° to the right (or to the left depending on the spectacles). (B) Stages at which prisms were placed on owl eyes. (C) Tuning curves showing responses to sounds with different interaural time intervals. In the normal juvenile owl at 60 days, the response of the neuron resembles that of Figure 26.17 in which an interaural time difference (ITD) value of 0 μs corresponds to the center of the visual field. The tuning curves of adult owls (120 days old) were shifted after rearing with prisms. Visual receptive fields were displaced to the left or the right by 23°. Now the best responses (small arrowheads) of the auditory neurons occurred at ITDs that corresponded to the displaced visual fields. The auditory and visual maps were again in register. (D) Time course of shift in auditory tuning curves in three immature owls. The adult owl (owl 222) showed no correction. (A kindly provided by E. Knudsen; B after Feldman and Knudsen, 1997; C,D after Brainard and Knudsen, 1998.)

alignment in register with the original visual map. In other tests, it was shown that owls that had once learned to adapt to prisms in early life could reestablish these connections as adults, unlike animals that had no such experience beforehand.[94]

Effects of Enriched Sensory Experience in Early Life

Since the flexibility of connections and structure in early life make the brain vulnerable to sensory deprivation, it is natural to wonder whether enrichment and a fuller early life during the critical period could enhance cortical function. Tests to assess this outcome are in practice difficult to devise. First, newborn animals need to be with their mothers for

[94] Knudsen, E. I. 1998. *Science* 279: 1531–1533.

much of the time. Second, it is hard to know for sure what would constitute an extra rich and pleasant set of stimuli for, say, baby birds, rats, mice, or monkeys in the wild.

An interesting and unexpected effect of enrichment in early life was found by Brainard and Knudsen[95] in barn owls, using the visual field displacements already mentioned. In a second set of experiments, they repeated their earlier procedures, with one important difference in the way the young owls were brought up. Instead of the young owl being confined on its own in a cage (as in the earlier experiments), it spent the first weeks of life in an aviary where it lived with other owls and was able to fly around. In these owls with enriched experience, they confirmed that, as before, the effect of a visual field displacement in early life was corrected in the tectum by realignment of the auditory field onto the new position of the visual field. And as before, when the prisms were removed from the young animals the auditory field moved back to its original mapping in the cortex, thereby allowing auditory and visual responses to be matched. However, in contrast to the earlier experiment, now this realignment could occur after periods longer than 200 days, at times when the owl had matured and become adult. Thus, the richer early environment lengthened the period of adaptability in the auditory system.

Behavioral relevance also can have profound influences on brain plasticity, even in adults.[96] Ordinarily, adult owls never adapt to the effect of displacing prisms. However, adult owls that had to hunt live mice (rather than simply receiving dead mice) adapted partially to prisms over the eyes that displaced the visual fields (Figure 26.19). Interaural time differences (ITDs) measured in the optic tectum were essentially unchanged after 10 weeks of displacing prisms in adult owls that were fed dead mice. However, when adult owls with displacing prisms were required to hunt live mice for 10 weeks, interaural time differences in the optic tectum were shifted approximately half the expected maximum (somewhat less than in juveniles). What is particularly striking here is that all the adult birds had otherwise identical experiences, including flight in an open aviary. The additional impact of hunting was to couple a reward (food) with the desired change in circuitry, emphasizing that motivation is an important parameter in brain plasticity.

The impact of behavioral conditioning has also been shown in mapping of the rat auditory cortex. The usual sensitive period for frequency representation in rat primary auditory cortex is from postnatal days 11 to 13.[97] However, reward conditioning can alter frequency mapping in the adult.[98] When adult rats were trained to distinguish either specific frequencies or specific intensities of sounds, substantial shifts were produced in the representation of those parameters in auditory cortex (Figure 26.20). Although all animals received identical sound exposures, only the rewarded (i.e., motivated) component gained cortical representation.

[95] Brainard, M. S., and Knudsen, E. I. 1998. *J. Neurosci.* 18: 3929–3942.

[96] Bergan, J. F. et al. 2005. *J. Neurosci.* 25: 9816–9820.

[97] de Villers-Sidani, E. et al. 2007. *J. Neurosci.* 27: 180–189.

[98] Polley, D. B., Steinberg, E. E., and Merzenich, M. M. 2006. *J. Neurosci.* 26: 4970–4982.

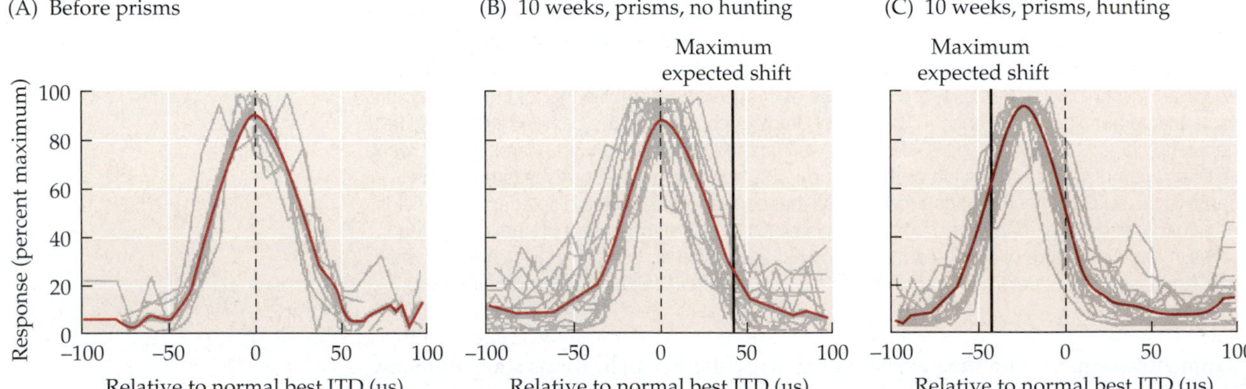

FIGURE 26.19 Effect of Prism Experience. With and without hunting, on interaural time difference (ITD) tuning in the optic tectum of an adult barn owl, negative ITDs designate left-ear leading ITDs. Shown are ITD tuning curves derived from all sites sampled in the same bird. Gray lines are curves from individual sites; red lines are population averages. (A) Data collected before prism experience. (B) Data collected after 10 weeks of experience with rightward shifting prisms and no hunting. (C) Data collected from the same owl after 10 weeks of experience with leftward shifting prisms and hunting. The shifts in ITD tuning were larger and more consistent when the owl had to hunt live prey. (After Bergan et al., 2005.)

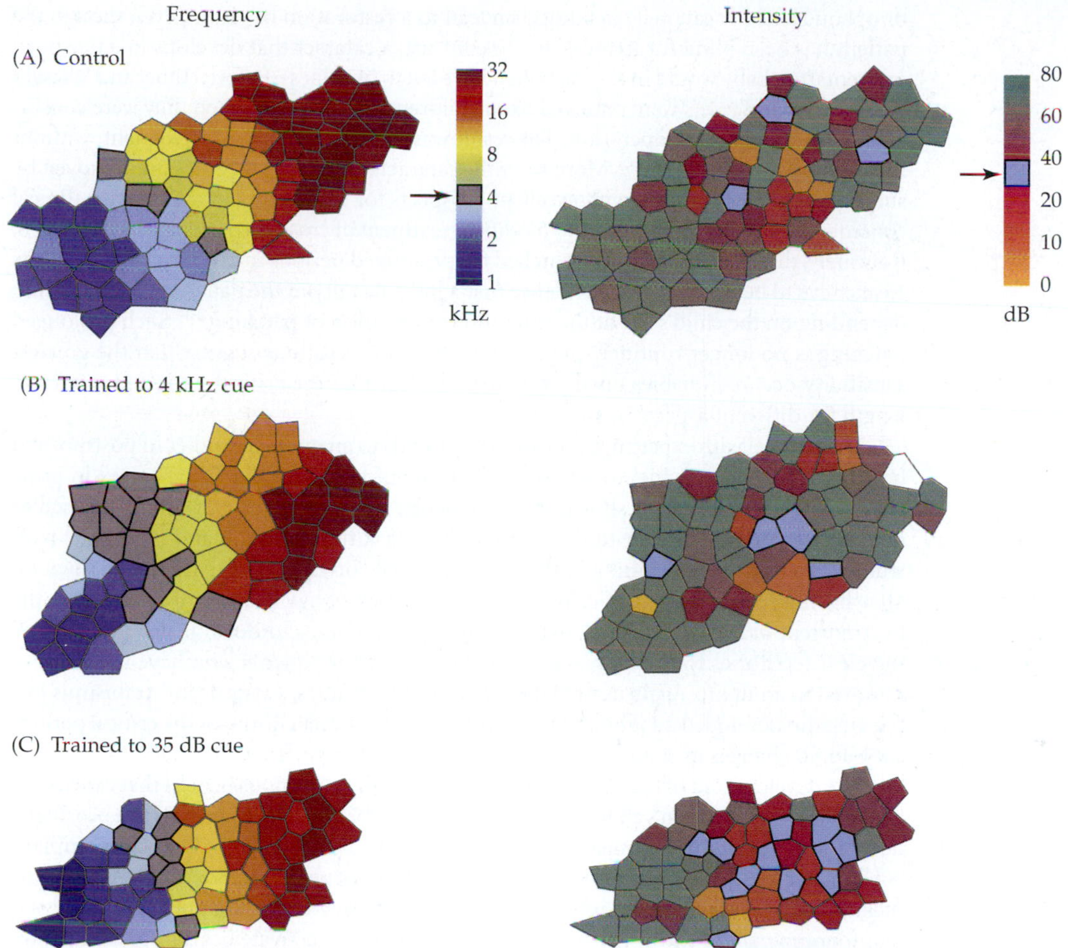

Figure 26.20 Task-Specific Reorganization of Cortical Maps in Primary Auditory Cortex of Rats. Adult rats were trained to respond to either a specific tone frequency or specific tone intensity. Multi-unit recordings were made to characterize the best frequency and best intensity level throughout primary auditory cortex. Each polygon represents a single recording site. The resulting control maps are shown in (A) for frequency (left) or intensity (right). Four gray polygons responded to 4 kHz, three blue polygons were selective for the 35-decibel (dB) intensity. Color scales are shown on the right. (B) Rats trained to the 4-kHz frequency cue showed expanded representation of that frequency (left, gray polygons), but no increase in representation of the 35 dB or best-intensity frequency (right, blue polygons). (C) Rats trained to the 35-dB intensity cue showed expansion of that best intensity (right, blue polygons) but no major change in representation of the 4-kHz frequency (left, gray polygons). Note as well that much of the cortex responds best to 80-dB sound (right panels, green polygons), while the effect of training to 35 dB has greatly reduced the 80-dB best area (right panel, bottom row). (After Polley et al., 2006.)

The ability to alter the adult cortex has potential significance for overcoming lesions or combating degenerative and developmental disorders. Ocular dominance plasticity in rodents can be promoted by various manipulations, including enzymatic degradation of extracellular matrix, pharmacological enhancement of inhibition, dark exposure, or modulation through serotonergic or cholinergic inputs (reviewed in[99]). Evidence continues to accumulate that experience-dependent plasticity extends to greater ages for higher-order cortical functions. One can expect that different molecular and cellular mechanisms will be responsible at different times.[88,100]

Critical Periods in Humans and Clinical Consequences

The susceptibility of cortical development in animals during early life is reminiscent of clinical observations made in humans. It has long been known that removal of a clouded

[99] Morishita, H., and Hensch, T. K. 2008. *Curr. Opin. Neurobiol.* 18: 101–107.

[100] Hooks, B. M., and Chen, C. 2007. *Neuron* 56: 312–326.

or opaque lens (or cataract) in adults can lead to a restoration of vision, even though the patient has been blind for many years. In contrast, a cataract that develops in a newborn or premature baby would in the past often have led to blindness. Before Hubel and Wiesel's experiments, cataracts were removed from children at a late stage, when they were considered to be ready for the operation. The result was that blindness was permanent, without the possibility of recovery.[101] More recently, cataracts in newborn babies are removed by surgery as early as possible, with excellent prospects for full vision.[102,103] A second clinical consequence is provided by the now-modified treatment of cross-eyed or wall-eyed children. Routinely, the good eye would be patched for prolonged periods in order to encourage the weaker eye to be used. There is evidence that a loss of acuity in the patched eye may result, depending on the child's age at the time and the duration of patching.[104] Such prolonged patching is no longer routinely practiced. Clinical observations suggest that the greatest sensitivity occurs in babies during the first year but that the critical period may vary in length for different aspects of vision.[105]

The parallel with experiments in animals extends to anatomical studies of postmortem human brain tissue. Cytochrome oxidase staining of layer 4 of the visual cortex in postmortem human brains has shown that monocular deprivation causes changes in ocular dominance columns similar to those in monkeys or kittens. Horton and Hocking[21] have studied the patterns of staining in the primary visual cortex of people that had grown up after having had one eye surgically removed at 1 week of age because of a tumor. In the postmortem brains, as expected, the staining in layer 4C was uniform instead of displaying clear territories for each eye, as in normal subjects or patients who have had one eye removed in adult life. Again as predicted from animal studies, a long-term strabismus in a patient who developed a squint in the second year of life (after closure of the critical period) showed no changes in column width postmortem at the age of 79.[106]

The development of cochlear implants illustrates the enormous benefit that can accrue by tapping into the brain's ability to reorganize with altered sensory input. These prosthetic devices can restore functional hearing after the loss of sensory hair cells. The implant consists of a series of electrodes inserted into the cochlear duct (see Box 26.1). An external pickup converts sound waves into their component frequencies and applies electrical signals **tonotopically** to the inserted electrodes, which in turn activate nearby spiral ganglion neurons. Although these devices continue to improve, even the optimum performance is exceedingly limited compared to the selective signaling of the thousands of hair cells they replace. Nonetheless, cochlear implants can enable speech comprehension (even on the telephone) for deaf adults with previously developed spoken language.[107] Still more impressive is the performance achieved by profoundly deaf children provided with implants early in life—some of whom achieve age-appropriate language skills.[108] Longitudinal studies of children given cochlear implants between 18 months and 5 years of age show that the earlier the implant is received the better is the development of language comprehension and expression. Also as predicted from animal studies, better outcomes are correlated with more extensive parent–child interactions and socioeconomic status (enrichment). Even apparently minor levels of hearing loss can have long-term consequences. For example, chronic middle ear infections in childhood can result in long-lasting binaural hearing deficits.

There exists a wealth of literature related to other complex behavioral processes in a variety of animals that shows periods of susceptibility. Behavioral studies in dogs indicate that if they are handled by humans during a critical period of 4 to 8 weeks after birth, they are far more tractable and tame than animals that have been isolated from human contact.[109] Imprinting is another well-known example. Lorenz[110] has shown that a duck will follow any moving object, provided that it is presented during the first day after hatching and will act throughout life as if that object were its mother. Similarly, birdsongs, such as produced by zebra finches, depend on the tutorial songs that the juvenile birds hear during a critical period.[87] Birds of a particular species will remember and reproduce appropriate (but not inappropriate) melodies that they hear in early life, although they might be exposed to multiple song models, try out different ones, and even preserve memories of discarded songs.[111]

The critical period in an animal's development seems to correspond to a time during which a significant sharpening of senses or faculties occurs. Why should plasticity in early life be so pronounced? The developing brain must not only form itself but must be able to

[101] Francois, J. 1979. *Ophthalmology* 86: 1586–1598.

[102] Lloyd, I. C. et al. 2007. *Eye (Lond).* 21: 1301–1309.

[103] Hamill, M. B., and Koch, D. D. 1999. *Curr. Opin. Ophthalmol.* 10: 4–9.

[104] Daw, N. W. 1998. *Arch. Ophthalmol.* 116: 502–505.

[105] Lewis, T. L., and Maurer, D. 2005. *Dev. Psychobiol.* 46: 163–183.

[106] Horton, J. C., and Hocking, D. R. 1996. *Vis. Neurosci.* 13: 787–795.

[107] Fallon, J. B., Irvine, D. R., and Shepherd, R. K. 2008. *Hear Res.* 238: 110–117.

[108] Niparko, J. K., and Marlowe, A. 2010. In *The Oxford University Handbook of Auditory Science: The Ear.* Oxford University Press, New York. pp. 409–436.

[109] Fuller, J. L. 1967. *Science* 158: 1645–1652.

[110] Lorenz, K. 1970. *Studies in Animal and Human Behavior.* Harvard Press, Cambridge, MA.

[111] Prather, J. F. et al. 2010. *J. Neurosci.* 30: 10586–10598.

represent the outside world, the body, and its movements.[112] The eye, for example, must grow to be the right size for distant objects to be in focus on the retina through the relaxed lens. And the two eyes, separated by different distances in different newborn babies, must act together. As if this were not enough, gross changes occur in limb length, skull diameter, and therefore body image in the first months and years of life. At all times, the maps in the brain for different functions must be in register, as in the owl experiments of Knudsen. Will lessons from animal studies of brain plasticity guide future therapeutic strategies? This is certainly to be hoped for, since deficits in communication and social behavior can have such debilitating consequences.[113] Indeed, it is tempting to speculate about the effects of deprivation on higher functions in humans. One can imagine, as Hubel has said,[13]

> Perhaps the most exciting possibility for the future is the extension of this type of work to other systems besides sensory. Experimental psychologists and psychiatrists both emphasize the importance of early experience on subsequent behavior patterns—could it be that deprivation of social contacts or the existence of other abnormal emotional situations early in life may lead to a deterioration or distortion of connections in some yet unexplored parts of the brain?

To find a physiological basis for such behavioral questions seems a distant, but not impossible, goal.

[112] Singer, W. 1995. *Science* 270: 758–764.

[113] Blakemore, S. J. 2010. *Neuron* 65: 744–747.

SUMMARY

- Receptive fields and cortical architecture in newborn monkeys and kittens resemble those of adults in many respects.
- In layer 4 of the cortex, however, incoming axons from the lateral geniculate overlap and cells are driven by both eyes instead of just one during the first 6 weeks of life.
- A critical period of about 3 months exists after birth, during which closure of the lids of one eye causes changes in structure and function.
- Closure of the lids of one eye leads to blindness in that eye. Cortical cells are no longer driven by the deprived eye, and its ocular dominance columns shrink.
- After the critical period, closure of lids or enucleation of an eye does not change cortical architecture.
- Binocular lid closure or induction of squint during the critical period does not cause changes in ocular dominance columns but does prevent binocular responses. Such results suggest that the two eyes compete for cells in the visual cortex.
- Endogenous spontaneous activity of the sensory periphery shapes cortical organization before the onset of sensory function.
- Inhibitory cortical interneurons are key players in establishing and closing cortical critical periods.
- The plasticity of visual–auditory integration in barn owls is enhanced by enrichment in early life or behavioral motivation.

Suggested Reading

General Reviews

Adams, D. L., and Horton, J. C. 2009. Ocular dominance columns: enigmas and challenges. *Neuroscientist* 15: 62–77.

Daw, N. 2006. *Visual development*, 2nd ed. Springer, New York.

Erzurumlu, R. 2010. Critical period for the whisker-barrel system. *Exp. Neurol.* 222: 10–12.

Hanganu-Opatz, I. 2010. Between molecules and experience: role of early patterns of coordinated activity for the development of cortical maps and sensory abilities. *Brain Res. Rev.* 64: 160–176.

Hensch, T. K. 2004. Critical period regulation. *Annu. Rev. Neurosci.* 27: 549–579.

Hooks, B. M., and Chen, C. 2007. Critical periods in the visual system: changing views for a model of experience-dependent plasticity. *Neuron* 56: 312–326.

Hubel, D. H., and Wiesel, T. N. 2005. *Brain and visual perception*. Oxford University Press, New York.

Keroughlian, A., and Knudsen, E. 2007. Adaptive auditory plasticity in developing and adult animals. *Prog. Neurobiol.* 82: 109–121.

Sanes, D., and Bao, S. 2009. Tuning up the developing auditory CNS. *Curr. Opin. Neurobiol.* 19: 188–199.

Tropea, D., Van Wart, A., and Sur, M. 2009. Molecular mechanisms of experience-dependent plasticity in visual cortex. *Philos. Trans. R. Soc. Lond., B, Biol. Sci.* 364: 341–355.

Wiesel, T. N. 1982. The postnatal development of the visual cortex and the influence of environment. *Nature* 299: 583–591.

Original Papers

Antonini, A., Gillespie, D. C., Crair, M. C., and Stryker, M. P. 1998. Morphology of single geniculocortical afferents and functional recovery of the visual cortex after reverse monocular deprivation in the kitten. *J. Neurosci.* 18: 9896–9909.

Bergan, J. F., Ro, P., Ro, D., and Knudsen, E. I. 2005. Hunting increases adaptive auditory map plasticity in adult barn owls. *J. Neurosci.* 25: 9816–9820.

Brainard, M. S., and Knudsen, E. I. 1998. Sensitive periods for visual calibration of the auditory space map in the barn owl optic tectum. *J. Neurosci.* 18: 3929–3942.

Carlson, M., Hubel, D. H., and Wiesel, T. N. 1986. Effects of monocular exposure to oriented lines on monkey striate cortex. *Brain Res.* 390: 71–81.

Chattopadhyaya, B., Di Cristo, G., Higashiyama, H., Knott, G. W., Kuhlman, S. J., Welker, E., and Huang, Z. J. 2004. Experience and activity-dependent maturation of perisomatic GABAergic innervation in primary visual cortex during a postnatal critical period. *J. Neurosci.* 24: 9598–9611.

de Villers-Sidani, E., Chang, E. F., Bao, S., and Merzenich, M. M. 2007. Critical period window for spectral tuning defined in the primary auditory cortex (A1) in the rat. *J. Neurosci.* 27: 180–189.

Hensch, T. K., and Stryker, M. P. 2004. Columnar architecture sculpted by GABA circuits in developing cat visual cortex. *Science* 303: 1678–1681.

Horton, J. C., and Hocking, D. R. 1996. An adult-like pattern of ocular dominance columns in striate cortex of newborn monkeys prior to visual experience. *J. Neurosci.* 16: 1791–1807.

Hubel, D. H., and Wiesel, T. N. 1965. Binocular interaction in striate cortex of kittens reared with artificial squint. *J. Neurophysiol.* 28: 1041–1059.

Hubel, D. H., Wiesel, T. N., and LeVay, S. 1977. Plasticity of ocular dominance columns in monkey striate cortex. *Philos. Trans. R. Soc. Lond., B, Biol. Sci.* 278: 377–409.

Knudsen, E. I., and Knudsen, P. F. 1990. Sensitive and critical periods for visual calibration of sound localization by barn owls. *J. Neurosci.* 10: 222–232.

LeVay, S., Wiesel, T. N., and Hubel, D. H. 1980. The development of ocular dominance columns in normal and visually deprived monkeys. *J. Comp. Neurol.* 191: 1–51.

Lebedev, M. A., Mirabella, G., Erchova, I., and Diamond, M. E. 2000. Experience-dependent plasticity of rat barrel cortex: redistribution of activity across barrel-columns. *Cereb. Cortex* 10: 23–31.

Meissirel, C., Wikler, K. C., Chalupa, L. M., and Rakic, P. 1997. Early divergence of magnocellular and parvocellular functional subsystems in the embryonic primate visual system. *Proc. Natl. Acad. Sci. USA* 94: 5900–5905.

Meister, M., Wong, R. O., Baylor, D. A., and Shatz, C. J. 1991. Synchronous bursts of action potentials in ganglion cells of the developing mammalian retina. *Science* 252: 939–943.

Popescu, M. V., and Polley, D. B. Monaural deprivation disrupts development of binaural selectivity in auditory midbrain and cortex. *Neuron* 65: 718–731.

Ryugo, D. K., Kretzmer, E. A., and Niparko, J. K. 2005. Restoration of auditory nerve synapses in cats by cochlear implants. *Science* 310: 1490–1492

Tritsch, N. X., Yi, E., Gale, J. E., Glowatzki, E., and Bergles, D. E. 2007. The origin of spontaneous activity in the developing auditory system. *Nature* 450: 50–55.

Wiesel, T. N., and Hubel, D. H. 1963. Single-cell responses in striate cortex of kittens deprived of vision in one eye. *J. Neurophysiol.* 26: 1003–1017.

Yang, J. W., Hanganu-Opatz, I. L., Sun, J. J., and Luhmann, H. J. 2009. Three patterns of oscillatory activity differentially synchronize developing neocortical networks in vivo. *J. Neurosci.* 29: 9011–9025.

CHAPTER 27

Regeneration of Synaptic Connections after Injury

When an axon in the vertebrate nervous system is severed, the distal portion degenerates. Over time, changes occur in its cell body and its targets. The changes result from the interruption of axonal transport of molecules that control neuronal differentiation and survival and from alterations in the pattern of electrical activity. After they have been denervated, vertebrate skeletal muscle fibers express acetylcholine (ACh) receptors over their entire surface and become more sensitive to acetylcholine. Direct electrical stimulation of denervated muscles causes the region sensitive to acetylcholine to shrink back to the original end plate. The effects of activity are mediated by calcium influx and activation of intracellular second messengers. Unlike innervated muscle fibers, denervated muscles accept new innervation anywhere along their length. Denervated muscle fibers also induce nearby, undamaged nerve terminals to sprout new branches. After nerve cells have been deprived of their presynaptic inputs, they too become supersensitive to transmitters and cause nearby nerve terminals to sprout. In the peripheral nervous system of adult mammals, damaged axons regrow to restore sensory and motor functions. They are guided back to skin and muscle by Schwann cells and the basal lamina. Agrin, a large protein secreted by motor nerve terminals during regeneration and during development, triggers differentiation of postsynaptic specializations in muscle cells.

The ability of damaged central nervous system (CNS) axons to grow and innervate appropriate targets varies widely among species. In invertebrates, such as the leech, and in lower vertebrates, such as frogs, axons in the CNS can regenerate and reconnect precisely with their original synaptic partners after injury. By contrast, in the adult mammalian central nervous system, regeneration and repair fail to occur. CNS neurons sprout new axons and form new synapses only over short distances. They can, however, extend axons for long distances through peripheral nerve grafts to establish synaptic connections. In fetal and neonatal mammals the central nervous system does regenerate, but after a critical period the capacity for repair is lost. Embryonic neurons and neural stem cells grafted into the adult CNS differentiate, extend axons, and can become integrated appropriately into the existing synaptic circuitry. Such transplantation techniques hold promise for the amelioration of functional deficits resulting from CNS lesions and neurodegenerative diseases.

In many animals the nervous system has a remarkable ability to reestablish with a high degree of specificity synaptic connections that have been disrupted by trauma. The regenerative powers of neurons in the CNS were first demonstrated by Matthey, who in the 1920s sectioned the optic nerve of a newt and found that vision was restored within a few weeks.[1] Beginning in the 1940s, Sperry, Stone, and their colleagues took advantage of this regenerative capacity to explore how specific connections form within the nervous system. Their experiments on regenerating retinotectal connections in frogs and fish provided support for the idea that neurons selectively reinnervate their targets during regeneration, rather than making connections at random that are subsequently reorganized[2] (see Chapter 25). Later, studies in leeches, crickets, and crayfish demonstrated that, after being severed, axons of individual identified neurons in invertebrates can reconnect precisely with their original synaptic partners, avoiding a multitude of other potential targets.[3] In contrast, regeneration of severed connections in the adult mammalian nervous system is typically incomplete or absent altogether.[4]

In this chapter we first describe the sequence of events that occurs after an axon in the periphery is damaged. Once again, the connections of motoneurons to skeletal muscle fibers are convenient for studying basic mechanisms of regeneration after injury. We then consider the problem of why regeneration typically fails in mammalian central nervous system.

Regeneration in the Peripheral Nervous System

Wallerian Degeneration and Removal of Debris

After a sensory or motor axon in a vertebrate peripheral nerve is severed, a characteristic sequence of changes occurs (Figure 27.1). The distal portion of the axon degenerates, as does a short length of the proximal portion. The Schwann cells that had formed the

[1]Matthey, R. 1925. *C. R. Soc. Biol.* 93: 904–906.

[2]Sperry, R. W. 1963. *Proc. Natl. Acad. Sci. USA* 50: 703–710.

[3]Anderson, H., Edwards, J. S., and Palka, J. 1980. *Annu. Rev. Neurosci.* 3: 97–139.

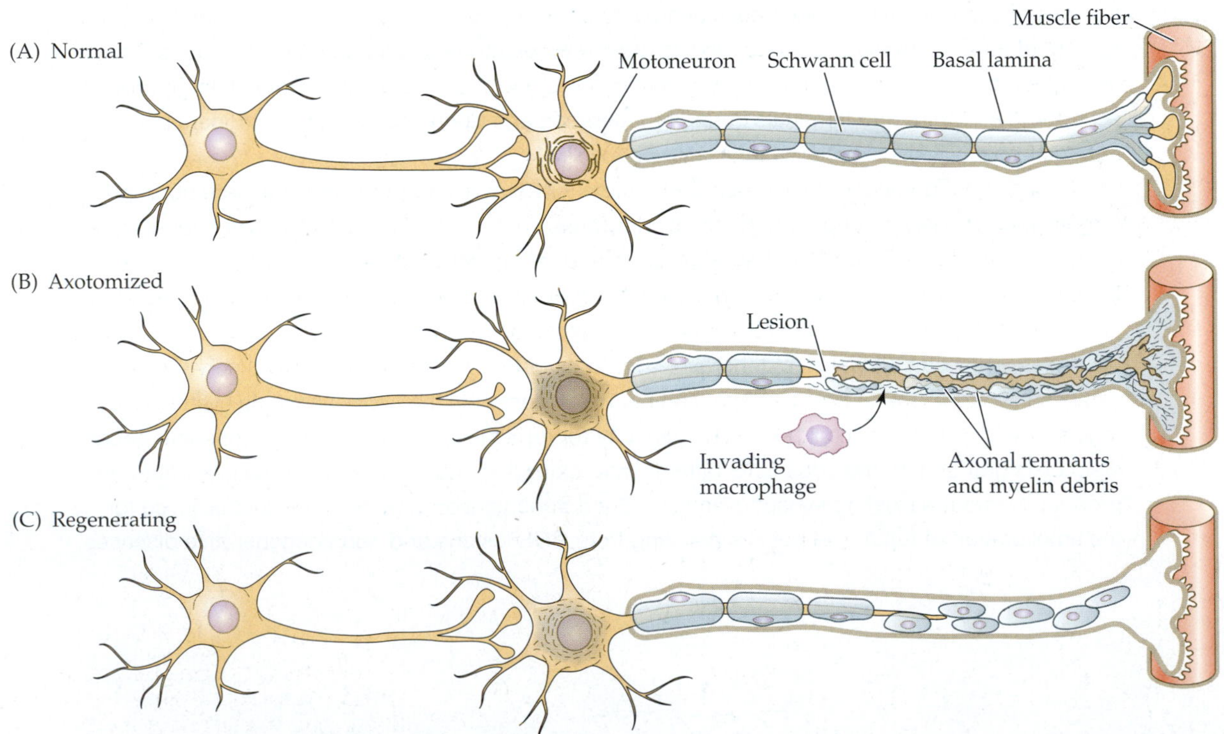

FIGURE 27.1 Degenerative Changes after Axotomy. (A) A typical motoneuron in an adult vertebrate. (B) After axotomy, both the nerve terminal (i.e., the distal segment of the axon) and a short length of the proximal segment of the axon degenerate. Schwann cells dedifferentiate, proliferate, and, together with invading microglial cells and macrophages, phagocytize the axonal and myelin remnants. The axotomized neuron undergoes chromatolysis, presynaptic terminals retract, and degenerative changes may occur in pre- and postsynaptic cells. (C) The axon regenerates along the column of Schwann cells within the endoneurial tube and sheath of basal lamina that had surrounded the original axon.

myelin sheath of the distal segment of the nerve de-differentiate and proliferate together with macrophages that are recruited from the blood stream to the site of injury. There they scavenge and remove debris, including myelin, as a first step in promoting regrowth of axons. The new axonal sprouts emerge from near the tip of the proximal stump and begin regenerating within a few hours. This response is called **Wallerian degeneration**, after the nineteenth-century anatomist Augustus Waller, who first described it.[5,6] The cell body and its nucleus swell, the nucleus moves from its typical position in the center of the cell soma to an eccentric location, and the ordered arrays of endoplasmic reticulum, called **Nissl substance**, disperse. Since the Nissl substance stains prominently with commonly used basic dyes, its dispersal following axotomy causes a decrease in intensity, which is referred to as **chromatolysis**. Chromatolysis also occurs after axons are severed in the CNS. After a regenerating peripheral axon successfully reestablishes contact with a target, the cell body regains its original appearance. If regeneration of the peripheral nerve fails, many motor and dorsal root ganglion sensory neurons die. Autonomic ganglion cells that survive become less sensitive to acetylcholine and shrink in size. In the adult mammalian central nervous system (see following section), the responses of neurons that fail to reestablish contact with their targets are variable. Thus, retinal ganglion cells rapidly die if their axons in the optic nerve are severed,[7,8] whereas motoneurons survive axotomy.

Retrograde Transsynaptic Effects of Axotomy

Axotomy can also cause changes in the presynaptic neurons that provide inputs to the damaged cell. For example, after axotomy of an autonomic ganglion cell in a chick, rat, mouse, or guinea pig, synaptic inputs onto the ganglion cell become less effective.[9,10] This is due in part to a decrease in the sensitivity of the axotomized cell to the neurotransmitter ACh. In addition, retrograde transsynaptic effects cause many of the presynaptic terminals to retract from the axotomized cell and the remaining terminals to release fewer quanta of transmitter (Figure 27.2).[11] Thus, damage to a neuron alters its ability to hold onto its presynaptic inputs. Rotshenker has shown an additional, retrograde transsynaptic effect in motoneurons in the frog and the mouse.[12] When a motor nerve is cut on one side of the animal, a signal

[4] Mladinic, M., Muller, K. J., and Nicholls, J. G. 2009. *J. Physiol.* 587: 2775–2782.

[5] Coleman, M. P., and Freeman, M. R. 2010. *Annu. Rev. Neurosci.* 33: 245–267.

[6] Rotshenker, S. 2009. *J. Mol. Neurosci.* 39: 99–103.

[7] Aguayo, A. J. et al. 1996. *Ciba Found. Symp.* 196: 135–144.

[8] McKernan, D. P., and Cotter, T. G. 2007. *J. Neurochem.* 102: 922–930.

[9] Brenner, H. R., and Martin, A. R. 1976. *J. Physiol.* 260: 159–175.

[10] Simões, G. F., and Oliveira, A. L. 2010. *Neuropathol. Appl. Neurobiol.* 36: 55–70.

[11] Matthews, M. R., and Nelson, V. H. 1975. *J. Physiol.* 245: 91–135.

[12] Rotshenker, S. 1988. *Trends Neurosci.* 11: 363–366.

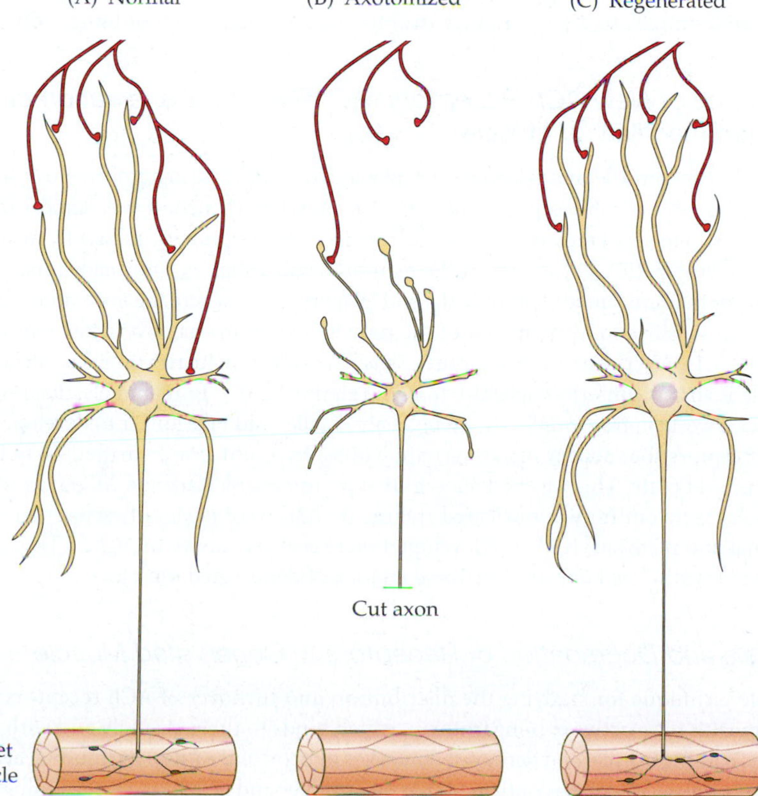

FIGURE 27.2 Axotomized Autonomic Ganglion Cells Atrophy and Lose Presynaptic Inputs. (A) Normal neuron. (B) Within a few days after axotomy, neurons atrophy and many dendrites show large varicosities. Many presynaptic terminals retract from dendrites; those that remain release less transmitter. (C) If the postganglionic axon regenerates and reinnervates its peripheral target, the cell and synaptic inputs recover. (After Purves, 1975.)

spreads from the axotomized neurons, crosses the spinal cord, and influences undamaged motoneurons on the other side of the animal. The axon terminals of those intact motor neurons sprout new branches and form additional synapses on the corresponding muscle. Motoneurons innervating other muscles are not affected.

Certain effects of axotomy—chromatolysis, atrophy, and cell death—result from the loss of trophic substances that are produced by the target tissue and transported retrogradely along the axon to the cell body.[13] Clear examples come from studies of the effects of nerve growth factor (NGF) on sensory and sympathetic neurons, as discussed in Chapter 25. Thus, in guinea pig autonomic ganglia the effects of axotomy are mimicked by injecting antibodies to NGF subcutaneously for several days or by blocking retrograde transport in postganglionic nerves. Conversely, the effects of axotomy are largely prevented by application of NGF to the ganglion.[14]

Effects of Denervation on Postsynaptic Cells

The Denervated Muscle Membrane

Toward the end of the nineteenth century, it was found that denervated skeletal muscles show certain clear-cut changes, such as spontaneous, asynchronous contractions called **fibrillation.** Fibrillation is initiated by the muscle membrane, not by ACh,[15] although most of the spontaneous action potentials producing fibrillation originate in the region of the former end plate.[16] The onset of fibrillation may be as early as 2 to 5 days after denervation in rats, guinea pigs, or rabbits and well over a week in monkeys and humans.

Before or at the start of fibrillation, mammalian muscle fibers become supersensitive to a variety of chemicals. This means that the concentration of a substance required to excite a muscle is reduced by a factor of several hundred to a thousand. For example, a denervated mammalian skeletal muscle is about 1000 times more sensitive to ACh, applied either directly in the bathing fluid or injected into an artery supplying the muscle, than is a normally innervated muscle.[17] The action potential in denervated muscles also changes, becoming more resistant to tetrodotoxin (TTX), the puffer fish poison that blocks sodium channels (see Chapter 7). This change is due to the reappearance of TTX-resistant sodium channels that are the prevailing form in immature muscle.[18] Other changes occur in denervated muscle, such as a gradual atrophy (i.e., wasting away) of muscle fibers.[19]

Appearance of New ACh Receptors (AChRs) after Denervation or Prolonged Inactivity of Muscle

Supersensitivity of muscle to acetylcholine is explained by an increase in number and a change in the distribution of ACh receptors in denervated muscles. This effect was demonstrated by applying ACh to small regions of the muscle surface by ionophoretic release from an extracellular micropipette while recording the membrane potential. In a normally innervated frog, snake, or mammalian muscle, only the end-plate region—where the nerve fiber makes a synapse—is sensitive to ACh; the rest of the muscle membrane has a very low sensitivity (see Chapter 11). After denervation, the area sensitive to ACh increases until the surface of the muscle is almost uniformly sensitive to ACh (Figure 27.3).[20] In mammals, this process takes about a week; in frog muscle, the changes are smaller and take longer to develop.[21]

The receptors that appear in extrasynaptic areas have not simply drifted away from the original end plate. This was first shown in experiments by Katz and Miledi in which frog muscles were cut in two; nucleated fragments that were physically separated from the original end plate survived and developed increased sensitivity to ACh.[22] Thus, new receptors are synthesized in extrajunctional regions of denervated muscles.

Synthesis and Degradation of Receptors in Denervated Muscle

A valuable technique for studying the distribution and turnover of ACh receptors is to label them with radioactive α-bungarotoxin, which binds to them strongly and with high specificity. Bathing normal and denervated muscles in toxin and measuring binding at end-plate and end-plate-free areas confirmed that the number and distribution of binding sites

[13] Campenot, R. B. 2009. *Results Probl. Cell. Differ.* 48: 141–158.
[14] Nja, A., and Purves, D. 1978. *J. Physiol.* 277: 55–75.
[15] Purves, D., and Sakmann, B. 1974. *J. Physiol.* 239: 125–153.
[16] Belmar, J., and Eyzaguirre, C. 1966. *J. Neurophysiol.* 29: 425–441.
[17] Brown, G. L. 1937. *J. Physiol.* 89: 438–461.
[18] Kallen, R. G. et al. 1990. *Neuron* 4: 233–342.
[19] Guth, L. 1968. *Physiol. Rev.* 48: 645–687.
[20] Axelsson, J., and Thesleff, S. 1959. *J. Physiol.* 147: 178–193.
[21] Miledi, R. 1960. *J. Physiol.* 151: 1–23.
[22] Katz, B., and Miledi, R. 1964. *J. Physiol.* 170: 389–396.

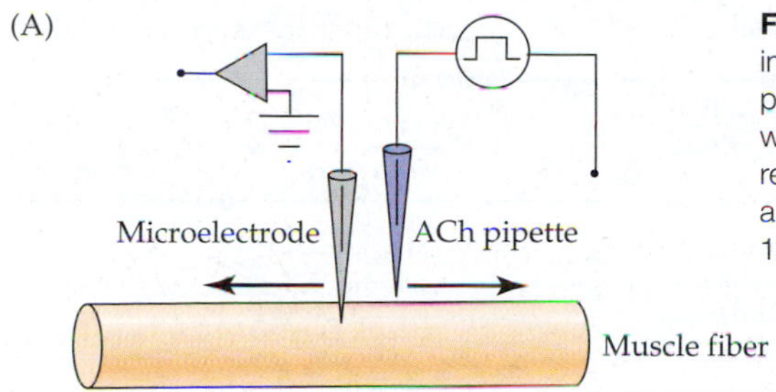

FIGURE 27.3 New Acetylcholine Receptors (AChR) Appear after Denervation in cat muscle. (A) Pulses of ACh are applied from an ACh-filled pipette at different positions along the surface of a muscle fiber, while the membrane potential is recorded with an intracellular microelectrode. (B) In a muscle fiber with intact innervation, a response is seen only in the vicinity of the end plate. (C) After 14 days of denervation, a muscle fiber responds to ACh along its entire length. (After Axelsson and Thesleff, 1959.)

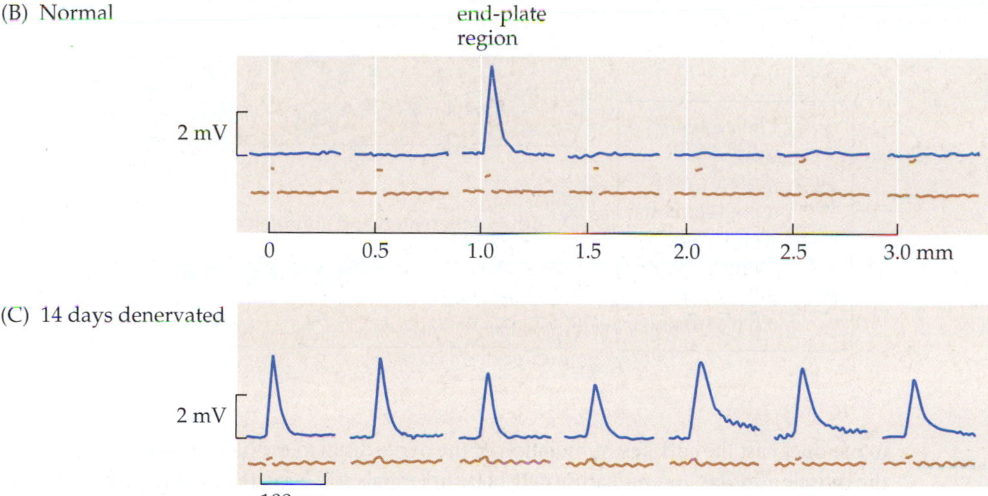

is changed after denervation.[23,24] In normal muscle, there are approximately 10^4 binding sites per square micrometer (μm^2) in the postsynaptic membrane, compared with fewer than $10/\mu m^2$ in end-plate-free areas. After denervation, receptor sites in the extrasynaptic regions increase to about $10^3/\mu m^2$, with little change in density in the synaptic region.

The increase in the number of receptors in denervated muscle is attributable to enhanced receptor synthesis.[23,25] Thus, the rate of appearance of new ACh receptors increases markedly after denervation, and substances that block protein synthesis (e.g., actinomycin and puromycin) prevent the increase in extrasynaptic receptor density in muscles maintained in organ culture. Northern blot analysis and in situ hybridization demonstrate that in normal muscle, only those few nuclei located immediately beneath the end plate are synthesizing ACh receptor subunit mRNAs; in contrast AChR genes are transcribed by nuclei all along the length of denervated muscle fibers (Figure 27.4).[26]

Denervation also affects the subunit composition and rate of degradation of ACh receptors. In mature muscle, junctional and extrajunctional AChRs contain an ε-subunit and have a half-life of about 10 days.[27,28] Following denervation, the half-life of ε-subunit–containing receptors remaining at the end plate decreases to 3 days. Turnover can be slowed again by reinnervation or by an increase in intracellular cAMP and consequent activation of protein kinase A.[29]

New receptors synthesized in denervated muscle (whether synaptic or extrasynaptic) resemble those in embryonic muscle (see below and Chapter 5). They contain a γ- instead of an ε-subunit and turn over with a half-life of 1 day.[30–32]

Role of Muscle Inactivity in Denervation Supersensitivity

How does sectioning of a nerve lead to the appearance of new receptors? Lømo and Rosenthal[33] investigated this problem by blocking conduction in rat nerves. A local anesthetic or diphtheria toxin was applied through a cuff to a short length of the nerve some distance from the muscle. With this technique, the muscles became inactive because motor impulses failed

[23] Fambrough, D. M. 1979. *Physiol. Rev.* 59: 165–227.

[24] Salpeter, M. M., and Loring, R. H. 1985. *Prog. Neurobiol.* 25: 297–325.

[25] Scheutze, S. M., and Role, L. M. 1987. *Annu. Rev. Neurosci.* 10: 403–457.

[26] Fontaine, B., and Changeux, J.-P. 1989. *J. Cell Biol.* 108: 1025–1037.

[27] Salpeter, M. M., and Marchaterre, M. 1992. *J. Neurosci.* 12: 35–38.

[28] Sala, C. et al. 1997. *J. Neurosci.* 17: 8937–8944.

[29] Xu, R., and Salpeter, M. M. 1995. *J. Cell. Physiol.* 165: 30–39.

[30] Mishina, M. et al. 1986. *Nature* 321: 406–411.

[31] Shyng, S.-L., Xu, R., and Salpeter, M. M. 1991. *Neuron* 6: 469–475.

[32] O'Malley, J., Moore, C. T., and Salpeter, M. M. 1997. *J. Cell Biol.* 138: 159–165.

[33] Lømo, T., and Rosenthal, J. 1972. *J. Physiol.* 221: 493–513.

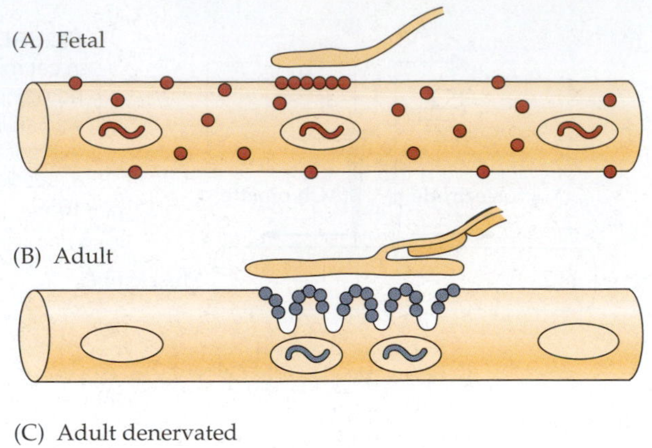

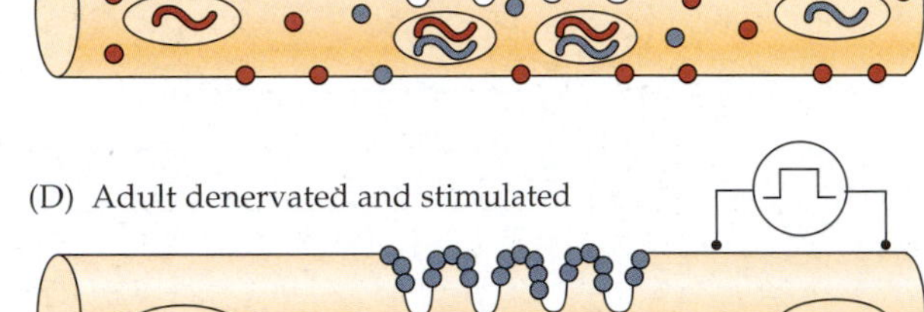

FIGURE 27.4 Synthesis and Distribution of Acetylcholine Receptors (AChRs) in rat muscle. (A) In fetal muscles, mRNAs for the α, β, γ, and δ-subunits of the AChR are expressed in nuclei all along the length of the myofiber. The embryonic $\alpha_2\beta\gamma\delta$ form of the receptor is found over the entire surface of the myofiber and accumulates at the site of innervation. (B) In adult muscles, mRNAs for the α, β, δ, and ε-subunits are expressed only in nuclei directly beneath the end plate. The adult form of the receptor is highly localized to the crests of the junctional folds. (C) In denervated adult muscles, nuclei directly beneath the end plate express α, β, γ, δ, and ε-subunits; all other nuclei re-express the fetal pattern of subunits. Embryonic AChRs are found all over the surface of the myofiber (producing denervation supersensitivity), including the postsynaptic membrane; the adult form of the receptor is restricted to the end-plate region. (D) If denervated muscles are stimulated directly, the pattern of AChR expression resembles that in innervated myofibers. (After Witzemann, Brenner, and Sakmann, 1991.)

to conduct past the cuff. Test stimulation of the nerve distal to the block produced a twitch of the muscle as usual, and miniature end-plate potentials still occurred normally, showing that synaptic transmission was intact. And yet, after 7 days of nerve block, the muscle had become supersensitive (Figure 27.5). Other experiments demonstrated that new extrajunctional receptors appeared when neuromuscular transmission was blocked by long-term application of curare or α-bungarotoxin to a muscle. These results show that denervation supersensitivity is produced by the loss of synaptic activation of the muscle.[34,35]

[34] Berg, D. K., and Hall, Z. W. 1975. *J. Physiol.* 244: 659–676.

[35] Witzemann, V., Brenner, H.-R., and Sakmann, B. 1991. *J. Cell Biol.* 114: 125–141.

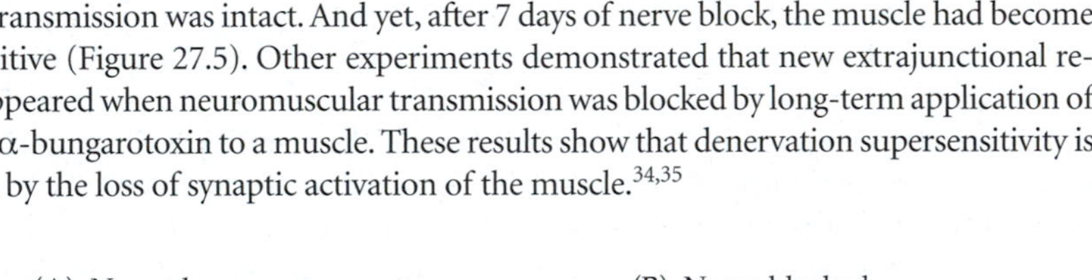

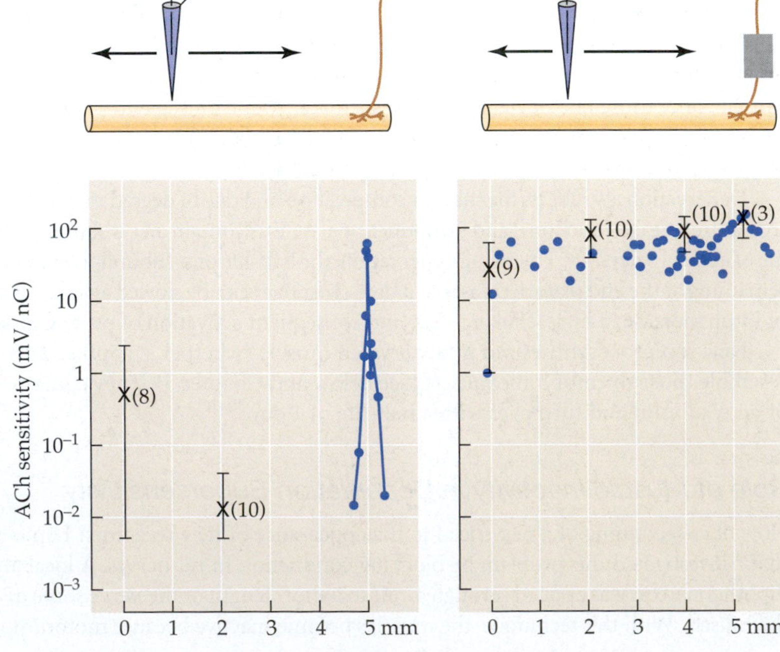

FIGURE 27.5 New Acetylcholine Receptors (AChRs). These receptors appear after block of nerve conduction in rat muscle. (A) In the normal muscle, ACh sensitivity is restricted to the end-plate region (near the 5-mm position). (B) After the nerve to the muscle was blocked for 7 days by a local anesthetic, the ACh sensitivity is distributed over the entire muscle fiber surface. Sensitivity is expressed numerically in millivolts of depolarization (mV) per nanocoulomb (nC) of charge ejected from the pipette (see Chapter 9). The crosses and bars represent the mean and range of sensitivities of a number (in parentheses) of adjacent muscle fibers. (From Lømo and Rosenthal, 1972.)

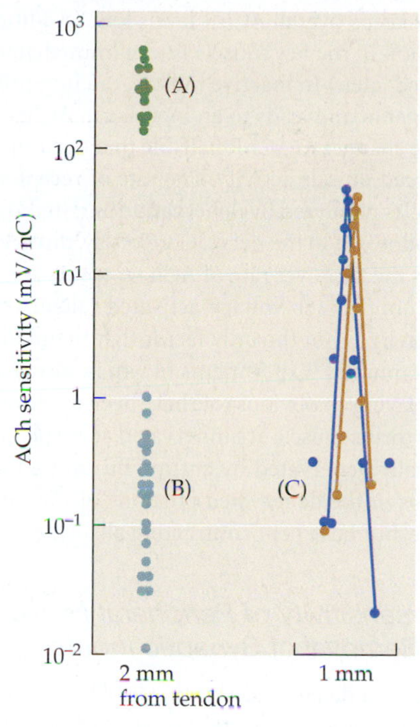

FIGURE 27.6 Reversal of Supersensitivity in a denervated rat muscle by direct stimulation of the muscle fibers. (A) Increased sensitivity in the extrasynaptic portion of a muscle fiber after 14 days of denervation. (B) Sensitivity in the extrasynaptic region of a muscle that had been denervated for 7 days without stimulation and then stimulated intermittently for another 7 days. This treatment reversed the denervation supersensitivity. (C) Acetylcholine (ACh) sensitivity in two stimulated fibers of the same muscle near their denervated end-plate regions. The high sensitivity is confined to this region in the stimulated muscle. (After Lømo and Rosenthal, 1972.)

The importance of muscle activity itself as a factor in controlling supersensitivity was confirmed in experiments in which denervated muscles in the rat were stimulated directly through permanently implanted electrodes. Repetitive, direct stimulation of muscles over several days caused the sensitive area to become restricted, so that once again only the synaptic region was sensitive to ACh (Figure 27.6; see also Figure 27.4D).[33] The frequency of spontaneous activity in fibrillating muscle is too low to reverse the effects of denervation on the distribution of AChRs.[15]

Role of Calcium in Development of Supersensitivity in Denervated Muscle

What is it about the lack of muscle activity that causes supersensitivity to develop? Changes in intracellular calcium appear to be the key factor (Figure 27.7).[36] Electrical activity of innervated muscle results in the influx of calcium through voltage-activated calcium channels. Increased intracellular calcium activates protein kinase C, which in turn phosphorylates and inhibits a molecule known as myogenin. Myogenin is a transcription factor that induces the expression of AChR-subunit genes and also regulates other aspects of muscle differen-

[36] Sakuma, K., and Yamaguchi, A. 2010. *J. Biomed. Biotechnol.* 2010: 721219.

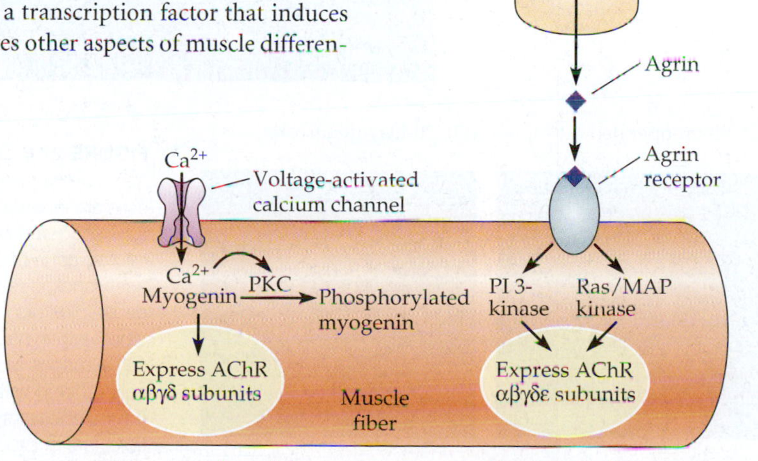

FIGURE 27.7 Control of Acetylcholine Receptor (AChR) Synthesis by Calcium and Neural Factors. In extrasynaptic regions of a vertebrate skeletal muscle fiber, influx of calcium through voltage-activated calcium channels activates protein kinase C (PKC), which phosphorylates and inactivates myogenin. At the synapse, agrin (see Figure 27.15) is released from nerve terminals and interacts with agrin receptors. This activates phosphatidylinositol 3-kinase (PI3K) and Ras/mitogen-activated protein kinase (Ras/MAPK) pathways, leading to expression of AChR α, β, γ, δ, and ε-subunits. Other neural signals suppress expression of the γ-subunit.

tiation.[37] Thus, in innervated muscle, calcium influx inhibits AChR gene expression and thereby keeps overall AChR levels low. (Additional signals that specifically induce AChR expression in the few muscle nuclei immediately beneath the postsynaptic membrane are discussed later.) In inactive muscle, calcium influx is reduced; this removes the inhibition of myogenin and leads to an increase in AChR expression.

The changes in AChR half-life that occur in denervated muscle are also a consequence of reduced muscle activity. The rate of receptor degradation increases to a similar extent in muscles paralyzed by denervation and in those made inactive by continuous application of tetrodotoxin to the nerve. Conversely, direct electrical stimulation of denervated muscle restores the turnover rate of ACh receptors at synaptic sites to normal levels. Again, influx of calcium through voltage-activated calcium channels plays an important role.[31,36,38]

Activity is not the only factor that maintains the normal complement of receptors in skeletal muscles. Experiments in which slowly developing changes occur without activity per se playing an obvious role have been made on partially denervated muscles. Fibers in the frog sartorius muscle are innervated at more than one site along their length. If the muscle is partially denervated by cutting intramuscular branches of the nerve, supersensitivity develops in the denervated portions of the muscle fibers. Yet these fibers have not been inactive, but have kept contracting all along.[21]

Supersensitivity of Peripheral Nerve Cells after Removal of Synaptic Inputs

The effect of denervation on the distribution of transmitter receptors in neurons has been studied in autonomic ganglion cells in frogs and chicks. In the living frog heart, parasympathetic neurons can be seen in the transparent interatrial septum, greatly facilitating application of ACh to discrete spots on the cell surface. The cells have no dendrites; synapses are made on the cell soma (Figure 27.8A). Like skeletal muscle fibers, these neurons are sensitive to the transmitter ACh at selected spots on their surfaces (i.e., immediately

[37] Macpherson, P. C., Cieslak, D., and Goldman, D. 2006. *Mol. Cell. Neurosci.* 31: 649–660.

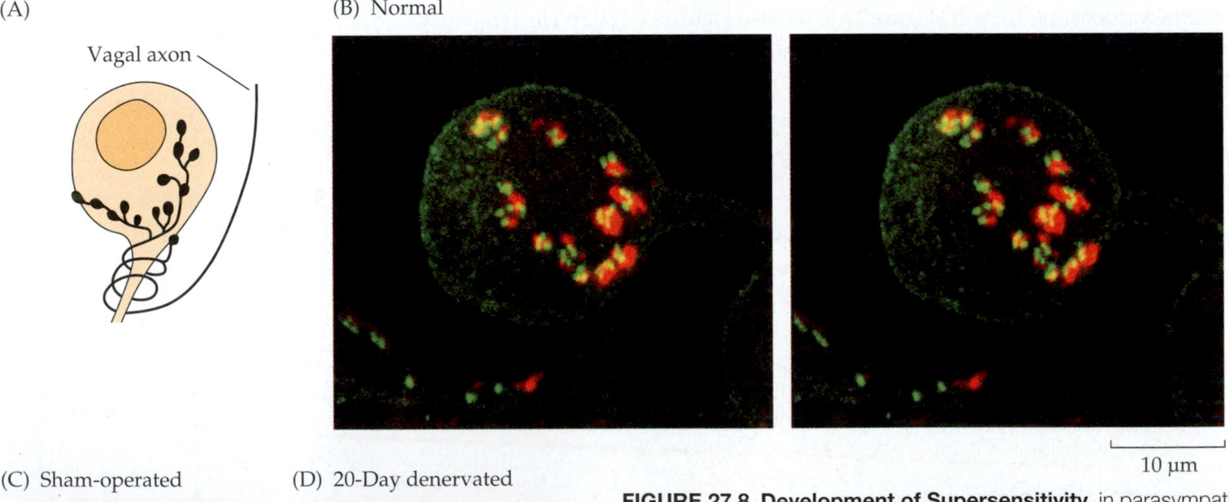

FIGURE 27.8 Development of Supersensitivity in parasympathetic nerve cells in frog heart after denervation. (A) Parasympathetic ganglion cells are innervated by axons in the vagus nerve, which form terminal boutons scattered over the cell surface. (B) Stereo pair immunofluorescence micrographs of a ganglion cell in a normal animal, labeled with antibodies to acetylcholine receptors (AChRs; green) and to synaptic vesicles (red). Large, dense clusters of AChRs are located at synaptic sites; more than a hundred small extrasynaptic clusters are spread over the rest of the cell surface. (C,D) Images of sham-operated (C) and 20-day denervated (D) ganglion cells labeled with antibodies to AChRs. Denervation causes a decrease in the number of synaptic clusters and a marked increase in small extrasynaptic clusters, producing supersensitivity. (From Wilson, Horch, and Sargent, 1996; micrographs kindly provided by P. B. Sargent.)

under the presynaptic terminals).[39] When the distribution of AChRs was assessed by immunofluorescence microscopy, each ganglion cell was found to have approximately 30 large, dense AChR clusters located at synaptic sites and more than 100 small extrasynaptic clusters spread over the cell surface (Figure 27.8B).[40] Approximately 20% of the receptors were at extrasynaptic sites.[41]

To study the effects of denervation, the two vagus nerves to the heart were cut, and the frog was left to recover.[42,43] Synaptic transmission between vagal nerve terminals and ganglion cells failed rapidly, starting on the second day after denervation. At the same time, the area of the neuronal surface membrane sensitive to ACh increased. By days 4 to 5, ACh caused a membrane depolarization when applied anywhere on the cell surface. In other respects, the cells were normal. It was not the number of AChRs that changed, rather it was the distribution (Figure 27.8C,D).[44] Denervation reduced the number of synaptic clusters by 90% and caused a two- to threefold increase in the number of small extrasynaptic clusters spread over the cell surface. Denervation also caused a decrease in the level of acetylcholinesterase. Thus, the sensitivity to ACh was increased. If the original nerve was allowed to grow back, the sensitive area became restricted once more to synaptic sites.[43] In certain other chick and frog ganglia, denervation has little or no effect on the number or distribution of surface AChRs.[45]

Susceptibility of Normal and Denervated Muscles to New Innervation

In adult mammals and frogs, an innervated muscle fiber will not accept innervation by an additional nerve.[46] Thus, if a cut motor nerve is placed on an innervated muscle, it will not form additional new end plates on the muscle fibers. In contrast, nerve fibers do grow out and form end plates on denervated or injured muscle fibers. Unlike the situation during development, in which growth cones contact muscle fibers at random sites[47] (but see also Kummer, Misgeld, and Sanes, 2005[48]), reinnervation usually occurs at the site of the original end plate. Regenerating axons appear to be guided to the original synaptic sites by the endoneurial tubes of the former axons (see Figure 27.1) and by processes extended by the Schwann cells that had capped the former axon terminal (see Chapter 10).[49] However, if a cut nerve is placed far enough away from the denervated end plate, then an entirely new end plate can be formed on a muscle. This means that nerve fibers can form synapses in a region that had never been innervated, and there they induce both pre- and postsynaptic specializations.

Similarly, after rat muscles are made supersensitive by blocking impulse transmission in the nerve or by application of botulinum toxin, foreign nerves are able to form additional distant synapses.[46,49] After release of the block, each of the two nerves can give rise to synaptic potentials and evoke contractions. Conversely, when a denervated muscle is stimulated directly, its ability to accept extra innervation is lost together with its supersensitivity. It is not, however, a prerequisite that the muscle be supersensitive to ACh for innervation to occur. Thus, reinnervation occurs in denervated rat and *Xenopus* muscles when ACh receptors are blocked by α-bungarotoxin or curare.[50,51]

Role of Schwann Cells and Microglia in Axon Outgrowth after Injury

Schwann cells of the peripheral nervous system provide an environment conducive to axon regeneration.[52] Moreover, they proliferate after injury. If a peripheral nerve is crushed rather than cut, the endoneurial tubes and Schwann cell basal lamina that surround the axons remain intact (see Figure 27.1). Under such conditions, axons regenerate within their parent tubes and are guided back to their original targets. If the endoneurial tubes are disrupted, as when a nerve is cut, then regenerating axons enter tubes in the distal portion of the nerve at random; frequently, the regenerating axons are guided to inappropriate targets. The growth-promoting activity of Schwann cells is due to secretion of trophic factors, surface expression of cell adhesion molecules and integrins, and production of extracellular matrix components such as laminin.[53] For example, experiments in which the sciatic nerve is lesioned have shown that as the peripheral portion of the axon degenerates, proliferating Schwann cells synthesize high levels of brain-derived neurotrophic factor (BDNF) and nerve growth factor (NGF) (Figure 27.9).[54] Thus, Schwann cells temporarily supply regenerating motor, sensory, and sympathetic axons with BDNF and NGF as they grow back to their peripheral targets. It is

[38] Caroni, P. et al. 1993. *J. Neurosci.* 13: 1315–1325.

[39] Harris, A. J., Kuffler, S. W., and Dennis, M. L. 1971. *Proc. R. Soc. Lond., B, Biol. Sci.* 177: 541–553.

[40] Wilson Horch, H. L., and Sargent, P. B. 1996. *J. Neurosci.* 16: 1720–1729.

[41] Sargent, P. B., and Pang, D. Z. 1989. *J. Neurosci.* 9: 1062–1072.

[42] Kuffler, S. W., Dennis, M. J., and Harris, A. J. 1971. *Proc. R. Soc. Lond., B, Biol. Sci.* 177: 555–563.

[43] Dennis, M. J., and Sargent, P. B. 1979. *J. Physiol.* 289: 263–275.

[44] Sargent, P. B. et al. 1991. *J. Neurosci.* 11: 3610–3623.

[45] Wilson Horch, H. L., and Sargent, P. B. 1995. *J. Neurosci.* 15: 7778–7795.

[46] Jansen, J. K. S. et al. 1973. *Science* 181: 559–561.

[47] Lin, S. et al. 2008. *J. Neurosci.* 28: 3333–3340.

[48] Kummer, T. T., Misgeld, T., and Sanes, J. R. 2006. *Curr. Opin. Neurobiol.* 16: 74–82.

[49] Son, Y. J., and Thompson, W. J. 1995. *Neuron* 14: 125–132.

[50] Cohen, M. W. 1972. *Brain Res.* 41: 457–463.

[51] Van Essen, D., and Jansen, J. K. 1974. *Acta Physiol. Scand.* 91: 571–573.

[52] Fawcett, J. W., and Keynes, R. J. 1990. *Annu. Rev. Neurosci.* 13: 43–60.

[53] Madduri, S., and Gander, B. 2010. *J. Peripher. Nerv. Syst.* 15: 93–103.

[54] Meyer, M. et al. 1992. *J. Cell Biol.* 119: 45–54.

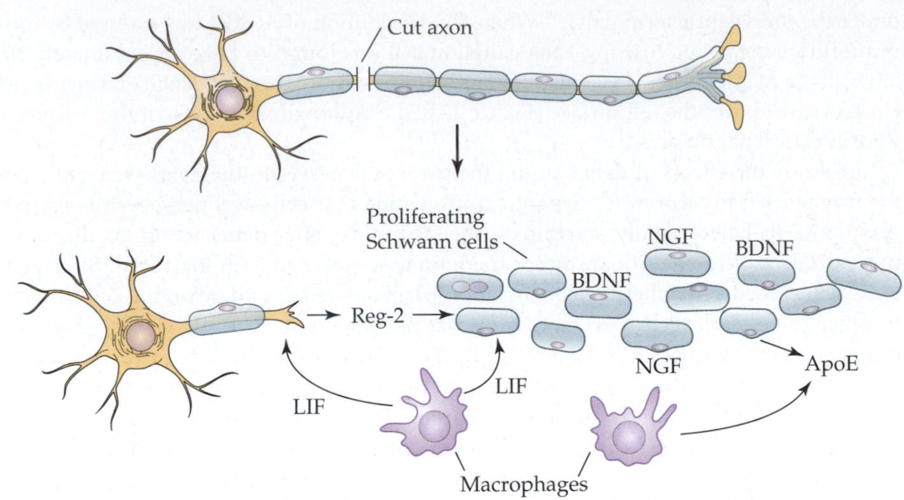

FIGURE 27.9 Schwann Cells Promote Axon Regrowth in the vertebrate peripheral nervous system. After axotomy, the distal portion of the axon and the myelin degenerate and are phagocytized. Schwann cell proliferation is stimulated by two cytokines: leukemia inhibitory factor (LIF) from macrophages and Reg-2 from axon terminals. Expression of Reg-2 is enhanced by LIF. Proliferating Schwann cells synthesize two neurotrophic factors, brain-derived neurotrophic factor (BDNF) and nerve growth factor (NGF), which are held on the cell surface by low-affinity BDNF/NGF receptors and help sustain regenerating axons and guide them to their targets. Schwann cells and macrophages also synthesize apolipoprotein E (ApoE), which may help promote neuron survival and axon regrowth.

interesting that such denervated Schwann cells also express large numbers of low-affinity NGF/BDNF receptors on their surface, perhaps to hold the NGF and BDNF they produce along the path that regenerating axons should take. As regeneration progresses, the Schwann cells cease production of NGF and BDNF and once again ensheathe the axons.

Apolipoprotein E (ApoE), synthesized by Schwann cells and macrophages, also accumulates in the distal portion of damaged peripheral nerves and becomes associated with the Schwann cell basement membrane (see Figure 27.11).[55,56] ApoE promotes the health and survival of neurons by virtue of its protective effects against oxidative damage; it also promotes neurite outgrowth and adhesion. Factors that promote Schwann cell proliferation include cytokines, leukemia inhibitory factor, and mitogens.[57] Microglial cells (discussed in Chapter 8) also contribute to repair after a lesion.[58] They migrate rapidly to the site of an injury, where they remove debris and provide growth-promoting molecules to axons.[59]

Denervation-Induced Axonal Sprouting

Not only are denervated muscles amenable to innervation, but also they actively induce undamaged nerves to sprout new terminal branches. After a muscle is partially denervated, the remaining axon terminals sprout and innervate the denervated muscle fibers (Figure 27.10).[60] As with regulation of ACh-receptor synthesis and degradation, muscle inactivity triggers this process. Sprouting and hyperinnervation occur if muscle activity is prevented (by blocking action potential propagation in the nerve with a cuff impregnated with TTX) or if neuromuscular transmission is blocked (with botulinum toxin or α-bungarotoxin). The terminal Schwann cell plays an important role in the regulation of axon terminal sprouting (see Chapter 10).[61–63] Axons of sensory cells other than motoneurons can also sprout to supply denervated territories. In the leech, for example, the killing of a particular sensory neuron by injecting it with pronase, induces axon sprouting into the denervated area of skin. However, only axons of cells that have the same sensory modality grow into the denervated area.[64]

Appropriate and Inappropriate Reinnervation

For complete recovery of function, regenerating axons must reestablish connections with their original targets. Classic experiments of Langley demonstrated that regenerating

[55] Skene, J. H. P., and Shooter, E. M. 1983. *Proc. Natl. Acad. Sci. USA* 80: 4169–4173.

[56] Fullerton, S. M., Strittmatter, W. J., and Matthew, W. D. 1998. *Exp. Neurol.* 153: 156–163.

[57] Banner, L. R., and Patterson, P. H. 1994. *Proc. Natl. Acad. Sci. USA* 91: 7109–7113.

[58] Gitik M., Reichert, F., and Rotshenker, S. 2010. *FASEB J.* 24: 2211–2221.

[59] Samuels, S. E. et al. 2010. *J. Gen. Physiol.* 136: 425–452.

[60] Brown, M. C., Holland, R. L., and Hopkins, W. G. 1981. *Annu. Rev. Neurosci.* 4: 17–42.

[61] Son, Y. J., and Thompson, W. J. 1995. *Neuron* 14: 133–141.

[62] Love, F. M., Son, Y. J., and Thompson, W. J. 2003. *J. Neurobiol.* 54: 566–576.

[63] Hayworth, C. R. et al. 2006. *J. Neurosci.* 26: 6873–6884.

[64] Blackshaw, S. E., Nicholls, J. G., and Parnas, I. 1982. *J. Physiol.* 326: 261–268.

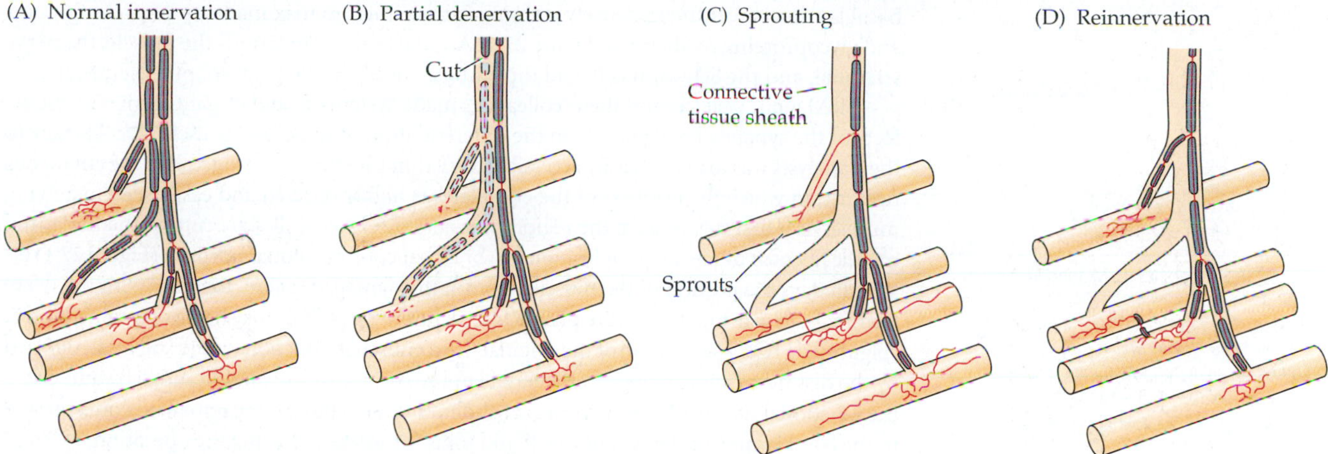

FIGURE 27.10 Nerve Terminals Sprout in Response to Partial Denervation of a mammalian skeletal muscle. (A) Normal pattern of innervation. (B) Some fibers are denervated by cutting a few of the axons innervating the muscle. (C) Axons sprout from the terminals and from nodes along the preterminal axons of undamaged motoneurons to innervate the denervated fibers. (D) After 1 or 2 months, sprouts that have contacted vacant end plates are retained, while other sprouts disappear. (After Brown, Holland, and Hopkins, 1981.)

mammalian preganglionic autonomic axons can reinnervate the appropriate postganglionic neurons. One mechanism for reestablishing connections selectively is competition between axons; in salamander muscles that have been innervated by inappropriate axons, the foreign synapses are eliminated after the normal nerve reestablishes its connection.[65] In adult mammals, foreign nerves can be as effective as the original ones in innervating muscle fibers, if they manage to reach them.[52,66] Observations concerning the consequences of inappropriate contacts date back to 1904, when Langley and Anderson made the remarkable observation that muscles of the cat could become innervated by cholinergic preganglionic sympathetic fibers,[67] which normally make synapses on nerve cells in ganglia (see Chapter 17). (At that time, it was of course not known that ACh was the transmitter used by preganglionic fibers or even that transmitters existed!) Similar synapses have been shown to be formed by autonomic nerves on frog and rat skeletal muscle.[68] In such experiments, many of the properties of the nerve and muscle were found to remain unchanged, despite the abnormal innervation.

By contrast, the properties of muscles can become markedly changed by foreign innervation. Slow and fast skeletal muscle fibers in the frog have quite different properties. Slow fibers are diffusely innervated, have a characteristic fine structure, and do not give regenerative impulses or twitches. After denervation, slow fibers can become reinnervated by nerves that normally innervate twitch muscles at discrete end plates. Under these conditions, slow fibers become able to give conducted action potentials and twitches.[69] Eccles, Eccles, Buller, and their colleagues[70] cut and interchanged the nerves to rapidly and slowly contracting skeletal muscles in kittens and rats. Both these types of mammalian muscle fibers produce propagating action potentials and are called slow-twitch and fast-twitch fibers. After the muscles were reinnervated by the inappropriate nerves, the slow-twitch muscles became faster and the fast-twitch ones slower.[71,72] A major factor in the transformation is the pattern of impulses in the nerve and the resulting muscle contractions—the motoneurons innervating slow- and fast-twitch muscle fibers tend to fire at different frequencies.[73]

Basal Lamina, Agrin, and the Formation of Synaptic Specializations

The only example at present available to show how a nerve causes a synapse to form on its target is provided by the neuromuscular junction. There it has been shown that a structure that plays a key role in the regeneration of neuromuscular synapses is the **synaptic basal lamina**, which lies between the nerve terminal and the muscle membrane. The synaptic

[65] Dennis, M. J., and Yip, J. W. 1978. *J. Physiol.* 274: 299–310.

[66] Bixby, J. L., and Van Essen, D. C. 1979. *Nature* 282: 726–728.

[67] Langley, J. N., and Anderson, H. K. 1904. *J. Physiol.* 31: 365–391.

[68] Grinnell, A. D., and Rheuben, M. B. 1979. *J. Physiol.* 289: 219–240.

[69] Miledi, R., Stefani, E., and Steinbach, A. B. 1971. *J. Physiol.* 217: 737–754.

[70] Buller, A. J., Eccles, J. C., and Eccles, R. M.1960. *J. Physiol.* 150: 417–439.

[71] Close, R. I. 1972. *Physiol. Rev.* 52: 129–197.

[72] Pette, D. 2001. *J. Appl. Physiol.* 90: 1119–1124.

[73] Salmons, S., and Sreter, F. A. 1975. *J. Anat.* 120: 412–415.

basal lamina constitutes a densely staining extracellular matrix made up of proteoglycans and glycoproteins. As shown in Figure 27.11A, basal lamina surrounds the muscle, the nerve terminal, and the Schwann cell, and dips into the folds in the postsynaptic membrane.

McMahan, Wallace and their colleagues made systematic and elegant studies of the effects of the synaptic basal lamina on the differentiation of nerve and muscle.[74–77] The key to their analysis was to use an easily accessible, very thin muscle in the frog, called the cutaneous pectoris, in which the position of the end plates is highly ordered and easily seen in a living muscle. As a first step, cells in the region of innervation were killed by cutting the nerve and muscle fibers or by repeated application of a brass bar cooled in liquid nitrogen (Figure 27.11B). Within days the portion of the muscle fibers in the damaged region, together with the nerve terminals, degenerated and were phagocytized, but the basal lamina sheaths remained intact (Figure 27.11C). The location of the original neuromuscular junctions could still be recognized by the distinctive morphology of the basal lamina sheaths of the muscle and Schwann cell at the junctional sites and by the presence of cholinesterase. The enzyme remained concentrated in the basal lamina of the synaptic cleft and folds for weeks following the operation.

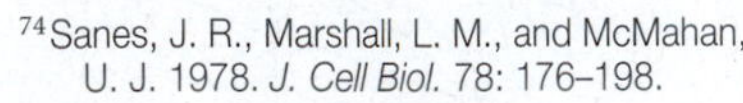

[74]Sanes, J. R., Marshall, L. M., and McMahan, U. J. 1978. *J. Cell Biol.* 78: 176–198.

[75]Burden, S. J., Sargent, P. B., and McMahan, U. J. 1979. *J. Cell Biol.* 82: 412–425.

[76]McMahan, U. J., and Slater, C. R. 1984. *J. Cell Biol.* 98: 1453–1473.

[77]Anglister, L., and McMahan, U. J. 1985. *J. Cell Biol.* 101: 735–743.

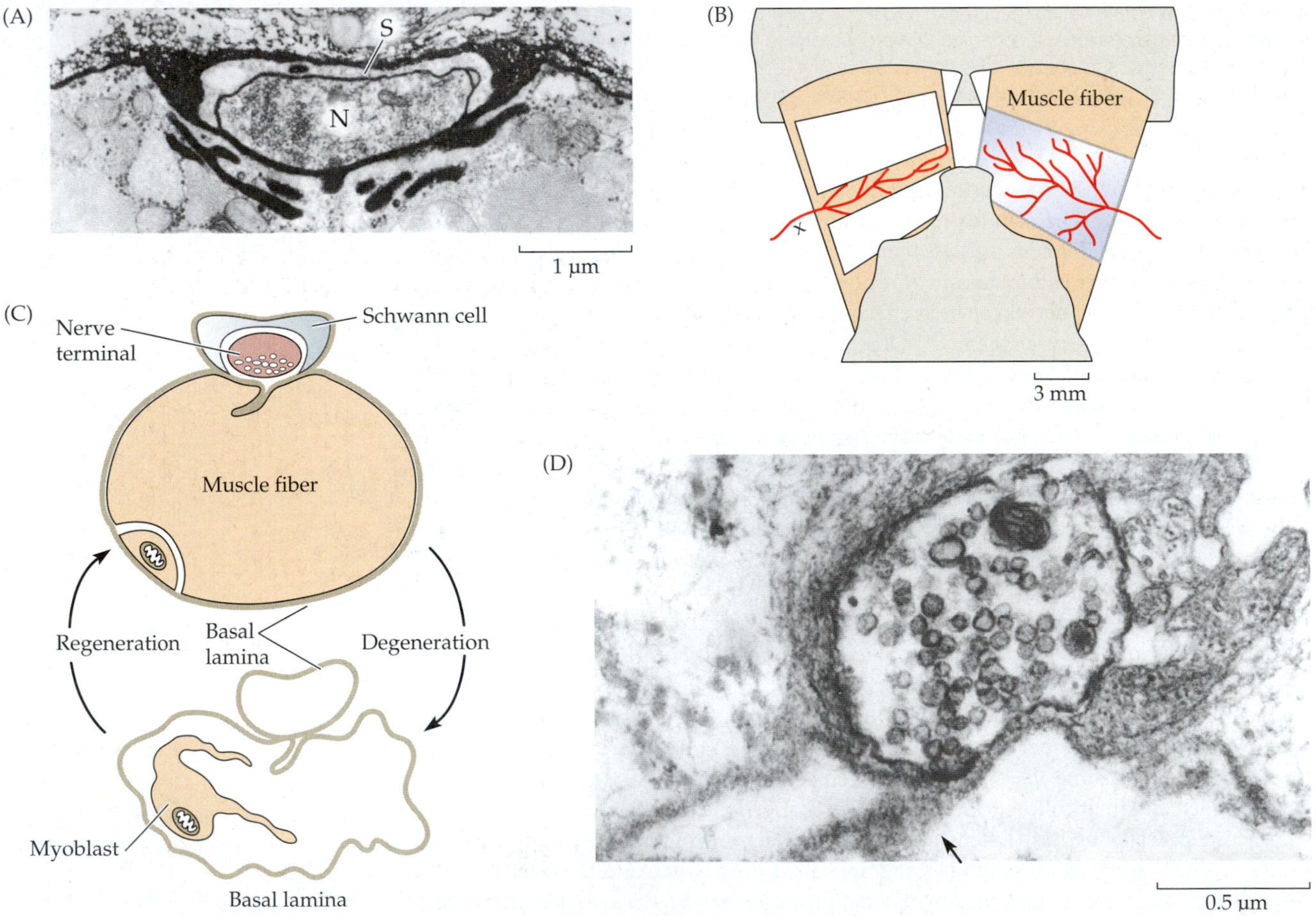

FIGURE 27.11 Basal Lamina and Regeneration of Synapses. (A) Electron micrograph of a normal neuromuscular synapse in the frog, stained with ruthenium red to show the basal lamina that dips into the postsynaptic folds and surrounds the Schwann cell (S) and nerve terminal (N). (B) Diagram of the cutaneous pectoris muscle, showing the region frozen (right) or cut away (left) to damage muscle fibers. (C) Freezing causes all cellular elements of the neuromuscular junction to degenerate and be phagocytized, leaving only the basal lamina sheath of the muscle fiber and Schwann cell intact. New neuromuscular junctions are restored by regenerating axons and muscle fibers. (D) Nerve and muscle were damaged, and regeneration of muscle fibers was prevented by X-irradiation. In the absence of muscle fibers, axons regenerated; contacted original synaptic sites, marked by the tongue of basal lamina that had extended into the junctional fold (arrow); and formed active zones. (After McMahan, Edgington, and Kuffler, 1980; micrographs kindly provided by U. J. McMahan.)

Two weeks after damage to the muscles, new myofibers had formed within the basal lamina sheaths and were contacted by regenerated axon terminals, which evoked muscle twitches when the nerves were stimulated. Nearly all of the regenerated synapses were located precisely at original synaptic sites, as marked by cholinesterase. Thus, signals associated with the synaptic basal lamina had specified where regenerating synapses were to be formed.

To investigate further the nature of the signals associated with the synaptic basal lamina, muscles were damaged and the nerve was crushed, but muscle fiber regeneration was prevented by X-irradiation. Regenerating axons grew to the former synaptic sites on the basal lamina, as marked by cholinesterase, and formed active zones for release precisely opposite portions of the basal lamina that had projected into the junctional folds—all this without a postsynaptic target (Figure 27.11D).

In a parallel series of experiments, it was shown that synaptic basal lamina in the adult frog contains a factor that triggers differentiation of postsynaptic specializations in regenerating myofibers. Muscles were damaged as described, but reinnervation was prevented by removing a long segment of the nerve. When new muscle fibers regenerated within the basal lamina sheaths, they formed junctional folds and aggregates of ACh receptors and acetylcholinesterase precisely at the point where they came in contact with the original synaptic basal lamina (Figure 27.12). Thus, signals stably associated with synaptic basal lamina can trigger the formation of synaptic specializations in both regenerating myofibers and nerve terminals.

Bruce Wallace, co-discoverer of agrin and co-author of previous editions of *From Neuron to Brain*.

U. J. McMahan

Identification of Agrin

In order to identify the molecule in synaptic basal lamina that triggers postsynaptic differentiation, McMahan, Wallace and their colleagues prepared basal lamina-containing extracts from the electric organ of the marine ray *Torpedo californica*.[78] The electric organ, which generates large currents that are used to stun animals, is a tissue derived embryologically from muscle that receives very dense cholinergic innervation. It resembles a giant array of motor end plates. There would be no chance of isolating the relevant molecule from normal skeletal muscle fibers because the area of the end plate is so small compared to the whole surface of the muscle. When added to myofibers in culture, purified extracts from torpedo mimicked the effects of synaptic basal lamina on regenerating muscle fibers; that

[78] McMahan, U. J., and Wallace, B. G. 1989. *Dev. Neurosci.* 11: 227–247.

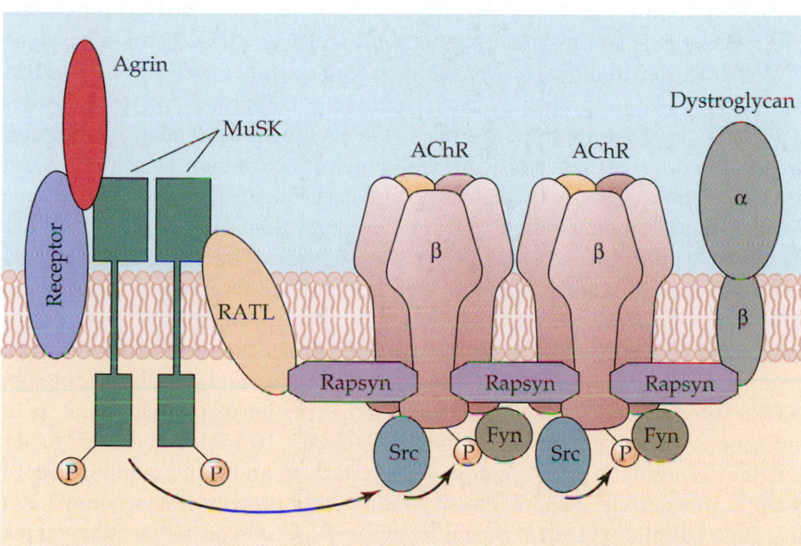

FIGURE 27.12 Interaction of Agrin with its receptor (LRP4) activates MuSK. MuSK then triggers differentiation of postsynaptic specializations in muscle cells at which acetylcholine receptors (AChRs), rapsyn, and dystroglycan accumulate. Binding of agrin also causes tyrosine autophosphorylation of MuSK and activation of intracellular tyrosine kinases Src and Fyn. Activated MuSK recruits rapsyn, via an unidentified transmembrane protein, RATL. Rapsyn, in turn, recruits ACh receptors, which become phosphorylated on tyrosine residues of the β-subunit and dystroglycan. Many additional postsynaptic components accumulate through interactions with dystroglycan (not shown).

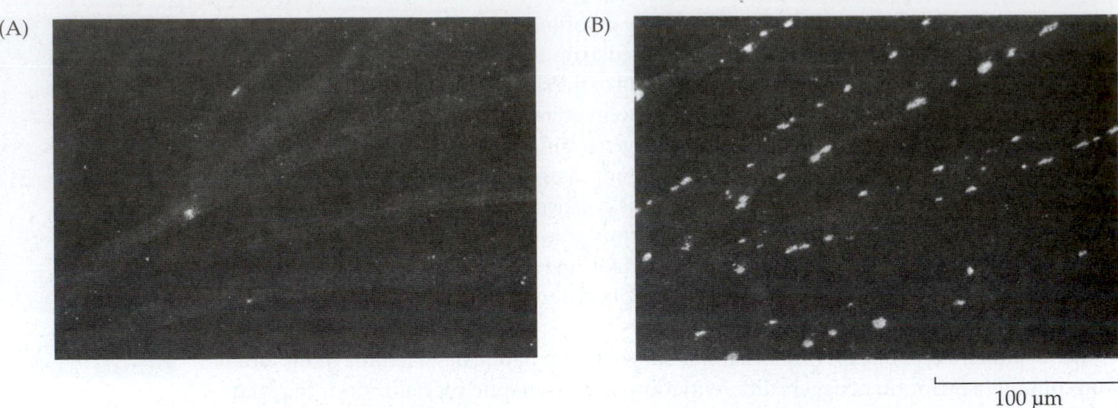

FIGURE 27.13 Agrin Causes Aggregation of Acetylcholine Receptors (AChR) seen here in chick myotubes in culture. Fluorescence micrographs of myotubes labeled with rhodamine-conjugated α-bungarotoxin to mark ACh receptors. (A) Receptors are distributed over the surface of control myotubes at low density. (B) Overnight incubation with agrin causes the formation of patches at which ACh receptors accumulate, together with other components of the postsynaptic apparatus. (After McMahan and Wallace, 1989.)

is, they induced the formation of specializations at which AChRs accumulated, together with several other components of the postsynaptic apparatus[78] (Figures 27.13 and 27.14; see also Figure 27.12). The active component in the extracts, called **agrin**, was purified and characterized, and cDNAs encoding it were cloned from chick, rat, and ray.[79]

Results of in situ hybridization and immunohistochemical studies demonstrate that agrin is synthesized by motor neurons, transported down their axons, and released to induce differentiation of the postsynaptic apparatus at developing neuromuscular junctions.[80] Agrin itself becomes incorporated into the synaptic basal lamina, where it helps maintain the postsynaptic apparatus in the adult and triggers its differentiation during regeneration.

The Role of Agrin in Synapse Formation

Additional evidence for the role of agrin in synapse formation is provided by comparisons made between developing and regenerating neuromuscular synapses. Early in development, AChRs are distributed diffusely over the surface of un-innervated myotubes, as in denervated muscle. When the growth cone of a motoneuron approaches a myotube, depolarizing potentials arise due to the release of ACh.[81,82] A functional synaptic connection is established within hours. AChRs and acetylcholinesterase accumulate beneath the axon terminal.[83] Detailed morphological and physiological experiments demonstrate that motor axons contact the developing muscle cells at random positions on their surface, rather than at preexisting AChR clusters.[47,84] Over time, the γ-subunit of the AChR is replaced by an ε-subunit, (see Figure 27.4 and Chapter 5), and the density of receptors in nonsynaptic portions of the muscle fiber decreases.[35]

The signal that produces these changes is not ACh. Acetylcholine receptors still accumulate beneath axon terminals in cultures grown in the presence of curare or α-bungarotoxin, which block the interaction of ACh with its receptor. During normal development, as in the reinnervation of denervated muscle, it is agrin, released by motor nerve terminals, that triggers the accumulation of receptors, cholinesterase, and other components of the postsynaptic apparatus at synaptic sites.[78] Among other proteins that accumulate in agrin-induced specializations is ARIA, a member of the neuregulin protein family, and the receptor proteins ERBB-2, ERBB-3, and ERBB-4.[85] It was originally proposed that activation of ERBB receptors in muscle might be the mechanism responsible for accumulation of AChR subunits. However, mice in which *Aria* genes have been deleted can still form normal neuromuscular synapses.[86,87]

In other experiments by McMahan and his colleagues, denervated rat soleus muscles were transfected with cDNA encoding neural agrin. After transfection, extrajunctional regions of the muscle fibers expressed and secreted neural agrin. Moreover, they formed

[79] Bowe, M. A., and Fallon, J. R. 1995. *Annu. Rev. Neurosci.* 18: 443–462.

[80] McMahan, U. J. 1990. *Cold Spring Harb. Symp. Quant. Biol.* 50: 407–418.

[81] Evers, J. et al. 1989. *J. Neurosci.* 9: 1523–1539.

[82] Dan, Y., Lo, Y., and Poo, M. M. 1995. *Prog. Brain Res.* 105: 211–215.

[83] Sanes, J. R., and Lichtman, J. W. 1999. *Annu. Rev. Neurosci.* 22: 389–442.

[84] Anderson, M. J., and Cohen, M. W. 1977. *J. Physiol.* 268: 757–773.

[85] Fischbach, G. D., and Rosen, K. M. 1997. *Annu. Rev. Neurosci.* 20: 429–458.

[86] Escher, P. et al. 2005. *Science* 308: 1920–1923.

[87] Rimer, M. 2010. *J. Biol. Chem.* 285: 32370–32377.

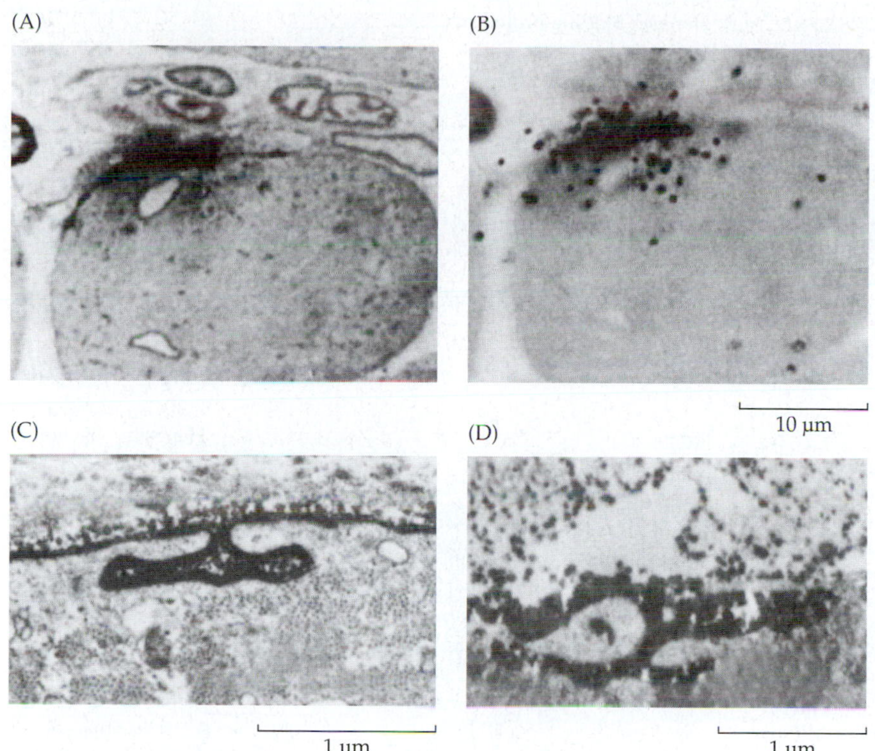

FIGURE 27.14 Accumulation of Acetylcholine Receptors and Acetylcholinesterase at original synaptic sites seen here on muscle fibers regenerating in the absence of nerve. The muscle was frozen as in Figure 27.11B, but the nerve was prevented from regenerating. New muscle fibers formed within the basal lamina sheaths. (A,B) Light-microscope autoradiography of a regenerated muscle stained for cholinesterase to mark the original synaptic site (in focus in [A]) and incubated with radioactive α-bungarotoxin to label ACh receptors (silver grains in focus in [B]). (C) Electron micrograph of the original synaptic site in a regenerated muscle labeled with horseradish peroxidase (HRP)–α-bungarotoxin. The distribution of ACh receptors is indicated by the dense HRP reaction product, which lines the muscle fiber surface and the junctional folds. (D) Electron micrograph of the original synaptic site in a regenerated muscle stained for cholinesterase. The original cholinesterase was permanently inactivated at the time the muscle was frozen. Thus, the dense reaction product is due to cholinesterase synthesized and accumulated at the original synaptic site by the regenerating muscle fiber. (A,B after McMahan, Edgington, and Kuffler, 1980; C after McMahan and Slater, 1984; D after Anglister and McMahan, 1985; micrographs kindly provided by U. J. McMahan.)

typical postsynaptic specializations, including membrane infoldings and aggregates of AChRs at sites far distant from the original synapse.[88]

Mechanism of Action of Agrin

Agrin occurs in several isoforms that arise from a single gene by alternative splicing.[81] Motoneurons, muscle cells, and Schwann cells all express agrin, but only motoneurons express the isoform that is potent in inducing postsynaptic differentiation. Agrin is a large heparan sulfate proteoglycan, with domains that interact with its receptor (LRP4), and with laminin, heparin-binding proteins, α-dystroglycan, heparin, and integrins (Figure 27.15).[89]

The ability to induce the formation of postsynaptic specializations resides in the C-terminal domain (Figure 27.16). The essential role of agrin in the formation of the neuromuscular junction is evident in mice in which agrin expression is prevented by homologous recombination.[90,91] In such agrin gene knockouts, myofibers appear normally and axons grow into the developing muscles, but neuromuscular junctions fail to form. (The agrin knockout animals die because of failure of respiration). A similar phenotype is seen in mice in which the muscle-specific receptor kinase (MuSK) is knocked out.[92] The lack of presynaptic specializations in agrin- and MuSK-deficient mutant mice suggests that

[88] Cohen, I. et al. 1997. *Mol. Cell. Neurosci.* 9: 237–253.

[89] Denzer, A. J. et al. 1998. *EMBO J.* 17: 335–343.

[90] Gautam, M. et al. 1996. *Cell* 85: 525–535.

[91] Burgess, R. W. et al. 1999. *Neuron* 23: 33–44.

[92] DeChiara, T. M. et al. 1995. *Cell* 83: 313–322.

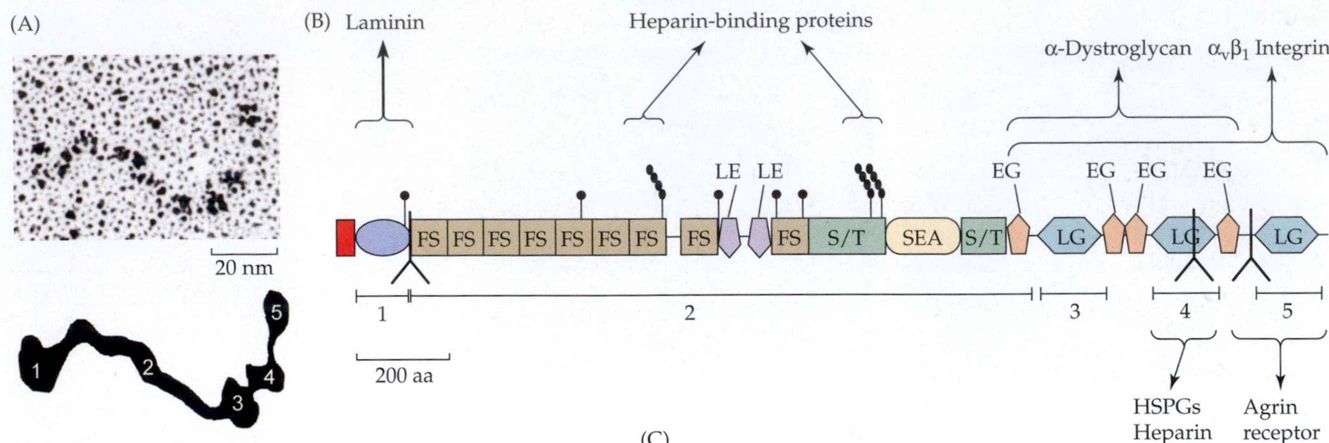

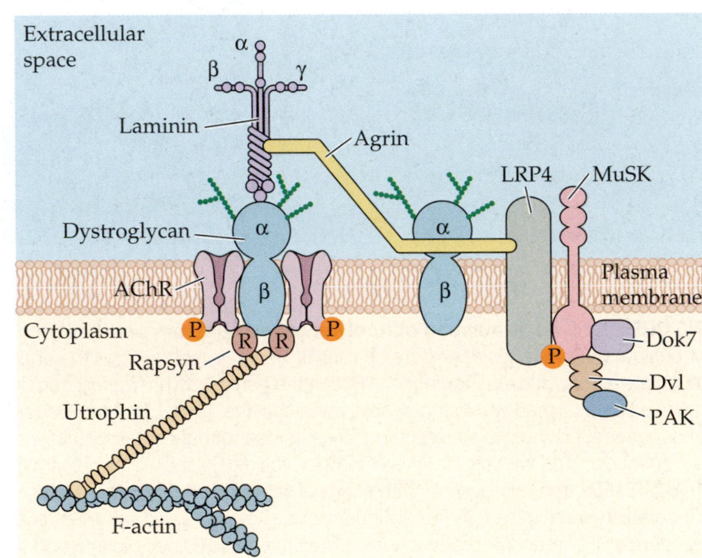

FIGURE 27.15 Agrin Is a Large Heparan Sulfate Proteoglycan (400 to 600 kDa) with domains that interact with laminin, heparan sulfate proteoglycans (HSPGs), heparin, α-dystroglycan, integrin, heparin-binding proteins, and the agrin receptor (LRP4) that causes acetylcholine receptor (AChR) aggregation. (A) Electron micrographs of agrin after rotary shadowing. (B) Schematic diagram of the structural and binding domains of chick agrin. Binding regions are indicated, as are globular (1, 3–5) and extended (2) regions of the molecule that can be recognized in part (A). (C) Diagram to show relation of agrin to the receptor LRP4 and downstream molecules. EG = epidermal growth-factor–like domain; FS = follistatin-like domain; LE = laminin EGF–like domain; LG = laminin G–like domain; SEA = motif found in sea urchin sperm protein, enterokinase, and agrin; S/T = serine and/or threonine-rich domain. (After Denzer et al., 1998; micrograph and C kindly provided by M. Ruegg.)

during development presynaptic differentiation is triggered by retrograde signals released by muscle cells in response to agrin.[93]

The sequence of events initiated by agrin is as follows: first, agrin binds to its LRP4 receptor and thus stimulates MuSK. This leads to phosphorylation of both proteins.[94,95] Phosphorylated MuSK then interacts with the adapter protein Dok7.[96] Mutation of the agrin LRP4 receptor, MuSK, or Dok7 prevents agrin from inducing AChR clusters to form in cultured muscle cells and prevents the formation of neuromuscular junctions in embryonic mice.[92,96,97] While there is evidence that another adapter protein, Crk, interacts with MuSK, the downstream signal pathways of the LRP4/MuSK/Dok7 complex remain unclear.[98] Ultimately, rapsyn[99] recruits the aggregation of AChRs. Much less is known about differentiation of the presynaptic nerve terminal in regenerating motor axons, even though

[93]Noakes, P. G. et al. 1995. *Nature* 374: 258–262.
[94]Kim, N. et al. 2008. *Cell* 135: 334–342.
[95]Zhang, B. et al. 2008. *Neuron* 60: 285–297.
[96]Okada, K. et al. 2006. *Science* 312: 1802–1805.
[97]Weatherbee, S. D., Anderson, K. V., and Niswander, L. A. 2006. *Development* 133: 4993–5000.
[98]Hallock, P. T. et al. 2010. *Genes Dev.* 24: 2451–2461.
[99]Moransard, M. et al. 2003. *J. Biol. Chem.* 278: 7350–7359.

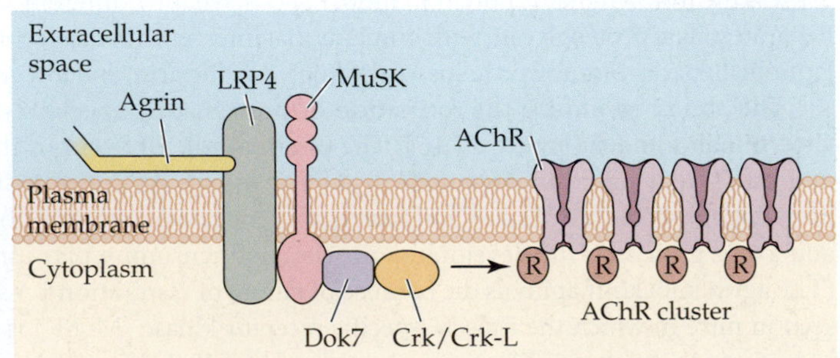

FIGURE 27.16 Molecules Genetically Proved to Be Critical for the formation of neuromuscular synaptic structure and for aggregation of acetylcholine receptors at the motor end plate. (Scheme kindly provided by M. Ruegg.)

it has been shown that molecules stably associated with the synaptic basal lamina in adult muscle can induce the formation of active zones in regenerating axons (see Figure 27.11).[78]

Regeneration in the Mammalian CNS

The adult mammalian central nervous system has limited capacity for regeneration. Transection of major axon tracts is not followed by axon regrowth and restitution of function. Nevertheless, it has become apparent that after tracts in the CNS are severed, axons can, under suitable circumstances, regrow for distances of several centimeters and form synapses with appropriate targets.[100–103]

Glial Cells and CNS Regeneration

Of importance for limiting axonal regeneration in the central nervous system is the immediate environment provided by CNS glial cells (see Chapter 10). Clues to an inhibitory role of CNS glial cells are provided by several experiments. First, although axons severed in the central nervous system typically do not regrow, motor neurons (whose cell bodies lie within the spinal cord) can regenerate severed peripheral axons (Figure 27.17). Likewise, axons of sensory neurons regrow to their targets in the periphery but fail to regenerate when severed within the CNS. Indeed, after a dorsal root is cut, sensory axons regenerate toward the spinal cord but stop growing when they reach the astrocytic processes that delimit the surface of the central nervous system. Moreover, axons in the periphery will not enter an optic nerve graft, which consists of CNS glial cells.[104] These findings suggest that CNS glial cells actively inhibit growth.

On the other hand, when dorsal root ganglion neurons are injected into CNS white matter tracts in such a way as to minimize trauma, they frequently extend axons for long distances in the white matter, invade gray matter, and form terminal arbors.[105] Thus, when there is no trauma-induced glial reaction, regeneration of axons by adult neurons is not prevented by contact with CNS glial cells.

When tracts in the CNS are lesioned, astrocytes, microglial cells, meningeal cells, and oligodendrocyte precursor cells accumulate at the site of the lesion to form a glial scar. These cells produce a variety of molecules that have been shown to inhibit axon growth, including free radicals, nitric oxide, arachidonic acid derivatives, and a variety of proteoglycans.[106,107]

In addition, it has been shown that oligodendrocytes from the mature central nervous system have a protein on their surface, known as Nogo-A, which is a member of the reticulon family of genes.[108] Nogo binds to a specific receptor (Nogo-66 receptor), induces long-lasting collapse of growth cones, and inhibits neurite outgrowth in vitro.[103,109] Application of a monoclonal antibody that neutralized this activity allowed axons to regenerate across a spinal cord lesion. Partial locomotor function could be restored, although the extent of regeneration under such conditions was still meager.[110]

Such observations led to the suggestion that failure of CNS neurons to regenerate might be due to the action of Nogo and other molecules (such as myelin associated glycoprotein [MAG] and myelin basic protein). It seems unlikely that inhibition on its own could be the

[100] Dusart, I. et al. 2005. *Brain Res. Brain Res. Rev.* 49: 300–316.

[101] Bunge, M. B. 2008. *J Spinal Cord Med.* 31: 262–269.

[102] Afshari, F. T., Kappagantula, S., and Fawcett, J. W. 2009. *Expert Rev. Mol. Med.* 11: e37.

[103] Huebner, E. A., and Strittmatter, S. M. 2009. *Results Probl. Cell Differ.* 48: 339–351.

[104] Aguayo, A. J. et al. 1978. *Neurosci. Lett.* 9: 97–104.

[105] Davies, S. J. A. et al. 1997. *Nature* 390: 680–683.

[106] Camand, E. et al. 2004. *Eur. J. Neurosci.* 20: 1161–1176.

[107] Kuzhandaivel, A. et al. 2010. *Neuroscience* 169: 325–338.

[108] Schwab, M. E., and Caroni, P. 1988. *J. Neurosci.* 8: 2381–2393.

[109] Fournier, A. E., GrandPre, T., and Strittmatter, S. M. 2001. *Nature.* 409: 341–346.

[110] Schnell, L., and Schwab, M. E. 1990. *Nature* 343: 269–272.

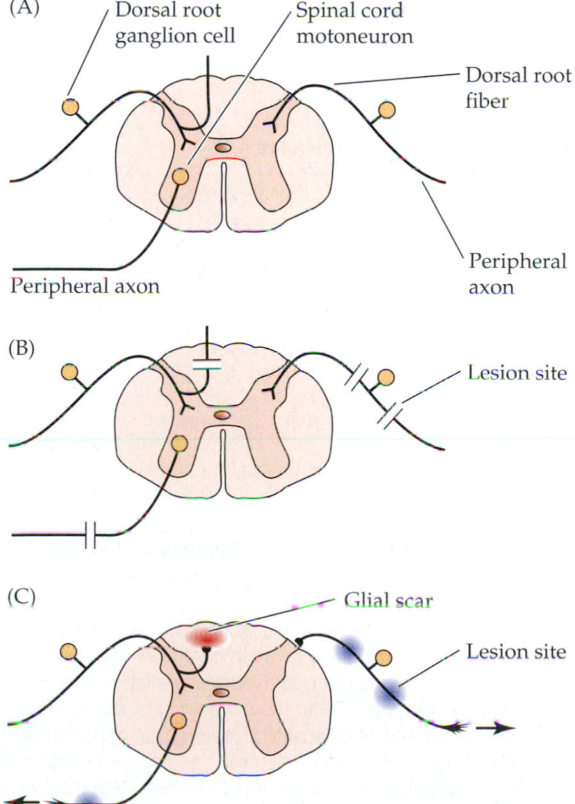

FIGURE 27.17 Axons of Sensory and Motor Neurons Regenerate in the Periphery but Not in the CNS. (A) Motoneurons, dorsal root ganglion sensory neurons, and their axonal processes in the mammalian nervous system. (B) Sites of axon lesions. (C) Extent of regeneration. Axons of dorsal root ganglion neurons and motoneurons regenerate through lesion sites in peripheral nerves and dorsal roots (blue). However, regenerating dorsal root fibers stop when they reach the astrocytic processes that delimit the surface of the spinal cord. Axons of dorsal root ganglion sensory neurons also do not regenerate through glial scars that form at lesion sites in the CNS (red).

complete explanation,[111] since unmyelinated axons (which are abundant in the spinal cord) also fail to regenerate.[112,113] In addition, regeneration still fails in transgenic animals in which growth inhibitory molecules have been deleted.

A further problem for regeneration is the glial scar that is produced by astrocytes days and weeks after a spinal cord lesion.[105,114] It is clear however that growth of damaged axons starts to occur within hours after the lesion has been made. In the CNS, this initial growth fizzles out within a few microns unless it is somehow enhanced. Hence, after spinal cord injuries, the scar might well be a *consequence* of failure of regeneration rather than being the main obstacle to growth.[115]

Schwann Cell Bridges and Regeneration

Schwann cells produce a favorable environment for the growth of axons of CNS neurons. For example, when segments of peripheral nerves are grafted between the cut ends of the spinal cord in a mouse or rat, fibers grow across and fill the gap.[116] The graft is composed of Schwann cells and connective tissue; the peripheral axons degenerate. Similarly, cultures of Schwann cells implanted into the spinal cord promote growth. This effect can be enhanced by genetically engineering the Schwann cells to produce supranormal amounts of neurotrophic factors.[117] Injection of ensheathing glial cells either into the stumps of a transected spinal cord or at the site of an electrolytic lesion in the corticospinal tract likewise enhances regeneration of axons.[118,119] Ensheathing glial cells are found only in the olfactory system, where new neurons are born and extend axons into the CNS throughout adulthood.

A dramatic effect is observed by the use of bridges of the type shown in Figure 27.18.[120] One end of a segment of sciatic nerve is implanted into the spinal cord, the other into a higher region of the nervous system (upper spinal cord, medulla, or thalamus). Bridges have even been made to extend from cortex to another part of the CNS or to muscle. After several weeks or months, the graft resembles a normal nerve trunk filled with myelinated and unmyelinated axons. These neurons fire impulses and are electrically excited or inhibited by stimuli applied above or below the sites of implantation. By cutting the bridge and dipping the cut ends into horseradish peroxidase or other markers, the cells of origin become labeled and their distribution can be mapped (Figure 27.18B). Such experiments show that axons in the bridge, which have grown over distances of several centimeters, arise from neurons whose cell bodies lie within the CNS. Usually only those neurons with somata that are not more than a few millimeters from the bridge send axons into it. Similarly, axons leaving the bridge to enter the CNS grow only a short distance before terminating.

Not all CNS neurons extend axons into permissive environments. For example, if the axons of cerebellar Purkinje cells are severed in the adult, the cells survive indefinitely but no axonal regrowth occurs,[121,122] even if pieces of embryonic cerebellum are grafted adjacent to the severed axons. Axons of other cerebellar cells readily innervate such grafts. Thus, regeneration depends both on growth-permissive or growth-promoting conditions and on the intrinsic properties of the neuron. The inability of Purkinje cells to regrow

[111] Silver, J. 2010. *Neuron* 66: 619–621.

[112] Hu, F., and Strittmatter, S. M. 2004. *Semin. Perinatol.* 28: 371–378.

[113] Griffin, J. W., and Thompson, W. J. 2008. *Glia* 56: 1518–1531.

[114] Fitch, M. T., and Silver, J. 1999. In *CNS Regeneration: Basic Science and Clinical Advances*. Academic Press, San Diego. pp. 55–88.

[115] Rolls, A., Shechter, R., and Schwartz, M. 2009. *Nat. Rev. Neurosci.* 10: 235–241.

[116] Richardson, P. M., McGuinness, U. M., and Aguayo, A. J. 1980. *Nature* 284: 264–265.

[117] Menei, P. et al. 1998. *Eur. J. Neurosci.* 10: 607–621.

[118] Li, Y., Field, P. M., and Raisman, G. 1998. *J. Neurosci.* 18: 10514–10524.

[119] Raisman, G. 2007. *C. R. Biol.* 330: 557–560.

[120] David, S., and Aguayo, A. J. 1981. *Science* 214: 931–933.

[121] Zagrebelsky, M. et al. 1998. *J. Neurosci.* 18: 7912–7929.

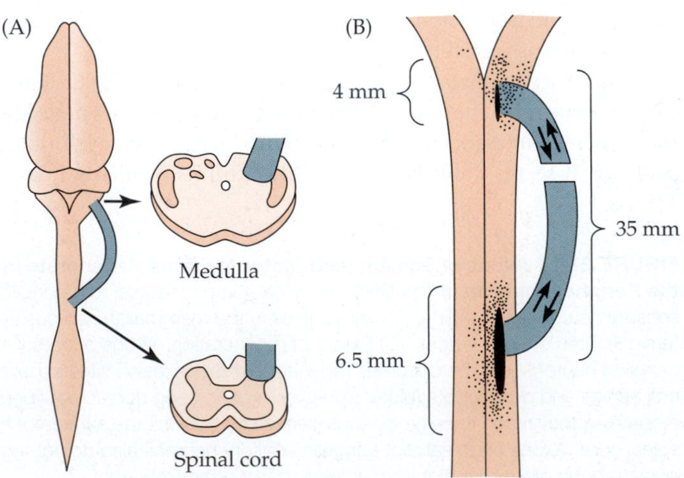

FIGURE 27.18 Bridges between Medulla and Spinal Cord enable CNS neurons to grow for prolonged distances. The grafted bridge consists of a segment of adult rat sciatic nerve in which axons have degenerated, leaving Schwann cells. The bridges act as a conduit along which central axons can grow. (A) Sites of insertion of the graft. (B) Neurons are labeled by cutting the graft and applying HRP to the cut ends. Positions of 1472 neuronal cell bodies were labeled by retrograde transport of HRP in seven grafted rats. Most of the cells sending axons into the graft are situated close to its points of insertion. (After David and Aguayo, 1981.)

(A)

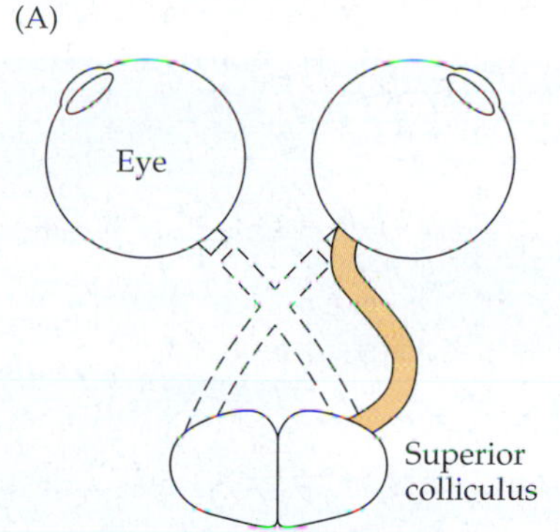

(B)

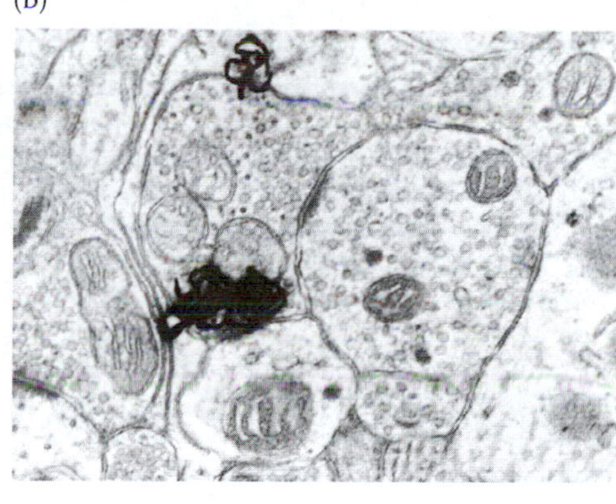

FIGURE 27.19 Reconnection of the Retina and Superior Colliculus through a peripheral nerve graft in an adult rat. (A) The optic nerves were severed, and one was replaced by a 3 to 4 cm segment of the peroneal nerve (yellow). Regeneration was tested by injecting anterograde tracers into the eye or by recording responses of superior colliculus neurons to light flashed onto the retina. (B) Electron microscope autoradiogram of a regenerated retinal ganglion cell axon terminal in the superior colliculus. [^{3}H]-labeled amino acids were injected into the eye 2 days before the brain was fixed and sectioned; silver grains exposed by radiolabeled proteins transported from the injected eye identify ganglion cell axon terminals. The regenerated terminal resembles those seen in control animals; it is filled with round synaptic vesicles and forms asymmetric synapses. (After Vidal-Sanz, Bray, and Aguayo, 1991; micrograph kindly provided by A. J. Aguayo.)

severed axons is correlated with their failure to upregulate proteins involved in axon growth in response to axotomy and not with the presence of myelin.[123]

Formation of Synapses by Axons Regenerating in the Mammalian CNS

Can axons regenerating in the CNS of mammals locate their correct targets and make functional synapses? Experiments on regenerating retinal ganglion cell axons indicate that the answer is yes.[124] If the optic nerve is cut and a peripheral nerve bridge is inserted between the eye and the superior colliculus, retinal ganglion cell axons that grow through the bridge can extend into the target, arborize, and form synapses (Figure 27.19). The regenerated synapses are formed on the correct regions of their target cells, have a normal structure when visualized by electron microscopy, and are functional in that the postsynaptic cells can be driven by illumination of the eye. Nevertheless, in spite of occasional reports to the contrary, very few axons actually succeed in growing through the graft to their targets.

Regeneration in Immature Mammalian CNS

Compared to the adult, the immature mammalian CNS provides a favorable environment for regeneration.[4,123] If the spinal cord of a neonatal opossum is crushed or cut, axons grow across the lesion and conduction through the damaged region is restored within a few days, even when the spinal cord is removed from the animal and maintained in culture (Figures 27.20 and 27.21).[125–127] Similar results have been obtained in embryonic rat or mouse spinal cord in culture. Even after complete transection of the spinal cord in a newborn opossum, prolonged survival leads to substantial and precise regeneration and excellent functional recovery. For example, sensory axons reestablish direct synaptic con-

[122] Carulli, D., Buffo, A., and Strata, P. 2004. *Prog. Neurobiol.* 72: 373–398.

[123] Bouslama-Oueghlani, L. et al. 2003. *J. Neurosci.* 23: 8318–8329.

[124] Bray, G. M. et al. 1991. *Ann. N Y Acad. Sci.* 633: 214–228.

[125] Nicholls, J., and Saunders, N. 1996. *Trends Neurosci.* 19: 229–234.

[126] Varga, Z. M. et al. 1995. *Eur. J. Neurosci.* 7: 2119–2129.

[127] Saunders, N. R. et al. 1998. *J. Neurosci.* 18: 339–355.

(A)

(B)

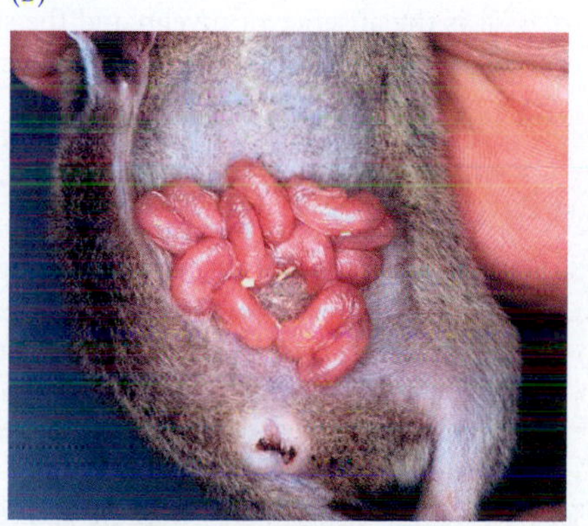

(C)

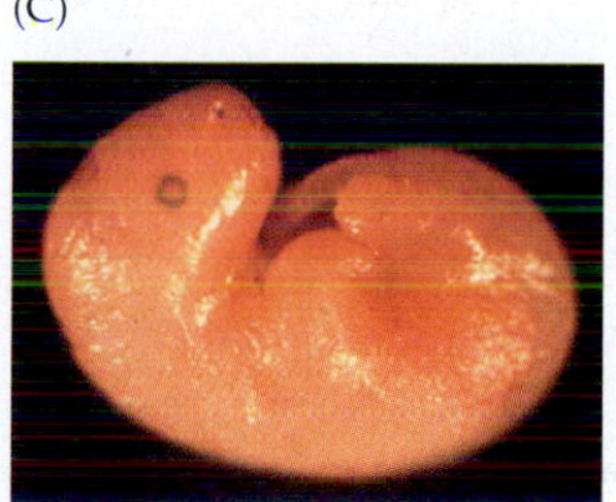

FIGURE 27.20 The South American Opossum (*Monodelphis domestica*) (A) is born in an immature state (B,C) corresponding roughly to a 15.16 day mouse embryo. The entire central nervous system (CNS) can be removed and maintained in culture for more than one week. (From Nicholls and Saunders, 1996.)

(A)

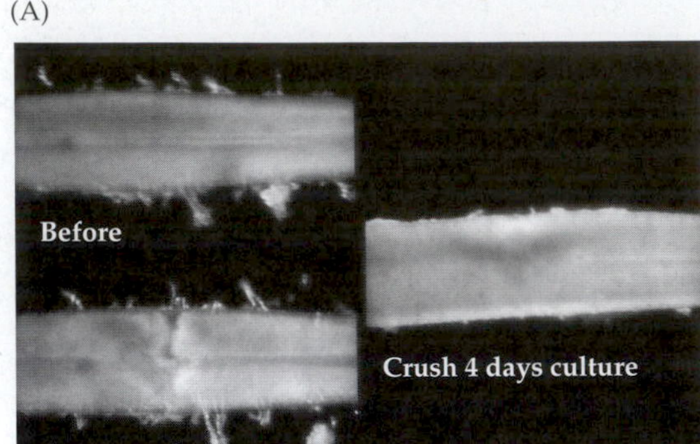

(B)

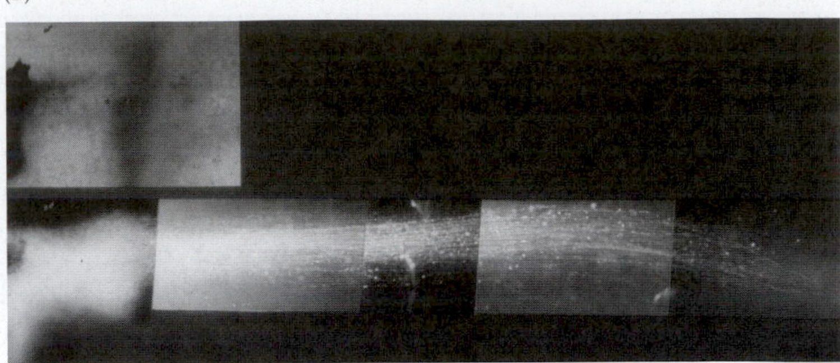

(C)

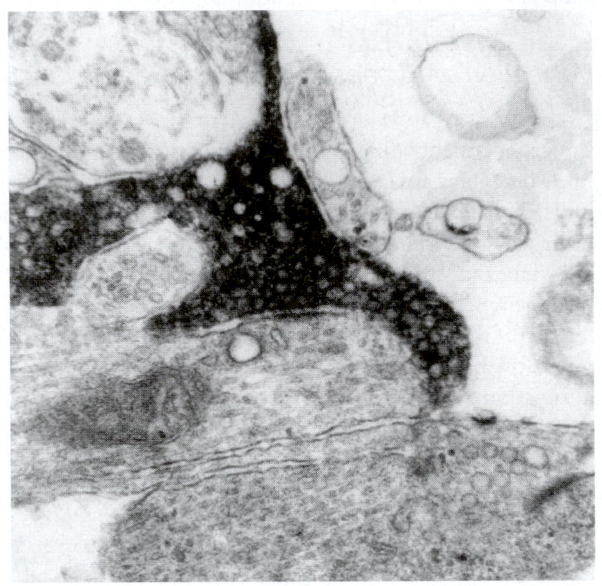

FIGURE 27.21 Regeneration of Axons after Spinal Cord Lesions in isolated CNS of 8-day-old opossum. (A) Whole mount of opossum spinal cord in culture before and after lesioning one side. After 4 days, the lesion becomes hard to detect. (B) Growth of axons labeled by the fluorescent dye DiI 5 days after injury, bright field (above) and fluorescence microscopy, below. Note the large number of fibers and their extensive and rapid growth through and beyond the lesion. (C) Horseradish peroxidase (HRP)-labeled sensory axon that formed a synapse on a motoneuron 6 days after injury. (From Nicholls and Saunders, 1996; C electron micrograph kindly provided by J. Fernandez.)

nections onto motor neurons (see Figure 27.21), and the animal can walk, swim, and climb in a coordinated manner.

There is a critical period in early life during which regeneration can occur. The spinal cord of a 9-day-old opossum regenerates after a lesion has been made, while that of a 12-day-old animal does not. A striking feature of the opossum spinal cord at 9 days of age is the absence of myelin and the small number of glial cells it contains. As in opossums, neurons in the CNS of embryonic chicks regenerate if the spinal cord is transected prior to the onset of myelination.[128]

The newborn opossum offers advantages for identifying molecules that are responsible for promoting or inhibiting spinal cord regeneration since RNA expression can be compared in spinal cords that can (9 days) and cannot (12 days) regenerate. Mladinic and her colleagues[129,130] have measured changes in RNA between 9 and 12 days by polymerase chain reaction, cDNA microarray, Northern blot, immunocytochemistry, and in situ hybridization. Growth-promoting molecules and their receptors are over expressed in cords that regenerate. By contrast, inhibitory molecules and their receptors appear at 12 days, when regeneration is no longer possible. The lists of molecules are long, which is not surprising since during the period between 9 and 12 days after birth extensive developmental changes are occurring.[131] Candidate molecules overexpressed at 9 days include cytokines, mitogen activated kinase, and laminin receptors, to name only a few, all of which can promote neurite survival or outgrowth. Those overexpressed at 12 days include inhibitory molecules or molecules inducing cell death, such as myelin basic protein, reticulon, annexins, semaphorins, and ephrins.[130] A next step will be to test these molecules separately or in combination to deter-

[128] Keirstead, H. S. et al. 1995. *J. Neurosci.* 15: 6963–6974.

[129] Mladinic, M. et al. 2005. *Cell. Mol. Neurobiol.* 25: 405–424.

[130] Mladinic, M. et al. 2010. *Brain Res.* 1363: 20–39.

[131] Lane, M. A. et al. 2007. *Eur. J. Neurosci.* 25: 1725–1742.

mine whether transfection with a growth-promoting molecule allows regeneration to occur after 12 days or, conversely, whether an inhibitor prevents it at less than 9 days. For such experiments, plasmids can be injected into cells in a selected region of the developing spinal cord by a simple technique known as electroporation. At present, it is not known which molecules or how many different molecules are involved in the initiation and prevention of neurite outgrowth across a lesion.

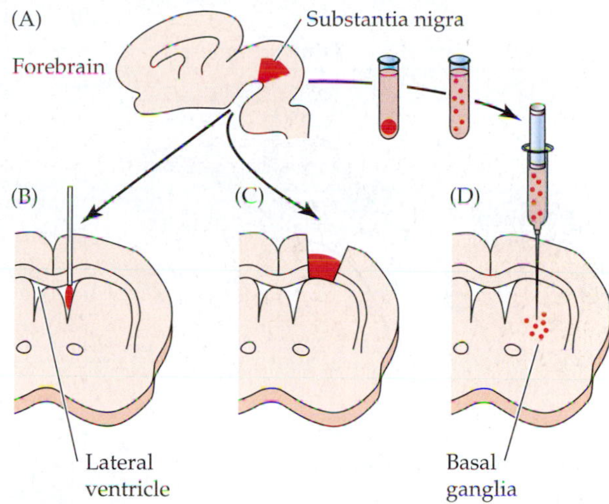

FIGURE 27.22 Procedures for Transplanting Embryonic Tissue into adult rat brain. Tissue rich in cells containing dopamine is dissected from the substantia nigra (A) and is injected into the lateral ventricle (B) or grafted into a cavity in the cortex overlying the basal ganglia (C). Alternatively, a suspension of dissociated substantia nigra cells can be injected directly into the basal ganglia (D). Such embryonic cells survive, sprout, and secrete transmitter. (After Dunnett, Björklund, and Stenevi, 1983.)

Neuronal Transplants

Among the most devastating of human diseases are those resulting from the spontaneous degeneration of CNS neurons, such as Parkinson's disease, Alzheimer's disease, and Huntington's disease. In the adult, most nerve cells are postmitotic; at present, no physiological mechanisms are known for replacing neurons that have been lost. One approach to cell replacement has been to transplant embryonic nerve cells into the adult brain.[132] Unlike neurons from the adult central nervous system, which die following transplantation, cells taken from fetal or neonatal animals can survive and grow after being inserted into the gray matter of the adult CNS (Figure 27.22). There they differentiate, extend axons, and release transmitters.

An example of such transplantation is provided by experiments in which neurons were transplanted into the basal ganglia of rats after destruction of dopamine-containing neurons in the substantia nigra—a loss that mimics in some ways the deficits caused by Parkinson's disease in humans (see also Chapters 14 and 24).[133] In normal animals, the dopaminergic neurons in the substantia nigra (a region in the midbrain) innervate cells in the basal ganglia (a region involved in programming movements; see Chapter 24 and Appendix C). If a lesion of this dopamine pathway is made on one side of a rat, a disorder of movement results in which the animal turns toward the side of the lesion in response to stress or certain drug treatments. This asymmetry of movement disappears after dopamine-containing neurons from the substantia nigra of *immature* animals are transplanted into the basal ganglia on the lesioned side.[133] Ultrastructural studies have shown that the transplanted neurons extend axons into the surrounding region and form synapses with host neurons. Many attempts are being made to repair spinal cord injury by implantation of stem cells.[134,135]

The degree of functional recovery following grafts depends on the extent to which synaptic connections are reestablished.[136,137] Grafts of embryonic entorhinal cortex have been made in adult rats with entorhinal cortex lesions. Neurons of the graft grow to reinnervate deafferented zones in the hippocampus, form synaptic connections, and partially ameliorate deficits in spatial memory.[138] As in partially denervated muscles (see above), reinnervation can occur after lesions are made in the adult hippocampus: surviving neurons sprout new collaterals to reconstitute functional circuits.[139]

A remarkable example of effective transplantation is the appropriate integration of transplanted embryonic cerebellar Purkinje cells in the adult *pcd* (*P*urkinje *c*ell *d*egeneration) mouse—a mutant whose cerebellar Purkinje cells degenerate shortly after birth (Figure 27.23).[140] Sotelo and his colleagues grafted either dissociated cells or solid pieces of the cerebellar primordium into the cerebellum of the adult mutant mouse. Donor Purkinje cells migrated out of the graft to the positions originally occupied by the degenerated Purkinje cells. They did so along the host Bergmann radial glial cells, which were induced by the graft to re-express proteins involved in guiding Purkinje cells. Within 2 weeks many transplanted cells formed dendritic arbors that resembled those of normal Purkinje cells, climbing fibers formed synapses (first on the cell body, then on the proximal dendrites), and parallel fibers innervated the distal dendrites. Characteristic synaptic potentials were recorded following stimulation of the climbing-fiber and mossy-fiber inputs. However, the implanted cells rarely succeeded in establishing synaptic connections with their normal targets in the deep cerebellar nuclei of the host, instead innervating nearby donor deep nuclear neurons that survived in the remnant of the graft. Nevertheless, such experiments

[132] Björklund, A. 2000. *Novartis Found. Symp.* 231: 7–15.

[133] Thompson, L. H., and Björklund, A. 2009. *Prog. Brain Res.* 175: 53–79.

[134] Sahni, V., and Kessler, J. A. 2010. *Nat. Rev. Neurol.* 6: 363–372.

[135] Rossi, S. L., and Keirstead, H. S. 2009. *Curr. Opin. Biotechnol.* 20: 552–562.

[136] Aimone, J. B., Deng, W., and Gage, F. H. 2010. *Trends Cogn. Sci.* 14: 325–337.

[137] Björklund, A. 1991. *Trends Neurosci.* 14: 319–322.

[138] Zhou, W., Raisman, G., and Zhou, C. 1998. *Brain Res.* 788: 202–206.

[139] Deller, T. et al. 2006. *Adv. Exp. Med. Biol.* 557: 101–121.

[140] Sotelo, C. et al. 1994. *J. Neurosci.* 14: 124–133.

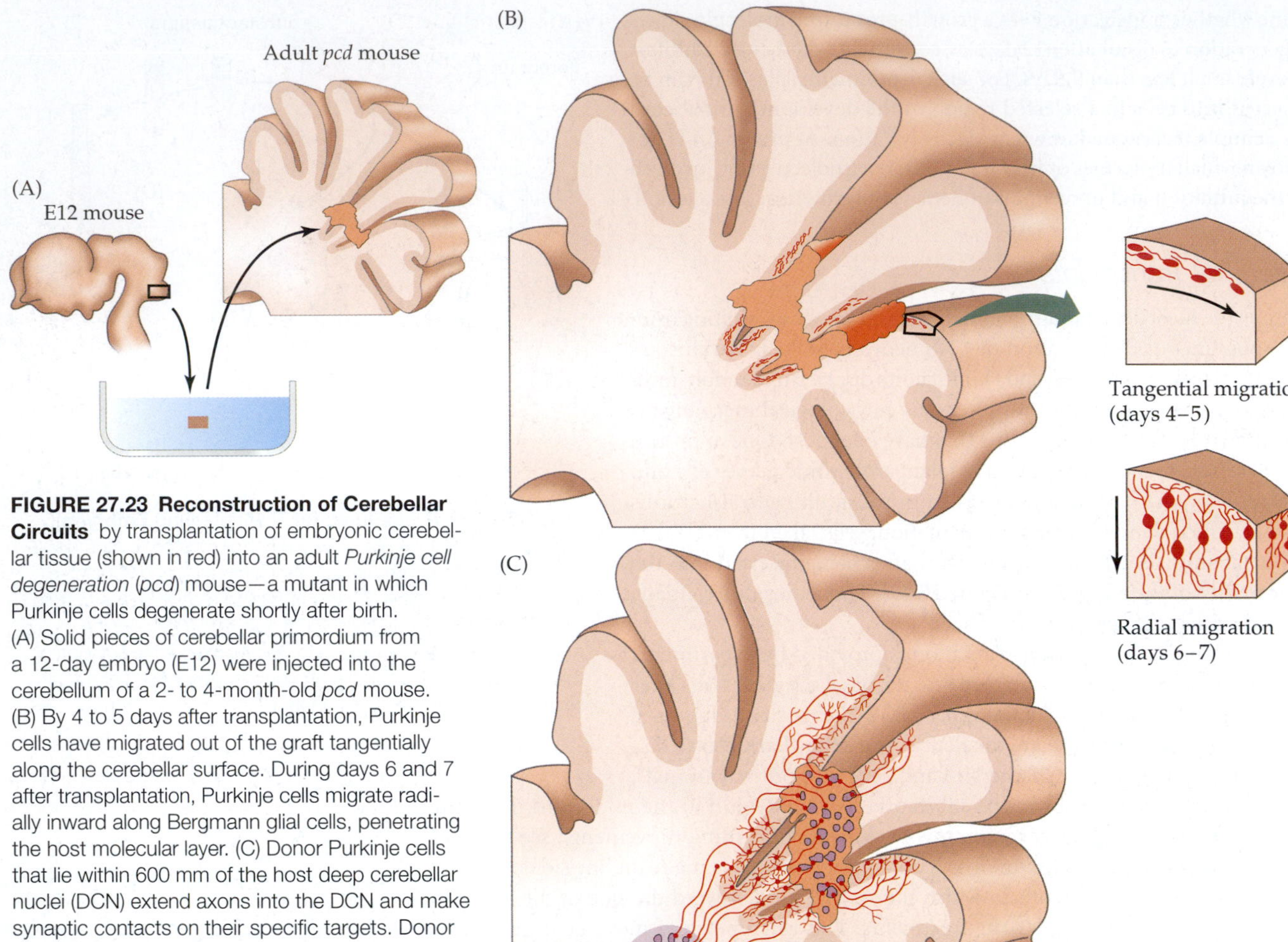

FIGURE 27.23 Reconstruction of Cerebellar Circuits by transplantation of embryonic cerebellar tissue (shown in red) into an adult *Purkinje cell degeneration* (*pcd*) mouse—a mutant in which Purkinje cells degenerate shortly after birth. (A) Solid pieces of cerebellar primordium from a 12-day embryo (E12) were injected into the cerebellum of a 2- to 4-month-old *pcd* mouse. (B) By 4 to 5 days after transplantation, Purkinje cells have migrated out of the graft tangentially along the cerebellar surface. During days 6 and 7 after transplantation, Purkinje cells migrate radially inward along Bergmann glial cells, penetrating the host molecular layer. (C) Donor Purkinje cells that lie within 600 mm of the host deep cerebellar nuclei (DCN) extend axons into the DCN and make synaptic contacts on their specific targets. Donor Purkinje cells farther from the host DCN make contact with donor DCN cells in the graft remnant. (After Sotelo and Alvarado-Mallart, 1991.)

demonstrate that transplanted cells can become incorporated into the synaptic circuitry of an adult host to a remarkable extent.

Prospects for Developing Treatment of Spinal Cord Injury in Humans

It is still not known what causes the failure of regeneration that is so evident after most CNS lesions. Identification of ways to suppress endogenous growth-inhibiting factors is an active area of research, as is the development of neural stem cell lines that offer the potential of providing a readily available source of glial cells and neurons whose properties can be manipulated by genetic engineering (see Chapter 25). Such advances, combined with improved transplantation techniques, may provide hope for the amelioration of functional deficits resulting from CNS lesions and neurodegenerative diseases.

Two important provisos must be borne in mind—important because they are of vital interest to patients who have suffered spinal cord injuries. We mention them here because many of the papers referred to in this chapter end on an optimistic note and suggest that new and effective treatments are just around the corner. First, if and when an effective therapy is developed, it will almost certainly be applicable only for repair of acute spinal cord injuries. It is beyond reasonable hope to imagine that walking and sensation could be restored in a person whose spinal cord had been transected say 2 years (or perhaps 2 months) earlier. By that time, the distal part of the spinal cord will have undergone major degenerative changes. Moreover, as with every therapy, there may well be side effects, perhaps serious ones. Second, while it is natural that optimism should drive research, no one can predict *how long* it will take until a reliable therapy becomes available. No one (not even the enthusiastic scientists who make the predictions) can say for sure whether it will take weeks, months, years, decades or longer. To repair a spinal cord is literally millions of times more difficult than repairing a computer. While time is not of the essence for a research worker, it is for the patient. False hopes aroused in paraplegics can (and often do) lead to failure to adapt themselves to their new circumstance as well as they could otherwise. Neurologists testify that, if a cure seems to be imminent, there is a temptation for spinal cord injury patients to avoid the intensive, continuous physical and mental rehabilitation that is required if the fullest possible life is to be led for all the years ahead.

SUMMARY

- When an axon is severed in the vertebrate nervous system, the distal portion degenerates. The axotomized cell body may undergo chromatolysis or die.
- Many of the presynaptic terminals innervating an axotomized neuron retract.
- In denervated skeletal muscle fibers, new ACh receptors are synthesized and inserted in extrasynaptic regions, making the muscle supersensitive to ACh. Denervated neurons also become supersensitive to the transmitters released by damaged presynaptic axons.
- Muscle activity is an important factor determining receptor number and distribution. Muscle activity also influences the rate at which ACh receptors are degraded and replaced.
- In adult mammals or frogs, an innervated muscle will not accept innervation by an additional nerve. In contrast, nerve fibers will form new synapses on denervated or injured muscle fibers.
- Partially denervated muscles and neurons cause nearby undamaged nerves to sprout new branches and form new synapses.
- In the peripheral nervous system, Schwann cells provide an environment conducive to axonal regrowth.
- The synaptic portion of the basal lamina sheath that surrounds muscle fibers has associated with a molecule known as agrin. Agrin is a proteoglycan synthesized by motor neurons and released from their axon terminals. It becomes associated with the synaptic basal lamina and induces the formation of postsynaptic specializations.
- The adult mammalian CNS has limited capacity for regeneration.
- Schwann cells, in the form of a peripheral nerve graft or injected as a cell suspension at the site of a lesion, produce a favorable environment for regrowth of axons of mammalian CNS neurons.
- Regeneration occurs in the CNS of immature mammals.
- Neurons from fetal or neonatal animals as well as neurons and glial cells derived from neural stem cell lines can survive and grow when transplanted into the adult mammalian CNS.
- Transplanted cells can become incorporated into the existing synaptic circuitry and partially restore normal function.

Suggested Reading

General Reviews

Aimone, J. B., Deng, W., and Gage, F. H. 2010. Adult neurogenesis: integrating theories and separating functions. *Trends Cogn. Sci.* 14: 325–337.

Bunge, M. B. 2008. Novel combination strategies to repair the injured mammalian spinal cord. *J. Spinal Cord Med.* 31: 262–269.

Deller, T, Haas, C. A., Freiman, T. M., Phinney, A., Jucker, M., and Frotscher, M. 2006. Lesion-induced axonal sprouting in the central nervous system. *Adv. Exp. Med. Biol.* 557: 101–121.

Dusart, I., Ghoumari, A., Wehrle, R., Morel, M. P., Bouslama-Oueghlani, L., Camand, E., and Sotelo, C. 2005. Cell death and axon regeneration of Purkinje cells after axotomy: challenges of classical hypotheses of axon regeneration. *Brain Res. Brain Res. Rev.* 49: 300–316.

Huebner, E. A., and Strittmatter, S. M. 2009. Axon regeneration in the peripheral and central nervous systems. *Results Probl. Cell Differ.* 48: 339–351.

Madduri, S., and Gander, B. 2010. Schwann cell delivery of neurotrophic factors for peripheral nerve regeneration. *J. Peripher. Nerv. Syst.* 15: 93–103.

Mladinic, M., Muller, K. J., and Nicholls. J. G. 2009. Central nervous system regeneration: from leech to opossum. *J. Physiol.* 587: 2775–2782.

McMahan, U. J. 1990. The agrin hypothesis. *Cold Spring Harb. Symp. Quant. Biol.* 50: 407–418.

Pette, D. 2001. Historical Perspectives: plasticity of mammalian skeletal muscle. *J. Appl. Physiol.* 90: 1119–1124.

Raisman, G. 2007. Repair of spinal cord injury by transplantation of olfactory ensheathing cells. *C. R. Biol.* 330: 557–560.

Silver, J. 2010. Much Ado about Nogo. *Neuron* 66: 619–621.

Original Papers

Björklund, A., Dunnett, S. B., Stenevi, U., Lewis, N. E., and Iversen, S. D. 1980. Reinnervation of the denervated striatum by substantia nigra transplants: Functional consequences as revealed by pharmacological and sensorimotor testing. *Brain Res.* 199: 307–333.

Burden, S. J., Sargent, P. B., and McMahan, U. J. 1979. Acetylcholine receptors in regenerating muscle accumulate at original synaptic sites in the absence of the nerve. *J. Cell Biol.* 82: 412–425.

David, S., and Aguayo, A. J. 1981. Axonal elongation into peripheral nervous system "bridges" after central nervous system injury in adult rats. *Science* 214: 931–933.

Kim, N., Stiegler, A. L., Cameron, T. O., Hallock, P. T., Gomez, A. M., Huang, J. H., Hubbard, S. R., Dustin, M. L., and Burden, S. J. 2008. Lrp4 is a receptor for agrin and forms a complex with MuSK. *Cell* 135: 334–342.

Fournier, A. E., GrandPre, T., and Strittmatter, S. M. 2001. Identification of a receptor mediating Nogo-66 inhibition of axonal regeneration. *Nature* 409: 341–346.

Lin, S., Landmann, L., Ruegg, M. A., and Brenner, H. R. 2008. The role of nerve- versus muscle-derived factors in mammalian neuromuscular junction formation. *J. Neurosci.* 28: 3333–3340.

Lømo, T., and Rosenthal, J. 1972. Control of ACh sensitivity by muscle activity in the rat. *J. Physiol.* 221: 493–513.

Love, F. M., Son, Y. J., and Thompson, W. J. 2003. Activity alters muscle reinnervation and terminal sprouting by reducing the number of Schwann cell pathways that grow to link synaptic sites. *J. Neurobiol.* 54: 566–576.

Rotshenker, S. 2009. The role of Galectin-3/MAC-2 in the activation of the innate-immune function of phagocytosis in microglia in injury and disease. *J. Mol. Neurosci.* 39: 99–103.

Samuels, S. E., Lipitz, J. B., Dahl, G., and Muller, K. J. 2010. Neuroglial ATP release through innexin channels controls microglial cell movement to a nerve injury. *J. Gen. Physiol.* 136: 425–452.

Saunders, N. R., Kitchener, P., Knott, G. W., Nicholls, J. G., Potter, A., and Smith, T. J. 1998. Development of walking, swimming and neuronal connections after complete spinal cord transection in the neonatal opossum, *Monodelphis domestica*. *J. Neurosci.* 18: 339–355.

Schwab, M. E., and Caroni, P. 1988. Oligodendrocytes and CNS myelin are nonpermissive substrates for neurite growth and fibroblast spreading in vitro. *J. Neurosci.* 8: 2381–2393.

PART VII

Conclusion

CHAPTER 28
Open Questions

With each new edition of *From Neuron to Brain*, our understanding of how nerve cells produce electrical signals, how they communicate with one another, how they act in concert, and how they become connected during development has become deeper. In the last years, major advances have been obtained through novel molecular biological, genetic, and imaging techniques. There seems, however, to be no way to guess what new techniques will become available in the future and what new questions will arise as a consequence. Thus, at the time of the first edition of this book in 1976, few could conceive of the use of site-directed mutagenesis to study gating currents, transgenic labeling of defined groups of neurons with fluorescent proteins, or the optical demonstration of functional columns in the living brain. Many problems that now seem unapproachable will require techniques and advances not yet imagined.

What can one predict today about novel concepts that might be incorporated into the next edition of this book? One reasonable guess is that more intensive collaboration among basic scientists working at the cellular and molecular levels, cognitive neuroscientists, and clinical neurologists will be important for understanding integrative and higher brain functions relating to perception, movement, sleep, and memory. One can hope also that an increase in fundamental knowledge of the nervous system will lead to the prevention and alleviation of diseases of the nervous system that arise from unknown causes and that cannot yet be treated effectively.

Open questions about the nervous system and the brain are very different from those in subjects such as physics, chemistry, or even biology in general. It is not only the reader of a book like this who can point to the important deficiencies in our knowledge and understanding. A layperson outside science is aware that we do not know the mechanisms of higher functions such as consciousness, learning, sleep, the production of coordinated movements, or even how one initiates the bending of a finger as an act of will. The same person, even if highly sophisticated and well educated, would probably have more difficulty in pointing out what things one still needs to know about relativity, particle physics, chemical reactions, or genetics. It is this wealth of obvious, unsolved, and important human questions that makes neuroscience so appealing today.

To illustrate one everyday example of our present ignorance of how the brain performs its functions, consider a sport such as tennis. An expert player—say, Roger Federer—sees his opponent hit the ball. He can compute rapidly where it will land and how high it will bounce. The ball may be traveling at 100 kilometers/hour, but he can rush to the right spot, arm extended, and hit the ball in the center of the racket, with exactly the right force to send it exactly onto the line in the other court (exploiting the remembered weakness of his opponent's backhand), all this in fractions of seconds. We could just as well have picked as examples the way in which a pelican dives for a fish, a frog catches a fly with its tongue, or a bee drinks from a particular flower. In each of these examples, objects must be recognized against a rich background and highly coordinated movements must be planned, initiated, regulated, and brought to fulfillment. And somehow, the necessary neuronal connections must have been formed. Moreover housekeeping functions, such as the control of respiration, heart rate, and the gut must match the needs of the body at all times.

In the following paragraphs we consider selected unsolved problems in neuroscience that might become approachable in the future, particularly in relation to the topics emphasized in this book.

Cellular and Molecular Studies of Neuronal Functions

So much new information has become available with such rapidity in the past few years about channels, receptors, transmitters, transporters, second messengers, and long-term changes at synapses that open questions posed today might already have solutions by the time you read this book.

Still not worked out completely are the intimate structural changes that mediate the opening, closing, and inactivation of channels. Another major problem, now in its early stages, concerns the way in which molecules are transported to precise regions of neurons, sodium channels to nodes, receptors to dendritic spines, and synaptic vesicles to active zones in presynaptic terminals. The formation of postsynaptic specializations at the neuromuscular junction constitutes an example in which the signal triggering the localization of key molecules has been identified (see the discussion of agrin in Chapter 27). Yet the cellular and molecular mechanisms by which neurons within the CNS form synapses are still not known.

In the pursuit of problems of high interest, such as long-term potentiation and long-term depression in relation to learning and memory, an enormous number of detailed experiments have been made to unravel the underlying mechanisms. What other mechanisms for storage and retrieval remain to be discovered?[1] As for retrieval, there are at present no serious hypotheses about the way in which you can remember the number of your hotel room when you need it or your mother's birthday. It goes without saying that we still have no idea of what neural mechanisms produce consciousness in a dog that begs for food or a person reading this book.

Functional Importance of Intercellular Transfer of Materials

Numerous experiments have demonstrated the transsynaptic transfer of amino acids or proteins from neuron to neuron—for example, from the retina through the lateral geniculate nucleus to the visual cortex. That such transfer occurs is certain, yet we do not have crucial information about the mechanisms for transfer or the functional significance. Intercellular transfer of small molecules also occurs between cells linked by gap junctions (see Chapter 8).

[1] Shallice,T., and Cooper, R.P. 2011. *The Organization of Mind*. Oxford University Press, Oxford England.

It has been suggested that intercellular transfer represents a mechanism for controlling growth and development. A related question concerns the role of glial cells in relation to neuronal signaling (see Chapter 10), particularly in relation to quantitative measures of exchange between glial cells and neurons and the importance of such exchange for function.

Development and Regeneration

In spite of remarkable progress, it is still not known how neurons select their precise targets. One can now approach, in molecular terms, problems such as directed neurite outgrowth toward targets, termination of growth, and the refinement of connections by selective pruning and cell death. At the same time, we can only wonder how the extraordinary precision of connections is achieved, for example, by terminals of muscle spindle afferent fibers on motoneurons in the spinal cord. With several thousand neurons in each cubic millimeter of tissue, how are the appropriate motor cells selected and innervated at the appropriate sites? And through what mechanisms does the same sensory cell form synapses with quite different release characteristics on specific neurons in the medulla? As for the failure of regeneration after injury to the mammalian central nervous system, the reasons are still not known, in spite of considerable advances in our understanding of molecular mechanisms that promote and inhibit neurite outgrowth.

Another major question concerns the development that proceeds in early life under the influence of experience. In particular, we know next to nothing at the cellular level about the effects of critical periods on the maturation of higher functions, including emotional states and personality.

Genetic Approaches to Understanding the Nervous System

It is hard to predict the consequences of the revolution in genetic techniques for understanding brain function. The present use of transgenic animals in which identified genes have been altered or deleted provides a powerful tool, but one still hampered by difficulties of interpretation owing to redundancy of function and unexpected side effects. With the completion of the Human Genome Project, candidate genes and molecules that are altered in disease and development are becoming known. To analyze this large array of information and separate the important from the incidental constitutes an immense task. The scope of the problem is illustrated by the study of inherited diseases, such as Huntington's disease, in which the altered gene can be identified by linkage analysis of the affected families.[2] Yet, although the altered sequences of the Huntington's disease gene were identified long ago, the function of the protein remains unknown. Similarly, mutations in genes coding for voltage-gated calcium channels are associated with familial hemiplegic migraine and cerebellar ataxia.[3] But again, there is no clear link in terms of mechanisms. Even for a molecule as important as the prion protein, which is abundant in the normal brain and which, when transformed, gives rise to transmissible spongiform encephalopathies (of which bovine spongiform encephalopathy, or mad cow disease, is the best known), there is no known normal function, nor is there complete information about the mechanism by which cortical tissue becomes infected through eating infected brains.[4] A long-term hope is that genetic therapies will be developed for those conditions as well as for certain conditions that cause degeneration of the retina. At present, genetic therapies are being clinically tested in patients suffering from macular degeneration and retinitis pigmentosa.[5] Nevertheless, it is worth commenting that the slow progress in devising treatments for Huntington's disease and other long-established monogenic diseases (such as cystic fibrosis,[6] a defect in an epithelial anion transporter, the genetic cause for which was discovered nearly 25 years ago[7]) suggest that to develop effective gene therapy may take a long time.

On the positive side, the study of human (and animal) genetic mutations, coupled with information yielded by the Human Genome Project, has provided important advances in basic neuroscience. An example is the discovery of the entire orexin/hypocretin system for controlling sleep and appetite described in Chapter 14. On a smaller scale, it was only through genetic analysis of inherited human epilepsies that the molecular structure of the M-channel became known.[8] It is clear that application of known genetic information will

[2] Ross, C. A., and Tabrizi, S. J. 2011. *Lancet Neurol.* 10: 83–98.

[3] Pietrobon, D. 2010. *Pflügers Arch.* 460: 375–393.

[4] Weissmann, C. 2009. *Folia Neuropathol.* 47: 104–113.

[5] Vugler, A. A. 2010. *Retina* 30: 983–1001.

[6] O'Sullivan, B. P., and Freedman, S. D. 2009. *Lancet* 373: 1891–1904.

[7] Kerem, B. et al. 1989. *Science* 245: 1073–1080.

[8] Jentsch, T. J. 2000. *Nat. Rev. Neurosci.* 1: 21–30.

continue to provide new information about the function of individual proteins, through the use of small-interfering and short-hairpin RNAs (siRNA, shRNA). Genetic knowledge can also yield big advances in technology, for example, the possibility of color-coding individual neurons,[9] stimulating or silencing them, and recording their activity in the brain in situ.[10] These techniques are already becoming valuable for sorting out the brain's functional wiring.[11]

Sensory and Motor Integration

A serious deficiency in our knowledge concerns the enormous numbers of neurons with no obvious function, particularly unmyelinated fibers that greatly outnumber myelinated fibers. One example from this book is how the various amacrine cell types (more than 20) contribute to processing in the retina. Another is the role of group II afferents from muscle spindles in spinal cord function.

Mechanisms for the initiation and control of coordinated movements represent problems that have seen progress but still remain open. Thanks to noninvasive techniques for imaging and stimulation, one can now obtain detailed images of brain activity. Yet more than 50 years ago, in remarkably prescient comments, Adrian pointed out (see Chapter 24) that once you have learned to write your name, you can do it at once by holding the pencil between your toes.[12] For our ability to transfer such programs from one effector system to another, we have no explanation.

Similarly, for sensory systems, the neural mechanisms for integrating the entire picture of, say, a bulldog or an artichoke, let alone a whole room, remain beyond our reach at present. In discussions of this type at this stage, the dreaded homunculus makes his appearance—the cell or little man in the brain who actually sees what we see. To ridicule this concept is both fashionable and a sure sign of sophistication. Nevertheless, the homunculus does have a useful function: He represents and continually reminds us of our ignorance about higher cortical function. As soon as answers are found, he will die a natural death, like phlogiston. We have no way as yet to replace him by a computer.

In addition to these obvious gaps in our knowledge, the mechanisms for the precise control of body temperature, blood pressure, and intestinal functions remain black boxes. Interactions of the brain with the immune system represent another major field of active research that is still at an early stage, with many open questions.

Mathematical modeling and computational neuroscience represent fields that depend critically on measurements made in channels, membranes, individual neurons, synapses, and networks. However, successes comparable to those of the Hodgkin–Huxley equations, which described comprehensively the permeability changes responsible for the action potential, have not been realized in other areas. One principal reason is the incompleteness of the facts available for modeling complex processes, such as synaptic plasticity and integration. For example, how fully could one have hoped to model cortical circuits before the discovery of *N*-methyl-D-aspartate (NMDA) receptors or conduction block, and how many more such mechanisms await discovery? (Important to bear in mind is that the Hodgkin-Huxley *model*, as it is often called, was in fact not a theoretical model: the equations were derived from accurate measurements and curve fitting).

Rhythmicity

Neuronal rhythms considered in this book include respiration and circadian rhythms as well as the periodicity of firing by neurons in the cerebellum, hippocampus, thalamus, and spinal cord. Except in a few examples, such as the stomatogastric ganglion of the lobster and the swimming of the leech, we have no detailed information about the mechanisms that underlie the genesis or the regularity of firing patterns. Moreover, it is not at all clear what functions are played by current oscillations in well-known phenomena such as theta waves or the alpha and delta waves of the electroencephalogram.[13] While information about circadian rhythms is becoming available at the molecular level (thanks largely to work on *Drosophila*), the precise role of sleep and the way it arises are still obscure, as are the mechanisms by which anesthetics produce their effects.

[9] Livet, J. et al. 2007. *Nature* 450: 56–62.

[10] Knöpfel, T. et al. 2010. *J. Neurosci* 30: 14998–15004.

[11] Gradinaru, V. et al. 2009. *Science* 324: 354–359.

[12] Adrian, E. D. 1946. *The Physical Background of Perception*. Clarendon, Oxford, England.

[13] Lisman, J. E. et al. 2010. *Biol. Psychiatry* 68: 17–24.

Input from Clinical Neurology to Studies of the Brain

For many years neurology was not only inseparable from neurobiology but provided the only method for studying higher functions in relation to brain structure. A triumph of the early neurologists was their application of nature's own experiments to describe functions of various brain areas from careful correlation of symptoms with lesions. Their achievements are all the more remarkable because the use of lesions to assess function is fraught with pitfalls. With the newer techniques now available, such as magnetic resonance imaging (fMRI) and positron emission tomography (PET), the neurologist is now able to locate and observe lesions directly, to follow their progress in the living brain, and to make inferences about areas of brain related to higher cortical functions.[1] Improvements in spatial and temporal resolution seem around the corner. This could allow one to follow in real time at the micro level the sequence of neuronal events that give rise to a decision, to perception, or to the laying down of a memory.

The dramatic story of Phineas Gage emphasizes the advantages and pitfalls of lesions and deficits as a means of analyzing brain function.[14] In 1848, at the age of 25, Phineas Gage suffered a massive lesion to the brain while working as a construction foreman on a railway in Vermont. As he pushed on a tamping iron to place a charge of gunpowder into a rock, the gunpowder exploded and blew the iron rod clear through the front of his skull. Gage lost consciousness only briefly and could soon sit up and speak. What astounded the doctor was that he recovered rapidly and was able to lead a relatively normal life for more than 12 years. Gage's personality, however, underwent a major change. From being a well-liked, quiet, sober, industrious, and careful worker, he changed into a loud-mouthed, boastful, impatient, and restless braggart. At a time when nothing was known of sensory, motor, visual, or auditory cortex, the neurological investigation showed that the prefrontal area was associated with the very highest functions of human conduct and personality.

Other examples of neurological observations made in the nineteenth century that defined specific brain areas involved in higher functions were those of Broca and Wernicke, who correlated defects in speech with the areas of cortex that had been damaged by vascular accidents or tumors. Even when precise areas are not known, clinicians and neuropsychologists can reliably define and separate processes such as long- and short-term memory.

Highly counterintuitive and difficult to comprehend at first sight are effects of lesions to the parietal lobe on one side, usually the right. Patients with such lesions may no longer recognize that there are two sides to the body and to the outside world. The left side of the body ceases to exist, so a patient will not recognize his left hand as his own. When such patients with a lesion of the right parietal lobe are asked to draw a daisy, all the petals are on the right, as are all the spokes of a bicycle wheel.[15] A drawing of a cat made by a right-handed, 61-year-old patient with a parietal lesion is shown in Figure 28.1. It is important to emphasize that these are true neurological defects, not hysterical reactions of the patient.

[14] Harlow, J. M. 1868. *Publ. Mass. Med. Soc.* 2: 328–334.

[15] Driver, J., and Halligan, P. W. 1991. *Cogn. Neuropsychol.* 8: 475–496.

Figure 28.1 Drawing of a Cat. The drawing (right) was made by a patient who had a large lesion of the right parietal lobe. Only the right side of the drawing was copied. All details on the left were overlooked. Such deficits are commonly seen following lesions of this type. (From Driver and Halligan, 1991; drawing kindly provided by J. Driver.)

Such clinical observations show that our inner world, which seems so complete, so unitary, and so perfect, is composed of elementary components welded together to form a continuum.

As information becomes available from brain scans and from sophisticated tests of language and performance, one can expect the exploration of higher functions to depend ever more on input from cognitive neuroscience and neurology. Moreover, even as knowledge of human brain advances, it is clear from the work of Kravitz and his colleagues that genetic studies of lower animals, such as lobsters and flies, can shed light on higher functions such as the role of specific molecules in aggressive behavior.[16]

Input from Basic Neuroscience to Neurology

There clearly exists a two-way street between basic and applied neuroscience. Molecular biological and genetic techniques are already beginning to play a role in diagnoses of such conditions as retinoblastoma[5] and Huntington's disease, and the possibility of treatment with genetically engineered cells is now being intensively investigated in muscular dystrophy, Parkinson's disease, and spinal cord lesions. Sophisticated electrophysiological techniques are used by neurosurgeons for recording from individual neurons, for implanting electrodes (as for control of the bladder), for noninvasive stimulation, and for devising prostheses to replace lost functions. And yet, the development of apparently rather simple techniques, such as the implantation of electrodes over long periods to selectively stimulate neurons for the relief of pain or for evoking movement, are fraught with difficulties. Surely in the future it will become possible to produce effective prostheses for the visually impaired in a manner resembling cochlear implants (see Chapter 27). Furthermore, it may not be too far-fetched to hope that, eventually, interception of signals from the central nervous system or the brain could be used to initiate coordinated movements in paralyzed muscles.

One example from experiments described in this book can illustrate how research in basic neuroscience can help to provide new treatments for serious conditions. As a result of the work of Hubel and Wiesel on sensory deprivation in newborn kittens and monkeys, it became evident that a newborn baby that suffers from a cataract should have it removed as soon as possible. Such procedures have prevented countless cases of blindness. This was not an outcome that the investigators had in mind at the time they were doing their initial experiments on receptive fields in the visual cortex.

For most diseases of the nervous system that afflict humanity (e.g., Alzheimer's disease or amyotrophic lateral sclerosis), we have little or no knowledge of the root cause and no effective treatment. It might be argued that it would be better to invest the money used for basic neuroscience in applied science or neurology. Surely it would be better to find cures for the diseases directly rather than trying to find out how the nervous system works? In instances in which applied research has been emphasized over basic biological research the results have been disappointing, to say the least. For example, the Soviet Union established and supported massive institutes for applied research in physiology and pharmacology, each with hundreds of research workers. (The type of research in which the investigator followed a scientific problem for its interest and its beauty was considered to be "bourgeois," and was not permitted.) Yet, during the existence of the Soviet Union, not one new drug was developed there that came into routine clinical use.

In fact, the best reason for studying neuroscience is really to find out how the nervous system works, in both man and in animals. If through this, we get a better understanding of what goes wrong in disease states and how to cure them, that is a wonderful bonus. Of course, a desire to treat horrible diseases is a noble motive for a scientist, but it is rarely successful in the absence of basic knowledge. To cite just one example, the revolutionary experiments of Katz and his colleagues on neuromuscular transmission in the frog were not initially stimulated by a desire to cure myasthenia gravis. Yet, without their work we would not have the faintest understanding or modern treatment of the disease. Helmholtz, often quoted in this book, stated in 1862, "Whoever in the pursuit of science, seeks after immediate practical utility may rest assured that he seeks in vain."[17]

[16] Fernández, M. de L. et al. 2010. *PLoS Biol.* 8: e1000541.

[17] Brasch, F. E. 1922. *Science* 55: 405–408.

The Rate of Progress

Although books on the brain and consciousness continue to appear at an alarming rate, it is surely a disservice to the field to pretend that answers to exceedingly difficult questions are just around the corner. For example, if one considers the development of the circuitry necessary for playing tennis, it seems somewhat rash to have prophesied in 1996 (in an editorial in *Science*!) that "the main principles of neural development will have been discovered by the end of this century."[18] There is a natural tendency for scientists and journalists to be optimistic and to offer hope that solutions for difficult problems are imminent. Thus, the time required for the cure for spinal cord injuries has been stated on occasion to be 7 or 10 years (more than 15 years ago!). Although such pronouncements might well encourage neuroscientists interested in the field, they can have a devastating effect on patients if they are not fulfilled in the time specified, as unfortunately is often the case.[19]

[18] Raff, M. 1996. *Science* 274: 1063.

[19] Krauthammer, C. 2000. *Time* February 14:76.

Conclusions

When one is faced with the fantastic range of animal behavior, from navigation by an ant to the reading of a textbook by a student, it is clear that understanding how the nervous system works is a fascinating, open-ended task of first-order interest in its own right.

An obvious inference from history is that approaches to the treatment of diseases often arise unexpectedly from experiments devoted to quite different questions. Further, an increase of natural knowledge on its own is a worthy objective, for without it, logical approaches to prevention and treatment of neurological problems can be only partially realized. In this context, it is almost impossible to define the "relevance" of any particular project at the time it is undertaken. Indeed, when asked about the "significance" of a research plan, complete honesty usually requires a very simple answer: "Don't know!"

Quite apart from the treatment of disease, the dividends that can accrue for society as a result of understanding the development and the functions of the nervous system are beyond today's imagination.

Suggested Reading

Adams, R. D., Victor, M., and Ropper, A. H. 2009. *Principles of Neurology*, 9th ed. McGraw-Hill, New York.

Fernández, M. de L., Chan, Y. B., Yew, J. Y., Billeter, J. C., Dreisewerd, K., Levine, J. D., and Kravitz, E. A. 2010. Pheromonal and behavioral cues trigger male-to-female aggression in *Drosophila*. *PLoS Biol.* 8: e1000541.

Hawkins, J., and Blakeslee, S. 2004. *On Intelligence*. Times Books, New York.

Pietrobon, D. 2010. CaV2.1 channelopathies. *Pflügers Arch.* 460: 375–393.

Ross, C. A., and Tabrizi, S. J. 2011. Huntington's disease: from molecular pathogenesis to clinical treatment. *Lancet Neurol.* 10: 83–98.

Shallice, T., and Cooper, R. P. 2011. *The Organization of Mind*. Oxford University Press, Oxford England.

Weissmann, C. 2009. Thoughts on mammalian prion strains. *Folia Neuropathol.* 47: 104–113.

APPENDIX A
Current Flow in Electrical Circuits

A few basic concepts are required to understand the electrical circuits used in this presentation. For our purposes it is sufficient to describe the properties of circuit elements and explain how they work when connected together in ways that correspond to the circuits described for nerves. The difficulties sometimes encountered on first reading accounts of electrical circuits often stem from the apparently abstract nature of the forces and movements involved. It is reassuring, therefore, to realize that many of the original pioneers in the field must have been faced with similar problems, since the terms devised in the last century are mainly related to the movement of fluids. Thus, the words *current*, *flow*, *potential*, *resistance*, and *capacitance* apply equally well to both electricity and hydraulics. The analogy between the two systems is illustrated by the fact that complex problems in hydraulics may be solved by using solutions to equivalent electrical circuits.

The analogy between a simple electrical circuit and its hydraulic equivalent is illustrated in Figure 1. The first point to be made is that a source of energy is required to keep the current flowing. In the hydraulic circuit, it is a pump; in the electrical circuit, a battery. The second point is that neither water nor electrical charge is created or lost within such a system. Thus, the flow rate of water is the same at points a, b, and c in the hydraulic circuit, since no water is added or removed between them. Similarly the electrical current in the equivalent circuit is the same at the three corresponding points. In both circuits, there are a number of *resistances* to current flow. In the hydraulic circuit, such resistance is offered by narrow tubes; similarly, thinner wires offer greater resistance to electrical current flow.

Terms and Units Describing Electrical Currents

The unit used to express rate of flow is to some extent a matter of choice; one can measure flow of water through a pipe in cubic feet per minute, for example, although in some other situation milliliters per hour might be more suitable. Electrical current flow is conventionally measured in **coulombs/sec** or **amperes** (abbreviated A). One coulomb is equal to the charge carried by 6.24×10^{18} electrons. In electrical circuits and equations, current is usually designated by I or i. As with flow of water, flow of current is a vector quantity, which is just a way of saying that it has a specified direction. The direction of flow is often indicated by arrows, as in Figure 1, current always being assumed to flow from the positive to the negative pole of a battery.

What do *positive* and *negative* mean with regard to current flow? Here the hydraulic analogy does not help. It is useful instead to consider the effects of passing current through a chemical solution. For example, suppose two copper wires are dipped into a solution of copper sulfate and connected to the positive and negative poles of a battery. Copper ions in solution are repelled from the positive wire, move through the solution, and are deposited from the solution onto the negative wire. In short, positive ions move in the direction conventionally designated for current: from positive to negative in the circuit. At the same time, sulfate ions move in the opposite direction and are deposited onto the positive wire. The direction specified for current, then, is the direction in which positive charges move in the circuit; negative charges move in the reverse direction.

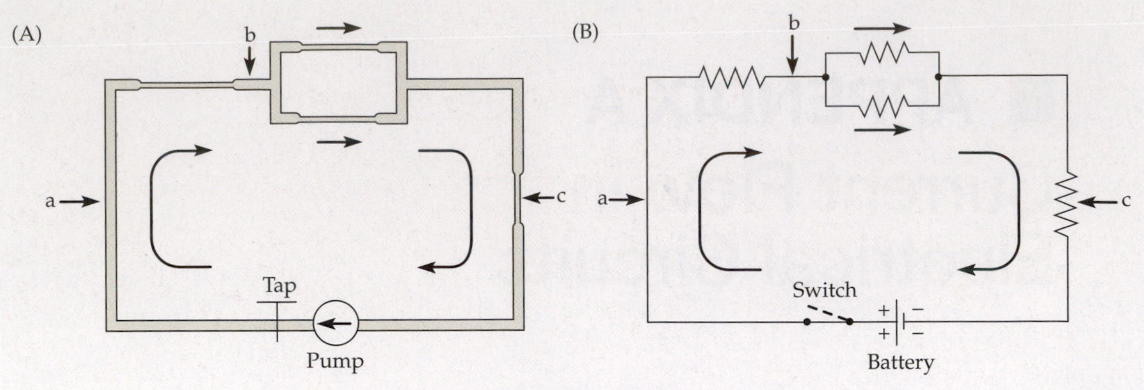

FIGURE 1 Hydraulic and Electrical Circuits. (A,B) Corresponding circuits for the flow of water and of electrical current. A battery is analogous to a pump that operates at constant pressure, the switch to a tap in the hydraulic line, and resistors to constrictions in the tubes. The letters "a," "b," and "c" indicate equivalent points in the two currents.

To explain the energy source for current flow and the meaning of electrical **potential**, the hydraulic analogy is again useful. The flow of fluid depicted in Figure 1 depends on a pressure difference, the direction of flow being from high to low. No net movement occurs between two parts of the circuit at the same pressure. The overall pressure in the circuit is supplied by expenditure of energy in driving the pump. In the electrical circuit shown here, the electrical "pressure" or **potential** is provided by a **battery** in which chemical energy is stored. Hydraulic pressure is measured in gm/cm^2; electrical potential is measured in **volts.**

Symbols used in electrical circuit diagrams and arrangements of circuit elements in series and in parallel are illustrated in Figure 2. As the names imply, a **voltmeter** measures electrical potential and is equivalent to a pressure gauge in hydraulics; an **ammeter** measures current flowing in a circuit and is equivalent to a flowmeter.

Ohm's Law and Electrical Resistance

In hydraulic systems, at least under ideal circumstances, the amount of current flowing through the system increases with pressure. The factor that determines the relation between pressure and flow rate is an inherent characteristic of the pipes, their **resistance.** Small-diameter, long pipes have greater resistances than large-diameter, short ones. Similarly, current flow in electrical circuits depends on the resistance in the circuit. Again, small, long wires have larger resistances than large, short ones. If current is being passed through an ionic solution, the resistance of the solution will increase as the solution is made more dilute. This is because there are fewer ions available to carry the current. In conductors

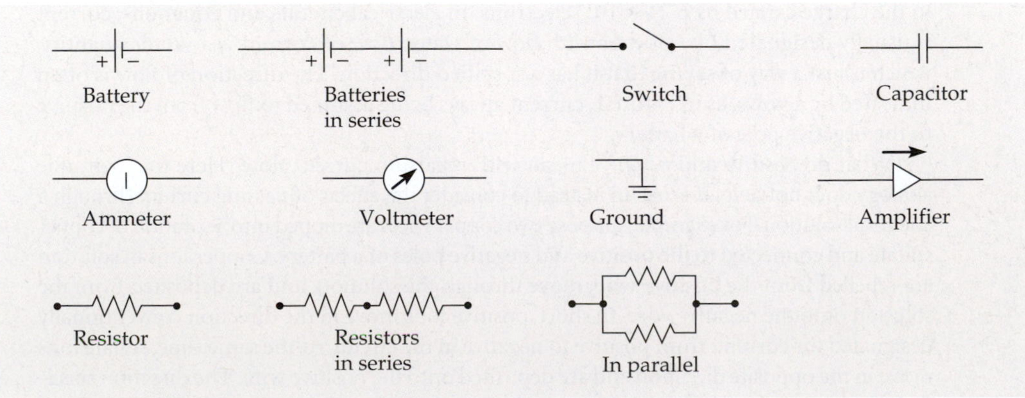

FIGURE 2 Symbols in Electrical Circuit Diagrams

such as wires, the relation between current and potential difference is described by Ohm's law, formulated by Ohm in the 1820s. The law says that the amount of current (I) flowing in a conductor is related to the potential difference (V) applied to it, $I = V/R$. The constant R is the resistance of the wire. If I is in amperes and V is in volts, then R is in units of **ohms** (Ω). The reciprocal of resistance is called conductance, and is a measure of the ease with which current flows through a conductor. Conductance is indicated by $g = 1/R$; the units of conductance are **siemens** (S). Thus, Ohm's law may also be written $I = gV$.

Use of Ohm's Law in Understanding Circuits

Ohm's law holds whenever the graph of current against potential is a straight line. In any circuit or part of a circuit for which this is true, any one of the three variables in the equation may be calculated if the other two are known. For example,

1. We can pass a known current through a nerve membrane, measure the change in potential, and then calculate the membrane resistance ($R = V/I$).
2. If we measure the potential difference produced by an unknown current and know the membrane resistance, we can calculate the applied current ($I = V/R$).
3. If we pass a known current through the membrane and know its resistance, then we can calculate the change in potential ($V = IR$).

Two additional simple, but important, rules (Kirchoff's laws) should be mentioned:

1. The algebraic sum of all the battery voltages is equal to the algebraic sum of all the IR voltage drops in a loop. An example of this is shown in Figure 3B: $V = IR_1 + IR_2$ (this is a statement of the conservation of energy).
2. The algebraic sum of all the currents flowing toward any junction is zero. For example, at point a in Figure 4, $I_{total} + I_{R1} + I_{R3} = 0$, which means that I_{total} (arriving) $= -I_{R1} - I_{R3}$ (leaving) (this is merely a statement that charge is neither created nor destroyed anywhere in the circuit).

We can now examine in more detail the circuits of Figures 3 and 4, which are needed to construct a model of the membrane. Figure 3A shows a battery (V) of 10 V connected to a resistance (R) of 10 Ω. The switch S can be opened or closed, thereby interrupting or establishing current flow. The voltage applied to R is 10 V; therefore the current measured by the ammeter, I, is, by Ohm's law, 1.0 A. In Figure 3B, the resistor is replaced by two resistors, R_1 and R_2, **in series**. By the first of Kirchoff's laws, the current flowing into point b must be equal to that leaving. Therefore, the same current, I, must flow through both the resistors. By the second of Kirchoff's laws, then, $IR_1 + IR_2 = V$ (10 V). It follows that the

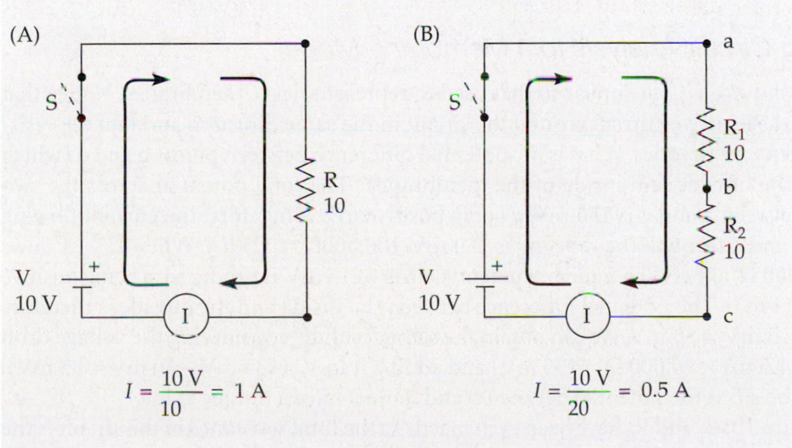

FIGURE 3 Ohm's Law Applied to Simple Circuits. (A) Current $I = (10\text{ V})/(10\ \Omega) = 1$ A. (B) Current = $(10\text{ V})/(20\ \Omega) = 0.5$ A, and the voltage across each resistor is 5 V.

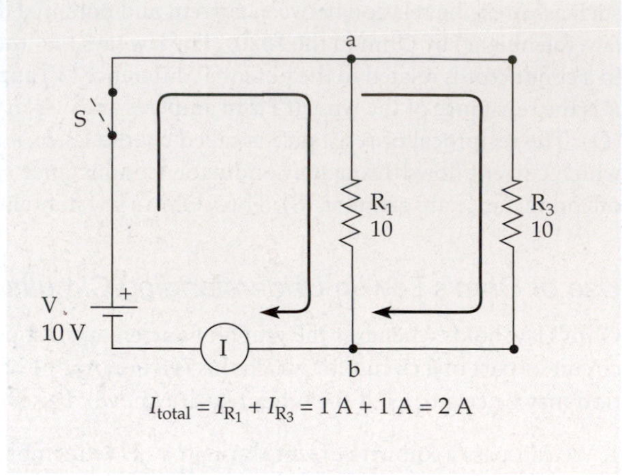

FIGURE 4 Parallel Resistors. When R_1 and R_3 are in parallel, the voltage drop across each resistor is 10 V and the total current is 2 A.

current, $I = V/(R_1 + R_2) = 0.5$ A. The voltage at b, then, is 5 V positive to that at c and a is 5 V positive to b. Note that because there is only one path for the current, the total resistance, R_{total}, seen by the battery is simply the sum of the two resistors; that is,

$$R_{total} = R_1 + R_2$$

What happens if, as shown in Figure 4, we add a second resistor, also of 10 Ω, **in parallel** rather than in series? In the circuit, the two resistors R_1 and R_3 provide two separate pathways for current. Both have a voltage V (10 V) across them, so the respective currents will be

$$I_{R_1} = V/R_1 = 1\text{ A}$$
$$I_{R_3} = V/R_3 = 1\text{ A}$$

Therefore, to satisfy the first of Kirchoff's laws, there must be 2 A arriving at point a and 2 A leaving point b. The ammeter, then, will read 2 A. Now the combined resistance of R_1 and R_3 is $R_{total} = V/I = (10\text{ V})/(2\text{ A}) = 5\ \Omega$, or half that of the individual resistors. This makes sense if one thinks of the hydraulic analogy: Two pipes in parallel will offer less resistance to flow than one pipe alone. In the parallel electrical circuit the *conductances* add: $g_{total} = g_1 + g_3$, or $1/R_{total} = 1/R_1 + 1/R_3$.

If we now generalize to any number (n) of resistors, resistances in series simply add:

$$R_{total} = R_1 + R_2 + R_3 + \cdots + R_n$$

and in parallel their reciprocals add:

$$1/R_{total} = 1/R_1 + 1/R_2 + 1/R_3 + \cdots + 1/R_n$$

Applying Circuit Analysis to a Membrane Model

Figure 5A shows a circuit similar to that used to represent nerve membranes. Notice that the two batteries drive current around the circuit in the same direction and that the resistors R_1 and R_2 are in series. What is the potential difference between points b and d (which represent the outside and inside of the membrane)? The total potential across the two resistors between a and c is 150 mV, a being positive to c. Therefore, the current flowing between a and c through the resistors is 150 mV/100,000Ω = 1.5 μA. When 1.5 μA flows across 10,000 Ω, as between a and b, a potential drop of 15 mV is produced, a being positive with respect to b. The potential difference between the inside and the outside is therefore 100 mV – 15 mV = 85 mV. We can obtain the same result by considering the voltage drop across R_2 (1.5 μA × 90,000 Ω = 135 mV) and adding it to V_2 (135 mV – 50 mV = 85 mV). This *must* be so, as the potential between b and d must have a unique value.

In Figure 5B, R_1 and R_2 have been exchanged. As the total resistance in the circuit is the same, the current must be the same, 1.5 μA. Now the potential drop across R_2, between a and b, is 90,000 Ω × 1.5 μA = 135 mV, a being positive to b. Now the potential across the membrane is 100 mV – 135 mV = –35 mV, **outside negative**; the same result can, of course,

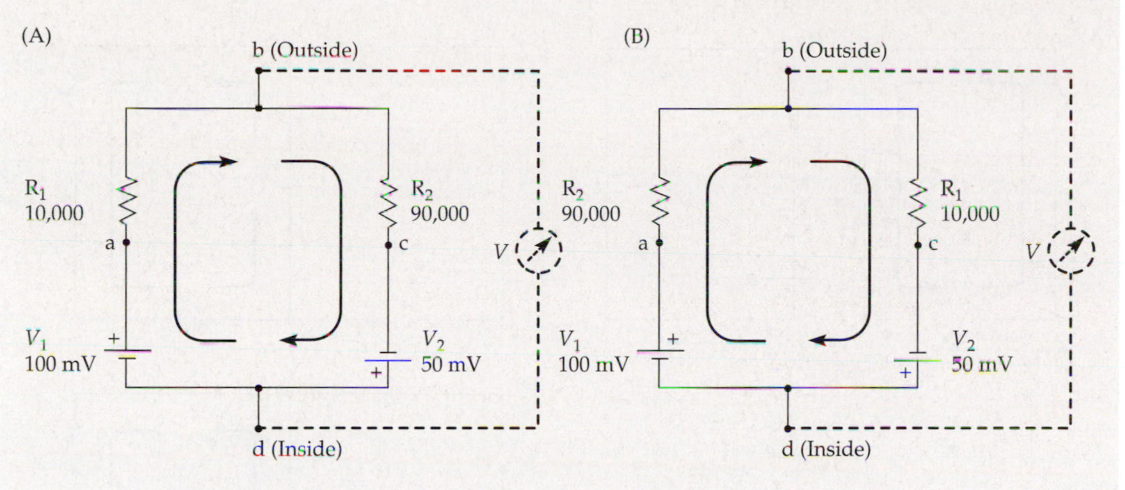

FIGURE 5 Analogue Circuits for Nerve Membranes. In A and B the resistors R_1 and R_2 are reversed; otherwise the circuits are the same. The batteries V_1 and V_2 are in series. In (A), point b (the "outside" of the membrane) is positive with respect to d (the "inside") by 85 mV; in (B) it is negative by 35 mV. These circuits illustrate how changes in resistance can give rise to membrane potential changes even though the batteries (which represent ionic equilibrium potentials) remain constant.

be obtained from the current through R_1. This simple circuit illustrates an important point about membrane physiology: *The potential across a membrane can change as a result of resistance changes while the batteries remain unchanged.* A general expression for the membrane potential in the circuit shown in Figure 5A can be derived simply, as follows:

$$V_m = V_1 - IR_1$$

As $I = (V_1 + V_2)/(R_1 + R_2)$:

$$V_m = V_1 - \frac{(V_1 + V_2)R_1}{R_1 + R_2}$$

On rearranging:

$$V_m = \frac{V_1 R_2/R_1 - V_2}{1 + R_2/R_1}$$

Electrical Capacitance and Time Constant

In the circuits described in Figures 3 and 4, closing or opening the switch produces instantaneous and simultaneous changes in current and potential. Capacitors introduce a time element into the consideration of current flow. They accumulate and store electrical charge, and, when they are present in a circuit, current and voltage changes are no longer simultaneous. A capacitor consists of two conducting plates (usually of metal) separated by an insulator (air, mica, oil, or plastic). When voltage is applied between the plates (Figure 6A), there is an instantaneous displacement of charge from one plate to the other through the external circuit. Once the capacitor is fully charged, however, there is no further current, as none can flow across the insulator. The **capacitance** (C) of a capacitor is defined by how much charge (q) it can store for each volt applied to it:

$$C = q/V$$

The units of capacitance are coulombs/volt or **farads** (F). The larger the plates of a capacitor and the closer together they are, the greater its capacitance. A one-farad capacitor is very large; capacitances in common use are in the range of microfarads (μF) or smaller.

When the switch in Figure 6A is closed, then, there is an instantaneous charge separation at the plates. The amount of charge stored in the capacitor is proportional to its capaci-

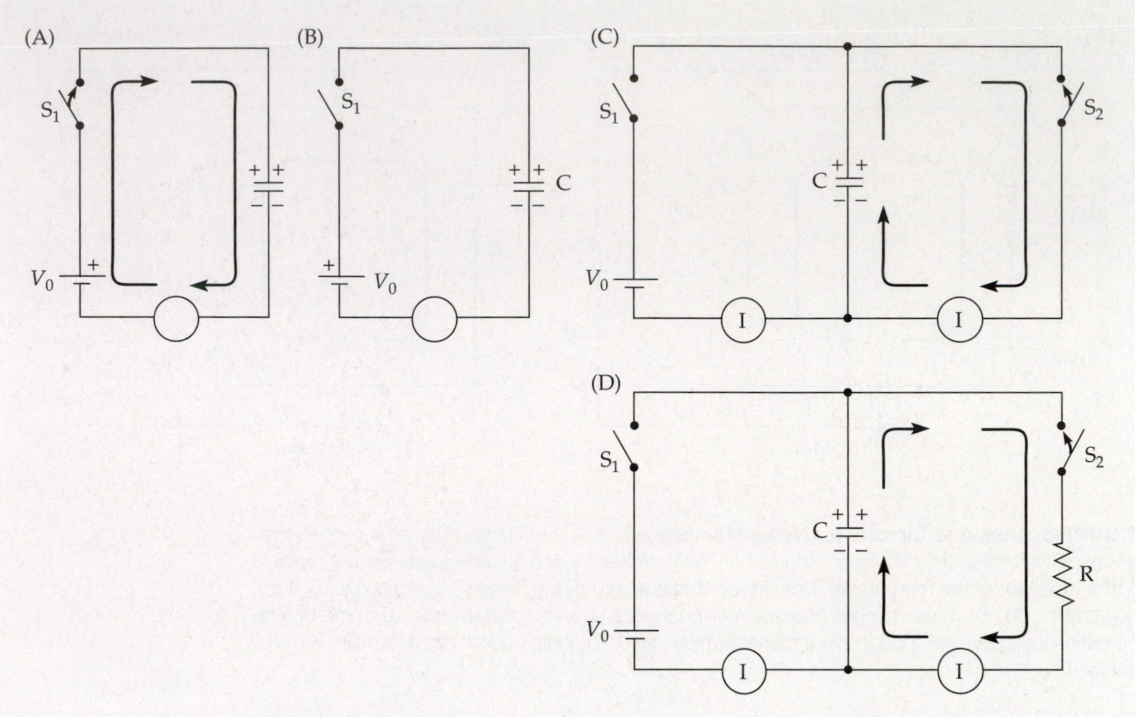

FIGURE 6 Capacitors in Electrical Circuits. A, B, and C are idealized circuits having no resistance. When S_1 is closed in (A), the capacitor is charged instantaneously to voltage V_0. If S_1 is then opened (B), the potential remains on the capacitor. Closing switch S_2 (C) discharges the capacitor instantaneously. In (D) the capacitor is discharged through resistor R. The maximum discharge current is $I = V_0/R$.

tance and to the magnitude of the applied voltage (V_0). When the switch is opened, as in B, the charge on the capacitor remains, as does the voltage (V) between the plates. (One can sometimes get a surprising shock from electronic apparatus after it has been turned off because some of the capacitors in the circuits may remain charged.) The capacitor can be discharged by shorting it with a second switch, as in Figure 6C. Again, the current flow is instantaneous, returning the charge and the voltage on the capacitor to zero. If, instead, the capacitor is discharged through a resistor (R, Figure 6D), the discharge is no longer instantaneous. This is because the resistor limits the current flow. If the voltage on the capacitor is V, then by Ohm's law the maximum current is $I = V/R$. With no resistor in the circuit, the current becomes infinitely large and the capacitor is discharged in an infinitesimal time period; if R is very large, the discharge process takes a very long time. The rate of discharge at any given time, dq/dt, is the current flowing at that particular time. In other words, $dq/dt = -V/R$ (negative because the charge is decreasing with time), where V initially is equal to the battery voltage and decreases as the capacitor is discharged. As $q = CV$, $dq/dt = CdV/dt$, and we can then write $CdV/dt = -V/R$, or

$$dV/dt = -V/RC$$

The equation says that the rate of loss of voltage from the capacitor is proportional to the voltage remaining. Thus, as the voltage decreases, the rate of discharge decreases. The constant of proportionality, $1/RC$, is the **rate constant** for the process: RC is its **time constant**. This kind of process arises over and over again in nature. For example, the rate at which water drains from a bathtub decreases as the depth, and hence the pressure at the drain, decreases. In this kind of situation, the discharge process is described by an exponential function:

$$V = V_0 e^{-t/\tau}$$

where V_0 is the initial charge on the capacitor and the time constant $\tau = RC$. Similarly, when the capacitor is charged through a resistor, as in Figure 7, the charging process takes

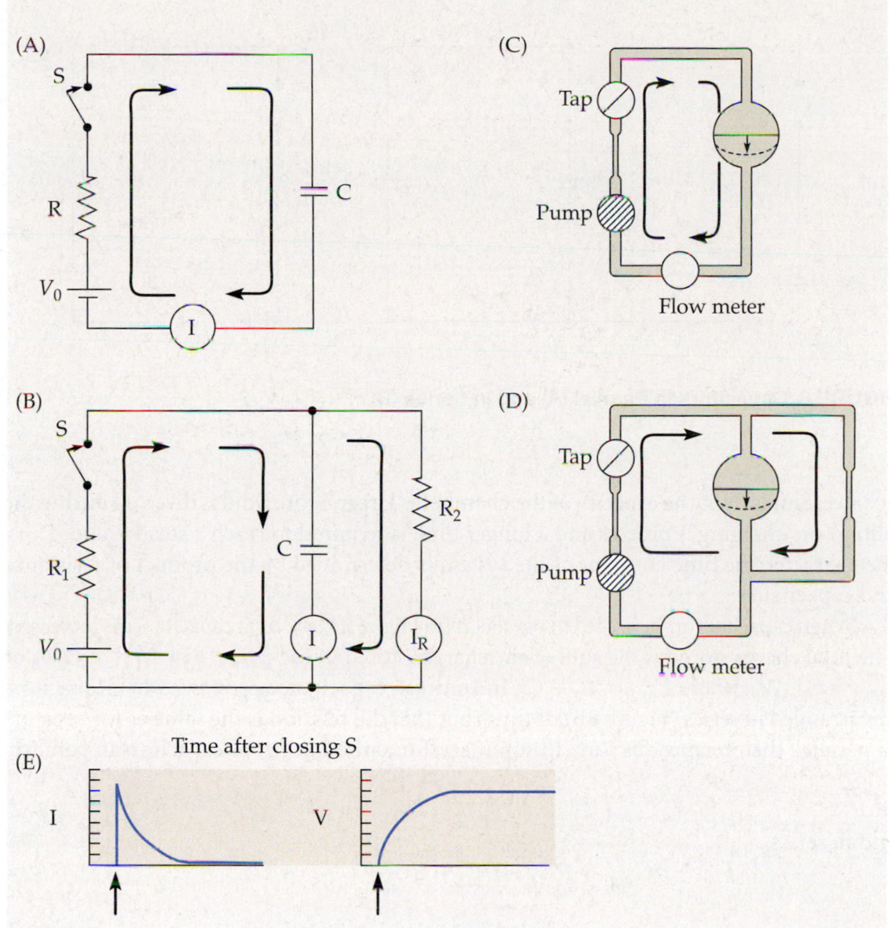

FIGURE 7 Charging a Capacitor. In (A) the capacitor is charged at a rate limited by the resistor, the initial rate being $I = V_0/R$. In (B) the charging rate depends on both resistors in the circuit. In (E) the capacitative current and the voltage across the capacitor are shown as functions of time. The voltage reaches its final value only when the capacitor is fully charged, i.e., when no more current flows into the capacitor. (C) and (D) are hydraulic analogues of the circuits in A and B.

a finite time. The voltage between the plates increases with time until the battery voltage is reached and no further current flows. The charging process is now a rising exponential, with a time constant $\tau = RC$:

$$V = V_0(1 - e^{-t/\tau})$$

These examples illustrate another property of a capacitor. Current flows into and out of the capacitor only when the potential is changing:

$$I_c = dQ/dt = CdV/dt$$

When the voltage across the capacitor is steady ($dV/dt = 0$), the capacitative current, I_c, is zero. In other words, the capacitance has an "infinite resistance" for a steady potential difference and a "low resistance" for a rapidly changing potential. Figure 7B shows a circuit in which current flows through a resistor and capacitor in parallel and Figure 7E the time courses of the capacitative current and voltage.

The properties of a capacitor in a circuit can be illustrated by the slightly more elaborate hydraulic analogy shown in Figure 7C. The capacitor is represented by an elastic diaphragm that forms a partition in a fluid-filled chamber. When the tap is opened, fluid is pumped from one side of the chamber to the other. The pressure generated by the pump causes the diaphragm to bulge. Fluid continues to flow until, because of its elasticity, the diaphragm provides an equal and opposite pressure; then there is no more fluid flow and the chamber is fully charged. If a tube is placed alongside, as in Figure 7D, some fluid flows through the tube and some is used to expand the diaphragm. The rate of expansion depends on the resistance of the tube, and on the capacity of the chamber. If the tube is of high resistance, then for a given flow the pressure difference between its two ends will be relatively large. In that case, the distention of the diaphragm will be large and take a relatively long time to

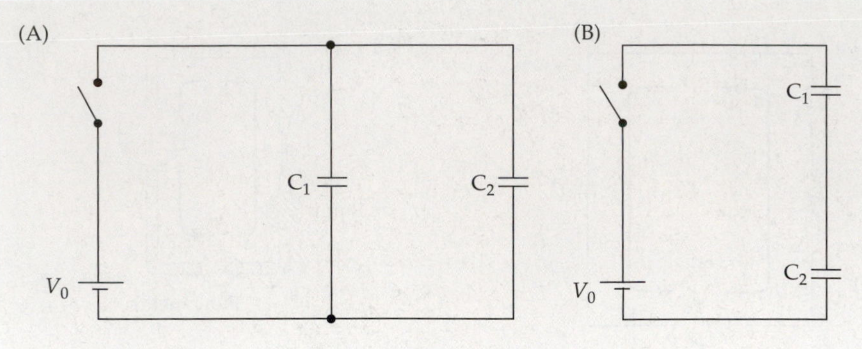

FIGURE 8 Capacitors in Parallel (A) and in Series (B)

achieve. Similarly, if the capacity of the chamber is larger, more fluid is diverted during the filling (or "charging") process and a longer time is required to reach a steady state. Thus, the characteristic time constant of the system is determined by the product of resistance and capacitance.

When capacitors are arranged in parallel, as in Figure 8A, the total capacitance is increased. The total charge stored is the sum of the charges stored in each: $q_1 + q_2 = C_1V_0 + C_2V_0$ or $q_{total} = C_{total}V_0$, where $C_{total} = C_1 + C_2$. In contrast, capacitance *decreases* when capacitors are arranged in series (Figure 8B). It turns out that the relation is the same as for resistors in parallel: their reciprocals sum. In summary, for a number (n) of capacitors in parallel:

$$C_{total} = C_1 + C_2 + C_3 + \cdot\ \cdot\ \cdot + C_n$$

and in series,

$$1/C_{total} = 1/C_1 + 1/C_2 + 1/C_3 + \cdot\ \cdot\ \cdot + 1/C_n$$

APPENDIX B

Metabolic Pathways for the Synthesis and Inactivation of Low-Molecular-Weight Transmitters

The figures on the following pages summarize the predominant metabolic pathways for the low-molecular-weight transmitters acetylcholine, GABA, glutamate, dopamine, norepinephrine, epinephrine, 5-HT, and histamine. Glycine, purines, NO, and CO are not included; there appear to be no special neuronal pathways for their synthesis or degradation. Pathways for endocannabinoid synthesis and degradation are still under investigation (see Chapter 15). For each metabolic step, the portion of the molecule being modified is highlighted in color. Further information can be found in several comprehensive texts:

Berg, J. M., Tymoczko, T., and Stryer, L. B. (eds.) 2011. *Biochemistry*, 7th ed. W. H. Freeman, New York.

Brunton, L. L., Chabner, B. S., and Knollmann, B. C. (eds.) 2011. *Goodman and Gilman's The Pharmacological Basis of Therapeutics*, 12th ed. McGraw-Hill, New York.

Siegel, J., Albers, R. W., Brady, S. T., and Price, D. L. (eds.) 2006. *Basic Neurochemistry: Molecular, Cellular, and Medical Aspects*, 7th ed. Elsevier Academic Press, Burlington, MA.

ACETYLCHOLINE (ACH)

Synthesis

$$\underset{\text{Acetyl-CoA}}{H_3C-\overset{\overset{O}{\|}}{C}-S-CoA} + \underset{\text{Choline}}{HO-CH_2-CH_2-\overset{+}{N}-(CH_3)_3} \xrightleftharpoons{\text{Choline acetyltransferase}} \underset{\text{CoA}}{HS-CoA} + \underset{\text{Acetylcholine}}{H_3C-\overset{\overset{O}{\|}}{C}-O-CH_2-CH_2-\overset{+}{N}-(CH_3)_3}$$

Degradation

$$\underset{\text{Acetylcholine}}{H_3C-\overset{\overset{O}{\|}}{C}-O-CH_2-CH_2-\overset{+}{N}-(CH_3)_3} + H_2O \xrightleftharpoons{\text{Acetylcholinesterase}} \underset{\text{Acetate}}{H_3C-C(=O)O^-} + \underset{\text{Choline}}{HO-CH_2-CH_2-\overset{+}{N}-(CH_3)_3} + H^+$$

γ-AMINOBUTYRIC ACID (GABA)

Synthesis

$$\underset{\text{Glutamate}}{{}^+H_3N-\underset{\underset{COO^-}{|}}{\overset{\overset{H}{|}}{C}}-CH_2-CH_2-COO^-} \xrightarrow[\text{CO}_2]{\text{Glutamic acid decarboxylase}} \underset{\gamma\text{-Aminobutyric acid (GABA)}}{{}^+H_3N-CH_2-CH_2-CH_2-COO^-}$$

Degradation

$$\underset{\gamma\text{-Aminobutyric acid (GABA)}}{{}^+H_3N-CH_2-CH_2-CH_2-COO^-} \xrightarrow[\alpha\text{-Ketoglutarate} \;\to\; \text{Glutamate}]{\text{GABA }\alpha\text{-oxoglutarate transaminase}} \underset{\text{Succinic semialdehyde}}{O{=}C(H)-CH_2-CH_2-COO^-} \xrightarrow[H_2O + NAD^+ \;\to\; H^+ + NADH]{\text{Succinic semialdehyde dehydrogenase}} \underset{\text{Succinate}}{{}^-OOC-CH_2-CH_2-COO^-}$$

Glutamate

Synthesis

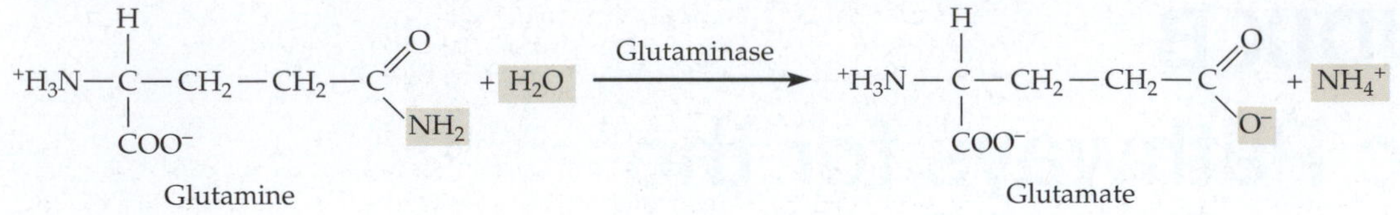

Glutamine + H_2O —Glutaminase→ Glutamate + NH_4^+

$^{+}H_3N-C(H)(COO^-)-CH_2-CH_2-C(=O)NH_2$ + H_2O → $^{+}H_3N-C(H)(COO^-)-CH_2-CH_2-C(=O)O^-$ + NH_4^+

Glutamine — Glutamate

Degradation

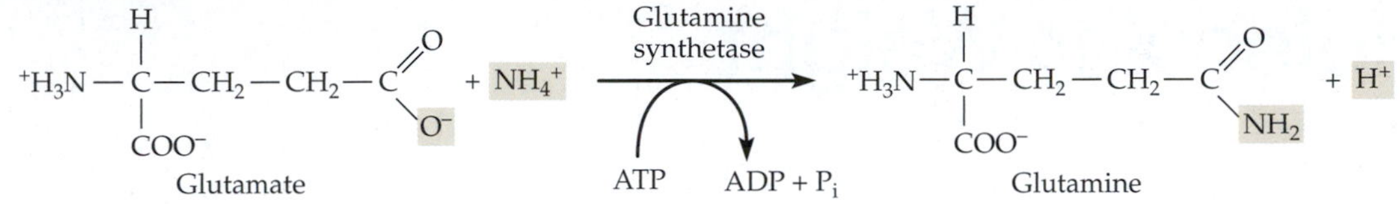

Glutamate + NH_4^+ —Glutamine synthetase (ATP → ADP + P_i)→ Glutamine + H^+

$^{+}H_3N-C(H)(COO^-)-CH_2-CH_2-C(=O)O^-$ + NH_4^+ → $^{+}H_3N-C(H)(COO^-)-CH_2-CH_2-C(=O)NH_2$ + H^+

Glutamate — Glutamine

Catecholamines: Dopamine

Synthesis

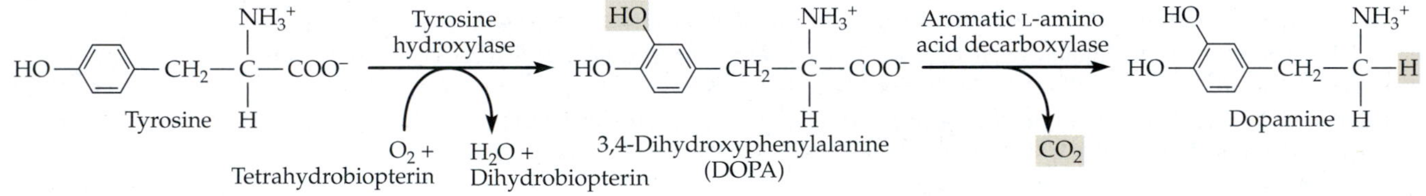

Tyrosine —Tyrosine hydroxylase (O_2 + Tetrahydrobiopterin → H_2O + Dihydrobiopterin)→ 3,4-Dihydroxyphenylalanine (DOPA) —Aromatic L-amino acid decarboxylase (→ CO_2)→ Dopamine

HO–C₆H₄–$CH_2-C(NH_3^+)(H)-COO^-$ (Tyrosine)

(HO)₂C₆H₃–$CH_2-C(NH_3^+)(H)-COO^-$ (3,4-Dihydroxyphenylalanine (DOPA))

(HO)₂C₆H₃–$CH_2-C(NH_3^+)(H)-H$ (Dopamine)

Degradation

Dopamine —COMT→ 3-Methoxytyramine

Dopamine —MAO→ 3,4-Dihydroxy-β-phenylacetaldehyde

3-Methoxytyramine —MAO→ 3-Methoxy-4-hydroxy-β-phenylacetaldehyde

3,4-Dihydroxy-β-phenylacetaldehyde —AR→ 3,4-Dihydroxy-β-phenylethanol

3,4-Dihydroxy-β-phenylethanol —COMT→ 3-Methoxy-4-hydroxy-β-phenylethanol

3-Methoxy-4-hydroxy-β-phenylacetaldehyde —AR→ 3-Methoxy-4-hydroxy-β-phenylethanol

3,4-Dihydroxy-β-phenylacetaldehyde —ADH→ 3,4-Dihydroxyphenylacetic acid (DOPAC)

3,4-Dihydroxyphenylacetic acid (DOPAC) —COMT→ 3-Methoxy-4-hydroxyphenylacetic acid (HVA)

3-Methoxy-4-hydroxy-β-phenylacetaldehyde —ADH→ 3-Methoxy-4-hydroxyphenylacetic acid (HVA)

Structures: Dopamine: (HO)₂C₆H₃–$CH_2-C(NH_3^+)(H)-H$; 3-Methoxytyramine: H_3CO(HO)C₆H₃–$CH_2-C(NH_3^+)(H)-H$; 3,4-Dihydroxy-β-phenylethanol: (HO)₂C₆H₃–CH_2-CH_2OH; 3-Methoxy-4-hydroxy-β-phenylethanol: H_3CO(HO)C₆H₃–CH_2-CH_2OH; 3,4-Dihydroxy-β-phenylacetaldehyde: (HO)₂C₆H₃–$CH_2-C(=O)H$; 3-Methoxy-4-hydroxy-β-phenylacetaldehyde: H_3CO(HO)C₆H₃–$CH_2-C(=O)H$; 3,4-Dihydroxyphenylacetic acid (DOPAC): (HO)₂C₆H₃–CH_2-COO^-; 3-Methoxy-4-hydroxyphenylacetic acid (HVA): H_3CO(HO)C₆H₃–CH_2-COO^-

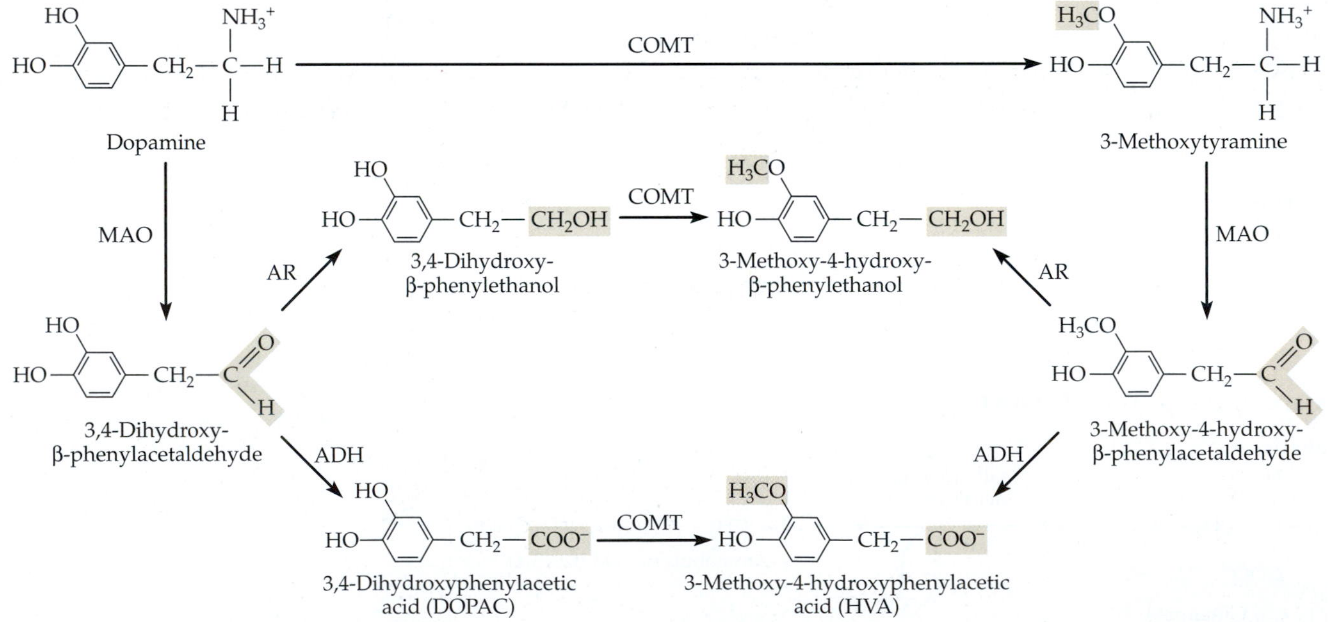

Catecholamines: Norepinephrine and Epinephrine

Synthesis

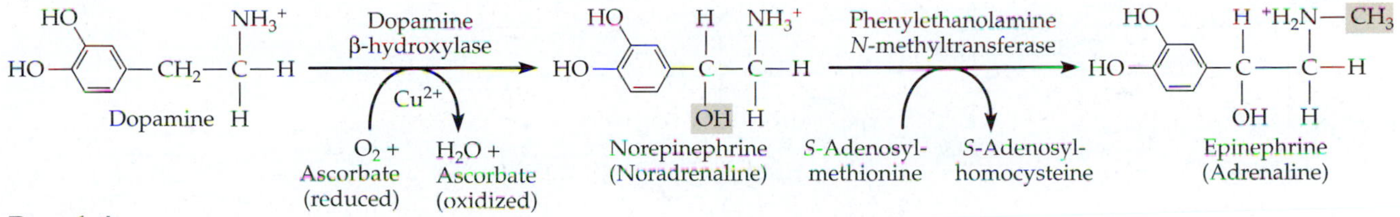

Dopamine

Dopamine β-hydroxylase

Cu^{2+}

O_2 + Ascorbate (reduced)

H_2O + Ascorbate (oxidized)

Norepinephrine (Noradrenaline)

Phenylethanolamine *N*-methyltransferase

S-Adenosyl-methionine

S-Adenosyl-homocysteine

Epinephrine (Adrenaline)

Degradation

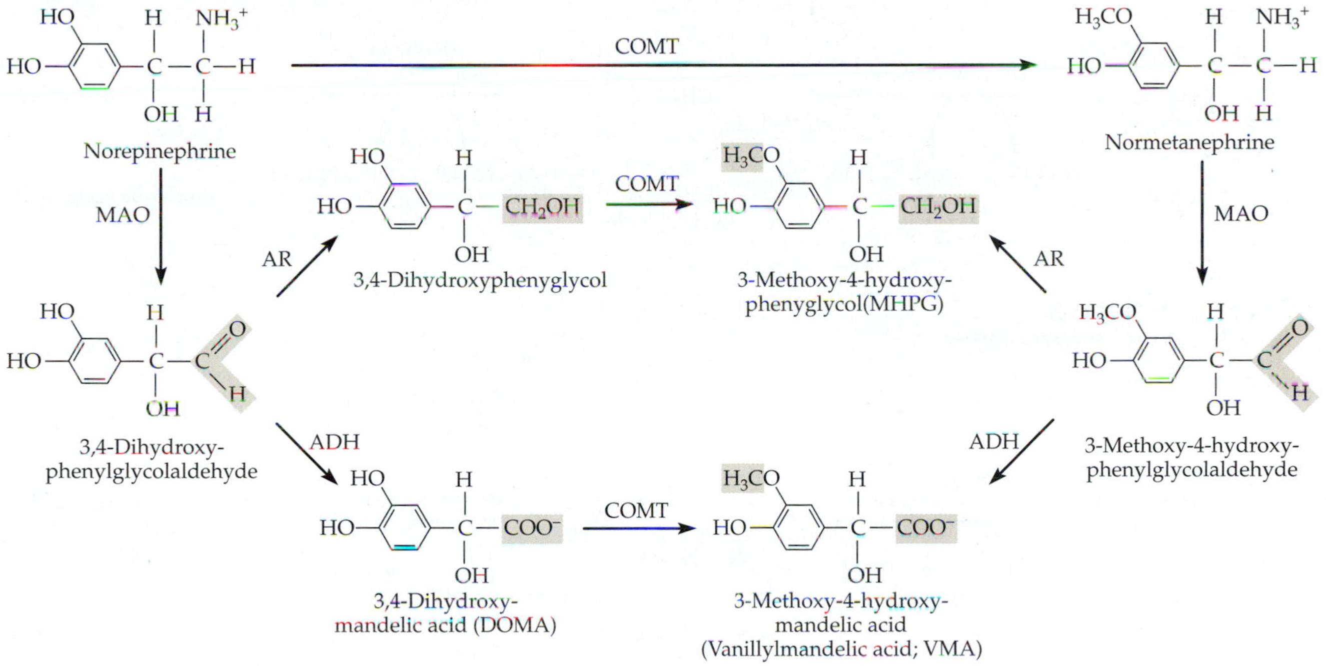

Norepinephrine

COMT

Normetanephrine

MAO

3,4-Dihydroxy-phenylglycolaldehyde

AR

3,4-Dihydroxyphenyglycol

COMT

3-Methoxy-4-hydroxy-phenyglycol(MHPG)

AR

MAO

3-Methoxy-4-hydroxy-phenylglycolaldehyde

ADH

3,4-Dihydroxy-mandelic acid (DOMA)

COMT

3-Methoxy-4-hydroxy-mandelic acid (Vanillylmandelic acid; VMA)

ADH

5-Hydroxytryptamine (5-HT; Serotonin)

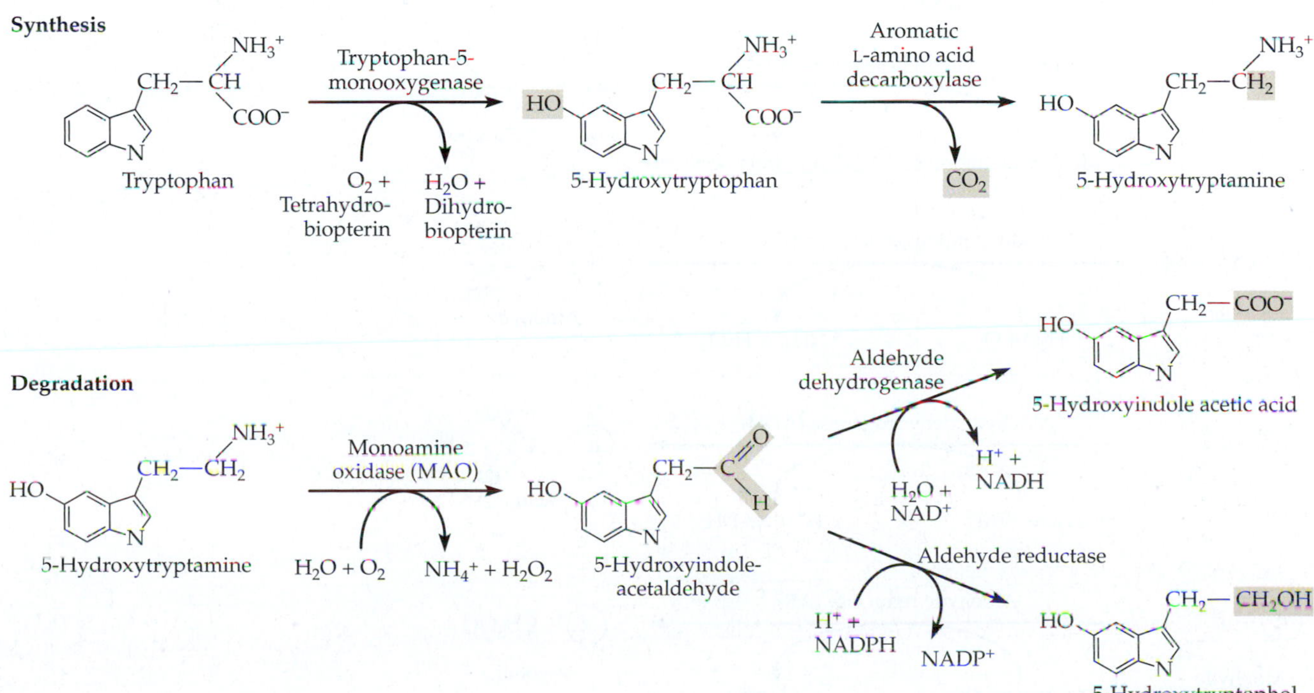

HISTAMINE

Synthesis

NH_3^+ — CH_2—C—COO^- — H — HN N

L-Histidine → Histidine decarboxylase → Histamine (CO_2 released)

Degradation

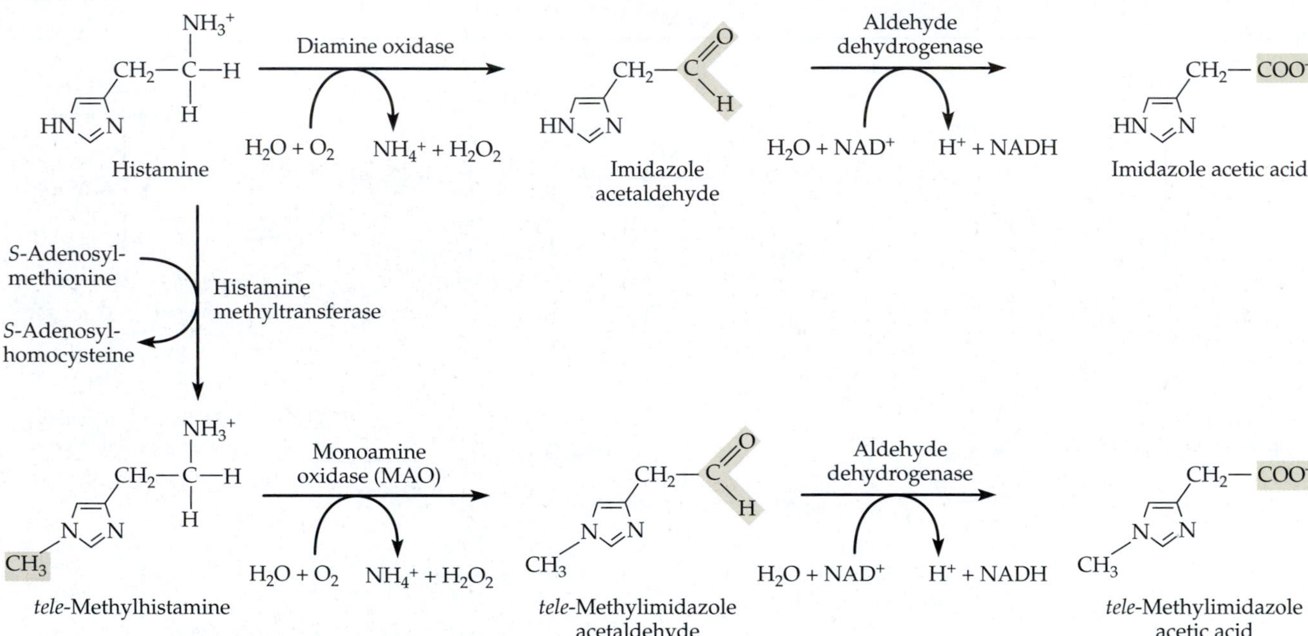

DEGRADATION OF BIOGENIC AMINES

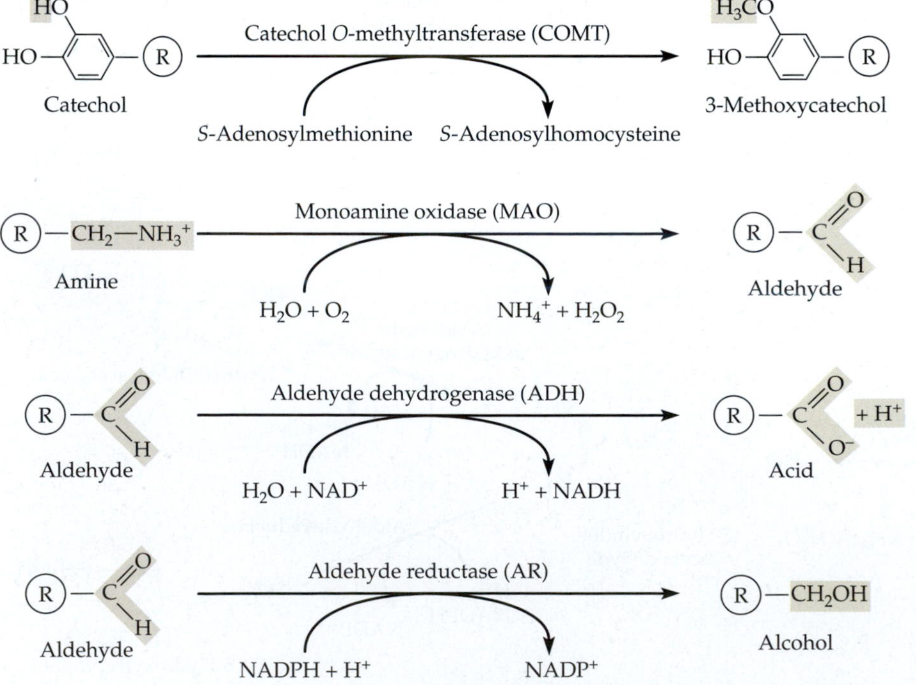

APPENDIX C
Structures and Pathways of the Brain

The following figures show the brain viewed from different aspects and cut in different sections. The aim is to provide a visual equivalent of a glossary relating to material in the text, rather than to present a full atlas. Consequently, only key landmarks and structures are illustrated. Further anatomical information can be found in a number of comprehensive texts:

Carpenter, M. B. 1991. *Core Text of Neuroanatomy*, 4th ed. Williams and Wilkins, Baltimore.

Martin, J. H. 2003. *Neuroanatomy: Text and Atlas*, 3rd ed. Mcgraw-Hill Medical, New York.

Nolte, J. 2008. *The Human Brain: An Introduction to Its Functional Anatomy*, 6th ed. Mosby-Year Book, St. Louis.

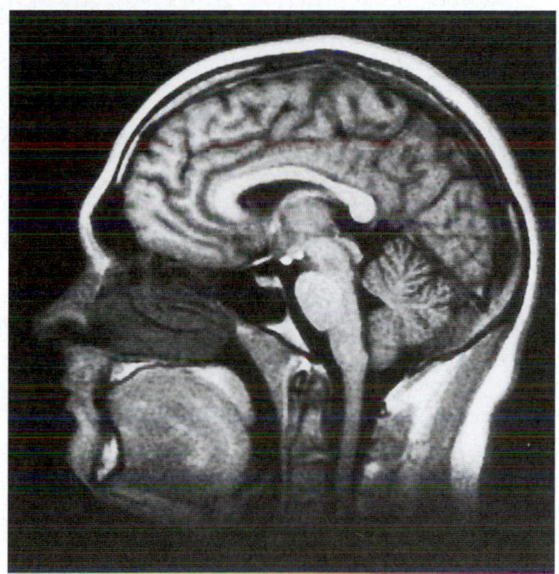

Magnetic resonance image of a living human brain (sagittal section). Copyright 1984 by the General Electric Company. Reproduced with permission.

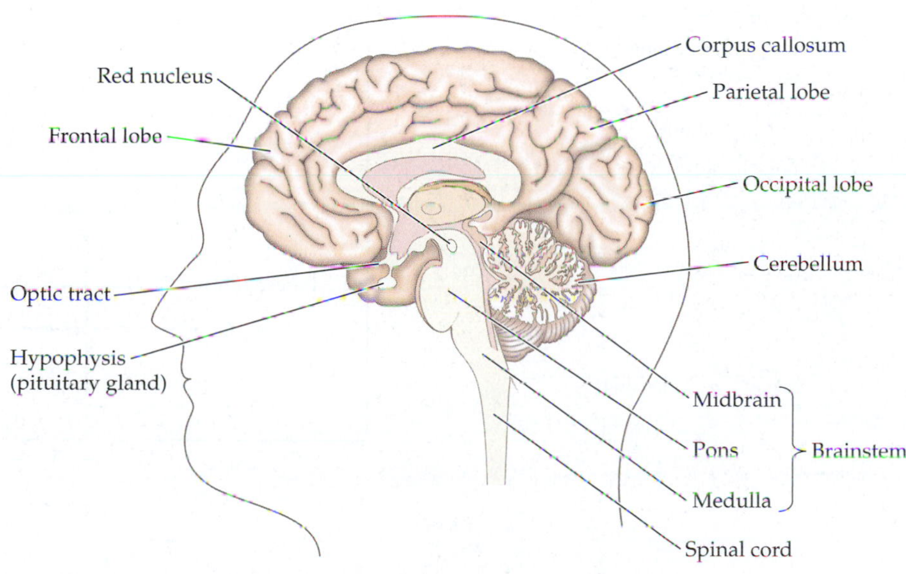

Side view

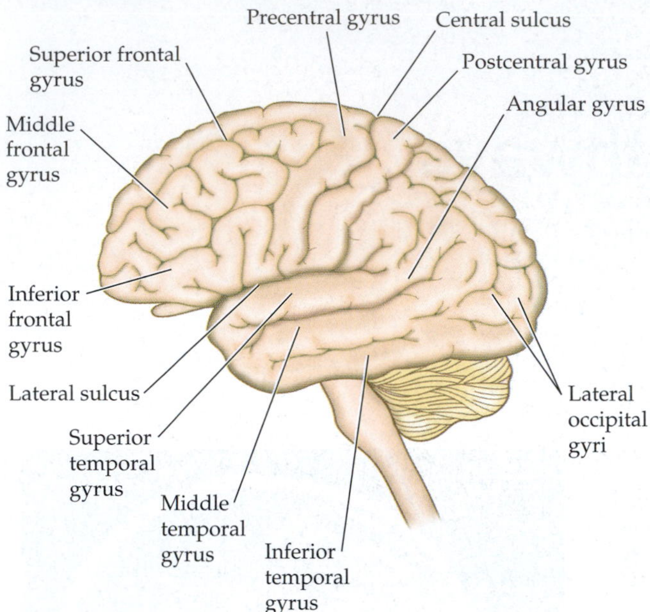
Precentral gyrus
Central sulcus
Superior frontal gyrus
Postcentral gyrus
Angular gyrus
Middle frontal gyrus
Inferior frontal gyrus
Lateral sulcus
Superior temporal gyrus
Middle temporal gyrus
Inferior temporal gyrus
Lateral occipital gyri

From above

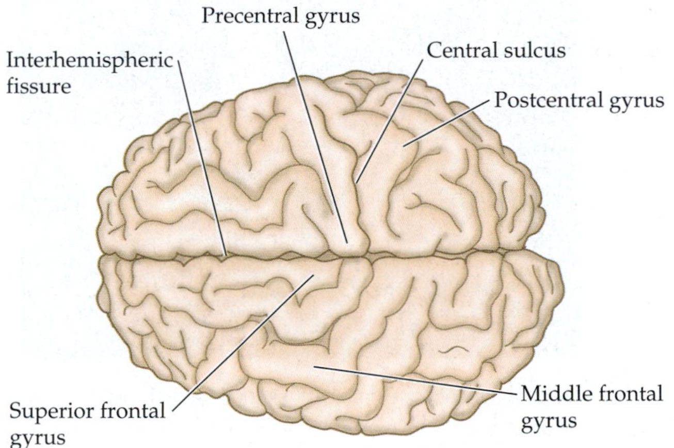
Precentral gyrus
Central sulcus
Interhemispheric fissure
Postcentral gyrus
Superior frontal gyrus
Middle frontal gyrus

From below

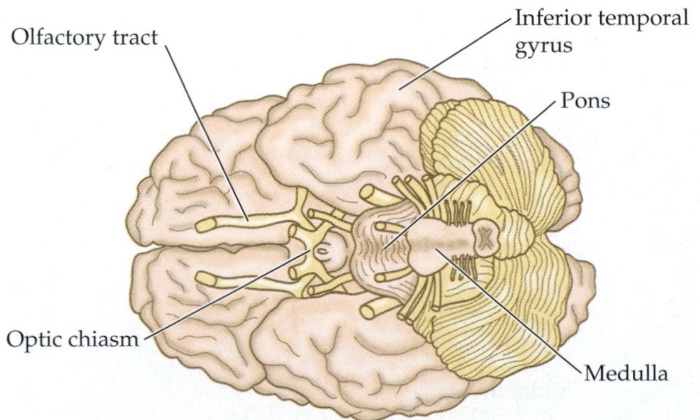
Olfactory tract
Inferior temporal gyrus
Pons
Optic chiasm
Medulla

Directional terms

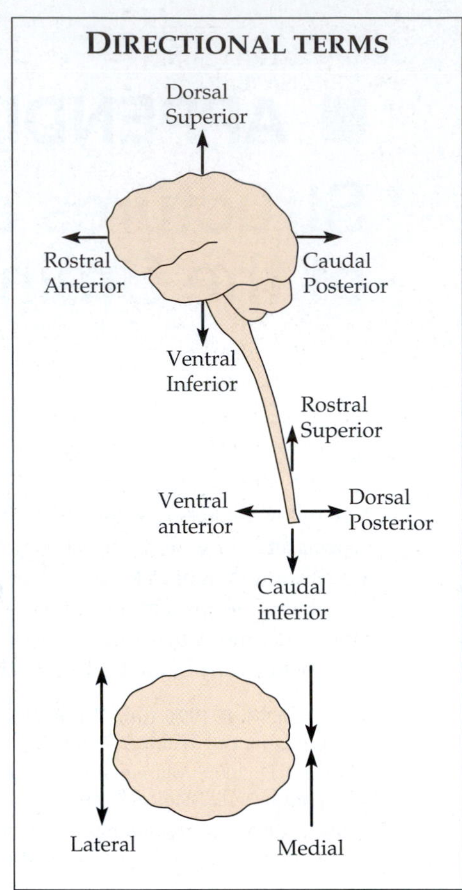
Dorsal
Superior
Rostral
Anterior
Caudal
Posterior
Ventral
Inferior
Rostral
Superior
Ventral
anterior
Dorsal
Posterior
Caudal
inferior
Lateral
Medial

Planes of section

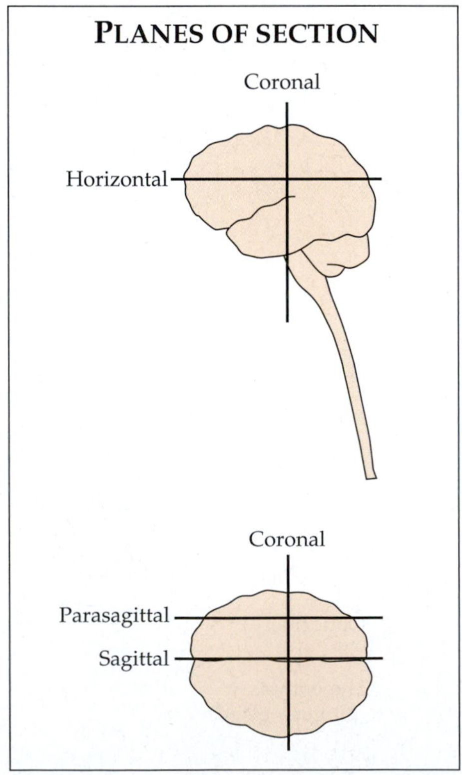
Coronal
Horizontal
Coronal
Parasagittal
Sagittal

NUMBERED ANATOMICAL AREAS OF THE CEREBRAL CORTEX (BRODMANN'S AREAS)

LOCALIZATION OF MOTOR AND SENSORY FUNCTIONS

LATERAL VIEW

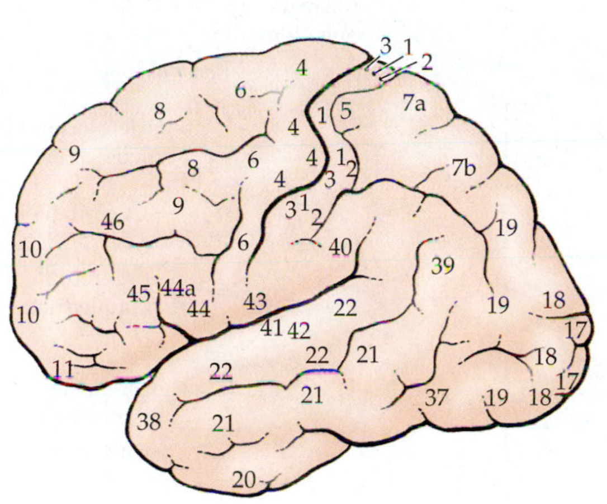

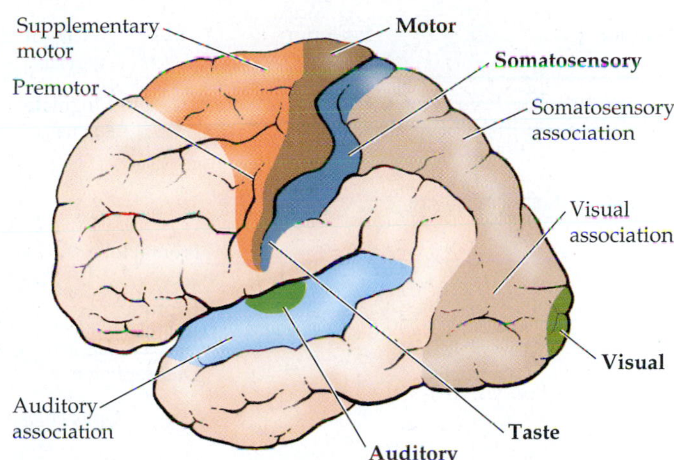

SAGITTAL VIEW

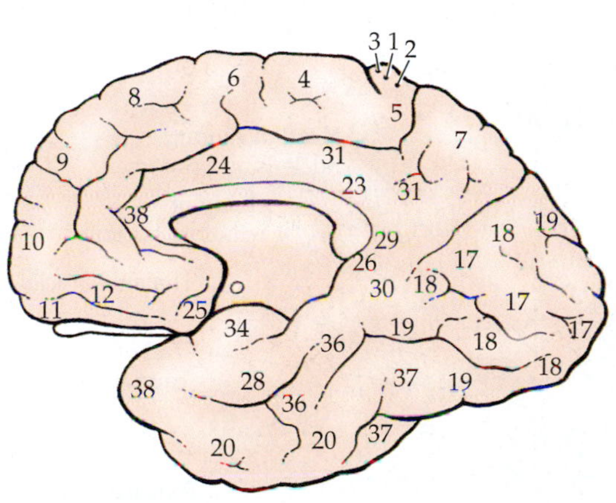

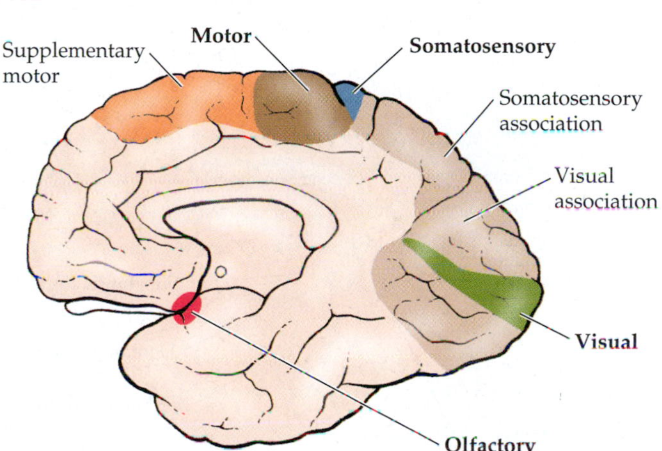

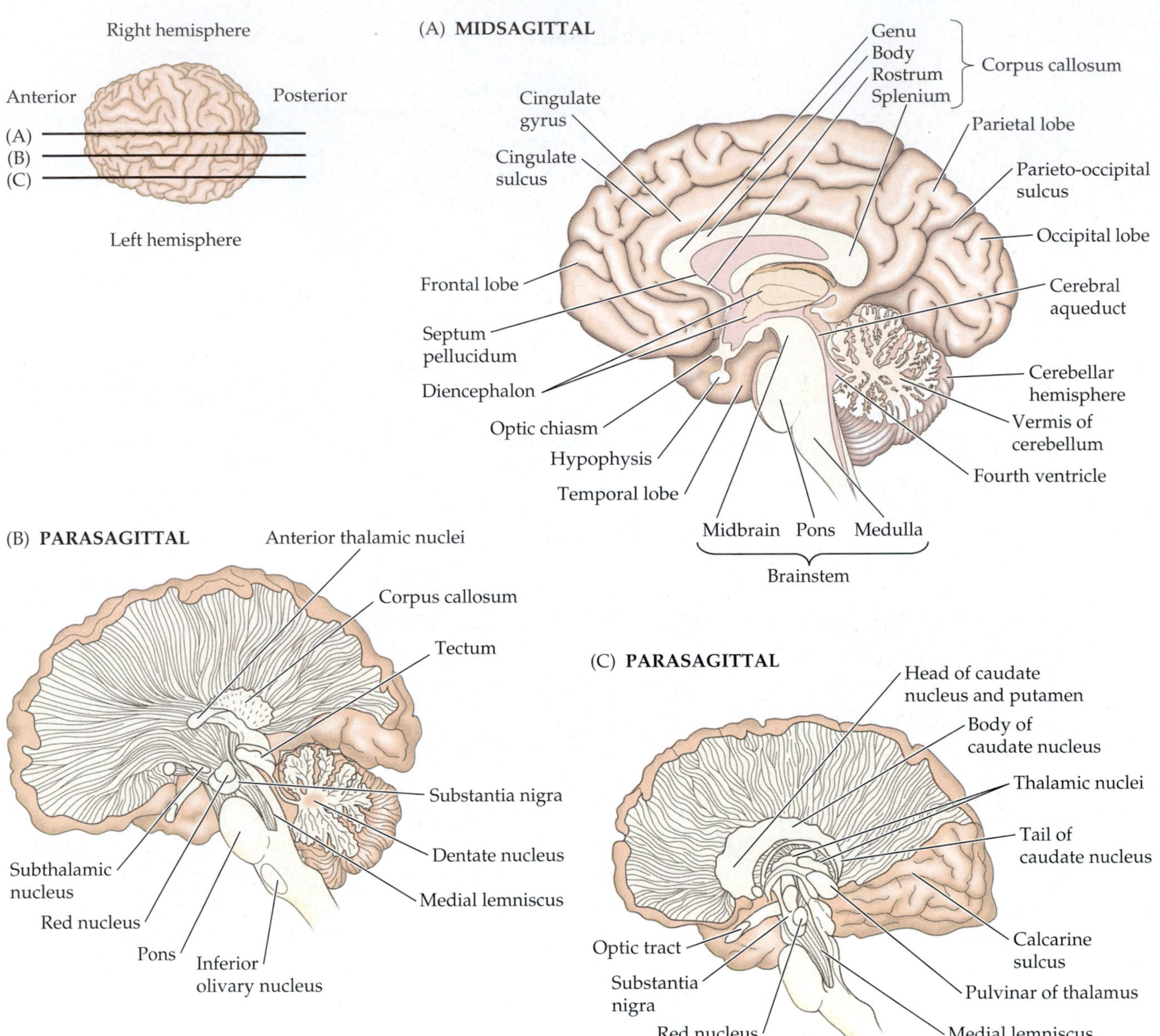
SAGITTAL SECTIONS
Right hemisphere
Anterior
Posterior
(A)
(B)
(C)
Left hemisphere
(A) MIDSAGITTAL
Genu
Body
Rostrum
Splenium
Corpus callosum
Cingulate gyrus
Cingulate sulcus
Parietal lobe
Parieto-occipital sulcus
Occipital lobe
Frontal lobe
Cerebral aqueduct
Septum pellucidum
Diencephalon
Cerebellar hemisphere
Vermis of cerebellum
Optic chiasm
Hypophysis
Fourth ventricle
Temporal lobe
Midbrain
Pons
Medulla
Brainstem
(B) PARASAGITTAL
Anterior thalamic nuclei
Corpus callosum
Tectum
Substantia nigra
Dentate nucleus
Subthalamic nucleus
Medial lemniscus
Red nucleus
Pons
Inferior olivary nucleus
(C) PARASAGITTAL
Head of caudate nucleus and putamen
Body of caudate nucleus
Thalamic nuclei
Tail of caudate nucleus
Calcarine sulcus
Optic tract
Substantia nigra
Pulvinar of thalamus
Red nucleus
Medial lemniscus

HORIZONTAL SECTIONS

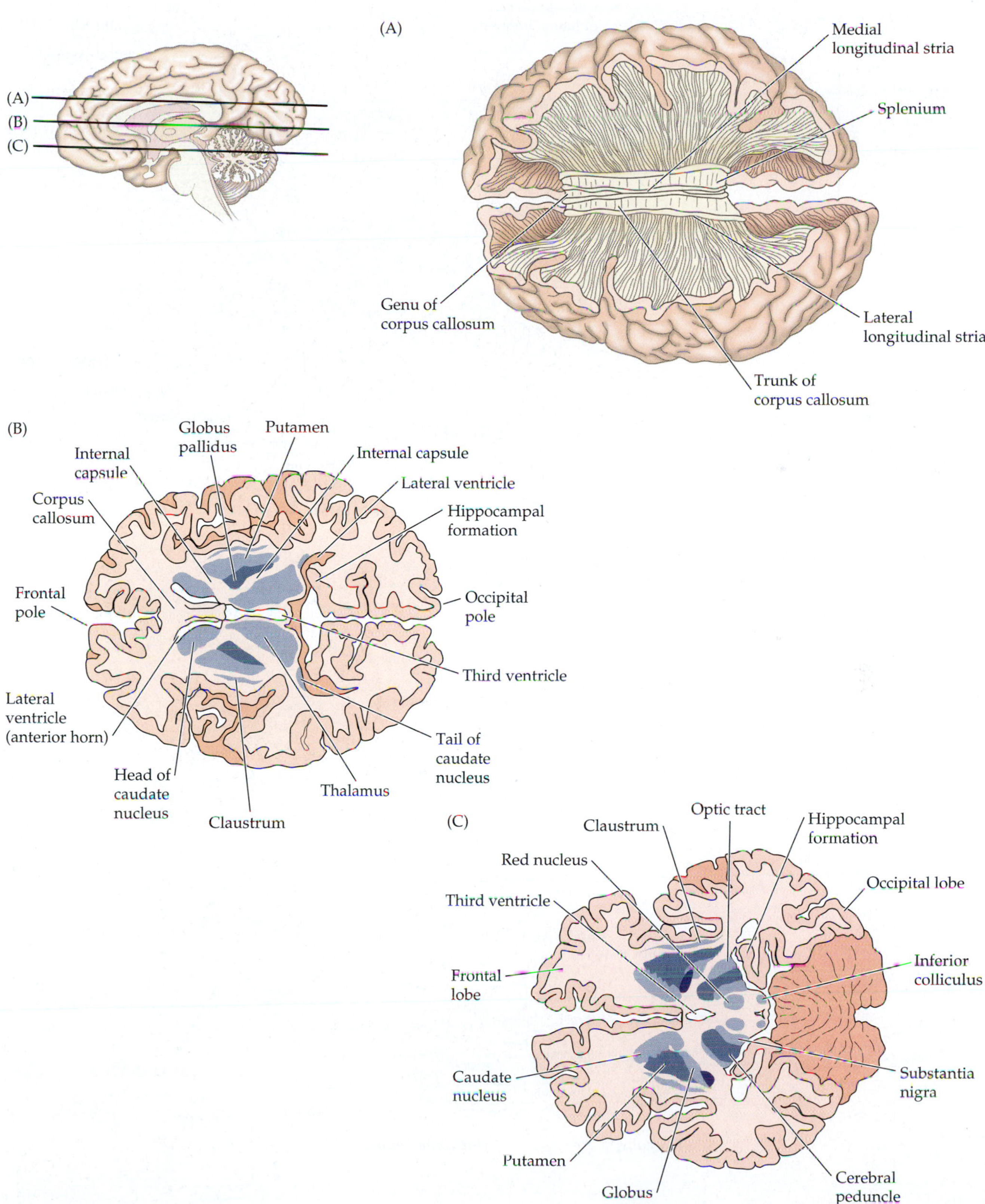
(A)
(B)
(C)
(A)
Medial
longitudinal stria
Splenium
Genu of
corpus callosum
Lateral
longitudinal stria
Trunk of
corpus callosum
(B)
Globus
pallidus
Putamen
Internal
capsule
Internal capsule
Lateral ventricle
Corpus
callosum
Hippocampal
formation
Frontal
pole
Occipital
pole
Third ventricle
Lateral
ventricle
(anterior horn)
Tail of
caudate
nucleus
Head of
caudate
nucleus
Thalamus
Claustrum
(C)
Claustrum
Optic tract
Hippocampal
formation
Red nucleus
Occipital lobe
Third ventricle
Frontal
lobe
Inferior
colliculus
Caudate
nucleus
Substantia
nigra
Putamen
Globus
pallidus
Cerebral
peduncle

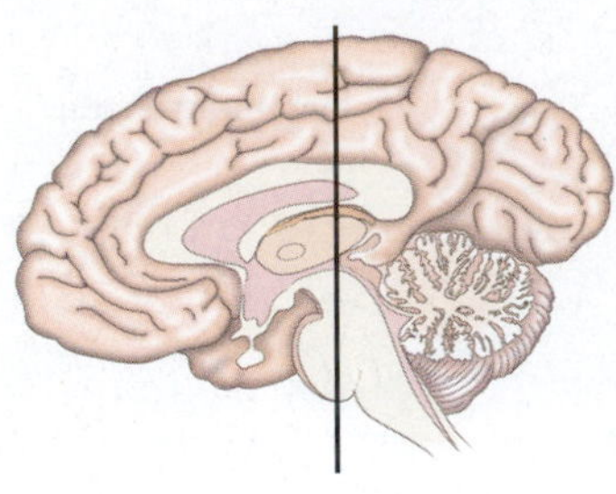

CORONAL SECTIONS

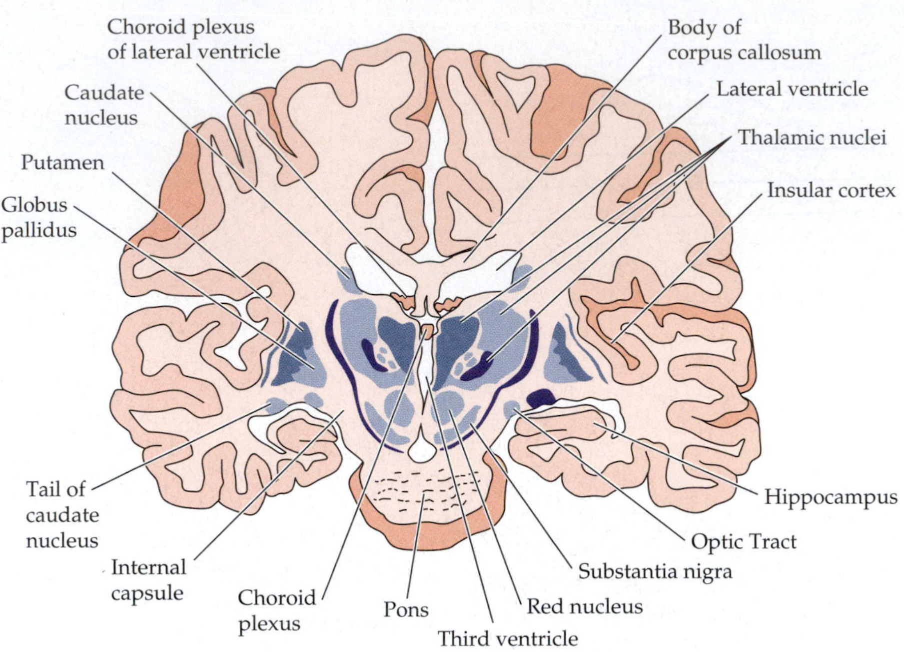
Choroid plexus of lateral ventricle
Caudate nucleus
Putamen
Globus pallidus
Tail of caudate nucleus
Internal capsule
Choroid plexus
Pons
Third ventricle
Red nucleus
Substantia nigra
Optic Tract
Hippocampus
Insular cortex
Thalamic nuclei
Lateral ventricle
Body of corpus callosum

THE CEREBELLUM

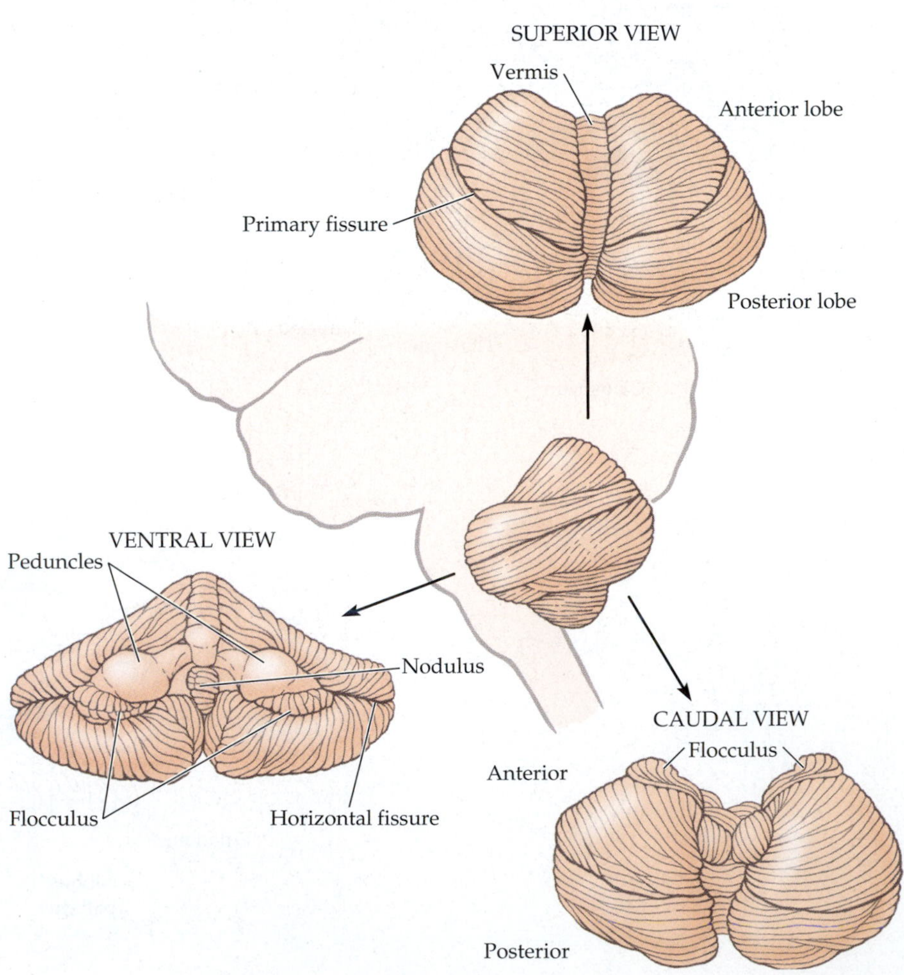
SUPERIOR VIEW
Vermis
Anterior lobe
Primary fissure
Posterior lobe
VENTRAL VIEW
Peduncles
Nodulus
Flocculus
Horizontal fissure
CAUDAL VIEW
Flocculus
Anterior
Posterior

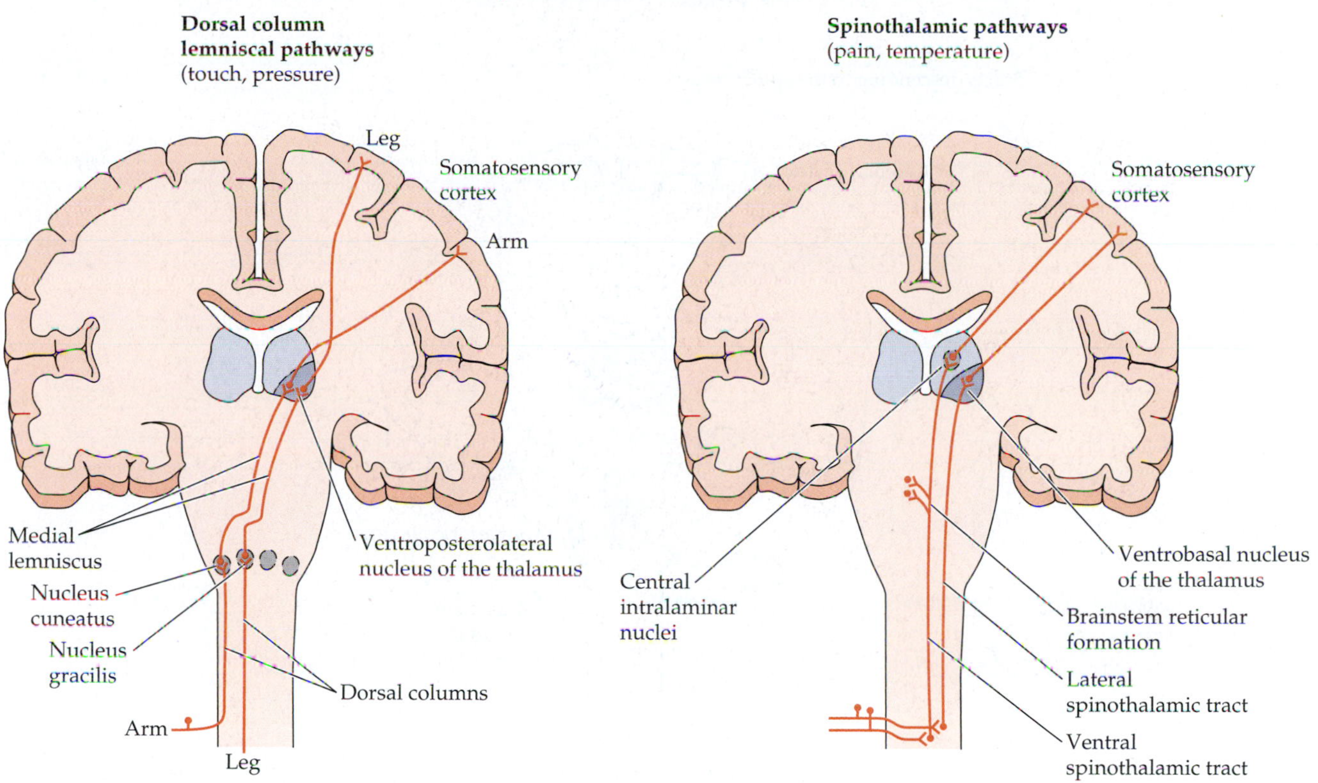
MAJOR SENSORY PATHWAYS
Dorsal column
lemniscal pathways
(touch, pressure)
Leg
Somatosensory
cortex
Arm
Medial
lemniscus
Nucleus
cuneatus
Nucleus
gracilis
Ventroposterolateral
nucleus of the thalamus
Dorsal columns
Arm
Leg
Spinothalamic pathways
(pain, temperature)
Somatosensory
cortex
Central
intralaminar
nuclei
Ventrobasal nucleus
of the thalamus
Brainstem reticular
formation
Lateral
spinothalamic tract
Ventral
spinothalamic tract

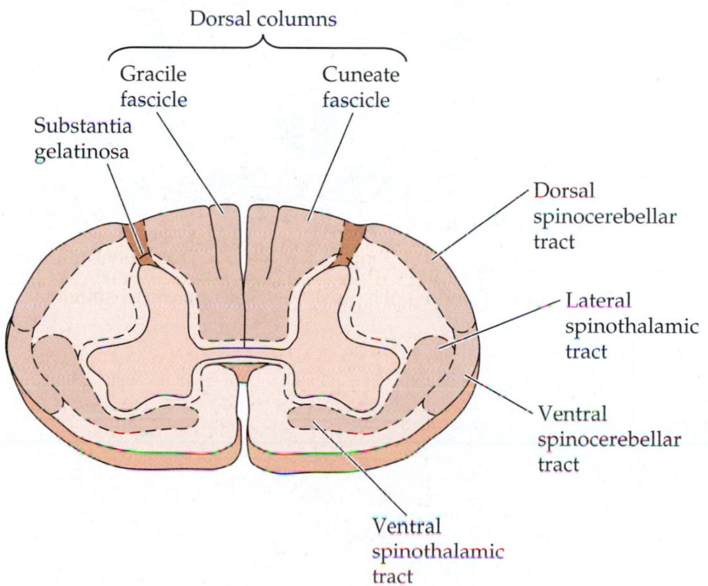
CROSS SECTION OF CERVICAL SPINAL CORD
Dorsal columns
Gracile
fascicle
Cuneate
fascicle
Substantia
gelatinosa
Dorsal
spinocerebellar
tract
Lateral
spinothalamic
tract
Ventral
spinocerebellar
tract
Ventral
spinothalamic
tract

MAJOR MOTOR PATHWAYS

Tracts descending to the spinal cord

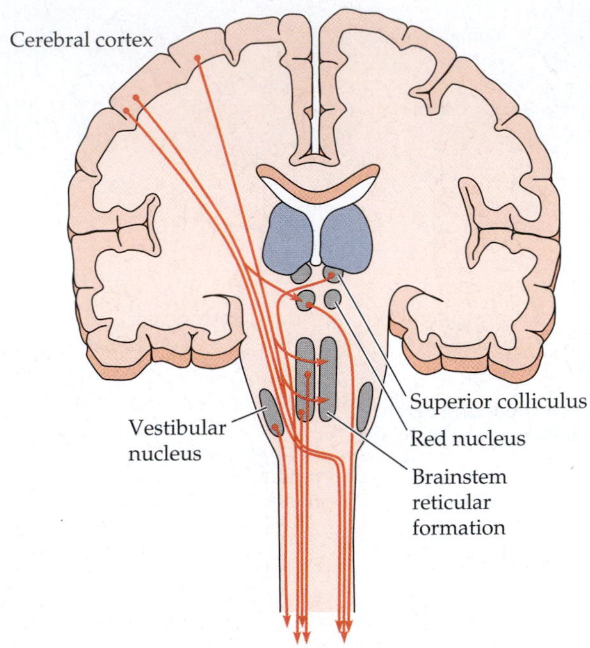

Cross section of cervical spinal cord

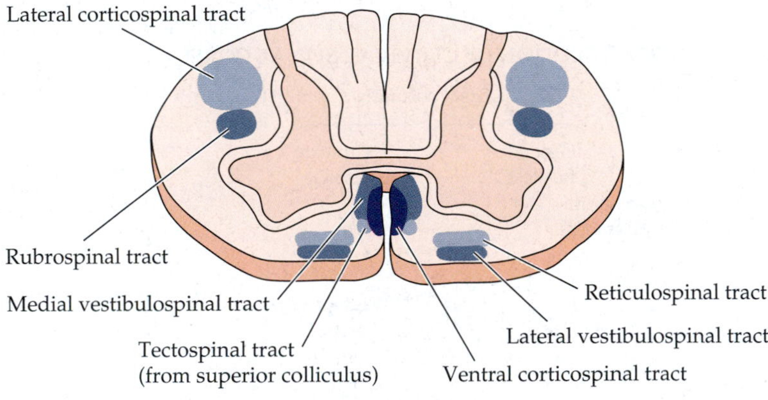

GLOSSARY

The definitions below apply to the terms used in the context of this book. Thus, excitation, adaptation, and inhibition all have additional meanings that are not included.

For structural formulae of transmitters see Appendix B.

For anatomical terms see Appendix C.

A

acetylcholine (ACh) Transmitter liberated by vertebrate motoneurons, preganglionic autonomic neurons, and in various central nervous system pathways.

acetylcholine receptor (ACh receptor) Membrane protein that binds ACh. Two different varieties:

nicotinic ACh receptor (nAChr) Activated by nicotine, consists of five polypeptide subunits that form a cation channel when activated.

muscarinic ACh receptor (mAChr) Activated by muscarine, contains a single protein molecule coupled by a G protein to one or more intracellular second messenger systems.

A channel A type of voltage-activated potassium channel.

action potential Brief regenerative, all-or-nothing electrical potential that propagates along an axon or muscle fiber.

activation 1. Initiation of an action potential. 2. Increase in the probability that an ion channel will open.

active transport Movement of ions or molecules against an electrochemical gradient.

primary active transport Utilizes metabolic energy.

secondary active transport Utilizes energy provided by the electrochemical gradient for another ion (usually sodium).

active zone Region in a presynaptic nerve terminal characterized by densely staining material on the cytoplasmic surface of the presynaptic membrane and a cluster of synaptic vesicles; believed to be the site of transmitter release.

activity The effective concentration of a substance in solution. The activity of a substance is smaller than its actual concentration because of interactions between the dissolved particles.

adaptation Decline in response of a sensory neuron to a maintained stimulus.

adenosine 5-triphosphate (ATP) A common metabolite; hydrolysis of the terminal phosphoester linkage provides energy for many cellular reactions. Serves as phosphate donor in phosphorylation reactions. Also found in adrenergic and cholinergic synaptic vesicles; acts as a transmitter at synapses made by vertebrate sympathetic neurons and some central neurons.

adenylyl cyclase Enzyme catalyzing the synthesis of cyclic AMP from ATP.

adequate stimulus The form of stimulus energy to which a sensory receptor neuron is most sensitive.

adrenergic Referring to neurons releasing norepinephrine as a transmitter.

afferent Axon conducting impulses toward the central nervous system. Also called primary afferent.

afterdepolarizing potential (ADP) Depolarization following one or more action potentials due to persistent changes in membrane conductance.

afterhyperpolarizing potential (AHP) Hyperpolarization following one or more action potentials due to persistent changes in membrane conductance.

agonist A molecule that activates a receptor.

agnosia Loss of ability to recognize real-world things (for instance, objects, persons, sounds, shapes, or smells) in spite of preserved capacities in sensing the elemental properties of things.

agrin A protein secreted by motoneurons that becomes embedded in extracellular matrix at the motor end plate and give rise to postsynaptic specialization.

angstrom Unit of length equal to 1×10^{-10} meter. Its symbol is the letter Å.

γ-aminobutyric acid (GABA) An inhibitory neurotransmitter.

amphipathic Containing separate regions of hydrophilic and hydrophobic amino acid groups.

ampulla The sensory region of a semicircular canal in the vestibular apparatus.

anion A negatively charged ion.

annulus Morphological subdivision of a segment of body wall, visible as a circumferential ring in leeches and other annelids.

antagonist A molecule that prevents activation of a receptor.

anterograde In the direction from the neuronal cell body toward the axon terminal. Compare with *retrograde*.

antibody An immunoglobulin molecule.

anticholinesterase Cholinesterase inhibitor (e.g., neostigmine, physostigmine); such agents prevent the hydrolysis of ACh and thereby allow its action to be prolonged.

apoptosis Also known as programmed cell death. Extracellular or intracellular stimuli trigger a cell to destroy itself by a defined cascade of proteases.

astrocyte A class of glial cells found in the vertebrate CNS.

asymmetry current See *gating current.*

area centralis In the cat, the area of retina with highest acuity, containing cones.

augmentation An increase in evoked transmitter release from nerve terminals, following a brief train of repetitive stimuli. Augmentation can last for several seconds.

autonomic nervous system Part of the vertebrate nervous system innervating viscera, skin, smooth muscle, glands, and heart, and consisting of two distinct divisions, parasympathetic and sympathetic.

axon The process or processes of a neuron conducting impulses, usually over long distances.

axon hillock Region of the cell body from which the axon originates; often the site of impulse initiation. See *initial segment.*

axonal transport Term for the movement of proteins, intracellular particles and organelles along axons.

axoplasm Intracellular constituents of an axon.

axoplasmic flow The bulk movement of axoplasm along the axon. One form of axonal transport.

axotomy Severing an axon.

B

ball and chain model Proposed mechanism for inactivation of voltage-activated channels whereby a ball of amino acids, tethered by an amino acid chain, swings into the mouth of an open channel to block its pore.

barrelette A cluster of neurons in the brainstem of rats and mice associated with a single whisker on the face. It receives input from sensory receptor neurons and projects to a barreloid in the thalamus.

barreloid A cluster of neurons in the somatosensory thalamus of rats and mice associated with a single whisker on the face. It receives input from a barrelette and projects to a cortical barrel. See *cortical barrel.*

basal lamina An extracellular, glycoprotein- and proteoglycan-containing matrix that ensheaths many tissues in the body, including nerves and muscle fibers.

basilar membrane The acellular sheet upon which rests the organ of Corti and whose vibration by sound drives motion of the hair cells stereocilia.

biogenic amine A general term referring to any of several bioactive amines, including norepinephrine, epinephrine, dopamine, serotonin, and histamine.

bipolar cell Neuron with two major processes arising from the cell body; in the vertebrate retina, interposed between receptors and ganglion cells.

blobs Small, regularly spaced assemblies of neurons in the visual cortex of the monkey; they are stained with cytochrome oxidase and respond mainly to color stimuli.

blood–brain barrier Term denoting restricted access of substances to neurons and glial cells within the brain.

botulinum toxin A bacterial toxin that blocks release of transmitter from vertebrate motor nerve terminals. Also called botulin toxin.

bouton Small terminal expansion of the presynaptic nerve fiber at a synapse; site of transmitter release.

brain–machine interface A direct communication channel between the nervous system and an external device or a computer with the aim of assisting, improving, or repairing human sensorimotor functions.

Brodmann's areas Regions of the cerebral cortex identified and numbered by the German anatomist K. Brodmann on the basis of their neuronal morphology and structure. Also see Appendix C.

α-bungarotoxin Toxin from venom of the snake *Bungarus multicinctus*; binds to the nicotinic ACh receptor with high affinity.

C

calcium ATPase A molecule that transports calcium across a cell membrane against its electrochemical gradient, utilizing energy derived from the hydrolysis of ATP.

calcium wave A depolarization involving an increase in intracellular calcium concentration, spreading along epithelial and glial cells and from cell to cell through gap junctions. Channels in the glial cell membranes allow ATP to leak out.

calmodulin (CaM) An intracellular protein that binds calcium ions to form calcium-calmodulin, which can then act as an intracellular messenger signalling to many other proteins.

cannabinoid A chemical compound in the *cannabis* plant that acts on cannabinoid receptors, a family of G protein–coupled receptors. Includes related compounds that imitate their action. See *endocannabinoid.*

capacitative current A transient current that flows across a cell membrane in response to a voltage change, thereby recharging the membrane capacitance to the new membrane potential.

carrier molecule A molecule involved in transporting ions or other molecules across cell membranes.

catecholamine A general term referring to molecules having both a catechol ring and an amino group, typically dopamine, norepinephrine, and epinephrine.

cation A positively charged ion.

cDNA (complementary DNA) DNA synthesized by reverse transcriptase using mRNA as a template.

cell fate The sum of all the measurable properties of a cell and its progeny, i.e., patterns of gene expression, differentiation, and proliferation.

cell lineage The pattern of cell divisions by which any given cell or groups of cells arise from precursor cells.

centrifugal control Regulation of performance of peripheral sense organs by axons coming from the central nervous system.

cerebrospinal fluid (CSF) Clear liquid filling the ventricles of the brain and spaces between the meninges. See *subarachnoid space; ventricles.*

charybdotoxin (CTX) A toxin obtained from scorpions that blocks potassium channels.

chemotransduction The generation of a voltage change in a sensory neuron by exposure to a chemical.

chimera An experimentally derived embryo or organ comprising cells arising from two or more genetically distinct sources.

choice probability index A method to quantify how accurately behavioral responses can be predicted from single neuronal responses. When the index measured for one neuron is high, it is likely that that neuron contributes to the behavioral outcome.

cholinergic Neuron releasing acetylcholine as a transmitter.

cholinesterase An enzyme that hydrolyzes acetylcholine into acetic acid and choline.

choroid plexus Folded processes that are rich in blood vessels and project into ventricles of the brain and secrete cerebrospinal fluid.

chromaffin granules Large vesicles found in cells of the adrenal medulla, containing epinephrine (often norepinephrine as well), ATP, dopamine β-hydroxylase and chromogranins.

clone 1. All the progeny of a single cell. 2. To isolate a cell or gene of interest from a diverse starting population of cells or genes.

cochlea The bony canal containing the sensory apparatus for hearing.

compound eye Type of eye found in invertebrates that consists of receptors, supporting cells and a lens, each forming a picture of the outside world. The number of such clusters varies from animal to animal and can reach 30,000 in moths

conductance (γ) Reciprocal of electrical resistance and thus a measure of the ability to conduct electricity; in cell membranes or ion channels, a measure of permeability to one or more ion species.

cone Retinal photoreceptor that discriminates colors.

connexin One of a family of proteins that can assemble to form a connexon.

connexon A membrane channel bridging the space between two adjacent cells, connecting the cytoplasm of one to that of the other. See *gap junction.*

contralateral Relating to the opposite side of the body.

co-transmission The release and action of two different chemical neurotransmitters from the same neuron or nerve terminal.

convergence The coming together of a group of presyn-aptic neurons to make synapses on one postsynaptic neuron.

cortical barrel Barrel-shaped aggregate of cortical neurons related to a specific peripheral sensory structure (e.g., a facial whisker, see *whisker barrels*).

cortical column An aggregate of cortical neurons, extending inward from the pial surface, that shares common properties (e.g., sensory modality, receptive field position, eye dominance, orientation, movement sensitivity).

coulomb Unit of electrical charge.

coupling potential A potential change in one cell produced by current spread from another through an electrical synapse.

critical period A time during nervous system maturation when activity permanently shapes connectivity.

crustacean Any member of the class Crustacea, which includes arthropods with hard shells, such as lobsters, crabs, barnacles, and shrimp.

curare A plant extract that blocks nicotinic ACh receptors. See also *tubocurarine.*

D

dalton (Da) A unit of molecular mass, numerically equal to 1/12 the mass of a carbon atom; often expressed in thousands, or kilodaltons (kDa).

deactivation The response of a channel to a stimulus that reduces the probability that the channel will open.

delayed rectifier A type of potassium channel activated by membrane depolarization after a brief delay.

dendrite Process of a neuron specialized to act as a postsyn aptic receptor region of a neuron.

depolarization Reduction in magnitude of the resting membrane potential toward zero.

desensitization Reduction of the response of a receptor to a ligand after prolonged or repeated exposure.

diacylglycerol (DAG) An intracellular second messenger produced by phospholipase C–catalyzed hydrolysis of phospholipids. DAG activates protein kinase C.

dishabituation A form of behavior, in which the recovery of sensitivity to a weak stimulus is brought about by exposure to a stronger stimulus.

divalent Having an electronic charge of +2 or –2.

divergence The branching of a nerve axon to form synapses with several other neurons.

domain 1. A particular region of a polypeptide; for example, one of the four repeating regions of the voltage-activated sodium channel. 2. The region occupied by a substance (such as Ca^{2+}ions) after entering a cell through membrane channels. Nanodomains extend for a few tens of nm, microdomains several hundred nm.

dopamine Transmitter liberated by neurons in some autonomic ganglia and in the central nervous system.

dorsal root ganglion An encapsulated cluster of sensory cell bodies situated along the vertebral column. The neurons in the ganglion transmit signals from the body into the spinal cord.

driving force The difference between the membrane potential and the equilibrium potential for movement of an ion species through a membrane channel.

E

ectoderm The outermost layer of cells in an embryo, arising from cell movements of early development (gastrulation), which typically gives rise to epidermal and neural tissues of the adult.

efferent Axon conducting impulses outward from the central nervous system.

γ-efferent fiber Small myelinated motor axon supplying intrafusal muscle fiber. See *fusimotor.*

EGTA Ethylene glycol bis-(β-aminoethyl ether) N,N,N′N′-tetraacetic acid. A high-affinity calcium-binding compound.

electrical coupling Spread of a potential change from one cell to another, usually by current flow through gap junctions.

electrochemical gradient The transmembrane difference in potential energy of an ion arising from the combined electrical and diffusional forces acting on it.

electroencephalogram (EEG) A recording of the electrical activity of the brain by external electrodes on the scalp.

electrogenesis The addition of an increment to the membrane potential by the active transport of ionic charges across the membrane.

electrogenic pump Active transport of ions across a cell membrane in which a net transfer of electrical charges contributes directly to the membrane potential.

electromotility The voltage-driven motion of outer hair cells of the cochlea.

electromyogram (EMG) A recording of muscular activity by external electrodes.

electroretinogram (ERG) Potential changes in response to light recorded by external electrodes on the eye.

electrotonic potentials Localized, graded, subthreshold potentials produced by artificially applied currents, and characterized by the passive electrical properties of cells.

equilibrium The condition in which there is no electrochemical gradient tending to move an ion across a cell membrane.

endocannabinoids Messengers in the nervous system that act as intercellular messengers by stimulating cannabinoid receptors. See *cannabinoids*.

endocytosis Process whereby membrane, together with some extracellular fluid, is internalized by the invagination and pinching off of a vesicle from the plasma membrane.

endoderm The innermost layer of cells in an embryo, arising in early development (gastrulation), which typically gives rise to gut and other internal tissues of the adult.

endorphins Endogenous peptides with chemical structure resembling that of opiates, that function as neurotransmitters.

endoplasmic reticulum A subcellular compartment which contains calcium that can be released by inositol-1,4,5-trisphosphate.

endothelial cells Layer of cells lining blood vessels.

end plate Postsynaptic area of vertebrate skeletal muscle fiber.

end plate potential (EPP) Postsynaptic potential change in a skeletal muscle fiber produced by ACh liberated from presynaptic terminals.

enkephalins Small neuropeptides comprising five amino acids that act on opioid receptors. See *opioids*.

ependyma Layer of cells lining cerebral ventricles and central canal of spinal cord.

epinephrine (adrenaline) Hormone secreted by the adrenal medulla; certain of its actions resemble those of sympathetic nerves.

equilibrium potential Membrane potential at which there is no net passive movement of a permeant ion species into or out of a cell.

eserine An anticholinesterase; also known as physostigmine.

excitation Process tending to produce action potentials.

excitatory postsynaptic potential (EPSP) Depolarization of the postsynaptic membrane of a neuron produced by an excitatory transmitter released from presynaptic terminals.

exocytosis Process whereby synaptic vesicles fuse with presynaptic terminal membrane and empty transmitter molecules into the synaptic cleft.

explant A piece of tissue placed in culture.

extracellular matrix A scaffolding of large glycoproteins and proteoglycans that surrounds and separates cells or tissues.

extrafusal Muscle fibers making up the mass of a skeletal muscle (i.e., not within the sensory muscle spindles).

facilitation An increase in evoked release of transmitter from nerve terminals, due to previous synaptic activity. Facilitation can last for several hundred ms.

F

farad (F) Unit of capacitance; more commonly used is microfarad ($\mu F = 10^{-6} F$).

faraday (F) The number of coulombs carried by 1 mole of a univalent ion (96,500).

fasciculation 1. The aggregation of neuronal processes to form a bundle. 2. The spontaneous contraction of muscle fibers in groups.

fate mapping The process of determining what structures or tissues specific cells in an embryo give rise to in later development.

fetus A relatively late-stage mammalian embryo.

field axis orientation For simple cortical neurons in the visual system, the angle of the long axis of the receptive field (e.g., horizontal, vertical, oblique).

formants Frequency components of basic speech elements.

filopodia Thin projections from the plasma membrane of a cell or growth cone, containing actin filaments, accessory proteins and membrane proteins that sense and respond to the extracellular environment.

fovea Central depression in the retinal composed of slender cones; area of greatest visual resolution.

free nerve ending A sensory fiber that terminates in the skin and is not encapsulated by any accessory structure.

functional magnetic resonance imaging (fMRI) Method developed in the 1990s for measuring the changes in blood oxygenation related to neuronal activity in the brain or spinal cord of humans or other animals.

fusimotor Motoneurons supplying muscle fibers in a muscle spindle.

G

G protein Receptor-coupled protein that binds guanine nucleotides and activates intracellular messenger systems.

ganglion A discrete collection of nerve cells.

gap junction The region of contact between two cells in which the space between adjacent membranes is reduced to about 2 nm and bridged by connecting channels. See *connexon*.

gate The mechanism whereby a channel is opened and closed.

gating current Current produced by movement of charge within a cell membrane, associated with the opening or closing of a channel.

gene expression The transcription of DNA into mRNA and the translation of mRNA into protein.

glia See *neuroglia*.

glioblast A dividing cell, the progeny of which will develop into glial cells.

glutamate An amino acid that serves as a major excitatory neurotransmitter.

glutamatergic Neurons that release the neurotransmitter glutamate.

glycine A transmitter liberated at many inhibitory synapses in the spinal cord and brain stem.

glycoprotein A protein containing carbohydrate residues.

golgi tendon organ Sensory element in muscle tendons activated by muscle stretch or contraction.

gray matter Part of the central nervous system composed predominantly of the cell bodies of neurons and fine terminals, as opposed to major axon tracts (white matter).

group I afferents Sensory fibers from muscle with conduction velocities in the range of 80–120 m/sec.

group Ia Arise from muscle spindles.

group Ib Arise from Golgi tendon organs.

group II afferents Sensory fibers from muscle spindles with conduction velocities in the range of 30–80 m/sec.

growth cone The expanded tip of a growing axon.

gustation The sense of taste.

H

habituation A form of behavior characterized by a decreasing response to repeated applications of a given stimulus.

hair cells Sensory cells in which bending of stereocilia ("hairs") causes a change in membrane potential; responsible for auditory transduction, transduction of vestibular stimuli, and vibratory transduction in lateral line organs of fish.

hair follicle Richly innervated skin structure in which a hair is rooted.

heterosynaptic Relating to the influence of activity at one synapse on the behavior of another synapse.

histamine A transmitter liberated by a small number of neurons in the vertebrate CNS.

homology The degree of identity of base pairs in two nucleotide sequences or of amino acids in two polypeptide sequences.

homosynaptic Relating to the influence of activity at a synapse on its own subsequent behavior.

horseradish peroxidase Enzyme used as a histochemical marker for tracing processes of neurons or spaces between cells.

hydropathy index A measure of insolubility of an amino acid, or amino acid sequence, in water, and hence its preference for a lipid environment.

hydrophilic Having a relatively high water solubility; polar.

hydrophobic Having a relatively low water solubility; nonpolar.

5-hydroxytryptamine (5-HT) See *serotonin*.

hyperpolarization Increased negativity of the resting membrane potential, tending to reduce excitability

hypocretins See *orexins*.

I

immunohistochemistry The use of antibodies to locate proteins in cells and tissues. The antibodies are tagged with a fluorophore, an electron-dense particle or an enzyme producing an electron-dense deposit so that they can be seen under the microscope. (Also called immunocytochemistry.)

impulse See *action potential*.

inactivation Removal of the ability of a voltage-activated channel to respond to a change in membrane potential.

inhibition Effect of one neuron upon another tending to prevent it from initiating impulses.

postsynaptic inhibition Mediated through a permeability change in the postsynaptic cell, holding the membrane potential away from threshold.

presynaptic inhibition Mediated by an inhibitory fiber ending on an excitatory terminal, reducing the release of transmitter.

initial segment The region of an axon close to the cell body; often the site of impulse initiation; see *axon hillock*.

input resistance (rinput) The resistance measured by injecting current into a cell or fiber; in a cylindrical fiber $r_{input} = 0.5\sqrt{(r_m r_i)}$.

in situ hybridization Technique for localizing a specific mRNA in cells or tissues by incubating them with a labeled oligonucleotide probe whose sequence is complementary to that of the target mRNA.

inositol-1,4,5-triphosphate (IP_3) An intracellular second messenger liberated by phospholipase C–catalyzed hydrolysis of phosphatidyl inositol. IP_3 triggers the release of calcium from intracellular stores.

integration The process whereby a neuron sums the various excitatory and inhibitory influences converging upon it and synthesizes a new output signal.

intercellular clefts Narrow fluid-filled spaces between membranes of adjacent cells; usually about 20 nm wide.

interneuron A neuron that is neither purely sensory nor purely motor but connects other neurons.

internode The myelinated portion of a nerve axon lying between two nodes of Ranvier.

intrafusal fiber Muscle fiber within a muscle spindle; its contraction initiates or modulates sensory discharge.

ion channel See *channel*.

ionophoresis Ejection of ions by passing current through a micropipette; used for applying charged molecules with a high degree of temporal and spatial resolution. Also spelled iontophoresis.

ionotropic receptor An ion channel that is opened by a neurotransmitter or other chemical ligand, for example the nicotinic acetylcholine receptor.

ipsilateral On the same side of the body.

inhibitory postsynaptic potential (IPSP) The potential change (usually hyperpolarizing) in a neuron produced by an inhibitory transmitter released from presynaptic terminals.

intracellular recording A recording of the membrane potential in an intact cell using a fine microelectrode inserted through the cell membrane.

invertebrate Any animal without a backbone.

in vitro Referring to a biological process studied outside an intact living organism.

in vivo Referring to a biological process studied within an intact living organism.

inward rectifier A type of potassium channel that allows potassium ions to move inward, but not outward, across the cell membrane.

isotypes Gene products of the same family, but with variations in amino acid sequence (e.g., voltage-activated sodium channels from brain and from muscle).

J

jellyfish green fluorescent protein (GFP) A protein from jellyfish that, when expressed in neurons, causes them to be fluorescent.

K

kilodalton (kDa) One thousand daltons.

kiss-and-run A mechanism of transmitter release in which synaptic vesicles combine briefly with the nerve terminal membrane to form a pore and then return directly to the cytoplasm.

L

laminin A prominent extracellular matrix glycoprotein; promotes neurite outgrowth in vitro.

lateral geniculate nucleus Small knee-shaped nucleus; part of posteroinferior aspect of thalamus acting as a relay in the visual pathway.

leak current A small steady current through resting ion channels, seen when the cell membrane is displaced from its resting potential.

length constant, $\lambda = \sqrt{(r_m r_i)}$ Distance (usually in millimeters) over which a localized graded potential decreases to $1/e$ of its original size in an axon or a muscle fiber.

ligand-activated channel A channel that is activated by binding an ion or molecule to an external or internal receptor region of the channel.

long-term depression (LTD) Decrease in size of a synaptic potential lasting hours or more, produced by previous synaptic activation.

long-term potentiation (LTP) Increase in size of a synaptic potential lasting hours or more, produced by previous synaptic activation.

M

magnocellular pathways In the visual system, large retinal ganglion cells and lateral geniculate cells that project to discrete cortical areas, particularly sensitive to movement and small changes in contrast.

Mauthner cell Large nerve cell in the mesencephalon of fishes and amphibians, up to 1 mm in length.

M-channel A voltage-activated potassium channel that is inactivated by acetylcholine through muscarinic receptors.

mechanotransduction The conversion of mechanical force into a voltage change in a sensory receptor neuron.

Meissner's corpuscle Rapidly adapting mechanoreceptor in superficial skin.

melanopsin Photopigment located in photosensitive ganglion cells of the retina that are involved in the regulation of w. f. circadian rhythms and the pupillary light reflex.

membrane capacitance (Cm) Property of the cell membrane enabling electrical charge to be stored and separated, usually measured in microfarads per square cm ($\mu F/cm^2$).

membrane channel An aqueous pathway through a membrane allowing the passage of ions or molecules, formed by a single large protein or an assembly of polypeptide subunits.

membrane resistance (Rm) Property of the cell membrane, measured in ohm cm^2, reflecting the difficulty ions encounter in moving across it. Inverse of conductance.

membrane time constant (τ) A measure of the rate of buildup or decay of a localized graded potential; equal numerically to the product of the resistance and the capacitance of the membrane.

Merkel's disc Slowly adapting mechanoreceptor in superficial skin.

mesoderm The intermediate layer of cells in an embryo, arising in early development (gastrulation), which gives rise to muscle and other internal tissues of the adult.

metabotropic receptor A neurotransmitter receptor that interacts with another membrane protein such as a G protein to produce its effect on the neuron.

microfilament Dynamic, actin-based polymer contributing to one of two main cytoskeletal systems in animal cells. See *microtubule*.

microneurography A method to measure electrical activity arising in the skin of human subjects by insertion of metal microelectrodes into nerve fascicles, usually in the arm.

microglia Wandering, macrophage-like cells in the central nervous system that accumulate at sites of injury and scavenge debris.

microtubule A component of the cytoskeleton prominent in axons, formed by polymerization of tubulin monomers.

miniature end-plate potential (MEPP) Small depolarization at a neuromuscular synapse caused by spontaneous release of a single quantum of transmitter from the presynaptic terminal.

modality Class of sensation (e.g., touch, vision, olfaction).

monoclonal antibody An antibody molecule raised from a clone of transformed lymphocytes.

monovalent Having an electronic charge of +1 or –1.

monosynaptic A direct pathway from one neuron to the next, involving only one synapse.

motoneuron (motor neuron) A neuron that innervates muscle fibers.

α-motoneuron Supplies extrafusal muscle fibers.

γ-motoneuron Supplies intrafusal muscle fibers.

motor unit A single motoneuron and the muscle fibers it innervates.

MRI Magnetic resonance imaging; provides high-resolution pictures of brain structures. Formerly known as nuclear magnetic resonance imaging.

mRNA (messenger RNA) Polymer of ribonucleic acids transcribed from DNA that serves as a template for protein synthesis.

multimeric Composed of more than one polypeptide subunit (e.g., the pentameric acetylcholine receptor).

homomultimeric Composed of identical subunits.

heteromultimeric Composed of nonidentical subunits.

muscarinic See *acetylcholine receptor*.

muscle spindle Fusiform (spindle-shaped) structure in skeletal muscles containing small muscle fibers and sensory receptors activated by stretch.

mutagenesis Alteration of a gene to produce a product different from the naturally occurring, or "wild-type," variety.

myelin Fused membranes of Schwann cells or glial cells forming a high-resistance sheath surrounding an axon.

myoblast A dividing cell, the progeny of which develop into muscle cells.

myotube A developing muscle fiber formed by the fusion of myoblasts.

N

neostigmine An anticholinesterase; also known as prostigmine.

neurite Any neuronal process (axon or dendrite), typically used to refer to the processes of neurons in cell culture.

neuroblast A dividing cell, the progeny of which develop into neurons.

neuroglia Non-neuronal satellite cells associated with neurons. In the mammalian CNS, the main groupings are astrocytes and oligodendrocytes; in peripheral nerves the satellite cells are called Schwann cells.

neurokinins Neuropeptides that act on neurokinin (NK) receptors. They include substance P.

neuromodulator A substance liberated from a neuron that modifies the efficacy of synaptic transmission.

neuropeptides Peptides present in the nervous system that act as neurotransmitters or neurohormones.

neuropil A network of axons, dendrites, and synapses.

neurotransmitter See *transmitter.*

nicotinic See *acetylcholine receptor.*

nocebo A sham or simulated medical intervention, opposite of the **placebo**, in which the patient, due to pessimistic expectations, experiences a substantial increase in pain sensation without administration of pharmacologically active substances.

nociception The signaling of noxious, painful stimuli by sensory neurons. See *placebo.*

nociceptive Responding to noxious (tissue-damaging or painful) stimulation.

node of Ranvier Localized area devoid of myelin occurring at intervals along a myelinated axon.

noise Fluctuations in membrane potential or current due to random opening and closing of ion channels.

norepinephrine (noradrenaline) Transmitter liberated by most sympathetic nerve terminals.

O

ocular dominance The greater effectiveness of one eye over the other for driving simple or complex cells in the visual cortex.

Ohm's law Relates current (I) to voltage (V) and resistance (R); $I = V/R$.

olfaction The sense of smell.

oligodendroglia Class of vertebrate CNS glial cells that forms myelin.

ommatidium An individual photo detector unit in an invertebrate compound eye, typically containing lens, photoreceptor and pigment cells.

open channel block Block of ion flow though an open channel by a physical obstruction, such as a large molecule in the channel mouth.

opiate Term denoting products derived from the juice of opium poppy seed capsules.

opioid Any directly acting compound whose actions are similar to those of opiates and are specifically antagonized by naloxone.

opioid Any directly acting compound whose actions are similar to those of opiates and act on opoid receptors (a family of G protein–coupled receptors that are stimulated by morphine and related compounds).

optic chiasm The point of crossing, or decussation, of the optic nerves. In cats and primates, fibers arising from the medial part of the retina cross to supply the opposite lateral geniculate nucleus.

optogenetics A method for selectively activating or inhibiting neurons with light by expressing a light-sensitive ion channel.

orexins Hypothalmic neuropeptides that regulate sleep and feeding. Also called hypocretins.

organ of Corti The sensory epithelium of the cochlea.

otolithic membrane An acellular gelatinous matrix enclosing calcium carbonate crystals to mass-load hair cells in the saccule and utricle.

ouabain G-strophanthidin, a glycoside that specifically blocks sodium–potassium ATPase.

oval window The membranous partition between the *scala vestibuli* and the middle ear that receives auditory vibrations from the ear drum through bony interconnections.

overshoot Reversal of the membrane potential during the peak of the action potential.

oxytocin Hypothalamic hormone that acts as a central neurotransmitter to regulate social behavior. When released into the bloodtsream during labor it induces uterine contraction.

P

Pacinian corpuscle A rapidly adapting mechanoreceptor sensitive to vibration; found in deep skin and other tissues.

parvocellular pathways In the visual system, smaller retinal ganglion cells and lateral geniculate cells that project to discrete areas of the visual cortex; concerned with color detection and fine discrimination.

parasympathetic nervous system A division of the autonomic nervous system arising from the cranial and sacral segments of the central nervous system.

patch clamp A technique whereby a small patch of membrane is sealed to the tip of a micropipette, enabling currents through single membrane channels to be recorded.

peptidase The name frequently given to the proteases that generate neuropeptides from their precursor proteins.

permeability Property of a membrane or channel allowing substances to pass into or out of the cell.

PET scan See *positron emission tomography.*

phagocytosis Endocytosis and degradation of foreign or degenerating material.

phenotype The physical characteristics of an animal.

phoneme Basic sound element of speech.

phosphatidylinositol-4,5-bisphosphate (also known as PIP_2) A phosphoinositide (q.v) that regulates the activity of many membrane ion channels and transporters. When hydrolysed by phospholipase C it forms diacylglycerol and inositol-1,4,5-trisphosphate.

phosphoinositides Membrane phospholipids that contain inositol phosphates.

phospholipase C; phospholipase A2 Enzymes that hydrolyze phospholipids.

phosphorylation The covalent addition of one or more phosphate ions to a molecule, for example, a channel protein.

placebo A sham or simulated medical intervention in which the patient, due to optimistic expectations, experiences a substantial relief of symptoms without administration of pharmacologically active drugs or effective surgery.

polar A molecule with separate positively charged and negatively charged regions. See *hydrophilic.*

polysynaptic A pathway involving a series of synaptic connections.

positron emission tomography A technique for mapping active areas of the brain. Glucose is labeled with isotopes that emit positrons; sites of glucose uptake are then located by detecting positron emissions.

postmitotic A cell that is no longer capable of undergoing cell division.

posttetanic potentiation (PTP) An increase in evoked transmitter release from nerve terminals following a train of repetitive stimuli. PTP can last for several minutes.

prosopagnosia A type of agnosia in which the ability to recognize faces is impaired, while the ability to recognize other objects remains normal. See *agnosia*.

protease An enzyme that hydrolyzes protein molecules by splitting the peptide (CO-NH) bond between two amino acids. Also spelled *proteinase*.

protein kinase An enzyme that phosphorylates proteins.

protein phosphatase An enzyme that cleaves phosphate residues from proteins.

pump Active transport mechanism.

pyramidal cell Any neuron with a long apical dendrite and shorter basal dendrites, a morphology characteristic of many cortical neurons.

Q

quantal release Secretion of multimolecular packets (quanta) of transmitter by the presynaptic nerve terminal.

quantal size The number of molecules of neurotransmitter in a quantum.

quantum The quantity of neurotransmitter substance stored in a presynaptic vesicle that, when released, produces one miniature synaptic potential.

quantum content The number of quanta in a synaptic response.

R

receptive field The area of the periphery whose stimulation influences the firing of a neuron. For cells in the visual pathway, the receptive field refers to an area on the retina whose illumination influences the activity of a neuron.

receptor 1. A nerve terminal or accessory cell, associated with sensory transduction. 2. A molecule in the cell, usually a trans membrane protein that initiates a response when combined with a specific chemical.

directly coupled receptors Molecules that form ion channels that span the membrane.

indirectly coupled (metabotropic) receptors Molecules that activate G proteins, which, in turn, modify the activity of channels or pumps either directly or through a second messenger pathway.

receptor potential Graded, localized potential change in a sensory receptor initiated by the appropriate stimulus; the electrical sign of the transduction process.

reciprocal innervation Interconnections of neurons arranged so that pathways exciting one group of muscles inhibit the antagonistic motoneurons.

rectification The property of a membrane, or membrane channel, that allows it to conduct ionic current more readily in one direction than in the other.

reflex Involuntary movement or other response elicited by activation of sensory receptors, and involving conduction through one or more synapses in the CNS.

refractory period

absolute refractory period The time following an action potential during which a stimulus cannot elicit a second action potential.

relative refractory period The time following the absolute refractory period when the threshold for initiation of a second action potential is increased.

resting potential The steady electrical potential across the membrane in the quiescent state.

retinotectal Referring to the projection of retinal ganglion cells to neurons in the optic tectum.

retrograde In the direction from the axon terminal toward the cell body.

retrovirus Virus using RNA instead of DNA to encode its heritable information. In a host cell, the retrovirus must first reverse transcribe its genome into DNA.

reversal potential The value of the membrane potential at which a chemical transmitter produces no change in potential.

ribbon synapse The active zone of hair cells and retinal cells specialized for continuous transmitter release.

rhodopsin The photopigment of retinal rods.

Ringer's solution A saline solution containing sodium chloride, potassium chloride, and calcium chloride; named after Sidney Ringer.

RNA editing Replacement of one codon by another in messenger RNA after transcription.

rod Retinal photoreceptors sensitive to dim light.

round window The membranous partition between the *scala tympani* and the middle ear.

Ruffini's capsule Slowly adapting mechanoreceptor found deep in skin.

S

saccule The part of the vestibular apparatus that responds to vertical acceleration of the head.

saltatory conduction Conduction along a myelinated axon whereby the leading edge of the propagating action potential leaps from node to node.

scala tympani, scala media, scala vestibuli Fluid-filled compartments of the cochlea.

Schwann cell Satellite cell in the peripheral nervous system, responsible for making the myelin sheath.

second messenger A molecule forming part of a second messenger system.

second messenger system A series of molecular reactions inside a cell initiated by occupation of extracellular receptor sites and leading to a functional response, such as opening or closing of membrane channels.

segmentation Organization of the body plan into repeating units (segments) that are more or less similar along the anterior-posterior axis of the animal.

semicircular canal Fluid-filled loop in the vestibular apparatus associated with detection of head rotation.

sensitization Increased responses to a standard weak stimulus following an initial exposure to a much stronger stimulus.

serotonin Also known as 5-hydroxytryptamine, or 5-HT; a neurotransmitter.

siemens (S) Unit of conductance; reciprocal of ohm.

size principle The orderly recruitment of motor units of increasing size as the strength of a muscle contraction increases.

site-directed mutagenesis The alteration of a gene to produce a product with one or more amino acid substitutions at specified locations.

sodium–potassium ATPase A molecule that transports sodium and potassium across a cell membrane against their electrochemical gradients, utilizing energy derived from the hydrolysis of ATP.

sodium–potassium exchange pump Sodium–potassium ATPase.

soma Cell body.

somatotopic Organized in an orderly manner in relation to an outline of the body, or a part of the body.

stem cell A cell that can divide to give rise to the parent cell and also a cell that differentiates further into one or more specialized types, for example neurons or blood cells.

stereocilia Specialized microvilli of graded length projecting from the apical surface of a hair cell.

specific resistance 1. The resistance of one square centimeter of cell membrane 2. The resistance of a volume of cytoplasm one square centimeter in cross section and one centimeter in length.

steady state The state of a cell in which passive gain or loss of solutes is exactly balanced by active transport processes, so that the composition of the cytoplasm remains constant.

striate cortex Also known as area 17 or visual I; primary visual region of occipital lobe marked by striation of Gennari, visible with the naked eye.

subarachnoid space The space filled by cerebrospinal fluid between the arachnoid and pia, two layers of connective tissue (meninges) surrounding the brain.

substance P An 11-amino acid member of the neurokinin (neuropeptide) family

subunit The basic structural "building block" of a multimeric protein, such as a membrane channel; usually a single polypeptide.

summation Addition of synaptic potentials.

temporal summation Addition in sequence, with each potential adding to the one preceding.

spatial summation Addition of potentials arising in different parts of a cell; for example, of potentials spreading to the axon hillock from different branches of a dendritic tree.

superfamily A group of gene families, each of whose ancestral genes arose by duplication and divergence from a yet more ancient ancestral gene.

supersensitivity An increase in responsiveness of target cells to chemical transmitters seen, for example, after denervation.

sympathetic nervous system A division of the autonomic nervous system arising from the thoracic and lumbar segments of the CNS.

synapse Site at which neurons make functional contact; a term coined by Sherrington.

synaptic cleft The space between the membranes of the pre- and postsynaptic cells at a chemical synapse across which transmitter diffuses.

synaptic delay The time between a presynaptic nerve impulse and a postsynaptic response.

synaptic vesicles Small membrane-enclosed sacs contained in presynaptic nerve terminals. Those with dense cores contain catecholamines and serotonin; clear vesicles store other transmitters, such as acetylcholine.

T

tectorial membrane An acellular gelatinous sheet that lies over the organ of Corti and contacts the hair cells stereocilia.

tetraethylammonium (TEA) Quaternary ammonium compound that selectively blocks certain voltage-gated potassium channels in neurons and muscle fibers.

tetanus A train of action potentials; also the resulting sustained muscular contraction.

tetrodotoxin (TTX) Toxin from puffer fish that selectively blocks the voltage-activated sodium channel in neurons and muscle fibers.

threshold 1. Critical value of membrane potential or depolarization at which an impulse is initiated. 2. Minimal stimulus required for a sensation.

tight junction Site at which fusion occurs between the outer leaflets of membranes of adjacent cells, resulting in a five-layered junction. It is called a *macula occludens* if the area is a spot and a *zonula occludens* if the junction is a circumferential ring. Such complete junctions prevent the movement of substances through the extracellular space between the cells.

tip link Fine extracellular fiber connecting the tips of adjacent stereocilia on hair cells.

tonotopy The frequency specific organization of the cochlea and higher centers of the auditory system.

topographic map A sensory representation in the brain in which nearby locations on the sensory apparatus project to nearby locations in the map.

transcription Synthesis of mRNA using DNA as a template.

transcription factor A protein that binds to specific regulatory sequences in the genomic DNA of a cell, thereby activating or repressing the transcription of specific target genes.

transducer Device for converting one form of energy into another (e.g., a microphone, photoelectric cell, loudspeaker, or light bulb).

transducin The G protein mediating phototransduction.

transgenesis The process of altering the properties of a cell or organism by the transient or permanent introduction of a new gene.

translation Synthesis of protein using mRNA as a template.

transmitter Chemical substance liberated by a presynaptic nerve terminal causing an effect on the membrane of the postsynaptic cell, usually an increase in permeability to one or more ions.

transporter A protein mediating transport of ions or molecules across a cell membrane.

tubocurarine An alkaloid isolated from curare.

tuning The property of a receptor (e.g., a cochlear hair cell) that restricts its response to a specified frequency range.

U

utricle Part of the vestibular apparatus that responds to horizontal acceleration of the head.

V

varicosity Swelling along an axon from which transmitter is liberated.

vasopressin Hypothalamic neuropeptide that regulates social behavior. When released into the bloodstream promotes water reabsorption by the kidney. (Also called antidiuretic hormone, ADH).

ventricles Cavities within the brain containing cerebrospinal fluid and lined by ependymal cells.

ventricular zone Region adjacent to the lumen of the neural tube (future ventricle) in the developing vertebrate neuroepithelium where cell proliferation occurs.

vertebrate An animal with a backbone.

vestibular ocular reflex A three neuron reflex from vestibular receptors to motoneurons controlling extraocular muscles that serves to maintain visual fixation during head movement.

voltage clamp Technique for displacing membrane potential abruptly to a desired value and keeping the potential constant while measuring currents across the cell membrane; devised by Cole and Marmont.

voltage-sensitive channel A channel that is activated or inactivated by changes in membrane potential.

W

whisker barrels Columnar organization of somatosensory cortex related to facial whiskers.

white matter The part of the CNS that appears white; consists of myelinated fiber tracts. See *gray matter.*

whole-cell patch recording Recording of membrane currents of an intact cell with a patch-clamp electrode, through an opening in the cell membrane.

working memory Information that is temporarily maintained by neuronal networks so that it can be monitored or manipulated to guide an ongoing task.

BIBLIOGRAPHY

Abbott, N. J., Patabendige, A. A., Dolman, D. E., Yusof, S. R., and Begley, D. J. 2010. Structure and function of the blood-brain barrier. *Neurobiol. Dis.* 37: 13–25. [10]

Abbracchino, M. P., Burnstock, G., Verkhratsky, A., and Zimmerman, H. 2009. Purinergic signalling in the nervous system: an overview. *Trends Neurosci.* 32: 19–29. [5, 14]

Abbracchio, M. P., Burnstock, G., Boeynaems, J.-M., Barnard, E. A., Boyer, J. L., Kennedy, C., Knight, G. E., Fumagalli, M., Gachet, C., Jacobson, K. A., and Weisman, G. A. 2006. International Union of Pharmacology. Update and subclassification of the P2Y G protein-coupled nucleotide receptors: from molecular mechanisms and pathophysiology to therapy. *Pharmacol. Rev.* 58: 281–341. [14]

Abe, H., Honma, S., Ohtsu, H., and Honma, K. 2004. Circadian rhythms in behavior and clock gene expressions in the brain of mice lacking histidine decarboxylase. *Brain Res. Mol. Brain Res.* 124: 178–187. [14]

Abraham, W. C., and Williams, J. M. 2003. Properties and mechanisms of LTP maintenance. *Neuroscientist* 9: 463–474. [16]

Accili, E. A., Proenza, C., Baruscotti, M., and DiFrancesco. D. 2002. From funny current to HCN channels: 20 years of excitation. *News Physiol. Sci.* 17: 32–37. [12]

Ache, B. W., and Zhainazarov, A. 1995. Dual second-messenger pathways in olfactory transduction. *Curr. Opin. Neurobiol.* 5: 461–466. [19]

Acklin, S. E. 1988. Electrical properties and anion permeability of doubly rectifying junctions in the leech central nervous system. *J. Exp. Biol.* 137: 1–11. [11]

Adams, D. J., Dwyer, T. M., and Hille, B. 1980. The permeability of endplate channels to monovalent and divalent metal cations. *J. Gen. Physiol.* 75: 493–510. [11]

Adams, D. L., and Zeki, S. 2001. Functional organization of macaque V3 for stereoscopic depth. *J. Neurophysiol.* 86: 2195–2203. [3]

Adams, D. L., Sincich, L. C., and Horton, J. C. 2007. Complete pattern of ocular dominance columns in human primary visual cortex. *J. Neurosci.* 27: 10391–10403. [3]

Adams, P. R., and Brown, D. A. 1975. Actions of gamma-aminobutyric acid on sympathetic ganglion cells. *J. Physiol.* 250: 85–120. [11]

Adams, P. R., and Brown, D. A. 1980. Luteinizing hormone-releasing factor and muscarinic agonists act on the same voltage-sensitive K^+-currents in bullfrog sympathetic neurones. *Brit. J. Pharmacol.* 68: 353–355. [12, 17]

Adams, P. R., Brown, D. A., and Jones, S. W. 1983. Substance P inhibits the M-current in bullfrog sympathetic neurones. *Brit. J. Pharmacol.* 79: 330–333. [19]

Adams, S. R., Kao, J. P. Y., Grynkiewicz, G., Minta, A., and Tsien, R. Y. 1988. Biologically useful chelators that release Ca^{2+} upon illumination. *J. Am. Chem. Soc.* 110: 3212–3220. [13]

Adar, E., Nottebohm, F., and Barnea, A. 2008. The relationship between nature of social change, age, and position of new neurons and their survival in adult zebra finch brain. *J. Neurosci.* 28: 5394–5400. [25]

Adler, E. M., Augustine, G. J., Duffy, S. N., and Charlton, M. P. 1991. Alien intracellular calcium chelators attenuate neurotransmitter release at the squid giant synapse. *J. Neurosci.* 11: 1496–1507. [13]

Adler, E., Hoon, M. A., Mueller, K. L., Chandrashekar, J., Ryba, N. J., and Zuker, C. S. 2000. A novel family of mammalian taste receptors. *Cell* 100: 693–702. [19]

Adrian, E. D. 1946. *The Physical Background of Perception.* Clarendon, Oxford, England. [1, 24, 28]

Adrian, E. D. 1953. Sensory messages and sensation; the response of the olfactory organ to different smells. *Acta Physiol. Scand.* 29: 5–14. [19]

Adrian, E. D. 1959. *The Mechanism of Nervous Action.* University of Pennsylvania Press, Philadelphia. [24]

Adrian, E. D., and Zotterman, Y. 1926. The impulses produced by sensory nerve-endings: Part II. The response of a Single End-Organ. *J. Physiol.* 61: 151–171. [19]

Afraz, S., Kiani, R., and Esteky, H. 2006. Microstimulation of inferotemporal cortex influences face categorization. *Nature* 442: 692–695. [23]

Afshari, F. T., Kappagantula, S., and Fawcett, J. W. 2009. Extrinsic and intrinsic factors controlling axonal regeneration after spinal cord injury. *Expert Rev. Mol. Med.* 11: e37. [27]

Aghajanian, G. K., and VanderMaelen, C. P. 1982. Alpha 2-adrenoceptor-mediated hyperpolarization of locus coeruleus neurons: intracellular studies in vivo. *Science* 215: 1394–1396. [14]

Aghajanian, G. K., Cedarbaum, J. M., and Wang, R. Y. 1977. Evidence for norepinephrine-mediated collateral inhibition of locus coeruleus neurons. *Brain Res.* 136: 570–577. [14]

Aguayo, A. J., Clarke, D. B., Jelsma, T. N., Kittlerova, P., Friedman, H. C., and Bray, G. M. 1996. Effects of neurotrophins on the survival and regrowth of injured retinal neurons. *Ciba Found. Symp.* 196: 135–144. [27]

Aguayo, A. J., Dickson, R., Trecarten, J., Attiwell, M., Bray, G. M., and Richardson, P. 1978. Ensheathment and myelination of regenerating PNS fibres by transplanted optic nerve glia. *Neurosci. Lett.* 9: 97–104. [27]

Agulhon, C., Fiacco, T. A., and McCarthy, K. D. 2010. Hippocampal short- and long-term plasticity are not modulated by astrocyte Ca^{2+} signaling. *Science.* 327: 1250–1254. [10]

Ahern, P., Klyachko, V. A., and Jackson, M. B. 2002. cGMP and S-nitrosylation: two routes for modulation of neuronal excitability by NO. *Trends Neurosci.* 25: 510–517. [12]

Ahlquist, R. P. 1948. A study of the adrenotropic receptors. *Am. J. Physiol.* 153: 586–600. [17]

Aiba, A., Chen, C., Herrup, K., Rosenmund, C., Stevens, C. F., and Tonegawa, S. 1994. Reduced hippocampal long-term potentiation and context-specific deficit in associative learning in mGluR1 mutant mice. *Cell* 79: 365–375. [14]

Aimone, J. B., Deng, W., and Gage, F. H. 2010. Adult neurogenesis: integrating theories and separating functions. *Trends Cogn. Sci.* 14: 325–337. [27]

Airaksinen, M. S., and Saarma, M. 2002. The GDNF family: signalling, biological functions and therapeutic value. *Nat. Rev. Neurosci.* 3: 383–394. [25]

Akaaboune, M., Grady, R. M., Turney, S., Sanes, J. R., and Lichtman, J. W. 2002. Neurotransmitter receptor dynamics studied in vivo by reversible photo-unbinding of fluorescent ligands. *Neuron* 34: 865–876. [11]

Albillos, A., Dernick, G., Horstmann, H., Almers, W., Alvarez de Toledo, G., and Lindau, M. 1997. The exocytotic event in chromaffin cells revealed by patch amperometry. *Nature* 389: 509–512. [13]

Albrecht, J., Sonnewald, U., Waagepetersen, H. S., and Schousboe, A. 2007. Glutamine in the central nervous system: function and dysfunction. *Front Biosci.* 12: 332–343. [15]

Albright, T. D. 1984. Direction and orientation selectivity of neurons in visual area MT of the macaque. *J. Neurophysiol.* 52: 1106–1130. [23]

Albus, H., Vansteensel, M. J., Michel, S., Block, G. D., and Meijer, J. H. 2005. A GABAergic mechanism is necessary for coupling dissociable ventral and dorsal regional oscillators within the circadian clock. *Current Biology* 15: 886–893. [17]

Alkadhi, K. A., Alzoubi, K. H., and Aleisa, A. M. 2005. Plasticity of synaptic transmission in autonomic ganglia. *Prog. Neurobiol.* 75: 83–108. [16]

Allen, N. J., and Barres, B. A. 2005. Signaling between glia and neurons: Focus on synaptic plasticity. *Curr. Opin. Neurobiol.* 15: 542–548. [10]

Allen, R. D., Allen, N. S., and Travis, J. L. 1981. Video-enhanced differential interference contrast (AVEC-DIC) microscopy: A new method capable of analyzing microtubule-related movement in the reticulopodial network of *Allogromia laticollaris*. *Cell Motil.* 1: 291–302. [15]

Allen, T. G., and Brown, D. A. 1993. M_2 muscarinic receptor-mediated inhibition of the Ca^{2+} current in rat magnocellular cholinergic basal forebrain neurones. *J. Physiol.* 466: 173–189. [14]

Allen, T. G., and Brown, D. A. 1996. Detection and modulation of acetylcholine release from neurites of rat basal forebrain cells in culture. *J. Physiol.* 492: 453–466. [14]

Allen, T. G., Abogadie, F. C., and Brown, D. A. 2006. Simultaneous release of glutamate and acetylcholine from single magnocellular "cholinergic" basal forebrain neurons. *J. Neurosci.* 26: 1588–1595. [14, 15]

Allen, T. J., Ansems, G. E., and Proske, U. 2008. Evidence from proprioception of fusimotor coactivation during voluntary contractions in humans. *Exp. Physiol.* 93: 391–398. [24]

Alonso, J. M. 2002. Neural connections and receptive field properties in the primary visual cortex. *Neuroscientist* 8: 443–456. [2]

Alonso, J. M. 2009. My recollections of Hubel and Wiesel and a brief review of functional circuitry in the visual pathway. *J. Physiol.* 587: 2783–2790. [2]

Alonso, J. M., Yeh, C. I., Weng, C., and Stoelzel, C. 2006. Retinogeniculate connections: A balancing act between connection specificity and receptive field diversity. *Prog. Brain Res.* 154: 3–13. [2]

Altaf, M. A., and Sood, M. R. 2008. The nervous system and gastrointestinal function. *Dev. Disabil. Res. Rev.* 14: 87–95. [17]

Amanzio, M., and Benedetti, F. 1999. Neuropharmacological dissection of placebo analgesia: expectation-activated opioid systems versus conditioning-activated specific subsystems. *J. Neurosci.* 19: 484–494. [21]

Amar, A. P., and Weiss, M. H. 2003. Pituitary anatomy and physiology. *Neurosurg. Clin. N. Am.* 14: 11–23. [17]

Amitai, Y., Gibson, J. R., Beierlein, M., Patrick, S. L., Ho, A. M., Connors, B. W., and Golomb, D. 2002. The spatial dimensions of electrically coupled networks of interneurons in the neocortex. *J. Neurosci.* 22: 4142–4152. [11]

Anderson, C. R., and Stevens, C. F. 1973. Voltage clamp analysis of acetylcholine-produced end-plate current fluctuations at frog neuromuscular junction. *J. Physiol.* 235: 665–691. [4]

Anderson, H., Edwards, J. S., and Palka, J. 1980. Developmental neurobiology of invertebrates. *Annu. Rev. Neurosci.* 3: 97–139. [27]

Anderson, M. J., and Cohen, M. W. 1977. Nerve-induced and spontaneous redistribution of acetylcholine receptors on cultured muscle cells. *J. Physiol.* 268: 757–773. [27]

Andres, F. L., and van der Loos, H. 1985. From sensory periphery to cortex: the architecture of the barrelfield as modified by various early manipulations of the mouse whiskerpad. *Anat. Embryol. (Berl.)* 172: 11–20. [21]

Angelaki, D. E., and Cullen, K. E. 2008. Vestibular system: the many facets of a multimodal sense. *Annu. Rev. Neurosci.* 31: 125–150. [22]

Angelaki, D. E., Klier, E. M., and Snyder, L. H. 2009. A vestibular sensation: probabilistic approaches to spatial perception. *Neuron* 64: 448–461. [22]

Angevine, J. B., Jr., and Sidman, R. L. 1961. Autoradiographic study of cell migration during histogenesis of cerebral cortex in the mouse. *Nature* 192: 766–768. [25]

Angleson, J. K., and Betz, W. J. 1997. Monitoring secretion in real time: Capacitance, amperometry and fluorescence compared. *Trends Neurosci.* 20: 281–287. [13]

Anglister, L., and McMahan, U. J. 1985. Basal lamina directs acetylcholinesterase accumulation at synaptic sites in regenerating muscle. *J. Cell Biol.* 101: 735–743. [27]

Annunziato, L., Pignatoro, G., and DiRenzo, G. F. 2004. Pharmacology of brain Na^+/Ca^{2+} exchanger: From molecular biology to therapeutic perspectives. *Pharmacol. Rev.* 56: 633–654. [9]

Antonini, A., and Stryker, M. P. 1993a. Development of individual geniculocortical arbors in cat striate cortex and effects of binocular impulse blockade. *J. Neurosci.* 13: 3549–3573. [26]

Antonini, A., and Stryker, M. P. 1993b. Rapid remodeling of axonal arbors in the visual cortex. *Vis. Neurosci.* 15: 401–409. [26]

Antonini, A., Gillespie, D. C., Crair, M. C., and Stryker, M. P. 1998. Morphology of single geniculocortical afferents and functional recovery of the visual cortex after reverse monocular deprivation in the kitten. *J. Neurosci.* 18: 9896–9909. [26]

Apkarian, A. V., Bushnell, M. C., Treede, R. D., and Zubieta, J. K. 2005. Human brain mechanisms of pain perception and regulation in health and disease. *Eur. J. Pain* 9: 463–484. [21]

Arabzadeh, E., Panzeri, S., and Diamond, M. E. 2006. Deciphering the spike train of a sensory neuron: counts and temporal patterns in the rat whisker pathway. *J. Neurosci.* 26: 9216–9226. [21]

Arabzadeh, E., Zorzin, E., and Diamond, M. E. 2005. Neuronal encoding of texture in the whisker sensory pathway. *PLoS Biol.* 3: e17. [21]

Arbuthnott, E. R., Boyd, I. A., and Kalu, K. U. 1980. Ultrastructural dimensions of myelinated peripheral nerve fibres in the cat and their relation to conduction velocity. *J. Physiol.* 308: 125–157. [8]

Armstrong-James, M., Fox, K., and Das-Gupta, A. 1992. Flow of excitation within rat barrel cortex on striking a single vibrissa. *J. Neurophysiol.* 68: 1345–1358. [21]

Armstrong, C. M. 1981. Sodium channels and gating currents. *Physiol. Rev.* 61: 644–683. [7]

Armstrong, C. M., and Bezanilla, F. 1974. Charge movements associated with the opening and closing of the activation gates of sodium channels. *J. Gen. Physiol.* 63: 533–552. [7]

Armstrong, C. M., and Bezanilla, F. 1977. Inactivation of the sodium channel II. Gating current experiments. *J. Gen. Physiol.* 70: 567–590. [7]

Armstrong, C. M., and Hille, B. 1972. The inner quaternary ammonium ion receptor in potassium channels of the node of Ranvier. *J. Gen. Physiol.* 59: 388–400. [7]

Armstrong, C. M., and Hille, B. 1998. Voltage-gated ion channels and electrical excitability. *Neuron* 20: 371–380. [7]

Arrang, J. M., Garbarg, M., and Schwartz, J. C. 1983. Auto-inhibition of brain histamine release mediated by a novel class (H_3) of histamine receptor. *Nature* 302: 832–837. [14]

Art, J. J., and Fettiplace, R. 1984. Efferent desensitization of auditory nerve fibre responses in the cochlea of the turtle *Pseudemys scripta elegans*. *J. Physiol.* 356: 507–523. [22]

Art, J. J., and Fettiplace, R. 1987. Variation of membrane properties in hair cells isolated from the turtle cochlea. *J. Physiol.* 385: 207–242. [22]

Art, J. J., Crawford, A. C., Fettiplace, R., and Fuchs, P. A. 1985. Efferent modulation of hair cell tuning in the cochlea of the turtle. *J. Physiol.* 360: 397–421. [22]

Art, J. J., Fettiplace, R., and Fuchs, P. A. 1984. Synaptic hyperpolarization and inhibition of turtle cochlear hair cells. *J. Physiol.* 356: 525–550. [22]

Art, J. J., Wu, Y. C., and Fettiplace, R. 1995. The calcium-activated potassium channels of turtle hair cells. *J. Gen. Physiol.* 105: 49–72. [22]

Arthur, B. J., Wyttenbach, R. A., Harrington, L. C., and Hoy, R. R. 2010. Neural responses to one- and two-tone stimuli in the hearing organ of the dengue vector mosquito. *J. Exp. Biol.* 213: 1376–1385. [18]

Arunlakshana, O., and Schild, H. O. 1959. Some quantitative uses of drug antagonists. *Brit. J. Pharmacol. Chemother.* 14: 48–58. [11]

Asada, H., Kawamura, Y., Maruyama, K., Kume, H., Ding, R. G., Kanbara, N., Kuzume, H., Sanbo, M., Yagi, T., and Obata, K. 1997. Cleft palate and decreased brain γ-aminobutyric acid in mice lacking the 67-kDa isoform of glutamic acid decarboxylase. *Proc. Natl. Acad. Sci. USA* 94: 6496–6499. [15]

Ashmore, J. 2008. Cochlear outer hair cell motility. *Physiol. Rev.* 88: 173–210. [22]

Ashmore, J. F. 1987. A fast motile response in guinea-pig outer hair cells: the cellular basis of the cochlear amplifier. *J. Physiol.* 388: 323–347. [22]

Assad, J. A., Shepherd, G. M., and Corey, D. P. 1991. Tip-link integrity and mechanical transduction in vertebrate hair cells. *Neuron* 7: 985–994. [19]

Attwell, D., Barbour, B., and Szatkowski, M. 1993. Nonvesicular release of neurotransmitter. *Neuron* 11: 401–407. [9]

Atwood, H. L., and Morin, W. A. 1970. Neuromuscular and axoaxonal synapses of the crayfish opener muscle. *J. Ultrastruct. Res.* 32: 351–369. [11]

Atwood, H. L., ed. 1982. *Biology of Crustacea*. Academic Press, New York. [18]

Aubert, A., Costalat, R., Magistretti, P. J., and Pellerin L. 2005. Brain lactate kinetics: Modeling evidence for neuronal lactate uptake upon activation. *Proc. Natl. Acad. Sci. USA* 102: 16448–16553. [10]

Augustine, G. J., Santamaria, F., and Tanaka, F. 2003. Local calcium signaling in neurons. *Neuron* 40: 331–346. [13]

Avenet, P., and Lindemann, B. 1991. Noninvasive recording of receptor cell action potentials and sustained currents from single taste buds maintained in the tongue: the response to mucosal NaCl and amiloride. *J. Membr. Biol.* 124: 33–41. [19]

Avidan, G., Levy, I., Hendler, T., Zohary, E., and Malach, R. 2003. Spatial vs. object specific attention in high-order visual areas. *Neuroimage* 19: 308–318. [23]

Awapara, J., Landua, A. J., Fuerst, R., and Seale, B. 1950. Free γ-aminobutyric acid in brain. *J. Biol. Chem.* 187: 35–39. [14]

Axelrod, J. 1971. Noradrenaline: fate and control of its biosynthesis. *Science* 173: 598–606. [15]

Axelsson, J., and Thesleff, S. 1959. A study of supersensitivity in denervated mammalian skeletal muscle. *J. Physiol.* 147: 178–193. [11, 27]

Azouz, R., Jensen, M. S., and Yaari, Y. 1996. Ionic basis of spike afterdepolarization and burst generation in adult rat hippocampal CA1 pyramidal cells. *J. Physiol.* 492: 211–223. [7]

Baader, A. P., and Kristan, W. B., Jr. 1995. Parallel pathways coordinate crawling in the medicinal leech, *Hirudo medicinalis*. *J. Comp. Physiol. A* 176: 715–726. [18]

Baas, P. W., and Brown, A. 1997. Slow axonal transport: The polymer transport model. *Trends Cell Biol.* 7: 380–384. [15]

Baca, S. M., Marin-Burgin, A., Wagenaar, D. A., and Kristan, W. B., Jr. 2008. Widespread inhibition proportional to excitation controls the gain of a leech behavioral circuit. *Neuron* 57: 276–289. [18]

Baccus, S. A. 2007. Timing and computation in inner retinal circuitry. *Annu. Rev. Physiol.* 69: 271–290. [20]

Baccus, S. A., and Meister, M. 2002. Fast and slow contrast adaptation in retinal circuitry. *Neuron* 36: 909–919. [20]

Baccus, S. A., Burrell, B. D., Sahley, C. L., and Muller, K. J. 2000. Action potential reflection and failure at axon branch points cause stepwise changes in EPSPs in a neuron essential for learning. *J. Neurophysiol.* 83: 1693–1700. [8]

Baccus, S. A., Olveczky, B. P., Manu, M., and Meister, M. 2008. A retinal circuit that computes object motion. *J. Neurosci.* 28: 6807–6817. [2, 20]

Bailey, C. H., Kandel, E. R., and Si, K. 2004. The persistence of long-term memory: a molecular approach to self-sustaining changes in learning-induced synaptic growth. *Neuron* 44: 49–57. [18]

Baird, R. A., Desmadryl, G., Fernandez, C., and Goldberg, J. M. 1988. The vestibular nerve of the chinchilla. II. Relation between afferent response properties and peripheral innervation patterns in the semicircular canals. *J. Neurophysiol.* 60: 182–203. [22]

Bakalyar, H. A., and Reed, R. R. 1990. Identification of a specialized adenylyl cyclase that may mediate odorant detection. *Science* 250: 1403–1406. [19]

Baker, M. W., Kauffman, B., Macagno, E. R., and Zipser, B. 2003. *In vivo* dynamics of CNS sensory arbor formation: a time-lapse study in the embryonic leech. *J. Neurobiol.* 56: 41–53. [18]

Baker, M. W., Peterson, S. M., and Macagno, E. R. 2008. The receptor phosphatase HmLAR2 collaborates with focal adhesion proteins in filopodial tips to control growth cone morphology. *Dev. Biol.* 320: 215–225. [25]

Baker, P. F., Blaustein, M. P., Keynes, R. D., Manil, J., Shaw, T. I., and Steinhardt, R. A., 1969. The ouabain-sensitive fluxes of sodium and potassium in squid giant axons. *J. Physiol.* 200: 459–496. [9]

Baker, P. F., Hodgkin, A. L., and Ridgeway, E. B. 1971. Depolarization and calcium entry in squid giant axons. *J. Physiol.* 218: 709–755. [6, 9]

Baker, P. F., Hodgkin, A. L., and Shaw, T. I. 1962. Replacement of the axoplasm of giant squid fibres with artificial solutions. *J. Physiol.* 164: 330–354. [6]

Balaban, C. D. 1999. Vestibular autonomic regulation (including motion sickness and the mechanism of vomiting). *Curr. Opin. Neurol.* 12: 29–33. [22]

Balasubramanian, V., and Sterling, P. 2009. Receptive fields and functional architecture in the retina. *J. Physiol.* 587: 2753–2767. [2, 20]

Balice-Gordon, R. J., and Lichtman, J. W. 1994. Long-term synapse loss induced by focal blockade of postsynaptic receptors. *Nature* 372: 519–524. [25]

Ballerini, L., Galante, M., Grandolfo, M., and Nistri, A. 1999. Generation of rhythmic patterns of activity by ventral interneurones in rat organotypic spinal slice culture. *J. Physiol.* 517: 459–475.Bampton, E. T., and Taylor, J. S. 2005. Effects of Schwann cell secreted factors on PC12 cell neuritogenesis and survival. *J. Neurobiol.* 63: 29–48. [10]

Bandettini, P. A. 2009. What's new in neuroimaging methods? *Ann. N Y Acad. Sci.* 1156: 260–293. [1]

Banks, G. B., Fuhrer, C., Adams, M. E., and Froehner, S. C. 2003. The postsynaptic submembrane machinery at the neuromuscular junction: requirement for rapsyn and the utrophin/dystrophin-associated complex. *J. Neurocytol.* 32: 709–726. [11]

Bannatyne, B. A., Liu, T. T., Hammar, I., Stecina, K., Jankowska, E., and Maxwell, D. J. 2009. Excitatory and inhibitory intermediate zone interneurons in pathways from feline group I and II afferents: differences in axonal projections and input. *J. Physiol.* 587: 379–399. [24]

Banner, L. R., and Patterson, P. H. 1994. Major changes in the expression of the mRNAs for cholinergic differentiation factor/ leukemia inhibitory factor and its receptor after injury to adult peripheral nerves and ganglia. *Proc. Natl. Acad. Sci. USA* 91: 7109–7113. [27]

Bao, H., Hakeem, A., Henteleff, M., Starkus, J. G., and Rayner, M. D. 1999. Voltage-insensitive gating after charge-neutralizing mutations in the S4 segment of *Shaker* channels. *J. Gen. Physiol.* 113: 139–151. [5]

Barchi, R. L. 1983. Protein components of the purified sodium channel from rat skeletal muscle sarcolemma. *J. Neurochem.* 40: 1377–1385. [5]

Barchi, R. L. 1997. Ion channel mutations and diseases of skeletal muscle. *Neurobiol. Dis.* 4: 254–264. [6]

Barde, Y. A. 1989. Trophic factors and neuronal survival. *Neuron* 2: 1525–1534. [25]

Barlow, H. B. 1953. Summation and inhibition in the frog's retina. *J. Physiol.* 119: 69–88. [2]

Barlow, H. B., Blakemore, C., and Pettigrew, J. D. 1967. The neural mechanism of binocular depth discrimination. *J. Physiol.* 193: 327–342. [2, 3]

Barlow, H. B., Hill, R. M., and Levick, W. R. 1964. Retinal ganglion cells responding selectively to direction and speed of image motion in the rabbit. *J. Physiol.* 173: 377–407. [2]

Barnes, N. M., and Sharp, T. 1999. A review of central 5-HT receptors and their function. *Neuropharmacology* 38: 1083–1152. [14]

Barnes, N. M., Hales, T. G., Lummis, S. C., and Peters, J. A. 2009. The 5-HT_3 receptor—the relationship between structure and function. *Neuropharmacology* 56: 273–284. [14]

Barres, B. A. 2008. The mystery and magic of glia: A perspective on their roles in health and disease. *Neuron* 60: 430–440. [10]

Barres, B. A., Chun, L. L., and Corey, D. P. 1988. Ion channel expression by white matter glia: I. Type 2 astrocytes and oligodendrocytes. *Glia* 1: 10–30. [10]

Barrett, E. F., and Barrett, J. N. 1976. Separation of two voltage-sensitive potassium currents, and demonstration of tetrodotoxin-resistant calcium current in frog motoneurones. *J. Physiol.* 255: 737–774. [7]

Barrionuevo, G., and Brown, T. H. 1983. Associative long-term potentiation in hippocampal slices. *Proc. Natl. Acad. Sci. USA* 80: 7347–7351. [16]

Barron, D. H., and Matthews, B. H. 1935. Intermittent conduction in the spinal cord. *J. Physiol.* 85: 73–103. [18]

Barry, M. F., Vickery, R. M., Bolsover, S. R., and Bindman, L. J. 1996. Intracellular studies of heterosynaptic long-term depression (LTD) in CA1 hippocampal slices. *Hippocampus* 6: 3–8. [16]

Bartlett, S. E., Reynolds, A. J., and Hendry, I. A. 1998. Retrograde axonal transport of neurotrophins: Differences between neuronal populations and implications for motor neuron disease. *Immunol. Cell Biol.* 76: 419–423. [15]

Bartol, T. M., Land, B. R., Salpeter, E. E., and Salpeter, M. M. 1991. Monte Carlo simulation of miniature endplate current generation in the vertebrate neuromuscular junction. *Biophys. J.* 59: 1290–1307. [15]

Basheer, R., Strecker, R. E., Thakkar, M. M., and McCarley, R. W. 2004. Adenosine and sleep-wake regulation. *Prog. Neurobiol.* 73: 379–396. [14]

Bateson, W. 1984. *Materials for the Study of Variation: Treated with Special Regard to Discontinuity in the Origin of Species.* Macmillan and Co., New York. [25]

Baumgartner, T., Heinrichs, M., Vonlanthen, A., Fischbacher, U., and Fehr, E. 2008. Oxytocin shapes the neural circuitry of trust and trust adaptation in humans. *Neuron.* 58: 639–650. [14]

Baylis, G. C., Rolls, E. T., and Leonard, C. M. 1987. Functional subdivisions of the temporal lobe neocortex. *J. Neurosci.* 7: 330–342. [23]

Baylor, D. 1996. How photons start vision. *Proc. Natl. Acad. Sci. USA* 93: 560–565. [20]

Baylor, D. A. 1987. Photoreceptor signals and vision. Proctor lecture. *Invest. Ophthalmol. Vis. Sci.* 28: 34–49. [20]

Baylor, D. A., and Burns, M. E. 1998. Control of rhodopsin activity in vision. *Eye (Lond).* 12 (Pt 3b): 521–525. [20]

Baylor, D. A., and Fettiplace, R. 1977. Transmission from photoreceptors to ganglion cells in turtle retina. *J. Physiol.* 271: 391–424. [1]

Baylor, D. A., and Fettiplace, R. 1979. Synaptic drive and impulse generation in ganglion cells of turtle retina. *J. Physiol.* 288: 107–127. [1]

Baylor, D. A., and Fuortes, M. G. F. 1970. Electrical responses of single cones in the retina of the turtle. *J. Physiol.* 207: 77–92. [6, 20]

Baylor, D. A., and Nicholls, J. G. 1969. Chemical and electrical synaptic connexions between cutaneous mechanoreceptor neurones in the central nervous system of the leech. *J. Physiol.* 203: 591–609. [11]

Baylor, D. A., Fuortes, M. G., and O'Bryan, P. M. 1971. Receptive fields of cones in the retina of the turtle. *J. Physiol.* 214: 265–294. [20]

Baylor, D. A., Lamb, T. D., and Yau, K. W. 1979. The membrane current of single rod outer segments. *J. Physiol.* 288: 589–611. [20]

Baylor, D. A., Nunn, B. J., and Schnapf, J. L. 1984. The photocurrent, noise and spectral sensitivity of rods of the monkey *Macaca fascicularis. J. Physiol.* 357: 575–607. [20]

Bazan, N. G. 2006. Eicosanoids, platelet-activating factor and inflammation. In: G. J. Siegel, B. W. Agranoff, R. W. Albers, S. K. Fisher, and M. D. Uhler (eds.), *Basic Neurochemistry: Molecular, Cellular and Medical Aspects*, 7th ed. Lippincott-Raven, Philadelphia, pp. 731–741. [12]

Bazemore, A., Elliott, K. A., and Florey, E. 1956. Factor I and γ-aminobutyric acid. *Nature* 178: 1052–1053. [14]

Beadle, D. J., Lees, G., and Kater, S. B. 1988. *Cell Culture Approaches to Invertebrate Neuroscience.* Academic Press, London. [18]

Bean, B. P. 1989. Neurotransmitter inhibition of neuronal calcium currents by changes in channel voltage dependence. *Nature* 340: 153–157. [12]

Bean, B. P. 2007. The action potential in mammalian central neurons. *Nat. Rev. Neurosci.* 8: 451–465. [7]

Beckers, J. M., and Stevens, C. F. 1990. Presynaptic mechanism for long-term potentiation in the hippocampus. *Nature* 346: 724–729. [16]

Beg, A. A., and Jorgensen, E. M. 2003. EXP-1 is an excitatory GABA-gated cation channel. *Nat. Neurosci.* 6: 1145–1152. [5]

Belenky, M. A., Sollars, P. J., Mount, D. B., Alper, S. L., Yarom, Y., and Pickard, G. E. 2010. Cell-type specific distribution of chloride transporters in the rat suprachiasmatic nucleus. *Neuroscience* 165: 1519–1537. [17]

Bell, J., Bolanowski, S., and Holmes, M. H. 1994. The structure and function of Pacinian corpuscles: a review. *Prog. Neurobiol.* 42: 79–128. [19]

Bellochio, E. E., Reimer, R. J., Fremeau, R. T., and Edwards, R. H. 2000. Uptake of glutamate into synaptic vesicles by an organic phosphate transporter. *Science* 289: 957–960. [9]

Belmar, J., and Eyzaguirre, C. 1966. Pacemaker site of fibrillation potentials in denervated mammalian muscle. *J. Neurophysiol.* 29: 425–441. [27]

Ben-Ari, Y., Gaiarsa, J. L., Tyzio, R., and Khazipov, R. 2007. GABA: a pioneer transmitter that excites immature neurons and generates primitive oscillations. *Physiol. Rev.* 87: 1215–1284. [11]

Ben-Chaim, Y., Chanda, B., Dascal, N., Bezanilla, F., Parnas, I., and Parnas, H. 2006. Movement of 'gating charge' is coupled to ligand binding in a G-protein-coupled receptor. *Nature* 444: 106–109. [13]

Benardete, E. A., Kaplan, E., and Knight, B. W. 1992. Contrast gain control in the primate retina: P cells are not X-like, some M cells are. *Vis. Neurosci.* 8: 483–486. [20]

Benardo, L. S. 1993. Characterization of cholinergic and noradrenergic slow excitatory postsynaptic potentials from rat cerebral cortical neurons. *Neuroscience* 53: 11–22. [14]

Benedetti, F., Amanzio, M., Vighetti, S., and Asteggiano, G. 2006. The biochemical and neuroendocrine bases of the hyperalgesic nocebo effect. *J. Neurosci.* 26: 12014–12022. [21]

Benians, A., Nobles, M., Hosny, S., and Tinker, A. 2005. Regulators of G-protein signaling form a quaternary complex with the agonist, receptor, and G-protein. A novel explanation for the acceleration of signaling activation kinetics. *J. Biol. Chem.* 280: 13383–13394. [12]

Bennett, M. V. 1997. Gap junctions as electrical synapses. *J. Neurocytol.* 26: 349–366. [11]

Bennett, M. V., and Zukin, R. S. 2004. Electrical coupling and neuronal synchronization in the mammalian brain. *Neuron* 41: 495–511. [11]

Bensmaia, S. J., and Hollins, M. 2003. The vibrations of texture. *Somatosens. Mot. Res.* 20: 33–43. [21]

Bentley, D., and Caudy, M. 1983. Navigational substrates for peripheral pioneer growth cones: limb-axis polarity cues, limb-segment boundaries, and guidepost neurons. *Cold Spring Harb. Symp. Quant. Biol.* 48 Pt 2: 573–585. [25]

Berardi, N., and Maffei, L. 1999. From visual experience to visual function: roles of neurotrophins. *J. Neurobiol.* 41: 119–126. [26]

Berg, D. K., and Hall, Z. W. 1975. Increased extrajunctional acetylcholine sensitivity produced by chronic postsynaptic neuromuscular blockade. *J. Physiol.* 244: 659–676. [27]

Bergan, J. F., Ro, P., Ro, D., and Knudsen, E. I. 2005. Hunting increases adaptive auditory map plasticity in adult barn owls. *J. Neurosci.* 25: 9816–9820. [26]

Berlucchi, G., and Rizzolatti, G. 1968. Binocularly driven neurons in visual cortex of split-chiasm cats. *Science* 159: 308–310. [3]

Berman, D. M., and Gilman, A. G. 1998. Mammalian RGS proteins: Barbarians at the gate. *J. Biol. Chem.* 273: 1269–1272. [12]

Berne, R. M., and Levy, M. N. (eds.). 1988. *Physiology*, 2nd ed. Mosby, St. Louis, MO. [22]

Bernstein, J. 1902. Untersuchungen zur Thermodynamik der bioelektrischen Strome. *Pflügers Arch.* 92: 521–562. [6]

Bernstein, M., and Lichtman, J. W. 1999. Axonal atrophy: the retraction reaction. *Curr. Opin. Neurobiol.* 9: 364–370. [26]

Berretta, N., and Cherubini, E. 1998. A novel form of long-term depression in the CA1 area of the adult rat hippocampus independent of glutamate receptor activation. *Eur. J. Neurosci.* 10: 2957–2963. [16]

Berridge, C. W., and Waterhouse, B. D. 2003. The locus coeruleus-noradrenergic system: modulation of behavioral state and state-dependent cognitive processes. *Brain Res. Brain Res. Rev.* 42: 33–84. [14]

Berridge, M. J., Lipp, P., and Bootman, M. D. 2000. The versatility and universality of calcium signalling. *Nat. Rev. Mol. Cell Biol.* 1: 11–21. [12]

Berry, M. J., 2nd, Brivanlou, I. H., Jordan, T. A., and Meister, M. 1999. Anticipation of moving stimuli by the retina. *Nature* 398: 334–338. [20]

Berson, D. M. 2003. Strange vision: ganglion cells as circadian photoreceptors. *Trends Neurosci.* 26: 314–320. [20]

Berson, D. M. 2007. Phototransduction in ganglion-cell photoreceptors. *Pflügers Arch.* 454: 849–855. [20]

Berson, D. M., Dunn, F. A., and Takao, M. 2002. Phototransduction by retinal ganglion cells that set the circadian clock. *Science* 295: 1070–1073. [17, 20]

Bestmann, S., Ruff, C. C., Blankenburg, F., Weiskopf, N., Driver, J., and Rothwell, J. C. 2008. Mapping causal interregional influences with concurrent TMS-fMRI. *Exp. Brain. Res.* 191: 383–402. [1]

Bettler, B., Kaupmann, K., Mosbacher, J., and Gassmann, M. 2004. Molecular structure and physiological functions of $GABA_B$ receptors. *Physiol. Rev.* 84: 835–867. [14]

Betz, H., and Laube, B. 2006. Glycine receptors: recent insights into their structural organization and functional diversity. *J. Neurochem.* 97: 1600–1610. [14]

Betz, W. J. 1970. Depression of transmitter release at the neuromuscular junction of the frog. *J. Physiol.* 206: 629–644. [16]

Betz, W. J., and Sakmann, B. 1973. Effects of proteolytic enzymes on function and structure of frog neuromuscular junctions. *J. Physiol.* 230: 673–688. [11]

Betz, W. J., Caldwell, J. H., and Ribchester, R. R. 1980. The effects of partial denervation at birth on the development of muscle fibres and motor units in rat lumbrical muscle. *J. Physiol.* 303: 265–279. [25]

Beurg, M., Fettiplace, R., Nam, J. H., and Ricci, A. J. 2009. Localization of inner hair cell mechanotransducer channels using high-speed calcium imaging. *Nat. Neurosci.* 12: 553–558. [19]

Bevan, S., and Yeats, J. 1991. Protons activate a cation conductance in a sub-population of rat dorsal root ganglion neurones. *J. Physiol.* 433: 145–161. [19]

Bezanilla, F. 2005. Voltage-gated ion channels. *IEEE Trans. Nanobiosci.* 4: 34–48. [7]

Bezanilla, F. 2008. How membrane proteins sense voltage. *Nat. Rev. Mol. Cell Biol.* 9: 323–332. [5, 7]

Bhardwaj, R. D., Curtis, M. A., Spalding, K. L., Buchholz, B. A., Fink, D., Bjork-Eriksson, T., Nordborg, C., Gage, F. H., Druid, H., Eriksson, P. S., and Frisen, J. 2006. Neocortical neurogenesis in humans is restricted to development. *Proc. Natl. Acad. Sci. USA* 103: 12564–12568. [25]

Bhattarai, J. P., Park, S. A., Park, J. B., Lee, S. Y., Herbison, A. E., Ryu, P. D., and Han, S. K. 2011. Tonic extrasynaptic GABA(A) receptor currents control gonadotropin-releasing hormone neuron excitability in the mouse. *Endocrinology* 152: 1551–1561. [17]

Bialek, W. 1987. Physical limits to sensation and perception. *Annu. Rev. Biophys. Biophys. Chem.* 16: 455–478. [19]

Bianchi, M. T., Botzolakis, E. J., Lagrange, A. H., and Macdonald, R. L. 2009. Benzodiazepine modulation of $GABA_A$ receptor opening frequency depends on activation context: a patch clamp and simulation study. *Epilepsy Res.* 85: 212–220. [14]

Bichet, D., Haase F. A., and Jan, L. Y. 2003. Merging functional studies with structures of inward-rectifier K(+) channels. *Nat. Rev. Neurosci.* 4: 957–967. [12]

Biel, M., Wahl-Schott, C., Michalakas, S., and Zong, X. 2009. Hyperpolarization-activated cation channels: From genes to function. *Physiol. Rev.* 89: 847–885. [6]

Bignami, A., and Dahl, D. 1974. Astrocyte-specific protein and neuroglial differentiation: An immunofluorescence study with antibodies to the glial fibrillary acidic protein. *J. Comp. Neurol.* 153: 27–38. [10]

Billups, B., and Attwell, D. 1996. Modulation of non-vesicular glutamate release by pH. *Nature* 379: 171–174. [10]

Bird, M. K., and Lawrence, A. J. 2009. The promiscuous mGlu5 receptor—a range of partners for therapeutic possibilities? *Trends Pharmacol. Sci.* 30: 617–623. [14]

Birder, L. A., and Perl, E. R. 1994. Cutaneous sensory receptors. *J. Clin. Neurophysiol.* 11: 534–552. [21]

Birkmayer, W., and Hornykiewicz, O. 1962. Der L-Dioxyphenylalanin (=L-DOPA)-Effekt beim Parkinson-Syndrom des Menschen: Zur Pathogenese und Behandlung def Parkinson-Akinese. *Arch. Psychiatr. Nervenkr.* 203: 560–574. [14]

Birks, R. I., and MacIntosh, F. C. 1961. Acetylcholine metabolism of a sympathetic ganglion. *Can. J. Biochem. Physiol.* 39: 787–827. [15]

Birks, R., Huxley, H. E., and Katz, B. 1960. The fine structure of the neuromuscular junction of the frog. *J. Physiol.* 150: 134–144. [13]

Birks, R., Katz, B., and Miledi, R. 1960. Physiological and structural changes at the amphibian myoneural junction in the course of nerve degeneration. *J. Physiol.* 150: 145–168. [13]

Bixby, J. L., and Van Essen, D. C. 1979. Competition between foreign and original nerves in adult mammalian skeletal muscle. *Nature* 282: 726–728. [27]

Björklund, A. 1991. Neural transplantation—An experimental tool with clinical possibilities. *Trends Neurosci.* 14: 319–322. [27]

Björklund, A. 2000. Cell replacement strategies for neurodegenerative disorders. *Novartis Found. Symp.* 231: 7–15. [27]

Black, J. A., and Waxman, S. G. 1988. The perinodal astrocyte. *Glia* 1: 169–183. [10]

Black, J. W., and Prichard, B. N. 1973. Activation and blockade of β adrenoceptors in common cardiac disorders. *Br. Med. Bull.* 29: 163–167. [17]

Blackman, J. G., and Purves, R. D. 1969. Intracellular recordings from the ganglia of the thoracic sympathetic chain of the guinea-pig. *J. Physiol.* 203: 173–198. [13]

Blackmer, T., Larsen, E. C., Bartleson, C., Kowalchyk, J. A., Yoon, E. J., Preininger, A. M., Alford, S., Hamm, H. E., and Martin, T. F. 2005. G protein βδ directly regulates SNARE protein fusion machinery for secretory granule exocytosis. *Nat. Neurosci.* 8: 421–425. [14]

Blackshaw, S. E. 1981. Morphology and distribution of touch cell terminals in the skin of the leech. *J. Physiol.* 320: 219–228. [18]

Blackshaw, S. E., and Nicholls, J. G. 1995. Neurobiology and development of the leech. *J. Neurobiol.* 27: 267–276. [18]

Blackshaw, S. E., and Thompson, S. W. 1988. Hyperpolarizing responses to stretch in sensory neurones innervating leech body wall muscle. *J. Physiol.* 396: 121–137. [18, 19]

Blackshaw, S. E., Nicholls, J. G., and Parnas, I. 1982. Expanded receptive fields of cutaneous mechanoreceptor cells after single neurone deletion in leech central nervous system. *J. Physiol.* 326: 261–268. [18, 27]

Blake, D. J., Weir, A., Newey, S. E., and Davies, K. E. 2002. Function and genetics of dystrophin and dystrophin-related proteins in muscle. *Physiol. Rev.* 82: 291–329. [11]

Blake, D. T., Byl, N. N., and Merzenich, M. M. 2002. Representation of the hand in the cerebral cortex. *Behav. Brain Res.* 135: 179–184. [24]

Blakemore, C. 1977. *Mechanics of the Mind.* Cambridge University Press, Cambridge. [21]

Blakemore, C., and Van Sluyters, R. C. 1974. Reversal of the physiological effects of monocular deprivation in kittens: further evidence for a sensitive period. *J. Physiol.* 237: 195–216. [26]

Blakemore, S. J. 2010. The developing social brain: implications for education. *Neuron* 65: 744–747. [26]

Blanchet, C., Erostegui, C., Sugasawa, M., and Dulon, D. 1996. Acetylcholine-induced potassium current of guinea pig outer hair cells: its dependence on a calcium influx through nicotinic-like receptors. *J. Neurosci.* 16: 2574–2584. [22]

Blankenburg, F., Ruben, J., Meyer, R., Schwiemann, J., and Villringer, A. 2003. Evidence for a rostral-to-caudal somatotopic organization in human primary somatosensory cortex with mirror-reversal in areas 3b and 1. *Cereb. Cortex* 13: 987–993. [21]

Blau, J., Blanchard, F., Collins, B., Dahdal, D., Knowles, A., Mizrak, D., and Ruben, M. 2007. What is there left to learn about the *Drosophila* clock? *Cold Spring Harb. Symp. Quant. Biol.* 72: 243–250. [17]

Blauert, J. 1982. Binaural localization. *Scand. Audiol. Suppl.* 15: 7–26. [22]

Blaustein, M. P., and Lederer, W. J. 1999. Sodium/calcium exchange: Its physiological implications. *Physiol. Rev.* 79: 763–854. [9]

Blaustein, M. P., Juhaszova, M., Golovina, V. A., Church, P. J., and Stanley E. F. 2002. Na/Ca exchanger and PMCA localization in neurons and astrocytes: Functional implications. *Ann. N Y Acad. Sci.* 976: 356–366. [10]

Blazquez, P. M., Hirata, Y., and Highstein, S. M. 2004. The vestibulo-ocular reflex as a model system for motor learning: what is the role of the cerebellum? *Cerebellum* 3: 188–192. [22]

Bliss, T. V. P., and Lømo, T. 1973. Long-lasting potentiation of synaptic transmission in the dentate of the anesthetized rabbit following stimulation of the perforant path. *J. Physiol.* 232: 331–356. [16]

Bloomfield, S. A., and Völgyi, B. 2009. The diverse functional roles and regulation of neuronal gap junctions in the retina. *Nat. Rev. Neurosci.* 10: 495–506. [11, 20]

Boadle-Biber, M. C. 1993. Regulation of serotonin synthesis. *Prog. Biophys. Mol. Biol.* 60: 1–15. [15]

Bockaert, J., Perroy, J., Bécamel, C., Marin, P., and Fagni, L. 2010. GPCR interacting proteins (GIPs) in the nervous system: Roles in physiology and pathologies. *Annu. Rev. Pharmacol. Toxicol.* 50: 89–109. [12]

Bocquet, N., Nury, H., Baaden, M., Le Poupon, C., Changeux, J. P., Delarue, M., and Corringer, P. J. 2009. X-ray structure of a pentameric ligand-gated ion channel in the apparently open conformation. *Nature* 457: 111–114. [5]

Boekhoff, I., Tareilus, E., Strotmann, J., and Breer, H. 1990. Rapid activation of alternative second messenger pathways in olfactory cilia from rats by different odorants. *EMBO J.* 9: 2453–2458. [19]

Bohlhalter, S., Fretz, C., and Weder, B. 2002. Hierarchical versus parallel processing in tactile object recognition: a behavioural-neuroanatomical study of aperceptive tactile agnosia. *Brain* 125: 2537–2548. [21]

Boistel, J., and Fatt, P. 1958. Membrane permeability change during inhibitory transmitter action in crustacean muscle. *J. Physiol.* 144: 176–191. [14]

Bolanowski, S. J., Jr., Gescheider, G. A., Verrillo, R. T., and Checkosky, C. M. 1988. Four channels mediate the mechanical aspects of touch. *J. Acoust. Soc. Am.* 84: 1680–1694. [21]

Bollman, J. H., and Sakmann, B. 2005. Control of synaptic strength and timing by the release-site Ca^{2+} signal. *Nat. Neurosci.* 8: 426–434. [13]

Bonanomi, D., and Pfaff, S. L. 2010. Motor axon pathfinding. *Cold Spring Harb. Perspect. Biol.* 2: a001735. [25]

Boncinelli, E., Mallamaci, A., and Broccoli, V. 1998. Body plan genes and human malformation. *Adv. Genet.* 38: 1–29. [25]

Bonhoeffer, T., and Grinvald, A. 1991. Iso-orientation domains in cat visual cortex are arranged in pin-wheel-like patterns. *Nature* 353: 429–431. [3]

Bonnavion, P., and de Lecea, L. 2010. Hypocretins in the control of sleep and wakefulness. *Curr. Neurol. Neurosci. Rep.* 10: 174–179. [14]

Booth M. C., and Rolls, E. T. 1998. View-invariant representations of familiar objects by neurons in the inferior temporal visual cortex. *Cereb. Cortex* 8: 510–523. [23]

Borghuis, B. G., Ratliff, C. P., Smith, R. G., Sterling, P., and Balasubramanian, V. 2008. Design of a neuronal array. *J. Neurosci.* 28: 3178–3189. [2]

Boring, E. G. 1942. *Sensation and Perception in the History of Experimental Psychology*. Appleton-Century, New York. [19]

Borst, J. G. G., and Sakmann, B. 1996. Calcium influx and transmitter release in a fast CNS synapse. *Nature* 383: 431–434. [13]

Borst, J. G. G., and Sakmann, B. 1998. Facilitation of presynaptic calcium currents in the rat brainstem. *J. Physiol.* 513: 149–155. [16]

Borst, J. G. G., Helmchen, F., and Sakmann, B. 1995. Pre- and postynaptic whole-cell recording in the medial nucleus of the trapezoid body of the rat. *J. Physiol.* 489: 825–840. [13, 16]

Boschat, C., Pelofi, C., Randin, O., Roppolo, D., Lüscher, C., Broillet, M. C., and Rodriguez, I. 2002. Pheromone detection mediated by a V1r vomeronasal receptor. *Nat. Neurosci.* 5: 1261–1262. [19]

Bostock, H., and Sears, T. A., and Sherratt, R. M. 1981. The effects of 4-aminopyradine and tetraethylammonium on normal and demyelinated nerve fibers. *J. Physiol.* 313: 301–315. [8]

Bouslama-Oueghlani, L., Wehrlé, R., Sotelo, C., and Dusart, I. 2003. The developmental loss of the ability of Purkinje cells to regenerate their axons occurs in the absence of myelin: an in vitro model to prevent myelination. *J. Neurosci.* 23: 8318–8329. [27]

Bowe, M. A., and Fallon, J. R. 1995. The role of agrin in synapse formation. *Annu. Rev. Neurosci.* 18: 443–462. [27]

Bowery, N. G., Hill, D. R., Hudson, A. L., Doble, A., Middlemiss, D. N., Shaw, J., and Turnbull, M. 1980. (–)Baclofen decreases neurotransmitter release in the mammalian CNS by an action at a novel GABA receptor. *Nature* 283: 92–94. [14]

Bowling, D. B., and Michael, C. R. 1980. Projection patterns of single physiologically characterized optic tract fibers in the cat. *Nature* 286: 899–902. [3]

Bowling, D., Nicholls, J., and Parnas, I. 1978. Destruction of a single cell in the central nervous system of the leech as a means of analysing its connexions and functional role. *J. Physiol.* 282: 169–80. [18]

Bowser, D. N., and Khakh, B. S. 2004. ATP excites interneurons and astrocytes to increase synaptic inhibition in neuronal networks. *J. Neurosci.* 24: 8606–8620. [14]

Boycott, B. B., and Dowling, J. E. 1969. Organization of primate retina: light microscopy. *Philos. Trans. R. Soc. Lond., B, Biol. Sci.* 255: 109–184. [20]

Boycott, B., and Wässle, H. 1999. Parallel processing in the mammalian retina: the Proctor Lecture. *Invest. Ophthalmol. Vis. Sci.* 40: 1313–1327. [20]

Boyd, I. A., and Martin, A. R. 1956. The end-plate potential in mammalian muscle. *J. Physiol.* 132: 74–91. [11, 13]

Bozza, T., Feinstein, P., Zheng, C., and Mombaerts, P. 2002. Odorant receptor expression defines functional units in the mouse olfactory system. *J. Neurosci.* 22: 3033–3043. [19]

Bradley, J., Resisert, J., and Frings, S. 2005. Regulation of cyclic nucleotide-gated channels. *Curr. Opin. Neurobiol.* 15: 343–349. [5]

Brady, S. T., Lasek, R. J., and Allen, R. D. 1982. Fast axonal transport in extruded axoplasm from squid giant axon. *Science* 218: 1129–1131. [15]

Brainard, M. S., and Knudsen, E. I. 1998. Sensitive periods for visual calibration of the auditory space map in the barn owl optic tectum. *J. Neurosci.* 18: 3929–3942. [26]

Brandstatter, J. H. 2002. Glutamate receptors in the retina: the molecular substrate for visual signal processing. *Curr. Eye Res.* 25: 327–331. [20]

Brannstrom, T. 1993. Quantitative synaptology of functionally different types of cat medial gastrocnemius α-motoneurons. *J. Comp. Neurol.* 330: 439–454. [24]

Brasch, F. E. 1922. History of Science. *Science* 55: 405–408. [28]

Bray, G. M., Villegas-Perez, M. P., Vidal-Sanz, M., Carter, D. A., and Aguayo, A. J. 1991. Neuronal and nonneuronal influences on retinal ganglion cell survival, axonal regrowth, and connectivity after axotomy. *Ann. N Y Acad. Sci.* 633: 214–228. [27]

Brecht, M., Preilowski, B., and Merzenich, M. M. 1997. Functional architecture of the mystacial vibrissae. *Behav. Brain Res.* 84: 81–97. [21]

Bredt, D. S., and Snyder, S. H. 1989. Nitric oxide mediates glutamate-linked enhancement of cGMP levels in the cerebellum. *Proc. Natl. Acad. Sci. USA* 86: 9030–9033. [12]

Breer, H., Boekhoff, I., and Tareilus, E. 1990. Rapid kinetics of second messenger formation in olfactory transduction. *Nature* 345: 65–68. [19]

Brefczynski, J. A., and DeYoe, E. A. 1999. A physiological correlate of the 'spotlight' of visual attention. *Nat. Neurosci.* 2: 370–374. [23]

Bregy, P., Sommer, S., and Wehner, R. 2008. Nest-mark orientation versus vector navigation in desert ants. *J. Exp. Biol.* 211: 1868–1873. [18]

Breitwieser, G. E., and Szabo, G. 1985. Uncoupling of cardiac muscarinic and β-adrenergic receptors from ion channels by a guanine nucleotide analogue. *Nature* 317: 538–540. [12]

Brejc, K., Van Dijk, W. J., Klaassen, R. V., Schuurmanns, M., van der Oost, J., Smit, A. B., and Sixma, T. K. 2001. Crystal structure of an ACh-binding protein reveals the ligand-binding domain of nicotinic receptors. *Nature* 411: 269–276. [5]

Brennan, P. A., and Kendrick, K. M. 2006. Mammalian social odours: attraction and individual recognition. *Philos. Trans. R. Soc. Lond., B, Biol. Sci.* 361: 2061–2078. [19]

Brenner S. 1974. The genetics of *Caenorhabditis elegans*. *Genetics* 77: 71–94. [25]

Brenner, H. R., and Martin, A. R. 1976. Reduction in acetylcholine sensitivity of axotomized ciliary ganglion cells. *J. Physiol.* 260: 159–175. [27]

Brew, H., Gray, P. T., Mobbs, P., and Attwell, D. 1986. Endfeet of retinal glial cells have higher densities of ion channels that mediate K^+ buffering. *Nature* 324: 466–468. [10]

Brichta, A. M., Aubert, A., Eatock, R. A., and Goldberg, J. M. 2002. Regional analysis of whole cell currents from hair cells of the turtle posterior crista. *J. Neurophysiol.* 88: 3259–3278. [22]

Briggman, K. L., and Kristan, W. B. 2008. Multifunctional pattern-generating circuits. *Annu. Rev. Neurosci.* 31: 271–294. [24]

Briggman, K. L., Abarbanel, H. D., and Kristan, W. B., Jr. 2005. Optical imaging of neuronal populations during decision-making. *Science* 307: 896–901. [18]

Briggman, K. L., Abarbanel, H. D., and Kristan, W. B., Jr. 2006. From crawling to cognition: analyzing the dynamical interactions among populations of neurons. *Curr. Opin. Neurobiol.* 16: 135–144. [18]

Brightman, M. W., and Reese, T. S. 1969. Junctions between intimately apposed cell membranes in the vertebrate brain. *J. Cell Biol.* 40: 668–677. [10]

Brightman, M. W., Reese, T. S., and Feder, N. 1970. Assessment with the electron microscope of the permeability to peroxidase of cerebral endothelium in mice and sharks. In E. H. Thaysen (ed.), *Capillary Permeability* (Alfred Benzon Symposium II). Munskgaard, Copenhagen, Denmark. [10]

Brincat, S. L., and Connor, C. E. 2004. Underlying principles of visual shape selectivity in posterior inferotemporal cortex. *Nat. Neurosci.* 7: 880–886. [23]

Briscoe, J. 2009. Making a grade: Sonic Hedgehog signalling and the control of neural cell fate. *EMBO J.* 28: 457–465. [25]

Brivanlou, I. H., Warland, D. K., and Meister, M. 1998. Mechanisms of concerted firing among retinal ganglion cells. *Neuron* 20: 527–539. [20, 26]

Brock, L. G., Coombs, J. S., and Eccles, J. C. 1952. The recording of potentials from motoneurones with an intracellular electrode. *J. Physiol.* 117: 431–460. [11]

Brodfuehrer, P. D., Debski, E. A., O'Gara, B. A., and Friesen, W. O. 1995. Neuronal control of leech swimming. *J. Neurobiol.* 27: 403–418. [18]

Bronner-Fraser, M. 1985. Alterations in neural crest migration by a monoclonal antibody that affects cell adhesion. *J. Cell Biol.* 101: 610–617. [25]

Brooks, V. B. 1956. An intracellular study of the action of repetitive nerve volleys and of botulinum toxin on miniature end-plate potentials. *J. Physiol.* 134: 264–277. [13]

Brose, K., Bland, K. S., Wang, K. H., Arnott, D., Henzel, W., Goodman, C. S., Tessier-Lavigne, M., and Kidd, T. 1999. Slit proteins bind Robo receptors and have an evolutionarily conserved role in repulsive axon guidance. *Cell* 96: 795–806. [25]

Brown, A. G., and Fyffe, R. E. W. 1981. Direct observations on the contacts made between Ia afferent fibers and -motoneurones in the cat's lumbosacral spinal cord. *J. Physiol.* 313: 121–140. [24]

Brown, A. M., and Ransom, B. R. 2007. Astrocyte glycogen and brain energy metabolism. *Glia* 55: 1263–1271. [10]

Brown, D. A. 2000. Neurobiology: the acid test for resting potassium channels. *Curr. Biol.* 10: R456–459. [6]

Brown, D. A. 2010. Muscarinic acetylcholine receptors (mAChRs) in the nervous system: some functions and mechanisms. *J. Mol. Neurosci.* 41: 340–346. [14]

Brown, D. A., and Adams, P. R. 1980. Muscarinic suppression of a novel voltage-sensitive K^+-current in a vertebrate neurone. *Nature* 283: 673–676. [12]

Brown, D. A., and Passmore, G. M. 2009. Neural KCNQ (Kv7) channels. *Brit. J. Pharmacol.* 156: 1185–1195. [12, 17]

Brown, D. A., and Selyanko, A. A. 1985. Membrane currents underlying the cholinergic slow excitatory post-synaptic potential in the rat sympathetic ganglion. *J. Physiol.* 365: 365–387. [11, 15]

Brown, D. A., Constanti, A., and Adams, P. R. 1983. Ca-activated potassium current in vertebrate sympathetic neurons. *Cell Calcium* 4: 407–420. [12]

Brown, D. A., Docherty, R. J., and Halliwell, J., V. 1983. Chemical transmission in the rat interpeduncular nucleus in vitro. *J. Physiol.* 341: 655–670. [14]

Brown, D. A., Hughes, S. A., Marsh, S. J., and Tinker, A. 2007. Regulation of M(Kv7.2/7.3) channels in neurons by PIP_2 and products of PIP_2 hydrolysis: significance for receptor-mediated inhibition. *J. Physiol.* 582: 917–925. [17]

Brown, G. L. 1937. The actions of acetylcholine on denervated mammalian and frog's muscle. *J. Physiol.* 89: 438–461. [27]

Brown, H. F., DiFrancesco, D., and Noble, D. 1979. How does adrenaline accelerate the heart? *Nature* 280: 235–236. [12]

Brown, H. M., Ottoson, D., and Rydqvist, B. 1978. Crayfish stretch receptor: an investigation with voltage-clamp and ion-sensitive electrodes. *J. Physiol.* 284: 155–179. [19]

Brown, M. C. 1987. Morphology of labeled afferent fibers in the guinea pig cochlea. *J. Comp. Neurol.* 260: 591–604. [22]

Brown, M. C., Holland, R. L., and Hopkins, W. G. 1981. Motor nerve sprouting. *Annu. Rev. Neurosci.* 4: 17–42. [27]

Brown, M. C., Hopkins, W. G., and Keynes, R. J. 1982. Short- and long-term effects of paralysis on the motor innervation of two different neonatal mouse muscles. *J. Physiol.* 329: 439–450. [25]

Brown, M. C., Jansen, J. K., and Van Essen, D. 1976. Polyneuronal innervation of skeletal muscle in new-born rats and its elimination during maturation. *J. Physiol.* 261: 387–422. [25]

Brown, N., Kerby, J., Bonnert, T. P., Whiting, P. J., and Wafford, K. A. 2002. Pharmacological characterization of a novel cell line expressing human $\alpha 4\beta 3\delta$ $GABA_A$ receptors. *Brit. J. Pharmacol.* 136: 965–974. [14]

Brown, P. K., and Wald, G. 1963. Visual pigments in human and monkey retinas. *Nature* 200: 37–43. [20]

Brown, S. P., Brenowitz, S. D., and Regehr, W. D. 2003. Brief presynaptic bursts evoke synapse-specific retrograde inhibition mediated by endogenous cannabinoids. *Nat. Neurosci.* 10: 1047–1058. [12]

Brown, T. G. 1911. The intrinsic factor in the act of progression in the mammal. *Proc. R. Soc. Lond., B, Biol. Sci.* 84: 308–319. [24]

Brownell, W. E., Bader, C. R., Bertrand, D., and de Ribaupierre, Y. 1985. Evoked mechanical responses of isolated cochlear outer hair cells. *Science* 227: 194–196. [22]

Bruening-Wright, A., Schumacher, M. A., Adelman, J. P., and Maylie, J. 2002. Localization of the activation gate for small conductance Ca^{2+}-activated K^+ channels. *J. Neurosci.* 22: 6499–6506. [5]

Brunet, L. J., Gold, G. H., and Ngai, J. 1996. General anosmia caused by a targeted disruption of the mouse olfactory cyclic nucleotide-gated cation channel. *Neuron* 17: 681–693. [19]

Brungart, D. S., Durlach, N. I., and Rabinowitz, W. M. 1999. Auditory localization of nearby sources. II. Localization of a broadband source. *J. Acoust. Soc. Am.* 106: 1956–1968. [22]

Brusa, R., Zimmermann, F., Koh, D. S., Feldmeyer, D., Gass, P., Seeburg, P. H., and Sprengel, R. 1995. Early-onset epilepsy and postnatal lethality associated with an editing-deficient GluR-B allele in mice. *Science* 270: 1677–1680. [25]

Bucher, D., Taylor, A. L., and Marder, E. 2006. Central pattern generating neurons simultaneously express fast and slow rhythmic activities in the stomatogastric ganglion. *J. Neurophysiol.* 95: 3617–3632. [18]

Buck, L., and Axel, R. 1991. A novel multigene family may encode odorant receptors: a molecular basis for odor recognition. *Cell* 65: 175–187. [19]

Buckley, C. E., Marguerie, A., Alderton, W. K., and Franklin, R. J. 2010. Temporal dynamics of myelination in the zebrafish spinal cord. *Glia* 58: 802–812. [10]

Buell, T. N., and Hafter, E. R. 1988. Discrimination of interaural differences of time in the envelopes of high-frequency signals: integration times. *J. Acoust. Soc. Am.* 84: 2063–2066. [22]

Buell, T. N., Trahiotis, C., and Bernstein, L. R. 1991. Lateralization of low-frequency tones: relative potency of gating and ongoing interaural delays. *J. Acoust. Soc. Am.* 90: 3077–3085. [22]

Buller, A. J., Eccles, J. C., and Eccles, R. M. 1960. Interactions between motoneurones and muscles in respect of the characteristic speeds of their responses. *J. Physiol.* 150: 417–439. [27]

Bullock, T. H., and Hagiwara, S. 1957. Intracellular recording from the giant synapse of the squid. *J. Gen. Physiol.* 40: 565–577. [13]

Bult, H., Boeckxstaens, G. E., Pelckmans, P. A., Jordaens, F. H., Van Maercke, Y. M., and Herman, A. G. 1990. Nitric oxide as an inhibitory non-adrenergic non-cholinergic neurotransmitter. *Nature* 345: 346–347. [12]

Bunge, M. B. 2008. Novel combination strategies to repair the injured mammalian spinal cord. *J. Spinal Cord Med.* 31: 262–269. [27]

Bunge, R. P. 1968. Glial cells and the central myelin sheath. *Physiol. Rev.* 48: 197–251. [10]

Burbach, J. P., Luckman, S. M., Murphy, D., and Gainer, H. 2001. Gene regulation in the magnocellular hypothalamo-neurohypophysial system. *Physiol. Rev.* 81: 1197–1267. [17]

Burdakov, D., Gerasimenko, O., and Verkhratsky, A. 2005. Physiological changes in glucose differentially modulate the excitability of hypothalamic melanin-concentrating hormone and orexin neurons in situ. *J. Neurosci.* 25: 2429–2433. [14]

Burdakov, D., Jensen, L. T., Alexopoulos, H., Williams, R. H., Fearon, I. M., O'Kelly, I., Gerasimenko, O., Fugger, L., and Verkhratsky, A. 2006. Tandem-pore K^+ channels mediate inhibition of orexin neurons by glucose. *Neuron* 50: 711–722. [14]

Burden, S. J., Sargent, P. B., and McMahan, U. J. 1979. Acetylcholine receptors in regenerating muscle accumulate at original synaptic sites in the absence of the nerve. *J. Cell Biol.* 82: 412–425. [11, 27]

Burgen, A. S. V., and Terroux, K. G. 1953. On the negative inotropic effect in the cat's auricle. *J. Physiol.* 120: 449–464. [12]

Burgess, G. M., Mullaney, I., McNeill, M., Dunn, P. M., and Rang, H. P. 1989. Second messengers involved in the mechanism of action of bradykinin in sensory neurons in culture. *J. Neurosci.* 9: 3314–3325. [19]

Burgess, R. W., Nguyen, Q. T., Son, Y. J., Lichtman, J. W., and Sanes, J. R. 1999. Alternatively spliced isoforms of nerve- and muscle-derived agrin: their roles at the neuromuscular junction. *Neuron* 23: 33–44. [27]

Burke, R. E., and Glenn, L. L. 1996. Horseradish peroxidase study of the spatial and electrotonic distribution of group Ia synapses on type-identified ankle extensor motoneurons in the cat. *J. Comp. Neurol.* 372: 465–485. [24]

Burnashev, N. 1996 Calcium permeability of glutamate-gated channels in the central nervous system. *Curr. Opin. Neurobiol.* 6: 311–317. [11]

Burnashev, N., Zhou, Z., Neher, E., and Sakmann, B. 1995. Fractional calcium currents through recombinant GluR channels of the NMDA, AMPA and kainate receptor subtypes. *J. Physiol.* 485: 403–418. [12]

Burnstock, G. 1972. Purinergic nerves. *Pharmacol. Rev.* 24: 509–581. [14]

Burnstock, G. 1976. Do some nerve cells release more than one transmitter? *Neuroscience* 1: 239–248. [15]

Burnstock, G. 1995. Noradrenaline and ATP: Cotransmitters and neuromodulators. *J. Physiol. Pharmacol.* 46: 365–384. [17]

Burnstock, G., ed. 1990–99. *The Autonomic Nervous System.* 8 vols. Harwood Academic, New Jersey. [17]

Burnstock, G. 2006. Historical review: ATP as a neurotransmitter. *Trends Pharmacol. Sci.* 27: 166–176. [17]

Burnstock, G., and Holman, M. E. 1961. The transmission of excitation from autonomic nerve to smooth muscle. *J. Physiol.* 155: 115–133. [17]

Burrell, B. D., and Crisp, K. M. 2008. Serotonergic modulation of afterhyperpolarization in a neuron that contributes to learning in the leech. *J. Neurophysiol.* 99: 605–616. [18]

Burrell, B. D., Sahley, C. L., and Muller, K. J. 2003. Progressive recovery of learning during regeneration of a single synapse in the medicinal leech. *J. Comp. Neurol.* 457: 67–74. [18]

Burton, H., and Sinclair, R. J. 1991. Second somatosensory cortical area in macaque monkeys: 2. Neuronal responses to punctate vibrotactile stimulation of glabrous skin on the hand. *Brain Res.* 538: 127–135. [21]

Buss, R. R., Sun, W., and Oppenheim, R. W. 2006. Adaptive roles of programmed cell death during nervous system development. *Annu. Rev. Neurosci.* 29: 1–35. [25]

Buszaki, G. 1984. Feed-forward inhibition in the hippocampal formation. *Prog. Neurobiol.* 22: 131–153. [14]

Butt, A. M., and Ransom, B. R. 1993. Morphology of astrocytes and oligodendrocytes during development in the intact rat optic nerve. *J. Comp. Neurol.* 338: 141–158. [10]

Butt, A. M., Hamilton, N., Hubbard, P., Pugh, M., and Ibrahim, M. 2005. Synantocytes: The fifth element. *J. Anat.* 207: 695–706. [10]

Buttner, N., Siegelbaum, S. A., and Volterra, A. 1989. Direct modulation of *Aplysia* S-K^+ channels by a 12-lipoxygenase metabolite of arachidonic acid. *Nature* 342: 553–555. [12]

Cain, D. P. 1998. Testing the NMDA, long-term potentiation, and cholinergic hypothesis of spatial learning. *Neurosci. Biobehav. Rev.* 22: 181–193. [16]

Caldwell, P. C., Hodgkin, A. L., Keynes, R. D., and Shaw, T. L. 1960. The effects of injecting "energy-rich" phosphate compounds on the active transport of ions in the giant axons of *Loligo*. *J. Physiol.* 152: 561–590. [9]

Callaway, E. M. 2005. Structure and function of parallel pathways in the primate early visual system. *J. Physiol.* 566: 13–19. [3]

Calvino, M. A., and Szczupak, L. 2008. Spatial-specific action of serotonin within the leech midbody ganglion. *J. Comp. Physiol. A* 194: 523–531. [18]

Camand, E., Morel, M. P., Faissner, A., Sotelo, C., and Dusart, I. 2004. Long-term changes in the molecular composition of the glial scar and progressive increase of serotoninergic fibre sprouting after hemisection of the mouse spinal cord. *Eur. J. Neurosci.* 20: 1161–1176. [27]

Cameron, O. G. 2009 Visceral brain–body information transfer. *Neuroimage* 47: 787–794. [17]

Cammack, J. N., and Schwartz, E. A. 1993. Ions required for the electrogenic transport of GABA by horizontal cells of the catfish retina. *J. Physiol.* 472: 81–102. [9, 13]

Cammack, J. N., Rakhilin, S. V., and Schwartz, E. A. 1994. A GABA transporter operates asymmetrically and with variable stoichiometry. *Neuron* 13: 949–960. [13]

Campbell, K., and Gotz, M. 2002. Radial glia: multi-purpose cells for vertebrate brain development. *Trends Neurosci.* 25: 235–238. [25]

Campenot, R. B. 1977. Local control of neurite development by nerve growth factor. *Proc. Natl. Acad. Sci. USA* 74: 4516–4519. [25]

Campenot, R. B. 1982. Development of sympathetic neurons in compartmentalized cultures. II. Local control of neurite survival by nerve growth factor. *Dev. Biol.* 93: 13–21. [25]

Campenot, R. B. 2009. NGF uptake and retrograde signaling mechanisms in sympathetic neurons in compartmented cultures. *Results Probl. Cell Differ.* 48: 141–158. [25, 27]

Canessa, C. M., Schild, L., Buell, G., Thorens, B., Gautschi, I., Horisberger, J. D., and Rossier, B. C. 1994. Amiloride-sensitive epithelial Na+ channel is made of three homologous subunits. *Nature* 367: 463–467. [19]

Cannon, S. C. 1996. Ion channel defects and aberrant excitability in myotonia and periodic paralysis. *Trends Neurosci.* 19: 3–10. [6]

Cant, N. B., and Casseday, J. H. 1986. Projections from the anteroventral cochlear nucleus to the lateral and medial superior olivary nuclei. *J. Comp. Neurol.* 247: 457–476. [22]

Cantino, D., and Mugnani, E. 1975. The structural basis for electrotonic coupling in the avian ciliary ganglion: A study with thin sectioning and freeze-fracturing. *J. Neurocytol.* 4: 505–536. [8]

Cao, Y. Q., Mantyh, P. W., Carlson, E. J., Gillespie, A. M., Epstein, C. J., and Basbaum, A. I. 1998. Primary afferent tachykinins are required to experience moderate to intense pain. *Nature* 392: 390–394. [14]

Capecchi, M. R. 1997. *Hox* genes and mammalian development. *Cold Spring Harb. Symp. Quant. Biol.* 62: 273–281. [25]

Capsoni, S., and Cattaneo, A. 2006. On the molecular basis linking nerve growth factor (NGF) to Alzheimer's disease. *Cell Mol. Neurobiol.* 26: 619–633. [25]

Capsoni, S., Tongiorgi, E., Cattaneo, A., and Domenici, L. 1999. Dark rearing blocks the developmental down-regulation of brain-derived neurotrophic factor messenger RNA expression in layers IV and V of the rat visual cortex. *Neuroscience* 88: 393–403. [26]

Carafoli, E., and Brini, M. 2000. Calcium pumps: Structural basis for and mechanisms of calcium transmembrane transport. *Curr. Opin. Chem. Biol.* 4: 152–161. [9]

Cariboni, A., Maggi, R., and Parnevalas, J. G. 2007. From nose to fertility: the long migratory journey of gonadotropin-releasing hormone neurons. *Trends Neurosci.* 30: 638–644. [17]

Carlson, M., Hubel, D. H., and Wiesel, T. N. 1986. Effects of monocular exposure to oriented lines on monkey striate cortex. *Brain Res.* 390: 71–81. [26]

Carlsson, A., Lindqvist, M., Magnusson, T., and Waldeck, B. 1958. On the presence of 3-hydroxytyramine in brain. *Science* 127: 471. [14]

Caroni, P., and Schwab, M. E. 1988. Two membrane protein fractions from rat central myelin with inhibitory properties for neurite growth and fibroblast spreading. *J. Cell Biol.* 106: 1281–1288. [10]

Caroni, P., Rotsler, S., Britt, J. C., and Brenner, H. R. 1993. Calcium influx and protein phosphorylation mediate the metabolic stabilization of synaptic acetylcholine receptors in muscle. *J. Neurosci.* 13: 1315–1325. [27]

Carrol, R. C., Lissin, D. V., von Zastrow, M., Nicoll, R. A., and Malenka, R. C. 1999. Rapid redistribution of glutamate receptors contributes to long-term depression in hippocampal cultures. *Nat. Neurosci.* 2: 454–460. [16]

Carulli, D., Buffo, A., and Strata, P. 2004. Reparative mechanisms in the cerebellar cortex. *Prog. Neurobiol.* 72: 373–398. [27]

Carvell, G. E., and Simons, D. J. 1990. Biometric analyses of vibrissal tactile discrimination in the rat. *J. Neurosci.* 10: 2638–2648. [21]

Casagrande, V. A., Yazar, F., Jones, K. D., and Ding, Y. 2007. The morphology of the koniocellular axon pathway in the macaque monkey. *Cereb. Cortex* 17: 2334–2345. [3]

Cascio, C. J., and Sathian, K. 2001. Temporal cues contribute to tactile perception of roughness. *J. Neurosci.* 21: 5289–5296. [21]

Cassell, J. F., and McLachlan, E. M. 1987. Muscarinic agonists block five different potassium conductances in guinea-pig sympathetic neurones. *Brit. J. Pharmacol.* 91: 259–261. [12]

Castellucci, V., and Kandel, E. R. 1974. A quantal analysis of the synaptic depression underlying habituation of the gill-withdrawal reflex in *Aplysia*. *Proc. Natl. Acad. Sci. USA* 71: 5004–5008. [16]

Castellucci, V., Pinsker, H., Kupfermann, I., and Kandel, E. R. 1970. Neuronal mechanisms of habituation and dishabituation of the gill-withdrawal reflex in *Aplysia*. *Science* 167: 1745–1748. [16]

Catania, K. C. 1999. A nose that looks like a hand and acts like an eye: the unusual mechanosensory system of the star-nosed mole. *J. Comp. Physiol. A* 185: 367–372. [21]

Catania, K. C., and Kaas, J. H. 1997. Somatosensory fovea in the star-nosed mole: behavioral use of the star in relation to innervation patterns and cortical representation. *J. Comp. Neurol.* 387: 215–233. [21]

Caterina, M. J., Schumacher, M. A., Tominaga, M., Rosen, T. A., Levine, J. D., and Julius, D. 1997. The capsaicin receptor: a heat-activated ion channel in the pain pathway. *Nature* 389: 816–824. [19]

Cator, L. J., Arthur, B. J., Harrington, L. C., and Hoy, R. R. 2009. Harmonic convergence in the love songs of the dengue vector mosquito. *Science* 323: 1077–1079. [18]

Cattaneo, L., and Rizzolatti, G. 2009. The mirror neuron system. *Arch. Neurol.* 66: 557–560. [24]

Catterall, W. A., and Few, A. P. 2008. Calcium channel regulation and presynaptic plasticity. *Neuron* 59: 882–901. [16]

Caulfield, M. P., and Brown, D. A. 1992. Cannabinoid receptor agonists inhibit Ca current in NG108-15 neuroblastoma cells via a pertussis toxin-sensitive mechanism. *Brit. J. Pharmacol.* 106: 231–232. [12]

Caviness, V. S., Jr. 1982. Neocortical histogenesis in normal and reeler mice: a developmental study based upon [3H]thymidine autoradiography. *Dev. Brain Res.* 4: 293–302. [25]

Ceccarelli, B., and Hurlbut, W. P. 1980. Vesicle hypothesis of the release of quanta of acetylcholine. *Physiol. Rev.* 60: 396–441. [13]

Celikel, T., Szostak, V. A., and Feldman, D. E. 2004. Modulation of spike timing by sensory deprivation during induction of cortical map plasticity. *Nat. Neurosci.* 7: 534–541. [21]

Cellerino, A., Arango-Gonzalez, B., Pinzon-Duarte, G., and Kohler, K. 2003. Brain-derived neurotrophic factor regulates expression of vasoactive intestinal polypeptide in retinal amacrine cells. *J. Comp. Neurol.* 467: 97–104. [20]

Cepko, C., and Pear, W. 1996. Transduction of genes using retrovirus vectors. *Curr. Protoc. Mol. Biol.* Sec. III, Unit 9.9. [25]

Cepko, C. L., Austin, C. P., Yang, X., Alexiades, M., and Ezzeddine, D. 1996. Cell fate determination in the vertebrate retina. *Proc. Natl. Acad. Sci. USA* 93: 589–595. [25]

Cesare, P., and McNaughton, P. A. 1996. A novel heat-activated current in nociceptive neurons and its sensitization by bradykinin. *Proc. Natl. Acad. Sci. USA* 93: 15435–15439. [12, 19]

Cha, A., Snyder, G. E., Selvin, P. R., and Bezanilla, F. 1999. Atomic scale movement of the voltage-sensing region in a potassium channel measured via spectroscopy. *Nature* 402: 809–813. [7]

Chalfie, M. 2009. Neurosensory mechanotransduction. *Nat. Rev. Mol. Cell Biol.* 10: 44–52. [19]

Chalfie, M., Tu, Y., Euskirchen, G., Ward, W. W., and Prasher, D. C. 1994. Green fluorescent protein as a marker for gene expression. *Science* 263: 802–805. [25]

Chameau, P., and van Hooft, J. A. 2006. Serotonin 5-HT_3 receptors in the central nervous system. *Cell Tissue Res.* 326: 573–581. [14]

Champagnat, J., Morin-Surun, M. P., Fortin, G., and Thoby-Brisson, M. 2009. Developmental basis of the rostro-caudal organization of the brainstem respiratory rhythm generator. *Philos. Trans. R. Soc. Lond., B, Biol. Sci.* 364: 2469–2476. [24]

Chan, J. R., Watkins, T. A., Cosgaya, J. M., Zhang, C., Chen, L., Reichardt, L. F., Shooter, E. M., and Barres, B. A. 2004. NGF controls axonal receptivity to myelination by Schwann cells or oligodendrocytes. *Neuron* 43: 183–191. [10]

Chanda, B., Asamoah, O. K., Blunck, R., Roux, B., and Bezanilla, F. 2005. Gating charge displacment in voltage-gated ion channels involves limited transmembrane movement. *Nature* 436: 852–856. [7]

Chandrashekar, J., Kuhn, C., Oka, Y., Yarmolinsky, D. A., Hummler, E., Ryba, N. J. P., and Zuker, C. S. 2010. The cells and peripheral representation of sodium taste in mice. *Nature* 464: 297–301. [19]

Chapman, B., Stryker, M. P., and Bonhoeffer, T. 1996. Development of orientation preference maps in ferret primary visual cortex. *J. Neurosci.* 16: 6443–6453. [26]

Chapman, C. E., and Beauchamp, E. 2006. Differential controls over tactile detection in humans by motor commands and peripheral reafference. *J. Neurophysiol.* 96: 1664–1675. [21]

Charnet, P., Labarca, C., Leonard, R. J., Vogelaar, N. J., Czyzyk, L., Gouin, A. Davidson, N., and Lester, H. A. 1990. An open-channel blocker interacts with adjacent turns of α-helices in the nicotinic acetylcholine receptor. *Neuron* 4: 87–85. [5]

Chattopadhyaya, B., Di Cristo, G., Higashiyama, H., Knott, G. W., Kuhlman, S. J., Welker, E., and Huang, J. 2004. Experience and activity-dependent maturation of perisomatic GABAergic innervation in primary visual cortex during a postnatal critical period. *J. Neurosci.* 24: 9598–9611. [26]

Chaudhari, N., Landin, A. M., and Roper, S. D. 2000. A metabotropic glutamate receptor variant functions as a taste receptor. *Nat. Neurosci.* 3: 113–119. [19]

Chemelli, R. M., Willie, J. T., Sinton, C. M., Elmquist, J. K., Scammell, T., Lee, C., Richardson, J. A., Williams, S. C., Xiong, Y., Kisanuki, Y., Fitch, T. E., Nakazato, M., Hammer, R. E., Saper, C. B., and Yanagisawa, M. 1999. Narcolepsy in orexin knockout mice: molecular genetics of sleep regulation. *Cell* 98: 437–451. [14]

Chen, A., Kumar, S. M., Sahley, C. L., and Muller, K. J. 2000. Nitric oxide influences injury-induced microglial migration and accumulation in the leech CNS. *J. Neurosci.* 20: 1036–1043. [10]

Chen, C. C., Akopian, A. N., Sivilotti, L., Colquhoun, D., and Wood, J. N. 1995. A P2X purinoceptor expressed by a subset of sensory neurons. *Nature* 377: 428–431. [19]

Chen, C. K. 2005. The vertebrate phototransduction cascade: amplification and termination mechanisms. *Rev. Physiol. Biochem. Pharmacol.* 154: 101–121. [20]

Chen, J., Makino, C. L., Peachey, N. S., Baylor, D. A., and Simon, M. I. 1995. Mechanisms of rhodopsin inactivation in vivo as revealed by a COOH-terminal truncation mutant. *Science* 267: 374–377. [20]

Chen, N-H., Reith, M. E. A., and Quick, M. W. 2004. Synaptic uptake and beyond: The sodium- and chloride-dependent neurotransmitter transporter family SLC6. *Pflügers Arch.* 447: 519–531. [9]

Chen, R., Tilley, M. R., Wei, H., Zhou, F., Zhou, F-M., Ching, S., Quan, N., Stephens, R. L., Hill, E. R., Nottoli, T., Han, D. D., and Gu, H. H. 2006. Abolished cocaine reward in mice with a cocaine-insensitive dopamine transporter. *Proc. Natl. Acad. Sci. USA* 103: 9333–9338. [14]

Chen, X., Tomochick, D. R., Kovrigin, E., Arac, D., Machius, M., Südhof, T., and Rizo, J. 2002. Three-dimensional structure of the complexin/SNARE complex. *Neuron* 33: 397–409. [13]

Chen, Z., Gore, B. B., Long, H., Ma, L., and Tessier-Lavigne, M. 2008. Alternative splicing of the Robo3 axon guidance receptor governs the midline switch from attraction to repulsion. *Neuron* 58: 325–532. [25]

Cheney, D. P., and Fetz, E. E. 1980. Functional classes of primate corticomotoneuronal cells and their relation to active force. *J. Neurophysiol.* 44: 773–791. [24]

Cheng, D., Hoogenraad, C. C., Rush, J., Ramm, E., Schlager, M. A., Duong, D. M., Xu, P., Wijayawardana, S. R., Hanfelt, J., Nakagawa, T., Sheng, M., and Peng, J. 2006. Relative and absolute quantification of postsynaptic density proteome isolated from rat forebrain and cerebellum. *Mol. Cell. Proteomics* 5: 1158–1170. [11]

Cheng, H., and Lederer, W. J. 2008. Calcium sparks. *Physiol. Rev.* 88: 1491–1545. [12]

Cheng, K., Narendra, A., Sommer, S., and Wehner, R. 2009. Traveling in clutter: navigation in the Central Australian desert ant *Melophorus bagoti. Behav. Processes* 80: 261–268. [18]

Cherubini, E., Gaiarsa, J. L., and Ben-Ari, Y. 1991. GABA: An excitatory transmitter in early postnatal life. *Trends Neurosci.* 14: 515–519. [17, 25]

Cherubini, E., Griguoli, M., Safiulina, V., and Lagostena, L. 2011. The depolarizing action of GABA controls early network activity in the developing hippocampus. *Mol. Neurobiol.* 43: 97–106. [25]

Chesler, A. T., Zou, D. J., Le Pichon, C. E., Peterlin, Z. A., Matthews, G. A., Pei, X., Miller, M. C., and Firestein, S. 2007. A G protein/cAMP signal cascade is required for axonal convergence into olfactory glomeruli. *Proc. Natl. Acad. Sci. USA* 104: 1039–1044. [26]

Chiu, S. Y., and Ritchie, J. M. 1981. Evidence for the presence of potassium channels in the paranodal region of acutely demyelinated mammalian nerve fibres. *J. Physiol.* 313: 415–437. [8]

Choi, H. J., Lee, C. J., Schroeder, A., Kim, Y. S., Jung, S. H., Kim, J. S., Kim, D. Y., Son, E. J., Han, H. C., Hong, S. K., Colwell, C. S., and Kim, Y. I. 2008. Excitatory actions of GABA in the suprachiasmatic nucleus. *J. Neurosci.* 28: 5450–5459. [17]

Choi, K. L., Aldrich, R. W., and Yellen, G. 1991. Tetraethylammonium blockade distinguishes two inactivation states in voltage-gated K^+ channels. *Proc. Natl. Acad. Sci. USA* 88: 5092–5095. [7]

Choquet, D., and Trille, A. 2003. The role of receptor diffusion in the organization of the postsynaptic membrane. *Nat. Rev. Neurosci.* 4: 251–265. [14]

Christie, B. R., and Abraham, W. C. 1994. L-type voltage-sensitive calcium channel antagonists block heterosynaptic long-term facilitation depression in the dentate gyrus of anesthetized rats. *Neurosci. Lett.* 167: 41–45. [16]

Christmann, C., Koeppe, C., Braus, D. F., Ruf, M., and Flor, H. 2007. A simultaneous EEG–fMRI study of painful electric stimulation. *Neuroimage* 34: 1428–1437. [21]

Cifra, A., Nani, F., Sharifullina, E., and Nistri, A. 2009. A repertoire of rhythmic bursting produced by hypoglossal motoneurons in physiological and pathological conditions. *Philos. Trans. R. Soc. Lond., B, Biol. Sci.* 364: 2493–2500. [24]

Cipelletti, B., Avanzini, G., Vitellaro-Zuccarello, L., Franceschetti, S., Sancini, G., Lavazza, T., Acampora, D., Simeone, A., Spreafico, R., and Frassoni, C. 2002. Morphological organization of somatosensory cortex in $Otx1^{-/-}$ mice. *Neuroscience* 115: 657–667. [25]

Civelli, O., Saito, Y., Wang, Z., Nothacker, H. P., and Reinscheid, R. K. 2006. Orphan GPCRs and their ligands. *Pharmacol. Ther.* 110: 525–532. [14]

Clapham, D. E. 2007. Calcium signaling. *Cell* 131: 1047–1058. [12]

Clapham, D. E., and Neer, E. J. 1997. G protein beta gamma subunits. *Annu. Rev. Pharmacol. Toxicol.* 37: 167–203. [12]

Clarey, J. C., Barone, P., and Imig, T. J. 1994. Functional organization of sound direction and sound pressure level in primary auditory cortex of the cat. *J. Neurophysiol.* 72: 2383–2405. [22]

Cline, H. T. 1986. Evidence for GABA as a neurotransmitter in the leech. *J. Neurosci.* 6: 2848–2856. [18]

Close, R. I. 1972. Dynamic properties of mammalian skeletal muscles. *Physiol. Rev.* 52: 129–197. [27]

Clowry, G., Molnár, Z., and Rakic, P. 2010. Renewed focus on the developing human neocortex. *J. Anat.* 217: 276–288. [25]

Cobb, S. R., Buhl, E. H., Halasy, K., Paulsen, O., and Somogyi, P. 1995. Synchronization of neuronal activity in hippocampus by individual GABAergic interneurons. *Nature* 378: 75–78. [14]

Cochilla, A. J., Angleson, J. K., and Betz, W. 1999. Monitoring secretory membrane with FM1-43 fluorescence. *Annu. Rev. Neurosci.* 22: 1–10. [13]

Coesmans, M., Weber, J. T., De Zeeuw, C. I., and Hansel, C. 2004. Bidirectional parallel fiber plasticity in the cerebellum under climbing fiber control. *Neuron* 44: 691–700. [24]

Coggeshall, R. E., and Fawcett, D. W. 1964. The fine structure of the central nervous system of the leech, *Hirudo Medicinalis. J. Neurophysiol.* 27: 229–289. [18]

Cohen, B., Maruta, J., and Raphan, T. 2001. Orientation of the eyes to gravitoinertial acceleration. *Ann. N Y Acad. Sci.* 942: 241–258. [22]

Cohen, I., Rimer, M., Lømo, T., and McMahan, U. J. 1997. Agrin-induced postsynaptic-like apparatus in skeletal muscle fibers in vivo. *Mol. Cell. Neurosci.* 9: 237–253. [27]

Cohen, M. R., and Newsome, W. T. 2004. What electrical microstimulation has revealed about the neural basis of cognition. *Curr. Opin. Neurobiol.* 14: 169–177. [3, 23]

Cohen, M. R., and Newsome, W. T. 2009. Estimates of the contribution of single neurons to perception depend on timescale and noise correlation. *J. Neurosci.* 29: 6635–6648. [3]

Cohen, M. W. 1972. The development of neuromuscular connexions in the presence of D-tubocurarine. *Brain Res.* 41: 457–463. [27]

Cohen, M. W., Jones, O. T., and Angelides, K. J. 1991. Distribution of Ca^{2+} channels on frog motor nerve terminals revealed by fluorescent omega-conotoxin. *J. Neurosci.* 11: 1032–1039. [13]

Cole, A. E., and Nicoll, R. A. 1984. Characterization of a slow cholinergic post-synaptic potential recorded in vitro from rat hippocampal pyramidal cells. *J. Physiol.* 352: 173–188. [14]

Coleman, G. T., Bahramali, H., Zhang, H. Q., and Rowe, M. J. 2001. Characterization of tactile afferent fibers in the hand of the marmoset monkey. *J. Neurophysiol.* 85: 1793–1804. [21]

Coleman, M. P., and Freeman, M. R. 2010. Wallerian degeneration, Wld^s, and Nmnat. *Annu. Rev. Neurosci.* 33: 245–267. [27]

Collett, M., Collett, T. S., and Wehner, R. 1999. Calibration of vector navigation in desert ants. *Curr. Biol.* 9: 1031–1034. [18]

Collett, M., Collett, T. S., Bisch, S., and Wehner, R. 1998. Local and global vectors in desert ant navigation. *Nature* 394: 269–272. [18]

Collett, T. S., and Baron, J. 1994. Biological compasses and the coordinate frame of landmark memories in honeybees. *Nature* 368: 137–140. [18]

Collett, T. S., and Graham, P. 2004. Animal navigation: path integration, visual landmarks and cognitive maps. *Curr. Biol.* 14: R475–477. [18]

Collin, E., Mauborgne, A., Bourgoin, S., Chantrel, D., Hamon, M., and Cesselin, F. 1991. *In vivo* tonic inhibition of spinal substance P (-like material) release by endogenous opioid(s) acting at δ receptors. *Neuroscience* 44: 725–731. [14]

Collingridge, G. L., Kehl, S. J., and McClennan, H. 1983. Excitatory amino acids in synaptic transmission in the Schaffer collateral-commissural pathway of the rat hippocampus. *J. Physiol.* 334: 33–46. [16]

Collingridge, G. L., Olsen, R. W. Peters, J., and Spedding, M. 2009. A nomenclature for ligand-gated ion channels. *Neuropharmacology* 56: 2–5. [5]

Collins, M. O., Husi, H., Yu, L., Brandon, J. M., Anderson, C. N., Blackstock, W. P., Choudhary, J. S., and Grant, S. G. 2006. Molecular characterization and comparison of the components and multiprotein complexes in the postsynaptic proteome. *J. Neurochem.* 97(Suppl. 1): 16–23. [11]

Colquhoun, D. 2007. What have we learned from single ion channels? *J. Physiol.* 581: 425–427. [11]

Colquhoun, D., and Sakmann, B. 1981. Fluctuations in the microsecond time range of the current through single acetylcholine receptor ion channels. *Nature* 294: 464–466. [11]

Colquhoun, D., and Sakmann, B. 1985. Fast events in single-channel currents activated by acetylcholine and its analogues at the frog muscle end-plate. *J. Physiol.* 369: 501–557. [11]

Colquhoun, D., Dreyer, F., and Sheridan, R. E. 1979. The actions of tubocurarine at the frog neuromuscular junction. *J. Physiol.* 293: 247–284. [11]

Colwell, C. S. 2011. Linking neural activity and molecular oscillations in the SCN. *Nat. Rev. Neurosci.* 12: 553–569. [17]

Comb, M., Hyman, S. E., and Goodman, H. M. 1987. Mechanisms of trans-synaptic regulation of gene expression. *Trends Neurosci.* 10: 473–478. [15]

Congar, P., Leinekugel, X., Ben-Ari, Y., and Crépel, V. 1997. A long-lasting calcium-activated nonselective cationic current is generated by synaptic stimulation or exogenous activation of group I metabotropic glutamate receptors in CA1 pyramidal neurons. *J. Neurosci.* 17: 5366–5379. [12]

Conley, M., Penny, G. R., and Diamond, I. T. 1987. Terminations of individual optic tract fibers in the lateral geniculate nuclei of *Galago crassicaudatus* and *Tupaia belangeri. J. Comp. Neurol.* 256: 71–87. [2]

Connor, C. E., and Johnson, K. O. 1992. Neural coding of tactile texture: comparison of spatial and temporal mechanisms for roughness perception. *J. Neurosci.* 12: 3414–3426. [21]

Connor, C. E., Hsiao, S. S., Phillips, J. R., and Johnson, K. O. 1990. Tactile roughness: neural codes that account for psychophysical magnitude estimates. *J. Neurosci.* 10: 3823–3836. [21]

Connors, B. W., and Long, M. A. 2004. Electrical synapses in the mammalian brain. *Annu. Rev. Neurosci.* 27: 393–418. [11]

Consiglio, J. F., Andalib, P., and Korn, S. J. 2003. Influence of pore residues on permeation properties in the Kv2.1 potassium channel. Evidence for a selective functional interaction of K^+ with the outer vestibule. *J. Gen. Physiol.* 121: 111–124. [5]

Constantinidis, C., Franowicz, M. N., and Goldman-Rakic, P. S. 2001. The sensory nature of mnemonic representation in the primate prefrontal cortex. *Nat. Neurosci.* 4: 311–316. [23]

Cook, S. P., Vulchanova, L., Hargreaves, K. M., Elde, R., and McCleskey, E. W. 1997. Distinct ATP receptors on pain-sensing and stretch-sensing neurons. *Nature* 387: 505–508. [19]

Coombs, J. S., Curtis, D. R., and Eccles, J. C. 1957. The generation of impulses in motoneurones. *J. Physiol.* 139: 232–249. [11]

Coombs, J. S., Eccles, J. C., and Fatt, P. 1955a. The electrical properties of the motoneuron membrane. *J. Physiol.* 130: 291–325. [8]

Coombs, J. S., Eccles, J. C., and Fatt, P. 1955b. The specific ionic conductances and the ionic movements across the motoneuronal membrane that produce the inhibitory post-synaptic potential. *J. Physiol.* 130: 326–373. [11]

Cooper, J. R., Bloom, F. E., and Roth, R. H. 2002. *The Biochemical Basis of Pharmacology*. Oxford University Press, New York. [17]

Cooper, J. R., Bloom, F. E., and Roth, R. H. 2003. *The Biochemical Basis of Neuropharmacology*, 8th ed. Oxford University Press. [14]

Copley, R. R. 2008. The animal in the genome: comparative genomics and evolution. *Philos. Trans. R. Soc. Lond., B, Biol. Sci.* 363: 1453–1461. [18]

Corey, D. P., and Hudspeth, A. J. 1979. Ionic basis of the receptor potential in a vertebrate hair cell. *Nature* 281: 675–677. [19]

Corey, D. P., and Hudspeth, A. J. 1983. Kinetics of the receptor current in bullfrog saccular hair cells. *J. Neurosci.* 3: 962–976. [19]

Couteaux, R., and Pecot-Déchavassine, M. 1970. Vésicules synaptiques et poches au niveau des zones actives de la jonction neuromusculaire. *C. R. Acad. Sci. (Paris)* 271: 2346–2349. [13]

Couve, A., Moss, S. J., and Pangalos, M. N. 2000. $GABA_B$ receptors: a new paradigm in G protein signaling. *Mol. Cell. Neuroscience* 16: 296–312. [14]

Cowen, P. J. 2008. Serotonin and depression: pathophysiological mechanism or marketing myth? *Trends Pharmacol. Sci.* 29: 433–436. [14]

Cox, E. C., Muller, B., and Bonhoeffer, F. 1990. Axonal guidance in the chick visual system: posterior tectal membranes induce collapse of growth cones from the temporal retina. *Neuron* 4: 31–37. [25]

Crago, P. E., Houk, J. C., and Rymer, W. Z. 1982. Sampling of total muscle force by tendon organs. *J. Neurophysiol.* 47: 1069–1083. [24]

Craig, A. M., and Kang, Y. 2007. Neurexin–neuroligin signaling in synapse development. *Curr. Opin. Neurobiol.* 17: 43–52. [25]

Crair, M. C., Gillespie, D. C., and Stryker, M. P. 1998. The role of visual experience in the development of columns in cat visual cortex. *Science* 279: 566–570. [26]

Crawford, A. C., and Fettiplace, R. 1981. An electrical tuning mechanism in turtle cochlear hair cells. *J. Physiol.* 312: 377–412. [22]

Crawford, A. C., and Fettiplace, R. 1985. The mechanical properties of ciliary bundles of turtle cochlear hair cells. *J. Physiol.* 364: 359–379. [19]

Crawford, A. C., Evans, M. G., and Fettiplace, R. 1989. Activation and adaptation of transducer currents in turtle hair cells. *J. Physiol.* 419: 405–434. [19]

Crawford, A. C., Evans, M. G., and Fettiplace, R. 1991. The actions of calcium on the mechano-electrical transducer current of turtle hair cells. *J. Physiol.* 434: 369–398. [19]

Crepel, F., and Jaillard, D. 1991. Pairing of pre- and postsynaptic activities in cerebellar Purkinje cells induces long-term changes in synaptic efficacy in vitro. *J. Physiol.* 432: 123–141. [16]

Critchlow, V., and von Euler, C. 1963. Intercostal muscle spindle activity and its γ–motor control. *J. Physiol.* 168: 820–847. [24]

Crivellato, E., Nico, B., and Ribatti, D. 2008. The chromaffin vesicle: advances in understanding the composition of a versatile, multifunctional secretory organelle. *Anat. Rec. (Hoboken)* 291: 1587–1602. [17]

Croner, L. J., and Kaplan, E. 1995. Receptive fields of P and M ganglion cells across the primate retina. *Vision Res.* 35: 7–24. [20]

Crowley, C., Spencer, S. D., Nishimura, M. C., Chen, K. S., Pitts-Meek, S., Armanini, M. P., Ling, L. H., McMahon, S. B., Shelton, D. L., Levinson, A. D., and Phillips, H. S. 1994. Mice lacking nerve growth factor display perinatal loss of sensory and sympathetic neurons yet develop basal forebrain cholinergic neurons. *Cell* 76: 1001–1011. [25]

Cull-Candy, S., Kelly, L., and Farrant, M. 2007. Regulation of Ca^{2+}-permeable AMPA receptors: synaptic plasticity and beyond. *Curr. Opin. Neurobiol.* 17: 277–280. [14]

Cull-Candy, S. G., Miledi, R., and Parker, I. 1980. Single glutamate-activated channels recorded from locust muscle fibres with perfused patch-clamp electrodes. *J. Physiol.* 321: 195–210. [4]

Cullen, K. E., Minor, L. B., Beraneck, M., and Sadeghi, S. G. 2009. Neural substrates underlying vestibular compensation: contribution of peripheral versus central processing. *J. Vestib. Res.* 19: 171–182. [22]

Cully, D. S., Vassilatis, D. K., Liu, K. K., Paress, P. S., Van Der Ploeg, L. H. T., Schaeffer, J. M., and Arena, J. P. 1994. Cloning of an avermectin-sensitive glutamate-gated chloride channel from *Caenorhabditis elegans*. *Nature* 371: 707–711. [5]

Curtis, B. M., and Catterall, W. A. 1986. Reconstitution of the voltage-sensitive calcium channel purified from skeletal muscle transverse tubules. *Biochemistry* 25: 3077–3083. [12]

Curtis, D. R., and Phillis, J. W. 1958. Gamma-amino-*n*-butyric acid and spinal synaptic transmission. *Nature* 182: 323. [14]

Curtis, D. R., Phillis, J. W., and Watkins, J. C. 1959. The depression of spinal neurones by γ-amino-*n*-butyric acid and β-alanine. *J. Physiol* 146: 185–203. [14]

Curtis, D. R., Phillis, J. W., and Watkins, J. C. 1960. The chemical excitation of spinal neurones by certain acidic amino acids. *J. Physiol.* 150: 656–682. [14]

Cuttle, M. F., Tsujimoto, T., Forsythe, I. D., and Takahashi, T. 1998. Facilitation of the presynaptic calcium current at an auditory synapse in rat brainstem. *J. Physiol.* 512: 723–729. [16]

Cynader, M., and Mitchell, D. E. 1980. Prolonged sensitivity to monocular deprivation in dark-reared cats. *J. Neurophysiol.* 43: 1026–1040. [26]

D'Antoni, S., Berretta, A., Bonaccorso, C. M., Bruno, V., Aronica, E., Nicoletti, F., and Catania, M. V. 2008. Metabotropic glutamate receptors in glial cells. *Neurochem. Res.* 33: 2436–2443. [10]

D'Arcangelo, G., Nakajima, K., Miyata, T., Ogawa, M., Mikoshiba, K., and Curran, T. 1997. Reelin is a secreted glycoprotein recognized by the CR-50 monoclonal antibody. *J. Neurosci.* 17: 23–31. [25]

d'Incamps, B. L., and Ascher, P. 2008. Four excitatory postsynaptic ionotropic receptors coactivated at the motoneuron-Renshaw cell synapse. *J. Neurosci.* 28: 14121–14131. [14]

Da Silva, K. M. C., Sayers, B. M., Sears, T. A., and Stagg, D. T. 1977. The changes in configuration of the rib cage and abdomen during breathing in the anaesthetized cat. *J. Physiol.* 266: 499–521. [24]

Dacke, M., and Srinivasan, M. V. 2007. Honeybee navigation: distance estimation in the third dimension. *J. Exp. Biol.* 210: 845–853. [18]

Dahlstom, A., and Fuxe, K. 1964. Evidence for the existence of monoamine-containing neurons in the central nervous system. I. Demonstration of monoamines in the cell bodies of brain stem neurons. *Acta Physiol. Scand. Suppl.* 232: 1–55. [14]

Dahlstrom, A. B. 2010. Fast intra-axonal transport: Beginning, development and post-genome advances. *Prog. Neurobiol.* 90: 119–145. [15]

Dale, H. H. 1914. The action of certain esters and ethers of choline, and their relation to muscarine. *J. Pharmacol. Exp. Ther.* 6: 147–190. [12]

Dale, H. H. 1933. Nomenclature of fibres in the autonomic nervous system and their effects. *J. Physiol.* 80: 10–11. [15]

Dale, H. H. 1953. *Adventures in Physiology*. Pergamon, London. [11]

Dale, H. H., Feldberg, W., and Vogt, M. 1936. Release of acetylcholine at voluntary motor nerve endings. *J. Physiol.* 86: 353–380. [11]

Dallos, P. 2008. Cochlear amplification, outer hair cells and prestin. *Curr. Opin. Neurobiol.* 18: 370–376. [22]

Dan, Y., Lo, Y., and Poo, M. M. 1995. Plasticity of developing neuromuscular synapses. *Prog. Brain Res.* 105: 211–215. [27]

Dani, J. A., and Bertrand, D. 2007. Nicotinic acetylcholine receptors and nicotinic cholinergic mechanisms of the central nervous system. *Annu. Rev. Pharmacol. Toxicol.* 47: 699–729. [14]

Daniel, P. 1946. Spinal nerve endings in the extrinsic eye muscles of man. *J. Anat.* 80: 189–193. [24]

Daniel, P. M., and Whitteridge, D. 1961. The representation of the visual field on the cerebral cortex in monkeys. *J. Physiol.* 159: 203–221. [3]

Danjo, T., Eiraku, M., Muguruma, K., Watanabe, K., Kawada, M., Yanagawa, Y., Rubenstein, J. L., and Sasai, Y. 2011. Subregional specification of embryonic stem cell-derived ventral telencephalic tissues by timed and combinatory treatment with extrinsic signals. *J. Neurosci.* 31: 1919–1933. [25]

Darbon, P., Tscherter, A., Yvon, C., and Streit, J. 2003. Role of the electrogenic Na/K pump in disinhibition-induced bursting in cultured spinal networks. *J. Neurophysiol.* 90: 3119–3129. [7]

Darland, T., Heinricher, M. M., and Grandy, D. K. 1998. Orphanin FQ/nociceptin: a role in pain and analgesia, but so much more. *Trends Neurosci.* 21: 215–221. [14]

Dartnall, H. J., Bowmaker, J. K., and Mollon, J. D. 1983. Human visual pigments: microspectrophotometric results from the eyes of seven persons. *Proc. R. Soc. Lond., B, Biol. Sci.* 220: 115–130. [20]

Dascal, N. 2001. Ion-channel regulation by G proteins. *Trends Endocrinol. Metab.* 12: 391–398. [12]

Dasen, J. S., and Jessell, T. M. 2009. Hox networks and the origins of motor neuron diversity. *Curr. Top. Dev. Biol.* 88: 169–200. [24]

DaSilva, A. F., Becerra, L., Makris, N., Strassman, A. M., Gonzalez, R. G., Geatrakis, N., and Borsook, D. 2002. Somatotopic activation in the human trigeminal pain pathway. *J. Neurosci.* 22: 8183–8192. [21]

David, S., and Aguayo, A. J. 1981. Axonal elongation into peripheral nervous system "bridges" after central nervous system injury in adult rats. *Science* 214: 931–933. [27]

Davidson, A. J., Yamazaki, S., and Menaker, M. 2003. SCN: ringmaster of the circadian circus or conductor of the circadian orchestra? *Novartis Found. Symp.* 253: 110–121. [17]

Davies, A. M., and Lumsden, A. 1990. Ontogeny of the somatosensory system: origins and early development of primary sensory neurons. *Annu. Rev. Neurosci.* 13: 61–73. [25]

Davies, C. H., and Collingridge, G. L. 1993. The physiological regulation of synaptic inhibition by GABAB autoreceptors in rat hippocampus. *J. Physiol.* 472: 245–265. [11]

Davies, J. G., Kirkwood, P. A., and Sears, T. A. 1985. The distribution of monosynaptic connexions from inspiratory bulbospinal neurones to inspiratory motoneurones in the cat. *J. Physiol.* 368: 63–87. [24]

Davies, K. E., and Nowak, K. J. 2006. Molecular mechanisms of muscular dystrophies: old and new players. *Nat. Rev. Mol. Cell Biol.* 7: 762–773. [11]

Davies, P. A., Wang, W., Hales, T. G., and Kirkness, E. F. 2003. A novel class of ligand-gated ion channel is activated by $Zn2^+$ *J. Biol. Chem.* 278: 712–717. [5]

Davies, S. J. A., Fitch, M. T., Memberg, S. P., Hall, A. K., Raisman, G., and Silver, J. 1997. Regeneration of adult axons in white matter tracts of the central nervous system. *Nature* 390: 680–683. [27]

Davis, K. D. 2000. The neural circuitry of pain as explored with functional MRI. *Neurol. Res.* 22: 313–317. [21]

Davis, R., and Koelle, G. B. 1978. Electron microscope localization of acetylcholinesterase and butyrylcholinesterase in the superior cervical ganglion of the cat. I. Normal ganglion. *J. Cell Biol.* 78: 785–809. [15]

Daw, N. W. 1998. Critical periods and amblyopia. *Arch. Ophthalmol.* 116: 502–505. [26]

Daw, N. W., Jensen, R. J., and Brunken, W. J. 1990. Rod pathways in mammalian retinae. *Trends Neurosci.* 13: 110–115. [2, 20]

Daw, N. W., Reid, S. N., Wang, X. F., and Flavin, H. J. 1995. Factors that are critical for plasticity in the visual cortex. *Ciba Found. Symp.* 193: 258–276; discussion 322–254. [26]

De Biase, L. M., Nishiyama, A., and Bergles, D. E. 2010. Excitability and synaptic communication within the oligodendrocyte lineage. *J. Neurosci.* 30: 3600–3611. [10]

De Boysson-Bardies, B., Halle, P., Sagart, L., and Durand, C. 1989. A crosslinguistic investigation of vowel formants in babbling. *J. Child Lang.* 16: 1–17. [22]

De Felipe, C., Herrero, J. F., O'Brien, J. A., Palmer, J. A., Doyle, C. A., Smith, A. J., Laird, J. M., Belmonte, C., Cervero, F., and Hunt, S. P. 1998. Altered nociception, analgesia and aggression in mice lacking the receptor for substance P. *Nature* 392: 394–397. [14]

de Lecea, L., Kilduff, T. S., Peyron, C., Gao, X., Foye, P. E., Danielson, P. E., Fukuhara, C., Battenberg, E. L., Gautvik, V. T., Bartlett, F. S. 2nd, Frankel, W. N., van den Pol, A. N., Bloom, F. E., Gautvik, K. M., and Sutcliffe, J. G. 1998. The hypocretins: hypothalamus-specific peptides with neuroexcitatory activity. *Proc. Natl. Acad. Sci. USA* 95: 322–327. [14]

De Potter, W. P., Smith, A. D., and De Schaepdryver, A. F. 1970. Subcellular fractionation of splenic nerve: ATP, chromogranin A, and dopamine β-hydroxylase in noradrenergic vesicles. *Tissue Cell* 2: 529–546. [15]

de Villers-Sidani, E., Chang, E. F., Bao, S., and Merzenich, M. M. 2007. Critical period window for spectral tuning defined in the primary auditory cortex (A1) in the rat. *J. Neurosci.* 27: 180–189. [26]

De Waard, M., Hering, J., Weiss, N., and Feltz, A. 2005. How do G proteins directly control neuronal Ca^{2+} channel function? *Trends Pharmacol. Sci.* 26: 427–436. [12]

De-Miguel, F. F., and Trueta, C. 2005. Synaptic and extrasynaptic secretion of serotonin. *Cell Mol. Neurobiol.* 25: 297–312. [15, 18]

DeAngelis, G. C., Cumming, B. G., and Newsome, W. T. 1998. Cortical area MT and the perception of stereoscopic depth. *Nature* 394: 677–680. [2]

Deans, M. R., Volgyi, B., Goodenough, D. A., Bloomfield, S. A., and Paul, D. L. 2002. Connexin36 is essential for transmission of rod-mediated visual signals in the mammalian retina. *Neuron* 36: 703–712. [20]

Debanne, D., and Thompson, S. M. 1996. Associative long-term depression in the hippocampus in vitro. *Hippocampus* 6: 9–16. [16]

Debby-Brafman, A., Burstyn-Cohen, T., Klar, A., and Kalcheim, C. 1999. F-Spondin, expressed in somite regions avoided by neural crest cells, mediates inhibition of distinct somite domains to neural crest migration. *Neuron* 22: 475–488. [25]

DeChiara, T. M., Vejsada, R., Poueymirou, W. T., Acheson, A., Suri, C., Conover, J. C., Friedman, B., McClain, J., Pan, L., Stahl, N., and Yancopolous, G. 1995. Mice lacking the CNTF receptor, unlike mice lacking CNTF, exhibit profound motor neuron deficits at birth. *Cell* 83: 313–322. [27]

Dedek, K., Kunath, B., Kananura, C., Reuner, U., Jentsch, T. J., and Steinlein, O. K. 2001. Myokymia and neonatal epilepsy caused by a mutation in the voltage sensor of KCNQ2 K^+ channel. *Proc. Natl. Acad. Sci. USA.* 98: 12272–12277. [8]

Deeb, S. S. 2006. Genetics of variation in human color vision and the retinal cone mosaic. *Curr. Opin. Genet. Dev.* 16: 301–307. [20]

Deeb, S. S., and Kohl, S. 2003. Genetics of color vision deficiencies. *Dev. Ophthalmol.* 37: 170–187. [20]

DeFazio, R. A., Dvoryanchikov, G., Maruyama, Y., Kim, J. W., Pereira, E., Roper, S. D., and Chaudhari, N. 2006. Separate populations of receptor cells and presynaptic cells in mouse taste buds. *J. Neurosci.* 26: 3971–3980. [19]

Del Bel, E. A., Guimarães, F. S., Bermúdez-Echeverry, M., Gomes, M. Z., Schiaveto-de-souza, A., Padovan-Neto, F. E., Tumas, V., Barion-Cavalcanti, A. P., Lazzarini, M., Nucci-da-Silva, L. P., and de Paula-Souza, D. 2005. Role of nitric oxide on motor behavior. *Cell Mol. Neurobiol.* 25: 371–392. [24]

del Castillo, J., and Katz, B. 1954a. Quantal components of the end-plate potential. *J. Physiol.* 124: 560–573. [13]

del Castillo, J., and Katz, B. 1954b. Statistical factors involved in neuromuscular facilitation and depression. *J. Physiol.* 124: 574–585. [16]

del Castillo, J., and Katz, B. 1954c. Changes in end-plate activity produced by presynaptic polarization. *J. Physiol.* 124: 586–604. [13]

del Castillo, J., and Katz, B. 1955. On the localization of end-plate receptors. *J. Physiol.* 128: 157–181. [11]

del Castillo, J., and Katz, B. 1956. Biophysical aspects of neuromuscular transmission. *Prog. Biophys. Biophys. Chem.* 6: 121–170. [13]

del Castillo, J., and Stark, L. 1952. The effect of calcium ions on the motor end-plate potentials. *J. Physiol.* 116: 507–515. [13]

Del Rio-Hortega, P. 1920. La microglia y su transformacion celulas en basoncito y cuerpos granulo-adiposos. *Trab. Lab. Invest. Biol. Madrid* 18: 37–82. [10]

Deller, T., Haas, C. A., Freiman, T. M., Phinney, A., Jucker, M., and Frotscher, M. 2006. Lesion-induced axonal sprouting in the central nervous system. *Adv. Exp. Med. Biol.* 557: 101–121. [27]

Delmas, P., and Brown, D. A. 2005. Pathways modulating neural KCNQ/M (Kv7) potassium channels. *Nat. Rev. Neurosci.* 6: 850–862. [17]

Delmas, P., Brown, D. A., Dayrell, M., Abogadie, F. C., Caulfield, M. P., and Buckley, N. J. 1998. On the role of endogenous G-protein beta gamma subunits in N-type Ca^{2+} current inhibition by neurotransmitters in rat sympathetic neurones. *J. Physiol.* 506: 319–329. [12]

Delmas, P., Crest, M., and Brown, D. A. 2004. Functional organization of PLC signaling microdomains in neurons. *Trends Neurosci.* 27: 41–47. [12]

Delmas, P., Wanaverbecq, N., Abogadie, F. C., Mistry, M., and Brown, D. A. 2002. Signaling microdomains define the specificity of receptor-mediated InsP(3) pathways in neurons. *Neuron* 34: 209–220. [12]

DeLong, M., and Wichmann, T. 2009. Update on models of basal ganglia function and dysfunction. *Parkinsonism Relat Disord.* 15(Suppl. 3): S247–240. [24]

Denda, S., and Reichardt, L. F. 2007. Studies on integrins in the nervous system. *Methods Enzymol.* 426: 203–221. [25]

Deniz, S., Wersinger, E., Schwab, Y., Mura, C., Erdelyi, F., Szabo, G., Rendon, A., Sahel, J. A., Picaud, S., and Roux, M. J. 2011. Mammalian retinal horizontal cells are unconventional GABAergic neurons. *J. Neurochem.* 116: 350–362. [20]

Denk, W., Holt, J. R., Shepherd, G. M., and Corey, D. P. 1995. Calcium imaging of single stereocilia in hair cells: localization of transduction channels at both ends of tip links. *Neuron* 15: 1311–1321. [19]

Denk, W., Sugimori, M., and Llinas, R. 1995. Two types of calcium response limited to single spines in cerebellar Purkinje cells. *Proc. Natl. Acad. Sci. USA* 92: 8279–8282. [12]

Dennis, M. J., and Miledi, R. 1974. Characteristics of transmitter release at regenerating frog neuromuscular junctions. *J. Physiol.* 239: 571–594. [13]

Dennis, M. J., and Sargent, P. B. 1979. Loss of extrasynaptic acetylcholine sensitivity upon reinnervation of parasympathetic ganglion cells. *J. Physiol.* 289: 263–275. [27]

Dennis, M. J., and Yip, J. W. 1978. Formation and elimination of foreign synapses on adult salamander muscle. *J. Physiol.* 274: 299–310. [27]

Dennis, M. J., Harris, A. J., and Kuffler, S. W. 1971. Synaptic transmission and its duplication by focally applied acetylcholine in parasympathetic neurons in the heart of the frog. *Proc. Roy. Soc. Lond., B, Biol. Sci.* 177: 509–539. [11]

Denzer, A. J., Schulthess, T., Fauser, C., Schumacher, B., Kammerer, R. A., Engel, J., and Ruegg, M. A. 1998. Electron microscopic structure of agrin and mapping of its binding site in laminin-1. *EMBO J.* 17: 335–343. [27]

Derijck, A. A., Van Erp, S., and Pasterkamp, R. J. 2010. Semaphorin signaling: molecular switches at the midline. *Trends Cell Biol.* 20: 568–576. [25]

Descarries, L., Gisiger, V., and Steriade, M. 1997. Diffuse transmission by acetylcholine in the CNS. *Prog. Neurobiol.* 53: 603–625. [14]

Desimone, R., Albright, T. D., Gross, C. G., and Bruce, C. 1984. Stimulus-selective properties of inferior temporal neurons in the macaque. J. Neurosci. 4: 2051–2062. [23]

Devaux, J. J., Kleopas, K. A., Cooper, E. C., and Scherer, S. S. 2004. KCNQ2 is a nodal K^+ channel. *J. Neurosci.* 24: 1236–1244. [8]

DeVries, S. H. 2000. Bipolar cells use kainate and AMPA receptors to filter visual information into separate channels. *Neuron* 28: 847–856. [20]

Deyoe, E. A., Hockfield, S., Garren, H., Van Essen, D. C. 1990. Antibody labeling of functional subdivisions in visual cortex: Cat-301 immunoreactivity in striate and extrastriate cortex of the macaque monkey. *Vis. Neurosci.* 5: 67–81. [3]

Dhallan, R. S., Yau, K. W., Schrader, K. A., and Reed, R. R. 1990. Primary structure and functional expression of a cyclic nucleotide-activated channel from olfactory neurons. *Nature* 347: 184–187. [19]

Diamond, M. E., Armstrong-James, M., and Ebner, F. F. 1993. Experience-dependent plasticity in adult rat barrel cortex. *Proc. Natl. Acad. Sci. USA* 90: 2082–2086. [21]

Diamond, M. E., Huang, W., and Ebner, F. F. 1994. Laminar comparison of somatosensory cortical plasticity. *Science* 265: 1885–1888. [21]

Diamond, M. E., von Heimendahl, M., Knutsen, P. M., Kleinfeld, D., and Ahissar, E. 2008. 'Where' and 'what' in the whisker sensorimotor system. *Nat. Rev. Neurosci.* 9: 601–612. [21]

Diana, M., and Tepper, J. M. 2002. Electrophysiological pharmacology of mesencephalic dopaminergic neurons. In *Handbook of Experimental Pharmacology*, vol. 154, part 1. Springer-Verlag, Berlin. pp. 1–62. [14]

Dibaj, P., Nadrigny, F., Steffens, H., Scheller, A., Hirrlinger, J., Schomburg, E. D., Neusch, C., and Kirchhoff, F. 2010. NO mediates microglial response to acute spinal cord injury under ATP control in vivo. *Glia* 58: 1133–1144. [10]

DiCarlo, J. J., and Cox, D. D. 2007. Untangling invariant object recognition. *Trends Cogn. Sci.* 11: 333–341. [23]

DiCarlo, J. J., and Johnson, K. O. 2000. Spatial and temporal structure of receptive fields in primate somatosensory area 3b: effects of stimulus scanning direction and orientation. *J. Neurosci.* 20: 495–510. [21]

DiCarlo, J. J., and Johnson, K. O. 2002. Receptive field structure in cortical area 3b of the alert monkey. *Behav. Brain Res.* 135: 167–178. [21]

DiCarlo, J. J., Johnson, K. O., and Hsiao, S. S. 1998. Structure of receptive fields in area 3b of primary somatosensory cortex in the alert monkey. *J. Neurosci.* 18: 2626–2645. [21]

Dickinson-Nelson, A., and Reese, T. S. 1983. Structural changes during transmitter release at synapses in the frog sympathetic ganglion. *J. Neurosci.* 3: 42–52. [13]

Dickson, B. J., and Gilestro, G. F. 2006. Regulation of commissural axon pathfinding by slit and its Robo receptors. *Annu. Rev. Cell. Dev. Biol.* 22: 651–675. [25]

Diesseroth, K. 2011. Optogenetics. *Nat. Methods* 8: 26–29. [14]

Dietrichs, E. 2008. Clinical manifestation of focal cerebellar disease as related to the organization of neural pathways. *Acta Neurol Scand. Suppl.* 188: 6–11. [24]

DiFrancesco, D., and Tortura, D. P. 1991. Direct activation of cardiac pacemaker channels by intracellular cyclic AMP. *Nature* 351: 145–147. [12]

Dingledine, R., Borges, K., Bowie, D., and Traynelis, S. F. 1999. The glutamate receptor ion channels. *Pharmacol. Rev.* 51: 7–61. [5]

Dionne, V. E., and Leibowitz, M. D. 1982. Acetylcholine receptor kinetics. A description from single-channel currents at the snake neuromuscular junction. *Biophys. J.* 39: 253–261. [11]

DiPaola, M., Czajkowski, C., and Karlin, A. 1989. The sidedness of the COOH terminus of the acetylcholine receptor δ subunit. *J. Biol. Chem.* 264: 15457–15463. [5]

Diss, J. K. J., Fraser, S. P., and Djamgoz, M. B. A. 2004. Voltage-gated Na^+ channels: multiplicity of expression, plasticity, functional implications and pathophysiological aspects. *Eur. Biophys. J.* 33: 180–193. [5]

Dittman, J. 2009. Worm watching: imaging nervous system structure and function in *Caenorhabditis elegans*. *Adv. Genet.* 65: 39–78. [18]

Do, M. T., Kang, S. H., Xue, T., Zhong, H., Liao, H. W., Bergles, D. E., and Yau, K. W. 2009. Photon capture and signalling by melanopsin retinal ganglion cells. *Nature* 457: 281–287. [17]

Dodd, J., and Horn, J. P. 1983a. A reclassification of B and C neurones in the ninth and tenth paravertebral sympathetic ganglia of the bullfrog. *J. Physiol.* 334: 255–269. [12]

Dodd, J., and Horn, J. P. 1983b. Muscarinic inhibition of sympathetic C cells in the bullfrog. *J Physiol.* 334: 271–291. [12]

Dodge, F. A., Jr., and Rahamimoff, R. 1967. Co-operative action of calcium ions in transmitter release at the neuromuscular junction. *J. Physiol.* 193: 419–432. [13, 16]

Doe, C. Q., Kuwada, J. Y., and Goodman, C. S. 1985. From epithelium to neuroblasts to neurons: the role of cell interactions and cell lineage during insect neurogenesis. *Philos. Trans. R. Soc. Lond., B, Biol. Sci.* 312: 67–81. [25]

Doherty, G. J., and McMahon, H. T. 2009. Mechanisms of endocytosis. *Annu. Rev. Biochem.* 78: 815–902. [13]

Domenici, M. R., Berretta, N., and Cherubini, E. 1998. Two distinct forms of long-term depression co-exist at the mossy fiber-CA3 synapse in the hippocampus during development. *Proc. Natl. Acad. Sci. USA* 95: 8310–8315. [16]

Doron, K. W., and Gazzaniga, M. S. 2008. Neuroimaging techniques offer new perspectives on callosal transfer and interhemispheric communication. *Cortex* 44: 1023–1029. [3]

Dougherty, K. J., and Kiehn, O. 2010. Functional organization of V2a-related locomotor circuits in the rodent spinal cord. *Ann. N Y Acad. Sci.* 1198: 85–93. [24]

Doupnik, C. A., Davidson, N., Lester, H. A., and Kofuji, P. 1997. RGS proteins reconstitute the rapid gating kinetics of gbetagamma-activated inwardly rectifying K+ channels. *Proc. Natl. Acad. Sci. USA* 94: 10461–10466. [12]

Dowdall, M. J., Boyne, A. F., and Whittaker, V. P. 1974. Adenosine triphosphate: A constituent of cholinergic synaptic vesicles. *Biochem. J.* 140: 1–12. [15]

Dowling, J. E. 1987. *The Retina: An Approachable Part of the Brain.* Harvard University Press, Cambridge, MA. [20]

Dowling, J. E., and Boycott, B. B. 1966. Organization of the primate retina: Electron microscopy. *Proc. R. Soc. Lond., B, Biol. Sci.* 166: 80–111. [2, 20]

Downing, P. E., Chan, A. W., Peelen, M. V., Dodds, C. M., and Kanwisher, N. 2006. Domain specificity in visual cortex. *Cereb. Cortex* 16: 1453–1461. [23]

Doyle, D. A., Cabral, J. M., Pfeutzner, A. K., Gulbis, J. M., Cohen, S. L., Chait, B. T., and McKinnon, R. 1998. The structure of the potassium channel: Molecular basis of K^+ conductance and selectivity. *Science* 280: 69–77. [5]

Drachman, D. B. 1994. Myasthenia gravis. *New England J. Med.* 330: 1797–1810. [13]

Drescher, U., Kremoser, C., Handwerker, C., Loschinger, J., Noda, M., and Bonhoeffer, F. 1995. In vitro guidance of retinal ganglion cell axons by RAGS, a 25 kDa tectal protein related to ligands for Eph receptor tyrosine kinases. *Cell* 82: 359–370. [25]

Driver, J., and Halligan, P. W. 1991. Can visual neglect operate in object-centred co-ordinates? An affirmative single case study. *Cogn. Neuropsychol.* 8: 475–496. [28]

Droz, B., and Leblond, C. P. 1963. Axonal migration of proteins in the central nervous system and peripheral nerves as shown by radioautography. *J. Comp. Neurol.* 121: 325–346. [15]

Du Bois-Reymond, E. 1848. *Untersuchungen über thierische Electricität.* Reimer, Berlin. [11]

Du Vigneaud, V. 1955. Hormones of the posterior pituitary gland: oxytocin and vasopressin. *Harvey Lect.* 50: 1–26. [14]

Duan, Y., Panoff, J., Burrell, B. D., Sahley, C. L., and Muller, K. J. 2005. Repair and regeneration of functional synaptic connections: cellular and molecular interactions in the leech. *Cell Mol. Neurobiol.* 25: 441–450. [18]

Duan, Y., Sahley, C. L., and Muller, K. J. 2009. ATP and NO dually control migration of microglia to nerve lesions. *Dev. Neurobiol.* 69: 60–72. [10]

Dubois, J. M. 1983. Potassium currents in the frog node of Ranvier. *Prog. Biophys. Mol. Biol.* 42: 1–20. [8]

Duboule, D. 2007. The rise and fall of *Hox* gene clusters. *Development* 134: 2549–2560. [25]

Dubreuil, V., Barhanin, J., Goridis, C., and Brunet, J. F. 2009 Breathing with *phox2b*. *Philos. Trans. R. Soc. Lond., B Biol. Sci.* 364: 2477–2483. [24]

Dudar, J. D., and Szerb, J. C. 1969. The effect of topically applied atropine on resting and evoked cortical acetylcholine release. *J. Physiol.* 203: 741–762. [14]

Dudek, S. M., and Bear, M. F. 1992. Homosynaptic long-term depression in area CA1 of hippocampus and the effects of *N*-methyl-D-aspartate receptor blockade. *Proc. Natl. Acad. Sci. USA* 89: 4363–4367. [16]

Dudel, J., and Kuffler, S. W. 1961. Presynaptic inhibition at the crayfish neuromuscular junction. *J. Physiol.* 155: 543–562. [11]

Dugué, G. P., Brunel, N., Hakim, V., Schwartz, E., Chat, M., Lévesque, M., Courtemanche, R., Léna, C., and Dieudonné, S. 2009. Electrical coupling mediates tunable low-frequency oscillations and resonance in the cerebellar Golgi cell network. *Neuron* 61: 126–139. [11]

Dulac, C., and Axel, R. 1995. A novel family of genes encoding putative pheromone receptors in mammals. *Cell* 83: 195–206. [19]

Dunnett, S. B., Björklund, A., and Stenevi, U. 1983. Dopamine-rich transplants in experimental parkinsonism. *Trends Neurosci.* 6: 266–270. [27]

Dunwiddie, T. V., and Masino, S. A. 2001. The role and regulation of adenosine in the central nervous system. *Annu. Rev. Neurosci.* 24: 31–55. [14, 15]

Dupin, E., Ziller, C., and Le Douarin, N. M. 1998. The avian embryo as a model in developmental studies: chimeras and *in vitro* clonal analysis. *Curr. Top. Dev. Biol.* 36: 1–35. [25]

Durbaba, R., Taylor, A., Ellaway, P. H., and Rawlinson, S. 2003. The influence of bag_2 and chain intrafusal muscle fibers on secondary spindle afferents in the cat. *J. Physiol.* 550: 263–278. [24]

Durbeej, M. 2010. Laminins. *Cell Tissue Res.* 339: 259–268. [25]

Dursteler, R. M., Wurtz, R. H., and Newsome, W. T. 1987. Directional pursuit deficits following lesions of the foveal representation within the superior temporal sulcus of the macaque monkey. *J. Neurophysiol.* 57: 1262–1287. [23]

Dusart, I., Ghoumari, A., Wehrle, R., Morel, M. P., Bouslama-Oueghlani, L., Camand, E., and Sotelo, C. 2005. Cell death and axon regeneration of Purkinje cells after axotomy: challenges of classical hypotheses of axon regeneration. *Brain Res. Brain Res. Rev.* 49: 300–316. [27]

Dwyer, T. M., Adams, D. J., and Hille, B. 1980. The permeability of the endplate channel to organic ions in frog muscle. *J. Gen. Physiol.* 75: 469–492. [5]

Dykes, I. M., Freeman, F. M., Bacon, J. P., and Davies, J. A. 2004. Molecular basis of gap junctional communication in the CNS of the leech *Hirudo medicinalis*. *J. Neurosci.* 24: 886–894. [11]

Eatock, R. A., Rusch, A., Lysakowski, A., Saeki, M. 1998. Hair cells in mammalian utricles. *Otolaryngol. Head Neck Surg.* 119: 172–181. [22]

Eatock, R. A., Xue, J., and Kalluri, R. 2008. Ion channels in mammalian vestibular afferents may set regularity of firing. *J. Exp. Biol.* 211: 1764–1774. [22]

Eccles, J. C. 1981. Physiology of motor control in man. *Appl. Neurophysiol.* 44: 5–15. [24]

Eccles, J. C., and O'Connor, W. J. 1939. Responses which nerve impulses evoke in mammalian striated muscles. *J. Physiol.* 97: 44–102. [11]

Eccles, J. C., and Sherrington, C. S. 1930. Numbers and contraction-values of individual motor-units examined in some muscles of the limb. *Proc. R. Soc. Lond., B, Biol. Sci.* 106: 326–357. [24]

Eccles, J. C., Eccles, R. M., and Magni, F. 1961. Central inhibitory action attributable to presynaptic depolarization produced by muscle afferent volleys. *J. Physiol.* 159: 147–166. [11]

Eccles, J. C., Fatt, P., and Koketsu, K. 1954. Cholinergic and inhibitory synapses in a pathway from motor-axon collaterals to motoneurones. *J. Physiol.* 126: 524–562. [14, 15]

Eccles, J. C., Katz, B., and Kuffler, S. W. 1941. Nature of the "endplate potential" in curarized muscle. *J. Neurophysiol.* 4: 362–387. [11]

Eccles, J. C., Katz, B., and Kuffler, S. W. 1942. Effects of eserine on neuromuscular transmission. *J. Neurophysiol.* 5: 211–230. [11]

Edin, B. B., and Vallbo, A. B. 1990. Muscle afferent responses to isometric contractions and relaxations in humans. *J. Neurophysiol.* 63: 1307–1313. [24]

Edmonds, B., Gibb, A. J., and Colquhoun, D. 1995. Mechanisms of activation of glutamate receptors and the time course of excitatory synaptic currents. *Annu. Rev. Physiol.* 57: 495–519. [11]

Edwards, C., Ottoson, D., Rydqvist, B., and Swerup, C. 1981. The permeability of the transducer membrane of the crayfish stretch receptor to calcium and other divalent cations. *Neuroscience* 6: 1455–1460. [19]

Edwards, F. A., Gibb, A. J., and Colquhoun, D. 1992. ATP receptor-mediated synaptic currents in the central nervous system. *Nature* 359: 144–147. [14]

Edwards, F. A., Konnerth, A., and Sakmann, B. 1990. Quantal analysis of inhibitory synaptic transmission in the dentate gyrus of rat hippocampal slices: A patch-clamp study. *J. Physiol.* 430: 213–249. [13]

Edwards, J. S. 1997. The evolution of insect flight: Implications for the evolution of the nervous system. *Brain Behav. Evol.* 50: 8–12. [18]

Edwards, R. H. 2007. The neurotransmitter cycle and quantal size. *Neuron* 55: 835–858. [9, 15]

Egan, T. M., Henderson, G., North, R. A., and, Williams, J. T. 1983. Noradrenaline-mediated synaptic inhibition in rat locus coeruleus neurones. *J. Physiol.* 345: 477–88. [14]

Eggermann, E., and Feldmeyer, D. 2009. Cholinergic filtering in the recurrent excitatory microcircuit of cortical layer 4. *Proc. Natl. Acad. Sci. USA* 106: 11753–11758. [14]

Ehringer, H., and Hornykiewicz, O. 1960. Verteilung von Noradrenalin und Dopamin (3 -Hydroxytyramin) im Gehirn des Menschen und ihr Verhalten bei Erkrankungen des extrapyramidalen Systems. *Klin. Wochenschr.* 38: 1236–1239. [14]

Eiden, L. E. 1998. The cholinergic gene locus. *J. Neurochem.* 70: 2227–2240. [15]

Eikeles, N., and Esler, M. 2005. The neurobiology of human obesity. *Exp. Physiol.* 90: 673–682. [17]

Eilers, J., Plant, T., and Konnerth, A. 1996. Localized calcium signaling and neural integration in cerebellar Purkinje neurons. *Cell Calcium* 20: 215–226. [12]

Elbert, T., Pantev, C., Wienbruch, C., Rockstroh, B., and Taub, E. 1995. Increased cortical representation of the fingers of the left hand in string players. *Science* 270: 305–307. [21]

Elden, L. E., Schäfer, M. K.-H., Weihe, E., and Schütz, B. 2004. The vesicular amine transporter family (SLC18): Amine/proton antiporters required for vesicular accumulation and regulated exocytotic secretion of monoamines and acetylcholine. *Pflügers Arch.* 447: 636–640. [9]

Elgersma, Y., and Silva, A. J. 1999. Molecular mechanisms of synaptic plasticity. *Curr. Opin. Neurobiol.* 9: 209–213. [16]

Elgoyhen, A. B., Johnson, D. S., Boulter, J., Vetter, D. E., and Heinemann, S. 1994. α9: an acetylcholine receptor with novel pharmacological properties expressed in rat cochlear hair cells. *Cell* 79: 705–715. [22]

Elgoyhen, A. B., Vetter, D. E., Katz, E., Rothlin, C. V., Heinemann, S. F., and Boulter, J. 2001. α10: a determinant of nicotinic cholinergic receptor function in mammalian vestibular and cochlear mechanosensory hair cells. *Proc. Natl. Acad. Sci. USA* 98: 3501–3506. [22]

Elhamdani, A., Azizi, F., and Artalejo, C. R. 2006. Double patch clamp reveals that transient fusion (kiss-and-run) is a major mechanism of secretion in calf adrenal chromaffin cells: high calcium shifts the mechanism from kiss-and-run to complete fusion. *J. Neurosci.* 26: 3030–3036. [13]

Elliot, T. R. 1904. On the action of adrenalin. *J. Physiol.* 31: (Proc.) xx–xxi. [11]

Elliott, E. J., and Muller, K. J. 1983. Sprouting and regeneration of sensory axons after destruction of ensheathing glial cells in the leech central nervous system. *J. Neurosci.* 3: 1994–2006. [18]

Ellis-Davies, G. C. R. 2008. Neurobiology with caged calcium. *Chem. Rev.* 108: 1603–1613. [13]

Engel, J. E., and Wu, C. F. 2009. Neurogenetic approaches to habituation and dishabituation in *Drosophila*. *Neurobiol. Learn. Mem.* 92: 166–175. [18]

Engel, J., Braig, C., Ruttiger, L., Kuhn, S., Zimmermann, U., Blin, N., Sausbier, M., Kalbacher, H., Münker, S., Rohbock, K., Ruth, P., Winter, H., and Knipper, M. 2006. Two classes of outer hair cells along the tonotopic axis of the cochlea. *Neuroscience* 143: 837–849. [22]

Engert, F., and Bonhoeffer, T. 1999. Dendritic spine changes associated with hippocampal long-term synaptic plasticity. *Nature* 399: 66–70. [16]

England, J. D., Levinson, S. R., and Shrager, P. 1996. Immunocytochemical investigations of sodium channels along nodal and internodal portions of demyelinating axons. *Microsc. Res. Tech.* 34: 445–451. [8]

Enoki, R., Hu, Y., Hamilton, D., and Fine, A. 2009. Expression of long-term plasticity at individual synapses in hippocampus is graded, bidirectional, and mainly presynaptic: Optical quantal analysis. *Neuron* 62: 242–253. [16]

Enroth-Cugell, C., and Robson, J. G. 1966. The contrast sensitivity of retinal ganglion cells of the cat. *J. Physiol.* 187: 517–552. [20]

Ercan-Sencicek, A. G., Stillman, A. A., Ghosh, A. K., Bilguvar, K., O'Roak, B. J., Mason, C. E., Abbott, T., Gupta, A., King, R. A., Pauls, D. L., Tischfield, J. A., Heiman, G. A., Singer, H. S., Gilbert, D. L., Hoekstra, P. J., Morgan, T. M., Loring, E., Yasuno, K., Fernandez, T., Sanders, S., Louvi, A., Cho, J. H., Mane, S., Colangelo, C. M., Biederer, T., Lifton, R. P., Gunel, M., and State, M. W. 2010. L-histidine decarboxylase and Tourette's syndrome. *New England J. Med.* 362: 1901–1908. [14]

Erickson, J. D., De Gois, S., Varoqui, H., Schafer, M. K., and Weihe, E. 2006. Activity-dependent regulation of vesicular glutamate and GABA transporters: a means to scale quantal size. *Neurochem. Int.* 48: 643–649. [15]

Erickson, J. D., Varoqui, H., Schäfer, M. K., Modi, W., Diebler, M. F., Weihe E., Rand, J., Eiden, L., Bonner, T. I., and Usdin, T. B. 1994. Functional identification of a vesicular acetylcholine transporter and its expression from a "cholinergic" gene locus. *J. Biol. Chem.* 269: 21929–21932. [15]

Ericson, J., Muhr, J., Placzek, M., Lints, T., Jessell, T. M., and Edlund, T. 1995. Sonic hedgehog induces the differentiation of ventral forebrain neurons: a common signal for ventral patterning within the neural tube. *Cell* 81: 747–756. [25]

Erlander, M. G., Tillakaratne, N. J. K., Feldblum, S., Patel, N., and Tobin, A. J. 1991. Two genes encode distinct glutamate decarboxylases. *Neuron* 7: 91–100. [15]

Ernsberger, U. 2009. Role of neurotrophin signalling in the differentiation of neurons from dorsal root ganglia and sympathetic ganglia. *Cell Tissue Res.* 336: 349–384. [25]

Ertel, E. A., Campbell, K. P., Harpold, M. M., Hofmann, F., Mori, Y., Perez-Reyes, E., Schwartz, A., Snutch, T. P., Tanabe, T., Birnbaumer, L., Tsien, R. W., and Catterall, W. A. 2000. Nomenclature of voltage-gated calcium channels. *Neuron* 25: 533–525. [5]

Erxleben, C. 1989. Stretch-activated current through single ion channels in the abdominal stretch receptor organ of the crayfish. *J. Gen. Physiol.* 94: 1071–1083. [19]

Erxleben, C. F. 1993. Calcium influx through stretch-activated cation channels mediates adaptation by potassium current activation. *Neuroreport* 4: 616–618. [19]

Erxleben, C., and Kriebel, M. E. 1988. Subunit composition of the spontaneous miniature end-plate currents at the mouse neuromuscular junction. *J. Physiol.* 400: 659–676Erzurumlu, R. S. 2010. Critical period for the whisker-barrel system. *Exp. Neurol.* 222: 10–12. [26]

Escher, P., Lacazette, E., Courtet, M., Blindenbacher, A., Landmann, L., Bezakova, G., Lloyd, K. C., Mueller, U., and Brenner, H. R. 2005. Synapses form in skeletal muscles lacking neuregulin receptors. *Science* 308: 1920–1923. [27]

Eugenin, J., and Nicholls, J. G. 1997. Chemosensory and cholinergic stimulation of fictive respiration in isolated CNS of neonatal opossum. *J. Physiol.* 501: 425–437. [24]

Eugenin, J., Nicholls, J. G., Cohen, L. B., and Muller, K. J. 2006. Optical recording from respiratory pattern generator of fetal mouse brainstem reveals a distributed network. *Neuroscience* 137: 1221–1227. [24]

Evans, M. G. 1996. Acetylcholine activates two currents in guinea-pig outer hair cells. *J. Physiol.* 491: 563–578. [22]

Evans, W. H., and Martin, P. E. M. 2002: Gap junctions: structure and function. *Mol. Membr. Biol.* 19: 121–136. [8]

Evarts, E. V. 1965. Relation of discharge frequency to conduction velocity in pyramidal neurons. *J. Neurophysiol.* 28: 216–228. [24]

Evarts, E. V. 1966. Pyramidal tract activity associated with a conditioned hand movement in the monkey. *J. Neurophysiol.* 29: 1011–1027. [24]

Evers, J., Laser, M., Sun, Y. A., Xie, Z. P., and Poo, M. M. 1989. Studies of nerve–muscle interactions in *Xenopus* cell culture: Analysis of early synaptic currents. *J. Neurosci.* 9: 1523–1539. [25, 27]

Eyzaguirre, C., and Kuffler, S. W. 1955. Processes of excitation in the dendrites and in the soma of single isolated sensory nerve cells of the lobster and crayfish. *J. Gen. Physiol.* 39: 87–119. [19]

Fagg, A. H., Hatsopoulos, N. G., de Lafuente, V., Moxon, K. A., Nemati, S., Rebesco, J. M., Romo, R., Solla, S. A., Reimer, J., Tkach, D., Pohlmeyer, E. A., and Miller, L. E. 2007. Biomimetic brain machine interfaces for the control of movement. *J. Neurosci.* 27: 11842–11846. [23]

Fain, G. L., Matthews, H. R., and Cornwall, M. C. 1996. Dark adaptation in vertebrate photoreceptors. *Trends Neurosci.* 19: 502–507. [20]

Faissner, A., and Steindler, D. 1995. Boundaries and inhibitory molecules in developing neural tissues. *Glia* 13: 233–254. [10]

Falck, B., Hillarp, N. A., Thieme, G., and Torp, A. 1962. Fluorescence of catacholamines and related compounds condensed with formaldehyde. *J. Histochem. Cytochem.* 10: 348–354. [14]

Falkenburger, B. H., Jensen, J. B., and Hille, B. 2010a. Kinetics of M1 muscarinic receptor and G protein signaling to phospholipase C in living cells. *J. Gen. Physiol.* 135: 81–97. [12]

Falkenburger, B. H., Jensen, J. B., and Hille, B. 2010b. Kinetics of PIP_2 metabolism and KCNQ2/3 channel regulation studied with a voltage-sensitive phosphatase in living cells. *J. Gen. Physiol.* 135: 99–114. [12]

Falker, B., and Adelman, J. P. 2008. Control of K_{Ca} channels by calcium nano-microdomains. *Neuron* 59: 873–881. [5]

Fallon, J. B., Irvine, D. R., and Shepherd, R. K. 2008. Cochlear implants and brain plasticity. *Hear Res.* 238: 110–117. [26]

Fambrough, D. M. 1979. Control of acetylcholine receptors in skeletal muscle. *Physiol. Rev.* 59: 165–227. [27]

Famiglietti, E. V., Jr., and Kolb, H. 1975. A bi-stratified amacrine cell and synaptic circuitry in the inner plexiform layer of the retina. *Brain Res.* 84: 293–300. [20]

Fanselow, M. S., and Kim, J. J. 1994. Acquisition of contextual Pavlovian fear conditioning is blocked by application of an NMDA receptor antagonist D,L-2-amino-5-Phosphonovaleric acid to the basolateral amygdala. *Behav. Neurosci.* 108: 210–212. [16]

Färber, K., and Kettenmann, H. 2005. Physiology of microglial cells. *Brain Res. Brain Res. Rev.* 48: 133–143. [10]

Farbman, A. I. 1994. Developmental biology of olfactory sensory neurons. *Semin. Cell Biol.* 5: 3–10. [19]

Farinas, I., Yoshida, C. K., Backus, C., and Reichardt, L. F. 1996. Lack of neurotrophin-3 results in death of spinal sensory neurons and premature differentiation of their precursors. *Neuron* 17: 1065–1078. [25]

Farrant, M., and Nusser, Z. 2005. Variations on an inhibitory theme: phasic and tonic activation of $GABA_A$ receptors. *Nat. Rev. Neurosci.* 6: 215–229. [14]

Fatt, P., and Ginsborg, B. L. 1958. The ionic requirements for the production of action potentials in crustacean muscle fibres. *J. Physiol.* 142: 516–543. [7]

Fatt, P., and Katz, B. 1951. An analysis of the end-plate potential recorded with an intra-cellular electrode. *J. Physiol.* 115: 320–370. [6, 11, 15]

Fatt, P., and Katz, B. 1952. Spontaneous subthreshold potentials at motor nerve endings. *J. Physiol.* 117: 109–128. [13]

Fatt, P., and Katz, B. 1953. The effect of inhibitory nerve impulses on a crustacean muscle fibre. *J. Physiol.* 121: 374–389. [11]

Fawcett, J. 2009. Molecular control of brain plasticity and repair. *Prog. Brain Res.* 175: 501–509. [26]

Fawcett, J. W., and Keynes, R. J. 1990. Peripheral nerve regeneration. *Annu. Rev. Neurosci.* 13: 43–60. [27]

Feinberg, K., Eshed–Eisenbach, Y., Frechter, S., Amor, V., Salomon, D., Sabanay, H., Dupree, J. L., Grumet, M., Brophy, P. J., Shrager, P., and Peles, E. 2010. A glial signal consisting of gliomedin and NrCAM clusters axonal Na^+ channels during the formation of nodes of Ranvier. *Neuron* 65: 490–502. [10]

Fekete, D. M., Rouiller, E. M., Liberman, M. C., and Ryugo, D. K. 1984. The central projections of intracellularly labeled auditory nerve fibers in cats. *J. Comp. Neurol.* 229: 432–450. [22]

Feldberg, W. 1945. Present views of the mode of action of acetylcholine in the central nervous system. *Physiol. Rev.* 25: 596–642. [11]

Feldberg, W. 1950. The role of acetylcholine in the central nervous system. *Br. Med. Bull.* 6: 312–321. [14]

Feldheim, D. A., and O'Leary, D. D. 2010. Visual map development: bidirectional signaling, bifunctional guidance molecules, and competition. *Cold Spring Harb. Perspect. Biol.* 2: a001768. [25]

Feldman, D. E., and Knudsen, E. I. 1997. An anatomical basis for visual calibration of the auditory space map in the barn owl's midbrain. *J. Neurosci.* 17: 6820–6837. [26]

Feldman, J. L., Mitchell, G. S., and Nattie, E. E. 2003. Breathing: rhythmicity, plasticity, chemosensitivity. *Annu. Rev. Neurosci.* 26: 249–266. [24]

Feller, M. B. 2009. Retinal waves are likely to instruct the formation of eye-specific retinogeniculate projections. *Neural Dev.* 4: 24. [26]

Fellman, D. J., and Van Essen, D. C. 1991. Distributed hierarchical processing in the primate cerebral cortex. *Cereb. Cortex* 1: 1–47. [23]

Felmy, F., Neher, E., and Schneggenberger, R. 2003. The timing of phasic transmitter release is Ca^{2+}-dependent and lacks direct influence of presynaptic terminal membrane potential. *Proc. Natl. Acad. Sci. USA* 100: 15200–15205. [13]

Feng, D., Kim, T., Ozkan, E., Light, M., Torkin, R., Teng, K. K., Hempstead, B. L., and Garcia, K. C. 2010. Molecular and structural insight into proNGF engagement of p75NTR and sortilin. *J. Mol. Biol.* 396: 967–984. [25]

Fernandez-Fernandez, J. M., Abogadie, F. C., Milligan, G., Delmas, P., and Brown, D. A. 2001. Multiple pertussis toxin-sensitive G-proteins can couple receptors to GIRK channels in rat sympathetic neurons when expressed heterologously, but only native G(i)-proteins do so in situ. *Eur. J. Neurosci.* 14: 283–292. [12]

Fernandez-Fernandez, J. M., Wanaverbecq, N., Halley, P., Caulfield, M. P., and Brown, D. A. 1999. Selective activation of heterologously expressed G protein-gated K^+ channels by M2 muscarinic receptors in rat sympathetic neurones. *J. Physiol.* 515: 631–637. [12]

Fernández, M. de L., Chan, Y. B., Yew, J. Y., Billeter, J. C., Dreisewerd, K., Levine, J. D., and Kravitz, E. A. 2010. Pheromonal and behavioral cues trigger male-to-female aggression in *Drosophila. PLoS Biol.* 8: e1000541. [28]

Ferrington, D. G., and Rowe, M. 1980. Differential contributions to coding of cutaneous vibratory information by cortical somatosensory areas I and II. *J. Neurophysiol.* 43: 310–331. [21]

Ferster, D., Chung, S., and Wheat, H. 1996. Orientation selectivity of thalamic input to simple cells of cat visual cortex. *Nature* 380: 249–252. [2]

Fertuck, H. C., and Salpeter, M. M. 1974. Localization of acetylcholine receptor by ^{125}I-labeled α-bungarotoxin binding at mouse motor endplates. *Proc. Natl. Acad. Sci. USA* 71: 1376–1378. [11]

Fesenko, E. E., Kolesnikov, S. S., and Lyubarsky, A. L. 1985. Induction by cyclic GMP of cationic conductance in plasma membrane of retinal rod outer segment. *Nature* 313: 310–313. [20]

Fettiplace, R. 1987. Electrical tuning of hair cells in the inner ear. *Trends Neurosci.* 10: 421–425. [22]

Fiacco, A., and McCarthy, K. D. 2004. Intracellular astrocyte calcium waves in situ increase the frequency of spontaneous AMPA receptor currents in CA1 pyramidal neurons. *J. Neurosci.* 24: 722–732. [10]

Fiacco, T. A., Agulhon, C., and McCarthy, K. D. 2009. Sorting out astrocyte physiology from pharmacology. *Annu. Rev. Pharmacol. Toxicol.* 49: 151–174. [10]

Fields, R. D., and Burnstock, G. 2006. Purinergic signaling in neuron glial interactions. *Nat. Rev. Neurosci.* 7: 423–436. [14]

Fierro, L., and Llano, I. 1996. High endogenous calcium buffering in Purkinje cells from rat cerebellar slices. *J. Physiol.* 496: 617–625. [12]

Filip, M., and Bader, M. 2009. Overview on 5-HT receptors and their role in physiology and pathology of the central nervous system. *Pharmacol Rep.* 61: 761–777. [14]

Filippov, A. K., Choi, R. C., Simon, J., Barnard, E. A., and Brown, D. A. 2006. Activation of $P2Y_1$ nucleotide receptors induces inhibition of the M-type K^+ current in rat hippocampal pyramidal neurons. *J. Neurosci.* 26: 9340–9348. [14]

Filippov, A. K., Couve, A., Pangalos, M. N., Walsh, F. S., Brown, D. A., and Moss, S. J. 2000. Heteromeric assembly of $GABA_BR1$ and $GABA_BR2$ receptor subunits inhibits Ca^{2+} current in sympathetic neurons. *J. Neurosci.* 20: 2867–2874. [14]

Fillenz, M. 2005a. In vivo neurochemical monitoring and the study of behaviour. *Neurosci. Biobehav. Rev.* 29: 949–962. [1]

Fillenz, M. 2005b. The role of lactate in brain metabolism. *Neurochem. Int.* 47: 413–417. [10]

Finger, T. E., Danilova, V., Barrows, J., Bartel, D. L., Vigers, A. J., Stone, L., Hellekant, G., and Kinnamon, C. 2005. ATP signaling is crucial for communication from taste buds to gustatory nerves. *Science* 310: 1495–1499. [19]

Finn, I. M., and Ferster, D. 2007. Neural connections and receptive field properties in the primary visual cortex. *J. Neurosci.* 27: 9638–9648. [2]

Finn, I. M., Priebe, N. J., and Ferster, D. 2007. The emergence of contrast-invariant orientation tuning in simple cells of cat visual cortex. *Neuron* 54: 137–152. [2]

Firestein, S., Shepherd, G. M., and Werblin, F. S. 1990. Time course of the membrane current underlying sensory transduction in salamander olfactory receptor neurones. *J. Physiol.* 430: 135–158. [19]

Fischbach, G. D., and Rosen, K. M. 1997. ARIA: A neuromuscular junction neuregulin. *Annu. Rev. Neurosci.* 20: 429–458. [27]

Fischer, W., Bjorklund, A., Chen, K., and Gage, F. H. 1991. NGF improves spatial memory in aged rodents as a function of age. *J. Neurosci.* 11: 1889–1906. [25]

Fischmeister, R., and Hartzell, H. C. 1986. Mechanism of action of acetylcholine on calcium current in single cells from frog ventricle. *J. Physiol.* 376: 183–202. [12]

Fischmeister, R., Castro, L. R., Abi-Gerges, A., Rochais, F., Jurevicius, J., Leroy, J., and Vandecasteele, G. 2006. Compartmentation of cyclic nucleotide signaling in the heart: the role of cyclic nucleotide phosphodiesterases. *Circ. Res.* 99: 816–828. [12]

Fisher, S. K., and Boycott, B. B. 1974. Synaptic connections made by horizontal cells within the outer plexiform layer of the retina of the cat and the rabbit. *Proc. R. Soc. Lond., B, Biol. Sci.* 186: 317–331. [1]

Fitch, M. T., and Silver, J. 1999. Beyond the glial scar: Cellular and molecular mechanisms by which glial cells contribute to CNS regenerative failure. In M. H. Tuszynski and J. H. Kordower (eds.), *CNS Regeneration: Basic Science and Clinical Advances.* Academic Press, San Diego. pp. 55–88. [27]

Fitzpatrick, J. S., Haggenston, A. S., Hertle, D. N., Gipson, K. E., Bertetto-d'Angelo, L., and Yekel, M. F. 2009. Inositol-1,4,5- trisphosphate receptor-mediated Ca^{2+} waves in pyramidal neuron dendrites propagate through hot spots and cold spots. *J. Physiol.* 587: 1439–1459. [12]

Flavell, S. W., and Greenberg, M. E. 2008. Signaling mechanisms linking neuronal activity to gene expression and plasticity of the nervous system. *Annu. Rev. Neurosci.* 31: 563–590. [12]

Fleishman, S. J., Ungar, V. M., Yeager, M., and Ben-Tal, N. 2004. A C^{α} model for the transmembrane α helices of gap junction intercellular channels. *Mol. Cell* 15: 879–888. [8]

Flock, A. 1965. Transducing mechanisms in the lateral line canal organ receptors. *Cold Spring Harb. Symp. Quant. Biol.* 30: 133–145. [19]

Flock, A., and Russell, I. 1976. Inhibition by efferent nerve fibres: action on hair cells and afferent synaptic transmission in the lateral line canal organ of the burbot *Lota lota. J. Physiol.* 257: 45–62. [22]

Flock, A., Flock, B., and Murray, E. 1977. Studies on the sensory hairs of receptor cells in the inner ear. *Acta Otolaryngol.* 83: 85–91. [19]

Flockerzi, V., Oeken, H-J., Hofmann, F., Pelzer, D., Cavalié, A., and Trautwein, W. 1986. Purified dihydropyridine-binding site from skeletal muscle t-tubules is a functional calcium channel. *Nature* 323: 66–68. [12]

Florey, E. 1954. An inhibitory and an excitatory factor of mammalian central nervous system, and their action on a single sensory neuron. *Arch. Int. Physiol.* 62: 33–53. [14]

Flynn, G. E., Johnson, J. P., Jr., and Zagotta, W. N. 2001. Cyclic nucleotide-gated channels: Shedding light on the openings of a channel pore. *Nat. Rev. Neurosci.* 2: 643–652. [5]

Fogassi, L., and Luppino, G. 2005. Motor functions of the parietal lobe. *Curr. Opin. Neurobiol.* 15: 626–631. [24]

Fontaine, B., and Changeux, J-P. 1989. Localization of nicotinic acetylcholine receptor α-subunit transcripts during myogenesis and motor endplate development in the chick. *J. Cell Biol.* 108: 1025–1037. [27]

Foote, S. L., Bloom, F. E., and Aston-Jones, G. 1983. Nucleus locus ceruleus: new evidence of anatomical and physiological specificity. *Physiol. Rev.* 63: 844–914. [14]

Forman, D. S., Padjen, A. L., and Siggins, G. R. 1977. Axonal transport of organelles visualized by light microscopy: cinemicrographic and computer analysis. *Brain Res.* 136: 197–213. [15]

Forscher, P., and Smith, S. J. 1988. Actions of cytochalasins on the organization of actin filaments and microtubules in a neuronal growth cone. *J. Cell Biol.* 4: 1505–1516. [25]

Förster, E., Bock, H. H., Herz, J., Chai, X., Frotscher, M., and Zhao, S. 2010. Emerging topics in Reelin function. *Eur. J. Neurosci.* 31: 1511–1518. [25]

Forsythe, I. D. 1994. Direct patch recording from identified presynaptic terminals mediating glutaminergic EPSCs in the rat CNS *in vitro*. *J. Physiol.* 479: 381–387. [13]

Forsythe, I. D., Tsujimoto, T., Barnes-Davies, M., Cuttle, M. F., and Takahashi, T. 1998. Inactivation of presynaptic calcium current contributes to synaptic depression at a fast central synapse. *Neuron* 20: 797–807. [16]

Foster, R. G., Provencio, I., Hudson, D., Fiske, S., De Grip, W., and Menaker, M. 1991. Circadian photoreception in the retinally degenerate mouse (rd/rd). *J. Comp. Physiol. A* 169: 39–50. [20]

Fournier, A. E., GrandPre, T., and Strittmatter, S. M. 2001. Identification of a receptor mediating Nogo-66 inhibition of axonal regeneration. *Nature* 409: 341–346. [27]

Foust, A., Popovic, M., Zecevic, D., and McCormick, D. A. 2009. Action potentials initiate in the axon initial segment and propagate through axon collaterals reliably in cerebellar Purkinje neurons. *Neuroscience* 162: 836–851. [24]

Fowler, C. E., Aryal, P., Suen, K. F., and Slesinger, P. A. 2007. Evidence for association of $GABA_B$ receptors with Kir3 channels and regulators of G protein signalling (RGS4) proteins. *J. Physiol.* 580: 51–65. [12]

Fowler, C. J., Griffiths, D., de Groat, W. C. 2008. The neural control of micturition. *Nat. Rev. Neurosci.* 9: 453–466. [17]

Franchini, L. F., and Elgoyhen, A. B. 2006. Adaptive evolution in mammalian proteins involved in cochlear outer hair cell electromotility. *Mol. Phylogenet. Evol.* 41: 622–635. [22]

Francis, N. J., and Landis, S. C. 1999. Cellular and molecular determinants of sympathetic neuron development. *Annu. Rev. Neurosci.* 22: 541–566. [25]

Francois, J. 1979. Late results of congenital cataract surgery. *Ophthalmology* 86: 1586–1598. [26]

Frank, K., and Fuortes, M. G. F. 1957. Presynaptic and postsynaptic inhibition of monosynaptic reflexes. *Fed. Proc.* 16: 39–40. [11]

Frank, T., Khimich, D., Neef, A., and Moser, T. 2009. Mechanisms contributing to synaptic Ca^{2+} signals and their heterogeneity in hair cells. *Proc. Natl. Acad. Sci. USA* 106: 4483–4488. [22]

Frankenhaeuser, B., and Hodgkin, A. L. 1957. The actions of calcium on the electrical properties of squid axons. *J. Physiol.* 137: 218–244. [7]

Franks, N. P. 2008. General anaesthesia: from molecular targets to neuronal pathways of sleep and arousal. *Nat. Rev. Neurosci.* 9: 370–386. [14]

Fraser, J. A., and Huang, C. L. H. 2004. A quantitative analysis of cell volume and resting potential determination and regulation of excitable cells. *J. Physiol.* 559: 459–478. [6]

Freedman, M. S., Lucas, R. J., Soni, B., von Schantz, M., Muñoz, M., David-Gray, Z., and Foster, R. 1999. Regulation of mammalian circadian behavior by non-rod, non-cone, ocular photoreceptors. *Science* 284: 502–504. [20]

Freeman, A. W., and Johnson, K. O. 1982. Cutaneous mechanoreceptors in macaque monkey: temporal discharge patterns evoked by vibration, and a receptor model. *J. Physiol.* 323: 21–41. [21]

Freiwald, W. A., Tsao, D. Y., and Livingstone, M. S. 2009. A face feature space in the macaque temporal lobe. *Nat. Neurosci.* 12: 1187–1196. [23]

Fremeau, Jr., R. T., Burman, J., Qureshi, T., Tran, C. H., Proctor, J., Johnson, J., Zhang, H., Sulzer, D., Copenhagen, D. R., Storm-Mathisen, J., Reimer, R. J., Chaudhry, F. A., and Edwards, R. H. 2002. The identification of vesicular glutamate transporter 3 suggests novel modes of signaling by glutamate. *Proc. Natl. Acad. Sci. USA* 99: 14488–14493. [9]

Fremeau, Jr., R. T., Kam, K., Qureshi, T., Johnson, J., Copenhagen, D. R., Storm-Mathisen, J., Chaudhry, F. A., Nicol, R. A., and Edwards, R. H. 2004. Vesicular glutamate transporters 1 and 2 target to functionally distinct synaptic release sites. *Science* 304: 1815–1819. [9]

French, C. R., Sah, P., Buckett, K. J., and Gage, P. W. 1990. A voltage-dependent persistent sodium current in mammalian hippocampal neurons. *J. Gen. Physiol.* 95: 1139–1157. [7]

French, K. A., and Muller, K. J. 1986. Regeneration of a distinctive set of axosomatic contacts in the leech central nervous system. *J. Neurosci.* 6: 318–324. [18]

French, R. J., and Zamponi, G. W. 2005. Voltage-gated sodium and calcium channels in nerve, muscle, and heart. *IEEE Trans. Nanobioscience* 4: 58–69. [5]

Fried, S. I., Munch, T. A., and Werblin, F. S. 2002. Mechanisms and circuitry underlying directional selectivity in the retina. *Nature* 420: 411–414. [20]

Friesen, W. O., and Kristan, W. B. 2007. Leech locomotion: swimming, crawling, and decisions. *Curr. Opin. Neurobiol.* 17: 704–711. [18]

Fritsch, G., and Hitzig, E. 1870. Ueber die electrische Erregbarkheit des Grosshirns. *Arch. Anat. Physiol. Wiss. Med.* 37: 300–332. [24]

Fritschy, J. M., Harvey, R. J., and Schwarz, G. 2008. Gephyrin: where do we stand, where do we go? *Trends Neurosci.* 31: 257–264. [11]

Frye, M. A., and Dickinson, M. H. 2001. Fly flight: a model for the neural control of complex behavior. *Neuron* 32: 385–388. [18]

Fu, Y., and Yau, K. W. 2007. Phototransduction in mouse rods and cones. *Pflügers Arch.* 454: 805–819. [20]

Fu, Y., Liao, H. W., Do, M. T., and Yau, K. W. 2005. Non-image forming ocular photoreception in vertebrates. 2005. *Curr. Opin. Neurobiol.* 15: 415–422. [2]

Fuchs, P. A., and Getting, P. A. 1980. Ionic basis of presynaptic inhibitory potentials at crayfish claw opener. *J. Neurophysiol.* 43: 1547–1557. [11]

Fuchs, P. A., and Murrow, B. W. 1992. Cholinergic inhibition of short (outer) hair cells of the chick's cochlea. *J. Neurosci.* 12: 800–809. [12, 22]

Fucile, S. 2004. Ca^{2+} permeability of nicotinic acetylcholine receptors. *Cell Calcium* 35: 1–8. [12]

Fujisawa, H. 2004. Discovery of semaphorin receptors, neuropilin and plexin, and their functions in neural development. *J. Neurobiol.* 59: 24–33. [25]

Fukami, Y., and Hunt, C. C. 1977. Structures in sensory region of snake spindles and their displacement during stretch. *J. Neurophysiol.* 40: 1121–1131. [19]

Fukuda, N., Shirasu, M., Sato, K., Ebisui, E., Touhara, K., and Mikoshiba, K. 2008. Decreased olfactory mucus secretion and nasal abnormality in mice lacking type 2 and type 3 IP3 receptors. *Eur. J. Neurosci.* 27: 2665–2675. [19]

Fuller, J. L. 1967. Experimental deprivation and later behavior. *Science* 158: 1645–1652. [26]

Fullerton, S. M., Strittmatter, W. J., and Matthew, W. D. 1998. Peripheral sensory nerve defects in apolipoprotein E knockout mice. *Exp. Neurol.* 153: 156–163. [27]

Fulop, T., Radabaugh, S., and Smith, C. 2005. Activity-dependent differential transmitter release in mouse adrenal chromaffin cells. *J. Neurosci.* 25: 7324–7332. [17]

Fuortes, M. G., and Poggio, G. F. 1963. Transient responses to sudden illumination in cells of the eye of *Limulus*. *J. Gen. Physiol.* 46: 435–452. [20]

Furchgott, R. F., and Zawadzki, J. V. 1980. The obligatory role of the endothelial cells in the relaxation of arterial smooth muscle by acetylcholine. *Nature* 288: 373–376. [12]

Furness, D. N., Dehnes, Y., Akhtar, A. Q., Rossi, D. J., Hamann, M., Grutle, N. J., Gundersen, V., Holmseth, S., Lehre, K. P., Ullensvang, K., Wojewodzic, M., Zhou, Y., Attwell, D., and Danbolt, N. C. 2008. A quantitative assessment of glutamate uptake into hippocampal synaptic terminals and astrocytes: New insights into a neuronal role for excitatory amino acid transporter 2 (EAAT2). *Neuroscience* 157: 80–94. [10]

Furness, J. B., Bornstein, J. C., Murphy, R., and Pompolo, S. 1992. Roles of peptides in transmission in the enteric nervous system. *Trends Neurosci.* 15: 66–71. [14]

Furshpan, E. J., and Potter, D. D. 1959. Transmission at the giant motor synapses of the crayfish. *J. Physiol.* 145: 289–325. [11]

Furshpan, E. J., MacLeish, P. R., O'Lague, P. H., and Potter, D. D. 1976. Chemical transmission between rat sympathetic neurons and cardiac myocytes developing in microcultures: evidence for cholinergic, adrenergic, and dual-function neurons. *Proc. Natl. Acad. Sci. USA* 73: 4225–4229. [25]

Fuxe, K., Dahlström, A. B., Jonsson, G., Marcellino, D., Guescini, M., Dam, M., Manger, P., and Agnati, L. 2010. The discovery of central monoamine neurons gave volume transmission to the wired brain. *Prog. Neurobiol.* 90: 82–100. [14]

Gaddum, J. H. 1943. Symposium on chemical constitution and pharmacological action. *Trans. Faraday Soc.* 39: 323–332. [11]

Gage, F. H., Armstrong, D. M., Williams, L. R., and Varon, S. 1988. Morphological response of axotomized septal neurons to nerve growth factor. *J. Comp. Neurol.* 269: 147–155. [25]

Gahwiler, B. H., and Brown, D. A. 1985a. Functional innervation of cultured hippocampal neurones by cholinergic afferents from co-cultured septal explants. *Nature* 313: 577–579. [14]

Gahwiler, B. H., and Brown, D. A. 1985b. $GABA_B$-receptor-activated K+ current in voltage-clamped CA3 pyramidal cells in hippocampal cultures. *Proc. Natl. Acad. Sci. USA* 82: 1558–1562. [14]

Galambos, R. 1956. Suppression of auditory nerve activity by stimulation of efferent fibers to the cochlea. *J. Neurophysiol.* 19: 424–437. [22]

Gallagher, J. P., Higashi, H., and Nishi, S. 1978. Characterization and ionic basis of GABA-induced depolarizations recorded in vitro from cat primary afferent neurones. *J. Physiol.* 275: 263–282. [11]

Gallo, G., and Letourneau, P. C. 2004. Regulation of growth cone actin filaments by guidance cues. *J. Neurobiol.* 58: 92–102. [25]

Galzi, J. L., Devillers-Thiery, A., Hussey, N., Bertrand, S., Changeux, J. P., and Bertrand, D. 1992. Mutations in the channel domain of a neuronal nicotinic receptor convert ion selectivity from cationic to anionic. *Nature* 359: 500–505. [5]

Gamal El-Din, T. M., Heldstab, H., Lehmann, C., and Greef, N. G. 2010. Double gaps along *Shaker* S4 demonstrate omega currents at three different closed states. *Channels* 4: 1–8. [7]

Gamper, N. S., and Shapiro, M. S. 2007. Regulation of ion transport proteins by membrane phosphoinositides. *Nat. Rev. Neurosci.* 8: 1–14. [12]

Gamzu, E., and Ahissar, E. 2001. Importance of temporal cues for tactile spatial- frequency discrimination. *J. Neurosci.* 21: 7416–7427. [21]

Gandhi, S. P., and Stevens, C. F. 2003. Three modes of synaptic vesicular recycling revealed by single-vesicle imaging. *Nature* 423: 607–613. [13]

Gao, T., Yatani, A., Dell'Acqua, M. L., Sako, H., Green, S. A., Dascal, N., Scott, J. D., and Hosey, M. M. 1997. cAMP-dependent regulation of cardiac L-type Ca^{2+} channels requires membrane targeting of PKA and phosphorylation of channel subunits. *Neuron* 19: 185–196. [12]

Garcia-Anoveros, J., and Corey, D. P. 1997. The molecules of mechanosensation. *Annu. Rev. Neurosci.* 20: 567–594. [19]

Gardner, E. P., Palmer, C. I., Hamalainen, H. A., and Warren, S. 1992. Simulation of motion on the skin. V. Effect of stimulus temporal frequency on the representation of moving bar patterns in primary somatosensory cortex of monkeys. *J. Neurophysiol.* 67: 37–63. [21]

Garfield, A. S., and Heisler, L. K. 2009. Pharmacological targeting of the serotonergic system for the treatment of obesity. *J. Physiol.* 587: 49–60. [14]

Garthwaite, J. 2008. Concepts of neural nitric oxide-mediated transmission. *Eur. J. Neurosci.* 27: 2783–2802. [12, 24]

Garthwaite, J., Charles, S. L., and Chess-Williams, R. 1988. Endothelium-derived relaxing factor release on activation of NMDA receptors suggests role as intercellular messenger in the brain. *Nature* 336: 385–388. [12]

Gasnier, B. 2004. The SLC32 transporter, a key protein for the synaptic release of inhibitory amino acids. *Pflügers Arch.* 447: 756–759. [9]

Gautam, M., Noakes, P. G., Moscoso, L., Rupp, F., Scheller, R. H., Merlie, J. P., and Sanes, J. R. 1996. Defective neuromuscular synaptogenesis in agrin-deficient mutant mice. *Cell* 85: 525–535. [27]

Gautvik, K. M., de Lecea, L., Gautvik, V. T., Danielson, P. E., Tranque, P., Dopazo, A., Bloom, F. E., and Sutcliffe, J. G. 1998. Overview of the most prevalent hypothalamus-specific mRNAs, as identified by directional tag PCR subtraction. *Proc. Natl. Acad. Sci. USA* 93: 8733–8738. [14]

Gazzaniga, M. S. 2005. Forty-five years of split-brain research and still going strong. *Nat. Rev. Neurosci.* 6: 653–659. [3]

Gehring, W. J., Kloter, U., and Suga, H. 2009. Evolution of the *Hox* gene complex from an evolutionary ground state. *Curr. Top. Dev. Biol.* 88: 35–61. [25]

Geiger, J. R., Lübke, J., Roth, A., Frotscher, M., and Jonas, P. 1997. Submillisecond AMPA receptor-mediated signaling at a principal neuron-interneuron synapse. *Neuron* 18: 1009–1023. [14]

Georgopoulos, A. P., Merchant, H., Naselaris, T., and Amirikian, B. 2007. Mapping of the preferred direction in the motor cortex. *Proc. Natl. Acad. Sci. USA* 104: 11068–1072. [24]

Georgopoulos, A. P., Schwartz, A. B., and Kettner, R. E. 1986. Neuronal population coding of movement direction. *Science* 243: 1416–1419. [24]

Gerencser, G. A., and Zhang, J. 2003. Existence and nature of the chloride pump. *Biochim. Biophys. Acta* 1618: 133–139. [9]

Gether, U., Andersen, P. H., Larsson, O. M., and Schousboe, A. 2006. Neurotransmitter transporters: molecular function of important drug targets. *Trends Pharmacol. Sci.* 27: 375–383. [9, 15]

Ghosh, A., and Greenberg, M. E. 1995. Calcium signaling in neurons: molecular mechanisms and cellular consequences. *Science* 268: 239–247. [12]

Gibbins, I. L., and Morris, J. L. 2006. Structure of peripheral synapses: autonomic ganglia. *Cell Tissue Res.* 326: 205–220. [17]

Giepmans, B. N., Adams, S. R., Ellisman, M. H., and Tsien, R. Y. 2006. The fluorescent toolbox for assessing protein location and function. *Science* 312: 217–224. [25]

Gil, J. M., and Rego, A. C. 2008. Mechanisms of neurodegeneration in Huntington's disease. *Eur. J. Neurosci.* 27: 2803–2820. [24]

Gilbert, C. D., and Wiesel, T. N. 1979. Morphology and intracortical projections of functionally characterised neurones in the cat visual cortex. *Nature* 280: 120–125. [2, 3]

Gilbert, C. D., and Wiesel, T. N. 1981. Laminar specialization and intracortical connections in cat primary visual cortex. In F. O. Schmitt, F. G. Worden, and F. Dennis (eds.), *The organization of the cerebral cortex.* MIT Press, Cambridge. pp. 163–198. [3]

Gilbert, C. D., and Wiesel, T. N. 1983. Cluster intrinsic connections in cat visual cortex. *J. Neurosci.* 3: 1116–1133. [3]

Gilbert, C. D., and Wiesel, T. N. 1989. Columnar specificity of intrinsic horizontal and corticocortical connections in cat visual cortex. *J. Neurosci.* 9: 2432–2442. [3]

Gilbert, S. F. 2000. *Developmental Biology*, 6th ed. Sinauer, Sunderland, MA. [25]

Gilbert, S. F. 2010. *Developmental Biology*, 9th ed. Sinauer, Sunderland, MA. [25]

Gilbertson, T. A., Roper, S. D., and Kinnamon, S. C. 1993. Proton currents through amiloride-sensitive Na^+ channels in isolated hamster taste cells: enhancement by vasopressin and cAMP. *Neuron* 10: 931–942. [19]

Gillespie, J. S., Liu, X. R., and Martin, W. 1989. The effects of L-arginine and NG-monomethyl L-arginine on the response of the rat anococcygeus muscle to NANC nerve stimulation. *Brit. J. Pharmacol.* 98: 1080–1082. [12]

Gillespie, P. G., and Muller, U. 2009. Mechanotransduction by hair cells: models, molecules, and mechanisms. *Cell* 139: 33–44. [19]

Gillespie, P. G., Wagner, M. C., and Hudspeth, A. J. 1993. Identification of a 120 kd hair-bundle myosin located near stereociliary tips. *Neuron* 11: 581–594. [19]

Gilman, A. G. 1987. G proteins: Transducers of receptor-generated signals. *Annu. Rev. Biochem.* 56: 615–649. [12]

Giniatullin, R., Nistri, A., and Fabbretti, E. 2008. Molecular mechanisms of sensitization of pain-transducing P2X3 receptors by the migraine mediators CGRP and NGF. *Mol. Neurobiol.* 37: 83–90. [17]

Giraudat, J., Dennis, M., Heidmann, T., Chang, J. Y., and Changeux, J. P. 1986. Structure of the high-affinity binding site for noncompetitive blockers of the acetylcholine receptor: serine 262 of the delta subunit is labeled by [3H] chlorpromazine. *Proc. Natl. Acad. Sci. USA* 83: 2719–2723. [5]

Giraudat, J., Dennis, M., Heidmann, T., Haumont, P. Y., Lederer, F., and Changeux, J. P. 1987. Structure of the high-affinity binding site for noncompetitive blockers of the acetylcholine receptor: [3H] chlorpromazine labels homologous residues in the beta and delta chains. *Biochemistry* 26: 2410–2418. [5]

Girouard, H., Bonev, A. D., Hannah, R. M., Meredith, A., Aldrich, R. W., and Nelson, M. T. 2010. Astrocytic endfoot Ca^{2+} and BK channels determine both arteriolar dilation and constriction. *Proc. Natl. Acad. Sci. USA*. 107: 3811–3816. [10]

Gitik, M., Reichert, F., and Rotshenker, S. 2010. Cytoskeleton plays a dual role of activation and inhibition in myelin and zymosan phagocytosis by microglia. *FASEB J.* 24: 2211–2221. [27]

Giuditta, A., Chun, J. T., Eyman, M., Cefaliello, C., Bruno, A. P., and Crispino, M. 2008. Local gene expression in axons and nerve endings: The glia-neuron unit. *Physiol. Rev.* 88: 515–555. [15]

Glanzman, D. L. 2009. Habituation in *Aplysia*: The cheshire cat of neurobiology. *Neurobiol. Learn. Mem.* 92: 147–154. [16, 18]

Glickstein, M., and Berlucchi, G. 2008. Classical disconnection studies of the corpus callosum. *Cortex* 44: 914–927. [3]

Glickstein, M., Strata, P., and Voogd, J. 2009. Cerebellum: history. *Neuroscience* 162: 549–559. [24]

Glickstein, M., Sultan, F., and Voogd, J. 2011. Functional localization in the cerebellum. *Cortex* 47: 59–80. [24]

Glover, J. C., and Kramer, A. P. 1982. Serotonin analog selectively ablates identified neurons in the leech embryo. *Science* 216: 317–319. [18]

Glover, J. C., Stuart, D. K., Cline, H. T., McCaman, R. E., Magill, C., and Stent, G. S. 1987. Development of neurotransmitter metabolism in embryos of the leech *Haementeria ghilianii*. *J. Neurosci.* 7: 581–594. [18]

Glowatzki, E., and Fuchs, P. A. 2002. Transmitter release at the hair cell ribbon synapse. *Nat. Neurosci.* 5: 147–154. [22]

Glowatzki, E., Cheng, N., Hiel, H., Yi, E., Tanaka, K., Ellis-Davies, G. C., Rothstein, J. D., and Bergles, D. E. 2006. The glutamate-aspartate transporter GLAST mediates glutamate uptake at inner hair cell afferent synapses in the mammalian cochlea. *J. Neurosci.* 26: 7659–7664. [9]

Goard, M., and Dan, Y. 2009. Basal forebrain activation enhances cortical coding of natural scenes. *Nat. Neurosci.* 12: 1444–1449. [14]

Godde, B., Diamond, M., and Braun, C. 2010. Feeling for space or for sime: task-dependent modulation of the cortical cepresentation of identical vibrotactile stimuli. *Neurosci. Lett.* 480: 143–147. [23]

Godecke, I., and Bonhoeffer, T. 1996. Development of identical orientation maps for two eyes without common visual experience. *Nature* 379: 251–254. [26]

Gold, J. I., and Knudsen, E. I. 1999. Hearing impairment induces frequency-specific adjustments in auditory spatial tuning in the optic tectum of young owls. *J. Neurophysiol.* 82: 2197–2209. [26]

Gold, M. R., and Martin, A. R. 1983a. Characteristics of inhibitory post-synaptic currents in brain-stem neurones of the lamprey. *J. Physiol.* 342: 85–98. [13]

Gold, M. R., and Martin, A. R. 1983b. Analysis of glycine-activated inhibitory post-synaptic channels in brain-stem neurones of the lamprey. *J. Physiol.* 342: 99–117. [6, 11]

Gold, M. R., and Martin, A. R. 1984. γ-Aminobutyric acid and glycine activate Cl^- channels having different characteristics in CNS neurones. *Nature* 308: 639–641. [14]

Gold, M. S., Reichling, D. B., Shuster, M. J., and Levine, J. D. 1996. Hyperalgesic agents increase a tetrodotoxin-resistant Na^+ current in nociceptors. *Proc. Natl. Acad. Sci. USA* 93: 1108–1112. [19]

Goldberg, J. M. 2000. Afferent diversity and the organization of central vestibular pathways. *Exp. Brain Res.* 130: 277–297. [22]

Goldin, A. L., Barchi, R. L., Caldwell, J. H., Hofmann, F., Howe, J. R., Hunter, J. C., Kallen, R. G., Mandel, G., Meisler, M. H., Netter, Y. B., Noda, M., Tamkun, M. M., Waxman, S. G., Wood, J. N., and Catterall, W. A. 2000. Nomenclature of sodium channels. *Neuron* 28: 365–368. [5]

Goldman, D. E. 1943. Potential, impedance and rectification in membranes. *J. Gen. Physiol.* 27: 37–60. [6]

Goldsmith, T. H., and Wehner, R. 1977. Restrictions on rotational and translational diffusion of pigment in the membranes of a rhabdomeric photoreceptor. *J. Gen. Physiol.* 70: 453–490. [18]

Goldstein, S. A. N., Bockenhauer, D., O'Kelly, I., and Zilberberg, N. 2001. Potassium leak channels and the KCNK family of two-P-domain subunits. *Nat. Rev. Neurosci.* 2: 175–184. [5, 6]

Golgi, C. 1903. *Opera Omnia*, vols. 1 and 2. U. Hoepli, Milan, Italy. [10]

Gollisch, T., and Meister, M. 2008. Rapid neural coding in the retina with relative spike latencies. *Science* 319: 1108–1111. [2]

Golomb, D., Yue, C., and Yaari, Y. 2006. Contribution of persistent Na^+ current and M-type K^+ current to somatic bursting in CA1 pyramidal cells: Combined experimental and modeling study. *J. Neurophysiol.* 96: 1912–1926. [7]

Gomeza, J., Hälsmann, S., Ohno, K., Eulenburg, V., Szöke, K., Richter, D., and Betz, H. 2003. Inactivation of the glycine transporter 1 gene discloses vital role of glial glycine uptake in glycinergic inhibition. *Neuron* 40: 785–796. [10]

Gorin, P. D., and Johnson, E. M. 1979. Experimental autoimmune model of nerve growth factor deprivation: effects on developing peripheral sympathetic and sensory neurons. *Proc. Natl. Acad. Sci. USA* 76: 5382–5386. [25]

Gotz, M., and Barde, Y. A. 2005. Radial glial cells defined and major intermediates between embryonic stem cells and CNS neurons. *Neuron* 46: 369–372. [10]

Gourine, A. V., Kasymov, V., Marina, N., Tang, F., Figueiredo, M. F., Lane, S., Teschemacher, A. G., Spyer, K. M., Deisseroth, K., and Kasparov, S. 2010. Astrocytes control breathing through pH-dependent release of ATP. *Science* 329: 571–575. [10, 14]

Govek, E. E., Hatten, M. E., and Van Aelst, L. 2011. The role of Rho GTPase proteins in CNS neuronal migration. *Dev. Neurobiol.* 71: 528–553. [25]

Gradinaru, V., Mogri, M., Thompson, K. R., Henderson, J. M., and Deisseroth, K. 2009. Optical deconstruction of parkinsonian neural circuitry. *Science* 324: 354–359. [28]

Gradinaru, V., Zhang, F., Ramakrishnan, C., Mattis. J., Prakash, R., Diester, I., Goshen, I., Thompson, K. R., and Deisseroth, K. 2010. Molecular and cellular approaches for diversifying and extending optogenetics. *Cell* 141: 154–165. [14]

Grafstein, B., and Forman, D. S. 1980. Intracellular transport in neurons. *Physiol. Rev.* 60: 1167–1283. [15]

Granseth, B., Odermaatt, B., Royle, S. J., and Lagnado, L. 2006. Clathrin-mediated endocytosis is the dominant mechanism of vesicle retrieval at hippocampal synapses. *Neuron* 51: 773–786. [13]

Grant, L., Yi, E., and Glowatzki, E. 2010. Two modes of release shape the postsynaptic response at the inner hair cell ribbon synapse. *J. Neurosci.* 30: 4210–4220. [22]

Grassi, F., and Lux, H. D. 1989. Voltage-dependent GABA-induced modulation of calcium currents in chick sensory neurons. *Neurosci. Lett.* 105: 113–119. [12]

Graybiel, A. M. 2008. Habits, rituals, and the evaluative brain. *Annu. Rev. Neurosci.* 31: 359–387. [24]

Greene, L. A., and Shooter, E. M. 1980. The nerve growth factor: biochemistry, synthesis, and mechanism of action. *Annu. Rev. Neurosci.* 3: 353–402. [25]

Grichtchenko, I. I., Choi, I., Zhong, X., Bray-Ward, P., Russell, J. M., and Boron, W. F. 2001. Cloning, characterization, and chromosomal mapping of a human electroneutral Na^+-driven Cl-$HC0_3$ exchanger. *J. Biol. Chem.* 276: 8358–8363. [9]

Griesinger, C. B., Richards, C. D., and Ashmore, J. F. 2005. Fast vesicle replenishment allows indefatigable signalling at the first auditory synapse. *Nature* 435: 212–215. [22]

Griffin, J. W., and Thompson, W. J. 2008. Biology and pathology of nonmyelinating Schwann cells. *Glia* 56: 1518–1531. [27]

Griffiths, T. D., Buchel, C., Frackowiak, R. S., and Patterson, R. D. 1998. Analysis of temporal structure in sound by the human brain. *Nat. Neurosci.* 1: 422–427. [22]

Grill-Spector, K., and Malach, R. 2004. The human visual cortex. *Annu. Rev. Neurosci.* 27: 649–677. [23]

Grill-Spector, K., Knouf, N., and Kanwisher, N. 2004. The fusiform face area subserves face perception, not generic within-category identification. *Nat. Neurosci.* 7: 555–562. [23]

Grill-Spector, K., Kushnir, T., Edelman, S., Avidan, G., Itzchak, Y., and Malach, R. 1999. Differential processing of objects under various viewing conditions in the human lateral occipital complex. *Neuron* 24: 187–203. [23]

Grill-Spector, K., Kushnir, T., Hendler, T., Edelman, S., Itzchak, Y., and Malach, R. 1998. A sequence of object-processing stages revealed by fMRI in the human occipital lobe. *Hum. Brain Mapp.* 6: 316–328. [23]

Grimes, W. N., Zhang, J., Graydon, C. W., Kachar, B., and Diamond, J. S. Retinal parallel processors: more than 100 independent microcircuits operate within a single interneuron. *Neuron* 65: 873–885. [20]

Grinnell, A. D., and Rheuben, M. B. 1979. The physiology, pharmacology and trophic effectiveness of synapses formed by autonomic preganglionic nerves on frog skeletal muscles. *J. Physiol.* 289: 219–240. [27]

Grinvald, A., Lieke, E., Frostig, R. D., Gilbert, C. D., and Wiesel, T. N. 1986. Functional architecture of cortex revealed by optical imaging of intrinsic signals. *Nature* 324: 361–364. [3]

Groh, J. M., Born, R. T., and Newsome, W. T. 1997. How is a sensory map read out? Effects of microstimulation in visual area MT on saccades and smooth pursuit eye movements. *J. Neurosci.* 17: 4312–4330. [23]

Gross, C. G., Rocha-Miranda, C. E., and Bender, D. B. 1972. Visual properties of neurons in inferotemporal cortex of the Macaque. *J. Neurophysiol.* 35: 96–111. [23]

Grossman, Y., Parnas, I., and Spira, M. E. 1979. Differential conduction block in branches of a bifurcating axon. *J. Physiol.* 295: 283–305. [18]

Grumbacher-Reinert, S., and Nicholls, J. 1992. Influence of substrate on retraction of neurites following electrical activity of leech Retzius cells in culture. *J. Exp. Biol.* 167: 1–14. [25]

Grynkiewicz, G., Poenie, M., and Tsien, R. Y. 1985. A new generation of Ca^{2+} indicators with greatly improved fluorescence properties. *J. Biol. Chem.* 260: 3440–3450. [12]

Gu, X. N. 1991. Effect of conduction block at axon bifurcations on synaptic transmission to different postsynaptic neurones in the leech. *J. Physiol.* 441: 755–778. [18]

Gu, X. N., Macagno, E. R., and Muller, K. J. 1989. Laser microbeam axotomy and conduction block show that electrical transmission at a central synapse is distributed at multiple contacts. *J. Neurobiol.* 20: 422–434. [8]

Gu, X. N., Muller, K. J., and Young, S. R. 1991. Synaptic integration at a sensory-motor reflex in the leech. *J. Physiol.* 441: 733–754. [18]

Guertin, P. A. 2009. The mammalian central pattern generator for locomotion. *Brain Res. Rev.* 62: 45–56. [24]

Guertin, P. A., and Hounsgaard, J. 1998. Chemical and electrical stimulation induce rhythmic motor activity in an in vitro preparation of the spinal cord from adult turtles. *Neurosci. Lett.* 245: 5–8. [24]

Guharay, F., and Sachs, F. 1984. Stretch-activated single ion channel currents in tissue-cultured embryonic chick skeletal muscle. *J. Physiol.* 352: 685–701. [19]

Guic-Robles, E., Jenkins, W. M., and Bravo, H. 1992. Vibrissal roughness discrimination is barrelcortex-dependent. *Behav. Brain Res.* 48: 145–152. [21]

Guidry, G., Willison, B. D., Blakely, R. D., Landis, S. C., and Habecker, B. A. 2005. Developmental expression of the high affinity choline transporter in cholinergic sympathetic neurons. *Auton. Neurosci.* 123: 54–61. [17]

Guillery, R. W. 1970. The laminar distribution of retinal fibers in the dorsal lateral geniculate nucleus of the rat: A new interpretation. *J. Comp. Neurol.* 138: 339–368. [3]

Guillery, R. W. 2005. Anatomical pathways that link perception and action. *Prog. Brain Res.* 149: 235–256. [2]

Guillery, R. W., and Stelzner, D. J. 1970. The differential effects of unilateral lid closure upon the monocular and binocular segments of the dorsal lateral geniculate nucleus in the cat. *J. Comp. Neurol.* 139: 413–421. [26]

Güler, A. D., Ecker, J. L., Lall, G. S., Haq, S., Altimus, C. M., Liao, H. W., Barnard, A. R., Cahill, H., Badea, T. C., Zhao, H., Hankins, M. W., Berson, D. M., Lucas, R. J., Yau, K. W., and Hattar, S. 2008. Melanopsin cells are the principal conduits for rod-cone input to non-image-forming vision. *Nature* 453: 102–105. [17]

Gulledge, A. T., and Stuart, G. J. 2005. Cholinergic inhibition of neonatal pyramidal neurons. *J. Neurosci.* 28: 10305–10320. [12]

Gulledge, A. T., Bucci, D. J., Zhang, S. S., Matsui, M., and Yeh, H. H. 2009. M1 receptors mediate cholinergic modulation of excitability in neocortical pyramidal neurons. *J. Neurosci.* 29: 9888–9902. [14]

Gunthorpe, M. J., and Lummis, S. C. R. 2001. Conversion of the ion selectivity of the 5-HT_{3A} receptor from cationic to anionic reveals a conserved feature of the ligand-gated ion channel superfamily. *J. Biol. Chem.* 276: 10977–10983. [5]

Gurdon, J. B., and Melton, D. A. 2008. Nuclear reprogramming in cells. *Science* 322: 1811–1815. [25]

Guth, L. 1968. "Trophic" influences of nerve. *Physiol. Rev.* 48: 645–687. [27]

Haas, H., and Panula, P. 2003. The role of histamine and the tuberomamillary nucleus in the nervous system. *Nat. Rev. Neurosci.* 4: 121–130. [14]

Haas, H. L., and Konnerth, A. 1983. Histamine and noradrenaline decrease calcium-activated potassium conductance in hippocampal pyramidal cells. *Nature* 302: 432–434. [14]

Haas, H. L., and Selbach, O. 2000. Functions of neuronal adenosine receptors. *Naunyn Schmiedebergs Arch. Pharmacol.* 362: 375–381. [14]

Haas, H. L., Sergeeva, O. A., and Selbach, O. 2008. Histamine in the nervous system. *Physiol. Rev.* 88: 1183–1241. [14]

Habas, C. 2010. Functional imaging of the deep cerebellar nuclei: a review. *Cerebellum* 9: 22–28. [24]

Habets, R. L., and Borst, J. G. 2005. Post-tetanic potentiation in the rat calyx of Held synapse. *J. Physiol.* 564: 173–187. [16]

Habets, R. L., and Borst, J. G. 2006. An increase in calcium influx contributes to post-tetanic potentiation at the rat calyx of Held synapse. *J. Neurophysiol.* 96: 2868–2876. [16]

Habets, R. L., and Borst, J. G. 2007. Dynamics of the readily releasable pool during post-tetanic potentiation in the rat calyx of Held synapse. *J. Physiol.* 581: 467–478. [16]

Hackett, T. A., Preuss, T. M., and Kaas, J. H. 2001. Architectonic identification of the core region in auditory cortex of macaques, chimpanzees, and humans. *J. Comp. Neurol.* 441: 197–222. [22]

Hagiwara, S., and Byerly, L. 1981. Calcium channel. *Annu. Rev. Neurosci.* 4: 69–125. [7]

Halassa, M. M., and Haydon, P. G. 2010. Integrated brain circuits: Astrocytic networks modulate neuronal activity and behavior. *Annu. Rev. Physiol.* 72: 335–55. [10]

Halder, G., Callaerts, P., and Gehring, W. J. 1995. Induction of ectopic eyes by targeted expression of the eyeless gene in *Drosophila*. *Science* 267: 1788–1792. [1]

Hall, Z. W., Bownds, M. D., and Kravitz, E. A. 1970. The metabolism of γ-aminobutyric acid in the lobster nervous system. *J. Cell Biol.* 46: 290–299. [15]

Halliwell, J. V., and Adams, P. R. 1982. Voltage-clamp analysis of muscarinic excitation in hippocampal neurons. *Brain Res.* 250: 71–92. [14]

Halliwell, J. V., and Horne, A. L. 1998. Evidence for enhancement of gap junctional coupling between rat island of Calleja granule cells *in vitro* by the activation of dopamine D_3 receptors. *J. Physiol.* 506: 175–194. [11]

Hallock, P. T., Xu, C. F., Park, T. J., Neubert, T. A., Curran, T., and Burden, S. J. 2010. Dok-7 regulates neuromuscular synapse formation by recruiting Crk and Crk-L. *Genes Dev.* 24: 2451–2461. [27]

Hamill, M. B., and Koch, D. D. 1999. Pediatric cataracts. *Curr. Opin. Ophthalmol.* 10: 4–9. [26]

Hamill, O. P., and Sakmann, B. 1981. Multiple conductance of single acetylcholine receptor channels in embryonic muscle cells. *Nature* 294: 462–464. [4]

Hamill, O. P., Marty, A., Neher, E., Sakmann, B., and Sigworth, J. 1981. Improved patch-clamp techniques for high-resolution current recording from cells and cell-free membrane patches. *Pflügers Arch.* 391: 85–100. [4]

Hamilton, N. B., and Attwell, D. 2010. Do astrocytes really exocytose neurotransmitters? *Nat. Rev. Neurosci.* 11: 227–238. [10]

Hamon, M., Bourgoin, S., Artaud, F., and El Mestikawy, S. 1981. The respective roles of tryptophan uptake and tryptophan hydroxylase in the regulation of serotonin synthesis in the central nervous system. *J. Physiol. (Paris)* 77: 269–279. [15]

Han, G. A., Malintan, N. T., Collins, B. M., Meunier, F. A., and Sugita, S. 2010. Munc18-1 as a key regulator of neurosecretion. *J. Neurochem.* 115: 1–10. [13]

Han, Y. K., Kover, H., Insanally, M. N., Semerdjian, J. H., and Bao, S. 2007. Early experience impairs perceptual discrimination. *Nat. Neurosci.* 10: 1191–1197. [26]

Hanlon, M. R., and Wallace, B. A. 2002. Structure and function of voltage-dependent ion channel regulatory β subunits. *Biochemistry* 41: 2886–2894. [5]

Hansen, D. V., Lui, J. H., Parker, P. R., and Kriegstein, A. R. 2010. Neurogenic radial glia in the outer subventricular zone of human neocortex. *Nature* 464: 554–561. [10, 25]

Hansen, H. H., Waroux, O., Seutin, V., Jentsch, T. J., Aznar, S., and Mikkelsen, J. D. 2008. Kv7 channels: interaction with dopaminergic and serotonergic neurotransmission in the CNS. *J. Physiol.* 586: 1823–1832. [17]

Hansen, S. M., Berezin, V., and Bock, E. 2008. Signaling mechanisms of neurite outgrowth induced by the cell adhesion molecules NCAM and N-cadherin. *Cell Mol. Life Sci.* 65: 3809–3821. [25]

Harik, S. I. 1984. Locus ceruleus lesion by local 6-hydroxydopamine infusion causes marked and specific destruction of noradrenergic neurons, long-term depletion of norepinephrine and the enzymes that synthesize it, and enhanced dopaminergic mechanisms in the ipsilateral cerebral cortex. *J. Neurosci.* 4: 699–707. [14]

Harlow, H. F., and Woolsey, C. N. 1958. Biological and biochemical bases of behavior. University of Wisconsin Press, Madison. [21]

Harlow, J. M. 1868. Recovery from passage of an iron bar through the head. *Publ. Mass. Med. Soc.* 2: 328–334. [28]

Harlow, M. L., Ress, D., Stoschek, A. Marshall, R. M., and McMahan, U. J. 2001. The architecture of active zone material at the frog's neuromuscular junction. *Nature* 409: 479–484. [13]

Harris, A. J., Kuffler, S. W., and Dennis, M. L. 1971. Differential chemosensitivity of synaptic and extrasynaptic areas on the neuronal surface membrane in parasympathetic neurones of the frog, tested by microapplication of acetylcholine. *Proc. R. Soc. Lond., B, Biol. Sci.* 177: 541–553. [27]

Harris, G. W., and Ruf, K. B. 1970. Luteinizing hormone releasing factor in rat hypophysial portal blood collected during electrical stimulation of the hypothalamus. *J. Physiol.* 208: 243–250. [17]

Harris, J. A., Harris, I. M., and Diamond, M. E. 2001. The topography of tactile learning in humans. *J. Neurosci.* 21: 1056–1061. [21]

Harris, J. A., Petersen, R. S., and Diamond, M. E. 1999. Distribution of tactile learning and its neural basis. *Proc. Natl. Acad. Sci. USA* 96: 7587–7591. [1, 21]

Harris, K. M., and Landis, D. M. M. 1986. Membrane structure at synaptic junctions in area CA1 of the rat hippocampus. *Neuroscience* 19: 857–872. [13]

Hartline, H. K. 1940. The receptive fields of optic nerve fibers. *Am. J. Physiol.* 130: 690–699. [2, 18]

Hartshorn, R. P., and Catterall, W. A. 1984. The sodium channel from rat brain: Purification and subunit composition. *J. Biol. Chem.* 259: 1667–1675. [5]

Hartzell, H. C., Kuffler, S. W., and Yoshikami, D. 1975. Post-synaptic potentiation: Interaction between quanta of acetylcholine at the skeletal neuromuscular synapse. *J. Physiol.* 251: 427–463. [13]

Hashimotodani, Y., Ohno-Shosaku, T., and Kano, M. 2007. Endocannabinoids and synaptic function in the CNS. *Neuroscientist* 13: 127–137. [12]

Hashimotodani, Y., Ohno-Shosaku, T., Tsubokawa, H., Ogata, H., Emoto, K., Maejima, T., Araishi, K., Shin, H. S., and Kano, M. 2005. Phospholipase Cbeta serves as a coincidence detector through its Ca^{2+} dependency for triggering retrograde endocannabinoid signal. *Neuron* 45: 257–268. [12]

Hasselmo, M. E. 2006. The role of acetylcholine in learning and memory. *Curr. Opin. Neurobiol.* 16: 710–715. [14]

Hasson, U., Hendler, T., Ben Bashat, D., and Malach, R. 2001. Vase or face? A neural correlate of shape-selective grouping processes in the human brain. *J. Cogn. Neurosci.* 13: 744–753. [23]

Hata, Y., Tsumoto, T., and Stryker, M. P. 1999. Selective pruning of more active afferents when cat visual cortex is pharmacologically inhibited. *Neuron* 22: 375–381. [26]

Hattar, S., Liao, H. W., Takao, M., Berson, D. M., and Yau, K.-W. 2002. Melanopsin-containing retinal ganglion cells: architecture, projections, and intrinsic photosensitivity. *Science* 295: 1065–1070. [17, 20]

Hatten, M. E. 1990. Riding the glial monorail: A common mechanism for glial-guided neuronal migration in different regions of the developing mammalian brain. *Trends Neurosci.* 13: 179–184. [10, 25]

Hatten, M. E. 1999. Central nervous system neuronal migration. *Annu. Rev. Neurosci.* 22: 511–539. [10]

Hatten, M. E., Liem, R. K., and Mason, C. A. 1986. Weaver mouse cerebellar granule neurons fail to migrate on wild-type astroglial processes *in vitro*. *J. Neurosci.* 6: 2676–2683. [25]

Hawkins, J., and Blakeslee, S. 2004. *On Intelligence*. Times Books, New York. [1]

Hayworth, C. R., Moody, S. E., Chodosh, L. A., Krieg, P., Rimer, M., and Thompson, W. J. 2006. Induction of neuregulin signaling in mouse Schwann cells in vivo mimics responses to denervation. *J. Neurosci.* 26: 6873–6884. [27]

He, L., and Wu, L-G. 2007. The debate on the kiss-and-run fusion at synapses. *Trends Neurosci.* 30: 447–455. [13]

He, L., Wu, X-S., Mohan, R., and Wu, L-G. 2006. Two modes of fusion pore opening revealed by cell-attached recordings at a synapse. *Nature* 444: 102–105. [13]

Hebb, D. O. 1949. *The Organization of Behavior: A Neuropsychological Theory.* Wiley, New York. [16]

Hecht, S., Shlaer, S., and Pirenne, M. H. 1942. Energy, quanta, and vision. *J. Gen. Physiol.* 25: 819–840. [20]

Hedgecock, E. M., Culotti, J. G., and Hall, D. H. 1990. The *unc-5*, *unc-6*, and *unc-40* genes guide circumferential migrations of pioneer axons and mesodermal cells on the epidermis in *C. elegans*. *Neuron* 4: 61–85. [25]

Hediger, M. A., Romero, M. F., Peng, J. B., Rolfs, A., Takanaga, H., and Bruford, E. A. 2004. The ABCs of solute carriers: physiological, pathological and therapeutic implications of human membrane transport proteins. *Pflügers Arch.* 447: 465–468. [15]

Heidelberger, R., and Matthews, G. 1992. Calcium influx and calcium current in single synaptic terminals of goldfish retinal bipolar neurons. *J. Physiol.* 447: 235–256. [13]

Heidelberger, R., Heinnemann, C., Neher, E., and Matthews, G. 1994. Calcium dependence of the rate of exocytosis in a synaptic terminal. *Nature* 371: 513–515. [13]

Heidelberger, R., Thoreson, W. B., and Witkovsky, P. 2005. Synaptic transmission at retinal ribbon synapses. *Prog. Retin. Eye Res.* 24: 682–720. [20]

Heil, P., Rajan, R., and Irvine, D. R. 1994. Topographic representation of tone intensity along the isofrequency axis of cat primary auditory cortex. *Hear. Res.* 76: 188–202. [22]

Heiligenberg, W. 1989. Coding and processing of electrosensory information in gymnotiform fish. *J. Exp. Biol.* 146: 255–275. [19]

Hein, P., Frank, M., Hoffmann, C., Lohse, M. J., and Bünemann, M. 2005. Dynamics of receptor/G protein coupling in living cells. *EMBO J.* 24: 4106–4114. [12]

Heinemann, S. H., Terlau, H., Stühmer, W., Imoto, K and Numa, S. 1992. Calcium channel characteristics conferred on the sodium channel by single mutations. *Nature* 356: 441–443. [5]

Helmholtz, H. von. 1889. *Popular Scientific Lectures*. Longmans, London. [1]

Helmholtz, H. von. 1962/1924 *Helmholtz's Treatise on Physiological Optics*. Dover, New York. [20]

Hemmati-Brivanlou, A., and Melton, D. 1997. Vertebrate neural induction. *Annu. Rev. Neurosci.* 20: 43–60. [25]

Henderson, T. A., Woolsey, T. A., and Jacquin, M. F. 1992. Infraorbital nerve blockade from birth does not disrupt central trigeminal pattern formation in the rat. *Brain Res. Dev. Brain Res.* 66: 146–152. [26]

Hendry, S. H. C., and Calkins, D. J. 1998. Neuronal chemistry and functional organization in the primate visual system. *Trends Neurosci.* 21: 344–349. [2, 3]

Henneberger, C., Papouin, T., Oliet, S. H., and Rusakov, D. A. 2010. Long-term potentiation depends on release of D-serine from astrocytes. *Nature* 463: 232–236. [10]

Henneman, E., Somjen, G., and Carpenter, D. O. 1965. Functional significance of cell size in spinal motoneurons. *J. Neurophysiol.* 28: 560–580. [24]

Hensch, T. K. 2004. Critical period regulation. *Annu. Rev. Neurosci.* 27: 549–579. [26]

Hensch, T. K., and Stryker, M. P. 2004. Columnar architecture sculpted by GABA circuits in developing cat visual cortex. *Science* 303: 1678–1681. [26]

Hensch, T. K., Fagiolini, M., Mataga, N., Stryker, M. P., Baekkeskov, S., and Kash, S. F. 1998. Local GABA circuit control of experience-dependent plasticity in developing visual cortex. *Science* 282: 1504–1508. [26]

Herbert, S. C., Mount, D. B., and Gamba, G. 2004. Molecular physiology of cation-coupled Cl^- cotransport. *Pflügers Arch.* 447: 580–593. [9]

Herbst, H., and Thier, P. 1996. Different effects of visual deprivation on vasoactive intestinal polypeptide (VIP)-containing cells in the retinas of juvenile and adult rats. *Exp. Brain Res.* 111: 345–355. [20]

Herlitze, S., Garcia, D. E., Mackie, K., Hille, B., Scheuer, T., and Catterall, W. A. 1996. Modulation of Ca^{2+} channels by G-protein beta gamma subunits. *Nature* 380: 258–262. [12]

Herlitze, S., Villarroel, A., Witzemann, V., Koenen, M., and Sakmann, B. 1996. Structural determinants of channel conductance in fetal and adult rat muscle acetylcholine receptors. *J. Physiol.* 492: 775–787. [11]

Hernandez, A., Salinas, E., Garcia, R., and Romo, R. 1997. Discrimination in the sense of flutter: new psychophysical measurements in monkeys. *J. Neurosci.* 17: 6391–6400. [23]

Hernandez, A., Zainos, A., and Romo, R. 2000. Neuronal correlates of sensory discrimination in the somatosensory cortex. *Proc. Natl. Acad. Sci. USA* 97: 6191–6196. [23]

Hernandez, A., Zainos, A., and Romo, R. 2002. Temporal evolution of a decision-making process in medial premotor cortex. *Neuron* 33: 959–972. [23]

Hernandez, C. C., Zaika, O., Tolstykh, G. P., and Shapiro, M. S. 2008. Regulation of neural KCNQ channels: signalling pathways, structural motifs and functional implications. *J. Physiol.* 586: 1811–1821. [17]

Hernandez, R. E., Rikhof, H. A., Bachmann, R., and Moens, C. B. 2004. vhnf1 integrates global RA patterning and local FGF signals to direct posterior hindbrain development in zebrafish. *Development* 131: 4511–4520. [25]

Herrada, G., and Dulac, C. 1997. A novel family of putative pheromone receptors in mammals with a topographically organized and sexually dimorphic distribution. *Cell* 90: 763–773. [19]

Herrero, J. L., Roberts, M. J., Delicato, L. S., Gieselmann, M. A., Dayan, P., and Thiele, A. 2008. Acetylcholine contributes through muscarinic receptors to attentional modulation in V1. *Nature* 454: 1110–1114. [14]

Hertting, G., and Axelrod, J. 1961. Fate of tritiated noradrenaline at the sympathetic nerve endings. *Nature* 192: 172–173. [15]

Hertting, G., Axelrod, J., Kopin, I. J., and Whitby, L. J. 1961. Lack of uptake of catecholamines after chronic denervation of sympathetic nerves. *Nature* 189: 66. [15]

Hertz L. 2004. Intercellular metabolic compartmentation in the brain: past, present and future. *Neurochem. Int.* 45: 285–296. [15]

Hestrin, S., and Galarreta, M. 2005. Electrical synapses define networks of neocortical GABAergic neurons. *Trends Neurosci.* 28: 304–309. [11]

Heuser, J. E. 1989. Review of electron microscopic evidence favouring vesicle exocytosis as the structural basis for quantal release during synaptic transmission. *Quart. J. Exp. Physiol.* 74: 1051–1069. [13]

Heuser, J. E., and Reese, T. S. 1973. Evidence for recycling of synaptic vesicle membrane during transmitter release at the frog neuromuscular junction. *J. Cell Biol.* 57: 315–344. [13]

Heuser, J. E., and Reese, T. S. 1981. Structural changes after transmitter release at the frog neuromuscular junction. *J. Cell Biol.* 88: 564–580. [13]

Heuser, J. E., Reese, T. S., and Landis, D. M. D. 1974. Functional changes in frog neuromuscular junction studied with freeze-fracture. *J. Neurocytol.* 3: 109–131. [13]

Heuser, J. E., Reese, T. S., Dennis, M. J., Jan, Y., Jan, L., and Evans, L. 1979. Synaptic vesicle exocytosis captured by quick freezing and correlated with quantal transmitter release. *J. Cell Biol.* 81: 275–300. [13]

Hevner, R. F. 2006. From radial glia to pyramidal-projection neuron: Transcription factor cascades in cerebral cortex development. *Mol. Neurobiol.* 33: 33–50. [10]

Hickie, C., Cohen, L. B., and Balaban, P. M. 1997. The synapse between LE sensory neurons and gill motoneurons makes only a small contribution to the *Aplysia* gill-withdrawal reflex. *Eur. J. Neurosci.* 9: 627–636. [18]

Higashida, H., Lopatina, O., Yoshihara, T., Pichugina, Y. A., Soumarokov, A. A., Munesue, T., Minabe, Y., Kikuchi, M., Ono, Y., Korshunova, N., and Salmina, A. B. 2010. Oxytocin signal and social behaviour: comparison among adult and infant oxytocin, oxytocin receptor and CD38 gene knockout mice. *J. Neuroendocrin.* 22: 373–379. [14]

Highstein, S. M. 1991. The central nervous system efferent control of the organs of balance and equilibrium. *Neurosci. Res.* 12: 13–30. [22]

Highstein, S. M., Rabbitt, R. D., Holstein, G. R., and Boyle, R. D. 2005. Determinants of spatial and temporal coding by semicircular canal afferents. *J. Neurophysiol.* 93: 2359–2370. [22]

Hilaire, G. G., Nicholls, J. G., and Sears, T. A. 1983. Central and proprioceptive influences on the activity of levator costae motoneurones in the cat. *J. Physiol.* 342: 527–548. [24]

Hilf, R. J., and Dutzler, R. 2008. X-ray structure of a prokaryotic pentameric ligand-gated ion channel. *Nature* 452: 375–379. [5]

Hilf, R. J., and Dutzler, R. 2009 Structure of a potentially open state of a proton-activated pentameric ligand-gated ion channel. *Nature* 457: 115–118. [Hill, R. 2000. NK1 (substance P) receptor antagonists—why are they not analgesic in humans? *Trends Pharmacol. Sci.* 21: 244–246. [14]

Hille, B. 1970. Ionic channels in nerve membranes. *Prog. Biophys. Mol. Biol.* 21: 1–32. [7]

Hille, B. 1994. Modulation of ion-channel function by G-protein-coupled receptors. *Trends Neurosci.* 17: 531–536. [12]

Hille, B. 2001. *Ion Channels in Excitable Membranes*, 3rd ed. Sinauer Associates, Sunderland MA. [4, 5, 7, 11]

Hirano, T. 1990a. Depression and potentiation of the synaptic transmission between a granule cell and a Purkinje cell in rat cerebellar culture. *Neurosci. Lett.* 119: 141–144. [16]

Hirano, T. 1990b. Effects of postsynaptic depolarization in the induction of synaptic depression between a granule cell and a Purkinje cell in rat cerebellar culture. *Neurosci. Lett.* 119: 145–147. [16]

Hirokawa, N. 1998. Kinesin and dynein superfamily proteins and the mechanism of organelle transport. *Science* 279: 519–552. [15]

Hirokawa, N., Niwa, S., and Tanaka, Y. 2010. Molecular motors in neurons: transport mechanisms and roles in brain function, development, and disease. *Neuron* 68: 610–638. [15]

Hirokawa, N., Terada, S., Funakoshi, T., and Takeda, S. 1997. Slow axonal transport: The subunit transport model. *Trends Cell Biol.* 7: 382–388. [15]

Hirsch, J. A., and Martinez, L. M. 2006. Circuits that build visual cortical receptive fields. *Trends Neurosci.* 29: 30–39. [2]

Hirsch, J. A., Gallagher, C. A., Alonso, J. M., and Martinez, L. M. 1998. Ascending projections of simple and complex cells in layer 6 of the cat striate cortex. *J. Neurosci.* 18: 8086–8094. [3]

Ho, R. K. 1992. Cell movements and cell fate during zebrafish gastrulation. *Dev. Suppl.* 65–73. [25]

Hobert, O. 2010. Neurogenesis in the nematode *Caenorhabditis elegans. WormBook.* 4: 1–24. [25]

Hodgkin, A. L. 1964. *The Conduction of the Nervous Impulse.* Liverpool University Press, Liverpool, England. [1]

Hodgkin, A. L. 1973. Presidential address. *Proc. R. Soc. Lond., B, Biol. Sci.* 183: 1–19. [6]

Hodgkin, A. L., and Huxley, A. F. 1952a. Currents carried by sodium and potassium ion through the membrane of the giant axon of *Loligo. J. Physiol.* 116: 449–472. [7]

Hodgkin, A. L., and Huxley, A. F. 1952b. The components of the membrane conductance in the giant axon of *Loligo. J. Physiol.* 116: 473–496. [7]

Hodgkin, A. L., and Huxley, A. F. 1952c. The dual effect of membrane potential on sodium conductance in the giant axon of *Loligo. J. Physiol.* 116: 497–506. [7]

Hodgkin, A. L., and Huxley, A. F. 1952d. A quantitative description of membrane current and its application to conduction and excitation in nerve. *J. Physiol.* 117: 500–544. [7]

Hodgkin, A. L., and Katz, B. 1949. The effect of sodium ions on the electrical activity of the giant axon of the squid. *J. Physiol.* 108: 37–77. [6]

Hodgkin, A. L., and Keynes, R. D. 1955. Active transport of cations in giant axons from *Sepia* and *Loligo. J. Physiol.* 128: 28–60. [9]

Hodgkin, A. L., and Keynes, R. D. 1956. Experiments on the injection of substances into squid giant axons by means of a microsyringe. *J. Physiol.* 131: 592–617. [6]

Hodgkin, A. L., and Keynes, R. D. 1957. Movements of labelled calcium in squid giant axons. *J Physiol.* 138: 253–281. [12]

Hodgkin, A. L., and Rushton, W. A. H. 1946. The electrical constants of a crustacean nerve fiber. *Proc. R. Soc. Lond., B, Biol. Sci.* 133: 444–479. [8]

Hodgkin, A. L., Huxley, A. F., and Katz, B. 1952. Measurement of current-voltage relations in the membrane of the giant axon of *Loligo. J. Physiol.* 116: 424–448. [7]

Hofmann, F., Biel, M., and Flockerzi, V. 1994. Molecular basis for Ca^{2+} channel diversity. *Annu. Rev. Neurosci.* 17: 399–418. [5]

Hökfelt, T., Broberger, C., Xu, Z. Q., Sergeyev, V., Ubink, R., and Diez, M. 2000. Neuropeptides—an overview. *Neuropharmacology* 39: 1337–1356. [14]

Hökfelt, T., Johansson, O., Llungdahl, A., Lundberg, M., and Schultzberg, M. 1980. Peptidergic neurons. *Nature* 284: 515–521. [17]

Hökfelt, T., Kellerth, J. O., Nilsson, G., and Pernow, B. 1975. Substance P; localization in the central nervous system and in some primary sensory neurons. *Science* 190: 889–890. [14]

Hökfelt, T., Ljungdahl, A., Terenius, L., Elde, R., and Nilsson, G. 1977. Immunohistochemical analysis of peptide pathways possibly related to pain and analgesia: enkephalin and substance P. *Proc. Natl. Acad. Sci. USA* 74: 3081–3085. [14]

Hollins, M., and Bensmaia, S. J. 2007. The coding of roughness. *Can. J. Exp. Psychol.* 61: 184–195. [21]

Hollins, M., Fox, A., and Bishop, C. 2000. Imposed vibration influences perceived tactile smoothness. *Perception* 29: 1455–1465. [21]

Hollman, M., and Heinemann, S. 1994. Cloned glutamate receptors. *Annu. Rev. Neurosci.* 17: 31–108. [14]

Hollmann, M., Hartley, M., and Heinemann, S. 1991. $Ca2^+$ permeability of KA–AMPA-gated glutamate receptor channels depends on subunit composition. *Science* 252: 851–853. [25]

Hollmann, M., Maron, C., and Heinemann, S. 1994. N-Glycosylation sit tagging suggests a three transmembrane domain topology for the glutamate receptor GluR1. *Neuron* 13: 1331–1343. [5]

Hollyday, M., and Hamburger, V. 1976. Reduction of the naturally occurring motor neuron loss by enlargement of the periphery. *J. Comp. Neurol.* 170: 311–320. [25]

Hölscher, C. 1999. Synaptic plasticity and learning and memory: LTP and beyond. *J. Neurosci. Res.* 58: 62–75. [16]

Holt, J. C., Chatlani, S., Lysakowski, A., and Goldberg, J. M. 2007. Quantal and nonquantal transmission in calyx-bearing fibers of the turtle posterior crista. *J. Neurophysiol.* 98: 1083–1101. [22]

Holton, P. 1959. The liberation of adenosine triphosphate on antidromic stimulation of sensory nerves. *J. Physiol.* 145: 494–504. [14]

Homma, Y., Baker, B. J., Jin, L., Garaschuk, O., Konnerth, A., Cohen, L. B., and Zecevic, D. 2009. Wide-field and two-photon imaging of brain activity with voltage- and calcium-sensitive dyes. *Philos. Trans. R. Soc. Lond., B, Biol. Sci.* 364: 2453–2467. [1]

Hooks, B. M., and Chen, C. 2007. Critical periods in the visual system: changing views for a model of experience-dependent plasticity. *Neuron* 56: 312–326. [26]

Hooper, S. L., and DiCaprio, R. A. 2004. Crustacean motor pattern generator networks. *Neurosignals* 13: 50–69. [18]

Horisberger, J. D. 2004. Recent insights into the structure and mechanism of the sodium pump. *Physiology* 19: 377–387. [9]

Horn, R., and Marty, A. 1988. Muscarinic activation of ionic currents measured by a new whole-cell recording method. *J. Gen. Physiol.* 92: 145–149. [4]

Horton, J. C., and Hocking, D. R. 1996. Pattern of ocular dominance columns in human striate cortex in strabismic amblyopia. *Vis. Neurosci.* 13: 787–795. [26]

Horton, J. C., and Hocking, D. R. 1997. Timing of the critical period for plasticity of ocular dominance columns in macaque striate cortex. *J. Neurosci.* 17: 3684–3709. [26]

Horton, J. C., and Hocking, D. R. 1998. Effect of early monocular enucleation upon ocular dominance columns and cytochrome oxidase activity in monkey and human visual cortex. *Vis. Neurosci.* 15: 289–303. [26]

Hoshi, T., Zagotta, W. N., and Aldrich, R. W. 1990. Biophysical and molecular mechanisms of *Shaker* potassium channel inactivation. *Science* 250: 533–550. [7]

Hoshi, T., Zagotta, W. N., and Aldrich, R. W. 1991. Two types of inactivation in *Shaker* K^+ channels: Effects of alterations in the carboxyl terminal region. *Neuron* 7: 547–566. [7]

Hosie, A. M., Wilkins, M. E., da Silva, H. M., and Smart, T. G., 2006. Endogenous neurosteroids regulate GABAA receptors through two discrete transmembrane sites. *Nature* 444: 486–489. [14]

Hosoya, T., Baccus, S. A., and Meister, M. 2005. Dynamic predictive coding by the retina. *Nature* 436: 71–77. [20]

Housley, G. D., and Ashmore, J. F. 1991. Direct measurement of the action of acetylcholine on isolated outer hair cells of the guinea pig cochlea. *Proc. R. Soc. Lond., B, Biol. Sci.* 244: 161–167. [22]

Howard, J., and Hudspeth, A. J. 1988. Compliance of the hair bundle associated with gating of mechanoelectrical transduction channels in the bullfrog's saccular hair cell. *Neuron* 1: 189–199. [19]

Howard, J., Hudspeth, A. J., and Vale, R. D. 1989. Movement of microtubules by single kinesin molecules. *Nature* 342: 154–158. [15]

Howes, O. D., and Kapur, S. 2009. The dopamine hypothesis of schizophrenia: version III—the final common pathway. *Schizophr. Bull.* 35: 549–562. [14]

Hu, F., and Strittmatter, S. M. 2004. Regulating axon growth within the postnatal central nervous system. *Semin. Perinatol.* 28: 371–378. [27]

Huang, C-L., Slesinger, P. A., Casey, P. J., Jan, Y. N., and Jan, L. Y. 1995. Evidence that direct binding of Gβγ to the GIRK1 G protein-gated inwardly rectifying K^+ channel is important for channel activation. *Neuron* 15: 1133–1143. [12]

Huang, Y., Jellies, J., Johansen, K. M., and Johansen, J. 1998. Development and pathway formation of peripheral neurons during leech embryogenesis. *J. Comp. Neurol.* 397: 394–402. [18]

Huang, Y. J., Maruyama, Y., Lu, K. S., Pereira, E., Plonsky, I., Baur, J. E., Wu, D., and Roper, S. D. 2005. Mouse taste buds use serotonin as a neurotransmitter. *J. Neurosci.* 25: 843–847. [19]

Huang, Z. J., Kirkwood, A., Pizzorusso, T., Porciatti, V., Morales, B., Bear, M. F., Maffei, L., and Tonegawa, S. 1999. BDNF regulates the maturation of inhibition and the critical period of plasticity in mouse visual cortex. *Cell* 98: 739–755. [26]

Huang, Z. L., Qu, W. M., Li, W. D., Mochizuki, T., Eguchi, N., Watanabe, T., Urade, Y., and Hayaishi, O. 2001. Arousal effect of orexin A depends on activation of the histaminergic system. *Proc. Natl. Acad. Sci. USA* 98: 9965–9970. [14]

Hubel, D. H. 1982. Exploration of the primary visual cortex. *Nature* 299: 515–524. [2]

Hubel, D. H. 1988. *Eye, Brain and Vision.* Scientific American Library, New York. [2, 3, 26]

Hubel, D. H., and Wiesel, T. N. 1959. Receptive fields of single neurons in the cat's striate cortex. *J. Physiol.* 148: 574–591. [1–3]

Hubel, D. H., and Wiesel, T. N. 1961. Integrative action in the cat's lateral geniculate body. *J. Physiol.* 155: 385–398. [2]

Hubel, D. H., and Wiesel, T. N. 1962. Receptive fields, binocular interaction and functional architecture in the cat's visual cortex. *J. Physiol.* 160: 106–154. [2, 3]

Hubel, D. H., and Wiesel, T. N. 1963. Receptive fields of cells in striate cortex of very young, visually inexperienced kittens. *J. Neurophysiol.* 26: 994–1002. [26]

Hubel, D. H., and Wiesel, T. N. 1965a. Binocular interaction in striate cortex of kittens reared with artificial squint. *J. Neurophysiol.* 28: 1041–1059. [26]

Hubel, D. H., and Wiesel, T. N. 1965b. Receptive fields and functional architecture in two non-striate visual areas (18 and 19) of the cat. *J. Neurophysiol.* 28: 229–289. [2, 3]

Hubel, D. H., and Wiesel, T. N. 1967. Cortical and callosal connections concerned with the vertical meridian of visual field in the cat. *J. Neurophysiol.* 30: 1561–1573. [3]

Hubel, D. H., and Wiesel, T. N. 1968. Receptive fields and functional architecture of monkey striate cortex. *J. Physiol.* 195: 215–243. [2, 3]

Hubel, D. H., and Wiesel, T. N. 1970. The period of susceptibility to the physiological effects of unilateral eye closure in kittens. *J. Physiol.* 206: 419–436. [26]

Hubel, D. H., and Wiesel, T. N. 1972. Laminar and columnar distribution of geniculo-cortical fibers in the macaque monkey. *J. Comp. Neurol.* 146: 421–450. [2, 3]

Hubel, D. H., and Wiesel, T. N. 1974. Sequence regularity and geometry of orientation columns in the monkey striate cortex. *J. Comp. Neurol.* 158: 267–294. [3]

Hubel, D. H., and Wiesel, T. N. 1977. Functional architecture of macaque monkey visual cortex (Ferrier Lecture). *Proc. R. Soc. Lond., B, Biol. Sci.* 198: 1–59. [1, 3, 26]

Hubel, D. H., Wiesel, T. N., and LeVay, S. 1977. Plasticity of ocular dominance columns in monkey striate cortex. *Philos. Trans. R. Soc. Lond., B, Biol. Sci.* 278: 377–409. [26]

Hubener, M., Shoham, D., Grinvald, A., and Bonhoeffer, T. 1997. Spatial relationships among three columnar systems in cat area 17. *J. Neurosci.* 17: 9270–9284. [3]

Hübner, C. A., Stein, V., Hermans-Borgmeyer, I., Meyer, T., Ballanyi, K., and Jentsch, T. J. 2001. Disruption of KCC2 reveals an essential role of K-Cl cotransport already in early synaptic inhibition. *Neuron* 30: 515–524. [11]

Huckstepp, R. T., id Bihi, R., Eason, R., Spyer, K. M., Dicke, N., Willecke, K., Marina, N., Gourine, A. V., and Dale, N. 2010. Connexin hemichannel-mediated CO_2-dependent release of ATP in the medulla oblongata contributes to central respiratory chemosensitivity. *J. Physiol.* 588: 3901–3920. [24]

Hudspeth, A. J. 1982. Extracellular current flow and the site of transduction by vertebrate hair cells. *J. Neurosci.* 2: 1–10. [19]

Hudspeth, A. J., and Corey, D. P. 1977. Sensitivity, polarity, and conductance change in the response of vertebrate hair cells to controlled mechanical stimuli. *Proc. Natl. Acad. Sci. USA* 74: 2407–2411. [19]

Hudspeth, A. J., and Gillespie, P. G. 1994. Pulling springs to tune transduction: adaptation by hair cells. *Neuron* 12: 1–9. [19]

Hudspeth, A. J., and Jacobs, R. 1979. Stereocilia mediate transduction in vertebrate hair cells (auditory system/cilium/vestibular system). *Proc. Natl. Acad. Sci. USA* 76: 1506–1509. [19]

Hudspeth, A. J., and Lewis, R. S. 1988. Kinetic analysis of voltage- and ion-dependent conductances in saccular hair cells of the bull-frog, *Rana catesbeiana. J. Physiol.* 400: 237–274. [22]

Hudspeth, A. J., Poo, M. M., and Stuart, A. E. 1977. Passive signal propagation and membrane properties in median photoreceptors of the giant barnacle. *J. Physiol.* 272: 25–43. [19]

Huebner, E. A., and Strittmatter, S. M. 2009. Axon regeneration in the peripheral and central nervous systems. *Results Probl. Cell Differ.* 48: 339–351. [27]

Hughes, J. 1975. Isolation of an endogenous compound from the brain with pharmacological properties similar to morphine. *Brain Res.* 88: 295–308. [14]

Hughes, J., Smith, T. W., Kosterlitz, H. W., Fothergill, L. A., Morgan, B. A., and Morris, H. R. 1975. Identification of two related pentapeptides from the brain with potent opiate agonist activity. *Nature* 258: 577–580. [14]

Hughes, S., Marsh, S. J., Tinker, A., and Brown, D. A. 2007. PIP_2-dependent inhibition of M-type (Kv7.2/7.3) potassium channels: direct on-line assessment of PIP_2 depletion by Gq-coupled receptors in single living neurons. *Pflügers Arch.* 455: 115–124. [12]

Hultborn, H. 2006. Spinal reflexes, mechanisms and concepts: from Eccles to Lundberg and beyond. *Prog. Neurobiol.* 78: 215–242. [24]

Humphrey, A. L., Sur, M., Uhlrich, D. J., and Sherman, S. M. 1985. Projection patterns of individual X- and Y-cell axons from the lateral geniculate nucleus to cortical area 17 in the cat. *J. Comp. Neurol.* 233: 159–189. [26]

Hunt, C. C., Wilkinson, R. S., and Fukami, Y. 1978. Ionic basis of the receptor potential in primary endings of mammalian muscle spindles. *J. Gen. Physiol.* 71: 683–698. [19]

Hurlemann, R., Patin, A., Onur, O. A., Cohen, M. X., Baumgartner, T., Metzler, S., Dziobek, I., Gallinat, J., Wagner, M., Maier, W., and Kendrick, K. M. 2010. Oxytocin enhances amygdala-dependent, socially reinforced learning and emotional empathy in humans. *J. Neurosci.* 30: 4999–5007. [14]

Hutchins, B. I., and Kalil, K. 2008. Differential outgrowth of axons and their branches is regulated by localized calcium transients. *J. Neurosci.* 28: 143–153. [25]

Hutter, O. F., and Trautwein, W. 1956. Vagal and sympathetic effects on the pacemaker fibers in the sinus venosus of the heart. *J. Gen. Physiol.* 39: 715–733. [12]

Huxley, A. 1928. *Point Counter Point.* Harper Collins, New York. [22]

Hyvarinen, J., Poranen, A., and Jokinen, Y. 1980. Influence of attentive behavior on neuronal responses to vibration in primary somatosensory cortex of the monkey. *J. Neurophysiol.* 43: 870–882. [21]

Hyvarinen, J., Sakata, H., Talbot, W. H., and Mountcastle, V. B. 1968. Neuronal coding by cortical cells of the frequency of oscillating peripheral stimuli. *Science* 162: 1130–1132. [21]

Ichida, J. M., and Casagrande, V. A. 2002. Organization of the feedback pathway from striate cortex (V1) to the LGN (LGN) in the owl monkey (*Aotus trivirgatus*). *J. Comp. Neurol.* 454: 272–283. [2]

Iggo, A., and Muir, A. R. 1969. The structure and function of a slowly adapting touch corpuscle in hairy skin. *J. Physiol.* 200: 763–796. [21]

Iglesias, R., Dahl, G., Qiu, F., Spray, D. C., and Scemes, E. 2009. Pannexin 1: The molecular substrate of astrocyte "hemichannels." *J. Neurosci.* 29: 7092–7097. [10]

Ignarro, J. 1990. Biosynthesis and metabolism of endothelium-derived nitric oxide. *Annu. Rev. Physiol.* 30: 535–560. [12]

Ikeda, S. R. 1996. Voltage-dependent modulation of N-type calcium channels by G-protein beta gamma subunits. *Nature* 380: 255–258. [12]

Ikoma, A., Rukwied, R., Stander, S., Steinhoff, M., Miyachi Y., and Schmelz, M. 2003. Neurophysiology of pruritus: interaction of itch and pain. *Arch. Dermatol.* 139: 1475–1478. [21]

Imoto, K., Busch, C., Sakmann, B., Mishina, M., Konno, T., Nakai, J., Bujo, H., Mori, Y., Fukada, K., and Numa, S. 1998. Rings of negatively charged amino acids determine the acetylcholine receptor conductance. *Nature* 335: 645–648. [5]

Imoto, K., Konno, T., Nakai, J., Wang, F., Mishina, M., and Numa, S. 1991. A ring of uncharged polar amino acids as a component of channel constriction in the nicotinic acetylcholine receptor. *FEBS Lett.* 289: 193–200. [5]

Ingber, D. E. 2006. Cellular mechanotransduction: putting all the pieces together again. *FASEB J.* 20: 811–827. [19]

Inoue, S. 1981. Video image processing greatly enhances contrast, quality and speed in polarization-based microscopy. *J. Cell Biol.* 89: 346–356. [15]

Insanally, M. N., Kover, H., Kim, H., and Bao, S. 2009. Feature-dependent sensitive periods in the development of complex sound representation. *J. Neurosci.* 29: 5456–5462. [26]

Isaac, J. T., Nicoll, R. A., and Malenka, R. C. 1995. Evidence for silent synapses: Implications for the expression of LTP. *Neuron* 15: 427–434. [16]

Isaac, R. E., Bland, N. D., and Shirras, A. D. 2009. Neuropeptidases and the metabolic inactivation of insect neuropeptides. *Gen. Comp. Endocrinol.* 162: 8–17. [15]

Isbister, C. M., and O'Connor, T. P. 1999. Filopodial adhesion does not predict growth cone steering events *in vivo*. *J. Neurosci.* 19: 2589–2600. [25]

Ishikawa, T., Sahara, Y., and Takahashi, T. 2002. A single packet of transmitter does not saturate postsynaptic glutamate receptors. *Neuron* 34: 613–621. [15]

Ito, M. 1972. Neural design of the cerebellar motor control system. *Brain Res.* 40: 81–84. [22]

Ito, M. 1984. *The Cerebellum and Neural Control.* Raven, New York. [24]

Ito, M., and Simpson, J. I. 1971. Discharges in Purkinje cell axons during climbing fiber activation. *Brain Res.* 31: 215–219. [24]

Ito, M., Sakurai, M., and Tongroach, P. 1982. Climbing fibre induced depression of both mossy fibre responsiveness and glutamate sensitivity of cerebellar Purkinje cells. *J. Physiol.* 324: 113–134. [16]

Ito, M., Tamura, H., Fujita, I., and Tanaka, K. 1995. Size and position invariance of neuronal responses in monkey inferotemporal cortex. *J. Neurophysiol.* 73: 218–226. [23]

Iversen, L. 2003. Cannabis and the brain. *Brain* 126: 1252–1270. [12]

Iversen, L. 2006. Neurotransmitter transporters and their impact on the development of psychopharmacology. *Brit. J. Pharmacol.* 147(Suppl. 1): S82–88. [15]

Iversen, L. L. 1971. Role of transmitter uptake mechanisms in synaptic neurotransmission. *Brit. J. Pharmacol.* 41: 571–591. [14]

Iversen, L. L., Iverson, S. D., Bloom, F. E., and Roth, R. H. 2009. *Introduction to Neuropsychopharmacology.* Oxford University Press, New York. [9]

Izquierdo, I., and Medina, J. H. 1995. Correlation between the pharmacology of long-term potentiation and the pharmacology of memory. *Neurobiol. Learn. Mem.* 63: 19–32. [16]

Jacobs, B. L., and Azmitia, E. C. 1992. Structure and function of the brain serotonin system. *Physiol. Rev.* 72: 165–229. [14]

Jacobs, G. A., Miller, J. P., and Aldworth, Z. 2008. Computational mechanisms of mechanosensory processing in the cricket. *J. Exp. Biol.* 211: 1819–1828. [18]

Jacobs, G. H. 2008. Primate color vision: a comparative perspective. *Vis. Neurosci.* 25: 619–633. [20]

Jan, Y. N., Jan, L. Y., and Kuffler, S. W. 1980. Further evidence for peptidergic transmission in sympathetic ganglia. *Proc. Natl. Acad. Sci. USA* 77: 5008–5012. [17]

Janig, W., and McLachlan, E. M. 1992. Characteristics of function-specific pathways in the sympathetic nervous system. *Trends Neurosci.* 15: 475–481. [17]

Jansen, J. K. S., and Matthews, P. B. C. 1962. The central control of the dynamic response of muscle spindle receptors. *J. Physiol.* 161: 357–378. [19]

Jansen, J. K. S., and Nicholls, J. G. 1973. Conductance changes, an electrogenic pump and the hyperpolarization of leech neurones following impulses. *J. Physiol.* 229: 635–655. [7]

Jansen, J. K. S., Lømo, T., Nicholaysen, K., and Westgaard, R. H. 1973. Hyperinnervation of skeletal muscle fibers: dependence on muscle activity. *Science* 181: 559–561. [27]

Jaramillo, F., and Hudspeth, A. J. 1991. Localization of the hair cell's transduction channels at the hair bundle's top by iontophoretic application of a channel blocker. *Neuron* 7: 409–420. [19]

Jarvilehto, T., Hamalainen, H., and Laurinen, P. 1976. Characteristics of single mechanoreceptive fibres innervating hairy skin of the human hand. *Exp. Brain Res.* 25: 45–61. [21]

Jarvis, M. F., and Khakh, B. S. 2009. ATP-gated P2X channels. *Neuropharmacology* 56: 208–215. [5]

Jasper, H., and Penfield, W. 1954. *Epilepsy and the Functional Anatomy of the Human Brain*, 2nd ed. Boston: Little, Brown and Co. [21]

Jasser, A., and Guth, P. S. 1973. The synthesis of acetylcholine by the olivo-cochlear bundle. *J. Neurochem.* 20: 45–53. [22]

Jeanmonod, D., Rice, F. L., and van Der Loos, H. 1977. Mouse somatosensory cortex: Development of the alterations in the barrel field which are caused by injury to the vibrissal follicles. *Neurosci. Lett.* 6: 151–156. [21]

Jenkinson, D. H. 1960. The antagonism between tubocurarine and substances which depolarize the motor end-plate. *J. Physiol.* 152: 309–324. [11]

Jenkinson, D. H. 2006. Potassium channels—multiplicity and challenges. *Brit. J. Pharmacol.* 147: S63–S71. [7]

Jenkinson, D. H. 2011. Classical approaches to the study of drug-receptor interactions. In J. C. Foreman, T. Johansen, and A. J. Gibb (ed.), *Textbook of Receptor Pharmacology*, 3rd ed. CRC Press, London, U. K. pp. 3–76. [11]

Jenkinson, D. H., and Nicholls, J. G. 1961. Contractures and permeability changes produced by acetylcholine in depolarized denervated muscle. *J. Physiol.* 159: 111–127. [11]

Jennings, E. A., Ryan, R. M., and Christie, M. J. 2004. Effects of sumatriptan on rat medullary dorsal horn neurons. *Pain* 111: 30–37. [14]

Jentsch, T. J. 2000. Neuronal KCNQ potassium channels: physiology and role in disease. *Nat. Rev. Neurosci.* 1: 21–30. [28]

Ji, T. H., Grossmann, M., and Ji, I. 1998. G protein–coupled receptors. I. Diversity of receptor-ligand interactions. *J. Biol. Chem.* 273: 17299–17302. [12]

Jiang, G. J., Zidanic, M., Michaels, R. L., Michael, T. H., Griguer, C., and Fuchs, P. A. 1997. CSlo encodes calcium-activated potassium channels in the chick's cochlea. *Proc. R. Soc. Lond., B, Biol. Sci.* 264: 731–737. [22]

Jiang, Y., Ruta, V., Chen, J., Lee, A., and MacKinnon, R. 2003. The principle of gating charge movement in a voltage-dependent K^+ channel. *Nature* 423: 42–48. [7]

Jo, Y. H., and Role, L. W. 2002. Coordinate release of ATP and GABA at *in vitro* synapses of lateral hypothalamic neurons. *J. Neurosci.* 22: 4794–4804. [14]

Joh, T. H., Park, D. H., and Reis, D. J. 1978. Direct phosphorylation of brain tyrosine hydroxylase by cyclic AMP-dependent protein kinase: Mechanism of enzyme activation. *Proc. Natl. Acad. Sci. USA* 75: 4744–4748. [15]

Johansson, P. A., Dziegielewska, K. M., Liddelow, S. A., and Saunders, N. R. 2008. The blood-CSF barrier explained: When development is not immaturity. *Bioessays* 30: 237–248. [10]

Johansson, R. S., and Vallbo, A. B. 1979. Detection of tactile stimuli. Thresholds of afferent units related to psychophysical thresholds in the human hand. *J. Physiol.* 297: 405–422. [21]

Johansson, R. S., and Vallbo, A. B. 1983. Tactile sensory coding in the glabrous skin of the human hand. *Trends Neurosci.* 6: 27–32. [21]

Johnson, E. W., and Wernig, A. 1971. The binomial nature of transmitter release at the crayfish neuromuscular junction. *J. Physiol.* 218: 757–767. [13]

Johnson, F. H., Eyring, H., and Polissar, M. J. 1954. *The Kinetic Basis of Molecular Biology*. Wiley, New York. [4]

Johnson, M. H. 2005. Subcortical face processing. *Nat. Rev. Neurosci.* 6: 766–774. [23]

Jomphe, C., Bourque, M. J., Fortin, G. D., St-Gelais, F., Okano, H., Kobayashi, K., and Trudeau, L. E. 2005. Use of TH-EGFP transgenic mice as a source of identified dopaminergic neurons for physiological studies in postnatal cell culture. *J. Neurosci. Methods* 146: 1–12. [14]

Jonas, P., Bischofberger, J., and Sandkühler, J. 1998. Corelease of two fast neurotransmitters at a central synapse. *Science* 281: 419–424. [14, 15]

Jonas, P., Major, G., and Sakmann, B. 1993. Quantal components of unitary EPSCs at the mossy fibre synapse on CA3 pyramidal cells of rat hippocampus. *J. Physiol.* 472: 615–663. [13]

Jones, D. T., and Reed, R. R. 1989. Golf: an olfactory neuron specific-G protein involved in odorant signal transduction. *Science* 244: 790–795. [19]

Jones, E. M., Gray-Keller, M., and Fettiplace, R. 1999. The role of Ca^{2+}-activated K^+ channel spliced variants in the tonotopic organization of the turtle cochlea. *J. Physiol.* 518: 653–665. [22]

Jones, E. M., Laus, C., and Fettiplace, R. 1998. Identification of Ca^{2+}-activated K^+ channel splice variants and their distribution in the turtle cochlea. *Proc. R. Soc. Lond., B, Biol. Sci.* 265: 685–692. [22]

Jones, S. W., and Adams, P. R. 1987. In *Neuromodulation: The Biochemical Control of Neuronal Excitability*. Oxford University Press, New York. pp. 159–186. [17]

Jones, T. A., Leake, P. A., Snyder, R. L., and Stakhovskaya, O., and Bonham, B. 2007. Spontaneous discharge patterns in cochlear spiral ganglion cells before the onset of hearing in cats. *J. Neurophysiol.* 98: 1898–1908. [26]

Jope, R. 1979. High-affinity choline uptake and acetylcholine production in the brain. Role in regulation of ACh synthesis. *Brain Res. Rev.* 1: 313–344. [15]

Jorgensen, P. L., Hakansson, K. O., and Karlish, S. J. D. 2003. Structure and mechanism of the Na,K-ATPases: Functional sites and their interactions. *Annu. Rev. Physiol.* 65: 817–849. [9]

Juge, N., Gray, J. A., Omote, H., Miyaji, T., Inoue, T., Hara, C., Uneyama, H., Edwards, R. H., Nicoll, R. A., and Moriyama, Y. 2010. Metabolic control of vesicular glutamate transport and release. *Neuron* 68: 99–112. [15]

Kaas, J. H. 1983. What, if anything, is, S. I.? Organization of first somatosensory area of cortex. *Physiol. Rev.* 63: 206–231. [21]

Kalb, R. 2005. The protean actions of neurotrophins and their receptors on the life and death of neurons. *Trends Neurosci.* 28: 5–11. [25]

Kalivas, P. W., and Duffy, P. 1990. Effect of acute and daily cocaine treatment on extracellular dopamine in the nucleus accumbens. *Synapse* 5: 48–58. [14]

Kallen, R. G., Sheng, Z-H., Yang, J., Chen, L., Rogart, R. B., and Barchi, R. L. 1990. Primary structure and expression of a sodium channel characteristic of denervated and immature rat skeletal muscle. *Neuron* 4: 233–342. [27]

Kalmijn, A. J. 1982. Electric and magnetic field detection in elasmobranch fishes. *Science* 218: 916–918. [19]

Kanai, Y., and Hediger, M. A. 2004. The glutamate/neutral amino acid transporter family SLC1: Molecular, physiological, and pharmacological aspects. *Pflügers Arch.* 447: 469–479. [9]

Kandel, E. R. 1979. *Behavioral Biology of Aplysia*. W. H. Freeman, San Francisco. [18]

Kandel, E. R. 2001. The molecular biology of memory storage: a dialogue between genes and synapses. *Science* 294: 1030–1038. [18]

Kaneko, A. 1970. Physiological and morphological identification of horizontal, bipolar and amacrine cells in goldfish retina. *J. Physiol.* 207: 623–633. [20]

Kaneko, A. 1971. Electrical connexions between horizontal cells in the dogfish retina. *J. Physiol.* 213: 95–105. [20]

Kaneko, A., and Hashimoto, H. 1969. Electrophysiological study of single neurons in the inner nuclear layer of the carp retina. *Vision Res.* 9: 37–55. [1, 20]

Kaneko, A., and Tachibana, M. 1986. Effects of gamma-aminobutyric acid on isolated cone photoreceptors of the turtle retina. *J. Physiol.* 373: 443–461. [20]

Kanning, K. C., Kaplan, A., and Henderson, C. E. 2010. Motor neuron diversity in development and disease. *Annu. Rev. Neurosci.* 33: 409–440. [24]

Kano, M., Ohno-Shosaku, T., Hashimotodani, Y., Uchigashima, M., and Watanabe, M. 2009. Endocannabinoid-mediated control of synaptic transmission. *Physiol. Rev.* 89: 309–380. [12]

Kanold, P. O., and Shatz, C. J. 2006. Subplate neurons regulate maturation of cortical inhibition and outcome of ocular dominance plasticity. *Neuron* 51: 627–638. [25]

Kanwal, J. S., and Rauschecker, J. P. 2007. Auditory cortex of bats and primates: managing species-specific calls for social communication. *Front. Biosci.* 12: 4621–4640. [22]

Kanwisher, N., and Yovel, G. 2006. The fusiform face area: a cortical region specialized for the perception of faces. *Philos. Trans. R. Soc. Lond., B, Biol. Sci.* 361: 2109–2128. [23]

Kaplan, E., and Shapley, R. M. 1986. The primate retina contains two types of ganglion cells, with high and low contrast sensitivity. *Proc. Natl. Acad. Sci. USA* 83: 2755–2757. [20]

Karadottir, R., Cavelier, P., Bergersen, L. H., and Attwell, D. 2005. NMDA receptors are expressed in oligodendrocytes and activated in ischaemia. *Nature* 438: 1162–1166. [10]

Karlin, A. 2002. Emerging structure of nicotinic acetylcholine receptors. *Nat. Rev. Neurosci.* 3: 102–114. [5]

Karplus, M., and Petsko, G. A. 1990. Molecular dynamics simulations in biology. *Nature* 347: 631–639. [4]

Kasai, H. 1999. Comparative biology of Ca^{2+}-dependent exocytosis: Implications of kinetic diversity for secretory function. *Trends Neurosci.* 22: 88–93. [13]

Kasakov, L., Ellis, J., Kirkpatrick, K., Milner, P., and Burnstock, G. 1988. Direct evidence for concomitant release of noradrenaline, adenosine 5′-triphosphate and neuropeptide Y from sympathetic nerve supplying the guinea-pig vas deferens. *J. Auton. Nerv. Syst.* 22: 75–82. [17]

Kask, K., Zamanillo, D., Rozov, A., Burnashev, N., Sprengel, R., and Seeburg, P. H. 1998. The AMPA receptor subunit GluR-B in its Q/R site-unedited form is not essential for brain development and function. *Proc. Natl. Acad. Sci. USA* 95: 13777–13782. [25]

Kaspar, J., Schor, R. H., and Wilson, V. J. 1988. Response of vestibular neurons to head rotations in vertical planes. II. Response to neck stimulation and vestibular–neck interactions. *J. Neurophysiol.* 60: 1765–1768. [24]

Kastner, S., Pinsk, M. A., De Weerd, P., Desimone, R., and Ungerleider, L. G. 1999. Increased activity in human visual cortex during directed attention in the absence of visual stimulation. *Neuron* 22: 751–761. [23]

Kastner, S., Schneider, K. A., and Wunderlich, K. 2006. Beyond a relay nucleus: neuroimaging views on the human LGN. *Prog. Brain Res.* 155: 125–143. [2]

Katona, I., Sperlágh, B., Maglóczky, Z., Sántha, E., Köfalvi, A., Czirják, S., Mackie, K., Vizi, E. S., and Freund, T. F. 2000. GABAergic interneurons are the targets of cannabinoid actions in the human hippocampus. *Neuroscience* 100: 797–804. [12]

Katsuki, Y. 1961. Neural mechanisms of auditory sensation in cats. In W. A. Rosenblith (ed.), *Sensory Communication*. MIT Press, Cambridge, MA. pp. 561–583. [22]

Katz, B. 1950. Depolarization of sensory terminals and the initiation of impulses in the muscle spindle. *J. Physiol.* 111: 261–282. [19]

Katz, B. 1971. Quantal mechanism of neural transmitter release. *Science* 173: 123–126. [1]

Katz, B., and Miledi, R. 1964. The development of acetylcholine sensitivity in nerve-free segments of skeletal muscle. *J. Physiol.* 170: 389–396. [27]

Katz, B., and Miledi, R. 1967a. The timing of calcium action during neuromuscular transmission. *J. Physiol.* 189: 535–544. [13]

Katz, B., and Miledi, R. 1967b. A study of synaptic transmission in the absence of nerve impulses. *J. Physiol.* 192: 407–436. [13]

Katz, B., and Miledi, R. 1968. The role of calcium in neuromuscular facilitation. *J. Physiol.* 195: 481–492. [16]

Katz, B., and Miledi, R. 1972. The statistical nature of the acetylcholine potential and its molecular components. *J. Physiol.* 224: 665–699. [4, 13]

Katz, B., and Miledi, R. 1973. The binding of acetylcholine to receptors and its removal from the synaptic cleft. *J. Physiol.* 231: 549–574. [11, 15]

Kaufman, C. M., and Menaker, M. J. 1993. Effect of transplanting suprachiasmatic nuclei from donors of different ages into completely SCN lesioned hamsters. *J. Neural Transplant. Plast.* 4: 257–265. [17]

Kaupp, U. B., and Seifert, R. 2002. Cyclic nucleotide-gated ion channels. *Physiol. Rev.* 82: 769–824. [20]

Kaur, G., Han, S. J., Yang, I., and Crane, C. 2010. Microglia and central nervous system immunity. *Neurosurg. Clin. N. Am.* 21: 43–51. [10]

Kazmierczak, P., Sakaguchi, H., Tokita, J., Wilson-Kubalek, E. M., Milligan, R. A., Müller, U., and Kachar, B. 2007. Cadherin 23 and protocadherin 15 interact to form tip-link filaments in sensory hair cells. *Nature* 449: 87–91. [19]

Keirstead, H. S., and Miller, R. F. 1997. Metabotropic glutamate receptor agonists evoke calcium waves in isolated Müller cells. *Glia* 21: 194–203. [10]

Keirstead, H. S., Dyer, J. K., Sholomenko, G. N., McGraw, J., Delaney, K. R., and Steeves, J. D. 1995. Axonal regeneration and physiological activity following transection and immunological disruption of myelin within the hatchling chick spinal cord. *J. Neurosci.* 15: 6963–6974. [27]

Kellenberger, S. West, J. W., Catterall W. A., and Scheuer, T. 1997. Molecular analysis of potential hinge residues in the inactivation gate of brain type IIA Na^+ channels. *J. Gen. Physiol.* 109: 607–617. [7]

Kellenberger, S., West, J. W., Scheuer, T., and Catterall, W. A. 1997. Molecular analysis of a putative inactivation particle in the inactivation gate of brain type IIA sodium channels *J. Gen Physiol.* 109: 589–605. [7]

Kelley, S. P., Dunlop, J. I., Kirkness, E. F., Lambert, J. J., and Peters, J. A. 2003. A cytoplasmic region determines single channel conductance in $5HT_3$ receptors. *Nature* 424: 321–324. [5]

Kelsch, W., Sim, S., and Lois, C. 2010. Watching synaptogenesis in the adult brain. *Annu. Rev. Neurosci.* 33: 131–149. [25]

Kemp, D. T. 1978. Stimulated acoustic emissions from within the human auditory system. *J. Acoust. Soc. Am.* 64: 1386–1391. [19]

Kennedy, P. R. 1990. Corticospinal, rubrospinal and rubro-olivary projections: A unifying hypothesis. *Trends Neurosci.* 13: 474–479. [24]

Kennedy, T. E., Serafini, T., de la Torre, J. R., and Tessier-Lavigne, M. 1994. Netrins are diffusible chemotropic factors for commissural axons in the embryonic spinal cord. *Cell* 78: 425–435. [25]

Kenshalo, D. R., Iwata, K., Sholas, M., and Thomas, D. A. 2000. Response properties and organization of nociceptive neurons in area 1 of monkey primary somatosensory cortex. *J. Neurophysiol.* 84: 719–729. [21]

Keramidas, A., Moorhouse, A. J., French, C. R., Schofield, P. R., and Barry, P. H. 2000. M2 pore mutations convert the glycine receptor channel from being anion- to cation-selective. *J. Biophys.* 78: 247–259. [5]

Keramidas, A., Moorhouse, A. J., Schofield, P. R., and Barry, P. H. 2004. Ligand-gated ion channels: mechanisms underlying ion selectivity. *Prog. Biophys. Mol. Biol.* 86: 161–204. [5]

Kerchner, G. A., and Nicoll, R. A. 2008. Silent synapses and the emergence of a postsynaptic mechanism for LTP. *Nat. Neurosci.* 9: 813–825. [16]

Kerem, B., Rommens, J. M., Buchanan, J. A., Markiewicz, D., Cox, T. K., Chakravarti, A., Buchwald, M., and Tsui, L. C. 1989. Identification of the cystic fibrosis gene: genetic analysis. *Science* 245: 1073–1080. [28]

Kessels, H. W., and Malinow, R. 2009. Synaptic AMPA receptor plasticity and behavior. *Neuron* 61: 340–350. [16]

Kettenmann, H., and Ransom, B. R. (eds.). 2005. *Neuroglia*, 2nd ed. Oxford University Press, New York. [10]

Keuroghlian, A. S., and Knudsen, E. I. 2007. Adaptive auditory plasticity in developing and adult animals. *Prog. Neurobiol.* 82: 109–121. [26]

Keynes, R. D., and Lumsden, A. 1990. Segmentation and the origin of regional diversity in the vertebrate central nervous system. *Neuron* 2: 1–9. [25]

Keynes, R. D., and Rojas, E. 1974. Kinetics and steady-state properties of the charged system controlling sodium conductance in the squid giant axon. *J. Physiol.* 239: 393–434. [7]

Khirug, S., Yamada, J., Afzalov, R., Voipio, J., Khiroug, L., and Kaila, K. 2008. GABAergic depolarization of the axon initial segment in cortical principal neurons is caused by the Na–K–2Cl cotransporter NKCC1. *J. Neurosci.* 28: 4635–4639. [11]

Kiang, N. Y., Rho, J. M., Northrop, C. C., Liberman, M. C., and Ryugo, D. K. 1982. Hair-cell innervation by spiral ganglion cells in adult cats. *Science* 217: 175–177. [22]

Kiani, R., Esteky, H., and Tanaka, K. 2004. Differences in onset latency of macaque inferotemporal neural responses to primate and non-primate faces. *J. Neurophysiol.* 94: 1587–1596. [23]

Kieffer, B. L., and Gavériaux-Ruff, C. 2002. Exploring the opioid system by gene knockout. *Prog. Neurobiol.* 66: 285–306. [14]

Kier, C. K., Buchsbaum, G., and Sterling, P. 1995. How retinal microcircuits scale for ganglion cells of different size. *J. Neurosci.* 15: 7673–7683. [20]

Kikkawa, S., Nakagawa, M., Iwasa, T., Kaneko, A., and Tsuda, M. 1993. GTP-binding protein couples with metabotropic glutamate receptor in bovine retinal on-bipolar cell. *Biochem. Biophys. Res. Commun.* 195: 374–379. [20]

Kim, D. S., and Bonhoeffer, T. 1994. Reverse occlusion leads to a precise restoration of orientation preference maps in visual cortex. *Nature* 370: 370–372. [26]

Kim, E., and Sheng, M. 2004. PDZ domain proteins of synapses. *Nat. Rev. Neurosci.* 5: 771–781. [11]

Kim, J. M., Beyer, R., Morales, M., Chen, S., Liu, L. Q., and Duncan, R. K. 2010. Expression of BK-type calcium-activated potassium channel splice variants during chick cochlear development. *J. Comp. Neurol.* 518: 2554–2569. [22]

Kim, N., Stiegler, A. L., Cameron, T. O., Hallock, P. T., Gomez, A. M., Huang, J. H., Hubbard, S. R., Dustin, M. L., and Burden, S. J. 2008. Lrp4 is a receptor for Agrin and forms a complex with MuSK. *Cell* 135: 334–342. [27]

Kinnamon, S. C., Dionne, V. E., and Beam, K. G. 1988. Apical localization of K+ channels in taste cells provides the basis for sour taste transduction. *Proc. Natl. Acad. Sci. USA* 85: 7023–7027. [19]

Kirkwood, P. A., and Sears, T. A. 1982. Excitatory postsynaptic potentials from single muscle spindle afferents in external intercostal motoneurones of the cat. *J. Physiol.* 322: 287–314. [24]

Kirsch, J., Wolters, I., Triller, A., and Betz, H. 1993. Gephyrin antisense oligonucleotides prevent glycine receptor clustering in spinal neurons. *Nature* 366: 745–748. [11]

Kirshner, N. 1969. Storage and secretion of adrenal catecholamines. *Adv. Biochem. Psychopharmacol.* 1: 71–89. [13]

Klausberger, T., and Somogyi, P. 2008. Neuronal diversity and temporal dynamics: the unity of hippocampal circuit operations. *Science* 321: 53–57. [14]

Kleene, S. J. 2008. The electrochemical basis of odor transduction in vertebrate olfactory cilia. *Chem. Senses.* 33: 839–859. [19]

Kleene, S. J., and Gesteland, R. C. 1991. Calcium-activated chloride conductance in frog olfactory cilia. *J. Neurosci.* 11: 3624–3629. [19]

Klyachko, V. A., and Jackson, M. B. 2002. Capacitance steps and fusion pores of small and large-dense-core vesicles in nerve terminals. *Nature* 418: 89–92. [13]

Kneussel, M., and Loebrich, S. 2007. Trafficking and synaptic anchoring of ionotropic inhibitory neurotransmitter receptors. *Biol. Cell* 99: 297–309. [11]

Knöpfel, T., Lin, M. Z., Levskaya, A., Tian, L., Lin, J. Y., and Boyden, E. S. 2010. Toward the second generation of optogenetic tools. *J. Neurosci.* 30: 14998–15004. [28]

Knudsen, E. I. 1998. Capacity for plasticity in the adult owl auditory system expanded by juvenile experience. *Science* 279: 1531–1533. [26]

Knudsen, E. I. 1999. Mechanisms of experience-dependent plasticity in the auditory localization pathway of the barn owl. *J. Comp. Physiol. A* 185: 305–321. [26]

Knudsen, E. I., and Knudsen, P. F. 1990. Sensitive and critical periods for visual calibration of sound localization by barn owls. *J. Neurosci.* 10: 222–232. [26]

Kodera, N., Yamamoto, D., Ishikawa, R., and Ando, T. 2010. Video imaging of walking myosin V by high-speed atomic force microscopy. *Nature* 468: 72–76. [15]

Koehler, R. C., Roman, R. J., and Harder, D. R. 2009. Astrocytes and the regulation of cerebral blood flow. *Trends Neurosci.* 32: 160–169. [10]

Koehnle, T. J., and Brown, A. 1999. Slow axonal transport of neurofilament protein in cultured neurons. *J. Cell Biol.* 144: 447–458. [15]

Koelle, G. B., and Friedenwald, J. S. 1949. A histochemical method for localizing cholinesterase activity. *Proc. Soc. Exp. Biol. Med.* 70: 617–622. [14]

Kofuji, P., and Newman, E. A. 2004. Potassium buffering in the central nervous system. *Neuroscience* 129: 1045–1056. [10]

Kofuji, P., Ceelen, P., Zahs, K. R., Surbeck, L. W., Lester, H. A., and Newman, E. A. 2000. Genetic inactivation of an inwardly rectifying potassium channel (Kir4.1 subunit) in mice: Phenotypic impact in retina. *J. Neurosci.* 20: 5733–5740. [10]

Köhler, M., Burnashev, N., Sakmann, B., and Seeburg, P. H. 1993. Determinants of Ca^{2+} permeability in both TM1 and TM2 of high-affinity kainate receptor channels: Diversity by RNA editing. *Neuron* 10: 491–500. [5]

Koike, C., Numata, T., Ueda, H., Mori, Y., and Furukawa, T. 2010. TRPM1: A vertebrate TRP channel responsible for retinal ON bipolar function. *Cell Calcium* 48: 95–101. [20]

Kolb, H. 1997. Amacrine cells of the mammalian retina: neurocircuitry and functional roles. *Eye (Lond).* 11 (Pt 6): 904–923. [20]

Kolodkin, A. L., and Tessier-Lavigne, M. 2011. Mechanisms and molecules of neuronal wiring: a primer. *Cold Spring Harb. Perspect. Biol.* 3: a001727. [25]

Kong, J-H., Adelman, J. P., and Fuchs, P. A. 2008. Expression of the SK2 calcium-activated potassium channel is required for cholinergic function in mouse cochlear hair cells *J. Physiol.* 586: 5471–5485. [12]

Konishi, M. 2004. The role of auditory feedback in birdsong. *Ann. N Y Acad. Sci.* 1016: 463–475. [24]

Kononenko, N. I., Kuehl-Kovarik, M. C., Partin, K. M., and Dudek, F. E. 2008. Circadian difference in firing rate of isolated rat suprachiasmatic nucleus neurons *Neurosci. Lett.* 436: 314–316. [17]

Kopin, I. J. 1968. False adrenergic transmitters. *Annu. Rev. Pharmacol.* 8: 377–394. [15]

Kopin, I. J., Breese, G. R., Krauss, K. R., and Weise, V. K. 1968. Selective release of newly synthesized norepinephrine from the cat spleen during sympathetic nerve stimulation. *J. Pharmacol. Exp. Ther.* 161: 271–278. [13]

Koppl, C. 1997. Phase locking to high frequencies in the auditory nerve and cochlear nucleus magnocellularis of the barn owl, *Tyto alba. J. Neurosci.* 17: 3312–3321. [22]

Korn, S. J., and Trapani, J. G. 2005. Potassium channels. *IEEE Trans. Nanobiosci.* 4: 21–33. [5]

Kornack, D. R., and Rakic, P. 1999. Continuation of neurogenesis in the hippocampus of the adult macaque monkey. *Proc. Natl. Acad. Sci. USA* 96: 5768–5773. [25]

Korogod, N., Lou, X., and Schneggenburger, R. 2005. Presynaptic Ca^{2+} requirements and developmental regulation of posttetanic potentiation in the calyx of Held. *J. Neurosci.* 25: 5127–5137. [16]

Kosslyn, S. M., Pascual-Leone, A., Felician, O., Camposano, S., Keenan, J. P., Thompson, W. L., Ganis, G., Sukel, K. E., and Alpert, N. M. 1999. The role of area 17 in visual imagery: convergent evidence from PET and rTMS. *Science* 284: 167–170. [23]

Kosterin, P., Kim, G. H., Muschol, M., Obaid, A. L., and Salzberg, B. M. 2005. Changes in FAD and NADH fluorescence in neurosecretory terminals are triggered by calcium entry and by ADP production. *J. Membr. Biol.* 208: 113–124. [17]

Koutalos, Y., and Yau, K. W. 1996. Regulation of sensitivity in vertebrate rod photoreceptors by calcium. *Trends Neurosci.* 19: 73–81. [20]

Kovalchuk, Y., Holthoff, K., and Konnerth, A. 2004. Neurotrophin action on a rapid timescale. *Curr. Opin. Neurobiol.* 14: 558–563. [25]

Kozlov, A., Huss, M., Lansner, A., Kotaleski, J. H., and Grillner, S. 2009. Simple cellular and network control principles govern complex patterns of motor behavior. *Proc. Natl. Acad. Sci. USA* 106: 20027–20032. [24]

Kramer, A. P., and Weisblat, D. A. 1985. Developmental neural kinship groups in the leech. *J. Neurosci.* 5: 388–407. [25]

Krauthammer, C. 2000. Restoration, reality and Christopher Reeve. *Time* February 14: 76. [28]

Kravitz, A. V., Freeze, B. S., Parker, P. R. L., Kay, K., Thwin, M. T., Deisseroth, K., and Kreitzer, A. C. 2010. Regulation of parkinsonian motor behaviours by optogenetic control of basal ganglia circuitry. *Nature* 466: 622–626. [14]

Kravitz, E. A. 2000. Serotonin and aggression: insights gained from a lobster model system and speculations on the role of amine neurons in a complex behavior. *J. Comp. Physiol. A* 186: 221–238. [14]

Kreitzer, A. C. 2009. Physiology and pharmacology of striatal neurons. *Annu. Rev. Neurosci.* 32: 127–147. [24]

Kreitzer, A. C., and Malenka, R. C. 2008. Striatal plasticity and basal ganglia circuit function. *Neuron* 60: 543–554. [24]

Kreitzer, A. C., and Regehr, W. G. 2001. Retrograde inhibition of presynaptic calcium influx by endogenous cannabinoids at excitatory synapses onto Purkinje cells. *Neuron* 29: 717–727. [12]

Kriebel, M. E., and Gross, C. E. 1974. Multimodal distribution of frog miniature endplate potentials in adult denervated and tadpole leg muscles. *J. Gen. Physiol.* 64: 85–103. [13]

Kriegstein, A., and Alvarez-Buylia, A. 2009. The glial nature of embryonic and adult neural stem cells. *Annu. Rev. Neurosci.* 32: 149–184. [25]

Kristan, W. B., Jr., Calabrese, R. L., and Friesen, W. O. 2005. Neuronal control of leech behavior. *Prog. Neurobiol.* 76: 279–327. [18]

Krnjevic, K., and Phillis, J. W. 1963. Iontophoretic studies of neurones in the mammalian cerebral cortex. *J. Physiol.* 165: 274–304. [14]

Krnjevic, K., and Schwartz, S. 1967. The action of γ-aminobutyric acid on cortical neurons. *Exp. Brain Res.* 3: 320–336. [14]

Kros, C. J. 2007. How to build an inner hair cell: challenges for regeneration. *Hear. Res.* 227: 3–10. [22]

Krug, M., Muller-Welde, P., Wagner, M., Ott, T., and Mathies, H. 1985. Functional plasticity in two afferent systems of the granule cells in the rat dentate area: Frequency-related changes, long-term potentiation and heterosynaptic depression. *Brain. Res.* 360: 264–272. [16]

Krüger, J., Caruana, F., Volta, R. D., and Rizzolatti, G. 2010. Seven years of recording from monkey cortex with a chronically implanted multiple microelectrode. *Front. Neuroeng.* 3: 6. [24]

Kubo, Y., Adelman, J. P., Clapham, D. E., Jan, L. Y. Karschin, A., Kurachi, Y., Lazdunski, M., Nichols, C. G., Seino, S., and Vandenberg, C. S. 2005. International Union of Pharmacology. LIV. Nomenclature and molecular relationships of inwardly rectifying potassium channels. *Pharmacol. Rev.* 57: 509–526. [5]

Kubota, Y., Ito, C., Sakurai, E., Sakurai, E., Watanabe, T., and Ohtsu, H. 2002. Increased methamphetamine-induced locomotor activity and behavioral sensitization in histamine-deficient mice. *J. Neurochem.* 83: 837–845. [14]

Kuffler, S. W. 1953. Discharge patterns and functional organization of the mammalian retina. *J. Neurophysiol.* 16: 37–68. [1, 2, 18, 20]

Kuffler, S. W. 1980. Slow synaptic responses in autonomic ganglia and the pursuit of a peptidergic transmitter. *J. Exp. Biol.* 89: 257–286. [14, 17]

Kuffler, S. W., and Edwards, C. 1958. Mechanism of gamma aminobutyric acid (GABA) action and its relation to synaptic inhibition. *J. Neurophysiol.* 21: 589–610. [14]

Kuffler, S. W., and Eyzaguirre, C. 1955a. Processes of excitation in the dendrites and in the soma of single isolated sensory nerve cells of the lobster and crayfish. *J. Gen. Physiol.* 39: 87–119. [8]

Kuffler, S. W., and Eyzaguirre, C. 1955b. Synaptic inhibition in an isolated nerve cell. *J. Gen. Physiol.* 39: 155–184. [11]

Kuffler, S. W., and Nicholls, J. G. 1966. The physiology of neuroglial cells. *Ergeb. Physiol.* 57: 1–90. [10]

Kuffler, S. W., and Potter, D. D. 1964. Glia in the leech central nervous system: Physiological properties and neuron-glia relationship. *J. Neurophysiol.* 27: 290–320. [10, 18]

Kuffler, S. W., and Yoshikami, D. 1975a. The distribution of acetylcholine sensitivity at the post-synaptic membrane of vertebrate skeletal twitch muscles: Iontophoretic mapping in the micron range. *J. Physiol.* 244: 703–730. [11]

Kuffler, S. W., and Yoshikami, D. 1975b. The number of transmitter molecules in a quantum: An estimate from iontophoretic application of acetylcholine at the neuromuscular synapse. *J. Physiol.* 251: 465–482. [13]

Kuffler, S. W., Dennis, M. J., and Harris, A. J. 1971. The development of chemosensitivity in extrasynaptic areas of the neuronal surface after denervation of parasympathetic ganglion cells in the heart of the frog. *Proc. R. Soc. Lond., B, Biol. Sci.* 177: 555–563. [27]

Kuffler, S. W., Hunt, C. C., and Quilliam, J. P. 1951. Function of medullated small-nerve fibers in mammalian ventral roots: Efferent muscle spindle innervation. *J. Neurophysiol.* 14: 29–51. [24]

Kuffler, S. W., Nicholls, J. G., and Orkand, R. K. 1966. Physiological properties of glial cells in the central nervous system of amphibia. *J. Neurophysiol.* 29: 768–787. [10]

Kuljis, R. O., and Rakic, P. 1990. Hypercolumns in primate visual cortex can develop in the absence of cues from photoreceptors. *Proc. Natl. Acad. Sci. USA* 87: 5303–5306. [26]

Kumar, S., Porcu, P., Werner, D. F., Matthews, D. B., Diaz-Granados, J. L., Helfand, R. S., and Morrow, A. L. 2009. The role of $GABA_A$ receptors in the acute and chronic effects of ethanol: a decade of progress. *Psychopharmacology* (*Berl.*) 205: 529–564. [14]

Kummer, T. T., Misgeld, T., and Sanes, J. R. 2006. Assembly of the postsynaptic membrane at the neuromuscular junction: paradigm lost. *Curr. Opin. Neurobiol.* 16: 74–82. [27]

Kuno, M. 1964a. Quantal components of excitatory synaptic potentials in spinal motoneurones. *J. Physiol.* 175: 81–99. [13]

Kuno, M. 1964b. Mechanism of facilitation and depression of the excitatory synaptic potential in spinal motoneurones. *J. Physiol.* 175: 100–112. [11, 16]

Kuno, M. 1971. Quantum aspects of central and ganglionic synaptic transmission in vertebrates. *Physiol. Rev.* 51: 647–678. [24]

Kupchik, Y. M., Rashkovan, G., Ohana, L., Keren-Raifman, T., Dascal, N., Parnas, H., and Parnas, I. 2008. Molecular mechanisms that control initiation and termination of physiological depolarization-evoked transmitter release. *Proc. Natl. Acad. Sci. USA* 105: 4435–4440. [12, 13]

Kupfermann, I. 1991. Functional studies of cotransmission. *Physiol. Rev.* 71: 683–732. [15]

Kurahashi, T., and Yau, K. W. 1993. Co-existence of cationic and chloride components in odorant-induced current of vertebrate olfactory receptor cells. *Nature* 363: 71–74. [19]

Kurth-Nelson, Z. L., Mishra, A., Newman, E. A. 2009. Spontaneous glial calcium waves in the retina develop over early adulthood. *J. Neurosci.* 29: 11339–11346. [10]

Kushmerick, C., Price, G. D., Taschenberger, H., Puente, N., Renden, R., Wadiche, J. I., Duvoisin, R. M., Grandes, P., and von Gersdorff, H. 2004. Retroinhibition of presynaptic Ca^{2+} currents by endocannabinoids released via postsynaptic mGluR activation at a calyx synapse. *J. Neurosci.* 24: 5955–5965. [12]

Kuwada, S., Fitzpatrick, D. C., Batra, R., and Ostapoff, E. M. 2006. Sensitivity to interaural time differences in the dorsal nucleus of the lateral lemniscus of the unanesthetized rabbit: comparison with other structures. *J. Neurophysiol.* 95: 1309–1322. [22]

Kuypers, H. G. J. M., and Ugolini, G. 1990. Viruses as trans-neuronal tracers. *Trends Neurosci.* 13: 71–75. [15]

Kuzhandaivel, A., Margaryan, G., Nistri, A., and Mladinic, M. 2010. Extensive glial apoptosis develops early after hypoxic-dysmetabolic insult to the neonatal rat spinal cord in vitro. *Neuroscience* 169: 325–338. [27]

Kwak, S., and Kawahara, Y. 2005. Deficient RNA editing of GluR2 and neuronal death in amyotropic lateral sclerosis. *J. Mol. Med.* 83: 110–120. [25]

Kwan, K. Y., and Corey, D. P. 2009. Burning cold: involvement of TRPA1 in noxious cold sensation. *J. Gen. Physiol.* 133: 251–256. [19]

Kwan, K. Y., Allchorne, A. J., Vollrath, M. A., Christensen, A. P., Zhang, D-S., Woolf, C. J., and Corey, D. P. 2006. TRPA1 contributes to cold, mechanical, and chemical nociception but is not essential for hair-cell transduction. *Neuron* 50: 277–289. [19]

Kyte, J., and Doolittle, R. F. 1982. A simple method for displaying the hydrophobic character of a protein. *J. Mol. Biol.* 157: 105–132. [5]

Labhart, T. 1988. Polarization-opponent interneurons in the insect visual system. *Nature* 331: 435–437. [18]

Labhart, T. 2000. Polarization-sensitive interneurons in the optic lobe of the desert ant *Cataglyphis bicolor*. *Naturwissenschaften* 87: 133–136. [18]

Labhart, T., Petzold, J., and Helbling, H. 2001. Spatial integration in polarization-sensitive interneurones of crickets: a survey of evidence, mechanisms and benefits. *J. Exp. Biol.* 204: 2423–2430. [18]

Lagostena, L., Rosato-Siri, M., D'Onofrio, M., Brandi, R., Arisi, I., Capsoni, S., Franzot, J., Cattaneo, A., and Cherubini, E. 2010. In the adult hippocampus, chronic nerve growth factor deprivation shifts GABAergic signaling from the hyperpolarizing to the depolarizing direction. *J. Neurosci.* 30: 885–893. [25]

Lak, A., Arabzadeh, E., and Diamond, M. E. 2008. Enhanced response of neurons in rat somatosensory cortex to stimuli containing temporal noise. *Cereb. Cortex* 18: 1085–1093. [23]

Lak, A., Arabzadeh, E., Harris, J., and Diamond, M. 2010. Correlated physiological and perceptual effects of noise in a tactile stimulus. *Proc. Natl. Acad. Sci. USA* 17: 7981–7986. [23]

Lamas, J. A., Reboreda, A., and Codesido, V. 2002. Ionic basis of the membrane potential in cultured rat sympathetic neurons. *Neuroreport* 13: 585–591. [6]

Lamb, T. D. 2009. Evolution of vertebrate retinal photoreception. *Philos. Trans. R. Soc. Lond., B, Biol. Sci.* 364: 2911–2924. [20]

Lamotte d'Incamps, B., and Ascher, P. 2008. Four excitatory postsynaptic ionotropic receptors coactivated at the motoneuron-Renshaw cell synapse. *J. Neurosci.* 28: 14121–14131. [15]

LaMotte, R. H., and Mountcastle, V. B. 1975. Capacities of humans and monkeys to discriminate vibratory stimuli of different frequency and amplitude: a correlation between neural events and psychological measurements. *J. Neurophysiol.* 38: 539–559. [21, 23]

Lampl, I., Anderson, J. S., Gillespie, D. C., and Ferster, D. 2001. Prediction of orientation selectivity from receptive field architecture in simple cells of cat visual cortex. *Neuron* 30: 263–274. [2]

Land, M. F. 2009. Vision, eye movements, and natural behavior. *Vis. Neurosci.* 26: 51–62. [22, 24]

Lane, M. A., Truettner, J. S., Brunschwig, J. P., Gomez, A., Bunge, M. B., Dietrich, W. D., Dziegielewska, K. M., Ek, C. J., Vandeberg, J. L., and Saunders, N. R. 2007. Age-related differences in the local cellular and molecular responses to injury in developing spinal cord of the opossum, *Monodelphis domestica*. *Eur. J. Neurosci.* 25: 1725–1742. [27]

Lang, T., Wacker, I., Steyer, J., Kaether, C., Wunderlich, I., Soldati, T., Gerdes, H-H., and Almers, W. 1997. Ca^{2+}-triggered peptide secretion in single cells imaged with green fluorescent protein and evanescent-wave microscopy. *Neuron* 18: 857–863. [13]

Langer, P., Grunder, S., and Rusch, A. 2003. Expression of Ca^{2+}-activated BK channel mRNA and its splice variants in the rat cochlea. *J. Comp. Neurol.* 455: 198–209. [22]

Langley, J. N. 1907. On the contraction of muscle, chiefly in relation to the presence of "receptive" substances. *J. Physiol.* 36: 347–384. [11]

Langley, J. N., and Anderson, H. K. 1904. The union of different kinds of nerve fibres. *J. Physiol.* 31: 365–391. [27]

Larkum, M. E., Zhu, J. J., and Sakmann, B. 1999. A new cellular mechanism for coupling inputs arriving at different cortical layers. *Nature* 398: 338–341. [8]

Larrson, H. P., Baker, O. S., Dhillon, D. S., and Isacoff, E. Y. 1996. Transmembrane movement of the *Shaker* K^+ channel S4. *Neuron* 16: 387–397. [5]

Lawrence, D. G., and Kuypers, H. G. J. M. 1968. The functional organization of the motor system in the monkey. I. The effects of bilateral pyramidal lesions. *Brain* 91: 1–14. [24]

Lawrence, J. J. 2008. Cholinergic control of GABA release: emerging parallels between neocortex and hippocampus. *Trends Neurosci.* 31: 317–327. [14]

Le Douarin, N. M. 1986. Cell line segregation during peripheral nervous system ontogeny. *Science* 231: 1515–1522. [25]

Lebedev, M. A., and Nelson, R. J. 1996. High-frequency vibratory sensitive neurons in monkey primary somatosensory cortex: entrained and nonentrained responses to vibration during the performance of vibratory-cued hand movements. *Exp. Brain Res.* 111: 313–325. [21]

Lebedev, M. A., Mirabella, G., Erchova, I., and Diamond, M. E. 2000. Experience-dependent plasticity of rat barrel cortex: redistribution of activity across barrel-columns. *Cereb. Cortex.* 10: 23–31. [26]

Ledent, C., Vaugeois, J. M., Schiffmann, S. N., Pedrazzini, T., El Yacoubi, M., Vanderhaeghen, J. J., Costentin, J., Heath, J. K., Vassart, G., and Parmentier, M. 1997. Aggressiveness, hypoalgesia and high blood pressure in mice lacking the adenosine A_{2a} receptor. *Nature* 388: 674–678. [14]

Lee, B. B., Martin, P. R., and Grunert, U. Retinal connectivity and primate vision. *Prog. Retin. Eye Res.* 29: 622–639. [20]

Lee, C. C., and Winer, J. A. 2005. Principles governing auditory cortex connections. *Cereb. Cortex* 15: 1804–1814. [22]

Lee, H-K., Kameyama, K., Huganir, R. L., and Bear, M. F. 1998. NMDA induces long-term synaptic depression and dephosphorylation of the GluR1 subunit of AMPA receptors in hippocampus. *Neuron* 21: 1151–1162. [16]

Lee, L. J., Chen, W. J., Chuang, Y. W., and Wang, Y. C. 2009. Neonatal whisker trimming causes long-lasting changes in structure and function of the somatosensory system. *Exp. Neurol.* 219: 524–532. [26]

Lee, R. G., Tonolli, I., Viallet, F., Aurenty, R., and Massion, J. 1995. Preparatory postural adjustments in parkinsonian patients with postural instability. *Can. J. Neurol. Sci.* 22: 126–135. [24]

Lee, S. J., Escobedo-Lozoya, Y., Szatmari, E. M., and Yasuda, R. 2009. Activation of CaMKII in single dendritic spines during long-term potentiation. *Nature* 458: 299–304. [12]

Lee, V. H., Lee, L. T., and Chow, B. K. 2008. Gonadotropin-releasing hormone: regulation of the GnRH gene. *FEBS J.* 275: 5458–5478. [17]

Lehrer, M., and Collett, T. S. 1994. Approaching and departing bees learn different cues to the distance of a landmark. *J. Comp. Physiol. A* 175: 171–177. [18]

Leksell, L. 1945. The action potential and excitatory effects of the small ventral root fibres to skeletal muscle. *Acta Physiol. Scand.* 10(Suppl. 31): 1–84. [24]

Lemon, R. N. 2008. Descending pathways in motor control. *Annu. Rev. Neurosci.* 31: 195–218. [24]

Lendvai, B., and Vizi, E. S. 2008. Nonsynaptic chemical transmission through nicotinic acetylcholine receptors. *Physiol. Rev.* 88: 333–349. [14]

Leonard, J. L., and Edstrom, J. P. 2004. Parallel processing in an identified neural circuit: the *Aplysia californica* gill-withdrawal response model system. *Biol. Rev. Camb. Philos. Soc.* 79: 1–59. [18]

Leonard, R. J., Labarca, C. G., Charnet, P., Davidson, N., and Lester, H. A. 1988. Evidence that the M2 membrane-spanning region lines the ion channel pore of a nicotinic receptor. *Science* 242: 1578–1581. [5]

Lerner, Y., Hendler, T., Ben-Bashat, D., Harel, M., and Malach. R. 2001. A hierarchical axis of object processing stages in the human visual cortex. *Cereb. Cortex* 11: 287–297. [23]

Lester, H. A., Dibas, M. I., Dahan, D. S., Leite, J. F., and Dougherty, D. A. 2004. Cys-loop receptors: new twists and turns. *Trends Neurosci.* 27: 329–336. [5]

LeVay, S., Connolly, M., Houde, J., and Van Essen, D. C. 1985. The complete pattern of ocular dominance stripes in the striate cortex and visual field of the macaque monkey. *J. Neurosci.* 5: 486–501. [3]

LeVay, S., Hubel, D. H., and Wiesel, T. N. 1975. The pattern of ocular dominance columns in macaque visual cortex revealed by a reduced silver stain. *J. Comp. Neurol.* 159: 559–576. [3]

LeVay, S., Stryker, M. P., and Shatz, C. J. 1978. Ocular dominance columns and their development in layer IV of the cat's visual cortex: a quantitative study. *J. Comp. Neurol.* 179: 223–244. [26]

LeVay, S., Wiesel, T. N., and Hubel, D. H. 1980. The development of ocular dominance columns in normal and visually deprived monkeys. *J. Comp. Neurol.* 191: 1–51. [26]

Levenes, C., Daniel, H., and Crepel, F. 1998. Long-term depression of synaptic transmission in the cerebellum: Cellular and molecular mechanisms revisited. *Prog. Neurobiol.* 55: 79–91. [16]

Levi-Montalcini, R. 1982. Developmental neurobiology and the natural history of nerve growth factor. *Annu. Rev. Neurosci.* 5: 341–362. [25]

Levick, W. R. 1967. Receptive fields and trigger features of ganglion cells in the visual streak of the rabbit's retina. *J. Physiol.* 188: 285–307. [20]

Levitan, I. B. 1994. Modulation of ion channels by protein phosphorylation and dephosphorylation. *Annu. Rev. Physiol.* 56: 193–212. [12]

Levitan, I. B. 2006. Signaling protein complexes associated with neuronal ion channels. *Nat. Neurosci.* 9: 305–310. [12]

Lewcock, J. W., and Reed, R. R. 2004. A feedback mechanism regulates monoallelic odorant receptor expression. *Proc. Natl. Acad. Sci. USA* 101: 1069–1074. [19]

Lewis, C., Neidhart, S., Holy, C., North, R. A., Buell, G., and Surprenant, A. 1995. Coexpression of P2X2 and P2X3 receptor subunits can account for ATP-gated currents in sensory neurons. *Nature* 377: 432–435. [19]

Lewis, J. E., and Kristan, W. B., Jr. 1998. A neuronal network for computing population vectors in the leech. *Nature* 391: 76–79. [18]

Lewis, K. E., and Eisen, J. S. 2003. From cells to circuits: Development of the zebrafish spinal cord. *Prog. Neurobiol.* 69: 419–449. [10]

Lewis, P. R., and Shute, C. C. 1967. The cholinergic limbic system: projections to hippocampal formation, medial cortex, nuclei of the ascending cholinergic reticular system, and the subfornical organ and supra-optic crest. *Brain* 90: 521–540. [14]

Lewis, R. S., and Hudspeth, A. J. 1983. Voltage- and ion-dependent conductances in solitary vertebrate hair cells. *Nature* 304: 538–541. [22]

Lewis, T. L., and Maurer, D. 2005. Multiple sensitive periods in human visual development: evidence from visually deprived children. *Dev. Psychobiol.* 46: 163–183. [26]

LeWitt, P. A. 2008. Levodopa for the treatment of Parkinson's disease. *New England J. Med.* 359: 2468–2476. [14]

Leyssen, M., and Hassan, B. A. 2007. A fruitfly's guide to keeping the brain wired. *EMBO Rep.* 8: 46–50. [18]

Li, C.-L., and Jasper, H. H. 1953. Microelectrode studies of the electrical activity in the cerebral cortex of the cat. *J. Physiol.* 121: 117–140. [8]

Li, N., and DiCarlo, J. J. 2008. Unsupervised natural experience rapidly alters invariant object representation in visual cortex. *Science* 321: 1502–1507. [23]

Li, Q., and Burrell, B. D. 2009. Two forms of long-term depression in a polysynaptic pathway in the leech CNS: one NMDA receptor-dependent and the other cannabinoid-dependent. *J. Comp. Physiol. A* 195: 831–841. [18]

Li, X. J., Blackshaw, S., and Snyder, S. H. 1994. Expression and localization of amiloride-sensitive sodium channel indicate a role for non-taste cells in taste perception. *Proc. Natl. Acad. Sci. USA* 91: 1814–1818. [19]

Li, Y., Atkin, G. M., Morales, M. M., Liu, L. Q., Tong, M., and Duncan, R. K. 2009. Developmental expression of BK channels in chick cochlear hair cells. *BMC Dev. Biol.* 9: 67. [22]

Li, Y., Field, P. M., and Raisman, G. 1998. Regeneration of adult corticospinal axons induced by transplanted olfactory ensheathing cells. *J. Neurosci.* 18: 10514–10524. [27]

Li, Y., Gamper, N., Hilgemann, D. W., and Shapiro, M. S. 2005. Regulation of Kv7 (KCNQ) K^+ channel open probability by phosphatidylinositol 4,5-bisphosphate. *J. Neurosci.* 25: 9825–9835. [12]

Liao, D., Hessler, N. A., and Malinow, R. 1995. Activation of postsynaptically silent synapses during pairing-induced LTP in the CA1 region of hippocampal slice. *Nature* 375: 400–404. [16]

Libby, R. T., and Steel, K. P. 2000. The roles of unconventional myosins in hearing and deafness. *Essays Biochem.* 35: 159–174. [19]

Liberles, S. D., and Buck, L. B. 2006. A second class of chemosensory receptors in the olfactory epithelium. *Nature* 442: 645–650. [19]

Liberman, M. C. 1982. Single-neuron labeling in the cat auditory nerve. *Science* 216: 1239–1241. [22]

Liberman, M. C. 1988. Response properties of cochlear efferent neurons: monaural vs. binaural stimulation and the effects of noise. *J. Neurophysiol.* 60: 1779–1798. [22]

Lichtneckert, R., and Reichert, H. 2008. Anteroposterior regionalization of the brain: genetic and comparative aspects. *Adv. Exp. Med. Biol.* 628: 32–41. [18]

Liley, A. W. 1956. The quantal components of the mammalian endplate potential. *J. Physiol.* 133: 571–587. [16]

Lim, M. M., Wang, Z., Olazábal, D. E., Ren, X., Terwilliger, E. F., and Young, L. J. 2004. Enhanced partner preference in a promiscuous species by manipulating the expression of a single gene. *Nature* 429: 754–757. [14]

Lin, D. M., Wang, F., Lowe, G., Gold, G. H., Axel, R., Ngai, J., and Brunet, L. 2000. Formation of precise connections in the olfactory bulb occurs in the absence of odorant-evoked neuronal activity. *Neuron* 26: 69–80. [26]

Lin, J. H., and Rydqvist, B. 1999. The mechanotransduction of the crayfish stretch receptor neurone can be differentially activated or inactivated by local anaesthetics. *Acta Physiol. Scand.* 166: 65–74. [19]

Lin, J. S. 2000. Brain structures and mechanisms involved in the control of cortical activation and wakefulness, with emphasis on the posterior hypothalamus and histaminergic neurons. *Sleep Med. Rev.* 4: 471–503. [14]

Lin, L., Faraco, J., Li, R., Kadotani, H., Rogers, W., Lin, X., Qiu, X., de Jong, P. J., Nishino, S., and Mignot, E. 1999. The sleep disorder canine narcolepsy is caused by a mutation in the hypocretin (orexin) receptor 2 gene. *Cell* 98: 365–376. [14]

Lin, S., Landmann, L., Ruegg, M. A., and Brenner, H. R. 2008. The role of nerve- versus muscle-derived factors in mammalian neuromuscular junction formation. *J. Neurosci.* 28: 3333–3340. [27]

Linden, D. J., and Connor, J. A. 1995. Long-term synaptic depression. *Annu. Rev. Neurosci.* 18: 319–357. [16]

Ling, G., and Gerard, R. W. 1949. The normal membrane potential of frog sartorius fibers. *J. Cell Comp. Physiol.* 34: 383–396. [4, 11]

Lippe, W. R. 1994. Rhythmic spontaneous activity in the developing avian auditory system. *J. Neurosci.* 14: 1486–1495. [26]

Lipscombe, D., Kongsamut, S., and Tsien, R. W. 1989. α-Adrenergic inhibition of sympathetic neurotransmitter release mediated by modulation of N-type calcium-channel gating. *Nature* 340: 639–642. [12]

Lisberger, S. G. 2009. Internal models of eye movement in the floccular complex of the monkey cerebellum. *Neuroscience* 162: 763–776. [22]

Lisman, J. E. 2009. The pre/post LTP debate. *Neuron* 63: 281–284. [16]

Lisman, J. E., Pi, H. J., Zhang, Y., and Otmakhova, N. A. 2010. A thalamo-hippocampal-ventral tegmental area loop may produce the positive feedback that underlies the psychotic break in schizophrenia. *Biol. Psychiatry* 68: 17–24. [28]

Little, S. C., and Mullins, M. C. 2006. Extracellular modulation of BMP activity in patterning the dorsoventral axis. *Birth Defects Res. C Embryo Today* 78: 224–242. [25]

Liu, B. P., Cafferty, W. B., Budel, S. O., and Strittmatter, S. M. 2006. Extracellular regulators of axonal growth in the adult central nervous system. *Philos. Trans. R. Soc. Lond., B, Biol. Sci.* 361: 1593–1610. [25]

Liu, C. R., Xu, L., Zhong, Y. M., Li, R. X., and Yang, X. L. 2009. Expression of connexin 35/36 in retinal horizontal and bipolar cells of carp. *Neuroscience* 164: 1161–1169. [20]

Liu, N., Varma, S., Shooter, E. M., and Tolwani, R. J. 2005. Enhancement of Schwann cell myelin formation by K252a in the Trembler-J mouse dorsal root ganglion explant culture. *J. Neurosci. Res.* 79: 310–317. [10]

Liu, Q., Tang, Z., Surdenikova, L., Kim, S., Patel, K. N., Kim, A., Ru, F., Guan, Y., Weng, H. J., Geng, Y., Undem, B. J., Kollarik, M., Chen, Z. F., Anderson, D. J., and Dong, X. 2009. Sensory neuron-specific GPCR Mrgprs are itch receptors mediating chloroquine-induced pruritus. *Cell* 139: 1353–1365. [19]

Liu, Y., and Joho, R. H. 1998. A side chain in S6 influences both open-state stability and ion permeation in a voltage-gated K^+ channel. *Pflügers Arch.* 435: 654–661. [5]

Liu, Y., Jurman, M. E., and Yellen, G. 1997. Gated access to the pore of a voltage-dependent K^+ channel. *Neuron* 19: 175–184. [5]

Livet, J., Weissman, T. A., Kang, H., Draft, R. W., Lu, J., Bennis, R. A., Sanes, J. R., and Lichtman, J. W. 2007. Transgenic strategies for combinatorial expression of fluorescent proteins in the nervous system. *Nature* 450: 56–62. [28]

Livingstone, M. S., and Hubel, D. H. 1984. Anatomy and physiology of a color system in the primate visual cortex. *J. Neurosci.* 4: 309–356. [3]

Livingstone, M. S., and Hubel, D. H. 1987. Connections between layer 4B of area 17 and the thick cytochrome oxidase stripes of area 18 in the squirrel monkey. *J. Neurosci.* 7: 3371–3377. [3]

Livingstone, M. S., and Hubel, D. H. 1988. Segregation of form, color, movement, and depth: Anatomy, physiology, and perception. *Science* 240: 740–749. [3]

Livingstone, P. D., and Wonnacott, S. 2009. Nicotinic acetylcholine receptors and the ascending dopamine pathways. *Biochem. Pharmacol.* 78: 744–755. [14]

Llano, I., Leresche, N., and Marty, A. 1991. Calcium entry increases the sensitivity of cerebellar Purkinje cells to applied GABA and decreases inhibitory synaptic currents. *Neuron* 6: 564–674. [12]

Lledo, P. M., Alonso, M., and Grubb, M. S. 2006. Adult neurogenesis and functional plasticity in neuronal circuits. *Nat. Rev. Neurosci.* 7: 179–193. [25]

Llewellyn, L. E. 2009. Sodium channel inhibiting marine toxins. *Prog. Mol. Subcell. Biol.* 46: 67–87. [7]

Llinás, R. 1982. Calcium in synaptic transmission. *Sci. Am.* 247(4): 56–65. [13]

Llinás, R., and Sugimori, M. 1980. Electrophysiological properties of *in vitro* Purkinje cell dendrites in mammalian cerebellar slices. *J. Physiol.* 305: 197–213. [7, 8]

Llinás, R., Leznik, E., and Makarenko, V. I. 2002. On the amazing olivocerebellar system. *Ann. N Y Acad. Sci.* 978: 258–272. [24]

Llinás, R., Sugimori, M., and Silver, R. B. 1992. Microdomains of high calcium concentration in a presynaptic terminal. *Science* 256: 677–679. [12, 13]

Llobet, A., Beaumont, V., and Lagnado, L. 2003. Real-time measurements of exocytosis at synapses and neuroendocrine cells using interference reflection microscopy. *Neuron* 40: 1075–1086. [13]

Lloyd, I. C., Ashworth, J., Biswas, S., and Abadi, R. V. 2007. Advances in the management of congenital and infantile cataract. *Eye (Lond)*. 21: 1301–1309. [26]

Lockery, S. R., and Kristan, W. B., Jr. 1990. Distributed processing of sensory information in the leech. II. Identification of interneurons contributing to the local bending reflex. *J. Neurosci.* 10: 1816–1829. [18]

Lodge, D. 2009. The history of the pharmacology and cloning of ionotropic glutamate receptors and the development of idiosyncratic nomenclature. *Neuropharmacology* 56: 6–21. [5, 14]

Loewenstein, O., and Wersall, J. 1959. A functional interpretation of the electron-microscopic structure of the sensory hairs in the cristæ of the elasmobranch Raja clavata in terms of directional sensitivity. *Nature* 184: 1807–1808. [19]

Loewenstein, W. 1981. Junctional intercellular communication: The cell-to-cell membrane channel. *Physiol. Rev.* 61: 829–913. [11]

Loewenstein, W. R. 1999. *The Touchstone of Life*. Oxford University Press, New York. [10]

Loewenstein, W. R., and Mendelson, M. 1965. Components of receptor adaptation in a Pacinian corpuscle. *J. Physiol.* 177: 377–397. [19]

Loewi, O. 1921. Über humorale Übertragbarkeit der Herznervenwirkung. *Pflügers Arch.* 189: 239–242. [11]

Logothetis, N. K., Pauls, J., and Poggio, T. 1995. Shape representation in the inferior temporal cortex of monkeys. *Curr. Biol.* 5: 552–563. [23]

Logothetis, N. K., Pauls, J., Bulthoff, H. H., and Poggio, T. 1994. View-dependent object recognition by monkeys. *Curr. Biol.* 4: 401–414. [23]

Lomeli, J., Quevedo, J., Linares, P., and Rudomin, P. 1998. Local control of information flow in segmental and ascending collaterals of single afferents. *Nature* 395: 600–604. [11]

Lømo T. 2003. The discovery of long-term potentiation. *Philos. Trans. R. Soc. Lond., B, Biol. Sci.* 358: 617–620. [16]

Lømo, T., and Rosenthal, J. 1972. Control of ACh sensitivity by muscle activity in the rat. *J. Physiol.* 221: 493–513. [27]

Long, S. B., Campbell, E. B., and MacKinnon, R. 2005a. Crystal structure of a mammalian voltage-dependent *Shaker* family K^+ channel. *Science* 309: 897–903. [5]

Long, S. B., Campbell, E. B., and MacKinnon, R. 2005b. Voltage sensor of Kv1.2: Structural basis of electromechanical coupling. *Science* 309: 903–908. [5]

Lonsbury-Martin, B. L., and Martin, G. K. 2003. Otoacoustic emissions. *Curr. Opin. Otolaryngol. Head Neck Surg.* 11: 361–366. [19]

Lorente de No, R. 1933. Vestibular ocular reflex arc. *Arch. Neurol. Psychiatry*. 30: 245–291. [22]

Lorenz, K. 1970. *Studies in Animal and Human Behavior*. Harvard Press, Cambridge, MA. [26]

Loring, R. H., and Salpeter, M. M. 1980. Denervation increases turnover rate of junctional acetylcholine receptors. *Proc. Natl. Acad. Sci. USA* 77: 2293–2297. [11]

Lottem, E., and Azouz, R. 2009. Mechanisms of tactile information transmission through whisker vibrations. *J. Neurosci.* 29: 11686–11697. [21]

Love, F. M., Son, Y. J., and Thompson, W. J. 2003. Activity alters muscle reinnervation and terminal sprouting by reducing the number of Schwann cell pathways that grow to link synaptic sites. *J. Neurobiol.* 54: 566–576. [10, 27]

Lovinger, D. M. 2010. Neurotransmitter roles in synaptic modulation, plasticity and learning in the dorsal striatum. *Neuropharmacology* 58: 951–961. [24]

Lowel, S., and Singer, W. 1992. Selection of intrinsic horizontal connections in the visual cortex by correlated neuronal activity. *Science* 255: 209–212. [26]

Lowenstein, O., Osborne, M. P., and Thornhill, R. A. 1968. The anatomy and ultrastructure of the labyrinth of the lamprey (*Lampetra fluviatilis* L.). *Proc. R. Soc. Lond., B, Biol. Sci.* 170: 113–134. [22]

Lowery, L. A., and Van Vactor, D. 2009. The trip of the tip: understanding the growth cone machinery. *Nat. Rev. Mol. Cell Biol.* 10: 332–343. [25]

Lowrey, P. L., and Takahashi, J. S. 2004. Mammalian circadian biology: elucidating genome-wide levels of temporal organization. *Annu. Rev. Genomics Hum. Genet.* 5: 407–441. [17]

Lu, B., Su, Y., Das, S., Liu, J., Xia, J., and Ren, D. 2007. The neuronal channel NALCN contributes resting sodium permeability and is required for normal respiratory rhythm. *Cell* 129: 371–383. [6]

Lu, H. D., and Roe, A. W. 2008. Functional organization of color domains in V1 and V2 of macaque monkey revealed by optical imaging. *Cereb. Cortex* 18: 516–533. [3]

Lu, J., Karadsheh, M., and Delpire, E. 1999. Developmental regulation of the neuronal-specific isoform of K-Cl cotransporter KCC2 in postnatal rat brains. *J. Neurobiol.* 39: 558–568. [9]

Lu, T., Nguyen, B., Zhang, X., and Yang, J. 1999. Architecture of a K^+ channel inner pore revealed by stoichiometric covalent modification. *Neuron* 22: 571–580. [5]

Lu, T., Ting, A. Y., Mainland, J., Jan, L. Y., Schultz, P. G., and Yang, J. 2001. Probing ion permeation and gating in a K^+ channel with backbone mutations in the selectivity filter. *Nat. Neurosci.* 4: 239–246. [5]

Lu, Z. 2004. Mechanism of rectification in inward-rectifier K^+ channels. *Annu. Rev. Physiol.* 66: 103–129. [5]

Lucas, S. M., and Binder, M. D. 1984. Topographic factors in distribution of homonymous group Ia-afferent input to cat medial gastrocnemius motoneurons. *J. Neurophysiol.* 51: 50–63. [24]

Luebke, A. E., and Robinson, D. A. 1994. Gain changes of the cat's vestibulo-ocular reflex after flocculus deactivation. *Exp. Brain Res.* 98: 379–390. [22]

Lumpkin, E. A., and Caterina, M. J. 2007. Mechanisms of sensory transduction in the skin. *Nature* 445: 858–865. [19]

Lumpkin, E. A., and Hudspeth, A. J. 1995. Detection of Ca^{2+} entry through mechanosensitive channels localizes the site of mechanoelectrical transduction in hair cells. *Proc. Natl. Acad. Sci. USA* 92: 10297–10301. [19]

Lumsden, A., and Krumlauf, R. 1996. Patterning the vertebrate neuroaxis. *Science* 274: 1109–1115. [25]

Lumsden, A. G., and Davies, A. M. 1986. Chemotropic effect of specific target epithelium in the developing mammalian nervous system. *Nature* 323: 538–539. [25]

Lundberg, A., and Quilisch, H. 1953. On the effect of calcium on presynaptic potentiation and depression at the neuromuscular junction. *Acta Physiol. Scand.* 30(Suppl. III): 121–129. [16]

Lundstrom, R. J. 1986. Responses of mechanoreceptive afferent units in the glabrous skin of the human hand to vibration. *Scand. J. Work Environ. Health.* 12: 413–416. [21]

Luo, D. G., Xue, T., and Yau, K. W. 2008. How vision begins: an odyssey. *Proc. Natl. Acad. Sci. USA* 105: 9855–9862. [20]

Luo, L., and O'Leary, D. D. 2005. Axon retraction and degeneration in development and disease. *Annu. Rev. Neurosci.* 28: 127–156. [25]

Luskin, M. B. 1993. Restricted proliferation and migration of postnatally generated neurons derived from the forebrain subventricular zone. *Neuron* 11: 173–189. [25]

Luskin, M. B. 1998. Neuroblasts of the postnatal mammalian forebrain: Their phenotype and fate. *J. Neurobiol.* 36: 221–233. [10]

Luskin, M. B., and Shatz, C. J. 1985a. Neurogenesis of the cat's primary visual cortex. *J. Comp. Neurol.* 242: 611–631. [25]

Luskin, M. B., and Shatz, C. J. 1985b. Studies of the earliest generated cells of the cat's visual cortex: cogeneration of subplate and marginal zones. *J. Neurosci.* 5: 1062–1075. [25]

Lustig, L. R. 2006. Nicotinic acetylcholine receptor structure and function in the efferent auditory system *Anat. Rec.* 288A: 424–234. [12]

Lustig, L. R., Leake, P. A., Snyder, R. L., and Rebscher, S. J. 1994. Changes in the cat cochlear nucleus following neonatal deafening and chronic intracochlear electrical stimulation. *Hear. Res.* 74: 29–37. [26]

Lyckman, A. W., Horng, S., Leamey, C. A., Tropea, D., Watakabe, A., van Wart, A., McCurry, C., Yamamori, T., and Sur, M. 2008. Gene expression patterns in visual cortex during the critical period: synaptic stabilization and reversal by visual deprivation. *Proc. Natl. Acad. Sci. USA* 105: 9409–9414. [26]

Lynch, G. S., Dunwiddie, T., and Gribkoff, V. 1977. Heterosynaptic depression: A postsynaptic correlate of long-term potentiation. *Nature* 266: 737–739. [16]

Lynch, J. W. 2009. Native glycine receptor subtypes and their physiological roles. *Neuropharmacol.* 56: 303–309. [5]

Lynch, M. A. 2004. Long-term potentiation and memory. *Physiol. Rev.* 84: 87–136. [16]

Ma, M., and Dahl, G. 2006. Cosegregation of permeability and single channel conductance in chimeric connexins. *Biophys. J.* 90: 151–163. [8]

Ma, W., Ribeiro-da-Silva, A., De Koninck, Y., Radhakrishnan, V., Cuello, A. C., and Henry, J. L. 1997. Substance P and enkephalin immunoreactivities in axonal boutons presynaptic to physiologically identified dorsal horn neurons. An ultrastructural multiple-labeling study in the cat. *Neuroscience* 77: 793–811. [14]

Mabb, A. M., and Ehlers, M. D. 2010. Ubiquitination in postsynaptic function and plasticity. *Annu. Rev. Cell Dev. Biol.* 26: 179–210. [25]

Macagno, E. R. 1980. Number and distribution of neurons in leech segmental ganglia. *J. Comp. Neurol.* 190: 283–302. [18]

Macagno, E. R., Muller, K. J., and Pitman, R. M. 1987. Conduction block silences parts of a chemical synapse in the leech central nervous system. *J. Physiol.* 387: 649–664. [18]

Macaluso, E., Frith, C. D., and Driver, J. 2000. Modulation of human visual cortex by crossmodal spatial attention. *Science* 289: 1206–1208. [23]

MacDermott, A. B., Connor, E. A., Dionne, V. E., and Parsons, R. L. 1980. Voltage clamp study of fast excitatory synaptic currents in bullfrog sympathetic ganglion cells. *J. Gen. Physiol.* 75: 39–60. [15]

MacKie, K., and Hille, B. 1992. Cannabinoids inhibit N-type calcium channels in neuroblastoma–glioma cells. *Proc. Natl. Acad. Sci. USA* 89: 3825–3829. [12]

MacKinnon, R., Cohen, S. L., Kuo, A., Lee, A., and Chait, B. T. 1998. Structural conservation in prokaryotic and eukaryotic potassium channels. *Science* 280: 106–109. [5]

MacLennan, D. H., Abu-Abed, M., and Kang, C. H. 2002. Structure–function relationships in Ca^{2+} cycling proteins. *J. Mol. Cell. Cardiol.* 34: 897–918. [9]

MacNeil, M. A., Heussy, J. K., Dacheux, R. F., Raviola, E., and Hasland, R. H. 1999. The shapes and numbers of amacrine cells: matching of photofilled with Golgi-stained cells in the rabbit retina and comparison with other mammalian species. *J. Comp. Neurol.* 413: 305–326. [20]

Macpherson, P. C., Cieslak, D., and Goldman, D. 2006. Myogenin-dependent nAChR clustering in aneural myotubes. *Mol. Cell. Neurosci.* 31: 649–660. [27]

Madduri, S., and Gander, B. 2010. Schwann cell delivery of neurotrophic factors for peripheral nerve regeneration. *J. Peripher. Nerv. Syst.* 15: 93–103. [27]

Madison, D. V., and Nicoll, R. A. 1986a. Actions of noradrenaline recorded intracellularly in rat hippocampal CA1 pyramidal neurones, in vitro. *J. Physiol.* 372: 221–244. [14]

Madison, D. V., and Nicoll, R. A. 1986b. Cyclic adenosine 3',5'-monophosphate mediates beta-receptor actions of noradrenaline in rat hippocampal pyramidal cells. *J. Physiol.* 372: 245–259. [14]

Madison, D. V., Lancaster, B., and Nicoll, R. A. 1987. Voltage clamp analysis of cholinergic action in the hippocampus. *J. Neurosci.* 7: 733–741. [14]

Maeda, S., Nakagawa, S., Suga, M., Yamashita, E., Oshima, A., Fujiyoshi, Y., and Tsukihara, T. 2009. Structure of the connexin 26 gap junction channel at 3.5 Å resolution. *Nature* 458: 597– 604. [8]

Maeno, T., Edwards, C., and Anraku, M. 1977. Permeability of the endplate membrane activated by acetylcholine to some organic cations. *J. Neurobiol.* 8: 173–184. [5]

Maffei, L., and Galli-Resta, L. 1990. Correlation in the discharges of neighboring rat retinal ganglion cells during prenatal life. *Proc. Natl. Acad. Sci. USA* 87: 2861–2864. [20, 26]

Magistretti, P. J. 2009. Role of glutamate in neuron-glia metabolic coupling. *Am. J. Clin. Nutr.* 90: 875–880. [10]

Magleby, K. L., and Stevens, C. F. 1972. The effect of voltage on the time course of end-plate currents. *J. Physiol.* 223: 151–171. [11]

Magleby, K. L., and Terrar, D. A. 1975. Factors affecting the time course of decay of end-plate currents: a possible cooperative action of acetylcholine on receptors at the frog neuromuscular junction. *J. Physiol.* 244: 467–495. [11]

Magleby, K. L., and Weinstock, M. M. 1980. Nickel and calcium ions modify the characteristics of the acetylcholine receptor-channel complex at the frog neuromuscular junction. *J. Physiol.* 299: 203–218. [13]

Magleby, K. L., and Zengel, J. E. 1976. Augmentation: A process that acts to increase transmitter release at the frog neuromuscular junction. *J. Physiol.* 257: 449–470. [16]

Mahaut-Smith, M. P., Martinez-Pinna, J., and Gurung, I. S. 2008. A role for membrane potential in regulating GPCRs? *Trends Pharmacol. Sci.* 29: 421–429. [12]

Mahns, D. A., Perkins, N. M., Sahai, V., Robinson, L., and Rowe, M. J. 2006. Vibrotactile frequency discrimination in human hairy skin. *J. Neurophysiol.* 95: 1442–1450. [21]

Maienschein, J. 1978. Cell lineage, ancestral reminiscence, and the biogenetic law. *J. Hist. Biol.* 11: 129–158. [18]

Mains, R. E., and Eipper, B. A. 1999. In *Basic Neurochemistry: Molecular, Cellular, and Medical Aspects*, 6th ed. Lippincott-Raven, Philadelphia. pp. 363–382. [15]

Mains, R. E., and Patterson, P. H. 1973. Primary cultures of dissociated sympathetic neurons. I. Establishment of long-term growth in culture and studies of differentiated properties. *J. Cell Biol.* 59: 329–345. [25]

Majdan, M., and Shatz, C. J. 2006. Effects of visual experience on activity-dependent gene regulation in cortex. *Nat. Neurosci.* 9: 650–659. [26]

Majewski, H., and Iannazzo, L. 1998. Protein kinase C: a physiological mediator of enhanced transmitter output. *Prog. Neurobiol.* 55: 463–476. [12]

Malach, R., Ebert, R., and Van Sluyters, R. C. 1984. Recovery from effects of brief monocular deprivation in the kitten. *J. Neurophysiol.* 51: 538–551. [26]

Malenka, R. C., and Nicoll, R. A. 1999. Long-term potentiation–A decade of progress? *Science* 285: 1870–1874. [16]

Malenka, R. C., Kauer, J. A., Perkel, D. J., Mank, M. D., Kelly, P. T., Nicoll, R. A., and Waxham, M. N. 1989. An essential role for postsynaptic calmodulin and protein kinase activity in long-term potentiation. *Nature* 340: 554–557. [16]

Malinow, R., and Malenka, R. C. 2002. AMPA receptor trafficking and synaptic plasticity. *Annu. Rev. Neurosci.* 25: 103–126. [16]

Malinow, R., and Tsien, R. W. 1990. Presynaptic enhancement shown by whole-cell recordings of long-term potentiation in hippocampal slices. *Nature* 346: 177–180. [16]

Malinow, R., Schulman, H., and Tsien, R. W. 1989. Inhibition of postsynaptic PKC or CaMKII blocks induction but not expression of LTP. *Science* 245: 862–866. [16]

Mallamaci, A. 2011. Molecular bases of cortico-cerebral regionalization. *Prog. Brain Res.* 189: 37–64. [25]

Mallart, A., and Martin, A. R. 1967. Analysis of facilitation of transmitter release at the neuromuscular junction of the frog. *J. Physiol.* 193: 679–697. [16]

Mallart, A., and Martin, A. R. 1968. The relation between quantum content and facilitation at the neuromuscular junction of the frog. *J. Physiol.* 196: 593–604. [16]

Malmierca, M. S. 2003. The structure and physiology of the rat auditory system: an overview. *Int. Rev. Neurobiol.* 56: 147–211. [22]

Malonek, D., Tootell, R. B. H., and Grinvald, A. 1994. Optical imaging reveals the functional architecture of neurons processing shape and motion in owl monkey area MT. *Proc. R. Soc. Lond., B, Biol. Sci.* 258: 109–119. [23]

Malpeli, J. G., Lee, D., and Baker, F. H. 1996. Laminar and retinotopic organization of the macaque LGN: magnocellular and parvocellular magnification functions. *J. Comp. Neurol.* 375: 363–377. [2]

Mandelkow, E., and Hoenger, A. 1999. Structures of kinesin and kinesin-microtubule interactions. *Curr. Opin. Cell Biol.* 11: 34–44. [15]

Manita, S., and Ross, W. N. 2009. Synaptic activation and membrane potential changes modulate the frequency of spontaneous elementary Ca^{2+} release events in the dendrites of pyramidal neurons. *J. Neurosci.* 29: 7833–7845. [12]

Manley, G. A. 2000. Cochlear mechanisms from a phylogenetic viewpoint. *Proc. Natl. Acad. Sci. USA* 97: 11736–11743. [22]

Mano, T., Iwase, S., and Toma, S. 2006. Microneurography as a tool in clinical neurophysiology to investigate peripheral neural traffic in humans. *Clin. Neurophysiol.* 117: 2357–2384. [21]

Mansour, A., Khachaturian, H., Lewis, M. E., Akil, H., and Watson, S. J. 1988. Anatomy of CNS opioid receptors. *Trends Neurosci.* 11: 308–314. [14]

Mansvelder, H. D., Keath, J. R., and McGehee, D. S. 2002. Synaptic mechanisms underlie nicotine-induced excitability of brain reward areas. *Neuron* 33: 905–919. [14]

Marban, E., Yamagishi, T., and Tomaselli, G. F. 1998. Structure and function of voltage-gated sodium channels. *J. Physiol.* 508: 647–657. [5]

Marcaggi, P., and Attwell, D. 2004. Role of glial amino acid transporters in synaptic transmission and brain energetics. *Glia* 47: 217–25. [10]

Marchand-Pauvert, V., Nicolas, G., Marque, P., Iglesias, C., and Pierrot-Deseilligny, E. 2005. Increase in group II excitation from ankle muscles to thigh motoneurones during human standing. *J. Physiol.* 566: 257–271. [24]

Marcus, D. C., Wu, T., Wangemann, P., and Kofuji, P. 2002. KCNJ10 (Kir4.1) potassium channel knockout abolishes endocochlear potential. *Am. J. Physiol. Cell Physiol.* 282: C403–407. [22]

Marder, E., and Bucher, D. 2007. Understanding circuit dynamics using the stomatogastric nervous system of lobsters and crabs. *Annu. Rev. Physiol.* 69: 291–316. [18]

Marin-Burgin, A., Kristan, W. B., Jr., and French, K. A. 2008. From synapses to behavior: development of a sensory-motor circuit in the leech. *Dev. Neurobiol.* 68: 779–787. [18]

Marín, O., and Rubenstein, J. L. 2003. Cell migration in the forebrain. *Annu. Rev. Neurosci.* 26: 441–483. [25]

Marín, O., Valiente, M., Ge, X., and Tsai, L. H. 2010. Guiding neuronal cell migrations. *Cold Spring Harb. Perspect. Biol.* 2: a001834. [25]

Marker, C. L., Luján, R., Colón, J., and Wickman, K. 2006. Distinct populations of spinal cord lamina II interneurons expressing G-protein-gated potassium channels. *J. Neurosci.* 26: 12251–12259. [14]

Markram, H., Lubke, J., Frotscher, M., and Sakmann, B. 1997. Regulation of synaptic efficacy by coincidence of postsynaptic APs and EPSPs. *Science* 275: 213–215. [8]

Marks, W. B., Dobelle, W. H., and Macnichol, E. F., Jr. 1964. Visual pigments of single primate cones. *Science* 143: 1181–1183. [20]

Marmont, G. 1949. Studies on the axon membrane; a new method. *J. Cell. Comp. Physiol.* 34: 351–382. [7]

Martin, A. R., and Fuchs, P. A. 1992. The dependence of calcium-activated potassium currents on membrane potential. *Proc. R. Soc. Lond., B, Biol. Sci.* 250: 71–76. [22]

Martin, A. R., and Levinson, S. R. 1985. Contribution of the Na^+-K^+ pump to membrane potential in familial periodic paralysis. *Muscle Nerve* 8: 354–362. [6]

Martin, A. R., and Pilar, G. 1963. Dual mode of synaptic transmission in the avian ciliary ganglion. *J. Physiol.* 168: 443–463. [11, 13]

Martin, A. R., and Pilar, G. 1964a. Quantal components of the synaptic potential in the ciliary ganglion of the chick. *J. Physiol.* 175: 1–16. [13]

Martin, A. R., and Pilar, G. 1964b. Presynaptic and postsynaptic events during post-tetanic potentiation and facilitation in the avian ciliary ganglion. *J. Physiol.* 175: 16–30. [16]

Martin, D. L. 1987. Regulatory properties of brain glutamate decarboxylase. *Cell. Mol. Neurobiol.* 7: 237–253. [15]

Martin, D. P., Schmidt, R. E., DiStefano, P. S., Lowry, O. H., Carter, J. G., and Johnson, E. M., Jr. 1988. Inhibitors of protein synthesis and RNA synthesis prevent neuronal death caused by nerve growth factor deprivation. *J. Cell Biol.* 106: 829–844. [25]

Martin, D. W. 2005. Structure-function relationships in the Na^+, K^+-pump. *Semin. Nephrol.* 25: 282–281. [9]

Martin, P., Mehta, A. D., and Hudspeth, A. J. 2000. Negative hair-bundle stiffness betrays a mechanism for mechanical amplification by the hair cell. *Proc. Natl. Acad. Sci. USA* 97: 12026–12031. [19]

Martinez, J. L., Jr., and Derrick, B. E. 1996. Long-term potentiation and learning. *Annu. Rev. Psychol.* 47: 173–203. [16]

Martinez, L. M., and Alonso, J. M. 2001. Construction of complex receptive fields in cat primary visual cortex. *Neuron* 32: 515–525. [2]

Marvizón, J. C., Chen, W., and Murphy, N. 2009. Enkephalins, dynorphins, and β-endorphin in the rat dorsal horn: an immunofluorescence colocalization study. *J. Comp. Neurol.* 517: 51–68. [14]

Masland, R. H. 1988. Amacrine cells. *Trends Neurosci.* 11: 405–410. [20]

Masland, R. H. 2001a. Neuronal diversity in the retina. *Curr. Opin. Neurobiol.* 11: 431–436. [20]

Masland, R. H. 2001b. The fundamental plan of the retina. *Nat. Neurosci.* 4: 877–886. [20]

Mason, A., and Muller, K. J. 1996. Accurate synapse regeneration despite ablation of the distal axon segment. *Eur. J. Neurosci.* 8: 11–20. [18]

Massey, P. V., and Bashir, Z. I. 2007. Long-term depression: Multiple forms and implications for brain functions. *Trends Neurosci.* 30: 176–184. [16]

Masu, M., Iwakabe, H., Tagawa, Y., Miyoshi, T., Yamashita, M., Fukuda, Y. Sasaki, H., Hiroi, K., Nakamura, Y., Shigemoto, R., Takada, M., Nakamura, K., Nakao, K., Katsuki, M., and Nakanishi, S. 1995. Specific deficit of the ON response in visual transmission by targeted disruption of the mGluR6 gene. *Cell* 80: 757–765. [20]

Matsuda, L., Lolait, S. J., Brownstein, M. J., Young, A. C., and Bonner, T. I. 1990. Structure of a cannabinoid receptor and functional expression of the cloned cDNA. *Nature* 346: 561–564. [12]

Matsunami, H., and Buck, L. B. 1997. A multigene family encoding a diverse array of putative pheromone receptors in mammals. *Cell* 90: 775–784. [19]

Matsunami, H., Montmayeur, J. P., and Buck, L. B. 2000. A family of candidate taste receptors in human and mouse. *Nature* 404: 601–604. [19]

Matsuzaki, M., Honkura, N., Ellis-Davies, G. C., and Kasai, H. 2004. Structural basis of long-term potentiation in single dendritic spines. *Nature* 429: 761– 766. [16]

Matthews, G., and Fuchs, P. 2010. The diverse roles of ribbon synapses in sensory neurotransmission. *Nat. Rev. Neurosci.* 11: 812–822. [13, 20, 22]

Matthews, G., and Wickelgren, W. O. 1979. Glycine, GABA and synaptic inhibition of reticulospinal neurones of the lamprey. *J. Physiol.* 293: 393–414. [6]

Matthews, M. R., and Nelson, V. H. 1975. Detachment of structurally intact nerve endings from chromatolytic neurones of rat superior cervical ganglion during the depression of synaptic transmission induced by postganglionic axotomy. *J. Physiol.* 245: 91–135. [27]

Matthews, P. B. 1981. Evolving views on the internal operation and functional role of the muscle spindle. *J. Physiol.* 320: 1–30. [19]

Matthews, P. B. C. 1964. Muscle spindles and their motor control. *Physiol. Rev.* 44: 219–288. [19]

Matthews, P. B. C. 1972. *Mammalian Muscle Receptors and Their Central Action*. Edward Arnold, London. [24]

Matthews, R. G., Hubbard, R., Brown, P. K., and Wald, G. 1963. Tautomeric forms of metarhodopsin. *J. Gen. Physiol.* 47: 215–240. [20]

Matthey, R. 1925. Récupération de la vue après résection des nerfs optiques chez le triton. *C. R. Soc. Biol.* 93: 904–906. [27]

Matulef, K., and Zagotta, W. N. 2003. Cyclic nucleotide-gated channels. *Annu. Rev. Cell. Dev. Biol.* 19: 23–44. [5]

Maturana, H. R., Lettvin, J. Y., McCulloch, W. S., and Pitts, W. H. 1960. Anatomy and physiology of vision in the frog (*Rana pipiens*). *J. Gen. Physiol.* 43: 129–175. [2]

Maue, R. A., and Dionne, V. E. 1987. Patch-clamp studies of isolated mouse olfactory receptor neurons. *J. Gen. Physiol.* 90: 95–125. [19]

Maunsell, J. H., and Newsome, W. T. 1987. Visual processing in monkey extrastriate cortex. *Annu. Rev. Neurosci.* 10: 363–401. [3, 23]

Maunsell, J. H. R., and Van Essen, D. C. 1983. Functional properties of neurons in the middle temporal visual area (MT) of the macaque monkey. I. Selectivity for stimulus direction, speed and orientation. *J. Neurophysiol.* 49: 1127–1147. [23]

May, B. J., and Huang, A. Y. 1996. Sound orientation behavior in cats. I. Localization of broadband noise. *J. Acoust. Soc. Am.* 100: 1059–1069. [22]

Mayer, M. L., Westbrook, G. L., and Guthrie, P. B. 1984. Voltage-dependent block by Mg^{2+} of NMDA responses in spinal cord neurones. *Nature* 309: 261–263. [11, 16]

Maylie, J., Bond, C. T., Herson, P. S., Lee, W. S., and Adelman, J. P. 2004. Small conductance Ca^{2+}-activated K^+ channels and calmodulin. *J. Physiol.* 554: 255–261. [12]

Mayser, W. Schloss, P., and Betz, H. 1992. Primary structure and functional expression of a choline transporter expressed in the rat nervous system. *FEBS Lett.* 305: 31–36. [9]

McConnell, S. K. 1995. Constructing the cerebral cortex: neurogenesis and fate determination. *Neuron* 15: 761–768. [25]

McCormick, D. A., and Williamson, A. 1989. Convergence and divergence of neurotransmitter action in human cerebral cortex. *Proc. Natl. Acad. Sci. USA* 86: 8098–8102. [14]

McCrea, P. D., Popot, J. L., and Engleman, D. M. 1987. Transmembrane topography of the nicotinic acetylcholine receptor subunit. *EMBO J.* 6: 3619–3626. [5]

McDonald, T. F., Pelzer, S., Trautwein, W., and Pelzer, D. J. 1994. Regulation and modulation of calcium channels in cardiac, skeletal, and smooth muscle cells. *Physiol. Rev.* 74: 365–507. [12]

McGehee, D. S., Heath, M. J., Gelber, S., Devay, P., and Role, L. W. 1995. Nicotine enhancement of fast excitatory synaptic transmission in CNS by presynaptic receptors. *Science* 269: 1692–1696. [14]

McGlade-McCulloh, E., Morrissey, A. M., Norona, F., and Muller, K. J. 1989. Individual microglia move rapidly and directly to nerve lesions in the leech central nervous system. *Proc. Natl. Acad. Sci. USA.* 86: 1093–1097. [10]

McGlone, F., Vallbo, A. B., Olausson, H., Loken, L., and Wessberg, J. 2007. Discriminative touch and emotional touch. *Can. J. Exp. Psychol.* 61: 173–183. [21]

McGraw, L. A., and Young, L. J. 2010. The prairie vole: an emerging model organism for understanding the social brain. *Trends Neurosci.* 33: 103–109. [14]

McIntire, S. L., Reimer, R. J., Schuske, K., Edwards, R. H., and Jorgensen, E. M. 1997. Identification and characterization of the vesicular GABA transporter. *Nature* 389: 870–876. [9]

McIntyre, A. 1980. Biological seismography. *Trends Neurosci.* 3: 202–205. [19]

McKenna, M. P., and Raper, J. A. 1988. Growth cone behavior on gradients of substratum bound laminin. *Dev. Biol.* 130: 232–236. [25]

McKernan, D. P., and Cotter, T. G. 2007. A critical role for Bim in retinal ganglion cell death. *J. Neurochem.* 102: 922–930. [27]

McKernan, M. G., and Shinnick-Gallagher, P. 1997. Fear conditioning induces a lasting potentiation of synaptic currents in vitro. *Nature* 390: 607–611. [16]

McLachlan, E. M., ed. 1995. *Autonomic Ganglia*. Gordon and Breach, London. [17]

McLaughlin, S. K., McKinnon, P. J., and Margolskee, R. F. 1992. Gustducin is a taste-cell-specific G protein closely related to the transducins. *Nature* 357: 563–569. [19]

McMahan, U. J. 1990. The agrin hypothesis. *Cold Spring Harb. Symp. Quant. Biol.* 55: 407–418. [11, 27]

McMahan, U. J., and Slater, C. R. 1984. The influence of basal lamina on the accumulation of acetylcholine receptors at synaptic sites in regenerating muscle. *J. Cell Biol.* 98: 1453–1473. [27]

McMahan, U. J., and Wallace, B. G. 1989. Molecules in basal lamina that direct the formation of synaptic specializations at neuromuscular junctions. *Dev. Neurosci.* 11: 227–247. [27]

McMahan, U. J., Edgington, D. R., and Kuffler, D. P. 1980. Factors that influence regeneration of the neuromuscular junction. *J. Exp. Biol.* 89: 31–42. [27]

McMahan, U. J., Sanes, J. R., and Marshall, L. M. 1978. Cholinesterase is associated with the basal lamina at the neuromuscular junction. *Nature* 271: 172–174. [15]

McMahan, U. J., Spitzer, N. C., and Peper, K. 1972. Visual identification of nerve terminals in living isolated skeletal muscle. *Proc. Roy. Soc. Lond., B, Biol. Sci.* 181: 421–430. [11]

Meier, J. C., Henneberger, C., Melnik, I., Racca, C., Harvey, R. J. Heinemann, U., Schmieden, V., and Grantyn, R. 2005. RNA editing produces glycine receptor alpha3 (P185L), resulting in high agonist potency. *Nature Neurosci.* 8: 736–744. [5]

Meier, J. D., Aflalo, T. N., Kastner, S., and Graziano, M. S. A. 2008. Complex organization of human primary motor cortex: a high-resolution fMRI study. *J. Neurophysiol.* 100: 1800–1812. [24]

Meis, S. 2003. Nociceptin/orphanin FQ: actions within the brain. *Neuroscientist* 9: 158–168. [14]

Meissirel, C., Wikler, K. C., Chalupa, L. M., and Rakic, P. 1997. Early divergence of magnocellular and parvocellular functional subsystems in the embryonic primate visual system. *Proc. Natl. Acad. Sci. USA* 94: 5900–5905. [26]

Meister, M., and Berry, M. J., 2nd. 1999. The neural code of the retina. *Neuron* 22: 435–450. [20]

Meister, M., Lagnado, L., and Baylor, D. A. 1995. Concerted signaling by retinal ganglion cells. *Science* 270: 1207–1210. [2, 20]

Meister, M., Wong, R. O., Baylor, D. A., and Shatz, C. J. 1991. Synchronous bursts of action potentials in ganglion cells of the developing mammalian retina. *Science* 252: 939–943. [26]

Meladinić, M., Bechetti, A., Didelon, F., Bradbury, A., and Cherubini, E. 1999. Low expression of the CLC-2 chloride channel during postnatal development: a mechanism for the paradoxical depolarizing action of GABA and glycine in the hippocampus. *Proc. Roy. Soc. Lond., B, Biol. Sci.* 266: 1207–1213. [6]

Melzack, R. 1973. *The Puzzle of Pain*. Harmondsworth: Penguin Books. [21]

Mendell, L. M., and Henneman, E. 1971. Terminals of single Ia fibers: Location, density, and distribution within a pool of 300 homonymous motoneurons. *J. Neurophysiol.* 34: 171–187. [24]

Menei, P., Montero-Menei, C., Whittemore, S. R., Bunge, R. P., and Bunge, M. B. 1998. Schwann cells genetically modified to secrete human BDNF promote enhanced axonal regrowth across transected adult rat spinal cord. *Eur. J. Neurosci.* 10: 607–621. [27]

Menini, A., Picco, C., and Firestein, S. 1995. Quantal-like current fluctuations induced by odorants in olfactory receptor cells. *Nature* 373: 435–437. [19]

Mercado, A., Mount, D. B., and Gamba, G. 2004. Electroneutral cation-chloride cotransporters in the central nervous system. *Neurochem. Res.* 29: 17–25. [9]

Merchant, H., Naselaris, T., and Georgopoulos, A. P. 2008. Dynamic sculpting of directional tuning in the primate motor cortex during three-dimensional reaching. *J. Neurosci.* 28: 9164–9172. [24]

Merigan, W. H., and Maunsell, J. H. R. 1993. How parallel are the primate visual pathways? *Annu. Rev. Neurosci.* 16: 369–402. [3, 23]

Merkle, T., and Wehner, R. 2008. Landmark guidance and vector navigation in outbound desert ants. *J. Exp. Biol.* 211: 3370–3377. [18]

Merzenich, M. M., Kaas, J. H., Sur, M., and Lin, C. S. 1978. Double representation of the body surface within cytoarchitectonic areas 3b and 1 in "SI" in the owl monkey (*Aotus trivirgatus*). *J. Comp. Neurol.* 181: 41–73. [21]

Merzenich, M. M., Knight, P. L., and Roth, G. L. 1975. Representation of cochlea within primary auditory cortex in the cat. *J. Neurophysiol.* 38: 231–249. [22]

Mesce, K. A., Esch, T., and Kristan, W. B., Jr. 2008. Cellular substrates of action selection: a cluster of higher-order descending neurons shapes body posture and locomotion. *J. Comp. Physiol. A.* 194: 469–481. [18]

Mesgarani, N., David, S. V., Fritz, J. B., and Shamma, S. A. 2008. Phoneme representation and classification in primary auditory cortex. *J. Acoust. Soc. Am.* 123: 899–909. [22]

Mesulam, M. 2004. The cholinergic lesion of Alzheimer's disease: pivotal factor or side show? *Learn. Mem.* 11: 43–49. [14]

Metea, M. R., and Newman, E. A. 2006a. Calcium signaling in specialized glial cells. *Glia* 54: 650–655. [10]

Metea, M. R., and Newman, E. A. 2006b. Glial cells dilate and constrict blood vessels: A mechanism of neurovascular coupling. *J. Neurosci.* 26: 2862–2870. [10]

Metea, M. R., Kofuji, P., and Newman, E. A. 2007. Neurovascular coupling is not mediated by potassium siphoning from glial cells. *J. Neurosci.* 27: 2468–2471. [10]

Meunier, J. C., Mollereau, C., Toll, L., Suaudeau, C., Moisand, C., Alvinerie, P., Butour, J. L., Guillemot, J. C., Ferrara, P., Monsarrat, B., Mazarguil, H., Vassart, G., Parmentier, M., and Costentin, J. 1995. Isolation and structure of the endogenous agonist of opioid receptor-like ORL_1 receptor. *Nature* 377: 532–535. [14]

Meves, H. 2008. Arachidonic acid and ion channels: an update. *Brit. J. Pharmacol.* 155: 4–16. [12]

Meyer, A. C., Frank, T., Khimich, D., Hoch, G., Riedel, D., Chapochnikov, N. M., Yarin, Y. M., Harke, B., Hell, S. W., Egner, A., and Moser, T. 2009. Tuning of synapse number, structure and function in the cochlea. *Nat. Neurosci.* 12: 444–453. [22]

Meyer, G. 2010. Building a human cortex: the evolutionary differentiation of Cajal-Retzius cells and the cortical hem. *J. Anat.* 217: 334–343. [25]

Meyer, M., Matsuoka, I., Wetmore, C., Olson, L., and Thoenen, H. 1992. Enhanced synthesis of brain-derived neurotrophic factor in the lesioned peripheral nerve: different mechanisms are responsible for the regulation of BDNF and NGF mRNA. *J. Cell Biol.* 119: 45–54. [27]

Middlebrooks, J. C., Dykes, R. W., and Merzenich, M. M. 1980. Binaural response-specific bands in primary auditory cortex (AI) of the cat: topographical organization orthogonal to isofrequency contours. *Brain. Res.* 181: 31–48. [22]

Mikoshiba, K. 2007. IP3 receptor/Ca^{2+} channel: from discovery to new signaling concepts. *J. Neurochem.* 102: 1426–1446. [12]

Miledi, R. 1960a. The acetylcholine sensitivity of frog muscle fibers after complete or partial denervation. *J. Physiol.* 151: 1–23. [27]

Miledi, R. 1960b. Junctional and extra-junctional acetylcholine receptors in skeletal muscle fibres. *J. Physiol.* 151: 24–30. [11]

Miledi, R., Parker, I., and Sumikawa, K. 1983. Recording single γ-aminobutyrate- and acetylcholine-activated receptor channels translated by exogenous mRNA in Xenopus oocytes. *Proc. R. Soc. Lond., B, Biol. Sci.* 218: 481–484. [5]

Miledi, R., Stefani, E., and Steinbach, A. B. 1971. Induction of the action potential mechanism in slow muscle fibres of the frog. *J. Physiol.* 217: 737–754. [27]

Millar, N. S., and Gotti, C. 2009. Diversity of vertebrate nicotinic acetylcholine receptors. *Neuropharmacology* 56: 237–246. [5, 14]

Miller, C. 1992. Hunting for the pore of voltage-gated channels. *Current Biol.* 2: 573–575. [5]

Miller, C., and White, M. M. 1984. Dimeric structure of single chloride channels from *Torpedo electroplax*. *Proc. Nat. Acad. Sci. USA* 81: 2772–2775. [5]

Miller, J., Agnew, W. S., and Levinson S. R. 1985. Principal glycopeptide of the tetrodotoxin/saxitoxin binding protein from *Electrophorus electricus*: isolation and partial chemical and physical purification. *Biochemistry* 22: 462–470. [5]

Miller, T. M., and Heuser, J. E. 1984. Endocytosis of synaptic vesicle membrane at the frog neuromuscular junction. *J. Cell Biol.* 98: 685–698. [13]

Milner, A. D., Perrett, D. I., Johnston, R. S., Benson, P. J., Jordan, T. R., Heeley, D. W., Bettucci, D., Mortara, F., Mutani, R., Terazzi, E., and Davidson, D. 1991. Perception and action in 'visual form agnosia'. *Brain* 114: 405–428. [23]

Mineo, M., Jolley, T., and Rodriguez, G. 2004. Leech therapy in penile replantation: a case of recurrent penile self-amputation. *Urology* 63: 981–983. [18]

Mink, J. W., and Thach, W. T. 1991. Basal ganglia motor control. III. Pallidal ablation: normal reaction time, muscle co-contraction, and slow movement. *J. Neurophysiol.* 65: 330–351. [24]

Mink, J. W., and Thach, W. T. 1993. Basal ganglia intrinsic circuits and their role in behavior. *Curr. Opin. Neurobiol.* 3: 950–957. [24]

Minor, L. B. 2005. Clinical manifestations of superior semicircular canal dehiscence. *Laryngoscope* 115: 1717–1727. [22]

Miserendino, M. J., Sananes, C. B., Melia, K. R., and Davis, M. 1990. Blocking acquisition but not expression of conditioned fear-potentiated startle by NMDA antagonists in the amygdala. *Nature* 345: 716–718. [16]

Mishina, M., Takai, T., Imoto, K., Noda, M., Takahashi, T., Numa, S., Methfessel, C., and Sakmann, B. 1986. Molecular distinction between fetal and adult forms of muscle acetylcholine receptor. *Nature* 321: 406–411. [11, 27]

Mitchell, E. A., Herd, M. B., Gunn, B. G., Lambert, J. J., and Belelli, D. 2008. Neurosteroid modulation of $GABA_A$ receptors: molecular determinants and significance in health and disease. *Neurochem. Int.* 52: 588–595. [14]

Miyakawa, H., Lev-Ram, V., Lasser-Ross, N., and Ross, W. N. 1992. Calcium transients evoked by climbing fiber and parallel fiber synaptic inputs in guinea pig cerebellar Purkinje neurons. *J. Neurophysiol.* 68: 1178–1189. [24]

Miyawaki, A., Llopis, J., Heim, R., McCaffery, J. M., Adams, J. A., Ikura, M., and Tsien, R. Y. 1997. Fluorescent indicators for Ca^{2+} based on green fluorescent proteins and calmodulin. *Nature* 388: 882–887. [12]

Miyazawa, A., Fujiyoshi, Y., and Unwin, N. 2003. Structure and gating mechanism of the acetylcholine receptor pore. *Nature* 423: 949–955. [5]

Mladinic, M., Lefèvre, C., Del Bel, E., Nicholls, J. G., and Digby, M. 2010. Developmental changes of gene expression after spinal cord injury in neonatal opossums. *Brain Res.* 1363: 20–39. [27]

Mladinic, M., Muller, K. J., and Nicholls. J. G. 2009. Central nervous system regeneration: from leech to opossum. *J. Physiol.* 587: 2775–2782. [27]

Mladinic, M., Wintzer, M., Casseler, C., Lazarevic, D., Crovella, S., Gustincich, S., Cattaneo, A., and Nicholls, J. 2005. Differential expression of genes at stages when regeneration can and cannot occur after injury to immature mammalian spinal cord. *Cell. Mol. Neurobiol.* 25: 405–424. [27]

Mochida, S., Few, A. P., Scheuer, T., and Catterall, W. A. 2008. Regulation of presynaptic $Ca_V2.1$ channels by Ca^{2+} sensor proteins mediates short-term synaptic plasticity. *Neuron* 57: 210–216. [16]

Modney, B. K., Sahley, C. L., and Muller, K. J. 1997. Regeneration of a central synapse restores nonassociative learning. *J. Neurosci.* 17: 6478–6482. [18]

Moeller, F. G., Dougherty, D. M., Swann, A. C., Collins, D., Davis, C. M., and Cherek, D. R. 1996. Tryptophan depletion and aggressive responding in healthy males. *Psychopharmacology (Berl)* 126: 96–103. [15]

Moens, C. B., and Prince, V. E. 2002. Constructing the hindbrain: insights from the zebrafish. *Dev. Dyn.* 224: 1–17. [25]

Möhler, H. 2006. GABA(A) receptor diversity and pharmacology. *Cell Tissue Res.* 326: 505–516. [14]

Moiseff, A. 1989. Bi-coordinate sound localization by the barn owl. *J. Comp. Physiol. A* 164: 637–644. [26]

Moller, A. R. 2006. History of cochlear implants and auditory brainstem implants. *Adv. Otorhinolaryngol.* 64: 1–10. [26]

Mombaerts, P., Wang, F., Dulac, C., Chao, S. K., Nemes, A., Mendelsohn, M., Edmonson, J., and Axel, R. 1996. Visualizing an olfactory sensory map. *Cell* 87: 675–686. [19]

Montal, M. O., Iwamoto, T., Tomich, J. M., and Montal, M. 1993. Design, synthesis, and functional characterization of a pentameric channel protein that mimics the presumed pore structure of the nicotinic cholinergic receptor. *FEBS Lett.* 320: 261–266. [5]

Montero, M., Brini, M., Marsault, R., Alvarez, J., Sitia, R., Pozzan, T., and Rizzuto, R. 1995. Monitoring dynamic changes in free Ca^{2+} concentration in the endoplasmic reticulum of intact cells. *EMBO J.* 14: 5467–5475. [12]

Mooney, R. 2009. Neurobiology of song learning. *Curr. Opin. Neurobiol.* 19: 654–660. [26]

Moore, M. J., and Caspary, D. M. 1983. Strychnine blocks binaural inhibition in lateral superior olivary neurons. *J. Neurosci.* 3: 237–242. [22]

Mora-Ferrer, C., and Neumeyer, C. 2009. Neuropharmacology of vision in goldfish: a review. *Vision Res.* 49: 960–969. [20]

Moransard, M., Borges, L. S., Willmann, R., Marangi, P. A., Brenner, H. R., Ferns, M. J., and Fuhrer, C. 2003. Agrin regulates rapsyn interaction with surface acetylcholine receptors, and this underlies cytoskeletal anchoring and clustering. *J. Biol. Chem.* 278: 7350–7359. [27]

Morgans, C. W., El Far, O., Berntson, A., Wässle, H., and Taylor, W. R. 1998. Calcium extrusion from mammalian photoreceptor terminals. *J. Neurosci.* 18: 2467–2474. [20]

Morishita, H., and Hensch, T. K. 2008. Critical period revisited: impact on vision. *Curr. Opin. Neurobiol.* 18: 101–107. [26]

Morita, K., and Barrett, E. F. 1990. Evidence for two calcium-dependent potassium conductances in lizard motor nerve terminals. *J. Neurosci.* 10: 2614–2625. [13]

Morrison, A. D. 1998. 1 + 1 = r4 and much much more. *Bioessays* 20: 794–797. [25]

Mörschel, M., and Dutschmann, M. 2009. Pontine respiratory activity involved in inspiratory/expiratory phase transition. *Philos. Trans. R. Soc. Lond., B, Biol. Sci.* 364: 2517–2526. [24]

Morton, N. E. 1991. Genetic epidemiology of hearing impairment. *Ann. N Y Acad. Sci.* 630: 16–31. [26]

Moss, S. J., and Smart, T. G. 2001. Constructing inhibitory synapses. *Nat. Rev. Neurosci.* 2: 240–250. [11]

Moulton, E. A., Pendse, G., Morris, S., Aiello-Lammens, M., Becerra, L., and Borsook, D. 2009. Segmentally arranged somatotopy within the face representation of human primary somatosensory cortex. *Hum. Brain Mapp.* 30: 757–765. [21]

Mountcastle, V. B. 1957. Modality and topographic properties of single neurons of cat's somatic sensory cortex. *J. Neurophysiol.* 20: 408–434. [3]

Mudge, A. W., Leeman, S. E., and Fischbach, G. D. 1979. Enkephalin inhibits release of substance P from sensory neurons in culture and decreases action potential duration. *Proc. Natl. Acad. Sci. USA* 76: 526–530. [14]

Mufson, E. J., Ginsberg, S. D., Ikonomovic, M. D., and DeKosky, S. T. 2003. Human cholinergic basal forebrain: chemoanatomy and neurologic dysfunction. *J. Chem. Neuroanat.* 26: 233–242. [25]

Mulkey, R. M., and Malenka, R. C. 1992. Mechanisms underlying induction of homosynaptic long-term depression in area CA1 of the hippocampus. *Neuron* 9: 967–975. [16]

Muller, D., Joly, M., and Lynch, G. 1988. Contributions of quisqualate and NMDA receptors to the induction and expression of LTP. *Science* 242: 1694–1697. [16]

Muller, K. J. 1973. Photoreceptors in the crayfish compound eye: electrical interactions between cells as related to polarized-light sensitivity. *J. Physiol.* 232: 573–595. [18]

Muller, K. J. 1981. Synapses and synaptic transmission. In K. J. Muller, J. G. Nicholls, and G. S. Stent (eds.), *Neurobiology of the Leech.* Cold Spring Harbor Laboratory, Cold Spring Harbor, NY. pp. 79–111. [18]

Muller, K. J., and Carbonetto, S. 1979. The morphological and physiological properties of a regenerating synapse in the C. N. S. of the leech. *J. Comp. Neurol.* 185: 485–516. [18]

Muller, K. J., and McMahan, U. J. 1976. The shapes of sensory and motor neurones and the distribution of their synapses in ganglia of the leech: a study using intracellular injection of horseradish peroxidase. *Proc. R. Soc. Lond., B, Biol. Sci.* 194: 481–499. [18]

Muller, K. J., and Nicholls, J. G. 1974. Different properties of synapses between a single sensory neurone and two different motor cells in the leech CNS. *J. Physiol.* 238: 357–369. [18]

Muller, K. J., Nicholls, J. G., and Stent, G. S., eds. 1981. *Neurobiology of the Leech.* Cold Spring Harbor Laboratory, Cold Spring Harbor, NY. [18]

Muller, K. J., Tsechpenakis, G., Homma, R., Nicholls, J. G., Cohen, L. B., and Eugenin, J. 2009. Optical analysis of circuitry for respiratory rhythm in isolated brainstem of foetal mice. *Philos. Trans. R. Soc. Lond., B, Biol. Sci.* 364: 2485–2491. [24]

Müller, M., and Schlue, W. R. 1998. Macroscopic and single-channel chloride currents in neuropile glial cells of the leech central nervous system. *Brain Res.* 781: 307–19. [10]

Müller, M., and Wehner, R. 1994. The hidden spiral: systematic search and path integration in desert ants, *Cataglyphis fortis. J. Comp. Physiol. A* 175: 525–530. [18]

Müller, M., and Wehner, R. 2007. Wind and sky as compass cues in desert ant navigation. *Naturwissenschaften* 94: 589–594. [18]

Müller, M., and Wehner, R. 2010. Path integration provides a scaffold for landmark learning in desert ants. *Curr. Biol.* 20: 1368–1371. [18]

Mullins, L. J. 1975. Ion selectivity of carriers and channels. *J. Biophys.* 15: 921–931. [5]

Mullins, L. J., and Noda, K. 1963. The influence of sodium-free solutions on membrane potential of frog muscle fibers. *J. Gen. Physiol.* 47: 117–132. [6]

Munch, T. A., and Werblin, F. S. 2006. Symmetric interactions within a homogeneous starburst cell network can lead to robust asymmetries in dendrites of starburst amacrine cells. *J. Neurophysiol.* 96: 471–477. [20]

Munk, H. 1881. *Ueber die Functionen der Grosshirnrinde; gesammelte Mittheilungen aus den Jahren 1877–80.* Hirschwald, Berlin. [23]

Murashima, M., and Hirano, T. 1999. Entire course and distinct phases of day-lasting depression of miniature EPSC amplitudes in cultured Purkinje neurons. *J. Neurosci.* 19: 7326–7333. [16]

Murphy, P. C., and Sillitoe, A. M. 1991. Cholinergic enhancement of direction sensitivity in the visual cortex of the cat. *Neuroscience* 40: 13–20. [14]

Nagatsu, T. 1995. Tyrosine hydroxylase: Human isoforms, structure and regulation in physiology and pathology. *Essays Biochem.* 30: 15–35. [15]

Nagatsu, T., and Ichinose, H. 1999. Regulation of pteridine-requiring enzymes by the cofactor tetrahydrobiopterin. *Mol. Neurobiol.* 19: 79–96. [15]

Nakajima, S., and Onodera, K. 1969a. Adaptation of the generator potential in the crayfish stretch receptors under constant length and constant tension. *J. Physiol.* 200: 187–204. [19]

Nakajima, S., and Onodera, K. 1969b. Membrane properties of the stretch receptor neurones of crayfish with particular reference to mechanisms of sensory adaptation. *J. Physiol.* 200: 161–185. [19]

Nakajima, S., and Takahashi, K. 1966. Post-tetanic hyperpolarization and electrogenic Na pump in stretch receptor neurone of crayfish. *J. Physiol.* 187: 105–127. [19]

Nakajima, Y., Tisdale, A. D., and Henkart, M. P. 1973. Presynaptic inhibition at inhibitory nerve terminals: A new synapse in the crayfish stretch receptor. *Proc. Natl. Acad. Sci. USA* 70: 2462–2466. [11]

Nakamura, T., and Gold, G. H. 1987. A cyclic nucleotide-gated conductance in olfactory receptor cilia. *Nature* 325: 442–444. [19]

Nakanishi, S., Nakajima, Y., Masu, M., Ueda, Y., Nakahar, K., Watanabe, D., Yamaguchi, S., Kawabata, S., and Okada, M. 1998. Glutamate receptors: brain function and signal transduction. *Brain Res. Brain Res. Rev.* 26: 230–235. [20]

Namer, B., and Handwerker, H. O. 2009. Translational nociceptor research as guide to human pain perceptions and pathophysiology. *Exp. Brain Res.* 196: 163–172. [21]

Nargeot, J., Nerbonne, J. M., Engels, J., and Lester, H. A. 1983. Time course of the increase in the myocardial slow inward current after a photochemically generated concentration jump of intracellular cAMP. *Proc. Natl. Acad. Sci. USA* 80: 2385–2399. [12]

Nastuk, W. L. 1953. Membrane potential changes at a single muscle end-plate produced by transitory application of acetylcholine with an electrically controlled microjet. *Fed. Proc.* 12: 102. [11]

Nathans, J. 1987. Molecular biology of visual pigments. *Annu. Rev. Neurosci.* 10: 163–194. [20]

Nathans, J. 1989. The genes for color vision. *Sci. Am.* 260: 42–49. [20]

Nathans, J. 1999. The evolution and physiology of human color vision: insights from molecular genetic studies of visual pigments. *Neuron* 24: 299–312. [20]

Nathans, J., and Hogness, D. S. 1984. Isolation and nucleotide sequence of the gene encoding human rhodopsin. *Proc. Natl. Acad. Sci. USA* 81: 4851–4855. [20]

Navaratnam, D. S., Bell, T. J., Tu, T. D., Cohen, E. L., and Oberholtzer, J. C. 1997. Differential distribution of Ca^{2+}-activated K^+ channel splice variants among hair cells along the tonotopic axis of the chick cochlea. *Neuron* 19: 1077–1085. [22]

Navarrete, M., and Araque, A. 2010. Endocannabinoids potentiate synaptic transmission through stimulation of astrocytes. *Neuron* 68: 113–126. [12]

Nave, K. A., and Trapp, B. D. 2008. Axon-glial signaling and the glial support of axon function. *Annu. Rev. Neurosci.* 31: 535–561. [10]

Naya, Y., Sakai, K., and Miyashita, Y. 1996. Activity of primate inferotemporal neurons related to a sought target in pair-association task. *Proc. Natl. Acad. Sci. USA* 93: 2664–2669. [23]

Naya, Y., Yoshida, M., and Miyashita, Y. 2001. Backward spreading of memory-retrieval signal in the primate temporal cortex. *Science* 291: 661–664. [23]

Neher, E., and Augustine, G. J. 1992. Calcium gradients and buffers in bovine chromaffin cells. *J. Physiol.* 450: 273–301. [12]

Neher, E., and Sakaba, T. 2008. Multiple roles of calcium ions in the regulation of neurotransmitter release. *Neuron* 59: 861–872. [16]

Neher, E., and Sakmann, B. 1976. Single channel currents recorded from membrane of denervated frog muscle fibres. *Nature* 260: 799–802. [11]

Neher, E., Sakmann, B., and Steinbach, J. H. 1978. The extracellular patch clamp: A method for resolving currents through individual open channels in biological membranes. *Pflügers Arch.* 375: 219–228. [4]

Neil, E. 1954. Reflexogenic areas of the circulation. *Arch. Middlesex Hosp.* 4: 16. (Modified from Berne, M., and Levy, M. N. 1990. *Principles of Physiology*. Wolfe, London.) [17]

Nelken, I. 2004. Processing of complex stimuli and natural scenes in the auditory cortex. *Curr. Opin. Neurobiol.* 14: 474–480. [22]

Nelken, I., and Bar-Yosef, O. 2008. Neurons and objects: the case of auditory cortex. *Front. Neurosci.* 2: 107–113. [22]

Nelken, I., Rotman, Y., and Bar Yosef, O. 1999. Responses of auditory-cortex neurons to structural features of natural sounds. *Nature* 397: 154–157. [22]

Nelson, N., and Lill, H. 1994. Porters and neurotransmitter transporters. *J. Exp. Biol.* 196: 213–228. [9]

Newman, E. A. 1986. High potassium conductance in astrocyte endfeet. *Science* 275: 844–847. [10]

Newman, E. A. 1987. Distribution of potassium conductance in mammalian Müller (glial) cells: A comparative study. *J. Neurosci.* 7: 2423–2432. [10]

Newman, E. A. 2003. Glial cell inhibition of neurons by release of ATP. *J. Neurosci.* 23: 1659–1666. [10]

Newman, E. A. 2004. A dialogue between glia and neurons in the retina: Modulation of neuronal excitability. *Neuron Glia Biol.* 1: 245–252. [10]

Newsome, W. T., and Pare, E. B. 1988. A selective impairment of motion perception following lesions of the middle temporal visual area (MT). *J. Neurosci.* 8: 2201–2211. [23]

Newsome, W. T., and Wurtz, R. H. 1988. Probing visual cortical function with discrete chemical lesions. *Trends Neurosci.* 11: 394–400. [23]

Nguyen, M. D., Mushynski, W. E., and Julien, J. P. 2002. Cycling at the interface between neurodevelopment and neurodegeneration. *Cell Death Differ.* 9: 1294–1306. [25]

Nicholls, C. G., and Lopatin, A. N. 1997. Inward rectifier potassium channels. *Annu. Rev. Physiol.* 59: 171–191. [5]

Nicholls, J. G. 2007. How acetylcholine gives rise to current at the motor end-plate. *J. Physiol.* 578: 621–622. [11]

Nicholls, J. G., and Baylor, D. A. 1968. Specific modalities and receptive fields of sensory neurons in the CNS of the leech. *J. Neurophysiol.* 31: 740–756. [6, 18]

Nicholls, J. G., and Purves, D. 1972. A comparison of chemical and electrical synaptic transmission between single sensory cells and a motoneurone in the central nervous system of the leech. *J. Physiol.* 225: 637–656. [11, 18]

Nicholls, J. G., and Saunders, N. 1996. Regeneration of immature mammalian spinal cord after injury. *Trends Neurosci.* 19: 229–234. [27]

Nicholls, J. G., and Wallace, B. G. 1978. Modulation of transmission at an inhibitory synapse in the central nervous system of the leech. *J. Physiol.* 281: 157–170. [11]

Nicholls, J. G., Stewart, R. R., Erulkar, S. D., and Saunders, N. R. 1990. Reflexes, fictive respiration, and cell division in the brain and spinal cord of the newborn opossum, *Monodelphis domestica*, isolated and maintained *in vitro*. *J. Exp. Biol.* 152: 1–15. [24]

Nicola, S. M., Surmeier, D. J., and Malenka, R. C. 2000. Dopaminergic modulation of neuronal excitability in the striatum and nucleus accumbens. *Annu. Rev. Neurosci.* 23: 185–215. [14]

Nicoll, R. A. 1985. The septo-hippocampal projection: a model cholinergic pathway. *Trends Neurosci.* 8: 533–536. [14]

Nicoll, R. A. 2004. My close encounters with $GABA_B$ receptors. *Biochem. Pharmacol.* 68: 1667–1674. [14]

Nicoll, R. A., and Schmitz, D. 2005. Synaptic plasticity at hippocampal mossy fibre synapses. *Nat. Rev. Neurosci.* 6: 863–876. [16]

Nicoll, R. A., Eccles, J. C., Oshima, T., and Rubia, F. 1975. Prolongation of hippocampal inhibitory postsynaptic potentials by barbiturates. *Nature* 258: 625–627. [14]

Nicoll, R. A., Tomita, S., and Bredt, D. S. 2006. Auxiliary subunits assist AMPA-type glutamate receptors. *Science* 311: 1253–1256. [14]

NIDCD. 2011. NIDCD Fact Sheet: *Cochlear Implants*. National Institute on Deafness and Other Communication Disorders. Publication No. 11-4798. //www.nidcd.nih.gov/ [26]

Nielsen, J. B., and Sinkjaer, T. 2002. Afferent feedback in the control of human gait. *J. Electromyogr. Kinesiol.* 12: 213–217. [24]

Niesler, B., Frank, B., Kapeller, J., and Rappold, G. A. 2003. Cloning, physical mapping and expression analysis of the human $5\text{-}HT_3$ serotonin receptor-like genes *HTR3C, HTR3D* and *HTR3E*. *Gene* 310: 101–111. [5]

Nilius, B., and Voets, T. 2005. TRP Channels: a TR(I)P through the world of multifunctional cation channels. *Pflügers Arch.* 451: 1–10. [5]

Nimigean, C. M., Chappie, J. S., and Miller, C. 2003. Electrostatic tuning of ion conductance in potassium channels. *Biochemistry* 42: 9263–9268. [5]

Niparko, J. K., and Marlowe, A. 2010. Hearing aids and cochlear implants. In P. A. Fuchs (ed.), *The Oxford University Handbook of Auditory Science: The Ear*. Oxford University Press, New York. pp. 409–436. [26]

Niswender, C., and Conn, P. J. 2010. Metabotropic glutamate receptors: physiology, pharmacology, and disease. *Annu. Rev. Pharmacol. Toxicol.* 50: 295–322. [14]

Nja, A., and Purves, D. 1978. The effects of nerve growth factor and its antiserum on synapses in the superior cervical ganglion of the guinea-pig. *J. Physiol.* 277: 55–75. [27]

Noakes, P. G., Gautam, M., Mudd, J., Sanes, J. R., and Merlie, J. P. 1995. Aberrant differentiation of neuromuscular junctions in mice lacking s-laminin/laminin β2. *Nature* 374: 258–262. [27]

Noctor, S. C., Flint, A. C., Weissman, T. A., Dammerman, R. S., and Kriegstein, A. R. 2001. Neurons derived from radial glial cells establish radial units in neocortex. *Nature* 409: 714–720. [25]

Noda, M., Shimizu, S., Tanabe, T., Takai, T., Kayano, T., Ikeda, T., Takahashi, H., Nakayama, H., Kanaoka, Y., Minamino, N., Kangawa, K., Matsuo, H., Raftery, M. A., Hirose, T., Inagama, S., Hayashida, H., Miyata, T., and Numa, S. 1984. Primary structure of *Electrophorus electricus* sodium channel deduced from cDNA sequence. *Nature* 312: 121–127. [5]

Noda, M., Takahashi, H., Tanabe, T., Toyosato, M., Furutani, Y., Hirose, T., Asai, M., Inayama, S., Miyata, T., and Numa, S. 1982. Primary structure of α-subunit precursor of *Torpedo californica* acetylcholine receptor deduced from cDNA sequence. *Nature* 299: 793–797. [5]

Noda, M., Takahashi, H., Tanabe, T., Toyosato, M., Kikyotani, S., Hirose, T., Asai, M., Takashima, H., Inayama, S., Miyata, T., and Numa, S. 1983a. Primary structure of β- and δ-subunit precursors of *Torpedo californica* acetylcholine receptor deduced from cDNA sequences. *Nature* 301: 251–255. [5]

Noda, M., Takahashi, H., Tanabe, T., Toyosato, M., Kikyotani, S., Furutani, Y., Hirose, T., Takashima, H., Inayama, S., Miyata, T., and Numa, S. 1983b. Structural homology of *Torpedo californica* acetylcholine receptor subunits. *Nature* 302: 528–532. [5]

Nolte, J. 1988. *The Human Brain*, 2nd ed. Mosby, St. Louis, Mo. [25]

Nomura, A., Shigemoto, R., Nakamura, Y., Okamoto, N., Mizuno, N., and Nakanishi, S. 1994. Developmentally regulated postsynaptic localization of a metabotropic glutamate receptor in rat rod bipolar cells. *Cell* 77: 361–369. [20]

North, R. A. 2002. Molecular physiology of P2X receptors. *Physiol. Rev.* 82: 1013–1067. [14]

North, R. A., Williams, J. T., Surprenant, A., and Christie, M. J. 1987. μ and δ receptors belong to a family of receptors that are coupled to potassium channels. *Proc. Natl. Acad. Sci. USA* 84: 5487–5491. [12]

Nottebohm, F. 1989. From bird song to neurogenesis. *Sci. Am.* 260: 74–79. [25]

Notterpek, L., Shooter, E. M., and Snipes, G. J. 1997. Up-regulation of the endosomal-lysosomal pathway in the trembler. *J. Neurosci.* 17: 4190–4200. [10]

Nowak, L., Bregestovski, P., Ascher, P., Herbet, A., and Prochiantz, A. 1984. Magnesium gates glutamate-activated channels in mouse central neurones. *Nature* 307: 462–465. [11, 16]

Numa, S., Noda, M., Takahashi, H., Tanabe, T., Toyosato, M., Furutani, Y., and Kikyotani, S. 1983. Molecular structure of the nicotinic acetylcholine receptor. *Cold Spring Harb. Symp. Quant. Biol.* 48: 57–69. [5]

Numano, R., Yamazaki, S., Umeda, N., Samura, T., Sujino, M., Takahashi, R., Ueda, M., Mori, A., Yamada, K., Sakaki, Y., Inouye, S. T., Menaker, M., and Tei, H. 2006. Constitutive expression of the *Period1* gene impairs behavioral and molecular circadian rhythms. *Proc. Natl. Acad. Sci. USA* 103: 3716–3721. [17]

Nussinovitch, I., and Rahamimoff, R. 1988. Ionic basis of tetanic and post-tetanic potentiation at a mammalian neuromuscular junction. *J. Physiol.* 396: 435–455. [16]

O'Craven, K. M., Downing, P. E., and Kanwisher, N. 1999. fMRI evidence for objects as the units of attentional selection. *Nature* 401: 584–587. [23]

O'Donnell, P., and Grace, A. A. 1993. Dopaminergic modulation of dye coupling between neurons in the core and shell regions of the nucleus accumbens. *J. Neurosci.* 13: 3456–3471. [11]

O'Malley, J., Moore, C. T., and Salpeter, M. M. 1997. Stabilization of acetylcholine receptors by exogenous ATP and its reversal by cAMP and calcium. *J. Cell Biol.* 138: 159–165. [27]

O'Rourke, N. A., Sullivan, D. P., Kaznowski, C. E., Jacobs, A. A., and McConnell, S. K. 1995. Tangential migration of neurons in the developing cerebral cortex. *Development* 121: 2165–2176. [25]

O'Sullivan, B. P., and Freedman, S. D. 2009. Cystic fibrosis. *Lancet* 373: 1891–1904. [28]

Obaid, A. L., Nelson, M. E., Lindstrom, J., and Salzberg, B. M. 2005. Optical studies of nicotinic acetylcholine receptor subtypes in the guinea- pig enteric nervous system. *J. Exp. Biol.* 208: 2891–3001. [17]

Obata, K., Ito, M., Ochi, R., and Sato, N. 1967. Pharmacological properties of the postsynaptic inhibition by Purkinje cell axons and the action of gamma-aminobutyric acid on Deiters NEURONES. *Exp. Brain Res.* 4: 43–57. [14]

Ochoa, J., and Torebjork, E. 1983. Sensations evoked by intraneural microstimulation of single mechanoreceptor units innervating the human hand. *J. Physiol.* 342: 633–654. [21]

Ochoa, J., and Torebjork, E. 1989. Sensations evoked by intraneural microstimulation of C nociceptor fibres in human skin nerves. *J. Physiol.* 415: 583–599. [21]

Odette, L. L., and Newman, E. A. 1988. Model of potassium dynamics in the central nervous system. *Glia* 1: 198–210. [10]

Oheim, M., Kirchhoff, F., and Stühmer, W. 2006 Calcium microdomains in regulated exocytosis. *Cell Calcium* 40: 423–439. [10, 12]

Ohki, K., Chung, S., Ch'ng, Y. H., Kara, P., and Reid, R. C. 2005. Functional imaging with cellular resolution reveals precise micro-architecture in visual cortex. *Nature* 433: 597–603. [3]

Ohki, K., Chung, S., Kara, P., Hübener, M., Bonhoeffer, T., and Reid, R. C. 2006. Highly ordered arrangement of single neurons in orientation pinwheels. *Nature* 442: 925–928. [3]

Ohmori, H. 1985. Mechano-electrical transduction currents in isolated vestibular hair cells of the chick. *J. Physiol.* 359: 189–217. [19]

Ohno-Shosaku, T., Maejima, T., and Kano, A. 2001 Endogenous cannabinoids mediate retrograde signals from depolarized postsynaptic neurons to presynaptic terminals. *Neuron* 29: 729–738. [12]

Okada, K., Inoue, A., Okada, M., Murata, Y., Kakuta, S., Jigami, T., Kubo, S., Shiraishi, H., Eguchi, K., Motomura, M., Akiyama, T., Iwakura, Y., Higuchi, O., and Yamanashi, Y. 2006. The muscle protein Dok-7 is essential for neuromuscular synaptogenesis. *Science* 312: 1802–1805. [27]

Okada, Y., Miyamoto, T., and Sato, T. 1994. Activation of a cation conductance by acetic acid in taste cells isolated from the bullfrog. *J. Exp. Biol.* 187: 19–32. [19]

Okita, K., Ichisaka, T., and Yamanaka, S. 2007. Generation of germline-competent induced pluripotent stem cells. *Nature* 448: 313–317. [25]

Oldham, W. H., and Hamm, H. E. 2008. Heterotrimeric G protein activation by G-protein-coupled receptors. *Nat. Rev. Mol. Cell. Biol.* 459: 356–363. [12]

Oliet, S., Malenka, R. C., and Nicoll, R. A. 1996. Bidirectional control of quantal size by synaptic activity in the hippocampus. *Science* 271: 1294–1297. [16]

Oliver, D., Klöcker, N., Schuck, J., Baukrowitz, T., Ruppersberg, J. P., and Fakler, B. 2000. Gating of Ca^{2+}-activated K^+ channels controls fast inhibitory synaptic transmission at auditory outer hair cells. *Neuron* 26: 595–601. [12]

Olivera, B. M., Miljanich, G. P., Ramachandran, J., and Adams, M. E. 1994. Calcium channel diversity and neurotransmitter release: The omega-conotoxins and omega-agatoxins. *Annu. Rev. Biochem.* 63: 823–867. [13]

Olsen, R. W., and Sieghart, W. 2008. International Union of Pharmacology. LXX. Subtypes of γ-aminobutyric $acid_A$ receptors: classification on the basis of subunit composition, pharmacology, and function. Update. *Pharmacol. Rev.* 60: 243–260. [14]

Olsen, R. W., and Sieghart, W. 2009. GABAA receptors: Subtypes provide diversity of function and pharmacology. *Neuropharmacology* 56: 141–148. [5]

Olveczky, B. P., Baccus, S. A., and Meister, M. 2007. Retinal adaptation to object motion. *Neuron* 56: 689–700. [20]

Orban, G. A., Van Essen, D., and Vanduffel, W. 2004. Comparative mapping of higher visual areas in monkeys and humans. *Trends. Cogn. Sci.* 8: 315–324. [3]

Orkand, R. K., Nicholls, J. G., and Kuffler, S. W. 1966. Effect of nerve impulses on the membrane potential of glial cells in the central nervous system of amphibia. *J. Neurophysiol.* 29: 788–806. [10]

Otsuka, M., and Yoshioka, K. 1993. Neurotransmitter functions of mammalian tachykinins. *Physiol. Rev.* 73: 229–308. [14]

Otsuka, M., Iversen, L. L., Hall, Z. W., and Kravitz, E. A. 1966. Release of γ-aminobutyric acid from inhibitory nerves of lobster. *Proc. Natl. Acad. Sci. USA* 56: 1110–1115. [14]

Ottoson, D. 1956. Analysis of the electrical activity of the olfactory epithelium. *Acta Physiol. Scand.* 35: 1–83. [19]

Ouyang, K., Zheng, H., Qin, X., Zhang, C., Yang, D., Wang, X., Wu, C., Zhou, Z., and Cheng, H. 2005. Ca^{2+} sparks and secretion in dorsal root ganglion neurons. *Proc. Natl. Acad. Sci. USA* 102: 12259–12264. [12]

Oyster, C. W., and Barlow, H. B. 1967. Direction-selective units in rabbit retina: distribution of preferred directions. *Science* 155: 841–842. [2]

Pace, U., Hanski, E., Salomon, Y., and Lancet, D. 1985. Odorant-sensitive adenylate cyclase may mediate olfactory reception. *Nature* 316: 255–258. [19]

Pack, C. C., Livingstone, M. S., Duffy, K. R., and Born, R. T. 2003. End-stopping and the aperture problem: two-dimensional motion signals in macaque V1. *Neuron* 39: 671–680. [2]

Palacin, M., Estévez, R., Bertran, J., and Zorzano, A. 1998. Molecular biology of mammalian plasma membrane amino acid transporters. *Physiol. Rev.* 78: 969–1054. [9]

Palmada, M., and Centelles, J. J. 1998. Excitatory amino acid neurotransmission. Pathways for metabolism, storage and reuptake of glutamate in brain. *Front. Biosci.* 3: 701–718. [15]

Palmer, A. M. 2010. The role of the blood-CNS barrier in CNS disorders and their treatment. *Neurobiol. Dis.* 37: 3–12. [10]

Palmer, A. R., and Russell, I. J. 1986. Phase-locking in the cochlear nerve of the guinea-pig and its relation to the receptor potential of inner hair-cells. *Hear. Res.* 24: 1–15. [22]

Palmer, R. M. J., Ferrige, J., and Moncada, S. 1987. Nitric oxide release accounts for the biological activity of endothelium-derived relaxing factor. *Nature* 324: 524–526. [12]

Pandi-Perumal, S. R., Srinivasan, V., Maestroni, G. J., Cardinali, D. P., Poeggeler, B., and Hardeland, R. 2006. Melatonin: Nature's most versatile biological signal? *FEBS J.* 273: 2813–2838. [17]

Pang, J. J., Gao, F., Lem, J., Bramblett, D. E., Paul, D. L., and Wu, S. M. 2010. Direct rod input to cone BCs and direct cone input to rod BCs challenge the traditional view of mammalian BC circuitry. *Proc. Natl. Acad. Sci. USA* 107: 395–400. [20]

Pang, Z. P., and Südhoff, T. C. 2010. Cell biology of Ca^{2+}-triggered exocytosis. *Curr. Opin. Cell Biol.* 22: 496–505. [13]

Pangrsic, T., Potokar, M., Stenovec, M., Kreft, M., Fabbretti, E., Nistri, A., Pryazhnikov, E., Khiroug, L., Giniatullin, R., and Zorec, R. 2007. Exocytotic release of ATP from cultured astrocytes. *J. Biol. Chem.* 282: 28749–28758. [10]

Pankratov, Y., Lalo, U., Verkhratsky, A., and North, R. A. 2006. Vesicular release of ATP at central synapses. *Pflügers Arch.* 452: 589–597. [14]

Papardia, S., and Hardingham, G. E. 2007. The dichotomy of NMDA receptor signalling. *Neuroscientist* 13: 572–579. [14]

Papazian, D. M., Timpe, L. C., Yan, Y. N., and Yan, L. Y. 1987. Cloning and complementary DNA from *Shaker,* a putative potassium channel gene from *Drosophila*. *Science* 237: 749–753. [5]

Pare, M., Smith, A. M., and Rice, F. L. 2002. Distribution and terminal arborizations of cutaneous mechanoreceptors in the glabrous finger pads of the monkey. *J. Comp. Neurol.* 445: 347–359. [21]

Pareek, S., Notterpek, L., Snipes, G. J., Naef, R., Sossin, W., Laliberte, J., Iacampo, S., Suter, U., Shooter, E. M., and Murphy, R. A. 1997. Neurons promote the translocation of peripheral myelin protein 22 into myelin. *J. Neurosci.* 17: 7754–7762. [10]

Parekh, A. B. 2008. Ca^{2+} microdomains near plasma membrane Ca^{2+} channels: impact on cell function. *J. Physiol.* 586: 3043–3054. [12]

Parker, A. J., and Newsome, W. T. 1998. Sense and the single neuron: probing the physiology of perception. *Annu. Rev. Neurosci.* 21: 227–277. [23]

Parnas, H., and Parnas, I. 2007. The chemical synapse goes electric: Ca^{2+}- and voltage-sensitive GPCRs control neurotransmitter release. *Trends Neurosci.* 30: 54–61. [12]

Parnas, I., and Parnas, H. 2010 Control of neurotransmitter release: From Ca^{2+} to voltage dependent G-protein coupled receptors. *Pflügers Arch.* 460: 975–990. [13]

Parsons, S. M., Prior, C., and Marshall, I. G. 1993. Acetylcholine transport, storage, and release. *Int. Rev. Neurobiol.* 35: 279–390. [15]

Paschal, B. M., and Vallee, R. B. 1987. Retrograde transport by the microtubule associated protein MAP 1C. *Nature* 330: 181–183. [15]

Pastor, J., Soria, B., and Belmonte, C. 1996. Properties of the nociceptive neurons of the leech segmental ganglion. *J. Neurophysiol.* 75: 2268–2279. [18]

Pastrana, E. 2011. Optogenetics: controlling cell function with light. *Nat. Methods* 8: 24–25. [14]

Patapoutian, A., Peier, A. M., Story, G. M., and Viswanath, V. 2003. ThermoTRP channels and beyond: Mechanisms of temperature sensation. *Nat. Rev. Neurosci.* 4: 529–539. [19]

Paulson, O. B., and Newman, E. A. 1987. Does the release of potassium from astrocyte endfeet regulate cerebral blood flow? *Science* 237: 896–898. [10]

Pavlos, N. J., Grønborg, M., Riedel, D., Chua, J. J. E., Boykken, J., Kloepper, T. H., Urlaub, H., Rizzoli, S. O., and Jahn, R. 2010. Quantitative analysis of synaptic vesicle Rabs uncovers distinct yet overlapping roles for Rab3a and Rab27b in Ca^{2+}-triggered exocytosis. *J. Neurosci.* 30: 13441–13453. [13]

Payne, J. A., Rivera, C., Voipio, J., and Kaila, K. 2003. Cation-chloride co-transporters in neuronal communication, development and trauma. *Trends Neurosci.* 26: 199–206. [11]

Payton, W. B. 1981. History of medicinal leeching and early medical references. In Muller, K. J., Nicholls, J. G., and Stent, G. S. (eds.) *Neurobiology of the Leech*. Cold Spring Harbor Laboratory, Cold Spring Harbor, NY. pp. 27–34. [18]

Pearson, K. 1976. The control of walking. *Sci. Am.* 245: 72–86. [24]

Pearson, K. G. 2008. Role of sensory feedback in the control of stance duration in walking cats. *Brain Res. Rev.* 57: 222–227. [24]

Peinado, A., Juste, R., and Kayz, L. C. 1993. Extensive dye coupling between rat neocortical neurons during the period of circuit formation. *Neuron* 10: 103–114. [11]

Peirson, S. N., Halford, S., and Foster, R. G. 2009. The evolution of irradiance detection: melanopsin and the non-visual opsins. *Philos. Trans. R. Soc. Lond., B, Biol. Sci.* 364: 2849–2865. [20]

Penfield, W. 1932. *Cytology and Cellular Pathology of the Nervous System*, Vol. 2. Hafner, New York. [10]

Penfield, W., and Rasmussen, T. 1950. *The Cerebral Cortex of Man. A Clinical Study of Localization of Function*. Macmillan, New York. [24]

Penington, N. J., Kelly, J. S., and Fox, A. P. 1993. Whole-cell recordings of inwardly rectifying K^+ currents activated by 5-HT_{1A} receptors on dorsal raphe neurones of the adult rat. *J. Physiol.* 469: 387–405. [14]

Penner, R., and Neher, E. 1988. The role of calcium in stimulus-secretion coupling in excitable and non-excitable cells. *J. Exp. Biol.* 139: 329–345. [13]

Penner, R., and Neher, E. 1989. The patch-clamp technique in the study of secretion. *Trends Neurosci.* 12: 159–163. [13]

Peper, K., and McMahan, U. J. 1972. Distribution of acetylcholine receptors in the vicinity of nerve terminals on skeletal muscle of the frog. *Proc. R. Soc. Lond., B, Biol. Sci.* 181: 431–440. [11]

Peper, K., Dreyer, F., Sandri, C., Akert, K., and Moore, H. 1974. Structure and ultrastructure of the frog motor end-plate: A freeze-etching study. *Cell Tissue Res.* 149: 437–455. [13]

Pepperberg, D. R., Okajima, T. L., Wiggert, B., Ripps, H., Crouch, R. K., and Chader, G. J. 1993. Interphotoreceptor retinoid-binding protein (IRBP). Molecular biology and physiological role in the visual cycle of rhodopsin. *Mol, Neurobiol.* 7: 61–85. [20]

Perea, G., and Araque, A. 2010. Glia modulates synaptic transmission. *Brain Res. Rev.* 63: 93–102. [10]

Perea, G., Navarrete, M., and Araque, A. 2009. Tripartite synapses: Astrocytes process and control synaptic information. *Trends Neurosci.* 32: 421–431. [10]

Peretz, I. 2006. The nature of music from a biological perspective. *Cognition* 100: 1–32. [23]

Pernía-Andrade, A. J., Kato, A., Witschi, R., Nyilas, R., Katona, I., Freund, T. F., Watanabe, M., Filitz, J., Koppert, W., Schüttler, J., Ji, G., Neugebauer, V., Marsicano, G., Lutz, B., Vanegas, H., and Zeilhofer, H. U. 2009. Spinal endocannabinoids and CB1 receptors mediate C-fiber-induced heterosynaptic pain sensitization. *Science* 325: 760–764. [12]

Perrett, D. I., Rolls, E. T., and Caan, W. 1982. Visual neurones responsive to faces in the monkey temporal cortex. *Exp. Brain Res.* 47: 329–342. [23]

Perrett, D. I., Smith, P. A., Potter, D. D., Mistlin, A. J., Head, A. S., Milner, A. D., and Jeeves, M. A. 1984. Neurones responsive to faces in the temporal cortex: studies of functional organization, sensitivity to identity and relation to perception. *Hum. Neurobiol.* 3: 197–208. [23]

Perry, V. H., Nicoll, J. A., and Holmes, C. 2010. Microglia in neurodegenerative disease. *Nat. Rev. Neurol.* 6: 193–201. [10]

Pert, C. B., and Snyder, S. H. 1973. Opiate receptor: demonstration in nervous tissue. *Science* 179: 1011–1014. [21]

Peters, A., Palay, S. L., and Webster, H. de F. 1991. *The Fine Structure of the Nervous System: Neurons and Their Supporting Cells,* 3rd ed. Oxford University Press, New York. [10]

Peters, J. A., Hales, T. G., and Lambert, J. J. 2006. Molecular determinants of single-channel conductance and ion selectivity in the Cys-loop family: insights from the 5-HT_3 receptor. *Trends Pharmacol. Sci.* 26: 587–594. [5]

Petersen, R. S., and Diamond, M. E. 2000. Spatial–temporal distribution of whisker-evoked activity in rat somatosensory cortex and the coding of stimulus location. *J. Neurosci.* 20: 6135–6143. [21]

Petrovic, P., Dietrich, T., Fransson, P., Andersson, J., Carlsson, K., and Ingvar, M. 2005. Placebo in emotional processing—induced expectations of anxiety relief activate a generalized modulatory network. *Neuron* 46: 957–969. [21]

Pette, D. 2001. Historical perspectives: plasticity of mammalian skeletal muscle. *J. Appl. Physiol.* 90: 1119–1124. [27]

Pevsner, J., Reed, R. R., Feinstein, P. G., and Snyder, S. H. 1988. Molecular cloning of odorant-binding protein: member of a ligand carrier family. *Science* 241: 336–339. [19]

Pfaffinger, P. J., Martin, J. M., Hunter, D. D., Nathanson, N. M., and Hille, B. 1985. GTP-binding proteins couple cardiac muscarinic receptors to a K channel. *Nature* 317: 536–538. [12]

Pfrieger, F. W., and Barres, B. A. 1996. New views on synapse-glia interactions. *Curr. Opin. Neurobiol.* 6: 615–621. [10]

Phelan, K. A., and Hollyday, M. 1990. Axon guidance in muscleless chick wings: the role of muscle cells in motoneuronal pathway selection and muscle nerve formation. *J. Neurosci.* 10: 2699–2716. [25]

Phelan, P., Bacon, J. P., Davies, J. A., Stebbings, L. A., Todman, M. G., Avery, L., Baines, R. A., Barnes, T. M., Ford, C., Hekimi, S., Lee, R., Shaw, J. E., Starich, T. A., Curtin, K. D., Sun, Y. A., and Wyman, R. J. 1998. Innexins: a family of invertebrate gap-junction proteins. *Trends Genet.* 14: 348–349. [11]

Phelan, P., Goulding, L. A., Tam, J. L., Allen, M. J., Dawber, R. J., Davies, J. A., and Bacon, J. P. 2008. Molecular mechanism of rectification at identified electrical synapses in the *Drosophila* giant fiber system. *Curr. Biol.* 18: 1955–1960. [11]

Philipson, K. D., and Nicoll, D. A. 2000. Sodium–calcium exchange: A molecular perspective. *Annu. Rev. Physiol.* 62: 111–133. [9]

Pickles, J. O., Comis, S. D., and Osborne, M. P. 1984. Cross-links between stereocilia in the guinea pig organ of Corti, and their possible relation to sensory transduction. *Hear. Res.* 15: 103–112. [19]

Pierrot-Deseilligny, C. 2009. Effect of gravity on vertical eye position. *Ann. N Y Acad. Sci.* 1164: 155–165. [24]

Pietrobon, D. 2010. CaV2.1 channelopathies. *Pflügers Arch.* 460: 375–393. [28]

Pifferi, S., Boccaccio, A., and Menini, A. 2006. Cyclic nucleotide-gated ion channels in sensory transduction. *FEBS Lett.* 580: 2853–2859. [19]

Pifferi, S., Dibattista, M., Sagheddu, C., Boccaccia, A., Al Qteishat, A., Ghirardi, F., Tirindelli, R., and Menini, A. 2009. Calcium-activated chloride currents in olfactory sensory neurons from mice lacking bestrophin-2. *J. Physiol.* 587: 4265–4279. [19]

Piggins, H. D., and London, A. 2005. Circadian Biology: Clocks within Clocks. *Curr. Biol.* 15: 455–457. [17]

Piñeyro, G., and Blier, P. 1999. Autoregulation of serotonin neurons: role in antidepressant actions. *Pharmacol. Rev.* 51: 533–591. [14]

Pinheiro, P. S., and Mulle, C. 2008. Presynaptic glutamate receptors: physiological functions and mechanisms of action. *Nat. Rev. Neurosci.* 9: 423–436. [11]

Piomelli, D. 2001. The ligand that came from within. *Trends Neurosci.* 22: 17–19. [12]

Piomelli, D., Volterra, A., Dale, N., Siegelbaum, S. A., Kandel, E. R., Schwartz, J. H., and Belardetti, F. 1987. Lipoxygenase metabolites of arachidonic acid as second messengers for presynaptic inhibition of Aplysia sensory cells. *Nature* 328: 38–43. [12]

Pitler, T. A., and Alger, B. E. 1992. Postsynaptic spike firing reduces synaptic $GABA_A$ responses in hippocampal pyramidal cells. *J. Neurosci.* 12: 4122–4132. [12]

Pizzorusso, T., Fagiolini, M., Fabris, M., Ferrari, G., and Maffei, L. 1994. Schwann cells transplanted in the lateral ventricles prevent the functional and anatomical effects of monocular deprivation in the rat. *Proc. Natl. Acad. Sci. USA* 91: 2572–2576. [26]

Pizzorusso, T., Medini, P., Berardi, N., Chierzi, S., Fawcett, J. W., and Maffei, L. 2002. Reactivation of ocular dominance plasticity in the adult visual cortex. *Science* 298: 1248–1251. [26]

Placzek, M., Yamada, T., Tessier-Lavigne, M., Jessell, T., and Dodd, J. 1991. Control of dorsoventral pattern in vertebrate neural development: induction and polarizing properties of the floor plate. *Development* 113(Suppl. 2): 105–122. [25]

Plotkin, M. D., Kaplan, M. R., Peterson, L. N., Gullans, S. R., Hebert, S. C., and Delpire, E. 1997. Expression of the Na^+-K^+-$2Cl^-$ cotransporter BSC2 in the nervous system. *Am. J. Physiol. Cell Physiol.* 272: C173–C183. [9]

Polley, D. B., Steinberg, E. E., and Merzenich, M. M. 2006. Perceptual learning directs auditory cortical map reorganization through top-down influences. *J. Neurosci.* 26: 4970–4982. [26]

Pollo, A., Amanzio, M., Arslanian, A., Casadio, C., Maggi, G., and Benedetti, F. 2001. Response expectancies in placebo analgesia and their clinical relevance. *Pain* 93: 77–84. [21]

Poritsky, R. 1969. Two- and three-dimensional ultrastructure of boutons and glial cells on the motoneural surface in the cat spinal cord. *J. Comp. Neurol.* 135: 423–452. [1]

Porro, C. A., Francescato, M. P., Cettolo, V., Diamond, M. E., Baraldi, P., Zuiani, C., Bazzocchi, M., and di Prampero, P. E. 1996. Primary motor and sensory cortex activation during motor performance and motor imagery: a functional magnetic resonance imaging study. *J. Neurosci.* 16: 7688–7698. [24]

Port, F., and Basler, K. 2010. Wnt trafficking: new insights into Wnt maturation, secretion and spreading. *Traffic* 11: 1265–1271. [25]

Porter, C. W., and Barnard, E. A. 1975. The density of cholinergic receptors at the postsynaptic membrane: Ultrastructural studies in two mammalian species. *J. Membr. Biol.* 20: 31–49. [13]

Potter, L. T. 1970. Synthesis, storage, and release of [^{14}C]acetylcholine in isolated rat diaphragm muscles. *J. Physiol.* 206: 145–166. [13, 15]

Pouille, F., and Scanziani, M. 2001. Enforcement of temporal fidelity in pyramidal cells by somatic feed-forward inhibition. *Science* 293: 1159–1163. [14]

Powell, T. P., and Mountcastle, V. B. 1959 Some aspects of the functional organization of the cortex of the postcentral gyrus of the monkey: a correlation of findings obtained in a single unit analysis with cytoarchitecture. *Bull. Johns Hopkins Hosp.* 105: 133–162. [21, 23]

Prado, M. A., Reis, R. A., Prado, V. F., de Mello, M. C., Gomez, M. V., and de Mello, F. G. 2002. Regulation of acetylcholine synthesis and storage. *Neurochem. Int.* 41: 291–299. [15]

Prather, J. F., Peters, S., Nowicki, S., and Mooney, R. 2010. Persistent representation of juvenile experience in the adult songbird brain. *J. Neurosci.* 30: 10586–10598. [26]

Preuss, T. M., and Coleman, G. Q. 2002. Human-specific organization of primary visual cortex: alternating compartments of dense Cat-301 and calbindin immunoreactivity in layer 4A. *Cereb. Cortex* 12: 671–691. [3]

Preuss, T. M., Stepniewska, I., and Kaas, J. H. 1996. Movement representation in the dorsal and ventral premotor areas of owl monkeys: A microstimulation study. *J. Comp. Neurol.* 371: 649–676. [24]

Price, D. J., Jarman, A. P., Mason, J. O., and Kind, P. C. 2011. *Building Brains: An Introduction to Neural Development*, Wiley Blackwell, Oxford, UK. [25]

Price, G. D., and Trussell, L. O. 2006. Estimate of the chloride concentration in a central glutamatergic terminal: a gramicidin perforated-patch study on the calyx of Held. *J. Neurosci.* 26: 11432–11436. [11]

Priebe, N. J., and Ferster, D. 2008. Inhibition, spike threshold, and stimulus selectivity in primary visual cortex. *Neuron* 57: 482–497. [2]

Prinz, A. A., Bucher, D., and Marder, E. 2004. Similar network activity from disparate circuit parameters. *Nat. Neurosci.* 7: 1345–1352. [18]

Provencio, I., Jiang, G., De Grip, W. J., Hayes, W. P., and Rollag, M. D. 1998. Melanopsin: an opsin in melanophores, brain, and eye. *Proc. Natl. Acad. Sci. USA* 95: 340–345. [20]

Prud'homme, M. J., Houdeau, E., Serghini, R., Tillet, Y., Schemann, M., and Rousseau, J. P. 1999. Small intensely fluorescent cells of the rat paracervical ganglion synthesize adrenaline, receive afferent innervation from postganglionic cholinergic neurones, and contain muscarinic receptors. *Brain Res.* 821: 141–149. [17]

Pun, R. Y. K., and Lecar, H. 2001. Patch clamp techniques and analysis. In: N. Sperelakis (ed.), *Cell Physiology Source Book*, 3rd ed. Academic Press, San Diego. pp. 441–453. [4]

Purali, N. 2005. Structure and function relationship in the abdominal stretch receptor organs of the crayfish. *J. Comp. Neurol.* 488: 369–383. [19]

Purves, D. 1975. Functional and structural changes in mammalian sympathetic neurones following interruption of their axons. *J. Physiol.* 252: 429–463. [27]

Purves, D., and Sakmann, B. 1974. Membrane properties underlying spontaneous activity of denervated muscle fibers. *J. Physiol.* 239: 125–153. [27]

Purves, D., Augustine, G. J., Fitzpatrick, D. Hall, W. C., LaMantia, A. S., McNamara, J. O., and White, L. E. 2008. *Neuroscience*, 4th ed. Sinauer Associates, Sunderland MA. [8]

Pusch, M., and Jentsch, T. J. 2005. Unique structure and function of chloride transporting CLC proteins. *IEEE Trans. Nanobiosci.* 4: 49–57. [5]

Puzzolo, E., and Mallamaci, A. 2010. Cortico-cerebral histogenesis in the opossum *Monodelphis domestica*: generation of a hexalaminar neocortex in the absence of a basal proliferative compartment. *Neural Development*. 5: 8. [1, 25]

Qian, H., Malchow, R. P., Chappell, R. L., and Ripps, H. 1996. Zinc enhances ionic currents induced in skate Müller (glial) cells by the inhibitory neurotransmitter GABA. *Proc. R. Soc. Lond., B, Biol. Sci.* 263: 791–796. [10]

Quilliam, T. A., and Armstrong, J. 1963. Mechanoreceptors. *Endeavour* 22: 55–60. [19]

Quiroga, R. Q., Reddy, L., Kreiman, G., Koch, C., and Fried, I. 2005. Invariant visual representation by single neurons in the human brain. *Nature* 435: 1102–1107. [23]

Raff, M. 1996. Neural development: Mysterious no more? *Science* 274: 1063. [28]

Raftery, M. A., Hunkapiller, M. W., Strader, C. D., and Hood, L. E. 1980. Acetylcholine receptor: Complex of homologous subunits. *Science* 208: 1454–1457. [5]

Raggenbass, M. 2001. Vasopressin- and oxytocin-induced activity in the central nervous system: electrophysiological studies using in-vitro systems. *Prog. Neurobiol.* 64: 307–326. [14]

Rahmouni, K., Haynes, W. G., and Mark, A. L. 2004. In D. Robertson (ed.), *Primer on the Autonomic Nervous system*. Academic Press, London. pp 86–89. [17]

Raisman, G. 2007. Repair of spinal cord injury by transplantation of olfactory ensheathing cells. *C. R. Biol.* 330: 557–560. [27]

Rajan, R. 1995. Frequency and loss dependence of the protective effects of the olivocochlear pathways in cats. *J. Neurophysiol.* 74: 598–615. [22]

Rakic P. 2003. Elusive radial glial cells: Historical and evolutionary perspective. *Glia* 43: 19–32. [10]

Rakic, P. 1974. Neurons in rhesus monkey visual cortex: systematic relation between time of origin and eventual disposition. *Science* 183: 425–427. [25]

Rakic, P. 1977. Prenatal development of the visual system in rhesus monkey. *Philos. Trans. R. Soc. Lond., B, Biol. Sci.* 278: 245–260. [26]

Rakic, P. 1981. Neuronal-glial interaction during brain development. *Trends Neurosci.* 4: 184–187. [10, 25]

Rakic, P. 2002. Adult neurogenesis in mammals: an identity crisis. *J. Neurosci.* 22: 614–618. [25]

Rakic, P. 2003. Developmental and evolutionary adaptations of cortical radial glia. *Cereb. Cortex.* 13: 541–549. [10]

Rakic, P. 2006a. A century of progress in corticoneurogenesis: from silver impregnation to genetic engineering. *Cereb. Cortex.* 16(Suppl. 1): i3–i17. [26]

Rakic, P. 2006b. Neuroscience. No more cortical neurons for you. *Science* 313: 928–929. [25]

Ralph, M. R., Foster, R., Davis, F. C., and Menaker, M. 1990. Transplanted suprachiasmatic nucleus determines circadian period. *Science* 247: 975–978. [17]

Ramachandran, R., Davis, K. A., and May, B. J. 1999. Single-unit responses in the inferior colliculus of decerebrate cats. I. Classification based on frequency response maps. *J. Neurophysiol.* 82: 152–163. [22]

Raman, I. M., Sprunger, L. K., Meisler, M. H., and Bean, B. P. 1997. Altered subthreshold sodium currents and disrupted firing patterns in Purkinje neurons of *Scn8a* mutant mice. *Neuron* 19: 881–891. [7]

Ramanathan, K., Michael, T. H., Jiang, G. J., Hiel, H., and Fuchs, P. A. 1999. A molecular mechanism for electrical tuning of cochlear hair cells. *Science* 283: 215–217. [22]

Ramón y Cajal, S. [1909–1911] 1995. *Histology of the Nervous System*, 2 vols. Translated by Neely Swanson and Larry Swanson. Oxford University Press, New York. [1, 10, 24]

Ramón y Cajal, S. 1955. *Histologie du Système Nerveux*, vol. 2. C.S.I.C., Madrid. [3]

Randall, A., and Tsien, R. W. 1995. Pharmacological dissection of multiple types of calcium channel currents in rat cerebellar granule neurons. *J. Neurosci.* 15: 2995–3012. [5]

Randlett, O., Norden, C., and Harris, W. A. 2011. The vertebrate retina: A model for neuronal polarization *in vivo*. *Dev. Neurobiol.* 71: 567–583. [25]

Randolph, M., and Semmes, J. 1974. Behavioral consequences of selective subtotal ablations in the postcentral gyrus of *Macaca mulatta*. *Brain Res.* 70: 55–70. [21]

Rang, H. P. 1981. The characteristics of synaptic currents and responses to acetylcholine of rat submandibular ganglion cells. *J. Physiol.* 311: 23–55. [15]

Ranganathan, R., Cannon, S. C., and Horvitz, H. R. 2000. MOD-1 is a serotonin-gated chloride channel that modulates locomotory behaviour in *C. elegans*. *Nature* 408: 470–475. [5]

Ransohoff, R. M., and Perry, V. H. 2009. Microglial physiology: Unique stimuli, specialized responses. *Annu. Rev. Immunol.* 27: 119–145. [10]

Ransom, B. R., and Goldring, S. 1973. Slow depolarization in cells presumed to be glia in cerebral cortex of cat. *J. Neurophysiol.* 36: 869–878. [10]

Ransom, B. R., and Sontheimer, H. 1992. The neurophysiology of glial cells. *J. Clin. Neurophysiol.* 9: 224–251. [10]

Rasband, M. N., Trimmer, J. S., Schwartz, T. L., Levinson, S. R., Ellisman, M. H., Schachner, M., and Shrager, P. 1998. Potassium channel distribution, clustering, and function in remyelinating rat axons. *J. Neurosci.* 18: 36–47. [8]

Rasmussen, G. 1946. The olivary peduncle and other fiber projections of the superior olivary complex. *J. Comp. Neurol.* 84: 141–219. [22]

Rav-Acha, M., Sagiv, N., Segev, I., Bergman, H., and Yarom, Y. 2005. Dynamic and spatial features of the inhibitory pallidal GABAergic synapses. *Neuroscience* 135: 791–802. [24]

Raveh, A., Riven, I., and Reuvenny, E. 2009. Elucidation of the gating of the GIRK channel using a spectroscopic approach. *J. Physiol.* 587: 5331–5335. [12]

Raymond, C. R. 2007. LTP forms 1, 2, and 3: Different mechanisms for the "long" in long-term potentiation. *Trends Neurosci.* 30: 167–175. [16]

Raymond, C. R., and Redman S. J. 2006. Spatial segregation of neuronal calcium signals encodes different forms of LTP in rat hippocampus. *J. Physiol.* 570: 97–111. [16]

Razak, K. A., Richardson, M. D., and Fuzessery, Z. M. 2008. Experience is required for the maintenance and refinement of FM sweep selectivity in the developing auditory cortex. *Proc. Natl. Acad. Sci. USA* 105: 4465–4470. [26]

Ready, D. F., Hanson, T. E., and Benzer, S. 1976. Development of the *Drosophila* retina, a neurocrystalline lattice. *Dev. Biol.* 53: 217–240. [25]

Reale, R. A., and Imig, T. J. 1980. Tonotopic organization in auditory cortex of the cat. *J. Comp. Neurol.* 192: 265–291. [22]

Rebsam, A., Seif, I., and Gaspar, P. 2005. Dissociating barrel development and lesion-induced plasticity in the mouse somatosensory cortex. *J. Neurosci.* 25: 706–710. [26]

Recanzone, G. H., Merzenich, M. M., and Schreiner, C. E. 1992. Changes in the distributed temporal response properties of S. I. cortical neurons reflect improvements in performance on a temporally based tactile discrimination task. *J. Neurophysiol.* 67: 1071–1091. [21]

Recanzone, G. H., Merzenich, M. M., Jenkins, W. M., Grajski, K. A., and Dinse, H. R. 1992. Topographic reorganization of the hand representation in cortical area 3b owl monkeys trained in a frequency-discrimination task. *J. Neurophysiol.* 67: 1031–1056. [21]

Redfern, P. A. 1970. Neuromuscular transmission in new-born rats. *J. Physiol.* 209: 701–709. [25]

Redman, S. 1990. Quantal analysis of synaptic potentials in neurons of the central nervous system. *Physiol. Rev.* 70: 165–198. [13]

Reed, R. R. 2004. After the Holy Grail: establishing a molecular basis for mammalian olfaction. *Cell* 116: 329–336. [19]

Reger, J. F. 1958. The fine structure of neuromuscular synapses of gastrocnemii from mouse and frog. *Anat. Rec.* 130: 7–23. [13]

Reichardt, L. F. 2006. Neurotrophin-regulated signalling pathways. *Philos. Trans. R. Soc. Lond., B, Biol. Sci.* 361: 1545–1564. [25]

Reid, G. 2005. ThermoTRP channels and cold sensing: what are they really up to? *Pflügers Arch.* 451: 250–263. [19]

Reinscheid, R. K., Nothacker, H. P., Bourson, A., Ardati, A., Henningsen, R. A., Bunzow, J. R., Grandy, D. K., Langen, H., Monsma, F. J. Jr., and Civelli, O. 1995. Orphanin FQ: a neuropeptide that activates an opioidlike G protein-coupled receptor. *Science* 270: 792–794. [14]

Reiser, G., and Miledi, R. 1988. Characteristics of Schwann-cell miniature end-plate currents in denervated frog muscle. *Pflügers Arch.* 412: 22–28. [10]

Reiser, G., and Miledi, R. 1989. Changes in the properties of synaptic channels opened by acetylcholine in denervated frog muscle. *Brain Res.* 479: 83–97. [13]

Reist, N. E., Werle, M. J., and McMahan, U. J. 1992. Agrin released by motor neurons induces the aggregation of acetylcholine receptors at neuromuscular junctions. *Neuron* 8: 865–868. [25]

Reiter, H. O., Waitzman, D. M., and Stryker, M. P. 1986. Cortical activity blockade prevents ocular dominance plasticity in the kitten visual cortex. *Exp. Brain Res.* 65: 182–188. [26]

Ress, D., Backus, B. T., and Heeger, D. J. 2000. Activity in primary visual cortex predicts performance in a visual detection task. *Nat. Neurosci.* 3: 940–945. [23]

Ressler, K. J., Sullivan, S. L., and Buck, L. B. 1993. A zonal organization of odorant receptor gene expression in the olfactory epithelium. *Cell* 73: 597–609. [19]

Restrepo, D., Miyamoto, T., Bryant, B. P., and Teeter, J. H. 1990. Odor stimuli trigger influx of calcium into olfactory neurons of the channel catfish. *Science* 249: 1166–1168. [19]

Reuter, H. 1974. Localization of β adrenergic receptors, and effects of noradrenaline and cyclic nucleotides on action potentials, ionic currents and tension in mammalian cardiac muscle. *J. Physiol.* 242: 429–451. [12]

Reuter, H., Cachelin, A. B., DePeyer, J. E., and Kokubun, S. 1983. Modulation of calcium channels in cultured cardiac cells by isoproternenol and 8-bromo-cAMP. *Cold Spring Harb. Symp. Quant. Biol.* 48: 193–200. [12]

Reuveny, E., Slesinger, P. A., Inglese, J., Morales, J. M., Iñiguez-Lluhi, J. A., Lefkowitz, R. J., Bourne, H. R., Jan, Y. N., and Jan, L. Y. 1994. Activation of the cloned muscarinic potassium channel by G protein βγ subunits. *Nature* 370: 143–146. [12]

Reynolds, B. A., and Weiss, S. 1996. Clonal and population analyses demonstrate that an EGF-responsive mammalian embryonic CNS precursor is a stem cell. *Dev. Biol.* 175: 1–13. [25]

Rhinn, M., Lun, K., Luz, M., Werner, M., and Brand, M. 2005. Positioning of the midbrain-hindbrain boundary organizer through global posteriorization of the neuroectoderm mediated by Wnt8 signaling. *Development* 132: 1261–1272. [25]

Ribchester, R. R., and Taxt, T. 1983. Motor unit size and synaptic competition in rat lumbrical muscles reinnervated by active and inactive motor axons. *J. Physiol.* 344: 89–111. [25]

Ricci, A. J., and Fettiplace, R. 1997. The effects of calcium buffering and cyclic AMP on mechano-electrical transduction in turtle auditory hair cells. *J. Physiol.* 501(Pt 1): 111–124. [19]

Rice, F. L., Mance, A., and Munger, B. L. 1986. A comparative light microscopic analysis of the sensory innervation of the mystacial pad. I. Innervation of vibrissal follicle-sinus complexes. *J. Comp. Neurol.* 252: 154–174. [21]

Rice, J. J., May, B. J., Spirou, G. A., Young, E. D. 1992. Pinna-based spectral cues for sound localization in cat. *Hear. Res.* 58: 132–152. [22]

Richards, D. A. 2009. Vesicular release mode shapes the postsynaptic response at hippocampal synapses. *J. Physiol.* 587: 5073–5080. [13]

Richards, D. A., Gautimosime, D., Rizzoli, S., and Betz, W. J. 2003. Synaptic vesicle pools at the frog neuromuscular junction. *Neuron* 39: 529–541. [13]

Richardson, P. M., McGuinness, U. M., and Aguayo, A. J. 1980. Axons from CNS neurones regenerate into PNS grafts. *Nature* 284: 264–265. [27]

Rieke, F., and Baylor, D. A. 1998. Origin of reproducibility in the responses of retinal rods to single photons. *Biophys. J.* 75: 1836–1857. [20]

Riesen, A. H., and Aarons, L. 1959. Visual movement and intensity discrimination in cats after early deprivation of pattern vision. *J. Comp. Physiol. Psychol.* 52: 142–149. [26]

Rijntjes, M., Dettmers, C., Buchel, C., Kiebel, S., Frackowiak, R. S., and Weiller, C. 1999. A blueprint for movement: Functional and anatomical representations in the human motor system. *J. Neurosci.* 19: 8043–8048. [24]

Rimer, M. 2010. Modulation of agrin-induced acetylcholine receptor clustering by extracellular signal-regulated kinases 1 and 2 in cultured myotubes. *J. Biol. Chem.* 285: 32370–32377. [27]

Ritchie, J. M. 1987. Voltage-gated cation and anion channels in mammalian Schwann cells and astrocytes. *J. Physiol.* (*Paris*) 82: 248–257. [10]

Ritchie, J. M., Black, J. A., Waxman, S. G., and Angelides, K. J. 1990. Sodium channels in the cytoplasm of Schwann cells. *Proc. Natl. Acad. Sci. USA.* 87: 9290–9294. [10]

Rivera, C., Voipio, J., Payne, J. A., Ruusuvuori, E., Lahtinen, H., Lamsa, K., Pirvola, U., Saarma, M., and Kaila, K. 1999. The K^+/Cl^- co-transporter KCC2 renders GABA hyperpolarizing during neuronal maturation. *Nature.* 397: 251–255. [9]

Riviere, S., Challet, L., Fluegge, D., Spehr, M., and Rodriguez, I. 2009. Formyl peptide receptor-like proteins are a novel family of vomeronasal chemosensors. *Nature* 459: 574–577. [19]

Rizzolatti, G., and Wolpert, D. M. 2005. Motor systems. *Curr. Opin. Neurobiol.* 15: 624–625. [24]

Rizzuto, R., and Pozzan, T. 2006. Microdomains of intracellular Ca^{2+}: Molecular determinants and functional consequences. *Physiol. Rev.* 86: 369–408. [9]

Roberts, A., and Bush, B. M. 1971. Coxal muscle receptors in the crab: the receptor current and some properties of the receptor nerve fibres. *J. Exp. Biol.* 54: 515–524. [19]

Roberts, E., and Frankel, S. 1950. γ-Aminobutyric acid in brain: its formation from glutamic acid. *J. Biol. Chem.* 187: 55–63. [14]

Robertson, D., ed. 2004. *Primer on the Autonomic Nervous system.* Academic Press, London. [17]

Robertson, S. J., and Edwards, F. A. 1998. ATP and glutamate are released from separate neurones in the rat medial habenula nucleus: frequency dependence and adenosine-mediated inhibition of release. *J. Physiol.* 508: 691–701. [14]

Robinson, D. A., and Fuchs, A. F. 1969. Eye movement evoked by stimulation of frontal eye fields. *J. Neurophysiol.* 32: 637–648. [23]

Robitaille, R., Adler, E. M., and Charlton, M. P. 1990. Strategic location of calcium channels at release sites of frog neuromuscular synapses. *Neuron* 5: 773–779. [13]

Rodieck, R. W. 1989. Starburst amacrine cells of the primate retina. *J. Comp. Neurol.* 285: 18–37. [20]

Rodieck, R. W., and Marshak, D. W. 1992. Spatial density and distribution of choline acetyltransferase immunoreactive cells in human, macaque, and baboon retinas. *J. Comp. Neurol.* 321: 46–64. [20]

Rodriguez, M. J., Perez-Etchegoyen, C. B., and Szczupak, L. 2009. Premotor nonspiking neurons regulate coupling among motoneurons that innervate overlapping muscle fiber population. *J. Comp. Physiol. A* 195: 491–500. [18]

Roehm, P. C., Xu, N., Woodson, E. A., Green, S. H., and Hansen, M. R. 2008. Membrane depolarization inhibits spiral ganglion neurite growth via activation of multiple types of voltage sensitive calcium channels and calpain. *Mol. Cell Neurosci.* 37: 376–387. [25]

Roelink, H., Porter, J. A., Chiang, C., Tanabe, Y., Chang, D. T., Beachy, P. A., and Jessell, T. M. 1995. Floor plate and motor neuron induction by different concentrations of the amino-terminal cleavage product of Sonic hedgehog autoproteolysis. *Cell* 81: 445–455. [25]

Rogan, M. T., Staubil, U. V., and LeDoux, J. E. 1997. Fear conditioning induces associative long-term potentiation in the amygdala. *Nature* 390: 604–607. [16]

Rogers, M., and Sargent, P. B. 2003. Rapid activation of presynaptic nicotinic acetylcholine receptors by nerve-released transmitter. *Eur. J. Neurosci.* 18: 2946–2956. [17]

Rojas, L., and Orkand, R. K. 1999. K^+ channel density increases selectively in the endfoot of retinal glial cells during development of Rana catesbiana. *Glia* 25: 199–203. [10]

Rokni, D., Llinas, R., and Yarom, Y. 2008. The morpho/functional discrepancy in the cerebellar cortex: looks alone are deceptive. *Front. Syst. Neurosci.* 2: 192–198. [24]

Role, L. W., and Berg, D. K. 1996. Nicotinic receptors in the development and modulation of CNS synapses. *Neuron* 16: 1077–1085. [14]

Rollema, H., Coe, J. W., Chambers, L. K., Hurst, R. S., Stahl, S. M., and Williams, K. E. 2007. Rationale, pharmacology and clinical efficacy of partial agonists of $\alpha_4\beta_2$ nACh receptors for smoking cessation. *Trends Pharmacol. Sci.* 28: 316–325. [14]

Rolls, A., Shechter, R., and Schwartz, M. 2009. The bright side of the glial scar in CNS repair. *Nat. Rev. Neurosci.* 10: 235–241. [27]

Rolls, E. T. 1984. Neurons in the cortex of the temporal lobe and in the amygdala of the monkey with responses selective for faces. *Hum. Neurobiol.* 3: 209–222. [23]

Romanski, L. M., and Averbeck, B. B. 2009. The primate cortical auditory system and neural representation of conspecific vocalizations. *Annu. Rev. Neurosci.* 32: 315–346. [22]

Romero, M. F., Fulton, C. M., and Boron, W. F. 2004. The SLC4 family of HCO_3^- transporters. *Pflügers Arch.* 447: 495–509. [9]

Romo, R., and Salinas, E. 2003. Flutter discrimination: neural codes, perception, memory and decision making. *Nat. Rev. Neurosci.* 4: 203–218. [23]

Romo, R., Hernández, A., and Zainos, A. 2004. Neuronal correlates of a perceptual decision in ventral premotor cortex. *Neuron* 41: 165–173. [23, 24]

Romo, R., Hernández, A., Zainos, A., Brody, C. D., and Lemus, L. 2000. Sensing without touching: psychophysical performance based on cortical microstimulation. *Neuron* 26: 273–278. [23]

Romo, R., Merchant, H., Zainos, A., and Hernández, A. 1997. Categorical perception of somesthetic stimuli: psychophysical measurements correlated with neuronal events in primate medial premotor cortex. *Cereb. Cortex* 7: 317–326. [23]

Roper, J., and Schwarz, J. R. 1989. Heterogeneous distribution fast and slow potassium channels in myelinated rat nerve. *J. Physiol.* 416: 93–110. [8]

Roper, S. D. 2007. Signal transduction and information processing in mammalian taste buds. *Pflügers Arch.* 454: 759–776. [19]

Rose, C. R., Ransom, B. R., and Waxman, S. G. 1997. Pharmacological characterization of Na^+ influx via voltage-gated Na^+ channels in spinal cord astrocytes. *J. Neurophysiol.* 78: 3249–3258. [10]

Rosenbaum, D. M., Rasmussen, S. G., and Kobilka, B. K. 2009. The structure and function of G-protein-coupled receptors. *Nature* 459: 356–363. [12]

Rosenblatt, K. P., Sun, Z. P., Heller, S., and Hudspeth, A. J. 1997. Distribution of Ca^{2+}-activated K^+ channel isoforms along the tonotopic gradient of the chicken's cochlea. *Neuron* 19: 1061–1075. [22]

Rosenthal, J. L. 1969. Post-tetanic potentiation at the neuromuscular junction of the frog. *J. Physiol.* 203: 121–133. [16]

Roska, B., and Werblin, F. 2003. Rapid global shifts in natural scenes block spiking in specific ganglion cell types. *Nat. Neurosci.* 6: 600–608. [20]

Ross, C. A., and Tabrizi, S. J. 2011. Huntington's disease: from molecular pathogenesis to clinical treatment. *Lancet Neurol.* 10: 83–98. [28]

Ross, W. N., and Werman, R. 1987. Mapping calcium transients in the dendrites of Purkinje cells form the guinea pig cerebellum in vitro. *J. Physiol.* 389: 319–336. [16]

Ross, W. N., Arechiga, H., and Nicholls, J. G. 1988. Influence of substrate on the distribution of calcium channels in identified leech neurons in culture. *Proc. Natl. Acad. Sci. USA* 85: 4075–4078. [12]

Ross, W. N., Lasser-Ross, N., and Werman, R. 1990. Spatial and temporal analysis of calcium-dependent electrical activity in guinea pig Purkinje cell dendrites. *Proc. R. Soc. Lond., B, Biol. Sci.* 240: 173–185. [7]

Rossant, J. 1985. Interspecific cell markers and lineage in mammals. *Philos. Trans. R. Soc. Lond., B, Biol. Sci.* 312: 91–100. [25]

Rossi, D. J., Oshima, T., and Attwell, D. 2000. Glutamate release in severe brain ischemia is mainly by reversed uptake. *Nature.* 403: 316–321. [9]

Rossi, S. L., and Keirstead, H. S. 2009. Stem cells and spinal cord regeneration. *Curr. Opin. Biotechnol.* 20: 552–562. [27]

Rothwell, J. C., Traub, M. M., Day, B. L., Obeso, J. A., Thomas, P. K., and Marsden, C. D. 1982. Manual motor performance in a deafferented man. *Brain* 105: 515–542. [24]

Rotshenker, S. 1988. Multiple modes and sites for the induction of axonal growth. *Trends Neurosci.* 11: 363–366. [27]

Rotshenker, S. 2009 The role of Galectin-3/MAC-2 in the activation of the innate-immune function of phagocytosis in microglia in injury and disease. *J. Mol. Neurosci.* 39: 99–103. [10, 27]

Rouiller, E. M., Moret, V., Tanne, J., and Boussaoud, D. 1996. Evidence for direct connections between the hand region of the supplementary motor area and motoneurons in the macaque monkey. *Eur. J. Neurosci.* 8: 1055–1059. [24]

Roux, I., Safieddine, S., Nouvian, R., Grati, M., Simmler, M. C., Bahloul, A., Perfettini, I., Le Gall, M., Rostaing, P., Hamard, G., Triller, A., Avan, P., Moser, T., and Petit, C. 2006. Otoferlin, defective in a human deafness form, is essential for exocytosis at the auditory ribbon synapse. *Cell* 127: 277–289. [22]

Rovainen, C. M. 1967. Physiological and anatomical studies on large neurons of the central nervous system of the sea lamprey (*Petromyzon marinus*). II. Dorsal cells and giant interneurons. *J. Neurophysiol.* 30: 1024–1042. [11]

Roy, S., Jayakumar, J., Martin, P. R., Dreher, B., Saalmann, Y. B., Hu, D., and Vidyasagar, T. R. 2009. Segregation of short-wavelength-sensitive (S) cone signals in the macaque dorsal lateral geniculate nucleus. *Eur. J. Neurosci.* 30: 1517–1526. [3]

Rubin, E. 1915. *Synsoplevede figurer, studier i psykologisk analyse.* København og Kristiania, Gyldendal, Nordisk forlag. [23]

Rudolf, R., Mongillo, M., Rizutto, R., and Pozzan, T. 2003. Looking forward to seeing calcium. *Nat. Rev. Mol. Cell Biol.* 4: 579–586. [13]

Rudomin, P. 2009. In search of lost presynaptic inhibition. *Exp. Brain Res.* 196: 139–151. [11]

Rueter, S. M., Burns, C. M., Coode, S. A., Mookherjee, P., and Emeson, R. B. 1995. Glutamate receptor RNA editing in vitro by enzymatic conversion of adenosine to inosine. *Science* 267: 1491–1494. [25]

Ruiz, M. L., and Karpen, J. W. 1997. Single cyclic nucleotide-gated channels locked in different ligand-bound states. *Nature* 389: 389–392. [20]

Russell, I. J., and Murugasu, E. 1997. Medial efferent inhibition suppresses basilar membrane responses to near characteristic frequency tones of moderate to high intensities. *J. Acoust. Soc. Am.* 102: 1734–1738. [22]

Russell, J. M. 2000. Sodium–potassium–chloride cotransport. *Physiol. Rev.* 80: 211–276. [9]

Russell, J. M., and Boron, W. F. 1976. Role of chloride transport in regulation of intracellular pH. *Nature* 264: 73–74. [9]

Rutherford, M. A., and Roberts, W. M. 2006. Frequency selectivity of synaptic exocytosis in frog saccular hair cells. *Proc. Natl. Acad. Sci. USA* 103: 2898–2903. [22]

Ryan, C. A., and Salvesen, G. S. 2003. Caspases and neuronal development. *Biol. Chem.* 384: 855–861. [25]

Ryba, N. J., and Tirindelli, R. 1997. A new multigene family of putative pheromone receptors. *Neuron* 19: 371–379. [19]

Rydqvist, B., and Purali, N. 1993. Transducer properties of the rapidly adapting stretch receptor neurone in the crayfish (*Pacifastacus leniusculus*). *J. Physiol.* 469: 193–211. [19]

Rymer, J., and Wildsoet, C. F. 2005. The role of the retinal pigment epithelium in eye growth regulation and myopia: a review. *Vis. Neurosci.* 22: 251–261. [20]

Ryugo, D. K., Kretzmer, E. A., and Niparko, J. K. 2005. Restoration of auditory nerve synapses in cats by cochlear implants. *Science* 310: 1490–1492. [26]

Sachdev, R. N., and Catania, K. C. 2002. Receptive fields and response properties of neurons in the star-nosed mole's somatosensory fovea. *J. Neurophysiol.* 87: 2602–2611. [21]

Sachs, F. 1988. Mechanical transduction in biological systems. *Crit. Rev. Biomed. Eng.* 16: 141–169. [19]

Sachs, M. B. 1984. Neural coding of complex sounds: speech. *Annu. Rev. Physiol.* 46: 261–273. [22]

Sadagopan, S., and Wang, X. 2008. Level invariant representation of sounds by populations of neurons in primary auditory cortex. *J. Neurosci.* 28: 3415–3426. [22]

Saez, J. C., Berthoud, V. M., Branes, M. C., Martinez, A. D., and Beyer, E. C. 2003. Plasma membrane channels formed by connexins: their regulation and functions. *Physiol. Rev.* 83: 1359–1400. [11]

Saez, L., Meyer, P., and Young, M. W. 2007. A PER/TIM/DBT interval timer for *Drosophila*'s circadian clock. *Cold Spring Harb. Symp. Quant. Biol.* 72: 69–74. [17]

Safieddine, S., and Wenthold, R. J. 1999. SNARE complex at the ribbon synapses of cochlear hair cells: analysis of synaptic vesicle- and synaptic membrane-associated proteins. *Eur. J. Neurosci.* 11: 803–812. [22]

Sagne, C., El Mestikaway, S., Isambert, M. F., Hamon, M., Henry, J. P., Giros, B., and Gasnier, B. 1997. Cloning of a functional vesicular GABA and glycine transporter by screening genome databases. *FEBS Lett.* 417: 177–183. [9]

Sah, P., Hestrin, S., and Nicoll, R. A. 1990. Properties of excitatory postsynaptic currents recorded in vitro from rat hippocampal interneurones. *J. Physiol.* 430: 605–616. [11]

Sahley, C. L., Modney, B. K., Boulis, N. M., and Muller, K. J. 1994. The S cell: an interneuron essential for sensitization and full dishabituation of leech shortening. *J. Neurosci.* 14: 6715–6721. [18]

Sahni, V., and Kessler, J. A. 2010. Stem cell therapies for spinal cord injury. *Nat. Rev. Neurol.* 6: 363–372. [27]

Sakmann, B. 1992. Elementary steps in synaptic transmission revealed by currents through single ion channels. *Neuron* 8: 613–629. [11]

Sakmann, B., Noma, A., and Trautwein, W. 1983. Acetylcholine activation of single muscarinic K^+ channels in isolated pacemaker cells of the mammalian heart. *Nature* 303: 250–253. [12]

Sakuma, K., and Yamaguchi, A. 2010. The functional role of calcineurin in hypertrophy, regeneration, and disorders of skeletal muscle. *J. Biomed. Biotechnol.* 2010: 721219. [27]

Sakurai, M. 1987. Synaptic modification of parallel fibre-Purkinje cell transmission in *in vitro* guinea-pig cerebellar slices. *J. Physiol.* 394: 463–480. [16]

Sakurai, M. 1990. Calcium is an intracellular mediator of the climbing fiber induction of cerebellar long-term depression. *Proc. Natl. Acad. Sci. USA* 87: 3383–3385. [16]

Sakurai, T., Amemiya, A., Ishii, M., Matsuzaki, I., Chemelli, R. M., Tanaka, H., Williams, S. C., Richardson, J. A., Kozlowski, G. P., Wilson, S., Arch, J. R., Buckingham, R. E., Haynes, A. C., Carr, S. A., Annan, R. S., McNulty, D. E., Liu, W. S., Terrett, J. A., Elshourbagy, N. A., Bergsma, D. J., and Yanagisawa, M. 1998. Orexins and orexin receptors: a family of hypothalamic neuropeptides and G protein-coupled receptors that regulate feeding behavior. *Cell* 92: 573–585. [14]

Sala, C., O'Malley, J., Xu, R., Fumagalli, G., and Salpeter, M. M. 1997. ε subunit-containing acetylcholine receptors in myotubes belong to the slowly degrading population. *J. Neurosci.* 17: 8937–8944. [27]

Salinas, E., Hernández, A., Zainos, A., and Romo, R. 2000. Periodicity and firing rate as candidate neural codes for the frequency of vibrotactile stimuli. *J. Neurosci.* 20: 5503–5515. [23]

Salio, C., Lossi, L., Ferrini, F., and Merighi, A. 2006. Neuropeptides as synaptic transmitters. *Cell Tissue Res.* 326: 583–598. [14, 15]

Salkoff, L., Baker, K., Butler, A., Covarrubius, M., Pak, M. D., and Wei, A. 1992. An essential 'set' of K^+ channels conserved in flies, mice and humans. *Trends Neurosci.* 15: 161–166. [5]

Salmons, S., and Sreter, F. A. 1975. The role of impulse activity in the transformation of skeletal muscle by cross innervation. *J. Anat.* 120: 412–415. [27]

Salpeter, M. M. 1987. Vertebrate neuromuscular junctions: General morphology, molecular organization, and functional consequences. In M. M. Salpeter (ed.), *The Vertebrate Neuromuscular Junction*. Alan R. Liss, New York, pp. 1–54. [11, 13, 15]

Salpeter, M. M., and Loring, R. H. 1985. Nicotinic acetylcholine receptors in vertebrate muscle: properties, distribution and neural control. *Prog. Neurobiol.* 25: 297–325. [27]

Salpeter, M. M., and Marchaterre, M. 1992. Acetylcholine receptors in extrajunctional regions of innervated muscle have a slow degradation rate. *J. Neurosci.* 12: 35–38. [27]

Samuels, S. E., Lipitz, J. B., Dahl, G., and Muller, K. J. 2010. Neuroglial ATP release through innexin channels controls microglial cell movement to a nerve injury. *J. Gen. Physiol.* 136: 425–442. [10, 14, 27]

Sandler, V. M., and Ross, W. N. 1999. Serotonin modulates spike backpropagation and associated $[Ca^{2+}]_i$ changes in the apical dendrites of hippocampal CA1 pyramidal neurons. *J. Neurophysiol.* 81: 216–224. [8]

Sandyk, R. 1992. L-Tryptophan in neuropsychiatric disorders: A review. *Int. J. Neurosci.* 67: 127–144. [15]

Sanes, D. H., and Bao, S. 2009. Tuning up the developing auditory CNS. *Curr. Opin. Neurobiol.* 19: 188–199. [26]

Sanes, D. H., Reh, T. A., and Harris, W. A. 2011. *Development of the Nervous System*, 3rd ed. Academic Press. Burlington, VT. [25]

Sanes, J. R., and Lichtman, J. W. 1999. Development of the vertebrate neuromuscular junction. *Annu. Rev. Neurosci.* 22: 389–442. [27]

Sanes, J. R., and Lichtman, J. W. 2001. Induction, assembly, maturation and maintenance of a postsynaptic apparatus. *Nat. Rev. Neurosci.* 2: 791–805. [11]

Sanes, J. R., and Zipursky, S. L. 2010. Design principles of insect and vertebrate visual systems. *Neuron* 66: 15–36. [25]

Sanes, J. R., Marshall, L. M., and McMahan, U. J. 1978. Reinnervation of muscle fiber basal lamina after removal of muscle fibers. *J. Cell Biol.* 78: 176–198. [27]

Sankaranarayanan, S., and Ryan, T. A. 2000. Real time measurements of vesicle SNARE recycling in synapses of the central nervous system. *Nat. Cell Biol.* 2: 197–204. [13]

Saper, C. B. 2002. The central autonomic nervous system: conscious visceral perception and autonomic pattern generation. *Annu. Rev. Neurosci.* 25: 433–469. [17]

Saper, C. B., and Fuller, P. M. 2007. Inducible clocks: living in an unpredictable world. *Cold Spring Harb. Symp. Quant. Biol.* 72: 543–550. [17]

Sara, S. J. 2008. The locus coeruleus and noradrenergic modulation of cognition. *Nat. Rev. Neurosci.* 10: 211–223. [14]

Sarfati, J., Dodé, C., and Young, J. 2010. Kallmann syndrome caused by mutations in the PROK2 and PROKR2 genes: pathophysiology and genotype-phenotype correlations. *Front. Horm. Res.* 39: 121–32. [25]

Sargent, P. B. 1993. The diversity of neuronal nicotinic acetylcholine receptors. *Annu. Rev. Neurosci.* 16: 403–443. [14]

Sargent, P. B., and Pang, D. Z. 1989. Acetylcholine receptor-like molecules are found in both synaptic and extrasynaptic clusters on the surface of neurons in the frog cardiac ganglion. *J. Neurosci.* 9: 1062–1072. [27]

Sargent, P. B., Bryan, G. K., Streichert, L. C., and Garrett, E. N. 1991. Denervation does not alter the number of neuronal bungarotoxin binding sites on autonomic neurons in the frog cardiac ganglion. *J. Neurosci.* 11: 3610–3623. [27]

Sarter, M., Hasselmo, M. E., Bruno, J. P., and Givens, B. 2005. Unraveling the attentional functions of cortical cholinergic inputs: interactions between signal-driven and cognitive modulation of signal detection. *Brain Res. Brain Res. Rev.* 48: 98–111. [14]

Sasai, Y. 1998. Identifying the missing links: genes that connect neural induction and primary neurogenesis in vertebrate embryos. *Neuron* 21: 455–458. [25]

Saunders, N. R., Ek, C. J., Habgood, M. D., and Dziegielewska, K. M. 2008. Barriers in the brain: A renaissance? *Trends Neurosci.* 31: 279–286. [10]

Saunders, N. R., Kitchener, P., Knott, G. W., Nicholls, J. G., Potter, A., and Smith, T. J. 1998. Development of walking, swimming and neuronal connections after complete spinal cord transection in the neonatal opossum, *Monodelphis domestica*. *J. Neurosci.* 18: 339–355. [27]

Savchenko, A., Barnes, S., and Kramer, R. H. 1997. Cyclic-nucleotide-gated channels mediate synaptic feedback by nitric oxide. *Nature* 390: 694–698. [12, 20]

Sawada, K., Echigo, N., Juge, N., Miyaji, T., Otsuka, M., Omote, H., Yamamoto, A., and Moriyama, Y. 2008. Identification of a vesicular nucleotide transporter. *Proc. Natl. Acad. Sci. USA* 105: 5683–5686. [15]

Scanziani, M. 2000. GABA spillover activates postsynaptic $GABA_B$ receptors to control rhythmic hippocampal activity. *Neuron* 25: 673–681. [14]

Scemes, E., Suadicani, S. O., Dahl, G., and Spray, D. C. 2007. Connexin and pannexin mediated cell-cell communication. *Neuron Glia Biol.* 3: 199–208. [10]

Schaller, K. L., Krzemien, D. M., Yarowsky, P. J., Krueger, B. K., and Caldwell, J. H. 1995. A novel, abundant sodium channel expressed in neurons and glia. *J. Neurosci.* 15: 3231–3242. [7]

Schalling, M., Stieg, P. E., Lindquist, C., Goldstein, M., and Hokfelt, T. 1989. Rapid increase in enzyme and peptide mRNA in sympathetic ganglia after electrical stimulation in humans. *Proc. Natl. Acad. Sci. USA* 86: 4302–4305. [15]

Schecterson, L. C., and Bothwell, M. 2010. Neurotrophin receptors: Old friends with new partners. *Dev. Neurobiol.* 70: 332–338. [25]

Scheutze, S. M., and Role, L. M. 1987. Developmental regulation of nicotinic acetylcholine receptors. *Annu. Rev. Neurosci.* 10: 403–457. [27]

Schikorski, T., and Stevens, C. F. 1997. Quantitative ultrastructural analysis of hippocampal excitatory synapses. *J. Neurosci.* 17: 5858–5867. [13]

Schiller, P. H. Parallel information processing channels created in the retina. *Proc. Natl. Acad. Sci. USA* 107: 17087–17094. [20]

Schliebs, R., and Arendt, T. 2006. The significance of the cholinergic system in the brain during aging and in Alzheimer's disease. *J. Neural Transm.* 113: 1625–1644. [14]

Schmelz, M., Schmidt, R., Weidner, C., Hilliges, M., Torebjork, H. E., and Handwerker, H. O. 2003. Chemical response pattern of different classes of C-nociceptors to pruritogens and algogens. *J. Neurophysiol.* 89: 2441–2448. [21]

Schmidt, R. F. 1971. Presynaptic inhibition in the vertebrate central nervous system. *Ergeb. Physiol.* 63: 20–101. [11]

Schmitt, S. 2003. Homeosis and atavistic regeneration: the 'biogenetic law' in Entwicklungsmechanik. *Hist. Philos. Life Sci.* 25: 193–210. [25]

Schmitz, F., Konigstorfer, A., and Sudhof, T. C. 2000. RIBEYE, a component of synaptic ribbons: a protein's journey through evolution provides insight into synaptic ribbon function. *Neuron* 28: 857–872. [20]

Schmolesky, M. T., Wang, Y., Hanes, D. P., Thompson, K. G., Leutgeb, S., Schall, J. D., and Leventhal, A. G. 1998. Signal timing across the macaque visual system. *J. Neurophysiol.* 79: 3272–3278. [23]

Schnapf, J. L., and Baylor, D. A. 1987. How photoreceptor cells respond to light. *Sci. Am.* 256: 40–47. [20]

Schnapf, J. L., Kraft, T. W., Nunn, B. J., and Baylor, D. A. 1988. Spectral sensitivity of primate photoreceptors. *Vis. Neurosci.* 1: 255–261. [20]

Schnapp, B. J., Vale, R. D., Sheetz, M. P., and Reese, T. S. 1985. Single microtubules from squid axoplasm support bidirectional movement of organelles. *Cell* 40: 455–462. [15]

Schnell, L., and Schwab, M. E. 1990. Axonal regeneration in the rat spinal cord produced by an antibody against myelin-associated neurite growth inhibitors. *Nature* 343: 269–272. [27]

Schnetkamp, P. P. 2004. The *SLC24* Na^+/Ca^{2+}- K^+ exchanger family: Vision and beyond. *Pflügers Arch.* 447: 683–688. [9]

Schnitzer, M. J., and Meister, M. 2003. Multineuronal firing patterns in the signal from eye to brain. *Neuron* 37: 499–511. [20]

Schnupp, J. W., Hall, T. M., Kokelaar, R. F., and Ahmed, B. 2006. Plasticity of temporal pattern codes for vocalization stimuli in primary auditory cortex. *J. Neurosci.* 26: 4785–4795. [22]

Scholfield, C. N. 1978. A barbiturate induced intensification of the inhibitory potential in slices of guinea-pig olfactory cortex. *J. Physiol.* 275: 559–566. [14]

Scholfield, C. N. 1980. Potentiation of inhibition by general anaesthetics in neurones of the olfactory cortex in vitro. *Pflügers Arch.* 383: 249–255. [14]

Schonweiler, R., Ptok, M., and Radu, H. J. 1998. A cross-sectional study of speech- and language-abilities of children with normal hearing, mild fluctuating conductive hearing loss, or moderate to profound sensoneurinal hearing loss. *Int. J. Pediatr. Otorhinolaryngol.* 44: 251–258. [26]

Schramm, M., and Selinger, Z. 1984. Message transmission: Receptor controlled adenylate cyclase system. *Science* 225: 1350–1356. [12]

Schreiner, C. E., and Winer, J. A. 2007. Auditory cortex mapmaking: principles, projections, and plasticity. *Neuron* 56: 356–365. [22]

Schuldiner, S., Schirvan, A., and Linial, M. 1995. Vesicular neurotransmitter transporters: From bacteria to humans. *Physiol. Rev.* 75: 369–392. [9]

Schulman, H. 1995. Protein phosphorylation in neuronal plasticity and gene expression. *Curr. Opin. Neurobiol.* 5: 375–381. [16]

Schultz, W., Dayan, P., and Montague, R. R. 1997. A neural substrate of prediction and reward. *Science* 275: 1593–1599. [14]

Schummers, J., Yu, H., and Sur, M. 2008. Tuned responses of astrocytes and their influence on hemodynamic signals in the visual cortex. *Science* 320: 1638–1643. [10]

Schwab, M. E. 2004. Nogo and axon regeneration. *Curr. Opin. Neurobiol.* 14: 118–1124. [10]

Schwab, M. E., and Caroni, P. 1988. Oligodendrocytes and CNS myelin are nonpermissive substrates for neurite growth and fibroblast spreading in vitro. *J. Neurosci.* 8: 2381–2393. [27]

Schwartz, E. A. 1987. Depolarization without calcium can release gamma-aminobutyric acid from a retinal neuron. *Science* 238: 350–355. [13, 20]

Schwartzman, R. J., and Semmes, J. 1971. The sensory cortex and tactile sensitivity. *Exp. Neurol.* 33: 147–158. [21]

Schwarz, J. R., Glassmeier, G., Cooper, E. C., Kao, T.-C., Nodera, H., Tabuena, D., Kaji, R., and Bostock, H. 2006. KCNQ channels mediate I_{Ks}, a slow K^+ current regulating excitability in the rat node of Ranvier. *J. Physiol.* 573: 17–34. [8]

Scott, S. H. 2008. Inconvenient truths about neural processing in primary motor cortex. *J. Physiol.* 586: 1217–1224. [24]

Seal, R. P., and Edwards, R. H. 2006. Functional implications of neurotransmitter co-release: glutamate and GABA share the load. *Curr. Opin. Pharmacol.* 6: 114–119. [15]

Seal, R. P., Akil, O., Yi, E., Weber, C. M., Grant, L., Yoo, J., Clause, A., Kandler, K., Nobels, J. L., Glowatzki, E., Lustig, L. R., and Edwards, R. H. 2008. Sensineuronal deafness and seizures in mice lacking vesicular glutamate transporter 3. *Neuron* 24: 173–174. [9]

Seamans, J. K., and Yang, C. R. 2004. The principal features and mechanisms of dopamine modulation in the prefrontal cortex. *Prog. Neurobiol.* 74: 1–58. [14]

Sears, T. A. 1964. Efferent discharges in alpha and fusimotor fibers of intercostal nerves of the cat. *J. Physiol.* 174: 295–315. [24]

Seeburg, P. H., and Hartner, J. 2003. Regulation of ion channel/neurotransmitter receptor function by RNA editing. *Curr. Opin. Neurobiol.* 13: 279–283. [14]

Seeburg, P. H., Single, F., Kuner, T., Higuchi, M., and Sprengel, R. 2001. Genetic manipulation of key determinants of ion flow in glutamate receptor channels in the mouse. *Brain Res.* 907: 233–243. [25]

Seeger, M., Tear, G., Ferres-Marco, D., and Goodman, C. S. 1993. Mutations affecting growth cone guidance in *Drosophila*: Genes necessary for guidance toward or away from the midline. *Neuron* 10: 409–426. [25]

Seidah, N. G., and Chretien, M. 1997. Eukaryotic protein processing: endoproteolysis of precursor proteins. *Curr. Opin. Biotechnol.* 8: 602–607. [15]

Selverston, A. I., and Ayers, J. 2006. Oscillations and oscillatory behavior in small neural circuits. *Biol. Cybern.* 95: 537–554. [17]

Semmes, J., Porter, L., and Randolph, M. C. 1974. Further studies of anterior postcentral lesions in monkeys. *Cortex* 10: 55–68. [21]

Sengpiel, F., Stawinski, P., and Bonhoeffer, T. 1999. Influence of experience on orientation maps in cat visual cortex. *Nat Neurosci.* 2: 727–732. [26]

Serizawa, S., Miyamichi, K., Nakatani, H., Suzuki, M., Saito, M., Yoshihara, Y., and Sakano, H. 2003. Negative feedback regulation ensures the one receptor-one olfactory neuron rule in mouse. *Science* 302: 2088–2094. [19]

Severini, C., Improta, G., Falconieri-Erspamer, G., Salvadori, S., and Erspamer, V. 2002. The tachykinin peptide family. *Pharmacol. Rev.* 54: 285–322. [14]

Shackleton, T. M., Skottun, B. C., Arnott, R. H., and Palmer, A. R. 2003. Interaural time difference discrimination thresholds for single neurons in the inferior colliculus of Guinea pigs. *J. Neurosci.* 23: 716–724. [22]

Shah, M. M., Migliore, M., Valencia, I., Cooper, E. C., and Brown, D. A. 2008. Functional significance of K_V7 channels in hippocampal pyramidal neurons. *Proc. Natl. Acad. Sci. USA.* 22: 7869–7874. [8]

Shah, M. M., Migliore, M., Valencia, I., Cooper, E. C., and Brown, D. A. 2008. Functional significance of axonal Kv7 channels in hippocampal pyramidal neurons. *Proc. Natl. Acad. Sci. USA* 105: 7869–7874. [14]

Shain, D. H., ed. 2009. *Annelids in Modern Biology*. Wiley-Blackwell, Hoboken, NJ. [18]

Shain, D. H., Stuart, D. K., Huang, F. Z., and Weisblat, D. A. 2000. Segmentation of the central nervous system in leech. *Development* 127: 735–744. [25]

Shallice, T., and Cooper, R. P. 2011. *The Organization of Mind*. Oxford University Press, Oxford England. [28]

Shapiro, L., Love, J., and Colman, D. R. 2007. Adhesion molecules in the nervous system: structural insights into function and diversity. *Annu. Rev. Neurosci.* 30: 451–474. [25]

Shapovalov, A. I., and Shiriaev, B. I. 1980. Dual mode of junctional transmission at synapses between single primary afferent fibres and motoneurones in the amphibian. *J. Physiol.* 306: 1–15. [11]

Sharuna, N., Harlow, M. L., Jung, J. H., Szule, J. A., Ress, D., Xu, J., Marshall, R. M., and McMahan, U. J. 2009. Macromolecular connections of active zone material to docked synaptic vesicles and presynaptic membrane at neuromuscular junctions of the mouse. *J. Comp. Neurol.* 513: 457–468. [13]

Shatz, C. J. 1977. Anatomy of interhemispheric connections in the visual system. *J. Comp. Neurol.* 173: 497–518. [3]

Shatz, C. J. 1996. Emergence of order in visual system development. *Proc. Natl. Acad. Sci. USA* 93: 602–608. [26]

Shatz, C. J., and Luskin, M. B. 1986. The relationship between the geniculocortical afferents and their cortical target cells during development of the cat's primary visual cortex. *J. Neurosci.* 6: 3655–3668. [25]

Sheetz, M. P. 1999. Motor and cargo interactions. *Eur. J. Biochem.* 262: 19–25. [15]

Shen, K., Fetter, R. D., and Bargmann, C. I. 2004. Synaptic specificity is generated by the synaptic guidepost protein SYG-2 and its receptor, SYG-1. *Cell* 116: 869–881. [25]

Shen, W., Hamilton, S. E., Nathanson, N. M., and Surmeier, D. J. 2005. Cholinergic suppression of KCNQ channel currents enhances excitability of striatal medium spiny neurons. *J. Neurosci.* 25: 7449–7458. [12, 14]

Shen, Y., Liu, X. L., and Yang, X. L. 2006. *N*-methyl-D-aspartate receptors in the retina. *Mol. Neurobiol.* 34: 163–179. [20]

Sheng, M., and Lee, S. H. 2000. Growth of the NMDA receptor industrial complex. *Nat. Neurosci.* 3: 633–635. [11]

Shepard, R. N. 1990. *Mind Sights: Original Visual Illusions, Ambiguities, and other Anomalies*. W. H. Freeman, New York. [3]

Sherman, S. M. 2007. The thalamus is more than just a relay. *Curr. Opin. Neurobiol.* 17: 417–422. [2]

Sherrington, C. S. 1906. *The Integrative Action of the Nervous System*. Reprint, Yale University Press, New Haven, CT, 1966. [1]

Sherrington, C. S. 1906. *The Integrative Action of the Nervous System*, 1961 ed. Yale University Press, New Haven, CT. [24]

Sherrington, C. S. 1910. Flexor-reflex of the limb, crossed extension reflex, and reflex stepping and standing (cat and dog). *J. Physiol.* 40: 28–121. [24]

Sherrington, C. S. 1933. *The brain and Its Mechanism*. Cambridge University Press, London. [24]

Sherrington, C. S. 1951. *Man on His Nature*. Cambridge University Press, Cambridge. [2, 20]

Shi, S.-H., Hayashi, Y., Esteban, J. A., and Malinowm R. Subunit-specific rules governing AMPA receptor trafficking to synapses in hippocampal pyramidal neurons. *Cell* 105: 331–343. [16]

Shi, S-H., Hayashi, Y., Petralia, R. S., Zaman, S. H., Wenthold, R. J., Svoboda, K., and Malinow, R. 1999. Rapid spine delivery and redistribution of AMPA receptors after synaptic NMDA receptor activation. *Science* 284: 1811–1816. [16]

Shik, M. L., and Orlovsky, G. N. 1976. Neurophysiology of locomotor automatism. *Physiol. Rev.* 56: 465–501. [24]

Shimomura, O., and Johnson, F. H. 1970. Calcium binding, quantum yield, and emitting molecule in aequorin bioluminescence. *Nature* 227: 1356–1357. [12]

Shlens, J., Rieke, F., and Chichilnisky, E. 2008. Synchronized firing in the retina. *Curr. Opin. Neurobiol.* 18: 396–402. [20]

Shmuel, A., Korman, M., Sterkin, A., Harel, M., Ullman, S., Malach, R., and Grinvald, A. 2005. Retinotopic axis specificity and selective clustering of feedback projections from V2 to V1 in the owl monkey. *J. Neurosci.* 25: 2117–2131. [3]

Shooter, E. M. 2001. Early days of the nerve growth factor proteins. *Annu. Rev. Neurosci.* 24: 601–629. [25]

Shotwell, S. L., Jacobs, R., and Hudspeth, A. J. 1981. Directional sensitivity of individual vertebrate hair cells to controlled deflection of their hair bundles. *Ann. N Y Acad. Sci.* 374: 1–10. [19]

Shoykhet, M., Doherty, D., and Simons, D. J. 2000. Coding of deflection velocity and amplitude by whisker primary afferent neurons: implications for higher level processing. *Somatosens. Mot. Res.* 17: 171–180. [21]

Shrager, P., Chiu, S. Y., and Ritchie, J. M. 1985. Voltage-dependent sodium and potassium channels in mammalian cultured Schwann cells. *Proc. Natl. Acad. Sci. USA.* 82: 948–952. [10]

Shyng, S.-L., Xu, R., and Salpeter, M. M. 1991. Cyclic AMP stabilizes the degradation of original junctional acetylcholine receptors in denervated muscle. *Neuron* 6: 469–475. [27]

Si, K., Lindquist, S., and Kandel, E. 2004. A possible epigenetic mechanism for the persistence of memory. *Cold Spring Harb. Symp. Quant. Biol.* 69: 497–498. [18]

Siepka, S. M., Yoo, S. H., Park, J., Lee, C., and Takahashi, J. S. 2007. Genetics and neurobiology of circadian clocks in mammals. *Cold Spring Harb. Symp. Quant. Biol.* 72: 251–259. [17]

Sigworth, F. J. 1994. Voltage gating of ion channels. *Q. Rev. Biophys.* 27: 1–40. [5]

Sigworth, F. J., and Neher, E. 1980. Single Na^+ channel currents observed in cultured rat muscle cells. *Nature* 287: 447–449. [7]

Silinsky, E. M., and Redman, R. S. 1996. Synchronous release of ATP and neurotransmitter within milliseconds of a motor-nerve impulse in the frog. *J. Physiol.* 492: 815–822. [13]

Silver, J. 2010. Much Ado about Nogo. *Neuron* 66: 619–621. [27]

Simões, G. F., and Oliveira, A. L. 2010. Alpha motoneurone input changes in dystrophic MDX mice after sciatic nerve transection. *Neuropathol. Appl. Neurobiol.* 36: 55–70. [27]

Simon, E. J., Hiller, J. M., and Edelman, I. 1973. Stereospecific binding of the potent narcotic analgesic [^{3}H]Etorphine to rat-brain homogenate. *Proc. Natl. Acad. Sci. USA* 70: 1947–1949. [21]

Simon, H., Hornbruch, A., and Lumsden, A. 1995. Independent assignment of antero-posterior and dorso-ventral positional values in the developing chick hindbrain. *Curr. Biol.* 5: 205–214. [25]

Simons, D. J. 1978. Response properties of vibrissa units in rat S. I. somatosensory neocortex. *J. Neurophysiol.* 41: 798–820. [21]

Simpson, E. H., Kellendonk, C., and Kandel, E. 2010. A possible role for the striatum in the pathogenesis of the cognitive symptoms of schizophrenia. *Neuron* 65: 585–596. [24]

Sims, T. J., Waxman, S. G., Black, J. A., and Gilmore, S. A. 1985. Perinodal astrocytic processes at notes of Ranvier in developing normal and glial cell deficient rat spinal cord. *Brain Res.* 337: 321–331. [10]

Sincich, L. C., and Horton, J. C. 2005. The circuitry of V_1 AND V_2: Integration of color, form, and motion. *Ann. Rev. Neurosci.* 28: 303–326. [3]

Sincich, L. C., Jocson, C. M., and Horton, J. C. 2007. Neurons in V1 patch columns project to V2 thin stripes. *Cereb. Cortex* 17: 935–941. [3]

Sincich, L. C., Park, K. F., Wohlgemuth, M. J., and Horton, J. C. 2004. Bypassing V1: a direct geniculate input to area MT. *Nat. Neurosci.* 7: 1123–1128. [3]

Sinclair, R. J., and Burton, H. 1993. Neuronal activity in the second somatosensory cortex of monkeys (*Macaca mulatta*) during active touch of gratings. *J. Neurophysiol.* 70: 331–350. [21]

Singer, W. 1995. Development and plasticity of cortical processing architectures. *Science* 270: 758–764. [26]

Sive, H. L., Grainger, R. M., and Harland, R. M. 2010. Microinjection of *Xenopus* embryos. *Cold Spring Harb. Protoc.* 2010: pdb ip81. [25]

Sivilotti, L. 2010. What single-channel analysis tells us of the activation mechanism of ligand-gated channels: the case of the glycine receptor. *J. Physiol.* 588: 45–58. [14]

Sivilotti, L., and Colquhoun, D. 1995. Acetylcholine receptors: too many channels, too few functions. *Science* 269: 1681–1682. [14]

Skene, J. H. P., and Shooter, E. M. 1983. Denervated sheath cells secrete a new protein after nerve injury. *Proc. Natl. Acad. Sci. USA* 80: 4169–4173. [27]

Skou, J. C. 1957. The influence of some cations on an adenosine triphosphatase from peripheral nerves. *Biochim. Biophys. Acta* 23: 394–401. [9]

Skou, J. C. 1988. Overview: The Na,K pump. *Methods Enzymol.* 156: 1–25. [9]

Slobodov, U., Reichert, F., Mirski, R., and Rotshenker, S. 2009. The role of Galectin-3/MAC-2 in the activation of the innate-immune function of phagocytosis in microglia in injury and disease. *J. Mol. Neurosci.* 39: 99–103. [10]

Smart, T. G., Hosie, A. M., and Miller, P. 2004. Zn^{2+} ions: modulators of excitatory and inhibitory synaptic activity. *Neuroscientist* 10: 432–442. [14]

Smeyne, R. J., Klein, R., Schnapp, A., Long, L. K., Bryant, S., Lewin, A., Lira, S. A., and Barbacid, M. 1994. Severe sensory and sympathetic neuropathies in mice carrying a disrupted Trk/NGF receptor gene. *Nature* 368: 246–249. [25]

Smith, A. D., de Potter, W. P., Moerman, E. J., and Schaepdryver, A. F. 1970. Release of dopamine β-hydroxylase and chromogranin A upon stimulation of the splenic nerve. *Tissue Cell* 2: 547–568. [13]

Smith, J. L., Crawford, M., Proske, U., Taylor, J. L., and Gandevia, S. C. 2009. Signals of motor command bias joint position sense in the presence of feedback from proprioceptors. *J. Appl. Physiol.* 106: 950–958. [24]

Smith, R. J., Bale, J. F., Jr., and White, K. R. 2005. Sensorineural hearing loss in children. *Lancet* 365: 879–890. [26]

Snyder, S. H., Jaffrey, S. R., and Zakhary, R. 1998. Nitric oxide and carbon monoxide: parallel roles as neural messengers. *Brain Res. Brain Res. Rev.* 26: 167–175. [12]

So, I., Ashmole, I., Davies, N. W., Sutcliffe, M. J., and Stanfield, P. R. 2001. The K^+ channel signature sequence of murine Kir 2.1: Mutations that affect microscopic gating but not ion selectivity. *J. Physiol.* 531.1: 37–50. [5]

Sobkowicz, H. M., Rose, J. E., Scott, G. L., and Slapnick, S. M. 1982. Ribbon synapses in the developing intact and cultured organ of Corti in the mouse. *J. Neurosci.* 2: 942–957. [22]

Soderling, T. 2000. CaM-kinases: modulators of synaptic plasticity. *Curr Opin. Neurobiol.* 10: 375–380. [12]

Soejima, M., and Noma, A. 1984. Mode of regulation of the ACh-sensitive K-channel by the muscarinic receptor in rabbit atrial cells. *Pflügers Arch.* 400: 424–431. [12]

Soghomonian, J. J., and Martin, D. L. 1998. Two isoforms of glutamate decarboxylase: why? *Trends Pharmacol. Sci.* 19: 500–505. [15]

Söhl, G., Maxeiner, S., and Willecke, K. 2005. Expression and functions of neuronal gap junctions. *Nat. Rev. Neurosci.* 6: 191–200. [11]

Sokoloff, L. 1977. Relation between physiological function and energy metabolism in the central nervous system. *J. Neurochem.* 29: 13–26. [3]

Sokolov, S., Scheuer, T., and Catterall, W. A. 2005. Ion permeation through a voltage-sensitive gating pore in brain sodium channels having voltage sensor mutations. *Neuron* 47: 183–189. [7]

Sokolov, S., Scheuer, T., and Catterall, W. A. 2007. Gating pore currents in an inherited ion channelopathy. *Nature* 446: 76–78. [6]

Sokolove, P. G., and Cooke, I. M. 1971. Inhibition of impulse activity in a sensory neuron by an electrogenic pump. *J. Gen. Physiol.* 57: 125–163. [19]

Sokolowski, B., Duncan, R. K., Chen, S., Karolat, J., Kathiresan, T., and Harvey, M. 2009. The large-conductance Ca^{2+}-activated K^{+} channel interacts with the apolipoprotein ApoA1. *Biochem. Biophys. Res. Commun.* 387: 671–675. [22]

Solecki, D. J., Model, L., Gaetz, J., Kapoor, T. M., and Hatten, M. E. 2004. Par6alpha signaling controls glial-guided neuronal migration. *Nat. Neurosci.* 7: 1169–1170. [10]

Sommer, B., Kohler, M., Sprengle, R., and Seeburg, P. H. 1991. RNA editing in brain controls a determinant of ion flow in glutamate-gated channels. *Cell* 67: 11–19. [5]

Sommer, M. A., and Wurtz, R. H. 2008. Brain circuits for the internal monitoring of movements. *Annu. Rev. Neurosci.* 31: 317–338. [24]

Son, Y. J., and Thompson, W. J. 1995a. Schwann cell processes guide regeneration of peripheral axons. *Neuron* 14: 125–132. [10, 27]

Son, Y. J., and Thompson, W. J. 1995b. Nerve sprouting in muscle is induced and guided by processes extended by Schwann cells. *Neuron* 14: 133–141. [10, 27]

Son, Y. J., Trachtenberg, J. T., and Thompson, W. J. 1996. Schwann cells induce and guide sprouting and reinnervation of neuromuscular junctions. *Trends Neurosci.* 19: 280–285. [10]

Soshnikova, N., and Duboule, D. 2009. Epigenetic regulation of vertebrate *Hox* genes: a dynamic equilibrium. *Epigenetics* 4: 537–540. [25]

Sosinsky, G. E., and Nicholson, B. J. 2005. Structural organization of gap junction channels. *Biochim. Biophys. Acta* 1711: 99–125. [8]

Sossin, W. S., Fisher, J. M., and Scheller, R. H. 1989. Cellular and molecular biology of neuropeptide processing and packaging. *Neuron* 2: 1407–1417. [15]

Sotelo, C., and Alvarado-Mallart, R. M. 1991. The reconstruction of cerebellar circuits. *Trends Neurosci.* 14: 350–355. [27]

Sotelo, C., Alvarado-Mallart, R. M., Frain, M., and Vernet, M. 1994. Molecular plasticity of adult Bergmann fibers is associated with radial migration of grafted Purkinje cells. *J. Neurosci.* 14: 124–133. [27]

Sotelo, C., Llinás, R., and Baker, R. 1974. Structural study of inferior olivary nucleus of the cat: Morphological correlates of electrotonic coupling. *J. Neurophysiol.* 37: 541–559. [8]

Soto, F., Garcia-Guzman, M., and Stühmer, W. 1997. Cloned ligand-gated channels activated by extracellular ATP (P2X receptors). *J. Membr. Biol.* 160: 91–100. [17]

Soucy, E., Wang, Y., Nirenberg, S., Nathans, J., and Meister, M. 1998. A novel signaling pathway from rod photoreceptors to ganglion cells in mammalian retina. *Neuron* 21: 481–493. [20]

Southwell, D. G., Froemke, R. C., Alvarez-Buylla, A., Stryker, M. P., and Gandhi, S. P. 2010. Cortical plasticity induced by inhibitory neuron transplantation. *Science* 327: 1145–1148. [26]

Spacek, J., and Harris, K. M. 1998. Three-dimensional organization of cell adhesion junctions at synapses and dendritic spines in area CA1 of the rat hippocampus. *J. Comp. Neurol.* 393: 58–68. [11]

Specht, S., and Grafstein, B. 1973. Accumulation of radioactive protein in mouse cerebral cortex after injection of [3H]-fucose into the eye. *Exp. Neurol.* 41: 705–722. [3]

Sperry, R. W. 1963. Chemoaffinity in the orderly growth of nerve fiber patterns and connections. *Proc. Natl. Acad. Sci. USA* 50: 703–710. [27]

Sperry, R. W. 1970. Perception in the absence of neocortical commissures. *Proc. Res. Assoc. Nerv. Ment. Dis.* 48: 123–138. [3]

Spoendlin, H. 1969. Innervation patterns in the organ of corti of the cat. *Acta Otolaryngol.* 67: 239–254. [22]

Spruston, N. 2008. Pyramidal neurons: dendritic structure and synaptic integration. *Nat. Rev. Neurosci.* 9: 206–221. [11]

Spyer, K. M., and Gourine, A. V. 2009. Chemosensory pathways in the brainstem controlling cardiorespiratory activity. *Philos. Trans. R. Soc. Lond., B, Biol. Sci.* 364: 2603–2610. [17, 24]

Sretavan, D. W., Shatz, C. J., and Stryker, M. P. 1988. Modification of retinal ganglion cell axon morphology by prenatal infusion of tetrodotoxin. *Nature* 336: 468–471. [26]

Srinivasan, M., Zhang, S., Lehrer, M., and Collett, T. 1996. Honeybee navigation en route to the goal: visual flight control and odometry. *J. Exp. Biol.* 199: 237–244. [18]

St. John, W. M. 2009. Noeud vital for breathing in the brainstem: gasping—yes, eupnoea—doubtful. *Philos. Trans. R. Soc. Lond., B, Biol. Sci.* 364: 2625–2633. [24]

Staley, K., Smith, R., Schaak, J., Wilcox, C., and Jentsch, T. J. 1996. Alteration of $GABA_A$ receptor function following gene transfer of the CLC-2 chloride channel. *Neuron* 17: 543–551. [6]

Stanke, M., Duong, C. V., Pape, M., Geissen, M., Burbach, G., Deller, T., Gascan, H., Otto, C., Parlato, R., Schutz, G., and Rohrer, H. 2006. Target-dependent specification of the neurotransmitter phenotype: cholinergic differentiation of sympathetic neurons is mediated in vivo by gp 130 signaling. *Development* 133: 141–150. [25]

Starling, E. H. 1941. *Starling's Principles of Human Physiology.* Churchill, London. [17]

Steinbach, J. H., and Akk, G. 2001. Modulation of $GABA_A$ receptor channel gating by pentobarbital. *J. Physiol.* 537: 715–733. [14]

Steindler, D. A., Cooper, N. G., Faissner, A., and Schachner, M. 1989. Boundaries defined by adhesion molecules during development of the cerebral cortex: the J1/tenascin glycoprotein in the mouse somatosensory cortical barrel field. *Dev. Biol.* 131: 243–260. [10]

Steiner, D. F., 1998. The proprotein convertases. *Curr. Opin. Chem. Biol.* 2: 31–39. [15]

Steinert, J. R., Kopp-Scheinpflug, C., Baker, C., Challiss, R. A., Mistry, R., Haustein, M. D., Griffin, S. J., Tong, H., Graham, B. P., and Forsythe, I. D. 2008. Nitric oxide is a volume transmitter regulating postsynaptic excitability at a glutamatergic synapse. *Neuron* 60: 642–656. [12]

Stent, G. S., and Weisblat, D. A. 1985. Cell lineage in the development of invertebrate nervous systems. *Annu. Rev. Neurosci.* 8: 45–70. [10]

Stent, G. S., Kristan, W. B., Jr., Friesen, W. O., Ort, C. A., Poon, M., and Calabrese, R. L. 1978. Neuronal generation of the leech swimming movement. *Science* 200: 1348–1357. [18]

Stephan, A. B., Shum, E. Y., Hirsh, S., Cygnar, K. D., Reisert, J., and Zhao, H. 2009. ANO2 is the cilial calcium-activated chloride channel that may mediate olfactory amplification. *Proc. Natl. Acad. Sci. USA* 106: 11776–11781. [12]

Sterling, P., and Demb, J. B. 2003. In G. M. Shepherd, *Synaptic Organization of the Brain.* Oxford University Press, New York. [20]

Stern, K., and McClintock, M. K. 1998. Regulation of ovulation by human pheromones. *Nature* 392: 177–179. [19]

Stettler, D. D., Das A., Bennett, J., and Gilbert, C. D. 2002. Lateral connectivity and contextual interactions in macaque primary visual cortex. *Neuron* 36: 739–750. [2, 3]

Stevens, C. F., and Wang, Y. 1993. Reversal of long-term potentiation by inhibitors of haem oxygenase. *Nature* 364: 147–149. [12]

Stevens, C. F., and Wesseling, J. F. 1999. Augmentation is a potentiation of the exocytotic process. *Neuron* 22: 139–146. [16]

Steyer, J. A., Horstmann, H., and Almers, W. 1997. Transport, docking and exocytosis of single secretory granules in live chromaffin cells. *Nature* 388: 474–478. [13]

Stoney, S. D., Jr., Thompson, W. D., and Asanuma, H. 1968. Excitation of pyramidal tract cells by intracortical microstimulation: effective extent of stimulating current. *J. Neurophysiol.* 31: 659–669. [23]

Streb, H., Irvine, R. F., Berridge, M. J., and Schulz, I. 1983. Release of Ca^{2+} from a nonmitochondrial intracellular store in pancreatic acinar cells by inositol-1,4,5-trisphosphate. *Nature* 306: 67–69. [12]

Streisinger, G., Walker, C., Dower, N., Knauber, D., and Singer, F. 1981. Production of clones of homozygous diploid zebra fish (*Brachydanio rerio*). *Nature* 291: 293–296. [25]

Strick, P. L., Dum, R. P., and Fiez, J. A. 2009. Cerebellum and nonmotor function. *Annu. Rev. Neurosci.* 32: 413–434. [24]

Stryer, L. 1987. The molecules of visual excitation. *Sci. Am.* 257: 42–50. [20]

Stryer, L., and Bourne, H. R. 1986. G proteins: a family of signal transducers. *Annu. Rev. Cell Biol.* 2: 391–419. [20]

Stryker, M. P., and Harris, W. A. 1986. Binocular impulse blockade prevents the formation of ocular dominance columns in cat visual cortex. *J. Neurosci.* 6: 2117–2133. [26]

Stuart, A. E. 1970. Physiological and morphological properties of motoneurones in the central nervous system of the leech. *J. Physiol.* 209: 627–646. [18]

Stuart, A. E., Borycz, J., and Meinertzhagen, I. A. 2007. The dynamics of signaling at the histaminergic photoreceptor synapse of arthropods. *Prog. Neurobiol.* 82: 202–227. [14]

Stuart, G., Schiller, J., and Sakmann, B. 1997. Action potential initiation and propagation in rat neocortical pyramidal neurons. *J. Physiol.* 505: 617–632. [8]

Stuart, G., Spruston, N., Sakmann, B., and Hausser, M. 1997. Action potential initiation and backpropagation in neurons of the mammalian CNS. *Trends Neurosci.* 20: 125–131. [8]

Stuber, G. D., Hnasko, T. S., Britt, J. P., Edwards, R. H., and Bonci, A. 2010. Dopaminergic terminals in the nucleus accumbens but not the dorsal striatum corelease glutamate. *J. Neurosci.* 30: 8229–8233. [14]

Sudhof, T. C. 2008. Neuroligins and neurexins link synaptic function to cognitive disease. *Nature* 455: 903–911. [25]

Sugiura, Y., Woppmann, A., Miljanich, G. P., and Ko, C-P. 1995. A novel omega-conopeptide for the presynaptic localization of calcium channels at the mammalian neuromuscular junction. *J. Neurocytol.* 24: 15–27. [13]

Suh, B. C., and Hille, B. 2008. PIP_2 is a necessary cofactor for ion channel function: how and why? *Annu. Rev. Biophys.* 37: 175–195. [17]

Suh, B. C., Horowitz, L. F., Hirdes, W., Mackie, K., and Hille, B. 2004. Regulation of KCNQ2/KCNQ3 current by G protein cycling: the kinetics of receptor-mediated signaling by Gq. *J. Gen. Physiol.* 123: 663–683. [12]

Suh, B. C., Inoue, T., Meyer, T., and Hille, B. 2006. Rapid chemically induced changes of PtdIns(4,5)P2 gate KCNQ ion channels. *Science* 314: 1454–1457. [12]

Sullivan, J. M. 2007. A simple depletion model of the readily releasable pool of synaptic vesicles cannot account for paired-pulse depression. *J. Neurophysiol.* 97: 948–950. [16]

Sulzer, D., Joyce, M. P., Lin, L., Geldwert, D., Haber, S. N., Hattori, T., and Rayport, S. 1998. Dopamine neurons make glutamatergic synapses in vitro. *J. Neurosci.* 18: 4588–4602. [15]

Surmeier, D. J., Ding, J., Day, M., Wang, Z., and Shen, W. 2007. D1 and D2 dopamine-receptor modulation of striatal glutamatergic signaling in striatal medium spiny neurons. *Trends Neurosci.* 30: 228–245. [24]

Surprenant, A., and North, R. A. 2009. Signaling at purinergic P2X receptors. *Annu. Rev. Physiol.* 71: 333–359. [14]

Susuki, K., and Rasband, M. N. 2008. Molecular mechanisms of node of Ranvier formation. *Curr. Opin. Cell Biol.* 20: 616–623. [8, 10]

Suter, D. M., Errante, L. D., Belotserkovsky, V., and Forscher, P. 1998. The Ig superfamily cell adhesion molecule, apCAM, mediates growth cone steering by substrate-cytoskeletal coupling. *J. Cell Biol.* 141: 227–240. [25]

Sutherland, E. W. 1972. Studies on the mechanism of hormone action. *Science* 177: 401–408. [12]

Sutter, M. L., and Schreiner, C. E. 1995. Topography of intensity tuning in cat primary auditory cortex: single-neuron versus multiple-neuron recordings. *J. Neurophysiol.* 73: 190–204. [22]

Suzuki, R., Hunt, S. P., and Dickenson, A. H. 2003. The coding of noxious mechanical and thermal stimuli of deep dorsal horn neurones is attenuated in NK1 knockout mice. *Neuropharmacology* 45: 1093–1100. [14]

Svoboda, K., Helmchen, F., Denk, W., and Tank, D. W. 1999. Spread of dendritic excitation in layer 2/3 pyramidal neurons in rat barrel cortex in vivo. *Nat. Neurosci.* 2: 65–73. [8]

Svoboda, K., Schmidt, C. F., Schnapp, B. J., and Block, S. M. 1993. Direct observation of kinesin stepping by optical trapping interferometry. *Nature* 365: 721–727. [15]

Swindale, N. V., Matsubara, J. A., and Cynader, M. S. 1987. Surface organization of orientation and direction selectivity in cat area 18. *J. Neurosci.* 7: 1414–1427. [3]

Szallasi, A., and Blumberg, P. M. 1996. Vanilloid receptors: new insights enhance potential as a therapeutic target. *Pain* 68: 195–208. [19]

Taccola, G., and Nistri, A. 2006. Oscillatory circuits underlying locomotor networks in the rat spinal cord. *Crit. Rev. Neurobiol.* 18: 25–36. [24]

Tadenev, A. L., Kulaga, H. M., May-Simera, H. L., Kelley, M. W., Katsanis, N., and Reed. R, R. 2011. Loss of Bardet-Biedl syndrome protein-8 (BBS8) perturbs olfactory function, protein localization, and axon targeting. *Proc. Natl. Acad. Sci. USA* 108(25): 10320–10325.

Takahashi, K., and Yamanaka, S. 2006. Induction of pluripotent stem cells from mouse embryonic and adult fibroblast cultures by defined factors. *Cell* 126: 663–676. [25]

Takahashi, K., Lin, J. S., and Sakai, K. 2006. Neuronal activity of histaminergic tuberomammillary neurons during wake-sleep states in the mouse. *J. Neurosci.* 26: 10292–10298. [14]

Takahashi, T., Kajikawa, Y., and Tsujimoto, T. 1998. G-protein-coupled modulation of presynaptic calcium currents and transmitter release by a $GABA_B$ receptor. *J. Neurosci.* 18: 3138–3146. [12, 14]

Takai, T., Noda, M., Mishina, M., Shimizu, S., Furutani, Y., Kayano, T., Ikeda, T., Kubo, T., Takahashi, H., Takahashi, T., Kuno, M., and Numa, S. 1985. Cloning, sequencing, and expression of cDNA for a novel subunit of acetylcholine receptor from calf muscle. *Nature* 315: 761–764. [5]

Takamori, S. 2006. VGLUTs: "Exciting" times for glutamate research? *Neurosci. Res.* 55: 343–351. [9]

Takeda, H., Inazu, M., and Matsumiya, T. 2002. Astroglial dopamine transport is mediated by norepinephrine transporter. *Naunyn Schmiedebergs Arch Pharmacol.* 366: 620–623. [10]

Takeuchi, A., and Takeuchi, N. 1959. Active phase of frog's end-plate potential. *J. Neurophysiol.* 22: 395–411. [11]

Takeuchi, A., and Takeuchi, N. 1960. On the permeability of the end plate membrane during the action of transmitter. *J. Physiol.* 154: 52–67. [11]

Takeuchi, A., and Takeuchi, N. 1964. The effect on crayfish muscle of iontophoretically applied glutamate. *J. Physiol.* 170: 296–317. [14]

Takeuchi, A., and Takeuchi, N. 1966. On the permeability of the presynaptic terminal of the crayfish neuromuscular junction during synaptic inhibition and the action of γ-aminobutyric acid. *J. Physiol.* 183: 433–449. [11]

Takeuchi, A., and Takeuchi, N. 1967. Anion permeability of the inhibitory post-synaptic membrane of the crayfish neuromuscular junction. *J. Physiol.* 191: 575–590. [11]

Takeuchi, N. 1963. Some properties of conductance changes at the end-plate membrane during the action of acetylcholine. *J. Physiol.* 167: 128–140. [11]

Takumi, Y., Matsubara, A., Rinvik, E., and Otterson, O. P. 1999. The arrangement of glutamate receptors in excitatory synapses. *Ann. N Y Acad. Sci.* 868: 474–481. [16]

Takumi, Y., Ramirez-Leon, V., Laake, P., Rinvik, E., and Otterson, O. P. 1999. Different modes of expression of AMPA and NMDA receptors in hippocampal synapses. *Nat. Neurosci.* 2: 618–624. [16]

Talbot, S. A., and Marshall, W. H. 1941. Physiological studies on neural mechanisms of visual localization and discrimination. *Am. J. Ophthalmol.* 24: 1255–1264. [3]

Tan, J., Epema, A. H., and Voogd, J. 1995. Zonal organization of the flocculovestibular nucleus projection in the rabbit: A combined axonal tracing and acetylcholinesterase study. *J. Comp. Neurol.* 356: 51–71. [24]

Tanabe, T. Takashima, H., Mikami, A., Flockerzi, V., Takahashi, H., Kangawa, K., Kojima, M., Matsuo, H., Hirose, T., and Numa, S. 1987. Primary structure of receptors for calcium channel blockers from skeletal muscle. *Nature* 328: 313–318. [5]

Tang, Y-G., and Zucker, R. S. 1997. Mitochondrial involvement in post-tetanic potentiation of synaptic transmission. *Neuron* 18: 483–491. [16]

Tank, D. W., Huganir, R. L., Greengard, P., and Webb, W. W. 1983. Patch-recorded single- channel currents of the purified and reconstituted *Torpedo* acetylcholine receptor. *Proc. Natl. Acad. Sci. USA.* 80: 5129–5133. [5]

Tao-Cheng, J. H., Nagy, Z., and Brightman, M. W. 1987. Tight junctions of brain endothelium in vitro are enhanced by astroglia. *J. Neurosci.* 7: 3293–3299. [10]

Taranda, J., Maison, S. F., Ballestero, J. A., Katz, E., Savino, J., Vetter, D. E., Boulter, J., Liberman, M. C., Fuchs, P. A., and Elgoyhen, A. B. 2009. A point mutation in the hair cell nicotinic cholinergic receptor prolongs cochlear inhibition and enhances noise protection. *PLoS Biol.* 7: e18. [22]

Tardin, C., Cognet, L., Bats, C., Lounis, B., and Choquet, D. 2003. Direct imaging of lateral movements of AMPA receptors inside synapses. *EMBO J.* 22: 4656–4665. [14]

Tarozzo, G., De Andrea, M., Feuilloley, M., Vaudry, H., and Fasolo, A. 1998. Molecular and cellular guidance of neuronal migration in the developing olfactory system of rodents. *Ann. N Y Acad. Sci.* 839: 196–200. [17]

Tarozzo, G., Peretto, P., Biffo, S., Varga, Z., Nicholls, J., and Fasolo, A. 1995. Development and migration of olfactory neurones in the nervous system of the neonatal opossum. *Proc. R. Soc. Lond., B, Biol. Sci.* 262: 95–101. [17]

Taylor, A. R., Gifondorwa, D. J., Newbern, J. M., Robinson, M. B., Strupe, J. L., Prevette, D., Oppenheim, R. W., and Milligan, C. E. 2007. Astrocyte and muscle-derived secreted factors differentially regulate motoneuron survival. *J. Neurosci.* 27: 634–644. [25]

Taylor, W. R., and Baylor, D. A. 1995. Conductance and kinetics of single cGMP-activated channels in salamander rod outer segments. *J. Physiol.* 483 (Pt 3): 567–582. [20]

Teillet, M. A., Ziller, C., and Le Douarin, N. M. 2008. Quail–chick chimeras. *Methods Mol. Biol.* 461: 337–350. [25]

Terenius, L. 1973. Stereospecific interaction between narcotic analgesics and a synaptic plasm a membrane fraction of rat cerebral cortex. *Acta Pharmacol. Toxicol. (Copenh.)* 32: 317–320. [21]

Terlau, H., and Olivera, B. M. 2004. *Conus* Venoms: A rich source of novel ion channel-targeted peptides. *Physiol. Rev.* 84: 41– 68. [7]

Tessier-Lavigne, M., Placzek, M., Lumsden, A. G., Dodd, J., and Jessell, T. M. 1988. Chemotropic guidance of developing axons in the mammalian central nervous system. *Nature* 336: 775–778. [25]

Tettamanti, G., Cattaneo, A. G., Gornati, R., de Eguileor, M., Bernardini, G., and Binelli, G. 2010. Phylogenesis of brain-derived neurotrophic factor (BDNF) in vertebrates. *Gene* 450: 85–93. [25]

Teune, T. M., van der Burg, J., de Zeeuw, C. I., Voogd, J., and Ruigrok, T. J. 1998. Single Purkinje cell can innervate multiple classes of projection neurons in the cerebellar nuclei of the rat: A light microscopic and ultrastructural triple-tracer study in the rat. *J. Comp. Neurol.* 392: 164–178. [15]

Thach, W. T. 2007. On the mechanism of cerebellar contributions to cognition *Cerebellum* 6: 163–167. [24]

Thach, W. T., Goodkin, H. G., and Keating, J. G. 1992. The cerebellum and the adaptive coordination of movement. *Annu. Rev. Neurosci.* 15: 403–442. [24]

Thoenen, H., Mueller, R. A., and Axelrod, J. 1969. Increased tyrosine hydroxylase activity after drug-induced alteration of sympathetic transmission. *Nature* 221: 1264. [15]

Thoenen, H., Otten, U., and Schwab, M. 1979. Orthograde and retrograde signals for the regulation of neuronal gene expression: The peripheral sympathetic nervous system as a model. In *The Neurosciences: Fourth Study Program*. MIT Press, Cambridge, MA. pp. 911–928. [15]

Thomas, R. C. 1969. Membrane currents and intracellular sodium changes in a snail neurone during extrusion of injected sodium. *J. Physiol.* 201: 495–514. [9]

Thomas, R. C. 1972. Intracellular sodium activity and the sodium pump in snail neurones. *J. Physiol.* 220: 55–71. [9]

Thomas, R. C. 1977. The role of bicarbonate, chloride and sodium ions in the regulation of intracellular pH in snail neurones. *J. Physiol.* 273: 317–338. [9]

Thompson, L. H., and Björklund, A. 2009. Transgenic reporter mice as tools for studies of transplantability and connectivity of dopamine neuron precursors in fetal tissue grafts. *Prog. Brain Res.* 175: 53–79. [27]

Thompson, W. 1983. Synapse elimination in neonatal rat muscle is sensitive to pattern of muscle use. *Nature* 302: 614–616. [25]

Thoreson, W. B. 2007. Kinetics of synaptic transmission at ribbon synapses of rods and cones. *Mol. Neurobiol.* 36: 205–223. [20]

Thorogood, M. S., Almeida, V. W., and Brodfuehrer, P. D. 1999. Glutamate receptor 5/6/7-like and glutamate transporter-1-like immunoreactivity in the leech central nervous system. *J. Comp. Neurol.* 405: 334–344. [18]

Thorpe, S. J., Fize, D., and Marlot, C. 1996. Speed of processing in the human visual system. *Nature* 381: 520–522. [23]

Thyssen, A., Hirnet, D., Wolburg, H., Schmalzing, G., Deitmer, J. W., and Lohr, C. 2010. Ectopic vesicular neurotransmitter release along sensory axons mediates neurovascular coupling via glial calcium signaling. *Proc. Natl. Acad. Sci. USA* 107: 15258–15263. [14]

Timpe, L. C., Schwartz, T. L., Tempel, B. L., Papazian, D. M., Jan, Y. N., and Jan, L. Y. 1988. Expression of functional potassium channels from *Shaker* cDNA in *Xenopus* oocytes. *Nature* 331: 143–145. [5]

Tirindelli, R., Dibattista, M., Pifferi, S., and Menini, A. 2009. From pheromones to behavior. *Physiol. Rev.* 89: 921–956. [19]

Tkatchenko, A. V., Walsh, P. A., Tkatchenko, T. V., Gustincich, S., and Raviola, E. 2006. Form deprivation modulates retinal neurogenesis in primate experimental myopia. *Proc. Natl. Acad. Sci. USA* 103: 4681–4686. [20]

Toda, M., and Okamura, T. 2003. The pharmacology of nitric oxide in the peripheral nervous system of blood vessels. *Pharmacol. Rev.* 55: 271–324. [17]

Togashi, H., Sakisaka, T., and Takai, Y. 2009. Cell adhesion molecules in the central nervous system. *Cell Adh. Migr.* 3: 29–35. [25]

tom Dieck, S., and Brandstatter, J. H. 2006. Ribbon synapses of the retina. *Cell Tissue Res.* 326: 339–346. [20]

Tombola, F., Pathak, M. M., and Isacoff, E. E. 2005. Voltage-sensing arginines in a potassium channel permeate and occlude cation-selective pores. *Neuron* 45: 379–388. [7]

Tomita, H., Ohbayashi, M., Nakahara, K., Hasegawa, I., and Miyashita, Y. 1999. Top-down signal from prefrontal cortex in executive control of memory retrieval. *Nature* 401: 699–703. [23]

Tongiorgi, E. 2008. Activity-dependent expression of brain-derived neurotrophic factor in dendrites: facts and open questions. *Neurosci. Res.* 61: 335–346. [25]

Tootell, R. B., and Hadjikhani, N. 2000. Attention—brains at work! *Nat. Neurosci.* 3: 206–208. [23]

Torebjork, E. 1985. Nociceptor activation and pain. *Philos. Trans. R. Soc. Lond., B, Biol. Sci.* 308: 227–234. [21]

Tornqvist, K., Yang, X. L., and Dowling, J. E. 1988. Modulation of cone horizontal cell activity in the teleost fish retina. III. Effects of prolonged darkness and dopamine on electrical coupling between horizontal cells. *J. Neurosci.* 8: 2279–2288. [20]

Torre, V., Ashmore, J. F., Lamb, T. D., and Menini, A. 1995. Transduction and adaptation in sensory receptor cells. *J. Neurosci.* 15: 7757–7768. [20]

Torres, G. E., and Amara, S. G. 2007. Glutamate and monamine transporters: New visions of form and function. *Curr. Opin. Neurobiol.* 17: 304–312. [9, 15]

Toyoshima, C., and Unwin, N. 1988. Ion channel of acetylcholine receptor reconstituted from images of postsynaptic membranes. *Nature* 336: 247–250. [5]

Toyoshima, C., Nakasako, M., Nomura, H., and Ogawa, H. 2000. Crystal structure of the calcium pump of the sarcoplasmic reticulum at 2.6 A resolution. *Nature* 405: 647–655. [9]

Trautwein, W., Osterrieder, W., and Noma, A. 1981. Potassium channels and the muscarinic receptor in the sino-atrial node of the heart. In N. J. M. Birdsall (ed.), in *Drug Receptors and their Effectors*. London: MacMillan. pp. 5–22. [12]

Trautwein, W., Taniguchi, J., and Noma, A. 1982. The effect of intracellular cyclic nucleotides and calcium on the action potential and acetylcholine response of isolated cardiac cells. *Pflügers Arch.* 392: 307–314. [12]

Trigo, F. F., Marty, A., and Stell, B. M. 2008. Axonal $GABA_A$ receptors. *Eur. J. Neurosci.* 28: 841–848. [11]

Tritsch, N. X., and Bergles, D. E. 2010. Developmental regulation of spontaneous activity in the mammalian cochlea. *J. Neurosci.* 30: 1539–1550. [26]

Tritsch, N. X., Yi, E., Gale, J. E., Glowatzki, E., and Bergles, D. E. 2007. The origin of spontaneous activity in the developing auditory system. *Nature* 450: 50–55. [26]

Tropea, D., Van Wart, A., and Sur, M. 2009. Molecular mechanisms of experience-dependent plasticity in visual cortex. *Philos. Trans. R. Soc. Lond., B, Biol. Sci.* 364: 341–355. [26]

Trouslard, J., Marsh, S. J., and Brown, D. A. 1993. Calcium entry through nicotinic receptor channels and calcium channels in cultured rat superior cervical ganglion cells. *J. Physiol.* 481: 251–271. [12]

Trueta, C., Mendez, B., and De-Miguel, F. F. 2003. Somatic exocytosis of serotonin mediated by L-type calcium channels in cultured leech neurones. *J. Physiol.* 547: 405–416. [18]

Truman, J. W., Thorn, R. S., and Robinow, S. 1992. Programmed neuronal death in insect development. *J. Neurobiol.* 23: 1295–1311. [25]

Ts'o, D. Y., and Gilbert, C. D. 1988. The organization of chromatic and spatial interactions in the primate striate cortex. *J. Neurosci.* 8: 1712–1727. [3]

Ts'o, D. Y., Frostig, R. D., Lieke, E. E., and Grinvald, A. 1990. Functional organization of primate visual cortex revealed by high resolution optical imaging. *Science* 249: 417–420. [3]

Ts'o, D. Y., Gilbert, C. D., and Wiesel, T. N. 1986. Relationships between horizontal interactions and functional architecture in cat striate cortex as revealed by cross-correlation analysis. *J. Neurosci.* 6: 1160–1170. [3]

Ts'o, D. Y., Roe, A. W., and Gilbert, C. D. 2001. A hierarchy of the functional organization for color, form and disparity in primate visual area V2. *Vision Res.* 41: 1333–1349. [3]

Ts'o, D. Y., Zarella, M., and Burkitt, G. 2009. Whither the hypercolumn? *J. Physiol.* 587: 2791–2805. [3]

Tsao, D. Y., and Livingstone, M. S. 2008. Mechanisms of face perception. *Annu. Rev. Neurosci.* 31: 411–437. [3]

Tsao, D. Y., Freiwald, W. A., Tootell, R. B., and Livingstone, M. S. 2006. A cortical region consisting entirely of face-selective cells. *Science* 311: 670–674. [3, 23]

Tschopp, P., and Duboule, D. 2011. A regulatory 'landscape effect' over the *HoxD* cluster. *Dev. Biol.* 351: 288–296. [25]

Tsetlin, V., and Hucho, F. 2009. Nicotinic acetylcholine receptors at atomic resolution. *Curr. Opin. Pharmacol.* 9: 306–310. [5]

Tsien, R. W. 1987. Calcium currents in heart cells and neurons. In *Neuromodulation: The Biochemical Control of Neuronal Excitability*. Oxford University Press, New York, pp. 206–242. [12]

Tsien, R. Y. 1989. Fluorescent probes of cell signaling. *Annu. Rev. Neurosci.* 12: 227–253. [13]

Tsubokawa, H., and Ross, W. N. 1996. IPSPs modulate spike backpropagation and associated $[Ca^{2+}]_i$ changes in the dendrites of hippocampal CA1 pyramidal neurons. *J. Neurophysiol.* 76: 2896–2906. [8]

Tsubokawa, H., and Ross, W. N. 1997. Muscarinic modulation of spike backpropagation in the apical dendrites of hippocampal CA1 pyramidal neurons. *J Neurosci* 17: 5782–5791. [18]

Tsuda, M., Tozaki-Saitoh, H., and Inoue, K. 2010. Pain and purinergic signaling. *Brain Res. Rev.* 63: 222–232. [14]

Tsujino, N., and Sakurai, T. 2009. Orexin/hypocretin: a neuropeptide at the interface of sleep, energy homeostasis, and reward system. *Pharmacol. Rev.* 61: 162–176. [14]

Tsunoo, A., Yoshii, M., and Narahashi, T. 1986: Block of calcium channels by enkephalin and somatostatin in neuroblastoma-glioma hybrid NG108-15 cells. *Proc. Nat. Acad. Sci. USA* 83: 9832–9836. [12]

Tsuzuki, K., and Suga, N. 1988. Combination-sensitive neurons in the ventroanterior area of the auditory cortex of the mustached bat. *J. Neurophysiol.* 60: 1908–1923. [22]

Tucker, K. L., Meyer, M., and Barde, Y. A. 2001. Neurotrophins are required for nerve growth during development. *Nat. Neurosci.* 4: 29–37. [25]

Turner, D. L., and Cepko, C. L. 1987. A common progenitor for neurons and glia persists in rat retina late in development. *Nature* 328: 131–136. [25]

Turrini, P., Casu, M. A., Wong, T. P., De Koninck, Y., Ribeiro-da-Silva, A., and Cuello, A. C. 2001. Cholinergic nerve terminals establish classical synapses in the rat cerebral cortex: synaptic pattern and age-related atrophy. *Neuroscience* 105: 277–285. [14]

Tyrrell, L., Renganathan, M., Dib-Hajj, S. D., and Waxman, S. G. 2001. Glycosylation alters steady-state inactivation of sodium channel $Na_v1.9$/NaN in dorsal root ganglion neurons and is developmentally regulated. *J. Neurosci.* 21: 9629–9637. [5]

Tyzio, R., Allene, C., Nardou, R., Picardo, M. A., Yamamoto, S., Sivakumaran, S., Caiati, M. D., Rheims, S., Minlebaev, M., Milh, M., Ferré, P., Khazipov, R., Romette, J-L., Lorquin, J., Cossart, R., Khalilov, I., Nehlig, A., Cherubini, E., and Ben-Ari, Y. 2011. Depolarizing actions of GABA in immature neurons depend neither on ketone bodies nor on pyruvate. *J. Neurosci.* 31: 34–45. [25]

Udenfriend, S. 1950. Identification of γ-aminobutyric acid in brain by the isotope derivative method. *J. Biol. Chem.* 187: 65–69. [14]

Uğurbil, K., Hu, X., Chen, W., Zhu, Z-H., Kim, S-G., and Georgopolos, A. 1999. Functional mapping in the human brain using high magnetic fields. *Philos. Trans. R. Soc. Lond., B, Biol. Sci.* 354: 1195–1213. [1]

Ullian, E. M., McIntosh, J. M., and Sargent, P. B. 1997. Rapid synaptic transmission in the avian ciliary ganglion is mediated by two distinct classes of nicotinic receptors. *J. Neurosci.* 17: 7210–7219. [17]

Umbriaco, D., Watkins, K. C., Descarries, L., Cozzari, C., and Hartman, B. K. 1994. Ultrastructural and morphometric features of the acetylcholine innervation in adult rat parietal cortex: an electron microscopic study in serial sections. *J. Comp. Neurol.* 348: 351–373. [14]

Ungerstedt, U. 1971. Stereotaxic mapping monoamine pathways in the rat brain. *Acta Physiol. Scand. Suppl.* 367: 1–49. [14]

Unwin, N. 1995. Acetylcholine receptor imaged in the open state. *Nature* 373: 37–43. [5]

Unwin, N. 2005. Refined structure of the nicotinic acetylcholine receptor at 4Å resolution. *J. Mol. Biol.* 346: 967–989. [5]

Unwin, N., Miyazawa, A., Li, J., and Fujiyoshi, Y. 2002. Activation of the nicotinic acetylcholine receptor involves a switch in conformation of the α subunits. *J. Mol. Biol.* 319: 1165–1176. [5]

Unwin, N., Toyoshima, C., and Kubalek, E. 1988. Arrangement of acetylcholine receptor subunits in the resting and desensitized states, determined by cryoelectron microscopy of crystallized

Torpedo postsynaptic membranes. *J. Cell Biol.* 107: 1123–1138. [5]

Vabnick, I., and Shrager, P. 1998. Ion channel redistribution and function during development of the myelinated axon. *J. Neurobiol.* 37: 80–96. [8]

Vale, R. D., and Fletterick, R. J. 1997. The design plan of kinesin motors. *Annu. Rev. Cell Dev. Biol.* 13: 745–777. [15]

Vallbo, A. B., and Hagbarth, K. E. 1968. Mechnoreceptor activity recorded from human peripheral nerves. *Electroencephalogr. Clin. Neurophysiol.* 25: 407. [21]

Vallee, R. B., and Bloom, G. S. 1991. Mechanisms of fast and slow axonal transport. *Annu. Rev. Neurosci.* 14: 59–92. [15]

Vallee, R. B., and Gee, M. A. 1998. Make room for dynein. *Trends Cell Biol.* 8: 490–494. [15]

Vallee, R. B., Shpetner, H. S., and Paschal, B. M. 1989. The role of dynein in retrograde axonal transport. *Trends Neurosci.* 12: 66–70. [15]

Van Boven, R. W., Hamilton, R. H., Kauffman, T., Keenan, J. P., and Pascual-Leone, A. 2000. Tactile spatial resolution in blind Braille readers. *Neurology* 54: 2230–2236. [21]

van den Pol, A. N. 2010. Excitatory neuromodulator reduces dopamine release, enhancing prolactin secretion. *Neuron* 65: 147–149. [14]

van der Loos, H., and Dorfl, J. 1978. Does the skin tell the somatosensory cortex how to construct a map of the periphery? *Neurosci. Lett.* 7: 23–30. [21]

van der Loos, H., and Woolsey, T. A. 1973. Somatosensory cortex: structural alterations following early injury to sense organs. *Science* 179: 395–398. [26]

van der Loos, H., Dorfl, J., and Welker, E. 1984. Variation in pattern of mystacial vibrissae in mice. A quantitative study of ICR stock and several inbred strains. *J. Hered.* 75: 326–336. [21]

Van Essen, D. C. 1997. A tension based theory of morphogenesis and compact wiring in the central nervous system. *Nature* 385: 313–318. [3]

Van Essen, D. C., and Drury, H. A. 1997. Structural and functional analyses of human cerebral cortex using a surface-based atlas. *J. Neurosci.* 17: 7079–7102. [3]

Van Essen, D., and Jansen, J. K. 1974. Reinnervation of rat diaphragm during perfusion with α-bungarotoxin. *Acta Physiol. Scand.* 91: 571–573. [27]

Van Essen, D., and Kelly, J. 1973. Correlation of cell shape and function in the visual cortex of the cat. *Nature* 241: 403–405. [10]

van Hooft, J. A., Spier, A. D., Yakel, J. L., Lummis, S. C., and Vijverberg, H. P. 1998. Promiscuous coassembly of serotonin 5-HT_3 and nicotinic α4 receptor subunits into Ca^{2+}-permeable ion channels. *Proc. Natl. Acad. Sci. USA* 95: 11456–11461. [14]

van Westen, D., Fransson, P., Olsrud, J., Rosen, B., Lundborg, G., and Larsson, E. M. 2004. Fingersomatotopy in area 3b: an fMRI-study. *BMC Neurosci.* 5: 28. [21]

Vandermaelen, C. P., and Aghajanian, G. K. 1983. Electrophysiological and pharmacological characterization of serotonergic dorsal raphe neurons recorded extracellularly and intracellularly in rat brain slices. *Brain Res.* 289: 109–119. [14]

Varga, Z. M., Bandtlow, C. E., Erulkar, S. D., Schwab, M. E., and Nicholls, J. G. 1995. The critical period for repair of CNS of neonatal opossum (*Monodelphis domestica*) in culture: correlation with development of glial cells, myelin and growth-inhibitory molecules. *Eur. J. Neurosci.* 7: 2119–2129. [27]

Varma, N., Carlson, G. C., Ledent, C., and Alger, B. E. 2001. Metabotropic glutamate receptors drive the endocannabinoid system in hippocampus. *J. Neurosci.* 21: RC188 (1–5). [12]

Vassar, R., Ngai, J., and Axel, R. 1993. Spatial segregation of odorant receptor expression in the mammalian olfactory epithelium. *Cell* 74: 309–318. [19]

Vautrin, J., and Kriebel, M. E. 1991. Characteristics of slow-miniature end plate currents show a subunit composition. *Neuroscience* 41: 71–88. [13]

Vergara, C., Latorre, R. Marrion, N. V., and Adelman, J. P. 1998. Calcium-activated potassium channels. *Curr. Opin. Neurobiol.* 8: 321–329. [5]

Verkhratsky, A., Krishtal, O. A., and Burnstock, G. 2009. Purinoceptors on neuroglia. *Mol. Neurobiol.* 39: 190–208. [10]

Veronin, L. L., and Cherubini, E. 2004. "Deaf, mute and whispering" silent synapses: their role in synaptic plasticity. *J. Physiol.* 557: 3–12. [16]

Vessal, M., and Darian-Smith, C. 2010. Adult neurogenesis occurs in primate sensorimotor cortex following cervical dorsal rhizotomy. *J. Neurosci.* 30: 8613–8623. [25]

Vidal-Sanz, M., Bray, G. M., and Aguayo, A. J. 1991. Regenerated synapses in the superior colliculus after the regrowth of retinal ganglion cell axons. *J. Neurocytol.* 20: 940–952. [27]

Viktorin, G., Chiuchitu, C., Rissler, M., Varga, Z. M., and Westerfield, M. 2009. Emx3 is required for the differentiation of dorsal telencephalic neurons. *Dev. Dyn.* 238: 1984–1998. [25]

Villanueva, S., Fiedler, J., and Orrego, F. 1990. A study in rat brain cortex synaptic vesicles of endogenous ligands for *N*-methyl-D-aspartate receptors. *Neuroscience* 37: 23–30. [13]

Villarroel, A., Herlitze, S., Koenen, M., and Sakmann, B. 1991. Location of threonine residue in the α-subunit M2 transmembrane segment that determines the ion flow through the acetylcholine receptor channel. *Proc. R. Soc. Lond., B, Biol. Sci.* 243: 69–74. [5]

Vincent, S. B. 1912. The function of the vibrissae in the behavior of the white rat. *Behav. Monogr:* 1–82. [21]

Virchow, R. 1959. *Cellularpathologie.* (F. Chance, trans.) Hirschwald, Berlin. (Excerpts are from pp. 310, 315, and 317.) [10]

Virkki, L. V., Choi, I., Davis, B. A., and Boron, W. F. 2003. Cloning of a Na^+-driven Cl-HCO_3 exchanger from squid giant fiber lobe. *Am. J. Physiol.* 285: C771–C780. [9]

Voderholzer, U., Hornyak, M., Thiel, B., Huwig-Poppe, C., Kiemen, A., Konig, A., Backhaus, J., Riemann, D., Berger, M., and Hohagen, F. 1998. Impact of experimentally induced serotonin deficiency by tryptophan depletion on sleep EEG in healthy subjects. *Neuropsychopharmacology* 18: 112–124. [15]

Vollrath, M. A., Kwan, K. Y., and Corey, D. P. 2007. The micromachinery of mechanotransduction in hair cells. *Annu. Rev. Neurosci.* 30: 339–365. [19]

von Bekesy, G. 1960. *Experiments in Hearing*. McGraw-Hill, New York. [22]

von Euler, U. S. 1956. *Noradrenaline.* Charles Thomas, Springfield, IL. [17]

Von Euler, U. S., and Gaddum, J. H. 1931. An unidentified depressor substance in certain tissue extracts. *J. Physiol.* 72: 74–87. [14]

von Heimendahl, M., Itskov, P. M., Arabzadeh, E., and Diamond, M. E. 2007. Neuronal activity in rat barrel cortex underlying texture discrimination. *PLoS Biol* 5: e305. [21]

Vugler, A. A. 2010. Progress toward the maintenance and repair of degenerating retinal circuitry. *Retina* 30: 983–1001. [28]

Vyskoĉil, F., Malomouzh, A. I., and Nikolsky, E. E. 2009. Non-quantal acetylcholine release at the neuromuscular junction. *Physiol. Res.* 58: 763–784. [13]

Wada, Y., and Yamamoto, T. 2001. Selective impairment of facial recognition due to a haematoma restricted to the right fusiform and lateral occipital region. *J. Neurol. Neurosurg. Psychiatry* 71: 254–257. [23]

Wager, T. D., Rilling, J. K., Smith, E. E., Sokolik, A., Casey, K. L., Davidson, R. J., Kosslyn, S. M., Rose, R. M., and Cohen, J. D. 2004. Placebo-induced changes in fMRI in the anticipation and experience of pain. *Science* 303: 1162–1167. [21]

Wagner, J. A., Carlson, S. S., and Kelly, R. B. 1978. Chemical and physical characterization of cholinergic synaptic vesicles. *Biochemistry* 17: 1199–1206. [13]

Wagner, S., Castel, M., Gainer, H., and Yarom, Y. 1997. GABA in the mammalian suprachiasmatic nucleus and its role in diurnal rhythmicity. *Nature* 387: 598–603. [17]

Wagner, S., Sagiv, N., and Yarom, Y. 2001. GABA-induced current and circadian regulation of chloride in neurones of the rat suprachiasmatic nucleus. *J. Physiol.* 537: 853–869. [17]

Wahl-Schott, C., and Biel, M. 2009. HCN channels: structure, cellular regulation and physiological function. *Cell. Mol. Life Sci.* 66: 470–494. [12]

Wainer, B. H., Levey, A. I., Mufson, E. J., and Mesulam, M. M. 1984. Cholinergic systems in mammalian brain identified with antibodies against choline acetyltransferase. *Neurochem. Int.* 6: 163–182. [14]

Waite, A., Tinsley, C. L., Locke, M., and Blake, D. J. 2009. The neurobiology of the dystrophin-associated glycoprotein complex. *Ann. Med.* 41: 344–359. [11]

Waldmann, R., Champigny, G., Bassilana, F., Heurteaux, C., and Lazdunski, M. 1997. A proton-gated cation channel involved in acid-sensing. *Nature* 386: 173–177. [19]

Walker, D., and De Waard, M. 1998. Subunit interaction sites in voltage-dependent Ca^{2+} channels: role in channel function. *Trends Neurosci.* 21: 148–154. [5]

Wallner, M., Hanchar, H. J., and Olsen, R. W. 2003. Ethanol enhances $\alpha_4\beta_3\delta$ and $\alpha_6\beta_3\delta$ γ-aminobutyric acid type A receptors at low concentrations known to affect humans. *Proc. Natl. Acad. Sci. USA* 100: 15218–1522. [14]

Walsh, E. J., and McGee, J. 1987. Postnatal development of auditory nerve and cochlear nucleus neuronal responses in kittens. *Hear Res.* 28: 97–116. [26]

Walter, J., Henke-Fahle, S., and Bonhoeffer, F. 1987. Avoidance of posterior tectal membranes by temporal retinal axons. *Development* 101: 909–913. [25]

Walter, J., Kern-Veits, B., Huf, J., Stolze, B., and Bonhoeffer, F. 1987. Recognition of position-specific properties of tectal cell membranes by retinal axons *in vitro*. *Development* 101: 685–696. [25]

Walters, E. T., and Cohen, L. B. 1997. Functions of the LE sensory neurons in *Aplysia*. *Invert. Neurosci.* 3: 15–25. [18]

Walum, H., Westberg, L., Henningsson, S., Neiderhiser, J. M., Reiss, D., Igl, W., Ganiban, J. M., Spotts, E. L., Pedersen, N. L., Eriksson, E., and Lichtenstein, P. 2008. Genetic variation in the vasopressin receptor 1a gene (*AVPR1A*) associates with pair-bonding behavior in humans. *Proc. Natl. Acad. Sci. USA* 105: 14153–14156. [14]

Wang, D. O., Kim, S. M., Zhao, Y., Hwang, H., Miura, S. K., Sossin, W. S., and Martin, K. C. 2009. Synapse- and stimulus-specific local translation during long-term neuronal plasticity. *Science* 324: 1536–1540. [12]

Wang, H-S., and McKinnon, D. 1995. Potassium currents in rat prevertebral and paravertebral sympathetic neurones: control of firing properties. *J. Physiol.* 485: 319–325. [17]

Wang, H-S., Pan, Z., Shi, W., Brown, B. S., Wymore, R. S., Cohen, I. S., Dixon, J. E., and McKinnon, D. 1998. KCNQ2 and KCNQ3 potassium channel subunits: molecular correlates of the M-channel. *Science* 282: 1890–1893. [12]

Wang, H., and Macagno, E. R. 1997. The establishment of peripheral sensory arbors in the leech: *in vivo* time-lapse studies reveal a highly dynamic process. *J. Neurosci.* 17: 2408–2419. [18]

Wang, H., and Macagno, E. R. 1998. A detached branch stops being recognized as self by other branches of a neuron. *J. Neurobiol.* 35: 53–64. [18]

Wang, H., Kunkel, D. D., Martin, T. M., Schwartzkroin, P. A., and Tempel, B. L. 1993. Heteromultimeric K^+ channels in terminal and juxtaparanodal regions of neurons. *Nature* 365: 75–79. [8]

Wang, R. Y., and Aghajanian, G. K. 1977. Antidromically identified serotonergic neurons in the rat midbrain raphe: evidence for collateral inhibition. *Brain Res.* 132: 186–193. [14]

Wang, T., and Montell, C. 2007. Phototransduction and retinal degeneration in *Drosophila*. *Pflugers Arch.* 454: 821–847. [20]

Wang, X., and Kadia, S. C. 2001. Differential representation of species-specific primate vocalizations in the auditory cortices of marmoset and cat. *J. Neurophysiol.* 86: 2616–2620. [22]

Wangemann, P. 2006. Supporting sensory transduction: cochlear fluid homeostasis and the endocochlear potential. *J. Physiol.* 576: 11–21. [22]

Ward, N. S., and Frackowiak, R. S. 2004. Towards a new mapping of brain cortex function. *Cerebrovasc. Dis.* 17(Suppl 3): 35–38. [1, 24]

Warr, W. B. 1975. Olivocochlear and vestibular efferent neurons of the feline brain stem: their location, morphology and number determined by retrograde axonal transport and acetylcholinesterase histochemistry. *J. Comp. Neurol.* 161: 159–181. [22]

Warr, W. B., Guinan J. J., and White J. S. 1986. Organization of the efferent fibers: the later and medial olivocochlear systems. In Altschuler, R. A., Hoffman, D. W., and Bobbin, R. P. (eds.), *Neurobiology of Hearing: The Cochlea*. Raven Press, New York. pp. 333–348. [22]

Waselus, M., Van Bockstaele, E. J. 2007. Co-localization of corticotropin-releasing factor and vesicular glutamate transporters within axon terminals of the rat dorsal raphe nucleus. *Brain Res.* 1174: 53–65. [14]

Wässle, H. 2004. Parallel processing in the mammalian retina. *Nat. Rev. Neurosci.* 5: 747–757. [20]

Wässle, H., Koulen, P., Brandstatter, J. H., Fletcher, E. L., and Becker, C. M. 1998. Glycine and GABA receptors in the mammalian retina. *Vision Res.* 38: 1411–1430. [20]

Waters, J., Schaefer, A., and Sakmann, B. 2005. Backpropagating action potentials in neurones: Measurement, mechanisms, and potential functions. *Prog. Biophys. Mol. Biol.* 87: 145–170. [8]

Watkins, J. C., and Evans, R. H. 1981. Excitatory amino acid transmitters. *Annu. Rev. Pharmacol. Toxicol.* 21: 165–204. [11]

Watson, T. L., and Krekelberg, B. 2009. The relationship between saccadic suppression and perceptual stability. *Curr. Biol.* 19: 1040–1043. [24]

Watters, M. R. 2005 Tropical marine neurotoxins: Venoms to drugs. *Semin. Neurol.* 25: 278–289. [7]

Weatherbee, S. D., Anderson, K. V., and Niswander, L. A. 2006. LDL-receptor-related protein 4 is crucial for formation of the neuromuscular junction. *Development* 133: 4993–5000. [27]

Weber, E., Evans, C. J., and Barchas, J. D. 1983. Multiple endogenous ligands for opioid receptors. *Trends Neurosci.* 6: 333–336. [14]

Webster, H., and Aström, K. E. 2009. Gliogenesis: Historical perspectives, 1839–1985. *Adv. Anat. Embryol. Cell Biol.* 202: 1–109. [10]

Wehner, R. 1989. Neurobiology of polarization vision. *Trends Neurosci.* 12: 353–359. [18]

Wehner, R. 1994a. Himmelsbild und kompassauge—neurobiologie eines Navigationssystems, *Verhand. Deutsch. Zool. Ges.* 87: 9–37. [18]

Wehner, R. 1994b. The polarization-vision project: championing organismic biology. *Fortschr. Zool.* 31: 11–53. [18]

Wehner, R. 1996. Polarisationsmuster analyse bei Insekten. *Nova Acta Leoplodina NF* 72: 159–183. [18]

Wehner, R. 1997. The ant's celestial compass: Spectral and polarization channels. In Lehrer, M. (ed.) *Orientation and Communication in Arthropods*. Birkhauser, Basel, Switzerland. pp. 145–185. [18]

Wehner, R., and Bernard, G. D. 1993. Photoreceptor twist: a solution to the false-color problem. *Proc. Natl. Acad. Sci. USA* 90: 4132–4135. [18]

Wehner, R., and Muller, M. 2006. The significance of direct sunlight and polarized skylight in the ant's celestial system of navigation. *Proc. Natl. Acad. Sci. USA* 103: 12575–12579. [18]

Wei, A. D., Gutman, G. A., Aldrich, R., Chandy, K. G., Grissmer, S., and Wulff, H. 2005. International Union of Pharmacology. LII. Nomenclature and molecular relationships of calcium-activated potassium channels. *Pharmacol. Rev.* 57: 463–472. [5]

Weiner, N., and Rabadjija, M. 1968. The effect of nerve stimulation on the synthesis and metabolism of norepinephrine from cat spleen during sympathetic nerve stimulation. *J. Pharmacol. Exp. Ther.* 160: 61–71. [15]

Weinfeld, A. B., Yuksel, E., Boutros, S., Gura, D. H., Akyurek, M., and Friedman, J. D. 2000. Clinical and scientific considerations in leech therapy for the management of acute venous congestion: an updated review. *Ann. Plast. Surg.* 45: 207–212. [18]

Weinreich, D. 1970. Ionic mechanisms of post-tetanic potentiation at the neuromuscular junction of the frog. *J. Physiol.* 212: 431–446. [16]

Weinstein, S., Semmes, J., Ghent, L., and Teuber, H. L. 1958. Roughness discrimination after penetrating brain injury in man: analysis according to locus of lesion. *J. Comp. Physiol. Psychol.* 51: 269–275. [21]

Weisblat, D. A., and Kuo, D.-H. 2009. *Helobdella* (Leech), a model for developmental studies. In *Emerging Model Organisms, a Laboratory Manual.* Cold Spring Harbor Laboratory, Cold Spring Harbor, NY. pp. 245–274. [18]

Weisblat, D. A., and Shankland, M. 1985. Cell lineage and segmentation in the leech. *Philos. Trans. R. Soc. Lond., B, Biol. Sci.* 312: 39–56. [25]

Weisblat, D. A., Zackson, S. L., Blair, S. S., and Young, J. D. 1980. Cell lineage analysis by intracellular injection of fluorescent tracers. *Science* 209: 1538–1541. [25]

Weiss, P., and Hiscoe, H. B. 1948. Experiments of the mechanism of nerve growth. *J. Exp. Zool.* 107: 315–395. [15]

Weissman, T. A., Riquelme, P. A., Ivic, L., Flint, A. C., and Kriegstein, A. R. 2004. Calcium waves propagate through radial glial cells and modulate proliferation in the developing neocortex. *Neuron* 43: 647–661. [10]

Weissmann, C. 2009. Thoughts on mammalian prion strains. *Folia Neuropathol.* 47: 104–113. [28]

Weitsen, H. A., and Weight, F. F. 1977. Synaptic innervation of sympathetic ganglion cells in the bullfrog. *Brain Res.* 128: 197–211. [15]

Welker, C. 1976. Receptive fields of barrels in the somatosensory neocortex of the rat. *J. Comp. Neurol.* 166: 173–189. [21]

Welker, E., and van der Loos, H. 1986. Quantitative correlation between barrel-field size and the sensory innervation of the whiskerpad: a comparative study in six strains of mice bred for different patterns of mystacial vibrissae. *J. Neurosci.* 6: 3355–3373. [21]

Welsh, J. P., Yamaguchi, H., Zeng, X. H., Kojo, M., Nakada, Y., Takagi, A., Sugimori, M., and Llinás, R. R. 2005. Normal motor learning during pharmacological prevention of Purkinje cell long-term depression. *Proc. Natl. Acad. Sci. USA* 102: 17166–17171. [24]

Werker, J. F., and Tees, R. C. 2005. Speech perception as a window for understanding plasticity and commitment in language systems of the brain. *Dev. Psychobiol.* 46: 233–251. [26]

Werman, R., Davidoff, R. A., and Aprison, M. H. 1968. Inhibitory action of glycine on spinal neurons in the cat. *J. Neurophysiol.* 31: 81–95. [14]

Wersinger, E., McLean, W. J., Fuchs, P. A., and Pyott, S. J. 2010. BK channels mediate cholinergic inhibition of high frequency cochlear hair cells. *PLoS One* 5: e13836. [22]

Wessberg, J., Olausson, H., Fernström, K. W., and Vallbo, A. B. 2003. Receptive field properties of unmyelinated tactile afferents in the human skin. *J. Neurophysiol.* 89: 1567–1575. [21]

Westheimer, G. 2009. The third dimension in the primary visual cortex. *J. Physiol.* 587: 2807–2816. [3]

Weston, J. A. 1970. The migration and differentiation of neural crest cells. *Adv. Morphog.* 8: 41–114. [25]

Whim, M. D., Church, P. J., and Lloyd, P. E. 1993. Functional roles of peptide cotransmitters at neuromuscular synapses in *Aplysia*. *Mol. Neurobiol.* 7: 335–347. [15]

Whitfield, I. C. 1979. The object of the sensory cortex. *Brain Behav. Evol.* 16: 129–154. [23]

Whittington, M. A., and Traub, R. D. 2003. Interneuron diversity series: inhibitory interneurons and network oscillations *in vitro*. *Trends Neurosci.* 26: 676–682. [11, 14]

Wickelgren, W. O., Leonard, J. P., Grimes, M. J., and Clark, R. D. 1985. Ultrastructural correlates of transmitter release in presynaptic areas of lamprey reticulospinal axons. *J. Neurosci.* 5: 1188–1201. [13]

Wickman, K. D., Iñiguez-Lluhi, J. A., Davenport, P. A., Taussig, R., Krapivinsky, G. B., Linder, M. E., Gilman, A. G., and Clapham, D. E. 1994. Recombinant G-protein βγ subunits activate the muscarinic-gated atrial potassium channel. *Nature* 368: 255–257. [12]

Wiech, K., Preissl, H., and Birbaumer, N. 2001. Neural networks and pain processing. New insights from imaging techniques. *Anaesthesist* 50: 2–12. [21]

Wiederhold, M. L., and Kiang, N. Y. 1970. Effects of electric stimulation of the crossed olivocochlear bundle on single auditory-nerve fibers in the cat. *J. Acoust. Soc. Am.* 48: 950–965. [22]

Wiese, K., ed. 2002. *The Crustacean Nervous System.* Berlin, Germany. [18]

Wiesel, T. N. 1982. Postnatal development of the visual cortex and the influence of environment. *Nature* 299: 583–591. [25, 26]

Wiesel, T. N., and Hubel, D. H. 1963a. Effects of visual deprivation on morphology and physiology of cells in the cats lateral geniculate body. *J. Neurophysiol.* 26: 978–993. [26]

Wiesel, T. N., and Hubel, D. H. 1963b. Single-cell responses in striate cortex of kittens deprived of vision in one eye. *J. Neurophysiol.* 26: 1003–1017. [26]

Wiesel, T. N., and Hubel, D. H. 1965. Comparison of the effects of unilateral and bilateral eye closure on cortical unit responses in kittens. *J. Neurophysiol.* 28: 1029–1040. [26]

Wiesel, T. N., and Hubel, D. H. 1974. Ordered arrangement of orientation columns in monkeys lacking visual experience. *J. Comp. Neurol.* 158: 307–318. [26]

Willard, A. L. 1981. Effects of serotonin on the generation of the motor program for swimming by the medicinal leech. *J. Neurosci.* 1: 936–944. [18]

Williams, J. T., North, R. A., Shefner, S. A., Nishi, S., and Egan, T. M. 1984. Membrane properties of rat locus coeruleus neurones. *Neurosience* 13: 137–156. [14]

Williams, K. W., Scott, M. M., and Elmquist, J. K. 2009. From observation to experimentation: leptin action in the mediobasal hypothalamus. *Am. J. Clin. Nutr.* 89: 985S–990S. [17]

Williams, P. R., Suzuki, S. C., Yoshimatsu, T., Lawrence, O. T., Waldron, S. J., Parsons, M. J., Nonet, M. L., and Wong, R. O. 2010. In vivo development of outer retinal synapses in the absence of glial contact. *J. Neurosci.* 30: 11951–11961. [10]

Williamson, R., and Chrachri, A. 2007. A model biological neural network: the cephalopod vestibular system. *Philos. Trans. R. Soc. Lond., B, Biol. Sci.* 362: 473–481. [22]

Wilson Horch, H. L., and Sargent, P. B. 1995. Perisynaptic surface distribution of multiple classes of nicotinic acetylcholine receptors on neurons in the chicken ciliary ganglion. *J. Neurosci.* 15: 7778–7795. [27]

Wilson Horch, H. L., and Sargent, P. B. 1996. Effects of denervation on acetylcholine receptor clusters on frog cardiac ganglion neurons as revealed by quantitative laser scanning confocal microscopy. *J. Neurosci.* 16: 1720–1729. [27]

Wilson, L., and Maden, M. 2005. The mechanisms of dorsoventral patterning in the vertebrate neural tube. *Dev. Biol.* 282: 1–13. [25]

Wilson, P. M., Fryer, R. H., Fang, Y., and Hatten, M. E. 2010. *Astn2*, a novel member of the astrotactin gene family, regulates the trafficking of ASTN1 during glial-guided neuronal migration. *J. Neurosci.* 30: 8529–8540. [25]

Wilson, R. I., and Nicoll, R. A. 2001. Endogenous cannabinoids mediate retrograde signalling at hippocampal synapses. *Nature* 410: 588–592. [12]

Wilson, R. I., and Nicoll, R. A. 2002. Endocannabinoid signaling in the brain. *Science* 296: 678–682. [12]

Windhorst, U. 2007. Muscle proprioceptive feedback and spinal networks. *Brain. Res. Bull.* 73: 155–202. [24]

Winer, J. A., and Lee, C. C. 2007. The distributed auditory cortex. *Hear. Res.* 229: 3–13. [22]

Winks, J. S., Hughes, S., Filippov, A. K., Tatulian, L., Abogadie, F. C., Brown, D. A., and Marsh, S. J. 2005. Relationship between membrane phosphatidylinositol-4,5-bisphosphate and receptor-mediated inhibition of native neuronal M channels. *J. Neurosci.* 25: 3400–3413. [12]

Winslow, J. T., Hastings, N., Carter, C. S., Harbaugh, C. R., and Insel, T. R. 1993. A role for central vasopressin in pair bonding in monogamous prairie voles. *Nature* 365: 545–548. [14]

Winslow, R. L., and Sachs, M. B. 1987. Effect of electrical stimulation of the crossed olivocochlear bundle on auditory nerve response to tones in noise. *J. Neurophysiol.* 57: 1002–1021. [22]

Wise, A., Jupe, S. C., and Rees, S. 2004. The identification of ligands at orphan G-protein coupled receptors. *Annu. Rev. Pharmacol. Toxicol.* 44: 43–66. [14]

Wise, D. S., Schoenborn, B. P., and Karlin, A. 1981. Structure of acetylcholine receptor dimer determined by neutron scattering and electron microscopy. *J. Biol. Chem.* 256: 4124–4126. [5]

Wittlinger, M., Wehner, R., and Wolf, H. 2006. The ant odometer: stepping on stilts and stumps. *Science* 312: 1965–1967. [18]

Wittlinger, M., Wehner, R., and Wolf, H. 2007. The desert ant odometer: a stride integrator that accounts for stride length and walking speed. *J. Exp. Biol.* 210: 198–207. [18]

Witzemann, V., Brenner, H-R., and Sakmann, B. 1991. Neural factors regulate AChR subunit mRNAs at rat neuromuscular synapses. *J. Cell Biol.* 114: 125–141. [11, 27]

Wojciulik, E., Kanwisher, N., and Driver, J. 1998. Covert visual attention modulates face-specific activity in the human fusiform gyrus: fMRI study. *J. Neurophysiol.* 79: 1574–1578. [23]

Wolburg, H., and Paulus, W. 2010. Choroid plexus: Biology and pathology. *Acta. Neuropathol.* 119: 75–88. [10]

Wolburg, H., Noell, S., Mack, A., Wolburg-Buchholz, K., and Fallier-Becker, P. 2009. Brain endothelial cells and the glio-vascular complex. *Cell Tissue Res.* 335: 75–96. [10]

Wolfe, J., Hill, D. N., Pahlavan, S., Drew, P. J., Kleinfeld, D., and Feldman, D. E. 2008. Texture coding in the rat whisker system: slip-stick versus differential resonance. *PLoS Biol* 6: e215. [21]

Wollmuth, L. P., and Sobolevsky, A. I. 2004. Structure and gating of the glutamate receptor ion channel. *Trends Neurosci.* 27: 321–328. [5, 11]

Womac, A. D., Burkeen, J. F., Neuendorff, N., Earnest, D. J., and Zoran, M. J. 2009. Circadian rhythms of extracellular ATP accumulation in suprachiasmatic nucleus cells and cultured astrocytes. *Eur. J. Neurosci.* 30: 869–876. [10]

Wong-Riley, M. 1989. Cytochrome oxidase: An endogenous metabolic marker for neuronal activity. *Trends Neurosci.* 12: 94–101. [3]

Wong, R. O. 1999. Retinal waves and visual system development. *Annu. Rev. Neurosci.* 22: 29–47. [26]

Woolf, N. J. 1991. Cholinergic systems in mammalian brain and spinal cord. *Prog. Neurobiol.* 37: 475–524. [14]

Woolsey, T. A., and van der Loos, H. 1970. The structural organization of layer IV in the somatosensory region (SI) of mouse cerebral cortex. The description of a cortical field composed of discrete cytoarchitectonic units. *Brain Res.* 17: 205–242. [21]

Wotring, V. E., Miller, T. S., and Weiss, D. S. 2003. Mutations in the GABA receptor selectivity filter: a possible role for effective charges. *J. Physiol.* 548: 527–540. [5]

Wray, S. 2010. From nose to brain: development of gonadotrophin-releasing hormone-1 neurones. *J. Neuroendocrinol.* 22: 743–753. [25]

Wray, S., Grant, P., and Gainer, H. 1989. Evidence that cells expressing luteinizing hormone-releasing hormone mRNA in mouse are derived from progenitor cells in the olfactory placode. *Proc. Natl. Acad. Sci. USA* 86: 8132–8136. [17, 25]

Wu, L-G., and Saggau, P. 1997. Presynaptic inhibition of elicited neurotransmitter release. *Trends Neurosci.* 20: 204–212. [14]

Wu, L-G., Ryan, T. A., and Lagnado, L. 2007. Modes of synaptic vesicle retrieval at ribbon synapses, calyx-type synapses, and small central synapses. *J. Neuosci.* 27: 11793–11802. [13]

Wu, L-J. Sweet, T.-B., and Clapham, D. E. 2010. International Union of Basic and Clinical Pharmacology. LXXVI. Current Progress in the Mammalian TRP Ion Channel Family. *Pharmacol. Rev.* 62: 381–404. [5]

Wu, S. M. Synaptic organization of the vertebrate retina: general principles and species-specific variations: the Friedenwald lecture. *Invest. Ophthalmol. Vis. Sci.* 51: 1263–1274. [20]

Wu, W., and Wu, L. G. 2007. Rapid bulk endocytosis and its kinetics of fission pore closure at a central synapse. *Proc. Natl. Acad. Sci. USA* 104: 10234–10239. [13]

Wu, W., Li, L., Yick, L. W., Chai, H., Xie, Y., Yang, Y., Prevette, D. M., and Oppenheim, R. W. 2003. GDNF and BDNF alter the expression of neuronal NOS, c-Jun, and p75 and prevent motoneuron death following spinal root avulsion in adult rats. *J. Neurotrauma* 20: 603–612. [25]

Wurmser, A. E., Nakashima, K., Summers, R. G., Toni, N., D'Amour, K. A., Lie, D. C., and Gage, F. H. 2004. Cell fusion-independent differentiation of neural stem cells to the endothelial lineage. *Nature* 430: 350–356. [25]

Wurtz, R. H. 2008. Neuronal mechanisms of visual stability. *Vision Res.* 48: 2070–2089. [24]

Xerri, C., Stern, J. M., and Merzenich, M. M. 1994. Alterations of the cortical representation of the rat ventrum induced by nursing behavior. *J. Neurosci.* 14: 1710–1721. [21]

Xiao, C., Nashmi, R., McKinney, S., Cai, H., McIntosh, J. M., and Lester, H. A. 2009. Chronic nicotine selectively enhances $\alpha_4\beta_2^*$ nicotinic acetylcholine receptors in the nigrostriatal dopamine pathway. *J Neurosci.* 29: 12428–12439. [14]

Xiao, J., Kilpatrick, T. J., and Murray, S. S. 2009. The role of neurotrophins in the regulation of myelin development. *Neurosignals* 17: 265–276. [10]

Xu, J., and Wu, L-G. 2005. The decrease in the presynaptic calcium current is a major cause of short-term depression at a calyx-type synapse. *Neuron* 46: 633–645. [16]

Xu, J., He, L., and Wu, L-G. 2007. Role of calcium channels in short-term synaptic plasticity. *Curr. Opin. Neurobiol.* 17: 352–359. [16]

Xu, K., and Terakawa, S. 1999. Fenestration nodes and the wide submyelinic space form the basis for unusually fast impulse conduction of shrimp myelinated axons. *J. Exp. Biol.* 202: 1979–1989. [8]

Xu, R., and Salpeter, M. M. 1995. Protein kinase A regulates the degradation rate of Rs acetylcholine receptors. *J. Cell. Physiol.* 165: 30–39. [27]

Yamada, T., Placzek, M., Tanaka, H., Dodd, J., and Jessell, T. M. 1991. Control of cell pattern in the developing nervous system: polarizing activity of the floor plate and notochord. *Cell* 64: 635–647. [25]

Yamamori, T., Fukada, K., Aebersold, R., Korsching, S., Fann, M. J., and Patterson, P. H. 1989. The cholinergic neuronal differentiation factor from heart cells is identical to leukemia inhibitory factor. *Science* 246: 1412–1416. [25]

Yamane, Y., Carlson, E. T., Bowman, K. C., Wang, Z., and Connor, C. E. 2008. A neural code for three-dimensional object shape in macaque inferotemporal cortex. *Nat. Neurosci.* 11: 1352–1360. [23]

Yamasaki, M., Matsui, M., and Watanabe, M. 2010. Preferential localization of muscarinic M_1 receptor on dendritic shaft and spine of cortical pyramidal cells and its anatomical evidence for volume transmission. *J. Neurosci.* 30: 4408–4418. [14]

Yamashita, A., Singh, S. K., Kawate, T., Jin, Y., and Gouaux, E. 2005. Crystal structure of a bacterial homologue of Na^+/Cl^--dependent neurotransmitter transporters. *Nature* 437: 215–223. [9]

Yamashita, M., and Wässle, H. 1991. Responses of rod bipolar cells isolated from the rat retina to the glutamate agonist 2-amino-4-phosphonobutyric acid (APB). *J. Neurosci.* 11: 2372–2382. [20]

Yan, T. C., Hunt, S. P., and Stanford, S. C. 2009. Behavioural and neurochemical abnormalities in mice lacking functional tachykinin-1 (NK1) receptors: a model of attention deficit hyperactivity disorder. *Neuropharmacology* 57: 627–635. [14]

Yang, J., Ellinor, P. T., Sather, W. A., Zhang, J. F., and Tsien, R. W. 1993. Molecular determinants of Ca^{2+} selectivity and ion permeation in L-type calcium channels. *Nature* 366: 158–161. [5]

Yang, N., George, A. L., and Horn, R. 1996. Molecular basis of charge movement in voltage-gated sodium channels. *Neuron* 16: 113–122. [5]

Yang, X. F., Miao, Y., Ping, Y., Wu, H. J., Yang, X. L., and Wang, Z. 2011. Melatonin inhibits tetraethylammonium-sensitive potassium channels of rod ON type bipolar cells via MT2 receptors in rat retina. *Neuroscience* 173: 19–29. [20]

Yang, X. L., Gao, F., and Wu, S. M. 1999. Modulation of horizontal cell function by $GABA_A$ and $GABA_C$ receptors in dark- and light-adapted tiger salamander retina. *Vis. Neurosci.* 16: 967–979. [20]

Yarom, Y., Sugimori, M., and Llinás, R. 1985. Ionic currents and firing patterns or mammalian vagal motoneurons *in vitro*. *Neuroscience* 16: 719–737. [7]

Yates, B. J., and Miller, A. D. 1994. Properties of sympathetic reflexes elicited by natural vestibular stimulation: implications for cardiovascular control. *J. Neurophysiol.* 71: 2087–2092. [22]

Yau, K. W. 1976a. Physiological properties and receptive fields of mechanosensory neurones in the head ganglion of the leech: comparison with homologous cells in segmental ganglia. *J. Physiol.* 263: 489–512. [18]

Yau, K. W. 1976b. Receptive fields, geometry and conduction block of sensory neurones in the central nervous system of the leech. *J. Physiol.* 263: 513–538. [18]

Yau, K. W., and Nakatani, K. 1985. Light-suppressible, cyclic GMP-sensitive conductance in the plasma membrane of a truncated rod outer segment. *Nature* 317: 252–255. [20]

Ye, W., Shimamura, K., Rubenstein, J. L., Hynes, M. A., and Rosenthal, A. 1998. FGF and Shh signals control dopaminergic and serotonergic cell fate in the anterior neural plate. *Cell* 93: 755–766. [25]

Yeh, C. I., Stoelzel, C. R., Weng, C., and Alonso, J. M. 2009. Functional consequences of neuronal divergence within the retinogeniculate pathway. *J. Neurophysiol.* 101: 2166–2185. [2]

Yellen, G., Jurman, M. E., Abramson, T., and MacKinnon, R. 1991. Mutations affecting internal TEA blockade identify the probable pore-forming region of a K^+ channel. *Science* 251: 939–942. [5]

Yernool, D., Boudker, O., Jin, Y., and Gouaux, E. 2004. Structure of a glutamate transporter homologue from *Pyrococcus horikoshil*. *Nature* 431: 811–818. [9]

Yool, A. J., and Schwartz, T. L. 1991. Alterations in ionic selectivity of a K^+ channel by mutation of the H5 region. *Nature* 349: 700–704. [5]

Yoshimura, M., and Furue, H. 2006. Mechanisms for the anti-nociceptive actions of the descending noradrenergic and serotonergic systems in the spinal cord. *J. Pharmacol. Sci.* 101: 107–117. [14]

Young, E. D. 2008. Neural representation of spectral and temporal information in speech. *Philos. Trans. R. Soc. Lond., B, Biol. Sci.* 363: 923–945. [22]

Young, E. D., and Sachs, M. B. 1979. Representation of steady-state vowels in the temporal aspects of the discharge patterns of populations of auditory-nerve fibers. *J. Acoust. Soc. Am.* 66: 1381–1403. [22]

Young, J. Z. 1936. The giant nerve fibres and epistellar body of cephalopods. *Q. J. Microsc. Sci.* 78: 367–386. [6]

Young, L. J. 2009. Love: neuroscience reveals all. *Nature* 457: 148. [14]

Young, L. J., Nilsen, R., Waymire, K. G., MacGregor, G. R., and Insel, T. R. 1999. Increased affiliative response to vasopressin in mice expressing the V_{1a} receptor from a monogamous vole. *Nature* 400: 766–768. [14]

Yu, F. H., and Catterall, W. A. 2003. Overview of the voltage-gated sodium channel family. *Genome Biol.* 4: 207.1–207.7. [5]

Yu, F. H., Westenbroek, R. E., Silos-Santiago, I., McCormick, K. A., Lawson, D., Ge, P., Ferriera, H., Lilly, J., DiStefano, P. S. Catterall, W. A., Scheuer, T., and Curtis, R. 2003. Sodium channel β4, a new disulfide-linked auxiliary subunit with similarity to β2. *J. Neurosci.* 23: 7577–7585. [5]

Yu, W. M., Yu, H., and Chen, Z. L. 2007. Laminins in peripheral nerve development and muscular dystrophy. *Mol. Neurobiol.* 35: 288–297. [10]

Yurek, D. M., and Sladek, J. R., Jr. 1990. Dopamine cell replacement: Parkinson's disease. *Annu. Rev. Neurosci.* 13: 415–440. [14]

Zagha, E., Manita, S., Ross, W. N., and Rudy, B. 2010. Dendritic Kv3.3 potassium channels in cerebellar purkinje cells regulate generation and spatial dynamics of dendritic Ca^{2+} spikes. *J. Neurophysiol.* 103: 3516–3525. [24]

Zagotta, W. N., Hoshi, T., and Aldrich, R. W. 1990. Restoration of inactivation in mutants of *Shaker* potassium channels by a peptide derived from ShB. *Science* 250: 568–571. [7]

Zagrebelsky, M., Buffo, A., Skerra, A., Schwab, M. E., Strata, P., and Rossi, F. 1998. Retrograde regulation of growth-associated gene expression in adult rat Purkinje cells by myelin-associated neurite growth inhibitory proteins. *J. Neurosci.* 18: 7912–7929. [27]

Zanazzi, G., and Matthews, G. 2009. The molecular architecture of ribbon presynaptic terminals. *Mol. Neurobiol.* 39: 130–148. [13, 20]

Zehring, W. A., Wheeler, D. A., Reddy, P., Konopka, R. J., Kyriacou, C. P., Rosbash, M., and Hall, J. C. 1984. P-element transformation with period locus DNA restores rhythmicity to mutant, arrhythmic *Drosophila melanogaster*. *Cell* 39: 369–376. [17]

Zeki, S. 1990. Colour visions and functional specialization in the visual cortex. *Disc. Neurosci.* 6: 1–64. [3]

Zeki, S. 2007. The neurobiology of love. *FEBS Lett.* 581: 2575–2579. [14]

Zeki, S., Watson, J. D., Lueck, C. J., Friston, K. J., Kennard, C., and Frackowiak, R. S. 1991. A direct demonstration of functional specialization in human visual cortex. *J. Neurosci.* 11: 641–649. [3]

Zeki, S. M. 1974. Functional organization of a visual area in the posterior bank of the superior temporal sulcus of the rhesus monkey. *J. Physiol.* 236: 549–573. [23]

Zelenin, P. V., Beloozerova, I. N., Sirota, M. G., Orlovsky, G. N., and Deliagina, T. G. 2010. Activity of red nucleus neurons in the cat during postural corrections. *J. Neurosci.* 30: 14533–14542. [24]

Zenisek, D., Horst, N. K., Merrifield, C., Sterling, P., and Matthews, G. 2004. Visualizing synaptic ribbons in the living cell. *J. Neurosci.* 24: 9752–9759. [20]

Zenisek, D., Steyer, J. A., Feldman, M. E., and Almers, W. 2002. A membrane marker leaves synaptic vesicles in milliseconds in retinal bipolar cells. *Neuron* 35: 1085–1097. [13]

Zhang, B., Luo, S., Wang, Q., Suzuki, T., Xiong, W. C., and Mei, L. 2008. LRP4 serves as a coreceptor of agrin. *Neuron* 60: 285–297. [27]

Zhang, H. Q., Murray, G. M., Coleman, G. T., Turman, A. B., Zhang, S. P., and Rowe, M. J. 2001. Functional characteristics of the parallel SI- and SII-projecting neurons of the thalamic ventral posterior nucleus in the marmoset. *J. Neurophysiol.* 85: 1805–1822. [21]

Zhang, Q., Li, Y., and Tsien, R. W. 2009. The dynamic control of kiss-and-run and vesicle reuse probed with single nanoparticles. *Science* 323: 1449–1453. [13]

Zhang, W., Basile, A. S., Gomeza, J., Volpicelli, L. A., Levey, A. I., and Wess, J. 2002. Characterization of central inhibitory muscarinic autoreceptors by the use of muscarinic acetylcholine receptor knock-out mice. *J. Neurosci.* 22: 1709–1717. [14]

Zhang, Y., Hoon, M. A., Chandrashekar, J., Mueller, K. L., Cook, B., Wu, D., Zuker, C. S., and Ryba, N. J. 2003. Coding of sweet, bitter, and umami tastes: different receptor cells sharing similar signaling pathways. *Cell* 112: 293–301. [19]

Zhang, Z., Zhao, Z., Margolskee, R., and Liman, E. 2007. The transduction channel TRPM5 is gated by intracellular calcium in taste cells. *J. Neurosci.* 27: 5777–5786. [19]

Zhao, S., and Frotscher, M. 2010. Go or stop? Divergent roles of Reelin in radial neuronal migration. *Neuroscientist* 16: 421–434. [25]

Zheng, C., Heintz, N., and Hatten, M. E. 1996. CNS gene encoding astrotactin, which supports neuronal migration along glial fibers. *Science* 272: 417–419. [25]

Zheng, J., Shen, W., He, D. Z., Long, K. B., Madison, L. D., and Dallos, P. 2000. Prestin is the motor protein of cochlear outer hair cells. *Nature* 405: 149–155. [22]

Zhou, F. M., Liang, Y., Salas, R., Zhang, L., De Biasi, M., and Dani, J. A. 2005. Corelease of dopamine and serotonin from striatal dopamine terminals. *Neuron* 46: 65–74. [15]

Zhou, F. M., Wilson, C. J., and Dani, J. A. 2002. Cholinergic interneuron characteristics and nicotinic properties in the striatum. *J. Neurobiol.* 53: 590–605. [14]

Zhou, J., Shapiro, M. S., and Hille, B. 1997. Speed of Ca^{2+} channel modulation by neurotransmitters in rat sympathetic neurons. *J. Neurophysiol.* 77: 2040–2048. [12]

Zhou, W., Raisman, G., and Zhou, C. 1998. Transplanted embryonic entorhinal neurons make functional synapses in adult host hippocampus. *Brain Res.* 788: 202–206. [27]

Zhou, Z. J. 1998. Direct participation of starburst amacrine cells in spontaneous rhythmic activities in the developing mammalian retina. *J. Neurosci.* 18: 4155–4165. [26]

Zhuo, M., Small, S. A., Kandel, E. R., and Hawkins, R. D. 1993. Nitric oxide and carbon monoxide produce activity-dependent long-term synaptic enhancement in hippocampus. *Science* 260: 1946–1950. [12]

Zigmond, R. E., Schwarzchild, M. A., and Rittenhouse, A. R. 1989. Acute regulation of tyrosine hydroxylase by nerve activity and by neurotransmitters via phosphorylation. *Annu. Rev. Neurosci.* 12: 415–461. [15]

Zimmermann, H. 2006. Ectonucleotidases in the nervous system. In *Novartis Foundation Symposium 276: Purinergic Signalling in Neuron-Glial Interactions.* Wiley, Chichester, UK. pp. 113–128. [15]

Zimmermann, H., and Denston, C. R. 1977a. Recycling of synaptic vesicles in the cholinergic synapses of the *Torpedo* electric organ during induced transmitter release. *Neuroscience* 2: 695–714. [13]

Zimmermann, H., and Denston, C. R. 1977b. Separation of synaptic vesicles of different functional states from the cholinergic synapses of the *Torpedo* electric organ. *Neuroscience* 2: 715–730. [13]

Zito, K., Knott, G., Shepherd, G. M., Shenolikar, S., and Svoboda, K. 2004. Induction of spine growth and synapse formation by regulation of the spine actin cytoskeleton. *Neuron* 44: 321–334. [11]

Zoccolan, D., Kouh, M., Poggio, T., and DiCarlo, J. J. 2007. Trade-off between object selectivity and tolerance in monkey inferotemporal cortex. *J. Neurosci.* 27: 12292–12307. [23]

Zollikofer, C., Wehner, R., and Fukushi, T. 1995. Optical scaling in conspecific *Cataglyphis* ants. *J. Exp. Biol.* 198: 1637–1646. [18]

Zou, D. J., Feinstein, P., Rivers, A. L., Mathews, G. A., Kim, A., Greer, C. A., Mombaerts, P., and Firestein, S. 2004. Postnatal refinement of peripheral olfactory projections. *Science* 304: 1976–1979. [26]

Zubieta, J. K., and Stohler, C. S. 2009. Neurobiological mechanisms of placebo responses. *Ann. N Y Acad. Sci.* 1156: 198–210. [21]

Zubieta, J. K., Bueller, J. A., Jackson, L. R., Scott, D. J., Xu, Y., Koeppe, R. A., Nichols, T. E., and Stohler, C. S. 2005. Placebo effects mediated by endogenous opioid activity on μ-opioid receptors. *J. Neurosci.* 25: 7754–7762. [21]

Zucker, R. S. 1993. Calcium and transmitter release. *J. Physiol. (Paris)* 87: 25–36. [13]

Zucker, R. S., and Regehr, W. G. 2002. Short-term synaptic plasticity. *Annu. Rev. Physiol.* 64: 355–405. [16]

Zuo, Y., Lubischer, J. L., Kang, H., Tian, L., Mikesh, M., Marks, A., Scofield, V. L., Maika, S., Newman, C., Krieg, P., and Thompson, W. J. 2004. Fluorescent proteins expressed in mouse transgenic lines mark subsets of glia, neurons, macrophages, and dendritic cells for vital examination. *J. Neurosci.* 24: 10999–11009. [10]

Zuo, Y., Perkon, I., and Diamond, M. E. 2011. Whisking and whisker kinematics during a texture classification task. *Philos. Trans. R. Soc. Lond., B, Biol. Sci.* 366: 3058–3069. [21]

INDEX

D

H

N

W

X

Y

Z

About This Book

Editor: Sydney Carroll
Project Editor: Kathaleen Emerson and Chelsea Holabird
Copy Editor: Lucy Anderson
Production Manager: Christopher Small
Book and Cover Design: Joan Gemme
Book Production: Joan Gemme
Illustration Program: Elizabeth Morales Scientific Illustration
Book and Cover Manufacture: Courier Companies, Inc.